Nonlinear Dynamical Systems and Control

Nonlinear Dynamical Systems and Control

A Lyapunov-Based Approach

Wassim M. Haddad

VijaySekhar Chellaboina

PRINCETON UNIVERSITY PRESS

PRINCETON AND OXFORD

Published by Princeton University Press
41 William Street, Princeton, New Jersey 08540

In the United Kingdom: Princeton University Press
6 Oxford St, Woodstock, Oxfordshire OX20 1TW

Library of Congress Cataloging-in-Publication Data

Haddad, Wassim M., 1961–
Nonlinear dynamical systems and control : a Lyapunov-based approach / Wassim M. Haddad, VijaySekhar Chellaboina.
p. cm.
Includes bibliographical references and index.
ISBN-9780691133294 (alk. paper)
1. Nonlinear control theory. 2. Automatic control. 3. Lyapunov functions. I. Chellaboina, VijaySekhar, 1970- II. Title.
QA402.35 .H33 2008
629.8′36–dc22 2007061026

British Library Cataloging-in-Publication Data is available

This book has been composed in LaTeX

The publisher would like to acknowledge the authors of this volume for providing the camera-ready copy from which this book was printed.

Printed on acid-free paper. ∞

press.princeton.edu

Printed in the United States of America

10 9 8 7 6 5 4 3 2 1

To my parents Mikhael and Sofia Haddad, with gratitude, appreciation, and love. Through their actions they inspired me to passionately pursue knowledge throughout my life because it is a more lasting possession than anything else

W. M. H.

To my wife Padma, who preserves the stability of my family life despite life's nonlinearities, turbulence, and uncertainties while I am deeply entrenched in my own world

V. C.

Αρχήν τε καί τέλος έχον ουδέν, ούτε αΐδιον ούτε άπειρον.

Anything that has a beginning and an end cannot at the same time be infinite and everlasting.

—Anaximander of Miletus, Ionia, Greece

Κόσμον (τόνδε), τὸν αὐτὸν ἁπάντων, οὔτε τις θεῶν, οὔτε ἀνθρώπων ἐποίησεν, ἀλλ' ἦν ἀεὶ καὶ ἔστιν καὶ ἔσται πῦρ ἀείζωον, ἁπτόμενον μέτρα καὶ ἀποσβεννύμενον μέτρα.

This universe which is the same everywhere, and which no one god or man has made, existed, exists, and will continue to exist as an eternal source of energy set on fire by its own natural laws, and will dissipate under its own laws.

Τα πάντα ρεί καί ούδέν μένει.

Everything is in a state of flux and nothing is stationary.

Ποταμείς τοίς αυτοίς εμβαίνομεν τε καί ουκ εμβαίνομεν, είμεν τε καί ουκ είμεν.

Man cannot step into the same river twice, because neither the man nor the river is the same.

—Herakleitos of Ephesus, Ionia, Greece

Δός μοι πού στώ και ταν γάν κινάσω.

Give me a place to stand and I will move the Earth.

—Archimedes of Syracuse, Sicily, Greater Greece

Contents

Conventions and Notation

In a definition or when a word is defined in the text, the concept defined is italicized. Italics in the running text is also used for emphasis. The definition of a word, phrase, or symbol is to be understood as an "if and only if" statement. Lower-case letters such as x denote vectors, upper-case letters such as A denote matrices, upper-case script letters such as $\mathcal{S}$ denote sets, and lower-case Greek letters such as α denote scalars; however, there are a few exceptions to this convention. The notation $\mathcal{S}_1 \subset \mathcal{S}_2$ means that $\mathcal{S}_1$ is a proper subset of $\mathcal{S}_2$, whereas $\mathcal{S}_1 \subseteq \mathcal{S}_2$ means that either $\mathcal{S}_1$ is a proper subset of $\mathcal{S}_2$ or $\mathcal{S}_1$ is equal to $\mathcal{S}_2$. Throughout the book we use two basic types of mathematical statements, namely, *existential* and *universal* statements. An existential statement has the form: there exists $x \in \mathcal{X}$ such that a certain condition C is satisfied; whereas a universal statement has the form: condition C holds for all $x \in \mathcal{X}$. For universal statements we often omit the words "for all" and write: condition C holds, $x \in \mathcal{X}$. The notation used in this book is fairly standard. The reader is urged to glance at the notation below before starting to read the book.

$\mathbb{Z}$	set of integers
$\overline{\mathbb{Z}}_+$, $\mathbb{Z}_+$, $\overline{\mathbb{Z}}_-$, $\mathbb{Z}_-$	set of nonnegative, positive, nonpositive, negative integers
$\mathbb{R}$	set of real numbers
$\mathbb{R}^{n\times m}$	set of $n \times m$ real matrices
$\mathbb{R}^n$	$\mathbb{R}^{n\times 1}$ (real column vectors)
$\overline{\mathbb{R}}_+$, $\mathbb{R}_+$, $\overline{\mathbb{R}}_-$, $\mathbb{R}_-$	set of nonnegative, positive, nonpositive, negative real numbers
$\mathbb{C}$	set of complex numbers
$\mathbb{C}^{n\times m}$	set of $n \times m$ complex matrices
$\mathbb{C}^n$	$\mathbb{C}^{n\times 1}$ (complex column vectors)
$\overline{\mathbb{C}}_+$, $\mathbb{C}_+$, $\overline{\mathbb{C}}_-$, $\mathbb{C}_-$	set of complex numbers with nonnegative, positive, nonpositive, negative real parts
$\mathbb{F}$, $\mathbb{F}^n$, $\mathbb{F}^{n\times m}$	$\mathbb{R}$ or $\mathbb{C}$, $\mathbb{R}^n$ or $\mathbb{C}^n$, $\mathbb{R}^{n\times m}$ or $\mathbb{C}^{n\times m}$
CLHP, OLHP	closed, open left half plane

CRHP, ORHP	closed, open right half plane
$\jmath$	$\sqrt{-1}$
$\jmath\mathbb{R}$	imaginary numbers
Re z, Im z	real part; imaginary part of a complex number z
$\triangleq$	equals by definition
$\emptyset$	empty set
$\{\ \}$, $\{\ \}_{\mathrm{m}}$	set, multiset
$\cup$, $\cap$	union, intersection
$\in$, $\notin$	is an element of, is not an element of
$\subseteq$, $\subset$	is a subset of, is a proper subset of
$\to$	approaches
0, $0_{n\times m}$, 0_n	zero matrix, $n\times m$ zero matrix, $0_{n\times n}$
I_n, I	$n\times n$ identity matrix
$\mathcal{R}(A)$, $\mathcal{N}(A)$	range space of A, null space of A
x_i, $x_{(i)}$	ith component of vector $x\in\mathbb{R}^n$
$A_{(i,j)}$	(i,j) entry of A
$\mathrm{col}_i(A)$, $\mathrm{row}_i(A)$	ith column of A, ith row of A
$\mathrm{diag}[A_{(1,1)},\ldots,A_{(n,n)}]$	diagonal matrix $\begin{bmatrix} A_{(1,1)} & & 0 \\ & \ddots & \\ 0 & & A_{(n,n)} \end{bmatrix}$
block-diag$[A_1,\ldots,A_k]$	block-diagonal matrix $\begin{bmatrix} A_1 & & 0 \\ & \ddots & \\ 0 & & A_k \end{bmatrix}$, $A_i\in\mathbb{R}^{n_i\times m_i}$, $i=1,\ldots,k$
A^{T}	transpose of A
$\bar{A}$	complex conjugate of A
A^*	$\bar{A}^{\mathrm{T}}$
A^{-1}	inverse of A
A^{+}	Moore-Penrose generalized inverse of A
$A^{\#}$	group generalized inverse of A
$A^{-\mathrm{T}}$, A^{-*}	$(A^{\mathrm{T}})^{-1}$, $(A^*)^{-1}$
tr A	trace of A
det A	determinant of A
rank A	rank of A
$\mathbb{S}^n$	set of $n\times n$ symmetric matrices
$\mathbb{N}^n$	set of $n\times n$ nonnegative-definite matrices
$\mathbb{P}^n$	set of $n\times n$ positive-definite matrices
$A\geq\geq 0$ $(A>>0)$	$A_{(i,j)}\geq 0$ $(A_{(i,j)}>0)$ for all i and j
$A\geq\geq B$ $(A>>B)$	$A_{(i,j)}\geq B_{(i,j)}$ $(A_{(i,j)}>B_{(i,j)})$, where A and B are matrices with identical dimensions
$A\geq 0$ $(A>0)$	nonnegative (respectively, positive) definite matrix; that is, symmetric matrix with

	nonnegative (respectively, positive) eigenvalues
$A \geq B$	$A - B \in \mathbb{N}^n$
$A > B$	$A - B \in \mathbb{P}^n$
$\mathbb{R}_+^n$, $\overline{\mathbb{R}}_+^n$	$\{x \in \mathbb{R}^n :\ x >> 0\}$, $\{x \in \mathbb{R}^n :\ x \geq\geq 0\}$
$\otimes$, $\oplus$	Kronecker product, Kronecker sum
$x^{[k]}$	$x \otimes \cdots \otimes x$ (k times)
$\overset{k}{\oplus} A$	$A \oplus A \oplus \cdots \oplus A$ (k times)
$\mathcal{N}^{(k,n)}$	$\{\Psi \in \mathbb{R}^{1\times n^k} : \Psi x^{[k]} \geq 0,\ x \in \mathbb{R}^n\}$
vec	column-stacking operator
spec(A)	spectrum of A including multiplicity
$\rho(A)$	spectral radius of A
$\alpha(A)$	spectral abscissa of A
$\lvert\alpha\rvert$	absolute value of α
$\sigma_i(A)$	ith singular value of A
$\sigma_{\min}(A)$, $\sigma_{\max}(A)$	minimum, maximum singular value of A
$\lVert\cdot\rVert$, $\vert\vert\vert\cdot\vert\vert\vert$	vector or matrix norm, vector or matrix operator norm
$\lVert x\rVert_2$	Euclidean norm of x ($=\sqrt{x^*x}$)
$\lVert x\rVert_p$	Hölder vector norms, $[\sum_{i=1}^n \lvert x_i\rvert^p]^{1/p}$, $1 \leq p < \infty$
$\lVert x\rVert_\infty$	$\max_i \lvert x_{(i)}\rvert$
$\lVert A\rVert_p$	Hölder matrix norms, $\left[\sum_{i=1}^m \sum_{j=1}^n \lvert A_{(i,j)}\rvert^p\right]^{1/p}$, $1 \leq p < \infty$
$\lVert A\rVert_\infty$	$\max_{i,j} \lvert A_{(i,j)}\rvert$
$\lVert A\rVert_{\sigma p}$	$\left[\sum_{i=1}^r \sigma_i^p(A)\right]^{1/p}$, $1 \leq p < \infty$, $r = \operatorname{rank} A$
$\lVert A\rVert_{\sigma\infty}$	$\sigma_{\max}(A)$
$\lVert A\rVert_{\mathrm{s}}$	spectral norm of A ($=\sigma_{\max}(A)$)
$\lVert A\rVert_{\mathrm{F}}$	Frobenius matrix norm of A ($=(\operatorname{tr} AA^*)^{1/2}$)
$\lVert A\rVert_{q,p}$	induced matrix norm
$\lambda_i(A)$	ith eigenvalue of $A \in \mathbb{R}^{n\times n}$
$\lambda_{\min}(A)$, $\lambda_{\max}(A)$	minimum, maximum eigenvalues of the Hermitian matrix A
He A, Sh A	$\frac{1}{2}(A + A^*)$, $\frac{1}{2}(A - A^*)$
$E_{(i,j)}$	elementary matrix with unity in the (i,j) entry and zeros elsewhere
$\log_e$	logarithm with base $e = 2.71828\cdots$
e_i	vector with unity in the ith component and zeros elsewhere
$\mathbf{e}$	$[1,\ 1,\ \ldots,\ 1]^{\mathrm{T}}$
$\mathcal{L}_p$	Lebesgue space, $1 \leq p \leq \infty$
$\mathcal{L}_2$	space of square-integrable Lebesgue measurable functions on $[0,\infty)$
$\mathcal{L}_\infty$	space of bounded Lebesgue measurable functions

	on $[0, \infty)$
$\lVert\lvert f \rvert\rVert_{p,q}$	$\{\int_0^\infty \lVert f(t)\rVert_q^p \mathrm{d}t\}^{1/p}$, $1 \le p < \infty$
$\lVert\lvert f \rvert\rVert_{\infty,q}$	$\operatorname{ess\,sup}_{t\ge 0} \lVert f(t)\rVert_q$
$\langle f, g\rangle$	$\int_0^\infty f^{\mathrm{T}}(t)g(t)\mathrm{d}t$
ℓ_p	sequence space, $1 \le p \le \infty$
ℓ_2	space of square-summable sequences on $\overline{\mathbb{Z}}_+$
ℓ_∞	space of bounded sequences on $\overline{\mathbb{Z}}_+$
$\mathcal{H}_p$	analytic function space
$\mathcal{H}_2$	Hardy space of real-rational transfer function matrices square-integrable on the imaginary axis (unit disk) with analytic continuation in the right half plane (outside the unit disk)
$\mathcal{H}_\infty$	Hardy space of real-rational transfer function matrices bounded on the imaginary axis (unit disk) with analytic continuation in the right half plane (outside the unit disk)
$\Re\mathcal{H}_2$	real-rational subspace of $\mathcal{H}_2$
$\Re\mathcal{H}_\infty$	real-rational subspace of $\mathcal{H}_\infty$
$[a, b]$	closed interval
(a, b)	open interval
$\mathcal{X} \times \mathcal{Y}$	Cartesian product of $\mathcal{X}$ and $\mathcal{Y}$
$f : \mathcal{X} \to \mathcal{Y}$	function f with domain $\mathcal{X}$ and codomain $\mathcal{Y}$
$\frac{\partial f}{\partial x_i}(x_0)$	partial derivative of f with respect to x_i at x_0
$f'(x_0)$	Fréchet derivative of f at x_0
$f^{(k)}(x_0)$	kth Fréchet derivative of f at x_0
$D^+f(x_0)$	upper right Dini derivative of f at x_0
$D_+f(x_0)$	lower right Dini derivative of f at x_0
$f^{-1}(\mathcal{D})$	inverse image of the set $\mathcal{D}$
$f_2 \circ f_1$	composition of two functions; $(f_2 \circ f_1)(\cdot) = f_2(f_1(\cdot))$
$\mathcal{L}[z(t)]$	Laplace transform of $z(\cdot)$
$G(s) \sim \left[\begin{array}{c\|c} A & B \\ \hline C & D \end{array}\right]$	state space realization of transfer function $G(s) = C(sI - A)^{-1}B + D$
$G(s) \overset{\min}{\sim} \left[\begin{array}{c\|c} A & B \\ \hline C & D \end{array}\right]$	minimal state space realization of $G(s)$
$\mathcal{B}_\varepsilon(\alpha)$	$\{x \in \mathbb{R}^n : \lVert x - \alpha\rVert < \varepsilon\}$
$\mathcal{B}_\varepsilon[\alpha]$	$\{x \in \mathbb{R}^n : \lVert x - \alpha\rVert \le \varepsilon\}$
$\mathcal{X} \backslash \mathcal{Y}$	$\{x \in \mathcal{X} : x \notin \mathcal{Y}\}$ for sets $\mathcal{X}$ and $\mathcal{Y}$
$\partial\mathcal{S}$	boundary of the set $\mathcal{S}$
$\overset{\circ}{\mathcal{S}}$	interior of the set $\mathcal{S}$
$\overline{\mathcal{S}}$	closure of the set $\mathcal{S}$
$\mathcal{S}^{\mathrm{c}}$ or $\mathcal{S}^\sim$	complement of the set $\mathcal{S}$

inf	infimum; greatest lower bound
sup	supremum; least upper bound
$\liminf_{n\to\infty} f(x_n)$	limit inferior of $f(x_n)$; $\liminf_{n\to\infty} f(x_n) = \sup_n \inf_{k\geq n} f(x_k)$
$\limsup_{n\to\infty} f(x_n)$	limit superior of $f(x_n)$, $\limsup_{n\to\infty} f(x_n) = \inf_n \sup_{k\geq n} f(x_k)$
min, max	minimum, maximum
C^0	continuous functions
C^r	functions with r-continuous derivatives
C^∞	infinitely differentiable functions
$\mathcal{C}[a, b]$	space of continuous functions
$\mathbb{E}$	expectation
a.e.	almost everywhere
$\triangle$	end of example
$\square$	quod erat demonstrandum or end of proof

Preface

Dynamical system theory provides a paradigm for modeling and studying phenomena that undergo spatial and temporal evolution. A dynamical system consists of three elements—namely, a setting (called the *state space*) in which the dynamical behavior takes place, such as a torus, topological space, manifold, or locally compact metric space; a mathematical rule or *dynamic* which specifies the evolution of the system over time; and an initial condition or *state* from which the system starts at some initial time. Ever since its inception, the basic questions concerning dynamical system theory have involved qualitative solutions for the properties of a dynamical system; questions such as: For a particular initial system state, does the dynamical system have at least one solution? What are the asymptotic properties of the system solutions? How are the system solutions dependent on the system initial conditions? How are the system solutions dependent on the form of the mathematical description of the dynamic of the system? How do system solutions depend on system parameters? And how do system solutions depend on the properties of the state space on which the system is defined?

Even though qualitative properties of solutions to dynamical systems were first pioneered by Henri Poincaré, mathematical dynamical system theory can be traced back to Isaac Newton. Newton was the first to model the motion of physical systems with differential equations. However, the development of dynamical system theory is a natural outcome of a much broader theme which is as old as science itself and can be traced back to the great cosmic theorists of ancient Greece—namely, the universe is in a constant state of flux (τα πάντα ρεί). Ancient Greek astronomers and mathematicians such as Eudoxus, Ptolemy, and Archimedes were the first to use abstract mathematical models and attach them to the physical world. More importantly, using abstract thought and theoretical calculations they were able to deduce something that is true about the physical world. Newton's greatest achievement, however, was the discovery that the motion of the planets and moons of the solar system resulted from a single fundamental source—the gravitational attraction of the heavenly bodies. As a consequence, Newtonian physics developed into the first field of modern science—dynamical systems as a branch of mathematical physics—wherein the circular, elliptical, and parabolic orbits of the heavenly bodies of our solar system were no longer fundamental determinants of motion, but rather approximations of the universal laws of the cosmos specified by governing differential equations of motion.

In the past century, dynamical system theory quickly spread from the study of astronomical bodies to modeling and studying numerous physical phenomena occurring in nature involving changes over time, and it has become one of the most important and fundamental fields of modern science and engineering. The application of dynamical systems has crossed interdisciplinary boundaries from chemistry to biochemistry to chemical kinetics, from medicine to biology to population genetics, from economics to sociology to psychology, and from physics to mechanics to engineering. The increasingly complex nature of engineering systems requiring feedback control to obtain a desired system behavior also gives rise to dynamical systems. Feedback control theory involves the analysis and synthesis of a feedback controller that manipulates system inputs to obtain a desired effect on the output of the system in the face of system uncertainty and system disturbances. Furthermore, since most physical and engineering systems are inherently *nonlinear*, the resulting feedback dynamical system can exhibit a very rich dynamical behavior.

One of the most important concepts in the study of dynamical system theory and control theory is the concept of stability. Stability theory concerns the behavior of the system trajectories of a dynamical system when the system initial state is near an equilibrium state. Since exogenous disturbances and system component uncertainty are always present in every actual system, stability theory plays a central role in dynamical systems and control. As in the study of dynamical system theory, the origins of stability theory can be traced back to the study of idealizations of astronomical problems involving persistent small oscillations (librations) about a state of motion. One of the first modern treatments of stability theory was given by Joseph-Louis Lagrange wherein he concluded that for a conservative mechanical system an isolated minimum of the system potential energy corresponds to a stable equilibrium system state. The most complete stability analysis framework for dynamical systems was developed by Aleksandr Lyapunov. Lyapunov's method is based on construction of a function of the system state coordinates that serves as a generalized norm of the solution of the dynamical system. Its appeal comes from the fact that conclusions about the behavior of the dynamical system can be drawn without actually computing the system solution trajectories. As a result, Lyapunov stability theory has become one of the cornerstones of systems and control theory.

Aleksandr Mikhailovich Lyapunov was born on the 6th of June, 1857 to the family of the prominent astronomer Mikhail Vasilyevich Lyapunov in Yaroslavl, Russia. After completing the gymnasium (high school) in Nizhny Novogorod where he was taught Greek, Latin, and basic science and mathematics, in 1876 he entered the Physics and Mathematics Department of Saint Petersburg University. Upon graduating from Saint Petersburg University in 1880, he remained in the department of mechanics at the university to prepare for his academic career which was greatly influenced by Chebyshev. His master's work concentrated on the problem of a homogeneous, incompressible fluid mass held together by gravitational forces of its particles and rotating about a fixed axis. Chebyshev posed the question, Under what conditions will the motion of the fluid be stable and what possible equilibrium forms can this stable rotating fluid take? This famous question of the piriform body, which was of considerable importance to cosmogony and had been studied by prominent mathematicians such as Newton, MacLaurin, Jacobi, and Poincaré, was eventually settled by Lyapunov in 1905.

After receiving his Master's degree in applied mathematics in 1884 for his work *On the Stability of Ellipsoidal Forms in the Equilibrium of a Rotating Fluid*, Lyapunov moved to Kharkov University as a *Privatdozent* (private reader or lecturer) in mechanics where he continued his research towards his doctoral thesis. In 1892 the Kharkov Mathematical Society published Lyapunov's seminal work on *The General Problem of the Stability of Motion* which he defended as his doctoral thesis at Moscow State University (Moskovskii Gosudarstvennii Universitet) later that year. His paper "Sur le Problème General de la Stabilité du Mouvement" published in the *Annales de la Faculté des Sciences de l'Université de Toulouse* in 1907 marks the beginning of modern stability theory in the West. In 1893 Lyapunov was appointed as a professor at Kharkov University and in 1902 he returned to Saint Petersburg University as Chair of Applied Mathematics. In 1900 he was elected as a corresponding member of the Russian Academy of Sciences, and in 1901 as a full member of the Academy in the field of applied mathematics.

Even though Lyapunov is predominantly known for his work on the stability of equilibria and motion of mechanical systems with a finite number of degrees of freedom in the dynamical systems and control community, his scientific work includes fundamental contributions to stability of equilibrium figures of rotating fluids; equilibrium figures of uniformly rotating fluids; mathematical physics; probability theory; and theoretical mechanics. In all these disciplines Lyapunov established a deep level of accuracy and rigor that resulted in fundamental and classical mathematical results. Lyapunov tragically died on the 3rd of November, 1918 in Odessa, Russia (now Ukraine) as a result of a voluntary act three days after the death of his wife. His wish, which he had stated in an ante mortem note, was to be buried with his wife.

The main objective of this book is to present and develop necessary mathematical tools for stability analysis and control design of nonlinear dynamical systems, with an emphasis on Lyapunov-based methods. This book is intended to be useful to applied mathematicians, dynamical system theorists, control theorists, and engineers. Since dynamical system theory and Lyapunov stability theory lie at the heart of mathematical sciences and engineering, researchers and graduate students in these fields who seek a fundamental understanding of the rich behavior of nonlinear dynamical systems and control will also find this textbook useful. The appropriate background for this book is a first course in linear system theory and a first course in advanced (multivariable) calculus.

After a brief introduction on dynamical systems in Chapter 1, a systematic development of nonlinear ordinary differential equations is given in Chapter 2. In Chapters 3 and 4, we present fundamental stability theory as well as advanced stability theory for nonlinear dynamical systems. Chapter 5 provides a treatment of classical dissipativity theory and absolute stability theory. A detailed treatment of stability of feedback interconnections, control Lyapunov functions, feedback linearization, zero dynamics, and stability margins for nonlinear regulators is given in Chapter 6. Chapter 7 focuses on input-output stability and dissipativity theory. In Chapter 8, we present the nonlinear optimal control problem, while stability and optimality results for backstepping control problems are given in Chapter 9. Chapters 10–12 provide novel extensions to disturbance rejection and robust control of nonlinear dynamical systems. Finally, Chapters 13 and 14 present discrete-time extensions of the aforementioned topics.

The authors would like to thank Dennis S. Bernstein, Sanjay P. Bhat, Tomohisa Hayakawa, V. Lakshmikantham, and Frank L. Lewis for their constructive comments and feedback. In some parts of the book we have relied on work we have done jointly with Chaouki T. Abdallah, Dennis S. Bernstein, Sanjay P. Bhat, Jerry L. Fausz, Qing Hui, Vikram Kapila, Alexander Leonessa, Sergey G. Nersesov, and David A. Wilson; it is a pleasure to acknowledge their contributions. In addition, we thank Sanjay P. Bhat, Tomohisa Hayakawa, Qing Hui, Vikram Kapila, and Sergey G. Nersesov for graciously agreeing to read chapters of the book for clarity and technical accuracy. We are also grateful to our students for their helpful comments, insightful suggestions, and corrections, all of which greatly improved the exposition of the book.

The authors would also like to thank Paul Katinas for providing a translation of Anaximander's, Herakleitos', and Archimedes' statements quoted in ancient Greek on page vii. In addition, we thank Paul for several insightful and enlightening philosophical discussions on the ramifications of these statements to cosmology, mathematics, science, and engineering. Anaximander's aphorism introduces for the first time the concepts of beginning (αρχήν) and infinity (άπειρον) which are central to the study of cosmology and modern mathematics. A common misconception among scholars has been that modern mathematicians and not ancient Greek mathematicians were the first to be able to address notions of infinity in a rigorous and precise manner. However, as has been recently discovered in the Archimedes Palimpsest, Archimedes was the first to rigorously address the *science of infinity* with infinitely large sets using precise mathematical proofs in his work on *The Method of Mechanical Theorems.*

Heraklitos' profound statements created the foundation for all physics and metaphysics. His first statement marks the beginning of science and postulates the big bang theory as the origin of the universe as well as the heat death of the universe. He further postulates that the universe evolves in accordance with its own laws which are the only unchangeable things in the universe (i.e., universal conservation and nonconservation laws). His second and third statements give the earliest perception of irreversibility of nature and the universe along with time's arrow. The idea that the universe is in constant change (τα πάντα ρεί) and there is an underlying order to this change—the *Logos* (Λόγος)—postulates the existence of entropy as a physical property of matter permeating the whole of nature and the universe. In addition, postulating that the fundamental uniform fact in nature is constant change and everything both is and is not at the same time (είμεν τε καί ουκ είμεν), as well as that energy is change and substance is change (Μὲν οὖν φησιν εἶναι τὸ πᾶν διαιρετὸν ἀδιαίρετον, γενητὸν ἀγένητον, θνητὸν ἀθάνατον, λὸγον αἰῶνα, πατέρα υἱὸν,... ἐστίν ἕν πάντα εἶναι.), he arrives at a precursor to the principle of relativity and the mass-energy equivalence.

Finally, Archimedes' statement leads to the foundation of mathematical mechanics. His law of the lever—involving the earliest treatment of a constrained mechanical system—associated with this statement led to the idea of energy as the product of force and distance, to the concept of the conservation of energy, and to the principle of virtual velocities. In his treatise on *The Method of Mechanical Theorems* he established the foundations of integral calculus using infinitesimals, as well as the foundations of mathematical mechanics. In addition, in one of his problems he constructed the tangent at any given point for a spiral, establishing the origins of differential calculus.

Among the greatest gifts of Greece to humankind and world civilization are philosophy, mathematics, science, and engineering. Monumental scholars in these disciplines inspired a deep love for creative work and an unparalleled thirst for knowledge (γνώση). These modern minds in ancient bodies stood out in bold relief as towering talents throughout the centuries, paving the way for modern mathematics, science, and engineering. Consider, for example, how intense Archimedes' passion for study (μελέτη) must have been. He was deeply immersed in his mathematics when he delivered his legendary last words, Μη μου τους κύκλους τάραττε! (Do not disturb my circles!), before being murdered by a Roman soldier. The golden age of Greece, which set the standard for western civilization and whose glory has transcended time itself, was superseded by the Roman imperialists who brought nothing more than sterility to science and mathematics. The derailment of the pursuit of abstract science and mathematics in favor of practicality obfuscated scientific scholarship. This, along with religious fundamentalism, plunged civilization into the dark ages. The intellectual light (φως) that was gifted to the world by the ancient Greeks and provided the pathway for unlocking the most intriguing mysteries of our world lay idle for over one thousand years. It was not until the fall of Byzantium to Ottoman tribes from Mongolia that caused Greek scholars to flee to Florence, Venice, and Rome, sparking a revival in learning and humanism. This renaissance or rebirth (αναγέννησης) led to the scientific revolution which further led to the marvels of modern-day science and engineering.

Atlanta, Georgia, USA, October 2007, *Wassim M. Haddad*

Knoxville, Tennessee, USA, October 2007, *VijaySekhar Chellaboina*

Chapter One

Introduction

A *system* is a combination of components or parts that is perceived as a single entity. The parts making up the system may be clearly or vaguely defined. These parts are related to each other through a particular set of variables, called the *states* of the system, that completely determine the behavior of the system at any given time. A *dynamical system* is a system whose state changes with time. Specifically, the state of a dynamical system can be regarded as an information storage or memory of past system events. The set of (internal) states of a dynamical system must be sufficiently rich to completely determine the behavior of the system for any future time. Hence, the state of a dynamical system at a given time is uniquely determined by the state of the system at the initial time and the present input to the system. In other words, the state of a dynamical system in general depends on both the present input to the system and the past history of the system. Even though it is often assumed that the state of a dynamical system is the *least* set of state variables needed to completely predict the effect of the past upon the future of the system, this is often a convenient simplifying assumption.

We regard a dynamical system $\mathcal{G}$ as a mathematical model structure involving an input, state, and output that can capture the dynamical description of a given class of physical systems. Specifically, at each moment of time $t \in \mathbb{T}$, where $\mathbb{T}$ denotes a time-ordered subset of the reals, the dynamical system $\mathcal{G}$ receives an input $u(t)$ (e.g., matter, energy, information) and generates an output $y(t)$. The values of the input are taken from the fixed set U. Furthermore, over a time segment the input function $u : [t_1, t_2) \to U$ is not arbitrary but belongs to the admissible input class $\mathcal{U}$, that is, for every $u(\cdot) \in \mathcal{U}$ and $t \in \mathbb{T}$, $u(t) \in U$. The input class $\mathcal{U}$ depends on the physical description of the system. In addition, each system output $y(t)$ belongs to the fixed set Y with $y(\cdot) \in \mathcal{Y}$ over a given time segment, where $\mathcal{Y}$ denotes an output space. In general, the output of $\mathcal{G}$ depends on both the present input of $\mathcal{G}$ and the past history of $\mathcal{G}$. Thus, the state, and hence the output at some time $t \in \mathbb{T}$, depends on both the initial state $x(t_0) = x_0$ and the input segment $u : [t_0, t) \to U$. In other words, knowledge of both x_0 and $u \in \mathcal{U}$ is necessary and sufficient to determine the present and future state $x(t) = s(t, t_0, x_0, u)$ of $\mathcal{G}$.

In light of the above discussion, we view a dynamical system as a precise mathematical object defined on a time set as a mapping between vector spaces satisfying a set of axioms. A mathematical dynamical system thus consists of the space of states $\mathcal{D}$ of the system together with a rule or *dynamic* that determines the state of the system at a given future time from a given present state. This is formalized by the following definition. For this definition $\mathbb{T} = \mathbb{R}$ for continuous-time systems and $\mathbb{T} = \mathbb{Z}$ for discrete-time systems.

Definition 1.1. A *dynamical system* $\mathcal{G}$ on $\mathcal{D}$ is the octuple $(\mathcal{D}, \mathcal{U}, U, \mathcal{Y}, Y, \mathbb{T}, s, h)$, where $s : \mathbb{T} \times \mathbb{T} \times \mathcal{D} \times \mathcal{U} \to \mathcal{D}$ and $h : \mathbb{T} \times \mathcal{D} \times U \to Y$ are such that the following axioms hold:

i) (Continuity): For every $t_0 \in \mathbb{T}$, $x_0 \in \mathcal{D}$, and $u \in \mathcal{U}$, $s(\cdot, t_0, x_0, u)$ is continuous for all $t \in \mathbb{T}$.

ii) (Consistency): For every $x_0 \in \mathcal{D}$, $u \in \mathcal{U}$, and $t_0 \in \mathbb{T}$, $s(t_0, t_0, x_0, u) = x_0$.

iii) (Determinism): For every $t_0 \in \mathbb{T}$ and $x_0 \in \mathcal{D}$, $s(t, t_0, x_0, u_1) = s(t, t_0, x_0, u_2)$ for all $t \in \mathbb{T}$ and $u_1, u_2 \in \mathcal{U}$ satisfying $u_1(\tau) = u_2(\tau)$, $\tau \in [t_0, t]$.

iv) (Group property): $s(t_2, t_0, x_0, u) = s(t_2, t_1, s(t_1, t_0, x_0, u), u)$ for all $t_0, t_1, t_2 \in \mathbb{T}$, $t_0 \le t_1 \le t_2$, $x_0 \in \mathcal{D}$, and $u \in \mathcal{U}$.

v) (Read-out map): There exists $y \in \mathcal{Y}$ such that $y(t) = h(t, s(t, t_0, x_0, u), u(t))$ for all $x_0 \in \mathcal{D}$, $u \in \mathcal{U}$, $t_0 \in \mathbb{T}$, and $t \in \mathbb{T}$.

We denote the dynamical system $(\mathcal{D}, \mathcal{U}, U, \mathcal{Y}, Y, \mathbb{T}, s, h)$ by $\mathcal{G}$ and we refer to the map $s(\cdot, t_0, \cdot, u)$ as the *flow* or *trajectory* corresponding to $x_0 \in \mathcal{D}$, $t_0 \in \mathbb{T}$, and $u \in \mathcal{U}$; and for a given trajectory $s(t, t_0, x_0, u)$, $t \in \mathbb{T}$, we refer to $t_0 \in \mathbb{T}$ as an *initial time* of $\mathcal{G}$, $x_0 \in \mathcal{D}$ as an *initial condition* of $\mathcal{G}$, and $u \in \mathcal{U}$ as an *input* to $\mathcal{G}$. The dynamical system $\mathcal{G}$ is *isolated* if the input space consists of one element only, that is, $u(t) = u^*$, and the dynamical system is *undisturbed* if $u^* = 0$. If $\mathcal{G}$ is isolated, then $\mathcal{G}$ is isolated from any inputs and the environment is the only input acting on the system. This, for example, would correspond to a conservative mechanical system wherein the only external force acting on the system is gravity. In general, the output of $\mathcal{G}$ depends on both the present input of $\mathcal{G}$ and the past history of $\mathcal{G}$. Hence, the output of the dynamical system at some time $t \in \mathbb{T}$ depends on the state $s(t, t_0, x_0, u)$ of $\mathcal{G}$, which effectively serves as an information storage (memory) of past history. Furthermore, the determinism axiom ensures that the state, and hence the output, before some time $t \in \mathbb{T}$ is not influenced by the values of the output after time t. Thus, future inputs to $\mathcal{G}$ do not affect past and present outputs of $\mathcal{G}$. This is simply a statement of causality that holds for all physical systems. The notion of a dynamical system as defined in Definition 1.1 is far too general to develop useful practical deductions for dynamical systems. This notion of a dynamical system is introduced here to develop terminology and to introduce certain key concepts. As we will see in the next chapter, under additional regularity conditions the flow of a dynamical system describing the motion of the system as a function of time can generate a differential equation on the state space, allowing for the development of a large array of mathematical results leading to useful and practical analysis and control synthesis tools.

Determining the rule or dynamic that defines the state of physical and engineering systems at a given future time from a given present state is one of the central problems of science and engineering. Once the flow of a dynamical system describing the motion of the system starting from a given initial state is given, dynamical system theory can be used to describe the behavior of the system states over time for different initial conditions. Throughout the centuries—from the great cosmic theorists of ancient Greece to the present-day quest for a unified field theory—the most important dynamical system is our universe. By using abstract mathematical models and attaching them to the physical world, astronomers, mathematicians, and physicists have used abstract thought to deduce something that is true about the natural system of the cosmos.

The quest by scientists, such as Brahe, Kepler, Galileo, Newton, Huygens, Euler, Lagrange, Laplace, and Maxwell, to understand the regularities inherent in the distances of the planets from the sun and their periods and velocities of revolution around the sun led to the science of dynamical systems as a branch of mathematical physics. One of the most basic issues in dynamical system theory that was spawned from the study of mathematical models of our solar system is the stability of dynamical systems. System stability involves the investigation of small deviations from a system's steady state of motion. In particular, a dynamical system is *stable* if the system is allowed to perform persistent small oscillations about a system equilibrium, or about a state of motion. Among the first investigations of the stability of a given state of motion is by Isaac Newton. In particular, in his *Principia Mathematica* [335] Newton investigated whether a small perturbation would make a particle moving in a plane around a center of attraction continue to move near the circle, or diverge from it. Newton used his analysis to analyze the motion of the moon orbiting the Earth.

Numerous astronomers and mathematicians who followed made significant contributions to dynamical stability theory in an effort to show that the observed deviations of planets and satellites from fixed elliptical orbits were in agreement with Newton's principle of universal gravitation. Notable contributions include the work of Torricelli [431], Euler [116], Lagrange [252], Laplace [257], Dirichlet [106], Liouville [283], Maxwell [310], and Routh [369]. The most complete contribution to the stability analysis of dynamical systems was introduced in the late nineteenth century by the Russian mathematician Aleksandr Mikhailovich Lyapunov in his seminal work entitled *The General Problem of the Stability of Motion* [293–295]. Lyapunov's *direct method* states that if a positive-definite function (now called a Lyapunov function) of the state coordinates of a dynamical system can be constructed for which its time rate of change following small perturbations from the system equilibrium is always negative or zero, then the system equilibrium state is stable. In other words, Lyapunov's method is based on the construction of a Lyapunov function that serves as a generalized norm of the solution of a dynamical system. Its appeal comes from the fact that stability properties of the system solutions are derived directly from the governing dynamical system equations; hence the name, Lyapunov's direct method.

Dynamical system theory grew out of the desire to analyze the mechanics of heavenly bodies and has become one of the most fundamental fields of modern science as it provides the foundation for unlocking many of the mysteries in nature and the universe that involve the evolution of time. Dynamical system theory is used to study ecological systems, geological systems, biological systems, economic systems, neural systems, and physical systems (e.g., mechanics, thermodynamics, fluids, magnetic fields, galaxies, etc.), to cite but a few examples. Dynamical system theory has also played a crucial role in the analysis and control design of numerous complex engineering systems. In particular, advances in feedback control theory have been intricately coupled to progress in dynamical system theory, and conversely, dynamical system theory has been greatly advanced by the numerous challenges posed in the analysis and control design of increasingly complex feedback control systems.

Since most physical and engineering systems are inherently nonlinear, with system nonlinearities arising from numerous sources including, for example, friction (e.g., Coulomb, hysteresis), gyroscopic effects (e.g., rotational motion), kinematic effects (e.g., backlash), input constraints (e.g., saturation, deadband), and geometric constraints, system nonlinearities must be accounted for in system analysis and control design. Nonlinear systems, however, can exhibit a very rich dynamical behavior, such as multiple equilibria, limit cycles, bifurcations, jump resonance phenomena, and chaos, which can make general nonlinear system analysis and control notoriously difficult. Lyapunov's results provide a powerful framework for analyzing the stability of nonlinear dynamical systems. Lyapunov-based methods have also been used by control system designers to obtain stabilizing feedback controllers for nonlinear systems. In particular, for smooth feedback, Lyapunov-based methods were inspired by Jurdjevic and Quinn [224], who give sufficient conditions for smooth stabilization based on the ability to construct a Lyapunov function for the closed-loop system. More recently, Artstein [13] introduced the notion of a control Lyapunov function whose existence guarantees a feedback control law which globally stabilizes a nonlinear dynamical system. Even though for certain classes of nonlinear dynamical systems a universal construction of a feedback stabilizer can be obtained using control Lyapunov functions [13, 406], there does not exist a unified procedure for finding a Lyapunov function that will stabilize the closed-loop system for general nonlinear systems. In light of this, advances in Lyapunov-based methods have been developed for analysis and control design for numerous classes of nonlinear dynamical systems. As a consequence, Lyapunov's direct method has become one of the cornerstones of systems and control theory.

The main objective of this book is to present necessary mathematical tools for stability analysis and control design of nonlinear systems, with an emphasis on Lyapunov-based methods. The main contents of the book are as follows. In Chapter 2, we provide a systematic development of nonlinear ordinary differential equations, which is central to the study of nonlinear dynamical system theory. Specifically, we develop qualitative solutions properties, existence of solutions, uniqueness of solutions, continuity of solutions, and continuous dependence of solutions on system initial conditions for nonlinear dynamical systems.

In Chapter 3, we develop stability theory for nonlinear dynamical systems. Specifically, Lyapunov stability theorems are developed for time-invariant nonlinear dynamical systems. Furthermore, invariant set stability theorems, converse Lyapunov theorems, and Lyapunov instability theorems are also considered. Finally, we present several systematic approaches for constructing Lyapunov functions as well as stability of linear systems and Lyapunov's linearization method. Chapter 4 provides an advanced treatment of stability theory including partial stability, stability theory for time-varying systems, Lagrange stability, boundedness, ultimate boundedness, input-to-state stability, finite-time stability, semistability, and stability theorems via vector Lyapunov functions. In addition, Lyapunov and asymptotic stability of sets as well as stability of periodic orbits are also systematically addressed. In particular, local and global stability theorems are given using lower semicontinuous Lyapunov functions. Furthermore, generalized invariant set theorems are derived wherein system trajectories converge to a union of largest invariant sets contained on the boundary of the intersections over finite intervals of the closure of generalized Lyapunov level surfaces. These results provide transparent generalizations to standard Lyapunov and invariant set theorems.

In Chapter 5, using generalized notions of system energy storage and external energy supply, we present a systematic treatment of dissipativity theory [456]. Dissipativity theory provides a fundamental framework for the analysis and design of nonlinear dynamical systems using an input, state, and output system description based on system-energy-related considerations. As a direct application of dissipativity theory, absolute stability theory involving the stability of a feedback system whose forward path contains a dynamic linear time-invariant system and whose feedback path contains a memoryless (possibly time-varying) nonlinearity is also addressed. The Aizerman conjecture and the Luré problem, as well as the circle and Popov criteria, are extensively developed.

Using the concepts of dissipativity theory, in Chapter 6 we present feedback interconnection stability results for nonlinear dynamical systems. General stability criteria are given for Lyapunov, asymptotic, and exponential stability of feedback dynamical systems. Using quadratic supply rates corresponding to net system power and weighted input-output energy, we specialize these results to the classical positivity and small gain theorems. In addition, notions of a control Lyapunov function, feedback linearization, zero dynamics, minimum-phase systems, and stability margins for nonlinear feedback systems are also introduced. Finally, to address optimality issues within nonlinear control-system design we consider an optimal control problem in which a performance function is minimized over all possible closed-loop system trajectories. The value of the performance function is given by a solution to the Hamilton-Jacobi-Bellman equation. In Chapter 7, we provide a brief treatment of input-output stability and dissipativity theory. In particular, we introduce input-output system models as well as $\mathcal{L}_p$ stability. In addition, we develop connections between dissipativity theory of input, state, and output systems, and input-output dissipativity theory.

In Chapter 8, we develop a unified framework to address the problem of optimal nonlinear analysis and feedback control. Asymptotic stability of the closed-loop nonlinear system is guaranteed by means of a Lyapunov function which can clearly be seen to be the solution to the steady-state form of the Hamilton-Jacobi-Bellman equation, and hence guarantees both stability and optimality. The overall framework provides the foundation for extending linear-quadratic controller synthesis to nonlinear-nonquadratic problems. Guaranteed stability margins for nonlinear optimal and inverse optimal regulators that minimize a nonlinear-nonquadratic performance criterion are also established. Using the optimal control framework of Chapter 8, in Chapter 9, we give a unification between nonlinear-nonquadratic optimal control and backstepping control. Backstepping control has received a great deal of attention in the nonlinear control literature [247,395]. The popularity of this control methodology is due in large part to the fact that it provides a systematic procedure for finding a Lyapunov function for nonlinear closed-loop cascade systems.

In Chapter 10, we develop an optimality-based framework to address the problem of nonlinear-nonquadratic control for disturbance rejection of nonlinear systems with bounded exogenous disturbances. Specifically, using dissipativity theory with appropriate storage functions and supply rates we transform the nonlinear disturbance rejection problem into an optimal control problem by modifying a nonlinear-nonquadratic cost functional to account for the exogenous disturbances. As a consequence, the resulting solution to the modified optimal control problem guarantees disturbance rejection for nonlinear systems with bounded input disturbances. Furthermore, it is shown that the Lyapunov function guaranteeing closed-loop stability is a solution to the steady-state Hamilton-Jacobi-Isaacs equation for the controlled system. The overall framework generalizes the Hamilton-Jacobi-Bellman conditions developed in Chapter 8 to address the design of optimal controllers for nonlinear systems with exogenous disturbances.

In Chapter 11, we concentrate on developing a unified framework to address the problem of optimal nonlinear robust control. As in the disturbance rejection problem, we transform the given robust control problem into an optimal control problem by properly modifying the cost functional to account for the system uncertainty. As a consequence, the resulting solution to the modified optimal control problem guarantees robust stability and performance for a class of nonlinear uncertain systems. In Chapter 12, we extend the framework developed in Chapter 11 to nonlinear systems with nonlinear time-invariant real parameter uncertainty. Robust stability of the closed-loop nonlinear system is guaranteed by means of a parameter-dependent Lyapunov function composed of a fixed (parameter-independent) and variable (parameter-dependent) part. The fixed part of the Lyapunov function can be seen to be the solution to the steady-state Hamilton-Jacobi-Bellman equation for the nominal system. Finally, in Chapters 13 and 14 we give a condensed presentation of the continuous-time analysis and control synthesis results developed in Chapters 2–12 for discrete-time systems.

Chapter Two

Dynamical Systems and Differential Equations

2.1 Introduction

In this chapter, we provide a thorough treatment of some of the basic results in nonlinear differential equations. The study of differential equations in dynamical systems and control is of the highest importance since differential equations arise in nearly all disciplines of science, engineering, medicine, economics, biocenology, and demography. In as much as differential equations are central to these fields, their study has additionally led to the development of entire fields of abstract mathematics. In particular, algebraic topology and group theory were developed to solve problems in dynamical system theory, which was originally a branch of differential equations. Fourier analysis was developed for analyzing the heat equation, while topology and set theory were essential developments for the study of convergence problems for differential equations. And of course most recently control theory, whose strength has been its effective use of differential equations, has significantly contributed to the theory of differential equations. The study of differential equations is usually divided into two parts; qualitative theory and quantitative theory. In this chapter, we will focus almost exclusively on qualitative analysis of differential equations. The notions of openness, convergence, continuity, and compactness that we will use throughout this book will refer to the topology generated on the n-dimensional Euclidean space $\mathbb{R}^n$ by the vector norm $\|\cdot\|$.

As discussed in Chapter 1, a nonlinear dynamical system consists of a set of possible states such that the knowledge of these states at some time $t = t_0$, together with the knowledge of an external input to the system for $t \geq t_0$, completely determines the behavior of the dynamical system at any time $t \geq t_0$. Hence, a dynamical system can generate a differential equation of the form

$$\dot{x}(t) = F(t, x(t), u(t)), \quad x(t_0) = x_0, \quad t \in [t_0, t_1], \tag{2.1}$$

where $x(t) \in \mathcal{D}$, $t \in [t_0, t_1]$, is the *state* of the dynamical system, $\mathcal{D}$ is an open subset of $\mathbb{R}^n$ with $0 \in \mathcal{D}$, $u(t) \in U \subseteq \mathbb{R}^m$, $t \in [t_0, t_1]$, is the *input* or *control* signal to the system, and $F : [t_0, t_1] \times \mathcal{D} \times U \to \mathbb{R}^n$ is piecewise continuous in t and continuous in x and u on $[t_0, t_1] \times \mathcal{D} \times U$. The input or control u is assumed to be a piecewise continuous function of time with values in U, that is, $u : [t_0, t_1] \to U$. Here, the time index t runs over the set $\overline{\mathbb{R}}_+ = [0, \infty)$, or sometimes the set $\mathbb{R} = (-\infty, \infty)$ of all reals. Even though for physical dynamical systems $t \in \overline{\mathbb{R}}_+$, on some occasions involving the mathematical analysis of differential equations it becomes necessary to assume that $t \in \mathbb{R}$. In studying dynamical systems of the form (2.1), we can distinguish between system *analysis* and system *synthesis*. Specifically, in analyzing (2.1) we evaluate the behavior of the system for a specified fixed input or control signal. Alternatively, synthesis refers to a design procedure which yields a control input that achieves a desired behavior of the system state. In this book we consider both the analysis and control synthesis problems for nonlinear dynamical systems.

In the remainder of this section we provide several classifications of (2.1) along with some definitions used throughout the book. First, we refer to (2.1) as a nonlinear *disturbed* or *controlled time-varying* dynamical system. Alternatively, in the case where $u(t) \equiv 0$, we let $f(t, x) = F(t, x, 0)$ so that (2.1) becomes

$$\dot{x}(t) = f(t, x(t)), \quad x(t_0) = x_0, \quad t \in [t_0, t_1], \tag{2.2}$$

where $f : [t_0, t_1] \times \mathcal{D} \to \mathbb{R}^n$ is piecewise continuous in t and continuous in x on $[t_0, t_1] \times \mathcal{D}$. In this case, we refer to (2.2) as a nonlinear *undisturbed* or *uncontrolled* time-varying dynamical system. Note that if the input $u(\cdot)$ is *a priori* uniquely specified, then defining $f(t, x) = F(t, x, u(t))$, (2.1) can be written as (2.2). Hence, in this case, the distinction between a disturbed and undisturbed system is not precise since an undisturbed system can describe the case where $u(t) \equiv 0$ or the case when $u(t)$ is specified over $[t_0, t_1]$. Note that if $u : [t_0, t_1] \to U$ is piecewise continuous, then $f(t, x) = F(t, x, u(t))$ is also piecewise continuous in t over $[t_0, t_1]$ for all $x \in \mathcal{D}$. The following definitions provide several classifications of the nonlinear dynamical system (2.2). Analogous classifications also hold for (2.1).

Definition 2.1. Consider the nonlinear dynamical system (2.2). If $f(t, x) = f(t_0, x)$ for all $(t, x) \in [t_0, t_1] \times \mathcal{D}$, then (2.2) is called a *time-invariant* or *autonomous* dynamical system.

Definition 2.2. Consider the nonlinear dynamical system (2.2) with

$f : [t_0, \infty) \times \mathcal{D} \to \mathbb{R}^n$. If there exists $T > 0$ such that $f(t, x) = f(t + T, x)$ for all $(t, x) \in [t_0, \infty) \times \mathcal{D}$, then (2.2) is called a *periodic* dynamical system. Furthermore, the minimum time $T > 0$ for which $f(t, x) = f(t + T, x)$ is called the *period*.

Definition 2.3. Consider the dynamical system (2.2) with $\mathcal{D} = \mathbb{R}^n$. If $f(t, x) = A(t)x$, where $A : [t_0, t_1] \to \mathbb{R}^{n\times n}$ is piecewise continuous on $[t_0, t_1]$ and $x \in \mathbb{R}^n$, then (2.2) is called a *linear time-varying* dynamical system. If, alternatively, for all $t \in [t_0, \infty)$ there exists $T > 0$ such that $A(t) = A(t + T)$, then (2.2) is called a *linear periodic* dynamical system. Finally, if $f(t, x) = Ax$, where $A \in \mathbb{R}^{n\times n}$ and $x \in \mathbb{R}^n$, then (2.2) is called a *linear autonomous* dynamical system.

Next, we introduce the concept of an equilibrium point of a nonlinear dynamical system.

Definition 2.4. Consider the nonlinear dynamical system (2.2) with $f : [t_0, \infty) \times \mathcal{D} \to \mathbb{R}^n$. A point $x_e \in \mathcal{D}$ is said to be an *equilibrium point* of (2.2) at time $t_e \in [t_0, \infty)$ if $f(t, x_e) = 0$ for all $t \geq t_e$.

Note that in the case of autonomous systems and periodic systems, $x_e \in \mathbb{R}^n$ is an equilibrium point at time t_e if and only if x_e is an equilibrium point at all times. Furthermore, if x_e is an equilibrium point at time t_e, then defining $\tau \triangleq t - t_e$ and $\tau_0 \triangleq t_0 - t_e$ yields

$$\dot{x}(\tau + t_e) = f(\tau + t_e, x(\tau + t_e)), \quad x(\tau_0 + t_e) = x_0, \quad \tau \in [\tau_0, \infty), \tag{2.3}$$

where $\dot{x}(\cdot)$ denotes differentiation with respect to τ, and hence x_e is an equilibrium point of (2.3) at $\tau = 0$ since $f(\tau + t_e, x_e) = 0$ for all $\tau \geq 0$. Hence, we assume without loss of generality that $t_e = 0$. Furthermore, if $f(\cdot, \cdot)$ has at least one equilibrium point $x_e \in \mathbb{R}^n$ then, without loss of generality, we can assume that $x_e = 0$ so that $f(t, 0) = 0$, $t \geq 0$. To see this, define $x_s \triangleq x - x_e$ and note that

$$\begin{aligned} \dot{x}_s(t) = \dot{x}(t) = f(t, x(t)) &= f(t, x_s(t) + x_e) \triangleq f_s(t, x_s(t)), \\ & x_s(t_0) = x_0 - x_e, \quad t \in [t_0, \infty). \end{aligned} \tag{2.4}$$

Now, the claim follows by noting that $f_s(t, 0) = f(t, x_e) = 0$, $t \geq t_0$. Similar observations hold for the forced dynamical system (2.1) by noting that a point $x_e \in \mathbb{R}^n$ is an equilibrium point of (2.1) if and only if there exists $u_e \in \mathbb{R}^m$ such that $F(t, x_e, u_e) = 0$ for all $t \geq 0$. Finally, we note that if $x_e \in \mathbb{R}^n$ is an equilibrium point of (2.2) and $x(t_0) = x_e$, then $x(t) = x_e$, $t \geq t_0$, is a solution (see Section 2.7) to (2.2).

Example 2.1. Consider the nonlinear dynamical system describing the motion of a simple pendulum with viscous damping given by

$$\dot{x}_1(t) = x_2(t), \quad x_1(0) = x_{10}, \quad t \geq 0, \tag{2.5}$$
$$\dot{x}_2(t) = -cx_2(t) - \tfrac{g}{l}\sin x_1(t), \quad x_2(0) = x_{20}, \tag{2.6}$$

where $c > 0$ is the viscous damping coefficient, g is the acceleration due to gravity, and l is the length of the pendulum. Note that (2.5) and (2.6) have the form of (2.2) with $f(t,x) = f(x)$. Even though the physical pendulum has two equilibrium points, namely, $(0,0)$ and $(\pi, 0)$, the mathematical model of the pendulum given by (2.5) and (2.6) has countably infinitely many equilibrium points in $\mathbb{R}^2$ given by $f(x_e) = 0$. In particular, $(x_{1e}, x_{2e}) = (n\pi, 0)$, $n = 0, \pm 1, \pm 2, \ldots$. △

Example 2.2. There are dynamical systems that have no equilibrium points. In particular, consider the system

$$\dot{x}_1(t) = \alpha + \sin[x_1(t) + x_2(t)] + x_1(t), \quad x_1(0) = x_{10}, \quad t \geq 0, \tag{2.7}$$
$$\dot{x}_2(t) = \alpha + \sin[x_1(t) + x_2(t)] - x_1(t), \quad x_2(0) = x_{20}, \tag{2.8}$$

where $\alpha > 1$. For this system there does not exist an $x_e = [x_{1e}\ x_{2e}]^{\mathrm{T}}$ such that $f(x_e) = 0$. Hence, (2.7) and (2.8) does not possess any equilibrium points in $\mathbb{R}^2$. △

As seen in Examples 2.1 and 2.2, the nonlinear algebraic equation $f(x) = 0$ may have multiple solutions or no solutions. However, $f(x) = 0$ can also have a continuum of solutions. To see this, let $f(x) = Ax$, where $A \in \mathbb{R}^{n \times n}$, which corresponds to the dynamics of a linear autonomous system. Now, $Ax = 0$ has a unique solution $x_e = 0$ if and only if $\det A \neq 0$. However, if $\det A = 0$, then $x_e \in \mathcal{N}(A) \triangleq \{x \in \mathbb{R}^n : Ax = 0\}$, which corresponds to a continuum of equilibria in $\mathbb{R}^n$. This observation leads to the following definition. For this definition we define the open ball $\mathcal{B}_\varepsilon(x_e) \triangleq \{x \in \mathbb{R}^n : \|x - x_e\| < \varepsilon\}$, where $\|\cdot\|$ denotes a vector norm in $\mathbb{R}^n$ (see Section 2.2).

Definition 2.5. An equilibrium point $x_e \in \mathbb{R}^n$ of (2.2) is said to be an *isolated equilibrium point* if there exists $\varepsilon > 0$ such that $\mathcal{B}_\varepsilon(x_e)$ contains no equilibrium points other than x_e.

The following proposition provides sufficient conditions for the existence of isolated equilibria of the dynamical system (2.2). For this result we assume that $f : [t_0, t_1] \times \mathcal{D} \to \mathbb{R}^n$ is continuously differentiable on $\mathcal{D}$ (see Section 2.3) and we use the notions of vector norms as defined in Section 2.2.

Proposition 2.1. Consider the nonlinear dynamical system (2.2). Assume $f(t, \cdot) : \mathcal{D} \to \mathbb{R}^n$ is continuously differentiable on $\mathcal{D}$ for every $t \in [t_0, t_1]$, $f(t, x_e) = 0$, $t \in [t_0, t_1]$, where $x_e \in \mathbb{R}^n$, and $f(\cdot, x) : [t_0, t_1] \to \mathbb{R}^n$

is piecewise continuous on $[t_0, t_1]$ for every $x \in \mathcal{D}$. If for every $t \in [t_0, t_1]$,

$$A(t) \triangleq \frac{\partial f}{\partial x}(t,x)\bigg|_{x=x_e} \tag{2.9}$$

is nonsingular, then x_e is an isolated equilibrium point.

Proof. Let $t \in [t_0, t_1]$. Since $\det A(t) \neq 0$, there exists $\varepsilon > 0$ such that $\|A(t)x\| \geq \varepsilon\|x\|$, $x \in \mathcal{D}$ (see Problem 2.1). Next, expanding $f(t,x)$ via a Taylor series expansion about the equilibrium point $x = x_e$ and using the fact that $f(t, x_e) = 0$ yields[1]

$$\begin{aligned} f(t,x) &= f(t,x_e) + \frac{\partial f}{\partial x}(t,x_e)(x - x_e) + \mathcal{O}(t,x) \\ &= A(t)(x - x_e) + \mathcal{O}(t,x). \end{aligned} \tag{2.10}$$

Now, since $\frac{\partial f}{\partial x}(t,x)$ is continuous on $\mathcal{D}$ it follows that $\lim_{\|x-x_e\|\to 0} \frac{\|\mathcal{O}(t,x)\|}{\|x-x_e\|} = 0$. Hence, for every $\hat{\varepsilon} > 0$ there exists $\delta > 0$ such that

$$\|\mathcal{O}(t,x)\| \leq \hat{\varepsilon}\|x - x_e\|, \quad \|x - x_e\| < \delta. \tag{2.11}$$

Setting $\hat{\varepsilon} = \varepsilon/2$, using (2.11), and using the fact that $\|A(t)x\| \geq \varepsilon\|x\|$, $x \in \mathcal{D}$, it follows that

$$\begin{aligned} \|f(t,x)\| &= \|A(t)(x - x_e) + \mathcal{O}(t,x)\| \\ &\geq \|A(t)(x - x_e)\| - \|\mathcal{O}(t,x)\| \\ &\geq \tfrac{\varepsilon}{2}\|x - x_e\|, \quad \|x - x_e\| < \delta, \end{aligned} \tag{2.12}$$

which implies that $\|f(t,x)\| > 0$ for all $x \in \mathcal{B}_\delta(x_e)$, $x \neq x_e$. □

Example 2.3. Consider the nonlinear dynamical system

$$\dot{x}_1(t) = -x_1(t) + 2x_1(t)x_2(t), \quad x_1(0) = x_{10}, \quad t \geq 0, \tag{2.13}$$

$$\dot{x}_2(t) = -2x_1(t)x_2(t), \quad x_2(0) = x_{20}. \tag{2.14}$$

Note that with $x = [x_1, x_2]^{\mathrm{T}}$, $f(x) = 0$ implies that $x_e = [x_{1e}\ x_{2e}]^{\mathrm{T}} = [0\ x_{2e}]^{\mathrm{T}}$, where $x_{2e} \in \mathbb{R}$, which shows that every point on the x_2 axis is an equilibrium point of (2.13) and (2.14). △

In the remainder of this chapter we address the problems of existence and uniqueness of solutions to the nonlinear dynamical system (2.2) along with continuous dependence of solutions on the system initial conditions and system parameters. First, however, we review some basic definitions and results on vector and matrix norms, topology and analysis, and vector and Banach spaces.

---

[1] $\mathcal{O}(t,x)$ denotes the *Landau order*, which means that $\|\mathcal{O}(t,x)\|/\|x - x_e\|$ is bounded as $\|x - x_e\| \to 0$.

2.2 Vector and Matrix Norms

In this section the concepts of vector and matrix norms are introduced. Given a number $\alpha \in \mathbb{R}$, the absolute value $|\alpha|$ of α denotes the magnitude of α while discarding the sign of α. For $x \in \mathbb{R}^n$ and $A \in \mathbb{R}^{n\times m}$ we can extend the definition of absolute value by replacing each component of x and each entry of A by its absolute value, that is, by defining $|x| \in \mathbb{R}^n$ and $|A| \in \mathbb{R}^{n\times m}$ by

$$|x|_{(i)} \triangleq |x_{(i)}|, \quad |A|_{(i,j)} \triangleq |A_{(i,j)}|$$

for all $i = 1, \ldots, n$ and $j = 1, \ldots, m$. For many applications it is more useful to have a scalar measure of the magnitude of x or A. *Vector* and *matrix norms* provide such measures.

Definition 2.6. A *vector norm* $\|\cdot\|$ on $\mathbb{R}^n$ is a function $\|\cdot\| : \mathbb{R}^n \to \mathbb{R}$ that satisfies the following axioms:

i) $\|x\| \geq 0$, $x \in \mathbb{R}^n$.

ii) $\|x\| = 0$ if and only if $x = 0$.

iii) $\|\alpha x\| = |\alpha|\|x\|$, $\alpha \in \mathbb{R}$, $x \in \mathbb{R}^n$.

iv) $\|x+y\| \leq \|x\| + \|y\|$, $x, y \in \mathbb{R}^n$.

Condition *iv)* is known as the *triangle inequality.* It is important to note that if there exists $\alpha \geq 0$ such that $x = \alpha y$ or $y = \alpha x$, then *iv)* holds as an equality. There are many different norms. The most useful class of vector norms are the *p-norms* (also called *Hölder norms*).

Proposition 2.2. For $1 \leq p \leq \infty$, $\|\cdot\|_p$ defined by

$$\|x\|_p \triangleq \left[\sum_{i=1}^{n} |x_{(i)}|^p\right]^{1/p}, \quad 1 \leq p < \infty,$$

$$\|x\|_\infty \triangleq \max_{i=1,\ldots,n} |x_{(i)}|,$$

is a vector norm on $\mathbb{R}^n$.

Proof. The only difficulty is in proving the triangle inequality, which in this case is known as Minkowski's inequality. If $\|x+y\|_p = 0$, then obviously $\|x+y\|_p \leq \|x\|_p + \|y\|_p$ since $\|x\|_p$ and $\|y\|_p$ are nonnegative for $1 \leq p \leq \infty$. Suppose $\|x+y\|_p > 0$. The cases $p = 1$ and $p = \infty$ are immediate. Hence, assume $1 < p < \infty$ and note that

$$\sum_{i=1}^{n} |x_{(i)} + y_{(i)}|^p \leq \sum_{i=1}^{n} |x_{(i)} + y_{(i)}|^{p-1}|x_{(i)}| + \sum_{i=1}^{n} |x_{(i)} + y_{(i)}|^{p-1}|y_{(i)}|. \quad (2.15)$$

Now, letting $q = p/(p-1)$ and applying Hölder's inequality (2.18) to each of the summations on the right-hand side of (2.15) yields

$$\begin{aligned}\sum_{i=1}^{n} |x_{(i)} + y_{(i)}|^p &\\ &\le \left(\sum_{i=1}^{n} |x_{(i)} + y_{(i)}|^{(p-1)q}\right)^{1/q} \left[\left(\sum_{i=1}^{n} |x_{(i)}|^p\right)^{1/p} + \left(\sum_{i=1}^{n} |y_{(i)}|^p\right)^{1/p}\right] \\ &= \left(\sum_{i=1}^{n} |x_{(i)} + y_{(i)}|^p\right)^{1/q} \left[\left(\sum_{i=1}^{n} |x_{(i)}|^p\right)^{1/p} + \left(\sum_{i=1}^{n} |y_{(i)}|^p\right)^{1/p}\right]. \quad (2.16)\end{aligned}$$

Next, dividing both sides of (2.16) by $\left(\sum_{i=1}^{n} |x_{(i)} + y_{(i)}|^p\right)^{1/q}$ and using the fact that $1 - 1/q = 1/p$ we obtain

$$\left(\sum_{i=1}^{n} |x_{(i)} + y_{(i)}|^p\right)^{1/p} \le \left(\sum_{i=1}^{n} |x_{(i)}|^p\right)^{1/p} + \left(\sum_{i=1}^{n} |y_{(i)}|^p\right)^{1/p}, \quad (2.17)$$

which proves the result. □

The notation $\|\cdot\|_p$ does not explicitly refer to the dimension of $\mathbb{R}^n$ to which it is applied. We assume this is implied by context. In the case where $p = 1$,

$$\|x\|_1 = \sum_{i=1}^{n} |x_{(i)}|$$

is called the *absolute sum* norm; the case $p = 2$,

$$\|x\|_2 = \left[\sum_{i=1}^{n} x_{(i)}^2\right]^{1/2} = (x^{\mathrm{T}}x)^{1/2}$$

is called the *Euclidean* norm; and the case $p = \infty$,

$$\|x\|_\infty = \max_{i=1,\ldots,n} |x_{(i)}|$$

is called the *infinity* norm.

The next result is known as Hölder's inequality. For this result (and henceforth) we interpret $1/\infty = 0$.

Proposition 2.3. Let $x, y \in \mathbb{R}^n$, and let $1 \le p \le \infty$ and $1 \le q \le \infty$ satisfy $1/p + 1/q = 1$. Then

$$|x^{\mathrm{T}}y| \le \|x\|_p \|y\|_q. \quad (2.18)$$

Proof. The cases $p = 1, \infty$ and $q = \infty, 1$ are straightforward and are left as an exercise for the reader. Assume $1 < p < \infty$ so that $1 < q < \infty$. For $u \geq 0$, $v \geq 0$, and $\alpha \in (0,1)$ note that

$$u^\alpha v^{(1-\alpha)} \leq \alpha u + (1-\alpha)v, \tag{2.19}$$

with equality holding in (2.19) if and only if $u = v$. To see this, let $f : \overline{\mathbb{R}}_+ \to \mathbb{R}$ be defined by $f(z) \triangleq z^\alpha - \alpha z + \alpha - 1$ and note that $f'(z) = \alpha(z^{(\alpha-1)} - 1)$. Since $\alpha \in (0,1)$ it follows that $f'(z) > 0$ for all $z \in (0,1)$ and $f'(z) < 0$ for all $z > 1$. Now, it follows that for $z \geq 0$, $f(z) \leq f(1) = 0$ with equality holding for $z = 1$. Hence, $z^\alpha \leq \alpha z + 1 - \alpha$ with equality holding for $z = 1$. Taking $z = u/v$, $v \neq 0$, yields (2.19), while for $v = 0$ (2.19) is trivially satisfied.

Now, for each $i \in \{1, \ldots, n\}$, $\alpha = 1/p$, and $1 - \alpha = 1/q$, on setting

$$u = \left(\frac{|x_{(i)}|}{\|x\|_p}\right)^p, \quad v = \left(\frac{|y_{(i)}|}{\|y\|_q}\right)^q, \tag{2.20}$$

(2.19) yields

$$\frac{|x_{(i)}y_{(i)}|}{\|x\|_p\|y\|_q} \leq \frac{1}{p}\left(\frac{|x_{(i)}|}{\|x\|_p}\right)^p + \frac{1}{q}\left(\frac{|y_{(i)}|}{\|y\|_q}\right)^q. \tag{2.21}$$

Summing over $[1, n]$ yields

$$\frac{1}{\|x\|_p\|y\|_q}\sum_{i=1}^n |x_{(i)}y_{(i)}| \leq \frac{1}{p} + \frac{1}{q} = 1, \tag{2.22}$$

which proves (2.18). □

The case $p = q = 2$ is known as the *Cauchy-Schwarz inequality*. Since this result is important, we state it as a corollary.

Corollary 2.1. Let $x, y \in \mathbb{R}^n$. Then

$$|x^\mathrm{T} y| \leq \|x\|_2\|y\|_2. \tag{2.23}$$

Proof. Note that for $x = 0$ or $y = 0$ the inequality is immediate. Next, note that for $\lambda \in \mathbb{R}$,

$$\begin{aligned}(x - \lambda y)^\mathrm{T}(x - \lambda y) &= x^\mathrm{T} x - 2\lambda x^\mathrm{T} y + \lambda^2 y^\mathrm{T} y \\ &= \|x\|_2^2 - 2\lambda x^\mathrm{T} y + \lambda^2\|y\|_2^2 \\ &\geq 0, \qquad x, y \in \mathbb{R}^n. \end{aligned} \tag{2.24}$$

Now, taking $\lambda = x^\mathrm{T} y/\|y\|_2^2$ yields the result. □

Two norms $\|\cdot\|$ and $\|\cdot\|'$ on $\mathbb{R}^n$ are *equivalent* if there exist positive constants α, β such that

$$\alpha\|x\| \leq \|x\|' \leq \beta\|x\|, \tag{2.25}$$

for all $x \in \mathbb{R}^n$. Note that (2.25) can be written as

$$\frac{1}{\beta}\|x\|' \leq \|x\| \leq \frac{1}{\alpha}\|x\|'.$$

Hence, the word "equivalent" is justified. The next result shows that all norms are equivalent in finite-dimensional spaces.

Theorem 2.1. If $\|\cdot\|$ and $\|\cdot\|'$ are vector norms on $\mathbb{R}^n$, then $\|\cdot\|$ and $\|\cdot\|'$ are equivalent.

Proof. Let $x, y \in \mathbb{R}^n$ and note that

$$\begin{aligned}\|x-y\| &= \left\|\sum_{i=1}^n (x_i - y_i)e_i\right\| \\ &\leq \sum_{i=1}^n |x_i - y_i|\|e_i\| \\ &\leq C\|x-y\|_\infty,\end{aligned}$$

where $C \triangleq \max_{i=1,\ldots,n}\|e_i\|$. Similarly, it can be shown that $\|x-y\|' \leq C'\|x-y\|_\infty$, where $C' \triangleq \max_{i=1,\ldots,n}\|e_i\|'$. Hence, it follows that $\|\cdot\|$ and $\|\cdot\|'$ are continuous on $\mathbb{R}^n$ with respect to $\|\cdot\|_\infty$.

Next, since $\{x \in \mathbb{R}^n : \|x\|_\infty = 1\}$ is a compact set (see Definition 2.14) it follows from Weierstrass' theorem (Theorem 2.13) that there exists $\alpha, \beta > 0$ such that $\alpha = \min_{\{x\in\mathbb{R}^n:\|x\|_\infty=1\}}\|x\|$ and $\beta = \max_{\{x\in\mathbb{R}^n:\|x\|_\infty=1\}}\|x\|$. Now it follows that

$$\alpha = \min_{\{x\in\mathbb{R}^n:\|x\|_\infty=1\}}\|x\| = \min_{\{x\in\mathbb{R}^n:\|x\|_\infty\neq 0\}}\frac{\|x\|}{\|x\|_\infty} \leq \frac{\|x\|}{\|x\|_\infty}, \quad x \neq 0, \tag{2.26}$$

and

$$\beta = \max_{\{x\in\mathbb{R}^n:\|x\|_\infty=1\}}\|x\| = \max_{\{x\in\mathbb{R}^n:\|x\|_\infty\neq 0\}}\frac{\|x\|}{\|x\|_\infty} \geq \frac{\|x\|}{\|x\|_\infty}, \quad x \neq 0, \tag{2.27}$$

which implies that

$$\alpha\|x\|_\infty \leq \|x\| \leq \beta\|x\|_\infty, \quad x \in \mathbb{R}^n. \tag{2.28}$$

Similarly, it can be shown that

$$\alpha'\|x\|_\infty \leq \|x\|' \leq \beta'\|x\|_\infty, \quad x \in \mathbb{R}^n, \tag{2.29}$$

where $\alpha' = \min_{\{x\in\mathbb{R}^n:\|x\|_\infty=1\}}\|x\|'$ and $\beta' = \max_{\{x\in\mathbb{R}^n:\|x\|_\infty=1\}}\|x\|'$. Now it follows from (2.28) and (2.29) that

$$\frac{\alpha}{\beta'}\|x\|' \leq \|x\| \leq \frac{\beta}{\alpha'}\|x\|', \quad x \in \mathbb{R}^n, \tag{2.30}$$

which proves the equivalence of $\|\cdot\|$ and $\|\cdot\|'$. □

The notion of vector norms allows us to define a *matrix* norm by viewing a matrix $A \in \mathbb{R}^{n\times m}$ as a vector in $\mathbb{R}^{nm}$, for example, as vec A, where vec$(\cdot)$ denotes the column stacking operator.

Definition 2.7. A *matrix norm* $\|\cdot\|$ on $\mathbb{R}^{n\times m}$ is a function $\|\cdot\| : \mathbb{R}^{n\times m} \to \mathbb{R}$ that satisfies the following axioms:

i) $\|A\| \geq 0$, $A \in \mathbb{R}^{n\times m}$.

ii) $\|A\| = 0$ if and only if $A = 0$.

iii) $\|\alpha A\| = |\alpha|\|A\|$, $\alpha \in \mathbb{R}$.

iv) $\|A + B\| \leq \|A\| + \|B\|$, $A, B \in \mathbb{R}^{n\times m}$.

If $\|\cdot\|$ is a vector norm on $\mathbb{R}^{nm}$, then $\|\cdot\|'$ defined by $\|A\|' = \|\text{vec } A\|$ is a matrix norm on $\mathbb{R}^{n\times m}$. For example, the p-norms can be extended to matrices in this way. Hence, for $A \in \mathbb{R}^{n\times m}$ define

$$\|A\|_p \triangleq \left[\sum_{i=1}^{n}\sum_{j=1}^{m} |A_{(i,j)}|^p\right]^{1/p}, \quad 1 \leq p < \infty,$$

$$\|A\|_\infty \triangleq \max_{i=1,\ldots,n,j=1,\ldots,m} |A_{(i,j)}|.$$

Note that we use the same symbol $\|\cdot\|_p$ to denote the p-norm for both vectors and matrices. This notation is consistent since if $A \in \mathbb{R}^{n\times 1}$, then $\|A\|_p$ coincides with the vector p-norm. Furthermore, if $A \in \mathbb{R}^{n\times m}$, then

$$\|A\|_p = \|\text{vec } A\|_p. \tag{2.31}$$

The matrix p-norms in the cases $p = 1, 2, \infty$ are the most commonly used. In particular, when $p = 2$ we write $\|A\|_{\mathrm{F}} \triangleq \|A\|_2$, where $\|\cdot\|_{\mathrm{F}}$ is called the *Frobenius* norm. Since $\|A\|_2 = \|\text{vec } A\|_2$, we have

$$\|A\|_{\mathrm{F}} = \|A\|_2 = \|\text{vec } A\|_2 = \|\text{vec } A\|_{\mathrm{F}}. \tag{2.32}$$

It is easy to see that $\|A\|_{\mathrm{F}} = (\text{tr } AA^{\mathrm{T}})^{1/2}$ for all $A \in \mathbb{R}^{n\times m}$.

Let $\|\cdot\|$, $\|\cdot\|'$, and $\|\cdot\|''$ denote matrix norms on $\mathbb{R}^{n\times l}$, $\mathbb{R}^{n\times m}$, and $\mathbb{R}^{m\times l}$, respectively. We say $(\|\cdot\|, \|\cdot\|', \|\cdot\|'')$ is a *submultiplicative triple* of matrix norms with constant $\delta > 0$ if

$$\|AB\| \leq \delta\|A\|'\|B\|'', \tag{2.33}$$

for all $A \in \mathbb{R}^{n\times m}$ and $B \in \mathbb{R}^{m\times l}$. If $\delta = 1$, then we may omit the phrase "with constant 1." If $\|\cdot\|$ is a norm on $\mathbb{R}^{n\times n}$ and

$$\|AB\| \leq \|A\|\|B\|, \tag{2.34}$$

for all $A, B \in \mathbb{R}^{n\times n}$, then we say that $\|\cdot\|$ is *submultiplicative*. That is, a matrix norm $\|\cdot\|$ on $\mathbb{R}^{n\times n}$ is submultiplicative if $(\|\cdot\|, \|\cdot\|, \|\cdot\|)$ is a submultiplicative triple of matrix norms with constant 1. We see in (2.34) with $n = 2$, for example, that $\|\cdot\|_p$, $1 \le p \le 2$, is submultiplicative. If a submultiplicative matrix norm $\|\cdot\|$ on $\mathbb{R}^{n\times n}$ satisfies $\|I_n\| = 1$, then we say $\|\cdot\|$ is *normalized*.

A special case of submultiplicative triples is the case in which B is a vector, that is, $l = 1$. Hence, for $A \in \mathbb{R}^{n\times m}$ and $x \in \mathbb{R}^m$ consider (setting $\delta = 1$)

$$\|Ax\| \le \|A\|'\|x\|'', \tag{2.35}$$

where $\|\cdot\|$, $\|\cdot\|'$, and $\|\cdot\|''$ are norms on $\mathbb{R}^n$, $\mathbb{R}^{n\times m}$, and $\mathbb{R}^m$, respectively. If $n = m$ and $\|\cdot\| = \|\cdot\|''$, that is,

$$\|Ax\| \le \|A\|'\|x\|,$$

then we say that $\|\cdot\|$ and $\|\cdot\|'$ are *compatible*.

We now consider a very special class of matrix norms, namely, the *induced matrix norms* wherein (2.35) holds as an equality.

Definition 2.8. Let $\|\cdot\|$ and $\|\cdot\|'$ be vector norms on $\mathbb{R}^m$ and $\mathbb{R}^n$, respectively. Then the function $\|\cdot\|'' : \mathbb{R}^{n\times m} \to \mathbb{R}$ defined by

$$\|A\|'' = \max_{\{x\in\mathbb{R}^m:\ x\neq 0\}} \frac{\|Ax\|'}{\|x\|} \tag{2.36}$$

is called an *induced matrix norm*.

It can be shown that (2.36) is equivalent to each of the expressions

$$\|A\|'' = \max_{\{x\in\mathbb{R}^m:\ \|x\|\le 1\}} \|Ax\|', \tag{2.37}$$

$$\|A\|'' = \max_{\{x\in\mathbb{R}^m:\ \|x\|=1\}} \|Ax\|'. \tag{2.38}$$

In this case, we say $\|\cdot\|''$ is *induced by* $\|\cdot\|$ and $\|\cdot\|'$. If $m = n$ and $\|\cdot\| = \|\cdot\|'$, then we say that $\|\cdot\|''$ is induced by $\|\cdot\|$ and we call $\|\cdot\|''$ an *equi-induced norm*. It remains to be shown, however, that $\|\cdot\|''$ defined by (2.36) is indeed a matrix norm.

Theorem 2.2. Every induced matrix norm is a matrix norm.

Proof. Let $\|\cdot\|$ and $\|\cdot\|'$ be vector norms on $\mathbb{R}^m$ and $\mathbb{R}^n$, respectively, and let $\|\cdot\|'' : \mathbb{R}^{n\times m} \to \mathbb{R}$ be defined as in (2.36) or, equivalently, (2.38). Clearly, $\|\cdot\|''$ satisfies Axioms *i*)–*iii*) of Definition 2.7. To show that Axiom *iv*) holds, let $A, B \in \mathbb{R}^{n\times m}$ and note that

$$\begin{aligned}
\|A+B\|'' &= \max_{\{x\in\mathbb{R}^m:\ \|x\|=1\}} \|(A+B)x\|' \\
&= \max_{\{x\in\mathbb{R}^m:\ \|x\|=1\}} \|Ax+Bx\|' \\
&\le \max_{\{x\in\mathbb{R}^m:\ \|x\|=1\}} (\|Ax\|'+\|Bx\|') \\
&\le \max_{\{x\in\mathbb{R}^m:\ \|x\|=1\}} \|Ax\|' + \max_{\{x\in\mathbb{R}^m:\ \|x\|=1\}} \|Bx\|' \\
&= \|A\|''+\|B\|''.
\end{aligned}$$

Hence, $\|\cdot\|'' : \mathbb{R}^{n\times m} \to \mathbb{R}$ is a matrix norm on $\mathbb{R}^{n\times m}$. □

In the notation of (2.36),

$$\|A\|'' = \max_{\{x\in\mathbb{R}^m:\ x\ne 0\}} \frac{\|Ax\|'}{\|x\|} \ge \frac{\|Ax\|'}{\|x\|}, \quad x \ne 0.$$

Hence,

$$\|Ax\|' \le \|A\|''\|x\|.$$

In this case, $(\|\cdot\|', \|\cdot\|'', \|\cdot\|)$ is a submultiplicative triple. If $m = n$ and $\|\cdot\| = \|\cdot\|'$, then the induced norm $\|\cdot\|''$ is compatible with $\|\cdot\|$. The next result shows that submultiplicative triples of matrix norms can be obtained from induced matrix norms.

Proposition 2.4. Let $\|\cdot\|$, $\|\cdot\|'$, and $\|\cdot\|''$ denote vector norms on $\mathbb{R}^l$, $\mathbb{R}^m$, and $\mathbb{R}^n$, respectively. Let $\|\cdot\|'''$ be the matrix norm on $\mathbb{R}^{m\times l}$ induced by $\|\cdot\|$ and $\|\cdot\|'$, let $\|\cdot\|''''$ be the matrix norm on $\mathbb{R}^{n\times m}$ induced by $\|\cdot\|'$ and $\|\cdot\|''$, and let $\|\cdot\|'''''$ be the matrix norm on $\mathbb{R}^{n\times l}$ induced by $\|\cdot\|$ and $\|\cdot\|''$. If $A \in \mathbb{R}^{n\times m}$ and $B \in \mathbb{R}^{m\times l}$, then

$$\|AB\|''''' \le \|A\|''''\|B\|'''. \tag{2.39}$$

Proof. Note that for all $x \in \mathbb{R}^l$ and $y \in \mathbb{R}^m$, $\|Bx\|' \le \|B\|'''\|x\|$ and $\|Ay\|'' \le \|A\|''''\|y\|'$. Hence, for all $x \in \mathbb{R}^l$,

$$\|ABx\|'' \le \|A\|''''\|Bx\|' \le \|A\|''''\|B\|'''\|x\|, \tag{2.40}$$

which implies (2.39). □

The following corollary to Proposition 2.4 is immediate.

Corollary 2.2. Every equi-induced matrix norm is a normalized submultiplicative matrix norm.

2.3 Set Theory and Topology

The notion of a norm on $\mathbb{R}^n$ discussed in the previous section provides the basis for the development of some basic results in topology and analysis. In this section, we restrict our attention to the Euclidean space $\mathbb{R}^n$. If $\mathcal{X}$ is a *set*, then $x \in \mathcal{X}$ denotes that x is an *element* of $\mathcal{X}$. The notation $x \notin \mathcal{X}$ denotes that x is not an element of $\mathcal{X}$. A set may be specified by listing its elements such as $\mathcal{S} = \{x_1, x_2, x_3\}$ or, alternatively, a set may be specified as consisting of all elements of set $\mathcal{X}$ having a certain property P. In particular, $\mathcal{S} = \{x \in \mathcal{X} : P(x)\}$. The *union* of two sets is denoted by $\mathcal{X} \cup \mathcal{Y}$ and consists of those elements that are in either $\mathcal{X}$ or $\mathcal{Y}$. Hence,

$$\mathcal{X} \cup \mathcal{Y} \triangleq \{x : x \in \mathcal{X} \text{ or } x \in \mathcal{Y}\} = \mathcal{Y} \cup \mathcal{X}.$$

The *intersection* of two sets $\mathcal{X}$ and $\mathcal{Y}$ is denoted by $\mathcal{X} \cap \mathcal{Y}$ and consists of those elements that are common to both $\mathcal{X}$ and $\mathcal{Y}$. Hence,

$$\begin{aligned} \mathcal{X} \cap \mathcal{Y} &= \{x : x \in \mathcal{X},\ x \in \mathcal{Y}\} \\ &= \{x \in \mathcal{X},\ x \in \mathcal{Y}\} \\ &= \{x \in \mathcal{Y},\ x \in \mathcal{X}\} \\ &= \mathcal{Y} \cap \mathcal{X}. \end{aligned}$$

The *complement* of a set $\mathcal{X}$ relative to another set $\mathcal{Y}$ is defined by $\mathcal{Y} \setminus \mathcal{X} \triangleq \{x \in \mathcal{Y} : x \notin \mathcal{X}\}$, and consists of those elements of $\mathcal{Y}$ that are not in $\mathcal{X}$. When it is clear that the complement is with respect to a *universal* set $\mathcal{Y}$, that is, a fixed set from which we take elements and subsets, then we write $\mathcal{X}^\sim \triangleq \mathcal{Y} \setminus \mathcal{X}$. If $x \in \mathcal{X}$ implies $x \in \mathcal{Y}$, then $\mathcal{X}$ is *contained* in $\mathcal{Y}$ or, equivalently, $\mathcal{X}$ is a subset of $\mathcal{Y}$. In this case, we write $\mathcal{X} \subseteq \mathcal{Y}$. Clearly, $\mathcal{X} = \mathcal{Y}$ is equivalent to the validity of both $\mathcal{X} \subseteq \mathcal{Y}$ and $\mathcal{Y} \subseteq \mathcal{X}$. If $\mathcal{X} \subseteq \mathcal{Y}$ and $\mathcal{X} \neq \mathcal{Y}$, then $\mathcal{X}$ is said to be a proper subset of $\mathcal{Y}$, and is written as $\mathcal{X} \subset \mathcal{Y}$. The set with no elements is called the *empty set* and is denoted by $\emptyset$. Two sets $\mathcal{X}$ and $\mathcal{Y}$ are *disjoint* if their intersection is empty, that is, $\mathcal{X} \cap \mathcal{Y} = \emptyset$. Finally, a collection of pairwise disjoint subsets of a set $\mathcal{X}$ whose union is equal to $\mathcal{X}$ is called a *partition* of $\mathcal{X}$.

Example 2.4. Let $\mathcal{X} = (0,3)$ and $\mathcal{Y} = [2,4)$. Then $\mathcal{X} \setminus \mathcal{Y} = (0,2)$, $\mathcal{X} \cup \mathcal{Y} = (0,4)$, and $\mathcal{X} \cap \mathcal{Y} = [2,3)$. With $\mathbb{R}$ as the universal set, $\mathcal{Y}^\sim = (-\infty, 2) \cup [4, \infty)$. Note that $\mathcal{X} \cap \mathcal{Y}^\sim = (0,3) \cap ((-\infty,2) \cup [4,\infty)) = (0,2) = \mathcal{X} \setminus \mathcal{Y}$. △

The *Cartesian product* $\mathcal{X}_1 \times \mathcal{X}_2 \times \cdots \times \mathcal{X}_n$ of n sets $\mathcal{X}_1, \ldots, \mathcal{X}_n$ is the set consisting of *ordered* elements of the form $(x_1, x_2, \ldots, x_n)$, where $x_i \in \mathcal{X}_i$, $i = 1, \ldots, n$. Hence,

$$\mathcal{X}_1 \times \cdots \times \mathcal{X}_n = \{(x_1, \ldots, x_n) : x_1 \in \mathcal{X}_1, \ldots, x_n \in \mathcal{X}_n\}.$$

We call $x_i \in \mathcal{X}_i$ the ith *element of the ordered set* $\mathcal{X}_1 \times \cdots \times \mathcal{X}_n$. The definition of the union, intersection, and product of sets can be extended from two sets (or n sets in the case of products) to a finite or infinite collection of sets $\{\mathcal{S}_i : i \in \mathcal{I}\}$, where i runs over some set $\mathcal{I}$ of indices. We call $\{\mathcal{S}_i : i \in \mathcal{I}\}$ an *indexed family of sets*. The definition of the union and intersection of a collection of sets is

$$\bigcup_{i \in \mathcal{I}} \mathcal{S}_i \triangleq \{x : x \in \mathcal{S}_i \text{ for at least one } i \in \mathcal{I}\},$$

$$\bigcap_{i \in \mathcal{I}} \mathcal{S}_i \triangleq \{x : x \in \mathcal{S}_i \text{ for every } i \in \mathcal{I}\}.$$

If the collection is finite, the set of indices $\mathcal{I}$ is usually $\{1, \ldots, n\}$, and if the collection is countably infinite $\mathcal{I}$ is usually $\mathbb{Z}_+$, although there are occasional exceptions to this.

To begin let $\|\cdot\|$ be a norm on $\mathbb{R}^n$ and define the *open ball* $\mathcal{B}_\varepsilon(x) \subset \mathbb{R}^n$ with *radius* ε and *center* x by $\mathcal{B}_\varepsilon(x) \triangleq \{y \in \mathbb{R}^n : \|x - y\| < \varepsilon\}$. It is important to note that $\mathcal{B}_\varepsilon(x)$ is not necessarily an open sphere or an open ball; its shape will depend on the norm chosen. For example, if $x \in \mathbb{R}^2$ and $\|\cdot\| = \|\cdot\|_\infty$, then $\mathcal{B}_1(0)$, for which the point x is the origin (0,0), is the interior of the square $\max\{|y_1|, |y_2|\} < 1$.

Definition 2.9. Let $\mathcal{S} \subseteq \mathbb{R}^n$. Then $x \in \mathcal{S}$ is an *interior point* of $\mathcal{S}$ if there exists $\varepsilon > 0$ such that $\mathcal{B}_\varepsilon(x) \subseteq \mathcal{S}$. The *interior* of $\mathcal{S}$ is the set

$$\overset{\circ}{\mathcal{S}} \triangleq \{x \in \mathcal{S} : x \text{ is an interior point of } \mathcal{S}\}.$$

Finally, the set $\mathcal{S}$ is *open* if every element of $\mathcal{S}$ is an interior point, that is, if $\mathcal{S} = \overset{\circ}{\mathcal{S}}$.

Note that $\overset{\circ}{\mathcal{S}} \subseteq \mathcal{S}$ and every open ball is an open set.

Example 2.5. Let $\mathcal{S} \triangleq \{(x_1, x_2) \in \mathbb{R}^2 : x_1^2 + x_2^2 < 1\}$. $\mathcal{S}$ is an open set in $\mathbb{R}^2$ since given any $x \in \mathcal{S}$, there exists $\varepsilon > 0$ such that $\mathcal{B}_\varepsilon(0) \subseteq \mathcal{S}$. $\triangle$

Definition 2.10. Let $x \in \mathbb{R}^n$. A *neighborhood* of x is an open subset $\mathcal{N} \subseteq \mathbb{R}^n$ containing x.

In certain parts of the book we denote a neighborhood of $x \in \mathbb{R}^n$ by $\mathcal{N}_\varepsilon(x)$ to refer to $\mathcal{B}_\varepsilon(x)$ (or an ε-neighborhood of x).

Definition 2.11. Let $\mathcal{S} \subseteq \mathbb{R}^n$. A vector $x \in \mathbb{R}^n$ is a *closure point* of $\mathcal{S}$ if, for every $\varepsilon > 0$, the set $\mathcal{S} \cap \mathcal{B}_\varepsilon(x)$ is not empty. A closure point of $\mathcal{S}$ is an *isolated point* of $\mathcal{S}$ if there exists $\varepsilon > 0$ such that $\mathcal{S} \cap \mathcal{B}_\varepsilon(x) = \{x\}$. The *closure* of $\mathcal{S}$ is the set

$$\overline{\mathcal{S}} \triangleq \{x \in \mathbb{R}^n : x \text{ is a closure point of } \mathcal{S}\}.$$

The set $\mathcal{S}$ is *closed* if every closure point of $\mathcal{S}$ is an element of $\mathcal{S}$, that is, $\mathcal{S} = \overline{\mathcal{S}}$. Finally, the *boundary* of $\mathcal{S}$ is the set $\partial\mathcal{S} \triangleq \overline{\mathcal{S}} \setminus \overset{\circ}{\mathcal{S}} = \overline{\mathcal{S}} \cap \overline{(\mathbb{R}^n \setminus \mathcal{S})}$.

A closure point is sometimes referred to as an *adherent point* or *contact point* in the literature. Note that $\overset{\circ}{\mathcal{S}} \subseteq \mathcal{S} \subseteq \overline{\mathcal{S}}$. A closure point should be carefully distinguished from an accumulation point.

Definition 2.12. Let $\mathcal{S} \subseteq \mathbb{R}^n$. A vector $x \in \mathbb{R}^n$ is an *accumulation point* of $\mathcal{S}$ if, for every $\varepsilon > 0$, the set $\mathcal{S} \cap (\mathcal{B}_\varepsilon(x) \setminus \{x\})$ is not empty. The set of all accumulation points of $\mathcal{S}$ is called a *derived set* and is denoted by $\mathcal{S}'$.

Note that if $\mathcal{S} \subset \mathbb{R}^n$, then the vector $x \in \mathbb{R}^n$ is an accumulation point of $\mathcal{S}$ if and only if every open ball centered at x contains a point of $\mathcal{S}$ distinct from x or, equivalently, every open ball centered at x contains an infinite number of points of $\mathcal{S}$. Note that if x is an accumulation point of $\mathcal{S}$, then x is a closure point of $\mathcal{S}$. Clearly, for every $\mathcal{S} \subset \mathbb{R}^n$, $\overline{\mathcal{S}} = \mathcal{S} \cup \mathcal{S}'$, and hence $\mathcal{S}$ is closed if and only if $\mathcal{S}' \subseteq \mathcal{S}$. Thus, a set $\mathcal{S} \subseteq \mathbb{R}^n$ is closed if and only if it contains all of its accumulation points. An accumulation point is sometimes referred to as a *cluster point* or *limit point* in the literature.

Example 2.6. Let $\mathcal{S} \subset \mathbb{R}$ be given by $\mathcal{S} = (0, 1]$. Note that every point $\alpha \in [0, 1]$ is an accumulation point of $\mathcal{S}$ since every open ball (open interval on $\mathbb{R}$ in this case) containing α will contain points of $(0, 1]$ distinct from α. Note that the accumulation point 0 is not an element of $\mathcal{S}$. Since 0 is the only accumulation point that does not belong to $\mathcal{S}$, the closure of $\mathcal{S}$ is given by $\overline{\mathcal{S}} = \mathcal{S} \cup \mathcal{S}' = (0, 1] \cup [0, 1] = (0, 1] \cup \{0\} = [0, 1]$. Finally, note that the boundary of $\mathcal{S}$ is $\partial\mathcal{S} = \overline{\mathcal{S}} \setminus \overset{\circ}{\mathcal{S}} = \overline{\mathcal{S}} \cap \overline{(\mathbb{R}^n \setminus \mathcal{S})} = [0, 1] \cap ((-\infty, 0] \cup [1, \infty)) = \{0, 1\}$. △

Note that for a given set $\mathcal{S} \subseteq \mathbb{R}^n$, closure points can be isolated points of $\mathcal{S}$, which always belong to $\mathcal{S}$, accumulation points of $\mathcal{S}$ belonging to $\mathcal{S}$, and accumulation points of $\mathcal{S}$ which do not belong to $\mathcal{S}$. For example, let $\mathcal{S} \triangleq \{x \in \mathbb{R} : 0 < x < 1, x = 2\}$. Clearly, $x = 2$ is an isolated closure point of $\mathcal{S}$, $x = 0$ and $x = 1$ are accumulation points, and hence, closure points of $\mathcal{S}$ which do not belong to $\mathcal{S}$, and every point x in the set $\{x \in \mathbb{R} : 0 < x < 1\}$ is an accumulation point, and hence, a closure point of $\mathcal{S}$ belonging to $\mathcal{S}$.

The next proposition shows that the complement of an open set is closed and the complement of a closed set is open.

Proposition 2.5. Let $\mathcal{S} \subseteq \mathbb{R}^n$. Then $\mathcal{S}$ is closed if and only if $\mathbb{R}^n \setminus \mathcal{S}$

is open.

Proof. $\mathcal{S}$ is closed if and only if, for every point $x \in \mathbb{R}^n \setminus \mathcal{S}$, there exists $\varepsilon > 0$ such that $\mathcal{B}_\varepsilon(x) \cap \mathcal{S} = \emptyset$. This holds if and only if, for every $x \in \mathbb{R}^n \setminus \mathcal{S}$, there exists an $\varepsilon > 0$ such that $\mathcal{B}_\varepsilon(x) \subset \mathbb{R}^n \setminus \mathcal{S}$, which is true if and only if $\mathbb{R}^n \setminus \mathcal{S}$ is open. □

The empty set $\emptyset$ and $\mathbb{R}^n$ are the only subsets of $\mathbb{R}^n$ that are both open and closed. To see this, note that for every $x \in \mathbb{R}^n$, every open ball $\mathcal{B}_\varepsilon(x) \subset \mathbb{R}^n$. Hence, every point in $\mathbb{R}^n$ is an interior point. Thus, $\mathbb{R}^n$ is open. Alternatively, note that $\emptyset$ has no points, and hence, every point of $\emptyset$ is an interior point of $\emptyset$. Hence, $\emptyset$ is an open subset of $\mathbb{R}^n$. The converse follows immediately by noting that $\mathcal{S} \subseteq \mathbb{R}^n$ is closed if and only if $\mathbb{R}^n \setminus \mathcal{S}$ is open (see Proposition 2.5). Alternatively, $\mathbb{R}^n$ is closed since it contains all points, and hence, all accumulation points. The empty set is closed since it has no accumulation points. Hence, $\partial\mathcal{S} = \emptyset$ if and only if $\mathcal{S} = \emptyset$ or $\mathcal{S} = \mathbb{R}^n$. Finally, note that since

$$\partial\mathcal{S} = [\overset{\circ}{\mathcal{S}} \cup (\overset{\circ}{\mathcal{S}^{\sim}})]^{\sim},$$

$\partial\mathcal{S}$ is always a closed set and is given by the elements in $\mathbb{R}^n$ that belong to the set $\overline{\mathcal{S}}$ but not $\overset{\circ}{\mathcal{S}}$.

Example 2.7. Let $\mathcal{X} = \{(x_1, x_2) \in \mathbb{R}\times\mathbb{R} : -1 < x_1 \leq 1, -1 < x_2 \leq 1\}$. Note that $\mathcal{X}$ is neither open nor closed since the boundary point $(x_1, x_2) = (1, 1)$ is contained in $\mathcal{X}$ and the boundary point $(x_1, x_2) = (-1, -1)$ is not in $\mathcal{X}$. The closure of $\mathcal{X}$ is $\overline{\mathcal{X}} = \{(x_1, x_2) \in \mathbb{R}\times\mathbb{R} : -1 \leq x_1 \leq 1, -1 \leq x_2 \leq 1\}$, whereas the interior is $\overset{\circ}{\mathcal{X}} = \{(x_1, x_2) \in \mathbb{R}\times\mathbb{R} : -1 < x_1 < 1, -1 < x_2 < 1\}$. The boundary of $\mathcal{X}$ is formed by the four segments $\{(x_1, x_2) \in \mathbb{R}\times\mathbb{R} : x_1 = 1, -1 \leq x_2 \leq 1\}$, $\{(x_1, x_2) \in \mathbb{R}\times\mathbb{R} : x_1 = -1, -1 \leq x_2 \leq 1\}$, $\{(x_1, x_2) \in \mathbb{R}\times\mathbb{R} : x_2 = 1, -1 \leq x_1 \leq 1\}$, and $\{(x_1, x_2) \in \mathbb{R}\times\mathbb{R} : x_2 = -1, -1 \leq x_1 \leq 1\}$, and hence, is given by the union of these four sets. △

Definition 2.13. Let $\mathcal{S} \subseteq \mathbb{R}^n$. Then $\mathcal{Q} \subseteq \mathcal{S}$ is *dense* in $\mathcal{S}$ if $\mathcal{S} \subseteq \overline{\mathcal{Q}}$. The set $\mathcal{Q}$ is *nowhere dense* in $\mathcal{S}$ if $\mathcal{S}\setminus\overline{\mathcal{Q}}$ is dense in $\mathcal{S}$.

Definition 2.13 implies that if $\mathcal{Q}$ is dense in $\mathcal{S}$, then every element of $\mathcal{S}$ is a closure point of $\mathcal{Q}$, or an element of $\mathcal{Q}$, or both. If $\mathcal{S}$ is closed, then $\mathcal{Q} \subseteq \mathcal{S}$ is dense if and only if $\overline{\mathcal{Q}} = \mathcal{S}$. Note that the set of rational numbers $\mathbb{Q}$ is a dense subset of $\mathbb{R}$, whereas the set of integers $\mathbb{Z}$ is a nowhere dense subset of $\mathbb{R}$.

Definition 2.14. Let $\mathcal{S} \subset \mathbb{R}^n$. Then $\mathcal{S}$ is *bounded* if there exists $\varepsilon > 0$ such that $\|x - y\| < \varepsilon$ for all $x, y \in \mathcal{S}$. Furthermore, $\mathcal{S}$ is *compact* if it is closed and bounded.

Definition 2.15. Let $\mathcal{S} \subseteq \mathbb{R}^n$. The set $\mathcal{Q} \subseteq \mathcal{S}$ is *open* (respectively, *closed*) *relative to* $\mathcal{S}$ if there exists an open (respectively, closed) set $\mathcal{R} \subseteq \mathbb{R}^n$ such that $\mathcal{Q} = \mathcal{S} \cap \mathcal{R}$.

Note that since $\mathcal{S} = \mathcal{S} \cap \mathbb{R}^n$, it follows that any set is open relative to itself.

Example 2.8. Let $\mathcal{Q} = (0, 1]$ and $\mathcal{S} = (-1, 1]$. Note that $\mathcal{Q}$ is open relative to $\mathcal{S}$. In particular, for all $\alpha > 1$, $(0, 1] = (-1, 1] \cap (0, \alpha)$. Similarly, $(0, 1]$ is closed relative to $(0, 2)$ since $(0, 1] = (0, 2) \cap [\alpha, 1]$ for all $\alpha < 0$. △

A sequence of scalars $\{x_n\}_{n=0}^{\infty} \subset \mathbb{R}$ is said to *converge* to a scalar $x \in \mathbb{R}$ if for every $\varepsilon > 0$, there exists $N = N(\varepsilon)$ such that $|x_n - x| < \varepsilon$ for every $n > N$. If a sequence $\{x_n\}_{n=0}^{\infty}$ converges to some $x \in \mathbb{R}$, we say x is the *limit* of $\{x_n\}_{n=0}^{\infty}$, that is, $\lim_{n\to\infty} x_n = x$. A sequence $\{x_n\}_{n=0}^{\infty}$ is a *Cauchy sequence* if for every $\varepsilon > 0$, there exists $N = N(\varepsilon)$ such that $|x_n - x_m| < \varepsilon$ for all $n, m > N$.

A scalar sequence $\{x_n\}_{n=0}^{\infty} \subset \mathbb{R}$ is said to be *bounded above* (respectively, *bounded below*) if there exists $\alpha \in \mathbb{R}$ such that $x_n \leq \alpha$ (respectively, $x_n \geq \alpha$) for all $n \in \overline{\mathbb{Z}}_+$. A scalar sequence $\{x_n\}_{n=0}^{\infty}$ is said to be *nonincreasing* (respectively, *nondecreasing*) if $x_{n+1} \leq x_n$ (respectively, $x_{n+1} \geq x_n$) for all $n \in \overline{\mathbb{Z}}_+$.

Example 2.9. Consider the sequence $\{\frac{1}{n}\}_{n=1}^{\infty} = \{1, \frac{1}{2}, \frac{1}{3}, \ldots\}$. To see that this sequence converges to $x = 0$, let $\varepsilon > 0$. In this case, $|1/n - 0| = |1/n| = 1/n < \varepsilon$ if and only if $n > 1/\varepsilon$. Hence, letting $N = N(\varepsilon) > 1/\varepsilon$, it follows that $|1/n - 0| < \varepsilon$ for every $n > N$. △

If a finite limit $x \in \mathbb{R}$ does not exist for a given scalar sequence, then the sequence is said to be *divergent*. In particular, the sequence $\{n\}_{n=0}^{\infty}$ diverges as $n \to \infty$. In addition, the sequence $\{(-1)^n\}_{n=0}^{\infty}$, though bounded, is also divergent since it yields the oscillating sequence $\{-1, 1, -1, 1, \ldots\}$, and hence does not converge to a finite limit.

The *supremum* of a nonempty set $\mathcal{S} \subset \mathbb{R}$ of scalars, denoted by $\sup \mathcal{S}$, is defined to be the smallest scalar x such that $x \geq y$ for all $y \in \mathcal{S}$. If no such scalar exists, then $\sup \mathcal{S} \triangleq \infty$. Similarly, the *infimum* of $\mathcal{S}$, denoted by $\inf \mathcal{S}$, is defined to be the largest scalar x such that $x \leq y$ for all $y \in \mathcal{S}$. If no such scalar exists, then $\inf \mathcal{S} \triangleq -\infty$. Given a scalar sequence $\{x_n\}_{n=0}^{\infty} \subset \mathbb{R}$, the supremum of the sequence, denoted by $\sup_n x_n$, is defined as $\sup\{x_n : n = 1, 2, \ldots\}$. Similarly, the infimum of the sequence, denoted by $\inf_n x_n$, is defined as $\inf\{x_n : n = 1, 2, \ldots\}$. Finally, given a sequence $\{x_n\}_{n=0}^{\infty}$, let $y_m = \inf\{x_n : n \geq m\}$ and $z_m = \sup\{x_n : n \geq m\}$. Since the sequences $\{y_m\}_{m=0}^{\infty}$ and $\{z_m\}_{m=0}^{\infty}$ are nondecreasing and nonincreasing, respectively, it follows from Proposition 2.6 that if these sequences are bounded, then they have finite limits. The limit of $\{y_m\}_{m=0}^{\infty}$ is denoted by $\liminf_{m\to\infty} x_m$ and is called the *limit inferior* of x_m, and the limit of $\{z_m\}_{m=0}^{\infty}$ is denoted by $\limsup_{m\to\infty} x_m$ and is called the *limit superior* of x_m. If $\{y_m\}_{m=0}^{\infty}$ and $\{z_m\}_{m=0}^{\infty}$ are unbounded, then $\liminf_{m\to\infty} x_m = -\infty$ and $\limsup_{m\to\infty} x_m = \infty$.

An equivalent definition for the limit superior of an infinite sequence of real numbers $\{x_n\}_{n=0}^{\infty}$ is the existence of a real number S such that for every $\varepsilon > 0$ there exists an integer N such that for all $n > N$, $x_n < S + \varepsilon$, and for every $\varepsilon > 0$ and $M > 0$ there exists an $n > M$ such that $x_n > S - \varepsilon$. Then S is the limit superior of $\{x_n\}_{n=0}^{\infty}$. The limit inferior of x_n is simply $\liminf_{n\to\infty} x_n = -\limsup_{n\to\infty}(-x_n)$. Finally, note that $\{x_n\}_{n=0}^{\infty}$ converges if and only if $\limsup_{n\to\infty} x_n = \liminf_{n\to\infty} x_n = \lim_{n\to\infty} x_n$. (See Problem 2.18.)

Proposition 2.6. Let $\{x_n\}_{n=0}^{\infty} \subset \mathbb{R}$ be a nonincreasing or nondecreasing scalar sequence. If $\{x_n\}_{n=0}^{\infty}$ is bounded, then $\{x_n\}_{n=0}^{\infty}$ converges to a finite real number.

Proof. Let $\{x_n\}_{n=0}^{\infty} \subset \mathbb{R}$ be a nondecreasing scalar sequence that is bounded above. Then, by the completeness axiom,[2] the supremum $\alpha \triangleq \sup\{x_n : n \in \overline{\mathbb{Z}}_+\}$ exists and is finite. Now, let $\varepsilon > 0$ and choose $N \in \overline{\mathbb{Z}}_+$ such that $\alpha - \varepsilon < x_N \le \alpha$. Since $x_N \le x_n$ for $n \ge N$ and $x_n \le \alpha$ for all $n \in \overline{\mathbb{Z}}_+$, it follows that $\alpha - \varepsilon < x_n \le \alpha$ for all $n \ge N$, and hence, $\lim_{n\to\infty} x_n = \alpha$. If, alternatively, $\{x_n\}_{n=0}^{\infty} \subset \mathbb{R}$ is nonincreasing with infimum $\beta \triangleq \inf\{x_n : n \in \overline{\mathbb{Z}}_+\}$, then $\{-x_n\}_{n=0}^{\infty}$ is nondecreasing with supremum $-\beta$. Hence, $\beta = -(-\beta) = -\lim_{n\to\infty}(-x_n) = \lim_{n\to\infty} x_n$. □

Example 2.10. Consider the scalar sequence $\mathcal{S} = \{\frac{1}{n}\}_{n=1}^{\infty} \subset \mathbb{R}$. Clearly, $\mathcal{S}$ is bounded below, and $\inf \mathcal{S} = 0$. However, $0 \notin \mathcal{S}$. Alternatively, $\sup \mathcal{S} = 1$ and $1 \in \mathcal{S}$. Hence, the maximum element of the set $\mathcal{S}$ is attained and is given by $\sup \mathcal{S}$, whereas the minimum element of the set $\mathcal{S}$ does not exist. Finally, note that $\liminf_{n\to\infty} \frac{1}{n} = \limsup_{n\to\infty} \frac{1}{n} = \lim_{n\to\infty} \frac{1}{n} = 0$. △

Norms can be used to define convergent sequences in $\mathbb{R}^n$. For the next definition, a *sequence of vectors* $\{x_1, x_2, \ldots\}$ is defined as an ordered multiset with countably infinite elements.

Definition 2.16. A sequence of vectors $\{x_n\}_{n=0}^{\infty} \subset \mathbb{R}^n$ *converges* to $x \in \mathbb{R}^n$ if $\lim_{n\to\infty} \|x - x_n\| = 0$. In this case, we write $\lim_{n\to\infty} x_n = x$ or,

[2]The completeness axiom states that every nonempty set $\mathcal{S}$ of real numbers which is bounded above has a supremum, that is, there exists a real number α such that $\alpha = \sup \mathcal{S}$. As a consequence of this axiom it follows that every nonempty set of real numbers which is bounded below has an infimum.

equivalently, $x_n \to x$ as $n \to \infty$. The sequence $\{x_n\}_{n=0}^{\infty} \subset \mathbb{R}^n$ is called a *Cauchy sequence* if, for every $\varepsilon > 0$, there exists an integer $N = N(\varepsilon)$ such that $\|x_n - x_m\| < \varepsilon$, whenever $n, m > N$.

Note that convergence guarantees that for every $\varepsilon > 0$, there exists $N = N(\varepsilon)$ such that $\|x_n - x\| < \varepsilon$ for all $n > N$. The next proposition shows that if a sequence converges, its limit is unique.

Proposition 2.7. Let $\{x_n\}_{n=0}^{\infty} \subset \mathbb{R}^n$ be a convergent sequence. Then its limit is unique.

Proof. Suppose, *ad absurdum*, $x_n \to x$ and $x_n \to y$ as $n \to \infty$, where $x \neq y$. Then it follows from the triangle inequality for norms that

$$\|x - y\| = \|x - x_n + x_n - y\| \leq \|x - x_n\| + \|x_n - y\|.$$

Hence, as $n \to \infty$, $\|x - y\| \to 0$. Thus, $x = y$. □

It is always possible to construct a *subsequence* from any sequence by selecting arbitrary elements from that sequence. However, in this construction the order of the terms must be preserved; that is, no interchange of terms is permissible. More precisely, given a sequence $\{x_n\}_{n=1}^{\infty} \subset \mathbb{R}^n$ and a sequence of positive integers $\{n_k\}_{k=1}^{\infty}$ such that $n_1 < n_2 < \ldots$, then the sequence $\{x_{n_k}\}_{k=1}^{\infty}$ is a subsequence of $\{x_n\}_{n=1}^{\infty}$. Sequences in general do not converge, but they often have subsequences that do. The next proposition gives a characterization of a closure point in terms of sequences.

Proposition 2.8. Let $\mathcal{D} \subseteq \mathbb{R}^n$. Then $x \in \mathbb{R}^n$ is a closure point of $\mathcal{D}$ if and only if there exists a sequence $\{x_n\}_{n=1}^{\infty} \subseteq \mathcal{D}$ such that $x = \lim_{n\to\infty} x_n$.

Proof. Let $x \in \mathbb{R}^n$ be a closure point of $\mathcal{D}$. Then, for all $n \in \mathbb{Z}_+$, there exists $x_n \in \mathcal{D}$ such that $\|x - x_n\| < 1/n$. Hence, $x_n \to x$ as $n \to \infty$. Conversely, suppose $\{x_n\}_{n=1}^{\infty} \subseteq \mathcal{D}$ is such that $x = \lim_{n\to\infty} x_n$ and let $\varepsilon > 0$. Then, there exists $k \in \mathbb{Z}_+$ such that $\|x - x_n\| < \varepsilon$ for $n > k$. Thus, $x_{k+1} \in \mathcal{D} \cap \mathcal{B}_\varepsilon(x)$, which implies that $\mathcal{D} \cap \mathcal{B}_\varepsilon(x)$ is not empty. Hence, x is a closure point of $\mathcal{D}$. □

The following theorem is due to Bolzano and Weierstrass and forms the cornerstone of mathematical analysis.

Theorem 2.3 (Bolzano-Weierstrass). Let $\mathcal{S} \subset \mathbb{R}^n$ be a bounded set that contains infinitely many points. Then there exists at least one point $p \in \mathbb{R}^n$ that is an accumulation point of $\mathcal{S}$.

Proof. See [371]. □

The next result shows an equivalence between *sequential compactness* and compactness; that is, every sequence in a compact set has a convergent subsequence and conversely, every compact set is sequentially compact. This result is known as the Bolzano-Lebesgue theorem.

Theorem 2.4 (Bolzano-Lebesgue). Let $\mathcal{D}_{\rm c} \subset \mathbb{R}^n$. For every sequence $\{x_n\}_{n=1}^{\infty} \subseteq \mathcal{D}_{\rm c}$ there exists a convergent subsequence $\{x_{n_k}\}_{k=1}^{\infty} \subseteq \{x_n\}_{n=1}^{\infty}$ such that $\lim_{k\to\infty} x_{n_k} \in \mathcal{D}_{\rm c}$ if and only if $\mathcal{D}_{\rm c}$ is compact.

Proof. To show necessity, let $\mathcal{D}_{\rm c}$ be compact, let $\{x_n\}_{n=1}^{\infty} \subseteq \mathcal{D}_{\rm c}$, and let $\mathcal{Q} = \{x_1, x_2, \ldots\}$. Note that $\mathcal{Q}$ is an infinite set contained in $\mathcal{D}_{\rm c}$. Now, since $\mathcal{D}_{\rm c}$ is bounded, $\mathcal{Q}$ is bounded, and hence, by the Bolzano-Weierstrass theorem (Theorem 2.3), $\mathcal{Q}$ has an accumulation point x. Since $\mathcal{D}_{\rm c}$ is closed, x is also an accumulation point of $\mathcal{D}_{\rm c}$, and hence, $x \in \mathcal{D}_{\rm c}$. Thus, for each $k \in \mathbb{Z}_+$ we can find an $n_k > n_{k-1}$ such that $x_{n_k} \in \mathcal{B}_{1/k}(x)$. Hence, $\lim_{k\to\infty} x_{n_k} = x \in \mathcal{D}_{\rm c}$.

To show sufficiency, assume that every sequence contained in $\mathcal{D}_{\rm c}$ has a convergent subsequence and suppose, *ad absurdum*, that $\mathcal{D}_{\rm c}$ is not bounded. Then there exists a sequence $\{x_n\}_{n=1}^{\infty} \subseteq \mathcal{D}_{\rm c}$ such that $\|x_n\| \geq n$, $n \in \mathbb{Z}_+$, and there exists a convergent subsequence $\{x_{n_k}\}_{k=1}^{\infty} \subseteq \{x_n\}_{n=1}^{\infty}$ such that $\lim_{k\to\infty} x_{n_k} = x \in \mathcal{D}_{\rm c}$. Now, for every $k \in \mathbb{Z}_+$ such that $n_k > 1 + \|x\|$,

$$\|x_{n_k} - x\| \geq \|x_{n_k}\| - \|x\| \geq n_k - \|x\| > 1,$$

which is a contradiction. Hence, $\mathcal{D}_{\rm c}$ is bounded.

Next, suppose, *ad absurdum*, that $\mathcal{D}_{\rm c}$ is not closed. Let $\{x_n\}_{n=1}^{\infty} \subseteq \mathcal{D}_{\rm c}$ be such that $\lim_{n\to\infty} x_n = x \notin \mathcal{D}_{\rm c}$. By assumption, there exists a subsequence $\{x_{n_k}\}_{k=1}^{\infty} \subseteq \{x_n\}_{n=1}^{\infty}$ such that $\lim_{k\to\infty} x_{n_k} = y \in \mathcal{D}_{\rm c}$. Now, since $\lim_{n\to\infty} x_n = x$ it follows that for every $\varepsilon > 0$, there exists $N \in \mathbb{Z}_+$ such that $\|x_n - x\| < \varepsilon$, $n \geq N$. Furthermore, there exists $k \in \mathbb{Z}_+$ such that $n_k \geq N$, which implies that $\|x_{n_k} - x\| < \varepsilon$. Hence, $\lim_{k\to\infty} x_{n_k} = x$. Now, it follows from the uniqueness of limits of sequences (see Proposition 2.7) that $x = y$, which is a contradiction. Hence, $\mathcal{D}_{\rm c}$ is compact. □

The next lemma is needed to show that a sequence in $\mathbb{R}^n$ converges if and only if it is a Cauchy sequence.

Lemma 2.1. A Cauchy sequence in $\mathbb{R}^n$ is bounded.

Proof. Let $\{x_n\}_{n=0}^{\infty} \subset \mathbb{R}^n$ be a Cauchy sequence and let $N \in \overline{\mathbb{Z}}_+$ be such that $\|x_n - x_N\| < 1$ for all $n > N$. Now, for $n > N$,

$$\|x_n\| = \|x_n - x_N + x_N\| \leq \|x_n - x_N\| + \|x_N\| < 1 + \|x_N\|, \tag{2.41}$$

which proves the result. □

Theorem 2.5. Let $\{x_n\}_{n=0}^{\infty} \subset \mathbb{R}^n$. Then $\{x_n\}_{n=0}^{\infty}$ is convergent if and only if $\{x_n\}_{n=0}^{\infty}$ is a Cauchy sequence.

Proof. Assume $\{x_n\}_{n=0}^{\infty}$ is convergent, and hence, $x_n \to x$ as $n \to \infty$ for some $x \in \mathbb{R}^n$. Given $\varepsilon > 0$, let $N \in \overline{\mathbb{Z}}_+$ be such that $\|x_n - x\| < \varepsilon/2$ whenever $n > N$. Now, if $m > N$, then $\|x_m - x\| < \varepsilon/2$. If $n > N$ and $m > N$ it follows that

$$\begin{aligned}
\|x_n - x_m\| &= \|x_n - x + x - x_m\| \\
&\leq \|x_n - x\| + \|x - x_m\| \\
&< \frac{\varepsilon}{2} + \frac{\varepsilon}{2} \\
&= \varepsilon, \qquad (2.42)
\end{aligned}$$

which shows that $\{x_n\}_{n=0}^{\infty}$ is a Cauchy sequence. Conversely, assume that $\{x_n\}_{n=0}^{\infty}$ is a Cauchy sequence. In this case, it follows from Lemma 2.1 that $\mathcal{S} \triangleq \{x_n\}_{n=0}^{\infty}$ is bounded, and hence, by the Bolzano-Weierstrass theorem (Theorem 2.3), $\mathcal{S}$ has an accumulation point $p \in \mathbb{R}^n$. Now, since $\{x_n\}_{n=0}^{\infty}$ is Cauchy, given $\varepsilon > 0$, there exists N such that $\|x_n - x_m\| < \varepsilon/2$ whenever $n, m > N$. Since p is an accumulation point of $\mathcal{S}$ it follows that the open ball $\mathcal{B}_{\varepsilon/2}(p)$ contains a point x_m with $m > N$. Hence, if $m > N$ then it follows that

$$\begin{aligned}
\|x_n - p\| &= \|x_n - x_m + x_m - p\| \\
&\leq \|x_n - x_m\| + \|x_m - p\| \\
&< \varepsilon/2 + \varepsilon/2 \\
&= \varepsilon, \qquad (2.43)
\end{aligned}$$

and hence, $x_n \to p$ as $n \to \infty$. □

It follows from Proposition 2.8 that the definition of convergence can be used to characterize closed sets.

Proposition 2.9. Let $\mathcal{D} \subseteq \mathbb{R}^n$. Then $\mathcal{D}$ is closed if and only if every convergent sequence $\{x_n\}_{n=0}^{\infty} \subseteq \mathcal{D}$ has its limit point in $\mathcal{D}$.

Proof. The result is a direct consequence of Proposition 2.8. □

The next results show that the finite intersection and infinite union of open sets is open, whereas the finite union and infinite intersection of closed sets is closed.

Theorem 2.6. The following statements hold:

i) Let $\mathcal{S}_1, \ldots, \mathcal{S}_n$ be open subsets of $\mathbb{R}^n$. Then $\mathcal{S} = \mathcal{S}_1 \cap \cdots \cap \mathcal{S}_n$ is an open set.

ii) Let $\{\mathcal{S}_i : i \in \mathcal{I}\}$ be an indexed family of open sets contained in $\mathbb{R}^n$ with the index set $\mathcal{I}$ being finite or infinite. Then $\mathcal{S} = \bigcup_{i\in\mathcal{I}} \mathcal{S}_i$ is an open set.

Proof. *i)* The result is obvious if $\mathcal{S} = \emptyset$. Assume $\mathcal{S} \neq \emptyset$, let $x \in \mathcal{S}$, and note that $x \in \mathcal{S}_i$ for all $i = 1, \ldots, n$. Since for all $i = 1, \ldots, n$, $\mathcal{S}_i$ is an open set, there exists $\varepsilon_i > 0$, $i = 1, \ldots, n$, such that $\mathcal{B}_{\varepsilon_i}(x) \subseteq \mathcal{S}_i$, $i = 1, \ldots, n$. Now, taking $\varepsilon = \min \varepsilon_i$, it follows that $\mathcal{B}_\varepsilon(x) \subseteq \mathcal{S}_i$, $i = 1, \ldots, n$, and hence, $\mathcal{B}_\varepsilon(x) \subseteq \mathcal{S}$, which proves $\mathcal{S}$ is open.

ii) Let $x \in \mathcal{S}$ and note that since $\mathcal{S} = \bigcup_{i\in\mathcal{I}} \mathcal{S}_i$ it follows that $x \in \mathcal{S}_j$ for some $j \in \mathcal{I}$. Since $\mathcal{S}_j$ is open, there exists $\varepsilon > 0$ such that $\mathcal{B}_\varepsilon(x) \subseteq \mathcal{S}_j$, which implies that $\mathcal{B}_\varepsilon(x) \subseteq \mathcal{S}$. Thus, $x \in \overset{\circ}{\mathcal{S}}$, and hence, for every $x \in \mathcal{S}$, $x \in \overset{\circ}{\mathcal{S}}$; that is, $\overset{\circ}{\mathcal{S}} = \mathcal{S}$, which proves that $\mathcal{S}$ is open. □

Theorem 2.7. The following statements hold:

i) Let $\mathcal{S}_1, \ldots, \mathcal{S}_n$ be closed subsets of $\mathbb{R}^n$. Then $\mathcal{S} = \mathcal{S}_1 \cup \cdots \cup \mathcal{S}_n$ is a closed set.

ii) Let $\{\mathcal{S}_i : i \in \mathcal{I}\}$ be an indexed family of closed sets contained in $\mathbb{R}^n$ with the index set $\mathcal{I}$ being finite or infinite. Then $\mathcal{S} = \bigcap_{i\in\mathcal{I}} \mathcal{S}_i$ is a closed set.

Proof. *i)* Let $x \in \mathbb{R}^n$ be an accumulation point of $\mathcal{S}$. Then, for every $\varepsilon > 0$, the set $\mathcal{S} \cap (\mathcal{B}_\varepsilon(x) \setminus \{x\})$ is not empty or, equivalently, there exists a sequence $\{x_n\}_{n=1}^\infty \subseteq \mathcal{S}$ such that $x = \lim_{n\to\infty} x_n$. Now, since $\mathcal{S}$ is a finite collection of closed sets, at least one, say $\mathcal{S}_1$, contains an infinite subsequence $\{x_{n_k}\}_{k=1}^\infty \subseteq \{x_n\}_{n=1}^\infty$ such that $x = \lim_{k\to\infty} x_{n_k}$. (This follows from the fact that if $x = \lim_{n\to\infty} x_n$, then $x = \lim_{k\to\infty} x_{n_k}$ for every subsequence $\{x_{n_k}\}_{k=1}^\infty$ of $\{x_n\}_{n=1}^\infty$.) Then, $x \in \mathbb{R}^n$ is an accumulation point of $\mathcal{S}_1$, and, since $\mathcal{S}_1$ is closed, $x \in \mathcal{S}_1$. Hence, $x \in \mathcal{S}$.

ii) Let $x \in \mathbb{R}^n$ be an accumulation point of $\mathcal{S}$ so that, for every $\varepsilon > 0$, the set $\mathcal{S} \cap (\mathcal{B}_\varepsilon(x) \setminus \{x\})$ is not empty. Then every open ball centered at x contains an infinite number of points of $\mathcal{S}$, and hence, since for every $i \in \mathcal{I}$, $\mathcal{S} \subseteq \mathcal{S}_i$, contains an infinite number of points of $\mathcal{S}_i$. Hence, x is an accumulation point of $\mathcal{S}_i$, and hence, $x \in \mathcal{S}_i$ for each $i \in \mathcal{I}$ since the sets $\mathcal{S}_i$, $i \in \mathcal{I}$, are all closed. Thus, $x \in \mathcal{S}$, and hence, $\mathcal{S}$ is closed. □

Definition 2.17. Let $\mathcal{S} \subseteq \mathbb{R}^n$. Then $\mathcal{S}$ is *connected* if there do not exist open sets $\mathcal{O}_1$ and $\mathcal{O}_2$ in $\mathbb{R}^n$ such that $\mathcal{S} \subset \mathcal{O}_1 \cup \mathcal{O}_2$, $\mathcal{S} \cap \mathcal{O}_1 \neq \emptyset$,

$\mathcal{S} \cap \mathcal{O}_2 \neq \emptyset$, and $\mathcal{S} \cap \mathcal{O}_1 \cap \mathcal{O}_2 = \emptyset$. $\mathcal{S}$ is *arcwise connected* if for every two points $x, y \in \mathcal{S}$ there exists a continuous function $g : [0, 1] \to \mathcal{S}$ such that $g(0) = x$ and $g(1) = y$. A *connected component* of the set $\mathcal{S}$ is a connected subset of $\mathcal{S}$ that is not properly contained in any connected subset of $\mathcal{S}$.

Note that $\mathcal{S}$ is connected if and only if it is not the union of two disjoint, nonempty, relatively open subsets of $\mathcal{S}$. Recall that $\mathcal{S}$ is a connected subset of $\mathbb{R}$ if and only if $\mathcal{S}$ is either an interval or a single point. $\mathcal{S}$ is arcwise connected if any two points in $\mathcal{S}$ can be joined by a continuous curve that lies in $\mathcal{S}$. Arcwise connectedness is an important notion in dynamical system theory since the state of a dynamical system taking values in $\mathbb{R}^n$ is typically a continuous function of time. Hence, in order to control the system state from any point $x \in \mathcal{D} \subseteq \mathbb{R}^n$ to any other point in $\mathcal{D}$, the subset of the state space $\mathcal{D}$ must be arcwise connected.

Example 2.11. The set of rational numbers $\mathbb{Q} \subset \mathbb{R}$ is disconnected. To see this, note that $\mathbb{Q} = \mathcal{X} \cap \mathcal{Y}$, where $\mathcal{X} = \{x \in \mathbb{Q} : x < \sqrt{3}\}$ and $\mathcal{Y} = \{y \in \mathbb{Q} : y > \sqrt{3}\}$. Since $\mathcal{X} = (-\infty, \sqrt{3}) \cap \mathbb{Q}$ and $\mathcal{Y} = (\sqrt{3}, \infty) \cap \mathbb{Q}$, $\mathcal{X}$ and $\mathcal{Y}$ are nonempty, disjoint, and open relative to $\mathbb{Q}$. Hence, $\mathbb{Q}$ is disconnected. △

Definition 2.18. Let $\mathcal{S} \subseteq \mathbb{R}^n$. Then $\mathcal{S}$ is *convex* if $\mu x + (1 - \mu)y \in \mathcal{S}$ for all $0 \leq \mu \leq 1$ and $x, y \in \mathcal{S}$. The *convex hull* of $\mathcal{S}$, denoted by $\operatorname{co} \mathcal{S}$, is the intersection of all convex sets containing $\mathcal{S}$, that is, the smallest convex set that contains $\mathcal{S}$.

Proposition 2.10. Let $\mathcal{B}_\varepsilon(x) \subset \mathbb{R}^n$. Then $\mathcal{B}_\varepsilon(x)$ is convex.

Proof. Without loss of generality, consider the open ball with radius 1 and center 0, that is, $\mathcal{B}_1(0) = \{x \in \mathbb{R}^n : \|x\| < 1\}$. Now, note that if $x_0, y_0 \in \mathcal{B}_1(0)$, then $\|x_0\| < 1$ and $\|y_0\| < 1$. Next, let $\mu \in [0, 1]$ and note that

$$\begin{aligned} \|\mu x_0 + (1 - \mu)y_0\| &\leq \|\mu x_0\| + \|(1 - \mu)y_0\| \\ &= \mu\|x_0\| + (1 - \mu)\|y_0\| \\ &< 1, \quad x_0, y_0 \in \mathcal{B}_1(0). \end{aligned} \tag{2.44}$$

Hence, $\mu x_0 + (1 - \mu)y_0 \in \mathcal{B}_1(0)$ for all $x_0, y_0 \in \mathcal{B}_1(0)$ and $\mu \in [0, 1]$. □

Proposition 2.11. Let $\mathcal{C} \subseteq \mathbb{R}^n$ be a convex set. Then $\overline{\mathcal{C}}$ and $\overset{\circ}{\mathcal{C}}$ are convex.

Proof. If $\overline{\mathcal{C}} = \emptyset$, then $\overline{\mathcal{C}}$ is convex. Let $x_0, y_0 \in \overline{\mathcal{C}}$ and choose $\mu = (0, 1)$. Now, for a given $\varepsilon > 0$, let $x, y \in \mathcal{C}$ be such that $\|x - x_0\| < \varepsilon$ and $\|y - y_0\| < \varepsilon$. In this case, $\|\mu x + (1 - \mu)y - \mu x_0 - (1 - \mu)y_0\| < \varepsilon$, and hence,

$\|z - z_0\| < \varepsilon$, where $z \triangleq \mu x + (1-\mu)y$ and $z_0 \triangleq \mu x_0 + (1-\mu)y_0$, and $z \in \mathcal{C}$. Since $\varepsilon > 0$ is arbitrary, it follows that z_0 is a closure point of $\mathcal{C}$, and hence, $\overline{\mathcal{C}}$ is convex.

If $\overset{\circ}{\mathcal{C}} = \emptyset$, then $\overset{\circ}{\mathcal{C}}$ is convex. Let $x_0, y_0 \in \overset{\circ}{\mathcal{C}}$ and choose $\mu \in (0,1)$. Since $x_0, y_0 \in \overset{\circ}{\mathcal{C}}$, there exists $\varepsilon > 0$ such that $\mathcal{B}_\varepsilon(x_0) \subset \mathcal{C}$ and $\mathcal{B}_\varepsilon(y_0) \in \mathcal{C}$. Now, since $z_0 + u = \mu(x_0 + u) + (1-\mu)(y_0 + u)$, it follows that $z_0 + u \in \mathcal{C}$ for all $\|u\| < \varepsilon$. Hence, z_0 is an interior point of $\mathcal{C}$, and hence, $\overset{\circ}{\mathcal{C}}$ is convex. □

Note that in the light of Proposition 2.11 the closed ball $\overline{\mathcal{B}}_\varepsilon(x)$ is also convex. The next result shows that the intersection of an arbitrary number of convex sets is convex.

Proposition 2.12. Let $\mathcal{S}$ be an arbitrary collection of convex sets. Then $\mathcal{Q} = \bigcap_{\mathcal{C}\in\mathcal{S}} \mathcal{C}$ is convex.

Proof. If $\mathcal{Q} = \emptyset$, then the result is immediate. Now, assume $x_1, x_2 \in \mathcal{Q}$ and choose $\mu \in [0,1]$. Then $x_1, x_2 \in \mathcal{C}$ for all $\mathcal{C} \in \mathcal{S}$, and since $\mathcal{C}$ is convex, $\mu x_1 + (1-\mu)x_2 \in \mathcal{C}$ for all $\mathcal{C} \in \mathcal{S}$. Thus, $\mu x_1 + (1-\mu)x_2 \in \mathcal{Q}$, and hence, $\mathcal{Q}$ is convex. □

2.4 Analysis in $\mathbb{R}^n$

If $\mathcal{X}$ and $\mathcal{Y}$ are sets, then a function $f(\cdot)$ (or f) that maps $\mathcal{X}$ into $\mathcal{Y}$ is a rule $f : \mathcal{X} \to \mathcal{Y}$ that assigns a unique element $f(x)$ of $\mathcal{Y}$ to each element x in $\mathcal{X}$. Equivalently, a function f can be viewed as a subset $\mathcal{F}$ of $\mathcal{X} \times \mathcal{Y}$ such that if $(x_1, y_1) \in \mathcal{F}$, $(x_2, y_2) \in \mathcal{F}$, and $x_1 = x_2$, then $y_1 = y_2$. This connection is denoted by $\mathcal{F} = \mathrm{graph}(f)$. The set $\mathcal{X}$ is called the *domain* of f while the set $\mathcal{Y}$ is called the *codomain* of f. If $\mathcal{X}_1 \subseteq \mathcal{X}$, then it is convenient to define $f(\mathcal{X}_1) \triangleq \{f(x) : x \in \mathcal{X}_1\}$. The set $f(\mathcal{X})$ is called the *range* of f. If, in addition, $\mathcal{Z}$ is a set and $g : \mathcal{Y} \to \mathcal{Z}$, then $g \circ f : \mathcal{X} \to \mathcal{Z}$ is defined by $(g \circ f)(x) \triangleq g(f(x))$. If $x_1, x_2 \in \mathcal{X}$ and $f(x_1) = f(x_2)$ implies that $x_1 = x_2$, then f is *one-to-one* (or *injective*); if $f(\mathcal{X}) = \mathcal{Y}$, then f is *onto* (or *surjective*). If f is one-to-one and onto then f is *bijective*. The function $I_{\mathcal{X}} : \mathcal{X} \to \mathcal{X}$ defined by $I_{\mathcal{X}}(x) = x$, $x \in \mathcal{X}$, is called the *identity* on $\mathcal{X}$. The function $f : \mathcal{X} \to \mathcal{Y}$ is called *left invertible* if there exists a function $g : \mathcal{Y} \to \mathcal{X}$ (called a *left inverse* of f) such that $g \circ f = I_{\mathcal{X}}$, while f is *right invertible* if there exists a function $h : \mathcal{Y} \to \mathcal{X}$ (called a *right inverse* of f) such that $f \circ h = I_{\mathcal{Y}}$. In addition, the function $f : \mathcal{X} \to \mathcal{Y}$ is *invertible* if there exists $f^{-1} : \mathcal{Y} \to \mathcal{X}$ such that $f^{-1} \circ f = I_{\mathcal{X}}$ and $f \circ f^{-1} = I_{\mathcal{Y}}$.

A function $f : \mathcal{X} \to \mathcal{Y}$ is said to be *into* if and only if $f(\mathcal{X}) \subset \mathcal{Y}$. For example, if $\mathcal{X} = \mathcal{Y} = \mathbb{Z}$, then $f(x) = x^2$ is an into function since $f(x)$

contains only integers that can be expressed as x^2 for $x \in \mathbb{Z}$, and hence, $f(\mathcal{X}) \subset \mathcal{Y} = \mathbb{Z}$. Alternatively, if $f(x) = 2x$, then f is injective since for every $x_1, x_2 \in \mathbb{Z}$ such that $x_1 \neq x_2$, $f(x_1) \neq f(x_2)$. For $\mathcal{X} = \mathbb{R}$ and $\mathcal{Y} = \mathbb{R}_+$, the function $f(x) = x^2$ is surjective since $f(\mathcal{X}) = \mathcal{Y} = \mathbb{R}_+$. Finally, for $\mathcal{X} = \mathcal{Y} = \mathbb{Z}$, the function $f(x) = -x$ is bijective since for every $x_1, x_2 \in \mathbb{Z}$ such that $x_1 \neq x_2$, $f(x_1) = -x_1 \neq f(x_2) = -x_2$, and hence, f is injective. In addition, since for every $-x \in \mathbb{Z}$, $x \in \mathbb{Z}$ it follows that $f(\mathcal{X}) = \mathcal{Y} = \mathbb{Z}$, and hence, f is also surjective.

Two functions f and g are *equal*, that is, $f = g$, if and only if they have the same domain, the same codomain, and for all x in the common domain, $f(x) = g(x)$. For example, given $f : \mathbb{R} \to \mathbb{R}_+$ defined by $f(x) = |x|$ and $g : \mathbb{R} \to \mathbb{R}_+$ defined by $g(x) = \sqrt{x^2}$, then $f(x) = g(x)$ for all $x \in \mathbb{R}$.

The notion of bijection is used to determine set equivalence. In particular, given two sets $\mathcal{X}$ and $\mathcal{Y}$ we say that $\mathcal{X}$ and $\mathcal{Y}$ are *equivalent*, denoted as $\mathcal{X} \sim \mathcal{Y}$, if and only if there exists a bijective mapping $f : \mathcal{X} \to \mathcal{Y}$. Since f is one-to-one and onto, it follows that there exists a one-to-one correspondence between the elements of the equivalent sets $\mathcal{X}$ and $\mathcal{Y}$. We say a set $\mathcal{X}$ is *finite* if and only if $\mathcal{X} = \varnothing$ or $\mathcal{X}$ is equivalent to the set $\{1, \ldots, n\}$ for some $n \in \mathbb{Z}_+$; otherwise $\mathcal{X}$ is *infinite*. If a set $\mathcal{X}$ is equivalent to $\mathbb{Z}_+$, then $\mathcal{X}$ is called *countably infinite* (or *denumerable*). A set $\mathcal{X}$ is *countable* if and only if it is either finite or countably infinite.

Example 2.12. The set $\mathbb{Z}$ of all integers is countable. In particular, the function $f : \mathbb{Z} \to \mathbb{Z}_+$ defined by

$$f(n) = \begin{cases} 2n+1, & n \in \overline{\mathbb{Z}}_+, \\ -2n, & n \in \mathbb{Z}_-, \end{cases} \tag{2.45}$$

is bijective, and hence, $\mathbb{Z}$ is a countable set. The set $\mathbb{E} = \{n \in \mathbb{Z}_+ : n \text{ is even}\}$ of all positive even numbers is countable. Specifically, the function $f : \mathbb{Z}_+ \to \mathbb{E}$ defined by $f(n) = 2n$ is bijective. Finally, the set $\mathbb{P} = \{2, 4, 8, \ldots, 2^n, \ldots\}$ of powers of 2 is countable as shown by the obvious one-to-one correspondence of $f : \mathbb{Z}_+ \to \mathbb{P}$ given by $f(n) = 2^n$. △

Definition 2.19. Let $\mathcal{D} \subseteq \mathbb{R}^n$ and $f : \mathcal{D} \to \mathbb{R}^n$. Then the *image* of $\mathcal{X} \subset \mathcal{D}$ under f is the set

$$f(\mathcal{X}) = \{y : \ y = f(x) \text{ for some } x \in \mathcal{X}\}.$$

The *inverse image* of $\mathcal{Y} \subset \mathbb{R}^n$ under f is the set

$$f^{-1}(\mathcal{Y}) = \{x \in \mathcal{D} : \ f(x) \in \mathcal{Y}\}.$$

Note that $f^{-1}(\mathcal{Y})$ may be empty even when $\mathcal{Y} \neq \varnothing$. In particular, an inverse function f^{-1} does not exist for the function $f : \mathbb{R} \to \mathbb{R}_+$ defined by $y = f(x) = x^2$ since $x = \pm\sqrt{y}$, and hence, there always exist two image points $\sqrt{y}$ and $-\sqrt{y}$ for every $y > 0$. This violates the uniqueness property of a function. However, $f : \mathbb{R}_+ \to \mathbb{R}_+$ and $g : \mathbb{R}_- \to \mathbb{R}_+$ defined by $y = f(x) = g(x) = x^2$, $f^{-1} = \mathbb{R}_+ \to \mathbb{R}_+$ and $g^{-1} : \mathbb{R}_+ \to \mathbb{R}_-$ exist and are given by $x = \sqrt{y}$ and $x = -\sqrt{y}$, respectively.

The next result summarizes some of the important properties of images and inverse images of functions.

Theorem 2.8. Let $\mathcal{X}$ and $\mathcal{Y}$ be sets, let $f : \mathcal{X} \to \mathcal{Y}$, let $\mathcal{A}$, $\mathcal{A}_1$, and $\mathcal{A}_2$ be subsets of $\mathcal{X}$, and let $\mathcal{B}_1$ and $\mathcal{B}_2$ be subsets of $\mathcal{Y}$. Then the following statements hold:

i) If $\mathcal{A}_1 \subseteq \mathcal{A}$, then $f(\mathcal{A}_1) \subseteq f(\mathcal{A})$.

ii) $f(\mathcal{A}_1 \cup \mathcal{A}_2) = f(\mathcal{A}_1) \cup f(\mathcal{A}_2)$.

iii) $f^{-1}(\mathcal{B}_1 \cup \mathcal{B}_2) = f^{-1}(\mathcal{B}_1) \cup f^{-1}(\mathcal{B}_2)$.

iv) $f^{-1}(\mathcal{B}_1 \cap \mathcal{B}_2) = f^{-1}(\mathcal{B}_1) \cap f^{-1}(\mathcal{B}_2)$.

Proof. *i)* Let $y \in f(\mathcal{A}_1)$. In this case, there exists $x \in \mathcal{A}_1$ such that $y = f(x)$. Since $\mathcal{A}_1 \subseteq \mathcal{A}$, $x \in \mathcal{A}$. Hence, $f(x) = y \in f(\mathcal{A})$, which proves $f(\mathcal{A}_1) \subseteq f(\mathcal{A})$.

ii) Let $y \in f(\mathcal{A}_1 \cup \mathcal{A}_2)$ and note that $y = f(x)$, where $x \in \mathcal{A}_1$ or $x \in \mathcal{A}_2$. Hence, $y \in f(\mathcal{A}_1)$ or $y \in f(\mathcal{A}_2)$ or, equivalently, $y \in f(\mathcal{A}_1) \cup f(\mathcal{A}_2)$. Conversely, let $y \in f(\mathcal{A}_1) \cup f(\mathcal{A}_2)$ and note that $y = f(x)$, where $x \in \mathcal{A}_1$ or $x \in \mathcal{A}_2$. Thus, $x \in \mathcal{A}_1 \cup \mathcal{A}_2$, and hence, $y = f(x) \in f(\mathcal{A}_1 \cup \mathcal{A}_2)$.

iii) Let $x \in f^{-1}(\mathcal{B}_1 \cup \mathcal{B}_2)$ and note that in this case $f(x) \in \mathcal{B}_1 \cup \mathcal{B}_2$. Hence, $f(x) \in \mathcal{B}_1$ or $f(x) \in \mathcal{B}_2$. This implies that $x \in f^{-1}(\mathcal{B}_1)$ or $x \in f^{-1}(\mathcal{B}_2)$ or, equivalently, $x \in f^{-1}(\mathcal{B}_1) \cup f^{-1}(\mathcal{B}_2)$. Conversely, let $x \in f^{-1}(\mathcal{B}_1) \cup f^{-1}(\mathcal{B}_2)$. In this case, $x \in f^{-1}(\mathcal{B}_1)$ or $x \in f^{-1}(\mathcal{B}_2)$, and hence, $f(x) \in \mathcal{B}_1$ or $f(x) \in \mathcal{B}_2$. Thus, $f(x) \in \mathcal{B}_1 \cup \mathcal{B}_2$, which implies that $x \in f^{-1}(\mathcal{B}_1 \cup \mathcal{B}_2)$.

iv) Let $x \in f^{-1}(\mathcal{B}_1 \cap \mathcal{B}_2)$ and note that in this case $f(x) \in \mathcal{B}_1 \cap \mathcal{B}_2$. Hence, $f(x) \in \mathcal{B}_1$ and $f(x) \in \mathcal{B}_2$. This implies that $x \in f^{-1}(\mathcal{B}_1)$ and $x \in f^{-1}(\mathcal{B}_2)$ or, equivalently, $x \in f^{-1}(\mathcal{B}_1) \cap f^{-1}(\mathcal{B}_2)$. Conversely, let $x \in f^{-1}(\mathcal{B}_1) \cap f^{-1}(\mathcal{B}_2)$. In this case, $x \in f^{-1}(\mathcal{B}_1)$ and $x \in f^{-1}(\mathcal{B}_2)$, and hence, $f(x) \in \mathcal{B}_1$ and $f(x) \in \mathcal{B}_2$. Thus, $f(x) \in \mathcal{B}_1 \cap \mathcal{B}_2$, which implies that $x \in f^{-1}(\mathcal{B}_1 \cap \mathcal{B}_2)$. $\square$

Note that in general the image of the intersection of two sets does not necessarily equal the intersection of the images of the sets. However, this statement is true if and only if $f : \mathcal{X} \to \mathcal{Y}$ is injective. (See Problem 2.26.) Theorem 2.8 also holds for unions and intersections of an arbitrary number (infinite or finite) of sets.

Theorem 2.9. Let $\mathcal{X}$ and $\mathcal{Y}$ be sets, let $f : \mathcal{X} \to \mathcal{Y}$, let $\{\mathcal{A}_\alpha : \alpha \in \mathcal{I}\}$ be an indexed family of sets in $\mathcal{X}$, and let $\{\mathcal{B}_\alpha : \alpha \in \mathcal{J}\}$ be an indexed family of sets in $\mathcal{Y}$. Then the following statements hold:

i) $f(\bigcup_{\alpha\in\mathcal{I}} \mathcal{A}_\alpha) = \bigcup_{\alpha\in\mathcal{I}} f(\mathcal{A}_\alpha)$.

ii) $f^{-1}(\bigcup_{\alpha\in\mathcal{J}} \mathcal{B}_\alpha) = \bigcup_{\alpha\in\mathcal{J}} f^{-1}(\mathcal{B}_\alpha)$.

iii) $f^{-1}(\bigcap_{\alpha\in\mathcal{J}} \mathcal{B}_\alpha) = \bigcap_{\alpha\in\mathcal{J}} f^{-1}(\mathcal{B}_\alpha)$.

Proof. The proof is similar to the proof of Theorem 2.8 and is left as an exercise for the reader. □

The next definition introduces the concept of convex functions defined on convex sets.

Definition 2.20. Let $\mathcal{C} \subseteq \mathbb{R}^n$ be a convex set and let $f : \mathcal{C} \to \mathbb{R}$. Then f is *convex* if

$$f(\mu x_1 + (1-\mu)x_2) \le \mu f(x_1) + (1-\mu) f(x_2), \tag{2.46}$$

for all $x_1, x_2 \in \mathcal{C}$ and $\mu \in [0,1]$. f is *strictly convex* if inequality (2.46) is strict for all $x_1, x_2 \in \mathcal{C}$ such that $x_1 \ne x_2$ and $\mu \in (0,1)$.

Definition 2.21. Let $\mathcal{D} \subseteq \mathbb{R}^n$ and let $f : \mathcal{D} \to \mathbb{R}$. For $\alpha \in \mathbb{R}$, the set $f^{-1}(\alpha) \triangleq \{x \in \mathcal{D} : f(x) = \alpha\}$ is called the *α-level set* of f. The set $f^{-1}((-\infty, \alpha]) \triangleq \{x \in \mathcal{D} : f(x) \le \alpha\}$ is called the *α-sublevel set* of f. For $\beta \in \mathbb{R}$, $\alpha \le \beta$, the set $f^{-1}([\alpha, \beta]) \triangleq \{x \in \mathcal{D} : \alpha \le f(x) \le \beta\}$ is called the *$[\alpha, \beta]$-sublevel set* of f.

Note that if $f(x) = x^2$, then $f^{-1}(0) = \{0\}$, and $f^{-1}(4) = \{-2, 2\}$, whereas $f^{-1}((-\infty, 4]) = [-2, 2]$. The following proposition states that every sublevel set of a convex function defined on a convex set is convex.

Proposition 2.13. Let $\mathcal{C} \subseteq \mathbb{R}^n$ be convex and let $f : \mathcal{C} \to \mathbb{R}$ be convex. Then $f^{-1}((-\infty, \alpha])$, $\alpha \in \mathbb{R}$, is convex.

Proof. Let $x_1, x_2 \in f^{-1}((-\infty, \alpha])$, $\alpha \in \mathbb{R}$, so that $f(x_1) \le \alpha$ and

$f(x_2) \leq \alpha$. Since f is convex in $\mathcal{C}$, it follows that

$$f(\mu x_1 + (1-\mu)x_2) \leq \mu f(x_1) + (1-\mu)f(x_2) \leq \alpha, \tag{2.47}$$

for all $x_1, x_2 \in f^{-1}((-\infty, \alpha])$, $\alpha \in \mathbb{R}$, and $\mu \in [0,1]$. Hence, $\mu x_1 + (1-\mu)x_2 \in f^{-1}((-\infty, \alpha])$, $\alpha \in \mathbb{R}$, and hence, $f^{-1}((-\infty, \alpha])$, $\alpha \in \mathbb{R}$, is convex. □

Definition 2.22. Let $\mathcal{D} \subseteq \mathbb{R}^n$ and let $f : \mathcal{D} \to \mathbb{R}$. The *graph* of f is defined by

$$\mathcal{F} \triangleq \{(x, y) \in \mathcal{D} \times \mathbb{R} : \ y = f(x)\}.$$

The *epigraph* of f is defined by

$$\mathcal{E} \triangleq \{(x, y) \in \mathcal{D} \times \mathbb{R} : \ y \geq f(x)\}.$$

Definition 2.23. Let $\mathcal{D} \subset \mathbb{R}^n$. Then $\mathcal{D}$ is a *hyperplane* if there exists $\mu \in \mathbb{R}^n$, $\mu \neq 0$, such that $\mathcal{D} = \{x \in \mathbb{R}^n : \ \mu^{\mathrm{T}} x = 0\}$.

The following two definitions introduce the notions of convergent sequences of functions.

Definition 2.24. Let $\mathcal{D} \subseteq \mathbb{R}^n$, $f : \mathcal{D} \to \mathbb{R}^n$, and $f_n : \mathcal{D} \to \mathbb{R}^n$, $n = 1, \ldots$. A sequence of functions $\{f_n\}_{n=0}^{\infty}$ *converges* to f if, for every $x \in \mathcal{D}$, $\lim_{n\to\infty} \|f(x) - f_n(x)\| = 0$ or, equivalently, for every $\varepsilon > 0$, there exists $N = N(\varepsilon, x)$ such that $\|f(x) - f_n(x)\| < \varepsilon$ for all $n \geq N$.

Definition 2.25. Let $\mathcal{D} \subseteq \mathbb{R}^n$, $f : \mathcal{D} \to \mathbb{R}^n$, and $f_n : \mathcal{D} \to \mathbb{R}^n$, $n = 1, \ldots$. A sequence of functions $\{f_n\}_{n=0}^{\infty}$ *converges uniformly* to f if, for every $\varepsilon > 0$, there exists $N = N(\varepsilon)$ such that $\|f(x) - f_n(x)\| < \varepsilon$ for all $x \in \mathcal{D}$ and $n \geq N$.

Example 2.13. Consider the infinite sequence of functions $\{f_n\}_{n=1}^{\infty}$, where $f_n : [0,1] \to \mathbb{R}$ is given by $f_n(x) = x^n$. This sequence has the form $\{x, x^2, x^3, \ldots\}$ and its limit depends on the value of x. In particular, for $0 \leq x < 1$, the sequence converges to zero as $n \to \infty$. For example, for $x = 1/2$, $\{f_n\}_{n=1}^{\infty} = \{\frac{1}{2}, \frac{1}{4}, \frac{1}{8}, \ldots, \frac{1}{2^n}, \ldots\}$, which converges to zero as $n \to \infty$. For $x = 1$, however, $\{f_n\}_{n=1}^{\infty} = \{1, 1, 1, \ldots\}$, which converges to 1. Hence, given $\varepsilon > 0$, there exists $N = N(\varepsilon, x)$ such that $|f(x) - x^n| < \varepsilon$ for all $n \geq N$ and $0 \leq x \leq 1$, where $f(x) = \lim_{n\to\infty} x^n$. Thus, the sequence converges pointwise to $f(x)$. However, the sequence does not converge uniformly to f. To see this, note that for every $x \in [0, 1)$ we have $|f(x) - x^n| = |0 - x^n| = x^n$. Choosing $x \approx 1$, we can ensure that $x^n > \varepsilon$ no matter how large $N = N(\varepsilon) \neq N(\varepsilon, x)$ is chosen. In particular, $x^n > \varepsilon$ implies that $x > \sqrt[n]{\varepsilon}$, and hence, any $x \in (\sqrt[n]{\varepsilon}, 1)$ will yield $|f(x) - x^n| > \varepsilon$, $n \geq N$, for $N = N(\varepsilon)$. Hence, the sequence of functions $f_n(x)$ do not converge uniformly to f. See Figure 2.1. △

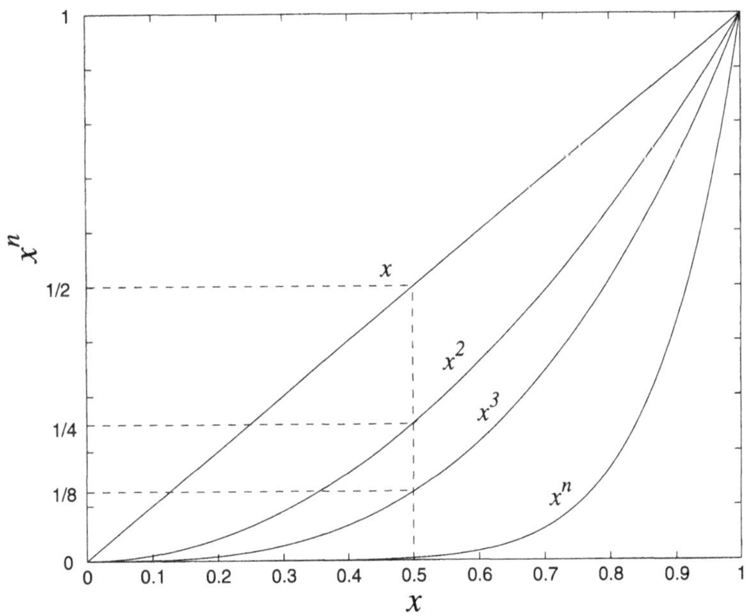

Figure 2.1 Graphs for x^n.

The next definition introduces the notion of monotonic functions defined on an interval $\mathcal{I} \subseteq \mathbb{R}$.

Definition 2.26. Let $\mathcal{I} \subseteq \mathbb{R}$ and let $f : \mathcal{I} \to \mathbb{R}$. f is *strictly increasing* on $\mathcal{I}$ if for every $x, y \in \mathcal{I}$, $x < y$ implies $f(x) < f(y)$. f is *increasing* (or *nondecreasing*) on $\mathcal{I}$ if for every $x, y \in \mathcal{I}$, $x < y$ implies $f(x) \leq f(y)$. f is *strictly decreasing* on $\mathcal{I}$ if for every $x, y \in \mathcal{I}$, $x < y$ implies $f(x) > f(y)$. f is *decreasing* (or *nonincreasing*) on $\mathcal{I}$ if for every $x, y \in \mathcal{I}$, $x < y$ implies $f(x) \geq f(y)$. f is *monotonic* on $\mathcal{I}$ if it is increasing on $\mathcal{I}$ or decreasing on $\mathcal{I}$.

Note that if f is an increasing function, then $-f$ is a decreasing function. Hence, in many situations involving monotonic functions it suffices to consider only the case of increasing or decreasing functions.

Definition 2.27. Let $\mathcal{D} \subseteq \mathbb{R}^n$ and let $f : \mathcal{D} \to \mathbb{R}^n$. f is *bounded on* $\mathcal{D}$ if there exists $\alpha > 0$ such that $\|f(x)\| \leq \alpha$ for all $x \in \mathcal{D}$.

The next theorem shows that if a monotone function on $\mathbb{R}$ is bounded, then its limit exists and is finite.

Theorem 2.10 (Monotone Convergence Theorem). Let $f : \mathbb{R} \to \mathbb{R}$ be decreasing (respectively, increasing) on $\mathbb{R}$ and assume that there exists

$\gamma \in \mathbb{R}$ such that $f(x) \geq \gamma$, $x \in \mathbb{R}$ (respectively, $f(x) \leq \gamma$, $x \in \mathbb{R}$). Then $\lim_{x\to\infty} f(x)$ exists.

Proof. Assume f is decreasing and bounded below, that is, there exists $\gamma \in \mathbb{R}$ such that $f(x) \geq \gamma$, $x \in \mathbb{R}$. Let $\alpha = \inf_{x\in\mathbb{R}} f(x)$ and note that $\alpha \geq \gamma$ since $f(x) \geq \gamma$, $x \in \mathbb{R}$. It follows from the definition of infimum that for every $\varepsilon > 0$ there exists $x_\varepsilon \in \mathbb{R}$ such that $\alpha \leq f(x_\varepsilon) < \alpha + \varepsilon$, which implies that $\alpha \leq f(x) < \alpha + \varepsilon$, $x \geq x_\varepsilon$. Hence, for every $\varepsilon > 0$, there exists $x_\varepsilon \in \mathbb{R}$ such that $0 \leq f(x) - \alpha < \varepsilon$, $x \geq x_\varepsilon$, which implies that $\lim_{x\to\infty} f(x) = \alpha$. The proof for the case where f is increasing and bounded above follows using identical arguments and, hence, is omitted. □

The next definition introduces the very important notion of continuity of a function at a point. The notion of continuity allows one to deduce the behavior of a function in a neighborhood of a given point by using knowledge of the value of the function at that point.

Definition 2.28. Let $\mathcal{D} \subseteq \mathbb{R}^n$, $f : \mathcal{D} \to \mathbb{R}^n$, and $x \in \mathcal{D}$. Then f is *continuous* at $x \in \mathcal{D}$ if, for every $\varepsilon > 0$, there exists $\delta = \delta(\varepsilon, x) > 0$ such that $\|f(x) - f(y)\| < \varepsilon$ for all $y \in \mathcal{D}$ satisfying $\|x - y\| < \delta$. f is *discontinuous* at x if f is not continuous at x.

Definition 2.28 is equivalent to

$$f(\mathcal{B}_\delta(x)) \subset \mathcal{B}_\varepsilon(f(x)). \tag{2.48}$$

In particular, for every open ball $\mathcal{B}_\varepsilon(f(x))$ there exists an open ball $\mathcal{B}_\delta(x)$ such that the points of $\mathcal{B}_\delta(x)$ are mapped into $\mathcal{B}_\varepsilon(f(x))$. It follows from (2.48) that

$$\mathcal{B}_\delta(x) \subset f^{-1}(\mathcal{B}_\varepsilon(f(x))), \tag{2.49}$$

that is, the open ball $\mathcal{B}_\delta(x)$ of radius $\delta > 0$ and center x lies in the subset $f^{-1}(\mathcal{B}_\varepsilon(f(x)))$ of $\mathcal{D}$. (See Figure 2.2.)

An equivalent statement for continuity at a point can be given in terms of convergent sequences. Specifically, $f : \mathcal{D} \to \mathbb{R}^n$ is continuous at x if, for every sequence $\{x_n\}_{n=1}^\infty \subset \mathcal{D}$ such that $\lim_{n\to\infty} x_n = x$, $\lim_{n\to\infty} f(x_n) = f(x)$ or, equivalently, $\lim_{n\to\infty} f(x_n) = f(\lim_{n\to\infty} x_n)$. To see the equivalence between this definition and Definition 2.28, assume f is continuous at x_0 so that for every $\varepsilon > 0$ there exists $\delta > 0$ such that $\|x - x_0\| < \delta$ implies $\|f(x) - f(x_0)\| < \varepsilon$, $x \in \mathcal{D}$. Now, take any sequence $\{x_n\}_{n=1}^\infty \subset \mathcal{D}$ which converges to $x_0 \in \mathcal{D}$. Hence, for the given $\delta > 0$, there exists an integer N such that $n > N$ implies $\|x_n - x_0\| < \delta$. Since $\|f(x) - f(x_0)\| < \varepsilon$ whenever $x \in \mathcal{D}$ and $\|x - x_0\| < \delta$, it follows that $\|f(x_n) - f(x_0)\| < \varepsilon$ for $n > N$, and hence, $\{f(x_n)\}_{n=1}^\infty$ converges to $f(x_0)$. This shows sufficiency. To show necessity, assume for every sequence $\{x_n\}_{n=1}^\infty \subset \mathcal{D}$ such that $x_n \to$

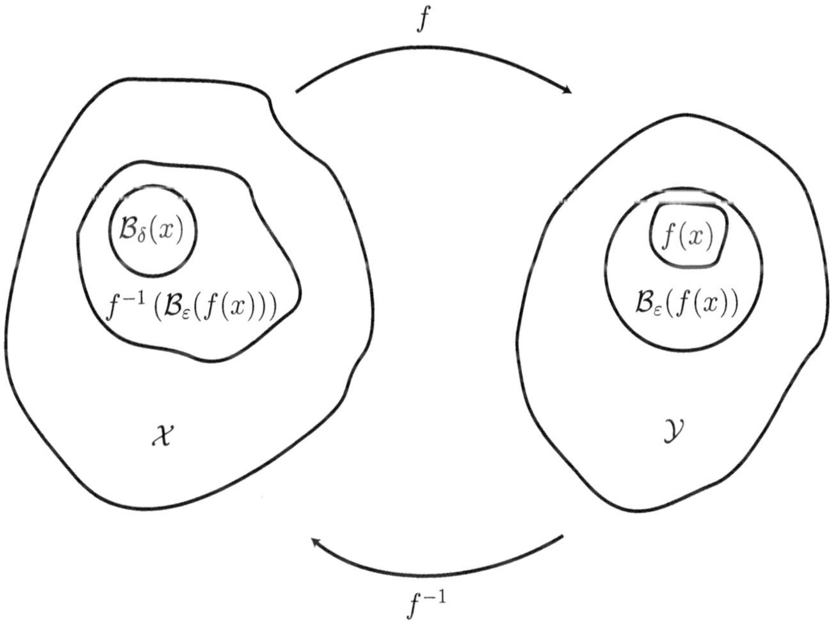

Figure 2.2 Continuity of $f : \mathcal{X} \to \mathcal{Y}$ at x.

x_0, $f(x_n) \to f(x_0)$ as $n \to \infty$. Now, suppose, *ad absurdum*, for some $\varepsilon > 0$ and every $\delta > 0$ there exists $x \in \mathcal{D}$ such that $\|x - x_0\| < \delta$ and $\|f(x) - f(x_0)\| \geq \varepsilon$. Choose $\delta = 1/n$, $n = 1, 2, \ldots$. This implies there exists a corresponding sequence of points $\{x_n\}_{n=1}^{\infty} \subset \mathcal{D}$ such that $\|x_n - x_0\| < 1/n$ and $\|f(x_n) - f(x_0)\| \geq \varepsilon$. Clearly, the sequence $\{x_n\}_{n=1}^{\infty}$ converges to x_0; however, the sequence $\{f(x_n)\}_{n=1}^{\infty}$ does not converge to $f(x_0)$, leading to a contradiction. This proves necessity.

The function f is said to be *continuous on* $\mathcal{D}$ if f is continuous at every point $x \in \mathcal{D}$. The next proposition establishes the fact that the continuous image of a compact set is compact.

Proposition 2.14. Let $\mathcal{D}_\mathrm{c} \subset \mathbb{R}^m$ be a compact set and let $f : \mathcal{D}_\mathrm{c} \to \mathbb{R}^n$ be continuous on $\mathcal{D}_\mathrm{c}$. Then the image of $\mathcal{D}_\mathrm{c}$ under f is compact.

Proof. Let $\{y_n\}_{n=1}^{\infty}$ be a sequence in the range of f. In this case, there exists a corresponding sequence $\{x_n\}_{n=1}^{\infty} \subset \mathcal{D}_\mathrm{c}$ such that $y_n = f(x_n)$. Since $\mathcal{D}_\mathrm{c}$ is compact, it follows from Theorem 2.4 that there exists a subsequence $\{x_{n_k}\}_{k=1}^{\infty} \subseteq \{x_n\}_{n=1}^{\infty}$ such that $\lim_{k\to\infty} x_{n_k} = x \in \mathcal{D}_\mathrm{c}$. Now, since f is continuous it follows that $\lim_{k\to\infty} f(x_{n_k}) = f(\lim_{k\to\infty} x_{n_k}) = f(x) \in f(\mathcal{D}_\mathrm{c})$. Hence, $\{y_n\}_{n=1}^{\infty}$ has a convergent subsequence in $f(\mathcal{D}_\mathrm{c})$, and hence, by Theorem 2.4, $f(\mathcal{D}_\mathrm{c})$ is compact. □

It is important to note that the existence of δ in Definition 2.28 is in general dependent on both ε and x. In the case where δ is not a function of the point x we have the stronger notion of uniform continuity over the set $\mathcal{D}$.

Definition 2.29. Let $\mathcal{D} \subseteq \mathbb{R}^n$ and $f : \mathcal{D} \to \mathbb{R}^n$. Then f is *uniformly continuous* on $\mathcal{D}$ if, for every $\varepsilon > 0$, there exists $\delta = \delta(\varepsilon) > 0$ such that $\|f(x) - f(y)\| < \varepsilon$ for all $x, y \in \mathcal{D}$ satisfying $\|x - y\| < \delta$.

Note that if $f : \mathcal{D} \to \mathbb{R}^n$ is uniformly continuous on $\mathcal{D}$, then f is continuous at every $x \in \mathcal{D}$. The converse of this statement, however, is not true.

Example 2.14. The function $f : (0, \infty) \to \mathbb{R}$ defined by $f(x) = 1/x$ is continuous at every point in $(0, \infty)$ but is not uniformly continuous on $(0, \infty)$. For example, let $\mathcal{D} = (0, 1)$. Clearly, f is continuous on $\mathcal{D}$ but not uniformly continuous on $\mathcal{D}$. To see this, let $\varepsilon > 1$ be such that $1/\varepsilon < \delta$, and let $x = 1/\varepsilon$ and $y = 1/(\varepsilon + 1)$. Note that $x, y \in \mathcal{D}$. In this case,

$$|x - y| = \left|\frac{1}{\varepsilon} - \frac{1}{\varepsilon + 1}\right| = \frac{1}{\varepsilon(\varepsilon + 1)} < \frac{1}{\varepsilon} < \delta.$$

However,

$$|f(x) - f(y)| = |\varepsilon - (\varepsilon + 1)| = 1 > \varepsilon.$$

Hence, for these two points we have $|f(x) - f(y)| > \varepsilon$ whenever $|x - y| < \delta$, contradicting the definition of uniform continuity. △

Example 2.15. The function $f : \mathcal{D} \to \mathbb{R}$ defined by $f(x) = x^2$ is uniformly continuous on $\mathcal{D} = (0, 1]$. To see this, note that for all $x, y \in \mathcal{D}$,

$$|f(x) - f(y)| = |x^2 - y^2| = |(x - y)(x + y)| \leq 2|x - y|.$$

If $|x - y| < \delta$, then $|f(x) - f(y)| < 2\delta$. Hence, given $\varepsilon > 0$, we need only take $\delta = \varepsilon/2$ to guarantee that $|f(x) - f(y)| < \varepsilon$ for every $x, y \in \mathcal{D}$ such that $|x - y| < \delta$. Since δ is independent of x, this shows uniform continuity on $\mathcal{D}$. △

Example 2.16. The function $f : \mathcal{D} \to \mathbb{R}$ defined by $f(x) = 1/(x^2+1)$ is uniformly continuous on $\mathcal{D} = [-1, 1]$. To see this, note that for all $x, y \in \mathcal{D}$,

$$\begin{aligned}
|f(x) - f(y)| &= \left|\frac{1}{x^2 + 1} - \frac{1}{y^2 + 1}\right| \\
&= \left|\frac{(y - x)(y + x)}{(x^2 + 1)(y^2 + 1)}\right| \\
&= \frac{|y - x||y + x|}{|x^2 + 1||y^2 + 1|} \\
&\leq |y - x||y + x|
\end{aligned}$$

$$\leq (|y| + |x|)|y - x|$$
$$\leq 2|y - x|.$$

Hence, given $\varepsilon > 0$ we need only take $\delta = \varepsilon/2$ to guarantee that $|f(x) - f(y)| < \varepsilon$ for every $x, y \in \mathcal{D}$ whenever $|x - y| < \delta$. In particular, $|f(x) - f(y)| \leq 2|x - y| < 2\delta = \varepsilon$. Since δ is independent of x, this shows uniform continuity on $\mathcal{D}$. △

Example 2.17. The signum function sgn : $\mathbb{R} \to \{-1, 0, 1\}$ is defined as $\operatorname{sgn} x \triangleq x/|x|$, $x \neq 0$, and $\operatorname{sgn}(0) \triangleq 0$. Since $\lim_{x\to 0^+} \operatorname{sgn}(x) = 1 \neq \operatorname{sgn}(0)$, the signum function is discontinuous at $x = 0$. △

Example 2.16 shows that if a continuous function is defined on a compact set, then the function is uniformly continuous. In other words, for compact sets, continuity implies uniform continuity. This fact is established in the following proposition.

Proposition 2.15. Let $\mathcal{D}_c \subset \mathbb{R}^m$ be a compact set and let $f : \mathcal{D}_c \to \mathbb{R}^n$ be continuous on $\mathcal{D}_c$. Then f is uniformly continuous on $\mathcal{D}_c$.

Proof. Suppose, *ad absurdum*, that f is not uniformly continuous on $\mathcal{D}_c$. In this case, there exists $\varepsilon > 0$ for which no $\delta = \delta(\varepsilon) > 0$ exists such that $\|f(x) - f(y)\| < \varepsilon$ for all $x, y \in \mathcal{D}_c$ satisfying $\|x - y\| < \delta$. In particular, none of the δ's given by $\delta = 1/2, 1/3, \ldots, 1/n, \ldots$ can be used. Hence, for each $n \in \mathbb{Z}_+$, there exist $x_n, y_n \in \mathcal{D}_c$ such that $\|x_n - y_n\| < 1/n$ and $\|f(x_n) - f(y_n)\| \geq \varepsilon$. Since $\mathcal{D}_c$ is compact, it follows from Theorem 2.4 that there exists a convergent subsequence $\{x_{n_k}\}_{k=1}^{\infty} \subseteq \{x_n\}_{n=1}^{\infty} \subseteq \mathcal{D}_c$ such that

$$\lim_{k\to\infty} x_{n_k} = x, \quad x \in \mathcal{D}_c. \tag{2.50}$$

Now, using the fact that $\|x_n - y_n\| < 1/n$ and $\|f(x_n) - f(y_n)\| \geq \varepsilon$, it follows that

$$\lim_{k\to\infty} (x_{n_k} - y_{n_k}) = 0. \tag{2.51}$$

Next, subtracting (2.51) from (2.50) yields $\lim_{k\to\infty} y_{n_k} = x$. Since f is continuous it follows that

$$\lim_{k\to\infty} [f(x_{n_k}) - f(y_{n_k})] = f(x) - f(x) = 0,$$

which implies that, for large enough $k \in \mathbb{Z}_+$, $\|f(x_{n_k}) - f(y_{n_k})\| < \varepsilon$. This leads to a contradiction. □

Continuity can alternatively be characterized by the following proposition.

Proposition 2.16. Let $\mathcal{D} \subseteq \mathbb{R}^n$ and let $f : \mathcal{D} \to \mathbb{R}^n$. Then f is continuous on $\mathcal{D}$ if and only if, for every open (respectively, closed) set $\mathcal{Q} \subseteq \mathbb{R}^n$, the inverse image $f^{-1}(\mathcal{Q}) \subseteq \mathcal{D}$ of $\mathcal{Q}$ is open (respectively, closed) relative to $\mathcal{D}$.

Proof. Let f be continuous on $\mathcal{D}$, let $\mathcal{Q} \subseteq \mathbb{R}^n$ be open, and let $x \in f^{-1}(\mathcal{Q})$, that is, $y \triangleq f(x) \in \mathcal{Q}$. Since $\mathcal{Q}$ is open it follows that there exists $\varepsilon > 0$ such that $\mathcal{B}_\varepsilon(y) \subseteq \mathcal{Q}$, and since f is continuous at x it follows that there exists $\delta > 0$ such that $f(\hat{x}) \in \mathcal{B}_\varepsilon(y)$ for all $\hat{x} \in \mathcal{B}_\delta(x)$ or, equivalently, $f(\mathcal{B}_\delta(x)) \subseteq \mathcal{B}_\varepsilon(y)$. Hence,

$$\mathcal{B}_\delta(x) \subseteq f^{-1}(f(\mathcal{B}_\delta(x))) \subseteq f^{-1}(\mathcal{B}_\varepsilon(y)) \subseteq f^{-1}(\mathcal{Q}),$$

which shows that $f^{-1}(\mathcal{Q})$ is open relative to $\mathcal{D}$.

Conversely, assume that $f^{-1}(\mathcal{Q})$ is open relative to $\mathcal{D}$ for every open $\mathcal{Q} \subseteq \mathbb{R}^n$. Let $x \in \mathcal{D}$ and let $y = f(x)$. Since for every $\varepsilon > 0$, $\mathcal{B}_\varepsilon(y)$ is open in $\mathbb{R}^n$ it follows that $f^{-1}(\mathcal{B}_\varepsilon(y))$ is open relative to $\mathcal{D}$. Now, since $x \in f^{-1}(\mathcal{B}_\varepsilon(y))$, there exists $\delta > 0$ such that $\mathcal{B}_\delta(x) \subseteq f^{-1}(\mathcal{B}_\varepsilon(y))$, which proves that f is continuous at x.

Finally, the proof of the statement f is continuous on $\mathcal{D}$ if and only if, for every closed set $\mathcal{Q} \subseteq \mathbb{R}^n$, the inverse image $f^{-1}(\mathcal{Q}) \subseteq \mathcal{D}$ of $\mathcal{Q}$ is closed relative to $\mathcal{D}$ is analogous to the proof given above and, hence, is omitted.

□

It follows from Proposition 2.16 that every level set and every sublevel set of a continuous scalar-valued function is closed relative to the domain of the function.

It is important to note that boundedness is not preserved by a continuous mapping. For example, consider the continuous function $f(x) = 1/x$ and consider the bounded set $\mathcal{Q} \triangleq (0, 1)$. Then $f(\mathcal{Q}) = (1, \infty)$ is not bounded. However, as shown in the next proposition, boundedness is preserved under a uniformly continuous mapping.

Proposition 2.17. Let $\mathcal{D} \subseteq \mathbb{R}^n$, let $\mathcal{Q} \subseteq \mathcal{D}$ be bounded, and let $f : \mathcal{D} \to \mathbb{R}^n$ be uniformly continuous on $\mathcal{Q}$. Then $f(\mathcal{Q})$ is bounded.

Proof. Let $\varepsilon > 0$ be given. Since f is uniformly continuous on $\mathcal{Q}$, there exists $\delta > 0$ such that $\|f(x) - f(y)\| < \varepsilon$ for all $x, y \in \mathcal{Q}$ such that $\|x - y\| < \delta$. Now, since $\mathcal{Q}$ is bounded, there exists a finite subset $\hat{\mathcal{Q}} = \{x_1, \ldots, x_q\}$ of $\mathcal{Q}$ such that, for each $x \in \mathcal{Q}$, there exists $k \in \{1, 2, \ldots, q\}$ such that $\|x - x_k\| < \delta$. Next, define

$$\beta \triangleq \max_{1 \le i,j \le q} \|f(x_i) - f(x_j)\|.$$

Now, let $x, y \in \mathcal{Q}$ and $x_k, x_l \in \hat{\mathcal{Q}}$ be such that $\|x - x_k\| < \delta$ and $\|y - x_l\| < \delta$. Then,

$$\begin{aligned}\|f(x) - f(y)\| &\leq \|f(x) - f(x_k)\| + \|f(x_k) - f(x_l)\| + \|f(x_l) - f(y)\| \\ &\leq 2\varepsilon + \beta.\end{aligned}$$

Hence, $f(\mathcal{Q})$ is bounded. □

Even though the image of a bounded set under a continuous mapping is not necessarily bounded, the image of a compact set under a continuous function is compact. An identical statement is true for connected sets. For details see Problem 2.54.

Example 2.18. Let $\mathcal{D} \subset \mathbb{R}$ and let $f : \mathcal{D} \to \mathbb{R}$ be given by $f(x) = 1/x$. For $\mathcal{D} = (1,2)$, f is bounded on $\mathcal{D}$ since $\inf\{1/x : 1 < x < 2\} = 1/2$ and $\sup\{1/x : 1 < x < 2\} = 1$. For $\mathcal{D} = [2, \infty)$, f is also bounded on $\mathcal{D}$ since $\inf\{1/x : x \geq 2\} = 0$ and $\sup\{1/x : x \geq 2\} = 1/2$. In this case, however, the infimum of zero is not attained since the set $\{1/x : x \geq 2\}$ does not contain zero. Finally, note that for $\mathcal{D} = (0,1)$, f is not bounded on $\mathcal{D}$ since $\sup\{1/x : 0 < x < 1\} = \infty$. △

The next definition introduces the notion of lower semicontinuous and upper semicontinuous functions at $x \in \mathcal{D}$.

Definition 2.30. Let $\mathcal{D} \subseteq \mathbb{R}^n$, $f : \mathcal{D} \to \mathbb{R}$, and $x \in \mathcal{D}$. f is *lower semicontinuous* at $x \in \mathcal{D}$ if for every sequence $\{x_n\}_{n=0}^{\infty} \subset \mathcal{D}$ such that $\lim_{n\to\infty} x_n = x$, $f(x) \leq \liminf_{n\to\infty} f(x_n)$.

Note that a function $f : \mathcal{D} \to \mathbb{R}$ is lower semicontinuous at $x \in \mathcal{D}$ if and only if for each $\alpha \in \mathbb{R}$ the set $\{x \in \mathcal{D} : f(x) > \alpha\}$ is open. Alternatively, a bounded function $f : \mathcal{D} \to \mathbb{R}$ is lower semicontinuous at $x \in \mathcal{D}$ if and only if for each $\varepsilon > 0$ there exists $\delta > 0$ such that $\|x - y\| < \delta$, $y \in \mathcal{D}$, implies $f(x) - f(y) \leq \varepsilon$.

Definition 2.31. Let $\mathcal{D} \subseteq \mathbb{R}^n$, $f : \mathcal{D} \to \mathbb{R}$, and $x \in \mathcal{D}$. f is *upper semicontinuous* at $x \in \mathcal{D}$ if for every sequence $\{x_n\}_{n=0}^{\infty} \subset \mathcal{D}$ such that $\lim_{n\to\infty} x_n = x$, $f(x) \geq \limsup_{n\to\infty} f(x_n)$, or, equivalently, for each $\alpha \in \mathbb{R}$ the set $\{x \in \mathcal{D} : f(x) < \alpha\}$ is open.

As in the case of continuous functions, a function f is said to be *lower* (respectively, *upper*) *semicontinuous on* $\mathcal{D}$ if f is lower (respectively, upper) semicontinuous at every point $x \in \mathcal{D}$. Clearly, if f is both lower and upper semicontinuous, then f is continuous. The function $f(x) = -1$ for $x < 0$ and $f(x) = 1$ for $x \geq 0$ is upper semicontinuous at $x = 0$, but not lower semicontinuous at $x = 0$. The *floor function* $f(x) = \lfloor x \rfloor$, which returns the greatest integer less than or equal to a given x, is upper semicontinuous on $\mathbb{R}$. Similarly, the *ceiling function* $f(x) = \lceil x \rceil$, which returns the smallest integer greater than or equal to a given x, is lower semicontinuous on $\mathbb{R}$.

Next, we present three key theorems due to Weierstrass involving the existence of global minimizers and maximizers of lower and upper semicontinuous functions on compact sets.

Theorem 2.11 (Weierstrass Theorem). Let $\mathcal{D}_c \subset \mathbb{R}^n$ be compact and let $f : \mathcal{D}_c \to \mathbb{R}$ be lower semicontinuous on $\mathcal{D}_c$. Then there exists $x^* \in \mathcal{D}_c$ such that $f(x^*) \leq f(x)$, $x \in \mathcal{D}_c$.

Proof. Let $\{x_n\}_{n=0}^{\infty} \subset \mathcal{D}_c$ be a sequence such that $\lim_{n\to\infty} f(x_n) = \inf_{x\in\mathcal{D}_c} f(x)$. Since $\mathcal{D}_c$ is bounded, it follows from the Bolzano-Weierstrass theorem (Theorem 2.3) that every sequence in $\mathcal{D}_c$ has at least one accumulation point x^*. Now, since $\mathcal{D}_c$ is closed, $x^* \in \mathcal{D}_c$, and since $f(\cdot)$ is lower semicontinuous on $\mathcal{D}_c$, it follows that $f(x^*) \leq \lim_{n\to\infty} f(x_n) = \inf_{x\in\mathcal{D}_c} f(x)$. Hence, $f(x^*) = \inf_{x\in\mathcal{D}_c} f(x)$. □

Theorem 2.12. Let $\mathcal{D}_c \subset \mathbb{R}^n$ be compact and let $f : \mathcal{D}_c \to \mathbb{R}$ be upper semicontinuous on $\mathcal{D}_c$. Then there exists $x^* \in \mathcal{D}_c$ such that $f(x^*) \geq f(x)$, $x \in \mathcal{D}_c$.

Proof. The proof is identical to the proof of Theorem 2.11. □

The following theorem combines Theorems 2.11 and 2.12 to show that continuous functions on compact sets attain a minimum and maximum.

Theorem 2.13. Let $\mathcal{D}_c \subset \mathbb{R}^n$ be compact and let $f : \mathcal{D}_c \to \mathbb{R}$ be continuous on $\mathcal{D}_c$. Then there exist $x_{\min} \in \mathcal{D}_c$ and $x_{\max} \in \mathcal{D}_c$ such that $f(x_{\min}) \leq f(x)$, $x \in \mathcal{D}_c$, and $f(x_{\max}) \geq f(x)$, $x \in \mathcal{D}_c$.

Proof. Since f is continuous on $\mathcal{D}_c$ if and only if f is both lower and upper semicontinuous on $\mathcal{D}_c$, the result is a direct consequence of Theorems 2.11 and 2.12. □

A slightly weaker notion of a continuous function on an interval $\mathcal{I} \subset \mathbb{R}$ is a piecewise continuous function on $\mathcal{I}$, wherein the function is continuous everywhere on an interval except possibly at a finite number of points in every finite interval. For the next definition, recall that an interval is a connected subset of $\mathbb{R}$ and for an interval $\mathcal{I} \subseteq \mathbb{R}$, the *endpoints* of $\mathcal{I}$ are the boundary points of $\mathcal{I}$ that belong to $\mathcal{I}$.

Definition 2.32. Let $\mathcal{I} \subseteq \mathbb{R}$ and let $f : \mathcal{I} \to \mathbb{R}^n$. Then f is *piecewise continuous* on $\mathcal{I}$ if for every bounded subinterval $\mathcal{I}_0 \subset \mathcal{I}$, f is continuous for all $t \in \mathcal{I}_0$ except at a finite number of points at which the left-side limit $f(t^-) \triangleq \lim_{h\to 0, h>0} f(t+h)$ and the right-side limit $f(t^+) \triangleq \lim_{h\to 0, h>0} f(t-h)$ exist. f is *left continuous at* t if $f(t^-) = f(t)$, and *right continuous at* t if $f(t^+) = f(t)$. The function is *left* (respectively, *right*) *piecewise continuous* if f is left (respectively, right) continuous at every point in the interior of $\mathcal{I}$ and if $\mathcal{I}$ has a left endpoint t_0, then $f(t_0^+) = f(t_0)$, while, if $\mathcal{I}$ has a right endpoint $t_{\rm f}$, then $f(t_{\rm f}^-) = f(t_{\rm f})$. The function $f : \mathcal{I} \to \mathbb{R}^n$ is *piecewise continuously differentiable* on $\mathcal{I}$ if f is piecewise continuous and its first derivative exists and is piecewise continuous on each subinterval $\mathcal{I}_k = \{t : t_{k-1} < t < t_k\}$, $k = 1, \ldots, n$, of $\mathcal{I}$, where $t_n = t_{\rm f}$.

Note that a function f is piecewise continuous on $[t_0, t_1]$ *i*) if it is continuous on $[t_0, t_1]$ at all but a finite number of points on (t_0, t_1); *ii*) if f is discontinuous at $t^* \in (t_0, t_1)$, then the left and right limits of $f(t)$ exist as t approaches t^* from the left and the right; and *iii*) if f is discontinuous at t^*, then $f(t^*)$ is equal to either the left or the right limit of $f(t)$.

Example 2.19. Consider the function $f : \mathbb{R} \to \{0, 1\}$ given by

$$f(x) = \begin{cases} 1, & x \neq 0, \\ 0, & x = 0. \end{cases} \tag{2.52}$$

Clearly, f is not continuous at $x = 0$. The left-side limit $f(0^-)$ and right-side limit of f exist and satisfy $f(0^-) = f(0^+) = 1$. Hence, f is piecewise continuous on $\mathbb{R}$. However, f is neither left piecewise continuous nor right piecewise continuous on $\mathbb{R}$. △

Proposition 2.18. Let $\mathcal{I} \subset \mathbb{R}$ be a compact interval and let $f : \mathcal{I} \to \mathbb{R}^n$ be piecewise continuous on $\mathcal{I}$. Then $f(\mathcal{I})$ is bounded.

Proof. Since f is piecewise continuous on a compact interval $\mathcal{I} \subseteq \mathbb{R}$, by definition, f is continuous on a finite number of intervals $[t_0, t_1)$, (t_1, t_2), $\ldots$, $(t_{n-1}, t_n]$, where $\cup_{k=1}^n [t_{k-1}, t_k] = \mathcal{I}$. Furthermore, for every $k \in \{1, \ldots, n\}$, the left-side and the right-side limits of $f(t)$ exist as t approaches t_k from the left and the right. Hence, $\eta_k \triangleq \sup_{t \in (t_{k-1}, t_k)} |f(t)| < \infty$, which implies that

$$|f(t)| \le \max\left\{ \max_{k\in\{1,\ldots,n\}} \eta_k, \max_{k\in\{0,1,\ldots,n\}} f(t_k) \right\},$$

proving that $f(\mathcal{I})$ is bounded. □

The next definition introduces the concept of *Lipschitz continuity*.

Definition 2.33. Let $\mathcal{D} \subseteq \mathbb{R}^m$ and let $f : \mathcal{D} \to \mathbb{R}^n$. Then f is *Lipschitz continuous at* $x_0 \in \mathcal{D}$ if there exists a *Lipschitz constant* $L = L(x_0) > 0$ and a neighborhood $\mathcal{N} \subset \mathcal{D}$ of x_0 such that

$$\|f(x) - f(y)\| \le L\|x - y\|, \quad x, y \in \mathcal{N}. \tag{2.53}$$

f is *Lipschitz continuous on* $\mathcal{D}$ if f is Lipschitz continuous at every point in $\mathcal{D}$. f is *uniformly Lipschitz continuous on* $\mathcal{D}$ if there exists $L > 0$ such that

$$\|f(x) - f(y)\| \le L\|x - y\|, \quad x, y \in \mathcal{D}. \tag{2.54}$$

Finally, f is *globally Lipschitz continuous* if f is uniformly Lipschitz continuous on $\mathcal{D} = \mathbb{R}^m$.

Note that a function can be Lipschitz continuous on $\mathcal{D}$ but not uniformly Lipschitz continuous on $\mathcal{D}$ since the Lipschitz condition might not hold for the same Lipschitz constant on $\mathcal{D}$. For example, $f(x) = x^2$ is Lipschitz continuous on $\mathbb{R}$ but *not* uniformly Lipschitz continuous on $\mathbb{R}$. However, if we restrict the domain to the closed interval $[0, 1]$, then $f(x) = x^2$ is uniformly Lipschitz continuous over $[0, 1]$ with Lipschitz constant $L = 2$. To see this, note that $|x^2 - y^2| = |x + y|\,|x - y| \le 2|x - y|$ for all $x, y \in [0, 1]$. For $f : \mathbb{R} \to \mathbb{R}$ the Lipschitz condition gives

$$\frac{|f(x) - f(y)|}{|x - y|} \le L, \tag{2.55}$$

which implies that any line segment joining two arbitrary points on the graph of f cannot have a slope greater than $\pm L$. Hence, any function having an infinite slope at a point $x_0 \in \mathcal{D}$ is not Lipschitz continuous at x_0. This of course rules out all discontinuous functions from being Lipschitz continuous at the point of discontinuity. Alternatively, if f is Lipschitz continuous at a point $x_0 \in \mathcal{D}$, then f is clearly continuous at x_0. The converse, however, is not true. The same remarks hold for uniform Lipschitz continuity. In particular, if f is uniformly Lipschitz continuous on $\mathcal{D}$, then f is uniformly continuous on $\mathcal{D}$. Once again, the converse of this statement is not true.

Example 2.20. Consider the function $f : \mathbb{R} \to \mathbb{R}$ given by

$$f(x) = \begin{cases} x^2 \sin(1/x), & x \ne 0, \\ 0, & x = 0. \end{cases} \tag{2.56}$$

For $x \ne 0$ and $y = 0$, it follows that

$$\begin{aligned} |f(x) - f(y)| &= |f(x)| \\ &= |x^2 \sin(1/x)| \\ &\le x^2 \\ &\le \rho|x| \\ &= \rho|x - y|, \qquad |x| \le \rho. \end{aligned} \tag{2.57}$$

Similarly, for $x = 0$ and $y \neq 0$, it follows that

$$|f(x) - f(y)| \leq \rho|x - y|, \qquad |y| \leq \rho. \tag{2.58}$$

For $x \neq 0$ and $y \neq 0$, it follows that

$$\begin{aligned}|f(x) - f(y)| &= |x^2 \sin(1/x) - y^2 \sin(1/y)| \\ &= |(x^2 - y^2)\sin(1/x) + y^2(\sin(1/x) - \sin(1/y))| \\ &\leq |(x^2 - y^2)\sin(1/x)| + |y^2(\sin(1/x) - \sin(1/y))| \\ &\leq (|x + y| + |y/x|)|x - y|.\end{aligned}$$

Hence, f is Lipschitz continuous on $\mathbb{R}$. △

In light of the above discussion we have the following propositions.

Proposition 2.19. Let $\mathcal{D} \subseteq \mathbb{R}^m$ and $f : \mathcal{D} \to \mathbb{R}^n$. If f is Lipschitz continuous at $x_0 \in \mathcal{D}$, then f is continuous at x_0.

Proof. Since f is Lipschitz continuous at x_0 there exists $L = L(x_0) > 0$ and a neighborhood $\mathcal{N} \subset \mathcal{D}$ of x_0 such that

$$\|f(x) - f(y)\| \leq L\|x - y\|, \qquad x, y \in \mathcal{N}.$$

Now, for every $\varepsilon > 0$ let $\delta = \delta(\varepsilon, x_0) = \varepsilon/L$. Hence, for all $\|x_0 - y\| < \delta$, it follows that $\|f(x_0) - f(y)\| < L\delta = \varepsilon$. The result is now immediate from the definition of continuity. □

The following result shows that uniform Lipschitz continuity implies uniform continuity.

Proposition 2.20. Let $\mathcal{D} \subseteq \mathbb{R}^n$ and let $f : \mathcal{D} \to \mathbb{R}^n$. If f is uniformly Lipschitz continuous on $\mathcal{D}$, then f is uniformly continuous on $\mathcal{D}$.

Proof. Since f is uniformly Lipschitz continuous on $\mathcal{D}$ it follows that there exists $L > 0$ such that

$$\|f(x) - f(y)\| \leq L\|x - y\|, \quad x, y \in \mathcal{D}. \tag{2.59}$$

Now, for a given $\varepsilon > 0$ let $\delta = \varepsilon/L$. Hence, for all $\|x - y\| < \delta$, it follows that $\|f(x) - f(y)\| < L\delta = \varepsilon$. The result is now immediate from the definition of uniform continuity. □

Lipschitz continuity is essentially a smoothness property almost everywhere on $\mathcal{D}$. Specifically, if f is continuously differentiable at $x_0 \in \mathcal{D}$, then f is Lipschitz continuous at x_0. The converse, however, is not true. A simple counterexample is $f(x) = |x|$, $x \in \mathbb{R}$, which is Lipschitz continuous at $x = 0$ but not continuously differentiable at $x = 0$. Alternatively, if the (left or right) derivative exists at some point x_0, then it does not automatically imply that f is Lipschitz continuous at x_0 since the existence of the derivative of f at x_0 does not automatically imply differentiability of f at x_0. Of course, if $f'(x)$ is continuously differentiable and bounded on $\mathcal{D}$, then f is uniformly Lipschitz continuous on $\mathcal{D}$.

Next, we extend the above observations to vector-valued functions. However, first we need the following definitions and results from mathematical analysis.

Definition 2.34. Let $\mathcal{D} \subseteq \mathbb{R}^m$ be an open set. The function $f : \mathcal{D} \to \mathbb{R}^n$ is *differentiable* at $x_0 \in \mathcal{D}$ if there exists a linear transformation $Df(x_0) \in \mathbb{R}^{n\times m}$ satisfying

$$\lim_{\|h\|\to 0} \frac{\|f(x_0+h) - f(x_0) - Df(x_0)h\|}{\|h\|} = 0, \tag{2.60}$$

where $x_0 + h \in \mathcal{D}$ and $h \neq 0$.

The linear transformation $Df(x_0)$ is called the *derivative* of f at x_0. We usually denote $Df(x_0)$ by $\frac{\partial f}{\partial x}(x_0)$ or $f'(x_0)$. A function $f : \mathbb{R}^m \to \mathbb{R}^n$ is *continuously differentiable* at x_0 if $Df(x_0)$ exists and is continuous at x_0. A function f is continuously differentiable on $\mathcal{D} \subseteq \mathbb{R}^m$ if f is continuously differentiable at all $x \in \mathcal{D}$. A function $f : \mathbb{R}^m \to \mathbb{R}^n$ is *differentiable* on $\mathbb{R}^m$ if f is differentiable at every point $x \in \mathbb{R}^m$. If f is differentiable on $\mathbb{R}^m$, then f is continuous on $\mathbb{R}^m$. To see this, let $x_0 \in \mathbb{R}^m$ and assume f is differentiable at x_0. In this case, $\|f(x_0+h) - f(x_0)\|/\|h\|$ tends to a definite limit $Df(x_0)$ as $\|h\| \to 0$, and hence, $\|f(x_0+h) - f(x_0)\| \to 0$ as $\|h\| \to 0$, which implies that f is continuous at x_0. However, not every continuous function has a derivative at every point in $\mathcal{D} \subseteq \mathbb{R}^m$.

Example 2.21. Consider the function $f : \mathbb{R} \to \mathbb{R}$ given by $f(x) = |x|$. Clearly, f is continuous at $x = 0$. However, for $h > 0$,

$$\lim_{h\to 0^+} \frac{f(h) - f(0)}{h} = \frac{h-0}{h} = 1, \tag{2.61}$$

whereas for $h < 0$,

$$\lim_{h\to 0^-} \frac{f(h) - f(0)}{h} = \frac{-h-0}{h} = -1. \tag{2.62}$$

Hence, $f'(0^-) \neq f'(0^+)$, and hence, f is not differentiable at $x = 0$. △

Another important distinction is between differentiability and continuous differentiability. In particular, if a function f is continuous on $\mathcal{D}$ and its derivative exists at every point on $\mathcal{D}$, then f is not necessarily continuously differentiable on $\mathcal{D}$. That is, it is not necessarily true that f' is continuous on $\mathcal{D}$.

Example 2.22. Consider the function $f : \mathbb{R} \to \mathbb{R}$ given in Example 2.20. The derivative of f is given by

$$f'(x) = \begin{cases} 2x\sin(1/x) - \cos(1/x), & x \neq 0, \\ 0, & x = 0, \end{cases} \tag{2.63}$$

and hence, f is differentiable everywhere on $\mathbb{R}$. However, $\lim_{x\to 0} f'(x)$ does not exist since f' is unbounded in every neighborhood of the origin. Thus, f' is not continuous at $x = 0$, and hence, although f is differentiable everywhere on $\mathbb{R}$, it is not continuously differentiable at $x = 0$. △

Definition 2.35. Let $\mathcal{X}$ and $\mathcal{Y}$ be open subsets of $\mathbb{R}^n$ and let $f : \mathcal{X} \to \mathcal{Y}$. Then f is a *homeomorphism* or a *topological mapping* of $\mathcal{X}$ onto $\mathcal{Y}$ if f is a one-to-one continuous map with a continuous inverse. Furthermore, $\mathcal{X}$ and $\mathcal{Y}$ are said to be *homeomorphic* or *topologically equivalent*. f is a *diffeomorphism* of $\mathcal{X}$ onto $\mathcal{Y}$ if f is a homeomorphism and f and f^{-1} are continuously differentiable. In this case, $\mathcal{X}$ and $\mathcal{Y}$ are said to be *diffeomorphic*.

Example 2.23. Consider the function $f : (-1,1) \to \mathbb{R}$ defined by $f(x) = \tan(\frac{\pi x}{2})$. Clearly, f is one-to-one and continuous on $(-1,1)$. The inverse function $f^{-1} : \mathbb{R} \to (-1,1)$ given by $x = \frac{2}{\pi}\tan^{-1} f(x)$ is also continuous, and hence, f is a homeomorphism and $(-1,1)$ and $\mathbb{R}$ are homeomorphic. △

If $\|h\|$ in (2.60) is sufficiently small, then $x + h \in \mathcal{D}$, since $\mathcal{D}$ is open. Hence, $f(x+h)$ is defined, $f(x+h) \in \mathbb{R}^n$, and, since $Df(x) : \mathbb{R}^m \to \mathbb{R}^n$ is a linear operator, $Df(x)h \in \mathbb{R}^n$. The next proposition shows that the linear transformation $Df(x_0) \in \mathbb{R}^{n\times m}$ satisfying (2.60) is unique.

Proposition 2.21. Let $\mathcal{D}$ and $f : \mathcal{D} \to \mathbb{R}^n$ be as in Definition 2.34 and let $x \in \mathcal{D}$. Then $Df(x) : \mathbb{R}^m \to \mathbb{R}^n$ satisfying (2.60) is unique.

Proof. Suppose, *ad absurdum*, that (2.60) holds with $Df(x) = X$ and $Df(x) = Y$, where $X \neq Y$. Define $Z \triangleq X - Y$ and note that

$$\begin{aligned} \|Zh\| &= \|f(x+h) - f(x) + Yh - f(x+h) + f(x) - Xh\| \\ &\leq \|f(x+h) - f(x) - Xh\| + \| - f(x+h) + f(x) + Yh\| \\ &= \|f(x+h) - f(x) - Xh\| + \|f(x+h) - f(x) - Yh\|, \end{aligned}$$

which implies that $\|Zh\|/\|h\| \to 0$ as $\|h\| \to 0$. Now, for fixed $h \neq 0$, it follows that $\|\varepsilon Zh\|/\|\varepsilon h\| \to 0$ as $\varepsilon \to 0$, and hence, $Zh = 0$ for every $h \in \mathbb{R}^m$. Thus, $Z = X - Y = 0$, which leads to a contradiction. □

Note that (2.60) can be rewritten as

$$f(x+h) - f(x) = f'(x)h + r(h), \tag{2.64}$$

where the *remainder* $r(h)$ satisfies

$$\lim_{\|h\|\to 0} \frac{\|r(h)\|}{\|h\|} = 0. \tag{2.65}$$

Hence, for a fixed $x \in \mathcal{D} \subseteq \mathbb{R}^m$ and sufficiently small h, the left-hand side of (2.64) is approximately equal to $f'(x)h$. If $f : \mathcal{D} \to \mathbb{R}^n$ and $\mathcal{D}$ is as in Definition 2.34, and if f is differentiable on $\mathcal{D}$, then for every $x \in \mathcal{D}$, $f'(x)$ is a linear transformation on $\mathbb{R}^m$ into $\mathbb{R}^n$. Note that if follows from (2.64) that f is continuous at any point at which f is differentiable.

Next, we introduce the notion of a partial derivative of the function $f = [f_1, \ldots, f_n]^{\mathrm{T}} : \mathcal{D} \to \mathbb{R}^n$. Let $\{e_1, \ldots, e_m\}$ and $\{v_1, \ldots, v_n\}$ be normalized bases of $\mathbb{R}^m$ and $\mathbb{R}^n$, respectively. In this case,

$$f(x) = \sum_{i=1}^{n} f_i(x) v_i, \quad x \in \mathcal{D}, \tag{2.66}$$

or, equivalently, $f_i(x) = f^{\mathrm{T}}(x) v_i$, $i = 1, \ldots, n$. Now, for $x \in \mathcal{D}$ and $i = 1, \ldots, n$, $j = 1, \ldots, m$, define

$$D_j f_i(x) \triangleq \lim_{\varepsilon \to 0} \frac{f_i(x + \varepsilon e_j) - f_i(x)}{\varepsilon}, \tag{2.67}$$

whenever the limit on the right-hand side exists. Hence, noting that $f_i(x) = f_i(x_1, \ldots, x_m)$ it follows that $D_j f_i(x)$ is the derivative of f_i with respect to x_j, keeping the other variables fixed. We usually denote $D_j f_i(x)$ by $\frac{\partial f_i}{\partial x_j}$ and call $\frac{\partial f_i}{\partial x_j}$ a *partial derivative*. The next theorem shows that if f is differentiable at $x \in \mathcal{D}$, then its partial derivatives exist at x, and they completely determine $f'(x)$.

Theorem 2.14. Let $\mathcal{D} \subseteq \mathbb{R}^m$ be open and $f = [f_1, \ldots, f_n]^{\mathrm{T}} : \mathcal{D} \to \mathbb{R}^n$, and suppose f is differentiable at $x \in \mathcal{D}$. Then $D_j f_i(x)$, $i = 1, \ldots, n$, $j = 1, \ldots, m$, exist, and

$$f'(x) e_j = \sum_{i=1}^{n} D_j f_i(x) v_i, \quad j = 1, \ldots, m. \tag{2.68}$$

Proof. Let $j \in \{1, \ldots, m\}$ and note that since f is differentiable at x,

$$f(x + \varepsilon e_j) - f(x) = f'(x)(\varepsilon e_j) + r(\varepsilon e_j), \tag{2.69}$$

where $\|r(\varepsilon e_j)\|/\varepsilon \to 0$ as $\varepsilon \to 0$. Since $f'(x)$ is linear it follows that

$$\lim_{\varepsilon\to 0} \frac{f(x+\varepsilon e_j) - f(x)}{\varepsilon} = f'(x)e_j, \tag{2.70}$$

or, equivalently,

$$\lim_{\varepsilon\to 0} \sum_{i=1}^{n} \frac{f_i(x+\varepsilon e_j) - f_i(x)}{\varepsilon} v_i = f'(x)e_j, \tag{2.71}$$

where $\{e_1, \ldots, e_m\}$ and $\{v_1, \ldots, v_n\}$ are normalized bases of $\mathbb{R}^m$ and $\mathbb{R}^n$, respectively. Now, since each quotient in (2.71) has a limit as $\varepsilon \to 0$, it follows that $D_j f_i(x)$, $i = 1, \ldots, n$, $j = 1, \ldots, m$, exist. Finally, (2.68) follows from (2.71). □

The next theorem gives necessary and sufficient conditions for a function to be continuously differentiable on $\mathcal{D} \subseteq \mathbb{R}^m$.

Theorem 2.15. Let $\mathcal{D} \subseteq \mathbb{R}^m$ be open and $f = [f_1, \ldots, f_n]^{\mathrm{T}} : \mathcal{D} \to \mathbb{R}^n$. Then f is continuously differentiable on $\mathcal{D}$ if and only if $\frac{\partial f_i}{\partial x_j}$, $i = 1, \ldots, n$, $j = 1, \ldots, m$, exist and are continuous on $\mathcal{D}$.

Proof. Assume f is continuously differentiable on $\mathcal{D}$ and note that it follows from (2.68) that $D_j f_i(x) = (f'(x)e_j)^{\mathrm{T}} v_i$ for all $i = 1, \ldots, n$, $j = 1, \ldots, m$, and all $x \in \mathcal{D}$, where $\{e_1, \ldots, e_m\}$ and $\{v_1, \ldots, v_n\}$ are normalized bases of $\mathbb{R}^m$ and $\mathbb{R}^n$, respectively. Hence, for $i = 1, \ldots, n$, $j = 1, \ldots, m$, and $x, y \in \mathcal{D}$,

$$D_j f_i(y) - D_j f_i(x) = [(f'(y) - f'(x))e_j]^{\mathrm{T}} v_i. \tag{2.72}$$

Now, using the Cauchy-Schwarz inequality and the fact that $\|v_i\| = \|e_j\| = 1$ it follows from (2.72) that for $i = 1, \ldots, n$, $j = 1, \ldots, m$, and $x, y \in \mathcal{D}$,

$$\begin{aligned} |D_j f_i(y) - D_j f_i(x)| &= |[(f'(y) - f'(x))e_j]^{\mathrm{T}} v_i| \\ &\leq \|[f'(y) - f'(x)]e_j\| \\ &\leq \|f'(y) - f'(x)\|, \end{aligned} \tag{2.73}$$

which shows that $D_j f_i(x)$, $i = 1, \ldots, n$, $j = 1, \ldots, m$, is continuous on $\mathcal{D}$.

To prove the converse, it suffices to consider the case where $n = 1$. Let $x \in \mathcal{D}$ and $\varepsilon > 0$, and note that since $\mathcal{D}$ is open, there exists $r > 0$ such that $\mathcal{B}_r(x) \subset \mathcal{D}$. Now, it follows from the continuity of $D_j f(x)$, $j = 1, \ldots, m$, that r can be chosen so that $|D_j f(y) - D_j f(x)| < \varepsilon/m$, $y \in \mathcal{B}_r(x)$, and $j = 1, \ldots, m$.

Next, let $h = \sum_{j=1}^{m} e_j$ be such that $\|h\| < r$ and let $u_k = h_1 e_1 + \cdots + h_k e_k$, $k \in \{1, \ldots, m\}$, with $u_0 \triangleq 0$. Now, writing the difference $f(x+h) -$

$f(x)$ as a telescoping sum yields

$$f(x+h) - f(x) = \sum_{j=1}^{m} [f(x+u_j) - f(x+u_{j-1})]. \tag{2.74}$$

Since $\|u_k\| < r$ for $k \in \{1, \ldots, m\}$ and since $\mathcal{B}_r(x)$ is convex, it follows that $\{z_j : z_j = \mu_j(x+u_{j-1}) + (1-\mu_j)(x+u_j), \mu_j \in (0,1)\} \subset \mathcal{B}_r(x)$. Using the mean value theorem from calculus and noting that $u_j = u_{j-1} + h_j e_j$, it follows that the jth summand of (2.74) is equal to $h_j D_j f(x+u_{j-1}+\mu_j h_j e_j)$. Hence, for $j = 1, \ldots, m$,

$$|h_j D_j f(x+u_{j-1}+\mu_j h_j e_j) - h_j D_j f(x)| < |h_j|\varepsilon/m.$$

Now, it follows from (2.74) that

$$\left| f(x+h) - f(x) - \sum_{j=1}^{m} h_j D_j f(x) \right| \le \frac{1}{m} \sum_{j=1}^{m} |h_j|\varepsilon \le \|h\|_1 \varepsilon \tag{2.75}$$

for all h such that $\|h\|_1 < r$.

Equation (2.75) implies that f is differentiable at x and $f'(x)$ is a linear map assigning the number $\sum_{j=1}^{m} h_j D_j f(x)$ to the vector $h = \sum_{j=1}^{m} h_j e_j$. Now, since the gradient $f'(x) = [D_1 f(x), \ldots, D_m f(x)]$ and $D_j f(x)$, $j = 1, \ldots, m$, are continuous functions on $\mathcal{D}$, it follows that f is continuously differentiable on $\mathcal{D}$. □

In the case where $f : \mathbb{R}^n \to \mathbb{R}$ is continuously differentiable, $\frac{\partial f}{\partial x} = f'(x) = [\frac{\partial f}{\partial x_1}, \ldots, \frac{\partial f}{\partial x_n}] \in \mathbb{R}^{1\times n}$ is called the *gradient* of f at x. Alternatively, for a continuously differentiable function $f : \mathbb{R}^m \to \mathbb{R}^n$, $\frac{\partial f}{\partial x} = f'(x) \in \mathbb{R}^{n\times m}$ is called the *Jacobian* of f at x and is given by

$$f'(x) = \begin{bmatrix} \frac{\partial f_1}{\partial x_1}(x) & \frac{\partial f_1}{\partial x_2}(x) & \cdots & \frac{\partial f_1}{\partial x_m}(x) \\ \frac{\partial f_2}{\partial x_1}(x) & \frac{\partial f_2}{\partial x_2}(x) & & \vdots \\ \vdots & \vdots & \ddots & \vdots \\ \frac{\partial f_n}{\partial x_1}(x) & \cdots & \cdots & \frac{\partial f_n}{\partial x_m}(x) \end{bmatrix}. \tag{2.76}$$

Higher-order derivatives $f^{(r)}(x_0)$ of a function $f : \mathcal{D} \to \mathbb{R}^n$ are defined in a similar way, and it can be shown that similar theorems to Theorem 2.15 hold for a function to be r-times continuously differentiable on $\mathcal{D}$. For details the interested reader is referred to [373, p. 235].

Next, we present the *chain rule* for vector-valued functions. In particular, let $\mathcal{D} \subseteq \mathbb{R}^m$, let $f : \mathcal{D} \to \mathbb{R}^n$ be continuously differentiable at $x_0 \in \mathcal{D}$, and let $g : f(\mathcal{D}) \to \mathbb{R}^l$, where $f(\mathcal{D})$ is open and g is continuously differentiable at $f(x_0)$. Then the function $h : \mathcal{D} \to \mathbb{R}^l$ given by

$h(x) = g(f(x))$ is continuously differentiable at x_0, and $\frac{\partial h}{\partial x}(x_0)$ is given by

$$\frac{\partial h}{\partial x}(x_0) = \frac{\partial g}{\partial f}(f(x_0))\frac{\partial f}{\partial x}(x_0). \tag{2.77}$$

Finally, we present three important results from mathematical analysis. The proofs of these theorems can be found in any analysis or advanced calculus textbook, and hence, are not given here.

Theorem 2.16 (Mean Value Theorem). Let $\mathcal{D} \subseteq \mathbb{R}^n$ be open, let $f : \mathcal{D} \to \mathbb{R}^m$ be continuously differentiable in $\mathcal{D}$, and assume there exist $x, y \in \mathcal{D}$ such that $\mathcal{L} \triangleq \{z : z = \mu x + (1-\mu)y, \ \mu \in (0,1)\} \subset \mathcal{D}$. Then for every $v \in \mathbb{R}^m$ there exists $z \in \mathcal{L}$ such that

$$v^{\mathrm{T}}[f(y) - f(x)] = v^{\mathrm{T}}[f'(z)(y-x)]. \tag{2.78}$$

Theorem 2.17 (Inverse Function Theorem). Let $\mathcal{D} \subseteq \mathbb{R}^n$ be open, let $f : \mathcal{D} \to \mathbb{R}^n$ be continuously differentiable on $\mathcal{D}$, and assume there exists $x_0 \in \mathcal{D}$ such that $\det f'(x_0) \neq 0$. Then there exists a neighborhood $\mathcal{N}$ of $f(x_0)$ and a unique function $g : \mathcal{N} \to \mathcal{D}$ such that $f(g(y)) = y$ for all $y \in \mathcal{N}$. Furthermore, the function g is continuously differentiable on $\mathcal{N}$.

It follows from the inverse function theorem that if $f : \mathbb{R}^n \to \mathbb{R}^n$ is a bijection with $\det f'(x) = 0$ for some $x \in \mathbb{R}^n$, then f^{-1} is not differentiable at $f(x)$. To see this, assume, *ad absurdum*, that f^{-1} is differentiable at $f(x)$ and note that $(f^{-1} \circ f)(x) = x$. Now, it follows from the chain rule (2.77) that

$$\begin{aligned}\frac{\partial f^{-1}}{\partial f}(f(x))\frac{\partial f}{\partial x}(x) &= Df^{-1}(f(x)) \circ Df(x) \\ &= D(f^{-1} \circ f)(x) \\ &= I_n,\end{aligned} \tag{2.79}$$

and hence, $\det Df^{-1}(f(x)) \det Df(x) = 1$. This contradicts the fact that $\det Df(x) = 0$.

Theorem 2.18 (Implicit Function Theorem). Let $\mathcal{D} \subseteq \mathbb{R}^n$ and $\mathcal{Q} \subseteq \mathbb{R}^m$ be open, and let $f : \mathcal{D} \times \mathcal{Q} \to \mathbb{R}^n$. Suppose $f(x_0, y_0) = 0$ for $(x_0, y_0) \in \mathcal{D} \times \mathcal{Q}$ and suppose the Jacobian matrix $\frac{\partial f}{\partial x}(x_0, y_0)$ is nonsingular. Then there exist neighborhoods $\mathcal{U} \subset \mathcal{D}$ of x_0 and $\mathcal{V} \subset \mathcal{Q}$ of y_0 such that $f(x, y) = 0$, $y \in \mathcal{V}$, has a unique solution $x \in \mathcal{U}$. Moreover, there exists a continuously differentiable function $g : \mathcal{V} \to \mathcal{U}$ at $y = y_0$ such that $x = g(y)$.

The implicit function theorem can be used to guarantee the existence of a feedback controller that provides a characterization of an equilibrium manifold. Specifically, consider the dynamical system (2.1) with $F(t, x, u) = F(x, u)$, where $F : \mathcal{D} \times U \to \mathbb{R}^n$ and $\mathcal{D} \subseteq \mathbb{R}^n$ and $U \subseteq \mathbb{R}^m$. If there exists

$u_e \in U$ such that $F(x_e, u_e) = 0$, then (x_e, u_e) is an equilibrium point of (2.1). Since F is an *implicit function*, it follows that to find a feedback control law to solve the *implicit equation* $F(x, u) = 0$ we require the existence of a function $\phi : \mathcal{D} \to \mathbb{R}^m$ such that the *explicit equation* $u = \phi(x)$ is satisfied. If such a function exists, then $F(x, \phi(x)) = 0$. Now, suppose there exists $(x_e, u_e) \in \mathcal{D} \times U$ such that $F(x_e, u_e) = 0$. If $\det \frac{\partial F}{\partial u}(x_e, u_e) \neq 0$, then it follows from the implicit function theorem that there exist open neighborhoods $\mathcal{X}$ and $\mathcal{U}$ of x_e and u_e, respectively, such that, for each $x \in \mathcal{X}$, there exists a unique feedback control law $\phi : \mathcal{X} \to \mathbb{R}^m$ such that *i)* $\phi(x_e) = u_e$, *ii)* ϕ is continuously differentiable in $\mathcal{X}$, and *iii)* $F(x, \phi(x)) = 0$ for all $x \in \mathcal{X}$.

Next, using the above results we concretize our previous observations on scalar Lipschitz continuous functions to vector-valued functions. Our first result shows that continuous differentiability implies Lipschitz continuity.

Proposition 2.22. Let $\mathcal{D} \subseteq \mathbb{R}^n$ be open and let $f : \mathcal{D} \to \mathbb{R}^n$. If f is continuously differentiable on $\mathcal{D}$, then f is Lipschitz continuous on $\mathcal{D}$.

Proof. Since $\mathcal{D}$ is open, for a given $x_0 \in \mathcal{D}$ there exists $\varepsilon > 0$ such that $\mathcal{B}_\varepsilon(x_0) \subset \mathcal{D}$. Now, since $f'(x)$ is continuous on $\mathcal{D}$ it follows from Theorem 2.13 that $L = \max_{\|x-x_0\| \leq \varepsilon} \|f'(x)\|'$ exists, where $\|\cdot\|'$ is the equi-induced matrix norm generated by the vector norm $\|\cdot\|$. Next, for some $x, y \in \mathcal{B}_\varepsilon(x_0)$, set $z = y - x$. Since $\mathcal{B}_\varepsilon(x_0)$ is a convex set, it follows that $x + \mu z \in \mathcal{B}_\varepsilon(x_0)$, $\mu \in [0, 1]$. Now, define the function $g : [0, 1] \to \mathbb{R}^n$ by $g(\mu) \triangleq f(x + \mu z)$. Using the chain rule, it follows that $g'(\mu) = f'(x + \mu z)z$, and hence,

$$f(y) - f(x) = g(1) - g(0) = \int_0^1 g'(\mu)\mathrm{d}\mu = \int_0^1 f'(x + \mu z)z\mathrm{d}\mu. \tag{2.80}$$

Thus,

$$\begin{aligned} \|f(y) - f(x)\| &\leq \int_0^1 \|f'(x + \mu z)z\|\mathrm{d}\mu \\ &\leq \int_0^1 \|f'(x + \mu z)\|'\|z\|\mathrm{d}\mu \\ &\leq L\|z\| \\ &= L\|y - x\|, \end{aligned} \tag{2.81}$$

for all $x, y \in \mathcal{B}_\varepsilon(x_0)$. □

The next result shows that if $\mathcal{D}$ is compact, then Lipschitz continuity implies uniform Lipschitz continuity.

Proposition 2.23. Let $\mathcal{D} \subseteq \mathbb{R}^n$, let $\mathcal{D}_c \subset \mathcal{D}$ be a compact set, and let $f : \mathcal{D} \to \mathbb{R}^n$. If f is Lipschitz continuous on $\mathcal{D}$, then f is uniformly Lipschitz continuous on $\mathcal{D}_c$.

Proof. Since $f : \mathcal{D} \to \mathbb{R}^n$ is continuous on the compact set $\mathcal{D}_c$ it follows that $\alpha = \max_{x \in \mathcal{D}_c} f(x)$ exists. Now, suppose, *ad absurdum*, that f is not uniformly Lipschitz continuous on $\mathcal{D}_c$. Then, for every $L > 0$, there exists $x, y \in \mathcal{D}_c$ such that $\|f(y) - f(x)\| > L\|y - x\|$. In particular, there exist sequences $\{x_n\}_{n=1}^{\infty}$ and $\{y_n\}_{n=1}^{\infty} \subseteq \mathcal{D}_c$ such that

$$\|f(y_n) - f(x_n)\| > n\|y_n - x_n\|, \quad n \in \mathbb{Z}_+. \tag{2.82}$$

Now, since $\mathcal{D}_c$ is a compact set, it follows from the Bolzano-Lebesgue theorem (Theorem 2.4) that there exist subsequences $\{x_{n_k}\}_{k=1}^{\infty}$ and $\{y_{n_k}\}_{k=1}^{\infty}$ such that $x_{n_k} \to x^* \in \mathcal{D}_c$ and $y_{n_k} \to y^* \in \mathcal{D}_c$ as $k \to \infty$. Since $n_k \in \mathbb{Z}_+$, it follows that

$$\|y^* - x^*\| = \lim_{k \to \infty} \|y_{n_k} - x_{n_k}\| \le \frac{1}{n_k}\|f(y_{n_k}) - f(x_{n_k})\| \le \frac{2\alpha}{n_k}, \tag{2.83}$$

and hence, $x^* = y^*$. Now, since f is Lipschitz continuous on $\mathcal{D}$, there exists a neighborhood $\mathcal{N} \subset \mathcal{D}$ of x^* and a Lipschitz constant $L = L(x^*) > 0$ such that

$$\|f(y) - f(x)\| \le L\|y - x\|, \quad x, y \in \mathcal{N}. \tag{2.84}$$

Since $x_{n_k} \to x^*$ and $y_{n_k} \to y^*$ as $k \to \infty$ it follows that for sufficiently large n, x_{n_k} and $y_{n_k} \in \mathcal{N}$, and hence,

$$\|f(y_{n_k}) - f(x_{n_k})\| \le L\|y_{n_k} - x_{n_k}\|. \tag{2.85}$$

For $n \ge L$, (2.85) contradicts (2.82). Hence, f is uniformly Lipschitz continuous on $\mathcal{D}$. □

The following proposition is a direct consequence of Proposition 2.22.

Proposition 2.24. Let $\mathcal{D} \subseteq \mathbb{R}^n$ and let $f : \mathcal{D} \to \mathbb{R}^n$ be continuous on $\mathcal{D}$. If f' exists and is continuous on $\mathcal{D}$, then f is Lipschitz continuous on $\mathcal{D}$.

Proof. It follows from Theorem 2.15 that the existence and continuity of f' on $\mathcal{D}$ implies that f is continuously differentiable on $\mathcal{D}$. Now, the result is a direct consequence of Proposition 2.22. □

Proposition 2.25. Let $\mathcal{D} \subseteq \mathbb{R}^n$, $\mathcal{C} \subset \mathcal{D}$ be a convex set, and let $f : \mathcal{D} \to \mathbb{R}^n$. Suppose f' exists and is continuous on $\mathcal{D}$. If there exists $L > 0$ such that $\|f'(x)\| \le L$ for all $x \in \mathcal{C}$, then f is uniformly Lipschitz continuous on $\mathcal{C}$.

Proof. It follows from Theorem 2.15 that the existence and continuity of f' on $\mathcal{D}$ implies that f is continuously differentiable on $\mathcal{D}$. Now, it follows as in the proof of Proposition 2.22 with $\mathcal{B}_\varepsilon(x_0)$ replaced by $\mathcal{C}$ that

$$\|f(x) - f(y)\| \le L\|x - y\|, \qquad x, y \in \mathcal{C},$$

which implies the result since L is independent of x and y. □

Finally, we provide necessary and sufficient conditions for a function f to be globally Lipschitz continuous.

Proposition 2.26. Let $f : \mathbb{R}^n \to \mathbb{R}^n$ be continuously differentiable on $\mathbb{R}^n$. Then f is globally Lipschitz continuous if and only if there exists $L > 0$ such that $\|f'(x)\| \le L$ for all $x \in \mathbb{R}^n$.

Proof. Sufficiency is a direct consequence of Proposition 2.25 with $\mathcal{C} = \mathbb{R}^n$. To show necessity suppose f is globally Lipschitz continuous with the Lipschitz constant L. Now, since $f(\cdot)$ is continuously differentiable on $\mathbb{R}^n$, it follows from Definition 2.34 and Theorem 2.15 that for every $x \in \mathbb{R}^n$ and $\varepsilon > 0$, there exists $\delta > 0$ such that

$$\left\|f(y) - f(x) - \frac{\partial f}{\partial x}(x)(y-x)\right\| \le \varepsilon\|y-x\|, \qquad y \in \mathcal{B}_\delta(x). \tag{2.86}$$

Next, note that

$$\begin{aligned}\left\|\frac{\partial f}{\partial x}(x)(y-x)\right\| &\le \|f(y) - f(x)\| + \left\|f(y) - f(x) - \frac{\partial f}{\partial x}(x)(y-x)\right\| \\ &\le (L+\varepsilon)\|y-x\|, \quad y \in \mathcal{B}_\delta(x), \end{aligned}\tag{2.87}$$

and hence,

$$\left\|\frac{\partial f}{\partial x}(x)z\right\| \le (L+\varepsilon)\|z\|, \quad z \in \mathcal{B}_\delta(0). \tag{2.88}$$

Thus,

$$\left\|\frac{\partial f}{\partial x}(x)\right\| = \max_{z\in\mathcal{B}_\delta(0)} \frac{\|\frac{\partial f}{\partial x}(x)z\|}{\|z\|} \le (L+\varepsilon), \tag{2.89}$$

where $\|\cdot\| : \mathbb{R}^{n\times n} \to \mathbb{R}$ is the equi-induced matrix norm induced by the vector norm $\|\cdot\| : \mathbb{R}^n \to \mathbb{R}$. Finally, since ε is arbitrary it follows that $\|\frac{\partial f}{\partial x}(x)\| \le L$, which proves the result. □

Note that $\|f'(x)\| \le L$, $x \in \mathbb{R}^n$, is equivalent to saying that f' is *uniformly bounded*, that is, the bound is independent of x in $\mathbb{R}^n$.

2.5 Vector Spaces and Banach Spaces

In this section, we introduce concepts involving linear vector spaces, normed linear spaces, and Banach spaces. First, however, we introduce the concepts of binary operations, semigroups, groups, rings, and fields. Let $\mathcal{V}$ be a set which may be finite or infinite. A *binary operation* $\circ$ takes any two elements $x, y \in \mathcal{V}$ and transforms them to $z = x \circ y$, not necessarily in $\mathcal{V}$. A binary operation $\circ$ is said to be *closed* if, for every $x, y \in \mathcal{V}$, $x \circ y \in \mathcal{V}$. Hence, a closed binary operation is a mapping $\phi : \mathcal{V}\times\mathcal{V} \to \mathcal{V}$ defined by $\phi(x,y) = x\circ y$.

An *algebraic system* $(\mathcal{V}, \circ)$ is a nonempty set $\mathcal{V}$ with the binary operation $\circ : \mathcal{V} \times \mathcal{V} \to \mathcal{V}$.

A *semigroup* $(\mathcal{S}, \circ)$ is a nonempty set $\mathcal{S}$ with associativity of the binary operation $\circ$ such that $\mathcal{S}$ is closed. That is, $x \circ y \in \mathcal{S}$ for all $x, y \in \mathcal{S}$, and $x \circ (y \circ z) = (x \circ y) \circ z$ for all $x, y, z \in \mathcal{S}$. A *group* $(\mathfrak{G}, \circ)$ is a nonempty set $\mathfrak{G}$ with a binary operation $\circ$ such that *i)* $\mathfrak{G}$ is closed under $\circ$, *ii)* $\circ$ is associative in $\mathfrak{G}$, *iii)* there exists an identity element $1 \in \mathfrak{G}$ such that $x \circ 1 = x = 1 \circ x$ for all $x \in \mathfrak{G}$, and *iv)* for each $x \in \mathfrak{G}$, there exists a unique inverse $x^{-1} \in \mathfrak{G}$ such that $x \circ x^{-1} = 1 = x^{-1} \circ x$. An *Abelian* group is a group $(\mathfrak{G}, \circ)$ with commutativity of the binary operation, that is, $x \circ y = y \circ x$ for all $x, y \in \mathfrak{G}$.

A *ring* $(\mathcal{R}, +, \cdot)$ is a nonempty set $\mathcal{R}$ with the two binary operations of addition $(+)$ and multiplication $(\cdot)$ connected by distributive laws. That is, $x \cdot (y + z) = x \cdot y + x \cdot z$ and $(y + z) \cdot x = y \cdot x + y \cdot z$ for all $x, y, z \in \mathcal{R}$. Here we assume that $(\mathcal{R}, +)$ is an Abelian group with the identity element denoted by $0 \in \mathcal{R}$ and $(\mathcal{R}, \cdot)$ is a semigroup with respect to multiplication with the identity element $1 \in \mathcal{R}$. Furthermore, 0 is referred to as the zero element and the elements $x \in \mathcal{R}$, $x \neq 0$, are referred to as nonzero elements. Finally, a *field* $\mathbb{F}$ is a commutative ring with $1 \in \mathbb{F}$ and with the nonzero elements in $\mathbb{F}$ forming a group under the binary operation of multiplication.

For the following definition we let $\mathbb{F}$ denote a field. For example, $\mathbb{F}$ can denote the field of real numbers, the field of complex numbers, the binary field, the field of rational functions, etc.

Definition 2.36. A *linear vector space* $(\mathcal{V}, +, \cdot)$ over a field $\mathbb{F}$ is a set $\mathcal{V}$ with the (addition and multiplication) operations $+ : \mathcal{V} \times \mathcal{V} \to \mathcal{V}$ and $\cdot : \mathbb{F} \times \mathcal{V} \to \mathcal{V}$ such that the following axioms hold:

i) (Commutativity of addition): $x + y = y + x$ for all $x, y \in \mathcal{V}$.

ii) (Associativity of addition): $x + (y + z) = (x + y) + z$ for all $x, y, z \in \mathcal{V}$.

iii) (Existence of additive identity): For every $x \in \mathcal{V}$, there exists a unique element $0 \in \mathcal{V}$ such that $0 + x = x + 0 = x$.

iv) (Existence of additive inverse): For every $x \in \mathcal{V}$, there exists $-x \in \mathcal{V}$ such that $x + (-x) = 0$.

v) $\alpha(\beta x) = (\alpha\beta)x$ for all $\alpha, \beta \in \mathbb{F}$ and $x \in \mathcal{V}$.

vi) $\alpha(x + y) = \alpha x + \alpha y$ for all $\alpha \in \mathbb{F}$ and $x, y \in \mathcal{V}$.

vii) $(\alpha + \beta)x = \alpha x + \beta x$ for all $\alpha, \beta \in \mathbb{F}$ and $x \in \mathcal{V}$.

viii) For every $x \in \mathcal{V}$ and the identity element $1 \in \mathbb{F}$, $1 \cdot x = x$.

Henceforth, unless otherwise stated, we choose $\mathbb{F} = \mathbb{R}$. There are many examples of linear vector spaces. As a first example note that the set $\mathbb{R}^n$ constitutes a linear vector space with addition and scalar multiplication defined in the usual way. In particular, for $x = [x_1, \ldots, x_n]^{\mathrm{T}}$, $y = [y_1, \ldots, y_n]^{\mathrm{T}}$, and $\alpha \in \mathbb{R}$, with addition and scalar multiplication defined as

$$x + y \triangleq [x_1 + y_1, \ldots, x_n + y_n]^{\mathrm{T}}, \tag{2.90}$$

$$\alpha x \triangleq [\alpha x_1, \ldots, \alpha x_n]^{\mathrm{T}}, \tag{2.91}$$

x, y satisfy Axioms *i)*–*viii)* of Definition 2.36, and hence, $\mathbb{R}^n$ is a real linear vector space. As another example consider vector-valued functions $f(\cdot)$ defined on $[a, b]$, that is, $f : [a, b] \to \mathbb{R}^n$. Then the set of such functions constitutes a linear vector space since $\alpha f(\cdot)$ can be defined by

$$(\alpha f)(t) \triangleq \alpha f(t), \quad t \in [a, b], \tag{2.92}$$

and $(f + g)(\cdot)$ can be defined by

$$(f + g)(t) \triangleq f(t) + g(t), \quad t \in [a, b], \tag{2.93}$$

where addition and scalar multiplication in the right-hand side of (2.92) and (2.93) are defined as in (2.90) and (2.91). Many different linear vector spaces can be created by specifying which functions are allowed. For examples, we may admit continuous functions, differentiable functions, or bounded functions. Each of these cases gives a different vector space.

Definition 2.37. Let $\mathcal{V}$ be a linear vector space and let $\mathcal{S} \subset \mathcal{V}$. Then $\mathcal{S}$ is a *subspace* if $\alpha x + \beta y \in \mathcal{S}$ for all $x, y \in \mathcal{S}$ and $\alpha, \beta \in \mathbb{F}$.

Example 2.24. Consider the subset of $\mathbb{R}^2$ given by $\mathcal{H} = \{(x_1, x_2) \in \mathbb{R}^2 : x_2 \geq 0\}$. Note that $\mathcal{H}$ is not a subspace since $\mathcal{H}$ is not an additive group, that is, $\mathcal{H}$ is not closed on taking negatives. However, the boundary of $\mathcal{H}$ given by $\partial\mathcal{H} = \{(x, 0) : x \in \mathbb{R}\} \subset \mathbb{R}^2$ is a subspace of $\mathbb{R}^2$. △

Let $\{x_1, \ldots, x_m\} \subset \mathbb{R}^n$. Then the set of all vectors of the form $\alpha_1 x_1 + \cdots + \alpha_m x_m$, where $\alpha_i \in \mathbb{R}$, $i = 1, \ldots, m$, is a subspace of $\mathbb{R}^n$. This subspace is called the *span* of $\{x_1, \ldots, x_m\}$ and is denoted by $\mathrm{span}\{x_1, \ldots, x_m\}$.

Proposition 2.27. Let $\mathcal{M}$ and $\mathcal{N}$ be subspaces of a linear vector space $\mathcal{V}$. Then $\mathcal{M} \cap \mathcal{N}$ is a subspace of $\mathcal{V}$.

Proof. Note that $0 \in \mathcal{M} \cap \mathcal{N}$ since $0 \in \mathcal{M}$ and $0 \in \mathcal{N}$. Hence, $\mathcal{M} \cap \mathcal{N} \neq \emptyset$. Now, if $x, y \in \mathcal{M} \cap \mathcal{N}$, then $x, y \in \mathcal{M}$ and $x, y \in \mathcal{N}$. Next, for every $\alpha, \beta \in \mathbb{F}$ the vector $\alpha x + \beta y \in \mathcal{M}$ and $\alpha x + \beta y \in \mathcal{N}$ since $\mathcal{M}$ and $\mathcal{N}$ are subspaces. Hence, $\alpha x + \beta y \in \mathcal{M} \cap \mathcal{N}$. □

Definition 2.38. Let $\mathcal{V}$ be a linear vector space and let $\mathcal{M} \subset \mathcal{V}$ and $\mathcal{N} \subset \mathcal{V}$. Then $\mathcal{M} + \mathcal{N} = \{x + y \in \mathcal{V} : x \in \mathcal{M}, y \in \mathcal{N}\}$ is called the *sum* of $\mathcal{M}$ and $\mathcal{N}$.

Proposition 2.28. Let $\mathcal{M}$ and $\mathcal{N}$ be subspaces of a linear vector space $\mathcal{V}$. Then $\mathcal{M} + \mathcal{N}$ is a subspace.

Proof. First, note that $0 \in \mathcal{M} + \mathcal{N}$ since $0 \in \mathcal{M}$ and $0 \in \mathcal{N}$. Next, let $x, y \in \mathcal{M} + \mathcal{N}$. In this case, there exist vectors $u, v \in \mathcal{M}$ and $w, z \in \mathcal{N}$ such that $x = u + w$ and $y = v + z$. Now, for every $\alpha, \beta \in \mathbb{F}$, $\alpha x + \beta y = (\alpha u + \beta v) + (\alpha w + \beta z)$, and hence, $\alpha x + \beta y \in \mathcal{M} + \mathcal{N}$ for all $x, y \in \mathcal{M} + \mathcal{N}$ and $\alpha, \beta \in \mathbb{F}$. □

Next, we introduce the notion of distance for a linear vector space. As seen in Section 2.2, since norms provide a natural measure of distance, endowing a linear vector space with a norm gives a *normed linear space.*

Definition 2.39. A *normed linear space* over a field $\mathbb{F}$ is a linear vector space $\mathcal{X}$ with a norm $|||\cdot||| : \mathcal{X} \to \mathbb{R}$ such that the following axioms hold:

i) $|||x||| \geq 0$, $x \in \mathcal{X}$.

ii) $|||x||| = 0$ if and only if $x = 0$.

iii) $|||\alpha x||| = |\alpha| |||x|||$, $x \in \mathcal{X}$, $\alpha \in \mathbb{F}$.

iv) $|||x + y||| \leq |||x||| + |||y|||$, $x, y \in \mathcal{X}$.

In Definition 2.39 the notation $|||\cdot|||$ is used to denote a vector or matrix operator (function) norm as opposed to $\|\cdot\|$, which is used to denote a vector or matrix norm. In the case where $\mathcal{X} = \mathbb{R}^n$ or $\mathcal{X} = \mathbb{C}^n$, $|||\cdot|||$ should be interpreted as $\|\cdot\|$. (The reason for this change in notation will become clear later.)

We can create a normed function space by assigning a norm to a linear vector space. For example, for $f : [0, \infty) \to \mathbb{R}$ define

$$|||f||| = \sup_{t \in [0,\infty)} |f(t)|, \tag{2.94}$$

where we allow only continuous functions $f(\cdot)$. Then it can be easily verified that the conditions in Definition 2.39 with the norm $|||\cdot|||$ are satisfied. It is important to note that in general many different norms can be assigned to the same linear space.

Example 2.25. Consider the linear vector space $\mathbb{R}^n$ with norm $\|\cdot\|_\infty : \mathbb{R}^n \to \mathbb{R}$ defined by $\|x\|_\infty = \max_{i=1,\ldots,n} |x_{(i)}|$. Note that $\|\cdot\|_\infty$ satisfies

Axioms *i)*–*iii)* of Definition 2.39 trivially. To show *iv)* let $x = [x_1, \ldots, x_n]^{\mathrm{T}}$ and $y = [y_1, \ldots, y_n]^{\mathrm{T}}$. Now,

$$\|x+y\|_\infty = \max_{i=1,\ldots,n} |x_{(i)} + y_{(i)}| \leq \max_{i=1,\ldots,n} |x_{(i)}| + \max_{i=1,\ldots,n} |y_{(i)}| = \|x\|_\infty + \|y\|_\infty, \tag{2.95}$$

and hence, *iv)* holds. Hence, the linear vector space $\mathbb{R}^n$ endowed with the infinity norm is a normed linear space. In a similar fashion it can be shown that $\mathbb{R}^n$ endowed with the absolute sum norm and the Euclidean norm are normed linear spaces. △

In dynamical system theory, and in particular controlled dynamical systems, the system input vector $u(t) \in \mathbb{R}^m$, $t \geq 0$, is often constrained to lie in some compact set $U \subset \mathbb{R}^m$. In addition, in certain applications, some or all of the components of $u(t)$ may be discontinuous with respect to t. In this case, there might exist a finite or infinite number of times t where the vector field $F(\cdot,\cdot)$, and hence, the derivative of the state vector $\dot{x}$ in (2.1), is discontinuous. To formally address such cases it is useful to quantify the size of a set.

Let $\mathcal{X}$ be a normed vector space over $\mathbb{R}^m$ and let $\mathcal{S} \subset \mathcal{X}$. In addition, let $\{\mathcal{A}_i : i \in \mathcal{I}\}$ be a (possibly infinite) set of subsets such that $\mathcal{S} \subseteq \bigcup_{i\in\mathcal{I}} \mathcal{A}_i$. If, for each $i \in \mathcal{I}$, $\mathcal{A}_i$ is an open set, then $\{\mathcal{A}_i : i \in \mathcal{I}\}$ is called a *cover* (or an *open cover*) of $\mathcal{S}$. Furthermore, if $\{\mathcal{A}_i : i \in \mathcal{I}\} = \{\mathcal{A}_1, \ldots, \mathcal{A}_n\}$ is a finite set, then $\{\mathcal{A}_1, \ldots, \mathcal{A}_n\}$ is called a *finite cover* of $\mathcal{S}$.[3] For a collection S of bounded subsets of $\mathbb{R}^n$, the *measure* $\mu : S \to [0,\infty)$ assigns to each set in S a nonnegative number. (See [288] and Problem 2.71 for the properties of $\mu(\mathcal{S})$.) A subset $\mathcal{S} \subset \mathbb{R}^n$ is a *set of measure zero* if and only if for all $\varepsilon > 0$, there exists a cover $\{\mathcal{A}_1, \mathcal{A}_2, \ldots\}$ of $\mathcal{S}$, where $\mathcal{A}_i$ is an open (or closed) n-dimensional symmetric polytope (or cube), such that the sum of their volumes $\mathrm{vol}(\mathcal{A}_i)$ is less than ε, that is, $\sum_{i=1}^{\infty} \mathrm{vol}(\mathcal{A}_i) < \varepsilon$. Any finite or countably infinite set of points in $\mathbb{R}^n$ has measure zero. A statement about $\mathcal{S} \subseteq \mathbb{R}^n$ holds *almost everywhere* (a.e.) in $\mathcal{S}$ if and only if the statement holds for every element in $\mathcal{S}$ except for a subset of $\mathcal{S}$ with zero measure.

A measure on $\mathbb{R}^n$ is a natural generalization of the notion of length of a line segment for $n = 1$, the area of a rectangle for $n = 2$, and the volume of a parallelepiped for $n = 3$. In particular, for a closed interval $\mathcal{I} = [a, b]$, the *length* $\ell(\mathcal{I})$ of the set $\mathcal{I}$ is given by $\ell(\mathcal{I}) = b - a$. Analogously, the *volume* of an n-dimensional interval $\mathcal{I} = [a_1, b_1] \times \cdots \times [a_n, b_n] \subset \mathbb{R}^n$ is given by $\mathrm{vol}(\mathcal{I}) = \Pi_{i=1}^{n}(b_i - a_i)$. For $n = 1$, $\ell(\mathcal{I}) = \mathrm{vol}(\mathcal{I})$.

---

[3]If $\mathcal{S} \subset \mathcal{X}$, where $\mathcal{X}$ is a normed space, then $\mathcal{S}$ is *compact* if and only if every open cover of $\mathcal{S}$ contains a finite subcollection of open sets that also covers $\mathcal{S}$. In the case where $\mathcal{S} \subset \mathbb{R}^n$, this definition is equivalent to the compactness definition given in Definition 2.14.

Definition 2.40. Let $\mathcal{S} \subseteq \mathbb{R}^n$. The *outer measure* of $\mathcal{S}$ is defined by

$$\mu^*(\mathcal{S}) = \inf_{\mathcal{I}} \sum_{I_j \in \mathcal{I}} \mathrm{vol}(I_j), \tag{2.96}$$

where the infimum in (2.96) is taken over all countable collections $\mathcal{I} = \{I_j\}_{j=1}^{\infty}$ of closed intervals such that $\mathcal{S} \subset \bigcup_{j=1}^{\infty} I_j$.

Note that $\mu^*(\mathbb{R}^n) = \infty$ while $\mu^*(\emptyset) = 0$.

Definition 2.41. Let $\mathcal{S} \subseteq \mathbb{R}^n$. $\mathcal{S}$ is *Lebesgue measurable*, or simply *measurable*, if, for each $\varepsilon > 0$, there exists an open set $\mathcal{Q} \subseteq \mathbb{R}^n$ such that $\mathcal{S} \subset \mathcal{Q}$ and

$$\mu^*(\mathcal{Q} \backslash \mathcal{S}) < \varepsilon. \tag{2.97}$$

If $\mathcal{S}$ is measurable, then the *Lebesgue measure* (or *measure*) of $\mathcal{S}$, denoted by $\mu(\mathcal{S})$, is given by $\mu(\mathcal{S}) \triangleq \mu^*(\mathcal{S})$.

It follows from Definition 2.41 that a measurable set can be approximated arbitrarily closely, in terms of its outer measure, by open subsets of $\mathbb{R}^n$. Hence, every open set in $\mathbb{R}^n$ is measurable. More precisely, the measurable sets of $\mathbb{R}^n$ are defined as the members of the smallest family of sets of $\mathbb{R}^n$ containing all open, closed, and measure-zero sets of $\mathbb{R}^n$, as well as every difference, countable union, and countable intersection of its members. Sets that are not measurable are exceptional and seldom encountered in applications.

As noted in Section 2.4, if $f : \mathcal{D} \to \mathbb{R}$ is continuous on $\mathcal{D} \subseteq \mathbb{R}^n$, then $\{x \in \mathcal{D} : f(x) > \alpha\}$ is open for every $\alpha \in \mathbb{R}$. Using the notions of measurable sets, one can define measurable functions as a generalization of continuous functions.

Definition 2.42. Let $\mathcal{D} \subseteq \mathbb{R}^n$ be measurable. The function $f : \mathcal{D} \to \mathbb{R}$ is *measurable* on $\mathcal{D}$ if, for every $\alpha \in \mathbb{R}$, $\{x \in \mathcal{D} : f(x) > \alpha\}$ is measurable.

It can be shown that a measurable function is continuous almost everywhere in its domain of definition. On an interval $\mathcal{I} \subset \mathbb{R}$, a real-valued function $f : \mathcal{I} \to \mathbb{R}^n$ is a measurable function on $\mathcal{I}$ if and only if it is the pointwise limit, except for a set of measure zero, of a sequence of piecewise constant functions on $\mathcal{I}$. That is, if there exists a sequence of step functions $\{s_n\}_{n=0}^{\infty}$ on $\mathcal{I}$ such that $\lim_{n\to\infty} s_n(x) = f(x)$ almost everywhere on $\mathcal{I}$, then f is a measurable function on $\mathcal{I}$.

In the theory of Riemann integration, a piecewise continuous function over the compact interval $\mathcal{I} = [a, b] \subset \mathbb{R}$, a continuous function over the interval $\mathcal{I}$, or a bounded function over the interval $\mathcal{I}$ with at most countable many discontinuities in every finite subinterval of $\mathcal{I}$ is integrable. Specifically, we can ignore sets of measure zero points at which the function is discontinuous since the value of the function can be altered arbitrarily at these points without altering the value of the integral. That is, if $f : \mathcal{I} \subset \mathbb{R} \to \mathbb{R}^n$ and

$$\int_a^b f(t)\mathrm{d}t = \int_{\mathcal{I}} f(t)\mathrm{d}t \tag{2.98}$$

exists and is finite, then

$$\int_{\mathcal{I}} f(t)\mathrm{d}t = \int_{\mathcal{I}\backslash\mathcal{S}} f(t)\mathrm{d}t, \tag{2.99}$$

where $\mathcal{S} \subset \mathcal{I} = [a, b]$ has measure zero. If $f : \mathcal{I} \to \mathbb{R}^n$ and $g : \mathcal{I} \to \mathbb{R}^n$ are equal almost everywhere in $\mathcal{I}$, that is, $f(t) = g(t)$ for all $t \in \mathcal{I}$ except on a set of measure zero, then

$$\int_{\mathcal{I}} f(t)\mathrm{d}t = \int_{\mathcal{I}} g(t)\mathrm{d}t. \tag{2.100}$$

In light of the above discussion, the following theorem, due to Lebesgue, presents necessary and sufficient conditions for Riemann integrability of a bounded function.

Theorem 2.19. Let $\mathcal{I} \subset \mathbb{R}$ be a compact interval and assume $f : \mathcal{I} \to \mathbb{R}^n$ is bounded. Then f is Riemann integrable on $\mathcal{I}$ if and only if the set of points at which f is discontinuous has zero measure.

It follows from Theorem 2.19 that a bounded function f on a compact interval $\mathcal{I}$ is Riemann integrable on $\mathcal{I}$ if and only if f is continuous almost everywhere on $\mathcal{I}$.

Example 2.26. The functions $f : \mathbb{R} \to \{0, 1\}$ and $g : \mathbb{R} \to \{0, 1\}$ defined by

$$f(x) = \begin{cases} 1, & x \in \mathbb{Z}, \\ 0, & \text{otherwise}, \end{cases}$$

and

$$g(x) = \begin{cases} 1, & x \in \mathbb{Q}, \\ 0, & \text{otherwise}, \end{cases}$$

where $\mathbb{Q} \triangleq \{x : x = p/q, p, q \in \mathbb{Z}, q \neq 0\}$, are zero almost everywhere in $\mathbb{R}$. In addition, $f(x) = g(x)$ almost everywhere in $\mathbb{R}$. Since a set consisting of a single point has measure zero (see Problem 2.72), it follows that every countable subset of $\mathbb{R}$ has measure zero. In particular, the set of integers $\mathbb{Z}$ and the set of rational numbers $\mathbb{Q}$ have measure zero. However, even though $\mathbb{Q}$ is countable, it is not the set of discontinuities of g, whereas $\mathbb{Z}$ is the set of discontinuities of f. Since $\mathbb{Q}$ is dense in $\mathbb{R}$ it follows that g is discontinuous at every point in $\mathbb{R}$. Hence, it follows from Theorem 2.19 that the Riemann integral of g does not exist on every compact interval of $\mathbb{R}$, whereas the Riemann integral of f does exist. △

As shown by Theorem 2.19, the concept of the Riemann integral applies to piecewise continuous, continuous, and bounded functions with at most countable many discontinuities over an interval $\mathcal{I} \subset \mathbb{R}$. However, the Riemann integral of a general measurable function f, wherein f is discontinuous everywhere or when f is defined on an abstract *measurable space* (that is, an arbitrary set $\mathcal{X}$ equipped with a measure), cannot be formed. For such functions, one has to use the more general notion of Lebesgue integration [288]. The *Lebesgue integral* allows the integration of more general (i.e., measurable) functions. In particular, for a function f defined on a closed interval $\mathcal{I} \subset \mathbb{R}$, the Riemann integral of f is formed by dividing $\mathcal{I}$ into subintervals, thereby grouping together neighboring points of $\mathbb{R}$. In contrast, the Lebesgue integral of f is formed by grouping together points of $\mathbb{R}$ at which the function f takes neighboring values; that is, the range of the function f rather than its domain is partitioned. Hence, the Lebesgue integral is an extension of the Riemann integral in the sense that whenever the Riemann integral exists its value is equal to the Lebesgue integral.

For a given function $f : \mathcal{D} \to \mathbb{R}$, where $\mathcal{D} \subseteq \mathbb{R}^n$, the Lebesgue integral of f is denoted by

$$\int_{\mathcal{D}} f(x)\mathrm{d}x. \tag{2.101}$$

If $f : \mathcal{D} \to \mathbb{R}$ is measurable, where $\mathcal{D} \subseteq \mathbb{R}^n$ is measurable, and if (2.101) exists and is finite, then we say that f is *Lebesgue integrable* on $\mathcal{D}$. In this case, the set of integrable functions on $\mathcal{D}$ is denoted by

$$\mathcal{L}(\mathcal{D}) \triangleq \left\{ f : \int_{\mathcal{D}} f(x)\mathrm{d}x < \infty \right\}. \tag{2.102}$$

The following theorem gives necessary and sufficient conditions for Lebesgue integrability.

Theorem 2.20. Let $\mathcal{D}$ be measurable and assume $f : \mathcal{D} \to \mathbb{R}$ is measurable on $\mathcal{D}$. Then f is Lebesgue integrable over $\mathcal{D}$ if and only if $|f|$ is Lebesgue integrable over $\mathcal{D}$.

If (2.101) exists where $\mathcal{D} \subseteq \mathbb{R}^n$ is measurable and $f : \mathcal{D} \to \mathbb{R}$ is measurable on $\mathcal{D}$, then

$$\left| \int_{\mathcal{D}} f(x)\mathrm{d}x \right| \le \int_{\mathcal{D}} |f(x)|\mathrm{d}x. \tag{2.103}$$

To see this, first note that, by Theorem 2.20, $|f(x)|$, $x \in \mathcal{D}$, is Lebesgue integrable over $\mathcal{D}$. Next, let $y = \int_{\mathcal{D}} f(x)\mathrm{d}x$ and note that $\alpha y = |y|$, where $|\alpha| = 1$. Now, since $|\alpha f| = |f|$,

$$\left|\int_{\mathcal{D}} f(x)\mathrm{d}x\right| = \alpha \int_{\mathcal{D}} f(x)\mathrm{d}x = \int_{\mathcal{D}} \alpha f(x)\mathrm{d}x \leq \int_{\mathcal{D}} |\alpha f(x)|\mathrm{d}x = \int_{\mathcal{D}} |f(x)|\mathrm{d}x,$$

which proves (2.103). Inequality (2.103) is known as the *absolute value theorem for integrals*. In light of (2.103), the set of Lebesgue integrable functions on $\mathcal{D}$ defined by (2.102) becomes

$$\mathcal{L}(\mathcal{D}) = \left\{f :\ f \text{ is measurable and } \int_{\mathcal{D}} |f(x)|\mathrm{d}x < \infty\right\}. \tag{2.104}$$

Note that the measurability of f in (2.104) is included since it is possible for $|f|$ to be integrable but not measurable.

Finally, if $\mathcal{D} = \mathcal{I}$, where $\mathcal{I}$ is a measurable subset of $\mathbb{R}$, $f : \mathcal{I} \to \mathbb{R}$ is measurable on $\mathcal{I}$, and $f \in \mathcal{L}(\mathcal{I})$, then

$$\int_{\mathcal{I}} f(t)\mathrm{d}t = \int_{\mathcal{I}\backslash\mathcal{S}} f(t)\mathrm{d}t, \tag{2.105}$$

where $\mathcal{S} \subset \mathcal{I}$ has measure zero. Hence, the value of the integral over $\mathcal{I}$ is not affected by removing the set $\mathcal{S}$ of measure zero from $\mathcal{I}$. In addition, if $f(t) = g(t)$ almost everywhere in $\mathcal{I}$, where $g : \mathcal{I} \to \mathbb{R}$ is measurable on $\mathcal{I}$ and $g \in \mathcal{L}(\mathcal{I})$, then

$$\int_{\mathcal{I}} f(t)\mathrm{d}t = \int_{\mathcal{I}} g(t)\mathrm{d}t. \tag{2.106}$$

Analogous results hold for functions f and g defined on $\mathcal{D} \subseteq \mathbb{R}^n$.

Example 2.27. Consider the *Dirichlet function* $g : [0,1] \to \{0,1\}$ defined in Example 2.26. Since the set of rational numbers $\mathbb{Q}$ is countable and therefore has measure zero, it follows that $g(x) = 0$ almost everywhere on $[0,1]$. Since the function 0 is Riemann integrable on $[0,1]$, it is also Lebesgue integrable on $[0,1]$. Hence, by (2.106),

$$\int_{[0,1]} g(x)\mathrm{d}x = \int_0^1 0\mathrm{d}x = 0,$$

and therefore g is Lebesgue integrable on $[0,1]$. △

Next, we consider some special normed vector spaces. Specifically, we consider the normed function space of all functions $f : [0,\infty) \to \mathbb{R}^n$ and with norm defined by

$$|||f|||_{p,q} \triangleq \left[\int_0^\infty \|f(t)\|_q^p \mathrm{d}t\right]^{1/p}, \qquad 1 \leq p < \infty, \tag{2.107}$$

$$|||f|||_{\infty,q} \triangleq \sup_{t\in[0,\infty)} \|f(t)\|_q. \tag{2.108}$$

Here, the vector norm (q part) is the *spatial* norm; the p part is the *temporal* norm. Note that the norm $|||\cdot|||_{p,q}$ satisfies Axioms *i*) and *iii*) of Definition 2.39. However, since there exist many nonzero functions $f : [0,\infty) \to \mathbb{R}^n$ such that $|||f|||_{p,q} = 0$, $p \in [1,\infty)$, it follows that Axiom *ii*) of Definition 2.39 can be violated with the norm $|||\cdot|||_{p,q}$. Specifically, if $f(\cdot)$ is a function such that it is zero almost everywhere or, equivalently, $f(\cdot)$ is zero everywhere on $[0,\infty)$ except on a set of measure zero [288], then $f(\cdot)$ satisfies $|||f|||_{p,q} = 0$. To address such functions the concepts of Lebesgue measure and measurable spaces [288] can be used with the integral in (2.107) denoting the Lebesgue integral. In this case, the function $|||\cdot|||_{p,q}$ is a valid norm on the space of measurable functions $f : [0,\infty) \to \mathbb{R}^n$ with the interpretation that if $|||f|||_{p,q} = 0$, then $f(t) = 0$ almost everywhere on $[0,\infty)$. We denote this normed linear space by $\mathcal{L}_p$ or, equivalently, the set of measurable functions such that $|||f|||_{p,q} < \infty$, $q \in [1,\infty]$, that is,

$$\mathcal{L}_p \triangleq \{f : [0,\infty) \to \mathbb{R}^n : \ f \text{ is measurable and } |||f|||_{p,q} < \infty,\ q \in [1,\infty]\}.$$

In this case, $|||f|||_{\infty,q}$ given by (2.108) is replaced by

$$|||f|||_{\infty,q} \triangleq \operatorname*{ess\,sup}_{t\in[0,\infty)} \|f(t)\|_q, \tag{2.109}$$

where "ess" denotes essential.

Example 2.28. Let $\mathcal{C}[a,b]$ denote the set of continuous functions mapping the bounded interval $[a,b]$ into $\mathbb{R}^n$. Furthermore, for $f,g \in \mathcal{C}[a,b]$ and $\alpha \in \mathbb{R}$, let $(f+g)(t) \triangleq f(t)+g(t)$ and $(\alpha f)(t) = \alpha f(t)$. Now, define the norm $|||\cdot||| : \mathbb{R}^n \to \mathbb{R}$ by $|||f|||_{\infty,q} = \max_{t\in[a,b]} \|f(t)\|_q$. Note that since f is defined on the bounded interval $[a,b]$ and f is continuous, $|||\cdot|||$ is a well-defined norm and is finite for every $f(\cdot) \in \mathcal{C}[a,b]$. Clearly, $|||f|||_{\infty,q} \geq 0$ and is zero only for $f(t) \equiv 0$. To show that Axioms *iii*) and *iv*) of Definition 2.39 hold note that for $f(\cdot), g(\cdot) \in \mathcal{C}[a,b]$ it follows that

$$\begin{aligned} |||f+g|||_{\infty,q} &= \max_{t\in[a,b]} \|f(t)+g(t)\|_q \\ &\leq \max_{t\in[a,b]} [\|f(t)\|_q + \|g(t)\|_q] \\ &= |||f|||_{\infty,q} + |||g|||_{\infty,q}. \end{aligned} \tag{2.110}$$

Furthermore,

$$\begin{aligned} |||\alpha f|||_{\infty,q} &= \max_{t\in[a,b]} \|\alpha f(t)\|_q \\ &= \max_{t\in[a,b]} |\alpha|\|f(t)\|_q \\ &= |\alpha| \max_{t\in[a,b]} \|f(t)\|_q \end{aligned}$$

$$= |\alpha| \|f\|_{\infty,q}. \tag{2.111}$$

Hence, $C[a,b]$ endowed with the norm $\|\cdot\|_{\infty,q}$ is a normed linear space. △

Note that in the above discussion we can also allow $f : [0,\infty) \to \mathbb{R}^{n\times m}$ in a similar way where the spatial norm is a matrix norm. Furthermore, in the matrix case, we can define induced norms for linear operators $\mathcal{G} : \mathcal{L}_p \to \mathcal{L}_q$ of the form $y(t) = \mathcal{G}[u](t)$. One such induced norm $\|\mathcal{G}\|_{(q,s),(p,r)}$ is defined by

$$\|\mathcal{G}\|_{(q,s),(p,r)} \triangleq \sup_{\|u\|_{p,r}=1} \|\mathcal{G}[u]\|_{q,s}, \tag{2.112}$$

which corresponds to an induced operator norm from an input signal $u(t)$ with p temporal norm and r spatial norm to an output signal $y(t)$ with q temporal norm and s spatial norm. For further details, see Chapter 7.

Some normed vector spaces are not large enough to permit limit operations. In particular, if $\{f_n\}_{n=1}^{\infty}$ denotes a sequence of functions belonging to a normed linear space such that $\lim_{n\to\infty} \|f - f_n\| = 0$, then a natural question is whether f is a member of the normed linear space. As an example, consider the space of continuous functions on $[0,1]$ with the norm defined by $\|f\| = \int_0^1 |f(t)|\mathrm{d}t$. Now, consider the sequence of continuous functions

$$f_n(t) = \begin{cases} 1 - nt, & t \in [0, 1/n), \\ 0, & t > 1/n, \end{cases} \tag{2.113}$$

and define f by

$$f(t) = \begin{cases} 1, & t = 0, \\ 0, & t > 0. \end{cases} \tag{2.114}$$

Then it can be easily seen that $\lim_{n\to\infty} \|f_n - f\| = 0$. However, the limit f is not continuous, that is, f is not a member of the vector space. In other words, our linear space was too small for the norm we defined on it. In order to address the above observations we introduce the concept of a *Cauchy sequence.*

Definition 2.43. A sequence $\{x_n\}_{n=1}^{\infty}$ in a normed linear space is called a *Cauchy sequence* if, for every $\varepsilon > 0$, there exists an integer $N = N(\varepsilon)$ such that $\|x_n - x_m\| < \varepsilon$, whenever $n, m > N$.

Note that an equivalent condition for a Cauchy sequence is $\lim_{n,m\to\infty} \|x_n - x_m\| = 0$. Thus, the main difference between a convergent sequence and a Cauchy sequence is that in the former case the terms in the sequence approach a limit point, whereas in the latter case the terms in the sequence approach each other. The following proposition shows that every convergent sequence in a normed linear space is a Cauchy sequence.

Proposition 2.29. Assume that the sequence $\{x_n\}_{n=1}^{\infty}$ converges in a normed linear space. Then for every $\varepsilon > 0$ there exists an integer N such that $\|x_n - x_m\| < \varepsilon$ whenever $n > N$ and $m > N$.

Proof. Let $x = \lim_{n\to\infty} x_n$. Given $\varepsilon > 0$, let N be such that $\|x_n - x\| < \varepsilon/2$, whenever $n > N$. Now, if $m > N$, then $\|x_m - x\| < \varepsilon/2$. If $n > N$ and $m > N$ it follows from the triangle inequality that

$$\|x_n - x_m\| = \|x_n - x + x - x_m\| \leq \|x_n - x\| + \|x - x_m\| < \frac{\varepsilon}{2} + \frac{\varepsilon}{2} = \varepsilon.$$

Hence, $\{x_n\}_{n=1}^{\infty}$ is a Cauchy sequence. □

Even though every convergent sequence is Cauchy, the converse is not necessarily true in a normed linear space. In particular, if the elements of a sequence get closer to each other, that does not imply that the sequence is convergent. However, there exist normed linear spaces where this is true. These spaces are known as *Banach spaces.*

Definition 2.44. A normed linear space is called a *complete space* or a *Banach space* if every Cauchy sequence converges to an element in the space.

Example 2.29. Once again, we consider the set $\mathcal{C}[a,b]$ of continuous functions mapping the bounded interval $[a,b]$ into $\mathbb{R}^n$ addressed in Example 2.28. To show that $\mathcal{C}[a,b]$ is a Banach space let $\{f_n\}_{n=0}^{\infty}$ be a Cauchy sequence in $\mathcal{C}[a,b]$. Now, for every fixed $t \in [a,b]$,

$$\|f_n(t) - f_m(t)\|_q \leq \|f_n - f_m\|_{\infty,q} \to 0 \text{ as } n, m \to \infty, \tag{2.115}$$

and hence the real vectors $\{f_n(t)\}_{n=0}^{\infty}$ form a Cauchy sequence in $\mathbb{R}^n$. Since $\mathbb{R}^n$ with $\|\cdot\|_q$, $q \in [1,\infty]$, is a complete space [373] it follows that there exists $f(t)$ such that $f_n(t) \to f(t)$ as $n \to \infty$. To show that the sequence $\{f_n\}_{n=0}^{\infty}$ converges to f uniformly in t, let $\varepsilon > 0$ and choose N such that $\|f_n - f_m\|_{\infty,q} < \varepsilon/2$ for all $n, m > N$. Now, let $t \in [a,b]$ and let $m > N$ such that $\|f_m(t) - f(t)\|_q < \varepsilon/2$. Next, it follows from the triangle inequality that for all $n > N$,

$$\begin{aligned}\|f_n(t) - f(t)\|_q &= \|f_n(t) - f_m(t) + f_m(t) - f(t)\|_q \\ &\leq \|f_n(t) - f_m(t)\|_q + \|f_m(t) - f(t)\|_q \\ &\leq \|f_n - f_m\|_{\infty,q} + \|f_m(t) - f(t)\|_q \\ &< \varepsilon. \end{aligned} \tag{2.116}$$

Hence, since $t \in [a,b]$ is arbitrary, it follows that $\{f_n\}_{n=0}^{\infty}$ converges to f uniformly in t. Now, since $f_n(\cdot)$ is continuous and convergence is uniform, the limit function $f(\cdot)$ is also continuous.

To see this, let $t_0 \in [a,b]$ be an arbitrary time in $[a,b]$. Since the convergence of $\{f_n\}_{n=0}^{\infty}$ is uniform, given $\varepsilon > 0$, there exists sufficiently large n such that

$$\|f_n(t) - f(t)\|_q < \varepsilon/3, \qquad t \in [a,b]. \tag{2.117}$$

However, $f_n(t)$ is continuous at t_0, and hence, there exists some $\delta > 0$ such that

$$\|f_n(t) - f_n(t_0)\|_q < \varepsilon/3 \tag{2.118}$$

whenever $|t - t_0| < \delta$. Thus, if $|t - t_0| < \delta$, it follows that

$$\begin{aligned} \|f(t) - f(t_0)\|_q &= \|f(t) - f_n(t) + f_n(t) - f_n(t_0) + f_n(t_0) - f(t_0)\|_q \\ &\le \|f(t) - f_n(t)\|_q + \|f_n(t) - f_n(t_0)\|_q + \|f_n(t_0) - f(t_0)\|_q \\ &< \varepsilon/3 + \varepsilon/3 + \varepsilon/3 \\ &= \varepsilon, \end{aligned} \tag{2.119}$$

which shows that f is continuous at t_0. Now, since t_0 is an arbitrary point in $[a,b]$, it follows that f is continuous on $[a,b]$. Hence, since a sequence of functions $\{f_n\}_{n=0}^{\infty}$ in $\mathcal{C}[a,b]$ converges to a function $f(\cdot) \in \mathcal{C}[a,b]$ if and only if $\{f_n(t)\}_{n=0}^{\infty}$ converges to $f(t)$ uniformly on $[a,b]$, it follows that $\mathcal{C}[a,b]$ endowed with the norm $|\!|\!|\cdot|\!|\!|_{\infty,q}$ is a Banach space. △

We close this section with a very important theorem from mathematical analysis needed to derive existence and uniqueness results for nonlinear differential equations. The theorem is known as the *contraction mapping theorem* or the *Banach fixed point theorem*. The following definition is needed for this result.

Definition 2.45. Let $\mathcal{V}$ be a linear vector space and let $T : \mathcal{V} \to \mathcal{V}$. The point $x^* \in \mathcal{V}$ is a *fixed point* of $T(\cdot)$ if $T(x^*) = x^*$.

Theorem 2.21 (Banach Fixed Point Theorem). Let $\mathcal{X}$ be a Banach space with norm $|\!|\!|\cdot|\!|\!| : \mathcal{X} \to \mathbb{R}$ and let $T : \mathcal{X} \to \mathcal{X}$. Suppose there exists a constant $\rho \in [0,1)$ such that

$$|\!|\!|T(x) - T(y)|\!|\!| \le \rho |\!|\!|x - y|\!|\!|, \qquad x, y \in \mathcal{X}. \tag{2.120}$$

Then there exists a unique $x^* \in \mathcal{X}$ such that $T(x^*) = x^*$. Furthermore, for each $x_0 \in \mathcal{X}$, the sequence $\{x_n\}_{n=0}^{\infty} \subset \mathcal{X}$ defined by

$$x_{n+1} = T(x_n) \tag{2.121}$$

converges to x^*. Finally,

$$|\!|\!|x^* - x_n|\!|\!| \le \frac{\rho^n}{1-\rho} |\!|\!|T(x_0) - x_0|\!|\!|, \quad n \ge 0. \tag{2.122}$$

Proof. Let $x_0 \in \mathcal{X}$ be arbitrary and define the sequence $\{x_n\}_{n=0}^{\infty}$ by

$x_{n+1} = T(x_n)$. Now, for each $n \geq 0$ it follows from (2.120) that

$$\begin{aligned}
\|x_{n+1} - x_n\| &= \|T(x_n) - T(x_{n-1})\| \\
&\leq \rho\|x_n - x_{n-1}\| \\
&= \rho\|T(x_{n-1}) - T(x_{n-2})\| \\
&\leq \rho^2\|x_{n-1} - x_{n-2}\| \\
&\ \ \vdots \\
&\leq \rho^n\|x_1 - x_0\| \\
&= \rho^n\|T(x_0) - x_0\|. \qquad (2.123)
\end{aligned}$$

Next, let $m = n + r$, $r \geq 0$, be given. Then (2.123) implies that

$$\begin{aligned}
\|x_m - x_n\| &= \|x_{n+r} - x_n\| \\
&= \|x_{n+r} - x_{n+r-1} + x_{n+r-1} - x_{n+r-2} + \cdots + x_{n+1} - x_n\| \\
&\leq \|x_{n+r} - x_{n+r-1}\| + \|x_{n+r-1} - x_{n+r-2}\| + \cdots + \|x_{n+1} - x_n\| \\
&\leq (\rho^{n+r-1} + \rho^{n+r-2} + \cdots + \rho^n)\|x_1 - x_0\| \\
&= \frac{\rho^n(1-\rho^r)}{1-\rho}\|T(x_0) - x_0\|. \qquad (2.124)
\end{aligned}$$

Now, since $\rho < 1$, $\rho^n \to 0$ as $n \to \infty$. Hence, given $\varepsilon > 0$ and choosing a large enough N, it follows that $\|x_m - x_n\| < \varepsilon$ for all $m > n \geq N$. This proves that the sequence $\{x_n\}_{n=0}^{\infty}$ is Cauchy. Furthermore, since $\mathcal{X}$ is a Banach space, the sequence converges to an element $x^* \in \mathcal{X}$. Noting that $T(\cdot)$ is a uniformly continuous function, it follows that

$$T(x^*) = T(\lim_{n\to\infty} x_n) = \lim_{n\to\infty} T(x_n) = \lim_{n\to\infty} x_{n+1} = x^*, \qquad (2.125)$$

and hence, x^* is a fixed point.

Next, suppose, *ad absurdum*, that x^* is not unique. That is, suppose there exists $x \in \mathcal{X}$, $x \neq x^*$, such that $T(x) = x$. Then, it follows from (2.120) that

$$\|x^* - x\| = \|T(x^*) - T(x)\| \leq \rho\|x^* - x\|. \qquad (2.126)$$

Since $\rho < 1$, (2.126) holds only if $\|x^* - x\| = 0$, and hence, $x^* = x$ leading to a contradiction. Thus, x^* is a unique fixed point.

Finally, to show (2.122), let $r \to \infty$ in (2.124) and recall that $\|\cdot\| : \mathcal{X} \to \mathbb{R}$ is continuous on $\mathcal{X}$ (see Problem 2.86) so that

$$\begin{aligned}
\|x^* - x_n\| &= \|\lim_{r\to\infty} x_{n+r} - x_n\| \\
&= \lim_{r\to\infty} \|x_{n+r} - x_n\| \\
&\leq \lim_{r\to\infty} \frac{\rho^n(1-\rho^r)}{1-\rho}\|T(x_0) - x_0\|
\end{aligned}$$

$$= \frac{\rho^n}{1-\rho} |||T(x_0) - x_0|||, \tag{2.127}$$

where the last inequality in (2.127) follows from the fact that $\rho < 1$. □

It is important to note that it is not possible to replace (2.120) with the weaker hypothesis

$$|||T(x) - T(y)||| < |||x - y|||, \qquad x, y \in \mathcal{X}. \tag{2.128}$$

See Problem 2.91. Finally, we give a slightly different version of Theorem 2.21 for the case where T maps a closed subset $\mathcal{S}$ of $\mathcal{X}$ into itself. This version of the theorem is useful in applications.

Theorem 2.22. Let $\mathcal{X}$ be a Banach space with norm $|||\cdot||| : \mathcal{X} \to \mathbb{R}$, let $\mathcal{S}$ be a closed subset of $\mathcal{X}$, and let $T : \mathcal{S} \to \mathcal{S}$. Suppose there exists a constant $\rho \in [0, 1)$ such that

$$|||T(x) - T(y)||| \le \rho |||x - y|||, \qquad x, y \in \mathcal{S}. \tag{2.129}$$

Then there exists a unique $x^* \in \mathcal{S}$ such that $T(x^*) = x^*$. Furthermore, for each $x_0 \in \mathcal{S}$, the sequence $\{x_n\}_{n=0}^{\infty} \subset \mathcal{S}$ defined by $x_{n+1} = T(x_n)$ converges to x^*. Finally,

$$|||x^* - x_n||| \le \frac{\rho^n}{1-\rho} |||T(x_0) - x_0|||, \quad n \ge 0. \tag{2.130}$$

Proof. The fact that $\mathcal{S}$ is closed guarantees that $x^* \in \mathcal{S}$. Now, the proof is identical to the proof of Theorem 2.21. □

Finally, we present a key result connecting continuity and boundedness of a linear operator defined on normed linear spaces. First, however, the following definition is needed.

Definition 2.46. Let $\mathcal{X}$ and $\mathcal{Y}$ be normed linear spaces with norms $|||\cdot||| : \mathcal{X} \to \mathbb{R}$ and $|||\cdot|||' : \mathcal{Y} \to \mathbb{R}$. A linear operator $T : \mathcal{X} \to \mathcal{Y}$ is *bounded* if there exists $\alpha \ge 0$ such that $|||Tx|||' \le \alpha |||x|||$ for all $x \in \mathcal{X}$.

Theorem 2.23. Let $\mathcal{X}$ and $\mathcal{Y}$ be normed linear spaces with norms $|||\cdot||| : \mathcal{X} \to \mathbb{R}$ and $|||\cdot|||' : \mathcal{Y} \to \mathbb{R}$, and let $T : \mathcal{X} \to \mathcal{Y}$ be a linear operator. Then T is uniformly continuous on $\mathcal{X}$ if and only if T is bounded on $\mathcal{X}$.

Proof. Assume $T(\cdot)$ is bounded. Since $T(\cdot)$ is a linear operator, it follows that

$$|||T(x) - T(y)|||' = |||Tx - Ty|||' = |||T(x - y)|||' \le \alpha |||x - y|||$$

for all $x, y \in \mathcal{X}$ and some $\alpha \ge 0$. Now, choosing $\delta = \varepsilon/\alpha$, it follows that given $\varepsilon > 0$ there exists $\delta = \delta(\varepsilon)$ such that $|||T(x) - T(y)|||' < \varepsilon$ for all $x, y \in \mathcal{X}$

satisfying $\|x - y\| < \delta$. Hence, $T(\cdot)$ is uniformly continuous on $\mathcal{X}$.

Conversely, assume $T(\cdot)$ is uniformly continuous on $\mathcal{X}$. Then $T(\cdot)$ is continuous at $x = 0$. Hence, given $\varepsilon > 0$ there exists $\delta = \delta(\varepsilon)$ such that $\|x\| \leq \varepsilon$ implies $\|T(x)\|' = \|Tx\|' \leq \delta$. Next, for $x \in \mathcal{X}$, $x \neq 0$, let $y = \beta x$, where $\beta = \varepsilon/\|x\|$. Now, since $\|y\| = \|\beta x\| = \varepsilon$ it follows that $\|Ty\|' \leq \delta$. Hence,

$$\|Ty\|' = \|T\beta x\|' = \beta\|Tx\|' = \varepsilon\frac{\|Tx\|'}{\|x\|} \leq \delta,$$

which implies that

$$\|Tx\|' \leq \frac{\delta}{\varepsilon}\|x\|,$$

and hence, $T(\cdot)$ is bounded on $\mathcal{X}$. □

2.6 Dynamical Systems, Flows, and Vector Fields

As discussed in Chapter 1, a system is a combination of components or parts which is perceived as a single entity. The parts making up the system are typically clearly defined with a particular set of variables, called the states of the system, that completely determine the behavior of the system at a given time. Hence, a dynamical system consists of a set of possible states in a given space, together with a rule that determines the present state of the system in terms of past states. Thus, a dynamical system on $\mathcal{D} \subseteq \mathbb{R}^n$ tells us for a specific time $t = t_0$ and state x in the space $\mathcal{D}$ where the system state x will be at time $t \geq t_0$. In this book, we view a dynamical system as a precise mathematical object defined on a time interval as a mapping between vector spaces satisfying a set of axioms. For this definition $\mathcal{D}$ is an open subset of $\mathbb{R}^n$.

Definition 2.47. A *dynamical system* on $\mathcal{D}$ is the triple $(\mathcal{D}, \mathbb{R}, s)$, where $s : \mathbb{R} \times \mathcal{D} \to \mathcal{D}$ is such that the following axioms hold:

i) (Continuity): $s(\cdot, \cdot)$ is continuous on $\mathcal{D} \times \mathbb{R}$ and for every $t \in \mathbb{R}$, $s(\cdot, x)$ is continuously differentiable on $\mathcal{D}$.

ii) (Consistency): $s(0, x_0) = x_0$ for all $x_0 \in \mathcal{D}$.

iii) (Group property): $s(\tau, s(t, x_0)) = s(t + \tau, x_0)$ for all $x_0 \in \mathcal{D}$ and $t, \tau \in \mathbb{R}$.

Henceforth, we denote the dynamical system $(\mathcal{D}, \mathbb{R}, s)$ by $\mathcal{G}$ and we refer to the map $s(\cdot, \cdot)$ as the *flow* or *trajectory* of $\mathcal{G}$ corresponding to $x_0 \in \mathcal{D}$, and for a given $s(t, x_0)$, $t \geq 0$, we refer to $x_0 \in \mathcal{D}$ as an *initial condition* of $\mathcal{G}$. Given $t \in \mathbb{R}$ we denote the map $s(t, \cdot) : \mathcal{D} \to \mathcal{D}$ by $s_t(x_0)$ or s_t. Hence, for

$t \in \mathbb{R}$ the set of mappings defined by $s_t(x_0) = s(t, x_0)$ for every $x_0 \in \mathcal{D}$ give the *flow* of $\mathcal{G}$. In particular, if $\mathcal{D}_0$ is a collection of initial conditions such that $\mathcal{D}_0 \subset \mathcal{D}$, then the flow $s_t : \mathcal{D}_0 \to \mathcal{D}$ is nothing more than the motion of all points $x_0 \in \mathcal{D}_0$ or, equivalently, the image of $\mathcal{D}_0 \subset \mathcal{D}$ under the flow s_t, that is, $s_t(\mathcal{D}_0) \subset \mathcal{D}$ (see Figure 2.3(a)). Alternatively, if the initial condition $x_0 \in \mathcal{D}$ is fixed and we let $[\alpha, \beta] \subset \mathbb{R}$, then the mapping $s(\cdot, x_0) : [\alpha, \beta] \to \mathcal{D}$ defines the *solution curve* or *trajectory* of the dynamical system $\mathcal{G}$. Hence, the mapping $s(\cdot, x_0)$ generates a graph in $[\alpha, \beta] \times \mathcal{D}$ identifying the trajectory corresponding to motion along a curve $\mathcal{C}$ through the point x_0 in a subset $\mathcal{D}$ of the state space (see Figure 2.3(b)). Given $x \in \mathcal{D}$ we denote the map $s(\cdot, x) : \mathbb{R} \to \mathcal{D}$ by $s^x(t)$ or s^x.

If we think of a dynamical system $\mathcal{G}$ as describing the motion of a fluid, then the flow of $\mathcal{G}$ describes the motion of the entire fluid and is consistent with an Eulerian description of the dynamical system wherein the motion is analyzed over a continuous medium (control volume). Alternatively, the trajectory of $\mathcal{G}$ describes the motion of an individual particle in the fluid and is consistent with a Lagrangian formulation of the dynamical system which describes the motion (position) of a particle as a function of time.

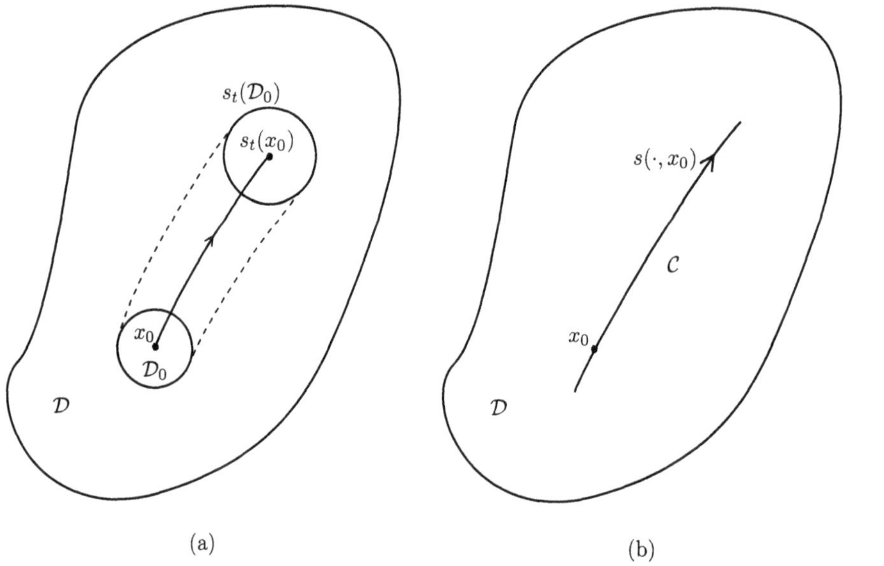

Figure 2.3 (a) Flow of a dynamical system. (b) Solution curve of a dynamical system.

In terms of the map $s_t : \mathcal{D} \to \mathcal{D}$ Axioms *ii*) and *iii*) can be equivalently written as *ii*)$'$ $s_0(x_0) = x_0$ and *iii*)$'$ $(s_\tau \circ s_t)(x_0) = s_\tau(s_t(x_0)) = s_{t+\tau}(x_0)$. Note that it follows from *i*) and *iii*) that the map $s_t : \mathcal{D} \to \mathcal{D}$ is a continuous function with a continuous inverse s_{-t}. Thus, s_t with $t \in \mathbb{R}$ generates a one-parameter family of homeomorphisms on $\mathcal{D}$ forming a commutative group under composition. To see that s_t is one-to-one note that if $s(t, y) = s(t, z)$, then $y = z$ follows from

$$\begin{aligned} y &= s(0, y) \\ &= s(-t + t, y) \\ &= s(-t, s(t, y)) \\ &= s(-t, s(t, z)) \\ &= s(-t + t, z) \\ &= s(0, z) \\ &= z. \end{aligned} \tag{2.131}$$

Also note that if $y \in \mathcal{D}$, then $s_t(x) = y$ for $x = s(-t, y)$, and hence, s_t is onto. Finally, to see that s_t has a continuous inverse we need only show that s_{-t} is the inverse of s_t. To see this, note that for any two flows s_t and s_τ, $s_t \circ s_\tau = s_{t+\tau}$ since, for every $x \in \mathcal{D}$,

$$(s_t \circ s_\tau)(x) = s_t(s_\tau(x)) = s_t(s(\tau, x)) = s(t + \tau, x) = s_{t+\tau}(x). \tag{2.132}$$

In addition, note that for every $x \in \mathcal{D}$, $s_0(x) = s(0, x) = x$ is the identity operator on $\mathcal{D}$. Hence, $s_{-t} \circ s_t = s_{t-t} = s_0$, which establishes that s_{-t} is the inverse of s_t.

Since a dynamical system $\mathcal{G}$ involves the function $s(\cdot, \cdot)$ describing the motion of $x \in \mathcal{D}$ for all $t \in \mathbb{R}$, it generates a differential equation on $\mathcal{D}$. In particular, the function $f : \mathcal{D} \to \mathbb{R}^n$ given by

$$f(x) \triangleq \frac{\mathrm{d}}{\mathrm{d}t} s(t, x)\bigg|_{t=0} \tag{2.133}$$

defines a continuous *vector field* on $\mathcal{D}$. For $x \in \mathcal{D}$, $f(x)$ belongs to $\mathbb{R}^n$ and corresponds to the tangent vector to the curve $s_t(x)$ at $t = 0$. Hence, for $s_t : \mathcal{D} \to \mathcal{D}$ satisfying Axioms *i*)–*iii*) of Definition 2.47, letting $x(t) = s(t, x_0)$ and defining $f : \mathcal{D} \to \mathbb{R}^n$ as in (2.133) it follows that

$$\dot{x}(t) = f(x(t)), \qquad x(0) = x_0, \qquad t \in \mathbb{R}. \tag{2.134}$$

In this book, we use the notation $s(t, x_0)$, $t \in \mathbb{R}$, and $x(t)$, $t \in \mathbb{R}$, interchangeably to denote the solution of the nonlinear dynamical system (2.134) with initial condition $x(0) = x_0$. Even though for physical dynamical systems $t \geq 0$, in this chapter we allow $t \in \mathbb{R}$ in order to develop general analysis results for (2.134) possessing *reversible* flows. In the later chapters we consider dynamical systems on the semi-infinite interval $[0, \infty)$.

Example 2.30. In this example we analyze the solution curves and flow of a linear system. Specifically, define $s : [0, \infty) \times \mathbb{R}^n \to \mathbb{R}^n$ by $s(t, x) = e^{At}x$, where $A \in \mathbb{R}^{n \times n}$. Hence, $s_t : \mathbb{R}^n \to \mathbb{R}^n$ is represented by $e^{At} \in \mathbb{R}^{n \times n}$ so that $s_t(x) = e^{At}x$. Note that since $e^{0_{n \times n}} = I_n$ and $e^{A(t+\tau)} = e^{At}e^{A\tau}$ it follows that $s_0(x_0) = x_0$ and $(s_\tau \circ s_t)(x_0) = s_\tau(s_t(x_0)) = e^{A\tau}e^{At}x_0$, and hence,

Axioms *ii*) and *iii*) of Definition 2.47 are satisfied. Axiom *i*) is trivially satisfied. Now, for a given time t, if $x \in \mathbb{R}^n$ the flow $s_t(x)$ is the image $e^{At}x$ of x and is given by $e^{At}x = \sum_{n=1}^{n} x_{(i)}\mathrm{col}_i(e^{At})$, where $\mathrm{col}_i(e^{At})$ denotes the ith column of e^{At}. Hence, the flow is given by $\mathcal{R}(e^{At}) \triangleq \{y \in \mathbb{R}^n : y = e^{At}x \text{ for some } x \in \mathbb{R}^n\} \subseteq \mathbb{R}^n$ for each fixed time. Alternatively, given $x_0 \in \mathcal{D}$, $s^{x_0} : [0, \infty) \to \mathbb{R}^n$ defines the system trajectory. Hence, $s^{x_0}(\cdot)$ is represented by $e^{At}x_0 : [0, \infty) \to \mathbb{R}^n$ so that $s^{x_0} = e^{At}x_0$. Finally, to show that $s(t, x) = e^{At}x$ generates a linear differential equation on $\mathbb{R}^n$ note that $f(x) = \frac{\mathrm{d}}{\mathrm{d}t}s(t, x)|_{t=0} = \frac{\mathrm{d}}{\mathrm{d}t}e^{At}x|_{t=0} = Ax$. Hence, with $x(t) = s(t, x_0)$ it follows that

$$\dot{x}(t) = Ax(t), \qquad x(0) = x_0, \qquad t \geq 0, \tag{2.135}$$

which defines a linear, time-invariant dynamical system. △

2.7 Nonlinear Differential Equations

In Section 2.6 we saw that a nonlinear dynamical system $\mathcal{G}$ as defined by Definition 2.47 gives rise to a nonlinear differential equation. In this section, we present several general results on nonlinear dynamical systems characterized by differential equations of the form

$$\dot{x}(t) = f(x(t)), \quad x(t_0) = x_0, \quad t \in \mathcal{I}_{x_0}, \tag{2.136}$$

where $x(t) \in \mathcal{D}$, $t \in \mathcal{I}_{x_0}$, $\mathcal{D}$ is an open subset of $\mathbb{R}^n$ with $0 \in \mathcal{D}$, $f : \mathcal{D} \to \mathbb{R}^n$ is continuous on $\mathcal{D}$, and $\mathcal{I}_{x_0} = (\tau_{\min}, \tau_{\max})$ is the *maximal interval of existence* for the solution $x(\cdot)$ of (2.136). A continuously differentiable function $x : \mathcal{I}_{x_0} \to \mathcal{D}$ is said to be a *solution* to (2.136) on the interval $\mathcal{I}_{x_0} \subseteq \mathbb{R}$ with initial condition $x(t_0) = x_0$, if $x(t)$ satisfies (2.136) for all $t \in \mathcal{I}_{x_0}$. Unlike linear differential equations, the existence and uniqueness of solutions of (2.136) are not guaranteed. In addition, a solution may only exist on some proper subinterval $(\tau_{\min}, \tau_{\max}) \subset \mathbb{R}$ (maximal interval of existence). Note that if $x(\cdot)$ is a solution to (2.136) and $f : \mathcal{D} \to \mathbb{R}^n$ is continuous, then $x(\cdot)$ satisfies the *integral equation*

$$x(t) = x_0 + \int_{t_0}^{t} f(x(s))\mathrm{d}s, \qquad t \in \mathcal{I}_{x_0}. \tag{2.137}$$

Conversely, if $x(\cdot)$ is continuous on $\mathcal{I}_{x_0}$ and satisfies (2.137), then $x(\cdot)$ is continuously differentiable on $\mathcal{I}_{x_0}$ and satisfies (2.136). Hence, (2.136) and (2.137) are equivalent in the sense that $x(\cdot)$ is a solution to (2.136) if and only if $x(\cdot)$ is a solution to (2.137).

Since the existence and uniqueness of solutions for nonlinear differential equations are not always guaranteed, in this and the next section we address the following questions:

i) Under what conditions does (2.136) have at *least one* solution for a given $x_0 \in \mathcal{D}$?

ii) Under what conditions does (2.136) have a unique solution for a given $x_0 \in \mathcal{D}$?

iii) What is the maximal interval of existence over which one or more solutions to (2.136) exist?

iv) What is the sensitivity of the solutions to (2.136) to initial data and/or parameter perturbations?

Before addressing each of the above questions, we present a series of examples that demonstrate the need for providing conditions that guarantee the existence and uniqueness of solutions to nonlinear differential equations.

Example 2.31. Consider the scalar nonlinear dynamical system

$$\dot{x}(t) = -\mathrm{sign}(x(t)), \qquad x(0) = 0, \qquad t \geq 0, \tag{2.138}$$

where

$$\mathrm{sign}(x) \triangleq \begin{cases} 1, & x \geq 0, \\ -1, & x < 0. \end{cases} \tag{2.139}$$

Now, note that since $\frac{\mathrm{d}}{\mathrm{d}t}x^2(t) = -2x(t)\mathrm{sign}(x(t)) = -2|x(t)| \leq 0$, it follows that $x^2(\cdot)$ is a decreasing function of time. Hence, if $x(0) = 0$, then $x(t) = 0$ for all $t \geq 0$, which implies $\dot{x}(t) = 0$, $t \geq 0$. In this case, $\mathrm{sign}(0) = 0$, which leads to a contradiction. Hence, there does not exist a continuously differentiable function $x(\cdot)$ that satisfies (2.138). Note that the function $f(x) = -\mathrm{sign}(x)$ is not continuous at $x = 0$. △

Example 2.32. Consider the scalar nonlinear dynamical system

$$\dot{x}(t) = 3x^{2/3}(t), \quad x(0) = 0, \quad t \geq 0. \tag{2.140}$$

This system has two solutions given by $x(t) = t^3$ and $x(t) = 0$, $t \geq 0$. Note that the function $f(x) = 3x^{2/3}$ is continuous at $x = 0$; however, it is not Lipschitz continuous at $x = 0$. △

Example 2.33. Consider the scalar nonlinear dynamical system

$$\dot{x}(t) = x^2(t), \quad x(0) = 1, \quad t \geq 0. \tag{2.141}$$

Using separation of variables, the solution to (2.141) is given by

$$x(t) = \frac{1}{1-t}. \tag{2.142}$$

The solution (2.142) is defined for $t \in [0, 1)$ and becomes unbounded as $t \to 1$. The solution has a *finite escape time* and, hence, exists only locally (see Figure 2.4). Note that even though (2.142) has another branch defined on the interval $(1, \infty)$, this branch is *not* considered part of the solution to (2.141) since the initial time $t_0 = 0 \notin (1, \infty)$. Finally, note that $f(x) = x^2$ is continuous and Lipschitz continuous at $x = 0$. However, $f(x)$ is not globally Lipschitz continuous on $\mathbb{R}$. △

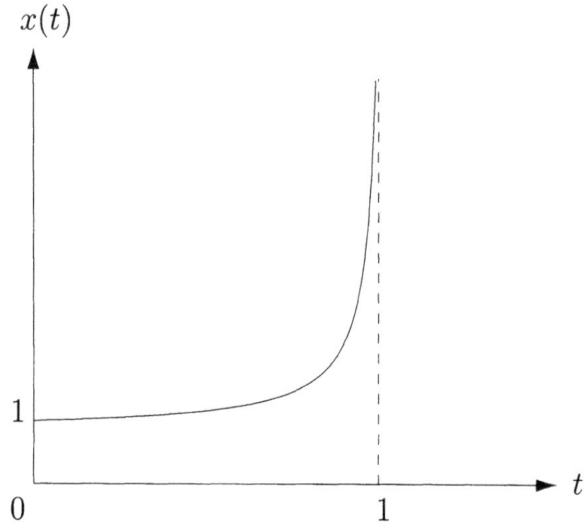

Figure 2.4 Solution exhibiting finite escape time.

Example 2.34. Consider the scalar nonlinear dynamical system

$$\dot{x}(t) = \sqrt{|x(t)|}, \quad x(0) = 0, \quad t \geq 0. \tag{2.143}$$

Clearly, $x(t) = 0$, $t \geq 0$, as well as $x(t) = \frac{1}{4}t^2$ are solutions to (2.143). In fact, there are infinitely many solutions to (2.143), parameterized by the arbitrary time T, and are given by (see Figure 2.5)

$$x(t) = \begin{cases} 0, & 0 \leq t \leq T, \\ \frac{(t-T)^2}{4}, & t > T. \end{cases} \tag{2.144}$$

Note that the function $f(x) = \sqrt{|x|}$ is continuous at $x = 0$; however, $f(x)$ is not Lipschitz continuous at $x = 0$. △

Examples 2.32–2.34 analyze nonlinear differential equations having multiple solutions as well as finite escape time. However, the nonlinear differential equation in Example 2.31 does not have a solution in the sense of the definition given in this section. The question of existence of solutions is fundamental in the study of dynamical systems. The following result provides a sufficient condition for the existence of solutions of (2.136) over an interval $[t_0, \tau]$ for sufficiently small $\tau \in (t_0, t_1)$.

Theorem 2.24 (Peano). Consider the nonlinear dynamical system

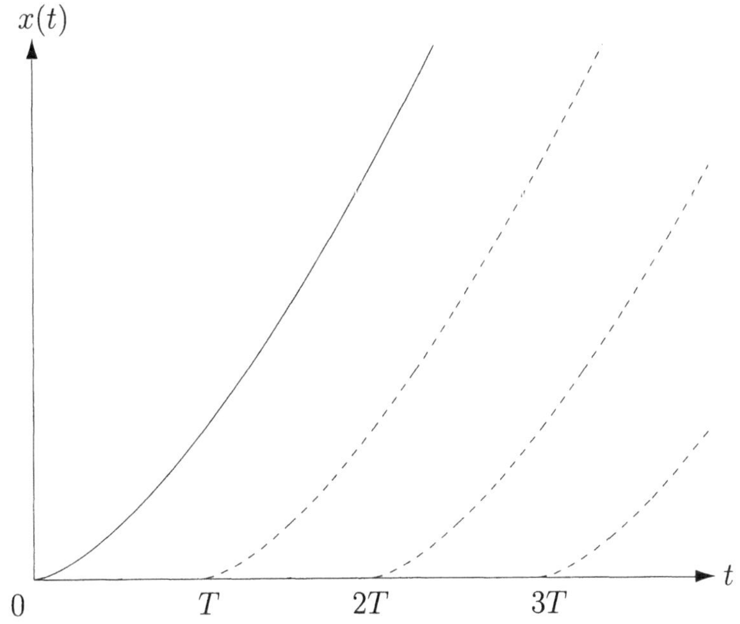

Figure 2.5 Multiple solutions.

(2.136). Assume $f : \mathcal{D} \to \mathbb{R}^n$ is continuous on $\mathcal{D}$. Then for every $x_0 \in \mathcal{D}$ and $t_0 > 0$ there exists $\tau > t_0$ such that (2.136) has a continuously differentiable solution $x : [t_0, \tau] \to \mathbb{R}^n$.

Proof. Let $\eta > 0$ be such that $\mathcal{B}_\eta(x_0) \subseteq \mathcal{D}$ and let $M \triangleq \sup\{\|f(x)\| : x \in \mathcal{B}_\eta(x_0)\}$. Now, let $\alpha, \beta > 0$ be such that $M\alpha \le \beta \le \varepsilon$ and let

$$\mathcal{S} \triangleq \{x(\cdot) \in \mathcal{C}[t_0, \tau] : \ |\!|\!|x - x_0|\!|\!| \le \eta, \ x(t_0) = x_0, \ t \in [t_0, \tau]\},$$

where $\tau \triangleq t_0 + \alpha$. Note that $\mathcal{S}$ is convex, closed, and bounded. Next, let $P : \mathcal{C}[t_0, \tau] \to \mathcal{C}[t_0, \tau]$ be given by

$$(Px)(t) \triangleq x_0 + \int_{t_0}^{t} f(x(s))\mathrm{d}s, \quad t \in [t_0, \tau]. \tag{2.145}$$

Now, it follows that

$$\begin{aligned} \|(Px)(t) - x_0\| &= \left\| \int_{t_0}^{t} f(x(s))\mathrm{d}s \right\| \\ &\le \int_{t_0}^{t} \|f(x(s))\|\mathrm{d}t \\ &\le M|t - t_0| \\ &\le M\alpha \end{aligned}$$

$$\leq \beta, \quad t \in [t_0, \tau],$$

which implies that $P(\mathcal{S})$ is bounded. Furthermore, since f is continuous on $\mathcal{D}$ it follows that for every $\varepsilon > 0$, there exists $\delta > 0$ such that

$$\begin{aligned}
\|(Px)(t) - (P\hat{x})(t)\| &= \left\| \int_{t_0}^{t} [f(x(s)) - f(\hat{x}(s))] \mathrm{d}s \right\| \\
&\leq \int_{t_0}^{t} \|f(x(s)) - f(\hat{x}(s))\| \mathrm{d}t \\
&\leq \varepsilon\alpha, \quad t \in [t_0, \tau], \quad \sup_{t \in [t_0, \tau]} \|x(t) - \hat{x}(t)\| < \delta.
\end{aligned}$$

The result now is a direct consequence of the Schauder fixed point theorem (see Problem 2.93) and the fact that $x(t) = (Px)(t)$ is a solution to (2.136) if and only if $x(t)$ is a fixed point of P. □

In order to guarantee the existence and uniqueness of solutions to (2.136) we strengthen the hypothesis in Theorem 2.24. Specifically, we assume that f is Lipschitz continuous on $\mathcal{D}$. The following theorem gives sufficient conditions for existence and uniqueness of solutions to (2.136). We state the result for the right-hand limit; the result for the left-hand limit is analogous.

Theorem 2.25. Consider the nonlinear dynamical system (2.136). Assume that $f : \mathcal{D} \to \mathbb{R}^n$ is Lipschitz continuous on $\mathcal{D}$. Then, for all $x_0 \in \mathcal{D}$, there exists $\tau \in (t_0, t_1)$ such that (2.136) has a unique solution $x : [t_0, \tau] \to \mathbb{R}^n$ over the interval $[t_0, \tau]$.

Proof. It follows from Theorem 2.24 that there exists $\tau > 0$ and $x : [t_0, \tau] \to \mathbb{R}^n$ such that $x(\cdot)$ is a continuous solution to (2.136). Now, let

$$(Px)(t) \triangleq x_0 + \int_{t_0}^{t} f(x(s)) \mathrm{d}s, \quad t \in [t_0, \tau], \tag{2.146}$$

so that $x(t) = (Px)(t)$. Note that $P : C[t_0, \tau] \to C[t_0, \tau]$. Furthermore, define $\mathcal{S} \triangleq \{x(\cdot) \in C[t_0, \tau] : \|\!|x - x_0|\!\| \leq r\}$, where $r > 0$ and $x_0 : C[t_0, \tau] \to C[t_0, \tau]$ is such that $x_0(t) = x_0$ for all $t \in [t_0, \tau]$. Note that $x(\cdot)$ is a solution to (2.136) over $[t_0, \tau]$ if and only if $(Px)(\cdot) = x(\cdot)$, that is, $x(\cdot)$ is a fixed point of P. Hence, using the fact that $f : \mathcal{D} \to \mathbb{R}^n$ is Lipschitz continuous on $\mathcal{D}$ and $\|x(t) - x_0\| \leq r$, $t \in [t_0, \tau]$, it follows that, for each $x(\cdot) \in \mathcal{S}$ and each $t \in [t_0, \tau]$,

$$\begin{aligned}
\|(Px)(t) - x_0\| &= \left\| \int_{t_0}^{t} f(x(s)) \mathrm{d}s \right\| \\
&= \left\| \int_{t_0}^{t} [f(x(s)) - f(x_0) + f(x_0)] \mathrm{d}s \right\|
\end{aligned}$$

$$
\begin{aligned}
&\leq \int_{t_0}^{t} [\|f(x(s)) - f(x_0)\| + \|f(x_0)\|]\mathrm{d}s \\
&\leq \int_{t_0}^{t} [L\|x(s) - x_0\| + \alpha]\mathrm{d}s \\
&\leq \int_{t_0}^{t} (Lr + \alpha)\mathrm{d}s \\
&= (t - t_0)(Lr + \alpha) \\
&\leq (\tau - t_0)(Lr + \alpha), \qquad (2.147)
\end{aligned}
$$

where $L > 0$ is the Lipschitz constant of f with respect to x on $\mathcal{B}_r(x_0) \subset \mathcal{D}$ and $\alpha = \alpha(x_0) = \|f(x_0)\|$, and hence,

$$
|||Px - x_0||| = \max_{t\in[t_0,\tau]} \|(Px)(t) - x_0\| \leq (\tau - t_0)(Lr + \alpha). \qquad (2.148)
$$

Now, choosing $\tau - t_0 \leq \frac{r}{Lr+\alpha}$ ensures that $|||Px - x_0||| \leq r$, and hence, $P : \mathcal{S} \to \mathcal{S}$.

To show that $P : \mathcal{S} \to \mathcal{S}$ is a contraction, let $x(\cdot)$ and $y(\cdot) \in \mathcal{S}$ so that $x(t), y(t) \in \mathcal{B}_r(x_0)$ for all $t \in [t_0, \tau]$. Hence,

$$
(Px)(t) - (Py)(t) = \int_{t_0}^{t} [f(x(s)) - f(y(s))]\mathrm{d}s, \qquad (2.149)
$$

which implies

$$
\begin{aligned}
\|(Px)(t) - (Py)(t)\| &\leq \int_{t_0}^{t} \|f(x(s)) - f(y(s))\|\mathrm{d}s \\
&\leq \int_{t_0}^{t} L\|x(s) - y(s)\|\mathrm{d}s \\
&\leq L|||x - y||| \int_{t_0}^{t} \mathrm{d}s \\
&= L(t - t_0)|||x - y||| \\
&\leq \rho|||x - y|||,
\end{aligned}
$$

where $|||x - y||| = \max_{t\in[t_0,\tau]} \|x(t) - y(t)\|$ and $L(t - t_0) \leq L(\tau - t_0) \leq \rho$. Thus,

$$
|||Px - Py||| = \max_{t\in[t_0,\tau]} \|(Px)(t) - (Py)(t)\| \leq \rho|||x - y|||. \qquad (2.150)
$$

Now, choosing $(\tau - t_0) \leq \rho/L$ and $\rho < 1$ ensures that $P : \mathcal{S} \to \mathcal{S}$ is a contraction. Hence, it follows from Theorem 2.22 that for

$$
\tau - t_0 \leq \min\left\{t_1 - t_0, \frac{r}{Lr + \alpha}, \frac{\rho}{L}\right\},
$$

(2.136) has a unique solution in $\mathcal{S}$.

Finally, we show that P has a unique fixed point in $\mathcal{C}[t,\tau]$. Suppose $x(\cdot) \in \mathcal{C}[t_0,\tau]$ satisfies (2.137). Then, $x(0) = x_0 \in \mathcal{B}_r(x_0)$. Now, since $x(\cdot)$ is continuous, it follows for sufficiently small t, $x(t) \in \mathcal{B}_r(x_0)$. Suppose, *ad absurdum*, that there exists $\hat{\tau} \in (t_0,\tau)$ such that $x(\hat{\tau}) \notin \mathcal{B}_r(x_0)$, that is, $\|x(\hat{\tau}) - x_0\| > r$. Since $\|x(t) - x_0\|$ is continuous in t and $\|x(0) - x_0\| = 0$, it follows that there exists $\tau^* < \hat{\tau} < \tau$ such that $\|x(t) - x_0\| < r$ for all $t \in [t_0, \tau^*)$ and $\|x(\tau^*) - x_0\| = r$. Now, for all $t \leq \tau^*$,

$$\begin{aligned}
\|x(t) - x_0\| &= \left\| \int_{t_0}^{t} f(x(s))\mathrm{d}s \right\| \\
&= \left\| \int_{t_0}^{t} [f(x(s)) - f(x_0) + f(x_0)]\mathrm{d}s \right\| \\
&\leq \int_{t_0}^{t} L\|x(s) - x_0\|\mathrm{d}s + \alpha(t - t_0) \\
&\leq (t - t_0)(Lr + \alpha) \\
&\leq (\tau^* - t_0)(Lr + \alpha). \qquad (2.151)
\end{aligned}$$

Since, $t \leq \tau^*$ it follows that $r = \|x(\tau^*) - x_0\| \leq (\tau^* - t_0)(Lr + \alpha)$. Hence, $(\tau^* - t_0) \geq \frac{r}{Lr+\alpha} \geq \tau - t_0$, which is a contradiction. Hence, if $x(\cdot) \in \mathcal{C}[t_0,\tau]$ satisfies (2.136) such that $\tau - t_0 \leq \min\{t_1 - t_0, \frac{r}{Lr+\alpha}, \frac{\rho}{L}\}$, then $x(\cdot) \in \mathcal{S}$. Thus, if $x^*(\cdot)$ is a fixed point of $\mathcal{C}[t_0,\tau]$, then $x^* \in \mathcal{S}$, which shows that (2.136) has a unique solution over the interval $[t_0,\tau]$. □

The following corollary to Theorem 2.25 is immediate.

Corollary 2.3. Consider the nonlinear dynamical system (2.136). Assume that $f : \mathcal{D} \to \mathbb{R}^n$ is continuously differentiable on $\mathcal{D}$. Then, for all $x_0 \in \mathcal{D}$, there exists $\tau \in (t_0, t_1)$ such that (2.136) has a unique solution $x : [t_0,\tau] \to \mathbb{R}^n$ over the interval $[t_0,\tau]$.

Proof. The proof follows by noting that continuous differentiability on $\mathcal{D}$ implies Lipschitz continuity on $\mathcal{D}$. □

Next, we state and prove a very important theorem regarding the sensitivity of solutions of (2.136) to the system initial data and system parameters. For this result we require a key lemma due to Gronwall that converts an implicit bound to an explicit bound.

Lemma 2.2 (Gronwall Lemma). Assume there exists a continuous function $x : \mathbb{R} \to \mathbb{R}$ such that

$$x(t) \leq \alpha + \int_{t_0}^{t} \beta x(s)\mathrm{d}s, \quad t \geq t_0, \qquad (2.152)$$

where $\alpha \in \mathbb{R}$ and $\beta \geq 0$. Then

$$x(t) \leq \alpha e^{\beta(t-t_0)}, \quad t \geq t_0. \tag{2.153}$$

Proof. Let $\alpha \in \mathbb{R}$ and let

$$y(t) = \alpha + \int_{t_0}^{t} \beta x(s)\mathrm{d}s, \quad t \geq t_0. \tag{2.154}$$

Now, it follows from (2.152) that $x(t) \leq y(t)$, $t \geq t_0$, and hence, using (2.154),

$$\dot{y}(t) = \beta x(t) \leq \beta y(t), \quad y(t_0) = \alpha, \quad t \geq t_0. \tag{2.155}$$

Next, define $z(t) \triangleq \dot{y}(t) - \beta y(t)$ and note that (2.155) implies $z(t) \leq 0$, $t \geq t_0$. Hence,

$$y(t) = y(t_0)e^{\beta(t-t_0)} + \int_{t_0}^{t} e^{\beta(t-\sigma)} z(\sigma)\mathrm{d}\sigma,$$

which implies

$$y(t) \leq y(t_0)e^{\beta(t-t_0)} = \alpha e^{\beta(t-t_0)}, \quad t \geq t_0.$$

The result is now immediate by noting that $x(t) \leq y(t)$, $t \geq t_0$. □

Theorem 2.26. Consider the nonlinear dynamical system (2.136). Assume that $f : \mathcal{D} \to \mathbb{R}^n$ is uniformly Lipschitz continuous on $\mathcal{D}$. Furthermore, let $x(t)$ and $y(t)$ be solutions to (2.136) with initial conditions $x(t_0) = x_0$ and $y(t_0) = y_0$ over the closed interval $[t_0, t_1]$. Then, for each $\varepsilon > 0$ and $t \in [t_0, t_1]$, there exists $\delta = \delta(\varepsilon, t - t_0) > 0$ such that if $\|x_0 - y_0\| < \delta$, then $\|x(t) - y(t)\| \leq \varepsilon$.

Proof. Note that $x(t)$ and $y(t)$, $t \in [t_0, t_1]$, satisfy

$$x(t) = x_0 + \int_{t_0}^{t} f(x(s))\mathrm{d}s, \tag{2.156}$$

$$y(t) = y_0 + \int_{t_0}^{t} f(y(s))\mathrm{d}s. \tag{2.157}$$

Subtracting (2.157) from (2.156) yields

$$x(t) - y(t) = x_0 - y_0 + \int_{t_0}^{t} [f(x(s)) - f(y(s))]\mathrm{d}s. \tag{2.158}$$

Now, (2.158) implies

$$\begin{aligned}\|x(t) - y(t)\| &\leq \|x_0 - y_0\| + \int_{t_0}^{t} \|f(x(s)) - f(y(s))\|\mathrm{d}s \\ &\leq \|x_0 - y_0\| + L\int_{t_0}^{t} \|x(s) - y(s)\|\mathrm{d}s, \end{aligned} \tag{2.159}$$

where $L > 0$ is the Lipschitz constant of f. Using Lemma 2.2 it follows that $\|x(t) - y(t)\| \leq \|x_0 - y_0\| e^{L(t-t_0)}$, $t \in [t_0, t_1]$. Hence, for every $\varepsilon > 0$, choosing $\delta = \delta(\varepsilon, t - t_0) = \frac{\varepsilon}{e^{L(t-t_0)}}$ yields the result. □

Theorem 2.26 shows that the solution $x(t)$, $t \in [t_0, t_1]$, of (2.136) depends continuously on the initial condition $x(0)$ over a *finite* time interval. This is not true in general over the semi-infinite interval $[t_0, \infty)$. If this were the case over the semi-infinite interval and $\delta(\varepsilon, t)$ could be chosen independent of t, then continuous dependence of solutions uniformly in t for all $t \geq 0$ would imply *Lyapunov stability of the solutions*; a concept introduced in Chapter 3. Furthermore, since Theorem 2.26 implies

$$\|x(t) - y(t)\| \leq \|x_0 - y_0\| e^{L(t-t_0)}, \qquad t \in [t_0, t_1], \tag{2.160}$$

it follows that for each $t \in [t_0, t_1]$,

$$\lim_{y_0 \to x_0} s(t, y_0) = s(t, x_0). \tag{2.161}$$

In addition, (2.160) implies that this limit is uniform for all $t \in [t_0, t_1]$. It is important to note that Theorem 2.26 also holds for the case where $f : \mathcal{D} \to \mathbb{R}^n$ is Lipschitz continuous on $\mathcal{D}$. In this case, however, continuous dependence on the initial conditions of $s(t, y_0)$ holds for $s(\cdot, \cdot) \in \mathcal{Q}$, where $\mathcal{Q} = [t_0, t_1] \times \mathcal{N}_\delta(x_0)$ and $\mathcal{N}_\delta(x_0) \subset \mathcal{D}$. Finally, it is important to note that Gronwall's lemma can be used to give an alternative proof of Theorem 2.25. In particular, if, *ad absurdum*, we assume that $x(t)$ and $y(t)$ are two solutions to (2.136) with initial conditions $x(t_0) = x_0$ and $y(t_0) = x_0$ over the closed interval $[t_0, t_1]$, then it follows from (2.160) that $\|x(t) - y(t)\| \leq 0$, $t \in [t_0, t_1]$. This of course implies that $x(t) = y(t)$, $t \in [t_0, t_1]$, establishing uniqueness of solutions.

The next result presents a more general version of Theorem 2.26 involving continuous dependence on initial conditions and system parameters. For this result, consider the nonlinear dynamical system

$$\dot{x}(t) = f(x(t), \lambda), \quad x(t_0) = x_0, \quad t \in \mathcal{I}_{x_0,\lambda}, \tag{2.162}$$

where $x(t) \in \mathcal{D}$, $t \in \mathcal{I}_{x_0,\lambda}$, $\mathcal{D}$ is an open subset of $\mathbb{R}^n$, $\lambda \in \mathbb{R}^m$ is a system parameter vector, $f : \mathcal{D} \times \mathbb{R}^m \to \mathbb{R}^n$ is such that $f(\cdot, \lambda)$ is Lipschitz continuous on $\mathcal{D}$, $f(x, \cdot)$ is uniformly Lipschitz continuous on $\mathbb{R}^m$, and $\mathcal{I}_{x_0,\lambda} = (\tau_{\min}, \tau_{\max}) \subset \mathbb{R}$ is the maximal interval of existence for the solution $x(\cdot)$ of (2.162).

Theorem 2.27. Consider the nonlinear dynamical system (2.162). Assume that $f : \mathcal{D} \times \mathbb{R}^m \to \mathbb{R}^n$ is such that for every $\lambda \in \mathbb{R}^m$, $f(\cdot, \lambda)$ is Lipschitz continuous on $\mathcal{D}$ and for every $x \in \mathcal{D}$, $f(x, \cdot)$ is globally Lipschitz continuous on $\mathbb{R}^m$. Furthermore, let $x(t)$ and $y(t)$ be solutions to (2.162) with system parameters λ and μ, respectively, and initial conditions

$x(t_0) = x_0$ and $y(t_0) = y_0$ over the closed interval $[t_0, t_1]$. Then, for each $\varepsilon > 0$ and $t \in [t_0, t_1]$, there exists $\delta = \delta(\varepsilon, t-t_0) > 0$ such that if $\|x_0 - y_0\| < \delta$ and $\|\lambda - \mu\| < \delta$, then $\|x(t) - y(t)\| \leq \varepsilon$.

Proof. The proof is a direct consequence of Theorem 2.26 with x_0, x, y_0, and y replaced by $[x_0^{\mathrm{T}}\ \lambda^{\mathrm{T}}]^{\mathrm{T}}$, $[x^{\mathrm{T}}\ \lambda^{\mathrm{T}}]^{\mathrm{T}}$, $[y_0^{\mathrm{T}}\ \mu^{\mathrm{T}}]^{\mathrm{T}}$, and $[y^{\mathrm{T}}\ \mu^{\mathrm{T}}]^{\mathrm{T}}$, respectively. $\square$

As in the case of Theorem 2.26, Theorem 2.27 implies that

$$\|s(t, x_0, \lambda) - s(t, y_0, \mu)\| \leq (\|x_0 - y_0\| + \|\lambda - \mu\|)e^{L(t-t_0)}, \quad t \in [t_0, t_1], \tag{2.163}$$

and hence,

$$\lim_{(y_0,\mu)\to(x_0,\lambda)} s(t, y_0, \mu) = s(t, x_0, \lambda), \tag{2.164}$$

uniformly for all $t \in [t_0, t_1]$.

Next, we present a different kind of continuity of parameter result. In particular, we show that given two dynamical systems

$$\dot{x}(t) = f(x(t)), \quad x(t_0) = x_0, \quad t \in [t_0, t_1], \tag{2.165}$$

$$\dot{y}(t) = g(y(t)), \quad y(t_0) = y_0, \quad t \in [t_0, t_1], \tag{2.166}$$

where $f : \mathcal{D} \to \mathbb{R}^n$ and $g : \mathcal{D} \to \mathbb{R}^n$ are both Lipschitz continuous on $\mathcal{D}$, and $\|f(x) - g(x)\| \leq \varepsilon$, $x \in \mathcal{D}$, the solutions to (2.165) and (2.166) with $\|x_0 - y_0\| \leq \gamma$ remain nearby over a finite time interval.

Theorem 2.28. Consider the nonlinear dynamical system (2.165) and (2.166). Assume that $f : \mathcal{D} \to \mathbb{R}^n$ is uniformly Lipschitz continuous on $\mathcal{D}$ with Lipschitz constant L and $g : \mathcal{D} \to \mathbb{R}^n$ is Lipschitz continuous on $\mathcal{D}$. Furthermore, suppose that

$$\|f(x) - g(x)\| \leq \varepsilon, \quad x \in \mathcal{D}. \tag{2.167}$$

If $x(t)$ and $y(t)$ are solutions to (2.165) and (2.166) on some time interval $\mathcal{I} \subset \mathbb{R}$ with $\|x_0 - y_0\| \leq \gamma$, then

$$\|x(t) - y(t)\| \leq \gamma e^{L|t-t_0|} + \frac{\varepsilon}{L}(e^{L|t-t_0|} - 1). \tag{2.168}$$

Proof. For $t \in \mathcal{I}$ it follows that

$$x(t) - y(t) = x_0 - y_0 + \int_{t_0}^{t} [f(x(s)) - g(y(s))]\mathrm{d}s. \tag{2.169}$$

Now, (2.169) implies

$$\|x(t) - y(t)\| \leq \|x_0 - y_0\| + \int_{t_0}^{t} \|f(x(s)) - g(y(s))\|\mathrm{d}s$$

$$\leq \gamma + \int_{t_0}^{t} \|f(x(s)) - f(y(s)) + f(y(s)) - g(y(s))\|\mathrm{d}s$$

$$\leq \gamma + \int_{t_0}^{t} \|f(x(s)) - f(y(s))\|\mathrm{d}s + \int_{t_0}^{t} \|f(y(s)) - g(y(s))\|\mathrm{d}s$$

$$\leq \gamma + \int_{t_0}^{t} L\|x(s) - y(s)\|\mathrm{d}s + \int_{t_0}^{t} \varepsilon \mathrm{d}s, \quad t \in \mathcal{I}. \tag{2.170}$$

Next, defining $q(t) \triangleq \|x(t) - y(t)\|$, (2.170) implies

$$q(t) \leq \gamma + L \int_{t_0}^{t} \left[q(s) + \frac{\varepsilon}{L} \right] \mathrm{d}s, \quad t \in \mathcal{I}, \tag{2.171}$$

or, equivalently,

$$q(t) + \frac{\varepsilon}{L} \leq \gamma + \frac{\varepsilon}{L} + L \int_{t_0}^{t} \left[q(s) + \frac{\varepsilon}{L} \right] \mathrm{d}s, \quad t \in \mathcal{I}. \tag{2.172}$$

Now, using Gronwall's lemma it follows that

$$q(t) + \frac{\varepsilon}{L} \leq \left(\frac{\varepsilon}{L} + \gamma \right) e^{L|t-t_0|}, \quad t \in \mathcal{I}, \tag{2.173}$$

which implies (2.168). □

Theorem 2.28 shows that, given two solutions to a dynamical system with initial conditions that are close at the same value of time, these solutions will remain close over the entire time interval $\mathcal{I}$ and not just at the initial time. Furthermore, it is clear from (2.168) that bounds on the initial condition errors and the solution errors grow exponentially in time, with the Lipschitz constant L controlling the growth rate. In the study of *robustness*, that is, sensitivity to system parameter variations, (2.168) tells us that a solution to an approximate dynamical system will serve as an approximate solution to the actual system over a finite time interval. Furthermore, (2.168) clearly shows why the qualitative study of differential equations is necessary and why numerical methods (which are approximate by definition) are inherently fragile over long periods of time and can be untrustworthy.

Finally, we close this section with a proposition that shows that the solution of a nonlinear system satisfying a uniform Lipschitz continuity condition is exponentially bounded from above and below.

Proposition 2.30. Consider the nonlinear dynamical system (2.136). Assume $f : \mathcal{D} \to \mathbb{R}^n$ is uniformly Lipschitz continuous on $\mathcal{D}$ with Lipschitz constant L and $f(0) = 0$. Then the solution $x(t)$ of (2.136) over the closed interval $[t_0, t_1]$ satisfies

$$\|x_0\|_2 e^{-L|t-t_0|} \leq \|x(t)\|_2 \leq \|x_0\|_2 e^{L|t-t_0|}, \quad t \in [t_0, t_1]. \tag{2.174}$$

Proof. Since $\|x\|_2^2 = x^{\mathrm{T}}x$ it follows that

$$\frac{\mathrm{d}}{\mathrm{d}t}x^{\mathrm{T}}(t)x(t) = 2x^{\mathrm{T}}(t)\dot{x}(t) = 2x^{\mathrm{T}}(t)f(x(t)), \quad t \in [t_0, t_1]. \tag{2.175}$$

Now, since $f(0) = 0$ and $f : \mathcal{D} \to \mathbb{R}^n$ is uniformly Lipschitz continuous on $\mathcal{D}$ it follows that $\|f(x)\|_2 \le L\|x\|_2$. Hence, using the Cauchy-Schwarz inequality it follows that

$$\left|\frac{\mathrm{d}}{\mathrm{d}t}x^{\mathrm{T}}(t)x(t)\right| \le 2\|x(t)\|_2\|f(x(t))\|_2 \le 2L\|x(t)\|_2^2, \quad t \in [t_0, t_1]. \tag{2.176}$$

Now, defining $q(t) \triangleq x^{\mathrm{T}}(t)x(t)$ and $q_0 \triangleq q(0) = x_0^{\mathrm{T}}x_0$, it follows from (2.176) that

$$-2Lq(t) \le \dot{q}(t) \le 2Lq(t), \quad t \in [t_0, t_1], \tag{2.177}$$

which implies

$$-\int_{t_0}^{t} 2L\mathrm{d}s \le \int_{q_0}^{q} \frac{\mathrm{d}q}{q} \le \int_{t_0}^{t} 2L\mathrm{d}s. \tag{2.178}$$

Hence,

$$q_0 e^{-2L|t-t_0|} \le q(t) \le q_0 e^{2L|t-t_0|}, \quad t \in [t_0, t_1]. \tag{2.179}$$

The result is now immediate by noting that $q(t) = \|x(t)\|_2^2$ and $q_0 = \|x_0\|_2^2$.
□

2.8 Extendability of Solutions

In Section 2.7 we showed that under the conditions of Lipschitz continuity on $\mathcal{D}$, the nonlinear dynamical system (2.136) is guaranteed to have a unique solution over the closed interval $[t_0, \tau]$. In this section, we consider the question of whether it is possible to extend the solution $x : [t_0, \tau] \to \mathbb{R}^n$ of (2.136) to a larger interval. This question can be intuitively answered by noting that if (2.136) has a unique solution over $[t_0, \tau]$, then by re-applying Theorem 2.25 with τ serving as the initial time and $x(\tau)$ as the initial condition to (2.136), it follows that there exists a unique solution to (2.136) over the interval $[\tau, \tau_1]$. By iteratively repeating this process and concatenating all solutions, it follows that there does not exist a largest closed bounded interval over which (2.136) has a unique solution. Hence, the solution will exist over the semiopen interval $[t_0, \tau_{\max})$. This implies that the solution $x(t)$ to (2.136) must approach the boundary of $\mathcal{D}$, tend to infinity, or both. For negative time, an identical argument for the left-hand limit can be used to show that the maximal interval of existence for a solution to (2.136) is $(\tau_{\min}, \tau_{\max}) \subseteq \mathbb{R}$. Before summarizing the above discussion we consider a simple example.

Example 2.35. Consider the scalar nonlinear dynamical system

$$\dot{x}(t) = 1 + x^2(t), \quad x(0) = 0, \quad t \in \mathbb{R}. \tag{2.180}$$

This system has a unique solution given by $x(t) = \tan t$. Since $x(t) \to \pm\infty$ as $t \to \pm\pi/2$, the maximal interval of existence of (2.180) is $(-\pi/2, \pi/2) \subset \mathbb{R}$. △

The following theorem shows that if a solution to (2.136) cannot be extended, it must leave any given compact set of the state space.

Theorem 2.29. Consider the nonlinear dynamical system (2.136). Assume $f : \mathcal{D} \to \mathbb{R}^n$ is Lipschitz continuous on $\mathcal{D}$. Furthermore, let $x(t)$ denote the solution of (2.136) on the maximal interval of existence $\mathcal{I}_{x_0} = (\tau_{\min}, \tau_{\max}) \subset \mathbb{R}$ with $\tau_{\max} < \infty$. Then given any compact set $\mathcal{D}_c \subset \mathcal{D}$, there exist $t_1 \in (\tau_{\min}, t_0)$ and $t_2 \in (t_0, \tau_{\max})$ such that $x(t_1) \notin \mathcal{D}_c$ and $x(t_2) \notin \mathcal{D}_c$.

Proof. Suppose, *ad absurdum*, that there does not exist $t \in \mathcal{I}_{x_0}$ such that $x(t) \notin \mathcal{D}_c$. Then, $x(t) \in \mathcal{D}_c$, $t \in (\tau_{\min}, \tau_{\max})$. Since $f : \mathcal{D} \to \mathbb{R}^n$ is continuous on $\mathcal{D}_c$, it follows from Theorem 2.13 that there exists $\alpha > 0$ such that $\|f(x)\| \leq \alpha$ for all $x \in \mathcal{D}_c$. Next, let $\sigma \in (t_0, \tau_{\max})$ and note that for $\tau_{\min} < t_0 < t < \tau_{\max}$,

$$\begin{aligned} \|x(t) - x(t_0)\| &= \left\| \int_{t_0}^{t} f(x(s))\mathrm{d}s \right\| \\ &\leq \int_{t_0}^{t} \|f(x(s))\|\mathrm{d}s \\ &\leq \alpha(t - t_0), \end{aligned} \tag{2.181}$$

which shows that $x(\cdot)$ is uniformly continuous on $\mathcal{I}_{x_0}$.

Now, since $x(\cdot)$ is uniformly continuous on $\mathcal{I}_{x_0}$, it follows that

$$\begin{aligned} x(\tau_{\max}) &= x(\sigma) + \lim_{t \to \tau_{\max}} \int_{\sigma}^{t} f(x(s))\mathrm{d}s \\ &= x(\sigma) + \int_{\sigma}^{\tau_{\max}} f(x(s))\mathrm{d}s. \end{aligned} \tag{2.182}$$

Thus,

$$x(t) = x(\sigma) + \int_{\sigma}^{t} f(x(s))\mathrm{d}s, \quad t \in [\sigma, \tau_{\max}], \tag{2.183}$$

which implies that $\dot{x}(\tau_{\max}) = f(x(\tau_{\max}))$, and hence, $x(\cdot)$ is (left) differentiable at $\tau_{\max}$. Hence, $x : [\sigma, \tau_{\max}] \to \mathbb{R}^n$ is a solution to (2.136). Now, using Theorem 2.25 it follows that the solution $x(t)$ to (2.136) can be extended to the interval $(\tau_{\min}, \tau^*)$, $\tau^* > \tau_{\max}$, which contradicts the claim that $\mathcal{I}_{x_0} = (\tau_{\min}, \tau_{\max})$ is the maximal interval of existence of (2.136). Hence, there exists $t_2 \in (t_0, \tau_{\max})$ such that $x(t_2) \notin \mathcal{D}_c$. The existence of $t_1 \in (\tau_{\min}, t_0)$ such that $x(t_1) \notin \mathcal{D}_c$ follows in a similar manner. □

As mentioned earlier, in most physical dynamical systems and especially feedback control systems we are more interested in the future behavior of the system as opposed to the past. Hence, we give a version of Theorem 2.29 wherein $t \in [0, \tau_{\max})$. This is sometimes referred to in the literature as the *right maximal interval of existence*.

Theorem 2.30. Consider the nonlinear dynamical system (2.136). Assume $f : \mathcal{D} \to \mathbb{R}^n$ is Lipschitz continuous on $\mathcal{D}$. Furthermore, let $x(t)$ denote the solution to (2.136) on the maximal interval of existence $\mathcal{I}_{x_0} = [0, \tau_{\max})$ with $\tau_{\max} < \infty$. Then given any compact set $\mathcal{D}_c \subset \mathcal{D}$, there exists $t \in \mathcal{I}_{x_0}$ such that $x(t) \notin \mathcal{D}_c$.

Proof. The proof is virtually identical to the proof of Theorem 2.29 and, hence, is omitted. □

The next result shows that if $\tau_{\max} < \infty$ and if $\lim_{t\to\tau_{\max}} x(t)$ exists, then $x(t) \to \partial\mathcal{D}$ as $t \to \tau_{\max}$.

Corollary 2.4. Consider the nonlinear dynamical system (2.136). Assume $f : \mathcal{D} \to \mathbb{R}^n$ is Lipschitz continuous on $\mathcal{D}$. Furthermore, let $x(t)$ denote the solution to (2.136) on the maximal interval of existence $[0, \tau_{\max})$ with $\tau_{\max} < \infty$. If $\lim_{t\to\tau_{\max}} x(t)$ exists, then $\lim_{t\to\tau_{\max}} x(t) \in \partial\mathcal{D}$.

Proof. Assume $\lim_{t\to\tau_{\max}} x(t)$ exists and is given by x^*. Then,

$$s(t, x_0) \triangleq \begin{cases} x(t), & t \in [0, \tau_{\max}), \\ x^*, & t = \tau_{\max}, \end{cases} \tag{2.184}$$

is continuous on $[0, \tau_{\max}]$. Next, let

$$\mathcal{D}_c \triangleq \{y \in \mathbb{R}^n : \ y = s(t, x_0) \text{ for some } t \in [0, \tau_{\max}]\}. \tag{2.185}$$

Since $\mathcal{D}_c$ is the image of the compact set $[0, \tau_{\max}]$ under the continuous map $s(\tau, x_0)$, $\mathcal{D}_c$ is a compact subset of $\mathcal{D}$. Now, suppose, *ad absurdum*, that $x^* \in \mathcal{D}$. Then $\mathcal{D}_c \subset \mathcal{D}$ and by Theorem 2.30 there must exist $t \in (0, \tau_{\max})$ such that $x(t) \notin \mathcal{D}_c$, which is a contradiction. Hence, $x^* \notin \mathcal{D}$. However, since $x(t) \in \mathcal{D}$ for all $t \in [0, \tau_{\max})$, it follows that $x^* = \lim_{t\to\tau_{\max}} x(t) \in \overline{\mathcal{D}}$, and hence, $x^* \in \overline{\mathcal{D}} \setminus \mathcal{D} = \partial\mathcal{D}$. □

Example 2.36. Consider the scalar nonlinear dynamical system

$$\dot{x}(t) = -[2x(t)]^{-1}, \quad x(0) = 1, \quad t \geq 0. \tag{2.186}$$

This system has a solution given by $x(t) = \sqrt{1-t}$ on the maximal interval of existence $\mathcal{I}_{x_0} = [0, 1)$. Note that on $\mathcal{D} = (0, \infty)$, the function $f(x) = -1/2x$ is continuously differentiable. Now, noting that $\partial\mathcal{D} = \{0\}$, it follows that $\lim_{t\to 1} x(t) = 0 \in \partial\mathcal{D}$. △

An immediate and very important corollary to Theorem 2.30 is the contrapositive statement of the theorem, which states that if a solution to (2.136) lies entirely in a compact set, then $\tau_{\max} = \infty$, and hence, there exists a unique solution to (2.136) for all $t \geq 0$.

Corollary 2.5. Consider the nonlinear dynamical system (2.136). Assume $f : \mathcal{D} \to \mathbb{R}^n$ is Lipschitz continuous on $\mathcal{D}$. Furthermore, let $\mathcal{D}_c \subset \mathcal{D}$ be compact and suppose for $x_0 \in \mathcal{D}_c$, the solution $x : [0, \tau] \to \mathcal{D}$ lies entirely in $\mathcal{D}_c$. Then there exists a unique solution $x : [0, \infty) \to \mathcal{D}$ to (2.136) for all $t \geq 0$.

Proof. Let $[0, \tau)$ be the maximal interval of existence of (2.136) with the solution $x(t)$, $t \in [0, \tau)$, lying entirely in $\mathcal{D}_c$. Then, by assumption $\{y \in \mathcal{D} : \ y = x(t) \text{ for some } t \in [0, \tau)\} \subset \mathcal{D}_c$ or, equivalently, $x([0, \tau)) \subset \mathcal{D}_c$. Hence, it follows from Theorem 2.30 that $\tau = \infty$. □

2.9 Global Existence and Uniqueness of Solutions

In this section, we give sufficient conditions for global existence and uniqueness of solutions over all time. First, however, we present a theorem that strengthens the results of Theorem 2.26 on uniform convergence with respect to initial conditions.

Theorem 2.31. Consider the nonlinear dynamical system (2.136). Assume that $f : \mathcal{D} \to \mathbb{R}^n$ is Lipschitz continuous on $\mathcal{D}$. Furthermore, let $x(t)$ denote a solution to (2.136) with initial condition $x(t_0) = x_0$ defined on the closed interval $[t_0, t_1]$. Then there exists a neighborhood $\mathcal{N} \subset \mathcal{D}$ of x_0 and a constant $L > 0$ such that for all $y(t_0) = y_0 \in \mathcal{N}$, there exists a unique solution $y(t)$ of (2.136) defined over the closed interval $[t_0, t_1]$ satisfying

$$\|x(t) - y(t)\| \leq \|x_0 - y_0\| e^{L|t - t_0|}, \tag{2.187}$$

for all $t \in [t_0, t_1]$. Moreover,

$$\lim_{y_0 \to x_0} s(t, y_0) = s(t, x_0), \tag{2.188}$$

uniformly for all $t \in [t_0, t_1]$, where $s(t, x_0)$ denotes the solution to (2.136) with initial condition x_0.

Proof. Since $[t_0, t_1]$ is compact and $s(\cdot, x_0)$ is continuous on $[t_0, t_1]$, it follows that $\{x \in \mathbb{R}^n : \ x = s(t, x_0), \ t \in [t_0, t_1]\}$ is a compact subset of $\mathcal{D}$. Now, since $\mathcal{D}$ is an open set, there exists $\varepsilon > 0$ such that the compact set

$$\mathcal{D}_c = \{x \in \mathbb{R}^n : \ \|x - s(t, x_0)\| \leq \varepsilon, \ t \in [t_0, t_1]\} \subset \mathcal{D}.$$

Next, since by assumption $f : \mathcal{D} \to \mathbb{R}^n$ is Lipschitz continuous on $\mathcal{D}$, it follows from Proposition 2.23 that there exists $L > 0$ such that $\|f(y) -$

$f(x)\| \leq L\|y - x\|$, $x, y \in \mathcal{D}_\mathrm{c}$. Let $\delta > 0$ be such that $\delta \leq \varepsilon$ and $\delta \leq \varepsilon e^{-L(t_1 - t_0)}$. Now, let $y_0 \in \mathcal{N}_\delta(x_0)$ and let $s(t, y_0)$ be the solution to (2.136) over the maximal interval of existence $(\tau_{\min}, \tau_{\max})$. Suppose, *ad absurdum*, that $\tau_{\max} \leq t_1$. Then it follows that $s(t, y_0) \in \mathcal{D}_\mathrm{c}$ for all $t \in (\tau_{\min}, \tau_{\max})$. To see this, suppose, *ad absurdum*, that there exists $\tau \in (\tau_{\min}, \tau_{\max})$ such that $s(t, x_0) \in \mathcal{D}_\mathrm{c}$ for some $t \in (\tau_{\min}, \tau]$ and $s(\tau, y_0) \in \partial\mathcal{D}_\mathrm{c}$. In this case,

$$\begin{aligned} \|s(t, y_0) - s(t, x_0)\| &\leq \|y_0 - x_0\| + \int_{t_0}^{t} \|f(y(\sigma)) - f(x(\sigma))\| \mathrm{d}\sigma \\ &\leq \|y_0 - x_0\| + L \int_{t_0}^{t} \|s(\sigma, y_0) - s(\sigma, x_0)\| \mathrm{d}\sigma, \quad (2.189) \end{aligned}$$

for all $t \in (\tau_{\min}, \tau]$. Now, since $\tau < \tau_{\max} \leq t_1$, it follows from Gronwall's lemma that

$$\|s(\tau, y_0) - s(\tau, x_0)\| \leq \|y_0 - x_0\| e^{L|\tau - t_0|} < \delta e^{L(\tau - t_0)} \leq \varepsilon. \quad (2.190)$$

Hence, $s(\tau, y_0) \in \mathring{\mathcal{D}}_\mathrm{c}$, which leads to a contradiction. Thus, $s(t, y_0) \in \mathcal{D}_\mathrm{c}$ for all $t \in (\tau_{\min}, \tau_{\max})$. However, by Theorem 2.30, $(\tau_{\min}, \tau_{\max})$ cannot be the maximal interval of existence of $s(t, y_0)$, leading to a contradiction. Thus, $t_1 < \tau_{\max}$. Similarly, it can be shown that $\tau_{\min} < t_0$, and hence, $[t_0, t_1] \subset (\tau_{\min}, \tau_{\max})$. Hence, for all $y_0 \in \mathcal{N}_\delta(x_0)$, $y(t)$ is the unique solution to (2.136) defined over the closed interval $[t_0, t_1]$.

Next, to show that $s(t, y_0) \in \mathcal{D}_\mathrm{c}$ for all $t \in [t_0, t_1]$, suppose, *ad absurdum*, that there exists $\tau \in [t_0, t_1)$ such that $s(t, y_0) \in \mathcal{D}_\mathrm{c}$ for all $t \in [t_0, \tau)$ and $s(\tau, y_0) \in \partial\mathcal{D}_\mathrm{c}$. Repeating the identical steps as above, it can be shown that this leads to a contradiction, and hence, $s(t, y_0) \in \mathcal{D}_\mathrm{c}$ for all $t \in [t_0, t_1]$. Now, a reapplication of Gronwall's lemma over the closed interval $[t_0, t_1]$ yields

$$\|s(t, y_0) - s(t, x_0)\| \leq \|y_0 - x_0\| e^{L|t - t_0|}, \quad t \in [t_0, t_1]. \quad (2.191)$$

Finally, (2.188) is now immediate. □

The main difference between Theorem 2.26 and Theorem 2.31 is that in Theorem 2.26 we assumed that both solutions to (2.136) are defined on the same closed interval. Alternatively, Theorem 2.31 shows that solutions to (2.136) that start nearby will be defined on the same interval and remain nearby over a finite time. An extension of Theorem 2.31 involving continuous dependence on initial conditions and system parameters can be easily addressed as in Theorem 2.31. Hence, the following result is immediate.

Theorem 2.32. Consider the nonlinear dynamical system (2.162). Assume $f : \mathcal{D} \times \mathbb{R}^m \to \mathbb{R}^n$ is such that for every $\lambda \in \mathbb{R}^m$, $f(\cdot, \lambda)$ is Lipschitz continuous on $\mathcal{D}$ and for every $x \in \mathcal{D}$, $f(x, \cdot)$ is Lipschitz continuous on $\mathbb{R}^m$. Furthermore, let $x(t)$ denote a solution to (2.162) with system parameter λ and initial condition $x(t_0) = x_0$ defined over the closed interval $[t_0, t_1]$. Then there exist neighborhoods $\mathcal{N}_\delta(x_0) \subset \mathcal{D}$ and $\mathcal{N}_\delta(\mu) \subset \mathbb{R}^m$, and a constant $L > 0$ such that for all $y(t_0) = y_0 \in \mathcal{N}_\delta(x_0)$ and $\lambda \in \mathcal{N}_\delta(\mu)$ there exists a unique solution $y(t)$ of (2.162) defined over the closed interval $[t_0, t_1]$ such that

$$\|x(t) - y(t)\| \leq [\|x_0 - y_0\| + \|\lambda - \mu\|]e^{L|t-t_0|}, \tag{2.192}$$

for all $t \in [t_0, t_1]$. Moreover,

$$\lim_{(y_0,\mu)\to(x_0,\lambda)} s(t, y_0, \mu) = s(t, x_0, \lambda), \tag{2.193}$$

uniformly for all $t \in [t_0, t_1]$, where $s(t, x_0, \lambda)$ denotes the solution to (2.162) with initial condition x_0 and parameters λ.

Next, we turn our attention to the question of global existence and uniqueness of solutions to (2.136). For this result the following lemma is needed.

Lemma 2.3. Let $y : \mathbb{R} \to \mathbb{R}^n$ be a continuously differentiable function on (α, β). If

$$\|\dot{y}(t)\| \leq q(t), \quad y(t_0) = y_0, \quad t \in (\alpha, \beta), \tag{2.194}$$

then

$$\|y(t)\| \leq \|y(t_0)\| + \int_{t_0}^{t} q(s)\mathrm{d}s, \quad t \in (\alpha, \beta). \tag{2.195}$$

Proof. Note that for $t \in (\alpha, \beta)$,

$$\left\|\int_{t_0}^{t} \dot{y}(s)\mathrm{d}s\right\| \leq \int_{t_0}^{t} \|\dot{y}(s)\|\mathrm{d}s \leq \int_{t_0}^{t} q(s)\mathrm{d}s, \tag{2.196}$$

which implies

$$\|y(t) - y(t_0)\| \leq \int_{t_0}^{t} q(s)\mathrm{d}s, \quad t \in (\alpha, \beta).$$

Hence,

$$\|y(t)\| = \|y(t) - y(t_0) + y(t_0)\| \leq \|y(t_0)\| + \int_{t_0}^{t} q(s)\mathrm{d}s,$$

for all $t \in (\alpha, \beta)$. □

Theorem 2.33. Consider the nonlinear dynamical system (2.136). Assume that $f : \mathbb{R}^n \to \mathbb{R}^n$ is globally Lipschitz continuous with Lipschitz constant L. Then, for all $x_0 \in \mathbb{R}^n$, (2.136) has a unique solution $x : (-\infty, \infty) \to \mathbb{R}^n$ over all $t \in \mathbb{R}$.

Proof. Let $x(t)$ be the solution to (2.136) on the maximal interval of existence $\mathcal{I}_{x_0} = (\tau_{\min}, \tau_{\max})$. Then it follows from the triangle inequality for vector norms and the globally Lipschitz condition on f that

$$\begin{aligned} \|\dot{x}(t)\| &= \|f(x(t))\| \\ &= \|f(x(t)) - f(x_0) + f(x_0)\| \\ &\leq \|f(x(t)) - f(x_0)\| + \|f(x_0)\| \\ &\leq L\|x(t) - x_0\| + \alpha, \quad t \in (\tau_{\min}, \tau_{\max}), \end{aligned} \tag{2.197}$$

where $\alpha \triangleq \|f(x_0)\|$. Now, suppose, *ad absurdum*, that $\tau_{\max} < \infty$. In this case, it follows from Lemma 2.3, with $y(t) = x(t) - x_0$ and $q(t) = L\|x(t) - x_0\| + \alpha$, that

$$\begin{aligned} \|x(t) - x_0\| &\leq \int_{t_0}^{t} [L\|x(s) - x_0\| + \alpha]\mathrm{d}s \\ &\leq \alpha(\tau_{\max} - t_0) + L\int_{t_0}^{t} \|x(s) - x_0\|\mathrm{d}s, \end{aligned} \tag{2.198}$$

for all $t \in (t_0, \tau_{\max})$. Next, using Gronwall's lemma (2.198) implies

$$\|x(t) - x_0\| \leq \alpha(\tau_{\max} - t_0)e^{L(t-t_0)}, \quad t \in [t_0, \tau_{\max}), \tag{2.199}$$

which further implies that the solution to (2.136) through the point x_0 at time $t = t_0$ is contained in the compact set

$$\mathcal{D}_{\rm c} = \{x \in \mathbb{R}^n : \ \|x - x_0\| \leq \alpha(\tau_{\max} - t_0)e^{L(\tau_{\max} - t_0)}\} \subset \mathbb{R}^n.$$

Hence, it follows from Theorem 2.30 that the supposition $\tau_{\max} < \infty$ leads to a contradiction. Thus, $\tau_{\max} = \infty$. Similarly, it can be shown that $\tau_{\min} = -\infty$. Hence, for all $x_0 \in \mathbb{R}^n$, (2.136) has a unique solution $x(t)$ over all $t \in \mathbb{R}$, where uniqueness follows from Theorem 2.25. □

Example 2.37. Consider the linear dynamical system

$$\dot{x}(t) = Ax(t), \quad x(t_0) = x_0, \quad t \in [t_0, t_1], \tag{2.200}$$

where $x(t) \in \mathbb{R}^n$, $t \in [t_0, t_1]$, and $A \in \mathbb{R}^{n\times n}$. Since A is constant, $\|A\| \leq \alpha$, where $\|\cdot\| : \mathbb{R}^{n\times n} \to \mathbb{R}$ is the equi-induced matrix norm induced by the vector norm $\|\cdot\|$. Now, with $f(x) = Ax$ it follows that

$$\|f(x) - f(y)\| = \|A(x - y)\| \leq \|A\|\|x - y\| \leq \alpha\|x - y\|, \tag{2.201}$$

for all $x, y \in \mathbb{R}^n$, and hence, (2.136) is globally Lipschitz on $\mathbb{R}^n$. Hence, it follows from Theorem 2.33 that linear systems have unique solutions over all $t \in \mathbb{R}$. △

The following corollary to Theorem 2.33 is immediate.

Corollary 2.6. Consider the nonlinear dynamical system (2.136).

Assume that $f : \mathbb{R}^n \to \mathbb{R}^n$ is continuously differentiable and its derivative is uniformly bounded on $\mathbb{R}^n$. Then, for all $x_0 \in \mathbb{R}^n$, (2.136) has a unique solution $x : (-\infty, \infty) \to \mathbb{R}^n$ over all $t \in \mathbb{R}$.

Proof. It follows from Proposition 2.26 that if $f : \mathbb{R}^n \to \mathbb{R}^n$ is continuously differentiable on $\mathbb{R}^n$ and its derivative is uniformly bounded on $\mathbb{R}^n$, then $f(\cdot)$ is globally Lipschitz continuous. Existence and uniqueness of solutions over $\mathbb{R}$ now follows from Theorem 2.33. □

2.10 Flows and Dynamical Systems

As shown in Section 2.6, a dynamical system defines a continuous flow in the state space. This flow is a one-parameter family of homeomorphisms and is one possible mathematical expression of the behavior of the dynamical system. In this section, we provide some additional properties of flows defined by differential equations. In particular, we consider the nonlinear dynamical system

$$\dot{x}(t) = f(x(t)), \quad x(0) = x_0, \quad t \in \mathcal{I}_{x_0}, \tag{2.202}$$

where $x(t) \in \mathcal{D}$, $t \in \mathcal{I}_{x_0}$, $\mathcal{D}$ is an open subset of $\mathbb{R}^n$, $f : \mathcal{D} \to \mathbb{R}^n$ is Lipschitz continuous on $\mathcal{D}$, and $\mathcal{I}_{x_0} = (\tau_{\min}, \tau_{\max}) \subseteq \mathbb{R}$ is the maximal interval of existence for the solution $x(\cdot)$ of (2.202). As in Section 2.6, for $t \in \mathcal{I}_{x_0}$, the set of mappings $s : \mathcal{I}_{x_0} \times \mathcal{D} \to \mathcal{D}$ defined by $s_t(x_0) = s(t, x_0)$ will denote the flow of (2.202). Furthermore, we define the set $\mathcal{S} \triangleq \{(t, x_0) \in \mathbb{R} \times \mathcal{D} : \ t \in \mathcal{I}_{x_0}\}$ so that $s : \mathcal{S} \to \mathcal{D}$.

Theorem 2.34. Let $s : \mathcal{S} \to \mathcal{D}$ be the flow generated by (2.202) and let $x_0 \in \mathcal{D}$. If $t \in \mathcal{I}_{x_0}$ and $\tau \in \mathcal{I}_{s_t(x_0)}$, then $\tau + t \in \mathcal{I}_{x_0}$ and

$$s_{t+\tau}(x_0) = s_\tau(s_t(x_0)). \tag{2.203}$$

Proof. Let $t \in \mathcal{I}_{x_0}$ and let $\tau \in \mathcal{I}_{s_t(x_0)}$ be such that $\tau > 0$. Furthermore, let $\mathcal{I}_{x_0} = (\tau_{\min}, \tau_{\max})$ and define the solution curves $s^{x_0} : (\tau_{\min}, \tau + t] \to \mathcal{D}$ by

$$s^{x_0}(\sigma) = \begin{cases} s(\sigma, x_0), & \tau_{\min} < \sigma \le t, \\ s(\sigma - t, s_t(x_0)), & t \le \sigma \le \tau + t. \end{cases} \tag{2.204}$$

Clearly, $s^{x_0}(\sigma)$ is the solution curve of (2.202) over the interval $(\tau_{\min}, \tau + t]$ with initial condition $s^{x_0}(0) = x_0$. Hence, $\tau + t \in \mathcal{I}_{x_0}$. Furthermore, since $f : \mathcal{D} \to \mathbb{R}^n$ is Lipschitz continuous on $\mathcal{D}$ it follows from the uniqueness property of solutions that

$$s_{t+\tau}(x_0) = s^{x_0}(t + \tau) = s(\tau, s_t(x_0)) = s_\tau(s_t(x_0)), \tag{2.205}$$

which proves (2.203) for the case $\tau > 0$. Next, in the case where $\tau = 0$, (2.203) is immediate. Finally, let $\tau < 0$ and define the solution curve s^{x_0} :

$[\tau + t, \tau_{\max}) \to \mathcal{D}$ by

$$s^{x_0}(\sigma) = \begin{cases} s(\sigma, x_0), & t \le \sigma < \tau_{\max}, \\ s(\sigma - t, s_t(x_0)), & \tau + t \le \sigma \le t. \end{cases} \quad (2.206)$$

Note that $s^{x_0}(\sigma)$ is the solution curve of (2.206) over the interval $[\tau+t, \tau_{\max})$ with initial condition $s^{x_0}(0) = x_0$. Now, using an identical argument as above shows (2.203). □

Theorem 2.34 shows that $s_{\tau+t}(x_0)$ is defined for $\tau + t \in \mathcal{I}_{x_0}$ if and only if $s_\tau(s_t(x_0))$ is defined for $t \in \mathcal{I}_{x_0}$ and $\tau \in \mathcal{I}_{s_t(x_0)}$.

Theorem 2.35. Let $s : \mathcal{S} \to \mathcal{D}$ be the flow generated by (2.202) and let $x_0 \in \mathcal{D}$. Then $\mathcal{S}$ is an open subset of $\mathbb{R} \times \mathcal{D}$ and $s : \mathcal{S} \to \mathcal{D}$ is continuous on $\mathcal{S}$.

Proof. To show that $\mathcal{S}$ is open, let $(t_1, x_0) \in \mathcal{S}$ and $t_1 \ge 0$. In this case, the solution $x(t) = s(t, x_0)$ to (2.202) is defined over $[0, t_1]$, and hence, by Theorem 2.25, can be extended on $[0, t_1 + \tau]$ for some $\tau > 0$. Hence, repeating this argument for the left-side interval, $s(t, x_0)$ is defined over $[t_1 - \tau, t_1 + \tau]$ for each $x_0 \in \mathcal{D}$.

Now, it follows from Theorem 2.31 that there exists $\delta > 0$ and $L > 0$ such that for all $y_0 \in \mathcal{B}_\delta(x_0)$, $s(t, y_0)$ is defined on $[t_1 - \tau, t_1 + \tau] \times \mathcal{D}_\delta(x_0)$ and satisfies

$$\|s(t, y_0) - s(t, x_0)\| \le \|y_0 - x_0\| e^{L|t-t_0|}, \quad t \in [t_1 - \tau, t_1 - \tau]. \quad (2.207)$$

Thus, $(t_1 - \tau, t_1 + \tau) \times \mathcal{D}_\delta(x_0) \subset \mathcal{S}$, and hence, $\mathcal{S}$ is open in $\mathbb{R} \times \mathcal{D}$.

To show that $s : \mathcal{S} \to \mathcal{D}$ is continuous at (t_1, x_0) note that since $f : \mathcal{D} \to \mathbb{R}^n$ is Lipschitz continuous on $\mathcal{D}$ and $\mathcal{D}_c = s([t_1 - \tau, t_1 - \tau] \times \overline{\mathcal{B}}_\delta(x_0)) \subset \mathcal{D}$ is a compact set, it follows from Proposition 2.23 that $f : \mathcal{D} \to \mathbb{R}^n$ is uniformly Lipschitz continuous on $\mathcal{D}_c$. Next, let $\alpha = \max\{\|f(x)\| : x \in \mathcal{D}_c\}$ and let $\delta > 0$ be such that $\delta < \tau$ and $\mathcal{B}_\delta(x_0) \subset \mathcal{D}_c$. Furthermore, let y be such that $y \in \mathcal{B}_\delta(x_0)$. Now, suppose $|t^* - t_1| < \delta$ and $\|y - x_0\| < \delta$. Then,

$$\begin{aligned} \|s(t^* y) - s(t_1, x_0)\| &= \|s(t^*, y) - s(t^*, x_0) + s(t^*, x_0) - s(t_1, x_0)\| \\ &\le \|s(t^*, y) - s(t^*, x_0)\| + \|s(t^*, x_0) - s(t_1, x_0)\|. \end{aligned} \quad (2.208)$$

Since $s(\cdot, x_0)$ is continuous in t, it follows that $\|s(t^*, x_0) - s(t_1, x_0)\| \to 0$ as $\delta \to 0$. Alternatively, it follows from (2.207) that $\|s(t^*, y) - s(t^*, x_0)\| < \delta e^{L\delta}$, and hence, $\|s(t^*, y) - s(t^*, x_0)\| \to 0$ as $\delta \to 0$. Hence, $s(\cdot, \cdot)$ is continuous on $\mathcal{S}$. An analogous proof holds for the case where $t_1 < 0$. □

Finally, we show that $s_t : \mathcal{D} \to \mathcal{D}$ is a one-to-one map and has a continuous inverse.

Theorem 2.36. Let $s_t : \mathcal{D} \to \mathcal{D}$ be the flow generated by (2.202) and let $x_0 \in \mathcal{D}$. If $(t, x_0) \in \mathcal{S}$ then there exists a neighborhood $\mathcal{N}$ of x_0 such that $\{t\} \times \mathcal{N} \subset \mathcal{S}$. Furthermore, $U = s_t(\mathcal{N}) \subset \mathcal{D}$ is open, $\{-t\} \times U \subset \mathcal{S}$, and

$$s_{-t}(s_t(x)) = x, \quad x \in \mathcal{N}, \tag{2.209}$$
$$s_t(s_{-t}(y)) = y, \quad y \in U. \tag{2.210}$$

Proof. Let $(t, x_0) \in \mathcal{S}$. Using similar arguments as in the proof of Theorem 2.35 it follows that there exists a neighborhood $\mathcal{N}$ of x_0 and $\tau > 0$ such that $(t-\tau, t+\tau) \times \mathcal{N} \subset \mathcal{S}$, and hence, $\{t\} \times \mathcal{N} \subset \mathcal{S}$. Next, to show that s_t has a continuous inverse, let $y = s_t(x)$ for $x \in \mathcal{N}$ and $t \in \mathcal{I}_x$. Since $\phi(\tau) = s(\tau + t, y)$ is a solution of (2.202) over the interval $[-t, 0]$ with initial condition $\phi(-t) = y$, it follows that $-t \in \mathcal{I}_y$. Hence, s_{-t} is defined on $U = s_t(\mathcal{N})$. Now, Theorem 2.35 yields $s_{-t}(s_t(x)) = s_0(x) = x$ for all $x \in \mathcal{N}$ and $s_t(s_{-t}(y)) = s_0(y) = y$ for all $y \in U$. Finally, to show that U is open let $s_{-t} : U_{\max} \to \mathcal{D}$, where $U_{\max} \supset U$ is such that $\mathbb{R} \times U_{\max} = \mathcal{S}$. Since $\mathcal{S}$ is open, $U_{\max}$ is open. Furthermore, since, by Theorem 2.35, s_t is continuous it follows that $s_{-t} : U_{\max} \to \mathcal{D}$ is also continuous. Now, since the inverse image of the open set $\mathcal{N}$ under the continuous map s_{-t} is open, and since this inverse image is U, that is, $s_{-t}(U) = \mathcal{N}$, it follows that U is open. □

2.11 Time-Varying Nonlinear Dynamical Systems

In Section 2.6 we defined a dynamical system as a precise mathematical object satisfying a set of axioms. A key implicit assumption in Definition 2.47 was that the solution curve of the dynamical system $\mathcal{G}$ remained unchanged under translation of time. In many dynamical systems this assumption does not hold, giving rise to time-varying or nonautonomous differential equations. Even though this allows for a more general class of dynamical systems, the fundamental theory developed in Sections 2.7–2.10 does not change significantly for the class of time-varying systems. In this section, we outline the salient changes for time-varying dynamical systems as compared to time-invariant dynamical systems. Since these results very closely parallel the results on time-invariant systems, we present the most important theorems needed for later developments. Furthermore, we leave the proofs of these theorems as exercises for the reader. For the following definition let $\mathcal{D}$ be an open subset of $\mathbb{R}^n$.

Definition 2.48. A *time-varying dynamical system* on $\mathcal{D}$ is the triple $(\mathcal{D}, \mathbb{R}, s)$, where $s : \mathbb{R} \times \mathbb{R} \times \mathcal{D} \to \mathcal{D}$ is such that the following axioms hold:

i) (Continuity): $s(t, t_0, \cdot)$ is continuous and $s(\cdot, t_0, x_0)$ is continuously differentiable in t for all $x_0 \in \mathcal{D}$ and $t, t_0 \in \mathbb{R}$.

ii) (Consistency): For every $x_0 \in \mathcal{D}$ and $t_0 \in \mathbb{R}$, $s(t_0, t_0, x_0) = x_0$.

iii) (Group property): $s(t_2, t_0, x_0) = s(t_2, t_1, s(t_1, t_0, x_0))$ for all $t_0, t_1, t_2 \in \mathbb{R}$ and $x_0 \in \mathcal{D}$.

As for the time-invariant case, we denote the time-varying dynamical system $(\mathcal{D}, \mathbb{R}, s)$ by $\mathcal{G}$. Note that if for every $t_0, t \in \mathbb{R}$ such that $t \geq t_0$, and $\tau \in \mathbb{R}$, $x_0 \in \mathcal{D}$, $s(t + \tau, t_0 + \tau, x_0) = s(t, t_0, x_0)$, then the function $s(\cdot, \cdot, \cdot)$ remains unchanged under time translation. Hence, without loss of generality we can set $t_0 = 0$. In this case, Definition 2.48 collapses to Definition 2.47 with $s(t, 0, x_0) = s(t, x_0)$. As in the time-invariant case, since a time-varying dynamical system involves the function $s(\cdot, \cdot, \cdot)$ describing the motion of $x \in \mathcal{D}$ for all $t \in \mathbb{R}$, it generates a time-varying differential equation on $\mathcal{D}$. In particular, the function $f : \mathbb{R} \times \mathcal{D} \to \mathbb{R}^n$ given by

$$f(t, x) \triangleq \lim_{\tau \to t} \frac{1}{t - \tau} [s(t, t_0, x) - s(\tau, t_0, x)] \tag{2.211}$$

defines a continuous vector field on $\mathcal{D}$. Hence, the dynamical system $\mathcal{G}$ defined by Definition 2.48 gives rise to the time-varying differential equation

$$\dot{x}(t) = f(t, x(t)), \quad x(t_0) = x_0, \quad t \in \mathcal{I}_{x_0}. \tag{2.212}$$

Existence and uniqueness of solutions, extendability of solutions, and continuous dependence on initial conditions and system parameter theorems can now be developed for (2.212) as in the case for time-invariant systems. To see that these results will not significantly differ from the results already developed note that by defining $x_1(\tau) \triangleq x(t)$ and $x_2(\tau) \triangleq t$, where $\tau = t - t_0$, the solution $x(t)$, $t \in \mathcal{I}_{x_0}$, to the nonlinear time-varying dynamical system (2.212) can be equivalently characterized by the solution $x_1(\tau)$, $\tau \in \mathcal{I}_{x_0,t_0}$, to the nonlinear time-invariant system

$$\dot{x}_1(\tau) = f(x_2(\tau), x_1(\tau)), \quad x_1(0) = x_0, \quad \tau \in \mathcal{I}_{x_0,t_0}, \tag{2.213}$$

$$\dot{x}_2(\tau) = 1, \quad x_2(0) = t_0, \tag{2.214}$$

where $\dot{x}_1(\cdot)$ and $\dot{x}_2(\cdot)$ in (2.213) and (2.214) denote differentiation with respect to τ. Next, we state the key results on the existence and uniqueness of solutions for time-varying dynamical systems needed for later developments. The proofs of these results are similar to the proofs given in earlier sections and are left as exercises for the reader.

Theorem 2.37. Consider the nonlinear dynamical system (2.212). Assume $f(t, \cdot) : \mathcal{D} \to \mathbb{R}^n$ is continuous on $\mathcal{D}$ for all $t \in [t_0, t_1]$ and $f(\cdot, x) : [t_0, t_1] \to \mathbb{R}^n$ is piecewise continuous on $[t_0, t_1]$ for all $x \in \mathcal{D}$. Then for every

$x_0 \in \mathcal{D}$ there exists $\tau > t_0$ such that (2.212) has a piecewise continuously differentiable solution $x : [t_0, \tau] \to \mathbb{R}^n$. Furthermore, if $f(\cdot, x) : [t_0, t_1] \to \mathbb{R}^n$ is continuous on $[t_0, t_1]$ for all $x \in \mathcal{D}$, then $x : [t_0, \tau] \to \mathbb{R}^n$ is continuously differentiable on $[t_0, \tau]$.

Theorem 2.38. Consider the nonlinear dynamical system (2.212). Assume $f(t, \cdot) : \mathcal{D} \to \mathbb{R}^n$ is Lipschitz continuous on $\mathcal{D}$ for all $t \in [t_0, t_1]$ and $f(\cdot, x) : [t_0, t_1] \to \mathbb{R}^n$ is piecewise continuous on $[t_0, t_1]$ for all $x \in \mathcal{D}$. Then, for every $x_0 \in \mathcal{D}$, there exists $\tau \in (t_0, t_1)$ such that (2.212) has a unique solution $x : [t_0, \tau] \to \mathbb{R}^n$ over the interval $[t_0, \tau]$.

Theorem 2.39. Consider the nonlinear dynamical system (2.212). Assume $f(t, \cdot) : \mathcal{D} \to \mathbb{R}^n$ is Lipschitz continuous on $\mathcal{D}$ for all $t \in [t_0, t_1]$ and $f(\cdot, x) : [t_0, t_1] \to \mathbb{R}^n$ is piecewise continuous on $[t_0, t_1]$ for all $x \in \mathcal{D}$. Furthermore, let $\mathcal{D}_c \subset \mathcal{D}$ be compact and suppose for $x_0 \in \mathcal{D}_c$ every solution $x : [t_0, \tau] \to \mathcal{D}$ lies entirely in $\mathcal{D}_c$. Then there exists a unique solution $x : [t_0, \infty) \to \mathbb{R}^n$ to (2.212) for all $t \geq t_0$.

Theorem 2.40. Consider the nonlinear dynamical system (2.212). Assume $f(t, \cdot) : \mathcal{D} \to \mathbb{R}^n$ is Lipschitz continuous on $\mathcal{D}$ for all $t \in [t_0, t_1]$ and $f(\cdot, x) : [t_0, t_1] \to \mathbb{R}^n$ is piecewise continuous on $[t_0, t_1]$ for all $x \in \mathcal{D}$. In addition, assume there exists $\alpha = \alpha(x_0)$ such that $\max_{t \in \mathbb{R}} \|f(t, x_0)\| \leq \alpha$. Then, for all $x_0 \in \mathbb{R}^n$, (2.212) has a unique solution $x : (-\infty, \infty) \to \mathbb{R}^n$ over all $t \in \mathbb{R}$.

If we assume that $f(t, 0) = 0$, $t \in \mathbb{R}$, then the assumption $\|f(t, x_0)\| \leq \alpha$ in Theorem 2.40 follows as a direct consequence of Lipschitz continuity condition on $f(t, \cdot) : \mathcal{D} \to \mathbb{R}^n$ and, hence, is automatic. To see this, note that in this case Lipschitz continuity implies that $\|f(t, x_0)\| \leq L\|x_0\|$. In this book we will assume that $f(\cdot, \cdot)$ has at least one equilibrium point in $\mathcal{D}$ so that, without loss of generality, $f(t, 0) = 0$, $t \in \mathbb{R}$. However, as seen in Example 2.2, there are systems that have no equilibrium points. For these systems, the assumption $\|f(t, x_0)\| \leq \alpha$ in Theorem 2.40 is necessary.

2.12 Limit Points, Limit Sets, and Attractors

In this section, we introduce the notion of invariance with respect to the flow $s_t(x_0)$ of a nonlinear dynamical system. Consider the nonlinear dynamical system

$$\dot{x}(t) = f(x(t)), \quad x(0) = x_0, \quad t \in \mathbb{R}, \tag{2.215}$$

where $x(t) \in \mathcal{D}$, $t \in \mathbb{R}$, $\mathcal{D}$ is an open subset of $\mathbb{R}^n$, and $f : \mathcal{D} \to \mathbb{R}^n$ is Lipschitz continuous on $\mathcal{D}$. Recall that (2.215) defines a dynamical system in the sense of Definition 2.47 with flow $s : \mathbb{R} \times \mathcal{D} \to \mathcal{D}$. In addition, recall that for $x \in \mathcal{D}$, the map $s(\cdot, x) : \mathbb{R} \to \mathcal{D}$ defines the solution curve or trajectory of (2.215) through the point x in $\mathcal{D}$. Identifying $s(\cdot, x)$ with its graph, the trajectory or *orbit* of a point $x_0 \in \mathcal{D}$ is defined as the motion along the curve

$$\mathcal{O}_{x_0} \triangleq \{x \in \mathcal{D}: \ x = s(t, x_0), \ t \in \mathbb{R}\}. \tag{2.216}$$

For $t \geq 0$, we define the *positive orbit* through the point $x_0 \in \mathcal{D}$ as the motion along the curve

$$\mathcal{O}_{x_0}^{+} \triangleq \{x \in \mathcal{D}: \ x = s(t, x_0), \ t \geq 0\}. \tag{2.217}$$

Similarly, the *negative orbit* through the point $x_0 \in \mathcal{D}$ is defined as

$$\mathcal{O}_{x_0}^{-} \triangleq \{x \in \mathcal{D}: \ x = s(t, x_0), \ t \leq 0\}. \tag{2.218}$$

Hence, the orbit $\mathcal{O}_x$ of a point $x \in \mathcal{D}$ is given by $\mathcal{O}_x^+ \cup \mathcal{O}_x^- = \{s(t,x): \ t \geq 0\} \cup \{s(t,x): \ t \leq 0\}$.

Definition 2.49. A point $p \in \mathcal{D}$ is a *positive limit point* of the trajectory $s(\cdot, x)$ of (2.215) if there exists a monotonic sequence $\{t_n\}_{n=0}^{\infty}$ of positive numbers, with $t_n \to \infty$ as $n \to \infty$, such that $s(t_n, x) \to p$ as $n \to \infty$. A point $q \in \mathcal{D}$ is a *negative limit point* of the trajectory $s(\cdot, x)$ of (2.215) if there exists a monotonic sequence $\{t_n\}_{n=0}^{\infty}$ of negative numbers, with $t_n \to -\infty$ as $n \to \infty$, such that $s(t_n, x) \to q$ as $n \to \infty$. The set of all positive limit points of $s(t, x)$, $t \geq 0$, is the *positive limit set* $\omega(x)$ of $s(\cdot, x)$ of (2.215). The set of all negative limit points of $s(t, x)$, $t \leq 0$, is the *negative limit set* $\alpha(x)$ of $s(\cdot, x)$ of (2.215).

An equivalent definition of $\omega(x_0)$ and $\alpha(x_0)$ which is geometrically easier to see is $\omega(x_0) = \cap_{t \geq 0} \overline{\mathcal{O}_{x_0}^+}$ and $\alpha(x_0) = \cap_{t \leq 0} \overline{\mathcal{O}_{x_0}^-}$ (see Problem 2.114). In the literature, the positive limit set is often referred to as the *ω-limit set* while the negative limit set is referred to as the *α-limit set*. Note that if $p \in \mathcal{D}$ is a positive limit point of the trajectory $s(\cdot, x)$, then for all $\varepsilon > 0$ and finite time $T > 0$ there exists $t > T$ such that $\|x(t) - p\| < \varepsilon$. This follows from the fact that $\|x(t) - p\| < \varepsilon$ for all $\varepsilon > 0$ and some $t > T > 0$ is equivalent to the existence of a sequence $\{t_n\}_{n=0}^{\infty}$, with $t_n \to \infty$ as $n \to \infty$, such that $x(t_n) \to p$ as $n \to \infty$. An analogous observation holds for negative limit points.

Definition 2.50. A set $\mathcal{M} \subset \mathcal{D} \subseteq \mathbb{R}^n$ is a *positively invariant set* with respect to the nonlinear dynamical system (2.215) if $s_t(\mathcal{M}) \subseteq \mathcal{M}$ for all $t \geq 0$, where $s_t(\mathcal{M}) \triangleq \{s_t(x) : \ x \in \mathcal{M}\}$. A set $\mathcal{M} \subset \mathcal{D} \subseteq \mathbb{R}^n$ is a *negatively invariant set* with respect to the nonlinear dynamical system (2.215) if $s_t(\mathcal{M}) \subseteq \mathcal{M}$ for all $t \leq 0$. A set $\mathcal{M} \subseteq \mathcal{D}$ is an *invariant set* with respect to the dynamical system (2.215) if $s_t(\mathcal{M}) = \mathcal{M}$ for all $t \in \mathbb{R}$.

In the case where $t \geq 0$ in (2.215), note that a set $\mathcal{M} \subseteq \mathcal{D}$ is a negatively invariant set with respect to the nonlinear dynamical system (2.215) if, for every $y \in \mathcal{M}$ and every $t \geq 0$, there exists $x \in \mathcal{M}$ such that $s(t,x) = y$ and $s(\tau, x) \in \mathcal{M}$ for all $\tau \in [0,t]$. Hence, if $\mathcal{M}$ is negatively invariant, then $\mathcal{M} \subseteq s_t(\mathcal{M})$ for all $t \geq 0$; the converse, however, is not generally true. Furthermore, a set $\mathcal{M} \subset \mathcal{D}$ is an invariant set with respect to (2.215) (defined over $t \geq 0$) if $s_t(\mathcal{M}) = \mathcal{M}$ for all $t \geq 0$. Note that a set $\mathcal{M}$ is invariant if and only if $\mathcal{M}$ is positively and negatively invariant.

The following propositions give several properties of positively invariant, negatively invariant, and invariant sets.

Proposition 2.31. Consider the nonlinear dynamical system $\mathcal{G}$ given by (2.215). Let $\{\mathcal{M}_i : i \in \mathcal{I}\}$ be a collection of positively invariant (respectively, negatively invariant or invariant) sets with respect to $\mathcal{G}$. Then $\mathcal{M} = \bigcap_{i \in \mathcal{I}} \mathcal{M}_i$ and $\mathcal{N} = \bigcup_{i \in \mathcal{I}} \mathcal{M}_i$ are positively invariant (respectively, negatively invariant or invariant) sets with respect to $\mathcal{G}$.

Proof. Let the sets $\mathcal{M}_i$ be positively invariant. For every $x \in \mathcal{N}$, it follows that $x \in \mathcal{M}_i$ for some $i \in \mathcal{I}$. Since $\mathcal{M}_i$ is positively invariant, $s(t,x) \in \mathcal{M}_i$ for all $t \geq 0$. Thus, since $\mathcal{M}_i \subseteq \mathcal{N}$, $s(t,x) \in \mathcal{N}$ for all $t \geq 0$, and hence, $\mathcal{N}$ is positively invariant. Next, let $x \in \mathcal{M}$. In this case, $x \in \mathcal{M}_i$ for every $i \in \mathcal{I}$ and since each $\mathcal{M}_i$ is positively invariant, $s(t,x) \in \mathcal{M}_i$ for each $i \in \mathcal{I}$ and each $t \geq 0$. Hence, $s(t,x) \in \bigcap_{i \in \mathcal{I}} \mathcal{M}_i = \mathcal{M}$ for each $t \geq 0$, which proves positive invariance of $\mathcal{M}$. The proofs for negative invariance and invariance are analogous and are left as exercises for the reader. □

Proposition 2.32. Consider the nonlinear dynamical system $\mathcal{G}$ given by (2.215). Let $\mathcal{M} \subset \mathcal{D}$ be a positively invariant (respectively, negatively invariant or invariant) set with respect to $\mathcal{G}$. Then $\overline{\mathcal{M}}$ is positively invariant (respectively, negatively invariant or invariant) with respect to $\mathcal{G}$.

Proof. Let $\mathcal{M}$ be invariant and let $x \in \overline{\mathcal{M}}$ and $t \in \mathbb{R}$. In this case, there exists a sequence $\{x_n\}_{n=0}^{\infty} \subseteq \mathcal{M}$ such that $x_n \to x$ as $n \to \infty$. Since $\mathcal{M}$ is invariant, $s(t, x_n) \in \mathcal{M}$ for each n. Furthermore, since $s(t,x_n) \to s(t,x)$ as $n \to \infty$ it follows that $s(t,x) \in \overline{\mathcal{M}}$, and hence, $\overline{\mathcal{M}}$ is invariant. The proofs of positive and negative invariance are identical and, hence, are omitted. □

Proposition 2.33. Consider the nonlinear dynamical system $\mathcal{G}$ given by (2.215). Then $\mathcal{M} \subset \mathcal{D}$ is positively invariant with respect to $\mathcal{G}$ if and only if $\mathcal{D} \setminus \mathcal{M}$ is negatively invariant with respect to $\mathcal{G}$. Furthermore, $\mathcal{M}$ is invariant with respect to $\mathcal{G}$ if and only if $\mathcal{D} \setminus \mathcal{M}$ is invariant with respect to $\mathcal{G}$.

Proof. Let $\mathcal{M}$ be positively invariant and suppose, *ad absurdum*, that if $x \in \mathcal{D} \setminus \mathcal{M}$ and $t \leq 0$, then $s(t,x) \notin \mathcal{D} \setminus \mathcal{M}$. Hence, $s(t,x) \in \mathcal{M}$ and since

$-t \geq 0$, $s(-t, s(t,x)) = s(-t+t, x) = s(0,x) = x \in \mathcal{M}$ by positive invariance of $\mathcal{M}$. This contradiction shows that $\mathcal{D} \setminus \mathcal{M}$ is negatively invariant. The converse statement follows by retracing the steps above. Finally, the second part of the theorem follows as a direct consequence from the first part. □

Proposition 2.34. Consider the nonlinear dynamical system $\mathcal{G}$ given by (2.215). Let $\mathcal{M} \subset \mathcal{D}$ be invariant with respect to $\mathcal{G}$. Then $\partial\mathcal{M}$ and $\overset{\circ}{\mathcal{M}}$ are invariant with respect to $\mathcal{G}$.

Proof. It follows from Proposition 2.33 that if $\mathcal{M}$ is invariant with respect to $\mathcal{G}$, then $\mathcal{D} \setminus \mathcal{M}$ is invariant with respect to $\mathcal{G}$. Hence, by Proposition 2.32, $\overline{\mathcal{M}}$ and $\overline{\mathcal{D} \setminus \mathcal{M}}$ are invariant with respect to $\mathcal{G}$. Now, it follows from Propositions 2.31 and 2.33 that $\partial\mathcal{M} = \overline{\mathcal{M}} \cap (\overline{\mathcal{D} \setminus \mathcal{M}})$ and $\overset{\circ}{\mathcal{M}} = \mathcal{D} \setminus (\overline{\mathcal{D} \setminus \mathcal{M}})$ are invariant with respect to $\mathcal{G}$. □

The converse of Proposition 2.34 holds whenever $\mathcal{M}$ is open or closed. To see this, note that if $\partial\mathcal{M}$ is invariant with respect to $\mathcal{G}$, then $\overset{\circ}{\mathcal{M}}$ is invariant with respect to $\mathcal{G}$. This simple fact can be shown by a contradiction argument. Specifically, suppose, *ad absurdum*, that there exist $x \in \overset{\circ}{\mathcal{M}}$ and $t \in \mathbb{R}$ such that $s(t,x) \notin \overset{\circ}{\mathcal{M}}$. Now, since the set $\mathcal{Q} \triangleq \{y \in \mathcal{D} : y = s(t,x), t \in [0, \tau]\}$ is connected it follows that $\mathcal{Q} \cap \partial\mathcal{M} \neq \emptyset$, where we have assumed $t \geq 0$ for convenience. Hence, there exists $t^* \in (0, t]$ such that $s(t^*, x) \in \partial\mathcal{M}$. However, in this case, $x = s(-t^*, s(t^*, x)) = s(-t^* + t^*, x) \in \partial\mathcal{M}$, which by invariance of $\partial\mathcal{M}$ contradicts $x \in \overset{\circ}{\mathcal{M}}$. This shows that if $\partial\mathcal{M}$ is invariant with respect to $\mathcal{G}$, then $\overset{\circ}{\mathcal{M}}$ is invariant with respect to $\mathcal{G}$. Thus, if $\mathcal{M}$ is closed then $\mathcal{M} = \partial\mathcal{M} \cup \overset{\circ}{\mathcal{M}}$ is invariant, whereas if $\mathcal{M}$ is open then $\mathcal{M} = \overset{\circ}{\mathcal{M}}$ is invariant. Similarly, it can be shown that if $\overset{\circ}{\mathcal{M}}$ is nonempty and invariant, then $\mathcal{M}$ is also invariant whenever $\mathcal{M}$ is closed or open.

Proposition 2.35. Consider the nonlinear dynamical system $\mathcal{G}$ given by (2.215). Let $\mathcal{M} \subset \mathcal{D}$ be positively (respectively, negatively) invariant with respect to $\mathcal{G}$. Then $\overset{\circ}{\mathcal{M}}$ is positively (respectively, negatively) invariant with respect to $\mathcal{G}$.

Proof. Assume $\mathcal{M}$ is positively invariant. Then by Propositions 2.32 and 2.33 $\mathcal{D} \setminus \mathcal{M}$ and $\overline{\mathcal{D} \setminus \mathcal{M}}$ are negatively invariant. Hence, $\overset{\circ}{\mathcal{M}} = \mathcal{D} \setminus (\overline{\mathcal{D} \setminus \mathcal{M}})$ is positively invariant. The case where $\mathcal{M}$ is negatively invariant follows using identical arguments. □

Definition 2.51. The trajectory $s(\cdot, x)$ of (2.215) is *bounded* if there exists $\gamma > 0$ such that $\|s(t,x)\| < \gamma$, $t \in \mathbb{R}$.

Next, we state and prove a key theorem involving positive limit sets. An analogous result holds for negative limit sets and is left as an exercise for the reader (see Problem 2.113). In the case where $t \geq 0$, we refer to the group property in Definition 2.47 as the semigroup property. Furthermore, we use the notation $x(t) \to \mathcal{M} \subseteq \mathcal{D}$ as $t \to \infty$ to denote that $x(t)$ approaches $\mathcal{M}$, that is, for each $\varepsilon > 0$ there exists $T > 0$ such that $\operatorname{dist}(x(t), \mathcal{M}) < \varepsilon$ for all $t > T$, where $\operatorname{dist}(p, \mathcal{M}) \triangleq \inf_{x \in \mathcal{M}} \|p - x\|$.

Theorem 2.41. Suppose the solution $x(t)$ to (2.215) corresponding to an initial condition $x(0) = x_0$ is bounded for all $t \geq 0$. Then the positive limit set $\omega(x_0)$ of $x(t)$, $t \geq 0$, is a nonempty, compact, invariant, and connected set. Furthermore, $x(t) \to \omega(x_0)$ as $t \to \infty$.

Proof. Let $x(t)$, $t \geq 0$, or, equivalently, $s(t, x_0)$, $t \geq 0$, denote the solution to (2.215) corresponding to the initial condition $x(0) = x_0$. Next, since $x(t)$ is bounded for all $t \geq 0$, it follows from the Bolzano-Weierstrass theorem (Theorem 2.3) that every sequence in the positive orbit $\mathcal{O}^+_{x_0} \triangleq \{s(t,x) : t \in [0,\infty)\}$ has at least one accumulation point $p \in \mathcal{D}$ as $t \to \infty$, and hence, $\omega(x_0)$ is nonempty. Next, let $p \in \omega(x_0)$ so that there exists an increasing unbounded sequence $\{t_n\}_{n=0}^{\infty}$, with $t_0 = 0$, such that $\lim_{n\to\infty} x(t_n) = p$. Now, since $x(t_n)$ is uniformly bounded in n it follows that the limit point p is bounded, which implies that $\omega(x_0)$ is bounded. To show that $\omega(x_0)$ is closed let $\{p_i\}_{i=0}^{\infty}$ be a sequence contained in $\omega(x_0)$ such that $\lim_{i\to\infty} p_i = p$. Now, since $p_i \to p$ as $i \to \infty$ for every $\varepsilon > 0$, there exists an i such that $\|p - p_i\| < \varepsilon/2$. Next, since $p_i \in \omega(x_0)$, there exists $t \geq T$, where T is arbitrary and finite, such that $\|p_i - x(t)\| < \varepsilon/2$. Now, since $\|p - p_i\| < \varepsilon/2$ and $\|p_i - x(t)\| < \varepsilon/2$ it follows that $\|p - x(t)\| \leq \|p_i - x(t)\| + \|p - p_i\| < \varepsilon$, and hence, $p \in \omega(x_0)$. Thus, every accumulation point of $\omega(x_0)$ is an element of $\omega(x_0)$ so that $\omega(x_0)$ is closed. Hence, since $\omega(x_0)$ is closed and bounded it is compact.

To show positive invariance of $\omega(x_0)$ let $p \in \omega(x_0)$ so that there exists an increasing unbounded sequence $\{t_n\}_{n=0}^{\infty}$ such that $x(t_n) \to p$ as $n \to \infty$. Now, let $s(t_n, x_0)$ denote the solution $x(t_n)$ of (2.215) with initial condition $x(0) = x_0$ and note that since $f : \mathcal{D} \to \mathbb{R}^n$ in (2.215) is Lipschitz continuous on $\mathcal{D}$, $x(t)$, $t \geq 0$, is the unique solution to (2.215) so that by the semigroup property $s(t + t_n, x_0) = s(t, s(t_n, x_0)) = s(t, x(t_n))$. Now, since $x(t)$, $t \geq 0$, is continuous it follows that, for $t + t_n \geq 0$, $\lim_{n\to\infty} s(t + t_n, x_0) = \lim_{n\to\infty} s(t, x(t_n)) = s(t,p)$, and hence, $s(t,p) \in \omega(x_0)$. Hence, $s_t(\omega(x_0)) \subseteq \omega(x_0)$, $t \geq 0$, establishing positive invariance of $\omega(x_0)$.

To show invariance of $\omega(x_0)$ let $y \in \omega(x_0)$ so that there exists an increasing unbounded sequence $\{t_n\}_{n=0}^{\infty}$ such that $s(t_n, x_0) \to y$ as $n \to \infty$. Next, let $t \in [0, \infty)$ and note that there exists N such that $t_n > t$, $n \geq N$. Hence, it follows from the semigroup property that $s(t, s(t_n - t, x_0)) = s(t_n, x_0) \to y$ as $n \to \infty$. Now, it follows from the Bolzano-Lebesgue theorem (Theorem 2.4) that there exists a subsequence $\{z_{n_k}\}_{k=1}^{\infty}$ of the sequence $z_n = s(t_n - t, x_0)$, $n = N, N+1, \ldots$, such that $z_{n_k} \to z \in \mathcal{D}$ as $k \to \infty$ and, by definition, $z \in \omega(x_0)$. Next, it follows from the continuous dependence property that $\lim_{k\to\infty} s(t, z_{n_k}) = s(t, \lim_{k\to\infty} z_{n_k})$, and hence, $y = s(t, z)$, which implies that $\omega(x_0) \subseteq s_t(\omega(x_0))$, $t \in [0, \infty)$. Now, using positive invariance of $\omega(x_0)$ it follows that $s_t(\omega(x_0)) = \omega(x_0)$, $t \geq 0$, establishing invariance of the positive limit set $\omega(x_0)$.

To show connectedness of $\omega(x_0)$, suppose, *ad absurdum*, that $\omega(x_0)$ is not connected. In this case, there exist two nonempty closed sets $\mathcal{P}_1^+$ and $\mathcal{P}_2^+$ such that $\mathcal{P}_1^+ \cap \mathcal{P}_2^+ = \emptyset$ and $\omega(x_0) = \mathcal{P}_1^+ \cup \mathcal{P}_2^+$. Since $\mathcal{P}_1^+$ and $\mathcal{P}_2^+$ are closed and disjoint there exist two open sets $\mathcal{S}_1$ and $\mathcal{S}_2$ such that $\mathcal{S}_1 \cap \mathcal{S}_2 = \emptyset$, $\mathcal{P}_1^+ \subset \mathcal{S}_1$, and $\mathcal{P}_2^+ \subset \mathcal{S}_2$. Next, since $f : \mathcal{D} \to \mathbb{R}$ is Lipschitz continuous on $\mathcal{D}$ it follows that the solution $x(t)$, $t \geq 0$, to (2.215) is a continuous function of t. Hence, there exist sequences $\{t_n\}_{n=0}^{\infty}$ and $\{\tau_n\}_{n=0}^{\infty}$ such that $x(t_n) \in \mathcal{S}_1$, $x(\tau_n) \in \mathcal{S}_2$, and $t_n < \tau_n < t_{n+1}$, which implies that there exists a sequence $\{\tau_n\}_{n=0}^{\infty}$, with $t_n < \tau_n < \tau_{n+1}$, such that $x(\tau_n) \notin \mathcal{S}_1 \cup \mathcal{S}_2$. Next, since $x(t)$ is bounded for all $t \geq 0$, it follows that $x(\tau_n) \to \hat{p} \notin \omega(x_0)$ as $n \to \infty$, leading to a contradiction. Hence, $\omega(x_0)$ is connected.

Finally, to show $x(t) \to \omega(x_0)$ as $t \to \infty$, suppose, *ad absurdum*, $x(t) \not\to \omega(x_0)$ as $t \to \infty$. In this case, there exists a sequence $\{t_n\}_{n=0}^{\infty}$, with $t_n \to \infty$ as $n \to \infty$, such that

$$\inf_{p \in \omega(x_0)} \|x(t_n) - p\| > \varepsilon, \quad n \in \overline{\mathbb{Z}}_+. \tag{2.219}$$

However, since $x(t)$, $t \geq 0$, is bounded, the bounded sequence $\{x(t_n)\}_{n=0}^{\infty}$ contains a convergent subsequence $\{x(t_n^*)\}_{n=0}^{\infty}$ such that $x(t_n^*) \to p^* \in \omega(x_0)$ as $n \to \infty$, which contradicts (2.219). Hence, $x(t) \to \omega(x_0)$ as $t \to \infty$. □

It is important to note that Theorem 2.41 holds for time-invariant nonlinear dynamical systems (2.215) possessing unique solutions forward in time with the solutions being continuous functions of the initial conditions. More generally, letting $s(\cdot, x_0)$ denote the solution of a dynamical system with initial condition $x(0) = x_0$, Theorem 2.41 holds if $s(t + \tau, x_0) = s(t, s(\tau, x_0))$, $t, \tau \geq 0$, and $s(\cdot, x_0)$ is a continuous function of $x_0 \in \mathcal{D}$. Of course, if $f(\cdot)$ is Lipschitz continuous on $\mathcal{D}$ then there exists a unique solution to (2.215), and hence, the required semigroup property $s(t + \tau, x_0) = s(t, s(\tau, x_0))$, $t, \tau \geq 0$, and the continuity of $s(t, \cdot)$ on $\mathcal{D}$, $t \geq 0$, hold. Alternatively, uniqueness of solutions in forward time along with the continuity of $f(\cdot)$ ensure that the solutions to (2.215) satisfy the semigroup property and are continuous functions of the initial conditions $x_0 \in \mathcal{D}$ even when $f(\cdot)$ is not Lipschitz continuous on $\mathcal{D}$ (see [96, Theorem 4.3, p. 59]). More generally, $f(\cdot)$ need not be continuous. In particular, if $f(\cdot)$ is discontinuous but bounded and $x(\cdot)$ is the unique solution to (2.215) in the sense of Filippov [118], then the required semigroup property along with the continuous dependence of solutions on initial conditions hold [118].

Note that Theorem 2.41 implies that if $p \in \mathcal{D}$ is an ω-limit point of a trajectory $s(\cdot, x)$ of (2.215), then all other points of the trajectory $s(\cdot, p)$ of (2.215) through the point p are also ω-limit points of $s(\cdot, x)$, that is, if $p \in \omega(x)$ then $\mathcal{O}_p^+ \subset \omega(x)$. Furthermore, since every equilibrium point $x_\mathrm{e} \in \mathcal{D}$ of (2.215) satisfies $s(t, x_\mathrm{e}) = x_\mathrm{e}$ for all $t \in \mathbb{R}$, all equilibrium points $x_\mathrm{e} \in \mathcal{D}$ of (2.215) are their own α- and ω-limit sets. If a trajectory of (2.215) possesses a unique ω-limit point x_e, then it follows from Theorem 2.41 that since $\omega(x_\mathrm{e})$ is invariant with respect to the flow s_t of (2.215), x_e is an equilibrium point of (2.215).

Next, we establish an important property of limit sets with orbit closures.

Theorem 2.42. Consider the nonlinear dynamical system $\mathcal{G}$ given by (2.215) and let $\mathcal{O}_x^+$ and $\mathcal{O}_x^-$ denote, respectively, the positive and negative orbits of $\mathcal{G}$ through the point $x \in \mathcal{D}$. Then $\overline{\mathcal{O}_x^+} = \mathcal{O}_x^+ \cup \omega(x)$ and $\overline{\mathcal{O}_x^-} = \mathcal{O}_x^- \cup \alpha(x)$.

Proof. Recall that $\mathcal{O}_x^+ = \{y \in \mathcal{D} : y = s(t, x), t \geq 0\}$. Now, it follows from the definition of $\omega(x)$ that $\overline{\mathcal{O}_x^+} \supseteq \mathcal{O}_x^+ \cup \omega(x)$. To show that $\overline{\mathcal{O}_x^+} \subseteq \mathcal{O}_x^+ \cup \omega(x)$, let $z \in \overline{\mathcal{O}_x^+}$. In this case, there exists a sequence $\{z_n\}_{n=0}^\infty \subseteq \overline{\mathcal{O}_x^+}$ such that $z_n \to z$ as $n \to \infty$. Next, let $z_n = s(t_n, x)$ for $t_n \in \mathbb{R}_+$. Now, either there exists a subsequence $\{t_{n_k}\}_{k=0}^\infty$ such that $t_{n_k} \to \infty$ as $k \to \infty$, in which case $z \in \omega(x)$, or since $\overline{\mathbb{R}}_+$ is closed it follows from the Bolzano-Lebesgue theorem (Theorem 2.4) that there exists a subsequence $\{t_{n_k}\}_{k=0}^\infty$ such that $t_{n_k} \to t \in \overline{\mathbb{R}}_+$. In the second case, $s(t_{n_k}, x) \to s(t, x) \in \mathcal{O}_x^+$ as $k \to \infty$, and since $s(t_{n_k}, x) \to z$ as $k \to \infty$ it follows that $z = s(t, x) \in \mathcal{O}_x^+$. Hence, $\overline{\mathcal{O}_x^+} \subseteq \mathcal{O}_x^+ \cup \omega(x)$. The proof of $\overline{\mathcal{O}_x^-} = \mathcal{O}_x^- \cup \alpha(x)$ follows using identical arguments. □

The next definition introduces the notions of attracting sets and attractors. For this definition recall that a neighborhood $\mathcal{N}$ of a set $\mathcal{M} \subset \mathcal{D}$ is defined as $\mathcal{N} \triangleq \{x \in \mathcal{D} : \|x - y\| < \varepsilon,\ y \in \mathcal{M}\}$ for some small $\varepsilon > 0$.

Definition 2.52. A closed invariant set $\mathcal{M} \subset \mathcal{D}$ is an *attracting set* of the dynamical system (2.215) if there exists a neighborhood $\mathcal{N}$ of $\mathcal{M}$ such that, for all $x \in \mathcal{N}$, $s_t(x) \in \mathcal{N}$ for all $t \geq 0$ and $s_t(x) \to \mathcal{M}$ as $t \to \infty$. A set $\mathcal{M} \subset \mathcal{D}$ is an *attractor* of the dynamical system (2.215) if $\mathcal{M}$ is an attracting set of (2.215) and contains a dense orbit.

Example 2.38. Consider the nonlinear dynamical system

$$\dot{x}_1(t) = -x_2(t) + x_1(t)[1 - x_1^2(t) - x_2^2(t)], \quad x_1(0) = x_{10}, \quad t \geq 0, \tag{2.220}$$

$$\dot{x}_2(t) = x_1(t) + x_2(t)[1 - x_1^2(t) - x_2^2(t)], \quad x_2(0) = x_{20}. \tag{2.221}$$

Rewriting (2.220) and (2.221) in terms of the polar coordinates $r = \sqrt{x_1^2 + x_2^2}$ and $\theta = \tan^{-1}(x_2/x_1)$ as

$$\dot{r}(t) = r(t)[1 - r^2(t)], \quad r(0) = r_0 \triangleq \sqrt{x_{10}^2 + x_{20}^2}, \quad t \geq 0, \tag{2.222}$$

$$\dot{\theta}(t) = 1, \quad \theta(0) = \theta_0 \triangleq \tan^{-1}(x_{20}/x_{10}), \tag{2.223}$$

it can be shown that the set of equilibria $f^{-1}(0)$ consists of the origin $x = [x_1\ x_2]^{\mathrm{T}} = 0$. All solutions of the system starting from nonzero initial conditions $x(0)$ that are not on the unit circle $\mathcal{C} = \{x \in \mathbb{R}^2 : x_1^2 + x_2^2 = 1\}$ approach the unit circle. In particular, since $\dot{r} > 0$ for $r \in (0, 1)$, all solutions starting inside the unit circle spiral counterclockwise toward the unit circle. Alternatively, since $\dot{r} < 0$ for $r > 1$, all solutions starting outside the unit circle spiral inward counterclockwise (see Figure 2.6). Hence, all solutions to (2.220) and (2.221) are bounded and converge to $\mathcal{C}$. Note that the counterclockwise flow on the unit circle characterizes a trajectory $\mathcal{O}_{x_0}^+$ of (2.220) and (2.222) since $\dot{r} = 0$ on $r = 1$. Furthermore, $\mathcal{O}_{x_0}^+$ is the ω-limit set of every trajectory of (2.220) and (2.221) with the exception of the equilibrium point $x = 0$. Clearly, $\mathcal{O}_{x_0}^+$ is an attractor.

Since the vector field (2.222) and (2.223) are Lipschitz continuous, the solutions to (2.222) and (2.223) are unique and all solutions are defined on $\mathbb{R}$. The system trajectories of (2.222) and (2.223) are shown in Figure 2.6 and consist of *i*) an equilibrium point corresponding to the origin, *ii*) a periodic trajectory coinciding with the limit cycle $\mathcal{C}$ (see Definition 2.54), and *iii*) spiraling trajectories passing through each point $p = (r, \theta)$ with $r \neq 0$ and $r \neq 1$. For points p such that $r \in (0, 1)$, $\omega(r, \theta) = \mathcal{C}$ and $\alpha(r, \theta) = \{0\}$. For points p such that $r > 1$, $\omega(r, \theta) = \mathcal{C}$ and $\alpha(r, \theta) = \emptyset$. For points p such that $r = 1$, $\omega(r, \theta) = \alpha(r, \theta) = \mathcal{C}$. Hence, $\omega(r, \theta) = \mathcal{C} \cup \{0\}$ and $\alpha(r, \theta) = \mathcal{C} \cup \{0\} \cup \emptyset = \mathcal{C} \cup \{0\}$, $(r, \theta) \in \mathbb{R} \times \mathbb{R}$. △

2.13 Periodic Orbits, Limit Cycles, and Poincaré-Bendixson Theorems

In this section we discuss periodic orbits and limit cycles. In particular, we present several key techniques for predicting the existence of periodic orbits

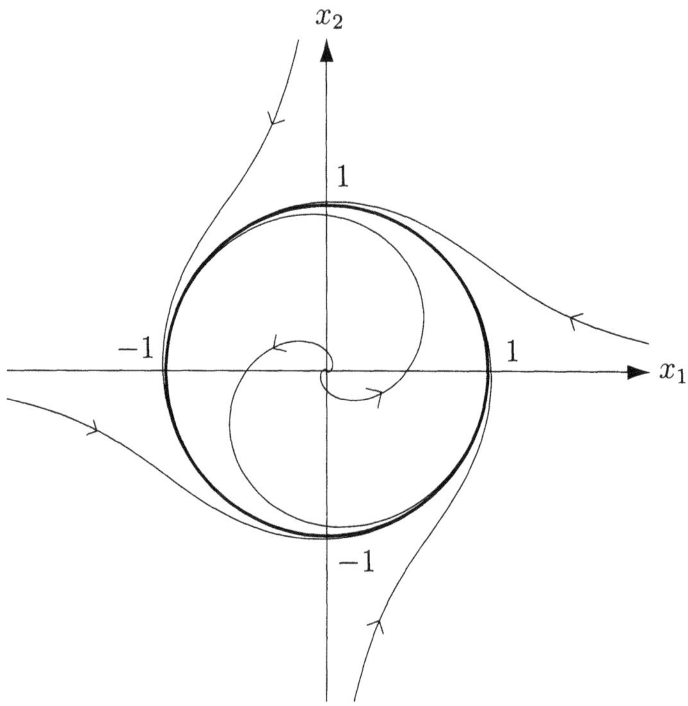

Figure 2.6 Attractor for Example 2.38.

and limit cycles in nonlinear dynamical systems. Here we limit our attention to planar (second-order) systems, whereas in Section 4.10 we address n-dimensional periodic dynamical systems. Once again, we consider dynamical systems of the form

$$\dot{x}(t) = f(x(t)), \quad x(0) = x_0, \quad t \in \mathbb{R}, \tag{2.224}$$

where $x(t) \in \mathcal{D}$, $t \in \mathbb{R}$, $\mathcal{D}$ is an open subset of $\mathbb{R}^n$, and $f : \mathcal{D} \to \mathbb{R}^n$ is Lipschitz continuous on $\mathcal{D}$. The following definition introduces the notion of periodic solutions and periodic orbits for (2.224).

Definition 2.53. A solution $s(t, x_0)$ of (2.224) is *periodic* if there exists a finite time $T > 0$ such that $s(t+T, x_0) = s(t, x_0)$ for all $t \in \mathbb{R}$. The minimal T for which the solution $s(t, x_0)$ of (2.224) is periodic is called the *period*. A set $\mathcal{O} \subset \mathcal{D}$ is a *periodic orbit* of (2.224) if $\mathcal{O} = \{x \in \mathcal{D} : x = s(t, x_0), t \in \mathbb{R}\}$ for some periodic solution $s(t, x_0)$ of (2.224).

Note that for every $x \in \mathcal{D}$, $s(t, x)$ has a period $T = 0$; however, $s(t, x)$ need not be periodic. Furthermore, if $x_e \in \mathcal{D}$ is an equilibrium point, then every $T \in \mathbb{R}$ is a period of $s(t, x_e)$, and hence, $s(t, x_e)$ is periodic. The next proposition gives a characterization of a periodic point of a nonlinear dynamical system.

Proposition 2.36. Consider the nonlinear dynamical system $\mathcal{G}$ given by (2.224). Then $x \in \mathcal{D}$ is a periodic point of $\mathcal{G}$ if and only if there exists $T \neq 0$ such that $x = s(T, x)$.

Proof. Necessity is immediate. To show sufficiency let $t \in \mathbb{R}$ and note that $s(t, x) = s(t, s(T, x)) = s(t + T, x)$, which proves the result. □

The next theorem establishes an important relationship between periodic points, ω-limit sets, and positive orbits.

Theorem 2.43. Consider the nonlinear dynamical system $\mathcal{G}$ given by (2.224). Then $x \in \mathcal{D}$ is a periodic point of $\mathcal{G}$ if and only if $\mathcal{O}_x^+ = \omega(x)$.

Proof. Let $\mathcal{O}_x^+ = \omega(x)$. In this case, $x \in \omega(x)$ and since $\omega(x)$ is invariant if follows that $\mathcal{O}_x^+ = \omega(x) = \mathcal{O}_x$. Hence, $s(\tau, x) \in \mathcal{O}_x^+$ for each $\tau < 0$, and hence, there exists $\mu \geq 0$ such that $s(\tau, x) = s(\mu, x)$. Now, it follows from the group axiom that $s(t, x) = s(t+\mu-\tau, x)$ for all $t \in \mathbb{R}$, which establishes that x is periodic with period $T = \mu - \tau > 0$. The converse is immediate. □

For finite-dimensional dynamical systems wherein $\mathcal{D} \subseteq \mathbb{R}^n$, it can be shown that $x \in \mathcal{D}$ is a periodic point of $\mathcal{G}$ if and only if $\omega(x) = \alpha(x) = \mathcal{O}_x$. (See Problem 2.127.)

It is important to distinguish between *trivial periodic orbits* and *nontrivial periodic orbits* or *limit cycles*. To see the distinction consider the dynamical equations for the simple harmonic oscillator given by

$$\dot{x}_1(t) = x_2(t), \quad x_1(0) = x_{10}, \quad t \geq 0, \tag{2.225}$$
$$\dot{x}_2(t) = -x_1(t), \quad x_1(0) = x_{20}, \tag{2.226}$$

where the solution is characterized by

$$x_1(t) = A\cos(t - \theta_0), \quad x_2(t) = -A\sin(t - \theta_0), \tag{2.227}$$

where $A = \sqrt{x_{10}^2 + x_{20}^2}$ and $\theta_0 = \tan^{-1}(\frac{x_{20}}{x_{10}})$. Note that the solution to (2.225) and (2.226) is periodic for all initial conditions $x_{10}, x_{20} \in \mathbb{R}$. That is, given any initial condition in the plane x_1-x_2, one can find a periodic solution passing through this point. This is called a trivial periodic orbit. This is in contrast to a nontrivial periodic orbit or limit cycle wherein a dynamical system possesses an isolated periodic orbit, that is, there exists a neighborhood of the periodic orbit that does not contain any other periodic solution. To see this, consider once again (2.220) and (2.221) given in Example 2.38. In particular, using separation of variables the solution to

(2.220) and (2.221) is given by

$$r(t) = \left[1 + \left(\frac{1}{A^2} - 1\right) e^{-2t}\right]^{-1/2}, \quad (2.228)$$

$$\theta(t) = t + \theta_0. \quad (2.229)$$

Clearly, (2.228) shows that (2.220) and (2.221) has only one nontrivial periodic solution corresponding to $r = 1$, that is, $x_{10}^2 + x_{20}^2 = 1$. In light of the above observations we have the following definition for a nontrivial periodic orbit or limit cycle.

Definition 2.54. Consider the nonlinear dynamical system (2.224). A *limit cycle* of (2.224) is a closed curve[4] $\Gamma \subset \mathbb{R}^n$ such that Γ is the positive limit set of the positive orbit $\mathcal{O}_x^+$ of (2.224) or the negative limit set of the negative orbit $\mathcal{O}_x^-$ of (2.224) for $x \notin \Gamma$.

Note that it follows from Definition 2.54 that a limit cycle is compact and invariant. Furthermore, if a nontrivial periodic solution Γ is isolated such that it is the positive limit set of the positive orbit $\mathcal{O}_x^+$ of (2.224) or the negative limit set of the negative orbit $\mathcal{O}_x^-$ of (2.224) for $x \notin \Gamma$, then $\mathcal{G}$ is a limit cycle.

Next, we present a key result due to Ivar Bendixson [39] that guarantees the absence of limit cycles for planar systems. For this result, we consider second-order nonlinear dynamical systems of the form

$$\dot{x}_1(t) = f_1(x_1(t), x_2(t)), \quad x_1(0) = x_{10}, \quad t \in \mathbb{R}, \quad (2.230)$$

$$\dot{x}_2(t) = f_2(x_1(t), x_2(t)), \quad x_2(0) = x_{20}, \quad (2.231)$$

where $f_1(x_1, x_2)$ and $f_2(x_1, x_2)$ are continuously differentiable. Furthermore, define the divergence operator

$$\nabla f(x) \triangleq \frac{\partial f_1}{\partial x_1}(x_1, x_2) + \frac{\partial f_2}{\partial x_2}(x_1, x_2). \quad (2.232)$$

The following definition is required for the statement of Bendixson's theorem.

Definition 2.55. A subset $\mathcal{D} \subseteq \mathbb{R}^2$ is *simply connected* if there exist $y \in \mathcal{D}$ and a continuous function $g : [0, 1] \times \mathcal{D} \to \mathcal{D}$ such that $g(0, x) = x$ and $g(1, x) = y$ for all $x \in \mathcal{D}$.

Theorem 2.44 (Bendixson's Theorem). Consider the second-order nonlinear dynamical system (2.230) and (2.231) with vector field $f : \mathcal{D} \to \mathbb{R}^2$, where $\mathcal{D}$ is a simply connected region in $\mathbb{R}^2$ such that there are no

[4] Γ is a *closed curve* contained in $\mathbb{R}^n$ if there exists a continuous mapping $\gamma : [0, 1] \to \mathbb{R}^n$ such that $\gamma(0) = \gamma(1)$, $\Gamma = \{\gamma(\sigma) : \sigma \in [0, 1]\}$, and $\Gamma \neq \{\gamma(0)\}$.

equilibrium points of (2.230) and (2.231) in $\mathcal{D}$. If there exists $x \in \mathcal{D}$ such that $\nabla f(x) \neq 0$ and $\nabla f(x)$ does not change sign in $\mathcal{D}$, then (2.230) and (2.231) has no periodic orbits lying entirely in $\mathcal{D}$.

Proof. Suppose, *ad absurdum*, that $\Gamma = \{x \in \mathcal{D} : x = x(t), 0 \leq t \leq T\}$ is a closed periodic orbit lying entirely in $\mathcal{D}$. Then, for each $x = (x_1, x_2) \in \Gamma$, $f(x)$ is tangent to Γ, that is, $f(x) \cdot n(x) = 0$ for all $x \in \Gamma$, where $n(x)$ denotes the outward normal vector to Γ at x (see Figure 2.7). Hence, it follows from Green's theorem that

$$\oint_\Gamma f(x) \cdot n(x) \mathrm{d}\Gamma = \int\int_\mathcal{S} \nabla f(x) \mathrm{d}\mathcal{S} = 0, \tag{2.233}$$

or, equivalently,

$$\begin{aligned} &\oint_\Gamma (f_2(x_1, x_2)\mathrm{d}x_1 - f_1(x_1, x_2)\mathrm{d}x_2) \\ &\quad = \int\int_\mathcal{S} \left(\frac{\partial f_1}{\partial x_1}(x_1, x_2) + \frac{\partial f_2}{\partial x_2}(x_1, x_2) \right) \mathrm{d}x_1 \mathrm{d}x_2 \\ &\quad = 0, \end{aligned} \tag{2.234}$$

where $\mathcal{S}$ is the area enclosed by Γ. Now, if there exists $x \in \mathcal{D}$ such that $\nabla f(x) \neq 0$ and $\nabla f(x)$ does not change sign in $\mathcal{S}$, then it follows from the continuity of the divergence operator $\nabla f(x)$ in $\mathcal{D}$ that (2.233) is either positive or negative, which leads to a contradiction. Hence, $\mathcal{D}$ contains no periodic solutions of (2.230) and (2.231). □

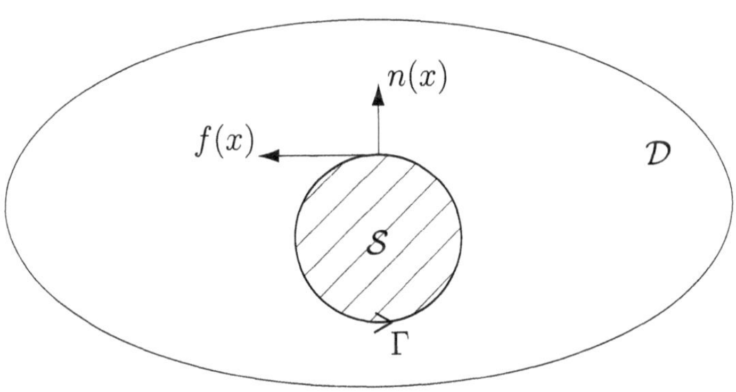

Figure 2.7 Visualization of sets and vector field used in the proof of Theorem 2.30.

Example 2.39. Consider the second-order nonlinear dynamical system

$$\ddot{x}(t) + \alpha\dot{x}(t) + g(x(t)) = 0, \quad x(0) = x_0, \quad \dot{x}(0) = \dot{x}_0, \quad t \geq 0, \tag{2.235}$$

where $\alpha \neq 0$ and $g : \mathbb{R} \to \mathbb{R}$ is continuous and satisfies $g(0) = 0$. Note that with $x_1 = x$ and $x_2 = \dot{x}$, (2.235) can be equivalently written as

$$\dot{x}_1(t) = x_2(t), \quad x_1(0) = x_0, \quad t \geq 0, \tag{2.236}$$

$$\dot{x}_2(t) = -g(x_1(t)) - \alpha x_2(t), \quad x_2(0) = \dot{x}_0. \tag{2.237}$$

Now, computing $\nabla f(x_1, x_2)$ yields

$$\nabla f(x_1, x_2) = \frac{\partial f_1}{\partial x_1}(x_1, x_2) + \frac{\partial f_2}{\partial x_2}(x_1, x_2) = -\alpha \neq 0, \quad (x_1, x_2) \in \mathbb{R} \times \mathbb{R}. \tag{2.238}$$

Hence, it follows from Bendixson's theorem that (2.235) has no limit cycles in $\mathbb{R}^2$. It is interesting to note that for the special case where $g(x) = -x + x^3$ and $\alpha > 0$, (2.235) is known as *Duffing's equation.* △

Example 2.40. Consider the *van der Pol oscillator* given by

$$\ddot{x}(t) - \varepsilon[1 - x^2(t)]\dot{x}(t) + x(t) = 0, \quad x(0) = x_0, \quad \dot{x}(0) = \dot{x}_0, \quad t \geq 0, \tag{2.239}$$

where $\varepsilon > 0$. Now, with $x_1 = x$ and $x_2 = \dot{x}$, (2.239) can be equivalently written as

$$\dot{x}_1(t) = x_2(t), \quad x_1(0) = x_0, \quad t \geq 0, \tag{2.240}$$

$$\dot{x}_2(t) = -x_1(t) + \varepsilon[1 - x_1^2(t)]x_2(t), \quad x_2(0) = \dot{x}_0. \tag{2.241}$$

Computing $\nabla f(x_1, x_2)$ yields

$$\nabla f(x_1, x_2) = \frac{\partial f_1}{\partial x_1}(x_1, x_2) + \frac{\partial f_2}{\partial x_2}(x_1, x_2) = \varepsilon(1 - x_1^2). \tag{2.242}$$

Hence, it follows from Bendixson's theorem that if the van der Pol oscillator were to possess a periodic solution it would have to intersect $\{(x_1, x_2) : x_1 = 1\}$ or $\{(x_1, x_2) : x_1 = -1\}$ or both sets. △

Example 2.41. Consider the linear dynamical system

$$\dot{x}(t) = Ax(t), \quad x(0) = x_0, \quad t \geq 0, \tag{2.243}$$

where $x(t) \in \mathbb{R}^2$, $t \geq 0$, and $A \in \mathbb{R}^{2\times 2}$. From linear system theory it follows that a necessary and sufficient condition for the existence of periodic solutions to (2.243) is $\mathrm{Re}\,\lambda = 0$ and $\mathrm{Im}\,\lambda \neq 0$, where $\lambda \in \mathrm{spec}(A)$. Recall that the eigenvalues of A for a second-order linear system are given by

$$\det(\lambda I_2 - A) = \lambda^2 - \mathrm{tr}\, A\lambda + \det A = 0. \tag{2.244}$$

Hence, (2.243) possesses periodic solutions if and only if $\mathrm{tr}\, A = 0$ and $\det A > 0$. A contrapositive statement to the above yields that the absence of periodic solutions will be guaranteed if and only if $\mathrm{tr}\, A \neq 0$ or $\det A \leq 0$. Using Bendixson's theorem it follows that $\nabla f(x) = \mathrm{tr}\, A$, $x \in \mathbb{R}^2$. Requiring $\mathrm{tr}\, A \neq 0$ thus guarantees the absence of periodic solutions to second-order linear systems. △

A more general criterion for ruling out closed orbits in a simply connected region of $\mathbb{R}^2$ is due to Dulac.

Theorem 2.45 (Dulac's Theorem). Consider the second-order nonlinear dynamical system (2.230) and (2.231) with vector field $f : \mathcal{D} \to \mathbb{R}^2$, where $\mathcal{D}$ is a simply connected region in $\mathbb{R}^2$ such that there are no equilibrium points of (2.230) and (2.231) in $\mathcal{D}$. If there exists $x \in \mathcal{D}$ and a continuously differentiable function $g : \mathcal{D} \to \mathbb{R}$ such that $\nabla(g(x)f(x)) \neq 0$ and $\nabla(g(x)f(x))$ does not change sign in $\mathcal{D}$, then (2.230) and (2.231) has no closed orbits lying entirely in $\mathcal{D}$.

Proof. The proof of this result is also based on Green's theorem and is virtually identical to the proof of Theorem 2.44. □

Example 2.42. Consider the second-order nonlinear dynamical system

$$\dot{x}_1(t) = x_1(t)[2 - x_1(t) - x_2(t)], \quad x_1(0) = x_{10}, \quad t \geq 0, \tag{2.245}$$

$$\dot{x}_2(t) = x_2(t)[4x_1(t) - x_1^2(t) - 3], \quad x_2(0) = x_{20}. \tag{2.246}$$

To show that (2.245) and (2.246) has no closed orbits in the positive orthant $\mathbb{R}_+^2$, consider the Dulac function $g(x_1, x_2) = 1/x_1x_2$. Computing $\nabla(g(x_1, x_2)f(x_1, x_2))$ yields

$$\begin{aligned}
\nabla(g(x_1, x_2)f(x_1, x_2)) &= \frac{\partial}{\partial x_1}(g(x_1, x_2)f_1(x_1, x_2)) + \frac{\partial}{\partial x_2}(g(x_1, x_2)f_2(x_1, x_2)) \\
&= \frac{\partial}{\partial x_1}\left(\frac{2 - x_1 - x_2}{x_2}\right) + \frac{\partial}{\partial x_2}\left(\frac{4x_1 - x_1^2 - 3}{x_1}\right) \\
&= -1/x_2 \\
&< 0, \quad x_1 > 0, \quad x_2 > 0.
\end{aligned} \tag{2.247}$$

Since $\{(x_1, x_2) \in \mathbb{R} \times \mathbb{R} : x_1 > 0 \text{ and } x_2 > 0\}$ is simply connected in $\mathbb{R}^2$ and $g(\cdot)$ is continuously differentiable it follows from Dulac's theorem that (2.245) and (2.246) has no closed orbits in $\mathbb{R}_+^2$. △

The next theorem is due to Henri Poincaré [358–360] and Ivar Bendixson [39] and gives conditions for the existence of periodic orbits for planar systems.

Theorem 2.46 (Poincaré-Bendixson Theorem). Consider the second-order nonlinear dynamical system (2.230) and (2.231). Let $\mathcal{O}_{x_0}^+$ be a positive orbit of (2.230) and (2.231) with positive limit set $\omega(x_0)$. If $\mathcal{O}_{x_0}^+$ is contained in a compact subset $\mathcal{M}$ of $\mathcal{D} \subset \mathbb{R}^2$ and $\omega(x_0)$ contains no equilibria of (2.230) and (2.231), then $\omega(x_0)$ is a periodic orbit of (2.230) and (2.231).

An analogous result is valid for negative orbits $\mathcal{O}_x^-$ with negative limit set $\alpha(x_0)$. The proof of the Poincaré-Bendixson theorem is based on a series of technical lemmas from algebraic topology involving positive limit sets for two-dimensional systems and will not be given here. The interested reader is referred to [349, 454] for the proof. Nevertheless, the result itself is quite intuitive. In particular, the theorem states that if a compact set $\mathcal{M} \subset \mathbb{R}^2$ can be constructed that does not contain any equilibrium points of (2.230) and (2.231) and all limit points of (2.230) and (2.231) are contained in $\mathcal{M}$, then $\mathcal{M}$ must contain at least one periodic orbit. Of course, it is necessary to find such a region $\mathcal{M}$. One way to do this is to construct $\mathcal{M}$ such that $\mathcal{M}$ does not contain any equilibria of (2.230) and (2.231) and is positively invariant. In this case, $\mathcal{O}_x^+ \subseteq \mathcal{M}$ for $x \in \mathcal{M}$ and since $\mathcal{M}$ is compact it contains all its limit points, and hence, $\omega(x_0) \subseteq \mathcal{M}$ for $x_0 \in \mathcal{M}$. Hence, every compact, nonempty positively invariant set $\mathcal{M}$ contains either an equilibrium point or a periodic orbit. This result is stated as a theorem below.

Theorem 2.47. Consider the second-order nonlinear dynamical system (2.230) and (2.231). Let $\mathcal{M}$ be a compact, positively invariant set with respect to (2.230) and (2.231). Then $\mathcal{M}$ contains an equilibrium point or a periodic orbit.

Proof. If $x \in \mathcal{M}$, then $\omega(x)$ is a nonempty subset of $\mathcal{M}$, that is, $\omega(x) \subset \mathcal{M}$. The result now follows as a direct consequence of the Poincaré-Bendixson theorem. □

An identical result is true for compact, negatively invariant sets.

Example 2.43. To illustrate the Poincaré-Bendixson theorem we consider once again the simple harmonic oscillator given by (2.225) and (2.226). Now, letting $V(x_1, x_2) = x_1^2 + x_2^2$ it follows that $\dot{V}(x_1, x_2) = 2x_1\dot{x}_1 + 2x_2\dot{x}_2 = 0$. Hence, the system trajectories of (2.225) and (2.226) cannot cross the α-level set of V, that is, $V^{-1}(\alpha) = \{(x_1, x_2) \in \mathbb{R}^2 : V(x_1, x_2) = \alpha\}$, for every $\alpha > 0$. Now, define the $[\alpha, \beta]$-sublevel set $\mathcal{M} \triangleq V^{-1}([\alpha, \beta]) = \{(x_1, x_2) \in \mathbb{R}^2 : \alpha \leq V(x) \leq \beta\}$, where $\beta > \alpha > 0$. Note that $\mathcal{M}$ is compact and, since $\dot{V}(x_1, x_2) = 0$, positively invariant. Hence, since $\mathcal{M}$ contains no equilibrium points it follows from the Poincaré-Bendixson theorem that $\mathcal{M}$ contains a periodic orbit. △

Example 2.44. Once again, consider the nonlinear dynamical system (2.220) and (2.221) given in Example 2.38. Letting $V(x_1, x_2) = x_1^2 + x_2^2$ it follows that

$$\begin{aligned}\dot{V}(x_1, x_2) &= 2x_1\dot{x}_1 + 2x_2\dot{x}_2 \\ &= -2x_1x_2 + 2x_1^2(1 - x_1^2 - x_2^2) + 2x_1x_2 + 2x_2^2(1 - x_1^2 - x_2^2) \\ &= 2V(x_1, x_2)[1 - V(x_1, x_2)]. \end{aligned} \tag{2.248}$$

Note that $\dot{V}(x_1, x_2) > 0$ for $V(x_1, x_2) < 1$ and $\dot{V}(x_1, x_2) < 0$ for $V(x_1, x_2) > 1$. Hence, on the α-level set $V^{-1}(\alpha)$, where $0 < \alpha < 1$, all the system trajectories of (2.220) and (2.221) are moving out toward the circle $x_1^2 + x_2^2 = 1$. Alternatively, on the β-level set $V^{-1}(\beta)$ for $\beta > 1$, all system trajectories of (2.220) and (2.221) are moving in toward the circle $x_1^2+x_2^2=1$. Hence, $\mathcal{M} = \{(x_1,x_2) \in \mathbb{R}^2 : \alpha \le V(x) \le \beta\}$ is positively invariant. Since $\mathcal{M}$ is compact and does not contain any equilibrium points, it follows from the Poincaré-Bendixson theorem that $\mathcal{M}$ contains a periodic orbit. Furthermore, since, for every $0 < \alpha < 1$ and $\beta > 1$, $\mathcal{M}$ contains a periodic orbit it follows that $\mathcal{C} = \{(x_1,x_2) \in \mathbb{R}^2 : x_1^2+x_2^2 = 1\}$ contains a periodic orbit. △

2.14 Problems

Problem 2.1. Let $A \in \mathbb{R}^{n\times n}$ be such that $\det A \neq 0$. Show that there exists $\varepsilon > 0$ such that $\|Ax\| \ge \varepsilon\|x\|$, $x \in \mathbb{R}^n$, where $\|\cdot\|$ is a vector norm on $\mathbb{R}^n$.

Problem 2.2. Let $p \in [1,\infty]$, $\bar{p} = p/(p-1)$, and let $A \in \mathbb{R}^{m\times n}$. Show that:

i) $\|A\|_{2,2} = \sigma_{\max}(A)$.

ii) $\|A\|_{p,1} = \max_{i=1,\ldots,n} \|\mathrm{col}_i(A)\|_p$.

iii) $\|A\|_{\infty,p} = \max_{i=1,\ldots,m} \|\mathrm{row}_i(A)\|_{\bar{p}}$.

Problem 2.3. Using the results of Problem 2.2 show that for $A \in \mathbb{R}^{m\times n}$ the following hold:

i) $\|A\|_{1,1} = \max_{i=1,\ldots,n} \|\mathrm{col}_i(A)\|_1$.

ii) $\|A\|_{\infty,\infty} = \max_{i=1,\ldots,m} \|\mathrm{row}_i(A)\|_1$.

iii) $\|A\|_{\infty,1} = \max_{1,\ldots,m} \|A\|_\infty$.

iv) $\|A\|_{2,1} = \mathrm{d}_{\max}^{1/2}(A^\mathrm{T}A)$.

v) $\|A\|_{\infty,2} = \mathrm{d}_{\max}^{1/2}(AA^\mathrm{T})$.

In *iv)* and *v)* $\mathrm{d}_{\max}(X) \triangleq \max_{i=1,\ldots,n} X_{(i,i)}$ for $X \in \mathbb{R}^{n\times n}$.

Problem 2.4 (Pythagorean Theorem). Let $x, y \in \mathbb{R}^n$. Show that if $x^\mathrm{T}y = 0$, then $\|x+y\|_2^2 = \|x\|_2^2 + \|y\|_2^2$.

Problem 2.5. Show that Axiom *iv)* of Definition 2.6 holds as an equality if and only if $x, y \in \mathbb{R}^n$ are linearly dependent.

Problem 2.6. Show that the Euclidean norm satisfies Axioms *i*)–*iv*) of Definition 2.6.

Problem 2.7. Show that equality holds in (2.18) if and only if $|x^{\mathrm{T}}y| = |x|^{\mathrm{T}}|y|$ and

$$\begin{aligned} |x| \circ |y| &= \|y\|_\infty |x|, \quad p = 1, \\ \|y\|_q^{1/p} |x|^{\{1/q\}} &= \|x\|_p^{1/q} |y|^{\{1/p\}}, \quad 1 < p < \infty, \\ |x| \circ |y| &= \|x\|_\infty |y|, \quad p = \infty, \end{aligned}$$

where $|z| \in \mathbb{R}^n$ is a vector whose components are the absolute values of $z \in \mathbb{R}^n$, $\circ$ denotes the Hadamard product (i.e., component-by-component product), and $z^{\{\alpha\}} \triangleq [z_1^\alpha, \ldots, z_n^\alpha]^{\mathrm{T}}$.

Problem 2.8. Let $\mathcal{S} \triangleq \{x \in \mathbb{R}^2 : -1 \le x_1 < 1, -1 \le x_2 < 1\}$. Obtain the closure, interior, and boundary of $\mathcal{S}$. Is $\mathcal{S}$ open? Is $\mathcal{S}$ closed?

Problem 2.9. Show that the union of a finite number of bounded sets is bounded.

Problem 2.10. Let $\mathcal{S} \subseteq \mathbb{R}^n$. Show that $x \in \overline{\mathcal{S}}$ if and only if $\mathcal{B}_\varepsilon(x) \cap \mathcal{S} \neq \emptyset$ for every $\varepsilon > 0$.

Problem 2.11. Show that the sequence $\{x_n\}_{n=1}^\infty$, where $x_n = (1 + 1/n)^n$, is a convergent sequence.

Problem 2.12. Let $\{x_n\}_{n=1}^\infty \subset \mathbb{R}$. For *i*) $x_n = (-1)^n(1 + 1/n)$, *ii*) $x_n = (-1)^n$, *iii*) $x_n = (-1)^n n$, and *iv*) $x_n = n^2 \sin^2(\frac{n\pi}{2})$, compute the limit inferior and limit superior of x_n.

Problem 2.13. Consider the sequence of functions $\{f_n\}_{n=0}^\infty$, where $f_n : \mathbb{R} \to \mathbb{R}$ is given by $f_n(x) = 1/(1 + x^{2n})$. Show that the sequence $\{f_n\}_{n=0}^\infty$ exhibits point wise convergence on $\mathbb{R}$.

Problem 2.14. Consider the sequence of functions $\{f_n\}_{n=1}^\infty$, where $f_n(x) = \ln(1 + x) = x - \frac{1}{2}x^2 + \frac{1}{3}x^3 - \cdots + \frac{(-1)^{n-1}}{n}x^n + \cdots$. Show that this sequence converges for $|x| < 1$ and diverges for $|x| > 1$.

Problem 2.15. Consider the sequence of functions $\{f_n\}_{n=0}^\infty$, where $f_n : [0, 1] \to \mathbb{R}$ is given by $f_n(x) = \sum_{n=0}^\infty \frac{1}{n!}(x^2)^n$. Show that the sequence $\{f_n\}_{n=1}^\infty$ converges uniformly to $f(x) = e^{x^2}$.

Problem 2.16. Consider the sequence of functions $\{f_n\}_{n=1}^\infty$, where $f_n : [0, 1] \to \mathbb{R}$ is given by $f_n(x) = n^2 x(1 - x)^n$. Show that even though the sequence $\{f_n\}_{n=1}^{\infty}$ converges,

$$\lim_{n\to\infty}\int_0^1 f_n(x)\mathrm{d}x \neq \int_0^1 \lim_{n\to\infty} f_n(x)\mathrm{d}x.$$

Problem 2.17. Construct an example of a convergent sequence of differentiable functions $\{f_n(x)\}_{n=1}^{\infty}$ such that, for all $x \in \mathbb{R}$, $\lim_{n\to\infty} f_n'(x)$ does not exist.

Problem 2.18. Let $\{x_n\}_{n=0}^{\infty} \subset \mathbb{R}$ be a scalar sequence. Show that:

i) $\inf x_n \leq \liminf_{n\to\infty} x_n \leq \limsup_{n\to\infty} x_n \leq \sup x_n$.

ii) $\{x_n\}_{n=0}^{\infty}$ converges if and only if $\liminf_{n\to\infty} x_n = \limsup_{n\to\infty} x_n = \lim_{n\to\infty} x_n$.

iii) If $\limsup_{n\to\infty} x_n$ (respectively, $\liminf_{n\to\infty} x_n$) is finite, then show that $\limsup_{n\to\infty} x_n$ (respectively, $\liminf_{n\to\infty} x_n$) is the largest (respectively, smallest) limit point of $\{x_n\}_{n=0}^{\infty}$.

Problem 2.19. Explain why *i)* of Theorem 2.6 may not be true for infinite intersections.

Problem 2.20. Let $f : \mathbb{R} \to \mathbb{R}$ and $g : \mathbb{R} \to \mathbb{R}$ be given by $f(x) = x - 1$ and $g(x) = \sin x$. Find the composite maps $f \circ g$ and $g \circ f$. Is the composition of functions commutative?

Problem 2.21. Let $\mathcal{X} \subseteq \mathbb{R}$ and $\mathcal{Y} \subseteq \mathbb{R}$, and let $f : \mathcal{X} \to \mathcal{Y}$ be given by $f(x) = \sin x$. What is the inverse image of $\mathcal{Y}$ under f if *i)* $\mathcal{X} = \mathcal{Y} = \mathbb{R}$ and *ii)* $\mathcal{X} = [-\pi/2, \pi/2]$ and $\mathcal{Y} = \mathbb{R}$.

Problem 2.22. Let $A \in \mathbb{R}^{n\times n}$ and define the map $f : \mathbb{R}^{n\times n} \to \mathbb{R}$ by $f(A) = \det A$. Show that f is surjective. Is f bijective? What is $f^{-1}(\mathbb{R}\setminus\{0\})$?

Problem 2.23. Let $f : \mathbb{R} \to \mathbb{R}$ be given by $f(x) = x^2 - 1$. Find the image of $\mathcal{X}$ under f, where $\mathcal{X} = [-1, 2]$ and $\mathcal{X} = \mathbb{R}$. Is f surjective, injective, or bijective?

Problem 2.24. Let $f : \mathcal{X} \to \mathcal{Y}$ and let $\mathcal{U} \subseteq \mathcal{X}$ be an open set. Prove or refute that $f(\mathcal{U})$ is open in $\mathcal{Y}$.

Problem 2.25. Let $\mathcal{X} \subseteq \mathbb{R}^n$ and $\mathcal{Y} \subseteq \mathbb{R}^n$, and let $f : \mathcal{X} \to \mathcal{Y}$. Show that f is injective if and only if $f(\mathcal{Q}) \cap f(\mathcal{X}\setminus\mathcal{Q}) = \emptyset$ for all $\mathcal{Q} \subseteq \mathcal{X}$.

Problem 2.26. Let $\mathcal{X}$ and $\mathcal{Y}$ be sets, let $\mathcal{A}$, $\mathcal{A}_1$, and $\mathcal{A}_2$ be subsets of

$\mathcal{X}$, and let $\mathcal{B}$ be a subset of $\mathcal{Y}$. Show that the following statements hold:

i) $f(\mathcal{A}_1 \cap \mathcal{A}_2) \subseteq f(\mathcal{A}_1) \cap f(\mathcal{A}_2)$.

ii) $f^{-1}(f(\mathcal{A})) \supseteq \mathcal{A}$.

iii) $f(f^{-1}(\mathcal{B})) \subseteq \mathcal{B}$.

iv) $f(\mathcal{A}_1 \cap \mathcal{A}_2) = f(\mathcal{A}_1) \cap f(\mathcal{A}_2)$ if and only if f is injective.

v) $f^{-1}(f(\mathcal{A})) = \mathcal{A}$ if and only if f is injective.

vi) $f(f^{-1}(\mathcal{B})) = \mathcal{B}$ if and only if f is surjective.

Problem 2.27. Prove or refute that the union of convex sets is convex.

Problem 2.28. Let $\mathcal{C} \subset \mathbb{R}^n$ and $\mathcal{Q} \subset \mathbb{R}^n$ be convex sets. Show that:

i) $\alpha\mathcal{C} = \{x \in \mathbb{R}^n : x = \alpha y, y \in \mathcal{C}\}$ is convex, where $\alpha \in \mathbb{R}$.

ii) $\mathcal{C} + \mathcal{Q}$ is convex.

Problem 2.29. Let $\|\cdot\| : \mathbb{R}^n \to \mathbb{R}$ be a vector norm on $\mathbb{R}^n$. Show that $\|\cdot\|$ is convex.

Problem 2.30. Let $\mathcal{I}$ be an index set, $\mathcal{C} \subset \mathbb{R}^n$ be a convex set, and $f_i : \mathcal{C} \to \mathbb{R}$ be a convex function for each $i \in \mathcal{I}$. Show that $g : \mathcal{C} \to \mathbb{R} \cup \{\infty\}$ given by $g(x) = \sup_{i \in \mathcal{I}} f_i(x)$ is convex.

Problem 2.31. Let $\mathcal{C} \subset \mathbb{R}^n$ be a convex set and let $f : \mathcal{C} \to \mathbb{R}$ be continuously differentiable. Show that:

i) f is convex if and only if

$$f(y) \geq f(x) + (y - x)^{\mathrm{T}} f'(x), \qquad x, y \in \mathcal{C}. \tag{2.249}$$

ii) If the inequality in (2.249) is strict whenever $x \neq y$, then f is strictly convex.

Problem 2.32. Let $\mathcal{C} \subset \mathbb{R}^n$ be a convex set, let $f : \mathcal{C} \to \mathbb{R}$ be two-times continuously differentiable, and let P be an $n \times n$ symmetric matrix. Show that:

i) If $f''(x) \geq 0$, $x \in \mathcal{C}$, then f is convex.

ii) If $f''(x) > 0$, $x \in \mathcal{C}$, $x \neq 0$, then f is strictly convex.

iii) If $\mathcal{C} = \mathbb{R}^n$ and f is convex, then $f''(x) \geq 0$, $x \in \mathcal{C}$.

iv) $f(x) = x^{\mathrm{T}} Px$ is convex if and only if $P \geq 0$.

v) $f(x) = x^{\mathrm{T}} Px$ is strictly convex if and only if $P > 0$.

Problem 2.33. Let $f : \mathbb{R}^n \to \mathbb{R}^n$ be continuously differentiable and let $\alpha > 0$. Show that if

$$[f'(x) - f'(y)]^{\mathrm{T}}(x - y) \geq \alpha \|x - y\|^2, \qquad x, y \in \mathbb{R}^n, \tag{2.250}$$

then f is strictly convex. Furthermore, show that if f is two-times continuously differentiable, then (2.250) is equivalent to $f''(x) \geq \alpha I_n$, for every $x \in \mathbb{R}^n$.

Problem 2.34. Let $\mathcal{C} \subset \mathbb{R}^n$ be a convex set and let $f : \mathcal{C} \to \mathbb{R}$ be a convex function. Show that if $x \in \mathcal{C}$ is a local minimum of f, then $x \in \mathcal{C}$ is also a global minimum of f. If, in addition, f is strictly convex, show that there exists at most one global minimum of f.

Problem 2.35. Show that the function $f : \mathbb{R} \to \mathbb{R}$ given by

$$f(x) = \begin{cases} x^2 \sin(\frac{1}{x}), & x \neq 0, \\ 0, & x = 0, \end{cases}$$

is continuous at $x = 0$.

Problem 2.36. Show that the function $f : \mathbb{R}^2 \to \mathbb{R}$ given by

$$f(x_1, x_2) = \begin{cases} x_1^2 x_2^2/(x_1^4 + x_2^4), & (x_1, x_2) \neq (0, 0), \\ 0, & (x_1, x_2) = (0, 0), \end{cases}$$

is discontinuous at $(x_1, x_2) = (0, 0)$.

Problem 2.37. Show that the function $f : \mathbb{R} \to \mathbb{R}$ given by $f(x) = \sin(\frac{1}{x})$, $x \neq 0$, and $f(0) = \alpha$, $\alpha \in \mathbb{R}$, is discontinuous at $x = 0$.

Problem 2.38. Show that the function $f : \mathbb{R}^2 \to \mathbb{R}$ given by

$$f(x_1, x_2) = \begin{cases} (x_1^2 + x_2)^4/(x_1^2 + x_2^2), & (x_1, x_2) \neq (0, 0), \\ 0, & (x_1, x_2) = (0, 0), \end{cases}$$

is continuous at $(x_1, x_2) = (0, 0)$.

Problem 2.39. Let $f : \mathbb{R}^m \to \mathbb{R}^n$ be given by $f(x) = Ax$, where $x \in \mathbb{R}^m$ and $A \in \mathbb{R}^{n \times m}$. Show that f is continuous on $\mathbb{R}^m$. Furthermore, for $m = n$ and $\det A \neq 0$, show that f is a diffeomorphism.

Problem 2.40. Let $f : \mathbb{R} \to \mathbb{R}$ be given by

$$f(x) = \begin{cases} x\sin(\frac{1}{x}), & x \neq 0, \\ 0, & x = 0. \end{cases}$$

Show that f is not differentiable at $x = 0$.

Problem 2.41. Let $f : [a,b] \to \mathbb{R}$, where $[a,b] \subset \mathbb{R}$, be such that f is differentiable on (a,b) and $f'(x) > 0$, $x \in (a,b)$. Show that $f^{-1} : [f(a), f(b)] \to \mathbb{R}$ is differentiable on $(f(a), f(b))$.

Problem 2.42. Let $f : (0,\infty) \to \mathbb{R}$ be given by

$$f(x) = \begin{cases} 1/x, & 0 < x \leq 1, \\ 1/2, & x > 1. \end{cases}$$

Show that f is left continuous, piecewise continuous, left piecewise continuous, and upper semicontinuous on $(0,\infty)$.

Problem 2.43. Let $f : \mathbb{R} \to \mathbb{R}$ be given by

$$f(x) = \begin{cases} \sin(1/x), & x < 0, \\ 0, & x \geq 0. \end{cases}$$

Show that f is right continuous on $\mathbb{R}$. Is f piecewise continuous on $\mathbb{R}$?

Problem 2.44. Let $f : \mathbb{R} \to \mathbb{R}$ be given by $f(x) = |x|$. Show that f is continuous on $\mathbb{R}$ and differentiable at all $x \in \mathbb{R}$, $x \neq 0$. Furthermore, show that f is piecewise continuously differentiable on $\mathbb{R}$. Does $f'(0)$ exist?

Problem 2.45. Let $f : \mathbb{R} \to \mathbb{R}$ and $g : \mathbb{R} \to \mathbb{R}$ be such that f and g are Lipschitz continuous on $\mathbb{R}$. Show that $f + g$, fg, and $g \circ f$ are Lipschitz continuous on $\mathbb{R}$.

Problem 2.46. Let $\|\cdot\| : \mathbb{R}^n \to \mathbb{R}$ be a vector norm on $\mathbb{R}^n$. Show that $\|\cdot\|$ is continuous on $\mathbb{R}^n$. Is $\|\cdot\|$ uniformly continuous on $\mathbb{R}^n$?

Problem 2.47. Let $\mathcal{X} \subseteq \mathbb{R}^n$, $\mathcal{Y} \subseteq \mathbb{R}^m$, and $\mathcal{Z} \subseteq \mathbb{R}^p$, and let $f : \mathcal{X} \to \mathcal{Y}$ and $g : \mathcal{Y} \to \mathcal{Z}$ be continuous functions. Show that $g \circ f : \mathcal{X} \to \mathcal{Z}$ is continuous on $\mathcal{X}$.

Problem 2.48. Let $\mathcal{D} \subset \mathbb{R}^n$ and let $\|\cdot\| : \mathcal{D} \to \mathbb{R}$ and $\|\cdot\|' : \mathcal{D} \to \mathbb{R}$ be vector norms on $\mathcal{D}$. Show that if $\mathcal{D}$ is open (respectively, closed, bounded, compact) under $\|\cdot\|$, then $\mathcal{D}$ is open (respectively, closed, bounded, compact) under $\|\cdot\|'$.

Problem 2.49. A function $f : \mathbb{R}^m \to \mathbb{R}^n$ is called *homogeneous* (respectively, *positively homogeneous*) *of degree* r if $f(\lambda x) = \lambda^r f(x)$ for all $\lambda \in \mathbb{R}$ (respectively, $\lambda > 0$). Show that if $0 < r < 1$, then f is not Lipschitz continuous at $x = 0$.

Problem 2.50. Let $f : \mathbb{R}^n \to \mathbb{R}$. Then f is *Hölder continuous* with exponent $\alpha > 0$ at $x \in \mathbb{R}^n$ if there exist a constant $k > 0$ (called the *Hölder constant*) and an open neighborhood $\mathcal{D}$ of x such that

$$|f(x) - f(y)| \leq k\|x - y\|^\alpha, \qquad x \in \mathcal{D}. \tag{2.251}$$

f is simply said to be Hölder continuous at x if f is Hölder continuous at x with some exponent $\alpha > 0$. Show that Hölder continuity at x implies continuity at x. In addition, show that if $\alpha > 1$, then Hölder continuity at x implies differentiability at x.

Problem 2.51. Let $f : \mathbb{R}^n \to \mathbb{R}$ be two-times continuously differentiable on $\mathcal{B}_\varepsilon(x) \subset \mathbb{R}^n$. Show that:

i) For all $y \in \mathbb{R}^n$ such that $x + y \in \mathcal{B}_\varepsilon(x)$,

$$f(x+y) = f(x) + y^\mathrm{T} f'(x) + \frac{1}{2} y^\mathrm{T} \left[\int_0^1 \left(\int_0^t f''(x + \sigma y) \mathrm{d}\sigma \right) \mathrm{d}t \right] y. \tag{2.252}$$

ii) For all $y \in \mathbb{R}^n$ such that $x + y \in \mathcal{B}_\varepsilon(x)$, there exists $\alpha \in [0, 1]$ such that

$$f(x+y) = f(x) + y^\mathrm{T} f'(x) + \frac{1}{2} y^\mathrm{T} f''(x + \alpha y) y. \tag{2.253}$$

iii) For all $y \in \mathbb{R}^n$ such that $x + y \in \mathcal{B}_\varepsilon(x)$,

$$f(x+y) = f(x) + y^\mathrm{T} f'(x) + \frac{1}{2} y^\mathrm{T} f''(x) y + \mathcal{O}(\|y\|^2). \tag{2.254}$$

Problem 2.52. Let $\mathcal{D} \subset \mathbb{R}^n$ be bounded, let $\varepsilon > 0$ be such that $\|x - y\| < \varepsilon$ for all $x, y \in \mathcal{D}$, and let $z \in \mathcal{D}$. Show that $\mathcal{D} \subseteq \mathcal{B}_\varepsilon(z)$.

Problem 2.53. Let $\mathcal{C} \subseteq \mathbb{R}^n$ be a convex set and let $f : \mathcal{C} \to \mathbb{R}$ be a convex function. Show that the $[0, \beta]$-sublevel set of f is convex.

Problem 2.54. Let $\mathcal{D} \subseteq \mathbb{R}^m$ be a connected set and let $f : \mathcal{D} \to \mathbb{R}^n$ be continuous. Show that the image of $\mathcal{D}$ under f is connected.

Problem 2.55. Let $\mathcal{C} \subseteq \mathbb{R}^n$ be a convex set and let $f : \mathcal{C} \to \mathbb{R}$ be a convex function. Show that f is continuous on $\overset{\circ}{\mathcal{C}}$.

Problem 2.56. Show that every $[\alpha, \beta]$-sublevel set of a continuous function is closed.

Problem 2.57. Show that every hyperplane is a subspace.

Problem 2.58. Let $\mathcal{D} \subseteq \mathbb{R}^n$ and let $f : \mathcal{D} \to \mathbb{R}^n$. Show that the following statements are equivalent:

i) f is continuous.

ii) For all closed $\mathcal{Q} \subseteq \mathbb{R}^n$, the inverse image of $\mathcal{Q}$ under f is closed relative to $\mathcal{D}$.

Problem 2.59. Let $\mathcal{D} \subseteq \mathbb{R}^m$, let $f : \mathcal{D} \to \mathbb{R}^n$ be continuous on $\mathcal{Q} \subseteq \mathcal{D}$, and assume that $\mathcal{Q}$ is bounded and $\partial\mathcal{Q} \subseteq \mathcal{D}$. Show that f is uniformly continuous on $\mathcal{Q}$.

Problem 2.60. Let $f : (\alpha, \beta) \to \mathbb{R}$, where $\alpha, \beta \in \mathbb{R}$. Show that the following statements hold:

i) If f is strictly convex, then f is continuous.

ii) If f is strictly increasing or strictly decreasing, then f is injective.

iii) If f is injective and strictly convex, then f is strictly increasing or strictly decreasing.

Problem 2.61. Let $f : [0, \infty) \to [0, \infty)$. Assume that f is differentiable, $f \in \mathcal{L}([0, \infty))$, and assume that there exists $\alpha \in [0, \infty)$ such that $|f'(x)| \leq \alpha$ for all $x \in [0, \infty)$. Show that $\lim_{x\to\infty} f(x) = 0$.

Problem 2.62. Let $\mathcal{C} \subseteq \mathbb{R}^n$ be a convex set and $f : \mathcal{C} \to \mathbb{R}^n$ be continuously differentiable. Furthermore, suppose $0 \in \mathcal{C}$ and $f(0) = 0$. Show that

$$f(x) = \int_0^1 \frac{\partial f}{\partial x}(\sigma x)x\mathrm{d}\sigma, \qquad x \in \mathcal{C}. \tag{2.255}$$

Problem 2.63. Show that if $f : \mathbb{R}^n \to \mathbb{R}^n$ is Lipschitz continuous on $\mathcal{B}_\varepsilon(0) = \{x \in \mathbb{R}^n : \|x\| < \varepsilon\}$ and $f(0) = 0$, then there exists $L > 0$ such that $\|f(x)\| \leq L\|x\|$ for all $x \in \mathcal{B}_\varepsilon(0)$.

Problem 2.64. Let $\|\cdot\|_p$ and $\|\cdot\|_q$, where $p, q \in [1, \infty]$, denote two vector norms on $\mathbb{R}^n$. Show that $f : \mathbb{R}^n \to \mathbb{R}^n$ is Lipschitz continuous under $\|\cdot\|_p$ if and only if $f(\cdot)$ is Lipschitz continuous under $\|\cdot\|_q$.

Problem 2.65. Show that if $f : \mathcal{D} \to \mathbb{R}^n$ is continuously differentiable at $x_0 \in \mathcal{D}$, then there exists $\delta > 0$ and $L > 0$ such that $\|f(x) - f(x_0)\| < L\|x - x_0\|$ for all $x \in \mathcal{B}_\delta(x_0)$.

Problem 2.66. Let $(\mathcal{C}[0,T], ||| \cdot |||_{2,2})$ denote the space of continuous functions $f : [0,T] \to \mathbb{R}$ with norm

$$|||f|||_{2,2} = \left(\int_0^T |f(t)|^2 \mathrm{d}t \right)^{1/2}.$$

Show that $(\mathcal{C}[0,T], ||| \cdot |||_{2,2})$ is not a Banach space. (**Hint:** Consider the sequence of functions $\{f_n\}_{n=1}^{\infty} \subset \mathcal{C}[0,T]$, where $f_n(t) = t/(|t| + 1/n)$.)

Problem 2.67. Let $\mathcal{X} \neq \emptyset$ be a set. Show that $\mathcal{F} = \{f : \mathcal{X} \to \mathcal{X}\}$ is a semigroup with identity under operation of composition $g \circ f$, where $f, g \in \mathcal{F}$.

Problem 2.68. Let $\mathbb{Q} = \{x : x = p/q,\ p, q \in \mathbb{Z},\ q \neq 0\}$. Show that $\mathbb{Q}$ is an Abelian group with respect to addition $(+)$ on $\mathbb{R}$, and $\mathbb{Q} \setminus \{0\}$ is an Abelian group with respect to multiplication $(\cdot)$ on $\mathbb{R}$.

Problem 2.69. Show that every subspace of $\mathbb{R}^n$ is closed.

Problem 2.70. Let $x, y \in \mathbb{R}^n$ and $\mathcal{S} \subset \mathbb{R}^n$. Then x and y are *orthogonal* if $x^{\mathrm{T}} y = 0$. Furthermore, the *orthogonal complement* of $\mathcal{S}$ is defined as $\mathcal{S}^{\perp} \triangleq \{x \in \mathbb{R}^n : x^{\mathrm{T}} y = 0, y \in \mathcal{S}\}$. Show that if $\mathcal{S}_1$ and $\mathcal{S}_2$ are subspaces of $\mathbb{R}^n$ with $\mathcal{S}_1 \subseteq \mathcal{S}_2$, then $\mathcal{S}_2^{\perp} \subseteq \mathcal{S}_1^{\perp}$.

Problem 2.71. Let $\mathcal{S}$ be a collection of bounded, measurable subsets of $\mathbb{R}$ with measure $\mu : \mathcal{S} \to [0, \infty)$. Show that the following statements hold:

i) $\mu(\emptyset) = 0$.

ii) $\mu([a,b)) = \mu([a,b]) = \mu((a,b)) = \mu((a,b]) = b - a$.

iii) If $\mathcal{S}_1, \mathcal{S}_2 \in \mathcal{S}$, then

$$\mu(\mathcal{S}_1 \cup \mathcal{S}_2) = \mu(\mathcal{S}_1) + \mu(\mathcal{S}_2) - \mu(\mathcal{S}_1 \cap \mathcal{S}_2).$$

iv) If $\{\mathcal{S}_i\}_{i=1}^{\infty} \subseteq \mathbb{R}$ is such that $\mathcal{S}_{i+1} \subseteq \mathcal{S}_i$ and $\mathcal{S}_i \in \mathcal{S}$, then

$$\mu\left(\bigcap_{i=1}^{\infty} \mathcal{S}_i \right) = \mu(\lim_{i \to \infty} \mathcal{S}_i) = \lim_{i \to \infty} \mu(\mathcal{S}_i).$$

v) If $\{\mathcal{S}_i\}_{i=1}^{\infty} \subseteq \mathbb{R}$ is such that $\mathcal{S}_i \cap \mathcal{S}_j = \emptyset$ and $\mathcal{S}_i \in \mathcal{S}$ for $i, j = 1, 2, \ldots,$ $i \neq j$, then

$$\mu\left(\bigcup_{i=1}^{\infty} \mathcal{S}_i \right) = \sum_{i=1}^{\infty} \mu(\mathcal{S}_i).$$

Problem 2.72. Let $\mathcal{S} = \{a\}$, where $a \in \mathbb{R}$. Show that $\mu(\mathcal{S}) = 0$. (**Hint:** Use the properties in Problem 2.71.)

Problem 2.73. Let $\mathcal{S} = \bigcup_i \mathcal{S}_i$ be a finite or countable union of sets in $\mathbb{R}^n$ where $\mu(\mathcal{S}_i) = 0$ for each i. Show that $\mu(\mathcal{S}) = 0$.

Problem 2.74. Show that the following statements hold:

i) If $\mathcal{S}_1 \subset \mathcal{S}_2$, then $\mu^*(\mathcal{S}_1) \leq \mu^*(\mathcal{S}_2)$.

ii) If $\mathcal{S} = \bigcup_{i=1}^{\infty} \mathcal{S}_i$, then $\mu^*(\mathcal{S}) \leq \sum_{i=1}^{\infty} \mu^*(\mathcal{S}_i)$.

iii) A countable union of measurable sets is measurable.

iv) Closed sets are measurable.

v) The complement of a measurable set is measurable.

vi) A countable intersection of measurable sets is measurable.

vii) Let $\{\mathcal{S}_i\}_{i=1}^{\infty} \subseteq \mathbb{R}^n$ be such that $\mathcal{S}_i \cap \mathcal{S}_j = \emptyset$ for $i, j = 1, 2, \ldots, j$, $i \neq j$, and $\mathcal{S} = \bigcup_{i=1}^{\infty} \mathcal{S}_i$. Then $\mu(\mathcal{S}) = \sum_{i=1}^{\infty} \mu(\mathcal{S}_i)$.

Problem 2.75. Give a function that is Riemann integrable but not Lebesgue integrable. (**Hint:** Consider improper integrals.)

Problem 2.76. Let $\mathcal{D} \subseteq \mathbb{R}^n$ be measurable, and let $f : \mathcal{D} \to \mathbb{R}$. Show that the following statements are equivalent:

i) f is measurable.

ii) $\{x \in \mathcal{D} : f(x) > \alpha\}$ is measurable for every $\alpha \in \mathbb{R}$.

iii) $\{x \in \mathcal{D} : f(x) \geq \alpha\}$ is measurable for every $\alpha \in \mathbb{R}$.

iv) $\{x \in \mathcal{D} : f(x) < \alpha\}$ is measurable for every $\alpha \in \mathbb{R}$.

v) $\{x \in \mathcal{D} : f(x) \leq \alpha\}$ is measurable for every $\alpha \in \mathbb{R}$.

Problem 2.77. Let $\mathcal{D} \subseteq \mathbb{R}^n$ be measurable, and let $\mathcal{I}$ be a dense subset of $\mathbb{R}$. Show that if $\{x \in \mathcal{D} : f(x) > \alpha\}$ is measurable for all $\alpha \in \mathcal{I}$, then $f : \mathcal{D} \to \mathbb{R}$ is measurable.

Problem 2.78. Let $f : \mathcal{D} \to \mathbb{R}$ and $g : \mathcal{D} \to \mathbb{R}$ be measurable, where $\mathcal{D} \subseteq \mathbb{R}^n$ is measurable. Show that the following statements hold:

i) $(f + g) : \mathcal{D} \to \mathbb{R}$ is measurable.

ii) $fg : \mathcal{D} \to \mathbb{R}$ is measurable.

iii) If $g(x) \neq 0$ for almost all $x \in \mathcal{D}$, then $(f/g) : \mathcal{D}\backslash\{x \in \mathcal{D} : g(x) = 0\} \to \mathbb{R}$ is measurable.

iv) f^p, where $p \in \mathbb{Z}_+$, is measurable.

Problem 2.79. Let $\{f_n\}_{n=1}^{\infty}$ be a sequence of measurable functions on $\mathcal{D} \subseteq \mathbb{R}^n$. Show that the following statements hold:

i) $\sup_n f_n(x)$ is measurable.

ii) $\inf_n f_n(x)$ is measurable.

iii) $\limsup_{n\to\infty} f_n(x)$ is measurable.

iv) $\liminf_{n\to\infty} f_n(x)$ is measurable.

v) If $\lim_{n\to\infty} f_n(x) = f(x)$, then $f(x)$ is measurable.

Problem 2.80. Prove or refute that the composition of two measurable functions is measurable.

Problem 2.81 (Bounded Convergence Theorem). Let $\{f_n\}_{n=1}^{\infty}$ be a sequence of measurable functions on $\mathcal{D} \subset \mathbb{R}^n$, where $\mathcal{D}$ is measurable with $\mu(\mathcal{D}) < \infty$, such that $\lim_{n\to\infty} f_n = f$ almost everywhere on $\mathcal{D}$. Assume there exists $\alpha \geq 0$ such that $|f_n(x)| \leq \alpha$ for all $x \in \mathcal{D}$ and $n \in \mathbb{Z}_+$. Show that $f \in \mathcal{L}(\mathcal{D})$ and

$$\lim_{n\to\infty} \int_{\mathcal{D}} f_n(x)\mathrm{d}x = \int_{\mathcal{D}} f(x)\mathrm{d}x.$$

Problem 2.82 (Monotone Convergence Theorem). Let $\{f_n\}_{n=1}^{\infty}$ be a sequence of measurable functions on $\mathcal{D} \subset \mathbb{R}^n$ such that $f_{n+1}(x) \geq f_n(x)$ for all $x \in \mathcal{D}$ and $n \in \mathbb{Z}_+$. Assume that $\lim_{n\to\infty} f_n(x) = f$ almost everywhere on $\mathcal{D}$. Show that $f \in \mathcal{L}(\mathcal{D})$ and

$$\lim_{n\to\infty} \int_{\mathcal{D}} f_n(x)\mathrm{d}x = \int_{\mathcal{D}} f(x)\mathrm{d}x.$$

Problem 2.83 (Dominated Convergence Theorem). Let $g \in \mathcal{L}(\mathcal{D})$ and let $\{f_n\}_{n=1}^{\infty}$ be a sequence of measurable functions on $\mathcal{D}$ such that $|f_n(x)| \leq g(x)$ for all $x \in \mathcal{D}$. Assume that $\lim_{n\to\infty} f_n = f$ almost everywhere on $\mathcal{D}$. Show that $f \in \mathcal{L}(\mathcal{D})$ and

$$\lim_{n\to\infty} \int_{\mathcal{D}} f_n(x)\mathrm{d}x = \int_{\mathcal{D}} f(x)\mathrm{d}x.$$

Problem 2.84. Let $\mathcal{X}$ be a normed linear space with norm $\|\cdot\| : \mathcal{X} \to \mathbb{R}$. Show that $\|x\| - \|y\| \leq \|x - y\|$, $x, y \in \mathcal{X}$.

Problem 2.85. Show that every Cauchy sequence in a normed linear space is bounded.

Problem 2.86. Let $\mathcal{X}$ be a normed linear space with norm $\|\cdot\| : \mathcal{X} \to \mathbb{R}$. Show that $\|\cdot\|$ is uniformly continuous.

Problem 2.87. An *inner product space* is a linear vector space $\mathcal{X}$ with associated field $\mathbb{F}$ and mapping $<\cdot,\cdot>: \mathcal{X} \times \mathcal{X} \to \mathbb{F}$ such that the following axioms hold:

i) $< x, x > \geq 0$, $x \in \mathcal{X}$.

ii) $< x, x > = 0$ if and only if $x = 0$.

iii) $< x, \alpha y > = \alpha < x, y >$, $x, y \in \mathcal{X}$ and $\alpha \in \mathbb{F}$.

iv) $< x, y + z > = < x, y > + < x, z >$, $x, y, z \in \mathcal{X}$.

v) $< x, y > = \overline{< y, x >}$, $x, y \in \mathcal{X}$.

For $\|\cdot\| : \mathcal{X} \to \mathbb{R}$ defined by $\|x\| = < x, x >^{1/2}$ show that $\|\cdot\|$ is a norm on $\mathcal{X}$, and hence, $\mathcal{X}$ is a normed linear space. (**Hint:** First show that $| < x, y > | \leq \|x\| \|y\|$, where x and y belong to the inner product space.)

Problem 2.88. Let $\mathcal{X}$ be an inner product space (see Problem 2.87) with inner product $<\cdot,\cdot>: \mathcal{X} \times \mathcal{X} \to \mathbb{R}$. Show that for each $y \in \mathcal{X}$, the function mapping x into $< x, y >$ is uniformly continuous.

Problem 2.89. A complete inner product space (see Problem 2.87) with norm defined by the inner product is called a *Hilbert space*. Discuss the Hilbert space $\mathbb{R}^n$. What are the inner product and corresponding norm?

Problem 2.90. Let $f : [0, \infty) \to \mathbb{R}^{n \times m}$. Which matrix norm (as a spatial norm) makes $\mathcal{L}_2$ a Hilbert space (see Problem 2.89)?

Problem 2.91. Let $\mathcal{X} = \mathbb{R}$ and define $T : \mathbb{R} \to \mathbb{R}$ by $T(x) = x - \tan^{-1}(x) + \pi/2$. Show that $T(\cdot)$ satisfies (2.128) but T has no fixed point in $\mathbb{R}$.

Problem 2.92. Let $\mathcal{X}$ be a Banach space with norm $\|\cdot\| : \mathcal{X} \to \mathbb{R}$, let $\mathcal{S}$ be a subset of $\mathcal{X}$, and let $T : \mathcal{S} \to \mathcal{X}$. Suppose there exists a constant $\rho \in [0, 1)$ such that

$$\|T(x) - T(y)\| \leq \rho \|x - y\|, \quad x, y \in \mathcal{S}, \tag{2.256}$$

and suppose there exists $x_0 \in \mathcal{S}$ such that

$$\mathcal{B} = \left\{ x \in \mathcal{X} : \ \|x - x_0\| \leq \frac{\|T(x) - T(x_0)\|}{1 - \rho} \right\} \subset \mathcal{S}. \tag{2.257}$$

Show that there exists unique $x^* \in \mathcal{S}$ such that $T(x^*) = x^*$. Furthermore, show that for each $x_0 \in \mathcal{S}$, the sequence $\{x_n\}_{n=0}^{\infty} \subset \mathcal{S}$ defined by $x_{n+1} = T(x_n)$ converges to x^*. Finally, show that

$$\|x^* - x_n\| \leq \frac{\rho^n}{1-\rho} \|T(x_0) - x_0\|, \quad n \geq 0. \tag{2.258}$$

Problem 2.93 (Schauder Fixed Point Theorem). Let $\mathcal{C} \subseteq \mathbb{R}^n$ be a nonempty, convex, and closed set, let $f : \mathcal{C} \to \mathcal{C}$ be continuous, and assume $f(\mathcal{C})$ is bounded. Show that there exists $x \in \mathcal{C}$ such that $f(x) = x$.

Problem 2.94 (Brouwer Fixed Point Theorem). Let $\mathcal{C} \subseteq \mathbb{R}^n$ be a nonempty, compact, and convex set, and let $f : \mathcal{C} \to \mathcal{C}$ be continuous. Show that there exists $x \in \mathcal{C}$ such that $f(x) = x$.

Problem 2.95. Let $\mathcal{X}$ be a Banach space with norm $\|\cdot\|' : \mathcal{X} \to \mathbb{R}$, let $T : \mathcal{X} \to \mathcal{X}$, let $I : \mathcal{X} \to \mathcal{X}$ be the identity operator, and let $\|T\| < 1$, where

$$\|T\| \triangleq \sup_{x \in \mathcal{X},\, x \neq 0} \frac{\|Tx\|'}{\|x\|'}.$$

Show that the range of $I - T$ is $\mathcal{X}$ and $(I - T)^{-1}$ exists and is bounded, and satisfies

$$\|(I - T)^{-1}\| \leq \frac{1}{1 - \|T\|}.$$

Furthermore, show that

$$\sum_{k=0}^{\infty} T^k = (I - T)^{-1}.$$

Problem 2.96. Consider the nonlinear dynamical system given by

$$\dot{x}_1(t) = x_2(t), \quad x_1(0) = x_{10}, \quad t \geq 0, \tag{2.259}$$
$$\dot{x}_2(t) = -\mathrm{sgn}[x_1(t) + x_2(t)], \quad x_2(0) = x_{20}, \tag{2.260}$$

where $\mathrm{sgn}(x) \triangleq x/|x|$, $x \neq 0$, and $\mathrm{sgn}(0) \triangleq 0$. Show that the dynamical system (2.259) and (2.260) has a unique solution for all initial conditions. Furthermore, show that the solution reaches the manifold $\{(x_1, x_2) \in \mathbb{R} \times \mathbb{R} : x_1 + x_2 = 0\}$ in finite time and slides on the manifold towards the origin for large values of time.

Problem 2.97. Consider the scalar nonlinear dynamical system

$$\dot{x}(t) = f(x(t)), \quad x(0) = 0, \quad t \geq 0, \tag{2.261}$$

where

$$f(x) \triangleq \begin{cases} -1 - x^2, & x \geq 0, \\ 1 + x^2, & x < 0. \end{cases} \tag{2.262}$$

Show that there does not exist a continuously differentiable function $x(\cdot)$ satisfying (2.261).

Problem 2.98. Consider the nonlinear dynamical system (2.136) and let $s(t, x)$, $t \geq 0$, denote the solution to (2.136) with initial condition $x(0) = x_0$. Show that if $f : \mathcal{D} \to \mathbb{R}^n$ is Lipschitz continuous on $\mathcal{D}$, then for every $\varepsilon, T > 0$, and $x_0 \in \mathcal{D}$, there exists $\delta(\varepsilon, T, x_0) > 0$ such that if $\|x_0 - y\| < \delta(\varepsilon, T, x_0)$, $y \in \mathcal{D}$, then $\|s(t, x_0) - s(t, y)\| < \varepsilon$ for all $t \in [0, T]$.

Problem 2.99. Let $x : (\alpha, \beta) \to \mathbb{R}^n$ be a continuously differentiable function on (α, β). Show that $\frac{\mathrm{d}}{\mathrm{d}t}\|x(t)\|_2 \leq \|\dot{x}(t)\|_2$, $t \in (\alpha, \beta)$.

Problem 2.100. Consider the nonlinear dynamical system (2.136) with $f : \mathcal{D} \to \mathbb{R}^n$ continuously differentiable on $\mathcal{D}$. Let $x_1(t)$ and $x_2(t)$ be two solutions of (2.136) over intervals $\mathcal{I}_1$ and $\mathcal{I}_2$, respectively. Show that $t_0 \in \mathcal{I}_1 \cap \mathcal{I}_2$ and if $\mathcal{I} \subset \mathcal{I}_1 \cap \mathcal{I}_2$ is any open interval containing t_0, show that $x_1(t) = x_2(t)$ for all $t \in \mathcal{I}$.

Problem 2.101. Consider the nonlinear dynamical system (2.136). *Picard's method* of successive approximations is based on the fact that $x(t)$ is a solution to (2.136) if and only if $x(t)$ is a solution to (2.137). In particular, the successive approximations of the solution to (2.137) are given by the sequence of functions

$$y_0(t) = x_0, \tag{2.263}$$

$$y_{n+1}(t) = x_0 + \int_{t_0}^{t} f(y_n(s))\mathrm{d}s, \tag{2.264}$$

for $n = 0, 1, \ldots$. Use Picard's method of successive approximations to show that successive approximations for the linear dynamical system

$$\dot{x}(t) = Ax(t), \quad x(0) = x_0, \quad t \geq 0, \tag{2.265}$$

where $x(t) \in \mathbb{R}^n$, $t \geq 0$, and $A \in \mathbb{R}^{n \times n}$, converge to $x(t) = e^{At}x_0$, $t \geq 0$.

Problem 2.102. Prove Theorem 2.24 using Picard's method of successive approximations (see Problem 2.101).

Problem 2.103. Consider the nonlinear dynamical system (2.136). Show that if $\lim_{t \to \tau_{\max}} x(t) = x^*$ exists and $x^* \in \mathcal{D}$, then $\tau_{\max} = \infty$.

Moreover, show that x^* is an equilibrium point of (2.136).

Problem 2.104 (Gronwall-Bellman Lemma). Assume there exist continuous functions $\alpha, x : \mathbb{R} \to \mathbb{R}$ and $\beta : \mathbb{R} \to \overline{\mathbb{R}}_+$ such that

$$x(t) \le \alpha(t) + \int_{t_0}^{t} \beta(s)x(s)\mathrm{d}s, \quad t \ge t_0 \tag{2.266}$$

Show that

$$x(t) \le \alpha(t) + \int_{t_0}^{t} \alpha(s)\beta(s)e^{\int_s^t \beta(\tau)\mathrm{d}\tau}\mathrm{d}s, \quad t \ge t_0. \tag{2.267}$$

Furthermore, if $\alpha(t) \equiv \alpha \in \mathbb{R}$, then show that

$$x(t) \le \alpha e^{\int_{t_0}^t \beta(s)\mathrm{d}s}, \quad t \ge t_0. \tag{2.268}$$

Finally, if, in addition, $\beta(t) \equiv \beta \ge 0$ show that $x(t)$, $t \ge t_0$, satisfies (2.153).

Problem 2.105. Consider the nonlinear dynamical system

$$\dot{x}(t) = f(x(t)), \quad x(0) = x_0, \quad t \ge 0, \tag{2.269}$$
$$\dot{x}(t) = g(x(t)), \quad x(0) = x_0, \quad t \ge 0, \tag{2.270}$$

where $f, g : \mathcal{D} \to \mathbb{R}^n$ are continuously differentiable functions on $\mathcal{D}$. The dynamical systems (2.269) and (2.270) are *topologically equivalent* on $\mathcal{D}$ if there exists a homeomorphism $\mathcal{H} : \mathcal{D} \to \mathcal{D}$ which maps trajectories of (2.269) onto trajectories of (2.270) and preserves their orientation in time. Show that the nonlinear dynamical system

$$\dot{x}(t) = f(x(t))[1 + \|f(x(t))\|]^{-1}, \quad x(0) = x_0, \quad t \ge 0, \tag{2.271}$$

where $f : \mathcal{D} \to \mathbb{R}^n$ is continuously differentiable on $\mathcal{D}$, has a unique solution defined for all $t \ge 0$. Moreover, show that (2.271) is topologically equivalent to (2.269) on $\mathbb{R}^n$.

Problem 2.106. Consider the nonlinear time-varying dynamical system (2.211) where $f(t, \cdot) : \mathcal{D} \to \mathbb{R}^n$ is Lipschitz continuous on $\mathcal{D}$ for all $t \in [t_0, t_1]$ and $f(\cdot, x) : [t_0, t_1] \to \mathbb{R}^n$ is piecewise continuous on $[t_0, t_1]$ for all $x \in \mathcal{D}$. Furthermore, assume there exist constants $\alpha, \beta \in \mathbb{R}$ such that

$$\|f(t, x)\| \le \alpha + \beta\|x\|, \quad (t, x) \in [0, \infty) \times \mathcal{D}. \tag{2.272}$$

Show that the solution $x(t)$, $t \in \mathcal{I}_{x_0}$, to (2.211) satisfies

$$\|x(t)\| \le \|x_0\|e^{\beta(t-t_0)} + \frac{\alpha}{\beta}[e^{\beta(t-t_0)} - 1], \quad t \in \mathcal{I}_{x_0}, \tag{2.273}$$

and hence, cannot exhibit finite escape time.

Problem 2.107 (Comparison Principle). Consider the scalar nonlinear dynamical system

$$\dot{x}(t) = f(t, x(t)), \quad x(t_0) = x_0, \quad t \geq t_0, \tag{2.274}$$

where $f(t, \cdot) : \mathcal{D} \to \mathbb{R}$ is Lipschitz continuous on $\mathcal{D}$ for all $t \in [t_0, t_1]$ and $f(\cdot, x) : [t_0, t_1] \to \mathbb{R}$ is continuous on $[t_0, t_1]$ for all $x \in \mathcal{D}$. Let $v : [t_0, t_1] \to \mathcal{D}$ be such that

$$\dot{v}(t) \leq f(t, v(t)), \quad v(t_0) \leq x_0, \quad t \geq t_0. \tag{2.275}$$

Show that $v(t) \leq x(t)$ for all $t \in [t_0, t_1]$.

Problem 2.108. Prove Theorem 2.25 using Picard's method of successive approximations (see Problem 2.101).

Problem 2.109. Prove Theorem 2.37.

Problem 2.110. Prove Theorem 2.38.

Problem 2.111. Prove Theorem 2.39.

Problem 2.112. Consider the nonlinear dynamical system (2.136). Let $f : \mathcal{D} \to \mathbb{R}^n$ be continuously differentiable on $\mathcal{D}$ and let $s : \mathcal{S} \to \mathcal{D}$ be the flow generated by (2.136) and $\mathcal{S} \triangleq \{(t, x_0) \in \mathbb{R} \times \mathcal{D} : t \in \mathcal{I}_{x_0}\}$. Show that $s : \mathcal{S} \to \mathcal{D}$ is continuously differentiable on $\mathcal{S}$.

Problem 2.113. Show that if the solution $x(t)$ to (2.136) corresponding to an initial condition $x(0) = x_0$ is bounded for all $t \leq 0$, then the negative limit set $\alpha(x_0)$ of $x(t)$, $t \leq 0$, is a nonempty, compact, invariant, and connected set. Furthermore, show that $x(t) \to \alpha(x_0)$ as $t \to -\infty$.

Problem 2.114. Consider the nonlinear dynamical system (2.136) and let $s(t, x)$, $t \geq 0$, denote the solution to (2.136) with the initial condition $x(0) = x$. Show that the positive limit set of (2.136) is given by $\omega(x_0) = \bigcap_{t \geq 0} \overline{\mathcal{O}_x^+}$, where $\mathcal{O}_x^+ \triangleq \{s(t, x) : t \in [0, \infty)\}$ denotes the positive orbit of (2.136).

Problem 2.115. Consider the nonlinear dynamical system (2.136). Let $\mathcal{R} \subset \mathcal{D}$ and let $\mathcal{M}$ be the largest invariant set in $\mathcal{R}$. Show that:

i) $\mathcal{M}$ is the union of all motions defined on $(-\infty, \infty)$ that remain in $\mathcal{R}$ for all $t \in \mathbb{R}$.

ii) $x \in \mathcal{M}$ if and only if $\mathcal{I}_x = (-\infty, \infty)$ and $s(t, x) \in \mathcal{R}$ for all $t \in \mathbb{R}$.

iii) If $\mathcal{R}$ is compact, then $\mathcal{M}$ is compact.

Problem 2.116. Consider the nonlinear dynamical system $\mathcal{G}$ given by (2.215). Show that $\mathcal{M} \subset \mathcal{D}$ is invariant with respect to $\mathcal{G}$ if and only if $\mathcal{M}$ is positively and negatively invariant with respect to $\mathcal{G}$.

Problem 2.117. Consider the nonlinear dynamical system $\mathcal{G}$ given by (2.215) with maximum interval of existence $\mathcal{I}_{x_0} = \mathbb{R}$ for all $x_0 \subset \mathbb{R}^n$. Show that the sets $\mathcal{D}$ and $\emptyset$ are positively and negatively invariant with respect to $\mathcal{G}$.

Problem 2.118. Consider the nonlinear dynamical system $\mathcal{G}$ given by (2.215). Show that for every $x \in \mathcal{D}$, $\mathcal{O}_x$, $\mathcal{O}_x^+$, and $\mathcal{O}_x^-$ are, respectively, invariant, positively invariant, and negatively invariant with respect to $\mathcal{G}$.

Problem 2.119. Consider the nonlinear dynamical system $\mathcal{G}$ given by (2.215), let $\mathcal{M} \subset \mathcal{D}$, and let $\mathcal{O}_x(\mathcal{M}) = \cup\{\mathcal{O}_x : x \in \mathcal{M}\}$. Show that $\mathcal{M}$ is invariant, positively invariant, or negatively invariant with respect to $\mathcal{G}$ if and only if, respectively, $\mathcal{O}_x(\mathcal{M}) \subseteq \mathcal{M}$, $\mathcal{O}_x^+(\mathcal{M}) \subseteq \mathcal{M}$, or $\mathcal{O}_x^-(\mathcal{M}) \subseteq \mathcal{M}$.

Problem 2.120. Consider the nonlinear dynamical system $\mathcal{G}$ given by (2.215) and let $\mathcal{M} \subset \mathcal{D}$. Show that $\mathcal{M}$ is invariant, positively invariant, or negatively invariant with respect to $\mathcal{G}$ if and only if, respectively, each of the connected components of $\mathcal{M}$ is invariant, positively invariant, or negatively invariant with respect to $\mathcal{G}$.

Problem 2.121. Consider the nonlinear dynamical system $\mathcal{G}$ given by (2.215). Show that if $x \in \mathcal{D}$ is such that $x = s(t, x)$ for some $t \in \mathbb{R}$, then $x = s(nt, x)$ for all $n \in \mathbb{Z}$.

Problem 2.122. Consider the nonlinear dynamical system $\mathcal{G}$ given by (2.215). Let $x \in \mathcal{D}$. Show that the following statements are equivalent:

i) x is an equilibrium point.

ii) $\{x\} = \mathcal{O}_x$.

iii) $\{x\} = \mathcal{O}_x^+$.

iv) $\{x\} = \mathcal{O}_x^-$.

v) $\{x\} = \{y \in \mathcal{D} : y = s(t, x), t \in [a, b]\}$ for $a < b$.

vi) There exists a sequence $\{t_n\}_{n=0}^{\infty}$, with $t_n > 0$ and $t_n \to 0$ as $n \to \infty$, such that $x = s(t_n, x)$ for each n.

(**Hint:** Use Problem 2.121 to establish the equivalence of *i)* with *vi)*.)

Problem 2.123. Consider the nonlinear dynamical system $\mathcal{G}$ given by (2.215). Show that if $x \neq s(t,x)$ for some $x \in \mathcal{D}$ and $t \in \mathbb{R}$, then there exists open neighborhoods $\mathcal{U}$ of x and $\mathcal{V}$ of $s(t,x)$ such that $\mathcal{V} = \{s(t,y) : y \in \mathcal{U} \text{ and } t \in \{t\}\}$ and $\mathcal{U} \cap \mathcal{V} = \emptyset$.

Problem 2.124. Consider the nonlinear dynamical system $\mathcal{G}$ given by (2.215) with $t \geq 0$. Show that $x \in \mathcal{D}$ is an equilibrium point of $\mathcal{G}$ if and only if every neighborhood of x contains a positive orbit.

Problem 2.125. Consider the nonlinear dynamical system $\mathcal{G}$ given by (2.215). Show that the set of all equilibrium points $x \in \mathcal{D}$ of $\mathcal{G}$ is closed.

Problem 2.126. Consider the nonlinear dynamical system $\mathcal{G}$ given by (2.215). Show that for every $x \in \mathcal{D}$:

i) $\omega(x) = \bigcap\{\overline{\mathcal{O}_y^+} : y \in \mathcal{O}_x^+\} = \bigcap\{\overline{\mathcal{O}_{x_n}^+} : n \in \mathbb{Z}\}$.

ii) $\omega(x) = \omega(s(t,x))$ for $t \in \mathbb{R}$.

Problem 2.127. Consider the nonlinear dynamical system $\mathcal{G}$ given by (2.215). Show that $x \in \mathcal{D}$ is a periodic point of $\mathcal{G}$ if and only if $\omega(x) = \alpha(x) = \mathcal{O}_x$.

Problem 2.128. Consider the nonlinear dynamical system

$$\dot{x}_1(t) = x_2(t) + x_1(t)x_2^2(t), \quad x_1(0) = x_{10}, \quad t \geq 0, \tag{2.276}$$
$$\dot{x}_2(t) = -x_1(t) + x_1^2(t)x_2(t), \quad x_2(0) = x_{20}. \tag{2.277}$$

Show that the linearization of (2.276) and (2.277) at $(x_1, x_2) = (0,0)$ possesses a continuum of periodic solutions whereas (2.276) and (2.277) has no periodic solutions.

Problem 2.129. Consider the nonlinear dynamical system given by (2.220) and (2.221) in Example 2.38. Show that the divergence $\nabla f(x) < 0$ in the annular region $\mathcal{D} = \{x \in \mathbb{R}^2 : 2/3 < x_1^2 + x_2^2 < 2\}$ and yet as shown in Example 2.38, (2.220) and (2.221) possesses a limit cycle in $\mathcal{D}$. Why does this contradict Bendixson's theorem?

Problem 2.130. Consider the second-order nonlinear mechanical system consisting of a unit mass with a nonlinear spring and a nonlinear damper given by

$$\ddot{x}(t) + f(\dot{x}(t)) + g(x(t)) = 0, \quad x(0) = x_0, \quad \dot{x}(0) = \dot{x}_0, \quad t \geq 0, \tag{2.278}$$

where $f(\cdot)$ is the damping force exerted by the damper and $g(\cdot)$ is the restoring force of the nonlinear spring. Suppose $f(\cdot)$ and $g(\cdot)$ are continuously differentiable.

i) Show that this system has no periodic solutions if $f'(\dot{x}) \neq 0$ for all $\dot{x} \in \mathbb{R}$. What does this mean physically?

ii) Use the Poincaré-Bendixson theorem to show that the undamped nonlinear mechanical system ($f(\dot{x}) \equiv 0$) always has a continuum of periodic solutions if $xg(x) > 0$, $x \neq 0$.

Problem 2.131. Consider the nonlinear dynamical system

$$\dot{x}_1(t) = x_2(t), \quad x_1(0) = x_{10}, \quad t \geq 0, \tag{2.279}$$
$$\dot{x}_2(t) = x_1(t) - x_1^3(t) - \alpha x_2(t) + x_1^2(t)x_2(t), \quad x_2(0) = x_{20}, \tag{2.280}$$

where $\alpha > 0$. Does (2.279) and (2.280) possess a periodic solution?

Problem 2.132. Show that if $\nabla f(x) = 0$ for all $x \in \mathcal{D}$, where $f : \mathbb{R}^2 \to \mathbb{R}^2$, then Bendixson's theorem implies that there may be a trivial periodic orbit in $\mathcal{D}$ but there is no limit cycle in $\mathcal{D}$.

Problem 2.133. Consider the nonlinear dynamical system

$$\dot{x}_1(t) = x_2(t)\cos x_1(t), \quad x_1(0) = x_{10}, \quad t \geq 0, \tag{2.281}$$
$$\dot{x}_2(t) = \sin x_1(t), \quad x_2(0) = x_{20}. \tag{2.282}$$

Show that (2.281) and (2.282) does not possess a limit cycle.

Problem 2.134. Consider the nonlinear dynamical system

$$\dot{x}_1(t) = x_2(t), \quad x_1(0) = x_{10}, \quad t \geq 0, \tag{2.283}$$
$$\dot{x}_2(t) = -x_1(t) - x_2(t) + x_1^2(t) + x_2^2(t), \quad x_2(0) = x_{20}. \tag{2.284}$$

Use Dulac's theorem to show that (2.283) and (2.284) does not possess a periodic solution. (**Hint:** Use the Dulac function $g(x_1, x_2) = e^{-2x_1}$.)

Problem 2.135. Consider the nonlinear dynamical system

$$\dot{x}_1(t) = x_1(t) - x_1^2(t) - x_1(t)x_2(t), \quad x_1(0) = x_{10}, \quad t \geq 0, \tag{2.285}$$
$$\dot{x}_2(t) = x_2(t) - x_2^2(t) - x_1(t)x_2(t), \quad x_2(0) = x_{20}. \tag{2.286}$$

Use Dulac's theorem to show that (2.285) and (2.286) does not possess a periodic solution in the nonnegative orthant $\overline{\mathbb{R}}_+^2$. (**Hint:** Use the Dulac function $g(x_1, x_2) = 1/x_1x_2$, $x_1 \neq 0$ and $x_2 \neq 0$.)

Problem 2.136. Consider the second-order nonlinear dynamical system (2.230) and (2.231) with vector field $f : \mathcal{D} \to \mathbb{R}^2$, where $\mathcal{D}$ is not a simply connected region in $\mathbb{R}^2$. Show that in this case Dulac's theorem is no longer valid.

Problem 2.137. Consider Dulac's theorem with $\mathcal{D}$ being topologically equivalent to an annulus, that is, $\mathcal{D}$ has exactly one hole in it. Show that in this case (2.230) and (2.231) possesses at most one closed orbit in $\mathcal{D}$. (**Hint:** Use Green's theorem.)

Problem 2.138. Prove Dulac's theorem.

Problem 2.139. Consider the nonlinear dynamical system

$$\dot{x}_1(t) = x_1(t) - x_2(t) - x_1^3(t), \quad x_1(0) = x_{10}, \quad t \geq 0, \tag{2.287}$$
$$\dot{x}_2(t) = x_1(t) + x_2(t) - x_2^3(t), \quad x_2(0) = x_{20}. \tag{2.288}$$

Use the Poincaré-Bendixson theorem to show that (2.287) and (2.288) has a periodic orbit.

Problem 2.140. Consider the nonlinear dynamical system

$$\dot{x}_1(t) = x_2(t),\ x_1(0) = x_{10},\ t \geq 0, \tag{2.289}$$
$$\dot{x}_2(t) = -x_1(t) + x_2(t)[1 - x_1^2(t) - x_2^2(t)],\ x_2(0) = x_{20}. \tag{2.290}$$

Use the Poincaré-Bendixson theorem to show that (2.289) and (2.290) has a unique limit cycle. Furthermore, show that every trajectory of (2.289) and (2.290), except for the equilibrium point, approaches this limit cycle.

Problem 2.141. Consider the nonlinear dynamical system

$$\dot{x}_1(t) = -x_2(t) + x_1(t)[1 - x_1^2(t) - x_2^2(t)],\ x_1(0) = x_{10},\ t \geq 0, \tag{2.291}$$
$$\dot{x}_2(t) = x_1(t) + x_2(t)[1 - x_1^2(t) - x_2^2(t)],\ x_2(0) = x_{20} \tag{2.292}$$
$$\dot{x}_3(t) = \alpha x_3(t),\ x_3(0) = x_{30}, \tag{2.293}$$

where $\alpha \in \mathbb{R}$. Use the Poincaré-Bendixson theorem to show that (2.291)–(2.293) has a periodic orbit.

Problem 2.142. Consider the nonlinear dynamical system

$$\dot{x}_1(t) = x_2(t) + \alpha x_1(t)[\beta^2 - x_1(t) - x_2^2(t)]^\gamma,\ x_1(0) = x_{10},\ t \geq 0, \tag{2.294}$$
$$\dot{x}_2(t) = -x_1(t) + \alpha x_2(t)[\beta^2 - x_1(t) - x_2^2(t)]^\gamma,\ x_2(0) = x_{20}, \tag{2.295}$$

where $\alpha, \beta \in \mathbb{R}$ and $\gamma \geq 1$. For $\gamma = 1$, find the solution of (2.294) and (2.295). Does (2.294) and (2.295) exhibit periodic solutions for $\gamma = 1$? Repeat the problem for $\gamma = 2$.

Problem 2.143. Consider the nonlinear dynamical system

$$\dot{x}_1(t) = x_1(t) + x_2(t) - x_1(t)[|x_1(t)| + |x_2(t)|],\ x_1(0) = x_{10},\ t \geq 0, \tag{2.296}$$
$$\dot{x}_2(t) = -2x_1(t) + x_2(t) - x_2(t)[|x_1(t)| + |x_2(t)|],\ x_2(0) = x_{20}. \tag{2.297}$$

Use the Poincaré-Bendixson theorem to show that (2.296) and (2.297) has a periodic orbit.

Problem 2.144. Consider the nonlinear dynamical system

$$\dot{r}(t) = r(t)[1 - r^2(t)] + \mu r(t)\cos\theta(t), \quad r(0) = r_0, \quad t \geq 0, \quad (2.298)$$
$$\dot{\theta}(t) = 1, \quad \theta(0) = \theta_0. \quad (2.299)$$

As shown in Example 2.38, when $\mu = 0$ (2.298) and (2.299) possesses a stable limit cycle. Show that a closed orbit still exists for sufficiently small $\mu > 0$.

Problem 2.145. Consider the nonlinear dynamical system

$$\dot{x}_1(t) = x_1(t) - x_2(t) - x_1(t)[x_1^2(t) + 5x_2^2(t)], \; x_1(0) = x_{10}, \; t \geq 0, \quad (2.300)$$
$$\dot{x}_2(t) = x_1(t) + x_2(t) - x_2(t)[x_1^2(t) + 5x_2^2(t)], \; x_2(0) = x_{20}. \quad (2.301)$$

Use the Poincaré-Bendixson theorem to show that (2.300) and (2.301) has a periodic orbit. (**Hint:** Rewrite the system in polar coordinates using $r\dot{r} = x_1\dot{x}_1 + x_2\dot{x}_2$ and $\dot{\theta} = (x_1\dot{x}_2 - x_2\dot{x}_1)/r^2$.)

Problem 2.146. Consider the nonlinear dynamical system

$$\dot{x}_1(t) = x_1(t) - x_2(t) - x_1^3(t), \; x_1(0) = x_{10}, \; t \geq 0, \quad (2.302)$$
$$\dot{x}_2(t) = x_1(t) + x_2(t) - x_2^3(t), \; x_2(0) = x_{20}. \quad (2.303)$$

Use the Poincaré-Bendixson theorem to show that (2.302) and (2.303) has a periodic orbit.

Problem 2.147. Consider the nonlinear dynamical system

$$\dot{x}_1(t) = x_1(t)[1 - 4x_1^2(t) - x_2^2(t)] - \tfrac{1}{2}x_2(t)[1 + x_1(t)], \; x_1(0) = x_{10}, \; t \geq 0, \quad (2.304)$$
$$\dot{x}_2(t) = x_2(t)[1 - 4x_1^2(t) - x_2^2(t)] + 2x_1(t)[1 + x_1(t)], \; x_2(0) = x_{20}. \quad (2.305)$$

Use the Poincaré-Bendixson theorem to show that all trajectories of (2.304) and (2.305) approach the ellipse $\mathcal{E} \triangleq \{(x_1, x_2) \in \mathbb{R}^2 : 4x_1^2 + x_2^2 = 1\}$ as $t \to \infty$.

Problem 2.148. Consider the nonlinear dynamical system

$$\dot{x}_1(t) = -x_1(t) - x_2(t) + x_1(t)[x_1^2(t) + 2x_2^2(t)], \quad x_1(0) = x_{10}, \quad t \geq 0, \quad (2.306)$$
$$\dot{x}_2(t) = x_1(t) - x_2(t) + x_2(t)[x_1^2(t) + 2x_2^2(t)], \quad x_2(0) = x_{20}. \quad (2.307)$$

Use the Poincaré-Bendixson theorem to show that (2.306) and (2.307) has a periodic orbit.

Problem 2.149. Consider the second-order dynamical system (2.230) and (2.231). Let $V : \mathcal{D} \subseteq \mathbb{R}^2 \to \mathbb{R}$ be conserved along the flow of (2.230) and (2.231), that is,

$$\dot{V}(x_1, x_2) = \frac{\partial V}{\partial x_1} f_1(x_1, x_2) + \frac{\partial V}{\partial x_2} f_2(x_1, x_2) = 0. \tag{2.308}$$

Show that if V is not constant on any open set $U \subset \mathcal{D}$, then (2.230) and (2.231) does not possess any limit cycle.

Problem 2.150 (Isocline Method). The method of isoclines is a technique for generating solution curves in the two-dimensional state space (that is, the *phase plane*) for various initial conditions. In particular, given the second-order nonlinear dynamical system (2.230) and (2.231), and assuming $f_1(x_1, x_2) \neq 0$, it follows that the slope of the system trajectories of (2.230) and (2.231) in the phase plane passing through a point (x_1, x_2) is given by

$$\frac{\mathrm{d}x_2}{\mathrm{d}x_1} = \frac{f_2(x_1, x_2)}{f_1(x_1, x_2)} = s(x_1, x_2). \tag{2.309}$$

Isoclines correspond to curves $x_2 = g(x_1)$ in the phase plane on which the slope $s(x_1, x_2) = c$, where c is a constant. These curves can be obtained by solving the equation

$$f_2(x_1, x_2) = c f_1(x_1, x_2). \tag{2.310}$$

A family of phase plane trajectories corresponding to various initial conditions is called a *phase portrait* of (2.230) and (2.231). If the slope at points (x_1, x_2) and $(x_1, -x_2)$ is such that $s(x_1, x_2) = -s(x_1, -x_2)$, then the phase portrait is symmetric about the x_1-axis. Similarly, if $s(x_1, x_2) = -s(-x_1, x_2)$, then the phase portrait is symmetric about the x_2-axis. Finally, if $s(x_1, x_2) = s(-x_1, -x_2)$, then the phase portrait is symmetric about the origin. Using the method of isoclines, sketch the phase portraits of the following dynamical systems.

$$\ddot{x}(t) + x^3(t) - x(t) = 0, \quad x(0) = x_0, \quad \dot{x}(0) = \dot{x}_0, \quad t \geq 0, \tag{2.311}$$

$$\ddot{x}(t) + \sin x(t) = 0, \quad x(0) = x_0, \quad \dot{x}(0) = \dot{x}_0, \quad t \geq 0, \tag{2.312}$$

$$\ddot{x}(t) - 0.2[1 - x^2(t)]\dot{x}(t) + x(t) = 0, \; x(0) = x_0, \; \dot{x}(0) = \dot{x}_0, \; t \geq 0, \tag{2.313}$$

$$\ddot{x}(t) + \tfrac{1}{2}\dot{x}(t) + 2x(t) + x^2(t) = 0, \; x(0) = x_0, \; \dot{x}(0) = \dot{x}_0, \; t \geq 0. \tag{2.314}$$

Verify your sketches by simulations.

Problem 2.151. Consider the nonlinear *Lienard system* given by

$$\ddot{x}(t) + f(x(t))\dot{x}(t) + g(x(t)) = 0, \quad x(0) = x_0, \quad \dot{x}(0) = \dot{x}_0, \quad t \geq 0. \tag{2.315}$$

Suppose that f and g satisfy the following conditions:

i) $f(x)$ and $g(x)$ are continuously differentiable for all $x \in \mathbb{R}$.

ii) $g(-x) = -g(x)$ for all $x \in \mathbb{R}$.

iii) $xg(x) > 0$, $x \in \mathbb{R}$, $x \neq 0$.

iv) $f(-x) = f(x)$ for all $x \in \mathbb{R}$.

v) $F(x) = \int_0^x f(\sigma)\mathrm{d}\sigma$ is such that $F(0) = 0$, $F'(0) < 0$, $F(\cdot)$ has a single positive zero at $x = a$, and $F(x) \to \infty$ for $x \geq a$ as $x \to \infty$.

Show that (2.315) has a unique, stable limit cycle.

2.15 Notes and References

The qualitative analysis of differential equations was first developed by Henri Poincaré [358–360] with further developments given by Birkhoff [60,61]. The material on existence, uniqueness, and continuity of solutions with respect to system initial conditions of nonlinear differential equations is standard and can be found in most textbooks on differential equations. Notable textbooks include those by Hartman [185], Coddington and Levinson [96], Hirsch and Smale [196], Agarwal and Lakshmikantham [4], Lefschetz [264], Nemytskii and Stepanov [333], Hale [179], and Miller and Michel [315]. The topics on matrix analysis are also standard and can be found in Bellman [37], Horn and Johnson [201,202], Lancaster and Tismenetsky [256], Gantmacher [132], Stewart and Sun [417], and Bernstein [45]. For a thorough presentation on advanced calculus and mathematical analysis the reader is referred to Rudin [373], Royden [371], Apostol [12], Graves [139], Edwards [114], Naylor and Sell [332], Fleming [120], Munkres [323], Hoffman [198], and Bartle [30]. See also Halmos [182] and Luenberger [288]. Finally, the study of the existence and absence of periodic orbits in nonlinear dynamical systems was fathered by Poincaré [358–360] and further developed by Bendixson [39]. For a modern treatment of these results, see Hirsch and Smale [196], Hale and Kocak [180], Wiggins [454], and Perko [349].

Chapter Three

Stability Theory for Nonlinear Dynamical Systems

3.1 Introduction

One of the most basic issues in system theory is stability of dynamical systems. System stability is characterized by analyzing the response of a dynamical system to small perturbations in the system states. Specifically, an equilibrium point of a dynamical system is said to be *stable* if, for sufficiently small values of initial disturbances, the perturbed motion remains in an arbitrarily prescribed small region of the state space. More precisely, stability is equivalent to continuity of solutions as a function of the system initial conditions over a neighborhood of the equilibrium point uniformly in time. If, in addition, all solutions of the dynamical system approach the equilibrium point for large values of time, then the equilibrium point is said to be *asymptotically stable.* The most complete contribution to the stability analysis of nonlinear dynamical systems was introduced in the late nineteenth century by the Russian mathematician A. M. Lyapunov in his seminal work entitled *The General Problem of the Stability of Motion* [293–295]. Lyapunov's results which include the direct and indirect methods, along with the Barbashin-Krasovskii-LaSalle invariance principle [23, 245, 258, 260], provide a powerful framework for analyzing the stability of nonlinear dynamical systems as well as designing feedback controllers that guarantee closed-loop system stability.

Lyapunov's *direct method* states that if a continuously differentiable positive-definite function of the states of a given dynamical system can be constructed for which its time rate of change due to perturbations in a neighborhood of the system's equilibrium is always negative or zero, then the system's equilibrium point is stable or, equivalently, *Lyapunov stable.* Alternatively, if the time rate of change of the positive-definite function is strictly negative, then the system's equilibrium point is asymptotically stable. Unlike Lyapunov's direct method, which can provide global stability conclusions for an equilibrium point for a nonlinear dynamical system,

Lyapunov's *indirect method* draws conclusions about local stability of the equilibrium point by examining the stability of the linearized nonlinear system about the equilibrium point in question. Since the analysis and controller design frameworks presented in this book are predominantly based on Lyapunov stability theory, in this chapter we present the main Lyapunov stability results needed for developing these frameworks.

3.2 Lyapunov Stability Theory

In this section, we develop the fundamental results of Lyapunov stability theory. We begin by considering the general nonlinear autonomous dynamical system

$$\dot{x}(t) = f(x(t)), \qquad x(0) = x_0, \qquad t \in \mathcal{I}_{x_0}, \tag{3.1}$$

where $x(t) \in \mathcal{D} \subseteq \mathbb{R}^n$, $t \in \mathcal{I}_{x_0}$, is the system state vector, $\mathcal{D}$ is an open set with $0 \in \mathcal{D}$, $f : \mathcal{D} \to \mathbb{R}^n$ is continuous on $\mathcal{D}$, and $\mathcal{I}_{x_0} = [0, \tau_{x_0})$, $0 \leq \tau_{x_0} \leq \infty$. We assume that for every initial condition $x(0) \in \mathcal{D}$ and every $\tau_{x_0} > 0$, the dynamical system (3.1) possesses a unique solution $x : [0, \tau_{x_0}) \to \mathcal{D}$ on the interval $[0, \tau_{x_0})$. We denote the solution to (3.1) with initial condition $x(0) = x_0$ by $s(\cdot, x_0)$, so that the flow of the dynamical system (3.1) given by the map $s : [0, \tau_{x_0}) \times \mathcal{D} \to \mathcal{D}$ is continuous in x and continuously differentiable in t and satisfies the consistency property $s(0, x_0) = x_0$ and the semigroup property $s(\tau, s(t, x_0)) = s(t + \tau, x_0)$, for all $x_0 \in \mathcal{D}$ and $t, \tau \in [0, \tau_{x_0})$ such that $t + \tau \in [0, \tau_{x_0})$. Unless otherwise stated, we assume $f(0) = 0$ and $f(\cdot)$ is Lipschitz continuous on $\mathcal{D}$. The following definition introduces several types of stability corresponding to the zero solution $x(t) \equiv 0$ of (3.1) for $\mathcal{I}_{x_0} = [0, \infty)$.

Definition 3.1. *i*) The zero solution $x(t) \equiv 0$ to (3.1) is *Lyapunov stable* if, for all $\varepsilon > 0$, there exists $\delta = \delta(\varepsilon) > 0$ such that if $\|x(0)\| < \delta$, then $\|x(t)\| < \varepsilon$, $t \geq 0$ (see Figure 3.1).

ii) The zero solution $x(t) \equiv 0$ to (3.1) is *(locally) asymptotically stable* if it is Lyapunov stable and there exists $\delta > 0$ such that if $\|x(0)\| < \delta$, then $\lim_{t\to\infty} x(t) = 0$ (see Figure 3.2).

iii) The zero solution $x(t) \equiv 0$ to (3.1) is *(locally) exponentially stable* if there exist positive constants α, β, and δ such that if $\|x(0)\| < \delta$, then $\|x(t)\| \leq \alpha \|x(0)\| e^{-\beta t}$, $t \geq 0$.

iv) The zero solution $x(t) \equiv 0$ to (3.1) is *globally asymptotically stable* if it is Lyapunov stable and for all $x(0) \in \mathbb{R}^n$, $\lim_{t\to\infty} x(t) = 0$.

v) The zero solution $x(t) \equiv 0$ to (3.1) is *globally exponentially stable* if there exist positive constants α and β such that $\|x(t)\| \leq \alpha\|x(0)\|e^{-\beta t}$, $t \geq 0$, for all $x(0) \in \mathbb{R}^n$.

vi) Finally, the zero solution $x(t) \equiv 0$ to (3.1) is *unstable* if it is not Lyapunov stable.

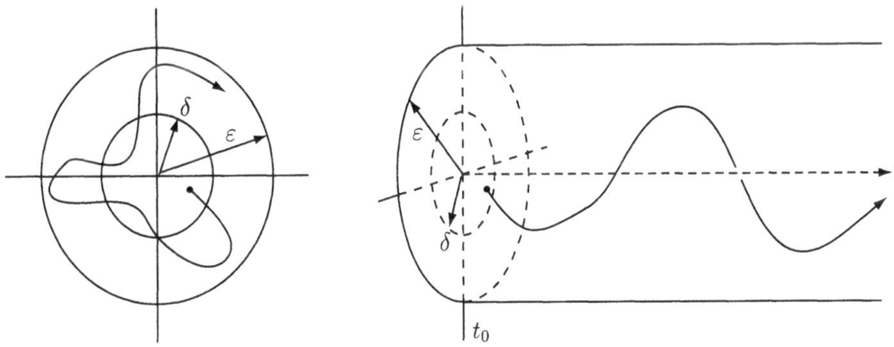

Figure 3.1 Lyapunov stability of an equilibrium point.

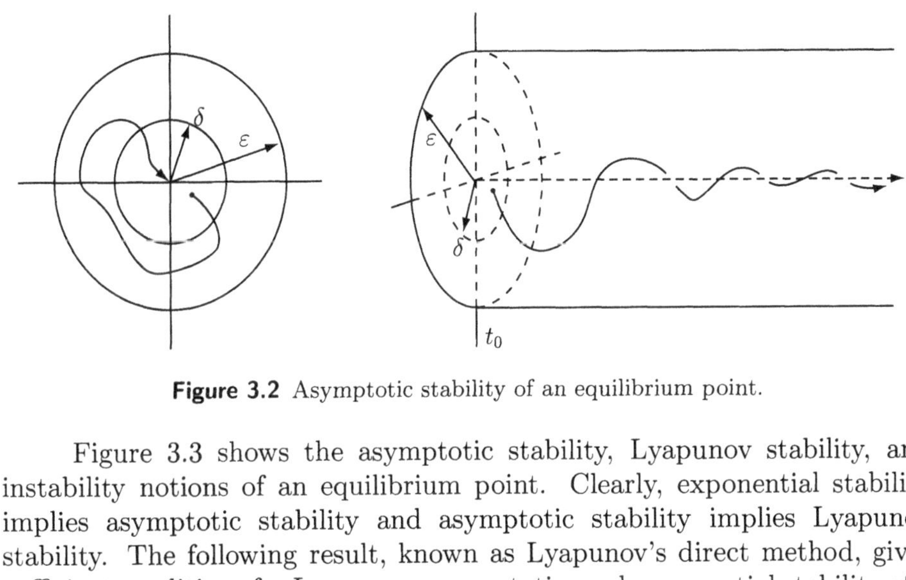

Figure 3.2 Asymptotic stability of an equilibrium point.

Figure 3.3 shows the asymptotic stability, Lyapunov stability, and instability notions of an equilibrium point. Clearly, exponential stability implies asymptotic stability and asymptotic stability implies Lyapunov stability. The following result, known as Lyapunov's direct method, gives sufficient conditions for Lyapunov, asymptotic, and exponential stability of a nonlinear dynamical system. For this result, let $V : \mathcal{D} \to \mathbb{R}$ be a continuously differentiable function with derivative *along the trajectories* of (3.1) given by $\dot{V}(x) \triangleq V'(x)f(x)$. Note that $\dot{V}(x)$ is dependent on the system dynamics (3.1). Since, using the chain rule, $\dot{V}(x) = \frac{\mathrm{d}}{\mathrm{d}t}V(s(t,x))\big|_{t=0} = V'(x)f(x)$ it follows that if $\dot{V}(x)$ is negative, then $V(x)$ decreases along the solution $s(t, x_0)$ of (3.1) through $x_0 \in \mathcal{D}$ at $t = 0$.

Theorem 3.1 (Lyapunov's Theorem). Consider the nonlinear dynamical system (3.1) and assume that there exists a continuously differentiable

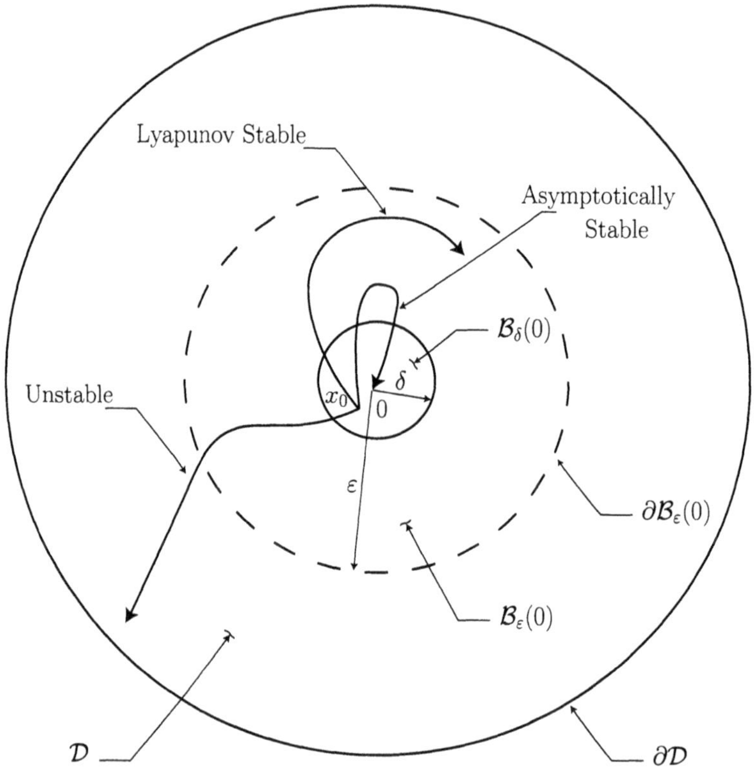

Figure 3.3 Asymptotically stable, Lyapunov stable, and unstable equilibrium point.

function $V\colon \mathcal{D} \to \mathbb{R}$ such that

$$V(0) = 0, \tag{3.2}$$

$$V(x) > 0, \quad x \in \mathcal{D}, \quad x \neq 0, \tag{3.3}$$

$$V'(x)f(x) \leq 0, \quad x \in \mathcal{D}. \tag{3.4}$$

Then the zero solution $x(t) \equiv 0$ to (3.1) is Lyapunov stable. If, in addition,

$$V'(x)f(x) < 0, \quad x \in \mathcal{D}, \quad x \neq 0, \tag{3.5}$$

then the zero solution $x(t) \equiv 0$ to (3.1) is asymptotically stable. Finally, if there exist scalars $\alpha, \beta, \varepsilon > 0$, and $p \geq 1$, such that $V : \mathcal{D} \to \mathbb{R}$ satisfies

$$\alpha\|x\|^p \leq V(x) \leq \beta\|x\|^p, \quad x \in \mathcal{D}, \tag{3.6}$$

$$V'(x)f(x) \leq -\varepsilon V(x), \quad x \in \mathcal{D}, \tag{3.7}$$

then the zero solution $x(t) \equiv 0$ to (3.1) is exponentially stable.

Proof. Let $\varepsilon > 0$ be such that $\mathcal{B}_\varepsilon(0) \subseteq \mathcal{D}$. Since $\partial\mathcal{B}_\varepsilon(0)$ is compact and $V(x)$, $x \in \mathcal{D}$, is continuous, it follows that $V(\partial\mathcal{B}_\varepsilon(0))$ is compact (see Proposition 2.14), and hence, by Theorem 2.13, $\alpha \triangleq \min_{x\in\partial\mathcal{B}_\varepsilon(0)} V(x)$ exists. Note $\alpha > 0$ since $0 \notin \partial\mathcal{B}_\varepsilon(0)$ and $V(x) > 0$, $x \in \mathcal{D}$, $x \neq 0$. Next, let

$\beta \in (0, \alpha)$ and define $\mathcal{D}_\beta$ to be the *arcwise connected component* of $\{x \in \mathcal{D} : V(x) \le \beta\}$ containing the origin; that is, $\mathcal{D}_\beta$ is the set of all $x \in \mathcal{D}$ such that there exists a continuous function $\psi : [0,1] \to \mathcal{D}$ such that $\psi(0) = x$, $\psi(1) = 0$, and $V(\psi(\mu)) \le \beta$, for all $\mu \in [0,1]$.[1] Note that $\mathcal{D}_\beta \subset \mathcal{B}_\varepsilon(0)$ (see Figure 3.4). To see this, suppose, *ad absurdum*, that $\mathcal{D}_\beta \not\subset \mathcal{B}_\varepsilon(0)$. In this case, there exists a point $p \in \mathcal{D}_\beta$ such that $p \in \partial\mathcal{B}_\varepsilon(0)$, and hence, $V(p) \ge \alpha > \beta$, which is a contradiction. Now, since $\dot{V}(x) \triangleq V'(x)f(x) \le 0$, $x \in \mathcal{D}_\beta$, it follows that $V(x(t))$ is a nonincreasing function of time, and hence, $V(x(t)) \le V(x(0)) \le \beta$, $t \ge 0$. Hence, $\mathcal{D}_\beta$ is a positively invariant set with respect to (3.1). Furthermore, since $\mathcal{D}_\beta$ is compact, it follows from Corollary 2.5 that for all $x(0) \in \mathcal{D}_\beta$, (3.1) has a unique solution defined for all $t \ge 0$.

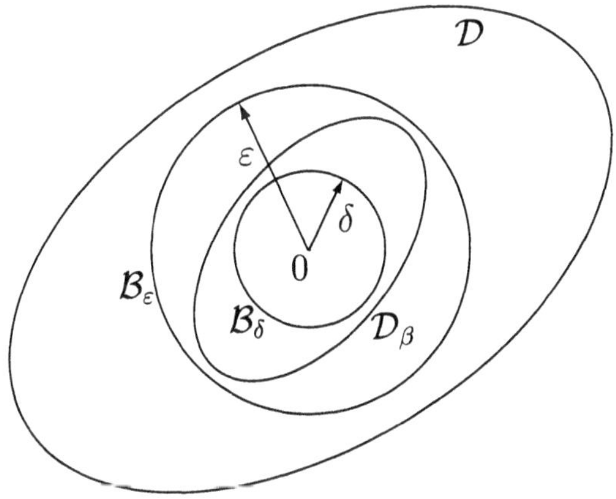

Figure 3.4 Visualization of sets used in the proof of Theorem 3.1.

Next, since $V(\cdot)$ is continuous and $V(0) = 0$, there exists $\delta = \delta(\varepsilon) \in (0, \varepsilon)$ such that $V(x) < \beta$, $x \in \mathcal{B}_\delta(0)$. Now, let $x(t)$, $t \ge 0$, satisfy (3.1) with $\|x(0)\| < \delta$. Since, $\mathcal{B}_\delta(0) \subset \mathcal{D}_\beta \subset \mathcal{B}_\varepsilon(0) \subseteq \mathcal{D}$ and $V'(x)f(x) \le 0$, $x \in \mathcal{D}$, it follows that

$$V(x(t)) - V(x(0)) = \int_0^t V'(x(s))f(x(s))\mathrm{d}s \le 0, \qquad t \ge 0,$$

and hence, for all $x(0) \in \mathcal{B}_\delta(0)$,

$$V(x(t)) \le V(x(0)) < \beta, \qquad t \ge 0.$$

Now, since $V(x) \ge \alpha$, $x \in \partial\mathcal{B}_\varepsilon(0)$, and $\beta \in (0, \alpha)$, it follows that $x(t) \notin \partial\mathcal{B}_\varepsilon(0)$, $t \ge 0$. Hence, for all $\varepsilon > 0$ there exists $\delta = \delta(\varepsilon) > 0$ such that if $\|x(0)\| < \delta$, then $\|x(t)\| < \varepsilon$, $t \ge 0$, which proves Lyapunov stability of the zero solution $x(t) \equiv 0$ to (3.1).

[1] Unless otherwise stated, in the remainder of the book we assume that sets of the form $\mathcal{D}_\beta = \{x \in \mathcal{D} : V(x) \le \beta\}$ correspond to the arcwise connected component of $\{x \in \mathcal{D} : V(x) \le \beta\}$ containing the origin. This minor abuse of notation considerably simplifies the presentation.

To prove asymptotic stability of the zero solution $x(t) \equiv 0$ to (3.1) suppose that $V'(x)f(x) < 0$, $x \in \mathcal{D}$, $x \neq 0$, and $x(0) \in \mathcal{B}_\delta(0)$. Then it follows that $x(t) \in \mathcal{B}_\varepsilon(0)$, $t \geq 0$. However, $V(x(t))$, $t \geq 0$, is decreasing and bounded from below by zero. Now, *ad absurdum*, suppose $x(t)$, $t \geq 0$, does not converge to zero. This implies that $V(x(t))$, $t \geq 0$, is lower bounded, that is, there exists $L > 0$ such that $V(x(t)) \geq L > 0$, $t \geq 0$. Hence, by continuity of $V(\cdot)$ there exists $\delta' > 0$ such that $V(x) < L$ for $x \in \mathcal{B}_{\delta'}(0)$, which further implies that $x(t) \notin \mathcal{B}_{\delta'}(0)$ for all $t \geq 0$. Next, define $L_1 \triangleq \min_{\delta' \leq \|x\| \leq \varepsilon} -V'(x)f(x)$. Now, (3.5) implies $-V'(x)f(x) \geq L_1$, $\delta' \leq \|x\| \leq \varepsilon$ or, equivalently,

$$V(x(t)) - V(x(0)) = \int_0^t V'(x(s))f(x(s))\mathrm{d}s \leq -L_1 t,$$

and hence, for all $x(0) \in \mathcal{B}_\delta(0)$,

$$V(x(t)) \leq V(x(0)) - L_1 t.$$

Letting $t > \frac{V(x(0))-L}{L_1}$, it follows that $V(x(t)) < L$, which is a contradiction. Hence, $x(t) \to 0$ as $t \to \infty$, establishing asymptotic stability.

Finally, to prove exponential stability of the zero solution $x(t) \equiv 0$ to (3.1) note that (3.7) implies

$$V(x(t)) \leq V(x(0))e^{-\varepsilon t}, \qquad t \geq 0. \tag{3.8}$$

Now, since by assumption $V(x(0)) \leq \beta \|x(0)\|^p$ and $\alpha \|x(t)\|^p \leq V(x(t))$, it follows that

$$\alpha \|x(t)\|^p \leq \beta \|x(0)\|^p e^{-\varepsilon t}, \qquad t \geq 0, \tag{3.9}$$

which implies that

$$\|x(t)\| \leq \left(\frac{\beta}{\alpha}\right)^{1/p} \|x(0)\| e^{(-\varepsilon/p)t}, \qquad t \geq 0, \tag{3.10}$$

establishing exponential stability. □

If $x_{\mathrm{e}} \neq 0$ is an equilibrium point of (3.1), then Theorem 3.1 holds with $V(0) = 0$ and $x \neq 0$ replaced by $V(x_{\mathrm{e}}) = 0$ and $x \neq x_{\mathrm{e}}$ (see Problem 3.39 for details). A continuously differentiable function $V(\cdot)$ satisfying (3.2) and (3.3) is called a *Lyapunov function candidate* for the nonlinear dynamical system (3.1). If, additionally, $V(\cdot)$ satisfies (3.4), $V(\cdot)$ is called a *Lyapunov function* for the nonlinear dynamical system (3.1).

In light of conditions (3.2)–(3.5), $V(\cdot)$ can be regarded as a generalized *energy function* for the nonlinear dynamical system (3.1). In particular, viewing the *Lyapunov level surfaces* $V(x) = \alpha$, for sufficiently small constants $\alpha > 0$, as constant energy surfaces covering the neighborhood $\mathcal{D}_\beta \subset \mathcal{B}_\varepsilon(0)$, where $V(x_0) = \beta$, of the nonlinear dynamical system (3.1), it follows from Theorem 3.1 that requiring the Lyapunov derivative $\dot{V}(x) \triangleq V'(x)f(x)$ to be negative, the system trajectories of (3.1) move from one energy surface to an inner, or lower, energy surface. Of course, if condition (3.4) is satisfied, the system trajectories will approach the origin and remain within a ball of radius $\varepsilon > 0$ of the origin for every system initial condition lying inside a constant energy surface contained in the ball of radius ε. Alternatively, if condition (3.5) is satisfied, then the system's energy surfaces shrink to zero so that the system trajectory approaches zero asymptotically (see Figure 3.5).

For planar ($\mathbb{R}^2$) systems the Lyapunov function provides the following interpretation. For a given solution $s(t, x_0)$ of (3.1) to cross the constant energy surface $V(x) = \beta$, where $\beta = V(x_0)$, the angle between the outward normal of the gradient vector $V(x_0)$ and the derivative of $s(t, x_0)$ at $t = t_0$ must be greater than $\pi/2$, that is, $\dot{V}(x_0) = V'(x_0)f(x_0) < 0$. For this to occur at all points, we require $\dot{V}(x) = V'(x)f(x) < 0$, $x \in \mathcal{D}_\beta$. Hence, $V(s(t, x_0))$ is a decreasing function of time. This of course implies that $\dot{V}(s(t, x))$ along the solution $s(t, x_0)$ of (3.1) must be negative in $\mathcal{D}_\beta$.

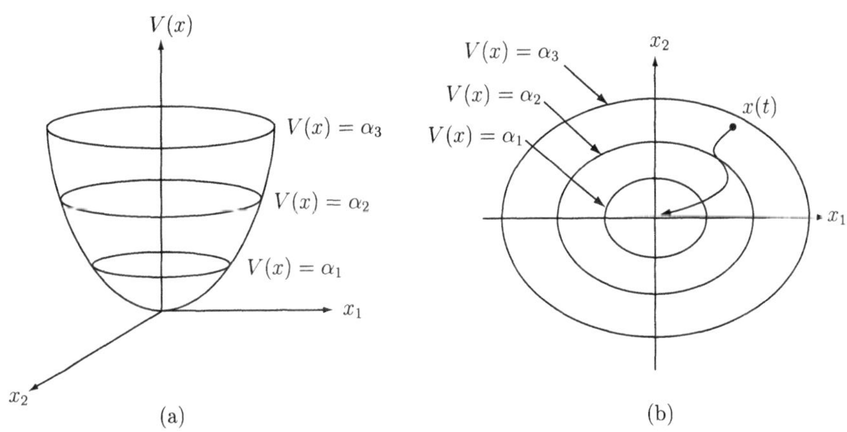

Figure 3.5 (a) Typical Lyapunov function candidate. (b) Constant Lyapunov energy surfaces. For both figures $\alpha_1 < \alpha_2 < \alpha_3$.

Example 3.1. Consider the nonlinear dynamical system representing a rigid spacecraft given by

$$\dot{x}_1(t) = I_{23}x_2(t)x_3(t), \quad x_1(0) = x_{10}, \quad t \geq 0, \tag{3.11}$$

$$\dot{x}_2(t) = I_{31}x_3(t)x_1(t), \quad x_2(0) = x_{20}, \tag{3.12}$$

$$\dot{x}_3(t) = I_{12}x_1(t)x_2(t), \quad x_3(0) = x_{30}, \tag{3.13}$$

where $I_{23} = (I_2 - I_3)/I_1$, $I_{31} = (I_3 - I_1)/I_2$, $I_{12} = (I_1 - I_2)/I_3$, and I_1, I_2, and I_3 are the principal moments of inertia of the spacecraft such that $I_1 > I_2 > I_3 > 0$. To examine the stability of this system consider the

Lyapunov function candidate $V(x_1,x_2,x_3) = \frac{1}{2}(\alpha_1 x_1^2 + \alpha_2 x_2^2 + \alpha_3 x_3^2)$, where $\alpha_1, \alpha_2, \alpha_3 > 0$. Now, the Lyapunov derivative is given by

$$\dot{V}(x_1,x_2,x_3) = \alpha_1 x_1 \dot{x}_1 + \alpha_2 x_2 \dot{x}_2 + \alpha_3 x_3 \dot{x}_3 = x_1 x_2 x_3 (\alpha_1 I_{23} + \alpha_2 I_{31} + \alpha_3 I_{12}). \tag{3.14}$$

Since $I_{31} < 0$ it follows that there exist $\alpha_1, \alpha_2, \alpha_3 > 0$ such that $\alpha_1 I_{23} + \alpha_2 I_{31} + \alpha_3 I_{12} = 0$. In this case, $\dot{V}(x_1,x_2,x_3) = 0$, and hence, the zero solution to (3.11)–(3.13) is Lyapunov stable. Show that the zero solution to (3.11)–(3.13) is Lyapunov stable without the constraint $I_1 > I_2 > I_3 > 0$ so long as $I_1, I_2, I_3 > 0$. △

Example 3.2. Consider the nonlinear dynamical system describing the motion of a simple pendulum with viscous damping given by

$$\ddot{\theta}(t) + \dot{\theta}(t) + \tfrac{g}{l}\sin\theta(t) = 0, \quad \theta(0) = \theta_0, \quad \dot{\theta}(0) = \dot{\theta}_0, \quad t \ge 0, \tag{3.15}$$

where g is the acceleration due to gravity and l is the length of the pendulum. To analyze the stability of this system consider the Lyapunov function candidate

$$V(\theta,\dot{\theta}) = \tfrac{1}{2}\dot{\theta}^2 + \tfrac{1}{2}(\theta+\dot{\theta})^2 + \tfrac{2g}{l}(1-\cos\theta). \tag{3.16}$$

Now, the Lyapunov derivative is given by

$$\dot{V}(\theta,\dot{\theta}) = \dot{\theta}\ddot{\theta} + (\theta+\dot{\theta})(\dot{\theta}+\ddot{\theta}) + \tfrac{2g}{l}\dot{\theta}\sin\theta = -(\dot{\theta}^2 + \tfrac{g}{l}\theta\sin\theta), \tag{3.17}$$

which is locally negative definite, and hence, the simple pendulum with viscous damping is locally asymptotically stable. Show that the *energy* Lyapunov function $V(\theta,\dot{\theta}) = \frac{1}{2}\dot{\theta}^2 + \frac{g}{l}(1-\cos\theta)$ fails to show that the Lyapunov derivative is strictly decreasing and, hence, cannot be used in conjunction with Theorem 3.1 to establish asymptotic stability of (3.15). △

Theorem 3.1 can be used to provide an estimate of the *domain of attraction* for the nonlinear dynamical system (3.1); that is, finding an open connected set $\mathcal{D}_0$ in $\mathbb{R}^n$ containing the origin with the property that every trajectory starting in $\mathcal{D}_0$ converges to the origin as time approaches infinity. It is important to note here that (3.3) and (3.5) are not sufficient to guarantee that every solution that starts in $\mathcal{D}$ will remain in $\mathcal{D}$ for all time. However, if (3.3) and (3.5) hold, then every *invariant set* of the nonlinear dynamical system (3.1) contained in $\mathcal{D}$ is also contained in the domain of attraction $\mathcal{D}_0$ of (3.1). As shown in the proof of Theorem 3.1 if there exists a Lyapunov function $V : \mathcal{D} \to \mathbb{R}$ such that (3.2), (3.3), and (3.5) hold, and if $\mathcal{D}_\beta = \{x \in \mathcal{D} : V(x) \le \beta\}$ is bounded, then $\mathcal{D}_\beta$ is a positively invariant set with respect to (3.1) and every system trajectory starting in $\mathcal{D}_\beta$ converges to the origin as time approaches infinity. Hence, $\mathcal{D}_\beta$ is an estimate of the domain of attraction. The question is then how large can we take $\beta > 0$ such that $\mathcal{D}_\beta$ remains bounded? Since $V(\cdot)$ is continuous and positive definite, there always exists a small enough $\beta > 0$ such that the Lyapunov level surface $V(x) \equiv \beta$ is bounded, and hence, $\mathcal{D}_\beta$ is bounded since $\mathcal{D}_\beta \subset \mathcal{B}_\varepsilon(0)$

for some $\varepsilon > 0$. However, depending on the structure of $V(\cdot)$, as β increases the Lyapunov level surface $V(x) = \beta$ can be unbounded, and hence, $\mathcal{D}_\beta$ becomes unbounded. For $\mathcal{D}_\beta \subset \mathcal{B}_\varepsilon(0)$, β must satisfy $\beta < \inf_{\|x\|\geq\varepsilon} V(x)$. Hence, if $\beta < \gamma$, where

$$\gamma = \lim_{\varepsilon\to\infty} \inf_{\|x\|\geq\varepsilon} V(x) < \infty \tag{3.18}$$

and $\gamma > 0$, then $\mathcal{D}_\beta$ is guaranteed to be bounded.

Next, we give a precise definition for the domain, or region, of attraction of the zero solution $x(t) \equiv 0$ of the nonlinear dynamical system (3.1).

Definition 3.2. Suppose the zero solution $x(t) \equiv 0$ to (3.1) is asymptotically stable. Then the *domain of attraction* $\mathcal{D}_0 \subseteq \mathcal{D}$ of (3.1) is given by

$$\mathcal{D}_0 \triangleq \{x_0 \in \mathcal{D} : \text{ if } x(0) = x_0, \text{ then } \lim_{t\to\infty} x(t) = 0\}. \tag{3.19}$$

As discussed above, the motivation for constructing the domain of attraction of a nonlinear dynamical system follows from the fact that there is no guarantee that a system trajectory starting in a subset $\mathcal{D}$ of the state space will remain in $\mathcal{D}$ even though the Lyapunov derivative is negative in $\mathcal{D}$, that is, the system trajectories move from one energy level to an inner energy level (see Figure 3.6). The problem of constructing the domain of attraction for locally stable nonlinear dynamical systems has received considerable attention in the literature [102, 174, 178, 285, 314, 443, 481]. Since, however, constructing the actual domain of attraction of a nonlinear dynamical system is system trajectory dependent, most of the techniques proposed in the literature provide a guaranteed subset of the domain of attraction.

To estimate a subset of the domain of attraction of the dynamical system (3.1) assume there exists a continuously differentiable function $V : \mathcal{D} \to \mathbb{R}$ such that (3.2), (3.3), and (3.5) are satisfied. Next, let $\mathcal{D}_\beta$ be the connected component of $\{x \in \mathcal{D} : V(x) \leq \beta\}$ containing the origin. Note that every trajectory starting in $\mathcal{D}_\beta$ will move to an inner energy surface and, hence, cannot escape $\mathcal{D}_\beta$. Hence, $\mathcal{D}_\beta$ is an estimate of the domain of attraction of the nonlinear dynamical system (3.1). Now, to maximize this estimate of the domain of attraction we maximize β such that $\mathcal{D}_\beta \subseteq \mathcal{D}$. Hence, define $V_\Gamma \triangleq \sup\{\beta > 0 : \mathcal{D}_\beta \subseteq \mathcal{D}\}$ so that

$$\mathcal{D}_{\mathrm{A}} \triangleq \{x \in \mathcal{D} : V(x) \leq V_\Gamma\}, \tag{3.20}$$

is a subset of the domain of attraction for (3.1) since $\dot{V}(x) < 0$ for all $x \in \mathcal{D}_{\mathrm{A}} \setminus \{0\} \subseteq \mathcal{D} \setminus \{0\}$.

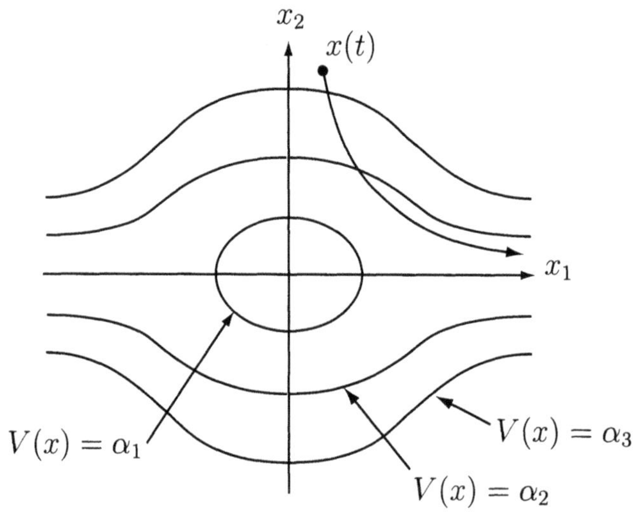

Figure 3.6 Visualization of the radial unboundedness requirement; $\alpha_1 < \alpha_2 < \alpha_3$.

The above discussion raises an interesting question; namely, under what conditions will the domain of attraction correspond to the entire state space $\mathbb{R}^n$? Of course, this corresponds to the case where the trajectory $s(t, x_0)$ of (3.1) approaches the origin as $t \to \infty$ for all $x_0 \in \mathbb{R}^n$ or, equivalently, the dynamical system is globally asymptotically stable. It follows from the proof of Theorem 3.1 that global asymptotic stability will hold if every point $x \in \mathbb{R}^n$ is contained in the compact set $\mathcal{D}_\beta$ for some $\beta > 0$. Clearly, in this case we require $\mathcal{D} = \mathbb{R}^n$ and that for every $\beta > 0$, $\mathcal{D}_\beta$ is bounded, that is, for every $\beta > 0$ there exists $r > 0$ such that if $x \in \mathcal{D}_\beta$, then $x \in \mathcal{B}_r(0)$ or, equivalently, $V(x) > \beta$ for all $x \notin \mathcal{B}_r(0)$. This condition ensuring that $\mathcal{D}_\beta$ is bounded for all $\beta > 0$ is implied by

$$V(x) \to \infty \quad \text{as} \quad \|x\| \to \infty. \tag{3.21}$$

A function $V(\cdot)$ satisfying (3.21) is called *proper* or *radially unbounded.*

Next, we state the global Lyapunov stability theorem where the domain of attraction of (3.1) is the entire state space.

Theorem 3.2. Consider the nonlinear dynamical system (3.1) and assume there exists a continuously differentiable function $V\colon \mathbb{R}^n \to \mathbb{R}$ such that

$$V(0) = 0, \tag{3.22}$$
$$V(x) > 0, \quad x \in \mathbb{R}^n, \quad x \neq 0, \tag{3.23}$$
$$V'(x)f(x) < 0, \quad x \in \mathbb{R}^n, \quad x \neq 0, \tag{3.24}$$
$$V(x) \to \infty \text{ as } \|x\| \to \infty. \tag{3.25}$$

Then the zero solution $x(t) \equiv 0$ to (3.1) is globally asymptotically stable. If, alternatively, there exist scalars $\alpha, \beta, \varepsilon > 0$, and $p \geq 1$, such that $V : \mathbb{R}^n \to \mathbb{R}$ satisfies

$$\alpha\|x\|^p \leq V(x) \leq \beta\|x\|^p, \qquad x \in \mathbb{R}^n \tag{3.26}$$

$$V'(x)f(x) \leq \ cV(x), \qquad x \in \mathbb{R}^n, \tag{3.27}$$

then the zero solution $x(t) \equiv 0$ to (3.1) is globally exponentially stable.

Proof. Let $x(0) \in \mathbb{R}^n$, and let $\beta \triangleq V(x_0)$. Now, the radial unboundedness condition (3.25) implies that there exists $\varepsilon > 0$ such that $V(x) > \beta$ for all $x \in \mathbb{R}^n$ such that $\|x\| \geq \varepsilon$. Hence, it follows from (3.24) that $V(x(t)) \leq V(x_0) = \beta$, $t \geq 0$, which implies that $x(t) \in \mathcal{B}_\varepsilon(0)$, $t \geq 0$. Now, the proof follows as in the proof of Theorem 3.1. □

The following example adopted from [178] shows the motivation for the radial unboundedness condition in ensuring global asymptotic stability.

Example 3.3. Consider the nonlinear dynamical system

$$\dot{x}_1(t) = -\frac{6x_1(t)}{[1+x_1^2(t)]^2} + 2x_2(t), \quad x_1(0) = x_{10}, \quad t \geq 0, \tag{3.28}$$

$$\dot{x}_2(t) = -\frac{2[x_1(t)+x_2(t)]}{[1+x_1^2(t)]^2}, \quad x_2(0) = x_{20}. \tag{3.29}$$

To examine the stability of this system, consider the Lyapunov function candidate $V(x_1, x_2) = x_1^2/(1+x_1^2) + x_2^2$. Note that $V(0,0) = 0$ and $V(x_1, x_2) > 0$, $(x_1, x_2) \in \mathbb{R} \times \mathbb{R}$, $(x_1, x_2) \neq (0,0)$; however, $V(x_1, x_2)$ is not radially unbounded. The Lyapunov derivative $\dot{V}(x_1, x_2)$ is given by

$$\begin{aligned}
\dot{V}(x_1, x_2) &= \left[\frac{2x_1(1+x_1^2) - 2x_1^3}{(1+x_1^2)^2}\right]\dot{x}_1 + 2x_2\dot{x}_2 \\
&= \frac{2x_1}{(1+x_1^2)^2}\left[-\frac{6x_1}{(1+x_1^2)^2} + 2x_2\right] - \frac{4x_2(x_1+x_2)}{(1+x_1^2)^2} \\
&= -\frac{12x_1^2}{(1+x_1^2)^4} - \frac{4x_2^2}{(1+x_1^2)^2} \\
&< 0, \quad (x_1, x_2) \in \mathbb{R} \times \mathbb{R}, \quad (x_1, x_2) \neq (0,0).
\end{aligned} \tag{3.30}$$

Clearly, the Lyapunov derivative is negative definite in all of $\mathbb{R}^2$, and hence, the zero solution $(x_1(t), x_2(t)) \equiv (0,0)$ to (3.28) and (3.29) is asymptotically stable. However, to show that (3.28) and (3.29) is not globally asymptotically stable, consider the hyperbola $x_2 = 2/(x_1 - \sqrt{2})$ in the x_1-x_2 plane. Now, on the hyperbola, $\dot{x}_2/\dot{x}_1$ is given by

$$\left.\frac{\dot{x}_2}{\dot{x}_1}\right|_{x_2=\frac{2}{x_1-\sqrt{2}}} = \left.\frac{-2(x_1+x_2)}{-6x_1+2x_2(1+x_1^2)^2}\right|_{x_2=\frac{2}{x_1-\sqrt{2}}} = -\frac{1}{2x_1^2+2\sqrt{2}x_1+1}, \tag{3.31}$$

whereas the slope of the hyperbola is given by

$$\frac{\mathrm{d}x_2}{\mathrm{d}x_1} = -\frac{2}{(x_1-\sqrt{2})^2} = -\frac{1}{\frac{1}{2}x_1^2-\sqrt{2}x_1+1}. \tag{3.32}$$

Next, for $x_1 > \sqrt{2}$, it follows from (3.31) and (3.32) that

$$2x_1^2+2\sqrt{2}x_1+1 > \frac{1}{2}x_1^2-\sqrt{2}x_1+1,$$

and hence,

$$\left.\frac{\dot{x}_2}{\dot{x}_1}\right|_{x_2=\frac{2}{x_1-\sqrt{2}}} > \frac{\mathrm{d}x_2}{\mathrm{d}x_1}, \tag{3.33}$$

for $x_1 > \sqrt{2}$. Furthermore, note that for $x_1 > \sqrt{2}$,

$$\begin{aligned}\dot{x}_1|_{x_2=\frac{2}{x_1-\sqrt{2}}} &= -\frac{6x_1}{(1+x_1^2)^2}+\frac{4}{x_1-\sqrt{2}}\\ &= \frac{4+6\sqrt{2}x_1+2x_1^2+4x_1^4}{(1+x_1^2)^2(x_1-\sqrt{2})}\\ &> 0.\end{aligned} \tag{3.34}$$

Since on the hyperbola $\dot{x}_1 > 0$ for $x_1 > \sqrt{2}$, it follows that the trajectories of (3.28) and (3.29) cannot cut the branch of the hyperbola lying in the first quadrant in the x_1-x_2 plane, and in the direction toward the x_1-x_2 axes (see Figure 3.7). Hence, since trajectories starting to the right of the hyperbola cannot reach the origin, it follows that the zero solution $(x_1(t), x_2(t)) \equiv (0,0)$ to (3.28) and (3.29) is not globally asymptotically stable. △

The radial unboundedness condition (3.25) ensures that the constant energy surfaces $V(x) = \alpha$, $\alpha > 0$, are hypersurfaces, and hence, since the system trajectories move from one energy surface to an inner energy surface, the system trajectories cannot drift away from the system equilibrium.

Example 3.4. Consider the nonlinear dynamical system

$$\dot{x}_1(t) = -x_1(t)+x_2^3(t), \quad x_1(0)=x_{10}, \quad t \geq 0, \tag{3.35}$$

$$\dot{x}_2(t) = -x_1(t)-x_2(t), \quad x_2(0)=x_{20}. \tag{3.36}$$

To examine the stability of this system, consider the Lyapunov function candidate $V(x_1, x_2) = \frac{1}{2}x_1^2 + \frac{1}{4}x_2^4$, which is positive definite and radially

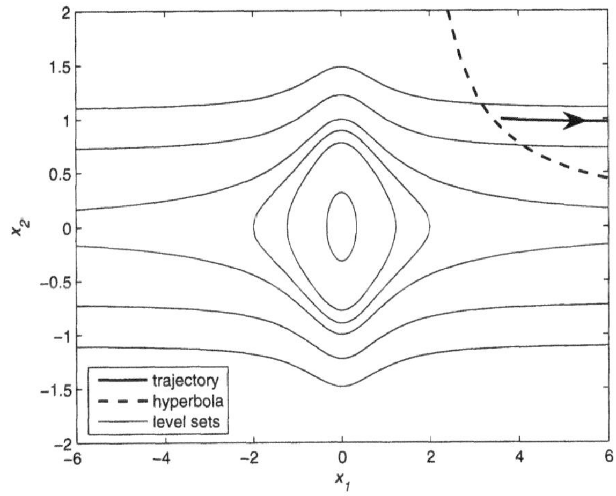

Figure 3.7 Level sets and an unbounded trajectory for Example 3.3.

unbounded in $\mathbb{R}^2$. Now, the Lyapunov derivative is given by

$$\dot{V}(x_1, x_2) = x_1\dot{x}_1 + x_2^3\dot{x}_2 = -x_1^2 - x_2^4 < 0, \qquad (x_1, x_2) \neq (0, 0), \qquad (3.37)$$

which implies that the zero solution $(x_1(t), x_2(t)) \equiv (0, 0)$ to (3.35) and (3.36) is globally asymptotically stable. Is this system globally exponentially stable? △

3.3 Invariant Set Stability Theorems

In this section, we introduce the Barbashin-Krasovskii-LaSalle invariance principle to relax one of the conditions on the Lyapunov function $V(\cdot)$ in the theorems given in Section 3.2. In particular, the strict negative-definiteness condition on the Lyapunov derivative can be relaxed while ensuring system asymptotic stability. Specifically, if a continuously differentiable function defined on a compact invariant set with respect to the nonlinear dynamical system (3.1) can be constructed whose derivative along the system's trajectories is negative semidefinite and no system trajectories can stay indefinitely at points where the function's derivative vanishes, then the system's equilibrium point is asymptotically stable. This result follows from the Barbashin-Krasovskii-LaSalle invariance principle for nonlinear dynamical systems, which we now state and prove.

Theorem 3.3 (Barbashin-Krasovskii-LaSalle Theorem). Consider the nonlinear dynamical system (3.1), assume that $\mathcal{D}_c \subset \mathcal{D}$ is a compact positively invariant set with respect to (3.1), and assume there exists a continuously differentiable function $V\colon \mathcal{D}_c \to \mathbb{R}$ such that $V'(x)f(x) \leq 0$, $x \in \mathcal{D}_c$. Let $\mathcal{R} \triangleq \{x \in \mathcal{D}_c : V'(x)f(x) = 0\}$ and let $\mathcal{M}$ be the largest invariant set contained in $\mathcal{R}$. If $x(0) \in \mathcal{D}_c$, then $x(t) \to \mathcal{M}$ as $t \to \infty$.

Proof. Let $x(t)$, $t \geq 0$, be a solution to (3.1) with $x(0) \in \mathcal{D}_c$. Since $V'(x)f(x) \leq 0$, $x \in \mathcal{D}_c$, it follows that

$$V(x(t)) - V(x(\tau)) = \int_\tau^t V'(x(s))f(x(s))\mathrm{d}s \leq 0, \qquad t \geq \tau,$$

and hence, $V(x(t)) \leq V(x(\tau))$, $t \geq \tau$, which implies that $V(x(t))$ is a nonincreasing function of t. Next, since $V(\cdot)$ is continuous on the compact set $\mathcal{D}_c$, there exists $\beta \in \mathbb{R}$ such that $V(x) \geq \beta$, $x \in \mathcal{D}_c$. Hence, $\gamma_{x_0} \triangleq \lim_{t\to\infty} V(x(t))$ exists. Now, for all $p \in \omega(x_0)$ there exists an increasing unbounded sequence $\{t_n\}_{n=0}^\infty$, with $t_0 = 0$, such that $x(t_n) \to p$ as $n \to \infty$. Since $V(x)$, $x \in \mathcal{D}_c$, is continuous, $V(p) = V(\lim_{n\to\infty} x(t_n)) = \lim_{n\to\infty} V(x(t_n)) = \gamma_{x_0}$, and hence, $V(x) = \gamma_{x_0}$ on $\omega(x_0)$. Now, since $\mathcal{D}_c$ is compact and positively invariant it follows that $x(t)$, $t \geq 0$, is bounded, and hence, it follows from Theorem 2.41 that $\omega(x_0)$ is a nonempty, compact invariant set. Hence, it follows that $V'(x)f(x) = 0$ on $\omega(x_0)$ and thus $\omega(x_0) \subset \mathcal{M} \subset \mathcal{R} \subset \mathcal{D}_c$. Finally, since $x(t) \to \omega(x_0)$ as $t \to \infty$, it follows that $x(t) \to \mathcal{M}$ as $t \to \infty$. □

The construction of $V(\cdot)$ in Theorem 3.3 can be used to guarantee the existence of the compact positively invariant set $\mathcal{D}_c$. Specifically, if $\mathcal{D}_\beta = \{x \in \mathcal{D} : V(x) \leq \beta\}$, where $\beta > 0$, is bounded and $\dot{V}(x) \leq 0$, $x \in \mathcal{D}_\beta$, then we can always take $\mathcal{D}_c = \mathcal{D}_\beta$. As discussed in Section 3.1, if $V(\cdot)$ is positive definite, then $\mathcal{D}_\beta$ is bounded for sufficiently small $\beta > 0$. Alternatively, if $V(\cdot)$ is radially unbounded, then $\mathcal{D}_\beta$ is bounded for every $\beta > 0$ irrespective of whether $V(\cdot)$ is positive definite or not.

Example 3.5. The Barbashin-Krasovskii-LaSalle invariant set theorem can be used to examine the stability of limit cycles. To see this, consider the nonlinear dynamical system

$$\dot{x}_1(t) = 4x_1^2(t)x_2(t) - g_1(x_1(t))[x_1^2(t) + 2x_2^2(t) - 4], \quad x_1(0) = x_{10}, \quad t \geq 0, \tag{3.38}$$

$$\dot{x}_2(t) = -2x_1^3(t) - g_2(x_2(t))[x_1^2(t) + 2x_2^2(t) - 4], \quad x_2(0) = x_{20}, \tag{3.39}$$

where $x_1 g_1(x_1) > 0$, $x_1 \neq 0$, $x_2 g_2(x_2) > 0$, $x_2 \neq 0$, $g_1(0) = 0$, and $g_2(0) = 0$. Now, note that the set defined by the ellipse $\mathcal{E} \triangleq \{(x_1, x_2) \in \mathbb{R} \times \mathbb{R} : x_1^2 + 2x_2^2 - 4 = 0\}$ is invariant since

$$\begin{aligned}\frac{\mathrm{d}}{\mathrm{d}t}[x_1^2(t) + 2x_2^2(t) - 4] &= -[2x_1(t)g_1(x_1(t)) + 4x_2(t)g_2(x_2(t))] \\ &\quad \cdot[x_1^2(t) + 2x_2^2(t) - 4] \\ &= 0, \qquad t \geq 0,\end{aligned} \tag{3.40}$$

and hence, if $(x_1(0), x_2(0)) \in \mathcal{E}$, then $(x_1(t), x_2(t)) \in \mathcal{E}$, $t \geq 0$. The motion on $\mathcal{E}$ is characterized by either of the equations

$$\dot{x}_1(t) = 4x_1^2(t)x_2(t), \tag{3.41}$$
$$\dot{x}_2(t) = -2x_1^3(t), \tag{3.42}$$

which shows that $\mathcal{E}$ is a limit cycle for (3.38) and (3.39) where the state vector moves clockwise.

To examine whether this limit cycle is attractive, define the function $V : \mathbb{R}^2 \to \mathbb{R}$ representing a measure of the distance to the limit cycle by $V(x_1, x_2) = (x_1^2 + 2x_2^2 - 4)^2$ and note that $V(0,0) = 16$. Next, let $\beta > 0$ and define

$$\begin{aligned}\mathcal{D}_{\rm c} &\triangleq \{(x_1, x_2) \in \mathbb{R} \times \mathbb{R} : \ V(x_1, x_2) \le \beta\} \\ &= \{(x_1, x_2) \in \mathbb{R} \times \mathbb{R} : \ (x_1^2 + 2x_2^2 - 4)^2 \le \beta\}.\end{aligned} \tag{3.43}$$

Now, note that

$$\begin{aligned}\dot{V}(x_1, x_2) &= 2(x_1^2 + 2x_2^2 - 4)\frac{\mathrm{d}}{\mathrm{d}t}(x_1^2 + 2x_2^2 - 4) \\ &= -2(x_1^2 + 2x_2^2 - 4)^2[2x_1 g_1(x_1) + 4x_2 g_2(x_2)] \\ &\le 0, \qquad (x_1, x_2) \in \mathcal{D}_{\rm c}.\end{aligned} \tag{3.44}$$

Next, defining $\mathcal{R} \triangleq \{(x_1, x_2) \in \mathbb{R} \times \mathbb{R} : \ \dot{V}(x_1, x_2) = 0\}$ it follows that the largest invariant set $\mathcal{M} \subseteq \mathcal{R}$ is given by

$$\mathcal{M} = \{(0,0)\} \cup \{(x_1, x_2) \in \mathbb{R} \times \mathbb{R} : \ x_1^2 + 2x_2^2 - 4 = 0\}. \tag{3.45}$$

Hence, it follows from Theorem 3.3 that all system trajectories starting in $\mathcal{D}_{\rm c}$ go to $(0,0)$ or $\mathcal{E}$. However, for $\beta < 16$, since $V(0,0) = 16$ and $V(1,0) = 9 < 16$, it follows that $(0,0) \notin \mathcal{D}_{\rm c}$ and $(0,0)$ corresponds to a local maximum of $V(x_1, x_2)$. Hence, $(0,0)$ is unstable. (Can this result be obtained by Lyapunov's indirect method? See Theorem 3.19.) Thus, if $(x_1(0), x_2(0)) \in \mathcal{D}_{\rm c}$, then $(x_1(t), x_2(t)) \to \mathcal{E}$ as $t \to \infty$, establishing that the limit cycle characterized by $\mathcal{E}$ is attractive. △

Next, using Theorem 3.3 we provide a generalization of Theorem 3.1 for local asymptotic stability of a nonlinear dynamical system.

Corollary 3.1. Consider the nonlinear dynamical system (3.1), assume that $\mathcal{D}_{\rm c} \subset \mathcal{D}$ is a compact positively invariant set with respect to (3.1) such that $0 \in \overset{\circ}{\mathcal{D}}_{\rm c}$, and assume that there exists a continuously differentiable function $V\colon \mathcal{D}_{\rm c} \to \mathbb{R}$ such that $V(0) = 0$, $V(x) > 0$, $x \ne 0$, and $V'(x)f(x) \le 0$, $x \in \mathcal{D}_{\rm c}$. Furthermore, assume that the set $\mathcal{R} \triangleq \{x \in \mathcal{D}_{\rm c}\colon \ V'(x)f(x) = 0\}$ contains no invariant set other than the set $\{0\}$. Then the zero solution $x(t) \equiv 0$ to (3.1) is asymptotically stable and $\mathcal{D}_{\rm c}$ is a subset of the domain of attraction of (3.1).

Proof. Lyapunov stability of the zero solution $x(t) \equiv 0$ to (3.1) follows from Theorem 3.1 since $V'(x)f(x) \leq 0$, $x \in \mathcal{D}_c$. Now, it follows from Theorem 3.3 that if $x_0 \in \mathcal{D}_c$, then $\omega(x_0) \subseteq \mathcal{M}$, where $\mathcal{M}$ denotes the largest invariant set contained in $\mathcal{R}$, which implies that $\mathcal{M} = \{0\}$. Hence, $x(t) \to \mathcal{M} = \{0\}$ as $t \to \infty$, establishing asymptotic stability of the zero solution $x(t) \equiv 0$ to (3.1). □

Example 3.6. To see the utility of Corollary 3.1, consider the nonlinear dynamical system describing the notion of a simple pendulum given in Example 3.2. To examine the stability of this system consider the more natural *energy* Lyapunov function candidate given by

$$V(\theta, \dot{\theta}) = \tfrac{1}{2}\dot{\theta}^2 + \tfrac{g}{l}(1 - \cos\theta), \tag{3.46}$$

with Lyapunov derivative

$$\dot{V}(\theta, \dot{\theta}) = \dot{\theta}\ddot{\theta} + \tfrac{g}{l}\dot{\theta}\sin\theta = -\dot{\theta}^2 \leq 0, \quad (\theta, \dot{\theta}) \in \mathbb{R} \times \mathbb{R}, \tag{3.47}$$

establishing Lyapunov stability of the zero solution $(\theta(t), \dot{\theta}(t)) \equiv (0, 0)$. Next, let $\beta > 0$ be such that $\mathcal{D}_c \triangleq \{(\theta, \dot{\theta}) : V(\theta, \dot{\theta}) \leq \beta\}$ is compact. Note that $\mathcal{D}_c$ is positively invariant. Now, to show asymptotic stability for this system let $\mathcal{R} \triangleq \{(\theta, \dot{\theta}) \in \mathbb{R} \times \mathbb{R} : \dot{V}(\theta, \dot{\theta}) = 0\} = \{(\theta, \dot{\theta}) \in \mathbb{R} \times \mathbb{R} : \dot{\theta} = 0\}$ and note that $\dot{V}(\theta, \dot{\theta}) < 0$ everywhere except on the line $\dot{\theta} = 0$, where $\dot{V}(\theta, \dot{\theta}) = 0$. Now, let $\mathcal{M}$ be the largest invariant set contained in $\mathcal{R}$ and note that $(0, 0) \in \mathcal{M} \subset \mathcal{R}$ since $(0, 0)$ is an equilibrium point. Furthermore, note that for the system to guarantee the condition $\dot{V}(\theta, \dot{\theta}) = 0$, the trajectory of the system must lie on the line $\dot{\theta} = 0$. Since $\dot{\theta}(t) \equiv 0$ implies $\ddot{\theta}(t) \equiv 0$ which, using (3.15), further implies $\sin\theta(t) \equiv 0$, it follows that, for $\theta \in (-\pi, \pi)$, $\mathcal{M} = \{(0, 0)\}$ is the largest invariant set contained in $\mathcal{R}$, and hence, by Corollary 3.1, $(\theta(t), \dot{\theta}(t)) \to (0, 0)$ as $t \to \infty$, establishing local asymptotic stability. △

Example 3.7. Corollary 3.1 can also be used to characterize the domain of attraction of a nonlinear dynamical system. To see this, consider the nonlinear dynamical system

$$\dot{x}_1(t) = x_1(t)[x_1^2(t) + x_2^2(t) - 1] - x_2(t), \quad x_1(0) = x_{10}, \quad t \geq 0, \tag{3.48}$$

$$\dot{x}_2(t) = x_1(t) + x_2(t)[x_1^2(t) + x_2^2(t) - 1], \quad x_2(0) = x_{20}, \tag{3.49}$$

with Lyapunov function candidate $V(x_1, x_2) = x_1^2 + x_2^2$. Now, the Lyapunov derivative is given by

$$\dot{V}(x_1, x_2) = 2x_1\dot{x}_1 + 2x_2\dot{x}_2 = 2(x_1^2 + x_2^2)(x_1^2 + x_2^2 - 1), \tag{3.50}$$

which is strictly negative if $0 < x_1^2 + x_2^2 < 1$. Next, define $\mathcal{D}_\beta \triangleq \{(x_1, x_2) \in \mathbb{R} \times \mathbb{R} : x_1^2 + x_2^2 \leq \beta\}$, where $\beta \in (0, 1)$, and note that $\mathcal{R} \triangleq \{(x_1, x_2) \in$

$\mathcal{D}_\beta : \dot{V}(x_1, x_2) = 0\} = \{(0,0)\} = \mathcal{M}$. Now, it follows from Corollary 3.1 that the zero solution $(x_1(t), x_2(t)) \equiv (0,0)$ to (3.48) and (3.49) is locally asymptotically stable with $\mathcal{D}_\beta$ being a subset of the domain of attraction for every $\beta \in (0,1)$. Hence, $\mathcal{D}_\mathrm{c} \triangleq \{(x_1, x_2) \in \mathbb{R} \times \mathbb{R} : x_1^2 + x_2^2 < 1\}$ is contained in the domain of attraction of (3.48) and (3.49). △

In Theorem 3.3 and Corollary 3.1 we explicitly assumed that there exists a compact invariant set $\mathcal{D}_\mathrm{c} \subset \mathcal{D}$ of (3.1). Next, we provide a result that does not require the explicit assumption of the existence of a compact invariant $\mathcal{D}_\mathrm{c}$.

Theorem 3.4. Consider the nonlinear dynamical system (3.1) and assume that there exists a continuously differentiable function $V\colon \mathbb{R}^n \to \mathbb{R}$ such that

$$V(0) = 0, \tag{3.51}$$
$$V(x) > 0, \quad x \in \mathbb{R}^n, \quad x \neq 0, \tag{3.52}$$
$$V'(x)f(x) \leq 0, \quad x \in \mathbb{R}^n. \tag{3.53}$$

Let $\mathcal{R} \triangleq \{x \in \mathbb{R}^n \colon V'(x)f(x) = 0\}$ and let $\mathcal{M}$ be the largest invariant set contained in $\mathcal{R}$. Then all solutions $x(t)$, $t \geq 0$, of (3.1) that are bounded approach $\mathcal{M}$ as $t \to \infty$.

Proof. Let $x \in \mathbb{R}^n$ be such that trajectory $s(t, x)$, $t \geq 0$, of (3.1) is bounded. Now, with $\mathcal{D}_\mathrm{c} = \overline{\mathcal{O}_x^+}$, it follows from Theorem 3.3 that $s(t, x) \to \mathcal{M}$ as $t \to \infty$. □

Next, we present the global invariant set theorem for guaranteeing global asymptotic stability of a nonlinear dynamical system.

Theorem 3.5. Consider the nonlinear dynamical system (3.1) and assume there exists a continuously differentiable function $V\colon \mathbb{R}^n \to \mathbb{R}$ such that

$$V(0) = 0, \tag{3.54}$$
$$V(x) > 0, \quad x \in \mathbb{R}^n, \quad x \neq 0, \tag{3.55}$$
$$V'(x)f(x) \leq 0, \quad x \in \mathbb{R}^n, \tag{3.56}$$
$$V(x) \to \infty \text{ as } \|x\| \to \infty. \tag{3.57}$$

Furthermore, assume that the set $\mathcal{R} \triangleq \{x \in \mathbb{R}^n\colon V'(x)f(x) = 0\}$ contains no invariant set other than the set $\{0\}$. Then the zero solution $x(t) \equiv 0$ to (3.1) is globally asymptotically stable.

Proof. Since (3.54)–(3.56) hold, it follows from Theorem 3.1 that the zero solution $x(t) \equiv 0$ to (3.1) is Lyapunov stable while the radial unboundedness condition (3.57) implies that all solutions to (3.1) are bounded. Now, Theorem 3.4 implies that $x(t) \to \mathcal{M}$ as $t \to \infty$. However, since $\mathcal{R}$ contains no invariant set other than the set $\{0\}$, the set $\mathcal{M}$ is $\{0\}$, and hence, global asymptotic stability is immediate. □

Theorems 3.3, 3.4, and 3.5 are known as invariant set theorems. Since for local asymptotic stability $V(x)$ is defined on a compact positively invariant set $\mathcal{D}_{\rm c}$, $\mathcal{D}_{\rm c}$ provides an estimate of the domain of attraction for the nonlinear dynamical system (3.1) which is not necessarily of the form given by (3.20). Finally, unlike Lyapunov's theorem, the Barbashin-Krasovskii-LaSalle invariant set theorem relaxes the strict negative-definiteness condition on the Lyapunov derivative $\dot{V}(x)$, $x \in \mathcal{D}_{\rm c}$, for both local and global asymptotic stability.

3.4 Construction of Lyapunov Functions

As shown in Sections 3.2 and 3.3 the key requirement in applying Lyapunov's direct method and the Barbashin-Krasovskii-LaSalle invariance principle for examining the stability of nonlinear systems is the construction of a Lyapunov function. For certain systems, especially ones that have physical energy interpretations, constructing a system Lyapunov function can be straightforward. In other cases, however, that can be a difficult task. In this section, we present four systematic approaches for constructing Lyapunov functions; namely, the *variable gradient*, *Krasovskii's*, *Zubov's*, and the *Energy-Casimir* methods.

The variable gradient method assumes a certain form for the gradient of an unknown Lyapunov function and then by integrating the assumed gradient one can often arrive at a Lyapunov function. To see this, let $V : \mathcal{D} \to \mathbb{R}$ be a continuously differentiable function and let $g(x) = \left(\frac{\partial V}{\partial x}\right)^{\rm T}$. Now, the derivative of $V(x)$ along the trajectories of (3.1) is given by

$$\begin{aligned} \dot{V}(x) &= \sum_{i=1}^{n} \frac{\partial V}{\partial x_i} \dot{x}_i \\ &= \sum_{i=1}^{n} \frac{\partial V}{\partial x_i} f_i(x) \\ &= \left[\begin{array}{cccc} \frac{\partial V}{\partial x_1}, & \frac{\partial V}{\partial x_2}, & \ldots, & \frac{\partial V}{\partial x_n} \end{array} \right] \left[\begin{array}{c} f_1(x) \\ f_2(x) \\ \vdots \\ f_n(x) \end{array} \right] \\ &= \frac{\partial V}{\partial x} f(x) \end{aligned}$$

$$= g^{\mathrm{T}}(x)f(x). \tag{3.58}$$

Next, construct $g(x)$ such that $g(x)$ is a gradient for a positive-definite function and $\dot{V}(x) = g^{\mathrm{T}}(x)f(x) < 0$, $x \in \mathcal{D}$, $x \neq 0$. Specifically, it follows from (3.58) that the function $V(x)$ can be computed from the line integral

$$V(x) = \int_0^x g^{\mathrm{T}}(s)\mathrm{d}s = \int_0^x \sum_{i=1}^{n} g_i(s)\mathrm{d}s_i. \tag{3.59}$$

Recall that the line integral of a gradient vector $g : \mathbb{R}^n \to \mathbb{R}^n$ is path independent [12, Theorem 10-37], and hence, the integration in (3.59) can be taken along any path joining the origin to $x \in \mathbb{R}^n$. Choosing a path made up of line segments parallel to the coordinate axes, (3.59) becomes

$$\begin{aligned} V(x) &= \int_0^{x_1} g_1(s_1, 0, \ldots, 0)\mathrm{d}s_1 + \int_0^{x_2} g_2(x_1, s_2, \ldots, 0)\mathrm{d}s_2 \\ &\quad + \cdots + \int_0^{x_n} g_n(x_1, x_2, \ldots, x_{n-1}, s_n)\mathrm{d}s_n. \end{aligned} \tag{3.60}$$

Alternatively, using the transformation $s = \sigma x$, where $\sigma \in [0, 1]$, (3.59) can be rewritten as

$$V(x) = \int_0^1 g^{\mathrm{T}}(\sigma x)x\mathrm{d}\sigma = \int_0^1 \sum_{i=1}^{n} g_i(\sigma x)x_i\mathrm{d}\sigma. \tag{3.61}$$

The following result shows that $g(x)$ is a gradient of a real valued function $V : \mathbb{R}^n \to \mathbb{R}$ if and only if the Jacobian matrix $\partial g/\partial x$ is symmetric.

Proposition 3.1. The function $g : \mathbb{R}^n \to \mathbb{R}^n$ is the gradient vector of a scalar-valued function $V : \mathbb{R}^n \to \mathbb{R}$ if and only if

$$\frac{\partial g_i}{\partial x_j} = \frac{\partial g_j}{\partial x_i}, \quad i, j = 1, \ldots, n. \tag{3.62}$$

Proof. If $g^{\mathrm{T}}(x) = \frac{\partial V}{\partial x}$, then $g_i(x) = \frac{\partial V}{\partial x_i}$. Since $\frac{\partial g_i}{\partial x_j} = \frac{\partial^2 V}{\partial x_j \partial x_i} = \frac{\partial g_j}{\partial x_i}$, $i, j = 1, \ldots, n$, necessity is immediate. To show sufficiency, assume $\frac{\partial g_i}{\partial x_j} = \frac{\partial g_j}{\partial x_i}$, $i, j = 1, \ldots, n$, and define $V(x)$ as the line integral

$$V(x) = \int_0^1 g^{\mathrm{T}}(\sigma x)x\mathrm{d}\sigma = \int_0^1 \sum_{j=1}^{n} g_j(\sigma x)x_j\mathrm{d}\sigma. \tag{3.63}$$

Hence,

$$\frac{\partial V}{\partial x_i} = \int_0^1 \sum_{j=1}^{n} \frac{\partial g_j}{\partial x_i}(\sigma x)x_j\sigma\mathrm{d}\sigma + \int_0^1 g_i(\sigma x)\mathrm{d}\sigma$$

$$
\begin{aligned}
&= \int_0^1 \sum_{j=1}^{n} \frac{\partial g_i}{\partial x_j}(\sigma x)x_j\sigma \mathrm{d}\sigma + \int_0^1 g_i(\sigma x)\mathrm{d}\sigma \\
&= \int_0^1 \frac{\mathrm{d}(\sigma g_i(\sigma x))}{\mathrm{d}\sigma}\mathrm{d}\sigma \\
&= g_i(x), \qquad i = 1, \ldots, n, \qquad (3.64)
\end{aligned}
$$

which implies that $g^{\mathrm{T}}(x) = \frac{\partial V}{\partial x}$. □

Choosing $g(x)$ such that $g^{\mathrm{T}}(x)f(x) < 0$, $x \in \mathcal{D}$, $x \neq 0$, it follows from Proposition 3.1 that $V(x)$ can be computed from the line integral

$$
V(x) = \int_0^1 g^{\mathrm{T}}(\sigma x)x\mathrm{d}\sigma = \int_0^1 \sum_{j=1}^{n} g_j(\sigma x)x_j\mathrm{d}\sigma. \qquad (3.65)
$$

Once arriving at $V(x)$, $x \in \mathcal{D}$, it is important to check whether $V(\cdot)$ is positive definite.

Example 3.8. Consider the nonlinear dynamical system

$$
\begin{aligned}
\dot{x}_1(t) &= x_2(t), \qquad x_1(0) = x_{10}, \qquad t \geq 0, \qquad (3.66) \\
\dot{x}_2(t) &= -[x_1(t) + x_2(t)] - \sin(x_1(t) + x_2(t)), \qquad x_2(0) = x_{20}. \qquad (3.67)
\end{aligned}
$$

To construct a Lyapunov function for (3.66) and (3.67) let $g(x) = [a_{11}x_1 + a_{12}x_2,\ a_{21}x_1 + a_{22}x_2]^{\mathrm{T}}$ and let $a_{12} = a_{21} = \beta$ so that the symmetry requirement (3.62) holds. Now,

$$
\begin{aligned}
\dot{V}(x) &= g^{\mathrm{T}}(x)f(x) \\
&= (a_{11}x_1 + \beta x_2)x_2 - (\beta x_1 + a_{22}x_2)[(x_1 + x_2) + \sin(x_1 + x_2)]. \qquad (3.68)
\end{aligned}
$$

Taking $a_{11} = 2\beta$, $a_{22} = \beta$, and $\beta > 0$ it follows that

$$
\dot{V}(x) = -\beta x_1^2 - \beta(x_1 + x_2)\sin(x_1 + x_2) < 0, \qquad (x_1, x_2) \in \mathcal{D}, \qquad (3.69)
$$

where $\mathcal{D} \triangleq \{(x_1, x_2) : |x_1 + x_2| < \pi\}$. Hence,

$$
\begin{aligned}
V(x) &= \int_0^1 [g_1(\sigma x_1, \sigma x_2)x_1 + g_2(\sigma x_1, \sigma x_2)x_2]\mathrm{d}\sigma \\
&= \int_0^1 \beta[2x_1^2 + 2x_1x_2 + x_2^2]\sigma\mathrm{d}\sigma \\
&= \beta x_1^2 + \beta x_1x_2 + \tfrac{1}{2}\beta x_2^2. \qquad (3.70)
\end{aligned}
$$

Note that $V(0,0) = 0$ and $V(x_1, x_2) = \frac{1}{2}\beta x_1^2 + \frac{1}{2}\beta(x_1 + x_2)^2 > 0$, $(x_1, x_2) \in \mathbb{R} \times \mathbb{R}$, $(x_1, x_2) \neq (0, 0)$, and hence, $V(x)$, $x \in \mathcal{D}$, is a Lyapunov function for (3.66) and (3.67). △

Next, we present Krasovskii's method for constructing a Lyapunov function for a given nonlinear system. First, however, the following proposition is needed.

Proposition 3.2. Let $f, g : \mathbb{R}^n \to \mathbb{R}^n$ be continuously differentiable functions such that $f(0) = 0$. Then for every $x \in \mathbb{R}^n$ there exists $\alpha \in [0,1]$ such that

$$g^{\mathrm{T}}(x)f(x) = g^{\mathrm{T}}(x)\frac{\partial f}{\partial x}(\alpha x)x. \tag{3.71}$$

Proof. Let $p \in \mathbb{R}^n$ and note that it follows from the mean value theorem (Theorem 2.16) that for every $x \in \mathbb{R}^n$ there exists $\alpha \in [0,1]$ such that

$$\begin{aligned} g^{\mathrm{T}}(p)f(x) &= g^{\mathrm{T}}(p)[f(x) - f(0)] \\ &= g^{\mathrm{T}}(p)\left[\frac{\partial f}{\partial x}(\alpha x)x\right]. \end{aligned}$$

Hence, there exists $\alpha \in [0,1]$ such that

$$g^{\mathrm{T}}(p)f(p) = g^{\mathrm{T}}(p)\left[\frac{\partial f}{\partial x}(\alpha p)p\right]. \tag{3.72}$$

Now, since p is arbitrary, the result follows. □

Theorem 3.6 (Krasovskii's Theorem). Let $x(t) \equiv 0$ be an equilibrium point for the nonlinear dynamical system

$$\dot{x}(t) = f(x(t)), \qquad x(0) = x_0, \qquad t \geq 0, \tag{3.73}$$

where $f : \mathcal{D} \to \mathbb{R}^n$ is continuously differentiable and $\mathcal{D}$ is an open set with $0 \in \mathcal{D}$. Assume there exist positive definite matrices $P \in \mathbb{R}^{n\times n}$ and $R \in \mathbb{R}^{n\times n}$ such that

$$\left[\frac{\partial f}{\partial x}(x)\right]^{\mathrm{T}} P + P\left[\frac{\partial f}{\partial x}(x)\right] \leq -R, \qquad x \in \mathcal{D}, \qquad x \neq 0. \tag{3.74}$$

Then, the zero solution $x(t) \equiv 0$ to (3.73) is a unique asymptotically stable equilibrium with Lyapunov function $V(x) = f^{\mathrm{T}}(x)Pf(x)$. If, in addition, $\mathcal{D} = \mathbb{R}^n$, then the zero solution $x(t) \equiv 0$ to (3.73) is a unique globally asymptotically stable equilibrium.

Proof. Suppose, *ad absurdum*, that there exists $x_{\mathrm{e}} \in \mathcal{D}$ such that $x_{\mathrm{e}} \neq 0$ and $f(x_{\mathrm{e}}) = 0$. In this case, it follows from Proposition 3.2 that for every $x_{\mathrm{e}} \in \mathcal{D}$ there exists $\alpha \in [0,1]$ such that $x_{\mathrm{e}}^{\mathrm{T}}Pf(x_{\mathrm{e}}) = x_{\mathrm{e}}^{\mathrm{T}}P\frac{\partial f}{\partial x}(\alpha x_{\mathrm{e}})x_{\mathrm{e}}$. Hence,

$$x_{\mathrm{e}}^{\mathrm{T}}\left\{\left[\frac{\partial f}{\partial x}(\alpha x_{\mathrm{e}})\right]^{\mathrm{T}} P + P\left[\frac{\partial f}{\partial x}(\alpha x_{\mathrm{e}})\right]\right\} x_{\mathrm{e}} = 0, \tag{3.75}$$

which contradicts (3.74). Hence, there does not exist $x_{\mathrm{e}} \in \mathcal{D}$, $x_{\mathrm{e}} \neq 0$, such that $f(x_{\mathrm{e}}) = 0$. Next, note that $V(x) = f^{\mathrm{T}}(x)Pf(x) \geq \lambda_{\min}(P)\|f(x)\|_2^2 \geq 0$,

$x \in \mathcal{D}$, which implies that $V(x) = 0$ if and only if $f(x) = 0$ or, equivalently, $V(x) = 0$ if and only if $x = 0$. Hence, $V(x) = f^{\mathrm{T}}(x)Pf(x) > 0$, $x \in \mathcal{D}$, $x \neq 0$. Next, computing the derivative of $V(x)$ along the trajectories of (3.73) and using (3.74) yields

$$\begin{aligned}
\dot{V}(x) &= V'(x)f(x) \\
&= 2f^{\mathrm{T}}(x)P\frac{\partial f(x)}{\partial x}f(x) \\
&= f^{\mathrm{T}}(x)\left\{\left[\frac{\partial f}{\partial x}(x)\right]^{\mathrm{T}} P + P\left[\frac{\partial f}{\partial x}(x)\right]\right\} f(x) \\
&\leq -f^{\mathrm{T}}(x)Rf(x) \\
&\leq -\lambda_{\min}(R)\|f(x)\|_2^2 \\
&\leq 0, \qquad x \in \mathcal{D}.
\end{aligned} \tag{3.76}$$

Now, since $f(x) = 0$ if and only if $x = 0$, it follows that $\dot{V}(x) < 0$, $x \in \mathcal{D}$, $x \neq 0$, which proves that the zero solution $x(t) \equiv 0$ to (3.73) is a unique asymptotically stable equilibrium with Lyapunov function $V(x) = f^{\mathrm{T}}(x)Pf(x)$.

Finally, in the case where $\mathcal{D} = \mathbb{R}^n$ we need only show that $V(x) \to \infty$ as $\|x\| \to \infty$. Note that it follows from Proposition 3.2 that for every $x \in \mathbb{R}^n$ and some $\alpha \in (0,1)$,

$$\begin{aligned}
x^{\mathrm{T}}Pf(x) &= x^{\mathrm{T}}P\frac{\partial f}{\partial x}(\alpha x)x \\
&= \tfrac{1}{2}x^{\mathrm{T}}\left\{\left[\frac{\partial f}{\partial x}(\alpha x)\right]^{\mathrm{T}} P + P\left[\frac{\partial f}{\partial x}(\alpha x)\right]\right\} x \\
&\leq -\tfrac{1}{2}x^{\mathrm{T}}Rx \\
&\leq -\tfrac{1}{2}\lambda_{\min}(R)\|x\|_2^2, \qquad x \in \mathbb{R}^n,
\end{aligned} \tag{3.77}$$

which implies

$$\frac{|x^{\mathrm{T}}Pf(x)|}{\|x\|_2^2} \geq \tfrac{1}{2}\lambda_{\min}(R), \qquad x \in \mathbb{R}^n, \qquad x \neq 0. \tag{3.78}$$

Hence, since $|x^{\mathrm{T}}Pf(x)| \leq \lambda_{\max}(P)\|x\|\|f(x)\|$, $x \in \mathbb{R}^n$, it follows from (3.78) that $\|f(x)\| \geq \frac{\lambda_{\min}(R)}{2\lambda_{\max}(P)}\|x\|$, which implies that $V(x) \to \infty$ as $\|x\| \to \infty$. The result now is immediate by repeating the steps of the first part of the proof. □

Example 3.9. Consider the nonlinear dynamical system

$$\dot{x}_1(t) = -3x_1(t) + x_2(t), \qquad x_1(0) = x_{10}, \qquad t \geq 0, \tag{3.79}$$

$$\dot{x}_2(t) = x_1(t) - x_2(t) - x_2^3(t), \qquad x_2(0) = x_{20}. \tag{3.80}$$

Note that $(0,0)$ is the only equilibrium point of (3.79) and (3.80). Next, computing

$$\frac{\partial f}{\partial x} = \begin{bmatrix} \frac{\partial f_1(x)}{\partial x_1} & \frac{\partial f_1(x)}{\partial x_2} \\ \frac{\partial f_2(x)}{\partial x_1} & \frac{\partial f_2(x)}{\partial x_2} \end{bmatrix} = \begin{bmatrix} -3 & 1 \\ 1 & -1-3x_2^2 \end{bmatrix}, \tag{3.81}$$

it can easily be shown that (3.74) holds with $P = I_2$ and $R = I_2$ so that all the conditions of Theorem 3.6 are satisfied. Hence, the zero solution $(x_1(t), x_2(t)) \equiv (0,0)$ to (3.79) and (3.80) is globally asymptotically stable with Lyapunov function $V(x) = f^{\mathrm{T}}(x)Pf(x) = f^{\mathrm{T}}(x)f(x) = (-3x_1+x_2)^2 + (x_1 - x_2 - x_2^3)^2$. △

Next, we present Zubov's method for constructing Lyapunov functions for nonlinear systems. Unlike the variable gradient method and Krasovskii's method, Zubov's method additionally characterizes a domain of attraction for a given nonlinear system.

Theorem 3.7 (Zubov's Theorem). Consider the nonlinear dynamical system (3.1) with $f(0) = 0$. Let $\mathcal{D} \subset \mathbb{R}^n$ be bounded and assume there exist a continuously differentiable function $V : \mathcal{D} \to \mathbb{R}$ and a continuous function $h : \mathbb{R}^n \to \mathbb{R}$ such that $V(0) = 0$, $h(0) = 0$, and

$$0 < V(x) < 1, \qquad x \in \mathcal{D}, \qquad x \neq 0, \tag{3.82}$$

$$V(x) \to 1 \text{ as } x \to \partial\mathcal{D}, \tag{3.83}$$

$$h(x) > 0, \qquad x \in \mathbb{R}^n, \qquad x \neq 0, \tag{3.84}$$

$$V'(x)f(x) = -h(x)[1 - V(x)]. \tag{3.85}$$

Then, the zero solution $x(t) \equiv 0$ to (3.1) is asymptotically stable with domain of attraction $\mathcal{D}$.

Proof. It follows from (3.82), (3.84), and (3.85) that in a neighborhood $\mathcal{B}_\varepsilon(0)$ of the origin $V(x) > 0$ and $\dot{V}(x) < 0$, $x \in \mathcal{B}_\varepsilon(0)$. Hence, the origin is locally asymptotically stable. Now, to show that $\mathcal{D}$ is the domain of attraction we need to show that $x(0) \in \mathcal{D}$ implies $x(t) \to 0$ as $t \to \infty$ and $x(0) \notin \mathcal{D}$ implies $x(t) \not\to 0$ as $t \to \infty$. Let $x(0) \in \mathcal{D}$. Then, by (3.82), $V(x(0)) < 1$. Next, let $\beta > 0$ be such that $V(x(0)) \leq \beta < 1$ and define $\mathcal{D}_\beta \triangleq \{x \in \mathcal{D} : V(x) \leq \beta\}$. Note that $\mathcal{D}_\beta \subset \mathcal{D}$ and $\mathcal{D}_\beta$ is bounded since $\lim_{x\to\partial\mathcal{D}} V(x) = 1$ and $\beta < 1$. Furthermore, since $\dot{V}(x) < 0$, $x \in \mathcal{D}_\beta$, it follows that $\mathcal{D}_\beta$ is a positively invariant set. Now, using (3.82) it follows that $\dot{V}(x) = 0$, $x \in \mathcal{D}$, implies that $h(x) = 0$, $x \in \mathcal{D}$, which further implies $x = 0$. Hence, it follows from Theorem 3.3 that $x(t) \to 0$ as $t \to \infty$.

Next, let $x(0) \notin \mathcal{D}$ and assume, *ad absurdum*, that $x(t) \to 0$ as $t \to \infty$. In this case, $x(t) \to \mathcal{D}$ for some $t \geq 0$. Hence, there exist finite times t_1 and t_2 such that $x(t_1) \in \partial\mathcal{D}$ and $x(t) \in \mathcal{D}$ for all $t \in (t_1, t_2]$. Next, define

$W(x) \triangleq 1 - V(x)$ and note that $\dot{W}(x) = h(x)W(x)$ or, equivalently,

$$\int_{W_0}^{W(x(t))} \frac{\mathrm{d}w}{w} = \int_{t_0}^{t} h(x(s))\mathrm{d}s, \tag{3.86}$$

where $W_0 \triangleq W(x(t_0))$. Integrating (3.86) and rearranging terms yields

$$1 - V(x(t_0)) = [1 - V(x(t))]e^{-\int_{t_0}^{t} h(x(s))\mathrm{d}s}. \tag{3.87}$$

Now, taking $t = t_2$, letting $t_0 \to t_1$, and using (3.83) it follows that $\lim_{t_0 \to t_1}[1 - V(x(t_0))] = 0$ and $\lim_{t_0 \to t_1}[1 - V(x(t_2))]e^{-\int_{t_0}^{t} h(x(s))\mathrm{d}s} > 0$, which is a contradiction. Hence, for $x(0) \notin \mathcal{D}$, $x(t) \not\to 0$ as $t \to \infty$. $\square$

Note that in the case where $\mathcal{D}$ is unbounded or $\mathcal{D} = \mathbb{R}^n$, (3.83) becomes $V(x) \to 1$ as $\|x\| \to \infty$. In latter case, the conditions of Theorem 3.7, with (3.83) replaced by $V(x) \to 1$ as $\|x\| \to \infty$, guarantee that the zero solution $x(t) \equiv 0$ to (3.1) is globally asymptotically stable.

Example 3.10. Consider the second-order nonlinear dynamical system adopted from [178] given by

$$\dot{x}_1(t) = -f_1(x_1(t)) + f_2(x_2(t)), \qquad x_1(0) = x_{10}, \qquad t \geq 0, \tag{3.88}$$

$$\dot{x}_2(t) = -f_3(x_1(t)), \qquad x_2(0) = x_{20}, \tag{3.89}$$

where $f_i(0) = 0$, $\sigma f_i(\sigma) > 0$, $\sigma \in (-a_i, b_i)$, for $i = 1, 2, 3$, $a_1 = a_3$, $b_1 = b_3$, and $\int_0^y f_i(s)\mathrm{d}s \to \infty$ as $y \to -a_i$ or $y \to b_i$, for $i = 2, 3$. Next, let $h(x) = f_1(x_1)f_3(x_1)$ and let $V(x)$ be of the form $V(x) = 1 - V_1(x_1)V_2(x_2)$, where $V_i : \mathbb{R} \to \mathbb{R}$, $V_i(0) = 1$, $i = 1, 2$. Now, it follows from (3.85) that

$$\begin{aligned}
&[V_1'(x_1) + f_3(x_1)V_1(x_1)]V_2(x_2)f_1(x_1) \\
&\qquad + [V_2'(x_2)V_1(x_1)f_3(x_1) - V_1'(x_1)V_2(x_2)f_2(x_1)] = 0,
\end{aligned} \tag{3.90}$$

which can be satisfied by setting $V_1'(x_1) = -f_3(x_1)V_1(x_1)$ and $V_2'(x_2) = -f_2(x_2)V_2(x_2)$. Hence, (3.85) holds with

$$V(x) = 1 - e^{-[\int_0^{x_1} f_3(s)\mathrm{d}s + \int_0^{x_2} f_2(s)\mathrm{d}s]}. \tag{3.91}$$

Note that (3.91) satisfies $V(0) = 0$, (3.82), and (3.83) for all $x \in \mathcal{D} = \{x \in \mathbb{R}^2 : -a_i < x_i < b_i\}$, $i = 1, 2$. Furthermore, it can be easily shown that $\dot{V}(x) = -f_1(x_1)f_3(x_1)[1 - V(x)] \leq 0$, which proves Lyapunov stability of (3.88) and (3.89). To show asymptotic stability note that $\dot{V}(x) = 0$ implies $f_1(x_1)f_3(x_1) = 0$, which further implies $x_1 = 0$. Furthermore, $x_1(t) \equiv 0$ implies $f_2(x_2(t)) \equiv 0$, which further implies $x_2(t) \equiv 0$. Hence, with $\mathcal{D}_{\mathrm{c}} = \overline{\mathcal{D}}$, it follows from Theorem 3.3 that the zero solution $(x_1(t), x_2(t)) \equiv (0, 0)$ to (3.88) and (3.89) is asymptotically stable with domain of attraction $\overline{\mathcal{D}}$. $\triangle$

Finally, we present the energy-Casimir method for constructing Lyapunov functions for nonlinear dynamical systems. This method exploits the existence of *dynamical invariants*, or *integrals of motion*, called *Casimir functions* of the nonlinear dynamical system (3.1). In particular, a function $C : \mathcal{D} \to \mathbb{R}$ is an integral of motion of (3.1) if it is conserved along the flow of (3.1), that is, $C'(x)f(x) = 0$. For the statement of our next result let $r \geq 2$ and let $C_i : \mathcal{D} \to \mathbb{R}$, $i = 1, \ldots, r$, be two-times continuously differentiable Casimir functions. Furthermore, define

$$E(x) \triangleq \sum_{i=1}^{r} \mu_i C_i(x), \tag{3.92}$$

for $\mu_i \in \mathbb{R}$, $i = 1, \ldots, r$.

Theorem 3.8 (Energy-Casimir Theorem). Consider the nonlinear dynamical system (3.1) where $f : \mathcal{D} \to \mathbb{R}^n$ is Lipschitz continuous on $\mathcal{D}$. Let $x_\mathrm{e} \in \mathcal{D}$ be an equilibrium point of (3.1) and let $C_i : \mathcal{D} \to \mathbb{R}$, $i = 1, \ldots, r$, be Casimir functions of (3.1). Assume that the vectors $C_i'(x_\mathrm{e})$, $i = 2, \ldots, r$, are linearly independent, and suppose there exists $\mu = [\mu_1, \ \mu_2, \ \ldots, \mu_r]^\mathrm{T} \in \mathbb{R}^r$ such that $\mu_1 \neq 0$, $E'(x_\mathrm{e}) = 0$, and $x^\mathrm{T} E''(x_\mathrm{e}) x > 0$, $x \in \mathcal{M}$, where $\mathcal{M} \triangleq \{x \in \mathcal{D} : \ C_i'(x_\mathrm{e})x = 0, \ i = 2, \ldots, r\}$. Then, there exists $\alpha \geq 0$ such that

$$E''(x_\mathrm{e}) + \alpha \sum_{i=2}^{r} \left(\frac{\partial C_i}{\partial x}(x_\mathrm{e})\right)^\mathrm{T} \left(\frac{\partial C_i}{\partial x}(x_\mathrm{e})\right) > 0. \tag{3.93}$$

Furthermore, the equilibrium solution $x(t) \equiv x_\mathrm{e}$ of (3.1) is Lyapunov stable with Lyapunov function

$$V(x) = E(x) - E(x_\mathrm{e}) + \frac{\alpha}{2} \sum_{i=2}^{r} [C_i(x) - C_i(x_\mathrm{e})]^2. \tag{3.94}$$

Proof. Note that

$$\begin{aligned}
\dot{V}(x) &= V'(x) f(x) \\
&= E'(x) f(x) + \alpha \sum_{i=2}^{r} [C_i(x) - C_i(x_\mathrm{e})] C_i'(x) f(x) \\
&= \sum_{i=1}^{r} \mu_i C_i'(x) f(x) + \alpha \sum_{i=2}^{r} [C_i(x) - C_i(x_\mathrm{e})] C_i'(x) f(x) \\
&= 0, \quad x \in \mathcal{D}.
\end{aligned} \tag{3.95}$$

Now, it need only be shown that $V(x_\mathrm{e}) = 0$ and $V(x) > 0$, $x \in \mathcal{D}$, $x \neq x_\mathrm{e}$. Clearly, $V(x_\mathrm{e}) = 0$. Furthermore,

$$V'(x) = E'(x) + \alpha \sum_{i=2}^{r} [C_i(x) - C_i(x_\mathrm{e})] C_i'(x), \tag{3.96}$$

and hence, $V'(x_e) = 0$.

Next, note that

$$V''(x) = E''(x) + \alpha \sum_{i=2}^{r} \{[C_i'(x)]^T C_i'(x) + [C_i(x) - C_i(x_e)]C_i''(x)\}, \quad (3.97)$$

and hence,

$$V''(x_e) = E''(x_e) + \alpha \sum_{i=2}^{r} \left(\frac{\partial C_i}{\partial x}(x_e)\right)^T \left(\frac{\partial C_i}{\partial x}(x_e)\right). \quad (3.98)$$

Now, to show the existence of α satisfying (3.93), note that since $C_i'(x_e)x = 0$, $i = 2, \ldots, r$, $x \in \mathcal{M}$, it follows that there exists an invertible matrix $S \in \mathbb{R}^{n \times n}$ such that $[0,\ I_{r-1}]Sx = 0$, $x \in \mathcal{M}$, and $[I_{n-(r-1)},\ 0]Sx = 0$, $x \in \mathcal{M}^c$ [288]. Hence, for $x_e \in \mathcal{D}$,

$$E''(x_e) = S^T \begin{bmatrix} E_1 & E_{12} \\ E_{12}^T & E_2 \end{bmatrix} S \quad (3.99)$$

and

$$\sum_{i=2}^{r} \left(\frac{\partial C_i}{\partial x}(x_e)\right)^T \left(\frac{\partial C_i}{\partial x}(x_e)\right) = S^T \begin{bmatrix} 0 & 0 \\ 0 & N \end{bmatrix} S, \quad (3.100)$$

where $E_1 \in \mathbb{R}^{(n-r+1)\times(n-r+1)}$, $E_{12} \in \mathbb{R}^{(n-r+1)\times(r-1)}$, $E_2 \in \mathbb{R}^{(r-1)\times(r-1)}$, and $N \in \mathbb{R}^{(r-1)\times(r-1)}$ are symmetric, and E_1 and N are positive definite. Next, substituting (3.99) and (3.100) into (3.98) yields

$$V''(x_e) = S^T \begin{bmatrix} E_1 & E_{12} \\ E_{12}^T & E_2 + \alpha N \end{bmatrix} S \triangleq S^T Q S. \quad (3.101)$$

Now, choosing $\alpha \geq 0$ such that $Q > 0$, (3.93) is satisfied, and hence, $V''(x_e) > 0$. Since $V(\cdot)$ is two-times continuously differentiable, it follows that (see Problem 3.2) $V(x)$, $x \in \mathcal{D}$, is positive definite in the neighborhood of x_e. □

It is clear from Theorem 3.8 that the existence of energy-Casimir functions for (3.1) can be used to construct Lyapunov functions for (3.1). In particular, suppose we can construct a function $H : \mathcal{D} \to \mathbb{R}$ such that $\dot{H}(x) = 0$ along the trajectories of the nonlinear dynamical system (3.1). If $C_1, \ldots, C_r$ are Casimir functions for (3.1), then

$$\frac{\mathrm{d}}{\mathrm{d}t}[H + E(C_1, \ldots, C_r)](x(t)) \equiv 0 \quad (3.102)$$

for *every* function $E : \mathbb{R}^r \to \mathbb{R}$. Hence, even if H is not positive definite at the equilibrium $x_e \in \mathcal{D}$, the function $V(x) = H(x) + E(C_1(x), \ldots, C_r(x))$ can be made positive definite at $x_e \in \mathcal{D}$ by properly choosing E so that $V(x)$ is a Lyapunov function for (3.1).

Example 3.11. This example is adopted from [450] and considers the nonlinear dynamical system representing a rigid spacecraft given by (3.11)–(3.13) in Example 3.1. To show that the equilibrium solution $x(t) \equiv x_{\rm e}$, where $x_{\rm e} = [0, 0, x_{3\rm e}]^{\rm T}$, to (3.11)–(3.13) is Lyapunov stable note that

$$C_1(x) = \tfrac{1}{2}(I_1 x_1^2 + I_2 x_2^2 + I_3 x_3^2), \tag{3.103}$$
$$C_2(x) = \tfrac{1}{2}(I_1^2 x_1^2 + I_2^2 x_2^2 + I_3^2 x_3^2), \tag{3.104}$$

are Casimir functions for (3.11)–(3.13). Now, letting $E(x) = \mu_1 C_1(x) + \mu_2 C_2(x)$ it follows that $E'(x_{\rm e}) = 0$ and $x^{\rm T} E''(x_{\rm e}) x > 0$, $x \in \mathcal{M}$, $x \neq 0$, are satisfied with $\mu_1 = -I_3$ and $\mu_2 = 1$. Next, using

$$V(x_1, x_2, x_3) = E(x_1, x_2, x_3) - E(0, 0, x_{3\rm e}) + \tfrac{\alpha}{2}[C_2(x_1, x_2, x_3) - C_2(0, 0, x_{3\rm e})]^2, \tag{3.105}$$

it follows that Q in (3.101) is given by

$$Q = \begin{bmatrix} I_1(I_1 - I_3) & 0 & 0 \\ 0 & I_2(I_2 - I_3) & 0 \\ 0 & 0 & \alpha I_3^4 x_{3\rm e}^2 \end{bmatrix}. \tag{3.106}$$

Note that $Q > 0$ for every $\alpha > 0$. Hence, it follows from Theorem 3.8 that the equilibrium solution $x(t) \equiv x_{\rm e}$ to (3.11)–(3.13) is Lyapunov stable with Lyapunov function (3.105). △

3.5 Converse Lyapunov Theorems

In the previous sections the existence of a Lyapunov function is assumed while stability properties of a nonlinear dynamical system are deduced. This raises the question of whether or not there always exists a Lyapunov function for a Lyapunov stable and an asymptotically stable nonlinear dynamical system. A number of results concerning the existence of continuously differentiable Lyapunov functions for Lyapunov stable and asymptotically stable *time-varying* nonlinear systems known as *converse Lyapunov theorems* address this problem [178, 298, 306, 307, 445, 474]. However, unlike converse theorems for time-varying systems where the existence of a continuously differentiable Lyapunov function for a Lyapunov stable system is ensured, in the time-invariant case Lyapunov stability does *not* in general imply the existence of a continuously differentiable or even a continuous *time-independent* Lyapunov function (see Example 4.15). For further discussion on this often overlooked fact the interested reader is referred to [134, 178]. However, the existence of a lower semicontinuous Lyapunov function for a Lyapunov stable system is guaranteed (see Section 4.8). As shown below, however, there always exists a continuously differentiable time-independent Lyapunov function for asymptotically stable, time-invariant nonlinear dynamical systems. In order to state and prove the converse Lyapunov theorems we need several definitions and one key lemma.

Definition 3.3. A continuous function $\gamma : [0,a) \to [0,\infty)$, where $a \in (0,\infty]$, is of *class* $\mathcal{K}$ if it is strictly increasing and $\gamma(0) = 0$. A continuous function $\gamma : [0,\infty) \to [0,\infty)$ is of *class* $\mathcal{K}_\infty$ if it is strictly increasing, $\gamma(0) = 0$, and $\gamma(s) \to \infty$ as $s \to \infty$. A continuous function $\gamma : [0,\infty) \to [0,\infty)$ is of *class* $\mathcal{L}$ if it is strictly decreasing and $\gamma(s) \to 0$ as $s \to \infty$. Finally, a continuous function $\gamma : [0,a) \times [0,\infty) \to [0,\infty)$ is of *class* $\mathcal{KL}$ if, for each fixed s, $\gamma(r,s)$ is of class $\mathcal{K}$ with respect to r and, for each fixed r, $\gamma(r,s)$ is of class $\mathcal{L}$ with respect to s.

The following key lemma due to Massera [306] is needed for the main result of this section.

Lemma 3.1 (Massera's Lemma). Let $\sigma : [0,\infty) \to [0,\infty)$ be a class $\mathcal{L}$ function and let λ be a given positive scalar. Then there exists an infinitely differentiable function $\gamma : [0,\infty) \to [0,\infty)$ such that $\gamma(\cdot), \gamma'(\cdot) \in \mathcal{K}$, $\int_0^\infty \gamma[\sigma(t)]\mathrm{d}t < \infty$, and $\int_0^\infty \gamma'[\sigma(t)]e^{\lambda t}\mathrm{d}t < \infty$.

Proof. Since $\sigma(\cdot)$ is strictly decreasing there exists an unbounded sequence $\{t_n\}_{n=1}^\infty$ such that

$$\sigma(t_n) \le \frac{1}{n+1}, \qquad n = 1,2,\ldots.$$

Now, define $\eta : \mathbb{R}_+ \to \mathbb{R}_+$ by

$$\eta(t) \triangleq \begin{cases} \left(\frac{t_1}{t}\right)^p, & 0 < t \le t_1, \\ \frac{(n+1)t_{n+1} - nt_n - t}{n(n+1)(t_{n+1} - t_n)}, & t_n \le t \le t_{n+1}, \quad n = 1,2,\ldots, \end{cases}$$

where $p > \frac{t_1^2}{2(t_2 - t_1)}$. Note that $\eta(\cdot)$ is strictly decreasing, $\sigma(t) < \eta(t)$, $t \in [t_1,\infty)$, and $\lim_{t\to 0, t>0} \eta(t) = \infty$. Since $\eta(\cdot)$ is an infinitely differentiable function except at a countable number of values of t, it can be approximated by an infinitely differentiable function on $[0,\infty)$. Next, it can be shown that $\eta^{-1}(\cdot)$ is a strictly decreasing function such that $\lim_{s\to 0, s>0} \eta^{-1}(s) = \infty$ and $\lim_{s\to\infty} \eta^{-1}(s) = 0$. Now, let $\mu : \mathbb{R}_+ \to \mathbb{R}_+$ be such that $\mu(0) = 0$ and $\mu(s) = e^{-(\lambda+1)\eta^{-1}(s)}$, $s > 0$. Note that $\mu(\cdot)$ is infinitely differentiable on $(0,\infty)$ and $\mu(\cdot)$ is strictly increasing, which implies that $\mu(\cdot)$ is a class $\mathcal{K}$ function.

Next, define $\gamma : [0,\infty) \to [0,\infty)$ by

$$\gamma(r) \triangleq \int_0^r \mu(s)\mathrm{d}s.$$

Note that $\gamma(\cdot)$ and $\gamma'(\cdot) = \mu(\cdot)$ are class $\mathcal{K}$ functions. Furthermore, note that

$$\int_0^\infty \gamma'[\sigma(t)]e^{\lambda t}\mathrm{d}t = \int_0^{t_1} \mu[\sigma(t)]e^{\lambda t}\mathrm{d}t + \int_{t_1}^\infty \mu[\sigma(t)]e^{\lambda t}\mathrm{d}t$$

$$
\begin{aligned}
&< \int_0^{t_1} \mu[\sigma(t)]e^{\lambda t}\mathrm{d}t + \int_{t_1}^{\infty} \mu[\eta(t)]e^{\lambda t}\mathrm{d}t \\
&= \int_0^{t_1} \mu[\sigma(t)]e^{\lambda t}\mathrm{d}t + \int_{t_1}^{\infty} e^{-t}\mathrm{d}t \\
&< \infty.
\end{aligned}
$$

Next, since for all $t \geq t_1$ and $s \in (0, \eta(t)]$, $t \leq \eta^{-1}(s)$, it follows that

$$
\begin{aligned}
\int_0^{\eta(t)} \mu(s)\mathrm{d}s &= \int_0^{\eta(t)} e^{-(\lambda+1)\eta^{-1}(s)}\mathrm{d}s \\
&\leq \int_0^{\eta(t)} e^{-(\lambda+1)t}\mathrm{d}s \\
&= e^{-(\lambda+1)t}\eta(t) \\
&\leq e^{-(\lambda+1)t}.
\end{aligned}
$$

Hence,

$$
\begin{aligned}
\int_{t_1}^{\infty} \gamma[\sigma(t)]\mathrm{d}t &= \int_{t_1}^{\infty}\int_0^{\sigma(t)} \mu(s)\mathrm{d}s\mathrm{d}t \\
&< \int_{t_1}^{\infty}\int_0^{\eta(t)} \mu(s)\mathrm{d}s\mathrm{d}t \\
&\leq \int_{t_1}^{\infty} e^{-(\lambda+1)t}\mathrm{d}t \\
&< \infty.
\end{aligned}
$$

Now, the result follows from the fact that $\int_0^{\infty} \gamma[\sigma(t)]\mathrm{d}t < \infty$ if and only if $\int_{t_1}^{\infty} \gamma[\sigma(t)]\mathrm{d}t < \infty$. □

Theorem 3.9. Assume that the zero solution $x(t) \equiv 0$ to (3.1) is asymptotically stable, $f : \mathcal{D} \to \mathbb{R}^n$ is continuously differentiable, and let $\delta > 0$ be such that $\mathcal{B}_\delta(0) \subset \mathcal{D}$ is contained in the domain of attraction of (3.1). Then there exists a continuously differentiable function $V : \mathcal{B}_\delta(0) \to \mathbb{R}$ such that $V(0) = 0$, $V(x) > 0$, $x \in \mathcal{B}_\delta(0)$, $x \neq 0$, and $V'(x)f(x) < 0$, $x \in \mathcal{B}_\delta(0)$, $x \neq 0$.

Proof. Since the zero solution $x(t) \equiv 0$ to (3.1) is, by assumption, asymptotically stable it follows that (see Problem 3.76) there exist class $\mathcal{K}$ and $\mathcal{L}$ functions $\alpha(\cdot)$ and $\beta(\cdot)$, respectively, such that if $\|x_0\| < \delta$ then $\|x(t)\| \leq \alpha(\|x_0\|)\beta(t)$, $t \geq 0$, where $\|\cdot\|$ is the Euclidean norm. Next, let $x \in \mathcal{B}_\delta(0)$ and let $s(t, x)$ denote the solution to (3.1) with the initial condition $x(0) = x$ so that for all $\|x\| < \delta$, $\|s(t,x)\| \leq \alpha(\|x\|)\beta(t)$, $t \geq 0$.

Furthermore, note that $s(t,x) = x + \int_0^t f(s(\tau,x))\mathrm{d}\tau$, which implies that

$$\frac{\partial s(t,x)}{\partial x} = I + \int_0^t f'(s(\tau,x))\frac{\partial s(\tau,x)}{\partial x}\mathrm{d}\tau. \tag{3.107}$$

Hence,

$$\left\|\frac{\partial s(t,x)}{\partial x}\right\| \leq 1 + \int_0^t \lambda \left\|\frac{\partial s(\tau,x)}{\partial x}\right\| \mathrm{d}\tau, \tag{3.108}$$

where $\lambda \triangleq \max_{x\in\mathcal{B}_\delta(0)} \|f'(x)\|$. Furthermore, it follows from Lemma 2.2 that $\|\frac{\partial s(t,x)}{\partial x}\| \leq e^{\lambda t}$. Now, it follows from Lemma 3.1 that there exists an infinitely differentiable function $\gamma(\cdot)$ such that $\gamma(\cdot)$ and $\gamma'(\cdot)$ are class $\mathcal{K}$ functions, $\int_0^\infty \gamma(\alpha(\delta)\beta(t))\mathrm{d}t < \infty$, and $\int_0^\infty \gamma'(\alpha(\delta)\beta(t))e^{\lambda t}\mathrm{d}t < \infty$. Since $\|s(t,x)\| \leq \alpha(\|x\|)\beta(t) \leq \alpha(\delta)\beta(t)$, $t \geq 0$, it follows that the function

$$V(x) = \int_0^\infty \gamma(\|s(t,x)\|)\mathrm{d}t \tag{3.109}$$

is well defined on $\mathcal{B}_\delta(0)$. Furthermore, note that $V(0) = 0$.

Next, since $f(\cdot)$ in (3.1) is uniformly Lipschitz continuous on $\mathcal{B}_\delta(0)$ it follows from Proposition 2.30 that

$$V(x) = \int_0^\infty \gamma(\|s(t,x)\|)\mathrm{d}t \geq \int_0^\infty \gamma(\|x\|e^{-Lt})\mathrm{d}t,$$

where L denotes the Lipschitz constant of $f(\cdot)$ on $\mathcal{B}_\delta(0)$, which implies that $V(x) > 0$, $x \in \mathcal{D}_0$, $x \neq 0$. Now, note that since $\|s(t,x)\| = \sqrt{s^{\mathrm{T}}(t,x)s(t,x)}$,

$$V'(x) = \int_0^\infty \gamma'(\|s(t,x)\|)\frac{s^{\mathrm{T}}(t,x)}{\|s(t,x)\|}\frac{\partial s(t,x)}{\partial x}\mathrm{d}t, \tag{3.110}$$

and hence,

$$\begin{aligned}\|V'(x)\| &\leq \int_0^\infty \gamma'(\|s(t,x)\|)\left\|\frac{\partial s(t,x)}{\partial x}\right\|\mathrm{d}t \\ &\leq \int_0^\infty \gamma'(\alpha(\delta)\beta(t))e^{\lambda t}\mathrm{d}t \\ &< \infty, \qquad x \in \mathcal{B}_\delta(0), \end{aligned} \tag{3.111}$$

which proves that $V'(\cdot)$ is bounded. Now, (3.110) implies that $V'(\cdot)$ is continuous, and hence, $V(\cdot)$ is continuously differentiable.

Finally, note that since $f(\cdot)$ is Lipschitz continuous on $\mathcal{D}$ the trajectory $s(t,x)$, $t \geq 0$, is unique, and hence,

$$\begin{aligned}V(s(t,x)) &= \int_0^\infty \gamma(\|s(\tau,s(t,x))\|)\mathrm{d}\tau \\ &= \int_0^\infty \gamma(\|s(\tau+t,x)\|)\mathrm{d}\tau\end{aligned}$$

$$= \int_t^\infty \gamma(\|s(\tau, x)\|)\mathrm{d}\tau, \tag{3.112}$$

which implies that $\dot{V}(s(t,x)) = -\gamma(\|s(t,x)\|) < 0$, $s(t,x) \neq 0$. The result is now immediate by noting that $V'(x)f(x) = \dot{V}(s(0,x)) = -\gamma(\|x\|) < 0$, $x \in \mathcal{B}_\delta(0)$, $x \neq 0$. □

The next result gives a converse Lyapunov theorem for exponential stability.

Theorem 3.10. Assume that the zero solution $x(t) \equiv 0$ to (3.1) is exponentially stable, $f : \mathcal{D} \to \mathbb{R}^n$ is continuously differentiable, and let $\delta > 0$ be such that $\mathcal{B}_\delta(0) \subset \mathcal{D}$ is contained in the domain of attraction of (3.1). Then, for every $p > 1$, there exists a continuously differentiable function $V : \mathcal{D} \to \mathbb{R}$ and scalars α, β, and $\varepsilon > 0$ such that

$$\alpha\|x\|^p \leq V(x) \leq \beta\|x\|^p, \quad x \in \mathcal{B}_\delta(0), \tag{3.113}$$

$$V'(x)f(x) \leq -\varepsilon V(x), \quad x \in \mathcal{B}_\delta(0). \tag{3.114}$$

Proof. Since the zero solution $x(t) \equiv 0$ to (3.1) is, by assumption, exponentially stable it follows that there exist positive scalars α_1 and β_1 such that if $\|x_0\| < \delta$, then $\|x(t)\| \leq \alpha_1\|x_0\|e^{-\beta_1 t}$, $t \geq 0$. Next, let $x \in \mathcal{B}_\delta(0)$ and let $s(t,x)$, $t \geq 0$, denote the solution to (3.1) with the initial condition $x(0) = x$ so that for all $\|x\| < \delta$, $\|s(t,x)\| \leq \alpha_1\|x\|e^{-\beta_1 t}$, $t \geq 0$. Now, using identical arguments as in the proof of Theorem 3.9 it follows that for $\alpha(\delta) = \alpha_1$, $\beta(t) = e^{-\beta_1 t}$, $\gamma(\sigma) = \sigma^p$, and $(p-1)\beta_1 > \lambda$, where $\lambda = \max_{x \in \mathcal{B}_\delta(0)} \|f'(x)\|$,

$$\int_0^\infty \alpha_1^p e^{-\beta_1 p t}\mathrm{d}t < \infty \tag{3.115}$$

and

$$\int_0^\infty p\alpha_1^{p-1} e^{-\beta_1(p-1)t} e^{\lambda t}\mathrm{d}t < \infty. \tag{3.116}$$

Hence,

$$V(x) = \int_0^\infty \|s(t,x)\|^p \mathrm{d}t \tag{3.117}$$

is a continuously differentiable Lyapunov function candidate for (3.1).

To show that (3.113) holds, note that

$$\begin{aligned} V(x) &= \int_0^\infty \|s(t,x)\|^p \mathrm{d}t \\ &\leq \int_0^\infty \alpha_1^p \|x\|^p e^{-\beta_1 p t}\mathrm{d}t \end{aligned}$$

$$= \frac{\alpha_1^p}{p\beta_1}\|x\|^p, \qquad x \in \mathcal{D}, \tag{3.118}$$

which proves the upper bound in (3.113). Next, since $\|f'(x)\| \le \lambda$, $x \in \mathcal{B}_\delta(0)$, and $f(\cdot)$ is uniformly Lipschitz continuous on $\mathcal{B}_\delta(0)$ with Lipschitz constant $L = \lambda$ it follows that

$$\|f(s(t,x))\| \le \lambda\|s(t,x)\| \le \lambda\alpha_1\|x\|e^{-\beta_1 t} \le \lambda\alpha_1\|x\|, \quad t \ge 0. \tag{3.119}$$

Hence,

$$s(t,x) = x + \int_0^t f(s(\tau,x))\mathrm{d}\tau \tag{3.120}$$

implies

$$\begin{aligned} \|s(t,x)\| &\ge \|x\| - \|x\|\lambda\alpha_1 t \\ &\ge \frac{\|x\|}{2}, \quad t \in [0, \tfrac{1}{2\lambda\alpha_1}]. \end{aligned} \tag{3.121}$$

Thus, it follows from (3.117) that

$$V(x) \ge \int_0^{\frac{1}{2\lambda\alpha_1}} \frac{\|x\|^p}{2^p}\mathrm{d}t = \frac{1}{2^{p+1}\lambda\alpha_1}\|x\|^p, \quad x \in \mathcal{B}_\delta(0), \tag{3.122}$$

which proves the lower bound in (3.113).

Finally, to show (3.114) note that with $\gamma(\sigma) = \sigma^p$ it follows as in the proof of Theorem 3.9 that $\dot{V}(x) = -\gamma(\|x\|) = -\|x\|^p$. Now, the result is immediate from (3.113) by noting that

$$\dot{V}(x) = -\|x\|^p \le -\tfrac{1}{\beta}V(x), \quad x \in \mathcal{B}_\delta(0), \tag{3.123}$$

which proves (3.114). □

Next, we present a corollary to Theorem 3.10 that shows that p in Theorem 3.10 can be taken to be equal to 2 without loss of generality.

Corollary 3.2. Assume that the zero solution $x(t) \equiv 0$ to (3.1) is exponentially stable, $f : \mathcal{D} \to \mathbb{R}^n$ is continuously differentiable, and let $\delta > 0$ be such that $\mathcal{B}_\delta(0) \subset \mathcal{D}$ is contained in the domain of attraction of (3.1). Then there exists a continuously differentiable function $V : \mathcal{D} \to \mathbb{R}$ and scalars α, β, and $\varepsilon > 0$ such that

$$\alpha\|x\|^2 \le V(x) \le \beta\|x\|^2, \quad x \in \mathcal{B}_\delta(0), \tag{3.124}$$

$$V'(x)f(x) \le -\varepsilon V(x), \quad x \in \mathcal{B}_\delta(0). \tag{3.125}$$

Proof. The proof is a direct consequence of Theorem 3.10 with $p = 2$. □

Finally, we present a converse theorem for global exponential stability.

Theorem 3.11. If the zero solution $x(t) \equiv 0$ to (3.1) is globally exponentially stable and $f : \mathbb{R}^n \to \mathbb{R}^n$ is continuously differentiable and globally Lipschitz, then there exists a continuously differentiable function $V : \mathcal{D} \to \mathbb{R}$ and scalars α, β, and $\varepsilon > 0$, such that

$$\alpha\|x\|^2 \le V(x) \le \beta\|x\|^2, \quad x \in \mathbb{R}^n, \tag{3.126}$$
$$V'(x)f(x) \le -\varepsilon V(x), \quad x \in \mathbb{R}^n. \tag{3.127}$$

Proof. By Proposition 2.26 $f(\cdot)$ is globally Lipschitz if and only if $\|f'(x)\| \le \lambda$, $x \in \mathbb{R}^n$. Now, the proof is identical to the proof of Theorem 3.10 by replacing $\mathcal{B}_\delta(0)$ by $\mathbb{R}^n$ and $p = 2$. □

Finally, it is important to note that even though we assumed that the vector field f is continuously differentiable in order to establish converse Lyapunov theorems for asymptotic stability, these results also hold for vector fields that are locally Lipschitz continuous as well as continuous. In particular, Massera [307] proved a converse theorem involving the existence of a smooth (i.e., infinitely differentiable) Lyapunov function for locally Lipschitz continuous vector fields. In addition, Kurzweil [250] proved the existence of a smooth Lyapunov function for asymptotic stability under the assumption of f only being continuous. The proofs of these results, however, are considerably more involved than the converse proofs given in this section and involve concepts not introduced in this book. For further details, see [250, 460].

3.6 Lyapunov Instability Theorems

In the preceding sections we established sufficient conditions for Lyapunov and asymptotic stability of nonlinear dynamical systems. In this section, we provide three key instability theorems based on Lyapunov's direct method for proving that the zero solution $x(t) \equiv 0$ to (3.1) is unstable. The main utility of these instability theorems are when Lyapunov's indirect method (see Theorem 3.19) fails to provide any information about the stability of a nonlinear dynamical system.

Theorem 3.12 (Lyapunov's First Instability Theorem). Consider the nonlinear dynamical system (3.1). Assume that there exist a continuously differentiable function $V : \mathcal{D} \to \mathbb{R}$ and a scalar $\varepsilon > 0$ such that

$$V(0) = 0, \tag{3.128}$$
$$V'(x)f(x) > 0, \quad x \in \mathcal{B}_\varepsilon(0), \quad x \neq 0. \tag{3.129}$$

Furthermore, assume that for every sufficiently small $\delta > 0$ there exists

$x_0 \in \mathcal{D}$ such that $\|x_0\| < \delta$ and $V(x_0) > 0$. Then, the zero solution $x(t) \equiv 0$ to (3.1) is unstable.

Proof. Suppose, *ad absurdum*, that there exists $\delta > 0$ such that if $x_0 \in \mathcal{B}_\delta(0)$, then $x(t) \in \mathcal{B}_\varepsilon(0)$, $t \geq 0$. By assumption, there exists $x_0 \in \mathcal{D}$ such that $V(x_0) = c > 0$ and $\|x_0\| < \delta$. In this case, it follows from (3.129) that $V'(x(t))f(x(t)) \geq 0$, $t \geq 0$, and hence, $V(x(t)) \geq c > 0$, $t \geq 0$. Thus, on the trajectory, $V'(x(t))f(x(t)) > 0$, $t \geq 0$, where $x(t)$, $t \geq 0$, denotes the solution of (3.1) with initial condition x_0. Now, consider the set $\mathcal{S} \triangleq \{y \in \mathbb{R}^n : y = x(t),\ t \geq 0\}$ and note that $\overline{\mathcal{S}}$ is compact. Since $V'(x(t))f(x(t)) > 0$, $t \geq 0$, it follows that there exists $d = \min_{y \in \overline{\mathcal{S}}} V'(y)f(y) > 0$. Next, since $V(\cdot)$ is continuous on $\overline{\mathcal{B}}_\varepsilon(0)$ it follows that there exists $\alpha > 0$ such that $V(x) \leq \alpha$, $x \in \mathcal{B}_\varepsilon(0) \cap \mathcal{S}$. Hence, it follows that

$$\alpha \geq V(x(t)) = V(x(t_1)) + \int_{t_1}^{t} V'(x(s))f(x(s))\mathrm{d}s \geq c + (t - t_1)d, \quad t \geq t_1. \tag{3.130}$$

Since the right-hand side of (3.130) is unbounded it follows that there exists $t \geq t_1$ such that $\alpha < c + (t - t_1)d$, which contradicts (3.130). Hence, there does not exist $\delta > 0$ such that if $x_0 \in \mathcal{B}_\delta(0)$, then $x(t) \in \mathcal{B}_\varepsilon(0)$, $t \geq 0$. Thus, the zero solution $x(t) \equiv 0$ to (3.1) is unstable. □

It is interesting to note that the function $V(\cdot)$ in Theorem 3.12 can be positive as well as negative in $\mathcal{D}$. However, $V(\cdot)$ is required to be positive for some points $x_0 \neq 0$ arbitrarily close to the origin of the nonlinear dynamical system. A more restrictive version of Theorem 3.12 is the case where $V : \mathcal{D} \to \mathbb{R}$ is positive definite for all $x \in \mathcal{B}_\varepsilon(0)$. In this case, the zero solution $x(t) \equiv 0$ to (3.1) is *completely unstable* in the sense that there exists $\varepsilon > 0$ such that *every* trajectory starting in $\mathcal{B}_\varepsilon(0)$, other than the trivial trajectory, eventually leaves $\mathcal{B}_\varepsilon(0)$. Similar remarks hold for the next instability theorem.

Example 3.12. Consider the nonlinear dynamical system

$$\dot{x}_1(t) = -x_1(t) + x_2^6(t), \quad x_1(0) = x_{10}, \quad t \geq 0, \tag{3.131}$$

$$\dot{x}_2(t) = x_2^3(t) + x_1^6(t), \quad x_2(0) = x_{20}. \tag{3.132}$$

To examine the stability of this system consider the function $V(x_1, x_2) = -\frac{1}{6}x_1^6 + \frac{1}{4}x_2^4$. Note that on the line $x_1 = 0$, $V(x_1, x_2) > 0$ at points arbitrarily close to the origin. Evaluating $\dot{V}(x_1, x_2)$ yields

$$\dot{V}(x_1, x_2) = -x_1^5\dot{x}_1 + x_2^3\dot{x}_2 = x_1^6 + x_2^6 - x_1^5x_2^6 + x_2^3x_1^6. \tag{3.133}$$

Now, there exist a neighborhood $\mathcal{N}$ of the origin and $\delta \in (0, 1)$ such that $|-x_1^5x_2^6 + x_2^3x_1^6| \leq \delta(x_1^6 + x_2^6)$, $(x_1, x_2) \in \mathcal{N}$, which implies that $\dot{V}(x_1, x_2) \geq (1 - \delta)(x_1^6 + x_2^6) > 0$, $(x_1, x_2) \in \mathcal{N}$. Hence, it follows from Theorem 3.12 that the zero solution $(x_1(t), x_2(t)) \equiv (0, 0)$ to (3.131) and (3.132) is unstable. $\triangle$

Theorem 3.13 (Lyapunov's Second Instability Theorem). Consider the nonlinear dynamical system (3.1). Assume that there exist a continuously differentiable function $V : \mathcal{D} \to \mathbb{R}$, a function $W : \mathcal{D} \to \mathbb{R}$, and scalars $\varepsilon, \lambda > 0$, such that

$$\begin{aligned} V(0) &= 0, && (3.134) \\ W(x) &\geq 0, \quad x \in \mathcal{B}_\varepsilon(0), && (3.135) \\ V'(x)f(x) &= \lambda V(x) + W(x). && (3.136) \end{aligned}$$

Furthermore, assume that for every sufficiently small $\delta > 0$ there exists $x_0 \in \mathcal{D}$ such that $\|x_0\| < \delta$ and $V(x_0) > 0$. Then the zero solution $x(t) \equiv 0$ to (3.1) is unstable.

Proof. Suppose *ad absurdum*, that there exists $\delta > 0$ such that if $x_0 \in \mathcal{B}_\delta(0)$, then $x(t) \in \mathcal{B}_\varepsilon(0)$, $t \geq 0$. By assumption, there exists $x_0 \in \mathcal{B}_\delta(0)$ such that $V(x_0) > 0$. It follows from (3.135) and (3.136) that

$$V'(x)f(x) \geq \lambda V(x), \quad x \in \mathcal{B}_\varepsilon(0),$$

or, equivalently,

$$V'(x)f(x) - \lambda V(x) \geq 0, \quad x \in \mathcal{B}_\varepsilon(0). \tag{3.137}$$

Next, forming $e^{-\lambda t}$ (3.137), $t \geq 0$, yields

$$e^{-\lambda t}V'(x(t))f(x(t)) - \lambda e^{-\lambda t}V(x(t)) \geq 0, \quad t \geq 0, \tag{3.138}$$

and hence,

$$\frac{\mathrm{d}}{\mathrm{d}t}[e^{-\lambda t}V(x(t))] \geq 0, \quad t \geq 0, \tag{3.139}$$

where $x(t)$, $t \geq 0$, denotes the solution to (3.1) with initial condition x_0. Integrating both sides of (3.139) yields $e^{-\lambda t}V(x(t)) - V(x(0)) \geq 0$, $t \geq 0$, and hence, $V(x(t)) \geq e^{\lambda t}V(x(0))$, $t \geq 0$. Thus, since $V(x_0) > 0$, $x(t) \notin \mathcal{B}_\varepsilon(0)$ as $t \to \infty$, which is a contradiction. Hence, the zero solution $x(t) \equiv 0$ to (3.1) is unstable since there does not exist $\delta > 0$ such that if $x_0 \in \mathcal{B}_\delta(0)$, then $x(t) \in \mathcal{B}_\varepsilon(0)$, $t \geq 0$. $\square$

Example 3.13. Consider the nonlinear dynamical system

$$\begin{aligned} \dot{x}_1(t) &= x_1(t) + 2x_2(t) + x_1(t)x_2^2(t), \quad x_1(0) = x_{10}, \quad t \geq 0, && (3.140) \\ \dot{x}_2(t) &= 2x_1(t) + x_2(t) - x_1^2(t)x_2(t), \quad x_2(0) = x_{20}. && (3.141) \end{aligned}$$

To examine the stability of this system consider the function $V(x_1, x_2) = x_1^2 - x_2^2$. Note that on the line $x_2 = 0$, $V(x_1, x_2) > 0$ at points arbitrarily close to the origin. Evaluating $\dot{V}(x_1, x_2)$ yields

$$\dot{V}(x_1, x_2) = 2x_1\dot{x}_1 - 2x_2\dot{x}_2$$

$$\begin{aligned} &= 2x_1^2 - 2x_2^2 + 4x_1^2x_2^2 \\ &= 2V(x_1, x_2) + 4x_1^2x_2^2. \end{aligned} \tag{3.142}$$

Now, with $W(x_1, x_2) \triangleq 4x_1^2x_2^2 \geq 0$, $(x_1, x_2) \in \mathbb{R} \times \mathbb{R}$, it follows that the conditions of Theorem 3.13 are satisfied, and hence, the zero solution $(x_1(t), x_2(t)) \equiv (0, 0)$ to (3.140) and (3.141) is unstable. △

The final instability theorem is due to Chetaev [92] and is appropriately known as Chetaev's instability theorem.

Theorem 3.14 (Chetaev's Instability Theorem). Consider the nonlinear dynamical system (3.1) and assume that there exist a continuously differentiable function $V : \mathcal{D} \to \mathbb{R}$, a scalar $\varepsilon > 0$, and an open set $\mathcal{Q} \subseteq \mathcal{B}_\varepsilon(0)$ such that

$$V(x) > 0, \quad x \in \mathcal{Q}, \tag{3.143}$$

$$\sup_{x \in \mathcal{Q}} V(x) < \infty, \tag{3.144}$$

$$0 \in \partial\mathcal{Q}, \tag{3.145}$$

$$V(x) = 0, \quad x \in \partial\mathcal{Q} \cap \mathcal{B}_\varepsilon(0), \tag{3.146}$$

$$V'(x)f(x) > 0, \quad x \in \mathcal{Q}. \tag{3.147}$$

Then the zero solution $x(t) \equiv 0$ to (3.1) is unstable.

Proof. Let $x_0 \in \mathcal{Q}$ and suppose, *ad absurdum*, that there exists a closed set $\mathcal{P}$ such that $x(t) \in \mathcal{P} \subset \mathcal{Q} \subseteq \mathcal{B}_\varepsilon(0)$, $t \geq 0$. Hence, it follows from (3.147) that

$$\begin{aligned} V(x(t)) &= V(x(0)) + \int_0^t \dot{V}(x(s))\mathrm{d}s \\ &= V(x(0)) + \int_0^t V'(x(s))f(x(s))\mathrm{d}s \\ &\geq V(x(0)) + \alpha t, \end{aligned}$$

where $\alpha = \min_{x \in \mathcal{P}} V'(x)f(x) > 0$, which implies that $V(x(t)) \to \infty$ as $t \to \infty$, contradicting (3.144). Thus, there exists $T > 0$ such that either $x(T) \in \partial\mathcal{Q}$ or $x(t) \to \partial\mathcal{Q}$ as $t \to \infty$. First, consider the case in which $x(T) \in \partial\mathcal{Q}$. In this case, since (3.147) implies that $V(x(t))$, $t \geq 0$, is strictly increasing for all $x(t) \in \mathcal{Q}$ it follows that $V(x(T)) > 0$. Hence, since $V(x) = 0$, $x \in \partial\mathcal{Q} \cap \mathcal{B}_\varepsilon(0)$, it follows that $x(T) \notin \partial\mathcal{Q} \cap \mathcal{B}_\varepsilon(0)$, which implies that $x(T) \in \partial\mathcal{Q} \backslash \mathcal{B}_\varepsilon(0)$. Next, since $\partial\mathcal{Q} = \overline{\mathcal{Q}} \cap \partial\mathcal{Q}$ and $\overline{\mathcal{Q}} \subseteq \overline{\mathcal{B}}_\varepsilon(0)$ it follows that $\partial\mathcal{Q} \backslash \mathcal{B}_\varepsilon(0) = (\overline{\mathcal{Q}} \cap \partial\mathcal{Q}) \backslash \mathcal{B}_\varepsilon(0) \subseteq (\overline{\mathcal{B}}_\varepsilon(0) \cap \partial\mathcal{Q}) \backslash \mathcal{B}_\varepsilon(0) = \partial\mathcal{Q} \cap \partial\mathcal{B}_\varepsilon(0)$. Hence, $x(T) \in \partial\mathcal{B}_\varepsilon(0)$. Similarly, if $x(t) \to \partial\mathcal{Q}$ as $t \to \infty$, then $x(t) \to \partial\mathcal{B}_\varepsilon(0)$ as $t \to \infty$. Thus, there does not exist $\delta > 0$ such that if $x_0 \in \mathcal{B}_\delta(0)$, then $x(t) \in \mathcal{B}_\varepsilon(0)$, $t \geq 0$. Hence, the zero solution $x(t) \equiv 0$ to (3.1) is unstable. □

Unlike Theorems 3.12 and 3.13 requiring that $V(\cdot)$ and $\dot{V}(\cdot)$ satisfy certain conditions at all points in a neighborhood of $\mathcal{D}$, Chetaev's instability theorem requires that $V(\cdot)$ and $\dot{V}(\cdot)$ satisfy certain conditions in a subregion $\mathcal{Q}$ of $\mathcal{D}$.

Example 3.14. Consider the nonlinear dynamical system

$$\dot{x}_1(t) = x_1^3(t) + x_2(t)x_1^2(t), \quad x_1(0) = x_{10}, \quad t \geq 0, \tag{3.148}$$
$$\dot{x}_2(t) = -x_2(t) + x_1^2(t), \quad x_2(0) = x_{20}. \tag{3.149}$$

To examine the stability of this system consider the function $V(x_1, x_2) = \frac{1}{2}x_1^2 - \frac{1}{2}x_2^2$ and the open set $\mathcal{Q} \triangleq \{(x_1, x_2) \in \mathcal{B}_\varepsilon(0) : x_1 > x_2 > -x_1\}$. Note that $V(x_1, x_2) > 0$, $(x_1, x_2) \in \mathcal{Q}$, and $V(x_1, x_2) = 0$, $x \in \partial\mathcal{Q} \cap \mathcal{B}_\varepsilon(0)$, where $\mathcal{B}_\varepsilon(0)$ is a sufficiently small neighborhood of the origin. Next, evaluating $\dot{V}(x_1, x_2)$ yields

$$\dot{V}(x_1, x_2) = x_1\dot{x}_1 - x_2\dot{x}_2 = x_1^4 - x_2(x_1^2 - x_1^3) + x_2^2. \tag{3.150}$$

Now, there exist a neighborhood $\mathcal{N}$ of the origin and $\delta \in [0, 1)$ such that

$$\dot{V}(x_1, x_2) \geq x_1^4 - (1+\delta)|x_2|x_1^2 + x_2^2, \quad (x_1, x_2) \in \mathcal{N} \tag{3.151}$$

which shows that $\dot{V}(x_1, x_2) > 0$, $(x_1, x_2) \in \mathcal{Q} \cap \mathcal{N}$. Hence, it follows from Theorem 3.14 that the zero solution $(x_1(t), x_2(t)) \equiv (0, 0)$ to (3.148) and (3.149) is unstable. △

3.7 Stability of Linear Systems and Lyapunov's Linearization Method

In this section, we consider linear time-invariant dynamical systems of the form

$$\dot{x}(t) = Ax(t), \qquad x(0) = x_0, \qquad t \geq 0, \tag{3.152}$$

where $x(t) \in \mathbb{R}^n$, $t \geq 0$, and $A \in \mathbb{R}^{n \times n}$. Note that the equilibrium solution $x(t) \equiv x_\mathrm{e}$ to (3.152) corresponds to $x_\mathrm{e} \in \mathcal{N}(A)$, and hence, every point in the null space of A is an equilibrium point for the linear dynamical system (3.152). In the case where $\det A \neq 0$, A has a trivial null space, and hence, $x_\mathrm{e} = 0$. To examine the stability of (3.152) recall that the solution to (3.152) is given by $x(t) = e^{At}x(0)$, $t \geq 0$. The structure of e^{At} can be understood by considering the Jordan form of A. In particular, it follows from the Jordan decomposition [201] that $A = SJS^{-1}$, where $S \in \mathbb{C}^{n \times n}$ is a nonsingular matrix and $J = \text{block-diag}[J_1, \ldots, J_m]$ is the Jordan form of A. Hence, $e^{At} = Se^{Jt}S^{-1}$, where $e^{Jt} = \text{block-diag}[e^{J_1 t}, \ldots, e^{J_m t}]$. The structure of e^{Jt} can thus be determined by considering each Jordan block J_i. Recall that J_i is of the form $J_i = \lambda_i I_{n_i} + N_{n_i}$, where, for $i \in \{1, \ldots, m\}$, n_i is the order of the ith Jordan block, $\lambda_i \in \operatorname{spec}(A)$, and N_{n_i} is an $n_i \times n_i$ nilpotent matrix which has ones on the superdiagonal and zeros elsewhere. By convention, $N_1 \triangleq 0_{1\times 1}$. Also note that $N_{n_i}^{n_i} = 0$.

Now, since $\lambda_i I_{n_i}$ and N_{n_i} commute, it follows that

$$e^{J_i t} = e^{(\lambda_i I_{n_i} + N_{n_i})t} = e^{\lambda_i I_{n_i} t} e^{N_{n_i} t} = e^{\lambda_i t} e^{N_{n_i} t}. \tag{3.153}$$

Furthermore, since $N_{n_i}^{n_i} = 0$, it follows that $e^{N_{n_i} t}$ is a finite sum of powers of $N_{n_i} t$. Specifically,

$$\begin{aligned} e^{N_{n_i} t} &= I_{n_i} + N_{n_i} t + \tfrac{1}{2} N_{n_i}^2 t^2 + \cdots + [(n_i - 1)]!]^{-1} N_{n_i}^{n_i - 1} t^{n_i - 1} \\ &= \begin{bmatrix} 1 & t & \frac{t^2}{2} & \cdots & \frac{t^{n_i-1}}{(n_i-1)!} \\ 0 & 1 & t & & \frac{t^{n_i-2}}{(n_i-2)!} \\ \vdots & & & & \vdots \\ 0 & \cdots & & \cdots & 1 \end{bmatrix}. \end{aligned} \tag{3.154}$$

Hence,

$$e^{At} = S e^{Jt} S^{-1} = \sum_{i=1}^{m} \sum_{j=1}^{n_i} t^{j-1} e^{\lambda_i t} p_{ij}(A), \tag{3.155}$$

where m is the number of distinct eigenvalues of A and $p_{ij}(A)$ are constant matrices. The following theorem gives necessary and sufficient conditions for Lyapunov stability as well as global asymptotic stability of (3.152).

Theorem 3.15. Consider the linear dynamical system (3.152). The zero solution $x(t) \equiv 0$ to (3.152) is Lyapunov stable if and only if for every $\lambda \in \mathrm{spec}(A)$, either $\mathrm{Re}\,\lambda < 0$, or both $\mathrm{Re}\,\lambda = 0$ and λ is semisimple. Alternatively, the zero solution $x(t) \equiv 0$ to (3.152) is globally asymptotically stable if and only if $\mathrm{Re}\,\lambda < 0$.

Proof. Since $x(t) = e^{At} x(0)$, $t \geq 0$, it follows that the zero solution to (3.152) is Lyapunov stable if and only if there exists $\alpha > 0$ such that $\|e^{At}\| < \alpha$, $t \geq 0$. Now, it follows from (3.155) that e^{At}, $t \geq 0$, is bounded if and only if either $\mathrm{Re}\,\lambda < 0$, or both $\mathrm{Re}\,\lambda = 0$ and λ is semisimple. Alternatively, asymptotic stability is immediate since it follows from (3.155) that $e^{At} \to 0$ as $t \to \infty$ if and only if $\mathrm{Re}\,\lambda < 0$. Finally, since $x(t)$, $t \geq 0$, depends linearly on $x(0)$, asymptotic stability is global. □

Theorem 3.15 gives necessary and sufficient conditions for Lyapunov and asymptotic stability for the zero solution $x(t) \equiv 0$ of the linear system (3.152) by examining the eigenvalues of A. In this book we say A is *Lyapunov stable* if and only if every eigenvalue of A has nonpositive real part and every eigenvalue of A with zero real part is semisimple. In addition, we say A is *asymptotically stable* or *Hurwitz* if and only if every eigenvalue of A has negative real part. Lyapunov's stability theorem can also be used to develop necessary and sufficient conditions for the stability of the zero solution $x(t) \equiv 0$ to (3.152). In particular, to apply Theorem 3.1 to (3.152), consider the Lyapunov function candidate $V(x) = x^{\mathrm{T}}Px$, where $x \in \mathbb{R}^n$ and $P \in \mathbb{R}^{n\times n}$ is positive definite. Next, note that

$$\dot{V}(x) = V'(x)f(x) = 2x^{\mathrm{T}}PAx = x^{\mathrm{T}}(A^{\mathrm{T}}P + PA)x. \tag{3.156}$$

Now, (3.4) will be satisfied if there exists a nonnegative-definite matrix $R \in \mathbb{R}^{n\times n}$ such that $A^{\mathrm{T}}P + PA = -R$ so that $V'(x)f(x) = -x^{\mathrm{T}}Rx \leq 0$, $x \in \mathbb{R}^n$. Hence, we wish to determine the existence of a positive-definite matrix P satisfying

$$0 = A^{\mathrm{T}}P + PA + R. \tag{3.157}$$

Equation (3.157) is appropriately called a *Lyapunov equation*. The next result addresses existence and uniqueness of solutions for the Lyapunov equation (3.157).

Lemma 3.2. Let $A \in \mathbb{R}^{n\times n}$. Then there exists a unique matrix $P \in \mathbb{R}^{n\times n}$ satisfying (3.157) if and only if

$$\lambda_i(A) + \lambda_j(A) \neq 0, \qquad i,j = 1,\ldots,n, \tag{3.158}$$

where $\lambda_k(A) \in \mathrm{spec}(A)$, $k = 1,\ldots,n$.

Proof. Rewriting (3.157) as

$$\mathrm{vec}(A^{\mathrm{T}}P) + \mathrm{vec}(PA) = -\mathrm{vec}(R), \tag{3.159}$$

where vec: $\mathbb{R}^{n\times n} \to \mathbb{R}^{n^2}$ denotes the column stacking operator, and using the identities $\mathrm{vec}(XYZ) = (Z^{\mathrm{T}} \otimes X)\mathrm{vec}\, Y$ and $I \otimes X + Y^{\mathrm{T}} \otimes I = Y^{\mathrm{T}} \oplus X$ [45, pp. 249 and 251], it follows that $(A^{\mathrm{T}} \oplus A^{\mathrm{T}})\mathrm{vec}\, P = -\mathrm{vec}\, R$. Hence, there exists a unique solution $P \in \mathbb{R}^{n\times n}$ satisfying (3.157) if and only if $\det(A^{\mathrm{T}} \oplus A^{\mathrm{T}}) \neq 0$. Now, the result follows from the fact that $\mathrm{spec}(X \oplus Y) = \{\lambda + \mu :\ \lambda \in \mathrm{spec}(X),\ \mu \in \mathrm{spec}(Y)\}$ [45, p. 251]. □

It follows from Lemma 3.2 that if for some $R \in \mathbb{R}^{n\times n}$, (3.157) does not have a unique solution, then the zero solution $x(t) \equiv 0$ to (3.152) is not an asymptotically stable equilibrium. The following theorem gives necessary and sufficient conditions for global asymptotic stability of (3.152) in terms of the solution of the Lyapunov equation (3.157).

Theorem 3.16. Consider the linear dynamical system (3.152). The zero solution $x(t) \equiv 0$ to (3.152) is globally asymptotically stable if and only if for every positive-definite matrix $R \in \mathbb{R}^{n\times n}$ there exists a unique positive-definite matrix $P \in \mathbb{R}^{n\times n}$ satisfying (3.157).

Proof. Sufficiency follows from Theorem 3.1 by noting that if there exists a positive-definite matrix $P \in \mathbb{R}^{n\times n}$ and a positive-definite matrix $R \in$

$\mathbb{R}^{n\times n}$ such that (3.157) holds, then global asymptotic stability is immediate with Lyapunov function $V(x) = x^{\mathrm{T}}Px$. To show necessity, suppose A is asymptotically stable, that is, Re $\lambda < 0$, $\lambda \in \mathrm{spec}(A)$, and define

$$P \triangleq \int_0^\infty e^{A^{\mathrm{T}}t}Re^{At}\mathrm{d}t. \tag{3.160}$$

Note that the integral in (3.160) is well defined since the integrand involves a sum of terms of the from $t^{j-1}e^{\lambda_i t}$, where Re $\lambda_i < 0$, and $i = 1,\ldots,m$, $j = 1,\ldots,n_i$. Next, note that

$$\begin{aligned} A^{\mathrm{T}}P + PA &= \int_0^\infty A^{\mathrm{T}}e^{A^{\mathrm{T}}t}Re^{At}\mathrm{d}t + \int_0^\infty e^{A^{\mathrm{T}}t}Re^{At}A\mathrm{d}t \\ &= \int_0^\infty \frac{\mathrm{d}}{\mathrm{d}t}(e^{A^{\mathrm{T}}t}Re^{At})\mathrm{d}t \\ &= e^{A^{\mathrm{T}}t}Re^{At}\Big|_0^\infty \\ &= -R, \end{aligned} \tag{3.161}$$

which shows that there exists a matrix $P \in \mathbb{R}^{n\times n}$ satisfying (3.157).

To show $P > 0$, let $C \in \mathbb{R}^{p\times n}$ be such that $R = C^{\mathrm{T}}C$ and note that for $x \in \mathbb{R}^n$,

$$\begin{aligned} x^{\mathrm{T}}Px &= \int_0^\infty x^{\mathrm{T}}e^{A^{\mathrm{T}}t}C^{\mathrm{T}}Ce^{At}x\mathrm{d}t \\ &= \int_0^\infty \|Ce^{At}x\|_2^2\mathrm{d}t, \end{aligned} \tag{3.162}$$

which shows that $P = P^{\mathrm{T}}$ and $P \geq 0$. Now, suppose, *ad absurdum*, that P is not positive definite. Then, there exists $x \in \mathbb{R}^n$, $x \neq 0$, such that $x^{\mathrm{T}}Px = 0$, which implies that $Ce^{At}x = 0$ for all $t \geq 0$. For $t = 0$ it follows that $Cx = 0$, which implies $C^{\mathrm{T}}Cx = 0$, and hence, $x = 0$, which is a contradiction. Hence, $P > 0$.

Finally, uniqueness of P follows from Lemma 3.2. Alternatively, suppose, *ad absurdum*, that there exist two solutions P_1 and P_2 to (3.157). Then

$$0 = A^{\mathrm{T}}P_1 + P_1A + R, \tag{3.163}$$
$$0 = A^{\mathrm{T}}P_2 + P_2A + R. \tag{3.164}$$

Subtracting (3.164) from (3.163) yields

$$0 = A^{\mathrm{T}}(P_1 - P_2) + (P_1 - P_2)A. \tag{3.165}$$

Next, forming $e^{A^{\mathrm{T}}t}[(3.165)]e^{At}$, $t \geq 0$, yields

$$0 = e^{A^{\mathrm{T}}t}[A^{\mathrm{T}}(P_1 - P_2) + (P_1 - P_2)A]e^{At}$$

$$= \frac{\mathrm{d}}{\mathrm{d}t} e^{A^{\mathrm{T}}t}(P_1 - P_2)e^{At}, \quad t \geq 0, \tag{3.166}$$

which further implies

$$\begin{aligned} 0 &= \int_0^\infty \frac{\mathrm{d}}{\mathrm{d}t} e^{A^{\mathrm{T}}t}(P_1 - P_2)e^{At}\mathrm{d}t \\ &= e^{A^{\mathrm{T}}t}(P_1 - P_2)e^{At}\Big|_0^\infty \\ &= -(P_1 - P_2). \end{aligned} \tag{3.167}$$

This leads to the contradiction $P_1 = P_2$, and hence, there exists a unique solution $P \in \mathbb{R}^{n\times n}$ satisfying (3.157). □

To apply Theorem 3.16 we choose *any* positive-definite matrix $R \in \mathbb{R}^{n\times n}$ and solve the Lyapunov equation for $P \in \mathbb{R}^{n\times n}$. If (3.157) has no solution or has multiple solutions, then the zero solution $x(t) \equiv 0$ to (3.152) is not asymptotically stable. Alternatively, if P is a unique positive-definite solution to (3.157), then the zero solution $x(t) \equiv 0$ to (3.152) is globally asymptotically stable.

Next, we use the Barbashin-Krasovskii-LaSalle invariant set theorem to weaken the conditions in Theorem 3.16.

Theorem 3.17. Consider the linear dynamical system (3.152). Let $R = C^{\mathrm{T}}C$, where $C \in \mathbb{R}^{l\times n}$, and assume (A, C) is observable. Then the zero solution $x(t) \equiv 0$ to (3.152) is globally asymptotically stable if and only if there exists a unique positive-definite matrix $P \in \mathbb{R}^{n\times n}$ satisfying

$$0 = A^{\mathrm{T}}P + PA + C^{\mathrm{T}}C. \tag{3.168}$$

Proof. To show sufficiency, consider the Lyapunov function candidate $V(x) = x^{\mathrm{T}}Px$, where $x \in \mathbb{R}^n$ and P satisfies (3.168). Thus, the corresponding Lyapunov derivative is given by

$$V'(x)f(x) = 2x^{\mathrm{T}}PAx = x^{\mathrm{T}}(A^{\mathrm{T}}P + PA)x = -x^{\mathrm{T}}C^{\mathrm{T}}Cx \leq 0, \quad x \in \mathbb{R}^n, \tag{3.169}$$

which proves Lyapunov stability. Next, note that $V'(x)f(x) = 0$ if and only if $Cx = 0$. Let $\mathcal{R} \triangleq \{x \in \mathbb{R}^n : V'(x)f(x) = 0\} = \{x \in \mathbb{R}^n : Cx = 0\}$. Now, let $\mathcal{M}$ denote the largest invariant set contained in $\mathcal{R}$ and note that since $Cx = 0$ it follows that for all $x \in \mathcal{M}$, $C\dot{x} = 0$, $C\ddot{x} = 0, \ldots, Cx^{(n-1)} = 0$, and hence, for all $x \in \mathcal{M}$, $Cx = 0$, $CAx = 0$, $\ldots$, $CA^{n-1}x = 0$ or, equivalently, $\mathcal{O}x = 0$, where $\mathcal{O}$ denotes the observability matrix. Next, since (A, C) is observable it follows that $\mathcal{O}x = 0$ if and only if $x = 0$, and hence, $\mathcal{M} = \{0\}$. Now, it follows from Theorem 3.3 that $x(t) \to \mathcal{M} = \{0\}$ as $t \to \infty$, establishing asymptotic stability. Furthermore, global asymptotic stability is a direct consequence of the fact that $V(x) \to \infty$ as $\|x\| \to \infty$.

To show necessity, suppose A is asymptotically stable. Then existence and uniqueness of a nonnegative-definite $P \in \mathbb{R}^{n\times n}$ satisfying (3.168) follows as in the proof of Theorem 3.16. To show $P > 0$, suppose, *ad absurdum*, that P is not positive definite. Then there exists $x \in \mathbb{R}^n$, $x \neq 0$, such that $x^{\mathrm{T}}Px = 0$, which, by (3.162), implies $Ce^{At}x = 0$ for all $t \geq 0$. Next, let $g(t) = Ce^{At}x$ and note that $g(t) = 0$, $\dot{g}(t) = 0$, $\ddot{g}(t) = 0$, ..., $g^{(n-1)}(t) = 0$, $t \geq 0$, or, equivalently, $Ce^{At}x = 0$, $CAe^{At}x = 0$, $CA^2e^{At}x = 0$,..., $CA^{n-1}e^{At}x = 0$, $t \geq 0$. Now, for $t = 0$, it follows that $\mathcal{O}x = 0$, which, since (A, C) is observable, implies that $x = 0$ and, hence, leads to a contradiction. Hence, $P > 0$. □

Finally, we give necessary and sufficient conditions for Lyapunov stability of the zero solution $x(t) \equiv 0$ to (3.152) in terms of the existence of a positive-definite solution to the Lyapunov equation (3.157).

Theorem 3.18. Consider the linear dynamical system (3.152). The zero solution $x(t) \equiv 0$ to (3.152) is Lyapunov stable if and only if there exists a positive-definite matrix $P \in \mathbb{R}^{n\times n}$ and a nonnegative-definite matrix $R \in \mathbb{R}^{n\times n}$ such that (3.157) holds.

Proof. Sufficiency is immediate from Theorem 3.1 with the Lyapunov function candidate $V(x) = x^{\mathrm{T}}Px$. To show necessity note that if the zero solution $x(t) \equiv 0$ to (3.152) is Lyapunov stable, then it follows from Theorem 3.15 that $\mathrm{Re}\,\lambda < 0$, or $\mathrm{Re}\,\lambda = 0$ and λ is semisimple, where $\lambda \in \mathrm{spec}(A)$. Hence, it follows from the real Jordan decomposition [201] that $A = SJS^{-1}$, where $S \in \mathbb{R}^{n\times n}$ is a nonsingular matrix and $J = \text{block–diag}[J_1, J_2, \ldots, J_m]$ is the real Jordan form of A. Now, without loss of generality, assume that J_1, J_2, ..., J_r correspond to Jordan blocks of A with $\mathrm{Re}\,\lambda < 0$, and $J_{r+1}, \ldots, J_m$ correspond to Jordan blocks of A with $\mathrm{Re}\,\lambda = 0$ and λ semisimple. Next, it follows from Theorem 3.16 that there exists a positive-definite matrix P_1 such that $J_{\mathrm{a}}^{\mathrm{T}}P_1 + P_1J_{\mathrm{a}} < 0$, where $J_{\mathrm{a}} \triangleq \text{block–diag}[J_1, \ldots, J_r]$. Furthermore, since $J_{r+1}, \ldots, J_m$ correspond to Jordan blocks of A with $\mathrm{Re}\,\lambda = 0$ and λ semisimple, it follows that $J_{\mathrm{s}} = \text{block–diag}[J_{r+1}, \ldots, J_m]$ is skew-symmetric, and hence, $J_{\mathrm{s}}^{\mathrm{T}} + J_{\mathrm{s}} = 0$, which implies that there exists a nonnegative-definite matrix $\hat{R} \in \mathbb{R}^{n\times n}$ such that

$$0 = J^{\mathrm{T}}\hat{P} + \hat{P}J + \hat{R}, \tag{3.170}$$

where $\hat{P} = \text{block–diag}[P_1, I]$. Now, forming $S^{-\mathrm{T}}(3.170)S^{-1}$ yields (3.157) with $P = S^{-\mathrm{T}}\hat{P}S^{-1}$ and $R = S^{-\mathrm{T}}\hat{R}S^{-1}$. □

Next, using the results of Section 3.2 and this section we provide a key result on linearization of nonlinear systems. Specifically, we present *Lyapunov's indirect method* to draw conclusions about local stability of an equilibrium point of a nonlinear system by examining the stability of the linearization of the nonlinear system about the equilibrium point in question. We begin by considering (3.1) with $f(0) = 0$ and assume that $f : \mathcal{D} \to \mathbb{R}^n$ is continuously differentiable. Next, expanding $f(x)$ via a Taylor expansion about the equilibrium point $x = 0$ it follows that (3.1) can be written as

$$\dot{x}(t) = f(0) + \left.\frac{\partial f}{\partial x}\right|_{x=0} x(t) + g(x(t)), \qquad x(0) = x_0, \qquad t \geq 0, \qquad (3.171)$$

where $g(x) \triangleq f(x) - \left.\frac{\partial f}{\partial x}\right|_{x=0} x$. Note that since $\frac{\partial f}{\partial x}$ is continuous it follows that $\lim_{\|x\|\to 0} \frac{\|g(x)\|}{\|x\|} = 0$. This shows that in a small neighborhood of the origin a nonlinear system can be approximated by its linearization about the origin. The following theorem known as Lyapunov's indirect method gives sufficient conditions for stability of an equilibrium point of a nonlinear system by examining the equilibrium point of the linearized system.

Theorem 3.19 (Lyapunov's Indirect Theorem). Let $x(t) \equiv 0$ be an equilibrium point for the nonlinear dynamical system

$$\dot{x}(t) = f(x(t)), \qquad x(0) = x_0, \qquad t \geq 0, \qquad (3.172)$$

where $f : \mathcal{D} \to \mathbb{R}^n$ is continuously differentiable and $\mathcal{D}$ is an open set with $0 \in \mathcal{D}$. Furthermore, let

$$A = \left.\frac{\partial f}{\partial x}\right|_{x=0}.$$

Then the following statements hold:

i) If Re $\lambda < 0$, where $\lambda \in \operatorname{spec}(A)$, then the zero solution $x(t) \equiv 0$ to (3.172) is exponentially stable.

ii) If there exists $\lambda \in \operatorname{spec}(A)$ such that Re $\lambda > 0$, then the zero solution $x(t) \equiv 0$ to (3.172) is unstable.

Proof. *i)* If Re $\lambda < 0$, $\lambda \in \operatorname{spec}(A)$, A is asymptotically stable, and hence, it follows from Theorem 3.16 that there exists a unique positive-definite matrix $P \in \mathbb{R}^{n\times n}$ satisfying (3.157). Next, choose the Lyapunov function candidate $V(x) = x^{\mathrm{T}}Px$ for the nonlinear system (3.172) and note that the Lyapunov derivative along the system trajectories of (3.172) is given by

$$\begin{aligned} \dot{V}(x) &= V'(x)f(x) \\ &= 2x^{\mathrm{T}}P[Ax + g(x)] \\ &= x^{\mathrm{T}}(A^{\mathrm{T}}P + PA)x + 2x^{\mathrm{T}}Pg(x) \\ &= -x^{\mathrm{T}}Rx + 2x^{\mathrm{T}}Pg(x), \qquad x \in \mathcal{D}. \end{aligned} \qquad (3.173)$$

Now, noting $-x^{\mathrm{T}}Rx \leq -\lambda_{\min}(R)\|x\|_2^2$ and using the Cauchy-Schwarz inequality it follows that

$$\dot{V}(x) \leq -\lambda_{\min}(R)\|x\|_2^2 + 2\lambda_{\max}(P)\|x\|_2\|g(x)\|_2, \quad x \in \mathcal{D}. \tag{3.174}$$

Next, since $g(x)$, $x \in \mathcal{D}$, is such that $\lim_{\|x\|_2 \to 0} \frac{\|g(x)\|_2}{\|x\|_2} = 0$ it follows that for every $\gamma > 0$, there exists $\varepsilon > 0$ such that $\|g(x)\|_2 < \gamma\|x\|_2$ for all $x \in \mathcal{B}_\varepsilon(0)$. Hence, (3.174) implies

$$\dot{V}(x) < -[\lambda_{\min}(R) - 2\gamma\lambda_{\max}(P)]\|x\|_2^2, \quad x \in \mathcal{B}_\varepsilon(0). \tag{3.175}$$

Now, choosing $\gamma \leq \lambda_{\min}(R)/2\lambda_{\max}(P)$ it follows that $\dot{V}(x) < 0$, $x \in \mathcal{B}_\varepsilon(0)$, $x \neq 0$, which proves local asymptotic stability. To show local exponential stability it need only be noted that $\lambda_{\min}(P)\|x\|_2^2 \leq V(x) \leq \lambda_{\max}(P)\|x\|_2^2$.

ii) First, assume that A is such that (3.158) holds. Hence, it follows from Lemma 3.2 that

$$A^{\mathrm{T}}P + PA = I_n \tag{3.176}$$

has a unique solution with $P \in \mathbb{R}^{n\times n}$. By assumption, there exists $\lambda \in \operatorname{spec}(A)$ such that $\operatorname{Re}\lambda > 0$. Next, assume that the remaining eigenvalues λ of A are such that $\operatorname{Re}\lambda < 0$; that is, the eigenvalues of A cluster into a group of eigenvalues in the open right half plane and a group of eigenvalues in the open left half plane. Now, it follows from the real Jordan decomposition [201] that $A = SJS^{-1}$, where $S \in \mathbb{R}^{n\times n}$ is a nonsingular matrix and $J = \text{block-diag}[J_1, J_2, \ldots, J_m]$ is the real Jordan form of A. Now, without loss of generality, let $J = \text{block-diag}[J_+, J_-]$, where J_+ corresponds to a block diagonal matrix having Jordan blocks with $\operatorname{Re}\lambda > 0$ and J_- corresponds to a block diagonal matrix having Jordan blocks with $\operatorname{Re}\lambda < 0$. Next, since $-J_+$ and J_- are asymptotically stable, it follows from Theorem 3.16 that there exist positive-definite matrices P_+ and P_- such that $-J_+^{\mathrm{T}}P_+ - P_+J_+ + I = 0$ and $J_-^{\mathrm{T}}P_- + P_-J_- + I = 0$, which implies that

$$0 = J^{\mathrm{T}}\hat{P} + \hat{P}J + I, \tag{3.177}$$

where $\hat{P} = \text{block-diag}[-P_+, P_-]$. Now, forming $S^{-\mathrm{T}}(3.170)S^{-1}$ yields (3.176) with $P = S^{-\mathrm{T}}\hat{P}S^{-1}$. Hence, the solution P to (3.176) has at least one positive eigenvalue. Now, using identical arguments as in *i*) it can be shown that for every sufficiently small $\delta > 0$ there exists $x_0 \in \mathcal{D}$ such that $\|x_0\| < \delta$, $V(x_0) > 0$, and $V'(x)f(x) > 0$, $x \in \mathcal{B}_\varepsilon(0)$, $x \neq 0$, where $\varepsilon > 0$ and $V(x) = x^{\mathrm{T}}Px$. Hence, it follows from Theorem 3.12 that the zero solution $x(t) \equiv 0$ to (3.172) is unstable. The proof of the case in which there do not exist $\lambda \in \operatorname{spec}(A)$ such that $\operatorname{Re}\lambda < 0$ follows identically with $J = J_+$. Finally, in the case where some of the eigenvalues of A lie on the $\jmath\omega$-axis with at least one eigenvalue in the right half plane, the proof follows by shifting the imaginary axis and using the fact that the eigenvalues of a matrix are continuous with respect to the entries of the matrix. □

Theorem 3.19 gives a straightforward procedure for examining the stability of a nonlinear system by examining the stability of the linearized system. Furthermore, it follows from the proof of Theorem 3.19 that if the linearized system is asymptotically stable, then we can always construct a quadratic Lyapunov function that guarantees *local* exponential stability of the nonlinear system. To maximize the domain of attraction with a quadratic Lyapunov function it is also clear from the proof of Theorem 3.19 that the larger the ratio $\lambda_{\min}(R)/\lambda_{\max}(P)$ the larger the possible choice of ε. Finally, it should be noted that if $x_{\rm e} \neq 0$ is an equilibrium point of (3.172), then Theorem 3.19 holds with $A = \frac{\partial f}{\partial x}\big|_{x=0}$ replaced by $A = \frac{\partial f}{\partial x}\big|_{x=x_{\rm e}}$.

Example 3.15. Consider the dynamical system given in Example 3.2 describing the motion of a simple pendulum with viscous damping given by

$$\dot{x}_1(t) = x_2(t), \qquad x_1(0) = x_{10}, \qquad t \geq 0, \tag{3.178}$$
$$\dot{x}_2(t) = -\frac{g}{l}\sin x_1(t) - x_2(t), \qquad x_2(0) = x_{20}. \tag{3.179}$$

This model has countably infinitely many equilibrium points in $\mathbb{R}^2$ given by $(x_{1\rm e}, x_{2\rm e}) = (n\pi, 0)$, $n = 0, \pm 1, \pm 2, \ldots$. Here, we examine the stability of the physical equilibria given by $(0,0)$ and $(\pi, 0)$ using Lyapunov's linearization method. For (3.178) and (3.179) the Jacobian matrix $\frac{\partial f}{\partial x}$ is given by

$$\frac{\partial f}{\partial x} = \begin{bmatrix} \frac{\partial f_1}{\partial x_1} & \frac{\partial f_1}{\partial x_2} \\ \frac{\partial f_2}{\partial x_1} & \frac{\partial f_2}{\partial x_2} \end{bmatrix} = \begin{bmatrix} 0 & 1 \\ -\frac{g}{l}\cos x_1 & -1 \end{bmatrix}. \tag{3.180}$$

Note that

$$A_1 = \frac{\partial f}{\partial x}\bigg|_{x=(0,0)} = \begin{bmatrix} 0 & 1 \\ -\frac{g}{l} & -1 \end{bmatrix}, \qquad A_2 = \frac{\partial f}{\partial x}\bigg|_{x=(\pi,0)} = \begin{bmatrix} 0 & 1 \\ \frac{g}{l} & -1 \end{bmatrix}. \tag{3.181}$$

Furthermore, since $\operatorname{tr} A_1 < 0$, $\det A_1 > 0$, $\operatorname{tr} A_2 < 0$, and $\det A_2 < 0$, it follows from Theorem 3.19 that the zero solution $(x_1(t), x_2(t)) \equiv (0,0)$ to (3.178) and (3.179) is locally exponentially stable while the solution $(x_1(t), x_2(t)) \equiv (\pi, 0)$ to (3.178) and (3.179) is unstable. △

Example 3.16. Consider the Van der Pol oscillator given by

$$\dot{x}_1(t) = x_2(t), \qquad x_1(0) = x_{10}, \qquad t \geq 0, \tag{3.182}$$
$$\dot{x}_2(t) = -\mu(1 - x_1^2(t))x_2(t) - x_1(t), \qquad x_2(0) = x_{20}, \tag{3.183}$$

where $\mu \in \mathbb{R}$. Note that (3.182) and (3.183) has a unique equilibrium at $(x_{1\rm e}, x_{2\rm e}) = (0,0)$. Furthermore, the Jacobian matrix $\frac{\partial f}{\partial x}$ evaluated at $(0,0)$ is given by

$$A = \frac{\partial f}{\partial x}\bigg|_{x=(0,0)} = \begin{bmatrix} 0 & 1 \\ -1 & -\mu \end{bmatrix}. \tag{3.184}$$

Now, it follows from Theorem 3.19 that if $\mu < 0$, then the zero solution $(x_1(t), x_2(t)) \equiv (0,0)$ to (3.182) and (3.183) is unstable and if $\mu > 0$, then the zero solution $(x_1(t), x_2(t)) \equiv (0,0)$ to (3.182) and (3.183) is locally exponentially stable. △

Finally, we apply Lyapunov's indirect method to the stabilization of a nonlinear dynamical system. Specifically, consider the nonlinear controlled system

$$\dot{x}(t) = F(x(t), u(t)), \qquad x(0) = x_0, \qquad t \geq 0, \tag{3.185}$$

where $x(t) \in \mathbb{R}^n$, $u(t) \in \mathbb{R}^m$, $t \geq 0$, and $F : \mathbb{R}^n \times \mathbb{R}^m \to \mathbb{R}^n$. Here, we seek *feedback controllers* of the form $u(t) = \phi(x(t))$, where $\phi : \mathbb{R}^n \to \mathbb{R}^m$ and $\phi(0) = 0$, so that the zero solution $x(t) \equiv 0$ of the closed-loop system given by

$$\dot{x}(t) = F(x(t), \phi(x(t))), \qquad x(0) = x_0, \qquad t \geq 0, \tag{3.186}$$

is asymptotically stable.

Theorem 3.20. Consider the nonlinear controlled system (3.185) where $F : \mathbb{R}^n \times \mathbb{R}^m \to \mathbb{R}^n$ is continuously differentiable and $F(0,0) = 0$. Furthermore, define

$$A \triangleq \left.\frac{\partial F}{\partial x}\right|_{(x,u)=(0,0)}, \qquad B \triangleq \left.\frac{\partial F}{\partial u}\right|_{(x,u)=(0,0)}, \tag{3.187}$$

and assume (A, B) is stabilizable. Then, there exists a matrix $K \in \mathbb{R}^{m \times n}$ such that $\mathrm{spec}(A + BK) \subset \mathbb{C}_-$. In addition, with the linear control law $u(t) = Kx(t)$, the zero solution $x(t) \equiv 0$ of the closed-loop nonlinear system

$$\dot{x}(t) = F(x(t), Kx(t)), \qquad x(0) = x_0, \qquad t \geq 0, \tag{3.188}$$

is locally exponentially stable.

Proof. First, note that the closed-loop system (3.188) has the form

$$\dot{x}(t) = F(x(t), Kx(t)) = f(x(t)), \qquad x(0) = x_0, \qquad t \geq 0. \tag{3.189}$$

Next, it follows that

$$\left.\frac{\partial f}{\partial x}\right|_{x=0} = \left.\left[\frac{\partial F}{\partial x}(x, Kx) + \frac{\partial F}{\partial u}(x, Kx)K\right]\right|_{x=0} = A + BK, \tag{3.190}$$

and hence, since (A, B) is stabilizable, K can be chosen so that $A + BK$ is asymptotically stable. Now, the result is a direct consequence of *i*) of Theorem 3.19. □

It follows from Theorem 3.20 that if the linearization of (3.185) is stabilizable, that is, (A, B) is stabilizable, then we can always construct a Lyapunov function for the nonlinear closed-loop system (3.186). In particular, since there exists $K \in \mathbb{R}^{m\times n}$ such that $A+BK$ is asymptotically stable, then for every $R > 0$ the Lyapunov equation

$$0 = (A+BK)^{\mathrm{T}}P + P(A+BK) + R, \tag{3.191}$$

is guaranteed to have a unique positive-definite solution by Theorem 3.16. Hence, the quadratic Lyapunov function $V(x) = x^{\mathrm{T}}Px$ is a Lyapunov function for the closed-loop *nonlinear* system (3.186) that guarantees local asymptotic stability.

A similar approach can be used to design dynamic output feedback controllers for the nonlinear system (3.185). Specifically, consider the nonlinear dynamical system

$$\dot{x}(t) = F(x(t), u(t)), \quad x(0) = x_0, \qquad t \geq 0, \tag{3.192}$$
$$y(t) = h(x(t), u(t)), \tag{3.193}$$

where $x(t) \in \mathbb{R}^n$, $u(t) \in \mathbb{R}^m$, $y(t) \in \mathbb{R}^l$, $t \geq 0$, $F : \mathbb{R}^n \times \mathbb{R}^m \to \mathbb{R}^n$ is such that $F(0,0) = 0$, and $h : \mathbb{R}^n \times \mathbb{R}^m \to \mathbb{R}^l$ is such that $h(0,0) = 0$. Next, linearizing (3.192) and (3.193) about $x = 0$ and $u = 0$ yields

$$\dot{x}(t) = Ax(t) + Bu(t), \quad x(0) = x_0, \qquad t \geq 0, \tag{3.194}$$
$$y(t) = Cx(t) + Du(t), \tag{3.195}$$

where A and B are given by (3.187) and

$$C \triangleq \left.\frac{\partial h}{\partial x}\right|_{(x,u)=(0,0)}, \qquad D \triangleq \left.\frac{\partial h}{\partial u}\right|_{(x,u)=(0,0)}. \tag{3.196}$$

Assuming (A, B) is stabilizable and (A, C) is detectable, then we can design a full-order dynamic compensator of the form

$$\dot{x}_{\mathrm{c}}(t) = A_{\mathrm{c}}x_{\mathrm{c}}(t) + B_{\mathrm{c}}y(t), \quad x_{\mathrm{c}}(0) = x_{\mathrm{c}0}, \qquad t \geq 0, \tag{3.197}$$
$$u(t) = C_{\mathrm{c}}x_{\mathrm{c}}(t), \tag{3.198}$$

where $x_{\mathrm{c}}(t) \in \mathbb{R}^n$, $t \geq 0$, $A_{\mathrm{c}} \in \mathbb{R}^{n\times n}$, $B_{\mathrm{c}} \in \mathbb{R}^{n\times l}$, and $C_{\mathrm{c}} \in \mathbb{R}^{m\times n}$, such that with $A_{\mathrm{c}} = A + BC_{\mathrm{c}} - B_{\mathrm{c}}C - B_{\mathrm{c}}DC_{\mathrm{c}}$,

$$\tilde{A} = \begin{bmatrix} A & BC_{\mathrm{c}} \\ B_{\mathrm{c}}C & A_{\mathrm{c}} + B_{\mathrm{c}}DC_{\mathrm{c}} \end{bmatrix} \tag{3.199}$$

is asymptotically stable. The asymptotic stability of $\tilde{A}$ follows from the fact that (A, B) is stabilizable and (A, C) is detectable, and hence, there exist $B_{\mathrm{c}} \in \mathbb{R}^{n\times l}$ and $C_{\mathrm{c}} \in \mathbb{R}^{m\times n}$ such that $A + BC_{\mathrm{c}}$ and $A - B_{\mathrm{c}}C$ are asymptotically stable.

Now, applying the controller (3.197) and (3.198) to the nonlinear system (3.192) and (3.193) gives the closed-loop system

$$\dot{x}(t) = F(x(t), C_{\mathrm{c}}x_{\mathrm{c}}(t)), \quad x(0) = x_0, \quad t \geq 0, \tag{3.200}$$

$$\dot{x}_{\rm c}(t) = A_{\rm c}x_{\rm c}(t) + B_{\rm c}h(x, C_{\rm c}x_{\rm c}(t)), \quad x_{\rm c}(0) = x_{\rm c0}. \tag{3.201}$$

Note that since $F(0,0) = 0$ and $h(0,0) = 0$, $(x, x_{\rm c}) = (0,0)$ is an equilibrium point of (3.200) and (3.201). Furthermore, linearizing (3.200) and (3.201) about $x = 0$ and $x_{\rm c} = 0$ yields

$$\dot{\tilde{x}}(t) = \tilde{A}\tilde{x}(t), \quad \tilde{x}(0) = \tilde{x}_0, \qquad t \geq 0, \tag{3.202}$$

where $\tilde{x} \triangleq [x^{\rm T},\ x_{\rm c}^{\rm T}]^{\rm T}$ and $\tilde{A}$ is given by (3.199). Hence, choosing the compensator triple $(A_{\rm c}, B_{\rm c}, C_{\rm c})$ such that the zero solution $\tilde{x}(t) \equiv 0$ to (3.202) is asymptotically stable, it follows from Theorem 3.19 that the nonlinear closed-loop system (3.200) and (3.201) is locally exponentially stable. Furthermore, a Lyapunov function guaranteeing local exponential stability of (3.200) and (3.201) is given by $V(\tilde{x}) = \tilde{x}^{\rm T}\tilde{P}\tilde{x}$, where $\tilde{P}$ is the $2n \times 2n$ positive definite solution to the Lyapunov equation

$$0 = \tilde{A}^{\rm T}\tilde{P} + \tilde{P}\tilde{A} + \tilde{R}, \tag{3.203}$$

where $\tilde{R}$ is any $2n \times 2n$ positive-definite matrix.

3.8 Problems

Problem 3.1. Give an example of a function that is positive definite but not radially unbounded. Alternatively, given an example of a function that is radially unbounded but not positive definite.

Problem 3.2. Let $V : \mathcal{D} \subseteq \mathbb{R}^n \to \mathbb{R}$ be such that $V(\cdot)$ is two-times continuously differentiable, $V(0) = 0$, $V'(0) = 0$, and $V''(0) > 0$. Show that there exists an open set $\mathcal{D}_0 \subset \mathcal{D}$ such that $0 \in \mathcal{D}_0$ and $V(x) > 0$, $x \in \mathcal{D}_0$, $x \neq 0$.

Problem 3.3. Let $V : \mathbb{R}^n \to \mathbb{R}$ be such that $V(\cdot)$ is continuously differentiable, $V(0) = 0$, $V(x) > 0$, $x \in \mathcal{D} \subset \mathbb{R}^n$, $x \neq 0$, and $\|V'(x)\| > 0$, $x \in \mathbb{R}^n$, $x \neq 0$. Show that $V(\cdot)$ is globally positive definite with a unique global minimum at $x = 0$.

Problem 3.4. Consider the dynamical system (3.1) with $f(x) = Ax$, where $A \in \mathbb{R}^{n \times n}$. Show that the following statements are equivalent:

i) The zero solution $x(t) \equiv 0$ to (3.1) is Lyapunov stable.

ii) For every initial condition $x_0 \in \mathbb{R}^n$, $x(t)$, $t \geq 0$, is bounded.

iii) If $\lambda \in \operatorname{spec}(A)$, then either $\operatorname{Re} \lambda < 0$, or both $\operatorname{Re} \lambda = 0$ and λ is semisimple.

iv) There exists $\alpha > 0$ such that $\|e^{At}\| < \alpha$, $t \geq 0$.

Problem 3.5. Consider the dynamical system (3.1) with $f(x) = Ax$, where $A \in \mathbb{R}^{n\times n}$. Show that the following statements are equivalent:

i) The zero solution $x(t) \equiv 0$ to (3.1) is globally asymptotically stable.

ii) For every initial condition $x_0 \in \mathbb{R}^n$, $\lim_{t\to\infty} x(t) = 0$.

iii) If $\lambda \in \operatorname{spec}(A)$, then $\operatorname{Re}\,\lambda < 0$.

iv) $\lim_{t\to\infty} e^{At} = 0$.

Problem 3.6. Consider the dynamical system (3.1) with $f(x) = Ax$, where $A \in \mathbb{R}^{n\times n}$. Show that if there exist a positive-definite matrix $P \in \mathbb{R}^{n\times n}$ and a scalar $\alpha > 0$ such that

$$0 = (A + \alpha I_n)^{\mathrm{T}} P + P(A + \alpha I_n) + R, \tag{3.204}$$

where R is a positive-definite matrix, then the eigenvalues of A have real part less than $-\alpha$.

Problem 3.7. Let $A \in \mathbb{R}^{n\times n}$. A is *essentially nonnegative* if $A_{(i,j)} \geq 0$, $i, j = 1, \ldots, n$, $i \neq j$. Consider the dynamical system (3.1) with $f(x) = Ax$, where $A \in \mathbb{R}^{n\times n}$. Show that $\overline{\mathbb{R}}_+^n$ is an invariant set with respect to (3.1) if and only if A is essentially nonnegative.

Problem 3.8. The nonlinear dynamical system (3.1) is *nonnegative* if for every $x(0) \in \overline{\mathbb{R}}_+^n$, the solution $x(t)$, $t \geq 0$, to (3.1) is nonnegative, that is, $x(t) \geq\geq 0$, $t \geq 0$. The equilibrium solution $x(t) \equiv x_{\mathrm{e}}$ of a nonnegative dynamical system is *Lyapunov stable* if, for every $\varepsilon > 0$, there exists $\delta = \delta(\varepsilon) > 0$ such that if $x_0 \in \mathcal{B}_\delta(x_{\mathrm{e}}) \cap \overline{\mathbb{R}}_+^n$, then $x(t) \in \mathcal{B}_\varepsilon(x_{\mathrm{e}}) \cap \overline{\mathbb{R}}_+^n$, $t \geq 0$. The equilibrium solution $x(t) \equiv x_{\mathrm{e}}$ of a nonnegative dynamical system is *asymptotically stable* if it is Lyapunov stable and there exists $\delta > 0$ such that if $x_0 \in \mathcal{B}_\delta(x_{\mathrm{e}}) \cap \overline{\mathbb{R}}_+^n$, then $\lim_{t\to\infty} x(t) = x_{\mathrm{e}}$. Consider the dynamical system (3.1) with $f(x) = Ax$, where $A \in \mathbb{R}^{n\times n}$ is essentially nonnegative (see Problem 3.7). Show that the following statements hold:

i) If there exist vectors $p, r \in \mathbb{R}^n$ such that $p >> 0$ and $r \geq\geq 0$ satisfy

$$0 = A^{\mathrm{T}} p + r, \tag{3.205}$$

then A is Lyapunov stable.

ii) If there exist vectors $p, r \in \mathbb{R}^n$ such that $p \geq\geq 0$ and $r \geq\geq 0$ satisfy (3.205) and (A, r^{T}) is observable, then $p >> 0$ and A is asymptotically stable.

Furthermore, show that the following statements are equivalent:

iii) A is asymptotically stable.

iv) There exist vectors $p, r \in \mathbb{R}^n$ such that $p >> 0$ and $r >> 0$ satisfy (3.205).

v) There exist vectors $p, r \in \mathbb{R}^n$ such that $p \geq\geq 0$ and $r >> 0$ satisfy (3.205).

vi) For every $r \in \mathbb{R}^n$ such that $r >> 0$, there exists $p \in \mathbb{R}^n$ such that $p >> 0$ satisfies (3.205).

(**Hint:** Use the Lyapunov function candidate $V(x) = p^{\mathrm{T}}x$ in your analysis and show that the Lyapunov stability theorem of Section 3.2 and the invariant set theorems of Section 3.3 can be used directly for nonnegative systems with the required sufficient conditions verified on $\overline{\mathbb{R}}_+^n$.)

Problem 3.9. Consider the nonlinear dynamical system

$$\dot{x}_1(t) = x_2(t), \quad x_1(0) = x_{10}, \quad t \geq 0, \tag{3.206}$$
$$\dot{x}_2(t) = -k \sin x_1(t), \quad x_2(0) = x_{20}, \tag{3.207}$$

where $k > 0$. Show that the zero solution $(x_1(t), x_2(t)) \equiv (0,0)$ to (3.206) and (3.207) is Lyapunov stable.

Problem 3.10. Consider the nonlinear dynamical system

$$\dot{x}_1(t) = x_2(t) + \alpha x_1(t)[x_1^2(t) + x_2^2(t)], \quad x_1(0) = x_{10}, \quad t \geq 0, \tag{3.208}$$
$$\dot{x}_2(t) = -x_1(t) + \alpha x_2(t)[x_1^2(t) + x_2^2(t)], \quad x_2(0) = x_{20}, \tag{3.209}$$

where $\alpha < 0$. Show that the zero solution $(x_1(t), x_2(t)) \equiv (0,0)$ to (3.208) and (3.209) is globally asymptotically stable.

Problem 3.11. Consider the nonlinear dynamical system

$$\dot{x}_1(t) = [x_1(t) - \alpha_2 x_2(t)][x_1^2(t) + x_2^2(t) - 1], \; x_1(0) = x_{10}, \; t \geq 0, \tag{3.210}$$
$$\dot{x}_2(t) = [\alpha_1 x_1(t) + x_2(t)][x_1^2(t) + x_2^2(t) - 1], \; x_2(0) = x_{20}, \tag{3.211}$$

where $\alpha_1 > 0$ and $\alpha_2 > 0$. Show that the zero solution $(x_1(t), x_2(t)) \equiv (0,0)$ to (3.210) and (3.211) is asymptotically stable. Is the result global? If not, give a characterization for the domain of attraction.

Problem 3.12. Consider the nonlinear dynamical system

$$\dot{x}_1(t) = -2x_1(t) - x_2^2(t), \quad x_1(0) = x_{10}, \quad t \geq 0, \tag{3.212}$$
$$\dot{x}_2(t) = -x_1^2(t) - x_2(t), \quad x_2(0) = x_{20}. \tag{3.213}$$

Using the quadratic Lyapunov function candidate $V(x_1, x_2) = x_1^2 + x_2^2$ show that the zero solution $(x_1(t), x_2(t)) \equiv (0,0)$ to (3.212) and (3.213) is asymptotically stable. Find the maximum domain of attraction $\mathcal{D}_{\mathrm{A}}$ for

(3.212) and (3.213) predicated on $V(x_1, x_2)$. Plot $\mathcal{D}_{\rm A}$ in the x_1-x_2 plane and show the region where $\dot{V}(x_1, x_2) < 0$, $(x_1, x_2) \in \mathbb{R} \times \mathbb{R}$.

Problem 3.13. Consider the nonlinear system

$$\ddot{x}(t) + g(x(t)) = 0, \quad x(0) = x_0, \quad \dot{x}(0) = \dot{x}_0, \quad t \geq 0, \tag{3.214}$$

where $xg(x) > 0$, $x \neq 0$, and $g(0) = 0$. Using $x_1 = x$ and $x_2 = \dot{x}$ show that the trajectories of (3.214) satisfy $V(x_1, x_2) = k^2$, where $V(x_1, x_2) = \frac{1}{2}x_2^2 + \int_0^{x_1} g(s)\mathrm{d}s$ and k is a constant, and the origin is Lyapunov stable. Plot sample trajectories for the cases *i)* $g(x) = x$, *ii)* $g(x) = \frac{1}{2}x$, and *iii)* $g(x) = x$, $x > 0$, and $g(x) = \frac{1}{2}x$, $x < 0$.

Problem 3.14. Show that the rigid spacecraft dynamical system in Example 3.1 is at an equilibrium if and only if at least two of the quantities (x_1, x_2, x_3) are equal to zero, and hence, the set of equilibria consists of the union of the x_1, x_2, and x_3 axes. Furthermore, show that equilibria of the form $(x_{1\rm e}, 0, 0)$, $x_{1\rm e} \neq 0$, and $(0, 0, x_{3\rm e})$, $x_{3\rm e} \neq 0$, are Lyapunov stable while equilibria of the form $(0, x_{2\rm e}, 0)$, $x_{2\rm e} \neq 0$, are unstable. What does this result imply in regards to the spacecraft spinning about its major, minor, and intermediate axes?

Problem 3.15. Consider the controlled rigid spacecraft given by

$$\dot{x}_1(t) = I_{23}x_2(t)x_3(t) + u_1(t), \quad x_1(0) = x_{10}, \quad t \geq 0, \tag{3.215}$$
$$\dot{x}_2(t) = I_{31}x_3(t)x_1(t) + u_2(t), \quad x_2(0) = x_{20}, \tag{3.216}$$
$$\dot{x}_3(t) = I_{12}x_1(t)x_2(t) + u_3(t), \quad x_3(0) = x_{30}, \tag{3.217}$$

where $I_{23} = (I_2 - I_3)/I_1$, $I_{31} = (I_3 - I_1)/I_2$, $I_{12} = (I_1 - I_2)/I_3$, and I_1, I_2, and I_3 are the principal moments of inertia of the spacecraft such that $I_1 > I_2 > I_3 > 0$. Assume that the attitude control torques u_1, u_2, and u_3 are given by $u_1(t) = \alpha_1 x_1(t)$, $u_2(t) = \alpha_2 x_2(t)$, and $u_3(t) = \alpha_3 x_3(t)$. Using Lyapunov's direct method determine sufficient conditions on α_1, α_2, and α_3 such that the zero solution $x(t) \equiv 0$ to (3.215)–(3.217) is globally asymptotically stable.

Problem 3.16. Consider the controlled rigid spacecraft given in Problem 3.15 with the *bang-bang control* law

$$u_i(t) = f_i(x_i) \triangleq U_i \mathrm{sgn}(x_i) = \begin{cases} U_i, & x_i > 0 \\ -U_i, & x_i < 0, \end{cases} \tag{3.218}$$

where $U_i > 0$, $i = 1, 2, 3$. Show that the feedback control law (3.218) globally stabilizes the zero solution $(x_1(t), x_2(t), x_3(t)) \equiv (0, 0, 0)$ to (3.215)–(3.217). Furthermore, show that the trajectories of the controlled spacecraft reach the origin in finite time $T \leq \|L(x_0)\|_2/U$, where $L(x_0)$ is the initial angular momentum vector of the rigid spacecraft and $U = \min\{U_1, U_2, U_3\}$.

Problem 3.17. Consider the controlled linearized approximation of the attitude motion of a satellite given by

$$I_1\ddot{\theta}(t)+3\omega^2(I_2-I_3)\theta(t)=u(t),\quad \theta(0)=\theta_{10},\quad \dot{\theta}(0)=\theta_{20},\quad t\geq 0, \tag{3.219}$$

where I_1, I_2, and I_3 are the moments of inertia of the satellite about the pitch, roll, and yaw axis, respectively; ω is the satellite orbiting angular rate; and θ is the pitch plane angular deviation from the local gravitational vertical. Show that the nonlinear control torque $u=-f(\alpha\theta+\beta\dot{\theta})$, where f is continuous, $f(0)=0$, and $zf(z)>0$, $z\in\mathbb{R}$, $z\neq 0$, can render the zero solution $(\theta(t),\dot{\theta}(t)\equiv(0,0)$ to (3.219) asymptotically stable. To analyze the stability of the closed-loop system use the Lyapunov function candidate

$$V(\theta,\dot{\theta})=\frac{1}{2}(\theta^2+\dot{\theta}^2)+b\int_0^{\alpha\theta+\beta\dot{\theta}}f(\sigma)\mathrm{d}\sigma, \tag{3.220}$$

where α, β, a, and b are parameters to be chosen. Can the feedback control law $u=-f(\beta\dot{\theta})$ asymptotically stabilize (3.219)? Under what conditions on $f(\cdot)$ will the origin be globally asymptotically stable?

Problem 3.18. Consider the nonlinear Lienard system given by

$$\ddot{x}(t)+f(x(t))\dot{x}(t)+g(x(t))=0,\quad x(0)=x_0,\quad \dot{x}(0)=\dot{x}_0,\quad t\geq 0, \tag{3.221}$$

where $g(0)=0$ and f and g are continuously differentiable, and define $F(x)=\int_0^x f(s)\mathrm{d}s$. Show that the system can be written as

$$\dot{x}_1(t) = x_2(t),\quad x_1(0)=x_{10},\quad t\geq 0, \tag{3.222}$$

$$\dot{x}_2(t) = -f(x_1(t))x_2(t)-g(x_1(t)),\quad x_2(0)=x_{20}, \tag{3.223}$$

or, equivalently,

$$\dot{x}_1(t) = x_2(t)-F(x_1(t)),\quad x_1(0)=x_{10},\quad t\geq 0, \tag{3.224}$$

$$\dot{x}_2(t) = -g(x_1(t)),\quad x_2(0)=x_{20}. \tag{3.225}$$

Analyze the stability of both forms using the Lyapunov function candidates

$$V(x_1,x_2)=\tfrac{1}{2}x_2^2+\int_0^{x_1}g(s)\mathrm{d}s \tag{3.226}$$

and

$$V(x_1,x_2)=\tfrac{1}{2}[x_2+F(x_1)]^2+\int_0^{x_1}g(s)\mathrm{d}s. \tag{3.227}$$

Make sure you specify under what conditions (3.226) and (3.227) are valid Lyapunov function candidates.

Problem 3.19. Consider the second-order nonlinear mechanical system consisting of a unit mass with a nonlinear spring and a nonlinear damper given by

$$\ddot{x}(t)+f(\dot{x}(t))+g(x(t))=0,\quad x(0)=x_0,\quad \dot{x}(0)=\dot{x}_0,\quad t\geq 0, \tag{3.228}$$

where $f(\cdot)$ is the force due to friction and $g(\cdot)$ is the restoring force of the nonlinear spring. Suppose that $f(\cdot)$ and $g(\cdot)$ are continuous and $f(0) = 0$ and $g(0) = 0$. In addition, suppose there exists $\varepsilon > 0$ such that $zf(z) > 0$, $z \in \mathcal{B}_\varepsilon(0)$, $z \neq 0$, and $zg(z) > 0$, $z \in \mathcal{B}_\varepsilon(0)$, $z \neq 0$.

i) Show that the origin is an equilibrium point.

ii) Show that the zero solution $(x(t), \dot{x}(t)) \equiv (0, 0)$ to (3.228) is asymptotically stable.

iii) Is the zero solution $(x(t), \dot{x}(t)) \equiv (0, 0)$ to (3.228) globally asymptotically stable? If not, give conditions on $f(\cdot)$ and $g(\cdot)$ that guarantee global asymptotic stability.

(**Hint:** Select as your Lyapunov function $V(x, \dot{x})$ the total mechanical energy in the system and show that the α-sublevel set of $V(x, \dot{x})$ is bounded for every $\alpha > 0$. Clearly state any necessary assumptions needed to show this.)

Problem 3.20. Consider the nonlinear dynamical system

$$\dot{x}_1(t) = x_2(t), \quad x_1(0) = x_{10}, \quad t \geq 0, \tag{3.229}$$
$$\dot{x}_2(t) = x_3(t), \quad x_2(0) = x_{20}, \tag{3.230}$$
$$\dot{x}_3(t) = -f(x_1(t)) - g(x_2(t)) - ax_3(t), \quad x_3(0) = x_{30}, \tag{3.231}$$

where $f(0) = g(0) = 0$ and f, g are continuously differentiable functions. Show that the zero solution $(x_1(t), x_2(t), x_3(t)) = (0, 0, 0)$ to (3.229)–(3.231) is globally asymptotically stable if *i)* $a > 0$, *ii)* $f(x_1)/x_1 \geq \varepsilon_1 > 0$, $x_1 \neq 0$, and *iii)* $ag(x_2)/x_2 - f'(x_1) \geq \varepsilon_2 > 0$, $x_1 \in \mathbb{R}$, $x_2 \neq 0$. (**Hint:** Use the Lyapunov function candidate $V(x_1, x_2, x_3) = aF(x_1) + f(x_1)x_2 + G(x_2) + \frac{1}{2}(ax_2 + x_3)^2$, where $F(x_1) = \int_0^{x_1} f(s)\mathrm{d}s$ and $G(x_2) = \int_0^{x_2} g(s)\mathrm{d}s$, to analyze (3.229)–(3.231).)

Problem 3.21. Consider the nonlinear dynamical system

$$\dot{x}_1(t) = x_2(t), \quad x_1(0) = x_{10}, \quad t \geq 0, \tag{3.232}$$
$$\dot{x}_2(t) = x_3(t), \quad x_2(0) = x_{20}, \tag{3.233}$$
$$\dot{x}_3(t) = -f(x_2(t))x_3(t) - ax_2(t) - bx_1(t), \quad x_3(0) = x_{30}, \tag{3.234}$$

where $a > 0$, $b > 0$, and f is continuous. Determine conditions on the damping function $f(\cdot)$ so that the zero solution $(x_1(t), x_2(t), x_3(t)) \equiv (0, 0, 0)$ to (3.232)–(3.234) is globally asymptotically stable. (**Hint:** Consider the Lyapunov function candidate

$$V(x_1, x_2, x_3) = \frac{1}{2}(bx_1 + ax_2)^2 + \frac{a}{2}x_3^2 + bx_2x_3 + b\int_0^{x_2} f(\sigma)\sigma\mathrm{d}\sigma, \tag{3.235}$$

and show that (3.235) is a valid candidate.)

Problem 3.22. Consider the nonlinear dynamical system

$$\dot{x}_1(t) = x_2(t), \quad x_1(0) = x_{10}, \quad t \geq 0, \tag{3.236}$$
$$\dot{x}_2(t) = -2\beta x_1(t) - 3x_1^2(t) - \alpha x_2(t), \quad x_2(0) = x_{20}, \tag{3.237}$$

where $\alpha > 0$ and $\beta > 0$. Show that $(x_{e1}, x_{e2}) = (0,0)$ and $(x_{e1}, x_{e2}) = (-\frac{2}{3}\beta, 0)$ are equilibrium points of (3.236) and (3.237). Using the function $V(x_1, x_2) = \beta x_1^2 + x_1^3 + \frac{1}{2}x_2^2$ show that the region $V(x_1, x_2) < \frac{4}{27}\beta^3$ is the union of two components $\mathcal{Q}_1$ and $\mathcal{Q}_2$, where $\mathcal{Q}_1$ is a bounded component containing the origin to the right of $(-\frac{2}{3}\beta, 0)$ and $\mathcal{Q}_2$ is an unbounded component to the left of $(-\frac{2}{3}\beta, 0)$. Finally, show that every solution starting in $\mathcal{Q}_1$ approaches the origin and hence $\mathcal{Q}_1$ is the domain of attraction of the equilibrium point $(0,0)$, and every solution starting in $\mathcal{Q}_2$ approaches infinity and hence $(-\frac{2}{3}\beta, 0)$ is unstable. Is the equilibrium point $(0,0)$ asymptotically stable?

Problem 3.23. Consider the nonlinear dynamical system

$$\dot{x}_1(t) = -x_1(t) + x_2^2(t), \quad x_1(0) = x_{10}, \quad t \geq 0, \tag{3.238}$$
$$\dot{x}_2(t) = -2x_2(t) + 3x_1^2(t), \quad x_2(0) = x_{20}. \tag{3.239}$$

Using the Lyapunov function candidate $V(x_1, x_2) = \frac{1}{2}x_1^2 + \frac{1}{4}x_2^2$ show that the zero solution $(x_1(t), x_2(t)) \equiv (0,0)$ to (3.238) and (3.239) is asymptotically stable and determine the largest ellipse contained in the domain of attraction of (3.238) and (3.239).

Problem 3.24. Consider the nonlinear dynamical system

$$\dot{x}_1(t) = x_2(t) - x_1(t)f(x_1(t), x_2(t)), \quad x_1(0) = x_{10}, \quad t \geq 0, \tag{3.240}$$
$$\dot{x}_2(t) = -x_1(t) - x_2(t)f(x_1(t), x_2(t)), \quad x_2(0) = x_{20}, \tag{3.241}$$

where $f : \mathbb{R}^2 \to \mathbb{R}$ is continuously differentiable and $f(0,0) = 0$. Using Lyapunov's theorem obtain conditions on $f(x_1, x_2)$ such that the zero solution $(x_1(t), x_2(t)) \equiv (0,0)$ to (3.240) and (3.241) is Lyapunov stable, asymptotically stable, and unstable.

Problem 3.25. Consider the nonlinear dynamical system

$$\dot{x}_1(t) = -x_1(t) + 2x_2^3(t) - 2x_2^4(t), \quad x_1(0) = x_{10}, \quad t \geq 0, \tag{3.242}$$
$$\dot{x}_2(t) = -x_1(t) - x_2(t) + x_1(t)x_2(t), \quad x_2(0) = x_{20}. \tag{3.243}$$

Analyze the stability of the zero solution $(x_1(t), x_2(t)) \equiv (0,0)$ to (3.242) and (3.243) using the Lyapunov function candidate $V(x_1, x_2) = x_1^p + \alpha x_2^q$, where α, p, and q are parameters to be chosen.

Problem 3.26. Consider the nonlinear dynamical system

$$\dot{x}_1(t) = -x_1(t) + 4x_2(t), \quad x_1(0) = x_{10}, \quad t \geq 0, \tag{3.244}$$

$$\dot{x}_2(t) = -x_1(t) - x_2^3(t), \quad x_2(0) = x_{20}. \tag{3.245}$$

Analyze the stability of the zero solution $(x_1(t), x_2(t)) \equiv (0, 0)$ to (3.244) and (3.245) using the Lyapunov function candidate $V(x_1, x_2) = x_1^2 + \alpha x_2^2$, where $\alpha > 0$ is a parameter to be chosen.

Problem 3.27. Consider the nonlinear dynamical system

$$\dot{x}_1(t) = x_2(t) - x_1^3(t), \quad x_1(0) = x_{10}, \quad t \geq 0, \tag{3.246}$$
$$\dot{x}_2(t) = -x_1(t) - x_2^3(t), \quad x_2(0) = x_{20}. \tag{3.247}$$

Analyze the stability of the zero solution $(x_1(t), x_2(t)) \equiv (0, 0)$ to (3.246) and (3.247) using the Lyapunov function candidate $V(x_1, x_2) = \alpha x_1^2 + \beta x_2^2$, where $\alpha > 0$ and $\beta > 0$ are parameters to be chosen.

Problem 3.28. Consider the nonlinear dynamical system

$$\ddot{x}(t) + |x^2(t) - 1| x^3(t) + x(t) - \sin\left(\frac{\pi x(t)}{2}\right) = 0,$$
$$x(0) = x_0, \quad \dot{x}(0) = \dot{x}_0, \quad t \geq 0. \tag{3.248}$$

Show that $(x_{e1}, \dot{x}_{e1}) = (1, 0)$, $(x_{e2}, \dot{x}_{e2}) = (-1, 0)$, and $(x_{e3}, \dot{x}_{e3}) = (0, 0)$ are equilibrium points of (3.248). Use Lyapunov's method to investigate the stability of each of these equilibrium points.

Problem 3.29. Consider the scalar uncertain system

$$\dot{x}(t) = \alpha x(t) + u(t), \quad x(0) = x_0, \quad t \geq 0, \tag{3.249}$$

where α is an unknown constant parameter. Show that the *adaptive controller* $u(t) = -k(t)x(t)$ with *update law*

$$\dot{k}(t) = \gamma x^2(t), \quad k(0) = k_0, \quad t \geq 0, \tag{3.250}$$

where $\gamma > 0$, guarantees that the equilibrium solution $(x(t), k(t)) = (0, k^*)$, where $k^* > \alpha$, of the closed-loop system (3.249) and (3.250) is Lyapunov stable and $\lim_{t\to\infty} x(t) = 0$ for all $x_0 \in \mathbb{R}$. (**Hint:** Consider the Lyapunov function candidate $V(x, k) = \frac{1}{2}x^2 + \frac{1}{2\gamma}(k - k^*)^2$.)

Problem 3.30. Consider the nonlinear dynamical system

$$\dot{x}_1(t) = -x_2(t) + x_1(t)[x_1^2(t) + x_2^2(t) - 1], \quad x_1(0) = x_{10}, \quad t \geq 0, \tag{3.251}$$
$$\dot{x}_2(t) = x_1(t) + x_2(t)[x_1^2(t) + x_2^2(t) - 1], \quad x_2(0) = x_{20}. \tag{3.252}$$

Is the zero solution $(x_1(t), x_2(t)) \equiv (0, 0)$ to (3.251) and (3.252) asymptotically stable? Is the zero solution $(x_1(t), x_2(t)) \equiv (0, 0)$ to (3.251) and (3.252) globally asymptotically stable?

Problem 3.31. Consider the nonlinear dynamical system

$$\dot{x}_1(t) = -x_1(t) + g(x_3(t)), \quad x_1(0) = x_{10}, \quad t \geq 0, \tag{3.253}$$
$$\dot{x}_2(t) = -g(x_3(t)), \quad x_2(0) = x_{20}, \tag{3.254}$$
$$\dot{x}_3(t) = -\alpha x_1(t) + \beta x_2(t) - \gamma g(x_3(t)), \quad x_3(0) = x_{30}, \tag{3.255}$$

where $\alpha, \beta, \gamma > 0$ and g is continuously differentiable and satisfies $g(0) = 0$ and $xg(x) > 0$, $0 < |x| < \sigma$, $\sigma > 0$. Show that the origin $(0, 0, 0)$ is an isolated equilibrium point of (3.253)–(3.255). Using the Lyapunov function candidate

$$V(x_1, x_2, x_3) = \tfrac{1}{2}\alpha x_1^2 + \tfrac{1}{2}\beta x_2^2 + \int_0^{x_3} g(s)\mathrm{d}s, \tag{3.256}$$

show that the zero solution $(x_1(t), x_2(t), x_3(t)) \equiv (0, 0, 0)$ to (3.253)–(3.255) is asymptotically stable. If g is such that $xg(x) > 0$ for all $x \in \mathbb{R}$, $x \neq 0$, is the zero solution $(x_1(t), x_2(t), x_3(t)) \equiv (0, 0, 0)$ globally asymptotically stable?

Problem 3.32. Consider the nonlinear *port-controlled Hamiltonian* system given by

$$\dot{x}(t) = [\mathcal{J}(x(t)) - \mathcal{R}(x(t))]\left(\frac{\partial \mathcal{H}}{\partial x}(x(t))\right)^{\mathrm{T}} + G(x(t))u(t), \; x(0) = x_0, \; t \geq 0, \tag{3.257}$$

$$y(t) = G^{\mathrm{T}}(x(t))\left(\frac{\partial \mathcal{H}}{\partial x}(x(t))\right)^{\mathrm{T}}, \tag{3.258}$$

where $x(t) \in \mathcal{D} \subseteq \mathbb{R}^n$, $\mathcal{D}$ is an open set with $0 \in \mathcal{D}$, $u(t) \in U \subseteq \mathbb{R}^m$, $y(t) \in Y \subseteq \mathbb{R}^m$, $\mathcal{H} : \mathcal{D} \to \mathbb{R}$ is a continuously differentiable lower bounded Hamiltonian (energy) function, $\mathcal{J} : \mathcal{D} \to \mathbb{R}^{n \times n}$ is a skew-symmetric function, $\mathcal{R} : \mathcal{D} \to \mathbb{R}^{n \times n}$ is a nonnegative-definite dissipation function, and $G : \mathcal{D} \to \mathbb{R}^{n \times m}$.

i) Show that the zero solution $x(t) \equiv 0$ of the uncontrolled ($u(t) \equiv 0$) system (3.257) is Lyapunov stable.

ii) With $u(t) = -Ky(t)$, where K is a positive-definite gain matrix, show that the zero solution $x(t) \equiv 0$ to (3.257) is asymptotically stable. Here, assume that the only solution of

$$\dot{x}(t) = \mathcal{J}(x(t))\left(\frac{\partial \mathcal{H}}{\partial x}(x(t))\right)^{\mathrm{T}} \tag{3.259}$$

that can stay in the set

$$\left\{x \in \mathcal{D} : \mathcal{R}(x)\left(\frac{\partial \mathcal{H}}{\partial x}(x)\right)^{\mathrm{T}} = 0\right\} \tag{3.260}$$

is the trivial solution $x(t) \equiv 0$.

iii) Under what conditions will the origin be globally asymptotically stable?

Problem 3.33. Consider the controlled nonlinear dynamical system given by

$$\dot{x}(t) = f(x(t)) + G(x(t))u(t), \quad x(0) = x_0, \quad t \geq 0, \tag{3.261}$$
$$u(t) = \phi(x(t)), \tag{3.262}$$

where $x(t) \in \mathcal{D} \subseteq \mathbb{R}^n$, $u(t) \in U \subseteq \mathbb{R}^m$, $t \geq 0$, $f : \mathcal{D} \to \mathbb{R}^n$, $G : \mathcal{D} \to \mathbb{R}^{n\times m}$, and $\phi : \mathcal{D} \to \mathbb{R}^m$. Assume that $f(0) = 0$, $\phi(0) = 0$, and $f(\cdot)$, $G(\cdot)$, and $\phi(\cdot)$ are continuously differentiable. Furthermore, let the controller $u = \phi(x)$ be given by

$$u = \phi(x) = -\alpha R_2^{-1}(x) G^{\mathrm{T}}(x) \left(\frac{\partial V}{\partial x}\right)^{\mathrm{T}}, \tag{3.263}$$

where $\alpha \in \mathbb{R}$ and $V : \mathcal{D} \to \mathbb{R}$ is continuously differentiable, $V(0) = 0$, $V(x) > 0$, $x \in \mathcal{D}$, $x \neq 0$, and $\frac{\partial V}{\partial x}$ satisfies the *Hamilton-Jacobi-Bellman* equation

$$0 = \frac{\partial V}{\partial x} f(x) + L(x) - \frac{1}{4}\frac{\partial V}{\partial x} G(x) R_2^{-1}(x) G^{\mathrm{T}}(x) \left(\frac{\partial V}{\partial x}\right)^{\mathrm{T}}, \quad x \in \mathcal{D}, \tag{3.264}$$

where $L : \mathcal{D} \to \mathbb{R}$ and $R_2 : \mathcal{D} \to \mathbb{R}^{m\times m}$ is a positive-definite matrix-valued function.

i) Show that if $L(x) > 0$, $x \in \mathcal{D}$, $x \neq 0$, and $\alpha \geq \frac{1}{4}$, then the zero solution $x(t) \equiv 0$ of the closed-loop system (3.261) and (3.263) is asymptotically stable.

ii) Show that if $L(x) \geq 0$, $x \in \mathcal{D}$, $\alpha > \frac{1}{4}$, and the only solution $\dot{x} = f(x)$ that can stay identically in the set $\{x \in \mathcal{D} : L(x) = 0\}$ is the trivial solution $x(t) \equiv 0$, then the zero solution $x(t) \equiv 0$ of the closed-loop system (3.261) and (3.263) is asymptotically stable.

iii) Under what conditions will the origin be globally asymptotically stable?

Problem 3.34. Consider the nonlinear dynamical system

$$\dot{x}_1(t) = \frac{x_1^2(t)[x_2(t) - x_1(t)] + x_2^5(t)}{[x_1^2(t) + x_2^2(t)][1 + [x_1^2(t) + x_2^2(t)]^2]}, \quad x_1(0) = x_{10}, \quad t \geq 0, \tag{3.265}$$
$$\dot{x}_2(t) = \frac{x_2^2(t)[x_2(t) - 2x_1(t)]}{[x_1^2(t) + x_2^2(t)][1 + [x_1^2(t) + x_2^2(t)]^2]}, \quad x_2(0) = x_{20}. \tag{3.266}$$

Show that even though $\lim_{t\to\infty}(x_1(t), x_2(t)) = 0$, the zero solution $(x_1(t), x_2(t)) \equiv (0,0)$ to (3.265) and (3.266) is unstable.

Problem 3.35. Consider the nonlinear dynamical system

$$\dot{x}_1(t) = [x_1(t) - x_2(t)][1 - a^2x_1^2(t) - b^2x_2^2(t)], \; x_1(0) = x_{10}, \; t \geq 0, \quad (3.267)$$
$$\dot{x}_2(t) = [x_1(t) + x_2(t)][1 - a^2x_1^2(t) - b^2x_2^2(t)], \; x_2(0) = x_{20}. \quad (3.268)$$

Analyze the stability of (3.267) and (3.268) using the Lyapunov function candidate

$$V(x_1, x_2) = x_1^2 + x_2^2.$$

Is the ellipse $\mathcal{E} = \{(x_1, x_2) \in \mathbb{R} \times \mathbb{R} : \; a^2x_1^2 + b^2x_2^2 = 1\}$ a limit cycle or a continuum of equilibrium points?

Problem 3.36. Consider the nonlinear dynamical system

$$\dot{x}_1(t) = x_2(t) - x_1^7(t)[x_1^4(t) + 2x_2^2(t) - 10], \; x_1(0) = x_{10}, \; t \geq 0, \quad (3.269)$$
$$\dot{x}_2(t) = -x_1^3(t) - 3x_2^5(t)[x_1^4(t) + 2x_2^2(t) - 10], \; x_2(0) = x_{20}. \quad (3.270)$$

i) Show that the set defined by $\mathcal{E} \triangleq \{(x_1, x_2) \in \mathbb{R}\times\mathbb{R} : \; x_1^4+2x_2^2-10 = 0\}$ is invariant.

ii) What is the motion on $\mathcal{E}$ characterized by? Is this a limit cycle or a continuum of equilibrium points?

iii) Examine the stability of $\mathcal{E}$ using the Barbashin-Krasovskii-LaSalle theorem.

Problem 3.37. Consider the nonlinear dynamical system

$$\dot{x}_1(t) = x_2(t), \quad x_1(0) = x_{10}, \quad t \geq 0, \quad (3.271)$$
$$\dot{x}_2(t) = -x_1(t) + x_2(t)[1 - x_1^2(t) - x_2^2(t)], \quad x_2(0) = x_{20}. \quad (3.272)$$

i) Show that the set defined by $\mathcal{E} \triangleq \{(x_1, x_2) \in \mathbb{R} \times \mathbb{R} : x_1^2 + x_2^2 - 1 = 0\}$ is invariant.

ii) What is the motion on $\mathcal{E}$ characterized by? Is this a limit cycle or a continuum of equilibrium points?

iii) Examine the stability of $\mathcal{E}$ using the Barbashin-Krasovskii-LaSalle theorem.

Problem 3.38. Consider the nonlinear dynamical system (3.1) and assume that there exist a continuously differentiable function $V : \mathbb{R}^n \to \mathbb{R}$ and positive scalars α, β such that

$$V(0) = 0, \quad (3.273)$$

$$V(x) > 0, \quad x \in \mathbb{R}^n, \quad x \neq 0, \tag{3.274}$$
$$V(x) \to \infty \text{ as } \|x\| \to \infty, \tag{3.275}$$
$$V'(x(t))f(x(t)) \leq \alpha e^{-\beta t}, \quad t \geq 0. \tag{3.276}$$

Show that $x(t)$, $t \geq 0$, is bounded.

Problem 3.39. Consider the nonlinear dynamical system (3.1) with $f(x_e) = 0$. Show that if there exists a continuously differentiable function $V : \mathcal{D} \to \mathbb{R}$ such that

$$V(x_e) = 0, \tag{3.277}$$
$$V(x) > 0, \quad x \in \mathcal{D}, \quad x \neq x_e, \tag{3.278}$$
$$V'(x)f(x) \leq 0, \quad x \in \mathcal{D}, \tag{3.279}$$

then the equilibrium solution $x(t) \equiv x_e$ to (3.1) is Lyapunov stable. If, in addition,

$$V'(x)f(x) < 0, \quad x \in \mathcal{D}, \quad x \neq x_e, \tag{3.280}$$

show that the equilibrium solution $x(t) \equiv x_e$ to (3.1) is asymptotically stable. Finally, show that if there exist scalars α, β, $\varepsilon > 0$, and $p \geq 1$ such that $V : \mathcal{D} \to \mathbb{R}$ satisfies

$$\alpha\|x - x_e\|^p \leq V(x) \leq \beta\|x - x_e\|^p, \quad x \in \mathcal{D}, \tag{3.281}$$
$$V'(x)f(x) \leq -\varepsilon V(x), \quad x \in \mathcal{D}, \tag{3.282}$$

then the equilibrium solution $x(t) \equiv x_e$ to (3.1) is exponentially stable.

Problem 3.40. Consider the nonlinear dynamical system (3.1), assume $\mathcal{D}_c \subset \mathcal{D}$ is a compact invariant set with respect to (3.1), and assume there exists a continuously differentiable function $V : \mathcal{D}_c \to \mathbb{R}$ such that $V'(x)f(x) \leq 0$, $x \in \mathcal{D}_c$. Furthermore, suppose that the largest invariant set $\mathcal{M}$ contained in $\mathcal{R} \triangleq \{x \in \mathcal{D}_c : V'(x)f(x) = 0\}$ consists of a finite collection of isolated points. Show that $\lim_{t\to\infty} x(t)$ exists and is equal to one of the isolated points contained in $\mathcal{M}$.

Problem 3.41. Consider the nonlinear dynamical system

$$\dot{x}_1(t) = -x_1(t)|x_2(t)|^\alpha, \quad x_1(0) = x_{10}, \quad t \geq 0, \tag{3.283}$$
$$\dot{x}_2(t) = -x_2(t)|x_1(t)|^\beta, \quad x_2(0) = x_{20}, \tag{3.284}$$

where $\alpha > 0$ and $\beta > 0$. Show that $\lim_{t\to\infty}(x_1(t), x_2(t))$ exists and is equal to one of the points $(0, a)$, $(0, -a)$, $(a, 0)$, or $(-a, 0)$, where $a \in \mathbb{R}$.

Problem 3.42. A *dynamical system* on $\mathcal{D} \subseteq \mathbb{R}^n$ is the triple $(\mathcal{D}, [0, \infty), s)$, where $s : [0, \infty) \times \mathcal{D} \to \mathcal{D}$ is such that the following axioms hold:

i) (Continuity): $s(\cdot,\cdot)$ is jointly continuous on $[0,\infty)\times\mathcal{D}$.

ii) (Consistency): For every $x_0\in\mathcal{D}$, $s(0,x_0)=x_0$.

iii) (Semigroup property): $s(\tau,s(t,x_0))=s(t+\tau,x_0)$ for all $x_0\in\mathcal{D}$ and $t,\tau\in[0,\infty)$.

Let $\mathcal{G}$ denote the dynamical system $(\mathcal{D},[0,\infty),s)$ and let $s(t,x_0)$, $t\geq 0$, denote the trajectory of $\mathcal{G}$ corresponding to the initial condition $x_0\in\mathcal{D}$. Furthermore, let $\mathcal{D}_{\rm c}\subset\mathcal{D}$ be a compact invariant set with respect to $\mathcal{G}$ and define $V^{-1}(\gamma)\triangleq\{x\in\mathcal{D}_{\rm c}: V(x)=\gamma\}$, where $\gamma\in\mathbb{R}$ and $V:\mathcal{D}_{\rm c}\to\mathbb{R}$ is a continuous function. Show that if $V(s(t,x_0))\leq V(s(\tau,x_0))$, for all $0\leq\tau\leq t$ and $x_0\in\mathcal{D}_{\rm c}$, then $s(t,x_0)\to\mathcal{M}\triangleq\cup_{\gamma\in\mathbb{R}}\mathcal{M}_\gamma$ as $t\to\infty$, where $\mathcal{M}_\gamma$ denotes the largest invariant set (with respect to the dynamical system $\mathcal{G}$) contained in $V^{-1}(\gamma)$.

Problem 3.43. Consider the nonlinear dynamical system (3.1) where $f:\mathcal{D}\to\mathbb{R}^n$ is continuous and (3.1) possesses unique solutions in forward time; that is, if the solutions agree at some time t_0, then they agree at any time $t\geq t_0$. The zero solution $x(t)\equiv 0$ to (3.1) is *finite-time stable* if the origin is Lyapunov stable and there exists an open neighborhood $\mathcal{N}\subseteq\mathcal{D}$ of the origin and a function $T:\mathcal{N}\setminus\{0\}\to(0,\infty)$, called a *settling-time function*, such that for every $x_0\in\mathcal{N}\setminus\{0\}$, $s(\cdot,x_0):[0,T(x_0))\to\mathcal{N}\setminus\{0\}$ and $s(t,x_0)\to 0$ as $t\to T(x_0)$. The zero solution $x(t)\equiv 0$ is *globally finite-time stable* if it is finite-time stable with $\mathcal{D}=\mathcal{N}=\mathbb{R}^n$. Show that if there exist a continuously differentiable function $V:\mathcal{D}\to\mathbb{R}$, positive scalars $\alpha\in(0,1)$ and $\beta>0$, and an open neighborhood $\mathcal{N}\subseteq\mathcal{D}$ of the origin such that

$$V(0)=0, \tag{3.285}$$
$$V(x)>0,\quad x\in\mathcal{D},\quad x\neq 0, \tag{3.286}$$
$$\dot V(x)\leq -\beta V(x)^\alpha,\quad x\in\mathcal{N}\setminus\{0\}, \tag{3.287}$$

then the zero solution to (3.1) is finite-time stable. Furthermore, show that $T:\mathcal{N}\setminus\{0\}\to(0,\infty)$ is continuous on $\mathcal{N}$ and

$$T(x_0)\leq\frac{1}{\beta(1-\alpha)}V(x)^{1-\alpha},\quad x\in\mathcal{N}. \tag{3.288}$$

Finally, show that if $\mathcal{D}=\mathbb{R}^n$, $V(x)\to\infty$ as $\|x\|\to\infty$, and (3.287) is satisfied on $\mathbb{R}^n\setminus\{0\}$, then the zero solution $x(t)\equiv 0$ to (3.1) is globally finite-time stable. (**Hint:** Use the comparison principle (see Problem 2.107) to construct a comparison system whose solution is finite-time stable and relate this finite-time stability property to the stability property of (3.1).)

Problem 3.44. Consider the nonlinear dynamical system (3.1). An equilibrium point $x\in\mathcal{D}$ is *semistable* if $x\in\mathcal{D}$ is Lyapunov stable and there exists $\delta > 0$ such that if $x_0 \in \mathcal{B}_\delta(x)$, then $\lim_{t\to\infty} s(t, x_0)$ exists and $s(t, x_0)$, $t \geq 0$, converges to a Lyapunov stable equilibrium point. The nonlinear dynamical system (3.1) is *semistable* if every equilibrium point of (3.1) is semistable. Suppose the orbit $\mathcal{O}_x$ of (3.1) is bounded for all $x \in \mathcal{D}$ and assume that there exists a continuously differentiable function $V : \mathcal{D} \to \mathbb{R}$ such that

$$V'(x)f(x) \leq 0, \quad x \in \mathcal{D}. \tag{3.289}$$

Let $\mathcal{M}$ denote the largest invariant set contained in $\mathcal{R} \triangleq \{x \in \mathcal{D} : V'(x)f(x) = 0\}$. Show that if $\mathcal{M} \subseteq \{x \in \mathcal{D} : f(x) = 0 \text{ and } x \text{ is Lyapunov stable}\}$, then (3.1) is semistable. (**Hint:** First show that if the positive orbit $\mathcal{O}_x^+$ of (3.1) contains a Lyapunov stable equilibrium point y, then $y = \lim_{t\to\infty} s(t, x)$ for all $x \in \mathcal{D}_c$, where $\mathcal{D}_c \subseteq \mathcal{D}$ is a positively invariant set with respect to (3.1).)

Problem 3.45. Consider the dynamical system (3.1) with $f(x) = Ax$, where $A \in \mathbb{R}^n$. Show that if there exists $n \times n$ matrices $P \geq 0$ and $R \geq 0$ such that

$$0 = A^{\mathrm{T}}P + PA + R, \tag{3.290}$$

$$\mathcal{N}\left(\begin{bmatrix} R \\ RA \\ \vdots \\ RA^{n-1} \end{bmatrix}\right) = \mathcal{N}(A), \tag{3.291}$$

then the zero solution $x(t) \equiv 0$ to (3.1) is semistable (see Problem 3.44). (**Hint:** First show that $\mathcal{N}(P) \subseteq \mathcal{N}(A) \subseteq \mathcal{N}(R)$ and $\mathcal{N}(A) \cap \mathcal{R}(A) = \{0\}$.)

Problem 3.46. Consider the dynamical system (3.1) with $f(x) = Ax$, where $A \in \mathbb{R}^{n\times n}$. Using the results of Problem 3.44 show that the following statements are equivalent:

i) The zero solution $x(t) \equiv 0$ to (3.1) is semistable.

ii) For every initial condition $x_0 \in \mathbb{R}^n$, $\lim_{t\to\infty} x(t)$ exists.

iii) If $\lambda \in \operatorname{spec}(A)$, then either $\operatorname{Re}\lambda < 0$, or both $\lambda = 0$ and λ is semisimple.

iv) $\lim_{t\to\infty} e^{At}$ exists.

Problem 3.47. Consider the nonlinear dynamical system

$$\dot{x}_1(t) = -2x_1(t), \quad x_1(0) = x_{10}, \quad t \geq 0, \tag{3.292}$$

$$\dot{x}_2(t) = -2x_2(t) + 2x_1(t)x_2^2, \quad x_2(0) = x_{20}. \tag{3.293}$$

Use the variable gradient method to construct a Lyapunov function for (3.292) and (3.293). Is the zero solution $(x_1(t), x_2(t)) \equiv (0,0)$ to (3.292) and (3.293) globally asymptotically stable?

Problem 3.48. Consider the nonlinear dynamical system

$$\dot{x}_1(t) = \frac{-6x_1(t)}{[1+x_2^2(t)]^2} + 2x_2(t), \quad x_1(0) = x_{10}, \quad t \geq 0, \tag{3.294}$$

$$\dot{x}_2(t) = -2x_1(t) - \frac{2x_2(t)}{[1+x_1^2(t)]^2}, \quad x_2(0) = x_{20}. \tag{3.295}$$

Use the variable gradient method to construct a Lyapunov function for (3.294) and (3.295). Is the zero solution $(x_1(t), x_2(t)) \equiv (0,0)$ to (3.294) and (3.295) globally asymptotically stable?

Problem 3.49. Consider the nonlinear dynamical system

$$\dot{x}_1(t) = -x_1(t) + x_1^2(t), \quad x_1(0) = x_{10}, \quad t \geq 0, \tag{3.296}$$

$$\dot{x}_2(t) = -x_2(t) + x_3^2(t), \quad x_2(0) = x_{20}, \tag{3.297}$$

$$\dot{x}_3(t) = \alpha x_3(t) - x_1^2(t), \quad x_3(0) = x_{30}. \tag{3.298}$$

Using Lyapunov's indirect method, investigate the stability of the zero solution $(x_1(t), x_2(t), x_3(t)) \equiv (0,0,0)$ to (3.296)–(3.298) for $\alpha > 0$, $\alpha < 0$, and $\alpha = 0$.

Problem 3.50. Consider the nonlinear Lienard system given in Problem 3.18 with $g(x) = x$ and $f(0) > 0$. Use Lyapunov's indirect theorem (Theorem 3.19) to draw conclusions about the local stability of the zero solution $(x(t), \dot{x}(t)) \equiv (0,0)$ to (3.221).

Problem 3.51. Consider the nonlinear dynamical system

$$\dot{x}_1(t) = f_1(x_1(t)) + f_2(x_2(t)), \quad x_1(0) = x_{10}, \quad t \geq 0, \tag{3.299}$$

$$\dot{x}_2(t) = x_1(t) + ax_2(t), \quad x_2(0) = x_{20}, \tag{3.300}$$

where $f_1(0) = f_2(0) = 0$, $a \in \mathbb{R}$, and f_1, f_2 are continuously differentiable functions. Using Krasovskii's theorem (Theorem 3.6) find conditions on $f_1'(x_1)$ and $f_2'(x_2)$ for all $(x_1, x_2) \in \mathbb{R} \times \mathbb{R}$ such that $V(x_1, x_2) = f_1^2(x_1) + f_2^2(x_2)$ serves as a Lyapunov function for (3.299) and (3.300).

Problem 3.52. Prove *i*) of Theorem 3.19 using Krasovskii's theorem (Theorem 3.6).

Problem 3.53. Consider the nonlinear dynamical system (3.1) with $f(x) = A(x)x$, where $A : \mathbb{R}^n \to \mathbb{R}^{n\times n}$ is a continuous matrix-valued function. Show that if there exist positive definite matrices $P \in \mathbb{R}^{n\times n}$ and $R \in \mathbb{R}^{n\times n}$

such that

$$0 = A^{\mathrm{T}}(x)P + PA(x) + R, \quad x \in \mathbb{R}^n, \quad x \neq 0, \tag{3.301}$$

then the zero solution $x(t) \equiv 0$ to (3.1) is globally asymptotically stable. Furthermore, prove or refute that if, for each $x \in \mathbb{R}^n$, all of the eigenvalues of $A(x)$ lie in the open left half complex plane, then the zero solution $x(t) \equiv 0$ to (3.1) is globally asymptotically stable.

Problem 3.54. Consider the scalar nonlinear system

$$\dot{x}(t) = x^3(t) - x(t), \qquad x(0) = 0, \qquad t \geq 0. \tag{3.302}$$

Using Zubov's method construct a Lyapunov function for (3.302) and provide a characterization for the domain of attraction of the zero solution $x(t) \equiv 0$ to (3.302).

Problem 3.55. Consider the nonlinear dynamical system representing a rigid spacecraft given by (3.11)–(3.13) in Example 3.1. Use the energy-Casimir method to show that the equilibrium solution $x(t) \equiv x_{\mathrm{e}}$, where $x_{\mathrm{e}} = [x_{1\mathrm{e}},\ 0,\ 0]^{\mathrm{T}}$, to (3.11)–(3.13) is Lyapunov stable.

Problem 3.56. Consider the controlled rigid spacecraft given by (3.215) –(3.217) in Problem 3.15 with $u_1(t) \equiv 0$ and $u_2(t) \equiv 0$. Show that with $u_3(t) = -k\frac{I_1 I_2}{I_3}x_1(t)x_2(t)$, where $k > c$ and $c \triangleq (I_1 - I_2)/(I_1 I_2)$, the closed-loop system can be written as

$$\begin{aligned}
\dot{z}_1(t) &= a_0 a z_2(t) z_3(t), \quad z_1(0) = z_{10}, \quad t \geq 0, && (3.303)\\
\dot{z}_2(t) &= a_0 b z_3(t) z_1(t), \quad z_2(0) = z_{20}, && (3.304)\\
\dot{z}_3(t) &= c z_1(t) z_2(t), \quad z_3(0) = z_{30}, && (3.305)
\end{aligned}$$

where $z_1 = I_1 x_1$, $z_2 = I_2 x_2$, $z_3 = (c/(c-k))I_3 x_3$, $a_0 \triangleq (c-k)/c < 0$, $a \triangleq (I_2 - I_3)/(I_2 I_3)$, and $b \triangleq (I_3 - I_1)/(I_3 I_1)$. Furthermore, use the energy-Casimir method to show that the equilibrium solution $z(t) \equiv z_{\mathrm{e}}$, where $z_{\mathrm{e}} = [0,\ z_{2\mathrm{e}},\ 0]^{\mathrm{T}}$ and $z_{2\mathrm{e}} = I_2 x_{2\mathrm{e}}$, to (3.303)–(3.305) is Lyapunov stable. (**Hint:** Show that $C_1(z) = \frac{1}{2}(\frac{z_1^2}{I_1} + \frac{z_2^2}{I_2} + a_0\frac{z_3^2}{I_3})$ and $C_2(z) = \frac{1}{2}(z_1^2 + z_2^2 + a_0 z_3^2)$ are Casimir functions for (3.303)–(3.305).)

Problem 3.57. In the spread of epidemics in large populations the basic *susceptible-infected-removed* (SIR) model has been proposed in the literature. In particular, the model of a SIR epidemic is given by

$$\begin{aligned}
\dot{x}_1(t) &= -\tfrac{\lambda}{N}x_1(t)x_2(t) - \mu x_1(t) + u, \quad x_1(0) = x_{10}, \quad t \geq 0, && (3.306)\\
\dot{x}_2(t) &= \tfrac{\lambda}{N}x_1(t)x_2(t) - (\gamma + \mu)x_2(t), \quad x_2(0) = x_{20}, && (3.307)\\
\dot{x}_3(t) &= \gamma x_2(t) - \mu x_3(t), \quad x_3(0) = x_{30}, && (3.308)
\end{aligned}$$

where x_1 denotes the number of susceptibles, x_2 denotes the number of infectives, x_3 denotes the number of immunes, $\mu > 0$ is the death rate coefficient, N is a constant denoting the total size of the population, that is, $x_1(t) + x_2(t) + x_3(t) = N$, u is the rate of recruitment of new members into the susceptible pool and is assumed to be a constant that just makes up for the deaths, that is, $u = \mu N$, $\gamma > 0$ is a rate constant for recovery, and $\lambda > 0$ is the mean constant rate per person for contacts that transmit the disease. For $\alpha \leq 1$ and $\alpha > 1$, where $\alpha \triangleq \lambda/(\gamma+\mu)$, compute all equilibria of (3.306)–(3.308). Using Lyapunov's linearization method, analyze the stability of these equilibria. Repeat the above analysis for the case corresponding to a zero death rate, that is, $\mu = 0$. (**Hint:** Note that since $x_1(t)+x_2(t)+x_3(t) = N$, (3.308) is superfluous and need not be considered in the analysis.)

Problem 3.58. Consider the model of two populations which interact in a predator-prey relationship given by the *Lotka-Volterra* equations

$$\dot{x}_1(t) = \alpha x_1(t) - \beta x_1(t)x_2(t), \quad x_1(0) = x_{10}, \quad t \geq 0, \tag{3.309}$$

$$\dot{x}_2(t) = -\gamma x_2(t) + \delta x_1(t)x_2(t), \quad x_2(0) = x_{20}, \tag{3.310}$$

where α, β, γ, $\delta > 0$, x_1 denotes the number of prey, and x_2 denotes the number of predators. Compute all equilibria of (3.309) and (3.310). Analyze the stability of these equilibria. (**Hint:** In the case where Lyapunov's linearization method provides no stability information use the Lyapunov function candidate $V(x) = E(x) - E(x_{\rm e})$, where $E(x) = \delta x_1 + \beta x_2 - \gamma \ln x_1 - \alpha \ln x_2$. Make sure you justify that $V(x)$, $x \in \mathbb{R}^2$, is a valid Lyapunov function candidate.)

Problem 3.59. Consider the nonlinear dynamical system (3.1). Show that if the zero solution $x(t) \equiv 0$ to (3.1) is asymptotically stable and $f : \mathcal{D} \to \mathbb{R}^n$ is continuously differentiable, then $V : \mathcal{D} \to \mathbb{R}$ satisfying the conditions of Theorem 3.9 additionally satisfies $\|V'(x)\| \leq \nu(\|x\|)$, $x \in \mathcal{B}_\delta(0)$, where $\nu : [0, \delta] \to [0, \infty)$ is a class $\mathcal{K}$ function.

Problem 3.60. Consider the nonlinear dynamical system (3.1). Show that if the zero solution $x(t) \equiv 0$ to (3.1) is exponentially stable and $f : \mathcal{D} \to \mathbb{R}^n$ is continuously differentiable, then $V : \mathcal{D} \to \mathbb{R}$ satisfying the conditions of Theorem 3.10 additionally satisfies $\|V'(x)\| \leq \nu \|x\|^{p-1}$, $x \in \mathcal{B}_\delta(0)$, where $\nu > 0$. Can $p = 2$ without loss of generality? Justify your answer.

Problem 3.61. Let $f = [f_1, \ldots, f_n]^{\rm T} : \mathcal{D} \to \mathbb{R}^n$, where $\mathcal{D}$ is an open subset of $\mathbb{R}^n$ that contains $\overline{\mathbb{R}}_+^n$. Then f is *essentially nonnegative* if $f_i(x) \geq 0$, for all $i = 1, \ldots, n$, and $x \in \overline{\mathbb{R}}_+^n$ such that $x_i = 0$, where x_i denotes the ith component of x. Consider the nonlinear dynamical system (3.1). Show that the following statements hold:

i) Suppose $\overline{\mathbb{R}}_+^n \subset \mathcal{D}$. Then $\overline{\mathbb{R}}_+^n$ is an invariant set with respect to (3.1) if and only if $f : \mathcal{D} \to \mathbb{R}^n$ is essentially nonnegative.

ii) Suppose $f(0) = 0$ and $f : \mathcal{D} \to \mathbb{R}^n$ is essentially nonnegative and continuously differentiable on $\overline{\mathbb{R}}^n_+$. Then $A \triangleq \frac{\partial f}{\partial x}\Big|_{x=0}$ is essentially nonnegative (see Problem 3.7).

iii) If $f(x) = Ax$, where $A \in \mathbb{R}^{n\times n}$, then f is essentially nonnegative if and only if A is essentially nonnegative.

Problem 3.62. Let $f : [0, \infty) \to \mathbb{R}$. Show that if $f(\cdot) \in \mathcal{L}_2 \cap \mathcal{L}_\infty$ and $\dot{f}(\cdot)$ is bounded, then $\lim_{t\to\infty} f(t) = 0$.

Problem 3.63. Consider the nonlinear *gradient dynamical system* given by

$$\dot{x}(t) = -\left[\frac{\partial V}{\partial x}(x(t))\right]^{\mathrm{T}}, \qquad x(0) = x_0, \qquad t \geq 0, \tag{3.311}$$

where $x \in \mathcal{D} \subseteq \mathbb{R}^n$ and $V : \mathcal{D} \to \mathbb{R}$ is a two-times continuously differentiable function. Show that:

i) $\dot{V}(x) \leq 0$, $x \in \mathcal{D}$, and $\dot{V}(x) = 0$ if and only if x is an equilibrium point of (3.311).

ii) Let $\mathcal{D}_{\mathrm{c}} \triangleq \{x \in \mathbb{R}^n : V(x) \leq c\}$ be compact for every $c > 0$ and let $x_0 \in \mathcal{D}_{\mathrm{c}}$. Then there exists a unique solution to (3.311) that is defined for all $t \geq 0$.

iii) Let $\frac{\partial V}{\partial x} = 0$ for a finite number of points $p_1, \ldots, p_q$. Then for every solution $x(t)$, $t \geq 0$, of (3.311), $\lim_{t\to\infty} x(t) = p_i$, $i \in \{1, \ldots, q\}$.

iv) If $x_{\mathrm{e}} \in \mathcal{D}$ is an isolated minimizer of $V(\cdot)$, then x_{e} is an asymptotically stable equilibrium point of (3.311).

Problem 3.64. Consider the nonlinear dynamical system

$$\dot{x}_1(t) = f_1(x_1(t), x_2(t)), \quad x_1(0) = x_{10}, \quad t \geq 0, \tag{3.312}$$

$$\dot{x}_2(t) = f_2(x_1(t), x_2(t)), \quad x_2(0) = x_{20}, \tag{3.313}$$

where $f_1(\cdot,\cdot)$ and $f_2(\cdot,\cdot)$ are continuously differentiable functions. Show that a necessary condition for (3.312) and (3.313) to be a gradient system (see Problem 3.63) is $\frac{\partial f_1}{\partial x_2} = \frac{\partial f_2}{\partial x_1}$. Is this condition also sufficient? For $f_1(x_1, x_2) = x_2 + 2x_1x_2$ and $f_2(x_1, x_2) = x_1 + x_1^2 - x_2^2$ show that (3.312) and (3.313) is a gradient system and find its potential function $V(x_1, x_2)$.

Problem 3.65. Consider the nonlinear *functional dynamical system*

$$\dot{x}(t) = f(x(t), x(t-\tau_{\mathrm{d}})), \quad x(\theta) = \phi(\theta), \quad -\tau_{\mathrm{d}} \leq \theta \leq 0, \tag{3.314}$$

where $x \in \mathbb{R}^n$, $f : \mathbb{R}^n \times \mathbb{R}^n \to \mathbb{R}^n$ is Lipschitz continuous and satisfies $f(0,0) = 0$, and $\phi \in \mathcal{C}([-\tau_\mathrm{d}, 0], \mathbb{R}^n)$, where $\mathcal{C}([-\tau_\mathrm{d}, 0], \mathbb{R}^n)$ is a Banach space of continuous functions mapping the interval $[-\tau_\mathrm{d}, 0]$ into $\mathbb{R}^n$ with topology of uniform convergence and designated norm given by $|||\phi||| = \sup_{-\tau_\mathrm{d} \le \theta \le 0} \|\phi(\theta)\|$. Let $\phi : [-\tau_\mathrm{d}, 0] \to \mathbb{R}^n$ be a continuous vector-valued function specifying the initial state of the system. Furthermore, let $x_t \in \mathcal{C}([-\tau_\mathrm{d}, 0], \mathbb{R}^n)$ defined by $x_t(\theta) \triangleq x(t+\theta)$, $\theta \in [-\tau_\mathrm{d}, 0]$, denote the (infinite-dimensional) state of (3.314) at time t corresponding to the *piece of trajectories* x between $t - \tau_\mathrm{d}$ and t or, equivalently, the *element* x_t in the space of continuous functions defined on the interval $[-\tau_\mathrm{d}, 0]$ and taking values in $\mathbb{R}^n$. The zero solution $x_t \equiv 0$ of (3.314) is *Lyapunov stable* if, for all $\varepsilon > 0$, there exists $\delta > 0$ such that $\|\phi(\theta)\| < \delta$, $\theta \in [-\tau_\mathrm{d}, 0]$, implies $\|x(t, \phi(\theta))\| < \varepsilon$ for $t \ge 0$, where $x(t, \phi(\theta))$ denotes the solution to (3.314). The zero solution $x_t \equiv 0$ of (3.314) is *asymptotically stable* if it is Lyapunov stable and there exists $\delta > 0$ such that $\|\phi(\theta)\| < \delta$, $\theta \in [-\tau_\mathrm{d}, 0]$, implies that $\lim_{t\to\infty} \|x(t, \phi(\theta))\| = 0$. Show that if there exists a continuously differentiable *Lyapunov functional* $V : \mathcal{C}([-\tau_\mathrm{d}, 0], \mathbb{R}^n) \to \mathbb{R}$ such that

$$\alpha(\|\psi(0)\|) \le V(\psi), \tag{3.315}$$

$$\dot{V}(\psi) \le 0, \tag{3.316}$$

where $\psi \in \mathcal{C}([-\tau_\mathrm{d}, 0], \mathbb{R}^n)$, $\alpha(\cdot)$ is a class $\mathcal{K}_\infty$ function, and $\dot{V}(\psi) \triangleq \lim_{h\to 0} [V(x_h(\psi)) - V(\psi)]$, then the zero solution $x_t \equiv 0$ to (3.314) is Lyapunov stable. If, in addition, there exists a class $\mathcal{K}_\infty$ function $\gamma(\cdot)$ such that

$$\dot{V}(\psi) \le -\gamma(\|\psi(0)\|), \tag{3.317}$$

show that the zero solution $x_t \equiv 0$ to (3.314) is asymptotically stable.

Problem 3.66. A mapping $\mathcal{T} : \mathcal{D} \to \mathbb{R}^n$ is a diffeomorphism on $\mathcal{D} \subseteq \mathbb{R}^n$ if it is invertible on $\mathcal{D}$, that is, there exists a function $\mathcal{T}^{-1}(x)$ such that $\mathcal{T}^{-1}(\mathcal{T}(x)) = x$, $x \in \mathcal{D}$, and $\mathcal{T}(x)$ and $\mathcal{T}^{-1}(x)$ are continuously differentiable. Consider the nonlinear dynamical system (3.1) and let $\mathcal{T} : \mathbb{R}^n \to \mathbb{R}^n$ be a diffeomorphism in the neighborhood of the origin such that $z = \mathcal{T}(x)$ and $\mathcal{T}(0) = 0$. Show that the zero solution $z(t) \equiv 0$ of the transformed nonlinear system $\dot{z}(t) = \hat{f}(z(t))$, $z(0) = z_0$, where $\hat{f}(z) = \frac{\partial \mathcal{T}(x)}{\partial x} f(x)\Big|_{x=\mathcal{T}^{-1}(z)}$, is Lyapunov stable (respectively, asymptotically stable) if and only if the zero solution $x(t) \equiv 0$ to (3.1) is Lyapunov stable (respectively, asymptotically stable).

Problem 3.67. Consider the nonlinear dynamical system

$$\dot{x}(t) = f(x(t)), \quad x(0) = x_0, \quad t \ge 0, \tag{3.318}$$

where $f : \mathcal{D} \subseteq \mathbb{R}^n \to \mathbb{R}$ is continuously differentiable with $0 \in \mathcal{D}$. Assume there exist continuous model validity functions $w_i : \mathbb{R}^n \to \mathbb{R}$, $i = 1, \ldots, r$, such that (3.318) can be represented by

$$\dot{x}(t) = \sum_{i=1}^{r} w_i(x(t))A_i x(t), \tag{3.319}$$

where $w_i(x) > 0$, $i = 1, \ldots, r$, and $A_i = \left.\frac{\partial f}{\partial x}\right|_{x=x_i}$. Show that if there exists a single positive-definite matrix $P \in \mathbb{R}^{n\times n}$ satisfying

$$A_i^{\mathrm{T}} P + PA_i < 0, \quad i = 1, \ldots, r, \tag{3.320}$$

then the zero solution $x(t) \equiv 0$ to (3.318) is globally asymptotically stable.

Problem 3.68. Show that Theorem 3.7 also holds in the case where (3.85) is replaced by

$$\frac{V'(x)f(x)}{[1 - V(x)][1 + \|f(x)\|^2]^{1/2}} < 0, \qquad x \in \mathcal{D}. \tag{3.321}$$

Problem 3.69 (Convergence Lemma). Let $V : [0, \infty) \to \mathbb{R}$ be a continuously differentiable function such that

$$\dot{V}(t) + \alpha V(t) \le 0, \qquad t \ge 0, \tag{3.322}$$

where $\alpha \in \mathbb{R}$. Show that $V(t) \le V(0)e^{-\alpha t}$, $t \ge 0$.

Problem 3.70. Consider the nonlinear dynamical system given in Example 3.7. Show that if $\|x(0)\|_2^2 < 1$, then $\|x(t)\|_2^2 < \alpha e^{-2t}$, $t \ge 0$, $\alpha > 0$. Alternatively, show that if $\|x(0)\|_2^2 > 1$, then $\|x(t)\|$ tends to infinity in finite time. (**Hint:** Use the convergence lemma of Problem 3.69.)

Problem 3.71. Let $\gamma_1 : [0, a) \to [0, \infty)$ and $\gamma_2 : [0, a) \to [0, \infty)$ be class $\mathcal{K}$ functions, $\gamma_3 : [0, \infty) \to [0, \infty)$ and $\gamma_4 : [0, \infty) \to [0, \infty)$ be class $\mathcal{K}_\infty$ functions, and $\beta : [0, a) \times [0, \infty) \to [0, \infty)$ be a class $\mathcal{KL}$ function. Show that the following statements hold:

i) $\gamma_1(\sigma_1 + \sigma_2) \le \gamma_1(2\sigma_1) + \gamma_1(2\sigma_2)$ for all $\sigma_1, \sigma_2 \in [0, a)$.

ii) $\gamma^{-1} : [0, \gamma_1(a)) \to [0, \infty)$ and $\gamma^{-1}(\cdot) \in \mathcal{K}$.

iii) $\gamma_3^{-1} : [0, \infty) \to [0, \infty)$ and $\gamma_3^{-1}(\cdot) \in \mathcal{K}_\infty$.

iv) $\gamma_1 \circ \gamma_2 \in \mathcal{K}$.

v) $\gamma_3 \circ \gamma_4 \in \mathcal{K}_\infty$.

vi) $\gamma_1(\beta(\gamma_2(\cdot), \cdot)) \in \mathcal{KL}$.

Problem 3.72. Let $V : \mathcal{D} \subseteq \mathbb{R}^n \to \mathbb{R}$ be a continuous function such that $V(0) = 0$, $V(x) > 0$, $x \in \mathcal{D}$, $x \neq 0$. Show that there exist class $\mathcal{K}$ functions $\alpha(\cdot)$ and $\beta(\cdot)$ defined on $[0, \varepsilon]$ such that

$$\alpha(\|x\|) \leq V(x) \leq \beta(\|x\|), \qquad x \in \mathcal{B}_\varepsilon(0), \tag{3.323}$$

where $\mathcal{B}_\varepsilon(0) \triangleq \{x \in \mathcal{D} : \ \|x\| < \varepsilon\}$. Furthermore, show that if $\mathcal{D} = \mathbb{R}^n$ and $V(x) \to \infty$ as $\|x\| \to \infty$, then $\alpha(\cdot)$ and $\beta(\cdot)$ can be chosen to be class $\mathcal{K}_\infty$ and (3.323) holds for all $x \in \mathbb{R}^n$.

Problem 3.73. Show that if the zero solution $x(t) \equiv 0$ of the nonlinear dynamical system (3.1) is asymptotically stable with Lyapunov function $V(x)$, $x \in \mathcal{D}$, then there exist class $\mathcal{K}$ functions $\alpha(\cdot)$, $\beta(\cdot)$, and $\gamma(\cdot)$ defined on $[0, \varepsilon]$ such that

$$\alpha(\|x\|) \leq V(x) \leq \beta(\|x\|), \quad x \in \mathcal{B}_\varepsilon(0), \tag{3.324}$$
$$V'(x)f(x) \leq -\gamma(\|x\|) < 0, \quad x \in \mathcal{B}_\varepsilon(0), \quad x \neq 0. \tag{3.325}$$

Problem 3.74. Suppose the zero solution $x(t) \equiv 0$ to (3.1) is asymptotically stable. Show that the domain of attraction $\mathcal{D}_0$ of (3.1) is an open, connected invariant set.

Problem 3.75. Show that the zero solution $x(t) \equiv 0$ to (3.1) is Lyapunov stable if and only if there exist $\delta > 0$ and a class $\mathcal{K}$ function $\alpha(\cdot)$ such that if $\|x_0\| < \delta$, then $\|x(t)\| \leq \alpha(\|x_0\|)$, $t \geq 0$.

Problem 3.76. Show that the zero solution $x(t) \equiv 0$ to (3.1) is asymptotically stable (respectively, globally asymptotically stable) if and only if there exist $\delta > 0$ and class $\mathcal{K}$ and $\mathcal{L}$ functions $\alpha(\cdot)$ and $\beta(\cdot)$, respectively, such that if $\|x_0\| < \delta$ (respectively, $x_0 \in \mathbb{R}^n$), then $\|x(t)\| \leq \alpha(\|x_0\|)\beta(t)$, $t \geq 0$.

Problem 3.77. Show that the zero solution $x(t) \equiv 0$ to (3.1) with $f(x) = Ax$, where $A \in \mathbb{R}^{n \times n}$, is asymptotically stable if and only if it is exponentially stable.

Problem 3.78. Consider the nonlinear dynamical system

$$\dot{x}_1(t) = x_1(t) + x_2(t) + x_1(t)x_2^4(t), \quad x_1(0) = x_{10}, \quad t \geq 0, \tag{3.326}$$
$$\dot{x}_2(t) = x_1(t) + x_2(t) - x_1^2(t)x_2(t), \quad x_2(0) = x_{20}. \tag{3.327}$$

Show that the zero solution $(x_1(t), x_2(t)) \equiv (0, 0)$ to (3.326) and (3.327) is unstable.

Problem 3.79. Consider the nonlinear dynamical system

$$\dot{x}_1(t) = x_1(t) - x_2(t) + x_1(t)x_2(t), \quad x_1(0) = x_{10}, \quad t \geq 0, \tag{3.328}$$

$$\dot{x}_2(t) = -x_2(t) - x_2^2(t), \quad x_2(0) = x_{20}. \tag{3.329}$$

Show that the zero solution $(x_1(t), x_2(t)) \equiv (0,0)$ to (3.328) and (3.329) is unstable. (**Hint:** Consider the function $V(x_1, x_2) = (2x_1 - x_2)^2 - x_2^2$.)

Problem 3.80. Consider the nonlinear dynamical system

$$\dot{x}_1(t) = \alpha x_2(t), \quad x_1(0) = x_{10}, \quad t \geq 0, \tag{3.330}$$
$$\dot{x}_2(t) = -\beta x_1(t) + \mu[\gamma - x_1^4(t)]^5 x_2^7(t), \quad x_2(0) = x_{20}, \tag{3.331}$$

where α, β, γ, and μ are positive constants. Use Theorem 3.12 to show that the zero solution $(x_1(t), x_2(t)) \equiv (0,0)$ to (3.330) and (3.331) is unstable. (**Hint:** Use Bendixson's theorem to show that no limit cycles exist in the simply connected region $-\gamma^{1/4} < x_1 < \gamma^{1/4}$.)

Problem 3.81. Consider the nonlinear dynamical system

$$\dot{x}_1(t) = x_1(t) + x_2(t), \quad x_1(0) = x_{10}, \quad t \geq 0, \tag{3.332}$$
$$\dot{x}_2(t) = x_1(t) - x_2(t) + x_1(t)x_2(t), \quad x_2(0) = x_{20}. \tag{3.333}$$

Use Chetaev's instability theorem to show that the zero solution $(x_1(t), x_2(t)) \equiv (0,0)$ to (3.332) and (3.333) is unstable. (**Hint:** Consider the function $V(x_1, x_2) = x_1 x_2$ and the subregion $\mathcal{Q} = \{(x_1, x_2) \in \mathbb{R}^2 : x_1 > 0,\ x_2 > 0, \text{ and } x_1^2 + x_2^2 < 1\}$.)

Problem 3.82. Consider the nonlinear dynamical system

$$\dot{x}_1(t) = x_1^3(t) + x_1(t)x_2(t), \quad x_1(0) = x_{10}, \quad t \geq 0, \tag{3.334}$$
$$\dot{x}_2(t) = -x_1(t) + x_2^2(t) + x_1(t)x_2(t) - x_1^3(t), \quad x_2(0) = x_{20}. \tag{3.335}$$

Using Chetaev's instability theorem show that the zero solution $(x_1(t), x_2(t)) \equiv (0,0)$ to (3.334) and (3.335) is unstable. (**Hint:** Consider the function $V(x_1, x_2) = \frac{1}{4}x_1^4 - \frac{1}{2}x_2^2$.)

Problem 3.83. Consider the nonlinear controlled dynamical system

$$\dot{x}_1(t) = x_1(t)x_2(t), \quad x_1(0) = x_{10}, \quad t \geq 0, \tag{3.336}$$
$$\dot{x}_2(t) = x_1^2(t) + u(t), \quad x_2(0) = x_{20}. \tag{3.337}$$

Show that there does not exist a linear feedback control law (that is, $u(t) = -k_1 x_1(t) - k_2 x_2(t)$) that renders the zero solution $(x_1(t), x_2(t)) \equiv (0,0)$ to (3.336) and (3.337) locally asymptotically stable. (**Hint:** First try using Lyapunov's indirect method. If this fails to yield any information, use Chetaev's instability theorem with function $V(x_1, x_2) = \frac{1}{2}(x_1^2 - x_2^2)$.)

Problem 3.84. Consider the nonlinear dynamical system

$$\dot{x}_1(t) = 2x_1(t) + x_2(t) + x_1^2(t), \quad x_1(0) = x_{10}, \quad t \geq 0, \tag{3.338}$$
$$\dot{x}_2(t) = x_1(t) - x_2(t) + x_1(t)x_2(t), \quad x_2(0) = x_{20}. \tag{3.339}$$

Use Chetaev's instability theorem to show that the zero solution $(x_1(t), x_2(t)) \equiv (0,0)$ to (3.338) and (3.339) is unstable. (**Hint:** Consider the function $V(x_1, x_2) = x_1 x_2$ and the subregion $\mathcal{Q} = \{(x_1, x_2) \in \mathbb{R}^2 : x_1 > 0,\ x_2 > 0, \text{ and } x_1^2 + x_2^2 < 1\}$.)

Problem 3.85. Use Chetaev's instability theorem to show that the rigid spacecraft dynamical system spinning about its intermediate axis considered in Problem 3.14 is unstable. (**Hint:** Translate the system coordinates so that the equilibrium $(0, x_{2e}, 0)$, $x_{2e} \neq 0$, becomes the origin. Then use the function $V(x_1, x_2, x_3) = x_1 x_3$ and the subregion $\mathcal{Q} = \{(x_1, x_2 - x_{2e}, x_3) \in \mathcal{B}_{\varepsilon/2}(0) : x_1 > 0, x_3 > 0\}$.)

Problem 3.86. Prove Theorem 3.12 using Chetaev's instability theorem. (**Hint:** Consider the subregion $\mathcal{Q} = \{x \in \mathcal{B}_\varepsilon(0) : V(x) > 0\}$, where $\mathcal{B}_\varepsilon(0) \subset \mathcal{D}$.)

Problem 3.87. Prove Theorem 3.13 using Chetaev's instability theorem.

3.9 Notes and References

The original work on Lyapunov stability theory is due to the Russian mathematician Aleksandr Mikhailovich Lyapunov [293]; it was translated into French in 1907 [294] and English in 1992 [295]. Lyapunov stability theory is extensively developed in the classical textbooks by Hahn [178], Krasovskii [245], LaSalle and Lefschetz [260], and Yoshizawa [474]. A more modern textbook treatment is given by Vidyasagar [445] and Khalil [235]. See also the papers by Kalman and Bertram [228] and LaSalle [258]. The invariant set stability theorems are due to Barbashin and Krasovskii [23,245] and LaSalle [258]. These theorems were first proved by Barbashin and Krasovskii [23] for autonomous systems, and by Krasovskii [245] for periodic systems. The invariant set theorem for autonomous systems was later rediscovered by LaSalle [258]. In the western literature the invariant set theorem is often incorrectly referred to as LaSalle's theorem. The systematic construction of Lyapunov functions was originally studied by Krasovskii [245] and Zubov [481]. The earliest result on stability using integrals of motion is due to Routh [370] with a Lyapunov function foundation given by Rouche, Habets, and Laloy [368]. A tutorial exposition of these results is given by Wan, Coppola, and Bernstein [450]. The treatment here is adopted from Wan, Coppola, and Bernstein [450]. Converse Lyapunov theorems are due to Persidskii [350], Malkin [298], and Massera [306, 307]. For an excellent textbook treatment see Vidyasagar [445]. The Lyapunov instability theorems can be found in Chetaev [92] and Vidyasagar [445].

Finally, the concept of semistability introduced in Problem 3.44 is due to Bhat and Bernstein [54] for nonlinear systems and Campbell and Rose [81] for linear systems. See also Bernstein and Bhat [46]. Finite-time stability as introduced in Problem 3.43 is due to Bhat and Bernstein [55]. Stability theory for functional retarded systems as introduced in Problem 3.65 is due to Krasovskii [245] with an excellent textbook treatment given by Hale and Lunel [181].

Chapter Four

Advanced Stability Theory

4.1 Introduction

In Chapter 3, we developed the basic concepts and mathematical tools for Lyapunov stability theory. In this chapter, we present several generalizations and extensions of Lyapunov stability theory. In particular, partial stability theorems are presented and derived, wherein stability with respect to part of the system state is addressed. In addition, we present stability theorems for time-varying nonlinear dynamical systems as a special case of partial stability. Lagrange stability, boundedness, ultimate boundedness, input-to-state stability, finite-time stability, and semistability notions are also presented. Finally, advanced stability theorems involving generalized Lyapunov functions, stability of sets, stability of periodic orbits, and stability theorems via vector Lyapunov functions are also established.

4.2 Partial Stability of Nonlinear Dynamical Systems

In many engineering applications, *partial stability*, that is, stability with respect to part of the system's states, is often necessary. In particular, partial stability arises in the study of electromagnetics [482], inertial navigation systems [403], spacecraft stabilization via gimballed gyroscopes and/or flywheels [448], combustion systems [16], vibrations in rotating machinery [289], and biocenology [368], to cite but a few examples. For example, in the field of biocenology involving Lotka-Volterra predator-prey models of population dynamics with age structure, if some of the species preyed upon are left alone, then the corresponding population increases without bound while a subset of the prey species remains stable [368, pp. 260–269]. The need to consider partial stability in the aforementioned systems arises from the fact that stability notions involve equilibrium coordinates as well as a hyperplane of coordinates that is closed but *not* compact. Hence, partial stability involves motion lying in a subspace instead of an equilibrium point.

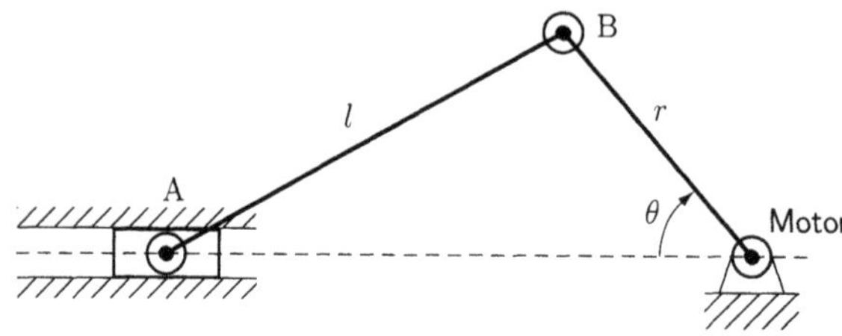

Figure 4.1 Slider-crank mechanism.

Additionally, *partial stabilization*, that is, closed-loop stability with respect to part of the closed-loop system's state, also arises in many engineering applications [289, 448]. Specifically, in spacecraft stabilization via gimballed gyroscopes asymptotic stability of an equilibrium position of the spacecraft is sought while requiring Lyapunov stability of the axis of the gyroscope relative to the spacecraft [448]. Alternatively, in the control of rotating machinery with mass imbalance, spin stabilization about a nonprincipal axis of inertia requires motion stabilization with respect to a subspace instead of the origin [289]. Perhaps the most common application where partial stabilization is necessary is adaptive control, wherein asymptotic stability of the closed-loop plant states is guaranteed without necessarily achieving parameter error convergence.

To further demonstrate the utility and need for partial stability theory, we consider two simple examples. Specifically, consider the equation of motion for the slider-crank mechanism shown in Figure 4.1 given by [44, 199]

$$m(\theta(t))\ddot{\theta}(t) + c(\theta(t))\dot{\theta}^2(t) = u(t), \quad \theta(0) = \theta_0, \quad \dot{\theta}(0) = \dot{\theta}_0, \quad t \geq 0, \tag{4.1}$$

where

$$m(\theta) = m_{\mathrm{B}}r^2 + m_{\mathrm{A}}r^2\left(\sin\theta + \frac{r\cos\theta\sin\theta}{\sqrt{l^2 - r^2\sin^2\theta}}\right)^2, \tag{4.2}$$

$$\begin{aligned} c(\theta) = {} & m_{\mathrm{A}}r^2\left(\sin\theta + \frac{r\cos\theta\sin\theta}{\sqrt{l^2 - r^2\sin^2\theta}}\right) \\ & \cdot\left[\cos\theta + r\frac{l^2(1 - 2\sin^2\theta) + r^2\sin^4\theta}{(l^2 - r^2\sin^2\theta)^{3/2}}\right], \end{aligned} \tag{4.3}$$

and m_{A} and m_{B} are point masses, r and l are the lengths of the rods, and $u(\cdot)$ is the control torque applied by the motor. Now, suppose we choose the feedback control law $u = \phi(\theta, \dot{\theta})$ so that the angular velocity of the crank is constant, that is, $\dot{\theta}(t) \to \Omega$ as $t \to \infty$, where $\Omega > 0$. This implies that $\theta(t) \approx \Omega t \to \infty$ as $t \to \infty$. Furthermore, since $m(\theta)$ and $c(\theta)$ are functions of θ we cannot ignore the angular position θ. Hence, since θ does *not* converge, it is clear that (4.1) is unstable in the standard sense but partially asymptotically stable with respect to $\dot{\theta}$ (see Definition 4.1).

Our next example involves a nonlinear system originally studied as a simplified model of a dual-spin spacecraft to investigate the resonance capture phenomenon [365] and more recently was studied to investigate the utility of a rotational/translational proof-mass actuator for stabilizing translational motion [73]. The system (see Figure 4.2) involves an eccentric rotational inertia on a translational oscillator giving rise to nonlinear coupling between the undamped oscillator and the rotational rigid body mode. The oscillator cart of mass M is connected to a fixed support via a linear spring of stiffness k. The cart is constrained to one-dimensional motion and the rotational proof-mass actuator consists of a mass m and mass moment of inertia I located at a distance e from the cart's center of mass. Letting q, $\dot{q}$, θ, and $\dot{\theta}$ denote the translational position and velocity of the cart and the angular position and velocity of the rotational proof mass, respectively, the dynamic equations of motion are given by

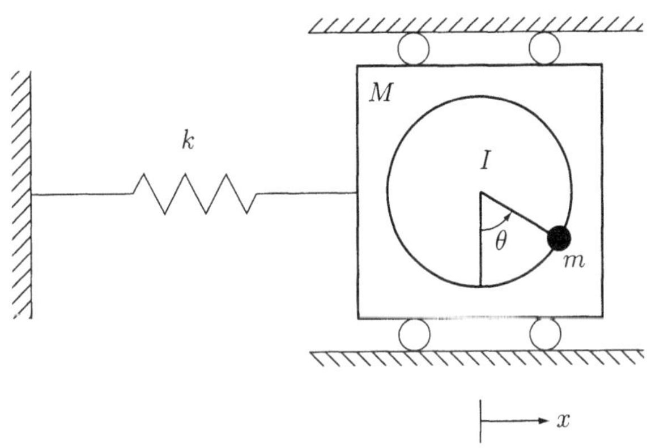

Figure 4.2 Rotational/translational proof-mass actuator.

$$(M+m)\ddot{q}(t)+me[\ddot{\theta}(t)\cos\theta(t)-\dot{\theta}^2(t)\sin\theta(t)]+kq(t) = 0, \quad (4.4)$$

$$(I+me^2)\ddot{\theta}(t)+me\ddot{q}(t)\cos\theta(t) = 0, \quad (4.5)$$

where $t \geq 0$, $q(0) = q_0$, $\dot{q}(0) = \dot{q}_0$, $\theta(0) = \theta_0$, and $\dot{\theta}(0) = \dot{\theta}_0$. Note that since the motion is constrained to the horizontal plane, the gravitational forces are not considered in the dynamic analysis. Analyzing (4.4) and (4.5) (see Example 4.1 for details), it follows that the zero solution $(q(t), \dot{q}(t), \theta(t), \dot{\theta}(t)) \equiv (0,0,0,0)$ to (4.4) and (4.5) is unstable in the standard sense but partially Lyapunov stable with respect to q, $\dot{q}$, and $\dot{\theta}$ (see Definition 4.1). Once again, standard Lyapunov stability theory cannot be used to arrive at this result since the angular position θ of the rotational proof mass cannot be ignored from (4.4) and (4.5) and $\theta(t) \to \infty$ as $t \to \infty$.

Another application of partial stability theory is the extra flexibility it provides in constructing Lyapunov functions for nonlinear dynamical systems. Specifically, generalizing Lyapunov's stability theorem to include partial stability weakens the hypotheses on the Lyapunov function (see Theorem 4.1) thus enlarging the class of allowable functions that can be used in analyzing system stability. Perhaps the clearest example of this is the *Lagrange-Dirichlet stability problem* [368] involving the conservative Euler-Lagrange system with a nonnegative-definite kinetic energy function T and a positive-definite potential function U. In this case, the Lagrange-Dirichlet energy function $V = T + U$ is only *nonnegative definite* and, hence, cannot be used as a Lyapunov function candidate to analyze the stability of the system using standard Lyapunov theory. However, the Lagrange-Dirichlet energy function can be used as a valid Lyapunov function within partial stability theory to guarantee Lyapunov stability of the Lagrange-Dirichlet problem (see Example 4.2).

In this section, we present partial stability theorems for nonlinear dynamical systems. Specifically, consider the nonlinear autonomous dynamical system

$$\dot{x}_1(t) = f_1(x_1(t), x_2(t)), \quad x_1(0) = x_{10}, \quad t \in \mathcal{I}_{x_0}, \tag{4.6}$$

$$\dot{x}_2(t) = f_2(x_1(t), x_2(t)), \quad x_2(0) = x_{20}, \tag{4.7}$$

where $x_1 \in \mathcal{D}$, $\mathcal{D} \subseteq \mathbb{R}^{n_1}$ is an open set such that $0 \in \mathcal{D}$, $x_2 \in \mathbb{R}^{n_2}$, $f_1 : \mathcal{D} \times \mathbb{R}^{n_2} \to \mathbb{R}^{n_1}$ is such that, for every $x_2 \in \mathbb{R}^{n_2}$, $f_1(0, x_2) = 0$ and $f_1(\cdot, x_2)$ is locally Lipschitz in x_1, $f_2 : \mathcal{D} \times \mathbb{R}^{n_2} \to \mathbb{R}^{n_2}$ is such that, for every $x_1 \in \mathcal{D}$, $f_2(x_1, \cdot)$ is locally Lipschitz in x_2, and $\mathcal{I}_{x_0} \triangleq [0, \tau_{x_0})$, $0 < \tau_{x_0} \leq \infty$, is the maximal interval of existence for the solution $(x_1(t), x_2(t))$, $t \in \mathcal{I}_{x_0}$, to (4.6) and (4.7). Note that under the above assumptions the solution $(x_1(t), x_2(t))$ to (4.6) and (4.7) exists and is unique over $\mathcal{I}_{x_0}$. The following definition introduces eight types of partial stability, that is, stability with respect to x_1, for the nonlinear dynamical system (4.6) and (4.7).

Definition 4.1. *i*) The nonlinear dynamical system (4.6) and (4.7) is *Lyapunov stable with respect to* x_1 if, for every $\varepsilon > 0$ and $x_{20} \in \mathbb{R}^{n_2}$, there exists $\delta = \delta(\varepsilon, x_{20}) > 0$ such that $\|x_{10}\| < \delta$ implies that $\|x_1(t)\| < \varepsilon$ for all $t \geq 0$ (see Figure 4.3(a)).

ii) The nonlinear dynamical system (4.6) and (4.7) is *Lyapunov stable with respect to* x_1 *uniformly in* x_{20} if, for every $\varepsilon > 0$, there exists $\delta = \delta(\varepsilon) > 0$ such that $\|x_{10}\| < \delta$ implies that $\|x_1(t)\| < \varepsilon$ for all $t \geq 0$ and for all $x_{20} \in \mathbb{R}^{n_2}$.

iii) The nonlinear dynamical system (4.6) and (4.7) is *asymptotically stable with respect to* x_1 if it is Lyapunov stable with respect to x_1 and, for every $x_{20} \in \mathbb{R}^{n_2}$, there exists $\delta = \delta(x_{20}) > 0$ such that $\|x_{10}\| < \delta$ implies

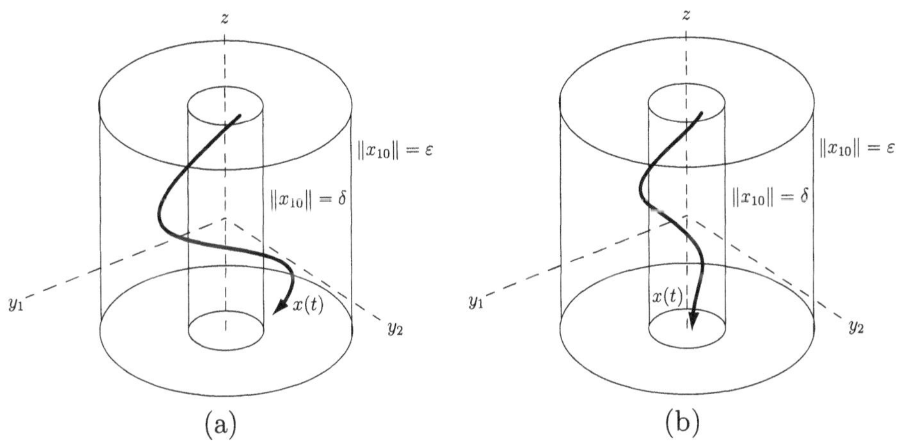

Figure 4.3 (a) Partial Lyapunov stability with respect to x_1. (b) Partial asymptotic stability with respect to x_1. $x_1 = [y_1\ y_2]^{\mathrm{T}}$, $x_2 = z$, and $x = [x_1^{\mathrm{T}}\ x_2]^{\mathrm{T}}$.

that $\lim_{t\to\infty} x_1(t) = 0$ (see Figure 4.3(b)).

iv) The nonlinear dynamical system (4.6) and (4.7) is *asymptotically stable with respect to* x_1 *uniformly in* x_{20} if it is Lyapunov stable with respect to x_1 uniformly in x_{20} and there exists $\delta > 0$ such that $\|x_{10}\| < \delta$ implies that $\lim_{t\to\infty} x_1(t) = 0$ uniformly in x_{10} and x_{20} for all $x_{20} \in \mathbb{R}^{n_2}$.

v) The nonlinear dynamical system (4.6) and (4.7) is *globally asymptotically stable with respect to* x_1 if it is Lyapunov stable with respect to x_1 and $\lim_{t\to\infty} x_1(t) = 0$ for all $x_{10} \in \mathbb{R}^{n_1}$ and $x_{20} \in \mathbb{R}^{n_2}$.

vi) The nonlinear dynamical system (4.6) and (4.7) is *globally asymptotically stable with respect to* x_1 *uniformly in* x_{20} if it is Lyapunov stable with respect to x_1 uniformly in x_{20} and $\lim_{t\to\infty} x_1(t) = 0$ uniformly in x_{10} and x_{20} for all $x_{10} \in \mathbb{R}^{n_1}$ and $x_{20} \in \mathbb{R}^{n_2}$.

vii) The nonlinear dynamical system (4.6) and (4.7) is *exponentially stable with respect to* x_1 *uniformly in* x_{20} if there exist scalars $\alpha, \beta, \delta > 0$ such that $\|x_{10}\| < \delta$ implies that $\|x_1(t)\| \leq \alpha \|x_{10}\| e^{-\beta t}$, $t \geq 0$, for all $x_{20} \in \mathbb{R}^{n_2}$.

viii) The nonlinear dynamical system (4.6) and (4.7) is *globally exponentially stable with respect to* x_1 *uniformly in* x_{20} if there exist scalars $\alpha, \beta > 0$ such that $\|x_1(t)\| \leq \alpha \|x_{10}\| e^{-\beta t}$, $t \geq 0$, for all $x_{10} \in \mathbb{R}^{n_1}$ and $x_{20} \in \mathbb{R}^{n_2}$.

Next, we present sufficient conditions for partial stability of the nonlinear dynamical system (4.6) and (4.7). For the following result define $\dot{V}(x_1, x_2) \triangleq V'(x_1, x_2) f(x_1, x_2)$, where $f(x_1, x_2) \triangleq [f_1^{\mathrm{T}}(x_1, x_2)\ f_2^{\mathrm{T}}(x_1, x_2)]^{\mathrm{T}}$, for a given continuously differentiable function $V : \mathcal{D} \times \mathbb{R}^{n_2} \to \mathbb{R}$. Furthermore, we assume that the solution $(x_1(t), x_2(t))$ to (4.6) and (4.7) exists and is unique for all $t \geq 0$. It is important to note that unlike standard theory (see Corollary 2.5) the existence of a Lyapunov function $V(x_1, x_2)$ satisfying the conditions in Theorem 4.1 below is not sufficient to ensure that all solutions of (4.6) and (4.7) starting in $\mathcal{D} \times \mathbb{R}^{n_2}$ can be extended to infinity, since none of the states of (4.6) and (4.7) serve as an independent variable. We do note, however, that Lipschitz continuity of $f_1(\cdot,\cdot)$ and $f_2(\cdot,\cdot)$ provides a sufficient condition for the existence and uniqueness of solutions to (4.6) and (4.7) over a forward time interval.

Theorem 4.1. Consider the nonlinear dynamical system (4.6) and (4.7). Then the following statements hold:

i) If there exist a continuously differentiable function $V : \mathcal{D} \times \mathbb{R}^{n_2} \to \mathbb{R}$ and a class $\mathcal{K}$ function $\alpha(\cdot)$ such that

$$V(0, x_2) = 0, \quad x_2 \in \mathbb{R}^{n_2}, \tag{4.8}$$

$$\alpha(\|x_1\|) \leq V(x_1, x_2), \quad (x_1, x_2) \in \mathcal{D} \times \mathbb{R}^{n_2}, \tag{4.9}$$

$$\dot{V}(x_1, x_2) \leq 0, \quad (x_1, x_2) \in \mathcal{D} \times \mathbb{R}^{n_2}, \tag{4.10}$$

then the nonlinear dynamical system given by (4.6) and (4.7) is Lyapunov stable with respect to x_1.

ii) If there exist a continuously differentiable function $V : \mathcal{D} \times \mathbb{R}^{n_2} \to \mathbb{R}$ and class $\mathcal{K}$ functions $\alpha(\cdot), \beta(\cdot)$ satisfying (4.9), (4.10), and

$$V(x_1, x_2) \leq \beta(\|x_1\|), \quad (x_1, x_2) \in \mathcal{D} \times \mathbb{R}^{n_2}, \tag{4.11}$$

then the nonlinear dynamical system given by (4.6) and (4.7) is Lyapunov stable with respect to x_1 uniformly in x_{20}.

iii) If there exist continuously differentiable functions $V : \mathcal{D} \times \mathbb{R}^{n_2} \to \mathbb{R}$ and $W : \mathcal{D} \times \mathbb{R}^{n_2} \to \mathbb{R}$, and class $\mathcal{K}$ functions $\alpha(\cdot)$, $\beta(\cdot)$, and $\gamma(\cdot)$ such that $\dot{W}(x_1(\cdot), x_2(\cdot))$ is bounded from below or above, (4.8) and (4.9) hold, and

$$W(0, x_2) = 0, \quad x_2 \in \mathbb{R}^{n_2}, \tag{4.12}$$

$$\beta(\|x_1\|) \leq W(x_1, x_2), \quad (x_1, x_2) \in \mathcal{D} \times \mathbb{R}^{n_2}, \tag{4.13}$$

$$\dot{V}(x_1, x_2) \leq -\gamma(W(x_1, x_2)), \quad (x_1, x_2) \in \mathcal{D} \times \mathbb{R}^{n_2}, \tag{4.14}$$

then the nonlinear dynamical system given by (4.6) and (4.7) is asymptotically stable with respect to x_1.

iv) If there exist a continuously differentiable function $V : \mathcal{D} \times \mathbb{R}^{n_2} \to \mathbb{R}$ and class $\mathcal{K}$ functions $\alpha(\cdot), \beta(\cdot), \gamma(\cdot)$ satisfying (4.9), (4.11), and

$$\dot{V}(x_1, x_2) \leq -\gamma(\|x_1\|), \quad (x_1, x_2) \in \mathcal{D} \times \mathbb{R}^{n_2}, \tag{4.15}$$

then the nonlinear dynamical system given by (4.6) and (4.7) is asymptotically stable with respect to x_1 uniformly in x_{20}.

v) If $\mathcal{D} = \mathbb{R}^{n_1}$ and there exist continuously differentiable functions $V : \mathbb{R}^{n_1} \times \mathbb{R}^{n_2} \to \mathbb{R}$ and $W : \mathbb{R}^{n_1} \times \mathbb{R}^{n_2} \to \mathbb{R}$, class $\mathcal{K}$ functions $\beta(\cdot)$, $\gamma(\cdot)$, and a class $\mathcal{K}_\infty$ function $\alpha(\cdot)$ such that $\dot{W}(x_1(\cdot), x_2(\cdot))$ is bounded from below or above, and (4.8), (4.9), and (4.12)–(4.14) hold, then the nonlinear dynamical system given by (4.6) and (4.7) is globally asymptotically stable with respect to x_1.

vi) If $\mathcal{D} = \mathbb{R}^{n_1}$ and there exist a continuously differentiable function $V : \mathbb{R}^{n_1} \times \mathbb{R}^{n_2} \to \mathbb{R}$, a class $\mathcal{K}$ function $\gamma(\cdot)$, and class $\mathcal{K}_\infty$ functions $\alpha(\cdot), \beta(\cdot)$ satisfying (4.9), (4.11), and (4.15), then the nonlinear dynamical system given by (4.6) and (4.7) is globally asymptotically stable with respect to x_1 uniformly in x_{20}.

vii) If there exist a continuously differentiable function $V : \mathcal{D} \times \mathbb{R}^{n_2} \to \mathbb{R}$ and positive constants α, β, γ, $p \geq 1$ satisfying

$$\alpha \|x_1\|^p \leq V(x_1, x_2) \leq \beta \|x_1\|^p, \qquad (x_1, x_2) \in \mathcal{D} \times \mathbb{R}^{n_2}, \quad (4.16)$$
$$\dot{V}(x_1, x_2) \leq -\gamma \|x_1\|^p, \qquad (x_1, x_2) \in \mathcal{D} \times \mathbb{R}^{n_2}, \quad (4.17)$$

then the nonlinear dynamical system given by (4.6) and (4.7) is exponentially stable with respect to x_1 uniformly in x_{20}.

viii) If $\mathcal{D} = \mathbb{R}^{n_1}$ and there exist a continuously differentiable function $V : \mathbb{R}^{n_1} \times \mathbb{R}^{n_2} \to \mathbb{R}$ and positive constants α, β, γ, $p \geq 1$ satisfying (4.16) and (4.17), then the nonlinear dynamical system given by (4.6) and (4.7) is globally exponentially stable with respect to x_1 uniformly in x_{20}.

Proof. *i)* Let $x_{20} \in \mathbb{R}^{n_2}$, let $\varepsilon > 0$ be such that $\mathcal{B}_\varepsilon(0) \triangleq \{x_1 \in \mathbb{R}^{n_1} : \|x_1\| < \varepsilon\} \subset \mathcal{D}$, define $\eta \triangleq \alpha(\varepsilon)$, and define $\mathcal{D}_\eta \triangleq \{x_1 \in \mathcal{B}_\varepsilon(0) : V(x_1, x_{20}) < \eta\}$. Since $V(\cdot, \cdot)$ is continuous and $V(0, x_{20}) = 0$ it follows that $\mathcal{D}_\eta$ is nonempty and there exists $\delta = \delta(\varepsilon, x_{20}) > 0$ such that $V(x_1, x_{20}) < \eta$, $x_1 \in \mathcal{B}_\delta(0)$. Hence, $\mathcal{B}_\delta(0) \subseteq \mathcal{D}_\eta$. Next, since $\dot{V}(x_1, x_2) \leq 0$ it follows that $V(x_1(t), x_2(t))$ is a nonincreasing function of time, and hence, for every $x_{10} \in \mathcal{B}_\delta(0) \subseteq \mathcal{D}_\eta$ it follows that

$$\alpha(\|x_1(t)\|) \leq V(x_1(t), x_2(t)) \leq V(x_{10}, x_{20}) < \eta = \alpha(\varepsilon).$$

Thus, for every $x_{10} \in \mathcal{B}_\delta(0) \subseteq \mathcal{D}_\eta$, $x_1(t) \in \mathcal{B}_\varepsilon(0)$, $t \geq 0$, establishing Lyapunov stability with respect to x_1.

ii) Let $\varepsilon > 0$ and let $\mathcal{B}_\varepsilon(0)$ and η be given as in the proof of *i)*. Now, let $\delta = \delta(\varepsilon) > 0$ be such that $\beta(\delta) = \alpha(\varepsilon)$. Then it follows from (4.11) that for all $(x_{10}, x_{20}) \in \mathcal{B}_\delta(0) \times \mathbb{R}^{n_2}$,

$$\alpha(\|x_1(t)\|) \le V(x_1(t), x_2(t)) \le V(x_{10}, x_{20}) < \beta(\delta) = \alpha(\varepsilon),$$

and hence, $x_1(t) \in \mathcal{B}_\varepsilon(0)$, $t \ge 0$.

iii) Lyapunov stability follows from *i)*. To show asymptotic stability suppose, *ad absurdum*, that $W(x_1(t), x_2(t)) \nrightarrow 0$ as $t \to \infty$ or, equivalently, $\limsup_{t\to\infty} W(x_1(t), x_2(t)) > 0$. In addition, suppose $\liminf_{t\to\infty} W(x_1(t), x_2(t)) > 0$, which implies that there exist constants $T > 0$ and $k > 0$ such that $W(x_1(t), x_2(t)) \ge k$, $t \ge T$. Then it follows from (4.14) that $V(x_1(t), x_2(t)) \to -\infty$ as $t \to \infty$, which contradicts (4.9), and hence, $\liminf_{t\to\infty} W(x_1(t), x_2(t)) = 0$. Now, since $\limsup_{t\to\infty} W(x_1(t), x_2(t)) > 0$ and $\liminf_{t\to\infty} W(x_1(t), x_2(t)) = 0$ it follows that there exist two increasing sequences $\{t_i\}_{i=0}^\infty$ and $\{t_i'\}_{i=0}^\infty$, and a constant $k > 0$ such that $t_i < t_i' < t_{i+1}$, $i = 0, 1, \ldots$, $t_i \to \infty$ as $i \to \infty$, and

$$W(x_1(t_i), x_2(t_i)) = \frac{k}{2} < W(x_1(t), x_2(t)) < k = W(x_1(t_i'), x_2(t_i')),$$
$$t_i < t < t_i', \quad i = 0, 1, \ldots. \tag{4.18}$$

Furthermore, since $\dot{W}(x_1(\cdot), x_2(\cdot))$ is upper (respectively, lower) bounded there exists $\eta > 0$ (respectively, $\eta < 0$) such that $\dot{W}(x_1(t), x_2(t)) \le \eta$, $t \ge 0$ (respectively, $\dot{W}(x_1(t), x_2(t)) \ge \eta$, $t \ge 0$). In the case where $\dot{W}(x_1(\cdot), x_2(\cdot))$ is upper bounded it follows from (4.18) that $t_i' - t_i \ge \frac{k}{2\eta}$, $i = 1, 2, \ldots$, and hence, $\int_{t_i}^{t_i'} \gamma(W(x_1(t), x_2(t)))\mathrm{d}t \ge \gamma(\frac{k}{2})\frac{k}{2\eta}$. Now, using (4.14), it follows that

$$\begin{aligned} V(x_1(t_n'), x_2(t_n')) &= V(x_{10}, x_{20}) + \int_0^{t_1} \dot{V}(x_1(t), x_2(t))\mathrm{d}t \\ &\quad + \sum_{i=1}^{n} \int_{t_i}^{t_i'} \dot{V}(x_1(t), x_2(t))\mathrm{d}t \\ &\quad + \sum_{i=1}^{n-1} \int_{t_i'}^{t_{i+1}} \dot{V}(x_1(t), x_2(t))\mathrm{d}t \\ &\le V(x_{10}, x_{20}) + \sum_{i=1}^{n} \int_{t_i}^{t_i'} \dot{V}(x_1(t), x_2(t))\mathrm{d}t \\ &\le V(x_{10}, x_{20}) - \sum_{i=1}^{n} \int_{t_i}^{t_i'} \gamma(W(x_1(t), x_2(t)))\mathrm{d}t \\ &\le V(x_{10}, x_{20}) - n\gamma\left(\frac{k}{2}\right)\frac{k}{2\eta}. \end{aligned} \tag{4.19}$$

Hence, for sufficiently large n the right-hand side of (4.19) becomes negative, which contradicts (4.9). The case where $\dot{W}(x_1(\cdot), x_2(\cdot))$ is lower bounded leads to a similar contradiction using identical arguments. Thus, in either case, $W(x_1(t), x_2(t)) \to 0$ as $t \to \infty$, and hence, it follows from (4.13) that $x_1(t) \to 0$ as $t \to \infty$, proving asymptotic stability of (4.6) and (4.7) with respect to x_1.

iv) Lyapunov stability uniformly in x_{20} follows from *ii)*. Next, let $\varepsilon > 0$ and $\delta = \delta(\varepsilon) > 0$ be such that for every $x_{10} \in \mathcal{B}_\delta(0)$, $x_1(t) \in \mathcal{B}_\varepsilon(0)$, $t \geq 0$, (the existence of such a (δ, ε) pair follows from uniform Lyapunov stability), and assume that (4.15) holds. Since (4.15) implies (4.10) it follows that for every $x_{10} \in \mathcal{B}_\delta(0)$, $V(x_1(t), x_2(t))$ is a nonincreasing function of time and, since $V(\cdot, \cdot)$ is bounded from below, it follows from the monotone convergence theorem (Theorem 2.10) that there exists $L \geq 0$ such that $\lim_{t\to\infty} V(x_1(t), x_2(t)) = L$. Now, suppose that for some $x_{10} \in \mathcal{B}_\delta(0)$, *ad absurdum*, $L > 0$ so that $\mathcal{D}_L \triangleq \{x_1 \in \mathcal{B}_\varepsilon(0) : V(x_1, x_2) \leq L \text{ for all } x_2 \in \mathbb{R}^{n_2}\}$ is nonempty and $x_1(t) \notin \mathcal{D}_L$, $t \geq 0$. Thus, as in the proof of *i)*, there exists $\hat{\delta} > 0$ such that $\mathcal{B}_{\hat{\delta}}(0) \subset \mathcal{D}_L$. Hence, it follows from (4.15) that for the given $x_{10} \in \mathcal{B}_\delta(0) \setminus \mathcal{D}_L$ and $t \geq 0$,

$$\begin{aligned} V(x_1(t), x_2(t)) &= V(x_{10}, x_{20}) + \int_0^t \dot{V}(x_1(s), x_2(s))\mathrm{d}s \\ &\leq V(x_{10}, x_{20}) - \int_0^t \gamma(\|x_1(s)\|)\mathrm{d}s \\ &\leq V(x_{10}, x_{20}) - \gamma(\hat{\delta})t. \end{aligned}$$

Letting $t > \frac{V(x_{10}, x_{20}) - L}{\gamma(\hat{\delta})}$, it follows that $V(x_1(t), x_2(t)) < L$, which is a contradiction. Hence, $L = 0$, and, since $x_{10} \in \mathcal{B}_\delta(0)$ was chosen arbitrarily, it follows that $V(x_1(t), x_2(t)) \to 0$ as $t \to \infty$ for all $x_{10} \in \mathcal{B}_\delta(0)$. Now, since $V(x_1(t), x_2(t)) \geq \alpha(\|x_1(t)\|) \geq 0$, it follows that $\alpha(\|x_1(t)\|) \to 0$ or, equivalently, $x_1(t) \to 0$ $t \to \infty$, establishing asymptotic stability with respect to x_1 uniformly in x_{20}.

v) Let $\delta > 0$ be such that $\|x_{10}\| < \delta$. Since $\alpha(\cdot)$ is a class $\mathcal{K}_\infty$ function, it follows that there exists $\varepsilon > 0$ such that $V(x_{10}, x_{20}) < \alpha(\varepsilon)$. Now, (4.14) implies that $V(x_1(t), x_2(t))$ is a nonincreasing function of time, and hence, it follows from (4.9) that $\alpha(\|x_1(t)\|) \leq V(x_1(t), x_2(t)) \leq V(x_{10}, x_{20}) < \alpha(\varepsilon)$, $t \geq 0$. Hence, $x_1(t) \in \mathcal{B}_\varepsilon(0)$, $t \geq 0$. Now, the proof follows as in the proof of *iii)*.

vi) Let $\delta > 0$ be such that $\|x_{10}\| < \delta$. Since $\alpha(\cdot)$ is a class $\mathcal{K}_\infty$ function, it follows that there exists $\varepsilon > 0$ such that $\beta(\delta) \leq \alpha(\varepsilon)$. Now, (4.15) implies that $V(x_1(t), x_2(t))$ is a nonincreasing function of time, and hence, it follows from (4.11) that $\alpha(\|x_1(t)\|) \leq V(x_1(t), x_2(t)) \leq V(x_{10}, x_{20}) < \beta(\delta) \leq \alpha(\varepsilon)$, $t \geq 0$. Hence, $x_1(t) \in \mathcal{B}_\varepsilon(0)$, $t \geq 0$. Now, the proof follows as in the proof of *iv)*.

vii) Let $\varepsilon > 0$ and $\mathcal{B}_\varepsilon(0)$ be given as in the proof of *i*) and let $\eta \triangleq \alpha\varepsilon^p$ and $\delta = \left(\frac{\eta}{\beta}\right)^{1/p}$. Now, (4.17) implies that $\dot{V}(x_1, x_2) \leq 0$, and hence, as in the proof of *ii*), it follows that for all $(x_{10}, x_{20}) \in \mathcal{B}_\delta(0) \times \mathbb{R}^{n_2}$, $x_1(t) \in \mathcal{B}_\varepsilon(0)$, $t \geq 0$. Furthermore, it follows from (4.16) and (4.17) that for all $t \geq 0$ and $(x_{10}, x_{20}) \in \mathcal{B}_\delta(0) \times \mathbb{R}^{n_2}$,

$$\dot{V}(x_1(t), x_2(t)) \leq -\gamma\|x_1(t)\|^p \leq -\frac{\gamma}{\beta}V(x_1(t), x_2(t)),$$

which implies that

$$V(x_1(t), x_2(t)) \leq V(x_{10}, x_{20})e^{-\frac{\gamma}{\beta}t}.$$

It now follows from (4.16) that

$$\alpha\|x_1(t)\|^p \leq V(x_1(t), x_2(t)) \leq V(x_{10}, x_{20})e^{-\frac{\gamma}{\beta}t} \leq \beta\|x_{10}\|^p e^{-\frac{\gamma}{\beta}t}, \quad t \geq 0,$$

and hence,

$$\|x_1(t)\| \leq \left(\frac{\beta}{\alpha}\right)^{1/p} \|x_{10}\|e^{-\frac{\gamma}{\beta p}t}, \quad t \geq 0,$$

establishing exponential stability with respect to x_1 uniformly in x_{20}.

viii) The proof follows as in *vi*) and *vii*). □

By setting $n_1 = n$ and $n_2 = 0$, Theorem 4.1 specializes to the case of nonlinear autonomous systems of the form $\dot{x}_1(t) = f_1(x_1(t))$. In this case, Lyapunov (respectively, asymptotic) stability with respect to x_1 and Lyapunov (respectively, asymptotic) stability with respect to x_1 uniformly in x_{20} are equivalent to the classical Lyapunov (respectively, asymptotic) stability of nonlinear autonomous systems presented in Section 3.2. In particular, note that it follows from Problems 3.75 and 3.76 that there exists a continuously differentiable function $V : \mathcal{D} \to \mathbb{R}$ such that (4.9), (4.11), and (4.15) hold if and only if $V(0) = 0$, $V(x_1) > 0$, $x_1 \neq 0$, $V'(x_1)f_1(x_1) < 0$, $x_1 \neq 0$. In addition, if $\mathcal{D} = \mathbb{R}^{n_1}$ and there exist class $\mathcal{K}_\infty$ functions $\alpha(\cdot), \beta(\cdot)$ and a continuously differentiable function $V(\cdot)$ such that (4.9), (4.11), and (4.15) hold if and only if $V(0) = 0$, $V(x_1) > 0$, $x_1 \neq 0$, $V'(x_1)f_1(x_1) < 0$, $x_1 \neq 0$, and $V(x_1) \to \infty$ as $\|x_1\| \to \infty$. Hence, in this case, Theorem 4.1 collapses to the classical Lyapunov stability theorem for autonomous systems given in Section 3.2.

It is important to note that there is a key difference between the partial stability definitions given in Definition 4.1 and the definitions of partial stability given in [374, 448]. In particular, the partial stability definitions given in [374, 448] require that both the initial conditions x_{10} and x_{20} lie in a neighborhood of the origin, whereas in Definition 4.1 x_{20} can be arbitrary. As will be seen in the next section, this difference allows us to unify autonomous partial stability theory with time-varying stability theory. Lyapunov (respectively, asymptotic) stability with respect to x_1 given in Definition 4.1 is referred to in [448] as x_1-stability (respectively, x_1-asymptotic stability) for large x_2 while Lyapunov (respectively, asymptotic) stability with respect to x_1 uniformly in x_{20} given in Definition 4.1 is referred to in [448] as x_1-stability (respectively, x_1-asymptotic stability) with respect to the whole of x_2. Note that if a nonlinear dynamical system is Lyapunov (respectively, asymptotically) stable with respect to x_1 in the sense of Definition 4.1, then the system is x_1-stable (respectively, x_1-asymptotically stable) in the sense of the definition given in [374, 448]. Furthermore, if there exist a continuously differentiable function $V : \mathcal{D} \times \mathbb{R}^{n_2} \to \mathbb{R}$ and a class $\mathcal{K}$ function $\alpha(\cdot)$ (respectively, class $\mathcal{K}$ functions $\beta(\cdot)$ and $\gamma(\cdot)$) such that $V(0,0) = 0$ and (4.9) and (4.10) (respectively, (4.9), (4.11), and (4.15)) hold, then the nonlinear dynamical system (4.6) and (4.7) is x_1-stable (respectively, uniformly asymptotically x_1-stable with respect to x_{20}) in the sense of the definition given in [448]. It is important to note that the condition $V(0,x_2) = 0$, $x_2 \in \mathbb{R}^{n_2}$, allows us to prove partial stability in the sense of Definition 4.1. Finally, an additional difference between our formulation of the partial stability problem and the partial stability problem considered in [374, 448] is in the treatment of the equilibrium of (4.6) and (4.7). Specifically, in our formulation, we require the partial equilibrium condition $f_1(0,x_2) = 0$ for every $x_2 \in \mathbb{R}^{n_2}$, whereas in [374,448], the authors require the equilibrium condition $f_1(0,0) = 0$ and $f_2(0,0) = 0$.

Example 4.1. In this example, we use Theorem 4.1 to show that the rotational/translational proof-mass model (4.4) and (4.5) is partially Lyapunov stable with respect to q, $\dot{q}$, and $\dot{\theta}$. To show this, let $x_1 = q$, $x_2 = \dot{q}$, $x_3 = \theta$, $x_4 = \dot{\theta}$ and consider the Lyapunov function candidate

$$V(x_1,x_2,x_3,x_4) = \tfrac{1}{2}[kx_1^2 + (M+m)x_2^2 + (I+me^2)x_4^2 + 2mex_2x_4\cos x_3]. \tag{4.20}$$

Note that $V(x_1,x_2,x_3,x_4) = \frac{1}{2}kx_1^2 + \frac{1}{2}\tilde{x}^{\mathrm{T}}P(x_3)\tilde{x}$, where $\tilde{x} = [x_2\ x_4]^{\mathrm{T}}$ and

$$P(x_3) = \begin{bmatrix} M+m & me\cos x_3 \\ me\cos x_3 & I+me^2 \end{bmatrix}. \tag{4.21}$$

Since

$$2\lambda_{\min}(P(x_3)) = M+m+I+me^2 - \sqrt{(M+m-I-me^2)^2 + 4m^2e^2\cos^2 x_3}, \tag{4.22}$$

$$2\lambda_{\max}(P(x_3)) = M+m+I+me^2 + \sqrt{(M+m-I-me^2)^2 + 4m^2e^2\cos^2 x_3}, \tag{4.23}$$

it follows that $\alpha_{\min}I_2 \le 2P(x_3) \le \alpha_{\max}I_2$, $x_3 \in \mathbb{R}$, where

$$\alpha_{\min} \triangleq M+m+I+me^2 - \sqrt{(M+m-I-me^2)^2 + 4m^2e^2}, \tag{4.24}$$

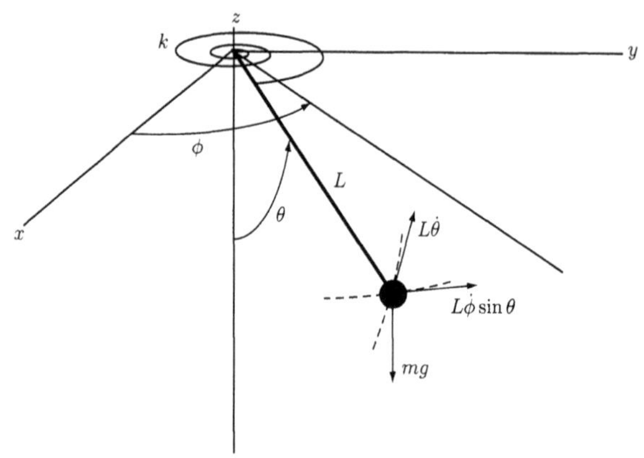

Figure 4.4 Spherical pendulum.

$$\alpha_{\max} \triangleq M + m + I + me^2 + \sqrt{(M + m - I - me^2)^2 + 4m^2e^2}. \quad (4.25)$$

Hence, it follows that $\frac{1}{2}x_1^2 + \frac{\alpha_{\min}}{4}(x_2^2 + x_4^2) \le V(x_1, x_2, x_3, x_4) \le \frac{1}{2}x_1^2 + \frac{\alpha_{\max}}{4}(x_2^2 + x_4^2)$, which implies that $V(\cdot)$ satisfies (4.9) and (4.11). Now, since $\dot{V}(x_1, x_2, x_3, x_4) = 0$, it follows from *ii*) of Theorem 4.1 that (4.4) and (4.5) is Lyapunov stable with respect to x_1, x_2, and x_4 uniformly in x_{30}. Furthermore, since the system involves a nonlinear coupling of an undamped oscillator with a rotational rigid body mode it follows that $x_3(t)$ does not converge as $t \to \infty$. △

Example 4.2. In this example, we apply Theorem 4.1 to a Lagrange-Dirichlet problem involving a conservative Euler-Lagrange system with a nonnegative-definite kinetic energy function T and a positive-definite potential function U. Specifically, we consider the motion of the spherical pendulum shown in Figure 4.4, where θ denotes the angular position of the pendulum with respect to vertical z-axis, ϕ denotes the angular position of the pendulum in the x-y plane, m denotes the mass of the pendulum, L denotes the length of the pendulum, k denotes the torsional spring stiffness, and g denotes the gravitational acceleration. Defining $q \triangleq [\theta\ \phi]^{\mathrm{T}}$ to be the generalized system positions and $\dot{q} \triangleq [\dot{\theta}\ \dot{\phi}]^{\mathrm{T}}$ to be the generalized system velocities, it follows that the governing equations of motion are given by the Euler-Lagrange equation

$$\frac{\mathrm{d}}{\mathrm{d}t}\left(\frac{\partial \mathcal{L}}{\partial \dot{q}}(q(t), \dot{q}(t))\right) - \left(\frac{\partial \mathcal{L}}{\partial q}(q(t), \dot{q}(t))\right) = 0, \ q(0) = q_0, \ \dot{q}(0) = \dot{q}_0, \ t \ge 0, \quad (4.26)$$

where $\mathcal{L}(q, \dot{q}) = T(q, \dot{q}) - U(q)$ denotes the system Lagrangian, $T(q, \dot{q}) \triangleq \frac{1}{2}m[(L\dot{\theta})^2 + (\dot{\phi}L\sin\theta)^2]$ denotes the system kinetic energy, and $U(q) \triangleq mgL(1 - \cos\theta) + \frac{1}{2}k\phi^2$ denotes the system potential energy. Equivalently,

(4.26) can be rewritten as

$$\ddot{\theta}(t) - \sin\theta(t)\cos\theta(t)\dot{\phi}^2(t) + (g/L)\sin\theta(t) = 0,$$
$$\theta(0) = \theta_0, \quad \dot{\theta}(0) = \dot{\theta}_0, \quad t \geq 0, \qquad (4.27)$$
$$\sin^2\theta(t)\ddot{\phi}(t) + 2\sin\theta(t)\cos\theta(t)\dot{\phi}(t)\dot{\theta}(t) + (k/mL^2)\phi(t) = 0,$$
$$\phi(0) = \phi_0, \quad \dot{\phi}(0) = \dot{\phi}_0. \qquad (4.28)$$

Next, consider the Lagrange-Dirichlet energy function $V(q,\dot{q}) = T(q, \dot{q}) + U(q)$ and note that since the system kinetic energy function $T(q,\dot{q})$ is not positive definite in $\dot{q}$ at θ such that $\sin\theta = 0$, the function $V(q,\dot{q})$ cannot be used as a Lyapunov function candidate to analyze the stability of the system using standard Lyapunov theory. However, the Lagrange-Dirichlet energy function $V(q,\dot{q})$ can be used as a valid Lyapunov function within partial stability theory to guarantee partial Lyapunov stability with respect to $[\theta\ \phi\ \dot{\theta}]^{\mathrm{T}}$. Specifically, let $x_1 = [\theta\ \phi\ \dot{\theta}]^{\mathrm{T}}$, let $x_2 = \dot{\phi}$, and let $\alpha(s) = \max\{mgL(1-\cos(s)), \frac{1}{2}ks^2, \frac{1}{2}mL^2s^2\}$ and $\|x_1\| = \max\{|\theta|, |\phi|, |\dot{\theta}|\}$. Now, note that $\alpha(\cdot)$ is a class $\mathcal{K}$ function and $V(x_1,x_2) = V(q,\dot{q}) \geq \alpha(\|x_1\|)$. Furthermore, note that $V(0,x_2) = 0$, $x_2 \in \mathbb{R}$, and $\dot{V}(x_1,x_2) = 0$. Now, it follows from *i*) of Theorem 4.1 that the Euler-Lagrange system given by (4.27) and (4.28) is partially Lyapunov stable with respect to x_1. Finally, it can be easily shown via simulations that the Euler-Lagrange system given by (4.27) and (4.28) is not Lyapunov stable in the standard sense. △

Another important application of partial stability theory is the unification it provides between time-invariant stability theory and stability theory for time-varying systems. To see this, consider the time-varying nonlinear dynamical system given by

$$\dot{x}(t) = f(t, x(t)), \qquad x(t_0) = x_0, \qquad t \geq t_0, \qquad (4.29)$$

where $x(t) \in \mathbb{R}^n$, $t \geq t_0$, and $f : [t_0, t_1) \times \mathbb{R}^n \to \mathbb{R}^n$. Now, define $x_1(\tau) \triangleq x(t)$ and $x_2(\tau) \triangleq t$, where $\tau \triangleq t - t_0$, and note that the solution $x(t)$, $t \geq t_0$, to the nonlinear time-varying dynamical system (4.29) can be equivalently characterized by the solution $x_1(\tau)$, $\tau \geq 0$, to the nonlinear autonomous dynamical system

$$\dot{x}_1(\tau) = f(x_2(\tau), x_1(\tau)), \qquad x_1(0) = x_0, \qquad \tau \geq 0, \qquad (4.30)$$
$$\dot{x}_2(\tau) = 1, \qquad x_2(0) = t_0, \qquad (4.31)$$

where $\dot{x}_1(\cdot)$ and $\dot{x}_2(\cdot)$ denote differentiation with respect to τ. However, in this case, stability results for time-invariant systems do not apply to the augmented system (4.30) and (4.31) since one of the states, namely, the state x_2 representing time, is unbounded. However, writing the time-varying nonlinear system (4.29) as (4.30) and (4.31), it is clear that partial stability theory provides a natural formulation for addressing stability theory for autonomous and nonautonomous systems within a unified framework. The following example demonstrates the utility of Theorem 4.1 to the stability of time-varying systems.

Example 4.3. Consider the spring-mass-damper system with time-varying damping coefficient given by

$$\ddot{q}(t) + c(t)\dot{q}(t) + kq(t) = 0, \quad q(0) = q_0, \quad \dot{q}(0) = \dot{q}_0, \quad t \geq 0. \tag{4.32}$$

This is an interesting system to analyze since physical intuition would lead one to surmise that if $c(t) \geq \alpha > 0$, $t \geq 0$, then the zero solution $(q(t), \dot{q}(t)) \equiv (0,0)$ to (4.32) is asymptotically stable since we have constant dissipation of energy. However, this is *not* the case. A simple counterexample (see [424]) is $c(t) = 2 + e^t$, $k = 1$, with $q(0) = 2$ and $\dot{q}(0) = -1$, which gives $q(t) = 1 + e^{-t}$, $t \geq 0$, and hence, $q(t) \to 1$ as $t \to \infty$. This is due to the fact that damping increases so fast that the system halts at $q = 1$.

To analyze (4.32) using Theorem 4.1, we consider $c(t) = 3 + \sin t$ and $k > 1$. Now, (4.32) can be equivalently written as

$$\dot{z}_1(t) = z_2(t), \qquad z_1(0) = q_0, \qquad t \geq 0, \tag{4.33}$$

$$\dot{z}_2(t) = -kz_1(t) - c(t)z_2(t), \qquad z_2(0) = \dot{q}_0, \tag{4.34}$$

where $z_1 \triangleq q$ and $z_2 \triangleq \dot{q}$. Next, let $n_1 = 2$, $n_2 = 1$, $x_1 = [z_1, \ z_2]^{\mathrm{T}}$, $x_2 = t$, $f_1(x_1, x_2) = [x_1^{\mathrm{T}}v, \ -x_1^{\mathrm{T}}h(x_2)]^{\mathrm{T}}$, and $f_2(x_1, x_2) = 1$, where $h(x_2) = [k, \ c(x_2)]^{\mathrm{T}}$ and $v = [0, \ 1]^{\mathrm{T}}$. Now, the solution $(z_1(t), z_2(t))$, $t \geq 0$, to the nonlinear time-varying dynamical system (4.33) and (4.34) is equivalently characterized by the solution $x_1(t)$, $t \geq 0$, to the nonlinear autonomous dynamical system

$$\dot{x}_1(t) = f_1(x_1(t), x_2(t)), \quad x_1(0) = [q_0, \ \dot{q}_0]^{\mathrm{T}}, \quad t \geq 0, \tag{4.35}$$

$$\dot{x}_2(t) = 1, \quad x_2(0) = 0. \tag{4.36}$$

To examine the stability of this system, consider the Lyapunov function candidate $V(x_1, x_2) = x_1^{\mathrm{T}} P(x_2) x_1$, where

$$P(x_2) = \begin{bmatrix} k + 3 + \sin(x_2) & 1 \\ 1 & 1 \end{bmatrix}.$$

Note that since

$$x_1^{\mathrm{T}} P_1 x_1 \leq V(x_1, x_2) \leq x_1^{\mathrm{T}} P_2 x_1, \qquad (x_1, x_2) \in \mathbb{R}^2 \times \mathbb{R}, \tag{4.37}$$

where

$$P_1 = \begin{bmatrix} k+2 & 1 \\ 1 & 1 \end{bmatrix}, \qquad P_2 = \begin{bmatrix} k+4 & 1 \\ 1 & 1 \end{bmatrix},$$

it follows that $V(x_1, x_2)$ satisfies (4.16) with $\mathcal{D} = \mathbb{R}^2$ and $p = 2$. Next, since

$$\begin{aligned}\dot{V}(x_1, x_2) &= -2x_1^{\mathrm{T}}[R + R_1(x_2)]x_1 \\ &\le -2x_1^{\mathrm{T}} R x_1 \\ &\le -\min\{k-1, 1\}\|x_1\|_2^2, \end{aligned} \tag{4.38}$$

where $R = \mathrm{diag}[k-1, 1] > 0$ and $R_1(x_2) = \mathrm{diag}[1 - \frac{1}{2}\cos(x_2),\ 1 + \sin(x_2)]$, it follows from $v)$ and $vi)$ of Theorem 4.1 that the dynamical system (4.35) and (4.36) is globally exponentially stable with respect to x_1 uniformly in x_{20}. △

In the case of time-invariant systems the Barbashin-Krasovskii-LaSalle invariance theorem (Theorem 3.3) shows that bounded system trajectories of a nonlinear dynamical system approach the largest invariant set $\mathcal{M}$ characterized by the set of all points in a compact set $\mathcal{D}$ of the state space where the Lyapunov derivative identically vanishes. In the case of partially stable systems, however, it is not generally clear how to define the set $\mathcal{M}$ since $\dot{V}(x_1, x_2)$ is a function of both x_1 and x_2. However, if $\dot{V}(x_1, x_2) \le -W(x_1) \le 0$, where $W : \mathcal{D} \to \mathbb{R}$ is continuous and nonnegative definite, then a set $\mathcal{R} \supset \mathcal{M}$ can be defined as the set of points where $W(x_1)$ identically vanishes, that is, $\mathcal{R} = \{x_1 \in \mathcal{D} : \ W(x_1) = 0\}$. In this case, as shown in the next theorem, the partial system trajectories $x_1(t)$ approach $\mathcal{R}$ as t tends to infinity. For this result, the following lemma, known as Barbalat's lemma, is necessary.

Lemma 4.1 (Barbalat's Lemma). Let $\sigma : [0, \infty) \to \mathbb{R}$ be a uniformly continuous function and suppose that $\lim_{t\to\infty} \int_0^t \sigma(s)\mathrm{d}s$ exists and is finite. Then, $\lim_{t\to\infty} \sigma(t) = 0$.

Proof. Suppose, *ad absurdum*, that $\limsup_{t\to\infty} |\sigma(t)| > 0$ and let $\alpha_1 \in \mathbb{R}$ be such that $0 < \alpha_1 < \limsup_{t\to\infty} |\sigma(t)|$. In this case, for every $T_1 > 0$, there exists $t_1' \ge T_1$ such that $|\sigma(t_1')| \ge \alpha_1$. Next, since $\sigma : [0, \infty) \to \mathbb{R}$ is uniformly continuous, it follows that there exists $\alpha_2 = \alpha_2(\alpha_1) > 0$ such that $|\sigma(t + \tau) - \sigma(t)| < \alpha_1/2$ for all $t \ge 0$ and $\tau \in [0, \alpha_2]$. Hence,

$$\begin{aligned}|\sigma(t)| &= |\sigma(t) - \sigma(t_1') + \sigma(t_1')| \\ &\ge |\sigma(t_1')| - |\sigma(t) - \sigma(t_1')| \\ &> \alpha_1/2, \qquad t \in [t_1', t_1' + \alpha_2]. \end{aligned} \tag{4.39}$$

Now, since $\sigma : [0, \infty) \to \mathbb{R}$ is continuous and $|\sigma(t)| > \alpha_1/2 > 0$, $t \in [t_1', t_1' + \alpha_2]$, it follows that the sign of $\sigma(t)$ is constant over $t \in [t_1', t_1' + \alpha_2]$. Hence,

$$\left| \int_{t_1'}^{t_1'+\alpha_2} \sigma(t)\mathrm{d}t \right| = \int_{t_1'}^{t_1'+\alpha_2} |\sigma(t)|\mathrm{d}t > \tfrac{1}{2}\alpha_1\alpha_2. \tag{4.40}$$

Now, repeating the above argument it can be shown that there exists a sequence $\{t_i'\}_{i=1}^{\infty}$ such that $t_i' + \alpha_2 \le t_{i+1}'$, $i = 1, 2, \ldots$, and $\int_{t_i'}^{t_i'+\alpha_2} |\sigma(t)\mathrm{d}t > \frac{1}{2}\alpha_1\alpha_2$, $i = 1, 2, \ldots$. Hence,

$$\int_0^{\infty} |\sigma(t)|\mathrm{d}t \ge \sum_{i=1}^{\infty} \int_{t_i'}^{t_i'+\alpha_2} |\sigma(t)|\mathrm{d}t > \alpha_1 \sum_{i=1}^{\infty} \alpha_2 = \infty, \tag{4.41}$$

which is a contradiction. □

Theorem 4.2. Consider the nonlinear dynamical system (4.6) and (4.7), and assume $\mathcal{D} \times \mathbb{R}^{n_2}$ is a positively invariant set with respect to (4.6) and (4.7) where $f_1(\cdot, x_2)$ is Lipschitz continuous in x_1, uniformly in x_2. Furthermore, assume there exist functions $V : \mathcal{D} \times \mathbb{R}^{n_2} \to \mathbb{R}$, $W, W_1, W_2 : \mathcal{D} \to \mathbb{R}$ such that $V(\cdot, \cdot)$ is continuously differentiable, $W_1(\cdot)$ and $W_2(\cdot)$ are continuous and positive definite, $W(\cdot)$ is continuous and nonnegative definite, and, for all $(x_1, x_2) \in \mathcal{D} \times \mathbb{R}^{n_2}$,

$$W_1(x_1) \le V(x_1, x_2) \le W_2(x_1), \tag{4.42}$$

$$\dot{V}(x_1, x_2) \le -W(x_1). \tag{4.43}$$

Then there exists $\mathcal{D}_0 \subseteq \mathcal{D}$ such that for all $(x_{10}, x_{20}) \in \mathcal{D}_0 \times \mathbb{R}^{n_2}$, $x_1(t) \to \mathcal{R} \triangleq \{x_1 \in \mathcal{D} : W(x_1) = 0\}$ as $t \to \infty$. If, in addition, $\mathcal{D} = \mathbb{R}^{n_1}$ and $W_1(\cdot)$ is radially unbounded, then for all $(x_{10}, x_{20}) \in \mathbb{R}^{n_1} \times \mathbb{R}^{n_2}$, $x_1(t) \to \mathcal{R} \triangleq \{x_1 \in \mathbb{R}^{n_1} : W(x_1) = 0\}$ as $t \to \infty$.

Proof. Assume (4.42) and (4.43) hold. Then it follows from Theorem 4.1 that the nonlinear dynamical system given by (4.6) and (4.7) is Lyapunov stable with respect to x_1 uniformly in x_{20}. Let $\varepsilon > 0$ be such that $\mathcal{B}_\varepsilon(0) \subset \mathcal{D}$ and let $\delta = \delta(\varepsilon) > 0$ be such that if $x_{10} \in \mathcal{B}_\delta(0)$, then $x_1(t) \in \mathcal{B}_\varepsilon(0)$, $t \ge 0$. Now, since $V(x_1(t), x_2(t))$ is nonincreasing and bounded from below by zero, it follows from the monotone convergence theorem (Theorem 2.10) that $\lim_{t\to\infty} V(x_1(t), x_2(t))$ exists and is finite. Hence, since for every $t \ge 0$,

$$\int_0^t W(x_1(\tau))\mathrm{d}\tau \le -\int_0^t \dot{V}(x_1(\tau), x_2(\tau))\mathrm{d}\tau = V(x_{10}, x_{20}) - V(x_1(t), x_2(t)),$$

it follows that $\lim_{t\to\infty} \int_0^t W(x_1(\tau))\mathrm{d}\tau$ exists and is finite.

Next, since by (4.42) $\|x_1(t)\| \le \varepsilon$, $t \ge 0$, and, for every $x_2 \in \mathbb{R}^{n_2}$, $f_1(\cdot, x_2)$ is Lipschitz continuous on $\mathcal{D}$ uniformly in x_2, it follows that

$$\begin{aligned} \|x_1(t_2) - x_1(t_1)\| &= \left\| \int_{t_1}^{t_2} f_1(x_1(\tau), x_2(\tau))\mathrm{d}\tau \right\| \\ &\le L_1 \int_{t_1}^{t_2} \|x_1(\tau)\|\mathrm{d}\tau \\ &\le L_1\varepsilon(t_2 - t_1), \quad t_2 \ge t_1, \end{aligned} \tag{4.44}$$

where L_1 is the Lipschitz constant of $f_1(\cdot,\cdot)$ on $\{x_1 \in \mathcal{D} : \|x_1\| \le \varepsilon\}$. Now, for every $\hat{\varepsilon} > 0$, letting $\delta = \delta(\hat{\varepsilon}) = \frac{\hat{\varepsilon}}{L_1 \varepsilon}$ yields

$$\|x_1(t_2) - x_1(t_1)\| < \hat{\varepsilon}, \qquad t_2 - t_1 \le \delta,$$

which shows that $x_1(\cdot)$ is uniformly continuous. Next, since $x_1(\cdot)$ is uniformly continuous and $W(\cdot)$ is continuous on a compact set $\overline{\mathcal{B}}_\varepsilon(0)$, it follows that $W(x_1(t))$ is uniformly continuous at every $t \ge 0$. Now, it follows from Barbalat's lemma (Lemma 4.1) that $W(x_1(t)) \to 0$ as $t \to \infty$. Finally, if, in addition, $\mathcal{D} = \mathbb{R}^{n_1}$ and $W_1(\cdot)$ is radially unbounded, then, as in the proof of *iv*) of Theorem 4.1, for every $x_{10} \in \mathbb{R}^{n_1}$ there exists $\varepsilon, \delta > 0$ such that $x_{10} \in \mathcal{B}_\delta(0)$ and $x_1(t) \in \mathcal{B}_\varepsilon(0)$, $t \ge 0$. Now, the proof follows by repeating the above arguments. □

Theorem 4.2 shows that the partial system trajectories $x_1(t)$ approach $\mathcal{R}$ as t tends to infinity. However, since the positive limit set of the partial trajectory $x_1(t)$ is a subset of $\mathcal{R}$, Theorem 4.2 is a weaker result than the standard invariance principle wherein one would conclude that the partial trajectory $x_1(t)$ approaches the *largest invariant set* $\mathcal{M}$ contained in $\mathcal{R}$. This is not true in general for partially stable systems since the positive limit set of a partial trajectory $x_1(t)$, $t \ge 0$, is not an invariant set. However, in the case where $f_1(\cdot, x_2)$ is periodic, almost periodic, or asymptotically independent of x_2, then an invariance principle for partially stable systems can be derived. This result is left as an exercise for the reader.

Next, we state two converse theorems for partial stability. The proofs of these theorems are virtually identical to the proofs of the converse theorems given in Section 3.5 and are left as exercises for the reader.

Theorem 4.3. Assume that the nonlinear dynamical system (4.6) and (4.7) is asymptotically stable with respect to x_1 uniformly in x_{20} and $f_1 : \mathcal{D} \times \mathbb{R}^{n_2} \to \mathbb{R}^{n_1}$ and $f_2 : \mathcal{D} \times \mathbb{R}^{n_2} \to \mathbb{R}^{n_2}$ are continuously differentiable. Let $\delta > 0$ be such that $\mathcal{B}_\delta(0) \subset \mathcal{D}$ and $x_{10} \in \mathcal{B}_\delta(0)$ implies that $\lim_{t\to\infty} x_1(t) = 0$ uniformly in x_{10} and x_{20} for all $x_{20} \in \mathbb{R}^{n_2}$, and assume $\frac{\partial f}{\partial x_1}$ is bounded on $\mathcal{B}_\delta(0)$ uniformly in x_2. Then there exist a continuously differentiable function $V : \mathcal{B}_\delta(0) \times \mathbb{R}^{n_2} \to \mathbb{R}$ and class $\mathcal{K}$ functions $\alpha(\cdot)$, $\beta(\cdot)$, and $\gamma(\cdot)$ such that

$$\alpha(\|x_1\|) \le V(x_1, x_2) \le \beta(\|x_1\|), \quad (x_1, x_2) \in \mathcal{B}_\delta(0) \times \mathbb{R}^{n_2}, \tag{4.45}$$

$$\dot{V}(x_1, x_2) \le -\gamma(\|x_1\|), \quad (x_1, x_2) \in \mathcal{B}_\delta(0) \times \mathbb{R}^{n_2}. \tag{4.46}$$

Theorem 4.4. Assume that the nonlinear dynamical system (4.6) and (4.7) is exponentially stable with respect to x_1 uniformly in x_{20} and $f_1 : \mathcal{D} \times \mathbb{R}^{n_2} \to \mathbb{R}^{n_1}$ and $f_2 : \mathcal{D} \times \mathbb{R}^{n_2} \to \mathbb{R}^{n_2}$ are continuously differentiable. Let $\delta > 0$ be such that $\mathcal{B}_\delta(0) \subset \mathcal{D}$ and $x_{10} \in \mathcal{B}_\delta(0)$ implies that $\|x_1(t)\| \le \alpha \|x_{10}\| e^{-\beta t}$ for some $\alpha, \beta > 0$ and for all $t \ge 0$ and $x_{20} \in \mathbb{R}^{n_2}$, and assume $\frac{\partial f}{\partial x_1}$ is bounded on $\mathcal{B}_\delta(0)$ uniformly in x_2. Then, for every $p > 1$, there exist a continuously differentiable function $V : \mathcal{B}_\delta(0) \times \mathbb{R}^{n_2} \to \mathbb{R}$ and positive constants α, β, and γ such that

$$\alpha\|x_1\|^p \le V(x_1, x_2) \le \beta\|x_1\|^p, \quad (x_1, x_2) \in \mathcal{B}_\delta(0) \times \mathbb{R}^{n_2}, \tag{4.47}$$

$$\dot{V}(x_1, x_2) \le -\gamma\|x_1\|^p, \quad (x_1, x_2) \in \mathcal{B}_\delta(0) \times \mathbb{R}^{n_2}. \tag{4.48}$$

Finally, we close this section by addressing a partial stability notion wherein both initial conditions x_{10} and x_{20} lie in a neighborhood of the origin. For this result we modify Definition 4.1 to reflect the fact that the entire initial state $x_0 = [x_{10}^{\mathrm{T}},\ x_{20}^{\mathrm{T}}]^{\mathrm{T}}$ lies in the neighborhood of the origin so that $\|x_{10}\| < \delta$ is replaced by $\|x_0\| < \delta$ in the definition. Furthermore, for this result we assume $f_1 : \mathcal{D} \times \mathbb{R}^{n_2} \to \mathbb{R}^{n_1}$ and $f_2 : \mathcal{D} \times \mathbb{R}^{n_2} \to \mathbb{R}^{n_2}$ are such that $f_1(0, 0) = 0$ and $f_2(0, 0) = 0$.

Theorem 4.5. Consider the nonlinear dynamical system (4.6) and (4.7). Then the following statements hold:

i) If there exist a continuously differentiable function $V : \mathcal{D} \times \mathbb{R}^{n_2} \to \mathbb{R}$ and a class $\mathcal{K}$ function $\alpha(\cdot)$ such that $V(0, 0) = 0$,

$$\alpha(\|x_1\|) \le V(x_1, x_2), \qquad (x_1, x_2) \in \mathcal{D} \times \mathbb{R}^{n_2}, \tag{4.49}$$

$$\dot{V}(x_1, x_2) \le 0, \qquad (x_1, x_2) \in \mathcal{D} \times \mathbb{R}^{n_2}, \tag{4.50}$$

then the nonlinear dynamical system given by (4.6) and (4.7) is Lyapunov stable with respect to x_1.

ii) If there exist a continuously differentiable function $V : \mathcal{D} \times \mathbb{R}^{n_2} \to \mathbb{R}$ and class $\mathcal{K}$ functions $\alpha(\cdot), \beta(\cdot)$ satisfying (4.49), (4.50), and

$$V(x_1, x_2) \le \beta(\|x\|), \qquad (x_1, x_2) \in \mathcal{D} \times \mathbb{R}^{n_2}, \tag{4.51}$$

where $x \triangleq [x_1^{\mathrm{T}},\ x_2^{\mathrm{T}}]^{\mathrm{T}}$, then the nonlinear dynamical system given by (4.6) and (4.7) is Lyapunov stable with respect to x_1 uniformly in x_{20}.

iii) If there exist continuously differentiable functions $V : \mathcal{D} \times \mathbb{R}^{n_2} \to \mathbb{R}$ and $W : \mathcal{D} \times \mathbb{R}^{n_2} \to \mathbb{R}$, and class $\mathcal{K}$ functions $\alpha(\cdot)$, $\beta(\cdot)$, and $\gamma(\cdot)$ such that $\dot{W}(x_1(\cdot), x_2(\cdot))$ is bounded from below or above, (4.49) holds, and

$$\beta(\|x\|) \le W(x_1, x_2), \qquad (x_1, x_2) \in \mathcal{D} \times \mathbb{R}^{n_2}, \tag{4.52}$$

$$\dot{V}(x_1, x_2) \le -\gamma(W(x_1, x_2)), \qquad (x_1, x_2) \in \mathcal{D} \times \mathbb{R}^{n_2}, \tag{4.53}$$

then the nonlinear dynamical system given by (4.6) and (4.7) is asymptotically stable with respect to x_1.

iv) If there exist a continuously differentiable function $V : \mathcal{D} \times \mathbb{R}^{n_2} \to \mathbb{R}$ and class $\mathcal{K}$ functions $\alpha(\cdot), \beta(\cdot), \gamma(\cdot)$ satisfying (4.49), (4.51), and

$$\dot{V}(x_1, x_2) \le -\gamma(\|x\|), \qquad (x_1, x_2) \in \mathcal{D} \times \mathbb{R}^{n_2}, \tag{4.54}$$

then the nonlinear dynamical system given by (4.6) and (4.7) is asymptotically stable with respect to x_1 uniformly in x_{20}.

v) If $\mathcal{D} = \mathbb{R}^{n_1}$ and there exist continuously differentiable functions $V : \mathbb{R}^{n_1} \times \mathbb{R}^{n_2} \to \mathbb{R}$ and $W : \mathbb{R}^{n_1} \times \mathbb{R}^{n_2} \to \mathbb{R}$, class $\mathcal{K}$ functions $\beta(\cdot)$, $\gamma(\cdot)$, and a class $\mathcal{K}_\infty$ function $\alpha(\cdot)$ such that $\dot{W}(x_1(\cdot), x_2(\cdot))$ is bounded from below or above, and (4.49), (4.52), and (4.53) hold, then the nonlinear dynamical system given by (4.6) and (4.7) is globally asymptotically stable with respect to x_1.

vi) If $\mathcal{D} = \mathbb{R}^{n_1}$ and there exist a continuously differentiable function $V : \mathbb{R}^{n_1} \times \mathbb{R}^{n_2} \to \mathbb{R}$, a class $\mathcal{K}$ function $\gamma(\cdot)$, and class $\mathcal{K}_\infty$ functions $\alpha(\cdot), \beta(\cdot)$ satisfying (4.49), (4.51), and (4.54), then the nonlinear dynamical system given by (4.6) and (4.7) is globally asymptotically stable with respect to x_1 uniformly in x_{20}.

vii) If there exist a continuously differentiable function $V : \mathcal{D} \times \mathbb{R}^{n_2} \to \mathbb{R}$ and positive constants α, β, γ, $p \geq 1$ satisfying

$$\alpha\|x_1\|^p \leq V(x_1, x_2) \leq \beta\|x\|^p, \qquad (x_1, x_2) \in \mathcal{D} \times \mathbb{R}^{n_2}, \quad (4.55)$$

$$\dot{V}(x_1, x_2) \leq -\gamma\|x\|^p, \qquad (x_1, x_2) \in \mathcal{D} \times \mathbb{R}^{n_2}, \quad (4.56)$$

then the nonlinear dynamical system given by (4.6) and (4.7) is exponentially stable with respect to x_1 uniformly in x_{20}.

viii) If $\mathcal{D} = \mathbb{R}^{n_1}$ and there exist a continuously differentiable function $V : \mathbb{R}^{n_1} \times \mathbb{R}^{n_2} \to \mathbb{R}$ and positive constants α, β, γ, $p \geq 1$ satisfying (4.55) and (4.56), then the nonlinear dynamical system given by (4.6) and (4.7) is globally exponentially stable with respect to x_1 uniformly in x_{20}.

Proof. The proof is virtually identical to the proof of Theorem 4.1 and is left as an exercise for the reader. □

4.3 Stability Theory for Nonlinear Time-Varying Systems

In this section, we use the results of Section 4.2 to extend Lyapunov's direct method to nonlinear time-varying systems, thereby providing a unification between partial stability theory for autonomous systems and stability theory for time-varying systems. Specifically, we consider the nonlinear time-varying dynamical system

$$\dot{x}(t) = f(t, x(t)), \quad x(t_0) = x_0, \quad t \geq t_0, \qquad (4.57)$$

where $x(t) \in \mathcal{D}$, $t \geq t_0$, $\mathcal{D} \subseteq \mathbb{R}^n$ such that $0 \in \mathcal{D}$, $f : [t_0, t_1) \times \mathcal{D} \to \mathbb{R}^n$ is such that $f(\cdot, \cdot)$ is jointly continuous in t and x, and for every $t \in [t_0, t_1)$, $f(t, 0) = 0$ and $f(t, \cdot)$ is locally Lipschitz in x uniformly in t for all t in compact subsets of $[0,\infty)$. Note that under the above assumptions the solution $x(t)$, $t \geq t_0$, to (4.57) exists and is unique over the interval $[t_0, t_1)$. The following definition provides eight types of stability for the nonlinear time-varying dynamical system (4.57).

Definition 4.2. *i*) The nonlinear time-varying dynamical system (4.57) is *Lyapunov stable* if, for every $\varepsilon > 0$ and $t_0 \in [0,\infty)$, there exists $\delta = \delta(\varepsilon, t_0) > 0$ such that $\|x_0\| < \delta$ implies that $\|x(t)\| < \varepsilon$ for all $t \geq t_0$.

ii) The nonlinear time-varying dynamical system (4.57) is *uniformly Lyapunov stable* if, for every $\varepsilon > 0$, there exists $\delta = \delta(\varepsilon) > 0$ such that $\|x_0\| < \delta$ implies that $\|x(t)\| < \varepsilon$ for all $t \geq t_0$ and for all $t_0 \in [0,\infty)$.

iii) The nonlinear time-varying dynamical system (4.57) is *asymptotically stable* if it is Lyapunov stable and for every $t_0 \in [0,\infty)$, there exists $\delta = \delta(t_0) > 0$ such that $\|x_0\| < \delta$ implies that $\lim_{t\to\infty} x(t) = 0$.

iv) The nonlinear time-varying dynamical system (4.57) is *uniformly asymptotically stable* if it is uniformly Lyapunov stable and there exists $\delta > 0$ such that $\|x_0\| < \delta$ implies that $\lim_{t\to\infty} x(t) = 0$ uniformly in t_0 and x_0 for all $t_0 \in [0,\infty)$.

v) The nonlinear time-varying dynamical system (4.57) is *globally asymptotically stable* if it is Lyapunov stable and $\lim_{t\to\infty} x(t) = 0$ for all $x_0 \in \mathbb{R}^n$ and $t_0 \in [0,\infty)$.

vi) The nonlinear time-varying dynamical system (4.57) is *globally uniformly asymptotically stable* if it is uniformly Lyapunov stable and $\lim_{t\to\infty} x(t) = 0$ uniformly in t_0 and x_0 for all $x_0 \in \mathbb{R}^n$ and $t_0 \in [0,\infty)$.

vii) The nonlinear time-varying dynamical system (4.57) is (*uniformly*) *exponentially stable* if there exist scalars $\alpha, \beta, \delta > 0$ such that $\|x_0\| < \delta$ implies that $\|x(t)\| \leq \alpha \|x_0\| e^{-\beta t}$, $t \geq t_0$ and $t_0 \in [0,\infty)$.

viii) The nonlinear time-varying dynamical system (4.57) is *globally* (*uniformly*) *exponentially stable* if there exist scalars $\alpha, \beta > 0$ such that $\|x(t)\| \leq \alpha \|x_0\| e^{-\beta t}$, $t \geq t_0$, for all $x_0 \in \mathbb{R}^n$ and $t_0 \in [0,\infty)$.

Note that uniform asymptotic stability is equivalent to uniform Lyapunov stability and the existence of $\delta > 0$ such that, for every $\varepsilon > 0$, there exists $T = T(\varepsilon) > 0$ such that $\|x_0\| < \delta$ and $t_0 \geq 0$ implies that $\|x(t)\| < \varepsilon$ for all $t \geq t_0 + T(\varepsilon)$. That is, for every initial condition in a ball of radius δ at time $t = t_0$, the graph of the solution of (4.57) is guaranteed to be inside a given cylinder for all $t > t_0 + T(\varepsilon)$ (see Figure 4.5). Uniform Lyapunov stability and uniform asymptotic stability can be additionally characterized by class $\mathcal{K}$ and class $\mathcal{KL}$ functions. For details, see Problems 4.5 and 4.6.

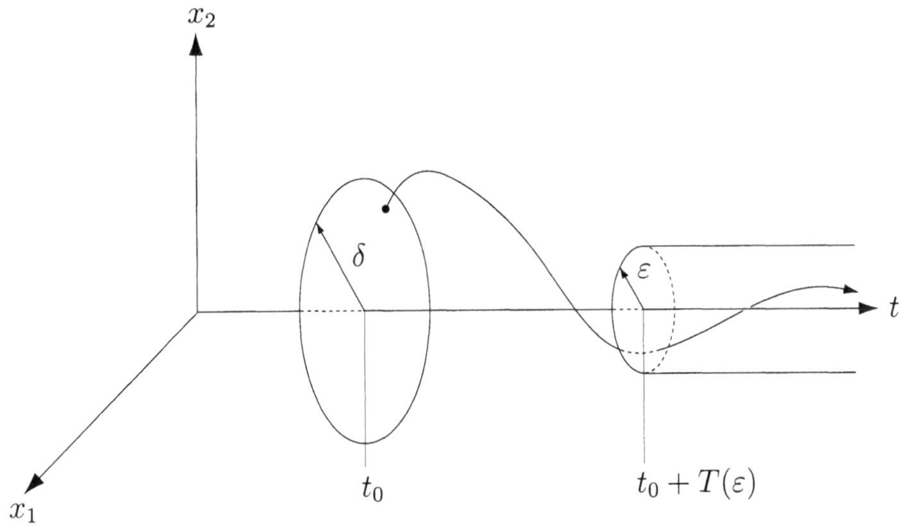

Figure 4.5 Uniform asymptotic stability.

Example 4.4. To elucidate the difference between Lyapunov stability and uniform Lyapunov stability consider the scalar linear dynamical system adopted from [307] given by

$$\dot{x}(t) = (6t \sin t - 2t)x(t), \quad x(t_0) = x_0, \quad t \geq t_0. \tag{4.58}$$

The solution of (4.58) is given by

$$\begin{aligned} x(t) &= x(t_0)e^{\int_{t_0}^{t}(6s\sin s - 2s)ds} \\ &= x(t_0)e^{[6\sin t - 6t\cos t - t^2 - 6\sin t_0 + 6t_0 \cos t_0 + t_0^2]}. \end{aligned} \tag{4.59}$$

Clearly, for every fixed t_0, the term $-t^2$ in (4.59) will eventually dominate, which shows that the exponential term in (4.59) is bounded by a constant $\sigma(t_0)$ dependent on t_0 for all $t \geq t_0$. Hence, $|x(t)| < |x(t_0)|\sigma(t_0)$, $t \geq t_0$. Now, given $\varepsilon > 0$, we can choose $\delta = \delta(\varepsilon, t_0) = \varepsilon/\sigma(t_0)$, which shows that $|x(t_0)| < \delta$ implies $|x(t)| < \varepsilon$, $t \geq t_0$, and hence, the zero solution $x(t) \equiv 0$ to (4.58) is Lyapunov stable.

Next, suppose t_0 takes the successive values $t_0 = 2n\pi$, $n = 0, 1, 2, \ldots$, and $x(t)$ is evaluated π seconds later for each n. In this case,

$$x(t_0 + \pi) = x(t_0)e^{[(4n+1)(6-\pi)\pi]}, \tag{4.60}$$

which implies that, for $x(t_0) \neq 0$,

$$\frac{x(t_0+\pi)}{x(t_0)} = e^{[(4n+1)(6-\pi)\pi]} \to \infty \text{ as } n \to \infty. \tag{4.61}$$

Since $\sigma(t_0) \geq e^{[(4n+1)(6-\pi)\pi]}$ it follows that $\sigma(t_0) \to \infty$ as $n \to \infty$. Thus, given $\varepsilon > 0$, for stability we need $\delta = \varepsilon/\sigma(t_0) \to 0$ as $t_0 \to \infty$, which shows that we cannot choose δ independent of t_0 to satisfy the requirement for uniform Lyapunov stability. △

Next, using Theorem 4.1 we present sufficient conditions for stability of the nonlinear time-varying dynamical system (4.57). For the following result define

$$\dot{V}(t,x) \triangleq \frac{\partial V}{\partial t}(t,x) + \frac{\partial V}{\partial x}(t,x)f(t,x),$$

for a given continuously differentiable function $V : [0,\infty) \times \mathcal{D} \to \mathbb{R}$.

Theorem 4.6. Consider the time-varying dynamical system given by (4.57). Then the following statements hold:

i) If there exist a continuously differentiable function $V : [0,\infty) \times \mathcal{D} \to \mathbb{R}$ and a class $\mathcal{K}$ function $\alpha(\cdot)$ such that

$$V(t,0) = 0, \qquad t \in [0,\infty), \tag{4.62}$$
$$\alpha(\|x\|) \leq V(t,x), \qquad (t,x) \in [0,\infty) \times \mathcal{D}, \tag{4.63}$$
$$\dot{V}(t,x) \leq 0, \qquad (t,x) \in [0,\infty) \times \mathcal{D}, \tag{4.64}$$

then the nonlinear time-varying dynamical system given by (4.57) is Lyapunov stable.

ii) If there exist a continuously differentiable function $V : [0,\infty) \times \mathcal{D} \to \mathbb{R}$ and class $\mathcal{K}$ functions $\alpha(\cdot), \beta(\cdot)$ satisfying (4.63), (4.64), and

$$V(t,x) \leq \beta(\|x\|), \qquad (t,x) \in [0,\infty) \times \mathcal{D}, \tag{4.65}$$

then the nonlinear time-varying dynamical system given by (4.57) is uniformly Lyapunov stable.

iii) If there exist continuously differentiable functions $V : [0,\infty) \times \mathcal{D} \to \mathbb{R}$ and $W : [0,\infty) \times \mathcal{D} \to \mathbb{R}$, and class $\mathcal{K}$ functions $\alpha(\cdot)$, $\beta(\cdot)$, and $\gamma(\cdot)$ such that $\dot{W}(\cdot, x(\cdot))$ is bounded from below or above, (4.62) and (4.63) hold, and

$$W(t,0) = 0, \qquad t \in [0,\infty), \tag{4.66}$$
$$\beta(\|x\|) \leq W(t,x), \qquad (t,x) \in [0,\infty) \times \mathcal{D}, \tag{4.67}$$
$$\dot{V}(t,x) \leq -\gamma(W(t,x)), \qquad (t,x) \in [0,\infty) \times \mathcal{D}, \tag{4.68}$$

then the zero solution $x(t) \equiv 0$ to (4.57) is asymptotically stable.

iv) If there exist a continuously differentiable function $V : [0,\infty) \times \mathcal{D} \to \mathbb{R}$ and class $\mathcal{K}$ functions $\alpha(\cdot), \beta(\cdot), \gamma(\cdot)$ satisfying (4.63), (4.65), and

$$\dot{V}(t,x) \le -\gamma(\|x\|), \qquad (t,x) \in [0,\infty) \times \mathcal{D}, \tag{4.69}$$

then the nonlinear time-varying dynamical system given by (4.57) is uniformly asymptotically stable.

v) If $\mathcal{D} = \mathbb{R}^n$ and there exist continuously differentiable functions $V : [0,\infty) \times \mathcal{D} \to \mathbb{R}$ and $W : [0,\infty) \times \mathcal{D} \to \mathbb{R}$, class $\mathcal{K}$ functions $\beta(\cdot)$, $\gamma(\cdot)$, and a class $\mathcal{K}_\infty$ function $\alpha(\cdot)$ such that $\dot{W}(\cdot, x(\cdot))$ is bounded from below or above, and (4.62), (4.63), and (4.66)–(4.68) hold, then the zero solution $x(t) \equiv 0$ to (4.57) is globally asymptotically stable.

vi) If $\mathcal{D} = \mathbb{R}^n$ and there exist a continuously differentiable function $V : [0,\infty) \times \mathbb{R}^n \to \mathbb{R}$, a class $\mathcal{K}$ function $\gamma(\cdot)$, class $\mathcal{K}_\infty$ functions $\alpha(\cdot), \beta(\cdot)$ satisfying (4.63), (4.65), and (4.69), then the nonlinear time-varying dynamical system given by (4.57) is globally uniformly asymptotically stable.

vii) If there exist a continuously differentiable function $V : [0,\infty) \times \mathcal{D} \to \mathbb{R}$ and positive constants α, β, γ, p such that $p \ge 1$ and

$$\alpha\|x\|^p \le V(t,x) \le \beta\|x\|^p, \qquad (t,x) \in [0,\infty) \times \mathcal{D}, \tag{4.70}$$

$$\dot{V}(t,x) \le -\gamma\|x\|^p, \qquad (t,x) \in [0,\infty) \times \mathcal{D}, \tag{4.71}$$

then the nonlinear time-varying dynamical system given by (4.57) is (uniformly) exponentially stable.

viii) If $\mathcal{D} = \mathbb{R}^n$ and there exist a continuously differentiable function $V : [0,\infty) \times \mathbb{R}^n \to \mathbb{R}$ and positive constants α, β, γ, $p \ge 1$ satisfying (4.70) and (4.71), then the nonlinear time-varying dynamical system given by (4.57) is globally (uniformly) exponentially stable.

Proof. First note that, requiring the existence of a Lyapunov function $V : [0,\infty) \times \mathcal{D} \to \mathbb{R}$ satisfying the conditions above, it follows from Theorem 2.39 that there exists a unique solution to (4.57) for all $t \ge t_0$. Next, let $n_1 = n$, $n_2 = 1$, $x_1(t - t_0) = x(t)$, $x_2(t - t_0) = t$, $f_1(x_1, x_2) = f(x_2, x_1)$, and $f_2(x_1, x_2) = 1$. Now, note that with $\tau = t - t_0$, the solution $x(t)$, $t \ge t_0$, to the nonlinear time-varying dynamical system (4.57) is equivalently characterized by the solution $x_1(\tau)$, $\tau \ge 0$, to the nonlinear autonomous dynamical system

$$\begin{aligned} \dot{x}_1(\tau) &= f_1(x_1(\tau), x_2(\tau)), \quad x_1(0) = x_0, \quad \tau \ge 0, \\ \dot{x}_2(\tau) &= 1, \quad x_2(0) = t_0, \end{aligned}$$

where $\dot{x}_1(\cdot)$ and $\dot{x}_2(\cdot)$ denote differentiation with respect to τ. Furthermore, note that since $f(t, 0) = 0$, $t \ge 0$, it follows that $f_1(0, x_2) = 0$, for every $x_2 \in \mathbb{R}^{n_2}$. Now, the result is a direct consequence of Theorem 4.1. □

It is important to note that (4.63) along with (4.69) are not sufficient to guarantee asymptotic stability of a nonlinear time-varying dynamical system. This is shown by the following counterexample adopted from [306].

Example 4.5. Consider the scalar nonlinear dynamical system given by

$$\dot{x}(t) = \frac{\dot{g}(t)}{g(t)}x(t), \qquad x(0) = x_0, \qquad t \geq 0, \tag{4.72}$$

where

$$g(t) = \sum_{n=1}^{\infty} \frac{1}{1+n^4(t-n)^2}, \quad t \geq 0. \tag{4.73}$$

Note that it can be shown that for every $t \geq 0$, $g(t)$ is well defined and $g(t) > 0$, $t \geq 0$, and hence, the right-hand side of (4.72) is well defined. Furthermore, note that the solution to (4.72) is given by $x(t) = \frac{g(t)}{g(0)}x_0$.

Next, note that for all $m \in \mathbb{Z}_+$,

$$g(m) = \sum_{n=1}^{\infty} \frac{1}{1+n^4(m-n)^2} > 1.$$

Now, let $t \in [0, \infty)$ and let $n_1, n_2 \in \overline{\mathbb{Z}}_+$ be such that $n_1 \leq t \leq n_2$. In this case, for all $n \in \overline{\mathbb{Z}}_+ \setminus \{n_1, n_2\}$,

$$\frac{1}{1+n^4(t-n)^2} \leq \frac{1}{n^4}$$

and

$$\frac{1}{1+n_i^4(t-n_i)^2} \leq 1, \quad i = 1, 2.$$

Hence, it follows that

$$g(t) < 2 + \sum_{n=1}^{\infty} \frac{1}{n^4} < \frac{10}{3}, \qquad t \geq 0, \tag{4.74}$$

which implies that $|||g|||_\infty < 10/3$.

Next, note that

$$\begin{aligned}
\int_0^\infty g(t)\mathrm{d}t &= \int_0^\infty \sum_{n=1}^{\infty} \frac{1}{1+n^4(t-n)^2}\mathrm{d}t \\
&= \sum_{n=1}^{\infty} \int_0^\infty \frac{1}{1+n^4(t-n)^2}\mathrm{d}t \\
&= \sum_{n=1}^{\infty} \int_{-n^3}^\infty \frac{1}{n^2(1+t^2)}\mathrm{d}t
\end{aligned}$$

$$\begin{aligned} &< \sum_{n=1}^{\infty} \int_{-\infty}^{\infty} \frac{1}{n^2(1+t^2)} \mathrm{d}t \\ &= \sum_{n=1}^{\infty} \frac{\pi}{n^2} \\ &< 2\pi, \end{aligned}$$

which implies that $\|g\|_1 < \infty$. Furthermore, note that

$$\int_0^\infty g^2(t)\mathrm{d}t \le \|g\|_\infty \int_0^\infty g(t)\mathrm{d}t = \|g\|_\infty \|g\|_1 < \frac{20\pi}{3}.$$

Now, consider the Lyapunov function candidate

$$V(t,x) = \frac{x^2}{g^2(t)}\left[\gamma^2 + \int_t^\infty g^2(s)\mathrm{d}s\right], \tag{4.75}$$

and note that $V(t,0) = 0$, $t \ge 0$. Since $g(t) \le \gamma$, $t \ge 0$,

$$\begin{aligned} V(t,x) &\ge \gamma^2 \frac{x^2}{g^2(t)} \\ &\ge x^2, \qquad t \in [0,\infty) \qquad x \in \mathbb{R}, \end{aligned}$$

and since $\int_0^\infty g^2(t)\mathrm{d}t < 20\pi/3$, $V(t,x)$ is a well-defined function. Now, it is easy to verify that $\dot{V}(t,x) = -x^2$. Hence, $V(t,x)$ satisfies (4.62), (4.63), and (4.69), which implies that the zero solution $x(t) \equiv 0$ is Lyapunov stable. However, since $\lim_{t\to\infty} g(t) \ne 0$ (since $g(m) > 1$, $m \in \mathbb{Z}_+$) it follows that the zero solution $x(t) \equiv 0$ to (4.72) is not asymptotically stable. △

In light of Theorem 4.6 it follows that Theorem 4.1 can be trivially extended to address partial stability for time-varying dynamical systems. Specifically, consider the nonlinear time-varying dynamical system

$$\dot{x}_1(t) = f_1(t, x_1(t), x_2(t)), \quad x_1(t_0) = x_{10}, \quad t \ge t_0, \tag{4.76}$$

$$\dot{x}_2(t) = f_2(t, x_1(t), x_2(t)), \quad x_2(t_0) = x_{20}, \tag{4.77}$$

where $x_1 \in \mathcal{D}$, $x_2 \in \mathbb{R}^{n_2}$, $f_1 : [t_0, t_1) \times \mathcal{D} \times \mathbb{R}^{n_2} \to \mathbb{R}^{n_1}$ is such that, for every $t \in [t_0, t_1)$ and $x_2 \in \mathbb{R}^{n_2}$, $f_1(t, 0, x_2) = 0$ and $f_1(t, \cdot, x_2)$ is locally Lipschitz in x_1, and $f_2 : [t_0, t_1) \times \mathcal{D} \times \mathbb{R}^{n_2} \to \mathbb{R}^{n_2}$ is such that, for every $x_1 \in \mathcal{D}$, $f_2(\cdot, x_1, \cdot)$ is locally Lipschitz in x_2. Next, let $\hat{x}_1(t-t_0) = x_1(t)$, $\hat{x}_2(t-t_0) = [x_2^{\mathrm{T}}(t)\ t]^{\mathrm{T}}$, $\hat{f}_1(\hat{x}_1, \hat{x}_2) = f_1(t, x_1, x_2)$, and $\hat{f}_2(\hat{x}_1, \hat{x}_2) = [f_2^{\mathrm{T}}(t, x_1, x_2)\ 1]^{\mathrm{T}}$. Now, note that with $\tau = t - t_0$, the solution $(x_1(t), x_2(t))$, $t \ge t_0$, to the nonlinear time-varying dynamical system (4.76) and (4.77) is equivalently characterized by the solution $(\hat{x}_1(\tau), \hat{x}_2(\tau))$, $\tau \ge 0$, to the nonlinear autonomous dynamical system

$$\dot{\hat{x}}_1(\tau) = \hat{f}_1(\hat{x}_1(\tau), \hat{x}_2(\tau)), \quad \hat{x}_1(0) = x_{10}, \quad \tau \ge 0, \tag{4.78}$$

$$\dot{\hat{x}}_2(\tau) = \hat{f}_2(\hat{x}_1(\tau), \hat{x}_2(\tau)), \quad \hat{x}_2(0) = [x_{20}^{\mathrm{T}}\ t_0]^{\mathrm{T}}, \tag{4.79}$$

where $\dot{x}_1(\cdot)$ and $\dot{x}_2(\cdot)$ denote differentiation with respect to τ. Hence, Theorem 4.1 can be used to derive sufficient conditions for partial stability results for the nonlinear time-varying dynamical systems of the form (4.76) and (4.77). Of course, in this case it is important to note that partial stability may be uniform with respect to either or both of x_{20} and t_0.

Using Theorems 4.1 and 4.6 we present some insight on the complexity of analyzing stability of linear, time-varying systems. Specifically, let $f(t,x) = A(t)x$, where $A : [0,\infty) \to \mathbb{R}^{n\times n}$ is continuous, so that (4.57) becomes

$$\dot{x}(t) = A(t)x(t), \quad x(t_0) = x_0, \quad t \geq t_0. \tag{4.80}$$

Now, in order to analyze the stability of (4.80), let $n_1 = n$, $n_2 = 1$, $x_1(t - t_0) = x(t)$, $x_2(t - t_0) = t$, $f_1(x_1, x_2) = A(t)x$, and $f_2(x_1, x_2) = 1$. Hence, the solution $x(t)$, $t \geq t_0$, to (4.80) can be equivalently characterized by the solution $x_1(\tau)$, $\tau \geq 0$, to the *nonlinear* autonomous dynamical system

$$\dot{x}_1(\tau) = A(x_2(\tau))x_1(\tau), \quad x_1(0) = x_0, \quad \tau \geq 0, \tag{4.81}$$
$$\dot{x}_2(\tau) = 1, \quad x_2(0) = t_0, \tag{4.82}$$

where $\dot{x}_1(\cdot)$ and $\dot{x}_2(\cdot)$ denote differentiation with respect to τ. It is clear from (4.81) and (4.82) that in spite of the fact that (4.80) is a linear system, its solutions are inherently characterized by a nonlinear system of the form (4.81) and (4.82).

Example 4.6. Consider the linear time-varying dynamical system

$$\dot{x}_1(t) = -x_1(t) - e^{-t}x_2(t), \quad x_1(0) = x_{10}, \quad t \geq 0, \tag{4.83}$$
$$\dot{x}_2(t) = x_1(t) - x_2(t), \quad x_2(0) = x_{20}. \tag{4.84}$$

To examine stability of this system, consider the Lyapunov function candidate $V(t,x) = x_1^2 + (1 + e^{-t})x_2^2$. Note that since

$$x_1^2 + x_2^2 \leq V(t,x) \leq x_1^2 + 2x_2^2, \quad (x_1, x_2) \in \mathbb{R} \times \mathbb{R}, \quad t \geq 0, \tag{4.85}$$

it follows that $V(t,x)$ is positive definite, radially unbounded and satisfies (4.63) and (4.65). Next, since

$$\begin{aligned} \dot{V}(t,x) &= -2x_1^2 + 2x_1x_2 - 2x_2^2 - 3e^{-t}x_2^2 \\ &\leq -2x_1^2 + 2x_1x_2 - 2x_2^2 \\ &= -x^{\mathrm{T}}Rx \\ &\leq -\lambda_{\min}(R)\|x\|_2^2, \end{aligned} \tag{4.86}$$

where

$$R = \begin{bmatrix} 2 & -1 \\ -1 & 2 \end{bmatrix} > 0$$

and $x = [x_1 \;\; x_2]^{\mathrm{T}}$, it follows from *vi*) and *vii*) of Theorem 4.6 with $p = 2$ that the zero solution $(x_1(t), x_2(t)) \equiv (0,0)$ to (4.83) and (4.84) is globally exponentially stable. △

Example 4.7. It is important to note that it is not always necessary to construct a time-varying Lyapunov function to show stability for a nonlinear time-varying dynamical system. In particular, consider the nonlinear time-varying dynamical system

$$\dot{x}_1(t) = -x_1^3(t) + (\sin\omega t)x_2(t), \quad x_1(0) = x_{10}, \quad t \geq 0, \tag{4.87}$$
$$\dot{x}_2(t) = -(\sin\omega t)x_1(t) - x_2^3(t), \quad x_2(0) = x_{20}. \tag{4.88}$$

To show that the origin is globally uniformly asymptotically stable, consider the *time-invariant* Lyapunov function candidate $V(x) = \frac{1}{2}(x_1^2+x_2^2)$. Clearly, $V(x)$, $x \in \mathbb{R}^2$, is positive definite and radially unbounded. Furthermore,

$$\begin{aligned}\dot{V}(x) &= x_1[-x_1^3 + (\sin\omega t)x_2] + x_2[-(\sin\omega t)x_1 - x_2^3] \\ &= -x_1^4 - x_2^4 \\ &< 0, \quad (x_1, x_2) \in \mathbb{R}\times\mathbb{R}, \quad (x_1,x_2) \neq (0,0),\end{aligned} \tag{4.89}$$

which shows that the zero solution $(x_1(t), x_2(t)) \equiv (0,0)$ to (4.87) and (4.88) is globally uniformly asymptotically stable. △

Example 4.8. Consider the linear time-varying dynamical system

$$\dot{x}(t) = A(t)x(t), \quad x(t_0) = x_0, \quad t \geq t_0, \tag{4.90}$$

where $A : [0,\infty) \to \mathbb{R}^{n\times n}$ is continuous. To examine the stability of this system, consider the quadratic Lyapunov function candidate $V(t,x) = x^{\mathrm{T}}P(t)x$, where $P : [0,\infty) \to \mathbb{R}^{n\times n}$ is a continuously differentiable, uniformly bounded, and positive-definite matrix function (that is, $P(t)$ is positive definite for every $t \geq t_0$) satisfying

$$-\dot{P}(t) = A^{\mathrm{T}}(t)P(t) + P(t)A(t) + R(t), \tag{4.91}$$

where $R(\cdot)$ is continuous such that $R(t) \geq \gamma I_n > 0$, $\gamma > 0$, for all $t \geq t_0$. Now, since $P(\cdot)$ is continuously differentiable, uniformly bounded, and positive definite it follows that there exist $\alpha, \beta > 0$ such that

$$0 < \alpha I_n \leq P(t) \leq \beta I_n, \quad t \geq t_0, \tag{4.92}$$

and hence,

$$\alpha\|x\|_2^2 \leq V(t,x) \leq \beta\|x\|_2^2. \tag{4.93}$$

Thus, $V(t,x)$ is positive definite and radially unbounded. Next, since

$$\begin{aligned}\dot{V}(t,x) &= x^{\mathrm{T}}\dot{P}(t)x + 2x^{\mathrm{T}}P(t)A(t)x \\ &= x^{\mathrm{T}}[\dot{P}(t) + A^{\mathrm{T}}(t)P(t) + P(t)A(t)]x \\ &= -x^{\mathrm{T}}R(t)x \\ &\leq -\gamma\|x\|_2^2,\end{aligned} \tag{4.94}$$

it follows from *vi*) and *vii*) of Theorem 4.6 with $p = 2$ that the zero solution $x(t) \equiv 0$ to (4.90) is globally exponentially stable. Hence, a sufficient condition for global exponential stability for a linear time-varying system is the existence of a continuously differentiable, bounded, and positive-definite matrix function $P : [0, \infty) \to \mathbb{R}^{n\times n}$ satisfying (4.91). $\triangle$

In Chapter 3 it was shown that for time-invariant dynamical systems the invariance principle can be used to relax the strict negative-definiteness condition on the Lyapunov derivative while still ensuring asymptotic stability of the origin by showing that the set $\mathcal{R}$ where the Lyapunov derivative vanishes contains no invariant set other than the origin. For time-varying systems, there does not exist an analogous result for establishing uniform asymptotic stability since, in general, positive limit sets for time-varying systems are not invariant. However, the following theorem provides a similar, albeit weaker, result for time-varying systems.

Theorem 4.7 (LaSalle-Yoshizawa Theorem). Consider the time-varying dynamical system (4.57) and assume $[0, \infty) \times \mathcal{D}$ is a positively invariant set with respect to (4.57) where $f(t, \cdot)$ is Lipschitz in x, uniformly in t. Furthermore, assume there exist functions $V : [0, \infty) \times \mathcal{D} \to \mathbb{R}$ and $W, W_1, W_2 : \mathcal{D} \to \mathbb{R}$ such that $V(\cdot, \cdot)$ is continuously differentiable, $W_1(\cdot)$ and $W_2(\cdot)$ are continuous and positive definite, $W(\cdot)$ is continuous and nonnegative definite, and, for all $(t, x) \in [0, \infty) \times \mathcal{D}$,

$$W_1(x) \le V(t, x) \le W_2(x), \tag{4.95}$$

$$\dot{V}(t, x) \le -W(x). \tag{4.96}$$

Then there exists $\mathcal{D}_0 \subseteq \mathcal{D}$ such that for all $(t_0, x_0) \in [0, \infty) \times \mathcal{D}_0$, $x(t) \to \mathcal{R} \triangleq \{x \in \mathcal{D} : W(x) = 0\}$ as $t \to \infty$. If, in addition, $\mathcal{D} = \mathbb{R}^n$ and $W_1(\cdot)$ is radially unbounded, then for all $(t_0, x_0) \in [0, \infty) \times \mathbb{R}^n$, $x(t) \to \mathcal{R} \triangleq \{x \in \mathbb{R}^n : W(x) = 0\}$ as $t \to \infty$.

Proof. The proof is a direct consequence of Theorem 4.2 with $n_1 = n$, $n_2 = 1$, $x_1(t - t_0) = x(t)$, $x_2(t - t_0) = t$, $f_1(x_1, x_2) = f(t, x)$, $f_2(x_1, x_2) = 1$, and $V(x_1, x_2) = V(t, x)$. $\square$

If W is a homeomorphism and $W^{-1}(0) = \{0\}$, then Theorem 4.7 establishes attraction to the origin. Alternatively, if, in place of (4.96), $\dot{V}(t, x) \le 0$, $(t, x) \in [0, \infty) \times \mathcal{D}$, and the integral of $\dot{V}(t, x)$ satisfies the inequality

$$\int_t^{t+\varepsilon} \dot{V}(\tau, s(\tau, t, x))\mathrm{d}\tau \le -\alpha V(t, x), \tag{4.97}$$

where $\alpha \in (0, 1)$, for all $t \ge 0$ and $x \in \mathcal{D}$, and some $\varepsilon > 0$, then uniform asymptotic stability of the zero solution $x(t) \equiv 0$ to (4.57) can be established. This result is left as an exercise for the reader.

Next, we provide two converse theorems for stability of time-varying systems.

Theorem 4.8. Assume that the nonlinear time-varying dynamical system (4.57) is uniformly asymptotically stable and $f : [0,\infty) \times \mathcal{D} \to \mathbb{R}^n$ is continuously differentiable. Let $\delta > 0$ be such that $\mathcal{B}_\delta(0) \subset \mathcal{D}$ is contained in the domain of attraction of (4.57) and assume $\frac{\partial f}{\partial x}$ is bounded on $\mathcal{B}_\delta(0)$ uniformly in t. Then there exist a continuously differentiable function $V : [0,\infty) \times \mathcal{B}_\delta(0) \to \mathbb{R}$ and class $\mathcal{K}$ functions $\alpha(\cdot)$, $\beta(\cdot)$, and $\gamma(\cdot)$ such that

$$\alpha(\|x\|) \le V(t,x) \le \beta(\|x\|), \quad (t,x) \in [0,\infty) \times \mathcal{B}_\delta(0), \tag{4.98}$$
$$\dot{V}(t,x) \le -\gamma(\|x\|), \quad (t,x) \in [0,\infty) \times \mathcal{B}_\delta(0). \tag{4.99}$$

Proof. The proof is a direct consequence of Theorem 4.3 with $n_1 = n$, $n_2 = 1$, $x_1(t-t_0) = x(t)$, $x_2(t-t_0) = t$, $f_1(x_1,x_2) = f(t,x)$, $f_2(x_1,x_2) = 1$, and $V(x_1,x_2) = V(t,x)$. □

Theorem 4.9. Assume that the nonlinear time-varying dynamical system (4.57) is (uniformly) exponentially stable and $f : [0,\infty) \times \mathcal{D} \to \mathbb{R}^n$ is continuously differentiable. Let $\delta > 0$ be such that $\mathcal{B}_\delta(0) \subset \mathcal{D}$ is contained in the domain of attraction of (4.57) and assume $\frac{\partial f}{\partial x}$ is bounded on $\mathcal{B}_\delta(0)$ uniformly in t. Then, for every $p > 1$, there exist a continuously differentiable function $V : [0,\infty) \times \mathcal{B}_\delta(0) \to \mathbb{R}$ and positive constants α, β, and γ such that

$$\alpha\|x\|^p \le V(t,x) \le \beta\|x\|^p, \quad (t,x) \in [0,\infty) \times \mathcal{B}_\delta(0), \tag{4.100}$$
$$\dot{V}(t,x) \le -\gamma\|x\|^p, \quad (t,x) \in [0,\infty) \times \mathcal{B}_\delta(0). \tag{4.101}$$

Proof. The proof is a direct consequence of Theorem 4.4 with $n_1 = n$, $n_2 = 1$, $x_1(t-t_0) = x(t)$, $x_2(t-t_0) = t$, $f_1(x_1,x_2) = f(t,x)$, $f_2(x_1,x_2) = 1$, and $V(x_1,x_2) = V(t,x)$. □

Finally, we state Lyapunov's and Chetaev's instability theorems for time-varying systems. The proofs of these results are virtually identical to the proofs of their time-invariant counterparts given in Section 3.6, and hence, are left as an exercise for the reader.

Theorem 4.10. Consider the nonlinear dynamical system (4.57). Assume that there exist a continuously differentiable function $V : [0,\infty) \times \mathcal{D} \to \mathbb{R}$ and class $\mathcal{K}$ functions $\beta(\cdot)$ and $\gamma(\cdot)$ such that

$$V(t,0) = 0, \qquad t \in [0,\infty), \tag{4.102}$$
$$V(t,x) \le \beta(\|x\|), \qquad (t,x) \in [0,\infty) \times \mathcal{D}, \tag{4.103}$$

$$\dot{V}(t,x) \geq \gamma(\|x\|), \qquad (t,x) \in [0,\infty) \times \mathcal{D}. \tag{4.104}$$

Furthermore, assume that for every sufficiently small $\delta > 0$ there exist a time $t_0 \geq 0$ and $x_0 \in \mathcal{D}$ such that $\|x_0\| < \delta$ and $V(t_0, x_0) > 0$. Then the zero solution $x(t) \equiv 0$ to (4.57) is unstable.

Theorem 4.11. Consider the nonlinear dynamical system (4.57). Assume that there exist a continuously differentiable function $V : [0,\infty) \times \mathcal{D} \to \mathbb{R}$, a class $\mathcal{K}$ function $\beta(\cdot)$, a function $W : [0,\infty) \times \mathcal{D} \to \mathbb{R}$, a time $t_0 \geq 0$, and scalars $\varepsilon, \lambda > 0$ such that

$$V(t_0, 0) = 0, \tag{4.105}$$
$$V(t,x) \leq \beta(\|x\|), \qquad (t,x) \in [0,\infty) \times \mathcal{D}, \tag{4.106}$$
$$W(t,x) \geq 0, \qquad (t,x) \in [0,\infty) \times \mathcal{B}_\varepsilon(0), \tag{4.107}$$
$$\dot{V}(t,x) \geq \lambda V(t,x) + W(t,x). \tag{4.108}$$

Furthermore, assume that for every sufficiently small $\delta > 0$ there exists $x_0 \in \mathcal{D}$ such that $\|x_0\| < \delta$ and $V(t_0, x_0) > 0$. Then the zero solution $x(t) \equiv 0$ to (4.57) is unstable.

Theorem 4.12. Consider the nonlinear dynamical system (4.57). Assume that there exist a continuously differentiable function $V : [0,\infty) \times \mathcal{D} \to \mathbb{R}$, a class $\mathcal{K}$ function $\gamma(\cdot)$, a scalar $\varepsilon > 0$, and an open set $\mathcal{Q} \subseteq \mathcal{B}_\varepsilon(0)$ such that

$$V(t,x) > 0, \qquad (t,x) \in [0,\infty) \times \mathcal{Q} \tag{4.109}$$
$$\sup_{t\geq 0} \sup_{x\in\mathcal{Q}} V(t,x) < \infty, \tag{4.110}$$
$$0 \in \partial\mathcal{Q}, \tag{4.111}$$
$$V(t,x) = 0, \quad (t,x) \in [0,\infty) \times (\partial\mathcal{Q} \cap \mathcal{B}_\varepsilon(0)), \tag{4.112}$$
$$\dot{V}(t,x) \geq \gamma(\|x\|), \quad (t,x) \in [0,\infty) \times \mathcal{Q}. \tag{4.113}$$

Then the zero solution $x(t) \equiv 0$ to (4.57) is unstable.

4.4 Lagrange Stability, Boundedness, and Ultimate Boundedness

In the previous sections we introduced the concepts of stability and partial stability for nonlinear dynamical systems. In certain engineering applications, however, it is more natural to ascertain whether for every system initial condition in a ball of radius δ the solution of the nonlinear dynamical system is bounded. This leads to the notions of *Lagrange stability, boundedness*, and *ultimate boundedness*. These notions are closely related to what is known in the literature as *practical stability*. In this section, we present Lyapunov-like theorems for boundedness and ultimate boundedness of nonlinear dynamical systems.

Definition 4.3. *i*) The nonlinear dynamical system (4.6) and (4.7) is *Lagrange stable with respect to* x_1 if, for every $x_{10} \in \mathcal{D}$ and $x_{20} \in \mathbb{R}^{n_2}$, there exists $\varepsilon = \varepsilon(x_{10}, x_{20}) > 0$ such that $\|x_1(t)\| < \varepsilon$, $t \geq 0$.

ii) The nonlinear dynamical system (4.6) and (4.7) is *bounded with respect to* x_1 *uniformly in* x_2 if there exists $\gamma > 0$ such that, for every $\delta \in (0, \gamma)$, there exists $\varepsilon = \varepsilon(\delta) > 0$ such that $\|x_{10}\| < \delta$ implies $\|x_1(t)\| < \varepsilon$, $t \geq 0$. The nonlinear dynamical system (4.6) and (4.7) is *globally bounded with respect to* x_1 *uniformly in* x_2 if, for every $\delta \in (0, \infty)$, there exists $\varepsilon = \varepsilon(\delta) > 0$ such that $\|x_{10}\| < \delta$ implies $\|x_1(t)\| < \varepsilon$, $t \geq 0$.

iii) The nonlinear dynamical system (4.6) and (4.7) is *ultimately bounded with respect to* x_1 *uniformly in* x_2 *with bound* ε if there exists $\gamma > 0$ such that, for every $\delta \in (0, \gamma)$, there exists $T = T(\delta, \varepsilon) > 0$ such that $\|x_{10}\| < \delta$ implies $\|x_1(t)\| < \varepsilon$, $t \geq T$. The nonlinear dynamical system (4.6) and (4.7) is *globally ultimately bounded with respect to* x_1 *uniformly in* x_2 *with bound* ε if, for every $\delta \in (0, \infty)$, there exists $T = T(\delta, \varepsilon) > 0$ such that $\|x_{10}\| < \delta$ implies $\|x_1(t)\| < \varepsilon$, $t \geq T$.

Note that if a nonlinear dynamical system is globally bounded with respect to x_1 uniformly in x_2, then it is Lagrange stable with respect to x_1. Alternatively, if a nonlinear dynamical system is (globally) bounded with respect to x_1 uniformly in x_2, then there exists $\varepsilon > 0$ such that it is (globally) ultimately bounded with respect to x_1 uniformly in x_2 with a bound ε. Conversely, if a nonlinear dynamical system is (globally) ultimately bounded with respect to x_1 uniformly in x_2 with a bound ε, then it is (globally) bounded with respect to x_1 uniformly in x_2. The following results present Lyapunov-like theorems for boundedness and ultimate boundedness. For these results define $\dot{V}(x_1, x_2) \triangleq V'(x_1, x_2) f(x_1, x_2)$, where $f(x_1, x_2) \triangleq [f_1^{\mathrm{T}}(x_1, x_2),\ f_2^{\mathrm{T}}(x_1, x_2)]^{\mathrm{T}}$ and $V : \mathcal{D} \times \mathbb{R}^{n_2} \to \mathbb{R}$ is a given continuously differentiable function.

Theorem 4.13. Consider the nonlinear dynamical system (4.6) and (4.7). Assume that there exist a continuously differentiable function $V : \mathcal{D} \times \mathbb{R}^{n_2} \to \mathbb{R}$ and class $\mathcal{K}$ functions $\alpha(\cdot)$, $\beta(\cdot)$ such that

$$\alpha(\|x_1\|) \leq V(x_1, x_2) \leq \beta(\|x_1\|), \quad x_1 \in \mathcal{D}, \quad x_2 \in \mathbb{R}^{n_2}, \tag{4.114}$$

$$\dot{V}(x_1, x_2) \leq 0, \quad x_1 \in \mathcal{D}, \quad \|x_1\| \geq \mu, \quad x_2 \in \mathbb{R}^{n_2}, \tag{4.115}$$

where $\mu > 0$ is such that $\mathcal{B}_{\alpha^{-1}(\eta)}(0) \subset \mathcal{D}$ for some $\eta \geq \beta(\mu)$. Then the nonlinear dynamical system (4.6) and (4.7) is bounded with respect to x_1 uniformly in x_2. Furthermore, for every $\delta \in (0, \gamma)$, $x_{10} \in \overline{\mathcal{B}}_\delta(0)$ implies that $\|x_1(t)\| \leq \varepsilon$, $t \geq 0$, where

$$\varepsilon = \varepsilon(\delta) \triangleq \begin{cases} \alpha^{-1}(\beta(\delta)), & \delta \in (\mu, \gamma), \\ \alpha^{-1}(\eta), & \delta \in (0, \mu], \end{cases} \tag{4.116}$$

and $\gamma \triangleq \sup\{r > 0 : \mathcal{B}_{\alpha^{-1}(\beta(r))}(0) \subset \mathcal{D}\}$. If, in addition, $\mathcal{D} = \mathbb{R}^{n_1}$ and $\alpha(\cdot)$ is a class $\mathcal{K}_\infty$ function, then the nonlinear dynamical system (4.6) and (4.7) is globally bounded with respect to x_1 uniformly in x_2 and for every $x_{10} \in \mathbb{R}^{n_1}$, $\|x_1(t)\| \leq \varepsilon$, $t \geq 0$, where ε is given by (4.116) with $\delta = \|x_{10}\|$.

Proof. First, let $\delta \in (0, \mu]$ and assume $\|x_{10}\| \leq \delta$. If $\|x_1(t)\| \leq \mu$, $t \geq 0$, then it follows from (4.114) that $\|x_1(t)\| \leq \mu \leq \alpha^{-1}(\beta(\mu)) \leq \alpha^{-1}(\eta)$, $t \geq 0$. Alternatively, if there exists $T > 0$ such that $\|x_1(T)\| > \mu$, then it follows from the continuity of $x_1(\cdot)$ that there exists $\tau < T$ such that $\|x_1(\tau)\| = \mu$ and $\|x_1(t)\| \geq \mu$, $t \in [\tau, T]$. Hence, it follows from (4.114) and (4.115) that

$$\alpha(\|x_1(T)\|) \leq V(x_1(T), x_2(T)) \leq V(x_1(\tau), x_2(\tau)) \leq \beta(\mu) \leq \eta,$$

which implies that $\|x_1(T)\| \leq \alpha^{-1}(\eta)$. Next, let $\delta \in (\mu, \gamma)$ and assume $x_{10} \in \overline{\mathcal{B}}_\delta(0)$ and $\|x_{10}\| > \mu$. Now, for every $\hat{t} > 0$ such that $\|x_1(t)\| \geq \mu$, $t \in [0, \hat{t}]$, it follows from (4.114) and (4.115) that

$$\alpha(\|x_1(t)\|) \leq V(x_1(t), x_2(t)) \leq V(x_{10}, x_{20}) \leq \beta(\delta), \qquad t \geq 0,$$

which implies that $\|x_1(t)\| \leq \alpha^{-1}(\beta(\delta))$, $t \in [0, \hat{t}]$. Next, if there exists $T > 0$ such that $\|x_1(T)\| \leq \mu$, then it follows as in the proof of the first case that $\|x_1(t)\| \leq \alpha^{-1}(\eta)$, $t \geq T$. Hence, if $x_{10} \in \mathcal{B}_\delta(0) \setminus \mathcal{B}_\mu(0)$, then $\|x_1(t)\| \leq \alpha^{-1}(\beta(\delta))$, $t \geq 0$. Finally, if $\mathcal{D} = \mathbb{R}^{n_1}$ and $\alpha(\cdot)$ is a class $\mathcal{K}_\infty$ function it follows that $\beta(\cdot)$ is a class $\mathcal{K}_\infty$ function, and hence, $\gamma = \infty$. Hence, the nonlinear dynamical system (4.6) and (4.7) is globally bounded with respect to x_1 uniformly in x_2. □

Theorem 4.14. Consider the nonlinear dynamical system (4.6) and (4.7). Assume that there exist a continuously differentiable function $V : \mathcal{D} \times \mathbb{R}^{n_2} \to \mathbb{R}$ and class $\mathcal{K}$ functions $\alpha(\cdot)$, $\beta(\cdot)$ such that (4.114) holds. Furthermore, assume that there exists a continuous function $W : \mathcal{D} \to \mathbb{R}$ such that $W(x_1) > 0$, $\|x_1\| > \mu$, and

$$\dot{V}(x_1, x_2) \leq -W(x_1), \quad x_1 \in \mathcal{D}, \quad \|x_1\| > \mu, \quad x_2 \in \mathbb{R}^{n_2}, \tag{4.117}$$

where $\mu > 0$ is such that $\mathcal{B}_{\alpha^{-1}(\eta)}(0) \subset \mathcal{D}$ for some $\eta > \beta(\mu)$. Then the nonlinear dynamical system (4.6) and (4.7) is ultimately bounded with respect to x_1 uniformly in x_2 with bound $\varepsilon \triangleq \alpha^{-1}(\eta)$. Furthermore, $\limsup_{t\to\infty} \|x_1(t)\| \leq \alpha^{-1}(\beta(\mu))$. If, in addition, $\mathcal{D} = \mathbb{R}^{n_1}$ and $\alpha(\cdot)$ is a class $\mathcal{K}_\infty$ function, then the nonlinear dynamical system (4.6) and (4.7) is globally ultimately bounded with respect to x_1 uniformly in x_2 with bound ε.

Proof. First, let $\delta \in (0, \mu]$ and assume $\|x_{10}\| \leq \delta$. As in the proof of Theorem 4.13, it follows that $\|x_1(t)\| \leq \alpha^{-1}(\eta) = \varepsilon$, $t \geq 0$. Next, let $\delta \in (\mu, \gamma)$, where $\gamma \triangleq \sup\{r > 0 : \mathcal{B}_{\alpha^{-1}(\beta(r))}(0) \subset \mathcal{D}\}$ and assume $x_{10} \in \mathcal{B}_\delta(0)$

and $\|x_{10}\| > \mu$. In this case, it follows from Theorem 4.13 that $\|x_1(t)\| \le \alpha^{-1}(\beta(\delta))$, $t \ge 0$. Suppose, *ad absurdum*, that $\|x_1(t)\| \ge \beta^{-1}(\eta)$, $t \ge 0$, or, equivalently, $x_1(t) \in \mathcal{O} \triangleq \mathcal{B}_{\alpha^{-1}(\beta(\delta))}(0) \setminus \mathcal{B}_{\beta^{-1}(\eta)}(0)$, $t \ge 0$. Since $\overline{\mathcal{O}}$ is compact and $W(\cdot)$ is continuous and $W(x_1) > 0$, $\|x_1\| \ge \beta^{-1}(\eta) > \mu$, it follows from Theorem 2.13 that $k \triangleq \min_{x_1 \in \overline{\mathcal{O}}} W(x_1) > 0$ exists. Hence, it follows from (4.117) that

$$V(x_1(t), x_2(t)) \le V(x_{10}, x_{20}) - kt, \qquad t \ge 0, \tag{4.118}$$

which implies that

$$\alpha(\|x_1(t)\|) \le \beta(\|x_{10}\|) - kt \le \beta(\delta) - kt, \qquad t \ge 0. \tag{4.119}$$

Now, letting $t > \beta(\delta)/k$ it follows that $\alpha(\|x_1(t)\|) < 0$, which is a contradiction. Hence, there exists $T = T(\delta, \eta) > 0$ such that $\|x_1(T)\| < \beta^{-1}(\eta)$. Thus, it follows from Theorem 4.13 that $\|x_1(t)\| \le \alpha^{-1}(\beta(\beta^{-1}(\eta))) = \alpha^{-1}(\eta)$, $t \ge T$, which proves that the nonlinear dynamical system (4.6) and (4.7) is ultimately bounded with respect to x_1 uniformly in x_2 with bound $\varepsilon = \alpha^{-1}(\eta)$. Furthermore, $\limsup_{t\to\infty} \|x_1(t)\| \le \alpha^{-1}(\beta(\mu))$. Finally, if $\mathcal{D} = \mathbb{R}^{n_1}$ and $\alpha(\cdot)$ is a class $\mathcal{K}_\infty$ function it follows that $\beta(\cdot)$ is a class $\mathcal{K}_\infty$ function, and hence, $\gamma = \infty$. Hence, the nonlinear dynamical system (4.6) and (4.7) is globally ultimately bounded with respect to x_1 uniformly in x_2 with bound ε. □

Next, we specialize Theorems 4.13 and 4.14 to nonlinear time-varying dynamical systems. The following definition is needed for these results.

Definition 4.4. *i*) The nonlinear time-varying dynamical system (4.57) is *Lagrange stable* if, for every $x_0 \in \mathbb{R}^n$ and $t_0 \in \mathbb{R}$, there exists $\varepsilon = \varepsilon(t_0, x_0) > 0$ such that $\|x(t)\| < \varepsilon$, $t \ge t_0$.

ii) The nonlinear time-varying dynamical system (4.57) is *uniformly bounded* if there exists $\gamma > 0$ such that, for every $\delta \in (0, \gamma)$, there exists $\varepsilon = \varepsilon(\delta) > 0$ such that $\|x_0\| < \delta$ implies $\|x(t)\| < \varepsilon$, $t \ge t_0$. The nonlinear time-varying dynamical system (4.57) is *globally uniformly bounded* if, for every $\delta \in (0, \infty)$, there exists $\varepsilon = \varepsilon(\delta) > 0$ such that $\|x_0\| < \delta$ implies $\|x(t)\| < \varepsilon$, $t \ge t_0$.

iii) The nonlinear time-varying dynamical system (4.57) is *uniformly ultimately bounded with bound* ε if there exists $\gamma > 0$ such that, for every $\delta \in (0, \gamma)$, there exists $T = T(\delta, \varepsilon) > 0$ such that $\|x_0\| < \delta$ implies $\|x(t)\| < \varepsilon$, $t \ge t_0 + T$. The nonlinear time-varying dynamical system (4.57) is *globally uniformly ultimately bounded with bound* ε if, for every $\delta \in (0, \infty)$, there exists $T = T(\delta, \varepsilon) > 0$ such that $\|x_0\| < \delta$ implies $\|x(t)\| < \varepsilon$, $t \ge t_0 + T$.

For the following result define

$$\dot{V}(t,x) \triangleq \frac{\partial V}{\partial t} + \frac{\partial V}{\partial x}(t,x)f(t,x),$$

where $V : \mathbb{R} \times \mathcal{D} \to \mathbb{R}$ is a given continuously differentiable function.

Corollary 4.1. Consider the nonlinear time-varying dynamical system (4.57). Assume that there exist a continuously differentiable function $V : \mathbb{R} \times \mathcal{D} \to \mathbb{R}$ and class $\mathcal{K}$ functions $\alpha(\cdot)$, $\beta(\cdot)$ such that

$$\alpha(\|x\|) \le V(t,x) \le \beta(\|x\|), \quad x \in \mathcal{D}, \quad t \in \mathbb{R}, \tag{4.120}$$

$$\dot{V}(t,x) \le 0, \quad x \in \mathcal{D}, \quad \|x\| \ge \mu, \quad t \in \mathbb{R}, \tag{4.121}$$

where $\mu > 0$ is such that $\mathcal{B}_{\alpha^{-1}(\eta)}(0) \subset \mathcal{D}$ for some $\eta \ge \beta(\mu)$. Then the nonlinear time-varying dynamical system (4.57) is uniformly bounded. If, in addition, $\mathcal{D} = \mathbb{R}^n$ and $\alpha(\cdot)$ is a class $\mathcal{K}_\infty$ function, then the nonlinear time-varying dynamical system (4.57) is globally uniformly bounded.

Proof. The result is a direct consequence of Theorem 4.13. Specifically, let $n_1 = n$, $n_2 = 1$, $x_1(t-t_0) = x(t)$, $x_2(t-t_0) = t$, $f_1(x_1, x_2) = f(t,x)$, and $f_2(x_1, x_2) = 1$. Now, note that with $\tau = t - t_0$, the solution $x(t)$, $t \ge t_0$, to the nonlinear time-varying dynamical system (4.57) is equivalently characterized by the solution $x_1(\tau)$, $\tau \ge 0$, to the nonlinear autonomous dynamical system

$$\dot{x}_1(\tau) = f_1(x_1(\tau), x_2(\tau)), \quad x_1(0) = x_0, \quad \tau \ge 0,$$

$$\dot{x}_2(\tau) = 1, \quad x_2(0) = t_0,$$

where $\dot{x}_1(\cdot)$ and $\dot{x}_2(\cdot)$ denote differentiation with respect to τ. Now, the result is a direct consequence of Theorem 4.13. □

Corollary 4.2. Consider the nonlinear time-varying dynamical system (4.57). Assume that there exist a continuously differentiable function $V : \mathbb{R} \times \mathcal{D} \to \mathbb{R}$ and class $\mathcal{K}$ functions $\alpha(\cdot)$, $\beta(\cdot)$ such that (4.120) holds. Furthermore, assume that there exists a continuous function $W : \mathcal{D} \to \mathbb{R}$ such that $W(x) > 0$, $\|x\| > \mu$, and

$$\dot{V}(t,x) \le -W(x), \quad x \in \mathcal{D}, \quad \|x\| > \mu, \quad t \in \mathbb{R}, \tag{4.122}$$

where $\mu > 0$ is such that $\mathcal{B}_{\alpha^{-1}(\eta)}(0) \subset \mathcal{D}$ for some $\eta > \beta(\mu)$. Then the nonlinear time-varying dynamical system (4.57) is uniformly ultimately bounded with bound $\varepsilon \triangleq \alpha^{-1}(\eta)$. Furthermore, $\limsup_{t\to\infty} \|x(t)\| \le \alpha^{-1}(\beta(\mu))$. If, in addition, $\mathcal{D} = \mathbb{R}^n$ and $\alpha(\cdot)$ is a class $\mathcal{K}_\infty$ function, then the nonlinear time-varying dynamical system (4.57) is globally uniformly ultimately bounded with bound ε.

Proof. The proof is a direct consequence of Theorem 4.14 using similar arguments as in the proof of Corollary 4.1 and, hence, is omitted. □

Finally, we specialize Corollaries 4.1 and 4.2 to nonlinear time-invariant dynamical systems. For these results we need the following specialization of Definition 4.4.

Definition 4.5. *i*) The nonlinear dynamical system (3.1) is *Lagrange stable* if, for every $x_0 \in \mathbb{R}^n$, there exists $\varepsilon = \varepsilon(x_0) > 0$ such that $\|x(t)\| < \varepsilon$, $t \geq 0$.

ii) The nonlinear dynamical system (3.1) is *bounded* if there exists $\gamma > 0$ such that, for every $\delta \in (0, \gamma)$, there exists $\varepsilon = \varepsilon(\delta) > 0$ such that $\|x_0\| < \delta$ implies $\|x(t)\| < \varepsilon$, $t \geq 0$. The nonlinear dynamical system (3.1) is *globally bounded* if, for every $\delta \in (0, \infty)$, there exists $\varepsilon = \varepsilon(\delta) > 0$ such that $\|x_0\| < \delta$ implies $\|x(t)\| < \varepsilon$, $t \geq 0$.

iii) The nonlinear dynamical system (3.1) is *ultimately bounded with bound* ε if there exists $\gamma > 0$ such that, for every $\delta \in (0, \gamma)$, there exists $T = T(\delta, \varepsilon) > 0$ such that $\|x_0\| < \delta$ implies $\|x(t)\| < \varepsilon$, $t \geq T$. The nonlinear dynamical system (3.1) is *globally ultimately bounded with bound* ε if, for every $\delta \in (0, \infty)$, there exists $T = T(\delta, \varepsilon) > 0$ such that $\|x_0\| < \delta$ implies $\|x(t)\| < \varepsilon$, $t \geq T$.

Corollary 4.3. Consider the nonlinear dynamical system (3.1). Assume that there exist a continuously differentiable function $V : \mathcal{D} \to \mathbb{R}$ and class $\mathcal{K}$ functions $\alpha(\cdot)$ and $\beta(\cdot)$ such that

$$\alpha(\|x\|) \leq V(x) \leq \beta(\|x\|), \quad x \in \mathcal{D}, \tag{4.123}$$

$$V'(x)f(x) \leq 0, \quad x \in \mathcal{D}, \quad \|x\| \geq \mu, \tag{4.124}$$

where $\mu > 0$ is such that $\mathcal{B}_{\alpha^{-1}(\eta)}(0) \subset \mathcal{D}$ for some $\eta \geq \beta(\mu)$. Then the nonlinear dynamical system (3.1) is bounded. If, in addition, $\mathcal{D} = \mathbb{R}^n$ and $V(x) \to \infty$ as $\|x\| \to \infty$, then the nonlinear dynamical system (3.1) is globally bounded.

Proof. The result is a direct consequence of Corollary 4.1. □

Corollary 4.4. Consider the nonlinear dynamical system (3.1). Assume that there exist a continuously differentiable function $V : \mathcal{D} \to \mathbb{R}$ and class $\mathcal{K}$ functions $\alpha(\cdot)$ and $\beta(\cdot)$ such that (4.123) holds and

$$V'(x)f(x) < 0, \quad x \in \mathcal{D}, \quad \|x\| > \mu, \tag{4.125}$$

where $\mu > 0$ is such that $\mathcal{B}_{\alpha^{-1}(\eta)}(0) \subset \mathcal{D}$ for some $\eta > \beta(\mu)$. Then the nonlinear dynamical system (3.1) is ultimately bounded with bound $\varepsilon \triangleq \alpha^{-1}(\eta)$. Furthermore, $\limsup_{t\to\infty} \|x(t)\| \leq \alpha^{-1}(\beta(\mu))$. If, in addition, $\mathcal{D} =$

$\mathbb{R}^n$ and $V(x) \to \infty$ as $\|x\| \to \infty$, then the nonlinear dynamical system (3.1) is globally ultimately bounded with bound ε.

Proof. The proof is a direct consequence of Corollary 4.2. □

Corollaries 4.3 and 4.4 present Lyapunov-like theorems for boundedness and ultimate boundedness of a nonlinear dynamical system. To further elucidate these results, consider the nonlinear dynamical system (3.1) and assume that there exist a positive-definite, radially unbounded, continuously differentiable function $V : \mathbb{R}^n \to \mathbb{R}$ and scalar $\mu > 0$ such that

$$\dot{V}(x) \leq 0, \quad x \in \mathbb{R}^n, \quad \|x\| \geq \mu, \quad t \geq 0. \tag{4.126}$$

Furthermore, let $\alpha(\cdot)$ and $\beta(\cdot)$ be class $\mathcal{K}_\infty$ functions such that

$$\alpha(\|x\|) \leq V(x) \leq \beta(\|x\|), \quad x \in \mathbb{R}^n. \tag{4.127}$$

In this case, it can be shown that the set $\mathcal{D}_\mu \triangleq \{x \in \mathbb{R}^n : V(x) \leq \beta(\mu)\}$ is positively invariant. To see this, suppose, *ad absurdum*, that there exists a trajectory $x(t)$, $t \geq 0$, such that $x(0) \in \mathcal{D}_\mu$ and $x(T) \notin \mathcal{D}_\mu$ for some $T > 0$. Now, note that if $x \in \mathcal{B}_\mu(0)$, then $V(x) \leq \beta(\|x\|) \leq \beta(\mu)$. Next, since $x(t)$, $t \geq 0$, is continuous it follows that there exists $\hat{t} > 0$ such that $V(x(\hat{t})) = \beta(\mu)$ and $x(t) \notin \mathcal{D}_\mu$, $t \in (\hat{t}, T]$, and hence, $\|x(t)\| > \mu$, $t \in (\hat{t}, T]$. Now, it follows from (4.126) that

$$\beta(\mu) < V(x(T)) = V(x(\hat{t})) + \int_{\hat{t}}^{T} \dot{V}(x(t))\mathrm{d}t \leq V(x(\hat{t}) = \beta(\mu),$$

which is a contradiction. Hence, if $x(0) \in \mathcal{D}_\mu$, then $x(t) \in \mathcal{D}_\mu$, $t \geq 0$. Similarly, for $\delta > \mu$, if $x(0) \in \mathcal{B}_\delta(0)$, then it can be shown that $x(t) \in \mathcal{D}_\delta \triangleq \{x \in \mathbb{R}^n : V(x) \leq \beta(\delta)\}$. Next, if $x \in \mathcal{D}_\delta$, then $\alpha(\|x\|) \leq V(x) \leq \beta(\delta)$, which implies that $x \in \mathcal{B}_\varepsilon(0)$, where $\varepsilon \triangleq \alpha^{-1}(\beta(\delta))$. Hence, if $x(0) \in \mathcal{B}_\delta(0)$, then $x(t) \in \mathcal{B}_\varepsilon(0)$, $t \geq 0$ (see Figure 4.6).

If (4.126) is replaced by

$$\dot{V}(x) < 0, \quad x \in \mathbb{R}^n, \quad \|x\| > \mu, \quad t \geq 0, \tag{4.128}$$

then, using identical arguments as above, it can be shown that if $x(0) \in \mathcal{B}_\delta(0)$, then $x(t) \in \mathcal{D}_\delta(0)$, $t \geq 0$. Furthermore, it can be shown that for every $\eta > \mu$, the trajectory starting in $\mathcal{B}_\delta(0)$ enters $\mathcal{D}_\eta \triangleq \{x \in \mathbb{R}^n : V(x) \leq \beta(\eta)\}$ in finite time. Hence, the trajectory either enters $\mathcal{D}_\mu$ in finite time or approaches $\mathcal{D}_\mu$ as $t \to \infty$, which, since $\mathcal{D}_\mu$ is positively invariant, implies that the trajectory ultimately enters $\mathcal{B}_\varepsilon(0)$, where $\varepsilon \triangleq \alpha^{-1}(\beta(\mu))$ (see Figure 4.7).

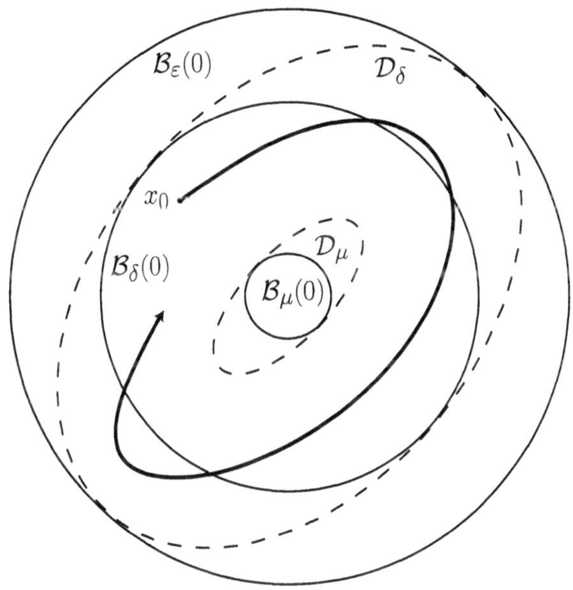

Figure 4.6 Visualizations of the sets $\mathcal{B}_\mu(0) \subset \mathcal{D}_\mu \subset \mathcal{B}_\delta(0) \subset \mathcal{D}_\delta \subset \mathcal{D}_\varepsilon$ and a bounded trajectory.

Example 4.9. Consider the nonlinear dynamical system

$$\dot{x}_1(t) = -x_1(t) + x_2^2(t), \quad x_1(0) = x_{10}, \quad t \geq 0, \tag{4.129}$$
$$\dot{x}_2(t) = -x_2(t) + u, \quad x_2(0) = x_{20}, \tag{4.130}$$

where $u \in \mathbb{R}$. To show that (4.129) and (4.130) is globally bounded consider the radially unbounded, positive definite function $V(x_1, x_2) = \frac{1}{2}x_1^2 + \frac{k}{4}x_2^4$, $k > 0$, and note that

$$\begin{aligned}
\dot{V}(x_1, x_2) &= -x_1(-x_1 + x_2^2) + kx_2^3(-x_2 + u) \\
&= -x_1^2 + x_1x_2^2 - k(1-\varepsilon)x_2^4 - k\varepsilon x_2^4 + kx_2^3 u, \quad \varepsilon \in (0,1), \\
&\leq -x_1^2 + x_1x_2^2 - k(1-\varepsilon)x_2^4, \quad |x_2| \geq \tfrac{|u|}{\varepsilon}, \\
&\leq -\begin{bmatrix} x_1 & x_2^2 \end{bmatrix} R_1 \begin{bmatrix} x_1 \\ x_2^2 \end{bmatrix}, \quad |x_2| \geq \tfrac{|u|}{\varepsilon},
\end{aligned} \tag{4.131}$$

where

$$R_1 = \begin{bmatrix} 1 & -1/2 \\ -1/2 & k(1-\varepsilon) \end{bmatrix}.$$

Now, choosing $k \geq 1/4(1-\varepsilon)$ ensures that $R_1 \geq 0$, and hence, $\dot{V}(x_1, x_2) \leq 0$, $|x_2| \geq \frac{|u|}{\varepsilon}$.

Next, for $|x_2| \leq \frac{|u|}{\varepsilon}$, it follows that

$$\dot{V}(x_1, x_2) \leq -x_1^2 + x_1x_2^2 - kx_2^4 + k|u|^4/\varepsilon^3$$

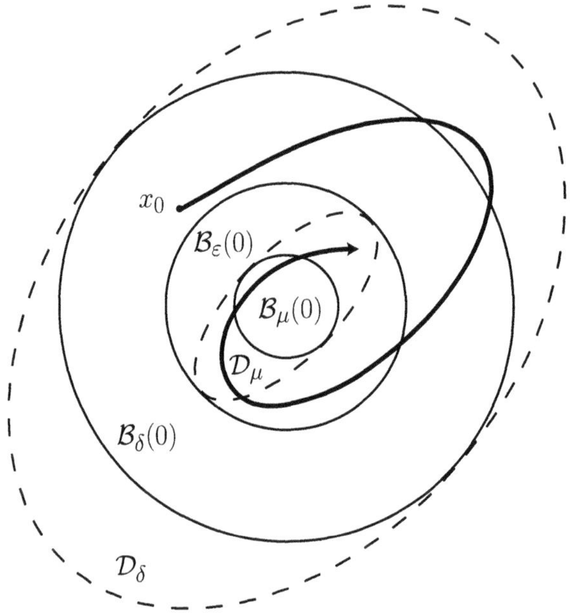

Figure 4.7 Visualizations of the sets $\mathcal{B}_\mu(0) \subset \mathcal{D}_\mu \subset \mathcal{B}_\varepsilon(0) \subset \mathcal{B}_\delta(0) \subset \mathcal{D}_\delta$ and an ultimate bounded trajectory.

$$
\begin{aligned}
&= -(1-\varepsilon)x_1^2 + x_1x_2^2 - kx_2^4 - \varepsilon x_1^2 + k|u|^4/\varepsilon^3 \\
&\le -\left[\ x_1 \quad x_2^2\ \right] R_2 \begin{bmatrix} x_1 \\ x_2^2 \end{bmatrix}, \quad |x_1| \ge k^{1/2}\tfrac{|u|^2}{\varepsilon^2},
\end{aligned} \tag{4.132}
$$

where

$$
R_2 = \begin{bmatrix} 1-\varepsilon & -1/2 \\ -1/2 & k \end{bmatrix}.
$$

Note that with $k \ge 1/4(1-\varepsilon)$, $R_2 \ge 0$, and hence, $\dot{V}(x_1, x_2) \le 0$, $|x_2| \le \frac{|u|}{\varepsilon}$. Next, let $\delta = \min\{\lambda_{\min}(R_1), \lambda_{\min}(R_2)\}$ and note that (4.131) and (4.132) imply $\dot{V}(x_1, x_2) \le -\delta(x_1^2 + x_2^4)$, $\|x\|_\infty \ge k^{1/2}|u|^2/\varepsilon^2$, and hence, it follows from Corollary 4.3 that (4.129) and (4.130) is globally bounded.

Next, to show that (4.129) and (4.130) is globally ultimately bounded, choose $k > 1/4(1-\varepsilon)$ so that $\delta > 0$. Now, with $W(x) = \delta(x_1^2 + x_2^4)$, $\alpha(\theta) = \min_{\|x\|_\infty=\theta} V(x)$, $\beta(\theta) = \max_{\|x\|_\infty=\theta} V(x)$, and $\eta > \beta(k^{1/2}|u|^2/\varepsilon^2)$, it follows from Corollary 4.4 that (4.129) and (4.130) is globally ultimately bounded with bound $\alpha^{-1}(\eta)$. Next, note that $\alpha(\theta) = \min\{\frac{1}{2}\theta^2, \frac{k}{2}\theta^4\}$ and $\beta(\theta) = \frac{1}{2}\theta^2 + \frac{k}{2}\theta^4$, which implies that

$$
\alpha^{-1}(\beta(k^{1/2}|u|^2/\varepsilon^2)) = \max\{\gamma^{1/2}(|u|), (\gamma(|u|)/k)^{1/4}\},
$$

where $\gamma(|u|) \triangleq \frac{k|u|^4}{\varepsilon^4} + \frac{k^3|u|^8}{\varepsilon^8}$. Now, it follows from Corollary 4.4 that $\limsup_{t\to\infty} \|x(t)\|_\infty \le \max\{\gamma^{1/2}(|u|), (\gamma(|u|)/k)^{1/4}\}$, which implies that, with $u = 0$, the zero solution $(x_1(t), x_2(t)) \equiv 0$ to (4.129) and (4.130) is globally attractive, that is, $\lim_{t\to\infty}(x_1(t), x_2(t)) = (0,0)$ for all $(x_{10}, x_{20}) \in \mathbb{R}^2$. △

4.5 Input-to-State Stability

In the previous sections we examined the stability and boundedness of undisturbed dynamical systems. In this section, we introduce the notion of *input-to-state stability* involving the stability of disturbed nonlinear dynamical systems. In particular, we consider nonlinear dynamical systems of the form

$$\dot{x}(t) = F(x(t), u(t)), \quad x(0) = x_0, \quad t \geq 0, \tag{4.133}$$

where $x(t) \in \mathbb{R}^n$, $t \geq 0$, $u(t) \in \mathbb{R}^m$, $t \geq 0$, and $F : \mathbb{R}^n \times \mathbb{R}^m \to \mathbb{R}^n$ is Lipschitz continuous on $\mathbb{R}^n \times \mathbb{R}^m$. The input u is assumed to be a piecewise continuous function of time with values in $\mathbb{R}^m$, that is, $u : [0, \infty) \to \mathbb{R}^m$. Now, suppose that (4.133) with $u(t) \equiv 0$ is globally asymptotically stable, then we are interested in whether a bounded input $u(t)$, $t \geq 0$, implies that the state $x(t)$, $t \geq 0$, is bounded. This is precisely the notion of input-to-state stability.

For the linear dynamical system

$$\dot{x}(t) = Ax(t) + Bu(t), \quad x(0) = x_0, \quad t \geq 0, \tag{4.134}$$

where A is Hurwitz, it follows that

$$x(t) = e^{At}x(0) + \int_0^t e^{A(t-\tau)}Bu(\tau)\mathrm{d}\tau. \tag{4.135}$$

Using the fact that for an asymptotically stable matrix $A \in \mathbb{R}^{n\times n}$, $\|e^{At}\| \leq \gamma e^{-\beta t}$, $t \geq 0$, where $\|\cdot\|$ is a submultiplicative matrix norm, $\gamma > 0$, and $0 < \beta < -\alpha(A)$, where $\alpha(A) \triangleq \max\{\mathrm{Re}\ \lambda : \lambda \in \mathrm{spec}(A)\}$, it follows that

$$\begin{aligned} \|x(t)\| &\leq \gamma e^{-\beta t}\|x_0\| + \int_0^t \gamma e^{-\beta(t-\tau)}\|B\|\|u(\tau)\|\mathrm{d}\tau \\ &\leq \gamma e^{-\beta t}\|x_0\| + \frac{\gamma}{\beta}\|B\| \sup_{0\leq\tau\leq t} \|u(\tau)\|, \end{aligned} \tag{4.136}$$

which shows that if $u(t)$, $t \geq 0$, is bounded, then $x(t)$, $t \geq 0$, is also bounded. Note that in the case where $x_0 = 0$, the state response is proportional to the input bound. However, for nonlinear asymptotically stable dynamical systems a bounded input does not necessarily imply that the state is bounded. To see this, consider

$$\dot{x}(t) = -x(t) + x^2(t)u(t), \quad x(0) = x_0, \quad t \geq 0, \tag{4.137}$$

which is globally asymptotically stable for $u(t) \equiv 0$. However, in the case where $x(0) = 2$ and $u(t) \equiv 1$, the solution to (4.137) is given by $x(t) = 1/(1-0.5e^t)$, which is unbounded and, in fact, possesses a finite escape time. In light of the above, we introduce the notion of input-to-state stability for nonlinear dynamical systems.

Definition 4.6. A nonlinear dynamical system given by (4.133) is said to be *input-to-state stable* if for every $x_0 \in \mathbb{R}^n$ and every continuous and bounded input $u(t) \in \mathbb{R}^m$, $t \geq 0$, the solution $x(t)$, $t \geq 0$, of (4.133) exists and satisfies

$$\|x(t)\| \leq \eta(\|x_0\|, t) + \gamma\left(\sup_{0\leq\tau\leq t} \|u(\tau)\|\right), \qquad t \geq 0, \tag{4.138}$$

where $\eta(s,t)$, $s > 0$, is a class $\mathcal{KL}$ function and $\gamma(s)$, $s > 0$, is a class $\mathcal{K}$ function.

The input-to-state inequality (4.138) guarantees that for a bounded input $u(t) \in \mathbb{R}^m$, $t \geq 0$, the state $x(t)$, $t \geq 0$, is bounded. In particular, as t increases, the state $x(t)$, $t \geq 0$, is bounded by a class $\mathcal{K}$ function of $\sup_{t\geq 0} \|u(t)\|$. In the case where $u(t) \equiv 0$, (4.138) reduces to

$$\|x(t)\| \leq \eta(\|x(t_0)\|, t), \tag{4.139}$$

and hence, input-to-state stability implies that the zero solution $x(t) \equiv 0$ of (4.133) (with $u(t) \equiv 0$) is globally asymptotically stable. Additionally, (4.138) implies that if $\lim_{t\to\infty} u(t) = 0$, then $\lim_{t\to\infty} x(t) = 0$. To see this, let $\mu > 0$ be such that $\gamma(\mu) \leq \varepsilon/2$ for a given $\varepsilon > 0$. Since $\lim_{t\to\infty} u(t) = 0$, it follows that there exists $t_1 > 0$ such that $\|u(t)\| \leq \mu$, $t \geq t_1$. Now, since $x(t)$, $t \geq 0$, is bounded, it follows that

$$\|x(t)\| \leq \eta(\|x(t_1)\|, t - t_1) + \gamma(\mu) \leq \eta(\delta, t - t_1) + \varepsilon/2, \quad t \geq t_1, \tag{4.140}$$

for some $\delta > 0$. Since $\eta(\delta, t-t_1) \to 0$ as $t \to \infty$, there exists $t_2 > 0$ such that $\eta(\delta, t) \leq \varepsilon/2$, $t \geq t_2$. Thus, it follows from (4.140) that $\|x(t)\| \leq \varepsilon$, $t \geq \tau$, where $\tau = \max(t_1, t_2)$, which implies that $\lim_{t\to\infty} x(t) = 0$. The following theorem gives necessary and sufficient conditions for input-to-state stability of a nonlinear dynamical system.

Theorem 4.15. The nonlinear dynamical system (4.133) is input-to-state stable if and only if there exist a continuously differentiable radially unbounded, positive-definite function $V : \mathbb{R}^n \to \mathbb{R}$ and continuous functions $\gamma_1, \gamma_2 \in \mathcal{K}$ such that for every $u \in \mathbb{R}^m$,

$$V'(x)F(x,u) \leq -\gamma_1(\|x\|), \quad \|x\| \geq \gamma_2(\|u\|). \tag{4.141}$$

Proof. Assume (4.141) holds and let $u(\cdot)$ be such that $u(t) \in \mathbb{R}^m$, $t \geq 0$. With $f(t,x) = F(x,u)$, $V(t,x) = V(x)$, $W(x) = \gamma_1(\|x\|)$, and $\mu = \gamma_2(|||u|||)$, where $|||u||| \triangleq \sup_{t\in[0,\infty)} \|u(t)\|$, it follows from Corollary 4.2

that there exists $t_1 > 0$ such that

$$\|x(t)\| \leq \gamma(\|u\|), \quad t \geq t_1, \tag{4.142}$$

where $\gamma = \alpha^{-1} \circ \beta \circ \gamma_2$ and where $\alpha(\cdot)$, $\beta(\cdot)$ are class $\mathcal{K}_\infty$ functions such that $\alpha(\|x\|) \leq V(x) \leq \beta(\|x\|)$, $x \in \mathbb{R}^n$ (see Problem 3.72). Next, without loss of generality, let $t_1 > 0$ be such that $\|x(t)\| \geq \gamma_2(\|u\|)$, $t \leq t_1$. Thus,

$$\begin{aligned} \frac{\mathrm{d}V(x(t))}{\mathrm{d}t} &= V'(x)F(x, u(t))|_{x=x(t)} \\ &\leq -\gamma_1(\|x(t)\|) \\ &\leq -\gamma_1 \circ \beta^{-1}(V(x(t)), \quad \text{a.e.} \quad t \leq t_1. \end{aligned} \tag{4.143}$$

Note that (4.143) guarantees that $x(t)$ is defined for all $t \geq 0$. Furthermore, it follows from the comparison principle (see Problem 2.107) that (4.143) implies that there exists a class $\mathcal{KL}$ function $\hat{\eta}$ such that $V(x(t)) \leq \hat{\eta}(V(x_0), t)$, $t \leq t_2$. Hence,

$$\|x(t)\| \leq \eta(\|x_0\|, t), \qquad t \leq t_1, \tag{4.144}$$

where $\eta(s, t) = \alpha^{-1}\hat{\eta}(\beta(s), t)$. Now, it follows from (4.142) and (4.144) that $\|x(t)\| \leq \eta(\|x_0\|, t) + \gamma(\|u\|)$, $t \geq 0$. Since x_0 and $u(\cdot)$ are arbitrary, it follows that (4.133) is input-to-state stable.

Necessity is considerably more involved and requires concepts not introduced in this book. For details, see [409]. □

The next result provides sufficient conditions for input-to-state stability as an immediate consequence of the converse global exponential stability theorem.

Proposition 4.1. Consider the nonlinear dynamical system (4.133) where $F : \mathbb{R}^n \times \mathbb{R}^m \to \mathbb{R}^n$ is continuously differentiable and globally Lipschitz continuous on $\mathbb{R}^n \times \mathbb{R}^m$. If the zero solution $x(t) \equiv 0$ of the undisturbed $(u(t) \equiv 0)$ system (4.133) is globally exponentially stable, then (4.133) is input-to-state stable.

Proof. Since the zero solution $x(t) \equiv 0$ of the undisturbed $(u(t) \equiv 0)$ system (4.133) is globally exponentially stable, it follows from Theorem 3.11 that there exist a continuously differentiable positive-definite function $V : \mathbb{R}^n \to \mathbb{R}$ and scalars α, β, and $\varepsilon > 0$ such that

$$\alpha\|x\|^2 \leq V(x) \leq \beta\|x\|^2, \quad x \in \mathbb{R}^n, \tag{4.145}$$

$$V'(x)F(x, 0) \leq -\varepsilon V(x), \quad x \in \mathbb{R}^n. \tag{4.146}$$

Next, using the fact that $\|V'(x)\| \leq \nu\|x\|$, $x \in \mathbb{R}^n$, $\nu > 0$, (see Problem 3.60) and the global Lipschitz continuity of $F(\cdot, \cdot)$ on $\mathbb{R}^n \times \mathbb{R}^m$, for every $u \in \mathbb{R}^m$, it follows that the derivative of $V(x)$ along the trajectories of (4.133) through

$x \in \mathbb{R}^n$ at $t = 0$ is given by

$$\begin{aligned}\dot{V}(x) &= V'(x)F(x,u)\\ &= V'(x)F(x,0) + V'(x)[F(x,u) - F(x,0)]\\ &\leq -\varepsilon\alpha\|x\|^2 + \nu L\|x\|\|u\|, \end{aligned} \qquad (4.147)$$

where $L > 0$ denotes the Lipschitz constant of $F(x,\cdot)$. Rewriting (4.147) as

$$\dot{V}(x) \leq -\varepsilon\alpha(1-\mu)\|x\|^2 - \varepsilon\alpha\mu\|x\|^2 + \nu L\|x\|\|u\|, \qquad (4.148)$$

where $\mu \in (0,1)$, it follows that

$$\dot{V}(x) \leq -\varepsilon\alpha(1-\mu)\|x\|^2, \quad \|x\| \geq \frac{\nu L}{\alpha\varepsilon\mu}\|u\|. \qquad (4.149)$$

Now, the result is a direct consequence of Theorem 4.15 with $\gamma_1(\sigma) = \varepsilon\alpha(1-\mu)\sigma^2$ and $\gamma_2(\sigma) = \frac{\nu L}{\alpha\varepsilon\mu}\sigma$. □

Example 4.10. Consider the nonlinear dynamical system

$$\dot{x}_1(t) = -x_1(t) + x_2^2(t), \quad x_1(0) = x_{10}, \quad t \geq 0, \qquad (4.150)$$
$$\dot{x}_2(t) = -x_2(t) + u(t), \quad x_2(0) = x_{20}. \qquad (4.151)$$

To show that (4.150) and (4.151) is input-to-state stable consider the radially unbounded, positive definite function $V(x_1,x_2) = \frac{1}{2}x_1^2 + \frac{\alpha}{4}x_2^4$, $\alpha > 0$, and note that

$$\begin{aligned}\dot{V}(x_1,x_2) &= x_1(-x_1 + x_2^2) + \alpha x_2^3(-x_2 + u)\\ &= -x_1^2 + x_1x_2^2 - \alpha(1-\varepsilon)x_2^4 - \alpha\varepsilon x_2^4 + \alpha x_2^3 u, \quad \varepsilon \in (0,1),\\ &\leq -x_1^2 + x_1x_2^2 - \alpha(1-\varepsilon)x_2^4, \quad |x_2| \geq \tfrac{|u|}{\varepsilon},\\ &\leq -\left[\begin{array}{cc} x_1 & x_2^2 \end{array}\right] R_1 \left[\begin{array}{c} x_1 \\ x_2^2 \end{array}\right], \quad |x_2| \geq \tfrac{|u|}{\varepsilon}, \end{aligned} \qquad (4.152)$$

where

$$R_1 = \left[\begin{array}{cc} 1 & -1/2 \\ -1/2 & \alpha(1-\varepsilon) \end{array}\right].$$

Now, choosing $\alpha > 1/4(1-\varepsilon)$ ensures that $R_1 > 0$, and hence, $\dot{V}(x_1,x_2) < 0$, $|x_2| \geq \frac{|u|}{\varepsilon}$. Next, for $|x_2| \leq \frac{|u|}{\varepsilon}$, it follows that

$$\begin{aligned}\dot{V}(x_1,x_2) &\leq -x_1^2 + x_1x_2^2 - \alpha x_2^4 + \alpha|u|^4/\varepsilon^3\\ &= -(1-\varepsilon)x_1^2 + x_1x_2^2 - \alpha x_2^4 - \varepsilon x_1^2 + \alpha|u|^4/\varepsilon^3\\ &\leq -\left[\begin{array}{cc} x_1 & x_2^2 \end{array}\right] R_2 \left[\begin{array}{c} x_1 \\ x_2^2 \end{array}\right], \quad |x_1| \geq \alpha^{1/2}\tfrac{|u|^2}{\varepsilon^2}, \end{aligned} \qquad (4.153)$$

where

$$R_2 = \left[\begin{array}{cc} 1-\varepsilon & -1/2 \\ -1/2 & \alpha \end{array}\right].$$

Note that with $\alpha > 1/4(1-\varepsilon)$, $R_2 > 0$, and hence, $\dot{V}(x_1, x_2) < 0$, $|x_2| \leq \frac{|u|}{\varepsilon}$. Next, let $\delta = \min\{\lambda_{\min}(R_1), \lambda_{\min}(R_2)\}$ and let $\gamma_2(s) = \frac{\alpha^{1/2}s^2}{\varepsilon^2}$, $s \geq 0$. Note that $\gamma_2 \in \mathcal{K}_\infty$. Now, (4.152) and (4.153) imply $\dot{V}(x_1, x_2) \leq -\delta(x_1^2 + x_2^4)$, $\|x\|_\infty \geq \gamma_2(\|u\|)$, and hence, it follows from Theorem 4.15 that (4.150) and (4.151) is input-to-state stable. △

The following propositions address stability of cascade and interconnected dynamical systems. Here, we provide the global versions of these propositions; the local cases are identical except for restricting the domain of analysis.

Proposition 4.2. Consider the nonlinear cascade dynamical system

$$\dot{x}_1(t) = f_1(x_1(t), x_2(t)), \quad x_1(0) = x_{10}, \quad t \geq 0, \tag{4.154}$$

$$\dot{x}_2(t) = f_2(x_2(t)), \quad x_2(0) = x_{20}, \tag{4.155}$$

where $f_1 : \mathbb{R}^{n_1} \times \mathbb{R}^{n_2} \to \mathbb{R}^{n_1}$ and $f_2 : \mathbb{R}^{n_2} \to \mathbb{R}^{n_2}$ are Lipschitz continuous and satisfy $f_1(0,0) = 0$ and $f_2(0) = 0$. If (4.154), with x_2 viewed as the input, is input-to-state stable and the zero solution $x_2(t) \equiv 0$ to (4.155) is globally asymptotically stable, then the zero solution $(x_1(t), x_2(t)) \equiv (0,0)$ of the cascade dynamical system (4.154) and (4.155) is globally asymptotically stable.

Proof. Since (4.154) is input-to-state stable and the zero solution $x_2(t) \equiv 0$ of (4.155) is globally asymptotically stable it follows that there exist $\mathcal{KL}$ functions $\eta_1(\cdot,\cdot)$ and $\eta_2(\cdot,\cdot)$ and a class $\mathcal{K}$ function $\gamma(\cdot)$ such that

$$\|x_1(t)\| \leq \eta_1(\|x_1(s)\|, t-s) + \gamma\left(\sup_{s\leq\tau\leq t} \|x_2(\tau)\|\right), \tag{4.156}$$

$$\|x_2(t)\| \leq \eta_2(\|x_2(s)\|, t-s), \tag{4.157}$$

where $t \geq s \geq 0$. Setting $s = t/2$ in (4.157) yields

$$\|x_1(t)\| \leq \eta_1(\|x_1(t/2)\|, t/2) + \gamma\left(\sup_{t/2\leq\tau\leq t} \|x_2(\tau)\|\right). \tag{4.158}$$

Next, setting $s = 0$ and replacing t by $t/2$ in (4.156) yields

$$\|x_1(t/2)\| \leq \eta_1(\|x_1(0)\|, t/2) + \gamma\left(\sup_{0\leq\tau\leq t/2} \|x_2(\tau)\|\right). \tag{4.159}$$

Now, using (4.157), it follows that

$$\sup_{0\leq\tau\leq t/2} \|x_2(\tau)\| \leq \eta_2(\|x_2(0)\|, 0), \tag{4.160}$$

$$\sup_{t/2\leq\tau\leq t} \|x_2(\tau)\| \leq \eta_2(\|x_2(0)\|, t/2). \tag{4.161}$$

Next, substituting (4.159)–(4.161) into (4.158) and using the fact that $\|x_1(0)\| \le \|x(0)\|$, $\|x_2(0)\| \le \|x(0)\|$, and $\|x(t)\| \le \|x_1(t)\| + \|x_2(t)\|$, $t \ge 0$, where $x(t) \triangleq [x_1^{\mathrm{T}}(t),\ x_2^{\mathrm{T}}(t)]^{\mathrm{T}}$, it follows that

$$\|x(t)\| \le \eta(\|x(0)\|, t), \tag{4.162}$$

where

$$\eta(r,s) = \eta_1(\eta_1(r, s/2) + \gamma(\eta_2(r,0)), s/2) + \gamma(\eta_2(r, s/2)) + \eta_2(r,s). \tag{4.163}$$

Finally, since $\eta(\cdot, t)$ is a strictly increasing function with $\eta(0,t) = 0$ and $\eta(r, \cdot)$ is a decreasing function of time such that $\lim_{t\to\infty} \eta(r,t) = 0$, $r > 0$, it follows that the zero solution $x(t) \equiv 0$ of the interconnected dynamical system (4.154) and (4.155) is globally asymptotically stable. □

Proposition 4.3. Consider the nonlinear interconnected dynamical system

$$\dot{x}_1(t) = f_1(x_1(t), x_2(t)), \quad x_1(0) = x_{10}, \quad t \ge 0, \tag{4.164}$$
$$\dot{x}_2(t) = f_2(x_1(t), x_2(t)), \quad x_2(0) = x_{20}, \tag{4.165}$$

where $f_1 : \mathbb{R}^{n_1} \times \mathbb{R}^{n_2} \to \mathbb{R}^{n_1}$ is such that, for every $x_2 \in \mathbb{R}^{n_2}$, $f_1(\cdot, x_2)$ is Lipschitz continuous in x_1, and $f_2 : \mathbb{R}^{n_1} \times \mathbb{R}^{n_2} \to \mathbb{R}^{n_2}$ is such that for every $x_1 \in \mathbb{R}^{n_1}$, $f_2(x_1, \cdot)$ is Lipschitz continuous in x_2. If (4.165) is input-to-state stable with x_1 viewed as the input and (4.164) and (4.165) is globally asymptotically stable with respect to x_1 uniformly in x_{20}, then the zero solution $(x_1(t), x_2(t)) \equiv (0,0)$ of the interconnected dynamical system (4.164) and (4.165) is globally asymptotically stable.

Proof. Since (4.165) is input-to-state stable with x_1 viewed as the input and (4.164) and (4.165) is globally asymptotically stable with respect to x_1 uniformly in x_{20}, it follows that there exist class $\mathcal{KL}$ functions $\eta_1(\cdot,\cdot)$ and $\eta_2(\cdot,\cdot)$ and a class $\mathcal{K}$ function $\gamma(\cdot)$ (see Problem 4.6) such that

$$\|x_1(t)\| \le \eta_1(\|x_1(s)\|, t-s), \tag{4.166}$$
$$\|x_2(t)\| \le \eta_2(\|x_2(s)\|, t-s) + \gamma\left(\sup_{s\le\tau\le t} \|x_1(\tau)\|\right), \tag{4.167}$$

where $t \ge s \ge 0$. Setting $s = t/2$ in (4.167) yields

$$\|x_2(t)\| \le \eta_2(\|x_2(t/2)\|, t/2) + \gamma\left(\sup_{t/2\le\tau\le t} \|x_1(\tau)\|\right). \tag{4.168}$$

Next, setting $s = 0$ and replacing t by $t/2$ in (4.167) yields

$$\|x_2(t/2)\| \le \eta_2(\|x_2(0)\|, t/2) + \gamma\left(\sup_{0\le\tau\le t/2} \|x_1(\tau)\|\right). \tag{4.169}$$

Now, using (4.166), it follows that

$$\sup_{0\le\tau\le t/2} \|x_1(\tau)\| \le \eta_1(\|x_1(0)\|, 0), \tag{4.170}$$

$$\sup_{t/2\le\tau\le t} \|x_1(\tau)\| \le \eta_1(\|x_1(0)\|, t/2). \tag{4.171}$$

Next, substituting (4.169)–(4.171) into (4.168) and using the fact that $\|x_1(0)\| \le \|x(0)\|$, $\|x_2(0)\| \le \|x(0)\|$, and $\|x(t)\| \le \|x_1(t)\| + \|x_2(t)\|$, $t \ge 0$, where $x(t) \triangleq [x_1^{\mathrm{T}}(t), x_2^{\mathrm{T}}(t)]^{\mathrm{T}}$, it follows that

$$\|x(t)\| \le \eta(\|x(0)\|, t), \tag{4.172}$$

where

$$\eta(r,s) = \eta_1(r,s) + \eta_2(\eta_2(r, s/2) + \gamma(\eta_1(r,0)), s/2) + \gamma(\eta_1(r, s/2)). \tag{4.173}$$

Finally, since $\eta(\cdot, t)$ is a strictly increasing function with $\eta(0,t) = 0$ and $\eta(r,\cdot)$ is a decreasing function such that $\lim_{t\to\infty} \eta(r,t) = 0$, $r > 0$, it follows that the zero solution $x(t) \equiv 0$ of the interconnected dynamical system (4.164) and (4.165) is globally asymptotically stable. □

Example 4.11. To illustrate the utility of Proposition 4.2 for feedback stabilization, consider the nonlinear system with a linear input subsystem

$$\dot{x}(t) = f(x(t)) + G(x(t))\hat{x}(t), \quad x(0) = x_0, \quad t \ge 0, \tag{4.174}$$

$$\dot{\hat{x}}(t) = A\hat{x}(t) + Bu(t), \quad \hat{x}(0) = \hat{x}_0, \tag{4.175}$$

where $x \in \mathbb{R}^n$, $\hat{x} \in \mathbb{R}^{\hat{n}}$, $f : \mathbb{R}^n \to \mathbb{R}^n$ and satisfies $f(0) = 0$, $G : \mathbb{R}^n \to \mathbb{R}^{n\times\hat{n}}$, $u \in \mathbb{R}^m$, $A \in \mathbb{R}^{\hat{n}\times\hat{n}}$, $B \in \mathbb{R}^{\hat{n}\times m}$, and (A, B) is controllable. Furthermore, assume (4.174) is input-to-state stable with $\hat{x}$ viewed as the input. Now, since (A, B) is controllable, there exists $K \in \mathbb{R}^{m\times\hat{n}}$ such that with $u = K\hat{x}$, (4.175) is asymptotically stable, that is, the zero solution of $\dot{\hat{x}}(t) = (A + BK)\hat{x}(t)$, $\hat{x}(0) = \hat{x}_0$, $t \ge 0$, is globally asymptotically stable. Hence, it follows from Proposition 4.2 that the zero solution $(x(t), \hat{x}(t)) \equiv (0, 0)$ of the cascade connection (4.174) and (4.175) is globally asymptotically stable. △

Finally, we present a proposition on ultimate boundedness of interconnected systems.

Proposition 4.4. Consider the nonlinear interconnected dynamical system (4.164) and (4.165). If (4.165) is input-to-state stable with x_1 viewed as the input and (4.164) and (4.165) is ultimately bounded with respect to x_1 uniformly in x_{20}, then the solution $(x_1(t), x_2(t))$, $t \ge 0$, of the interconnected dynamical system (4.164) and (4.165) is ultimately bounded.

Proof. Let $\delta > 0$. Since (4.164) and (4.165) is ultimately bounded with respect to x_1 (uniformly in x_{20}), for every $\|x_{10}\| < \delta$, there exist positive constants δ, ε and $T = T(\delta, \varepsilon)$ such that if $\|x_0\| < \delta$, then $\|x_1(t)\| < \varepsilon$, $t \geq T$. Furthermore, since (4.165) is input-to-state stable with x_1 viewed as the input, it follows that $x_2(T)$ is finite, and hence, there exist a class $\mathcal{KL}$ function $\eta(\cdot, \cdot)$ and a class $\mathcal{K}$ function $\gamma(\cdot)$ such that

$$\begin{aligned} \|x_2(t)\| &\leq \eta(\|x_2(T)\|, t - T) + \gamma\left(\sup_{T \leq \tau \leq t} \|x_1(\tau)\|\right) \\ &\leq \eta(\|x_2(T)\|, t - T) + \gamma(\varepsilon) \\ &\leq \eta(\|x_2(T)\|, 0) + \gamma(\varepsilon), \quad t \geq T, \end{aligned} \tag{4.176}$$

which proves that the solution $(x_1(t), x_2(t))$, $t \geq 0$, to (4.164) and (4.165) is ultimately bounded. □

4.6 Finite-Time Stability of Nonlinear Dynamical Systems

The notions of asymptotic and exponential stability in dynamical system theory imply convergence of the system trajectories to an equilibrium state over the infinite horizon. In many applications, however, it is desirable that a dynamical system possesses the property that trajectories that converge to a Lyapunov stable equilibrium state must do so in finite time rather than merely asymptotically. The stability theorems presented in Chapter 3 involve system dynamics with Lipschitz continuous vector fields, which implies uniqueness of system solutions in forward and backward times. Hence, convergence to an equilibrium state is achieved over an infinite time interval. In order to achieve convergence in finite time, the system dynamics need to be non-Lipschitzian, giving rise to nonuniqueness of solutions in backward time. Uniqueness of solutions in forward time, however, can be preserved in the case of finite-time convergence. Sufficient conditions that ensure uniqueness of solutions in forward time in the absence of Lipschitz continuity are given in [4, 118, 232, 474]. In addition, it is shown in [96, Theorem 4.3, p. 59] that uniqueness of solutions in forward time along with continuity of the system dynamics ensure that the system solutions are continuous functions of the system initial conditions even when the dynamics are not Lipschitz continuous.

In this section, we develop Lyapunov and converse Lyapunov theorems for finite-time stability of autonomous systems. Specifically, consider the nonlinear dynamical system given by

$$\dot{x}(t) = f(x(t)), \quad x(t_0) = x_0, \quad t \in \mathcal{I}_{x_0}, \tag{4.177}$$

where $x(t) \in \mathcal{D} \subseteq \mathbb{R}^n$, $t \in \mathcal{I}_{x_0}$, is the system state vector, $\mathcal{I}_{x_0}$ is the maximal interval of existence of a solution $x(t)$ of (4.177), $\mathcal{D}$ is an open set, $0 \in \mathcal{D}$, $f(0) = 0$, and $f(\cdot)$ is continuous on $\mathcal{D}$. We assume that (4.177) possesses unique solutions in forward time for all initial conditions except possibly the origin in the following sense. For every $x \in \mathcal{D}\backslash\{0\}$ there exists $\tau_x > 0$ such that, if $y_1 : [0, \tau_1) \to \mathcal{D}$ and $y_2 : [0, \tau_2) \to \mathcal{D}$ are two solutions of (4.177) with $y_1(0) = y_2(0) = x$, then $\tau_x \leq \min\{\tau_1, \tau_2\}$ and $y_1(t) = y_2(t)$ for all $t \in [0, \tau_x)$. Without loss of generality, we assume that for each x, τ_x is chosen to be the largest such number in $\overline{\mathbb{R}}_+$. Sufficient conditions for forward uniqueness in the absence of Lipschitz continuity can be found in [4, 118, 232, 474].

The next result presents the classical comparison principle for nonlinear dynamical systems.

Theorem 4.16. Consider the nonlinear dynamical system (4.177). Assume there exists a continuously differentiable function $V : \mathcal{D} \to \mathbb{R}$ such that

$$V'(x)f(x) \leq w(V(x)), \quad x \in \mathcal{D}, \tag{4.178}$$

where $w : \mathbb{R} \to \mathbb{R}$ is a continuous function and

$$\dot{z}(t) = w(z(t)), \quad z(t_0) = z_0, \quad t \in \mathcal{I}_{z_0}, \tag{4.179}$$

has a unique solution $z(t)$, $t \in \mathcal{I}_{z_0}$. If $[t_0,\, t_0 + \tau] \subseteq \mathcal{I}_{x_0} \cap \mathcal{I}_{z_0}$ is a compact interval and $V(x_0) \leq z_0$, $z_0 \in \mathbb{R}$, then $V(x(t)) \leq z(t)$, $t \in [t_0,\, t_0 + \tau]$.

Proof. Consider the family of dynamical systems given by

$$\dot{z}(t) = w(z(t)) + \frac{1}{n}, \quad z(t_0) = z_0, \tag{4.180}$$

where $n \in \mathbb{Z}_+$ and $t \in \mathcal{I}_{z_0,n}$, and denote the solution to (4.180) by $s_{(n)}(t, z_0)$, $t \in \mathcal{I}_{z_0,n}$. Now, it follows from [98, p. 17, Theorem 3] that there exists a compact interval $[t_0, t_0 + \tau] \subseteq \mathcal{I}_{x_0} \cap \mathcal{I}_{z_0}$ such that $s_{(n)}(t, z_0)$, $t \in [t_0, t_0 + \tau]$, is defined for all sufficiently large n. Furthermore, it follows from Lemma 3.1 of [179] that $s_{(n)}(t, z_0) \to z(t)$ as $n \to \infty$ uniformly on $[t_0, t_0 + \tau]$, where $z(t)$, $t \in \mathcal{I}_{z_0}$, is a solution to (4.179). Next, we show that $V(x(t)) \leq s_{(n)}(t, z_0)$, $n > m$, $t \in [t_0, t_0 + \tau]$, where m is sufficiently large so that $s_{(n)}(t, z_0)$ is well defined on $[t_0, t_0 + \tau]$ for all $n > m$. Note that at $t = t_0$, $V(x_0) = V(x(t_0)) \leq z_0 = s_{(n)}(t_0, z_0)$. Now, suppose, *ad absurdum*, that there exist $t_1, t_2 \in [t_0, t_0 + \tau]$ such that $V(x(t)) > s_{(n)}(t, z_0)$, $t \in (t_1, t_2]$, and $V(x(t_1)) = s_{(n)}(t_1, z_0)$ for some $n > m$. Since $V(\cdot)$ is continuously differentiable, it follows that

$$\begin{aligned}
\dot{V}(x(t_1)) &\geq \dot{s}_{(n)}(t_1, z_0) \\
&= w(s_{(n)}(t_1, z_0)) + \frac{1}{n} \\
&= w(V(x(t_1))) + \frac{1}{n} \\
&> w(V(x(t_1))),
\end{aligned} \tag{4.181}$$

which is a contradiction. Thus, $V(x(t)) \leq s_{(n)}(t, z_0)$, $t \in [t_0, t_0 + \tau]$, $n > m$.

Since $s_{(n)}(t, z_0) \to z(t)$ uniformly on $[t_0, t_0 + \tau]$, this proves the result. $\square$

The next definition introduces the notion of finite-time stability. For this definition and the remainder of the section we assume, without loss of generality, that $t_0 = 0$.

Definition 4.7. Consider the nonlinear dynamical system (4.177). The zero solution $x(t) \equiv 0$ to (4.177) is *finite-time stable* if there exist an open neighborhood $\mathcal{N} \subseteq \mathcal{D}$ of the origin and a function $T : \mathcal{N}\backslash\{0\} \to (0, \infty)$, called the *settling-time function*, such that the following statements hold:

i) Finite-time convergence. For every $x \in \mathcal{N}\backslash\{0\}$, $s^x(t)$ is defined on $[0, T(x))$, $s^x(t) \in \mathcal{N}\backslash\{0\}$ for all $t \in [0, T(x))$, and $\lim_{t\to T(x)} s(x, t) = 0$.

ii) Lyapunov stability. For every $\varepsilon > 0$ there exists $\delta > 0$ such that $\mathcal{B}_\delta(0) \subset \mathcal{N}$ and for every $x \in \mathcal{B}_\delta(0)\backslash\{0\}$, $s(t, x) \in \mathcal{B}_\varepsilon(0)$ for all $t \in [0, T(x))$.

The zero solution $x(t) \equiv 0$ of (4.177) is *globally finite-time stable* if it is finite-time stable with $\mathcal{N} = \mathcal{D} = \mathbb{R}^n$.

Note that if the zero solution $x(t) \equiv 0$ to (4.177) is finite-time stable, then it is asymptotically stable, and hence, finite-time stability is a stronger notion than asymptotic stability. Next, we show that if the zero solution $x(t) \equiv 0$ to (4.177) is finite-time stable, then (4.177) has a unique solution $s(\cdot, \cdot)$ defined on $\mathbb{R}_+ \times \mathcal{N}$ for every initial condition in an open neighborhood of the origin, including the origin, and $s(t, x) = 0$ for all $t \geq T(x)$, $x \in \mathcal{N}$, where $T(0) \triangleq 0$.

Proposition 4.5. Consider the nonlinear dynamical system (4.177). Assume that the zero solution $x(t) \equiv 0$ to (4.177) is finite-time stable and let $\mathcal{N} \subseteq \mathcal{D}$ and $T : \mathcal{N}\backslash\{0\} \to (0, \infty)$ be as in Definition 4.7. Then, $s(\cdot, \cdot)$ is a unique solution of (4.177) and is defined on $\mathbb{R}_+ \times \mathcal{N}$, and $s(t, x) = 0$ for all $t \geq T(x)$, $x \in \mathcal{N}$, where $T(0) \triangleq 0$.

Proof. It follows from Lyapunov stability of the origin that $x(t) \equiv 0$ is the unique solution $x(\cdot)$ of (4.177) satisfying $x(0) = 0$. This proves that $\mathbb{R}_+ \times \{0\}$ is contained in the domain of the definition of $s(\cdot, \cdot)$ and $s^0(t) \equiv 0$.

Next, let $\mathcal{N} \subseteq \mathcal{D}$ and $T(\cdot)$ be as in Definition 4.7, and let $x_0 \in \mathcal{N}\backslash\{0\}$. Define

$$x(t) \triangleq \begin{cases} s(t, x_0), & 0 \leq t \leq T(x_0), \\ 0, & T(x_0) \leq t. \end{cases} \tag{4.182}$$

Note that by construction, $x(\cdot)$ is continuously differentiable on $\mathbb{R}_+\backslash\{T(x_0)\}$ and satisfies (4.177) on $\mathbb{R}_+\backslash\{T(x_0)\}$. Furthermore, since $f(\cdot)$ is continuous,

$$\lim_{t\to T(x_0)^-} \dot{x}(t) = \lim_{t\to T(x_0)^-} f(x(t)) = 0 = \lim_{t\to T(x_0)^+} \dot{x}(t), \tag{4.183}$$

and hence, $x(\cdot)$ is continuously differentiable at $T(x_0)$ and $x(\cdot)$ satisfies (4.177). Hence, $x(\cdot)$ is a solution of (4.177) on $\mathbb{R}_+$.

To show uniqueness, assume $y(\cdot)$ is a solution of (4.177) on $\mathbb{R}_+$ satisfying $y(0) = x_0$. In this case, $x(\cdot)$ and $y(\cdot)$ agree on $[0, T(x_0))$, and by continuity, $x(\cdot)$ and $y(\cdot)$ must also agree on $[0, T(x_0)]$, and hence, $y(T(x_0)) = 0$. Now, Lyapunov stability implies that $y(t) = 0$ for $t > T(x)$, which proves uniqueness. Finally, by definition, $s^{x_0}(t) = x(t)$, and hence, $s^{x_0}(\cdot)$ is defined on $\mathbb{R}_+$ and satisfies $s^{x_0}(t) = 0$ on $[T(x_0), \infty)$ for every $x_0 \in \mathcal{N}$. This proves the result. □

It follows from Proposition 4.5 that if the zero solution $x(t) \equiv 0$ to (4.177) is finite-time stable, then the solutions of (4.177) define a continuous *global semiflow* on $\mathcal{N}$; that is, $s : \mathbb{R}_+ \times \mathcal{N} \to \mathcal{N}$ is jointly continuous and satisfies $s(0, x) = x$ and $s(t, s(\tau, x)) = s(t + \tau, x)$ for every $x \in \mathcal{N}$ and $t, \tau \in \mathbb{R}_+$. Furthermore, $s(\cdot, \cdot)$ satisfies $s(T(x) + t, x) = 0$ for all $x \in \mathcal{N}$ and $t \in \mathbb{R}_+$. Finally, it also follows from Proposition 4.5 that we can extend $T(\cdot)$ to all of $\mathcal{N}$ by defining $T(0) \triangleq 0$. It is easy to see from Definition 4.7 that

$$T(x) = \inf\{t \in \mathbb{R}_+ : s(t, x) = 0\}, \quad x \in \mathcal{N}. \tag{4.184}$$

The following example adopted from [55] presents a finite-time stable system with a continuous but non-Lipschitzian vector field.

Example 4.12. Consider the scalar nonlinear dynamical system given by

$$\dot{x}(t) = -k\,\mathrm{sign}(x(t))|x(t)|^\alpha, \quad x(0) = x_0, \quad t \geq 0, \tag{4.185}$$

where $x_0 \in \mathbb{R}$, $\mathrm{sign}(x) \triangleq \frac{x}{|x|}$, $x \neq 0$, $\mathrm{sign}(0) \triangleq 0$, $k > 0$, and $\alpha \in (0, 1)$. The right-hand side of (4.185) is continuous everywhere and locally Lipschitz everywhere except the origin. Hence, every initial condition in $\mathbb{R}\backslash\{0\}$ has a unique solution in forward time on a sufficiently small time interval. The solution to (4.185) is obtained by direct integration and is given by

$$s(t, x_0) = \begin{cases} \mathrm{sign}(x_0)\left[|x_0|^{1-\alpha} - k(1-\alpha)t\right]^{\frac{1}{1-\alpha}}, & t < \frac{1}{k(1-\alpha)}|x_0|^{1-\alpha}, \quad x_0 \neq 0, \\ 0, & t \geq \frac{1}{k(1-\alpha)}|x_0|^{1-\alpha}, \quad x_0 \neq 0, \\ 0, & t \geq 0, \quad x_0 = 0. \end{cases} \tag{4.186}$$

It is clear from (4.186) that i) in Definition 4.7 is satisfied with $\mathcal{N} = \mathcal{D} = \mathbb{R}$ and with the settling-time function $T : \mathbb{R} \to \mathbb{R}_+$ given by

$$T(x_0) = \frac{1}{k(1-\alpha)}|x_0|^{1-\alpha}, \quad x_0 \in \mathbb{R}. \tag{4.187}$$

Lyapunov stability follows by considering the Lyapunov function $V(x) = x^2$, $x \in \mathbb{R}$. Thus, the zero solution $x(t) \equiv 0$ to (4.185) is globally finite-time stable. △

The next proposition shows that the settling-time function of a finite-time stable system is continuous on $\mathcal{N}$ if and only if it is continuous at the origin.

Proposition 4.6. Consider the nonlinear dynamical system (4.177). Assume that the zero solution $x(t) \equiv 0$ to (4.177) is finite-time stable, let $\mathcal{N} \subseteq \mathcal{D}$ be as in Definition 4.7, and let $T : \mathcal{N} \to \mathbb{R}_+$ be the settling-time function. Then $T(\cdot)$ is continuous on $\mathcal{N}$ if and only if $T(\cdot)$ is continuous at $x = 0$.

Proof. Necessity is immediate. To show sufficiency, suppose $T(\cdot)$ is continuous at $x = 0$, let $y \in \mathcal{N}$, and consider the sequence $\{y_n\}_{n=0}^\infty$ in $\mathcal{N}$ converging to y. Let $\tau^- = \liminf_{n\to\infty} T(y_n)$ and $\tau^+ = \limsup_{n\to\infty} T(y_n)$. Note that $\tau^-, \tau^+ \in \mathbb{R}_+$ and

$$\tau^- \leq \tau^+. \tag{4.188}$$

Next, let $\{y_m^+\}_{m=0}^\infty$ be a subsequence of $\{y_n\}_{n=0}^\infty$ such that $T(y_m^+) \to \tau^+$ as $m \to \infty$. The sequence $\{T(y), y_m^+\}_{m=0}^\infty$ converges in $\mathbb{R}_+ \times \mathcal{N}$ to $(T(y), y)$. Now, it follows from continuity and $s(T(x)+t, x) = 0$ for all $x \in \mathcal{N}$ and $t \in \mathbb{R}_+$ that $s(T(y), y_m^+) \to s(T(y), y) = 0$ as $m \to \infty$. Since, by assumption, $T(\cdot)$ is continuous at $x = 0$, $T(s(T(y), y_m^+)) \to T(0) = 0$ as $m \to \infty$. Next, using (4.184), the semigroup property $s(t, s(\tau, x)) = s(t+\tau, x)$, $x \in \mathcal{N}$ and $t, \tau \in \mathbb{R}_+$, and $s(T(x)+t, x) = 0$, $x \in \mathcal{N}$ and $t, \tau \in \mathbb{R}_+$, it follows that

$$T(s(t,x)) = \max\{T(x) - t, 0\}. \tag{4.189}$$

Now, with $t = T(y)$ and $x = y_m^+$, it follows from (4.189) that $\max\{T(y_m^+) - T(y), 0\} \to 0$ as $m \to \infty$. Hence, $\max\{\tau^+ - T(y), 0\} = 0$, that is,

$$\tau^+ \leq T(y). \tag{4.190}$$

Finally, let $\{y_m^-\}_{m=0}^\infty$ be a subsequence of $\{y_m\}_{n=0}^\infty$ such that $T(y_m^-) \to \tau^-$ as $m \to \infty$. Now, it follows from (4.188) and (4.190) that $\tau^- \in \mathbb{R}_+$, and hence, the sequence $\{T(y_m^-), y_m^-\}_{m=0}^\infty$ converges in $\mathbb{R}_+ \times \mathcal{N}$ to (τ^-, y). Since $s(\cdot,\cdot)$ is jointly continuous, it follows that $s(T(y_m^-), y_m^-) \to s(\tau^-, y)$ as $m \to \infty$. Now, $s(T(x)+t, x) = 0$ for all $x \in \mathcal{N}$ and $t \in \mathbb{R}_+$ implies that

$s(T(y_m^-), y_m^-) = 0$ for each m. Hence, $s(\tau^-, y) = 0$ and, by (4.184),

$$T(y) \leq \tau^-. \tag{4.191}$$

Now, it follows from (4.188), (4.190), and (4.191) that $\tau^- = \tau^+ = T(y)$, and hence, $T(y_n) \to T(y)$ as $n \to \infty$, which proves that $T(\cdot)$ is continuous on $\mathcal{N}$. □

Next, we present sufficient conditions for finite-time stability using a Lyapunov function involving a scalar differential inequality.

Theorem 4.17. Consider the nonlinear dynamical system (4.177). Assume there exist a continuously differentiable function $V : \mathcal{D} \to \overline{\mathbb{R}}_+$, real numbers $c > 0$ and $\alpha \in (0, 1)$, and a neighborhood $\mathcal{M} \subseteq \mathcal{D}$ of the origin such that

$$V(0) = 0, \tag{4.192}$$

$$V(x) > 0, \quad x \in \mathcal{M}\backslash\{0\}, \tag{4.193}$$

$$V'(x)f(x) \leq -c(V(x))^\alpha, \quad x \in \mathcal{M}\backslash\{0\}. \tag{4.194}$$

Then the zero solution $x(t) \equiv 0$ to (4.177) is finite-time stable. Moreover, there exist an open neighborhood $\mathcal{N}$ of the origin and a settling-time function $T : \mathcal{N} \to [0, \infty)$ such that

$$T(x_0) \leq \frac{1}{c(1-\alpha)}(V(x_0))^{1-\alpha}, \quad x_0 \in \mathcal{N}, \tag{4.195}$$

and $T(\cdot)$ is continuous on $\mathcal{N}$. If, in addition, $\mathcal{D} = \mathbb{R}^n$, $V(\cdot)$ is radially unbounded, and (4.194) holds on $\mathbb{R}^n$, then the zero solution $x(t) \equiv 0$ to (4.177) is globally finite-time stable.

Proof. Since $V(\cdot)$ is positive definite and $\dot{V}(\cdot)$ takes negative values on $\mathcal{M}\backslash\{0\}$, it follows that $x(t) \equiv 0$ is the unique solution of (4.177) for $t \geq 0$ satisfying $x(0) = 0$ [4, Section 3.15], [474, Theorem 1.2, p. 5]. Thus, for every initial condition in $\mathcal{D}$, (4.177) has a unique solution in forward time.

Let $\mathcal{V} \subseteq \mathcal{M}$ be a bounded open set such that $0 \in \mathcal{V}$ and $\overline{\mathcal{V}} \subset \mathcal{D}$. Then $\partial\mathcal{V}$ is compact and $0 \notin \partial\mathcal{V}$. Now, it follows from Weierstrass' theorem (Theorem 2.13) that the continuous function $V(\cdot)$ attains a minimum on $\partial\mathcal{V}$ and since $V(\cdot)$ is positive definite, $\min_{x\in\partial\mathcal{V}} V(x) > 0$. Let $0 < \beta < \min_{x\in\partial\mathcal{V}} V(x)$ and $\mathcal{D}_\beta \triangleq \{x \in \mathcal{V} : V(x) \leq \beta\}$. It follows from (4.194) that $\mathcal{D}_\beta \subset \mathcal{M}$ is positively invariant with respect to (4.177). Furthermore, it follows from (4.194), the positive definiteness of $V(\cdot)$, and standard Lyapunov arguments that, for every $\varepsilon > 0$, there exists $\delta > 0$ such that $\mathcal{B}_\delta(0) \subset \mathcal{D}_\beta \subset \mathcal{M}$ and

$$\|x(t)\| \leq \varepsilon, \quad \|x_0\| < \delta, \quad t \in \mathcal{I}_{x_0}. \tag{4.196}$$

Moreover, since the solution $x(t)$ to (4.177) is bounded for all $t \in \mathcal{I}_{x_0}$, it can be extended on the semi-infinite interval $[0, \infty)$, and hence, $x(t)$ is defined for all $t \geq 0$. Furthermore, it follows from Theorem 4.16, with $w(y) = -cy^{\alpha}$ and $z(t) = s(t, V(x_0))$, where $\alpha \in (0, 1)$, that

$$V(x(t)) \leq s(t, V(x_0)), \quad x_0 \in \mathcal{B}_\delta(0), \quad t \in [0, \infty), \tag{4.197}$$

where $s(\cdot, \cdot)$ is given by (4.186) with $k = c$. Now, it follows from (4.186), (4.197), and the positive definiteness of $V(\cdot)$ that

$$x(t) = 0, \quad t \geq \frac{1}{c(1-\alpha)}(V(x_0))^{1-\alpha}, \quad x_0 \in \mathcal{B}_\delta(0), \tag{4.198}$$

which implies finite-time convergence of the trajectories of (4.177) for all $x_0 \in \mathcal{B}_\delta(0)$. This along with (4.196) implies finite-time stability of the zero solution $x(t) \equiv 0$ to (4.177) with $\mathcal{N} \triangleq \mathcal{B}_\delta(0)$.

Since $s(0, x) = x$ and $s(\cdot, \cdot)$ is continuous, $\inf\{t \in \mathbb{R}_+ : s(t, x) = 0\} > 0$, $x \in \mathcal{N} \backslash \{0\}$. Furthermore, it follows from (4.198) that $\inf\{t \in \mathbb{R}_+ : s(t, x) = 0\} < \infty$, $x \in \mathcal{N}$. Now, defining $T : \mathcal{N} \to \mathbb{R}_+$ by using (4.184), (4.195) is immediate from (4.198). Finally, the right-hand side of (4.195) is continuous at the origin, and hence, by Proposition 4.6, continuous on $\mathcal{N}$.

Finally, if $\mathcal{D} = \mathbb{R}^n$ and $V(\cdot)$ is radially unbounded, then global finite-time stability follows using standard arguments. □

Example 4.13. Consider the nonlinear dynamical system $\mathcal{G}$ given by

$$\dot{x}(t) = -(x(t))^{\frac{1}{3}} - (x(t))^{\frac{1}{5}}, \quad x(0) = x_0, \quad t \geq 0, \tag{4.199}$$

where $x \in \mathbb{R}$. For this system, we show that the zero solution $x(t) \equiv 0$ to (4.199) is globally finite-time stable. To see this, consider $V(x) = x^{\frac{4}{3}}$ and let $\mathcal{D} = \mathbb{R}$. Then, $\dot{V}(x) = -\frac{4}{3}(x^{\frac{2}{3}} + x^{\frac{8}{15}}) \leq -\frac{4}{3}x^{\frac{2}{3}} = -\frac{4}{3}(V(x))^{\frac{1}{2}}$ for all $x \in \mathbb{R}$. Hence, it follows from Theorem 4.17 that the zero solution $x(t) \equiv 0$ to (4.199) is globally finite-time stable. Figure 4.8 shows the state trajectory versus time of $\mathcal{G}$ with $x_0 = 1$. △

Finally, we present a converse theorem for finite-time stability in the case where the settling-time function is continuous. For the statement of this result, define

$$\dot{V}(x) \triangleq \lim_{h \to 0^+} \frac{1}{h}[V(s(h, x)) - V(x)], \quad x \in \mathcal{D}, \tag{4.200}$$

for a given continuous function $V : \mathcal{D} \to \mathbb{R}$ and for every $x \in \mathcal{D}$ such that the limit in (4.200) exists.

Theorem 4.18. Let $\alpha \in (0, 1)$ and let $\mathcal{N}$ be as in Definition 4.7. If the zero solution $x(t) \equiv 0$ to (4.177) is finite-time stable and the settling-time

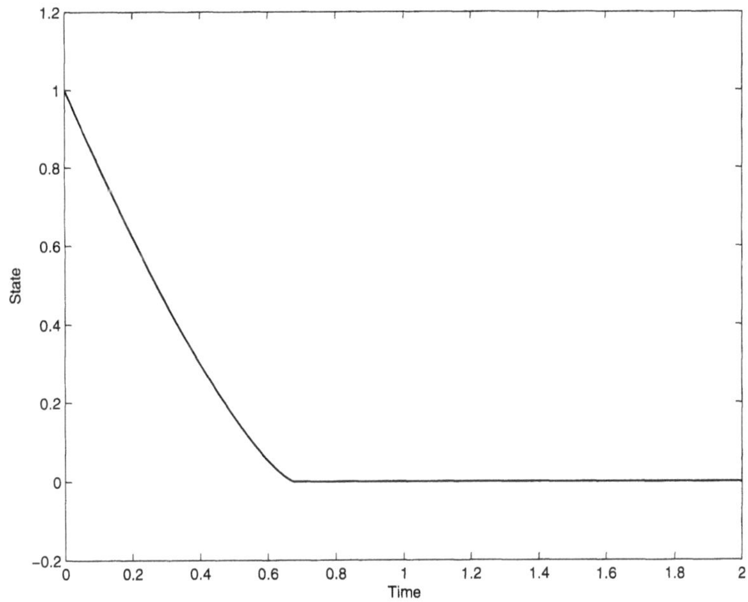

Figure 4.8 State trajectory versus time for Example 4.13.

function $T(\cdot)$ is continuous at $x = 0$, then there exist a continuous function $V : \mathcal{N} \to \mathbb{R}$ and a scalar $c > 0$ such that $V(0) = 0$, $V(x) > 0$, $x \in \mathcal{N}$, $x \neq 0$, and $\dot{V}(x) \leq -c(V(x))^{\alpha}$, $x \in \mathcal{N}$.

Proof. First, it follows from Proposition 4.6 that the settling-time function $T : \mathcal{N} \to \mathbb{R}_+$ is continuous. Next, define $V : \mathcal{N} \to \mathbb{R}_+$ by $V(x) = (T(x))^{\frac{1}{1-\alpha}}$. Note that $V(\cdot)$ is continuous and positive definite and, by $s(T(x)+t, x) = 0$ for all $x \in \mathcal{N}$ and $t \in \mathbb{R}_+$, $\dot{V}(0) = 0$. Now, (4.189) implies that $V(s^x(t))$ is continuously differentiable on $[0, T(x))$, and hence, (4.200) yields

$$\dot{V}(x) = -\frac{1}{1-\alpha}(T(x))^{\frac{\alpha}{1-\alpha}} = -\frac{1}{1-\alpha}(V(x))^{\alpha}. \tag{4.201}$$

Hence, $\dot{V}(\cdot)$ is continuous and negative definite on $\mathcal{N}$ and satisfies $\dot{V}(x) + c(V(x))^{\alpha} = 0$ for all $x \in \mathcal{N}$ with $c = \frac{1}{1-\alpha}$. □

4.7 Semistability of Nonlinear Dynamical Systems

In this section, we develop a stability analysis framework for systems having a continuum of equilibria. Since every neighborhood of a nonisolated equilibrium contains another equilibrium, a nonisolated equilibrium cannot be asymptotically stable. Hence, asymptotic stability is not the appropriate notion of stability for systems having a continuum of equilibria. Two notions that are of particular relevance to such systems are *convergence* and *semistability*. Convergence is the property whereby every system solution converges to a limit point that may depend on the system initial condition. Semistability is the additional requirement that all solutions converge to limit points that are Lyapunov stable. Semistability for an equilibrium thus implies Lyapunov stability, and is implied by asymptotic stability.

It is important to note that semistability is not merely equivalent to asymptotic stability of the set of equilibria. Indeed, it is possible for a trajectory to converge to the set of equilibria without converging to any one equilibrium point (see Problem 4.32). Conversely, semistability does not imply that the equilibrium set is asymptotically stable in any accepted sense. This is because stability of sets (see Section 4.9) is defined in terms of distance (especially in case of noncompact sets), and it is possible to construct examples in which the dynamical system is semistable, but the domain of semistability (see Definition 4.9) contains no ε-neighborhood (defined in terms of the distance) of the (noncompact) equilibrium set, thus ruling out asymptotic stability of the equilibrium set. Hence, semistability and set stability of the equilibrium set are independent notions.

The dependence of the limiting state on the initial state is seen in numerous dynamical systems including compartmental systems [220] which arise in chemical kinetics [47], biomedical [219], environmental [338], economic [40], power [400], and thermodynamic systems [167]. For these systems, every trajectory that starts in a neighborhood of a Lyapunov stable equilibrium converges to a (possibly different) Lyapunov stable equilibrium, and hence these systems are semistable. Semistability is especially pertinent to networks of dynamic agents which exhibit convergence to a state of consensus in which the agents agree on certain quantities of interest [208]. Semistability was first introduced in [81] for linear systems, and applied to matrix second-order systems in [46]. References [57] and [56] consider semistability of nonlinear systems, and give several stability results for systems having a continuum of equilibria based on nontangency and arc length of trajectories, respectively.

In this section, we develop necessary and sufficient conditions for semistability. Specifically, we consider nonlinear dynamical systems $\mathcal{G}$ of the form

$$\dot{x}(t) = f(x(t)), \quad x(0) = x_0, \quad t \in \mathcal{I}_{x_0}, \tag{4.202}$$

where $x(t) \in \mathcal{D} \subseteq \mathbb{R}^n$, $t \in \mathcal{I}_{x_0}$, is the system state vector, $\mathcal{D}$ is an open set, $f : \mathcal{D} \to \mathbb{R}^n$ is Lipschitz continuous on $\mathcal{D}$, $f^{-1}(0) \triangleq \{x \in \mathcal{D} : f(x) = 0\}$ is nonempty, and $\mathcal{I}_{x_0} = [0, \tau_{x_0})$, $0 \leq \tau_{x_0} \leq \infty$, is the maximal interval of existence for the solution $x(\cdot)$ of (4.202). Here, we assume that for every initial condition $x_0 \in \mathcal{D}$, (4.202) has a unique solution defined on $[0, \infty)$, and hence, the solutions of (4.202) define a continuous global semiflow on $\mathcal{D}$. We say that the dynamical system (4.202) is *convergent* with respect to the closed set $\mathcal{D}_c \subseteq \mathcal{D}$ if $\lim_{t\to\infty} s(t,x)$ exists for every $x \in \mathcal{D}_c$.

The following proposition gives a sufficient condition for a trajectory of (4.202) to converge to a limit. For this result, $\mathcal{D}_c \subseteq \mathcal{D}$ denotes a positively invariant set with respect to (4.202) so that the orbit $\mathcal{O}_x$ of (4.202) is contained in $\mathcal{D}_c$ for all $x \in \mathcal{D}_c$.

Proposition 4.7. Consider the nonlinear dynamical system (4.202) and let $x \in \mathcal{D}_c$. If the positive orbit $\mathcal{O}_x^+$ of (4.202) contains a Lyapunov stable equilibrium point y, then $y = \lim_{t\to\infty} s(t,x)$, that is, $\mathcal{O}_x^+ = \{y\}$.

Proof. Suppose $y \in \mathcal{O}_x^+$ is Lyapunov stable and let $\mathcal{N}_\varepsilon \subseteq \mathcal{D}_c$ be an open neighborhood of y. Since y is Lyapunov stable, there exists an open neighborhood $\mathcal{N}_\delta \subset \mathcal{D}_c$ of y such that $s_t(\mathcal{N}_\delta) \subseteq \mathcal{N}_\varepsilon$ for every $t \geq 0$. Now, since $y \in \mathcal{O}_x^+$, it follows that there exists $\tau \geq 0$ such that $s(\tau, x) \in \mathcal{N}_\delta$. Hence, $s(t+\tau, x) = s_t(s(\tau,x)) \in s_t(\mathcal{N}_\delta) \subseteq \mathcal{N}_\varepsilon$ for every $t > 0$. Since $\mathcal{N}_\varepsilon \subseteq \mathcal{D}_c$ is arbitrary, it follows that $y = \lim_{t\to\infty} s(t,x)$. Thus, $\lim_{n\to\infty} s(t_n, x) = y$ for every sequence $\{t_n\}_{n=1}^{\infty}$, and hence, $\mathcal{O}_x^+ = \{y\}$. $\square$

The following definitions and key proposition are necessary for the main results of this section.

Definition 4.8. An equilibrium point $x \in \mathcal{D}$ of (4.202) is *Lyapunov stable* if for every open subset $\mathcal{N}_\varepsilon$ of $\mathcal{D}$ containing x, there exists an open subset $\mathcal{N}_\delta$ of $\mathcal{D}$ containing x such that $s_t(\mathcal{N}_\delta) \subset \mathcal{N}_\varepsilon$ for all $t \geq 0$. An equilibrium point $x \in \mathcal{D}$ of (4.202) is *semistable* if it is Lyapunov stable and there exists an open subset $\mathcal{Q}$ of $\mathcal{D}$ containing x such that for all initial conditions in $\mathcal{Q}$, the trajectory of (4.202) converges to a Lyapunov stable equilibrium point, that is, $\lim_{t\to\infty} s(t,x) = y$, where $y \in \mathcal{D}$ is a Lyapunov stable equilibrium point of (4.202) and $x \in \mathcal{Q}$. If, in addition, $\mathcal{Q} = \mathcal{D} = \mathbb{R}^n$, then an equilibrium point $x \in \mathcal{D}$ of (4.202) is a *globally semistable equilibrium*. The system (4.202) is said to be *Lyapunov stable* if every equilibrium point of (4.202) is Lyapunov stable. The system (4.202) is said to be *semistable* if every equilibrium point of (4.202) is semistable. Finally, (4.202) is said to be *globally semistable* if (4.202) is semistable and $\mathcal{Q} = \mathcal{D} = \mathbb{R}^n$.

Definition 4.9. The *domain of semistability* is the set of points $x_0 \in \mathcal{D}$ such that if $x(t)$ is a solution to (4.202) with $x(0) = x_0$, $t \geq 0$, then $x(t)$ converges to a Lyapunov stable equilibrium point in $\mathcal{D}$.

Note that if (4.202) is semistable, then its domain of semistability contains the set of equilibria in its interior. Next, we present alternative equivalent characterizations of semistability of (4.202).

Proposition 4.8. Consider the nonlinear dynamical system $\mathcal{G}$ given by (4.202). Then the following statements are equivalent:

i) $\mathcal{G}$ is semistable.

ii) For each $x_{\rm e} \in f^{-1}(0)$, there exist class $\mathcal{K}$ and $\mathcal{L}$ functions $\alpha(\cdot)$ and $\beta(\cdot)$, respectively, and $\delta = \delta(x_{\rm e}) > 0$, such that if $\|x_0 - x_{\rm e}\| < \delta$, then $\|x(t) - x_{\rm e}\| \le \alpha(\|x_0 - x_{\rm e}\|)$, $t \ge 0$, and $\operatorname{dist}(x(t), f^{-1}(0)) \le \beta(t)$, $t \ge 0$.

iii) For each $x_{\rm e} \in f^{-1}(0)$, there exist class $\mathcal{K}$ functions $\alpha_1(\cdot)$ and $\alpha_2(\cdot)$, a class $\mathcal{L}$ function $\beta(\cdot)$, and $\delta = \delta(x_{\rm e}) > 0$, such that if $\|x_0 - x_{\rm e}\| < \delta$, then $\operatorname{dist}(x(t), f^{-1}(0)) \le \alpha_1(\|x(t) - x_{\rm e}\|)\beta(t) \le \alpha_2(\|x_0 - x_{\rm e}\|)\beta(t)$, $t \ge 0$.

Proof. To show that *i)* implies *ii)*, suppose (4.202) is semistable and let $x_{\rm e} \in f^{-1}(0)$. It follows from Problem 3.75 that there exists $\delta = \delta(x_{\rm e}) > 0$ and a class $\mathcal{K}$ function $\alpha(\cdot)$ such that if $\|x_0 - x_{\rm e}\| \le \delta$, then $\|x(t) - x_{\rm e}\| \le \alpha(\|x_0 - x_{\rm e}\|)$, $t \ge 0$. Without loss of generality, we may assume that δ is such that $\overline{\mathcal{B}_\delta(x_{\rm e})}$ is contained in the domain of semistability of (4.202). Hence, for every $x_0 \in \overline{\mathcal{B}_\delta(x_{\rm e})}$, $\lim_{t\to\infty} x(t) = x^* \in f^{-1}(0)$ and, consequently, $\lim_{t\to\infty} \operatorname{dist}(x(t), f^{-1}(0)) = 0$.

For each $\varepsilon > 0$ and $x_0 \in \overline{\mathcal{B}_\delta(x_{\rm e})}$, define $T_{x_0}(\varepsilon)$ to be the infimum of T with the property that $\operatorname{dist}(x(t), f^{-1}(0)) < \varepsilon$ for all $t \ge T$, that is, $T_{x_0}(\varepsilon) \triangleq \inf\{T : \operatorname{dist}(x(t), f^{-1}(0)) < \varepsilon, t \ge T\}$. For each $x_0 \in \overline{\mathcal{B}_\delta(x_{\rm e})}$, the function $T_{x_0}(\varepsilon)$ is nonnegative and nonincreasing in ε, and $T_{x_0}(\varepsilon) = 0$ for sufficiently large ε.

Next, let $T(\varepsilon) \triangleq \sup\{T_{x_0}(\varepsilon) : x_0 \in \overline{\mathcal{B}_\delta(x_{\rm e})}\}$. We claim that T is well defined. To show this, consider $\varepsilon > 0$ and $x_0 \in \overline{\mathcal{B}_\delta(x_{\rm e})}$. Since $\operatorname{dist}(s(t, x_0), f^{-1}(0)) < \varepsilon$ for every $t > T_{x_0}(\varepsilon)$, it follows from the continuity of s that, for every $\eta > 0$, there exists an open neighborhood $\mathcal{U}$ of x_0 such that $\operatorname{dist}(s(t, z), f^{-1}(0)) < \varepsilon$ for every $z \in \mathcal{U}$. Hence, $\limsup_{z\to x_0} T_z(\varepsilon) \le T_{x_0}(\varepsilon)$ implying that the function $x_0 \mapsto T_{x_0}(\varepsilon)$ is upper semicontinuous at the arbitrarily chosen point x_0, and hence on $\overline{\mathcal{B}_\delta(x_{\rm e})}$. Since an upper semicontinuous function defined on a compact set achieves its supremum, it follows that $T(\varepsilon)$ is well defined. The function $T(\cdot)$ is the pointwise supremum of a collection of nonnegative and nonincreasing functions, and is hence nonnegative and nonincreasing. Moreover, $T(\varepsilon) = 0$ for every $\varepsilon > \max\{\alpha(\|x_0 - x_{\rm e}\|) : x_0 \in \overline{\mathcal{B}_\delta(x_{\rm e})}\}$.

Let $\psi(\varepsilon) \triangleq \frac{2}{\varepsilon}\int_{\varepsilon/2}^{\varepsilon} T(\sigma)\mathrm{d}\sigma + \frac{1}{\varepsilon} \geq T(\varepsilon) + \frac{1}{\varepsilon}$. The function $\psi(\varepsilon)$ is positive, continuous, strictly decreasing, and $\psi(\varepsilon) \to 0$ as $\varepsilon \to \infty$. Choose $\beta(\cdot) = \psi^{-1}(\cdot)$. Then $\beta(\cdot)$ is positive, continuous, strictly decreasing, and $\beta(\sigma) \to 0$ as $\sigma \to \infty$. Furthermore, $T(\beta(\sigma)) < \psi(\beta(\sigma)) = \sigma$. Hence, $\mathrm{dist}(x(t), f^{-1}(0)) \leq \beta(t)$, $t \geq 0$.

Next, to show that *ii*) implies *iii*), suppose *ii*) holds and let $x_{\mathrm{e}} \in f^{-1}(0)$. Then it follows from Problem 3.75 that x_{e} is Lyapunov stable. Choosing x_0 sufficiently close to x_{e}, it follows from the inequality $\|x(t) - x_{\mathrm{e}}\| \leq \alpha(\|x_0 - x_{\mathrm{e}}\|)$, $t \geq 0$, that trajectories of (4.202) starting sufficiently close to x_{e} are bounded, and hence, the positive limit set of (4.202) is nonempty. Since $\lim_{t\to\infty} \mathrm{dist}(x(t), f^{-1}(0)) = 0$, it follows that the positive limit set is contained in $f^{-1}(0)$. Now, since every point in $f^{-1}(0)$ is Lyapunov stable, it follows from Proposition 4.7 that $\lim_{t\to\infty} x(t) = x^*$, where $x^* \in f^{-1}(0)$ is Lyapunov stable. If $x^* = x_{\mathrm{e}}$, then it follows using similar arguments as above that there exists a class $\mathcal{L}$ function $\hat{\beta}(\cdot)$ such that $\mathrm{dist}(x(t), f^{-1}(0)) \leq \|x(t) - x_{\mathrm{e}}\| \leq \hat{\beta}(t)$ for every x_0 satisfying $\|x_0 - x_{\mathrm{e}}\| < \delta$ and $t \geq 0$. Hence, $\mathrm{dist}(x(t), f^{-1}(0)) \leq \sqrt{\|x(t) - x_{\mathrm{e}}\|}\sqrt{\hat{\beta}(t)}$, $t \geq 0$. Next, consider the case where $x^* \neq x_{\mathrm{e}}$ and let $\alpha_1(\cdot)$ be a class $\mathcal{K}$ function. In this case, note that $\lim_{t\to\infty} \mathrm{dist}(x(t), f^{-1}(0))/\alpha_1(\|x(t) - x_{\mathrm{e}}\|) = 0$, and hence, it follows using similar arguments as above that there exists a class $\mathcal{L}$ function $\beta(\cdot)$ such that $\mathrm{dist}(x(t), f^{-1}(0)) \leq \alpha_1(\|x(t) - x_{\mathrm{e}}\|)\beta(t)$, $t \geq 0$. Finally, note that $\alpha_1 \circ \alpha$ is of class $\mathcal{K}$ (by Problem 3.71), and hence, *iii*) follows immediately.

Finally, to show that *iii*) implies *i*), suppose *iii*) holds and let $x_{\mathrm{e}} \in f^{-1}(0)$. Then it follows that $\alpha_1(\|x(t) - x_{\mathrm{e}}\|) \leq \alpha_2(\|x(0) - x_{\mathrm{e}}\|)$, $t \geq 0$, that is, $\|x(t) - x_{\mathrm{e}}\| \leq \alpha(\|x(0) - x_{\mathrm{e}}\|)$, where $t \geq 0$ and $\alpha = \alpha_1^{-1} \circ \alpha_2$ is of class $\mathcal{K}$ (by Problem 3.71). It now follows from Problem 3.75 that x_{e} is Lyapunov stable. Since x_{e} was chosen arbitrarily, it follows that every equilibrium point is Lyapunov stable. Furthermore, $\lim_{t\to\infty} \mathrm{dist}(x(t), f^{-1}(0)) = 0$. Choosing x_0 sufficiently close to x_{e}, it follows from the inequality $\|x(t) - x_{\mathrm{e}}\| \leq \alpha(\|x_0 - x_{\mathrm{e}}\|)$, $t \geq 0$, that trajectories of (4.202) starting sufficiently close to x_{e} are bounded, and hence, the positive limit set of (4.202) is nonempty. Since every point in $f^{-1}(0)$ is Lyapunov stable, it follows from Proposition 4.7 that $\lim_{t\to\infty} x(t) = x^*$, where $x^* \in f^{-1}(0)$ is Lyapunov stable. Hence, by definition, (4.202) is semistable. □

Next, we present a sufficient condition for semistability.

Theorem 4.19. Consider the nonlinear dynamical system (4.202). Let $\mathcal{Q}$ be an open neighborhood of $f^{-1}(0)$ and assume that there exists a continuously differentiable function $V : \mathcal{Q} \to \mathbb{R}$ such that

$$V'(x)f(x) < 0, \quad x \in \mathcal{Q}\backslash f^{-1}(0). \tag{4.203}$$

If (4.202) is Lyapunov stable, then (4.202) is semistable.

Proof. Since (4.202) is Lyapunov stable by assumption, for every $z \in f^{-1}(0)$, there exists an open neighborhood $\mathcal{V}_z$ of z such that $s([0,\infty) \times \mathcal{V}_z)$ is bounded and contained in $\mathcal{Q}$. The set $\mathcal{V} \triangleq \bigcup_{z\in f^{-1}(0)} \mathcal{V}_z$ is an open neighborhood of $f^{-1}(0)$ contained in $\mathcal{Q}$. Consider $x \in \mathcal{V}$ so that there exists $z \in f^{-1}(0)$ such that $x \in \mathcal{V}_z$ and $s(t,x) \in \mathcal{V}_z$, $t \geq 0$. Since $\mathcal{V}_z$ is bounded it follows that the positive limit set of x is nonempty and invariant. Furthermore, it follows from (4.203) that $\dot{V}(s(t,x)) \leq 0$, $t \geq 0$, and hence, it follows from Theorem 3.3 that $s(t,x) \to \mathcal{M}$ as $t \to \infty$, where $\mathcal{M}$ is the largest invariant set contained in the set $\mathcal{R} = \{y \in \mathcal{V}_z : V'(y)f(y) = 0\}$. Note that $\mathcal{R} = f^{-1}(0)$ is invariant, and hence, $\mathcal{M} = \mathcal{R}$, which implies that $\lim_{t\to\infty} \operatorname{dist}(s(t,x), f^{-1}(0)) = 0$. Finally, since every point in $f^{-1}(0)$ is Lyapunov stable, it follows from Proposition 4.7 that $\lim_{t\to\infty} s(t,x) = x^*$, where $x^* \in f^{-1}(0)$ is Lyapunov stable. Hence, by definition, (4.202) is semistable. □

Next, we present a slightly more general theorem for semistability wherein we do not assume that all points in $\dot{V}^{-1}(0)$ are Lyapunov stable but rather we assume that all points in the largest invariant subset of $\dot{V}^{-1}(0)$ are Lyapunov stable.

Theorem 4.20. Consider the nonlinear dynamical system (4.202) and let $\mathcal{Q}$ be an open neighborhood of $f^{-1}(0)$. Suppose the orbit $\mathcal{O}_x$ of (4.202) is bounded for all $x \in \mathcal{Q}$ and assume that there exists a continuously differentiable function $V : \mathcal{Q} \to \mathbb{R}$ such that

$$V'(x)f(x) \leq 0, \quad x \in \mathcal{Q}. \tag{4.204}$$

If every point in the largest invariant subset $\mathcal{M}$ of $\{x \in \mathcal{Q} : V'(x)f(x) = 0\}$ is Lyapunov stable, then (4.202) is semistable.

Proof. Since every solution of (4.202) is bounded, it follows from the hypotheses on $V(\cdot)$ that, for every $x \in \mathcal{Q}$, the positive limit set $\omega(x)$ of (4.202) is nonempty and contained in the largest invariant subset $\mathcal{M}$ of $\{x \in \mathcal{Q} : V'(x)f(x) = 0\}$. Since every point in $\mathcal{M}$ is a Lyapunov stable equilibrium, it follows from Proposition 4.7 that $\omega(x)$ contains a single point for every $x \in \mathcal{Q}$ and $\lim_{t\to\infty} s(t,x)$ exists for every $x \in \mathcal{Q}$. Now, since $\lim_{t\to\infty} s(t,x) \in \mathcal{M}$ is Lyapunov stable for every $x \in \mathcal{Q}$, semistability is immediate. □

Example 4.14. Consider the nonlinear dynamical system given by

$$\dot{x}_1(t) = \sigma_{12}(x_2(t)) - \sigma_{21}(x_1(t)), \quad x_1(0) = x_{10}, \quad t \geq 0, \tag{4.205}$$
$$\dot{x}_2(t) = \sigma_{21}(x_1(t)) - \sigma_{12}(x_2(t)), \quad x_2(0) = x_{20}, \tag{4.206}$$

where $x_1, x_2 \in \mathbb{R}$, $\sigma_{ij}(\cdot)$, $i, j = 1, 2$, $i \neq j$, are Lipschitz continuous, $\sigma_{12}(x_2) - \sigma_{21}(x_1) = 0$ if and only if $x_1 = x_2$, and $(x_1 - x_2)(\sigma_{12}(x_2) - \sigma_{21}(x_1)) \leq 0$, $x_1, x_2 \in \mathbb{R}$. Note that $f^{-1}(0) = \{(x_1, x_2) \in \mathbb{R}^2 : x_1 = x_2 = \alpha, \alpha \in \mathbb{R}\}$. To show that (4.205) and (4.206) is semistable, consider the Lyapunov function candidate $V(x_1, x_2) = \frac{1}{2}(x_1 - \alpha)^2 + \frac{1}{2}(x_2 - \alpha)^2$, where $\alpha \in \mathbb{R}$. Now, it follows that

$$\begin{aligned} \dot{V}(x_1, x_2) &= (x_1 - \alpha)[\sigma_{12}(x_2) - \sigma_{21}(x_1)] + (x_2 - \alpha)[\sigma_{21}(x_1) - \sigma_{12}(x_2)] \\ &= x_1[\sigma_{12}(x_2) - \sigma_{21}(x_1)] + x_2[\sigma_{21}(x_1) - \sigma_{12}(x_2)] \\ &= (x_1 - x_2)[\sigma_{12}(x_2) - \sigma_{21}(x_1)] \\ &\leq 0, \quad (x_1, x_2) \in \mathbb{R} \times \mathbb{R}, \end{aligned} \tag{4.207}$$

which implies that $x_1 = x_2 = \alpha$ is Lyapunov stable.

Next, let $\mathcal{R} \triangleq \{(x_1, x_2) \in \mathbb{R}^2 : \dot{V}(x_1, x_2) = 0\} = \{(x_1, x_2) \in \mathbb{R}^2 : x_1 = x_2 = \alpha, \alpha \in \mathbb{R}\}$. Since $\mathcal{R}$ consists of equilibrium points, it follows that $\mathcal{M} = \mathcal{R}$. Hence, for every $x_1(0), x_2(0) \in \mathbb{R}$, $(x_1(t), x_2(t)) \to \mathcal{M}$ as $t \to \infty$. Hence, it follows from Theorem 4.20 that $x_1 = x_2 = \alpha$ is semistable for all $\alpha \in \mathbb{R}$. △

Finally, we provide a converse Lyapunov theorem for semistability. For this result, recall that for a given continuous function $V : \mathcal{D} \to \mathbb{R}$, the *upper right Dini derivative* of V along the solution of (4.202) is defined by

$$\dot{V}(s(t, x)) \triangleq \limsup_{h \to 0^+} \frac{1}{h}[V(s(t + h, x)) - V(s(t, x))]. \tag{4.208}$$

It is easy to see that $\dot{V}(x_{\rm e}) = 0$ for every $x_{\rm e} \in f^{-1}(0)$. Also note that it follows from (4.208) that $\dot{V}(x) = \dot{V}(s(0, x))$.

Theorem 4.21. Consider the nonlinear dynamical system (4.202). Suppose (4.202) is semistable with the domain of semistability $\mathcal{D}_0$. Then there exist a continuous nonnegative function $V : \mathcal{D}_0 \to \overline{\mathbb{R}}_+$ and a class $\mathcal{K}$ function $\alpha(\cdot)$ such that *i)* $V(x) = 0$, $x \in f^{-1}(0)$, *ii)* $V(x) \geq \alpha(\mathrm{dist}(x, f^{-1}(0)))$, $x \in \mathcal{D}_0$, and *iii)* $\dot{V}(x) < 0$, $x \in \mathcal{D}_0 \backslash f^{-1}(0)$.

Proof. Define the function $V : \mathcal{D}_0 \to \overline{\mathbb{R}}_+$ by

$$V(x) \triangleq \sup_{t \geq 0} \left\{ \frac{1 + 2t}{1 + t} \mathrm{dist}(s(t, x), f^{-1}(0)) \right\}, \quad x \in \mathcal{D}_0. \tag{4.209}$$

Note that $V(\cdot)$ is well defined since (4.202) is semistable. Clearly, *i)* holds.

Furthermore, since $V(x) \geq \text{dist}(x, f^{-1}(0))$, $x \in \mathcal{D}_0$, it follows that *ii*) holds.

To show that $V(\cdot)$ is continuous on $\mathcal{D}_0 \backslash f^{-1}(0)$, define $T : \mathcal{D}_0 \backslash f^{-1}(0) \to [0, \infty)$ by $T(z) \triangleq \inf\{h : \text{dist}(s(t,z), f^{-1}(0)) < \text{dist}(z, f^{-1}(0))/2 \text{ for all } t \geq h > 0\}$, and denote

$$\mathcal{W}_\varepsilon \triangleq \{x \in \mathcal{D}_0 : \text{dist}(s(t,x), f^{-1}(0)) < \varepsilon, t \geq 0\}. \tag{4.210}$$

Note that $\mathcal{W}_\varepsilon \supset f^{-1}(0)$ is open and positively invariant, and contains an open neighborhood of $f^{-1}(0)$. Consider $z \in \mathcal{D}_0 \backslash f^{-1}(0)$ and define $\lambda \triangleq \text{dist}(z, f^{-1}(0)) > 0$. Then it follows from semistability of (4.202) that there exists $h > 0$ such that $s(h,z) \in \mathcal{W}_{\lambda/2}$. Consequently, $s(h+t, z) \in \mathcal{W}_{\lambda/2}$ for all $t \geq 0$, and hence, it follows that $T(z)$ is well defined. Since $\mathcal{W}_{\lambda/2}$ is open, there exists a neighborhood $\mathcal{B}_\sigma(s(T(z),z)) \subset \mathcal{W}_{\lambda/2}$. Hence, $\mathcal{N} \triangleq s_{-T(z)}(\mathcal{B}_\sigma(s(T(z),z)))$ is a neighborhood of z and $\mathcal{N} \subset \mathcal{D}_0$. Choose $\eta > 0$ such that $\eta < \lambda/2$ and $\mathcal{B}_\eta(z) \subset \mathcal{N}$. Then, for every $t > T(z)$ and $y \in \mathcal{B}_\eta(z)$, $[(1+2t)/(1+t)]\text{dist}(s(t,y), f^{-1}(0)) \leq 2\text{dist}(s(t,y), f^{-1}(0)) \leq \lambda$. Therefore, for each $y \in \mathcal{B}_\eta(z)$,

$$\begin{aligned} V(z) - V(y) &= \sup_{t \geq 0} \left\{ \frac{1+2t}{1+t} \text{dist}(s(t,z), f^{-1}(0)) \right\} \\ &\quad - \sup_{t \geq 0} \left\{ \frac{1+2t}{1+t} \text{dist}(s(t,y), f^{-1}(0)) \right\} \\ &= \sup_{0 \leq t \leq T(z)} \left\{ \frac{1+2t}{1+t} \text{dist}(s(t,z), f^{-1}(0)) \right\} \\ &\quad - \sup_{0 \leq t \leq T(z)} \left\{ \frac{1+2t}{1+t} \text{dist}(s(t,y), f^{-1}(0)) \right\}. \end{aligned} \tag{4.211}$$

Hence,

$$\begin{aligned} &|V(z) - V(y)| \\ &\quad \leq \sup_{0 \leq t \leq T(z)} \left| \frac{1+2t}{1+t} \left(\text{dist}(s(t,z), f^{-1}(0)) - \text{dist}(s(t,y), f^{-1}(0)) \right) \right| \\ &\quad \leq 2 \sup_{0 \leq t \leq T(z)} \left| \text{dist}(s(t,z), f^{-1}(0)) - \text{dist}(s(t,y), f^{-1}(0)) \right| \\ &\quad \leq 2 \sup_{0 \leq t \leq T(z)} \text{dist}(s(t,z), s(t,y)), \quad z \in \mathcal{D}_0 \backslash f^{-1}(0), \quad y \in \mathcal{B}_\eta(z). \end{aligned} \tag{4.212}$$

Now, it follows from continuous dependence of solutions $s(\cdot,\cdot)$ on system initial conditions (Theorem 2.26) and (4.212) that $V(\cdot)$ is continuous on $\mathcal{D}_0 \backslash f^{-1}(0)$.

To show that $V(\cdot)$ is continuous on $f^{-1}(0)$, consider $x_e \in f^{-1}(0)$. Let $\{x_n\}_{n=1}^{\infty}$ be a sequence in $\mathcal{D}_0 \backslash f^{-1}(0)$ that converges to x_e. Since x_e is Lyapunov stable, it follows that $x(t) \equiv x_e$ is the unique solution to (4.202) with $x_0 = x_e$. By continuous dependence of solutions $s(\cdot,\cdot)$ on system initial conditions (Theorem 2.26), $s(t, x_n) \to s(t, x_e) = x_e$ as $n \to \infty$, $t \geq 0$.

Let $\varepsilon > 0$ and note that it follows from *ii*) of Proposition 4.8 that there exists $\delta = \delta(x_e) > 0$ such that for every solution of (4.202) in $\mathcal{B}_\delta(x_e)$ there exists $\hat{T} = \hat{T}(x_e, \varepsilon) > 0$ such that $s_t(\mathcal{B}_\delta(x_e)) \subset \mathcal{W}_\varepsilon$ for all $t \geq \hat{T}$. Next, note that there exists a positive integer N_1 such that $x_n \in \mathcal{B}_\delta(x_e)$ for all $n \geq N_1$. Now, it follows from (4.209) that

$$V(x_n) \leq 2 \sup_{0 \leq t \leq \hat{T}} \operatorname{dist}(s(t, x_n), f^{-1}(0)) + 2\varepsilon, \quad n \geq N_1. \tag{4.213}$$

Next, it follows from Lemma 3.1 of Chapter I of [179] that $s(\cdot, x_n)$ converges to $s(\cdot, x_e)$ uniformly on $[0, \hat{T}]$. Hence,

$$\begin{aligned} \lim_{n\to\infty} \sup_{0 \leq t \leq \hat{T}} \operatorname{dist}(s(t, x_n), f^{-1}(0)) &= \sup_{0 \leq t \leq \hat{T}} \operatorname{dist}(\lim_{n\to\infty} s(t, x_n), f^{-1}(0)) \\ &= \sup_{0 \leq t \leq \hat{T}} \operatorname{dist}(x_e, f^{-1}(0)) \\ &= 0, \end{aligned} \tag{4.214}$$

which implies that there exists a positive integer $N_2 = N_2(x_e, \varepsilon) \geq N_1$ such that $\sup_{0 \leq t \leq \hat{T}} \operatorname{dist}(s(t, x_n), f^{-1}(0)) < \varepsilon$ for all $n \geq N_2$. Combining (4.213) with the above result yields $V(x_n) < 4\varepsilon$ for all $n \geq N_2$, which implies that $\lim_{n\to\infty} V(x_n) = 0 = V(x_e)$.

Next, we show that $V(x(t))$ is strictly decreasing along the solution of (4.202) on $\mathcal{D} \backslash f^{-1}(0)$. Note that for every $x \in \mathcal{D}_0 \backslash f^{-1}(0)$ and $0 < h \leq 1/2$ such that $s(h, x) \in \mathcal{D}_0 \backslash f^{-1}(0)$, it follows from the definition of $T(\cdot)$ that $V(s(h, x))$ is reached at some time $\hat{t}$ such that $0 \leq \hat{t} \leq T(x)$. Hence,

$$\begin{aligned} V(s(h, x)) &= \operatorname{dist}(s(\hat{t} + h, x), f^{-1}(0)) \frac{1 + 2\hat{t}}{1 + \hat{t}} \\ &= \operatorname{dist}(s(\hat{t} + h, x), f^{-1}(0)) \frac{1 + 2\hat{t} + 2h}{1 + \hat{t} + h} \left[1 - \frac{h}{(1 + 2\hat{t} + 2h)(1 + \hat{t})}\right] \\ &\leq V(x) \left[1 - \frac{h}{2(1 + T(x))^2}\right], \end{aligned} \tag{4.215}$$

which implies that $\dot{V}(x) \leq -\frac{1}{2} V(x)(1 + T(x))^{-2} < 0$, $x \in \mathcal{D}_0 \backslash f^{-1}(0)$, and hence, *iii*) holds. □

It is important to note that a converse Lyapunov theorem for semistability involving a smooth (i.e., infinitely differentiable) Lyapunov function for dynamical systems with continuous vector fields can also be established. For details, see [209].

4.8 Generalized Lyapunov Theorems

Lyapunov's results, along with the Barbashin-Krasovskii-LaSalle invariance principle, provide a powerful framework for analyzing the stability of nonlinear dynamical systems as well as designing feedback controllers that guarantee closed-loop system stability. In particular, as discussed in Sections 3.2 and 3.3, Lyapunov's direct method can provide local and global stability conclusions of an equilibrium point of a nonlinear dynamical system if a continuously differentiable positive-definite function of the nonlinear system states (Lyapunov function) can be constructed for which its time rate of change due to perturbations in a neighborhood of the system's equilibrium is always negative or zero, with strict negative-definiteness ensuring asymptotic stability. Alternatively, using the Barbashin-Krasovskii-LaSalle invariance principle the strict negative-definiteness condition on the Lyapunov derivative can be relaxed while ensuring asymptotic stability. In particular, if a continuously differentiable function defined on a compact invariant set with respect to the nonlinear dynamical system can be constructed whose derivative along the system's trajectories is negative semidefinite, and no system trajectories can stay indefinitely at points where the function's derivative identically vanishes, then the system's equilibrium is asymptotically stable.

Most Lyapunov stability and invariant set theorems presented in the literature require that the Lyapunov function candidate for a nonlinear dynamical system be a continuously differentiable function with a negative-definite derivative (see [178, 228, 235, 260, 445, 474] and the numerous references therein). This is due to the fact that the majority of the dynamical systems considered are systems possessing continuous motions, and hence, Lyapunov theorems provide stability conditions that do not require knowledge of the system trajectories. However, in light of the increasingly complex nature of dynamical systems such as biological systems [284], hybrid systems [473], sampled-data systems [176], discrete-event systems [346], gain scheduled systems [271,299,348], constrained mechanical systems [18], and impulsive systems [20], system discontinuities arise naturally. Even though standard Lyapunov theory is applicable for systems with discontinuous system dynamics and continuous motions, it might be simpler to construct discontinuous "Lyapunov" functions to establish system stability. For example, in gain scheduling control it is not uncommon to use several different controllers designed over several fixed operating points covering the system's operating range and to switch between them over this range. Even though for each operating range one can construct a continuously differentiable Lyapunov function, to show closed-loop system stability over the whole system operating envelope for a given switching control strategy, a *generalized* Lyapunov function involving combinations of the Lyapunov functions for each operating range can be constructed [269–271, 299, 348]. However, in this case, it can be shown that the generalized Lyapunov function is nonsmooth and noncontinuous [269–271, 299, 348].

In this section, we develop generalized Lyapunov and invariant set theorems for nonlinear dynamical systems wherein all regularity assumptions on the Lyapunov function and the system dynamics are removed. In particular, local and global stability theorems are presented using generalized Lyapunov functions that are lower semicontinuous. Furthermore, generalized invariant set theorems are derived wherein system trajectories converge to a union of largest invariant sets contained in intersections over finite intervals of the closure of generalized Lyapunov level surfaces. In the case where the generalized Lyapunov function is taken to be a continuously differentiable function, the results collapse to the standard Lyapunov stability and invariant set theorems presented earlier. Lower semicontinuous Lyapunov functions have been considered in [15] in the context of viability theory and differential inclusions. However, the present formulation provides invariant set stability theorem generalizations not considered in [15].

To present the main results of this section recall that a continuous function $x : \mathcal{I}_{x_0} \to \mathcal{D}$ is said to be a solution to (3.1) on the interval $\mathcal{I}_{x_0} \subseteq \mathbb{R}$ if $x(t)$ satisfies (3.1) for all $t \in \mathcal{I}_{x_0}$. Furthermore, unless otherwise stated, in this section we do not assume any regularity conditions on the system dynamics $f(\cdot)$. However, we do assume that $f(\cdot)$ is such that the solution $x(t)$, $t \geq 0$, to (3.1) is well defined on the time interval $\mathcal{I}_{x_0} = [0, \infty)$. That is, we assume that for every $y \in \mathcal{D}$ there exists a unique solution $x(\cdot)$ of (3.1) defined on $[0, \infty)$ satisfying $x(0) = y$. Furthermore, we assume that all the solutions $x(t)$, $t \geq 0$, to (3.1) are continuous functions of the initial conditions $x_0 \in \mathcal{D}$.

The following result presents sufficient conditions for Lyapunov and asymptotic stability of a nonlinear dynamical system, wherein the assumption of continuous differentiability on the Lyapunov function with a negative-definite derivative is relaxed.

Theorem 4.22. Consider the nonlinear dynamical system (3.1) and let $x(t)$, $t \geq 0$, denote the solution to (3.1). Assume that there exists a lower semicontinuous function $V : \mathcal{D} \to \mathbb{R}$ such that $V(\cdot)$ is continuous at the origin and

$$V(0) = 0, \tag{4.216}$$

$$V(x) > 0, \quad x \in \mathcal{D}, \quad x \neq 0, \tag{4.217}$$

$$V(x(t)) \leq V(x(\tau)), \qquad 0 \leq \tau \leq t. \tag{4.218}$$

Then the zero solution $x(t) \equiv 0$ to (3.1) is Lyapunov stable. If, in addition, there exists an increasing unbounded sequence $\{t_n\}_{n=0}^{\infty}$, with $t_0 = 0$, such that

$$V(x(t_{n+1})) < V(x(t_n)), \quad n = 0, 1, \ldots, \tag{4.219}$$

then the zero solution $x(t) \equiv 0$ to (3.1) is asymptotically stable.

Proof. To show Lyapunov stability, let $\varepsilon > 0$ be such that $\mathcal{B}_\varepsilon(0) \subseteq \mathcal{D}$. Since $\partial\mathcal{B}_\varepsilon(0)$ is compact and $V(x)$, $x \in \mathcal{D}$, is lower semicontinuous it follows from Theorem 2.11 that there exists $\alpha = \min_{x\in\partial\mathcal{B}_\varepsilon(0)} V(x)$. Note that $\alpha > 0$ since $0 \notin \partial\mathcal{B}_\varepsilon(0)$ and $V(x) > 0$, $x \in \mathcal{D}$, $x \neq 0$. Next, since $V(0) = 0$ and $V(\cdot)$ is continuous at the origin it follows that there exists $\delta \in (0, \varepsilon]$ such that $V(x) < \alpha$, $x \in \mathcal{B}_\delta(0)$. Now, it follows from (4.218) that for all $x(0) \in \mathcal{B}_\delta(0)$,

$$V(x(t)) \leq V(x(0)) < \alpha, \qquad t \geq 0,$$

which, since $V(x) \geq \alpha$, $x \in \partial\mathcal{B}_\varepsilon(0)$, implies that $x(t) \notin \partial\mathcal{B}_\varepsilon(0)$, $t \geq 0$. Hence, for all $\varepsilon > 0$ such that $\mathcal{B}_\varepsilon(0) \subseteq \mathcal{D}$ there exists $\delta > 0$ such that if $\|x(0)\| < \delta$, then $\|x(t)\| < \varepsilon$, $t \geq 0$, which proves Lyapunov stability.

To prove asymptotic stability let $x_0 \in \mathcal{B}_\delta(0)$ and suppose there exists an increasing unbounded sequence $\{t_n\}_{n=0}^{\infty}$, with $t_0 = 0$, such that (4.219) holds. Then it follows that $x(t) \in \mathcal{B}_\varepsilon(0)$, $t \geq 0$, and hence, it follows from Theorem 2.41 that the positive limit set $\omega(x_0)$ of $x(t)$, $t \geq 0$, is a nonempty, compact, invariant connected set. Furthermore, $x(t) \to \omega(x_0)$ as $t \to \infty$. Now, since $V(x(t))$, $t \geq 0$, is nonincreasing and bounded from below by zero it follows that $\beta \triangleq \lim_{t\to\infty} V(x(t)) \geq 0$ is well defined. Furthermore, since $V(\cdot)$ is lower semicontinuous it can be shown that $V(y) \leq \beta$, $y \in \omega(x_0)$, and hence, since $V(\cdot)$ is nonnegative, $\beta = 0$ if and only if $\omega(x_0) = \{0\}$ or, equivalently, $x(t) \to 0$ as $t \to \infty$. Now, suppose, *ad absurdum*, that $x(t)$, $t \geq 0$, does not converge to zero or, equivalently, $\beta > 0$. Furthermore, let $y \in \omega(x_0)$ and let $\{\tau_n\}_{n=0}^{\infty}$ be an increasing unbounded sequence such that $\lim_{n\to\infty} x(\tau_n) = y$. Since $\{V(x(\tau_n))\}_{n=0}^{\infty}$ is a lower bounded nonincreasing sequence, $\lim_{n\to\infty} V(x(\tau_n))$ exists and is equal to β. Hence, $y \neq 0$, which implies that $0 \notin \omega(x_0)$. Next, let $\gamma \triangleq \min_{x\in\omega(x_0)} V(x) > 0$ and let $x_\gamma \in \omega(x_0)$ be such that $V(x_\gamma) = \gamma$. Now, since $\omega(x_0)$ is an invariant set it follows that for all $x(0) \in \omega(x_0)$, $x(t) \in \omega(x_0)$, $t \geq 0$, and hence, $V(x(t)) \geq \gamma$, $t \geq 0$. However, since for all $x_0 \in \mathcal{D}$ there exists an increasing unbounded sequence $\{t_n\}_{n=1}^{\infty}$ such that (4.219) holds, it follows that if $x(0) = x_\gamma \in \omega(x_0)$, there exists $t > 0$ such that $V(x(t)) < V(x(0)) = \gamma$, which is a contradiction. Hence, $x(t) \to 0$ as $t \to \infty$, establishing asymptotic stability. □

A lower semicontinuous function $V(\cdot)$, with $V(\cdot)$ being continuous at the origin, satisfying (4.216) and (4.217) is called a *generalized Lyapunov function candidate* for the nonlinear dynamical system (3.1). If, additionally, $V(\cdot)$ satisfies (4.218), $V(\cdot)$ is called a *generalized Lyapunov function* for the nonlinear dynamical system (3.1). Note that if the function $V(\cdot)$ is continuously differentiable on $\mathcal{D}$ in Theorem 4.22 and $V'(x)f(x) \leq 0$, $x \in \mathcal{D}$ (respectively, $V'(x)f(x) < 0$, $x \in \mathcal{D}$, $x \neq 0$), then $V(x(t))$ is a nonincreasing function of time (respectively, there exists an increasing unbounded sequence $\{t_n\}_{n=0}^{\infty}$, with $t_0 = 0$, such that $V(x(t_{n+1})) < V(x(t_n))$, $n = 0, 1, \ldots$). In this case, Theorem 4.22 specializes to Theorem 3.1.

Next, we provide a partial converse to Theorem 3.1. Specifically, we show that if the nonlinear dynamical system (3.1) is Lyapunov stable, then there exists a generalized Lyapunov function for (3.1).

Theorem 4.23. Consider the nonlinear dynamical system (3.1) and let $s(t, x_0)$, $t \geq 0$, denote the solution to (3.1) with initial condition x_0. Assume that the zero solution $x(t) \equiv 0$ to (3.1) is Lyapunov stable. Then there exists a lower semicontinuous function $V : \mathcal{D}_0 \to \mathbb{R}$, where $\mathcal{D}_0 \subseteq \mathcal{D}$, such that $0 \in \overset{\circ}{\mathcal{D}}_0$, $V(\cdot)$ is continuous at the origin, and

$$V(0) = 0, \tag{4.220}$$
$$V(x) > 0, \quad x \in \mathcal{D}_0, \quad x \neq 0, \tag{4.221}$$
$$V(s(t,x)) \leq V(s(\tau,x)), \quad 0 \leq \tau \leq t, \quad x \in \mathcal{D}_0. \tag{4.222}$$

Proof. Let $\varepsilon > 0$. Since the zero solution $x(t) \equiv 0$ to (3.1) is Lyapunov stable it follows that there exists $\delta > 0$ such that if $x_0 \in \mathcal{B}_\delta(0)$, then $s(t, x_0) \in \mathcal{B}_\varepsilon(0)$, $t \geq 0$. Now, let $\mathcal{D}_0 = \{y \in \mathcal{B}_\varepsilon(0) :$ there exists $t \geq 0$ and $x_0 \in \mathcal{B}_\delta(0)$ such that $y = s(t, x_0)\}$, that is, $\mathcal{D}_0 = \cup_{t \geq 0} s_t(\mathcal{B}_\delta(0))$. Note that $\mathcal{D}_0 \subseteq \mathcal{B}_\varepsilon(0)$, $\mathcal{D}_0$ is positively invariant, and $\mathcal{B}_\delta(0) \subseteq \mathcal{D}_0$. Hence, $0 \in \overset{\circ}{\mathcal{D}}_0$. Next, define $V(x) \triangleq \sup_{t \geq 0} \|s(t,x)\|$, $x \in \mathcal{D}_0$, and since $\mathcal{D}_0$ is positively invariant and bounded it follows that $V(\cdot)$ is well defined on $\mathcal{D}_0$. Now, $x = 0$ implies $s(t, x) \equiv 0$, and hence, $V(0) = 0$. Furthermore, $V(x) \geq \|s(0,x)\| = \|x\| > 0$, $x \in \mathcal{D}_0$, $x \neq 0$.

Next, since $f(\cdot)$ in (3.1) is such that for every $x \in \mathcal{D}_0$, $s(t,x)$, $t \geq 0$, is the unique solution to (3.1), it follows that $s(t, x) = s(t - \tau, s(\tau, x))$, $0 \leq \tau \leq t$. Hence, for every $t, \tau \geq 0$, such that $t \geq \tau$,

$$\begin{aligned} V(s(\tau,x)) &= \sup_{\theta \geq 0} \|s(\theta, s(\tau,x))\| \\ &= \sup_{\theta \geq 0} \|s(\tau + \theta, x)\| \\ &\geq \sup_{\theta \geq t-\tau} \|s(\tau + \theta, x)\| \end{aligned}$$

$$
\begin{aligned}
&= \sup_{\theta \geq t-\tau} \|s(\theta - (t-\tau), s(t,x))\| \\
&= \sup_{\theta \geq 0} \|s(\theta, s(t,x))\| \\
&= V(s(t,x)), \qquad (4.223)
\end{aligned}
$$

which proves (4.222). Next, since the zero solution $x(t) \equiv 0$ to (3.1) is Lyapunov stable it follows that for every $\hat{\varepsilon} > 0$ there exists $\hat{\delta} > 0$ such that if $x_0 \in \mathcal{B}_{\hat{\delta}}(0)$, then $s(t, x_0) \in \mathcal{B}_{\hat{\varepsilon}/2}(0)$, $t \geq 0$, which implies that $V(x_0) = \sup_{t\geq 0} \|s(t,x_0)\| \leq \hat{\varepsilon}/2$. Hence, for every $\hat{\varepsilon} > 0$ there exists $\hat{\delta} > 0$ such that if $x_0 \in \mathcal{B}_{\hat{\delta}}(0)$, then $V(x_0) < \hat{\varepsilon}$, establishing that $V(\cdot)$ is continuous at the origin.

Finally, to show that $V(\cdot)$ is lower semicontinuous everywhere on $\mathcal{D}_0$, let $x \in \mathcal{D}_0$ and let $\hat{\varepsilon} > 0$, and note that since $V(x) = \sup_{t\geq 0} \|s(t,x)\|$ there exists $T = T(x, \hat{\varepsilon}) > 0$ such that $V(x) - \|s(T,x)\| < \hat{\varepsilon}$. Now, consider a sequence $\{x_i\}_{i=1}^{\infty} \in \mathcal{D}_0$ such that $x_i \to x$ as $i \to \infty$. Next, since by assumption $s(t, \cdot)$ is continuous for every $t \geq 0$ and $\|\cdot\| : \mathcal{D}_0 \to \mathbb{R}$ is continuous, it follows that $\|s(T,x)\| = \lim_{i\to\infty} \|s(T,x_i)\|$. Next, note that $\|s(T,x_i)\| \leq \sup_{t\geq 0} \|s(t,x_i)\|$, $i = 1, 2, \ldots$, and hence,

$$
\liminf_{i\to\infty} \sup_{t\geq 0} \|s(t,x_i)\| \geq \liminf_{i\to\infty} \|s(T,x_i)\| = \lim_{i\to\infty} \|s(T,x_i)\|, \quad i = 1, 2, \ldots,
$$

which implies that

$$
\begin{aligned}
V(x) &< \|s(T,x)\| + \hat{\varepsilon} \\
&= \lim_{i\to\infty} \|s(T,x_i)\| + \hat{\varepsilon} \\
&\leq \liminf_{i\to\infty} \sup_{t\geq 0} \|s(t,x_i)\| + \hat{\varepsilon} \\
&= \liminf_{i\to\infty} V(x_i) + \hat{\varepsilon}. \qquad (4.224)
\end{aligned}
$$

Now, since $\hat{\varepsilon} > 0$ is arbitrary, (4.224) implies that $V(x) \leq \liminf_{i\to\infty} V(x_i)$. Thus, since $\{x_i\}_{i=1}^{\infty}$ is an arbitrary sequence converging to x, it follows that $V(\cdot)$ is lower semicontinuous on $\mathcal{D}_0$. □

In the following example we show that there exist Lyapunov stable systems with continuously differentiable vector fields that do not posses continuous Lyapunov functions.

Example 4.15. Consider the scalar nonlinear dynamical system

$$
\dot{x}(t) = x^4(t) \sin^2\left(\frac{1}{x(t)}\right), \quad x(0) = x_0, \quad t \geq 0. \qquad (4.225)
$$

Note that the set of equilibria for the dynamical system (4.225) is characterized by

$$\mathcal{E} \triangleq \{0\} \cup \left\{ \frac{\pm 1}{\pi}, \frac{\pm 1}{2\pi}, \frac{\pm 1}{3\pi}, \ldots \right\}.$$

Next, let $\varepsilon > 0$ and let $\delta = \frac{1}{n\pi} > 0$, where $n \in \{1, 2, \ldots\}$, such that $\frac{1}{n\pi} < \varepsilon$. Now, for every $x_0 \in [0, \delta)$, since $\dot{x}(t) \geq 0$, $t \geq 0$, it follows that $x(t) \geq x_0$, $t \geq 0$. Furthermore, since there exists at least one equilibrium point $x_\mathrm{e} \in [x_0, \delta]$ it follows that $0 \leq x(t) \leq x_\mathrm{e} \leq \delta < \varepsilon$, $t \geq 0$. Similarly, for every $x_0 \in (-\delta, 0]$, since $\dot{x}(t) \geq 0$, $t \geq 0$, it follows that $x(t) \geq -\delta$, $t \geq 0$, and since there exists at least one equilibrium point $x_\mathrm{e} \in [x_0, 0]$, it follows that $-\varepsilon < -\delta < x_0 \leq x(t) \leq x_\mathrm{e} \leq 0$, $t \geq 0$. Hence, for every $\varepsilon > 0$ there exists $\delta > 0$ such that if $|x_0| < \delta$, then $|x(t)| < \varepsilon$ or, equivalently, the zero solution $x(t) \equiv 0$ to (4.225) is Lyapunov stable.

Next, we show that there does not exist a continuous Lyapunov function that proves Lyapunov stability of the zero solution $x(t) \equiv 0$ to (4.225). To see this, let $\mathcal{D} \subset \mathbb{R}$ be an open interval such that $0 \in \mathcal{D}$ and suppose, *ad absurdum*, there exists a continuous function $V : \mathcal{D} \to \mathbb{R}$ such that $V(0) = 0$, $V(x) > 0$, $x \in \mathcal{D}$, $x \neq 0$, and $V(x(t))$, $t \geq 0$, is a nonincreasing function of time. Now, let $n \in \{1, 2, \ldots\}$ be such that $\frac{1}{n\pi} \in \mathcal{D}$. Next, note that, for every $x_0 \in (\frac{1}{(n+1)\pi}, \frac{1}{n\pi})$, $x(t) \to \frac{1}{n\pi}$ as $t \to \infty$. Now, since $V(x(t))$, $t \geq 0$, is nonincreasing and $V(\cdot)$ is continuous on $\mathcal{D}$ it follows that $V(x_0) \geq \lim_{t\to\infty} V(x(t)) = V(\lim_{t\to\infty} x(t)) = V(\frac{1}{n\pi}) > 0$. Hence, since $x_0 \in (\frac{1}{(n+1)\pi}, \frac{1}{n\pi})$ is arbitrary it follows from the continuity of $V(\cdot)$ that $V(\frac{1}{(n+1)\pi}) \geq V(\frac{1}{n\pi}) > 0$. Repeating these arguments iteratively it follows that

$$V(\tfrac{1}{N\pi}) \geq V(\tfrac{1}{n\pi}), \qquad N \in \{n+1, n+2, \ldots\},$$

and hence, it follows from the continuity of $V(\cdot)$ that

$$\begin{aligned} V(0) &= V\left(\lim_{N\to\infty} \tfrac{1}{N\pi}\right) \\ &= \lim_{N\to\infty} V(\tfrac{1}{N\pi}) \\ &\geq V(\tfrac{1}{n\pi}) \\ &> 0, \end{aligned}$$

which is a contradiction. Hence, there does not exist a continuous Lyapunov function that proves Lyapunov stability of the zero solution $x(t) \equiv 0$ to (4.225). △

Next, we generalize the invariant set stability theorems of Section 3.3 to the case in which the function $V(\cdot)$ is lower semicontinuous. For the remainder of the results of this section we define the notation

$$\mathcal{R}_\gamma \triangleq \bigcap_{c>\gamma} \overline{V^{-1}([\gamma, c])}, \tag{4.226}$$

for arbitrary $V : \mathcal{D} \subseteq \mathbb{R}^n \to \mathbb{R}$ and $\gamma \in \mathbb{R}$, and let $\mathcal{M}_\gamma$ denote the largest invariant set (with respect to (3.1)) contained in $\mathcal{R}_\gamma$.

Theorem 4.24. Consider the nonlinear dynamical system (3.1), let $x(t)$, $t \geq 0$, denote the solution to (3.1), and let $\mathcal{D}_c \subset \mathcal{D}$ be a compact invariant set with respect to (3.1). Assume that there exists a lower semicontinuous function $V : \mathcal{D}_c \to \mathbb{R}$ such that $V(x(t)) \leq V(x(\tau))$, $0 \leq \tau \leq t$, for all $x_0 \in \mathcal{D}_c$. If $x_0 \in \mathcal{D}_c$, then $x(t) \to \mathcal{M} \triangleq \cup_{\gamma\in\mathbb{R}} \mathcal{M}_\gamma$ as $t \to \infty$.

Proof. Let $x(t)$, $t \geq 0$, be the solution to (3.1) with $x_0 \in \mathcal{D}_c$. Since $V(\cdot)$ is lower semicontinuous on the compact set $\mathcal{D}_c$, there exists $\beta \in \mathbb{R}$ such that $V(x) \geq \beta$, $x \in \mathcal{D}_c$. Hence, since $V(x(t))$, $t \geq 0$, is nonincreasing, $\gamma_{x_0} \triangleq \lim_{t\to\infty} V(x(t))$, $x_0 \in \mathcal{D}_c$, exists. Now, for every $p \in \omega(x_0)$ there exists an increasing unbounded sequence $\{t_n\}_{n=0}^\infty$, with $t_0 = 0$, such that $x(t_n) \to p$ as $n \to \infty$. Next, since $V(x(t_n))$, $n \geq 0$, is nonincreasing it follows that for all $N \geq 0$, $\gamma_{x_0} \leq V(x(t_n)) \leq V(x(t_N))$, $n \geq N$, or, equivalently, since $\mathcal{D}_c$ is invariant, $x(t_n) \in V^{-1}([\gamma_{x_0}, V(x(t_N))])$, $n \geq N$. Now, since $\lim_{n\to\infty} x(t_n) = p$ it follows that $p \in \overline{V^{-1}([\gamma_{x_0}, V(x(t_n))])}$, $n \geq 0$. Furthermore, since $\lim_{n\to\infty} V(x(t_n)) = \gamma_{x_0}$ it follows that for every $c > \gamma_{x_0}$, there exists $n \geq 0$ such that $\gamma_{x_0} \leq V(x(t_n)) \leq c$, which implies that for every $c > \gamma_{x_0}$, $p \in \overline{V^{-1}([\gamma_{x_0}, c])}$. Hence, $p \in \mathcal{R}_{\gamma_{x_0}}$, which implies that $\omega(x_0) \subseteq \mathcal{R}_{\gamma_{x_0}}$. Now, since $\mathcal{D}_c$ is compact and invariant it follows that the solution $x(t)$, $t \geq 0$, to (3.1) is bounded for all $x_0 \in \mathcal{D}_c$, and hence, it follows from Theorem 2.41 that $\omega(x_0)$ is a nonempty compact invariant set which further implies that $\omega(x_0)$ is a subset of the largest invariant set contained in $\mathcal{R}_{\gamma_{x_0}}$, that is, $\omega(x_0) \subseteq \mathcal{M}_{\gamma_{x_0}}$. Hence, for all $x_0 \in \mathcal{D}_c$, $\omega(x_0) \subseteq \mathcal{M}$. Finally, since $x(t) \to \omega(x_0)$ as $t \to \infty$ it follows that $x(t) \to \mathcal{M}$ as $t \to \infty$. □

If in Theorem 4.24 $\mathcal{M}$ contains no invariant set other than the set $\{0\}$, then the zero solution $x(t) \equiv 0$ to (3.1) is attractive and $\mathcal{D}_c$ is a subset of the domain of attraction. Furthermore, note that if $V : \mathcal{D}_c \to \mathbb{R}$ is a lower semicontinuous function such that all the conditions of Theorem 4.24 are satisfied, then for every $x_0 \in \mathcal{D}_c$ there exists $\gamma_{x_0} \leq V(x_0)$ such that $\omega(x_0) \subseteq \mathcal{M}_{\gamma_{x_0}} \subseteq \mathcal{M}$. In addition, since $V^{-1}([\gamma, c]) = \{x \in \mathcal{D}_c : V(x) \geq \gamma\} \cap \{x \in \mathcal{D}_c : V(x) \leq c\}$ and $\{x \in \mathcal{D}_c : V(x) \leq c\}$ is a closed set, it follows that $\hat{\mathcal{R}}_{\gamma,c} \subset \{x \in \mathcal{D}_c : V(x) < \gamma\}$, where $\hat{\mathcal{R}}_{\gamma,c} \triangleq \overline{V^{-1}([\gamma, c])} \setminus V^{-1}([\gamma, c])$,

$c > \gamma$, for a fixed $\gamma \in \mathbb{R}$. Hence,

$$\mathcal{R}_\gamma = \bigcap_{c>\gamma} \left(V^{-1}([\gamma, c]) \cup \hat{\mathcal{R}}_{\gamma,c} \right) = V^{-1}(\gamma) \cup \hat{\mathcal{R}}_\gamma,$$

where $\hat{\mathcal{R}}_\gamma \triangleq \cap_{c>\gamma} \hat{\mathcal{R}}_{\gamma,c}$, is such that $V(x) < \gamma$, $x \in \hat{\mathcal{R}}_\gamma$. Finally, if $V(\cdot)$ is continuous, then $\hat{\mathcal{R}}_{\gamma,c} = \emptyset$, $\gamma \in \mathbb{R}$, $c > \gamma$, and hence, $\mathcal{R}_\gamma = V^{-1}(\gamma)$.

It is important to note that as in standard Lyapunov and invariant set theorems involving continuously differentiable functions, Theorem 4.24 allows one to characterize the invariant set $\mathcal{M}$ *without* knowledge of the system trajectories $x(t)$, $t \geq 0$. Similar remarks hold for Theorems 4.25–4.27 given below.

Example 4.16. To illustrate the utility of Theorem 4.24 consider the simple scalar nonlinear dynamical system given by

$$\dot{x}(t) = -x(t)(x(t) - 1)(x(t) + 2), \quad x(0) = x_0, \quad t \geq 0, \tag{4.227}$$

with generalized Lyapunov function candidate

$$V(x) = \begin{cases} (x+2)^2, & x < 0, \\ (x-1)^2, & x \geq 0. \end{cases}$$

Now, note that for all $x \in \mathbb{R}$,

$$\dot{V}(x) \triangleq D^+V(x)[-x(x-1)(x+2)] = \begin{cases} -2x(x-1)(x+2)^2, & x < 0, \\ -2x(x-1)^2(x+2), & x \geq 0, \end{cases} \leq 0,$$

which implies that $V(x(t))$, $t \geq 0$, is nonincreasing along the system trajectories. Next, note that $\mathcal{R}_\gamma = V^{-1}(\gamma)$, $\gamma \in \mathbb{R} \setminus \{4\}$, and $\mathcal{R}_4 = V^{-1}(4) \cup \{0\}$. Since the only invariant sets in $\mathcal{R}_\gamma$, $\gamma \in \mathbb{R}$, for the dynamical system (4.227) are the equilibrium points $x_{\rm e1} = -2$, $x_{\rm e2} = 0$, $x_{\rm e3} = 1$, it follows that $\mathcal{M}_\gamma = \emptyset$, $\gamma \notin \{0, 1, 4\}$, $\mathcal{M}_0 = \{-2, 1\}$, $\mathcal{M}_1 = \{0\}$, and $\mathcal{M}_4 = \{0\}$, which implies that $\mathcal{M} = \{-2, 0, 1\}$. Hence, it follows from Theorem 4.24 that for every $x_0 \in \mathbb{R}$ the solution to (4.227) approaches the invariant set $\mathcal{M} = \{-2, 0, 1\}$ as $t \to \infty$ which can be easily verified. △

The following corollary to Theorem 4.24 presents sufficient conditions that guarantee local asymptotic stability of the nonlinear dynamical system (3.1).

Corollary 4.5. Consider the nonlinear dynamical system (3.1), let $x(t)$, $t \geq 0$, denote the solution to (3.1), and let $\mathcal{D}_{\rm c} \subset \mathcal{D}$ with $0 \in \overset{\circ}{\mathcal{D}}_{\rm c}$ be a compact invariant set with respect to (3.1). Assume that there exists a lower semicontinuous positive-definite function $V : \mathcal{D}_{\rm c} \to \mathbb{R}$ such that $V(\cdot)$ is continuous at the origin and $V(x(t)) \leq V(x(\tau))$, $0 \leq \tau \leq t$, for all $x_0 \in \mathcal{D}_{\rm c}$. Furthermore, assume that $\mathcal{M} \triangleq \cup_{\gamma \geq 0} \mathcal{M}_\gamma$ contains no invariant set other than the set $\{0\}$. Then the zero solution $x(t) \equiv 0$ to (3.1) is asymptotically stable and $\mathcal{D}_c$ is a subset of the domain of attraction of (3.1).

Proof. The result follows as a direct consequence of Theorems 4.22 and 4.24. □

Next, we specialize Theorem 4.24 to the Barbashin-Krasovskii-LaSalle invariant set theorem wherein $V(\cdot)$ is a continuously differentiable function.

Corollary 4.6. Consider the nonlinear dynamical system (3.1), assume that $\mathcal{D}_c \subset \mathcal{D}$ is a compact invariant set with respect to (3.1), and assume that there exists a continuously differentiable function $V : \mathcal{D}_c \to \mathbb{R}$ such that $V'(x)f(x) \le 0$, $x \in \mathcal{D}_c$. Let $\mathcal{R} \triangleq \{x \in \mathcal{D}_c : V'(x)f(x) = 0\}$ and let $\mathcal{M}$ be the largest invariant set contained in $\mathcal{R}$. If $x_0 \in \mathcal{D}_c$, then $x(t) \to \mathcal{M}$ as $t \to \infty$.

Proof. The result follows from Theorem 4.24. Specifically, since $V'(x)f(x) \le 0$, $x \in \mathcal{D}_c$, it follows that

$$V(x(t)) - V(x(\tau)) = \int_{\tau}^{t} V'(x(s))f(x(s))\mathrm{d}s \le 0, \quad t \ge \tau,$$

and hence, $V(x(t)) \le V(x(\tau))$, $t \ge \tau$. Now, since $V(\cdot)$ is continuously differentiable it follows that $\mathcal{R}_\gamma = V^{-1}(\gamma)$, $\gamma \in \mathbb{R}$. In this case, it follows from Theorem 4.24 that for every $x_0 \in \mathcal{D}_c$ there exists $\gamma_{x_0} \in \mathbb{R}$ such that $\omega(x_0) \subseteq \mathcal{M}_{\gamma_{x_0}}$, where $\mathcal{M}_{\gamma_{x_0}}$ is the largest invariant set contained in $\mathcal{R}_{\gamma_{x_0}} = V^{-1}(\gamma_{x_0})$, which implies that $V(x) = \gamma_{x_0}$, $x \in \omega(x_0)$. Hence, since $\mathcal{M}_{\gamma_{x_0}}$ is an invariant set it follows that for all $x(0) \in \mathcal{M}_{\gamma_{x_0}}$, $x(t) \in \mathcal{M}_{\gamma_{x_0}}$, $t \ge 0$, and thus, $\dot{V}(x(0)) \triangleq \left.\frac{\mathrm{d}V(x(t))}{\mathrm{d}t}\right|_{t=0} = V'(x(0))f(x(0)) = 0$, which implies the $\mathcal{M}_{\gamma_{x_0}}$ is contained in $\mathcal{M}$ which is the largest invariant set contained in $\mathcal{R}$. Hence, since $x(t) \to \omega(x_0) \subseteq \mathcal{M}$ as $t \to \infty$, it follows that $x(t) \to \mathcal{M}$ as $t \to \infty$. □

Next, we sharpen the results of Theorem 4.24 by providing a refined construction of the invariant set $\mathcal{M}$. In particular, we show that the system trajectories converge to a union of largest invariant sets contained in intersections over the largest limit value of $V(\cdot)$ at the origin of the closure of generalized Lyapunov surfaces. First, however, the following key lemma is needed.

Lemma 4.2. Let $\mathcal{Q} \subseteq \mathbb{R}^n$, let $V : \mathcal{Q} \to \mathbb{R}$, and let $\gamma_0 \triangleq \limsup_{x\to 0} V(x)$. If $0 \in \mathcal{R}_\gamma$ for some $\gamma \in \mathbb{R}$, then $\gamma \le \gamma_0$.

Proof. If $0 \in \mathcal{R}_\gamma$ for $\gamma \in \mathbb{R}$, then there exists a sequence $\{x_n\}_{n=0}^{\infty} \subset \mathcal{R}_\gamma$ such that $\lim_{n\to\infty} x_n = 0$. Now, since $\gamma_0 = \limsup_{x\to 0} V(x)$, it follows that $\limsup_{n\to\infty} V(x_n) \le \gamma_0$. Next, note that $x_n \in \overline{V^{-1}([\gamma, c])}$, $c > \gamma$, $n = 0, 1, \ldots$, which implies that $V(x_n) \ge \gamma$, $n = 0, 1, \ldots$. Thus, using the fact that $\limsup_{n\to\infty} V(x_n) \le \gamma_0$ it follows that $\gamma \le \gamma_0$. □

Note that if in Lemma 4.2 $V(\cdot)$ is continuous at the origin, then $\gamma_0 = V(0)$.

Theorem 4.25. Consider the nonlinear dynamical system (3.1), let $x(t)$, $t \ge 0$, denote the solution to (3.1), and let $\mathcal{D}_c \subset \mathcal{D}$ with $0 \in \overset{\circ}{\mathcal{D}}_c$ be a compact invariant set with respect to (3.1). Assume that there exists a lower semicontinuous positive-definite function $V : \mathcal{D}_c \to \mathbb{R}$ such that $V(x(t)) \le V(x(\tau))$, $0 \le \tau \le t$, for all $x_0 \in \mathcal{D}_c$. Furthermore, assume that for all $x_0 \in \mathcal{D}_c$, $x_0 \ne 0$, there exists an increasing unbounded sequence $\{t_n\}_{n=0}^{\infty}$, with $t_0 = 0$, such that

$$V(x(t_{n+1})) < V(x(t_n)), \quad n = 0, 1, \ldots. \tag{4.228}$$

If $x_0 \in \mathcal{D}_c$, then $x(t) \to \hat{\mathcal{M}} \triangleq \cup_{\gamma \in \mathcal{G}} \mathcal{M}_\gamma$ as $t \to \infty$, where $\mathcal{G} \triangleq \{\gamma \in [0, \gamma_0] : 0 \in \mathcal{R}_\gamma\}$ and $\gamma_0 \triangleq \limsup_{x\to 0} V(x)$. If, in addition, $V(\cdot)$ is continuous at the origin, then the zero solution $x(t) \equiv 0$ to (3.1) is locally asymptotically stable and $\mathcal{D}_c$ is a subset of the domain of attraction.

Proof. It follows from Theorem 4.24 and the fact that $V(\cdot)$ is positive definite that, for every $x_0 \in \mathcal{D}_c$, there exists $\gamma_{x_0} \ge 0$ such that $\omega(x_0) \subseteq \mathcal{M}_{\gamma_{x_0}} \subseteq \mathcal{R}_{\gamma_{x_0}}$. Furthermore, since all solutions $x(t)$, $t \ge 0$, to (3.1) are bounded, it follows from Theorem 2.41 that $\omega(x_0)$ is a nonempty, compact, invariant set. Now, *ad absurdum*, suppose $0 \notin \omega(x_0)$. Since $V(\cdot)$ is lower semicontinuous it follows from Theorem 2.11 that $\alpha \triangleq \min_{x\in\omega(x_0)} V(x)$ exists. Furthermore, there exists $\hat{x} \in \omega(x_0)$ such that $V(\hat{x}) = \alpha$. Now, with $x(0) = \hat{x} \ne 0$ it follows from (4.228) that there exists an increasing unbounded sequence $\{t_n\}_{n=0}^{\infty}$, with $t_0 = 0$, such that $V(x(t_{n+1})) < V(x(t_n))$, $n = 0, 1, \ldots$, which implies that there exists $t > 0$ such that $V(x(t)) < \alpha$, and hence, $x(t) \notin \omega(x_0)$, contradicting the fact that $\omega(x_0)$ is an invariant set. Hence, $0 \in \omega(x_0) \subseteq \mathcal{R}_{\gamma_{x_0}}$, which, using Lemma 4.2, implies that $\gamma_{x_0} \le \gamma_0$ for all $x_0 \in \mathcal{D}_c$, and which further implies that $\omega(x_0) \subseteq \hat{\mathcal{M}}$. Now, since $x(t) \to \omega(x_0) \subseteq \hat{\mathcal{M}}$ as $t \to \infty$ it follows that $x(t) \to \hat{\mathcal{M}}$ as $t \to \infty$.

Finally, if $V(\cdot)$ is continuous at the origin then Lyapunov stability follows from Theorem 4.22. Furthermore, in this case, $\gamma_0 = V(0) = 0$, which implies that $\hat{\mathcal{M}} \equiv \{0\}$. Hence, $x(t) \to 0$ as $t \to \infty$ for all $x_0 \in \mathcal{D}_c$, establishing local asymptotic stability with a subset of the domain of attraction given by $\mathcal{D}_c$. □

In all the above results we explicitly assumed that there exists a compact invariant set $\mathcal{D}_c \subset \mathcal{D}$ of (3.1). Next, we provide a result that does not require the existence of such a compact invariant $\mathcal{D}_c$.

Theorem 4.26. Consider the nonlinear dynamical system (3.1) and let $x(t)$, $t \geq 0$, denote the solution to (3.1). Assume that there exists a lower semicontinuous function $V : \mathbb{R}^n \to \mathbb{R}$ such that

$$V(0) = 0, \tag{4.229}$$
$$V(x) > 0, \quad x \in \mathbb{R}^n, \quad x \neq 0, \tag{4.230}$$
$$V(x(t)) \leq V(x(\tau)), \quad 0 \leq \tau \leq t. \tag{4.231}$$

Then all solutions $x(t)$, $t \geq 0$, to (3.1) that are bounded approach $\mathcal{M} \triangleq \cup_{\gamma \geq 0} \mathcal{M}_\gamma$ as $t \to \infty$. If, in addition, for all $x_0 \in \mathbb{R}^n$, $x_0 \neq 0$, there exists an increasing unbounded sequence $\{t_n\}_{n=0}^{\infty}$, with $t_0 = 0$, such that

$$V(x(t_{n+1})) < V(x(t_n)), \quad n = 0, 1, \ldots, \tag{4.232}$$

then all solutions $x(t)$, $t \geq 0$, to (3.1) that are bounded approach $\hat{\mathcal{M}} \triangleq \cup_{\gamma \in \mathcal{G}} \mathcal{M}_\gamma$, where $\mathcal{G} \triangleq \{\gamma \in [0, \gamma_0] : 0 \in \mathcal{R}_\gamma\}$ and $\gamma_0 \triangleq \limsup_{x \to 0} V(x)$.

Proof. Let $x_0 \in \mathbb{R}^n$ be such that the trajectory $x(t)$, $t \geq 0$, is bounded. Now, the proof is a direct consequence of Theorems 4.24 and 4.25 with $\mathcal{D}_c = \overline{\mathcal{O}_{x_0}^+}$. □

Next, we present a generalized global invariant set theorem for guaranteeing global attraction and global asymptotic stability of a nonlinear dynamical system.

Theorem 4.27. Consider the nonlinear dynamical system (3.1) with $\mathcal{D} = \mathbb{R}^n$ and let $x(t)$, $t \geq 0$, denote the solution to (3.1). Assume that there exists a lower semicontinuous function $V : \mathbb{R}^n \to \mathbb{R}$ such that

$$V(0) = 0, \tag{4.233}$$
$$V(x) > 0, \quad x \in \mathbb{R}^n, \quad x \neq 0, \tag{4.234}$$
$$V(x(t)) \leq V(x(\tau)), \quad 0 \leq \tau \leq t, \tag{4.235}$$
$$V(x) \to \infty \text{ as } \|x\| \to \infty. \tag{4.236}$$

Then for all $x_0 \in \mathbb{R}^n$, $x(t) \to \mathcal{M} \triangleq \cup_{\gamma \geq 0} \mathcal{M}_\gamma$ as $t \to \infty$. If, in addition, for all $x_0 \in \mathbb{R}^n$, $x_0 \neq 0$, there exists an increasing unbounded sequence $\{t_n\}_{n=0}^{\infty}$, with $t_0 = 0$, such that

$$V(x(t_{n+1})) < V(x(t_n)), \quad n = 0, 1, \ldots, \tag{4.237}$$

then $x(t) \to \hat{\mathcal{M}} \triangleq \cup_{\gamma \in \mathcal{G}} \mathcal{M}_\gamma$ as $t \to \infty$, where $\mathcal{G} \triangleq \{\gamma \in [0, \gamma_0] : 0 \in \mathcal{R}_\gamma\}$ and $\gamma_0 \triangleq \limsup_{x \to 0} V(x)$. Finally, if $V(\cdot)$ is continuous at the origin, then the zero solution $x(t) \equiv 0$ to (3.1) is globally asymptotically stable.

Proof. Note that since $V(x) \to \infty$ as $\|x\| \to \infty$ it follows that for every $\beta > 0$ there exists $r > 0$ such that $V(x) > \beta$ for all $x \notin \mathcal{B}_r(0)$ or, equivalently, $V^{-1}([0,\beta]) \subseteq \overline{\mathcal{B}}_r(0)$, which implies that $V^{-1}([0,\beta])$ is bounded for all $\beta > 0$. Hence, for all $x_0 \in \mathbb{R}^n$, $V^{-1}([0,\beta_{x_0}])$ is bounded, where $\beta_{x_0} \triangleq V(x_0)$. Furthermore, since $V(\cdot)$ is a positive-definite lower semicontinuous function it follows that $V^{-1}([0,\beta_{x_0}])$ is closed, and since $V(x(t))$, $t \geq 0$, is nonincreasing it follows that $V^{-1}([0,\beta_{x_0}])$ is an invariant set. Hence, for every $x_0 \in \mathbb{R}^n$, $V^{-1}([0,\beta_{x_0}])$ is a compact invariant set. Now, with $\mathcal{D}_c = V^{-1}([0,\beta_{x_0}])$ it follows from Theorem 4.24 that there exists $0 \leq \gamma_{x_0} \leq \beta_{x_0}$ such that $\omega(x_0) \subseteq \mathcal{M}_{\gamma_{x_0}} \subseteq \mathcal{R}_{\gamma_{x_0}}$, which implies that $x(t) \to \mathcal{M}$ as $t \to \infty$. If, in addition, for all $x_0 \in \mathbb{R}^n$, $x_0 \neq 0$, there exists an increasing unbounded sequence $\{t_n\}_{n=0}^{\infty}$, with $t_0 = 0$, such that (4.237) holds, then it follows from Theorem 4.25 that $x(t) \to \hat{\mathcal{M}}$ as $t \to \infty$.

Finally, if $V(\cdot)$ is continuous at the origin, then Lyapunov stability follows from Theorem 4.22. Furthermore, in this case, $\gamma_0 = V(0) = 0$, which implies that $\hat{\mathcal{M}} \equiv \{0\}$. Hence, $x(t) \to 0$ as $t \to \infty$, establishing global asymptotic stability. □

If in Theorems 4.25 and 4.27 the function $V(\cdot)$ is continuously differentiable on $\mathcal{D}_c$ and $\mathbb{R}^n$, respectively, and $V'(x)f(x) < 0$, $x \in \mathbb{R}^n$, $x \neq 0$, then every increasing unbounded sequence $\{t_n\}_{n=0}^{\infty}$, with $t_0 = 0$, is such that $V(x(t_{n+1})) < V(x(t_n))$, $n = 0, 1, \ldots$. In this case Theorems 4.25 and 4.27 specialize to the standard Lyapunov stability theorems for local and global asymptotic stability, respectively, presented in Section 3.2.

Example 4.17. To illustrate the generalized stability theorems presented in this section we consider the stability analysis of a nonlinear dynamical system given by

$$\dot{x}_1(t) = \sigma(x_1(t)) + u(t), \quad x_1(0) = x_{10}, \quad t \geq 0, \tag{4.238}$$
$$\dot{x}_2(t) = \sigma(x_2(t)) + u(t), \quad x_2(0) = x_{20}, \tag{4.239}$$
$$\dot{x}_3(t) = \sigma(x_3(t)) + u(t), \quad x_3(0) = x_{30}, \tag{4.240}$$

where $\sigma : \mathbb{R} \to \mathbb{R}$ is given by $\sigma(y) = -3my^2 - y^3$, $m > 0$, and

$$u = \begin{cases} 0, & x \in \mathcal{D}_I, \\ -\sigma(\alpha m), & x \notin \mathcal{D}_I, \end{cases} \tag{4.241}$$

where $x \triangleq [x_1 \ x_2 \ x_3]^{\mathrm{T}}$, $\alpha > 1$, and $\mathcal{D}_I \triangleq \{x \in \mathbb{R}^3 : x_1 \geq 0,\ x_2 \geq 0,\ x_3 \geq 0\}$. Next, consider the radially unbounded generalized Lyapunov function candidate given by

$$V(x) = \begin{cases} \frac{1}{3}\sum_{i=1}^{3} x_i^2, & x \in \mathcal{D}_I, \\ \sum_{i=1}^{3}(x_i - \alpha m)^2, & x \notin \mathcal{D}_I, \end{cases} \tag{4.242}$$

and note that $V(x)$ is continuous on $\mathbb{R}^3 \setminus \partial\mathcal{D}_I$.

Next, note that for $i = \{1,2,3\}$,

$$q_i^- \triangleq \lim_{x_i \to 0^-} V(x) = \alpha^2 m^2 + \sum_{j=1, j\neq i}^{3} (x_i - \alpha m)^2,$$

$$q_i^+ \triangleq \lim_{x_i \to 0^+} V(x) = \sum_{j=1, j\neq i}^{3} \tfrac{1}{3}x_i^2, \quad q_i \triangleq V(x)|_{x_i=0} = \sum_{j=1, j\neq i}^{3} \tfrac{1}{3}x_i^2,$$

and hence, $q_i^+ - q_i = 0$ and $q_i^- - q_i = \frac{2}{3}\sum_{j=1, j\neq i}^{3}(x_i - \frac{3}{2}\alpha m)^2$, which implies that $V(x)$ is lower semicontinuous on $\partial\mathcal{D}_I$, and hence, $V(\cdot)$ is lower semicontinuous on $\mathbb{R}^3$. Now, note that

$$\begin{aligned} \dot{V}(x) &\triangleq D^+V(x)f(x) \\ &= \begin{cases} -\frac{2}{3}\sum_{i=1}^{3} x_i(3mx_i^2 + x_i^3), & x \in \mathcal{D}_I, \\ -2\sum_{i=1}^{3}(x_i - \alpha m)[3m(x_i - \alpha m)^2 + x_i^3 - \alpha^3 m^3], & x \notin \mathcal{D}_I, \end{cases} \\ &= \begin{cases} -\frac{2}{3}\sum_{i=1}^{3} x_i(3mx_i^2 + x_i^3), & x \in \mathcal{D}_I, \\ -2\sum_{i=1}^{3}(x_i - \alpha m)^2[(x_i + \frac{m(\alpha+3)}{2})^2 + \frac{3m^2}{4}(\alpha+3)(\alpha-1)], & \\ \quad x \notin \mathcal{D}_I, \end{cases} \end{aligned} \tag{4.243}$$

where $f : \mathbb{R}^3 \to \mathbb{R}^3$ denotes the right-hand side of (4.238)–(4.240). Next, (4.243) implies that $V(x(t))$, $t \geq 0$, is nonincreasing along the system trajectories. Furthermore, since $\dot{V}(x) < 0$, $x \in \mathbb{R}^3$, $x \neq 0$, it follows that for all $(x_{10}, x_{20}, x_{30}) \in \mathbb{R}^3 \setminus \{0\}$ there exists an increasing unbounded sequence $\{t_n\}_{n=0}^{\infty}$, with $t_0 = 0$, such that $V(x(t_{j+1})) < V(x(t_j))$, $j = 0, 1, \ldots$, so that all the conditions of Theorem 4.27 are satisfied. Hence, it follows from Theorem 4.27 that $x(t) \to \hat{\mathcal{M}}$ as $t \to \infty$, where $\hat{\mathcal{M}}$ is defined as in Theorem 4.27 with $\mathcal{G} = \{0, 3\alpha^2 m^2\}$ and $\gamma_0 = \limsup_{x \to 0} V(x) = 3\alpha^2 m^2$.

Next, note that $\mathcal{R}_0 = \{0\}$ and

$$\mathcal{R}_{3\alpha^2 m^2} = \left\{ x \in \mathcal{D}_I : \sum_{i=1}^{3} x_i^2 = 9\alpha^2 m^2 \right\} \cup \{0\}. \tag{4.244}$$

Now, it follows that $\mathcal{M}_\gamma = \{0\}$, $\gamma \in \mathcal{G} = \{0, 3\alpha^2 m^2\}$, which implies that $\hat{\mathcal{M}} = \{0\}$. Hence, it follows from Theorem 4.27 that $x(t) \to 0$ as $t \to \infty$. Finally, it can be shown that the nonlinear dynamical system (4.238)–(4.240) is Lyapunov stable so that the asymptotic stability can be established. $\triangle$

Next, we remove the lower semicontinuity assumption on the function $V(\cdot)$ and show that the zero solution $x(t) \equiv 0$ to (3.1) is asymptotically stable if there simply exists an increasing unbounded sequence $\{t_n\}_{n=0}^{\infty}$, with

$t_0 = 0$, such that for $T \geq t_{n+1} - t_n > 0$, $n \in \mathbb{Z}_+$, and all $x(0) \in \mathcal{D}$, $V(x(T)) - V(x(0)) \leq -\gamma(\|x(0)\|) < 0$, where $\gamma(\cdot)$ is a class $\mathcal{K}$ function. However, for the remainder of the results of this section we assume that $f(\cdot)$ is Lipschitz continuous on $\mathcal{D}$ with maximum Lipschitz constant $L > 0$ over $\mathcal{D}$. The following lemma is needed for this result.

Lemma 4.3. Consider the nonlinear dynamical system (3.1). Assume $f : \mathcal{D} \to \mathbb{R}^n$ is Lipschitz continuous on $\mathcal{D}$ with maximum Lipschitz constant $L > 0$ over $\mathcal{D}$. If $x_0 \in \mathcal{B}_\delta(0)$, $\delta = \varepsilon e^{-LT}$, where $\varepsilon > 0$ and $T > 0$, then $x(t) \in \mathcal{B}_\varepsilon(0)$ for all $t \in [0, T]$.

Proof. It follows from (3.1) that

$$x(t) = x(0) + \int_0^t f(x(s))\mathrm{d}s, \quad t \geq 0. \tag{4.245}$$

Now, using the fact that $f : \mathcal{D} \to \mathbb{R}^n$ is Lipschitz continuous on $\mathcal{D}$, it follows that

$$\begin{aligned} \|x(t)\| &\leq \|x_0\| + \int_0^t \|f(x(s))\|\mathrm{d}s, \quad t \geq 0 \\ &\leq \|x_0\| + \int_0^t L\|x(s)\|\mathrm{d}s, \end{aligned} \tag{4.246}$$

where $L > 0$ is the maximum Lipschitz constant over $\mathcal{D}$. Using Lemma 2.2, (4.246) implies

$$\|x(t)\| \leq \|x_0\|e^{Lt}, \quad t \geq 0. \tag{4.247}$$

Next, let $x_0 \in \mathcal{B}_\delta(0) \subset \mathcal{B}_\varepsilon(0)$ and let $\tau > 0$ be such that $x(t) \in \mathcal{B}_\varepsilon(0)$, $t \in [0, \tau)$, and $\|x(\tau)\| = \varepsilon$. In this case, it follows from (4.247) with $t = \tau$ that $\|x(\tau)\| \leq \|x_0\|e^{L\tau}$. Now, since by assumption $\|x(\tau)\| = \varepsilon$, $\|x_0\| < \delta$, and $\delta = \varepsilon e^{-LT}$, it follows that $\varepsilon < \varepsilon e^{L(\tau - T)}$, and hence, $\tau > T$. Hence, since $x(t) \in \mathcal{B}_\varepsilon(0)$, $t \in [0, \tau)$, and $\tau > T$ it follows that $x(t) \in \mathcal{B}_\varepsilon(0)$, $t \in [0, T]$. □

Theorem 4.28. Consider the nonlinear dynamical system (3.1) and let $x(t)$, $t \geq 0$, denote the solution to (3.1). Assume there exist a function $V : \mathcal{D} \to \mathbb{R}$, an increasing unbounded sequence $\{t_n\}_{n=0}^{\infty}$, with $t_0 = 0$, and $T > 0$ such that $0 < t_{n+1} - t_n \leq T$, $n = 0, 1, \ldots$, and

$$\alpha(\|x\|) \leq V(x) \leq \beta(\|x\|), \quad x \in \mathcal{D}, \tag{4.248}$$

$$V(x(t_{n+1})) - V(x(t_n)) \leq 0, \quad x(t_n) \in \mathcal{D}, \quad n = 0, 1, \ldots, \tag{4.249}$$

where $\alpha(\cdot)$ and $\beta(\cdot)$ are class $\mathcal{K}$ functions defined on $[0, \varepsilon)$ for all $\varepsilon > 0$. Then the zero solution $x(t) \equiv 0$ to (3.1) is Lyapunov stable. If, in addition, there exists a class $\mathcal{K}$ function $\gamma(\cdot)$ defined on $[0, \infty)$ such that

$$V(x(t_{n+1})) - V(x(t_n)) \leq -\gamma(\|x(t_n)\|) < 0, \quad x(t_n) \in \mathcal{D}, \quad n = 0, 1, \ldots, \tag{4.250}$$

then the zero solution $x(t) \equiv 0$ to (3.1) is asymptotically stable.

Proof. To show Lyapunov stability, define $\delta \triangleq \varepsilon e^{-LT}$ and define $\eta \triangleq \beta^{-1}(\alpha(\delta))$, where L denotes the maximum Lipschitz constant over $\mathcal{D}$ for the nonlinear dynamical system (3.1). Now, for all $x_0 \in \mathcal{B}_\eta(0)$ it follows that $\beta(\|x_0\|) < \alpha(\delta)$, and hence, (4.248) implies $V(x_0) < \alpha(\delta)$. Next, (4.249) implies that $V(x(t_1)) \leq V(x_0) < \alpha(\delta)$, and hence, (4.248) implies $\alpha(\|x(t_1)\|) \leq V(x(t_1)) < \alpha(\delta)$ so that $x(t_1) \in \mathcal{B}_\delta(0)$. Now, using (4.249) with $n = 2$ it follows that $V(x(t_2)) \leq V(x(t_1)) < \alpha(\delta)$, and hence, repeating this procedure, $x(t_n) \in \mathcal{B}_\delta(0)$, $n = 0, 1, \ldots$. Since $\{t_n\}_{n=0}^{\infty}$ is an increasing sequence with $t_0 = 0$ and $t_n \to \infty$ as $n \to \infty$, there exists an integer $\hat{n}$ such that $t_{\hat{n}} \leq t \leq t_{\hat{n}+1}$, $t \geq 0$. Now, using the fact that $x(t_n) \in \mathcal{B}_\delta(0)$, it follows that $\|x(t_{\hat{n}})\| < \delta = \varepsilon e^{-LT}$. Hence, Lemma 4.3 implies that $\|x(t)\| < \varepsilon$, $t \geq 0$, and hence, for all $\varepsilon > 0$ there exists $\eta = \beta^{-1}(\alpha(\delta)) > 0$ such that if $\|x(0)\| < \eta$, then $\|x(t)\| < \varepsilon$, $t \geq 0$, which proves Lyapunov stability.

To show asymptotic stability, let $0 < \hat{\eta} < \eta(\varepsilon) = \beta^{-1}(\alpha(\varepsilon e^{-LT}))$ and let $x_0 \in \mathcal{B}_{\hat{\eta}}(0)$ so that $V(x_0) \leq \beta(\|x_0\|) < \beta(\hat{\eta})$. Now, suppose, *ad absurdum*, that $\|x(t_n)\| \geq l > 0$, $n = 0, 1, \ldots$. In this case, it follows from (4.250) that

$$V(x(t_{n+1})) - V(x(t_n)) \leq -\gamma(\|x(t_n)\|) \leq -\gamma(l), \quad n = 0, 1, \ldots,$$

and hence, $V(x(t_{n+1})) - V(x_0) \leq -(n+1)\gamma(l)$, $n = 0, 1, \ldots$. Now, since $0 \leq \alpha(\|x\|) \leq V(x)$, $x \in \mathcal{D}$, it follows that

$$0 \leq V(x(t_{n+1})) \leq V(x_0) - (n+1)\gamma(l) < \beta(\hat{\eta}) - (n+1)\gamma(l),$$

for all $x_0 \in \mathcal{B}_{\hat{\eta}}(0)$ and $n = 0, 1, \ldots$. Since $\beta(\cdot)$ and $\gamma(\cdot)$ are class $\mathcal{K}$ functions it follows that there exists an integer $\hat{n}$ such that $\beta(\hat{\eta}) - (\hat{n}+1)\gamma(l) < 0$, which leads to a contradiction, and hence, $\|x(t_{\hat{n}})\| < l$. Thus, it follows from (4.248) that $V(x(t_{\hat{n}})) \leq \beta(\|x(t_{\hat{n}})\|) < \beta(l)$. Next, define $\hat{l} \triangleq \alpha^{-1}(\beta(l))$ so that $V(x(t_{\hat{n}})) < \beta(l) = \alpha(\hat{l})$. Now, it follows from (4.248) and (4.250) that $\alpha(\|x(t_{n^*})\|) \leq V(x(t_{n^*})) < \alpha(\hat{l})$, $n^* \geq \hat{n}$, and hence, $\|x(t_{n^*})\| < \hat{l}$. Letting $t \geq t_{\hat{n}}$ and choosing $n^* \geq \hat{n}$ such that $t_{n^*} \leq t \leq t_{n^*+1}$, it follows from Lemma 4.3 that $x(t) \in \mathcal{B}_{\hat{\varepsilon}}(0)$, $\hat{\varepsilon} \triangleq \hat{l}e^{-LT}$, $t \geq t_{\hat{n}}$. Hence, for $x_0 \in \mathcal{B}_{\hat{\eta}}(0)$, $0 < \hat{\eta} < \eta(\varepsilon)$, there exists $t_{\hat{n}}$ such that $x(t) \in \mathcal{B}_{\hat{\varepsilon}}(0)$, $\hat{\varepsilon} > 0$, $t > t_{\hat{n}}$. Now, letting $\hat{n} \to \infty$ so that $t_{\hat{n}} \to \infty$, it follows that $\hat{\varepsilon} \to 0$, and hence, $x(t) \to 0$ as $t \to \infty$, which proves asymptotic stability. □

The results of this section provide sufficient conditions for Lyapunov and asymptotic stability of the nonlinear dynamical system (3.1) with no regularity assumptions on the function $V(\cdot)$. It is important to note that even though the stability conditions appearing in Theorems 4.24–4.27 are system trajectory dependent, the invariant set $\mathcal{M}$ can be characterized without knowledge of the system trajectories. Furthermore, as shown in Example 4.16, these theorems allow for a systematic way of constructing system Lyapunov functions by piecing together a collection of functions. Alternatively, $V(\cdot)$ in Theorem 4.28 is not required to be continuous, differentiable, or decreasing. In the case where $V(\cdot)$ is continuously differentiable, condition (4.250) implies that the time derivative of $V(\cdot)$ along the trajectories of the nonlinear dynamical system (3.1) may have negative and positive values. This allows the consideration of Lyapunov function candidates for proving asymptotic stability for nonlinear dynamical systems that might otherwise be excluded as valid Lyapunov function candidates using classical Lyapunov stability theory.

4.9 Lyapunov and Asymptotic Stability of Sets

In this section, we extend the results of Section 4.8 to address stability and attraction of dynamical systems with respect to compact positively invariant sets. These results are used in the next section to provide necessary and sufficient conditions for stability of periodic orbits and limit cycles. We begin by considering the nonlinear dynamical system

$$\dot{x}(t) = f(x(t)), \quad x(0) = x_0, \quad t \in \mathcal{I}_{x_0}, \tag{4.251}$$

where $x(t) \in \mathcal{D} \subseteq \mathbb{R}^n$, $t \in \mathcal{I}_{x_0}$, is the system state vector, $\mathcal{D}$ is an open set, $f : \mathcal{D} \to \mathbb{R}^n$, and $\mathcal{I}_{x_0} = [0, \tau_{x_0})$, $0 < \tau_{x_0} \leq \infty$, is the maximal interval of existence for the solution $x(\cdot)$ of (4.251). We assume that the dynamics $f(\cdot)$ are such that the solution $s(t, x_0)$ to (4.251) is unique for every initial condition in $\mathcal{D}$ and jointly continuous in t and x_0. A sufficient condition ensuring this is Lipschitz continuity of $f(\cdot)$. Furthermore, we assume that all solutions to (4.251) are bounded over $\mathcal{I}_{x_0}$, and hence, by Corollary 2.5 can be extended to infinity. The following definition introduces three types of stability notions as well as attraction of (4.251) with respect to a compact positively invariant set for $\mathcal{I}_{x_0} = [0, \infty)$.

Definition 4.10. Let $\mathcal{D}_0 \subset \mathcal{D}$ be a compact positively invariant set for the nonlinear dynamical system (4.251). $\mathcal{D}_0$ is *Lyapunov stable* if, for every open neighborhood $\mathcal{O}_1 \subseteq \mathcal{D}$ of $\mathcal{D}_0$, there exists an open neighborhood $\mathcal{O}_2 \subseteq \mathcal{O}_1$ of $\mathcal{D}_0$ such that $x(t) \in \mathcal{O}_1$, $t \geq 0$, for all $x_0 \in \mathcal{O}_2$. Equivalently, $\mathcal{D}_0$ is Lyapunov stable if, for all $\varepsilon > 0$, there exists $\delta = \delta(\varepsilon) > 0$ such that if $\operatorname{dist}(x_0, \mathcal{D}_0) < \delta$, then $\operatorname{dist}(s(t, x_0), \mathcal{D}_0) < \varepsilon$, $t \geq 0$. $\mathcal{D}_0$ is *attractive* if there exists an open neighborhood $\mathcal{O}_3 \subseteq \mathcal{D}$ of $\mathcal{D}_0$ such that $\omega(x_0) \subseteq \mathcal{D}_0$ for all $x_0 \in \mathcal{O}_3$. $\mathcal{D}_0$ is *asymptotically stable* if it is Lyapunov stable and attractive. Equivalently, $\mathcal{D}_0$ is asymptotically stable if $\mathcal{D}_0$ is Lyapunov stable and there exists $\varepsilon > 0$ such that if $\operatorname{dist}(x_0, \mathcal{D}_0) < \varepsilon$, then $\operatorname{dist}(s(t, x_0), \mathcal{D}_0) \to 0$ as $t \to \infty$. $\mathcal{D}_0$ is *globally asymptotically stable* if it is Lyapunov stable and $\omega(x_0) \subseteq \mathcal{D}_0$ for all $x_0 \in \mathbb{R}^n$. Finally, $\mathcal{D}_0$ is *unstable* if it is not Lyapunov stable.

Next, we give a set theoretic definition involving the domain, or region, of attraction of the compact positively invariant set $\mathcal{D}_0$ of (4.251).

Definition 4.11. Suppose the compact positively invariant set $\mathcal{D}_0 \subset \mathcal{D}$ of (4.251) is attractive. Then the *domain of attraction* $\mathcal{D}_{\mathrm{A}}$ of $\mathcal{D}_0$ is defined as

$$\mathcal{D}_{\mathrm{A}} \triangleq \{x_0 \in \mathcal{D} : \omega(x_0) \subseteq \mathcal{D}_0\}. \tag{4.252}$$

The following result gives sufficient conditions for Lyapunov and asymptotic stability of a compact positively invariant set with respect to the nonlinear dynamical system (4.251).

Theorem 4.29. Consider the nonlinear dynamical system (4.251), let $\mathcal{D}_0$ be a compact positively invariant set with respect to (4.251) such that $\mathcal{D}_0 \subset \mathcal{D}$, and let $x(t)$, $t \geq 0$, denote the solution to (4.251) with $x_0 \in \mathcal{D}$. Assume that there exists a lower semicontinuous function $V : \mathcal{D} \to \mathbb{R}$ such that $V(\cdot)$ is continuous on $\mathcal{D}_0$ and

$$V(x) = 0, \qquad x \in \mathcal{D}_0, \tag{4.253}$$

$$V(x) > 0, \qquad x \in \mathcal{D}, \qquad x \notin \mathcal{D}_0, \tag{4.254}$$

$$V(x(t)) \leq V(x(\tau)), \qquad 0 \leq \tau \leq t. \tag{4.255}$$

Then $\mathcal{D}_0$ is Lyapunov stable. If, in addition, for all $x_0 \in \mathcal{D}$, $x_0 \notin \mathcal{D}_0$, there exists an increasing unbounded sequence $\{t_n\}_{n=0}^{\infty}$, with $t_0 = 0$, such that

$$V(x(t_{n+1})) < V(x(t_n)), \quad n = 0, 1, \ldots, \tag{4.256}$$

then $\mathcal{D}_0$ is asymptotically stable.

Proof. Let $\mathcal{O}_1 \subseteq \mathcal{D}$ be a bounded open neighborhood of $\mathcal{D}_0$. Since $\partial\mathcal{O}_1$ is compact and $V(x)$, $x \in \mathcal{D}$, is lower semicontinuous, it follows from Theorem 2.11 that there exists $\alpha = \min_{x \in \partial\mathcal{O}_1} V(x)$. Note that $\alpha > 0$ since $\mathcal{D}_0 \cap \partial\mathcal{O}_1 = \emptyset$ and $V(x) > 0$, $x \in \mathcal{D}$, $x \notin \mathcal{D}_0$. Next, using the facts that $V(x) = 0$, $x \in \mathcal{D}_0$, and $V(\cdot)$ is continuous on $\mathcal{D}_0$, it follows that the set $\mathcal{O}_2 \triangleq \{x \in \mathcal{O}_1 : V(x) < \alpha\}^{\circ}$ is not empty. Now, it follows from (4.255) that for all $x(0) \in \mathcal{O}_2$,

$$V(x(t)) \leq V(x(0)) < \alpha, \qquad t \geq 0,$$

which, since $V(x) \geq \alpha$, $x \in \partial\mathcal{O}_1$, implies that $x(t) \notin \partial\mathcal{O}_1$, $t \geq 0$. Hence, for every open neighborhood $\mathcal{O}_1 \subseteq \mathcal{D}$ of $\mathcal{D}_0$, there exists an open neighborhood $\mathcal{O}_2 \subseteq \mathcal{O}_1$ of $\mathcal{D}_0$ such that, if $x(0) \in \mathcal{O}_2$, then $x(t) \in \mathcal{O}_1$, $t \geq 0$, which proves Lyapunov stability of the compact positively invariant set $\mathcal{D}_0$ of (4.251).

To prove asymptotic stability let $x_0 \in \mathcal{O}_2$ and suppose there exists an increasing unbounded sequence $\{t_n\}_{n=0}^{\infty}$, with $t_0 = 0$, such that (4.256) holds. Then it follows that $x(t) \in \mathcal{O}_1$, $t \geq 0$, and hence, it follows from Theorem 2.41 that the positive limit set $\omega(x_0)$ of $x(t)$, $t \geq 0$, is a nonempty, compact, invariant, and connected set. Furthermore, $x(t) \to \omega(x_0)$ as $t \to \infty$. Now, since $V(x(t))$, $t \geq 0$, is nonincreasing and bounded from below by zero it follows that $\beta \triangleq \lim_{t\to\infty} V(x(t)) \geq 0$ is well defined. Furthermore, since $V(\cdot)$ is lower semicontinuous it can be shown that $V(y) \leq \beta$, $y \in \omega(x_0)$, and hence, since $V(\cdot)$ is nonnegative, $\beta = 0$ if and only if $\omega(x_0) = \mathcal{D}_0$ or, equivalently, $x(t) \to \mathcal{D}_0$ as $t \to \infty$. Now, suppose, *ad absurdum*, that $x(t)$, $t \geq 0$, does not converge to $\mathcal{D}_0$ or, equivalently, $\beta > 0$. Furthermore, let $y \in \omega(x_0)$ and let $\{\tau_n\}_{n=0}^{\infty}$ be an increasing unbounded sequence such that $\lim_{n\to\infty} x(\tau_n) = y$. Since $\{V(x(\tau_n))\}_{n=0}^{\infty}$ is a lower bounded nonincreasing sequence, $\lim_{n\to\infty} V(x(\tau_n))$ exists and is equal to β. Hence, $y \notin \mathcal{D}_0$ and since $y \in \omega(x_0)$ is arbitrary, it follows that $\omega(x_0) \cap \mathcal{D}_0 = \emptyset$. Next, let $\gamma \triangleq \min_{x\in\omega(x_0)} V(x) > 0$ and let $x_\gamma \in \omega(x_0)$ be such that $V(x_\gamma) = \gamma$. Now, since $\omega(x_0)$ is an invariant set it follows that for all $x(0) \in \omega(x_0)$, $x(t) \in \omega(x_0)$, $t \geq 0$, and hence, $V(x(t)) \geq \gamma$, $t \geq 0$. However, since for all $x(0) \in \mathcal{D}$, $x(0) \notin \mathcal{D}_0$, there exists an increasing unbounded sequence $\{t_n\}_{n=1}^{\infty}$ such that (4.256) holds, it follows that if $x(0) = x_\gamma \in \omega(x_0)$, there exists $t > 0$ such that $V(x(t)) < V(x(0)) = \gamma$, which is a contradiction. Hence, $\omega(x_0) \subseteq \mathcal{D}_0$ and $x(t) \to \mathcal{D}_0$ as $t \to \infty$, establishing asymptotic stability. □

Note that in the case where the function $V(\cdot)$ is continuously differentiable on $\mathcal{D}$ in Theorem 4.29, it follows that $V(x(t)) \leq V(x(\tau))$, for all $t \geq \tau \geq 0$, is equivalent to $\dot{V}(x) \triangleq V'(x)f(x) \leq 0$, $x \in \mathcal{D}$. Furthermore, if $\dot{V}(x) = V'(x)f(x) < 0$, $x \in \mathcal{D}$, $x \notin \mathcal{D}_0$, then every increasing unbounded sequence $\{t_n\}_{n=0}^{\infty}$, with $t_0 = 0$, is such that $V(x(t_{n+1})) < V(x(t_n))$, $n = 0, 1, \ldots$. Hence, the following corollary to Theorem 4.29 is immediate.

Corollary 4.7. Consider the nonlinear dynamical system (4.251) and let $\mathcal{D}_0$ be a compact positively invariant set with respect to (4.251) such that $\mathcal{D}_0 \subset \mathcal{D}$. Assume that there exists a continuously differentiable function $V : \mathcal{D} \to \mathbb{R}$ such that

$$V(x) = 0, \quad x \in \mathcal{D}_0, \tag{4.257}$$

$$V(x) > 0, \quad x \in \mathcal{D}, \quad x \notin \mathcal{D}_0, \tag{4.258}$$

$$V'(x)f(x) \leq 0, \quad x \in \mathcal{D}. \tag{4.259}$$

Then $\mathcal{D}_0$ is Lyapunov stable. If, in addition,

$$V'(x)f(x) < 0, \quad x \in \mathcal{D}, \quad x \notin \mathcal{D}_0, \tag{4.260}$$

then $\mathcal{D}_0$ is asymptotically stable.

The following theorem provides a converse to Theorem 4.29. For this result define the notation $\mathcal{D}_r \triangleq \{x \in \mathcal{D} : \operatorname{dist}(x, \mathcal{D}_0) < r\}$, $r > 0$, to denote an r open neighborhood of $\mathcal{D}_0$.

Theorem 4.30. Consider the nonlinear dynamical system (4.251). If $\mathcal{D}_0$ is Lyapunov stable, then there exist $\mathcal{D}_a \subseteq \mathcal{D}$ and a lower semicontinuous, positive-definite (on $\mathcal{D}_a \setminus \mathcal{D}_0$) function $V : \mathcal{D}_a \to \mathbb{R}$ such that $\mathcal{D}_0 \subset \overset{\circ}{\mathcal{D}}_a$, $V(\cdot)$ is continuous on $\mathcal{D}_0$, and

$$V(s(t,x)) \leq V(s(\tau,x)), \quad 0 \leq \tau \leq t, \quad x \in \mathcal{D}_a. \tag{4.261}$$

If $\mathcal{D}_0$ is asymptotically stable, then there exists a continuous, positive-definite function $V : \mathcal{D}_a \to \mathbb{R}$ such that inequality (4.261) is strictly satisfied.

Proof. Let $\varepsilon > 0$. Since $\mathcal{D}_0$ is Lyapunov stable it follows that there exists $\delta = \delta(\varepsilon) > 0$ such that if $x_0 \in \mathcal{D}_\delta$, then $s(t,x_0) \in \mathcal{D}_\varepsilon$, $t \geq 0$. Now, let $\mathcal{D}_a \triangleq \{y \in \mathcal{D}_\varepsilon :$ there exists $t \geq 0$ and $x_0 \in \mathcal{D}_\delta$ such that $y = s(t,x_0)\}$. Note that $\mathcal{D}_a \subseteq \mathcal{D}_\varepsilon$, $\mathcal{D}_a$ is a positively invariant set, and $\mathcal{D}_\delta \subseteq \mathcal{D}_a$. Hence, $\mathcal{D}_0 \subset \overset{\circ}{\mathcal{D}}_a$. Next, define $V(x) \triangleq \sup_{t\geq 0} \operatorname{dist}(s(t,x),\mathcal{D}_0)$, $x \in \mathcal{D}_a$, and since $\mathcal{D}_a$ is positively invariant and bounded it follows that $V(\cdot)$ is well defined on $\mathcal{D}_a$. Now, since $\mathcal{D}_0$ is invariant, $x \in \mathcal{D}_0$ implies $V(x) = 0$, $x \in \mathcal{D}_0$. Furthermore, $V(x) \geq \operatorname{dist}(s(0,x),\mathcal{D}_0) > 0$, $x \in \mathcal{D}_a$, $x \notin \mathcal{D}_0$.

Next, since $f(\cdot)$ in (4.251) is such that for every $x \in \mathcal{D}_a$, $s(t,x)$, $t \geq 0$, is the unique solution to (4.251), it follows that $s(t,x) = s(t-\tau, s(\tau,x))$, $0 \leq \tau \leq t$. Hence, for every $t, \tau \geq 0$, such that $t \geq \tau$,

$$\begin{aligned} V(s(\tau,x)) &= \sup_{\theta\geq 0} \operatorname{dist}(s(\theta, s(\tau,x)),\mathcal{D}_0) \\ &= \sup_{\theta\geq 0} \operatorname{dist}(s(\tau+\theta,x),\mathcal{D}_0) \\ &\geq \sup_{\theta\geq t-\tau} \operatorname{dist}(s(\tau+\theta,x),\mathcal{D}_0) \\ &= \sup_{\theta\geq t-\tau} \operatorname{dist}(s(\theta-(t-\tau), s(t,x)),\mathcal{D}_0) \\ &= \sup_{\theta\geq 0} \operatorname{dist}(s(\theta, s(t,x)),\mathcal{D}_0) \\ &= V(s(t,x)), \end{aligned} \tag{4.262}$$

which proves (4.261). Next, let $p_{\mathcal{D}_0} \in \mathcal{D}_0$. Since $\mathcal{D}_0$ is Lyapunov stable it follows that for every $\hat{\varepsilon} > 0$ there exists $\hat{\delta} = \hat{\delta}(\varepsilon) > 0$ such that if $x_0 \in \mathcal{B}_{\hat{\delta}}(p_{\mathcal{D}_0})$, then $s(t,x_0) \in \mathcal{D}_{\hat{\varepsilon}/2}$, $t \geq 0$, which implies that $V(x_0) = \sup_{t\geq 0} \operatorname{dist}(s(t,x_0),\mathcal{D}_0) \leq \frac{\hat{\varepsilon}}{2}$. Hence, for every point $p_{\mathcal{D}_0} \in \mathcal{D}_0$ and $\hat{\varepsilon} > 0$ there exists $\hat{\delta} = \hat{\delta}(\hat{\varepsilon})$ such that if $x_0 \in \mathcal{B}_{\hat{\delta}}(p_{\mathcal{D}_0})$, then $V(x_0) < \hat{\varepsilon}$, establishing that $V(\cdot)$ is continuous on $\mathcal{D}_0$.

Finally, to show that $V(\cdot)$ is lower semicontinuous everywhere on $\mathcal{D}_a$, let $x \in \mathcal{D}_a$, let $\hat{\varepsilon} > 0$, and note that since $V(x) \triangleq \sup_{t\geq 0} \operatorname{dist}(s(t,x),\mathcal{D}_0)$, there exists $T = T(x,\hat{\varepsilon}) > 0$ such that $V(x) - \operatorname{dist}(s(T,x),\mathcal{D}_0) < \hat{\varepsilon}$. Now, consider a sequence $\{x_i\}_{i=1}^{\infty} \in \mathcal{D}_a$ such that $x_i \to x$ as $i \to \infty$. Next, since

$s(t,\cdot)$ is continuous for every $t \geq 0$ and $\mathrm{dist}(\cdot, \mathcal{D}_0) : \mathcal{D}_\mathrm{a} \to \mathbb{R}$ is continuous, it follows that $\mathrm{dist}(s(T,x), \mathcal{D}_0) = \lim_{i\to\infty} \mathrm{dist}(s(T,x_i), \mathcal{D}_0)$. Next, note that $\mathrm{dist}(s(T,x_i), \mathcal{D}_0) \leq \sup_{t\geq 0}\mathrm{dist}(s(t,x_0), \mathcal{D}_0)$, $i = 1, 2, \ldots$, and hence,

$$\begin{aligned}\liminf_{i\to\infty} \sup_{t\geq 0} \mathrm{dist}(s(t,x_i), \mathcal{D}_0) &\geq \liminf_{i\to\infty} \mathrm{dist}(s(T,x_i), \mathcal{D}_0)\\ &= \lim_{i\to\infty} \mathrm{dist}(s(T,x_i), \mathcal{D}_0), \quad i = 1, 2, \ldots,\end{aligned}$$

which implies that

$$\begin{aligned}V(x) &< \mathrm{dist}(s(T,x), \mathcal{D}_0) + \hat{\varepsilon}\\ &= \lim_{i\to\infty} \mathrm{dist}(s(T,x_i), \mathcal{D}_0) + \hat{\varepsilon}\\ &\leq \liminf_{i\to\infty} \sup_{t\geq 0} \mathrm{dist}(s(t,x_i), \mathcal{D}_0) + \hat{\varepsilon}\\ &= \liminf_{i\to\infty} V(x_i) + \hat{\varepsilon}. \end{aligned} \tag{4.263}$$

Now, since $\hat{\varepsilon} > 0$ is arbitrary, (4.263) implies that $V(x) \leq \liminf_{i\to\infty} V(x_i)$. Thus, since $\{x_i\}_{i=1}^{\infty}$ is an arbitrary sequence converging to x, it follows that $V(\cdot)$ is lower semicontinuous on $\mathcal{D}_\mathrm{a}$.

To show the existence of a continuous, positive-definite, strictly decreasing function along the system trajectories of (4.251) in the case where $\mathcal{D}_0$ is asymptotically stable, consider the function $V : \mathcal{D}_\mathrm{A} \to \mathbb{R}$ given by

$$V(x) \triangleq \int_0^\infty \sup_{t\geq 0} \mathrm{dist}(s(t, s(\sigma, x)), \mathcal{D}_0) e^{-\sigma} \mathrm{d}\sigma, \quad x \in \mathcal{D}_\mathrm{A}, \tag{4.264}$$

where $\mathcal{D}_\mathrm{A} \subseteq \mathcal{D}_\mathrm{a}$ is a domain of attraction of $\mathcal{D}_0$. Since $\mathcal{D}_0$ is asymptotically stable, it follows that $\hat{V}(x) \triangleq \sup_{t\geq 0} \mathrm{dist}(s(t,x), \mathcal{D}_0)$, $x \in \mathcal{D}_\mathrm{A}$, is continuous on $\mathcal{D}_\mathrm{A}$ (see [58, p. 67]), which implies that $V(\cdot)$ is continuous on $\mathcal{D}_\mathrm{A}$ and positive definite on $\mathcal{D}_\mathrm{A} \setminus \mathcal{D}_0$. Next, suppose, *ad absurdum*, that

$$V(s(t,x)) = V(s(\tau,x)), \tag{4.265}$$

for some $0 \leq \tau < t$ and $x \in \mathcal{D}_\mathrm{A}$. Now, since $\hat{V}(\cdot)$ is positive definite (on $\mathcal{D}_\mathrm{a} \setminus \mathcal{D}_0$) and (4.261) holds for $\hat{V}(\cdot)$, (4.265) implies that $\hat{V}(s(\sigma + t, x)) = \hat{V}(s(\sigma + \tau, x))$ for all $\sigma \geq 0$. However, since $\mathcal{D}_0$ is asymptotically stable it follows that $\mathrm{dist}(s(t,x), \mathcal{D}_0) \to 0$ as $t \to \infty$ for $x \in \mathcal{D}_\mathrm{A}$, and hence, for sufficiently large $\sigma^* > 0$, $\mathrm{dist}(s(\sigma^* + t, x), \mathcal{D}_0) < \mathrm{dist}(s(\sigma^* + \tau, x), \mathcal{D}_0)$. Thus, $\hat{V}(s(\sigma^* + t, x)) < \hat{V}(s(\sigma^* + \tau, x))$, which leads to a contradiction. Hence, inequality (4.261) is strictly satisfied for $V(\cdot)$ which completes the proof. □

Next, we generalize the Barbashin-Krasovskii-LaSalle invariant set theorems to the case in which the function $V(\cdot)$ is lower semicontinuous. In particular, we show that the system trajectories converge to a union of largest invariant sets contained on the boundary of the intersections over *finite* intervals of the closure of generalized Lyapunov level surfaces. For the remainder of the results of this section define the notation

$$\mathcal{R}_\gamma \triangleq \bigcap_{c>\gamma} \overline{V^{-1}([\gamma, c])}, \tag{4.266}$$

for arbitrary $V : \mathcal{D} \subseteq \mathbb{R}^n \to \mathbb{R}$ and $\gamma \in \mathbb{R}$, and let $\mathcal{M}_\gamma$ denote the largest invariant set (with respect to (4.251)) contained in $\mathcal{R}_\gamma$.

Theorem 4.31. Consider the nonlinear dynamical system (4.251), let $\mathcal{D}_\mathrm{c}$ and $\mathcal{D}_0$ be compact positively invariant sets with respect to (4.251) such that $\mathcal{D}_0 \subset \mathcal{D}_\mathrm{c} \subset \mathcal{D}$, and let $x(t)$, $t \geq 0$, denote the solution to (4.251) corresponding to $x_0 \in \mathcal{D}_\mathrm{c}$. Assume that there exists a lower semicontinuous function $V : \mathcal{D}_\mathrm{c} \to \mathbb{R}$ such that

$$V(x) = 0, \qquad x \in \mathcal{D}_0, \tag{4.267}$$

$$V(x) > 0, \qquad x \in \mathcal{D}_\mathrm{c}, \qquad x \notin \mathcal{D}_0, \tag{4.268}$$

$$V(x(t)) \leq V(x(\tau)), \qquad 0 \leq \tau \leq t. \tag{4.269}$$

Furthermore, assume that for all $x_0 \in \mathcal{D}_\mathrm{c}$, $x_0 \notin \mathcal{D}_0$, there exists an increasing unbounded sequence $\{t_n\}_{n=0}^\infty$, with $t_0 = 0$, such that

$$V(x(t_{n+1})) < V(x(t_n)), \qquad n = 0, 1, \ldots. \tag{4.270}$$

Then, either $\mathcal{M}_\gamma \subset \hat{\mathcal{R}}_\gamma \triangleq \mathcal{R}_\gamma \setminus V^{-1}(\gamma)$, or $\mathcal{M}_\gamma = \emptyset$, $\gamma > 0$. Furthermore, if $x_0 \in \mathcal{D}_\mathrm{c}$, then $x(t) \to \hat{\mathcal{M}} \triangleq \cup_{\gamma \in \mathcal{G}} \mathcal{M}_\gamma$ as $t \to \infty$, where $\mathcal{G} \triangleq \{\gamma \geq 0 : \mathcal{R}_\gamma \cap \mathcal{D}_0 \neq \emptyset\}$. If, in addition, $\mathcal{D}_0 \subset \overset{\circ}{\mathcal{D}}_\mathrm{c}$ and $V(\cdot)$ is continuous on $\mathcal{D}_0$, then $\mathcal{D}_0$ is locally asymptotically stable and $\mathcal{D}_\mathrm{c}$ is a subset of the domain of attraction.

Proof. Since $\mathcal{D}_\mathrm{c}$ is a compact positively invariant set, it follows that for all $x_0 \in \mathcal{D}_\mathrm{c}$, the forward solution $x(t)$, $t \geq 0$, to (4.251) is bounded. Hence, it follows from Theorem 2.41 that, for all $x_0 \in \mathcal{D}_\mathrm{c}$, $\omega(x_0)$ is a nonempty, compact, connected invariant set. Next, it follows from Theorem 4.24 and the fact that $V(\cdot)$ is positive definite (with respect to $\mathcal{D}_\mathrm{c} \setminus \mathcal{D}_0$), that for every $x_0 \in \mathcal{D}_\mathrm{c}$ there exists $\gamma_{x_0} \geq 0$ such that $\omega(x_0) \subseteq \mathcal{M}_{\gamma_{x_0}} \subseteq \mathcal{R}_{\gamma_{x_0}}$. Now, given $x(0) \in V^{-1}(\gamma_{x_0})$, $\gamma_{x_0} > 0$, (4.270) implies that there exists $t_1 > 0$ such that $V(x(t_1)) < \gamma_{x_0}$ and $x(t_1) \notin V^{-1}(\gamma_{x_0})$. Hence, $V^{-1}(\gamma_{x_0}) \subset \mathcal{R}_{\gamma_{x_0}}$ does not contain any invariant set. Alternatively, if $x(0) \in \hat{\mathcal{R}}_{\gamma_{x_0}}$, then $V(x(0)) < \gamma_{x_0}$ and (4.270) implies that $x(t) \notin V^{-1}(\gamma_{x_0})$, $t \geq 0$. Hence, any invariant set contained in $\mathcal{R}_{\gamma_{x_0}}$ is a subset of $\hat{\mathcal{R}}_{\gamma_{x_0}}$, which implies that $\mathcal{M}_{\gamma_{x_0}} \subset \hat{\mathcal{R}}_{\gamma_{x_0}}$, $\gamma_{x_0} > 0$. If $\hat{\gamma} > 0$ is such that $\hat{\gamma} \neq \gamma_{x_0}$ for all $x_0 \in \mathcal{D}_\mathrm{c}$, then there does not exist $x_0 \in \mathcal{D}_\mathrm{c}$ such that $\omega(x_0) \subseteq \mathcal{R}_{\hat{\gamma}}$, and hence, $\mathcal{M}_{\hat{\gamma}} = \emptyset$. Now, *ad absurdum*, suppose $\mathcal{D}_0 \cap \omega(x_0) = \emptyset$. Since $V(\cdot)$ is lower semicontinuous it follows from Theorem 2.11 that there exists $\hat{x} \in \omega(x_0)$ such that $\alpha = V(\hat{x}) \leq V(x)$, $x \in \omega(x_0)$. Now, with $x(0) = \hat{x} \notin \mathcal{D}_0$ it follows from (4.270) that there exists an increasing unbounded sequence $\{t_n\}_{n=0}^\infty$, with $t_0 = 0$, such that $V(x(t_{n+1})) < V(x(t_n))$, $n = 0, 1, \ldots$, which implies that there exists $t > 0$ such that $V(x(t)) < \alpha$, and hence, $x(t) \notin \omega(x_0)$, contradicting the fact that $\omega(x_0)$ is an invariant set. Hence, there exists $q \in \mathcal{D}_0$ such that $q \in \omega(x_0) \subseteq \mathcal{R}_{\gamma_{x_0}}$, which implies that $\mathcal{R}_{\gamma_{x_0}} \cap \mathcal{D}_0 \neq \emptyset$. Thus, $\gamma_{x_0} \in \mathcal{G}$ for all $x_0 \in \mathcal{D}_c$, which further implies that $\omega(x_0) \subseteq \hat{\mathcal{M}}$. Now, since $x(t) \to \omega(x_0) \subseteq \hat{\mathcal{M}}$ as $t \to \infty$ it follows that $x(t) \to \hat{\mathcal{M}}$ as $t \to \infty$.

If $V(\cdot)$ is continuous on $\mathcal{D}_0 \subset \overset{\circ}{\mathcal{D}}_c$, then Lyapunov stability of the compact positively invariant set $\mathcal{D}_0$ follows from Theorem 4.29. Furthermore, from the continuity of $V(\cdot)$ on $\mathcal{D}_0$ and the fact that $V(x) = 0$ for all $x \in \mathcal{D}_0$, it follows that $\mathcal{G} = \{0\}$ and $\hat{\mathcal{M}} \equiv \mathcal{M}_0$. Hence, $\omega(x_0) \subseteq \mathcal{D}_0$ for all $x_0 \in \mathcal{D}_c$, establishing local asymptotic stability of the compact positively invariant set $\mathcal{D}_0$ of (4.251) with a subset of the domain of attraction given by $\mathcal{D}_c$. □

Finally, we present a generalized global invariant set theorem for guaranteeing global attraction and global asymptotic stability of a compact positively invariant set of a nonlinear dynamical system.

Theorem 4.32. Consider the nonlinear dynamical system (4.251) with $\mathcal{D} = \mathbb{R}^n$ and let $x(t)$, $t \geq 0$, denote the solution to (4.251) corresponding to $x_0 \in \mathbb{R}^n$. Assume that there exists a compact positively invariant set $\mathcal{D}_0$ with respect to (4.251) and a lower semicontinuous function $V : \mathbb{R}^n \to \mathbb{R}$ such that

$$V(x) = 0, \quad x \in \mathcal{D}_0, \tag{4.271}$$

$$V(x) > 0, \quad x \in \mathbb{R}^n, \quad x \notin \mathcal{D}_0, \tag{4.272}$$

$$V(x(t)) \leq V(x(\tau)), \quad 0 \leq \tau \leq t, \tag{4.273}$$

$$V(x) \to \infty \text{ as } \|x\| \to \infty. \tag{4.274}$$

Then for all $x_0 \in \mathbb{R}^n$, $x(t) \to \mathcal{M} \triangleq \cup_{\gamma \geq 0} \mathcal{M}_\gamma$, as $t \to \infty$. If, in addition, for all $x_0 \in \mathbb{R}^n$, $x_0 \notin \mathcal{D}_0$, there exists an increasing unbounded sequence $\{t_n\}_{n=0}^{\infty}$, with $t_0 = 0$, such that

$$V(x(t_{n+1})) < V(x(t_n)), \quad n = 0, 1, \ldots, \tag{4.275}$$

then, either $\mathcal{M}_\gamma \subset \hat{\mathcal{R}}_\gamma \triangleq \mathcal{R}_\gamma \setminus V^{-1}(\gamma)$, or $\mathcal{M}_\gamma = \emptyset$, $\gamma > 0$. Furthermore, $x(t) \to \hat{\mathcal{M}} \triangleq \cup_{\gamma \in \mathcal{G}} \mathcal{M}_\gamma$ as $t \to \infty$, where $\mathcal{G} \triangleq \{\gamma \geq 0 : \mathcal{R}_\gamma \cap \mathcal{D}_0 \neq \emptyset\}$. Finally, if $V(\cdot)$ is continuous on $\mathcal{D}_0$ then the compact positively invariant set $\mathcal{D}_0$ of (4.251) is globally asymptotically stable.

Proof. Note that since $V(x) \to \infty$ as $\|x\| \to \infty$ it follows that for every $\beta > 0$ there exists $r > 0$ such that $V(x) > \beta$ for all $\|x\| > r$ or, equivalently, $V^{-1}([0, \beta]) \subseteq \{x : \|x\| \leq r\}$, which implies that $V^{-1}([0, \beta])$ is bounded for all $\beta > 0$. Hence, for all $x_0 \in \mathbb{R}^n$, $V^{-1}([0, \beta_{x_0}])$ is bounded, where $\beta_{x_0} \triangleq$

$V(x_0)$. Furthermore, since $V(\cdot)$ is a positive-definite lower semicontinuous function, it follows that $V^{-1}([0, \beta_{x_0}])$ is closed and, since $V(x(t))$, $t \geq 0$, is nonincreasing, $V^{-1}([0, \beta_{x_0}])$ is positively invariant. Hence, for every $x_0 \in \mathbb{R}^n$, $V^{-1}([0, \beta_{x_0}])$ is a compact positively invariant set. Now, with $\mathcal{D}_c = V^{-1}([0, \beta_{x_0}])$ it follows from Theorem 4.24 that there exists $\gamma_{x_0} \in [0, \beta_{x_0}]$ such that $\omega(x_0) \subseteq \mathcal{M}_{\gamma_{x_0}} \subset \hat{\mathcal{R}}_{\gamma_{x_0}}$, which implies that $x(t) \to \mathcal{M}$ as $t \to \infty$. If, in addition, for all $x_0 \in \mathbb{R}^n$, $x_0 \notin \mathcal{D}_0$, there exists an increasing unbounded sequence $\{t_n\}_{n=0}^{\infty}$, with $t_0 = 0$, such that (4.275) holds, then it follows from Theorem 4.31 that $x(t) \to \hat{\mathcal{M}}$ as $t \to \infty$.

Finally, if $V(\cdot)$ is continuous on $\mathcal{D}_0$ then Lyapunov stability follows as in the proof of Theorem 4.31. Furthermore, in this case, $\mathcal{G} = \{0\}$, which implies that $\hat{\mathcal{M}} = \mathcal{M}_0$. Hence, $\omega(x_0) \subseteq \mathcal{D}_0$, establishing global asymptotic stability of the compact positively invariant set $\mathcal{D}_0$ of (4.251). □

4.10 Poincaré Maps and Stability of Periodic Orbits

Poincaré's theorem [358] provides a powerful tool in analyzing the stability properties of periodic orbits and limit cycles of n-dimensional dynamical systems in the case where the trajectory of the system can be relatively easily integrated. Specifically, Poincaré's theorem provides necessary and sufficient conditions for stability of periodic orbits based on the stability properties of a fixed point of a discrete-time dynamical system constructed from a Poincaré return map. In particular, for a given candidate periodic trajectory, an $(n-1)$-dimensional hyperplane is constructed that is transversal to the periodic trajectory and which defines the Poincaré return map. Trajectories starting on the hyperplane which are sufficiently close to a point on the periodic orbit will intersect the hyperplane after a time approximately equal to the period of the periodic orbit. This mapping traces the system trajectory from a point on the hyperplane to its next corresponding intersection with the hyperplane. Hence, the Poincaré return map can be used to establish a relationship between the stability properties of a dynamical system with periodic solutions and the stability properties of an equilibrium point of an $(n-1)$-dimensional discrete-time system. In this section, using the notions of Lyapunov and asymptotic stability of sets developed in Section 4.9, we construct lower semicontinuous Lyapunov functions to provide a Lyapunov function proof of Poincaré's theorem. To begin, we introduce the notions of Lyapunov and asymptotic stability of a periodic orbit of the nonlinear dynamical system (4.251). For this definition recall the definitions of periodic solutions and periodic orbits of (4.251) given in Definition 2.53.

Definition 4.12. A periodic orbit $\mathcal{O}$ of (4.251) is *Lyapunov stable* if, for all $\varepsilon > 0$, there exists $\delta = \delta(\varepsilon) > 0$ such that if $\text{dist}(x_0, \mathcal{O}) < \delta$, then $\text{dist}(s(t, x_0), \mathcal{O}) < \varepsilon$, $t \geq 0$. A periodic orbit $\mathcal{O}$ of (4.251) is *asymptotically* *stable* if it is Lyapunov stable and there exists $\delta > 0$ such that if $\mathrm{dist}(x_0, \mathcal{O}) < \delta$, then $\mathrm{dist}(s(t, x_0), \mathcal{O}) \to 0$ as $t \to \infty$.

To proceed, we assume that for the point $p \in \mathcal{D}$, the dynamical system (4.251) has a periodic solution $s(t,p), t \geq 0$, with period $T > 0$ that generates the periodic orbit $\mathcal{O} \triangleq \{x \in \mathcal{D} : x = s(t,p),\ 0 \leq t \leq T\}$. Note that $\mathcal{O}$ is a compact invariant set. Furthermore, we assume that there exists a continuously differentiable function $\mathcal{X} : \mathcal{D} \to \mathbb{R}$ such that the $(n-1)$-dimensional hyperplane defined by $\mathcal{H} \triangleq \{x \in \mathcal{D} : \mathcal{X}(x) = 0\}$ contains the point $x = p$ and $\mathcal{X}'(p) \neq 0$. In addition, we assume that the hyperplane $\mathcal{H}$ is not tangent to the periodic orbit $\mathcal{O}$ at $x = p$, that is, $\mathcal{X}'(p)f(p) \neq 0$. Next, define the local section $\mathcal{S} \subset \mathcal{H}$ such that $p \in \mathcal{S}$, $\mathcal{X}'(x) \neq 0$, $x \in \mathcal{S}$, and no trajectory of (4.251) starting in $\mathcal{S}$ is tangent to $\mathcal{H}$, that is, $\mathcal{X}'(x)f(x) \neq 0, x \in \mathcal{S}$. Note that a trajectory $s(t,p)$ will intersect $\mathcal{S}$ at p in T seconds. Furthermore, let

$$\begin{aligned}\mathcal{U} \triangleq \{x \in \mathcal{S} : &\text{there exists } \hat{\tau} > 0 \text{ such that } s(\hat{\tau}, x) \in \mathcal{S} \\ &\text{and } s(t,x) \notin \mathcal{S},\ 0 < t < \hat{\tau}\},\end{aligned} \tag{4.276}$$

and let $\tau : \mathcal{U} \to \mathbb{R}_+$ be defined by

$$\tau(x) \triangleq \{\hat{\tau} > 0 : s(\hat{\tau}, x) \in \mathcal{S} \text{ and } s(t,x) \notin \mathcal{S},\ 0 < t < \hat{\tau}\}. \tag{4.277}$$

Finally, define the Poincaré return map $P : \mathcal{U} \to \mathcal{S}$ by

$$P(x) \triangleq s(\tau(x), x), \quad x \in \mathcal{U}. \tag{4.278}$$

Figure 4.9 gives a visualization of the Poincaré return map construction.

Next, define $\mathcal{D}_1 \triangleq \{x \in \mathcal{D} : \text{there exists } \tau(x) > 0 \text{ such that } s(\tau(x), x) \in \mathcal{S}\}$ and note that, for every $x \in \mathcal{O}$, there exists $\delta = \delta(x) > 0$ such that $\mathcal{B}_\delta(x) \subset \mathcal{D}_1$, and hence, $\mathcal{O} \subset \overset{\circ}{\mathcal{D}}_1$. Similarly, define $\mathcal{O}_\alpha \triangleq \{x \in \mathcal{D}_1 : s(\tau(x), x) \in \mathcal{S}_\alpha\}$ and $\mathcal{U}_\alpha \triangleq \{x \in \mathcal{S}_\alpha : s(\tau(x), x) \in \mathcal{S}_\alpha\}$, where $\mathcal{S}_\alpha \triangleq \mathcal{B}_\alpha(p) \cap \mathcal{S}$, $\alpha > 0$, and $\mathcal{O} \subset \overset{\circ}{\mathcal{O}}_\alpha \subseteq \mathcal{D}_1$. The function $\tau : \mathcal{D}_1 \to \overline{\mathbb{R}}_+$ defines the minimum time required for the trajectory $s(t,x)$, $x \in \mathcal{D}_1$, to return to the local section $\mathcal{S}$. Note that $\tau(x) > 0$, $x \in \mathcal{U}$. The following lemma shows that $\tau(\cdot)$ is continuous on $\overset{\circ}{\mathcal{D}}_1 \backslash \mathcal{H}$.

Lemma 4.4. Consider the nonlinear dynamical system (4.251). Assume that the point $p \in \mathcal{D}_1$ generates the periodic orbit $\mathcal{O} \triangleq \{x \in \mathcal{D}_1 : x = s(t,p),\ 0 \leq t \leq T\}$, where $s(t,p), t \geq 0$, is the periodic solution with period $T \equiv \tau(p)$. Then the function $\tau : \overset{\circ}{\mathcal{D}}_1 \to \overline{\mathbb{R}}_+$ is continuous on $\overset{\circ}{\mathcal{D}}_1 \backslash \mathcal{H}$.

Proof. Let $\varepsilon > 0$ and $x \in \overset{\circ}{\mathcal{D}}_1 \backslash \mathcal{H}$. Note that $x^* \triangleq s(\tau(x), x) \in \mathcal{S}$, and hence, $\mathcal{X}'(x^*) \neq 0$ and $\mathcal{X}'(x^*)f(x^*) \neq 0$. Now, since $s(\cdot, x)$ is continuous in

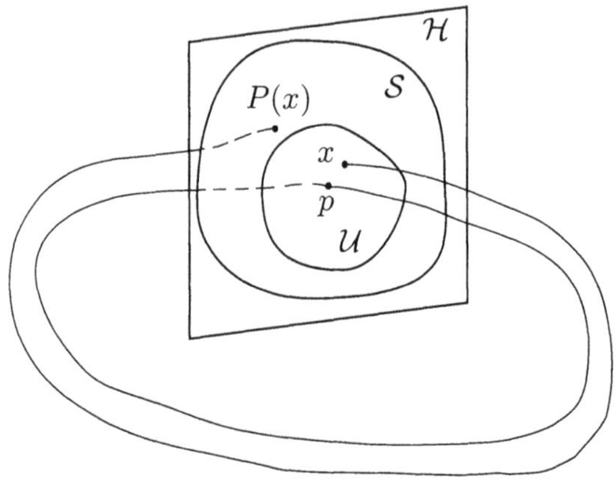

Figure 4.9 Visualization of the Poincaré return map.

t, $[0, t_1]$ is a compact interval it follows from the definition of $\tau(\cdot)$ that there exists $\hat{t} > 0$ such that for every $t_1 \in (\hat{t}, \tau(x))$,

$$\sigma(t_1) \triangleq \inf_{0 \le t \le t_1} \operatorname{dist}(s(t,x), \mathcal{S}) > 0. \tag{4.279}$$

Next, for sufficiently small $\hat{\varepsilon} > 0$, define $t_2 \triangleq \tau(x) + \frac{\hat{\varepsilon}}{2}$ and $x_2 \triangleq s(t_2, x)$. Since $\mathcal{X}'(x^*) \neq 0$ and $\mathcal{X}'(x^*) f(x^*) \neq 0$, it follows that $\operatorname{dist}(x_2, \mathcal{S}) > 0$. Now, define $t_1 \triangleq \tau(x) - \frac{\hat{\varepsilon}}{2}$. Then it follows from the continuous dependence of solutions to (4.251) in time and initial data that there exists $\delta > 0$ such that for all $y \in \mathcal{B}_\delta(x)$, $\sup_{\hat{t} \le t \le t_2} \|s(t,y) - s(t,x)\| < \min\{\operatorname{dist}(x_2, \mathcal{S}), \sigma(t_1)\}$. Hence, for all $y \in \mathcal{B}_\delta(x)$, it follows that $t_1 < \tau(y) < t_2$. Now, taking $\hat{\varepsilon} < \varepsilon$, it follows that $|\tau(y) - \tau(x)| < \varepsilon$, establishing the continuity of $\tau(\cdot)$ at $x \in \overset{\circ}{\mathcal{D}}_1 \backslash \mathcal{H}$. □

Finally, define the *discrete-time* dynamical system given by

$$z(k+1) = P(z(k)), \quad z(0) \in \mathcal{U}, \quad k \in \overline{\mathbb{Z}}_+. \tag{4.280}$$

Clearly $x = p$ is a fixed point of (4.280) since $T = \tau(p)$, and hence, $p = P(p)$. Since Poincaré's theorem provides necessary and sufficient conditions for stability of periodic orbits based on the stability properties of a fixed point of the discrete-time dynamical system (4.280), stability notions of discrete-time systems are required. Chapter 13 develops stability theory for discrete-time systems. Rather than embarking on a lengthy discussion of discrete-time stability theory, here we give two key necessary results for developing Poincaré's theorem. For these results, the definitions of Lyapunov and asymptotic stability of a discrete-time system are analogous to their continuous-time counterparts and are given in Chapter 13 (see Definition 13.1). First, we note that if $\rho(P'(z(p))) < 1$, where $\rho(\cdot)$ denotes the spectral radius, then the fixed point $x = p$ of the nonlinear discrete-time dynamical system (4.280) is asymptotically stable. This is Lyapunov's indirect method for nonlinear discrete-time systems. For details, see Problem 13.10. Second, the following theorem is a direct application of the standard discrete-time Lyapunov stability theorem (see Theorems 13.2 and 13.6) for general nonlinear dynamical systems to the dynamical system (4.280).

Theorem 4.33. The equilibrium solution $z(k) \equiv p$ to (4.280) is Lyapunov (respectively, asymptotically) stable if and only if there exist a scalar $\alpha > 0$ and a lower semicontinuous (respectively, continuous) function $V : \mathcal{S}_\alpha \to \mathbb{R}$ such that $V(\cdot)$ is continuous at $x = p$, $V(p) = 0$, $V(x) > 0$, $x \in \mathcal{S}_\alpha$, $x \neq p$, and $V(P(x)) - V(x) \leq 0$, $x \in \mathcal{U}_\alpha$ (respectively, $V(P(x)) - V(x) < 0$, $x \in \mathcal{U}_\alpha$, $x \neq p$).

Next, we present Poincaré's stability theorem.

Theorem 4.34. Consider the nonlinear dynamical system (4.251) with the Poincaré map defined by (4.278). Assume that the point $p \in \mathcal{D}$ generates the periodic orbit $\mathcal{O} \triangleq \{x \in \mathcal{D} : x = s(t,p), 0 \leq t \leq T\}$, where $s(t,p), t \geq 0$, is the periodic solution with period $T \equiv \tau(p)$. Then the following statements hold:

i) $p \in \mathcal{D}$ is a Lyapunov stable fixed point of (4.280) if and only if the periodic orbit $\mathcal{O}$ generated by p is Lyapunov stable.

ii) $p \in \mathcal{D}$ is an asymptotically stable fixed point of (4.280) if and only if the periodic orbit $\mathcal{O}$ generated by p is asymptotically stable.

Proof. *i)* To show necessity, assume that $x = p$ is a Lyapunov stable fixed point of (4.280). Then it follows from Theorem 4.33 that for sufficiently small $\alpha > 0$ there exists a lower semicontinuous function $V_{\mathrm{d}} : \mathcal{S}_\alpha \to \mathbb{R}$ such that $V_{\mathrm{d}}(\cdot)$ is continuous at $x = p$, $V_{\mathrm{d}}(p) = 0$, $V_{\mathrm{d}}(x) > 0$, $x \in \mathcal{S}_\alpha$, $x \neq p$, and $V_{\mathrm{d}}(P(x)) - V_{\mathrm{d}}(x) \leq 0$, $x \in \mathcal{U}_\alpha$. Next, define a function $V : \mathcal{O}_\alpha \to \mathbb{R}$ such that $V(x) = V_{\mathrm{d}}(s(\tau(x), x))$, $x \in \mathcal{O}_\alpha$. It follows from the definition of $\tau(\cdot)$, Lemma 4.4, and the joint continuity of solutions of (4.251) that $V(\cdot)$ is a lower semicontinuous function on $\mathcal{O}_\alpha$ and $V(x) = 0$, $x \in \mathcal{O}$, $V(x) > 0$, $x \in \mathcal{O}_\alpha \backslash \mathcal{O}$, where $\mathcal{O} \subset \overset{\circ}{\mathcal{O}}_\alpha$. Alternatively, it follows from the Lyapunov stability of the fixed point $x = p$ of (4.280) that for $\varepsilon > 0$ such that $\mathcal{B}_\varepsilon(p) \cap \mathcal{S} \subset \mathcal{U}_\alpha$, there exists $\delta = \delta(\varepsilon) > 0$ such that $z(k) \in \mathcal{B}_\varepsilon(p) \cap \mathcal{S}$, $k \in \overline{\mathbb{Z}}_+$, for all $z(0) \in \mathcal{B}_\delta(p) \cap \mathcal{S}$, where $z(k) \in \mathcal{N}$ satisfies (4.280). Now, define $\mathcal{O}_\delta \triangleq \{x \in \mathcal{O}_\alpha : s(\tau(x), x) \in \mathcal{B}_\delta(p) \cap \mathcal{S}\}$ and note that $\mathcal{O} \subset \overset{\circ}{\mathcal{O}}_\delta$. Hence, $V(s(t,x)) \leq V(s(\tau, x))$, $0 \leq \tau \leq t$, for every $x \in \mathcal{O}_\delta \subseteq \mathcal{O}_\alpha$. Now, to show that $V(\cdot)$ is continuous on $\mathcal{O}$ let $p_{\mathcal{O}} \in \mathcal{O}$ be such that $p_{\mathcal{O}} \neq p$ and consider any arbitrary sequence $\{x_n\}_{n=0}^{\infty} \in \mathcal{O}_\alpha$ such that $\lim_{n\to\infty} x_n = p_{\mathcal{O}}$. Then, since $\tau(\cdot)$ is continuous on $\mathcal{D}_1 \setminus \mathcal{H}$, it follows that $\lim_{n\to\infty} \tau(x_n) = \tau(p_{\mathcal{O}})$ and, by joint continuity of solutions of (4.251), $s(\tau(x_n), x_n) \to s(\tau(p_{\mathcal{O}}), p_{\mathcal{O}}) = p$ as $n \to \infty$. Now, since $V_{\rm d}(\cdot)$ is continuous at $x = p$, it follows that $\lim_{n\to\infty} V(x_n) = \lim_{n\to\infty} V_{\rm d}(s(\tau(x_n), x_n)) = V_{\rm d}(p) = 0$, which, since $\{x_n\}_{n=0}^{\infty}$ is arbitrary, implies continuity of $V(\cdot)$ at any point $p_{\mathcal{O}} \in \mathcal{O}$, $p_{\mathcal{O}} \neq p$.

Next, we show the continuity of $V(\cdot)$ at $x = p$. Note, that $V(\cdot)$ is not necessarily continuous at every point of $\mathcal{S}$ but $x = p$. Consider any arbitrary sequence $\{x_n\}_{n=0}^{\infty} \in \mathcal{O}_\alpha$ such that $\lim_{n\to\infty} x_n = p$. For this sequence we have one of the following three cases: either $\lim_{n\to\infty} \tau(x_n) = 0$, $\lim_{n\to\infty} \tau(x_n) = T$, or there exist subsequences $\{x_{n_k}\}_{k=0}^{\infty}$ and $\{x_{n_m}\}_{m=0}^{\infty}$ such that $\{x_{n_k}\}_{k=0}^{\infty} \cup \{x_{n_m}\}_{m=0}^{\infty} = \{x_n\}_{n=0}^{\infty}$, $x_{n_k} \to p$, $\tau(x_{n_k}) \to 0$, as $k \to \infty$, and $x_{n_m} \to p$, $\tau(x_{n_m}) \to T$, as $m \to \infty$. We assume the latter case, since the analysis for the first two cases follows immediately from the arguments for $p_{\mathcal{O}} \in \mathcal{O}$, $p_{\mathcal{O}} \neq p$, presented above. The characterization of both subsequences and joint continuity of solutions of (4.251) yield $s(\tau(x_{n_k}), x_{n_k}) \to s(0, p) = p$ and $s(\tau(x_{n_m}), x_{n_m}) \to s(T, p) = p$, as $k \to \infty$, and $m \to \infty$, respectively. Now, since $V_{\rm d}(\cdot)$ is continuous at $x = p$, it follows that $\lim_{k\to\infty} V(x_{n_k}) = \lim_{k\to\infty} V_{\rm d}(s(\tau(x_{n_k}), x_{n_k})) = V_{\rm d}(p) = 0$ and $\lim_{m\to\infty} V(x_{n_m}) = \lim_{m\to\infty} V_{\rm d}(s(\tau(x_{n_m}), x_{n_m})) = V_{\rm d}(p) = 0$, and thus, $\lim_{n\to\infty} V(x_n) = V(p) = 0$, which implies that $V(\cdot)$ is continuous at $x = p$, and hence, $V(\cdot)$ is continuous on $\mathcal{O}$. Finally, since all the assumptions of Theorem 4.29 hold, the periodic orbit $\mathcal{O}$ is Lyapunov stable.

To show sufficiency, assume that the periodic orbit $\mathcal{O}$ generated by the point $p \in \mathcal{D}$ is Lyapunov stable. Then it follows from the Theorem 4.30 that there exists a lower semicontinuous, positive-definite (on $\mathcal{D}_{\rm a} \setminus \mathcal{O}$) function $V : \mathcal{D}_{\rm a} \to \mathbb{R}$ such that (4.261) is satisfied. Now, for sufficiently small $\alpha > 0$, construct a function $V_{\rm d} : \mathcal{S}_\alpha \to \mathbb{R}$ such that $V_{\rm d}(x) = V(x)$, $x \in \mathcal{S}_\alpha$. Thus, in this case the sufficient conditions of Theorem 4.33 are satisfied for $V_{\rm d}(\cdot)$, which implies that the point $x = p$ is a Lyapunov stable fixed point of (4.280).

ii) To show necessity, assume that $x = p$ is an asymptotically stable fixed point of (4.280). Then it follows from Theorem 4.33 that there exists a continuous function $V_{\rm d} : \mathcal{S}_\alpha \to \mathbb{R}$ such that $V_{\rm d}(p) = 0$, $V_{\rm d}(x) > 0$, $x \in \mathcal{S}_\alpha$, $x \neq p$, and $V_{\rm d}(P(x)) - V_{\rm d}(x) < 0$, $x \in \mathcal{U}_\alpha$, $x \neq p$. Next, as in *i*), construct the lower semicontinuous function $V : \mathcal{O}_\alpha \to \mathbb{R}$ such that $V(x) = V_{\rm d}(s(\tau(x), x))$, $x \in \mathcal{O}_\alpha$, $V(x) = 0$, $x \in \mathcal{O}$, $V(x) > 0$, $x \in \mathcal{O}_\alpha$, $x \notin \mathcal{O}$, $V(s(t, x)) \leq V(s(\tau, x))$, $0 \leq \tau \leq t$, for every $x \in \mathcal{O}_\delta \subseteq \mathcal{O}_\alpha$, and $V(\cdot)$ is continuous on $\mathcal{O}$. Furthermore, for every $x \in \mathcal{O}_\delta$ define an increasing unbounded sequence $\{t_n\}_{n=0}^{\infty}$ such that $t_0 = 0$ and $t_k = \tau(z(k-1))$, $k = 1, 2, \ldots$, where $z(\cdot)$ satisfies (4.280) with $z(0) = x$. Then it follows from the definition of the function $V(\cdot)$ that $V(s(t_{n+1},x)) < V(s(t_n,x))$, $n \in \overline{\mathbb{Z}}_+$, establishing that all the conditions of the Theorem 4.29 hold, and hence, the periodic orbit $\mathcal{O}$ is asymptotically stable.

To show sufficiency, assume that the periodic orbit $\mathcal{O}$ generated by the point $p \in \mathcal{D}$ is asymptotically stable. Then it follows from the Theorem 4.30 that there exists a continuous, positive-definite (on $\mathcal{D}_\mathrm{a} \setminus \mathcal{O}$) function $V : \mathcal{D}_\mathrm{a} \to \mathbb{R}$ such that (4.261) is strictly satisfied. Now, for sufficiently small $\alpha > 0$, construct a function $V_\mathrm{d} : \mathcal{S}_\alpha \to \mathbb{R}$ such that $V_\mathrm{d}(x) = V(x)$, $x \in \mathcal{S}_\alpha$. Thus, the sufficient conditions of Theorem 4.33 are satisfied for $V_\mathrm{d}(\cdot)$ which implies that the point $x = p$ is an asymptotically stable fixed point of (4.280).

□

Theorem 4.34 is a restatement of the classical Poincaré theorem. However, in proving necessary and sufficient conditions for Lyapunov and asymptotic stability of the periodic orbit $\mathcal{O}$, we constructed explicit Lyapunov functions in the proof of Theorem 4.34. Specifically, in order to show necessity of Poincaré's theorem via Lyapunov's second method we constructed the lower semicontinuous (respectively, continuous), positive definite (on $\mathcal{O}_\alpha \setminus \mathcal{O}$) Lyapunov function

$$V(x) = V_\mathrm{d}(s(\tau(x), x)), \quad x \in \mathcal{O}_\alpha, \tag{4.281}$$

where the existence of the lower semicontinuous (respectively, continuous), positive definite (on $\mathcal{S}_\alpha \setminus p$) function $V_\mathrm{d} : \mathcal{S}_\alpha \to \mathbb{R}$ is guaranteed by the Lyapunov (respectively, asymptotic) stability of a fixed point $p \in \mathcal{D}$ of (4.280). Alternatively, in the proof of sufficiency, Lyapunov (respectively, asymptotic) stability of the periodic orbit $\mathcal{O}$ implies the existence of the lower semicontinuous (respectively, continuous), positive definite (on $\mathcal{D}_\mathrm{a} \setminus \mathcal{O}$) Lyapunov function given by

$$V(x) = \sup_{t \geq 0} \mathrm{dist}(s(t,x), \mathcal{O}), \quad x \in \mathcal{D}_\mathrm{a}, \tag{4.282}$$

and

$$V(x) = \int_0^\infty \sup_{t \geq 0} \mathrm{dist}(s(t, s(\tau, x)), \mathcal{O}) e^{-\tau} \mathrm{d}\tau, \quad x \in \mathcal{D}_\mathrm{A}, \tag{4.283}$$

respectively. Using the Lyapunov function $V_\mathrm{d}(x) = V(x)$, $x \in \mathcal{S}_\alpha$, we showed the stability of a fixed point $p \in \mathcal{D}$ of (4.280).

Theorem 4.34 presents necessary and sufficient conditions for Lyapunov and asymptotic stability of a periodic orbit of the nonlinear dynamical system (4.251) based on the stability properties of a fixed point of the n-dimensional discrete-time dynamical system (4.280) involving the Poincaré map (4.278). Next, we present a classical corollary to Poincaré's theorem that allows us to analyze the stability of periodic orbits by replacing the

nth-order nonlinear dynamical system by an $(n-1)$th-order discrete-time system. To present this result assume, without loss of generality, that $\frac{\partial \mathcal{X}(x)}{\partial x_n} \neq 0, x \in \mathcal{S}_\alpha$, where $x = [x_1, \ldots, x_n]^{\mathrm{T}}$ and $\alpha > 0$ is sufficiently small. Then it follows from the implicit function theorem (see Theorem 2.18) that $x_n = g(x_1, \ldots, x_{n-1})$, where $g(\cdot)$ is a continuously differentiable function at $x_{\mathrm{r}} \triangleq [x_1, \ldots, x_{n-1}]^{\mathrm{T}}$ such that $[x_{\mathrm{r}}^{\mathrm{T}}, g(x_{\mathrm{r}})]^{\mathrm{T}} \in \mathcal{S}_\alpha$. Note that in this case $P : \mathcal{U}_\alpha \to \mathcal{S}_\alpha$ in (4.280) is given by $P(x) \triangleq [P_1(x), \cdots, P_n(x)]^{\mathrm{T}}$, where

$$P_n(x_{\mathrm{r}}, g(x_{\mathrm{r}})) = g(P_1(x_{\mathrm{r}}, g(x_{\mathrm{r}})), \ldots, P_{n-1}(x_{\mathrm{r}}, g(x_{\mathrm{r}}))). \tag{4.284}$$

Hence, we can reduce the n-dimensional discrete-time system (4.280) to the $(n-1)$-dimensional discrete-time system given by

$$z_{\mathrm{r}}(k+1) = \mathcal{P}_{\mathrm{r}}(z_{\mathrm{r}}(k)), \quad k \in \overline{\mathbb{Z}}_+, \tag{4.285}$$

where $z_{\mathrm{r}} \in \mathbb{R}^{n-1}$, $[z_{\mathrm{r}}^{\mathrm{T}}(\cdot), g(z_{\mathrm{r}}(\cdot))]^{\mathrm{T}} \in \mathcal{S}_\alpha$, and

$$\mathcal{P}_{\mathrm{r}}(x_{\mathrm{r}}) \triangleq \begin{bmatrix} P_1(x_{\mathrm{r}}, g(x_{\mathrm{r}})) \\ \vdots \\ P_{n-1}(x_{\mathrm{r}}, g(x_{\mathrm{r}})) \end{bmatrix}. \tag{4.286}$$

Note that it follows from (4.284) and (4.286) that $p \triangleq [p_{\mathrm{r}}^{\mathrm{T}}, g(p_{\mathrm{r}})]^{\mathrm{T}} \in \mathcal{S}_\alpha$ is a fixed point of (4.280) if and only if p_{r} is a fixed point of (4.285). To present the following result define $\mathcal{S}_{\mathrm{r}\alpha} \triangleq \{x_{\mathrm{r}} \in \mathbb{R}^{n-1} : [x_{\mathrm{r}}^{\mathrm{T}}, g(x_{\mathrm{r}})]^{\mathrm{T}} \in \mathcal{S}_\alpha\}$ and $\mathcal{U}_{\mathrm{r}\alpha} \triangleq \{x_{\mathrm{r}} \in \mathcal{S}_{\mathrm{r}\alpha} : [x_{\mathrm{r}}^{\mathrm{T}}, g(x_{\mathrm{r}})]^{\mathrm{T}} \in \mathcal{U}_\alpha\}$.

Corollary 4.8. Consider the nonlinear dynamical system (4.251) with the Poincaré return map defined by (4.278). Assume that $\frac{\partial \mathcal{X}(x)}{\partial x_n} \neq 0$, $x \in \mathcal{S}_\alpha$, and the point $p \in \mathcal{S}_\alpha$ generates the periodic orbit $\mathcal{O} \triangleq \{x \in \mathcal{D} : x = s(t,p),\ 0 \leq t \leq T\}$, where $s(t,p)$, $t \geq 0$, is the periodic solution with the period $T = \tau(p)$ such that $s(\tau(p), p) = p$. Then the following statements hold:

i) For $p = [p_{\mathrm{r}}^{\mathrm{T}}, g(p_{\mathrm{r}})]^{\mathrm{T}} \in \mathcal{S}_\alpha$, p_{r} is a Lyapunov stable fixed point of (4.285) if and only if the periodic orbit $\mathcal{O}$ is Lyapunov stable.

ii) For $p = [p_{\mathrm{r}}^{\mathrm{T}}, g(p_{\mathrm{r}})]^{\mathrm{T}} \in \mathcal{S}_\alpha$, p_{r} is an asymptotically stable fixed point of (4.285) if and only if the periodic orbit $\mathcal{O}$ is asymptotically stable.

Proof. *i)* To show necessity, assume that $p_{\mathrm{r}} \in \mathcal{S}_{\mathrm{r}\alpha}$ is a Lyapunov stable fixed point of (4.285). Then it follows from Theorem 4.33 that there exists a lower semicontinuous function $V_{\mathrm{r}} : \mathcal{S}_{\mathrm{r}\alpha} \to \mathbb{R}$ such that $V_{\mathrm{r}}(\cdot)$ is continuous at p_{r}, $V_{\mathrm{r}}(p_{\mathrm{r}}) = 0$, $V_{\mathrm{r}}(x_{\mathrm{r}}) > 0$, $x_{\mathrm{r}} \neq p_{\mathrm{r}}$, $x_{\mathrm{r}} \in \mathcal{S}_{\mathrm{r}\alpha}$, and $V_{\mathrm{r}}(\mathcal{P}_{\mathrm{r}}(x_{\mathrm{r}})) - V_{\mathrm{r}}(x_{\mathrm{r}}) \leq 0$, $x_{\mathrm{r}} \in \mathcal{U}_{\mathrm{r}\alpha}$. Define $V : \mathcal{S}_\alpha \to \mathbb{R}$ such that $V(x) = V_{\mathrm{r}}(x_{\mathrm{r}})$, $x_{\mathrm{r}} \in \mathcal{S}_{\mathrm{r}\alpha}$. To show that $V(\cdot)$ is continuous at $p \in \mathcal{S}_\alpha$, consider an arbitrary sequence $\{x_k\}_{k=1}^\infty$ such that $x_k \in \mathcal{S}_\alpha$ and $x_k \to p$

as $k \to \infty$. Then, $x_{\mathrm{r}k} \to p_{\mathrm{r}}$ as $k \to \infty$ and, since $V_{\mathrm{r}}(\cdot)$ is continuous at p_{r}, $\lim_{k\to\infty} V(x_k) = \lim_{k\to\infty} V_{\mathrm{r}}(x_{\mathrm{r}k}) = V_{\mathrm{r}}(p_{\mathrm{r}}) = V(p)$. Hence, $V(\cdot)$ is continuous at $p \in \mathcal{S}_\alpha$. Similarly, for the sequence defined above $V(x) = V_{\mathrm{r}}(x_{\mathrm{r}}) \le \liminf_{k\to\infty} V_{\mathrm{r}}(x_{\mathrm{r}k}) = \liminf_{k\to\infty} V(x_k)$, $x \in \mathcal{S}_\alpha$, which implies that $V(\cdot)$ is lower semicontinuous. Next, note that $V(p) = V_{\mathrm{r}}(p_{\mathrm{r}}) = 0$ and suppose, *ad absurdum*, that there exists $x \neq p$ such that $x \in \mathcal{S}_\alpha$ and $V(x) = 0$. Then, $V_{\mathrm{r}}(x_{\mathrm{r}}) = 0$ and $x_{\mathrm{r}} = p_{\mathrm{r}}$, which implies that $x_n = g(p_{\mathrm{r}})$ and $x = p$, leading to a contradiction. Hence, $V(x) > 0$, $x \neq p$, $x \in \mathcal{S}_\alpha$. Next, note that

$$\begin{aligned} V(P(x)) - V(x) &= V_{\mathrm{r}}(P_1(x), \ldots, P_{n-1}(x)) - V_{\mathrm{r}}(x_{\mathrm{r}}) \\ &= V_{\mathrm{r}}(P_1(x_{\mathrm{r}}, g(x_{\mathrm{r}})), \ldots, P_{n-1}(x_{\mathrm{r}}, g(x_{\mathrm{r}}))) - V_{\mathrm{r}}(x_{\mathrm{r}}) \\ &= V_{\mathrm{r}}(\mathcal{P}_{\mathrm{r}}(x_{\mathrm{r}})) - V_{\mathrm{r}}(x_{\mathrm{r}}) \\ &\le 0, \qquad x \in \mathcal{U}_\alpha, \end{aligned} \tag{4.287}$$

and hence, by Theorem 4.33 the point $p \in \mathcal{S}_\alpha$ is a Lyapunov stable fixed point of (4.280). Finally, Lyapunov stability of the periodic orbit $\mathcal{O}$ follows from Theorem 4.34.

To show sufficiency, assume that the periodic orbit $\mathcal{O}$ is Lyapunov stable. Then, it follows from Theorem 4.34 that the point $p \in \mathcal{S}_\alpha$ is a Lyapunov stable fixed point of (4.280). Hence, it follows from Theorem 4.33 that there exists a lower semicontinuous function $V : \mathcal{S}_\alpha \to \mathbb{R}$ such that $V(\cdot)$ is continuous at $p \in \mathcal{S}_\alpha$, $V(p) = 0$, $V(x) > 0$, $x \neq p$, $x \in \mathcal{S}_\alpha$, and $V(P(x)) - V(x) \le 0$, $x \in \mathcal{U}_\alpha$. Next, define $V_{\mathrm{r}} : \mathcal{S}_{\mathrm{r}\alpha} \to \mathbb{R}$ such that $V_{\mathrm{r}}(x_{\mathrm{r}}) = V(x_{\mathrm{r}}, g(x_{\mathrm{r}}))$. The proofs of continuity of $V_{\mathrm{r}}(\cdot)$ at $p_{\mathrm{r}} \in \mathcal{S}_{\mathrm{r}\alpha}$ and lower semicontinuity of $V_{\mathrm{r}}(\cdot)$ follow similarly as in the proof of necessity. Next, note that $V_{\mathrm{r}}(p_{\mathrm{r}}) = V(p_{\mathrm{r}}, g(p_{\mathrm{r}})) = V(p) = 0$ and suppose, *ad absurdum*, that there exists $x_{\mathrm{r}} \neq p_{\mathrm{r}}$ such that $x_{\mathrm{r}} \in \mathcal{S}_{\mathrm{r}\alpha}$ and $V_{\mathrm{r}}(x_{\mathrm{r}}) = 0$. Then, $V(x_{\mathrm{r}}, g(x_{\mathrm{r}})) = 0$ and $x = [x_{\mathrm{r}}^{\mathrm{T}}, g(x_{\mathrm{r}})]^{\mathrm{T}} = p$, which implies that $x_{\mathrm{r}} = p_{\mathrm{r}}$, leading to a contradiction. Hence, $V_{\mathrm{r}}(x_{\mathrm{r}}) > 0$, $x_{\mathrm{r}} \neq p_{\mathrm{r}}$, $x_{\mathrm{r}} \in \mathcal{S}_{\mathrm{r}\alpha}$. Finally, using (4.284), it follows that

$$\begin{aligned} V_{\mathrm{r}}(\mathcal{P}_{\mathrm{r}}(x_{\mathrm{r}})) - V_{\mathrm{r}}(x_{\mathrm{r}}) &= V(\mathcal{P}_{\mathrm{r}}(x_{\mathrm{r}}), g(\mathcal{P}_{\mathrm{r}}(x_{\mathrm{r}}))) - V(x_{\mathrm{r}}, g(x_{\mathrm{r}})) \\ &= V(\mathcal{P}_{\mathrm{r}}(x_{\mathrm{r}}), P_n(x_{\mathrm{r}}, g(x_{\mathrm{r}}))) - V(x) \\ &= V(P(x)) - V(x) \\ &\le 0, \qquad x_{\mathrm{r}} \in \mathcal{U}_{\mathrm{r}\alpha}, \end{aligned} \tag{4.288}$$

and hence, by Theorem 4.33 the point $p_{\mathrm{r}} \in \mathcal{S}_{\mathrm{r}\alpha}$ is a Lyapunov stable fixed point of (4.285).

ii) The proof is analogous to that of *i*) and, hence, is omitted. □

Example 4.18. Once again consider the nonlinear dynamical system (2.220) and (2.221) given in Example 2.38 or, equivalently, in polar coordinates,

$$\dot{r}(t) = r(t)\left[1 - r^2(t)\right], \quad r(0) = r_0, \quad t \geq 0, \tag{4.289}$$
$$\dot{\theta}(t) = 1, \quad \theta(0) = \theta_0. \tag{4.290}$$

As shown in Example 2.38, the unit circle $r = 1$ is a periodic orbit for (4.289) and (4.290). Using separation of variables the solution to (4.289) and (4.290) is given by

$$r(t) = \left[1 + \left(\frac{1}{r_0^2} - 1\right)e^{-2t}\right]^{-1/2}, \tag{4.291}$$
$$\theta(t) = \theta_0 + t. \tag{4.292}$$

Note that if $\mathcal{S}$ is the ray $\theta = \theta_0$ through the origin of the x_1-x_2 plane, that is, $\mathcal{S} = \{(r,\theta) : \theta = \theta_0 \text{ and } r > 0\}$, then $\mathcal{S}$ is perpendicular to the periodic orbit $\mathcal{O}$ and the trajectory passing through the point $(r_0, \theta_0) \in \mathcal{S} \cap \mathcal{O}$ at $t = 0$ intersects the ray $\theta = \theta_0$ again at $T = 2\pi$ (see Figure 4.10). Thus, the Poincaré map is given by

$$P(r) = \left[1 + \left(\frac{1}{r^2} - 1\right)e^{-4\pi}\right]^{-1/2}. \tag{4.293}$$

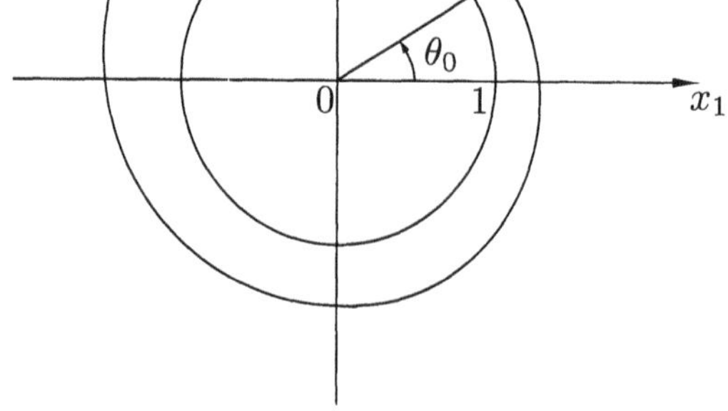

Figure 4.10 Poincaré map for Example 4.18.

Since (4.293) is a one-dimensional map, no reduction procedure is necessary and Corollary 4.8 can be used directly. Since $P(1) = 1$, $r = 1$ is a fixed point. Now, with $z = r$ we examine the stability of the discrete-time system

$$z(k+1) = \left[1 + \left(\frac{1}{z^2(k)} - 1\right) e^{-4\pi}\right]^{-1/2}, \quad z(0) \in \mathcal{B}_\delta(1), \qquad (4.294)$$

where $\delta > 0$ is sufficiently small. Specifically, evaluating $P'(z)$ yields

$$P'(z) = e^{-4\pi} z^{-3} \left[1 + \left(\frac{1}{z^2} - 1\right) e^{-4\pi}\right]^{-3/2}, \qquad (4.295)$$

and hence, $P'(1) = e^{-4\pi} < 1$. Thus, it follows from Corollary 4.8 that the periodic orbit $\mathcal{O} = \{(x_1, x_2) \in \mathbb{R} \times \mathbb{R} : x_1^2 + x_2^2 = 1\}$ is asymptotically stable.

$\triangle$

Example 4.19. Consider the nonlinear dynamical system

$$\dot{x}_1(t) = -x_2(t) + x_1(t)\left[1 - x_1^2(t) - x_2^2(t)\right], \quad x_1(0) = x_{10}, \quad t \geq 0, \qquad (4.296)$$

$$\dot{x}_2(t) = x_1(t) + x_2(t)\left[1 - x_1^2(t) - x_2^2(t)\right], \quad x_2(0) = x_{20}, \qquad (4.297)$$

$$\dot{x}_3(t) = -x_3(t), \quad x_3(0) = x_{30}. \qquad (4.298)$$

It can easily be shown that $x(t) = [x_1(t), x_2(t), x_3(t)]^{\mathrm{T}} = [\cos t, \sin t, 0]^{\mathrm{T}}$ is a periodic solution of (4.296)–(4.298) with period $T = 2\pi$ (see Problem 2.141). To examine the stability of this periodic solution we rewrite (4.296)–(4.298) in terms of the cylindrical coordinates $r = \sqrt{x_1^2 + x_2^2}$, $\theta = \tan^{-1}\left(\frac{x_2}{x_1}\right)$, and $\hat{z} = x_3$ as

$$\dot{r}(t) = r(t)[1 - r^2(t)], \quad r(0) = r_0, \quad t \geq 0, \qquad (4.299)$$

$$\dot{\theta}(t) = 1, \quad \theta(0) = \theta_0, \qquad (4.300)$$

$$\dot{\hat{z}}(t) = -\hat{z}(t), \quad \hat{z}(0) = \hat{z}_0. \qquad (4.301)$$

Note that the solution to (4.299) and (4.300) are given by (4.298) and (4.299), respectively, and the solution to (4.301) is given by $\hat{z}(t) = \hat{z}_0 e^{-t}$. Furthermore, since the periodic orbit of (4.296)–(4.298) lies in the x_1-x_2 plane we take $\mathcal{S} = \{(r, \theta, \hat{z}) : \theta = \theta_0, \ r > 0, \text{ and } \hat{z} \in \mathbb{R}\}$. Note that $\mathcal{S}$ is perpendicular to the periodic orbit and the trajectory passing through the point $(r_0, \theta_0, \hat{z}_0) \in \mathcal{S} \cap \mathcal{O}$ at $t = 0$ intersects the plane again at $T = 2\pi$. Thus, the Poincaré map is given by

$$P(r, \hat{z}) = \begin{bmatrix} \left(1 + \left(\frac{1}{r^2} - 1\right) e^{-4\pi}\right)^{-1/2} \\ \hat{z} e^{-2\pi} \end{bmatrix}. \qquad (4.302)$$

Clearly, $P(1, 0) = (1, 0)$, and hence, $[1, 0]^{\mathrm{T}}$ is a fixed point. Now, with

$z_1 = r$ and $z_2 = \hat{z}$ we examine the stability of the discrete-time system

$$\begin{bmatrix} z_1(k+1) \\ z_2(k+1) \end{bmatrix} = \begin{bmatrix} \left(1 + \left(\frac{1}{z_1^2(k)} - 1\right) e^{-4\pi}\right)^{-1/2} \\ z_2(k)e^{-2\pi} \end{bmatrix},$$
$$[z_1(0), z_2(0)]^{\mathrm{T}} \in \mathcal{B}_\delta([1,0]^{\mathrm{T}}), \qquad (4.303)$$

where $\delta > 0$ is sufficiently small. Specifically, evaluating $P'(z)$ yields

$$P'(z_1, z_2) = \begin{bmatrix} e^{-4\pi} z_1^{-3} \left(1 + \left(\frac{1}{z_1^2} - 1\right) e^{-4\pi}\right)^{-3/2} & 0 \\ 0 & e^{-2\pi} \end{bmatrix} \qquad (4.304)$$

so that $\rho(P'(1,0)) = e^{-2\pi} < 1$. Hence, it follows from Corollary 4.8 that the periodic orbit $\mathcal{O} = \{(x_1, x_2, x_3) \in \mathbb{R}^3 : x_1^2 + x_2^2 = 1 \text{ and } x_3 = 0\}$ is asymptotically stable. △

4.11 Stability Theory via Vector Lyapunov Functions

In this section, we introduce the notion of vector Lyapunov functions for stability analysis of nonlinear dynamical systems. The use of vector Lyapunov functions in dynamical system theory offers a very flexible framework since each component of the vector Lyapunov function can satisfy less rigid requirements as compared to a single scalar Lyapunov function. Specifically, since for many nonlinear dynamical systems constructing a system Lyapunov function can be a difficult task, weakening the hypothesis on the Lyapunov function enlarges the class of Lyapunov functions that can be used for analyzing system stability. Moreover, in certain applications, such as the analysis of large-scale nonlinear dynamical systems, several Lyapunov functions arise naturally from the stability properties of each individual subsystem. To develop the theory of vector Lyapunov functions, we first introduce some results on vector differential inequalities and the *vector comparison principle.* The following definition introduces the notion of class $\mathcal{W}$ functions involving *quasimonotone* increasing functions.

Definition 4.13. A function $w = [w_1, \ldots, w_q]^{\mathrm{T}} : \mathbb{R}^q \to \mathbb{R}^q$ is of *class* $\mathcal{W}$ if $w_i(z') \le w_i(z'')$, $i = 1, \ldots, q$, for all $z', z'' \in \mathbb{R}^q$ such that $z'_j \le z''_j$, $z'_i = z''_i$, $j = 1, \ldots, q$, $i \ne j$, where z_i denotes the ith component of z.

If $w(\cdot) \in \mathcal{W}$ we say that w satisfies the *Kamke condition*. Note that if $w(z) = Wz$, where $W \in \mathbb{R}^{q \times q}$, then the function $w(\cdot)$ is of class $\mathcal{W}$ if and only if W is *essentially nonnegative*, that is, all the off-diagonal entries of the matrix function W are nonnegative. Furthermore, note that it follows from Definition 4.13 that any scalar ($q = 1$) function $w(z)$ is of class $\mathcal{W}$.

Next, we introduce the notion of class $\mathcal{W}_{\rm d}$ functions involving *nondecreasing* functions.

Definition 4.14. A function $w = [w_1, \ldots, w_q]^{\rm T} : \mathbb{R}^q \to \mathbb{R}^q$ is of *class* $\mathcal{W}_{\rm d}$ if $w(z') \leq\leq w(z'')$ for all $z', z'' \in \mathbb{R}^q$ such that $z' \leq\leq z''$.

Note that if $w(\cdot) \in \mathcal{W}_{\rm d}$, then $w(\cdot) \in \mathcal{W}$. Next, we consider the nonlinear comparison system given by

$$\dot{z}(t) = w(z(t)), \quad z(t_0) = z_0, \quad t \in \mathcal{I}_{z_0}, \tag{4.305}$$

where $z(t) \in \mathcal{Q} \subseteq \mathbb{R}^q$, $t \in \mathcal{I}_{z_0}$, is the comparison system state vector, $\mathcal{I}_{z_0} \subseteq \mathcal{T} \subseteq \overline{\mathbb{R}}_+$ is the maximal interval of existence of a solution $z(t)$ of (4.305), $\mathcal{Q}$ is an open set, $0 \in \mathcal{Q}$, and $w : \mathcal{Q} \to \mathbb{R}^q$. We assume that $w(\cdot)$ satisfies the Lipschitz condition

$$\|w(z') - w(z'')\| \leq L\|z' - z''\|, \tag{4.306}$$

for all $z', z'' \in \mathcal{B}_\delta(z_0)$, where $\delta > 0$ and $L > 0$ is a Lipschitz constant. Hence, it follows from Theorem 2.25 that there exists $\tau > 0$ such that (4.305) has a unique solution over the time interval $[t_0, t_0 + \tau]$.

Theorem 4.35. Consider the nonlinear comparison system (4.305). Assume that the function $w : \mathcal{Q} \to \mathbb{R}^q$ is continuous and $w(\cdot)$ is of class $\mathcal{W}$. If there exists a continuously differentiable vector function $V = [v_1, \ldots, v_q]^{\rm T} : \mathcal{I}_{z_0} \to \mathcal{Q}$ such that

$$\dot{V}(t) << w(V(t)), \quad t \in \mathcal{I}_{z_0}, \tag{4.307}$$

then $V(t_0) << z_0$, $z_0 \in \mathcal{Q}$, implies

$$V(t) << z(t), \quad t \in \mathcal{I}_{z_0}, \tag{4.308}$$

where $z(t)$, $t \in \mathcal{I}_{z_0}$, is the solution to (4.305).

Proof. Since $V(t)$, $t \in \mathcal{I}_{z_0}$, is continuous it follows that for sufficiently small $\tau > 0$,

$$V(t) << z(t), \quad t \in [t_0, t_0 + \tau]. \tag{4.309}$$

Now, suppose, *ad absurdum*, that inequality (4.308) does not hold on the entire interval $\mathcal{I}_{z_0}$. Then there exists $\hat{t} \in \mathcal{I}_{z_0}$ such that $V(t) << z(t)$, $t \in [t_0, \hat{t})$, and for at least one $i \in \{1, \ldots, q\}$,

$$v_i(\hat{t}) = z_i(\hat{t}) \tag{4.310}$$

and

$$v_j(\hat{t}) \leq z_j(\hat{t}), \quad j \neq i, \quad j = 1, \ldots, q. \tag{4.311}$$

Since $w(\cdot) \in \mathcal{W}$, it follows from (4.307), (4.310), and (4.311) that

$$\dot{v}_i(\hat{t}) < w_i(V(\hat{t})) \leq w_i(z(\hat{t})) = \dot{z}_i(\hat{t}), \tag{4.312}$$

which, along with (4.310), implies that for sufficiently small $\hat{\tau} > 0$, $v_i(t) > z_i(t)$, $t \in [\hat{t} - \hat{\tau}, \hat{t})$. This contradicts the fact that $V(t) << z(t)$, $t \in [t_0, \hat{t})$, and establishes (4.308). □

Next, we present a stronger version of Theorem 4.35 where the strict inequalities are replaced by soft inequalities.

Theorem 4.36. Consider the nonlinear comparison system (4.305). Assume that the function $w : \mathcal{Q} \to \mathbb{R}^q$ is continuous and $w(\cdot)$ is of class $\mathcal{W}$. Let $z(t)$, $t \in \mathcal{I}_{z_0}$, be the solution to (4.305) and $[t_0, t_0 + \tau] \subseteq \mathcal{I}_{z_0}$ be a compact interval. If there exists a continuously differentiable vector function $V : [t_0, t_0 + \tau] \to \mathcal{Q}$ such that

$$\dot{V}(t) \leq\leq w(V(t)), \quad t \in [t_0, t_0 + \tau], \tag{4.313}$$

then $V(t_0) \leq\leq z_0$, $z_0 \in \mathcal{Q}$, implies

$$V(t) \leq\leq z(t), \quad t \in [t_0, t_0 + \tau]. \tag{4.314}$$

Proof. Consider the family of comparison systems given by

$$\dot{z}(t) = w(z(t)) + \tfrac{\varepsilon}{n}\mathbf{e}, \quad z(t_0) = z_0 + \tfrac{\varepsilon}{n}\mathbf{e}, \tag{4.315}$$

where $\varepsilon > 0$, $n \in \mathbb{Z}_+$, and $t \in \mathcal{I}_{z_0 + \frac{\varepsilon}{n}\mathbf{e}}$, and let the solution to (4.315) be denoted by $s_{(n)}(t, z_0 + \frac{\varepsilon}{n}\mathbf{e})$, $t \in \mathcal{I}_{z_0 + \frac{\varepsilon}{n}\mathbf{e}}$. Now, it follows from Theorem 3 of [98, p. 17] that $s_{(n)}(t, z_0 + \frac{\varepsilon}{n}\mathbf{e})$, $t \in [t_0, t_0 + \tau]$, is defined for all sufficiently large n. Moreover, it follows from Theorem 4.35 that

$$V(t) << s_{(n)}(t, z_0 + \tfrac{\varepsilon}{n}\mathbf{e}) << s_{(m)}(t, z_0 + \tfrac{\varepsilon}{m}\mathbf{e}), \quad n > m, \quad t \in [t_0, t_0 + \tau], \tag{4.316}$$

for all sufficiently large $m \in \mathbb{Z}_+$. Since the functions $s_{(n)}(t, z_0 + \frac{\varepsilon}{n}\mathbf{e})$, $t \in [t_0, t_0 + \tau]$, $n \in \mathbb{Z}_+$, are continuous in t, decreasing in n, and bounded from below, it follows that the sequence of functions $s_{(n)}(\cdot, z_0 + \frac{\varepsilon}{n}\mathbf{e})$ converges uniformly on the compact interval $[t_0, t_0 + \tau]$ as $n \to \infty$, that is, there exists a continuous function $\hat{z} : [t_0, t_0 + \tau] \to \mathcal{Q}$ such that

$$s_{(n)}(t, z_0 + \tfrac{\varepsilon}{n}\mathbf{e}) \to \hat{z}(t), \quad n \to \infty, \tag{4.317}$$

uniformly on $[t_0, t_0 + \tau]$. Hence, it follows from (4.316) and (4.317) that

$$V(t) \leq\leq \hat{z}(t), \quad t \in [t_0, t_0 + \tau]. \tag{4.318}$$

Next, note that it follows from (4.315) that

$$s_{(n)}(t, z_0 + \tfrac{\varepsilon}{n}\mathbf{e}) = z_0 + \tfrac{\varepsilon}{n}\mathbf{e} + \int_{t_0}^{t} w(s_{(n)}(\sigma, z_0 + \tfrac{\varepsilon}{n}\mathbf{e}))\mathrm{d}\sigma, \quad t \in [t_0,\, t_0 + \tau], \qquad (4.319)$$

which implies that $\hat{z}(t_0) = z_0$ and, since $w(\cdot)$ is continuous, $w(s_{(n)}(t, z_0 + \frac{\varepsilon}{n}\mathbf{e})) \to w(\hat{z}(t))$ as $n \to \infty$ uniformly on $[t_0,\, t_0 + \tau]$. Hence, taking the limit as $n \to \infty$ on both sides of (4.319) yields

$$\hat{z}(t) = z_0 + \int_{t_0}^{t} w(\hat{z}(\sigma))\mathrm{d}\sigma, \quad t \in [t_0,\, t_0 + \tau], \qquad (4.320)$$

which implies that $\hat{z}(t)$ is the solution to (4.305) on the interval $[t_0,\, t_0 + \tau]$. Hence, by uniqueness of solutions of (4.305) we obtain that $\hat{z}(t) = z(t)$, $[t_0,\, t_0 + \tau]$. This, along with (4.318), proves the result. □

Next, consider the nonlinear dynamical system given by

$$\dot{x}(t) = f(x(t)), \quad x(t_0) = x_0, \quad t \in \mathcal{I}_{x_0}, \qquad (4.321)$$

where $x(t) \in \mathcal{D} \subseteq \mathbb{R}^n$, $t \in \mathcal{I}_{x_0}$, is the system state vector, $\mathcal{I}_{x_0}$ is the maximal interval of existence of a solution $x(t)$ of (4.321), $\mathcal{D}$ is an open set, $0 \in \mathcal{D}$, and $f(\cdot)$ is Lipschitz continuous on $\mathcal{D}$. The following result is a direct consequence of Theorem 4.36.

Corollary 4.9. Consider the nonlinear dynamical system (4.321). Assume there exists a continuously differentiable vector function $V : \mathcal{D} \to \mathcal{Q} \subseteq \mathbb{R}^q$ such that

$$V'(x)f(x) \leq\leq w(V(x)), \quad x \in \mathcal{D}, \qquad (4.322)$$

where $w : \mathcal{Q} \to \mathbb{R}^q$ is a continuous function, $w(\cdot) \in \mathcal{W}$, and

$$\dot{z}(t) = w(z(t)), \quad z(t_0) = z_0, \quad t \in \mathcal{I}_{z_0}, \qquad (4.323)$$

has a unique solution $z(t)$, $t \in \mathcal{I}_{z_0}$. If $[t_0,\, t_0 + \tau] \subseteq \mathcal{I}_{x_0} \cap \mathcal{I}_{z_0}$ is a compact interval, then $V(x_0) \leq\leq z_0$, $z_0 \in \mathcal{Q}$, implies $V(x(t)) \leq\leq z(t)$, $t \in [t_0,\, t_0 + \tau]$.

Proof. For every $x_0 \in \mathcal{D}$, the solution $x(t)$, $t \in \mathcal{I}_{x_0}$, to (4.321) is a well defined function of time. Hence, define $\eta(t) \triangleq V(x(t))$, $t \in \mathcal{I}_{x_0}$, and note that (4.322) implies

$$\dot{\eta}(t) \leq\leq w(\eta(t)), \quad t \in \mathcal{I}_{x_0}. \qquad (4.324)$$

Moreover, if $[t_0,\, t_0 + \tau] \subseteq \mathcal{I}_{x_0} \cap \mathcal{I}_{z_0}$ is a compact interval, then it follows from Theorem 4.36, with $V(x_0) = \eta(t_0) \leq\leq z_0$, that $V(x(t)) = \eta(t) \leq\leq z(t)$, $t \in [t_0,\, t_0 + \tau]$, which establishes the result. □

If in (4.321) $f : \mathbb{R}^n \to \mathbb{R}^n$ is globally Lipschitz continuous, then (4.321) has a unique solution $x(t)$ for all $t \geq t_0$. A more restrictive sufficient condition for global existence and uniqueness of solutions to (4.321) is continuous differentiability of $f : \mathbb{R}^n \to \mathbb{R}^n$ and uniform boundedness of $f'(x)$ on $\mathbb{R}^n$. Note that if the solutions to (4.321) and (4.323) are globally defined for all $x_0 \in \mathcal{D}$ and $z_0 \in \mathcal{Q}$, then the result of Corollary 4.9 holds for every arbitrarily large but compact interval $[t_0, t_0 + \tau] \subset \overline{\mathbb{R}}_+$. For the remainder of this section we assume that the solutions to the systems (4.321) and (4.323) are defined for all $t \geq t_0$.

Consider the nonlinear comparison system given by

$$\dot{z}(t) = w(z(t)), \quad z(t_0) = z_0, \quad t \geq t_0, \tag{4.325}$$

and the nonlinear dynamical system given by

$$\dot{x}(t) = f(x(t)), \quad x(t_0) = x_0, \quad t \geq t_0, \tag{4.326}$$

where $z_0 \in \mathcal{Q} \subseteq \overline{\mathbb{R}}_+^q$, $x_0 \in \mathcal{D} \subseteq \mathbb{R}^n$, $w : \mathcal{Q} \to \mathbb{R}^q$ is continuous, $w(\cdot) \in \mathcal{W}$, $w(0) = 0$, $f : \mathcal{D} \to \mathbb{R}^n$ is Lipschitz continuous on $\mathcal{D}$, and $f(0) = 0$. Note that since $w(\cdot) \in \mathcal{W}$ and $w(0) = 0$, then for every $z \in \mathcal{Q} \cap \overline{\mathbb{R}}_+^q$ such that $z_i = 0$ it follows that $w_i(z) \geq 0$, $i = 1, \ldots, q$, which implies that for every $z_0 \in \mathcal{Q} \cap \overline{\mathbb{R}}_+^q$ the solution $z(t)$, $t \geq t_0$, remains in $\overline{\mathbb{R}}_+^q$ (see Problem 3.61).

Theorem 4.37. Consider the nonlinear dynamical system (4.326). Assume that there exist a continuously differentiable vector function $V : \mathcal{D} \to \mathcal{Q} \cap \overline{\mathbb{R}}_+^q$ and a positive vector $p \in \mathbb{R}_+^q$ such that $V(0) = 0$, the scalar function $v : \mathcal{D} \to \overline{\mathbb{R}}_+$ defined by $v(x) \triangleq p^{\mathrm{T}} V(x)$, $x \in \mathcal{D}$, is such that $v(x) > 0$, $x \neq 0$, and

$$V'(x)f(x) \leq\leq w(V(x)), \quad x \in \mathcal{D}, \tag{4.327}$$

where $w : \mathcal{Q} \to \mathbb{R}^q$ is continuous, $w(\cdot) \in \mathcal{W}$, and $w(0) = 0$. Then the following statements hold:

i) If the zero solution $z(t) \equiv 0$ to (4.325) is Lyapunov stable, then the zero solution $x(t) \equiv 0$ to (4.326) is Lyapunov stable.

ii) If the zero solution $z(t) \equiv 0$ to (4.325) is asymptotically stable, then the zero solution $x(t) \equiv 0$ to (4.326) is asymptotically stable.

iii) If $\mathcal{D} = \mathbb{R}^n$, $\mathcal{Q} = \mathbb{R}^q$, $v : \mathbb{R}^n \to \overline{\mathbb{R}}_+$ is radially unbounded, and the zero solution $z(t) \equiv 0$ to (4.325) is globally asymptotically stable, then the zero solution $x(t) \equiv 0$ to (4.326) is globally asymptotically stable.

iv) If there exist constants $\nu \geq 1$, $\alpha > 0$, and $\beta > 0$ such that $v : \mathcal{D} \to \overline{\mathbb{R}}_+$ satisfies

$$\alpha\|x\|^\nu \leq v(x) \leq \beta\|x\|^\nu, \quad x \in \mathcal{D}, \tag{4.328}$$

and the zero solution $z(t) \equiv 0$ to (4.325) is exponentially stable, then the zero solution $x(t) \equiv 0$ to (4.326) is exponentially stable.

v) If $\mathcal{D} = \mathbb{R}^n$, $\mathcal{Q} = \mathbb{R}^q$, there exist constants $\nu \geq 1$, $\alpha > 0$, and $\beta > 0$ such that $v : \mathbb{R}^n \to \overline{\mathbb{R}}_+$ satisfies (4.328), and the zero solution $z(t) \equiv 0$ to (4.325) is globally exponentially stable, then the zero solution $x(t) \equiv 0$ to (4.326) is globally exponentially stable.

Proof. Assume that there exist a continuously differentiable vector function $V : \mathcal{D} \to \mathcal{Q} \cap \overline{\mathbb{R}}^q_+$ and a positive vector $p \in \mathbb{R}^q_+$ such that $v(x) = p^{\mathrm{T}} V(x)$, $x \in \mathcal{D}$, is positive definite, that is, $v(0) = 0$ and $v(x) > 0$, $x \neq 0$. Note that since $v(x) = p^{\mathrm{T}} V(x) \leq \max_{i=1,\ldots,q}\{p_i\} \mathbf{e}^{\mathrm{T}} V(x)$, $x \in \mathcal{D}$, the function $\mathbf{e}^{\mathrm{T}} V(x)$, $x \in \mathcal{D}$, is also positive definite. Thus, there exist $r > 0$ and class $\mathcal{K}$ functions α, $\beta : [0, r] \to \overline{\mathbb{R}}_+$ such that $\mathcal{B}_r(0) \subset \mathcal{D}$ and

$$\alpha(\|x\|) \leq \mathbf{e}^{\mathrm{T}} V(x) \leq \beta(\|x\|), \quad x \in \mathcal{B}_r(0). \tag{4.329}$$

i) Let $\varepsilon > 0$ and choose $0 < \hat{\varepsilon} < \min\{\varepsilon, r\}$. It follows from Lyapunov stability of the nonlinear comparison system (4.325) that there exists $\mu = \mu(\hat{\varepsilon}) = \mu(\varepsilon) > 0$ such that if $\|z_0\|_1 < \mu$, where $\|\cdot\|_1$ denotes the absolute sum norm, then $\|z(t)\|_1 < \alpha(\hat{\varepsilon})$, $t \geq t_0$. Now, choose $z_0 = V(x_0) \geq\geq 0$, $x_0 \in \mathcal{D}$. Since $V(x)$, $x \in \mathcal{D}$, is continuous, the function $\mathbf{e}^{\mathrm{T}} V(x)$, $x \in \mathcal{D}$, is also continuous. Hence, for $\mu = \mu(\hat{\varepsilon}) > 0$ there exists $\delta = \delta(\mu(\hat{\varepsilon})) = \delta(\varepsilon) > 0$ such that $\delta < \hat{\varepsilon}$, and if $\|x_0\| < \delta$, then $\mathbf{e}^{\mathrm{T}} V(x_0) = \mathbf{e}^{\mathrm{T}} z_0 = \|z_0\|_1 < \mu$, which implies that $\|z(t)\|_1 < \alpha(\hat{\varepsilon})$, $t \geq t_0$. Now, with $z_0 = V(x_0) \geq\geq 0$, $x_0 \in \mathcal{D}$, and the assumption that $w(\cdot) \in \mathcal{W}$, $x \in \mathcal{D}$, it follows from (4.327) and Corollary 4.9 that $0 \leq\leq V(x(t)) \leq\leq z(t)$ on any compact interval $[t_0, t_0 + \tau]$, and hence, $\mathbf{e}^{\mathrm{T}} z(t) = \|z(t)\|_1$, $t \in [t_0, t_0 + \tau]$. Let $\tau > t_0$ be such that $x(t) \in \mathcal{B}_r(0)$, $t \in [t_0, t_0 + \tau]$, for all $x_0 \in \mathcal{B}_\delta(0)$. Thus, using (4.329), if $\|x_0\| < \delta$, then

$$\alpha(\|x(t)\|) \leq \mathbf{e}^{\mathrm{T}} V(x(t)) \leq \mathbf{e}^{\mathrm{T}} z(t) < \alpha(\hat{\varepsilon}), \quad t \in [t_0, t_0 + \tau], \tag{4.330}$$

which implies $\|x(t)\| < \hat{\varepsilon} < \varepsilon$, $t \in [t_0, t_0 + \tau]$. Now, suppose, *ad absurdum*, that for some $x_0 \in \mathcal{B}_\delta(0)$ there exists $\hat{t} > t_0 + \tau$ such that $\|x(\hat{t})\| = \hat{\varepsilon}$. Then, for $z_0 = V(x_0)$ and the compact interval $[t_0, \hat{t}]$ it follows from (4.327) and Corollary 4.9 that $V(x(\hat{t})) \leq\leq z(\hat{t})$, which implies that $\alpha(\hat{\varepsilon}) = \alpha(\|x(\hat{t})\|) \leq \mathbf{e}^{\mathrm{T}} V(x(\hat{t})) \leq \mathbf{e}^{\mathrm{T}} z(\hat{t}) < \alpha(\hat{\varepsilon})$. This is a contradiction, and hence, for a given $\varepsilon > 0$ there exists $\delta = \delta(\varepsilon) > 0$ such that for all $x_0 \in \mathcal{B}_\delta(0)$, $\|x(t)\| < \varepsilon$, $t \geq t_0$, which implies Lyapunov stability of the zero solution $x(t) \equiv 0$ to (4.326).

ii) It follows from i) and the asymptotic stability of the nonlinear comparison system (4.325) that the zero solution to (4.326) is Lyapunov stable and there exists $\mu > 0$ such that if $\|z_0\|_1 < \mu$, then $\lim_{t\to\infty} z(t) = 0$ for every $x_0 \in \mathcal{D}$. As in i), choose $z_0 = V(x_0) \geq\geq 0$, $x_0 \in \mathcal{D}$. It follows from Lyapunov stability of the zero solution to (4.326) and the continuity of $V : \mathcal{D} \to \mathcal{Q} \cap \overline{\mathbb{R}}_+^q$ that there exists $\delta = \delta(\mu) > 0$ such that if $\|x_0\| < \delta$, then $\|x(t)\| < r$, $t \geq t_0$, and $\mathbf{e}^{\mathrm{T}} V(x_0) = \mathbf{e}^{\mathrm{T}} z_0 = \|z_0\|_1 < \mu$. Thus, by asymptotic stability of (4.325) for every arbitrary $\varepsilon > 0$ there exists $T = T(\varepsilon) > t_0$ such that $\|z(t)\|_1 < \alpha(\varepsilon)$, $t \geq T$. Thus, it follows from (4.327) and Corollary 4.9 that $0 \leq\leq V(x(t)) \leq\leq z(t)$ on any compact interval $[t_0, T + \tau]$, and hence, $\mathbf{e}^{\mathrm{T}} z(t) = \|z(t)\|_1$, $t \in [t_0, T + \tau]$, and, by (4.329),

$$\alpha(\|x(t)\|) \leq \mathbf{e}^{\mathrm{T}} V(x(t)) \leq \mathbf{e}^{\mathrm{T}} z(t) < \alpha(\varepsilon), \quad t \in [T, T + \tau]. \tag{4.331}$$

Now, suppose, *ad absurdum*, that for some $x_0 \in \mathcal{B}_\delta(0)$, $\lim_{t\to\infty} x(t) \neq 0$, that is, there exists a sequence $\{t_k\}_{k=1}^{\infty}$, with $t_k \to \infty$ as $k \to \infty$, such that $\|x(t_k)\| \geq \hat{\varepsilon}$, $k \in \overline{\mathbb{Z}}_+$, for some $0 < \hat{\varepsilon} < r$. Choose $\varepsilon = \hat{\varepsilon}$ and the interval $[T, T + \tau]$ such that at least one $t_k \in [T, T + \tau]$. Then it follows from (4.331) that $\alpha(\varepsilon) \leq \alpha(\|x(t_k)\|) < \alpha(\varepsilon)$, which is a contradiction. Hence, there exists $\delta > 0$ such that for all $x_0 \in \mathcal{B}_\delta(0)$, $\lim_{t\to\infty} x(t) = 0$ which, along with Lyapunov stability, implies asymptotic stability of the zero solution $x(t) \equiv 0$ to (4.326).

iii) Suppose $\mathcal{D} = \mathbb{R}^n$, $\mathcal{Q} = \mathbb{R}^q$, $v : \mathbb{R}^n \to \overline{\mathbb{R}}_+$ is a radially unbounded function, and the nonlinear comparison system (4.325) is globally asymptotically stable. In this case, for $V : \mathbb{R}^n \to \overline{\mathbb{R}}_+^q$ the inequality (4.329) holds for all $x \in \mathbb{R}^n$, where the functions $\alpha, \beta : \overline{\mathbb{R}}_+ \to \overline{\mathbb{R}}_+$ are of class $\mathcal{K}_\infty$. Furthermore, Lyapunov stability of the zero solution $x(t) \equiv 0$ to (4.326) follows from *i*). Next, for every $x_0 \in \mathbb{R}^n$ and $z_0 = V(x_0) \in \overline{\mathbb{R}}_+^q$, identical arguments as in *ii*) can be used to show that $\lim_{t\to\infty} x(t) = 0$, which proves global asymptotic stability of the zero solution $x(t) \equiv 0$ to (4.326).

iv) Suppose (4.328) holds. Since $p \in \mathbb{R}_+^q$, then

$$\hat{\alpha}\|x\|^\nu \leq \mathbf{e}^{\mathrm{T}} V(x) \leq \hat{\beta}\|x\|^\nu, \quad x \in \mathcal{D}, \tag{4.332}$$

where $\hat{\alpha} \triangleq \alpha / \max_{i=1,\ldots,q}\{p_i\}$ and $\hat{\beta} \triangleq \beta / \min_{i=1,\ldots,q}\{p_i\}$. It follows from the exponential stability of the nonlinear comparison system (4.325) that there exist positive constants γ, μ, and η such that if $\|z_0\|_1 < \mu$, then

$$\|z(t)\|_1 \leq \gamma\|z_0\|_1 e^{-\eta(t-t_0)}, \quad t \geq t_0, \tag{4.333}$$

for all $x_0 \in \mathcal{D}$. Let $x_0 \in \mathcal{D}$ and $z_0 = V(x_0) \geq\geq 0$. By continuity of $V : \mathcal{D} \to \mathcal{Q} \cap \overline{\mathbb{R}}_+^q$, there exists $\delta = \delta(\mu) > 0$ such that for all $x_0 \in \mathcal{B}_\delta(0)$, $\mathbf{e}^{\mathrm{T}} V(x_0) = \mathbf{e}^{\mathrm{T}} z_0 = \|z_0\|_1 < \mu$. Furthermore, it follows from (4.327), (4.332), (4.333), and Corollary 4.9 that, for all $x_0 \in \mathcal{B}_\delta(0)$, the inequality

$$\hat{\alpha}\|x(t)\|^\nu \leq \mathbf{e}^{\mathrm{T}} V(x(t)) \leq \mathbf{e}^{\mathrm{T}} z(t) \leq \gamma\|z_0\|_1 e^{-\eta(t-t_0)} \leq \gamma\hat{\beta}\|x_0\|^\nu e^{-\eta(t-t_0)} \tag{4.334}$$

holds on any compact interval $[t_0, t_0+\tau]$. This in turn implies that, for every

$x_0 \in \mathcal{B}_\delta(0)$,

$$\|x(t)\| \le \left(\frac{\gamma\hat{\beta}}{\hat{\alpha}}\right)^{\frac{1}{\nu}} \|x_0\| e^{-\frac{\eta}{\nu}(t-t_0)}, \quad t \in [t_0, t_0+\tau]. \tag{4.335}$$

Now, suppose, *ad absurdum*, that for some $x_0 \in \mathcal{B}_\delta(0)$ there exists $\hat{t} > t_0 + \tau$ such that $\|x(\hat{t})\| > \left(\frac{\gamma\hat{\beta}}{\hat{\alpha}}\right)^{\frac{1}{\nu}} \|x_0\| e^{-\frac{\eta}{\nu}(\hat{t}-t_0)}$. Then for the compact interval $[t_0, \hat{t}]$, it follows from (4.335) that $\|x(\hat{t})\| \le \left(\frac{\gamma\hat{\beta}}{\hat{\alpha}}\right)^{\frac{1}{\nu}} \|x_0\| e^{-\frac{\eta}{\nu}(\hat{t}-t_0)}$, which is a contradiction. Thus, inequality (4.335) holds for all $t \ge t_0$, establishing exponential stability of the zero solution $x(t) \equiv 0$ to (4.326).

v) The proof is identical to the proof of *iv*). □

If $V : \mathcal{D} \to \mathcal{Q} \cap \overline{\mathbb{R}}_+^q$ satisfies the conditions of Theorem 4.37 we say that $V(x)$, $x \in \mathcal{D}$, is a *vector Lyapunov function*. Note that for stability analysis each component of a vector Lyapunov function need not be positive definite with a negative-definite or negative-semidefinite time derivative along the trajectories of (4.326). This provides more flexibility in searching for a vector Lyapunov function as compared to a scalar Lyapunov function for addressing the stability of nonlinear dynamical systems.

Example 4.20. Consider the nonlinear dynamical system given by

$$\dot{x}_1(t) = -x_1(t) - x_1^2(t)x_2^3(t), \quad x_1(0) = x_{10}, \quad t \ge 0, \tag{4.336}$$
$$\dot{x}_2(t) = -x_2^3(t) + x_1^2(t)x_2^2(t), \quad x_2(0) = x_{20}. \tag{4.337}$$

Note that Lyapunov's indirect method fails to yield any information on the stability of the zero solution $x(t) \triangleq [x_1(t), x_2(t)]^{\mathrm{T}} \equiv 0$ of (4.336) and (4.337). To examine the stability of (4.336) and (4.337) consider the vector Lyapunov function candidate $V(x) = [v_1(x), v_2(x)]^{\mathrm{T}}$, $x \in \mathbb{R}^2$, with $v_1(x) = \frac{1}{2}x_1^2$ and $v_2(x) = \frac{1}{4}x_2^4$. Clearly, $V(0) = 0$ and $\mathbf{e}^{\mathrm{T}}V(x)$, $x \in \mathbb{R}^2$, is a positive-definite function. Next, consider the domain $\mathcal{D} \triangleq \{x \in \mathbb{R}^2 : |x_1| \le 1, |x_2| \le c_2\}$, where $c_2 > 0$, and note that

$$\begin{aligned}\dot{v}_1(x(t)) &= x_1(t)(-x_1(t) - x_1^2(t)x_2^3(t)) \\ &\le -x_1^2(t) + |x_1^3(t)x_2^3(t)| \\ &\le (-2 + 2c_2^3)v_1(x(t)),\end{aligned} \tag{4.338}$$

$$\begin{aligned}\dot{v}_2(x(t)) &= x_2^3(t)(-x_2^3(t) + x_1^2(t)x_2^2(t)) \\ &\le -x_2^6(t) + |x_1^2(t)x_2^5(t)| \\ &\le 2c_2^5 v_1(x(t)) - 8v_2^{\frac{3}{2}}(x(t)),\end{aligned} \tag{4.339}$$

for all $x(t) \in \mathcal{D}$, $t \geq 0$. Thus, the comparison system (4.325) is given by

$$\dot{z}_1(t) = (-2 + 2c_2^3)z_1(t), \quad z_1(0) = z_{10}, \quad t \geq 0, \tag{4.340}$$

$$\dot{z}_2(t) = 2c_2^5 z_1(t) - 8z_2^{\frac{3}{2}}(t), \quad z_2(0) = z_{20}, \tag{4.341}$$

where $(z_{10}, z_{20}) \in \overline{\mathbb{R}}_+ \times \overline{\mathbb{R}}_+$. Note that $w(z) \triangleq [(-2+2c_2^3)z_1, 2c_2^5 z_1 - 8z_2^{\frac{3}{2}}]^{\mathrm{T}} \in \mathcal{W}$, where $z \triangleq [z_1, z_2]^{\mathrm{T}}$.

Next, to show stability of the zero solution $z(t) \equiv 0$ to the comparison system, consider the linear Lyapunov function candidate $v(z) = z_1 + z_2$, $z \in \overline{\mathbb{R}}_+^2$ (see Problem 3.8). Clearly, $v(0) = 0$ and $v(z) > 0$, $z \in \overline{\mathbb{R}}_+^2 \setminus \{0\}$. Moreover,

$$\dot{v}(z(t)) = 2(-1 + c_2^3 + c_2^5)z_1(t) - 8z_2^{\frac{3}{2}}(t), \quad t \geq 0. \tag{4.342}$$

In order to ensure asymptotic stability of the zero solution $z(t) \equiv 0$, it suffices to take $c_2 = 0.83$. In this case, it follows from Theorem 4.37 that the zero solution $x(t) \equiv 0$ to (4.336) and (4.337) is asymptotically stable. △

Next, we present a convergence result via vector Lyapunov functions that allows us to establish asymptotic stability of the nonlinear dynamical system (4.326).

Theorem 4.38. Consider the nonlinear dynamical system (4.326), assume that there exist a continuously differentiable vector function $V = [v_1, \ldots, v_q]^{\mathrm{T}} : \mathcal{D} \to \mathcal{Q} \cap \overline{\mathbb{R}}_+^q$ and a positive vector $p \in \mathbb{R}_+^q$ such that $V(0) = 0$, the scalar function $v : \mathcal{D} \to \overline{\mathbb{R}}_+$ defined by $v(x) \triangleq p^{\mathrm{T}} V(x)$, $x \in \mathcal{D}$, is such that $v(x) > 0$, $x \neq 0$, and

$$V'(x)f(x) \leq\leq w(V(x)), \quad x \in \mathcal{D}, \tag{4.343}$$

where $w : \mathcal{Q} \to \mathbb{R}^q$ is continuous, $w(\cdot) \in \mathcal{W}$, and $w(0) = 0$, such that the nonlinear comparison system (4.325) is Lyapunov stable. Let $\mathcal{R}_i \triangleq \{x \in \mathcal{D} : v_i'(x)f(x) - w_i(V(x)) = 0\}$, $i = 1, \ldots, q$. Then there exists $\mathcal{D}_{\mathrm{c}} \subset \mathcal{D}$ such that $x(t) \to \mathcal{R} \triangleq \cap_{i=1}^q \mathcal{R}_i$ as $t \to \infty$ for all $x(t_0) = x_0 \in \mathcal{D}_{\mathrm{c}}$. Moreover, if $\mathcal{R}$ contains no trajectory other than the trivial trajectory, then the zero solution $x(t) \equiv 0$ to (4.326) is asymptotically stable.

Proof. Since the nonlinear comparison system (4.325) is Lyapunov stable, it follows that there exists $\hat{\delta} > 0$ such that, if $\|z_0\|_1 < \hat{\delta}$, then the system trajectories $z(t)$, $t \geq t_0$, of (4.325) are bounded. Furthermore, since $V(x)$, $x \in \mathcal{D}$, is continuous, it follows that there exists $\delta_1 = \delta_1(\hat{\delta}) > 0$ such that $\mathbf{e}^{\mathrm{T}} V(x_0) < \hat{\delta}$ for all $x_0 \in \mathcal{B}_{\delta_1}(0)$. In addition, it follows from Theorem 4.37 that the zero solution $x(t) \equiv 0$ to (4.326) is Lyapunov stable, and hence, for a given $\varepsilon > 0$ such that $\mathcal{B}_\varepsilon(0) \subset \overset{\circ}{\mathcal{D}}$, there exists $\delta_2 = \delta_2(\varepsilon) > 0$ such that, if $x_0 \in \mathcal{B}_{\delta_2}(0)$, then $x(t) \in \mathcal{B}_\varepsilon(0)$, $t \geq t_0$, where $x(t)$, $t \geq t_0$, is the solution to (4.326). Choose $\delta = \min\{\delta_1, \delta_2\}$ and define $\mathcal{D}_c \triangleq \mathcal{B}_\delta(0) \subset \mathcal{D}$. Then for every $x_0 \in \mathcal{D}_c$ and $z_0 = V(x_0)$, it follows that $x(t) \in \mathcal{B}_\varepsilon(0)$, $t \geq t_0$, and $z(t)$, $t \geq t_0$, is bounded.

Next, consider the function

$$W_i(t) \triangleq v_i(x(t)) - \int_{t_0}^{t} w_i(V(x(s)))\mathrm{d}s, \ t \geq t_0, \ x \in \mathcal{D}, \ i = 1, \ldots, q.$$

It follows from (4.343) that

$$\dot{W}_i(t) = v_i'(x(t))f(x(t)) - w_i(V(x(t))) \leq 0, \ t \geq t_0, \ x_0 \in \mathcal{D}, \quad (4.344)$$

which implies that $W_i(t)$, $i \in \{1, \ldots, q\}$, is a nonincreasing function of time, and hence, $\lim_{t\to\infty} W_i(t)$, $i \in \{1, \ldots, q\}$, exists. Moreover, $W_i(t_0) = v_i(x(t_0)) < \infty$, $i \in \{1, \ldots, q\}$. Now suppose, *ad absurdum*, that for some initial condition $x(t_0) = x_0 \in \mathcal{D}_c$, $\lim_{t\to\infty} W_i(t) = -\infty$ for some $i \in \{1, \ldots, q\}$. Since the function $v_i(x)$, $x \in \mathcal{D}$, is continuous on the compact set $\overline{\mathcal{B}_\varepsilon(0)}$, it follows that $v_i(x(t))$, $t \geq t_0$, is uniformly bounded, and hence, $\lim_{t\to\infty} \int_{t_0}^{t} w_i(V(x(s)))\mathrm{d}s = \infty$. Now, it follows from (4.343) and Corollary 4.9 that $V(x(t)) \leq\leq z(t)$, $t \geq t_0$, for $z(t_0) = V(x(t_0))$. Note that since $x_0 \in \mathcal{D}_c$ it follows that $z(t)$, $t \geq t_0$, is bounded. Furthermore, since $w(\cdot) \in \mathcal{W}$ it follows that

$$v_i(x(t)) \leq v_i(x(t_0)) + \int_{t_0}^{t} w_i(V(x(s)))\mathrm{d}s \leq z_i(t_0) + \int_{t_0}^{t} w_i(z(s))\mathrm{d}s = z_i(t)$$

for all $t \geq t_0$. Since $z(t)$ and $v_i(x(t))$ are bounded for all $t \geq t_0$ it follows that there exists $M > 0$ such that $|\int_{t_0}^{t} w_i(V(x(s)))\mathrm{d}s| < M$, $t \geq t_0$. This is a contradiction, and hence, $\lim_{t\to\infty} W_i(t)$, $i \in \{1, \ldots, q\}$, exists and is finite for every $x_0 \in \mathcal{D}_c$. Thus, for every $x_0 \in \mathcal{D}_c$, it follows that

$$\begin{aligned}\int_{t_0}^{t} \dot{W}_i(s)\mathrm{d}s &= \int_{t_0}^{t} [v_i'(x(s))f(x(s)) - w_i(V(x(s)))]\mathrm{d}s \\ &= W_i(t) - W_i(t_0), \quad t \geq t_0, \end{aligned} \quad (4.345)$$

and hence, $\lim_{t\to\infty} \int_{t_0}^{t} [v_i'(x(s))f(x(s)) - w_i(V(x(s)))]\mathrm{d}s$, $i \in \{1, \ldots, q\}$, exists and is finite.

Next, since $f(\cdot)$ is Lipschitz continuous on $\mathcal{D}$ and $x(t) \in \mathcal{B}_\varepsilon(0)$ for all $x_0 \in \mathcal{D}_c$ and $t \geq t_0$ it follows that

$$\begin{aligned}\|x(t_2) - x(t_1)\| &= \left\| \int_{t_1}^{t_2} f(x(s))\mathrm{d}s \right\| \\ &\leq L \int_{t_1}^{t_2} \|x(s)\|\mathrm{d}s\end{aligned}$$

$$\leq L\varepsilon(t_2 - t_1), \qquad t_2 \geq t_1 \geq t_0, \qquad (4.346)$$

where L is the Lipschitz constant on $\mathcal{D}_c$. Thus, it follows from (4.346) that for every $\gamma > 0$ there exists $\mu = \mu(\gamma) = \frac{\gamma}{L\varepsilon}$ such that $\|x(t_2) - x(t_1)\| < \gamma$, $|t_2 - t_1| < \mu$, which shows that $x(t)$, $t \geq t_0$, is uniformly continuous. Next, since $x(t)$ is uniformly continuous and $v_i'(x)f(x) - w_i(V(x))$, $x \in \mathcal{D}$, $i \in \{1, \ldots, q\}$, is continuous, it follows that $v_i'(x)f(x) - w_i(V(x))$ is uniformly continuous on $\overline{\mathcal{B}}_\varepsilon(0)$, and hence, $v_i'(x(t))f(x(t)) - w_i(V(x(t)))$, $i \in \{1, \ldots, q\}$, is uniformly continuous at every $t \geq t_0$. Hence, it follows from Barbalat's lemma (Lemma 4.1) that $v_i'(x(t))f(x(t)) - w_i(V(x(t))) \to 0$ as $t \to \infty$ for all $x_0 \in \mathcal{D}_c$ and $i \in \{1, \ldots, q\}$. Repeating the above analysis for all $i = 1, \ldots, q$, it follows that $x(t) \to \mathcal{R} = \cap_{i=1}^q \mathcal{R}_i$ for all $x_0 \in \mathcal{D}_c$. Finally, if $\mathcal{R}$ contains no trajectory other than the trivial trajectory, then $\mathcal{R} = \{0\}$, and hence, $x(t) \to 0$ as $t \to \infty$ for all $x_0 \in \mathcal{D}_c$, which proves asymptotic stability of the zero solution $x(t) \equiv 0$ to (4.326). □

Note that $\mathcal{R} = \cap_{i=1}^q \mathcal{R}_i \neq \emptyset$ since $0 \in \mathcal{R}$. Furthermore, recall that for every bounded solution $x(t)$, $t \geq t_0$, to (4.326) with initial condition $x(t_0) = x_0$, the positive limit set $\omega(x_0)$ of (4.326) is a nonempty, compact, invariant, and connected set with $x(t) \to \omega(x_0)$ as $t \to \infty$. If $q = 1$ and $w(V(x)) \equiv 0$, then it can be shown that the Lyapunov derivative $\dot{V}(x)$ vanishes on the positive limit set $\omega(x_0)$, $x_0 \in \mathcal{D}_c$, so that $\omega(x_0) \in \mathcal{R}$. Moreover, since $\omega(x_0)$ is a positively invariant set with respect to (4.326), it follows that for all $x_0 \in \mathcal{D}_c$, the trajectory of (4.326) converges to the largest invariant set $\mathcal{M}$ contained in $\mathcal{R}$. In this case, Theorem 4.38 specializes to Theorem 3.3.

If for some $k \in \{1, \ldots, q\}$, $w_k(V(x)) \equiv 0$ and $v_k'(x)f(x) < 0$, $x \in \mathcal{D}$, $x \neq 0$, then $\mathcal{R} = \mathcal{R}_k = \{0\}$. In this case, it follows from Theorem 4.38 that the zero solution $x(t) \equiv 0$ to (4.326) is asymptotically stable. Note that even though for $k \in \{1, \ldots, q\}$ the time derivative $\dot{v}_k(x)$, $x \in \mathcal{D}$, is negative definite, the function $v_k(x)$, $x \in \mathcal{D}$, can be nonnegative definite, in contrast to classical Lyapunov stability theory, to ensure asymptotic stability of (4.326).

Next, we present a converse Lyapunov theorem that establishes the existence of a vector Lyapunov function for an asymptotically stable nonlinear dynamical system.

Theorem 4.39. Consider the nonlinear dynamical system (4.326). Assume that $f : \mathcal{D} \to \mathbb{R}^n$ is continuously differentiable, the zero solution $x(t) \equiv 0$ to (4.326) is asymptotically stable, and let $\delta > 0$ be such that $\mathcal{B}_\delta(0) \subset \mathcal{D}$ is contained in the domain of attraction of (4.326). Then there exist a continuously differentiable componentwise positive-definite vector function $V = [v_1, \ldots, v_q]^{\mathrm{T}} : \mathcal{B}_\delta(0) \to \overline{\mathbb{R}}_+^q$ and a continuous function $w = [w_1, \ldots, w_q]^{\mathrm{T}} : \overline{\mathbb{R}}_+^q \to \mathbb{R}^q$ such that $V(0) = 0$, $w(\cdot) \in \mathcal{W}$, $w(0) = 0$,

$V'(x)f(x) \leq\leq w(V(x))$, $x \in \mathcal{B}_\delta(0)$, and the zero solution $z(t) \equiv 0$ to

$$\dot{z}(t) = w(z(t)), \quad z(t_0) = z_0, \quad t \geq t_0, \tag{4.347}$$

where $z_0 \in \overline{\mathbb{R}}_+^q$, is asymptotically stable.

Proof. Since the zero solution $x(t) \equiv 0$ to (4.326) is asymptotically stable it follows from Theorem 3.9 that there exist a continuously differentiable positive definite function $\tilde{v} : \mathcal{B}_\delta(0) \to \overline{\mathbb{R}}_+$ and class $\mathcal{K}$ functions $\alpha(\cdot)$, $\beta(\cdot)$, and $\gamma(\cdot)$ such that

$$\alpha(\|x\|) \leq \tilde{v}(x) \leq \beta(\|x\|), \quad x \in \mathcal{B}_\delta(0), \tag{4.348}$$

$$\tilde{v}'(x)f(x) \leq -\gamma(\|x\|), \quad x \in \mathcal{B}_\delta(0). \tag{4.349}$$

Furthermore, it follows from (4.348) and (4.349) that

$$\tilde{v}'(x)f(x) \leq -\gamma \circ \beta^{-1}(\tilde{v}(x)), \quad x \in \mathcal{B}_\delta(0), \tag{4.350}$$

where "$\circ$" denotes the composition operator and $\beta^{-1} : [0, \beta(\delta)] \to \overline{\mathbb{R}}_+$ is the inverse function of $\beta(\cdot)$, and hence, $\beta^{-1}(\cdot)$ and $\gamma\circ\beta^{-1}(\cdot)$ are class $\mathcal{K}$ functions. Next, define $V = [v_1, \ldots, v_q]^{\mathrm{T}} : \mathcal{B}_\delta(0) \to \overline{\mathbb{R}}_+^q$ such that $v_i(x) \triangleq \tilde{v}(x)$, $x \in \mathcal{B}_\delta(0)$, $i = 1, \ldots, q$. Then it follows that $V(0) = 0$ and $V'(x)f(x) \leq\leq w(V(x))$, $x \in \mathcal{B}_\delta(0)$, where $w = [w_1, \ldots, w_q]^{\mathrm{T}} : \overline{\mathbb{R}}_+^q \to \mathbb{R}^q$ is such that $w_i(V(x)) = -\gamma \circ \beta^{-1}(v_i(x))$, $x \in \mathcal{B}_\delta(0)$. Note that $w(\cdot) \in \mathcal{W}$ and $w(0) = 0$. To show that the zero solution $z(t) \equiv 0$ to (4.347) is asymptotically stable, consider the Lyapunov function candidate $\hat{v}(z) \triangleq \mathbf{e}^{\mathrm{T}} z$, $z \in \overline{\mathbb{R}}_+^q$. Note that $\hat{v}(0) = 0$, $\hat{v}(z) > 0$, $z \in \overline{\mathbb{R}}_+^q$, $z \neq 0$, and $\dot{\hat{v}}(z) = -\sum_{i=1}^q \gamma \circ \beta^{-1}(z_i) < 0$, $z \in \overline{\mathbb{R}}_+^q$, $z \neq 0$. Thus, the zero solution $z(t) \equiv 0$ to (4.347) is asymptotically stable. □

Next, we provide a time-varying extension of Theorem 4.37. In particular, we consider the nonlinear time-varying dynamical system

$$\dot{x}(t) = f(t, x(t)), \quad x(t_0) = x_0, \quad t \geq t_0, \tag{4.351}$$

where $x(t) \in \mathcal{D} \subseteq \mathbb{R}^n$, $0 \in \mathcal{D}$, $f : [t_0, \infty) \times \mathcal{D} \to \mathbb{R}^n$ is such that $f(\cdot, \cdot)$ is jointly continuous in t and x, for every $t \in [t_0, \infty)$, $f(t, 0) = 0$, and $f(t, \cdot)$ is locally Lipschitz in x uniformly in t for all t in compact subsets of $[0, \infty)$.

Theorem 4.40. Consider the nonlinear time-varying dynamical system (4.351). Assume that there exist a continuously differentiable vector function $V : [0, \infty) \times \mathcal{D} \to \mathcal{Q} \cap \overline{\mathbb{R}}_+^q$, a positive vector $p \in \mathbb{R}_+^q$, and class $\mathcal{K}$ functions $\alpha, \beta : [0, r] \to \overline{\mathbb{R}}_+$ such that $V(t, 0) = 0$, $t \in [0, \infty)$, the scalar function $v : [0, \infty) \times \mathcal{D} \to \overline{\mathbb{R}}_+$ defined by $v(t, x) \triangleq p^{\mathrm{T}} V(t, x)$, $(t, x) \in [0, \infty) \times \mathcal{D}$, is such that

$$\alpha(\|x\|) \leq v(t, x) \leq \beta(\|x\|), \quad (t, x) \in [0, \infty) \times \mathcal{B}_r(0), \quad \mathcal{B}_r(0) \subseteq \mathcal{D}, \tag{4.352}$$

and

$$\frac{\partial V}{\partial t}(t,x) + \frac{\partial V}{\partial x}(t,x)f(t,x) \leq\leq w(t, V(t,x)), \quad (t,x) \in [0,\infty) \times \mathcal{D}, \quad (4.353)$$

where $w : [0,\infty) \times \mathcal{Q} \to \mathbb{R}^q$ is continuous, $w(t,\cdot) \in \mathcal{W}$, and $w(t,0) = 0$, $t \in [0,\infty)$. Then the stability properties of the zero solution $z(t) \equiv 0$ to

$$\dot{z}(t) = w(t, z(t)), \quad z(t_0) = z_0, \quad t \geq t_0, \quad (4.354)$$

where $z_0 \in \mathcal{Q} \cap \overline{\mathbb{R}}_+^q$, imply the corresponding stability properties of the zero solution $x(t) \equiv 0$ to (4.351). That is, if the zero solution $z(t) \equiv 0$ to (4.354) is uniformly Lyapunov (respectively, uniformly asymptotically) stable, then the zero solution $x(t) \equiv 0$ to (4.351) is uniformly Lyapunov (respectively, uniformly asymptotically) stable. If, in addition, $\mathcal{D} = \mathbb{R}^n$, $\mathcal{Q} = \mathbb{R}^q$, and $\alpha(\cdot), \beta(\cdot)$ are class $\mathcal{K}_\infty$ functions, then global uniform asymptotic stability of the zero solution $z(t) \equiv 0$ to (4.354) implies global uniform asymptotic stability of the zero solution $x(t) \equiv 0$ to (4.351). Moreover, if there exist constants $\nu \geq 1$, $\alpha > 0$, and $\beta > 0$ such that $v : [0,\infty) \times \mathcal{D} \to \overline{\mathbb{R}}_+$ satisfies

$$\alpha \|x\|^\nu \leq v(t,x) \leq \beta \|x\|^\nu, \quad (t,x) \in [0,\infty) \times \mathcal{D}, \quad (4.355)$$

then exponential stability of the zero solution $z(t) \equiv 0$ to (4.354) implies exponential stability of the zero solution $x(t) \equiv 0$ to (4.351). Finally, if $\mathcal{D} = \mathbb{R}^n$, $\mathcal{Q} = \mathbb{R}^q$, there exist constants $\nu \geq 1$, $\alpha > 0$, and $\beta > 0$ such that $v : [0,\infty) \times \mathbb{R}^n \to \overline{\mathbb{R}}_+$ satisfies (4.355), then global exponential stability of the zero solution $z(t) \equiv 0$ to (4.354) implies global exponential stability of the zero solution $x(t) \equiv 0$ to (4.351).

Proof. The proof is similar to the proof of Theorem 4.37 and is left as an exercise for the reader. □

Finally, to elucidate how to use the vector Lyapunov function framework to address the problem of control design for nonlinear dynamical systems consider the controlled nonlinear dynamical system given by

$$\dot{x}(t) = F(x(t), u(t)), \quad x(t_0) = x_0, \quad t \geq t_0, \quad (4.356)$$

where $x_0 \in \mathcal{D}$, $\mathcal{D} \subseteq \mathbb{R}^n$ is an open set, $0 \in \mathcal{D}$, $u(t) \in U \subseteq \mathbb{R}^m$ is the control input, U is the set of all admissible control inputs, $F : \mathcal{D} \times U \to \mathbb{R}^n$ is Lipschitz continuous for all $(x,u) \in \mathcal{D} \times U$, and $F(0,0) = 0$. Moreover, assume that for every $x_0 \in \mathcal{D}$ and $u(t) \in U$, $t \geq t_0$, the solution $x(t)$ to (4.356) is unique and defined for all $t \geq t_0$. Now, assume there exist a continuously differentiable vector function $V : \mathcal{D} \to \mathcal{Q} \cap \overline{\mathbb{R}}_+^q$ and a positive vector $p \in \mathbb{R}_+^q$ such that $V(0) = 0$, $v(x) \triangleq p^{\mathrm{T}} V(x)$, $x \in \mathcal{D}$, is positive definite, and

$$V'(x)F(x,u) \leq\leq w(V(x), u), \quad x \in \mathcal{D}, \quad u \in U, \quad (4.357)$$

where $w : \mathcal{Q} \times U \to \mathbb{R}^q$ is continuous. Furthermore, define the feedback control law $\phi : \mathcal{Q} \to U$ given by $u = \phi(V(x))$, $x \in \mathcal{D}$, so that $\phi(0) = 0$ and (4.356) is given by

$$\dot{x}(t) = F(x(t), \phi(V(x(t)))), \quad x(t_0) = x_0, \quad t \geq t_0. \tag{4.358}$$

Now, if $\phi(\cdot)$ is such that the zero solution $z(t) \equiv 0$ to

$$\dot{z}(t) = \tilde{w}(z(t)), \quad z(t_0) = z_0, \quad t \geq t_0, \tag{4.359}$$

where $\tilde{w}(z) \triangleq w(z, \phi(z))$, $z \in \mathcal{Q}$, $\tilde{w}(\cdot) \in \mathcal{W}$, $\tilde{w}(0) = 0$, $z_0 \in \mathcal{Q} \cap \overline{\mathbb{R}}_+^q$, is asymptotically stable then the zero solution $x(t) \equiv 0$ of the closed-loop system (4.358) is asymptotically stable.

4.12 Problems

Problem 4.1. Consider the nonlinear dynamical system

$$\dot{x}_1(t) = f_1(x_1(t), x_2(t)), \quad x_1(0) = x_{10}, \quad t \geq 0, \tag{4.360}$$

$$\dot{x}_2(t) = f_2(x_1(t), x_2(t)), \quad x_2(0) = x_{20}, \tag{4.361}$$

where $x_1 \in \mathbb{R}^{n_1}$, $x_2 \in \mathbb{R}^{n_2}$, $f_1 : \mathbb{R}^{n_1} \times \mathbb{R}^{n_2} \to \mathbb{R}^{n_1}$ is Lipschitz continuous on $\mathbb{R}^{n_1} \times \mathbb{R}^{n_2}$ and satisfies $f_1(0, x_2) = 0$ for all $x_2 \in \mathbb{R}^{n_2}$, and $f_2 : \mathbb{R}^{n_1} \times \mathbb{R}^{n_2} \to \mathbb{R}^{n_2}$ is Lipschitz continuous on $\mathbb{R}^{n_1} \times \mathbb{R}^{n_2}$. Show that if there exists a continuously differentiable function $V : \mathbb{R}^{n_1} \to \mathbb{R}$ such that

$$V(0) = 0, \tag{4.362}$$

$$V(x_1) > 0, \quad x_1 \in \mathbb{R}^n, \quad x_1 \neq 0, \tag{4.363}$$

$$V'(x_1) f_1(x_1, x_2) \leq 0, \quad (x_1, x_2) \in \mathbb{R}^{n_1} \times \mathbb{R}^{n_2}, \tag{4.364}$$

then the system (4.360) and (4.361) is Lyapunov stable with respect to x_1. If, in addition, there exists a class $\mathcal{K}_\infty$ function $\gamma : [0, \infty) \to [0, \infty)$ such that

$$V'(x_1) f_1(x_1, x_2) \leq -\gamma(\|x_1\|), \quad (x_1, x_2) \in \mathbb{R}^{n_1} \times \mathbb{R}^{n_2}, \tag{4.365}$$

show that (4.360) and (4.361) is asymptotically stable with respect to x_1. Finally, if $V(\cdot)$ is radially unbounded and satisfies (4.362), (4.363), and (4.365) show that (4.360) and (4.361) is globally asymptotically stable with respect to x_1.

Problem 4.2. Consider the nonlinear dynamical system representing a rigid spacecraft given by

$$\dot{x}_1(t) = I_{23} x_2(t) x_3(t) + \alpha_1 x_1(t), \quad x_1(0) = x_{10}, \quad t \geq 0, \tag{4.366}$$

$$\dot{x}_2(t) = I_{31} x_3(t) x_1(t) + \alpha_2 x_2(t), \quad x_2(0) = x_{20}, \tag{4.367}$$

$$\dot{x}_3(t) = I_{12} x_1(t) x_2(t), \quad x_3(0) = x_{30}, \tag{4.368}$$

where $I_{23} = (I_2 - I_3)/I_1$, $I_{31} = (I_3 - I_1)/I_2$, $I_{12} = (I_1 - I_2)/I_3$, I_1, I_2, and I_3 are the principal moments of inertia of the spacecraft such that

$I_1 > I_2 > I_3 > 0$, and $\alpha_1 < 0$ and $\alpha_2 < 0$ reflect dissipation in the x_1 and x_2 coordinates of the spacecraft. Show that (4.366)–(4.368) is exponentially stable with respect to (x_1, x_2) uniformly in x_{30}. (**Hint:** Use the Lyapunov function candidate $V(x_1, x_2, x_3) = \frac{1}{2}(-I_{31}x_1^2 + I_{23}x_2^2)$.)

Problem 4.3. Analyze the partial stability conditions given in Problem 4.1 and Theorem 4.1 and comment on their interrelationship. Are the conditions equivalent? Is one set of conditions implied by the other? Are the partial stability conditions in Problem 4.1 and Theorem 4.1 a special case of the stability conditions with respect to compact positively invariant sets discussed in Section 4.9? Explain your answers.

Problem 4.4. Consider the nonlinear dynamical system

$$\dot{x}_1(t) = \alpha_1 x_1(t) - \beta x_1(t)x_2(t)\cos x_3(t), \quad x_1(0) = x_{10}, \quad t \geq 0, \tag{4.369}$$
$$\dot{x}_2(t) = \alpha_2 x_2(t) + \beta x_1^2(t)\cos x_3(t), \quad x_2(0) = x_{20}, \tag{4.370}$$
$$\dot{x}_3(t) = 2\theta_1 - \theta_2 - \beta\left(\frac{x_1^2(t)}{x_2(t)} - 2x_2(t)\right)\sin x_3(t), \quad x_3(0) = x_{30}, \tag{4.371}$$

representing a time-averaged, two-mode thermoacoustic combustion model where $\alpha_1 < 0$, $\alpha_2 < 0$, and $\beta, \theta_1, \theta_2 \in \mathbb{R}$. Show that (4.369)–(4.371) is globally asymptotically stable with respect to $[x_1\ x_2]^{\mathrm{T}}$. (**Hint:** Use Problem 4.1 with the *partial* Lyapunov function candidate $V(x_1, x_2) = \frac{1}{2}x_1^2 + \frac{1}{2}x_2^2$.)

Problem 4.5. Show that the zero solution $x(t) \equiv 0$ to (4.57) is uniformly Lyapunov stable if and only if there exists a class $\mathcal{K}$ function $\alpha(\cdot)$ and a positive constant δ, independent of t_0, such that $\|x(t)\| \leq \alpha(\|x(t_0)\|)$ for all $t \geq t_0$ and $\|x(t_0)\| < \delta$.

Problem 4.6. Show that the zero solution $x(t) \equiv 0$ to (4.57) is uniformly asymptotically stable if and only if there exists a class $\mathcal{KL}$ function $\beta(\cdot,\cdot)$ and a positive constant δ, independent of t_0, such that $\|x(t)\| \leq \beta(\|x(t_0)\|, t - t_0)$ for all $t \geq t_0$ and $\|x(t_0)\| < \delta$. Additionally, show that if $x(t_0) \in \mathbb{R}^n$, then the above equivalence holds for global uniform asymptotic stability.

Problem 4.7. Consider the scalar linear equation

$$\dot{x}(t) = -\frac{1}{t+1}x(t), \quad x(t_0) = x_0, \quad t \geq t_0. \tag{4.372}$$

Show that the zero solution $x(t) \equiv 0$ to (4.372) is uniformly Lyapunov stable and globally asymptotically stable, but not uniformly asymptotically stable.

Problem 4.8. Consider the linear dynamical system

$$\dot{x}_1(t) = x_2(t), \quad x_1(0) = x_{10}, \quad t \geq 0, \tag{4.373}$$

$$\dot{x}_2(t) = -x_2(t) - e^{-t}x_1(t), \quad x_2(0) = x_{20}. \tag{4.374}$$

Show that the zero solution $(x_1(t), x_2(t)) \equiv (0,0)$ to (4.373) and (4.374) is Lyapunov stable.

Problem 4.9. Consider the linear dynamical system

$$\dot{x}_1(t) = -a(t)x_1(t) - bx_2(t), \quad x_1(0) = x_{10}, \quad t \geq 0, \tag{4.375}$$

$$\dot{x}_2(t) = bx_1(t) - c(t)x_2(t), \quad x_2(0) = x_{20}, \tag{4.376}$$

where $b \in \mathbb{R}$, $a(t) \geq \alpha > 0$, $t \geq 0$, and $c(t) \geq \beta > 0$, $t \geq 0$. Show that the zero solution $(x_1(t), x_2(t)) \equiv (0,0)$ to (4.375) and (4.376) is globally exponentially stable.

Problem 4.10. Consider the damped *Mathieu equation* given by

$$\ddot{x}(t) + \dot{x}(t) + (2 + \sin t)x(t) = 0, \quad x(0) = x_0, \quad \dot{x}(0) = \dot{x}_0, \quad t \geq 0. \tag{4.377}$$

Show that the zero solution $(x(t), \dot{x}(t)) \equiv (0,0)$ to (4.377) is uniformly Lyapunov stable.

Problem 4.11. Consider the spring-mass-damper system with a time-varying damping coefficient given by

$$\ddot{x}(t) + b(t)\dot{x}(t) + kx(t) = 0, \quad x(0) = x_0, \quad \dot{x}(0) = \dot{x}_0, \quad t \geq 0, \tag{4.378}$$

where $b(t)$ is such that

$$b(t) > \alpha > 0, \quad \dot{b}(t) \leq \beta < 2k, \quad t \geq 0. \tag{4.379}$$

Assuming $b(t)$ is upper bounded, show that the zero solution $(x(t), \dot{x}(t)) \equiv (0,0)$ to (4.378) is uniformly asymptotically stable. Are either or both of the conditions in (4.379) necessary to establish uniform asymptotic stability? (**Hint:** Consider the Lyapunov function candidate $V(t,x) = \frac{1}{2}(\dot{x} + \alpha x)^2 + \frac{1}{2}\gamma(t)x^2$, where $0 < \alpha < \sqrt{k}$ and $\gamma(t) = k - \alpha^2 + \alpha b(t)$.)

Problem 4.12. Consider the nonlinear dynamical system

$$\dot{x}_1(t) = \alpha(t)x_2(t) + \beta(t)x_1(t)[x_1^2(t) + x_2^2(t)], \quad x_1(0) = x_{10}, \quad t \geq 0, \tag{4.380}$$

$$\dot{x}_2(t) = -\alpha(t)x_1(t) + \beta(t)x_2(t)[x_1^2(t) + x_2^2(t)], \quad x_2(0) = x_{20}, \tag{4.381}$$

where $\alpha(\cdot)$ and $\beta(\cdot)$ are continuous functions. Analyze the stability of the zero solution $(x_1(t), x_2(t)) \equiv (0,0)$ to (4.380) and (4.381) using Lyapunov's direct method.

Problem 4.13. Consider the nonlinear dynamical system

$$\dot{x}_1(t) = x_2(t), \quad x_1(0) = x_{10}, \quad t \geq 0, \tag{4.382}$$

$$\dot{x}_2(t) = -\sin x_1(t) - a(t)x_2(t), \quad x_2(0) = x_{20}, \tag{4.383}$$

where $a(\cdot)$ is continuously differentiable and satisfies $0 < \alpha < a(t) \leq \beta < \infty$, $t \geq 0$, and $\dot{\alpha}(t) \leq \gamma < 2$, $t \geq 0$. Show that the zero solution $(x_1(t), x_2(t)) \equiv (0, 0)$ to (4.382) and (4.383) is uniformly asymptotically stable. (**Hint:** Consider the Lyapunov function candidate

$$V(t,x) = \frac{1}{2}(\alpha \sin x_1 + x_2)^2 + [1 + \alpha a(t) - \alpha^2](1 - \cos x_1), \tag{4.384}$$

and show that (4.384) is a valid candidate.)

Problem 4.14. Consider the nonlinear dynamical system

$$\dot{x}_1(t) = x_2(t), \quad x_1(0) = x_{10}, \quad t \geq 0, \tag{4.385}$$
$$\dot{x}_2(t) = -a(t)x_2(t) - e^{-t}x_1(t), \quad x_2(0) = x_{20}, \tag{4.386}$$

where $a(\cdot)$ is a continuous function. Analyze the stability of the zero solution $(x_1(t), x_2(t)) \equiv (0, 0)$ to (4.385) and (4.386) by using the Lyapunov function candidate $V(t, x_1, x_2) = x_1^2 + e^t x_2^2$.

Problem 4.15. Show that if $V : [0, \infty) \times \mathcal{D} \subseteq \mathbb{R}^n \to \mathbb{R}$ is continuously differentiable, lower bounded, $\dot{V}(t,x) \leq 0$, $t \in [0, \infty) \times \mathcal{D}$, and $\dot{V}(t,x)$ is uniformly continuous in time, then $\dot{V}(t,x) \to 0$ as $t \to \infty$. (**Hint:** Use Barbalat's lemma to prove the result.)

Problem 4.16. Consider the dynamical system

$$\dot{x}_1(t) = -x_1(t) + x_2(t)w(t), \quad x_1(0) = x_{10}, \quad t \geq 0, \tag{4.387}$$
$$\dot{x}_2(t) = -x_1(t)w(t), \quad x_2(0) = x_{20}, \tag{4.388}$$

where $w(t)$, $t \geq 0$, is a bounded continuous disturbance. Show that $x_1(t) \to 0$ as $t \to \infty$ and $x_2(t)$, $t \geq 0$, is bounded. (**Hint:** Use the result in Problem 4.15.)

Problem 4.17. Consider the time-varying nonlinear dynamical system (4.57). Show that if there exist a continuously differentiable function $V : [0, \infty) \times \mathcal{D} \to \mathbb{R}$ and class $\mathcal{K}$ functions $\alpha(\cdot)$ and $\gamma(\cdot)$ such that

$$V(t, 0) = 0, \quad t \in [0, \infty), \tag{4.389}$$
$$\alpha(\|x\|) \leq V(t,x), \quad (t,x) \in [0, \infty) \times \mathcal{D}, \tag{4.390}$$
$$\dot{V}(t,x) \leq -\gamma(V(t,x)), \quad (t,x) \in [0, \infty) \times \mathcal{D}, \tag{4.391}$$

then the zero solution $x(t) \equiv 0$ to (4.57) is asymptotically stable. If, in addition, $\mathcal{D} = \mathbb{R}^n$ and $\alpha(\cdot)$ is a class $\mathcal{K}_\infty$ function, show that the zero solution $x(t) \equiv 0$ to (4.57) is globally asymptotically stable.

Problem 4.18. Consider the time-varying nonlinear dynamical system (4.57). Show that if there exist continuously differentiable functions $V : [0, \infty) \times \mathcal{D} \to \mathbb{R}$ and $W : \mathcal{D} \to \mathbb{R}$, and class $\mathcal{K}$ functions $\alpha(\cdot)$ and $\gamma(\cdot)$ such that $W(0) = 0$, $W(x) > 0$, $x \in \mathcal{D}$, $x \neq 0$, $\dot{W}(x(\cdot))$ is bounded from below or above, and

$$\begin{aligned} V(t,0) &= 0, \qquad t \in [0,\infty), && (4.392)\\ \alpha(\|x\|) &\leq V(t,x), \qquad (t,x) \in [0,\infty) \times \mathcal{D}, && (4.393)\\ \dot{V}(t,x) &\leq -\gamma(W(x)), \qquad (t,x) \in [0,\infty) \times \mathcal{D}, && (4.394) \end{aligned}$$

then the zero solution $x(t) \equiv 0$ to (4.57) is asymptotically stable. If, in addition, $\mathcal{D} = \mathbb{R}^n$ and $\alpha(\cdot)$ is a class $\mathcal{K}_\infty$ function show that the zero solution $x(t) \equiv 0$ to (4.57) is globally asymptotically stable.

Problem 4.19. Prove Theorem 4.10.

Problem 4.20. Prove Theorem 4.11.

Problem 4.21. Prove Theorem 4.12.

Problem 4.22. Consider the nonlinear dynamical system

$$\begin{aligned} \dot{x}_1(t) &= -ax_1^3(t) + bx_2(t), \quad x_1(0) = x_{10}, \quad t \geq 0, && (4.395)\\ \dot{x}_2(t) &= -cx_2^3(t), \quad x_2(0) = x_{20}, && (4.396) \end{aligned}$$

where $a, b, c > 0$. Use Proposition 4.2 to show that the zero solution $(x_1(t), x_2(t)) \equiv (0,0)$ of (4.395) and (4.396) is globally asymptotically stable.

Problem 4.23. Consider the nonlinear dynamical system

$$\begin{aligned} \dot{x}_1(t) &= -x_1^2(t) - 2x_1^3(t) + x_2(t), \quad x_1(0) = x_{10}, \quad t \geq 0, && (4.397)\\ \dot{x}_2(t) &= -x_2^3(t), \quad x_2(0) = x_{20}. && (4.398) \end{aligned}$$

Use Proposition 4.2 to show that the zero solution $(x_1(t), x_2(t)) \equiv (0,0)$ is globally asymptotically stable.

Problem 4.24. Consider the nonlinear dynamical system

$$\begin{aligned} \dot{x}_1(t) &= -x_1^2 + \tfrac{1}{2}x_1^2(t)x_2(t), \quad x_1(0) = x_{10}, \quad t \geq 0, && (4.399)\\ \dot{x}_2(t) &= -\tfrac{1}{2}x_1^3(t) - x_2(t) + \tfrac{1}{2}u(t), \quad x_2(0) = x_{20}. && (4.400) \end{aligned}$$

Show that (4.399) and (4.400) is input-to-state stable.

Problem 4.25. Consider the nonlinear perturbed dynamical system

$$\dot{x}(t) = f(t, x(t)) + g(t, x(t)), \quad x(t_0) = x_0, \quad t \geq t_0, \qquad (4.401)$$

where $x(t) \in \mathcal{D}$, $t \geq 0$, $\mathcal{D} \subseteq \mathbb{R}$ such that $0 \in \mathcal{D}$, $f(t,\cdot) : \mathcal{D} \to \mathbb{R}^n$ and $g(t,\cdot) : \mathcal{D} \to \mathbb{R}^n$ are Lipschitz continuous on $\mathcal{D}$ for all $t \in [0,\infty)$, and $f(\cdot,x) : [0,\infty) \to \mathbb{R}^n$ and $g(\cdot,x) : [0,\infty) \to \mathbb{R}^n$ are piecewise continuous on $[0,\infty)$. Assume that the zero solution $x(t) \equiv 0$ to the *nominal* system (4.401), that is, (4.401) with $g(t,x) \equiv 0$, is exponentially stable. Furthermore, assume there exists a continuously differentiable function $V : [0,\infty) \times \mathcal{D} \to \mathbb{R}$ such that the conditions of Corollary 4.2 are satisfied for $\mathcal{D} = \{x \in \mathbb{R}^n : \|x\| < \varepsilon\}$. Finally, suppose $g(t,x)$ satisfies

$$\|g(t,x)\| \le \delta < \frac{\alpha_1}{\alpha_2}\sqrt{\frac{\alpha_3}{\alpha_4}}\mu\varepsilon, \quad (t,x) \in [0,\infty) \times \mathcal{D}, \tag{4.402}$$

where α_1, α_2, α_3, α_4, and μ are positive constants with $\mu < 1$. Show that, for all $\|x(t_0)\| < \sqrt{\frac{\alpha_3}{\alpha_4}}\varepsilon$,

$$\|x(t)\| \le \sigma\|x(t_0)\|e^{-\gamma(t-t_0)}, \quad t \le t < t_0 + T, \tag{4.403}$$

and

$$\|x(t)\| \le \beta, \qquad t \ge t_0 + T, \tag{4.404}$$

where $T > 0$ is a finite time, $\sigma \triangleq \sqrt{\frac{\alpha_3}{\alpha_4}}$, $\gamma \triangleq \frac{(1-\mu)\alpha_1}{2\alpha_4}$, and $\beta \triangleq \frac{\alpha_2}{\alpha_1}\frac{\delta}{\mu}\sqrt{\frac{\alpha_3}{\alpha_4}}$.

Problem 4.26. Consider the nonlinear dynamical system (4.177) with $\mathcal{D} = \{x \in \mathbb{R} : |x| < 1\}$ and $f : \mathcal{D} \to \mathbb{R}$ given by

$$f(x) = \begin{cases} -x(\ln|x|)^2, & x \in \mathcal{D}\backslash\{0\}, \\ 0, & x = 0. \end{cases} \tag{4.405}$$

Show that this system is finite-time stable with settling-time function

$$T(x) = \begin{cases} -\frac{1}{\ln|x|}, & x \in \mathcal{D}\backslash\{0\}, \\ 0, & x = 0. \end{cases} \tag{4.406}$$

Problem 4.27. Consider the nonlinear time-varying dynamical system (4.57) where $f(\cdot,\cdot)$ is continuous on $[0,\infty) \times \mathcal{D}$ and (4.57) possesses unique solutions in forward time for all $x_0 \in \mathcal{D}$ and $t_0 \in [0,\infty)$. The zero solution $x(t) \equiv 0$ to (4.57) is *finite-time stable* if the origin is Lyapunov stable and there exists an open neighborhood $\mathcal{N} \subseteq \mathcal{D}$ of the origin and a function $T : [0,\infty) \times \mathcal{N}\backslash\{0\} \to (0,\infty)$, called a *settling-time function*, such that for every $x_0 \in \mathcal{N}\backslash\{0\}$, $s(\cdot, t_0, x_0) : [t_0, T(t_0,x_0)) \to \mathcal{N}\backslash\{0\}$ and $s(t,t_0,x_0) \to 0$ as $t \to T(t_0,x_0)$. The zero solution to (4.57) is *uniformly finite-time stable* if the origin is uniformly Lyapunov stable and $s(t,t_0,x_0) \to 0$ as $t \to T(t_0,x_0)$ for every $x_0 \in \mathcal{N}\backslash\{0\}$. The zero solution is *globally uniformly finite-time stable* if it is uniformly finite-time stable with $\mathcal{D} = \mathcal{N} = \mathbb{R}^n$. Show that the following statements hold:

i) If there exist a continuously differentiable function $V : [0,\infty)\times\mathcal{D} \to \mathbb{R}$, class $\mathcal{K}$ function $\alpha(\cdot)$, a continuous function $c : [0,\infty) \to \mathbb{R}_+$ with $c(t) > 0$ for almost all $t \in [0,\infty)$, a real number $\lambda \in (0,1)$, and an open neighborhood $\mathcal{M} \subseteq \mathcal{D}$ of the origin such that

$$V(t,0) = 0, \quad t \in [0,\infty), \tag{4.407}$$

$$\alpha(\|x\|) \le V(t,x), \quad x \in \mathcal{M}, \quad t \in [0,\infty), \tag{4.408}$$
$$\dot{V}(t,x) \le -c(t)(V(t,x))^{\lambda}, \quad x \in \mathcal{M}, \quad t \in [0,\infty), \tag{4.409}$$

then the zero solution $x(t) \equiv 0$ to (4.57) is finite-time stable.

ii) If there exist a continuously differentiable function $V : [0,\infty)\times\mathcal{D} \to \mathbb{R}$, class $\mathcal{K}$ functions $\alpha(\cdot)$ and $\beta(\cdot)$, a continuous function $c : [0,\infty) \to \mathbb{R}_+$ with $c(t) > 0$ for almost all $t \in [0,\infty)$, a real number $\lambda \in (0,1)$, and an open neighborhood $\mathcal{M} \subseteq \mathcal{D}$ of the origin such that (4.408) and (4.409) hold, and

$$V(t,x) \le \beta(\|x\|), \quad x \in \mathcal{M}, \quad t \in [0,\infty), \tag{4.410}$$

then the zero solution $x(t) \equiv 0$ to (4.57) is uniformly finite-time stable.

iii) If $\mathcal{D} = \mathcal{M} = \mathbb{R}^n$ and there exist a continuously differentiable function $V : [0,\infty) \times \mathcal{D} \to \mathbb{R}$, class $\mathcal{K}_\infty$ functions $\alpha(\cdot)$ and $\beta(\cdot)$, a continuous function $c : [0,\infty) \to \mathbb{R}_+$ with $c(t) > 0$ for almost all $t \in [0,\infty)$, a real number $\lambda \in (0,1)$, and an open neighborhood $\mathcal{M} \subseteq \mathcal{D}$ of the origin such that (4.408)–(4.410) hold, then the zero solution $x(t) \equiv 0$ to (4.57) is globally uniformly finite-time stable.

Problem 4.28. Consider the nonlinear time-varying dynamical system (4.57). Assume that the zero solution $x(t) \equiv 0$ to (4.57) is finite-time stable and let $\mathcal{N} \subseteq \mathcal{D}$ and $T : [0,\infty) \times \mathcal{N}\backslash\{0\} \to (0,\infty)$ be defined as in Problem 4.27. Show that, for every $t_0 \in [0,\infty)$ and $x_0 \in \mathcal{N}$, there exists a unique solution $s(t,t_0,x_0)$, $t \ge t_0$, to (4.57) such that $s(t,t_0,x_0) \in \mathcal{N}$, $t \in [t_0, T(t_0,x_0))$, and $s(t,t_0,x_0) = 0$ for all $t \ge T(t_0,x_0)$, where $T(t_0,0) \triangleq t_0$.

Problem 4.29. Consider the nonlinear time-varying dynamical system (4.57). Assume that the zero solution $x(t) \equiv 0$ to (4.57) is finite-time stable and let $\mathcal{N} \subseteq \mathcal{D}$ and $T : [0,\infty) \times \mathcal{N}\backslash\{0\} \to (0,\infty)$ be defined as in Problem 4.27. Show that the following statements hold:

i) If $t_0 \in [0,\infty)$, $t \ge t_0$, and $x \in \mathcal{N}$, then the settling-time function $T(t, s(t,t_0,x_0)) = \max\{T(t_0,x), t\}$.

ii) $T(\cdot,\cdot)$ is jointly continuous on $[0,\infty) \times \mathcal{N}$ if and only if $T(\cdot,\cdot)$ is continuous at $(t,0)$ for all $t \ge 0$.

Problem 4.30. Consider the nonlinear time-varying dynamical system (4.57). Let $\lambda \in (0,1)$, let $\mathcal{N}$ be as in Problem 4.27, and assume that there exists a class $\mathcal{K}$ function $\mu : [0,r] \to [0,\infty)$, where $r > 0$, such that $\mathcal{B}_r(0) \subseteq \mathcal{N}$ and

$$\|f(t,x)\| \le \mu(\|x\|), \quad t \in [0,\infty), \quad x \in \mathcal{B}_r(0). \tag{4.411}$$

Show that if the zero solution $x(t) \equiv 0$ to (4.57) is uniformly finite-time stable and the settling-time function $T(\cdot,\cdot)$ is continuous at $(t,0)$ for all $t \geq 0$, then there exist a class $\mathcal{K}$ function $\alpha(\cdot)$, a positive constant $c > 0$, a continuous function $V : [0,\infty) \times \mathcal{N} \to \mathbb{R}$, and a neighborhood $\mathcal{M} \subseteq \mathcal{N}$ of the origin such that

$$\alpha(\|x\|) \leq V(t,x), \quad (t,x) \in [0,\infty) \times \mathcal{M}, \tag{4.412}$$
$$\dot{V}(t,x) \leq -c(V(t,x))^{\lambda}, \quad (t,x) \in [0,\infty) \times \mathcal{M}. \tag{4.413}$$

(**Hint:** Consider the Lyapunov function candidate $V(t,x) = [T(t,x)-t]^{\frac{1}{1-\lambda}}$.)

Problem 4.31. Consider the nonlinear time-varying dynamical system (4.57) and assume $f(t,\cdot)$ is Lipschitz continuous in x on $\mathcal{D}$ uniformly in t for all t in compact subsets of $[0,\infty)$. Furthermore, assume that the Lipschitz constant of $f(t,\cdot)$ is bounded for all $t \geq 0$ with maximum Lipschitz constant $L > 0$ over $\mathcal{D}$. Show that if there exist a function $V : [0,\infty) \times \mathcal{D} \to \mathbb{R}$ and a finite time $T > 0$ such that

$$\alpha(\|x\|) \leq V(t,x) \leq \beta(\|x\|), \quad (t,x) \in [0,\infty) \times \mathcal{D}, \tag{4.414}$$
$$V(t+T, x(t+T)) - V(t,x) \leq -\gamma(\|x\|) < 0, \quad (t,x) \in [0,\infty) \times \mathcal{D}, \tag{4.415}$$

where $\alpha(\cdot)$, $\beta(\cdot)$, and $\gamma(\cdot)$ are class $\mathcal{K}$ functions, then the zero solution $x(t) \equiv 0$ to (4.57) is uniformly asymptotically stable. (**Hint:** Show that (4.415) implies that there exists an increasing unbounded sequence $\{t_n\}_{n=0}^{\infty}$, with $t_0 = 0$, such that $T \geq t_{n+1} - t_n > 0$, $n = 0,1,\ldots$, and

$$V(t_{n+1}, x(t_{n+1})) - V(t_n, x(t_n)) \leq -\gamma(\|x(t_n)\|), \; x(t_n) \in \mathcal{D}, n = 0,1,\ldots. \tag{4.416}$$

Here, the function $V(\cdot,\cdot)$ does not satisfy any regularity assumptions.)

Problem 4.32. Consider the nonlinear dynamical system in polar coordinates given by

$$\dot{r}(t) = -r(t)\text{sign}[r^2(t)-1]|r^2(t)-1|^{\alpha}, \quad r(0) = r_0, \quad t \geq 0, \tag{4.417}$$
$$\dot{\theta}(t) = -\text{sign}[r^2(t)-1]|r^2(t)-1|^{\beta}, \quad \theta(0) = \theta_0, \tag{4.418}$$

where $\alpha, \beta \in \mathbb{R}$. Show that the set of equilibria of (4.417) and (4.418) consists of the origin $(r,\theta) = (0,0)$ and the unit circle $\mathcal{C} = \{(r,\theta) \in \mathbb{R} \times \mathbb{R} : r^2 = 1\}$. In addition, show that all solutions of (4.417) and (4.418) starting from nonzero initial conditions that are not on the unit circle $\mathcal{C}$ approach the unit circle, and hence, all solutions are bounded, and, for every choice of α and β, all solutions converge to the set of equilibria. However, show that if $\alpha \geq \beta+1$, then the dynamical system (4.417) and (4.418) is not convergent. (**Hint:** Use (4.417) and (4.418) to obtain $\frac{\mathrm{d}r}{\mathrm{d}\theta} = r|r^2-1|^{\alpha-\beta}$.)

Problem 4.33. Consider the nonlinear dynamical system

$$\dot{x}_1(t) = -4x_2(t) + x_1(t)[1 - \tfrac{1}{4}x_1^2(t) - x_2^2(t)], \quad x_1(0) = x_{10}, \quad t \geq 0, \tag{4.419}$$

$$\dot{x}_2(t) = x_1(t) + x_2(t)[1 - \tfrac{1}{4}x_1^2(t) - x_2^2(t)],\ x_2(0) = x_{20}. \quad (4.420)$$

Show that (4.419) and (4.420) has a limit cycle that lies in the ellipse $\mathcal{E} = \{(x_1, x_2) \in \mathbb{R}^2 : \frac{1}{2}x_1^2 + x_2^2 = 1\}$. Using Poincaré maps, examine the stability of this limit cycle.

Problem 4.34. Consider the nonlinear dynamical system

$$\dot{x}_1(t) = x_1(t) - 4x_2(t) - \tfrac{1}{4}x_1^3(t) - x_1(t)x_2^2(t),\ x_1(0) = x_{10},\ t \geq 0, \quad (4.421)$$

$$\dot{x}_2(t) = x_1(t) + x_2(t) - \tfrac{1}{4}x_1^2(t)x_2(t) - x_2^3(t),\ x_2(0) = x_{20}, \quad (4.422)$$

$$\dot{x}_3(t) = x_3(t),\ x_3(0) = x_{30}. \quad (4.423)$$

Show that (4.421)–(4.423) has a periodic orbit given by $x(t) = [2\cos 2t,\ 2\sin 2t,\ 0]^{\mathrm{T}}$. Using Poincaré maps examine the stability of this orbit.

Problem 4.35. Consider the nonlinear dynamical system

$$\dot{x}_1(t) = -x_2(t) + x_1(t)x_3^2(t), \quad x_1(0) = x_{10}, \quad t \geq 0, \quad (4.424)$$

$$\dot{x}_2(t) = x_1(t) + x_2(t)x_3^2(t), \quad x_2(0) = x_{20}, \quad (4.425)$$

$$\dot{x}_3(t) = -x_3(t)[x_1^2(t) + x_2^2(t)], \quad x_3(0) = x_{30}. \quad (4.426)$$

Show that (4.424)–(4.426) has a periodic orbit given by $x(t) = [\cos t,\ \sin t,\ 0]^{\mathrm{T}}$. Using Poincaré maps examine the stability of this orbit. (**Hint:** Use $V(x_1, x_2, x_3) = x_1^2 + x_2^2 + x_3^2$ to show that the trajectories of (4.424)–(4.426) lie on spheres $x_1^2 + x_2^2 + x_3^2 = k^2$, $k \in \mathbb{R}$.)

Problem 4.36. Consider the nonlinear dynamical system

$$\dot{x}_1(t) = -2x_2(t) + 4x_1[4 - 4x_1^2(t) - x_2^2(t) + x_3(t)],\ x_1(0) = x_{10},\ t \geq 0, \quad (4.427)$$

$$\dot{x}_2(t) = 8x_1(t) + 4x_2(t)[4 - 4x_1^2(t) - x_2^2(t) + x_3(t)],\ x_2(0) = x_{20}, \quad (4.428)$$

$$\dot{x}_3(t) = x_3(t)[x_1(t) - 16], \quad x_3(0) = x_{30}. \quad (4.429)$$

Show that (4.427)–(4.429) has a periodic orbit given by $x(t) = [\cos 4t,\ 2\sin 4t,\ 0]^{\mathrm{T}}$. Using Poincaré maps examine the stability of this orbit.

Problem 4.37. Consider the nonlinear dynamical system

$$\dot{x}_1(t) = -x_1(t) + x_1(t)[1 - x_1^2(t) - x_2^2(t)][16 - x_1^2(t) - x_2^2(t)],$$
$$x_1(0) = x_{10}, \quad t \geq 0, \quad (4.430)$$

$$\dot{x}_2(t) = x_1(t) + x_2(t)[1 - x_1^2(t) - x_2^2(t)][16 - x_1^2(t) - x_2^2(t)],$$
$$x_2(0) = x_{20}, \quad (4.431)$$

$$\dot{x}_3(t) = x_3(t), \quad x_3(0) = x_{30}. \quad (4.432)$$

Show that (4.430)–(4.432) has two periodic orbits. Using Poincaré maps examine the stability of these orbits.

Problem 4.38. Prove Theorem 4.40.

Problem 4.39. Consider the nonlinear dynamical system (4.177). Assume that there exist a continuously differentiable vector function $V : \mathcal{D} \to \mathcal{Q} \cap \overline{\mathbb{R}}_+^q$, where $\mathcal{Q} \subset \mathbb{R}^q$ and $0 \in \mathcal{Q}$, and a positive vector $p \in \mathbb{R}_+^q$ such that $V(0) = 0$, the scalar function $p^{\mathrm{T}} V(x)$, $x \in \mathcal{D}$, is positive definite, and

$$V'(x)f(x) \leq\leq w(V(x)), \quad x \in \mathcal{D}, \tag{4.433}$$

where $w : \mathcal{Q} \to \mathbb{R}^q$ is continuous, $w(\cdot) \in \mathcal{W}$, and $w(0) = 0$. In addition, assume that the vector comparison system

$$\dot{z}(t) = w(z(t)), \quad z(0) = z_0, \quad t \in \mathcal{I}_{z_0}, \tag{4.434}$$

has a unique solution in forward time $z(t)$, $t \in \mathcal{I}_{z_0}$, and there exist a continuously differentiable function $v : \mathcal{Q} \to \mathbb{R}$, real numbers $c > 0$ and $\alpha \in (0, 1)$, and a neighborhood $\mathcal{M} \subseteq \mathcal{Q}$ of the origin such that $v(\cdot)$ is positive definite and

$$v'(z)w(z) \leq -c(v(z))^{\alpha}, \quad z \in \mathcal{M}. \tag{4.435}$$

Show that the zero solution $x(t) \equiv 0$ to (4.177) is finite-time stable. Moreover, if $\mathcal{N}$ is as in Definition 4.7 and $T : \mathcal{N} \to [0, \infty)$ is the settling-time function, show that

$$T(x_0) \leq \frac{1}{c(1-\alpha)}(v(V(x_0)))^{1-\alpha}, \quad x_0 \in \mathcal{N}, \tag{4.436}$$

and $T(\cdot)$ is continuous on $\mathcal{N}$. Finally, show that if $\mathcal{D} = \mathbb{R}^n$, $v(\cdot)$ is radially unbounded, and (4.433) and (4.435) hold on $\mathbb{R}^n$, then the zero solution $x(t) \equiv 0$ to (4.177) is globally finite-time stable.

Problem 4.40. Consider the nonlinear dynamical system (4.177). Assume there exist a continuously differentiable vector function $V : \mathcal{D} \to \mathcal{Q} \cap \overline{\mathbb{R}}_+^q$, where $\mathcal{Q} \subset \mathbb{R}^q$ and $0 \in \mathcal{Q}$, and a positive vector $p \in \mathbb{R}_+^q$ such that $V(0) = 0$, the scalar function $p^{\mathrm{T}} V(x)$, $x \in \mathcal{D}$, is positive definite, and

$$V'(x)f(x) \leq\leq W(V(x))^{\{\alpha\}}, \quad x \in \mathcal{D}, \tag{4.437}$$

where $\alpha \in (0, 1)$, $W \in \mathbb{R}^{q \times q}$ is essentially nonnegative (see Problem 3.7) and Hurwitz, and $(V(x))^{\{\alpha\}} \triangleq [(V_1(x))^{\alpha}, \ldots, (V_q(x))^{\alpha}]^{\mathrm{T}}$. Show that the zero solution $x(t) \equiv 0$ to (4.177) is finite-time stable. (**Hint:** Use Problems 3.8 and 4.39.)

4.13 Notes and References

The concept of partial stability is due to Rumyantsev [374] with a thorough treatment given by Vorotnikov [448]. See also Rouche, Habets, and Laloy [368] and Chellaboina and Haddad [87]. The concept of input-to-state stability was introduced by Sontag [405] with key results given in Sontag [407] and Sontag and Wang [409]. A rigorous foundation for the theory of finite-time stability was first given by Bhat and Bernstein [55]. The presentation here is adopted from Bhat and Bernstein [55]. Semistability was first introduced by Campbell and Rose [81] for linear systems, and applied to matrix second order system by Bernstein and Bhat [46]. Bhat and Bernstein [54,56,57] also consider semistability of nonlinear systems, and give several stability results for systems having a continuum of equilibria based on nontangency and arc length of trajectories. Semistability was also addressed by Hui, Haddad, and Bhat [209] for consensus protocols in nonlinear dynamical networks.

The generalized Lyapunov and invariant set theorems predicated on lower semicontinuous Lyapunov functions presented in Section 4.8 are due to Chellaboina, Leonessa, and Haddad [91] while Theorem 4.28 is due to Aeyels and Peuteman [3]. Lyapunov and asymptotic stability of sets were first introduced by Zubov [481] and further developed by Yoshizawa [474] and Bhatia and Szegö [58]. The treatment here is adopted from Leonessa, Haddad, and Chellaboina [272,273]. Poincaré maps and stability of periodic orbits are due to Poincaré [358] with a Lyapunov function proof given by Haddad, Nersesov, and Chellaboina [175]. Vector Lyapunov functions were first introduced by Bellman [38] and Matrosov [308] with further developments given by Lakshmikantham, Matrosov, and Sivasundaram [255] and Nersesov and Haddad [334]. See also Siljak [400,402], Lakshmikantham and Leela [253], and Lakshmikantham, Leela, and Martynyuk [254].

Chapter Five

Dissipativity Theory for Nonlinear Dynamical Systems

5.1 Introduction

In control engineering, dissipativity theory provides a fundamental framework for the analysis and control design of dynamical systems using an input, state, and output system description based on system-energy-related considerations. The notion of energy here refers to abstract energy notions for which a physical system energy interpretation is not necessary. The dissipation hypothesis on dynamical systems results in a fundamental constraint on their dynamic behavior, wherein a dissipative dynamical system can deliver only a fraction of its energy to its surroundings and can store only a fraction of the work done to it. Many of the great landmarks of feedback control theory are associated with dissipativity theory. In particular, dissipativity theory provides the foundation for absolute stability theory which in turn forms the basis of the Luré problem, as well as the circle and Popov criteria, which are extensively developed in the classical monographs of Aizerman and Gantmacher [5], Lefschetz [265], and Popov [364]. Since absolute stability theory concerns the stability of a dynamical system for classes of feedback nonlinearities which, as noted in [147, 148], can readily be interpreted as an uncertainty model, it is not surprising that absolute stability theory (and, hence, dissipativity theory) also forms the basis of modern-day robust stability analysis and synthesis [147, 151, 172].

The key foundation in developing dissipativity theory for general nonlinear dynamical systems was presented by J. C. Willems [456, 457] in his seminal two-part paper on dissipative dynamical systems. In particular, Willems [456] introduced the definition of dissipativity for general dynamical systems in terms of a *dissipation inequality* involving a generalized system power input, or *supply rate*, and a generalized energy function, or *storage function.* The dissipation inequality implies that the increase in generalized system energy over a given time interval cannot exceed the generalized energy supply delivered to the system during this time interval. The set of all possible system storage functions is convex and every system storage function is bounded from below by the available system storage and bounded from above by the required supply. The available storage is the amount of internal generalized stored energy which can be extracted from the dynamical system and the required supply is the amount of generalized energy which can be delivered to the dynamical system to transfer it from a state of minimum potential to a given state. Hence, as noted above, a dissipative dynamical system can deliver only a fraction of its stored generalized energy to its surroundings and can store only a fraction of generalized work done to it.

Dissipativity theory is a system theoretic concept that provides a powerful framework for the analysis and control design of dynamical systems based on generalized energy considerations. In particular, dissipativity theory exploits the notion that numerous physical dynamical systems have certain input-output system properties related to conservation, dissipation, and transport of mass and energy. Such conservation laws are prevalent in dynamical systems such as mechanical systems, fluid systems, electromechanical systems, electrical systems, combustion systems, structural systems, biological systems, physiological systems, biomedical systems, ecological systems, economic systems, as well as feedback control systems. On the level of analysis, dissipativity can involve conditions on system parameters that render an input, state, output system dissipative. Or, alternatively, analyzing system stability robustness by viewing a dynamical system as an interconnection of dissipative dynamical subsystems. On the synthesis level, dissipativity can be used to design feedback controllers that add dissipation and guarantee stability robustness allowing stabilization to be understood in physical terms.

To elucidate the notion of dissipativity of an input, state, and output system, consider the single-degree-of-freedom spring-mass-damper mechanical system given by

$$M\ddot{x}(t) + C\dot{x}(t) + Kx(t) = u(t), \quad x(0) = x_0, \quad \dot{x}(0) = \dot{x}_0, \quad t \geq 0, \tag{5.1}$$

where $M > 0$ is the system mass, $C \geq 0$ is the system damping constant, $K \geq 0$ is the system stiffness, $x(t)$, $t \geq 0$, is the position of the mass M, and $u(t)$, $t \geq 0$, is an external force acting on the mass M. The energy of this system is given by

$$V_{\rm s}(x, \dot{x}) = \tfrac{1}{2}M\dot{x}^2 + \tfrac{1}{2}Kx^2. \tag{5.2}$$

Now, assuming that the measured output of this system is the system velocity, that is, $y(t) = \dot{x}(t)$, it follows that the time rate of change of the system energy along the system trajectories is given by

$$\dot{V}_{\rm s}(x, \dot{x}) = M\ddot{x}\dot{x} + Kx\dot{x} = uy - C\dot{x}^2. \tag{5.3}$$

Integrating (5.3) over the time interval $[0, T]$, it follows that

$$V_s(x(T), \dot{x}(T)) = V_s(x(0), \dot{x}(0)) + \int_0^T u(t)y(t)dt - \int_0^T C\dot{x}^2(t)dt, \quad (5.4)$$

which shows that the system energy at time $t = T$ is equal to the initial energy stored in the system plus the energy supplied to the system via the external force u minus the energy dissipated by the system damper. Equivalently, it follows from (5.3) that the rate of change in the system energy, or system power, is equal to the external supplied system power through the input port u minus the internal system power dissipated by the viscous damper.

Note that in the case where the external input force u is zero and $C = 0$, that is, no system supply or dissipation is present, (5.3) or, equivalently, (5.4) shows that the system energy is constant. Furthermore, note that since $C \geq 0$ and $V(x(T), \dot{x}(T)) \geq 0$, $T \geq 0$, it follows from (5.4) that

$$\int_0^T u(t)y(t)dt \geq -V_s(x_0, \dot{x}_0), \quad (5.5)$$

or, equivalently,

$$-\int_0^T u(t)y(t)dt \leq V_s(x_0, \dot{x}_0). \quad (5.6)$$

Equation (5.6) shows that the energy that can be *extracted* from the system through its input-output ports is less than or equal to the initial energy stored in the system. As will be seen in this chapter, this is precisely the notion of dissipativity.

Since, as discussed in Chapter 3, Lyapunov functions can be viewed as generalizations of energy functions for nonlinear dynamical systems, the notion of dissipativity, with appropriate storage functions and supply rates, can be used to construct Lyapunov functions for nonlinear feedback systems by appropriately combining storage functions for each subsystem. Even though the original work on dissipative dynamical systems was formulated in the state space setting, describing the system dynamics in terms of continuous flows on appropriate manifolds, an input-output formulation for dissipative dynamical systems extending the notions of passivity [476], nonexpansivity [477], and conicity [377,476] was presented in [188,191,320]. In this chapter, we introduce precise mathematical definitions for dissipativity as well as develop a general framework for characterizing system dissipativity in terms of system storage functions and supply rates.

5.2 Dissipative and Exponentially Dissipative Dynamical Systems

In this section, we introduce the definition of dissipativity for general dynamical systems in terms of an inequality involving generalized system power input, or supply rate, and a generalized energy function, or storage function. In Chapters 2–4, we considered *closed* dynamical systems wherein each system trajectory is determined by the system initial conditions and driven by the internal dynamics of the system without any influence from the environment. Alternatively, in this chapter we consider *open* dynamical systems wherein the system interaction with the environment is explicitly taken into account through the system inputs and outputs. Specifically, the environment acts on the dynamical system through the system inputs, and the dynamical system reacts through the system outputs.

We begin by considering nonlinear dynamical systems $\mathcal{G}$ of the form

$$\dot{x}(t) = F(x(t), u(t)), \quad x(t_0) = x_0, \quad t \geq t_0, \tag{5.7}$$

$$y(t) = H(x(t), u(t)), \tag{5.8}$$

where $x(t) \in \mathcal{D} \subseteq \mathbb{R}^n$, $\mathcal{D}$ is an open set with $0 \in \mathcal{D}$, $u(t) \in U \subseteq \mathbb{R}^m$, $y(t) \in Y \subseteq \mathbb{R}^l$, $F : \mathcal{D} \times U \to \mathbb{R}^n$, and $H : \mathcal{D} \times U \to Y$. For the dynamical system $\mathcal{G}$ given by (5.7) and (5.8) defined on the state space $\mathcal{D} \subseteq \mathbb{R}^n$, $\mathcal{U}$ and $\mathcal{Y}$ define an input and output space, respectively, consisting of continuous bounded U-valued and Y-valued functions on the semi-infinite interval $[0, \infty)$. The set U contains the set of input values, that is, for every $u(\cdot) \in \mathcal{U}$ and $t \in [0, \infty)$, $u(t) \in U$. The set Y contains the set of output values, that is, for every $y(\cdot) \in \mathcal{Y}$ and $t \in [0, \infty)$, $y(t) \in Y$. The spaces $\mathcal{U}$ and $\mathcal{Y}$ are assumed to be closed under the shift operator, that is, if $u(\cdot) \in \mathcal{U}$ (respectively, $y(\cdot) \in \mathcal{Y}$), then the function defined by $u_T \triangleq u(t+T)$ (respectively, $y_T \triangleq y(t+T)$) is contained in $\mathcal{U}$ (respectively, $\mathcal{Y}$) for all $T \geq 0$. We assume that $F(\cdot,\cdot)$ and $H(\cdot,\cdot)$ are continuously differentiable mappings in (x, u) and $F(\cdot,\cdot)$ has at least one equilibrium so that, without loss of generality, $F(0,0) = 0$ and $H(0,0) = 0$. Furthermore, for the nonlinear dynamical system $\mathcal{G}$ we assume that the required properties for the existence and uniqueness of solutions are satisfied, that is, $u(\cdot)$ satisfies sufficient regularity conditions such that the system (5.7) has a unique solution forward and backward in time. For the dynamical system $\mathcal{G}$ given by (5.7) and (5.8), a function $r : U \times Y \to \mathbb{R}$ such that $r(0,0) = 0$ is called a *supply rate* if $r(u, y)$ is locally integrable for all input-output pairs satisfying (5.7) and (5.8), that is, for all input-output pairs $u(\cdot) \in \mathcal{U}$ and $y(\cdot) \in \mathcal{Y}$ satisfying (5.7) and (5.8), $r(\cdot,\cdot)$ satisfies $\int_{t_1}^{t_2} |r(u(s), y(s))| \mathrm{d}s < \infty$, $t_1, t_2 \geq 0$.

Definition 5.1. A dynamical system $\mathcal{G}$ of the form (5.7) and (5.8) is *dissipative with respect to the supply rate* $r(u, y)$ if the *dissipation inequality*

$$0 \leq \int_{t_0}^{t} r(u(s), y(s))\mathrm{d}s \tag{5.9}$$

is satisfied for all $t \geq t_0$ and all $u(\cdot) \in \mathcal{U}$ with $x(t_0) = 0$ along the trajectories of $\mathcal{G}$. A dynamical system $\mathcal{G}$ of the form (5.7) and (5.8) is *exponentially dissipative with respect to the supply rate* $r(u, y)$ if there exists a constant $\varepsilon > 0$ such that the *exponential dissipation inequality*

$$0 \leq \int_{t_0}^{t} e^{\varepsilon s} r(u(s), y(s))\mathrm{d}s \tag{5.10}$$

is satisfied for all $t \geq t_0$ and all $u(\cdot) \in \mathcal{U}$ with $x(t_0) = 0$ along the trajectories of $\mathcal{G}$. A dynamical system $\mathcal{G}$ of the form (5.7) and (5.8) is *lossless with respect to the supply rate* $r(u, y)$ if $\mathcal{G}$ is dissipative with respect to the supply rate $r(u, y)$ and the dissipation inequality (5.9) is satisfied as an equality for all $t \geq t_0$ and all $u(\cdot) \in \mathcal{U}$ with $x(t_0) = x(t) = 0$ along the trajectories of $\mathcal{G}$.

In the following we shall use either 0 or t_0 to denote the initial time for $\mathcal{G}$. Next, define the *available storage* $V_{\mathrm{a}}(x_0)$ of the nonlinear dynamical system $\mathcal{G}$ by

$$\begin{aligned} V_{\mathrm{a}}(x_0) &\triangleq -\inf_{u(\cdot),\, T \geq 0} \int_0^T r(u(t), y(t))\mathrm{d}t \\ &= \sup_{u(\cdot),\, T \geq 0} \left[-\int_0^T r(u(t), y(t))\mathrm{d}t \right], \end{aligned} \tag{5.11}$$

where $x(t)$, $t \geq 0$, is the solution to (5.7) with $x(0) = x_0$ and admissible input $u(\cdot) \in \mathcal{U}$. The supremum in (5.11) is taken over all admissible inputs $u(\cdot)$, all time $t \geq 0$, and all system trajectories with initial value $x(0) = x_0$ and terminal value left free. Note that $V_{\mathrm{a}}(x) \geq 0$ for all $x \in \mathcal{D}$ since $V_{\mathrm{a}}(x)$ is the supremum over a set of numbers containing the zero element ($T = 0$). When the final state is not free but rather constrained to $x(t) = 0$ corresponding to the equilibrium of the uncontrolled system, then $V_{\mathrm{a}}(x_0)$ corresponds to the *virtual available storage*. For details, see Problem 5.3. It follows from (5.11) that the available storage of a nonlinear dynamical system $\mathcal{G}$ is the maximum amount of storage, or generalized stored energy, which can be extracted from the nonlinear dynamical system $\mathcal{G}$ at any time T. Similarly, define the *available exponential storage* $V_{\mathrm{a}}(x_0)$ of the nonlinear dynamical system $\mathcal{G}$ by

$$V_{\mathrm{a}}(x_0) \triangleq -\inf_{u(\cdot),\, T \geq 0} \int_0^T e^{\varepsilon t} r(u(t), y(t))\mathrm{d}t, \tag{5.12}$$

where $x(t)$, $t \geq 0$, is the solution to (5.7) with $x(0) = x_0$ and admissible input $u(\cdot) \in \mathcal{U}$. Note that if we define the available exponential storage as the time-varying function

$$\hat{V}_{\rm a}(x_0,t_0) = -\inf_{u(\cdot),\,T\geq t_0}\int_{t_0}^{T} e^{\varepsilon t} r(u(t),y(t)){\rm d}t, \tag{5.13}$$

where $x(t)$, $t\geq t_0$, is the solution to (5.7) with $x(t_0)=x_0$ and admissible input $u(\cdot)$, it follows that, since $\mathcal{G}$ is time invariant,

$$\hat{V}_{\rm a}(x_0,t_0) = -e^{\varepsilon t_0}\inf_{u(\cdot),\,T\geq 0}\int_{0}^{T} e^{\varepsilon t} r(u(t),y(t)){\rm d}t = e^{\varepsilon t_0}V_{\rm a}(x_0). \tag{5.14}$$

Hence, an alternative expression for the available exponential storage function $V_{\rm a}(x_0)$ is given by

$$V_{\rm a}(x_0) = -e^{-\varepsilon t_0}\inf_{u(\cdot),\,T\geq t_0}\int_{t_0}^{T} e^{\varepsilon t} r(u(t),y(t)){\rm d}t. \tag{5.15}$$

$\hat{V}_{\rm a}(x_0,t_0)$ given by (5.13) defines the available storage function for nonstationary (time-varying) dynamical systems [191,456]. As shown above, in the case of exponentially dissipative systems, $\hat{V}_{\rm a}(x_0,t_0)=e^{\varepsilon t_0}V_{\rm a}(x_0)$.

Next, we show that the available storage (respectively, available exponential storage) is finite and zero at the origin if and only if $\mathcal{G}$ is dissipative (respectively, exponentially dissipative). For this result we require three more definitions.

Definition 5.2. A nonlinear dynamical system $\mathcal{G}$ is *completely reachable* if for all $x_0\in\mathcal{D}\subseteq\mathbb{R}^n$ there exist a finite time $t_{\rm i}<t_0$ and a square integrable input $u(t)$ defined on $[t_{\rm i},t_0]$ such that the state $x(t)$, $t\geq t_{\rm i}$, can be driven from $x(t_{\rm i})=0$ to $x(t_0)=x_0$. $\mathcal{G}$ is *completely null controllable* if for all $x_0\in\mathcal{D}\subseteq\mathbb{R}^n$ there exist a finite time $t_{\rm f}>t_0$ and a square integrable input $u(t)$ defined on $[t_0,t_{\rm f}]$ such that $x(t)$, $t\geq t_0$, can be driven from $x(t_0)=x_0$ to $x(t_{\rm f})=0$.

Definition 5.3. Consider the nonlinear dynamical system $\mathcal{G}$ given by (5.7) and (5.8). A continuous, nonnegative-definite function $V_{\rm s}:\mathcal{D}\to\mathbb{R}$ satisfying $V_{\rm s}(0)=0$ and

$$V_{\rm s}(x(t)) \leq V_{\rm s}(x(t_0)) + \int_{t_0}^{t} r(u(s),y(s)){\rm d}s, \qquad t\geq t_0, \tag{5.16}$$

for all $t_0,t\geq 0$, where $x(t)$, $t\geq t_0$, is the solution of (5.7) with $u(\cdot)\in\mathcal{U}$, is called a *storage function* for $\mathcal{G}$.

Inequality (5.16) is known as the *dissipation inequality* and reflects the fact that some of the supplied generalized energy to the open dynamical system $\mathcal{G}$ is stored, and some is dissipated. The dissipated generalized energy is nonnegative and is given by the difference of what is supplied and what is stored. In addition, the amount of generalized stored energy is a function of the state of the dynamical system.

Definition 5.4. Consider the nonlinear dynamical system $\mathcal{G}$ given by (5.7) and (5.8). A continuous, nonnegative-definite function $V_{\rm s} : \mathcal{D} \to \mathbb{R}$ satisfying $V_{\rm s}(0) = 0$ and

$$e^{\varepsilon t}V_{\rm s}(x(t)) \le e^{\varepsilon t_0}V_{\rm s}(x(t_0)) + \int_{t_0}^{t} e^{\varepsilon s}r(u(s), y(s)){\rm d}s, \qquad t \ge t_0, \qquad (5.17)$$

for all $t_0, t \ge 0$, where $x(t)$, $t \ge t_0$, is the solution of (5.7) with $u(\cdot) \in \mathcal{U}$, is called an *exponential storage function* for $\mathcal{G}$.

Theorem 5.1. Consider the nonlinear dynamical system $\mathcal{G}$ given by (5.7) and (5.8), and assume that $\mathcal{G}$ is completely reachable. Then $\mathcal{G}$ is dissipative (respectively, exponentially dissipative) with respect to the supply rate $r(u, y)$ if and only if the available system storage $V_{\rm a}(x_0)$ given by (5.11) (respectively, the available exponential storage $V_{\rm a}(x)$ given by (5.12)) is finite for all $x_0 \in \mathcal{D}$ and $V_{\rm a}(0) = 0$. Moreover, if $V_{\rm a}(0) = 0$ and $V_{\rm a}(x_0)$ is finite for all $x_0 \in \mathcal{D}$, then $V_{\rm a}(x)$, $x \in \mathcal{D}$, is a storage function (respectively, exponential storage function) for $\mathcal{G}$. Finally, all storage functions (respectively, exponential storage functions) $V_{\rm s}(x)$, $x \in \mathcal{D}$, for $\mathcal{G}$ satisfy

$$0 \le V_{\rm a}(x) \le V_{\rm s}(x), \quad x \in \mathcal{D}. \qquad (5.18)$$

Proof. Suppose $V_{\rm a}(0) = 0$ and $V_{\rm a}(x_0)$, $x_0 \in \mathcal{D}$, is finite. Now, it follows from (5.11) (with $T = 0$) that $V_{\rm a}(x_0) \ge 0$, $x_0 \in \mathcal{D}$. Next, let $x(t)$, $t \ge t_0$, satisfy (5.7) with admissible input $u(t)$, $t \in [t_0, T]$. Since $-V_{\rm a}(x_0)$, $x_0 \in \mathcal{D}$, is given by the infimum over all admissible inputs $u(\cdot)$ in (5.11), it follows that for all admissible inputs $u(\cdot) \in \mathcal{U}$ and $T > t_0$,

$$\begin{aligned} -V_{\rm a}(x(t_0)) &\le \int_{t_0}^{T} r(u(t), y(t)){\rm d}t \\ &= \int_{t_0}^{t_{\rm f}} r(u(t), y(t)){\rm d}t + \int_{t_{\rm f}}^{T} r(u(t), y(t)){\rm d}t, \end{aligned}$$

which implies

$$-V_{\rm a}(x(t_0)) - \int_{t_0}^{t_{\rm f}} r(u(t), y(t)){\rm d}t \le \int_{t_{\rm f}}^{T} r(u(t), y(t)){\rm d}t.$$

Hence,

$$\begin{aligned} V_{\rm a}(x(t_0)) + \int_{t_0}^{t_{\rm f}} r(u(t), y(t)){\rm d}t &\ge -\inf_{u(\cdot),\, T\ge t_{\rm f}} \int_{t_{\rm f}}^{T} r(u(t), y(t)){\rm d}t \\ &= V_{\rm a}(x(t_{\rm f})) \end{aligned}$$

$$\geq 0, \tag{5.19}$$

which implies that

$$\int_{t_0}^{t_f} r(u(t), y(t))\mathrm{d}t \geq -V_{\mathrm{a}}(x(t_0)). \tag{5.20}$$

Hence, since by assumption $V_{\mathrm{a}}(0) = 0$, $\mathcal{G}$ is dissipative with respect to the supply rate $r(u, y)$. Furthermore, $V_{\mathrm{a}}(x)$, $x \in \mathcal{D}$, is a storage function for $\mathcal{G}$.

Conversely, suppose $\mathcal{G}$ is dissipative with respect to the supply rate $r(u, y)$. Since $\mathcal{G}$ is completely reachable it follows that for every $x_0 \in \mathcal{D}$ such that $x(t_0) = x_0$, there exist $\hat{t} \leq t < t_0$ and an admissible input $u(\cdot) \in \mathcal{U}$ defined on $[\hat{t}, t_0]$ such that $x(\hat{t}) = 0$ and $x(t_0) = x_0$. Now, since $\mathcal{G}$ is dissipative with respect to the supply rate and $x(\hat{t}) = 0$ it follows that

$$\int_{\hat{t}}^{T} r(u(t), y(t))\mathrm{d}t \geq 0, \qquad T > \hat{t},$$

or, equivalently,

$$\int_{t_0}^{T} r(u(t), y(t))\mathrm{d}t \geq -\int_{\hat{t}}^{t_0} r(u(t), y(t))\mathrm{d}t, \qquad T > t_0,$$

which implies that there exists a function $W : \mathcal{D} \to \mathbb{R}$ such that

$$\int_{t_0}^{T} r(u(t), y(t))\mathrm{d}t \geq W(x_0) > -\infty, \qquad T > t_0. \tag{5.21}$$

Now, it follows from (5.21) that for all $x \in \mathcal{D}$,

$$\begin{aligned} V_{\mathrm{a}}(x) &= -\inf_{u(\cdot),\, T \geq t_0} \int_{t_0}^{T} r(u(t), y(t))\mathrm{d}t \\ &\leq -W(x), \end{aligned} \tag{5.22}$$

and hence, the available storage $V_{\mathrm{a}}(x) < \infty$, $x \in \mathcal{D}$. Furthermore, with $x(t_0) = 0$, it follows that for all admissible $u(t)$, $t \geq t_0$,

$$\int_{t_0}^{T} r(u(t), y(t))\mathrm{d}t \geq 0, \qquad T \geq t_0, \tag{5.23}$$

which implies that

$$\sup_{u(\cdot),\, T \geq t_0} \left[-\int_{t_0}^{T} r(u(t), y(t))\mathrm{d}t \right] \leq 0, \tag{5.24}$$

or, equivalently, $V_{\mathrm{a}}(x(t_0)) = V_{\mathrm{a}}(0) \leq 0$. However, since $V_{\mathrm{a}}(x) \geq 0$, $x \in \mathcal{D}$, it follows that $V_{\mathrm{a}}(0) = 0$.

Moreover, if $V_a(x)$ is finite for all $x \in \mathcal{D}$, it follows from (5.19) that $V_a(x)$, $x \in \mathcal{D}$, is a storage function for $\mathcal{G}$. Next, if $V_s(x)$, $x \in \mathcal{D}$, is a storage function, then it follows that, for all $T > 0$ and $x_0 \in \mathcal{D}$,

$$\begin{aligned} V_s(x_0) &\geq V_s(x(T)) - \int_0^T r(u(t), y(t))\mathrm{d}t \\ &\geq -\int_0^T r(u(t), y(t))\mathrm{d}t, \end{aligned}$$

which implies

$$\begin{aligned} V_s(x_0) &\geq -\inf_{u(\cdot),\, T\geq 0} \int_0^T r(u(t), y(t))\mathrm{d}t \\ &= V_a(x_0), \end{aligned}$$

yielding (5.18).

Finally, the proof for the exponentially dissipative case follows an identical construction and, hence, is omitted. □

Theorem 5.1 presents necessary and sufficient conditions for the existence of an available storage function for a nonlinear dynamical system $\mathcal{G}$. The following result presents sufficient conditions for guaranteeing that the system storage function is a continuous function. First, however, the following definition is needed.

Definition 5.5. Consider the nonlinear dynamical system $\mathcal{G}$ given by (5.7) and (5.8) and let $\hat{x} \in \mathbb{R}^n$ and $\hat{u} \in \mathbb{R}^m$ be such that $x(t) \equiv \hat{x}$ and $u(t) \equiv \hat{u}$, $t \geq 0$, satisfy (5.7). $\mathcal{G}$ is *locally controllable at* $\hat{x}$ if, for every $T > 0$ and $\varepsilon > 0$, the set of points that can be reached from and to $\hat{x}$ in finite time T using admissible inputs $u : [0, T] \to U$, satisfying $\|u(t) - \hat{u}\| < \varepsilon$, contains a neighborhood of $\hat{x}$.

Consider the linearization of (5.7) at $x = \hat{x}$ and $u = \hat{u}$ given by

$$\dot{x}(t) = A(x(t) - \hat{x}) + B(u(t) - \hat{u}), \quad x(t_0) = x_0, \quad t \geq 0, \tag{5.25}$$

where $A = \frac{\partial F}{\partial x}\big|_{x=\hat{x},u=\hat{u}}$ and $B = \frac{\partial F}{\partial u}\big|_{x=\hat{x},u=\hat{u}}$. Now, it follows from Proposition 3.3 of [336] that if the pair (A, B) is controllable, then (5.7) is locally controllable.

Theorem 5.2. Consider the nonlinear dynamical system $\mathcal{G}$ given by (5.7) and (5.8), and assume that $\mathcal{G}$ is completely reachable. Furthermore, assume that for every $\hat{x} \in \mathcal{D}$, there exists $\hat{u} \in \mathbb{R}^m$ such that $x(t) \equiv \hat{x}$ and $u(t) \equiv \hat{u}$, $t \geq 0$, satisfy (5.7), and $\mathcal{G}$ is locally controllable at every $\hat{x} \in \mathcal{D}$. If $\mathcal{G}$ is dissipative (respectively, exponentially dissipative) with respect to the supply rate $r(u, y)$, then every storage function (respectively, exponential storage function) $V_{\rm s}(x)$, $x \in \mathcal{D}$, is continuous on $\mathcal{D}$.

Proof. Let $\hat{x} \in \mathcal{D}$ and $\hat{u} \in \mathbb{R}^m$ be such that $x(t) \equiv \hat{x}$ and $u(t) \equiv \hat{u}$, $t \geq 0$, satisfy (5.7). Now, let $\delta > 0$ and note that it follows from the continuity of $F(\cdot,\cdot)$ that there exist $T > 0$ and $\varepsilon > 0$ such that for every $u : [0,T) \to U$ and $\|u(t) - \hat{u}\| < \varepsilon$, $\|x(t) - \hat{x}\| < \delta$, $t \in [0,T)$, where $u(\cdot) \in \mathcal{U}$ and $x(t)$, $t \in [0,T)$, denotes the solution to (5.7) with the initial condition $\hat{x}$. Furthermore, it follows from the local controllability assumption that for every $\hat{T} \in (0,T]$, there exists a strictly increasing, continuous function $\gamma : \mathbb{R} \to \mathbb{R}$ such that $\gamma(0) = 0$, and for every $x_0 \in \mathcal{D}$ such that $\|x_0 - \hat{x}\| \leq \gamma(\hat{T})$, there exists $\hat{t} \in [0,\hat{T})$ and an input $u : [0,\hat{T}] \to \mathbb{R}^m$ such that $\|u(t) - \hat{u}\| < \varepsilon$, $t \in [0,\hat{t})$, and $x(\hat{t}) = x_0$. Hence, there exists $\beta > 0$ such that for every $x_0 \in \mathcal{D}$ such that $\|x_0 - \hat{x}\| \leq \beta$, there exists $\hat{t} \in [0, \gamma^{-1}(\|x_0 - \hat{x}\|)]$ and input $u : [0,\hat{t}] \to \mathbb{R}^m$ such that $\|u(t) - \hat{u}\| < \varepsilon$, $t \in [0,\hat{t}]$, and $x(\hat{t}) = x_0$.

Next, since $r(\cdot,\cdot)$ is locally integrable for all input-output pairs satisfying (5.7) and (5.8), it follows that there exists $M \in (0,\infty)$ such that

$$\sup_{\|x-\hat{x}\|<\delta, \|u-\hat{u}\|<\varepsilon} |r(u,y)| = M, \tag{5.26}$$

and hence, it follows that

$$\begin{aligned} \left| \int_0^{\hat{t}} r(u(s), y(s)) \mathrm{d}s \right| &\leq \int_0^{\hat{t}} |r(u(s), y(s))| \, \mathrm{d}s \\ &\leq M\hat{t} \\ &\leq M\gamma^{-1}(\|x_0 - \hat{x}\|). \end{aligned} \tag{5.27}$$

Now, if $V_{\rm s}(\cdot)$ is a storage function of $\mathcal{G}$, then

$$V_{\rm s}(x(\hat{t})) \leq V_{\rm s}(\hat{x}) + \int_0^{\hat{t}} r(u(s), y(s)) \mathrm{d}s, \tag{5.28}$$

or, equivalently,

$$-\int_0^{\hat{t}} r(u(s), y(s)) \mathrm{d}s \leq V_{\rm s}(\hat{x}) - V_{\rm s}(x(\hat{t})). \tag{5.29}$$

If $V_{\rm s}(\hat{x}) \leq V_{\rm s}(x(\hat{t}))$, then combining (5.27) and (5.29) yields

$$|V_{\rm s}(\hat{x}) - V_{\rm s}(x(\hat{t}))| \leq M\gamma^{-1}(\|x_0 - \hat{x}\|). \tag{5.30}$$

Alternatively, if $V_{\rm s}(x(\hat{t})) \geq V_{\rm s}(\hat{x})$, then (5.30) can be derived by reversing the roles of $\hat{x}$ and $x(\hat{t})$. Hence, since $\gamma(\cdot)$ is continuous and $x(\hat{t})$ is arbitrary, it follows that $V_{\rm s}(\cdot)$ is continuous on $\mathcal{D}$.

Finally, the proof for the exponentially dissipative case follows an identical construction and, hence, is omitted. □

The following corollary to Theorem 5.1 shows that the nonlinear dynamical system $\mathcal{G}$ is dissipative (respectively, exponentially dissipative) with respect to the supply rate $r(\cdot,\cdot)$ if and only if there exists a continuous storage function (respectively, exponential storage function) $V_{\rm s}(\cdot)$ satisfying (5.16) (respectively, (5.17)).

Corollary 5.1. Consider the nonlinear dynamical system $\mathcal{G}$ given by (5.7) and (5.8) and assume that $\mathcal{G}$ is completely reachable. Furthermore, assume that, for every $\hat{x} \in \mathcal{D}$, there exists $\hat{u} \in \mathbb{R}^m$ such that $x(t) \equiv \hat{x}$ and $u(t) \equiv \hat{u}$, $t \geq t_0$, satisfy (5.7), and $\mathcal{G}$ is locally controllable at every $\hat{x} \in \mathcal{D}$. Then $\mathcal{G}$ is dissipative (respectively, exponentially dissipative) with respect to the supply rate $r(u, y)$ if and only if there exists a continuous storage function (respectively, exponential storage function) $V_{\rm s}(x)$, $x \in \mathcal{D}$, satisfying (5.16) (respectively, (5.17)).

Proof. The result is immediate from Theorems 5.1 and 5.2 with $V_{\rm s}(x) = V_{\rm a}(x)$. □

In this book, we assume that $\mathcal{G}$ is locally controllable at every $\hat{x} \in \mathcal{D}$ and, for every $\hat{x} \in \mathcal{D}$, there exists $\hat{u} \in \mathbb{R}^m$ such that $x(t) \equiv \hat{x}$ and $u(t) \equiv \hat{u}$, $t \geq t_0$, satisfy (5.7), and hence, all storage functions of $\mathcal{G}$ are continuous on $\mathcal{D}$. The following theorem provides conditions for guaranteeing that all storage functions (respectively, exponential storage functions) of a given dissipative (respectively, exponentially dissipative) nonlinear dynamical system are positive definite. For this result we require the following definition.

Definition 5.6. A nonlinear dynamical system $\mathcal{G}$ is *zero-state observable* if $u(t) \equiv 0$ and $y(t) \equiv 0$ implies $x(t) \equiv 0$.

Theorem 5.3. Consider the nonlinear dynamical system $\mathcal{G}$ given by (5.7) and (5.8), and assume that $\mathcal{G}$ is completely reachable and zero-state observable. Furthermore, assume that $\mathcal{G}$ is dissipative (respectively, exponentially dissipative) with respect to the supply rate $r(u, y)$ and there exists a function $\kappa : Y \to U$ such that $\kappa(0) = 0$ and $r(\kappa(y), y) < 0$, $y \neq 0$. Then all the storage functions (respectively, exponential storage functions) $V_{\rm s}(x)$, $x \in \mathcal{D}$, for $\mathcal{G}$ are positive definite, that is, $V_{\rm s}(0) = 0$ and $V_{\rm s}(x) > 0$, $x \in \mathcal{D}$, $x \neq 0$.

Proof. It follows from Theorem 5.1 that the available storage $V_{\rm a}(x)$, $x \in \mathcal{D}$, is a storage function for $\mathcal{G}$. Next, suppose there exists $x \in \mathcal{D}$ such that $V_{\rm a}(x) = 0$, which implies that $r(u(t), y(t)) = 0$ almost everywhere $t \geq 0$, for all admissible inputs $u(\cdot) \in \mathcal{U}$. Since there exists a function $\kappa : Y \to U$ such that $\kappa(0) = 0$ and $r(\kappa(y), y) < 0$, $y \neq 0$, it follows that $y(t) = 0$

almost everywhere $t \geq 0$. Now, since $\mathcal{G}$ is zero-state observable it follows that $x = 0$, and hence, $V_{\rm a}(x) = 0$ if and only if $x = 0$. The result now follows from (5.18). Finally, the proof for the exponentially dissipative case is identical. □

If $V_{\rm s}(\cdot)$ is continuously differentiable in Corollary 5.1, then an equivalent statement for the dissipativeness of $\mathcal{G}$ with respect to the supply rate $r(u, y)$ is

$$\dot{V}_{\rm s}(x(t)) \leq r(u(t), y(t)), \qquad t \geq 0, \tag{5.31}$$

or, equivalently, $\dot{V}_{\rm s}(x) \leq r(u, y)$, where $\dot{V}_{\rm s}(x) = \frac{\rm d}{{\rm d}t}V(s(t, x, u))\big|_{t=0}$ denotes the total derivative of $V_{\rm s}(x)$ along the state trajectories $s(t, x, u)$ of (5.7) through $x \in \mathcal{D}$ with $u(\cdot) \in \mathcal{U}$ at $t = 0$. Alternatively, an equivalent statement for exponential dissipativeness of $\mathcal{G}$ with respect to the supply rate $r(u, y)$ is

$$\dot{V}_{\rm s}(x(t)) + \varepsilon V_{\rm s}(x(t)) \leq r(u(t), y(t)), \qquad t \geq 0. \tag{5.32}$$

Furthermore, a system $\mathcal{G}$ with storage function $V_{\rm s}(\cdot)$ is *strictly dissipative* with respect to the supply rate $r(u, y)$ if and only if

$$V_{\rm s}(x(t)) < V_{\rm s}(x(t_0)) + \int_{t_0}^{t} r(u(s), y(s)){\rm d}s, \qquad t > t_0. \tag{5.33}$$

Note that exponential dissipativity implies strict dissipativity; however, the converse is not necessarily true.

Next, we introduce the concept of a required supply of a nonlinear dynamical system. Specifically, define the *required supply* $V_{\rm r}(x_0)$ of the nonlinear dynamical system $\mathcal{G}$ by

$$V_{\rm r}(x_0) = \inf_{u(\cdot),\, T \geq 0} \int_{-T}^{0} r(u(t), y(t)){\rm d}t, \tag{5.34}$$

where $x(t)$, $t \geq -T$, is the solution to (5.7) with $x(-T) = 0$ and $x(0) = x_0$. The infimum in (5.34) is taken over all system trajectories starting from $x(-T) = 0$ and time $t = -T$ and ending at $x(0) = x_0$ at time $t = 0$, and all times $t \geq 0$ or, equivalently, over all admissible inputs $u(\cdot)$ which drive the dynamical system $\mathcal{G}$ from the origin to x_0 over the time interval $[-T, 0]$. If the system is not reachable from the origin, then we define $V_{\rm r}(x_0) = \infty$. It follows from (5.34) that the required supply of a nonlinear dynamical system is the minimum amount of generalized energy that has to be delivered to the dynamical system in order to transfer it from an initial state $x(-T) = 0$ to a given state $x(0) = x_0$. Similarly, define the *required exponential supply* of the nonlinear dynamical system $\mathcal{G}$ by

$$V_{\rm r}(x_0) = \inf_{u(\cdot),\, T \geq 0} \int_{-T}^{0} e^{\varepsilon t} r(u(t), y(t)){\rm d}t, \tag{5.35}$$

where $x(t)$, $t \geq -T$, is the solution to (5.7) with $x(-T) = 0$ and $x(0) = x_0$. Note that since, with $x(0) = 0$, the infimum in (5.34) is zero it follows that $V_{\rm r}(0) = 0$.

Next, using the notion of a required supply, we show that all storage functions are bounded from above by the required supply and bounded from below by the available storage, and hence, a dissipative dynamical system can deliver to its surroundings only a fraction of its generalized stored energy and can store only a fraction of the generalized work done to it.

Theorem 5.4. Consider the nonlinear dynamical system $\mathcal{G}$ given by (5.7) and (5.8), and assume that $\mathcal{G}$ is completely reachable. Then $\mathcal{G}$ is dissipative (respectively, exponentially dissipative) with respect to the supply rate $r(u, y)$ if and only if $0 \leq V_{\rm r}(x) < \infty$, $x \in \mathcal{D}$. Moreover, if $V_{\rm r}(x)$ is finite and nonnegative for all $x \in \mathcal{D}$, then $V_{\rm r}(x)$, $x \in \mathcal{D}$, is a storage function (respectively, exponential storage function) for $\mathcal{G}$. Finally, all storage functions (respectively, exponential storage functions) $V_{\rm s}(x)$, $x \in \mathcal{D}$, for $\mathcal{G}$ satisfy

$$0 \leq V_{\rm a}(x) \leq V_{\rm s}(x) \leq V_{\rm r}(x) < \infty, \qquad x \in \mathcal{D}. \tag{5.36}$$

Proof. Suppose $0 \leq V_{\rm r}(x) < \infty$, $x \in \mathcal{D}$. Next, let $x(t)$, $t \in \mathbb{R}$, satisfy (5.7) and (5.8) with admissible inputs $u(t)$, $t \in \mathbb{R}$, and $x(0) = x_0$. Since $V_{\rm r}(x)$, $x \in \mathcal{D}$, is given by the infimum over all admissible inputs $u(\cdot) \in \mathcal{U}$ and $T > 0$ in (5.34), it follows that for all admissible inputs $u(\cdot)$ and $T \leq t \leq 0$,

$$V_{\rm r}(x_0) \leq \int_{-T}^{0} r(u(t), y(t)){\rm d}t = \int_{-T}^{t} r(u(s), y(s)){\rm d}s + \int_{t}^{0} r(u(s), y(s)){\rm d}s,$$

and hence,

$$\begin{aligned} V_{\rm r}(x_0) &\leq \inf_{u(\cdot),\, T \geq 0} \left[\int_{-T}^{t} r(u(s), y(s)){\rm d}s \right] + \int_{t}^{0} r(u(s), y(s)){\rm d}s \\ &= V_{\rm r}(x(t)) + \int_{t}^{0} r(u(s), y(s)){\rm d}s, \end{aligned} \tag{5.37}$$

which shows that $V_{\rm r}(x)$, $x \in \mathcal{D}$, is a storage function for $\mathcal{G}$, and hence, Corollary 5.1 implies that $\mathcal{G}$ is dissipative.

Conversely, suppose $\mathcal{G}$ is dissipative with respect to the supply rate $r(u, y)$ and let $x_0 \in \mathcal{D}$. Since $\mathcal{G}$ is completely reachable it follows that there exist $T > 0$ and $u(t)$, $t \in [-T, 0]$, such that $x(-T) = 0$ and $x(0) = x_0$. Hence, since $\mathcal{G}$ is dissipative with respect to the supply rate $r(u, y)$ it follows that, for all $T \geq 0$,

$$0 \leq \int_{-T}^{0} r(u(t), y(t)){\rm d}t, \tag{5.38}$$

and hence,

$$0 \leq \inf_{u(\cdot),\, T \geq 0} \left[\int_{-T}^{0} r(u(t), y(t)) \mathrm{d}t \right], \tag{5.39}$$

which implies that

$$0 \leq V_{\mathrm{r}}(x_0) < \infty, \qquad x_0 \in \mathcal{D}. \tag{5.40}$$

Next, if $V_{\mathrm{s}}(\cdot)$ is a storage function for $\mathcal{G}$, then it follows from Theorem 5.1 that

$$0 \leq V_{\mathrm{a}}(x) \leq V_{\mathrm{s}}(x), \qquad x \in \mathcal{D}. \tag{5.41}$$

Furthermore, for all $T \geq 0$ such that $x(T) = 0$ it follows that

$$V_{\mathrm{s}}(x_0) \leq V_{\mathrm{s}}(0) + \int_{-T}^{0} r(u(t), y(t)) \mathrm{d}t, \tag{5.42}$$

and hence,

$$V_{\mathrm{s}}(x_0) \leq \inf_{u(\cdot),\, T \geq 0} \left[\int_{-T}^{0} r(u(t), y(t)) \mathrm{d}t \right] = V_{\mathrm{r}}(x_0) < \infty,$$

which implies (5.36).

Finally, the proof for the exponentially dissipative case follows a similar construction and, hence, is omitted. □

As a direct consequence of Theorems 5.1 and 5.4, we show that the set of all possible storage functions of a dynamical system forms a convex set parameterized by the system available storage and the system required supply. An identical result holds for exponential storage functions.

Proposition 5.1. Consider the nonlinear dynamical system $\mathcal{G}$ given by (5.7) and (5.8) with available storage $V_{\mathrm{a}}(x)$, $x \in \mathcal{D}$, and required supply $V_{\mathrm{r}}(x)$, $x \in \mathcal{D}$, and assume $\mathcal{G}$ is completely reachable. Then for every $\alpha \in [0, 1]$,

$$V_{\mathrm{s}}(x) = \alpha V_{\mathrm{a}}(x) + (1 - \alpha) V_{\mathrm{r}}(x), \qquad x \in \mathcal{D}, \tag{5.43}$$

is a storage function for $\mathcal{G}$.

Proof. The result is a direct consequence of the dissipation inequality (5.16) by noting that if $V_{\mathrm{a}}(x)$ and $V_{\mathrm{r}}(x)$ satisfy (5.16), then $V_{\mathrm{s}}(x)$ satisfies (5.16). □

In light of Theorems 5.1 and 5.4 we have the following result on lossless dynamical systems.

Theorem 5.5. Consider the nonlinear dynamical system $\mathcal{G}$ given by (5.7) and (5.8), and assume $\mathcal{G}$ is completely reachable to and from the origin.

Then $\mathcal{G}$ is lossless with respect to the supply rate $r(u, y)$ if and only if there exists a continuous storage function $V_s(x)$, $x \in \mathcal{D}$, satisfying (5.16) as an equality. Furthermore, if $\mathcal{G}$ is lossless with respect to the supply rate $r(u, y)$, then $V_a(x) = V_r(x)$, and hence, the storage function $V_s(x)$, $x \in \mathcal{D}$, is unique and is given by

$$V_s(x_0) = -\int_0^{T_+} r(u(t), y(t))dt = \int_{-T_-}^{0} r(u(t), y(t))dt, \tag{5.44}$$

where $x(t)$, $t \geq 0$, is the solution to (5.7) with admissible $u(\cdot) \in \mathcal{U}$ and $x(0) = x_0$, $x_0 \in \mathcal{D}$, for every $T_-, T_+ > 0$ such that $x(-T_-) = 0$ and $x(T_+) = 0$.

Proof. Suppose $\mathcal{G}$ is lossless with respect to the supply rate $r(u, y)$. Since $\mathcal{G}$ is completely reachable to and from the origin it follows that, for every $x_0 \in \mathcal{D}$, there exist $T_-, T_+ > 0$, and $u(t) \in U$, $t \in [-T_-, T_+]$, such that $x(-T_-) = 0$, $x(T_+) = 0$, and $x(0) = x_0$. Now, it follows that

$$\begin{aligned}
0 &= \int_{-T_-}^{T_+} r(u(t), y(t))dt \\
&= \int_{-T_-}^{0} r(u(t), y(t))dt + \int_0^{T_+} r(u(t), y(t))dt \\
&\geq \inf_{u(\cdot),\, T\geq 0} \int_{-T}^{0} r(u(t), y(t))dt + \inf_{u(\cdot),\, T\geq 0} \int_0^{T} r(u(t), y(t))dt \\
&= V_r(x_0) - V_a(x_0),
\end{aligned} \tag{5.45}$$

which implies that $V_r(x_0) \leq V_a(x_0)$, $x_0 \in \mathcal{D}$. However, since by definition $\mathcal{G}$ is dissipative with respect to the supply rate $r(u, y)$ it follows from Theorem 5.4 that $V_a(x_0) \leq V_r(x_0)$, $x_0 \in \mathcal{D}$, and hence, every storage function $V_s(x_0)$, $x_0 \in \mathcal{D}$, satisfies $V_a(x_0) = V_s(x_0) = V_r(x_0)$. Furthermore, it follows that the inequality in (5.45) is indeed an equality, which implies (5.44).

Next, let $t_0, t, T \geq 0$ be such that $t_0 < t < T$, $x(T) = 0$. Hence, it follows from (5.44) that

$$\begin{aligned}
0 &= V_s(x(t_0)) + \int_{t_0}^{T} r(u(s), y(s))ds \\
&= V_s(x(t_0)) + \int_{t_0}^{t} r(u(s), y(s))ds + \int_t^T r(u(s), y(s))ds \\
&= V_s(x(t_0)) + \int_{t_0}^{t} r(u(s), y(s))ds - V_s(x(t)),
\end{aligned}$$

which implies that (5.16) is satisfied as an equality.

Conversely, if there exists a storage function $V_s(x)$, $x \in \mathcal{D}$, satisfying

(5.16) as an equality, it follows from Corollary 5.1 that $\mathcal{G}$ is dissipative with respect to the supply rate $r(u, y)$. Furthermore, for every $u(\cdot) \in \mathcal{U}$, $t \geq 0$, and $x(t_0) = x(t) = 0$, it follows from (5.16) (with an equality) that

$$\int_{t_0}^{t} r(u(s), y(s))\mathrm{d}s = 0,$$

which implies that $\mathcal{G}$ is lossless with respect to the supply rate $r(u, y)$. □

Example 5.1. Consider the integrator scalar system

$$\dot{x}(t) = u(t), \qquad x(0) = x_0, \qquad t \geq 0, \tag{5.46}$$

$$y(t) = x(t). \tag{5.47}$$

To show that an integrator is the simplest storage element note that with $V_{\mathrm{s}}(x) = \frac{1}{2}x^2$ and $r(u, y) = uy$ it follows that $\dot{V}_{\mathrm{s}}(x(t)) = u(t)y(t)$, $t \geq 0$, and hence, $\int_0^t u(s)y(s)\mathrm{d}s = 0$ for all $t \geq 0$ with $x(0) = x(t) = 0$. Hence, (5.46) and (5.47) is lossless with respect to the supply rate $r(u, y) = uy$. Furthermore, the available storage for (5.46) and (5.47) is given by

$$V_{\mathrm{a}}(x_0) = \sup_{u(\cdot),\, T \geq 0} \left[-\int_0^T u(t)y(t)\mathrm{d}t \right] \geq \int_0^\infty y^2(t)\mathrm{d}t = x_0^2 \int_0^\infty e^{-2t}\mathrm{d}t = \tfrac{1}{2}x_0^2, \tag{5.48}$$

where the above inequality follows by choosing $u = -y$ and $T = \infty$. Now, since $V_{\mathrm{s}}(x_0) \geq V_{\mathrm{a}}(x_0)$, it follows that $V_{\mathrm{a}}(x_0) = \frac{1}{2}x_0^2$. △

5.3 Lagrangian and Hamiltonian Dynamical Systems

In this section, we show that two of the fundamental dynamical system formulations of analytical mechanics, namely, *Lagrangian* and *Hamiltonian* dynamical systems, can be formulated as special cases of dissipative dynamical system theory. In particular, we show that conservation of energy, internal system interaction, and interaction with the environment through input-output ports, inherent in Lagrangian and Hamiltonian dynamical system formulations, can be captured as a special case of dissipative dynamical system theory with appropriate storage functions and supply rates corresponding to physical system energy and supplied system power, respectively.

To begin, consider the governing equations of motion of an n degree-of-freedom dynamical system given by the *Euler-Lagrange equation*

$$\frac{\mathrm{d}}{\mathrm{d}t}\left[\frac{\partial \mathcal{L}}{\partial \dot{q}}(q, \dot{q})\right]^{\mathrm{T}} - \left[\frac{\partial \mathcal{L}}{\partial q}(q, \dot{q})\right]^{\mathrm{T}} = u, \tag{5.49}$$

where $q \in \mathbb{R}^n$ represents the generalized system positions, $\dot{q} \in \mathbb{R}^n$ represents the generalized system velocities, $\mathcal{L} : \mathbb{R}^n \times \mathbb{R}^n \to \mathbb{R}$ denotes the system

Lagrangian given by $\mathcal{L}(q,\dot{q}) = T(q,\dot{q}) - V(q)$, where $T : \mathbb{R}^n \times \mathbb{R}^n \to \mathbb{R}$ is the system kinetic energy and $V : \mathbb{R}^n \to \mathbb{R}$ is the system potential energy, and $u \in \mathbb{R}^n$ is the vector of generalized forces acting on the system. Furthermore, let $\mathcal{H} : \mathbb{R}^n \times \mathbb{R}^n \to \mathbb{R}$ denote the *Legendre transformation* of the Lagrangian function $\mathcal{L}(q,\dot{q})$ with respect to the generalized velocity $\dot{q}$ defined by

$$\mathcal{H}(q,p) \triangleq \dot{q}^{\mathrm{T}} p - \mathcal{L}(q,\dot{q})\big|_{\dot{q}=\Phi^{-1}(p)}. \tag{5.50}$$

Here p denotes the vector of *generalized momenta* given by

$$p(q,\dot{q}) = \left[\frac{\partial \mathcal{L}}{\partial \dot{q}}(q,\dot{q})\right]^{\mathrm{T}} \tag{5.51}$$

and $\Phi : \mathbb{R}^n \to \mathbb{R}^n$ is a bijective map from the generalized velocities $\dot{q}$ to the generalized momenta p. Now, if $\mathcal{H}(q,p)$ is lower bounded, then we can always shift $\mathcal{H}(q,p)$ so that, with minor abuse of notation, $\mathcal{H}(q,p) \geq 0$, $(q,p) \in \mathbb{R}^n \times \mathbb{R}^n$, and $\mathcal{H}(0,0) = 0$.

In this case, using (5.49) and the fact that

$$\frac{\mathrm{d}}{\mathrm{d}t}[\mathcal{L}(q,\dot{q})] = \frac{\partial \mathcal{L}}{\partial q}(q,\dot{q})\dot{q} + \frac{\partial \mathcal{L}}{\partial \dot{q}}(q,\dot{q})\ddot{q}, \tag{5.52}$$

it follows that

$$\begin{aligned}
u^{\mathrm{T}}\dot{q} &= \left\{\frac{\mathrm{d}}{\mathrm{d}t}\left[\frac{\partial \mathcal{L}}{\partial \dot{q}}(q,\dot{q})\right] - \frac{\partial \mathcal{L}}{\partial q}(q,\dot{q})\right\}\dot{q} \\
&= \frac{\mathrm{d}}{\mathrm{d}t}[p(q,\dot{q})]^{\mathrm{T}}\dot{q} - \frac{\partial \mathcal{L}}{\partial q}(q,\dot{q})\dot{q} \\
&= \frac{\mathrm{d}}{\mathrm{d}t}[p^{\mathrm{T}}(q,\dot{q})\dot{q}] - p^{\mathrm{T}}(q,\dot{q})\ddot{q} + \frac{\partial \mathcal{L}}{\partial \dot{q}}(q,\dot{q})\ddot{q} - \frac{\mathrm{d}}{\mathrm{d}t}\mathcal{L}(q,\dot{q}) \\
&= \frac{\mathrm{d}}{\mathrm{d}t}[p^{\mathrm{T}}(q,\dot{q})\dot{q} - \mathcal{L}(q,\dot{q})] \\
&= \frac{\mathrm{d}}{\mathrm{d}t}\mathcal{H}(q,p).
\end{aligned} \tag{5.53}$$

Hence, with $V_{\mathrm{s}}(q,\dot{q}) = \mathcal{H}(q,p(q,\dot{q}))$ and $y = \dot{q}$, it follows from Theorem 5.5 that the Euler-Lagrange system (5.49) is lossless with respect to the supply rate $r(u,y) = u^{\mathrm{T}}y$. This gives a power balance equation that states that the increase in system energy $\mathcal{H}(q,p)$ is equal to the supplied work due to the generalized force u.

Alternatively, if the n degree-of-freedom dynamical system possesses internal dissipation, then the Euler-Lagrange equation takes the form

$$\frac{\mathrm{d}}{\mathrm{d}t}\left[\frac{\partial \mathcal{L}}{\partial \dot{q}}(q,\dot{q})\right]^{\mathrm{T}} - \left[\frac{\partial \mathcal{L}}{\partial q}(q,\dot{q})\right]^{\mathrm{T}} + \left[\frac{\partial \mathcal{R}}{\partial \dot{q}}(\dot{q})\right]^{\mathrm{T}} = u, \tag{5.54}$$

where $\mathcal{R} : \mathbb{R}^n \to \mathbb{R}$ represents the *Rayleigh dissipation function* satisfying

$\frac{\partial \mathcal{R}}{\partial \dot{q}}(\dot{q})\dot{q} \geq 0$, $\dot{q} \in \mathbb{R}^n$. In this case, using (5.52) and (5.54), it follows that

$$\begin{aligned} u^{\mathrm{T}}\dot{q} &= \left\{\frac{\mathrm{d}}{\mathrm{d}t}\left[\frac{\partial \mathcal{L}}{\partial \dot{q}}(q,\dot{q})\right] - \frac{\partial \mathcal{L}}{\partial q}(q,\dot{q}) + \frac{\partial \mathcal{R}}{\partial \dot{q}}(\dot{q})\right\}\dot{q} \\ &= \frac{\mathrm{d}}{\mathrm{d}t}[p(q,\dot{q})]^{\mathrm{T}}\dot{q} - \frac{\partial \mathcal{L}}{\partial q}(q,\dot{q})\dot{q} + \frac{\partial \mathcal{R}}{\partial \dot{q}}(\dot{q})\dot{q} \\ &= \frac{\mathrm{d}}{\mathrm{d}t}[p^{\mathrm{T}}(q,\dot{q})\dot{q}] - p^{\mathrm{T}}(q,\dot{q})\ddot{q} + \frac{\partial \mathcal{L}}{\partial \dot{q}}(q,\dot{q})\ddot{q} - \frac{\mathrm{d}}{\mathrm{d}t}\mathcal{L}(q,\dot{q}) + \frac{\partial \mathcal{R}}{\partial \dot{q}}(\dot{q})\dot{q} \\ &= \frac{\mathrm{d}}{\mathrm{d}t}[p^{\mathrm{T}}(q,\dot{q})\dot{q} - \mathcal{L}(q,\dot{q})] + \frac{\partial \mathcal{R}}{\partial \dot{q}}(\dot{q})\dot{q} \\ &= \frac{\mathrm{d}}{\mathrm{d}t}\mathcal{H}(q,p) + \frac{\partial \mathcal{R}}{\partial \dot{q}}(\dot{q})\dot{q}, \end{aligned} \tag{5.55}$$

or, equivalently, integrating over the interval $[0,T]$,

$$\mathcal{H}(q(T),p(T)) = \mathcal{H}(q(0),p(0)) + \int_0^T u^{\mathrm{T}}(s)\dot{q}(s)\mathrm{d}s - \int_0^T \frac{\partial \mathcal{R}}{\partial \dot{q}}(\dot{q}(s))\dot{q}(s)\mathrm{d}s. \tag{5.56}$$

Equation (5.56) shows that the system energy at time $t = T$ is equal to the initial energy stored in the system plus the energy supplied to the system via the external force u minus the internal energy dissipated. Since $\frac{\partial \mathcal{R}}{\partial \dot{q}}(\dot{q})\dot{q} \geq 0$, $\dot{q} \in \mathbb{R}^n$, it follows from (5.56) and Corollary 5.1 that the Euler-Lagrange system (5.54) is dissipative with respect to the supply rate $r(u,y) = u^{\mathrm{T}}y$, where $y = \dot{q}$, and with storage function $V_{\mathrm{s}}(q,\dot{q}) = \mathcal{H}(q,p(q,\dot{q}))$.

Next, we compute the available storage and the required supply for the Euler-Lagrange dynamical system (5.54). Specifically, using (5.11) and the fact that the map between the generalized velocities to the generalized momenta is bijective we obtain

$$\begin{aligned} &V_{\mathrm{a}}(q_0,\dot{q}_0) \\ &\quad= \sup_{u(\cdot),\,T\geq 0}\left[-\int_0^T u^{\mathrm{T}}(t)\dot{q}(t)\mathrm{d}t\right] \\ &\quad= \sup_{u(\cdot),\,T\geq 0}\left\{-\int_0^T \left[\frac{\mathrm{d}}{\mathrm{d}t}\mathcal{H}(q(t),p(t)) + \frac{\partial \mathcal{R}}{\partial \dot{q}}(\dot{q}(t))\dot{q}(t)\right]\mathrm{d}t\right\} \\ &\quad= \sup_{u(\cdot),\,T\geq 0}\left[-\mathcal{H}(q(T),p(T)) + \mathcal{H}(q(0),p(0)) - \int_0^T \frac{\partial \mathcal{R}}{\partial \dot{q}}(\dot{q}(t))\dot{q}(t)\mathrm{d}t\right] \\ &\quad\leq \mathcal{H}(q(0),p(0)). \end{aligned} \tag{5.57}$$

Note that in the case where $\frac{\partial \mathcal{R}}{\partial \dot{q}}(\dot{q}) \equiv 0$ it follows that $V_{\mathrm{a}}(q_0,\dot{q}_0) = \mathcal{H}(q(0), p(0))$, which shows that the available storage is simply the total initial system energy.

Alternatively, using (5.34) we obtain

$$
\begin{aligned}
V_{\mathrm{r}}(q_0, \dot{q}_0) \\
&= \inf_{u(\cdot),\, T\geq 0} \int_{-T}^{0} u^{\mathrm{T}}(t)\dot{q}(t)\mathrm{d}t \\
&= \inf_{u(\cdot),\, T\geq 0} \left\{ \int_{-T}^{0} \left[\frac{\mathrm{d}}{\mathrm{d}t}\mathcal{H}(q(t),p(t)) + \frac{\partial \mathcal{R}}{\partial \dot{q}}(\dot{q}(t))\dot{q}(t) \right] \mathrm{d}t \right\} \\
&= \inf_{u(\cdot),\, T\geq 0} \left[\mathcal{H}(q(0),p(0)) - \mathcal{H}(q(-T),p(-T)) + \int_{-T}^{0} \frac{\partial \mathcal{R}}{\partial \dot{q}}(\dot{q}(t))\dot{q}(t)\mathrm{d}t \right] \\
&\geq \mathcal{H}(q(0),p(0)),
\end{aligned} \tag{5.58}
$$

where $(q(-T),p(-T)) = (0,0)$. Hence, $V_{\mathrm{a}}(q_0,\dot{q}_0) \leq \mathcal{H}(q(0),p(0)) \leq V_{\mathrm{r}}(q_0, \dot{q}_0)$. Note that in the case where $\frac{\partial \mathcal{R}}{\partial \dot{q}}(\dot{q}) \equiv 0$, it follows from (5.58) that

$$
V_{\mathrm{r}}(q_0,\dot{q}_0) = \mathcal{H}(q(0),p(0)) - \mathcal{H}(q(-T),p(-T)) = \mathcal{H}(q(0),p(0)), \tag{5.59}
$$

and hence, $V_{\mathrm{r}}(q_0,\dot{q}_0) = V_{\mathrm{a}}(q_0,\dot{q}_0) = \mathcal{H}(q(0),p(0))$.

Next, we transform the Euler-Lagrange equations to the *Hamiltonian equations* of motion. These equations provide a fundamental structure of the mathematical description of numerous physical dynamical systems by capturing energy conservation as well as internal interconnection structural properties of physical dynamical systems. To reduce the Euler-Lagrange equations (5.54) to a Hamiltonian system of equations consider the Legendre transformation $\mathcal{H}(q,p)$ given by (5.50) of the Lagrangian function $\mathcal{L}(q,\dot{q})$, called the *Hamiltonian function*, and the vector of generalized momenta given by (5.51). Now, it follows from (5.50), (5.51), and (5.54) that

$$
\begin{aligned}
\frac{\mathrm{d}}{\mathrm{d}t}p(q,\dot{q}) &= \frac{\mathrm{d}}{\mathrm{d}t}\left[\frac{\partial \mathcal{L}}{\partial \dot{q}}(q,\dot{q})\right]^{\mathrm{T}} \\
&= \left[\frac{\partial \mathcal{L}}{\partial q}(q,\dot{q})\right]^{\mathrm{T}} - \left[\frac{\partial \mathcal{R}}{\partial \dot{q}}(\dot{q})\right]^{\mathrm{T}} + u \\
&= -\left[\frac{\partial \mathcal{H}}{\partial q}(q,p)\right]^{\mathrm{T}} - \left[\frac{\partial \mathcal{R}}{\partial \dot{q}}(\dot{q})\right]^{\mathrm{T}} + u.
\end{aligned} \tag{5.60}
$$

Assuming that the Rayleigh dissipation function $\mathcal{R} : \mathbb{R}^n \to \mathbb{R}$ is quadratic, that is, $\mathcal{R}(\dot{q}) = \frac{1}{2}\dot{q}^{\mathrm{T}}D\dot{q}$, where D is a nonnegative-definite system dissipation matrix, (5.60) implies

$$
\dot{p} = -\left(\frac{\partial \mathcal{H}}{\partial q}(q,p)\right)^{\mathrm{T}} - D\dot{q} + u. \tag{5.61}
$$

Next, it follows from (5.50) that

$$\dot{q} = \left(\frac{\partial \mathcal{H}}{\partial p}(q,p)\right)^{\mathrm{T}}. \tag{5.62}$$

Hence, (5.61) and (5.62) can be equivalently written as

$$\dot{x}(t) = [\mathcal{J} - \mathcal{R}]\left(\frac{\partial \mathcal{H}}{\partial x}(x(t))\right)^{\mathrm{T}} + Gu(t), \quad x(0) = x_0, \quad t \geq 0, \tag{5.63}$$

where $x = [q^{\mathrm{T}}, p^{\mathrm{T}}]^{\mathrm{T}}$ and

$$\mathcal{J} \triangleq \begin{bmatrix} 0_n & I_n \\ -I_n & 0_n \end{bmatrix}, \quad \mathcal{R} \triangleq \begin{bmatrix} 0 & 0 \\ 0 & D \end{bmatrix}, \quad G \triangleq \begin{bmatrix} 0 \\ I_n \end{bmatrix}. \tag{5.64}$$

The dynamical system (5.63) with outputs $y(t) = G^{\mathrm{T}}(x(t))\left(\frac{\partial \mathcal{H}}{\partial x}(x(t))\right)^{\mathrm{T}}$ is called a *Hamiltonian dynamical system.*

The Hamiltonian dynamical system (5.63) can be further generalized to what is commonly referred to as *port-controlled Hamiltonian dynamical systems* described in local coordinates $x \in \mathcal{D} \subseteq \mathbb{R}^n$. In particular, port-controlled Hamiltonian dynamical systems are given by

$$\dot{x}(t) = [\mathcal{J}(x(t)) - \mathcal{R}(x(t))]\left(\frac{\partial \mathcal{H}}{\partial x}(x(t))\right)^{\mathrm{T}} + G(x(t))u(t),$$

$$x(0) = x_0, \quad t \geq 0, \tag{5.65}$$

$$y(t) = G^{\mathrm{T}}(x(t))\left(\frac{\partial \mathcal{H}}{\partial x}(x(t))\right)^{\mathrm{T}}, \tag{5.66}$$

where $x(t) \in \mathcal{D} \subseteq \mathbb{R}^n$, $\mathcal{D}$ is an open set with $0 \in \mathcal{D}$, $u(t) \in U \subseteq \mathbb{R}^m$, $y(t) \in Y \subseteq \mathbb{R}^l$, $\mathcal{H} : \mathcal{D} \to \mathbb{R}$ is a continuously differentiable *Hamiltonian function* for the system (5.65) and (5.66), $\mathcal{J} : \mathcal{D} \to \mathbb{R}^{n\times n}$ is such that $\mathcal{J}(x) = -\mathcal{J}^{\mathrm{T}}(x)$, $\mathcal{R} : \mathcal{D} \to \mathbb{S}^n$ is such that $\mathcal{R}(x) \geq 0$, $x \in \mathcal{D}$, $[\mathcal{J}(x) - \mathcal{R}(x)]\left(\frac{\partial \mathcal{H}}{\partial x}(x)\right)^{\mathrm{T}}$, $x \in \mathcal{D}$, is Lipschitz continuous on $\mathcal{D}$, and $G : \mathcal{D} \to \mathbb{R}^{n\times m}$. The skew-symmetric matrix function $\mathcal{J}(x)$, $x \in \mathcal{D}$, captures the internal system interconnection structure, the input matrix function $G(x)$, $x \in \mathcal{D}$, captures interconnections with the environment, and the symmetric nonnegative-definite matrix function $\mathcal{R}(x)$, $x \in \mathcal{D}$, captures system dissipation. Note, that the inputs and outputs are dual (conjugated) variables. Here, we assume that $u(\cdot)$ is restricted to the class of *admissible* inputs consisting of measurable functions such that $u(t) \in U$ for all $t \geq 0$.

Assuming that the Hamiltonian energy function $\mathcal{H}(\cdot)$ is lower bounded, it can be shown that port-controlled Hamiltonian systems provide an energy balance in terms of the stored or accumulated energy, supplied system energy, and dissipated energy. Specifically, computing the rate of change of the Hamiltonian along the system state trajectories $x(t)$, $t \geq 0$, yields the energy conservation equation

$$\dot{\mathcal{H}}(x(t)) = u^{\mathrm{T}}(t)y(t) - \frac{\partial \mathcal{H}}{\partial x}(x(t))\mathcal{R}(x(t))\left(\frac{\partial \mathcal{H}}{\partial x}(x(t))\right)^{\mathrm{T}}, \quad t \geq 0, \quad (5.67)$$

which shows that the rate of change in energy, or power, is equal to the system power input minus the internal system power dissipated. Note that (5.67) can be equivalently written as

$$\begin{aligned}\mathcal{H}(x(t)) - \mathcal{H}(x(0)) &= \int_0^t u^{\mathrm{T}}(s)y(s)\mathrm{d}s \\ &\quad - \int_0^t \frac{\partial \mathcal{H}}{\partial x}(x(s))\mathcal{R}(x(s))\left(\frac{\partial \mathcal{H}}{\partial x}(x(s))\right)^{\mathrm{T}}\mathrm{d}s. \quad (5.68)\end{aligned}$$

Equation (5.68) shows that the stored or accumulated system energy is equal to the energy supplied to the system via the external input u minus the energy dissipated over the time interval $[0, t]$. Since $\mathcal{R}(x)$ is nonnegative definite for all $x \in \mathcal{D}$, it follows from (5.68) that

$$-\int_0^t u^{\mathrm{T}}(s)y(s)\mathrm{d}s \leq \mathcal{H}(x(0)), \quad (5.69)$$

which shows that the energy that can be extracted from the port-controlled Hamiltonian system through the input-output ports is less than or equal to the initial energy stored in the system. Hence, port-controlled Hamiltonian systems are dissipative with respect to the (power) supply rate $r(u, y) = u^{\mathrm{T}}y$. Furthermore, note that in the case where no system dissipation is present, that is, $\mathcal{R}(x) \equiv 0$, then port-controlled dynamical systems are lossless with respect to the supply rate $r(u, y) = u^{\mathrm{T}}y$.

Finally, we exploit the skew-symmetric structure of the internal interconnection matrix function $\mathcal{J}(x)$, $x \in \mathcal{D}$, of (5.65) to establish the existence of Casimir functions for port-controlled Hamiltonian systems. Specifically, let $C : \mathcal{D} \to \mathbb{R}$ be such that

$$\frac{\partial C}{\partial x}(x)[\mathcal{J}(x) - \mathcal{R}(x)] = 0, \quad x \in \mathcal{D}. \quad (5.70)$$

In this case, it follows that

$$\begin{aligned}\dot{C}(x) &= \frac{\partial C}{\partial x}(x)[\mathcal{J}(x) - \mathcal{R}(x)]\left(\frac{\partial \mathcal{H}}{\partial x}(x)\right) + \frac{\partial C}{\partial x}(x)G(x)u \\ &= \frac{\partial C}{\partial x}(x)G(x)u. \quad (5.71)\end{aligned}$$

Now, if $u(t) \equiv 0$ or $\frac{\partial C}{\partial x}(x)G(x) = 0$, $x \in \mathcal{D}$, then $C : \mathcal{D} \to \mathbb{R}$ is conserved along the flow of (5.65) irrespective of the form of the system Hamiltonian.

Note that (5.70) is also implied by the stronger conditions

$$\frac{\partial C}{\partial x}(x)\mathcal{J}(x) = 0, \quad x \in \mathcal{D}, \tag{5.72}$$

$$\frac{\partial C}{\partial x}(x)\mathcal{R}(x) = 0, \quad x \in \mathcal{D}. \tag{5.73}$$

In addition, since $\mathcal{J}(x) = -\mathcal{J}^{\mathrm{T}}(x)$ and $\mathcal{R}(x) = \mathcal{R}^{\mathrm{T}}(x)$, (5.72) and (5.73) imply

$$[\mathcal{J}(x) - \mathcal{R}(x)]\left(\frac{\partial C}{\partial x}(x)\right)^{\mathrm{T}} = 0, \quad x \in \mathcal{D}. \tag{5.74}$$

Now, using the fact that the sum of a skew-symmetric matrix and a symmetric matrix is zero if and only if the individual matrices are zero, it follows that (5.72) and (5.73) hold if and only if (5.70) and (5.74) hold. Next, assuming $u(t) \equiv 0$, it follows from (5.71) that the α-level set of $C(x)$ given by $C^{-1}(\alpha) = \{x \in \mathcal{D} : C(x) = \alpha\}$, where $\alpha \in \mathbb{R}$, is invariant with respect to the port-controlled Hamiltonian system (5.65). Hence, if the system Hamiltonian $\mathcal{H}(\cdot)$ is not positive definite at an equilibrium point $x_{\mathrm{e}} \in \mathcal{D}$, then, constructing a *shaped Hamiltonian* $\mathcal{H}_{\mathrm{s}}(x) = \mathcal{H}(x) + \mathcal{H}_{\mathrm{c}}(C(x))$ such that $\mathcal{H}_{\mathrm{s}}(x)$ is positive definite at x_{e} by properly choosing $\mathcal{H}_{\mathrm{c}}$, it follows that $\mathcal{H}_{\mathrm{s}}(x)$ serves as a Lyapunov function candidate for (5.65) with $u(t) \equiv 0$. Theorem 3.8 can be used to construct such a shaped Hamiltonian. Now, (5.62) and (5.74) imply that $\dot{\mathcal{H}}_{\mathrm{s}}(x) \leq 0$, $x \in \mathcal{D}$, establishing Lyapunov stability of (5.65). More generally, in an identical fashion as above one can construct r independent two-times continuously differentiable Casimir functions and use Theorem 3.8 to construct shaped Hamiltonians as Lyapunov functions for (5.65) with $u(t) \equiv 0$.

5.4 Extended Kalman-Yakubovich-Popov Conditions for Nonlinear Dynamical Systems

In this section, we show that dissipativeness, exponential dissipativeness, and losslessness of nonlinear affine dynamical systems $\mathcal{G}$ of the form

$$\dot{x}(t) = f(x(t)) + G(x(t))u(t), \quad x(t_0) = x_0, \quad t \geq t_0, \tag{5.75}$$

$$y(t) = h(x(t)) + J(x(t))u(t), \tag{5.76}$$

where $x(t) \in \mathcal{D} \subseteq \mathbb{R}^n$, $\mathcal{D}$ is an open set with $0 \in \mathcal{D}$, $u(t) \in U \subseteq \mathbb{R}^m$, $y(t) \in Y \subseteq \mathbb{R}^l$, $f : \mathcal{D} \to \mathbb{R}^n$, $G : \mathcal{D} \to \mathbb{R}^{n \times m}$, $h : \mathcal{D} \to Y$, and $J : \mathcal{D} \to \mathbb{R}^{l \times m}$, can be characterized in terms of the system functions $f(\cdot)$, $G(\cdot)$, $h(\cdot)$, and $J(\cdot)$. We assume that $f(\cdot)$, $G(\cdot)$, $h(\cdot)$, and $J(\cdot)$ are continuously differentiable mappings and $f(\cdot)$ has at least one equilibrium so that, without loss of generality, $f(0) = 0$ and $h(0) = 0$. Furthermore, for the nonlinear dynamical system $\mathcal{G}$ we assume that the required properties for the existence and uniqueness of solutions in forward and backward time are satisfied. For the following result we consider the special case of dissipative systems with quadratic supply rates [457]. Specifically, set $\mathcal{D} = \mathbb{R}^n$, $U = \mathbb{R}^m$, $Y = \mathbb{R}^l$, let $Q \in \mathbb{S}^l$, $R \in \mathbb{S}^m$, and $S \in \mathbb{R}^{l\times m}$ be given, and assume $r(u,y) = y^{\mathrm{T}}Qy + 2y^{\mathrm{T}}Su + u^{\mathrm{T}}Ru$. Furthermore, we assume that there exists a function $\kappa : \mathbb{R}^l \to \mathbb{R}^m$ such that $\kappa(0) = 0$ and $r(\kappa(y), y) < 0$, $y \neq 0$, and the available storage $V_{\mathrm{a}}(x)$, $x \in \mathbb{R}^n$, for $\mathcal{G}$ is a continuously differentiable function.

Theorem 5.6. Let $Q \in \mathbb{S}^l$, $S \in \mathbb{R}^{l\times m}$, $R \in \mathbb{S}^m$, and let $\mathcal{G}$ be zero-state observable and completely reachable. $\mathcal{G}$ is dissipative with respect to the quadratic supply rate $r(u,y) = y^{\mathrm{T}}Qy + 2y^{\mathrm{T}}Su + u^{\mathrm{T}}Ru$ if and only if there exist functions $V_{\mathrm{s}} : \mathbb{R}^n \to \mathbb{R}$, $\ell : \mathbb{R}^n \to \mathbb{R}^p$, and $\mathcal{W} : \mathbb{R}^n \to \mathbb{R}^{p\times m}$ such that $V_{\mathrm{s}}(\cdot)$ is continuously differentiable and positive definite, $V_{\mathrm{s}}(0) = 0$, and, for all $x \in \mathbb{R}^n$,

$$0 = V_{\mathrm{s}}'(x)f(x) - h^{\mathrm{T}}(x)Qh(x) + \ell^{\mathrm{T}}(x)\ell(x), \tag{5.77}$$

$$0 = \tfrac{1}{2}V_{\mathrm{s}}'(x)G(x) - h^{\mathrm{T}}(x)(QJ(x) + S) + \ell^{\mathrm{T}}(x)\mathcal{W}(x), \tag{5.78}$$

$$0 = R + S^{\mathrm{T}}J(x) + J^{\mathrm{T}}(x)S + J^{\mathrm{T}}(x)QJ(x) - \mathcal{W}^{\mathrm{T}}(x)\mathcal{W}(x). \tag{5.79}$$

If, alternatively,

$$\mathcal{N}(x) \triangleq R + S^{\mathrm{T}}J(x) + J^{\mathrm{T}}(x)S + J^{\mathrm{T}}(x)QJ(x) > 0, \quad x \in \mathbb{R}^n, \tag{5.80}$$

then $\mathcal{G}$ is dissipative with respect to the quadratic supply rate $r(u,y) = y^{\mathrm{T}}Qy + 2y^{\mathrm{T}}Su + u^{\mathrm{T}}Ru$ if and only if there exists a continuously differentiable function $V_{\mathrm{s}} : \mathbb{R}^n \to \mathbb{R}$ such that $V_{\mathrm{s}}(\cdot)$ is positive definite, $V_{\mathrm{s}}(0) = 0$, and, for all $x \in \mathbb{R}^n$,

$$\begin{aligned} 0 \geq{} & V_{\mathrm{s}}'(x)f(x) - h^{\mathrm{T}}(x)Qh(x) + [\tfrac{1}{2}V_{\mathrm{s}}'(x)G(x) - h^{\mathrm{T}}(x)(QJ(x) + S)] \\ & \cdot\mathcal{N}^{-1}(x)[\tfrac{1}{2}V_{\mathrm{s}}'(x)G(x) - h^{\mathrm{T}}(x)(QJ(x) + S)]^{\mathrm{T}}. \end{aligned} \tag{5.81}$$

Proof. First, suppose that there exist functions $V_{\mathrm{s}} : \mathbb{R}^n \to \mathbb{R}$, $\ell : \mathbb{R}^n \to \mathbb{R}^p$, and $\mathcal{W} : \mathbb{R}^n \to \mathbb{R}^{p\times m}$ such that $V_{\mathrm{s}}(\cdot)$ is continuously differentiable and positive definite, and (5.77)–(5.79) are satisfied. Then for every admissible input $u(\cdot) \in \mathcal{U}$, $t_1, t_2 \in \mathbb{R}$, $t_2 \geq t_1 \geq 0$, it follows from (5.77)–(5.79) that

$$\begin{aligned} \int_{t_1}^{t_2} r(u,y)\mathrm{d}t &= \int_{t_1}^{t_2} [y^{\mathrm{T}}Qy + 2y^{\mathrm{T}}Su + u^{\mathrm{T}}Ru]\mathrm{d}t \\ &= \int_{t_1}^{t_2} [h^{\mathrm{T}}(x)Qh(x) + 2h^{\mathrm{T}}(x)(S + QJ(x))u \\ &\quad + u^{\mathrm{T}}(J^{\mathrm{T}}(x)QJ(x) + S^{\mathrm{T}}J(x) + J^{\mathrm{T}}(x)S + R)u]\mathrm{d}t \\ &= \int_{t_1}^{t_2} [V_{\mathrm{s}}'(x)(f(x) + G(x)u) + \ell^{\mathrm{T}}(x)\ell(x) + 2\ell^{\mathrm{T}}(x)\mathcal{W}(x)u \\ &\quad + u^{\mathrm{T}}\mathcal{W}^{\mathrm{T}}(x)\mathcal{W}(x)u]\mathrm{d}t \end{aligned}$$

$$= \int_{t_1}^{t_2} \Big[\dot{V}_s(x) + [\ell(x) + \mathcal{W}(x)u]^T[\ell(x) + \mathcal{W}(x)u]\Big] dt$$
$$\geq V_s(x(t_2)) - V_s(x(t_1)),$$

where $x(t)$, $t \geq 0$, satisfies (5.75) and $\dot{V}_s(\cdot)$ denotes the total derivative of the storage function along the trajectories $x(t)$, $t \geq 0$, of (5.75). Now, the result is immediate from Corollary 5.1.

Conversely, suppose that $\mathcal{G}$ is dissipative with respect to a quadratic supply rate $r(u, y)$. Now, it follows from Theorem 5.1 that the available storage $V_a(x)$ of $\mathcal{G}$ is finite for all $x \in \mathbb{R}^n$, $V_a(0) = 0$, and

$$V_a(x(t_2)) \leq V_a(x(t_1)) + \int_{t_1}^{t_2} r(u(t), y(t))dt, \quad t_2 \geq t_1, \tag{5.82}$$

for all admissible $u(\cdot) \in \mathcal{U}$. Dividing (5.82) by $t_2 - t_1$ and letting $t_2 \to t_1$ it follows that

$$\dot{V}_a(x(t)) \leq r(u(t), y(t)), \qquad t \geq 0, \tag{5.83}$$

where $x(t)$, $t \geq 0$, satisfies (5.75) and $\dot{V}_a(x(t)) \triangleq V_a'(x(t))(f(x(t)) + G(x(t)) \cdot u(t))$ denotes the total derivative of the available storage function along the trajectories $x(t)$, $t \geq 0$. Now, with $t = 0$, it follows from (5.83) that

$$V_a'(x_0)(f(x_0) + G(x_0)u) \leq r(u, y(0)), \qquad u \in \mathbb{R}^m. \tag{5.84}$$

Next, let $d : \mathbb{R}^n \times \mathbb{R}^m \to \mathbb{R}$ be such that

$$d(x, u) \triangleq -\dot{V}_a(x) + r(u, y) = -V_a'(x)(f(x) + G(x)u) + r(u, h(x) + J(x)u). \tag{5.85}$$

Now, it follows from (5.83) that $d(x, u) \geq 0$, $x \in \mathbb{R}^n$, $u \in \mathbb{R}^m$. Furthermore, note that $d(x, u)$ given by (5.85) is quadratic in u, and hence, there exist functions $\ell : \mathbb{R}^n \to \mathbb{R}$ and $\mathcal{W} : \mathbb{R}^n \to \mathbb{R}^{p\times m}$ such that

$$\begin{aligned} d(x, u) &= [\ell(x) + \mathcal{W}(x)u]^T[\ell(x) + \mathcal{W}(x)u] \\ &= -V_a'(x)(f(x) + G(x)u) + r(u, h(x) + J(x)u) \\ &= -V_a'(x)(f(x) + G(x)u) + (h(x) + J(x)u)^T Q(h(x) + J(x)u) \\ &\quad +2(h(x) + J(x)u)^T Su + u^T Ru. \end{aligned}$$

Now, equating coefficients of equal powers yields (5.77)–(5.79) with $V_s(x) = V_a(x)$ and the positive definiteness of $V_s(x)$, $x \in \mathbb{R}^n$, follows from Theorem 5.3.

Finally, to show (5.81) note that (5.77)–(5.79) can be equivalently written as

$$\begin{bmatrix} \mathcal{A}(x) & \mathcal{B}(x) \\ \mathcal{B}^T(x) & \mathcal{C}(x) \end{bmatrix} = - \begin{bmatrix} \ell^T(x) \\ \mathcal{W}^T(x) \end{bmatrix} \begin{bmatrix} \ell(x) & \mathcal{W}(x) \end{bmatrix} \leq 0, \quad x \in \mathbb{R}^n, \tag{5.86}$$

where $\mathcal{A}(x) \triangleq V_{\rm s}'(x)f(x) - h^{\rm T}(x)Qh(x)$, $\mathcal{B}(x) \triangleq \frac{1}{2}V_{\rm s}'(x)G(x) - h^{\rm T}(x)(QJ(x) + S)$, and $\mathcal{C}(x) \triangleq -(R + S^{\rm T}J(x) + J^{\rm T}(x)S + J^{\rm T}(x)QJ(x))$. Now, for all invertible $\mathcal{T} \in \mathbb{R}^{(m+1)\times(m+1)}$ (5.86) holds if and only if $\mathcal{T}^{\rm T}$(5.86)$\mathcal{T}$ holds. Hence, the equivalence of (5.77)–(5.79) to (5.81) in the case when (5.80) holds follows from the (1,1) block of $\mathcal{T}^{\rm T}$(5.86)$\mathcal{T}$, where

$$\mathcal{T} \triangleq \begin{bmatrix} 1 & 0 \\ -\mathcal{C}^{-1}(x)\mathcal{B}^{\rm T}(x) & I \end{bmatrix}.$$

This completes the proof. □

Note that the assumption of complete reachability in Theorem 5.6 is needed to establish the existence of a nonnegative-definite storage function $V_{\rm s}(\cdot)$ while zero-state observability ensures that $V_{\rm s}(\cdot)$ is positive definite. In the case where the existence of a continuously differentiable positive-definite storage function $V_{\rm s}(\cdot)$ is assumed for $\mathcal{G}$, then $\mathcal{G}$ is dissipative with respect to the quadratic supply rate $r(u, y)$ with storage function $V_{\rm s}(\cdot)$ if and only if (5.77)–(5.79) are satisfied.

Next, we provide necessary and sufficient conditions for exponential dissipativeness with respect to quadratic supply rates. Here, once again we assume that there exists a function $\kappa : \mathbb{R}^l \to \mathbb{R}^m$ such that $\kappa(0) = 0$ and $r(\kappa(y), y) < 0$, $y \neq 0$, and the available exponential storage $V_{\rm a}(x)$, $x \in \mathbb{R}^n$, for $\mathcal{G}$ is a continuously differentiable function.

Theorem 5.7. Let $Q \in \mathbb{S}^l$, $S \in \mathbb{R}^{l\times m}$, $R \in \mathbb{S}^m$, and let $\mathcal{G}$ be zero-state observable and completely reachable. $\mathcal{G}$ is exponentially dissipative with respect to the quadratic supply rate $r(u,y) = y^{\rm T}Qy + 2y^{\rm T}Su + u^{\rm T}Ru$ if and only if there exist functions $V_{\rm s} : \mathbb{R}^n \to \mathbb{R}$, $\ell : \mathbb{R}^n \to \mathbb{R}^p$, and $\mathcal{W} : \mathbb{R}^n \to \mathbb{R}^{p\times m}$ and a scalar $\varepsilon > 0$ such that $V_{\rm s}(\cdot)$ is continuously differentiable and positive definite, $V_{\rm s}(0) = 0$, and, for all $x \in \mathbb{R}^n$,

$$0 = V_{\rm s}'(x)f(x) + \varepsilon V_{\rm s}(x) - h^{\rm T}(x)Qh(x) + \ell^{\rm T}(x)\ell(x), \tag{5.87}$$
$$0 = \tfrac{1}{2}V_{\rm s}'(x)G(x) - h^{\rm T}(x)(QJ(x) + S) + \ell^{\rm T}(x)\mathcal{W}(x), \tag{5.88}$$
$$0 = R + S^{\rm T}J(x) + J^{\rm T}(x)S + J^{\rm T}(x)QJ(x) - \mathcal{W}^{\rm T}(x)\mathcal{W}(x). \tag{5.89}$$

If, alternatively, $\mathcal{N}(x) > 0$, $x \in \mathbb{R}^n$, then $\mathcal{G}$ is exponentially dissipative with respect to the quadratic supply rate $r(u,y) = y^{\rm T}Qy + 2y^{\rm T}Su + u^{\rm T}Ru$ if and only if there exists a continuously differentiable function $V_{\rm s} : \mathbb{R}^n \to \mathbb{R}$ and a scalar $\varepsilon > 0$ such that $V_{\rm s}(\cdot)$ is positive definite, $V_{\rm s}(0) = 0$, and, for all $x \in \mathbb{R}^n$,

$$\begin{aligned} 0 \geq\ & V_{\rm s}'(x)f(x) + \varepsilon V_{\rm s}(x) - h^{\rm T}(x)Qh(x) + [\tfrac{1}{2}V_{\rm s}'(x)G(x) - h^{\rm T}(x)(QJ(x) + S)] \\ & \cdot\mathcal{N}^{-1}(x)[\tfrac{1}{2}V_{\rm s}'(x)G(x) - h^{\rm T}(x)(QJ(x) + S)]^{\rm T}. \end{aligned} \tag{5.90}$$

Proof. The proof is analogous to the proof of Theorem 5.6. □

Finally, we provide necessary and sufficient conditions for the case where $\mathcal{G}$ given by (5.75) and (5.76) is lossless with respect to a quadratic supply rate $r(u, y)$.

Theorem 5.8. Let $Q \in \mathbb{S}^l$, $S \in \mathbb{R}^{l\times m}$, $R \in \mathbb{S}^m$, and let $\mathcal{G}$ be zero-state observable and completely reachable. $\mathcal{G}$ is lossless with respect to the quadratic supply rate $r(u, y) = y^{\mathrm{T}}Qy + 2y^{\mathrm{T}}Su + u^{\mathrm{T}}Ru$ if and only if there exists a function $V_{\mathrm{s}} : \mathbb{R}^n \to \mathbb{R}$ such that $V_{\mathrm{s}}(\cdot)$ is continuously differentiable and positive definite, $V_{\mathrm{s}}(0) = 0$, and, for all $x \in \mathbb{R}^n$,

$$0 = V_{\mathrm{s}}'(x)f(x) - h^{\mathrm{T}}(x)Qh(x), \tag{5.91}$$

$$0 = \tfrac{1}{2}V_{\mathrm{s}}'(x)G(x) - h^{\mathrm{T}}(x)(QJ(x) + S), \tag{5.92}$$

$$0 = R + S^{\mathrm{T}}J(x) + J^{\mathrm{T}}(x)S + J^{\mathrm{T}}(x)QJ(x). \tag{5.93}$$

Proof. The proof is analogous to the proof of Theorem 5.6. □

For particular examples of dynamical systems with force inputs and velocity outputs we can associate the storage function with the stored or available energy in the system and the supply rate with the net flow of energy or power into the system. However, as discussed in [320, 456], the concepts of supply rates and storage functions also apply to more general systems for which a physical energy interpretation is no longer valid. Specifically, using (5.77)–(5.79) it follows that

$$\int_{t_0}^{t} r(u(s), y(s))\mathrm{d}s = V_{\mathrm{s}}(x(t)) - V_{\mathrm{s}}(x(t_0)) + \int_{t_0}^{t} [\ell(x(s)) + \mathcal{W}(x(s))u(s)]^{\mathrm{T}}[\ell(x(s)) + \mathcal{W}(x(s))u(s)]\mathrm{d}s, \tag{5.94}$$

which can be interpreted as a *generalized* energy balance equation, where $V_{\mathrm{s}}(x(t)) - V_{\mathrm{s}}(x(t_0))$ is the stored or accumulated generalized energy of the system and the second path-dependent term on the right corresponds to the dissipated generalized energy of the system. Rewriting (5.94) as

$$\dot{V}_{\mathrm{s}}(x) = r(u, y) - [\ell(x) + \mathcal{W}(x)u]^{\mathrm{T}}[\ell(x) + \mathcal{W}(x)u], \tag{5.95}$$

yields a generalized energy conservation equation which shows that the rate of change in generalized system energy, or generalized power, is equal to the external generalized system power input minus the internal generalized system power dissipated.

Note that if $\mathcal{G}$ with a continuously differentiable positive-definite storage function is dissipative with respect to the quadratic supply rate

$r(u,y) = y^{\mathrm{T}}Qy + 2y^{\mathrm{T}}Su + u^{\mathrm{T}}Ru$, and if $Q \leq 0$ and $u(t) \equiv 0$, then it follows that

$$\dot{V}_{\mathrm{s}}(x(t)) \leq y^{\mathrm{T}}(t)Qy(t) \leq 0, \qquad t \geq 0. \tag{5.96}$$

Hence, the zero solution $x(t) \equiv 0$ of the undisturbed ($u(t) \equiv 0$) nonlinear system (5.75) is Lyapunov stable. Alternatively, if $\mathcal{G}$ with a continuously differentiable positive-definite storage function is exponentially dissipative with respect to the quadratic supply rate $r(u,y) = y^{\mathrm{T}}Qy + 2y^{\mathrm{T}}Su + u^{\mathrm{T}}Ru$, and if $Q \leq 0$ and $u(t) \equiv 0$, then it follows that

$$\dot{V}_{\mathrm{s}}(x(t)) \leq -\varepsilon V_{\mathrm{s}}(x(t)) + y^{\mathrm{T}}(t)Qy(t) \leq -\varepsilon V_{\mathrm{s}}(x(t)), \qquad t \geq 0. \tag{5.97}$$

Hence, the zero solution $x(t) \equiv 0$ of the undisturbed ($u(t) \equiv 0$) nonlinear system (5.75) is asymptotically stable. If, in addition, there exist scalars $\alpha, \beta > 0$ and $p \geq 1$ such that

$$\alpha\|x\|^p \leq V_{\mathrm{s}}(x) \leq \beta\|x\|^p, \qquad x \in \mathbb{R}^n, \tag{5.98}$$

then the zero solution $x(t) \equiv 0$ of the undisturbed ($u(t) \equiv 0$) nonlinear dynamical system (5.75) is exponentially stable.

Next, we provide several definitions of nonlinear dynamical systems which are dissipative or exponentially dissipative with respect to supply rates of a specific form.

Definition 5.7. A dynamical system $\mathcal{G}$ of the form (5.7) and (5.8) with $m = l$ is *passive* if $\mathcal{G}$ is dissipative with respect to the supply rate $r(u,y) = 2u^{\mathrm{T}}y$.

Definition 5.8. A dynamical system $\mathcal{G}$ of the form (5.7) and (5.8) with $m = l$ is *strictly passive* if $\mathcal{G}$ is strictly dissipative with respect to the supply rate $r(u,y) = 2u^{\mathrm{T}}y$.

Definition 5.9. A dynamical system $\mathcal{G}$ of the form (5.7) and (5.8) is *input strict passive* if there exists $\varepsilon > 0$ such that $\mathcal{G}$ is dissipative with respect to the supply rate $r(u,y) = 2u^{\mathrm{T}}y - \varepsilon u^{\mathrm{T}}u$.

Definition 5.10. A dynamical system $\mathcal{G}$ of the form (5.7) and (5.8) is *output strict passive* if there exists $\varepsilon > 0$ such that $\mathcal{G}$ is dissipative with respect to the supply rate $r(u,y) = 2u^{\mathrm{T}}y - \varepsilon y^{\mathrm{T}}y$.

Definition 5.11. A dynamical system $\mathcal{G}$ of the form (5.7) and (5.8) is *input-output strict passive* if there exist $\varepsilon, \hat{\varepsilon} > 0$ such that $\mathcal{G}$ is dissipative with respect to the supply rate $r(u,y) = 2u^{\mathrm{T}}y - \varepsilon u^{\mathrm{T}}u - \hat{\varepsilon}y^{\mathrm{T}}y$.

Definition 5.12. A dynamical system $\mathcal{G}$ of the form (5.7) and (5.8) is *nonexpansive* if $\mathcal{G}$ is dissipative with respect to the supply rate $r(u,y) =$

$\gamma^2 u^{\mathrm{T}} u - y^{\mathrm{T}} y$, where $\gamma > 0$ is given.

Definition 5.13. A dynamical system $\mathcal{G}$ of the form (5.7) and (5.8) with $m = l$ is *exponentially passive* if $\mathcal{G}$ is exponentially dissipative with respect to the supply rate $r(u, y) = 2u^{\mathrm{T}} y$.

Definition 5.14. A dynamical system $\mathcal{G}$ of the form (5.7) and (5.8) is *exponentially nonexpansive* if $\mathcal{G}$ is exponentially dissipative with respect to the supply rate $r(u, y) = \gamma^2 u^{\mathrm{T}} u - y^{\mathrm{T}} y$, where $\gamma > 0$ is given.

In light of the above definitions the following result is immediate.

Proposition 5.2. Consider the nonlinear dynamical system $\mathcal{G}$ given by (5.7) and (5.8). Then the following statements hold:

i) If $\mathcal{G}$ is passive with a continuously differentiable positive-definite storage function $V_{\mathrm{s}}(\cdot)$, then the zero solution $x(t) \equiv 0$ of the undisturbed ($u(t) \equiv 0$) system $\mathcal{G}$ is Lyapunov stable.

ii) If $\mathcal{G}$ is exponentially passive with a continuously differentiable positive-definite storage function $V_{\mathrm{s}}(\cdot)$, then the zero solution $x(t) \equiv 0$ of the undisturbed ($u(t) \equiv 0$) system $\mathcal{G}$ is asymptotically stable. If, in addition, $V_{\mathrm{s}}(\cdot)$ satisfies (5.98), then the zero solution $x(t) \equiv 0$ of the undisturbed ($u(t) \equiv 0$) system $\mathcal{G}$ is exponentially stable.

iii) If $\mathcal{G}$ is zero-state observable and nonexpansive with a continuously differentiable positive-definite storage function $V_{\mathrm{s}}(\cdot)$, then the zero solution $x(t) \equiv 0$ of the undisturbed ($u(t) \equiv 0$) system $\mathcal{G}$ is asymptotically stable.

iv) If $\mathcal{G}$ is exponentially nonexpansive with a continuously differentiable positive-definite storage function $V_{\mathrm{s}}(\cdot)$, then the zero solution $x(t) \equiv 0$ of the undisturbed ($u(t) \equiv 0$) system $\mathcal{G}$ is asymptotically stable. If, in addition, $V_{\mathrm{s}}(\cdot)$ satisfies (5.98), then the zero solution $x(t) \equiv 0$ of the undisturbed ($u(t) \equiv 0$) system $\mathcal{G}$ is exponentially stable.

v) If $\mathcal{G}$ is strictly passive with a continuously differentiable positive-definite storage function $V_{\mathrm{s}}(\cdot)$, then the zero solution $x(t) \equiv 0$ of the undisturbed ($u(t) \equiv 0$) system $\mathcal{G}$ is asymptotically stable.

vi) If $\mathcal{G}$ is input strict passive with a continuously differentiable positive-definite storage function $V_{\mathrm{s}}(\cdot)$, then the zero solution $x(t) \equiv 0$ of the undisturbed ($u(t) \equiv 0$) system $\mathcal{G}$ is Lyapunov stable.

vii) If $\mathcal{G}$ is zero-state observable and output strict passive with a continuously differentiable positive-definite storage function $V_{\mathrm{s}}(\cdot)$, then the zero solution $x(t) \equiv 0$ of the undisturbed ($u(t) \equiv 0$) system $\mathcal{G}$ is asymptotically stable.

viii) If $\mathcal{G}$ is zero-state observable and input-output passive with a continuously differentiable positive-definite storage function $V_s(\cdot)$, then the zero solution $x(t) \equiv 0$ of the undisturbed ($u(t) \equiv 0$) system $\mathcal{G}$ is asymptotically stable.

ix) If $\mathcal{G}$ is output strict passive with a continuously differentiable positive-definite storage function $V_s(\cdot)$, then $\mathcal{G}$ is nonexpansive.

x) If $\mathcal{G}$ is input strict passive and nonexpansive with a continuously differentiable positive-definite storage function $V_s(\cdot)$, then $\mathcal{G}$ is input-output strict passive.

Proof. Statements *i)*–*viii)* are immediate and follow from (5.31)–(5.33) using Lyapunov and invariant set stability arguments. To show *ix)*, note that if $\mathcal{G}$ is output strict passive with a continuously differentiable positive-definite storage function $V_s(\cdot)$ it follows that, for $\varepsilon > 0$,

$$\begin{aligned}\dot{V}_s(x) &\leq 2u^{\mathrm{T}}y - \varepsilon y^{\mathrm{T}}y \\ &= -\tfrac{1}{2\varepsilon}(2u - \varepsilon y)^{\mathrm{T}}(2u - \varepsilon y) + \tfrac{2}{\varepsilon}u^{\mathrm{T}}u - \tfrac{\varepsilon}{2}y^{\mathrm{T}}y \\ &\leq \tfrac{2}{\varepsilon}u^{\mathrm{T}}u - \tfrac{\varepsilon}{2}y^{\mathrm{T}}y, \end{aligned} \tag{5.99}$$

which implies that $\mathcal{G}$ is nonexpansive. Finally, to show *x)*, it follows that if $\mathcal{G}$ is input strict passive and nonexpansive with a continuously differentiable positive-definite storage function $V_s(\cdot)$ it follows that

$$\dot{V}_s(x) \leq 2u^{\mathrm{T}}y - \varepsilon u^{\mathrm{T}}u, \qquad \varepsilon > 0, \tag{5.100}$$

and

$$\dot{V}_s(x) \leq \gamma^2 u^{\mathrm{T}}u - y^{\mathrm{T}}y, \qquad \gamma > 0, \tag{5.101}$$

which implies that for all $\alpha > 0$,

$$(1+\alpha)\dot{V}_s(x) \leq 2u^{\mathrm{T}}y - \hat{\varepsilon}u^{\mathrm{T}}u - \alpha y^{\mathrm{T}}y, \tag{5.102}$$

where $\hat{\varepsilon} = \varepsilon - \alpha\gamma^2$. Now, the result is immediate by choosing $\alpha > 0$ such that $\hat{\varepsilon} > 0$. □

Example 5.2. Consider the matrix second-order nonlinear dynamical system of an n-link robot given by [342]

$$M(q(t))\ddot{q}(t) + C(q(t),\dot{q}(t))\dot{q}(t) + g(q(t)) = u(t),$$
$$q(0) = q_0, \quad \dot{q}(0) = \dot{q}_0, \quad t \geq 0, \tag{5.103}$$
$$y(t) = \dot{q}(t), \tag{5.104}$$

where q, $\dot{q}$, $\ddot{q} \in \mathbb{R}^n$ represent generalized position, velocity, and acceleration coordinates, respectively, $u \in \mathbb{R}^n$ is a force input, $y = \dot{q} \in \mathbb{R}^n$ is a velocity measurement, $M(q)$ is a positive-definite inertia matrix function for all $q \in \mathbb{R}^n$, $C(q, \dot{q})$ is an $n \times n$ matrix function accounting for centrifugal and Coriolis forces and has the property that $\frac{\mathrm{d}}{\mathrm{d}t}M(q(t)) - 2C(q(t), \dot{q}(t))$ is skew-symmetric for all $q, \dot{q} \in \mathbb{R}^n$, and $g(q)$ is an n-dimensional vector accounting for gravity forces and is given by $g(q) = \left[\frac{\partial V(q)}{\partial q}\right]^{\mathrm{T}}$, where $V(q)$ is the system potential. Here, we assume $V(0) = 0$, $V(q)$ is positive definite, and $g(q) = 0$ has an isolated root at $q = 0$.

To show that (5.103) and (5.104) is lossless, consider the energy storage function $V_{\mathrm{s}}(q, \dot{q}) = \dot{q}^{\mathrm{T}} M(q)\dot{q} + 2V(q)$. Now,

$$\begin{aligned} \dot{V}_{\mathrm{s}}(q, \dot{q}) &= 2\dot{q}^{\mathrm{T}} M(q)\ddot{q} + \dot{q}^{\mathrm{T}} \dot{M}(q)\dot{q} + 2\frac{\partial V(q)}{\partial q}\dot{q} \\ &= 2\dot{q}^{\mathrm{T}}[u - C(q, \dot{q})\dot{q} - g(q)] + \dot{q}^{\mathrm{T}} \dot{M}(q)\dot{q} + 2g^{\mathrm{T}}(q)\dot{q} \\ &= 2y^{\mathrm{T}} u, \end{aligned} \tag{5.105}$$

and hence, the system is lossless. Alternatively, with $u = -K_{\mathrm{p}}\dot{q} + v$, where K_{p} is a positive-definite matrix, it follows that

$$\dot{V}_{\mathrm{s}}(q, \dot{q}) = 2y^{\mathrm{T}} v - 2y^{\mathrm{T}} K_{\mathrm{p}} y, \tag{5.106}$$

and hence, the input-output map from v to y is output strict passive. Note that in the case where $v = 0$, $\dot{V}_{\mathrm{s}}(q, \dot{q}) = -2\dot{q}^{\mathrm{T}} K_{\mathrm{p}} \dot{q} \le 0$, and hence, $\mathcal{R} \triangleq \{(q, \dot{q}) : \dot{V}_{\mathrm{s}}(q, \dot{q}) = 0\} = \{(q, \dot{q}) : \dot{q} = 0\}$. Now, since $\dot{q}(t) \equiv 0$ it follows that $\ddot{q}(t) \equiv 0$ which, using (5.103), implies $g(q(t)) \equiv 0$, and hence, $q(t) \equiv 0$. Thus, $\mathcal{M} = \{(0, 0)\}$ is the largest invariant set contained in $\mathcal{R}$ so that the zero solution $(q(t), \dot{q}(t)) \equiv (0, 0)$ is asymptotically stable. Finally, we note that if $V(q)$ is radially unbounded, then global asymptotic stability is ensured. △

Example 5.3. Consider the nonlinear mass-spring-damper dynamical system

$$m\ddot{x}(t) + x^2(t)\dot{x}^3(t) + x^7(t) = 2u(t), \quad x(0) = x_0, \quad \dot{x}(0) = \dot{x}_0, \quad t \ge 0, \tag{5.107}$$

$$y(t) = \dot{x}(t). \tag{5.108}$$

To show that (5.107) and (5.108) is passive consider the energy storage function $V_{\mathrm{s}}(x, \dot{x}) = \frac{1}{2}m\dot{x}^2 + \frac{1}{8}x^8$. Now, $\dot{V}_{\mathrm{s}}(x, \dot{x}) = 2\dot{x}u - x^2\dot{x}^4 \le 2yu$, and hence, (5.107) and (5.108) is passive. △

Example 5.4. Consider the nonlinear dynamical system

$$\dot{x}_1(t) = x_2(t), \qquad x_1(0) = x_{10}, \qquad t \ge 0, \tag{5.109}$$

$$\dot{x}_2(t) = -g(x_1(t)) - ax_1(t) + u(t), \qquad x_2(0) = x_{20}, \tag{5.110}$$

$$y(t) = bx_1(t) + x_2(t), \tag{5.111}$$

where $0 < b < a$, $x_1 g(x_1) > 0$, $x_1 \in \mathbb{R}$, $x_1 \neq 0$, and $g(0) = 0$. To examine the passivity of (5.109)–(5.111) we consider the storage function

$$V_{\rm s}(x_1, x_2) = \tfrac{\alpha}{2}[\beta a^2 x_1^2 + 2\beta a x_1 x_2 + x_2^2] + \alpha \int_0^{x_1} g(\sigma){\rm d}\sigma, \tag{5.112}$$

where $\alpha > 0$ and $\beta \in (0, 1)$. Note that $V_{\rm s}(x_1, x_2)$ is positive definite and radially unbounded. Now, computing $\dot{V}_{\rm s}(x_1, x_2)$ yields

$$\begin{aligned}\dot{V}_{\rm s}(x_1, x_2) &= \alpha[\beta a^2 x_1 + \beta a x_2 + g(x_1)]x_2 + \alpha(\beta a x_1 + x_2)[-g(x_1) - a x_2 + u] \\ &= -\alpha\beta a x_1 g(x_1) + \alpha(\beta - 1)a x_2^2 + \alpha(\beta a x_1 + x_2)u. \end{aligned} \tag{5.113}$$

Setting $\alpha = 1$ and $\beta = b/a < 1$ if follows that

$$\dot{V}_{\rm s}(x_1, x_2) = uy - b x_1 g(x_1) - (a - b)x_2^2, \tag{5.114}$$

which shows that (5.109)–(5.111) is strictly passive. △

The following results present the nonlinear versions of the Kalman-Yakubovich-Popov *positive real lemma* and the *bounded real lemma.*

Corollary 5.2. Let $\mathcal{G}$ be zero-state observable and completely reachable. $\mathcal{G}$ is passive if and only if there exist functions $V_{\rm s} : \mathbb{R}^n \to \mathbb{R}$, $\ell : \mathbb{R}^n \to \mathbb{R}^p$, and $\mathcal{W} : \mathbb{R}^n \to \mathbb{R}^{p \times m}$ such that $V_{\rm s}(\cdot)$ is continuously differentiable and positive definite, $V_{\rm s}(0) = 0$, and, for all $x \in \mathbb{R}^n$,

$$0 = V_{\rm s}'(x)f(x) + \ell^{\rm T}(x)\ell(x), \tag{5.115}$$

$$0 = \tfrac{1}{2}V_{\rm s}'(x)G(x) - h^{\rm T}(x) + \ell^{\rm T}(x)\mathcal{W}(x), \tag{5.116}$$

$$0 = J(x) + J^{\rm T}(x) - \mathcal{W}^{\rm T}(x)\mathcal{W}(x). \tag{5.117}$$

If, alternatively,

$$J(x) + J^{\rm T}(x) > 0, \quad x \in \mathbb{R}^n, \tag{5.118}$$

then $\mathcal{G}$ is passive if and only if there exists a continuously differentiable function $V_{\rm s} : \mathbb{R}^n \to \mathbb{R}$ such that $V_{\rm s}(\cdot)$ is positive definite, $V_{\rm s}(0) = 0$, and, for all $x \in \mathbb{R}^n$,

$$\begin{aligned}0 \geq\ & V_{\rm s}'(x)f(x) + [\tfrac{1}{2}V_{\rm s}'(x)G(x) - h^{\rm T}(x)] \\ & \cdot[J(x) + J^{\rm T}(x)]^{-1}[\tfrac{1}{2}V_{\rm s}'(x)G(x) - h^{\rm T}(x)]^{\rm T}. \end{aligned} \tag{5.119}$$

Proof. The result is a direct consequence of Theorem 5.6 with $l = m$, $Q = 0$, $S = I_m$, and $R = 0$. Specifically, with $\kappa(y) = -y$ it follows that $r(\kappa(y), y) = -2y^{\rm T}y < 0$, $y \neq 0$, so that all the assumptions of Theorem 5.6 are satisfied. □

Example 5.5. Consider the nonlinear controlled Lienard system

$$\ddot{x}(t) + \hat{f}(x(t))\dot{x}(t) + g(x(t)) = u(t), \quad x(0) = x_0, \quad \dot{x}(0) = \dot{x}_0, \quad t \geq 0, \tag{5.120}$$

with output $y(t) = \frac{1}{2}\dot{x}(t)$ or, equivalently,

$$\dot{x}_1(t) = x_2(t) - F(x_1(t)), \quad x_1(0) = x_{10}, \quad t \geq 0, \tag{5.121}$$
$$\dot{x}_2(t) = -g(x_1(t)) + u(t), \quad x_2(0) = x_{20}, \tag{5.122}$$
$$y(t) = \tfrac{1}{2}x_2(t), \tag{5.123}$$

where $x_1 = x$, $x_2 = \dot{x} + F(x)$, $F(x_1) = \int_0^{x_1} \hat{f}(s)\mathrm{d}s$, $\hat{f}(0) = g(0) = 0$, and $\hat{f}$ and g are continuously differentiable. To examine the passivity of (5.121)–(5.123) we consider the energy function

$$V_{\mathrm{s}}(x_1, x_2) = \tfrac{1}{2}x_2^2 + \int_0^{x_1} g(s)\mathrm{d}s, \tag{5.124}$$

where $x_1 g(x_1) \geq 0$, $x_1 \in \mathbb{R}$, so that $V_{\mathrm{s}}(x_1, x_2) \geq 0$, $(x_1, x_2) \in \mathbb{R} \times \mathbb{R}$, is a candidate storage function. Now, we use Corollary 5.2 to determine conditions on $g(x_1)$ and $F(x_1)$ that guarantee (5.121)–(5.123) is passive. Note that (5.121)–(5.123) can be written in the state space form (5.75) and (5.76) with $x = [x_1,\ x_2]^{\mathrm{T}}$, $f(x) = [x_2 - F(x_1),\ -g(x_1)]^{\mathrm{T}}$, $G(x) = [0,\ \ 1]^{\mathrm{T}}$, $h(x) = \frac{1}{2}x_2$, and $J(x) = 0$. Now, using (5.115) it follows that

$$0 = \begin{bmatrix} g(x_1) & x_2 \end{bmatrix} \begin{bmatrix} x_2 - F(x_1) \\ -g(x_1) \end{bmatrix} + \ell^{\mathrm{T}}(x)\ell(x), \tag{5.125}$$

which requires that $g(x_1)F(x_1) \geq 0$. If $\hat{f}(x_1) \geq 0$, $x_1 \in \mathbb{R}$, then $x_1 F(x_1) \geq 0$, $x_1 \in \mathbb{R}$. Furthermore, since x_1 and $g(x_1)$ have the same sign it follows that $g(x_1)F(x_1) \geq 0$, $x_1 \in \mathbb{R}$. Next, note that (5.116) is automatically satisfied since

$$0 = \tfrac{1}{2}\begin{bmatrix} g(x_1) & x_2 \end{bmatrix} \begin{bmatrix} 0 \\ 1 \end{bmatrix} - \tfrac{1}{2}x_2. \tag{5.126}$$

Hence, if $\hat{f}(x_1) \geq 0$, $x_1 \in \mathbb{R}$, then the controlled Lienard system (5.121)–(5.123) is passive. △

Corollary 5.3. Let $\mathcal{G}$ be zero-state observable and completely reachable. $\mathcal{G}$ is nonexpansive if and only if there exist functions $V_{\mathrm{s}} : \mathbb{R}^n \to \mathbb{R}$, $\ell : \mathbb{R}^n \to \mathbb{R}^p$, and $\mathcal{W} : \mathbb{R}^n \to \mathbb{R}^{p \times m}$ such that $V_{\mathrm{s}}(\cdot)$ is continuously differentiable and positive definite, $V_{\mathrm{s}}(0) = 0$, and, for all $x \in \mathbb{R}^n$,

$$0 = V_{\mathrm{s}}'(x)f(x) + h^{\mathrm{T}}(x)h(x) + \ell^{\mathrm{T}}(x)\ell(x), \tag{5.127}$$
$$0 = \tfrac{1}{2}V_{\mathrm{s}}'(x)G(x) + h^{\mathrm{T}}(x)J(x) + \ell^{\mathrm{T}}(x)\mathcal{W}(x), \tag{5.128}$$
$$0 = \gamma^2 I_m - J^{\mathrm{T}}(x)J(x) - \mathcal{W}^{\mathrm{T}}(x)\mathcal{W}(x), \tag{5.129}$$

where $\gamma > 0$. If, alternatively,

$$\gamma^2 I_m - J^{\mathrm{T}}(x)J(x) > 0, \quad x \in \mathbb{R}^n, \tag{5.130}$$

then $\mathcal{G}$ is nonexpansive if and only if there exists a continuously differentiable function $V_{\mathrm{s}} : \mathbb{R}^n \to \mathbb{R}$ such that $V_{\mathrm{s}}(\cdot)$ is positive definite, $V_{\mathrm{s}}(0) = 0$, and, for all $x \in \mathbb{R}^n$,

$$0 \geq V_s'(x)f(x) + h^T(x)h(x) + [\tfrac{1}{2}V_s'(x)G(x) + h^T(x)J(x)] \cdot[\gamma^2 I_m - J^T(x)J(x)]^{-1}[\tfrac{1}{2}V_s'(x)G(x) + h^T(x)J(x)]^T. \quad (5.131)$$

Proof. The result is a direct consequence of Theorem 5.6 with $Q = -I_l$, $S = 0$, and $R = \gamma^2 I_m$. Specifically, with $\kappa(y) = -\frac{1}{2\gamma}y$ it follows that $r(\kappa(y), y) = -\frac{3}{4}y^T y < 0$, $y \neq 0$, so that all the assumptions of Theorem 5.6 are satisfied. □

Example 5.6. Consider the nonlinear controlled dynamical system

$$\dot{x}_1(t) = x_2(t), \quad x_1(0) = x_{10}, \quad t \geq 0, \quad (5.132)$$

$$\dot{x}_2(t) = -a \sin x_1(t) - bx_2(t) + u(t), \quad x_2(0) = x_{20}, \quad (5.133)$$

$$y(t) = x_2(t), \quad (5.134)$$

where $a, b > 0$. Note that (5.132)–(5.134) can be written in the state space form (5.75) and (5.76) with $x = [x_1,\ x_2]^T$, $f(x) = [x_2,\ -a\sin x_1 - bx_2]^T$, $G(x) = [0,\ 1]^T$, $h(x) = x_2$, and $J(x) = 0$. To examine the nonexpansivity of (5.132)–(5.134) we consider the storage function $V_s(x) = a(1 - \cos x_1) + \frac{1}{2}x_2^2$ satisfying $V_s(x) \geq 0$, $x \in \mathbb{R}^2$. Now, using Corollary 5.3 it follows from (5.131) that

$$0 \geq \begin{bmatrix} a\sin x_1 & x_2 \end{bmatrix} \begin{bmatrix} x_2 \\ -a\sin x_1 - bx_2 \end{bmatrix} + h^2(x) + \frac{1}{4\gamma^2}\begin{bmatrix} a\sin x_1 & x_2 \end{bmatrix}\begin{bmatrix} 0 \\ 1 \end{bmatrix}\begin{bmatrix} 0 & 1 \end{bmatrix}\begin{bmatrix} a\sin x_1 \\ x_2 \end{bmatrix}, \quad (5.135)$$

or, equivalently,

$$0 \geq (1-b)h^2(x) + \frac{1}{4\gamma^2}h^2(x). \quad (5.136)$$

Hence, (5.136) is satisfied if $\gamma \geq \frac{1}{2\sqrt{b-1}}$. △

Finally, the following results present the nonlinear versions of the Kalman-Yakubovich-Popov *strict positive real lemma* and *strict bounded real lemma* for exponentially passive and exponentially nonexpansive systems, respectively.

Corollary 5.4. Let $\mathcal{G}$ be zero-state observable and completely reachable. $\mathcal{G}$ is exponentially passive if and only if there exist functions $V_s : \mathbb{R}^n \to \mathbb{R}$, $\ell : \mathbb{R}^n \to \mathbb{R}^p$, and $\mathcal{W} : \mathbb{R}^n \to \mathbb{R}^{p\times m}$ and a scalar $\varepsilon > 0$ such that $V_s(\cdot)$ is continuously differentiable and positive definite, $V_s(0) = 0$, and, for all $x \in \mathbb{R}^n$,

$$0 = V_s'(x)f(x) + \varepsilon V_s(x) + \ell^T(x)\ell(x), \quad (5.137)$$

$$0 = \tfrac{1}{2}V_s'(x)G(x) - h^T(x) + \ell^T(x)\mathcal{W}(x), \quad (5.138)$$

$$0 = J(x) + J^{\mathrm{T}}(x) - \mathcal{W}^{\mathrm{T}}(x)\mathcal{W}(x). \tag{5.139}$$

If, alternatively, (5.118) holds, then $\mathcal{G}$ is exponentially passive if and only if there exist a continuously differentiable function $V_{\mathrm{s}} : \mathbb{R}^n \to \mathbb{R}$ and a scalar $\varepsilon > 0$ such that $V_{\mathrm{s}}(\cdot)$ is positive definite, $V_{\mathrm{s}}(0) = 0$, and, for all $x \in \mathbb{R}^n$,

$$\begin{aligned} 0 \geq\ & V_{\mathrm{s}}'(x)f(x) + \varepsilon V_{\mathrm{s}}(x) + [\tfrac{1}{2}V_{\mathrm{s}}'(x)G(x) - h^{\mathrm{T}}(x)] \\ & \cdot[J(x) + J^{\mathrm{T}}(x)]^{-1}[\tfrac{1}{2}V_{\mathrm{s}}'(x)G(x) - h^{\mathrm{T}}(x)]^{\mathrm{T}}. \end{aligned} \tag{5.140}$$

Proof. The result is a direct consequence of Theorem 5.7 with $l = m$, $Q = 0$, $S = I_m$, and $R = 0$. Specifically, with $\kappa(y) = -y$ it follows that $r(\kappa(y), y) = -2y^{\mathrm{T}}y < 0$, $y \neq 0$, so that all the assumptions of Theorem 5.7 are satisfied. □

Corollary 5.5. Let $\mathcal{G}$ be zero-state observable and completely reachable. $\mathcal{G}$ is exponentially nonexpansive if and only if there exist functions $V_{\mathrm{s}} : \mathbb{R}^n \to \mathbb{R}$, $\ell : \mathbb{R}^n \to \mathbb{R}^p$, and $\mathcal{W} : \mathbb{R}^n \to \mathbb{R}^{p\times m}$ and a scalar $\varepsilon > 0$ such that $V_{\mathrm{s}}(\cdot)$ is continuously differentiable and positive definite, $V_{\mathrm{s}}(0) = 0$, and, for all $x \in \mathbb{R}^n$,

$$0 = V_{\mathrm{s}}'(x)f(x) + \varepsilon V_{\mathrm{s}}(x) + h^{\mathrm{T}}(x)h(x) + \ell^{\mathrm{T}}(x)\ell(x), \tag{5.141}$$

$$0 = \tfrac{1}{2}V_{\mathrm{s}}'(x)G(x) + h^{\mathrm{T}}(x)J(x) + \ell^{\mathrm{T}}(x)\mathcal{W}(x), \tag{5.142}$$

$$0 = \gamma^2 I_m - J^{\mathrm{T}}(x)J(x) - \mathcal{W}^{\mathrm{T}}(x)\mathcal{W}(x), \tag{5.143}$$

where $\gamma > 0$. If, alternatively, (5.130) holds, then $\mathcal{G}$ is exponentially nonexpansive if and only if there exist a continuously differentiable function $V_{\mathrm{s}} : \mathbb{R}^n \to \mathbb{R}$ and a scalar $\varepsilon > 0$ such that $V_{\mathrm{s}}(\cdot)$ is positive definite, $V_{\mathrm{s}}(0) = 0$, and, for all $x \in \mathbb{R}^n$,

$$\begin{aligned} 0 \geq\ & V_{\mathrm{s}}'(x)f(x) + \varepsilon V_{\mathrm{s}}(x) + h^{\mathrm{T}}(x)h(x) + [\tfrac{1}{2}V_{\mathrm{s}}'(x)G(x) + h^{\mathrm{T}}(x)J(x)] \\ & \cdot[\gamma^2 I_m - J^{\mathrm{T}}(x)J(x)]^{-1}[\tfrac{1}{2}V_{\mathrm{s}}'(x)G(x) + h^{\mathrm{T}}(x)J(x)]^{\mathrm{T}}. \end{aligned} \tag{5.144}$$

Proof. The result is a direct consequence of Theorem 5.7 with $Q = -I_l$, $S = 0$, and $R = \gamma^2 I_m$. Specifically, with $\kappa(y) = -\frac{1}{2\gamma}y$ it follows that $r(\kappa(y), y) = -\frac{3}{4}y^{\mathrm{T}}y < 0$, $y \neq 0$, so that all the assumptions of Theorem 5.7 are satisfied. □

5.5 Linearization of Dissipative Dynamical Systems

In this section, we present several key results on linearization of dissipative, exponentially dissipative, passive, exponentially passive, nonexpansive, and exponentially nonexpansive dynamical systems. For these results, we assume that there exists a function $\kappa : \mathbb{R}^l \to \mathbb{R}^m$ such that $\kappa(0) = 0$ and $r(\kappa(y), y) <$

0, $y \neq 0$, and the available storage function (respectively, exponentially storage function) $V_{\rm a}(x)$, $x \in \mathbb{R}^n$, is a smooth function.

Theorem 5.9. Let $Q \in \mathbb{S}^l$, $S \in \mathbb{R}^{l\times m}$, $R \in \mathbb{S}^m$, and suppose $\mathcal{G}$ given by (5.75) and (5.76) is completely reachable and dissipative with respect to the quadratic supply rate $r(u,y) = y^{\rm T}Qy + 2y^{\rm T}Su + u^{\rm T}Ru$. Then, there exist matrices $P \in \mathbb{R}^{n\times n}$, $L \in \mathbb{R}^{p\times n}$, and $W \in \mathbb{R}^{p\times m}$, with P nonnegative definite, such that

$$0 = A^{\rm T}P + PA - C^{\rm T}QC + L^{\rm T}L, \tag{5.145}$$

$$0 = PB - C^{\rm T}(QD+S) + L^{\rm T}W, \tag{5.146}$$

$$0 = R + S^{\rm T}D + D^{\rm T}S + D^{\rm T}QD - W^{\rm T}W, \tag{5.147}$$

where

$$A = \left.\frac{\partial f}{\partial x}\right|_{x=0}, \quad B = G(0), \quad C = \left.\frac{\partial h}{\partial x}\right|_{x=0}, \quad D = J(0). \tag{5.148}$$

If, in addition, (A,C) is observable, then $P > 0$.

Proof. First note that since $\mathcal{G}$ is completely reachable and dissipative with respect to a quadratic supply rate there exist functions $V_{\rm s} : \mathbb{R}^n \to \mathbb{R}$, $\ell : \mathbb{R}^n \to \mathbb{R}^p$, and $\mathcal{W} : \mathbb{R}^n \to \mathbb{R}^{p\times m}$, with $V_{\rm s}(\cdot)$ nonnegative definite, such that (5.77)–(5.79) are satisfied. Now, expanding $V_{\rm s}(\cdot)$ via a Taylor series expansion about $x = 0$ and using the fact that $V_{\rm s}(\cdot)$ is nonnegative definite and $V_{\rm s}(0) = 0$ it follows that there exists a nonnegative-definite matrix $P \in \mathbb{R}^{n\times n}$ such that

$$V_{\rm s}(x) = x^{\rm T}Px + V_{\rm sr}(x),$$

where $V_{\rm sr} : \mathbb{R}^n \to \mathbb{R}$ contains the higher-order terms of $V_{\rm s}(x)$.

Next, let $f(x) = Ax + f_{\rm r}(x)$, $\ell(x) = Lx + \ell_{\rm r}(x)$, and $h(x) = Cx + h_{\rm r}(x)$, where $f_{\rm r}(x)$, $\ell_{\rm r}(x)$, and $h_{\rm r}(x)$ contain the nonlinear terms of $f(x)$, $\ell(x)$, and $h(x)$, respectively, and let $G(x) = B + G_{\rm r}(x)$, $J(x) = D + J_{\rm r}(x)$, and $\mathcal{W}(x) = W + \mathcal{W}_{\rm r}(x)$, where $W \triangleq \mathcal{W}(0)$ and $G_{\rm r}(x)$, $J_{\rm r}(x)$, and $\mathcal{W}_{\rm r}(x)$ contain the nonconstant terms of $G(x)$, $J(x)$, and $\mathcal{W}(x)$, respectively. Using the above expressions (5.77) and (5.78) can be written as

$$0 = x^{\rm T}(A^{\rm T}P + PA - C^{\rm T}QC + L^{\rm T}L)x + \gamma(x), \tag{5.149}$$

$$0 = x^{\rm T}(PB - C^{\rm T}(QD+S) + L^{\rm T}W) + \Gamma(x), \tag{5.150}$$

where

$$\begin{aligned}\gamma(x) &\triangleq V_{\rm sr}'(x)f(x) + \ell_{\rm r}^{\rm T}(x)\ell_{\rm r}(x) + 2x^{\rm T}(Pf_{\rm r}(x) - C^{\rm T}Qh_{\rm r}(x) + L^{\rm T}\ell_{\rm r}(x)) \\ &\quad -h_{\rm r}^{\rm T}(x)Qh_{\rm r}(x), \\ \Gamma(x) &\triangleq \tfrac{1}{2}V_{\rm sr}'(x)G(x) - h_{\rm r}^{\rm T}(x)(QJ(x)+S) - x^{\rm T}C^{\rm T}QJ_{\rm r}(x) + x^{\rm T}PG_{\rm r}(x) \\ &\quad +x^{\rm T}L^{\rm T}\mathcal{W}_{\rm r}(x) + \ell_{\rm r}^{\rm T}(x)W.\end{aligned}$$

Now, viewing (5.149) and (5.150) as the Taylor series expansion of (5.77) and (5.78), respectively, about $x = 0$, and noting that

$$\lim_{\|x\|\to 0} \frac{|\gamma(x)|}{\|x\|^2} = 0, \qquad \lim_{\|x\|\to 0} \frac{\|\Gamma(x)\|}{\|x\|} = 0,$$

where $\|\cdot\|$ denotes the Euclidean vector norm, it follows that P satisfies (5.145) and (5.146). Now, (5.147) follows from (5.79) by setting $x = 0$.

Next, it follows from Theorem 5.6 and (5.145)–(5.147) that the linear system

$$\begin{aligned} \dot{x}(t) &= Ax(t) + Bu(t), \qquad x(0) = x_0, \qquad t \geq 0, \\ y(t) &= Cx(t) + Du(t), \end{aligned}$$

with a storage function $V_s(x) = x^T P x$ is dissipative with respect to the quadratic supply rate $r(u, y)$. Now, the positive definiteness of P follows from Theorem 5.3. □

The following corollaries are immediate from Theorem 5.9 and provide linearization results for passive and nonexpansive dynamical systems, respectively.

Corollary 5.6. Suppose the nonlinear dynamical system $\mathcal{G}$ given by (5.75) and (5.76) is completely reachable and passive. Then there exist matrices $P \in \mathbb{R}^{n\times n}$, $L \in \mathbb{R}^{p\times n}$, and $W \in \mathbb{R}^{p\times m}$, with P nonnegative definite, such that

$$0 = A^T P + PA + L^T L, \tag{5.151}$$
$$0 = PB - C^T + L^T W, \tag{5.152}$$
$$0 = D + D^T - W^T W, \tag{5.153}$$

where A, B, C, and D are given by (5.148). If, in addition, (A, C) is observable, then $P > 0$.

Corollary 5.7. Suppose the nonlinear dynamical system $\mathcal{G}$ given by (5.75) and (5.76) is completely reachable and nonexpansive. Then there exist matrices $P \in \mathbb{R}^{n\times n}$, $L \in \mathbb{R}^{p\times n}$, and $W \in \mathbb{R}^{p\times m}$, with P nonnegative definite, such that

$$0 = A^T P + PA + C^T C + L^T L, \tag{5.154}$$
$$0 = PB + C^T D + L^T W, \tag{5.155}$$
$$0 = \gamma^2 I_m - D^T D - W^T W, \tag{5.156}$$

where A, B, C, and D are given by (5.148) and $\gamma > 0$. If, in addition, (A, C) is observable, then $P > 0$.

Next, we present a linearization theorem for exponentially dissipative systems.

Theorem 5.10. Let $Q \in \mathbb{S}^l$, $S \in \mathbb{R}^{l\times m}$, $R \in \mathbb{S}^m$, and suppose $\mathcal{G}$ given by (5.75) and (5.76) is completely reachable and exponentially dissipative with respect to the quadratic supply rate $r(u,y) = y^{\mathrm{T}}Qy + 2y^{\mathrm{T}}Su + u^{\mathrm{T}}Ru$. Then, there exist matrices $P \in \mathbb{R}^{n\times n}$, $L \in \mathbb{R}^{p\times n}$, and $W \in \mathbb{R}^{p\times m}$, with P nonnegative definite, and a scalar $\varepsilon > 0$ such that

$$0 = A^{\mathrm{T}}P + PA + \varepsilon P - C^{\mathrm{T}}QC + L^{\mathrm{T}}L, \tag{5.157}$$
$$0 = PB - C^{\mathrm{T}}(QD+S) + L^{\mathrm{T}}W, \tag{5.158}$$
$$0 = R + S^{\mathrm{T}}D + D^{\mathrm{T}}S + D^{\mathrm{T}}QD - W^{\mathrm{T}}W, \tag{5.159}$$

where A, B, C, and D are given by (5.148). If, in addition, (A,C) is observable, then $P > 0$.

Proof. The proof is analogous to the proof of Theorem 5.9. □

Linearization results for exponentially passive and exponentially nonexpansive dynamical systems follow immediately from Theorem 5.10.

5.6 Positive Real and Bounded Real Dynamical Systems

In this section, we specialize the results of Section 5.4 to the case of linear systems and provide connections to the frequency domain versions of passivity, exponential passivity, nonexpansivity, and exponential nonexpansivity. Specifically, we consider linear systems

$$\mathcal{G} = G(s) \sim \left[\begin{array}{c|c} A & B \\ \hline C & D \end{array}\right]$$

with a state space representation

$$\dot{x}(t) = Ax(t) + Bu(t), \qquad x(0) = 0, \qquad t \geq 0, \tag{5.160}$$
$$y(t) = Cx(t) + Du(t), \tag{5.161}$$

where $x \in \mathbb{R}^n$, $u \in \mathbb{R}^m$, $y \in \mathbb{R}^l$, $A \in \mathbb{R}^{n\times n}$, $B \in \mathbb{R}^{n\times m}$, $C \in \mathbb{R}^{l\times n}$, and $D \in \mathbb{R}^{l\times m}$. To present the main results of this section we first give several key definitions.

Definition 5.15. A function $p : \mathbb{C} \to \mathbb{C}$ of the form

$$p(s) = \alpha_n s^n + \alpha_{n-1}s^{n-1} + \cdots + \alpha_1 s + \alpha_0, \tag{5.162}$$

where $\alpha_0, \ldots, \alpha_n$ are real numbers, is called a *polynomial*. If the leading coefficient α_n is nonzero, then the *degree* of $p(s)$, denoted by deg $p(s)$, is n, whereas if $p(s) = 0$, then deg $p(s) = -\infty$. If $\alpha_n = 1$, then $p(s)$ is *monic.*

Definition 5.16. A function $f : \mathbb{C} \to \mathbb{C}$ is *analytic at a point* $z_0 \in \mathbb{C}$ if, for every $\varepsilon > 0$, there exists $\delta > 0$ such that $f'(z)$ exists for all $z \in \mathcal{B}_\delta(z_0)$. f is said to be *analytic* in $\mathbb{C}$ if f is analytic at each point in $\mathbb{C}$. $f(s)$ is a *rational function* if there exist polynomials $n(s)$ and $d(s)$ such that $f(s) = n(s)/d(s)$. $f(s)$ is called *strictly proper* (respectively, *proper*) if deg $n(s) <$ deg $d(s)$ (respectively, deg $n(s) \leq$ deg $d(s)$). The *relative degree* of $f(s)$ denoted by r, is $r \triangleq \deg d(s) - \deg n(s)$.

Definition 5.17. An $l \times m$ matrix transfer function $G(s)$ is a *rational transfer function* if every entry of $G(s)$ is a rational function. $G(s)$ is *strictly proper* (respectively, *proper*) if every entry of $G(s)$ is strictly proper (respectively, proper).

In this book all transfer functions are assumed to be rational proper transfer functions.

Definition 5.18. A square transfer function $G(s)$ is *positive real* if *i*) all the entries of $G(s)$ are analytic in $\mathrm{Re}[s] > 0$ and *ii*) He $G(s) \geq 0$, $\mathrm{Re}[s] > 0$. A square transfer function $G(s)$ is *strictly positive real* if there exists $\varepsilon > 0$ such that $G(s - \varepsilon)$ is positive real. Finally, a square transfer function $G(s)$ is *strongly positive real* if it is strictly positive real and $D + D^{\mathrm{T}} > 0$, where $D \triangleq G(\infty)$.

Definition 5.19. A transfer function $G(s)$ is *bounded real* if *i*) all the entries of $G(s)$ are analytic in $\mathrm{Re}[s] > 0$ and *ii*) $\gamma^2 I_m - G^*(s)G(s) \geq 0$, $\mathrm{Re}[s] > 0$, where $\gamma > 0$. A transfer function $G(s)$ is *strictly bounded real* if there exists $\varepsilon > 0$ such that $G(s - \varepsilon)$ is bounded real. Finally, a transfer function $G(s)$ is *strongly bounded real* if it is strictly bounded real and $\gamma^2 I_m - D^{\mathrm{T}} D > 0$, where $D \triangleq G(\infty)$.

It is interesting to note that *ii*) in Definition 5.19 implies that $G(s)$ is analytic in $\mathrm{Re}[s] \geq 0$, and hence, a bounded real transfer function is asymptotically stable. To see this, note that $\gamma^2 I_m - G^*(s)G(s) \geq 0$, $\mathrm{Re}[s] > 0$, implies that

$$[\gamma^2 I_m - G^*(s)G(s)]_{(i,i)} = \gamma^2 - \sum_{j=1}^{m} |G_{(j,i)}(s)|^2 \geq 0, \quad \mathrm{Re}[s] > 0, \qquad (5.163)$$

and hence, $|G_{(i,j)}(s)|$ is bounded by γ^2 at every point in $\mathrm{Re}[s] > 0$. Hence, $G_{(i,j)}(s)$ cannot possess a pole in $\mathrm{Re}[s] = 0$ since in this case $|G_{(i,j)}(s)|$ would take on arbitrary large values in $\mathrm{Re}[s] > 0$ in the vicinity of this pole. Hence, $G(\jmath\omega) = \lim_{\sigma \to 0, \sigma > 0} G(\sigma + \jmath\omega)$ exists for all $\omega \in \mathbb{R}$ and $\gamma^2 I_m - G^*(\jmath\omega)G(\jmath\omega) \geq 0$, $\omega \in \mathbb{R}$. Now, since $G^*(\jmath\omega)G(\jmath\omega) \leq \gamma^2 I_m$, $\omega \in \mathbb{R}$, is equivalent to $\sup_{\omega \in \mathbb{R}} \sigma_{\max}[G(\jmath\omega)] \leq \gamma$, it follows that $G(s)$ is bounded real if and only if $G(s)$ is asymptotically stable and $|||G(s)|||_\infty \leq \gamma$. Similarly, it can be shown that strict bounded realness is equivalent to $G(s)$ asymptotically stable and $G^*(\jmath\omega)G(\jmath\omega) < \gamma^2 I_m$, $\omega \in \mathbb{R}$, or, equivalently, $|||G(s)|||_\infty < \gamma$.

Alternatively, for a positive real transfer function it follows from *ii)* of Definition 5.19, using a limiting argument, that He $G(\jmath\omega) \geq 0$ for all $\omega \in \mathbb{R}$ such that $\jmath\omega$ is not a pole of any entry of $G(s)$. However, unlike bounded real transfer functions, there are positive real transfer functions possessing poles on the $\jmath\omega$ axis. A simple example is $G(s) = 1/s$ which is analytic in $\mathrm{Re}[s] > 0$ and He $G(s) = 2\mathrm{Re}[s]/|s|^2 > 0$, $\mathrm{Re}[s] > 0$. Hence, He $G(\jmath\omega) \geq 0$, $\omega \in \mathbb{R}$, does *not* provide a frequency domain test for positive realness. In the case where $G(s)$ is analytic in $\mathrm{Re}[s] > 0$ and He $G(\jmath\omega) \geq 0$ holds for all $\omega \in \mathbb{R}$ for which $\jmath\omega$ is not a pole of any entry of $G(s)$, one might surmise that $G(s)$ is positive real. Once again this is not true. A simple counterexample is $G(s) = -1/s$ which is analytic in $\mathrm{Re}[s] > 0$ and satisfies He $G(\jmath\omega) = 0$, $\omega \in \mathbb{R}$. However, He $G(s) \geq 0$, $\mathrm{Re}[s] > 0$, is *not* satisfied. The following theorem gives a frequency domain test for positive realness.

Theorem 5.11. Let $G(s)$ be a square, real rational transfer function. $G(s)$ is positive real if and only if the following conditions hold:

i) No entry of $G(s)$ has a pole in $\mathrm{Re}[s] > 0$.

ii) He $G(\jmath\omega) \geq 0$ for all $\omega \in \mathbb{R}$, with $\jmath\omega$ not a pole of any entry of $G(s)$.

iii) If $\jmath\hat{\omega}$ is a pole of any entry of $G(s)$ it is at most a simple pole, and the residue matrix $G_0 \triangleq \lim_{s\to\jmath\hat{\omega}}(s - \jmath\hat{\omega})G(s)$ is nonnegative-definite Hermitian. Alternatively, if the limit $G_\infty \triangleq \lim_{\omega\to\infty} G(\jmath\omega)/\jmath\omega$ exists, then G_∞ is nonnegative-definite Hermitian.

Proof. The proof follows from the maximum modulus theorem of complex variable theory by forming a Nyquist-type closed contour Γ in $\mathrm{Re}[s] > 0$ and analyzing the function $f(s) = x^*G(s)x$, $x \in \mathbb{C}^m$, on Γ. For details see [11]. □

Next, we present the key results of this section for characterizing positive realness, strict positive realness, bounded realness, and strict bounded realness of a linear dynamical system in terms of the system matrices A, B, C, and D. First, however, we present a key theorem due to Parseval.

Theorem 5.12 (Parseval's Theorem). Let $u : [0,\infty) \to \mathbb{R}^m$ and $y : [0,\infty) \to \mathbb{R}^l$ be in $\mathcal{L}_p$, $p \in [0,\infty)$, and let $u(s)$ and $y(s)$ denote their Laplace transforms, respectively. Then

$$\int_0^\infty u^{\mathrm{T}}(t)y(t)\mathrm{d}t = \frac{1}{2\pi}\int_{-\infty}^{\infty} u^*(\jmath\omega)y(\jmath\omega)\mathrm{d}\omega. \tag{5.164}$$

Proof. Since $y(\cdot) \in \mathcal{L}_p$, $p \in [0, \infty)$, it follows that the inverse Fourier transform of $y(t)$ is given by

$$y(t) = \frac{1}{2\pi}\int_{-\infty}^{\infty} y(\jmath\omega)e^{\jmath\omega t}\mathrm{d}\omega. \tag{5.165}$$

Now, for all $s \in \mathbb{C}$, $\mathrm{Re}[s] \geq 0$,

$$\begin{aligned}\int_0^\infty [u^{\mathrm{T}}(\tau)y(\tau)]e^{-s\tau}\mathrm{d}\tau &= \int_0^\infty e^{-s\tau}u^{\mathrm{T}}(\tau)\left[\tfrac{1}{2\pi}\int_{-\infty}^{\infty} y(\jmath\omega)e^{\jmath\omega\tau}\mathrm{d}\omega\right]\mathrm{d}\tau \\ &= \frac{1}{2\pi}\int_{-\infty}^{\infty}\left[\int_0^\infty u^{\mathrm{T}}(\tau)e^{-(s-\jmath\omega)\tau}\mathrm{d}\tau\right]y(\jmath\omega)\mathrm{d}\omega \\ &= \frac{1}{2\pi}\int_{-\infty}^{\infty} u^{\mathrm{T}}(s-\jmath\omega)y(\jmath\omega)\mathrm{d}\omega. \end{aligned} \tag{5.166}$$

Setting $s = 0$ yields (5.164). □

Theorem 5.13 (Positive Real Lemma). Consider the linear dynamical system

$$G(s) \overset{\min}{\sim} \left[\begin{array}{c|c} A & B \\ \hline C & D \end{array}\right]$$

with input $u(\cdot) \in \mathcal{U}$ and output $y(\cdot) \in \mathcal{Y}$. Then the following statements are equivalent:

i) $G(s)$ is positive real.

ii) $\int_0^T u^{\mathrm{T}}(t)y(t)\mathrm{d}t \geq 0, \quad T \geq 0.$

iii) There exist matrices $P \in \mathbb{R}^{n\times n}$, $L \in \mathbb{R}^{p\times n}$, and $W \in \mathbb{R}^{p\times m}$, with P positive definite, such that

$$0 = A^{\mathrm{T}}P + PA + L^{\mathrm{T}}L, \tag{5.167}$$

$$0 = PB - C^{\mathrm{T}} + L^{\mathrm{T}}W, \tag{5.168}$$

$$0 = D + D^{\mathrm{T}} - W^{\mathrm{T}}W. \tag{5.169}$$

If, alternatively, $D + D^{\mathrm{T}} > 0$, then $G(s)$ is positive real if and only if there exists an $n \times n$ positive-definite matrix P such that

$$0 \geq A^{\mathrm{T}}P + PA + (B^{\mathrm{T}}P - C)^{\mathrm{T}}(D + D^{\mathrm{T}})^{-1}(B^{\mathrm{T}}P - C). \tag{5.170}$$

Proof. First, we show that i) implies ii). Suppose $G(s)$ is positive real. Then it follows from Parseval's theorem that, for all $T \geq 0$ and truncated input function

$$u_T(t) = \begin{cases} u(t), & 0 \leq t \leq T, \\ 0, & t < 0, t > T, \end{cases} \tag{5.171}$$

$$\begin{aligned} \int_0^T y^{\mathrm{T}}(t)u(t)\mathrm{d}t &= \int_{-\infty}^{\infty} y^{\mathrm{T}}(t)u_T(t)\mathrm{d}t \\ &= \frac{1}{2\pi}\int_{-\infty}^{\infty} y^*(\jmath\omega)u_T(\jmath\omega)\mathrm{d}\omega \\ &= \frac{1}{4\pi}\int_{-\infty}^{\infty} u_T^*(\jmath\omega)[G(\jmath\omega)+G^*(\jmath\omega)]u_T(\jmath\omega)\mathrm{d}\omega \\ &\geq 0, \end{aligned}$$

which implies that $G(s)$ is passive.

Next, we show that ii) implies iii). If $G(s)$ is passive, then it follows from Corollary 5.6 with $f(x) = Ax$, $G(x) = B$, $h(x) = Cx$, $J(x) = D$, that there exist matrices $P \in \mathbb{R}^{n\times n}$, $L \in \mathbb{R}^{p\times n}$, and $W \in \mathbb{R}^{p\times m}$, with P positive definite, such that (5.167)–(5.169) are satisfied.

Next, to show that iii) implies i), note that if there exist matrices $P \in \mathbb{R}^{n\times n}$, $L \in \mathbb{R}^{p\times n}$, and $W \in \mathbb{R}^{p\times m}$, with P positive definite, such that (5.167)–(5.169) are satisfied, then, for all $\mathrm{Re}[s] > 0$,

$$\begin{aligned} &G(s) + G^*(s) \\ &\quad = C(sI_n - A)^{-1}B + D + B^{\mathrm{T}}(s^*I_n - A)^{-\mathrm{T}}C^{\mathrm{T}} + D^{\mathrm{T}} \\ &\quad = W^{\mathrm{T}}W + (B^{\mathrm{T}}P + W^{\mathrm{T}}L)(sI_n - A)^{-1}B \\ &\qquad + B^{\mathrm{T}}(s^*I_n - A)^{-\mathrm{T}}(PB + L^{\mathrm{T}}W) \\ &\quad = W^{\mathrm{T}}W + W^{\mathrm{T}}L(sI_n - A)^{-1}B + B^{\mathrm{T}}(s^*I_n - A)^{-\mathrm{T}}L^{\mathrm{T}}W \\ &\qquad + B^{\mathrm{T}}(s^*I_n - A)^{-\mathrm{T}}[(s^*I_n - A)^{\mathrm{T}}P + P(sI_n - A)](sI_n - A)^{-1}B \\ &\quad = W^{\mathrm{T}}W + W^{\mathrm{T}}L(sI_n - A)^{-1}B + B^{\mathrm{T}}(s^*I_n - A)^{-\mathrm{T}}L^{\mathrm{T}}W \\ &\qquad + B^{\mathrm{T}}(s^*I_n - A)^{-\mathrm{T}}[L^{\mathrm{T}}L + 2\mathrm{Re}[s]P](sI_n - A)^{-1}B \\ &\quad \geq [W + L(sI_n - A)^{-1}B]^*[W + L(sI_n - A)^{-1}B] \\ &\quad \geq 0. \end{aligned}$$

To show analyticity of the entries of $G(s)$ in $\mathrm{Re}[s] > 0$ note that an entry of $G(s)$ will have a pole at $s = \lambda$ only if $\lambda \in \mathrm{spec}(A)$. Now, it follows from (5.167) and the fact that $P > 0$ that all eigenvalues of A have nonpositive real parts. Hence, $G(s)$ is analytic in $\mathrm{Re}[s] > 0$, which implies that $G(s)$ is positive real. Finally, (5.170) follows from (5.119) with the linearization given above. □

Theorem 5.14 (Strict Positive Real Lemma). Consider the linear dynamical system

$$G(s) \overset{\min}{\sim} \left[\begin{array}{c|c} A & B \\ \hline C & D \end{array}\right]$$

with input $u(\cdot) \in \mathcal{U}$ and output $y(\cdot) \in \mathcal{Y}$. Then the following statements are equivalent:

i) $G(s)$ is strictly positive real.

ii) $\int_0^T e^{\varepsilon t} u^{\mathrm{T}}(t) y(t) \mathrm{d}t \geq 0, \qquad T \geq 0 \qquad \varepsilon > 0.$

iii) There exist matrices $P \in \mathbb{R}^{n \times n}$, $L \in \mathbb{R}^{p \times n}$, and $W \in \mathbb{R}^{p \times m}$, with P positive definite, and a scalar $\varepsilon > 0$ such that

$$0 = A^{\mathrm{T}} P + PA + \varepsilon P + L^{\mathrm{T}} L, \tag{5.172}$$

$$0 = PB - C^{\mathrm{T}} + L^{\mathrm{T}} W, \tag{5.173}$$

$$0 = D + D^{\mathrm{T}} - W^{\mathrm{T}} W. \tag{5.174}$$

Furthermore, $G(s)$ is strongly positive real if and only if there exist $n \times n$ positive-definite matrices P and R such that

$$0 = A^{\mathrm{T}} P + PA + (B^{\mathrm{T}} P - C)^{\mathrm{T}} (D + D^{\mathrm{T}})^{-1} (B^{\mathrm{T}} P - C) + R. \tag{5.175}$$

Proof. The equivalence of *i)* and *iii)* is a direct consequence of Theorem 5.13 by noting that $G(s)$ is strictly positive real if and only if there exists $\varepsilon > 0$ such that

$$G(s - \varepsilon/2) \overset{\min}{\sim} \left[\begin{array}{c|c} A + \frac{1}{2}\varepsilon I_n & B \\ \hline C & D \end{array}\right]$$

is positive real. The fact that *iii)* implies *ii)* follows from Corollary 5.4 with $f(x) = Ax$, $G(x) = B$, $h(x) = Cx$, $J(x) = D$, $V_{\mathrm{s}}(x) = x^{\mathrm{T}} P x$, $\ell(x) = Lx$, and $\mathcal{W}(x) = W$. To show that *ii)* implies *iii)*, note that if $G(s)$ is exponentially passive, then it follows from Theorem 5.10 with $f(x) = Ax$, $G(x) = B$, $h(x) = Cx$, $J(x) = D$, $Q = 0$, $S = I_m$, and $R = 0$, that there exist matrices $P \in \mathbb{R}^{n \times n}$, $L \in \mathbb{R}^{p \times n}$, and $W \in \mathbb{R}^{p \times m}$, with P positive definite, such that (5.172)–(5.174) are satisfied.

Finally, with the linearization given above, it follows from (5.140) that $G(s)$ is strongly positive real if and only if there exist a scalar $\varepsilon > 0$ and a positive-definite matrix $P \in \mathbb{R}^{n \times n}$ such that

$$0 \geq A^{\mathrm{T}} P + PA + \varepsilon P + (B^{\mathrm{T}} P - C)^{\mathrm{T}} (D + D^{\mathrm{T}})^{-1} (B^{\mathrm{T}} P - C). \tag{5.176}$$

Now, if there exist a scalar $\varepsilon > 0$ and a positive-definite matrix $P \in \mathbb{R}^{n \times n}$ such that (5.176) is satisfied, then there exists an $n \times n$ positive-definite matrix R such that (5.175) is satisfied. Conversely, if there exists an $n \times n$ positive-definite matrix R such that (5.175) is satisfied then, with $\varepsilon = \sigma_{\min}(R)/\sigma_{\max}(P)$, (5.175) implies (5.176). Hence, $G(s)$ is strongly positive real if and only if there exist $n \times n$ positive-definite matrices P and R such that (5.175) is satisfied. □

Next, we present analogous results for bounded real systems.

Theorem 5.15 (Bounded Real Lemma). Consider the linear dynamical system

$$G(s) \overset{\min}{\sim} \left[\begin{array}{c|c} A & B \\ \hline C & D \end{array}\right]$$

with input $u(\cdot) \in \mathcal{U}$ and output $y(\cdot) \in \mathcal{Y}$. Then the following statements are equivalent:

i) $G(s)$ is bounded real.

ii) $\int_0^T y^{\mathrm{T}}(t)y(t)\mathrm{d}t \leq \gamma^2 \int_0^T u^{\mathrm{T}}(t)u(t)\mathrm{d}t, \quad T \geq 0, \quad \gamma > 0.$

iii) There exist matrices $P \in \mathbb{R}^{n\times n}$, $L \in \mathbb{R}^{p\times n}$, and $W \in \mathbb{R}^{p\times m}$, with P positive definite, such that

$$0 = A^{\mathrm{T}}P + PA + C^{\mathrm{T}}C + L^{\mathrm{T}}L, \tag{5.177}$$
$$0 = PB + C^{\mathrm{T}}D + L^{\mathrm{T}}W, \tag{5.178}$$
$$0 = \gamma^2 I_m - D^{\mathrm{T}}D - W^{\mathrm{T}}W. \tag{5.179}$$

If, alternatively, $\gamma^2 I_m - D^{\mathrm{T}}D > 0$, then $G(s)$ is bounded real if and only if there exists an $n \times n$ positive-definite matrix P such that

$$0 \geq A^{\mathrm{T}}P + PA + C^{\mathrm{T}}C + (B^{\mathrm{T}}P + D^{\mathrm{T}}C)^{\mathrm{T}}(\gamma^2 I_m - D^{\mathrm{T}}D)^{-1}(B^{\mathrm{T}}P + D^{\mathrm{T}}C). \tag{5.180}$$

Proof. First, we show that *i)* implies *ii)*. Suppose $G(s)$ is bounded real. Then it follows from Parseval's theorem that, for all $T \geq 0$ and truncated input function $u_T(\cdot)$ given by (5.171),

$$\begin{aligned}
\int_0^T y^{\mathrm{T}}(t)y(t)\mathrm{d}t &= \int_{-\infty}^{\infty} y^{\mathrm{T}}(t)y(t)\mathrm{d}t \\
&= \frac{1}{2\pi}\int_{-\infty}^{\infty} y^*(\jmath\omega)y(\jmath\omega)\mathrm{d}\omega \\
&= \frac{1}{2\pi}\int_{-\infty}^{\infty} u_T^*(\jmath\omega)G^*(\jmath\omega)G(\jmath\omega)u_T(\jmath\omega)\mathrm{d}\omega \\
&\leq \frac{\gamma^2}{2\pi}\int_{-\infty}^{\infty} u_T^*(\jmath\omega)u_T(\jmath\omega)\mathrm{d}\omega
\end{aligned}$$

$$= \gamma^2 \int_0^T u^{\mathrm{T}}(t)u(t)\mathrm{d}t,$$

which implies that $G(s)$ is nonexpansive.

Next, we show that *ii*) implies *iii*). If $G(s)$ is nonexpansive, then it follows from Corollary 5.7 with $f(x) = Ax$, $G(x) = B$, $h(x) = Cx$, $J(x) = D$, that there exist matrices $P \in \mathbb{R}^{n\times n}$, $L \in \mathbb{R}^{p\times n}$, and $W \in \mathbb{R}^{p\times m}$, with P positive definite, such that (5.177)–(5.179) are satisfied.

Now, to show that *iii*) implies *i*), note that if there exist matrices $P \in \mathbb{R}^{n\times n}$, $L \in \mathbb{R}^{p\times n}$, and $W \in \mathbb{R}^{p\times m}$, with P positive definite, such that (5.177)–(5.179) are satisfied, then, for all $\mathrm{Re}[s] \geq 0$,

$$\begin{aligned}
\gamma^2 I_m - G^*(s)G(s)&\\
&= \gamma^2 I_m - [C(sI_n - A)^{-1}B + D]^*[C(sI_n - A)^{-1}B + D]\\
&= [\gamma^2 I_m - D^{\mathrm{T}}D] - [C(sI_n - A)^{-1}B]^*[C(sI_n - A)^{-1}B]\\
&\quad -[C(sI_n - A)^{-1}B]^*D - D^{\mathrm{T}}[C(sI_n - A)^{-1}B]\\
&= W^{\mathrm{T}}W + [B^{\mathrm{T}}P + W^{\mathrm{T}}L](sI_n - A)^{-1}B + B^{\mathrm{T}}(sI_n - A)^{-*}\\
&\quad \cdot[PB + L^{\mathrm{T}}W] - B^{\mathrm{T}}(sI_n - A)^{-*}C^{\mathrm{T}}C(sI_n - A)^{-1}B\\
&= W^{\mathrm{T}}W + W^{\mathrm{T}}L(sI_n - A)^{-1}B + B^{\mathrm{T}}(sI_n - A)^{-*}L^{\mathrm{T}}W\\
&\quad -B^{\mathrm{T}}(sI_n - A)^{-*}C^{\mathrm{T}}C(sI_n - A)^{-1}B\\
&\quad +B^{\mathrm{T}}(sI_n - A)^{-*}[(sI_n - A)^*P + P(sI_n - A)](sI_n - A)^{-1}B\\
&= W^{\mathrm{T}}W + W^{\mathrm{T}}L(sI_n - A)^{-1}B + B^{\mathrm{T}}(sI_n - A)^{-*}L^{\mathrm{T}}W\\
&\quad +B^{\mathrm{T}}(sI_n - A)^{-*}L^{\mathrm{T}}L(sI_n - A)^{-1}B\\
&\quad +2\mathrm{Re}[s]B^{\mathrm{T}}(sI_n - A)^{-*}P(sI_n - A)^{-1}B\\
&\geq [W + L(sI_n - A)^{-1}B]^*[W + L(sI_n - A)^{-1}B]\\
&\geq 0.
\end{aligned}$$

To show analyticity of the entries of $G(s)$ in $\mathrm{Re}[s] \geq 0$ note that it follows from (5.177) and the fact that $P > 0$ and (A, C) is observable that all the eigenvalues of A have negative real parts. Hence, $G(s)$ is analytic in $\mathrm{Re}[s] \geq 0$, which implies that $G(s)$ is bounded real. Finally, (5.180) follows from (5.131) with the linearization given above. □

Theorem 5.16 (Strict Bounded Real Lemma). Consider the linear dynamical system

$$G(s) \overset{\min}{\sim} \left[\begin{array}{c|c} A & B \\ \hline C & D \end{array}\right]$$

with input $u(\cdot) \in \mathcal{U}$ and output $y(\cdot) \in \mathcal{Y}$. Then the following statements are equivalent:

i) $G(s)$ is strictly bounded real.

ii) $\int_0^T e^{\varepsilon t} y^{\mathrm{T}}(t) y(t) \mathrm{d}t \leq \gamma^2 \int_0^T e^{\varepsilon t} u^{\mathrm{T}}(t) u(t) \mathrm{d}t, \quad T \geq 0, \quad \gamma > 0, \quad \varepsilon > 0.$

iii) There exist matrices $P \in \mathbb{R}^{n \times n}$, $L \in \mathbb{R}^{p \times n}$, and $W \in \mathbb{R}^{p \times m}$, with P positive definite, such that

$$0 = A^{\mathrm{T}} P + PA + \varepsilon P + C^{\mathrm{T}} C + L^{\mathrm{T}} L, \tag{5.181}$$

$$0 = PB + C^{\mathrm{T}} D + L^{\mathrm{T}} W, \tag{5.182}$$

$$0 = \gamma^2 I_m - D^{\mathrm{T}} D - W^{\mathrm{T}} W. \tag{5.183}$$

Furthermore, $G(s)$ is strongly bounded real if and only if there exist $n \times n$ positive-definite matrices P and R such that

$$0 = A^{\mathrm{T}} P + PA + C^{\mathrm{T}} C + (B^{\mathrm{T}} P + D^{\mathrm{T}} C)^{\mathrm{T}} (\gamma^2 I_m - D^{\mathrm{T}} D)^{-1} (B^{\mathrm{T}} P + D^{\mathrm{T}} C) + R. \tag{5.184}$$

Proof. The equivalence of *i)* and *iii)* is a direct consequence of Theorem 5.15 by noting that $G(s)$ is strictly bounded real if and only if there exists $\varepsilon > 0$ such that

$$G(s - \varepsilon/2) \overset{\min}{\sim} \left[\begin{array}{c|c} A + \frac{1}{2}\varepsilon I_n & B \\ \hline C & D \end{array}\right]$$

is bounded real. The fact that *iii)* implies *ii)* follows from Corollary 5.5 with $f(x) = Ax$, $G(x) = B$, $h(x) = Cx$, $J(x) = D$, $V_{\mathrm{s}}(x) = x^{\mathrm{T}} P x$, $\ell(x) = Lx$, and $\mathcal{W}(x) = W$. To show that *ii)* implies *iii)*, note that if $G(s)$ is exponentially nonexpansive, then it follows from Theorem 5.10 with $f(x) = Ax$, $G(x) = B$, $h(x) = Cx$, $J(x) = D$, $Q = -I_p$, $S = 0$, and $R = \gamma^2 I_m$, that there exist matrices $P \in \mathbb{R}^{n \times n}$, $L \in \mathbb{R}^{p \times n}$, and $W \in \mathbb{R}^{p \times m}$, with P positive definite, such that (5.181)–(5.183) are satisfied.

Finally, with the linearization given above, it follows from (5.144) that $G(s)$ is strongly bounded real if and only if there exist a scalar $\varepsilon > 0$ and a positive-definite matrix $P \in \mathbb{R}^{n \times n}$ such that

$$0 \geq A^{\mathrm{T}} P + PA + \varepsilon P + (B^{\mathrm{T}} P - C)^{\mathrm{T}} (D + D^{\mathrm{T}})^{-1} (B^{\mathrm{T}} P - C). \tag{5.185}$$

Now, if there exist a scalar $\varepsilon > 0$ and a positive-definite matrix $P \in \mathbb{R}^{n \times n}$ such that (5.185) is satisfied, then there exists an $n \times n$ positive-definite matrix R such that (5.184) is satisfied. Conversely, if there exists an $n \times n$ positive-definite matrix R such that (5.184) is satisfied then, with $\varepsilon = \sigma_{\min}(R)/\sigma_{\max}(P)$, (5.184) implies (5.185). Hence, $G(s)$ is strongly bounded real if and only if there exist $n \times n$ positive-definite matrices P and R such that (5.184) is satisfied. □

As noted earlier, strict bounded realness is equivalent to $G(s)$ asymptotically stable and $G^*(\jmath\omega) G(\jmath\omega) < \gamma^2 I_m$, $\omega \in \mathbb{R}$. However, as in the positive real case, the frequency domain test for strict positive realness is subtle and does not simply involve checking He $G(\jmath\omega) > 0$, $\omega \in \mathbb{R}$. To see this, consider the transfer function

$$G(s) = \frac{s + (\alpha + \beta)}{(s + \alpha)(s + \beta)}, \tag{5.186}$$

where $\alpha, \beta > 0$. Noting that

$$\text{He } G(\jmath\omega) = \frac{\alpha\beta(\alpha + \beta)}{(\omega^2 + \alpha^2)(\omega^2 + \beta^2)}, \tag{5.187}$$

it follows that He $G(\jmath\omega) > 0$, $\omega \in \mathbb{R}$. Now, it can be easily shown, using the definition of positive realness, that $G(s)$ is positive real. However, for $\varepsilon > 0$ it follows that

$$\text{He } G(\jmath\omega - \varepsilon) = \frac{-\varepsilon\omega^2 + (\alpha - \varepsilon)(\beta - \varepsilon)(\alpha + \beta - \varepsilon)}{(\omega^2 + (\alpha - \varepsilon)^2)(\omega^2 - (\beta - \varepsilon)^2)}, \tag{5.188}$$

which, for sufficiently large ω, is negative, and hence, $G(s - \varepsilon)$ is not positive real. Hence, even though He $G(\jmath\omega) > 0$, $\omega \in \mathbb{R}$, $G(s)$ is not strictly positive real. In light of the above, we characterize necessary and sufficient conditions for a transfer function to be strictly positive real in terms of a frequency domain test. For the statement of the next result recall that a rational transfer function $G(s)$ is *nonsingular* if and only if det $G(s)$ is not the zero polynomial. Equivalently, det $G(s)$ is not zero if and only if the *normal rank* of $G(s) \in \mathbb{C}^{m \times m}$ over the field of rational functions of s is m, that is, nrank $G(s) \triangleq \max_{s \in \mathbb{C}} \text{rank}_{\mathbb{C}} G(s) = m$.

Theorem 5.17. Let

$$G(s) \sim \left[\begin{array}{c|c} A & B \\ \hline C & D \end{array}\right]$$

be an $m \times m$ rational transfer function and suppose that $G(s)$ is not singular. Then, $G(s)$ is strictly positive real if and only if the following conditions hold:

i) No entry of $G(s)$ has a pole in $\text{Re}[s] \geq 0$.

ii) He $G(\jmath\omega) > 0$ for all $\omega \in \mathbb{R}$.

iii) Either $D + D^{\mathrm{T}} > 0$ or both $D + D^{\mathrm{T}} \geq 0$ and $\lim_{\omega \to \infty} \omega^2 Q^{\mathrm{T}}[\text{He } G(\jmath\omega)]Q > 0$ for every $Q \in \mathbb{R}^{m \times (m-q)}$, where $q = \text{rank}(D + D^{\mathrm{T}})$, such that $Q^{\mathrm{T}}(D + D^{\mathrm{T}})Q = 0$.

Proof. Suppose *i)*–*iii)* hold. Let $\varepsilon > 0$ and note that

$$\begin{aligned} G(s - \varepsilon) &= C((s - \varepsilon)I - A)^{-1}B + D \\ &= C(sI - A)(sI - A)^{-1}(sI - (A + \varepsilon I))^{-1}B + D \\ &= C(sI - A)^{-1}(sI - (A + \varepsilon I) + \varepsilon I)(sI - (A + \varepsilon I))^{-1}B + D \end{aligned}$$

$$= G(s) + \varepsilon G_\varepsilon(s), \tag{5.189}$$

where $G_\varepsilon(s) \triangleq C(sI - A)^{-1}(sI - (A + \varepsilon I))^{-1}B$. Since A is Hurwitz, $sI - A$ is nonsingular for all $s = \jmath\omega$ and there exists $\hat{\varepsilon}$ such that $sI - (A + \varepsilon I)$ is nonsingular for all $\varepsilon \in [0, \hat{\varepsilon}]$ and for all $s = \jmath\omega$. Hence, $f(\omega, \varepsilon) \triangleq \max_i |\lambda_i[\mathrm{He}\ G_\varepsilon(\jmath\omega)]|$ is finite for all $\varepsilon \in [0, \hat{\varepsilon}]$ and for all $\omega \in \mathbb{R}$. Since $\lim_{\omega\to\infty} f(\omega, \varepsilon) = 0$ for all $\varepsilon \in [0, \hat{\varepsilon}]$, it follows that there exists $k_1 > 0$ such that $f(\omega, \varepsilon) < k_1$ for all $\varepsilon \in [0, \hat{\varepsilon}]$ and for all $\omega \in \mathbb{R}$, and hence, $-k_1 I_m \leq \mathrm{He}\ G_\varepsilon(\sigma) \leq k_1 I_m$ for all $\varepsilon \in [0, \hat{\varepsilon}]$ and for all $s = \jmath\omega$. Now, if $D + D^{\mathrm{T}} > 0$, and since $\mathrm{He}\ G(\jmath\omega) > 0$ for all $\omega \in \mathbb{R}$ and $\lim_{\omega\to\infty} \mathrm{He}\ G(\jmath\omega) = D + D^{\mathrm{T}} > 0$, it follows that there exists $k_2 > 0$ such that $\mathrm{He}\ G(\jmath\omega) \geq k_2 I_m > 0$, $\omega \in \mathbb{R}$. Choosing $0 < \varepsilon < \min\{\hat{\varepsilon}, \frac{k_2}{k_1}\}$, it follows that

$$\mathrm{He}\ G(s - \varepsilon) = \mathrm{He}\ G(s) + \varepsilon \mathrm{He}\ G_\varepsilon(s) \geq k_2 I_m - \varepsilon k_1 I_m > 0, \tag{5.190}$$

for all $s = \jmath\omega$. Hence, $G(s - \varepsilon)$ is positive real, and thus, by definition, $G(s)$ is strictly positive real.

Alternatively, if $\det(D + D^{\mathrm{T}}) = 0$, it follows from *iii*) that $\mathrm{He}\ G(\jmath\omega)$ has q eigenvalues $\lambda(\omega)$ satisfying $\lim_{\omega\to\infty} \lambda(\omega) > 0$ and $m - q$ eigenvalues satisfying $\lim_{\omega\to\infty} \lambda(\omega) = 0$ and $\lim_{\omega\to\infty} \omega^2\lambda(\omega) > 0$. Hence, there exists $k_3 > 0$ for some $\hat{\omega} > 0$ such that $\omega^2 \lambda_{\min}[\mathrm{He}\ G(\jmath\omega)] \geq k_3$ for all $|\omega| \geq \hat{\omega}$. Furthermore, since $\lim_{\omega\to\infty} \omega^2 G_\varepsilon(\jmath\omega)$ exists and $-k_1 I_m \leq \mathrm{He}\ G_\varepsilon(\jmath\omega) \leq k_1 I_m$ for all $\omega \in \mathbb{R}$ and $\varepsilon \in [0, \hat{\varepsilon}]$, it follows that there exist $k_4 > 0$ and $\omega^* > 0$ such that $-k_4 I_m \leq \mathrm{He}\ G_\varepsilon(\jmath\omega) \leq k_4 I_m$ for all $|\omega| \geq \omega^*$ and $\varepsilon \in [0, \hat{\varepsilon}]$. Hence,

$$\begin{aligned}\omega^2 \mathrm{He}(\jmath\omega - \varepsilon) &= \omega^2 \mathrm{He}\ G(\jmath\omega) + \omega^2 \varepsilon \mathrm{He}\ G_\varepsilon(\jmath\omega) \geq k_3 I_m - \varepsilon k_4 I_m, \\ &\qquad |\omega| \geq \omega_m,\end{aligned} \tag{5.191}$$

where $\omega_m = \max\{\hat{\omega}, \omega^*\}$. Since $\mathrm{He}\ G(\jmath\omega) > 0$, $\omega \in \mathbb{R}$, it follows that $\lambda_{\min}[\mathrm{He}\ G(\jmath\omega)] \geq k_5 > 0$ for all $\omega \in [-\omega_m, \omega_m]$. Hence, using the fact that $-k_1 I_m \leq \mathrm{He}\ G_\varepsilon(\jmath\omega) \leq k_1 I_m$ for all $\omega \in \mathbb{R}$ it follows that $\mathrm{He}\ G(\jmath\omega - \varepsilon) \geq k_5 - \varepsilon k_1$ for all $|\omega| \leq \omega_m$. Now, taking $0 < \varepsilon < \min\{\frac{k_3}{k_4}, \frac{k_5}{k_1}\}$, it follows that $\mathrm{He}\ G(\jmath\omega - \varepsilon) > 0$, $\omega \in \mathbb{R}$. Hence, $G(s - \varepsilon)$ is positive real, and hence, $G(s)$ is strictly positive real.

Conversely, suppose $G(s)$ is strictly positive real and let $\varepsilon > 0$ be such that $G(s - \varepsilon)$ is positive real. Note that $G(s)$ is asymptotically stable and positive real. Hence, $\mathrm{He}\ G(\jmath\omega) \geq 0$, $\omega \in \mathbb{R}$, and $\mathrm{He}\ G(\infty) = D + D^{\mathrm{T}} \geq 0$. Now, let (A, B, C, D) be a minimal realization of $G(s)$. Using Theorem 5.14, it follows from (5.172)–(5.173) that

$$\begin{aligned}G(s) + G^*(s) &= D + D^{\mathrm{T}} + C(sI - A)^{-1}B + B^{\mathrm{T}}(s^*I - A)^{-\mathrm{T}}C^{\mathrm{T}} \\ &= W^{\mathrm{T}}W + (W^{\mathrm{T}}L + B^{\mathrm{T}}P)(sI - A)^{-1}B \\ &\quad + B^{\mathrm{T}}(s^*I - A)^{-\mathrm{T}}(PB + L^{\mathrm{T}}W)\end{aligned}$$

$$\begin{aligned}
&= W^{\mathrm{T}}W + W^{\mathrm{T}}L(sI-A)^{-1}B + B^{\mathrm{T}}(s^*I-A)^{-\mathrm{T}}L^{\mathrm{T}}W \\
&\quad +B^{\mathrm{T}}(s^*I-A)^{-\mathrm{T}}[(s^*+s)P - A^{\mathrm{T}}P - PA](sI-A)^{-1}B \\
&= W^{\mathrm{T}}W + +W^{\mathrm{T}}L(sI-A)^{-1}B + B^{\mathrm{T}}(s^*I-A)^{-\mathrm{T}}L^{\mathrm{T}}W \\
&\quad +B^{\mathrm{T}}(s^*I-A)^{-\mathrm{T}}(L^{\mathrm{T}}L+\varepsilon P)(sI-A)^{-1}B \\
&\quad +(s^*+s)B^{\mathrm{T}}(s^*I-A)^{-\mathrm{T}}P(sI-A)^{-1}B \\
&= [W+L(sI-A)^{-1}B]^*[W+L(sI-A)^{-1}B] \\
&\quad +(2\mathrm{Re}[s]+\varepsilon)B^{\mathrm{T}}(s^*I-A)^{-\mathrm{T}}P(sI-A)^{-1}B. \qquad (5.192)
\end{aligned}$$

Now, suppose, *ad absurdum*, that He $G(\jmath\omega)$ is not positive definite for all $\omega \in \mathbb{R}$. Then, for some $\omega = \hat{\omega}$ there exists $x \in \mathbb{C}^m$, $x \neq 0$, such that $x^*[\text{He } G(\jmath\omega)]x = 0$. In this case, it follows from (5.192) that $Bx = 0$ and $Wx = 0$. Hence, $x^*[\text{He } G(s)]x = 0$ for all $s \in \mathbb{C}$, and hence, $\det[\text{He } G(s)] \equiv 0$, which leads to a contradiction. Thus, He $G(\jmath\omega) > 0$, $\omega \in \mathbb{R}$. Now, if He $G(\infty) = D + D^{\mathrm{T}} > 0$ the result is immediate. Alternatively, let $Q \in \mathbb{R}^{m\times(m-q)}$ be a full row rank matrix such that $Q^{\mathrm{T}}(D+D^{\mathrm{T}})Q = Q^{\mathrm{T}}W^{\mathrm{T}}WQ = 0$. Hence, $WQ = 0$ and by (5.192),

$$Q^{\mathrm{T}}[\text{He } G(\jmath\omega)]Q = Q^{\mathrm{T}}B^{\mathrm{T}}(\jmath\omega I-A)^{-*}[L^{\mathrm{T}}L+\varepsilon P](\jmath\omega I-A)^{-1}BQ. \quad (5.193)$$

Now, since He $G(\jmath\omega) > 0$, $\omega \in \mathbb{R}$, and Q is full row rank it follows that BQ is full column rank. Using (5.193) it follows that

$$\lim_{\omega\to\infty} \omega^2 Q^{\mathrm{T}}[\text{He } G(\jmath\omega)]Q = Q^{\mathrm{T}}B^{\mathrm{T}}(L^{\mathrm{T}}L+\varepsilon P)BQ > 0, \quad (5.194)$$

which proves the result. □

Note that if $D + D^{\mathrm{T}} = 0$ then we can take $Q = I_m$ in Theorem 5.17. It follows from Theorem 5.17 that a necessary and sufficient frequency domain test for a scalar strictly positive real system is Re $G(\jmath\omega) > 0$, $\omega \in \mathbb{R}$, if $D > 0$, and Re $G(\jmath\omega) > 0$, $\omega \in \mathbb{R}$, and $\lim_{\omega\to\infty} \omega^2 \mathrm{Re}\ G(\jmath\omega) > 0$ if $D = 0$.

5.7 Absolute Stability Theory

Absolute stability theory guarantees stability of feedback systems whose forward path contains a dynamic linear time-invariant system and whose feedback path contains a memoryless (possibly time-varying) nonlinearity (see Figure 5.1). These stability criteria are generally stated in terms of the linear system and apply to every element of specified class of nonlinearities. Hence, absolute stability theory provides sufficient conditions for *robust stability* with a given class of uncertain elements.

The literature on absolute stability is extensive. Two of the most fundamental results concerning the stability of feedback systems with memoryless nonlinearities are the circle criterion, which includes the positivity and

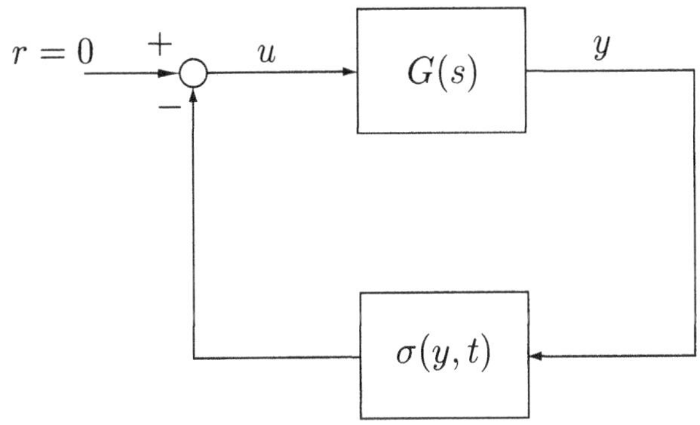

Figure 5.1 Feedback system with a static nonlinearity.

small gain theorems as special cases, and the Popov criterion. A convenient way of distinguishing these results is to focus on the allowable class of feedback nonlinearities. Specifically, the small gain, positivity, and circle theorems guarantee stability for arbitrarily time-varying nonlinearities, whereas the Popov criterion does not. This is not surprising since as will be shown the Lyapunov function upon which the small gain, positivity, and circle theorems are based is a fixed quadratic Lyapunov function, which permits arbitrary time variation of the nonlinearity. Alternatively, the Popov criterion is based on a *Luré-Postnikov Lyapunov function* which explicitly depends on the feedback nonlinearity thereby restricting its allowable time variation.

In this section, we present the absolute stability problem for feedback systems with time-varying memoryless input nonlinearities. Specifically, the forward path of the dynamical system is described by

$$\dot{x}(t) = Ax(t) + Bu(t), \qquad x(0) = x_0, \qquad t \geq 0, \tag{5.195}$$

$$y(t) = Cx(t) + Du(t), \tag{5.196}$$

where $x(t) \in \mathbb{R}^n$, $u(t) \in \mathbb{R}^m$, $y(t) \in \mathbb{R}^l$, and the feedback interconnection is given by

$$u(t) = -\sigma(y(t), t), \tag{5.197}$$

where

$$\sigma(\cdot,\cdot) \in \Phi \triangleq \{\sigma : \mathbb{R}^l \times \overline{\mathbb{R}}_+ \to \mathbb{R}^m : \ \sigma(0,\cdot) = 0, \ [\sigma(y,t) - M_1 y]^{\mathrm{T}}[\sigma(y,t) - M_2 y] \leq 0, \ y \in \mathbb{R}^l, \text{ a.e. } t \geq 0, \text{ and } \sigma(y,\cdot) \text{ is Lebesgue measurable for all } y \in \mathbb{R}^l\}, \tag{5.198}$$

where $M_1, M_2 \in \mathbb{R}^{m \times l}$. We say that $\sigma(\cdot,\cdot)$ *belongs to the sector* $[M_1, M_2]$ if and only if $\sigma(\cdot,\cdot) \in \Phi$. Note that in the single-input, single-output case $m = l = 1$, the sector condition characterizing Φ is equivalent to

$$M_1 y^2 \leq \sigma(y,t) y \leq M_2 y^2, \quad y \in \mathbb{R}, \quad \text{a.e.} \quad t \geq 0. \tag{5.199}$$

In this case, for each $t \in \overline{\mathbb{R}}_+$, the graph of $\sigma(y,t)$ lies in between two straight lines of slopes M_1 and M_2, respectively (see Figure 5.2).

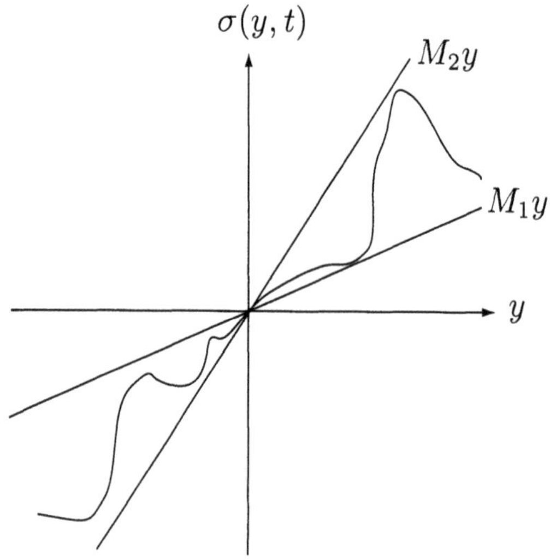

Figure 5.2 Sector-bounded nonlinearity $\sigma(y,t)$.

Now, the absolute stability problem, also known as the *Luré problem*, can be stated as follows. Given the dynamical system (5.195) and (5.196) where (A, B, C) is minimal and given M_1, $M_2 \in \mathbb{R}^{m\times l}$, derive conditions involving *only* the forward path (5.195) and (5.196) and the sector bounds M_1, M_2, such that the zero solution $x(t) \equiv 0$ of the feedback interconnection (5.195)–(5.197) is *globally* uniformly asymptotically stable for all $\sigma(\cdot,\cdot) \in \Phi$.

In an attempt to solve the absolute stability problem by examining the stability of all *linear time-invariant* systems within the family of nonlinear time-varying systems, in 1949 the Russian mathematician M. A. Aizerman made the following conjecture.

Conjecture 5.1 (Aizerman's Conjecture). Consider the nonlinear dynamical system (5.195)–(5.197) with $\sigma(y,t) = \sigma(y) \in \Phi$, $D = 0$, and $m = l = 1$. If the zero solution $x(t) \equiv 0$ to (5.195)–(5.197) with $\sigma(y) = Fy$, where $F \in [M_1, M_2]$, is asymptotically stable, then the zero solution $x(t) \equiv 0$ to (5.195)–(5.197) is globally asymptotically stable for all $\sigma(\cdot) \in \Phi$.

This conjecture is false. Motivated by Aizerman's conjecture, in 1957 R. E. Kalman made a refinement to this conjecture.

Conjecture 5.2 (Kalman's Conjecture). Consider the nonlinear dynamical system (5.195)–(5.197) with $\sigma(y,t) = \sigma(y)$, $D = 0$, and $m = l = 1$. Furthermore, assume $\sigma : \mathbb{R} \to \mathbb{R}$ is such that $M_1 \le \sigma'(y) \le M_2$ for all $y \in \mathbb{R}$. If the zero solution $x(t) \equiv 0$ to (5.195)–(5.197) with $\sigma(y) = Fy$, where

$F \in [M_1, M_2]$, is asymptotically stable, then the zero solution $x(t) \equiv 0$ to (5.195)–(5.197) is globally asymptotically stable for all $\sigma : \mathbb{R} \to \mathbb{R}$ such that $M_1 \leq \sigma'(y) \leq M_2$.

Note that since $\sigma : \mathbb{R} \to \mathbb{R}$ in Kalman's conjecture is such that $M_1 \leq \sigma'(y) \leq M_2$, it follows from the mean value theorem that $\sigma(\cdot) \in \Phi$. The converse, of course, is not necessarily true. Hence, Kalman's conjecture considers a refined class of feedback nonlinearities as compared to Aizerman's conjecture. Nevertheless, Kalman's conjecture is also false. However, it is important to note that if the zero solution $x(t) \equiv 0$ to (5.195)–(5.197) with $\sigma(y) = Fy$, where $F \in [M_1, M_2]$, is asymptotically stable, then the zero solution $x(t) \equiv 0$ to (5.195)–(5.197) is *locally* asymptotically stable for all $\sigma : \mathbb{R} \to \mathbb{R}$ such that $M_1 \leq \sigma'(y) \leq M_2$. This follows as a direct consequence of Lyapunov's indirect method.

5.8 The Positivity Theorem and the Circle Criterion

In this section, we present sufficient conditions for an absolute stability problem involving a dynamical system with memoryless, time-varying feedback nonlinearities. In the case where $m = l = 1$, these sufficient conditions involve inclusion/exclusion of the Nyquist plot of (5.195) and (5.197) in/from a disk region that encompasses the critical point, and hence, ensures stability. Appropriately, this result is known as the circle criterion or circle theorem. First, we present the simplest version of the circle criterion known as the positivity theorem involving a half-plane exclusion of the Nyquist plot of (5.195) and (5.196). For this result we assume that $\sigma(\cdot, \cdot) \in \Phi_{\mathrm{pr}}$, where

$$\Phi_{\mathrm{pr}} \triangleq \{\sigma : \mathbb{R}^l \times \overline{\mathbb{R}}_+ \to \mathbb{R}^m : \ \sigma(0, \cdot) = 0, \ \sigma^{\mathrm{T}}(y,t)[\sigma(y,t) - My] \leq 0, \ y \in \mathbb{R}^l, \\ \text{a.e. } t \geq 0, \text{ and } \sigma(y, \cdot) \text{ is Lebesgue measurable for all } y \in \mathbb{R}^l\}, \tag{5.200}$$

where $M \in \mathbb{R}^{m \times l}$.

Theorem 5.18 (Positivity Theorem). Consider the nonlinear dynamical system (5.195)–(5.197). Suppose

$$G(s) \sim \left[\begin{array}{c|c} A & B \\ \hline C & D \end{array}\right]$$

is minimal and $I_m + MG(s)$ is strictly positive real. Then the zero solution $x(t) \equiv 0$ of the negative feedback interconnection of (5.195)–(5.197) is globally exponentially stable for all $\sigma(\cdot, \cdot) \in \Phi_{\mathrm{pr}}$.

Proof. First note that the negative feedback interconnection of (5.195)–(5.197) has the state space representation

$$\dot{x}(t) = Ax(t) - B\sigma(y(t), t), \qquad x(0) = x_0, \qquad t \geq 0, \tag{5.201}$$
$$y(t) = Cx(t) - D\sigma(y(t), t). \tag{5.202}$$

Now, it follows from Theorem 5.14 that since $I_m + MG(s)$ is strictly positive real and minimal, there exist matrices $P \in \mathbb{R}^{n \times n}$, $L \in \mathbb{R}^{p \times n}$, and $W \in \mathbb{R}^{p \times m}$, with P positive definite, and a scalar $\varepsilon > 0$ such that

$$0 = A^{\mathrm{T}}P + PA + \varepsilon P + L^{\mathrm{T}}L, \tag{5.203}$$
$$0 = B^{\mathrm{T}}P - MC + W^{\mathrm{T}}L, \tag{5.204}$$
$$0 = (I + MD) + (I + MD)^{\mathrm{T}} - W^{\mathrm{T}}W. \tag{5.205}$$

Next, consider the Lyapunov function candidate $V(x) = x^{\mathrm{T}}Px$, where P satisfies (5.203)–(5.205). The corresponding Lyapunov derivative is given by

$$\begin{aligned} \dot{V}(x) &= V'(x)[Ax - B\sigma(y, t)] \\ &= 2x^{\mathrm{T}}P[Ax - B\sigma(y, t)] \\ &= x^{\mathrm{T}}(A^{\mathrm{T}}P + PA)x - 2\sigma^{\mathrm{T}}(y, t)B^{\mathrm{T}}Px \\ &= -x^{\mathrm{T}}(\varepsilon P + L^{\mathrm{T}}L)x - \sigma^{\mathrm{T}}(y, t)B^{\mathrm{T}}Px - x^{\mathrm{T}}PB\sigma(y, t). \end{aligned} \tag{5.206}$$

Now, adding and subtracting $2\sigma^{\mathrm{T}}(y, t)(I_m + MD)\sigma(y, t)$ and $2\sigma^{\mathrm{T}}(y, t)MCx$ to and from (5.206) yields

$$\begin{aligned} \dot{V}(x) = {} & -\varepsilon x^{\mathrm{T}}Px - x^{\mathrm{T}}L^{\mathrm{T}}Lx - \sigma^{\mathrm{T}}(y, t)(B^{\mathrm{T}}P - MC)x - x^{\mathrm{T}}(B^{\mathrm{T}}P - MC)^{\mathrm{T}} \\ & \cdot\sigma(y, t) - \sigma^{\mathrm{T}}(y, t)[(I_m + MD) + (I_m + MD)^{\mathrm{T}}]\sigma(y, t) + 2\sigma^{\mathrm{T}}(y, t) \\ & \cdot[\sigma(y, t) - M(Cx - D\sigma(y, t))], \end{aligned} \tag{5.207}$$

or, equivalently,

$$\begin{aligned} \dot{V}(x) = {} & -\varepsilon x^{\mathrm{T}}Px - [Lx - W\sigma(y, t)]^{\mathrm{T}}[Lx - W\sigma(y, t)] \\ & + 2\sigma^{\mathrm{T}}(y, t)[\sigma(y, t) - My]. \end{aligned} \tag{5.208}$$

Since $\sigma^{\mathrm{T}}(y, t)[\sigma(y, t) - My] \leq 0$ for all $t \geq 0$ and $\sigma(\cdot, \cdot) \in \Phi_{\mathrm{pr}}$, it follows that $\dot{V}(x) \leq -\varepsilon V(x)$, $x \in \mathbb{R}^n$, which shows that the zero solution $x(t) \equiv 0$ of the negative feedback interconnection of (5.195)–(5.197) is globally exponentially stable for all $\sigma(\cdot, \cdot) \in \Phi_{\mathrm{pr}}$. □

In the single-input, single-output case, the frequency domain condition in Theorem 5.18 has an interesting geometric interpretation in the Nyquist plane. Specifically, setting $G(\jmath\omega) = x + \jmath y$ and requiring that $1 + MG(s)$ be strictly positive real if follows that

$$x = \mathrm{Re}\ G(\jmath\omega) > -\tfrac{1}{M}, \qquad \omega \in \mathbb{R}, \tag{5.209}$$

which is equivalent to the graphical condition that the Nyquist plot of $G(\jmath\omega)$ lies to the right of the vertical line defined by $\mathrm{Re}[s] = -1/M$.

Example 5.7. Consider the linear dynamical system $G(s) = \frac{1}{s(s+1)^2}$ with a saturation feedback nonlinearity belonging to the sector $[0, M]$ shown in Figure 5.3. Note that $\lim_{\omega\to 0} |G(\jmath\omega)| = \infty$, $\lim_{\omega\to 0} \angle G(\jmath\omega) = -90°$,

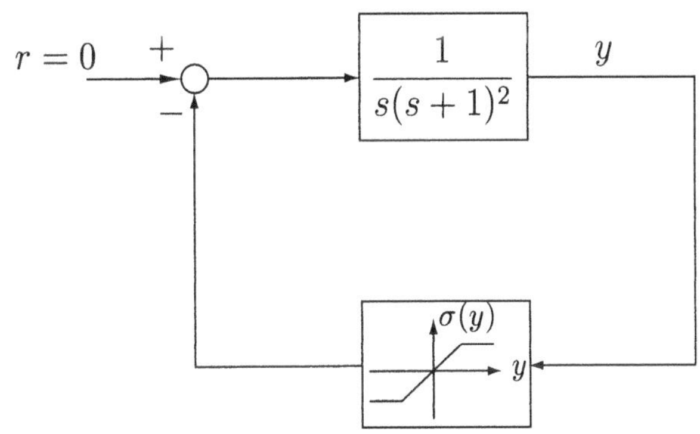

Figure 5.3 Feedback connection with saturation nonlinearity.

$\lim_{\omega\to\infty} |G(\jmath\omega)| = 0$, and $\lim_{\omega\to\infty} \angle G(\jmath\omega) = -270°$. Furthermore, the real axis crossing of the Nyquist plot of $G(\jmath\omega)$ corresponds to $\omega = 1$ rad/sec, and hence, $\mathrm{Re}[G(\jmath\omega)]|_{\omega=1} = -1/2$. Also note that $\mathrm{Re}[G(\jmath\omega)]|_{\omega=0} = -2$. Hence, the Nyquist plot of $G(\jmath\omega)$ lies to the right of the vertical line $\mathrm{Re}[s] = -2$ (see Figure 5.4). Thus, it follows from the positivity theorem that the system is globally exponentially stable for all nonlinearities in the sector $[0, 0.5]$. △

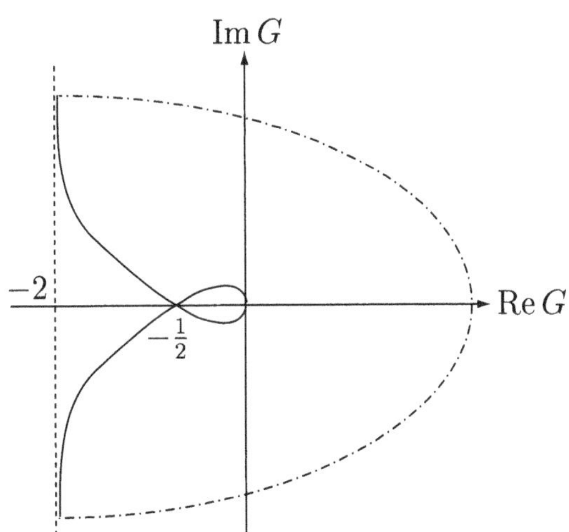

Figure 5.4 Nyquist plot for Example 5.7.

The positivity theorem applies to the case where the time-varying feedback nonlinearity belongs to the sector Φ_{pr}. Since in this case $\sigma(y, t) \equiv 0$

is an admissible feedback nonlinearity, it is clear that a necessary condition for guaranteeing absolute stability is that A be Hurwitz. Next, using loop shifting techniques, we consider the more general case where the feedback nonlinearity belongs to the general sector Φ. In this case, the restriction on A being Hurwitz can be removed. To consider double sector nonlinearities characterized by Φ note that if $\sigma(\cdot,\cdot) \in \Phi$, then the shifted nonlinearity $\sigma_{\mathrm{s}}(y,t) \triangleq \sigma(y,t) - M_1(y)$ belongs to Φ_{pr} with $M \triangleq M_2 - M_1$. Hence, transforming the forward path from $G(s)$ to $G_{\mathrm{s}}(s) \triangleq (I + G(s)M_1)^{-1}G(s)$ with feedback nonlinearity $\sigma_{\mathrm{s}}(\cdot,\cdot) \in \Phi_{\mathrm{pr}}$ gives an equivalent representation to the dynamical system $G(s)$ with feedback nonlinearity $\sigma(\cdot,\cdot) \in \Phi$. This equivalence is shown in Figure 5.5. Thus, the following result is a direct consequence of Theorem 5.18.

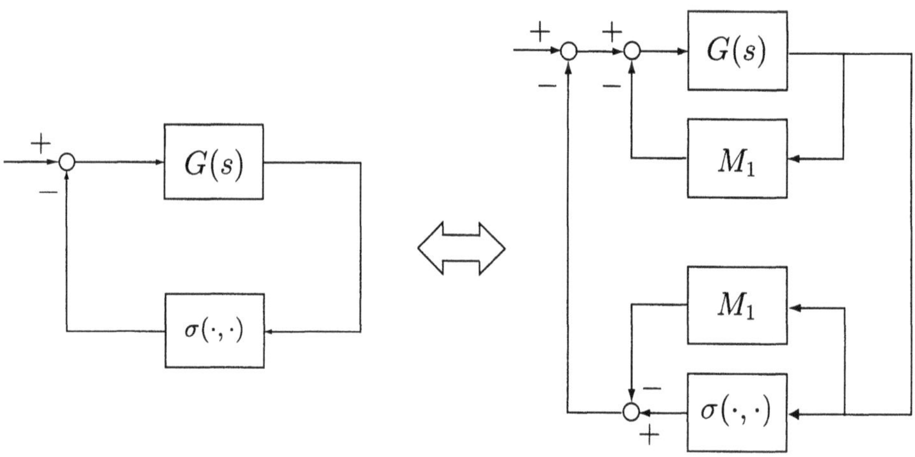

Figure 5.5 Equivalence via loop shifting.

Corollary 5.8 (Circle Theorem). Consider the nonlinear dynamical system (5.195)–(5.197). Suppose

$$G(s) \sim \left[\begin{array}{c|c} A & B \\ \hline C & D \end{array}\right]$$

is minimal, $\det[I_m + M_1G(s)] \neq 0$, $\mathrm{Re}[s] \geq 0$, and $[I_m + M_2G(s)][I_m + M_1G(s)]^{-1}$ is strictly positive real. Then, the zero solution $x(t) \equiv 0$ of the negative feedback interconnection of (5.195)–(5.197) is globally exponentially stable for all $\sigma(\cdot,\cdot) \in \Phi$.

Proof. It need only be shown that the transformed system satisfies the hypotheses of Theorem 5.18. First, it follows from simple algebraic manipulations that $[I + M_2G(s)][I + M_1G(s)]^{-1} = I + (M_2 - M_1)(I + G(s)M_1)^{-1}G(s) = I + MG_{\mathrm{s}}(s)$. Hence, by assumption, $I + MG_{\mathrm{s}}(s)$ is strictly positive real. In addition, the realization of $I + MG_{\rm s}(s)$ is given by

$$I + MG_{\rm s}(s) \sim \left[\begin{array}{c|c} A - B(I+M_1D)^{-1}M_1C & B(I+M_1D)^{-1} \\ \hline M(I+DM_1)^{-1}C & I + M(I+DM_1)^{-1}D \end{array}\right], \tag{5.210}$$

which is minimal if (A, B, C) is minimal. Hence, all the conditions of Theorem 5.18 are satisfied for the transformed system $G_{\rm s}(s)$ with feedback nonlinearities $\sigma_{\rm s}(\cdot,\cdot) \in \Phi_{\rm pr}$. □

Corollary 5.8 is known as the multivariable circle theorem or circle criterion. Note that in the case where $M_1 = 0$, Corollary 5.8 specializes to Theorem 5.18. Alternatively, in the case where $M_1 = -\gamma^{-1}I$ and $M_2 = \gamma^{-1}I$, where $\gamma > 0$, Corollary 5.8 reduces to the classical small gain theorem which states that if G is gain bounded, that is, $\|G(s)\|_\infty < \gamma$, then the zero solution of the feedback interconnection of G with a gain bounded nonlinearity $\|\sigma(y,t)\|_2 \leq \gamma^{-1}\|y\|_2$, $y \in \mathbb{R}^l$, almost everywhere $t \geq 0$, is uniformly asymptotically stable. To see this, first note that in the case where $M_1 = -\gamma^{-1}I$ and $M_2 = \gamma^{-1}I$, Φ specializes to

$$\Phi_{\rm br} \triangleq \{\sigma : \mathbb{R}^l \times \overline{\mathbb{R}}_+ \to \mathbb{R}^m : \sigma(0,\cdot) = 0,\ \|\sigma(y,t)\|_2 \leq \gamma^{-1}\|y\|_2,\ y \in \mathbb{R}^l, \text{ a.e. } t \geq 0, \text{ and } \sigma(y,\cdot) \text{ is Lebesgue measurable for all } y \in \mathbb{R}^l\}. \tag{5.211}$$

Furthermore, the strict positive real condition in Corollary 5.8 implies

$$[I + \gamma^{-1}G(\jmath\omega)][I - \gamma^{-1}G(\jmath\omega)]^{-1} + [I - \gamma^{-1}G^*(\jmath\omega)]^{-1}[I + \gamma^{-1}G^*(\jmath\omega)] > 0, \quad \omega \in \mathbb{R}, \tag{5.212}$$

or, equivalently, forming $[I - \gamma^{-1}G^*(\jmath\omega)](5.212)[I - \gamma^{-1}G(\jmath\omega)]$ yields

$$\begin{aligned} G^*(\jmath\omega)G(\jmath\omega) &< \gamma^2 I, \quad \omega \in \mathbb{R}, \\ \lambda_{\max}^{1/2}[G^*(\jmath\omega)G(\jmath\omega)] &< \gamma, \quad \omega \in \mathbb{R}, \\ \sigma_{\max}[G(\jmath\omega)] &< \gamma, \quad \omega \in \mathbb{R}, \\ \sup_{\omega\in\mathbb{R}} \sigma_{\max}[G(\jmath\omega)] &< \gamma, \\ \|G(s)\|_\infty &< \gamma. \end{aligned} \tag{5.213}$$

In the scalar case, the circle criterion provides a very interesting and elegant graphical interpretation in terms of an inclusion/exclusion of the Nyquist plot of $G(s)$ in/from a disk region in the Nyquist plane. To arrive at this result, let $M_1 \neq 0$, $M_2 \neq 0$, and $M_1 < M_2$, and note that setting $G(\jmath\omega) = x + \jmath y$, the strict positive real condition in Corollary 5.8 implies that

$$\operatorname{Re}\left[\frac{G(\jmath\omega)}{1 + M_1G(\jmath\omega)}\right] + \frac{1}{M_2 - M_1} > 0, \quad \omega \in \mathbb{R}, \tag{5.214}$$

or, equivalently,

$$\frac{(1+M_2x)(1+M_1x)+M_1M_2y^2}{(1+M_1x)^2+M_1^2y^2} > 0, \quad \omega \in \mathbb{R}. \tag{5.215}$$

Next, suppose A in the realization of $G(s)$ has ν eigenvalues with positive real parts. Then, it follows from the Nyquist criterion [445] that $G_{\rm s}(s)$ is asymptotically stable if and only if the number of counterclockwise encirclements of the $-1/M_1+\jmath 0$ point of the image of the clockwise Nyquist contour Γ under the mapping $G(s)$ equals the number of unstable poles of the loop gain transfer function $M_1G(s)$. Hence, it follows from Corollary 5.8 that the zero solution $x(t)\equiv 0$ of the negative feedback interconnection of (5.195)–(5.197) is globally exponentially stable for all $\sigma(\cdot,\cdot)\in\Phi$ if the Nyquist plot of $G(s)$ does not intersect the point $-1/M_1+\jmath 0$ and encircles it ν times counterclockwise, and (5.214) holds. However, (5.214) holds if and only if

$$(x+1/M_1)(x+1/M_2)+y^2>0, \qquad \omega\in\mathbb{R}, \tag{5.216}$$

for $M_1M_2>0$, and if and only if

$$(x+1/M_1)(x+1/M_2)+y^2<0, \qquad \omega\in\mathbb{R}, \tag{5.217}$$

for $M_1M_2<0$. Hence, (5.214) holds if and only if for all $\omega\in\mathbb{R}$, $G(\jmath\omega)$ lies *outside* the disk $\mathcal{D}(M_1,M_2)$ centered at $(M_1+M_2)/(2M_1M_2+\jmath 0)$ with radius $(M_2-M_1)/2|M_2M_1|$ in the case where $M_1M_2>0$, and lies in the *interior* of the disk $\mathcal{D}(M_1,M_2)$ in the case where $M_1M_2<0$. In the former case, if the Nyquist plot does not enter the disk $\mathcal{D}(M_1,M_2)$ and encircles the point $-1/M_1+\jmath 0$ ν times counterclockwise, then the Nyquist plot must encircle the disk $\mathcal{D}(M_1,M_2)$ ν times and vice versa. In light of the above the following theorem is immediate.

Theorem 5.19 (Circle Criterion). Consider the nonlinear dynamical system (5.195)–(5.197) with $m=l=1$ and $\sigma(\cdot,\cdot)\in\Phi$. Furthermore, suppose A has ν eigenvalues with positive real parts. Then, the zero solution $x(t)\equiv 0$ of the negative feedback interconnection of (5.195)–(5.197) is globally exponentially stable for all $\sigma(\cdot,\cdot)\in\Phi$ if one of the following conditions is satisfied, as appropriate:

i) If $0<M_1<M_2$, the Nyquist plot of $G(\jmath\omega)$ does not enter the disk $\mathcal{D}(M_1,M_2)$ and encircles it ν times counterclockwise.

ii) If $0=M_1<M_2$, A is Hurwitz, and the Nyquist plot of $G(\jmath\omega)$ lies in the half plane $\{s\in\mathbb{C}:\ \mathrm{Re}[s]>-1/M_2\}$.

iii) If $M_1<0<M_2$, A is Hurwitz, and the Nyquist plot of $G(\jmath\omega)$ lies in the interior of the disk $\mathcal{D}(M_1,M_2)$.

iv) If $M_1 < M_2 \le 0$, replace $G(\cdot)$ by $-G(\cdot)$, M_1 by $-M_2$, M_2 by $-M_1$, and apply *i)* or *ii)* as appropriate.

It is interesting to note that if $M_1 = M_2$, then the critical disk $\mathcal{D}(M_1, M_2)$ collapses to the critical point $-1/M_1 + \jmath 0$, and hence, the circle criterion reduces to the sufficiency portion of the Nyquist criterion. Figure 5.6 shows the five different cases addressed in Theorem 5.19 along with the associated forbidden regions in the Nyquist plane.

Example 5.8. Consider the linear dynamical system

$$G(s) = \frac{s+1}{s(0.1s+1)^2(s-1)}$$

with feedback nonlinearity shown in Figure 5.7. The Nyquist plot of $G(\jmath\omega)$ is shown in Figure 5.8. Since $G(s)$ has one pole in the open right half plane we use *i)* of the circle criterion. Specifically, we need to construct a disk $\mathcal{D}(M_1, M_2)$ such that the Nyquist plot does not enter $\mathcal{D}(M_1, M_2)$ and encircles it once counterclockwise. Inspecting the Nyquist plot of $G(\jmath\omega)$ shows that the disk $\mathcal{D}(1.85, 3.34)$ is encircled once in the counterclockwise direction by the left lobe of the Nyquist plot. Hence, we conclude that the system is exponentially stable for the nonlinearity shown in Figure 5.7 with slopes $M_1 = 1.85$ and $M_2 = 3.34$. △

5.9 The Popov Criterion

In this section, we present another absolute stability criterion known as the Popov criterion. Although often discussed in juxtaposition with the circle criterion, the Popov criterion is fundamentally distinct from the circle criterion in regard to its Lyapunov function foundation. Whereas the small gain, positivity, and circle results are based upon fixed quadratic Lyapunov functions, the Popov result is based upon a Lyapunov function that is a function of the sector-bounded nonlinearity. In particular, in the single-input single-output case, the Popov criterion is based upon the Luré-Postnikov Lyapunov function having the form

$$V(x) = x^{\mathrm{T}}Px + N\int_0^y \sigma(s)\mathrm{d}s, \tag{5.218}$$

where $P > 0$, $N > 0$, $y = Cx$, and $\sigma(\cdot)$ is a scalar memoryless *time-invariant* nonlinearity belonging to the sector $[0, M]$. Thus, in effect, the Popov result guarantees stability by means of a *family* of Lyapunov functions and, hence, does *not* in general apply to time-varying nonlinearities.

To present the multivariable Popov criterion, consider the dynamical

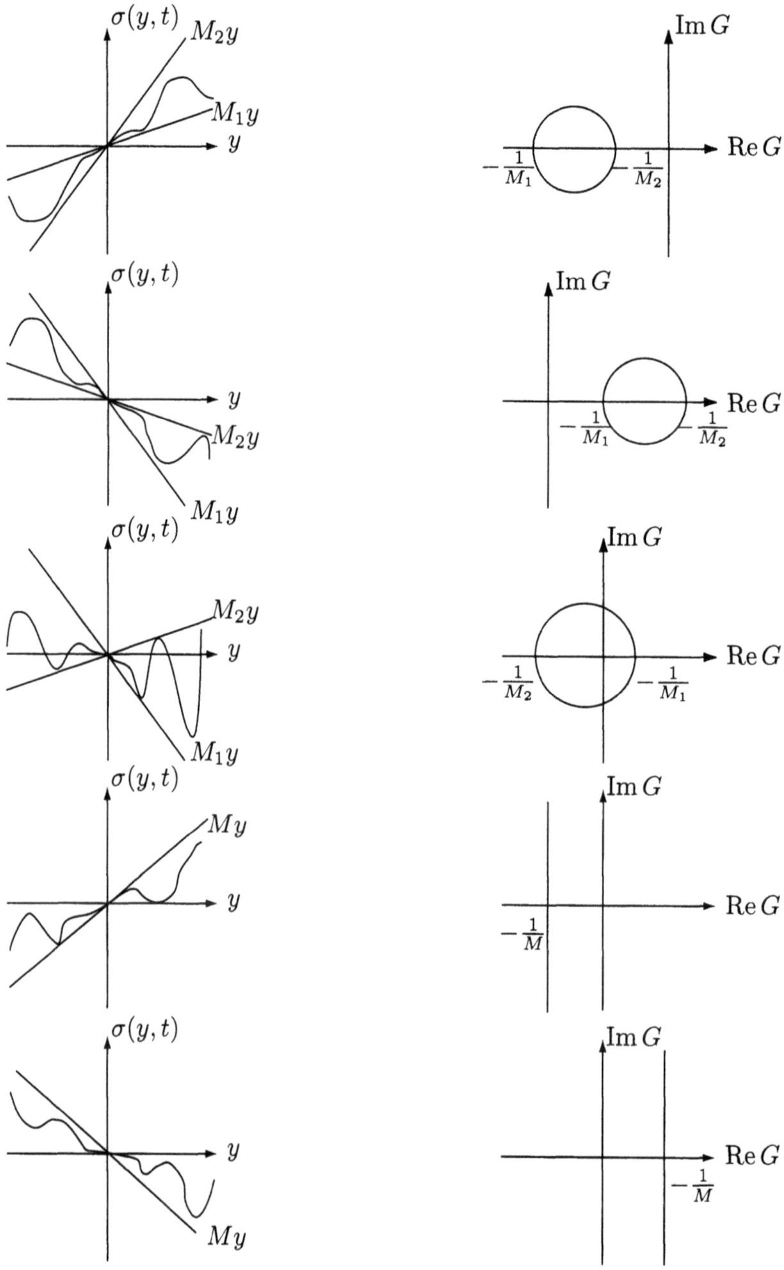

Figure 5.6 Sector nonlinearities and forbidden Nyquist regions.

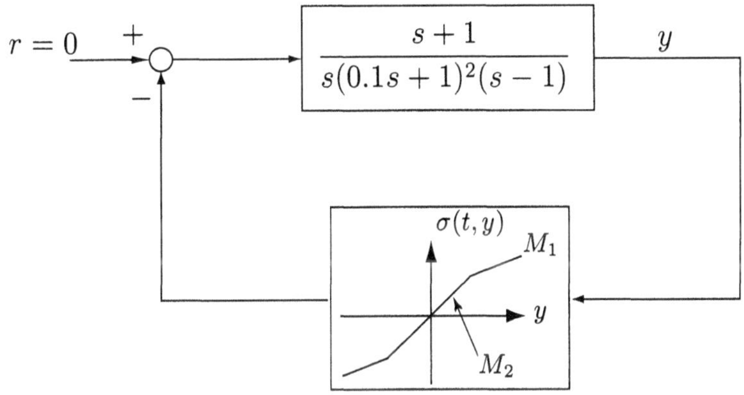

Figure 5.7 Feedback connection with multislope nonlinearity.

system

$$\dot{x}(t) = Ax(t) + Bu(t), \quad x(0) = x_0, \quad t \geq 0, \tag{5.219}$$
$$y(t) = Cx(t), \tag{5.220}$$

where $x(t) \in \mathbb{R}^n$, $u(t) \in \mathbb{R}^m$, and $y(t) \in \mathbb{R}^m$, with feedback nonlinearity

$$u(t) = -\sigma(y(t)), \tag{5.221}$$

where

$$\sigma(\cdot) \in \Phi_{\mathrm{P}} \triangleq \{\sigma : \mathbb{R}^m \to \mathbb{R}^m : \sigma(0) = 0,\ \sigma^{\mathrm{T}}(y)[\sigma(y) - My] \leq 0,\ y \in \mathbb{R}^m, \text{ and } \sigma(y) = [\sigma_1(y_1),\ \sigma_2(y_2), \ldots, \sigma_m(y_m)]^{\mathrm{T}}\}, \tag{5.222}$$

where $M \in \mathbb{R}^{m \times m}$ and $M > 0$. Note that the components of σ are assumed to be decoupled. If $M = \operatorname{diag}[M_1, \ldots, M_m]$, $M_i > 0$, $i = 1, \ldots, m$, then the sector condition characterizing Φ_{P} is implied by the scalar sector conditions

$$0 \leq \sigma_i(y_i)y_i \leq M_i y_i^2, \qquad y_i \in \mathbb{R}, \qquad i = 1, \ldots, m, \tag{5.223}$$

where $y_i \in \mathbb{R}$ denotes the ith component of $y \in \mathbb{R}^m$.

Theorem 5.20 (Popov Criterion). Consider the nonlinear dynamical system (5.219)–(5.221). Suppose

$$G(s) \sim \left[\begin{array}{c|c} A & B \\ \hline C & 0 \end{array}\right]$$

is minimal, and assume there exists $N = \operatorname{diag}[N_1, N_2, \ldots, N_m]$, $N_i \geq 0$, $i = 1, \ldots, m$, such that $I_m + M(I_m + Ns)G(s)$ is strictly positive real and $\det(I_m + \lambda N) \neq 0$, where $M = \operatorname{diag}[M_1, M_2, \ldots, M_m] > 0$ and $\lambda \in \operatorname{spec}(A)$. Then, the zero solution $x(t) \equiv 0$ of the negative feedback interconnection of (5.219)–(5.221) is globally asymptotically stable for all $\sigma(\cdot) \in \Phi_{\mathrm{P}}$.

Proof. First note that the negative feedback interconnection of

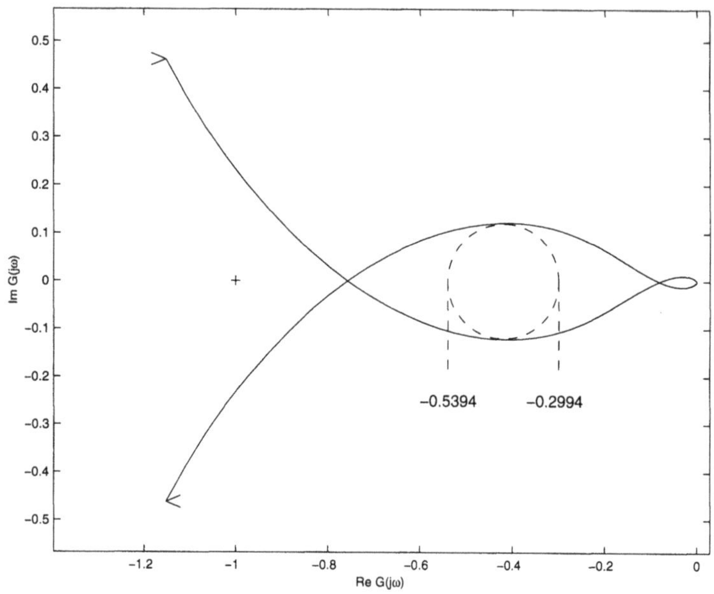

Figure 5.8 Nyquist plot for Example 5.8.

(5.219)–(5.221) has the state space representation

$$\dot{x}(t) = Ax(t) - B\sigma(y(t)), \qquad x(0) = x_0, \qquad t \geq 0, \tag{5.224}$$
$$y(t) = Cx(t). \tag{5.225}$$

Next, noting that $(I + Ns)G(s)$ has a differentiation effect on the plant transfer function $G(s)$ it follows that the realization of $I_m + M(I + Ns)G(s)$ is given by

$$I_m + M(I + Ns)G(s) \sim \left[\begin{array}{c|c} A & B \\ \hline MC + MNCA & I_m + MNCB \end{array}\right]. \tag{5.226}$$

Since $\det(I_m + \lambda N) \neq 0$, $\lambda \in \operatorname{spec}(A)$, it follows that the realization given in (5.226) is minimal. To see this, note that (A, B) is controllable by assumption. To show that $(A, MC + MNCA)$ is observable or, equivalently, $(A, C + NCA)$ is observable since $M > 0$, assume, *ad absurdum*, that $(A, C + NCA)$ is not observable. In this case, since $\det(I_m + \lambda N) \neq 0$, $\lambda \in \operatorname{spec}(A)$, it follows that there exists $\eta \in \mathbb{C}^m$, $\eta \neq 0$, such that $A\eta = \lambda\eta$ and $C\eta = 0$, which implies that (A, C) is not observable, and hence, contradicts the minimality assumption of $G(s)$.

Next, it follows from Theorem 5.14 that if $I_m + M(I_m + Ns)G(s)$ is strictly positive real, then there exists matrices $P \in \mathbb{R}^{n\times n}$, $L \in \mathbb{R}^{p\times n}$, and $W \in \mathbb{R}^{p\times m}$, with P positive definite, and a scalar $\varepsilon > 0$ such that

$$0 = A^{\mathrm{T}}P + PA + \varepsilon P + L^{\mathrm{T}}L, \tag{5.227}$$

$$0 = B^{\mathrm{T}}P - (MC + MNCA) + W^{\mathrm{T}}L, \quad (5.228)$$
$$0 = (I_m + MNCB) + (I_m + MNCB)^{\mathrm{T}} - W^{\mathrm{T}}W. \quad (5.229)$$

Now, consider the Luré-Postnikov Lyapunov function candidate

$$V(x) = x^{\mathrm{T}}Px + 2\sum_{i=1}^{m}\int_0^{y_i}\sigma_i(s)M_iN_i\mathrm{d}s, \quad (5.230)$$

where P satisfies (5.227)–(5.229). Note that since P is positive definite and $\sigma(\cdot) \in \Phi_{\mathrm{P}}$, $V(x)$ is positive definite for all nonzero $x \in \mathbb{R}^n$.

Using Leibnitz's integral rule,[1] the corresponding Lyapunov derivative of $V(\cdot)$ is given by

$$\dot{V}(x) = 2x^{\mathrm{T}}P[Ax - B\sigma(y)] + 2\sum_{i=1}^{m}\sigma_i(y_i)M_iN_i\dot{y}_i, \quad (5.231)$$

or, equivalently, using (5.227),

$$\dot{V}(x) = -x^{\mathrm{T}}(\varepsilon P + L^{\mathrm{T}}L)x - 2\sigma^{\mathrm{T}}(y)B^{\mathrm{T}}Px + 2\sigma^{\mathrm{T}}(y)MN\dot{y}. \quad (5.232)$$

Next, since $\dot{y} = C\dot{x} = CAx - CB\sigma(y)$, (5.232) becomes

$$\begin{aligned}\dot{V}(x) &= -\varepsilon x^{\mathrm{T}}Px - x^{\mathrm{T}}L^{\mathrm{T}}Lx - \sigma^{\mathrm{T}}(y)(B^{\mathrm{T}}P - MNCA)x \\ &\quad -x^{\mathrm{T}}(B^{\mathrm{T}}P - MNCA)^{\mathrm{T}}\sigma(y) - \sigma^{\mathrm{T}}(y)[MNCB + B^{\mathrm{T}}C^{\mathrm{T}}NM]\sigma(y).\end{aligned} \quad (5.233)$$

Adding and subtracting $2\sigma^{\mathrm{T}}(y)MCx$ and $2\sigma^{\mathrm{T}}(y)\sigma(y)$ to and from (5.233) and using (5.228) and (5.229) yields

$$\dot{V}(x) = -\varepsilon x^{\mathrm{T}}Px - [Lx - W\sigma(y)]^{\mathrm{T}}[Lx - W\sigma(y)] + 2\sigma^{\mathrm{T}}(y)[\sigma(y) - My]. \quad (5.234)$$

Since $\sigma^{\mathrm{T}}(y)[\sigma(y) - My] \leq 0$ for all $\sigma(\cdot) \in \Phi_{\mathrm{P}}$, it follows that $\dot{V}(x) < 0$, $x \in \mathbb{R}^n$, $x \neq 0$, which shows that the zero solution $x(t) \equiv 0$ of the negative feedback interconnection (5.219)–(5.221) is globally asymptotically stable for all $\sigma(\cdot) \in \Phi_{\mathrm{P}}$ □

As in the positivity case, the Popov criterion also has an interesting geometric interpretation in the Nyquist plane and in a modified plane called the *Popov plane*. To see this, let $m = 1$, set $G(\jmath\omega) = x + \jmath y$, and require that $1 + M(1 + Ns)G(s)$ be strictly positive real. In this case, we obtain

$$\omega y < \frac{1}{N}x + \frac{1}{NM}, \qquad \omega \in \mathbb{R}. \quad (5.235)$$

---

[1]Recall that for $f : \mathbb{R}^2 \to \mathbb{R}$, $g : \mathbb{R} \to \mathbb{R}$, and $h : \mathbb{R} \to \mathbb{R}$, Leibnitz's integral rule is

$$\frac{\mathrm{d}}{\mathrm{d}t}\int_{g(t)}^{h(t)} f(t,s)\mathrm{d}s = \int_{g(t)}^{h(t)}\frac{\partial}{\partial t}f(t,s)\mathrm{d}s + f(t,h(t))\frac{\mathrm{d}h(t)}{\mathrm{d}t} - f(t,g(t))\frac{\mathrm{d}g(t)}{\mathrm{d}t},$$

whenever the above integrals exist.

In the Nyquist plane, (5.235) requires the Nyquist plot at each $\omega \in \mathbb{R}$ be to the right of a rotating line that is a linear function of $\omega \in \mathbb{R}$. Alternatively, (5.235) is a frequency domain stability criterion with a graphical interpretation in a modified Nyquist plane, known as the *Popov plane* involving Re G and ωIm G, in terms of a fixed straight line (Popov line) with real axis intercept $-1/M$ and slope $1/N$ (see Figure 5.9). It is

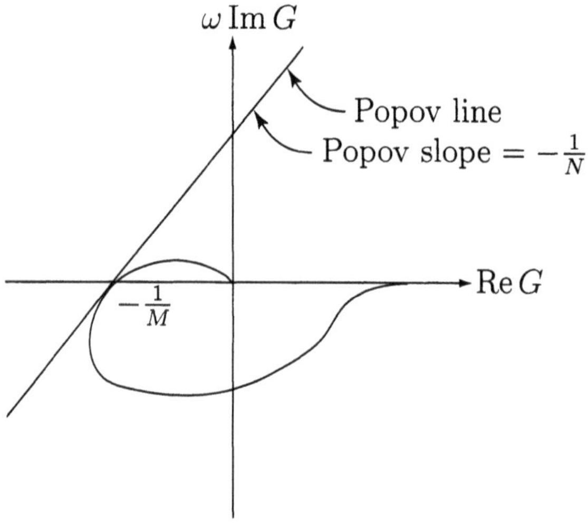

Figure 5.9 Popov plot.

interesting to note that, unlike the Nyquist plot, the Popov plot is an odd plot, and hence, we need only consider $\omega \in [0, \infty)$ in generating the plot. Furthermore, setting $N = 0$ the Popov criterion collapses to the positivity theorem. Finally, we note that as for the positivity theorem, the restriction on A being asymptotically stable can be removed using the loop sifting techniques as discussed in Section 5.8. However, in this case the resulting frequency domain stability conditions do not provide a simple graphical test as in the case of the single sector Popov criterion. See [331] and Problem 5.59 for further details.

Example 5.9. Consider the linear dynamical system with the saturation feedback nonlinearity addressed in Example 5.7. Figure 5.10 shows the Popov plot of $G(\jmath\omega)$. Note that the Popov plot lies to the right of any line of slope $1/N \leq 0.5$ that intersects the real axis to the left of $-1/M = -0.5$. Hence, the maximum possible value of M is 2, which is substantially less conservative than the result arrived at by the positivity theorem in Example 5.7. △

The strict positive real conditions appearing in Theorem 5.20 can be written as $I_m + MZ(s)G(s)$, where $Z(s) = Z_{\rm P}(s) \triangleq I_m + Ns$ is known as the *stability Popov multiplier*. Further refinements of the absolute stability

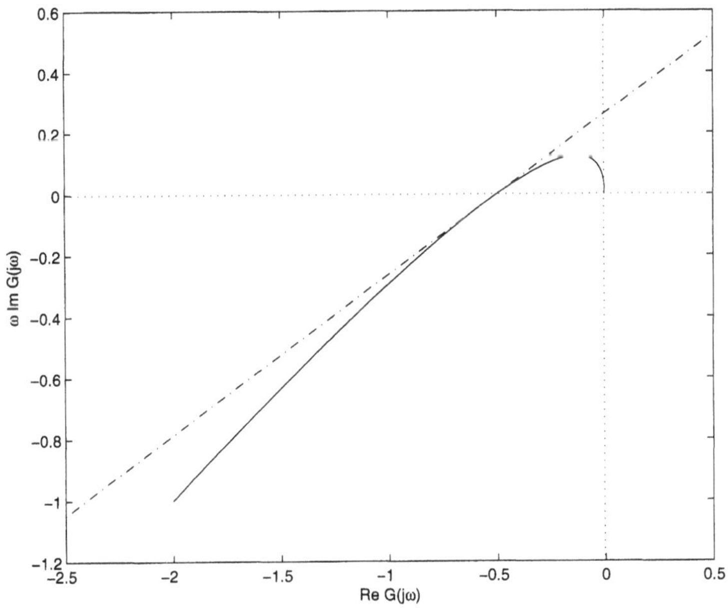

Figure 5.10 Popov plot for Example 5.9.

Popov criterion can be developed by considering extended Luré-Postnikov Lyapunov functions. Specifically, absolute stability criteria can be derived that extend the Popov criterion for sector-bounded, time-invariant nonlinear functions to monotonic and odd monotonic nonlinearities. In particular, suitable positive real stability multipliers $Z(s)$ can be constructed as driving-point impedances of passive electrical networks involving resistor-inductor (RL), resistor-capacitor (RC), and inductor-capacitor (LC) combinations which exhibit interlacing pole-zero patterns on the negative real axis and imaginary axis [71, 328, 330, 479]. These stability multipliers effectively place less restrictive conditions on the linear part of the system and more restrictive conditions on the allowable class of feedback nonlinearities. In addition, the stability criteria for these refined class of nonlinearities are predicated on extended Luré-Postnikov Lyapunov functions involving real signals obtained by passing the system outputs ($y = Cx$) through a parallel bank of decoupled low pass filters with specified time constants and positive gains corresponding to the RL, RC, and LC networks. However, as a result of the more involved multiplier construction, the resulting frequency domain conditions do not provide a simple graphical test involving fixed shapes in the Nyquist and Popov planes as in the case of the circle and Popov criteria.

5.10 Problems

Problem 5.1. Let $H(s) \in \mathbb{C}^{m\times m}$. Suppose $|||H(s)|||_\infty \le \gamma$, where $\gamma > 0$, and $\det[I - \gamma^{-1}H(s)] \neq 0$, $\mathrm{Re}[s] \ge 0$. Show that

$$G(s) = [I + \gamma^{-1}H(s)][I - \gamma^{-1}H(s)]^{-1}, \qquad (5.236)$$

is positive real. Conversely, show that if $G(s) \in \mathbb{C}^{m\times m}$ is positive real, then

$$\gamma^{-1}H(s) = [G(s) - I][G(s) + I]^{-1}, \qquad (5.237)$$

is bounded real (i.e., $|||H(s)|||_\infty \le \gamma$).

Problem 5.2. Consider the nonlinear dynamical system $\mathcal{G}$ given by (5.75) and (5.76). Show that:

i) If $\mathcal{G}$ is dissipative with respect to the supply rate $r(u,y) = u^{\mathrm{T}}y - \varepsilon y^{\mathrm{T}}y$, where $\varepsilon > 0$, then $\alpha\mathcal{G}$ is dissipative with respect to the supply rate $r(u,y) = \frac{1}{\sqrt{\alpha}}u^{\mathrm{T}}y - \frac{\varepsilon}{\alpha}y^{\mathrm{T}}y$, where $\alpha > 0$. Here, $\alpha\mathcal{G}$ denotes a nonlinear system with output $y_\alpha = \sqrt{\alpha}y$, where y is the output of $\mathcal{G}$.

ii) If $\mathcal{G}$ is dissipative with respect to the supply rate $r(u,y) = u^{\mathrm{T}}y - \varepsilon u^{\mathrm{T}}u$, where $\varepsilon > 0$, then $\alpha\mathcal{G}$ is dissipative with respect to the supply rate $r(u,y) = \sqrt{\alpha}u^{\mathrm{T}}y - \alpha\varepsilon u^{\mathrm{T}}u$, where $\alpha > 0$.

Problem 5.3. The nonlinear dynamical system $\mathcal{G}$ given by (5.7) and (5.8) is *cyclo-dissipative* (respectively, *exponentially cyclo-dissipative*) *with respect to the supply rate* $r(u,y)$ if (5.9) (respectively, (5.10)) is satisfied for all $t \ge t_0$ and all $u(\cdot) \in \mathcal{U}$ with $x(t_0) = x(t) = 0$. In this case, $V_{\mathrm{a}}(x_0)$ given by (5.11) (respectively, (5.12)) is called the *virtual available storage* (respectively, *virtual available exponential storage*) of $\mathcal{G}$. Show that $\mathcal{G}$ is cyclo-dissipative (respectively, exponentially cyclo-dissipative) with respect to the supply rate $r(u,y)$ if and only if $V_{\mathrm{a}}(x)$ is finite for all $x_0 \in \mathcal{D}$ and $V_{\mathrm{a}}(0) = 0$.

Problem 5.4. A function $V_{\mathrm{s}} : \mathcal{D} \to \mathbb{R}$ is a *virtual storage function* (respectively, *virtual exponential storage function*) of the nonlinear dynamical system $\mathcal{G}$ given by (5.7) and (5.8) if it satisfies $V_{\mathrm{s}}(0) = 0$ and (5.16) (respectively, (5.17)). Show that if the virtual available storage (respectively, virtual available exponential storage) $V_{\mathrm{a}}(x_0)$ is finite for all $x_0 \in \mathcal{D}$ and $V_{\mathrm{a}}(0) = 0$, then $V_{\mathrm{a}}(\cdot)$ is a virtual storage function (respectively, virtual exponential storage function) for $\mathcal{G}$. Furthermore, show that every virtual storage function (respectively, virtual exponential storage function) $V_{\mathrm{s}}(\cdot)$ for $\mathcal{G}$ satisfies $V_{\mathrm{a}}(x) \le V_{\mathrm{s}}(x)$, $x \in \mathcal{D}$.

Problem 5.5. The nonlinear dynamical system $\mathcal{G}$ given by (5.7) and

(5.8) is *cyclo-lossless with respect to the supply rate* $r(u,y)$ if (5.9) is satisfied as an equality for all $t \geq t_0$ and all $u(\cdot) \in \mathcal{U}$ with $x(t_0) = x(t) = 0$. Assume $\mathcal{G}$ is completely reachable to and from the origin. Show that $\mathcal{G}$ is cyclo-lossless with respect to the supply rate $r(u,y)$ if and only if there exists a virtual storage function $V_\mathrm{s}(x)$, $x \in \mathcal{D}$, satisfying (5.16) as an equality. Furthermore, show that if $\mathcal{G}$ is cyclo-lossless with respect to the supply rate $r(u,y)$, then the virtual storage function $V_\mathrm{s}(x)$, $x \in \mathcal{D}$, is unique and is given by (5.44) where $x(t)$, $t \geq 0$, is the solution to (5.7) with admissible $u(\cdot) \in \mathcal{U}$, $t \geq 0$, $x(-T) = 0$, $x(T) = 0$, and $x(0) = x_0$, $x_0 \in \mathcal{D}$.

Problem 5.6. Let $Q \in \mathbb{S}^l$, $S \in \mathbb{R}^{l \times m}$, $R \in \mathbb{S}^m$, and let $\mathcal{G}$ given by (5.75) and (5.76) be zero-state observable and completely reachable. Furthermore, assume that all virtual storage functions (see Problem 5.4) of $\mathcal{G}$ are continuously differentiable. Show that $\mathcal{G}$ is cyclo-dissipative (see Problem 5.3) with respect to the quadratic supply rate $r(u,y) = y^\mathrm{T}Qy + 2y^\mathrm{T}Su + u^\mathrm{T}Ru$ if and only if there exist functions $V_\mathrm{s} : \mathbb{R}^n \to \mathbb{R}$, $\ell : \mathbb{R}^n \to \mathbb{R}^p$, and $\mathcal{W} : \mathbb{R}^n \to \mathbb{R}^{p \times m}$ such that $V_\mathrm{s}(\cdot)$ is continuously differentiable, $V_\mathrm{s}(0) = 0$, and, for all $x \in \mathbb{R}^n$, (5.77)–(5.79) hold.

Problem 5.7. Let $Q \in \mathbb{S}^l$, $S \in \mathbb{R}^{l \times m}$, $R \in \mathbb{S}^m$, and let $\mathcal{G}$ given by (5.75) and (5.76) be zero-state observable and completely reachable. Furthermore, assume that all virtual exponential storage functions (see Problem 5.4) of $\mathcal{G}$ are continuously differentiable. Show that $\mathcal{G}$ is exponentially cyclo-dissipative (see Problem 5.3) with respect to the quadratic supply rate $r(u,y) - y^\mathrm{T}Qy + 2y^\mathrm{T}Su + u^\mathrm{T}Ru$ if and only if there exist functions $V_\mathrm{s} : \mathbb{R}^n \to \mathbb{R}$, $\ell : \mathbb{R}^n \to \mathbb{R}^p$, $\mathcal{W} : \mathbb{R}^n \to \mathbb{R}^{p \times m}$, and a scalar $\varepsilon > 0$ such that $V_\mathrm{s}(\cdot)$ is continuously differentiable, $V_\mathrm{s}(0) = 0$, and, for all $x \in \mathbb{R}^n$, (5.87)–(5.89) hold.

Problem 5.8. Consider the nonlinear dynamical system $\mathcal{G}$ given by (5.7) and (5.8). Define the *minimum input energy* of $\mathcal{G}$ by

$$V_\mathrm{e}(x_0) = \inf_{u(\cdot),\, T \geq 0} \int_0^T r(u(t), y(t))\mathrm{d}t, \tag{5.238}$$

where $x(t)$, $t \geq 0$, is the solution to (5.7) with $x(0) = 0$ and $x(T) = x_0$. The infimum in (5.238) is taken over all time $t \geq 0$ and all admissible inputs $u(\cdot)$ which drive $\mathcal{G}$ from $x(0) = 0$ to $x(T) = x_0$. It follows from (5.238) that the minimum input energy is the minimum energy it takes to drive $\mathcal{G}$ from the origin to a given state x_0. Assuming $\mathcal{G}$ is completely reachable, show that $V_\mathrm{a}(x_0) + V_\mathrm{e}(x_0) = V_\mathrm{r}(x_0)$.

Problem 5.9. The nonlinear dynamical system (5.75) and (5.76) is *nonnegative* if for every $x(0) \in \overline{\mathbb{R}}_+^n$ and $u(t) \geq\geq 0$, $t \geq 0$, the solution $x(t)$, $t \geq 0$, to (5.75) and the output $y(t)$, $t \geq 0$, are nonnegative, that is,

$x(t) \geq\geq 0$, $t \geq 0$, and $y(t) \geq\geq 0$, $t \geq 0$. Consider the nonlinear dynamical system $\mathcal{G}$ given by (5.75) and (5.76). Show that if $f : \mathcal{D} \to \mathbb{R}^n$ is essentially nonnegative (see Problem 3.8), $h(x) \geq\geq 0$, $G(x) \geq\geq 0$, and $J(x) \geq\geq 0$, $x \in \overline{\mathbb{R}}_+^n$, then $\mathcal{G}$ is nonnegative.

Problem 5.10. Let $q \in \mathbb{R}^l$ and $r \in \mathbb{R}^m$. Consider the nonlinear nonnegative dynamical system $\mathcal{G}$ given by (5.75) and (5.76) where $f : \mathcal{D} \to \mathbb{R}^n$ is essentially nonnegative (see Problem 3.8), $G(x) \geq\geq 0$, $h(x) \geq\geq 0$, and $J(x) \geq\geq 0$, $x \in \overline{\mathbb{R}}_+^n$. Show that $\mathcal{G}$ is exponentially dissipative (respectively, dissipative) with respect to the supply rate $s(u, y) = q^{\mathrm{T}}y + r^{\mathrm{T}}u$ if and only if there exist functions $V_{\mathrm{s}} : \overline{\mathbb{R}}_+^n \to \overline{\mathbb{R}}_+$, $\ell : \overline{\mathbb{R}}_+^n \to \overline{\mathbb{R}}_+$, and $\mathcal{W} : \overline{\mathbb{R}}_+^n \to \overline{\mathbb{R}}_+^m$, and a scalar $\varepsilon > 0$ (respectively, $\varepsilon = 0$) such that $V_{\mathrm{s}}(\cdot)$ is continuously differentiable and nonnegative definite, $V_{\mathrm{s}}(0) = 0$, and for all $x \in \overline{\mathbb{R}}_+^n$,

$$0 = V_{\mathrm{s}}'(x)f(x) + \varepsilon V_{\mathrm{s}}(x) - q^{\mathrm{T}}h(x) + \ell(x), \tag{5.239}$$
$$0 = V_{\mathrm{s}}'(x)G(x) - q^{\mathrm{T}}J(x) - r^{\mathrm{T}} + \mathcal{W}^{\mathrm{T}}(x). \tag{5.240}$$

(**Hint:** The definition of dissipativity and exponential dissipativity should be modified to reflect the fact that $x_0 \in \overline{\mathbb{R}}_+^n$, and $u(t)$, $t \geq 0$, and $y(t)$, $t \geq 0$, are nonnegative.)

Problem 5.11. Let $q \in \mathbb{R}^l$ and $r \in \mathbb{R}^m$ and consider the nonlinear nonnegative dynamical system $\mathcal{G}$ (see Problem 5.9) given by (5.75) and (5.76) where $f : \mathcal{D} \to \mathbb{R}^n$ is essentially nonnegative (see Problem 3.8), $G(x) \geq\geq 0$, $h(x) \geq\geq 0$, and $J(x) \geq\geq 0$, $x \in \overline{\mathbb{R}}_+^n$. Suppose $\mathcal{G}$ is exponentially dissipative (respectively, dissipative) with respect to the supply rate $s(u, y) = q^{\mathrm{T}}y + r^{\mathrm{T}}u$. Show that there exist $p \in \overline{\mathbb{R}}_+^n$, $l \in \overline{\mathbb{R}}_+^n$, and $w \in \overline{\mathbb{R}}_+^m$, and a scalar $\varepsilon > 0$ (respectively, $\varepsilon = 0$) such that

$$0 = A^{\mathrm{T}}p + \varepsilon p - C^{\mathrm{T}}q + l, \tag{5.241}$$
$$0 = B^{\mathrm{T}}p - D^{\mathrm{T}}q - r + w, \tag{5.242}$$

where

$$A = \left.\frac{\partial f}{\partial x}\right|_{x=0}, \quad B = G(0), \quad C = \left.\frac{\partial h}{\partial x}\right|_{x=0}, \quad D = J(0). \tag{5.243}$$

If, in addition, (A, C) is observable, show that $p >> 0$.

Problem 5.12. Let $q \in \mathbb{R}^l$ and $r \in \mathbb{R}^m$. Consider the nonnegative dynamical system $\mathcal{G}$ (see Problem 5.9) given by (5.160) and (5.161) where A is essentially nonnegative (see Problem 3.7), $B \geq\geq 0$, $C \geq\geq 0$, and $D \geq\geq 0$. Show that $\mathcal{G}$ is exponentially dissipative (respectively, dissipative) with respect to the supply rate $s(u, y) = q^{\mathrm{T}}y + r^{\mathrm{T}}u$ if and only if there exist $p \in \overline{\mathbb{R}}_+^n$, $l \in \overline{\mathbb{R}}_+^n$, and $w \in \overline{\mathbb{R}}_+^m$, and a scalar $\varepsilon > 0$ (respectively, $\varepsilon = 0$) such that

$$0 = A^{\mathrm{T}}p + \varepsilon p - C^{\mathrm{T}}q + l, \tag{5.244}$$
$$0 = B^{\mathrm{T}}p - D^{\mathrm{T}}q - r + w. \tag{5.245}$$

(**Hint:** Use Problems 5.10 and 5.11 to show this result.)

Problem 5.13. Consider the controlled rigid spacecraft given in Problem 3.15. Show that this system is a port-controlled Hamiltonian dynamical system. What does the output have to be in this case?

Problem 5.14. Consider the dynamical system

$$G(s) \overset{\min}{\sim} \left[\begin{array}{c|c} A & B \\ \hline C & D \end{array}\right]$$

with input $u(\cdot) \in \mathcal{U}$ and output $y(\cdot) \in \mathcal{Y}$. Let $Q \in \mathbb{R}^{l\times l}$, $R \in \mathbb{R}^{m\times m}$, and $S \in \mathbb{R}^{l\times m}$ be such that Q and R are symmetric. Show that the following statements are equivalent:

i) $G^*(s)QG(s) + G^*(s)S + S^{\mathrm{T}}G(s) + R \geq 0$, Re $[s] > 0$.

ii) $\int_0^T [y^{\mathrm{T}}(t)Qy(t) + 2y^{\mathrm{T}}(t)Su(t) + u^{\mathrm{T}}(t)Ru(t)]\mathrm{d}t \geq 0, \quad T \geq 0$.

iii) There exist matrices $P \in \mathbb{R}^{n\times n}$, $L \in \mathbb{R}^{p\times n}$, and $W \in \mathbb{R}^{p\times m}$, with P positive definite, such that (5.145)–(5.147) are satisfied.

If, alternatively, $R + S^{\mathrm{T}}D + D^{\mathrm{T}}S + D^{\mathrm{T}}QD > 0$, show that *i)* holds if and only if there exists an $n \times n$ positive-definite matrix P such that

$$\begin{aligned} 0 \geq\ & A^{\mathrm{T}}P + PA - C^{\mathrm{T}}QC + [B^{\mathrm{T}}P - (QD+S)^{\mathrm{T}}C]^{\mathrm{T}}(R + S^{\mathrm{T}}D + D^{\mathrm{T}}S \\ & + D^{\mathrm{T}}QD)^{-1}[B^{\mathrm{T}}P - (QD+S)^{\mathrm{T}}C]. \end{aligned} \tag{5.246}$$

Problem 5.15. Consider the dynamical system

$$G(s) \overset{\min}{\sim} \left[\begin{array}{c|c} A & B \\ \hline C & D \end{array}\right].$$

Let $Q \in \mathbb{R}^{l\times l}$, $R \in \mathbb{R}^{m\times m}$, and $S \in \mathbb{R}^{l\times m}$ be such that Q and R are symmetric, $Q > 0$, and $R < S^{\mathrm{T}}Q^{-1}S$. Show that the following statements are equivalent:

i) $G^*(s)QG(s) + G^*(s)S + S^{\mathrm{T}}G(s) + R \geq 0$, Re $[s] > 0$.

ii) $L^{1/2}[Q^{1/2}G(s) + Q^{-1/2}S]^{-1}$ is bounded real, where $L \triangleq S^{\mathrm{T}}Q^{-1}S - R$.

iii) $[QG(s) + S - Q^{1/2}L^{1/2}][QG(s) + S + Q^{1/2}L^{1/2}]^{-1}Q$ is positive real.

Problem 5.16. Consider the dynamical system

$$G(s) \overset{\min}{\sim} \left[\begin{array}{c|c} A & B \\ \hline C & D \end{array}\right].$$

Let $Q \in \mathbb{R}^{l\times l}$, $R \in \mathbb{R}^{m\times m}$, and $S \in \mathbb{R}^{l\times m}$ be such that Q and R are symmetric, $Q > 0$, and $R + S^{\mathrm{T}}Q^{-1}S > 0$. Show that the following statements are equivalent:

i) $-G^*(s)QG(s) + G^*(s)S + S^{\mathrm{T}}G(s) + R \geq 0$, $\mathrm{Re}\,[s] > 0$.

ii) $[Q^{1/2}G(s) - Q^{-1/2}S]L^{-1/2}$ is bounded real, where $L \triangleq R + S^{\mathrm{T}}Q^{-1}S$.

iii) $-[QG(s) - S - Q^{1/2}L^{1/2}][QG(s) - S + Q^{1/2}L^{1/2}]^{-1}Q$ is positive real.

Problem 5.17. Consider the dynamical system

$$G(s) \overset{\min}{\sim} \left[\begin{array}{c|c} A & B \\ \hline C & D \end{array}\right],$$

let $Q \in \mathbb{R}^{l\times l}$, $R \in \mathbb{R}^{m\times m}$, and $S \in \mathbb{R}^{l\times m}$ be such that Q and R are symmetric, $\det R \neq 0$, and either one of the following assumptions is satisfied:

i) A has no eigenvalues on the $\jmath\omega$-axis.

ii) Q is sign definite, that is, $Q \geq 0$ or $Q \leq 0$, (A, B) has no uncontrollable eigenvalues on the $\jmath\omega$-axis, and (A, C) has no unobservable eigenvalues on the $\jmath\omega$-axis.

Show that the following statements are equivalent:

i) $G^*(\jmath\hat{\omega})QG(\jmath\hat{\omega}) + G^*(\jmath\hat{\omega})S + S^{\mathrm{T}}G(\jmath\hat{\omega}) + R$ is singular for some $\hat{\omega} \in \mathbb{R}$.

ii) The Hamiltonian matrix

$$\mathcal{H} = \begin{bmatrix} A - B\hat{R}^{-1}(QD+S)^{\mathrm{T}}C & -B\hat{R}^{-1}B^{\mathrm{T}} \\ -[CQC^{\mathrm{T}} - C^{\mathrm{T}}(QD+S)\hat{R}^{-1}(QD+S)^{\mathrm{T}}C] & -[A - B\hat{R}^{-1}(QD+S)^{\mathrm{T}}C]^{\mathrm{T}} \end{bmatrix}, \tag{5.247}$$

where $\hat{R} = R + S^{\mathrm{T}}D + D^{\mathrm{T}}S + D^{\mathrm{T}}QD$, has no eigenvalues at $\jmath\hat{\omega}$.

Problem 5.18. Consider the dynamical system

$$G(s) \overset{\min}{\sim} \left[\begin{array}{c|c} A & B \\ \hline C & D \end{array}\right],$$

let $Q \in \mathbb{R}^{l\times l}$, $R \in \mathbb{R}^{m\times m}$, and $S \in \mathbb{R}^{l\times m}$ be such that Q and R are symmetric, $R > 0$, and either one of the assumptions *i)* or *ii)* of Problem 5.17 is satisfied. Show that the following statements are equivalent:

i) $G^*(\jmath\omega)QG(\jmath\omega) + G^*(\jmath\omega)S + S^{\mathrm{T}}G(\jmath\omega) + R > 0$, $\omega \in \mathbb{R}$.

ii) The Hamiltonian matrix (5.247) has no eigenvalue on the $\jmath\omega$-axis.

If, in addition, (A, B) is stabilizable show that:

iii) There exists a unique $P = P^{\mathrm{T}}$ such that

$$\begin{aligned} 0 = {} & [A - B\hat{R}^{-1}(QD+S)^{\mathrm{T}}C]^{\mathrm{T}}P + P[A - B\hat{R}^{-1}(QD+S)^{\mathrm{T}}C] \\ & + PB\hat{R}^{-1}B^{\mathrm{T}}P - C^{\mathrm{T}}QC + C^{\mathrm{T}}(QD+S)\hat{R}^{-1}(QD+S)^{\mathrm{T}}C, \end{aligned} \tag{5.248}$$

and $\mathrm{spec}(A - B\hat{R}^{-1}(QD+S)^{\mathrm{T}}C + B\hat{R}^{-1}B^{\mathrm{T}}P) \subset \mathbb{C}_-$, where $\hat{R} \triangleq R + S^{\mathrm{T}}D + D^{\mathrm{T}}S + D^{\mathrm{T}}QD$, if and only if *i)* holds.

Finally, show that the following statements are also equivalent:

iv) $G^*(\jmath\omega)QG(\jmath\omega) + G^*(\jmath\omega)S + S^{\mathrm{T}}G(\jmath\omega)R \geq 0$, $\omega \in \mathbb{R}$.

v) There exists a unique $P = P^{\mathrm{T}}$ such that (5.248) holds and $\mathrm{spec}(A - B\hat{R}^{-1}(QD+S)^{\mathrm{T}}C + B\hat{R}^{-1}B^{\mathrm{T}}P) \subset \overline{\mathbb{C}}_-$.

Problem 5.19. Let $\gamma > 0$,

$$G(s) \sim \left[\begin{array}{c|c} A & B \\ \hline C & D \end{array}\right] \in \mathcal{RH}_\infty,$$

and define

$$\mathcal{H} \triangleq \begin{bmatrix} A + B\hat{R}^{-1}D^{\mathrm{T}}C & B\hat{R}^{-1}B^{\mathrm{T}} \\ -C^{\mathrm{T}}(I + D\hat{R}^{-1}D^{\mathrm{T}})C & -(A + B\hat{R}^{-1}D^{\mathrm{T}}C)^{\mathrm{T}} \end{bmatrix}, \tag{5.249}$$

where $\hat{R} \triangleq \gamma^2 I_m - D^{\mathrm{T}}D$. Show that the following statements are equivalent:

i) $\|G(s)\|_\infty < \gamma$.

ii) $\sigma_{\max}(D) < \gamma$ and $\mathcal{H}$ has no eigenvalues on the $\jmath\omega$-axis.

iii) There exists $P \geq 0$ such that

$$0 = A^{\mathrm{T}}P + PA + C^{\mathrm{T}}C + (B^{\mathrm{T}}P + D^{\mathrm{T}}C)^{\mathrm{T}}\hat{R}^{-1}(B^{\mathrm{T}}P + D^{\mathrm{T}}C). \tag{5.250}$$

If, in addition, (A, C) is observable show that $P > 0$.

Problem 5.20. Let $\alpha \in (0,\pi]$, let $G(s)$ be a scalar transfer function, and let $q, r \in \mathbb{R}$ be such that $q, r < 0$ and $qr = \cos^2\alpha$. Show that if

$$qG^*(\jmath\omega)G(\jmath\omega) + G(\jmath\omega) + G^*(\jmath\omega) + r \geq 0, \qquad \omega \in \mathbb{R}, \tag{5.251}$$

then $\angle G(\jmath\omega) \in [-\alpha, \alpha]$, $\omega \in \mathbb{R}$.

Problem 5.21. Let $\alpha \in (0,\pi]$ and let

$$G(s) \sim \left[\begin{array}{c|c} A & B \\ \hline C & D \end{array}\right]$$

be a scalar transfer function, where $A \in \mathbb{R}^{n\times n}$, $B \in \mathbb{R}^{n\times 1}$, $C \in \mathbb{R}^{1\times n}$, and $D \in \mathbb{R}$. Show that if there exist a scalar $\mu > 0$ and matrices $P \in \mathbb{S}^n$, $L \in \mathbb{R}^{p\times n}$, and $W \in \mathbb{R}^{p\times 1}$ such that

$$0 = A^{\mathrm{T}}P + PA + \mu C^{\mathrm{T}}C + L^{\mathrm{T}}L, \tag{5.252}$$
$$0 = B^{\mathrm{T}}P + (\mu D - \lambda)C + W^{\mathrm{T}}L, \tag{5.253}$$
$$0 = 2D - \mu D^2 - \tfrac{1}{\mu}\cos^2\alpha - W^{\mathrm{T}}W, \tag{5.254}$$

then $\angle G(\jmath\omega) \in [-\alpha, \alpha]$, $\omega \in \mathbb{R}$.

Problem 5.22. Prove Theorem 5.7.

Problem 5.23. Prove Theorem 5.8.

Problem 5.24. Prove Theorem 5.10.

Problem 5.25. Consider the dynamical system $\mathcal{G}$ given by (5.75) and (5.76) with $f(x) = Ax$, $G(x) = B$, $h(x) = Cx$, and $J(x) \equiv 0$. Assume $\mathcal{G}$ is passive. Show that $\mathcal{G}$ is stabilizable if and only if $\mathcal{G}$ is detectable.

Problem 5.26. Consider the dynamical system $\mathcal{G}$ given by (5.75) and (5.76) with $f(x) = Ax$, $G(x) = B$, $h(x) = Cx$, and $J(x) = D$. Assume $\mathcal{G}$ is minimal and passive with storage function $V_{\mathrm{s}}(x)$, $x \in \mathbb{R}^n$. Show that there exists a constant $\varepsilon > 0$ such that $V_{\mathrm{s}}(x) > \varepsilon\|x\|^2 + \inf_{x\in\mathbb{R}^n} V_{\mathrm{s}}(x)$, $x \in \mathbb{R}^n$.

Problem 5.27. Consider the dynamical system $\mathcal{G}$ given by (5.75) and (5.76) with $f(x) = Ax$, $G(x) = B$, $h(x) = Cx$, and $J(x) = D$. Show that if $\mathcal{G}$ is passive, then the available storage is given by $V_{\mathrm{a}}(x) = x^{\mathrm{T}}Px$, where $P = \lim_{\varepsilon\to 0} P_\varepsilon$ and where P_ε is the nonnegative-definite solution to the Riccati equation

$$0 = A^{\mathrm{T}}P_\varepsilon + P_\varepsilon A + (B^{\mathrm{T}}P_\varepsilon - C)^{\mathrm{T}}(D + D^{\mathrm{T}} + \varepsilon I)^{-1}(B^{\mathrm{T}}P_\varepsilon - C). \tag{5.255}$$

Problem 5.28. Consider the dynamical system $\mathcal{G}$ given by (5.75) and (5.76) with $f(x) = Ax$, $G(x) = B$, $h(x) = Cx$, and $J(x) = D$. Assume $\mathcal{G}$ is minimal and passive. Show that the available storage is given by $V_{\mathrm{a}}(x) =$

$x^{\mathrm{T}}P_{-}x$ and the required supply is given by $V_{\mathrm{r}}(x) = x^{\mathrm{T}}P_{+}x$, where P_{-} and P_{+} are the minimal and maximal solutions to (5.255), respectively. Furthermore, show that there exist $\alpha > 0$ and $\beta > 0$ such that

$$\alpha\|x\|^2 \leq V_{\mathrm{a}}(x) \leq V_{\mathrm{s}}(x) \leq V_{\mathrm{r}}(x) \leq \beta\|x\|^2, \quad x \in \mathbb{R}^n. \tag{5.256}$$

Problem 5.29. Consider the nonlinear dynamical system $\mathcal{G}$ given by (5.75) and (5.76), and assume that $\mathcal{G}$ is zero-state observable and completely reachable. Show that if $\mathcal{G}$ is input strict passive, then $\det J(x) \neq 0$, $x \in \mathbb{R}^n$, and hence, $\mathcal{G}$ has relative degree zero.

Problem 5.30. Give a complete proof of Theorem 5.11.

Problem 5.31. Let $G(s)$ be a real rational matrix transfer function with input $u(\cdot) \in \mathcal{U}$ and output $y(\cdot) \in \mathcal{Y}$. Show that $G(s)$ is lossless with respect to the supply rate $r(u, y) = 2u^{\mathrm{T}}y$ if and only if $\mathrm{He}\, G(\jmath\omega) = 0$ for all $\omega \in \mathbb{R}$, with $\jmath\omega$ not a pole of any entry of $G(s)$, and if $\jmath\omega$ is a pole of any entry of $G(s)$ it is at most a simple pole and the residue matrix (Theorem 5.11) at $\jmath\omega$ is nonnegative definite Hermitian. Alternatively, show that $G(s)$ is lossless with respect to supply rate $r(u, y) = \gamma^2 u^{\mathrm{T}}u - y^{\mathrm{T}}y$, $\gamma > 0$, if and only if $\gamma^2 I - G^*(\jmath\omega)G(\jmath\omega) = 0$ for all $\omega \in \mathbb{R}$.

Problem 5.32. Let

$$G(s) \overset{\min}{\sim} \left[\begin{array}{c|c} A & B \\ \hline C & D \end{array}\right]$$

be positive real with $D + D^{\mathrm{T}} > 0$, and let P, L, and W, with $P > 0$, satisfy (5.151)–(5.153). Show that

$$G^{-1}(s) \overset{\min}{\sim} \left[\begin{array}{c|c} \hat{A} & \hat{B} \\ \hline \hat{C} & \hat{D} \end{array}\right],$$

where $\hat{A} = A - BD^{-1}C$, $\hat{B} = BD^{-1}$, $\hat{C} = -D^{-1}C$, and $\hat{D} = D^{-1}$, is also positive real, and P satisfies

$$0 = \hat{A}^{\mathrm{T}}P + P\hat{A} + \hat{L}^{\mathrm{T}}\hat{L}, \tag{5.257}$$

$$0 = P\hat{B} - \hat{C}^{\mathrm{T}} + \hat{L}^{\mathrm{T}}\hat{W}, \tag{5.258}$$

$$0 = \hat{D} + \hat{D}^{\mathrm{T}} - \hat{W}^{\mathrm{T}}\hat{W}, \tag{5.259}$$

where $\hat{L} = L - WD^{-1}C$ and $\hat{W} = WD^{-1}$.

Problem 5.33. Let

$$G(s) \overset{\min}{\sim} \left[\begin{array}{c|c} A & B \\ \hline C & 0 \end{array}\right]$$

be a positive real transfer function. $G(s)$ is a *self-dual* realization if $A + A^{\mathrm{T}} \leq$

0 and $B = C^{\mathrm{T}}$. Show that a self-dual realization can be obtained from the change of coordinates $z = P^{1/2}x$, where P satisfies (5.151) and (5.152), and x is the internal state of the realization of $G(s)$.

Problem 5.34. Consider the linear dynamical system

$$G(s) \overset{\min}{\sim} \left[\begin{array}{c|c} A & B \\ \hline C & D \end{array}\right]$$

with input $u(\cdot) \in \mathcal{U}$ and output $y(\cdot) \in \mathcal{Y}$. Assume that A is asymptotically stable. Show that the following statements are equivalent:

i) There exists $\varepsilon > 0$ such that $G(\jmath\omega) + G^*(\jmath\omega) \geq \varepsilon I$ for all $\omega \in \mathbb{R}$.

ii) There exists $\gamma > 0$ and a function $\beta : \mathbb{R}^n \to \mathbb{R}$, $\beta(0) = 0$, such that for all $T \geq 0$,

$$\int_0^T u^{\mathrm{T}}(t)y(t)\mathrm{d}t \geq \beta(x_0) + \gamma^2 \int_0^T u^{\mathrm{T}}(t)u(t)\mathrm{d}t. \tag{5.260}$$

iii) There exists $\varepsilon > 0$ such that $G(\jmath\omega - \varepsilon) + G^*(\jmath\omega - \varepsilon) \geq 0$ for all $\omega \in \mathbb{R}$.

iv) $G(\jmath\omega) + G^*(\jmath\omega) > 0$, for all $\omega \in \mathbb{R}$, and

$$\lim_{\omega \to \infty} \omega^2[G(\jmath\omega) + G^*(\jmath\omega)] > 0. \tag{5.261}$$

Problem 5.35. Consider the controlled nonlinear oscillator given by the undamped Duffing equation

$$\ddot{x}(t) + (2 + x^2(t))x(t) = u(t), \quad x(0) = x_0, \quad \dot{x}(0) = \dot{x}_0, \quad t \geq 0, \tag{5.262}$$

$$y(t) = \dot{x}(t). \tag{5.263}$$

Show that the input-output map from u to y is lossless with respect to the supply rate $r(u, y) = uy$.

Problem 5.36. Consider the controlled nonlinear damped oscillator given by

$$\ddot{x}(t) + \eta(x(t), \dot{x}(t))[\dot{x}(t) + x(t)] = u(t), \quad x(0) = x_0, \quad \dot{x}(0) = \dot{x}_0, \quad t \geq 0, \tag{5.264}$$

$$y(t) = x(t) + \dot{x}(t), \tag{5.265}$$

where $\eta(x, \dot{x}) = 2 + (x + \dot{x})^2$. Show that the input-output map from u to y is exponentially passive.

Problem 5.37. Consider the controlled nonlinear system

$$\dot{x}_1(t) = x_2(t), \quad x_1(0) = x_{10}, \quad t \geq 0, \tag{5.266}$$

$$\dot{x}_2(t) = -(1 + x_1^2(t))x_2(t) - x_1^3(t) + x_1(t)u(t), \quad x_2(0) = x_{20}, \tag{5.267}$$

$$y(t) = x_1(t)x_2(t). \tag{5.268}$$

Show that the input-output map from u to y is nonexpansive with $\gamma \geq 1$. (**Hint:** Use the storage function $V_{\rm s}(x_1, x_2) = \alpha x_1^4 + \beta x_2^2$, where $\alpha, \beta > 0$ are parameters to be chosen.)

Problem 5.38. Consider the linear dynamical system

$$\dot{x}(t) = -x(t) + u(t), \quad x(0) = x_0, \quad t \geq 0, \tag{5.269}$$
$$y(t) = x(t) + u(t). \tag{5.270}$$

Show that (5.269) and (5.270) is passive. Find $V_{\rm s}(x)$, $V_{\rm a}(x)$, and $V_{\rm r}(x)$ for (5.269) and (5.270).

Problem 5.39. Consider the controlled linear system

$$\dot{x}(t) = -x(t) + u(t), \quad x(0) = x_0, \quad t \geq 0, \tag{5.271}$$
$$y(t) = x(t). \tag{5.272}$$

Show that (5.271) and (5.272) is nonexpansive with $\gamma \geq 1$.

Problem 5.40. Consider the scalar dynamical system

$$\dot{x}(t) = -x(t) + 2u(t), \quad x(0) = x_0, \quad t \geq 0, \tag{5.273}$$
$$y(t) = \tan^{-1}(x(t)). \tag{5.274}$$

Show that (5.273) and (5.274) is output strict passive.

Problem 5.41. Consider the nonlinear dynamical system in polar coordinates given by

$$\dot{r}(t) = r(t)(r^2(t) - 1)(r^2(t) - 4) + r(t)(r^2(t) - 4)u(t),$$
$$r(0) = r_0, \quad t \geq 0, \tag{5.275}$$
$$\dot{\theta}(t) = 1, \quad \theta(0) = \theta_0, \tag{5.276}$$
$$y(t) = r^2(t) - 1. \tag{5.277}$$

Show that the set $\mathcal{D}_{\rm c} = \{(r, \theta) \in \mathbb{R} \times \mathbb{R} : r = 1\}$ is invariant under the uncontrolled (i.e., $u(t) \equiv 0$) dynamics and $\mathcal{D}_{\rm c}$ is asymptotically stable. In addition, show that the largest domain of attraction (with respect to $\mathcal{D}_{\rm c}$) of the uncontrolled system (5.275) and (5.276) is given by $\mathcal{D}_{\rm A} = \{(r, \theta) \in \mathbb{R} \times \mathbb{R} : 0 < r < 2\}$. Finally, show that (5.275)–(5.277) is nonexpansive with $\gamma \geq 1$. (**Hint:** Use the storage function $V_{\rm s}(r, \theta) = -\frac{1}{4}\ln r^2 - \frac{3}{4}\ln(4 - r^2) + \frac{3}{4}\ln 3$ with $(r, \theta) = (1, 0)$ being the equilibrium solution of (5.275) and (5.276).)

Problem 5.42. Consider the nonlinear dynamical system

$$\dot{x}(t) = x(t) + u(t), \quad x(0) = x_0, \quad t \geq 0, \tag{5.278}$$

$$y(t) = -\frac{\alpha x(t)}{1 + x^4(t)}, \tag{5.279}$$

where $\alpha > 0$. Show that $V_s(x) = \alpha(\frac{\pi}{2} - \tan^{-1}(x^2))$ satisfies the dissipation inequality (5.16) with $r(u, y) = 2uy$ and yet the zero solution $x(t) \equiv 0$ of the undisturbed system (5.278) is unstable. Why does this contradict *i*) of Proposition 5.2?

Problem 5.43. Consider the nonlinear dynamical system $\mathcal{G}$ given by (5.75) and (5.76), and assume $\mathcal{G}$ is completely reachable and zero state observable. Furthermore, assume $\mathcal{G}$ is passive and $J(x) + J^{\mathrm{T}}(x) > 0$, $x \in \mathbb{R}^n$. Show that $V_a(x)$ and $V_r(x)$ satisfy

$$\begin{aligned} 0 = {} & V'(x)f(x) + [\tfrac{1}{2}V'(x)G(x) - h^{\mathrm{T}}(x)] \\ & \cdot [J(x) + J^{\mathrm{T}}(x)]^{-1}[\tfrac{1}{2}V'(x)G(x) - h^{\mathrm{T}}(x)]^{\mathrm{T}}, \quad x \in \mathbb{R}^n, \end{aligned} \tag{5.280}$$

where $V(\cdot)$ is positive definite and $V(0) = 0$. (**Hint:** First show that

$$\begin{aligned} \int_{t_0}^{t} 2u^{\mathrm{T}}(s)y(s)\mathrm{d}s = {} & V(x(t)) - V(x(t_0)) + \\ & + \int_{t_0}^{t} \begin{bmatrix} 1 & u^{\mathrm{T}}(s) \end{bmatrix} \begin{bmatrix} \mathcal{A}(x(s)) & \mathcal{B}(x(s)) \\ \mathcal{B}^{\mathrm{T}}(x(s)) & \mathcal{C}(x(s)) \end{bmatrix} \begin{bmatrix} 1 \\ u(s) \end{bmatrix} \mathrm{d}s, \end{aligned}$$

where $\mathcal{A}(x) \triangleq -V'(x)f(x)$, $\mathcal{B}(x) \triangleq h^{\mathrm{T}}(x) - \frac{1}{2}V'(x)G(x)$, and $\mathcal{C}(x) \triangleq J(x) + J^{\mathrm{T}}(x)$.)

Problem 5.44. Consider the nonlinear time-varying dynamical system $\mathcal{G}$ given by

$$\dot{x}(t) = f(t, x(t)) + G(t, x(t))u(t), \quad x(t_0) = x_0, \quad t \geq t_0, \tag{5.281}$$

$$y(t) = h(t, x(t)) + J(t, x(t))u(t), \tag{5.282}$$

where $x(t) \in \mathcal{D} \subseteq \mathbb{R}^n$, $\mathcal{D}$ is an open set with $0 \in \mathcal{D}$, $u(t) \in U \subseteq \mathbb{R}^m$, $y(t) \in Y \subseteq \mathbb{R}^l$, $f : [t_0, \infty) \times \mathcal{D} \to \mathbb{R}^n$, $G : [t_0, \infty) \times \mathcal{D} \to \mathbb{R}^{n \times m}$, $h : [t_0, \infty) \times \mathcal{D} \to Y$, and $J : [t_0, \infty) \times \mathcal{D} \to \mathbb{R}^{l \times m}$. Here, assume that $f(\cdot, \cdot)$, $G(\cdot, \cdot)$, $h(\cdot, \cdot)$, and $J(\cdot, \cdot)$ are piecewise continuous in t and continuously differentiable in x on $[t_0, \infty) \times \mathcal{D}$. The available storage and required supply for nonlinear time-varying dynamical systems are defined as

$$V_a(t_0, x_0) \triangleq - \inf_{u(\cdot), T \geq t_0} \int_{t_0}^{T} r(u(t), y(t))\mathrm{d}t \tag{5.283}$$

and

$$V_r(t_0, x_0) \triangleq \inf_{u(\cdot), T \geq t_0} \int_{t_0}^{T} r(u(t), y(t))\mathrm{d}t. \tag{5.284}$$

Assuming that $\mathcal{G}$ is completely reachable and $V_a(t, x)$ and $V_r(t, x)$ are continuously differentiable on $[0, \infty) \times \mathcal{D}$, show that $\mathcal{G}$ is passive if and only if there exists an almost everywhere continuously differentiable function $V_s : [0, \infty) \times \mathcal{D} \to \mathbb{R}$ such that $V_s(t, x) \geq 0$, $(t, x) \in [0, \infty) \times \mathcal{D}$, $V_s(t, 0) = 0$, $t \in [0, \infty)$, and for all $(t, x) \in [0, \infty) \times \mathcal{D}$,

$$\begin{bmatrix} \frac{\partial V_s(t,x)}{\partial x} f(t,x) + \frac{\partial V_s(t,x)}{\partial t} & \frac{1}{2}\frac{\partial V_s(t,x)}{\partial x} G(t,x) - h^{\mathrm{T}}(t,x) \\ \frac{1}{2} G^{\mathrm{T}}(t,x) \frac{\partial V_s^{\mathrm{T}}(t,x)}{\partial x} & -(J(t,x) + J^{\mathrm{T}}(t,x)) \end{bmatrix} \leq 0. \quad (5.285)$$

Problem 5.45. Consider the nonlinear dynamical system $\mathcal{G}$ given by (5.7) and (5.8). Assume that $\mathcal{G}$ is passive with a two-times continuously differentiable storage function $V_s(\cdot)$. Show that

$$V_s'(x) F(x, 0) \leq 0, \quad x \in \mathcal{D}, \quad (5.286)$$

$$\tfrac{1}{2} V_s'(x) G(x, 0) = H^{\mathrm{T}}(x, 0), \quad x \in \mathcal{Q}, \quad (5.287)$$

$$\sum_{i=1}^{n} \frac{\partial^2 F_i}{\partial u^2}(x, 0) \frac{\partial V_s}{\partial x_i} \leq \frac{\partial H}{\partial u}(x, 0) + \left(\frac{\partial H}{\partial u}(x, 0) \right)^{\mathrm{T}}, \quad x \in \mathcal{Q}, \quad (5.288)$$

where $\mathcal{Q} = \{x \in \mathcal{D} : V_s'(x) F(x, 0) = 0\}$ and $F_i(x, u)$ denotes the ith component of $F(x, u)$.

Problem 5.46. Consider the governing equations of motion of an n-degree-of-freedom dynamical system given by the Euler-Lagrange equation

$$\frac{\mathrm{d}}{\mathrm{d}t} \left[\frac{\partial \mathcal{L}}{\partial \dot{q}}(q, \dot{q}) \right]^{\mathrm{T}} - \left[\frac{\partial \mathcal{L}}{\partial q}(q, \dot{q}) \right]^{\mathrm{T}} = u, \quad (5.289)$$

where $q \in \mathbb{R}^n$ represents the generalized system positions, $\dot{q} \in \mathbb{R}^n$ represents the generalized system velocities, $\mathcal{L} : \mathbb{R}^n \times \mathbb{R}^n \to \mathbb{R}$ denotes the system Lagrangian given by $\mathcal{L}(q, \dot{q}) = T(q, \dot{q}) - V(q)$, where $T : \mathbb{R}^n \times \mathbb{R}^n \to \mathbb{R}$ is the system kinetic energy and $V : \mathbb{R}^n \to \mathbb{R}$ is the system potential energy, and $u \in \mathbb{R}^n$ is the vector of generalized forces acting on the system.

i) Show that the Euler-Lagrange equation can be equivalently characterized by the state equations

$$\dot{q} = \left[\frac{\partial \mathcal{H}}{\partial p}(q, p) \right]^{\mathrm{T}}, \quad (5.290)$$

$$\dot{p} = -\left[\frac{\partial \mathcal{H}}{\partial q}(q, p) \right]^{\mathrm{T}} + u, \quad (5.291)$$

where $p \in \mathbb{R}^n$ represents the system generalized momenta and $\mathcal{H} : \mathbb{R}^n \times \mathbb{R}^n \to \mathbb{R}$ denotes the Legendre transformation given by $\mathcal{H}(q, p) = \dot{q}^{\mathrm{T}} p - \mathcal{L}(q, \dot{q})$.

ii) Show that the rate of change in system energy is equal to the external power input, that is,

$$\frac{\mathrm{d}}{\mathrm{d}t} \mathcal{H}(q, p) = \dot{q}^{\mathrm{T}} u, \quad (5.292)$$

and hence, $\mathcal{H}(q,p)$ is a storage function for (5.290) and (5.291).

iii) Show that if $V(q)$ is bounded from below, then the system input-output map from generalized forces u to generalized velocities $\dot{q}$ is lossless, that is,

$$0 = \int_0^T \dot{q}^{\mathrm{T}}(t)u(t)\mathrm{d}t, \tag{5.293}$$

for all $T \geq 0$ with $(q(0), \dot{q}(0)) = (q(T), \dot{q}(T)) = (0,0)$.

iv) For dynamical n degree-of-freedom systems with internal dissipation the Euler-Lagrange equations (5.289) take the form

$$\frac{\mathrm{d}}{\mathrm{d}t}\left[\frac{\partial \mathcal{L}}{\partial \dot{q}}(q,\dot{q})\right]^{\mathrm{T}} - \left[\frac{\partial \mathcal{L}}{\partial q}(q,\dot{q})\right]^{\mathrm{T}} + \left[\frac{\partial \mathcal{R}}{\partial \dot{q}}(\dot{q})\right]^{\mathrm{T}} = u, \tag{5.294}$$

where $\mathcal{R} : \mathbb{R}^n \to \mathbb{R}$ represents the Rayleigh dissipation function satisfying $\frac{\partial \mathcal{R}}{\partial \dot{q}}(\dot{q})\dot{q} \geq 0$, $\dot{q} \in \mathbb{R}^n$. Show that in this case (5.292) becomes

$$\frac{\mathrm{d}}{\mathrm{d}t}\mathcal{H}(q,p) = -\frac{\partial \mathcal{R}}{\partial \dot{q}}(\dot{q})\dot{q} + u^{\mathrm{T}}\dot{q}, \tag{5.295}$$

and the system input-output map from generalized forces u to generalized velocities $\dot{q}$ is passive, that is,

$$0 \leq \int_0^T \dot{q}^{\mathrm{T}}(t)u(t)\mathrm{d}t, \tag{5.296}$$

for all $T \geq 0$ with $(q(0), \dot{q}(0)) = (0,0)$.

v) Show that if (5.294) is fully damped, that is, there exists $\varepsilon > 0$ such that $\frac{\partial \mathcal{R}}{\partial \dot{q}}(\dot{q})\dot{q} \geq \varepsilon \dot{q}^{\mathrm{T}}\dot{q}$, $\dot{q} \in \mathbb{R}^n$, then (5.294) is output strict passive.

vi) Show that (5.294) can be interpreted as the negative feedback interconnection of two passive systems $\mathcal{G}_1$ and $\mathcal{G}_2$ with input-output pairs $(\hat{u}, \dot{q})$ and $(\dot{q}, \frac{\partial \mathcal{R}}{\partial \dot{q}})$, respectively, where $\hat{u} = u - \left(\frac{\partial \mathcal{R}}{\partial \dot{q}}\right)^{\mathrm{T}}$.

vii) Characterize the system dynamics for $\mathcal{G}_1$ and $\mathcal{G}_2$ in *vi)*.

viii) Show that the zero solution to (5.289) or, equivalently, (5.290) and (5.291) is asymptotically stable if $u = -K_{\mathrm{d}}\dot{q}$, where $K_{\mathrm{d}} \in \mathbb{R}^{n\times n}$ and satisfies $K_{\mathrm{d}} + K_{\mathrm{d}}^{\mathrm{T}} > 0$.

ix) Show that with $u = -K_{\mathrm{d}}\dot{q} + \hat{u}$, where $K_{\mathrm{d}} \in \mathbb{R}^{n\times n}$ and satisfies $K_{\mathrm{d}} + K_{\mathrm{d}}^{\mathrm{T}} > 0$, the input-output map from $\hat{u}$ to $\dot{q}$ is output strict passive.

x) Show that if $T(q,\dot{q}) = \frac{1}{2}\dot{q}^{\mathrm{T}}M(q)\dot{q}$, where $M(q) > 0$, $q \in \mathbb{R}^n$, is the system inertia matrix function, then (5.289) becomes

$$M(q)\ddot{q} + C(q,\dot{q})\dot{q} + g(q) = u, \tag{5.297}$$

where $g(q) = \left[\frac{\partial V}{\partial q}(q)\right]^{\mathrm{T}}$ and $C(q,\dot{q})_{(i,j)} = \sum_{k=1}^{n} \gamma_{ijk}(q)\dot{q}_k$, where $i,j = 1,\ldots,n$ and

$$\gamma_{ijk}(q) = \frac{1}{2}\left(\frac{\partial M_{(j,k)}(q)}{\partial q_i} + \frac{\partial M_{(k,i)}(q)}{\partial q_j} - \frac{\partial M_{(i,j)}(q)}{\partial q_k}\right). \tag{5.298}$$

xi) Show that $\frac{\mathrm{d}}{\mathrm{d}t}M(q) - 2C(q,\dot{q})$ is skew-symmetric for every $q, \dot{q} \in \mathbb{R}^n$.

xii) Show that $\frac{\mathrm{d}}{\mathrm{d}t}M(q) - 2C(q,\dot{q})$ is skew-symmetric if and only if

$$\frac{\mathrm{d}}{\mathrm{d}t}M(q) = C(q,\dot{q}) + C^{\mathrm{T}}(q,\dot{q}).$$

xiii) Show that the available storage of the dynamical system (5.297) is bounded above by $\frac{1}{2}\dot{q}(0)^{\mathrm{T}}M(q(0))\dot{q}(0) + V(q(0))$.

xiv) Show that the required supply of the dynamical system (5.297) is bounded below by $\frac{1}{2}\dot{q}(0)^{\mathrm{T}}M(q(0))\dot{q}(0) + V(q(0))$.

xv) For $C(q,\dot{q}) \equiv 0$ show that the available storage and the required supply of (5.297) are equal and is given by $\frac{1}{2}\dot{q}(0)^{\mathrm{T}}M(q(0))\dot{q}(0) + V(q(0))$.

Problem 5.47. Consider the governing equations of motion of an n-degree-of-freedom dynamical system given by Hamiltonian system

$$\dot{q} = \left[\frac{\partial \mathcal{H}}{\partial p}(q,p)\right]^{\mathrm{T}}, \tag{5.299}$$

$$\dot{p} = -\left[\frac{\partial \mathcal{H}}{\partial q}(q,p)\right]^{\mathrm{T}} + G(q)u, \tag{5.300}$$

$$y = G^{\mathrm{T}}(q)\left[\frac{\partial \mathcal{H}}{\partial p}(q,p)\right]^{\mathrm{T}}, \tag{5.301}$$

where $q \in \mathbb{R}^n$, $p \in \mathbb{R}^n$, and $\mathcal{H} : \mathbb{R}^n \times \mathbb{R}^n \to \mathbb{R}$ are as in Problem 5.46, and $u \in \mathbb{R}^m$, $y \in \mathbb{R}^m$, and $G : \mathbb{R}^n \to \mathbb{R}^{n\times m}$. Show that if (5.299)–(5.301) is zero-state observable, then the zero solution of (5.299) and (5.300), with $u = -y$, is asymptotically stable.

Problem 5.48. Consider the port-controlled Hamiltonian system

$$\dot{x}(t) = \mathcal{J}(x(t))\left[\frac{\partial \mathcal{H}}{\partial x}(x(t))\right]^{\mathrm{T}} + G(x(t))u(t), \quad x(0) = x_0, \quad t \geq 0, \tag{5.302}$$

$$y(t) = G^{\mathrm{T}}(x(t))\left[\frac{\partial \mathcal{H}}{\partial x}(x(t))\right]^{\mathrm{T}} \tag{5.303}$$

where $x \in \mathcal{D} \subseteq \mathbb{R}^n$, $u, y \in \mathbb{R}^m$, $\mathcal{H} : \mathcal{D} \to \mathbb{R}$, $G : \mathcal{D} \to \mathbb{R}^{n\times m}$, and $\mathcal{J} : \mathcal{D} \to \mathbb{R}^{n\times n}$ and satisfies $\mathcal{J}(x) = -\mathcal{J}^{\mathrm{T}}(x)$. Show that if (5.302) and (5.303) is zero-state observable, then the zero solution $x(t) \equiv 0$ to (5.302), with $u = -y$, is asymptotically stable.

Problem 5.49. Consider the rotational/translational nonlinear dynamical system shown in Figure 5.11 with oscillator cart mass M, linear spring stiffness k, rotational mass m, mass moment of inertia I located a distance e from the center of mass of the cart, and input torque N. Assume that the motion is constrained to the horizontal plane.

i) Using the Euler-Lagrange equations given by (5.289) show that the governing nonlinear dynamic equations of motion are given by

$$(M+m)\ddot{q} + kq = -me(\ddot{\theta}\cos\theta - \dot{\theta}^2\sin\theta), \quad (5.304)$$
$$(I+me^2)\ddot{\theta} = -me\ddot{q}\cos\theta + N, \quad (5.305)$$

where q, $\dot{q}$, θ, and $\dot{\theta}$ denote, respectively, the translational position and velocity of the cart and the angular position and velocity of the rotational mass.

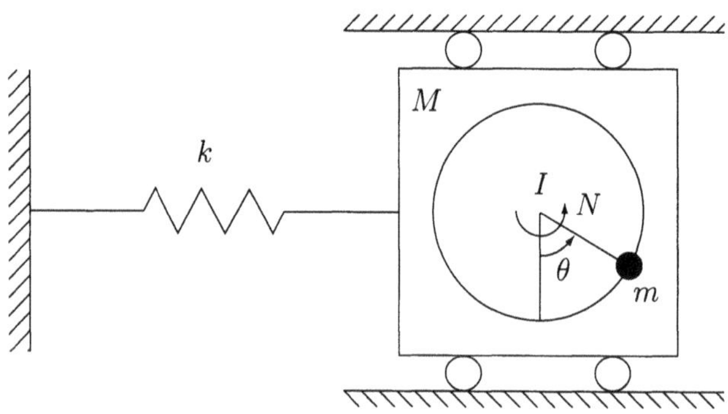

Figure 5.11 Rotational/translational nonlinear dynamical system.

ii) Show that with output $y = \dot{\theta}$ and input $u = N$ the system is passive but not zero-state observable.

iii) Show that with $u = -k_\theta\theta + \hat{u}$, where $k_\theta > 0$, the new system with output $y = \dot{\theta}$ and input $\hat{u}$ is passive and zero-state observable with positive-definite storage function

$$V_s(q,\dot{q},\theta,\dot{\theta}) = \tfrac{1}{2}[kq^2 + (M+m)\dot{q}^2 + k_\theta\theta^2 + (I+me^2)\dot{\theta}^2 + 2me\dot{q}\dot{\theta}\cos\theta]. \quad (5.306)$$

Problem 5.50. Consider the linear matrix second-order dynamical system given by

$$M\ddot{q}(t) + C\dot{q}(t) + Kq(t) = Bu(t), \; q(0) = q_0, \; \dot{q}(0) = \dot{q}_0, \; t \geq 0, \quad (5.307)$$
$$y(t) = B^{\mathrm{T}}\dot{q}(t), \quad (5.308)$$

where $q, \dot{q}, \ddot{q} \in \mathbb{R}^n$ represent generalized position, velocity, and acceleration coordinates, respectively, $u \in \mathbb{R}^m$ is a force input, $y \in \mathbb{R}^m$ is a velocity measurement, M, C, and K are inertia, damping, and stiffness matrices, respectively, and $B \in \mathbb{R}^{n\times m}$ is determined by the location of the system input-output topology. Assume that $M > 0$, $C \geq 0$, and $K \geq 0$.

i) Show that the input-output map from force inputs u to the velocity measurements y, with $q_0 = 0$ and $\dot{q}_0 = 0$, is given by $y(s) = G(s)u(s)$, where

$$G(s) \sim \left[\begin{array}{c|c} \begin{bmatrix} 0 & I \\ -M^{-1}K & -M^{-1}C \end{bmatrix} & \begin{bmatrix} 0 \\ M^{-1}B \end{bmatrix} \\ \hline \begin{bmatrix} 0 & B^{\mathrm{T}} \end{bmatrix} & 0 \end{array}\right]. \tag{5.309}$$

ii) Show that $G(s)$ is positive real.

iii) Show that (5.167)–(5.169) hold with

$$P = \begin{bmatrix} K & 0 \\ 0 & M \end{bmatrix}, \quad L = \begin{bmatrix} 0 & \sqrt{2}C^{1/2} \end{bmatrix}, \quad W = 0.$$

iv) Construct a Lyapunov function to show that if $C \geq 0$ and $K > 0$, then (5.307) with $u(t) \equiv 0$ is Lyapunov stable.

v) Construct a Lyapunov function to show that if $C > 0$ and $K \geq 0$, then (5.307) with $u(t) \equiv 0$ is semistable (see Problem 3.44).

vi) Construct a Lyapunov function to show that if $C > 0$ and $K > 0$, then (5.307) with $u(t) \equiv 0$ is asymptotically stable.

vii) Construct a Lyapunov function to show that if $K > 0$, $C \geq 0$, and $\operatorname{rank}[C \ KM^{-1}C \ \cdots \ (KM^{-1})^{n-1}C] = n$, then (5.307) with $u(t) \equiv 0$ is asymptotically stable.

viii) Show that if C in (5.307) is replaced by $C + G$, where $G = -G^{\mathrm{T}}$ captures system gyroscopic effects, and $C > 0$, $K > 0$, then (5.307) with $u(t) \equiv 0$ remains asymptotically stable.

ix) Show that with $q_0 = 0$ and $\dot{q}_0 = 0$, $u^{\mathrm{T}}y \geq 0$.

Problem 5.51. Consider the nonlinear dynamical system representing a controlled rigid spacecraft given by

$$\dot{x}_1(t) = I_{23}x_2(t)x_3(t) + \tfrac{1}{I_1}u_1(t), \quad x_1(0) = x_{10}, \quad t \geq 0, \tag{5.310}$$

$$\dot{x}_2(t) = I_{31}x_3(t)x_1(t) + \tfrac{1}{I_2}u_2(t), \quad x_2(0) = x_{20}, \tag{5.311}$$

$$\dot{x}_3(t) = I_{12}x_1(t)x_2(t) + \tfrac{1}{I_3}u_3(t), \quad x_3(0) = x_{30}, \tag{5.312}$$

where $I_{23} = (I_2 - I_3)/I_1$, $I_{31} = (I_3 - I_1)/I_2$, $I_{12} = (I_1 - I_2)/I_3$, and I_1, I_2, and I_3 are the principal moments of inertia of the spacecraft. Show that the input-output map from $u = [u_1,\ u_2,\ u_3]^{\mathrm{T}}$ to $y = x = [x_1,\ x_2,\ x_3]^{\mathrm{T}}$ is lossless with respect to the supply rate $r(u,y) = 2u^{\mathrm{T}}y$. Furthermore, show that the zero solution $x(t) \equiv 0$ to (5.310)–(5.312) is globally asymptotically stable if $u = -Kx$, where $K \in \mathbb{R}^{3\times 3}$ and satisfies $K + K^{\mathrm{T}} > 0$. Alternatively, show that if $u = -\phi(x)$ and satisfies $x^{\mathrm{T}}\phi(x) > 0$, $x \neq 0$, then the zero solution $x(t) \equiv 0$ to (5.310)–(5.312) is also globally asymptotically stable. Finally, if $\phi : \mathbb{R}^3 \to \mathbb{R}^3$ is such that $\phi(x) = [\phi_1(x_1),\ \phi_2(x_2),\ \phi_3(x_3)]^{\mathrm{T}}$, how would you pick $\phi_i(x_i)$, $i = 1, 2, 3$, so as to maximize the decay rate of the Lyapunov function candidate $V(x) = I_1x_1^2 + I_2x_2^2 + I_3x_3^2$?

Problem 5.52. Consider a thermodynamic system at a uniform temperature. The first law of thermodynamics states that during any cycle that a system undergoes, the cyclic integral of the heat is proportional to the cyclic integral of the work, that is,

$$\mathcal{J}\oint \mathrm{d}Q = \oint \mathrm{d}W, \tag{5.313}$$

where $\oint \mathrm{d}Q$ represents the net heat transfer during the cycle and $\oint \mathrm{d}W$ represents the net work during the cycle. $\mathcal{J}$ is a proportionality factor, which depends on the units used for work and heat. Here, assume SI units so that $\mathcal{J} = 1$. The second law of thermodynamics states that the transfer of heat from a lower temperature level (source) to a higher temperature level (sink) requires the input of additional work or energy, or, using Clausius' inequality,

$$\oint \frac{\mathrm{d}Q}{T} \leq 0, \tag{5.314}$$

where $\oint \frac{\mathrm{d}Q}{T}$ represents the system entropy and T represents the absolute system temperature. Writing the first and second laws as rate equations and assuming that every admissible system input and every initial system state yield locally integrable work and heat generation functions, show that the first and second laws of thermodynamics can be formulated using cyclo-dissipative system theoretic notions with appropriate virtual storage functions and supply rates (see Problems 5.3–5.5). Use the convention that the work done by the system and the heat delivered to the system are positive.

Problem 5.53. Let $\alpha, \beta \in \mathbb{R}$ be such that $\alpha \leq \beta$ and let $\sigma : \mathbb{R} \to \mathbb{R}$ with $\sigma(0) = 0$. Show that the following statements are equivalent:

i) $\alpha \leq \sigma(u)/u \leq \beta$, $u \in \mathbb{R}$, $u \neq 0$.

ii) $\alpha u^2 \leq \sigma(u)u \leq \beta u^2$, $u \in \mathbb{R}$.

iii) $(\sigma(u) - \alpha u)(\sigma(u) - \beta u) \leq 0$, $u \in \mathbb{R}$.

iv) $|\sigma(u) - cu|^2 \leq r^2|u|^2$, $u \in \mathbb{R}$, where $c \triangleq \frac{1}{2}(\alpha + \beta)$ and $r \triangleq \frac{1}{2}(\alpha - \beta)$.

Problem 5.54. Consider the damped Mathieu equation

$$\ddot{x}(t)+2\mu\dot{x}(t)+(\mu^2+a^2-q\cos\omega t)x(t)=0,\quad x(0)=x_0,\quad \dot{x}(0)=\dot{x}_0,\quad t\geq 0, \tag{5.315}$$

where q, μ, $a > 0$. Use the circle criterion to obtain sufficient conditions in terms of q, μ, and a such that the zero solution $x(t) \equiv 0$ to (5.315) is globally exponentially stable. For which values of ω does your result hold? Is stability guaranteed if $(\cos\omega t)x(t)$ is replaced by $\sigma(x, t)$, where $\sigma(\cdot,\cdot)$ is a memoryless time-varying nonlinearity belonging to the sector $[-1, 1]$? Explain your answer.

Problem 5.55. Consider the absolute stability problem with forward dynamics given by $G(s) = \frac{1}{(s+1)(s+2)(s+3)}$. Use the circle criterion to determine the largest range $[M_1, M_2]$ such that the feedback system is globally exponentially stable for all memoryless, time-varying nonlinearities $\sigma(\cdot,\cdot)$ belonging to the sector $[M_1, M_2]$. Repeat your analysis for $G(s) = \frac{1}{(s-1)(s+1)^2}$.

Problem 5.56. Consider the feedback system shown in Figure 5.3 with $G(s) = \frac{1}{(s+1)^3}$. Use the positivity and Popov theorems to determine the maximum allowable slope M on the saturation nonlinearity such that the feedback system is globally (uniformly) asymptotically stable. Repeat your analysis for $G(s) = \frac{s+3}{s^2+7s+10}$.

Problem 5.57. Consider the damped Matheiu equation

$$\ddot{x}(t) + a\dot{x}(t) + (A + B\cos t)x(t) = 0,\quad x(0) = x_0,\quad \dot{x}(0) = \dot{x}_0,\quad t \geq 0, \tag{5.316}$$

where $a > 0$. Give sufficient conditions such that the zero solution to (5.316) is asymptotically stable. Show that this condition determines a region in the (A, B) plane which is bounded by straight lines and a parabola. (**Hint:** Use Corollary 5.8 to obtain sufficient conditions for exponential stability.)

Problem 5.58 (Shifted Popov Criterion). Consider the controllable and observable dynamical system

$$\dot{x}(t) = Ax(t) + Bu(t),\quad x(0) = x_0,\quad t \geq 0, \tag{5.317}$$

$$y(t) = Cx(t), \tag{5.318}$$

where $x(t) \in \mathbb{R}^n$, $u(t) \in \mathbb{R}^m$, and $y(t) \in \mathbb{R}^m$, with feedback nonlinearity

$$u(t) = -\sigma(y(t)), \tag{5.319}$$

where

$$\sigma(\cdot) \in \Phi \triangleq \{\sigma : \mathbb{R}^m \to \mathbb{R}^m : \sigma(0) = 0,\ [\sigma(y) - M_1 y]^{\mathrm{T}}[\sigma(y) - M_2 y] \le 0, \\ y \in \mathbb{R}^m, \text{ and } \sigma(y) = [\sigma_1(y_1), \sigma_2(y_2), \ldots, \sigma_m(y_m)]^{\mathrm{T}}\},$$

$M_1 = \mathrm{diag}[\underline{M}_1, \underline{M}_2, \ldots, \underline{M}_m]$, $M_2 = \mathrm{diag}[\overline{M}_1, \overline{M}_2, \ldots, \overline{M}_m]$, and $M_2 - M_1 > 0$. Show that if there exists $N = \mathrm{diag}[N_1, N_2, \ldots, N_m]$, $N_i \ge 0$, $i = 1, \ldots, m$, such that $I_m + (M_2 - M_1)(I_m + Ns)(I_m + G(s)M_1)^{-1}G(s)$ is strictly positive real and $\det(I_m + \lambda N) \neq 0$, where $G(s) = C(sI_n - A)^{-1}B$ and $\lambda \in \mathrm{spec}(A)$, then the feedback interconnection (5.317)–(5.319) is asymptotically stable for all $\sigma(\cdot) \in \Phi$. (**Hint:** Use the *shifted Luré-Postnikov Lyapunov function* candidate

$$V(x) = x^{\mathrm{T}} P x + 2 \sum_{i=1}^{m} \int_0^{y_i} [\sigma_i(s) - \underline{M}_i s] N_i \mathrm{d}s, \tag{5.320}$$

where $y_i \in \mathbb{R}$ denotes the ith component of $y \in \mathbb{R}^m$, and state under what conditions this is a valid Lyapunov function.)

Problem 5.59 (Off-Axis Circle Criterion). Show that in the case where $m = 1$ the frequency domain condition for the shifted Popov criterion given in Problem 5.58 involves a frequency domain test in the Nyquist plane in terms of a family of frequency-dependent off-axis circles. In addition, show that the circle centers vary as a function of the phase of the Popov multiplier, but each circle has the same real axis intercepts. Give the real axis intercepts and the centers of the circles.

Problem 5.60 (Parabola Criterion). Show that in the case where $m = 1$ and $M_1 M_2 > 0$, the frequency domain condition for the shifted Popov criterion given in Problem 5.58 can be shown to provide a frequency domain test in the Popov plane in terms of a fixed parabola. Give the real axis intercepts of the parabola and show that the parabola will be tangent to straight lines drawn through the crossings of the real axis intercepts with slopes $\pm 1/N$.

Problem 5.61. Consider the nonlinear dynamical system $\mathcal{G}$ given by

$$\dot{x}(t) = f(x(t)) + G(x(t))u(t), \quad x(0) = x_0, \quad t \ge 0, \tag{5.321}$$

$$y(t) = h(x(t)) + J(x(t))u(t), \tag{5.322}$$

$$u(t) = -\sigma(y(t), t), \tag{5.323}$$

where $x \in \mathbb{R}^n$, $u, y \in \mathbb{R}^m$, $f : \mathbb{R}^n \to \mathbb{R}^n$, $G : \mathbb{R}^n \to \mathbb{R}^{n \times m}$, $h : \mathbb{R}^n \to \mathbb{R}^m$, $J : \mathbb{R}^n \to \mathbb{R}^{m \times m}$, and $\sigma(\cdot, \cdot) \in \Phi_{\mathrm{prg}}$, where

$$\Phi_{\mathrm{prg}} \triangleq \{\sigma : \mathbb{R}^m \times \overline{\mathbb{R}}_+ \to \mathbb{R}^m : \sigma(0, \cdot) = 0, \sigma^{\mathrm{T}}(y, t)y \ge 0, y \in \mathbb{R}^m, \text{ a.e. } \\ t \ge 0, \text{ and } \sigma(y, \cdot) \text{ is Lebesgue measurable for all } y \in \mathbb{R}^m\}.$$

Furthermore, suppose that $\mathcal{G}$ is zero-state observable and exponentially passive with a continuously differentiable, radially unbounded storage function $V_{\rm s}(\cdot)$. Show that the negative feedback interconnection of (5.321)–(5.323) is globally uniformly asymptotically stable for all $\sigma(\cdot,\cdot) \in \Phi_{\rm prg}$.

Problem 5.62. Consider the nonlinear dynamical system $\mathcal{G}$ given by

$$\dot{x}(t) = f(x(t)) + G(x(t))u(t), \quad x(0) = x_0, \quad t \geq 0, \tag{5.324}$$

$$y(t) = h(x(t)) + J(x(t))u(t), \tag{5.325}$$

$$u(t) = \sigma(y(t), t), \tag{5.326}$$

where $x \in \mathbb{R}^n$, $u \in \mathbb{R}^m$, $y \in \mathbb{R}^l$, $f : \mathbb{R}^n \to \mathbb{R}^n$, $G : \mathbb{R}^n \to \mathbb{R}^{n\times m}$, $h : \mathbb{R}^n \to \mathbb{R}^l$, $J : \mathbb{R}^n \to \mathbb{R}^{l\times m}$, and $\sigma(\cdot,\cdot) \in \Phi_{\rm br}$. Furthermore, suppose that $\mathcal{G}$ is zero-state observable and exponentially nonexpansive with a continuously differentiable, radially unbounded storage function $V_{\rm s}(\cdot)$. Show that the feedback interconnection of (5.324)–(5.326) is globally uniformly asymptotically stable for all $\sigma(\cdot,\cdot) \in \Phi_{\rm br}$.

Problem 5.63. Consider the nonlinear dynamical system $\mathcal{G}$ given by

$$\dot{x}(t) = f(x(t)) + G(x(t))u(t), \quad x(0) = x_0, \quad t \geq 0, \tag{5.327}$$

$$y(t) = h(x(t)) + J(x(t))u(t), \tag{5.328}$$

$$u(t) = -\sigma(y(t), t), \tag{5.329}$$

where $x \in \mathbb{R}^n$, $u, y \in \mathbb{R}^m$, $f : \mathbb{R}^n \to \mathbb{R}^n$, $G : \mathbb{R}^n \to \mathbb{R}^{n\times m}$, $h : \mathbb{R}^n \to \mathbb{R}^m$, $J : \mathbb{R}^n \to \mathbb{R}^{m\times m}$, and $\sigma(\cdot,\cdot) \in \Phi_{\rm pr}$. Furthermore, suppose that $\mathcal{G}$ is zero-state observable and exponentially dissipative with respect to the supply rate $r(u,y) = u^{\rm T}u + u^{\rm T}My$ and with a continuously differentiable, radially unbounded storage function $V_{\rm s}(\cdot)$, where $M = M^{\rm T} \in \mathbb{R}^{m\times m}$. Show that the feedback interconnection of (5.327)–(5.329) is globally uniformly asymptotically stable for all $\sigma(\cdot,\cdot) \in \Phi_{\rm pr}$. How would you modify $r(u,y)$ to address $\sigma(\cdot,\cdot) \in \Phi$?

Problem 5.64. Consider the nonlinear dynamical system $\mathcal{G}$ given by

$$\dot{x}(t) = f(x(t)) + G(x(t))u(t), \quad x(0) = x_0, \quad t \geq 0, \tag{5.330}$$

$$y(t) = h(x(t)), \tag{5.331}$$

$$u(t) = -\sigma(y(t)), \tag{5.332}$$

where $x \in \mathbb{R}^n$, $u, y \in \mathbb{R}^m$, $f : \mathbb{R}^n \to \mathbb{R}^n$, $G : \mathbb{R}^n \to \mathbb{R}^{n\times m}$, $h : \mathbb{R}^n \to \mathbb{R}^m$, and $\sigma(\cdot) \in \Phi_{\rm P}$. Furthermore, suppose that $\mathcal{G}$ is zero-state observable and exponentially dissipative with respect to the supply rate $r(u,y) = 2u^{\rm T}y + u^{\rm T}My + \dot{y}^{\rm T}Nu$ and with a continuously differentiable, radially unbounded storage function $V_{\rm s}(\cdot)$, where M is positive definite and N is a nonnegative-definite diagonal matrix. In addition, suppose that $y(t) + N\dot{y}(t) \neq 0$, $t \geq 0$, for $u(t) \equiv 0$. Show that the negative feedback interconnection of (5.327)–(5.329) is globally asymptotically stable for all $\sigma(\cdot) \in \Phi_{\rm P}$. (**Hint:** To address

$\dot{y}$ in the supply rate $r(u, y)$ form an augmented system with output $\hat{y} = [y^{\mathrm{T}}\ \dot{y}^{\mathrm{T}}]^{\mathrm{T}}$ and construct a supply rate of the form $r(u, \hat{y})$.)

Problem 5.65. Consider the nonnegative dynamical system $\mathcal{G}$ (see Problem 5.9) given by (5.160) and (5.161) and assume that (A, C) is observable and $\mathcal{G}$ is exponentially dissipative with respect to the supply rate $s(u, y) = \mathbf{e}^{\mathrm{T}}u - \mathbf{e}^{\mathrm{T}}My$, where $M >> 0$. Show that the positive feedback interconnection of $\mathcal{G}$ and $\sigma(\cdot, \cdot)$ is globally uniformly asymptotically stable for all $\sigma(\cdot, \cdot) \in \Phi$, where

$$\Phi \triangleq \{\sigma : \overline{\mathbb{R}}_+ \times \overline{\mathbb{R}}_+^l \to \overline{\mathbb{R}}_+^m : \sigma(\cdot, 0) = 0,\ 0 \leq\leq \sigma(t, y) \leq\leq My,\ y \in \overline{\mathbb{R}}_+^l, \text{ a.e. } t \geq 0, \text{ and } \sigma(\cdot, y) \text{ is Lebesgue measurable for all } y \in \overline{\mathbb{R}}_+^l\}, \tag{5.333}$$

$M >> 0$, and $M \in \mathbb{R}^{m \times l}$. (**Hint:** Use Problem 5.11 to show that if $\mathcal{G}$ is exponentially dissipative with respect to the supply rate $s(u, y) = \mathbf{e}^{\mathrm{T}}u - \mathbf{e}^{\mathrm{T}}My$, then there exists $p \in \mathbb{R}_+^n$, $l \in \overline{\mathbb{R}}_+^n$, and $w \in \overline{\mathbb{R}}_+^m$, and a scalar $\varepsilon > 0$ such that

$$0 = A^{\mathrm{T}}p + \varepsilon p + C^{\mathrm{T}}M^{\mathrm{T}}\mathbf{e} + l, \tag{5.334}$$

$$0 = B^{\mathrm{T}}p + D^{\mathrm{T}}M^{\mathrm{T}}\mathbf{e} - \mathbf{e} + w. \tag{5.335}$$

Now, use the Lyapunov function candidate $V(x) = p^{\mathrm{T}}x$.)

5.11 Notes and References

The original work on dissipative dynamical systems is due to J. C. Willems [456, 457]. Lagrangian and Hamiltonian dynamical systems arose from Euler's variational calculus and are due to the fundamental work of Joseph-Louis Lagrange [251] on analytical mechanics and William Rowan Hamilton's work on least action [183], developed in the eighteenth and nineteenth centuries, respectively. Port-controlled Hamiltonian systems were introduced by Maschke, van der Schaft, and Breedveld [305] and Maschke and van der Schaft [303, 304]. See also van der Schaft [441]. Theorem 5.6 presenting necessary and sufficient conditions for dissipativity with respect to quadratic supply rates is due to Hill and Moylan [188]. The concepts of exponential dissipativity, exponential passivity, and exponential nonexpansivity are due to Chellaboina and Haddad [88]. The concepts of input strict passivity, output strict passivity, and input-output strict passivity are also due to Hill and Moylan [189]. The classical concepts of passivity and nonexpansivity can be found in Popov [362, 364], Zames [476, 477], Sandberg [389], and Desoer and Vidyasagar [104]. Positive real and bounded real transfer functions are discussed in Anderson [7, 8] and Anderson and Vongpanitlerd [11]. The Kalman-Yakubovich-Popov lemma, also known as the positive real lemma, was discovered independently by

Kalman [226] and Yakubovich [468]. For an excellent treatment of the positive real lemma and the bounded real lemma see Anderson [7, 8] and Anderson and Vongpanitlerd [11]. The linearization results for dissipative and exponential dissipative dynamical systems are due to Haddad and Chellaboina [158] and Chellaboina and Haddad [88].

The absolute stability problem was first formulated in 1944 by A. I. Luré and V. N. Postnikov [292]. In particular, sufficient conditions for absolute stability in terms of a set of quadratic equations were derived using a Lyapunov function containing a quadratic plus an integral term involving the feedback nonlinearity [290,291]. This approach was further developed in a series of papers by Yakubovich [467], Malkin [297], and Rozenvasser [372]. The frequency domain approach to absolute stability was first developed by the Rumanian mathematician Vasile-Mikhai Popov [361]. Yakubovich was the first to show that the Luré-Postnikov Lyapunov function is necessary for proving the Popov criterion [468]. Ever since Popov derived a frequency domain condition for absolute stability, considerable effort by numerous researchers was devoted in deriving similar criteria for absolute stability [5, 64, 65, 67, 70, 71, 94, 326, 328–331, 361, 362, 364, 425, 427, 458, 476, 479]. The circle criterion evolved from this activity and was first derived by Bongiorno [64], with a Lyapunov function proof first given by Narendra and Goldwyn [327]. An extensive development of absolute stability theory is given in the classical monographs of Aizerman and Gantmacher [5], Lefschetz [265], Popov [364], and Narendra and Taylor [331].

Chapter Six

Stability and Optimality of Feedback Dynamical Systems

6.1 Introduction

In this chapter, we use the stability and dissipativity results developed in Chapters 3–5 to develop stability and optimality results for feedback dynamical systems. Specifically, general stability criteria are given for Lyapunov, asymptotic, and exponential stability of feedback dynamical systems. Furthermore, energy-based controllers for port-controlled Hamiltonian systems are established using passivity theory. In particular, constructive sufficient conditions for feedback stabilization are derived that provide a shaped energy function for the closed-loop system, while preserving the Hamiltonian structure at the closed-loop system level. In addition, relative stability margins are also derived for nonlinear regulators, and sector and disk margins are introduced as a generalization of classical gain and phase margins. The notion of a control Lyapunov function is also introduced, and it is shown that the existence of a smooth control Lyapunov function implies smooth (almost everywhere) stabilizability of a controlled nonlinear dynamical system. Next, a nonlinear control problem involving a notion of optimality with respect to a nonlinear-nonquadratic cost functional is introduced. Finally, we close the chapter by introducing the notions of feedback linearization, zero dynamics, and minimum-phase systems.

6.2 Feedback Interconnections of Dissipative Dynamical Systems

In this section, we consider feedback interconnections of dissipative dynamical systems. Specifically, using the notion of dissipative and exponentially dissipative dynamical systems, with appropriate storage functions and supply rates, we construct Lyapunov functions for interconnected dynamical systems by appropriately combining storage functions for each subsystem. The feedback system can be nonlinear and either dynamic or static. In the dynamic case, for generality, we allow the nonlinear feedback system (compensator) to be of fixed dimension n_c that may be less than the plant order n.

We begin by considering the nonlinear dynamical system $\mathcal{G}$ given by

$$\dot{x}(t) = f(x(t)) + G(x(t))u(t), \quad x(0) = x_0, \quad t \geq 0, \tag{6.1}$$
$$y(t) = h(x(t)) + J(x(t))u(t), \tag{6.2}$$

where $x \in \mathbb{R}^n$, $u \in \mathbb{R}^m$, $y \in \mathbb{R}^l$, $f : \mathbb{R}^n \to \mathbb{R}^n$ satisfies $f(0) = 0$, $G : \mathbb{R}^n \to \mathbb{R}^{n\times m}$, $h : \mathbb{R}^n \to \mathbb{R}^l$ satisfies $h(0) = 0$, and $J : \mathbb{R}^n \to \mathbb{R}^{l\times m}$, with the nonlinear feedback system $\mathcal{G}_c$ given by

$$\dot{x}_c(t) = f_c(x_c(t)) + G_c(u_c(t), x_c(t))u_c(t), \quad x_c(0) = x_{c0}, \quad t \geq 0, \tag{6.3}$$
$$y_c(t) = h_c(u_c(t), x_c(t)) + J_c(u_c(t), x_c(t))u_c(t), \tag{6.4}$$

where $x_c \in \mathbb{R}^{n_c}$, $u_c \in \mathbb{R}^l$, $y_c \in \mathbb{R}^m$, $f_c : \mathbb{R}^{n_c} \to \mathbb{R}^{n_c}$ satisfies $f_c(0) = 0$, $G_c : \mathbb{R}^l \times \mathbb{R}^{n_c} \to \mathbb{R}^{n_c\times l}$, $h_c : \mathbb{R}^l \times \mathbb{R}^{n_c} \to \mathbb{R}^m$ satisfies $h_c(0,0) = 0$, and $J_c : \mathbb{R}^l \times \mathbb{R}^{n_c} \to \mathbb{R}^{m\times l}$. We assume that $f(\cdot)$, $G(\cdot)$, $h(\cdot)$, $J(\cdot)$, $f_c(\cdot)$, $G_c(\cdot)$, $h_c(\cdot,\cdot)$, and $J_c(\cdot,\cdot)$ are continuous mappings and the required properties for the existence and uniqueness of solutions of the feedback interconnection of $\mathcal{G}$ and $\mathcal{G}_c$ are satisfied. Note that with the negative feedback interconnection given by Figure 6.1, $u_c = y$ and $y_c = -u$. Here and henceforth in the book

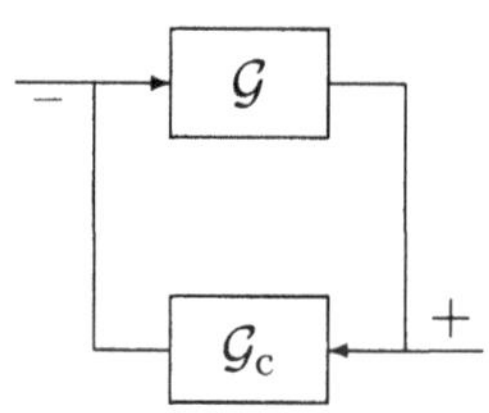

Figure 6.1 Feedback interconnection of $\mathcal{G}$ and $\mathcal{G}_c$.

we assume that the negative feedback interconnection of $\mathcal{G}$ and $\mathcal{G}_c$ is well posed, that is, $\det[I_m + J_c(y, x_c)J(x)] \neq 0$ for all y, x, and x_c.

The following results give sufficient conditions for Lyapunov, asymptotic, and exponential stability of the feedback interconnection given by Figure 6.1.

Theorem 6.1. Consider the closed-loop system consisting of the nonlinear dynamical systems $\mathcal{G}$ and $\mathcal{G}_c$ with input-output pairs (u, y) and (u_c, y_c), respectively, and with $u_c = y$ and $y_c = -u$. Assume $\mathcal{G}$ and $\mathcal{G}_c$ are zero-state observable and dissipative with respect to the supply rates $r(u, y)$ and $r_c(u_c, y_c)$ and with continuously differentiable, radially unbounded storage functions $V_s(\cdot)$ and $V_{sc}(\cdot)$, respectively, such that $V_s(0) = 0$ and $V_{sc}(0) = 0$. Furthermore, assume there exists a scalar $\sigma > 0$ such that

$r(u,y)+\sigma r_{\rm c}(u_{\rm c},y_{\rm c})\le 0$. Then the following statements hold:

i) The negative feedback interconnection of $\mathcal{G}$ and $\mathcal{G}_{\rm c}$ is Lyapunov stable.

ii) If $\mathcal{G}_{\rm c}$ is exponentially dissipative with respect to supply rate $r_{\rm c}(u_{\rm c},y_{\rm c})$ and $\operatorname{rank}[G_{\rm c}(u_{\rm c},0)] = m$, $u_{\rm c}\in\mathbb{R}^l$, then the negative feedback interconnection of $\mathcal{G}$ and $\mathcal{G}_{\rm c}$ is globally asymptotically stable.

iii) If $\mathcal{G}$ and $\mathcal{G}_{\rm c}$ are exponentially dissipative with respect to supply rates $r(u,y)$ and $r_{\rm c}(u_{\rm c},y_{\rm c})$, respectively, and $V_{\rm s}(\cdot)$ and $V_{\rm sc}(\cdot)$ are such that there exist constants $\alpha,\alpha_{\rm c},\beta$, and $\beta_{\rm c}>0$ such that

$$\alpha\|x\|^2\le V_{\rm s}(x)\le\beta\|x\|^2,\qquad x\in\mathbb{R}^n,\tag{6.5}$$

$$\alpha_{\rm c}\|x_{\rm c}\|^2\le V_{\rm sc}(x_{\rm c})\le\beta_{\rm c}\|x_{\rm c}\|^2,\qquad x_{\rm c}\in\mathbb{R}^{n_{\rm c}},\tag{6.6}$$

then the negative feedback interconnection of $\mathcal{G}$ and $\mathcal{G}_{\rm c}$ is globally exponentially stable.

Proof. *i)* Consider the Lyapunov function candidate $V(x,x_{\rm c})=V_{\rm s}(x)+\sigma V_{\rm sc}(x_{\rm c})$. Now, the corresponding Lyapunov derivative is given by

$$\dot V(x,x_{\rm c})=\dot V_{\rm s}(x)+\sigma\dot V_{\rm sc}(x_{\rm c})\le r(u,y)+\sigma r_{\rm c}(u_{\rm c},y_{\rm c})\le 0,\quad (x,x_{\rm c})\in\mathbb{R}^n\times\mathbb{R}^{n_{\rm c}},$$

which implies that the negative feedback interconnection of $\mathcal{G}$ and $\mathcal{G}_{\rm c}$ is Lyapunov stable.

ii) If $\mathcal{G}_{\rm c}$ is exponentially dissipative it follows that for some scalar $\varepsilon_{\rm c}>0$,

$$\begin{aligned}\dot V(x,x_{\rm c}) &= \dot V_{\rm s}(x)+\sigma\dot V_{\rm sc}(x_{\rm c})\\ &\le -\sigma\varepsilon_{\rm c}V_{\rm sc}(x_{\rm c})+r(u,y)+\sigma r_{\rm c}(u_{\rm c},y_{\rm c})\\ &\le -\sigma\varepsilon_{\rm c}V_{\rm sc}(x_{\rm c}),\qquad (x,x_{\rm c})\in\mathbb{R}^n\times\mathbb{R}^{n_{\rm c}}.\end{aligned}$$

Next, let $\mathcal{R}\triangleq\{(x,x_{\rm c})\in\mathbb{R}^n\times\mathbb{R}^{n_{\rm c}}:\ \dot V(x,x_{\rm c})=0\}$ and, since $V_{\rm sc}(x_{\rm c})$ is positive definite, note that $\dot V(x,x_{\rm c})=0$ only if $x_{\rm c}=0$. Now, since $\operatorname{rank}[G_{\rm c}(u_{\rm c},0)]=m$, $u_{\rm c}\in\mathbb{R}^l$, it follows that on every invariant set $\mathcal{M}$ contained in $\mathcal{R}$, $u_{\rm c}(t)=y(t)\equiv 0$, and hence, by (6.4), $u(t)\equiv 0$ so that $\dot x(t)=f(x(t))$. Now, since $\mathcal{G}$ is zero-state observable it follows that $\mathcal{M}=\{(0,0)\}$ is the largest invariant set contained in $\mathcal{R}$. Hence, it follows from Theorem 3.5 that $(x(t),x_{\rm c}(t))\to\mathcal{M}=\{(0,0)\}$ as $t\to\infty$. Now, global asymptotic stability of the negative feedback interconnection of $\mathcal{G}$ and $\mathcal{G}_{\rm c}$ follows from the fact that $V_{\rm s}(\cdot)$ and $V_{\rm sc}(\cdot)$ are, by assumption, radially unbounded.

iii) Finally, if $\mathcal{G}$ and $\mathcal{G}_{\rm c}$ are exponentially dissipative it follows that

$$\begin{aligned}\dot V(x,x_{\rm c}) &= \dot V_{\rm s}(x)+\sigma\dot V_{\rm sc}(x_{\rm c})\\ &\le -\varepsilon V_{\rm s}(x)-\sigma\varepsilon_{\rm c}V_{\rm sc}(x_{\rm c})+r(u,y)+\sigma r_{\rm c}(u_{\rm c},y_{\rm c})\end{aligned}$$

$$\leq -\min\{\varepsilon, \varepsilon_{\rm c}\}V(x, x_{\rm c}), \qquad (x, x_{\rm c}) \in \mathbb{R}^n \times \mathbb{R}^{n_{\rm c}},$$

which implies that the negative feedback interconnection of $\mathcal{G}$ and $\mathcal{G}_{\rm c}$ is globally exponentially stable. □

The next result presents Lyapunov, asymptotic, and exponential stability of dissipative feedback systems with quadratic supply rates.

Theorem 6.2. Let $Q \in \mathbb{S}^l$, $S \in \mathbb{R}^{l\times m}$, $R \in \mathbb{S}^m$, $Q_{\rm c} \in \mathbb{S}^m$, $S_{\rm c} \in \mathbb{R}^{m\times l}$, and $R_{\rm c} \in \mathbb{S}^l$. Consider the closed-loop system consisting of the nonlinear dynamical systems $\mathcal{G}$ given by (6.1) and (6.2) and $\mathcal{G}_{\rm c}$ given by (6.3) and (6.4), and assume $\mathcal{G}$ and $\mathcal{G}_{\rm c}$ are zero-state observable. Furthermore, assume $\mathcal{G}$ is dissipative with respect to the quadratic supply rate $r(u,y) = y^{\rm T}Qy + 2y^{\rm T}Su + u^{\rm T}Ru$ and has a continuously differentiable, radially unbounded storage function $V_{\rm s}(\cdot)$, and $\mathcal{G}_{\rm c}$ is dissipative with respect to the quadratic supply rate $r_{\rm c}(u_{\rm c}, y_{\rm c}) = y_{\rm c}^{\rm T}Q_{\rm c}y_{\rm c} + 2y_{\rm c}^{\rm T}S_{\rm c}u_{\rm c} + u_{\rm c}^{\rm T}R_{\rm c}u_{\rm c}$ and has a continuously differentiable, radially unbounded storage function $V_{\rm sc}(\cdot)$. Finally, assume there exists $\sigma > 0$ such that

$$\hat{Q} \triangleq \begin{bmatrix} Q + \sigma R_{\rm c} & -S + \sigma S_{\rm c}^{\rm T} \\ -S^{\rm T} + \sigma S_{\rm c} & R + \sigma Q_{\rm c} \end{bmatrix} \leq 0. \tag{6.7}$$

Then the following statements hold:

i) The negative feedback interconnection of $\mathcal{G}$ and $\mathcal{G}_{\rm c}$ is Lyapunov stable.

ii) If $\mathcal{G}_{\rm c}$ is exponentially dissipative with respect to supply rate $r_{\rm c}(u_{\rm c}, y_{\rm c})$ and $\operatorname{rank}[G_{\rm c}(u_{\rm c}, 0)] = m$, $u_{\rm c} \in \mathbb{R}^l$, then the negative feedback interconnection of $\mathcal{G}$ and $\mathcal{G}_{\rm c}$ is globally asymptotically stable.

iii) If $\mathcal{G}$ and $\mathcal{G}_{\rm c}$ are exponentially dissipative with respect to supply rates $r(u,y)$ and $r_{\rm c}(u_{\rm c}, y_{\rm c})$ and there exist constants $\alpha, \beta, \alpha_{\rm c}$, and $\beta_{\rm c} > 0$ such that (6.5) and (6.6) hold, then the negative feedback interconnection of $\mathcal{G}$ and $\mathcal{G}_{\rm c}$ is globally exponentially stable.

iv) If $\hat{Q} < 0$, then the negative feedback interconnection of $\mathcal{G}$ and $\mathcal{G}_{\rm c}$ is globally asymptotically stable.

Proof. Statements *i)*–*iii)* are a direct consequence of Theorem 6.1 by noting that

$$r(u,y) + \sigma r_{\rm c}(u_{\rm c}, y_{\rm c}) = \begin{bmatrix} y \\ y_{\rm c} \end{bmatrix}^{\rm T} \hat{Q} \begin{bmatrix} y \\ y_{\rm c} \end{bmatrix},$$

and hence, $r(u,y) + \sigma r_{\rm c}(u_{\rm c}, y_{\rm c}) \leq 0$.

To show *iv)* consider the Lyapunov function candidate $V(x, x_{\rm c}) = V_{\rm s}(x) + \sigma V_{\rm sc}(x_{\rm c})$. Noting that $u_{\rm c} = y$ and $y_{\rm c} = -u$ it follows that the corresponding Lyapunov derivative is given by

$$\begin{aligned}
\dot{V}(x,x_{\rm c}) &= \dot{V}_{\rm s}(x) + \sigma\dot{V}_{\rm sc}(x_{\rm c}) \\
&\le r(u,y) + \sigma r_{\rm c}(u_{\rm c},y_{\rm c}) \\
&= y^{\rm T}Qy + 2y^{\rm T}Su + u^{\rm T}Ru + \sigma(y_{\rm c}^{\rm T}Q_{\rm c}y_{\rm c} + 2y_{\rm c}^{\rm T}S_{\rm c}u_{\rm c} + u_{\rm c}^{\rm T}R_{\rm c}u_{\rm c}) \\
&= \begin{bmatrix} y \\ y_{\rm c} \end{bmatrix}^{\rm T} \hat{Q} \begin{bmatrix} y \\ y_{\rm c} \end{bmatrix} \\
&\le 0, \qquad (x,x_{\rm c}) \in \mathbb{R}^n \times \mathbb{R}^{n_{\rm c}},
\end{aligned}$$

which implies that the negative feedback interconnection of $\mathcal{G}$ and $\mathcal{G}_{\rm c}$ is Lyapunov stable. Next, let $\mathcal{R} \triangleq \{(x,x_{\rm c}) \in \mathbb{R}^n \times \mathbb{R}^{n_{\rm c}} : \dot{V}(x,x_{\rm c}) = 0\}$ and note that $\dot{V}(x,x_{\rm c}) = 0$ if and only if $(y,y_{\rm c}) = (0,0)$. Now, since $\mathcal{G}$ and $\mathcal{G}_{\rm c}$ are zero-state observable it follows that $\mathcal{M} = \{(0,0)\}$ is the largest invariant set contained in $\mathcal{R}$. Hence, it follows from Theorem 3.5 that $(x(t),x_{\rm c}(t)) \to \mathcal{M} = \{(0,0)\}$ as $t \to \infty$. Finally, global asymptotic stability follows from the fact that $V_{\rm s}(\cdot)$ and $V_{\rm sc}(\cdot)$ are, by assumption, radially unbounded, and hence, $V(x,x_{\rm c}) \to \infty$ as $\|(x,x_{\rm c})\| \to \infty$. □

The following two corollaries are a direct consequence of Theorem 6.2. For both results note that if a nonlinear dynamical system $\mathcal{G}$ is dissipative (respectively, exponentially dissipative) with respect to a supply rate $r(u,y) = u^{\rm T}y - \varepsilon u^{\rm T}u - \hat{\varepsilon}y^{\rm T}y$, where $\varepsilon, \hat{\varepsilon} \ge 0$, then with $\kappa(y) = ky$, where $k \in \mathbb{R}$ is such that $k(1-\varepsilon k) < \hat{\varepsilon}$, $r(u,y) = [k(1-\varepsilon k) - \hat{\varepsilon}]y^{\rm T}y < 0$, $y \neq 0$. Hence, if $\mathcal{G}$ is zero-state observable it follows from Theorem 5.6 that all storage functions (respectively, exponential storage functions) of $\mathcal{G}$ are positive definite. For the next result, we assume that all storage functions of $\mathcal{G}$ and $\mathcal{G}_{\rm c}$ are continuously differentiable.

Corollary 6.1. Consider the closed-loop system consisting of the nonlinear dynamical systems $\mathcal{G}$ given by (6.1) and (6.2) and $\mathcal{G}_{\rm c}$ given by (6.3) and (6.4), and assume $\mathcal{G}$ and $\mathcal{G}_{\rm c}$ are zero-state observable. Then the following statements hold:

i) If $\mathcal{G}$ is passive, $\mathcal{G}_{\rm c}$ is exponentially passive, and $\operatorname{rank}[G_{\rm c}(u_{\rm c},0)] = m$, $u_{\rm c} \in \mathbb{R}^l$, then the negative feedback interconnection of $\mathcal{G}$ and $\mathcal{G}_{\rm c}$ is asymptotically stable.

ii) If $\mathcal{G}$ and $\mathcal{G}_{\rm c}$ are exponentially passive with storage functions $V_{\rm s}(\cdot)$ and $V_{\rm sc}(\cdot)$, respectively, such that (6.5) and (6.6) hold, then the negative feedback interconnection of $\mathcal{G}$ and $\mathcal{G}_{\rm c}$ is exponentially stable.

iii) If $\mathcal{G}$ is nonexpansive with gain $\gamma > 0$, $\mathcal{G}_{\rm c}$ is exponentially nonexpansive with gain $\gamma_{\rm c} > 0$, $\operatorname{rank}[G_{\rm c}(u_{\rm c},0)] = m$, $u_{\rm c} \in \mathbb{R}^l$, and $\gamma\gamma_{\rm c} \le 1$, then the negative feedback interconnection of $\mathcal{G}$ and $\mathcal{G}_{\rm c}$ is asymptotically stable.

iv) If $\mathcal{G}$ and $\mathcal{G}_{\rm c}$ are exponentially nonexpansive with storage functions $V_{\rm s}(\cdot)$ and $V_{\rm sc}(\cdot)$, respectively, such that (6.5) and (6.6) hold, and with gains $\gamma > 0$ and $\gamma_{\rm c} > 0$, respectively, such that $\gamma\gamma_{\rm c} \le 1$, then the negative feedback interconnection of $\mathcal{G}$ and $\mathcal{G}_{\rm c}$ is exponentially stable.

v) If $\mathcal{G}$ is passive and $\mathcal{G}_{\rm c}$ is input-output strict passive, then the negative feedback interconnection of $\mathcal{G}$ and $\mathcal{G}_{\rm c}$ is asymptotically stable.

vi) If $\mathcal{G}$ and $\mathcal{G}_{\rm c}$ are input strict passive, then the negative feedback interconnection of $\mathcal{G}$ and $\mathcal{G}_{\rm c}$ is asymptotically stable.

vii) If $\mathcal{G}$ and $\mathcal{G}_{\rm c}$ are output strict passive, then the negative feedback interconnection of $\mathcal{G}$ and $\mathcal{G}_{\rm c}$ is asymptotically stable.

Proof. The proof is a direct consequence of Theorem 6.2. Specifically, *i)* and *ii)* follow from Theorem 6.2 with $Q = Q_{\rm c} = 0$, $S = S_{\rm c} = I_m$, and $R = R_{\rm c} = 0$, while *iii)* and *iv)* follow from Theorem 6.2 with $Q = -I_l$, $S = 0$, $R = \gamma^2 I_m$, $Q_{\rm c} = -I_{l_{\rm c}}$, $S_{\rm c} = 0$, and $R_{\rm c} = \gamma_{\rm c}^2 I_{m_{\rm c}}$. Statement *v)* follows from Theorem 6.2 with $Q = 0$, $S = I_m$, $R = 0$, $Q_{\rm c} = -\hat{\varepsilon} I_m$, $S_{\rm c} = I_m$, and $R_{\rm c} = -\varepsilon I_m$, where $\varepsilon, \hat{\varepsilon} > 0$. Statement *vi)* follows from Theorem 6.2 with $Q = 0$, $S = I_m$, $R = -\varepsilon I_m$, $Q_{\rm c} = 0$, $S_{\rm c} = I_m$, and $R_{\rm c} = -\hat{\varepsilon} I_m$, where $\varepsilon, \hat{\varepsilon} > 0$. Finally, *vii)* follows from Theorem 6.2 with $Q = -\varepsilon I_m$, $S = I_m$, $R = 0$, $Q_{\rm c} = -\hat{\varepsilon} I_m$, $S_{\rm c} = I_m$, and $R_{\rm c} = 0$, where $\varepsilon, \hat{\varepsilon} > 0$. □

Example 6.1. Consider the nonlinear dynamical system

$$\dot{x}(t) = Ax(t) - B\phi(Cx(t)), \quad x(0) = x_0, \quad t \ge 0, \tag{6.8}$$

where $x(t) \in \mathbb{R}^n$, $t \ge 0$, $A \in \mathbb{R}^{n\times n}$, $B \in \mathbb{R}^{n\times m}$, $C \in \mathbb{R}^{m\times n}$, (A, C) is observable, and $\phi(\cdot) \in \Phi_{\rm P}$, where

$$\Phi_{\rm P} \triangleq \{\phi : \mathbb{R}^m \to \mathbb{R}^m : \ \phi(0) = 0, \ \phi(y) = [\phi_1(y_1), \ \ldots, \ \phi_m(y_m)]^{\rm T}, \ y \in \mathbb{R}^m, \\ \phi_i(y_i)y_i > 0, y_i \neq 0, \ i = 1, \ldots, m\}. \tag{6.9}$$

Note that (6.8) can be rewritten as a negative feedback interconnection of a linear dynamical system given by the transfer function $G(s) = C(sI-A)^{-1}B$ and a memoryless feedback time-invariant nonlinearity $\phi(\cdot)$ (see Figure 6.2 (a)). Equivalently, we can rewrite the above feedback interconnection as a negative feedback interconnection of a linear dynamical system given by the transfer function $\tilde{G}(s)$ and a nonlinear dynamical system $\mathcal{G}_\phi$ (see Figure 6.2 (b)), where

$$\tilde{G}(s) \triangleq (M + Ns)G(s) \sim \left[\begin{array}{c|c} A & B \\ \hline MC + NCA & NCB \end{array}\right], \tag{6.10}$$

$M, N \in \mathbb{R}^{m\times m}$, $M, N > 0$ are diagonal, and $\mathcal{G}_\phi$ is given by

$$\dot{\hat{x}}(t) = -MN^{-1}\hat{x}(t) + N^{-1}\hat{u}(t), \quad \hat{x}(0) = 0, \quad t \ge 0, \tag{6.11}$$

$$\hat{y}(t) = \phi(\hat{x}(t)), \tag{6.12}$$

where $\hat{x}(t), \hat{u}(t), \hat{y}(t) \in \mathbb{R}^m$, $t \geq 0$, and $\hat{u}(t) = (MC + NCA)x(t)$.

Now, consider the function $\hat{V} : \mathbb{R}^m \to \mathbb{R}^m$ given by

$$\hat{V}(\hat{x}) = 2\sum_{i=1}^{m} \int_0^{\hat{x}_i} N_{(i,i)}\phi_i(\sigma)\mathrm{d}\sigma, \tag{6.13}$$

and note that $\hat{V}(\hat{x}) > 0$, $\hat{x} \in \mathbb{R}^m$, $\hat{x} \neq 0$, $\hat{V}(0) = 0$, and

$$\begin{aligned}
\dot{\hat{V}}(\hat{x}(t)) &= 2\sum_{i=1}^{m} N_{(i,i)}\phi(\hat{x}_i(t))\dot{\hat{x}}_i(t) \\
&= 2\dot{\hat{x}}^{\mathrm{T}}(t)N\phi(\hat{x}(t)) \\
&= -2\hat{x}^{\mathrm{T}}(t)M\phi(\hat{x}(t)) + 2\hat{u}^{\mathrm{T}}(t)\phi(\hat{x}(t)) \\
&\leq 2\hat{u}^{\mathrm{T}}(t)\hat{y}(t), \quad t \geq 0,
\end{aligned} \tag{6.14}$$

which implies that $\hat{V}(\hat{x})$ is a storage function for $\mathcal{G}_\phi$, and hence, $\mathcal{G}_\phi$ is a passive dynamical system. Now, it follows from *i*) of Corollary 6.1 that if $\tilde{G}(s)$ is strictly positive real, then the negative feedback interconnection of $\tilde{G}(s)$ and $\mathcal{G}_\phi$ is asymptotically stable or, equivalently, the zero solution $x(t) \equiv 0$ to (6.8) is asymptotically stable for all $\phi \in \Phi_{\mathrm{P}}$. Hence, it follows from Theorem 5.14 that $\tilde{G}(s)$ is strictly positive real if and only if there exist matrices $P \in \mathbb{R}^{n\times n}$, $P > 0$, $L \in \mathbb{R}^{p\times n}$, and $W \in \mathbb{R}^{p\times m}$, with P positive definite, and a scalar $\varepsilon > 0$ such that

$$0 = A^{\mathrm{T}}P + PA + \varepsilon P + L^{\mathrm{T}}L, \tag{6.15}$$
$$0 = B^{\mathrm{T}}P - MC - NCA + W^{\mathrm{T}}L, \tag{6.16}$$
$$0 = NCB + B^{\mathrm{T}}C^{\mathrm{T}}N - W^{\mathrm{T}}W. \tag{6.17}$$

Now, $V(x, \hat{x}) = x^{\mathrm{T}}Px + \hat{V}(\hat{x})$ or, equivalently,

$$V(x) = x^{\mathrm{T}}Px + 2\sum_{i=1}^{m} \int_0^{y_i} N_{(i,i)}\phi_i(\sigma)\mathrm{d}\sigma, \tag{6.18}$$

since $\hat{u} = y$, is a Lyapunov function for (6.8). △

Example 6.2. Consider the controlled undamped Duffing equation given by

$$\dot{x}_1(t) = x_2(t), \quad x_1(0) = x_{10}, \quad t \geq 0, \tag{6.19}$$
$$\dot{x}_2(t) = -[2 + x_1^2(t)]x_1(t) + u(t), \quad x_2(0) = x_{20}, \tag{6.20}$$
$$y(t) = x_2(t). \tag{6.21}$$

Defining $x = [x_1, \;\; x_2]^{\mathrm{T}}$, (6.19)–(6.21) can be written in state space form (6.1) and (6.2) with $f(x) = [x_2, \;\; -(2+x_1^2)x_1]^{\mathrm{T}}$, $G(x) = [0, \;\; 1]^{\mathrm{T}}$, $h(x) = x_2$,

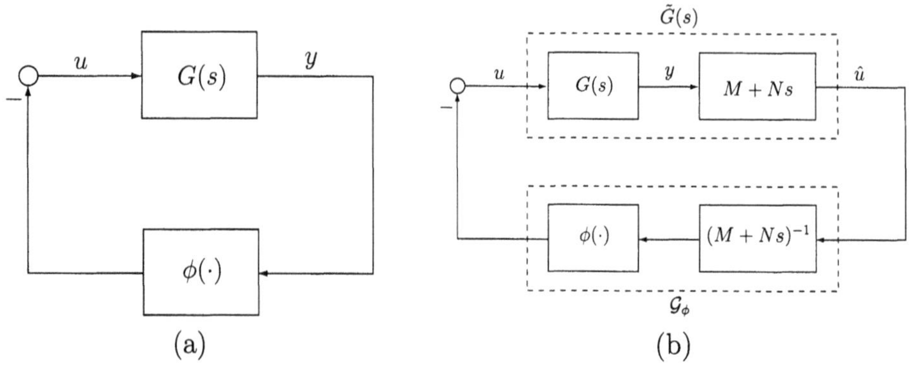

Figure 6.2 Feedback interconnection representation of an uncertain system.

and $J(x) = 0$. With $V_s(x) = x_1^2 + \frac{1}{4}x_1^4 + \frac{1}{2}x_2^2$, $\ell(x) \equiv 0$, and $\mathcal{W}(x) \equiv 0$, it follows from Corollary 5.2 that (6.19)–(6.21) is passive. Now, using Corollary 6.1 we can design a reduced-order linear dynamic compensator to asymptotically stabilize (6.19) and (6.20). Specifically, it follows from *i*) of Corollary 6.1 that if $\mathcal{G}_c$ given by (6.3) and (6.4) is exponentially passive with $\operatorname{rank}[G_c(0)] = 1$, then the negative feedback interconnection of $\mathcal{G}$ given by (6.19)–(6.21) and $\mathcal{G}_c$ is asymptotically stable. Here, we construct a reduced-order linear dynamic compensator $\mathcal{G}_c$ given by (6.3) and (6.4) with $f_c(x_c) = -10x_c$, $G_c(x_c) = 5$, $h_c(x_c) = 6x_c$, and $J_c(x_c) \equiv 0$. Note that with $V_s(x_c) = \frac{3}{5}x_c^2$, $\varepsilon = 20$, $\ell(x_c) \equiv 0$, and $\mathcal{W}(x_c) \equiv 0$, it follows from Corollary 5.4 that $\mathcal{G}_c$ is exponentially passive. Hence, Corollary 6.1 guarantees that the negative feedback interconnection of $\mathcal{G}$ and $\mathcal{G}_c$ is globally asymptotically stable. Figures 6.3 and 6.4 compare the time responses of the position x_1 and velocity x_2, respectively, for the open-loop and closed-loop systems for an initial condition $[x_{10},\ x_{20}]^{\mathrm{T}} = [1,\ 0.5]^{\mathrm{T}}$. Figure 6.5 compares the control effort versus time and Figure 6.6 gives the phase portraits of the open-loop and closed-loop systems. △

Corollary 6.2. Consider the closed-loop system consisting of the nonlinear dynamical systems $\mathcal{G}$ given by (6.1) and (6.2) and $\mathcal{G}_c$ given by (6.3) and (6.4). Let $a, b, a_c, b_c, \delta \in \mathbb{R}$ be such that $b > 0$, $0 < a + b$, $0 < 2\delta < b - a$, $a_c = a + \delta$, and $b_c = b + \delta$, let $M \in \mathbb{R}^{m \times m}$ be positive definite, and assume $\mathcal{G}$ and $\mathcal{G}_c$ are zero-state observable. If $\mathcal{G}$ is dissipative with respect to the supply rate $r(u, y) = u^{\mathrm{T}}My + \frac{ab}{a+b}y^{\mathrm{T}}My + \frac{1}{a+b}u^{\mathrm{T}}Mu$ and has a continuously differentiable radially unbounded storage function, and $\mathcal{G}_c$ is dissipative with respect to the supply rate $r_c(u_c, y_c) = u_c^{\mathrm{T}}My_c - \frac{1}{a_c+b_c}y_c^{\mathrm{T}}My_c - \frac{a_cb_c}{a_c+b_c}u_c^{\mathrm{T}}Mu_c$ and has a continuously differentiable radially unbounded storage function, then the negative feedback interconnection of $\mathcal{G}$ and $\mathcal{G}_c$ is globally asymptotically stable.

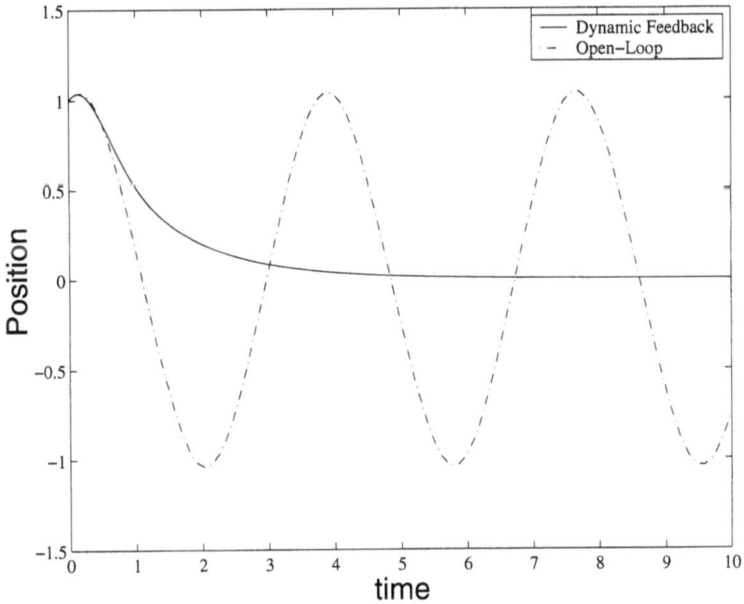

Figure 6.3 Position versus time.

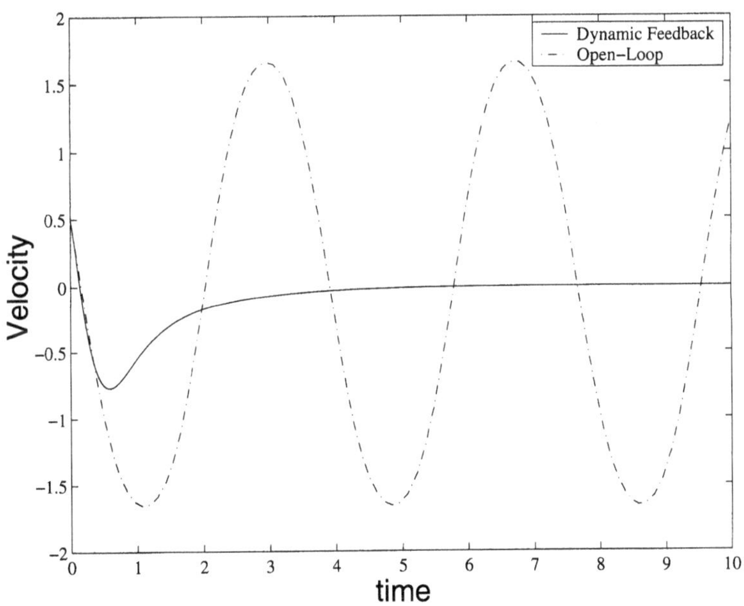

Figure 6.4 Velocity versus time.

Proof. The proof is a direct consequence of Theorem 6.2 with $Q = \frac{ab}{a+b}M$, $S = \frac{1}{2}M$, $R = \frac{1}{a+b}M$, $Q_\mathrm{c} = -\frac{1}{a_\mathrm{c}+b_\mathrm{c}}M$, $S_\mathrm{c} = \frac{1}{2}M$, and

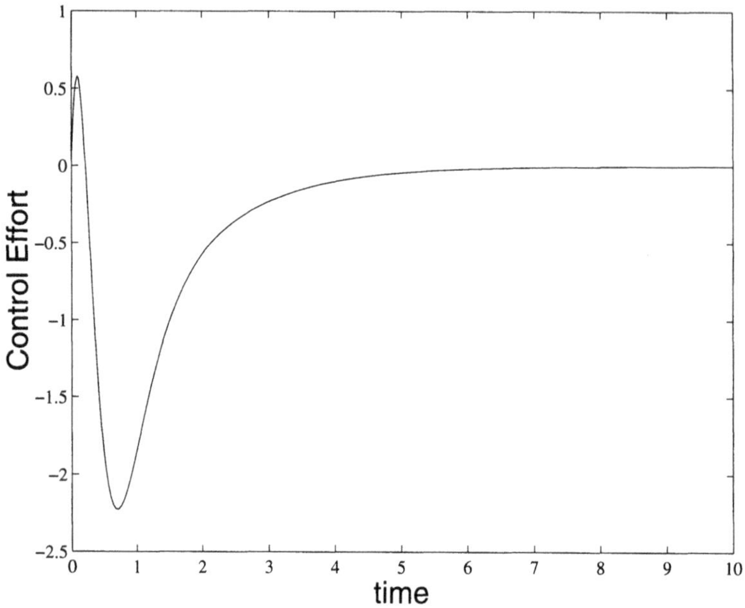

Figure 6.5 Control effort versus time.

$R_c = -\frac{a_c b_c}{a_c+b_c} M$. Specifically, let $\sigma > 0$ be such that

$$\sigma\left(\frac{\delta^2}{(a+b)^2} - \frac{1}{4}\right) + \frac{1}{4} > 0.$$

In this case, $\hat{Q}$ given by (6.7) satisfies

$$\hat{Q} = \begin{bmatrix} (\frac{ab}{a+b} - \frac{\sigma a_c b_c}{a_c+b_c})M & \frac{\sigma-1}{2}M \\ \frac{\sigma-1}{2}M & (\frac{1}{a+b} - \frac{\sigma}{a_c+b_c})M \end{bmatrix} < 0,$$

so that all the conditions of Theorem 6.2 are satisfied. □

6.3 Energy-Based Feedback Control

In this section, an energy-based control framework for port-controlled Hamiltonian systems is established. Specifically, we develop a controller design methodology that achieves stabilization via system passivation. In particular, the interconnection and damping matrix functions of the port-controlled Hamiltonian system are shaped so that the physical (Hamiltonian) system structure is preserved at the closed-loop level and the closed-loop energy function is equal to the difference between the physical energy of the system and the energy supplied by the controller. Since the Hamiltonian structure is preserved at the closed-loop level, the passivity-based controller is *robust* with respect to unmodeled passive dynamics. Furthermore, passivity-based control architectures are extremely appealing

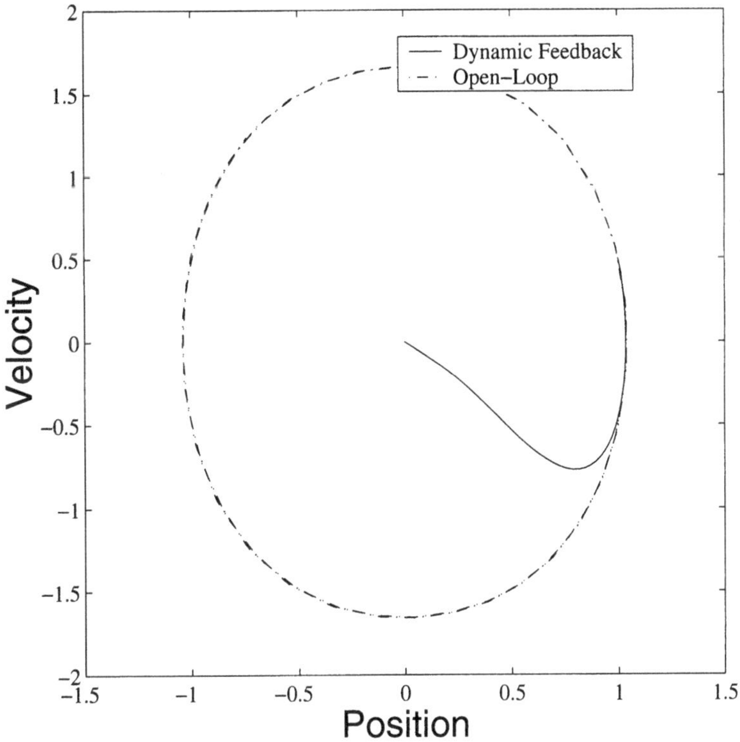

Figure 6.6 Phase portrait.

since the control action has a clear *physical* energy interpretation which can considerably simplify controller implementation.

We begin by considering the port-controlled Hamiltonian system given by

$$\dot{x}(t) = [\mathcal{J}(x(t)) - \mathcal{R}(x(t))]\left(\frac{\partial \mathcal{H}}{\partial x}(x(t))\right)^{\mathrm{T}} + G(x(t))u(t), \quad x(0) = x_0, \quad t \geq 0, \tag{6.22}$$

$$y(t) = G^{\mathrm{T}}(x(t))\left(\frac{\partial \mathcal{H}}{\partial x}(x(t))\right)^{\mathrm{T}}, \tag{6.23}$$

where $x(t) \in \mathcal{D} \subseteq \mathbb{R}^n$, $\mathcal{D}$ is an open set, $u(t) \in U \subseteq \mathbb{R}^m$, $y(t) \in Y \subseteq \mathbb{R}^l$, $\mathcal{H} : \mathcal{D} \to \mathbb{R}$ is a continuously differentiable *Hamiltonian function* for the system (6.22) and (6.23), $\mathcal{J} : \mathcal{D} \to \mathbb{R}^{n\times n}$ is such that $\mathcal{J}(x) = -\mathcal{J}^{\mathrm{T}}(x)$, $\mathcal{R} : \mathcal{D} \to \mathbb{S}^n$ is such that $\mathcal{R}(x) \geq 0$, $x \in \mathcal{D}$, $[\mathcal{J}(x) - \mathcal{R}(x)]\left(\frac{\partial \mathcal{H}}{\partial x}(x)\right)^{\mathrm{T}}$, $x \in \mathcal{D}$, is Lipschitz continuous on $\mathcal{D}$, and $G : \mathcal{D} \to \mathbb{R}^{n\times m}$. To address the energy-based feedback control problem let $\phi : \mathcal{D} \to U$. If $u(t) = \phi(x(t))$, $t \geq 0$, then $u(\cdot)$ is a *feedback control*. Next, we provide constructive sufficient conditions for energy-based feedback control of port-controlled Hamiltonian systems. Specifically, we seek feedback controllers $u(t) = \phi(x(t))$, $t \geq 0$, where $\phi : \mathcal{D} \to U$, such that the closed-loop system has the form

$$\begin{aligned}\dot{x}(t) &= [\mathcal{J}(x(t)) - \mathcal{R}(x(t))]\left(\frac{\partial \mathcal{H}}{\partial x}(x(t))\right)^{\mathrm{T}} + G(x(t))\phi(x(t)) \\ &= [\mathcal{J}_{\mathrm{s}}(x(t)) - \mathcal{R}_{\mathrm{s}}(x(t))]\left(\frac{\partial \mathcal{H}_{\mathrm{s}}}{\partial x}(x(t))\right)^{\mathrm{T}}, \quad x(0) = x_0, \quad t \geq 0, \quad (6.24)\end{aligned}$$

where $\mathcal{H}_{\mathrm{s}} : \mathcal{D} \to \mathbb{R}$ is a *shaped Hamiltonian function* for the closed-loop system (6.24), $\mathcal{J}_{\mathrm{s}} : \mathcal{D} \to \mathbb{R}^{n\times n}$ is a shaped interconnection matrix function for the closed-loop system and satisfies $\mathcal{J}_{\mathrm{s}}(x) = -\mathcal{J}_{\mathrm{s}}^{\mathrm{T}}(x)$, and $\mathcal{R}_{\mathrm{s}} : \mathcal{D} \to \mathbb{S}^n$ is a shaped dissipation matrix function for the closed-loop system and satisfies $\mathcal{R}_{\mathrm{s}}(x) \geq 0$, $x \in \mathcal{D}$.

Theorem 6.3. Consider the nonlinear port-controlled Hamiltonian system given by (6.22). Assume there exist functions $\phi : \mathcal{D} \to U$, $\mathcal{H}_{\mathrm{s}}, \mathcal{H}_{\mathrm{c}} : \mathcal{D} \to \mathbb{R}$, $\mathcal{J}_{\mathrm{s}}, \mathcal{J}_{\mathrm{a}} : \mathcal{D} \to \mathbb{R}^{n\times n}$, $\mathcal{R}_{\mathrm{s}}, \mathcal{R}_{\mathrm{a}} : \mathcal{D} \to \mathbb{R}^{n\times n}$ such that $\mathcal{H}_{\mathrm{s}}(x) = \mathcal{H}(x) + \mathcal{H}_{\mathrm{c}}(x)$ is continuously differentiable, $\mathcal{J}_{\mathrm{s}}(x) = \mathcal{J}(x) + \mathcal{J}_{\mathrm{a}}(x)$, $\mathcal{J}_{\mathrm{s}}(x) = -\mathcal{J}_{\mathrm{s}}^{\mathrm{T}}(x)$, $\mathcal{R}_{\mathrm{s}}(x) = \mathcal{R}(x) + \mathcal{R}_{\mathrm{a}}(x)$, $\mathcal{R}_{\mathrm{s}}(x) = \mathcal{R}_{\mathrm{s}}^{\mathrm{T}}(x) \geq 0$, $x \in \mathcal{D}$, and

$$\frac{\partial \mathcal{H}_{\mathrm{c}}}{\partial x}(x_{\mathrm{e}}) = -\frac{\partial \mathcal{H}}{\partial x}(x_{\mathrm{e}}), \quad x_{\mathrm{e}} \in \mathcal{D}, \quad (6.25)$$

$$\frac{\partial^2 \mathcal{H}_{\mathrm{c}}}{\partial x^2}(x_{\mathrm{e}}) > -\frac{\partial^2 \mathcal{H}}{\partial x^2}(x_{\mathrm{e}}), \quad x_{\mathrm{e}} \in \mathcal{D}, \quad (6.26)$$

$$\begin{aligned}[\mathcal{J}_{\mathrm{s}}(x) - \mathcal{R}_{\mathrm{s}}(x)]\left(\frac{\partial \mathcal{H}_{\mathrm{c}}}{\partial x}(x)\right)^{\mathrm{T}} &= -[\mathcal{J}_{\mathrm{a}}(x) - \mathcal{R}_{\mathrm{a}}(x)]\left(\frac{\partial \mathcal{H}}{\partial x}(x)\right)^{\mathrm{T}} \\ &\quad + G(x)\phi(x), \quad x \in \mathcal{D}. \quad (6.27)\end{aligned}$$

Then the equilibrium solution $x(t) \equiv x_{\mathrm{e}}$ of the closed-loop system (6.24) is Lyapunov stable. If, in addition, $\mathcal{D}_{\mathrm{c}} \subseteq \mathcal{D}$ is a compact positively invariant set with respect to (6.24) and the largest invariant set contained in $\mathcal{R} \triangleq \{x \in \mathcal{D}_{\mathrm{c}} : \frac{\partial \mathcal{H}_{\mathrm{s}}}{\partial x}(x)\mathcal{R}_{\mathrm{s}}(x)\left(\frac{\partial \mathcal{H}_{\mathrm{s}}}{\partial x}(x)\right)^{\mathrm{T}} = 0\}$ is $\mathcal{M} = \{x_{\mathrm{e}}\}$, then the equilibrium solution $x(t) \equiv x_{\mathrm{e}}$ of the closed-loop system (6.24) is locally asymptotically stable and $\mathcal{D}_{\mathrm{c}}$ is a subset of the domain of attraction of (6.24).

Proof. Condition (6.27) implies that with feedback controller $u(t) = \phi(x(t))$ the closed-loop system (6.22) has a Hamiltonian structure given by (6.24). Furthermore, it follows from (6.25) and (6.26) that the energy function $\mathcal{H}_{\mathrm{s}}(\cdot)$ has a local minimum at $x = x_{\mathrm{e}}$. Hence, $x = x_{\mathrm{e}}$ is an equilibrium point of the closed-loop system. Next, consider the Lyapunov function candidate for the closed-loop system (6.24) given by $V(x) = \mathcal{H}_{\mathrm{s}}(x) - \mathcal{H}_{\mathrm{s}}(x_{\mathrm{e}})$. Now, the corresponding Lyapunov derivative of $V(x)$ along the closed-loop state trajectories $x(t)$, $t \geq 0$, is given by

$$\dot{V}(x(t)) = \dot{\mathcal{H}}_{\mathrm{s}}(x(t)) = -\frac{\partial \mathcal{H}_{\mathrm{s}}}{\partial x}(x(t))\mathcal{R}_{\mathrm{s}}(x(t))\left(\frac{\partial \mathcal{H}_{\mathrm{s}}}{\partial x}(x(t))\right)^{\mathrm{T}} \leq 0, \quad t \geq 0. \tag{6.28}$$

Thus, it follows from Theorem 3.1 that the equilibrium solution $x(t) \equiv x_{\mathrm{e}}$ of (6.24) is Lyapunov stable. Asymptotic stability of the closed-loop system follows immediately from Corollary 3.1. □

Theorem 6.3 presents constructive sufficient conditions for feedback stabilization that preserve the physical Hamiltonian structure at the closed-loop level while providing a shaped Hamiltonian energy function as a Lyapunov function for the closed-loop system. These sufficient conditions consist of a partial differential equation parameterized by the auxiliary energy function $\mathcal{H}_{\mathrm{c}}$, the auxiliary interconnection matrix function $\mathcal{J}_{\mathrm{a}}$, and auxiliary dissipation matrix functions $\mathcal{R}_{\mathrm{a}}$, and whose solution characterizes the set of all desired shaped energy functions that can be assigned while preserving the system Hamiltonian structure at the closed-loop level. To apply Theorem 6.3, we fix the structure of the interconnection $\mathcal{J}_{\mathrm{s}}(\cdot)$ and dissipation $\mathcal{R}_{\mathrm{s}}(\cdot)$ matrix functions and solve for the closed-loop energy function $\mathcal{H}_{\mathrm{s}}(\cdot)$. Although in this case solving (6.27) appears formidable, it is in fact quite tractable since the partial differential equation (6.27) is parameterized via the interconnection and dissipation matrix functions which can be chosen by the control designer to satisfy system physical constraints. Alternatively, we can fix the shaped Hamiltonian $\mathcal{H}_{\mathrm{s}}$ and solve for the interconnection and dissipation matrix functions. In this case, we do not need to solve a partial differential equation but rather an algebraic equation.

If rank $G(x) = m$ and rank$[G(x)\ b(x)] = \operatorname{rank} G(x) = m$, where

$$b(x) = [\mathcal{J}_{\mathrm{s}}(x) - \mathcal{R}_{\mathrm{s}}(x)]\left(\frac{\partial \mathcal{H}_{\mathrm{c}}}{\partial x}(x)\right)^{\mathrm{T}} + [\mathcal{J}_{\mathrm{a}}(x) - \mathcal{R}_{\mathrm{a}}(x)]\left(\frac{\partial \mathcal{H}}{\partial x}(x)\right)^{\mathrm{T}}, \tag{6.29}$$

then an explicit expression for the stabilizing feedback controller satisfying (6.27) is given by $\phi(x) = (G^{\mathrm{T}}(x)G(x))^{-1}G^{\mathrm{T}}(x)b(x)$, $x \in \mathcal{D}$. Alternatively, if rank$[G(x)\ b(x)] = \operatorname{rank} G(x) < m$, $x \in \mathcal{D}$, then the feedback controller $\phi(x) = G^{\dagger}(x)b(x) + [I_m - G^{\dagger}(x)G(x)]z$, $x \in \mathcal{D}$, where $(\cdot)^{\dagger}$ denotes the Moore-Penrose generalized inverse and $z \in \mathbb{R}^m$, satisfies (6.27).

Under certain conditions on the system dissipation, the energy-based controller given by Theorem 6.3 provides an energy balance of the controlled system. To see this, let $\mathcal{R}_{\mathrm{a}}(x) \equiv 0$ and $\mathcal{R}(x)\left(\frac{\partial \mathcal{H}_{\mathrm{c}}}{\partial x}(x)\right)^{\mathrm{T}} = 0$, $x \in \mathcal{D}$. In this case, the closed-loop dynamics are given by

$$\dot{x}(t) = [\mathcal{J}_{\rm s}(x(t)) - \mathcal{R}(x(t))]\left(\frac{\partial \mathcal{H}_{\rm s}}{\partial x}(x(t))\right)^{\rm T}, \quad x(0) = x_0, \quad t \geq 0. \tag{6.30}$$

Along the trajectories $x(t)$, $t \geq 0$, it follows that

$$\begin{aligned}
&\dot{\mathcal{H}}_{\rm s}(x(t)) \\
&\quad = -\frac{\partial \mathcal{H}_{\rm s}}{\partial x}(x(t))\mathcal{R}(x(t))\left(\frac{\partial \mathcal{H}_{\rm s}}{\partial x}(x(t))\right)^{\rm T} \\
&\quad = -\left(\frac{\partial \mathcal{H}}{\partial x}(x(t)) + \frac{\partial \mathcal{H}_{\rm c}}{\partial x}(x(t))\right)\mathcal{R}(x(t))\left(\frac{\partial \mathcal{H}}{\partial x}(x(t)) + \frac{\partial \mathcal{H}_{\rm c}}{\partial x}(x(t))\right)^{\rm T} \\
&\quad = -\frac{\partial \mathcal{H}}{\partial x}(x(t))\mathcal{R}(x(t))\left(\frac{\partial \mathcal{H}}{\partial x}(x(t))\right)^{\rm T}, \quad t \geq 0,
\end{aligned} \tag{6.31}$$

or, equivalently using (5.67),

$$\dot{\mathcal{H}}_{\rm s}(x(t)) = \dot{\mathcal{H}}(x(t)) - u^{\rm T}(t)y(t), \quad t \geq 0. \tag{6.32}$$

Now, integrating (6.32) yields

$$\mathcal{H}_{\rm s}(x(t)) = \mathcal{H}(x(t)) - \int_{\hat{t}}^{t} u^{\rm T}(s)y(s){\rm d}s + \kappa, \quad 0 \leq \hat{t} \leq t, \tag{6.33}$$

where $\kappa \triangleq \mathcal{H}_{\rm s}(x(\hat{t})) - \mathcal{H}(x(\hat{t}))$, which shows that the closed-loop energy function $\mathcal{H}_{\rm s}(\cdot)$ is equal to the difference between the physical energy $\mathcal{H}(\cdot)$ of the system and the energy supplied by the controller modulo the constant κ.

Example 6.3. Consider the inverted pendulum shown in Figure 6.7, where $m = 1\,{\rm kg}$ and $L = 1\,{\rm m}$. The system is governed by the dynamic equation of motion

$$\ddot{\theta}(t) - g\sin\theta(t) = u(t), \quad \theta(0) = \theta_0, \quad \dot{\theta}(0) = \dot{\theta}_0, \quad t \geq 0, \tag{6.34}$$

where g denotes the gravitational acceleration and $u(\cdot)$ is a (thruster) control force. Defining $x_1 = \theta$ and $x_2 = \dot{\theta}$, we can rewrite the equation of motion in state space form (6.22) with $x \triangleq [x_1, x_2]^{\rm T}$,

$$\mathcal{J}(x) = \begin{bmatrix} 0 & 1 \\ -1 & 0 \end{bmatrix}, \quad \mathcal{R}(x) = 0, \quad G(x) = \begin{bmatrix} 0 \\ -1 \end{bmatrix}, \tag{6.35}$$

$\mathcal{D} = \mathbb{R}^2$, and Hamiltonian function $\mathcal{H}(\cdot)$ corresponding to the total energy in the system given by $\mathcal{H}(x) = \frac{x_2^2}{2} + g\cos x_1$.

Next, to stabilize the equilibrium point $x_{\rm e} = [\theta_{\rm e}, 0]^{\rm T}$ we assign the shaped Hamiltonian $\mathcal{H}_{\rm s}(x) = \frac{x_2^2}{2} + \frac{1}{2}(x_1 - \theta_{\rm e})^2$ function for the closed-loop

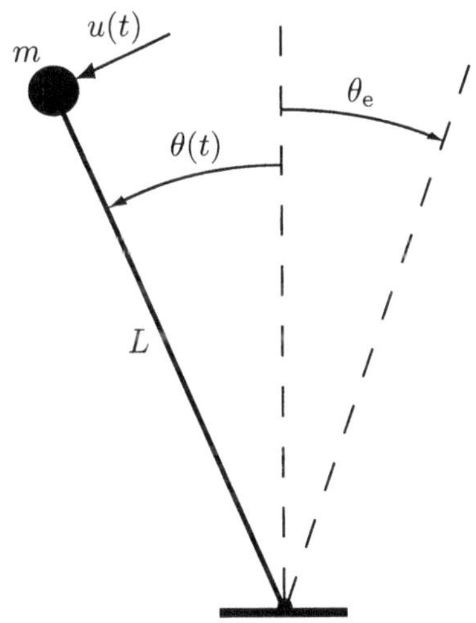

Figure 6.7 Inverted pendulum.

system. Furthermore, we set

$$\mathcal{J}_{\mathrm{a}}(x) = 0, \quad \mathcal{R}_{\mathrm{a}}(x) = \begin{bmatrix} 0 & 0 \\ 0 & 1 \end{bmatrix}, \quad x \in \mathcal{D}. \tag{6.36}$$

In this case, it follows from (6.27) that the feedback controller is given by $u = \phi(x) - x_2 + (x_1 - \theta_{\mathrm{e}}) + g \sin x_1$, $x \in \mathcal{D}$. Next, note that $\dot{\mathcal{H}}_{\mathrm{s}}(x) = -x_2^2 \leq 0$, $x \in \mathcal{D}$. Hence, $\mathcal{R} \triangleq \{x \in \mathcal{D} : \dot{\mathcal{H}}_{\mathrm{s}} = 0\} = \{x \in \mathcal{D} : x_2 = 0\}$. Finally, since for every $x \in \mathcal{R}$, $\dot{x}_2 \neq 0$ if and only if $x_1 \neq \theta_{\mathrm{e}}$, it follows that the largest invariant set contained in $\mathcal{R}$ is given by $\mathcal{M} = \{x_{\mathrm{e}}\}$, and hence, the equilibrium solution $x(t) \equiv [\theta_{\mathrm{e}}, 0]^{\mathrm{T}}$ is asymptotically stable. With $\theta_{\mathrm{e}} = 15^{\circ}$, Figure 6.8 shows the phase portrait of the port-controlled Hamiltonian system. Figure 6.9 shows the control force versus time and the shaped Hamiltonian versus time. △

Example 6.4. Consider the two-mass, two-spring system shown in Figure 6.10. A control force $\hat{u}(\cdot)$ acts on mass 2 with the goal to stabilize the position of the second mass. The system dynamics, with state variables defined in Figure 6.10, are given by

$$m_1\ddot{q}_1(t) + (k_1 + k_2)q_1(t) - k_2 q_2(t) = 0, \quad q_1(0) = q_{01}, \quad \dot{q}_1(0) = \dot{q}_{01}, \quad t \geq 0, \tag{6.37}$$

$$m_2\ddot{q}_2(t) - k_2 q_1(t) + k_2 q_2(t) = \hat{u}(t), \quad q_2(0) = q_{02}, \quad \dot{q}_2(0) = \dot{q}_{02}. \tag{6.38}$$

Defining $x_1 = q_1$, $x_2 = \dot{q}_1$, $x_3 = q_2$, and $x_4 = \dot{q}_2$, we can rewrite (6.37) and (6.38) in state space form (6.22) with $x = [x_1, x_2, x_3, x_4]^{\mathrm{T}}$, $\mathcal{R}(x) = 0$,

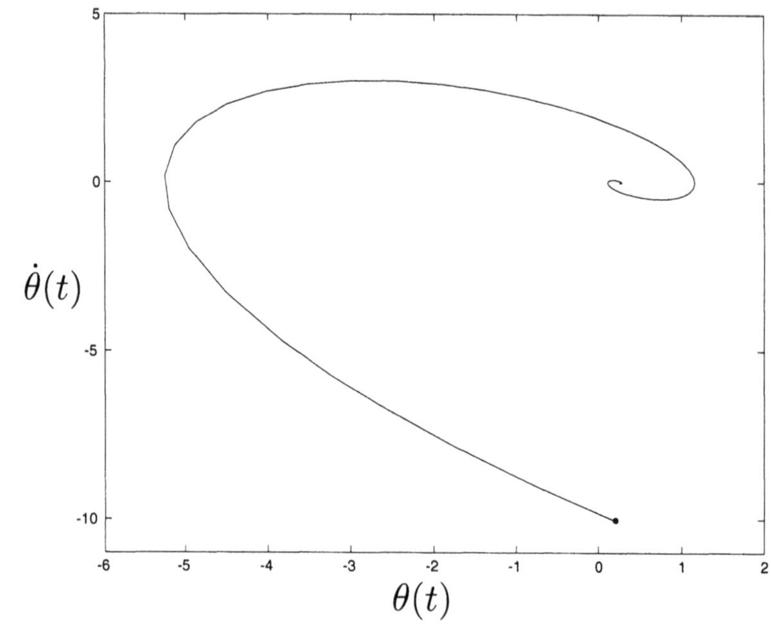

Figure 6.8 Phase portrait of the inverted pendulum.

$G(x) = [0,\, 0,\, 0,\, 1]^{\mathrm{T}}$, $u = \frac{\hat{u}}{m_2}$,

$$\mathcal{J}(x) = \begin{bmatrix} 0 & \frac{1}{m_1} & 0 & 0 \\ -\frac{1}{m_1} & 0 & 0 & 0 \\ 0 & 0 & 0 & \frac{1}{m_2} \\ 0 & 0 & -\frac{1}{m_2} & 0 \end{bmatrix}, \tag{6.39}$$

$\mathcal{D} = \{x \in \mathbb{R}^4 : x_1 \geq 0,\, x_3 \geq 0\}$, and Hamiltonian function $\mathcal{H}(\cdot)$ corresponding to the total energy in the system given by $\mathcal{H}(x) = \frac{m_1 x_2^2}{2} + \frac{m_2 x_4^2}{2} + \frac{k_1 x_1^2}{2} + \frac{k_2 (x_3 - x_1)^2}{2}$.

Next, to stabilize the equilibrium point $x_{\mathrm{e}} = [x_{1\mathrm{e}},\, 0,\, x_{3\mathrm{e}},\, 0]^{\mathrm{T}}$, where $x_{1\mathrm{e}} = \frac{k_2}{(k_1+k_2)} x_{3\mathrm{e}}$, with steady-state control value of $u_{\mathrm{c\,ss}} = \frac{k_1 k_2}{m_2(k_1+k_2)} x_{3\mathrm{e}}$, we assign the shaped Hamiltonian function $\mathcal{H}_{\mathrm{s}}(x) = \frac{m_1 x_2^2}{2} + \frac{m_2 x_4^2}{2} + \frac{k_1 x_1^2}{2} + \frac{k_2 (x_3 - x_1)^2}{2} - \frac{k_1 k_2}{(k_1+k_2)} x_{3\mathrm{e}} x_3$ for the closed-loop system. Furthermore, we set $\mathcal{J}_{\mathrm{a}}(x) \equiv 0$ and

$$\mathcal{R}_{\mathrm{a}}(x) = \begin{bmatrix} 0 & 0 & 0 & 0 \\ 0 & 0 & 0 & 0 \\ 0 & 0 & 0 & 0 \\ 0 & 0 & 0 & \frac{1}{m_2} \end{bmatrix}, \quad x \in \mathcal{D}. \tag{6.40}$$

In this case, it follows from (6.27) that the feedback controller is given by

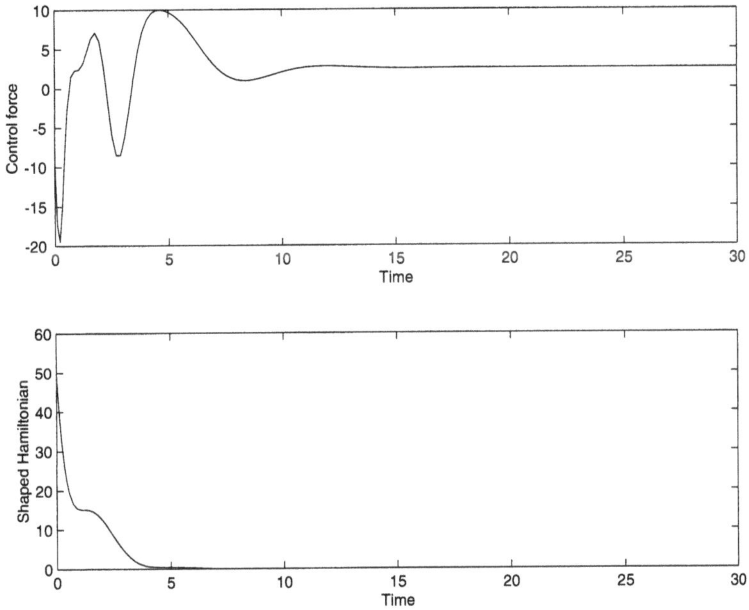

Figure 6.9 Control force and shaped Hamiltonian versus time.

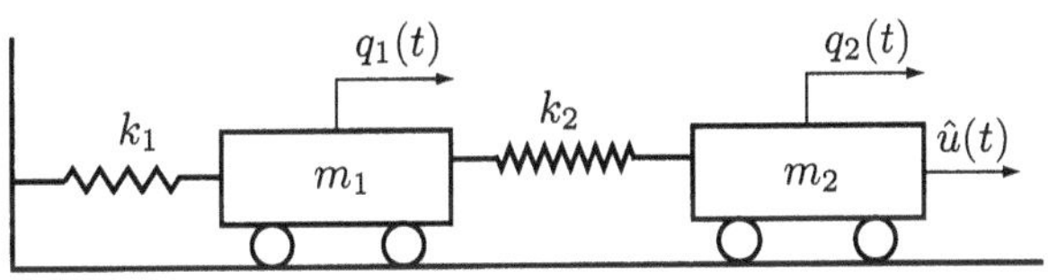

Figure 6.10 Two-mass, two-spring system.

$u = \phi(x) = \frac{k_1 k_2}{m_2(k_1+k_2)} x_{3\mathrm{e}} - x_4$, $x \in \mathcal{D}$. Next, note that $\dot{\mathcal{H}}_\mathrm{s}(x) = -m_2 x_4^2 \leq 0$, $x \in \mathcal{D}$. Hence, $\mathcal{R} \triangleq \{x \in \mathcal{D} : \dot{\mathcal{H}}_\mathrm{s} = 0\} = \{x \in \mathcal{D} : x_4 = 0\}$. Now, if $\mathcal{M} \subseteq \mathcal{R}$ is the largest invariant set contained in $\mathcal{R}$, then for every $x_0 \in \mathcal{M}$, $x_4(t) \equiv 0$, which implies that $x_1(t) - x_3(t) + \frac{k_1}{k_1+k_2} x_{3\mathrm{e}} = 0$ and $\dot{x}_3(t) = 0$, $t \geq 0$. In this case, it follows that $\dot{x}_1(t) = 0$, and hence, $\dot{x}_2(t) = 0$, $t \geq 0$. Hence, the only point that belongs to $\mathcal{M}$ is $x_\mathrm{e} = [\frac{k_2}{(k_1+k_2)} x_{3\mathrm{e}}, 0, x_{3\mathrm{e}}, 0]^\mathrm{T}$, which implies that x_e is an asymptotically stable equilibrium point of the closed-loop system. With $m_1 = 1.5\,\mathrm{kg}$, $m_2 = 0.8\,\mathrm{kg}$, $k_1 = 0.1\,\mathrm{N/m}$, $k_2 = 0.3\,\mathrm{N/m}$, $L = 0.4\,\mathrm{m}$, and $x_{3\mathrm{e}} = 3\,\mathrm{m}$, Figure 6.11 shows the phase portrait of x_2 versus x_4 of the port-controlled Hamiltonian system. Figures 6.12 and 6.13 show, respectively, the positions and velocities of the masses versus time. Finally, Figure 6.14 shows the control force versus time and the shaped Hamiltonian versus time. $\triangle$

Next, we consider energy-based dynamic control for port-controlled Hamiltonian systems wherein energy shaping is achieved by combining the physical energy of the plant and the emulated energy of the controller.

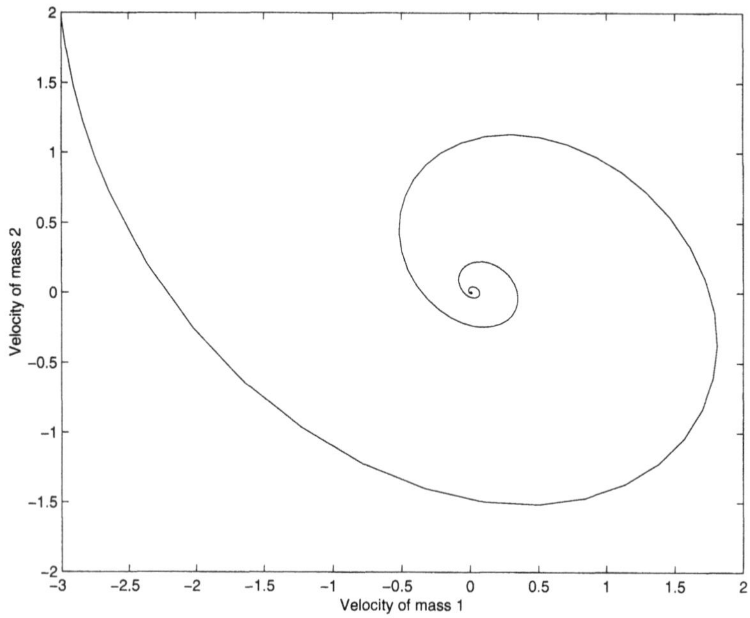

Figure 6.11 Phase portrait of x_2 versus x_4.

This approach has been extensively studied by Ortega *et al.* [340, 341] to design Euler-Lagrange controllers for potential energy shaping of mechanical systems.

We begin by considering the port-controlled Hamiltonian system $\mathcal{G}$ given by (6.22) and (6.23) with $m = l$. Furthermore, we consider the port-controlled Hamiltonian feedback control system $\mathcal{G}_c$ given by

$$\dot{x}_c(t) = [\mathcal{J}_c(x_c(t)) - \mathcal{R}_c(x_c(t))]\left(\frac{\partial \mathcal{H}_c}{\partial x_c}(x_c(t))\right)^{\rm T} + G_c(x_c(t))u_c(t), \quad x_c(0) = x_{c0}, \quad t \geq 0, \tag{6.41}$$

$$y_c(t) = G_c^{\rm T}(x_c(t))\left(\frac{\partial \mathcal{H}_c}{\partial x_c}(x_c(t))\right)^{\rm T}, \tag{6.42}$$

where $x_c(t) \in \mathbb{R}^{n_c}$, $u_c(t) \in U_c \subseteq \mathbb{R}^{m_c}$, $y_c(t) \in Y_c \subseteq \mathbb{R}^{l_c}$, $m_c = l_c$, $\mathcal{H}_c : \mathbb{R}^{n_c} \to \mathbb{R}$ is a continuously differentiable Hamiltonian function of the feedback control system $\mathcal{G}_c$, $\mathcal{J}_c : \mathbb{R}^{n_c} \to \mathbb{R}^{n_c \times n_c}$ is such that $\mathcal{J}_c(x_c) = -\mathcal{J}_c^{\rm T}(x_c)$, $\mathcal{R}_c : \mathbb{R}^{n_c} \to \mathbb{S}^{n_c}$ is such that $\mathcal{R}_c(x_c) \geq 0$, $x_c \in \mathbb{R}^{n_c}$, $[\mathcal{J}_c(x_c) - \mathcal{R}_c(x_c)]\left(\frac{\partial \mathcal{H}_c}{\partial x_c}(x_c)\right)^{\rm T}$, $x_c \in \mathbb{R}^{n_c}$, is Lipschitz continuous on $\mathbb{R}^{n_c}$, $G_c : \mathbb{R}^{n_c} \to \mathbb{R}^{n_c \times m_c}$, $m_c = l$, and $l_c = m$. Here, we assume that $u_c(\cdot)$ is restricted to the class of admissible inputs consisting of measurable functions such that $u_c(t) \in U_c$ for all $t \geq 0$. Note that with the feedback interconnection given by Figure 6.1, $u_c = y$ and $y_c = -u$. Hence, the closed-loop dynamics can be written in Hamiltonian

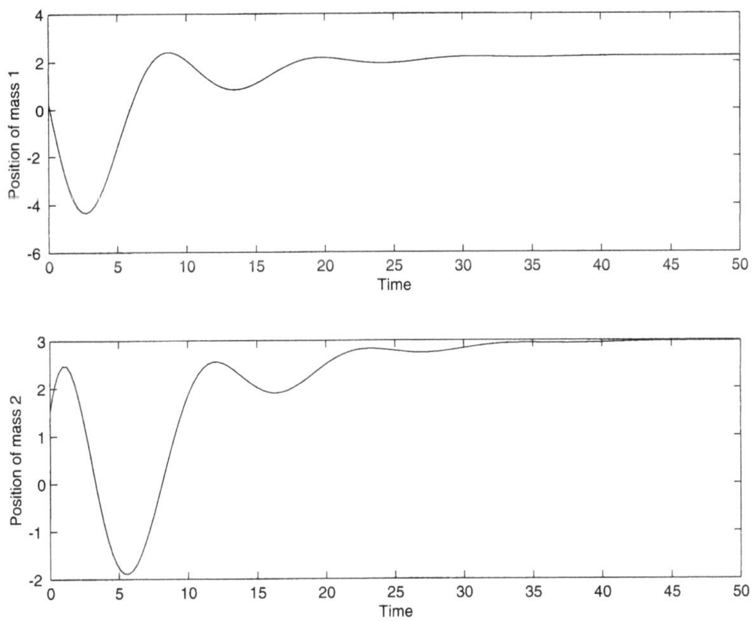

Figure 6.12 Mass positions versus time.

form given by

$$\dot{\tilde{x}}(t) = \left(\begin{bmatrix} \mathcal{J}(x(t)) & -G(x(t))G_{\mathrm{c}}^{\mathrm{T}}(x_{\mathrm{c}}(t)) \\ G_{\mathrm{c}}(x_{\mathrm{c}}(t))G^{\mathrm{T}}(x(t)) & \mathcal{J}_{\mathrm{c}}(x_{\mathrm{c}}(t)) \end{bmatrix} - \begin{bmatrix} \mathcal{R}(x(t)) & 0 \\ 0 & \mathcal{R}_{\mathrm{c}}(x_{\mathrm{c}}(t)) \end{bmatrix} \right) \begin{bmatrix} \left(\frac{\partial \mathcal{H}}{\partial x}(x(t))\right)^{\mathrm{T}} \\ \left(\frac{\partial \mathcal{H}_{\mathrm{c}}}{\partial x_{\mathrm{c}}}(x_{\mathrm{c}}(t))\right)^{\mathrm{T}} \end{bmatrix}, \quad \tilde{x}(0) = \tilde{x}_0, \quad t \geq 0, \tag{6.43}$$

where $\tilde{x} \triangleq [x^{\mathrm{T}}, x_{\mathrm{c}}^{\mathrm{T}}]^{\mathrm{T}}$.

It can be seen from (6.43) that by relating the controller state variables x_{c} to the plant state variables x, one can shape the Hamiltonian function $\mathcal{H}(\cdot) + \mathcal{H}_{\mathrm{c}}(\cdot)$ so as to preserve the Hamiltonian structure under dynamic feedback for part of the closed-loop system associated with the plant dynamics. Since the closed-loop dynamical system (6.43) is Hamiltonian involving skew-symmetric interconnection matrix function terms and nonnegative-definite dissipation matrix function terms, we can establish the existence of energy-Casimir functions [63, 441] (i.e., dynamical invariants) that are independent of the closed-loop Hamiltonian and relate the controller states to the plant states. Since, as shown in Section 3.4, energy-Casimir functions are composed of integrals of motion, it follows that these functions are constant along the trajectories of the closed-loop system (6.43). Furthermore, since the controller Hamiltonian $\mathcal{H}_{\mathrm{c}}(\cdot)$ can

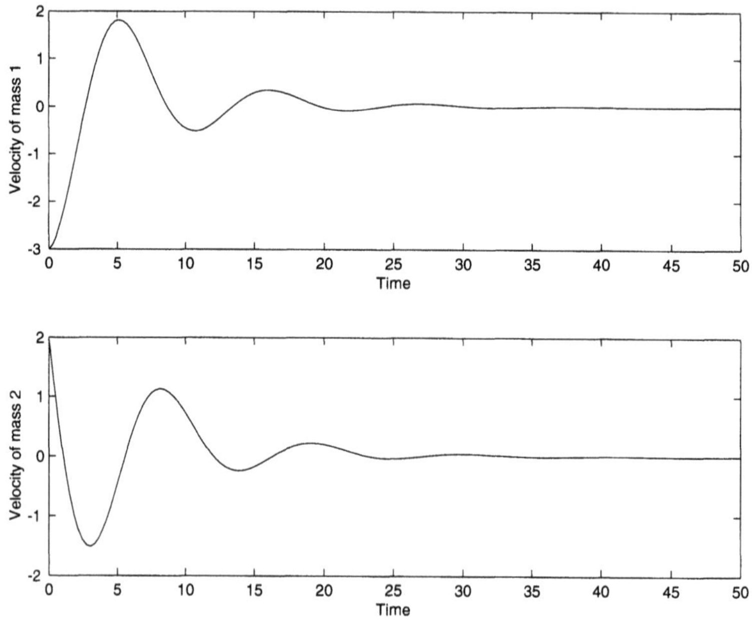

Figure 6.13 Mass velocities versus time.

be assigned, the energy-Casimir method can be used to construct suitable Lyapunov functions for the closed-loop system.

To proceed, consider the candidate vector energy-Casimir function $E : \mathcal{D} \times \mathbb{R}^{n_c} \to \mathbb{R}^{n_c}$, where $E(\cdot,\cdot)$ is continuously differentiable and has the form

$$E(x, x_c) = x_c - F(x), \quad (x, x_c) \in \mathcal{D} \times \mathbb{R}^{n_c}, \tag{6.44}$$

where $F : \mathcal{D} \to \mathbb{R}^{n_c}$ is a continuously differentiable function. To ensure that the candidate vector energy-Casimir function $E(\cdot,\cdot)$ is constant along the trajectories of (6.43) we require that

$$\dot{E}(x(t), x_c(t)) = \dot{x}_c(t) - \frac{\partial F}{\partial x}(x(t))\dot{x}(t) = 0, \quad t \geq 0. \tag{6.45}$$

Now, we can arrive at a set of *sufficient* conditions which guarantee that (6.45) holds. Specifically, it follows from (6.43) that (6.45) can be rewritten as

$$\dot{E}(x(t), x_c(t)) = \begin{bmatrix} \left[G_c(x_c)G^{\mathrm{T}}(x) - \frac{\partial F}{\partial x}(x)(\mathcal{J}(x) - \mathcal{R}(x))\right]^{\mathrm{T}} \\ \left[\mathcal{J}_c(x_c) - \mathcal{R}_c(x_c) + \frac{\partial F}{\partial x}(x)G(x)G_c^{\mathrm{T}}(x_c)\right]^{\mathrm{T}} \end{bmatrix}^{\mathrm{T}} \cdot \begin{bmatrix} \left(\frac{\partial \mathcal{H}}{\partial x}(x(t))\right)^{\mathrm{T}} \\ \left(\frac{\partial \mathcal{H}_c}{\partial x_c}(x_c(t))\right)^{\mathrm{T}} \end{bmatrix}. \tag{6.46}$$

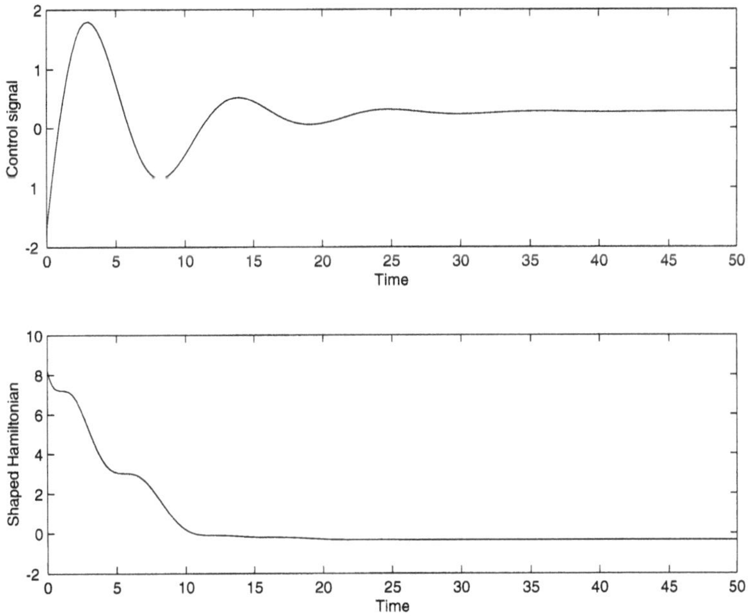

Figure 6.14 Control signal and shaped Hamiltonian versus time.

Hence, a set of sufficient conditions such that (6.45) holds is given by

$$G_c(x_c)G^{\mathrm{T}}(x) - \frac{\partial F}{\partial x}(x)(\mathcal{J}(x) - \mathcal{R}(x)) = 0, \ (x, x_c) \in \mathcal{D} \times \mathbb{R}^{n_c}, \quad (6.47)$$

$$\mathcal{J}_c(x_c) - \mathcal{R}_c(x_c) + \frac{\partial F}{\partial x}(x)G(x)G_c^{\mathrm{T}}(x_c) = 0, \ (x, x_c) \in \mathcal{D} \times \mathbb{R}^{n_c}. \quad (6.48)$$

The following proposition summarizes the above results.

Proposition 6.1. Consider the feedback interconnection of the port-controlled Hamiltonian systems $\mathcal{G}$ and $\mathcal{G}_c$ given by (6.22) and (6.23), and (6.41) and (6.42), respectively. If there exists a continuously differentiable function $F : \mathcal{D} \to \mathbb{R}^{n_c}$ such that for all $(x, x_c) \in \mathcal{D} \times \mathbb{R}^{n_c}$,

$$\frac{\partial F}{\partial x}(x)\mathcal{J}(x)\left(\frac{\partial F}{\partial x}(x)\right)^{\mathrm{T}} - \mathcal{J}_c(x_c) = 0, \quad (6.49)$$

$$\mathcal{R}_c(x_c) = 0, \quad (6.50)$$

$$\mathcal{R}(x)\left(\frac{\partial F}{\partial x}(x)\right)^{\mathrm{T}} = 0, \quad (6.51)$$

$$\frac{\partial F}{\partial x}(x)\mathcal{J}(x) - G_c(x_c)G^{\mathrm{T}}(x) = 0, \quad (6.52)$$

then

$$E(\tilde{x}(t)) = x_c(t) - F(x(t)) = c, \quad t \geq 0, \quad (6.53)$$

where $c \in \mathbb{R}^{n_c}$ and $\tilde{x}(t) = [x^{\mathrm{T}}(t),\ x_c^{\mathrm{T}}(t)]^{\mathrm{T}}$ satisfies (6.43).

Proof. Postmultiplying (6.47) by $\left(\frac{\partial F}{\partial x}(x)\right)^{\mathrm{T}}$, it follows from (6.47) and (6.48) that

$$\frac{\partial F}{\partial x}(x)[\mathcal{J}(x) - \mathcal{R}(x)]\left(\frac{\partial F}{\partial x}(x)\right)^{\mathrm{T}} = \mathcal{J}_{\mathrm{c}}(x_{\mathrm{c}}) + \mathcal{R}_{\mathrm{c}}(x_{\mathrm{c}}), \quad (x, x_{\mathrm{c}}) \in \mathcal{D} \times \mathbb{R}^{n_{\mathrm{c}}}. \tag{6.54}$$

Next, using the fact that the sum of a skew-symmetric and symmetric matrix is zero if and only if the individual matrices are zero, it follows that (6.54) is equivalent to

$$\frac{\partial F}{\partial x}(x)\mathcal{J}(x)\left(\frac{\partial F}{\partial x}(x)\right)^{\mathrm{T}} - \mathcal{J}_{\mathrm{c}}(x_{\mathrm{c}}) = 0, \quad (x, x_{\mathrm{c}}) \in \mathcal{D} \times \mathbb{R}^{n_{\mathrm{c}}}, \tag{6.55}$$

$$\mathcal{R}_{\mathrm{c}}(x_{\mathrm{c}}) + \frac{\partial F}{\partial x}(x)\mathcal{R}(x)\left(\frac{\partial F}{\partial x}(x)\right)^{\mathrm{T}} = 0, \quad (x, x_{\mathrm{c}}) \in \mathcal{D} \times \mathbb{R}^{n_{\mathrm{c}}}. \tag{6.56}$$

Now, since $\mathcal{R}(x) \geq 0$, $x \in \mathcal{D}$, and $\mathcal{R}_{\mathrm{c}}(x_{\mathrm{c}}) \geq 0$, $x \in \mathbb{R}^{n_{\mathrm{c}}}$, it follows that (6.55) and (6.56) are equivalent to (6.49)–(6.51). Hence, it follows that (6.47) can be rewritten as (6.52). The equivalence between (6.47)–(6.48) and (6.49)–(6.52) proves the result. □

Note that conditions (6.49)–(6.52) are necessary and sufficient for (6.47)–(6.48) to hold, which, in turn, provide sufficient conditions for guaranteeing that the vector energy-Casimir function $E(\cdot,\cdot)$ is constant along the trajectories of the closed-loop system (6.43). The constant vector $c \in \mathbb{R}^{n_{\mathrm{c}}}$ in (6.53) depends on the initial conditions for the plant and controller states. If conditions (6.49)–(6.52) are satisfied, then the controller state variables along the trajectories of the closed-loop system given by (6.43) can be represented in terms of the plant state variables as $x_{\mathrm{c}}(t) = F(x(t)) + c$, $t \geq 0$, $x(t) \in \mathcal{D}$, $c \in \mathbb{R}^{n_{\mathrm{c}}}$. In this case, it follows that the closed-loop system associated with the plant dynamics is given by

$$\begin{aligned}\dot{x}(t) &= [\mathcal{J}(x(t)) - \mathcal{R}(x(t))]\left(\frac{\partial \mathcal{H}}{\partial x}(x(t))\right)^{\mathrm{T}} \\ &\quad - G(x(t))G_{\mathrm{c}}^{\mathrm{T}}(x_{\mathrm{c}}(t))\left(\frac{\partial \mathcal{H}_{\mathrm{c}}}{\partial x_{\mathrm{c}}}(x_{\mathrm{c}}(t))\right)^{\mathrm{T}} \\ &= [\mathcal{J}(x(t)) - \mathcal{R}(x(t))]\left(\frac{\partial \mathcal{H}}{\partial x}(x(t)) + \frac{\partial \mathcal{H}_{\mathrm{c}}}{\partial x_{\mathrm{c}}}(x_{\mathrm{c}}(t))\frac{\partial F}{\partial x}(x(t))\right)^{\mathrm{T}} \\ &= [\mathcal{J}(x(t)) - \mathcal{R}(x(t))]\left(\frac{\partial \mathcal{H}_{\mathrm{s}}}{\partial x}(x(t))\right)^{\mathrm{T}}, \quad x(0) = x_0, \quad t \geq 0,\end{aligned} \tag{6.57}$$

where $\mathcal{H}_{\mathrm{s}}(x) = \mathcal{H}(x) + \mathcal{H}_{\mathrm{c}}(F(x) + c)$, $x \in \mathcal{D}$, is the *shaped* Hamiltonian function for the closed-loop system (6.57).

Next, we use the existence of the vector energy-Casimir function to construct stabilizing dynamic controllers that guarantee that the closed-loop system associated with the plant dynamics preserves the Hamiltonian structure without the need for solving a set of partial differential equations.

Theorem 6.4. Consider the feedback interconnection of the port-cont-rolled Hamiltonian systems $\mathcal{G}$ and $\mathcal{G}_c$ given by (6.22) and (6.23), and (6.41) and (6.42), respectively. Assume that there exists a continuously differentiable function $F : \mathcal{D} \to \mathbb{R}^{n_c}$ such that conditions (6.49)–(6.52) hold for all $(x, x_c) \in \mathcal{D} \times \mathbb{R}^{n_c}$, and assume that the Hamiltonian function $\mathcal{H}_c : \mathbb{R}^{n_c} \to \mathbb{R}$ of the feedback controller $\mathcal{G}_c$ is such that $\mathcal{H}_s : \mathcal{D} \to \mathbb{R}$ is given by $\mathcal{H}_s(x) = \mathcal{H}(x) + \mathcal{H}_c(F(x) + c)$, $x \in \mathcal{D}$. If

$$\frac{\partial \mathcal{H}_c}{\partial x}(F(x_e) + c) = -\frac{\partial \mathcal{H}}{\partial x}(x_e), \quad x_e \in \mathcal{D}, \tag{6.58}$$

$$\frac{\partial^2 \mathcal{H}_c}{\partial x^2}(F(x_e) + c) > -\frac{\partial^2 \mathcal{H}}{\partial x^2}(x_e), \quad x_e \in \mathcal{D}, \tag{6.59}$$

then the equilibrium solution $x(t) \equiv x_e$ of the system (6.57) is Lyapunov stable. If, in addition, $\mathcal{D}_c \subseteq \mathcal{D}$ is a compact positively invariant set with respect to (6.57) and the largest invariant set contained in $\mathcal{R} \triangleq \{x \in \mathcal{D}_c : \frac{\partial \mathcal{H}_s}{\partial x}(x)\mathcal{R}(x)\left(\frac{\partial \mathcal{H}_s}{\partial x}(x)\right)^{\mathrm{T}} = 0\}$ is $\mathcal{M} = \{x_e\}$, then the equilibrium solution $x(t) \equiv x_e$ of the closed-loop system (6.57) is locally asymptotically stable.

Proof. Conditions (6.49)–(6.52) imply that the closed-loop dynamics of the port-controlled Hamiltonian system $\mathcal{G}$ and the controller $\mathcal{G}_c$ associated with the plant states can be written in the form given by (6.57). Now, using identical arguments as in the proof of Theorem 6.3, conditions (6.58) and (6.59) guarantee the existence of the Lyapunov function candidate $V(x) = \mathcal{H}_s(x) - \mathcal{H}_s(x_e)$, $x \in \mathcal{D}$, which guarantees Lyapunov stability of the equilibrium solution $x(t) \equiv x_e$ of the closed-loop system (6.57). Asymptotic stability of $x(t) \equiv x_e$ follows from Corollary 3.1. □

As in the static controller case, the dynamic controller given by Theorem 6.4 also provides an energy balance interpretation over the trajectories of the controlled system. To see this, note that since by (6.50), $\mathcal{R}_c(x_c) = 0$, $x_c \in \mathbb{R}^{n_c}$, it follows that the controller Hamiltonian $\mathcal{H}_c(\cdot)$ satisfies

$$\dot{\mathcal{H}}_c(F(x(t)) + c) = y_c^{\mathrm{T}}(t)y(t) = -u^{\mathrm{T}}(t)y(t), \quad t \geq 0. \tag{6.60}$$

Now, it follows that

$$\begin{aligned} \dot{\mathcal{H}}_s(x(t)) &= \dot{\mathcal{H}}(x(t)) + \dot{\mathcal{H}}_c(F(x(t)) + c) \\ &= \dot{\mathcal{H}}(x(t)) - u^{\mathrm{T}}(t)y(t), \quad t \geq 0, \end{aligned} \tag{6.61}$$

which yields (6.33).

6.4 Stability Margins for Nonlinear Feedback Regulators

To develop relative stability margins for nonlinear regulators consider the nonlinear dynamical system $\mathcal{G}$ given by

$$\dot{x}(t) = f(x(t)) + G(x(t))u(t), \qquad x(0) = x_0, \qquad t \geq 0, \tag{6.62}$$
$$y(t) = -\phi(x(t)), \tag{6.63}$$

where $\phi : \mathbb{R}^n \to \mathbb{R}^m$ is such that $\mathcal{G}$ is asymptotically stable with $u = -y$. Furthermore, assume that the system $\mathcal{G}$ is zero-state observable. Next, we define the relative stability margins for $\mathcal{G}$ given by (6.62) and (6.63). Specifically, let $u_c \triangleq -y$, $y_c \triangleq u$, and consider the negative feedback interconnection $u = \Delta(-y)$ of $\mathcal{G}$ and $\Delta(\cdot)$ given in Figure 6.15, where $\Delta(\cdot)$ is either a linear operator $\Delta(u_c) = \Delta u_c$, a nonlinear static operator $\Delta(u_c) = \sigma(u_c)$, or a dynamic nonlinear operator $\Delta(\cdot)$ with input u_c and output y_c. Furthermore, we assume that in the nominal case $\Delta(\cdot) = I(\cdot)$ so that the nominal closed-loop system is asymptotically stable.

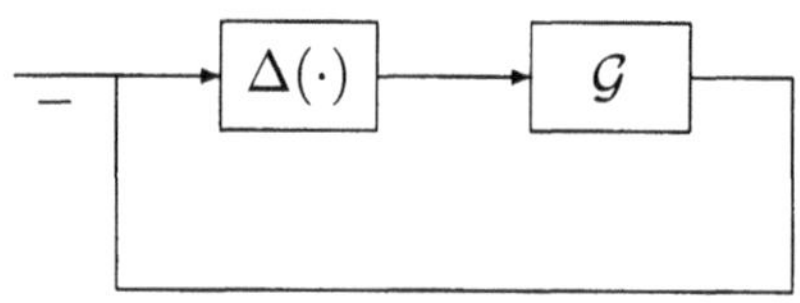

Figure 6.15 Multiplicative input uncertainty of $\mathcal{G}$ and input operator $\Delta(\cdot)$.

Definition 6.1. Let $\alpha, \beta \in \mathbb{R}$ be such that $0 < \alpha \leq 1 \leq \beta < \infty$. Then the nonlinear dynamical system $\mathcal{G}$ given by (6.62) and (6.63) is said to have a *gain margin* (α, β) if the negative feedback interconnection of $\mathcal{G}$ and $\Delta(u_c) = \Delta u_c$ is globally asymptotically stable for all $\Delta = \operatorname{diag}[k_1, \ldots, k_m]$, where $k_i \in (\alpha, \beta)$, $i = 1, \ldots, m$.

Definition 6.2. Let $\alpha, \beta \in \mathbb{R}$ be such that $0 < \alpha \leq 1 \leq \beta < \infty$. Then the nonlinear dynamical system $\mathcal{G}$ given by (6.62) and (6.63) is said to have a *sector margin* (α, β) if the negative feedback interconnection of $\mathcal{G}$ and $\Delta(u_c) = \sigma(u_c)$ is globally asymptotically stable for all nonlinearities $\sigma : \mathbb{R}^m \to \mathbb{R}^m$ such that $\sigma(0) = 0$, $\sigma(u_c) = [\sigma_1(u_{c1}), \ldots, \sigma_m(u_{cm})]^{\mathrm{T}}$, and $\alpha u_{ci}^2 < \sigma_i(u_{ci})u_{ci} < \beta u_{ci}^2$, for all $u_{ci} \neq 0$, $i = 1, \ldots, m$.

Definition 6.3. Let $\alpha, \beta \in \mathbb{R}$ be such that $0 < \alpha \leq 1 \leq \beta < \infty$. Then the nonlinear dynamical system $\mathcal{G}$ given by (6.62) and (6.63) is said to have a *disk margin* (α, β) if the negative feedback interconnection of $\mathcal{G}$ and $\Delta(\cdot)$ is globally asymptotically stable for all dynamic operators $\Delta(\cdot)$ such that $\Delta(\cdot)$ is zero-state observable and dissipative with respect to the supply rate

$r(u_c, y_c) = u_c^T y_c - \frac{1}{\hat{\alpha}+\hat{\beta}} y_c^T y_c - \frac{\hat{\alpha}\hat{\beta}}{\hat{\alpha}+\hat{\beta}} u_c^T u_c$, where $\hat{\alpha} = \alpha + \delta$, $\hat{\beta} = \beta - \delta$, and $\delta \in \mathbb{R}$ such that $0 < 2\delta < \beta - \alpha$.

Definition 6.4. Let $\alpha, \beta \in \mathbb{R}$ be such that $0 < \alpha \leq 1 \leq \beta < \infty$. Then the nonlinear dynamical system $\mathcal{G}$ given by (6.62) and (6.63) is said to have a *structured disk margin* (α, β) if the negative feedback interconnection of $\mathcal{G}$ and $\Delta(\cdot)$ is globally asymptotically stable for all dynamic operators $\Delta(\cdot)$ such that $\Delta(\cdot)$ is zero-state observable, $\Delta(u_c) = \operatorname{diag}[\delta_1(u_{c1}), \ldots, \delta_m(u_{cm})]$, and $\delta_i(\cdot)$, $i = 1, \ldots, m$, is dissipative with respect to the supply rate $r(u_{ci}, y_{ci}) = u_{ci} y_{ci} - \frac{1}{\hat{\alpha}+\hat{\beta}} y_{ci}^2 - \frac{\hat{\alpha}\hat{\beta}}{\hat{\alpha}+\hat{\beta}} u_{ci}^2$, where $\hat{\alpha} = \alpha + \delta$, $\hat{\beta} = \beta - \delta$, and $\delta \in \mathbb{R}$ such that $0 < 2\delta < \beta - \alpha$.

Note that if $\mathcal{G}$ has a disk margin (or structured disk margin) (α, β), then $\mathcal{G}$ has gain and sector margins (α, β). To see this, let $\Delta(u_c) = \sigma(u_c)$, where $\sigma : \mathbb{R}^m \to \mathbb{R}^m$ is such that $\sigma(0) = 0$ and

$$\sigma(u_c) = [\sigma_1(u_{c1}), \ldots, \sigma_m(u_{cm})]^T.$$

Now, if $\Delta(u_c)$ is dissipative with respect to the supply rate $r(u_c, y_c) = u_c^T y_c - \frac{1}{\hat{\alpha}+\hat{\beta}} y_c^T y_c - \frac{\hat{\alpha}\hat{\beta}}{\hat{\alpha}+\hat{\beta}} u_c^T u_c$, then $\sigma(\cdot)$ satisfies

$$0 \leq \sigma^T(u_c) u_c - \frac{1}{\hat{\alpha}+\hat{\beta}} \sigma^T(u_c) \sigma(u_c) - \frac{\hat{\alpha}\hat{\beta}}{\hat{\alpha}+\hat{\beta}} u_c^T u_c, \tag{6.64}$$

or, equivalently,

$$0 \geq \sum_{i=1}^{m} (\sigma_i(u_{ci}) - \hat{\alpha} u_{ci})(\sigma_i(u_{ci}) - \hat{\beta} u_{ci}). \tag{6.65}$$

Hence, if $\mathcal{G}$ has a disk margin (α, β), then $\mathcal{G}$ has a sector margin (α, β). Similarly, if $\Delta(\cdot)$ is a linear operator such that $\Delta(u_c) = \Delta u_c$, where $\Delta = \operatorname{diag}[k_1, \ldots, k_m]$, and $\Delta(u_c)$ is dissipative with respect to the supply rate $r(u_c, u_c) = u_c^T y_c - \frac{1}{\hat{\alpha}+\hat{\beta}} y_c^T y_c - \frac{\hat{\alpha}\hat{\beta}}{\hat{\alpha}+\hat{\beta}} u_c^T u_c$, it follows that

$$0 \geq \sum_{i=1}^{m} (k_i - \hat{\alpha})(k_i - \hat{\beta}). \tag{6.66}$$

Thus, if $\mathcal{G}$ has a disk margin (α, β), then $\mathcal{G}$ has a gain margin (α, β). Finally, letting $\beta \to \infty$ in Definition 6.3 we recover the definition of disk margins given in [395].

In the case where $\Delta(\cdot)$ is a dynamic operator we assume that $\Delta(\cdot)$ can be characterized as (6.3) and (6.4) where x_c denotes the internal state of the operator $\Delta(\cdot)$, u_c denotes the input, and y_c denotes the output of the operator. If $\Delta(\cdot)$ is a single-input/single-output, linear dynamic operator it follows from Parseval's theorem that if $\Delta(\cdot)$ is dissipative with respect to the supply rate $r(u_c, y_c) = u_c^T y_c - \frac{1}{\hat{\alpha}+\hat{\beta}} y_c^T y_c - \frac{\hat{\alpha}\hat{\beta}}{\hat{\alpha}+\hat{\beta}} u_c^T u_c$, then

$$|\Delta(\jmath\omega)|^2 - (\hat{\alpha}+\hat{\beta})\mathrm{Re}\ \Delta(\jmath\omega) + \hat{\alpha}\hat{\beta} < 0. \tag{6.67}$$

In this case, it can be easily shown that if $\mathcal{G}$ has a disk margin of (α, β), then $\mathcal{G}$ has a gain margin (α, β) and a phase margin φ, where $\cos(\varphi) = \frac{\alpha\beta}{\alpha+\beta}$. Hence, the concept of the disk margin for nonlinear systems provides a nonlinear hybrid analog to the concepts of gain and phase margins of linear systems.

The following results provide algebraic sufficient conditions that guarantee disk margins for the nonlinear dynamical system $\mathcal{G}$ given by (6.62) and (6.63).

Theorem 6.5. Consider the nonlinear dynamical system $\mathcal{G}$ given by (6.62) and (6.63). Let $\alpha, \beta \in \mathbb{R}$ be such that $0 < \alpha \leq 1 \leq \beta < \infty$. Suppose there exist a positive-definite diagonal matrix $Z \in \mathbb{R}^{m \times m}$ and functions $V_s : \mathbb{R}^n \to \mathbb{R}$, $\ell : \mathbb{R}^n \to \mathbb{R}^p$, and $\mathcal{W} : \mathbb{R}^n \to \mathbb{R}^{p \times m}$ such that $V_s(\cdot)$ is continuously differentiable, $V_s(0) = 0$, $V_s(x) > 0$, $x \in \mathbb{R}^n$, $x \neq 0$, and for all $x \in \mathbb{R}^n$,

$$0 = V_s'(x)f(x) - \frac{\alpha\beta}{\alpha+\beta}\phi^T(x)Z\phi(x) + \ell^T(x)\ell(x), \tag{6.68}$$

$$0 = V_s'(x)G(x) + \phi^T(x)Z + 2\ell^T(x)\mathcal{W}(x), \tag{6.69}$$

$$0 = \frac{1}{\alpha+\beta}Z - \mathcal{W}^T(x)\mathcal{W}(x). \tag{6.70}$$

Then the nonlinear system $\mathcal{G}$ has a structured disk margin (α, β). Alternatively, if (6.68)–(6.70) are satisfied with $Z = I_m$, then the nonlinear system $\mathcal{G}$ has a disk margin (α, β).

Proof. With $h(x) = -\phi(x)$, $J(x) \equiv 0$, $Q = \frac{\alpha\beta}{\alpha+\beta}Z$, $R = \frac{1}{\alpha+\beta}Z$, and $S = \frac{1}{2}Z$, it follows from Theorem 5.6 that the nonlinear system $\mathcal{G}$ is dissipative with respect to the supply rate $r(u, y) = u^T Z y + \frac{\alpha\beta}{\alpha+\beta} y^T Z y + \frac{1}{\alpha+\beta} u^T Z u$. Next, let $\Delta(\cdot)$ be zero-state observable and let $\Delta(u_c) = \mathrm{diag}[\delta_1(u_{c1}), \ldots, \delta_m(u_{cm})]$ be such that $\delta_i(\cdot)$, $i = 1, \ldots, m$, is dissipative with respect to the supply rate $r(u_{ci}, y_{ci}) = u_{ci}y_{ci} - \frac{1}{\hat{\alpha}+\hat{\beta}} y_{ci}^2 - \frac{\hat{\alpha}\hat{\beta}}{\hat{\alpha}+\hat{\beta}} u_{ci}^2$, where $\hat{\alpha} = \alpha + \delta$, $\hat{\beta} = \beta - \delta$, and $\delta \in \mathbb{R}$ is such that $0 < 2\delta < \beta - \alpha$. Now, noting that in this case $\Delta(\cdot)$ is dissipative with respect to the supply rate $r(u_c, y_c) = u_c^T Z y_c - \frac{1}{\hat{\alpha}+\hat{\beta}} y_c^T Z y_c - \frac{\hat{\alpha}\hat{\beta}}{\hat{\alpha}+\hat{\beta}} u_c^T Z u_c$, it follows from Definition 6.4 and Corollary 6.2 that the nonlinear system $\mathcal{G}$ has a structured disk margin (α, β). Finally, if (6.68)–(6.70) are satisfied with $Z = I_m$, then it follows from Theorem 5.6 that the nonlinear system $\mathcal{G}$ is dissipative with respect to the supply rate $r(u, y) = u^T y + \frac{\alpha\beta}{\alpha+\beta} y^T y + \frac{1}{\alpha+\beta} u^T u$. Hence, it follows from Definition 6.3 and Corollary 6.2 that the nonlinear system $\mathcal{G}$ has a disk margin (α, β). □

Corollary 6.3. Consider the nonlinear dynamical system $\mathcal{G}$ given by (6.62) and (6.63). Let $\alpha, \beta \in \mathbb{R}$ be such that $0 < \alpha \leq 1 \leq \beta < \infty$. Suppose there exist a positive-definite diagonal matrix $Z \in \mathbb{R}^{m \times m}$ and a continuously differentiable function $V_s : \mathbb{R}^n \to \mathbb{R}$ such that $V_s(0) = 0$, $V_s(x) > 0$, $x \in \mathbb{R}^n$, $x \neq 0$, and for all $x \in \mathbb{R}^n$,

$$0 \geq V_s'(x)f(x) - \tfrac{\alpha\beta}{\alpha+\beta}\phi^{\mathrm{T}}(x)Z\phi(x) + \tfrac{\alpha+\beta}{4}[\phi^{\mathrm{T}}(x)Z + V_s'(x)G(x)]Z^{-1}[\phi^{\mathrm{T}}(x)Z + V_s'(x)G(x)]^{\mathrm{T}}. \quad (6.71)$$

Then the nonlinear system $\mathcal{G}$ has a structured disk margin (α, β). Alternatively, if (6.71) is satisfied with $Z = I_m$, then the nonlinear system $\mathcal{G}$ has a disk margin (α, β).

Proof. It follows from Theorem 5.6 that (6.68)–(6.70) are equivalent to (6.71). Now, the result is a direct consequence of Theorem 6.5. □

The following theorem gives the nonlinear version of the results of [316].

Theorem 6.6. Consider the nonlinear dynamical system $\mathcal{G}$ given by (6.62) and (6.63). Let $\alpha, \beta \in \mathbb{R}$ be such that $0 < \alpha \leq 1 \leq \beta < \infty$. Suppose there exist a positive-definite diagonal matrix $Z \in \mathbb{R}^{m \times m}$, a continuously differentiable function $V : \mathbb{R}^n \to \mathbb{R}$, and a scalar $q > 0$ such that $V(0) = 0$, $V(x) > 0$, $x \in \mathbb{R}^n$, $x \neq 0$, and for all $x \in \mathbb{R}^n$,

$$0 \geq V'(x)f(x) - \tfrac{\alpha\beta}{q(\alpha+\beta)^2}V'(x)G(x)Z^{-1}G^{\mathrm{T}}(x)V'^{\mathrm{T}}(x). \quad (6.72)$$

Then, with $\phi(x) = -\frac{1}{q(\alpha+\beta)}Z^{-1}G^{\mathrm{T}}(x)V'^{\mathrm{T}}(x)$, the nonlinear system $\mathcal{G}$ has a structured disk margin (α, β). Alternatively, if (6.72) is satisfied with $Z = I_m$, then the nonlinear system $\mathcal{G}$ has a disk margin (α, β).

Proof. The result is a direct consequence of Corollary 6.3 with $V_s(x) = \frac{1}{q(\alpha+\beta)}V(x)$. Specifically, since $\phi(x) = -\frac{1}{q(\alpha+\beta)}Z^{-1}G^{\mathrm{T}}(x)V'^{\mathrm{T}}(x) = -Z^{-1} \cdot G^{\mathrm{T}}(x)V_s'^{\mathrm{T}}(x)$, it follows from (6.72) that for all $x \in \mathbb{R}^n$,

$$\begin{aligned} 0 &\geq V'(x)f(x) - \tfrac{\alpha\beta}{q(\alpha+\beta)^2}V'(x)G(x)Z^{-1}G^{\mathrm{T}}(x)V'^{\mathrm{T}}(x) \\ &= q(\alpha+\beta)\left(V_s'(x)f(x) - \tfrac{\alpha\beta}{\alpha+\beta}\phi^{\mathrm{T}}(x)Z\phi(x)\right) \\ &= q(\alpha+\beta)\Big(V_s'(x)f(x) - \tfrac{\alpha\beta}{\alpha+\beta}\phi^{\mathrm{T}}(x)Z\phi(x) \\ &\quad + \tfrac{\alpha+\beta}{4}(\phi(x)^{\mathrm{T}}Z + V_s'(x)G(x))Z^{-1}(\phi^{\mathrm{T}}(x)Z + V_s'(x)G(x))^{\mathrm{T}}\Big), \end{aligned}$$

which implies (6.71), so that the conditions of Corollary 6.3 are satisfied. □

6.5 Control Lyapunov Functions

In this section, we consider a feedback control problem and introduce the notion of *control Lyapunov functions.* Furthermore, using the concept of control Lyapunov functions we provide necessary and sufficient conditions for nonlinear system stabilization.

Consider the nonlinear controlled dynamical system given by

$$\dot{x}(t) = F(x(t), u(t)), \quad x(t_0) = x_0, \quad t \geq 0, \tag{6.73}$$

where $x(t) \in \mathcal{D} \subseteq \mathbb{R}^n$, $t \geq 0$, is the state vector, $\mathcal{D}$ is an open set with $0 \in \mathcal{D}$, $u(t) \in U \subseteq \mathbb{R}^m$ is the control input, and $F : \mathcal{D} \times U \to \mathbb{R}^n$ satisfies $F(0,0) = 0$. We assume that the control input $u(\cdot)$ in (6.73) is restricted to the class of *admissible controls* $\mathcal{U}$ consisting of measurable functions $u(\cdot)$ such that $u(t) \in U$ for all $t \geq 0$, where the constraint set U is given with $0 \in U$. A measurable mapping $\phi : \mathcal{D} \to U$ satisfying $\phi(0) = 0$ is called a *control law.* Furthermore, if $u(t) = \phi(x(t))$, where ϕ is a control law and $x(t)$, $t \geq 0$, satisfies (6.73), then $u(\cdot)$ is called a *feedback control law.* We assume that the mapping $\phi : \mathcal{D} \to U$ satisfies sufficient regularity conditions such that the resulting closed-loop system

$$\dot{x}(t) = F(x(t), \phi(x(t))), \quad x(0) = x_0, \quad t \geq 0, \tag{6.74}$$

has a unique solution forward in time. Specifically, we assume that $F(\cdot,\cdot)$ is Lipschitz continuous in a neighborhood of the origin in $\mathcal{D} \times U$.

The following two definitions are required for stating the results of this section.

Definition 6.5. Let $\phi : \mathcal{D} \to U$ be a mapping on $\mathcal{D}\backslash\{0\}$ with $\phi(0) = 0$. Then (6.73) is *feedback asymptotically stabilizable* if the zero solution $x(t) \equiv 0$ of the closed-loop system (6.74) is asymptotically stable.

Definition 6.6. Consider the controlled nonlinear dynamical system given by (6.73). A continuously differentiable positive-definite function $V : \mathcal{D} \to \mathbb{R}$ satisfying

$$\inf_{u \in U} V'(x)F(x,u) < 0, \quad x \in \mathcal{D}, \quad x \neq 0, \tag{6.75}$$

is called a *control Lyapunov function.*

Note that, if (6.75) holds, then there exists a feedback control law $\phi : \mathcal{D} \to U$ such that $V'(x)F(x,\phi(x)) < 0$, $x \in \mathcal{D}$, $x \neq 0$, and hence, Theorem 3.1 implies that if there exists a control Lyapunov function for the nonlinear dynamical system (6.73), then there exists a feedback control law $\phi(x)$ such that the zero solution $x(t) \equiv 0$ of the closed-loop nonlinear dynamical system (6.73) is asymptotically stable. Conversely, if there exists a feedback control law $u = \phi(x)$ such that the zero solution $x(t) \equiv 0$ of the nonlinear dynamical system (6.73) is asymptotically stable, then it follows from Theorem 3.9 that there exists a continuously differentiable positive-definite function $V : \mathcal{D} \to \mathbb{R}$ such that $V'(x)F(x, \phi(x)) < 0$, $x \in \mathcal{D}$, $x \neq 0$, or, equivalently, there exists a control Lyapunov function for the nonlinear dynamical system (6.73). Hence, a given nonlinear dynamical system of the form (6.73) is feedback asymptotically stabilizable if and only if there exists a control Lyapunov function satisfying (6.75). Finally, in the case where $\mathcal{D} = \mathbb{R}^n$ and $U = \mathbb{R}^m$ the zero solution $x(t) \equiv 0$ to (6.73) is globally asymptotically stabilizable if and only if $V(x) \to \infty$ as $\|x\| \to \infty$.

Next, we consider the special case of nonlinear affine systems in the control and construct state feedback controllers that globally asymptotically stabilize the zero solution of the nonlinear dynamical system under the assumption that the system has a radially unbounded control Lyapunov function. Specifically, we consider nonlinear affine systems of the form

$$\dot{x}(t) = f(x(t)) + G(x(t))u(t), \quad x(t_0) = x_0, \quad t \geq 0, \tag{6.76}$$

where $f : \mathbb{R}^n \to \mathbb{R}^n$ satisfies $f(0) = 0$ and $G : \mathbb{R}^n \to \mathbb{R}^{n \times m}$, and $f(\cdot)$ and $G(\cdot)$ are smooth functions (at least continuously differentiable mappings).

Theorem 6.7. Consider the controlled nonlinear system given by (6.76). Then a continuously differentiable positive-definite, radially unbounded function $V : \mathbb{R}^n \to \mathbb{R}$ is a control Lyapunov function of (6.76) if and only if

$$V'(x)f(x) < 0, \qquad x \in \mathcal{R}, \tag{6.77}$$

where $\mathcal{R} \triangleq \{x \in \mathbb{R}^n,\ x \neq 0 :\ V'(x)G(x) = 0\}$.

Proof. The proof is a direct consequence of the definition of a control Lyapunov function by noting that for systems of the form (6.76),

$$\inf_{u \in \mathbb{R}^m} V'(x)[f(x) + G(x)u] = -\infty, \quad x \notin \mathcal{R}, \quad x \neq 0.$$

Hence, (6.75) is equivalent to (6.77), which proves the result. □

Example 6.5. Consider the nonlinear controlled system

$$\dot{x}_1(t) = -x_1^3(t) + x_2(t)e^{x_1(t)}\cos(x_2(t)), \quad x_1(0) = x_{10}, \quad t \geq 0, \tag{6.78}$$

$$x_2(t) = x_1^5(t)\sin(x_2(t)) + u(t), \quad x_2(0) = x_{20}. \tag{6.79}$$

To show that $V(x_1, x_2) = \frac{1}{2}x_1^2 + \frac{1}{2}x_2^2$ is a control Lyapunov function for (6.78) and (6.79) note that

$$\inf_{u \in U} V'(x_1, x_2)F(x_1, x_2, u) = \inf_{u \in U}[-x_1^4 + x_1x_2e^{x_1}\cos(x_2)$$

$$+x_2 u + x_2 x_1^5 \sin(x_2)]$$
$$= \begin{cases} -x_1^4, & x_2 = 0, \\ -\infty, & x_2 \neq 0, \end{cases}$$

where $F(x_1, x_2, u)$ denotes the right-hand-side of (6.78) and (6.79). Thus, $V(x_1, x_2) = \frac{1}{2}x_1^2 + \frac{1}{2}x_2^2$ is a control Lyapunov function, and hence, (6.78) and (6.79) is globally asymptotically stabilizable. In particular, it is easily verifiable that the feedback control law

$$u(t) = -x_2(t) - x_1(t)e^{x_1(t)} \cos(x_2(t)) - x_1^5(t) \sin(x_2(t)) \tag{6.80}$$

globally stabilizes (6.78) and (6.79). △

It follows from Theorem 6.7 that the zero solution $x(t) \equiv 0$ of a nonlinear affine system of the form (6.76) is globally feedback asymptotically stabilizable if and only if there exists a continuously differentiable positive-definite, radially unbounded function $V : \mathbb{R}^n \to \mathbb{R}$ satisfying (6.77). Hence, Theorem 6.7 provides necessary and sufficient conditions for nonlinear system stabilization.

Next, using Theorem 6.7 we *construct* an explicit feedback control law that is a function of the control Lyapunov function $V(\cdot)$. Specifically, consider the feedback control law given by

$$\phi(x) = \begin{cases} -\left(c_0 + \frac{\alpha(x)+\sqrt{\alpha^2(x)+(\beta^{\mathrm{T}}(x)\beta(x))^2}}{\beta^{\mathrm{T}}(x)\beta(x)}\right)\beta(x), & \beta(x) \neq 0, \\ 0, & \beta(x) = 0, \end{cases} \tag{6.81}$$

where $\alpha(x) \triangleq V'(x)f(x)$, $\beta(x) \triangleq G^{\mathrm{T}}(x)V'^{\mathrm{T}}(x)$, and $c_0 \geq 0$. In this case, the control Lyapunov function $V(\cdot)$ of (6.76) is a Lyapunov function for the closed-loop system (6.76) with $u = \phi(x)$, where $\phi(x)$ is given by (6.81). In particular, the control Lyapunov derivative $\dot{V}(\cdot)$ along the trajectories of the nonlinear system (6.76) with $u = \phi(x)$ given by (6.81) is given by

$$\begin{aligned} \dot{V}(x) &\triangleq V'(x)[f(x) + G(x)\phi(x)] \\ &= \alpha(x) + \beta^{\mathrm{T}}(x)\phi(x) \\ &= \begin{cases} -c_0\beta^{\mathrm{T}}(x)\beta(x) - \sqrt{\alpha^2(x) + (\beta^{\mathrm{T}}(x)\beta(x))^2}, & \beta(x) \neq 0, \\ \alpha(x), & \beta(x) = 0, \end{cases} \\ &< 0, \quad x \in \mathbb{R}^n, \quad x \neq 0, \end{aligned} \tag{6.82}$$

which implies that $V(\cdot)$ is a Lyapunov function for the closed-loop system (6.76) guaranteeing global asymptotic stability with $u = \phi(x)$ given by (6.81).

Since $f(\cdot)$ and $G(\cdot)$ are smooth it follows that $\alpha(x)$ and $\beta(x)$, $x \in \mathbb{R}^n$, are smooth functions, and hence, $\phi(x)$ given by (6.81) is smooth for all $x \in \mathbb{R}^n$ if either $\beta(x) \neq 0$ or $\alpha(x) < 0$. Hence, the feedback control law given by (6.81) is smooth everywhere except for the origin. The following result provides necessary and sufficient conditions under which the feedback control law given by (6.81) is guaranteed to be continuous and Lipschitz continuous at the origin in addition to being smooth everywhere else.

Theorem 6.8. Consider the nonlinear dynamical system $\mathcal{G}$ given by (6.76) with a radially unbounded control Lyapunov function $V : \mathbb{R}^n \to \mathbb{R}$. Then the following statements hold:

i) The control law $\phi(x)$ given by (6.81) is continuous at $x = 0$ if and only if for every $\varepsilon > 0$, there exists $\delta > 0$ such that for all $0 < \|x\| < \delta$, there exists $u \in \mathbb{R}^m$ such that $\|u\| < \varepsilon$ and $\alpha(x) + \beta^{\mathrm{T}}(x)u < 0$.

ii) There exists a stabilizing control law $\hat{\phi}(x)$ such that $\alpha(x) + \beta^{\mathrm{T}}(x)\hat{\phi}(x) < 0$, $x \in \mathbb{R}^n$, $x \neq 0$, and $\hat{\phi}(x)$ is Lipschitz continuous at $x = 0$ if and only if the control law $\phi(x)$ given by (6.81) is Lipschitz continuous at $x = 0$.

Proof. Necessity of *i)* is trivial with $u = \phi(x)$. Conversely, assume that, for every $\varepsilon > 0$, there exists $\delta > 0$ such that for all $0 < \|x\| < \delta$, there exists $u \in \mathbb{R}^m$ such that $\|u\| < \varepsilon$ and $\alpha(x) + \beta^{\mathrm{T}}(x)u < 0$. In this case, since $\|u\| < \varepsilon$ it follows from the Cauchy-Schwarz inequality that $\alpha(x) < \varepsilon\|\beta(x)\|$. Furthermore, since $V(\cdot)$ is continuously differentiable and $G(\cdot)$ is continuous it follows that there exists $\hat{\delta} > 0$ such that for all $0 < \|x\| < \hat{\delta}$, $\|\beta(x)\| < \varepsilon$. Hence, for all $0 < \|x\| < \delta_{\min}$, where $\delta_{\min} \triangleq \min\{\delta, \hat{\delta}\}$, it follows that $\alpha(x) < \varepsilon\|\beta(x)\|$ and $\|\beta(x)\| < \varepsilon$. Furthermore, if $\beta(x) = 0$, then $\|\phi(x)\| = 0$, and if $\beta(x) \neq 0$, then it follows from (6.81) that

$$\begin{aligned}
\|\phi(x)\| &\leq c_0\|\beta(x)\| + \frac{|\alpha(x) + \sqrt{\alpha^2(x) + (\beta^{\mathrm{T}}(x)\beta(x))^2}|}{\|\beta(x)\|} \\
&\leq \frac{2\alpha(x) + (c_0 + 1)\|\beta(x)\|^2}{\|\beta(x)\|} \\
&\leq (c_0 + 3)\varepsilon, \quad 0 < \|x\| < \delta_{\min}, \quad \alpha(x) > 0,
\end{aligned}$$

and

$$\begin{aligned}
\|\phi(x)\| &\leq c_0\|\beta(x)\| + \frac{\alpha(x) + \sqrt{\alpha^2(x) + (\beta^{\mathrm{T}}(x)\beta(x))^2}}{\|\beta(x)\|} \\
&\leq c_0\|\beta(x)\| + \frac{\beta^{\mathrm{T}}(x)\beta(x)}{\|\beta(x)\|} \\
&= (c_0 + 1)\|\beta(x)\| < (c_0 + 1)\varepsilon, \quad 0 < \|x\| < \delta_{\min}, \quad \alpha(x) \leq 0.
\end{aligned}$$

Hence, it follows that for every $\hat{\varepsilon} \triangleq (c_0 + 3)\varepsilon > 0$, there exists $\delta_{\min} > 0$ such that for all $\|x\| < \delta_{\min}$, $\|\phi(x)\| < \hat{\varepsilon}$, which implies that $\phi(\cdot)$ is continuous at the origin.

Next, to show necessity of *ii)* assume that there exists a stabilizing control $\hat{\phi}(x)$ such that $\alpha(x) + \beta^{\mathrm{T}}(x)\hat{\phi}(x) < 0$, $x \in \mathbb{R}^n$, $x \neq 0$, and $\hat{\phi}(x)$ is Lipschitz continuous at $x = 0$ with a Lipschitz constant $\hat{L}$; that is, there exists $\delta > 0$ such that for all $x \in \mathcal{B}_\delta(0)$, $\|\hat{\phi}(x)\| \leq \hat{L}\|x\|$. Now, since $V(\cdot)$ is continuous and $V'(0) = 0$, it follows that there exists $K > 0$ such that $\|\beta(x)\| \leq K\|x\|$, $x \in \mathcal{B}_\delta(0)$. Hence,

$$
\begin{aligned}
\|\phi(x)\| &\leq c_0\|\beta(x)\| + \frac{|\alpha(x) + \sqrt{\alpha^2(x) + (\beta^{\mathrm{T}}(x)\beta(x))^2}|}{\|\beta(x)\|} \\
&\leq \frac{2\alpha(x) + (c_0+1)\|\beta(x)\|^2}{\|\beta(x)\|} \\
&\leq 2\|\hat{\phi}(x)\| + \|\beta(x)\| \\
&\leq (2\hat{L} + (c_0+1)K)\|x\|, \quad x \in \mathcal{B}_\delta(0), \quad \alpha(x) > 0,
\end{aligned}
$$

and

$$
\begin{aligned}
\|\phi(x)\| &\leq c_0\|\beta(x)\| + \frac{\alpha(x) + \sqrt{\alpha^2(x) + (\beta^{\mathrm{T}}(x)\beta(x))^2}}{\|\beta(x)\|} \\
&\leq c_0\|\beta(x)\| + \frac{\beta^{\mathrm{T}}(x)\beta(x)}{\|\beta(x)\|} \\
&= (c_0+1)\|\beta(x)\| \\
&< (c_0+1)K\|x\|, \quad x \in \mathcal{B}_\delta(0), \quad \alpha(x) \leq 0,
\end{aligned}
$$

which implies that for all $x \in \mathcal{B}_\delta(0)$, $\|\phi(x)\| \leq L\|x\|$, where $L \triangleq 2\hat{L} + (c_0+1)K$, and hence, $\phi(\cdot)$ is Lipschitz continuous. Finally, sufficiency of *ii)* follows immediately with $\hat{\phi}(x) = \phi(x)$. □

6.6 Optimal Control and the Hamilton-Jacobi-Bellman Equation

In this section, we consider a control problem involving a notion of optimality with respect to a nonlinear-nonquadratic cost functional. Specifically, we consider the following optimal control problem.

Optimal Control Problem. Consider the nonlinear controlled system given by

$$
\dot{x}(t) = F(x(t), u(t), t), \quad x(t_0) = x_0, \quad x(t_{\mathrm{f}}) = x_{\mathrm{f}}, \quad u(t) \in U, \quad t \geq t_0, \tag{6.83}
$$

where $x(t) \in \mathcal{D} \subseteq \mathbb{R}^n$, $t \geq 0$, is the state vector, $\mathcal{D}$ is an open set with $0 \in \mathcal{D}$, $u(t) \in U \subseteq \mathbb{R}^m$, $t \geq 0$, is the control input, $x(t_0) = x_0$ is given, $x(t_{\mathrm{f}}) = x_{\mathrm{f}}$ is fixed, and $F : \mathcal{D} \times U \times \mathbb{R} \to \mathbb{R}^n$ satisfies $F(0, 0, \cdot) = 0$. We assume that $u(\cdot)$ is restricted to the class of *admissible* controls $\mathcal{U}$ consisting of measurable functions $u(\cdot)$ such that $u(t) \in U$ for all $t \geq 0$, where the constraint set U is given with $0 \in U$. Furthermore, we assume that $F(\cdot, \cdot, \cdot)$ is Lipschitz continuous in a neighborhood of the origin in $\mathcal{D} \times U \times \mathbb{R}$. Then determine the control input $u(t) \in U$, $t \in [t_0, t_f]$, such that the cost functional

$$J(x_0, u(\cdot), t_0) = \int_{t_0}^{t_f} L(x(t), u(t), t)\mathrm{d}t, \tag{6.84}$$

is minimized, where $L : \mathcal{D} \times U \times \mathbb{R} \to \mathbb{R}$ is given.

To solve the optimal control problem we present *Bellman's principle of optimality*, which provides necessary and sufficient conditions, for a given control $u(t) \in U$, $t \geq 0$, for minimizing the cost functional (6.84).

Lemma 6.1. Let $u^*(\cdot) \in \mathcal{U}$ be an optimal control that generates the trajectory $x(t)$, $t \in [t_0, t_f]$, with $x(t_0) = x_0$. Then the trajectory $x(\cdot)$ from (t_0, x_0) to (t_f, x_f) is optimal if and only if for all $t_1, t_2 \in [t_0, t_f]$, the portion of the trajectory $x(\cdot)$ going from $(t_1, x(t_1))$ to $(t_2, x(t_2))$ optimizes the same cost functional over $[t_1, t_2]$, where $x(t_1) = x_1$ is a point on the optimal trajectory generated by $u^*(\cdot)$.

Proof. Let $u^*(\cdot) \in \mathcal{U}$ solve the optimal control problem and let $x(t)$, $t \in [t_0, t_f]$, be the solution to (6.83) generated by $u^*(\cdot)$. Next, suppose, *ad absurdum*, that there exist $t_1 \geq t_0$, $t_2 \leq t_f$, and $\hat{u}(t)$, $t \in [t_1, t_2]$, such that

$$\int_{t_1}^{t_2} L(\hat{x}(t), \hat{u}(t), t)\mathrm{d}t < \int_{t_1}^{t_2} L(x(t), u^*(t), t)\mathrm{d}t,$$

where $\hat{x}(t)$ solves (6.83) for all $t \in [t_1, t_2]$ with $u(t) = \hat{u}(t)$, $\hat{x}(t_1) = x(t_1)$, and $\hat{x}(t_2) = x(t_2)$. Now, define

$$u_0(t) \triangleq \begin{cases} u^*(t), & t \in [t_0, t_1), \\ \hat{u}(t), & t \in [t_1, t_2], \\ u^*(t), & t \in (t_2, t_f]. \end{cases}$$

Then,

$$\begin{aligned} J(x_0, u_0(\cdot), t_0) &= \int_{t_0}^{t_f} L(x(t), u_0(t), t)\mathrm{d}t \\ &= \int_{t_0}^{t_1} L(x(t), u^*(t), t)\mathrm{d}t + \int_{t_1}^{t_2} L(\hat{x}(t), \hat{u}(t), t)\mathrm{d}t \\ &\quad + \int_{t_2}^{t_f} L(x(t), u^*(t), t)\mathrm{d}t \\ &< \int_{t_0}^{t_1} L(x(t), u^*(t), t)\mathrm{d}t + \int_{t_1}^{t_2} L(x(t), u^*(t), t)\mathrm{d}t \end{aligned}$$

$$+\int_{t_2}^{t_{\rm f}} L(x(t), u^*(t), t)\mathrm{d}t$$
$$= J(x_0, u^*(\cdot), t_0),$$

which is a contradiction.

Conversely, if $u^*(\cdot)$ minimizes $J(\cdot,\cdot,\cdot)$ over $[t_1, t_2]$ for all $t_1 \geq t_0$ and $t_2 \leq t_{\rm f}$, then it minimizes $J(\cdot,\cdot,\cdot)$ over $[t_0, t_{\rm f}]$. □

Lemma 6.1 states that $u^*(\cdot)$ solves the optimal control problem over the time interval $[t_0, t_{\rm f}]$ if and only if $u^*(\cdot)$ solves the optimal control problem over every subset of the time interval $[t_0, t_{\rm f}]$. Next, let $u^*(\cdot) \in \mathcal{U}$ solve the optimal control problem and define the optimal cost $J^*(x_0, t_0) \triangleq J(x_0, u^*(\cdot), t_0)$. Furthermore, define, for $p \in \mathbb{R}^n$, the Hamiltonian $H(x, u, p, t) \triangleq L(x, u, t) + p^{\rm T} F(x, u, t)$. With these definitions we have the following result.

Theorem 6.9. Let $J^*(x, t)$ denote the minimal cost for the optimal control problem with $x_0 = x$ and $t_0 = t$, and assume that $J^*(\cdot,\cdot)$ is continuously differentiable in x. Then

$$0 = \frac{\partial J^*(x(t), t)}{\partial t} + \min_{u(t)\in U} H(x(t), u(t), p(x(t), t), t), \tag{6.85}$$

where $p(x(t), t) \triangleq \left(\frac{\partial J^*(x(t),t)}{\partial x}\right)^{\rm T}$. Furthermore, if $u^*(\cdot)$ solves the optimal control problem, then

$$0 = \frac{\partial J^*(x(t), t)}{\partial t} + H(x(t), u^*(t), p(x(t), t), t). \tag{6.86}$$

Proof. It follows from Lemma 6.1 that, for all $t_1 \geq t$,

$$\begin{aligned}
J^*(x(t), t) &= \min_{u(\cdot)\in\mathcal{U}} \int_t^{t_{\rm f}} L(x(s), u(s), s)\mathrm{d}s \\
&= \min_{u(\cdot)\in\mathcal{U}} \left\{ \int_t^{t_1} L(x(s), u(s), s)\mathrm{d}s + \int_{t_1}^{t_{\rm f}} L(x(s), u(s), s)\mathrm{d}s \right\} \\
&= \min_{u(\cdot)\in\mathcal{U}} \left\{ \int_t^{t_1} L(x(s), u(s), s)\mathrm{d}s + J^*(x(t_1), t_1) \right\},
\end{aligned}$$

or, equivalently,

$$\begin{aligned}
0 = \min_{u(\cdot)\in\mathcal{U}} &\left\{ \frac{1}{t_1 - t}(J^*(x(t_1), t_1) - J^*(x(t), t)) \right. \\
&\left. + \frac{1}{t_1 - t}\int_t^{t_1} L(x(s), u(s), s)\mathrm{d}s \right\}.
\end{aligned}$$

Now, letting $t_1 \to t$ yields

$$0 = \min_{u(t)\in U} \left\{ \frac{\mathrm{d}J^*(x(t),t)}{\mathrm{d}t} + L(x(t),u(t),t) \right\}. \tag{6.87}$$

Next, noting that

$$\frac{\mathrm{d}J^*(x(t),t)}{\mathrm{d}t} = \frac{\partial J^*(x(t),t)}{\partial t} + \frac{\partial J^*(x(t),t)}{\partial x} F(x(t),u(t),t),$$

(6.87) yields

$$0 = \min_{u\in U} \left\{ \frac{\partial J^*(x,t)}{\partial t} + \frac{\partial J^*(x,t)}{\partial x} F(x,u,t) + L(x,u,t) \right\}, \quad x \in \mathcal{D},$$

which is equivalent to (6.85). Finally, (6.86) can be proved in a similar manner by replacing $u(\cdot)$ with $u^*(\cdot)$, where $u^*(\cdot)$ is the optimal control. □

Next, we provide the converse result to Theorem 6.9.

Theorem 6.10. Suppose there exists a continuously differentiable function $V : \mathcal{D} \times \mathbb{R} \to \mathbb{R}$ and an optimal control $u^*(\cdot)$ such that $V(x(t_\mathrm{f}),t_\mathrm{f}) = 0$,

$$0 = \frac{\partial V(x,t)}{\partial t} + H\left(x, u^*(t), \left(\frac{\partial V(x,t)}{\partial x}\right)^\mathrm{T}, t\right), \quad x \in \mathcal{D}, \quad t \geq 0, \tag{6.88}$$

and

$$H\left(x, u^*(t), \left(\frac{\partial V(x,t)}{\partial x}\right)^\mathrm{T}, t\right) \leq H\left(x, u(t), \left(\frac{\partial V(x,t)}{\partial x}\right)^\mathrm{T}, t\right),$$
$$x \in \mathcal{D}, \quad u(t) \in U, \quad t \geq 0. \tag{6.89}$$

Then $u^*(\cdot)$ solves the optimal control problem, that is,

$$J^*(x_0,t_0) = J(x_0,u^*(\cdot),t_0) \leq J(x_0,u(\cdot),t_0), \qquad u(\cdot) \in \mathcal{U}, \tag{6.90}$$

and

$$J^*(x_0,t_0) = V(x_0,t_0). \tag{6.91}$$

Proof. Let $x(t)$, $t \geq 0$, satisfy (6.83) and, for all $t \in [t_0, t_\mathrm{f}]$, define

$$\dot{V}(x(t),t) \triangleq \frac{\partial V(x(t),t)}{\partial t} + \frac{\partial V(x(t),t)}{\partial x} F(x(t),u(t),t). \tag{6.92}$$

Then, with $u(\cdot) = u^*(\cdot)$, it follows from (6.88) that

$$0 = \dot{V}(x(t),t) + L(x(t),u^*(t),t).$$

Now, integrating over $[t_0, t_\mathrm{f}]$ and noting that $V(x_\mathrm{f},t_\mathrm{f}) = 0$ yields

$$V(x_0,t_0) = \int_{t_0}^{t_\mathrm{f}} L(x(t),u^*(t),t)\mathrm{d}t = J(x_0,u^*(\cdot),t_0) = J^*(x_0,t_0).$$

Next, for all $u(\cdot) \in \mathcal{U}$ it follows from (6.88) and (6.92) that

$$
\begin{aligned}
J(x_0, u(\cdot), t)& \\
&= \int_{t_0}^{t_f} L(x(t), u(t), t)\mathrm{d}t \\
&= \int_{t_0}^{t_f} \left\{ -\dot{V}(x(t), t) + L(x(t), u(t), t) + \frac{\partial V(x(t), t)}{\partial t} \right. \\
&\quad \left. + \frac{\partial V(x(t), t)}{\partial x} F(x(t), u(t), t) \right\} \mathrm{d}t \\
&= \int_{t_0}^{t_f} \left\{ -\dot{V}(x(t), t) + \frac{\partial V(x(t), t)}{\partial t} + H\left(x, u(t), \left(\frac{\partial V(x, t)}{\partial x}\right)^{\mathrm{T}}, t\right) \right\} \mathrm{d}t \\
&\geq \int_{t_0}^{t_f} \left\{ -\dot{V}(x(t), t) + \frac{\partial V(x(t), t)}{\partial t} + H\left(x, u^*(t), \left(\frac{\partial V(x, t)}{\partial x}\right)^{\mathrm{T}}, t\right) \right\} \mathrm{d}t \\
&= \int_{t_0}^{t_f} -\dot{V}(x(t), t)\mathrm{d}t \\
&= V(x_0, t_0) \\
&= J^*(x_0, t_0),
\end{aligned}
$$

which completes the proof. □

Note that (6.88) and (6.89) imply

$$
0 = \frac{\partial V(x(t), t)}{\partial t} + \min_{u(t) \in U} H\left(x, u(t), \left(\frac{\partial V(x, t)}{\partial x}\right)^{\mathrm{T}}, t\right), \tag{6.93}
$$

which is known as the *Hamilton-Jacobi-Bellman equation.* It follows from Theorems 6.9 and 6.10 that the Hamilton-Jacobi-Bellman equation provides necessary and sufficient conditions for characterizing the optimal control for time-varying nonlinear dynamical systems over a finite time interval or the infinite horizon. In the infinite-horizon, time-invariant case, $V(\cdot)$ is independent of t so that the Hamilton-Jacobi-Bellman equation reduces to the time-invariant partial differential equation

$$
0 = \min_{u \in U} H(x, u, V'^{\mathrm{T}}(x)), \qquad x \in \mathcal{D}. \tag{6.94}
$$

Example 6.6. Consider the controlled nonlinear scalar system

$$
\dot{x}(t) = x^2(t) + u(t), \quad x(0) = x_0, \quad t \geq 0, \tag{6.95}
$$

with performance functional

$$
J(x_0, u(\cdot)) = \int_0^{\infty} [x^2(t) + u^2(t)]\mathrm{d}t. \tag{6.96}
$$

To obtain the optimal control $u^*(t)$, $t \geq 0$, that minimizes (6.96) we use Theorem 6.10. Specifically, for (6.95) and (6.96) the Hamiltonian is given by $H(x, u, V'(x)) = x^2 + u^2 + V'(x)[x^2 + u]$. Now, if follows from (6.90) that the optimal control is given by $\frac{\partial H}{\partial u} = 2u + V'(x) = 0$ or, equivalently, $u = -\frac{1}{2}V'(x)$. Next, using (6.88) with $u^* = -\frac{1}{2}V'(x)$ it follows that

$$0 = x^2 + V'(x)x^2 - \tfrac{1}{4}[V'(x)]^2. \tag{6.97}$$

Solving (6.97) as a quadratic equation gives $V'(x) = 2x^2 + 2x\sqrt{x^2+1}$, which implies, with $V(0) = 0$, $V(x) = \frac{2}{3}(x^3 + (x^2+1)^{3/2} - 1)$. Hence, the optimal control is given by $u^*(t) = -\frac{1}{2}V'(x(t)) = -x^2(t) - x(t)\sqrt{x^2(t)+1}$. □

6.7 Feedback Linearization, Zero Dynamics, and Minimum-Phase Systems

Recent work involving differential geometric methods [75, 212, 336] has made the design of controllers for certain classes of nonlinear systems more methodical. Such frameworks include the concepts of zero dynamics and feedback linearization. Even though the nonlinear stabilization frameworks presented in this book are based on Lyapunov theory, in certain cases feedback linearization simplifies the construction of Lyapunov functions for nonlinear systems. Here, we present a brief introduction to feedback linearization needed to develop some of the results in this book. For an excellent treatment on this subject, the interested reader is referred to [212].

In this section, we consider square (i.e., $m = l$) nonlinear dynamical systems $\mathcal{G}$ of the form

$$\dot{x}(t) = f(x(t)) + G(x(t))u(t), \qquad x(0) = x_0, \qquad t \geq 0, \tag{6.98}$$

$$y(t) = h(x(t)), \tag{6.99}$$

where $x \in \mathbb{R}^n$, $u, y \in \mathbb{R}^m$, $f : \mathbb{R}^n \to \mathbb{R}^n$, $G : \mathbb{R}^n \to \mathbb{R}^{n\times m}$, and $h : \mathbb{R}^n \to \mathbb{R}^m$. We assume that $f(\cdot)$, $G(\cdot)$, and $h(\cdot)$ are smooth, that is, infinitely differentiable mappings, and $f(\cdot)$ has at least one equilibrium so that, without loss of generality, $f(0) = 0$ and $h(0) = 0$. Furthermore, for the nonlinear dynamical system $\mathcal{G}$ we assume that the required properties for the existence and uniqueness of solutions are satisfied, that is, $u(\cdot)$ satisfies sufficient regularity conditions such that the system (6.98) has a unique solution forward in time.

The controlled nonlinear system (6.98) is *feedback linearizable* [210,419] if there exist a global invertible state transformation $\mathcal{T} : \mathbb{R}^n \to \mathbb{R}^n$ and a nonlinear feedback control law $u = \alpha(x) + \beta(x)v$, where $\alpha : \mathbb{R}^n \to \mathbb{R}^m$ and $\beta : \mathbb{R}^n \to \mathbb{R}^{m\times m}$ satisfies $\det \beta(x) \neq 0$, $x \in \mathbb{R}^n$, that transforms (6.98) into a linear controllable companion form. In this case, standard linear control design techniques can be used to synthesize stabilizing linear feedback controllers for the linearized system. Since the original nonlinear system and the feedback linearized system are *feedback equivalent*, that is, the two systems exhibit the same input-state behavior [17, 69], the linear control law can be transformed through the nonlinear feedback and the state transformation to yield a stabilizing nonlinear controller for the original nonlinear system.

To elucidate the above discussion consider the nonlinear dynamical system (6.98) and suppose there exist an invertible state transformation $\mathcal{T} : \mathbb{R}^n \to \mathbb{R}^n$ such that $z = \mathcal{T}(x)$ $\mathcal{T}(0) = 0$, and a nonlinear feedback control law $u = \alpha(x) + \beta(x)v$, where $\alpha : \mathbb{R}^n \to \mathbb{R}^m$ and $\beta : \mathbb{R}^n \to \mathbb{R}^{m\times m}$ satisfies $\det\beta(x) \neq 0$, $x \in \mathbb{R}^n$, that transforms (6.98) into

$$\dot{z}(t) = Az(t) + B\beta^{-1}(x)[u(t) - \alpha(x(t))], \quad z(0) = z_0, \quad t \geq 0, \tag{6.100}$$

where $A \in \mathbb{R}^{n\times n}$, $B \in \mathbb{R}^{n\times m}$, and (A, B) is controllable. Furthermore, let $P \in \mathbb{R}^{n\times n}$ be a positive-definite matrix satisfying the algebraic Riccati equation

$$0 = A^{\mathrm{T}}P + PA + R_1 - PBR_2^{-1}B^{\mathrm{T}}P, \tag{6.101}$$

where $R_1 > 0$ and $R_2 > 0$. Now, the function $V(x) = \mathcal{T}^{\mathrm{T}}(x)P\mathcal{T}(x) = z^{\mathrm{T}}Pz$ satisfies

$$\begin{aligned}
&\inf_{u\in U} V'(x)[f(x) + G(x)u] \\
&\quad = \inf_{u\in U} \left[z^{\mathrm{T}}(A^{\mathrm{T}}P + PA)z + 2z^{\mathrm{T}}PB\beta^{-1}(x)[u - \alpha(x)]\right] \\
&\quad = \inf_{u\in U} \left[-z^{\mathrm{T}}R_1z + z^{\mathrm{T}}PB[R_2^{-1}B^{\mathrm{T}}Pz + 2\beta^{-1}(x)u \right. \\
&\qquad \left. -2\beta^{-1}(x)\alpha(x)]\right] \\
&\quad = \begin{cases} -z^{\mathrm{T}}R_1z, & z^{\mathrm{T}}PB = 0, \\ -\infty, & z^{\mathrm{T}}PB \neq 0, \end{cases}
\end{aligned} \tag{6.102}$$

and hence, $V(\cdot)$ is a control Lyapunov function for (6.100). Hence, it follows from the results of Section 6.5 that (6.100) is globally stabilizable. In particular, one such feedback controller is given by (6.81) with x replaced by z, and hence, a globally stabilizing feedback nonlinear controller for (6.98) is given by $u = \phi(\mathcal{T}(x))$.

The above discussion raises an interesting question, namely, under what conditions does there exist an invertible state transformation and a nonlinear control that feedback linearizes (6.98)? To present necessary and sufficient conditions for feedback linearization of the nonlinear system (6.98) the following definitions are needed. For the first definition, a k-dimensional vector field $f_1(x), f_2(x), \ldots, f_k(x)$, defined on an open subset $\mathcal{D}$, is a mapping that assigns a k-dimensional vector to each point x of $\mathcal{D}$.

Definition 6.7. A k-dimensional *distribution* $\mathbf{D}(\cdot)$ on $\mathcal{D}$ is a mapping that assigns, to each $x \in \mathcal{D}$, a k-dimensional subspace $\mathbf{D}(x)$ of $\mathbb{R}^n$ such that there exist smooth vector fields $f_1(x)$, $f_2(x)$, ..., $f_k(x)$, $x \in \mathcal{D}$, with $\{f_1(x), f_2(x), \ldots, f_k(x)\}$, $x \in \mathcal{D}$, forming a linearly independent set and $\mathbf{D}(x) = \operatorname{span}\{f_1(x), f_2(x), \ldots, f_k(x)\}$, $x \in \mathcal{D}$.

Definition 6.8. Let $f, G : \mathbb{R}^n \to \mathbb{R}^n$ be continuously differentiable functions. The *Lie bracket* of f and G is defined as

$$[f(x), G(x)] = \operatorname{ad}_f G(x) \triangleq \frac{\partial G}{\partial x} f(x) - \frac{\partial f}{\partial x} G(x).$$

Furthermore, the *zeroth-order* and *higher-order Lie brackets* are defined as, respectively,

$$\operatorname{ad}_f^0 G(x) \triangleq G(x), \quad \operatorname{ad}_f^k G(x) \triangleq [f(x), \operatorname{ad}_f^{k-1} G(x)],$$

where $k \geq 1$. Finally, define the *Lie derivative* of a scalar function $V(x)$ along the vector field of $f(x)$ by

$$L_f V(x) \triangleq \frac{\partial V(x)}{\partial x} f(x),$$

and define the *zeroth-order* and *higher-order Lie derivatives*, respectively, by

$$L_f^0 V(x) \triangleq V(x), \quad L_f^k V(x) \triangleq L_f(L_f^{k-1} V(x)),$$

where $k \geq 1$.

Definition 6.9. The distribution $\mathbf{D}(x)$, $x \in \mathcal{D}$, is *involutive* if $[f_i(x), f_j(x)] \in \mathbf{D}(x)$, for all $f_i(x), f_j(x) \in \mathbf{D}(x)$, and $i \neq j$, $i, j = 1, \ldots, k$, where $\mathbf{D}(x) = \operatorname{span}\{f_1(x), \ldots, f_k(x)\}$, $x \in \mathcal{D}$.

Example 6.7. Let $\mathcal{D} = \mathbb{R}^4$ and $\mathbf{D}(x) = \operatorname{span}\{f_1(x), f_2(x)\}$, where

$$f_1(x) = \begin{bmatrix} x_1 \\ 1 \\ 0 \\ x_3 \end{bmatrix}, \quad f_2(x) = \begin{bmatrix} -e^{x_2} \\ 0 \\ 0 \\ 0 \end{bmatrix}. \tag{6.103}$$

To show that $\mathbf{D}(x)$ is involutive note that $\frac{\partial f_2}{\partial x} f_1(x) - \frac{\partial f_1}{\partial x} f_2(x) = 0$, which shows that $\operatorname{rank}[f_1(x),\ f_2(x),\ [f_1(x), f_2(x)]] = 2$, $x \in \mathcal{D}$. △

For the statement of the next theorem define the distribution

$$\mathbf{D}_i(x) \triangleq \operatorname{span}\{\operatorname{ad}_f^k G_j(x),\ k = 0, 1, \ldots, i,\ j = 1, 2, \ldots, m\},$$

for $0 \leq i \leq n-1$, where $G_j(x)$ denotes the jth column of $G(x)$.

Theorem 6.11 ([210]). Suppose rank $G(0) = m$. Then (6.98) is feedback linearizable if and only if the following statements hold:

i) For each $i \in \{0, \ldots, n-1\}$, the distribution $\mathbf{D}_i(x)$ has a constant dimension in a neighborhood of the origin.

ii) For each $i \in \{0, \ldots, n-2\}$, the distribution $\mathbf{D}_i(x)$ is involutive in a neighborhood of the origin.

Example 6.8. The dynamics of an undamped single-link manipulator with flexible joints are given by (6.98) where ([415])

$$f(x) = \begin{bmatrix} x_2 \\ -\alpha \sin(x_1) - \beta(x_1 - x_3) \\ x_4 \\ \gamma(x_1 - x_3) \end{bmatrix}, \quad G(x) = \begin{bmatrix} 0 \\ 0 \\ 0 \\ \delta \end{bmatrix}, \tag{6.104}$$

and where α, β, γ, $\delta > 0$. The undisturbed system has an equilibrium at $x = 0$. To show that this system is feedback linearizable we compute the distribution $\mathbf{D}_i(x)$ for each $0 \le i \le 3$ and check conditions *i)* and *ii)* of Theorem 6.11. Specifically, it can be easily shown that at $x = 0$

$$\mathrm{ad}_f G(0) = \begin{bmatrix} 0 \\ 0 \\ -\delta \\ 0 \end{bmatrix}, \quad \mathrm{ad}_f^2 G(0) = \begin{bmatrix} 0 \\ \beta\delta \\ 0 \\ -\gamma\delta \end{bmatrix}, \quad \mathrm{ad}_f^3 G(0) = \begin{bmatrix} -\beta\delta \\ 0 \\ \gamma\delta \\ 0 \end{bmatrix}.$$

Now, since $\det[G(0), \mathrm{ad}_f G(0), \mathrm{ad}_f^2 G(0), \mathrm{ad}_f^3 G(0)] = \beta^2\gamma^4$ for all $x \in \mathbb{R}^4$ it follows that $\mathbf{D}_i(x)$, $0 \le i \le 3$, has constant dimension at $x = 0$. Furthermore, since $\mathrm{ad}_f^i G(0)$, $0 \le i \le 3$, are constant, it follows that the $\mathrm{span}\{G(0), \mathrm{ad}_f G(0), \mathrm{ad}_f^2 G(0)\}$ is involutive. Hence, it follows from Theorem 6.11 that the system is feedback linearizable. △

The *zero dynamics* of the nonlinear system (6.98) and (6.99) are the dynamics of the system subject to the constraint that the output $y(t)$, $t \ge 0$, is identically zero. The system (6.98) and (6.99) is said to be *minimum phase* if its zero dynamics are asymptotically stable, while the system is said to be *critically minimum phase* if its zero dynamics are Lyapunov stable. Furthermore, (6.98) and (6.99) is said to have *relative degree* $\{r_1, r_2, \ldots, r_m\}$ at a point x_0, if there exists a neighborhood $\mathcal{D}_0$ of x_0 such that, for all $x \in \mathcal{D}_0$,

$$L_{G_i} L_f^k h_j(x) = 0, \qquad 0 \le k < r_j - 1, \qquad 1 \le i, j \le m,$$

and the matrix

$$\mathcal{L}(x) \triangleq \begin{bmatrix} L_{G_1} L_f^{r_1-1} h_1(x) & \cdots & L_{G_m} L_f^{r_1-1} h_1(x) \\ \vdots & \ddots & \vdots \\ L_{G_1} L_f^{r_m-1} h_m(x) & \cdots & L_{G_m} L_f^{r_m-1} h_m(x) \end{bmatrix}$$

is nonsingular. In the above notation, $L_f h_j(x) \triangleq h_j'(x) f(x)$, $j \in \{1, \ldots, m\}$ and $L_f^k h_j(x) \triangleq L_f(L_f^{k-1} h_j(x))$, $2 \le k < r_j - 1$, $1 \le j \le m$, where $L_f^0 h_j(x) \triangleq$

$h_j(x)$, G_i, $i = 1, \ldots, m$, are the smooth column vector fields of G, and h_j, $j = 1, \ldots, m$, are the smooth components of h. In the case where the relative degree $\{r_1, r_2, \ldots, r_m\} = \{1, 1, \ldots, 1\}$,

$$\mathcal{L}(x) = L_G h(x) \triangleq \begin{bmatrix} L_{G_1} h_1(x) & \cdots & L_{G_m} h_1(x) \\ \vdots & \ddots & \vdots \\ L_{G_1} h_m(x) & \cdots & L_{G_m} h_m(x) \end{bmatrix},$$

which is nonsingular for all $x \in \mathcal{D}_0$.

Example 6.9. Consider the controlled Van der Pol oscillator

$$\dot{x}_1(t) = x_2(t), \quad x_1(0) = x_{10}, \quad t \geq 0, \tag{6.105}$$
$$\dot{x}_2(t) = -x_1(t) + \varepsilon(1 - x_1^2(t))x_2(t) + u(t), \quad x_2(0) = x_{20}, \tag{6.106}$$
$$y(t) = x_2(t), \tag{6.107}$$

where $\varepsilon > 0$. Note that (6.105)–(6.107) can be written in the state space form (6.98) and (6.99) with $x = [x_1,\ x_2]^{\mathrm{T}}$, $f(x) = [x_2,\ -x_1 + \varepsilon(1 - x_1^2)x_2]^{\mathrm{T}}$, $G(x) = [0,\ 1]^{\mathrm{T}}$, and $h(x) = x_2$. Now, computing $L_G h(x) = \frac{\partial h}{\partial x} G(x) = [0,\ 1][0,\ 1]^{\mathrm{T}} = 1$, and hence, the system has a relative degree 1 in $\mathbb{R}^2$. Next, setting $y(t) = x_2(t) = 0$, $t \geq 0$, it follows that the zero dynamics given by $\dot{x}_1(t) = 0$, $t \geq 0$, are not asymptotically stable, and hence, the system is not minimum phase. However, the system is critically minimum phase. △

For the nonlinear system $\dot{x} = f(x)$, $x \in \mathbb{R}^n$, we say that f is *complete* if f is an infinitely differentiable function defined on a manifold $\mathcal{M} \subset \mathbb{R}^n$ and the flow of f is defined on the whole Cartesian product $\mathbb{R} \times \mathcal{M}$. Furthermore, recall (see Definition 2.35) that a mapping $\mathcal{T} : \mathcal{D} \subseteq \mathbb{R}^n \to \mathbb{R}^n$ is a diffeomorphism on $\mathcal{D}$ if $\mathcal{T}(x)$ is invertible on $\mathcal{D}$ and $\mathcal{T}(x)$ and $\mathcal{T}^{-1}(x)$, $x \in \mathcal{D}$, are continuously differentiable. For nonlinear affine systems of the form (6.98) and (6.99), conditions for the existence of globally defined diffeomorphisms transforming (6.98) and (6.99) into several kinds of normal forms are given in [76]. Here, we only consider relative degree $\{1, 1, \ldots, 1\}$ systems with complete and involutive vector fields $G(L_G h)^{-1}$. The following result is used later in the book.

Lemma 6.2 ([76]). Assume (6.98) and (6.99) is minimum phase with relative degree $\{1, 1, \ldots, 1\}$. If the vector field $G(L_G h)^{-1}$ is complete and involutive, then there exists a global diffeomorphism $\mathcal{T} : \mathbb{R}^n \to \mathbb{R}^n$, an infinitely differentiable function $f_0 : \mathbb{R}^{n-m} \to \mathbb{R}^{n-m}$, and an infinitely differentiable function $r : \mathbb{R}^{n-m} \times \mathbb{R}^m \to \mathbb{R}^{(n-m)\times m}$ such that, in the coordinates

$$\begin{bmatrix} y \\ z \end{bmatrix} \triangleq \mathcal{T}(x), \tag{6.108}$$

(6.98) is equivalent to

$$\begin{bmatrix} \dot{y} \\ \dot{z} \end{bmatrix} = \begin{bmatrix} L_f h(x) \\ f_0(z) + r(z,y)y \end{bmatrix} + \begin{bmatrix} L_G h(x) \\ 0 \end{bmatrix} u. \tag{6.109}$$

As discussed in Section 5.2, dissipativity can be helpful in constructing Lyapunov functions for nonlinear dynamical systems. Next, using the results in this section, we present necessary and sufficient conditions for rendering a system passive and strictly passive via feedback. Specifically, suppose (6.98) and (6.99) is stabilizable. Then, we construct a feedback transformation

$$u = \alpha(x) + \beta(x)v, \tag{6.110}$$

where $\det \beta(x) \neq 0$, $x \in \mathbb{R}^n$, such that

$$\dot{x}(t) = f(x(t)) + G(x(t))\alpha(x(t)) + G(x(t))\beta(x(t))v(t), \quad x(0) = x_0, \quad t \geq 0, \tag{6.111}$$

$$y(t) = h(x(t)), \tag{6.112}$$

is passive. Now, since (6.111) and (6.112) is rendered passive via (6.110) or, equivalently, (6.98) and (6.99) is *feedback passive*, $v = -\phi(y)$, where $\phi : \mathbb{R}^m \to \mathbb{R}^m$ is such that $y^{\mathrm{T}}\phi(x) > 0$, $y \neq 0$, stabilizes (6.111) and (6.112).

First, we show that if (6.98) and (6.99) is passive, then (6.98) and (6.99) has relative degree $\{1, 1, \ldots, 1\}$ at $x = 0$.

Lemma 6.3. Assume (6.98) and (6.99) is passive with a two-times continuously differentiable storage function $V_{\mathrm{s}} : \mathbb{R}^n \to \mathbb{R}$. Suppose $\operatorname{rank} G(0) = \operatorname{rank} \frac{\partial h}{\partial x}(0) = m$. Then (6.98) and (6.99) has relative degree $\{1, 1, \ldots, 1\}$ at $x = 0$.

Proof. Since (6.98) and (6.99) is passive with a two-times continuously differentiable storage function $V_{\mathrm{s}} : \mathbb{R}^n \to \mathbb{R}$ it follows from (5.116) of Corollary 5.2 with $J(x) \equiv 0$ that

$$\tfrac{1}{2} G^{\mathrm{T}}(x) V_{\mathrm{s}}'^{\mathrm{T}}(x) = h(x). \tag{6.113}$$

Now, differentiating both sides of (6.113) with respect to x yields

$$\tfrac{1}{2} \frac{\partial}{\partial x} [G^{\mathrm{T}}(x) V_{\mathrm{s}}'^{\mathrm{T}}(x)] = \frac{\partial h}{\partial x}(x). \tag{6.114}$$

Forming (6.114)$G(x)$, it follows that

$$\tfrac{1}{2} \frac{\partial}{\partial x} [G^{\mathrm{T}}(x) V_{\mathrm{s}}'^{\mathrm{T}}(x)] G(x) = \frac{\partial h}{\partial x}(x) G(x). \tag{6.115}$$

Now, noting that at $x = 0$, $V_{\mathrm{s}}'(0) = 0$ and $V_{\mathrm{s}}''(0)$ is nonnegative definite, it follows from (6.115) that

$$\tfrac{1}{2}G^{\mathrm{T}}(0)V_{\mathrm{s}}''(0)G(0) = L_G h(0). \tag{6.116}$$

Since $V_{\mathrm{s}}''(0) \geq 0$ it follows from the Schur decomposition that there exists M such that $\frac{1}{2}V_{\mathrm{s}}''(0) = M^{\mathrm{T}}M$. Hence, $L_G h(0) = G^{\mathrm{T}}(0)M^{\mathrm{T}}MG(0)$. Next, evaluating (6.114) at $x = 0$ it follows that $\frac{\partial h}{\partial x}(0) = G^{\mathrm{T}}(0)M^{\mathrm{T}}M$. Now, since rank $G(0) =$ rank $\frac{\partial h}{\partial x}(0) = m$ it follows that rank $MG(0) = m$, and hence, $L_G h(0)$ is nonsingular. The result is now immediate from the definition of relative degree. □

Next, we show that if (6.98) and (6.99) is passive, then the zero dynamics of (6.98) and (6.99) are Lyapunov stable. Since, as discussed in Section 3.5, a Lyapunov stable time-invariant system does not ensure the existence of a continuously differentiable time-independent Lyapunov function, we require the following definition.

Definition 6.10. The system (6.98) and (6.99) is *weakly minimum phase* if its zero dynamics, evolving on the smooth $(n-m)$-dimensional submanifold $\mathcal{Z} \triangleq \{x \in \mathbb{R}^n : h(x) = 0\}$ described by $\dot{z} = f_0(z)$, are Lyapunov stable and there exists a continuously differentiable positive-definite function $V_0 : \mathbb{R}^{n-m} \to \mathbb{R}$ such that $V_0(0) = 0$ and $V_0'(z)f_0(z) \leq 0$, $z \in \mathcal{Z}$.

Lemma 6.4. Assume (6.98) and (6.99) is passive (respectively, strictly passive) with a continuously differentiable positive-definite storage function $V_{\mathrm{s}} : \mathbb{R}^n \to \mathbb{R}$. Then (6.98) and (6.99) is weakly minimum phase (respectively, minimum phase).

Proof. Since (6.98) and (6.99) is passive with a continuously differentiable positive-definite storage function $V_{\mathrm{s}} : \mathbb{R}^n \to \mathbb{R}$ it follows from (5.116) of Corollary 5.2 with $J(x) \equiv 0$ that

$$\tfrac{1}{2}V_{\mathrm{s}}'(x)G(x) = h^{\mathrm{T}}(x). \tag{6.117}$$

Next, since by definition the zero dynamics of the nonlinear system (6.98) and (6.99) are the dynamics subject to the external constraint $y(t) = h(x(t)) \equiv 0$, it follows from (6.117) that $V_{\mathrm{s}}'(x)G(x) = 0$, $x \in \mathbb{R}^n$. Now, using the above two facts it follows that

$$\dot{V}_{\mathrm{s}}(x) = V_{\mathrm{s}}'(x)[f(x) + G(x)u] = V_{\mathrm{s}}'(x)f(x) \leq 2u^{\mathrm{T}}y = 0,$$

which shows that $\dot{V}_{\mathrm{s}}(x) \leq 0$, $x \in \mathbb{R}^n$, along the submanifold $h(x) \equiv 0$. Thus, the zero dynamics of (6.98) and (6.99) are Lyapunov stable, and hence, (6.98) and (6.99) is weakly minimum phase. Finally, if (6.98) and (6.99) is strictly passive, then $\dot{V}_{\mathrm{s}}(x) < 2u^{\mathrm{T}}y$, and hence, it follows from (6.118) that $\dot{V}_{\mathrm{s}}(x) < 0$, $x \in \mathbb{R}^n$, $x \neq 0$, along the submanifold $h(x) \equiv 0$. Thus, the zero dynamics of (6.98) and (6.99) are asymptotically stable, and hence, (6.98) and (6.99) is minimum phase. □

Next, we present the main result of this section which gives necessary and sufficient conditions for rendering a nonlinear system feedback passive.

Theorem 6.12. Consider the nonlinear dynamical system $\mathcal{G}$ given by (6.98) and (6.99). Suppose the vector field $G(L_G h)^{-1}$ is complete and involutive. Then the system (6.98) and (6.99) is feedback passive (respectively, feedback strictly passive) with a two-times continuously differentiable positive-definite storage function $V_\mathrm{s} : \mathbb{R}^n \to \mathbb{R}$ if and only if (6.98) and (6.99) has relative degree $\{1, 1, \ldots, 1\}$ at $x = 0$ and is weakly minimum phase (respectively, minimum phase).

Proof. For both cases necessity is a direct consequence of Lemmas 6.3 and 6.4. Specifically, if $\mathcal{G}$ is feedback passive (respectively, feedback strictly passive), then there exist functions $\alpha : \mathbb{R}^n \to \mathbb{R}^m$ and $\beta : \mathbb{R}^n \to \mathbb{R}^{m\times m}$ such that $\det \beta(x) \neq 0$, $x \in \mathbb{R}^n$, and with $u = \alpha(x) + \beta(x)v$ the nonlinear system (6.111) and (6.112) is passive (respectively, strictly passive). That is,

$$\dot{x}(t) = \tilde{f}(x(t)) + \tilde{G}(x(t))v, \quad x(0) = x_0, \quad t \geq 0, \tag{6.118}$$

$$y(t) = h(x(t)), \tag{6.119}$$

is passive (respectively, strictly passive), where $\tilde{f}(x) \triangleq f(x) + G(x)\alpha(x)$ and $\tilde{G}(x) \triangleq G(x)\beta(x)$.

Next, noting that $\tilde{G}(L_{\tilde{G}}h)^{-1} = G(L_G h)^{-1}$ it follows that $\operatorname{rank} \tilde{G}(0) = \operatorname{rank} \frac{\partial h}{\partial x}(0) = m$, and hence, by Lemmas 6.3 and 6.4 it follows that the nonlinear system (6.118) and (6.119) has relative degree $\{1, 1, \ldots, 1\}$ at $x = 0$ and is weakly minimum phase (respectively, minimum phase). Now, it follows from Lemma 6.2 that there exists a global diffeomorphism $\mathcal{T} : \mathbb{R}^n \to \mathbb{R}^n$, an infinitely differentiable function $f_0 : \mathbb{R}^{n-m} \to \mathbb{R}^{n-m}$, and an infinitely differentiable function $r : \mathbb{R}^{n-m} \times \mathbb{R}^m \to \mathbb{R}^{(n-m)\times m}$ such that, in the coordinates

$$\begin{bmatrix} y \\ z \end{bmatrix} \triangleq \mathcal{T}(x), \tag{6.120}$$

(6.98) is equivalent to

$$\begin{aligned} \begin{bmatrix} \dot{y} \\ \dot{z} \end{bmatrix} &= \begin{bmatrix} L_{\tilde{f}}h(x) \\ f_0(z) + r(z,y)y \end{bmatrix} + \begin{bmatrix} L_{\tilde{G}}h(x) \\ 0 \end{bmatrix} v \\ &= \begin{bmatrix} L_f h(x) \\ f_0(z) + r(z,y)y \end{bmatrix} + \begin{bmatrix} L_G h(x) \\ 0 \end{bmatrix} u. \end{aligned} \tag{6.121}$$

The result now follows immediately from the structure of (6.121) by noting that relative degree and zero dynamics are invariant under a static state feedback transformation.

Conversely, assume (6.98) and (6.99) has relative degree $\{1, 1, \ldots, 1\}$ at $x = 0$ and is weakly minimum phase. In this case, it follows from Lemma 6.2 that (6.98) and (6.99) can be equivalently written as

$$\dot{y}(t) = L_f h(x(t)) + L_G h(x(t))u(t), \quad y(0) = y_0, \quad t \geq 0, \tag{6.122}$$
$$\dot{z}(t) = f_0(z(t)) + r(z(t), y(t))y(t), \quad z(0) = z_0. \tag{6.123}$$

Now, since by assumption (6.98) and (6.99) or, equivalently, (6.122) and (6.123), is weakly minimum phase there exists a continuously differentiable positive-definite function $V_0(z)$, $z \in \mathbb{R}^{n-m}$, such that

$$V_0'(z) f_0(z) \leq 0, \quad z \in \mathbb{R}^{n-m}. \tag{6.124}$$

Hence,

$$\dot{V}_0(z) = V_0'(z)[f_0(z) + r(z, y)y] \leq V_0'(z) r(z, y) y.$$

Next, using the state feedback transformation

$$u(z, y) = \tfrac{1}{2}(L_G h(x))^{-1}[-L_f h(x) - r^{\mathrm{T}}(z, y) V_0'^{\mathrm{T}}(z) + v],$$

and the positive-definite function

$$V_{\mathrm{s}}(z, y) = V_0(z) + y^{\mathrm{T}} y,$$

it follows that $\dot{V}_{\mathrm{s}}(z, y) \leq 2y^{\mathrm{T}} v$, which shows that (6.98) and (6.99) is feedback passive.

Finally, the proof of the equivalence between feedback strict passivity and relative degree $\{1, 1, \ldots, 1\}$ at $x = 0$ and minimum phase is identical with (6.124) replaced by

$$V_0'(z) f_0(z) < 0, \quad z \in \mathbb{R}^{n-m}, \quad z \neq 0,$$

since in this case the zero dynamics $\dot{z} = f_0(z)$ are asymptotically stable. □

To specialize Theorem 6.12 to linear dynamical systems let $f(x) = Ax$, $G(x) = B$, and $h(x) = Cx$ so that

$$\dot{x}(t) = Ax(t) + Bu(t), \quad x(0) = x_0, \quad t \geq 0, \tag{6.125}$$
$$y(t) = Cx(t). \tag{6.126}$$

Now, with rank $B = m$, it follows from Theorem 6.12 and Theorem 5.13 that (6.125) and (6.126) is feedback equivalent to a passive linear system with a quadratic positive-definite storage function $V_{\mathrm{s}}(\cdot)$ if and only if $\det(CB) \neq 0$ and (6.125) and (6.126) is weakly minimum phase. Since a minimal linear passive system is equivalent to a positive real system with a quadratic positive-definite storage function, it follows that any minimal linear system is feedback positive real if and only if $\det(CB) \neq 0$ and (6.125) and (6.126) is weakly minimum phase.

Next, we present sufficient conditions for which a given *block cascade system* is feedback equivalent to a passive system. Specifically, consider the nonlinear block cascade system

$$\dot{x}(t) = f(x(t)) + G(x(t))y(t), \quad x(0) = x_0, \quad t \geq 0, \tag{6.127}$$
$$\dot{\hat{x}}(t) = \hat{f}(\hat{x}(t)) + \hat{G}(\hat{x}(t))u(t), \quad \hat{x}(0) = \hat{x}_0, \tag{6.128}$$
$$y(t) = h(\hat{x}(t)), \tag{6.129}$$

where $x \in \mathbb{R}^n$, $\hat{x} \in \mathbb{R}^q$, $u, y \in \mathbb{R}^m$, $f : \mathbb{R}^n \to \mathbb{R}^n$ satisfies $f(0) = 0$, $G : \mathbb{R}^n \to \mathbb{R}^{n\times m}$, $\hat{f} : \mathbb{R}^q \to \mathbb{R}^q$ satisfies $\hat{f}(0) = 0$, $\hat{G} : \mathbb{R}^q \to \mathbb{R}^{q\times m}$, and $h : \mathbb{R}^q \to \mathbb{R}^m$ satisfies $h(0) = 0$.

Theorem 6.13. Consider the block cascade system (6.127)–(6.129). Assume that the input subsystem (6.128) and (6.129) is passive (respectively, strictly passive) with a continuously differentiable positive-definite storage function $V_{\rm s} : \mathbb{R}^q \to \mathbb{R}$, and suppose the zero solution $x(t) \equiv 0$ to (6.127) with $y(t) \equiv 0$ is globally asymptotically stable. Then (6.127)–(6.129) is feedback equivalent to a passive (respectively, strictly passive) system with a continuously differentiable positive-definite storage function $V : \mathbb{R}^{n+q} \to \mathbb{R}$.

Proof. Since the zero solution $x(t) \equiv 0$ to (6.127) with $y(t) \equiv 0$ is globally asymptotically stable it follows from Theorem 3.9 that there exists a continuously differentiable positive-definite function $V_{\rm sub} : \mathbb{R}^n \to \mathbb{R}$ such that $V_{\rm sub}(0) = 0$ and $V_{\rm sub}(x)f(x) < 0$, $x \in \mathbb{R}^n$, $x \neq 0$. Next, note that (6.127)–(6.129) can be equivalently written as

$$\dot{\tilde{x}}(t) = \tilde{f}(\tilde{x}(t)) + \tilde{G}(\tilde{x}(t))u(t), \quad \tilde{x}(0) = \tilde{x}_0, \quad t \geq 0, \tag{6.130}$$
$$y(t) = \tilde{h}(\tilde{x}(t)), \tag{6.131}$$

where

$$\tilde{x} \triangleq \begin{bmatrix} x \\ \hat{x} \end{bmatrix}, \quad \tilde{f}(\tilde{x}) \triangleq \begin{bmatrix} f(x) + G(x)y \\ \hat{f}(\hat{x}) \end{bmatrix},$$
$$\tilde{G}(\tilde{x}) \triangleq \begin{bmatrix} 0 \\ \hat{G}(\hat{x}) \end{bmatrix}, \quad \tilde{h}(\tilde{x}) \triangleq h(\hat{x}).$$

Now, let

$$V(x, \hat{x}) = V_{\rm sub}(x) + V_{\rm s}(\hat{x})$$

and note that since (6.128) and (6.129) is passive with a continuously differentiable positive-definite storage function $V_{\rm s} : \mathbb{R}^n \to \mathbb{R}$, it follows from (5.116) of Corollary 5.2 with $J(x) \equiv 0$ that

$$h^{\rm T}(\hat{x}) = \tfrac{1}{2}V_{\rm s}'(\hat{x})\hat{G}(\hat{x}) = \tfrac{1}{2}V'(\tilde{x})\tilde{G}(\tilde{x}) = \tilde{h}^{\rm T}(\tilde{x}). \tag{6.132}$$

Next, let $u = \alpha(\tilde{x}) + v$ and note that

$$V'(\tilde{x})[\tilde{f}(\tilde{x}) + \tilde{G}(\tilde{x})\alpha(\tilde{x})] = V_{\rm sub}'(x)f(x) + V_{\rm sub}'G(x)h(\hat{x})$$

$$+V_{\rm s}'(\hat{x})\hat{f}(\hat{x}) + 2h^{\rm T}(\hat{x})\alpha(\tilde{x}).$$

Now, choosing $\alpha(\tilde{x}) = -\frac{1}{2}V_{\rm sub}'^{\rm T}(x)G^{\rm T}(x)$ it follows that

$$V'(\tilde{x})[\tilde{f}(\tilde{x}) + \tilde{G}(\tilde{x})\alpha(\tilde{x})] = V_{\rm sub}'(x)f(x) + V_{\rm s}'(\hat{x})\hat{f}(\hat{x}) \leq 0, \quad (x, \hat{x}) \in \mathbb{R}^n \times \mathbb{R}^q, \tag{6.133}$$

which, with (6.132), shows that (6.127)–(6.129) is rendered passive with $u = \alpha(\tilde{x}) + v$. To show strict feedback passivity, it need only be noted that if (6.128) and (6.129) is strictly feedback passive, then $V_{\rm s}'(\hat{x})\hat{f}(\hat{x}) < 0$, $\hat{x} \in \mathbb{R}^q$, $\hat{x} \neq 0$, and hence, (6.133) holds with a strict inequality. □

Finally, it is important to note that since exponential passivity implies strict passivity, the results in this section also hold for exponentially passive systems.

6.8 Problems

Problem 6.1. Consider the nonlinear zero-state observable dissipative dynamical system $\mathcal{G}$ with inputs u_i, outputs y_i, $i = 1, \ldots, m$, and internal states x interconnected with m dissipative dynamical subsystems $\mathcal{G}_i$, $i = 1, \ldots, m$, given by

$$\dot{x}_i(t) = f_i(x_i(t)) + G_i(x_i(t))y_i(t), \quad x_i(0) = x_{0i}, \quad t \geq 0, \tag{6.134}$$

$$-u_i(t) = h(x_i(t)) + J_i(x_i(t))y_i(t). \tag{6.135}$$

Let $r_i(-u_i, y_i)$ and $V_{{\rm s}i}(x_i)$ denote the storage functions and supply rates of the $\mathcal{G}_i$th subsystem, respectively, and define the overall subsystem supply rate $R(u, y) \triangleq \sum_{i=1}^m \gamma_i r_i(-u_i, y_i)$, where $u = [u_1, \ldots, u_m]^{\rm T}$, $y = [y_1, \ldots, y_m]^{\rm T}$, and $\gamma_i > 0$, $i = 1, \ldots, m$. Show that if there exists a positive-definite storage function $V_{\mathcal{G}}(\cdot)$ such that $\dot{V}_{\mathcal{G}}(x) \leq -R(u, y)$, then the zero solution $(x(t), x_1(t), \ldots, x_m(t)) \equiv (0, 0, \ldots, 0)$ of the interconnected system $\mathcal{G}$ with the subsystems $\mathcal{G}_i$, $i = 1, \ldots, m$, is Lyapunov stable.

Problem 6.2. Let

$$G(s) \overset{\rm min}{\sim} \left[\begin{array}{c|c} A & B \\ \hline C & 0 \end{array}\right], \quad G_{\rm c}(s) \overset{\rm min}{\sim} \left[\begin{array}{c|c} A_{\rm c} & B_{\rm c} \\ \hline C_{\rm c} & 0 \end{array}\right]$$

be asymptotically stable transfer functions. If $G(s)$ and $G_{\rm c}(s)$ are bounded real, show that the feedback interconnection of $G(s)$ and $G_{\rm c}(s)$ is Lyapunov stable, that is, the linear system with dynamics matrix

$$\tilde{A} \triangleq \begin{bmatrix} A & BC_{\rm c} \\ B_{\rm c}C & A_{\rm c} \end{bmatrix}$$

is Lyapunov stable. If, alternatively, $G_{\rm c}(s)$ is strictly bounded real, then show that $\tilde{A}$ is asymptotically stable.

Problem 6.3. Let

$$G(s) \overset{\min}{\sim} \left[\begin{array}{c|c} A & B \\ \hline C & 0 \end{array}\right], \quad G_{\rm c}(s) \overset{\min}{\sim} \left[\begin{array}{c|c} A_{\rm c} & B_{\rm c} \\ \hline C_{\rm c} & 0 \end{array}\right]$$

be Lyapunov stable transfer functions. If $G(s)$ and $G_{\rm c}(s)$ are positive real, show that the negative feedback interconnection of $G(s)$ and $G_{\rm c}(s)$ is Lyapunov stable, that is, the linear system with dynamics matrix

$$\tilde{A} \triangleq \begin{bmatrix} A & -BC_{\rm c} \\ B_{\rm c}C & A_{\rm c} \end{bmatrix}$$

is Lyapunov stable. If, alternatively, $G_{\rm c}(s)$ is strictly positive real, then show that $\tilde{A}$ is asymptotically stable.

Problem 6.4. Consider two nonlinear dynamical systems $\mathcal{G}_1$ and $\mathcal{G}_2$ with input-output pairs (u_1, y_1) and (u_2, y_2), respectively. Assume $\mathcal{G}_1$ and $\mathcal{G}_2$ are passive. Show that:

i) The parallel interconnection of $\mathcal{G}_1$ and $\mathcal{G}_2$ is passive.

ii) The negative feedback interconnection of $\mathcal{G}_1$ and $\mathcal{G}_2$ is passive.

Problem 6.5. Consider two nonlinear dynamical system $\mathcal{G}_1$ and $\mathcal{G}_2$ with input-output pairs (u_1, y_1) and (u_2, y_2), respectively. Assume $\mathcal{G}_1$ and $\mathcal{G}_2$ are nonexpansive with gains $\gamma_1 > 0$ and $\gamma_2 > 0$, respectively. Show that the cascade interconnection of $\mathcal{G}_1$ and $\mathcal{G}_2$ is nonexpansive with gain $\gamma_1\gamma_2$.

Problem 6.6. Consider the two nonlinear dynamical systems $\mathcal{G}_1$ and $\mathcal{G}_2$ with input-output pairs (u_1, y_1) and (u_2, y_2), respectively. Show that:

i) If $\mathcal{G}_1$ and $\mathcal{G}_2$ are input strict passive, then the parallel interconnection of $\mathcal{G}_1$ and $\mathcal{G}_2$ is input strict passive.

ii) If $\mathcal{G}_1$ and $\mathcal{G}_2$ are output strict passive, then the parallel interconnection of $\mathcal{G}_1$ and $\mathcal{G}_2$ is output strict passive.

iii) If $\mathcal{G}_1$ and $\mathcal{G}_2$ are strictly passive, then the parallel interconnection of $\mathcal{G}_1$ and $\mathcal{G}_2$ is strictly passive.

Problem 6.7. Consider the closed-loop dynamical system consisting of the nonlinear system $\mathcal{G}$ given by (6.1) and (6.2), and the nonlinear compensator $\mathcal{G}_{\rm c}$ given by (6.3) and (6.4). Show that if there exist functions $V_{\rm s} : \mathbb{R}^n \to \mathbb{R}$, $V_{\rm sc} : \mathbb{R}^{n_{\rm c}} \to \mathbb{R}$, $\ell : \mathbb{R}^n \to \mathbb{R}^p$, $\ell_{\rm c} : \mathbb{R}^{n_{\rm c}} \to \mathbb{R}^{p_{\rm c}}$, $\mathcal{W} : \mathbb{R}^n \to \mathbb{R}^{p\times m}$, and $\mathcal{W}_{\rm c} : \mathbb{R}^m \times \mathbb{R}^{n_{\rm c}} \to \mathbb{R}^{p_{\rm c}\times m}$, such that $V_{\rm s}(\cdot)$ and $V_{\rm sc}(\cdot)$ are continuously differentiable and positive definite, $V_{\rm s}(0) = 0$, $V_{\rm sc}(0) = 0$, $V_{\rm s}(x) \to \infty$ as

$\|x\| \to \infty$, $V_{\mathrm{sc}}(x_{\mathrm{c}}) \to \infty$ as $\|x_{\mathrm{c}}\| \to \infty$, and

$$0 > V_{\mathrm{s}}'(x)f(x) + \ell^{\mathrm{T}}(x)\ell(x), \quad x \in \mathbb{R}^n, \quad x \neq 0, \tag{6.136}$$
$$0 = \tfrac{1}{2}V_{\mathrm{s}}'(x)G(x) - h^{\mathrm{T}}(x) + \ell^{\mathrm{T}}(x)\mathcal{W}(x), \tag{6.137}$$
$$0 = J(x) + J^{\mathrm{T}}(x) - \mathcal{W}^{\mathrm{T}}(x)\mathcal{W}(x), \tag{6.138}$$
$$0 > V_{\mathrm{sc}}'(x_{\mathrm{c}})f_{\mathrm{c}}(x_{\mathrm{c}}) + \ell_{\mathrm{c}}^{\mathrm{T}}(x_{\mathrm{c}})\ell_{\mathrm{c}}(x_{\mathrm{c}}), \quad x_{\mathrm{c}} \in \mathbb{R}^{n_{\mathrm{c}}}, \quad x_{\mathrm{c}} \neq 0, \tag{6.139}$$
$$0 = \tfrac{1}{2}V_{\mathrm{sc}}'(x_{\mathrm{sc}})G_{\mathrm{c}}(u_{\mathrm{c}}, x_{\mathrm{c}}) - h_{\mathrm{c}}^{\mathrm{T}}(u_{\mathrm{c}}, x_{\mathrm{c}}) + \ell_{\mathrm{c}}^{\mathrm{T}}(x_{\mathrm{c}})\mathcal{W}_{\mathrm{c}}(u_{\mathrm{c}}, x_{\mathrm{c}}), \tag{6.140}$$
$$0 = J_{\mathrm{c}}(u_{\mathrm{c}}, x_{\mathrm{c}}) + J_{\mathrm{c}}^{\mathrm{T}}(u_{\mathrm{c}}, x_{\mathrm{c}}) - \mathcal{W}_{\mathrm{c}}^{\mathrm{T}}(u_{\mathrm{c}}, x_{\mathrm{c}})\mathcal{W}_{\mathrm{c}}(u_{\mathrm{c}}, x_{\mathrm{c}}), \tag{6.141}$$

then the negative feedback interconnection of $\mathcal{G}$ and $\mathcal{G}_{\mathrm{c}}$ is globally asymptotically stable.

Problem 6.8. Consider the controlled undamped oscillator

$$\ddot{x}(t) + \tfrac{k}{m}x(t) = u(t), \quad x(0) = x_0, \quad \dot{x}(0) = \dot{x}_0, \quad t \geq 0, \tag{6.142}$$
$$y(t) = \dot{x}(t), \tag{6.143}$$

where $m, k > 0$. Show that the dynamic compensator

$$\dot{x}_{\mathrm{c}}(t) = A_{\mathrm{c}}x_{\mathrm{c}}(t) + B_{\mathrm{c}}y(t), \quad x_{\mathrm{c}}(0) = x_{\mathrm{c}0}, \quad t \geq 0, \tag{6.144}$$
$$-u(t) = C_{\mathrm{c}}x_{\mathrm{c}}(t) + D_{\mathrm{c}}y(t), \tag{6.145}$$

where

$$A_{\mathrm{c}} = \begin{bmatrix} 0 & 1 \\ -k_{\mathrm{a}}/m_{\mathrm{a}} & -c_{\mathrm{a}}/m_{\mathrm{a}} \end{bmatrix}, \quad B_{\mathrm{c}} = \begin{bmatrix} 0 \\ 1 \end{bmatrix},$$
$$C_{\mathrm{c}} = \begin{bmatrix} -c_{\mathrm{a}}k_{\mathrm{a}}/m_{\mathrm{a}} & k_{\mathrm{a}} - c_{\mathrm{a}}^2/m_{\mathrm{a}} \end{bmatrix}, \quad D_{\mathrm{c}} = c_{\mathrm{a}},$$

emulating a dynamic $(m_{\mathrm{a}}, k_{\mathrm{a}}, c_{\mathrm{a}})$ absorber with $m_{\mathrm{a}}, k_{\mathrm{a}}, c_{\mathrm{a}} > 0$ guarantees that the closed-loop system (6.142)–(6.145) is asymptotically stable.

Problem 6.9. Consider the nonlinear dynamical system

$$\dot{x}_1(t) = x_2(t), \quad x_1(0) = x_{10}, \quad t \geq 0, \tag{6.146}$$
$$\dot{x}_2(t) = -x_1(t) - f(x_2(t)) + u(t), \quad x_2(0) = x_{20}, \tag{6.147}$$
$$y(t) = x_2(t) + u(t), \tag{6.148}$$

where $zf(z) > 0$, $z \in \mathbb{R}$, $z \neq 0$, and $f(0) = 0$. Show that the linear dynamic compensator

$$\dot{x}_{\mathrm{c}1}(t) = x_{\mathrm{c}2}(t), \quad x_{\mathrm{c}1}(0) = x_{\mathrm{c}10}, \quad t \geq 0, \tag{6.149}$$
$$\dot{x}_{\mathrm{c}2}(t) = -x_{\mathrm{c}1}(t) + \frac{1}{4}x_{\mathrm{c}2}(t) + y(t), \quad x_{\mathrm{c}2}(0) = x_{\mathrm{c}20}, \tag{6.150}$$
$$u(t) = -x_{\mathrm{c}2}(t), \tag{6.151}$$

guarantees global stability of the closed-loop system (6.146)–(6.151).

Problem 6.10. Consider the rotational/translational nonlinear dynamical system $\mathcal{G}$ given in Problem 5.49 and the nonlinear controller $\mathcal{G}_c$ given by

$$\begin{aligned}\dot{x}_c(t) &= f_c(x_c(t)) + G_c(x_c(t), y(t))y(t), \quad x_c(0) = x_{c0}, \quad t \geq 0, && (6.152)\\ -N_c(t) &= h_c(x_c(t), y(t)) + J_c(x_c(t), y(t))y(t), && (6.153)\end{aligned}$$

where $x_c = [\theta \; \theta_c \; \dot{\theta}_c]^T$,

$$f_c(x_c) = \begin{bmatrix} 0 \\ \dot{\theta}_c \\ -\frac{\sin(\theta_c - \theta)}{m_c e_c^2} - \frac{g \sin \theta_c}{e_c} \end{bmatrix}, \quad G_c(x_c, y) = \begin{bmatrix} 1 \\ 0 \\ 0 \end{bmatrix},$$

$$h_c(x_c, y) = \kappa \sin(\theta - \theta_c), \quad J_c(x_c, y) = \frac{\alpha \tanh(\gamma y)}{y},$$

$\alpha, \gamma, \kappa, e_c, m_c > 0$, and $y = \dot{\theta}$. Show that the closed-loop system consisting of $\mathcal{G}$ and $\mathcal{G}_c$ with $N(t) = -mgl \sin \theta(t) - N_c(t)$ is globally asymptotically stable.

Problem 6.11. Let $\varepsilon_c, \delta_c > 0$ be such that $\varepsilon_c \delta_c < 1$. Consider the nonlinear dynamical system $\mathcal{G}$ given by (6.1) and (6.2) and the nonlinear controller $\mathcal{G}_c$ given by

$$\begin{aligned}\dot{x}_c(t) &= [A_c - P^{-1} M(u_c(t), x_c(t)] x_c(t) + (1 - 2\varepsilon_c \delta_c)\\ &\quad \cdot [B_c + N(u_c(t), x_c(t))] u_c(t), \quad x_c(0) = x_{c0}, \quad t \geq 0, && (6.154)\\ y_c(t) &= [C_c + N^T(u_c(t), x_c(t)) P] x_c(t) + \delta_c u_c(t), && (6.155)\end{aligned}$$

where $A_c \in \mathbb{R}^{n_c \times n_c}$, $B_c \in \mathbb{R}^{n_c \times m}$, $C_c \in \mathbb{R}^{m \times n_c}$, $N : \mathbb{R}^m \times \mathbb{R}^{n_c} \to \mathbb{R}^{n_c \times m}$, and $P \in \mathbb{R}^{n_c \times n_c}$, with $P > 0$, satisfying

$$\begin{aligned} 0 &> A_c^T P + P A_c, && (6.156)\\ C_c &= B_c^T P, && (6.157)\end{aligned}$$

and

$$M(u_c, x_c) \triangleq \varepsilon_c [C_c + N^T(u_c(t), x_c(t)) P]^T [C_c + N^T(u_c(t), x_c(t)) P], \quad (u_c, x_c) \in \mathbb{R}^m \times \mathbb{R}^{n_c}. \quad (6.158)$$

Show that if $\mathcal{G}$ is passive and zero-state observable, then the negative feedback interconnection of $\mathcal{G}$ and $\mathcal{G}_c$ is globally asymptotically stable.

Problem 6.12. Show that if in Problem 6.11 $\mathcal{G}$ is strictly passive, $M(u_c, x_c)$ is an arbitrary nonnegative-definite function, and $\delta_c = 0$, then the negative feedback interconnection of $\mathcal{G}$ and $\mathcal{G}_c$ is globally asymptotically stable.

Problem 6.13. Consider the square linear dynamical system

$$G(s) \sim \left[\begin{array}{c|c} A & B \\ \hline C & D \end{array}\right].$$

Let $R_1 \in \mathbb{R}^{n\times n}$ and $R_2 \in \mathbb{R}^{m\times m}$, with R_1 and R_2 positive definite, be such that

$$R_1 > C^{\mathrm{T}}[R_2 - (D + D^{\mathrm{T}})]C, \tag{6.159}$$
$$R_2 > D + D^{\mathrm{T}}. \tag{6.160}$$

Show that the linear dynamic controller

$$G_{\mathrm{c}}(s) \sim \left[\begin{array}{c|c} A - BR_2^{-1}[C + B^{\mathrm{T}}P - DR_2^{-1}B^{\mathrm{T}}P] & BR_2^{-1} \\ \hline R_2^{-1}B^{\mathrm{T}}P & 0 \end{array}\right], \tag{6.161}$$

where $P > 0$ satisfies

$$0 = A^{\mathrm{T}}P + PA + R_1 - PBR_2^{-1}B^{\mathrm{T}}P, \tag{6.162}$$

is strictly positive real. Alternatively, show that if $D = 0$, $G(s)$ is positive real, and R_1 satisfies

$$R_1 = L^{\mathrm{T}}L + C^{\mathrm{T}}R_2C, \tag{6.163}$$

where L satisfies (5.151), then the dynamic controller

$$G_{\mathrm{c}}(s) \sim \left[\begin{array}{c|c} A - 2BR_2^{-1}C & BR_2^{-1} \\ \hline R_2^{-1}C & 0 \end{array}\right] \tag{6.164}$$

is strictly positive real.

Problem 6.14. Consider the nonlinear dynamical system $\mathcal{G}$ given by (6.1) and (6.2) with $J(x) \equiv 0$ and $\mathrm{rank}[G(0)] = m$, and the nonlinear controller $\mathcal{G}_{\mathrm{c}}$ given by (6.3) and (6.4) with $J_{\mathrm{c}}(x) \equiv 0$. Assume that $\mathcal{G}$ is completely reachable, zero-state observable, and exponentially passive with continuously differentiable radially unbounded, positive-definite storage function $V_{\mathrm{s}} : \mathbb{R}^n \to \mathbb{R}$. Suppose there exists a continuously differentiable positive-definite function $V : \mathbb{R}^n \to \mathbb{R}$ such that $V(x) \to \infty$ as $\|x\| \to \infty$, and

$$0 = V'(x)f(x) + L_1(x) - \tfrac{1}{4}V'(x)G(x)R_2^{-1}G^{\mathrm{T}}(x)V'^{\mathrm{T}}(x), \quad x \in \mathbb{R}^n, \tag{6.165}$$

where $L_1 : \mathbb{R}^n \to \mathbb{R}$ is given by $L_1(x) = \ell^{\mathrm{T}}(x)\ell(x) + \varepsilon V_{\mathrm{s}}(x) + h^{\mathrm{T}}(x)R_2^{-1}h(x)$, $R_2 > 0$, and ε and $\ell(\cdot)$ satisfy (5.137). Show that with

$$f_{\mathrm{c}}(x_{\mathrm{c}}) = f(x_{\mathrm{c}}) - 2G(x_{\mathrm{c}})R_2^{-1}h(x_{\mathrm{c}}), \tag{6.166}$$
$$G_{\mathrm{c}}(x_{\mathrm{c}}) = G(x_{\mathrm{c}})R_2^{-1} \tag{6.167}$$
$$h_{\mathrm{c}}(x_{\mathrm{c}}) = R_2^{-1}h(x_{\mathrm{c}}), \tag{6.168}$$

the negative feedback interconnection of $\mathcal{G}$ and $\mathcal{G}_{\mathrm{c}}$ is globally asymptotically stable.

Problem 6.15. Consider the linear matrix second-order dynamical system in energy coordinates given by

$$\begin{bmatrix} \dot{x}_1(t) \\ \dot{x}_2(t) \end{bmatrix} = \begin{bmatrix} 0 & \Omega \\ -\Omega & -2\eta\Omega \end{bmatrix} \begin{bmatrix} x_1(t) \\ x_2(t) \end{bmatrix} + \begin{bmatrix} 0 \\ b \end{bmatrix} u(t),$$
$$\begin{bmatrix} x_1(0) \\ x_2(0) \end{bmatrix} = \begin{bmatrix} x_{10} \\ x_{20} \end{bmatrix}, \quad t \geq 0, \tag{6.169}$$
$$y(t) = b^{\mathrm{T}} x_2(t), \tag{6.170}$$

where $x_1 \in \mathbb{R}^n$, $x_2 \in \mathbb{R}^n$, $u, y, \in \mathbb{R}$, $\Omega = \mathrm{diag}[\omega_1, \ldots, \omega_n]$, $\omega_i > 0$, $i = 1, \ldots, n$, $\eta = \mathrm{diag}[\eta_1, \ldots, \eta_n]$, $\eta_i > 0$, $i = 1, \ldots, n$, and $b \in \mathbb{R}^n$. Show that the nonlinear dynamic compensator

$$\dot{x}_{\mathrm{c}}(t) = \left(\begin{bmatrix} 0 & \Omega_{\mathrm{c}} \\ -\Omega_{\mathrm{c}} & -2\eta_{\mathrm{c}}\Omega_{\mathrm{c}} \end{bmatrix} + 2\alpha[0 \ \mathbf{e}^{\mathrm{T}}]x_{\mathrm{c}}S \right) x_{\mathrm{c}}(t)$$
$$+\kappa \begin{bmatrix} 0 \\ e \end{bmatrix} y^2(t), \quad x_{\mathrm{c}}(0) = x_{\mathrm{c}0}, \quad t \geq 0, \tag{6.171}$$
$$u(t) = -\kappa([0 \ e^{\mathrm{T}}]x_{\mathrm{c}}(t))y(t), \tag{6.172}$$

where $x_{\mathrm{c}} \in \mathbb{R}^{2n_{\mathrm{c}}}$, $n_{\mathrm{c}} \leq n$, $\kappa > 0$, $\alpha > 0$, $\mathbf{e}^{\mathrm{T}} = [1, 1, \ldots, 1] \in \mathbb{R}^{1 \times n_{\mathrm{c}}}$, $S = -S^{\mathrm{T}}$, $\Omega_{\mathrm{c}} = \mathrm{diag}[\omega_{\mathrm{c}1}, \ldots, \omega_{\mathrm{c}n}]$, $\omega_{\mathrm{c}i} > 0$, $i = 1, \ldots, n$, and $\eta_{\mathrm{c}} = \mathrm{diag}[\eta_{\mathrm{c}1}, \ldots, \eta_{\mathrm{c}n}]$, $\eta_{\mathrm{c}i} > 0$, $i = 1, \ldots, n$, guarantees asymptotic stability of the closed-loop system (6.169)–(6.172).

Problem 6.16. Consider the nonlinear dynamical system $\mathcal{G}$ given by (6.1) and (6.2) and the nonlinear controller $\mathcal{G}_{\mathrm{c}}$ given by

$$\dot{x}_{\mathrm{c}}(t) = (A_{\mathrm{c}} + S)x_{\mathrm{c}}(t) + B_{\mathrm{c}}\,\mathrm{diag}(y(t))y(t), \quad x_{\mathrm{c}}(0) = x_{\mathrm{c}0}, \quad t \geq 0, \tag{6.173}$$
$$u(t) = -\mathrm{diag}(y(t))B_{\mathrm{c}}^{\mathrm{T}} x_{\mathrm{c}}(t), \tag{6.174}$$

where $A_{\mathrm{c}} \in \mathbb{R}^{n_{\mathrm{c}} \times n_{\mathrm{c}}}$, $B_{\mathrm{c}} \in \mathbb{R}^{n_{\mathrm{c}} \times m}$, $S \in \mathbb{R}^{n_{\mathrm{c}} \times n_{\mathrm{c}}}$ is skew symmetric, and $\mathrm{diag}(y)$ is a diagonal matrix whose entries on the diagonal are the components of y. Show that if $\mathcal{G}$ is passive and zero-state observable and the triple $(A_{\mathrm{c}}, B_{\mathrm{c}}, B_{\mathrm{c}}^{\mathrm{T}})$ is strictly positive real and self-dual (see Problem 5.33), then the negative feedback interconnection of $\mathcal{G}$ and $\mathcal{G}_{\mathrm{c}}$ is asymptotically stable.

Problem 6.17. Consider the nonlinear controlled oscillator given in Problem 5.35. Using Corollary 6.1 design a nonlinear reduced-order dynamic compensator that asymptotically stabilizes the closed-loop system. Compare your results with the static linear controller $u = -R_2^{-1}y$, where $R_2 = 1$, for the initial condition $[x_1(0), \ x_2(0)]^{\mathrm{T}} = [1, \ \ 0.5]^{\mathrm{T}}$.

Problem 6.18. Consider the controlled nonlinear damped oscillator given in Problem 5.36. Using Corollary 6.1 and Problem 6.13 design a nonlinear full-order dynamic compensator that asymptotically stabilizes the closed-loop system. Compare the open-loop and closed-loop responses for the initial condition $[x_1(0),\ x_2(0)]^{\mathrm{T}} = [0,\ 1]^{\mathrm{T}}$ and $R_2 = 0.01$.

Problem 6.19. Let $q \in \mathbb{R}^l$, $r \in \mathbb{R}^m$, $q_{\mathrm{c}} \in \mathbb{R}^{l_{\mathrm{c}}}$, and $r_{\mathrm{c}} \in \mathbb{R}^{m_{\mathrm{c}}}$. Consider the nonlinear nonnegative dynamical systems $\mathcal{G}$ and $\mathcal{G}_{\mathrm{c}}$ (see Problem 5.9) given by (6.1) and (6.2), and (6.3) and (6.4), respectively, where $f : \mathcal{D} \to \mathbb{R}^n$ is essentially nonnegative (see Problem 3.7), $G(x) \geq\geq 0$, $h(x) \geq\geq 0$, $J(x) \geq\geq 0$, $x \in \overline{\mathbb{R}}_+^n$, $f_{\mathrm{c}} : \mathbb{R}^{n_{\mathrm{c}}} \to \mathbb{R}^{n_{\mathrm{c}}}$ is essentially nonnegative, $G_{\mathrm{c}}(u_{\mathrm{c}}, x_{\mathrm{c}}) = G_{\mathrm{c}}(x_{\mathrm{c}}) \geq\geq 0$, $h_{\mathrm{c}}(u_{\mathrm{c}}, x_{\mathrm{c}}) = h_{\mathrm{c}}(x_{\mathrm{c}}) \geq\geq 0$, $x_{\mathrm{c}} \in \overline{\mathbb{R}}_+^{n_{\mathrm{c}}}$, and $J_{\mathrm{c}}(u_{\mathrm{c}}, x_{\mathrm{c}}) \equiv 0$. Assume that $\mathcal{G}$ is dissipative with respect to the linear supply rate $s(u, y) = q^{\mathrm{T}}y + r^{\mathrm{T}}u$ and with a continuously differentiable positive-definite storage function $V_{\mathrm{s}}(\cdot)$, and assume that $\mathcal{G}_{\mathrm{c}}$ is dissipative with respect to the linear supply rate $s_{\mathrm{c}}(u_{\mathrm{c}}, y_{\mathrm{c}}) = q_{\mathrm{c}}^{\mathrm{T}}y_{\mathrm{c}} + r_{\mathrm{c}}^{\mathrm{T}}u_{\mathrm{c}}$ and with a continuously differentiable positive-definite storage function $V_{\mathrm{sc}}(\cdot)$. Show that the following statements hold:

i) If there exists a scalar $\sigma > 0$ such that $q + \sigma r_{\mathrm{c}} \leq\leq 0$ and $r + \sigma q_{\mathrm{c}} \leq\leq 0$, then the positive feedback interconnection of $\mathcal{G}$ and $\mathcal{G}_{\mathrm{c}}$ is Lyapunov stable.

ii) If $\mathcal{G}$ and $\mathcal{G}_{\mathrm{c}}$ are zero-state observable and there exists a scalar $\sigma > 0$ such that $q + \sigma r_{\mathrm{c}} << 0$ and $r + \sigma q_{\mathrm{c}} << 0$, then the positive feedback interconnection of $\mathcal{G}$ and $\mathcal{G}_{\mathrm{c}}$ is asymptotically stable.

iii) If $\mathcal{G}$ is zero-state observable, rank $G_{\mathrm{c}}(0) = m_{\mathrm{c}}$, $\mathcal{G}_{\mathrm{c}}$ is exponentially dissipative with respect to the supply rate $s_{\mathrm{c}}(u_{\mathrm{c}}, y_{\mathrm{c}}) = q_{\mathrm{c}}^{\mathrm{T}}y_{\mathrm{c}} + r_{\mathrm{c}}^{\mathrm{T}}u_{\mathrm{c}}$, and there exists a scalar $\sigma > 0$ such that $q + \sigma r_{\mathrm{c}} \leq\leq 0$ and $r + \sigma q_{\mathrm{c}} \leq\leq 0$, then the positive feedback interconnection of $\mathcal{G}$ and $\mathcal{G}_{\mathrm{c}}$ is asymptotically stable.

iv) If $\mathcal{G}$ is exponentially dissipative with respect to the supply rate $s(u, y) = q^{\mathrm{T}}y + r^{\mathrm{T}}u$, $\mathcal{G}_{\mathrm{c}}$ is exponentially dissipative with respect to the supply rate $s_{\mathrm{c}}(u_{\mathrm{c}}, y_{\mathrm{c}}) = q_{\mathrm{c}}^{\mathrm{T}}y_{\mathrm{c}} + r_{\mathrm{c}}^{\mathrm{T}}u_{\mathrm{c}}$, and there exists a scalar $\sigma > 0$ such that $q + \sigma r_{\mathrm{c}} \leq\leq 0$ and $r + \sigma q_{\mathrm{c}} \leq\leq 0$, then the positive feedback interconnection of $\mathcal{G}$ and $\mathcal{G}_{\mathrm{c}}$ is asymptotically stable.

(**Hint:** First show that the positive feedback interconnection of $\mathcal{G}$ and $\mathcal{G}_{\mathrm{c}}$ gives a nonnegative closed-loop system.)

Problem 6.20. Consider the dynamical system $\mathcal{G}$ given by (6.62) and (6.63) with $f(x) = Ax$, $G(x) = B$, $\phi(x) = Kx$, where $A \in \mathbb{R}^{n\times n}$, $B \in \mathbb{R}^{n\times 1}$, and $K \in \mathbb{R}^{1\times n}$. Suppose that $\mathcal{G}$ has gain margin (α, β) and let $L(\jmath\omega) = -K(\jmath\omega I_n - A)^{-1}B$ denote the *loop gain* of (6.62) and (6.63). Using a Nyquist sketch show that the maximum amount by which the loop gain can be increased before instability is β, that is, the upward gain margin is β. Show the maximum amount by which the loop gain can be decreased before instability is $(1 - \alpha)100$, that is, the gain reduction tolerance is $(1 - \alpha)100$.

Finally, show that the maximum angle by which the loop gain can lag before instability is $\phi = \cos^{-1}\left(\frac{\alpha\beta+1}{\alpha+\beta}\right)$.

Problem 6.21. Consider the linear dynamical system

$$G(s) \overset{\min}{\sim} \left[\begin{array}{c|c} A & B \\ \hline C & D \end{array}\right].$$

Show that if $G(s)$ has a disk margin (α, β), then $[I + \beta G(s)][I + \alpha G(s)]^{-1}$ is positive real.

Problem 6.22. Consider the linear dynamical system

$$G(s) \overset{\min}{\sim} \left[\begin{array}{c|c} A & B \\ \hline C & D \end{array}\right]$$

with input $u(\cdot) \in \mathcal{U}$ and output $y(\cdot) \in \mathcal{Y}$. Show that $G(s)$ is dissipative with respect to the supply rate $r(u, y) = u^{\mathrm{T}}y + \alpha y^{\mathrm{T}}y$, $0 < \alpha < 1$, if and only if $G(s)$ has a disk margin (α, ∞).

Problem 6.23. Consider the linear dynamical system

$$G(s) \overset{\min}{\sim} \left[\begin{array}{c|c} A & B \\ \hline C & D \end{array}\right]$$

with input $u(\cdot) \in \mathcal{U}$ and output $y(\cdot) \in \mathcal{Y}$. Show that $G(s)$ is dissipative with respect to the supply rate $r(u, y) = u^{\mathrm{T}}y + \frac{1}{\beta}u^{\mathrm{T}}u$, $\beta > 1$, if and only if $G(s)$ has a disk margin $(0, \beta)$.

Problem 6.24. Consider the nonlinear controlled system (6.73). Show that if Condition (6.75) is satisfied, then there exists a *feedback* control law $\phi : \mathcal{D} \to U$, in general discontinuous, such that $V'(x)F(x, \phi(x)) < 0$, $x \in \mathcal{D}$, $x \neq 0$.

Problem 6.25. A smooth system of the form (6.76) with $f(0) = 0$ is a *Jurdjevic-Quinn* (J-Q) *type system* if there exists a continuously differentiable positive-definite radially unbounded function $V : \mathbb{R}^n \to \mathbb{R}$ such that:

i) $V'(x)f(x) \leq 0$, $x \in \mathbb{R}^n$.

ii) $\mathcal{W} \triangleq \{x \in \mathbb{R}^n : L_f^{k+1}V(x) = L_f^k L_{G_i} V(x) = 0,\ k = 0, 1, \ldots, i = 1, \ldots, m\}$
$= \{0\}$.

Show that $u = -[V'(x)G(x)]^{\mathrm{T}}$ is a globally stabilizing controller for a J-Q type system.

Problem 6.26. Show that for scalar stabilizable systems $V(x) = \frac{1}{2}x^2$ is always a control Lyapunov function. Using this fact, obtain a stabilizing controller for the dynamical system considered in Example 6.6 using the control Lyapunov function approach and compare this controller to the optimal Hamilton-Jacobi-Bellman controller.

Problem 6.27. Consider the nonlinear dynamical system (6.76). Assume there exists a continuously differentiable function $V : \mathcal{D} \to \overline{\mathbb{R}}_+$ such that $V(\cdot)$ is positive definite and

$$V'(x)f(x) \le -c(V(x))^\alpha, \quad x \in \mathcal{R}, \tag{6.175}$$

where $c > 0$, $\alpha \in (0,1)$, and $\mathcal{R} \triangleq \{x \in \mathbb{R}^n, x \neq 0 : V'(x)G(x) = 0\}$. Show that the zero solution $x(t) \equiv 0$ to (6.76) with the feedback controller $u = \phi(x)$, $x \in \mathbb{R}^n$, given by

$$\phi(x) = \begin{cases} -\left(c_0 + \frac{(\alpha(x)-w(V(x)))+\sqrt{(\alpha(x)-w(V(x)))^2+(\beta^{\mathrm{T}}(x)\beta(x))^2}}{\beta^{\mathrm{T}}(x)\beta(x)}\right)\beta(x), & \beta(x) \neq 0, \\ 0, & \beta(x) = 0, \end{cases} \tag{6.176}$$

where $c_0 > 0$, $\alpha(x) \triangleq V'(x)f(x)$, $x \in \mathbb{R}^n$, $\beta(x) \triangleq G^{\mathrm{T}}(x)V'^{\mathrm{T}}(x)$, $x \in \mathbb{R}^n$, and $w(V(x)) \triangleq -c(V(x))^\alpha$, $x \in \mathbb{R}^n$, is finite-time stable with the settling-time function $T(x_0) \le \frac{1}{c(1-\alpha)}(V(x_0))^{1-\alpha}$, $x_0 \in \mathbb{R}^n$. Furthermore, show that $V(\cdot)$ is a control Lyapunov function.

Problem 6.28. Consider the system (6.83) and (6.84) with $F(x,u,t) = Ax + Bu$ and $L(x,u,t) = x^{\mathrm{T}}R_1x + u^{\mathrm{T}}R_2u$, where $A \in \mathbb{R}^{n\times n}$, $B \in \mathbb{R}^{n\times m}$, $R_1 \in \mathbb{R}^{n\times n}$, and $R_2 \in \mathbb{R}^{m\times m}$, such that $R_1 \ge 0$, $R_2 > 0$. Show that the optimal control law characterized by the Hamilton-Jacobi-Bellman equation (6.93) is given by

$$u(t) = -R_2^{-1}B^{\mathrm{T}}P(t)x(t), \tag{6.177}$$

where

$$-\dot{P}(t) = A^{\mathrm{T}}P(t) + P(t)A + R_1 - P(t)BR_2^{-1}B^{\mathrm{T}}P(t), \qquad P(t_{\mathrm{f}}) = 0. \tag{6.178}$$

Problem 6.29. Consider the nonlinear dynamical system

$$\dot{x}(t) = f(x(t)) + G(x(t))u(t), \quad x(0) = x_0, \quad t \ge 0, \tag{6.179}$$

with performance functional

$$J(x_0, u(\cdot)) = \int_0^\infty [L_1(x(t)) + u^{\mathrm{T}}(t)R_2(x(t))u(t)]\mathrm{d}t, \tag{6.180}$$

where $x \in \mathbb{R}^n$, $u \in \mathbb{R}^m$, $f : \mathbb{R}^n \to \mathbb{R}^n$ satisfies $f(0) = 0$, $G : \mathbb{R}^n \to \mathbb{R}^{n\times m}$, $L_1 : \mathbb{R}^n \to \mathbb{R}$ is such that $L_1(x) > 0$, $x \in \mathbb{R}^n$, $x \neq 0$, and $R_2 : \mathbb{R}^n \to \mathbb{R}^{m\times m}$

is such that $R_2(x) > 0$, $x \in \mathbb{R}^n$. Show that the optimal control $u = \phi(x)$ characterized by the Hamilton-Jacobi-Bellman equation (6.93) is given by

$$\phi(x) = -\tfrac{1}{2}R_2^{-1}(x)G^{\mathrm{T}}(x)V'^{\mathrm{T}}(x), \tag{6.181}$$

where $V(0) = 0$ and

$$0 = L_1(x) + V'(x)f(x) - \tfrac{1}{4}V'(x)G(x)R_2^{-1}(x)G^{\mathrm{T}}(x)V'^{\mathrm{T}}(x), \quad x \in \mathbb{R}^n. \tag{6.182}$$

Also show that (6.181) is a stabilizing feedback control law.

Problem 6.30. Consider the nonlinear dynamical system (6.76). Show that the feedback control law (6.81) is an optimal stabilizing control law minimizing the performance functional

$$J = \tfrac{1}{2}\int_0^\infty [2\alpha(x(t)) - q(x(t))\beta^{\mathrm{T}}(x(t))\beta(x(t)) + q^{-1}(x(t))u^{\mathrm{T}}(t)u(t)]\mathrm{d}t, \tag{6.183}$$

where

$$q(x) \triangleq \begin{cases} c_0 + \frac{\alpha(x)+\sqrt{\alpha^2(x)+(\beta^{\mathrm{T}}(x)\beta(x))^2}}{\beta^{\mathrm{T}}(x)\beta(x)}, & \beta(x) \neq 0, \\ c_0, & \beta(x) = 0, \end{cases} \tag{6.184}$$

and $c_0 > 0$.

Problem 6.31. Consider the nonlinear dynamical system given by (6.76). Show that whenever the control Lyapunov function V satisfying (6.77) has the same level sets as the value function V^* satisfying the Hamilton-Jacobi-Bellman equation (6.182), the feedback control law (6.81) reduces to the optimal control law given by (6.181).

Problem 6.32. Consider the nonlinear dynamical system

$$\dot{x}(t) = F(x(t), u(t), t), \quad x(t_0) = x_0, \quad x(t_\mathrm{f}) = x_\mathrm{f}, \quad t \geq t_0, \tag{6.185}$$

where $x(t) \in \mathcal{D} \subseteq \mathbb{R}^n$, $t \in [t_0, t_\mathrm{f}]$, $x(t_0) = x_0$ is given, $x(t_\mathrm{f}) = x_\mathrm{f}$ is fixed, and $u(t) \in U \subseteq \mathbb{R}^m$, $t \in [t_0, t_\mathrm{f}]$, with performance functional

$$J(x_0, u(\cdot), t_0) = \int_{t_0}^{t_\mathrm{f}} L(x(t), u(t), t)\mathrm{d}t. \tag{6.186}$$

Let $J^*(x,t)$ and $u^*(t)$, $t \in [t_0, t_\mathrm{f}]$, denote the minimal cost and optimal control, respectively, for the optimal control problem given by (6.185) and (6.186) and assume that $J^*(\cdot,\cdot)$ is two-times continuously differentiable. Show that

$$\dot{p}(x(t),t) = -\left[\frac{\partial}{\partial x}H(x(t), u^*(t), p(x(t),t), t)\right]^{\mathrm{T}},$$

$$p(x(t_\mathrm{f}), t_\mathrm{f}) = \left[\frac{\partial}{\partial x} J^*(x(t_\mathrm{f}), t_\mathrm{f})\right]^\mathrm{T}, \quad (6.187)$$

where $p(x(t), t) = \left(\frac{\partial}{\partial x} J^*(x(t), t)\right)^\mathrm{T}$ and $H(x, u, p(x, t), t) = L(x, u, t) + p^\mathrm{T}(x, t)F(x, u, t)$.

Problem 6.33. Consider an n-degree-of-freedom dynamical system with *action integral*

$$J = \int_{t_0}^{t_\mathrm{f}} \mathcal{L}(q(t), \dot{q}(t))\mathrm{d}t, \quad (6.188)$$

where $q \in \mathbb{R}^n$ denotes generalized system positions, $\dot{q} \in \mathbb{R}^n$ denotes generalized system velocities, $\mathcal{L} : \mathbb{R}^n \times \mathbb{R}^n \to \mathbb{R}$ denotes the system Lagrangian given by $\mathcal{L}(q, \dot{q}) = T(q, \dot{q}) - V(q)$, where $T : \mathbb{R}^n \times \mathbb{R}^n \to \mathbb{R}$ is the system kinetic energy and $V : \mathbb{R}^n \to \mathbb{R}$ is the system potential energy. *Hamilton's principle of least action* states that among the set of all smooth kinematically possible paths that a dynamical system may move between two fixed end points over a specified time interval, the only dynamically possible system paths are those that render J stationary to all variations in the shape of the paths. Use the Hamilton-Jacobi-Bellman equation to show that Hamilton's principle for an n-degree-of-freedom dynamical system implies

$$\frac{\mathrm{d}}{\mathrm{d}t}\left[\frac{\partial \mathcal{L}}{\partial \dot{q}}(q, \dot{q})\right]^\mathrm{T} - \left[\frac{\partial \mathcal{L}}{\partial q}(q, \dot{q})\right]^\mathrm{T} = 0. \quad (6.189)$$

Problem 6.34. Consider the nonlinear dynamical system

$$\dot{x}_1(t) = x_1^2(t)x_2(t), \quad x_1(0) = x_{10}, \quad t \geq 0, \quad (6.190)$$

$$\dot{x}_2(t) = u(t), \quad x_2(0) = x_{20}, \quad (6.191)$$

$$y(t) = x_2(t). \quad (6.192)$$

Show that (6.190)–(6.192) is weakly minimum phase and $u = x_1^3 + v$ renders (6.190)–(6.192) passive. In addition, show that the output feedback controller $v = -y$ globally stabilizes (6.190)–(6.192).

Problem 6.35. Consider the nonlinear dynamical system

$$\dot{x}_1(t) = -x_1^3(t) + x_2(t), \quad x_1(0) = x_{10}, \quad t \geq 0, \quad (6.193)$$

$$\dot{x}_2(t) = x_1(t) + u(t), \quad x_2(0) = x_{20}, \quad (6.194)$$

$$y(t) = x_2(t). \quad (6.195)$$

Show that (6.193)–(6.195) has relative degree 1 and is minimum phase. Can this system be rendered passive via a static output feedback controller?

Problem 6.36. Consider the nonlinear dynamical system $\mathcal{G}$ given by (6.98) and (6.99) with $u, y \in \mathbb{R}$. Show that if $\mathcal{G}$ has relative degree r at a point x_0, then the vectors $\{\frac{\partial}{\partial x}h(x),\ \frac{\partial}{\partial x}[L_f h(x)], \ldots, \frac{\partial}{\partial x}[L_f^{r-1}h(x)]\}$ at $x = x_0$ are linearly independent.

Problem 6.37. Consider the nonlinear dynamical system $\mathcal{G}$ given by (6.98) and (6.99). Show that if $\mathcal{G}$ is input strictly passive, then $\mathcal{G}$ has relative degree zero.

Problem 6.38. Show that every minimum phase system of the form (6.98) and (6.99) with relative degree $\{1, 1, \ldots, 1\}$ and complete vector field $G(L_G h)^{-1}$ is feedback equivalent to a J-Q type system (see Problem 6.25).

Problem 6.39. Let

$$G(s) \sim \left[\begin{array}{c|c} A & B \\ \hline C & 0 \end{array}\right]$$

be an $m \times m$ rational transfer function and suppose $q \in \mathbb{C}$ is not a pole of $G(s)$. q is a *transmission zero* of $G(s)$ if rank $G(q) <$ nrank $G(s)$, where nrank $G(s) \triangleq \max_{s \in \mathbb{C}}$ rank $G(s)$. Show that if $\dot{z} = A_0 z$ denotes the zero dynamics of $G(s)$ and $G(s)$ is minimum phase and has relative degree $\{1, 1, \ldots, 1\}$, then the eigenvalues of A_0 are the transmission zeros of $G(s)$.

Problem 6.40. Consider the nonlinear dynamical system $\mathcal{G}$ given by (6.98) and (6.99). Show that if $\mathcal{G}$ is feedback linearizable, then the solutions to the partial differential equations

$$0 = L_{G_j} L_f^k h_i(x), \quad 0 \le k \le r_i - 2, \quad 1 \le j \le m,$$

exist, where G_j, $j = 1, \ldots, m$, are the smooth column vector fields of G, h_i, $i = 1, \ldots, m$, are the smooth components of h, $r_1 + r_2 + \cdots + r_m = n$, and r_i is the smallest integer such that at least one of the inputs u_i appears in $\frac{\mathrm{d}^{r_i}}{\mathrm{d}t^{r_i}} h_i(x)$. Also show that in the case where $f(x) = Ax$, $A \in \mathbb{R}^{n \times n}$, $G(x) = B$, $B \in \mathbb{R}^{n \times 1}$, $h(x) = Cx$, $C \in \mathbb{R}^{1 \times n}$, the integers $r_1, \ldots, r_m$, reduce to a single integer corresponding to the relative degree of the transfer function $G(s) = C(sI - A)^{-1}B$.

Problem 6.41. Consider the dynamical system (6.98) and (6.99) with $f(x) = Ax$, $G(x) = B$, and $h(x) = Cx$, where $A \in \mathbb{R}^{n \times n}$, $B \in \mathbb{R}^{n \times m}$, and $C \in \mathbb{R}^{l \times n}$. Show that in this case Condition *i*) of Theorem 6.11 reduces to

$$\text{rank}[B \ AB \ \cdots \ A^{n-1}B] = n.$$

Problem 6.42. Consider the nonlinear dynamical system (6.98). Show that (6.98) is feedback linearizable at $x_0 \in \mathbb{R}^n$ only if the linearized system is controllable at $x_0 \in \mathbb{R}^n$.

Problem 6.43. Let f, f_1, f_2, g, g_1, $g_2 : \mathbb{R}^n \to \mathbb{R}$ be continuously differentiable vector fields, let $h : \mathbb{R}^n \to \mathbb{R}$, and let α_1 and α_2 be real constants. Show that:

i) $[\alpha_1 f_1(x) + \alpha_2 f_2(x), g_1(x)] = \alpha_1[f_1(x), g_1(x)] + \alpha_2[f_2(x), g_1(x)]$.

ii) $[f_1(x), \alpha_1 g_1(x) + \alpha_2 g_2(x)] = \alpha_1[f_1(x), g_1(x)] + \alpha_2[f_1(x), g_2(x)]$.

iii) $[f(x), g(x)] = -[g(x), f(x)]$.

iv) $L_{[f,g]}h(x) = L_f L_g h(x) - L_g L_f h(x)$.

Problem 6.44. Let $\mathcal{T} : \mathcal{D} \to \mathbb{R}^n$. Show that if $\mathcal{T}'(x)$ is invertible at $x = x_0 \in \mathcal{D}$, then there exists a neighborhood $\mathcal{N}$ of x_0 such that $\mathcal{T}(x)$ is a diffeomorphism for all $x \in \mathcal{N}$.

Problem 6.45. $\mathcal{T} : \mathbb{R}^n \to \mathbb{R}^n$ is a *global diffeomorphism* if it is a diffeomorphism on $\mathbb{R}^n$ and $\mathcal{T}(\mathbb{R}^n) = \mathbb{R}^n$. Show that $\mathcal{T}(\cdot)$ is a global diffeomorphism if and only if $\mathcal{T}'(x)$ is invertible for all $x \in \mathbb{R}^n$ and $\lim_{\|x\|\to\infty} \|\mathcal{T}(x)\| = 0$.

Problem 6.46. Consider the controlled Van der Pol oscillator (6.105) and (6.106) given in Example 6.9 with (6.107) replaced by $y(t) = x_1(t)$. Calculate the relative degree of this system. Is this system minimum phase? Is this system feedback linearizable?

Problem 6.47. Consider the nonlinear dynamical system

$$\dot{x}_1(t) = x_1^2(t) + x_2(t), \quad x_1(0) = x_{10}, \quad t \geq 0, \tag{6.196}$$
$$\dot{x}_2(t) = u(t), \quad x_2(0) = x_{20}. \tag{6.197}$$

Using Theorem 6.11 show that (6.196) and (6.197) is feedback linearizable. Furthermore, using the diffeomorphism $\mathcal{T}(x) = [x_1 \ x_1^2 + x_2]^{\mathrm{T}}$, where $x = [x_1 \ x_2]^{\mathrm{T}}$, construct a stabilizing nonlinear controller that cancels out all nonlinearities in the system. Alternatively, construct a globally stabilizing controller based on the control Lyapunov function $V(x) = \mathcal{T}^{\mathrm{T}}(x)P\mathcal{T}(x)$, where $P > 0$ satisfies (6.101). Compare the state response and the control effort versus time of both designs.

6.9 Notes and References

For a treatment of feedback interconnections of dissipative systems, see Popov [364], Zames [476, 477], Desoer and Vidyasagar [104], Willems [456, 457], and Hill and Moylan [189]. Stability of feedback interconnections involving dissipative and exponentially dissipative systems was first introduced by Chellaboina and Haddad [88]. The treatment of stability for nonlinear dissipative feedback systems is adopted from Hill and Moylan [189] and

Chellaboina and Haddad [88]. Energy-based controllers for port-controlled Hamiltonian systems were first developed by Maschke, Ortega, and van der Schaft [301] and Maschke, Ortega, van der Schaft, and Escobar [302]. The treatment here is adopted from Ortega, van der Schaft, Maschke, and Escobar [343]. Gain and sector margins for nonlinear systems are motivated by work on Nyquist stability theory and absolute stability theory and can be found in Luré [291], Popov [364], Narendra and Taylor [331], and Safonov [377]. An operator approach is given by Zames [476, 477] while a state space approach can be found in Hill and Moylan [188,189]. The concept of disk margins can be found in the monograph by Sepulchre, Janković, and Kokotović [395].

The concept of control Lyapunov functions is due to Artstein [13], while the constructive feedback control law based on the control Lyapunov function given in Section 6.5 is due to Sontag [406] and is known as Sontag's universal formula. The principle of optimality is due to Bellman [36]. The Hamilton-Jacobi-Bellman equation was developed by Bellman [36] and can be viewed as an extension of Jacobi's work on conjugate points in fields of extremals and Hamilton's work on least action in mechanical systems; both of which were developed in the nineteenth century. For a modern textbook treatment see Kirk [239] and Bryson and Ho [72].

For a thorough textbook treatment of feedback linearization, zero dynamics, and minimum-phase systems the reader is referred to Isidori [212] and Nijmeijer and van der Schaft [336]. One of the first contributions to solving nonlinear control problems using differential geometric methods is due to Brockett [68]. The feedback linearization problem was first posed by Brockett [69] and solved by Su [419] in the single-input case and by Hunt, Su, and Meyer [210] in the multi-input case. The definitions in Section 6.7 are adopted from Isidori [212]. For an excellent treatment on zero dynamics and minimum-phase systems see Byrnes and Isidori [74–76] and Saberi, Kokotović, and Sussmann [376]. The concepts of feedback passivity equivalence and global stabilization of minimum phase systems are due to Kokotović and Sussmann [241] for linear systems and Byrnes, Isidori, and Willems [77] for nonlinear systems.

Chapter Seven

Input-Output Stability and Dissipativity

7.1 Introduction

As mentioned in Chapter 1, the main objective of this book is to present the necessary mathematical tools for stability and control design of nonlinear dynamical systems, with an emphasis on Lyapunov-based methods. For completeness of exposition, in this chapter we digress from our main objective to present an alternative approach to the mathematical modeling of nonlinear dynamical systems based on input-output notions. In contrast to the input, state, and output setting for nonlinear dynamical systems presented in the preceding chapters, an input-output representation of a dynamical system relates the output of the system directly to the input via an input-output mapping with no knowledge of the internal state of the system. There are numerous situations where an input-output system description is appropriate. These include, for example, intelligent machines such as computer hardwarc and signal processors, as well as computer software algorithm execution.

The principal advantage of the input-output modeling approach to dynamical systems is that it addresses infinite-dimensional systems almost as easily as finite-dimensional systems. This is also the case for continuous-time and discrete-time systems. In addition, the input-output approach leads to more powerful and general results. However, in most physical systems the output of the system also depends on the system's initial conditions. In addition, an input-output system description cannot deal with physical system interconnections. Hence, dynamical systems modeling predicated on state space models, wherein an internal state model is used to describe the system dynamics using physical laws and system interconnections, is of fundamental importance in the description of physical dynamical systems.

Input-output system descriptions can be traced back to the work of Heaviside on impedance circuit descriptions and the cybernetic force-response systems view of Wiener. This input-output systems approach led to the concept of input-output stability which was pioneered by Sandberg

[387, 388, 390] and Zames [476, 477]. In this chapter, we introduce input-output systems descriptions and define the concept of input-output $\mathcal{L}_p$ stability. In addition, we introduce the concepts of input-output finite-gain, dissipativity, passivity, and nonexpansivity. Furthermore, we develop connections between input-output stability and Lyapunov stability theory. Finally, we develop explicit formulas for induced convolution operator norms for linear input-output dynamical systems.

7.2 Input-Output Stability

In this section, we introduce the definition of input-output stability for general operator dynamical systems. Let $\mathcal{U}$ and $\mathcal{Y}$ define an input and an output spaces, respectively, consisting of continuous bounded U-valued and Y-valued functions on the semi-infinite interval $[0, \infty)$, where $U \subseteq \mathbb{R}^m$ and $Y \subseteq \mathbb{R}^l$. The set U contains the set of input values, that is, for every $u(\cdot) \in \mathcal{U}$ and $t \in [0, \infty)$, $u(t) \in U$. The set Y contains the set of output values, that is, for every $y(\cdot) \in \mathcal{Y}$ and $t \in [0, \infty)$, $y(t) \in Y$. The spaces $\mathcal{U}$ and $\mathcal{Y}$ are assumed to be closed under the shift operator, that is, if $u(\cdot) \in \mathcal{U}$ (respectively, $y(\cdot) \in \mathcal{Y}$), then the function defined by $u_{\mathrm{s}T} \triangleq u(t+T)$ (respectively, $y_{\mathrm{s}T} \triangleq y(t+T)$) is contained in $\mathcal{U}$ (respectively, $\mathcal{Y}$) for all $T \geq 0$.

In this chapter, we consider operator dynamical systems $\mathcal{G} : \mathcal{U} \to \mathcal{Y}$. For example, $\mathcal{G}$ may denote a linear, time-invariant, finite-dimensional dynamical system given by the transfer function

$$G(s) \sim \left[\begin{array}{c|c} A & B \\ \hline C & D \end{array}\right]$$

or, equivalently,

$$y(t) = \mathcal{G}[u](t) \triangleq \int_0^t H(t-\tau)u(\tau)\mathrm{d}\tau + Du(t), \qquad t \geq 0, \tag{7.1}$$

where $H(t) \triangleq Ce^{At}B$ is the *impulse response matrix function.* For notational convenience, we denote the functional dependence of $y \in \mathcal{Y}$ on $u \in \mathcal{U}$ given by (7.1) as $y = \mathcal{G}[u]$. Similarly, consider a nonlinear dynamical system described by

$$\dot{x}(t) = f(x(t)) + G(x(t))u(t), \qquad x(t_0) = 0, \qquad t \geq t_0, \tag{7.2}$$

$$y(t) = h(x(t)) + J(x(t))u(t), \tag{7.3}$$

where $x(t) \in \mathcal{D} \subseteq \mathbb{R}^n$, $\mathcal{D}$ is an open set with $0 \in \mathcal{D}$, $u(t) \in U \subseteq \mathbb{R}^m$, $y(t) \in Y \subseteq \mathbb{R}^l$, $f : \mathcal{D} \to \mathbb{R}^n$, $G : \mathcal{D} \to \mathbb{R}^{n\times m}$, $h : \mathcal{D} \to Y$, and $J : \mathcal{D} \to \mathbb{R}^{l\times m}$. The mapping from $u(\cdot)$ to $y(\cdot)$ is also an operator dynamical system. However, in this case, it is not possible in general to provide an explicit expression for the operator $\mathcal{G}[u]$.

In this book, we restrict our attention to *causal* operator dynamical systems. In order to provide the definition for causality we need the following definition.

Definition 7.1. Let $u(\cdot) \in \mathcal{U}$. Then for all $T \in [0, \infty)$, the function $u_T(\cdot)$ defined by

$$u_T(t) \triangleq \begin{cases} u(t), & 0 \leq t \leq T, \\ 0, & T < t, \end{cases}$$

is called the *truncation* of $u(\cdot)$ on the interval $[0, T]$.

Definition 7.2. An operator dynamical system $\mathcal{G} : \mathcal{U} \to \mathcal{Y}$ is said to be *causal* if for every $T \in [0, \infty)$ and $u(\cdot) \in \mathcal{U}$,

$$y_T = (\mathcal{G}[u])_T = (\mathcal{G}[u_T])_T.$$

Note that if $u(\cdot) \in \mathcal{U}$, then $u_T(\cdot) \in \mathcal{U}$, and hence, $\mathcal{G}[u_T]$ is well defined. Furthermore, for operator dynamical systems described by (7.1) or (7.2) and (7.3) it is easy to show that the $\mathcal{G}$ is a causal system. The following proposition is now immediate.

Proposition 7.1. Let $\mathcal{G} : \mathcal{U} \to \mathcal{Y}$. $\mathcal{G}$ is causal if and only if for every pair $u, v \in \mathcal{U}$ and for every $T > 0$ such that $u_T = v_T$, $(\mathcal{G}[u])_T = (\mathcal{G}[v])_T$.

Proof. Suppose $\mathcal{G}$ is causal and let $u, v \in \mathcal{U}$ be such that $u_T = v_T$ for some $T > 0$. Then, it follows that

$$(\mathcal{G}[u])_T = (\mathcal{G}[u_T])_T = (\mathcal{G}[v_T])_T = (\mathcal{G}[v])_T,$$

where the first and last equalities follow from the causality of $\mathcal{G}$. Conversely, suppose for every pair $u, v \in \mathcal{U}$ and for every $T > 0$ such that $u_T = v_T$, $\mathcal{G}$ satisfies $(\mathcal{G}[u])_T = (\mathcal{G}[v])_T$. Let $u \in \mathcal{U}$ and $v = u_T$ for some $T > 0$. Noting that $u_T = v_T$ and $(\mathcal{G}[u])_T = (\mathcal{G}[v])_T = (\mathcal{G}[u_T])_T$, the causality of $\mathcal{G}$ follows. □

Next, we define the notion of input-output stability. First, however, we restrict the input and output to belong to normed linear spaces. Specifically, let $u(\cdot) \in \mathcal{U}$ be measurable[1] and define the $\mathcal{L}_p^m$-*norm* as

$$|||u|||_p \triangleq \left[\int_0^\infty \|u(t)\|^p \mathrm{d}t \right]^{1/p}, \tag{7.4}$$

where $p \in [1, \infty)$ and $\|\cdot\|$ denotes the Euclidean vector norm defined on $\mathbb{R}^m$.

[1]A function $u : [0, \infty) \to \mathbb{R}^m$ is *measurable* if it is the pointwise limit, except for a set of Lebesgue measure zero, of a sequence of piecewise constant functions on $[0, \infty)$.

Furthermore, define the $\mathcal{L}_\infty^m$-*norm* as

$$|||u|||_\infty \triangleq \operatorname*{ess\ sup}_{t\geq 0} \|u(t)\|. \tag{7.5}$$

Now, for every $p \in [1,\infty]$, define the *Lebesgue normed space*[2] $\mathcal{L}_p^m$ as

$$\mathcal{L}_p^m \triangleq \{u(\cdot) \in \mathcal{U} : |||u|||_p < \infty\}. \tag{7.6}$$

The notion of input-output stability involves operator dynamical systems that map $\mathcal{L}_p^m$ to $\mathcal{L}_p^l$. Note that, in general, for an arbitrary operator dynamical system $\mathcal{G} : \mathcal{U} \to \mathcal{Y}$, $y = \mathcal{G}[u]$ need not belong to $\mathcal{L}_p^l$. In order to define input-output stability, we refer to an operator dynamical system $\mathcal{G} : \mathcal{U} \to \mathcal{Y}$ as *well behaved* if for every $u(\cdot) \in \mathcal{L}_p^m$, $\mathcal{G}[u]$ is measurable. Furthermore, define the *extended* $\mathcal{L}_p^m$ *space* by

$$\mathcal{L}_{\rm pe}^m \triangleq \{u(\cdot) \in \mathcal{U} : u_T \in \mathcal{L}_p^m \text{ for every } T > 0\}.$$

Note that the set $\mathcal{L}_{\rm pe}^m$ consists of all measurable functions whose truncations belong to $\mathcal{L}_p^m$. Hence, an operator dynamical system is well behaved if and only if $\mathcal{G} : \mathcal{L}_{\rm pe}^m \to \mathcal{L}_{\rm pe}^l$.

Definition 7.3. Let $\mathcal{G} : \mathcal{L}_{\rm pe}^m \to \mathcal{L}_{\rm pe}^l$. $\mathcal{G}$ is $\mathcal{L}_p$*-stable* if $\mathcal{G}[u] \in \mathcal{L}_p^l$ for every $u \in \mathcal{L}_p^m$. $\mathcal{G}$ is $\mathcal{L}_p$*-stable with finite gain* γ if there exist $\gamma, \beta > 0$ such that

$$|||\mathcal{G}[u]|||_p \leq \gamma |||u|||_p + \beta, \qquad u \in \mathcal{L}_p^m. \tag{7.7}$$

$\mathcal{G}$ is $\mathcal{L}_p$*-stable with finite gain and zero bias* if there exists $\gamma > 0$ such that

$$|||\mathcal{G}[u]|||_p \leq \gamma |||u|||_p, \qquad u \in \mathcal{L}_p^m. \tag{7.8}$$

Note that if $\mathcal{G}$ is $\mathcal{L}_p$-stable with finite gain and zero bias, then $\mathcal{G}$ is $\mathcal{L}_p$-stable with finite gain, and if $\mathcal{G}$ is $\mathcal{L}_p$-stable with finite gain, then $\mathcal{G}$ is $\mathcal{L}_p$-stable. Furthermore, note that if $\mathcal{G}$ is causal, then (7.7) is equivalent to

$$|||(\mathcal{G}[u])_T|||_p \leq \gamma |||u_T|||_p + \beta, \quad T > 0, \quad u \in \mathcal{L}_p^m, \tag{7.9}$$

and (7.8) is equivalent to

$$|||(\mathcal{G}[u])_T|||_p \leq \gamma |||u_T|||_p, \quad T > 0, \quad u \in \mathcal{L}_p^m. \tag{7.10}$$

Example 7.1. Consider the scalar linear dynamical system given by

$$\dot{x}(t) = x(t) + u(t), \quad x(0) = 0, \quad t \geq 0, \tag{7.11}$$

$$y(t) = x(t), \tag{7.12}$$

[2]For $p \in [1,\infty)$, the Lebesgue normed space $\mathcal{L}_p^m$ consists of all real-valued measurable functions $u : [0,\infty) \to \mathbb{R}^m$ for which $\|u\|^p$ is Lebesgue integrable with norm defined by (7.4). For $p = \infty$, the Lebesgue normed space $\mathcal{L}_\infty^m$ consists of all Lebesgue measurable functions on $[0,\infty)$ which are bounded, except possibly on a set of measure zero.

where $x(t)$, $u(t)$, and $y(t) \in \mathbb{R}$. Now, it follows that

$$y(t) = x(t) = \int_0^t e^{t-\tau} u(\tau) \mathrm{d}\tau,$$

which implies that for all $u(\cdot) \in \mathcal{L}_{2\mathrm{e}}$, $y(\cdot) \in \mathcal{L}_{2\mathrm{e}}$. Next, let $u(t) = e^{-t}$ so that $u(\cdot) \in \mathcal{L}_2$. In this case,

$$y(t) = \int_0^t e^{t-\tau} e^{-\tau} \mathrm{d}\tau = \sinh t,$$

which implies that $y(\cdot) \notin \mathcal{L}_2$. Hence, $\mathcal{G}$ given by (7.11) and (7.12) is not $\mathcal{L}_2$-stable. Note that the zero solution to (7.11) is unstable. △

Example 7.2. In this example, let $\mathcal{G}_i : \mathcal{L}_{\infty\mathrm{e}} \to \mathcal{L}_{\infty\mathrm{e}}$, $i = 1, 2, 3$, be given by

$$\begin{aligned} \mathcal{G}_1[u](t) &= u^2(t), \\ \mathcal{G}_2[u](t) &= u(t) + 1, \\ \mathcal{G}_3[u](t) &= \log_e(1 + u^2(t)). \end{aligned}$$

It is clear that $\mathcal{G}_1$ is $\mathcal{L}_\infty$-stable. However, note that there does not exist constants γ and β (with $p = \infty$) such that (7.7) holds which implies that $\mathcal{G}_1$ is not $\mathcal{L}_\infty$-stable with finite gain. Next, since $|||\mathcal{G}_2[u]|||_\infty \le |||u|||_\infty + 1$ it follows that $\mathcal{G}_2$ is $\mathcal{L}_\infty$-stable with finite gain but not $\mathcal{L}_\infty$-stable with finite gain and zero bias. Finally, since $|||\mathcal{G}_3[u]|||_\infty \le \sqrt{2}|||u|||_\infty$ it follows that $\mathcal{G}_3$ is $\mathcal{L}_\infty$-stable with finite gain and zero bias. △

Example 7.3. Consider the scalar operator dynamical system $\mathcal{G}$ given by the mapping

$$y(t) = \mathcal{G}[u](t) = \int_0^t e^{\alpha(t-\tau)} u(\tau) \mathrm{d}\tau, \quad t \ge 0,$$

where $\alpha \in \mathbb{R}$. Let $u(\cdot) \in \mathcal{L}_{\infty\mathrm{e}}$, let $T > 0$, and note that for all $t \le T$,

$$\begin{aligned} |y(t)| &\le \int_0^t e^{\alpha(t-\tau)} |u(\tau)| \mathrm{d}\tau \\ &\le |||u_T|||_\infty \int_0^t e^{\alpha(t-\tau)} \mathrm{d}\tau \\ &= \frac{e^{\alpha t} - 1}{\alpha} |||u_T|||_\infty \\ &\le \frac{e^{|\alpha| T} - 1}{|\alpha|} |||u_T|||_\infty, \end{aligned}$$

which implies that $|||y_T|||_\infty \le \frac{e^{|\alpha|T}-1}{|\alpha|} |||u_T|||_\infty$, establishing that $\mathcal{G} : \mathcal{L}_{\infty\mathrm{e}} \to \mathcal{L}_{\infty\mathrm{e}}$.

Now, let $\alpha < 0$ and let $u(\cdot) \in \mathcal{L}_\infty$. In this case, it follows that for all $t > 0$,

$$\begin{aligned} |y(t)| &\le \|u\|_\infty \int_0^t e^{\alpha(t-\tau)} \mathrm{d}\tau \\ &= \frac{e^{\alpha t} - 1}{\alpha} \|u\|_\infty \\ &\le \frac{1}{|\alpha|} \|u\|_\infty, \end{aligned}$$

which implies that $\mathcal{G}$ is $\mathcal{L}_\infty$-stable with finite gain and zero bias for all $\alpha < 0$.

Finally, let $\alpha > 0$ and let $u(t) \equiv 1$ so that $u(\cdot) \in \mathcal{L}_\infty$. In this case, $y(t) = \frac{e^{\alpha t}-1}{\alpha}$, which implies that $y(\cdot) \notin \mathcal{L}_\infty$. Hence, $\mathcal{G}$ is not $\mathcal{L}_\infty$-stable for all $\alpha > 0$. △

7.3 The Small Gain Theorem

In this section, we consider feedback interconnections of input-output stable systems and provide sufficient conditions for input-output stability of interconnected systems. Specifically, let $\mathcal{G}_1 : \mathcal{L}^m_{\mathrm{pe}} \to \mathcal{L}^l_{\mathrm{pe}}$ and $\mathcal{G}_2 : \mathcal{L}^l_{\mathrm{pe}} \to \mathcal{L}^m_{\mathrm{pe}}$, and consider the negative feedback interconnection given in Figure 7.1 where $(u_1, u_2) \in \mathcal{L}^m_{\mathrm{pe}} \times \mathcal{L}^l_{\mathrm{pe}}$ and $(y_1, y_2) \in \mathcal{L}^l_{\mathrm{pe}} \times \mathcal{L}^m_{\mathrm{pe}}$ are the input and the output signals of the closed-loop system, respectively, and $e_1 \triangleq u_1 - y_2 \in \mathcal{L}^m_{\mathrm{pe}}$ and $e_2 \triangleq u_2 + y_1 \in \mathcal{L}^l_{\mathrm{pe}}$ are the inputs to $\mathcal{G}_1$ and $\mathcal{G}_2$, respectively. Note that $y_1 = \mathcal{G}_1[e_1] = \mathcal{G}_1[u_1 - y_2]$ and $y_2 = \mathcal{G}_2[e_2] = \mathcal{G}_2[u_2 + y_1]$. Hence,

$$\begin{bmatrix} y_1 \\ y_2 \end{bmatrix} = \begin{bmatrix} \mathcal{G}_1[u_1 - y_2] \\ \mathcal{G}_2[u_2 + y_1] \end{bmatrix} \tag{7.13}$$

and

$$\begin{bmatrix} u_1 \\ u_2 \end{bmatrix} = \begin{bmatrix} e_1 + \mathcal{G}_2[e_2] \\ e_2 - \mathcal{G}_1[e_1] \end{bmatrix}. \tag{7.14}$$

In general, there may not exist a mapping $\tilde{\mathcal{G}} : \mathcal{L}^m_{\mathrm{pe}} \times \mathcal{L}^l_{\mathrm{pe}} \to \mathcal{L}^l_{\mathrm{pe}} \times \mathcal{L}^m_{\mathrm{pe}}$ such that $y = \tilde{\mathcal{G}}[u]$, where $y \triangleq [y_1^\mathrm{T}, y_2^\mathrm{T}]^\mathrm{T}$ and $u \triangleq [u_1^\mathrm{T}, u_2^\mathrm{T}]^\mathrm{T}$, and (7.13) holds. In this chapter, we restrict $\mathcal{G}_1$ and $\mathcal{G}_2$ such that there exists a mapping $\tilde{\mathcal{G}} : \mathcal{L}^m_{\mathrm{pe}} \times \mathcal{L}^l_{\mathrm{pe}} \to \mathcal{L}^l_{\mathrm{pe}} \times \mathcal{L}^m_{\mathrm{pe}}$ such that $y = \tilde{\mathcal{G}}[u]$ and (7.13) holds. In this case, the feedback interconnection given in Figure 7.1 is well defined. Note that such a $\tilde{\mathcal{G}}$ exists if and only if there exists a mapping $\hat{\mathcal{G}} : \mathcal{L}^m_{\mathrm{pe}} \times \mathcal{L}^l_{\mathrm{pe}} \to \mathcal{L}^l_{\mathrm{pe}} \times \mathcal{L}^m_{\mathrm{pe}}$ such that $e = \hat{\mathcal{G}}[u]$, where $e \triangleq [e_1^\mathrm{T}, e_2^\mathrm{T}]^\mathrm{T}$, and (7.14) holds.

Definition 7.4. Let the feedback interconnection of $\mathcal{G}_1$ and $\mathcal{G}_2$ given in Figure 7.1 be well defined and let $\tilde{\mathcal{G}} : \mathcal{L}^m_{\mathrm{pe}} \times \mathcal{L}^l_{\mathrm{pe}} \to \mathcal{L}^l_{\mathrm{pe}} \times \mathcal{L}^m_{\mathrm{pe}}$ be such that

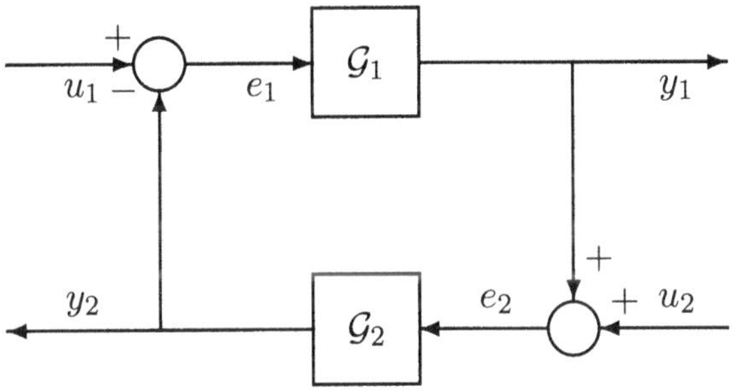

Figure 7.1 Feedback interconnection of $\mathcal{G}_1$ and $\mathcal{G}_2$.

$y = \tilde{\mathcal{G}}[u]$ and (7.13) holds. Then the feedback interconnection of $\mathcal{G}_1$ and $\mathcal{G}_2$ given in Figure 7.1 is, respectively, *$\mathcal{L}_p$-stable*, *$\mathcal{L}_p$-stable with finite gain*, and *$\mathcal{L}_p$-stable with finite gain and zero bias* if $\tilde{\mathcal{G}}$ is $\mathcal{L}_p$-stable, $\mathcal{L}_p$-stable with finite gain, and $\mathcal{L}_p$-stable with finite gain and zero bias.

The following result, known as *small gain theorem* provides a sufficient condition for $\mathcal{L}_p$-stability of the feedback interconnection given by Figure 7.1.

Theorem 7.1. Let $\mathcal{G}_1 : \mathcal{L}^m_{\mathrm{pe}} \to \mathcal{L}^l_{\mathrm{pe}}$ and $\mathcal{G}_2 : \mathcal{L}^l_{\mathrm{pe}} \to \mathcal{L}^m_{\mathrm{pe}}$ be causal operator dynamical systems such that the feedback interconnection given by Figure 7.1 is well defined. If $\mathcal{G}_1$ is $\mathcal{L}_p$-stable with finite gain γ_1 and $\mathcal{G}_2$ is $\mathcal{L}_p$-stable with finite gain γ_2 such that $\gamma_1\gamma_2 < 1$, then the feedback interconnection of $\mathcal{G}_1$ and $\mathcal{G}_2$ given in Figure 7.1 is $\mathcal{L}_p$-stable.

Proof. Since $\mathcal{G}_1$ and $\mathcal{G}_2$ are causal and $\mathcal{L}_p$-stable with finite gains γ_1 and γ_2, respectively, it follows that there exist $\beta_1, \beta_2 > 0$ such that for every $T > 0$, $e_1 \in \mathcal{L}^m_{\mathrm{pe}}$, and $e_2 \in \mathcal{L}^l_{\mathrm{pe}}$,

$$\|(\mathcal{G}_1[e_1])_T\|_p \le \gamma_1\|e_{1T}\|_p + \beta_1, \tag{7.15}$$

$$\|(\mathcal{G}_2[e_2])_T\|_p \le \gamma_2\|e_{2T}\|_p + \beta_2. \tag{7.16}$$

Hence, it follows that

$$\begin{aligned}\|e_{1T}\|_p &\le \|u_{1T}\|_p + \|(\mathcal{G}_2[e_2])_T\|_p \\ &\le \|u_{1T}\|_p + \gamma_2\|e_{2T}\|_p + \beta_2,\end{aligned} \tag{7.17}$$

and

$$\begin{aligned}\|e_{2T}\|_p &\le \|u_{2T}\|_p + \|(\mathcal{G}_1[e_1])_T\|_p \\ &\le \|u_{2T}\|_p + \gamma_1\|e_{1T}\|_p + \beta_1.\end{aligned} \tag{7.18}$$

Now, combining (7.17) and (7.18) yields

$$|||e_{1T}|||_p \leq |||u_{1T}|||_p + \beta_2 + \gamma_2 \left(|||u_{2T}|||_p + \gamma_1 |||e_{1T}|||_p + \beta_1\right), \quad (7.19)$$
$$|||e_{2T}|||_p \leq |||u_{2T}|||_p + \beta_1 + \gamma_1 \left(|||u_{1T}|||_p + \gamma_2 |||e_{2T}|||_p + \beta_2\right), \quad (7.20)$$

and hence,

$$|||e_{1T}|||_p \leq \frac{1}{1-\gamma_1\gamma_2} \left(|||u_{1T}|||_p + \gamma_2 |||u_{2T}|||_p + \beta_2 + \gamma_2\beta_1\right), \quad (7.21)$$
$$|||e_{2T}|||_p \leq \frac{1}{1-\gamma_1\gamma_2} \left(|||u_{2T}|||_p + \gamma_1 |||u_{1T}|||_p + \beta_1 + \gamma_1\beta_2\right). \quad (7.22)$$

Next, since $y_1 = \mathcal{G}_1[e_1]$ and $y_2 = \mathcal{G}_2[e_2]$ it follows from (7.15), (7.16), (7.21), and (7.22) that

$$|||y_{1T}|||_p \leq \frac{\gamma_1}{1-\gamma_1\gamma_2} \left(|||u_{1T}|||_p + \gamma_2 |||u_{2T}|||_p + \beta_2 + \gamma_2\beta_1\right) + \beta_1, \quad (7.23)$$
$$|||y_{2T}|||_p \leq \frac{\gamma_2}{1-\gamma_1\gamma_2} \left(|||u_{2T}|||_p + \gamma_1 |||u_{1T}|||_p + \beta_1 + \gamma_1\beta_2\right) + \beta_2. \quad (7.24)$$

Now, the result follows from the fact that $\mathcal{G}_1$ and $\mathcal{G}_2$ are causal dynamical systems. □

7.4 Input-Output Dissipativity Theory

In this section, we introduce the concept of dissipative systems within the context of operator dynamical systems. In order to do this, we restrict our attention to operator dynamical systems that map $\mathcal{L}_{2\mathrm{e}}^m$ to $\mathcal{L}_{2\mathrm{e}}^l$. Note that $\mathcal{L}_2^m$-space is a *Hilbert space* (see Problem 2.89) with the inner product

$$\langle u, y \rangle \triangleq \int_0^\infty u^{\mathrm{T}}(t) y(t) \mathrm{d}t, \qquad u(\cdot), y(\cdot) \in \mathcal{L}_2^m. \quad (7.25)$$

For $u, y \in \mathcal{L}_{2\mathrm{e}}^m$ define $\langle u, y \rangle_T \triangleq \langle u_T, y_T \rangle$. The following definition introduces the notion of dissipativity for operator dynamical systems.

Definition 7.5. Let $\mathcal{G} : \mathcal{L}_{2\mathrm{e}}^m \to \mathcal{L}_{2\mathrm{e}}^l$ be a causal operator dynamical system and let $Q \in \mathbb{S}^l$, $R \in \mathbb{S}^m$, and $S \in \mathbb{R}^{l\times m}$. $\mathcal{G}$ is *(Q,R,S)-dissipative* if for every $T > 0$ and for every $u \in \mathcal{L}_{2\mathrm{e}}^m$,

$$\langle y, Qy \rangle_T + 2\langle y, Su \rangle_T + \langle u, Ru \rangle_T \geq 0, \quad (7.26)$$

where $y = \mathcal{G}[u]$. $\mathcal{G}$ is *passive* if for every $T > 0$ and for every $u \in \mathcal{L}_{2\mathrm{e}}^m$, $2\langle y, u \rangle_T \geq 0$. $\mathcal{G}$ is *input strict passive* if there exists $\varepsilon > 0$ such that for every $T > 0$ and for every $u \in \mathcal{L}_{2\mathrm{e}}^m$, $2\langle y, u \rangle_T - \varepsilon\langle u, u \rangle_T \geq 0$. $\mathcal{G}$ is *output strict passive* if there exists $\varepsilon > 0$ such that for every $T > 0$ and for every $u \in \mathcal{L}_{2\mathrm{e}}^m$, $2\langle y, u \rangle_T - \varepsilon\langle y, y \rangle_T \geq 0$. $\mathcal{G}$ is *input-output strict passive* if there exist $\varepsilon_1, \varepsilon_2 > 0$ such that for every $T > 0$ and for every $u \in \mathcal{L}_{2\mathrm{e}}^m$, $2\langle y, u \rangle_T -$

$\varepsilon_1\langle y, y\rangle - \varepsilon_2\langle u, u\rangle_T \geq 0$. Finally, $\mathcal{G}$ is *nonexpansive with gain* $\gamma > 0$ if for every $T > 0$ and for every $u \in \mathcal{L}_{2\mathrm{e}}^m$, $\langle y, y\rangle_T - \gamma^2\langle u, u\rangle_T \leq 0$.

The following result connects (Q, R, S)-dissipativity with $\mathcal{L}_2$-stability.

Theorem 7.2. Let $\mathcal{G} : \mathcal{L}_{2\mathrm{e}}^m \to \mathcal{L}_{2\mathrm{e}}^l$ be a causal operator dynamical system and let $Q \in \mathbb{S}^l$, $R \in \mathbb{S}^m$, and $S \in \mathbb{R}^{l\times m}$ be such that $Q < 0$. If $\mathcal{G}$ is (Q, R, S)-dissipative, then $\mathcal{G}$ is $\mathcal{L}_2$-stable.

Proof. Let $M \triangleq -Q$ and let $\lambda > 0$ be such that $\lambda I_m + R > 0$ and $M - \frac{1}{\lambda}SS^{\mathrm{T}} \geq \varepsilon I_l$ for some $\varepsilon > 0$. (The existence of such a scalar λ can be easily established.) Since $\mathcal{G}$ is (Q, R, S)-dissipative it follows that for every $T > 0$ and $u \in \mathcal{L}_2^m$,

$$\begin{aligned}\langle y, My\rangle_T &\leq 2\langle y, Su\rangle_T + \langle u, Ru\rangle_T \\ &= \frac{1}{\lambda}\langle S^{\mathrm{T}}y, S^{\mathrm{T}}y\rangle_T + \lambda\langle u, u\rangle_T - \langle\frac{1}{\sqrt{\lambda}}S^{\mathrm{T}}y - \sqrt{\lambda}u, \frac{1}{\sqrt{\lambda}}S^{\mathrm{T}}y - \sqrt{\lambda}u\rangle_T \\ &\quad +\langle u, Ru\rangle_T \\ &\leq \frac{1}{\lambda}\langle S^{\mathrm{T}}y, S^{\mathrm{T}}y\rangle_T + \langle u, (R + \lambda I_m)u\rangle_T, \end{aligned} \tag{7.27}$$

which implies that

$$\begin{aligned}\varepsilon\langle y, y\rangle_T &\leq \langle y, (M - \frac{1}{\lambda}SS^{\mathrm{T}})y\rangle_T \\ &\leq \langle u, (R + \lambda I_m)u\rangle_T \\ &\leq (\lambda + \lambda_{\max}(R))\langle u, u\rangle_T. \end{aligned} \tag{7.28}$$

The result now follows immediately by noting that $\langle y, y\rangle_T = \|y_T\|_2$. □

Corollary 7.1. Let $\mathcal{G} : \mathcal{L}_{2\mathrm{e}}^m \to \mathcal{L}_{2\mathrm{e}}^l$ be a causal operator dynamical system. Then the following statements hold:

i) If $\mathcal{G}$ is output strict passive, then $\mathcal{G}$ is $\mathcal{L}_2$-stable.

ii) If $\mathcal{G}$ is input-output strict passive, then $\mathcal{G}$ is $\mathcal{L}_2$-stable.

iii) If $\mathcal{G}$ is nonexpansive with gain $\gamma > 0$, then $\mathcal{G}$ is $\mathcal{L}_2$-stable.

Proof. The proof is a direct consequence of Theorem 7.2 with *i)* $Q = -\varepsilon I_l$, $R = 0$, and $S = I_l$, *ii)* $Q = -\varepsilon_1 I_l$, $R = -\varepsilon_2 I_l$, and $S = I_l$, and *iii)* $Q = I_l$, $R = -\gamma^2 I_l$, and $S = 0$. □

Next, we provide a sufficient condition for input-output stability of a feedback interconnection of two dissipative dynamical systems $\mathcal{G}_1$ and $\mathcal{G}_2$.

Theorem 7.3. Let $\mathcal{G}_1 : \mathcal{L}_{2e}^m \to \mathcal{L}_{2e}^l$ and $\mathcal{G}_2 : \mathcal{L}_{2e}^l \to \mathcal{L}_{2e}^m$ be causal operator dynamical systems such that the feedback interconnection given by Figure 7.1 is well defined. Furthermore, let $Q_1, R_2 \in \mathbb{S}^l$, $Q_2, R_1 \in \mathbb{S}^m$, and $S_1, S_2^{\mathrm{T}} \in \mathbb{R}^{l\times m}$ be such that there exists a scalar $\sigma > 0$ such that

$$\hat{Q} \triangleq \begin{bmatrix} Q_1 + \sigma R_2 & -S_1 + \sigma S_2^{\mathrm{T}} \\ -S_1^{\mathrm{T}} + \sigma S_2 & R_1 + \sigma Q_2 \end{bmatrix} < 0.$$

If $\mathcal{G}_1$ is (Q_1, R_1, S_1)-dissipative and $\mathcal{G}_2$ is (Q_2, R_2, S_2)-dissipative, then the feedback interconnection of $\mathcal{G}_1$ and $\mathcal{G}_2$ given in Figure 7.1 is $\mathcal{L}_2$-stable.

Proof. Since $\mathcal{G}_1$ is (Q_1, R_1, S_1)-dissipative and $\mathcal{G}_2$ is (Q_2, R_2, S_2)-dissipative it follows that for every $T > 0$, $e_1 \in \mathcal{L}_{2e}^m$, and $e_2 \in \mathcal{L}_{2e}^l$,

$$\langle y_1, Q_1 y_1\rangle_T + 2\langle y_1, S_1 e_1\rangle_T + \langle e_1, R_1 e_1\rangle_T \geq 0 \tag{7.29}$$

and

$$\sigma\langle y_2, Q_2 y_2\rangle_T + 2\sigma\langle y_2, S_2 e_2\rangle_T + \sigma\langle e_2, R_2 e_2\rangle_T \geq 0. \tag{7.30}$$

Now, using $e_1 = u_1 - y_2$ and $e_2 = u_2 + y_1$, and combining (7.29) and (7.30) yields

$$\langle y, \hat{Q} y\rangle_T + 2\langle y, \hat{S} u\rangle_T + \langle u, \hat{R} u\rangle_T \geq 0, \tag{7.31}$$

where $u \triangleq [u_1^{\mathrm{T}}, u_2^{\mathrm{T}}]^{\mathrm{T}}$, $y \triangleq [y_1^{\mathrm{T}}, y_2^{\mathrm{T}}]^{\mathrm{T}}$, and

$$\hat{R} \triangleq \begin{bmatrix} R_1 & 0 \\ 0 & \sigma R_2 \end{bmatrix}, \qquad \hat{S} \triangleq \begin{bmatrix} S_1 & -\sigma R_2 \\ -R_1 & \sigma S_2 \end{bmatrix}.$$

The result now is an immediate consequence of Theorem 7.2. □

Corollary 7.2. Let $\mathcal{G}_1 : \mathcal{L}_{2e}^m \to \mathcal{L}_{2e}^l$ and $\mathcal{G}_2 : \mathcal{L}_{2e}^l \to \mathcal{L}_{2e}^m$ be causal operator dynamical systems such that the feedback interconnection given by Figure 7.1 is well defined. Then the following statements hold:

i) If $\mathcal{G}_1$ is input-output strict passive and $\mathcal{G}_2$ is passive, then the feedback interconnection of $\mathcal{G}_1$ and $\mathcal{G}_2$ given in Figure 7.1 is $\mathcal{L}_2$-stable.

ii) If $\mathcal{G}_1$ and $\mathcal{G}_2$ are input strict passive, then feedback interconnection of $\mathcal{G}_1$ and $\mathcal{G}_2$ given in Figure 7.1 is $\mathcal{L}_2$-stable.

iii) If $\mathcal{G}_1$ and $\mathcal{G}_2$ are output strict passive, then the feedback interconnection of $\mathcal{G}_1$ and $\mathcal{G}_2$ given in Figure 7.1 is $\mathcal{L}_2$-stable.

iv) If $\mathcal{G}_1$ is input strict passive and $\mathcal{L}_2$-stable with finite gain and $\mathcal{G}_2$ is passive, then the feedback interconnection of $\mathcal{G}_1$ and $\mathcal{G}_2$ given in Figure 7.1 is $\mathcal{L}_2$-stable.

v) If $\mathcal{G}_1$ is nonexpansive with gain γ_1 and $\mathcal{G}_2$ is nonexpansive with gain γ_2 such that $\gamma_1\gamma_2 < 1$, then the feedback interconnection of $\mathcal{G}_1$ and $\mathcal{G}_2$ given in Figure 7.1 is $\mathcal{L}_2$-stable.

Proof. *i*) The proof follows from Theorem 7.3 with $l = m$, $\sigma = 1$, $Q_1 = -\varepsilon_1 I_m$, $R_1 = -\varepsilon_2 I_m$, $S_1 = I_m$, $Q_2 = 0$, $R_2 = 0$, and $S_2 = I_m$ for some $\varepsilon_1, \varepsilon_2 > 0$.

ii) The proof follows from Theorem 7.3 with $l = m$, $\sigma = 1$, $Q_1 = 0$, $R_1 = -\varepsilon_1 I_m$, $S_1 = I_m$, $Q_2 = 0$, $R_2 = -\varepsilon_2 I_m$, and $S_2 = I_m$ for some $\varepsilon_1, \varepsilon_2 > 0$.

iii) The proof follows from Theorem 7.3 with $l = m$, $\sigma = 1$, $Q_1 = -\varepsilon_1 I_m$, $R_1 = 0$, $S_1 = I_m$, $Q_2 = -\varepsilon_2 I_m$, $R_2 = 0$, and $S_2 = I_m$ for some $\varepsilon_1, \varepsilon_2 > 0$.

iv) Note that if $\mathcal{G}_1$ is input strict passive and $\mathcal{L}_2$-stable with finite gain, then there exist scalars $\varepsilon > 0$ and $\gamma > 0$ such that for every $T > 0$ and $e_1 \in \mathcal{L}_{2\mathrm{e}}^m$,

$$2\langle y_1, e_1 \rangle_T - \varepsilon \langle e_1, e_1 \rangle_T \geq 0 \tag{7.32}$$

and

$$-\langle y_1, y_1 \rangle_T + \gamma \langle e_1, e_1 \rangle_T \geq 0. \tag{7.33}$$

Now, let $\sigma > 0$ be such that $\sigma\gamma < \varepsilon$. Adding (7.32) to σ(7.33) yields

$$-\sigma\langle y_1, y_1 \rangle_T + 2\langle y_1, e_1 \rangle_T - (\varepsilon - \sigma\gamma)\langle e_1, e_1 \rangle_T \geq 0,$$

which shows that $\mathcal{G}_1$ is input-output strict passive. The result now follows from *i*).

v) Since $\gamma_1\gamma_2 < 1$ there exists $\varepsilon > 0$ such that $(\gamma_1 + \varepsilon)\gamma_2 < 1$. The result now follows from Theorem 7.3 with $\sigma = \gamma_1 + \varepsilon$, $Q_1 = -I_l$, $R_1 = \gamma_1^2 I_m$, $S_1 = 0$, $Q_2 = -I_m$, $R_2 = \gamma_2^2 I_l$, and $S_2 = 0$. □

7.5 Input-Output Operator Dissipativity Theory

In this section, we extend the concept of dissipative systems introduced in Section 7.4. Once again, we restrict our attention to operator dynamical systems that map $\mathcal{L}_{2\mathrm{e}}^m$ to $\mathcal{L}_{2\mathrm{e}}^l$. The following definition introduces the notion of operator dissipativity for operator dynamical systems.

Definition 7.6. Let $\mathcal{G} : \mathcal{L}_{2\mathrm{e}}^m \to \mathcal{L}_{2\mathrm{e}}^l$ be a causal operator dynamical system and let $Q : \mathcal{L}_{2\mathrm{e}}^l \to \mathcal{L}_{2\mathrm{e}}^l$, $R : \mathcal{L}_{2\mathrm{e}}^m \to \mathcal{L}_{2\mathrm{e}}^m$, and $S : \mathcal{L}_{2\mathrm{e}}^m \to \mathcal{L}_{2\mathrm{e}}^l$ be causal operators such that Q and R are self-adjoint.[3] $\mathcal{G}$ is *(Q, R, S)-operator dissipative* if for every $T > 0$ and for every $u \in \mathcal{L}_{2\mathrm{e}}^m$,

$$\langle y, Qy \rangle_T + 2\langle y, Su \rangle_T + \langle u, Ru \rangle_T \geq 0, \tag{7.34}$$

[3] Given an operator $\mathcal{X} : \mathcal{L}_{2\mathrm{e}}^p \to \mathcal{L}_{2\mathrm{e}}^q$, the operator $\mathcal{X}^* : \mathcal{L}_{2\mathrm{e}}^q \to \mathcal{L}_{2\mathrm{e}}^p$ is the *adjoint operator* of $\mathcal{X}$ if for every $u \in \mathcal{L}_{2\mathrm{e}}^p$ and $y \in \mathcal{L}_{2\mathrm{e}}^q$, $\langle y, \mathcal{X}u \rangle = \langle \mathcal{X}^* y, u \rangle$. An operator $\mathcal{X} : \mathcal{L}_{2\mathrm{e}}^p \to \mathcal{L}_{2\mathrm{e}}^p$ is *self-adjoint* if for every $u \in \mathcal{L}_{2\mathrm{e}}^p$ and $y \in \mathcal{L}_{2\mathrm{e}}^q$, $\langle y, \mathcal{X}u \rangle = \langle \mathcal{X}y, u \rangle$, that is, if $\mathcal{X} = \mathcal{X}^*$.

where $y = \mathcal{G}[u]$.

The following result connects (Q, R, S)-operator dissipativity with $\mathcal{L}_2$-stability.

Theorem 7.4. Let $\mathcal{G} : \mathcal{L}_{2e}^m \to \mathcal{L}_{2e}^l$ be a causal operator dynamical system and let $Q : \mathcal{L}_{2e}^l \to \mathcal{L}_{2e}^l$, $R : \mathcal{L}_{2e}^m \to \mathcal{L}_{2e}^m$, and $S : \mathcal{L}_{2e}^m \to \mathcal{L}_{2e}^l$ be causal operators such that Q, R, and SS^* are bounded, that is, there exist positive scalars q, r, and s such that $\langle y, Qy\rangle \le q\langle y, y\rangle$, $y \in \mathcal{L}_{2e}^l$, $\langle u, Ru\rangle \le r\langle u, u\rangle$, $u \in \mathcal{L}_{2e}^m$, $\langle y, SS^*y\rangle = \langle S^*y, S^*y\rangle \le s\langle y, y\rangle$, $y \in \mathcal{L}_{2e}^l$, Q and R are self-adjoint, and there exists $\varepsilon > 0$ such that $\langle y, Qy\rangle < -\varepsilon\langle y, y\rangle$. If $\mathcal{G}$ is (Q, R, S)-operator dissipative, then $\mathcal{G}$ is $\mathcal{L}_2$-stable.

Proof. Let $M \triangleq -Q$ and let $\lambda > 0$ be such that $\lambda < s/\varepsilon$. Since $\mathcal{G}$ is (Q, R, S)-operator dissipative it follows that for every $T > 0$ and $u \in \mathcal{L}_2^m$,

$$\begin{aligned}
\varepsilon\langle y, y\rangle_T &\le \langle y, My\rangle_T \\
&\le 2\langle y, Su\rangle_T + \langle u, Ru\rangle_T \\
&= \frac{1}{\lambda}\langle S^*y, S^*y\rangle_T + \lambda\langle u, u\rangle_T - \langle \frac{1}{\sqrt{\lambda}}S^*y - \sqrt{\lambda}u, \frac{1}{\sqrt{\lambda}}S^*y - \sqrt{\lambda}u\rangle_T \\
&\quad + \langle u, Ru\rangle_T \\
&\le \frac{1}{\lambda}\langle S^*y, S^*y\rangle_T + (r+\lambda)\langle u, u\rangle_T \\
&\le \frac{s}{\lambda}\langle y, y\rangle_T + (r+\lambda)\langle u, u\rangle_T,
\end{aligned} \tag{7.35}$$

which implies that

$$\langle y, y\rangle_T \le \frac{\lambda(\lambda + r)}{\varepsilon\lambda - s}\langle u, u\rangle_T. \tag{7.36}$$

The result now follows immediately by noting that $\langle y, y\rangle_T = \|y_T\|_2$. □

Next, we provide a sufficient condition for input-output stability of a feedback interconnection of two operator dissipative dynamical systems $\mathcal{G}_1$ and $\mathcal{G}_2$.

Theorem 7.5. Let $\mathcal{G}_1 : \mathcal{L}_{2e}^m \to \mathcal{L}_{2e}^l$ and $\mathcal{G}_2 : \mathcal{L}_{2e}^l \to \mathcal{L}_{2e}^m$ be causal operator dynamical systems such that the feedback interconnection given by Figure 7.1 is well defined. Furthermore, let $Q_1, R_2 : \mathcal{L}_{2e}^l \to \mathcal{L}_{2e}^l$, $Q_2, R_1 : \mathcal{L}_{2e}^m \to \mathcal{L}_{2e}^m$, and $S_1, S_2^* : \mathcal{L}_{2e}^m \to \mathcal{L}_{2e}^l$ be bounded causal operators such that there exist scalar $\sigma, \varepsilon > 0$ and

$$\hat{Q} \triangleq \begin{bmatrix} Q_1 + \sigma R_2 & -S_1 + \sigma S_2^* \\ -S_1^* + \sigma S_2 & R_1 + \sigma Q_2 \end{bmatrix}$$

satisfies $\langle y, \hat{Q}y\rangle \le -\varepsilon\langle y, y\rangle$, $y \in \mathcal{L}_{2e}^{l+m}$. If $\mathcal{G}_1$ is (Q_1, R_1, S_1)-operator dissipative and $\mathcal{G}_2$ is (Q_2, R_2, S_2)-operator dissipative, then feedback interconnection of $\mathcal{G}_1$ and $\mathcal{G}_2$ given in Figure 7.1 is $\mathcal{L}_2$-stable.

Proof. Since $\mathcal{G}_1$ is (Q_1, R_1, S_1)-operator dissipative and $\mathcal{G}_2$ is (Q_2, R_2, S_2)-operator dissipative it follows that for every $T > 0$, $e_1 \in \mathcal{L}_{2e}^m$, and $e_2 \subset \mathcal{L}_{2e}^l$,

$$\langle y_1, Q_1 y_1\rangle_T + 2\langle y_1, S_1 e_1\rangle_T + \langle e_1, R_1 e_1\rangle_T \geq 0 \tag{7.37}$$

and

$$\sigma\langle y_2, Q_2 y_2\rangle_T + 2\sigma\langle y_2, S_2 e_2\rangle_T + \sigma\langle e_2, R_2 e_2\rangle_T \geq 0. \tag{7.38}$$

Now, using $e_1 = u_1 - y_2$ and $e_2 = u_2 + y_1$, and combining (7.37) and (7.38) yields

$$\langle y, \hat{Q} y\rangle_T + 2\langle y, \hat{S} u\rangle_T + \langle u, \hat{R} u\rangle_T \geq 0, \tag{7.39}$$

where $u \triangleq [u_1^{\mathrm{T}}, u_2^{\mathrm{T}}]^{\mathrm{T}}$, $y \triangleq [y_1^{\mathrm{T}}, y_2^{\mathrm{T}}]^{\mathrm{T}}$, and

$$\hat{R} \triangleq \begin{bmatrix} R_1 & 0 \\ 0 & \sigma R_2 \end{bmatrix}, \qquad \hat{S} \triangleq \begin{bmatrix} S_1 & -\sigma R_2 \\ -R_1 & \sigma S_2 \end{bmatrix}.$$

The result now is an immediate consequence of Theorem 7.4. □

7.6 Connections Between Input-Output Stability and Lyapunov Stability

In this section, we provide connections between input-output stability and Lyapunov stability. Since Lyapunov stability theory deals with state space dynamical systems we begin by considering nonlinear dynamical systems $\mathcal{G}$ of the form

$$\dot{x}(t) = F(x(t), u(t)), \qquad x(t_0) = x_0, \qquad t \geq t_0, \tag{7.40}$$

$$y(t) = H(x(t), u(t)), \tag{7.41}$$

where $x(t) \in \mathbb{R}^n$, $u(t) \in \mathbb{R}^m$, $y(t) \in \mathbb{R}^l$, $F : \mathbb{R}^n \times \mathbb{R}^m \to \mathbb{R}^n$, and $H : \mathbb{R}^n \times \mathbb{R}^m \to \mathbb{R}^l$. For the dynamical system $\mathcal{G}$ given by (7.40) and (7.41) defined on the state space $\mathbb{R}^n$, $\mathbb{R}^m$ and $\mathbb{R}^l$ define an input and output space, respectively, consisting of continuous bounded functions on the semi-infinite interval $[0, \infty)$. The input and output spaces $\mathcal{U}$ and $\mathcal{Y}$ are assumed to be closed under the shift operator, that is, if $u(\cdot) \in \mathcal{U}$ (respectively, $y(\cdot) \in \mathcal{Y}$), then the function defined by $u_T \triangleq u(t+T)$ (respectively, $y_T \triangleq y(t+T)$) is contained in $\mathcal{U}$ (respectively, $\mathcal{Y}$) for all $T \geq 0$. We assume that $F(\cdot,\cdot)$ and $H(\cdot,\cdot)$ are continuously differentiable mappings in (x, u) and $F(\cdot,\cdot)$ has at least one equilibrium so that, without loss of generality, $F(0,0) = 0$ and $H(0,0) = 0$.

Theorem 7.6. Consider the nonlinear dynamical system $\mathcal{G}$ given by (7.40) and (7.41). Assume that there exist a continuously differentiable function $V : \mathbb{R}^n \to [0,\infty)$ and positive scalars $\alpha, \beta, \gamma_1, \gamma_2$ such that

$$\alpha\|x\|^2 \le V(x) \le \beta\|x\|^2, \quad x \in \mathbb{R}^n, \tag{7.42}$$
$$V'(x)F(x,0) \le -\gamma_1\|x\|^2, \quad x \in \mathbb{R}^n, \tag{7.43}$$
$$\|V'(x)\| \le \gamma_2\|x\|, \quad x \in \mathbb{R}^n. \tag{7.44}$$

Furthermore, assume that there exist positive scalars L, η_1, and η_2 such that

$$\|F(x,u) - F(x,0)\| \le L\|u\|, \quad (x,u) \in \mathbb{R}^n \times \mathbb{R}^m, \tag{7.45}$$
$$\|H(x,u)\| \le \eta_1\|x\| + \eta_2\|u\|, \quad (x,u) \in \mathbb{R}^n \times \mathbb{R}^m. \tag{7.46}$$

Then, for every $x_0 \in \mathbb{R}^n$ and $p \in [1,\infty]$, $\mathcal{G}$ is $\mathcal{L}_p$-stable with finite gain.

Proof. It follows from (7.43)–(7.45) that

$$\begin{aligned} V'(x)F(x,u) &= V'(x)F(x,0) + V'(x)[F(x,u) - F(x,0)] \\ &\le -\gamma_1\|x\|^2 + \gamma_2 L\|x\|\|u\|, \quad x \in \mathbb{R}^n, \quad u \in \mathbb{R}^m. \end{aligned} \tag{7.47}$$

Now, let $u \in \mathcal{L}_p^m$ and let $x(t)$, $t \ge 0$, denote the solution to (7.40) with $x(0) = 0$ and define $W(t) \triangleq \sqrt{V(x(t))}$, $t \ge 0$. First, consider the case in which $W(t) \ne 0$, $t > 0$. In this case, it follows from (7.42) and (7.47) that

$$\begin{aligned} 2W(t)\dot{W}(t) &= \dot{V}(x(t)) \\ &= V'(x(t))F(x(t),u(t)) \\ &\le -\gamma_1\|x(t)\|^2 + \gamma_2 L\|x(t)\|\|u(t)\| \\ &\le -\tfrac{\gamma_1}{\beta}V(x(t)) + \tfrac{\gamma_2 L}{\sqrt{\alpha}}V^{1/2}(x(t))\|u(t)\| \\ &= -\tfrac{\gamma_1}{\beta}W^2(t) + \tfrac{\gamma_2 L}{\sqrt{\alpha}}W(t)\|u(t)\|, \quad t > 0, \end{aligned} \tag{7.48}$$

which implies that

$$\dot{W}(t) + \tfrac{\gamma_1}{2\beta}W(t) \le \tfrac{\gamma_2 L}{2\sqrt{\alpha}}\|u(t)\|, \quad t > 0. \tag{7.49}$$

Now, multiplying (7.49) by $e^{\frac{\gamma_1 t}{2\beta}}$, $t \ge 0$, yields

$$\frac{\mathrm{d}}{\mathrm{d}t}\left[e^{\frac{\gamma_1 t}{2\beta}}W(t)\right] \le e^{\frac{\gamma_1 t}{2\beta}}\tfrac{\gamma_2 L}{2\sqrt{\alpha}}\|u(t)\|, \quad t > 0. \tag{7.50}$$

Next, since $W(0) = V^{1/2}(x(0)) = 0$, integrating (7.50) yields

$$W(t) \le \tfrac{\gamma_2 L}{2\sqrt{\alpha}}\int_0^t e^{-\frac{\gamma_1}{2\beta}(t-\tau)}\|u(\tau)\|\mathrm{d}\tau, \quad t > 0. \tag{7.51}$$

Now, it follows from (7.51) that there exists a constant $\gamma > 0$ such that $|||W|||_p \le \gamma|||u|||_p$ and, since $\|x(t)\| \le \frac{1}{\sqrt{\alpha}}W(t)$, $t \ge 0$, it follows that $x(\cdot) \in \mathcal{L}_p^n$. The result is now a direct consequence of (7.46). The case in which $W(t) = 0$, $t \ge 0$, is a straightforward extension of the proof above since $W(t) = 0$, $t \ge 0$, if and only if $x(t) = 0$, $t \ge 0$. □

7.7 Induced Convolution Operator Norms of Linear Dynamical Systems

In the remainder of this chapter, we consider the dynamical system

$$\dot{x}(t) = Ax(t) + Bu(t), \qquad x(0) = 0, \qquad t \geq 0, \tag{7.52}$$

$$y(t) = Cx(t), \tag{7.53}$$

where $x(t) \in \mathbb{R}^n$, $u(t) \in \mathbb{R}^m$, $y(t) \in \mathbb{R}^l$, $t \in [0, \infty)$, $A \in \mathbb{R}^{n \times n}$ is asymptotically stable, $B \in \mathbb{R}^{n \times m}$, and $C \in \mathbb{R}^{l \times n}$. Here, $u(\cdot)$ is an input signal belonging to the class $\mathcal{L}_p$ of input signals and $y(\cdot)$ is an output signal belonging to the class $\mathcal{L}_r$ of output signals, where the notation $\mathcal{L}_p$ and $\mathcal{L}_r$ denote the set of functions in $\mathcal{L}_p^m$ and $\mathcal{L}_r^l$. In applications, (7.52) and (7.53) may denote a control system in closed-loop configuration where the objective is to determine the "size" of the output $y(\cdot)$ for a disturbance $u(\cdot)$. In the remainder of this chapter, we develop explicit formulas for convolution operator norms and their bounds induced by various norms on several classes of input-output signal pairs. These results generalize established induced convolution operator norms for linear dynamical systems.

If the input-output signals are constrained to be finite-energy signals so that $u, y \in \mathcal{L}_2$ then the equi-induced (that is, the domain and range spaces of the convolution operator are assigned the same temporal and spatial norms) signal norm is the $\mathcal{H}_\infty$ system norm [121, 478] given by

$$|||\mathcal{G}|||_{(2,2),(2,2)} \triangleq \sup_{u(\cdot) \in \mathcal{L}_2} \frac{|||y|||_{2,2}}{|||u|||_{2,2}} = \sup_{\omega \in \mathbb{R}} \sigma_{\max}[H(\jmath\omega)], \tag{7.54}$$

where the notation $|||\cdot|||_{p,q}$ denotes a signal norm with p temporal norm and q spatial norm, $\sigma_{\max}(\cdot)$ denotes maximum singular value, $\mathcal{G}$ denotes the convolution operator of (7.52) and (7.53), and $H(s) = C(sI - A)^{-1}B$ is the corresponding transfer function. Hence, the $\mathcal{H}_\infty$ system norm captures the supremum system energy gain.

Alternatively, if the input-output signals are constrained to be bounded amplitude signals so that $u, y \in \mathcal{L}_\infty$, then the equi-induced signal norm

$$|||\mathcal{G}|||_{(\infty,\infty),(\infty,\infty)} \triangleq \sup_{u(\cdot) \in \mathcal{L}_\infty} \frac{|||y|||_{\infty,\infty}}{|||u|||_{\infty,\infty}}, \tag{7.55}$$

is the $\mathcal{L}_1$ system norm[4] [101, 444]. Thus, the $\mathcal{L}_1$ system norm captures the worst-case amplification from input disturbance signals to output signals, where the signal size is taken to be the supremum over time of the signal's peak value pointwise in time [101, 444].

[4]In the single-input/single-output case, it is well known that the induced norm (7.55) corresponds to the $\mathcal{L}_1$ norm of the impulse response matrix function [101, 444].

Mixed input-output signals have also been considered. For example, if $u \in \mathcal{L}_2$ and $y \in \mathcal{L}_\infty$ then the resulting induced operator norm is ([461])

$$|||\mathcal{G}|||_{(\infty,2),(2,2)} \triangleq \sup_{u(\cdot)\in\mathcal{L}_2} \frac{|||y|||_{\infty,2}}{|||u|||_{2,2}} = \lambda_{\max}(CQC^{\mathrm{T}}), \tag{7.56}$$

where $\lambda_{\max}(\cdot)$ denotes the maximum eigenvalue and Q is the unique $n \times n$ nonnegative-definite solution to the Lyapunov equation

$$0 = AQ + QA^{\mathrm{T}} + BB^{\mathrm{T}}. \tag{7.57}$$

Hence, $|||\mathcal{G}|||_{(\infty,2),(2,2)}$ provides a worst-case measure of amplitude errors due to finite energy input signals. Alternatively, if the input and output signal norms are chosen as $|||\cdot|||_{2,2}$ and $|||\cdot|||_{\infty,\infty}$, respectively, then the resulting induced operator norm is ([461])

$$|||\mathcal{G}|||_{(\infty,\infty),(2,2)} \triangleq \sup_{u(\cdot)\in\mathcal{L}_2} \frac{|||y|||_{\infty,\infty}}{|||u|||_{2,2}} = \mathrm{d}_{\max}(CQC^{\mathrm{T}}), \tag{7.58}$$

where $\mathrm{d}_{\max}(\cdot)$ denotes the maximum diagonal entry. Hence, $|||\mathcal{G}|||_{(\infty,\infty),(2,2)}$ provides a worst-case peak excursion response due to finite energy disturbances.

It is clear from the above discussion that operator norms induced by classes of input-output signal pairs can be used to capture disturbance rejection performance objectives for controlled dynamical systems. In particular, $\mathcal{H}_\infty$ control theory [121, 478] has been developed to address the problem of disturbance rejection for systems with bounded energy $\mathcal{L}_2$ signal norms on the disturbance and performance variables. Since the induced $\mathcal{H}_\infty$ transfer function norm (7.54) corresponds to the worst-case disturbance attenuation, for systems with $\mathcal{L}_2$ disturbances which possess significant power within arbitrarily small bandwidths, $\mathcal{H}_\infty$ theory is clearly appropriate. Alternatively, to address pointwise in time the worst-case peak amplitude response due to bounded amplitude persistent $\mathcal{L}_\infty$ disturbances, $\mathcal{L}_1$ theory is appropriate [101, 444]. The problem of finding a stabilizing controller such that the closed-loop system gain from $|||\cdot|||_{2,2}$ to $|||\cdot|||_{\infty,q}$, where $q = 2$ or ∞, is below a specified level is solved in [367, 463]. In addition to the disturbance rejection problem, another application of induced operator norms is the problem of actuator amplitude and rate saturation [90, 105]. In particular, since the convolution operator norm $|||\mathcal{G}|||_{(\infty,\infty),(2,2)}$ given by (7.58) captures the worst-case peak amplitude response due to finite energy disturbances, defining the output (performance) variables y to correspond to the actuator amplitude and actuator rate signals, it follows that $|||\mathcal{G}|||_{(\infty,\infty),(2,2)}$ bounds actuator amplitude and actuator rate excursion. Furthermore, since uncertain signals can also be used to model uncertainty in a system, the treatment of certain classes of uncertain disturbances also enable the development of controllers that are robust with respect to input-output uncertainty blocks [101, 104].

In the papers [461, 462], Wilson developed explicit formulas for convolution operator norms induced by several classes of input-output signal pairs. In the remainder of this chapter we extend the results of [461, 462] to a larger class of input-output signal pairs and provide explicit formulas for induced convolution operator norms and operator norm bounds for linear dynamical systems. These results generalize several well-known induced convolution operator norm results in the literature including results on $\mathcal{L}_\infty$ equi-induced norms ($\mathcal{L}_1$ operator norms) and $\mathcal{L}_1$ equi-induced norms (resource norms). In cases where the induced convolution operator norm expressions are not finitely computable, we provide finitely computable norm bounds.

To develop induced convolution operator norms for linear systems, we first introduce some notation, definitions, and several key lemmas. Let $\|\cdot\|'$ and $\|\cdot\|''$ denote vector norms on $\mathbb{R}^n$ and $\mathbb{R}^m$, respectively, where $m, n \geq 1$. Then $\|\cdot\| : \mathbb{R}^{m\times n} \to \mathbb{R}$ defined by

$$\|A\| \triangleq \max_{\|x\|'=1} \|Ax\|''$$

is the *matrix norm induced by* $\|\cdot\|'$ and $\|\cdot\|''$. If $\|\cdot\|' = \|\cdot\|_p$ and $\|\cdot\|'' = \|\cdot\|_q$, where $p, q \in [1, \infty]$, then the matrix norm on $\mathbb{R}^{m\times n}$ induced by $\|\cdot\|_p$ and $\|\cdot\|_q$ is denoted by $\|\cdot\|_{q,p}$. Let $\|\cdot\|$ denote a vector norm on $\mathbb{R}^m$. Then the *dual norm* $\|\cdot\|_{\mathrm{D}}$ of $\|\cdot\|$ is defined by

$$\|y\|_{\mathrm{D}} \triangleq \max_{\|x\|=1} |y^{\mathrm{T}}x|,$$

where $y \in \mathbb{R}^m$ [417]. Note that $\|\cdot\|_{\mathrm{DD}} = \|\cdot\|$ [417]. Furthermore, if $p, q \in [1, \infty]$ satisfy $1/p + 1/q = 1$, then $\|\cdot\|_{p\mathrm{D}} = \|\cdot\|_q$ [417]. For $p \in [1, \infty]$ we denote the conjugate variable $q \in [1, \infty]$ satisfying $1/p + 1/q = 1$ by $\bar{p} = p/(p-1)$.

Let $\|\cdot\|$ denote a vector norm on $\mathbb{R}^n$. Then $\|\cdot\|$ is *absolute* if $\|x\| = \||x|\|$ for all $x \in \mathbb{R}^n$. Furthermore, $\|\cdot\|$ is *monotone* if $\|x\| \leq \|y\|$ for all $x, y \in \mathbb{R}^n$ such that $|x| \leq\leq |y|$. Note that $\|\cdot\|$ is absolute if and only if $\|\cdot\|$ is monotone [201, p. 285].

Lemma 7.1. Let $p \in [1, \infty]$ and let $A \in \mathbb{R}^{m\times n}$. Then

$$\|A\|_{2,2} = \sigma_{\max}(A), \tag{7.59}$$

$$\|A\|_{p,1} = \max_{i=1,\ldots,n} \|\mathrm{col}_i(A)\|_p, \tag{7.60}$$

and

$$\|A\|_{\infty,p} = \max_{i=1,\ldots,m} \|\mathrm{row}_i(A)\|_{\bar{p}}. \tag{7.61}$$

Proof. Expression (7.59) is standard; see [417] for a proof. To show (7.60), note that, for all $x \in \mathbb{R}^n$,

$$\|Ax\|_p = \left\|\sum_{i=1}^n x_i \mathrm{col}_i(A)\right\|_p \le \sum_{i=1}^n |x_i|\|\mathrm{col}_i(A)\|_p \le \max_{i=1,\ldots,n} \|\mathrm{col}_i(A)\|_p \|x\|_1,$$

and hence, $\|A\|_{p,1} \le \max_{i=1,\ldots,n} \|\mathrm{col}_i(A)\|_p$. Next, let $j \in \{1,\ldots,n\}$ be such that $\|\mathrm{col}_j(A)\|_p = \max_{i=1,\ldots,n} \|\mathrm{col}_i(A)\|_p$. Now, since $\|e_j\|_1 = 1$, it follows that $\|Ae_j\|_p = \|\mathrm{col}_j(A)\|_p$, which implies $\|A\|_{p,1} \ge \max_{i=1,\ldots,n} \|\mathrm{col}_i(A)\|_p$, and hence, (7.60) holds.

Finally, to show (7.61) note that, for all $x \in \mathbb{R}^n$, it follows from Hölder's inequality that

$$\|Ax\|_\infty = \max_{i=1,\ldots,m} |\mathrm{row}_i(A)x| \le \max_{i=1,\ldots,m} \|\mathrm{row}_i(A)\|_{\hat{p}} \|x\|_p,$$

which implies that $\|A\|_{\infty,p} \le \max_{i=1,\ldots,m} \|\mathrm{row}_i(A)\|_{\hat{p}}$. Next, let $j \in \{1,\ldots, n\}$ be such that $\|\mathrm{row}_j(A)\|_{\hat{p}} = \max_{i=1,\ldots,m} \|\mathrm{row}_i(A)\|_{\hat{p}}$ and let x be such that $|\mathrm{row}_j(A)x| = \|\mathrm{row}_j(A)\|_{\hat{p}}\|x\|_p$. Hence, $\|Ax\|_\infty = \max_{i=1,\ldots,m} |\mathrm{row}_i(A)x| \ge |\mathrm{row}_j(A)x| = \|\mathrm{row}_j(A)\|_{\hat{p}}\|x\|_p$, which implies that $\|A\|_{\infty,p} \ge \max_{i=1,\ldots,m} \|\mathrm{row}_i(A)\|_{\hat{p}}$, and hence, (7.61) holds. $\square$

Note that (7.60) and (7.61) generalize the well-known expressions $\|A\|_{1,1} = \max_{i=1,\ldots,n} \|\mathrm{col}_i(A)\|_1$ [201], $\|A\|_{\infty,\infty} = \max_{i=1,\ldots,m} \|\mathrm{row}_i(A)\|_1$ [201], and $\|A\|_{\infty,1} = \|A\|_\infty$ [225]. Furthermore, since $\max_{i=1,\ldots,n} \|\mathrm{col}_i(A)\|_2 = \mathrm{d}_{\max}^{1/2}(A^{\mathrm{T}}A)$ and

$$\max_{i=1,\ldots,m} \|\mathrm{row}_i(A)\|_2 = \mathrm{d}_{\max}^{1/2}(AA^{\mathrm{T}}),$$

it follows from (7.60) with $p = 2$ that $\|A\|_{2,1} = \mathrm{d}_{\max}^{1/2}(A^{\mathrm{T}}A)$ and from (7.61) with $p = 2$ that $\|A\|_{\infty,2} = \mathrm{d}_{\max}^{1/2}(AA^{\mathrm{T}})$.

Lemma 7.2. Let $\|\cdot\|'$ and $\|\cdot\|''$ denote absolute vector norms on $\mathbb{R}^n$ and $\mathbb{R}^m$, respectively, and let $\|\cdot\| : \mathbb{R}^{m\times n} \to \mathbb{R}$ be the matrix norm induced by $\|\cdot\|'$ and $\|\cdot\|''$. Then the following statements hold:

i) Let $A \in \mathbb{R}^{m\times n}$ be such that $A \ge\ge 0$. Then there exists $x \in \mathbb{R}^n$ such that $x \ge\ge 0$, $\|x\|' = 1$, and $\|A\| = \|Ax\|''$.

ii) Let $A, B \in \mathbb{R}^{m\times n}$ be such that $0 \le\le A \le\le B$. Then $\|A\| \le \|B\|$.

Proof. To prove *i)* let $y \in \mathbb{R}^n$ be such that $\|y\|' = 1$ and $\|A\| = \|Ay\|''$. Now, since $\|\cdot\|'$ and $\|\cdot\|''$ are absolute (and hence monotone) vector norms it follows that

$$\|A\| = \|Ay\|'' \le \|A|y|\|'' \le \|A\|\||y|\|' = \|A\|\|y\|' = \|A\|,$$

which implies that $\|A\| = \||A|y||\|''$, and hence, *i)* follows with $x = |y|$.

Next, to prove *ii)* let $x \in \mathbb{R}^n$ be such that $x \geq\geq 0$, $\|x\|' = 1$, and $\|A\| = \|Ax\|''$ (existence of such an x follows from *i)*). Hence, since $\|\cdot\|''$ is an absolute vector norm and $Ax \leq\leq Bx$ it follows that

$$\|A\| = \|Ax\|'' \leq \|Bx\|'' \leq \|B\|,$$

which implies *ii)*. □

The following result generalizes Hölder's inequality to mixed-signal norms.

Lemma 7.3. Let $p, r \in [1, \infty]$, and let $f \in \mathcal{L}_p$ and $g \in \mathcal{L}_{\bar{p}}$. Then

$$\langle f, g \rangle \leq |||f|||_{p,r} |||g|||_{\bar{p},\bar{r}}. \tag{7.62}$$

Finally, the following two results are needed for the results given in Section 7.8.

Lemma 7.4 ([461]). Let $p \in [1, \infty)$ and $r \in [1, \infty]$, and let $f \in \mathcal{L}_p$. Then

$$|||f|||_{p,r} = \sup_{g \in \mathfrak{G}} \langle f, g \rangle, \tag{7.63}$$

where $\mathfrak{G} \triangleq \{g \in \mathcal{L}_{\bar{p}} : |||g|||_{\bar{p},\bar{r}} \leq 1\}$.

Lemma 7.5 ([461]). Let $p \in [1, \infty)$, $r \in [1, \infty]$, and $f : [0, \infty) \times [0, \infty) \to \mathbb{R}^n$ be such that $f(t, \cdot)$ is integrable for almost all $t \in [0, \infty)$, $f(\cdot, \tau) \in \mathcal{L}_p$ for almost all $\tau \in [0, \infty)$, and $g \in \mathcal{L}_1$, where $g(\tau) \triangleq [\int_0^\infty \|f(t, \tau)\|_r^p \mathrm{d}t]^{1/p}$. Then

$$|||y|||_{p,r} \leq \int_0^\infty g(\tau) \mathrm{d}\tau, \tag{7.64}$$

where

$$y(t) = \int_0^\infty f(t, \tau) \mathrm{d}\tau, \qquad t \geq 0. \tag{7.65}$$

7.8 Induced Convolution Operator Norms of Linear Dynamical Systems and $\mathcal{L}_p$ Stability

In this section, we develop induced convolution operator norms. For the system (7.52) and (7.53), let $G : \mathbb{R} \to \mathbb{R}^{l \times m}$ denote the impulse response function

$$G(t) \triangleq \begin{cases} 0, & t < 0, \\ Ce^{At}B, & t \geq 0. \end{cases} \tag{7.66}$$

Next, let $\mathcal{G} : \mathcal{L}_p \to \mathcal{L}_q$ denote the convolution operator

$$y(t) = \mathcal{G}[u](t) = (\mathcal{G} * u)(t) \triangleq \int_0^\infty G(t-\tau)u(\tau)\mathrm{d}\tau, \tag{7.67}$$

and define the induced norm $|||\mathcal{G}|||_{(q,s),(p,r)}$ as

$$|||\mathcal{G}|||_{(q,s),(p,r)} \triangleq \sup_{|||u|||_{p,r}=1} |||\mathcal{G} * u|||_{q,s}. \tag{7.68}$$

First, note that if the induced norm $|||\mathcal{G}|||_{(p,s),(p,r)}$ is bounded for $p, r, s \in [1, \infty]$, then $\mathcal{G}$ is $\mathcal{L}_p$-stable with finite gain. The following result presents a sufficient condition for $\mathcal{L}_p$ stability of $\mathcal{G}$.

Proposition 7.2. Consider the linear dynamical system given by (7.52) and (7.53) where A is Hurwitz. Then, for every $p \in [1, \infty]$, $\mathcal{G}$ is $\mathcal{L}_p$-stable with finite gain.

Proof. Since A is Hurwitz it follows that there exist positive definite matrices $P \in \mathbb{R}^{n\times n}$ and $R \in \mathbb{R}^{n\times n}$ such that

$$0 = A^{\mathrm{T}}P + PA + R.$$

Now, the result is a direct consequence of Theorem 7.6. Specifically, with $F(x,u) = Ax + Bu$, $H(x,u) = Cx$, $V(x) = x^{\mathrm{T}}Px$, $\alpha = \sigma_{\min}(P)$, $\beta = \sigma_{\max}(P)$, $\gamma_1 = \sigma_{\min}(R)$, $\gamma_2 = 2\sigma_{\max}(P)$, $L = \sigma_{\max}(B)$, $\eta_1 = \sigma_{\max}(C)$, and $\eta_2 = 0$, all the conditions of Theorem 7.6 are satisfied. □

Note that if (A, B) is controllable and (A, C) is observable, then the converse of Proposition 7.2 is also true (see, for example [445]). In the remainder of the chapter we provide several explicit expressions for induced norms of $\mathcal{G}$.

The following lemma provides an explicit expression for $|||\mathcal{G}|||_{(\infty,\infty),(r,r)}$ for the case in which $\mathcal{G}$ is a single-input/single-output operator.

Lemma 7.6. Let $r \in [1, \infty]$ and let $l = m = 1$. Then $\mathcal{G} : \mathcal{L}_r \to \mathcal{L}_\infty$, there exists $u \in \mathcal{L}_r$ such that $\lim_{t\to\infty}(\mathcal{G} * u)(t) = |||G|||_{\bar{r},\bar{r}}|||u|||_{r,r}$, and

$$|||\mathcal{G}|||_{(\infty,\infty),(r,r)} = |||G|||_{\bar{r},\bar{r}}. \tag{7.69}$$

Proof. For $r = 1$ and $r = \infty$, (7.69) is standard; see [461] and [104, pp. 23-24], respectively. Next, let $r \in (1, \infty)$ and note for all $t \geq 0$, it follows from Lemma 7.3 with $p = r$ that

$$|y(t)| = \left| \int_0^\infty G(t-\tau)u(\tau)\mathrm{d}\tau \right|$$

$$
\begin{aligned}
&\leq \left[\int_0^\infty |G(t-\tau)|^{\bar{r}}\mathrm{d}\tau\right]^{1/\bar{r}} \|u\|_{r,r}\\
&= \left[\int_0^t |G(\tau)|^{\bar{r}}\mathrm{d}\tau\right]^{1/\bar{r}} \|u\|_{r,r}\\
&\leq \|G\|_{\bar{r},\bar{r}}\|u\|_{r,r}
\end{aligned}
$$

which implies

$$\|\mathcal{G}\|_{(\infty,\infty),(r,r)} \leq \|G\|_{\bar{r},\bar{r}}. \tag{7.70}$$

Next, let $T > 0$ and let $u(\cdot)$ be such that $u(t) = \mathrm{sgn}(G(T-t))|G(T-t)|^{1/(r-1)}$, $t \geq 0$, where $\mathrm{sgn}(\cdot)$ denotes the signum function. Now, since $\|u\|_{r,r} = \left[\int_0^\infty |G(T-t)|^{\bar{r}}\mathrm{d}t\right]^{1/r}$, it follows that

$$
\begin{aligned}
|y(T)| &= \left|\int_0^\infty G(T-\tau)u(\tau)\mathrm{d}\tau\right|\\
&= \int_0^\infty |G(T-\tau)|^{\bar{r}}\mathrm{d}\tau\\
&= \left[\int_0^\infty |G(T-\tau)|^{\bar{r}}\mathrm{d}\tau\right]^{1/\bar{r}} \|u\|_{r,r}\\
&= \left[\int_0^T |G(\tau)|^{\bar{r}}\mathrm{d}\tau\right]^{1/\bar{r}} \|u\|_{r,r}.
\end{aligned}
$$

Hence,

$$\|y\|_{\infty,\infty} \geq \lim_{T\to\infty} |y(T)| = \|G\|_{\bar{r},\bar{r}}\|u\|_{r,r},$$

which, with (7.70), proves the result. □

Note that it follows from Lemma 7.6 that there exists $u \in \mathcal{L}_r$ such that $\lim_{t\to\infty}(\mathcal{G} * u)(t) = \|G\|_{\bar{r},\bar{r}}\|u\|_{r,r}$.

Next, define $\mathcal{P} \in \mathbb{R}^{m\times m}$ and $\mathcal{Q} \in \mathbb{R}^{l\times l}$ by

$$\mathcal{P} \triangleq \int_0^\infty G^{\mathrm{T}}(t)G(t)\mathrm{d}t, \qquad \mathcal{Q} \triangleq \int_0^\infty G(t)G^{\mathrm{T}}(t)\mathrm{d}t. \tag{7.71}$$

Note that $\mathcal{P} = B^{\mathrm{T}}PB$ and $\mathcal{Q} = CQC^{\mathrm{T}}$, where the observability and controllability Gramians P and Q, respectively, are the unique $n \times n$ nonnegative-definite solutions to the Lyapunov equations

$$0 = A^{\mathrm{T}}P + PA + C^{\mathrm{T}}C, \qquad 0 = AQ + QA^{\mathrm{T}} + BB^{\mathrm{T}}. \tag{7.72}$$

Furthermore, let $G_{[p,q]}$ denote the $l \times m$ matrix whose (i,j)th entry is $\|G_{(i,j)}\|_{(p,p),(q,q)}$.

Theorem 7.7. The following statements hold:

i) $\mathcal{G}:\mathcal{L}_2\to\mathcal{L}_2$, and

$$|||\mathcal{G}|||_{(2,2),(2,2)}=\sup_{\omega\in\mathbb{R}}\ \sigma_{\max}(H(\jmath\omega)). \tag{7.73}$$

ii) Let $r\in[1,\infty]$. Then $\mathcal{G}:\mathcal{L}_1\to\mathcal{L}_2$, and

$$|||\mathcal{G}|||_{(2,2),(1,r)}=\|\mathcal{P}^{1/2}\|_{2,r}. \tag{7.74}$$

iii) Let $p\in[1,\infty]$. Then $\mathcal{G}:\mathcal{L}_2\to\mathcal{L}_\infty$, and

$$|||\mathcal{G}|||_{(\infty,p),(2,2)}=\|\mathcal{Q}^{1/2}\|_{2,\bar{p}}. \tag{7.75}$$

iv) Let $p,r\in[1,\infty]$. Then $\mathcal{G}:\mathcal{L}_1\to\mathcal{L}_\infty$, and

$$|||\mathcal{G}|||_{(\infty,p),(1,r)}=\sup_{t\ge0}\|G(t)\|_{p,r}. \tag{7.76}$$

v) Let $r\in[1,\infty]$. Then $\mathcal{G}:\mathcal{L}_r\to\mathcal{L}_\infty$, and

$$|||\mathcal{G}|||_{(\infty,\infty),(r,r)}=\max_{i=1,\ldots,l}\|\mathrm{row}_i(G_{[\bar{r},\bar{r}]})\|_{\bar{r}}. \tag{7.77}$$

vi) Let $p\in[1,\infty]$. Then $\mathcal{G}:\mathcal{L}_1\to\mathcal{L}_p$, and

$$|||\mathcal{G}|||_{(p,p),(1,1)}=\max_{j=1,\ldots,m}\|\mathrm{col}_j(G_{[p,p]})\|_{p}. \tag{7.78}$$

Proof. *i)* It follows from Theorem 5.12 that

$$\begin{aligned}
|||y|||_{2,2}^2 &= \int_0^\infty y^{\mathrm{T}}(t)y(t)\mathrm{d}t\\
&=\frac{1}{2\pi}\int_{-\infty}^\infty y^*(\jmath\omega)y(\jmath\omega)\mathrm{d}\omega\\
&=\frac{1}{2\pi}\int_{-\infty}^\infty u^*(\jmath\omega)H^*(\jmath\omega)H(\jmath\omega)u(\jmath\omega)\mathrm{d}\omega\\
&\le \sup_{\omega\in\mathbb{R}}\ \sigma_{\max}(H(\jmath\omega))\frac{1}{2\pi}\int_{-\infty}^\infty u^*(\jmath\omega)u(\jmath\omega)\mathrm{d}\omega\\
&=\sup_{\omega\in\mathbb{R}}\ \sigma_{\max}(H(\jmath\omega))|||u|||_{2,2}^2,
\end{aligned}$$

which implies that

$$|||\mathcal{G}|||_{(2,2),(2,2)}\le\sup_{\omega\in\mathbb{R}}\ \sigma_{\max}(H(\jmath\omega)).$$

Next, let $\gamma\triangleq|||\mathcal{G}|||_{(2,2),(2,2)}$ and note that since $\mathcal{G}$ is causal, for every $u\in\mathcal{L}_2$ and for every $T>0$, $|||y_T|||_{2,2}\le\gamma|||u_T|||_{2,2}$ or, equivalently,

$$\int_0^T y^{\mathrm{T}}(t)y(t)\mathrm{d}t\le\gamma^2\int_0^T u^{\mathrm{T}}(t)u(t)\mathrm{d}t,\quad T\ge0.$$

Hence, it follows from Theorem 5.15 that $H(s)$ is bounded real (with gain γ), that is,

$$\sup_{\omega\in\mathbb{R}} \sigma_{\max}(H(\jmath\omega)) \leq \gamma = |\!|\!|\mathcal{G}|\!|\!|_{(2,2),(2,2)},$$

which implies that

$$|\!|\!|\mathcal{G}|\!|\!|_{(2,2),(2,2)} = \sup_{\omega\in\mathbb{R}} \sigma_{\max}(H(\jmath\omega)).$$

ii) It follows from Lemma 7.5 that

$$\begin{aligned}
|\!|\!|y|\!|\!|_{2,2} &\leq \int_0^\infty |\!|\!|G(t-\tau)u(\tau)|\!|\!|_{2,2}\mathrm{d}\tau \\
&= \int_0^\infty \left\{\int_0^\infty u^{\mathrm{T}}(\tau)G^{\mathrm{T}}(t-\tau)G(t-\tau)u(\tau)\mathrm{d}t\right\}^{1/2}\mathrm{d}\tau \\
&= \int_0^\infty \|\mathcal{P}^{1/2}u(\tau)\|_2\mathrm{d}\tau \\
&\leq \int_0^\infty \|\mathcal{P}^{1/2}\|_{2,r}\|u(\tau)\|_r\mathrm{d}\tau \\
&= \|\mathcal{P}^{1/2}\|_{2,r}|\!|\!|u|\!|\!|_{1,r},
\end{aligned}$$

which implies that $|\!|\!|\mathcal{G}|\!|\!|_{(2,2),(1,r)} \leq \|\mathcal{P}^{1/2}\|_{2,r}$.

Next, let $u_k(\cdot) = \hat{u}v_k(\cdot)$, $k = 1, 2, \ldots$, where $\hat{u} \in \mathbb{R}^d$ is such that $\|\hat{u}\|_r = 1$, $\|\mathcal{P}^{1/2}\hat{u}\|_2 = \|\mathcal{P}^{1/2}\|_{2,r}\|\hat{u}\|_r$, and measurable $v_k : [0,\infty) \to \mathbb{R}$ is such that $|\!|\!|v_k|\!|\!|_{1,1} = 1$ and, as $k \to \infty$, $v_k(\cdot) \to \delta(\cdot)$, where $\delta(\cdot)$ is the Dirac delta function. Note that $|\!|\!|u_k|\!|\!|_{1,r} = 1$, $k = 1, 2, \ldots$, and $y_k(t) \to G(t)\hat{u}$, $t \geq 0$, as $k \to \infty$, where $y_k(t) \triangleq (\mathcal{G} * u_k)(t)$. Hence,

$$\begin{aligned}
|\!|\!|\mathcal{G}|\!|\!|_{(2,2),(1,r)} &\geq \lim_{k\to\infty} |\!|\!|y_k|\!|\!|_{2,2} \\
&= \left\{\int_0^\infty \|G(t)\hat{u}\|_2^2\mathrm{d}t\right\}^{1/2} \\
&= \left\{\int_0^\infty \hat{u}^{\mathrm{T}}G^{\mathrm{T}}(t)G(t)\hat{u}\mathrm{d}t\right\}^{1/2} \\
&= (\hat{u}^{\mathrm{T}}\mathcal{P}\hat{u})^{1/2} \\
&= \|\mathcal{P}^{1/2}\hat{u}\|_2 \\
&= \|\mathcal{P}^{1/2}\|_{2,r},
\end{aligned}$$

which implies that $|\!|\!|\mathcal{G}|\!|\!|_{(2,2),(1,r)} = \|\mathcal{P}^{1/2}\|_{2,r}$.

iii) With $p = r = 2$ it follows from Lemma 7.3 that for all $t \geq 0$,

$$\begin{aligned}
\|y(t)\|_p &= \left\| \int_0^\infty G(t-\tau)u(\tau)\mathrm{d}\tau \right\|_p \\
&= \max_{\{\hat{u}\in\mathbb{R}^n:\ \|\hat{u}\|_{\bar{p}}=1\}} \int_0^\infty \hat{u}^\mathrm{T} G(t-\tau)u(\tau)\mathrm{d}\tau \\
&\leq \max_{\{\hat{u}\in\mathbb{R}^n:\ \|\hat{u}\|_{\bar{p}}=1\}} \left[\int_0^\infty \|G(t-\tau)\hat{u}\|_2^2 \mathrm{d}\tau\right]^{1/2} |\!|\!| u |\!|\!|_{2,2} \\
&= \max_{\{\hat{u}\in\mathbb{R}^n:\ \|\hat{u}\|_{\bar{p}}=1\}} \left[\hat{u}^\mathrm{T} \int_0^t G(\tau)G^\mathrm{T}(\tau)\mathrm{d}\tau \hat{u}\right]^{1/2} |\!|\!| u |\!|\!|_{2,2} \\
&\leq \max_{\{\hat{u}\in\mathbb{R}^n:\ \|\hat{u}\|_{\bar{p}}=1\}} (\hat{u}^\mathrm{T} \mathcal{Q}\hat{u})^{1/2} |\!|\!| u |\!|\!|_{2,2} \\
&= \|\mathcal{Q}^{1/2}\|_{2,\bar{p}} |\!|\!| u |\!|\!|_{2,2},
\end{aligned}$$

which implies that $|\!|\!| y |\!|\!|_{\infty,p} \leq \|\mathcal{Q}^{1/2}\|_{2,\bar{p}} |\!|\!| u |\!|\!|_{2,2}$ for all $y \in \mathcal{L}_\infty$ and $u \in \mathcal{L}_2$, and hence, $|\!|\!| \mathcal{G} |\!|\!|_{(\infty,p),(2,2)} \leq \|\mathcal{Q}^{1/2}\|_{2,\bar{p}}$.

Next, let $\hat{u} \in \mathbb{R}^d$ be such that $\|\hat{u}\|_{\bar{p}} = 1$ and $\|\mathcal{Q}^{1/2}\hat{u}\|_2 = \|\mathcal{Q}^{1/2}\|_{2,\bar{p}}$, and let $T > 0$ and

$$u(t) = \frac{1}{\|\mathcal{Q}^{1/2}\|_{2,\bar{p}}} G^\mathrm{T}(T-t)\hat{u},$$

so that $|\!|\!| u |\!|\!|_{2,2} \leq 1$. Now, since $\|\cdot\|_{p\mathrm{D}} = \|\cdot\|_{\bar{p}}$, it follows that

$$\begin{aligned}
|\!|\!| y |\!|\!|_{\infty,p} &= \sup_{t\geq 0} \|y(t)\|_p \\
&= \sup_{t\geq 0} \max_{\{\hat{y}\in\mathbb{R}^n:\|\hat{y}\|_{\bar{p}}=1\}} \hat{y}^\mathrm{T} y(t) \\
&\geq \sup_{t\geq 0} \hat{u}^\mathrm{T} y(t) \\
&= \frac{1}{\|\mathcal{Q}^{1/2}\|_{2,\bar{p}}} \sup_{t\geq 0} \int_0^\infty \hat{u}^\mathrm{T} G(T-\tau)G^\mathrm{T}(T-\tau)\hat{u}\mathrm{d}\tau \\
&= \frac{1}{\|\mathcal{Q}^{1/2}\|_{2,\bar{p}}} \sup_{t\geq 0} \int_0^T \hat{u}^\mathrm{T} G(\tau)G^\mathrm{T}(\tau)\hat{u}\mathrm{d}\tau,
\end{aligned}$$

which implies that, for every $T > 0$, there exists $u \in \mathcal{L}_2$ such that $|\!|\!| u |\!|\!|_{2,2} \leq 1$ and

$$|\!|\!| y |\!|\!|_{\infty,p} \geq \frac{1}{\|\mathcal{Q}^{1/2}\|_{2,\bar{p}}} \sup_{t\geq 0} \int_0^T \hat{u}^\mathrm{T} G(\tau)G^\mathrm{T}(\tau)\hat{u}\mathrm{d}\tau,$$

or, equivalently,

$$|\!|\!| \mathcal{G} |\!|\!|_{(\infty,p),(2,2)} \geq \sup_{T>0} \frac{1}{\|\mathcal{Q}^{1/2}\|_{2,\bar{p}}} \sup_{t\geq 0} \int_0^T \hat{u}^\mathrm{T} G(\tau)G^\mathrm{T}(\tau)\hat{u}\mathrm{d}\tau$$

$$
\begin{aligned}
&= \frac{1}{\|Q^{1/2}\|_{2,\bar{p}}} \sup_{t\geq 0} \int_0^\infty \hat{u}^{\mathrm{T}} G(\tau) G^{\mathrm{T}}(\tau) \hat{u} \mathrm{d}\tau \\
&= \frac{1}{\|Q^{1/2}\|_{2,\bar{p}}} \hat{u}^{\mathrm{T}} Q \hat{u} \\
&= \|Q^{1/2}\|_{2,\bar{p}},
\end{aligned}
$$

which further implies that $|||\mathcal{G}|||_{(\infty,p),(2,2)} = \|Q^{1/2}\|_{2,\bar{p}}$.

iv) Note that for all $t \geq 0$,

$$
\begin{aligned}
\|y(t)\|_p &\leq \int_0^\infty \|G(t-\tau)u(\tau)\|_p \mathrm{d}\tau \\
&\leq \int_0^\infty \|G(t-\tau)\|_{p,r} \|u(\tau)\|_r \mathrm{d}\tau \\
&\leq \sup_{t\geq 0} \|G(t)\|_{p,r} \int_0^\infty \|u(\tau)\|_r \mathrm{d}\tau \\
&\leq \sup_{t\geq 0} \|G(t)\|_{p,r} |||u|||_{1,r},
\end{aligned}
$$

which implies that $|||\mathcal{G}|||_{(\infty,p),(1,r)} \leq \sup_{t\geq 0} \|G(t)\|_{p,r}$.

Next, let $\varepsilon > 0$ and $t_0 \in [0,\infty)$ be such that $\|G(t_0)\|_{p,r} > \sup_{t\geq 0} \|G(t)\|_{p,r} - \varepsilon$. In addition, let $u_k(\cdot) = v_k(\cdot)\hat{u}$, $k = 1, 2, \ldots$, where $\hat{u} \in \mathbb{R}^n$ is such that $\|\hat{u}\|_r = 1$, $\|G(t_0)\hat{u}\|_p = \|G(t_0)\|_{p,r}\|\hat{u}\|_r$, and measurable $v_k : [0,\infty) \to \mathbb{R}$ is such that $|||v_k|||_{1,1} = 1$ and, as $k \to \infty$, $v_k(\cdot) \to \delta(\cdot)$, where $\delta(\cdot)$ is the Dirac delta function. In this case, note that $|||u_k|||_{1,r} = 1$, $k = 1, 2, \ldots$, and $y_k(t) \to G(t)\hat{u}$, $t \geq 0$, as $k \to \infty$, where $y_k(t) \triangleq (\mathcal{G} * u_k)(t)$. Hence,

$$
\begin{aligned}
|||\mathcal{G}|||_{(\infty,p),(1,r)} &\geq \lim_{k\to\infty} \sup_{t\geq 0} \|y_k(t)\|_p \\
&= \sup_{t\geq 0} \|G(t)\hat{u}\|_p \\
&\geq \|G(t_0)\hat{u}\|_p \\
&= \|G(t_0)\|_{p,r} \\
&> \sup_{t\geq 0} \|G(t)\|_{p,r} - \varepsilon,
\end{aligned}
$$

which implies that

$$
\sup_{t\geq 0} \|G(t)\|_{p,r} - \varepsilon < |||\mathcal{G}|||_{(\infty,p),(1,r)} \leq \sup_{t\geq 0} \|G(t)\|_{p,r}, \qquad \varepsilon > 0,
$$

and hence, (7.76) holds.

v) Note that for all $u \in \mathcal{L}_r$ and $y \in \mathcal{L}_\infty$ it follows that $|||u|||_{r,r} = \|\bar{u}\|_r$ and $|||y|||_{\infty,\infty} = \|\bar{y}\|_\infty$, where $\bar{u} \in \mathbb{R}^m$ and $\bar{y} \in \mathbb{R}^l$ with $\bar{u}_i = |||u_i|||_{r,r}$, $i = 1, \ldots, m$, and $\bar{y}_i = |||y_i|||_{\infty,\infty}$, $i = 1, \ldots, l$. Next, it follows from Lemma 7.6

that $\|\mathcal{G}_{(i,j)}\|_{(\infty,\infty),(r,r)} = \|G_{(i,j)}\|_{\bar{r},\bar{r}}$, and hence,

$$\begin{aligned}
\|y_i\|_{\infty,\infty} &= \|\sum_{j=1}^{m} \mathcal{G}_{(i,j)} * u_j\|_{\infty,\infty} \\
&\le \sum_{j=1}^{m} \|\mathcal{G}_{(i,j)} * u_j\|_{\infty,\infty} \\
&\le \sum_{j=1}^{m} \|\mathcal{G}_{(i,j)}\|_{\bar{r},\bar{r}} \|u_j\|_{r,r} \\
&\le \|\mathrm{row}_i(G_{[\bar{r},\bar{r}]})\|_{\bar{r}} \|\bar{u}\|_r \\
&\le \max_{i=1,\ldots,l} \|\mathrm{row}_i(G_{[\bar{r},\bar{r}]})\|_{\bar{r}} \|\bar{u}\|_r,
\end{aligned}$$

which implies that $\|y\|_{\infty,\infty} = \|\bar{y}\|_\infty \le \max_{i=1,\ldots,l} \|\mathrm{row}_i(G_{[\bar{r},\bar{r}]})\|_{\bar{r}} \|u\|_{r,r}$, and hence,

$$\|\mathcal{G}\|_{(\infty,\infty),(r,r)} \le \max_{i=1,\ldots,l} \|\mathrm{row}_i(G_{[\bar{r},\bar{r}]})\|_{\bar{r}}. \tag{7.79}$$

Next, let $I \in \{1,\ldots,l\}$ be such that $\|\mathrm{row}_I(G_{[\bar{r},\bar{r}]})\|_{\bar{r}} = \max_{i=1,\ldots,l} \|\mathrm{row}_i(G_{[\bar{r},\bar{r}]})\|_{\bar{r}}$. Now, let $\hat{u} \in \mathbb{R}^m$ be such that $\|\hat{u}\|_r = 1$, let $\mathrm{row}_I(G_{[\bar{r},\bar{r}]})\hat{u} = \|\mathrm{row}_I(G_{[\bar{r},\bar{r}]})\|_{\bar{r}}$, and let $u_j \in \mathcal{L}_r$, $j = 1,\ldots,m$, be such that $\|u_j\|_{r,r} = \hat{u}_j$ and $\lim_{t\to\infty}(\mathcal{G}_{(I,j)} * u_j)(t) = \|G_{(I,j)}\|_{\bar{r},\bar{r}} \|u_j\|_{r,r}$. Note that existence of such a $u_j(\cdot)$ follows from Lemma 7.6. Now,

$$\begin{aligned}
\|y\|_{\infty,\infty} &\ge \|y_I\|_{\infty,\infty} \\
&\ge \lim_{t\to\infty} |y_I(t)| \\
&= \lim_{t\to\infty} |\sum_{j=1}^{m} (\mathcal{G}_{(I,j)} * u_j)(t)| \\
&= \sum_{j=1}^{m} \|G_{(I,j)}\|_{\bar{r},\bar{r}} \|u_j\|_{r,r} \\
&= \mathrm{row}_I(G_{[\bar{r},\bar{r}]})\bar{u} \\
&= \|\mathrm{row}_I(G_{[\bar{r},\bar{r}]})\|_{\bar{r}},
\end{aligned}$$

which, with (7.79), implies (7.77).

vi) For $p = \infty$, (7.78) is a direct consequence of *iv)* or *v)*. Now, let $p \in [1,\infty)$ and note that it follows from Lemma 7.4 that $\|y\|_{p,p} = \sup_{\{\hat{y}\in\mathcal{L}_{\bar{p}}\colon\ \|\hat{y}\|_{\bar{p},\bar{p}}=1\}} \langle y, \hat{y}\rangle$. Hence, with $p = r = 1$ it follows from Lemma 7.3 that

$$\|y\|_{p,p} = \sup_{\|\hat{y}\|_{\bar{p},\bar{p}}=1} \int_0^\infty y^{\mathrm{T}}(t)\hat{y}(t)\mathrm{d}t$$

$$\begin{aligned}
&= \sup_{\|\hat{y}\|_{\bar{p},\bar{p}}=1} \int_0^\infty \left(\int_0^\infty u^{\mathrm{T}}(\tau) G^{\mathrm{T}}(t-\tau) \mathrm{d}\tau \right) \hat{y}(t) \mathrm{d}t \\
&= \sup_{\|\hat{y}\|_{\bar{p},\bar{p}}=1} \int_0^\infty u^{\mathrm{T}}(\tau) \left(\int_0^\infty G^{\mathrm{T}}(t-\tau) \hat{y}(t) \mathrm{d}t \right) \mathrm{d}\tau \\
&= \sup_{\|\hat{y}\|_{\bar{p},\bar{p}}=1} \langle u, \hat{u} \rangle \\
&\leq \|u\|_{1,1} \sup_{\|\hat{y}\|_{\bar{p},\bar{p}}=1} \|\hat{u}\|_{\infty,\infty},
\end{aligned}$$

where $\hat{u}(t) \triangleq \int_0^\infty G^{\mathrm{T}}(\tau - t)\hat{y}(\tau)\mathrm{d}\tau$. Now, with $r = \bar{p}$, it follows from *v)* that

$$\|\mathcal{G}\|_{(p,p),(1,1)} \leq \sup_{\|\hat{y}\|_{\bar{p},\bar{p}}=1} \|\hat{u}\|_{\infty,\infty} = \max_{j=1,\ldots,m} \|\mathrm{col}_j(G_{[p,p]})\|_p. \tag{7.80}$$

Next, let $J \in \{1, \ldots, m\}$ be such that $\|\mathrm{col}_J(G_{[p,p]})\|_p = \max_{j=1,\ldots,m} \|\mathrm{col}_j(G_{[p,p]})\|_p$ and let $u_k(\cdot) \triangleq v_k(\cdot)e_J$, $k = 1, 2, \ldots$, where $v_k : [0, \infty) \to \mathbb{R}$ is a measurable function such that $\|v_k\|_{1,1} = 1$ and, as $k \to \infty$, $v_k(\cdot) \to \delta(\cdot)$, where $\delta(\cdot)$ is the Dirac delta function. In this case note that $\|u_k\|_{1,1} = 1$, $k = 1, 2, \ldots$, and $y_k(t) \to \mathrm{col}_J(G(t))$, $t \geq 0$, as $k \to \infty$, where $y_k(t) \triangleq (\mathcal{G} * u_k)(t)$. Hence,

$$\begin{aligned}
\|\mathcal{G}\|_{(p,p),(1,1)} &\geq \lim_{k\to\infty} \|y_k\|_{p,p} \\
&= \|\mathrm{col}_J(G)\|_{p,p} \\
&= \|\mathrm{col}_J(G_{[p,p]})\|_p \\
&= \max_{j=1,\ldots,m} \|\mathrm{col}_j(G_{[p,p]})\|_p,
\end{aligned}$$

which, with (7.80), implies (7.78). □

The following corollary specializes Theorem 7.7 to the results given in [461, 462] and [104, p. 26].

Corollary 7.3. The following statements hold:

i) $\mathcal{G} : \mathcal{L}_1 \to \mathcal{L}_2$, $\|\mathcal{G}\|_{(2,2),(1,2)} = \sigma_{\max}^{1/2}(\mathcal{P})$, and $\|\mathcal{G}\|_{(2,2),(1,1)} = \mathrm{d}_{\max}^{1/2}(\mathcal{P})$.

ii) $\mathcal{G} : \mathcal{L}_2 \to \mathcal{L}_\infty$, $\|\mathcal{G}\|_{(\infty,2),(2,2)} = \sigma_{\max}^{1/2}(\mathcal{Q})$, and $\|\mathcal{G}\|_{(\infty,\infty),(2,2)} = \mathrm{d}_{\max}^{1/2}(\mathcal{Q})$.

iii) $\mathcal{G} : \mathcal{L}_1 \to \mathcal{L}_\infty$, $\|\mathcal{G}\|_{(\infty,\infty),(1,1)} = \sup_{t\geq 0} \|G(t)\|_\infty$, and $\|\mathcal{G}\|_{(\infty,2),(1,2)} = \sup_{t\geq 0} \sigma_{\max}(G(t))$.

iv) $\mathcal{G} : \mathcal{L}_\infty \to \mathcal{L}_\infty$, and $\|\mathcal{G}\|_{(\infty,\infty),(\infty,\infty)} = \max_{i=1,\ldots,l} \|\mathrm{row}_i(G_{[1,1]})\|_1$.

v) $\mathcal{G} : \mathcal{L}_1 \to \mathcal{L}_1$, and $\|\mathcal{G}\|_{(1,1),(1,1)} = \max_{j=1,\ldots,m} \|\mathrm{col}_j(G_{[1,1]})\|_1$.

Recall that the $\mathcal{H}_2$ norm of the system (7.52) and (7.53) is given by $|||\mathcal{G}|||_{\mathcal{H}_2} = \|\mathcal{P}^{1/2}\|_{\mathrm{F}} = \|\mathcal{Q}^{1/2}\|_{\mathrm{F}}$. Hence, using the fact that $\|\cdot\|_{\mathrm{F}} = \sigma_{\max}(\cdot)$ for rank-one matrices, it follows from *i)* of Corollary 7.3 that if B (and hence $\mathcal{P}$) is a rank-one matrix then $|||\mathcal{G}|||_{\mathcal{H}_2} = |||\mathcal{G}|||_{(2,2),(1,2)}$. Similarly, it follows from *iii)* of Corollary 7.3 that if C (and hence $\mathcal{Q}$) is a rank-one matrix then $|||\mathcal{G}|||_{\mathcal{H}_2} = |||\mathcal{G}|||_{(\infty,2),(2,2)}$. Hence, in the single-input/multi-output and multi-input/single-output cases the $\mathcal{H}_2$ norm of a dynamical system is induced. In the multi-input/multi-output case, however, the $\mathcal{H}_2$ norm does not appear to be induced. For related details see [84].

Theorem 7.7 also applies to the more general case where $\mathcal{G}$ is a noncausal, time-invariant operator. In this case, the input-output spaces $\mathcal{L}_p$ and $\mathcal{L}_r$ are defined for $t \in (-\infty, \infty)$, $H(\jmath\omega)$ is the Fourier transform of $G(t)$, and the lower limit in the integrals defining $\mathcal{P}$ and $\mathcal{Q}$ is replaced by $-\infty$.

An alternative characterization of input-output properties is the Hankel norm which provides a mapping from past inputs $u(t)$, $t \in (-\infty, 0]$, to future outputs $y(t)$, $t \in [0, \infty)$ [136, 461]. For causal dynamical systems the Hankel operator $\Gamma : \mathcal{L}_p(-\infty, 0] \to \mathcal{L}_q$ is defined by

$$y(t) = (\Gamma * u)(t) \triangleq \int_0^\infty G(t+\tau)u(-\tau)\mathrm{d}\tau, \qquad t \in [0, \infty), \tag{7.81}$$

where $\mathcal{L}_p(-\infty, 0]$ denotes the set of functions in $\mathcal{L}_p$ on the time interval $(-\infty, 0]$, and the induced Hankel norm $|||\Gamma|||_{(q,s),(p,r)}$ is defined by

$$|||\Gamma|||_{(q,s),(p,r)} \triangleq \sup_{|||u|||_{p,r}=1} |||\Gamma * u|||_{q,s}. \tag{7.82}$$

Proposition 7.3. The following statements hold:

i) $\Gamma : \mathcal{L}_2(-\infty, 0] \to \mathcal{L}_2$, and

$$|||\Gamma|||_{(2,2),(2,2)} = \lambda_{\max}^{1/2}(PQ). \tag{7.83}$$

ii) Let $r \in [1, \infty]$. Then $\Gamma : \mathcal{L}_1(-\infty, 0] \to \mathcal{L}_2$, and

$$|||\Gamma|||_{(2,2),(1,r)} = \|\mathcal{P}^{1/2}\|_{2,r}. \tag{7.84}$$

iii) Let $p \in [1, \infty]$. Then $\Gamma : \mathcal{L}_2(-\infty, 0] \to \mathcal{L}_\infty$, and

$$|||\Gamma|||_{(\infty,p),(2,2)} = \|\mathcal{Q}^{1/2}\|_{2,\bar{p}}. \tag{7.85}$$

iv) Let $p, r \in [1, \infty]$. Then $\Gamma : \mathcal{L}_1(-\infty, 0] \to \mathcal{L}_\infty$, and

$$|||\Gamma|||_{(\infty,p),(1,r)} = \sup_{t \geq 0} \|G(t)\|_{p,r}. \tag{7.86}$$

v) Let $r \in [1, \infty]$. Then $\Gamma : \mathcal{L}_r(-\infty, 0] \to \mathcal{L}_\infty$, and

$$|\!|\!|\Gamma|\!|\!|_{(\infty,\infty),(r,r)} = \max_{i=1,\ldots,l} \|\mathrm{row}_i(G_{[\bar{r},\bar{r}]})\|_{\bar{r}}. \tag{7.87}$$

vi) Let $p \in [1, \infty]$. Then $\Gamma : \mathcal{L}_1(-\infty, 0] \to \mathcal{L}_p$, and

$$|\!|\!|\Gamma|\!|\!|_{(p,p),(1,1)} = \max_{j=1,\ldots,m} \|\mathrm{col}_j(G_{[p,p]})\|_p. \tag{7.88}$$

Proof. The proof of *i)* is standard; see [136] for a proof. The proof of *ii)*–*vi)* is similar to that of *ii)*–*vi)* of Theorem 7.7 with appropriate modifications to the time interval for the input space. □

7.9 Finitely Computable Upper Bounds for $|\!|\!|\mathcal{G}|\!|\!|_{(\infty,p),(1,r)}$

In this section, we obtain a finitely computable upper bound for (7.76). To do this we assume that there exist $H_\mathcal{L}(s), H_\mathcal{R}(s) \in \mathcal{RH}_2$ such that $H(s) = H_\mathcal{L}(s)H_\mathcal{R}(s)$, where $H(s) \in \mathcal{RH}_2$ denotes the Laplace transform of $G(t)$. Note that such a factorization exists only if $H(s)$ has relative degree two. Furthermore, note that the above factorization exists if and only if there exist linear, time-invariant asymptotically stable dynamical systems with impulse response functions $G_\mathcal{L} : \mathbb{R} \to \mathbb{R}^{l\times m_\mathrm{l}}$ and $G_\mathcal{R} : \mathbb{R} \to \mathbb{R}^{m_\mathrm{l}\times m}$ such that $G_\mathcal{L}(t) = 0$ and $G_\mathcal{R}(t) = 0$, $t < 0$, and

$$G(t) = \int_0^\infty G_\mathcal{L}(t-\tau)G_\mathcal{R}(\tau)\mathrm{d}\tau, \qquad t \geq 0. \tag{7.89}$$

Next, let $\mathcal{G}_\mathcal{L} : \mathcal{L}_2 \to \mathcal{L}_\infty$ and $\mathcal{G}_\mathcal{R} : \mathcal{L}_1 \to \mathcal{L}_2$ denote the convolution operators of $G_\mathcal{L}$ and $G_\mathcal{R}$, respectively, and define $\mathcal{P}_\mathcal{R} \in \mathbb{R}^{m\times m}$ and $\mathcal{Q}_\mathcal{L} \in \mathbb{R}^{l\times l}$ by

$$\mathcal{P}_\mathcal{R} \triangleq \int_0^\infty G_\mathcal{R}^\mathrm{T}(t)G_\mathcal{R}(t)\mathrm{d}t, \qquad \mathcal{Q}_\mathcal{L} \triangleq \int_0^\infty G_\mathcal{L}(t)G_\mathcal{L}^\mathrm{T}(t)\mathrm{d}t. \tag{7.90}$$

Finally, let $G_\mathcal{L}(t) = C_\mathcal{L}e^{A_\mathcal{L}t}B_\mathcal{L}$, $t \geq 0$, and $G_\mathcal{R}(t) = C_\mathcal{R}e^{A_\mathcal{R}t}B_\mathcal{R}$, $t \geq 0$, where $A_\mathcal{L} \in \mathbb{R}^{n_\mathrm{l}\times n_\mathrm{l}}$, $B_\mathcal{L} \in \mathbb{R}^{n_\mathrm{l}\times m_\mathrm{l}}$, $C_\mathcal{L} \in \mathbb{R}^{l\times n_\mathrm{l}}$, $A_\mathcal{R} \in \mathbb{R}^{n_\mathrm{r}\times n_\mathrm{r}}$, $B_\mathcal{R} \in \mathbb{R}^{n_\mathrm{r}\times m}$, and $C_\mathcal{R} \in \mathbb{R}^{m_\mathrm{l}\times n_\mathrm{r}}$, and let $P_\mathcal{R} \in \mathbb{R}^{n_\mathrm{r}\times n_\mathrm{r}}$ and $Q_\mathcal{L} \in \mathbb{R}^{n_\mathrm{l}\times n_\mathrm{l}}$ be the unique, nonnegative-definite solutions to the Lyapunov equations

$$0 = A_\mathcal{R}^\mathrm{T}P_\mathcal{R} + P_\mathcal{R}A_\mathcal{R} + C_\mathcal{R}^\mathrm{T}C_\mathcal{R}, \quad 0 = A_\mathcal{L}Q_\mathcal{L} + Q_\mathcal{L}A_\mathcal{L}^\mathrm{T} + B_\mathcal{L}B_\mathcal{L}^\mathrm{T}. \tag{7.91}$$

Note that $\mathcal{P}_\mathcal{R} = B_\mathcal{R}^\mathrm{T}P_\mathcal{R}B_\mathcal{R}$ and $\mathcal{Q}_\mathcal{L} = C_\mathcal{L}Q_\mathcal{L}C_\mathcal{L}^\mathrm{T}$.

Proposition 7.4. Let $p, r \in [1, \infty]$. If there exist $G_\mathcal{L} : \mathbb{R} \to \mathbb{R}^{l\times m_\mathrm{l}}$ and $G_\mathcal{R} : \mathbb{R} \to \mathbb{R}^{m_\mathrm{l}\times m}$ such that (7.89) holds, then

$$|\!|\!|\mathcal{G}|\!|\!|_{(\infty,p),(1,r)} \leq \|\mathcal{Q}_\mathcal{L}^{1/2}\|_{2,\bar{p}}\|\mathcal{P}_\mathcal{R}^{1/2}\|_{2,r}. \tag{7.92}$$

Proof. Note that $y(t) = (\mathcal{G}_{\mathcal{L}} * (\mathcal{G}_{\mathcal{R}} * u))(t)$. Now, since $\mathcal{G}_{\mathcal{L}} : \mathcal{L}_2 \to \mathcal{L}_\infty$ and $\mathcal{G}_{\mathcal{R}} : \mathcal{L}_1 \to \mathcal{L}_2$ it follows from Theorem 7.7 that

$$|||y|||_{\infty,p} \le \|\mathcal{Q}_{\mathcal{L}}^{1/2}\|_{2,\hat{p}} |||\mathcal{G}_{\mathcal{R}} * u|||_{2,2} \le \|\mathcal{Q}_{\mathcal{L}}^{1/2}\|_{2,\hat{p}} \|\mathcal{P}_{\mathcal{R}}^{1/2}\|_{2,r} |||u|||_{1,r},$$

which implies (7.92). □

The following corollary of Proposition 7.4 provides finitely computable bounds for the mixed-induced signal norm (7.76).

Corollary 7.4. Let $\mathcal{P}_{\mathcal{R}}$ and $\mathcal{Q}_{\mathcal{L}}$ be given by (7.90). Then the following inequalities hold:

i) $|||\mathcal{G}|||_{(\infty,\infty),(1,1)} \le \mathrm{d}_{\max}^{1/2}(\mathcal{Q}_{\mathcal{L}}) \mathrm{d}_{\max}^{1/2}(\mathcal{P}_{\mathcal{R}})$.

ii) $|||\mathcal{G}|||_{(\infty,2),(1,2)} \le \sigma_{\max}^{1/2}(\mathcal{Q}_{\mathcal{L}}) \sigma_{\max}^{1/2}(\mathcal{P}_{\mathcal{R}})$.

iii) $|||\mathcal{G}|||_{(\infty,\infty),(1,2)} \le \mathrm{d}_{\max}^{1/2}(\mathcal{Q}_{\mathcal{L}}) \sigma_{\max}^{1/2}(\mathcal{P}_{\mathcal{R}})$.

iv) $|||\mathcal{G}|||_{(\infty,2),(1,1)} \le \sigma_{\max}^{1/2}(\mathcal{Q}_{\mathcal{L}}) \mathrm{d}_{\max}^{1/2}(\mathcal{P}_{\mathcal{R}})$.

Proof. The results follow from Theorem 7.7 and Proposition 7.4. □

7.10 Upper Bounds for $\mathcal{L}_1$ Operator Norms

In this section, we provide upper bounds for the $\mathcal{L}_1$ operator norm $|||\mathcal{G}|||_{(\infty,p),(\infty,r)}$. For $\alpha > 0$, define the shifted impulse response function $G_\alpha : \mathbb{R} \to \mathbb{R}^{l \times m}$ by

$$G_\alpha(t) \triangleq \begin{cases} 0 & t < 0, \\ Ce^{(A+\frac{\alpha}{2}I)t}B, & t \ge 0, \end{cases} \tag{7.93}$$

and let $\mathcal{G}_\alpha$ denote its convolution operator

$$y(t) = (\mathcal{G}_\alpha * u)(t) \triangleq \int_0^\infty G_\alpha(t-\tau)u(\tau)\mathrm{d}\tau. \tag{7.94}$$

Furthermore, for some of the results in this section we assume there exist $H_{\mathcal{L}_\alpha}(s), H_{\mathcal{R}_\alpha}(s) \in \mathcal{RH}_2$ such that $H_\alpha(s) = H_{\mathcal{L}_\alpha}(s)H_{\mathcal{R}_\alpha}(s)$, where $H_\alpha(s) \in \mathcal{RH}_2$ denotes the Laplace transform of $G_\alpha(t)$. Note that the above factorization exists if and only if there exist linear time-invariant asymptotically stable dynamical systems with impulse response functions $G_{\mathcal{L}_\alpha} : \mathbb{R} \to \mathbb{R}^{l \times m_l}$ and $G_{\mathcal{R}_\alpha} : \mathbb{R} \to \mathbb{R}^{m_l \times m}$ such that $G_{\mathcal{L}_\alpha}(t) = 0$ and $G_{\mathcal{R}_\alpha}(t) = 0$, $t < 0$, and

$$G_\alpha(t) = \int_0^\infty G_{\mathcal{L}_\alpha}(t-\tau)G_{\mathcal{R}_\alpha}(\tau)\mathrm{d}\tau, \qquad t \ge 0. \tag{7.95}$$

Next, let $\mathcal{G}_{\mathcal{L}_\alpha} : \mathcal{L}_2 \to \mathcal{L}_\infty$ and $\mathcal{G}_{\mathcal{R}_\alpha} : \mathcal{L}_1 \to \mathcal{L}_2$ denote the convolution operators of $G_{\mathcal{L}_\alpha}$ and $G_{\mathcal{R}_\alpha}$, respectively, and define $\mathcal{P}_{\mathcal{R}_\alpha} \in \mathbb{R}^{m\times m}$ and $\mathcal{Q}_{\mathcal{L}_\alpha} \in \mathbb{R}^{l\times l}$ by

$$\mathcal{P}_{\mathcal{R}_\alpha} \triangleq \int_0^\infty G_{\mathcal{R}_\alpha}^{\mathrm{T}}(t)G_{\mathcal{R}_\alpha}(t)\mathrm{d}t, \qquad \mathcal{Q}_{\mathcal{L}_\alpha} \triangleq \int_0^\infty G_{\mathcal{L}_\alpha}(t)G_{\mathcal{L}_\alpha}^{\mathrm{T}}(t)\mathrm{d}t. \tag{7.96}$$

Theorem 7.8. Let $\alpha > 0$ be such that $A + \frac{\alpha}{2}I$ is asymptotically stable and let $Q_\alpha \in \mathbb{R}^{n\times n}$ be the unique, nonnegative-definite solution to the Lyapunov equation

$$0 = AQ_\alpha + Q_\alpha A^{\mathrm{T}} + \alpha Q_\alpha + BB^{\mathrm{T}}. \tag{7.97}$$

Furthermore, let $p, r \in [1,\infty]$. Then $\mathcal{G} : \mathcal{L}_\infty \to \mathcal{L}_\infty$,

$$|||\mathcal{G}|||_{(\infty,p),(\infty,2)} \le \frac{1}{\sqrt{\alpha}}|||\mathcal{G}_\alpha|||_{(\infty,p),(2,2)} = \frac{1}{\sqrt{\alpha}}\|(CQ_\alpha C^{\mathrm{T}})^{1/2}\|_{2,\bar{p}}, \tag{7.98}$$

and

$$|||\mathcal{G}|||_{(\infty,p),(\infty,r)} \le \frac{2}{\alpha}|||\mathcal{G}_\alpha|||_{(\infty,p),(1,r)} = \frac{2}{\alpha}\sup_{t\ge 0}\|G_\alpha(t)\|_{p,r}. \tag{7.99}$$

In addition, if there exist $G_{\mathcal{L}_\alpha} : \mathbb{R} \to \mathbb{R}^{l\times m_1}$ and $G_{\mathcal{R}_\alpha} : \mathbb{R} \to \mathbb{R}^{m_1\times m}$ such that (7.95) holds, then

$$|||\mathcal{G}|||_{(\infty,p),(\infty,r)} \le \frac{2}{\alpha}|||\mathcal{G}_\alpha|||_{(\infty,p),(1,r)} \le \frac{2}{\alpha}\|\mathcal{Q}_{\mathcal{L}_\alpha}^{1/2}\|_{2,\bar{p}}\|\mathcal{P}_{\mathcal{R}_\alpha}^{1/2}\|_{2,r}. \tag{7.100}$$

Proof. Let $T > 0$, $u \in \mathcal{L}_\infty$, and define

$$u_T(t) \triangleq \begin{cases} e^{\frac{\alpha}{2}(t-T)}u(t), & 0 \le t \le T, \\ 0, & t > T. \end{cases} \tag{7.101}$$

Now, note that

$$\begin{aligned} |||u_T|||_{2,2}^2 &= \int_0^\infty \|u_T(t)\|_2^2 \mathrm{d}t \\ &= \int_0^T e^{\alpha(t-T)}\|u(t)\|_2^2\mathrm{d}t \\ &\le |||u|||_{\infty,2}^2 \int_0^T e^{\alpha(t-T)}\mathrm{d}t \\ &= \frac{1}{\alpha}|||u|||_{\infty,2}^2, \end{aligned}$$

or, equivalently, $|||u_T|||_{2,2} \le \frac{1}{\sqrt{\alpha}}|||u|||_{\infty,2}$. Next, define $y_T(t) \triangleq e^{\frac{\alpha}{2}(t-T)}y(t)$ and, since $G(t) = 0$, $t < 0$, note that

$$y_T(t) = \int_0^\infty e^{\frac{\alpha}{2}(t-T)}G(t-\tau)u(\tau)\mathrm{d}\tau$$

$$
\begin{aligned}
&= \int_0^\infty e^{\frac{\alpha}{2}(t-\tau)} G(t-\tau) e^{\frac{\alpha}{2}(\tau-T)} u(\tau)\mathrm{d}\tau \\
&= \int_0^\infty G_\alpha(t-\tau) u_T(\tau)\mathrm{d}\tau \\
&= (\mathcal{G}_\alpha * u_T)(t).
\end{aligned}
$$

Next, it follows from (7.75) that

$$
\|y_T(t)\|_p \leq |||\mathcal{G}_\alpha|||_{(\infty,p),(2,2)} |||u_T|||_{2,2} \leq \frac{1}{\sqrt{\alpha}} |||\mathcal{G}_\alpha|||_{(\infty,p),(2,2)} |||u|||_{\infty,2}.
$$

Now, noting that $y(T) = y_T(T)$ it follows that

$$
\|y(T)\|_p \leq \frac{1}{\sqrt{\alpha}} |||\mathcal{G}_\alpha|||_{(\infty,p),(2,2)} |||u|||_{\infty,2}, \qquad T \geq 0,
$$

which implies (7.98).

Let $T > 0$, $u \in \mathcal{L}_\infty$, and let $u_T(\cdot)$ be given by (7.101). Then

$$
\begin{aligned}
|||u_T|||_{1,r} &= \int_0^\infty \|u_T(t)\|_r \mathrm{d}t \\
&= \int_0^T e^{\frac{\alpha}{2}(t-T)} \|u(t)\|_r \mathrm{d}t \\
&\leq |||u|||_{\infty,r} \int_0^T e^{\frac{\alpha}{2}(t-T)} \mathrm{d}t \\
&= \frac{2}{\alpha} |||u|||_{\infty,r}.
\end{aligned}
$$

Now, it follows from (7.76) that

$$
\|y_T(t)\|_p \leq |||\mathcal{G}_\alpha|||_{(\infty,p),(1,r)} |||u_T|||_{1,r} \leq \frac{2}{\alpha} |||\mathcal{G}_\alpha|||_{(\infty,p),(1,r)} |||u|||_{\infty,r}.
$$

Hence, since $y(T) = y_T(T)$,

$$
\|y(T)\|_p \leq \frac{2}{\alpha} |||\mathcal{G}_\alpha|||_{(\infty,p),(1,r)} |||u|||_{\infty,r}, \qquad T \geq 0,
$$

which implies (7.99). Finally, (7.100) follows from (7.99) and Proposition 7.4. □

Next we specialize Theorem 7.8 to Euclidean and infinity norms.

Corollary 7.5. Let $\alpha > 0$ be such that $A + \frac{\alpha}{2} I$ is asymptotically stable, let $G_\alpha(\cdot)$ be given by (7.93), and let $Q_\alpha \in \mathbb{R}^{n \times n}$ be the unique, nonnegative-definite solution to (7.97). Then the following statements hold:

i)

$$|||\mathcal{G}|||_{(\infty,2),(\infty,2)} \le \frac{1}{\sqrt{\alpha}}|||\mathcal{G}_\alpha|||_{(\infty,2),(2,2)} = \frac{1}{\sqrt{\alpha}}\sigma_{\max}^{1/2}(CQ_\alpha C^{\mathrm{T}}). \quad (7.102)$$

Furthermore, if there exist $G_{\mathcal{L}_\alpha} : \mathbb{R} \to \mathbb{R}^{l\times m_l}$ and $G_{\mathcal{R}_\alpha} : \mathbb{R} \to \mathbb{R}^{m_l\times m}$ such that (7.95) holds, then

$$|||\mathcal{G}|||_{(\infty,2),(\infty,2)} \le \frac{2}{\alpha}|||\mathcal{G}_\alpha|||_{(\infty,2),(1,2)} \le \frac{2}{\alpha}\sigma_{\max}^{1/2}(\mathcal{Q}_{\mathcal{L}_\alpha})\sigma_{\max}^{1/2}(\mathcal{P}_{\mathcal{R}_\alpha}). \quad (7.103)$$

ii)

$$|||\mathcal{G}|||_{(\infty,\infty),(\infty,2)} \le \frac{1}{\sqrt{\alpha}}|||\mathcal{G}_\alpha|||_{(\infty,\infty),(2,2)} = \frac{1}{\sqrt{\alpha}}\mathrm{d}_{\max}^{1/2}(CQ_\alpha C^{\mathrm{T}}). \quad (7.104)$$

Furthermore, if there exist $G_{\mathcal{L}_\alpha} : \mathbb{R} \to \mathbb{R}^{l\times m_l}$ and $G_{\mathcal{R}_\alpha} : \mathbb{R} \to \mathbb{R}^{m_l\times m}$ such that (7.95) holds, then

$$|||\mathcal{G}|||_{(\infty,\infty),(\infty,2)} \le \frac{2}{\alpha}|||\mathcal{G}_\alpha|||_{(\infty,\infty),(1,2)} \le \frac{2}{\alpha}\mathrm{d}_{\max}^{1/2}(\mathcal{Q}_{\mathcal{L}})\sigma_{\max}^{1/2}(\mathcal{P}_{\mathcal{R}}). \quad (7.105)$$

Proof. The proof is a direct consequence of Lemma 7.1 and Theorem 7.8. □

Using set theoretic arguments involving closed convex sets and support functions, the $\mathcal{L}_1$ norm bound in (7.102) was given by Schweppe [393]. Within the context of $\mathcal{L}_\infty$ equi-induced norms, this $\mathcal{L}_1$ norm bound is referred to as the star-norm in [324, 423]. The expression given by (7.103) provides an alternative finitely computable bound for the $\mathcal{L}_\infty$ equi-induced norm.

A summary of the results of Sections 7.8–7.10 is given in Table 7.1.

Next, we present an example to provide comparisons between the bounds (7.98)–(7.100) given in Theorem 7.8 for $|||\mathcal{G}|||_{(\infty,p),(\infty,r)}$, $p, r \in [1, \infty]$.

Example 7.4. Consider the system (7.52) and (7.53) with

$$A = \begin{bmatrix} -1 & 0 \\ 1 & -1 \end{bmatrix}, \qquad B = \begin{bmatrix} 1 \\ 0 \end{bmatrix}, \qquad C = \begin{bmatrix} 0 & 1 \end{bmatrix},$$

so that $G(t) = te^{-t}$, $t \ge 0$. Since $\mathcal{G}$ is a convolution operator for a single-input/single-output system it follows that $|||\mathcal{G}|||_{(\infty,\infty),(\infty,\infty)} = |||\mathcal{G}|||_{(\infty,p),(\infty,r)}$,

Table 7.1 Summary of induced operator norms for $p, r \in [1, \infty]$.

Input	Output	Induced Norm	Upper Bound
$\lVert\!\lvert \cdot \rVert\!\rvert_{2,2}$	$\lVert\!\lvert \cdot \rVert\!\rvert_{2,2}$	$\sup_{\omega\in\mathbb{R}} \sigma_{\max}(H(\jmath\omega))$	
$\lVert\!\lvert \cdot \rVert\!\rvert_{1,r}$	$\lVert\!\lvert \cdot \rVert\!\rvert_{2,2}$	$\lVert\mathcal{P}^{1/2}\rVert_{2,r}$	
$\lVert\!\lvert \cdot \rVert\!\rvert_{2,2}$	$\lVert\!\lvert \cdot \rVert\!\rvert_{\infty,p}$	$\lVert\mathcal{Q}^{1/2}\rVert_{2,\bar{p}}$	
$\lVert\!\lvert \cdot \rVert\!\rvert_{1,r}$	$\lVert\!\lvert \cdot \rVert\!\rvert_{\infty,p}$	$\sup_{t\geq 0} \lVert G(t)\rVert_{p,r}$	$\lVert\mathcal{Q}_{\mathcal{L}}^{1/2}\rVert_{2,\bar{p}}\lVert\mathcal{P}_{\mathcal{R}}^{1/2}\rVert_{2,r}$
$\lVert\!\lvert \cdot \rVert\!\rvert_{r,r}$	$\lVert\!\lvert \cdot \rVert\!\rvert_{\infty,\infty}$	$\max_{i=1,\ldots,l} \lVert\mathrm{row}_i(G_{[\bar{r},\bar{r}]})\rVert_{\bar{r}}$	
$\lVert\!\lvert \cdot \rVert\!\rvert_{1,1}$	$\lVert\!\lvert \cdot \rVert\!\rvert_{p,p}$	$\max_{j=1,\ldots,m} \lVert\mathrm{col}_j(G_{[p,p]})\rVert_{p}$	
$\lVert\!\lvert \cdot \rVert\!\rvert_{\infty,2}$	$\lVert\!\lvert \cdot \rVert\!\rvert_{\infty,p}$		$\frac{1}{\sqrt{\alpha}}\lVert CQ_\alpha C^{\mathrm{T}}\rVert_{2,\bar{p}}$
$\lVert\!\lvert \cdot \rVert\!\rvert_{\infty,r}$	$\lVert\!\lvert \cdot \rVert\!\rvert_{\infty,p}$		$\frac{2}{\alpha}\sup_{t\geq 0}\lVert G_\alpha(t)\rVert_{p,r}$
$\lVert\!\lvert \cdot \rVert\!\rvert_{\infty,r}$	$\lVert\!\lvert \cdot \rVert\!\rvert_{\infty,p}$		$\frac{2}{\alpha}\lVert\mathcal{Q}_{\mathcal{L}_\alpha}^{1/2}\rVert_{2,\bar{p}}\lVert\mathcal{P}_{\mathcal{R}_\alpha}^{1/2}\rVert_{2,r}$

$p, r \in [1, \infty]$. Hence, it follows from Lemma 7.6 with $r = \infty$ that

$$\lVert\!\lvert \mathcal{G} \rVert\!\rvert_{(\infty,\infty),(\infty,\infty)} = \int_0^\infty |G(t)|\mathrm{d}t = \int_0^\infty te^{-t}\mathrm{d}t = 1.$$

Now, with $p = 2$, it follows from (7.98) that

$$\lVert\!\lvert \mathcal{G} \rVert\!\rvert_{(\infty,\infty),(\infty,\infty)} \leq \frac{1}{\sqrt{\alpha}}\lVert\!\lvert \mathcal{G}_\alpha \rVert\!\rvert_{(\infty,2),(2,2)}$$

for all $0 < \alpha < 2$. Noting that

$$\lVert\!\lvert \mathcal{G}_\alpha \rVert\!\rvert^2_{(\infty,2),(2,2)} = \int_0^\infty G_\alpha^2(t)\mathrm{d}t = \int_0^\infty t^2 e^{-(2-\alpha)t}\mathrm{d}t = \frac{2}{(2-\alpha)^3},$$

it follows that

$$\lVert\!\lvert \mathcal{G} \rVert\!\rvert_{(\infty,\infty),(\infty,\infty)} \leq \inf_{0<\alpha<2} \frac{2}{\sqrt{\alpha(2-\alpha)^3}} = \sqrt{\frac{32}{27}} \approx 1.0887.$$

Next, using (7.99) to bound $\lVert\!\lvert \mathcal{G} \rVert\!\rvert_{(\infty,\infty),(\infty,\infty)}$ yields

$$\lVert\!\lvert \mathcal{G} \rVert\!\rvert_{(\infty,\infty),(\infty,\infty)} \leq \frac{2}{\alpha}\sup_{t\geq 0} G_\alpha(t) = \frac{4}{\alpha(2-\alpha)}e^{-1}, \qquad 0 < \alpha < 2,$$

which implies

$$\lVert\!\lvert \mathcal{G} \rVert\!\rvert_{(\infty,\infty),(\infty,\infty)} \leq \inf_{0<\alpha<2} \frac{4}{\alpha(2-\alpha)}e^{-1} = 4e^{-1} \approx 1.4715.$$

Finally, we compare (7.100) to (7.98) and (7.99). Since $H_\alpha(s) = \frac{1}{(s+1-\frac{\alpha}{2})^2}$

it follows from (7.100) with $H_{\mathcal{L}_\alpha}(s) = H_{\mathcal{R}_\alpha}(s) = \frac{1}{s+1-\frac{\alpha}{2}}$ that

$$|||\mathcal{G}|||_{(\infty,\infty),(\infty,\infty)} \leq \frac{2}{\alpha}|||\mathcal{G}_{\mathcal{L}_\alpha}|||_{(\infty,2),(2,2)}|||\mathcal{G}_{\mathcal{R}_\alpha}|||_{(2,2),(1,2)}.$$

Hence,

$$|||\mathcal{G}|||_{(\infty,\infty),(\infty,\infty)} \leq \frac{2}{\alpha}\int_0^\infty e^{-(2-\alpha)t}\mathrm{d}t = \frac{2}{\alpha(2-\alpha)}, \quad 0 < \alpha < 2,$$

which implies

$$|||\mathcal{G}|||_{(\infty,\infty),(\infty,\infty)} \leq \inf_{0<\alpha<2}\frac{2}{\alpha(2-\alpha)} = 2.$$

△

7.11 Problems

Problem 7.1. For each of the following functions $f_i : [0,\infty) \to \mathbb{R}$, $i = 1,\ldots,7$, state whether $f_i \in \mathcal{L}_1$, $f_i \in \mathcal{L}_2$, and $f_i \in \mathcal{L}_\infty$.

i) $f_1(t) = 1$.

ii) $f_2(t) = \frac{1}{1+t}$.

iii) $f_3(t) = \frac{1+t^{1/4}}{t^{1/4}(1+t)}$.

iv) $f_4(t) = e^{-t}$.

v) $f_5(t) = \frac{1+t^{1/4}}{t^{1/4}(1+t^2)}$.

vi) $f_6(t) = \frac{1+t^{1/2}}{t^{1/2}(1+t^2)}$.

vii) $f_7(t) = \sin t$.

Problem 7.2. Let $f : [0,\infty) \to \mathbb{R}$. Show that if $\int_0^t |f(s)|\mathrm{d}s$ converges, then $\int_0^t f(s)\mathrm{d}s$ converges.

Problem 7.3. Let $f : [0,\infty) \to \mathbb{R}$ and let $f \in \mathcal{L}_1$. Show that $\lim_{t\to\infty} f(t)$ exists.

Problem 7.4. Let $f : [0,\infty) \to \mathbb{R}$. Show that if $f \in \mathcal{L}_2$ and $\dot{f} \in \mathcal{L}_2$ then $f \in \mathcal{L}_\infty$ and $\lim_{t\to\infty} f(t) = 0$.

Problem 7.5. Let $f : [0,\infty) \to \mathbb{R}$. Show that if $f \in \mathcal{L}_1$ and $\dot{f} \in \mathcal{L}_1$, then $\lim_{t\to\infty} f(t) = 0$.

Problem 7.6. Let $f : [0, \infty) \to \mathbb{R}$. Show that if $\dot{f} \in \mathcal{L}_\infty$, then f is uniformly continuous on $\mathbb{R}$.

Problem 7.7. Let $f : [0, \infty) \to \mathbb{R}$. Show that if $f \in \mathcal{L}_2$ and f is Lipschitz continuous on $\mathbb{R}$, then $\lim_{t\to\infty} f(t) = 0$.

Problem 7.8. Let $f : [0, \infty) \to \mathbb{R}$ and $p \in [1, \infty]$. Show that if $f \in \mathcal{L}_p$ and f is uniformly continuous on $\mathbb{R}$, then $\lim_{t\to\infty} f(t) = 0$.

Problem 7.9. Let $f_1 : [0, \infty) \to \mathbb{R}$ and $f_2 : [0, \infty) \to \mathbb{R}$. Show that if $f_1 \in \mathcal{L}_2$ and $f_2 \in \mathcal{L}_2$, then $f_1 + f_2 \in \mathcal{L}_2$.

Problem 7.10. Let $f : [0, \infty) \to \mathbb{R}$ be continuously differentiable and square integrable on $[0, \infty)$. Furthermore, assume $\dot{f}$ is bounded on $[0, \infty)$. Show that $\lim_{t\to\infty} f(t) = 0$.

Problem 7.11. Let $f : [0, \infty) \to \mathbb{R}$ be continuously differentiable on $[0, \infty)$. Prove or refute that if $\lim_{t\to\infty} \dot{f}(t) = 0$, then $\lim_{t\to\infty} f(t)$ exists. Conversely, prove or refute that if $\lim_{t\to\infty} f(t)$ exists, then $\lim_{t\to\infty} \dot{f}(t) = 0$.

Problem 7.12. Consider the linear dynamical system given by

$$\dot{x}(t) = Ax(t) + Bu(t), \qquad x(0) = x_0, \qquad t \geq 0, \tag{7.106}$$

$$y(t) = Cx(t), \tag{7.107}$$

where $x(t) \in \mathbb{R}^n$, $u(t) \in \mathbb{R}^m$, $y(t) \in \mathbb{R}^l$, $t \geq 0$, and A is Hurwitz. Show that if $u(\cdot) \in \mathcal{L}_2$, then $x(\cdot) \in \mathcal{L}_2 \cap \mathcal{L}_\infty$, $\dot{x}(\cdot) \in \mathcal{L}_2$, and $\lim_{t\to\infty} x(t) = 0$. Furthermore, show that if $u(\cdot) \in \mathcal{L}_p$, then $y(\cdot) \in \mathcal{L}_p \cap \mathcal{L}_\infty$ and $\dot{y}(\cdot) \in \mathcal{L}_p$ for $p = 1, 2$, and ∞. In addition, for $p = 1$ and 2, show that $\lim_{t\to\infty} y(t) = 0$.

Problem 7.13. Consider the operator dynamical system $\mathcal{G} : \mathcal{L}_{pe} \to \mathcal{L}_{pe}$ given by

$$y(t) = \mathcal{G}[u](t) = \int_0^\infty h(t, \tau)u(\tau)\mathrm{d}\tau. \tag{7.108}$$

Show that $\mathcal{G}$ is causal if and only if $h(t, \tau) = 0$ whenever $t < \tau$.

Problem 7.14 (Minkowski's Inequality). Let $p \in [1, \infty]$ and suppose that $f, g \in \mathcal{L}_p$. Show that $f + g \in \mathcal{L}_p$ and $|||f + g|||_p \leq |||f|||_p + |||g|||_p$.

Problem 7.15. Show that for each $p \in [1, \infty]$, the pair $(\mathcal{L}_p, ||\cdot||_p)$ is a normed linear space. Furthermore, show that $(\mathcal{L}_p, ||\cdot||_p)$ is a Banach space. (**Hint:** Use Minkowski's inequality.)

Problem 7.16 (Hölder's Inequality). Let $p, q \in [1, \infty]$ be such that $1/p + 1/q = 1$. Suppose $f \in \mathcal{L}_p$ and $g \in \mathcal{L}_q$. Show that $h : [0, \infty \to \mathbb{R}$ defined by $h(t) \triangleq f(t)g(t)$ belongs to $\mathcal{L}_1$. In addition, show that $|||fg|||_1 \le |||f|||_p |||g|||_q$.

Problem 7.17. Consider the nonlinear feedback system

$$\dot{x}(t) = Ax(t) + Bu(t), \quad x(0) = x_0, \quad t \ge 0, \tag{7.109}$$
$$y(t) = Cx(t), \tag{7.110}$$
$$u(t) = r(t) - \sigma(t, y(t)), \tag{7.111}$$

where $x(t) \in \mathbb{R}^n$, $u(t) \in \mathbb{R}^n$, $y(t) \in \mathbb{R}^l$, $r(t) \in \mathbb{R}^m$, and $\sigma : \overline{\mathbb{R}}_+ \times \mathbb{R}^l \to \mathbb{R}^m$ satisfies $\sigma(t, 0) = 0$, $t \ge 0$. Show that if $\sigma(\cdot, \cdot)$ is globally Lipschitz continuous on $\mathbb{R}^l$ and the zero solution $x(t) \equiv 0$ of the undisturbed system (7.109) is globally exponentially stable, then (7.109) is $\mathcal{L}_p$-stable for all $p \in [1, \infty]$. Conversely, show that if (7.109) and (7.110) is minimal and (7.109) is $\mathcal{L}_2$ stable, then $\lim_{t\to\infty} x(t) = 0$ for all $x_0 \in \mathbb{R}^n$ and $u(t) \equiv 0$.

Problem 7.18. Consider the feedback system shown in Figure 7.1 where $\mathcal{G}_1$ is a linear single-input, single-output time-invariant dynamical system with transfer function $G(s) \in \mathcal{H}_\infty$ and $\mathcal{G}_2$ is a memoryless time-varying nonlinearity $\sigma(\cdot, \cdot) \in \Phi_{\rm br}$. Show that if $|||G(s)|||_\infty < \gamma$, where $\gamma > 0$, then the feedback interconnection of $\mathcal{G}_1$ and $\mathcal{G}_2$ is $\mathcal{L}_2$-stable with finite gain and zero bias.

Problem 7.19 (Circle Criterion). Consider the feedback system shown in Figure 7.1 where $\mathcal{G}_1$ is a linear single-input, single-output time-invariant dynamical system with transfer function $G(s) \in \mathcal{H}_\infty$ and $\mathcal{G}_2$ is a memoryless time-varying nonlinearity $\sigma(\cdot, \cdot) \in \Phi$. Show that if $|||G_{\rm s}(s)|||_\infty < r^{-1}$, where $G_{\rm s}(s) \triangleq G(s)/(1 + cG(s))$, $c \triangleq (M_1 + M_2)/2$, and $r \triangleq (M_2 - M_1)/2$, then the feedback interconnection of $\mathcal{G}_1$ and $\mathcal{G}_2$ is $\mathcal{L}_2$-stable with finite gain and zero bias. Connect this result to Theorem 5.19.

Problem 7.20 (Popov Criterion). Consider the feedback system shown in Figure 7.1 where $\mathcal{G}_1$ is a strictly proper linear single-input, single-output time-invariant dynamical system with transfer function $G(s) \in \mathcal{H}_\infty$ and $\mathcal{G}_2$ is a memoryless time-invariant nonlinearity $\sigma(\cdot) \in \Phi_{\rm P}$. Assume that $sG(s) \in \mathcal{H}_\infty$. Show that if there exists $N \ge 0$ such that

$$\beta \triangleq \inf_{\omega\in\mathbb{R}} \mathrm{Re}[(1 + \jmath\omega N)G(\jmath\omega)] + \frac{1}{M} > 0, \tag{7.112}$$

then there exists $\gamma > 0$ such that

$$|||y_i||| \le \gamma(|||u_1|||_2 + |||u_2|||_2 + |||\dot{u}_2|||_2), \quad i = 1, 2. \tag{7.113}$$

Problem 7.21. Consider the nonlinear dynamical system (7.40) and (7.41) where $F(\cdot, \cdot)$ is Lipschitz continuous in (x, u), $H(\cdot, \cdot)$ is continuous, $F(0, 0) = 0$, and $H(0, 0) = 0$. Assume there exists a continuously differentiable function $V : \mathbb{R}^n \to \mathbb{R}$ such that $V(0) = 0$, $V(x) > 0$, $x \in \mathbb{R}^n$,

$x \neq 0$, $V(x) \to \infty$ as $\|x\| \to \infty$, and

$$V'(x)F(x,u) \leq -W(x) + \phi(u), \qquad (x,u) \in \mathbb{R}^n \times \mathbb{R}^m, \tag{7.114}$$

where $W(x)$ is continuous on $\mathbb{R}^n$, $W(x) > 0$, $x \in \mathbb{R}^n$, $x \neq 0$, $W(x) \to \infty$ as $\|x\| \to \infty$, $\phi(\cdot)$ is continuous on $\mathbb{R}^n$, and $\phi(0) = 0$. Show that in this case (7.40) and (7.41) is $\mathcal{L}_\infty$-stable.

Problem 7.22. Consider the nonlinear dynamical system (7.2) and (7.3) where $f(\cdot)$ is Lipschitz continuous on $\mathbb{R}^n$, $G(\cdot)$, $h(\cdot)$, and $J(\cdot)$ are continuous on $\mathbb{R}^n$, and $f(0) = 0$ and $h(0) = 0$. Let $\gamma > 0$ and assume there exists a continuously differentiable function $V : \mathbb{R}^n \to \mathbb{R}$ such that $V(\cdot)$ is positive definite, $V(0) = 0$, and, for all $x \in \mathbb{R}^n$,

$$\begin{aligned} 0 \geq\ & V'(x)f(x) + h^{\mathrm{T}}(x)h(x) + [\tfrac{1}{2}V'(x)G(x) + h^{\mathrm{T}}(x)J(x)] \\ & \cdot[\gamma^2 I_m - J^{\mathrm{T}}(x)J(x)]^{-1}[\tfrac{1}{2}V'(x)G(x) + h^{\mathrm{T}}(x)J(x)]^{\mathrm{T}}. \end{aligned} \tag{7.115}$$

Show that in this case, for each $x_0 \in \mathbb{R}^n$, (7.2) and (7.3) is $\mathcal{L}_2$-stable and the $\mathcal{L}_2$ gain is less than or equal to γ.

Problem 7.23. Consider the finite gain operator dynamical systems $\mathcal{G}_1 : \mathcal{L}^m_{p\mathrm{e}} \to \mathcal{L}^l_{p\mathrm{e}}$ and $\mathcal{G}_1 : \mathcal{L}^l_{p\mathrm{e}} \to \mathcal{L}^q_{p\mathrm{e}}$. Show that $\mathcal{G}_2 \circ \mathcal{G}_1$ is a finite gain operator dynamical system.

Problem 7.24. Let $\mathcal{G}_i : \mathcal{L}^{m_i}_{2\mathrm{e}} \to \mathcal{L}^{l_i}_{2\mathrm{e}}$, $i = 1, \ldots, q$, be a collection of causal operator dynamical systems that are (Q_i, R_i, S_i)-dissipative for each $i \in \{1, \ldots, q\}$. Define the input vector $u \triangleq [u_1^{\mathrm{T}}, \ldots, u_q^{\mathrm{T}}]^{\mathrm{T}}$, the output vector $y \triangleq [y_1^{\mathrm{T}}, \ldots, y_q^{\mathrm{T}}]^{\mathrm{T}}$, and the system interconnection by

$$u = v - Hy, \tag{7.116}$$

where $H \in \mathbb{R}^{(m_1+\cdots+m_q)\times(l_1+\cdots+l_q)}$ and $v \in \mathcal{L}_{2\mathrm{e}}^{(m_1+\cdots+m_q)}$. In addition, define $Q = \mathrm{diag}[Q_1, \ldots, Q_q]$, $R = \mathrm{diag}[R_1, \ldots, R_q]$, and $S = \mathrm{diag}[S_1, \ldots, S_q]$. Show that if

$$\hat{Q} = Q + H^{\mathrm{T}}RH - SH - H^{\mathrm{T}}S^{\mathrm{T}} < 0, \tag{7.117}$$

then the interconnected system with input v and output y is finite-gain $\mathcal{L}_2$-stable.

Problem 7.25. For discrete-time nonlinear operator dynamical systems the input and output spaces are spaces of sequences, denoted by ℓ_p^m and ℓ_p^l, respectively. In particular, ℓ_p^n is the set of sequences such that $\|f\|_{p,q} < \infty$, $q \in [1, \infty]$, where $f : \overline{\mathbb{Z}}_+ \to \mathbb{R}^n$, that is,

$$\ell_p^n \triangleq \{f : \overline{\mathbb{Z}}_+ \to \mathbb{R}^n : \|f\|_{p,q} < \infty, q \in [1, \infty]\},$$

where

$$|||f|||_{p,q} \triangleq \left[\sum_{k=0}^{\infty} \|f(k)\|_q^p\right]^{1/p}, \quad 1 \le p < \infty, \tag{7.118}$$

$$|||f|||_{\infty,q} \triangleq \sup_{k \in \overline{\mathbb{Z}}_+} \|f(k)\|_q. \tag{7.119}$$

Show that the main results of this chapter including the small gain theorem and the (Q, R, S)-dissipativity theorem go through unchanged for discrete-time nonlinear operator dynamical systems.

7.12 Notes and References

Input-output system theory can be traced back to Norbert Wiener. In the late 1950s, Wiener used Volterra integral series to represent input-output maps of nonlinear dynamical systems [453]. The foundations of input-output stability theory were developed by Sandberg [387, 388, 390] and Zames [476, 477]. An excellent textbook treatment of input-output stability of feedback systems is given by Desoer and Vidyasagar [104]. Input-output dissipativity theory is due to Moylan and Hill [322], and Hill and Moylan [192]. Connections between input-output stability and Lyapunov stability theory were developed by Willems [455], Hill and Moylan [190], and Vidyasagar and Vannelli [446]. Finally, the induced convolution operator norms of linear dynamical systems presented in Sections 7.8–7.10 are due to Chellaboina, Haddad, Bernstein, and Wilson [89].

Chapter Eight

Optimal Nonlinear Feedback Control

8.1 Introduction

Under certain conditions nonlinear controllers offer significant advantages over linear controllers. In particular, if the plant dynamics and/or system measurements are nonlinear [34, 212, 398, 452], the plant/measurement disturbances are either nonadditive or non-Gaussian, the performance measure considered in nonquadratic [14, 33, 133, 244, 275, 366, 384, 391, 399, 411, 429], the plant model is uncertain [24, 28, 99, 142, 268, 351], or the control signals/state amplitudes are constrained [62, 125, 234, 375], then nonlinear controllers yield better performance than the best linear controllers. In a paper by Bernstein [43] the current status of continuous-time, nonlinear-nonquadratic problems was presented in a simplified and tutorial manner. The basic underlying ideas of the results in [43] are based on the fact that the steady-state solution of the Hamilton-Jacobi-Bellman equation is a Lyapunov function for the nonlinear system and thus guaranteeing both stability and optimality [217].

Building on the results of [43], in this chapter, we present a framework for analyzing and designing feedback controllers for nonlinear systems. Specifically, we consider a feedback control problem over an infinite horizon involving a nonlinear-nonquadratic performance functional. The performance functional can be evaluated in closed form as long as the nonlinear-nonquadratic cost functional considered is related in a specific way to an underlying Lyapunov function that guarantees asymptotic stability of the nonlinear closed-loop system. This Lyapunov function is shown to be the solution of the steady-state Hamilton-Jacobi-Bellman equation. The overall framework provides the foundation for extending linear-quadratic control to nonlinear-nonquadratic problems.

8.2 Stability Analysis of Nonlinear Systems

In this section, we present sufficient conditions for stability and performance for a given nonlinear system with a nonlinear-nonquadratic performance functional. For the class of nonlinear systems considered we assume that the required properties for the existence and uniqueness of solutions are satisfied. For the following result, let $\mathcal{D} \subseteq \mathbb{R}^n$ be an open set, assume $0 \in \mathcal{D}$, let $L : \mathcal{D} \to \mathbb{R}$, and let $f : \mathcal{D} \to \mathbb{R}^n$ be such that $f(0) = 0$.

Theorem 8.1. Consider the nonlinear dynamical system

$$\dot{x}(t) = f(x(t)), \qquad x(0) = x_0, \qquad t \geq 0, \tag{8.1}$$

with nonlinear-nonquadratic performance functional

$$J(x_0) \triangleq \int_0^\infty L(x(t))\, dt. \tag{8.2}$$

Furthermore, assume that there exists a continuously differentiable function $V : \mathcal{D} \to \mathbb{R}$ such that

$$V(0) = 0, \tag{8.3}$$

$$V(x) > 0, \quad x \in \mathcal{D}, \quad x \neq 0, \tag{8.4}$$

$$V'(x)f(x) < 0, \quad x \in \mathcal{D}, \quad x \neq 0, \tag{8.5}$$

$$L(x) + V'(x)f(x) = 0, \quad x \in \mathcal{D}. \tag{8.6}$$

Then the zero solution $x(t) \equiv 0$ to (8.1) is locally asymptotically stable and there exists a neighborhood of the origin $\mathcal{D}_0 \subseteq \mathcal{D}$ such that

$$J(x_0) = V(x_0), \qquad x_0 \in \mathcal{D}_0. \tag{8.7}$$

Finally, if $\mathcal{D} = \mathbb{R}^n$ and

$$V(x) \to \infty \text{ as } \|x\| \to \infty, \tag{8.8}$$

then the zero solution $x(t) \equiv 0$ to (8.1) is globally asymptotically stable.

Proof. Let $x(t)$, $t \geq 0$, satisfy (8.1). Then it follows from (8.5) that

$$\dot{V}(x(t)) = V'(x(t))f(x(t)) < 0, \quad t \geq 0, \quad x(t) \neq 0. \tag{8.9}$$

Thus, from (8.3), (8.4), and (8.9) it follows from Theorem 3.1 that $V(\cdot)$ is a Lyapunov function for the nonlinear dynamical system (8.1), which proves local asymptotic stability of the zero solution $x(t) \equiv 0$ to (8.1). Consequently, $x(t) \to 0$ as $t \to \infty$ for all initial conditions $x_0 \in \mathcal{D}_0$ for some neighborhood of the origin $\mathcal{D}_0 \subseteq \mathcal{D}$. Now, since

$$0 = -\dot{V}(x(t)) + V'(x(t))f(x(t)), \quad t \geq 0,$$

it follows from (8.6) that

$$L(x(t)) = -\dot{V}(x(t)) + L(x(t)) + V'(x(t))f(x(t)) = -\dot{V}(x(t)).$$

Now, integrating over $[0, t]$ yields

$$\int_0^t L(x(s))\mathrm{d}s = -V(x(t)) + V(x_0).$$

Letting $t \to \infty$ and noting that $V(x(t)) \to 0$ for all $x_0 \in \mathcal{D}_0$ yields $J(x_0) = V(x_0)$. Finally, for $\mathcal{D} = \mathbb{R}^n$ global asymptotic stability is a direct consequence of the radially unbounded condition (8.8) on $V(\cdot)$. □

It is important to note that if (8.6) holds, then (8.5) is equivalent to $L(x) > 0$, $x \in \mathcal{D}$, $x \neq 0$. Next, we specialize Theorem 8.1 to linear systems. For this result let $A \in \mathbb{R}^{n\times n}$ and let $R \in \mathbb{R}^{n\times n}$ be a positive-definite matrix.

Corollary 8.1. Consider the linear dynamical system

$$\dot{x}(t) = Ax(t), \quad x(0) = x_0, \quad t \geq 0, \tag{8.10}$$

with quadratic performance functional

$$J(x_0) \triangleq \int_0^\infty x^{\mathrm{T}}(t)Rx(t)\mathrm{d}t. \tag{8.11}$$

Furthermore, assume that there exists a positive-definite matrix $P \in \mathbb{R}^{n\times n}$ such that

$$0 = A^{\mathrm{T}}P + PA + R. \tag{8.12}$$

Then, the zero solution $x(t) \equiv 0$ to (8.10) is globally asymptotically stable and

$$J(x_0) = x_0^{\mathrm{T}}Px_0, \qquad x_0 \in \mathbb{R}^n. \tag{8.13}$$

Proof. The result is a direct consequence of Theorem 8.1 with $f(x) = Ax$, $L(x) = x^{\mathrm{T}}Rx$, $V(x) = x^{\mathrm{T}}Px$, and $\mathcal{D} = \mathbb{R}^n$. Specifically, conditions (8.3) and (8.4) are trivially satisfied. Now, $V'(x)f(x) = x^{\mathrm{T}}(A^{\mathrm{T}}P + PA)x$, and hence, it follows from (8.12) that $L(x) + V'(x)f(x) = 0$, $x \in \mathbb{R}^n$, so that all the conditions of Theorem 8.1 are satisfied. Finally, since $V(\cdot)$ is radially unbounded, the zero solution $x(t) \equiv 0$ to (8.10) is globally asymptotically stable. □

It follows from Corollary 8.1 that Theorem 8.1 is an extension of the $\mathcal{H}_2$ analysis framework to nonlinear systems. Recall that the $\mathcal{H}_2$ Hardy space consists of complex matrix-valued functions $G(s) \in \mathbb{C}^{l\times m}$ that are analytic in the open right half plane and satisfy

$$\sup_{\eta>0} \frac{1}{2\pi} \int_{-\infty}^\infty \|G(\eta + \jmath\omega)\|_{\mathrm{F}}^2 \mathrm{d}\omega < \infty. \tag{8.14}$$

The norm of an $\mathcal{H}_2$ function $G(s)$ is defined by

$$\|G\|_2 \triangleq \left[\frac{1}{2\pi}\int_{-\infty}^{\infty} \|G(\jmath\omega)\|_{\mathrm{F}}^2 \mathrm{d}\omega\right]^{1/2}. \tag{8.15}$$

Alternatively, using Parseval's theorem we can express the $\mathcal{H}_2$ norm as an $\mathcal{L}_2$ norm of the impulse response $H(t) = \frac{1}{2\pi}\int_{-\infty}^{\infty} G(\jmath\omega)e^{\jmath\omega t}\mathrm{d}\omega$. In particular, the $\mathcal{L}_2$ norm of the matrix-valued impulse response function $H(t) \in \mathbb{R}^{l\times m}$, $t \geq 0$, is defined by

$$\|H\|_2 \triangleq \left[\int_0^{\infty} \|H(t)\|_{\mathrm{F}}^2 \mathrm{d}t\right]^{1/2}. \tag{8.16}$$

Now, letting $R = E^{\mathrm{T}}E$ and defining the free response $z(t) \triangleq Ex(t) = Ee^{At}x_0$, $t \geq 0$, it follows that the performance functional (8.11) can be written as

$$\begin{aligned} J(x_0) &= \int_0^{\infty} z^{\mathrm{T}}(t)z(t)\mathrm{d}t \\ &= \int_0^{\infty} x_0^{\mathrm{T}}e^{A^{\mathrm{T}}t}E^{\mathrm{T}}Ee^{At}x_0\mathrm{d}t \\ &= x_0^{\mathrm{T}}Px_0 \\ &= \int_0^{\infty} \|H(t)\|_{\mathrm{F}}^2\mathrm{d}t \\ &= \frac{1}{2\pi}\int_{-\infty}^{\infty} \|G(\jmath\omega)\|_{\mathrm{F}}^2\mathrm{d}\omega \\ &= \|G\|_2^2, \end{aligned} \tag{8.17}$$

where $H(t) = Ee^{At}x_0$ and

$$G(s) \sim \left[\begin{array}{c|c} A & x_0 \\ \hline E & 0 \end{array}\right].$$

Alternatively, assuming $x_0x_0^{\mathrm{T}}$ has an expected value V, that is, $\mathbb{E}[x_0x_0^{\mathrm{T}}] = V$, where $\mathbb{E}$ denotes expectation, and letting $V = DD^{\mathrm{T}}$, it follows that the averaged performance functional is given by

$$\mathbb{E}[J(x_0)] = \mathbb{E}[x_0^{\mathrm{T}}Px_0] = \operatorname{tr} D^{\mathrm{T}}PD = \|G\|_2^2, \tag{8.18}$$

where P satisfies (8.12) and

$$G(s) \sim \left[\begin{array}{c|c} A & D \\ \hline E & 0 \end{array}\right].$$

Next, we specialize Theorem 8.1 to linear and nonlinear systems with multilinear cost functionals. First, however, we give several definitions involving multilinear functions and a key lemma establishing the existence and uniqueness of specific multilinear forms. A scalar function $\psi : \mathbb{R}^n \to \mathbb{R}$ is *q-multilinear* if q is a positive integer and $\psi(x)$ is a linear combination of terms of the form $x_1^{i_1} x_2^{i_2} \cdots x_n^{i_n}$, where i_j is a nonnegative integer for $j = 1, \ldots, n$, and $i_1 + i_2 + \cdots + i_n = q$. Furthermore, a q-multilinear function $\psi(\cdot)$ is *nonnegative definite* (respectively, *positive definite*) if $\psi(x) \geq 0$ for all $x \in \mathbb{R}^n$ (respectively, $\psi(x) > 0$ for all nonzero $x \in \mathbb{R}^n$). Note that if q is odd then $\psi(x)$ cannot be positive definite. If $\psi(\cdot)$ is a q-multilinear function, then $\psi(\cdot)$ can be represented by means of Kronecker products, that is, $\psi(x)$ is given by $\psi(x) = \Psi x^{[q]}$, where $\Psi \in \mathbb{R}^{1 \times n^q}$.

The following lemma due to Bernstein [43] is needed for several of the main results of this book. For this result define the *spectral abscissa* of $A \in \mathbb{R}^{n \times n}$ by

$$\alpha(A) \triangleq \max\{\mathrm{Re}\ \lambda : \ \lambda \in \mathrm{spec}(A)\}.$$

We say that the matrix $A \in \mathbb{R}^{n \times n}$ is *Hurwitz* if and only if $\alpha(A) < 0$.

Lemma 8.1. Let $A \in \mathbb{R}^{n \times n}$ be Hurwitz and let $h : \mathbb{R}^n \to \mathbb{R}$ be a q-multilinear function. Then there exists a unique q-multilinear function $g : \mathbb{R}^n \to \mathbb{R}$ such that

$$0 = g'(x)Ax + h(x), \qquad x \in \mathbb{R}^n. \tag{8.19}$$

Furthermore, if $h(x)$ is nonnegative (respectively, positive) definite, then $g(x)$ is nonnegative (respectively, positive) definite.

Proof. Let $h(x) = \Psi x^{[q]}$ and define $g(x) \triangleq \Gamma x^{[q]}$, where $\Gamma \triangleq -\Psi(\overset{q}{\oplus} A)^{-1}$. Note that $\overset{q}{\oplus} A$ is invertible since A is Hurwitz. Now, note that for all $x \in \mathbb{R}^n$,

$$\begin{aligned}
g'(x)Ax &= \Gamma \frac{\mathrm{d}}{\mathrm{d}x}(x^{[q]})Ax \\
&= \Gamma\left(x \otimes \cdots \otimes I + x \otimes \cdots \otimes I \otimes x + \cdots + I \otimes \cdots \otimes x\right)Ax \\
&= \Gamma\left(x \otimes \cdots \otimes Ax + x \otimes \cdots \otimes Ax \otimes x + \cdots + Ax \otimes \cdots \otimes x\right) \\
&= \Gamma\left(I \otimes \cdots \otimes A + I \otimes \cdots \otimes A \otimes I + \cdots + A \otimes I \cdots \otimes I\right)x^{[q]} \\
&= \Gamma\left(A \oplus A \oplus \cdots \oplus A\right)x^{[q]} \\
&= \Gamma(\overset{q}{\oplus} A)x^{[q]} \\
&= -\Psi x^{[q]} \\
&= -h(x).
\end{aligned}$$

To prove uniqueness, suppose that $\hat{g}(x) = \hat{\Gamma} x^{[q]}$ satisfies (8.19). Then it follows that

$$\Gamma(\overset{q}{\oplus} A)x^{[q]} = \hat{\Gamma}(\overset{q}{\oplus} A)x^{[q]}.$$

Since $\overset{q}{\oplus} A$ is Hurwitz and $e^{(\overset{q}{\oplus} A)t} = (e^{At})^{[q]}$, it follows that, for all $x \in \mathbb{R}^n$,

$$\begin{aligned}
\Gamma x^{[q]} &= \Gamma(\overset{q}{\oplus} A)(\overset{q}{\oplus} A)^{-1} x^{[q]} \\
&= -\Gamma(\overset{q}{\oplus} A) \int_0^\infty e^{(\overset{q}{\oplus} A)t} x^{[q]} \mathrm{d}t \\
&= -\Gamma(\overset{q}{\oplus} A) \int_0^\infty (e^{At})^{[q]} x^{[q]} \mathrm{d}t \\
&= -\Gamma(\overset{q}{\oplus} A) \int_0^\infty (e^{At}x)^{[q]} \mathrm{d}t \\
&= -\hat{\Gamma}(\overset{q}{\oplus} A) \int_0^\infty (e^{At}x)^{[q]} \mathrm{d}t \\
&= \hat{\Gamma} x^{[q]},
\end{aligned}$$

which shows that $g(x) = \hat{g}(x)$, $x \in \mathbb{R}^n$.

Finally, if $h(x)$ is nonnegative definite, then it follows that for all $x \in \mathbb{R}^n$,

$$g(x) = -\Psi(\overset{q}{\oplus} A)^{-1} x^{[q]} = \Psi \int_0^\infty e^{(\overset{q}{\oplus} A)t} x^{[q]} \mathrm{d}t = \Psi \int_0^\infty (e^{At}x)^{[q]} \mathrm{d}t \geq 0.$$

If, in addition, $x \neq 0$, then $e^{At}x \neq 0$, $t \geq 0$. Hence, if $h(x)$ is positive definite, then $g(x)$ is positive definite. □

Next, assume A is Hurwitz, let P be given by (8.12), and consider the case in which $L(\cdot)$, $f(\cdot)$, and $V(\cdot)$ are given by

$$L(x) = x^{\mathrm{T}} R x + h(x), \tag{8.20}$$

$$f(x) = Ax + N(x), \tag{8.21}$$

$$V(x) = x^{\mathrm{T}} P x + g(x), \tag{8.22}$$

where $h : \mathcal{D} \to \mathbb{R}$ and $g : \mathcal{D} \to \mathbb{R}$ are nonquadratic, and $N : \mathcal{D} \to \mathbb{R}^n$ is nonlinear. In this case, (8.6) holds if and only if

$$0 = x^{\mathrm{T}} R x + h(x) + x^{\mathrm{T}}(A^{\mathrm{T}}P + PA)x + 2x^{\mathrm{T}} P N(x) + g'(x)(Ax + N(x)), \quad x \in \mathcal{D}, \tag{8.23}$$

or, equivalently,

$$0 = x^{\mathrm{T}}(A^{\mathrm{T}}P + PA + R)x + g'(x)(Ax + N(x)) + h(x) + 2x^{\mathrm{T}} P N(x), \quad x \in \mathcal{D}. \tag{8.24}$$

Since A is Hurwitz, we can choose P to satisfy (8.12) as in the linear-quadratic $\mathcal{H}_2$ case. Now, suppose $N(x) \equiv 0$ and let P satisfy (8.12). Then (8.24) specializes to

$$0 = g'(x)Ax + h(x), \quad x \in \mathcal{D}. \tag{8.25}$$

Next, given $h(\cdot)$, we determine the existence of a function $g(\cdot)$ satisfying

(8.25). Here, we focus our attention on multilinear functionals for which (8.25) holds with $\mathcal{D} = \mathbb{R}^n$. Specifically, let $h(x)$ be a nonnegative-definite q-multilinear function, where q is necessarily even. Furthermore, let $g(x)$ be the nonnegative-definite q-multilinear function given by Lemma 8.1. Then, since $g'(x)Ax \leq 0$, $x \in \mathbb{R}^n$, it follows that $x^{\mathrm{T}}Px + g(x)$ is a Lyapunov function for (8.10). Hence, Lemma 8.1 can be used to generate Lyapunov functions of specific multilinear structures.

To demonstrate the above discussion suppose $h(x)$ in (8.20) is of the more general form given by

$$h(x) = \sum_{\nu=2}^{r} h_{2\nu}(x), \tag{8.26}$$

where, for $\nu = 2, 3, \ldots, r$, $h_{2\nu} : \mathbb{R}^n \to \mathbb{R}$ is a nonnegative-definite 2ν-multilinear function. Now, using Lemma 8.1, it follows that there exists a nonnegative-definite 2ν-multilinear function $g_{2\nu} : \mathbb{R}^n \to \mathbb{R}$ satisfying

$$0 = g'_{2\nu}(x)Ax + h_{2\nu}(x), \quad x \in \mathbb{R}^n, \quad \nu = 2, 3, \ldots, r. \tag{8.27}$$

Defining $g(x) \triangleq \sum_{\nu=2}^{r} g_{2\nu}(x)$ and summing (8.27) over ν yields (8.25). Since (8.6) is satisfied with $L(x)$ and $V(x)$ given by (8.20) and (8.22), respectively, (8.7) implies that

$$J(x_0) = x_0^{\mathrm{T}}Px_0 + g(x_0). \tag{8.28}$$

To illustrate condition (8.25) with quartic Lyapunov functions let

$$V(x) = x^{\mathrm{T}}Px + (x^{\mathrm{T}}Mx)^2, \tag{8.29}$$

where P satisfies (8.12) and assume M is an $n \times n$ symmetric matin. In this case, $g(x) = (x^{\mathrm{T}}Mx)^2$ is a nonnegative-definite 4-multilinear function and (8.25) yields

$$h(x) = -2(x^{\mathrm{T}}Mx)x^{\mathrm{T}}(A^{\mathrm{T}}M + MA)x. \tag{8.30}$$

Now, letting M satisfy

$$0 = A^{\mathrm{T}}M + MA + \hat{R}, \tag{8.31}$$

where $\hat{R}$ is an $n \times n$ symmetric matrix, it follows from (8.30) that $h(x)$ satisfying (8.25) is of the form

$$h(x) = 2(x^{\mathrm{T}}Mx)(x^{\mathrm{T}}\hat{R}x). \tag{8.32}$$

If $\hat{R}$ is nonnegative definite, then M is nonnegative definite, and hence, $h(x)$ is a nonnegative-definite 4-multilinear function. Thus, if $V(x)$ is a quartic Lyapunov function of the form given by (8.29), and $L(x)$ is given by

$$L(x) = x^{\mathrm{T}}Rx + 2(x^{\mathrm{T}}Mx)(x^{\mathrm{T}}\hat{R}x), \tag{8.33}$$

where M satisfies (8.31), then condition (8.25), and hence, (8.6), is satisfied. The following proposition generalizes the above results to general polynomial cost functionals.

Proposition 8.1. Let $A \in \mathbb{R}^{n\times n}$ be Hurwitz, $R \in \mathbb{R}^{n\times n}$, $R > 0$, and $\hat{R}_q \in \mathbb{R}^{n\times n}$, $\hat{R}_q \geq 0$, $q = 2, \ldots, r$. Consider the linear system (8.10) with performance functional

$$J(x_0) \triangleq \int_0^\infty \left\{ x^{\mathrm{T}}(t)Rx(t) + \sum_{q=2}^{r} \left[(x^{\mathrm{T}}(t)\hat{R}_q x(t))(x^{\mathrm{T}}(t)M_q x(t))^{q-1} \right] \right\} \mathrm{d}t, \tag{8.34}$$

where $M_q \in \mathbb{R}^{n\times n}$, $M_q \geq 0$, $q = 2, \ldots, r$, satisfy

$$0 = A^{\mathrm{T}}M_q + M_q A + \hat{R}_q. \tag{8.35}$$

Furthermore, assume there exists a positive-definite matrix $P \in \mathbb{R}^{n\times n}$ such that

$$0 = A^{\mathrm{T}}P + PA + R. \tag{8.36}$$

Then the zero solution $x(t) \equiv 0$ to (8.10) is globally asymptotically stable and

$$J(x_0) = x_0^{\mathrm{T}}Px_0 + \sum_{q=2}^{r} \tfrac{1}{q}(x_0^{\mathrm{T}}M_q x_0)^q, \qquad x_0 \in \mathbb{R}^n. \tag{8.37}$$

Proof. The result is a direct consequence of Theorem 8.1 with $f(x) = Ax$, $L(x) = x^{\mathrm{T}}Rx + \sum_{q=2}^{r}[(x^{\mathrm{T}}\hat{R}_q x)(x^{\mathrm{T}}M_q x)^{q-1}]$, $V(x) = x^{\mathrm{T}}Px + \sum_{q=2}^{r}\frac{1}{q}(x^{\mathrm{T}} M_q x)^q$, and $\mathcal{D} = \mathbb{R}^n$. Specifically, conditions (8.3) and (8.4) are trivially satisfied. Now,

$$V'(x)f(x) = x^{\mathrm{T}}(A^{\mathrm{T}}P + PA)x + \sum_{q=2}^{r}(x^{\mathrm{T}}M_q x)^{q-1}x^{\mathrm{T}}(A^{\mathrm{T}}, M_q + M_q A)x,$$

and hence, it follows from (8.35) and (8.36) that $L(x) + V'(x)f(x) = 0$, $x \in \mathbb{R}^n$, so that all the conditions of Theorem 8.1 are satisfied. Finally, since $V(\cdot)$ is radially unbounded (8.10) is globally asymptotically stable. □

Proposition 8.1 requires the solutions of $r - 1$ Lyapunov equations in (8.35) to obtain a closed-form expression for the nonlinear-nonquadratic cost functional (8.34). However, if $\hat{R}_q = \hat{R}_2$, $q = 3, \ldots, r$, then $M_q = M_2$, $q = 3, \ldots, r$, satisfies (8.35). In this case, the solution of only one Lyapunov equation in (8.35) is required.

8.3 Optimal Nonlinear Control

In this section, we consider a control problem involving a notion of optimality with respect to a nonlinear-nonquadratic cost functional. The optimal feedback controllers are derived as a direct consequence of Theorem 8.1. To address the optimal control problem let $\mathcal{D} \subseteq \mathbb{R}^n$ be an open set and let $U \subseteq \mathbb{R}^m$, where $0 \in \mathcal{D}$ and $0 \in U$. Furthermore, let $F : \mathcal{D} \times U \to \mathbb{R}^n$ be such that $F(0,0) = 0$. Next, consider the controlled nonlinear dynamical system $\mathcal{G}$ given by

$$\dot{x}(t) = F(x(t), u(t)), \qquad x(0) = x_0, \qquad t \geq 0, \tag{8.38}$$

where $u(\cdot)$ is restricted to the class of *admissible* controls consisting of measurable function $u(\cdot)$ such that $u(t) \in U$ for all $t \geq 0$, where the constraint set U is given. We assume $0 \in U$. Given a control law $\phi(\cdot)$ and a feedback control $u(t) = \phi(x(t))$, the closed-loop system shown in the Figure 8.1 has the form

$$\dot{x}(t) = F(x(t), \phi(x(t))), \qquad x(0) = x_0, \qquad t \geq 0. \tag{8.39}$$

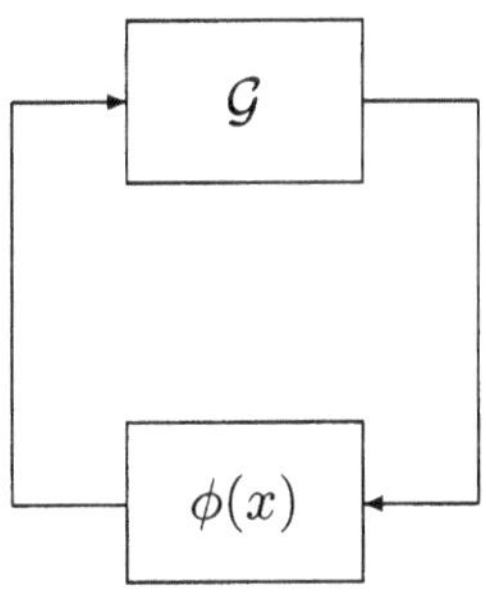

Figure 8.1 Nonlinear closed-loop feedback system.

Next, we present a main theorem due to Bernstein [43] for characterizing feedback controllers that guarantee stability and minimize a nonlinear-nonquadratic performance functional. For the statement of this result let $L : \mathcal{D} \times U \to \mathbb{R}$ and define the set of regulation controllers by

$$\mathcal{S}(x_0) \triangleq \{u(\cdot) : u(\cdot) \text{ is admissible and } x(\cdot) \text{ given by (8.38)} \\ \text{satisfies } x(t) \to 0 \text{ as } t \to \infty\}.$$

Theorem 8.2. Consider the nonlinear controlled dynamical system

(8.38) with performance functional

$$J(x_0, u(\cdot)) \triangleq \int_0^\infty L(x(t), u(t))\mathrm{d}t, \tag{8.40}$$

where $u(\cdot)$ is an admissible control. Assume that there exists a continuously differentiable function $V : \mathcal{D} \to \mathbb{R}$ and a control law $\phi : \mathcal{D} \to U$ such that

$$\begin{aligned}
V(0) &= 0, && && (8.41)\\
V(x) &> 0, && x \in \mathcal{D}, \quad x \neq 0, && (8.42)\\
\phi(0) &= 0, && && (8.43)\\
V'(x)F(x, \phi(x)) &< 0, && x \in \mathcal{D}, \quad x \neq 0, && (8.44)\\
H(x, \phi(x)) &= 0, && x \in \mathcal{D}, && (8.45)\\
H(x, u) &\geq 0, && x \in \mathcal{D}, \quad u \in U, && (8.46)
\end{aligned}$$

where

$$H(x, u) \triangleq L(x, u) + V'(x)F(x, u). \tag{8.47}$$

Then, with the feedback control $u(\cdot) = \phi(x(\cdot))$, the zero solution $x(t) \equiv 0$ of the closed-loop system (8.39) is locally asymptotically stable and there exists a neighborhood of the origin $\mathcal{D}_0 \subseteq \mathcal{D}$ such that

$$J(x_0, \phi(x(\cdot))) = V(x_0), \qquad x_0 \in \mathcal{D}_0. \tag{8.48}$$

In addition, if $x_0 \in \mathcal{D}_0$ then the feedback control $u(\cdot) = \phi(x(\cdot))$ minimizes $J(x_0, u(\cdot))$ in the sense that

$$J(x_0, \phi(x(\cdot))) = \min_{u(\cdot) \in \mathcal{S}(x_0)} J(x_0, u(\cdot)). \tag{8.49}$$

Finally, if $\mathcal{D} = \mathbb{R}^n$, $U = \mathbb{R}^m$, and

$$V(x) \to \infty \text{ as } \|x\| \to \infty, \tag{8.50}$$

then the zero solution $x(t) \equiv 0$ of the closed-loop system (8.39) is globally asymptotically stable.

Proof. Local and global asymptotic stability are a direct consequence of (8.41)–(8.44) and (8.50) by applying Theorem 8.1 to the closed-loop system (8.39). Furthermore, using (8.45), condition (8.48) is a restatement of (8.7) as applied to the closed-loop system. Next, let $x_0 \in \mathcal{D}_0$, let $u(\cdot) \in \mathcal{S}(x_0)$, and let $x(t)$, $t \geq 0$, be the solution of (8.38). Then it follows that

$$0 = -\dot{V}(x(t)) + V'(x(t))F(x(t), u(t)).$$

Hence,

$$\begin{aligned}
L(x(t), u(t)) &= -\dot{V}(x(t)) + L(x(t), u(t)) + V'(x(t))F(x(t), u(t))\\
&= -\dot{V}(x(t)) + H(x(t), u(t)).
\end{aligned}$$

Now, using (8.46) and the fact that $u(\cdot) \in \mathcal{S}(x_0)$, it follows that

$$\begin{aligned} J(x_0, u(\cdot)) &= \int_0^\infty [-\dot{V}(x(t)) + H(x(t), u(t))]\mathrm{d}t \\ &= -\lim_{t\to\infty} V(x(t)) + V(x_0) + \int_0^\infty H(x(t), u(t))\mathrm{d}t \\ &= V(x_0) + \int_0^\infty H(x(t), u(t))\mathrm{d}t \\ &\geq V(x_0) \\ &= J(x_0, \phi(x(\cdot)), \end{aligned}$$

which yields (8.49). □

Note that (8.45) is the steady-state Hamilton-Jacobi-Bellman equation for the nonlinear system $F(\cdot,\cdot)$ with the cost $J(x_0, u(\cdot))$. Furthermore, conditions (8.45) and (8.46) guarantee optimality with respect to the set of admissible stabilizing controllers $\mathcal{S}(x_0)$. However, it is important to note that an explicit characterization of $\mathcal{S}(x_0)$ is not required. In addition, the optimal stabilizing *feedback* control law $u = \phi(x)$ is independent of the initial condition x_0. Finally, in order to ensure asymptotic stability of the closed-loop system (8.38), Theorem 8.2 requires that $V(\cdot)$ satisfy (8.41), (8.42), and (8.44), which implies that $V(\cdot)$ is a Lyapunov function for the closed-loop system (8.38). However, for optimality $V(\cdot)$ need not satisfy (8.42) and (8.44). Specifically, if $V(\cdot)$ is a continuously differentiable function such that (8.41) is satisfied and $\phi(\cdot) \in \mathcal{S}(x_0)$, then (8.45) and (8.46) imply (8.48) and (8.49).

Next, we specialize Theorem 8.2 to linear systems and provide connections to the optimal linear-quadratic regulator problem. For the following result let $A \in \mathbb{R}^{n\times n}$, $B \in \mathbb{R}^{n\times m}$, $R_1 \in \mathbb{P}^n$, and $R_2 \in \mathbb{P}^m$ be given.

Corollary 8.2. Consider the linear controlled dynamical system

$$\dot{x}(t) = Ax(t) + Bu(t), \qquad x(0) = x_0, \qquad t \geq 0, \tag{8.51}$$

with quadratic performance functional

$$J(x_0, u(\cdot)) \triangleq \int_0^\infty [x^{\mathrm{T}}(t)R_1x(t) + u^{\mathrm{T}}(t)R_2u(t)]\mathrm{d}t, \tag{8.52}$$

where $u(\cdot)$ is an admissible control. Furthermore, assume that there exists a positive-definite matrix $P \in \mathbb{R}^{n\times n}$ such that

$$0 = A^{\mathrm{T}}P + PA + R_1 - PBR_2^{-1}B^{\mathrm{T}}P. \tag{8.53}$$

Then, with the feedback control $u = \phi(x) \triangleq -R_2^{-1}B^{\mathrm{T}}Px$, the zero solution

$x(t) \equiv 0$ to (8.51) is globally asymptotically stable and

$$J(x_0, \phi(x(\cdot))) = x_0^{\mathrm{T}} P x_0, \qquad x_0 \in \mathbb{R}^n. \tag{8.54}$$

Furthermore,

$$J(x_0, \phi(x(\cdot))) = \min_{u(\cdot) \in \mathcal{S}(x_0)} J(x_0, u(\cdot)), \tag{8.55}$$

where $\mathcal{S}(x_0)$ is the set of regulation controllers for (8.51) and $x_0 \in \mathbb{R}^n$.

Proof. The result is a direct consequence of Theorem 8.2 with $F(x,u) = Ax+Bu$, $L(x,u) = x^{\mathrm{T}}R_1x+u^{\mathrm{T}}R_2u$, $V(x) = x^{\mathrm{T}}Px$, $\mathcal{D} = \mathbb{R}^n$, and $U = \mathbb{R}^m$. Specifically, conditions (8.41) and (8.42) are trivially satisfied. Next, it follows from (8.53) that $H(x, \phi(x)) = 0$, and hence, $V'(x)F(x,\phi(x)) < 0$ for all $x \in \mathbb{R}^n$ and $x \neq 0$. Thus, $H(x,u) = H(x,u) - H(x,\phi(x)) = [u - \phi(x)]^{\mathrm{T}} R_2 [u - \phi(x)] \geq 0$ so that all the conditions of Theorem 8.2 are satisfied. Finally, since $V(\cdot)$ is radially unbounded the zero solution $x(t) \equiv 0$ to (8.51) with $u(t) = \phi(x(t)) = -R_2^{-1}B^{\mathrm{T}}Px(t)$ is globally asymptotically stable. □

The optimal feedback control law $\phi(x)$ in Corollary 8.2 is derived using the properties of $H(x,u)$ as defined in Theorem 8.2. Specifically, since $H(x,u) = x^{\mathrm{T}}R_1x + u^{\mathrm{T}}R_2u + x^{\mathrm{T}}(A^{\mathrm{T}}P + PA)x + 2x^{\mathrm{T}}PBu$ it follows that $\frac{\partial^2 H}{\partial u^2} = R_2 > 0$. Now, $\frac{\partial H}{\partial u} = 2R_2u + 2B^{\mathrm{T}}Px = 0$ gives the unique global minimum of $H(x,u)$. Hence, since $\phi(x)$ minimizes $H(x,u)$ it follows that $\phi(x)$ satisfies $\frac{\partial H}{\partial u} = 0$ or, equivalently, $\phi(x) = -R_2^{-1}B^{\mathrm{T}}Px$.

8.4 Inverse Optimal Control for Nonlinear Affine Systems

In this section, we specialize Theorem 8.2 to affine systems. Specifically, we construct nonlinear feedback controllers using an optimal control framework that minimizes a nonlinear-nonquadratic performance criterion. This is accomplished by choosing the controller such that the time derivative of the Lyapunov function is negative along the closed-loop system trajectories while providing sufficient conditions for the existence of asymptotically stabilizing solutions to the Hamilton-Jacobi-Bellman equation. Thus, these results provide a family of globally stabilizing controllers parameterized by the cost functional that is minimized.

The controllers obtained in this section are predicated on an *inverse optimal control problem* [10, 127, 135, 186, 217, 218, 227, 317, 320, 321, 449]. In particular, to avoid the complexity in solving the steady-state Hamilton-Jacobi-Bellman equation we do not attempt to minimize a *given* cost functional, but rather, we parameterize a family of stabilizing controllers that minimize some *derived* cost functional that provides flexibility in specifying the control law. The performance integrand is shown to explicitly depend on the nonlinear system dynamics, the Lyapunov function of the closed-loop system, and the stabilizing feedback control law, wherein the coupling is introduced via the Hamilton-Jacobi-Bellman equation. Hence, by varying parameters in the Lyapunov function and the performance integrand, the proposed framework can be used to characterize a class of globally stabilizing controllers that can meet closed-loop system response constraints.

Consider the nonlinear affine system given by

$$\dot{x}(t) = f(x(t)) + G(x(t))u(t), \qquad x(0) = x_0, \qquad t \geq 0, \tag{8.56}$$

where $f : \mathbb{R}^n \to \mathbb{R}^n$ satisfies $f(0) = 0$, $G : \mathbb{R}^n \to \mathbb{R}^{n\times m}$, $\mathcal{D} = \mathbb{R}^n$, and $U = \mathbb{R}^m$. Furthermore, we consider performance integrands $L(x, u)$ of the form

$$L(x,u) = L_1(x) + L_2(x)u + u^{\mathrm{T}} R_2(x)u, \tag{8.57}$$

where $L_1 : \mathbb{R}^n \to \mathbb{R}, L_2 : \mathbb{R}^n \to \mathbb{R}^{1\times m}$, and $R_2 : \mathbb{R}^n \to \mathbb{P}^m$ so that (8.40) becomes

$$J(x_0, u(\cdot)) = \int_0^\infty [L_1(x(t)) + L_2(x(t))u(t) + u^{\mathrm{T}}(t) R_2(x(t))u(t)]\mathrm{d}t. \tag{8.58}$$

Theorem 8.3. Consider the nonlinear controlled affine system (8.56) with performance functional (8.58). Assume that there exists a continuously differentiable function $V : \mathbb{R}^n \to \mathbb{R}$ and a function $L_2 : \mathbb{R}^n \to \mathbb{R}^{1\times m}$ such that

$$V(0) = 0, \tag{8.59}$$

$$L_2(0) = 0, \tag{8.60}$$

$$V(x) > 0, \qquad x \in \mathbb{R}^n, \qquad x \neq 0, \tag{8.61}$$

$$V'(x)[f(x) - \tfrac{1}{2}G(x)R_2^{-1}(x)L_2^{\mathrm{T}}(x) - \tfrac{1}{2}G(x)R_2^{-1}(x)G^{\mathrm{T}}(x)V'^{\mathrm{T}}(x)] < 0, \;\; x \in \mathbb{R}^n, \;\; x \neq 0, \tag{8.62}$$

and

$$V(x) \to \infty \text{ as } \|x\| \to \infty. \tag{8.63}$$

Then the zero solution $x(t) \equiv 0$ of the closed-loop system

$$\dot{x}(t) = f(x(t)) + G(x(t))\phi(x(t)), \qquad x(0) = x_0, \qquad t \geq 0, \tag{8.64}$$

is globally asymptotically stable with the feedback control law

$$\phi(x) = -\tfrac{1}{2}R_2^{-1}(x)[G^{\mathrm{T}}(x)V'^{\mathrm{T}}(x) + L_2^{\mathrm{T}}(x)], \tag{8.65}$$

and the performance functional (8.58), with

$$L_1(x) = \phi^{\mathrm{T}}(x)R_2(x)\phi(x) - V'(x)f(x), \tag{8.66}$$

is minimized in the sense that

$$J(x_0, \phi(x(\cdot))) = \min_{u(\cdot)\in\mathcal{S}(x_0)} J(x_0, u(\cdot)), \qquad x_0 \in \mathbb{R}^n. \tag{8.67}$$

Finally,

$$J(x_0, \phi(x(\cdot))) = V(x_0), \qquad x_0 \in \mathbb{R}^n. \tag{8.68}$$

Proof. The result is a direct consequence of Theorem 8.2 with $\mathcal{D} = \mathbb{R}^n$, $U = \mathbb{R}^m$, $F(x,u) = f(x)+G(x)u$, and $L(x,u) = L_1(x)+L_2(x)u+u^{\mathrm{T}}R_2(x)u$. Specifically, with (8.57) the Hamiltonian has the form

$$H(x,u) = L_1(x) + L_2(x)u + u^{\mathrm{T}}R_2(x)u + V'(x)(f(x) + G(x)u).$$

Now, the feedback control law (8.65) is obtained by setting $\frac{\partial H}{\partial u} = 0$. With (8.65), it follows that (8.59), (8.61), (8.62), and (8.63) imply (8.41), (8.42), (8.44), and (8.51), respectively. Next, since $V(\cdot)$ is continuously differentiable and $x = 0$ is a local minimum of $V(\cdot)$, it follows that $V'(0) = 0$, and hence, since by assumption $L_2(0) = 0$, it follows that $\phi(0) = 0$, which implies (8.43). Next, with $L_1(x)$ given by (8.66) and $\phi(x)$ given by (8.65) and (8.45) holds. Finally, since $H(x,u) = H(x,u) - H(x,\phi(x)) = [u - \phi(x)]^{\mathrm{T}}R_2(x)[u - \phi(x)]$ and $R_2(x)$ is positive definite for all $x \in \mathbb{R}^n$, condition (8.46) holds. The result now follows as a direct consequence of Theorem 8.2. □

Note that (8.62) is equivalent to

$$\dot{V}(x) \triangleq V'(x)[f(x) + G(x)\phi(x)] < 0, \qquad x \in \mathbb{R}^n, \qquad x \neq 0, \tag{8.69}$$

with $\phi(x)$ given by (8.65). Furthermore, conditions (8.59), (8.61), and (8.69) ensure that $V(\cdot)$ is a Lyapunov function for the closed-loop system (8.64). As discussed in [449], it is important to recognize that the function $L_2(x)$, which appears in the integrand of the performance functional (8.57) is an arbitrary function of $x \in \mathbb{R}^n$ subject to conditions (8.60) and (8.62). Thus, $L_2(x)$ provides flexibility in choosing the control law. This flexibility will be used in the following chapter to connect this inverse optimal control framework to backstepping control methods [247].

As noted in [449], with $L_1(x)$ given by (8.66) and $\phi(x)$ given by (8.65), $L(x,u)$ can be expressed as

$$\begin{aligned} L(x,u) &= u^{\mathrm{T}}R_2(x)u - \phi^{\mathrm{T}}(x)R_2(x)\phi(x) + L_2(x)(u - \phi(x)) \\ &\quad - V'(x)[f(x) + G(x)\phi(x)] \\ &= \left[u + \tfrac{1}{2}R_2^{-1}(x)L_2^{\mathrm{T}}(x)\right]^{\mathrm{T}} R_2(x)\left[u + \tfrac{1}{2}R_2^{-1}(x)L_2^{\mathrm{T}}(x)\right] - V'(x)[f(x) \\ &\quad + G(x)\phi(x)] - \tfrac{1}{4}V'(x)G(x)R_2^{-1}(x)G^{\mathrm{T}}(x)V'^{\mathrm{T}}(x). \end{aligned} \tag{8.70}$$

Since $R_2(x) > 0$, $x \in \mathbb{R}^n$, the first term on the right-hand side of (8.70) is nonnegative, while (8.69) implies that the second term is also nonnegative. Thus, it follows that

$$L(x,u) \geq -\tfrac{1}{4}V'(x)G(x)R_2^{-1}(x)G^{\mathrm{T}}(x)V'^{\mathrm{T}}(x), \tag{8.71}$$

which shows that $L(x,u)$ may be negative. As a result, there may exist a control input u for which the performance functional $J(x_0,u)$ is negative. However, if the control u is a regulation controller, that is, $u \in \mathcal{S}(x_0)$, then it follows from (8.67) and (8.68) that

$$J(x_0,u(\cdot)) \geq V(x_0) > 0, \qquad x_0 \in \mathbb{R}^n, \qquad u(\cdot) \in \mathcal{S}(x_0). \tag{8.72}$$

Furthermore, in this case, substituting $u = \phi(x)$ into (8.70) yields

$$L(x,\phi(x)) = -V'(x)[f(x) + G(x)\phi(x)], \tag{8.73}$$

which, by (8.69), is positive.

Example 8.1. To illustrate the utility of Theorem 8.3 we consider a simple example originally studied in [75] involving the two-state nonlinear controlled system given by

$$\dot{x}_1(t) = x_2^2(t) - x_1^5(t), \qquad x_1(0) = x_{10}, \qquad t \geq 0, \tag{8.74}$$

$$\dot{x}_2(t) = x_1^2(t) + u(t), \qquad x_2(0) = x_{20}. \tag{8.75}$$

To construct an inverse optimal globally stabilizing control law for (8.74) and (8.75) let $V(x)$ be a quadratic Lyapunov function of the form

$$V(x) = p_1 x_1^2 + p_2 x_2^2, \tag{8.76}$$

where $x \triangleq [x_1 \; x_2]^{\mathrm{T}}$, $p_1 > 0$, and $p_2 > 0$, and let $L(x,u) = L_1(x) + L_2(x)u + R_2u^2$, where $R_2 > 0$. Now, $L_2(x) = 2R_2(\frac{p_1}{p_2}x_1x_2 + x_1^2)$ satisfies (8.62) so that the inverse optimal control law (8.65) is given by

$$\phi(x) = -\tfrac{p_1}{p_2}x_1x_2 - x_1^2 - \tfrac{p_2}{R_2}x_2. \tag{8.77}$$

In this case, the performance functional (8.58), with

$$L_1(x) = R_2[\tfrac{p_1}{p_2}x_1x_2 + x_1^2 + \tfrac{p_2}{R_2}x_2]^2 - [2p_1x_1(x_2^2 - x_1^5) + 2p_2x_1^2x_2], \tag{8.78}$$

is minimized in the sense of (8.67). Furthermore, since $V(x)$ given by (8.76) is radially unbounded and

$$\dot{V}(x) = -2p_1x_1^6 - 2\tfrac{p_2^2}{R_2}x_2^2 < 0, \qquad x \in \mathbb{R}^2, \qquad x \neq 0, \tag{8.79}$$

the feedback control law (8.77) is globally stabilizing. △

Example 8.2. This example is adopted from [449] and considers the global stabilization of the Lorentz equations. These equations were proposed by Lorentz [286] to model fluid convection and can exhibit chaotic motion. To construct inverse optimal controllers for the controlled Lorentz dynamical system consider the system

$$\dot{x}_1(t) = -\sigma x_1(t) + \sigma x_2(t), \qquad x_1(0) = x_{10}, \qquad t \geq 0, \tag{8.80}$$

$$\dot{x}_2(t) = rx_1(t) - x_2(t) - x_1(t)x_3(t) + u(t), \qquad x_2(0) = x_{20}, \tag{8.81}$$

$$\dot{x}_3(t) = x_1(t)x_2(t) - bx_3(t), \qquad x_3(0) = x_{30}, \tag{8.82}$$

where $\sigma, r, b > 0$. Note that (8.80)–(8.82) can be written in the form of (8.56) with

$$f(x) = \begin{bmatrix} -\sigma x_1 + \sigma x_2 \\ r x_1 - x_2 - x_1 x_3 \\ x_1 x_2 - b x_3 \end{bmatrix}, \qquad G(x) = \begin{bmatrix} 0 \\ 1 \\ 0 \end{bmatrix}.$$

In order to design an inverse optimal control law for the controlled Lorentz dynamical system (8.80)–(8.82) consider the quadratic Lyapunov function candidate given by

$$V(x) = p_1 x_1^2 + p_2 x_2^2 + p_3 x_3^2, \tag{8.83}$$

where $x \triangleq [x_1, x_2, x_3]^{\mathrm{T}}$ and $p_1, p_2, p_3 > 0$. Now, letting $p_2 = p_3$ and $L(x, u) = L_1(x) + L_2(x)u + R_2 u^2$, where $R_2 > 0$, it follows that

$$L_2(x) = \tfrac{R_2}{p_2}(2p_1\sigma + 2p_2 r)x_1 - 2p_2 x_2, \tag{8.84}$$

satisfies (8.62); that is,

$$\begin{aligned} \dot{V}(x) &= V'(x)[f(x) - \tfrac{1}{2}G(x)R_2^{-1}L_2^{\mathrm{T}}(x) - \tfrac{1}{2}G(x)R_2^{-1}G^{\mathrm{T}}(x)V'^{\mathrm{T}}(x)] \\ &= -2(\sigma p_1 x_1^2 + p_2 x_2^2 + p_2 b x_3^2) \\ &< 0, \qquad x \in \mathbb{R}^3, \qquad x \neq 0. \end{aligned}$$

Hence, the *output* feedback control law $\phi(x) = -(\frac{p_1}{p_2}\sigma + r)x_1$ given by (8.65) globally stabilizes the controlled Lorentz dynamical system (8.80)–(8.82). Furthermore, the performance functional (8.58), with

$$L_1(x) = [2\sigma p_1 + R_2(\tfrac{p_1}{p_2}\sigma + r)^2]x_1^2 - 2(p_1\sigma + p_2 r)x_1 x_2 + 2p_2 x_2^2 + 2p_2 b x_3^2, \tag{8.85}$$

is minimized in the sense of (8.67). △

Next, we specialize Theorem 8.3 to linear systems controlled by nonlinear controllers that minimize a polynomial cost functional. For the following result let $R_1 \in \mathbb{P}^n$, $R_2 \in \mathbb{P}^m$, and $\hat{R}_q \in \mathbb{N}^n$, $q = 2, \ldots, r$, be given, where r is a positive integer, and define $S \triangleq BR_2^{-1}B^{\mathrm{T}}$.

Corollary 8.3. Consider the linear controlled dynamical system

$$\dot{x}(t) = Ax(t) + Bu(t), \qquad x(0) = x_0, \qquad t \geq 0, \tag{8.86}$$

where $u(\cdot)$ is admissible. Assume that there exist $P \in \mathbb{P}^n$ and $M_q \in \mathbb{N}^n$, $q = 2, \ldots, r$, such that

$$0 = A^{\mathrm{T}}P + PA + R_1 - PSP, \tag{8.87}$$

and

$$0 = (A - SP)^{\mathrm{T}}M_q + M_q(A - SP) + \hat{R}_q, \quad q = 2, \ldots, r. \tag{8.88}$$

Then the zero solution $x(t) \equiv 0$ of the closed-loop system

$$\dot{x}(t) = Ax(t) + B\phi(x(t)), \quad x(0) = x_0, \quad t \geq 0, \tag{8.89}$$

is globally asymptotically stable with the feedback control law

$$\phi(x) = -R_2^{-1}B^{\mathrm{T}}\left(P + \sum_{q=2}^{r}(x^{\mathrm{T}}M_q x)^{q-1}M_q\right)x, \tag{8.90}$$

and the performance functional (8.58) with $R_2(x) = R_2$, $L_2(x) = 0$, and

$$\begin{aligned} L_1(x) = x^{\mathrm{T}}\Bigg(R_1 &+ \sum_{q=2}^{r}(x^{\mathrm{T}}M_q x)^{q-1}\hat{R}_q + \left[\sum_{q=2}^{r}(x^{\mathrm{T}}M_q x)^{q-1}M_q\right]^{\mathrm{T}} S \\ &\cdot \left[\sum_{q=2}^{r}(x^{\mathrm{T}}M_q x)^{q-1}M_q\right]\Bigg)x, \end{aligned} \tag{8.91}$$

is minimized in the sense that

$$J(x_0, \phi(x(\cdot))) = \min_{u(\cdot)\in\mathcal{S}(x_0)} J(x_0, u(\cdot)), \qquad x_0 \in \mathbb{R}^n. \tag{8.92}$$

Finally,

$$J(x_0, \phi(x(\cdot))) = x_0^{\mathrm{T}}Px_0 + \sum_{q=2}^{r}\tfrac{1}{q}(x_0^{\mathrm{T}}M_q x_0)^q, \qquad x_0 \in \mathbb{R}^n. \tag{8.93}$$

Proof. The result is a direct consequence of Theorem 8.3 with $f(x) = Ax$, $G(x) = B$, $L_2(x) = 0$, $R_2(x) = R_2$, and

$$V(x) = x^{\mathrm{T}}Px + \sum_{q=2}^{r}\frac{1}{q}(x^{\mathrm{T}}M_q x)^q.$$

Specifically, (8.59)–(8.61) and (8.63) are trivially satisfied. Next, it follows from (8.87), (8.88), and (8.90) that

$$\begin{aligned} &V'(x)[f(x) - \tfrac{1}{2}G(x)R_2^{-1}(x)G^{\mathrm{T}}(x)V'^{\mathrm{T}}(x)] = \\ &\quad -x^{\mathrm{T}}R_1 x - \sum_{q=2}^{r}(x^{\mathrm{T}}M_q x)^{q-1}x^{\mathrm{T}}R_q x - \phi^{\mathrm{T}}(x)R_2\phi(x) \\ &\quad -x^{\mathrm{T}}\left[\sum_{q=2}^{r}(x^{\mathrm{T}}M_q x)^{q-1}M_q\right]^{\mathrm{T}} S \left[\sum_{q=2}^{r}(x^{\mathrm{T}}M_q x)^{q-1}M_q\right]x, \end{aligned}$$

which implies (8.62), so that all the conditions of Theorem 8.3 are satisfied.

□

Corollary 8.3 generalizes the deterministic version of the stochastic nonlinear-nonquadratic optimal control problem considered in [411] to polynomial performance criteria. Specifically, unlike the results of [411], Corollary 8.3 is not limited to sixth-order cost functionals and cubic nonlinear controllers since it addresses a polynomial performance criterion of an arbitrary even order. Corollary 8.3 requires the solutions of $r-1$ modified Riccati equations in (8.88) to obtain the optimal controller (8.90). However, if $\hat{R}_q = \hat{R}_2$, $q = 3, \ldots, r$, then $M_q = M_2$, $q = 3, \ldots, r$, satisfies (8.88). In this case, we require the solution of one modified Riccati equation in (8.88). Finally, it is important to note that the derived performance functional weighs the state variables by arbitrary even powers. Furthermore, $J(x_0, u(\cdot))$ has the form

$$J(x_0, u(\cdot)) = \int_0^\infty \Big[x^{\mathrm{T}}(R_1 + \sum_{q=2}^{r}(x^{\mathrm{T}} M_q x)^{q-1}\hat{R}_q)x + u^{\mathrm{T}} R_2 u + \phi_{\mathrm{NL}}^{\mathrm{T}}(x) R_2 \phi_{\mathrm{NL}}(x)\Big]\mathrm{d}t,$$

where $\phi_{\mathrm{NL}}(x)$ is the nonlinear part of the optimal feedback control

$$\phi(x) = \phi_{\mathrm{L}}(x) + \phi_{\mathrm{NL}}(x),$$

where $\phi_{\mathrm{L}}(x) \triangleq -R_2^{-1}B^{\mathrm{T}}Px$ and $\phi_{\mathrm{NL}}(x) \triangleq -R_2^{-1}B^{\mathrm{T}}\sum_{q=2}^{r}(x^{\mathrm{T}}M_q x)^{q-1}M_q x$.

Next, we specialize Theorem 8.3 to linear systems controlled by nonlinear controllers that minimize a multilinear cost functional. For the following result recall the definition of S and let $R_1 \in \mathbb{P}^n$, $R_2 \in \mathbb{P}^m$, and $\hat{R}_{2q} \in \mathcal{N}^{(2q,n)}$, $q = 2, \ldots, r$, be given, where r is a given integer.

Corollary 8.4. Consider the linear controlled dynamical system (8.86). Assume that there exist $P \in \mathbb{P}^n$ and $\hat{P}_q \in \mathcal{N}^{(2q,n)}$, $q = 2, \ldots, r$, such that

$$0 = A^{\mathrm{T}}P + PA + R_1 - PSP \tag{8.94}$$

and

$$0 = \hat{P}_q[\overset{2q}{\oplus}(A - SP)] + \hat{R}_{2q}, \qquad q = 2, \ldots, r. \tag{8.95}$$

Then the zero solution $x(t) \equiv 0$ of the closed-loop system (8.89) is globally asymptotically stable with the feedback control law

$$\phi(x) = -R_2^{-1}B^{\mathrm{T}}(Px + \tfrac{1}{2}g'^{\mathrm{T}}(x)), \tag{8.96}$$

where $g(x) \triangleq \sum_{q=2}^{r}\hat{P}_q x^{[2q]}$ and the performance functional (8.58) with $R_2(x) = R_2$, $L_2(x) = 0$, and

$$L_1(x) = x^{\mathrm{T}}R_1 x + \sum_{q=2}^{r}\hat{R}_{2q}x^{[2q]} + \tfrac{1}{4}g'(x)Sg'^{\mathrm{T}}(x), \tag{8.97}$$

is minimized in the sense that

$$J(x_0, \phi(x(\cdot))) = \min_{u(\cdot)\in\mathcal{S}(x_0)} J(x_0, u(\cdot)), \qquad x_0 \in \mathbb{R}^n. \tag{8.98}$$

Finally,

$$J(x_0, \phi(x(\cdot))) = x_0^{\mathrm{T}} P x_0 + \sum_{q=2}^{r} \hat{P}_q x_0^{[2q]}, \qquad x_0 \in \mathbb{R}^n. \tag{8.99}$$

Proof. The result is a direct consequence of Theorem 8.3 with $f(x) = Ax$, $G(x) = B$, $L_2(x) = 0$, $R_2(x) = R_2$, and $V(x) = x^{\mathrm{T}}Px + \sum_{q=2}^{r} \hat{P}_q x_0^{[2q]}$. Specifically, (8.59)–(8.61) and (8.63) are trivially satisfied. Next, it follows from (8.94)–(8.96) that

$$V'(x)[f(x) - \tfrac{1}{2}G(x)R_2^{-1}(x)G^{\mathrm{T}}(x)V'^{\mathrm{T}}(x)] = -x^{\mathrm{T}}R_1x - \sum_{q=2}^{r} \hat{R}_{2q}x^{[2q]}$$
$$-\phi^{\mathrm{T}}(x)R_2\phi(x) - \tfrac{1}{4}g'(x)Sg'^{\mathrm{T}}(x),$$

which implies (8.62) so that all the conditions of Theorem 8.3 are satisfied.

□

Note that since $g'(x)(A-SP)x = \sum_{q=2}^{r} \hat{P}_q[\overset{2q}{\oplus}(A-SP)]x^{[2q]}$ it follows that (8.95) can be equivalently written as

$$0 = g'(x)(A - SP)x + \sum_{q=2}^{r} \hat{R}_{2q}x^{[2q]},$$

for all $x \in \mathbb{R}^n$, and hence, it follows from Lemma 8.1 that there exists a unique $\hat{P}_q \in \mathcal{N}^{(2q,n)}$ such that (8.95) is satisfied.

8.5 Gain, Sector, and Disk Margins of Nonlinear-Nonquadratic Optimal Regulators

The gain and phase margins of state feedback linear-quadratic optimal regulators are well known [10, 227, 379, 466]. In particular, in terms of classical control relative stability notions, these controllers possess at least a $\pm 60^\circ$ phase margin, infinite gain margin, and 50 percent gain reduction for each control channel. Alternatively, in terms of absolute stability theory [10] these controllers guarantee sector margins in that the closed-loop system will remain asymptotically stable in the face of a memoryless static input nonlinearity each of whose components is contained in the conic sector $(\frac{1}{2}, \infty)$. In both cases, these results hold if the integrand of the quadratic performance criterion is chosen to be a quadratic nonnegative-definite function of the state and a quadratic positive-definite function of the control with a diagonal weighting matrix. Gain and phase margins of state feedback linear-quadratic optimal regulators involving cross-weighting terms in the quadratic performance criterion were obtained in [95]. Specifically, the authors in [95] provide explicit connections between relative stability margins and the selection of the state, control, and cross-weighting matrices. However, unlike the standard linear-quadratic case, no sector margin guarantees were shown in the linear-quadratic problem with cross-weighting terms.

The problem of guaranteed sector margins for state feedback nonlinear-nonquadratic *inverse* optimal regulators has also been considered in the literature [127, 217, 320, 321]. Specifically, nonlinear Hamilton-Jacobi-Bellman inverse optimal controllers that minimize a meaningful (in the terminology of [127, 395]) nonlinear-nonquadratic performance criterion involving a nonlinear-nonquadratic, nonnegative-definite function of the state and a quadratic positive-definite function of the feedback control are shown to possess sector margin guarantees to component decoupled input nonlinearities in the conic sector $(\frac{1}{2}, \infty)$. These results have been recently extended in [395] to disk margin guarantees where asymptotic stability of the closed-loop system is guaranteed in the face of a dissipative dynamic input operator.

In the remainder of this chapter we derive stability margins for the optimal and inverse optimal nonlinear regulators presented in Sections 8.3 and 8.4. Specifically, gain, sector, and disk margin guarantees are obtained for nonlinear dynamical systems controlled by nonlinear optimal and inverse optimal Hamilton-Jacobi-Bellman controllers that minimize a nonlinear-nonquadratic performance criterion *with* cross-weighting terms. In the case where the cross-weighting term in the performance criterion is deleted our results recover the gain, sector, and disk margins of [395]. Alternatively, retaining the cross-terms in the performance criterion and specializing the nonlinear-nonquadratic problem to a linear-quadratic problem our results recover the gain and phase margins of [95]. Finally, we note that even though the inclusion of cross-weighting terms in the performance criterion is shown to degrade gain, sector, and disk margins, the extra flexibility provided by the cross-weighting terms makes it possible to guarantee optimal and inverse optimal nonlinear controllers that may be far superior in terms of transient performance over meaningful inverse optimal controllers.

In this section, we start by deriving guaranteed gain, sector, and disk margins for nonlinear optimal and inverse optimal regulators that minimize a nonlinear-nonquadratic performance criterion. Specifically, sufficient conditions that guarantee gain, sector, and disk margins are given in terms of the state, control, and cross-weighting nonlinear-nonquadratic weighting functions. In particular, we consider the nonlinear system given by

$$\dot{x}(t) = f(x(t)) + G(x(t))u(t), \quad x(0) = x_0, \quad t \geq 0, \tag{8.100}$$
$$y(t) = -\phi(x(t)), \tag{8.101}$$

where $\phi : \mathbb{R}^n \to \mathbb{R}^m$, with a nonquadratic performance criterion

$$J(x_0, u(\cdot)) = \int_0^\infty [L_1(x(t)) + L_2(x(t))u(t) + u^{\mathrm{T}}(t)R_2(x(t))u(t)]\mathrm{d}t, \tag{8.102}$$

where $L_1 : \mathbb{R}^n \to \mathbb{R}$, $L_2 : \mathbb{R}^n \to \mathbb{R}^{1\times m}$, and $R_2 : \mathbb{R}^n \to \mathbb{R}^{m\times m}$ are given such that $R_2(x) > 0$, $x \in \mathbb{R}^n$, and $L_2(0) = 0$. In this case, the optimal nonlinear feedback controller $u = \phi(x)$ that minimizes the nonlinear-nonquadratic performance criterion (8.102) is given by the following result.

Theorem 8.4. Consider the nonlinear dynamical system (8.100) and (8.101) with performance functional (8.102). Assume that there exists a continuously differentiable function $V : \mathbb{R}^n \to \mathbb{R}$ such that

$$V(0) = 0, \tag{8.103}$$
$$V(x) > 0, \quad x \in \mathbb{R}^n, \quad x \neq 0, \tag{8.104}$$
$$L_2(0) = 0, \tag{8.105}$$
$$V'(x)[f(x) - \tfrac{1}{2}G(x)R_2^{-1}(x)L_2^{\mathrm{T}}(x) - \tfrac{1}{2}G(x)R_2^{-1}(x)G^{\mathrm{T}}(x)V'^{\mathrm{T}}(x)] < 0, \quad x \in \mathbb{R}^n, \quad x \neq 0, \tag{8.106}$$
$$0 = L_1(x) + V'(x)f(x) - \tfrac{1}{4}[V'(x)G(x) + L_2(x)] \cdot R_2^{-1}(x)[V'(x)G(x) + L_2(x)]^{\mathrm{T}}, \quad x \in \mathbb{R}^n, \tag{8.107}$$

and

$$V(x) \to \infty \text{ as } \|x\| \to \infty. \tag{8.108}$$

Then the zero solution $x(t) \equiv 0$ of the closed-loop system

$$\dot{x}(t) = f(x(t)) + G(x(t))\phi(x(t)), \quad x(0) = x_0, \quad t \geq 0, \tag{8.109}$$

is globally asymptotically stable with the feedback control law

$$\phi(x) = -\tfrac{1}{2}R_2^{-1}(x)[V'(x)G(x) + L_2(x)]^{\mathrm{T}}, \tag{8.110}$$

and the performance functional (8.102) is minimized in the sense that

$$J(x_0, \phi(x(\cdot))) = \min_{u(\cdot)\in\mathcal{S}(x_0)} J(x_0, u(\cdot)), \quad x_0 \in \mathbb{R}^n. \tag{8.111}$$

Finally,

$$J(x_0, \phi(x(\cdot))) = V(x_0), \quad x_0 \in \mathbb{R}^n. \tag{8.112}$$

Proof. The proof is identical to the proof of Theorem 8.3. □

Example 8.3. Consider the nonlinear dynamical system

$$\dot{x}_1(t) = x_1(t) + x_1(t)x_2^2(t), \quad x_1(0) = x_{10}, \quad t \geq 0, \tag{8.113}$$
$$\dot{x}_2(t) = -x_2(t) + x_1(t)u(t), \quad x_2(0) = x_{20}, \tag{8.114}$$

with performance functional

$$J(x_{10}, x_{20}, u(\cdot)) = \int_0^\infty [2x_1^2(t) + 2x_2^2(t) + \tfrac{1}{2}u^2(t)]\mathrm{d}t. \tag{8.115}$$

To design an optimal control law $\phi(x_1, x_2)$ that minimizes (8.115) we use Theorem 8.4 with $x = [x_1,\ x_2]^{\mathrm{T}}$, $f(x) = [-x_1 + x_1x_2^2,\ -x_2]^{\mathrm{T}}$, $G(x) = [0,\ x_1]^{\mathrm{T}}$, $L_1(x) = 2x^{\mathrm{T}}x$, $L_2(x) = 0$, and $R_2(x) = \frac{1}{2}$. In particular, it follows from (8.107) that

$$0 = V'(x)\begin{bmatrix} -x_1 + x_1x_2^2 \\ -x_2 \end{bmatrix} - \tfrac{1}{2}V'(x))\begin{bmatrix} 0 & 0 \\ 0 & x_1^2 \end{bmatrix} V'^{\mathrm{T}}(x) + 2(x_1^2 + x_2^2), \tag{8.116}$$

which implies that $V'(x) = [2x_1,\ 2x_2]$. Furthermore, since $V(0) = 0$, $V(x) = x_1^2 + x_2^2$. Hence, the optimal feedback control law is given by $\phi(x) = -\frac{1}{2}R_2^{-1}(x)G^{\mathrm{T}}(x)V'^{\mathrm{T}}(x) = -2x_1x_2$. Finally, note that (8.106) implies

$$\begin{bmatrix} 2x_1 & 2x_2 \end{bmatrix}\begin{bmatrix} -x_1 + x_1x_2^2 \\ -x_2 \end{bmatrix} - 2x_1^2x_2 = -2(x_1^2 + x_2^2) < 0, \tag{8.117}$$

for all $(x_1, x_2) \neq (0, 0)$, and hence, $\phi(x_1, x_2) = -2x_1x_2$ is a global stabilizer for (8.113) and (8.114). △

The following key lemma is needed for developing the main result of this section.

Lemma 8.2. Consider the nonlinear dynamical system $\mathcal{G}$ given by (8.100) and (8.101) where $\phi(x)$ is a stabilizing feedback control law given by (8.110) and where $V(x)$, $x \in \mathbb{R}^n$, satisfies

$$0 = V'(x)f(x) + L_1(x) - \tfrac{1}{4}[V'(x)G(x) + L_2(x)]R_2^{-1}(x)[V'(x)G(x) + L_2(x)]^{\mathrm{T}}. \tag{8.118}$$

Furthermore, suppose there exists $\theta \in \mathbb{R}$ such that $0 < \theta < 1$ and

$$(1 - \theta^2)L_1(x) - \tfrac{1}{4}L_2(x)R_2^{-1}(x)L_2^{\mathrm{T}}(x) \geq 0, \qquad x \in \mathbb{R}^n. \tag{8.119}$$

Then for all $u(\cdot) \in \mathcal{U}$ and $t_1, t_2 \geq 0$, $t_1 < t_2$, the solution $x(t)$, $t \geq 0$, to (8.100) satisfies

$$V(x(t_2)) \leq \int_{t_1}^{t_2} \{[u(t) + y(t)]^{\mathrm{T}}R_2(x(t))[u(t) + y(t)] -\theta^2 u^{\mathrm{T}}(t)R_2(x(t))u(t)\}\mathrm{d}t + V(x(t_1)). \tag{8.120}$$

Proof. Note that it follows from (8.118) and (8.119) that for all $x \in \mathbb{R}^n$

and $u \in \mathbb{R}^m$,

$$\begin{aligned}\theta^2 u^{\mathrm{T}} R_2(x) u &\leq \theta^2 u^{\mathrm{T}} R_2(x) u + \left[\frac{1}{2\sqrt{1-\theta^2}} L_2(x) R_2^{-1}(x) + \sqrt{1-\theta^2} u^{\mathrm{T}}\right] \\ &\quad \cdot R_2(x) \left[\frac{1}{2\sqrt{1-\theta^2}} L_2(x) R_2^{-1}(x) + \sqrt{1-\theta^2} u^{\mathrm{T}}\right]^{\mathrm{T}} \\ &= u^{\mathrm{T}} R_2(x) u + \frac{1}{4(1-\theta^2)} L_2(x) R_2^{-1}(x) L_2^{\mathrm{T}}(x) + L_2(x) u \\ &\leq u^{\mathrm{T}} R_2(x) u + L_2(x) u + L_1(x) \\ &= u^{\mathrm{T}} R_2(x) u + L_2(x) u - V'(x) f(x) + \phi^{\mathrm{T}}(x) R_2(x) \phi(x) \\ &= [u+y]^{\mathrm{T}} R_2(x) [u+y] - V'(x)[f(x) + G(x) u],\end{aligned}$$

which implies that, for all $u(\cdot) \in \mathcal{U}$ and $t \geq 0$,

$$\theta^2 u^{\mathrm{T}}(t) R_2(x(t)) u(t) \leq [u(t) + y(t)]^{\mathrm{T}} R_2(x(t)) [u(t) + y(t)] - \dot{V}(x(t)).$$

Now, integrating over $[t_1, t_2]$ yields (8.120). □

Next, we present disk margins for the nonlinear-nonquadratic optimal regulator given by Theorem 8.4. First, we consider the case in which $R_2(x)$, $x \in \mathbb{R}^n$, is a constant diagonal matrix.

Theorem 8.5. Consider the nonlinear dynamical system $\mathcal{G}$ given by (8.100) and (8.101) where $\phi(x)$ is a stabilizing feedback control law given by (8.110) and where $V(x)$, $x \in \mathbb{R}^n$, satisfies (8.107). If $R_2(x) \equiv \mathrm{diag}[r_1, \ldots, r_m]$, where $r_i > 0$, $i = 1, \ldots, m$, and there exists $\theta \in \mathbb{R}$ such that $0 < \theta < 1$ and (8.119) is satisfied, then the nonlinear system $\mathcal{G}$ has a structured disk margin $(\frac{1}{1+\theta}, \frac{1}{1-\theta})$. If, in addition, $R_2(x) \equiv I$ and there exists $\theta \in \mathbb{R}$ such that $0 < \theta < 1$ and (8.119) is satisfied, then the nonlinear system $\mathcal{G}$ has a disk margin $(\frac{1}{1+\theta}, \frac{1}{1-\theta})$.

Proof. Note that for all $u(\cdot) \in \mathcal{U}$ and $t_1, t_2 \geq 0$, $t_1 < t_2$, it follows from Lemma 8.2 that the solution $x(t)$, $t \geq 0$, to (8.100) satisfies

$$\begin{aligned}V(x(t_2)) - V(x(t_1)) \leq \int_{t_1}^{t_2} &\{[u(t) + y(t)]^{\mathrm{T}} R_2 [u(t) + y(t)] \\ &-\theta^2 u^{\mathrm{T}}(t) R_2 u(t)\} \, \mathrm{d}t.\end{aligned}$$

Hence, with the storage function $V_{\mathrm{s}}(x) = \frac{1}{2} V(x)$, $\mathcal{G}$ is dissipative with respect to the supply rate $r(u, y) = u^{\mathrm{T}} R_2 y + \frac{1-\theta^2}{2} u^{\mathrm{T}} R_2 u + y^{\mathrm{T}} R_2 y$. Now, the result is a direct consequence of Corollary 6.2 and Definitions 6.4 and 6.3 with $\alpha = \frac{1}{1+\theta}$ and $\beta = \frac{1}{1-\theta}$. □

Next, we consider the case in which $R_2(x)$, $x \in \mathbb{R}^n$, is not a diagonal constant matrix. For the following result define

$$\overline{\gamma} \triangleq \sup_{x\in\mathbb{R}^n} \sigma_{\max}(R_2(x)), \qquad \underline{\gamma} \triangleq \inf_{x\in\mathbb{R}^n} \sigma_{\min}(R_2(x)), \tag{8.121}$$

where $R_2(x)$ is such that $\overline{\gamma} < \infty$ and $\underline{\gamma} > 0$.

Theorem 8.6. Consider the nonlinear dynamical system $\mathcal{G}$ given by (8.100) and (8.101) where $\phi(x)$ is a stabilizing feedback control law given by (8.110) and where $V(x)$, $x \in \mathbb{R}^n$, satisfies (8.107). If there exists $\theta \in \mathbb{R}$ such that $0 < \theta < 1$ and (8.119) is satisfied, then the nonlinear system $\mathcal{G}$ has a disk margin $(\frac{1}{1+\eta\theta}, \frac{1}{1-\eta\theta})$, where $\eta \triangleq \sqrt{\underline{\gamma}/\overline{\gamma}}$.

Proof. Note that for all $u(\cdot) \in \mathcal{U}$ and $t_1, t_2 \geq 0$, $t_1 < t_2$, it follows from Lemma 8.2 that the solution $x(t)$, $t \geq 0$, to (8.100) satisfies

$$V(x(t_2)) - V(x(t_1)) \leq \int_{t_1}^{t_2} \{[u(t)+y(t)]^{\mathrm{T}} R_2(x(t))[u(t)+y(t)] \\ -\theta^2 u^{\mathrm{T}}(t) R_2(x(t)) u(t)\}\mathrm{d}t,$$

which implies that

$$V(x(t_2)) - V(x(t_1)) \leq \int_{t_1}^{t_2} \{\overline{\gamma}[u(t)+y(t)]^{\mathrm{T}}[u(t)+y(t)] - \underline{\gamma}\theta^2 u^{\mathrm{T}}(t)u(t)\}\mathrm{d}t.$$

Hence, with the storage function $V_{\mathrm{s}}(x) = \frac{1}{2\overline{\gamma}}V(x)$, $\mathcal{G}$ is dissipative with respect to the supply rate $r(u,y) = u^{\mathrm{T}}y + \frac{1-\eta^2\theta^2}{2}u^{\mathrm{T}}u + y^{\mathrm{T}}y$. Now, the result is a direct consequence of Corollary 6.2 and Definition 6.3 with $\alpha = \frac{1}{1+\eta\theta}$ and $\beta = \frac{1}{1-\eta\theta}$. □

Next, we provide an alternative result that guarantees sector and gain margins for the case in which $R_2(x)$, $x \in \mathbb{R}^n$, is diagonal.

Theorem 8.7. Consider the nonlinear dynamical system $\mathcal{G}$ given by (8.100) and (8.101) where $\phi(x)$ is a stabilizing feedback control law given by (8.110) and where $V(x)$, $x \in \mathbb{R}^n$, satisfies (8.107). Furthermore, let $R_2(x) = \mathrm{diag}\,[r_1(x), \ldots, r_m(x)]$, where $r_i : \mathbb{R}^n \to \mathbb{R}$, $r_i(x) > 0$, $i = 1, \ldots, m$. If $\mathcal{G}$ is zero-state observable and there exists $\theta \in \mathbb{R}$ such that $0 < \theta < 1$ and

$$(1-\theta^2)L_1(x) - \tfrac{1}{4}L_2(x)R_2^{-1}(x)L_2^{\mathrm{T}}(x) \geq 0, \qquad x \in \mathbb{R}^n, \tag{8.122}$$

then the nonlinear system $\mathcal{G}$ has a sector (and, hence, gain) margin $(\frac{1}{1+\theta}, \frac{1}{1-\theta})$.

Proof. Let $\Delta(-y) = \sigma(-y)$, where $\sigma : \mathbb{R}^m \to \mathbb{R}^m$ is a static nonlinearity such that $\sigma(0) = 0$, $\sigma(v) = [\sigma_1(v_1), \ldots, \sigma_m(v_m)]^{\mathrm{T}}$, and $\alpha v_i^2 < \sigma_i(v_i)v_i < \beta v_i^2$, for all $v_i \neq 0$, $i = 1, \ldots, m$, where $\alpha = \frac{1}{1+\theta}$ and $\beta = \frac{1}{1-\theta}$;

or, equivalently, $(\sigma_i(v_i) - \alpha v_i)(\sigma_i(v_i) - \beta v_i) < 0$, for all $v_i \neq 0$, $i = 1, \ldots, m$. In this case, the closed-loop system (8.100) and (8.101) with $u = \sigma(-y)$ is given by

$$\dot{x}(t) = f(x(t)) + G(x(t))\sigma(\phi(x(t))), \qquad x(0) = x_0, \qquad t \geq 0. \tag{8.123}$$

Next, consider the Lyapunov function candidate $V(x)$, $x \in \mathbb{R}^n$, satisfying (8.107) and let $\dot{V}(x)$ denote the Lyapunov derivative along the trajectories of the closed-loop system (8.123). Now, it follows from (8.107) and (8.122) that

$$\begin{aligned}
\dot{V}(x) &= V'(x)f(x) + V'(x)G(x)\sigma(\phi(x)) \\
&\leq V'(x)f(x) + V'(x)G(x)\sigma(\phi(x)) + L_1(x) \\
&\quad -\tfrac{1}{4(1-\theta^2)}L_2(x)R_2^{-1}(x)L_2^{\mathrm{T}}(x) \\
&\quad +(1-\theta^2)\left[\sigma(\phi(x)) + \tfrac{1}{2(1-\theta^2)}R_2^{-1}(x)L_2^{\mathrm{T}}(x)\right]^{\mathrm{T}} R_2(x) \\
&\quad \cdot\left[\sigma(\phi(x)) + \tfrac{1}{2(1-\theta^2)}R_2^{-1}(x)L_2^{\mathrm{T}}(x)\right] \\
&= V'(x)f(x) + L_1(x) + V'(x)G(x)\sigma(\phi(x)) \\
&\quad +(1-\theta^2)\sigma^{\mathrm{T}}(\phi(x))R_2(x)\sigma(\phi(x)) + L_2(x)\sigma(\phi(x)) \\
&= \phi^{\mathrm{T}}(x)R_2(x)\phi(x) - 2\phi^{\mathrm{T}}(x)R_2(x)\sigma(\phi(x)) \\
&\quad +(1-\theta^2)\sigma^{\mathrm{T}}(\phi(x))R_2(x)\sigma(\phi(x)) \\
&= \sum_{i=1}^{m} r_i(x)\left(\tfrac{1}{\beta}\sigma_i(-y_i) + y_i\right)\left(\tfrac{1}{\alpha}\sigma_i(-y_i) + y_i\right) \\
&= \tfrac{1}{\alpha\beta}\sum_{i=1}^{m} r_i(x)\left(\sigma_i(-y_i) + \alpha y_i\right)\left(\sigma_i(-y_i) + \beta y_i\right) \\
&\leq 0,
\end{aligned}$$

which implies that the closed-loop system (8.123) is Lyapunov stable.

Next, let $\mathcal{R} \triangleq \{x \in \mathbb{R}^n : \dot{V}(x) = 0\}$ and note that $\dot{V}(x) = 0$ if and only if $y = 0$. Now, since $\mathcal{G}$ is zero-state observable it follows that $\mathcal{M} \triangleq \{x \in \mathbb{R}^n : x = 0\}$ is the largest invariant set contained in $\mathcal{R}$. Hence, it follows from Theorem 3.5 that $x(t) \to \mathcal{M} = \{0\}$ as $t \to \infty$. Thus, the closed-loop system (8.123) is globally asymptotically stable for all $\sigma(\cdot)$ such that $\alpha v_i^2 < \sigma_i(v_i)v_i < \beta v_i^2$, $v_i \neq 0$, $i = 1, \ldots, m$, which implies that the nonlinear system $\mathcal{G}$ given by (8.100) and (8.101) has sector (and, hence, gain) margins (α, β). □

Note that in the case where $R_2(x)$, $x \in \mathbb{R}^n$, is diagonal, Theorem 8.7 guarantees larger gain and sector margins to the gain and sector margin guarantees provided by Theorem 8.6. However, Theorem 8.7 does not provide disk margin guarantees.

8.6 Inverse Optimality of Nonlinear Regulators

In this section, we give sufficient conditions that guarantee that a given nonlinear feedback controller has prespecified disk, sector, and gain margins. First, we present the following generalization of the results given in [321].

Proposition 8.2. Let $\theta \in (0,1)$ and let $R_2 \in \mathbb{R}^{m\times m}$ be a positive-definite matrix. Consider the nonlinear dynamical system $\mathcal{G}$ given by (8.100) and (8.101), where $\phi(x)$ is a stabilizing feedback control law. Then there exist functions $V : \mathbb{R}^n \to \mathbb{R}$, $L_1 : \mathbb{R}^n \to \mathbb{R}$, and $L_2 : \mathbb{R}^n \to \mathbb{R}^{1\times m}$ such that $\phi(x) = -\frac{1}{2}R_2^{-1}[V'(x)G(x) + L_2(x)]^{\mathrm{T}}$, $V(\cdot)$ is continuously differentiable, $V(0) = 0$, $V(x) > 0$, $x \in \mathbb{R}^n$, $x \neq 0$, and for all $x \in \mathbb{R}^n$,

$$0 = V'(x)f(x) + L_1(x) - [V'(x)G(x) + L_2(x)]R_2^{-1}[V'(x)G(x) + L_2(x)]^{\mathrm{T}}, \tag{8.124}$$

$$0 \leq (1-\theta^2)L_1(x) - \tfrac{1}{4}L_2(x)R_2^{-1}L_2^{\mathrm{T}}(x), \tag{8.125}$$

if and only if, for all $u(\cdot) \in \mathcal{U}$ and $t_1, t_2 \geq 0$, $t_1 < t_2$, there exists $V : \mathbb{R}^n \to \mathbb{R}$ such that $V(0) = 0$, $V(x) > 0$, $x \in \mathbb{R}^n$, $x \neq 0$, and the solution $x(t)$, $t \geq 0$, to (8.100) satisfies

$$V(x(t_2)) \leq \int_{t_1}^{t_2} \left\{[u(t)+y(t)]^{\mathrm{T}}R_2[u(t)+y(t)] - \theta^2 u^{\mathrm{T}}(t)R_2u(t)\right\}\mathrm{d}t + V(x(t_1)). \tag{8.126}$$

Proof. If there exist functions $V : \mathbb{R}^n \to \mathbb{R}$, $L_1 : \mathbb{R}^n \to \mathbb{R}$, and $L_2 : \mathbb{R}^n \to \mathbb{R}^{1\times m}$ such that $\phi(x) = -R_2^{-1}[V'(x)G(x) + L_2(x)]^{\mathrm{T}}$ and (8.124) and (8.125) are satisfied, then it follows from Lemma 8.2 that (8.126) is satisfied. Conversely, if for all $t_1, t_2 \geq 0$, $t_1 < t_2$, and $u(\cdot) \in \mathcal{U}$ the solution $x(t)$, $t \geq 0$, to (8.100) satisfies (8.126), then with $Q = R_2$, $S = R_2$, and $R = (1-\theta^2)R_2$, it follows from (5.81) of Theorem 5.6 that

$$0 \geq V'(x)f(x) - \phi^{\mathrm{T}}(x)R_2\phi(x) + \tfrac{1}{4(1-\theta^2)}[2\phi^{\mathrm{T}}(x)R_2 + V'(x)G(x)] \cdot R_2^{-1}[2\phi^{\mathrm{T}}(x)R_2 + V'(x)G(x)]^{\mathrm{T}}, \quad x \in \mathbb{R}^n.$$

The result now follows with $L_1(x) = -V'(x)f(x) + \phi^{\mathrm{T}}(x)R_2\phi(x)$ and $L_2(x) = -[2\phi^{\mathrm{T}}(x)R_2 + V'(x)G(x)]$. □

Note that if (8.124) and (8.125) are satisfied then it follows from Theorem 8.4 that the feedback control law $\phi(x) = -R_2^{-1}[V'(x)G(x) + L_2(x)]^{\mathrm{T}}$ minimizes the cost functional (8.58). Hence, Proposition 8.2 provides necessary and sufficient conditions for optimality of a given stabilizing feedback control law with prespecified disk margin guarantees. The following result presents specific disk margin guarantees for inverse optimal controllers.

Theorem 8.8. Let $\theta \in (0,1)$ be given. Consider the nonlinear dynamical system $\mathcal{G}$ given by (8.100) and (8.101) where $\phi(x)$ is a stabilizing feedback control law. Assume that there exist functions $V : \mathbb{R}^n \to \mathbb{R}$ and $R_2 : \mathbb{R}^n \to \mathbb{R}^{m\times m}$ such that $V(\cdot)$ is continuously differentiable, $R_2(x) > 0$, $x \in \mathbb{R}^n$, and

$$V(0) = 0, \tag{8.127}$$

$$V(x) > 0, \quad x \in \mathbb{R}^n, \quad x \neq 0, \tag{8.128}$$

$$V'(x)[f(x) + G(x)\phi(x)] < 0, \quad x \in \mathbb{R}^n, \quad x \neq 0, \tag{8.129}$$

$$V'(x)f(x) - \phi^{\mathrm{T}}(x)R_2^{-1}(x)\phi(x) + \tfrac{1}{1-\theta^2}\left(\phi^{\mathrm{T}}(x) + \tfrac{1}{2}V'(x)G(x)R_2^{-1}(x)\right) \cdot R_2(x)\left(\phi^{\mathrm{T}}(x) + \tfrac{1}{2}V'(x)G(x)R_2^{-1}(x)\right)^{\mathrm{T}} \leq 0, \quad x \in \mathbb{R}^n, \tag{8.130}$$

and

$$V(x) \to \infty \text{ as } \|x\| \to \infty. \tag{8.131}$$

Then the nonlinear dynamical system $\mathcal{G}$ has a disk margin $(\frac{1}{1+\eta\theta}, \frac{1}{1-\eta\theta})$, where $\eta = \sqrt{\underline{\gamma}/\overline{\gamma}}$ and $\underline{\gamma}$ and $\overline{\gamma}$ are given by (8.121). Furthermore, with the feedback control law $\phi(x)$ the performance functional

$$J(x_0, u(\cdot)) = \int_0^\infty [-V'(x(t))(f(x(t)) + G(x(t))u(t)) + (\phi(x(t)) - u(t))^{\mathrm{T}} R_2(x(t))(\phi(x(t)) - u(t))]\mathrm{d}t \tag{8.132}$$

is minimized in the sense that

$$J(x_0, \phi(x(\cdot))) = \min_{u(\cdot)\in\mathcal{S}(x_0)} J(x_0, u(\cdot)), \qquad x_0 \in \mathbb{R}^n. \tag{8.133}$$

Proof. The result is a direct consequence of Theorems 8.4 and 8.6 with $L_1(x) = -V'(x)f(x) + \phi^{\mathrm{T}}(x)R_2(x)\phi(x)$ and $L_2(x) = -(2\phi^{\mathrm{T}}(x)R_2(x) + V'(x)G(x))$. Specifically, in this case, all the conditions of Theorem 8.4 are trivially satisfied. Furthermore, note that (8.130) is equivalent to (8.119). The result is now immediate. □

The next result provides sufficient conditions that guarantee that a given nonlinear feedback controller has prespecified gain and sector margins.

Theorem 8.9. Let $\theta \in (0,1)$ be given. Consider the nonlinear dynamical system $\mathcal{G}$ given by (8.100) and (8.101) where $\phi(x)$ is a stabilizing feedback control law. Assume there exist functions $R_2(x) = \mathrm{diag}[r_1(x), \ldots, r_m(x)]$, where $r_i : \mathbb{R}^n \to \mathbb{R}$, $r_i(x) > 0$, $i = 1, \ldots, m$, and $V : \mathbb{R}^n \to \mathbb{R}$ such that $V(\cdot)$ is continuously differentiable and satisfies (8.127)–(8.131). Then the nonlinear dynamical system $\mathcal{G}$ has a disk margin $(\frac{1}{1+\theta}, \frac{1}{1-\theta})$. Furthermore, with the feedback control law $\phi(x)$ the performance functional (8.132) is minimized in the sense that

$$J(x_0, \phi(x(\cdot))) = \min_{u(\cdot)\in\mathcal{S}(x_0)} J(x_0, u(\cdot)), \qquad x_0 \in \mathbb{R}^n. \tag{8.134}$$

Proof. The result is a direct consequence of Theorems 8.4 and 8.7 with the proof being identical to the proof of Theorem 8.8. □

8.7 Linear-Quadratic Optimal Regulators

In this section, we specialize Theorems 8.5 and 8.6 to the case of linear systems. Specifically, consider the stabilizable linear system given by

$$\dot{x}(t) = Ax(t) + Bu(t), \qquad x(0) = x_0, \qquad t \geq 0, \tag{8.135}$$
$$y(t) = -Kx(t), \tag{8.136}$$

where $A \in \mathbb{R}^{n\times n}$, $B \in \mathbb{R}^{n\times m}$, and $K \in \mathbb{R}^{m\times n}$, and assume that (A, K) is detectable and the linear system (8.135) and (8.136) is asymptotically stable with the feedback $u = -y$ or, equivalently, $A + BK$ is Hurwitz. Furthermore, assume that K is an optimal regulator which minimizes the quadratic performance functional given by

$$J(x_0, u(\cdot)) = \int_0^\infty [x^{\mathrm{T}}(t)R_1x(t) + 2x^{\mathrm{T}}(t)R_{12}u(t) + u^{\mathrm{T}}(t)R_2u(t)]\mathrm{d}t, \tag{8.137}$$

where $R_1 \in \mathbb{R}^{n\times n}$, $R_{12} \in \mathbb{R}^{n\times m}$, and $R_2 \in \mathbb{R}^{m\times m}$ are such that $R_2 > 0$, $R_1 - R_{12}R_2^{-1}R_{12}^{\mathrm{T}} \geq 0$, and (A, R_1) is observable. In this case, it follows from Theorem 8.4 with $f(x) = Ax$, $G(x) = B$, $L_1(x) = x^{\mathrm{T}}R_1x$, $L_2(x) = 2x^{\mathrm{T}}R_{12}$, $R_2(x) = R_2$, $\phi(x) = Kx$, and $V(x) = x^{\mathrm{T}}Px$ that the optimal control law K is given by $K = -R_2^{-1}(B^{\mathrm{T}}P + R_{12})$, where $P > 0$ is the solution to the algebraic regulator Riccati equation given by

$$0 = (A - BR_2^{-1}R_{12}^{\mathrm{T}})^{\mathrm{T}}P + P(A - BR_2^{-1}R_{12}^{\mathrm{T}}) + R_1 - R_{12}R_2^{-1}R_{12}^{\mathrm{T}} - PBR_2^{-1}B^{\mathrm{T}}P. \tag{8.138}$$

The following results provide guarantees of disk, sector, and gain margins for the linear system (8.135) and (8.136).

Corollary 8.5. Consider the linear dynamical system (8.135) and (8.136) with performance functional (8.137) and let $\sigma_{\max}^2(R_{12}) < \sigma_{\min}(R_1) \cdot\sigma_{\min}(R_2)$. Then, with $K = -R_2^{-1}(B^{\mathrm{T}}P + R_{12})$, where $P > 0$ satisfies (8.138), the linear system (8.135) and (8.136) has disk margin (and, hence, sector and gain margins) $(\frac{1}{1+\eta\theta}, \frac{1}{1-\eta\theta})$, where

$$\eta = \frac{\sigma_{\min}(R_2)}{\sigma_{\max}(R_2)}, \qquad \theta = \left(1 - \frac{\sigma_{\max}^2(R_{12})}{\sigma_{\min}(R_1)\sigma_{\min}(R_2)}\right)^{1/2}. \tag{8.139}$$

Proof. The result is a direct consequence of Theorem 8.6 with $f(x) = Ax$, $G(x) = B$, $\phi(x) = Kx$, $V(x) = x^{\mathrm{T}}Px$, $L_1(x) = x^{\mathrm{T}}R_1x$, and $L_2(x) = 2x^{\mathrm{T}}R_{12}$. Specifically, note that (8.138) is equivalent to (8.107). Now, with θ given by (8.139), it follows that $(1-\theta^2)R_1 - R_{12}R_2^{-1}R_{12}^{\mathrm{T}} \geq 0$, and hence, (8.122) is satisfied so that all the conditions of Theorem 8.6 are satisfied. □

Corollary 8.6. Consider the linear dynamical system (8.135) and (8.136) with performance functional (8.137) and let $\sigma_{\max}^2(R_{12}) < \sigma_{\min}(R_1) \cdot\sigma_{\min}(R_2)$, where R_2 is diagonal. Then, with $K = -R_2^{-1}(B^{\mathrm{T}}P + R_{12})$, where $P > 0$ satisfies (8.138), the linear system (8.135) and (8.136) has structured disk margin (and, hence, gain and sector) margin $(\frac{1}{1+\theta}, \frac{1}{1-\theta})$, where

$$\theta = \left(1 - \frac{\sigma_{\max}^2(R_{12})}{\sigma_{\min}(R_1)\sigma_{\min}(R_2)}\right)^{1/2}. \tag{8.140}$$

Proof. The result is a direct consequence of Theorem 8.5 with $f(x) = Ax$, $G(x) = B$, $\phi(x) = Kx$, $V(x) = x^{\mathrm{T}}Px$, $L_1(x) = x^{\mathrm{T}}R_1x$, and $L_2(x) = 2x^{\mathrm{T}}R_{12}$. Specifically, note that (8.138) is equivalent to (8.107). Now, with θ given by (8.140), it follows that $(1-\theta^2)R_1 - R_{12}R_2^{-1}R_{12}^{\mathrm{T}} \geq 0$, and hence, (8.122) is satisfied so that all the conditions of Theorem 8.5 are satisfied. □

The gain margins obtained in Corollary 8.6 are precisely the gain margins given in [95] for linear-quadratic optimal regulators with cross-weighting terms in the performance criterion. Furthermore, since Corollary 8.6 guarantees structured disk margins of $(\frac{1}{1+\theta}, \frac{1}{1-\theta})$, it follows that the linear system has a phase margin ϕ given by (see Problem 6.20)

$$\cos(\phi) = 1 - \frac{\theta^2}{2}, \tag{8.141}$$

or, equivalently,

$$\sin\left(\frac{\phi}{2}\right) = \frac{\theta}{2}. \tag{8.142}$$

In the case where $R_{12} = 0$ it follows from (8.140) that $\theta = 1$, and hence, Corollary 8.6 guarantees a phase margin of $\pm 60°$ in each input-output channel. In addition, requiring that $R_1 \geq 0$, it follows from Corollary 8.6 that the linear system given by (8.135) and (8.136) has a gain and sector margin of $(\frac{1}{2}, \infty)$.

Finally, we specialize Proposition 8.2 to the case of linear systems to provide necessary and sufficient conditions for optimality of a given stabilizing linear feedback control law.

Proposition 8.3. Let $\theta \in (0,1)$ and let $R_2 \in \mathbb{R}^{m\times m}$ be a positive-definite matrix. Consider the linear system given by (8.135) and (8.136) where $K \in \mathbb{R}^{m\times n}$ is such that $A + BK$ is Hurwitz. Then the following statements are equivalent:

i) There exist matrices $P, R_1 \in \mathbb{R}^{n\times n}$, and $R_{12} \in \mathbb{R}^{n\times m}$ such that $K = -R_2^{-1}(B^{\mathrm{T}}P + R_{12}^{\mathrm{T}})$, P is positive definite, R_1 is nonnegative definite, and

$$0 = A^{\mathrm{T}}P + PA + R_1 - (PB + R_{12})R_2^{-1}(PB + R_{12})^{\mathrm{T}}, \quad (8.143)$$

$$0 \leq (1-\theta^2)R_1 - R_{12}R_2^{-1}R_{12}^{\mathrm{T}}. \quad (8.144)$$

ii) The transfer function $\theta R_2^{1/2}[I - G(s)]^{-1}R_2^{-1/2}$ is bounded real, where $G(s) \triangleq K(sI - A)^{-1}B$.

iii) The transfer function $R_2[(1+\theta)I - G(s)][(1-\theta)I - G(s)]^{-1}$ is positive real.

Proof. The result follows from Proposition 8.2 with $f(x) = Ax$, $G(x) = B$, $\phi(x) = Kx$, $L_1(x) = R_1$, $L_2(x) = 2x^{\mathrm{T}}R_{12}$, and $V(x) = x^{\mathrm{T}}Px$. Specifically, it follows from Proposition 8.2 that K, P, R_1, R_{12}, and R_2 satisfy (8.143) and (8.144) if and only if the linear system (8.135) and (8.136) is dissipative with respect to the supply rate $r(u, y) = [u + y]^{\mathrm{T}}R_2[u + y] - \theta^2 u^{\mathrm{T}}R_2 u$. Now, it follows from Problem 5.14 that the linear system (8.135) and (8.136) is dissipative with respect to the supply rate $r(u, y) = [u+y]^{\mathrm{T}}R_2[u+y] - \theta^2 u^{\mathrm{T}}R_2 u$ if and only if $G^*(s)R_2G(s) - G^*(s)R_2 - R_2G(s) + (1-\theta^2)R_2 \geq 0$, $\mathrm{Re}[s] > 0$. The result now follows as a direct consequence of Problem 5.15. □

8.8 Stability Margins, Meaningful Inverse Optimality, and Control Lyapunov Functions

In this section, we specialize the results of Section 8.5 to the case where $L(x, u)$ is nonnegative for all $(x, u) \in \mathbb{R}^n \times \mathbb{R}^m$. In the terminology of [128,395] this corresponds to a *meaningful cost functional*. Here, we assume $L_2(x) \equiv 0$ and $L_1(x) \geq 0$, $x \in \mathbb{R}^n$. In this case, we show that for every nonlinear dynamical system for which a control Lyapunov function can be constructed there exists an inverse optimal feedback control law with sector and gain margins of $(\frac{1}{2}, \infty)$. The first result specializes Theorem 8.4 to the case in which $L_2(x) \equiv 0$.

Theorem 8.10. Consider the nonlinear dynamical system (8.100) with performance functional (8.102) with $L_2(x) \equiv 0$ and $L_1(x) \geq 0$, $x \in \mathbb{R}^n$. Assume there exists a continuously differentiable function $V : \mathbb{R}^n \to \mathbb{R}$ such that

$$V(0) = 0, \tag{8.145}$$

$$V(x) > 0, \quad x \in \mathbb{R}^n, \quad x \neq 0, \tag{8.146}$$

$$0 = L_1(x) + V'(x)f(x) - \tfrac{1}{4}V'(x)G(x)R_2^{-1}(x)G^{\mathrm{T}}(x)V'^{\mathrm{T}}(x), \quad x \in \mathbb{R}^n, \tag{8.147}$$

and

$$V(x) \to \infty \text{ as } \|x\| \to \infty. \tag{8.148}$$

Furthermore, assume that the system (8.100) and (8.101) is zero-state observable with $y = L_1(x)$. Then the zero solution $x(t) \equiv 0$ of the closed-loop system

$$\dot{x}(t) = f(x(t)) + G(x(t))\phi(x(t)), \quad x(0) = x_0, \quad t \geq 0, \tag{8.149}$$

is globally asymptotically stable with the feedback control law

$$\phi(x) = -\tfrac{1}{2}R_2^{-1}(x)G^{\mathrm{T}}(x)V'^{\mathrm{T}}(x), \tag{8.150}$$

and the performance functional (8.102) is minimized in the sense that

$$J(x_0, \phi(x(\cdot))) = \min_{u(\cdot) \in \mathcal{S}(x_0)} J(x_0, u(\cdot)), \quad x_0 \in \mathbb{R}^n. \tag{8.151}$$

Finally,

$$J(x_0, \phi(x(\cdot))) = V(x_0), \quad x_0 \in \mathbb{R}^n. \tag{8.152}$$

Proof. The proof is similar to the proof of Theorem 8.3. □

Next, we show that for a given nonlinear dynamical system $\mathcal{G}$ given by (8.100) and (8.101), there exists an equivalence between optimality and passivity. For the following result we assume that for a given nonlinear system (8.100), if there exists a feedback control law $\phi(x)$ such that it minimizes the performance functional (8.102) with $R_2(x) \equiv I$, $L_2(x) \equiv 0$, and $L_1(x) \geq 0$, $x \in \mathbb{R}^n$, then there exists a continuously differentiable positive-definite function $V(x)$, $x \in \mathbb{R}^n$, such that (8.147) is satisfied.

Theorem 8.11. Consider the nonlinear dynamical system $\mathcal{G}$ given by (8.100) and (8.101). The feedback control law $u = \phi(x)$ is optimal with respect to a performance functional (8.102) with $R_2(x) \equiv I$, $L_2(x) \equiv 0$, and $L_1(x) \geq 0$, $x \in \mathbb{R}^n$, if and only if the nonlinear system $\mathcal{G}$ is dissipative with respect to the supply rate $r(u, y) = y^{\mathrm{T}}y + 2u^{\mathrm{T}}y$ and has a continuously differentiable positive-definite, radially unbounded storage function $V(x)$, $x \in \mathbb{R}^n$.

Proof. If the control law $\phi(x)$ is optimal with respect to a performance functional (8.102) with $R_2(x) \equiv I$, $L_2(x) \equiv 0$, and $L_1(x) \geq 0$, $x \in \mathbb{R}^n$, then, by assumption, there exists a continuously differentiable positive-definite function $V(x)$ such that (8.147) is satisfied. Hence, it follows from Proposition 8.2 that the solution $x(t)$, $t \geq 0$, to (8.100) satisfies

$$V(x(t_2)) \leq \int_{t_1}^{t_2} \{[u(t)+y(t)]^{\mathrm{T}}[u(t)+y(t)] - u^{\mathrm{T}}(t)u(t)\}\mathrm{d}t + V(x(t_1)), \quad 0 \leq t_1 \leq t_2,$$

which implies that $\mathcal{G}$ is dissipative with respect to the supply rate $r(u,y) = y^{\mathrm{T}}y + 2u^{\mathrm{T}}y$.

Conversely, if $\mathcal{G}$ is dissipative with respect to the supply rate $r(u,y) = y^{\mathrm{T}}y + 2u^{\mathrm{T}}y$ and has a continuously differentiable positive-definite storage function, then, with $h(x) = -\phi(x)$, $J(x) \equiv 0$, $Q = I$, $R = 0$, and $S = 2I$, it follows from Theorem 5.6 that there exists a function $\ell : \mathbb{R}^n \to \mathbb{R}^p$ such that $\phi(x) = -\frac{1}{2}G^{\mathrm{T}}(x)V'^{\mathrm{T}}(x)$ and, for all $x \in \mathbb{R}^n$,

$$0 = V'(x)f(x) - \tfrac{1}{4}V'(x)G(x)G^{\mathrm{T}}(x)V'^{\mathrm{T}}(x) + \ell^{\mathrm{T}}(x)\ell(x).$$

Now, the result follows from Theorem 8.10 with $L_1(x) = \ell^{\mathrm{T}}(x)\ell(x)$. □

Example 8.4. Consider the nonlinear scalar dynamical system discussed in Example 6.6 given by

$$\dot{x}(t) = x^2(t) + u(t), \quad x(0) = x_0, \quad t \geq 0. \tag{8.153}$$

Recall that the optimal controller $\phi(x) = -x^2 - x\sqrt{x^2+1}$ minimizes the cost functional $J(x_0, u(\cdot)) = \int_0^\infty [x^2(t) + u^2(t)]\mathrm{d}t$. Furthermore, recall that $V(x) = \frac{2}{3}(x^3 + (x^2+1)^{3/2} - 1)$ satisfies the Hamilton-Jacobi-Bellman equation. Now, it follows from Theorem 8.11 that (8.153) is dissipative with respect to the supply rate $r(u,y) = y^2 + 2uy$, where $y = -\phi(x) = x^2 + x\sqrt{x^2+1}$. To show this consider the storage function $V_{\mathrm{s}}(x) = V(x)$ and note that $\dot{V}_{\mathrm{s}}(x) = V'(x)(x^2+u) = 2(x^2 + x\sqrt{x^2+1})(x^2+u) = 2(yx^2 + yu)$. Now, noting that $2x^4 + 2x^3\sqrt{x^2+1} \leq (x^2 + x\sqrt{x^2+1})^2$ it follows that $2yx^2 \leq y^2$, and hence, $\dot{V}_{\mathrm{s}}(x) \leq y^2 + 2uy$. △

The next result gives disk and structured disk margins for the nonlinear dynamical system $\mathcal{G}$ given by (8.100) and (8.101).

Corollary 8.7. Consider the nonlinear dynamical system $\mathcal{G}$ given by (8.100) and (8.101) where $\phi(x)$ is a stabilizing feedback control law given by (8.110) with $L_2(x) \equiv 0$ and where $V(x)$, $x \in \mathbb{R}^n$, satisfies (8.107). Furthermore, assume $R_2(x) = \mathrm{diag}[r_1, \ldots, r_m]$, where $r_i > 0$, $i = 1, \ldots, m$, and $L_1(x) \geq 0$, $x \in \mathbb{R}^n$. Then the nonlinear dynamical system $\mathcal{G}$ has a structured disk margin $(\frac{1}{2}, \infty)$. If, in addition, $R_2(x) \equiv I_m$, then the nonlinear system $\mathcal{G}$ has a disk margin $(\frac{1}{2}, \infty)$

Proof. The result is a direct consequence of Theorem 8.5. Specifically, if $L_1(x) \geq 0$, $x \in \mathbb{R}^n$, and $L_2(x) \equiv 0$, then (8.119) is trivially satisfied for all $\theta \in (0,1)$. Now, the result follows immediately by letting $\theta \to 1$. □

Next, we provide sector and gain margins for the nonlinear dynamical system $\mathcal{G}$ given by (8.100) and (8.101).

Corollary 8.8. Consider the nonlinear dynamical system $\mathcal{G}$ given by (8.100) and (8.101) where $\phi(x)$ is a stabilizing feedback control law given by (8.65) with $L_2(x) \equiv 0$ and where $V(x)$, $x \in \mathbb{R}^n$, satisfies (8.107). Furthermore, assume $R_2(x) = \operatorname{diag}[r_1(x), \ldots, r_m(x)]$, where $r_i : \mathbb{R}^n \to \mathbb{R}$, $r_i(x) > 0$, $i = 1, \ldots, m$, and $L_1(x) \geq 0$, $x \in \mathbb{R}^n$. Then the nonlinear dynamical system $\mathcal{G}$ has a sector (and, hence, gain) margin $(\frac{1}{2}, \infty)$.

Proof. The result is a direct consequence of Theorem 8.7. Specifically, if $L_1(x) \geq 0$, $x \in \mathbb{R}^n$, and $L_2(x) \equiv 0$ then (8.119) is trivially satisfied for all $\theta \in (0,1)$. Now, the result follows immediately by letting $\theta \to 1$. □

Finally, we show that given a control Lyapunov function for a controlled nonlinear system, the feedback control law given by (6.81) guarantees sector and gain margins of $(\frac{1}{2}, \infty)$.

Theorem 8.12. Consider the nonlinear dynamical system $\mathcal{G}$ given by (8.100) and let the continuously differentiable positive-definite, radially unbounded function $V : \mathbb{R}^n \to \mathbb{R}$ be a control Lyapunov function of (8.100), that is,

$$V'(x)f(x) < 0, \qquad x \in \mathcal{R}, \tag{8.154}$$

where $\mathcal{R} \triangleq \{x \in \mathbb{R}^n : x \neq 0, V'(x)G(x) = 0\}$. Then with the feedback stabilizing control law given by

$$\phi(x) = \begin{cases} -\left(c_0 + \frac{\alpha(x)+\sqrt{\alpha^2(x)+(\beta^{\mathrm{T}}(x)\beta(x))^2}}{\beta^{\mathrm{T}}(x)\beta(x)}\right)\beta(x), & \beta(x) \neq 0, \\ 0, & \beta(x) = 0, \end{cases} \tag{8.155}$$

where $\alpha(x) \triangleq V'(x)f(x)$, $\beta(x) \triangleq G^{\mathrm{T}}(x)V'^{\mathrm{T}}(x)$, and $c_0 > 0$, the nonlinear dynamical system $\mathcal{G}$ given by (8.100) and (8.101) has a sector (and, hence, gain) margin $(\frac{1}{2}, \infty)$. Furthermore, with the feedback control law $u = \phi(x)$ the performance functional

$$J(x_0, u(\cdot)) = \int_0^\infty [\alpha(x(t)) - \tfrac{\gamma(x(t))}{2}\beta^{\mathrm{T}}(x(t))\beta(x(t)) + \tfrac{1}{2\gamma(x(t))}u^{\mathrm{T}}(t)u(t)]\mathrm{d}t, \tag{8.156}$$

where

$$\gamma(x) \triangleq \begin{cases} \left(c_0 + \frac{\alpha(x)+\sqrt{\alpha^2(x)+(\beta^{\mathrm{T}}(x)\beta(x))^2}}{\beta^{\mathrm{T}}(x)\beta(x)}\right), & \beta(x) \neq 0, \\ c_0, & \beta(x) = 0, \end{cases} \tag{8.157}$$

is minimized in the sense that

$$J(x_0, \phi(x(\cdot))) = \min_{u \in \mathcal{S}(x_0)} J(x_0, u(\cdot)), \quad x_0 \in \mathbb{R}^n. \tag{8.158}$$

Proof. The result is a direct consequence of Corollary 8.7 and Theorem 8.10 with $R_2(x) = \frac{1}{2\gamma(x)} I_m$ and $L_1(x) = -\alpha(x) + \frac{\gamma(x)}{2}\beta^{\mathrm{T}}(x)\beta(x)$. Specifically, it follows from (8.157) that $R_2(x) > 0$, $x \in \mathbb{R}^n$, and

$$\begin{aligned} L_1(x) &= -\alpha(x) + \tfrac{\gamma(x)}{2}\beta^{\mathrm{T}}(x)\beta(x) \\ &= \begin{cases} \frac{1}{2}\left(c_0\beta^{\mathrm{T}}(x)\beta(x) - \alpha(x) + \sqrt{\alpha^2(x) + (\beta^{\mathrm{T}}(x)\beta(x))^2}\right), & \beta(x) \neq 0, \\ -\alpha(x), & \beta(x) = 0. \end{cases} \end{aligned} \tag{8.159}$$

Now, it follows from (8.159) that $L_1(x) \geq 0$, $\beta(x) \neq 0$, and, since $V(\cdot)$ is a control Lyapunov function of (8.100), it follows from Theorem 6.7 that $L_1(x) = -\alpha(x) \geq 0$, for all $x \in \mathcal{R} = \{x \in \mathbb{R}^n : x \neq 0, \ \beta(x) = 0\}$. Hence, (8.159) yields $L_1(x) \geq 0$, $x \in \mathbb{R}^n$, so that all conditions of Corollary 8.7 are satisfied. □

Theorem 8.12 shows that given a nonlinear system for which a control Lyapunov function can be constructed, the feedback control law given by (8.155) is inverse optimal with respect to a meaningful cost functional and has a sector (and, hence, gain) margin $(\frac{1}{2}, \infty)$.

Example 8.5. In this example, we demonstrate the extra flexibility provided by inverse optimal controllers with performance criteria involving cross-weighting terms over inverse optimal controllers with meaningful cost functionals. Specifically, we consider the controlled Lorentz dynamical system discussed in Example 8.2. Our objective is to design and compare inverse optimal and meaningful inverse optimal controllers that stabilize the origin of the Lorentz equations (8.80)–(8.82). In particular, we compare the guaranteed gain and sector margins to input saturation-type nonlinearities and the transient performance in terms of maximum overshoot of both designs. Recall that the control law $\phi(x) = -(\frac{p_1}{p_2}\sigma + r)x_1$ stabilizes the nonlinear system (8.80)–(8.82) while minimizing the performance functional (8.102) with $L_1(x)$ given by (8.85) and $L_2(x)$ given by (8.84). In this case, (8.119) becomes

$$(1 - \theta^2)[\hat{x}^{\mathrm{T}} Q \hat{x} + 2p_2 b x_3^2] - \hat{x}^{\mathrm{T}} y y^{\mathrm{T}} \hat{x} \geq 0, \tag{8.160}$$

where $\hat{x} \triangleq [x_1, \ \ x_2]^{\mathrm{T}}$,

$$Q \triangleq R_2 \begin{bmatrix} 2\sigma p_1 + R_2(\frac{p_1}{p_2}\sigma + r)^2 & -(p_1\sigma + p_2 r) \\ -(p_1\sigma + p_2 r) & 2p_2 \end{bmatrix}, \tag{8.161}$$

and

$$y \triangleq \begin{bmatrix} R_2(\frac{p_1}{p_2}\sigma + r) & -p_2 \end{bmatrix}^{\mathrm{T}}. \tag{8.162}$$

Since $2p_2 b x_3^2$ is nonnegative for all $\theta \in (0,1)$ it follows that (8.160) is equivalent to $(1-\theta^2)Q \geq yy^{\mathrm{T}}$, which implies that, if Q is invertible, (8.119) is satisfied for all $\theta \in (0,1)$ such that $1-\theta^2 \geq y^{\mathrm{T}}Q^{-1}y$. Hence, the maximum possible θ such that (8.119) holds is given by

$$\theta = \sqrt{1 - y^{\mathrm{T}}Q^{-1}y}. \tag{8.163}$$

Now, it follows from Theorem 8.6 that for all $p_1, p_2, R_2 > 0$, such that $\det Q \neq 0$, with $\phi(x)$ given above the nonlinear system (8.80)–(8.82) has disk margins of $(\frac{1}{1+\theta}, \frac{1}{1-\theta})$, where θ is given by (8.163). Furthermore, for given p_1, p_2, and $R_2 > 0$ these disk margins are the maximum possible disk margins that are guaranteed by Theorem 8.6. Next, we vary p_1, p_2, and R_2 such that θ given by (8.163) is maximized. It can be shown that the maximum is achieved at $\frac{p_1}{p_2} = \frac{r}{\sigma}$ and $\frac{p_2}{R_2} = 0$ so that

$$\theta_{\max} = \frac{1}{\sqrt{r+1}}. \tag{8.164}$$

In this case, the control law $\phi(x)$ is given by $\phi(x) = -2rx_1$.

Next, using the control Lyapunov function $V(x) = p_1x_1^2 + p_2x_2^2 + p_2x_3^2$, where $p_1, p_2 > 0$ are such that $\frac{p_1}{p_2} = \frac{r}{\sigma}$, we design a meaningful inverse optimal controller using the feedback controller given by (8.155). Using the initial conditions $(x_{10}, x_{20}, x_{30}) = (-20, -20, 30)$, the data parameters $\sigma = 10$, $r = 15$, $b = 8/3$, and design parameters $p_1 = 1.5$ and $p_2 = 1$ the inverse optimal controller $\phi(x) = -2rx_1$ and the meaningful inverse optimal controller given by (8.155) were used to compare closed-loop system performance. First, we note that the downside disk and sector margins of the inverse optimal controller are 0.8 while the meaningful inverse optimal controller guarantees the standard 0.5 downside sector margin with *no* disk margin guarantees. Hence, both controllers have guaranteed robustness sector margins to actuator saturation nonlinearities with, as expected, the meaningful inverse optimal controller having a slightly larger guarantee. However, as shown in Figures 8.2 and 8.3, the inverse optimal controller with a cross-weighting term in the performance functional has better transient performance in terms of peak overshoot over the meaningful inverse optimal controller. (We note that the controlled third state is not shown since it is virtually identical for both designs.) Finally, Figure 8.4 compares the control effort versus time for both controllers. △

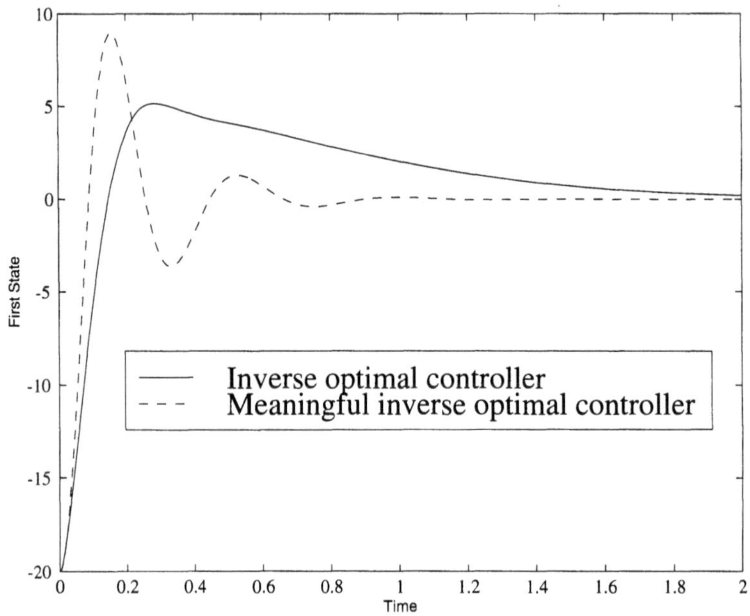

Figure 8.2 First state versus time.

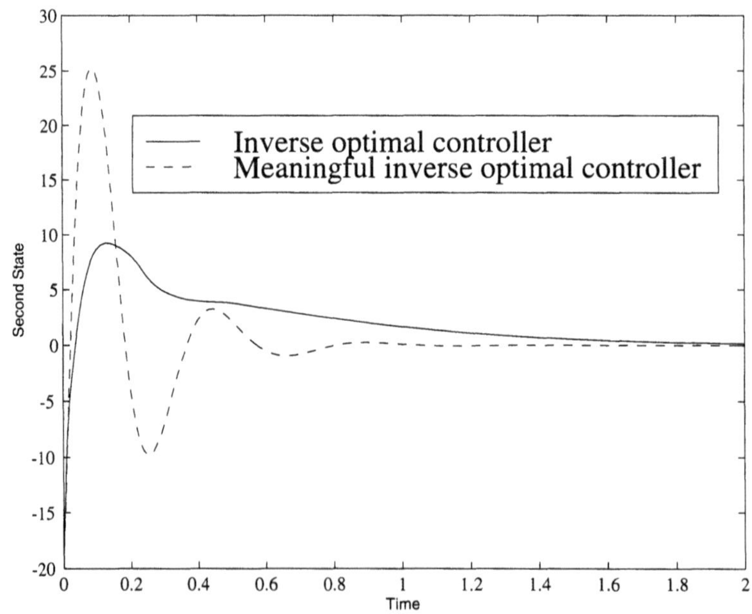

Figure 8.3 Second state versus time.

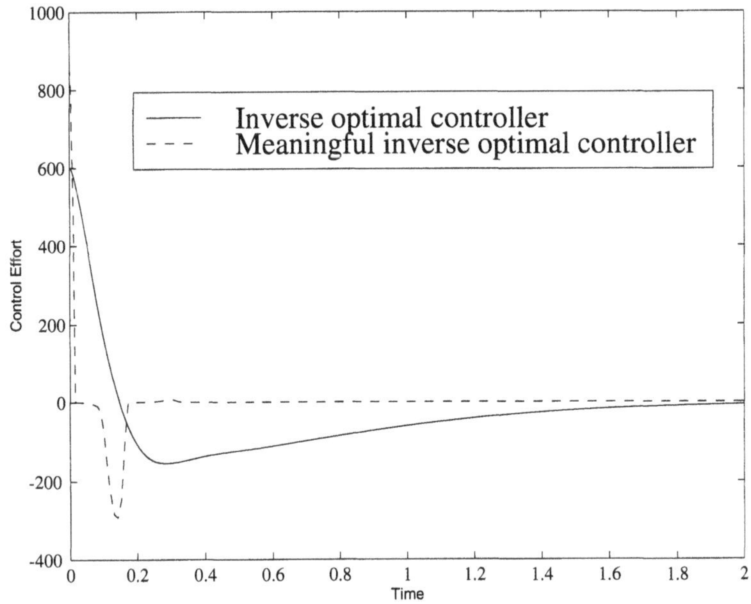

Figure 8.4 Control effort versus time.

8.9 Problems

Problem 8.1. Show that every $2n$th-order positive definite polynomial function of the form $V(x) = \sum_{q=1}^{n} (x^{\mathrm{T}} P_q x)^q$, where $P_1 > 0$ and $P_q \geq 0$, $q = 2, \ldots, n$, can be written as a q-multilinear function where q is even. Can every multilinear function be written as a polynomial function?

Problem 8.2. Let $A \in \mathbb{R}^{n\times n}$. Show that $e^{(\overset{q}{\oplus} A)} = (e^A)^{[q]}$. If, in addition, $\det A \neq 0$, show

$$\int_0^t e^{(\overset{q}{\oplus} As)} \mathrm{d}s = (\overset{q}{\oplus} A)^{-1}[e^{(\overset{q}{\oplus} At)} - I_{n^q}]. \tag{8.165}$$

Finally, if $\alpha(A) < 0$, show

$$\int_0^\infty e^{(\overset{q}{\oplus} As)} \mathrm{d}s = -(\overset{q}{\oplus} A)^{-1}. \tag{8.166}$$

Problem 8.3. Consider the nonlinear dynamical system (8.38) and let $V : \mathcal{D} \to \mathbb{R}$ be a continuously differentiable function satisfying (8.41) and (8.42). Then we say that the feedback control law $\phi : \mathcal{D} \to \Omega$ is *optimal with respect to V* if

$$V'(x)F(x, \phi(x)) \leq V'(x)F(x, u), \qquad x \in \mathcal{D}, \qquad u \in \Omega. \tag{8.167}$$

Now, consider the linear controlled system

$$\dot{x}(t) = Ax(t) + Bu(t), \qquad x(0) = x_0, \qquad t \geq 0, \tag{8.168}$$

where $x \in \mathbb{R}^n$ and $u \in \Omega \triangleq \{u \in \mathbb{R}^m : \ \|u\|_p \leq 1\}$, where $p = 1, 2, \infty$. Assume A is Hurwitz and define $V(x) = x^{\mathrm{T}}Px$, where $P \in \mathbb{R}^{n \times n}$ is the unique positive-definite solution to

$$0 = A^{\mathrm{T}}P + PA + R, \tag{8.169}$$

for a given $n \times n$ positive-definite matrix R. Furthermore, let $V(x) = x^{\mathrm{T}}Px$ be the Lyapunov function for the closed-loop system (8.168) with $u = \phi(x)$. For each value of $p \in \{1, 2, \infty\}$, construct a feedback control law $u = \phi(x)$ that is optimal with respect to $V(\cdot)$ or, equivalently, gives a maximum time decay rate of $V(x)$.

Problem 8.4. Consider the algebraic Riccati equation (8.53) with $R_1 = E_1^{\mathrm{T}}E_1 \geq 0$. Show that the following statements are equivalent:

i) (A, B) is stabilizable and (A, E_1) is detectable.

ii) There exists a nonnegative-definite solution P satisfying (8.53) and $A - BR_2^{-1}B^{\mathrm{T}}P$ is Hurwitz.

In addition, show that the following statements hold:

iii) If (A, E_1) is detectable, then P is the only nonnegative-definite solution to (8.53).

iv) If $R_1 > 0$, then $P > 0$.

Problem 8.5. Consider the linear controlled system (8.51) with quadratic performance functional (8.52) and $R_2 = I_m$. Show that if (A, B) is controllable, $-A$ is Hurwitz, and $R_1 = 0$, then

$$P = \left[\int_0^\infty e^{-At}BB^{\mathrm{T}}e^{-A^{\mathrm{T}}t}\mathrm{d}t\right]^{-1} \tag{8.170}$$

is a positive-definite solution to (8.53).

Problem 8.6. Consider the linear controlled system (8.51). Let $R_1 \in \mathbb{P}^n$, $R_2 \in \mathbb{P}^m$, $\alpha > 0$, and assume there exists a positive-definite matrix $P \in \mathbb{R}^{n \times n}$ such that

$$0 = (A + \alpha I_n)^{\mathrm{T}}P + P(A + \alpha I_n) + R_1 - PBR_2^{-1}B^{\mathrm{T}}P. \tag{8.171}$$

Show that the linear controlled system (8.51) with feedback control law

$$u(t) = -R_2^{-1}B^{\mathrm{T}}Px(t), \tag{8.172}$$

has the property that all the eigenvalues of the closed-loop system have real part less than $-\alpha$. Finally, show that the stabilizing controller (8.172) minimizes

$$J(x_0, u(\cdot)) = \int_0^\infty e^{2\alpha t}[x^{\mathrm{T}}(t)R_1x(t) + u^{\mathrm{T}}(t)R_2u(t)]\mathrm{d}t. \tag{8.173}$$

Problem 8.7. Consider the nonlinear dynamical system $\mathcal{G}$ given by

$$\dot{x}(t) = f(x(t)) + G(x(t))u(t), \quad x(0) = x_0, \quad t \geq 0, \tag{8.174}$$
$$y(t) = h(x(t)), \tag{8.175}$$

where $x \in \mathbb{R}^n$, $u, y \in \mathbb{R}^m$, $f : \mathbb{R}^n \to \mathbb{R}^n$ satisfies $f(0) = 0$, $G : \mathbb{R}^n \to \mathbb{R}^{n\times m}$, and $h : \mathbb{R}^n \to \mathbb{R}^m$ satisfies $h(0) = 0$. Assume that $\mathcal{G}$ is passive, zero-state observable, and completely reachable with a continuously differentiable radially unbounded storage function $V_{\mathrm{s}} : \mathbb{R}^n \to \mathbb{R}$. Furthermore, let $V : \mathbb{R}^n \to \mathbb{R}$ be a continuously differentiable function such that $V(0) = 0$, $V(x) > 0$, $x \in \mathbb{R}^n$, $x \neq 0$, $V(x) \to \infty$ as $\|x\| \to \infty$, and

$$0 = L_1(x) + V'(x)f(x) - \tfrac{1}{4}V'(x)G(x)R_2^{-1}G^{\mathrm{T}}(x)V'^{\mathrm{T}}(x), \quad x \in \mathbb{R}^n, \tag{8.176}$$

where $L_1(x) = \ell^{\mathrm{T}}(x)\ell(x) + h^{\mathrm{T}}(x)R_2^{-1}h(x)$ and $\ell(\cdot)$ satisfies (5.115). Show that the *output* feedback controller $u(t) = -R_2^{-1}y(t)$, where $R_2 > 0$, asymptotically stabilizes (8.174) and minimizes the performance criterion

$$J(x_0, u(\cdot)) = \int_0^\infty [\ell^{\mathrm{T}}(x(t))\ell(x(t)) + h^{\mathrm{T}}(x(t))R_2^{-1}h(x(t)) + u^{\mathrm{T}}(t)R_2u(t)]\mathrm{d}t, \tag{8.177}$$

in the sense that

$$J(x_0, \phi(x(\cdot))) = \min_{u(\cdot)\in\mathcal{S}(x_0)} J(x_0, u(\cdot)), \quad x_0 \in \mathbb{R}^n. \tag{8.178}$$

Furthermore, show that $J(x_0, \phi(x(\cdot))) = V(x_0)$, $x_0 \in \mathbb{R}^n$.

Problem 8.8. Consider the nonlinear oscillator

$$\dot{x}_1(t) = x_2(t), \quad x_1(0) = x_{10}, \quad t \geq 0, \tag{8.179}$$
$$\dot{x}_2(t) = -x_1(t) + x_2(t)\sinh(x_1^2(t) + x_2^2(t)) + u(t), \; x_2(0) = x_{20}. \tag{8.180}$$

Find a globally stabilizing feedback controller $u(t) = \phi(x(t))$ that minimizes

$$J(x_{10}, x_{20}, u(\cdot)) = \int_0^\infty [x_2^2(t) + u^2(t)]\mathrm{d}t. \tag{8.181}$$

Compare the state response and the control effort of the optimal controller to the feedback linearizing control law $u_{\mathrm{FL}}(t) = -x_2(t)[1 + \sinh(x_1^2(t) + x_2^2(t))]$.

Problem 8.9. Consider the nonlinear oscillator

$$\dot{x}_1(t) = x_2(t), \quad x_1(0) = x_{10}, \quad t \geq 0, \tag{8.182}$$

$$\dot{x}_2(t) = -e^{x_2(t)[x_1(t)+\frac{1}{2}x_2(t)]} + \tfrac{1}{2}x_2(t)e^{4x_1(t)+3x_2(t)} + e^{2x_1(t)+2x_2(t)}u(t),$$
$$x_2(0) = x_{20}. \quad (8.183)$$

Find a stabilizing feedback controller $u(t) = \phi(x(t))$ that minimizes

$$J(x_{10}, x_{20}, u(\cdot)) = \int_0^\infty [x_2^2(t) + u^2(t)]\mathrm{d}t. \quad (8.184)$$

Is your controller a global stabilizer?

Problem 8.10. Consider the nonlinear dynamical system

$$\dot{x}_1(t) = -x_1^3(t) + x_1(t)x_2^2(t), \quad x_1(0) = x_{10}, \quad t \geq 0, \quad (8.185)$$
$$\dot{x}_2(t) = -x_2(t) + x_1^2(t)u(t), \quad x_2(0) = x_{20}. \quad (8.186)$$

Find a stabilizing feedback controller $u(t) = \phi(x(t))$ that minimizes

$$J(x_{10}, x_{20}, u(\cdot)) = \int_0^\infty [2x_1^2(t) + 2x_2^2(t) + \tfrac{1}{2}u^2(t)]\mathrm{d}t. \quad (8.187)$$

Problem 8.11. Show that in the case where $0 < r \leq 1$, the uncontrolled ($u(t) \equiv 0$) Lorentz dynamical system (8.80)–(8.82) has only one equilibrium state, namely, $x_{\mathrm{e}1} \triangleq (0,0,0)^{\mathrm{T}}$. Alternatively, show that in the case where $r > 1$, the uncontrolled Lorentz dynamical system (8.80)–(8.82) has three equilibrium states, namely, $x_{\mathrm{e}1}$, $x_{\mathrm{e}2} \triangleq (\sqrt{b(r-1)}, \sqrt{b(r-1)}, r-1)^{\mathrm{T}}$, and $x_{\mathrm{e}3} \triangleq (-\sqrt{b(r-1)}, -\sqrt{b(r-1)}, r-1)^{\mathrm{T}}$. Finally, show that if $r \leq 1$, $x_{\mathrm{e}1}$ is a locally asymptotically stable equilibrium point while if $r > 1$, $x_{\mathrm{e}1}$ is unstable. What can you say about the stability of $x_{\mathrm{e}2}$ and $x_{\mathrm{e}3}$?

Problem 8.12. Using Theorem 8.3 construct a globally asymptotically stabilizing control law for the equilibrium state $x_{\mathrm{e}} \triangleq (\sqrt{b(r-1)}, \sqrt{b(r-1)}, r-1)^{\mathrm{T}}$, $r > 1$, of the controlled Lorentz dynamical system (8.80)–(8.82).

Problem 8.13. Consider the nonlinear controlled system (8.56) with performance functional (8.58). Assume there exists a continuously differentiable function $V : \mathbb{R}^n \to \mathbb{R}$ and a function $L_2 : \mathbb{R}^n \to \mathbb{R}^{1\times m}$ such that (8.59)–(8.61) and (8.63) hold, and

$$V'(x)f_{\mathrm{s}}(x) \leq 0, \quad x \in \mathbb{R}^n, \quad (8.188)$$

$$\mathcal{W} \triangleq \{x \in \mathbb{R}^n : L_{f_{\mathrm{s}}}^{k+1}V(x) = L_{f_{\mathrm{s}}}^{k}L_{G_i}V(x) = 0,\ k = 0,1,\ldots,\ i = 1,\ldots,m\}$$
$$= \{0\}, \quad (8.189)$$

where $f_{\mathrm{s}}(x) \triangleq f(x) - \frac{1}{2}G(x)R_2^{-1}(x)L_2^{\mathrm{T}}(x)$. Show that, with the feedback control law (8.65), the zero solution $x(t) \equiv 0$ of the closed-loop system (8.64) is globally asymptotically stable, and the performance functional (8.58) with $L_1(x)$ given by (8.66) is minimized in the sense that (8.67) and (8.68) hold.

Problem 8.14. Assume the controlled system (8.56) with output $y(t) = h(x(t))$, where $h : \mathbb{R}^n \to \mathbb{R}^m$, is minimum phase with relative degree $\{1, 1, \ldots, 1\}$ and complete and involutive vector field $G(L_G h)^{-1}$. Furthermore, let

$$V(x) = V_0(z) + y^{\mathrm{T}} P y, \tag{8.190}$$

$$L_2^{\mathrm{T}}(x) = R_2(x)[L_G h(x)]^{-1}[P^{-1} r^{\mathrm{T}}(z, y) \left(\frac{\partial V_0(z)}{\partial z}\right)^{\mathrm{T}} + 2L_f h(x)], \tag{8.191}$$

where $V_0 : \mathbb{R}^{n-m} \to \mathbb{R}$ is a continuously differentiable positive-definite function such that $V_0(0) = 0$, $V_0'(z) f_0(z) < 0$, $z \in \mathcal{Z} \triangleq \{x \in \mathbb{R}^n : h(x) = 0\}$, and $\dot{z} = f_0(z)$, and where P is an arbitrary $m \times m$ positive-definite matrix and $r(z, y)$ is defined as in Lemma 6.2. Show that the feedback control law

$$\begin{aligned}\phi(x) = & -\tfrac{1}{2}[L_G h(x)]^{-1}[P^{-1} r^{\mathrm{T}}(z, y) \left(\frac{\partial V_0(z)}{\partial z}\right)^{\mathrm{T}} + 2L_f h(x)] \\ & - R_2^{-1}(x)[L_G h(x)]^{\mathrm{T}} P h(x),\end{aligned} \tag{8.192}$$

globally asymptotically stabilizes (8.56) and minimizes the performance functional (8.58) with $L_1(x)$ given by (8.66) in the sense that (8.67) and (8.68) hold.

Problem 8.15. Consider the nonlinear dynamical system representing a controlled rigid spacecraft with one torque input given by

$$\dot{x}_1(t) = I_{23} x_2(t) x_3(t) + \tfrac{b_1}{I_1} u(t), \quad x_1(0) = x_{10}, \quad t \geq 0, \tag{8.193}$$

$$\dot{x}_2(t) = I_{31} x_3(t) x_1(t) + \tfrac{b_2}{I_2} u(t), \quad x_2(0) = x_{20}, \tag{8.194}$$

$$\dot{x}_3(t) = I_{12} x_1(t) x_2(t) + \tfrac{b_2}{I_3} u(t), \quad x_3(0) = x_{30}, \tag{8.195}$$

where $I_{23} = (I_2 - I_3)/I_1$, $I_{31} = (I_3 - I_1)/I_2$, $I_{12} = (I_1 - I_2)/I_3$, and I_1, I_2, and I_3 are principal moments of inertia of the spacecraft. Using Problem 8.13 with $V(x) = \frac{1}{2}(I_1 x_1^2 + I_2 x_2^2 + I_3 x_3^2)$ and $L_2(x) = 0$, where $x = [x_1 \; x_2 \; x_3]^{\mathrm{T}}$, construct a globally stabilizing controller for (8.193)–(8.195).

Problem 8.16. Consider the nonlinear dynamical system

$$\dot{x}_1(t) = u_1(t), \quad x_1(0) = x_{10}, \quad t \geq 0, \tag{8.196}$$

$$\dot{x}_2(t) = u_2(t), \quad x_2(0) = x_{20}, \tag{8.197}$$

$$\dot{x}_3(t) = x_1(t) x_2(t), \quad x_3(0) = x_{30}, \tag{8.198}$$

representing a controlled rigid spacecraft with two actuators along the principal axes and whose uncontrolled principal axis is not an axis of symmetry. Using Problem 8.14 with

$$y_1(t) = x_1(t) + \alpha x_3^k(t), \quad (8.199)$$
$$y_2(t) = x_2(t) + \beta x_3^{k+1}(t), \quad (8.200)$$

where k is a positive integer and α, β are arbitrary real numbers, construct a globally stabilizing controller for (8.196)–(8.198).

Problem 8.17. Prove Theorem 8.10.

Problem 8.18. Consider the port-controlled Hamiltonian system given by (6.22) with performance functional (8.58). Assume that there exist functions $\mathcal{H}_\mathrm{s}$, $\mathcal{H}_\mathrm{c} : \mathcal{D} \to \mathbb{R}$, $\mathcal{J}_\mathrm{s}$, $\mathcal{J}_\mathrm{a} : \mathcal{D} \to \mathbb{R}^{n\times n}$, $\mathcal{R}_\mathrm{s}$, $\mathcal{R}_\mathrm{a} : \mathcal{D} \to \mathbb{R}^{n\times n}$ such that $\mathcal{H}_\mathrm{s}(x) = \mathcal{H}(x) + \mathcal{H}_\mathrm{c}(x)$ is continuously differentiable, $\mathcal{J}_\mathrm{s}(x) = \mathcal{J}(x) + \mathcal{J}_\mathrm{a}(x)$, $\mathcal{J}_\mathrm{s}(x) = -\mathcal{J}_\mathrm{s}^\mathrm{T}(x)$, $\mathcal{R}_\mathrm{s}(x) = \mathcal{R}(x)+\mathcal{R}_\mathrm{a}(x)$, $\mathcal{R}_\mathrm{s}(x) = \mathcal{R}_\mathrm{s}^\mathrm{T}(x) \geq 0$, $x \in \mathcal{D}$, condition (6.25) is satisfied, and

$$\frac{\partial^2 \mathcal{H}_\mathrm{s}}{\partial x^2}(x) > 0, \quad x \in \mathcal{D}, \quad (8.201)$$

$$\begin{aligned}\left[\mathcal{J}_\mathrm{s}(x) - \mathcal{R}_\mathrm{s}(x) + \frac{1}{2}G(x)R_2^{-1}(x)G^\mathrm{T}(x)\right]\left(\frac{\partial \mathcal{H}_\mathrm{c}}{\partial x}(x)\right)^\mathrm{T} &= \left[-\left[\mathcal{J}_\mathrm{a}(x) - \mathcal{R}_\mathrm{a}(x)\right]\right.\\ \left. - \frac{1}{2}G(x)R_2^{-1}(x)G^\mathrm{T}(x)\right]\left(\frac{\partial \mathcal{H}}{\partial x}(x)\right)^\mathrm{T} &- \frac{1}{2}G(x)R_2^{-1}(x)L_2^\mathrm{T}(x), \quad x \in \mathcal{D}.\end{aligned} \quad (8.202)$$

Show that the equilibrium solution $x(t) \equiv x_\mathrm{e}$ of the closed-loop system given by (6.24) is Lyapunov stable with the feedback control law

$$\phi(x) = -\frac{1}{2}R_2^{-1}(x)\left(G^\mathrm{T}(x)\left(\frac{\partial \mathcal{H}_\mathrm{s}}{\partial x}(x)\right)^\mathrm{T} + L_2^\mathrm{T}(x)\right), \quad x \in \mathcal{D}. \quad (8.203)$$

If, in addition, $\mathcal{D}_\mathrm{c} \subseteq \mathcal{D}$ is a compact positively invariant set with respect to (6.24) and the largest invariant set contained in $\mathcal{R} \triangleq \{x \in \mathcal{D}_\mathrm{c} : \frac{\partial \mathcal{H}_\mathrm{s}}{\partial x}(x)\mathcal{R}_\mathrm{s}(x)\left(\frac{\partial \mathcal{H}_\mathrm{s}}{\partial x}(x)\right)^\mathrm{T} = 0\}$ is $\mathcal{M} = \{x_\mathrm{e}\}$, show that the equilibrium solution $x(t) \equiv x_\mathrm{e}$ of the closed-loop system (6.24) is locally asymptotically stable. Moreover, show that the performance functional (8.58), with

$$L_1(x) = \phi^\mathrm{T}(x)R_2(x)\phi(x) - \frac{\partial \mathcal{H}_\mathrm{s}}{\partial x}(x)\left[\mathcal{J}(x) - \mathcal{R}(x)\right]\left(\frac{\partial \mathcal{H}}{\partial x}(x)\right)^\mathrm{T}, \quad (8.204)$$

is minimized in the sense that

$$J(x_0, \phi(x(\cdot))) = \min_{u(\cdot)\in\mathcal{C}(x_0)} J(x_0, u(\cdot)), \quad x_0 \in \mathcal{D}_\mathrm{c}, \quad (8.205)$$

and $J(x_0, \phi(x(\cdot))) = \mathcal{H}_\mathrm{s}(x_0) - \mathcal{H}_\mathrm{s}(x_\mathrm{e})$, $x_0 \in \mathcal{D}_\mathrm{c}$.

Problem 8.19. Consider the nonlinear dynamical system $\mathcal{G}$ given by (8.100) and assume that $\mathcal{G}$ is a J-Q type system (see Problem 6.25) so that there exists a continuously differentiable positive-definite, radially unbounded function $V : \mathbb{R}^n \to \mathbb{R}$ satisfying *i*) and *ii*) of Problem 6.25. Show that with the feedback stabilizing control law

$$\phi(x) = -\kappa[V'(x)G(x)]^{\mathrm{T}}, \qquad \kappa > 0, \tag{8.206}$$

the nonlinear system (8.100) and (8.101) has a disk margin $(0, \infty)$. Furthermore, with the feedback control law $u = \phi(x)$, show that the performance functional

$$\begin{aligned} J(x_0, u(\cdot)) = \int_0^\infty & \{\tfrac{\kappa}{2}[V'(x(t))G(x(t))][V'(x(t))G(x(t))]^{\mathrm{T}} \\ & -V'(x(t))f(x(t)) + \tfrac{2}{\kappa}u^{\mathrm{T}}(t)u(t)\}\mathrm{d}t, \end{aligned} \tag{8.207}$$

is minimized in the sense that

$$J(x_0, \phi(x(\cdot))) = \min_{u(\cdot)\in\mathcal{S}(x_0)} J(x_0, u(\cdot)), \quad x_0 \in \mathbb{R}^n. \tag{8.208}$$

Problem 8.20. Consider the nonlinear dynamical system $\mathcal{G}$ given by (8.100) and let $V : \mathbb{R}^n \to \mathbb{R}$ be a continuously differentiable positive-definite, radially unbounded control Lyapunov function for (8.100). Let $\phi : \mathbb{R}^n \to \mathbb{R}^m$ satisfy

$$\|\phi(x)\| = \min\{\|u\| : V'(x)[f(x) + G(x)u] \le -\sigma(x),\ u \in \mathbb{R}^m\}, \qquad x \in \mathbb{R}^n, \tag{8.209}$$

where $\|\cdot\|$ denotes the Euclidean norm, $\sigma : \mathbb{R}^n \to \mathbb{R}$ is a continuous positive-definite function such that $V'(x)f(x) \le -\sigma(x)$, $x \in \mathcal{R} \triangleq \{x \in \mathbb{R}^n : V'(x)G(x) = 0\}$. Show that the control law $\phi(x)$ given by (8.209) is globally stabilizing and inverse optimal with respect to the cost functional

$$J(x_0, u(\cdot)) = \int_0^\infty [L_1(x(t)) + u^{\mathrm{T}}(t)R_2(x(t))u(t)]\mathrm{d}t, \tag{8.210}$$

where $L_1(x) \ge 0$ and $R_2(x) > 0$, $x \in \mathbb{R}^n$. If, in addition,

$$\sigma(x) = \sqrt{\alpha^2(x) + q(x)\beta^{\mathrm{T}}(x)\beta(x)},$$

where $\alpha(x) \triangleq V'(x)f(x)$, $\beta(x) \triangleq V'(x)G(x)$, and $q : \mathbb{R}^n \to \mathbb{R}$ is a nonnegative function, show that the control law $\phi(x)$ given by (8.209) specializes to

$$\phi(x) = \begin{cases} -\frac{\alpha(x)+\sqrt{\alpha^2(x)+q(x)\beta^{\mathrm{T}}(x)\beta(x)}}{\beta^{\mathrm{T}}(x)\beta(x)}\beta(x), & \beta(x) \neq 0, \\ 0, & \beta(x) = 0. \end{cases} \tag{8.211}$$

Finally, let $\kappa : \mathbb{R} \to \mathbb{R}$ be a continuously differentiable class $\mathcal{K}$ function and let $\hat{V} : \mathbb{R}^n \to \mathbb{R}$ be a continuously differentiable positive-definite, radially unbounded function such that $V(x) = \kappa(\hat{V}(x))$ and

$$0 = \hat{V}'(x)f(x) + q(x) - \tfrac{1}{4}\hat{V}'(x)G(x)G^{\mathrm{T}}(x)\hat{V}'^{\mathrm{T}}(x). \tag{8.212}$$

Show that the control law $u = \phi(x)$ given by (8.211) minimizes the cost functional

$$J(x_0, u(\cdot)) = \int_0^\infty [q(x(t)) + u^{\mathrm{T}}(t)u(t)]\mathrm{d}t. \tag{8.213}$$

Problem 8.21. Consider the port-controlled Hamiltonian system

$$\dot{x}(t) = \mathcal{J}(x(t))\left[\frac{\partial \mathcal{H}}{\partial x}(x(t))\right]^{\mathrm{T}} + G(x(t))u(t), \ x(0) = x_0, \ t \geq 0, \tag{8.214}$$

$$y(t) = G^{\mathrm{T}}(x(t))\left[\frac{\partial \mathcal{H}}{\partial x}(x(t))\right]^{\mathrm{T}}, \tag{8.215}$$

where $x \in \mathbb{R}^n$, $u, y \in \mathbb{R}^m$, $\mathcal{H} : \mathcal{D} \to \mathbb{R}$, $G : \mathbb{R}^n \to \mathbb{R}^{n\times m}$, and $\mathcal{J} : \mathbb{R}^n \to \mathbb{R}^{n\times n}$ satisfies $\mathcal{J}(x) = -\mathcal{J}^{\mathrm{T}}(x)$. Assume that (8.214) and (8.215) is zero-state observable. Show that the feedback control law $\phi(x) = -G^{\mathrm{T}}(x)\left[\frac{\partial \mathcal{H}}{\partial x}\right]^{\mathrm{T}}$ asymptotically stabilizes (8.214) and minimizes the performance functional

$$J(x_0, u(\cdot)) = \tfrac{1}{2}\int_0^\infty [y^{\mathrm{T}}(t)y(t) + u^{\mathrm{T}}(t)u(t)]\mathrm{d}t. \tag{8.216}$$

8.10 Notes and References

The results in Sections 8.2 and 8.3 on stability analysis and optimal control of nonlinear systems are due to Bernstein [43]. In particular, Bernstein [43] gives an excellent review of the nonlinear-nonquadratic control problem in a simplified and tutorial manner. Polynomial forms in the performance criterion were developed by Speyer [411] while multilinear forms were addressed by Bass and Webber [33]. A treatment of nonlinear-nonquadratic optimal control is also given by Jacobson [217]. The inverse optimal control problem has been studied by numerous authors including Kalman [227], Moylan and Anderson [321], Moylan [320], Molinari [317], Jacobson [217], Anderson and Moore [10], Wan and Bernstein [449], and more recently by Freeman and Kokotovic [127,128]. The presentation here parallels that given by Wan and Bernstein [449].

The equivalence between optimality and passivity is due to Kalman [227] for linear systems and Moylan [320] for nonlinear systems. Gain, sector, and disk margins of nonlinear-nonquadratic optimal regulators with performance measures involving cross-weighting terms are due to Chellaboina and Haddad [85]. An excellent treatment on meaningful inverse optimality is given by Freeman and Kokotović [128] and Sepulchre, Janković, and Kokotović [395].

Finally, the concept of optimality with respect to a Lyapunov function introduced in Problem 8.3 is due to Bernstein [43] while the inverse optimal controller for J-Q type systems introduced in Problem 8.13 and the inverse optimal controller for the minimum phase system introduced in Problem 8.14 is due to Wan and Bernstein [449]. The concept of the pointwise control minimization for generating control Lyapunov functions introduced in Problem 8.20 is due to Freeman and Primbs [129].

Chapter Nine

Inverse Optimal Control and Integrator Backstepping

9.1 Introduction

Control system designers have usually resorted to Lyapunov methods [445] in order to obtain stabilizing controllers for nonlinear systems. In particular, for smooth feedback, Lyapunov-based methods were inspired by Jurdjevic and Quinn [224] who give sufficient conditions for smooth stabilization based on the ability of constructing a Lyapunov function for the closed-loop system [434]. Unfortunately, however, there does not exist a unified procedure for finding a Lyapunov function candidate that will stabilize the closed-loop system for general nonlinear systems. Recent work involving differential geometric methods [212, 336] has made the design of controllers for certain classes of nonlinear systems more methodical. Such frameworks include the concepts of zero dynamics and feedback linearization and require that the system zero dynamics are asymptotically stable, ensuring the existence of globally defined diffeomorphisms to transform the nonlinear system into a normal form [212,336]. These techniques, however, usually rely on canceling out system nonlinearities using feedback and may therefore lead to inefficient designs since feedback linearizing controllers may generate unnecessarily large control effort to cancel beneficial system nonlinearities.

Backstepping control has recently received a great deal of attention in the nonlinear control literature [222, 230, 246, 249, 392]. The popularity of this control methodology can be explained in a large part due to the fact that it provides a framework for designing stabilizing nonlinear controllers for a large class of nonlinear dynamical cascade systems. This framework guarantees stability by providing a systematic procedure for finding a Lyapunov function for the closed-loop system and choosing the control such that the time derivative of the Lyapunov function along the trajectories of the closed-loop dynamical system is negative. Furthermore, the controller is obtained in such a way that the nonlinearities of the dynamical system, which may be useful in reaching performance objectives, need not be canceled as in state or output feedback linearization techniques. Using this framework, the control system designer has a significant amount of freedom in designing the controller to address specific performance objectives while guaranteeing closed-loop stability.

In [126, 127] optimal pointwise min-norm state tracking controllers are obtained for feedback linearizable systems by computing pointwise solutions of a static quadratic programming problem. A trade-off between control effort and tracking error is automatically taken into account in designing these controllers. The optimality of this control design method relies on the fact that every Lyapunov function solves the Hamilton-Jacobi-Bellman equation associated with a cost functional. However, this theory does not present a natural extension to the larger class of systems for which recursive backstepping is applicable. In particular, it is noted in [247] that for recursive control schemes such as backstepping, optimization of partial cost functionals at each step will by no means result in overall optimization. Backstepping does, however, utilize a Lyapunov function for the overall system based on stabilizing functions (virtual controls) which are defined at each recursion step. This Lyapunov function can be used to derive a performance criterion for which the overall control, consisting of the control law obtained at the final and intermediate steps, is optimal.

In this chapter, we extend the optimality-based nonlinear control framework developed in Chapter 8 to cascade and block cascade systems for which the backstepping control design methodology is applicable. The key motivation for developing an optimal and inverse optimal nonlinear backstepping control theory is that it provides a family of candidate backstepping controllers parameterized by the cost functional that is minimized. In order to address the optimality-based backstepping nonlinear control problem we use the nonlinear-nonquadratic optimal control framework developed in Chapter 8 to show that a particular controller derived via backstepping methods corresponds to the solution of an optimal control problem that minimizes an inverse nonlinear-nonquadratic performance criterion. This is accomplished by choosing the controller such that the Lyapunov derivative is negative along the closed-loop system trajectories while providing sufficient conditions for the existence of asymptotically stabilizing solutions to the Hamilton-Jacobi-Bellman equation. Thus, our results allow us to derive globally asymptotically stabilizing backstepping controllers for nonlinear systems that minimize a derived nonlinear-nonquadratic performance functional.

9.2 Cascade and Block Cascade Control Design

In this section, we consider the nonlinear cascade system

$$\dot{x}(t) = f(x(t)) + G(x(t))\hat{x}(t), \qquad x(0) = x_0, \qquad t \geq 0, \tag{9.1}$$
$$\dot{\hat{x}}(t) = u(t), \qquad \hat{x}(0) = \hat{x}_0, \tag{9.2}$$

where $x \in \mathbb{R}^n$, $\hat{x} \in \mathbb{R}^m$, $f : \mathbb{R}^n \to \mathbb{R}^n$ satisfies $f(0) = 0$, and $G : \mathbb{R}^n \to \mathbb{R}^m$. Here, we seek a globally stabilizing feedback controller for (9.1) and (9.2). To introduce the *integrator backstepping* approach note that (9.1) and (9.2) can be viewed as a cascade connection of two dynamical subsystems, as shown in Figure 9.1(a). Specifically, the first subsystem is (9.1) with input $\hat{x}$ and the second subsystem consists of m integrators. Next, we assume that there exists a continuously differentiable function $\alpha : \mathbb{R}^n \to \mathbb{R}^m$ such that the zero solution $x(t) \equiv 0$ of the first subsystem (9.1) is asymptotically stable with $\hat{x}$ replaced by $\alpha(x)$. In this case, it follows from Theorem 3.9 that there exists a continuously differentiable positive-definite function $V_{\rm sub} : \mathbb{R}^n \to \mathbb{R}$ such that

$$V'_{\rm sub}(x)[f(x) + G(x)\alpha(x)] < 0, \quad x \in \mathbb{R}^n, \quad x \neq 0. \tag{9.3}$$

Next, adding and subtracting $G(x)\alpha(x)$, $x \in \mathbb{R}^n$, to and from (9.1) yields the equivalent dynamical system

$$\dot{x}(t) = f(x(t)) + G(x(t))\alpha(x(t)) + G(x(t))[\hat{x}(t) - \alpha(x(t))],$$
$$x(0) = x_0, \ t \geq 0, \tag{9.4}$$
$$\dot{\hat{x}}(t) = u(t), \ \hat{x}(0) = \hat{x}_0, \tag{9.5}$$

shown in Figure 9.1(b). Introducing the change of variables $z(t) \triangleq \hat{x}(t) - \alpha(x(t))$ yields

$$\dot{x}(t) = f(x(t)) + G(x(t))\alpha(x(t)) + G(x(t))z(t), \ x(0) = x_0, \ t \geq 0, \tag{9.6}$$
$$\dot{z}(t) = u(t) - \dot{\alpha}(x(t)), \ z(0) = z_0. \tag{9.7}$$

As shown in Figure 9.1(c), transforming (9.1) and (9.2) to (9.6) and (9.7) can be viewed as "backstepping" $-\alpha(x)$ through the integrator subsystem. Now, with $v(t) \triangleq u(t) - \dot{\alpha}(x(t))$, (9.6) and (9.7) reduces to

$$\dot{x}(t) = f(x(t)) + G(x(t))\alpha(x(t)) + G(x(t))z(t), \ x(0) = x_0, \ t \geq 0, \tag{9.8}$$
$$\dot{z}(t) = v(t), \ z(0) = z_0, \tag{9.9}$$

which, when $z(t)$ is bounded and $z(t) \to 0$ as $t \to \infty$, is an asymptotically stable cascade system (see Proposition 4.2).

Exploiting this feature, we can stabilize the overall system by considering the Lyapunov function candidate

$$V(x, \hat{x}) = V_{\rm sub}(x) + (\hat{x} - \alpha(x))^{\rm T} P(\hat{x} - \alpha(x)), \tag{9.10}$$

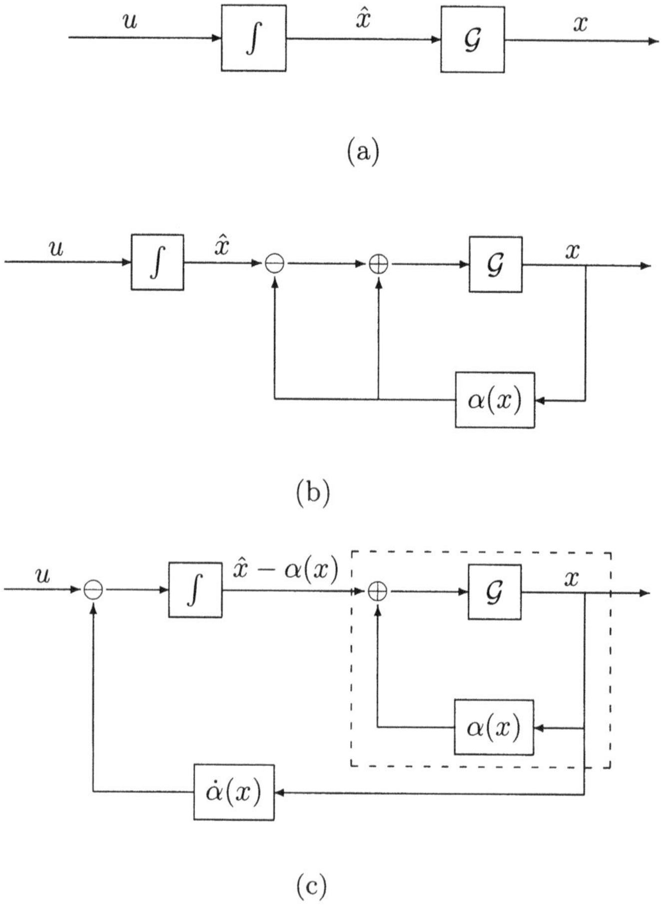

Figure 9.1 Visualization of integrator backstepping.

where $P \in \mathbb{R}^{m\times m}$ is an arbitrary positive-definite matrix. In this case, the Lyapunov derivative is given by

$$\dot{V}(x,\hat{x}) = V'_{\rm sub}(x)[f(x)+G(x)\alpha(x)] + V'_{\rm sub}(x)G(x)z + 2z^{\rm T}Pv. \qquad (9.11)$$

Letting $v = -\frac{1}{2}P^{-1}G^{\rm T}(x)V'^{\rm T}_{\rm sub}(x) - kz$, where $k>0$, yields

$$\begin{aligned}\dot{V}(x,\hat{x}) &= V'_{\rm sub}(x)[f(x)+G(x)\alpha(x)] - k(\hat{x}-\alpha(x))^{\rm T}P(\hat{x}-\alpha(x))\\ &< 0, \qquad (x,\hat{x})\in\mathbb{R}^n\times\mathbb{R}^m, \quad (x,\hat{x})\neq(0,0).\end{aligned} \qquad (9.12)$$

Hence, the control law

$$\begin{aligned}u &= v+\dot{\alpha}(x)\\ &= -k(\hat{x}-\alpha(x)) - \tfrac{1}{2}P^{-1}G^{\rm T}(x)V'^{\rm T}_{\rm sub}(x) + \alpha'(x)[f(x)+G(x)\hat{x}],\end{aligned} \qquad (9.13)$$

stabilizes the nonlinear cascade system (9.1) and (9.2). Using the above results the following proposition is immediate.

Proposition 9.1. Consider the nonlinear cascade system (9.1) and (9.2). Assume that there exist a continuously differentiable function $\alpha : \mathbb{R}^n \to \mathbb{R}^m$ and a continuously differentiable radially unbounded function $V_{\rm sub} : \mathbb{R}^n \to \mathbb{R}$ such that

$$\alpha(0) = 0, \tag{9.14}$$
$$V_{\rm sub}(0) = 0 \tag{9.15}$$
$$V_{\rm sub}(x) > 0, \quad x \in \mathbb{R}^n, \quad x \neq 0, \tag{9.16}$$
$$V'_{\rm sub}(x)[f(x) + G(x)\alpha(x)] < 0, \quad x \in \mathbb{R}^n, \quad x \neq 0. \tag{9.17}$$

Then, the zero solution $(x(t), \hat{x}(t)) \equiv (0, 0)$ of the cascade system (9.1) and (9.2) is globally asymptotically stable with the feedback control law (9.13).

Example 9.1. Consider the nonlinear cascade system

$$\dot{x}_1(t) = -\tfrac{3}{2}x_1^2(t) - \tfrac{1}{2}x_1^3(t) - x_2(t), \quad x_1(0) = x_{10}, \quad t \geq 0, \tag{9.18}$$
$$\dot{x}_2(t) = u(t), \quad x_2(0) = x_{20}. \tag{9.19}$$

Here, we seek a globally stabilizing controller for (9.18) and (9.19) using the integrator backstepping approach. Note that (9.18) and (9.19) has the correct form for the application of Proposition 9.1 where (9.18) makes up the nonlinear subsystem and x_2 is the integrator state. Specifically, (9.18) and (9.19) can be written in the form of (9.1) and (9.2) where $x = x_1$, $\hat{x} = x_2$, and

$$f(x) = -\tfrac{3}{2}x^2 - \tfrac{1}{2}x^3, \qquad G(x) = -1.$$

To apply Proposition 9.1 we require a stabilizing feedback for the subsystem (9.18) and a corresponding Lyapunov function $V_{\rm sub}(x)$ such that (9.14)–(9.17) are satisfied. For the nonlinear subsystem (9.18) we choose the Lyapunov function candidate

$$V_{\rm sub}(x) = x_1^2 \tag{9.20}$$

and the stabilizing feedback control

$$\alpha(x) = -\tfrac{3}{2}x^2. \tag{9.21}$$

It is straightforward to show that (9.20) and (9.21) satisfy conditions (9.14)–(9.17) of Proposition 9.1. Now, it follows from Proposition 9.1 that the control law given by (9.13), that is,

$$u = -k(x_2 + \tfrac{3}{2}x_1^2) + P^{-1}x_1 + 3x_1(\tfrac{3}{2}x_1^2 + \tfrac{1}{2}x_1^3 + x_2), \tag{9.22}$$

where k and P are positive constants, is a globally stabilizing feedback control law for the overall system (9.18) and (9.19). △

Example 9.2. Consider the nonlinear cascade system

$$\dot{x}_1(t) = -\tfrac{\sigma}{2}x_1(t)[\tfrac{1}{4}x_1^2(t) + 2x_2(t) + x_2^2(t)], \quad x_1(0) = x_{10}, \quad t \geq 0, \tag{9.23}$$

$$\dot{x}_2(t) = -\tfrac{3}{2}x_2^2(t) - \tfrac{1}{2}x_2^3(t) - \tfrac{3}{4}x_1^2(t)(1 + x_2(t)) - x_3(t), \ x_2(0) = x_{30}, \tag{9.24}$$

$$\dot{x}_3(t) = u(t), \ x_3(0) = x_{30}, \tag{9.25}$$

where $\sigma > 0$. Once again, we seek a globally stabilizing controller for (9.23)–(9.25) using Proposition 9.1. Note that (9.23)–(9.25) has the correct form for the application of Proposition 9.1 where (9.23) and (9.24) make up the nonlinear subsystem and x_3 is the integrator state. Specifically, (9.23)–(9.25) can be written in the form of (9.1) and (9.2) where $x = [x_1 \ \ x_2]^{\mathrm{T}}$, $\hat{x} = x_3$, and

$$f(x) = \begin{bmatrix} \frac{\sigma}{2}x_1(\frac{1}{4}x_1^2 + 2x_2 + x_2^2) \\ -\frac{3}{2}x_2^2 - \frac{1}{2}x_2^3 - \frac{3}{4}x_1^2(1 + x_2) \end{bmatrix}, \quad G(x) = \begin{bmatrix} 0 \\ -1 \end{bmatrix}.$$

To apply Proposition 9.1 we require a stabilizing feedback for the subsystem (9.23) and (9.24) and a corresponding Lyapunov function $V_{\mathrm{sub}}(x)$ such that (9.14)–(9.17) are satisfied. For the nonlinear subsystem (9.23) and (9.24) we choose the Lyapunov function candidate

$$V_{\mathrm{sub}}(x) = \varepsilon x_1^4 + x_2^2, \tag{9.26}$$

where $\varepsilon > 0$, and the stabilizing feedback control

$$\alpha(x) = -(2\varepsilon\sigma x_1^4 + \tfrac{3}{2}x_2^2 + \tfrac{3}{4}x_1^2). \tag{9.27}$$

It is straightforward to show that (9.26) and (9.27) satisfy conditions (9.14)–(9.17) of Proposition 9.1. Now, it follows from Proposition 9.1 that the control law given by (9.13), that is,

$$u = -k(x_3 - \alpha(x)) + P^{-1}x_2 + \alpha'(x)[f(x) + G(x)x_3], \tag{9.28}$$

where k and P are positive constants, is a globally stabilizing feedback control law for the overall system (9.23)–(9.25). △

Next, we consider the application of backstepping control design to more general *block cascade* systems of the form

$$\dot{x}(t) = f(x(t)) + G(x(t))\hat{x}(t), \quad x(0) = x_0, \quad t \geq 0, \tag{9.29}$$

$$\dot{\hat{x}}(t) = \hat{f}(x(t), \hat{x}(t)) + \hat{G}(x(t), \hat{x}(t))u(t), \quad \hat{x}(0) = \hat{x}_0, \tag{9.30}$$

where the subsystem (9.29) is as (9.1) and $\hat{x} \in \mathbb{R}^m$, $\hat{f} : \mathbb{R}^n \times \mathbb{R}^m \to \mathbb{R}^m$ satisfies $\hat{f}(0,0) = 0$, and $\hat{G} : \mathbb{R}^n \times \mathbb{R}^m \to \mathbb{R}^{m \times m}$. Now, suppose $\det \hat{G}(x, \hat{x}) \neq 0$, $(x, \hat{x}) \in \mathbb{R}^n \times \mathbb{R}^m$, and suppose there exists a continuously differentiable function $\alpha : \mathbb{R}^n \to \mathbb{R}^m$ such that the zero solution $x(t) \equiv 0$ of (9.29) is asymptotically stable with $\hat{x}$ replaced by $\alpha(x)$. In this case, it follows from Theorem 3.9 that there exists a continuously differentiable positive-definite function $V_{\mathrm{sub}} : \mathbb{R}^n \to \mathbb{R}$ such that (9.17) holds. Now, once again using the Lyapunov function candidate given by (9.10) for the overall system (9.29) and (9.30), it follows that

$$\begin{aligned}\dot V(x,\hat x) &= V'_{\rm sub}(x)[f(x)+G(x)\alpha(x)] + V'_{\rm sub}(x)G(x)[\hat x-\alpha(x)]\\ &\quad +2[\hat x-\alpha(x)]^{\rm T}P\left\{\hat f(x,\hat x)+\hat G(x,\hat x)u-\alpha'(x)[f(x)+G(x)\hat x]\right\}.\end{aligned} \tag{9.31}$$

Now, choosing

$$\begin{aligned}u &= \hat G^{-1}(x,\hat x)\Big\{\alpha'(x)[f(x)+G(x)\hat x]-\tfrac{1}{2}P^{-1}G^{\rm T}(x)V'^{\rm T}_{\rm sub}(x)-\hat f(x,\hat x)\\ &\quad -k(\hat x-\alpha(x))\Big\},\end{aligned} \tag{9.32}$$

where $k>0$, it follows that (9.12) is satisfied, and hence, (9.32) stabilizes the block cascade system (9.29) and (9.30). If, in addition, $V_{\rm sub}(\cdot)$ is radially unbounded, (9.32) is a global stabilizer.

Using the backstepping formulation discussed above, it follows that a recursive formulation of this approach can be used to stabilize *strict-feedback* nonlinear smooth systems of the form

$$\dot x(t) = f(x(t))+G(x(t))\hat x_1(t),\quad x(0)=x_0,\quad t\ge 0, \tag{9.33}$$

$$\dot{\hat x}_1(t) = \hat f_1(x(t),\hat x_1(t))+\hat G_1(x(t),\hat x_1(t))\hat x_2(t),\quad \hat x_1(0)=\hat x_{0_1}, \tag{9.34}$$

$$\dot{\hat x}_2(t) = \hat f_2(x(t),\hat x_1(t),\hat x_2(t))+\hat G_2(x(t),\hat x_1(t),\hat x_2(t))\hat x_3(t),\quad \hat x_3(0)=\hat x_{0_2}, \tag{9.35}$$

$$\vdots$$

$$\dot{\hat x}_m(t) = \hat f_m(x(t),\hat x_1(t),\ldots,\hat x_m(t))+\hat G_m(x(t),\hat x_1(t),\ldots,\hat x_m(t))u(t),\quad \hat x_m(0)=\hat x_{0_m}, \tag{9.36}$$

where $x\in\mathbb{R}^n$, $f:\mathbb{R}^n\to\mathbb{R}^n$ satisfies $f(0)=0$, $G:\mathbb{R}^n\to\mathbb{R}^m$, $\hat x_i\in\mathbb{R}$, $i=1,\ldots,m$, $\hat f_i:\mathbb{R}^n\times\mathbb{R}^i\to\mathbb{R}$ satisfies $\hat f_i(0)=0$, $i=1,\ldots,m$, and $\hat G_i:\mathbb{R}^n\times\mathbb{R}^i\to\mathbb{R}$, $i=1,\ldots,m$. Specifically, assuming $\hat G_i(x,\hat x_1,\ldots,\hat x_i)\neq 0$, $1\le i\le m$, and assuming the there exists an m-times continuously differentiable function $\alpha:\mathbb{R}^n\to\mathbb{R}$ such that the first subsystem (9.33) is asymptotically stable with $\hat x_1$ replaced by $\alpha(x)$, a recursive backstepping procedure can be used to stabilize (9.33)–(9.36).

In particular, considering the system

$$\dot x(t) = f(x(t))+G(x(t))\hat x_1(t),\quad x(0)=x_0,\quad t\ge 0, \tag{9.37}$$

$$\dot{\hat x}_1(t) = \hat f_1(x(t),\hat x_1(t))+\hat G_1(x(t),\hat x_1(t))\hat x_2(t),\quad \hat x_1(0)=\hat x_{0_1}, \tag{9.38}$$

it follows from (9.29)–(9.32), with $\hat x=\hat x_1$, $u=\hat x_2$, $\hat f(x,\hat x)=\hat f_1(x,\hat x_1)$, and

$\hat{G}(x,\hat{x}) = \hat{G}_1(x,\hat{x}_1)$, that the feedback control law

$$\alpha_1(x,\hat{x}_1) = \hat{G}_1^{-1}(x,\hat{x}_1)\Big\{\alpha'(x)[f(x)+G(x)\hat{x}_1] - \tfrac{1}{2}P_1^{-1}G^{\mathrm{T}}(x)V_{\mathrm{sub}}'^{\mathrm{T}}(x) - \hat{f}_1(x,\hat{x}_1) - k_1(\hat{x}_1-\alpha(x))\Big\}, \quad k_1>0, \tag{9.39}$$

asymptotically stabilizes (9.37) and (9.38) with Lyapunov function

$$V_1(x,\hat{x}_1) = V_{\mathrm{sub}}(x) + P_1(\hat{x}_1-\alpha(x))^2, \tag{9.40}$$

where $V_{\mathrm{sub}}(x)$ is such that (9.3) holds. Next, considering the system

$$\dot{x}(t) = f(x(t)) + \hat{G}(x(t))\hat{x}_1(t), \quad x(0)=x_0, \quad t\ge 0, \tag{9.41}$$

$$\dot{\hat{x}}_1(t) = \hat{f}_1(x(t),\hat{x}_1(t)) + \hat{G}_1(x(t),\hat{x}_1(t))\hat{x}_2(t), \quad \hat{x}_1(0)=\hat{x}_{0_1}, \tag{9.42}$$

$$\dot{\hat{x}}_2(t) = \hat{f}_2(x(t),\hat{x}_1(t),\hat{x}_2(t)) + \hat{G}_2(x(t),\hat{x}_1(t),\hat{x}_2(t))\hat{x}_3(t), \ \hat{x}_3(0)=\hat{x}_{0_2}, \tag{9.43}$$

it follows from (9.29)–(9.32), with $x = \begin{bmatrix} x^{\mathrm{T}} & \hat{x}_1^{\mathrm{T}} \end{bmatrix}^{\mathrm{T}}$, $\hat{x}=\hat{x}_2$, $u=\hat{x}_3$,

$$f(x) = \begin{bmatrix} f(x)+G(x)\hat{x}_1 \\ \hat{f}_1(x,\hat{x}_1) \end{bmatrix}, \quad G(x) = \begin{bmatrix} 0 \\ G_1(x,\hat{x}_1) \end{bmatrix},$$

$\hat{f}(x,\hat{x}) = \hat{f}_2(x,\hat{x}_1,\hat{x}_2)$, and $\hat{G}(x,\hat{x}) = \hat{G}_2(x,\hat{x}_1,\hat{x}_2)$, that the feedback control law

$$\begin{aligned}\alpha_2(x,\hat{x}_1,\hat{x}_2) = G_2^{-1}(x,\hat{x}_1,\hat{x}_2)\Bigg\{&\left[\frac{\partial\alpha_1(x,\hat{x}_1)}{\partial x}\right]^{\mathrm{T}}[f(x)+G(x)\hat{x}_1] \\ &+\left[\frac{\partial(x,\hat{x}_1)\alpha_1}{\partial\hat{x}_1}\right]^{\mathrm{T}}[\hat{f}_1(x,\hat{x}_1)+G_1(x,\hat{x}_1)\hat{x}_2] \\ &-\tfrac{1}{2}P_2^{-1}G_1(x,\hat{x}_1)\frac{\partial V_1(x,\hat{x}_1)}{\partial\hat{x}_1} \\ &-\hat{f}_2(x,\hat{x}_1)-k_2[\hat{x}_2-\alpha_1(x,\hat{x}_1)]\Bigg\}, \quad k_2>0,\end{aligned} \tag{9.44}$$

asymptotically stabilizes (9.41)–(9.43) with Lyapunov function

$$V_2(x,\hat{x}_1,\hat{x}_2) = V_1(x,\hat{x}_1) + P_2(\hat{x}_2-\alpha_1(x,\hat{x}_1))^2. \tag{9.45}$$

Repeating this procedure m-times, an overall state feedback controller of the form $u=\alpha_m(x,\hat{x}_1,\ldots,\hat{x}_m)$ with Lyapunov function $V_m(x,\hat{x}_1,\ldots,\hat{x}_m)$ can be obtained for the strict-feedback nonlinear system (9.34)–(9.36).

Example 9.3. To demonstrate the recursive backstepping procedure discussed above consider the nonlinear dynamical system

$$\dot{x}_1(t) = x_1^4(t) - x_1^5(t) + x_2(t), \quad x_1(0)=x_{10}, \quad t\ge 0, \tag{9.46}$$

$$\dot{x}_2(t) = x_3(t), \quad x_2(0)=x_{20}, \tag{9.47}$$

$$\dot{x}_3(t) = u(t), \quad x_3(0) = x_{30}. \tag{9.48}$$

We start by considering the system

$$\dot{x}_1(t) = x_1^4(t) - x_1^5(t) + x_2(t), \quad x_1(0) = x_{10}, \quad t \geq 0, \tag{9.49}$$
$$\dot{x}_2(t) = x_3(t), \quad x_2(0) = x_{20}, \tag{9.50}$$

which has the form of (9.1) and (9.2) with $x = x_1$, $\hat{x} = x_2$, $u = x_3$, $f(x_1) = x_1^4 - x_1^5$, and $G(x) = 1$. To apply Proposition 9.1 we require a stabilizing feedback for the subsystem (9.49) and a corresponding Lyapunov function $V_{\rm sub}(x_1)$ such that (9.14)–(9.17) are satisfied. For the nonlinear subsystem (9.49) we choose the Lyapunov function candidate $V_{\rm sub}(x_1) = \frac{1}{2}x_1^2$ and the stabilizing feedback control $\alpha(x_1) = -x_1^4 - x_1$. It is straightforward to show that $V_{\rm sub}(x_1)$ and $\alpha(x_1)$ given above satisfy the conditions of Proposition 9.1. Hence, it follows from Proposition 9.1 that

$$x_3 = -(x_2 + x_1^4 + x_1) - x_1 - (4x_1^3 + 1)(x_1^4 - x_1^5 + x_2) \tag{9.51}$$

is a globally stabilizing feedback control law for (9.49) and (9.50) with Lyapunov function

$$V_1(x_1, x_2) = V_{\rm sub}(x_1) + \tfrac{1}{2}(x_2 + x_1^4 + x_1)^2. \tag{9.52}$$

Next, we view (9.46)–(9.48) as (9.1) and (9.2) with $x = [x_1 \ x_2]^{\rm T}$, $\hat{x} = x_3$, $f(x_1, x_2) = [x_1^4 - x_1^5 + x_2 \ \ 0]^{\rm T}$, and $G(x_1, x_2) = [0 \ 1]^{\rm T}$. Now, with

$$\alpha_1(x_1, x_2) = -(x_2 + x_1^4 + x_1) - x_1 - (4x_1^3 + 1)(x_1^4 - x_1^5 + x_2), \tag{9.53}$$

and $V_1(x_1, x_2)$ given by (9.52) it follows that all the conditions of Proposition 9.1 are satisfied. Hence, it follows from Proposition 9.1 that

$$\begin{aligned} u = {} & -(x_3 - \alpha_1(x_1, x_2)) - G^{\rm T}(x_1, x_2)V_1'^{\rm T}(x_1, x_2) + \alpha'_1(x_1, x_2) \\ & \cdot [f(x_1, x_2) + G(x_1, x_2)x_3] \end{aligned} \tag{9.54}$$

globally stabilizes (9.46)–(9.49) with Lyapunov function

$$V(x_1, x_2, x_3) = V_1(x_1, x_2) + \tfrac{1}{2}(x_3 - \alpha_1(x_1, x_2))^2. \tag{9.55}$$

△

As shown in this section, backstepping control provides a systematic procedure for finding a Lyapunov function for cascade and block cascade systems while choosing a stabilizing feedback control law. Since the control designer has a significant amount of freedom in designing the controller, it is natural to ask the question: Does there exist analytical measures of performance or notions of optimality for backstepping controllers? This question is addressed in the remainder of this chapter.

9.3 Optimal Integrator Backstepping Controllers

In this section, we develop an optimality-based framework for backstepping controllers. The key motivation for developing an optimal nonlinear backstepping control theory is that it provides a family of candidate backstepping controllers parameterized by the cost functional that is minimized. In order to address the optimality-based backstepping nonlinear control problem we use the nonlinear-nonquadratic optimal control framework developed in Chapter 8. For our first result, define

$$L(x,\hat{x},u) \triangleq L_1(x,\hat{x}) + L_2(x,\hat{x})u + u^{\mathrm{T}}R_2(x,\hat{x})u, \tag{9.56}$$

where $L_1 : \mathbb{R}^n \times \mathbb{R}^m \to \mathbb{R}$, $L_2 : \mathbb{R}^n \times \mathbb{R}^m \to \mathbb{R}^{1\times m}$, and $R_2 : \mathbb{R}^n \times \mathbb{R}^m \to \mathbb{P}^m$, and define

$$\begin{aligned}\mathcal{S}(x_0,\hat{x}_0) \triangleq \{u(\cdot) :\ & u(\cdot) \in \mathcal{U} \text{ and } (x(\cdot),\hat{x}(\cdot)) \text{ given by} \\ & (9.1) \text{ and } (9.2) \text{ satisfies } (x(t),\hat{x}(t)) \to 0 \text{ as } t \to \infty\}.\end{aligned}$$

Theorem 9.1. Consider the nonlinear cascade system (9.1) and (9.2) with performance functional

$$J(x_0,\hat{x}_0,u(\cdot)) \triangleq \int_0^\infty L(x(t),\hat{x}(t),u(t))\mathrm{d}t, \tag{9.57}$$

where $u(\cdot)$ is admissible, $(x(t),\hat{x}(t))$, $t \geq 0$, satisfies (9.1) and (9.2), and $L(x,\hat{x},u)$ is given by (9.56). Assume that there exist continuously differentiable functions $\alpha : \mathbb{R}^n \to \mathbb{R}^m$ and $V_{\mathrm{sub}} : \mathbb{R}^n \to \mathbb{R}$ such that

$$\alpha(0) = 0, \tag{9.58}$$

$$V_{\mathrm{sub}}(0) = 0, \tag{9.59}$$

$$V_{\mathrm{sub}}(x) > 0, \quad x \in \mathbb{R}^n, \quad x \neq 0, \tag{9.60}$$

$$V'_{\mathrm{sub}}(x)[f(x) + G(x)\alpha(x)] < 0, \quad x \in \mathbb{R}^n, \quad x \neq 0. \tag{9.61}$$

Furthermore, let $L_2 : \mathbb{R}^n \times \mathbb{R}^m \to \mathbb{R}^{1\times m}$ be such that $L_2(0,0) = 0$ and

$$\begin{aligned}&(\hat{x} - \alpha(x))^{\mathrm{T}}\Big\{G^{\mathrm{T}}(x)V'^{\mathrm{T}}_{\mathrm{sub}}(x) - 2\hat{P}\Big[\alpha'(x)(f(x) + G(x)\hat{x}) \\ &+R_2^{-1}(x,\hat{x})[\hat{P}(\hat{x} - \alpha(x)) + \tfrac{1}{2}L_2^{\mathrm{T}}(x,\hat{x})]\Big]\Big\} < 0,\ (x,\hat{x}) \in \mathbb{R}^n \times \mathbb{R}^m,\ \hat{x} \neq \alpha(x),\end{aligned} \tag{9.62}$$

where $\hat{P} \in \mathbb{R}^{m\times m}$ is an arbitrary positive-definite matrix. Then the zero solution $(x(t),\hat{x}(t)) \equiv (0,0)$ of the cascade system (9.1) and (9.2) is globally asymptotically stable with the feedback control law

$$u = \phi(x,\hat{x}) = -R_2^{-1}(x,\hat{x})\hat{P}[\hat{x} - \alpha(x)] - \tfrac{1}{2}R_2^{-1}(x,\hat{x})L_2^{\mathrm{T}}(x,\hat{x}). \tag{9.63}$$

Furthermore,

$$J(x_0,\hat{x}_0,\phi(x(\cdot),\hat{x}(\cdot))) = V(x_0,\hat{x}_0), \qquad (x_0,\hat{x}_0)\in\mathbb{R}^n\times\mathbb{R}^m, \tag{9.64}$$

where

$$V(x,\hat{x}) = V_{\rm sub}(x) + [\hat{x}-\alpha(x)]^{\rm T}\hat{P}[\hat{x}-\alpha(x)], \tag{9.65}$$

and the performance functional (9.57), with

$$\begin{aligned} L_1(x,\hat{x}) &= \phi^{\rm T}(x,\hat{x})R_2(x,\hat{x})\phi(x,\hat{x}) - V'_{\rm sub}(x)[f(x)+G(x)\hat{x}] \\ &\quad +2[\hat{x}-\alpha(x)]^{\rm T}\hat{P}\alpha'(x)[f(x)+G(x)\hat{x}], \end{aligned} \tag{9.66}$$

is minimized in the sense that

$$J(x_0,\hat{x}_0,\phi(x(\cdot),\hat{x}(\cdot))) = \min_{u(\cdot)\in\mathcal{S}(x_0,\hat{x}_0)} J(x_0,\hat{x}_0,u(\cdot)). \tag{9.67}$$

Proof. The result follows as a direct consequence of Theorem 8.3 applied to the system

$$\dot{\tilde{x}}(t) = \tilde{f}(\tilde{x}(t)) + \tilde{G}(\tilde{x}(t))u(t), \qquad \tilde{x}(0)=\tilde{x}_0, \qquad t\geq 0, \tag{9.68}$$

where

$$\tilde{x} = \begin{bmatrix} x \\ \hat{x}\end{bmatrix}, \quad \tilde{f}(\tilde{x}) = \begin{bmatrix} f(x)+G(x)\hat{x} \\ 0\end{bmatrix}, \quad \tilde{G}(\tilde{x}) = \begin{bmatrix} 0 \\ I_m\end{bmatrix}.$$

Specifically, conditions (8.59)–(8.62) are trivially satisfied by (9.58)–(9.62) and (9.65). Next, using (9.63) and (9.68), $\phi(x,\hat{x})$ can be written as

$$\phi(\tilde{x}) = -\tfrac{1}{2}R_2^{-1}(\tilde{x})[L_2^{\rm T}(\tilde{x}) + \tilde{G}^{\rm T}(\tilde{x})V'^{\rm T}(\tilde{x})] \tag{9.69}$$

so that (8.65) is satisfied. Finally, using (9.65) and (9.68) it follows that (9.66) or, equivalently,

$$L_1(\tilde{x}) = \phi^{\rm T}(\tilde{x})R_2(\tilde{x})\phi(\tilde{x}) - V'(\tilde{x})\tilde{f}(\tilde{x}), \tag{9.70}$$

satisfies (8.66). □

Note that the only restrictions on the choice of $L_2(\cdot,\cdot)$ are the conditions $L_2(0,0)=0$ and (9.62) which must be satisfied. Therefore, a significant amount of freedom with regard to the form of the controller is available to the control designer. Furthermore, since $L_2(x,\hat{x})$ appears explicitly in both the feedback control law $\phi(x,\hat{x})$ and the cost functional $J(x_0,\hat{x}_0,u)$, it ties together the particular choice of control and the form of the performance criterion for which the control is optimal. A particular choice of $L_2(x,\hat{x})$ satisfying condition (9.62) is given by

$$L_2(x,\hat{x}) = \left[V'_{\rm sub}(x)G(x)\hat{P}^{-1} - 2[f(x)+G(x)\hat{x}]^{\rm T}\alpha'^{\rm T}(x)\right]R_2(x,\hat{x}). \tag{9.71}$$

In this case, for $u(t)\in\mathbb{R}$, $t\geq 0$, the feedback control law given by (9.63) specializes to the integrator backstepping controller given by Lemma 2.8

of [247] by setting $\hat{P} = \frac{1}{2}I_m$ and $R_2(x,\hat{x}) = 1/c$.

The next result provides inverse optimal controllers for cascade systems with guaranteed sector margins.

Theorem 9.2. Consider the nonlinear cascade system (9.1) and (9.2). Assume that there exist continuously differentiable functions $\alpha : \mathbb{R}^n \to \mathbb{R}^m$ and $V_{\rm sub} : \mathbb{R}^n \to \mathbb{R}$ such that

$$\alpha(0) = 0, \tag{9.72}$$

$$V_{\rm sub}(0) = 0, \tag{9.73}$$

$$V_{\rm sub}(x) > 0, \quad x \in \mathbb{R}^n, \quad x \neq 0, \tag{9.74}$$

$$V'_{\rm sub}(x)[f(x) + G(x)\alpha(x)] < 0, \quad x \in \mathbb{R}^n, \quad x \neq 0. \tag{9.75}$$

Then, with the feedback stabilizing control law given by

$$\phi(x,\hat{x}) = \begin{cases} -(c_0 + \rho(x,\hat{x}))\beta(x,\hat{x}), & \hat{x} \neq \alpha(x), \\ 0, & \hat{x} = \alpha(x), \end{cases} \tag{9.76}$$

where $\beta(x,\hat{x}) \triangleq 2\hat{P}(\hat{x} - \alpha(x))$, $\mu(x,\hat{x}) \triangleq 2\alpha'(x)[f(x) + G(x)\hat{x}] - \hat{P}^{-1}G^{\rm T}(x) \cdot V'^{\rm T}_{\rm sub}(x)$, $\hat{P} \in \mathbb{P}^m$,

$$\rho(x,\hat{x}) \triangleq \frac{\sqrt{(\beta^{\rm T}(x,\hat{x})\mu(x,\hat{x}))^2 + (\beta^{\rm T}(x,\hat{x})\beta(x,\hat{x}))^2} - \beta^{\rm T}(x,\hat{x})\mu(x,\hat{x})}{\beta^{\rm T}(x,\hat{x})\beta(x,\hat{x})},$$

and $c_0 > 0$, the cascade system (9.1) and (9.2) has a sector (and, hence, gain) margin $(\frac{1}{2}, \infty)$. Furthermore, with the feedback control law $u = \phi(x,\hat{x})$ the performance functional

$$\begin{aligned} J(x_0,\hat{x}_0,u(\cdot)) = \int_0^\infty & [\beta^{\rm T}(x(t),\hat{x}(t))\mu(x(t),\hat{x}(t)) - V'_{\rm sub}(x)[f(x) \\ & + G(x)\alpha(x)] - \tfrac{\eta(x(t),\hat{x}(t))}{2}\beta^{\rm T}(x(t),\hat{x}(t))\beta(x(t),\hat{x}(t)) \\ & + \tfrac{1}{2\eta(x(t),\hat{x}(t))}u^{\rm T}(t)u(t)]{\rm d}t, \end{aligned} \tag{9.77}$$

where

$$\eta(x,\hat{x}) \triangleq \begin{cases} c_0 + \rho(x,\hat{x}), & \hat{x} \neq \alpha(x), \\ c_0, & \hat{x} = \alpha(x), \end{cases} \tag{9.78}$$

is minimized in the sense that

$$J(x_0,\hat{x}_0,\phi(x(\cdot))) = \min_{u(\cdot)\in\mathcal{S}(x_0,\hat{x}_0)} J(x_0,\hat{x}_0,u(\cdot)), \qquad (x_0,\hat{x}_0) \in \mathbb{R}^n \times \mathbb{R}^m. \tag{9.79}$$

Proof. The result is a direct consequence of Corollary 8.8 and Theorem 8.10 with $R_2(x,\hat{x}) = \frac{1}{2\eta(x,\hat{x})}I_m$ and $L_1(x) = \beta^{\rm T}(x,\hat{x})\mu(x,\hat{x}) - V'_{\rm sub}(x)[f(x) + G(x)\alpha(x)] + \frac{\eta(x,\hat{x})}{2}\beta^{\rm T}(x,\hat{x})\beta(x,\hat{x})$. Specifically, it follows from (9.78) that

$R_2(x,\hat{x}) > 0$, $x \in \mathbb{R}^n$, $\hat{x} \in \mathbb{R}^m$, and

$$\begin{aligned} L_1(x,\hat{x}) &= \beta^{\mathrm{T}}(x,\hat{x})\mu(x,\hat{x}) - V'_{\mathrm{sub}}(x)[f(x)+G(x)\alpha(x)] \\ &\quad +\tfrac{\eta(x,\hat{x})}{2}\beta^{\mathrm{T}}(x,\hat{x})\beta(x,\hat{x}) \\ &= \begin{cases} -V'_{\mathrm{sub}}(x)[f(x)+G(x)\alpha(x)] \\ +\frac{1}{2}\beta^{\mathrm{T}}(x,\hat{x})\beta(x,\hat{x})(c_0+\rho(x,\hat{x})), & \beta(x) \neq 0, \\ -V'_{\mathrm{sub}}(x)[f(x)+G(x)\alpha(x)], & \beta(x) = 0. \end{cases} \end{aligned} \quad (9.80)$$

Now, it follows from (9.75) and (9.80) that $L_1(x,\hat{x}) \geq 0$, $(x,\hat{x}) \in \mathbb{R}^n \times \mathbb{R}^m$, so that all conditions of Corollary 8.8 are satisfied. □

9.4 Optimal Linear Block Backstepping Controllers

In this section, we generalize the results of Section 9.2 to cascade systems where the input subsystem is characterized by a linear time-invariant system. Specifically, we consider a nonlinear system with a linear square input subsystem having the form

$$\dot{x}(t) = f(x(t)) + G(x(t))y(t), \qquad x(0) = x_0, \qquad t \geq 0, \quad (9.81)$$

$$\dot{\hat{x}}(t) = A\hat{x}(t) + Bu(t), \qquad \hat{x}(0) = \hat{x}_0, \quad (9.82)$$

$$y(t) = C\hat{x}(t), \quad (9.83)$$

where $\hat{x} \in \mathbb{R}^q$, $u, y \in \mathbb{R}^m$, $A \in \mathbb{R}^{q\times q}$, $B \in \mathbb{R}^{q\times m}$, and $C \in \mathbb{R}^{m\times q}$.

First, we consider the case in which the linear input subsystem (9.82) is feedback strict positive real, that is, there exist matrices $K \in \mathbb{R}^{m\times q}$, $\hat{P} \in \mathbb{P}^q$, and $\hat{Q} \in \mathbb{N}^q$ such that

$$0 = (A+BK)^{\mathrm{T}}\hat{P} + \hat{P}(A+BK) + \hat{Q}, \quad (9.84)$$

$$0 = B^{\mathrm{T}}\hat{P} - C. \quad (9.85)$$

Recall that it follows from Theorem 6.12 that the input subsystem (9.82) and (9.83) is feedback strict positive real if and only if $\det(CB) \neq 0$ and (9.82) and (9.83) is minimum phase.

Theorem 9.3. Consider the nonlinear cascade system (9.81)–(9.83) with performance functional (9.57) where $L(x,\hat{x},u)$ is given by (9.56). Assume that the triple (A,B,C) is feedback strict positive real and the nonlinear subsystem (9.81) has a globally stable equilibrium at $x=0$ with $y=0$ and Lyapunov function $V_{\mathrm{sub}}: \mathbb{R}^n \to \mathbb{R}$ so that

$$V'_{\mathrm{sub}}(x)f(x) < 0, \qquad x \in \mathbb{R}^n, \qquad x \neq 0. \quad (9.86)$$

Furthermore, let $L_2: \mathbb{R}^n \times \mathbb{R}^q \to \mathbb{R}^{1\times m}$ be such that $L_2(0,0)=0$ and

$$y^{\mathrm{T}}[G^{\mathrm{T}}(x)V'^{\mathrm{T}}_{\mathrm{sub}}(x) - R_2^{-1}(x,\hat{x})L_2^{\mathrm{T}}(x,\hat{x}) - 2K\hat{x}] \leq 0, \quad (x,\hat{x}) \in \mathbb{R}^n \times \mathbb{R}^q. \quad (9.87)$$

Then the zero solution $(x(t), \hat{x}(t)) \equiv (0,0)$ of the cascade system (9.81)–(9.83) is globally asymptotically stable with the feedback control law

$$u = \phi(x,\hat{x}) = -\tfrac{1}{2}R_2^{-1}(x,\hat{x})[2B^{\mathrm{T}}\hat{P}\hat{x} + L_2^{\mathrm{T}}(x,\hat{x})], \tag{9.88}$$

where $\hat{P} \in \mathbb{P}^q$ satisfies (9.84) and (9.85). Furthermore,

$$J(x_0,\hat{x}_0,\phi(x(\cdot),\hat{x}(\cdot))) = V(x_0,\hat{x}_0), \qquad (x_0,\hat{x}_0) \in \mathbb{R}^n \times \mathbb{R}^q, \tag{9.89}$$

where

$$V(x,\hat{x}) = V_{\mathrm{sub}}(x) + \hat{x}^{\mathrm{T}}\hat{P}\hat{x}, \tag{9.90}$$

and the performance functional (9.57), with

$$\begin{aligned} L_1(x,\hat{x}) &= \phi^{\mathrm{T}}(x,\hat{x})R_2(x,\hat{x})\phi(x,\hat{x}) + \hat{x}^{\mathrm{T}}\hat{P}B(2K\hat{x} - G^{\mathrm{T}}(x)V_{\mathrm{sub}}'^{\mathrm{T}}(x)) \\ &\quad + \hat{x}^{\mathrm{T}}\hat{Q}\hat{x} - V_{\mathrm{sub}}'(x)f(x), \end{aligned} \tag{9.91}$$

is minimized in the sense that

$$J(x_0,\hat{x}_0,\phi(x(\cdot),\hat{x}(\cdot))) = \min_{u(\cdot)\in\mathcal{S}(x_0,\hat{x}_0)} J(x_0,\hat{x}_0,u(\cdot)), \quad (x_0,\hat{x}_0) \in \mathbb{R}^n \times \mathbb{R}^q. \tag{9.92}$$

Proof. The result follows as a direct consequence of Theorem 8.3 applied to the system

$$\dot{\tilde{x}}(t) = \tilde{f}(\tilde{x}(t)) + \tilde{G}(\tilde{x}(t))u(t), \qquad \tilde{x}(0) = \tilde{x}_0, \qquad t \geq 0, \tag{9.93}$$

where

$$\tilde{x} = \begin{bmatrix} x \\ \hat{x} \end{bmatrix}, \quad \tilde{f}(\tilde{x}) = \begin{bmatrix} f(x) + G(x)y \\ A\hat{x} \end{bmatrix}, \quad \tilde{G}(\tilde{x}) = \begin{bmatrix} 0 \\ B \end{bmatrix}.$$

Specifically, conditions (8.59)–(8.62) are trivially satisfied by (9.86), (9.87) and (9.90). Next, using (9.88) and (9.93), $\phi(x,\hat{x})$ can be written as

$$\phi(\tilde{x}) = -\tfrac{1}{2}R_2^{-1}(\tilde{x})[L_2^{\mathrm{T}}(\tilde{x}) + \tilde{G}^{\mathrm{T}}(\tilde{x})V'^{\mathrm{T}}(\tilde{x})] \tag{9.94}$$

so that (8.65) is satisfied. Finally, using (9.90) and (9.93) it follows that (9.91) or, equivalently,

$$L_1(\tilde{x}) = \phi^{\mathrm{T}}(\tilde{x})R_2(\tilde{x})\phi(\tilde{x}) - V'(\tilde{x})\tilde{f}(\tilde{x}) \tag{9.95}$$

satisfies (8.66). □

A particular choice of $L_2(x,\hat{x})$ satisfying condition (9.87) is given by

$$L_2(x,\hat{x}) = \left(V_{\mathrm{sub}}'(x)G(x) - 2\hat{x}^{\mathrm{T}}K^{\mathrm{T}}\right)R_2(x,\hat{x}). \tag{9.96}$$

In this case, the feedback control $\phi(x,\hat{x})$ is given by

$$\phi(x,\hat{x}) = K\hat{x} - R_2^{-1}(x,\hat{x})B^{\mathrm{T}}\hat{P}\hat{x} - \tfrac{1}{2}G^{\mathrm{T}}(x)V_{\mathrm{sub}}'^{\mathrm{T}}(x). \tag{9.97}$$

Alternatively, choosing $L_2(x,\hat{x})$ satisfying condition (9.87) by

$$L_2(x,\hat{x}) = V'_{\rm sub}(x)G(x)R_2(x,\hat{x}) - 2\hat{x}^{\rm T}(K^{\rm T}R_2(x,\hat{x}) + \hat{P}B), \tag{9.98}$$

the feedback control $\phi(x,\hat{x})$ given by (9.88) specializes to

$$\phi(x,\hat{x}) = K\hat{x} - \tfrac{1}{2}G^{\rm T}(x)V'^{\rm T}_{\rm sub}(x). \tag{9.99}$$

In the case where $m = 1$, (9.99) specializes to the control law obtained in Lemma 2.13 of [247].

Next, we consider a class of systems in which the subsystem (9.81) is assumed to be globally stabilizable at $x = 0$ through y instead of having a globally asymptotically stable equilibrium at $x = 0$ with $y = 0$.

Theorem 9.4. Consider the nonlinear cascade system (9.81)–(9.83) with performance functional

$$J(x_0,\hat{x}_0,u) = \int_0^\infty L(x(t),\hat{x}(t),u(t)){\rm d}t, \tag{9.100}$$

where $L(x,\hat{x},u)$ is given by

$$\begin{aligned} L(x,\hat{x},u) &= L_1(x,\hat{x}) + L_2(x,\hat{x})(CA\hat{x} + CBu) \\ &\quad +(CA\hat{x} + CBu)^{\rm T}R_2(x,\hat{x})(CA\hat{x} + CBu), \end{aligned} \tag{9.101}$$

where $u \in \mathbb{R}^m$, $L_1 : \mathbb{R}^n \times \mathbb{R}^q \to \mathbb{R}$, $L_2 : \mathbb{R}^n \times \mathbb{R}^q \to \mathbb{R}^{1\times m}$ satisfies $L_2(0,0) = 0$, and $R_2 : \mathbb{R}^n \times \mathbb{R}^q \to \mathbb{P}^m$. Assume that there exist continuously differentiable functions $\alpha : \mathbb{R}^n \to \mathbb{R}^m$ and $V_{\rm sub} : \mathbb{R}^n \to \mathbb{R}$ such that

$$\alpha(0) = 0, \tag{9.102}$$

$$V_{\rm sub}(0) = 0, \tag{9.103}$$

$$V_{\rm sub}(x) > 0, \quad x \in \mathbb{R}^n, \quad x \neq 0, \tag{9.104}$$

$$V'_{\rm sub}(x)[f(x) + G(x)\alpha(x)] < 0, \quad x \in \mathbb{R}^n, \quad x \neq 0. \tag{9.105}$$

Furthermore, let $L_2(x,\hat{x})$ satisfy

$$\begin{aligned} &(y-\alpha(x))^{\rm T}\Big\{G^{\rm T}(x)V'^{\rm T}_{\rm sub}(x) - 2\hat{P}\Big[\alpha'(x)(f(x) + G(x)y) \\ &+R_2^{-1}(x,\hat{x})[\hat{P}(y-\alpha(x)) + \tfrac{1}{2}L_2^{\rm T}(x,\hat{x})]\Big]\Big\} < 0,\ (x,y) \in \mathbb{R}^n \times \mathbb{R}^m,\ y \neq \alpha(x), \end{aligned} \tag{9.106}$$

where $\hat{P} \in \mathbb{R}^{m\times m}$ is an arbitrary positive-definite matrix, and assume that the linear subsystem (9.82) is minimum phase and has relative degree $\{1, 1, \ldots, 1\}$. Then the zero solution $(x(t),\hat{x}(t)) \equiv (0,0)$ of the cascade system (9.81)–(9.83) is globally asymptotically stable with the feedback control law

$$u = (CB)^{-1}(v - CA\hat{x}), \tag{9.107}$$

where

$$v = \phi(x, \hat{x}) = -R_2^{-1}(x, \hat{x})\hat{P}(y - \alpha(x)) - \tfrac{1}{2}R_2^{-1}(x, \hat{x})L_2^{\mathrm{T}}(x, \hat{x}). \quad (9.108)$$

Furthermore,

$$J(x_0, \hat{x}_0, (CB)^{-1}(\phi(x, \hat{x}) - CA\hat{x})) = V(x_0, C\hat{x}_0), \qquad (x_0, \hat{x}_0) \in \mathbb{R}^n \times \mathbb{R}^q, \quad (9.109)$$

where

$$V(x, y) = V_{\mathrm{sub}}(x) + [y - \alpha(x)]^{\mathrm{T}}\hat{P}[y - \alpha(x)], \quad (9.110)$$

and the performance functional (9.100), with

$$\begin{aligned} L_1(x, \hat{x}) &= \phi^{\mathrm{T}}(x, \hat{x})R_2(x, \hat{x})\phi(x, \hat{x}) - V'_{\mathrm{sub}}(x)(f(x) + G(x)y) \\ &\quad +2(y - \alpha(x))^{\mathrm{T}}\hat{P}\alpha'(x)(f(x) + G(x)y), \end{aligned} \quad (9.111)$$

is minimized in the sense that

$$J(x_0, \hat{x}_0, (CB)^{-1}(\phi(x, \hat{x}) - CA\hat{x})) = \min_{u(\cdot)\in\mathcal{S}(x_0,\hat{x}_0)} J(x_0, \hat{x}_0, u(\cdot)), \qquad (x_0, \hat{x}_0) \in \mathbb{R}^n \times \mathbb{R}^q. \quad (9.112)$$

Proof. Since the linear input subsystem (9.82) and (9.83) has relative degree $\{1, 1, \ldots, 1\}$, it follows from Lemma 6.2 that there exists a nonsingular transformation matrix $T \in \mathbb{R}^{q\times q}$ such that, in the coordinates

$$\begin{bmatrix} y \\ z \end{bmatrix} \triangleq T\hat{x}, \quad (9.113)$$

the linear differential equation (9.82) is equivalent to the normal form

$$\dot{y}(t) = CA\hat{x}(t) + CBu(t), \qquad y(0) = C\hat{x}_0, \qquad t \geq 0, \quad (9.114)$$

$$\dot{z}(t) = A_0 z(t) + B_0 y(t), \qquad z(0) = z_0, \quad (9.115)$$

where $A_0 \in \mathbb{R}^{(q-m)\times(q-m)}$ and $B_0 \in \mathbb{R}^{(q-m)\times m}$. Furthermore, the eigenvalues of A_0 are the transmission zeros of the transfer function matrix $H(s) = C(sI - A)^{-1}B$ of the linear input subsystem (9.82) and (9.83) (see Problem 6.39). Next, applying the feedback transformation (9.107) to (9.114) yields the transformed cascade system

$$\dot{x}(t) = f(x(t)) + G(x(t))y(t), \qquad x(0) = x_0, \qquad t \geq 0, \quad (9.116)$$

$$\dot{y}(t) = v(t), \qquad y(0) = C\hat{x}_0, \quad (9.117)$$

$$\dot{z}(t) = A_0 z(t) + B_0 y(t), \qquad z(0) = z_0. \quad (9.118)$$

Initially ignoring the zero dynamics (9.118), it follows from Theorem 9.1, with $\hat{x}$ replaced by y and u replaced by v, that the equilibrium $x = 0$, $y = 0$ of (9.116) and (9.117) with the feedback control law $v = \phi(x, \hat{x})$, where $\phi(x, \hat{x})$ is given by (9.108), is globally asymptotically stable. Now, since the linear subsystem (9.82) is minimum phase, A_0 is Hurwitz. Since the zero solution $y(t) \equiv 0$ to (9.117) can be made asymptotically stable, $y(t)$ is bounded and $\lim_{t\to\infty} y(t) = 0$. Hence, the zero solution $z(t) \equiv 0$ to (9.118) is globally asymptotically stable. Furthermore, since $\hat{x} = T^{-1}[y^{\mathrm{T}}\ z^{\mathrm{T}}]^{\mathrm{T}}$, global asymptotic stability of the zero solution $(y(t), z(t)) \equiv (0,0)$ to (9.117) and (9.118) implies global asymptotic stability of the zero solution $\hat{x}(t) \equiv 0$ to (9.82).

Next, with (9.116), (9.117), (9.101), and (9.110), the Hamiltonian has the form

$$\begin{aligned} H(x,\hat{x},u) &= L_1(x,\hat{x}) + L_2(x,\hat{x})(CA\hat{x}+CBu) \\ &\quad +(CA\hat{x}+CBu)^{\mathrm{T}}R_2(x,\hat{x})(CA\hat{x}+CBu) \\ &\quad +V'_{\mathrm{sub}}(x)(f(x)+G(x)y) \\ &\quad +2(y-\alpha(x))^{\mathrm{T}}\hat{P}[CA\hat{x}+CBu-\alpha'(x)(f(x)+G(x)y)]. \end{aligned} \tag{9.119}$$

Now, the feedback control law (9.107) is obtained by setting $\frac{\partial H}{\partial u} = 0$. Using $L_2(0,0) = 0$ it follows that $\phi(0,0) = 0$, which proves (8.43). Next, with $L_1(x,\hat{x})$ given by (9.111) it follows that (8.46) holds. Finally, since

$$H(x,\hat{x},u) = [v-\phi(x,\hat{x})]^{\mathrm{T}}R_2(x,\hat{x})[v-\phi(x,\hat{x})],$$

and $R_2(x,\hat{x}) > 0$, $(x,\hat{x}) \in \mathbb{R}^n \times \mathbb{R}^q$, (8.45) holds. The result now follows as a direct consequence of Theorem 8.3. □

A particular choice of $L_2(x,\ddot{x})$ satisfying condition (9.106) is given by

$$L_2(x,\hat{x}) = \left[\hat{P}^{-1}G^{\mathrm{T}}(x)V'^{\mathrm{T}}_{\mathrm{sub}}(x) - 2\alpha'(x)(f(x)+G(x)y)\right]^{\mathrm{T}} R_2(x,\hat{x}). \tag{9.120}$$

In this case, for $u(t) \in \mathbb{R}$, $t \geq 0$, the control law given by (9.107) specializes to the linear block backstepping controller obtained in Lemma 2.23 of [247] by setting $\hat{P} = \frac{1}{2}I_m$ and $R_2(x,\hat{x}) = 1/c$.

9.5 Optimal Nonlinear Block Backstepping Controllers

In this section, we generalize the results of Section 9.4 to the case where the input subsystem (9.82) and (9.83) is nonlinear. Specifically, consider the nonlinear cascade system

$$\dot{x}(t) = f(x(t)) + G(x(t))y(t), \quad x(0) = x_0, \quad t \geq 0, \tag{9.121}$$

$$\dot{\hat{x}}(t) = \hat{f}(\hat{x}(t)) + \hat{G}(\hat{x}(t))u(t), \quad \hat{x}(0) = \hat{x}_0, \tag{9.122}$$

$$y(t) = h(\hat{x}(t)), \tag{9.123}$$

where $\hat{x} \in \mathbb{R}^q$, $u, y \in \mathbb{R}^m$, $\hat{f} : \mathbb{R}^q \to \mathbb{R}^q$ satisfies $\hat{f}(0,0) = 0$, $\hat{G} : \mathbb{R}^q \to \mathbb{R}^{q\times m}$, and $h : \mathbb{R}^q \to \mathbb{R}^m$ satisfies $h(0) = 0$.

First, we consider the case in which the input subsystem (9.122) and (9.123) is feedback strictly passive. Specifically, we assume there exists a positive-definite storage function $V_{\rm s}:\mathbb{R}^q\to\mathbb{R}$ and a function $k:\mathbb{R}^q\to\mathbb{R}^m$ such that

$$0 > V_{\rm s}'(\hat{x})(\hat{f}(\hat{x})+\hat{G}(\hat{x})k(\hat{x})), \qquad \hat{x}\in\mathbb{R}^q, \qquad \hat{x}\neq 0, \qquad (9.124)$$
$$0 = \tfrac{1}{2}\hat{G}^{\rm T}(\hat{x})V_{\rm s}'^{\rm T}(\hat{x}) - h(\hat{x}). \qquad (9.125)$$

Recall that it follows from Theorem 6.12 that the nonlinear input subsystem (9.122) and (9.123) is feedback strictly passive if and only if (9.122) and (9.123) has relative degree $\{1,1,\ldots,1\}$ at $x=0$ and is minimum phase.

Theorem 9.5. Consider the nonlinear cascade system (9.121)–(9.123) with performance functional (9.57) where $L(x,\hat{x},u)$ is given by (9.56). Assume that the input subsystem (9.122) and (9.123) is feedback strictly passive with positive-definite storage function $V_{\rm s}:\mathbb{R}^q\to\mathbb{R}$ satisfying (9.124) and (9.125) and the subsystem (9.121) has a globally stable equilibrium at $x=0$ with $y=0$ and Lyapunov function $V_{\rm sub}:\mathbb{R}^n\to\mathbb{R}$ so that

$$V_{\rm sub}'(x)f(x)<0, \qquad x\in\mathbb{R}^n, \qquad x\neq 0. \qquad (9.126)$$

Furthermore, let $L_2:\mathbb{R}^n\times\mathbb{R}^q\to\mathbb{R}^{1\times m}$ be such that $L_2(0,0)=0$ and

$$y^{\rm T}[G^{\rm T}(x)V_{\rm sub}'^{\rm T}(x) - R_2^{-1}(x,\hat{x})L_2^{\rm T}(x,\hat{x}) - 2k(\hat{x})] \leq 0, \quad (x,\hat{x})\in\mathbb{R}^n\times\mathbb{R}^q. \qquad (9.127)$$

Then the zero solution $(x(t),\hat{x}(t))\equiv(0,0)$ of the cascade system (9.121)–(9.123) is globally asymptotically stable with the feedback control law

$$u=\phi(x,\hat{x}) = -\tfrac{1}{2}R_2^{-1}(x,\hat{x})[\hat{G}^{\rm T}(\hat{x})V_{\rm s}'^{\rm T}(\hat{x}) + L_2^{\rm T}(x,\hat{x})]. \qquad (9.128)$$

Furthermore,

$$J(x_0,\hat{x}_0,\phi(x(\cdot),\hat{x}(\cdot))) = V(x_0,\hat{x}_0), \qquad (x_0,\hat{x}_0)\in\mathbb{R}^n\times\mathbb{R}^q, \qquad (9.129)$$

where

$$V(x,\hat{x}) = V_{\rm sub}(x) + V_{\rm s}(\hat{x}), \qquad (9.130)$$

and the performance functional (9.57), with

$$L_1(x,\hat{x}) = \phi^{\rm T}(x,\hat{x})R_2(x,\hat{x})\phi(x,\hat{x}) + V_{\rm s}'(\hat{x})\hat{G}(\hat{x})(k(\hat{x}) - \tfrac{1}{2}G^{\rm T}(x)V_{\rm sub}'^{\rm T}(x))$$
$$-V_{\rm s}'(\hat{x})(\hat{f}(\hat{x})+\hat{G}(\hat{x})k(\hat{x})) - V_{\rm sub}'(x)f(x), \qquad (9.131)$$

is minimized in the sense that

$$J(x_0,\hat{x}_0,\phi(x(\cdot),\hat{x}(\cdot))) = \min_{u(\cdot)\in\mathcal{S}(x_0,\hat{x}_0)} J(x_0,\hat{x}_0,u(\cdot)), \qquad (x_0,\hat{x}_0)\in\mathbb{R}^n\times\mathbb{R}^q. \qquad (9.132)$$

Proof. The result follows as a direct consequence of Theorem 8.3

applied to the system

$$\dot{\tilde{x}}(t) = \tilde{f}(\tilde{x}(t)) + \tilde{G}(\tilde{x}(t))u(t), \qquad \tilde{x}(0) = \tilde{x}_0, \qquad t \geq 0, \tag{9.133}$$

where

$$\tilde{x} = \begin{bmatrix} x \\ \hat{x} \end{bmatrix}, \quad \tilde{f}(\tilde{x}) = \begin{bmatrix} f(x) + G(x)y \\ \hat{f}(\hat{x}) \end{bmatrix}, \quad \tilde{G}(\tilde{x}) = \begin{bmatrix} 0 \\ \hat{G}(\hat{x}) \end{bmatrix}.$$

Specifically, conditions (8.59)–(8.62) are trivially satisfied by (9.126), (9.127), and (9.130). Next, using (9.128) and (9.133), $\phi(x, \hat{x})$ can be written as

$$\phi(\tilde{x}) = -\tfrac{1}{2}R_2^{-1}(\tilde{x})(L_2^{\mathrm{T}}(\tilde{x}) + \tilde{G}^{\mathrm{T}}(\tilde{x})V'^{\mathrm{T}}(\tilde{x})) \tag{9.134}$$

so that (8.65) is satisfied. Finally, using (9.130) and (9.133) it follows that (9.131) or, equivalently,

$$L_1(\tilde{x}) = \phi^{\mathrm{T}}(\tilde{x})R_2(\tilde{x})\phi(\tilde{x}) - V'(\tilde{x})\tilde{f}(\tilde{x}) \tag{9.135}$$

satisfies (8.66). □

A particular choice of $L_2(x, \hat{x})$ satisfying condition (9.127) is given by

$$L_2(x, \hat{x}) = \left(V'_{\mathrm{sub}}(x)G(x) - 2k^{\mathrm{T}}(\hat{x})\right) R_2(x, \hat{x}). \tag{9.136}$$

In this case, the feedback control $\phi(x, \hat{x})$ is given by

$$\phi(x, \hat{x}) = k(\hat{x}) - \tfrac{1}{2}\left(R_2^{-1}(x, \hat{x})\hat{G}^{\mathrm{T}}(\hat{x})V_{\mathrm{s}}'^{\mathrm{T}}(\hat{x}) + G^{\mathrm{T}}(x)V_{\mathrm{sub}}'^{\mathrm{T}}(x)\right). \tag{9.137}$$

Alternatively, choosing $L_2(x, \hat{x})$ satisfying condition (9.127) by

$$L_2(x, \hat{x}) = V'_{\mathrm{sub}}(x)G(x)R_2(x, \hat{x}) - 2k^{\mathrm{T}}(\hat{x})R_2(x, \hat{x}) + V'_{\mathrm{s}}(\hat{x})\hat{G}(\hat{x}), \tag{9.138}$$

the feedback control $\phi(x, \hat{x})$ given by (9.128) specializes to

$$\phi(x, \hat{x}) = k(\hat{x}) - \tfrac{1}{2}G^{\mathrm{T}}(x)V_{\mathrm{sub}}'^{\mathrm{T}}(x). \tag{9.139}$$

In the case where $m = 1$, (9.139) specializes to the control law obtained in Lemma 2.17 of [247].

Next, we consider a class of systems given by (9.121)–(9.123) in which the subsystem (9.121) is assumed to be globally stabilizable at $x = 0$ through y instead of having a globally asymptotically stable equilibrium at $x = 0$ with $y = 0$. Furthermore, we assume that the zero dynamics of a nonlinear system are asymptotically stable with respect to the signals by which they are driven. Thus, the following result is a nonlinear analog to Theorem 9.4. Here, we assume that the vector field $\hat{G}(\mathrm{L}_{\hat{G}}\hat{h})^{-1}$ is complete and involutive.

Theorem 9.6. Consider the nonlinear cascade system (9.121)–(9.123) with performance functional (9.100) where $L(x, \hat{x}, u)$ is given by

$$L(x, \hat{x}, u) \;=\; L_1(x, \hat{x}) + L_2(x, \hat{x})(h'(\hat{x})\hat{f}(\hat{x}) + h'(\hat{x})\hat{G}(\hat{x})u) + (h'(\hat{x})\hat{f}(\hat{x})$$

$$+h'(\hat{x})\hat{G}(\hat{x})u)^{\mathrm{T}}R_2(x,\hat{x})(h'(\hat{x})\hat{f}(\hat{x})+h'(\hat{x})\hat{G}(\hat{x})u), \quad (9.140)$$

where $u \in \mathbb{R}^m$, $L_1 : \mathbb{R}^n \times \mathbb{R}^q \to \mathbb{R}$, $L_2 : \mathbb{R}^n \times \mathbb{R}^q \to \mathbb{R}^{1\times m}$ satisfies $L_2(0,0) = 0$, and $R_2 : \mathbb{R}^n \times \mathbb{R}^q \to \mathbb{P}^m$. Assume that there exist continuously differentiable functions $\alpha : \mathbb{R}^n \to \mathbb{R}^m$ and $V_{\mathrm{sub}} : \mathbb{R}^n \to \mathbb{R}$ such that

$$\alpha(0) = 0, \quad (9.141)$$

$$V_{\mathrm{sub}}(0) = 0, \quad (9.142)$$

$$V_{\mathrm{sub}}(x) > 0, \quad x \in \mathbb{R}^n, \quad x \neq 0, \quad (9.143)$$

$$V'_{\mathrm{sub}}(x)[f(x)+G(x)\alpha(x)] < 0, \quad x \in \mathbb{R}^n, \quad x \neq 0. \quad (9.144)$$

Furthermore, let $L_2(x,\hat{x})$ satisfy

$$(y-\alpha(x))^{\mathrm{T}}\Big\{G^{\mathrm{T}}(x)V'^{\mathrm{T}}_{\mathrm{sub}}(x) - 2\hat{P}\Big[\alpha'(x)(f(x)+G(x)y)$$
$$+R_2^{-1}(x,\hat{x})[\hat{P}(y-\alpha(x)) + \tfrac{1}{2}L_2^{\mathrm{T}}(x,\hat{x})]\Big]\Big\} < 0, \ (x,y) \in \mathbb{R}^n \times \mathbb{R}^m, \ y \neq \alpha(x), \quad (9.145)$$

where $\hat{P} \in \mathbb{R}^{m\times m}$ is an arbitrary positive-definite matrix, and assume that (9.122) and (9.123) has constant relative degree $\{1,1,\ldots,1\}$ globally defined uniformly in x with input-to-state stable zero dynamics. Then the zero solution $(x(t),\hat{x}(t)) \equiv (0,0)$ of the cascade system (9.121)–(9.123) is globally asymptotically stable with the feedback control law

$$u = (h'(\hat{x})\hat{G}(\hat{x}))^{-1}[v - h'(\hat{x})\hat{f}(\hat{x})], \quad (9.146)$$

where

$$v = \phi(x,\hat{x}) = -R_2^{-1}(x,\hat{x})\hat{P}(y-\alpha(x)) - \tfrac{1}{2}R_2^{-1}(x,\hat{x})L_2^{\mathrm{T}}(x,\hat{x}). \quad (9.147)$$

Furthermore,

$$J(x_0,\hat{x}_0,(h'(\hat{x})\hat{G}(\hat{x}))^{-1}(\phi(x,\hat{x}) - h'(\hat{x})\hat{f}(\hat{x}))) = V(x_0,h(\hat{x}_0)),$$
$$(x_0,\hat{x}_0) \in \mathbb{R}^n \times \mathbb{R}^q, \quad (9.148)$$

where

$$V(x,y) = V_{\mathrm{sub}}(x) + [y-\alpha(x)]^{\mathrm{T}}\hat{P}[y-\alpha(x)], \quad (9.149)$$

and the performance functional (9.140), with

$$L_1(x,\hat{x}) = \phi^{\mathrm{T}}(x,\hat{x})R_2(x,\hat{x})\phi(x,\hat{x}) - V'_{\mathrm{sub}}(x)(f(x)+G(x)y)$$
$$+2(y-\alpha(x))^{\mathrm{T}}\hat{P}\alpha'(x)(f(x)+G(x)y), \quad (9.150)$$

is minimized in the sense that

$$J(x_0,\hat{x}_0,(h'(\hat{x})\hat{G}(\hat{x}))^{-1}(\phi(x,\hat{x}) - h'(\hat{x})\hat{f}(\hat{x})))$$
$$= \min_{u(\cdot)\in\mathcal{S}(x_0,\hat{x}_0)} J(x_0,\hat{x}_0,u(\cdot)), \quad (x_0,\hat{x}_0) \in \mathbb{R}^n \times \mathbb{R}^q. \quad (9.151)$$

Proof. Since the nonlinear subsystem (9.122) and (9.123) has relative degree $\{1, 1, \ldots, 1\}$, it follows from Lemma 6.2 that there exists a global diffeomorphism $\mathcal{T} : \mathbb{R}^q \to \mathbb{R}^q$ such that, in the coordinates

$$\begin{bmatrix} y \\ z \end{bmatrix} \triangleq \mathcal{T}(\hat{x}), \tag{9.152}$$

the nonlinear differential equation (9.122) is equivalent to the normal form

$$\dot{y}(t) = h'(\hat{x}(t))[\hat{f}(\hat{x}(t)) + \hat{G}(\hat{x}(t))u(t)], \quad y(0) = h(\hat{x}_0), \quad t \geq 0, \tag{9.153}$$
$$\dot{z}(t) = f_0(z(t)) + r(z(t), y(t))y(t), \quad z(0) = z_0, \tag{9.154}$$

where $f_0 : \mathbb{R}^{(q-m)} \to \mathbb{R}^{(q-m)}$ and $r : \mathbb{R}^{(q-m)} \times \mathbb{R}^m \to \mathbb{R}^{(q-m)\times m}$. Next, applying the linearizing feedback transformation (9.146) to (9.153) yields the transformed cascade system

$$\dot{x}(t) = f(x(t)) + G(x(t))y(t), \qquad x(0) = x_0, \qquad t \geq 0, \tag{9.155}$$
$$\dot{y}(t) = v(t), \qquad y(0) = h(\hat{x}_0), \tag{9.156}$$
$$\dot{z}(t) = f_0(z(t)) + r(z(t), y(t))y(t), \qquad z(0) = z_0. \tag{9.157}$$

Initially ignoring the zero dynamics (9.157), it follows from Theorem 9.1, with $\hat{x}$ replaced by y and u replaced by v, that the equilibrium $x = 0$, $y = 0$ with the feedback control law $v = \phi(x, y)$, where $\phi(x, y)$ is given by (9.147), is asymptotically stable. Now, since the zero dynamics subsystem (9.157) is input-to-state stable and the zero solution $y(t) \equiv 0$ to (9.156) can be made asymptotically stable, $y(t)$ is bounded and $\lim_{t\to\infty} y(t) = 0$. Hence, it follows from Proposition 4.2 that the zero solution $z(t) \equiv 0$ to (9.157) is globally asymptotically stable. Furthermore, since $\hat{x} = \mathcal{T}^{-1}(y, z)$, global asymptotic stability of the zero solution $(y(t), z(t)) \equiv (0, 0)$ to (9.156) and (9.157) implies global asymptotic stability of the zero solution $\hat{x}(t) \equiv 0$ to (9.122).

Next, with (9.155), (9.156), (9.140), and (9.149), the Hamiltonian has the form

$$\begin{aligned} H(x, \hat{x}, u) = {} & L_1(x, \hat{x}) + L_2(x, \hat{x})(h'(\hat{x})\hat{f}(\hat{x}) + h'(\hat{x})\hat{G}(\hat{x})u) + (h'(\hat{x})\hat{f}(\hat{x}) \\ & + h'(\hat{x})\hat{G}(\hat{x})u)^{\mathrm{T}} R_2(x, \hat{x})(h'(\hat{x})\hat{f}(\hat{x}) + h'(\hat{x})\hat{G}(\hat{x})u) \\ & + V'_{\mathrm{sub}}(x)(f(x) + G(x)y) + 2(y - \alpha(x))^{\mathrm{T}}\hat{P}[h'(\hat{x})\hat{f}(\hat{x}) \\ & + h'(\hat{x})\hat{G}(\hat{x})u - \alpha'(x)(f(x) + G(x)y)]. \end{aligned} \tag{9.158}$$

Now, the feedback control law (9.146) is obtained by setting $\frac{\partial H}{\partial u} = 0$. Using $L_2(0, 0) = 0$ it follows that $\phi(0, 0) = 0$, which proves (8.43). Next, with $L_1(x, \hat{x})$ given by (9.150) it follows that (8.46) holds. Finally, since

$$H(x, \hat{x}, u) = [v - \phi(x, \hat{x})]^{\mathrm{T}} R_2(x, \hat{x})[v - \phi(x, \hat{x})],$$

and $R_2(x,\hat{x}) > 0$, $(x,\hat{x}) \in \mathbb{R}^n \times \mathbb{R}^q$, (8.45) holds. The result now follows as a direct consequence of Theorem 8.2. □

A particular choice of $L_2(x,\hat{x})$ satisfying condition (9.145) is given by

$$L_2(x,\hat{x}) = \left[\hat{P}^{-1}G^{\mathrm{T}}(x)V_{\mathrm{sub}}'^{\mathrm{T}}(x) - 2\alpha'(x)(f(x)+G(x)y)\right]^{\mathrm{T}} R_2(x,\hat{x}). \quad (9.159)$$

In this case, for $u(t) \in \mathbb{R}$, $t \geq 0$, the control law given by (9.146) specializes to the nonlinear backstepping controller obtained in Lemma 2.25 of [247] by setting $\hat{P} = \frac{1}{2}I_m$ and $R_2(x,\hat{x}) = 1/c$.

Finally, we consider nonlinear cascade systems of the form

$$\dot{x}(t) = f(x(t)) + G(x(t))y(t), \qquad x(0) = x_0, \qquad t \geq 0, \quad (9.160)$$
$$\dot{y}(t) = h'(\hat{x}(t))\hat{f}(\hat{x}(t)) + h'(\hat{x}(t))\hat{G}(\hat{x}(t))u(t), \qquad \hat{x}(0) = \hat{x}_0, \quad (9.161)$$
$$\dot{z}(t) = f_0(z(t)) + r(y(t),z(t))y(t), \qquad z(0) = z_0, \quad (9.162)$$

where the zero solution $z(t) \equiv 0$ to (9.162) is asymptotically stable, $[y^{\mathrm{T}},\ z^{\mathrm{T}}] \triangleq \mathcal{T}(\hat{x})$, and $\mathcal{T}(\cdot)$ is given by (9.152). The system (9.160)–(9.162) can be viewed as the normal form equivalent of (9.121)–(9.123), where the input subsystem (9.122) and (9.123) is minimum phase with relative degree $\{1,1,\ldots,1\}$, obtained by applying the global diffeomorphism (9.152). In this case, the stability assumption on (9.162) is implied by the minimum phase assumption on (9.122) and (9.123). For the cascade nonlinear system (9.160)–(9.162), we consider the performance functional

$$J(x_0, h(\hat{x}_0), z_0, u) = \int_0^\infty L(x(t), y(t), z(t), u(t))\mathrm{d}t, \quad (9.163)$$

where

$$L(x,y,z,u) = L_1(x,y,z) + L_2(x,y,z)u + u^{\mathrm{T}}R_2(x,y,z)u. \quad (9.164)$$

Furthermore, define

$$V_{\mathrm{c}}(y,z) \triangleq \int_0^\infty V_0'(\bar{z}(s))r(\bar{y}(s),\bar{z}(s))\bar{y}(s)\mathrm{d}s, \quad (9.165)$$

where $\bar{y}(\tau)$ and $\bar{z}(\tau)$, $\tau \geq 0$, are solutions of (9.160) and (9.161), respectively, with initial conditions $\bar{y}(0) = y$ and $\bar{z}(0) = z$, and $V_0(z)$ is a Lyapunov function for the asymptotically stable (by assumption) subsystem (9.162). Note that the time derivative of $V_{\mathrm{c}}(y,z)$ along the trajectories of y and z is given by $\dot{V}_{\mathrm{c}}(y,z) = -V_0'(z)r(y,z)y$.

Theorem 9.7. Consider the nonlinear cascade system (9.160)–(9.162) with performance functional (9.163), where $L(x,y,z,u)$ is given by (9.164) and where $u \in \mathbb{R}^m$, $L_1(x,y,z) : \mathbb{R}^n \times \mathbb{R}^m \times \mathbb{R}^{q-m} \to \mathbb{R}$, $L_2(x,y,z) : \mathbb{R}^n \times \mathbb{R}^m \times \mathbb{R}^{q-m} \to \mathbb{R}^{1\times m}$ satisfies $L_2(0,0,0) = 0$, and $R_2 : \mathbb{R}^n \times \mathbb{R}^m \times \mathbb{R}^{q-m} \to \mathbb{P}^m$.

Assume that there exist continuously differentiable functions $\alpha : \mathbb{R}^n \to \mathbb{R}^m$ and $V_{\rm sub} : \mathbb{R}^n \to \mathbb{R}$ such that

$$\begin{aligned} \alpha(0) &= 0, && && (9.166)\\ V_{\rm sub}(0) &= 0, && && (9.167)\\ V_{\rm sub}(x) &> 0, && x \in \mathbb{R}^n, \quad x \neq 0, && (9.168)\\ V'_{\rm sub}(x)[f(x) + G(x)\alpha(x)] &< 0, && x \in \mathbb{R}^n, \quad x \neq 0. && (9.169) \end{aligned}$$

Furthermore, let $L_2(x, y, z)$ satisfy

$$(y - \alpha(x))^{\rm T}\Big\{G^{\rm T}(x)V'^{\rm T}_{\rm sub}(x) + 2\hat{P}\Big[h'(\hat{x})\hat{f}(\hat{x}) - \alpha'(x)(f(x) + G(x)y)$$
$$-h'(\hat{x})\hat{G}(\hat{x})R_2^{-1}(x,y,z)((h'(\hat{x})\hat{G}(\hat{x}))^{\rm T}\hat{P}(y - \alpha(x)) + \tfrac{1}{2}L_2^{\rm T}(x,y,z))\Big]\Big\} < 0,$$
$$(x, y, z) \in \mathbb{R}^n \times \mathbb{R}^m \times \mathbb{R}^{q-m}, \qquad y \neq \alpha(x), \qquad (9.170)$$

where $\hat{P} \in \mathbb{R}^{m\times m}$ is an arbitrary positive-definite matrix, assume that the zero solution $z(t) \equiv 0$ to (9.162) with $y = 0$ is globally asymptotically stable with Lyapunov function $V_0(z)$, assume there exist continuously differentiable functions γ_1, $\gamma_2 \in \mathcal{K}_\infty$ such that

$$\|r(y, z)\| \leq \gamma_1(\|y\|) + \gamma_2(\|y\|)\|z\|, \qquad (9.171)$$

and assume there exist k, $c > 0$ such that $\|z\| > c$ implies

$$\left\|\frac{\partial V_0(z)}{\partial z}\right\| \|z\| \leq kV_0(z). \qquad (9.172)$$

Then the zero solution $(x(t), y(t), z(t)) \equiv (0, 0, 0)$ of the cascade system (9.160)–(9.162) is globally asymptotically stable with the feedback control law $u = \phi(x, y, z)$, where

$$\begin{aligned} \phi(x,y,z) &= -R_2^{-1}(x,y,z)(h'(\hat{x})\hat{G}(\hat{x}))^{\rm T}\hat{P}(y - \alpha(x))\\ &\quad -\tfrac{1}{2}R_2^{-1}(x,y,z)L_2^{\rm T}(x,y,z). \qquad (9.173) \end{aligned}$$

Furthermore,

$$J(x_0, h(\hat{x}_0), z_0, \phi(\cdot,\cdot,\cdot)) = V(x_0, h(\hat{x}_0), z_0), \ (x_0, \hat{x}_0, z_0) \in \mathbb{R}^n \times \mathbb{R}^q \times \mathbb{R}^{q-m}, \qquad (9.174)$$

where

$$V(x,y,z) = V_{\rm sub}(x) + (y - \alpha(x))^{\rm T}\hat{P}(y - \alpha(x)) + V_{\rm c}(y,z) + V_0(z), \quad (9.175)$$

and the performance functional (9.163), with

$$\begin{aligned} L_1(x,y,z) &= \phi^{\rm T}(x,y,z)R_2(x,y,z)\phi(x,y,z) - V'_{\rm sub}(x)(f(x) + G(x)y)\\ &\quad -V'_0(z)f_0(z) + (y - \alpha'(x))^{\rm T}[\alpha'(x)(f(x) + G(x)y) - h'(\hat{x})\hat{f}(\hat{x})], \qquad (9.176) \end{aligned}$$

is minimized in the sense that

$$J(x_0, h(\hat{x}_0), z_0, \phi(x,y,z)) = \min_{u(\cdot)\in\mathcal{S}(x_0,\hat{x}_0,z_0)} J(x_0, h(\hat{x}_0), z_0, u(\cdot)),$$
$$(x_0, \hat{x}_0, z_0) \in \mathbb{R}^n \times \mathbb{R}^q \times \mathbb{R}^{q-m}. \quad (9.177)$$

Proof. Since the subsystem (9.162) has an asymptotically stable equilibrium at $z(t) = 0$, $t \geq 0$, it follows from Theorem 3.9 that there exists a continuously differentiable positive-definite function $V_0(z)$ such that

$$V_0'(z)f_0(z) < 0, \qquad z \in \mathbb{R}^{q-m}, \qquad z \neq 0. \quad (9.178)$$

Next, since $r(y,z)$ and $V_0(z)$ satisfy (9.171) and (9.172), $V_{\mathrm{sub}}(x)$ satisfies (9.167) and (9.168), and, with $\hat{P} > 0$, it follows that the Lyapunov function candidate (9.175) is positive definite (see Problem 9.7). The result now follows as a direct consequence of Theorem 8.3 applied to the system

$$\dot{\tilde{x}}(t) = \tilde{f}(\tilde{x}(t)) + \tilde{G}(\tilde{x}(t))u(t), \quad \tilde{x}(0) = \tilde{x}_0, \quad t \geq 0, \quad (9.179)$$

where

$$\tilde{x} = \begin{bmatrix} x \\ y \\ z \end{bmatrix}, \quad \tilde{f}(\tilde{x}) = \begin{bmatrix} f(x) + G(x)y \\ h'(\hat{x})\hat{f}(\hat{x}) \\ f_0(z) + r(y,z)y \end{bmatrix}, \quad \tilde{G}(\tilde{x}) = \begin{bmatrix} 0 \\ h'(\hat{x})\hat{G}(\hat{x}) \\ 0 \end{bmatrix}.$$

Specifically, conditions (8.59)–(8.62) are satisfied by $L_2(0,0,0) = 0$, (9.166)–(9.170) and (9.175). Next, using (9.173) and (9.179), $\phi(x,y,z)$ can be written as

$$\phi(\tilde{x}) = -\tfrac{1}{2}R_2^{-1}(\tilde{x})[L_2^{\mathrm{T}}(\tilde{x}) + \tilde{G}^{\mathrm{T}}(\tilde{x})V'^{\mathrm{T}}(\tilde{x})] \quad (9.180)$$

so that (8.65) is satisfied. Finally, using (9.175) and (9.179) it follows that (9.176) or, equivalently,

$$L_1(\tilde{x}) = \phi^{\mathrm{T}}(\tilde{x})R_2(\tilde{x})\phi(\tilde{x}) - V'(\tilde{x})\tilde{f}(\tilde{x}) \quad (9.181)$$

satisfies (8.66). □

Assumption (9.171) requires that $r(y,z)$ of (9.162) has, at most, a linear growth in z. While this unavoidably narrows the class of admissible nonlinear systems which can be considered, it is not restrictive when the input subsystem (9.161) and (9.162) is linear so that the zero dynamics subsystem has the form (9.115). In this case, $r(y,z) = B_0$, where B_0 is a constant matrix, and hence, (9.171) is automatically satisfied. More generally, assumption (9.172) is satisfied by Lyapunov functions involving positive definite, radially unbounded polynomial functions of z (see Problem 9.8). Finally, a choice of $L_2(x,y,z)$ satisfying condition (9.170) is given by

$$L_2(x,y,z) = 2[h'(\hat{x})\hat{f}(\hat{x}) + \tfrac{1}{2}\hat{P}^{-1}G^{\mathrm{T}}(x)V_{\mathrm{sub}}'^{\mathrm{T}}(x)$$
$$-\alpha'(x)(f(x) + G(x)y)]^{\mathrm{T}}(h'(\hat{x})\hat{G}(\hat{x}))^{-\mathrm{T}}R_2(x,y,z). \quad (9.182)$$

9.6 Rotating Stall and Surge Control for Axial Compression Systems

The desire to develop an integrated control system design methodology for advanced propulsion systems has led to significant activity in modeling and control of flow compression systems in recent years (see, for example, [2, 19, 35, 103, 140, 141, 143, 248, 249, 277, 318, 319, 345, 451] and the numerous references therein). Two predominant aerodynamic instabilities in compression systems are rotating stall and surge. Rotating stall is an inherently three-dimensional[1] local compression system oscillation which is characterized by regions of flow that rotate at a fraction of the compressor rotor speed while surge is a one-dimensional axisymmetric global compression system oscillation, which involves axial flow oscillations and in some cases even axial flow reversal, which can damage engine components and cause flameout to occur.

Rotating stall and surge arise due to perturbations in stable system operating conditions involving steady, axisymmetric flow and can severely limit compressor performance. The transition from stable compressor operating conditions to rotating stall and surge is shown in Figure 9.2 representing a schematic of a compressor characteristic map where the

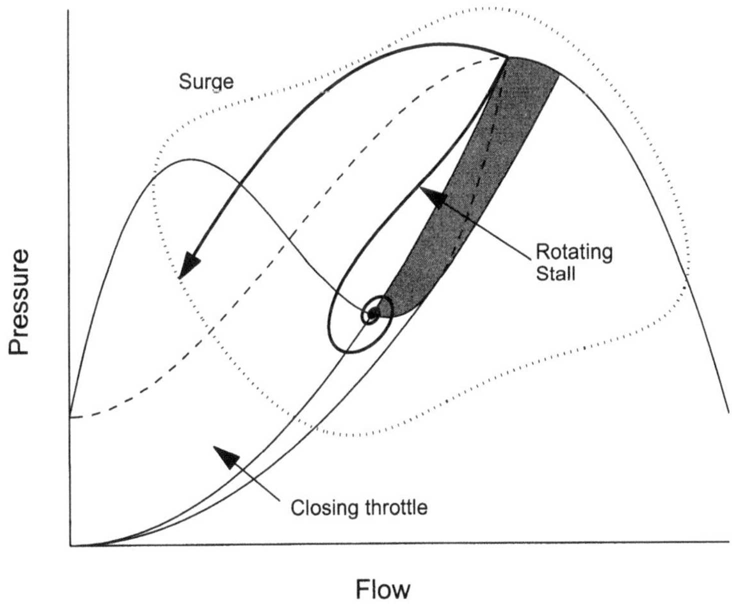

Figure 9.2 Schematic of compressor characteristic map for a typical compression system (——stable equilibria, – – – unstable equilibria).

[1]When analyzing high hub-to-tip ratio compressors, rotating stall can be approximated as a two-dimensional compression system oscillation.

abscissa corresponds to the circumferentially averaged mass flow through the compressor and the ordinate corresponds to the normalized total-to-static pressure rise in the compressor. For maximum compressor performance, operating conditions require that the pressure rise in the compressor correspond to the maximum pressure operating point on the stable axisymmetric branch for a given throttle opening. Here, we distinguish between compressor performance (pressure rise) and compressor efficiency (specific power consumption) where, depending on how the compressor is designed, the most *efficient* operating point may be to the right of the peak of the compressor characteristic map. In practice, however, compression system uncertainties and compression system disturbances can perturb the operating point into an unstable region driving the system to a stalled stable equilibrium, a stable limit cycle (surge), or both. In the case of rotating stall, an attempt to recover to a high pressure operating point by increasing the flow through the throttle traps the system within a flow range corresponding to two stable operating conditions involving steady axisymmetric flow and rotating stall resulting in severe hysteresis.

To avoid rotating stall and surge, traditionally system designers allow for a safety margin (rotating stall or surge margin) in compression system operation. However, to account for compression system uncertainties such as system modeling errors, in-service changes due to aging, etc., and compression system disturbances such as compressor speed fluctuations, combustion noise, etc., operating at or below the rotating stall/surge margin significantly reduces the efficiency of the compression system. In contrast, active control can enhance stable compression system operation to achieve peak compressor performance. However, compression system uncertainty and compression system disturbances are often significant and the need for robust disturbance rejection control is severe.

In order to develop control system design methodologies for compression systems, reliable models capturing the intricate physical phenomena of rotating stall and surge are necessary. A fundamental development in compression system modeling for low speed axial compressors is the Moore-Greitzer model given in [319]. Specifically, utilizing a one-mode expansion of the disturbance velocity potential in the compression system and assuming a nonlinear (cubic) characteristic for the compressor performance map, the authors in [319] develop a low-order, three-state nonlinear model involving the mean flow in the compressor, the pressure rise, and the amplitude of rotating stall. Starting from infinitesimal perturbations in the flow field the model captures the development of rotating stall and surge. In particular, the model predicts the experimentally verified subcritical pitchfork bifurcation at the onset of rotating stall [311]. Extensions to the Moore-Greitzer model that include blade row time lags and viscous transport terms have been reported in [187] and [1, 2, 451], respectively.

In this section, we apply the inverse optimal backstepping control framework to the control of rotating stall and surge in jet engine compression systems. To capture poststall transients in axial flow compression systems we use the one-mode Galerkin approximation model for the nonlinear partial differential equation characterizing the disturbance velocity potential at the compressor inlet proposed by Moore and Greitzer [319]. Specifically, we consider the basic compression system shown in Figure 9.3 consisting of an inlet duct, a compressor, an outlet duct, a plenum, and a control throttle. We assume that the plenum dimensions are large as compared to the compressor-duct dimensions so that the fluid velocity and acceleration in the plenum are negligible. In this case, the pressure in the plenum is spatially uniform. Furthermore, we assume that the flow is controlled by a throttle at the plenum exit. Finally, we assume a low-speed compression system with oscillation frequencies much lower than the acoustic resonance frequencies so that the flow can be considered incompressible. However, we do assume that the gas in the plenum is compressible, and therefore acts as a gas spring.

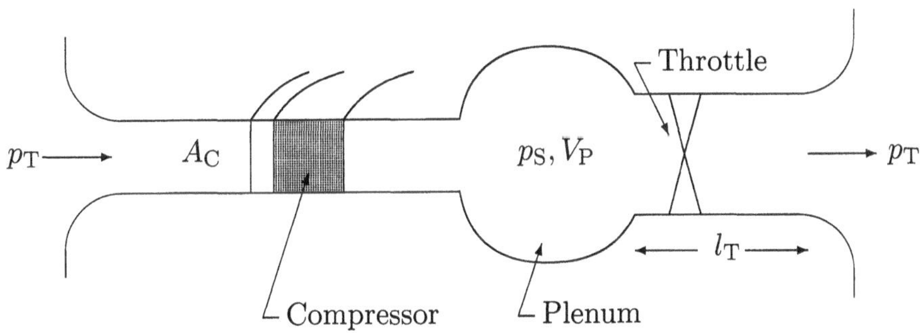

Figure 9.3 Compressor system geometry.

Invoking a momentum balance across the compression system, conservation of mass in the plenum, and using a Galerkin projection based on a one-mode circumferential spatial harmonic approximation for the non-axisymmetric flow disturbances yields [319]

$$\dot{A}(t) = \frac{\sigma}{3\pi}\int_0^{2\pi} \Psi_C(\Phi(t) + A(t)\sin\theta)\sin\theta\,d\theta,\ A(0) = A_0,\ t \geq 0, \quad (9.183)$$

$$\dot{\Phi}(t) = -\Psi(t) + \frac{1}{2\pi}\int_0^{2\pi} \Psi_C(\Phi(t) + A(t)\sin\theta)\,d\theta,\ \Phi(0) = \Phi_0, \quad (9.184)$$

$$\dot{\Psi}(t) = \frac{1}{\beta^2}(\Phi(t) - \Phi_T(t)),\ \Psi(0) = \Psi_0, \quad (9.185)$$

where Φ is the circumferentially averaged axial mass flow in the compressor, Ψ is the total-to-static pressure rise, A is the normalized stall cell amplitude of angular variation capturing a measure of nonuniformity in the flow, $\Phi_{\rm T}$ is the mass flow through the throttle, σ, β are positive constant parameters, and $\Psi_{\rm C}(\cdot)$ is a given compressor pressure-flow map. The compliance coefficient β is a function of the compressor rotor speed and plenum size. For large values of β a surge limit cycle can occur, while rotating stall can occur for any value of β.

Now, assuming that the compressor pressure-flow map $\Psi_{\rm C}(\cdot)$ is analytic, the integral terms in (9.183)–(9.185) can be expressed in terms of an infinite Taylor series expansion about the circumferentially averaged flow to give

$$\dot{A}(t) = \frac{2\sigma}{3}\sum_{k=1}^{\infty}\frac{1}{k!(k-1)!}\left.\frac{\mathrm{d}^{2k-1}\Psi_{\rm C}(\xi)}{\mathrm{d}\xi^{2k-1}}\right|_{\xi=\Phi(t)}\left(\frac{A(t)}{2}\right)^{2k-1}, \quad A(0) = A_0,\ t \geq 0, \tag{9.186}$$

$$\dot{\Phi}(t) = -\Psi(t) + \sum_{k=0}^{\infty}\frac{1}{(k!)^2}\left.\frac{\mathrm{d}^{2k}\Psi_{\rm C}(\xi)}{\mathrm{d}\xi^{2k}}\right|_{\xi=\Phi(t)}\left(\frac{A(t)}{2}\right)^{2k}, \quad \Phi(0) = \Phi_0, \tag{9.187}$$

$$\dot{\Psi}(t) = \frac{1}{\beta^2}(\Phi(t) - \Phi_{\rm T}(t)), \quad \Psi(0) = \Psi_0. \tag{9.188}$$

The specific compressor pressure-flow performance map $\Psi_{\rm C}$ which was considered in [319] is

$$\Psi_{\rm C}(\Phi) = \Psi_{\rm C0} + 1 + \tfrac{3}{2}\Phi - \tfrac{1}{2}\Phi^3, \tag{9.189}$$

where $\Psi_{\rm C0}$ is a constant parameter. In this case, (9.186)–(9.188) become

$$\dot{A}(t) = \tfrac{\sigma}{2}A(t)(1 - \Phi^2(t) - \tfrac{1}{4}A^2(t)), \quad A(0) = A_0, \quad t \geq 0, \tag{9.190}$$

$$\dot{\Phi}(t) = -\Psi(t) + \Psi_{\rm C}(\Phi(t)) - \tfrac{3}{4}\Phi(t)A^2(t), \quad \Phi(0) = \Phi_0, \tag{9.191}$$

$$\dot{\Psi}(t) = \tfrac{1}{\beta^2}(\Phi(t) - \Phi_{\rm T}(t)), \quad \Psi(0) = \Psi_0. \tag{9.192}$$

Next, define the control variable

$$u \triangleq \tfrac{1}{\beta^2}(\Phi_{\rm T} - \Phi) \tag{9.193}$$

so that for fixed values of flow through the throttle, $\Phi_{\rm T}(t) \equiv \Phi_{\rm Teq}$, (9.190)–(9.192) have an equilibrium point given by

$$(A_{\rm eq}, \Phi_{\rm eq}, \Psi_{\rm eq}) = (0, \Phi_{\rm Teq}, \Psi_{\rm C}(\Phi_{\rm eq})). \tag{9.194}$$

Defining the shifted state variables $x_1 \triangleq A$, $x_2 \triangleq \Phi - \Phi_{\rm eq}$, and $x_3 \triangleq \Psi - \Psi_{\rm eq}$, so that for a given equilibrium point on the axisymmetric branch of the compressor characteristic pressure-flow map the system equilibrium is translated to the origin, it follows that the translated nonlinear system is given by

$$\dot{x}_1(t) = \tfrac{\sigma}{2}x_1(t)(1-(x_2(t)+\lambda)^2-\tfrac{1}{4}x_1^2(t)), \quad x_1(0)=x_{10}, \quad t\geq 0, \tag{9.195}$$
$$\dot{x}_2(t) = -\tfrac{1}{2}x_2^3-\tfrac{3}{2}\lambda x_2^2-\tfrac{3}{2}(\lambda^2-1+\tfrac{1}{2}x_1^2)x_2-\tfrac{3}{4}\lambda x_1^2-x_3, \quad x_2(0)=x_{20}, \tag{9.196}$$
$$\dot{x}_3(t) = -u(t), \qquad x_3(0)=x_{30}, \tag{9.197}$$

where $\lambda \triangleq \Phi_{\mathrm{Teq}}$ and $u=\frac{1}{\beta^2}(\Phi_{\mathrm{T}}-\lambda-x_2)$.

Our objective is to stabilize the equilibrium ($A(t)=0$, $\Phi(t)=1$, $\Psi(t)=\Psi_{\mathrm{CO}}+2$) by controlling the throttle mass flow Φ_{T} which is related to the throttle opening γ_{throt} by $\Phi_{\mathrm{T}}=\gamma_{\mathrm{throt}}\sqrt{\Psi}$ [247]. To translate the desired equilibrium to the origin we apply the linear transformation $\Phi_{\mathrm{s}}\triangleq\Phi-1$ and $\Psi_{\mathrm{s}}\triangleq\Psi-\Psi_{\mathrm{CO}}-2$. Furthermore, in this case,

$$u=\tfrac{1}{\beta^2}(\Phi_{\mathrm{T}}-1-\Phi_{\mathrm{s}}), \tag{9.198}$$

and hence, with $\lambda=1$, (9.195)–(9.197) yield the transformed nonlinear system

$$\dot{A}(t) = -\tfrac{\sigma}{2}A(t)[\tfrac{1}{4}A^2(t)+2\Phi_{\mathrm{s}}(t)+\Phi_{\mathrm{s}}^2(t)], \quad A(0)=A_0, \quad t\geq 0, \tag{9.199}$$
$$\dot{\Phi}_{\mathrm{s}}(t) = -\tfrac{3}{2}\Phi_{\mathrm{s}}^2(t)-\tfrac{1}{2}\Phi_{\mathrm{s}}^3(t)-\tfrac{3}{4}A^2(t)[1+\Phi_{\mathrm{s}}(t)]-\Psi_{\mathrm{s}}(t), \quad \Phi_{\mathrm{s}}(0)=\Phi_{\mathrm{s}_0}, \tag{9.200}$$
$$\dot{\Psi}_{\mathrm{s}}(t) = -u(t), \quad \Psi_{\mathrm{s}}(0)=\Psi_{\mathrm{s}_0}. \tag{9.201}$$

Note that (9.199)–(9.201) has the correct form for the application of Theorem 9.1 where (9.199) and (9.200) make up the nonlinear subsystem and Ψ_{s} is the integrator state. Specifically, (9.199)–(9.201) can be written in the form of (9.1) and (9.2) where $x=[A\ \ \Phi_{\mathrm{s}}]^{\mathrm{T}}$, $\hat{x}=\Psi_{\mathrm{s}}$, and

$$f(A,\Phi_{\mathrm{s}})=\begin{bmatrix} -\frac{\sigma}{2}A(\frac{1}{4}A^2+2\Phi_{\mathrm{s}}+\Phi_{\mathrm{s}}^2) \\ -\Phi_{\mathrm{s}}(\frac{3}{2}\Phi_{\mathrm{s}}+\frac{1}{2}\Phi_{\mathrm{s}}^2+\frac{3}{4}A^2)-\frac{3}{4}A^2 \end{bmatrix}, \qquad G(A,\Phi_{\mathrm{s}})=\begin{bmatrix} 0 \\ -1 \end{bmatrix}.$$

To apply Theorem 9.1 we require a stabilizing feedback for the subsystem (9.199) and (9.200) and a corresponding Lyapunov function $V_{\mathrm{sub}}(A,\Phi_{\mathrm{s}})$ such that (9.59) and (9.60) are satisfied. For the nonlinear subsystem (9.199) and (9.200) we choose the Lyapunov function candidate

$$V_{\mathrm{sub}}(A,\Phi_{\mathrm{s}})=\varepsilon A^4+\Phi_{\mathrm{s}}^2, \tag{9.202}$$

where $\varepsilon>0$, and the stabilizing feedback control

$$\alpha(A,\Phi_{\mathrm{s}})=c_1\Phi_{\mathrm{s}}-\tfrac{3}{2}\Phi_{\mathrm{s}}^2-\tfrac{3}{4}A^2-2\varepsilon\sigma A^4, \tag{9.203}$$

where $c_1\geq 0$. It is straightforward to show that (9.202) and (9.203) satisfy conditions (9.59)–(9.61) of Theorem 9.1.

Applying Theorem 9.1 to the system (9.199)–(9.201) yields the family of control laws

$$u = -\phi(A, \Phi_s, \Psi_s) = R_2^{-1}\left[\hat{P}\left(\Psi_s - \alpha(A, \Phi_s)\right) + \tfrac{1}{2}L_2(A, \Phi_s, \Psi_s)\right], \quad (9.204)$$

with Lyapunov function

$$V(A, \Phi_s, \Psi_s) = \varepsilon A^4 + \Phi_s^2 + \hat{P}[\Psi_s - \alpha(A, \Phi_s)]^2, \quad (9.205)$$

where $R_2 > 0$ and $\hat{P} > 0$. Furthermore, the performance functional minimized by the control law (9.204) has the form

$$\begin{aligned} J(A_0, \Phi_{s_0}, \Psi_{s_0}, u) &= \int_0^\infty [L_1(A(t), \Phi_s(t), \Psi_s(t)) \\ &\quad + L_2(A(t), \Phi_s(t), \Psi_s(t))u(t) + R_2 u^2(t)]\mathrm{d}t, \end{aligned} \quad (9.206)$$

where

$$\begin{aligned} L_1(A, \Phi_s, \Psi_s) &\triangleq R_2\phi^2(A, \Phi_s, \Psi_s) - V'_{\mathrm{sub}}(A, \Phi_s)[f(A, \Phi_s) + G(A, \Phi_s)\Psi_s] \\ &\quad + 2[\Psi_s - \alpha(A, \Phi_s)]\hat{P}\alpha'(A, \Phi_s)[f(A, \Phi_s) + G(A, \Phi_s)\Psi_s]. \end{aligned} \quad (9.207)$$

Now $L_2(A, \Phi_s, \Psi_s)$ must be chosen to satisfy condition (9.62) or, equivalently,

$$\begin{aligned} (\Psi_s - \alpha(A, \Phi_s))\Big\{ &-2\Phi_s - 2\hat{P}\Big[\alpha'(A, \Phi_s)(f(A, \Phi_s) + G(A, \Phi_s)\Psi_s) \\ &+ R_2^{-1}[\hat{P}(\Psi_s - \alpha(A, \Phi_s)) + \tfrac{1}{2}L_2(A, \Phi_s, \Psi_s)]\Big]\Big\} < 0. \end{aligned} \quad (9.208)$$

A particular admissible choice for $L_2(A, \Phi_s, \Psi_s)$ satisfying (9.208) is given by

$$\begin{aligned} L_2(A, \Phi_s, \Psi_s) &= -2R_2\{\hat{P}^{-1}\Phi_s + \alpha'(A, \Phi_s)[f(A, \Phi_s) + G(A, \Phi_s)\Psi_s] \\ &\quad - c_3A^2[\Psi_s - \alpha(A, \Phi_s)]\}, \end{aligned} \quad (9.209)$$

where $c_3 \geq 0$. For this choice of $L_2(A, \Phi_s, \Psi_s)$ the feedback control (9.204) becomes

$$\begin{aligned} -\phi(A, \Phi_s, \Psi_s) &= R_2^{-1}\hat{P}[\Psi_s - \alpha(A, \Phi_s)] - \hat{P}^{-1}\Phi_s - \alpha'(A, \Phi_s)[f(A, \Phi_s) \\ &\quad + G(A, \Phi_s)\Psi_s] + c_3A^2(\Psi_s - \alpha(A, \Phi_s)), \end{aligned} \quad (9.210)$$

so that (9.208) satisfies

$$\begin{aligned} -2R_2^{-1}\hat{P}^2(\Psi_s - \alpha(A, \Phi_s))^2 - c_3\hat{P}A^2(\Psi_s - \alpha(A, \Phi_s))^2 &< 0, \\ (A, \Phi_s, \Psi_s) &\neq (0, 0, 0). \end{aligned} \quad (9.211)$$

Note that instead of using $L_2(A, \Phi_s, \Psi_s)$ to simply cancel the indefinite terms in (9.208), we have also added an extra term in which the stall cell squared amplitude A^2 is multiplied by the tracking error $\Psi_s - \alpha(A, \Phi_s)$ and a nonnegative constant c_3. This illustrates the flexibility available in choosing $L_2(A, \Phi_S, \Psi_S)$ in the control law. In the special case $\hat{P} = \frac{1}{2}$, $\varepsilon = 0$, $c_3 = 0$, and $R_2 = 1/c_2$ this control law specializes to the controller given in [247]. However, when ε vanishes the positive definiteness of the Lyapunov function (9.205) over the whole state space is destroyed, and hence, the optimality claims of Theorem 9.1 cannot be made for the controller given in [247]. Whereas by varying ε, c_3, and $\hat{P}$ in the control law (9.210), we can generate a family of controllers which guarantee global asymptotic stability and global optimality with respect to the performance functional (9.206).

Using the initial conditions $A_0 = 0.5$, $\Phi_{S_0} = 0$, $\Psi_{S_0} = 0$, and parameter values $\Psi_{C0} = 0.72$, $\sigma = 3.6$, $\beta = 0.356$, with $\hat{P} = 2.5$, $R_2 = 1$, $\varepsilon = 0.0625$, and $c_3 = 0.25$ the inverse optimal control law (9.210) and the controller given in [247] (with $c_1 = c_2 = 1$) were used to compare the closed-loop system response. The squared stall cell amplitude responses for the two controllers are compared in Figure 9.4, the compressor flow and pressure rise responses are compared in Figures 9.5 and 9.6, and the control efforts are compared in Figures 9.7 and 9.8. In Figure 9.9 a phase portrait is given comparing the overall system state trajectories for the two controllers. This comparison illustrates that the present framework allows the control designer to improve both the state response and the control effort using the state weights of the Lyapunov function, $\hat{P}$ and ε, and the control weight of the performance functional, R_2. Furthermore, the trade-off between achievable state response and allowable control effort is characterized by the performance functional (9.206). Finally, Figure 9.10 shows the compressor pressure-flow performance map parameterized as a function of the throttle opening which constitutes a coexistent set of stable and unstable equilibria. The controlled globally stable equilibrium point $(0, 1, 2.72)$ corresponds to the maximum pressure performance for the given compressor speed.

9.7 Surge Control for Centrifugal Compressors

While the literature on modeling and control of compression systems predominantly focuses on axial flow compression systems, the research literature on centrifugal flow compression systems is rather limited in comparison. Notable exceptions include [21,113,119,138,184,223,357] which address modeling and control of centrifugal compressors. In contrast to axial flow compression systems involving the aerodynamic instabilities of rotating stall and surge, a common feature of [21, 113, 119, 138, 184, 223, 357] is the realization that surge (and deep surge) is the predominant aerodynamic instability arising in centrifugal compression systems.

To address the problem of nonlinear stabilization for centrifugal compression systems we consider the basic centrifugal compression system

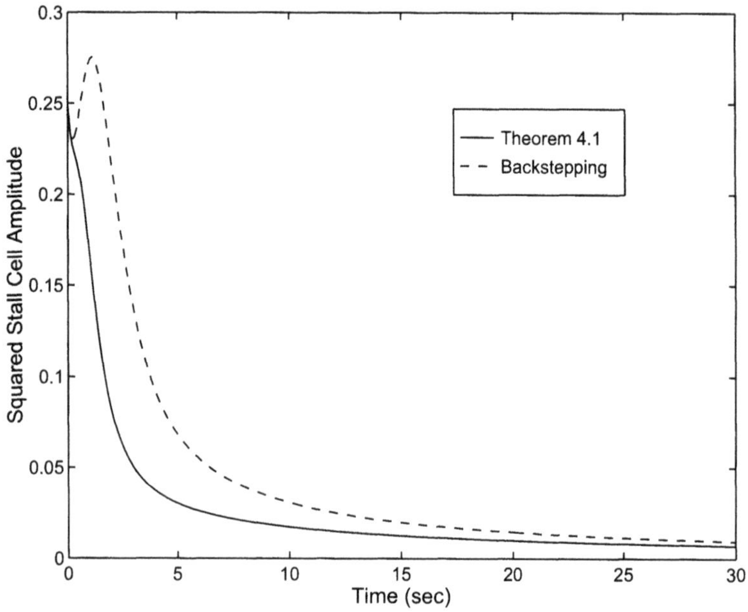

Figure 9.4 Squared stall cell amplitude versus time.

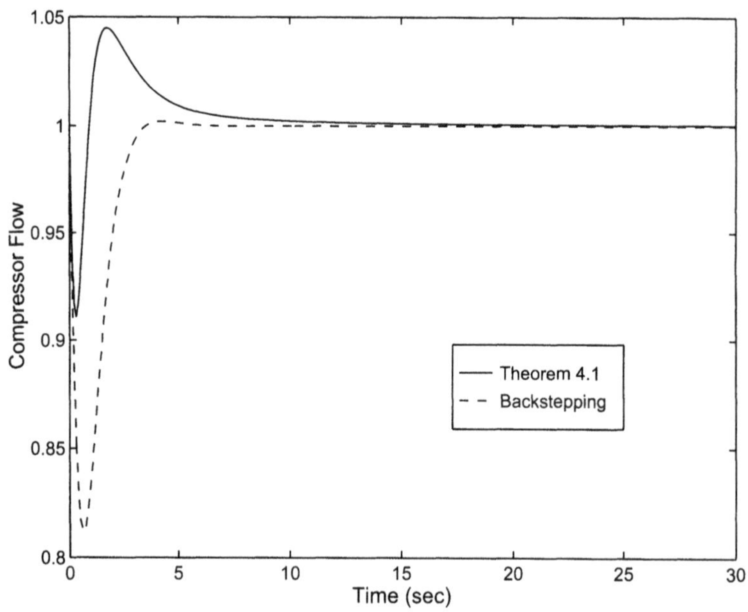

Figure 9.5 Compressor flow versus time.

shown in Figure 9.11, consisting of a short inlet duct, a compressor, an outlet duct, a plenum, an exit duct, and a control throttle. We assume

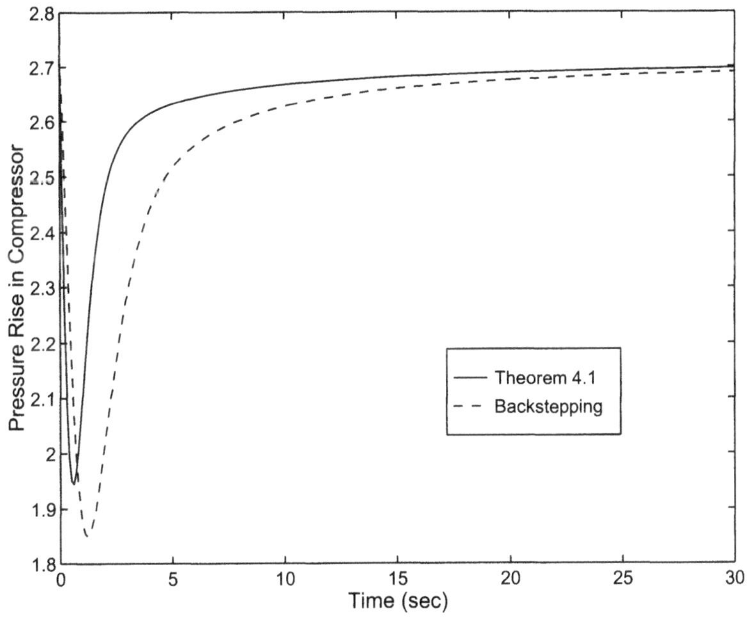

Figure 9.6 Pressure rise versus time.

that the plenum dimensions are large as compared to the compressor-duct dimensions so that the fluid velocity and acceleration in the plenum are negligible. In this case, the pressure in the plenum is spatially uniform. Furthermore, we assume that the flow is controlled by a throttle at the plenum exit. In addition, we assume a low-speed compression system with oscillation frequencies much lower than the acoustic resonance frequencies so that the flow can be considered incompressible. However, we do assume that the gas in the plenum is compressible and acts as a gas spring. Finally, we assume isentropic process dynamics in the plenum and negligible gas angular momentum in the compressor passages as compared to the impeller angular momentum.

To address the problem of nonlinear stabilization for centrifugal compression systems we use the three-state lumped parameter model for surge in centrifugal flow compression systems developed in [119, 138, 278]. Specifically, pressure and mass flow compression system dynamics are developed using principles of conservation of mass and momentum. Furthermore, in order to account for the influence of speed transients on the compression surge dynamics, turbocharger spool dynamics are also considered.

Using continuity it follows that mass conservation in the plenum is

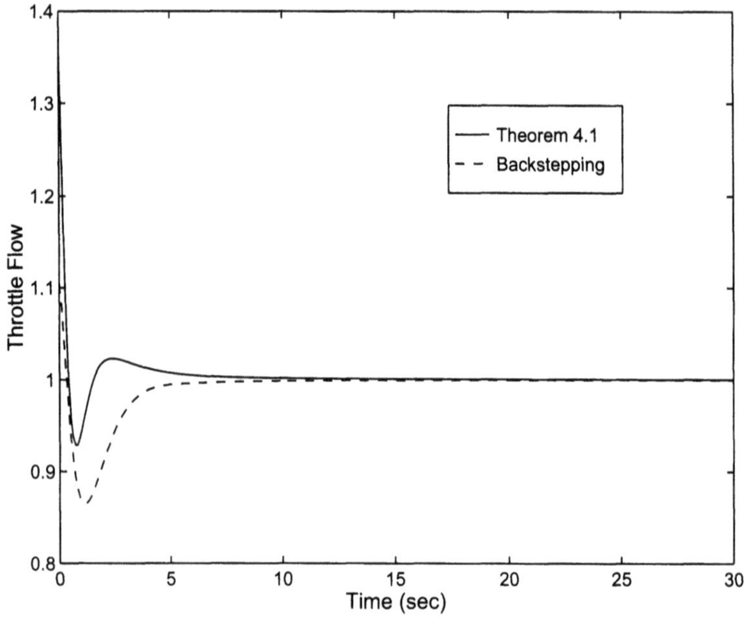

Figure 9.7 Throttle flow versus time.

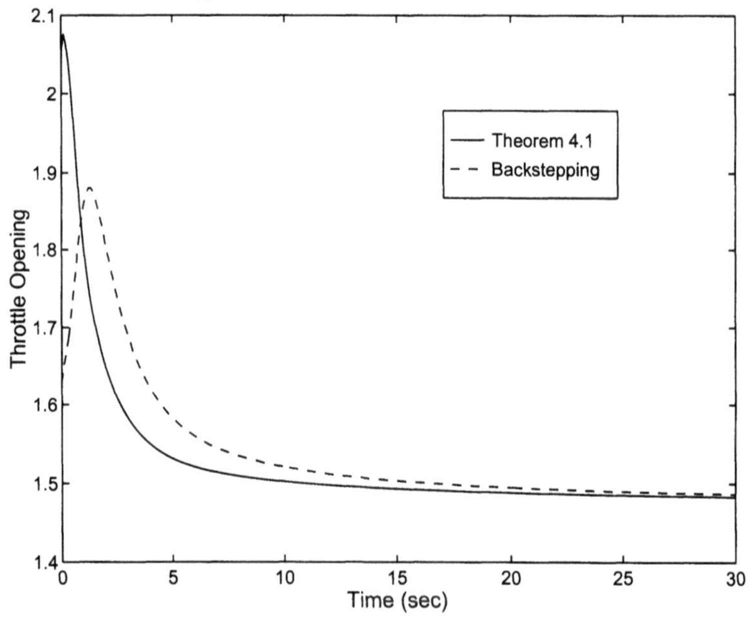

Figure 9.8 Throttle opening versus time.

given by

$$\dot{\Psi}(t) = a(\Phi(t) - \gamma_{\mathrm{throt}}\sqrt{\Psi(t)}), \quad \Psi(0) = \Psi_0, \quad t \geq 0, \tag{9.212}$$

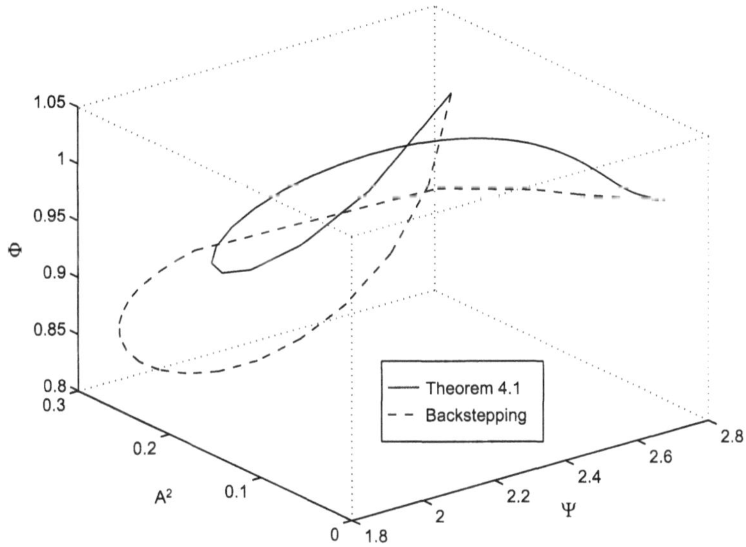

Figure 9.9 Phase portrait of state trajectories from $(0.5, 1, 2.72)^{\mathrm{T}}$.

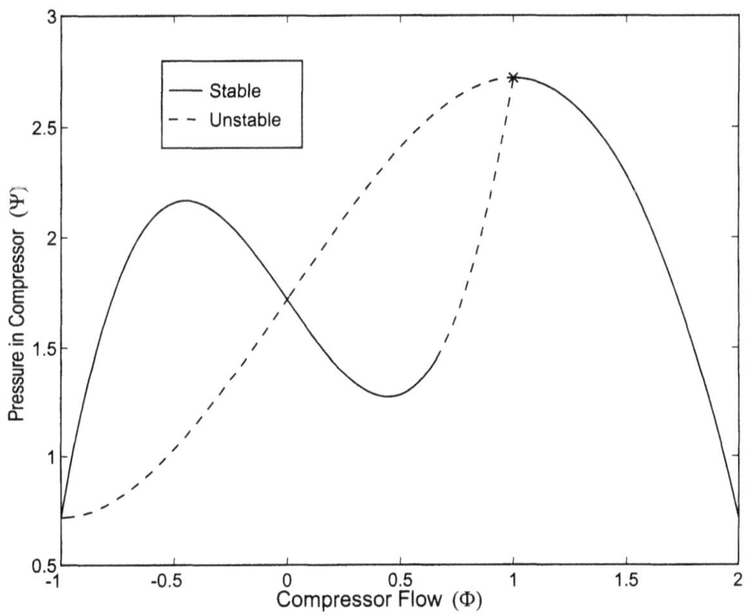

Figure 9.10 Compression system performance map.

where Φ is the nondimensional mass flow rate at the plenum entrance, Ψ is the total-to-static pressure ratio, γ_{throt} is a parameter proportional to the throttle opening, and a is a nondimensional parameter related to the compressor dimensions.

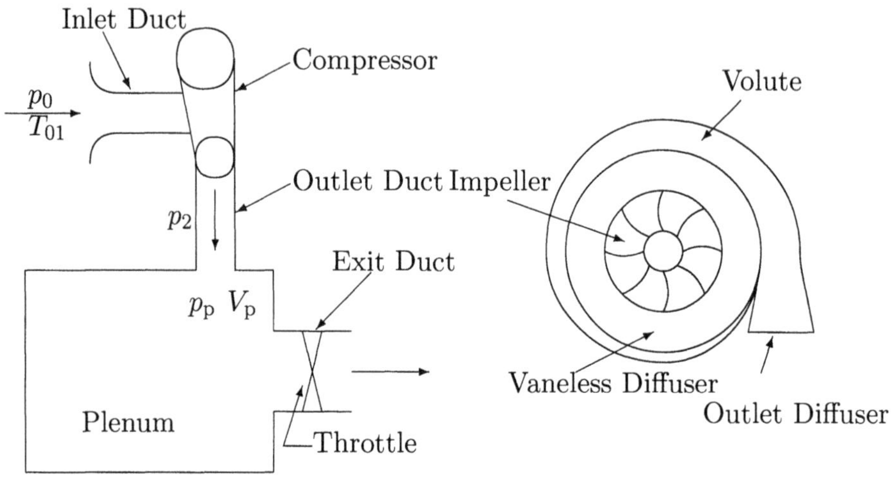

Figure 9.11 Centrifugal compressor system geometry.

Next, using a momentum balance with the assumption of incompressible flow, using the fact that the change in angular momentum of the fluid is equal to the compressor torque, assuming isentropic process dynamics with constant specific heat, and assuming an absence of prewhirl at the rotor inlet, it follows that [107, 193, 309]

$$\dot{\Phi}(t) = b(\Psi_c(\Phi(t), \Omega(t)) - \Psi(t)), \quad \Phi(0) = \Phi_0, \quad t \geq 0, \tag{9.213}$$

where Ω is the nondimensional angular velocity of compressor spool, $\Psi_c(\Phi, \Omega)$ is the compressor characteristic pressure-flow/angular velocity map given by

$$\Psi_c(\Phi, \Omega) \triangleq \left(1 + \eta_c(\Phi, \Omega)\sigma d\Omega^2\right)^{\frac{\gamma_{sh}}{\gamma_{sh}-1}} - 1, \tag{9.214}$$

where γ_{sh} is the specific heat ratio and $\eta_c(\Phi, \Omega)$ is the isentropic efficiency given by ([278])

$$\eta_c(\Phi, \Omega) \triangleq \frac{\sigma\Omega^2}{\sigma\Omega^2 + \frac{1}{2}(f_1\Omega - f_2\Phi)^2 + \frac{1}{2}(\sigma\Omega - f_3\Phi)^2 + \Phi^2(f_4 + f_5)}. \tag{9.215}$$

Here, σ is the slip factor and b, d, f_i, $i = 1, \ldots, 5$, are nondimensional parameters related to the compressor dimensions, the sound velocity in the plenum, the inducer and rotor geometry, and the friction coefficients, respectively.

It is important to note that the compressor characteristic map given by (9.214) holds for the case where the flow through the compressor is positive. In the case of deep surge involving negative mass flow, it is assumed that the pressure rise in the compressor is proportional to the square of the mass flow so that [184]

$$\Psi_{\rm c}(\Phi,\Omega) = \mu\Phi^2 + \Psi_{{\rm c}_0}(\Phi,\Omega), \qquad \Phi < 0, \tag{9.216}$$

where μ is a constant and

$$\Psi_{{\rm c}_0}(\Phi,\Omega) \triangleq \Psi_{\rm C}(\Phi,\Omega)\Big|_{\Phi=0} = (1+\sigma\eta_{\rm C0}d\Omega^2)^{\frac{\gamma_{\rm sh}}{\gamma_{\rm sh}-1}} - 1, \tag{9.217}$$

where

$$\eta_{{\rm C}_0} \triangleq \eta_{\rm C}(\Phi,\Omega)|_{\Phi=0} = \frac{2\sigma}{\sigma^2+2\sigma+f_1^2}. \tag{9.218}$$

It is shown in [278] that $\eta_{{\rm C}_{\max}}$ is constant for all spool speeds. This indicates that the compressor achieves the same maximum isentropic efficiency at each maximum pressure point for all spool speeds. However, since these points are critically stable, active control is needed to guarantee stable compression system operation for peak compressor performance. Figure 9.12 shows a typical family of compressor characteristic maps for different spool speeds along with the corresponding constant isentropic efficiency lines. The stone wall depicted in Figure 9.12 corresponds to choked flow at a given cross-section of the compression system [138, 278].

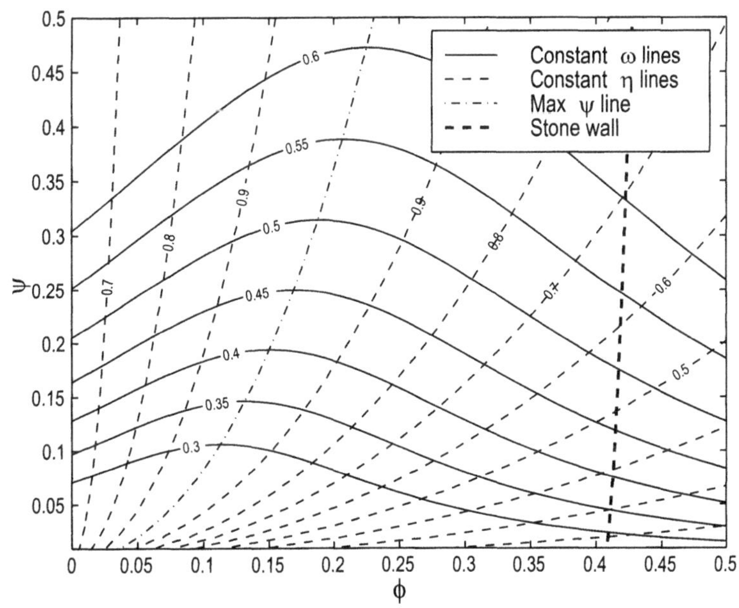

Figure 9.12 Compressor characteristic maps and efficiency lines for different spool speeds.

Finally, using conservation of angular momentum in the turbocharger spool it follows that the nondimensional spool dynamics are given by

$$\dot{\Omega}(t) = c(\tau(t) - \sigma\Phi(t)\Omega(t)), \quad \Omega(0) = \Omega_0, \quad t \geq 0, \tag{9.219}$$

where $\tau(\cdot)$ is the nondimensional driving torque and c is a nondimensional parameter related to the spool mass moment of inertia.

Next, we apply the inverse optimal backstepping control framework to control surge in centrifugal compression systems. First, we note that with control inputs $u_1 \triangleq \gamma_{\rm throt}\sqrt{\Psi}$ and $u_2 \triangleq \tau$ it follows from (9.212), (9.213), and (9.219), that a state space model for the centrifugal compressor is given by

$$\begin{aligned}\dot{\Psi}(t) &= a(\Phi(t) - u_1(t)), \quad \Psi(0) = \Psi_0, \quad t \geq 0, &(9.220)\\ \dot{\Phi}(t) &= b(\Psi_{\rm c}(\Phi(t), \Omega(t)) - \Psi(t)), \quad \Phi(0) = \Phi_0, &(9.221)\\ \dot{\Omega}(t) &= c(u_2(t) - \sigma\Phi(t)\Omega(t)), \quad \Omega(0) = \Omega_0. &(9.222)\end{aligned}$$

Note that for fixed values of the control inputs u_1 and u_2, (9.220), (9.221) and (9.222) give an equilibrium point $(\Psi_{\rm eq}, \Phi_{\rm eq}, \Omega_{\rm eq})$, where $(\Psi_{\rm eq}, \Phi_{\rm eq}, \Omega_{\rm eq})$ is given by

$$(\Psi_{\rm eq}, \Phi_{\rm eq}, \Omega_{\rm eq}) = \left(\Psi_{\rm c}(\Phi_{\rm eq}, \Omega_{\rm eq}),\ u_{1\rm eq},\ \frac{u_{2\rm eq}}{\sigma\Phi_{\rm eq}}\right). \qquad (9.223)$$

Defining the shifted state variables $x_1 \triangleq \Psi - \Psi_{\rm eq}$, $x_2 \triangleq \Phi - \Phi_{\rm eq}$, and $x_3 \triangleq \Omega - \Omega_{\rm eq}$, so that for a given equilibrium point on the compressor characteristic map the system equilibrium is translated to the origin, and defining the shifted controls $\tilde{u}_1 \triangleq u_1 - u_{1\rm eq}$ and $\tilde{u}_2 \triangleq u_2 - u_{2\rm eq}$, it follows that the translated nonlinear system is given by

$$\begin{aligned}\dot{x}_1(t) &= a(x_2(t) - \tilde{u}_1(t)), \quad x_1(0) = x_{10}, \quad t \geq 0, &(9.224)\\ \dot{x}_2(t) &= b(\Psi_{\rm Ceq}(x_2(t), x_3(t)) - x_1(t)), \quad x_2(0) = x_{20}, &(9.225)\\ \dot{x}_3(t) &= c(\tilde{u}_2(t) - f(x_2(t), x_3(t))), \quad x_3(0) = x_{30}, &(9.226)\end{aligned}$$

where

$$\begin{aligned}\Psi_{\rm Ceq}(x_2, x_3) &\triangleq \Psi_{\rm c}(\Phi_{\rm eq} + x_2, \Omega_{\rm eq} + x_3) - \Psi_{\rm c}(\Phi_{\rm eq}, \Omega_{\rm eq}), &(9.227)\\ f(x_2, x_3) &\triangleq \sigma(\Phi_{\rm eq}x_3 + \Omega_{\rm eq}x_2) + \sigma x_2 x_3. &(9.228)\end{aligned}$$

Now, setting

$$\begin{aligned}\hat{u} &= x_2 - \tilde{u}_1, &(9.229)\\ \tilde{u}_2 &= -k_3 x_3 + f(x_2, x_3), &(9.230)\end{aligned}$$

where $k_3 > 0$, and substituting (9.229) and (9.230) into (9.224)–(9.226) yields

$$\begin{aligned}\dot{x}_1(t) &= a\hat{u}(t), \quad x_1(0) = x_{10}, \quad t \geq 0, &(9.231)\\ \dot{x}_2(t) &= b(\Psi_{\rm Ceq}(x_2(t), x_3(t)) - x_1(t)), \quad x_2(0) = x_{20}, &(9.232)\\ \dot{x}_3(t) &= -k_3 c x_3(t), \quad x_3(0) = x_{30}. &(9.233)\end{aligned}$$

Note that (9.231)–(9.233) has the correct form for the application of

Theorem 9.1 where (9.232) and (9.233) make up the nonlinear subsystem and x_1 is the integrator state. Specifically, (9.231)–(9.233) can be written in the form of (9.1) and (9.2) where $x = [x_2 \;\; x_3]^{\mathrm{T}}$, $\hat{x} = x_1$, and

$$f(x_2, x_3) = \begin{bmatrix} b\Psi_{\mathrm{Ceq}}(x_2, x_3) \\ -k_3 c x_3 \end{bmatrix}, \qquad G(x_2, x_3) = \begin{bmatrix} -b \\ 0 \end{bmatrix}.$$

To apply Theorem 9.1 we require a stabilizing feedback for the subsystem (9.232) and (9.233) and a corresponding Lyapunov function $V_{\mathrm{sub}}(x_2, x_3)$ such that (9.59) and (9.60) are satisfied. For the nonlinear subsystem (9.232) and (9.233) we choose the Lyapunov function candidate

$$V_{\mathrm{sub}}(x_2, x_3) = \tfrac{1}{2}\alpha_2 x_2^2 + \tfrac{1}{2}\alpha_3 x_3^2, \tag{9.234}$$

where $\alpha_2, \alpha_3 > 0$, and the stabilizing feedback control

$$\alpha(x_2, x_3) = \Psi_{\mathrm{Ceq}}(x_2, x_3) + k_2 x_2, \tag{9.235}$$

where $k_2 > 0$. It is straightforward to show that (9.234) and (9.235) satisfy conditions (9.59)–(9.61) of Theorem 9.1.

Applying Theorem 9.1 to the system (9.231)–(9.233) yields the family of control laws

$$u = \phi(x_1, x_2, x_3) = -R_2^{-1}\left[\hat{P}\left(x_1 - \alpha(x_2, x_3)\right) + \tfrac{1}{2}L_2(x_1, x_2, x_3)\right], \tag{9.236}$$

with Lyapunov function

$$V(x_1, x_2, x_3) = \hat{P}[x_1 - \alpha(x_2, x_3)]^2 + \tfrac{1}{2}\alpha_2 x_2^2 + \tfrac{1}{2}\alpha_3 x_3^2, \tag{9.237}$$

where $R_2 > 0$ and $\hat{P} > 0$. Furthermore, the performance functional minimized by the control law (9.236) has the form

$$\begin{aligned} J(x_{10}, x_{20}, x_{30}, \hat{u}) = & \int_0^\infty [L_1(x_1(t), x_2(t), x_3(t)) \\ & + L_2(x_1(t), x_2(t), x_3(t))\hat{u}(t) + R_2\hat{u}^2(t)]\mathrm{d}t, \end{aligned} \tag{9.238}$$

where

$$\begin{aligned} L_1(x_1, x_2, x_3) \triangleq\; & R_2\phi^2(x_1, x_2, x_3) - V'_{\mathrm{sub}}(x_2, x_3)[f(x_2, x_3) + G(x_2, x_3)x_1] \\ & + 2[x_1 - \alpha(x_2, x_3)]\hat{P}\alpha'(x_2, x_3)[f(x_2, x_3) + G(x_2, x_3)x_1]. \end{aligned} \tag{9.239}$$

Now, $L_2(x_1, x_2, x_3)$ must be chosen to satisfy condition (9.62) or, equivalently,

$$\begin{aligned} (x_1 - \alpha(x_2, x_3)) \Big\{ & -2x_1 - 2\hat{P}\Big[\alpha'(x_2, x_3)(f(x_2, x_3) + G(x_2, x_3)x_1) \\ & + R_2^{-1}[\hat{P}(x_1 - \alpha(x_2, x_3) + \tfrac{1}{2}L_2^{\mathrm{T}}(x_1, x_2, x_3)]\Big]\Big\} < 0. \end{aligned} \tag{9.240}$$

A particular admissible choice for $L_2(x_1,x_2,x_3)$ satisfying (9.240) is given by

$$\begin{aligned} L_2(x_1,x_2,x_3) &= -\frac{\alpha_1}{k_1 a}\Big[b(k_2+\Psi_{\mathrm{c_{eq}},x_2}(x_2,x_3))(\Psi_{\mathrm{Ceq}}(x_2,x_3)-x_1) \\ &\quad -k_3 c\Psi_{\mathrm{c_{eq}},x_3}(x_2,x_3)x_3+\tfrac{\alpha_2}{\alpha_1}bx_2\Big], \end{aligned} \tag{9.241}$$

where $\alpha_1 \triangleq 2\hat{P}$ and $k_1 \triangleq R_2^{-1}\hat{P}$. For this choice of $L_2(x_1,x_2,x_3)$ the feedback control (9.236) becomes

$$\begin{aligned} \phi(x_1,x_2,x_3) &= -k_1(x_1-\Psi_{\mathrm{Ceq}}(x_2,x_3)-k_2x_2) \\ &\quad +\frac{1}{a}\Big[b(k_2+\Psi_{\mathrm{c_{eq}},x_2}(x_2,x_3))(\Psi_{\mathrm{Ceq}}(x_2,x_3)-x_1) \\ &\quad -k_3 c\Psi_{\mathrm{c_{eq}},x_3}(x_2,x_3)x_3+\frac{\alpha_2}{\alpha_1}bx_2\Big], \end{aligned} \tag{9.242}$$

where

$$\begin{aligned} \Psi_{\mathrm{c_{eq}},x_2}(x_2,x_3) &\triangleq \frac{\partial \Psi_{\mathrm{Ceq}}(x_2,x_3)}{\partial x_2} \\ &= \left.\frac{\partial \Psi_{\mathrm{c}}(x_2,x_3)}{\partial x_2}\right|_{(x_2,x_3)=(\Phi_{\mathrm{eq}}+x_2,\ \Omega_{\mathrm{eq}}+x_3)} \\ &= \frac{\gamma_{\mathrm{sh}}\sigma d x_3^2}{\gamma_{\mathrm{sh}}-1}\left[\Psi_{\mathrm{c}}(\Phi_{\mathrm{eq}}+x_2,\Omega_{\mathrm{eq}}+x_3)+1\right]^{\frac{1}{\gamma_{\mathrm{sh}}}} \left.\frac{\partial \eta_{\mathrm{c}}(x_2,x_3)}{\partial x_2}\right|_{(x_2,x_3)=(\Phi_{\mathrm{eq}}+x_2,\ \Omega_{\mathrm{eq}}+x_3)}, \end{aligned}$$

$$\begin{aligned} \Psi_{\mathrm{c_{eq}},x_3}(x_2,x_3) &\triangleq \frac{\partial \Psi_{\mathrm{c_{eq}}}(x_2,x_3)}{\partial x_3} \\ &= \left.\frac{\partial \Psi_{\mathrm{c}}(x_2,x_3)}{\partial x_3}\right|_{(x_2,x_3)=(\Phi_{\mathrm{eq}}+x_2,\ \Omega_{\mathrm{eq}}+x_3)} \\ &= \frac{\gamma_{\mathrm{sh}}\sigma d}{\gamma_{\mathrm{sh}}-1}\left[\Psi_{\mathrm{c}}(\Phi_{\mathrm{eq}}+x_2,\Omega_{\mathrm{eq}}+x_3)+1\right]^{\frac{1}{\gamma_{\mathrm{sh}}}} \left.\frac{\partial\left[x_3^2\eta_{\mathrm{c}}(x_2,x_3)\right]}{\partial x_3}\right|_{(x_2,x_3)=(\Phi_{\mathrm{eq}}+x_2,\ \Omega_{\mathrm{eq}}+x_3)}, \end{aligned}$$

and where

$$\frac{\partial \eta_{\mathrm{c}}(x_2,x_3)}{\partial x_2} = \frac{\sigma x_3^2(f_2(x_3f_1-x_2f_2)+f_3(\sigma x_3-x_2f_3)-2x_2(f_4+f_5))}{(\sigma x_3^2+\frac{1}{2}(x_3f_1-x_2f_2)^2+\frac{1}{2}(\sigma x_3-x_2f_3)^2+x_2^2(f_4+f_5))^2},$$

$$\begin{aligned} &\frac{\partial[x_3^2\eta_{\mathrm{c}}(x_2,x_3)]}{\partial x_3} \\ &\quad = \frac{\sigma x_3^3(2\sigma x_3^2+(x_3f_1-2x_2f_2)(x_3f_1-x_2f_2)+(\sigma x_3-2x_2f_3)(\sigma x_3-x_2f_3)+4x_2^2(f_4+f_5))}{(\sigma x_3^2+\frac{1}{2}(x_3f_1-x_2f_2)^2+\frac{1}{2}(\sigma x_3-x_2f_3)^2+x_2^2(f_4+f_5))^2}, \end{aligned}$$

so that (9.241) satisfies

$$-\alpha_1 k_1 a(x_1-\Psi_{\mathrm{Ceq}}(x_2,x_3)-k_2x_2)^2 < 0, \quad (x_1,x_2,x_3)\neq(0,0,0). \tag{9.243}$$

In this case, the overall nonlinear controller

$$\tilde{u} \triangleq \begin{bmatrix} \tilde{u}_1 \\ \tilde{u}_2 \end{bmatrix} = \begin{bmatrix} x_2 - \hat{u} \\ -k_3 x_3 + f(x_2, x_3) \end{bmatrix}, \tag{9.244}$$

guarantees that the closed-loop system (9.224)–(9.226) is globally asymptotically stable.

Next, with $(a, b, c, d, e) = (9.37, 310.81, 6.97, 0.38, 0.7)$, $(f_1, f_2, f_3, f_4, f_5) = (0.44, 1.07, 2.18, 0.17, 0.12)$, $\gamma_{\mathrm{sh}} = 1.4$, $\mu = 3$, and $\sigma = 0.9$, we design inverse optimal controllers for the three-state centrifugal compressor model discussed above. Since the torque dynamics given by (9.219) is for forward mass flow in the compressor, and since the compressor may enter deep surge, there is a need to derive an expression for the compressor torque for negative mass flow. Hence, assuming that a centrifugal compressor in reverse flow can be viewed as a throttling device and, hence, can be approximated as a turbine, it follows that [119]

$$\dot{\Omega}(t) = c(u_2(t) - \sigma|\Phi(t)|\Omega(t)), \quad \Omega(0) = \Omega_0, \quad t \geq 0. \tag{9.245}$$

Using the initial conditions $(\Psi_0, \Phi_0, \Omega_0) = (0.188, 0.141, 0.394)$ and the design parameters $(\alpha_1, \alpha_2, \alpha_3) = (1, 0.1, 1)$ and $(k_1, k_2, k_3) = (1, 3, 1)$, the closed-loop system response is compared to the open-loop response when the compression system is taken from an operating speed of 20,000 rpm to 25,000 rpm. Figure 9.13 shows the pressure-flow phase portrait of the state trajectories when the system is taken from an operating speed of 20,000 rpm to 25,000 rpm. The pressure rise, mass flow, and spool speed variations for the open-loop and controlled system are shown in Figure 9.14, 9.15, and 9.16, respectively. Figures 9.17 and 9.18 shows the control effort versus time. This comparison illustrates that open-loop control drives the compression system into deep surge, while the proposed globally stabilizing controller drives the system to the desired equilibrium point $(\Psi_{\mathrm{eq}}, \Phi_{\mathrm{eq}}, \Omega_{\mathrm{eq}}) = (0.304, 0.176, 0.493)$.

9.8 Problems

Problem 9.1. Consider the nonlinear dynamical system

$$\dot{x}_1(t) = 1 + x_2(t) + [x_1(t) - 1]^3, \; x_1(0) = x_{10}, \; t \geq 0, \tag{9.246}$$
$$\dot{x}_2(t) = x_1(t) + u(t), \; x_2(0) = x_{20}. \tag{9.247}$$

Using backstepping, find a globally stabilizing feedback controller $u(t) = \phi(x(t))$ for (9.246) and (9.247).

Problem 9.2. Consider the nonlinear dynamical system

$$\dot{x}_1(t) = x_1^2(t) - x_1^5(t) + x_2(t), \quad x_1(0) = x_{10}, \quad t \geq 0, \tag{9.248}$$

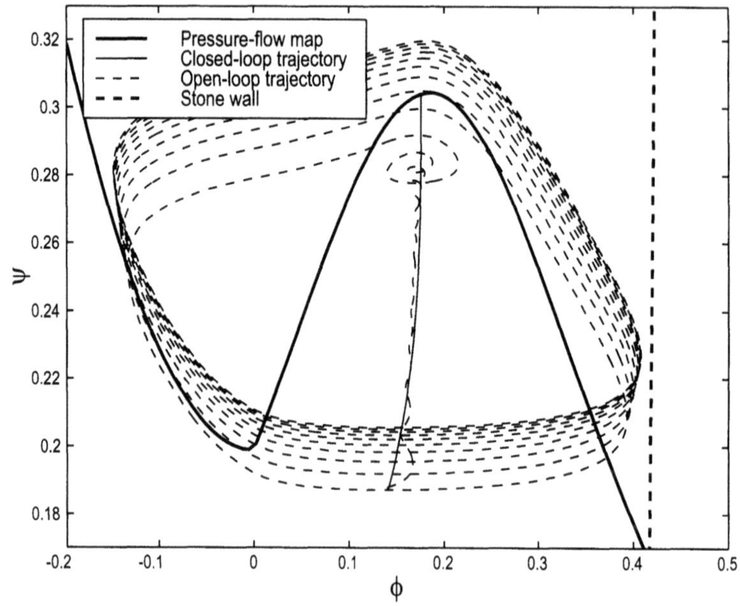

Figure 9.13 Controlled and uncontrolled phase portrait of pressure-flow state trajectories from 20,000 rpm to 25,000 rpm.

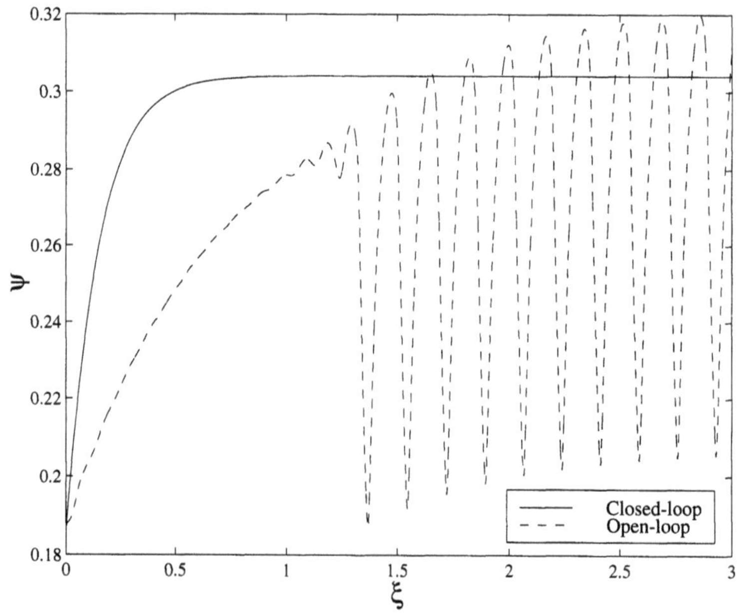

Figure 9.14 Pressure rise versus time.

$$\dot{x}_2(t) = x_3(t), \quad x_2(0) = x_{20}, \tag{9.249}$$

$$\dot{x}_3(t) = u(t), \quad x_3(0) = x_{30}. \tag{9.250}$$

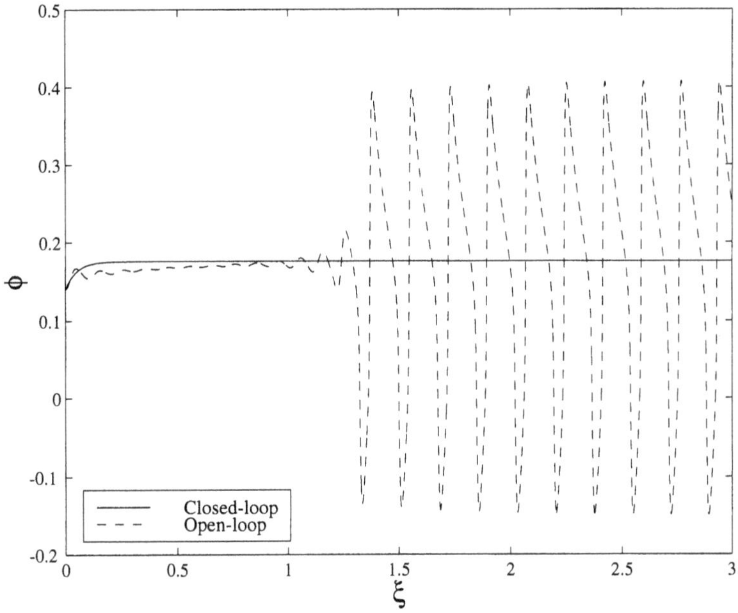

Figure 9.15 Mass flow versus time.

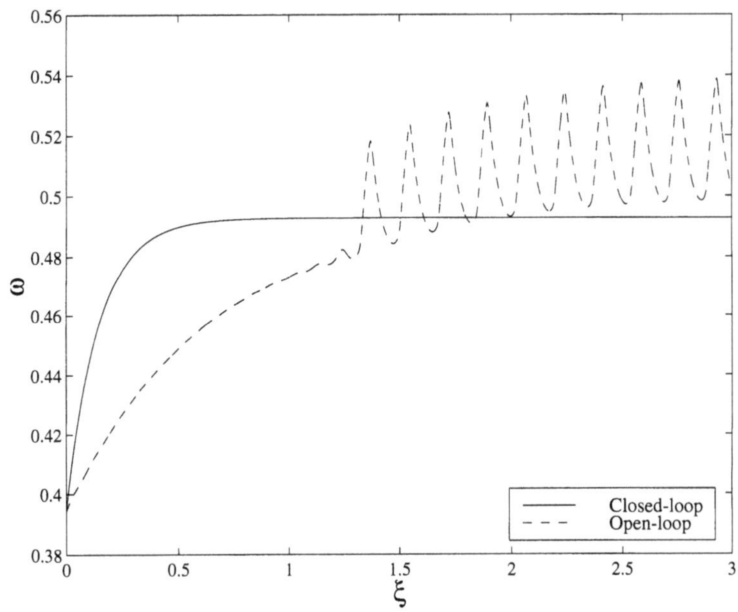

Figure 9.16 Compressor spool speed versus time.

Using backstepping, find a globally stabilizing feedback controller $u(t) = \phi(x(t))$ for (9.248)–(9.250).

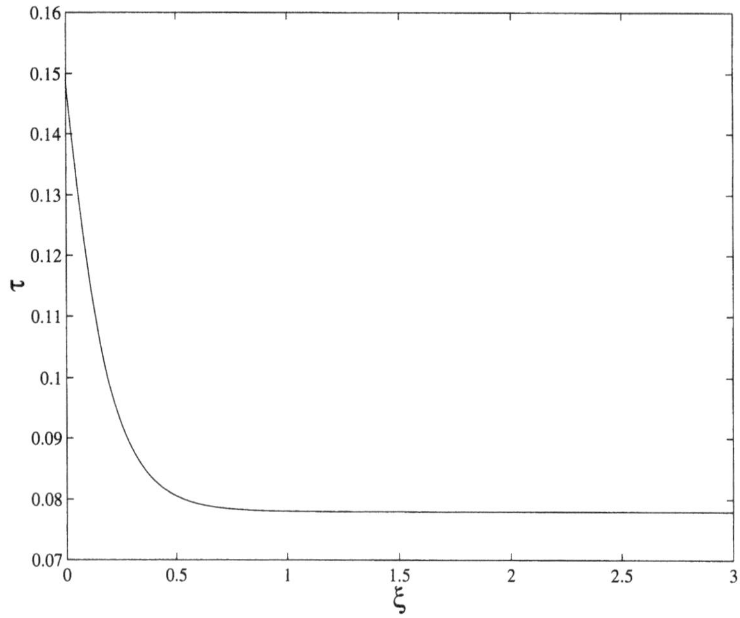

Figure 9.17 Driving torque versus time.

Problem 9.3. Consider the nonlinear dynamical system

$$\dot{x}_1(t) = x_2(t) - x_1^3(t), \quad x_1(0) = x_{10}, \quad t \geq 0, \tag{9.251}$$
$$\dot{x}_2(t) = u(t), \quad x_2(0) = x_{20}. \tag{9.252}$$

Using backstepping, find a stabilizing controller $u(t) = \phi(x(t))$ for (9.251) and (9.252).

Problem 9.4. Consider the nonlinear dynamical system

$$\dot{x}_1(t) = \cos x_1(t) - x_1^3(t) + x_2(t), \quad x_1(0) = x_{10}, \quad t \geq 0, \tag{9.253}$$
$$\dot{x}_2(t) = u(t), \quad x_2(0) = x_{20}. \tag{9.254}$$

Using backstepping, find a globally stabilizing controller $u(t) = \phi(x(t))$ for (9.253) and (9.254).

Problem 9.5. Consider the nonlinear dynamical system

$$\dot{x}_1(t) = -x_1(t) + x_2(t)x_1^2(t), \quad x_1(0) = x_{10}, \quad t \geq 0, \tag{9.255}$$
$$\dot{x}_2(t) = u(t), \quad x_2(0) = x_{20}. \tag{9.256}$$

Using backstepping, find a globally stabilizing controller $u(t) = \phi(x(t))$ for (9.255) and (9.256).

Problem 9.6. Consider the nonlinear dynamical system

$$\dot{x}_1(t) = -x_1^3(t) - [\alpha x_2(t) + \beta x_3(t)]x_1^3(t), \; x_1(0) = x_{10}, \; t \geq 0, \tag{9.257}$$

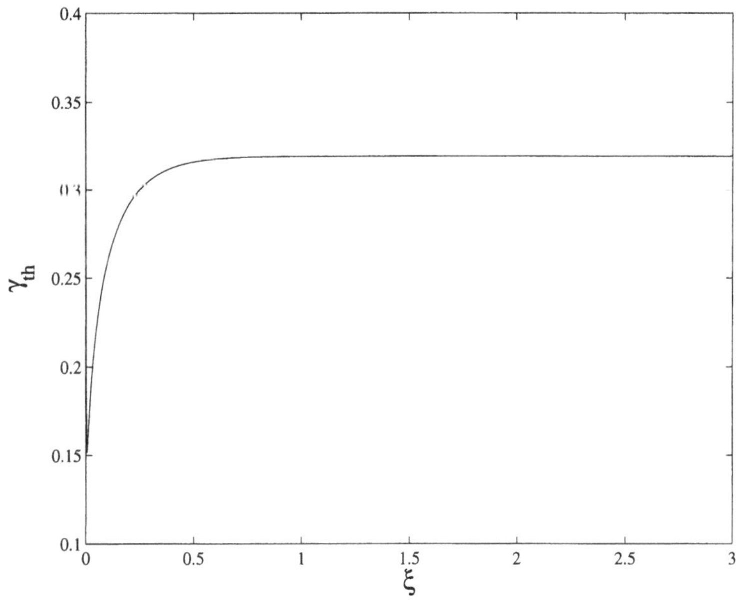

Figure 9.18 Throttle opening versus time.

$$\dot{x}_2(t) = x_3(t), \; x_2(0) = x_{20}, \tag{9.258}$$
$$\dot{x}_3(t) = u(t), \; x_3(0) = x_{30}, \tag{9.259}$$

where $\alpha\beta \geq 0$. Using backstepping, find a stabilizing controller $u(t) = \phi(x(t))$ for (9.257)–(9.259). Can the system be globally stabilized if $\alpha\beta < 0$?

Problem 9.7. Consider the nonlinear cascade system (9.160)–(9.162). Show that if (9.171) and (9.172) are satisfied then the following statements hold:

i) $V_{\rm c}(y, z)$ given by (9.165) exists and is continuous in $\mathbb{R}^m \times \mathbb{R}^{q-m}$.

ii) $V(x, y, z)$ given by (9.175) is positive definite in $\mathbb{R}^n \times \mathbb{R}^m \times \mathbb{R}^{q-m}$.

iii) $V(x, y, z)$ given by (9.175) is radially unbounded.

Problem 9.8. Show that if $V : \mathbb{R}^{q-m} \to \mathbb{R}$ is a radially unbounded, nonnegative-definite polynomial function, then condition (9.172) is satisfied.

Problem 9.9. Consider the three-state parameterized Moore-Greitzer model given by (9.195)–(9.197). Show that the linearized system is linearly stabilizable for $\lambda > 1$, while for $\lambda = 1$, corresponding to the maximum pressure rise equilibrium point, the linearized system is not linearly stabilizable.

Problem 9.10. Show that if $\Psi_{\rm C}(\cdot)$ in (9.183) and (9.184) is analytic, then the integral terms in (9.183) and (9.184) can be expressed in terms of an infinite Taylor expansion about the circumferentially average flow $\Phi(\cdot)$ to give (9.186) and (9.187).

Problem 9.11. To address only surge instabilities in axial flow compressor systems we restrict the three-state Moore-Greitzer model given by (9.199)–(9.201) to the invariant manifold where the rotating stall amplitude is zero, to obtain a two-state surge model. In this case,

$$\dot{\Phi}_{\rm s}(t) = -\tfrac{3}{2}\Phi_{\rm s}^2(t) - \tfrac{1}{2}\Phi_{\rm s}^3(t) - \Psi_{\rm s}(t), \quad \Phi_{\rm s}(0) = \Phi_{\rm s0}, \quad t \geq 0, \quad (9.260)$$

$$\dot{\Psi}_{\rm s}(t) = -u(t), \quad \Psi_{\rm s}(0) = \Psi_{\rm s0}, \quad (9.261)$$

where $\Phi_{\rm s}$ is the shifted axial average mass flow in the compressor, $\Psi_{\rm s}$ is the shifted total-to-static pressure rise, and u is the control input. Using Theorem 9.1, design a globally stabilizing controller for this system. Compare the state response and control effort to the feedback linearizing controller given by

$$u = \phi_{\rm FL}(\Phi_{\rm s}, \Psi_{\rm s}) = -100\Phi_{\rm s} + (5 - 3\Phi_{\rm s} - 1.5\Phi_{\rm s}^2)(\Psi_{\rm s} + 1.5\Phi_{\rm s}^2 + 0.15\Phi_{\rm s}^3). \quad (9.262)$$

9.9 Notes and References

Integrator backstepping for cascade and block cascade systems can be traced back to the works of Tsinias [433], Koditschek [240], Byrnes and Isidori [75], and Sontag and Sussmann [408]. Block cascade integrator backstepping via passivity notions was developed by Kokotović and Sussmann [241] and extended to nonlinear block cascade systems by Ortega [339] and Byrnes, Isidori, and Willems [77]. Further extensions were reported in Lozano, Brogliato, and Landau [287]. Recursive backstepping designs were reported in Saberi, Kokotović, and Sussmann [376] and Kanellakopoulos, Kokotović, and Morse [231]. For a textbook treatment of recursive integrator backstepping design see Krstić, Kanellakopoulos, and Kokotović [247].

The optimality and inverse optimality framework of integrator backstepping presented in Sections 9.3–9.5 were adopted from Haddad, Fausz, Chellaboina, and Abdallah [169]. Optimality issues for backstepping-like designs are also presented in Kolesnikov [242]. The Lyapunov function involving cross-terms in Section 9.5 and Problem 9.7 was introduced by Sepulchre, Janković, and Kokotović [395].

Chapter Ten

Disturbance Rejection Control for Nonlinear Dynamical Systems

10.1 Introduction

One of the fundamental problems in the analysis and feedback control design of linear systems is the ability of the control system to reject uncertain exogenous disturbances. To this end, $\mathcal{H}_\infty$ control theory has been developed to address the problem of disturbance rejection for linear systems with bounded energy (square-integrable) $\mathcal{L}_2$ signal norms on the disturbances and performance variables [121, 122, 478]. Since in this case the induced $\mathcal{H}_\infty$ transfer function norm corresponds to the worst-case disturbance attenuation, for systems with poorly modeled disturbances which possess significant power within arbitrarily small bandwidths $\mathcal{H}_\infty$ theory is clearly appropriate. For linear finite-dimensional, time-invariant systems the $\mathcal{H}_\infty$ control design problem has been thoroughly investigated in recent years (see, for example, [111, 237, 353, 355] and the numerous references therein). In particular, the $\mathcal{H}_\infty$ control design problem was formulated in the state space setting and was shown to correspond to a two-person zero-sum differential game problem, wherein the existence of an $\mathcal{H}_\infty$ (sub)optimal controller is equivalent to the existence of a solution to an algebraic Riccati equation arising in quadratic differential game theory [31, 32, 296].

Alternatively, the $\mathcal{H}_\infty$ analysis and synthesis control problem can also be formulated and solved in the state space setting using the notion of dissipativity theory [11]. In particular, using the bounded real Riccati equation it follows that the $\mathcal{H}_\infty$ norm of a (closed-loop) linear system is less than a prespecified positive number γ if and only if the (closed-loop) linear system is nonexpansive with respect to an appropriate quadratic supply rate involving the systems weighted input energy and output energy. Riccati-equation-based results for the mixed-norm $\mathcal{H}_2/\mathcal{H}_\infty$ problem [49, 112, 238] have also been developed using dissipativity notions to allow the trade-off between systems with stochastic white noise disturbance models ($\mathcal{H}_2$) possessing a fixed covariance (power spectral density) and deterministic bounded energy disturbance models. As in the pure $\mathcal{H}_\infty$ case, the mixed-norm $\mathcal{H}_2/\mathcal{H}_\infty$ problem can be compared to a game-theoretic framework involving a Nash differential game problem [279].

Using a nonlinear game-theoretic framework the authors in [22, 31] replace the algebraic Riccati equation arising in linear $\mathcal{H}_\infty$ theory with a particular Hamilton-Jacobi-Bellman equation (the *Isaacs equation*) to obtain a nonlinear equivalent to the $\mathcal{H}_\infty$ analysis and synthesis control problem. Sufficient conditions for the existence of stabilizing solutions of the Isaacs equation are given in [438–440] in terms of the existence of a linear (sub)optimal $\mathcal{H}_\infty$ controller for the linearized (about a given equilibrium point) nonlinear controlled system. In parallel research, the authors in [213–216] use nonlinear dissipativity theory [77,188,189,191,320,456,457] for nonlinear affine systems with appropriate storage functions and quadratic supply rates to obtain nonexpansive (gain bounded) closed-loop systems.

Although a nonlinear equivalent to $\mathcal{H}_\infty$ analysis and synthesis has been developed it is important to note that the methods and results discussed in [22,31,213–216,438–440] are independent of optimality considerations. In this chapter, we develop an optimality-based theory for disturbance rejection for nonlinear systems with bounded exogenous disturbances. The key motivation for developing an optimal and inverse optimal nonlinear control theory that additionally guarantees disturbance rejection is that it provides a class of candidate disturbance rejection controllers parameterized by the cost functional that is minimized. In the case of linear systems, optimality-based theories have proven extremely successful in numerous applications. Specifically, to fully address the trade-offs between $\mathcal{H}_2$ and $\mathcal{H}_\infty$ performance, the optimality-based linear-quadratic control problem was merged with $\mathcal{H}_\infty$ methods to address the mixed-norm $\mathcal{H}_2/\mathcal{H}_\infty$ control problem [49, 238].

In order to address the optimality-based disturbance rejection nonlinear control problem we extend the nonlinear-nonquadratic, continuous-time controller analysis and synthesis framework presented in Chapter 8. Specifically, using nonlinear dissipativity theory with appropriate storage functions and supply rates we transform the nonlinear disturbance rejection problem into an optimal control problem. This is accomplished by properly modifying the cost functional to account for exogenous disturbances so that the solution of the modified optimal nonlinear control problem serves as the solution to the disturbance rejection problem.

The framework guarantees that the closed-loop nonlinear input-output map is dissipative with respect to general supply rates. Specializing to quadratic supply rates involving net system energy flow and weighted input and output energy, the results guarantee passive and nonexpansive (gain bounded) closed-loop input-output maps, respectively. In the special case where the controlled system is linear the results, with appropriate quadratic supply rates, specialize to the mixed-norm $\mathcal{H}_2/\mathcal{H}_\infty$ framework developed in [49] and the mixed $\mathcal{H}_2$/positivity framework developed in [146, 151].

The main focus of this chapter is a methodology for designing optimal nonlinear controllers which guarantee disturbance rejection and minimize a (derived) performance functional that serves as an upper bound to a nonlinear-nonquadratic cost functional. In particular, the performance bound can be evaluated in closed form as long as the nonlinear-nonquadratic cost functional considered is related in a specific way to an underlying Lyapunov function that guarantees stability. This Lyapunov function is shown to be the solution to the steady-state form of the *Hamilton-Jacobi-Isaacs* equation for the controlled system and plays a key role in constructing the optimal nonlinear disturbance rejection control law. Furthermore, since the nonlinear-nonquadratic cost functional is closely related to the structure of the Lyapunov function the proposed framework provides a class of feedback stabilizing controllers that minimize a derived performance functional. Hence, the overall framework provides for a generalization of the Hamilton-Jacobi-Isaacs conditions for addressing the design of optimal and inverse optimal controllers for nonlinear systems with exogenous disturbances.

A key feature of the present chapter is that since the necessary and sufficient Hamilton-Jacobi-Isaacs optimality conditions are obtained for a modified nonlinear-nonquadratic performance functional rather than the original performance functional, *globally* optimal controllers are guaranteed to provide disturbance rejection. Of course, since the approach allows us to construct globally optimal controllers that minimize a given Hamiltonian, the resulting disturbance rejection controllers provide the best worst-case performance over the class of admissible input disturbances.

10.2 Nonlinear Dissipative Dynamical Systems with Bounded Disturbances

In this chapter, we consider nonlinear dynamical systems $\mathcal{G}$ of the form

$$\dot{x}(t) = f(x(t)) + J_1(x(t))w(t), \qquad x(0) = x_0, \qquad t \geq 0, \tag{10.1}$$
$$z(t) = h(x(t)) + J_2(x(t))w(t), \tag{10.2}$$

where $x \in \mathbb{R}^n$, $w \in \mathbb{R}^d$, $z \in \mathbb{R}^p$, $f : \mathbb{R}^n \to \mathbb{R}^n$, $J_1 : \mathbb{R}^n \to \mathbb{R}^{n\times d}$, $h : \mathbb{R}^n \to \mathbb{R}^p$, and $J_2 : \mathbb{R}^n \to \mathbb{R}^{p\times d}$. We assume that $f(\cdot)$, $J_1(\cdot)$, $h(\cdot)$, and $J_2(\cdot)$ are continuous mappings and $f(\cdot)$ has at least one equilibrium so that, without loss of generality, $f(0) = 0$ and $h(0) = 0$. Furthermore, for the nonlinear system $\mathcal{G}$ we assume that the required properties for the existence and uniqueness of solutions are satisfied, that is, $w(\cdot)$ satisfies sufficient regularity conditions such that (10.1) has a unique solution forward in time.

In this section, we present sufficient conditions for dissipativity for a class of nonlinear systems with bounded energy and bounded amplitude disturbances. In addition, we consider the problem of evaluating a performance bound for a nonlinear-nonquadratic cost functional. The cost bound is evaluated in closed form by relating the cost functional to an underlying Lyapunov function that guarantees asymptotic stability of the nonlinear system. Here, we restrict our attention to time-invariant, infinite-horizon systems. For the following result presenting sufficient conditions under which a nonlinear system is dissipative with respect to the supply rate $r(z,w)$, let $\mathcal{D} \subset \mathbb{R}^n$ be an open set, assume $0 \in \mathcal{D}$, let $f : \mathcal{D} \to \mathbb{R}^n$ be such that $f(0) = 0$, $h : \mathcal{D} \to \mathbb{R}^p$ be such that $h(0) = 0$, $J_1 : \mathcal{D} \to \mathbb{R}^{n\times d}$, and $J_2 : \mathcal{D} \to \mathbb{R}^{p\times d}$. Finally, let $\mathcal{W} \subset \mathbb{R}^d$ and let $r : \mathbb{R}^p \times \mathbb{R}^d \to \mathbb{R}$ be a given function.

Lemma 10.1. Consider the nonlinear dynamical system

$$\dot{x}(t) = f(x(t)) + J_1(x(t))w(t), \quad x(0) = x_0, \quad w(\cdot) \in \mathcal{L}_2, \quad t \geq 0, \tag{10.3}$$

$$z(t) = h(x(t)) + J_2(x(t))w(t). \tag{10.4}$$

Furthermore, assume that there exist functions $\Gamma : \mathcal{D} \to \mathbb{R}$ and $V : \mathcal{D} \to \mathbb{R}$, where $V(\cdot)$ is continuously differentiable, such that

$$V(0) = 0, \tag{10.5}$$

$$V(x) \geq 0, \quad x \in \mathcal{D}, \tag{10.6}$$

$$V'(x)J_1(x)w \leq r(z,w) + \Gamma(x), \quad x \in \mathcal{D}, \quad w \in \mathbb{R}^d, \tag{10.7}$$

$$V'(x)f(x) + \Gamma(x) \leq 0, \quad x \in \mathcal{D}. \tag{10.8}$$

Then the solution $x(t)$, $t \geq 0$, of (10.3) satisfies

$$V(x(T)) \leq \int_0^T r(z(t),w(t))\mathrm{d}t + V(x_0), \quad w(\cdot) \in \mathcal{L}_2, \quad T \geq 0. \tag{10.9}$$

Proof. Let $x(t)$, $t \geq 0$, satisfy (10.3) and let $w(\cdot) \in \mathcal{L}_2$. Then it follows from (10.7) and (10.8) that

$$\begin{aligned} \dot{V}(x(t)) &\triangleq \frac{\mathrm{d}V(x(t))}{\mathrm{d}t} \\ &= V'(x(t))[f(x(t)) + J_1(x(t))w(t)] \\ &\leq V'(x(t))f(x(t)) + \Gamma(x(t)) + r(z(t),w(t)) \\ &\leq r(z(t),w(t)), \quad t \geq 0. \end{aligned} \tag{10.10}$$

Now, integrating over $[0, T]$ yields

$$V(x(T)) - V(x_0) \leq \int_0^T r(z(t), w(t)) \mathrm{d}t, \quad w(\cdot) \in \mathcal{L}_2, \quad T \geq 0,$$

which proves the result. □

For the next result let $L : \mathcal{D} \to \mathbb{R}$ be given.

Theorem 10.1. Consider the nonlinear dynamical system given by (10.3) and (10.4) with performance functional

$$J(x_0) \triangleq \int_0^\infty L(x(t)) \mathrm{d}t, \tag{10.11}$$

where $x(t)$, $t \geq 0$, solves (10.3) with $w(t) \equiv 0$. Assume that there exist functions $\Gamma : \mathcal{D} \to \mathbb{R}$ and $V : \mathcal{D} \to \mathbb{R}$, where $V(\cdot)$ is continuously differentiable, such that $L(x) + \Gamma(x) \geq 0$, $x \in \mathcal{D}$,

$$V(0) = 0, \tag{10.12}$$

$$V(x) > 0, \quad x \in \mathcal{D}, \quad x \neq 0, \tag{10.13}$$

$$V'(x)f(x) < 0, \quad x \in \mathcal{D}, \quad x \neq 0, \tag{10.14}$$

$$V'(x)J_1(x)w \leq r(z, w) + L(x) + \Gamma(x), \quad x \in \mathcal{D}, \ w \in \mathcal{W}, \tag{10.15}$$

$$L(x) + V'(x)f(x) + \Gamma(x) = 0, \quad x \in \mathcal{D}. \tag{10.16}$$

Then the zero solution $x(t) \equiv 0$ of the undisturbed ($w(t) \equiv 0$) system (10.3) is locally asymptotically stable and there exists a neighborhood $\mathcal{D}_0 \subseteq \mathcal{D}$ of the origin such that if $\Gamma(x) \geq 0$, $x \in \mathcal{D}$, then

$$J(x_0) \leq \mathcal{J}(x_0) = V(x_0), \quad x_0 \in \mathcal{D}_0, \tag{10.17}$$

where

$$\mathcal{J}(x_0) \triangleq \int_0^\infty [L(x(t)) + \Gamma(x(t))] \mathrm{d}t \tag{10.18}$$

and where $x(t)$, $t \geq 0$, is a solution to (10.3) with $w(t) \equiv 0$. Furthermore, the solution $x(t)$, $t \geq 0$, to (10.3) satisfies the dissipativity constraint

$$\int_0^T r(z(t), w(t)) \mathrm{d}t + V(x_0) \geq 0, \quad w(\cdot) \in \mathcal{L}_2, \quad T \geq 0. \tag{10.19}$$

Finally, if $\mathcal{D} = \mathbb{R}^n$, $w(t) \equiv 0$, and

$$V(x) \to \infty \text{ as } \|x\| \to \infty, \tag{10.20}$$

then the zero solution $x(t) \equiv 0$ to (10.3) is globally asymptotically stable.

Proof. Let $x(t)$, $t \geq 0$, satisfy (10.3). Then

$$\dot{V}(x(t)) \triangleq \frac{\mathrm{d}}{\mathrm{d}t}V(x(t)) = V'(x(t))(f(x(t)) + J_1(x)w(t)), \qquad t \geq 0. \quad (10.21)$$

Hence, with $w(t) \equiv 0$, it follows from (10.14) that

$$\dot{V}(x(t)) < 0, \qquad t \geq 0, \qquad x(t) \neq 0. \quad (10.22)$$

Thus, from (10.12), (10.13), and (10.22) it follows that $V(\cdot)$ is a Lyapunov function for (10.3), which proves local asymptotic stability of the zero solution $x(t) \equiv 0$ of (10.3) with $w(t) \equiv 0$. Consequently, $x(t) \to 0$ as $t \to \infty$ for all initial conditions $x_0 \in \mathcal{D}_0$ for some neighborhood $\mathcal{D}_0 \subseteq \mathcal{D}$ of the origin.

Next, if $\Gamma(x) \geq 0$, $x \in \mathcal{D}$, and $w(t) \equiv 0$, (10.16) implies

$$\begin{aligned} L(x(t)) &= -\dot{V}(x(t)) + L(x(t)) + V'(x(t))f(x(t)) \\ &\leq -\dot{V}(x(t)) + L(x(t)) + V'(x(t))f(x(t)) + \Gamma(x(t)) \\ &= -\dot{V}(x(t)). \end{aligned}$$

Now, integrating over $[0,t)$ yields

$$\int_0^t L(x(s))\mathrm{d}s \leq -V(x(t)) + V(x_0).$$

Letting $t \to \infty$ and noting that $V(x(t)) \to 0$ for all $x_0 \in \mathcal{D}_0$ yields $J(x_0) \leq V(x_0)$.

Next, let $x(t)$, $t \geq 0$, satisfy (10.3) with $w(t) \equiv 0$. Then, with $L(x)$ replaced by $L(x) + \Gamma(x)$ and $J(x_0)$ replaced by $\mathcal{J}(x_0)$, it follows from Theorem 8.1 that $\mathcal{J}(x_0) = V(x_0)$. Finally, since $L(x) + \Gamma(x) \geq 0$, $x \in \mathcal{D}$, it follows that (10.12)–(10.16) implies (10.5)–(10.9), and hence, with $\Gamma(x)$ replaced by $L(x) + \Gamma(x)$, Lemma 10.1 yields

$$V(x(T)) \leq \int_0^T r(z(t), w(t))\mathrm{d}t + V(x_0), \qquad w(\cdot) \in \mathcal{L}_2, \qquad T \geq 0.$$

Now, (10.19) follows by noting that $V(x(T)) \geq 0$, $T \geq 0$. Finally, for $\mathcal{D} = \mathbb{R}^n$ global asymptotic stability of the zero solution $x(t) \equiv 0$ of the undisturbed ($w(t) \equiv 0$) system (10.3) is a direct consequence of the radially unbounded condition (10.20) on $V(x)$. □

10.3 Specialization to Dissipative Systems with Quadratic Supply Rates

In this section, we consider the special case in which $r(z, w)$ is a quadratic functional. Specifically, let $h : \mathcal{D} \to \mathbb{R}^p$, $J_2 : \mathcal{D} \to \mathbb{R}^{p \times d}$, $\hat{Q} \in \mathbb{S}^p$, $\hat{S} \in \mathbb{R}^{p \times d}$,

$\hat{R} \in \mathbb{S}^d$, and

$$r(z, w) = z^{\mathrm{T}}\hat{Q}z + 2z^{\mathrm{T}}\hat{S}w + w^{\mathrm{T}}\hat{R}w, \tag{10.23}$$

such that

$$N(x) \triangleq J_2^{\mathrm{T}}(x)\hat{Q}J_2(x) + J_2^{\mathrm{T}}(x)\hat{S} + \hat{S}^{\mathrm{T}}J_2(x) + \hat{R} > 0, \qquad x \in \mathcal{D}.$$

Furthermore, let $L(x) \geq 0$, $x \in \mathcal{D}$. Then

$$\begin{aligned}\Gamma(x) = {} & \left[\tfrac{1}{2}J_1^{\mathrm{T}}(x)V'^{\mathrm{T}}(x) - J_2^{\mathrm{T}}(x)\hat{Q}h(x) - \hat{S}^{\mathrm{T}}h(x)\right]^{\mathrm{T}} N^{-1}(x) \\ & \cdot \left[\tfrac{1}{2}J_1^{\mathrm{T}}(x)V'^{\mathrm{T}}(x) - J_2^{\mathrm{T}}(x)\hat{Q}h(x) - \hat{S}^{\mathrm{T}}h(x)\right] - h^{\mathrm{T}}(x)\hat{Q}h(x)\end{aligned}$$

satisfies (10.15) since in this case

$$\begin{aligned}& L(x) + \Gamma(x) - V'(x)J_1(x)w + r(z, w) \\ & = L(x) + \left[\tfrac{1}{2}J_1^{\mathrm{T}}(x)V'^{\mathrm{T}}(x) - J_2^{\mathrm{T}}(x)\hat{Q}h(x) - \hat{S}^{\mathrm{T}}(x)h(x) - N(x)w\right]^{\mathrm{T}} \\ & \quad \cdot N^{-1}(x)\left[\tfrac{1}{2}J_1^{\mathrm{T}}(x)V'^{\mathrm{T}}(x) - J_2^{\mathrm{T}}(x)\hat{Q}h(x) - \hat{S}^{\mathrm{T}}(x)h(x) - N(x)w\right] \\ & \geq 0.\end{aligned} \tag{10.24}$$

Corollary 10.1. Let $\gamma > 0$ and $L(x) \geq 0$, $x \in \mathcal{D}$, and consider the nonlinear dynamical system given by (10.3) and (10.4) with performance functional

$$J(x_0) \triangleq \int_0^\infty L(x(t))\mathrm{d}t, \tag{10.25}$$

where $x(t)$, $t \geq 0$, solves (10.3) with $w(t) \equiv 0$. Assume that there exists a continuously differentiable function $V : \mathcal{D} \to \mathbb{R}$ such that

$$V(0) = 0, \tag{10.26}$$

$$V(x) > 0, \quad x \in \mathcal{D}, \quad x \neq 0, \tag{10.27}$$

$$V'(x)f(x) < 0, \quad x \in \mathcal{D}, \quad x \neq 0, \tag{10.28}$$

$$L(x) + V'(x)f(x) + \Gamma(x) = 0, \quad x \in \mathcal{D}, \tag{10.29}$$

where

$$\begin{aligned}\Gamma(x) = {} & \left[\tfrac{1}{2}J_1^{\mathrm{T}}(x)V'^{\mathrm{T}}(x) + J_2^{\mathrm{T}}(x)h(x)\right]^{\mathrm{T}} \left[\gamma^2 I - J_2^{\mathrm{T}}(x)J_2(x)\right]^{-1} \\ & \cdot \left[\tfrac{1}{2}J_1^{\mathrm{T}}(x)V'^{\mathrm{T}}(x) + J_2^{\mathrm{T}}(x)h(x)\right] + h^{\mathrm{T}}(x)h(x).\end{aligned} \tag{10.30}$$

Then the zero solution $x(t) \equiv 0$ of the undisturbed ($w(t) \equiv 0$) system (10.3) is locally asymptotically stable and there exists a neighborhood $\mathcal{D}_0 \subseteq \mathcal{D}$ of the origin such that

$$J(x_0) \leq \mathcal{J}(x_0) = V(x_0), \quad x_0 \in \mathcal{D}_0, \tag{10.31}$$

where

$$\mathcal{J}(x_0) \triangleq \int_0^\infty [L(x(t)) + \Gamma(x(t))]\mathrm{d}t \tag{10.32}$$

and where $x(t)$, $t \geq 0$, solves (10.3) with $w(t) \equiv 0$. Furthermore, the solution $x(t)$, $t \geq 0$, of (10.3) satisfies the nonexpansivity constraint

$$\int_0^T z^{\mathrm{T}}(t)z(t)\mathrm{d}t \leq \gamma^2 \int_0^T w^{\mathrm{T}}(t)w(t)\mathrm{d}t + V(x_0), \qquad w(\cdot) \in \mathcal{L}_2, \qquad T \geq 0. \tag{10.33}$$

Finally, if $\mathcal{D} = \mathbb{R}^n$, $w(t) \equiv 0$, and

$$V(x) \to \infty \text{ as } \|x\| \to \infty, \tag{10.34}$$

then the zero solution $x(t) \equiv 0$ to (10.3) is globally asymptotically stable.

Proof. With $\hat{Q} = -I_p$, $\hat{S} = 0$, and $\hat{R} = \gamma^2 I_d$, it follows from (10.24) that $\Gamma(x)$ given by (10.30) satisfies (10.15). The result now follows as a direct consequence of Theorem 10.1. □

Note that if $L(x) = h^{\mathrm{T}}(x)h(x)$ in Corollary 10.1, then $\Gamma(x)$ can be chosen as

$$\begin{aligned}\Gamma(x) = {} & \left[\tfrac{1}{2}J_1^{\mathrm{T}}(x)V'^{\mathrm{T}}(x) + J_2^{\mathrm{T}}(x)h(x)\right]^{\mathrm{T}} \left[\gamma^2 I - J_2^{\mathrm{T}}(x)J_2(x)\right]^{-1} \\ & \cdot \left[\tfrac{1}{2}J_1^{\mathrm{T}}(x)V'^{\mathrm{T}}(x) + J_2^{\mathrm{T}}(x)h(x)\right].\end{aligned}$$

Example 10.1. Consider the nonlinear dynamical system

$$\dot{x}_1(t) = -x_2(t), \quad x_1(0) = x_{10}, \quad t \geq 0, \tag{10.35}$$

$$\dot{x}_2(t) = x_1(t) - x_2(t)\tanh(x_2^2(t) - x_3^2(t)) + x_2(t)w(t), \quad x_2(0) = x_{20}, \tag{10.36}$$

$$\dot{x}_3(t) = x_3(t)\tanh(x_2^2 - x_3^2(t)) - x_3(t)w(t), \quad x_3(0) = x_{30}, \tag{10.37}$$

$$z(t) = x_2^2(t) - x_3^2(t). \tag{10.38}$$

To show that (10.35)–(10.38) is nonexpansive with gain less than or equal to 1, note that (10.35)–(10.38) can be written in the state space form (10.3) and (10.4) with $x = [x_1\ x_2\ x_3]^{\mathrm{T}}$,

$$f(x) = \begin{bmatrix} -x_2 \\ x_1 - x_2\tanh(x_2^2 - x_3^2) \\ x_3\tanh(x_2^2 - x_3^2) \end{bmatrix}, \quad J_1(x) = \begin{bmatrix} 0 \\ x_2 \\ -x_3 \end{bmatrix}, \quad h(x) = x_2^2 - x_3^2, \tag{10.39}$$

and $J_2(x) = 0$. Now, with $V(x) = \frac{1}{2}x^{\mathrm{T}}x$ it follows that

$$\begin{aligned} V'(x)f(x) &= \begin{bmatrix} x_1 & x_2 & x_3 \end{bmatrix} \begin{bmatrix} -x_2 \\ x_1 - x_2\tanh(x_2^2 - x_3^2) \\ x_3\tanh(x_2^2 - x_3^2) \end{bmatrix} \\ &= -(x_2^2 - x_3^2)\tanh(x_2^2 - x_3^2) \\ &= -h(x)\tanh(h(x)) \end{aligned} \tag{10.40}$$

and

$$V'(x)J_1(x) = \begin{bmatrix} x_1 & x_2 & x_3 \end{bmatrix} \begin{bmatrix} 0 \\ x_2 \\ -x_3 \end{bmatrix} = h(x). \tag{10.41}$$

Next, let $\mathcal{D} \triangleq \{x \in \mathbb{R}^3 : |x_2^2 - x_3^2| \leq 1\}$. In this case, it can easily be shown that all conditions of Corollary 10.1 are satisfied with $\gamma = 1$ and $L(x) = h^{\mathrm{T}}(x)h(x)$, and hence, the nonlinear dynamical system (10.35)–(10.38) is nonexpansive with gain less than or equal to 1. △

The framework presented in Corollary 10.1 is an extension of the mixed-norm $\mathcal{H}_2/\mathcal{H}_\infty$ framework of Bernstein and Haddad [49] to nonlinear dynamical systems. Specifically, setting $f(x) = Ax$, $J_1(x) = D$, $h(x) = Ex$, $J_2(x) = 0$, $L(x) = x^{\mathrm{T}}Rx$, and $V(x) = x^{\mathrm{T}}Px$, where $A \in \mathbb{R}^{n\times n}$, $D \in \mathbb{R}^{n\times d}$, $E \in \mathbb{R}^{p\times n}$, $R \triangleq E^{\mathrm{T}}E > 0$, and $P \in \mathbb{P}^n$ satisfies

$$0 = A^{\mathrm{T}}P + PA + \gamma^{-2}PDD^{\mathrm{T}}P + R, \tag{10.42}$$

it follows from Corollary 10.1, with $L(x) = h^{\mathrm{T}}(x)h(x) = x^{\mathrm{T}}Rx$, $\Gamma(x) = \gamma^{-2}x^{\mathrm{T}}PDD^{\mathrm{T}}Px$, and $x_0 = 0$, that

$$\int_0^T x^{\mathrm{T}}(t)Rx(t)\mathrm{d}t \leq \gamma^2 \int_0^T w^{\mathrm{T}}(t)w(t)\mathrm{d}t, \qquad T \geq 0, \qquad w(\cdot) \in \mathcal{L}_2, \tag{10.43}$$

or, equivalently, the $\mathcal{H}_\infty$ norm of

$$G(s) \sim \left[\begin{array}{c|c} A & D \\ \hline E & 0 \end{array}\right]$$

satisfies

$$|||G|||_\infty \triangleq \sup_{\omega\in\mathbb{R}} \sigma_{\max}(G(\jmath\omega)) \leq \gamma. \tag{10.44}$$

Now, (10.31) implies

$$\begin{aligned}\int_0^\infty x^{\mathrm{T}}(t)Rx(t)\mathrm{d}t &\leq \int_0^\infty x^{\mathrm{T}}(t)(R + \gamma^{-2}PDD^{\mathrm{T}}P)x(t)\mathrm{d}t \\ &= \int_0^\infty x_0^{\mathrm{T}}e^{A^{\mathrm{T}}t}(R + \gamma^{-2}PDD^{\mathrm{T}}P)e^{At}x_0\mathrm{d}t,\end{aligned}$$

where $x(t)$, $t \geq 0$, solves (10.3) with $w(t) \equiv 0$.

As is common practice [274], to eliminate the explicit dependence of $J(x_0)$ on the initial condition x_0 we assume $x_0x_0^{\mathrm{T}}$ has expected value V, that is, $\mathbb{E}[x_0x_0^{\mathrm{T}}] = V$, where $\mathbb{E}$ denotes expectation. Invoking this step leads to

$$\mathbb{E}\left[\int_0^\infty x^{\mathrm{T}}(t)Rx(t)\mathrm{d}t\right] = \mathbb{E}\left[\int_0^\infty x_0^{\mathrm{T}}e^{A^{\mathrm{T}}t}Re^{At}x_0\mathrm{d}t\right] = \mathbb{E}[x_0^{\mathrm{T}}\hat{P}x_0] = \mathrm{tr}\,\hat{P}V,$$

where

$$0 = A^{\mathrm{T}}\hat{P} + \hat{P}A + R$$

and

$$\begin{aligned}\mathbb{E}\left[\int_0^\infty x^{\mathrm{T}}(t)(R+\gamma^{-2}PDD^{\mathrm{T}}P)x(t)\mathrm{d}t\right]\\ &= \mathbb{E}\left[\int_0^\infty x_0^{\mathrm{T}}e^{A^{\mathrm{T}}t}(R+\gamma^{-2}PDD^{\mathrm{T}}P)e^{At}x_0\mathrm{d}t\right]\\ &= \mathbb{E}\left[x_0^{\mathrm{T}}Px_0\right]\\ &= \mathrm{tr}\ PV,\end{aligned}$$

where P satisfies (10.42). Hence, $|||G|||_2^2 = \mathrm{tr}\ \hat{P}V \leq \mathrm{tr}\ PV$, which implies that $\mathcal{J}(x_0)$ given by (10.32) provides an upper bound to the $\mathcal{H}_2$ norm of $G(s)$.

Corollary 10.2. Let $L(x) \geq 0$, $x \in \mathcal{D}$, $p = d$, and consider the nonlinear dynamical system given by (10.3) and (10.4) with performance functional

$$J(x_0) \triangleq \int_0^\infty L(x(t))\mathrm{d}t, \tag{10.45}$$

where $x(t)$, $t \geq 0$, solves (10.3) with $w(t) \equiv 0$. Assume that there exists a continuously differentiable function $V : \mathcal{D} \to \mathbb{R}$ such that

$$V(0) = 0, \tag{10.46}$$

$$V(x) > 0, \quad x \in \mathcal{D}, \quad x \neq 0, \tag{10.47}$$

$$V'(x)f(x) < 0, \quad x \in \mathcal{D}, \quad x \neq 0, \tag{10.48}$$

$$L(x) + V'(x)f(x) + \Gamma(x) = 0, \quad x \in \mathcal{D}, \tag{10.49}$$

where

$$\begin{aligned}\Gamma(x) &= \left[\tfrac{1}{2}J_1^{\mathrm{T}}(x)V'^{\mathrm{T}}(x) - h(x)\right]^{\mathrm{T}}\left[J_2(x) + J_2^{\mathrm{T}}(x)\right]^{-1}\\ &\quad\cdot\left[\tfrac{1}{2}J_1^{\mathrm{T}}(x)V'^{\mathrm{T}}(x) - h(x)\right].\end{aligned} \tag{10.50}$$

Then the zero solution $x(t) \equiv 0$ of the undisturbed ($w(t) \equiv 0$) system (10.3) is locally asymptotically stable and there exists a neighborhood $\mathcal{D}_0 \subseteq \mathcal{D}$ of the origin such that

$$J(x_0) \leq \mathcal{J}(x_0) = V(x_0), \quad x_0 \in \mathcal{D}_0, \tag{10.51}$$

where

$$\mathcal{J}(x_0) \triangleq \int_0^\infty [L(x(t)) + \Gamma(x(t))]\mathrm{d}t \tag{10.52}$$

and where $x(t)$, $t \geq 0$, solves (10.3) with $w(t) \equiv 0$. Furthermore, the solution $x(t)$, $t \geq 0$, of (10.3) satisfies the passivity constraint

$$\int_0^T 2z^{\mathrm{T}}(t)w(t)\mathrm{d}t + V(x_0) \geq 0, \quad w(\cdot) \in \mathcal{L}_2, \quad T \geq 0. \tag{10.53}$$

Finally, if $\mathcal{D} = \mathbb{R}^n$, $w(t) \equiv 0$, and

$$V(x) \to \infty \text{ as } \|x\| \to \infty, \tag{10.54}$$

then the zero solution $x(t) \equiv 0$ to (10.3) is globally asymptotically stable.

Proof. With $\hat{Q} = 0$, $\hat{S} = I_d$, and $\hat{R} = 0$, it follows from (10.24) that $\Gamma(x)$ given by (10.50) satisfies (10.15). The result now follows as a direct consequence of Theorem 10.1. □

The framework presented in Corollary 10.2 is an extension of the $\mathcal{H}_2$/positivity framework of Haddad and Bernstein [146, 151] to nonlinear dynamical systems. Specifically, setting $f(x) = Ax$, $J_1(x) = D$, $h(x) = Ex$, $J_2(x) = E_\infty$, $L(x) = x^{\mathrm{T}}Rx$, $V(x) = x^{\mathrm{T}}Px$, and $\Gamma(x) = x^{\mathrm{T}}[D^{\mathrm{T}}P - E]^{\mathrm{T}}(E_\infty + E_\infty^{\mathrm{T}})^{-1}[D^{\mathrm{T}}P - E]x$, where $A \in \mathbb{R}^{n\times n}$, $D \in \mathbb{R}^{n\times d}$, $E \in \mathbb{R}^{d\times n}$, $E_\infty \in \mathbb{R}^{d\times d}$, $R \in \mathbb{P}^n$, and $P \in \mathbb{P}^n$ satisfies

$$0 = A^{\mathrm{T}}P + PA + (D^{\mathrm{T}}P - E)^{\mathrm{T}}(E_\infty + E_\infty^{\mathrm{T}})^{-1}(D^{\mathrm{T}}P - E) + R, \tag{10.55}$$

it follows from Corollary 10.2, with $x_0 = 0$, that

$$\int_0^T 2w^{\mathrm{T}}(t)z(t)\mathrm{d}t \geq 0, \qquad w(\cdot) \in \mathcal{L}_2, \qquad T \geq 0, \tag{10.56}$$

or, equivalently,

$$G_\infty(s) + G_\infty^*(s) \geq 0, \qquad \mathrm{Re}[s] > 0, \tag{10.57}$$

where

$$G_\infty(s) \sim \left[\begin{array}{c|c} A & D \\ \hline E & E_\infty \end{array}\right].$$

Now, using similar arguments as in the $\mathcal{H}_\infty$ case, (10.51) implies that

$$\begin{aligned}\mathrm{tr}\ \hat{P}V &= \mathbb{E}\left[\int_0^\infty x^{\mathrm{T}}(t)Rx(t)\mathrm{d}t\right] \\ &\leq \mathbb{E}\left[\int_0^\infty x^{\mathrm{T}}(t)[R + (D^{\mathrm{T}}P - E)^{\mathrm{T}}(E_\infty + E_\infty^{\mathrm{T}})^{-1}(D^{\mathrm{T}}P - E)]x(t)\mathrm{d}t\right],\end{aligned}$$

or, equivalently, since

$$\begin{aligned}&\mathbb{E}\left[\int_0^\infty x^{\mathrm{T}}(t)[R + (D^{\mathrm{T}}P - E)^{\mathrm{T}}(E_\infty + E_\infty^{\mathrm{T}})^{-1}(D^{\mathrm{T}}P - E)]x(t)\mathrm{d}t\right] \\ &= \mathbb{E}\left[\int_0^\infty x_0^{\mathrm{T}}e^{A^{\mathrm{T}}t}[R + (D^{\mathrm{T}}P - E)^{\mathrm{T}}(E_\infty + E_\infty^{\mathrm{T}})^{-1}(D^{\mathrm{T}}P - E)]e^{At}x_0\mathrm{d}t\right] \\ &= \mathbb{E}\left[x_0^{\mathrm{T}}Px_0\right] \\ &= \mathrm{tr}\ PV,\end{aligned}$$

$\|G\|_2^2 = \text{tr } \hat{P}V \leq \text{tr } PV$. Hence, $\mathcal{J}(x_0)$ given by (10.52) provides an upper bound to the $\mathcal{H}_2$ norm of $G(s)$.

Next, define the subset of square-integrable bounded disturbances

$$\mathcal{W}_\beta \triangleq \{w(\cdot) \in \mathcal{L}_2 : \int_0^\infty w^{\mathrm{T}}(t)w(t)\mathrm{d}t \leq \beta\}, \tag{10.58}$$

where $\beta > 0$. Furthermore, let $L : \mathcal{D} \to \mathbb{R}$ be given such that $L(x) \geq 0$, $x \in \mathcal{D}$.

Theorem 10.2. Let $\gamma > 0$ and consider the nonlinear dynamical system (10.3) with performance functional (10.11). Assume that there exists a continuously differentiable function $V : \mathcal{D} \to \mathbb{R}$ such that

$$V(0) = 0, \tag{10.59}$$

$$V(x) > 0, \quad x \in \mathcal{D}, \quad x \neq 0, \tag{10.60}$$

$$V'(x)f(x) < 0, \quad x \in \mathcal{D}, \quad x \neq 0, \tag{10.61}$$

$$L(x) + V'(x)f(x) + \frac{\beta}{4\gamma}V'(x)J_1(x)J_1^{\mathrm{T}}(x)V'^{\mathrm{T}}(x) = 0, \quad x \in \mathcal{D}. \tag{10.62}$$

Then then the zero solution $x(t) \equiv 0$ of the undisturbed ($w(t) \equiv 0$) system (10.3) is locally asymptotically stable and there exists a neighborhood $\mathcal{D}_0 \subseteq \mathcal{D}$ of the origin such that

$$J(x_0) \leq \mathcal{J}(x_0) = V(x_0), \quad x_0 \in \mathcal{D}_0, \tag{10.63}$$

where

$$\mathcal{J}(x_0) \triangleq \int_0^\infty [L(x(t)) + \Gamma(x(t))]\mathrm{d}t, \tag{10.64}$$

$$\Gamma(x) = \frac{\beta}{4\gamma}V'(x)J_1(x)J_1^{\mathrm{T}}(x)V'^{\mathrm{T}}(x), \tag{10.65}$$

and where $x(t)$, $t \geq 0$, solves (10.3) with $w(t) \equiv 0$. Furthermore, if $x_0 = 0$ then the solution $x(t)$, $t \geq 0$, of (10.3) satisfies

$$V(x(T)) \leq \gamma, \qquad w(\cdot) \in \mathcal{L}_2, \qquad T \geq 0. \tag{10.66}$$

Finally, if $\mathcal{D} = \mathbb{R}^n$, $w(t) \equiv 0$, and

$$V(x) \to \infty \text{ as } \|x\| \to \infty, \tag{10.67}$$

then the zero solution $x(t) \equiv 0$ to (10.3) is globally asymptotically stable.

Proof. The proofs for local and global asymptotic stability and the performance bound (10.63) are identical to the proofs of local and global asymptotic stability given in Theorem 10.1 and the performance bound (10.17). Next, with $r(z, w) = \frac{\gamma}{\beta}w^{\mathrm{T}}w$ and $\Gamma(x)$ given by (10.65) it follows from Lemma 10.1 that

$$V(x(T)) \le \frac{\gamma}{\beta}\int_0^T w^{\mathrm{T}}(t)w(t)\mathrm{d}t, \qquad w(\cdot)\in\mathcal{L}_2, \qquad T\ge 0,$$

which yields (10.66). □

10.4 A Riccati Equation Characterization for Mixed $\mathcal{H}_2/\mathcal{L}_1$ Performance

In this section, we consider the dynamical system (10.1) and (10.2) with $f(x) = Ax$, $J_1(x) = D$, $h(x) = Ex$, and $J_2(x) = E_\infty$, where $A \in \mathbb{R}^{n\times n}$, $D \in \mathbb{R}^{n\times d}$, $E \in \mathbb{R}^{p\times n}$, $E_\infty \in \mathbb{R}^{p\times d}$, and where A is asymptotically stable so that

$$\dot{x}(t) = Ax(t) + Dw(t), \quad x(0) = 0, \quad w(\cdot)\in\mathcal{L}_2, \quad t\ge 0, \tag{10.68}$$

$$z(t) = Ex(t) + E_\infty w(t), \tag{10.69}$$

where $\hat{\mathcal{W}}$ consists of unit-peak input signals defined by

$$\hat{\mathcal{W}} \triangleq \{w(\cdot) : w^{\mathrm{T}}(t)w(t) \le 1,\ t\ge 0\}. \tag{10.70}$$

The following result provides an upper bound to the $\mathcal{L}_1$ norm [104] ($\mathcal{L}_\infty$ equi-induced norm) of the convolution operator G of the linear, time-invariant system (10.68) and (10.69) given by

$$|\!|\!|G|\!|\!|_1 \triangleq \sup_{w(\cdot)\in\mathcal{L}_2}\left\{\sup_{t\ge 0}\|z(t)\|\right\},$$

where $\|\cdot\|$ denotes the Euclidean vector norm. From an input-output point of view the $\mathcal{L}_1$ norm captures the worst-case amplification from input disturbance signals to output signals, where the signal size is taken to be the supremum over time of the signal's pointwise-in-time Euclidean norm.

Theorem 10.3. Let $\alpha > 0$ and consider the linear dynamical system (10.68) and (10.69). Then

$$|\!|\!|G|\!|\!|_1 \le \sigma_{\max}^{1/2}(EP^{-1}E^{\mathrm{T}}) + \sigma_{\max}^{1/2}(E_\infty E_\infty^{\mathrm{T}}), \tag{10.71}$$

where $P > 0$ satisfies

$$0 \ge A^{\mathrm{T}}P + PA + \alpha P + \frac{1}{\alpha}PDD^{\mathrm{T}}P. \tag{10.72}$$

Proof. Let $T \ge 0$ and consider the shifted linear dynamical system

$$\dot{\tilde{x}}(t) = A_\alpha\tilde{x}(t) + Dv(t), \quad \tilde{x}(0) = 0, \quad t\ge 0, \tag{10.73}$$

$$z(t) = e^{-\frac{\alpha}{2}(t-T)}(E\tilde{x}(t) + E_\infty v(t)), \tag{10.74}$$

where $A_\alpha \triangleq A+\frac{\alpha}{2}I$, $\alpha > 0$, $\tilde{x}(t) \triangleq e^{\frac{\alpha}{2}(t-T)}x(t)$, and $v(t) \triangleq e^{\frac{\alpha}{2}(t-T)}w(t)$. Note that (10.73) and (10.74) are equivalent to (10.68) and (10.69). Furthermore, note that if $w(\cdot) \in \hat{\mathcal{W}}$ then $v(\cdot) \in \mathcal{V}$, where $\mathcal{V} \triangleq \{v(\cdot) : \int_0^T v^{\mathrm{T}}(t)v(t)\mathrm{d}t \leq \frac{1}{\alpha}\}$. Hence,

$$|\!|\!|G|\!|\!|_1^2 \leq \sup_{T\geq 0}\ \sup_{v(\cdot)\in\mathcal{V}} \|z(T)\|^2.$$

Next, with $f(x) = A_\alpha\tilde{x}(t)$, $J_1(x) = D$, $V(x) = \tilde{x}^{\mathrm{T}}P\tilde{x}$, $\mathcal{W} = \mathcal{V}$, $\beta = \frac{1}{\alpha}$, $\gamma = 1$, and $L(x) = \tilde{x}^{\mathrm{T}}R\tilde{x}$, where $R \in \mathbb{R}^{n\times n}$ is an arbitrary positive-definite matrix, it follows from Theorem 10.2 that if there exists $P > 0$ such that

$$0 = A_\alpha^{\mathrm{T}}P + PA_\alpha + \frac{1}{\alpha}PDD^{\mathrm{T}}P + R, \tag{10.75}$$

then

$$\tilde{x}^{\mathrm{T}}(T)P\tilde{x}(T) \leq 1,$$

and hence, for all $v(\cdot) \in \mathcal{V}$,

$$\|z(T)\| = \|E\tilde{x}(T) + E_\infty w(T)\| \leq \sigma_{\max}^{1/2}(EP^{-1}E^{\mathrm{T}}) + \sigma_{\max}^{1/2}(E_\infty E_\infty^{\mathrm{T}}).$$

The result is now immediate by noting that (10.75) is equivalent to (10.72). □

Note that we can replace the Riccati inequality (10.72) by the Riccati equation

$$0 = A^{\mathrm{T}}P + PA + \alpha P + \frac{1}{\alpha}PDD^{\mathrm{T}}P \tag{10.76}$$

in Theorem 10.3. In this case, if (A, D) is controllable and A_α is asymptotically stable, then there exists a positive-definite solution satisfying (10.76). In particular,

$$P = \left[\int_0^\infty e^{A_\alpha t}DD^{\mathrm{T}}e^{A_\alpha^{\mathrm{T}} t}\mathrm{d}t\right]^{-1}. \tag{10.77}$$

Now, letting $\mathcal{Q} = \alpha P^{-1}$ in Theorem 10.3 yields

$$|\!|\!|G|\!|\!|_1 \leq \frac{1}{\alpha}\sigma_{\max}^{1/2}(E\mathcal{Q}E^{\mathrm{T}}) + \sigma_{\max}^{1/2}(E_\infty E_\infty^{\mathrm{T}}), \tag{10.78}$$

where $\mathcal{Q} > 0$ satisfies

$$0 \geq A\mathcal{Q} + \mathcal{Q}A^{\mathrm{T}} + \alpha\mathcal{Q} + DD^{\mathrm{T}}. \tag{10.79}$$

Furthermore, it is interesting to note that in the case where $E_\infty = 0$ the solution $\mathcal{Q}$ to (10.79) satisfies the bound

$$Q \leq \mathcal{Q}, \tag{10.80}$$

where Q satisfies

$$0 = AQ + QA^{\mathrm{T}} + DD^{\mathrm{T}}, \tag{10.81}$$

and hence,

$$\|G\|_2^2 = \operatorname{tr}\, EQE^{\mathrm{T}} \leq \operatorname{tr}\, E\mathcal{Q}E^{\mathrm{T}}. \tag{10.82}$$

Thus, (10.79) can be used to provide a trade-off between $\mathcal{H}_2$ and mixed $\mathcal{H}_2/\mathcal{L}_1$ performance. For further details see [157].

10.5 Nonlinear-Nonquadratic Controllers for Systems with Bounded Disturbances

In this section, we consider a control problem involving a notion of optimality with respect to an *auxiliary cost* which guarantees a bound on the worst-case value of a nonlinear-nonquadratic cost functional over a prescribed set of bounded input disturbances. The optimal feedback controllers are derived as a direct consequence of Theorem 10.1 and provide a generalization of the Hamilton-Jacobi-Bellman conditions for time-invariant, infinite-horizon problems for addressing nonlinear feedback controllers for nonlinear systems with bounded energy disturbances that additionally minimize a nonlinear-nonquadratic cost functional.

To address the optimal control problem let $\mathcal{D} \subset \mathbb{R}^n$ be an open set and let $U \subset \mathbb{R}^m$, where $0 \in \mathcal{D}$ and $0 \in U$. Furthermore, let $\mathcal{W} \subset \mathbb{R}^d$ and let $r : \mathbb{R}^p \times \mathbb{R}^d \to \mathbb{R}$ be a given function. Next, consider the controlled dynamical system

$$\dot{x}(t) = F(x(t), u(t)) + J_1(x(t))w(t) \quad x(0) = x_0, \quad w(\cdot) \in \mathcal{L}_2, \quad t \geq 0, \tag{10.83}$$

with performance variables

$$z(t) = h(x(t), u(t)) + J_2(x(t))w(t), \tag{10.84}$$

where $F : \mathbb{R}^n \times \mathbb{R}^m \to \mathbb{R}^n$ satisfies $F(0,0) = 0$, $J_1 : \mathbb{R}^n \to \mathbb{R}^d$, $h : \mathbb{R}^n \times \mathbb{R}^m \to \mathbb{R}^p$ satisfies $h(0,0) = 0$, $J_2 : \mathbb{R}^n \to \mathbb{R}^{p\times d}$, and the control $u(\cdot)$ is restricted to the class of admissible controls consisting of measurable functions $u(\cdot) \in \mathcal{U}$ such that $u(t) \in U$ for all $t \geq 0$, where the control constraint set U is given. We assume $0 \in U$. Given a control law $\phi(\cdot)$ and a feedback control law $u(t) = \phi(x(t))$, the closed-loop system shown in Figure 10.1 has the form

$$\dot{x}(t) = F(x(t), \phi(x(t))) + J_1(x(t))w(t), \quad x(0) = x_0, \quad t \geq 0, \tag{10.85}$$

$$z(t) = h(x(t), \phi(x(t))) + J_2(x(t))w(t). \tag{10.86}$$

We assume that the mapping $\phi : \mathcal{D} \to U$ satisfies sufficient regularity conditions such that the resulting closed-loop system (10.85) has a unique solution forward in time.

Next, we present an extension of Theorem 8.2 for characterizing feedback controllers that guarantee stability, minimize an auxiliary performance functional, and guarantee that the input-output map of the closed-

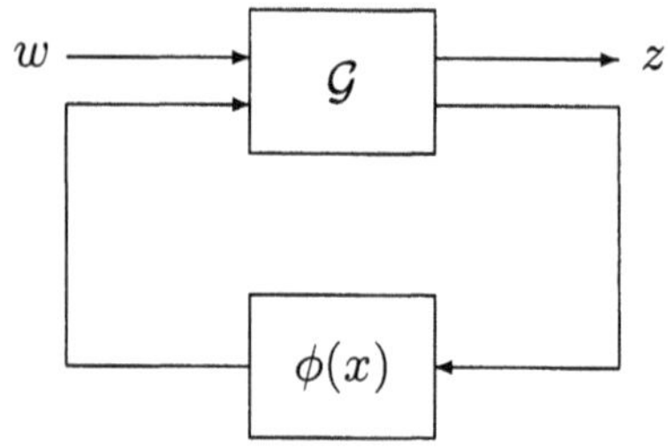

Figure 10.1 Disturbed nonlinear closed-loop feedback system.

loop system is dissipative, nonexpansive, or passive for bounded input disturbances. For the statement of these results let $L : \mathcal{D} \times U \to \mathbb{R}$ and define the set of regulation controllers for the nonlinear system with $w(t) \equiv 0$ by

$$\mathcal{S}(x_0) \triangleq \{u(\cdot) : u(\cdot) \text{ is admissible and } x(\cdot) \text{ given by (10.83)} \\ \text{satisfies } x(t) \to 0 \text{ as } t \to \infty \text{ with } w(t) \equiv 0\}.$$

Theorem 10.4. Consider the nonlinear controlled dynamical system (10.83) and (10.84) with performance functional

$$J(x_0, u(\cdot)) \triangleq \int_0^\infty L(x(t), u(t))\mathrm{d}t, \tag{10.87}$$

where $u(\cdot)$ is an admissible control. Assume that there exist a continuously differentiable function $V : \mathcal{D} \to \mathbb{R}$, a function $\Gamma : \mathcal{D} \times U \to \mathbb{R}$, and control law $\phi : \mathcal{D} \to U$ such that

$$V(0) = 0, \tag{10.88}$$
$$V(x) > 0, \quad x \in \mathcal{D}, \quad x \neq 0, \tag{10.89}$$
$$\phi(0) = 0, \tag{10.90}$$
$$V'(x)F(x,\phi(x)) < 0, \quad x \in \mathcal{D}, \quad x \neq 0, \tag{10.91}$$
$$V'(x)J_1(x)w \leq r(z,w) + L(x,\phi(x)) + \Gamma(x,\phi(x)), \; x \in \mathcal{D}, \; w \in \mathcal{W}, \tag{10.92}$$
$$H(x,\phi(x)) = 0, \quad x \in \mathcal{D}, \tag{10.93}$$
$$H(x,u) \geq 0, \quad x \in \mathcal{D}, \quad u \in U, \tag{10.94}$$

where

$$H(x,u) \triangleq V'(x)F(x,u) + L(x,u) + \Gamma(x,u). \tag{10.95}$$

Then, with the feedback control $u(\cdot) = \phi(x(\cdot))$, there exists a neighborhood $\mathcal{D}_0 \subseteq \mathcal{D}$ of the origin such that if $x_0 \in \mathcal{D}_0$ and $w(t) \equiv 0$, the zero solution $x(t) \equiv 0$ of the closed-loop system (10.85) is locally asymptotically stable. If, in addition, $\Gamma(x,\phi(x)) \geq 0$, $x \in \mathcal{D}$, then

$$J(x_0, \phi(x(\cdot))) \leq \mathcal{J}(x_0, \phi(x(\cdot))) = V(x_0), \tag{10.96}$$

where

$$\mathcal{J}(x_0, u(\cdot)) \triangleq \int_0^\infty [L(x(t), u(t)) + \Gamma(x(t), u(t))]\mathrm{d}t \tag{10.97}$$

and where $u(\cdot)$ is admissible and $x(t)$, $t \geq 0$, solves (10.83) with $w(t) \equiv 0$. In addition, if $x_0 \in \mathcal{D}_0$ then the feedback control $u(\cdot) = \phi(x(\cdot))$ minimizes $\mathcal{J}(x_0, u(\cdot))$ in the sense that

$$\mathcal{J}(x_0, \phi(x(\cdot))) = \min_{u(\cdot)\in\mathcal{S}(x_0)} \mathcal{J}(x_0, u(\cdot)). \tag{10.98}$$

Furthermore, the solution $x(t)$, $t \geq 0$, of (10.85) satisfies the dissipativity constraint

$$\int_0^T r(z(t), w(t)) + V(x_0) \geq 0, \qquad w(\cdot) \in \mathcal{L}_2, \qquad T \geq 0. \tag{10.99}$$

Finally, if $\mathcal{D} = \mathbb{R}^n$, $U = \mathbb{R}^m$, $w(t) \equiv 0$, and

$$V(x) \to \infty \text{ as } \|x\| \to \infty, \tag{10.100}$$

then the zero solution $x(t) \equiv 0$ of the closed-loop system (10.85) is globally asymptotically stable.

Proof. Local and global asymptotic stability is a direct consequence of (10.88)–(10.91) by applying Theorem 10.1 to the closed-loop system (10.85). Furthermore, using (10.93), the performance bound (10.96) is a restatement of (10.17) as applied to the closed-loop system. Next, let $u(\cdot) \in \mathcal{S}(x_0)$ and let $x(t)$, $t \geq 0$, be the solution of (10.83) with $w(t) \equiv 0$. Then (10.98) follows from Theorem 8.2 with $L(x, u)$ replaced by $L(x, u) + \Gamma(x, u)$ and $J(x_0, u(\cdot))$ replaced by $\mathcal{J}(x_0, u(\cdot))$. Finally, using (10.92), condition (10.99) is a restatement of (10.19) as applied to the closed-loop system. □

Next, we specialize Theorem 10.4 to linear systems with bounded energy disturbances and provide connections to the mixed-norm $\mathcal{H}_2/\mathcal{H}_\infty$ and mixed $\mathcal{H}_2$/positivity frameworks developed in [49] and [146, 420], respectively. Specifically, we consider the case in which $F(x, u) = Ax + Bu$, $J_1(x) = D$, $h(x, u) = E_1 x + E_2 u$, and $J_2(x) = E_\infty$, where $A \in \mathbb{R}^{n\times n}$, $B \in \mathbb{R}^{n\times m}$, $D \in \mathbb{R}^{n\times d}$, $E_1 \in \mathbb{R}^{p\times n}$, $E_2 \in \mathbb{R}^{p\times m}$, and $E_\infty \in \mathbb{R}^{p\times d}$. First, we consider the mixed-norm $\mathcal{H}_2/\mathcal{H}_\infty$ case where $r(z, w) = \gamma^2 w^\mathrm{T} w - z^\mathrm{T} z$, and where $\gamma > 0$ is given. For the following result define $R_1 \triangleq E_1^\mathrm{T} E_1 > 0$, $R_2 \triangleq E_2^\mathrm{T} E_2 > 0$, $S \triangleq B R_2^{-1} B^\mathrm{T}$, and assume $E_\infty = 0$ and $R_{12} \triangleq E_1^\mathrm{T} E_2 = 0$.

Corollary 10.3. Consider the linear controlled system

$$\dot{x}(t) = Ax(t) + Bu(t) + Dw(t), \quad x(0) = x_0, \quad w(\cdot) \in \mathcal{L}_2, \quad t \geq 0, \tag{10.101}$$

$$z(t) = E_1 x(t) + E_2 u(t), \tag{10.102}$$

with performance functional

$$J(x_0, u(\cdot)) = \int_0^\infty [x^{\mathrm{T}}(t)R_1x(t) + u^{\mathrm{T}}R_2u(t)]\mathrm{d}t, \tag{10.103}$$

where $u(\cdot)$ is admissible. Assume that there exists a positive-definite matrix $P \in \mathbb{R}^{n\times n}$ such that

$$0 = A^{\mathrm{T}}P + PA + R_1 + \gamma^{-2}PDD^{\mathrm{T}}P - PSP, \tag{10.104}$$

where $\gamma > 0$. Then, with the feedback control law $u = \phi(x) = -R_2^{-1}B^{\mathrm{T}}Px$, the zero solution $x(t) \equiv 0$ of the undisturbed $(w(t) \equiv 0)$ system (10.101) is globally asymptotically stable for all $x_0 \in \mathbb{R}^n$ and

$$J(x_0, \phi(x(\cdot))) \leq \mathcal{J}(x_0, \phi(x(\cdot))) = x_0^{\mathrm{T}}Px_0, \tag{10.105}$$

where

$$\mathcal{J}(x_0, u(\cdot)) = \int_0^\infty [x^{\mathrm{T}}(t)(R_1 + \gamma^{-2}PDD^{\mathrm{T}}P)x(t) + u^{\mathrm{T}}(t)R_2u(t)]\mathrm{d}t, \tag{10.106}$$

and where $u(\cdot)$ is admissible and $x(t)$, $t \geq 0$, solves (10.101) with $w(t) \equiv 0$. Furthermore,

$$\mathcal{J}(x_0, \phi(x(\cdot))) = \min_{u(\cdot)\in\mathcal{S}(x_0)} \mathcal{J}(x_0, u(\cdot)), \tag{10.107}$$

where $\mathcal{S}(x_0)$ is the set of regulation controllers for the system (10.101) with $w(t) \equiv 0$ and $x_0 \in \mathbb{R}^n$. Finally, if $x_0 = 0$ then, with $u = \phi(x)$, the solution $x(t)$, $t \geq 0$, of (10.101) satisfies the nonexpansivity constraint

$$\int_0^T z(t)^{\mathrm{T}}z(t)\mathrm{d}t \leq \gamma^2 \int_0^T w(t)^{\mathrm{T}}w(t)\mathrm{d}t, \qquad w(\cdot) \in \mathcal{L}_2, \qquad T \geq 0, \tag{10.108}$$

or, equivalently, $|||\tilde{G}(s)|||_\infty \leq \gamma$, where

$$\tilde{G}(s) \sim \left[\begin{array}{c|c} A + BK & D \\ \hline E_1 + E_2K & 0 \end{array}\right]$$

and $K \triangleq -R_2^{-1}B^{\mathrm{T}}P$.

Proof. The result is a direct consequence of Theorem 10.4 with $F(x,u) = Ax + Bu$, $J_1(x) = D$, $L(x,u) = x^{\mathrm{T}}R_1x + u^{\mathrm{T}}R_2u$, $V(x) = x^{\mathrm{T}}Px$, $\Gamma(x,u) = \gamma^{-2}x^{\mathrm{T}}PDD^{\mathrm{T}}Px$, $\mathcal{D} = \mathbb{R}^n$, and $U = \mathbb{R}^m$. Specifically, conditions (10.88)–(10.91) are trivially satisfied. Now, forming $x^{\mathrm{T}}(10.104)x$ it follows that, after some algebraic manipulations, $V'(x)J_1(x)w \leq r(z,w) + L(x,\phi(x),w) + \Gamma(x,\phi(x),w)$, for all $x \in \mathcal{D}$ and $w \in \mathcal{W}$. Furthermore, it follows from (10.104) that $H(x,\phi(x)) = 0$ and $H(x,u) = H(x,u) - H(x,\phi(x)) = [u-\phi(x)]^{\mathrm{T}}R_2[u-\phi(x)] \geq 0$ so that all conditions of Theorem 10.4 are satisfied. Finally, since $V(\cdot)$ is radially unbounded, (10.101), with $u(t) = \phi(x(t)) = -R_2^{-1}B^{\mathrm{T}}Px(t)$, is globally asymptotically stable. □

In the case where $U = \mathbb{R}^m$ the feedback control $u = \phi(x)$ is globally optimal since it minimizes $H(x,u)$ and satisfies (10.93). Specifically, setting

$$\frac{\partial}{\partial u} H(x,u) = 0, \tag{10.109}$$

yields the feedback control

$$\phi(x) = -R_2^{-1} B^{\mathrm{T}} P x. \tag{10.110}$$

Now, since

$$\frac{\partial^2}{\partial u^2} H(x,u) = R_2 > 0, \tag{10.111}$$

it follows that for all $x \in \mathbb{R}^n$ the feedback control given by (10.110) minimizes $H(x,u)$. In particular, the optimal feedback control law $\phi(x)$ in Corollary 10.3 is derived using the properties of $H(x,u)$ as defined in Theorem 10.4. Specifically, since $H(x,u) = x^{\mathrm{T}}(A^{\mathrm{T}}P + PA + R_1 + \gamma^{-2}PDD^{\mathrm{T}}P)x + u^{\mathrm{T}}R_2u + 2x^{\mathrm{T}}PBu$ it follows that $\partial^2 H/\partial u^2 > 0$. Now, $\partial H/\partial u = 2R_2u + 2B^{\mathrm{T}}Px = 0$ gives the unique global minimum of $H(x,u)$. Hence, since $\phi(x)$ minimizes $H(x,u)$ it follows that $\phi(x)$ satisfies $\partial H/\partial u = 0$ or, equivalently, $R_2\phi(x) + B^{\mathrm{T}}Px = 0$ so that $\phi(x)$ is given by (10.110). Similar remarks hold for the controllers developed in Corollary 10.4 and Section 10.8.

Next, we specialize Theorem 10.4 to provide connections to the mixed $\mathcal{H}_2$/positivity framework developed in [146, 420]. Specifically, we consider the case where $p = d$ and $r(z,w) = 2w^{\mathrm{T}}z$. For the following result define $R_0 \triangleq (E_\infty + E_\infty^{\mathrm{T}})^{-1}$, $R_{2\mathrm{s}} \triangleq R_2 + E_2^{\mathrm{T}}R_0E_2$, $R_{1\mathrm{s}} \triangleq E_1^{\mathrm{T}}(I + R_0)E_1 - E_1^{\mathrm{T}}R_0E_2R_{2\mathrm{s}}^{-1}E_2^{\mathrm{T}}R_0E_1$, $B_{\mathrm{s}} \triangleq B - DR_0E_2$, $A_{\mathrm{s}} \triangleq A - (B_{\mathrm{s}}R_{2\mathrm{s}}^{-1}E_2^{\mathrm{T}} + D)R_0E_1$, and $S_{\mathrm{s}} \triangleq B_{\mathrm{s}}R_{2\mathrm{s}}^{-1}B_{\mathrm{s}}^{\mathrm{T}}$. Furthermore, assume that $E_1^{\mathrm{T}}E_1 > 0$ and $E_2^{\mathrm{T}}E_2 > 0$. Note that using Schur complements it can be shown that $R_{1\mathrm{s}} > 0$.

Corollary 10.4. Consider the linear dynamical system (10.101) with performance variables

$$z(t) = E_1x(t) + E_2u(t) + E_\infty w(t) \tag{10.112}$$

and performance functional (10.103). Assume that there exists a positive-definite matrix $P \in \mathbb{R}^{n\times n}$ such that

$$0 = A_{\mathrm{s}}^{\mathrm{T}}P + PA_{\mathrm{s}} + R_{1\mathrm{s}} + PDR_0D^{\mathrm{T}}P - PS_{\mathrm{s}}P. \tag{10.113}$$

Then, with the feedback control law $u = \phi(x) = -R_{2\mathrm{s}}^{-1}(B_{\mathrm{s}}^{\mathrm{T}}P + E_2^{\mathrm{T}}R_0E_1)x$, the zero solution $x(t) \equiv 0$ of the undisturbed ($w(t) \equiv 0$) system (10.101) is globally asymptotically stable for all $x_0 \in \mathbb{R}^n$ and

$$J(x_0, \phi(x(\cdot))) \le \mathcal{J}(x_0, \phi(x(\cdot))) = x_0^{\mathrm{T}}Px_0, \tag{10.114}$$

where

$$\mathcal{J}(x_0, u(\cdot)) = \int_0^\infty [x^{\mathrm{T}}(t)(R_1 + (D^{\mathrm{T}}P - E_1)^{\mathrm{T}} R_0 (D^{\mathrm{T}}P - E_1))x(t) \\ +u^{\mathrm{T}}(t)R_{2\mathrm{s}}u(t) - 2x^{\mathrm{T}}(t)(D^{\mathrm{T}}P - E_1)^{\mathrm{T}} R_0 E_2 u(t)]\mathrm{d}t \tag{10.115}$$

and where $u(\cdot)$ is admissible and $x(t)$, $t \geq 0$, solves (10.101) with $w(t) \equiv 0$. Furthermore,

$$\mathcal{J}(x_0, \phi(x(\cdot))) = \min_{u(\cdot) \in \mathcal{S}(x_0)} \mathcal{J}(x_0, u(\cdot)), \tag{10.116}$$

where $\mathcal{S}(x_0)$ is the set of regulation controllers for the system (10.101) with $w(t) \equiv 0$ and $x_0 \in \mathbb{R}^n$. Finally, if $x_0 = 0$, then, with $u = \phi(x)$, the solution $x(t)$, $t \geq 0$, of (10.101) satisfies the passivity constraint

$$\int_0^T 2w^{\mathrm{T}}(t)z(t) \geq 0, \qquad w(\cdot) \in \mathcal{L}_2, \qquad T \geq 0, \tag{10.117}$$

or, equivalently, $\tilde{G}_\infty(s) + \tilde{G}_\infty^*(s) \geq 0$, $\mathrm{Re}[s] > 0$, where

$$\tilde{G}_\infty(s) \sim \left[\begin{array}{c|c} A + BK & D \\ \hline E_1 + E_2 K & E_\infty \end{array}\right]$$

and $K = -R_{2\mathrm{s}}^{-1}(B_{\mathrm{s}}^{\mathrm{T}}P + E_2^{\mathrm{T}} R_0 E_1)$.

Proof. The result is a direct consequence of Theorem 10.4 with $F(x,u) = Ax + Bu$, $J_1(x) = D$, $L(x,u) = x^{\mathrm{T}}R_1 x + u^{\mathrm{T}}R_2 u$, $V(x) = x^{\mathrm{T}}Px$, $\Gamma(x,u) = [(D^{\mathrm{T}}P - E_1)x - E_2 u]^{\mathrm{T}} R_0 [(D^{\mathrm{T}}P - E_1)x - E_2 u]$, $\mathcal{D} = \mathbb{R}^n$, and $U = \mathbb{R}^m$. Specifically, conditions (10.88)–(10.91) are trivially satisfied. Now, forming $x^{\mathrm{T}}(10.113)x$ it follows that, after some algebraic manipulations, $V'(x)J_1(x)w \leq r(z,w) + L(x,\phi(x),w) + \Gamma(x,\phi(x),w)$, for all $x \in \mathcal{D}$ and $w \in \mathcal{W}$. Furthermore, it follows from (10.113) that $H(x,\phi(x)) = 0$ and $H(x,u) = H(x,u) - H(x,\phi(x)) = [u - \phi(x)]^{\mathrm{T}} R_{2\mathrm{s}} [u - \phi(x)] \geq 0$ so that all conditions of Theorem 10.4 are satisfied. Finally, since $V(x)$ is radially unbounded, (10.101), with $u(t) = \phi(x(t)) = -R_{2\mathrm{s}}^{-1}(B_{\mathrm{s}}^{\mathrm{T}}P + E_2^{\mathrm{T}}R_0 E_1)x(t)$, is globally asymptotically stable. □

10.6 Optimal and Inverse Optimal Control for Affine Systems with $\mathcal{L}_2$ Disturbances

In this section, we specialize Theorem 10.4 to affine (in the control) systems of the form

$$\dot{x}(t) = f(x(t)) + G(x(t))u(t) + J_1(x(t))w(t), \; x(0) = x_0, \; w(\cdot) \in \mathcal{L}_2, \; t \geq 0, \tag{10.118}$$

with performance variables

$$z(t) = h(x(t)) + J(x(t))u(t), \tag{10.119}$$

where $f : \mathbb{R}^n \to \mathbb{R}^n$ satisfies $f(0) = 0$, $G : \mathbb{R}^n \to \mathbb{R}^{n\times m}$, $J : \mathbb{R}^n \to \mathbb{R}^{p\times m}$, $h : \mathbb{R}^n \to \mathbb{R}^p$ satisfies $h(0) = 0$, $\mathcal{D} = \mathbb{R}^n$, and $U = \mathbb{R}^m$. First, we consider the nonexpansivity case so that the supply rate $r(z, w)$ is given by $r(z, w) = \gamma^2 w^{\mathrm{T}} w - z^{\mathrm{T}} z$, where $\gamma > 0$. For the following result, we consider performance integrands $L(x, u)$ of the form

$$L(x, u) = [h(x) + J(x)u]^{\mathrm{T}}[h(x) + J(x)u], \tag{10.120}$$

where $J^{\mathrm{T}}(x)J(x) > 0$, $x \in \mathbb{R}^n$, so that (10.87) becomes

$$J(x_0, u(\cdot)) = \int_0^\infty [h(x(t)) + J(x(t))u(t)]^{\mathrm{T}}[h(x(t)) + J(x(t))u(t)]\mathrm{d}t. \tag{10.121}$$

Corollary 10.5. Consider the nonlinear controlled dynamical system (10.118) and (10.119) with performance functional (10.121). Assume that there exists a continuously differentiable function $V : \mathbb{R}^n \to \mathbb{R}$ such that

$$V(0) = 0, \tag{10.122}$$

$$V(x) > 0, \quad x \in \mathbb{R}^n, \quad x \neq 0, \tag{10.123}$$

$$V'(x)[f(x) - \tfrac{1}{2}G(x)R_2^{-1}(x)(V'(x)G(x) + 2h^{\mathrm{T}}(x)J(x))^{\mathrm{T}}] < 0, \quad x \in \mathbb{R}^n, \ x \neq 0, \tag{10.124}$$

$$\begin{aligned} 0 = {} & V'(x)f(x) + h^{\mathrm{T}}(x)h(x) + \tfrac{1}{4\gamma^2}V'(x)J_1(x)J_1^{\mathrm{T}}(x)V'^{\mathrm{T}}(x) \\ & -\tfrac{1}{4}[V'(x)G(x) + 2h^{\mathrm{T}}(x)J(x)]R_2^{-1}(x)[V'(x)G(x) + 2h^{\mathrm{T}}(x)J(x)]^{\mathrm{T}}, \quad x \in \mathbb{R}^n, \end{aligned} \tag{10.125}$$

and

$$V(x) \to \infty \text{ as } \|x\| \to \infty, \tag{10.126}$$

where $\gamma > 0$ and $R_2(x) \triangleq J^{\mathrm{T}}(x)J(x)$. Then the zero solution $x(t) \equiv 0$ of the undisturbed ($w(t) \equiv 0$) closed-loop system

$$\dot{x}(t) = f(x(t)) + G(x(t))\phi(x(t)), \qquad x(0) = x_0, \qquad t \geq 0, \tag{10.127}$$

is globally asymptotically stable with feedback control law

$$\phi(x) = -\tfrac{1}{2}R_2^{-1}(x)[V'(x)G(x) + 2h^{\mathrm{T}}(x)J(x)]^{\mathrm{T}}. \tag{10.128}$$

Furthermore, the performance functional (10.121) satisfies

$$J(x_0, \phi(x(\cdot))) \leq \mathcal{J}(x_0, \phi(x(\cdot))) = V(x_0), \tag{10.129}$$

where

$$\mathcal{J}(x_0, u(\cdot)) \triangleq \int_0^\infty [L(x(t), u(t)) + \Gamma(x(t), u(t))]\mathrm{d}t, \tag{10.130}$$

$$\Gamma(x,u) = \tfrac{1}{4\gamma^2}V'(x)J_1(x)J_1^{\mathrm{T}}(x)V'^{\mathrm{T}}(x). \tag{10.131}$$

In addition, the performance functional (10.130) is minimized in the sense that

$$\mathcal{J}(x_0,\phi(x(\cdot))) = \min_{u(\cdot)\in\mathcal{S}(x_0)} \mathcal{J}(x_0,u(\cdot)). \tag{10.132}$$

Finally, with $u(\cdot) = \phi(x(\cdot))$, the solution $x(t)$, $t \geq 0$, of the closed-loop system (10.118) satisfies the nonexpansivity constraint

$$\int_0^T z^{\mathrm{T}}(t)z(t)\mathrm{d}t \leq \gamma^2 \int_0^T w^{\mathrm{T}}(t)w(t)\mathrm{d}t + V(x_0), \quad T \geq 0, \quad w(\cdot) \in \mathcal{L}_2. \tag{10.133}$$

Proof. The result is a direct consequence of Theorem 10.4 with $F(x,u) = f(x)+G(x)u$, $z = h(x)+J(x)u$, $L(x,u) = [h(x)+J(x)u]^{\mathrm{T}}[h(x)+J(x)u]$, $J_2(x) = 0$, $\Gamma(x,u)$ given by (10.131), $\mathcal{D} = \mathbb{R}^n$, and $U = \mathbb{R}^m$. Specifically, with (10.120) and (10.131), the Hamiltonian has the form

$$\begin{aligned} H(x,u) &= [h(x) + J(x)u]^{\mathrm{T}}[h(x) + J(x)u] + V'(x)(f(x) + G(x)u) \\ &\quad + \tfrac{1}{4\gamma^2}V'(x)J_1(x)J_1^{\mathrm{T}}(x)V'^{\mathrm{T}}(x). \end{aligned}$$

Now, the feedback control law (10.128) is obtained by setting $\frac{\partial H}{\partial u} = 0$. With (10.128), it follows that (10.122)–(10.124) imply (10.88), (10.89), and (10.91). Next, since $V(\cdot)$ is continuously differentiable and $x = 0$ is a local minimum of $V(\cdot)$, it follows that $V'(0) = 0$, and hence, since by assumption $h(0) = 0$, it follows that $\phi(0) = 0$, which proves (10.90). Next, with $\phi(x)$ given by (10.128), it follows from (10.125) that (10.93) holds. Finally, since $H(x,u) = H(x,u) - H(x,\phi(x)) = [u - \phi(x)]^{\mathrm{T}}R_2(x)[u - \phi(x)]$, and $R_2(x)$ is positive definite for all $x \in \mathbb{R}^n$, condition (10.94) holds. The result now follows as a direct consequence of Theorem 10.4. □

Next, we consider performance integrands $L(x,u)$ of the form

$$L(x,u) = L_1(x) + L_2(x)u + u^{\mathrm{T}}R_2(x)u, \tag{10.134}$$

where $L_1 : \mathbb{R}^n \to \mathbb{R}$, $L_2 : \mathbb{R}^n \to \mathbb{R}^{1\times m}$, and $R_2 : \mathbb{R}^n \to \mathbb{P}^m$ so that (10.87) becomes

$$J(x_0,u(\cdot)) = \int_0^\infty [L_1(x(t)) + L_2(x(t))u(t) + u^{\mathrm{T}}(t)R_2(x)u(t)]\mathrm{d}t. \tag{10.135}$$

Corollary 10.6. Consider the nonlinear controlled dynamical system (10.118) and (10.119) with performance functional (10.135). Assume that there exist a continuously differentiable function $V : \mathbb{R}^n \to \mathbb{R}$ and a function $L_2 : \mathbb{R}^n \to \mathbb{R}^{1\times m}$ such that

$$V(0) = 0, \tag{10.136}$$

$$L_2(0) = 0, \tag{10.137}$$

$$V(x) > 0, \qquad x \in \mathbb{R}^n, \qquad x \neq 0, \tag{10.138}$$

$$V'(x)[f(x) - \tfrac{1}{2}G(x)R_{2a}^{-1}(x)(L_2^{\mathrm{T}}(x) + G^{\mathrm{T}}(x)V'^{\mathrm{T}}(x) + 2J^{\mathrm{T}}(x)h(x))] \\ +\Gamma(x, \phi(x)) < 0, \qquad x \in \mathbb{R}^n, \qquad x \neq 0, \tag{10.139}$$

and

$$V(x) \to \infty \text{ as } \|x\| \to \infty, \tag{10.140}$$

where $R_{2a}(x) \triangleq R_2(x)+J^{\mathrm{T}}(x)J(x)$, $\phi(x) = -\frac{1}{2}R_{2a}^{-1}(x)[L_2^{\mathrm{T}}(x)+G^{\mathrm{T}}(x)V'^{\mathrm{T}}(x) +2J^{\mathrm{T}}(x)h(x)]$, and

$$\Gamma(x,u) = \tfrac{1}{4\gamma^2}V'(x)J_1(x)J_1^{\mathrm{T}}(x)V'^{\mathrm{T}}(x) + [h(x) + J(x)u]^{\mathrm{T}}[h(x) + J(x)u], \tag{10.141}$$

where $\gamma > 0$, $u(\cdot)$ is admissible, and $x(t)$, $t \geq 0$, solves (10.118) with $w(t) \equiv 0$. Then the zero solution $x(t) \equiv 0$ of the undisturbed ($w(t) \equiv 0$) closed-loop system

$$\dot{x}(t) = f(x(t)) + G(x(t))\phi(x(t)), \qquad x(0) = x_0, \qquad t \geq 0, \tag{10.142}$$

is globally asymptotically stable with feedback control law

$$\phi(x) = -\tfrac{1}{2}R_{2a}^{-1}(x)[L_2^{\mathrm{T}}(x) + G^{\mathrm{T}}(x)V'^{\mathrm{T}}(x) + 2J^{\mathrm{T}}(x)h(x)]. \tag{10.143}$$

Furthermore, the performance functional (10.135) satisfies

$$J(x_0, \phi(x(\cdot))) \leq \mathcal{J}(x_0, \phi(x(\cdot))) = V(x_0), \tag{10.144}$$

where

$$\mathcal{J}(x_0, u(\cdot)) \triangleq \int_0^\infty [L(x(t), u(t)) + \Gamma(x(t), u(t))]\mathrm{d}t. \tag{10.145}$$

In addition, the performance functional (10.145), with

$$L_1(x) = \phi^{\mathrm{T}}(x)R_{2a}(x)\phi(x) - V'(x)f(x) - h^{\mathrm{T}}(x)h(x) \\ -\tfrac{1}{4\gamma^2}V'(x)J_1(x)J_1^{\mathrm{T}}(x)V'^{\mathrm{T}}(x), \tag{10.146}$$

is minimized in the sense that

$$\mathcal{J}(x_0, \phi(x(\cdot))) = \min_{u(\cdot)\in\mathcal{S}(x_0)} \mathcal{J}(x_0, u(\cdot)). \tag{10.147}$$

Finally, with $u(\cdot) = \phi(x(\cdot))$, the solution $x(t)$, $t \geq 0$, of the closed-loop system (10.118) satisfies the nonexpansivity constraint

$$\int_0^T z^{\mathrm{T}}(t)z(t)\mathrm{d}t \leq \gamma^2 \int_0^T w^{\mathrm{T}}(t)w(t)\mathrm{d}t + V(x_0), \qquad w(\cdot) \in \mathcal{L}_2, \qquad T \geq 0. \tag{10.148}$$

Proof. The result is a direct consequence of Theorem 10.4 with $F(x,u) = f(x) + G(x)u$, $z = h(x) + J(x)u$, $L(x,u) = L_1(x) + L_2(x)u + u^{\mathrm{T}}R_2(x)u$, $J_2(x) = 0$, $\Gamma(x,u)$ given by (10.141), $\mathcal{D} = \mathbb{R}^n$, and $U = \mathbb{R}^m$.

Specifically, with (10.118), (10.134), and (10.141), the Hamiltonian has the form

$$\begin{aligned} H(x,u) &= L_1(x) + L_2(x)u + u^{\mathrm{T}}R_2(x)u + V'(x)(f(x)+G(x)u) \\ &\quad + \tfrac{1}{4\gamma^2}V'(x)J_1(x)J_1^{\mathrm{T}}(x)V'^{\mathrm{T}}(x) + [h(x)+J(x)u]^{\mathrm{T}}[h(x)+J(x)u]. \end{aligned}$$

Now, the proof follows as the proof of Corollary 10.5. □

Note that since $\Gamma(x,\phi(x)) \geq 0$, $x \in \mathbb{R}^n$, (10.139) implies that

$$\dot{V}(x) \triangleq V'(x)[f(x)+G(x)\phi(x)] < 0, \quad x \in \mathbb{R}^n, \quad x \neq 0, \tag{10.149}$$

with $\phi(x)$ given by (10.143). Furthermore, (10.136), (10.138), and (10.149) ensure that $V(\cdot)$ is a Lyapunov function for the undisturbed closed-loop system (10.142). In addition, with $L_1(x)$ given by (10.146) and $\phi(x)$ given by (10.143), $L(x,u)+\Gamma(x,u)$ can be expressed as

$$\begin{aligned} &L(x,u)+\Gamma(x,u) \\ &\quad = [u-\phi(x)]^{\mathrm{T}}R_{2a}(x)[u-\phi(x)] - V'(x)[f(x)+G(x)u] \\ &\quad = [u+\tfrac{1}{2}R_{2a}^{-1}(x)(L_2^{\mathrm{T}}(x)+2J^{\mathrm{T}}(x)h(x)]^{\mathrm{T}}R_{2a}(x) \\ &\qquad [u+\tfrac{1}{2}R_{2a}^{-1}(x)(L_2^{\mathrm{T}}(x)+2J^{\mathrm{T}}(x)h(x)] \\ &\qquad -V'(x)[f(x)+G(x)\phi(x)] - \tfrac{1}{4}V'(x)G(x)R_{2a}^{-1}(x)G^{\mathrm{T}}(x)V'^{\mathrm{T}}(x). \end{aligned} \tag{10.150}$$

Since $R_{2a}(x) \geq R_2(x) > 0$ for all $x \in \mathbb{R}^n$ the first term of the right-hand side of (10.150) is nonnegative, while (10.149) implies that the second term is nonnegative. Thus, we have

$$L(x,u)+\Gamma(x,u) \geq -\tfrac{1}{4}V'(x)G(x)R_{2a}^{-1}(x)G^{\mathrm{T}}(x)V'^{\mathrm{T}}(x), \tag{10.151}$$

which shows that $L(x,u)+\Gamma(x,u)$ may be negative. As a result, there may exist a control input u for which the auxiliary performance functional $\mathcal{J}(x_0,u)$ is negative. However, if the disturbance rejection control u is a regulation controller, that is, $u \in \mathcal{S}(x_0)$, then it follows from (10.144) and (10.147) that

$$\mathcal{J}(x_0,u(\cdot)) \geq V(x_0) \geq 0, \qquad x_0 \in \mathbb{R}^n, \qquad u(\cdot) \in \mathcal{S}(x_0).$$

Furthermore, in this case substituting $u=\phi(x)$ into (10.150) yields

$$L(x,\phi(x)) + \Gamma(x,\phi(x)) = -V'(x)[f(x)+G(x)\phi(x)],$$

which, by (10.149), is positive.

Next, we specialize Theorem 10.4 to the passivity case. Specifically, we consider the case where $p=d$ and $r(z,w)=2z^{\mathrm{T}}w$. For the following result we consider performance variables

$$z(t) = h(x(t)) + J(x(t))u(t) + J_2(x(t))w(t), \tag{10.152}$$

where $J : \mathbb{R}^n \to \mathbb{R}^{p\times m}$, $J_2 : \mathbb{R}^n \to \mathbb{R}^{p\times p}$ satisfies $J_2(x) + J_2^{\mathrm{T}}(x) > 0$, $x \in \mathbb{R}^n$, and $h : \mathbb{R}^n \to \mathbb{R}^p$ satisfies $h(0) = 0$. Furthermore, we consider performance integrands $L(x, u)$ of the form given by (10.134).

Corollary 10.7. Consider the nonlinear controlled dynamical system (10.118) and (10.152) with performance functional (10.135). Assume that there exist a continuously differentiable function $V : \mathbb{R}^n \to \mathbb{R}$ and a function $L_2 : \mathbb{R}^n \to \mathbb{R}^{1\times m}$ such that

$$V(0) = 0, \tag{10.153}$$

$$L_2(0) = 0, \tag{10.154}$$

$$V(x) > 0, \quad x \in \mathbb{R}^n, \quad x \neq 0, \tag{10.155}$$

$$\begin{aligned} V'(x)[f(x) - \tfrac{1}{2}G(x)R_{2a}^{-1}(x)(L_2^{\mathrm{T}}(x) + J^{\mathrm{T}}(x)R_0(x)[2h(x) - J_1^{\mathrm{T}}(x)V'^{\mathrm{T}}(x)])] \\ +\Gamma(x, \phi(x)) < 0, \quad x \in \mathbb{R}^n, \quad x \neq 0, \end{aligned} \tag{10.156}$$

and

$$V(x) \to \infty \text{ as } \|x\| \to \infty, \tag{10.157}$$

where $R_0(x) \triangleq (J_2(x) + J_2^{\mathrm{T}}(x))^{-1}$, $R_{2a}(x) \triangleq R_2(x) + J^{\mathrm{T}}(x)R_0(x)J(x)$,

$$\begin{aligned} \phi(x) &= -\tfrac{1}{2}R_{2a}^{-1}(x)[G^{\mathrm{T}}(x)V'^{\mathrm{T}}(x) + L_2^{\mathrm{T}}(x) + J^{\mathrm{T}}(x)R_0(x)(2h(x) \\ &\quad -J_1^{\mathrm{T}}(x)V'^{\mathrm{T}}(x))], \end{aligned}$$

and

$$\begin{aligned} \Gamma(x, u) &= \left[\tfrac{1}{2}J_1^{\mathrm{T}}(x)V'^{\mathrm{T}}(x) - (h(x) + J(x)u)\right]^{\mathrm{T}} R_0(x) \\ &\quad \cdot \left[\tfrac{1}{2}J_1^{\mathrm{T}}(x)V'^{\mathrm{T}}(x) - (h(x) + J(x)u)\right], \end{aligned} \tag{10.158}$$

where $u(\cdot)$ is admissible, and $x(t)$, $t \geq 0$, solves (10.118) with $w(t) \equiv 0$. Then the zero solution $x(t) \equiv 0$ of the undisturbed ($w(t) \equiv 0$) closed-loop system (10.142) is globally asymptotically stable with feedback control law $\phi(x)$. Furthermore, the performance functional (10.135) satisfies

$$J(x_0, \phi(x(\cdot))) \leq \mathcal{J}(x_0, \phi(x(\cdot))) = V(x_0), \tag{10.159}$$

where

$$\mathcal{J}(x_0, u(\cdot)) \triangleq \int_0^\infty [L(x(t), u(t)) + \Gamma(x(t), u(t))]\mathrm{d}t. \tag{10.160}$$

In addition, the performance functional (10.160), with

$$\begin{aligned} L_1(x) &= \phi^{\mathrm{T}}(x)R_{2a}(x)\phi(x) - V'(x)f(x) - [\tfrac{1}{2}J_1^{\mathrm{T}}(x)V'^{\mathrm{T}}(x) - h(x)]^{\mathrm{T}}R_0(x) \\ &\quad \cdot[\tfrac{1}{2}J_1^{\mathrm{T}}(x)V'^{\mathrm{T}}(x) - h(x)]h^{\mathrm{T}}(x)h(x) - \tfrac{1}{4\gamma^2}V'(x)J_1(x)J_1^{\mathrm{T}}(x)V'^{\mathrm{T}}(x), \end{aligned} \tag{10.161}$$

is minimized in the sense that

$$\mathcal{J}(x_0, \phi(x(\cdot))) = \min_{u(\cdot)\in\mathcal{S}(x_0)} \mathcal{J}(x_0, u(\cdot)). \tag{10.162}$$

Finally, with $u(\cdot) = \phi(x(\cdot))$, the solution $x(t)$, $t \geq 0$, of the closed-loop system (10.118) satisfies the positivity constraint

$$\int_0^T 2z^{\mathrm{T}}(t)w(t)\mathrm{d}t + V(x_0) \geq 0, \qquad w(\cdot) \in \mathcal{L}_2, \qquad T \geq 0. \tag{10.163}$$

Proof. The result is a direct consequence of Theorem 10.4 with $F(x,u) = f(x) + G(x)u$, $z = h(x) + J(x)u + J_2(x)w$, $L(x,u) = L_1(x) + L_2(x)u + u^{\mathrm{T}}R_2(x)u$, $J_2(x) = 0$, $\Gamma(x,u)$ given by (10.158), $\mathcal{D} = \mathbb{R}^n$, and $U = \mathbb{R}^m$. Specifically, with (10.134) and (10.158), the Hamiltonian has the form

$$H(x,u) = L_1(x) + L_2(x)u + u^{\mathrm{T}}R_2(x)u + V'(x)(f(x) + G(x)u) + \Gamma(x,u).$$

The proof now follows as in the proof of Corollary 10.5. □

10.7 Stability Margins, Meaningful Inverse Optimality, and Nonexpansive Control Lyapunov Functions

In this section, we specialize the results of Section 10.6 to the case where $L(x,u)$ is nonnegative for all $(x,u) \in \mathbb{R}^n \times \mathbb{R}^m$. Here, we assume $L_2(x) \equiv 0$ and $L_1(x) = h^{\mathrm{T}}(x)h(x)$, $x \in \mathbb{R}^n$. We begin by specializing Corollary 10.5 to affine systems of the form (10.118) with performance variables (10.119) and we assume $h^{\mathrm{T}}(x)J(x) \equiv 0$ and $R_2(x) \triangleq J^{\mathrm{T}}(x)J(x) > 0$, $x \in \mathbb{R}^n$, so that the performance functional (10.121) becomes

$$J(x_0, u(\cdot)) = \int_0^\infty [L_1(x(t)) + u^{\mathrm{T}}(t)R_2(x(t))u(t)]\mathrm{d}t. \tag{10.164}$$

Corollary 10.8. Consider the nonlinear controlled dynamical system (10.118) and (10.119) with performance functional (10.164). Assume that there exists a continuously differentiable function $V : \mathbb{R}^n \to \mathbb{R}$ such that

$$V(0) = 0, \tag{10.165}$$

$$V(x) > 0, \quad x \in \mathbb{R}^n, \quad x \neq 0, \tag{10.166}$$

$$\begin{aligned} 0 = {} & V'(x)f(x) + h^{\mathrm{T}}(x)h(x) + \tfrac{1}{4\gamma^2}V'(x)J_1(x)J_1^{\mathrm{T}}(x)V'^{\mathrm{T}}(x) \\ & -\tfrac{1}{4}V'(x)G(x)R_2^{-1}(x)G^{\mathrm{T}}(x)V'^{\mathrm{T}}(x), \quad x \in \mathbb{R}^n, \end{aligned} \tag{10.167}$$

and

$$V(x) \to \infty \text{ as } \|x\| \to \infty, \tag{10.168}$$

where $\gamma > 0$. Furthermore, assume that the system (10.118) and (10.119) is zero-state observable. Then the zero solution $x(t) \equiv 0$ of the undisturbed ($w(t) \equiv 0$) closed-loop system

$$\dot{x}(t) = f(x(t)) + G(x(t))\phi(x(t)), \qquad x(0) = x_0, \qquad t \geq 0, \tag{10.169}$$

is globally asymptotically stable with feedback control law

$$\phi(x) = -\tfrac{1}{2}R_2^{-1}(x)G^{\mathrm{T}}(x)V'^{\mathrm{T}}(x). \tag{10.170}$$

Furthermore, the performance functional (10.164) satisfies

$$J(x_0, \phi(x(\cdot))) \le \mathcal{J}(x_0, \phi(x(\cdot))) = V(x_0), \tag{10.171}$$

where

$$\mathcal{J}(x_0, u(\cdot)) \triangleq \int_0^\infty [L(x(t), u(t)) + \Gamma(x(t), u(t))]\mathrm{d}t, \tag{10.172}$$

$$\Gamma(x, u) = \tfrac{1}{4\gamma^2}V'(x)J_1(x)J_1^{\mathrm{T}}(x)V'^{\mathrm{T}}(x), \tag{10.173}$$

and $\gamma > 0$. In addition, the performance functional (10.172) is minimized in the sense that

$$\mathcal{J}(x_0, \phi(x(\cdot))) = \min_{u(\cdot)\in\mathcal{S}(x_0)} \mathcal{J}(x_0, u(\cdot)). \tag{10.174}$$

Finally, with $u(\cdot) = \phi(x(\cdot))$, the solution $x(t)$, $t \ge 0$, of the closed-loop system (10.118) satisfies the nonexpansivity constraint

$$\int_0^T z^{\mathrm{T}}(t)z(t)\mathrm{d}t \le \gamma^2 \int_0^T w^{\mathrm{T}}(t)w(t)\mathrm{d}t + V(x_0), \quad w(\cdot) \in \mathcal{L}_2, \quad T \ge 0. \tag{10.175}$$

Proof. The result follows as a direct consequence of Corollary 10.5. □

Next, we provide sector and gain margins for the nonlinear dynamical system $\mathcal{G}$ given by (10.118) and (10.119). To consider relative stability margins for nonlinear nonexpansive regulators consider the nonlinear dynamical system given by (10.118) and (10.119), along with the output

$$y(t) = -\phi(x(t)), \tag{10.176}$$

where $\phi(\cdot)$ is such that the input-output map from w to z is nonexpansive with $u = \phi(x)$. Furthermore, assume that (10.118) and (10.176) is zero-state observable. For the next result define

$$\eta \triangleq \frac{\inf_{x\in\mathbb{R}^n} \sigma_{\min}(J_1(x)J_1^{\mathrm{T}}(x))}{\sup_{x\in\mathbb{R}^n} \sigma_{\max}(G(x)R_2^{-1}(x)G^{\mathrm{T}}(x))}. \tag{10.177}$$

Theorem 10.5. Let $\gamma > 0$ and $\rho \in (0, 1]$. Consider the nonlinear dynamical system $\mathcal{G}$ given by (10.118) and (10.119) where $\phi(x)$ is a nonexpansive feedback control law given by (10.170) and where $V(x)$, $x \in \mathbb{R}^n$, satisfies

$$V(0) = 0, \tag{10.178}$$

$$V(x) > 0, \quad x \in \mathbb{R}^n, \quad x \ne 0, \tag{10.179}$$

$$0 = V'(x)f(x) + h^{\mathrm{T}}(x)h(x) + \tfrac{1}{4\gamma^2}V'(x)J_1(x)J_1^{\mathrm{T}}(x)V'^{\mathrm{T}}(x)$$
$$-\tfrac{1}{4}V'(x)G(x)R_2^{-1}G^{\mathrm{T}}(x)V'^{\mathrm{T}}(x), \quad x \in \mathbb{R}^n. \tag{10.180}$$

Furthermore, assume $R_2(x) = \mathrm{diag}[r_1(x), \ldots, r_m(x)]$, where $r_i : \mathbb{R}^n \to \mathbb{R}$, $r_i(x) > 0$, $i = 1, \ldots, m$. Then the undisturbed ($w(t) \equiv 0$) nonlinear system $\mathcal{G}$ has a sector (and, hence, gain) margin $(\frac{\alpha}{2}, \infty)$, where

$$\alpha \triangleq 1 - \frac{\eta(1-\rho^2)}{\gamma^2}.$$

Finally, with $u = \sigma(\phi(x))$, the solution $x(t)$, $t \geq 0$, of the closed-loop system (10.118) satisfies the nonexpansivity constraint

$$\int_0^T z^{\mathrm{T}}(t)z(t)\mathrm{d}t \leq (\gamma/\rho)^2 \int_0^T w^{\mathrm{T}}(t)w(t)\mathrm{d}t + V(x_0), \quad w(\cdot) \in \mathcal{L}_2, \quad T \geq 0, \tag{10.181}$$

where $\sigma : \mathbb{R}^m \to \mathbb{R}^m$ is such that $\sigma(0) = 0$ and for every $u_{\mathrm{c}} \in \mathbb{R}^m$, $\sigma(u_{\mathrm{c}}) = [\sigma_1(u_{\mathrm{c}1}), \ldots, \sigma_m(u_{\mathrm{c}m})]^{\mathrm{T}}$ and $\alpha u_{\mathrm{c}i}^2 < 2\sigma_i(u_{\mathrm{c}i})u_{\mathrm{c}i}$, $u_{\mathrm{c}i} \neq 0$, $i = 1, \ldots, m$.

Proof. With $u = \sigma(\phi(x))$, the closed-loop system (10.118) and (10.176) is given by

$$\dot{x}(t) = f(x(t)) + G(x(t))\sigma(\phi(x(t)) + J_1(x(t))w(t),$$
$$x(0) = x_0, \quad w(\cdot) \in \mathcal{L}_2, \quad t \geq 0. \tag{10.182}$$

Next, consider the Lyapunov function candidate $V(x)$, $x \in \mathbb{R}^n$, satisfying (10.180) and let $\dot{V}(x)$ denote the Lyapunov derivative along the trajectories of the closed-loop system (10.182). Now, it follows from (10.180) that for all $w \in \mathbb{R}^d$,

$$\begin{aligned}
\dot{V}(x) &= V'(x)f(x) + V'(x)G(x)\sigma(\phi(x)) + V'(x)J_1(x)w \\
&\leq \phi^{\mathrm{T}}(x)R_2(x)\phi(x) + V'(x)G(x)\sigma(\phi(x)) + V'(x)J_1(x)w \\
&\quad -\tfrac{1}{4\gamma^2}V'(x)J_1(x)J_1^{\mathrm{T}}(x)V'^{\mathrm{T}}(x) - h^{\mathrm{T}}(x)h(x) \\
&\leq \alpha\phi^{\mathrm{T}}(x)R_2(x)\phi(x) + V'(x)G(x)\sigma(\phi(x)) + V'(x)J_1(x)w \\
&\quad -h^{\mathrm{T}}(x)h(x) - \tfrac{\rho^2}{4\gamma^2}V'(x)J_1(x)J_1^{\mathrm{T}}(x)V'^{\mathrm{T}}(x) \\
&\leq \sum_{i=1}^{m} r_i(x)y_i(\alpha y_i + 2\sigma_i(-y_i)) + (\gamma/\rho)^2 w^{\mathrm{T}}w - z^{\mathrm{T}}z \\
&\leq (\gamma/\rho)^2 w^{\mathrm{T}}w - z^{\mathrm{T}}z,
\end{aligned}$$

which proves the nonexpansivity property. The proof of sector (and, hence, gain) margin guarantees is similar to the proof of Theorem 8.7 and, hence, is omitted. □

The following specialization of Theorem 10.5 to linear dynamical systems is immediate.

Corollary 10.9. Let $\gamma > 0$ and $\rho \in (0,1]$. Consider the linear dynamical system $\mathcal{G}$ given by (10.101) and (10.102) and let $P \in \mathbb{P}^n$ satisfy (10.104). Furthermore, assume $R_2 = \operatorname{diag}[r_1, \ldots, r_m]$, where $r_i > 0$, $i = 1, \ldots, m$. Then with the nonexpansive feedback control law $\phi(x) = -R_2^{-1}B^{\mathrm{T}}Px$, the undisturbed ($w(t) \equiv 0$) linear dynamical system $\mathcal{G}$ has a sector (and, hence, gain) margin $(\frac{\alpha}{2}, \infty)$, where

$$\alpha \triangleq 1 - \frac{\eta(1-\rho^2)}{\gamma^2}.$$

Finally, with $u = \sigma(\phi(x))$, the solution $x(t)$, $t \geq 0$, of the closed-loop system satisfies the nonexpansivity constraint

$$\int_0^T z^{\mathrm{T}}(t)z(t)\mathrm{d}t \leq (\gamma/\rho)^2 \int_0^T w^{\mathrm{T}}(t)w(t)\mathrm{d}t + V(x_0), \quad w(\cdot) \in \mathcal{L}_2, \quad T \geq 0, \tag{10.183}$$

where $\sigma : \mathbb{R}^m \to \mathbb{R}^m$ is such that $\sigma(0) = 0$ and for every $u_{\mathrm{c}} \in \mathbb{R}^m$, $\sigma(u_{\mathrm{c}}) = [\sigma_1(u_{\mathrm{c}1}), \ldots, \sigma_m(u_{\mathrm{c}m})]^{\mathrm{T}}$ and $\alpha u_{\mathrm{c}i}^2 < 2\sigma_i(u_{\mathrm{c}i})u_{\mathrm{c}i}$, $u_{\mathrm{c}i} \neq 0$, $i = 1, \ldots, m$.

Next, we introduce the notion of *nonexpansive control Lyapunov functions* for the nonlinear dynamical system (10.118) and (10.119).

Definition 10.1. Consider the controlled nonlinear dynamical system given by (10.118) and (10.119). A continuously differentiable positive-definite function $V : \mathbb{R}^n \to \mathbb{R}$ satisfying

$$V'(x)f(x) + h^{\mathrm{T}}(x)h(x) + \frac{1}{4\gamma^2}V'(x)J_1(x)J_1^{\mathrm{T}}(x)V'^{\mathrm{T}}(x) < 0, \quad x \in \mathcal{R}, \tag{10.184}$$

where $\mathcal{R} \triangleq \{x \in \mathbb{R}^n : x \neq 0 : V'(x)G(x) = 0\}$, is called a *nonexpansive control Lyapunov function.*

Finally, we show that for every nonlinear dynamical system for which a nonexpansive control Lyapunov function can be constructed there exists an inverse optimal nonexpansive feedback control law with sector and gain margin guarantees of at least $(\frac{\alpha}{2}, \infty)$.

Theorem 10.6. Let $\gamma > 0$ and $\rho \in (0,1]$. Consider the nonlinear dynamical system $\mathcal{G}$ given by (10.118) and (10.119), and let the continuously differentiable positive-definite, radially unbounded function $V : \mathbb{R}^n \to \mathbb{R}$ be a nonexpansive control Lyapunov function of (10.118) and (10.119), that is,

$$V'(x)f(x) + h^{\mathrm{T}}(x)h(x) + \frac{1}{4\gamma^2}V'(x)J_1(x)J_1^{\mathrm{T}}(x)V'^{\mathrm{T}}(x) < 0, \quad x \in \mathcal{R}, \tag{10.185}$$

where $\mathcal{R} \triangleq \{x \in \mathbb{R}^n : x \neq 0 : V'(x)G(x) = 0\}$. Then, with the feedback control law given by

$$\phi(x) = \begin{cases} -\left(c_0 + \frac{\alpha(x)+\sqrt{\alpha^2(x)+(\beta^{\mathrm{T}}(x)\beta(x))^2}}{\beta^{\mathrm{T}}(x)\beta(x)}\right)\beta(x), & \beta(x) \neq 0, \\ 0, & \beta(x) = 0, \end{cases} \tag{10.186}$$

where $\alpha(x) \triangleq V'(x)f(x) + h^{\mathrm{T}}(x)h(x) + \frac{1}{4\gamma^2}V'(x)J_1(x)J_1^{\mathrm{T}}(x)V'^{\mathrm{T}}(x)$, $\beta(x) \triangleq G^{\mathrm{T}}(x)V'^{\mathrm{T}}(x)$, and $c_0 > 0$, the nonlinear system $\mathcal{G}$ given by (10.118) and (10.119) has a sector (and, hence, gain) margin $(\frac{\alpha}{2}, \infty)$, where

$$\alpha \triangleq 1 - \frac{\eta(1-\rho^2)}{\gamma^2}.$$

Finally, with $u = \sigma(\phi(x))$, the solution $x(t)$, $t \geq 0$, of the closed-loop system (10.118) satisfies the nonexpansivity constraint

$$\int_0^T z^{\mathrm{T}}(t)z(t)\mathrm{d}t \leq (\gamma/\rho)^2 \int_0^T w^{\mathrm{T}}(t)w(t)\mathrm{d}t + V(x_0), \quad w(\cdot) \in \mathcal{L}_2, \quad T \geq 0, \tag{10.187}$$

where $\sigma : \mathbb{R}^m \to \mathbb{R}^m$ is such that $\sigma(0) = 0$ and for every $u_{\mathrm{c}} \in \mathbb{R}^m$, $\sigma(u_{\mathrm{c}}) = [\sigma_1(u_{\mathrm{c}1}), \ldots, \sigma_m(u_{\mathrm{c}m})]^{\mathrm{T}}$ and $\alpha u_{\mathrm{c}i}^2 < 2\sigma_i(u_{\mathrm{c}i})u_{\mathrm{c}i}$, $u_{\mathrm{c}i} \neq 0$, $i = 1, \ldots, m$.

Proof. The result is a direct consequence of Corollary 10.8 and Theorem 10.5 with $R_2(x) = \frac{1}{2\eta(x)}I_m$ and $L_1(x) = -\alpha(x) + \frac{\eta(x)}{2}\beta^{\mathrm{T}}(x)\beta(x)$, where

$$\eta(x) = \begin{cases} -\left(c_0 + \frac{\alpha(x)+\sqrt{\alpha^2(x)+(\beta^{\mathrm{T}}(x)\beta(x))^2}}{\beta^{\mathrm{T}}(x)\beta(x)}\right), & \beta(x) \neq 0, \\ 0, & \beta(x) = 0. \end{cases} \tag{10.188}$$

Specifically, note that $R_2(x) > 0$, $x \in \mathbb{R}^n$, and

$$\begin{aligned} L_1(x) &= -\alpha(x) + \tfrac{\eta(x)}{2}\beta^{\mathrm{T}}(x)\beta(x) \\ &= \begin{cases} -\frac{1}{2}\Big(c_0\beta^{\mathrm{T}}(x)\beta(x) - \alpha(x) \\ \quad +\sqrt{\alpha^2(x) + (\beta^{\mathrm{T}}(x)\beta(x))^2}\Big), & \beta(x) \neq 0, \\ -\alpha(x), & \beta(x) = 0. \end{cases} \end{aligned} \tag{10.189}$$

Now, it follows from (10.189) that $L_1(x) \geq 0$, $\beta(x) \neq 0$, and since $V(\cdot)$ is a nonexpansive control Lyapunov function of (10.118), it follows that $L_1(x) = -\alpha(x) \geq 0$, for all $x \in \mathcal{R}$. Hence, (10.189) yields $L_1(x) \geq 0$, $x \in \mathbb{R}^n$, so that all the conditions of Corollary 10.9 are satisfied. □

10.8 Nonlinear Controllers with Multilinear and Polynomial Performance Criteria

In this section, we specialize the results of Section 10.5 to linear systems controlled by nonlinear controllers that minimize a multilinear cost functional. Specifically, we consider the linear system (10.101) controlled by nonlinear controllers. First, we consider the case in which $r(z,w) = \gamma^2 w^{\mathrm{T}} w - z^{\mathrm{T}} z$, where $\gamma > 0$. Recall the definitions of S, R_1, and R_2 and let $\ell : \mathbb{R}^n \to \mathbb{R}$ be a multilinear function such that $\ell(x) \geq 0$, $x \in \mathbb{R}^n$. Furthermore, assume that $\gamma^{-2} D D^{\mathrm{T}} \leq S$, where $\gamma > 0$ is given.

Proposition 10.1. Consider the linear dynamical system (10.101) and (10.102) and assume that there exist a positive-definite matrix $P \in \mathbb{R}^{n\times n}$ and a function $p : \mathbb{R}^n \to \mathbb{R}$ such that $p(x) \geq 0$, $x \in \mathbb{R}^n$,

$$0 = A^{\mathrm{T}} P + PA + R_1 + \gamma^{-2} P D D^{\mathrm{T}} P - PSP, \tag{10.190}$$

and

$$0 = p'(x)[A - (S - \gamma^{-2} D D^{\mathrm{T}})P]x + \ell(x), \tag{10.191}$$

where $\gamma > 0$. Then the zero solution $x(t) \equiv 0$ of the undisturbed ($w(t) \equiv 0$) system (10.101) is globally asymptotically stable for all $x_0 \in \mathbb{R}^n$ with the feedback control law $u = \phi(x) = -R_2^{-1} B^{\mathrm{T}}(Px + \frac{1}{2} p'^{\mathrm{T}}(x))$, and the performance functional (10.135) with $R_2(x) \equiv R_2$, $L_2(x) \equiv 0$, and

$$L_1(x) = x^{\mathrm{T}} R_1 x + \ell(x) + \tfrac{1}{4} p'(x)(S - \gamma^{-2} D D^{\mathrm{T}}) p'^{\mathrm{T}}(x), \tag{10.192}$$

satisfies

$$J(x_0, \phi(x(\cdot))) \leq \mathcal{J}(x_0, \phi(x(\cdot))) = x_0^{\mathrm{T}} P x_0 + p(x_0), \tag{10.193}$$

where

$$\mathcal{J}(x_0, u(\cdot)) = \int_0^\infty [L(x(t), u(t)) + \Gamma(x(t), u(t))]\mathrm{d}t \tag{10.194}$$

and where $u(\cdot)$ is admissible, $x(t)$, $t \geq 0$, solves (10.101) with $w(t) \equiv 0$, and

$$\Gamma(x,u) = \gamma^{-2}[Px + \tfrac{1}{2} p'^{\mathrm{T}}(x)]^{\mathrm{T}} D D^{\mathrm{T}} [Px + \tfrac{1}{2} p'^{\mathrm{T}}(x)]. \tag{10.195}$$

Furthermore,

$$\mathcal{J}(x_0, \phi(x(\cdot))) = \min_{u(\cdot) \in \mathcal{S}(x_0)} \mathcal{J}(x_0, u(\cdot)), \tag{10.196}$$

where $\mathcal{S}(x_0)$ is the set of regulation controllers for the system (10.101) with $w(t) \equiv 0$ and $x_0 \in \mathbb{R}^n$. Finally, with $u = \phi(x)$, the solution $x(t)$, $t \geq 0$, of (10.101) satisfies the nonexpansivity constraint

$$\int_0^T z^{\mathrm{T}}(t) z(t) \mathrm{d}t \leq \gamma^2 \int_0^T w(t)^{\mathrm{T}} w(t) \mathrm{d}t + V(x_0), \qquad w(\cdot) \in \mathcal{L}_2, \qquad T \geq 0. \tag{10.197}$$

Proof. The result is a direct consequence of Theorem 10.4 with $F(x,u) = Ax + Bu$, $J_1(x) = D$, $L(x,u) = x^{\mathrm{T}}R_1x + \frac{1}{2}p'(x)(S - \gamma^{-2}DD^{\mathrm{T}})$ $p'^{\mathrm{T}}(x) + \ell(x) + u^{\mathrm{T}}R_2u$, $V(x) = x^{\mathrm{T}}Px + p(x)$, $\mathcal{D} = \mathbb{R}^n$, and $U = \mathbb{R}^m$. Specifically, conditions (10.88)–(10.91) are trivially satisfied. Now, forming $x^{\mathrm{T}}(10.190)x + (10.191)$ it follows that, after some algebraic manipulations, $V'(x)J_1(x)w \le r(z,w) + L(x,\phi(x),w) + \Gamma(x,\phi(x),w)$ for all $x \in \mathcal{D}$ and $w \in \mathcal{W}$. Furthermore, it follows from (10.190) and (10.191) that $H(x,\phi(x)) = 0$ and $H(x,u) = H(x,u) - H(x,\phi(x)) = [u - \phi(x)]^{\mathrm{T}}R_2[u - \phi(x)] \ge 0$ so that all conditions of Theorem 10.4 are satisfied. Finally, since $V(\cdot)$ is radially unbounded, (10.101), with $u(t) = \phi(x(t)) = -R_2^{-1}(B^{\mathrm{T}}Px(t) + \frac{1}{2}p'^{\mathrm{T}}(x(t)))$, is globally asymptotically stable. □

Since $A - (S - \gamma^{-2}DD^{\mathrm{T}})P$ is Hurwitz and $\ell(x)$, $x \in \mathbb{R}^n$, is a nonnegative multilinear function, it follows from Lemma 8.1 that there exists a nonnegative $p(x)$, $x \in \mathbb{R}^n$, such that (10.191) is satisfied. Proposition 10.1 generalizes the classical results of Bass and Webber [33] to optimal control of nonlinear systems with bounded energy disturbances. As discussed in Chapter 8 the performance functional (10.192) is a derived performance functional in the sense that it cannot be arbitrarily specified. However, this performance functional does weigh the state variables by arbitrary even powers. Furthermore, (10.192) has the form

$$J(x_0,u(\cdot)) = \int_0^\infty \Big[x^{\mathrm{T}}R_1x + \ell(x) + u^{\mathrm{T}}R_2u - \tfrac{1}{4}\gamma^{-2}p'(x)DD^{\mathrm{T}}p'^{\mathrm{T}}(x) + \phi_{\mathrm{NL}}^{\mathrm{T}}(x)R_2\phi_{\mathrm{NL}}(x)\Big]\mathrm{d}t,$$

where $\phi_{\mathrm{NL}}(x) \triangleq -\frac{1}{2}R_2^{-1}B^{\mathrm{T}}p'^{\mathrm{T}}(x)$ is the nonlinear part of the optimal feedback control

$$\phi(x) = \phi_{\mathrm{L}}(x) + \phi_{\mathrm{NL}}(x),$$

where $\phi_{\mathrm{L}}(x) \triangleq -R_2^{-1}B^{\mathrm{T}}Px$.

If $p(x)$ is a polynomial function of the form $p(x) = \sum_{k=2}^{r}\frac{1}{k}(x^{\mathrm{T}}M_kx)^k$, then it follows from (10.191) that $\ell(x) = \sum_{k=2}^{r}(x^{\mathrm{T}}M_kx)^{k-1}x^{\mathrm{T}}\hat{R}_kx$, where $M_k, \hat{R}_k \in \mathbb{N}^n$, $k = 2,\ldots,r$, and M_k satisfies

$$0 = [A-(S-\gamma^{-2}DD^{\mathrm{T}}P)]^{\mathrm{T}}M_k + M_k[A-(S-\gamma^{-2}DD^{\mathrm{T}}P)] + \hat{R}_k,\quad k = 2,\ldots,r, \tag{10.198}$$

where P satisfies (10.190). In this case, the optimal control law $\phi(x)$ is given by

$$\phi(x) = -R_2^{-1}B^{\mathrm{T}}\left(Px + \sum_{k=2}^{r}(x^{\mathrm{T}}M_kx)^{k-1}M_kx\right),$$

the corresponding Lyapunov function guaranteeing closed-loop stability is given by

$$V(x) = x^{\mathrm{T}}Px + \sum_{k=2}^{r} \frac{1}{k}(x^{\mathrm{T}}M_k x)^k,$$

and $L(x, u)$ given by (10.192) becomes

$$\begin{aligned} L(x,u) &= x^{\mathrm{T}}\left(R_1 + \sum_{k=2}^{r}(x^{\mathrm{T}}M_k x)^{k-1}\hat{R}_k\right)x \\ &\quad + \left(\sum_{k=2}^{r}(x^{\mathrm{T}}M_k x)^{k-1}M_k x\right)^{\mathrm{T}}(S - \gamma^{-2}DD^{\mathrm{T}}) \\ &\quad \cdot \left(\sum_{k=2}^{r}(x^{\mathrm{T}}M_k x)^{k-1}M_k x\right) + u^{\mathrm{T}}R_2 u. \end{aligned}$$

Furthermore, if $\hat{R}_k = \hat{R}_2$, $k = 3, \ldots, r$, then $M_k = M_2$, $k = 3, \ldots, r$, satisfies (10.198). In this case, we require the solution of only one modified Riccati equation in (10.198). Proposition 10.1 generalizes the deterministic version of the stochastic nonlinear-nonquadratic optimal control problem considered in [411] to the disturbance rejection setting. Furthermore, unlike the results of [411], this result is not limited to sixth-order cost functionals and cubic nonlinear controllers since it addresses a polynomial nonlinear performance criterion.

Next, we consider the linear system (10.101) controlled by nonlinear controllers where $r(z, w) = 2z^{\mathrm{T}}w$. For the statement of the next result recall the definitions of R_0, S_{s}, $R_{1\mathrm{s}}$, $R_{2\mathrm{s}}$, A_{s}, B_{s} and let $\ell : \mathbb{R}^n \to \mathbb{R}$ be a multilinear function such that $\ell(x) \geq 0$, $x \in \mathbb{R}^n$. Furthermore, assume that $DR_0D^{\mathrm{T}} \leq S_{\mathrm{s}}$ and $p = d$.

Proposition 10.2. Consider the linear dynamical system (10.101) and (10.112) and assume that there exist a positive-definite matrix $P \in \mathbb{R}^{n\times n}$ and a function $p : \mathbb{R}^n \to \mathbb{R}$ such that $p(x) \geq 0$, $x \in \mathbb{R}^n$,

$$0 = A_{\mathrm{s}}^{\mathrm{T}}P + PA_{\mathrm{s}} + R_{1\mathrm{s}} + PDR_0D^{\mathrm{T}}P - PS_{\mathrm{s}}P, \tag{10.199}$$

and

$$0 = p'(x)[A_{\mathrm{s}} - (S_{\mathrm{s}} - DR_0D^{\mathrm{T}})P]x + \ell(x). \tag{10.200}$$

Then the zero solution $x(t) \equiv 0$ of the undisturbed ($w(t) \equiv 0$) system (10.101) is globally asymptotically stable for all $x_0 \in \mathbb{R}^n$ with the feedback control law $u = \phi(x) = -R_{2\mathrm{s}}^{-1}(B_{\mathrm{s}}^{\mathrm{T}}(Px + \frac{1}{2}p'^{\mathrm{T}}(x)) + E_2^{\mathrm{T}}R_0E_1x)$, and the performance functional (10.135) with $R_2(x) \equiv R_2$, $L_2(x) \equiv 0$, and

$$L_1(x) = x^{\mathrm{T}}R_1x + \ell(x) + \tfrac{1}{4}p'(x)(S_{\mathrm{s}} - DR_0D^{\mathrm{T}})p'^{\mathrm{T}}(x), \tag{10.201}$$

satisfies

$$J(x_0,\phi(x(\cdot))) \le \mathcal{J}(x_0,\phi(x(\cdot))) = x_0^{\mathrm{T}}Px_0 + p(x_0), \tag{10.202}$$

where

$$\mathcal{J}(x_0,u(\cdot)) = \int_0^\infty [L(x(t),u(t)) + \Gamma(x(t),u(t))]\mathrm{d}t \tag{10.203}$$

and where $u(\cdot)$ is admissible, $x(t)$, $t \ge 0$, solves (10.101) with $w(t) \equiv 0$, and

$$\begin{aligned}\Gamma(x,u) &= [D^{\mathrm{T}}(Px + \tfrac{1}{2}p'^{\mathrm{T}}(x)) - E_1x - E_2u]^{\mathrm{T}}R_0\\ &\quad\cdot[D^{\mathrm{T}}(Px + \tfrac{1}{2}p'^{\mathrm{T}}(x)) - E_1x - E_2u].\end{aligned} \tag{10.204}$$

Furthermore,

$$\mathcal{J}(x_0,\phi(x(\cdot))) = \min_{u(\cdot)\in\mathcal{S}(x_0)} \mathcal{J}(x_0,u(\cdot)), \tag{10.205}$$

where $\mathcal{S}(x_0)$ is the set of regulation controllers for the system (10.101) with $w(t) \equiv 0$ and $x_0 \in \mathbb{R}^n$. Finally, with $u = \phi(x)$, the solution $x(t)$, $t \ge 0$, of (10.101) satisfies the passivity constraint

$$\int_0^T 2w^{\mathrm{T}}(t)z(t) + V(x_0) \ge 0, \qquad w(\cdot) \in \mathcal{L}_2, \qquad T \ge 0. \tag{10.206}$$

Proof. The result is a direct consequence of Theorem 10.4 with $F(x,u) = Ax + Bu$, $J_1(x) = D$, $L(x,u) = x^{\mathrm{T}}R_1x + \frac{1}{2}p'(x)(S_{\mathrm{s}} - DR_0D^{\mathrm{T}})$ $p'^{\mathrm{T}}(x)$ $+\ell(x) + u^{\mathrm{T}}R_2u$, $V(x) = x^{\mathrm{T}}Px + p(x)$, $\mathcal{D} = \mathbb{R}^n$, and $U = \mathbb{R}^m$. Specifically, conditions (10.88)–(10.91) are trivially satisfied. Now, forming $x^{\mathrm{T}}(10.199)x + (10.200)$, it follows that, after some algebraic manipulations, $V'(x)J_1(x)w \le r(z,w) + L(x,\phi(x),w) + \Gamma(x,\phi(x),w)$ for all $x \in \mathcal{D}$ and $w \in \mathcal{W}$. Furthermore, it follows from (10.199) and (10.200) that $H(x,\phi(x)) = 0$ and $H(x,u) = H(x,u) - H(x,\phi(x)) = [u - \phi(x)]^{\mathrm{T}}R_{2\mathrm{s}}[u-\phi(x)] \ge 0$ so that all conditions of Theorem 10.4 are satisfied. Finally, since $V(\cdot)$ is radially unbounded, (10.101), with $u(t) = \phi(x(t)) = -R_{2\mathrm{s}}^{-1}(B_{\mathrm{s}}^{\mathrm{T}}(Px(t)+\frac{1}{2}p'^{\mathrm{T}}(x(t)))+E_2^{\mathrm{T}}R_0E_1x(t))$, is globally asymptotically stable. □

Since $A_{\mathrm{s}} - (S_{\mathrm{s}} - DR_0D^{\mathrm{T}})P$ is Hurwitz and $\ell(x)$, $x \in \mathbb{R}^n$, is a nonnegative multilinear function, it follows from Lemma 8.1 that there exists a nonnegative $p(x)$, $x \in \mathbb{R}^n$, such that (10.200) is satisfied.

Finally, if $p(x)$, $x \in \mathbb{R}^n$, is a polynomial function of the form $\sum_{k=2}^r \frac{1}{k}$ $(x^{\mathrm{T}}M_kx)^k$, then it follows from (10.200) that $\ell(x) = \sum_{k=2}^r (x^{\mathrm{T}}M_kx)^{k-1}x^{\mathrm{T}}$ $\cdot\hat{R}_kx$, where $M_k, \hat{R}_k \in \mathbb{N}^n$, $k = 2,\ldots,r$, and M_k satisfies

$$0 = [A_{\mathrm{s}}-(S_{\mathrm{s}}-DR_0D^{\mathrm{T}}P)]^{\mathrm{T}}M_k+M_k[A_{\mathrm{s}}-(S_{\mathrm{s}}-DR_0D^{\mathrm{T}}P)]+\hat{R}_k,\ k = 2,\ldots,r, \tag{10.207}$$

and where P satisfies (10.199). In this case, the optimal control law $\phi(x)$ is given by

$$\phi(x) = -R_{2\mathrm{s}}^{-1}\left[B_{\mathrm{s}}^{\mathrm{T}}\left(Px + \sum_{k=2}^{r}(x^{\mathrm{T}}M_k x)^{k-1}M_k x\right) + E_2^{\mathrm{T}}R_0 E_1 x\right],$$

the corresponding Lyapunov function guaranteeing closed-loop stability is given by

$$V(x) = x^{\mathrm{T}}Px + \sum_{k=2}^{r}\frac{1}{k}(x^{\mathrm{T}}M_k x)^k,$$

and $L(x,u)$ given by (10.201) becomes

$$\begin{aligned} L(x,u) &= x^{\mathrm{T}}\left(R_1 + \sum_{k=2}^{r}(x^{\mathrm{T}}M_k x)^{k-1}\hat{R}_k\right)x + u^{\mathrm{T}}R_2 u \\ &\quad + \left(\sum_{k=2}^{r}(x^{\mathrm{T}}M_k x)^{k-1}M_k x\right)^{\mathrm{T}}(S_{\mathrm{s}} - D_0 R_0 D^{\mathrm{T}}) \\ &\quad \cdot \left(\sum_{k=2}^{r}(x^{\mathrm{T}}M_k x)^{k-1}M_k x\right). \end{aligned}$$

10.9 Problems

Problem 10.1. Consider the linear dynamical system given by (10.68) and (10.69), where A is Hurwitz. Show that if (A, D) is controllable and $A_\alpha = A + \frac{\alpha}{2}I_n$, $\alpha > 0$, then the solution to the Riccati equation (10.76) is given by (10.77).

Problem 10.2. Let $X \in \mathbb{R}^{n\times n}$ be a nonnegative-definite matrix. Show that $\lim_{q\to\infty}[\mathrm{tr}\,(I_n \circ X)^q]^{1/q} = \mathrm{d}_{\max}(X)$ and $\lim_{q\to\infty}[\mathrm{tr}\,X^q]^{1/q} = \lambda_{\max}(X)$, where $\circ$ denotes the Hadamard product (i.e., entry-by-entry product), $\mathrm{d}_{\max}(X)$ denotes the maximum diagonal entry of X, and $\lambda_{\max}(X)$ denotes the maximum eigenvalue of X. Also show that $\mathrm{d}_{\max}(X) \le \lambda_{\max}(X) \le \mathrm{tr}(X)$. Use the above results to obtain differentiable bounds for $|\!|\!|G|\!|\!|_1$ given by (10.78).

Problem 10.3. Consider the linear time-invariant system given by (10.68) and (10.69). Assume A is Hurwitz. Show that

$$\sup_{\omega\in\mathbb{R}}\sigma_{\max}(G(\jmath\omega)) = \sup_{w(\cdot)\in\mathcal{L}_2}\frac{|\!|\!|z|\!|\!|_{2,2}}{|\!|\!|w|\!|\!|_{2,2}}, \qquad (10.208)$$

where $G(s) = E(sI - A)^{-1}D + E_\infty$.

Problem 10.4. Consider the linear dynamical system

$$G(s) \sim \left[\begin{array}{c|c} A & D \\ \hline E & 0 \end{array}\right],$$

where $G \in \mathcal{RH}_\infty$. Define the *entropy of G at infinity* by

$$I(G,\gamma) \triangleq \lim_{s_0 \to \infty} \left[-\frac{\gamma^2}{2\pi} \int_{-\infty}^{\infty} \ln \left| \det(I_m - \gamma^{-2} G^*(\jmath\omega) G(\jmath\omega)) \right| \left[\frac{s_0^2}{s_0^2 + \omega^2} \right] \mathrm{d}\omega \right], \tag{10.209}$$

and assume there exists a positive-definite matrix $P \in \mathbb{R}^{n \times n}$ satisfying

$$0 = A^\mathrm{T} P + PA + \gamma^{-2} PDD^\mathrm{T} P + E^\mathrm{T} E, \tag{10.210}$$

where $\gamma > 0$. Show that the following statements hold:

i) The transfer function G satisfies $|||G|||_\infty \leq \gamma$.

ii) If $|||G|||_\infty < \gamma$, then $I(G,\gamma) \leq \operatorname{tr} D^\mathrm{T} PD$.

iii) The $\mathcal{H}_2$ norm of G satisfies $|||G|||_2 \leq I(G,\gamma)$.

iv) All real symmetric solutions to (10.210) are nonnegative definite.

v) There exists a (unique) minimal solution to (10.210) in the class of real symmetric solutions.

vi) P is the minimal solution to (10.210) if and only if $\alpha(A + \gamma^{-2} DD^\mathrm{T} P) < 0$.

vii) $|||G|||_\infty < \gamma$ if and only if $A + \gamma^{-2} DD^\mathrm{T} P$ is Hurwitz, where P is the minimal solution to (10.210).

viii) If P is the minimal solution to (10.210) and $|||G|||_\infty < \gamma$, then $I(G,\gamma) = \operatorname{tr} D^\mathrm{T} PD$.

Problem 10.5. Consider the nonlinear dynamical system $\mathcal{G}$ given by (10.3) and (10.4) with $J_2(x) \equiv 0$ and $p = d$. Show that if there exists a continuously differentiable, positive-definite function $W : \mathcal{D} \to \mathbb{R}$ such that

$$W'(x) f(x) \leq -\delta h^\mathrm{T}(x) h(x), \quad \delta > 0, \tag{10.211}$$

$$W'(x) J_1(x) = h^\mathrm{T}(x), \tag{10.212}$$

then the nonlinear system $\mathcal{G}$ is nonexpansive with gain $1/\delta$.

Problem 10.6. A nonlinear dynamical system $\mathcal{G}$ given by (10.1) and (10.2) is said to be *input-to-output stable* if for every $x_0 \in \mathbb{R}^n$ and every continuous bounded input $w(t) \in \mathbb{R}^d$, $t \geq 0$, the solution $x(t)$, $t \geq 0$, to (10.1) exists and the output satisfies

$$\|z(t)\| \leq \eta(\|x_0\|, t) + \gamma \left(\sup_{0 \leq \tau \leq t} \|w(\tau)\| \right), \quad t \geq 0, \tag{10.213}$$

where $\eta(s,t)$, $s > 0$, is a class $\mathcal{KL}$ function and $\gamma(s)$, $s > 0$, is a class $\mathcal{K}$ function. Show that if $\mathcal{G}$ is input-to-state stable, then $\mathcal{G}$ is input-to-output stable. Alternatively, show that if

$$\|h(x) + J_2(x)w\| \leq \alpha_1(\|x\|) + \alpha_2(\|w\|), \tag{10.214}$$

where $\alpha_1(\cdot)$ and $\alpha_2(\cdot)$ are class $\mathcal{K}$ functions, then $\mathcal{G}$ is input-to-output stable.

Problem 10.7. Consider the linear dynamical system with state delay given by

$$\dot{x}(t) = Ax(t) + A_{\rm d}x(t-\tau_{\rm d}), \quad x(\theta) = \phi(\theta), \quad -\tau_{\rm d} \leq \theta \leq 0, \tag{10.215}$$

where $x(t) \in \mathbb{R}^n$, $A \in \mathbb{R}^{n\times n}$, $A_{\rm d} \in \mathbb{R}^{n\times n}$, and $\phi : [-\tau_{\rm d}, 0] \to \mathbb{R}^n$ is a continuous vector-valued function specifying the initial state of the system. Let $R > 0$ and suppose there exist an $n \times n$ positive-definite matrix P and a scalar $\alpha > 0$ such that

$$o = A^{\rm T}P + PA + \alpha^2 I_m + \alpha^{-2}PA_{\rm d}A_{\rm d}^{\rm T}P + R. \tag{10.216}$$

Show that the zero solution $x_t \equiv 0$ to (10.215) is globally asymptotically stable (in the sense of Problem 3.65) for all $\tau_{\rm d} \geq 0$. Furthermore, show that this problem can be formulated as a feedback involving the operator $\Delta x(t) \triangleq x(t-\tau_{\rm d})$ satisfying the nonexpansivity constraint $\|\Delta x(t)\|_2 \leq \|x(t)\|_2$. (**Hint:** Use the Lyapunov-Krasovskii functional candidate given by

$$V(\psi) = \psi^{\rm T}(0)P\psi(0) + \alpha^2 \int_{-\tau_{\rm d}}^{0} \psi^{\rm T}(\theta)\psi(\theta){\rm d}\theta, \quad \psi \in \mathcal{C}([-\tau_{\rm d}, 0], \mathbb{R}^n).) \tag{10.217}$$

Problem 10.8. Consider the linear controlled system with state delay given by

$$\dot{x}(t) = Ax(t) + A_{\rm d}x(t-\tau_{\rm d}) + Bu(t), \quad x(\theta) = \hat{\phi}(\theta), \quad -\tau_{\rm d} \leq \theta \leq 0, \tag{10.218}$$

where $x(t) \in \mathbb{R}^n$, $u(t) \in \mathbb{R}^m$, $A, A_{\rm d} \in \mathbb{R}^{n\times n}$, $B \in \mathbb{R}^{n\times m}$, and $\hat{\phi} : [-\tau_{\rm d}, 0] \to \mathbb{R}^n$ is a continuous vector-valued function specifying the initial state of the system. Let $\alpha > 0$, $R_1 > 0$, and $R_2 > 0$. Show that the zero solution $x_t \equiv 0$ to (10.218) is globally asymptotically stable (in the sense of Problem 3.65) for all $\tau_{\rm d} \geq 0$ with the feedback control $\phi(x) = -R_2^{-1}B^{\rm T}Px$, where $P > 0$ satisfies

$$0 = A^{\rm T}P + PA + R_1 + \alpha^2 I_n + \alpha^{-2}PA_{\rm d}A_{\rm d}^{\rm T}P - PBR_2^{-1}B^{\rm T}P. \tag{10.219}$$

Problem 10.9. Consider the nonlinear controlled system with state delay given by

$$\dot{x}(t) = Ax(t) + f_{\rm d}(x(t-\tau_{\rm d})) + Bu(t), \quad x(\theta) = \hat{\phi}(\theta), \quad -\tau_{\rm d} \leq \theta \leq 0,$$

(10.220)

where $x(t) \in \mathbb{R}^n$, $u(t) \in \mathbb{R}^m$, $A \in \mathbb{R}^{n\times n}$, $B \in \mathbb{R}^{n\times m}$, $f_{\rm d} : \mathbb{R}^n \to \mathbb{R}^n$ satisfies $f_{\rm d}(0) = 0$, and $\hat{\phi} : [-\tau_{\rm d}, 0] \to \mathbb{R}^n$ is a continuous vector-valued function specifying the initial state of the system. Let $\alpha > 0$, $R_1 > 0$, $R_2 > 0$, and assume that $\|f_{\rm d}(x)\| \le \gamma\|x\|_2$, $x \in \mathbb{R}^n$. Show that the zero solution $x_t \equiv 0$ to (10.218) is globally asymptotically stable (in the sense of Problem 3.65) for all $\tau_{\rm d} \ge 0$ with the feedback control $\phi(x) = -R_2^{-1}B^{\rm T}Px$, where $P > 0$ satisfies

$$0 = A^{\rm T}P + PA + R_1 + \alpha^{-2}P^2 - 2PBR_2^{-1}B^{\rm T}P. \tag{10.221}$$

(**Hint:** Use the Lyapunov-Krasovskii functional candidate given by

$$V(\psi) = \psi^{\rm T}(0)P\psi(0) + \alpha^2\int_{-\tau_{\rm d}}^{0} f_{\rm d}^{\rm T}(\psi(\theta))f_{\rm d}(\psi(\theta)){\rm d}\theta, \quad \psi \in \mathcal{C}([-\tau_{\rm d}, 0], \mathbb{R}^n).) \tag{10.222}$$

Problem 10.10. Consider the nonlinear scalar dynamical system

$$\dot{x}(t) = -x^3(t) + u(t) + x(t)w(t), \quad x(0) = x_0, \quad t \ge 0, \tag{10.223}$$

where $x(t), u(t) \in \mathbb{R}$ and $w(t) \in \mathcal{W} = [-1, 1]$. Find a stabilizing feedback controller $u(t) = \phi(x(t))$ that minimizes

$$J(x_0, u(\cdot)) = \int_0^\infty [x^2(t) + u^2(t)]{\rm d}t. \tag{10.224}$$

Compare your controller to the feedback linearization controller $u_{\rm FL}(t) = x^3(t) - 2x(t)$ by plotting $u(t)$ versus $x(t)$.

Problem 10.11. Consider the port-controlled Hamiltonian system given in Problem 8.21. Find an asymptotically stabilizing feedback control law of the form $u = \phi(x) = -\alpha(\gamma)G^{\rm T}(x)\left[\frac{\partial \mathcal{H}}{\partial x}(x)\right]^{\rm T}$, where $\alpha(\gamma) > 0$ and $\alpha(\gamma)\left[\frac{\partial \mathcal{H}}{\partial x}(x)\right]$ satisfies the Hamilton-Jacobi-Bellman equation.

Problem 10.12. Consider the nonlinear dynamical system $\mathcal{G}$ given by

$$\dot{x}(t) = f(x(t)) + G(x(t))u(t) + J_1(x(t))w(t), \; x(0) = x_0, \; w(\cdot) \in \mathcal{L}_2, \; t \ge 0, \tag{10.225}$$

$$y(t) = h(x(t)), \tag{10.226}$$

with performance variables

$$z(t) = E_{1\infty}(x(t)) + E_{2\infty}u(t), \tag{10.227}$$

where $x \in \mathbb{R}^n$, $u, y \in \mathbb{R}^m$, $w \in \mathbb{R}^d$, $z \in \mathbb{R}^p$, $f : \mathbb{R}^n \to \mathbb{R}^n$ satisfies $f(0) = 0$, $G : \mathbb{R}^n \to \mathbb{R}^{n\times m}$, $J_1 : \mathbb{R}^n \to \mathbb{R}^{n\times d}$, $h : \mathbb{R}^n \to \mathbb{R}^m$ satisfies $h(0) = 0$, $E_{1\infty} : \mathbb{R}^n \to \mathbb{R}^p$, $E_{2\infty} \in \mathbb{R}^{p\times m}$, $R_2 \triangleq E_{2\infty}^{\rm T}E_{2\infty} > 0$, and $E_{2\infty}^{\rm T}E_{1\infty}(x) = 0$,

$x \in \mathbb{R}^n$. Assume that, with $w(t) \equiv 0$, (10.225) and (10.226) is passive, zero-state observable, and completely reachable with a continuously differentiable radially unbounded storage function $V_{\rm s} : \mathbb{R}^n \to \mathbb{R}$. Furthermore, let $V : \mathbb{R}^n \to \mathbb{R}$ be a continuously differentiable function such that $V(0) = 0$, $V(x) > 0$, $x \in \mathbb{R}^n$, $x \neq 0$, $V(x) \to \infty$ as $\|x\| \to \infty$, and

$$\begin{aligned} 0 &= L_1(x) + V'(x)f(x) - \tfrac{1}{4}V'(x)G(x)R_2^{-1}G^{\rm T}(x)V'^{\rm T}(x) \\ &\quad + \tfrac{1}{4\gamma^2}V'(x)J_1(x)J_1^{\rm T}(x)V'^{\rm T}(x), \qquad x \in \mathbb{R}^n, \end{aligned} \tag{10.228}$$

where $\gamma > 0$, $L_1(x) = \ell^{\rm T}(x)\ell(x) + h^{\rm T}(x)R_2^{-1}h(x) - \frac{1}{4\gamma^2}V_{\rm s}'(x)J_1(x)J_1^{\rm T}(x) \cdot V_{\rm s}'^{\rm T}(x) \geq 0$, and $\ell(\cdot)$ satisfies (5.145). Show that the *output* feedback controller $u(t) = -R_2^{-1}y(t)$ asymptotically stabilizes the undisturbed ($w(t) \equiv 0$) nonlinear system (10.225) and minimizes the performance criterion

$$\mathcal{J}(x_0, u(\cdot)) = \int_0^\infty [\ell^{\rm T}(x(t))\ell(x(t)) + y^{\rm T}(t)R_2^{-1}y(t) + u^{\rm T}(t)R_2^{-1}u(t)]{\rm d}t, \tag{10.229}$$

where $x(t)$, $t \geq 0$, solves (10.225) with $w(t) \equiv 0$, in the sense that

$$\mathcal{J}(x_0, \phi(y(\cdot))) = \min_{u(\cdot)\in\mathcal{S}(x_0)} \mathcal{J}(x_0, u(\cdot)), \quad x_0 \in \mathbb{R}^n. \tag{10.230}$$

Furthermore, show that with $u(t) = -R_2^{-1}y(t)$, the solution $x(t)$, $t \geq 0$, of the closed-loop system (10.225) satisfies the nonexpansivity constraint

$$\int_0^T z^{\rm T}(t)z(t){\rm d}t \leq \gamma^2 \int_0^T w^{\rm T}(x)w(t){\rm d}t + V(x_0), \quad w(\cdot) \in \mathcal{L}_2, \quad T \geq 0. \tag{10.231}$$

Problem 10.13. Consider the nonlinear cascade dynamical system

$$\dot{x}(t) = f(x(t)) + G(x(t))\hat{x}(t) + J_1(x)w(t), \ x(0) = x_0, \ w(\cdot) \in \mathcal{L}_2, \ t \geq 0, \tag{10.232}$$

$$\dot{\hat{x}}(t) = u(t) + J_3(\hat{x})w(t), \ \hat{x}(0) = \hat{x}_0, \tag{10.233}$$

$$z(t) = h(x(t), \hat{x}(t)) + J(x(t), \hat{x}(t))u(t), \tag{10.234}$$

with performance functional

$$J(x_0, \hat{x}_0, u(\cdot)) \triangleq \int_0^\infty L(x(t), \hat{x}(t), u(t)){\rm d}t, \tag{10.235}$$

where $u(\cdot)$ is admissible and $(x(t), \hat{x}(t))$, $t \geq 0$, solves (10.232) and (10.233) and where

$$L(x, \hat{x}, u) \triangleq L_1(x, \hat{x}) + L_2(x, \hat{x})u + u^{\rm T}R_2(x, \hat{x})u, \tag{10.236}$$

where $L_1 : \mathbb{R}^n \times \mathbb{R}^m \to \mathbb{R}$, $L_2 : \mathbb{R}^n \times \mathbb{R}^m \to \mathbb{R}^{1\times m}$, and $R_2 : \mathbb{R}^n \times \mathbb{R}^m \to \mathbb{P}^m$. Assume there exist continuously differentiable functions $\alpha : \mathbb{R}^n \to \mathbb{R}^m$ and

$V_{\mathrm{sub}} : \mathbb{R}^n \to \mathbb{R}$ such that for all $x \in \mathbb{R}^n$ and $\hat{x} \in \mathbb{R}^m$,

$$\alpha(0) = 0, \tag{10.237}$$

$$V_{\mathrm{sub}}(0) = 0, \tag{10.238}$$

$$V_{\mathrm{sub}}(x) > 0, \quad x \neq 0, \tag{10.239}$$

$$V'_{\mathrm{sub}}(x)[f(x) + G(x)\alpha(x)] + \tfrac{1}{4\gamma^2} V'_{\mathrm{sub}}(x) J_1(x) J_1^{\mathrm{T}}(x) V'^{\mathrm{T}}_{\mathrm{sub}}(x) < 0, \quad x \neq 0, \tag{10.240}$$

where $R_{2a}(x, \hat{x}) \triangleq R_2(x, \hat{x}) + J^{\mathrm{T}}(x, \hat{x}) J(x, \hat{x})$ and $\gamma > 0$. Furthermore, assume that there exist a function $L_2 : \mathbb{R}^n \times \mathbb{R}^m \to \mathbb{R}^{1 \times m}$ such that $L_2(0, 0) = 0$ and a positive-definite matrix $\hat{P} \in \mathbb{R}^{m \times m}$ such that, for all $x \in \mathbb{R}^n$ and $\hat{x} \in \mathbb{R}^m$,

$$\begin{aligned} &2(\hat{x} - \alpha(x))^{\mathrm{T}} \hat{P} \Big[\tfrac{1}{2} \hat{P}^{-1} G^{\mathrm{T}}(x) V'^{\mathrm{T}}_{\mathrm{sub}}(x) - \alpha'(x)(f(x) + G(x)\hat{x}) \\ &- \tfrac{1}{2} R_{2a}^{-1}(x, \hat{x}) \{ 2\hat{P}(\hat{x} - \alpha(x)) + L_2^{\mathrm{T}}(x, \hat{x}) + 2J^{\mathrm{T}}(x, \hat{x}) h(x, \hat{x}) \} \\ &+ \tfrac{1}{2\gamma^2} \{ -\alpha'(x) J_1(x) J_1^{\mathrm{T}}(x) V'^{\mathrm{T}}_{\mathrm{sub}}(x) \\ &+ [\alpha'(x) J_1(x) J_1^{\mathrm{T}}(x) \alpha'^{\mathrm{T}}(x) + J_3(\hat{x}) J_3^{\mathrm{T}}(\hat{x})] \hat{P}(\hat{x} - \alpha(x)) \} \Big] \\ &+ [h(x, \hat{x}) + J(x, \hat{x}) \phi(x, \hat{x})]^{\mathrm{T}} [h(x, \hat{x}) + J(x, \hat{x}) \phi(x, \hat{x})] < 0, \quad \hat{x} \neq \alpha(x), \end{aligned} \tag{10.241}$$

where $\phi(x, \hat{x}) = -\frac{1}{2} R_{2a}^{-1}(x, \hat{x}) [L_2^{\mathrm{T}}(x, \hat{x}) + 2\hat{P}(\hat{x} - \alpha(x)) + 2J^{\mathrm{T}}(x, \hat{x}) h(x, \hat{x})]$. Show that, with the feedback control law

$$\phi(x, \hat{x}) = -\tfrac{1}{2} R_{2a}^{-1}(x, \hat{x}) [L_2^{\mathrm{T}}(x, \hat{x}) + 2\hat{P}(\hat{x} - \alpha(x)) + 2J^{\mathrm{T}}(x, \hat{x}) h(x, \hat{x})], \tag{10.242}$$

the zero solution $(x(t), \hat{x}(t)) \equiv (0, 0)$ of the undisturbed ($w(t) \equiv 0$) cascade system

$$\dot{x}(t) = f(x(t)) + G(x(t))\hat{x}(t), \quad x(0) = x_0, \quad t \geq 0, \tag{10.243}$$

$$\dot{\hat{x}}(t) = \phi(x(t), \hat{x}(t)), \quad \hat{x}(0) = \hat{x}_0, \tag{10.244}$$

is globally asymptotically stable. Furthermore, show that

$$J(x_0, \hat{x}_0, \phi(x, \hat{x})) \leq \mathcal{J}(x_0, \hat{x}_0, \phi(x, \hat{x})) = V(x_0, \hat{x}_0), \ (x_0, \hat{x}_0) \in \mathbb{R}^n \times \mathbb{R}^m, \tag{10.245}$$

where

$$\mathcal{J}(x_0, \hat{x}_0, u(\cdot)) \triangleq \int_0^\infty [L(x(t), \hat{x}(t), u(t)) + \Gamma(x(t), \hat{x}(t), u(t))] \mathrm{d}t, \tag{10.246}$$

$$V(x, \hat{x}) = V_{\mathrm{sub}}(x) + (\hat{x} - \alpha(x))^{\mathrm{T}} \hat{P}(\hat{x} - \alpha(x)), \tag{10.247}$$

and

$$\Gamma(x,\hat{x},u) = \tfrac{1}{4\gamma^2}V'(x,\hat{x})J_1(x,\hat{x})J_1^{\rm T}(x,\hat{x})V'^{\rm T}(x,\hat{x}) \\ +[h(x,\hat{x})+J(x,\hat{x})u]^{\rm T}[h(x,\hat{x})+J(x,\hat{x})u], \quad (10.248)$$

where $J_1(x,\hat{x}) \triangleq [J_1^{\rm T}(x),\ J_3^{\rm T}(\hat{x})]^{\rm T}$ and $(x(t),\hat{x}(t))$, $t > 0$, solves (10.232) and (10.233) with $w(t) \equiv 0$. In addition, show that the performance functional (10.246), with

$$L_1(x,\hat{x}) = \phi^{\rm T}(x,\hat{x})R_{2a}(x,\hat{x})\phi(x,\hat{x}) + 2(\hat{x}-\alpha(x))^{\rm T}\hat{P}\alpha'(x)(f(x)+G(x)\hat{x}) \\ -V'_{\rm sub}(x)(f(x)+G(x)\hat{x}) - h^{\rm T}(x,\hat{x})h(x,\hat{x}) \\ -\tfrac{1}{4\gamma^2}V'(x,\hat{x})J_1(x,\hat{x})J_1^{\rm T}(x,\hat{x})V'^{\rm T}(x,\hat{x}), \quad (10.249)$$

is minimized in the sense that

$$\mathcal{J}(x_0,\hat{x}_0,\phi(x(\cdot))) = \min_{u(\cdot)\in\mathcal{S}(x_0,\hat{x}_0)} \mathcal{J}(x_0,\hat{x}_0,u(\cdot)), \quad (10.250)$$

where $\mathcal{S}(x_0,\hat{x}_0)$ is the set of regulation controllers for the nonlinear system (10.232) and (10.233) with $w(t) \equiv 0$. With $u(\cdot) = \phi(x(\cdot),\ \hat{x}(\cdot))$, show that the solution $(x(t),\hat{x}(t))$, $t \geq 0$, of the cascade system (10.232) and (10.233) satisfies the nonexpansivity constraint

$$\int_0^T z^{\rm T}(t)z(t){\rm d}t \leq \gamma^2\int_0^T w^{\rm T}(t)w(t){\rm d}t + V(x_0,\hat{x}_0), \qquad w(\cdot)\in\mathcal{L}_2 \qquad T\geq 0. \quad (10.251)$$

Finally, for $J(x,\hat{x}) \equiv 0$ and $h(x,\hat{x}) = E(x,\hat{x})(\hat{x}-\alpha(x))$, where $E:\mathbb{R}^n\times\mathbb{R}^m \to \mathbb{R}^{p\times m}$, show that a particular choice satisfying (10.241) is given by

$$L_2(x,\hat{x}) = 2[\tfrac{1}{2}V'_{\rm sub}(x)G(x)\hat{P}^{-1} - [f(x)+G(x)\hat{x}]^{\rm T}\alpha'^{\rm T}(x) \\ -\tfrac{1}{2\gamma^2}\{V'_{\rm sub}(x)J_1(x)J_1^{\rm T}(x)\alpha'^{\rm T}(x) \\ -(\hat{x}-\alpha(x))^{\rm T}\hat{P}[\alpha'(x)J_1(x)J_1^{\rm T}(x)\alpha'^{\rm T}(x) + J_3(\hat{x})J_3^{\rm T}(\hat{x})]\} \\ +\tfrac{1}{2}(\hat{x}-\alpha(x))^{\rm T}E^{\rm T}(x,\hat{x})E(x,\hat{x})\hat{P}^{-1}]R_2(x,\hat{x}). \quad (10.252)$$

yielding the feedback control

$$\phi(x,\hat{x}) = -R_2^{-1}(x,\hat{x})\hat{P}(\hat{x}-\alpha(x)) - \tfrac{1}{2}\Big[\hat{P}^{-1}E^{\rm T}(x,\hat{x})E(x,\hat{x})\,(\hat{x}-\alpha(x)) \\ +\hat{P}^{-1}G^{\rm T}(x)V'^{\rm T}_{\rm sub}(x)\Big] + \alpha'(x)[f(x)+G(x)\hat{x}] \\ +\tfrac{1}{2\gamma^2}\Big\{\alpha'(x)J_1(x)J_1^{\rm T}(x)V'^{\rm T}_{\rm sub}(x) \\ -\Big[\alpha'(x)J_1(x)J_1^{\rm T}(x)\alpha'^{\rm T}(x) + J_3(\hat{x})J_3^{\rm T}(\hat{x})\Big]\,\hat{P}\,(\hat{x}-\alpha(x))\Big\}.$$

Problem 10.14. Consider the nonlinear dynamical system (10.232)–(10.234) with $J(x,\hat{x}) \equiv 0$. Assume that there exist continuously differentiable functions $\alpha : \mathbb{R}^n \to \mathbb{R}^m$ and $V_{\rm sub} : \mathbb{R}^n \to \mathbb{R}$ such that

$$\alpha(0) = 0, \tag{10.253}$$

$$V_{\rm sub}(0) = 0, \tag{10.254}$$

$$V_{\rm sub}(x) > 0, \quad x \in \mathbb{R}^n, \quad x \neq 0, \tag{10.255}$$

$$V'_{\rm sub}(x)[f(x) + G(x)\alpha(x)] + \tfrac{1}{4\gamma^2} V'_{\rm sub}(x) J_1(x) J_1^{\rm T}(x) V'^{\rm T}_{\rm sub}(x) < 0, \quad x \in \mathbb{R}^n, \quad x \neq 0, \tag{10.256}$$

and let $h(x, \hat{x}) = E(x, \hat{x})(\hat{x} - \alpha(x))$, where $E : \mathbb{R}^n \times \mathbb{R}^m \to \mathbb{R}^{p\times m}$. Show that, with the feedback stabilizing control law given by

$$\phi(x, \hat{x}) = \begin{cases} -(c_0 + \rho(x, \hat{x}))\beta(x, \hat{x}), & \hat{x} \neq \alpha(x), \\ 0, & \hat{x} = \alpha(x), \end{cases} \tag{10.257}$$

where $\beta(x, \hat{x}) \triangleq 2\hat{P}(\hat{x} - \alpha(x))$,

$$\rho(x, \hat{x}) \triangleq \frac{\sqrt{(\beta^{\rm T}(x, \hat{x})\mu(x, \hat{x}))^2 + (\beta^{\rm T}(x, \hat{x})\beta(x, \hat{x}))^2} - \beta^{\rm T}(x, \hat{x})\mu(x, \hat{x})}{\beta^{\rm T}(x, \hat{x})\beta(x, \hat{x})},$$

$\mu(x, \hat{x}) \triangleq 2\alpha'(x)[f(x) + G(x)\hat{x}] - \hat{P}^{-1}G^{\rm T}(x)V'^{\rm T}_{\rm sub}(x)$, $\hat{P} \in \mathbb{P}^m$, and $c_0 > 0$, the cascade system (10.232) and (10.233) has a sector (and, hence, gain) margin $(\frac{\alpha}{2}, \infty)$, where $\rho \in [0, 1]$ and $\alpha = 1 - \frac{\eta(1-\rho^2)}{\gamma^2}$. Finally, with $u = \sigma(\phi(x, \hat{x}))$, show that the solution $(x(t), \hat{x}(t))$, $t \geq 0$, of the closed-loop system (10.232) and (10.233) satisfies the nonexpansivity constraint

$$\int_0^T z^{\rm T}(t)z(t){\rm d}t \leq (\gamma/\rho)^2 \int_0^T w^{\rm T}(t)w(t){\rm d}t + V(x_0), \quad w(\cdot) \in \mathcal{L}_2, \quad T \geq 0, \tag{10.258}$$

where $\sigma : \mathbb{R}^m \to \mathbb{R}^m$ is such that $\sigma(0) = 0$ and for every $u_{\rm c} \in \mathbb{R}^m$, $\sigma(u_{\rm c}) = [\sigma_1(u_{\rm c1}), \ldots, \sigma_m(u_{{\rm c}m})]^{\rm T}$ and $\alpha u_{{\rm c}i}^2 < 2\sigma_i(u_{{\rm c}i})u_{{\rm c}i}$, $u_{{\rm c}i} \neq 0$, $i = 1, \ldots, m$.

Problem 10.15. Consider the nonlinear cascade dynamical system

$$\dot{x}(t) = f(x(t)) + G(x(t))y(t) + J_1(x)w(t), \quad x(0) = x_0, \quad w(\cdot) \in \mathcal{L}_2, \quad t \geq 0, \tag{10.259}$$

$$\dot{\hat{x}}(t) = \hat{f}(\hat{x}) + \hat{G}(\hat{x})u(t) + J_3(\hat{x})w(t), \quad \hat{x}(0) = \hat{x}_0, \tag{10.260}$$

$$y(t) = \hat{h}(\hat{x}), \tag{10.261}$$

$$z(t) = h(x(t), \hat{x}(t)) + J(x(t), \hat{x}(t))u(t), \tag{10.262}$$

with performance functional (10.235) where $L(x, \hat{x}, u)$ is given by (10.236) and $J_1(x)J_3^{\rm T}(\hat{x}) \equiv 0$. Assume that the input subsystem (10.260) and (10.261) is feedback strictly passive such that there exist a positive-definite storage function $V_{\rm s}(\hat{x})$ and a function $k : \mathbb{R}^q \to \mathbb{R}^m$ satisfying

$$0 > V'_{\rm s}(\hat{x})\left[\hat{f}(\hat{x}) + \hat{G}(\hat{x})k(\hat{x})\right], \quad \hat{x} \in \mathbb{R}^q, \quad \hat{x} \neq 0, \tag{10.263}$$

$$0 = \hat{G}^{\mathrm{T}}(\hat{x})V_{\mathrm{s}}'^{\mathrm{T}}(\hat{x}) - \hat{h}(\hat{x}), \tag{10.264}$$

and the subsystem (10.259) has a globally stable equilibrium at $x = 0$ with $y = 0$ and Lyapunov function $V_{\mathrm{sub}}(x)$ so that

$$V_{\mathrm{sub}}'(x)f(x) < 0, \quad x \in \mathbb{R}^n, \quad x \neq 0. \tag{10.265}$$

Furthermore, assume that there exists a function $L_2 : \mathbb{R}^n \times \mathbb{R}^q \to \mathbb{R}^{1\times m}$ such that $L_2(0,0) = 0$ and

$$\begin{aligned} &V_{\mathrm{sub}}'(x)f(x) + \tfrac{1}{4\gamma^2}V_{\mathrm{sub}}'(x)J_1(x)J_1^{\mathrm{T}}(x)V_{\mathrm{sub}}'^{\mathrm{T}}(x) < 0, \quad x \in \mathbb{R}^n, \quad x \neq 0,\\ &\quad y^{\mathrm{T}}[G^{\mathrm{T}}(x)V_{\mathrm{sub}}'^{\mathrm{T}}(x) - \tfrac{1}{2}R_{2a}^{-1}(x,\hat{x})\{L_2(x,\hat{x})\\ &\quad +V_{\mathrm{s}}'(\hat{x})\hat{G}(\hat{x}) + 2h^{\mathrm{T}}(x,\hat{x})J(x,\hat{x})\}^{\mathrm{T}} - k(\hat{x})]\\ &\quad +[h(x,\hat{x}) + J(x,\hat{x})\phi(x,\hat{x})]^{\mathrm{T}}[h(x,\hat{x}) + J(x,\hat{x})\phi(x,\hat{x})]\\ &\quad +\tfrac{1}{4\gamma^2}V_{\mathrm{s}}'(\hat{x})J_3(\hat{x})J_3^{\mathrm{T}}(\hat{x})V_{\mathrm{s}}'^{\mathrm{T}}(\hat{x}) \leq 0, \qquad (x,\hat{x}) \in \mathbb{R}^n \times \mathbb{R}^m, \end{aligned} \tag{10.266}$$

where $R_{2a}(x,\hat{x}) \triangleq R_2(x,\hat{x}) + J^{\mathrm{T}}(x,\hat{x})J(x,\hat{x})$, $\phi(x,\hat{x}) = -\frac{1}{2}R_{2a}^{-1}(x,\hat{x})[L_2(x,\hat{x}) + V_{\mathrm{s}}'(\hat{x})\hat{G}(\hat{x}) + 2h^{\mathrm{T}}(x,\hat{x})J(x,\hat{x})]^{\mathrm{T}}$, and $\gamma > 0$. Show that the zero solution $(x(t),\hat{x}(t)) \equiv (0,0)$ of the undisturbed ($w(t) \equiv 0$) cascade system

$$\dot{x}(t) = f(x(t)) + G(x(t))y(t), \qquad x(0) = x_0, \qquad t \geq 0, \tag{10.267}$$

$$\dot{\hat{x}}(t) = \hat{f}(\hat{x}) + \hat{G}(\hat{x})\phi(x(t),\hat{x}(t)), \qquad \hat{x}(0) = \hat{x}_0, \tag{10.268}$$

$$y(t) = \hat{h}(\hat{x}), \tag{10.269}$$

is globally asymptotically stable with the feedback control law

$$\phi(x,\hat{x}) = -\tfrac{1}{2}R_{2a}^{-1}(x,\hat{x})[L_2(x,\hat{x}) + V_{\mathrm{s}}'(\hat{x})\hat{G}(\hat{x}) + 2h^{\mathrm{T}}(x,\hat{x})J(x,\hat{x})]^{\mathrm{T}}. \tag{10.270}$$

Furthermore, show that the performance functional (10.235) satisfies

$$J(x_0,\hat{x}_0,\phi(x,\hat{x})) \leq \mathcal{J}(x_0,\hat{x}_0,\phi(x,\hat{x})) = V(x_0,\hat{x}_0), \quad (x_0,\hat{x}_0) \in \mathbb{R}^n \times \mathbb{R}^m, \tag{10.271}$$

where

$$\mathcal{J}(x_0,\hat{x}_0,u(\cdot)) \triangleq \int_0^\infty [L(x(t),\hat{x}(t),u(t)) + \Gamma(x(t),\hat{x}(t),u(t))]\mathrm{d}t, \tag{10.272}$$

$$V(x,\hat{x}) = V_{\mathrm{sub}}(x) + V_{\mathrm{s}}(\hat{x}), \tag{10.273}$$

and

$$\begin{aligned} \Gamma(x,\hat{x},u) = {} & \tfrac{1}{4\gamma^2}V'(x,\hat{x})J_1(x,\hat{x})J_1^{\mathrm{T}}(x,\hat{x})V'^{\mathrm{T}}(x,\hat{x})\\ & +[h(x,\hat{x}) + J(x,\hat{x})u]^{\mathrm{T}}[h(x,\hat{x}) + J(x,\hat{x})u], \end{aligned} \tag{10.274}$$

where $J_1(x,\hat{x}) \triangleq [J_1^{\mathrm{T}}(x),\ J_3^{\mathrm{T}}(\hat{x})]^{\mathrm{T}}$ and $(x(t),\hat{x}(t))$, $t \geq 0$, solves (10.259) and (10.260) with $w(t) \equiv 0$. In addition, show that the performance functional

(10.272), with

$$\begin{aligned} L_1(x,\hat{x}) &= \phi^{\mathrm{T}}(x,\hat{x})R_{2a}(x,\hat{x})\phi(x,\hat{x}) - V_{\mathrm{s}}'(\hat{x})\hat{f}(\hat{x}) - V_{\mathrm{sub}}'(x)(f(x)+G(x)\hat{h}(\hat{x})) \\ &\quad -h^{\mathrm{T}}(x,\hat{x})h(x,\hat{x}) - \tfrac{1}{4\gamma^2}V'(x,\hat{x})J_1(x,\hat{x})J_1^{\mathrm{T}}(x,\hat{x})V'^{\mathrm{T}}(x,\hat{x}), \qquad (10.275) \end{aligned}$$

is minimized in the sense that

$$\mathcal{J}(x_0,\hat{x}_0,\phi(x(\cdot))) = \min_{u(\cdot)\in\mathcal{S}(x_0,\hat{x}_0)} \mathcal{J}(x_0,\hat{x}_0,u(\cdot)). \qquad (10.276)$$

With $u(\cdot)=\phi(x(\cdot),\hat{x}(\cdot))$, show that the solution $(x(t),\hat{x}(t))$, $t\geq 0$, of the cascade system (10.259) and (10.260) satisfies the nonexpansivity constraint

$$\int_0^T z^{\mathrm{T}}(t)z(t)\mathrm{d}t \leq \gamma^2\int_0^T w^{\mathrm{T}}(t)w(t)\mathrm{d}t + V(x_0,\hat{x}_0), \quad w(\cdot)\in\mathcal{L}_2, \quad T\geq 0. \qquad (10.277)$$

Finally, for $J(x,\hat{x})\equiv 0$, $h(x,\hat{x}) = E(x,\hat{x})\hat{h}(x)$, and $J_3(x)=\hat{G}(\hat{x})$, where $E:\mathbb{R}^n\times\mathbb{R}^q\to\mathbb{R}^{p\times m}$, show that a particular choice satisfying (10.266) is given by

$$\begin{aligned} L_2(x,\hat{x}) &= 2[G^{\mathrm{T}}(x)V_{\mathrm{sub}}'^{\mathrm{T}}(x) + \tfrac{1}{4\gamma^2}\hat{h}(\hat{x}) \\ &\quad +E^{\mathrm{T}}(x,\hat{x})E(x,\hat{x})\hat{h}(\hat{x}) - k(\hat{x})]^{\mathrm{T}}R_2(x,\hat{x}), \end{aligned}$$

yielding a feedback control law

$$\begin{aligned} \phi(x,\hat{x}) &= k(\hat{x}) - \tfrac{1}{2}R_2^{-1}(x,\hat{x})\hat{G}^{\mathrm{T}}(\hat{x})V_{\mathrm{s}}'^{\mathrm{T}}(\hat{x}) - V_{\mathrm{sub}}'(x)G(x) - \tfrac{1}{4\gamma^2}\hat{h}(\hat{x}) \\ &\quad -E^{\mathrm{T}}(x,\hat{x})E(x,\hat{x})\hat{h}(\hat{x}). \end{aligned}$$

10.10 Notes and References

The disturbance rejection problem for analysis and feedback control design can be traced back to the 1960s and early 1970s with the work of Cruz [100] and Frank [124] giving a textbook treatment on the subject. During this period, the role of differential game theory was also recognized as a framework for disturbance rejection control; see for example Dorato and Drenick [108], Witsenhausen [464], and Salmon [385]. In the 1980s the disturbance rejection problem was formulated in the frequency domain by the pioneering work of Zames [478] using $\mathcal{H}_\infty$ theory.

For linear multivariable systems, the $\mathcal{H}_\infty$ disturbance rejection control problem was formulated in the frequency domain by Francis, Helton, and Zames [123], Chang and Pearson [82], Francis and Doyle [122], Francis [121], and Safonov, Jonckheere, Verma, and Limebeer [383]. A state-space Riccati equation approach to the $\mathcal{H}_\infty$ control problem was given by Petersen [353], Glover and Doyle [137], and Doyle, Glover, Khargonekar, and Francis [111].

See also Khargonekar, Petersen, and Rotea [236]. The mixed-norm $\mathcal{H}_2/\mathcal{H}_\infty$ control problem was first formulated by Bernstein and Haddad [49] and Haddad and Bernstein [144].

A nonlinear game-theoretic framework for the "nonlinear" $\mathcal{H}_\infty$ control problem is given by Ball and Helton [22] with a textbook treatment given in Basar and Bernhard [31]. In parallel research, Isidori [213, 214], Isidori and Astolfi [215, 216], and van der Schaft [438–441] addressed the "nonlinear" $\mathcal{H}_\infty$ control problem using dissipativity theory. The optimality-based nonlinear disturbance rejection framework presented in this chapter is adopted from Haddad and Chellaboina [159] and Haddad, Chellaboina, and Fausz [163].

Chapter Eleven

Robust Control for Nonlinear Uncertain Systems

11.1 Introduction

Unavoidable discrepancies between system models and real-world systems can result in degradation of control-system performance including instability [109, 410]. Thus, it is not surprising that one of the fundamental problems in feedback control design is the ability of the control system to guarantee robustness with respect to system uncertainties in the design model. Although the theory of linear robust control is highly mature, nonlinear robust control techniques remain relatively undeveloped. Traditionally, Lyapunov function theory along with Hamilton-Jacobi-Bellman theory have been instrumental in advancing nonlinear control theory by addressing control system stability and optimality. Unfortunately, however, there does not exist a unified procedure for finding Lyapunov function candidates that will stabilize closed-loop nonlinear systems. Furthermore, computational methods for establishing the existence of solutions to Hamilton-Jacobi-Bellman equations involving highly complex nonlinear partial differential equations have not been developed to a level comparable to the existence of solutions of Riccati equations arising in linear optimal control problems. These problems are further exacerbated when addressing robustness in uncertain nonlinear systems.

Differential geometric methods [212, 336, 445] have made the design of controllers for certain classes of nonlinear systems more methodical. These methods are predicated on feedback linearization, or dynamic inversion, wherein state feedback along with coordinate transformations are used to transform a class of nonlinear systems to a linear time-invariant system. In this case, the resulting linear system can be stabilized using standard linear-quadratic design techniques. However, a serious drawback of all feedback linearization techniques is the failure to account for system uncertainty since exact cancelation of the nonlinear dynamics via feedback is required, and hence, an exact knowledge of the dynamics is assumed resulting in non-robust designs. To see this, consider the scalar nonlinear system

$$\dot{x}(t) = x^2(t) + u(t), \quad x(0) = x_0, \quad t \geq 0. \tag{11.1}$$

Clearly, a feedback linearizing control law for this system is given by $u_{\rm FL} = -x^2 - x$ resulting in the globally asymptotically stable closed-loop system $\dot{x}(t) = -x(t)$. Now, suppose that (11.1) possesses an input uncertainty so that (11.1) is not valid. Rather, in place of (11.1) a more accurate model is given by

$$\dot{x}(t) = x^2(t) + (1+\varepsilon)u(t), \quad x(0) = x_0, \quad t \geq 0, \tag{11.2}$$

where $\varepsilon \neq 0$. Using the feedback linearizing control law $u_{\rm FL} = -x^2 - x$ on the actual model yields

$$\dot{x}(t) = -(1+\varepsilon)x(t) - \varepsilon x^2(t), \quad x(0) = x_0, \quad t \geq 0. \tag{11.3}$$

It can be easily shown that for every $\varepsilon \neq 0$ there exists an initial condition such that the solution to (11.3) has a finite escape time.

Even though robustness frameworks to parametric uncertainty via feedback linearization techniques involving a two stage design consisting of nominal feedback linearization followed by additional state feedback designed to guarantee robustness have been developed [386,404,413,414], the fact that such approaches do not *directly* account for system uncertainties can result in severe robustness problems with respect to nonlinear errors internal to the system dynamics. Furthermore, restrictive matching conditions are imposed to the structure of the uncertainty in order to address general feedback linearizable systems [412]. Finally, these techniques often lead to inefficient designs since feedback linearizing controllers may generate unnecessarily large control effort to cancel beneficial system nonlinearities [127, 247].

In this chapter, we extend the framework presented in Chapter 8 to develop an optimality-based framework for addressing the problem of nonlinear-nonquadratic optimal control for uncertain nonlinear systems with structured parametric uncertainty. Specifically, using a Lyapunov bounding framework, the robust nonlinear control problem is transformed into an optimal control problem by modifying a nonlinear-nonquadratic cost functional to account for system uncertainty. Furthermore, it is shown that the Lyapunov function guaranteeing closed-loop robust stability is a solution to the steady-state Hamilton-Jacobi-Bellman equation for the controlled nominal system.

The main focus of this chapter is to develop a methodology for designing nonlinear controllers which provide both robust stability and robust performance over a prescribed range of system uncertainty. The present framework extends the guaranteed cost control approach [50, 83] to nonlinear systems by utilizing a performance bound to provide robust performance in addition to robust stability. In particular, the performance bound can be evaluated in closed form as long as the nonlinear-nonquadratic cost functional considered is related in a specific way to an underlying Lyapunov function that guarantees robust stability over a prescribed uncertainty set. This Lyapunov function is shown to be the solution to the steady-state form of the Hamilton-Jacobi-Bellman equation for the nominal system and plays a key role in constructing the optimal nonlinear robust control law. Hence, the overall framework provides for a generalization of the Hamilton-Jacobi-Bellman conditions for addressing the design of robust optimal controllers for nonlinear uncertain systems.

A key feature of the present framework is that since the necessary and sufficient Hamilton-Jacobi-Bellman optimality conditions are obtained for a modified nonlinear-nonquadratic performance functional rather than the original performance functional, *globally* optimal controllers are guaranteed to provide both robust stability and performance. Of course, since our approach allows us to construct globally optimal controllers that minimize a given Hamiltonian, the resulting robust nonlinear controllers provide the best worst-case performance over the robust stability range.

11.2 Robust Stability Analysis of Nonlinear Uncertain Systems

In this section, we present sufficient conditions for robust stability for a class of nonlinear uncertain systems. Specifically, we extend the analysis framework of Chapter 8 in order to address robust stability of a class of nonlinear uncertain systems. In the present framework we consider the problem of evaluating a performance bound for a nonlinear-nonquadratic cost functional depending upon a class of nonlinear uncertain systems. It turns out that the cost bound can be evaluated in closed form as long as the cost functional is related in a specific way to an underlying Lyapunov function that guarantees *robust* stability over a prescribed uncertainty set. Hence, the overall framework provides for robust stability and performance where robust performance here refers to a guaranteed bound on the worst-case value of a nonlinear-nonquadratic cost criterion over a prescribed uncertainty set.

Once again we restrict our attention to time-invariant infinite horizon systems. Furthermore, for the class of nonlinear uncertain systems considered we assume that the required properties for the existence and uniqueness of solutions are satisfied. For the following result, let $\mathcal{D} \subset \mathbb{R}^n$ be an open set, assume $0 \in \mathcal{D}$, let $L : \mathcal{D} \to \mathbb{R}$, and let $\mathcal{F} \subset \{f : \mathcal{D} \to \mathbb{R}^n : f(0) = 0\}$ denote the class of uncertain nonlinear systems with $f_0(\cdot) \in \mathcal{F}$ defining the nominal nonlinear system. Within the context of robustness analysis, it is assumed that the zero solution $x(t) \equiv 0$ of the nominal nonlinear dynamical system $\dot{x}(t) = f_0(x(t))$, $x(0) = x_0$, is asymptotically stable.

Theorem 11.1. Consider the nonlinear uncertain dynamical system

$$\dot{x}(t) = f(x(t)), \qquad x(0) = x_0, \qquad t \geq 0, \tag{11.4}$$

where $f(\cdot) \in \mathcal{F}$, with performance functional

$$J_f(x_0) \triangleq \int_0^\infty L(x(t))\mathrm{d}t. \tag{11.5}$$

Furthermore, assume that there exist functions $\Gamma : \mathcal{D} \to \mathbb{R}$ and $V : \mathcal{D} \to \mathbb{R}$, where $V(\cdot)$ is a continuously differentiable function, such that

$$V(0) = 0, \tag{11.6}$$

$$V(x) > 0, \quad x \in \mathcal{D}, \quad x \neq 0, \tag{11.7}$$

$$V'(x)f(x) \leq V'(x)f_0(x) + \Gamma(x), \quad x \in \mathcal{D}, \quad f(\cdot) \in \mathcal{F}, \tag{11.8}$$

$$V'(x)f_0(x) + \Gamma(x) < 0, \quad x \in \mathcal{D}, \quad x \neq 0, \tag{11.9}$$

$$L(x) + V'(x)f_0(x) + \Gamma(x) = 0, \quad x \in \mathcal{D}, \tag{11.10}$$

where $f_0(\cdot) \in \mathcal{F}$ defines the nominal nonlinear system. Then there exists a neighborhood of the origin $\mathcal{D}_0 \subseteq \mathcal{D}$ such that if $x_0 \in \mathcal{D}_0$, then the zero solution $x(t) \equiv 0$ to (11.4) is locally asymptotically stable for all $f(\cdot) \in \mathcal{F}$, and

$$\sup_{f(\cdot)\in\mathcal{F}} J_f(x_0) \leq \mathcal{J}(x_0) = V(x_0), \tag{11.11}$$

where

$$\mathcal{J}(x_0) \triangleq \int_0^\infty [L(x(t)) + \Gamma(x(t))]\mathrm{d}t, \tag{11.12}$$

and where $x(t)$, $t \geq 0$, is the solution to (11.4) with $f(x(t)) = f_0(x(t))$. Finally, if $\mathcal{D} = \mathbb{R}^n$ and

$$V(x) \to \infty \text{ as } \|x\| \to \infty, \tag{11.13}$$

then the zero solution $x(t) \equiv 0$ to (11.4) is globally asymptotically stable for all $f(\cdot) \in \mathcal{F}$.

Proof. Let $f(\cdot) \in \mathcal{F}$ and $x(t)$, $t \geq 0$, satisfy (11.4). Then

$$\dot{V}(x(t)) \triangleq \frac{\mathrm{d}}{\mathrm{d}t}V(x(t)) = V'(x(t))f(x(t)), \qquad t \geq 0. \tag{11.14}$$

Hence, it follows from (11.8) and (11.9) that

$$\dot{V}(x(t)) < 0, \qquad t \geq 0, \qquad x(t) \neq 0. \tag{11.15}$$

Thus, from (11.6), (11.7), and (11.15) it follows that $V(\cdot)$ is a Lyapunov function for (11.4), which proves local asymptotic stability of the zero solution $x(t) \equiv 0$ to (11.4) for all $f(\cdot) \in \mathcal{F}$. Consequently, $x(t) \to 0$ as $t \to \infty$ for all initial conditions $x_0 \in \mathcal{D}_0$ for some neighborhood of the origin $\mathcal{D}_0 \subseteq \mathcal{D}$. Now, (11.14) implies that

$$0 = -\dot{V}(x(t)) + V'(x(t))f(x(t)), \qquad t \geq 0,$$

and hence, using (11.8) and (11.10),

$$\begin{aligned} L(x(t)) &= -\dot{V}(x(t)) + L(x(t)) + V'(x(t))f(x(t)) \\ &\leq -\dot{V}(x(t)) + L(x(t)) + V'(x(t))f_0(x(t)) + \Gamma(x(t)) \\ &= -\dot{V}(x(t)). \end{aligned}$$

Now, integrating over $[0, t)$ yields

$$\int_0^t L(x(s))\mathrm{d}s \leq -V(x(t)) + V(x_0).$$

Letting $t \to \infty$ and noting that $V(x(t)) \to 0$ for all $x_0 \in \mathcal{D}_0$ yields $J_f(x_0) \leq V(x_0)$.

Next, let $x(t)$, $t \geq 0$, satisfy (11.4) with $f(x(t)) = f_0(x(t))$. Then, with $L(x)$ replaced by $L(x) + \Gamma(x)$ and $J(x_0)$ replaced by $\mathcal{J}(x_0)$ it follows from Theorem 8.1 that $\mathcal{J}(x_0) = V(x_0)$. Finally, for $\mathcal{D} = \mathbb{R}^n$ and for all $f(\cdot) \in \mathcal{F}$, global asymptotic stability of the zero solution $x(t) \equiv 0$ to (11.4) is a direct consequence of the radially unbounded condition (11.13) on $V(x)$, $x \in \mathbb{R}^n$. □

Theorem 11.1 is an extension of Theorem 8.1. Note that conditions (11.6) and (11.7) ensure that $V(x)$ is a Lyapunov function candidate for the nonlinear uncertain system (11.4). Conditions (11.8) and (11.9) imply $\dot{V}(x(t)) < 0$, $t \geq 0$, for $x(\cdot)$ satisfying (11.4) for all $f(\cdot) \in \mathcal{F}$, and hence, $V(\cdot)$ is a Lyapunov function guaranteeing robust stability of the nonlinear uncertain system (11.4). It is important to note that condition (11.9) is a *verifiable* condition since it is independent of the uncertain system parameters $f(\cdot) \in \mathcal{F}$. To apply Theorem 11.1 we specify a bounding function $\Gamma(\cdot)$ for the uncertain set $\mathcal{F}$ such that $\Gamma(\cdot)$ bounds $\mathcal{F}$ (see Propositions 11.1 and 11.2). Also note that if $\mathcal{F}$ consists of only the nominal nonlinear system $f_0(\cdot)$, then $\Gamma(x) \equiv 0$ satisfies (11.8), and hence, $J_{f_0}(x_0) = \mathcal{J}(x_0)$. In this case, Theorem 11.1 specializes to Theorem 8.1. The key feature of Theorem 11.1 is that it provides sufficient conditions for robust stability of a class of nonlinear uncertain systems $f(\cdot) \in \mathcal{F}$. Furthermore, an upper bound to the nonlinear-nonquadratic performance functional is given in terms of a Lyapunov function which can be interpreted in terms of an auxiliary cost defined for the nominal system in the spirit of [48, 50, 51].

If $\Gamma(\cdot)$ bounds $\mathcal{F}$ then clearly $\Gamma(\cdot)$ bounds the convex hull of $\mathcal{F}$. Hence, only convex uncertainty sets $\mathcal{F}$ need be considered. Next, we use the obvious fact that if $\Gamma_1(\cdot)$ bounds $\mathcal{F}_1$ and $\Gamma_2(\cdot)$ bounds $\mathcal{F}_2$, then $\Gamma_1(\cdot)+\Gamma_2(\cdot)$ bounds $\mathcal{F}_1 + \mathcal{F}_2$. Hence, if $\mathcal{F}$ can be decomposed additively then it suffices to bound each component separately. Finally, if $\Gamma(\cdot)$ bounds $\mathcal{F}$ and there exists $\tilde{\Gamma} : \mathcal{D} \to \mathbb{R}$ such that $\Gamma(x) \leq \tilde{\Gamma}(x)$ for all $x \in \mathbb{R}^n$, then $\tilde{\Gamma}(\cdot)$ bounds $\mathcal{F}$. That is any *overbound* $\tilde{\Gamma}(\cdot)$ for $\Gamma(\cdot)$ also bounds $\mathcal{F}$. Of course, it is quite possible that an overbound $\tilde{\Gamma}(\cdot)$ for $\Gamma(\cdot)$ may actually bound a set $\tilde{\mathcal{F}}$ that is larger than the "original" uncertainty set $\mathcal{F}$.

Next, we specialize Theorem 11.1 to nonlinear uncertain systems of the form

$$\dot{x}(t) = f_0(x(t)) + \Delta f(x(t)), \quad x(0) = x_0, \quad t \geq 0, \tag{11.16}$$

where $f_0 : \mathcal{D} \to \mathbb{R}^n$ satisfies $f_0(0) = 0$ and $f_0 + \Delta f \in \mathcal{F}$. Here, $\mathcal{F}$ is such that

$$\mathcal{F} \subset \{f_0 + \Delta f : \mathcal{D} \to \mathbb{R}^n : \ \ \Delta f \in \boldsymbol{\Delta}\}, \tag{11.17}$$

where $\boldsymbol{\Delta}$ is a given nonlinear uncertainty set of nonlinear perturbations Δf of the nominal system dynamics $f_0(\cdot) \in \mathcal{F}$. Since $\mathcal{F} \subset \{f : \mathcal{D} \to \mathbb{R}^n : f(0) = 0\}$ it follows that $\Delta f(0) = 0$ for all $\Delta f \in \boldsymbol{\Delta}$.

Corollary 11.1. Consider the nonlinear uncertain dynamical system (11.16) with performance functional (11.5). Furthermore, assume that there exist functions $\Gamma : \mathcal{D} \to \mathbb{R}$ and $V : \mathcal{D} \to \mathbb{R}$, where $V(\cdot)$ is a continuously differentiable function, such that (11.6) and (11.7) hold and

$$V'(x)\Delta f(x) \leq \Gamma(x), \quad x \in \mathcal{D}, \quad \Delta f(\cdot) \in \boldsymbol{\Delta}, \tag{11.18}$$

$$V'(x)f_0(x) + \Gamma(x) < 0, \quad x \in \mathcal{D}, \quad x \neq 0, \tag{11.19}$$

$$L(x) + V'(x)f_0(x) + \Gamma(x) = 0, \quad x \in \mathcal{D}. \tag{11.20}$$

Then the zero solution $x(t) \equiv 0$ to (11.16) is locally asymptotically stable for all $\Delta f(\cdot) \in \boldsymbol{\Delta}$ and there exists a neighborhood of the origin $\mathcal{D}_0 \subseteq \mathcal{D}$ such that the performance functional (11.5) satisfies

$$\sup_{\Delta f(\cdot) \in \boldsymbol{\Delta}} J_{\Delta f}(x_0) \leq \mathcal{J}(x_0) = V(x_0), \quad x_0 \in \mathcal{D}_0, \tag{11.21}$$

where

$$\mathcal{J}(x_0) = \int_0^\infty [L(x(t)) + \Gamma(x(t))]\mathrm{d}t, \tag{11.22}$$

and where $x(t)$, $t \geq 0$, is the solution to (11.16) with $\Delta f(x) \equiv 0$. Finally, if $\mathcal{D} = \mathbb{R}^n$, and $V(x)$, $x \in \mathbb{R}^n$, satisfies (11.13), then the zero solution $x(t) \equiv 0$ to (11.16) is globally asymptotically stable for all $\Delta f(\cdot) \in \boldsymbol{\Delta}$.

Proof. The result is a direct consequence of Theorem 11.1 with $f(x) = f_0(x) + \Delta f(x)$. Specifically, in this case, it follows from (11.18) and (11.19) that $V'(x)f(x) \leq V'(x)f_0(x) + \Gamma(x)$ for all $x \in \mathcal{D}$ and $\Delta f(\cdot) \in \boldsymbol{\Delta}$. Hence, all the conditions of Theorem 11.1 are satisfied. □

There are many alternative definitions for the bound $\Gamma(\cdot)$. To illustrate some of these alternatives consider the nonlinear uncertain dynamical system (11.16) and assume the uncertainty set $\mathcal{F}$ to be of the form

$$\mathcal{F} = \{f_0 + \Delta f : \mathbb{R}^n \to \mathbb{R}^n : \ \Delta f(x) = G_\delta(x)\delta(h_\delta(x)), \ x \in \mathbb{R}^n, \ \delta(\cdot) \in \mathbf{\Delta}\}, \tag{11.23}$$

where

$$\mathbf{\Delta} = \{\delta : \mathbb{R}^{p_\delta} \to \mathbb{R}^{m_\delta} : \ \delta(0) = 0, \ \delta^{\mathrm{T}}(y)\delta(y) \le m^{\mathrm{T}}(y)m(y), \ y \in \mathbb{R}^{p_\delta}\}, \tag{11.24}$$

$G_\delta : \mathbb{R}^n \to \mathbb{R}^{n \times m_\delta}$ and $h_\delta : \mathbb{R}^n \to \mathbb{R}^{p_\delta}$ satisfying $h_\delta(0) = 0$ are fixed functions denoting the structure of the uncertainty, $\delta : \mathbb{R}^{p_\delta} \to \mathbb{R}^{m_\delta}$ is an uncertain function, and $m : \mathbb{R}^{p_\delta} \to \mathbb{R}^{m_\delta}$ is a given function such that $m(0) = 0$. The special case $m(y) = \gamma^{-1}y$, where $\gamma > 0$, is worth noting. Specifically, in this case, (11.24) specializes to

$$\mathbf{\Delta} = \{\delta : \mathbb{R}^{p_\delta} \to \mathbb{R}^{m_\delta} : \delta(0) = 0, \ \delta^{\mathrm{T}}(y)\delta(y) \le \gamma^{-2}y^{\mathrm{T}}y, \ y \in \mathbb{R}^{p_\delta}\}, \tag{11.25}$$

which corresponds to a nonlinear small gain-type norm bounded uncertainty characterization. For the structure of $\mathcal{F}$ as specified by (11.23) and $\mathbf{\Delta}$ given by (11.24), the bounding function $\Gamma(\cdot)$ satisfying (11.18) can now be given a concrete form.

Proposition 11.1. The function

$$\Gamma(x) = \tfrac{1}{4}V'(x)G_\delta(x)G_\delta^{\mathrm{T}}(x)V'^{\mathrm{T}}(x) + m^{\mathrm{T}}(h_\delta(x))m(h_\delta(x)) \tag{11.26}$$

satisfies (11.18) with $\mathcal{F}$ given by (11.23) and $\mathbf{\Delta}$ given by (11.24).

Proof. Note that

$$\begin{aligned}
0 &\le [\delta^{\mathrm{T}}(h_\delta(x)) - \tfrac{1}{2}V'(x)G_\delta(x)][\delta^{\mathrm{T}}(h_\delta(x)) - \tfrac{1}{2}V'(x)G_\delta(x)]^{\mathrm{T}} \\
&= \delta^{\mathrm{T}}(h_\delta(x))\delta(h_\delta(x)) + \tfrac{1}{4}V'(x)G_\delta(x)G_\delta^{\mathrm{T}}(x)V'^{\mathrm{T}}(x) - V'(x)G_\delta(x)\delta(h_\delta(x)) \\
&\le m^{\mathrm{T}}(h_\delta(x))m(h_\delta(x)) + \tfrac{1}{4}V'(x)G_\delta(x)G_\delta^{\mathrm{T}}(x)V'^{\mathrm{T}}(x) - V'(x)G_\delta(x)\delta(h_\delta(x)) \\
&= \Gamma(x) - V'(x)\Delta f(x),
\end{aligned}$$

which proves (11.18) with $\mathcal{F}$ given by (11.23) and $\mathbf{\Delta}$ given by (11.24). □

Alternatively, consider the nonlinear uncertain dynamical system given by (11.16) and assume the uncertainty set $\mathcal{F}$ is given by (11.23) with $\mathbf{\Delta}$ given by

$$\begin{aligned}
\mathbf{\Delta} = \{\delta : \mathbb{R}^{p_\delta} \to \mathbb{R}^{m_\delta} : \ \delta(0) = 0, \ [\delta(y) - m_1(y)]^{\mathrm{T}}[\delta(y) - m_2(y)] \le 0, \\
y \in \mathbb{R}^{p_\delta}\},
\end{aligned} \tag{11.27}$$

where $m_1, m_2 : \mathbb{R}^{p_\delta} \to \mathbb{R}^{m_\delta}$ are given functions such that $m_1(0) = 0$, $m_2(0) = 0$, and $m_1^{\mathrm{T}}(y)m_2(y) \le 0$, $y \in \mathbb{R}^{p_\delta}$. For the structure of $\mathcal{F}$ as specified by (11.23) with $\mathbf{\Delta}$ given by (11.27), the bounding function $\Gamma(\cdot)$ satisfying

(11.18) can now be given a concrete form. For this result define $m(y) \triangleq m_2(y) - m_1(y)$.

Proposition 11.2. The function

$$\Gamma(x) = \tfrac{1}{4}[m(h_\delta(x)) + G_\delta^{\mathrm{T}}(x)V'^{\mathrm{T}}(x)]^{\mathrm{T}}[m(h_\delta(x)) + G_\delta^{\mathrm{T}}(x)V'^{\mathrm{T}}(x)] + V'(x)G_\delta(x)m_1(h_\delta(x)) \quad (11.28)$$

satisfies (11.18) with $\mathcal{F}$ given by (11.23) and $\mathbf{\Delta}$ given by (11.27).

Proof. Note that

$$\begin{aligned}
0 &\leq [\tfrac{1}{2}m(h_\delta(x)) + \tfrac{1}{2}G_\delta^{\mathrm{T}}(x)V'^{\mathrm{T}}(x) - (\delta(h_\delta(x)) - m_1(h_\delta(x)))]^{\mathrm{T}} \\
&\quad \cdot[\tfrac{1}{2}m(h_\delta(x)) + \tfrac{1}{2}G_\delta^{\mathrm{T}}(x)V'^{\mathrm{T}}(x) - (\delta(h_\delta(x)) - m_1(h_\delta(x)))] \\
&= [\delta(h_\delta(x)) - m_1(h_\delta(x))]^{\mathrm{T}}[\delta(h_\delta(x)) - m_2(h_\delta(x))] - V'(x)G_\delta(x)\delta(h_\delta(x)) \\
&\quad + \tfrac{1}{4}[m(h_\delta(x)) + G_\delta^{\mathrm{T}}(x)V'^{\mathrm{T}}(x)]^{\mathrm{T}}[m(h_\delta(x)) + G_\delta^{\mathrm{T}}(x)V'^{\mathrm{T}}(x)] \\
&\quad + V'(x)G_\delta(x)m_1(h_\delta(x)) \\
&\leq \Gamma(x) - V'(x)\Delta f(x),
\end{aligned}$$

which proves (11.18) with $\mathcal{F}$ given by (11.23) and $\mathbf{\Delta}$ given by (11.27). □

Finally, consider the nonlinear uncertain dynamical system (11.16) and assume the uncertainty set $\mathcal{F}$ is given by (11.23) with $\mathbf{\Delta}$ given by

$$\mathbf{\Delta} = \{\delta : \mathbb{R}^{p_\delta} \to \mathbb{R}^{m_\delta} : \ \delta(0) = 0, \ \delta^{\mathrm{T}}(y)Q\delta(y) + 2\delta^{\mathrm{T}}(y)Sy + y^{\mathrm{T}}Ry \leq 0, \ y \in \mathbb{R}^{p_\delta}\}, \quad (11.29)$$

where $Q \in \mathbb{P}^{m_\delta}$, $-R \in \mathbb{N}^{p_\delta}$, and $S \in \mathbb{R}^{m_\delta \times p_\delta}$. For this uncertainty characterization, the bounding function $\Gamma(\cdot)$ satisfying (11.18) can now be given a concrete form.

Proposition 11.3. The function

$$\Gamma(x) = [\tfrac{1}{2}V'(x)G_\delta(x) - h_\delta^{\mathrm{T}}(x)S^{\mathrm{T}}]Q^{-1}[\tfrac{1}{2}V'(x)G_\delta(x) - h_\delta^{\mathrm{T}}(x)S^{\mathrm{T}}]^{\mathrm{T}} - h_\delta^{\mathrm{T}}(x)Rh_\delta(x) \quad (11.30)$$

satisfies (11.18) with $\mathcal{F}$ given by (11.23) and $\mathbf{\Delta}$ given by (11.29).

Proof. Note that

$$\begin{aligned}
0 &\leq [Q^{1/2}\delta(h_\delta(x)) + Q^{-1/2}(Sh_\delta(x) - \tfrac{1}{2}G_\delta^{\mathrm{T}}(x)V'^{\mathrm{T}}(x))]^{\mathrm{T}} \\
&\quad \cdot[Q^{1/2}\delta(h_\delta(x)) + Q^{-1/2}(Sh_\delta(x) - \tfrac{1}{2}G_\delta^{\mathrm{T}}(x)V'^{\mathrm{T}}(x))] \\
&= \delta^{\mathrm{T}}(h_\delta(x))Q\delta(h_\delta(x)) + 2\delta^{\mathrm{T}}(h_\delta(x))Sh_\delta(x) - V'(x)G_\delta(x)\delta(h_\delta(x)) \\
&\quad + [\tfrac{1}{2}V'(x)G_\delta(x) - h_\delta^{\mathrm{T}}(x)S^{\mathrm{T}}]Q^{-1}[\tfrac{1}{2}V'(x)G_\delta(x) - h_\delta^{\mathrm{T}}(x)S^{\mathrm{T}}]^{\mathrm{T}} \\
&= \delta^{\mathrm{T}}(h_\delta(x))Q\delta(h_\delta(x)) + 2\delta^{\mathrm{T}}(h_\delta(x))Sh_\delta(x) + h_\delta^{\mathrm{T}}(x)Rh_\delta(x)
\end{aligned}$$

$$-V'(x)\Delta f(x) + \Gamma(x)$$
$$\leq \Gamma(x) - V'(x)\Delta f(x),$$

which proves (11.18) with $\mathcal{F}$ given by (11.23) and $\mathbf{\Delta}$ given by (11.29). $\square$

Example 11.1. Consider the nonlinear uncertain dynamical system given by

$$\dot{x}_1(t) = x_2(t) + \delta_1 x_1(t)\cos(x_2(t) + \delta_2) + \delta_3 x_2(t)\sin(\delta_4 x_1(t) x_2(t)),$$
$$x_1(0) = x_{10}, \quad t \geq 0, \tag{11.31}$$
$$\dot{x}_2(t) = x_1(t)\sin(x_2(t)), \quad x_2(0) = x_{20}, \tag{11.32}$$

where $\delta_1 \in [-1, 1]$, $\delta_2 \in [-5, 50]$, $\delta_3 \in [0, 1]$, and $\delta_4 \in [-50, 0]$. Now, with $x = [x_1,\ \ x_2]^{\mathrm{T}}$, $f_0(x) = [x_2,\ \ x_1 \sin x_2]^{\mathrm{T}}$, $G_\delta(x) = [1,\ \ 0]^{\mathrm{T}}$, $h_\delta(x) = x$, and $\delta(x) = \delta_1 x_1 \cos(x_2 + \delta_2) + \delta_3 x_2 \sin(\delta_4 x_1 x_2)$, (11.31) and (11.31) can be written in the form of (11.16) with $\mathcal{F}$ given by (11.23). Furthermore, since $\delta^2(x) \leq x_1^2 + x_2^2$ it follows that $\delta(\cdot) \in \mathbf{\Delta}$, where $\mathbf{\Delta}$ is given by (11.24) with $m(x) = \sqrt{x_1^2 + x_2^2}$. $\triangle$

Example 11.2. Consider the nonlinear uncertain matrix second-order dynamical system of an n-link robot discussed in Example 5.2 given by

$$M(q(t))\ddot{q}(t) + W(q(t), \dot{q}(t)) = 0, \quad q(0) = q_0, \quad \dot{q}(0) = \dot{q}_0, \quad t \geq 0, \tag{11.33}$$

where q, $\dot{q}$, $\ddot{q} \in \mathbb{R}^n$ represent generalized position, velocity, and acceleration coordinates, respectively, $M(q)$ is a positive-definite inertia matrix function for all $q \in \mathbb{R}^n$, and $W(q, \dot{q})$ is a vector function lumping the centrifugal and Coriolis forces, dissipation forces due to friction, and gravitational forces. Here, we assume that $M_1(q) \leq M(q) \leq M_2(q)$, where $M_1(q) > 0$, $q \in \mathbb{R}^n$, and $\|W(q, \dot{q})\|_2 \leq \overline{w}(q, \dot{q})$. Now, with $x = [x_1^{\mathrm{T}},\ x_2^{\mathrm{T}}]^{\mathrm{T}}$, where $x_1 = q$ and $x_2 = \dot{q}$, $f_0(x) = [x_2^{\mathrm{T}},\ 0]^{\mathrm{T}}$, $G_\delta(x) = [0,\ I_n]^{\mathrm{T}}$, $h_\delta(x) = x$, and $\delta(x) = M^{-1}(x_1)W(x_1, x_2)$, (11.33) can be written in the form of (11.16) with $\mathcal{F}$ given by (11.23). Furthermore, since $\|\delta(x)\|_2 = \|M^{-1}(x_1)W(x_1, x_2)\|_2 \leq \sigma_{\max}(M^{-1}(x_1))\|W(x_1, x_2)\|_2 \leq \sigma_{\max}^{-1}(M_1(x_1))\overline{w}(x_1, x_2)$ it follows that $\delta(\cdot) \in \mathbf{\Delta}$, where $\mathbf{\Delta}$ is given by (11.24) with $\|m(x)\|_2 = \sigma_{\max}^{-1}(M_1(x_1))\overline{w}(x_1, x_2)$. $\triangle$

Example 11.3. Consider the nonlinear uncertain dynamical system given by

$$\dot{x}_1(t) = x_2(t) + x_1^2(t), \quad x_1(0) = x_{10}, \quad t \geq 0, \tag{11.34}$$
$$\dot{x}_2(t) = x_1^3(t) + \delta x_2(t)[0.55 + 0.44\delta\cos(x_2(t))], \quad x_2(0) = x_{20}, \tag{11.35}$$

where $\delta \in [-1, 1]$. Now, with $x = [x_1,\ x_2]^{\mathrm{T}}$, $f_0(x) = [x_2 + x_1^2,\ x_1^3]^{\mathrm{T}}$, $G_\delta(x) = [0,\ 1]^{\mathrm{T}}$, $h_\delta(x) = x_2$, and $\delta(x_2) = \delta x_2(0.55 + 0.44\delta\cos(x_2)$, (11.34) and (11.35) can be written in the form of (11.16) with $\mathcal{F}$ given by (11.23). Furthermore, since $\delta^2 x_2^2(0.55 + 0.44\delta\cos(x_2))^2 + 0.1x_2^2 \leq 1.1\delta x_2^2(0.55 + 0.44\delta\cos(x_2))$ it follows that $\delta(\cdot) \in \mathbf{\Delta}$, where $\mathbf{\Delta}$ is given by (11.27) with $m_1(x_2) = 0.1x_2$ and

$m_2(x_2) = x_2$. △

We now combine the results of Corollary 11.1 and Propositions 11.1–11.3 to obtain a series of conditions guaranteeing robust stability and performance for the nonlinear uncertain system (11.16).

Proposition 11.4. Consider the nonlinear uncertain system (11.16). Let $L(x) > 0$, $x \in \mathbb{R}^n$, and suppose there exists a continuously differentiable radially unbounded function $V : \mathbb{R}^n \to \mathbb{R}$ such that (11.6) and (11.7) hold and

$$0 = V'(x)f_0(x) + \tfrac{1}{4}V'(x)G_\delta(x)G_\delta^{\mathrm{T}}(x)V'^{\mathrm{T}}(x) + m^{\mathrm{T}}(h_\delta(x))m(h_\delta(x)) + L(x), \quad x \in \mathbb{R}^n. \tag{11.36}$$

Then the zero solution $x(t) \equiv 0$ to (11.16) is globally asymptotically stable for all $\Delta f(\cdot) \in \boldsymbol{\Delta}$ with $\boldsymbol{\Delta}$ given by (11.24), and the performance functional (11.5) satisfies

$$\sup_{\Delta f(\cdot) \in \boldsymbol{\Delta}} J_{\Delta f}(x_0) \le \mathcal{J}(x_0) = V(x_0), \tag{11.37}$$

where

$$\mathcal{J}(x_0) = \int_0^\infty [L(x(t)) + \tfrac{1}{4}V'(x(t))G_\delta(x(t))G_\delta^{\mathrm{T}}(x(t))V'^{\mathrm{T}}(x(t)) + m^{\mathrm{T}}(h_\delta(x(t)))m(h_\delta(x(t)))]\mathrm{d}t, \tag{11.38}$$

and $x(t)$, $t \ge 0$, is the solution of (11.16) with $\Delta f(x) \equiv 0$.

Proposition 11.5. Consider the nonlinear uncertain system (11.16). Let $L(x) > 0$, $x \in \mathbb{R}^n$, and suppose there exists a continuously differentiable radially unbounded function $V : \mathbb{R}^n \to \mathbb{R}$ such that (11.6) and (11.7) hold and

$$0 = V'(x)f_0(x) + \tfrac{1}{4}[m(h_\delta(x)) + G_\delta^{\mathrm{T}}(x)V'^{\mathrm{T}}(x)]^{\mathrm{T}}[m(h_\delta(x)) + G_\delta^{\mathrm{T}}(x)V'^{\mathrm{T}}(x)] + V'(x)G_\delta(x)m_1(h_\delta(x)) + L(x), \quad x \in \mathbb{R}^n. \tag{11.39}$$

Then the zero solution $x(t) \equiv 0$ to (11.16) is globally asymptotically stable for all $\Delta f(\cdot) \in \boldsymbol{\Delta}$ with $\boldsymbol{\Delta}$ given by (11.27), and the performance functional (11.5) satisfies

$$\sup_{\Delta f(\cdot) \in \boldsymbol{\Delta}} J_{\Delta f}(x_0) \le \mathcal{J}(x_0) = V(x_0), \tag{11.40}$$

where

$$\mathcal{J}(x_0) = \int_0^\infty [L(x(t)) + \tfrac{1}{4}[m(h_\delta(x(t))) + G_\delta^{\mathrm{T}}(x(t))V'^{\mathrm{T}}(x(t))]^{\mathrm{T}} \cdot [m(h_\delta(x(t))) + G_\delta^{\mathrm{T}}(x(t))V'^{\mathrm{T}}(x(t))] + V'(x(t))G_\delta(x(t))m_1(h_\delta(x(t)))]\mathrm{d}t, \tag{11.41}$$

and $x(t)$, $t \ge 0$, is the solution of (11.16) with $\Delta f(x) \equiv 0$.

Proposition 11.6. Consider the nonlinear uncertain system (11.16). Let $L(x) > 0$, $x \in \mathbb{R}^n$, and suppose there exists a continuously differentiable radially unbounded function $V : \mathbb{R}^n \to \mathbb{R}$ such that (11.6) and (11.7) hold and

$$0 = V'(x)f_0(x) + [\tfrac{1}{2}V'(x)G_\delta(x) - h_\delta^{\rm T}(x)S^{\rm T}]Q^{-1}[\tfrac{1}{2}V'(x)G_\delta(x) - h_\delta^{\rm T}(x)S^{\rm T}]^{\rm T} -h_\delta^{\rm T}(x)Rh_\delta(x) + L(x), \quad x \in \mathbb{R}^n. \tag{11.42}$$

Then the zero solution $x(t) \equiv 0$ to (11.16) is globally asymptotically stable for all $\Delta f(\cdot) \in \mathbf{\Delta}$ with $\mathbf{\Delta}$ given by (11.29), and the performance functional (11.5) satisfies

$$\sup_{\Delta f(\cdot) \in \mathbf{\Delta}} J_{\Delta f}(x_0) \le \mathcal{J}(x_0) = V(x_0), \tag{11.43}$$

where

$$\begin{aligned}\mathcal{J}(x_0) = & \int_0^\infty [L(x(t)) + [\tfrac{1}{2}V'(x(t))G_\delta(x(t)) - h_\delta^{\rm T}(x(t))S^{\rm T}]Q^{-1} \\ & \cdot[\tfrac{1}{2}V'(x(t))G_\delta(x(t)) - h_\delta^{\rm T}(x(t))S^{\rm T}]^{\rm T} \\ & -h_\delta^{\rm T}(x(t))Rh_\delta(x(t))]{\rm d}t,\end{aligned} \tag{11.44}$$

and $x(t)$, $t \ge 0$, is the solution of (11.16) with $\Delta f(x) \equiv 0$.

Example 11.4. Consider the nonlinear uncertain system

$$\dot{x}(t) = Ax(t) + B_0\delta(y(t)), \quad x(0) = x_0, \quad t \ge 0, \tag{11.45}$$

$$y(t) = C_0x(t), \tag{11.46}$$

where $A \in \mathbb{R}^{n\times n}$, $B_0 \in \mathbb{R}^{n\times m_\delta}$, $C_0 \in \mathbb{R}^{p_\delta \times n}$, and $\delta(y) \in \mathbf{\Delta}$ with $\mathbf{\Delta}$ given by (11.25). To obtain sufficient conditions for robust stability for the nonlinear uncertain system (11.45) and (11.46) we use Proposition 11.4 with $L(x) = x^{\rm T}Rx$, where $R > 0$, $m(h_\delta(x)) = \gamma^{-1}y = \gamma^{-1}C_0x$, and $V(x) = x^{\rm T}Px$, where $P > 0$. In this case, (11.36) yields

$$0 = A^{\rm T}P + PA + PB_0B_0^{\rm T}P + \gamma^{-2}C_0^{\rm T}C_0 + R, \tag{11.47}$$

or, equivalently,

$$0 = A^{\rm T}P + PA + \gamma^{-2}PB_0B_0^{\rm T}P + C_0^{\rm T}C_0 + R. \tag{11.48}$$

Now, it follows from Proposition 11.4 that if there exists a positive-definite $P \in \mathbb{R}^{n\times n}$ satisfying (11.48), then the zero solution $x(t) \equiv 0$ to (11.45) is globally asymptotically stable for all $\delta(\cdot) \in \mathbf{\Delta}$, where $\mathbf{\Delta}$ is given by (11.25).

Alternatively, suppose $\delta(y) \in \mathbf{\Delta}$, where $\mathbf{\Delta}$ is given by (11.27) with $m_1(t) = M_1y$ and $m_2(y) = M_2y$, where $M_1, M_2 \in \mathbb{R}^{m_\delta\times m_\delta}$ are symmetric matrices such that $M \triangleq M_2 - M_1$. In this case, to obtain sufficient conditions for robust stability for the nonlinear uncertain system (11.45) and (11.46) we use Proposition 11.5 with $L(x) = x^{\rm T}Rx$, where $R > 0$, $m(h_\delta(x)) = My =$

MC_0x, $m_1(h_\delta(x)) = M_1y = M_1C_0x$, and $V(x) = x^{\mathrm{T}}Px$, where $P > 0$. In this case, (11.39) yields

$$0 = (A + B_0M_1C_0)^{\mathrm{T}}P + P(A + B_0M_1C_0) \\ +(\tfrac{1}{2}MC_0 + B_0^{\mathrm{T}}P)^{\mathrm{T}}(\tfrac{1}{2}MC_0 + B_0^{\mathrm{T}}P) + R. \qquad (11.49)$$

Now, it follows from Proposition 11.5 that if there exists a positive-definite $P \in \mathbb{R}^{n\times n}$ satisfying (11.49), then the zero solution to (11.45) is globally asymptotically stable for all $\delta(\cdot) \in \boldsymbol{\Delta}$, where $\boldsymbol{\Delta}$ is given by (11.27) with $m_1(y) = M_1y$ and $m_2(y) = M_2y$. △

The following corollary specializes Theorem 11.1 to a class of linear uncertain systems which connects the framework of Theorem 11.1 to the quadratic Lyapunov bounding (Ω-bound) framework of Bernstein and Haddad [50]. Specifically, in this case, we consider $\mathcal{F}$ to be the set of uncertain linear systems given by

$$\mathcal{F} = \{(A + \Delta A)x : x \in \mathbb{R}^n, A \in \mathbb{R}^{n\times n}, \Delta A \in \boldsymbol{\Delta}_A\},$$

where $\boldsymbol{\Delta}_A \subset \mathbb{R}^{n\times n}$ is a given bounded uncertainty set of uncertain perturbations ΔA of the nominal asymptotically stable system matrix A such that $0 \in \boldsymbol{\Delta}_A$.

Corollary 11.2. Let $R \in \mathbb{P}^n$. Consider the linear uncertain dynamical system

$$\dot{x}(t) = (A + \Delta A)x(t), \qquad x(0) = x_0, \qquad t \geq 0, \qquad (11.50)$$

with performance functional

$$J_{\Delta A}(x_0) \triangleq \int_0^\infty x^{\mathrm{T}}(t)Rx(t)\mathrm{d}t, \qquad (11.51)$$

where $\Delta A \in \boldsymbol{\Delta}_A$. Let $\Omega : \mathcal{N} \subseteq \mathbb{S}^n \to \mathbb{N}^n$ be such that

$$\Delta A^{\mathrm{T}}P + P\Delta A \leq \Omega(P), \qquad \Delta A \in \boldsymbol{\Delta}_A, \qquad P \in \mathcal{N}. \qquad (11.52)$$

Furthermore, suppose there exist $P \in \mathbb{P}^n$ satisfying

$$0 = A^{\mathrm{T}}P + PA + \Omega(P) + R. \qquad (11.53)$$

Then the zero solution $x(t) \equiv 0$ to (11.50) is globally asymptotically stable for all $\Delta A \in \boldsymbol{\Delta}_A$, and

$$\sup_{\Delta A \in \boldsymbol{\Delta}_A} J_{\Delta A}(x_0) \leq \mathcal{J}(x_0) = x_0^{\mathrm{T}}Px_0, \qquad x_0 \in \mathbb{R}^n, \qquad (11.54)$$

where

$$\mathcal{J}(x_0) \triangleq \int_0^\infty x^{\mathrm{T}}(t)(\Omega(P) + R)x(t)\mathrm{d}t, \qquad (11.55)$$

and where $x(t)$, $t \geq 0$, solves (11.50) with $\Delta A = 0$.

Proof. The result is a direct consequence of Theorem 11.1 with $f(x) = (A + \Delta A)x$, $f_0(x) = Ax$, $L(x) = x^{\mathrm{T}}Rx$, $V(x) = x^{\mathrm{T}}Px$, $\Gamma(x) = x^{\mathrm{T}}\Omega(P)x$, and $\mathcal{D} = \mathbb{R}^n$. Specifically, conditions (11.6) and (11.7) are trivially satisfied. Now, $V'(x)f(x) = x^{\mathrm{T}}(A^{\mathrm{T}}P + PA)x + x^{\mathrm{T}}(\Delta A^{\mathrm{T}}P + P\Delta A)x$, and hence, it follows from (11.52) that $V'(x)f(x) \leq V'(x)f_0(x) + \Gamma(x) = x^{\mathrm{T}}(A^{\mathrm{T}}P + PA + \Omega(P))x$, for all $\Delta A \in \boldsymbol{\Delta}_A$. Furthermore, it follows from (11.53) that $L(x)+V'(x)f_0(x)+\Gamma(x) = 0$, and hence, $V'(x)f_0(x)+\Gamma(x) < 0$ for all $x \neq 0$, so that all the conditions of Theorem 11.1 are satisfied. Finally, since $V(x)$, $x \in \mathbb{R}^n$, is radially unbounded, (11.50) is globally asymptotically stable for all $\Delta A \in \boldsymbol{\Delta}_A$. □

Corollary 11.2 is the deterministic version of Theorem 4.1 of [50] involving quadratic Lyapunov bounds for addressing robust stability and performance analysis of linear uncertain systems. For convenience we shall say that $\Omega(\cdot)$ *bounds* $\boldsymbol{\Delta}_A$ if (11.52) is satisfied. To apply Corollary 11.2, we first specify a function $\Omega(\cdot)$ and an uncertainty set $\boldsymbol{\Delta}_A$ such that $\Omega(\cdot)$ bounds $\boldsymbol{\Delta}_A$. If the existence of a *positive-definite* solution P to (11.53) can be determined analytically or numerically, then robust stability is guaranteed and the performance bound (11.54) can be computed. We can then enlarge $\boldsymbol{\Delta}_A$, modify $\Omega(\cdot)$, and again attempt to solve (11.53). If, however, a positive-definite solution to (11.53) cannot be determined, then $\boldsymbol{\Delta}_A$ must be decreased in size until (11.53) is solvable. For example, $\Omega(\cdot)$ can be replaced by $\varepsilon\Omega(\cdot)$ to bound $\varepsilon\boldsymbol{\Delta}_A$, where $\varepsilon > 1$ enlarges $\boldsymbol{\Delta}_A$ and $\varepsilon < 1$ shrinks $\boldsymbol{\Delta}_A$. Of course, the actual range of uncertainty that can be bounded depends on the nominal matrix A, the function $\Omega(\cdot)$, and the structure of $\boldsymbol{\Delta}_A$.

Since the ordering induced by the cone of nonnegative-definite matrices is only a partial ordering, it should not be expected that there exists an operator $\Omega(\cdot)$ satisfying (11.52), which is a least upper bound. Next, we illustrate two bounding functions for two different uncertainty characterizations. Specifically, we assign explicit structure to the uncertainty set $\boldsymbol{\Delta}_A$ and the bound $\Omega(\cdot)$ satisfying (11.52). First, we assume that the uncertainty set $\boldsymbol{\Delta}_A$ to be of the form

$$\boldsymbol{\Delta}_A = \left\{ \Delta A \in \mathbb{R}^{n\times n} : \ \Delta A = \sum_{i=1}^{p} \delta_i A_i, \ \sum_{i=1}^{p} \frac{\delta_i^2}{\alpha_i^2} \leq 1 \right\}, \tag{11.56}$$

where for $i = 1, \ldots, p$, A_i is a fixed matrix denoting the structure of the parametric uncertainty; α_i is a given positive number; and δ_i is an uncertain parameter. Note that the uncertain parameters δ_i are assumed to lie in a specified ellipsoidal region in $\mathbb{R}^p$. For the structure $\boldsymbol{\Delta}_A$ as specified by (11.56), the bound $\Omega(\cdot)$ satisfying (11.52) can now be given a concrete form.

Proposition 11.7. Let $\alpha > 0$ and $\mathcal{N} = \mathbb{N}^n$. Then the function

$$\Omega(P) = \alpha P + \alpha^{-1} \sum_{i=1}^{p} \alpha_i^2 A_i^{\mathrm{T}} P A_i \tag{11.57}$$

satisfies (11.52) with $\mathbf{\Delta}_A$ given by (11.56).

Proof. Note that

$$\begin{aligned} 0 &\le \sum_{i=1}^{p} [(\tfrac{\alpha^{1/2}\delta_i}{\alpha_i}) I_n - (\tfrac{\alpha_1}{\alpha^{1/2}}) A_i]^{\mathrm{T}} P [(\tfrac{\alpha^{1/2}\delta_i}{\alpha_i}) I_n - (\tfrac{\alpha_1}{\alpha^{1/2}}) A_i] \\ &= \alpha \sum_{i=1}^{p} (\tfrac{\delta_i^2}{\alpha_i^2}) P + \alpha^{-1} \sum_{i=1}^{p} \alpha_i^2 A_i^{\mathrm{T}} P A_i - \sum_{i=1}^{p} \delta_i (A_i^{\mathrm{T}} P + P A_i), \end{aligned}$$

which, since $\sum_{i=1}^{p} \delta_i^2/\alpha_i^2 \le 1$, proves (11.52) with $\mathbf{\Delta}_A$ given by (11.56). □

Note that $\mathbf{\Delta}_A$ given by (11.56) includes repeated parameters without loss of generality. For example, if $\delta_1 = \delta_2$ then discard δ_2 and replace A_1 by $A_1 + A_2$. Furthermore, $\mathbf{\Delta}_A$ includes real full block uncertainty. For example if

$$\Delta A = \begin{bmatrix} \delta_1 & \delta_2 \\ \delta_3 & \delta_4 \end{bmatrix},$$

then $\Delta A = \sum_{i=1}^{4} \delta_i A_i$, where

$$A_1 = \begin{bmatrix} 1 & 0 \\ 0 & 0 \end{bmatrix},$$

and likewise for A_2, A_3, and A_4. Finally, for $i = 1, \ldots, p$, letting $A_i = B_i C_i$, where $B_i \in \mathbb{R}^{n \times q_i}$, $C_i \in \mathbb{R}^{q_i \times n}$, and $q_i \le n$, and defining $B_0 \triangleq [B_1 \cdots B_p]$ and $C_0 \triangleq [C_1^{\mathrm{T}} \cdots C_p^{\mathrm{T}}]^{\mathrm{T}}$, $\mathbf{\Delta}_A$ can be written as

$$\begin{aligned} \mathbf{\Delta}_A = \Big\{ &\Delta A \in \mathbb{R}^{n \times n} : \ \Delta A = B_0 \Delta C_0, \ \Delta = \text{block-diag}[\delta_1 I_{q_1}, \ldots, \delta_p I_{q_p}], \\ &\sum_{i=1}^{p} \tfrac{\delta_i^2}{\alpha_i^2} \le 1, \ i = 1, \ldots, p \Big\}. \end{aligned} \tag{11.58}$$

Since an uncertainty set of the form (11.58) can always be written in the form of (11.56) by partitioning B_0 and C_0 as above and defining $A_i = B_i C_i$, $i = 1, \ldots, p$, robust stability of $A + \Delta A$ for all $\Delta A \in \mathbf{\Delta}_A$ is equivalent to the robust stability of the feedback interconnection of $G(s) \triangleq C_0(sI - A)^{-1} B_0$ and Δ given in (11.58).

Next, assume the uncertainty set $\mathbf{\Delta}_A$ to be of the form

$$\mathbf{\Delta}_A = \{\Delta A \in \mathbb{R}^{n \times n} : \ \Delta A = B_0 F C_0, \ F^{\mathrm{T}} F \le N\}, \tag{11.59}$$

where $B_0 \in \mathbb{R}^{n\times r}$ and $C_0 \in \mathbb{R}^{s\times n}$ are fixed matrices denoting the structure of the uncertainty, $F \in \mathbb{R}^{r\times s}$ is an uncertain matrix, and $N \in \mathbb{N}^s$ is a given uncertainty bound. The special case $N = \gamma^{-2} I_s$, where $\gamma > 0$, is worth noting. Specifically, in this case, (11.59) specializes to

$$\mathbf{\Delta}_A = \{\Delta A \in \mathbb{R}^{n\times n} : \ \Delta A = B_0 F C_0, \ \sigma_{\max}(F) \le \gamma^{-1}\}, \tag{11.60}$$

which corresponds to a small gain norm bounded uncertainty characterization. For this uncertainty characterization, the bound $\Omega(\cdot)$ satisfying (11.52) can now be given a concrete form.

Proposition 11.8. Let $\alpha > 0$. Then the function

$$\Omega(P) = \alpha P B_0 B_0^{\mathrm{T}} P + \alpha^{-1} C_0^{\mathrm{T}} N C_0 \tag{11.61}$$

satisfies (11.52) with $\mathbf{\Delta}_A$ given by (11.59).

Proof. Note that

$$\begin{aligned} 0 &\le [\alpha^{1/2} B_0^{\mathrm{T}} P - \alpha^{-1/2} F C_0]^{\mathrm{T}} [\alpha^{1/2} B_0^{\mathrm{T}} P - \alpha^{-1/2} F C_0] \\ &= \alpha P B_0 B_0^{\mathrm{T}} P + \alpha^{-1} C_0^{\mathrm{T}} F^{\mathrm{T}} F C_0 - (C_0^{\mathrm{T}} F^{\mathrm{T}} B_0^{\mathrm{T}} P + P B_0 F C_0) \\ &\le \alpha P B_0 B_0^{\mathrm{T}} P + \alpha^{-1} C_0^{\mathrm{T}} N C_0 - (\Delta A^{\mathrm{T}} P + P \Delta A) \\ &= \Omega(P) - (\Delta A^{\mathrm{T}} P + P\Delta A), \end{aligned}$$

which proves (11.52) with $\mathbf{\Delta}_A$ given by (11.59). □

Finally, assume the uncertainty set $\mathbf{\Delta}_A$ to be of the form

$$\mathbf{\Delta}_A = \{\Delta A \in \mathbb{R}^{n\times n} : \ \Delta A = B_0 F C_0, \ F^{\mathrm{T}} Q F + F^{\mathrm{T}} S + S^{\mathrm{T}} F + R \le 0\}, \tag{11.62}$$

where $B_0 \in \mathbb{R}^{n\times r}$ and $C_0 \in \mathbb{R}^{s\times n}$ are fixed matrices denoting the structure of the uncertainty, $F \in \mathbb{R}^{r\times s}$ is an uncertain matrix, $Q \in \mathbb{P}^r$, $S \in \mathbb{R}^{r\times s}$, and $-R \in \mathbb{N}^s$. For this uncertainty characterization, the bound $\Omega(\cdot)$ satisfying (11.52) can now be given a concrete form.

Proposition 11.9. The function

$$\Omega(P) = [B_0^{\mathrm{T}} P - S C_0]^{\mathrm{T}} Q^{-1} [B_0^{\mathrm{T}} P - S C_0] - C_0^{\mathrm{T}} R C_0 \tag{11.63}$$

satisfies (11.52) with $\mathbf{\Delta}_A$ given by (11.62).

Proof. Note that

$$\begin{aligned} 0 &\le [Q^{1/2} F C_0 + Q^{-1/2}(S C_0 - B_0^{\mathrm{T}} P)]^{\mathrm{T}} [Q^{1/2} F C_0 + Q^{-1/2}(S C_0 - B_0^{\mathrm{T}} P)] \\ &= C_0^{\mathrm{T}} (F^{\mathrm{T}} Q F + F^{\mathrm{T}} S + S^{\mathrm{T}} F) C_0 - P B_0 F C_0 - C_0^{\mathrm{T}} F^{\mathrm{T}} B_0^{\mathrm{T}} P \\ &\quad + [B_0^{\mathrm{T}} P - S C_0]^{\mathrm{T}} Q^{-1} [B_0^{\mathrm{T}} P - S C_0] \\ &= C_0^{\mathrm{T}} (F^{\mathrm{T}} Q F + F^{\mathrm{T}} S + S^{\mathrm{T}} F + R) C_0 \\ &\quad - (\Delta A^{\mathrm{T}} P + P \Delta A) + \Omega(P) \end{aligned}$$

$$\leq \Omega(P) - (\Delta A^{\mathrm{T}} P + P\Delta A),$$

which proves (11.52) with $\boldsymbol{\Delta}_A$ given by (11.62). □

As in the nonlinear case, we now combine the results of Corollary 11.2 and Propositions 11.7–11.9 to obtain a series of conditions guaranteeing robust stability and performance for the linear uncertain system (11.50).

Proposition 11.10. Consider the linear uncertain system (11.50). Let $R \in \mathbb{P}^n$, α, $\alpha_1, \ldots, \alpha_p > 0$, and suppose there exists $P \in \mathbb{P}^n$ satisfying

$$0 = A_\alpha^{\mathrm{T}} P + PA_\alpha + \sum_{i=1}^{p} \tfrac{\alpha_i^2}{\alpha} A_i^{\mathrm{T}} PA_i + R, \tag{11.64}$$

where $A_\alpha = A + \frac{1}{2}\alpha I$. Then $A + \Delta A$ is asymptotically stable for all $\Delta A \in \boldsymbol{\Delta}_A$ given by (11.56), and

$$\sup_{\Delta A \in \boldsymbol{\Delta}_A} J_{\Delta A}(x_0) \leq x_0^{\mathrm{T}} P x_0, \quad x_0 \in \mathbb{R}^n. \tag{11.65}$$

Proposition 11.11. Consider the linear uncertain system (11.50). Let $R \in \mathbb{P}^n$, $\alpha > 0$, $N \in \mathbb{N}^n$, and suppose there exists $P \in \mathbb{P}^n$ satisfying

$$0 = A^{\mathrm{T}} P + PA + \alpha PB_0 B_0^{\mathrm{T}} P + \alpha^{-1} C_0^{\mathrm{T}} N C_0 + R. \tag{11.66}$$

Then $A + \Delta A$ is asymptotically stable for all $\Delta A \in \boldsymbol{\Delta}_A$ given by (11.59), and

$$\sup_{\Delta A \in \boldsymbol{\Delta}_A} J_{\Delta A}(x_0) \leq x_0^{\mathrm{T}} P x_0, \quad x_0 \in \mathbb{R}^n. \tag{11.67}$$

Proposition 11.12. Consider the linear uncertain system (11.50). Let $\hat{R} \in \mathbb{P}^n$, $Q \in \mathbb{P}^r$, $S \in \mathbb{R}^{r \times s}$, and $-R \in \mathbb{N}^s$, and suppose there exists $P \in \mathbb{P}^n$ satisfying

$$0 = A^{\mathrm{T}} P + PA + [B_0^{\mathrm{T}} P - SC_0]^{\mathrm{T}} Q^{-1} [B_0^{\mathrm{T}} P - SC_0] - C_0^{\mathrm{T}} R C_0 + \hat{R}. \tag{11.68}$$

Then $A + \Delta A$ is asymptotically stable for all $\Delta A \in \boldsymbol{\Delta}_A$ given by (11.62), and

$$\sup_{\Delta A \in \boldsymbol{\Delta}_A} J_{\Delta A}(x_0) \leq x_0^{\mathrm{T}} P x_0, \quad x_0 \in \mathbb{R}^n. \tag{11.69}$$

We conclude this section with several observations. First, note that $\sup_{\Delta A \in \boldsymbol{\Delta}_A} J_{\Delta A}(x_0)$ provides a worst-case characterization of the $\mathcal{H}_2$ norm of the uncertain system (11.50) with an averaged free response. Specifically, since $A + \Delta A$ is asymptotically stable for all $\Delta A \in \boldsymbol{\Delta}_A$ it follows that

$$\sup_{\Delta A \in \boldsymbol{\Delta}_A} \mathbb{E}[J_{\Delta A}(x_0)] = \sup_{\Delta A \in \boldsymbol{\Delta}_A} \mathbb{E}\left[\int_0^\infty x^{\mathrm{T}}(t) R x(t) \mathrm{d}t\right]$$

$$= \sup_{\Delta A \in \boldsymbol{\Delta}_A} \mathbb{E}\left[\int_0^{\mathrm{T}} x_0^{\mathrm{T}} e^{(A+\Delta A)^{\mathrm{T}} t} R e^{(A+\Delta A)t} x_0 \mathrm{d}t\right]$$
$$= \sup_{\Delta A \in \boldsymbol{\Delta}_A} \mathbb{E}[x_0^{\mathrm{T}} P_{\Delta A} x_0]$$
$$= \sup_{\Delta A \in \boldsymbol{\Delta}_A} \operatorname{tr} P_{\Delta A} V, \tag{11.70}$$

where

$$0 = (A + \Delta A)^{\mathrm{T}} P_{\Delta A} + P_{\Delta A}(A + \Delta A) + R \tag{11.71}$$

and $V = \mathbb{E}[x_0 x_0^{\mathrm{T}}]$. Now, since A is asymptotically stable,

$$\mathbb{E}[\mathcal{J}(x_0)] = \mathbb{E}\left[\int_0^{\infty} x^{\mathrm{T}}(t)(\Omega(P) + R)x(t)\mathrm{d}t\right]$$
$$= \mathbb{E}\left[\int_0^{\infty} x_0^{\mathrm{T}} e^{A^{\mathrm{T}} t}(\Omega(P) + R)e^{At} x_0 \mathrm{d}t\right]$$
$$= \mathbb{E}[x_0^{\mathrm{T}} P x_0]$$
$$= \operatorname{tr} PV, \tag{11.72}$$

where P satisfies (11.53). Hence, it follows from (11.54) that $\mathcal{J}(x_0)$, $x_0 \in \mathbb{R}^n$, provides an upper bound to the worst case, over the uncertainty set $\boldsymbol{\Delta}_A$, of the $\mathcal{H}_2$ norm of

$$G_{\Delta A}(s) \sim \left[\begin{array}{c|c} A + \Delta A & x_0 \\ \hline E & 0 \end{array}\right],$$

where E is such that $R = E^{\mathrm{T}} E$.

Example 11.5. Consider the linear uncertain dynamical system given by (11.50) and (11.51) with $n = 1$, $A < 0$, $R > 0$, $V = \mathbb{E}[x_0^2] > 0$, $A_1 = 1$, and $\boldsymbol{\Delta}_A = \{\Delta A : |\Delta A| \leq \alpha_1\}$. Note that for $\alpha_1 < -A$, $P_{\Delta A} = R/2(|A| - \Delta A)$ and $J_{\Delta A}(x_0) = R x_0^2 / 2(|A| - \alpha_1)$, where this worst-case performance is achieved for $\Delta A = \alpha_1$. Solving (11.64) with $\alpha = \alpha_1$ yields $P = R/2(|A| - \alpha_1)$, which is a nonconservative result for both robust stability and robust performance. To apply (11.66), set $\alpha_1 = \sqrt{N}$ and $B_0 = C_0 = 1$. Choosing $\alpha = 2\alpha_1(|A| - \alpha)NR$ again yields the nonconservative result $P = R/2(|A| - \alpha_1)$. △

Example 11.6. Consider the linear uncertain dynamical system given by (11.50) and (11.51) with $n = 2$, $A = -I_2$, $R = I_2$, $V = \mathbb{E}[x_0 x_0^{\mathrm{T}}] = I_2$, and $\boldsymbol{\Delta}_A = \{\Delta A : \Delta A = \delta_1 A_1, |\delta_1| \leq \alpha_1\}$, where

$$A_1 = \begin{bmatrix} 0 & 1 \\ 0 & 0 \end{bmatrix}. \tag{11.73}$$

Clearly, this perturbation is a nondestabilizing perturbation since $A + \Delta A$ remains asymptotically stable for all values of δ_1 since ΔA does not affect the nominal poles. Furthermore, note that

$$P_{\Delta A} = \begin{bmatrix} \frac{1}{2} & \frac{\delta_1}{4} \\ \frac{\delta_1}{4} & \frac{\delta_1^2}{4} + \frac{1}{2} \end{bmatrix} \tag{11.74}$$

and $\mathbb{E}[J_{\Delta A}(x_0)] = \frac{1}{4}\alpha_1^2 + 1$, where this worst case is achieved for $\delta_1 = \alpha_1$. In this case, (11.64) has the solution

$$P = \begin{bmatrix} (2 - \alpha\alpha_1)^{-1} & 0 \\ 0 & (2 - \alpha\alpha_1)^{-1} + \alpha^{-1}\alpha_1(2 - \alpha\alpha_1)^{-2} \end{bmatrix}, \tag{11.75}$$

which is positive definite for all α_1 so long as $\alpha < 2/\alpha_1$. Hence, (11.64) is nonconservative with respect to robust stability. For robust performance,

$$\mathbb{E}[\mathcal{J}(x_0)] = \operatorname{tr} PV = 2(2 - \alpha\alpha_1)^{-1} + \alpha^{-1}\alpha_1(2 - \alpha\alpha_1)^{-2}, \tag{11.76}$$

can be shown to be an upper bound for $\frac{1}{4}\delta_1^2 + 1$. Choosing, for example, $\alpha = \alpha_1^{-1}$ yields $\operatorname{tr} PV = \alpha_1^2 + 2$. △

11.3 A Dissipative Systems Perspective on Robust Stability

Although the Lyapunov-function-based robust analysis framework discussed in Section 11.2 applies to problems in which the nominal nonlinear system dynamics $f_0(\cdot)$ are perturbed by an uncertain function $\Delta f(\cdot) \in \boldsymbol{\Delta}$, a reinterpretation of these results yield standard nonlinear system theoretic criteria when viewed from the terminals of the uncertain parts of the system. For example, the bounding function

$$\Gamma(x) = \tfrac{1}{4}V'(x)G_\delta(x)G_\delta^{\mathrm{T}}(x)V'^{\mathrm{T}}(x) + m^{\mathrm{T}}(h_\delta(x))m(h_\delta(x))$$

in Proposition 11.1 forms the basis of nonlinear $\mathcal{H}_\infty$ theory while the bounding function

$$\begin{aligned}\Gamma(x) &= \tfrac{1}{4}[m(h_\delta(x)) + G_\delta^{\mathrm{T}}(x)V'^{\mathrm{T}}(x)]^{\mathrm{T}}[m(h_\delta(x)) + G_\delta^{\mathrm{T}}(x)V'^{\mathrm{T}}(x)] \\ &\quad + V'(x)G_\delta(x)m_1(h_\delta(x))\end{aligned}$$

in Proposition 11.2 forms the basis for nonlinear passivity and dissipativity theory. In particular, every operator $\delta(\cdot) \in \boldsymbol{\Delta}$, where $\boldsymbol{\Delta}$ is given by (11.25), is dissipative with respect to the supply rate $r(y, u_0) = y^{\mathrm{T}}y - \gamma^2 u_0^{\mathrm{T}}u_0$, where $u_0 = \delta(y)$.

Now, it follows from the results in Section 6.2 that the zero solution $x(t) \equiv 0$ of the nonlinear uncertain system (11.16) is globally asymptotically stable for all $\delta(\cdot) \in \boldsymbol{\Delta}$ if the nonlinear system $\mathcal{G}$ given by

$$\dot{x}(t) = f_0(x) + G_\delta(x)u_0(t), \quad x(0) = x_0, \quad t \geq 0, \tag{11.77}$$

$$y(t) = h_\delta(x(t)), \tag{11.78}$$

is dissipative with respect to the supply rate $r(u_0, y) = \gamma^2 u_0^{\mathrm{T}} u_0 - y^{\mathrm{T}} y$. Hence, it follows from Corollary 5.3 that a sufficient condition for robust stability of (11.16) is the existence of functions $\ell(\cdot)$, $\mathcal{W}(\cdot)$, and a continuously differentiable radially unbounded storage function $V(\cdot)$ such that

$$0 = V'(x) f_0(x) + h_\delta^{\mathrm{T}}(x) h_\delta(x) + \ell^{\mathrm{T}}(x)\ell(x), \quad (11.79)$$
$$0 = \tfrac{1}{2} V'(x) G_\delta(x) + \ell^{\mathrm{T}}(x)\mathcal{W}(x), \quad (11.80)$$
$$0 = \gamma^2 I - \mathcal{W}^{\mathrm{T}}(x)\mathcal{W}(x), \quad (11.81)$$

or, equivalently,

$$0 \geq V'(x) f_0(x) + \tfrac{1}{4}\gamma^{-2} V'(x) G_\delta(x) G_\delta^{\mathrm{T}}(x) V'^{\mathrm{T}}(x) + h_\delta^{\mathrm{T}}(x) h_\delta(x), \quad x \in \mathbb{R}^n, \quad (11.82)$$

which is identical to (11.36) with $m(y) = \gamma^{-1} y$ and $L_1(x) \geq 0$, $x \in \mathbb{R}^n$, for the uncertainty structure $\boldsymbol{\Delta}$ given by (11.24).

Similarly, every operator $\delta(\cdot) \in \boldsymbol{\Delta}$, where $\boldsymbol{\Delta}$ is given by (11.27) with $m_1(y) = M_1 y$, $m_2(y) = M_2 y$, and $M_1, M_2 \in \mathbb{S}^m$ such that $M_1 M_2 + M_2 M_1 \leq 0$, is dissipative with respect to the supply rate $r(y, u_0) = -(u_0 - M_1 y)^{\mathrm{T}}(u_0 - M_2 y)$, where $u_0 = \delta(y)$. Now, it follows from the results in Section 6.2 that the zero solution $x(t) \equiv 0$ of the nonlinear uncertain system (11.16) is globally asymptotically stable for all $\delta(\cdot) \in \boldsymbol{\Delta}$ if the nonlinear system $\mathcal{G}$ given by (11.77) and (11.78) is dissipative with respect to the supply rate $r(u_0, y) = (u_0 - M_1 y)^{\mathrm{T}}(u_0 - M_2 y)$. Hence, it follows from Theorem 5.6 that a sufficient condition for robust stability of (11.16) is the existence of functions $\ell(\cdot)$, $\mathcal{W}(x)$ and a continuously differentiable radially unbounded storage function $V(\cdot)$ such that

$$0 = V'(x) f_0(x) - h_\delta^{\mathrm{T}}(x) M_1 M_2 h_\delta(x) + \ell^{\mathrm{T}}(x)\ell(x), \quad (11.83)$$
$$0 = \tfrac{1}{2} V'(x) G_\delta(x) - h_\delta^{\mathrm{T}}(x)(M_1 + M_2) + \ell^{\mathrm{T}}(x)\mathcal{W}(x), \quad (11.84)$$
$$0 = I - \mathcal{W}^{\mathrm{T}}(x)\mathcal{W}(x), \quad (11.85)$$

or, equivalently,

$$0 \geq V'(x) f_0(x) + \tfrac{1}{4}[M h_\delta(x) + G_\delta^{\mathrm{T}}(x) V'^{\mathrm{T}}(x)]^{\mathrm{T}} \cdot [M h_\delta(x) + G_\delta^{\mathrm{T}}(x) V'^{\mathrm{T}}(x)] + V'(x) G_\delta(x) M_1 h_\delta(x), \quad x \in \mathbb{R}^n, \quad (11.86)$$

where $M \triangleq M_2 - M_1$, which is identical to (11.39) with $m_1(y) = M_1 y$, $m_2(y) = M_2 y$, and $L_1(x) \geq 0$, $x \in \mathbb{R}^n$, for the uncertainty structure $\boldsymbol{\Delta}$ given by (11.27). The above exposition demonstrates that dissipativity theory and nonlinear robustness theory are derivable from the same principles and are part of the same mathematical framework.

To see this for linear Ω-bound theory, rewrite (11.50) as

$$\dot{x}(t) = Ax(t) + \Delta A x(t), \qquad x(0) = x_0, \qquad t \geq 0. \tag{11.87}$$

Next, rewrite (11.87) as a negative feedback interconnection of a nominal system and an uncertain operator given by

$$\dot{x}(t) = Ax(t) - u(t), \qquad x(0) = x_0, \qquad t \geq 0, \tag{11.88}$$

$$y(t) = x(t), \qquad u_{\mathrm{c}}(t) = y(t), \tag{11.89}$$

$$y_{\mathrm{c}}(t) = \Delta A u_{\mathrm{c}}(t), \qquad u(t) = -y_{\mathrm{c}}(t), \tag{11.90}$$

or, equivalently, as a negative feedback interconnection given in Figure 11.1 where

$$\mathcal{G} = G(s) \sim \left[\begin{array}{c|c} A & -I_n \\ \hline I_n & 0 \end{array}\right]$$

and $\mathcal{G}_{\mathrm{c}} = \Delta A$.

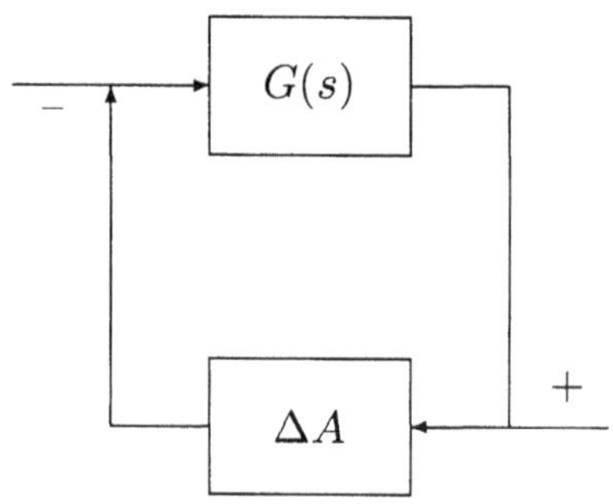

Figure 11.1 Feedback interconnection of $G(s)$ and ΔA.

The following two propositions show that Ω-bound theory for robust stability analysis is indeed a special case of dissipativity theory.

Proposition 11.13. Let $\Delta A \in \mathbf{\Delta}_A$ denote a linear operator $\Delta A : \mathbb{R}^n \to \mathbb{R}^n$ with input u_{c} and output y_{c}. Let $\Omega : \mathbb{S}^n \to \mathbb{S}^n$ be such that for every $P \in \mathbb{S}^n$,

$$\Delta A^{\mathrm{T}} P + P \Delta A \leq \Omega(P). \tag{11.91}$$

Then the linear operator $\Delta A(\cdot)$ is dissipative with respect to the supply rate $r_{\mathrm{c}}(u_{\mathrm{c}}, y_{\mathrm{c}}) = u_{\mathrm{c}}^{\mathrm{T}} \Omega(P) u_{\mathrm{c}} - 2 y_{\mathrm{c}}^{\mathrm{T}} P u_{\mathrm{c}}$.

Proof. The proof is a direct consequence of (11.91) by noting that $y_{\mathrm{c}} = \Delta A u_{\mathrm{c}}$, and hence, for every $u_{\mathrm{c}} \in \mathbb{R}^n$, $r_{\mathrm{c}}(u_{\mathrm{c}}, y_{\mathrm{c}}) = u_{\mathrm{c}}^{\mathrm{T}} \Omega(P) u_{\mathrm{c}} - 2 y_{\mathrm{c}}^{\mathrm{T}} P u_{\mathrm{c}} \geq 0$. $\square$

Proposition 11.14. Let

$$G(s) \sim \left[\begin{array}{c|c} A & -I_n \\ \hline I_n & 0 \end{array}\right]$$

and assume that there exists $P = P^{\mathrm{T}} > 0$ such that

$$0 > A^{\mathrm{T}}P + PA + \Omega(P). \tag{11.92}$$

Then the linear dynamical system given by the transfer function $G(s)$ with input u and output y is exponentially dissipative with respect to the supply rate $r(u, y) = -y^{\mathrm{T}}\Omega(P)y - 2y^{\mathrm{T}}Pu$.

Proof. It follows from (11.92) that there exists a scalar $\varepsilon > 0$ such that $0 \geq A^{\mathrm{T}}P + PA + \varepsilon P + \Omega(P)$. The result now follows immediately from Theorem 6.2 with $B = -I_n$, $C = I_n$, $D = 0$, $Q = -\Omega(P)$, $R = 0$, $S = -P$, $W = 0$, and $L = (-A^{\mathrm{T}}P - PA - \varepsilon P - \Omega(P))^{1/2}$. □

In light of Propositions 11.13 and 11.14, it follows from Theorem 6.2, with $\mathcal{G} = G(s)$, $\mathcal{G}_{\mathrm{c}} = \Delta A(\cdot)$, $Q = -R_{\mathrm{c}} = -\Omega(P)$, $R = Q_{\mathrm{c}} = 0$, $S = S_{\mathrm{c}} = -P$, and $\sigma = 1$, that if (11.91) and (11.92) hold, then $A + \Delta A$ is asymptotically stable for all $\Delta A \in \boldsymbol{\Delta}_A$. This of course establishes that Ω-bound theory for robust stability analysis is a special case of dissipativity theory. This exposition thus demonstrates that all (parameter-independent) guaranteed cost bounds developed in the literature including the bounded real bound [11, 147, 337, 356], the positive real bound [8, 147], the shifted bounded real bound [435], the shifted positive real bound [435], the implicit small gain bound [161], the absolute value bound [83], the linear bound [41,50,221,243], the inverse bound [50], the double commutator bound [436], the shifted linear bound [53], and the shifted inverse bound [53] are a special case of dissipativity theory.

11.4 Robust Optimal Control for Nonlinear Uncertain Systems

In this section, we consider a control problem for nonlinear uncertain dynamical systems involving a notion of optimality with respect to an *auxiliary cost* which guarantees a bound on the worst-case value of a nonlinear-nonquadratic cost criterion over a prescribed uncertainty set. The optimal robust feedback controllers are derived as a direct consequence of Theorem 11.1 and provide a generalization of the Hamilton-Jacobi-Bellman conditions for time-invariant, infinite-horizon problems for addressing robust feedback controllers of nonlinear uncertain systems. To address the robust optimal control problem let $\mathcal{D} \subset \mathbb{R}^n$ be an open set and let $U \subset \mathbb{R}^m$, where $0 \in \mathcal{D}$ and $0 \in U$. Furthermore, let $\mathcal{F} \subset \{F : \mathcal{D} \times U \to \mathbb{R}^n : F(0,0) = 0\}$.

Next, consider the controlled uncertain dynamical system

$$\dot{x}(t) = F(x(t), u(t)), \qquad x(0) = x_0, \qquad t \geq 0, \tag{11.93}$$

where $F(\cdot,\cdot) \in \mathcal{F}$ and the control $u(\cdot)$ is restricted to the class of admissible controls consisting of measurable functions $u(\cdot)$ such that $u(t) \in U$ for all $t \geq 0$, where the control constraint set U is given. We assume $0 \in U$. Given a control law $\phi(\cdot)$ and a feedback control law $u(t) = \phi(x(t))$, the closed-loop system has the form

$$\dot{x}(t) = F(x(t), \phi(x(t))), \qquad x(0) = x_0, \qquad t \geq 0, \tag{11.94}$$

for all $F(\cdot,\cdot) \in \mathcal{F}$.

Next, we present a generalization of Theorem 8.2 for characterizing robust feedback controllers that guarantee robust stability over a class of nonlinear uncertain systems and minimize an auxiliary performance functional. For the statement of this result let $L : \mathcal{D} \times U \to \mathbb{R}$ and define the set of regulation controllers for the nominal nonlinear system $F_0(\cdot,\cdot)$ by

$$\mathcal{S}(x_0) \triangleq \{u(\cdot) : u(\cdot) \text{ is admissible and } x(\cdot) \text{ given by (11.93)} \\ \text{satisfies } x(t) \to 0 \text{ as } t \to \infty \text{ with } F(\cdot,\cdot) = F_0(\cdot,\cdot)\}.$$

Theorem 11.2. Consider the nonlinear uncertain controlled system (11.93) with performance functional

$$J_F(x_0, u(\cdot)) \triangleq \int_0^\infty L(x(t), u(t))\mathrm{d}t, \tag{11.95}$$

where $F(\cdot,\cdot) \in \mathcal{F}$ and $u(\cdot)$ is an admissible control. Assume that there exist functions $V : \mathcal{D} \to \mathbb{R}$, $\Gamma : \mathcal{D} \times U \to \mathbb{R}$, and control law $\phi : \mathcal{D} \to U$, where $V(\cdot)$ is a continuously differentiable function, such that

$$V(0) = 0, \tag{11.96}$$

$$V(x) > 0, \quad x \in \mathcal{D}, \quad x \neq 0, \tag{11.97}$$

$$\phi(0) = 0, \tag{11.98}$$

$$V'(x)F(x, \phi(x)) \leq V'(x)F_0(x, \phi(x)) + \Gamma(x, \phi(x)), \ x \in \mathcal{D}, \ F(\cdot,\cdot) \in \mathcal{F}, \tag{11.99}$$

$$V'(x)F_0(x, \phi(x)) + \Gamma(x, \phi(x)) < 0, \quad x \in \mathcal{D}, \quad x \neq 0, \tag{11.100}$$

$$H(x, \phi(x)) = 0, \quad x \in \mathcal{D}, \tag{11.101}$$

$$H(x, u) \geq 0, \quad x \in \mathcal{D}, \quad u \in U, \tag{11.102}$$

where $F_0(\cdot,\cdot) \in \mathcal{F}$ defines the nominal system and

$$H(x, u) \triangleq L(x, u) + V'(x)F_0(x, u) + \Gamma(x, u). \tag{11.103}$$

Then, with the feedback control $u(\cdot) = \phi(x(\cdot))$, the zero solution $x(t) \equiv 0$ of the closed-loop system (11.94) is locally asymptotically stable for all

$F(\cdot,\cdot) \in \mathcal{F}$ and there exists a neighborhood of the origin $\mathcal{D}_0 \subseteq \mathcal{D}$ such that

$$\sup_{F(\cdot,\cdot)\in\mathcal{F}} J_F(x_0, \phi(x(\cdot))) \le \mathcal{J}(x_0, \phi(x(\cdot))) = V(x_0), \quad x_0 \in \mathcal{D}_0, \qquad (11.104)$$

where

$$\mathcal{J}(x_0, u(\cdot)) \triangleq \int_0^\infty [L(x(t), u(t)) + \Gamma(x(t), u(t))]\mathrm{d}t, \qquad (11.105)$$

and where $u(\cdot)$ is admissible and $x(t)$, $t \ge 0$, solves (11.93) with $F(x(t), u(t)) = F_0(x(t), u(t))$. In addition, if $x_0 \in \mathcal{D}_0$ then the feedback control $u(\cdot) = \phi(x(\cdot))$ minimizes $\mathcal{J}(x_0, u(\cdot))$ in the sense that

$$\mathcal{J}(x_0, \phi(x(\cdot))) = \min_{u(\cdot)\in\mathcal{S}(x_0)} \mathcal{J}(x_0, u(\cdot)). \qquad (11.106)$$

Finally, if $\mathcal{D} = \mathbb{R}^n$ and

$$V(x) \to \infty \text{ as } \|x\| \to \infty, \qquad (11.107)$$

then the zero solution $x(t) \equiv 0$ of the closed-loop system (11.94) is globally asymptotically stable for all $F(\cdot) \in \mathcal{F}$.

Proof. Local and global asymptotic stability are a direct consequence of (11.96)–(11.100) by applying Theorem 11.1 to the closed-loop system (11.94). Furthermore, using (11.101), condition (11.104) is a restatement of (11.11) as applied to the closed-loop system (11.94). Next, let $u(\cdot) \in \mathcal{S}(x_0)$ and let $x(\cdot)$ be the solution of (11.93) with $F(\cdot,\cdot) = F_0(\cdot,\cdot)$. Then (11.106) follows from Theorem 8.2 with $L(x,u)$ replaced by $L(x,u) + \Gamma(x,u)$ and $J(x_0, u(\cdot))$ replaced by $\mathcal{J}(x_0, u(\cdot))$. □

Note that conditions (11.101) and (11.102) correspond to the steady-state Hamilton-Jacobi-Bellman conditions for the nominal nonlinear system $F_0(\cdot,\cdot)$ with the auxiliary cost of $\mathcal{J}(x_0, u(\cdot))$. If $\mathcal{F}$ consists of only the nominal nonlinear closed-loop system $F_0(\cdot,\cdot)$, then $\Gamma(x,u) = 0$ for all $x \in \mathcal{D}$ and $u \in U$ satisfies (11.99), and hence, $J_F(x_0, u(\cdot)) = \mathcal{J}(x_0, u(\cdot))$. In this case, Theorem 11.2 specializes to Theorem 8.2.

Next, we specialize Theorem 11.2 to linear uncertain systems and provide connections to the quadratic Lyapunov bounding synthesis framework developed in [48, 51]. Specifically, we consider $\mathcal{F}$ to be the set of uncertain linear systems given by

$$\mathcal{F} = \{(A + \Delta A)x + (B + \Delta B)u : x \in \mathbb{R}^n, A \in \mathbb{R}^{n\times n}, B \in \mathbb{R}^{n\times m}, \\ (\Delta A, \Delta B) \in \mathbf{\Delta}_A \times \mathbf{\Delta}_B\}, \qquad (11.108)$$

where $\mathbf{\Delta}_A \subset \mathbb{R}^{n\times n}$ and $\mathbf{\Delta}_B \subset \mathbb{R}^{n\times m}$ are given bounded uncertainty sets of the uncertain perturbations $(\Delta A, \Delta B)$ of the nominal system (A, B) such that $(0,0) \in \mathbf{\Delta}_A \times \mathbf{\Delta}_B$. For the following result let $R_1 \in \mathbb{P}^n$ and $R_2 \in \mathbb{P}^m$

be given.

Corollary 11.3. Consider the linear uncertain controlled system

$$\dot{x}(t) = (A + \Delta A)x(t) + (B + \Delta B)u(t), \qquad x(0) = x_0, \qquad t \geq 0, \quad (11.109)$$

with performance functional

$$J_{\Delta \tilde{A}}(x_0, u(\cdot)) \triangleq \int_0^\infty [x^{\mathrm{T}}(t)R_1x(t) + u^{\mathrm{T}}(t)R_2u(t)]\mathrm{d}t, \quad (11.110)$$

where $u(\cdot)$ is admissible and $(\Delta A, \Delta B) \in \mathbf{\Delta}_A \times \mathbf{\Delta}_B$. Furthermore, assume there exist $P \in \mathbb{P}^n$, $\Omega_{xx} : \mathbb{P}^n \to \mathbb{N}^n$, $\Omega_{xu} : \mathbb{P}^n \to \mathbb{R}^{n\times m}$, and $\Omega_{uu} : \mathbb{P}^n \to \mathbb{N}^m$ such that

$$\begin{aligned}
&\Delta A^{\mathrm{T}}P + P\Delta A - P\Delta B(R_2 + \Omega_{uu}(P))^{-1}(B^{\mathrm{T}}P + \Omega_{xu}^{\mathrm{T}}(P))\\
&-(B^{\mathrm{T}}P + \Omega_{xu}^{\mathrm{T}}(P))^{\mathrm{T}}(R_2 + \Omega_{uu}(P))^{-1}\Delta B^{\mathrm{T}}P \leq \Omega_{xx}(P)\\
&-\Omega_{xu}(P)(R_2 + \Omega_{uu}(P))^{-1}(B^{\mathrm{T}}P + \Omega_{xu}^{\mathrm{T}}(P))\\
&-(B^{\mathrm{T}}P + \Omega_{xu}^{\mathrm{T}}(P))^{\mathrm{T}}(R_2 + \Omega_{uu}(P))^{-1}\Omega_{xu}^{\mathrm{T}}(P)\\
&+(B^{\mathrm{T}}P + \Omega_{xu}^{\mathrm{T}}(P))^{\mathrm{T}}(R_2 + \Omega_{uu}(P))^{-1}\Omega_{uu}(P)(R_2 + \Omega_{uu}(P))^{-1}\\
&\cdot(B^{\mathrm{T}}P + \Omega_{xu}^{\mathrm{T}}(P)), \ (\Delta A, \Delta B) \in \mathbf{\Delta}_A \times \mathbf{\Delta}_B, \quad (11.111)
\end{aligned}$$

and

$$\begin{aligned}
0 &= A^{\mathrm{T}}P + PA + R_1 + \Omega_{xx}(P)\\
&\quad -(B^{\mathrm{T}}P + \Omega_{xu}^{\mathrm{T}}(P))^{\mathrm{T}}(R_2 + \Omega_{uu}(P))^{-1}(B^{\mathrm{T}}P + \Omega_{xu}^{\mathrm{T}}(P)). \quad (11.112)
\end{aligned}$$

Then, with the feedback control $u = \phi(x) = -(R_2 + \Omega_{uu}(P))^{-1}(B^{\mathrm{T}}P + \Omega_{xu}^{\mathrm{T}}(P))x$, the zero solution $x(t) \equiv 0$ to (11.109) is globally asymptotically stable for all $x_0 \in \mathbb{R}^n$ and $(\Delta A, \Delta B) \in \mathbf{\Delta}_A \times \mathbf{\Delta}_B$, and

$$\sup_{(\Delta A,\Delta B)\in\mathbf{\Delta}_A\times\mathbf{\Delta}_B} J_{\Delta \tilde{A}}(x_0, \phi(x(\cdot))) \leq \mathcal{J}(x_0, \phi(x(\cdot))) = x_0^{\mathrm{T}}Px_0, \quad (11.113)$$

where

$$\begin{aligned}
\mathcal{J}(x_0, u(\cdot)) \triangleq \int_0^\infty &[x^{\mathrm{T}}(t)R_1x(t) + u^{\mathrm{T}}(t)R_2u(t) + x^{\mathrm{T}}(t)\Omega_{xx}(P)x(t)\\
&+2x^{\mathrm{T}}(t)\Omega_{xu}(P)u(t) + u^{\mathrm{T}}(t)\Omega_{uu}(P)u(t)]\mathrm{d}t, \quad (11.114)
\end{aligned}$$

and where $u(\cdot)$ is admissible and $x(t)$, $t \geq 0$, solves (11.109) with $(\Delta A, \Delta B) = (0, 0)$. Furthermore,

$$\mathcal{J}(x_0, \phi(x(\cdot))) = \min_{u(\cdot)\in\mathcal{S}(x_0)} \mathcal{J}(x_0, u(\cdot)), \quad (11.115)$$

where $\mathcal{S}(x_0)$ is the set of regulation controllers for the nominal system and $x_0 \in \mathbb{R}^n$.

Proof. The result is a direct consequence of Theorem 11.2 with $F(x,u) = (A+\Delta A)x+(B+\Delta B)u$, $F_0(x,u) = Ax+Bu$, $L(x,u) = x^{\mathrm{T}}R_1x+$

$u^{\mathrm{T}}R_2u$, $V(x) = x^{\mathrm{T}}Px$, $\Gamma(x,u) = x^{\mathrm{T}}\Omega_{xx}(P)x + 2x^{\mathrm{T}}\Omega_{xu}(P)u + u^{\mathrm{T}}\Omega_{uu}(P)u$, $\mathcal{D} = \mathbb{R}^n$, and $U = \mathbb{R}^m$. Specifically, conditions (11.96) and (11.97) are trivially satisfied. Now, forming $x^{\mathrm{T}}(11.111)x$ it follows that, after some algebraic manipulation, $V'(x)F(x,\phi(x)) \leq V'(x)F_0(x,\phi(x)) + \Gamma(x,\phi(x))$, for all $(\Delta A, \Delta B) \in \boldsymbol{\Delta}_A \times \boldsymbol{\Delta}_B$. Furthermore, it follows from (11.112) that $H(x,\phi(x)) = 0$, and hence, $V'(x)F_0(x,\phi(x)) + \Gamma(x,\phi(x)) < 0$ for all $x \neq 0$. Thus, $H(x,u) = H(x,u) - H(x,\phi(x)) = [u - \phi(x)]^{\mathrm{T}}(R_2 + \Omega_{uu}(P))[u-\phi(x)] \geq 0$ so that all the conditions of Theorem 11.2 are satisfied. Finally, since $V(\cdot)$ is radially unbounded, (11.109), with $u(t) = \phi(x(t)) = -(R_2+\Omega_{uu}(P))^{-1}(B^{\mathrm{T}}P+\Omega_{xu}^{\mathrm{T}}(P))x(t)$, is globally asymptotically stable for all $(\Delta A, \Delta B) \in \boldsymbol{\Delta}_A \times \boldsymbol{\Delta}_B$. □

Note that in the case where $U = \mathbb{R}^m$ the robust feedback control $u = \phi(x)$ is globally optimal since it minimizes $H(x,u)$ and satisfies (11.101). Specifically, setting

$$\frac{\partial}{\partial u}H(x,u) = 0, \tag{11.116}$$

yields the robust feedback control

$$\phi(x) = -(R_2 + \Omega_{uu}(P))^{-1}(B^{\mathrm{T}}P + \Omega_{xu}^{\mathrm{T}}(P))x. \tag{11.117}$$

Now, since

$$\frac{\partial^2}{\partial u^2}H(x,u) = R_2 + \Omega_{uu}(P) > 0, \tag{11.118}$$

it follows that for all $x \in \mathbb{R}^n$ the robust feedback control given by (11.117) minimizes $H(x,u)$. In particular, the optimal feedback control law $\phi(x)$ in Corollary 11.3 is derived using the properties of $H(x,u)$ as defined in Theorem 11.2. Specifically, since $H(x,u) = x^{\mathrm{T}}(A^{\mathrm{T}}P+PA+R_1+\Omega_{xx}(P))x+ u^{\mathrm{T}}(R_2+\Omega_{uu})u+2x^{\mathrm{T}}(B^{\mathrm{T}}P+\Omega_{xu}^{\mathrm{T}}(P))^{\mathrm{T}}u$ it follows that $\partial^2 H/\partial u^2 > 0$. Now, $\partial H/\partial u = 2(R_2+\Omega_{uu}(P))u+2(B^{\mathrm{T}}P+\Omega_{xu}^{\mathrm{T}}(P))x = 0$ gives the unique global minimum of $H(x,u)$. Hence, since $\phi(x)$ minimizes $H(x,u)$ it follows that $\phi(x)$ satisfies $\partial H/\partial u = 0$ or, equivalently, $(R_2 + \Omega_{uu}(P))\phi(x) + (B^{\mathrm{T}}P + \Omega_{xu}^{\mathrm{T}}(P))x = 0$ so that $\phi(x)$ is given by (11.117). Similar remarks hold for the nonlinear robust controllers developed in Sections 11.4 and 11.5.

In order to make explicit connections with linear robust control, we now assign explicit structure to the sets $\boldsymbol{\Delta}_A$ and $\boldsymbol{\Delta}_B$ and the bounding functions $\Omega_{xx}(\cdot)$, $\Omega_{xu}(\cdot)$, and $\Omega_{uu}(\cdot)$. First, the uncertainty set $\boldsymbol{\Delta}_A \times \boldsymbol{\Delta}_B$ is assumed to be of the form

$$\boldsymbol{\Delta}_A \times \boldsymbol{\Delta}_B \triangleq \left\{(\Delta A, \Delta B) : \Delta A = \sum_{i=1}^{p} \delta_i A_i,\ \Delta B = \sum_{i=1}^{p} \delta_i B_i,\ \sum_{i=1}^{p} \frac{\delta_i^2}{\alpha_i^2} \leq 1\right\}, \tag{11.119}$$

where for $i = 1,\ldots,p$: $A_i \in \mathbb{R}^{n\times n}$ and $B_i \in \mathbb{R}^{n\times m}$ are fixed matrices denoting the structure of the parametric uncertainty, α_i is a given positive number, and δ_i is an uncertain real parameter. As discussed in Section 11.2 the uncertain parameters δ_i are assumed to lie in a specified ellipsoidal region in $\mathbb{R}^p$ [50, 51]. In this case, let

$$\Omega_{xx}(P) = \sum_{i=1}^{p} \frac{\alpha_i^2}{\alpha} A_i^{\mathrm{T}} P A_i + \alpha P,$$

$$\Omega_{xu}(P) = \sum_{i=1}^{p} \frac{\alpha_i^2}{\alpha} A_i^{\mathrm{T}} P B_i,$$

$$\Omega_{uu}(P) = \sum_{i=1}^{p} \frac{\alpha_i^2}{\alpha} B_i^{\mathrm{T}} P B_i,$$

where α is an arbitrary positive scalar. Next, for notational convenience define $R_{2s} \triangleq R_2 + \sum_{i=1}^{p} \frac{\alpha_i^2}{\alpha} B_i^{\mathrm{T}} P B_i$ and $P_s \triangleq B^{\mathrm{T}} P + \sum_{i=1}^{p} \frac{\alpha_i^2}{\alpha} B_i^{\mathrm{T}} P A_i$. Now, note that

$$0 \le \sum_{i=1}^{p} \left[\frac{\alpha_i}{\alpha^{\frac{1}{2}}} P^{\frac{1}{2}} (A_i - B_i R_{2s}^{-1} P_s) - \frac{\delta_i \alpha^{\frac{1}{2}}}{\alpha_i} P^{\frac{1}{2}} \right]^{\mathrm{T}} \cdot \left[\frac{\alpha_i}{\alpha^{\frac{1}{2}}} P^{\frac{1}{2}} (A_i - B_i R_{2s}^{-1} P_s) - \frac{\delta_i \alpha^{\frac{1}{2}}}{\alpha_i} P^{\frac{1}{2}} \right],$$

or, equivalently,

$$\sum_{i=1}^{p} \delta_i [(A_i - B_i R_{2s}^{-1} P_s)^{\mathrm{T}} P + P(A_i - B_i R_{2s}^{-1} P_s)] \le \sum_{i=1}^{p} \frac{\alpha_i^2}{\alpha} (A_i - B_i R_{2s}^{-1} P_s)^{\mathrm{T}} P (A_i - B_i R_{2s}^{-1} P_s) + \frac{\delta_i^2 \alpha}{\alpha_i^2} P,$$

which, since $\sum_{i=1}^{p} \frac{\delta_i^2}{\alpha_i^2} \le 1$, implies

$$\sum_{i=1}^{p} \delta_i [(A_i - B_i R_{2s}^{-1} P_s)^{\mathrm{T}} P + P(A_i - B_i R_{2s}^{-1} P_s)] \le \sum_{i=1}^{p} \frac{\alpha_i^2}{\alpha} (A_i - B_i R_{2s}^{-1} P_s)^{\mathrm{T}} P (A_i - B_i R_{2s}^{-1} P_s) + \alpha P,$$

and hence, (11.111) holds. Furthermore, (11.112) specializes to

$$0 = A_\alpha^{\mathrm{T}} P + P A_\alpha + R_1 + \sum_{i=1}^{p} \frac{\alpha_i^2}{\alpha} A_i^{\mathrm{T}} P A_i - P_s^{\mathrm{T}} R_{2s}^{-1} P_s, \qquad (11.120)$$

where $A_\alpha \triangleq A + \frac{\alpha}{2} I_n$, and the optimal robust feedback law is given by $\phi(x) = -R_{2s}^{-1} P_s x$. This corresponds to the results obtained in [41, 243].

Alternatively, we can choose

$$\begin{aligned}\Omega_{xx}(P) &= \sum_{i=1}^{p} \frac{\alpha_i^2}{\alpha}(A_i - B_i R_2^{-1} B^{\mathrm{T}} P)^{\mathrm{T}} P (A_i - B_i R_2^{-1} B^{\mathrm{T}} P) + \alpha P, \\ \Omega_{xu}(P) &= 0, \\ \Omega_{uu}(P) &= 0.\end{aligned}$$

In this case, note that

$$\begin{aligned}0 \le \sum_{i=1}^{p} &\left[\frac{\alpha_i}{\alpha^{\frac{1}{2}}} P^{\frac{1}{2}} (A_i - B_i R_2^{-1} B^{\mathrm{T}} P) - \frac{\delta_i \alpha^{\frac{1}{2}}}{\alpha_i} P^{\frac{1}{2}}\right]^{\mathrm{T}} \\ &\cdot \left[\frac{\alpha_i}{\alpha^{\frac{1}{2}}} P^{\frac{1}{2}} (A_i - B_i R_2^{-1} B^{\mathrm{T}} P) - \frac{\delta_i \alpha^{\frac{1}{2}}}{\alpha_i} P^{\frac{1}{2}}\right],\end{aligned}$$

or, equivalently,

$$\begin{aligned}&\sum_{i=1}^{p} \delta_i [(A_i - B_i R_2^{-1} B^{\mathrm{T}} P)^{\mathrm{T}} P + P(A_i - B_i R_2^{-1} B^{\mathrm{T}} P)] \\ &\qquad \le \sum_{i=1}^{p} \frac{\alpha_i^2}{\alpha}(A_i - B_i R_2^{-1} B^{\mathrm{T}} P)^{\mathrm{T}} P (A_i - B_i R_2^{-1} B^{\mathrm{T}} P) + \frac{\delta_i^2 \alpha}{\alpha_i^2} P,\end{aligned}$$

which, since $\sum_{i=1}^{p} \frac{\delta_i^2}{\alpha_i^2} \le 1$, implies

$$\begin{aligned}&\sum_{i=1}^{p} \delta_i [(A_i - B_i R_2^{-1} B^{\mathrm{T}} P)^{\mathrm{T}} P + P(A_i - B_i R_2^{-1} B^{\mathrm{T}} P)] \\ &\qquad \le \sum_{i=1}^{p} \frac{\alpha_i^2}{\alpha}(A_i - B_i R_2^{-1} B^{\mathrm{T}} P)^{\mathrm{T}} P (A_i - B_i R_2^{-1} B^{\mathrm{T}} P) + \alpha P,\end{aligned}$$

and hence, (11.111) holds. Furthermore, (11.112) specializes to

$$0 = A_\alpha^{\mathrm{T}} P + P A_\alpha + R_1 + \sum_{i=1}^{p} \frac{\alpha_i^2}{\alpha}(A_i - B_i R_2^{-1} B^{\mathrm{T}} P)^{\mathrm{T}} P (A_i - B_i R_2^{-1} B^{\mathrm{T}} P), \tag{11.121}$$

and the optimal robust feedback law is given by $\phi(x) = -R_2^{-1} B^{\mathrm{T}} P x$. The robustified Riccati equation (11.121) does not appear to have been considered in the literature for the uncertainty structure given by (11.119).

Next, we consider the uncertainty set $\boldsymbol{\Delta}_A \times \boldsymbol{\Delta}_B$ given by

$$\boldsymbol{\Delta}_A \times \boldsymbol{\Delta}_B \triangleq \{(\Delta A, \Delta B) : \Delta A = B_0 F C_0,\ \Delta B = B_0 F D_0,\ F^{\mathrm{T}} F \le N\}, \tag{11.122}$$

where $B_0 \in \mathbb{R}^{n \times r}$, $C_0 \in \mathbb{R}^{s \times n}$, and $D_0 \in \mathbb{R}^{r \times m}$ are fixed matrices denoting the structure of the uncertainty, $F \in \mathbb{R}^{r \times s}$ is an uncertain matrix, and

$N \in \mathbb{N}^s$ is a given uncertainty bound [48]. In this case, let

$$\begin{aligned}\Omega_{xx}(P) &= C_0^{\mathrm{T}} N C_0 + P B_0 B_0^{\mathrm{T}} P,\\ \Omega_{xu}(P) &= C_0^{\mathrm{T}} N D_0,\\ \Omega_{uu}(P) &= D_0^{\mathrm{T}} N D_0.\end{aligned}$$

Next, for notational convenience define $R_{2a} \triangleq R_2 + D_0^{\mathrm{T}} N D_0$ and $P_a \triangleq B^{\mathrm{T}} P + D_0^{\mathrm{T}} N C_0$. Now, note that

$$0 \le \left[FC_0 - FD_0 R_{2a}^{-1} P_a - B_0^{\mathrm{T}} P\right]^{\mathrm{T}} \left[FC_0 - FD_0 R_{2a}^{-1} P_a - B_0^{\mathrm{T}} P\right],$$

or, equivalently,

$$\begin{aligned}&[FC_0 - FD_0 R_{2a}^{-1} P_a]^{\mathrm{T}} B_0^{\mathrm{T}} P + P B_0 [FC_0 - FD_0 R_{2a}^{-1} P_a]\\ &\qquad \le [C_0 - D_0 R_{2a}^{-1} P_a]^{\mathrm{T}} F^{\mathrm{T}} F [C_0 - D_0 R_{2a}^{-1} P_a] + P B_0 B_0^{\mathrm{T}} P,\end{aligned}$$

which, since $F^{\mathrm{T}} F \le N$, implies

$$\begin{aligned}&[FC_0 - FD_0 R_{2a}^{-1} P_a]^{\mathrm{T}} B_0^{\mathrm{T}} P + P B_0 [FC_0 - FD_0 R_{2a}^{-1} P_a]\\ &\qquad \le [C_0 - D_0 R_{2a}^{-1} P_a]^{\mathrm{T}} N [C_0 - D_0 R_{2a}^{-1} P_a] + P B_0 B_0^{\mathrm{T}} P,\end{aligned}$$

and hence, (11.111) holds. Furthermore, (11.112) specializes to

$$0 = A^{\mathrm{T}} P + PA + R_1 + C_0^{\mathrm{T}} N C_0 + P B_0 B_0^{\mathrm{T}} P - P_a^{\mathrm{T}} R_{2a}^{-1} P_a, \quad (11.123)$$

and the optimal robust feedback law is given by $\phi(x) = -R_{2a}^{-1} P_a x$. This corresponds to the results obtained in [41, 354, 356, 480].

Alternatively, we can choose

$$\begin{aligned}\Omega_{xx}(P) &= [C_0 - D_0 R_2^{-1} B^{\mathrm{T}} P]^{\mathrm{T}} N [C_0 - D_0 R_2^{-1} B^{\mathrm{T}} P] + P B_0 B_0^{\mathrm{T}} P,\\ \Omega_{xu}(P) &= 0,\\ \Omega_{uu}(P) &= 0.\end{aligned}$$

In this case, note that

$$0 \le \left[FC_0 - FD_0 R_2^{-1} B^{\mathrm{T}} P - B_0^{\mathrm{T}} P\right]^{\mathrm{T}} \left[FC_0 - FD_0 R_2^{-1} B^{\mathrm{T}} P - B_0^{\mathrm{T}} P\right],$$

or, equivalently,

$$\begin{aligned}&[FC_0 - FD_0 R_2^{-1} B^{\mathrm{T}} P] B_0^{\mathrm{T}} P + P B_0 [FC_0 - FD_0 R_2^{-1} B^{\mathrm{T}} P]\\ &\qquad \le [C_0 - D_0 R_2^{-1} B^{\mathrm{T}} P]^{\mathrm{T}} F^{\mathrm{T}} F [C_0 - D_0 R_2^{-1} B^{\mathrm{T}} P] + P B_0 B_0^{\mathrm{T}} P,\end{aligned}$$

which, since $F^{\mathrm{T}} F \le N$, implies

$$\begin{aligned}&[FC_0 - FD_0 R_2^{-1} B^{\mathrm{T}} P] B_0^{\mathrm{T}} P + P B_0 [FC_0 - FD_0 R_2^{-1} B^{\mathrm{T}} P]\\ &\qquad \le [C_0 - D_0 R_2^{-1} B^{\mathrm{T}} P]^{\mathrm{T}} N [C_0 - D_0 R_2^{-1} B^{\mathrm{T}} P] + P B_0 B_0^{\mathrm{T}} P,\end{aligned}$$

and hence, (11.111) holds. Furthermore, (11.112) becomes

$$0 = A^{\mathrm{T}}P + PA + R_1 - PSP + [C_0 - D_0R_2^{-1}B^{\mathrm{T}}P]^{\mathrm{T}}N[C_0 - D_0R_2^{-1}B^{\mathrm{T}}P] + PB_0B_0^{\mathrm{T}}P, \tag{11.124}$$

where $S \triangleq BR_2^{-1}B^{\mathrm{T}}$, and the optimal robust feedback law is given by

$$\phi(x) = -R_2^{-1}B^{\mathrm{T}}Px.$$

The robustified Riccati equation (11.124) does not appear to have been considered in the literature for the uncertainty structure given by (11.122).

11.5 Optimal and Inverse Optimal Robust Control for Nonlinear Uncertain Affine Systems

In this section, we specialize Theorem 11.2 to affine (in the control) uncertain systems having the form (see Figure 11.2)

$$\dot{x}(t) = f_0(x(t)) + \Delta f(x(t)) + [G_0(x(t)) + \Delta G(x(t))]\,u(t),\ x(0) = x_0,\ t \geq 0, \tag{11.125}$$

where $f_0 : \mathbb{R}^n \to \mathbb{R}^n$ satisfies $f_0(0) = 0$, $G_0 : \mathbb{R}^n \to \mathbb{R}^{n\times m}$, $\mathcal{D} = \mathbb{R}^n$, $U = \mathbb{R}^m$,

$$\mathcal{F} = \{f_0(x) + \Delta f(x) + [G_0(x) + \Delta G(x)]u :\ x \in \mathbb{R}^n, u \in \mathbb{R}^m, (\Delta f, \Delta G) \in \boldsymbol{\Delta}_f \times \boldsymbol{\Delta}_G\},\ \Delta f(\cdot) \in \boldsymbol{\Delta}_f \subset \{\Delta f : \mathbb{R}^n \to \mathbb{R}^n : \Delta f(0) = 0\},$$

and

$$\Delta G \in \boldsymbol{\Delta}_G \subset \{\Delta G : \mathbb{R}^n \to \mathbb{R}^{n\times m}\}.$$

In this section, no explicit structure is assumed for the elements of $\boldsymbol{\Delta}_f$ and $\boldsymbol{\Delta}_G$. In Section 11.6 the structure of variations in $\boldsymbol{\Delta}_f$ and $\boldsymbol{\Delta}_G$ will be specified. Furthermore, we consider performance integrands $L(x,u)$ of the form

$$L(x,u) = L_1(x) + L_2(x)u + u^{\mathrm{T}}R_2(x)u, \tag{11.126}$$

where $L_1 : \mathbb{R}^n \to \mathbb{R}$, $L_2 : \mathbb{R}^n \to \mathbb{R}^{1\times m}$, and $R_2 : \mathbb{R}^n \to \mathbb{P}^m$ so that (11.95) becomes

$$J(x_0, u(\cdot)) = \int_0^\infty [L_1(x(t)) + L_2(x(t))u(t) + u^{\mathrm{T}}(t)R_2(x)u(t)]\mathrm{d}t. \tag{11.127}$$

Corollary 11.4. Consider the nonlinear uncertain controlled affine system (11.125) with performance functional (11.127). Assume that there exist a continuously differentiable function $V : \mathbb{R}^n \to \mathbb{R}$ and functions $L_2 : \mathbb{R}^n \to \mathbb{R}^{1\times m}$, $\Gamma_{xx} : \mathbb{R}^n \to \mathbb{R}$, $\Gamma_{xu} : \mathbb{R}^n \to \mathbb{R}^{1\times m}$, and $\Gamma_{uu} : \mathbb{R}^n \to \mathbb{N}^m$ such that

$$V(0) = 0, \tag{11.128}$$

$$L_2(0) = 0, \tag{11.129}$$

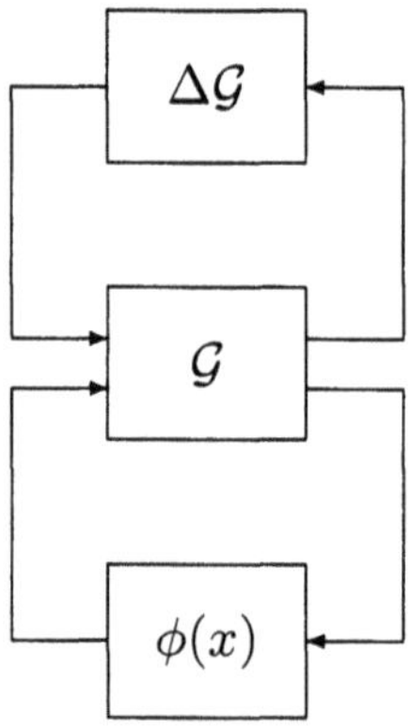

Figure 11.2 Uncertain nonlinear feedback system.

$$\Gamma_{xu}(0) = 0, \tag{11.130}$$

$$V(x) > 0, \quad x \in \mathbb{R}^n, \quad x \neq 0, \tag{11.131}$$

$$V'(x)\big(\Delta f(x) - \tfrac{1}{2}\Delta G(x) R_{2a}^{-1}(x) V_a(x)\big) \leq \Gamma_{xx}(x) - \tfrac{1}{2}\Gamma_{xu}(x) R_{2a}^{-1}(x) V_a(x) + \tfrac{1}{4} V_a^{\mathrm{T}}(x) R_{2a}^{-1}(x) \Gamma_{uu}(x) R_{2a}^{-1}(x) V_a(x), \quad (\Delta f, \Delta G) \in \boldsymbol{\Delta}_f \times \boldsymbol{\Delta}_G, \tag{11.132}$$

$$V'(x)[f_0(x) - \tfrac{1}{2} G_0(x) R_{2a}^{-1}(x) V_a(x)] + \Gamma_{xx}(x) - \tfrac{1}{2}\Gamma_{xu}(x) R_{2a}^{-1}(x) V_a(x) + \tfrac{1}{4} V_a^{\mathrm{T}}(x) R_{2a}^{-1}(x) \Gamma_{uu}(x) R_{2a}^{-1}(x) V_a(x) < 0, \quad x \in \mathbb{R}^n, \quad x \neq 0, \tag{11.133}$$

and

$$V(x) \to \infty \text{ as } \|x\| \to \infty, \tag{11.134}$$

where $R_{2a}(x) \triangleq R_2(x) + \Gamma_{uu}(x)$ and $V_a(x) \triangleq [L_2(x) + \Gamma_{xu}(x) + V'(x)G_0(x)]^{\mathrm{T}}$. Then the zero solution $x(t) \equiv 0$ of the nonlinear uncertain system (11.125) is globally asymptotically stable for all $(\Delta f, \Delta G) \in \boldsymbol{\Delta}_f \times \boldsymbol{\Delta}_G$ with the feedback control law

$$\phi(x) = -\tfrac{1}{2} R_{2a}^{-1}(x) V_a(x), \tag{11.135}$$

and the performance functional (11.127) satisfies

$$\sup_{(\Delta f, \Delta G) \in \boldsymbol{\Delta}_f \times \boldsymbol{\Delta}_G} J(x_0, \phi(x(\cdot))) \leq \mathcal{J}(x_0, \phi(x(\cdot))) = V(x_0), \tag{11.136}$$

where

$$\mathcal{J}(x_0, u(\cdot)) \triangleq \int_0^\infty [L(x(t), u(t)) + \Gamma(x(t), u(t))] \mathrm{d}t, \tag{11.137}$$

and

$$\Gamma(x, u) = \Gamma_{xx}(x) + \Gamma_{xu}(x) u + u^{\mathrm{T}} \Gamma_{uu}(x) u, \tag{11.138}$$

where $u(\cdot)$ is admissible, and $x(t)$, $t \geq 0$, solves (11.125) with $(\Delta f, \Delta G) = (0, 0)$. In addition, the performance functional (11.137), with

$$L_1(x) = \phi^{\mathrm{T}}(x) R_{2a}(x)\phi(x) - V'(x) f_0(x) - \Gamma_{xx}(x), \tag{11.139}$$

is minimized in the sense that

$$\mathcal{J}(x_0, \phi(x(\cdot))) = \min_{u(\cdot)\in\mathcal{S}(x_0)} \mathcal{J}(x_0, u(\cdot)). \tag{11.140}$$

Proof. The result is a direct consequence of Theorem 11.2 with $\mathcal{D} = \mathbb{R}^n$, $U = \mathbb{R}^m$, $F_0(x,u) = f_0(x) + G_0(x)u$, $F(x,u) = f_0(x) + \Delta f(x) + [G_0(x) + \Delta G(x)]u$, $\mathcal{F} = \{f_0(x) + \Delta f(x) + [G_0(x) + \Delta G(x)]u : x \in \mathbb{R}^n, u \in \mathbb{R}^m, (\Delta f, \Delta G) \in \mathbf{\Delta}_f \times \mathbf{\Delta}_G\}$, $L(x,u)$ given by (11.126), and $\Gamma(x,u)$ given by (11.138). Specifically, with (11.125), (11.126), and (11.138), the Hamiltonian has the form

$$\begin{aligned} H(x,u) &= L_1(x) + L_2(x)u + u^{\mathrm{T}} R_2(x) u + V'(x)(f_0(x) + G_0(x)u) \\ &\quad + \Gamma_{xx}(x) + \Gamma_{xu}(x)u + u^{\mathrm{T}}\Gamma_{uu}(x)u. \end{aligned}$$

Now, the feedback control law (11.135) is obtained by setting $\frac{\partial H}{\partial u} = 0$. With (11.135), it follows that (11.132) and (11.133) imply (11.99), and (11.100), respectively. Next, since $V(\cdot)$ is continuously differentiable and $x = 0$ is a local minimum of $V(\cdot)$, it follows that $V'(0) = 0$, and hence, since by assumption $L_2(0) = 0$ and $\Gamma_{xu}(0) = 0$, it follows that $\phi(0) = 0$, which proves (11.98). Next, with $L_1(x)$ given by (11.139) and $\phi(x)$ given by (11.135) it follows that (11.101) holds. Finally, since $H(x,u) = H(x,u) - H(x,\phi(x)) = [u - \phi(x)]^{\mathrm{T}} R_{2a}(x)[u - \phi(x)]$ and $R_{2a}(x)$ is positive definite for all $x \in \mathbb{R}^n$, condition (11.102) holds. The result now follows as a direct consequence of Theorem 11.2. □

Note that (11.133) implies

$$\begin{aligned} \dot{V}(x) &\triangleq V'(x)[f_0(x) + \Delta f(x) + (G_0(x) + \Delta G(x))\phi(x)] < 0, \\ & \qquad x \in \mathbb{R}^n, \quad x \neq 0, \quad (\Delta f, \Delta G) \in \mathbf{\Delta}_f \times \mathbf{\Delta}_G, \end{aligned} \tag{11.141}$$

with $\phi(x)$ given by (11.135). Furthermore, (11.128), (11.131), and (11.141) ensure that $V(x)$ is a Lyapunov function guaranteeing robust stability of the closed-loop system for all $(\Delta f, \Delta G) \in \mathbf{\Delta}_f \times \mathbf{\Delta}_G$. As noted in Chapter 9, it is important to recognize that the function $L_2(x)$ which appears in the integrand of the performance functional (11.127) is an arbitrary function of x subject to conditions (11.129), (11.132), and (11.133). Thus, $L_2(x)$ provides flexibility in choosing the control law.

With $L_1(x)$ given by (11.139) and $\phi(x)$ given by (11.135), $L(x,u) + \Gamma(x,u)$ can be expressed as

$$L(x,u) + \Gamma(x,u) = [u - \phi(x)]^{\mathrm{T}} R_{2a}(x)[u - \phi(x)] - V'(x)[f_0(x) + G_0(x)u]$$

$$= \left[u + \tfrac{1}{2}R_{2a}^{-1}(x)(L_2(x) + \Gamma_{xu}(x))^{\mathrm{T}}\right]^{\mathrm{T}} R_{2a}(x)$$
$$\cdot \left[u + \tfrac{1}{2}R_{2a}^{-1}(x)(L_2(x) + \Gamma_{xu}(x))^{\mathrm{T}}\right]$$
$$-V'(x)[f_0(x) + G_0(x)\phi(x)]$$
$$-\tfrac{1}{4}V'(x)G_0(x)R_{2a}^{-1}(x)G_0^{\mathrm{T}}(x)V'^{\mathrm{T}}(x). \tag{11.142}$$

Since $R_{2a}(x) \geq R_2(x) > 0$ for all $x \in \mathbb{R}^n$ the first term of the right-hand side of (11.142) is nonnegative, while (11.141) implies that the second term is nonnegative. Thus, we have

$$L(x,u) + \Gamma(x,u) \geq -\tfrac{1}{4}V'(x)G_0(x)R_{2a}^{-1}(x)G_0^{\mathrm{T}}(x)V'^{\mathrm{T}}(x), \tag{11.143}$$

which shows that $L(x,u) + \Gamma(x,u)$ may be negative. As a result, there may exist a control input u for which the auxiliary performance functional $\mathcal{J}(x_0, u)$ is negative. Note, however, if the control is a stabilizing feedback control, that is, $u \in \mathcal{S}(x_0)$, then it follows from (11.136) and (11.140) that

$$\mathcal{J}(x_0, u(\cdot)) \geq V(x_0) \geq 0, \quad x_0 \in \mathbb{R}^n, \quad u(\cdot) \in \mathcal{S}(x_0).$$

Furthermore, in this case, substituting $u = \phi(x)$ into (11.142) yields

$$L(x,\phi(x)) + \Gamma(x,\phi(x)) = -V'(x)[f_0(x) + G_0(x)\phi(x)],$$

which, by (11.141), is positive.

11.6 Nonlinear Guaranteed Cost Control

Having established the theoretical basis for our approach, we now assign explicit structure to the set $\boldsymbol{\Delta}_f \times \boldsymbol{\Delta}_G$ and the bounding functions $\Gamma_{xx}(x)$, $\Gamma_{xu}(x)$, and $\Gamma_{uu}(x)$, $x \in \mathbb{R}^n$. Even though both $\Delta f(x)$ and $\Delta G(x)$, $x \in \mathbb{R}^n$, uncertainties can be considered, for simplicity of exposition we assume that $\Delta G(x) = 0$, $x \in \mathbb{R}^n$ (see Problem 11.26 for the case where $\Delta G(x) \neq 0$, $x \in \mathbb{R}^n$). The uncertainty set $\mathcal{F}$ is assumed to be of the form given by (11.23) with $\boldsymbol{\Delta}$ given by (11.24).

Proposition 11.15. Consider the nonlinear uncertain controlled system (11.125) with performance functional (11.127). Assume that there exist a continuously differentiable function $V : \mathbb{R}^n \to \mathbb{R}$ and function $L_2 : \mathbb{R}^n \to \mathbb{R}^{1\times m}$ such that (11.128)–(11.131) are satisfied,

$$V'(x)[f_0(x) - \tfrac{1}{2}G_0(x)R_2^{-1}(x)V_a(x)] + \tfrac{1}{4}V'(x)G_\delta(x)G_\delta^{\mathrm{T}}(x)V'^{\mathrm{T}}(x)$$
$$+m^{\mathrm{T}}(h_\delta(x))m(h_\delta(x)) < 0, \quad x \in \mathbb{R}^n, \quad x \neq 0, \tag{11.144}$$

and

$$V(x) \to \infty \text{ as } \|x\| \to \infty, \tag{11.145}$$

where $V_a(x) \triangleq [L_2(x) + V'(x)G_0(x)]^{\mathrm{T}}$. Then the zero solution $x(t) \equiv 0$ of the nonlinear uncertain system (11.125) is globally asymptotically stable for all $\delta(\cdot) \in \mathbf{\Delta}$, where $\mathbf{\Delta}$ is given by (11.24), with the feedback control law

$$\phi(x) = -\tfrac{1}{2}R_2^{-1}(x)V_a(x). \tag{11.146}$$

Furthermore, the performance functional (11.127) satisfies

$$\sup_{\delta(\cdot)\in\mathbf{\Delta}} J(x_0,\phi(x(\cdot))) \leq \mathcal{J}(x_0,\phi(x(\cdot))) = V(x_0), \tag{11.147}$$

where

$$\begin{aligned}\mathcal{J}(x_0,u(\cdot)) \triangleq \int_0^\infty &[L_1(x) + L_2(x)u + u^{\mathrm{T}}R_2(x)u + \tfrac{1}{4}V'(x)G_\delta(x)G_\delta^{\mathrm{T}}(x)V'^{\mathrm{T}}(x)\\ &+m^{\mathrm{T}}(h_\delta(x))m(h_\delta(x))]\mathrm{d}t,\end{aligned} \tag{11.148}$$

where $u(\cdot)$ is admissible and $x(t)$, $t \geq 0$, solves (11.125) with $\delta(h_\delta(x)) \equiv 0$. In addition, the performance functional (11.148), with

$$\begin{aligned}L_1(x) &= \phi^{\mathrm{T}}(x)R_2(x)\phi(x) - V'(x)f_0(x) - \tfrac{1}{4}V'(x)G_\delta(x)G_\delta^{\mathrm{T}}(x)V'^{\mathrm{T}}(x)\\ &\quad -m^{\mathrm{T}}(h_\delta(x))m(h_\delta(x)),\end{aligned} \tag{11.149}$$

is minimized in the sense that

$$\mathcal{J}(x_0,\phi(x(\cdot))) = \min_{u(\cdot)\in\mathcal{S}(x_0)} \mathcal{J}(x_0,u(\cdot)). \tag{11.150}$$

Proof. The result is direct consequence of Corollary 11.4 with $\mathbf{\Delta}_f = \mathbf{\Delta}$, $\mathbf{\Delta}$ given by (11.24), $\Delta G = 0$, $\Gamma_{xx}(x) = \frac{1}{4}V'(x)G_\delta(x)G_\delta^{\mathrm{T}}(x)V'^{\mathrm{T}}(x) + m^{\mathrm{T}}(h_\delta(x))m(h_\delta(x))$, $\Gamma_{xu}(x) = 0$, and $\Gamma_{uu}(x) = 0$. Specifically, (11.128)–(11.131) are satisfied by assumption and (11.144) implies (11.133). Next, if $\delta(\cdot) \in \mathbf{\Delta}$ it follows that

$$\begin{aligned}V'(x)\Delta f(x) - \Gamma_{xx}(x) &= V'(x)G_\delta(x)\delta(h_\delta(x)) - \tfrac{1}{4}V'(x)G_\delta(x)G_\delta^{\mathrm{T}}(x)V'^{\mathrm{T}}(x)\\ &\quad -m^{\mathrm{T}}(h_\delta(x))m(h_\delta(x))\\ &\leq -[\tfrac{1}{2}G_\delta^{\mathrm{T}}(x)V'^{\mathrm{T}}(x) - \delta(h_\delta(x))]^{\mathrm{T}}\\ &\quad \cdot[\tfrac{1}{2}G_\delta^{\mathrm{T}}(x)V'^{\mathrm{T}}(x) - \delta(h_\delta(x))]\\ &\leq 0,\end{aligned}$$

which implies (11.132). The result now follows as a direct consequence of Corollary 11.4. □

We now assign a different structure to the uncertainty set $\mathbf{\Delta}$ and the bounding functions $\Gamma_{xx}(x)$, $\Gamma_{xu}(x)$, and $\Gamma_{uu}(x)$. Specifically, the uncertainty set $\mathbf{\Delta}$ is assumed to be of the form given by (11.27).

Proposition 11.16. Consider the nonlinear uncertain controlled system (11.125) with performance functional (11.127). Assume that there exist a continuously differentiable function $V : \mathbb{R}^n \to \mathbb{R}$ and function

$L_2 : \mathbb{R}^n \to \mathbb{R}^{1\times m}$ such that (11.128)–(11.131) are satisfied,

$$V'(x)[f_0(x) - \tfrac{1}{2}G_0(x)R_2^{-1}(x)V_a(x)] + \tfrac{1}{4}[m(h_\delta(x)) + G_\delta^{\mathrm{T}}(x)V'^{\mathrm{T}}(x)]^{\mathrm{T}} \cdot[m(h_\delta(x)) + G_\delta^{\mathrm{T}}(x)V'^{\mathrm{T}}(x)] + V'(x)G_\delta(x)m_1(h_\delta(x)) < 0, \ x \in \mathbb{R}^n, \ x \neq 0, \tag{11.151}$$

and

$$V(x) \to \infty \text{ as } \|x\| \to \infty, \tag{11.152}$$

where $V_a(x) \triangleq [L_2(x) + V'(x)G_0(x)]^{\mathrm{T}}$. Then the zero solution $x(t) \equiv 0$ of the nonlinear uncertain system (11.125) is globally asymptotically stable for all $\delta(\cdot) \in \mathbf{\Delta}$, where $\mathbf{\Delta}$ is given by (11.27), with the feedback control law

$$\phi(x) = -\tfrac{1}{2}R_2^{-1}(x)V_a(x). \tag{11.153}$$

Furthermore, the performance functional (11.127) satisfies

$$\sup_{\delta(\cdot)\in\mathbf{\Delta}} J(x_0, \phi(x(\cdot))) \leq \mathcal{J}(x_0, \phi(x(\cdot))) = V(x_0), \tag{11.154}$$

where

$$\mathcal{J}(x_0, u(\cdot)) \triangleq \int_0^\infty \Big[L_1(x) + L_2(x)u + u^{\mathrm{T}}R_2(x)u + \tfrac{1}{4}[m(h_\delta(x)) + G_\delta^{\mathrm{T}}(x)V'^{\mathrm{T}}(x)]^{\mathrm{T}}[m(h_\delta(x)) + G_\delta^{\mathrm{T}}(x)V'^{\mathrm{T}}(x)] + V'(x)G_\delta(x)m_1(h_\delta(x))\Big]\mathrm{d}t, \tag{11.155}$$

where $u(\cdot)$ is admissible, and $x(t)$, $t \geq 0$, solves (11.125) with $\delta(h_\delta(x)) \equiv 0$. In addition, the performance functional (11.155), with

$$L_1(x) = \phi^{\mathrm{T}}(x)R_2(x)\phi(x) - V'(x)f_0(x) - V'(x)G_\delta(x)m_1(h_\delta(x)) - \tfrac{1}{4}[m(h_\delta(x)) + G_\delta^{\mathrm{T}}(x)V'^{\mathrm{T}}(x)]^{\mathrm{T}}[m(h_\delta(x)) + G_\delta^{\mathrm{T}}(x)V'^{\mathrm{T}}(x)], \tag{11.156}$$

is minimized in the sense that

$$\mathcal{J}(x_0, \phi(x(\cdot))) = \min_{u(\cdot)\in\mathcal{S}(x_0)} \mathcal{J}(x_0, u(\cdot)). \tag{11.157}$$

Proof. The result is direct consequence of Corollary 11.4 with $\mathbf{\Delta}_f = \mathbf{\Delta}$, $\mathbf{\Delta}$ given by (11.27), $\Delta G = 0$, $\Gamma_{xx}(x) = \tfrac{1}{4}[m(h_\delta(x)) + G_\delta^{\mathrm{T}}(x)V'^{\mathrm{T}}(x)]^{\mathrm{T}}[m(h_\delta(x)) + G_\delta^{\mathrm{T}}(x)V'^{\mathrm{T}}(x)] + V'(x)G_\delta(x)m_1(h_\delta(x))$, $\Gamma_{xu}(x) = 0$, and $\Gamma_{uu}(x) = 0$. Specifically, (11.128)–(11.131) are satisfied by assumption and (11.151) implies (11.133). Next, if $\delta(\cdot) \in \mathbf{\Delta}$ it follows that

$$\begin{aligned} V'(x)\Delta f(x) - \Gamma_{xx}(x) &\leq -[\delta(h_\delta(x)) - m_1(h_\delta(x))]^{\mathrm{T}}[\delta(h_\delta(x)) - m_2(h_\delta(x))] \\ &\quad + V'(x)G_\delta(x)\delta(h_\delta(x)) \end{aligned}$$

$$
\begin{aligned}
& -\tfrac{1}{4}[m(h_\delta(x)) + G_\delta^{\mathrm{T}}(x)V'^{\mathrm{T}}(x)]^{\mathrm{T}}[m(h_\delta(x)) + G_\delta^{\mathrm{T}}(x)V'^{\mathrm{T}}(x)] \\
& -V'(x)G_\delta(x)m_1(h_\delta(x)) \\
= & -[\tfrac{1}{2}m(h_\delta(x)) + \tfrac{1}{2}G_\delta^{\mathrm{T}}(x)V'^{\mathrm{T}}(x) - (\delta(h_\delta(x) - m_1(h_\delta(x)))]^{\mathrm{T}} \\
& \cdot[\tfrac{1}{2}m(h_\delta(x)) + \tfrac{1}{2}G_\delta^{\mathrm{T}}(x)V'^{\mathrm{T}}(x) - (\delta(h_\delta(x) - m_1(h_\delta(x)))] \\
\leq & \ 0,
\end{aligned}
$$

which implies (11.132). The result now follows as a direct consequence of Corollary 11.4. □

11.7 Stability Margins, Meaningful Inverse Optimality, and Robust Control Lyapunov Functions

In this section, we specialize the results of Section 11.5 to the case where $L(x,u)$ is nonnegative for all $(x,u) \in \mathbb{R}^n \times \mathbb{R}^m$. Here, we assume that $L_2(x) \equiv 0$ and $L_1(x) \geq 0$, $x \in \mathbb{R}^n$. We begin by specializing Corollary 11.4 to the case where $L_2(x) \equiv 0$. In this case, the performance functional (11.127) becomes

$$
J(x_0, u(\cdot)) = \int_0^\infty [L_1(x(t)) + u^{\mathrm{T}}(t)R_2(x(t))u(t)]\mathrm{d}t. \qquad (11.158)
$$

For simplicity of exposition we assume $\Delta G(x) \equiv 0$, $\Gamma_{xu}(x) \equiv 0$, and $\Gamma_{uu}(x) \equiv 0$.

Corollary 11.5. Consider the nonlinear controlled dynamical system (11.125) with performance functional (11.158). Assume that there exist a continuously differentiable function $V : \mathbb{R}^n \to \mathbb{R}$ and a function $\Gamma_{xx} : \mathbb{R}^n \to \mathbb{R}$ such that

$$
V(0) = 0, \qquad (11.159)
$$

$$
V(x) > 0, \quad x \in \mathbb{R}^n, \quad x \neq 0, \qquad (11.160)
$$

$$
V'(x)\Delta f(x) \leq \Gamma_{xx}(x), \quad \Delta f(\cdot) \in \boldsymbol{\Delta}, \qquad (11.161)
$$

$$
0 = V'(x)f_0(x) + L_1(x) - \tfrac{1}{4}V'(x)G(x)R_2^{-1}(x)G^{\mathrm{T}}(x)V'^{\mathrm{T}}(x) + \Gamma_{xx}(x), \quad x \in \mathbb{R}^n, \qquad (11.162)
$$

and

$$
V(x) \to \infty \text{ as } \|x\| \to \infty. \qquad (11.163)
$$

Furthermore, assume that the system (11.125), with output $y = L_1(x)$, is zero-state observable. Then the zero solution $x(t) \equiv 0$ of the uncertain closed-loop system

$$
\dot{x}(t) = f_0(x(t)) + \Delta f(x) + G(x(t))\phi(x(t)), \qquad x(0) = x_0, \qquad t \geq 0, \qquad (11.164)
$$

is globally asymptotically stable for all $\Delta f(\cdot) \in \mathbf{\Delta}$ with feedback control law

$$\phi(x) = -\tfrac{1}{2}R_2^{-1}(x)G^{\mathrm{T}}(x)V'^{\mathrm{T}}(x). \tag{11.165}$$

Furthermore, the performance functional (11.158) satisfies

$$J(x_0, \phi(x(\cdot))) \le \mathcal{J}(x_0, \phi(x(\cdot))) = V(x_0), \tag{11.166}$$

where

$$\mathcal{J}(x_0, u(\cdot)) \triangleq \int_0^\infty [L(x(t), u(t)) + \Gamma_{xx}(x(t))]\mathrm{d}t. \tag{11.167}$$

In addition, the performance functional (11.167) is minimized in the sense that

$$\mathcal{J}(x_0, \phi(x(\cdot))) = \min_{u(\cdot)\in\mathcal{S}(x_0)} \mathcal{J}(x_0, u(\cdot)). \tag{11.168}$$

Proof. The result follows as a direct consequence of Corollary 11.4. □

Next, we provide sector and gain margins for the nonlinear dynamical system $\mathcal{G}$ given by (11.125). To consider relative stability margins for nonlinear robust regulators consider the nonlinear dynamical system given by (11.125) along with the output

$$y(t) = -\phi(x(t)), \tag{11.169}$$

where $\phi(\cdot)$ is such that $\mathcal{G}$ is robustly stable for all $\Delta f(\cdot) \in \mathbf{\Delta}$ with $u = \phi(x)$. Furthermore, assume that (11.125) and (11.169) is zero-state observable.

Theorem 11.3. Consider the nonlinear dynamical system $\mathcal{G}$ given by (11.125) and (11.169), where $\phi(x)$ is a feedback control law given by (11.165) and where $V(x)$, $x \in \mathbb{R}^n$, satisfies (11.159)–(11.162). Furthermore, assume $R_2(x) = \mathrm{diag}[r_1(x), \ldots, r_m(x)]$, where $r_i : \mathbb{R}^n \to \mathbb{R}$, $r_i(x) > 0$, $i = 1, \ldots, m$. Then the nonlinear system $\mathcal{G}$ has a sector (and, hence, gain) margin $(\frac{1}{2}, \infty)$.

Proof. Let $\Delta(u_{\mathrm{c}}) = \sigma(u_{\mathrm{c}})$, where $\sigma : \mathbb{R}^m \to \mathbb{R}^m$ is a static nonlinearity such that $\sigma(0) = 0$, $\sigma(u_{\mathrm{c}}) = [\sigma_1(u_{\mathrm{c}1}), \ldots, \sigma_m(u_{\mathrm{c}m})]^{\mathrm{T}}$, and $\frac{1}{2}u_{\mathrm{c}i}^2 < \sigma_i(u_{\mathrm{c}i})u_{\mathrm{c}i}$, for all $u_{\mathrm{c}i} \neq 0$, $i = 1, \ldots, m$. In this case, the closed-loop system (11.125) and (11.169) with $u = -\sigma(y)$ is given by

$$\begin{aligned} \dot{x}(t) &= f_0(x(t)) + \Delta f(x(t)) + G(x(t))\sigma(\phi(x(t))), \\ &\qquad x(0) = x_0, \quad \Delta f(\cdot) \in \mathbf{\Delta}, \quad t \ge 0. \end{aligned} \tag{11.170}$$

Next, consider the Lyapunov function candidate $V(x)$, $x \in \mathbb{R}^n$, satisfying (11.162) and let $\dot{V}(x)$ denote the Lyapunov derivative along with the trajectories of the closed-loop system (11.170). Now, it follows from (11.161) and (11.162) that for all $\Delta f(\cdot) \in \mathbf{\Delta}$,

$$\begin{aligned} \dot{V}(x) &= V'(x)f_0(x) + V'(x)\Delta f(x) + V'(x)G(x)\sigma(\phi(x)) \\ &\le \phi^{\mathrm{T}}(x)R_2^{-1}(x)\phi(x) + V'(x)G(x)\sigma(\phi(x)) + V'(x)\Delta f(x) \end{aligned}$$

$$
\begin{aligned}
&-L_1(x) - \Gamma_{xx}(x) \\
&\leq \phi^{\mathrm{T}}(x)R_2^{-1}(x)\phi(x) + V'(x)G(x)\sigma(\phi(x)) \\
&\leq \sum_{i=1}^{m} r_i(x)y_i(y_i + 2\sigma_i(-y_i)),
\end{aligned}
$$

which implies the result. □

The following result specializes Theorem 11.3 to linear uncertain systems.

Corollary 11.6. Consider the uncertain linear dynamical system $\mathcal{G}$ given by (11.109) and (11.169) with $\Delta B \equiv 0$, where $\phi(x) = -(R_2 + \Omega_{uu}(P))^{-1}(B^{\mathrm{T}}P + \Omega_{xu}^{\mathrm{T}}(P))x$ and where $P \in \mathbb{P}^n$, satisfies (11.111) and (11.112). Furthermore, assume $R_2 = \mathrm{diag}[r_1, \ldots, r_m]$, where $r_i > 0$, $i = 1, \ldots, m$. Then the linear system $\mathcal{G}$ has a sector (and, hence, gain) margin $(\frac{1}{2}, \infty)$.

Next, we introduce the notion of *robust control Lyapunov functions* for the nonlinear dynamical system (11.125).

Definition 11.1. Consider the controlled nonlinear dynamical system given by (11.125). A continuously differentiable positive-definite function $V : \mathbb{R}^n \to \mathbb{R}$ satisfying

$$
V'(x)f_0(x) + \Gamma_{xx}(x) < 0, \quad x \in \mathcal{R}, \tag{11.171}
$$

where $\mathcal{R} \triangleq \{x \in \mathbb{R}^n : x \neq 0 : V'(x)G(x) = 0\}$, is called a *robust control Lyapunov function.*

Finally, we show that for every nonlinear dynamical system for which a robust control Lyapunov function can be constructed there exists an inverse optimal robust feedback control law with sector and gain margin $(\frac{1}{2}, \infty)$.

Theorem 11.4. Consider the nonlinear dynamical system $\mathcal{G}$ given by (11.125) and let the continuously differentiable positive-definite, radially unbounded function $V : \mathbb{R}^n \to \mathbb{R}$ be a robust control Lyapunov function of (11.125), that is,

$$
V'(x)f_0(x) + \Gamma_{xx}(x) < 0, \quad x \in \mathcal{R}, \tag{11.172}
$$

where $\mathcal{R} \triangleq \{x \in \mathbb{R}^n : x \neq 0 : V'(x)G(x) = 0\}$. Then, with the feedback control law given by

$$
\phi(x) = \begin{cases} -\left(c_0 + \frac{\alpha(x) + \sqrt{\alpha^2(x) + (\beta^{\mathrm{T}}(x)\beta(x))^2}}{\beta^{\mathrm{T}}(x)\beta(x)}\right)\beta(x), & \beta(x) \neq 0, \\ 0, & \beta(x) = 0, \end{cases} \tag{11.173}
$$

where $\alpha(x) \triangleq V'(x)f_0(x) + \Gamma_{xx}(x)$, $\beta(x) \triangleq G^{\mathrm{T}}(x)V'^{\mathrm{T}}(x)$, and $c_0 > 0$, the nonlinear system $\mathcal{G}$ given by (11.125) has a sector (and, hence, gain) margin $(\frac{1}{2}, \infty)$.

Proof. The result is a direct consequence of Corollary 11.5 and Theorem 11.3 with $R_2(x) = \frac{1}{2\eta(x)}I_m$ and $L_1(x) = -\alpha(x) + \frac{\eta(x)}{2}\beta^{\mathrm{T}}(x)\beta(x)$, where

$$\eta(x) = \begin{cases} -\left(c_0 + \frac{\alpha(x)+\sqrt{\alpha^2(x)+(\beta^{\mathrm{T}}(x)\beta(x))^2}}{\beta^{\mathrm{T}}(x)\beta(x)}\right), & \beta(x) \neq 0, \\ 0, & \beta(x) = 0. \end{cases} \tag{11.174}$$

Specifically, note that $R_2(x) > 0$, $x \in \mathbb{R}^n$, and

$$\begin{aligned} L_1(x) &= -\alpha(x) + \tfrac{\eta(x)}{2}\beta^{\mathrm{T}}(x)\beta(x) \\ &= \begin{cases} -\frac{1}{2}\Big(c_0\beta^{\mathrm{T}}(x)\beta(x) - \alpha(x) \\ \quad +\sqrt{\alpha^2(x) + (\beta^{\mathrm{T}}(x)\beta(x))^2}\Big), & \beta(x) \neq 0, \\ -\alpha(x), & \beta(x) = 0. \end{cases} \end{aligned} \tag{11.175}$$

Now, it follows from (11.175) that $L_1(x) \geq 0$, $\beta(x) \neq 0$, and since $V(\cdot)$ is a robust control Lyapunov function of (11.125), it follows that $L_1(x) = -\alpha(x) \geq 0$, for all $x \in \mathcal{R}$. Hence, (11.175) yields $L_1(x) \geq 0$, $x \in \mathbb{R}^n$, so that all the conditions of Corollary 11.5 are satisfied. □

11.8 Robust Nonlinear Controllers with Polynomial Performance Criteria

In this section, we specialize the results of Section 11.5 to linear systems controlled by nonlinear controllers that minimize a polynomial cost functional. Specifically, assume $\mathcal{F}$ to be the set of uncertain systems given by

$$\mathcal{F} = \{(A+\Delta A)x + Bu : x \in \mathbb{R}^n, u \in \mathbb{R}^m, A \in \mathbb{R}^{n\times n}, B \in \mathbb{R}^{n\times m}, \Delta A \in \boldsymbol{\Delta}_A\}, \tag{11.176}$$

where $\boldsymbol{\Delta}_A \subset \mathbb{R}^{n\times n}$ is a given bounded uncertainty set of the uncertain perturbation ΔA of the nominal system A such that $0 \in \boldsymbol{\Delta}_A$. For simplicity of exposition here and in the remainder of the chapter we assume $\Delta B = 0$. For the following result recall the definition of S and let $R_1 \in \mathbb{P}^n$, $R_2 \in \mathbb{P}^m$, and $\hat{R}_k \in \mathbb{N}^n$, $k = 2, \ldots, r$, be given where r is a positive integer.

Theorem 11.5. Consider the linear uncertain controlled system

$$\dot{x}(t) = (A + \Delta A)x(t) + Bu(t), \qquad x(0) = x_0, \qquad t \geq 0, \tag{11.177}$$

where $u(\cdot) \in \mathcal{U}$ is admissible and $\Delta A \in \boldsymbol{\Delta}_A$. Assume there exists $\Omega : \mathbb{P}^n \to \mathbb{N}^n$ such that

$$\Delta A^{\mathrm{T}}P + P\Delta A \leq \Omega(P), \qquad \Delta A \in \boldsymbol{\Delta}_A, \quad P \in \mathbb{P}^n, \tag{11.178}$$

and there exist $P \in \mathbb{P}^n$ and $M_k \in \mathbb{N}^n$, $k = 2, \ldots, r$, such that

$$0 = A^{\mathrm{T}}P + PA + R_1 + \Omega(P) - PSP, \tag{11.179}$$

and

$$0 = (A - SP)^{\mathrm{T}}M_k + M_k(A - SP) + \hat{R}_k + \Omega(M_k), \quad k = 2, \ldots, r. \tag{11.180}$$

Then, with the feedback control law

$$u = \phi(x) = -R_2^{-1}B^{\mathrm{T}}\left(P + \sum_{k=2}^{r}(x^{\mathrm{T}}M_k x)^{k-1}M_k\right)x,$$

the zero solution $x(t) \equiv 0$ of the uncertain system (11.177) is globally asymptotically stable for all $x_0 \in \mathbb{R}^n$ and $\Delta A \in \boldsymbol{\Delta}_A$, and the performance functional (11.127) satisfies

$$\sup_{\Delta A \in \boldsymbol{\Delta}_A} J_{\Delta A}(x_0, \phi(x(\cdot))) \le \mathcal{J}(x_0, \phi(x(\cdot))) = x_0^{\mathrm{T}}Px_0 + \sum_{k=2}^{r}\frac{1}{k}(x_0^{\mathrm{T}}M_k x_0)^k, \tag{11.181}$$

where

$$\mathcal{J}(x_0, u(\cdot)) \triangleq \int_0^\infty [L(x, u) + \Gamma(x, u)]\mathrm{d}t, \tag{11.182}$$

and where $u(\cdot)$ is admissible, and $x(t)$, $t \ge 0$, solves (11.177) with $\Delta A = 0$, and

$$\Gamma(x, u) = x^{\mathrm{T}}(\Omega(P) + \sum_{k=2}^{r}(x^{\mathrm{T}}M_k x)^{k-1}\Omega(M_k))x,$$

where $u(\cdot)$ is admissible and $\Delta A \in \boldsymbol{\Delta}_A$. In addition, the performance functional (11.127), with $R_2(x) = R_2$, $L_2(x) = 0$, and

$$\begin{aligned} L_1(x) = x^{\mathrm{T}}\Bigg(&R_1 + \sum_{k=2}^{r}(x^{\mathrm{T}}M_k x)^{k-1}\hat{R}_k + \left[\sum_{k=2}^{r}(x^{\mathrm{T}}M_k x)^{k-1}M_k\right]^{\mathrm{T}} \\ &\cdot S\left[\sum_{k=2}^{r}(x^{\mathrm{T}}M_k x)^{k-1}M_k\right]\Bigg)x, \end{aligned}$$

is minimized in the sense that

$$\mathcal{J}(x_0, \phi(x(\cdot))) = \min_{u(\cdot) \in \mathcal{S}(x_0)} \mathcal{J}(x_0, u(\cdot)), \tag{11.183}$$

where $\mathcal{S}(x_0)$ is the set of regulation controllers for the nominal system and $x_0 \in \mathbb{R}^n$.

Proof. The result is a direct consequence of Corollary 11.4 with $f_0(x) = Ax$, $G_0(x) = B$, $\Delta f(x) = \Delta Ax$, $\Delta G(x) = 0$, $\boldsymbol{\Delta} = \boldsymbol{\Delta}_A$, and $V(x) = x^{\mathrm{T}}Px + \sum_{k=2}^{r} \frac{1}{k}(x^{\mathrm{T}}M_kx)^k$. Specifically, conditions (11.128)–(11.162) are trivially satisfied. Now, it follows from (11.178) that

$$0 \geq x^{\mathrm{T}}(\Delta A^{\mathrm{T}}P + P\Delta A - \Omega(P))x$$
$$+\sum_{k=2}^{r}(x^{\mathrm{T}}M_kx)^{k-1}x^{\mathrm{T}}(\Delta A^{\mathrm{T}}M_k + M_k\Delta A - \Omega(M_k))x,$$

$x \in \mathbb{R}^n$, which implies (11.133) for all $\Delta A \in \mathbf{\Delta}_A$ so that all the conditions of Corollary 11.4 are satisfied. □

Theorem 11.5 generalizes the deterministic version of the stochastic nonlinear-nonquadratic optimal control problem considered in [411] to the robustness setting. Furthermore, unlike the results of [411], Theorem 11.5 is not limited to sixth order cost functionals and cubic nonlinear controllers since it addresses a polynomial nonlinear performance criterion. Theorem 11.5 requires the solutions of $r-1$ modified Riccati equations in (11.180) to obtain the optimal robust controller. However, if $\hat{R}_k = \hat{R}_2$, $k = 3, \ldots, r$, then $M_k = M_2$, $k = 3, \ldots, r$, satisfies (11.180). In this case, we require the solution of one modified Riccati equation in (11.180). This special case is considered in Propositions 11.17 and 11.18 below.

As discussed in Chapter 9, the performance functional (11.182) is a derived performance functional in the sense that it cannot be arbitrarily specified. However, this performance functional does weight the state variables by arbitrary even powers. Furthermore, (11.182) has the form

$$J_{\Delta A}(x_0, u(\cdot)) = \int_0^\infty \Big[x^{\mathrm{T}}(R_1 + \sum_{k=2}^{r}(x^{\mathrm{T}}M_kx)^{k-1}\hat{R}_k)x + u^{\mathrm{T}}R_2u$$
$$+\phi_{\mathrm{NL}}^{\mathrm{T}}(x)R_2\phi_{\mathrm{NL}}(x)\Big]\mathrm{d}t,$$

where $\phi_{\mathrm{NL}}(x)$ is the nonlinear part of the optimal feedback control

$$\phi(x) = \phi_{\mathrm{L}}(x) + \phi_{\mathrm{NL}}(x),$$

where $\phi_{\mathrm{L}}(x) \triangleq -R_2^{-1}B^{\mathrm{T}}Px$ and $\phi_{\mathrm{NL}}(x) \triangleq -R_2^{-1}B^{\mathrm{T}}\sum_{k=2}^{r}(x^{\mathrm{T}}M_kx)^{k-1}M_kx$.

Next, we consider the special case in which $r = 2$. In this case, note that if there exist $P \in \mathbb{P}^n$ and $M_2 \in \mathbb{N}^n$ such that

$$0 = A^{\mathrm{T}}P + PA + R_1 + \Omega(P) - PSP$$

and

$$0 = (A - SP)^{\mathrm{T}}M_2 + M_2(A - SP) + \hat{R}_2 + \Omega(M_2),$$

then (11.177), with the performance functional

$$J_{\Delta A}(x_0, u(\cdot)) = \int_0^\infty [x^{\mathrm{T}}R_1x + u^{\mathrm{T}}R_2u + (x^{\mathrm{T}}M_2x)(x^{\mathrm{T}}\hat{R}_2x)$$

$$+(x^{\mathrm{T}}M_2x)^2(x^{\mathrm{T}}M_2SM_2x)]\mathrm{d}t,$$

is globally asymptotically stable for all $x_0 \in \mathbb{R}^n$ and $\Delta A \in \mathbf{\Delta}_A$ with the feedback control law $u = \phi(x) = -R_2^{-1}B^{\mathrm{T}}\left(P + (x^{\mathrm{T}}M_2x)M_2\right)x$.

Finally, using the explicit uncertainty characterizations given by (11.121) and (11.122) (with $\Delta B = 0$) we present two specializations of Theorem 11.5.

Proposition 11.17. Consider the linear uncertain controlled system

$$\dot{x}(t) = (A + \Delta A)x(t) + Bu(t), \qquad x(0) = x_0, \qquad t \geq 0, \tag{11.184}$$

where $u(\cdot) \in \mathcal{U}$ is admissible and $\Delta A \in \mathbf{\Delta}_A$, where $\mathbf{\Delta}_A$ is given by (11.121). Assume that there exist $P \in \mathbb{P}^n$ and $M_2 \in \mathbb{N}^n$ such that

$$0 = A_\alpha^{\mathrm{T}}P + PA_\alpha + R_1 + \sum_{i=1}^{p}\frac{\alpha_i^2}{\alpha}A_i^{\mathrm{T}}PA_i - PSP, \tag{11.185}$$

$$0 = (A_\alpha - SP)^{\mathrm{T}}M_2 + M_2(A_\alpha - SP) + \sum_{i=1}^{p}\frac{\alpha_i^2}{\alpha}A_i^{\mathrm{T}}M_2A_i + \hat{R}_2. \tag{11.186}$$

Then, with the feedback control

$$u = \phi(x) = -R_2^{-1}B^{\mathrm{T}}\left(P + \sum_{k=2}^{r}(x^{\mathrm{T}}M_2x)^{k-1}M_2\right)x,$$

the zero solution $x(t) \equiv 0$ to the uncertain system (11.184) is globally asymptotically stable for all $x_0 \in \mathbb{R}^n$ and $\Delta A \in \mathbf{\Delta}_A$, and the performance functional (11.127) satisfies

$$\sup_{\Delta A \in \mathbf{\Delta}_A} J_{\Delta A}(x_0, \phi(x(\cdot))) \leq \mathcal{J}(x_0, \phi(x(\cdot))) = x_0^{\mathrm{T}}Px_0 + \sum_{k=2}^{r}\frac{1}{k}(x_0^{\mathrm{T}}M_2x_0)^k, \tag{11.187}$$

where

$$\mathcal{J}(x_0, u(\cdot)) \triangleq \int_0^\infty [L(x,u) + \Gamma(x,u)]\mathrm{d}t, \tag{11.188}$$

and where $u(\cdot)$ is admissible, and $x(t)$, $t \geq 0$, solves (11.184) with $\Delta A = 0$ and

$$\begin{aligned}\Gamma(x,u) &= x^{\mathrm{T}}\left[\sum_{i=1}^{p}\frac{\alpha_i^2}{\alpha}A_i^{\mathrm{T}}PA_i + \alpha P\right.\\ &\quad\left. + \sum_{k=2}^{r}(x^{\mathrm{T}}M_2x)^{k-1}\left(\sum_{i=1}^{p}\frac{\alpha_i^2}{\alpha}A_i^{\mathrm{T}}M_2A_i + \alpha M_2\right)\right]x.\end{aligned}$$

In addition, the performance functional (11.127), with $R_2(x) = R_2$, $L_2(x) =$

0, and

$$L_1(x) = x^{\mathrm{T}} \left[R_1 + \sum_{k=2}^{r} (x^{\mathrm{T}} M_2 x)^{k-1} \hat{R}_2 + \left(\sum_{k=2}^{r} (x^{\mathrm{T}} M_2 x)^{k-1} \right)^2 M_2 S M_2 \right] x, \tag{11.189}$$

is minimized in the sense that

$$\mathcal{J}(x_0, \phi(x(\cdot))) = \min_{u(\cdot) \in \mathcal{S}(x_0)} \mathcal{J}(x_0, u(\cdot)), \tag{11.190}$$

where $\mathcal{S}(x_0)$ is the set of regulation controllers for the nominal system and $x_0 \in \mathbb{R}^n$.

Proof. It need only be noted that $\Delta A^{\mathrm{T}} P + P \Delta A \leq \sum_{i=1}^{p} \frac{\alpha_i^2}{\alpha} A_i^{\mathrm{T}} P A_i + \alpha P$ for all $\Delta A \in \mathbf{\Delta}_A$ and $P \in \mathbb{P}^n$. The result now is a direct consequence of Theorem 11.5 with $\Omega(P) = \sum_{i=1}^{p} \frac{\alpha_i^2}{\alpha} A_i^{\mathrm{T}} P A_i + \alpha P$. □

Proposition 11.18. Consider the linear uncertain controlled system

$$\dot{x}(t) = (A + \Delta A) x(t) + B u(t), \qquad x(0) = x_0, \qquad t \geq 0, \tag{11.191}$$

where $u(\cdot) \in \mathcal{U}$ is admissible and $\Delta A \in \mathbf{\Delta}_A$, where $\mathbf{\Delta}_A$ is given by (11.122). Assume that there exist $P \in \mathbb{P}^n$ and $M_2 \in \mathbb{N}^n$ such that

$$0 = A^{\mathrm{T}} P + P A + R_1 + C_0^{\mathrm{T}} N C_0 + P B_0 B_0^{\mathrm{T}} P - P S P, \tag{11.192}$$

$$0 = (A - SP)^{\mathrm{T}} M_2 + M_2 (A - SP) + \hat{R}_2 + C_0^{\mathrm{T}} N C_0 + M_2 B_0 B_0^{\mathrm{T}} M_2. \tag{11.193}$$

Then, with the feedback control

$$u = \phi(x) = -R_2^{-1} B^{\mathrm{T}} \left(P + \sum_{k=2}^{r} (x^{\mathrm{T}} M_2 x)^{k-1} M_2 \right) x,$$

the zero solution $x(t) \equiv 0$ to the uncertain system (11.191) is globally asymptotically stable for all $x_0 \in \mathbb{R}^n$ and $\Delta A \in \mathbf{\Delta}_A$, and the performance functional (11.127) satisfies

$$\sup_{\Delta A \in \mathbf{\Delta}_A} J_{\Delta A}(x_0, \phi(x(\cdot))) \leq \mathcal{J}(x_0, \phi(x(\cdot))) = x_0^{\mathrm{T}} P x_0 + \sum_{k=2}^{r} \frac{1}{k} (x_0^{\mathrm{T}} M_2 x_0)^k, \tag{11.194}$$

where

$$\mathcal{J}(x_0, u(\cdot)) \triangleq \int_0^{\infty} [L(x, u) + \Gamma(x, u)] \mathrm{d}t, \tag{11.195}$$

and where $u(\cdot)$ is admissible, and $x(t)$, $t \geq 0$, solves (11.191) with $\Delta A = 0$ and

$$\Gamma(x, u) = x^{\mathrm{T}} \left[C_0^{\mathrm{T}} N C_0 + P B_0 B_0^{\mathrm{T}} P \right.$$

$$+\sum_{k=2}^{r}(x^{\mathrm{T}}M_2x)^{k-1}(C_0^{\mathrm{T}}NC_0+M_2B_0B_0^{\mathrm{T}}M_2)\Bigg]x.$$

In addition, the performance functional (11.127), with $R_2(x)=R_2$, $L_2(x)=0$, and

$$L_1(x)=x^{\mathrm{T}}\left[R_1+\sum_{k=2}^{r}(x^{\mathrm{T}}M_2x)^{k-1}\hat{R}_2+\left(\sum_{k=2}^{r}(x^{\mathrm{T}}M_2x)^{k-1}\right)^2M_2SM_2\right]x, \tag{11.196}$$

is minimized in the sense that

$$\mathcal{J}(x_0,\phi(x(\cdot)))=\min_{u(\cdot)\in\mathcal{S}(x_0)}\mathcal{J}(x_0,u(\cdot)), \tag{11.197}$$

where $\mathcal{S}(x_0)$ is the set of regulation controllers for the nominal system and $x_0\in\mathbb{R}^n$.

Proof. It need only be noted that $\Delta A^{\mathrm{T}}P+P\Delta A\leq C_0^{\mathrm{T}}NC_0+PB_0B_0^{\mathrm{T}}P$ for all $\Delta A\in\mathbf{\Delta}_A$ and $P\in\mathbb{P}^n$. The result now is a direct consequence of Theorem 11.5 with $\Omega(P)=C_0^{\mathrm{T}}NC_0+PB_0B_0^{\mathrm{T}}P$. □

Propositions 11.17 and 11.18 are generalizations of results given in [48, 50] to nonlinear polynomial performance criteria.

Example 11.7. Consider the three-mass, two-spring system shown in Figure 11.3 with $m_1=m_2=m_3=1$ and uncertain spring stiffnesses k_1 and k_2. A control force acts on mass 2. The nominal dynamics, with state variables defined in Figure 11.3, are given by

$$A=\begin{bmatrix}0&0&0&1&0&0\\0&0&0&0&1&0\\0&0&0&0&0&1\\-k_{1\mathrm{nom}}&k_{1\mathrm{nom}}&0&0&0&0\\k_{1\mathrm{nom}}&-(k_{1\mathrm{nom}}+k_{2\mathrm{nom}})&k_{2\mathrm{nom}}&0&0&0\\0&k_{2\mathrm{nom}}&-k_{2\mathrm{nom}}&0&0&0\end{bmatrix},\quad B=\begin{bmatrix}0\\0\\0\\0\\1\\0\end{bmatrix}. \tag{11.198}$$

The actual spring stiffnesses can be written as $k_i=k_{i\mathrm{nom}}+\Delta k_i$, where $k_{i\mathrm{nom}}=1$, $i=1,2$, so that the actual dynamics of the system are given by $A_{\mathrm{actual}}=A+\sum_{i=1}^{2}\Delta k_iA_i$, where

$$A_1=\begin{bmatrix}0&0&0&0&0&0\\0&0&0&0&0&0\\0&0&0&0&0&0\\-1&1&0&0&0&0\\1&-1&0&0&0&0\\0&0&0&0&0&0\end{bmatrix},\qquad A_2=\begin{bmatrix}0&0&0&0&0&0\\0&0&0&0&0&0\\0&0&0&0&0&0\\0&0&0&0&0&0\\0&-1&1&0&0&0\\0&1&-1&0&0&0\end{bmatrix}.$$

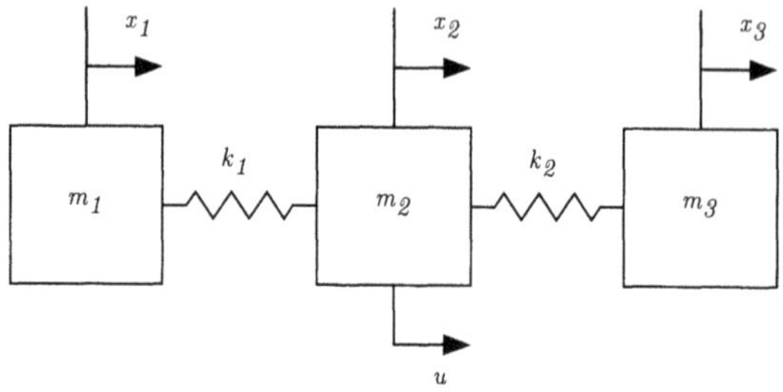

Figure 11.3 Three-mass oscillator.

Choosing the performance variables

$$z \triangleq E_1 x + E_2 u,$$

where

$$E_1 = \begin{bmatrix} 0 & 0 & 1 & 0 & 0 & 0 \\ 0 & 0 & 0 & 0 & 0 & 1 \end{bmatrix}, \qquad E_2 = \begin{bmatrix} 0 \\ 1 \end{bmatrix},$$

expresses the desire to regulate the displacement and velocity of mass 3. Furthermore, let the design parameters $R_1 = E_1^{\mathrm{T}} E_1$ and $R_2 = \rho E_2^{\mathrm{T}} E_2$, where $\rho = 0.001$. For the nonlinear control design, the additional design parameter $\hat{R}_2$ in (11.186) is taken as $\hat{R}_2 = 10 \cdot R_1$. The design goal of this problem is to achieve good nominal performance and demonstrate robust stability and performance for perturbed spring stiffness values in the range $0.75 \le k_i \le 1.25$, $i = 1, 2$.

Figures 11.4 and 11.5 compares the linear LQR controller to the nonlinear Speyer [411] controller (Proposition 11.17 with $\Delta A = 0$ and $r = 2$) for the nominal plant subject to an initial displacement of mass 1. Note that the nonlinear controller achieves better performance in the sense of state trajectory regulation for each of the states. This is primarily due to the relatively large initial displacement of mass 1 which allowed the nonlinear part of the control to initially have a significant impact on the response causing the position of mass 1 to approach zero faster than the linear controller design. The action of the nonlinear part of the controller also reduced the overshoot since the nonlinear control contribution becomes greater as the position deviates farther from the equilibrium thus causing the mass to recover quicker.

Next, using Proposition 11.17 (with $r = 2$) a robustified nonlinear controller was designed for the uncertain system with multivariable uncertainty in the stiffness values. This controller is compared to the robustified linear LQR controller given by (11.120), the Speyer [411] nonlinear controller, and the LQR controller. Figures 11.6 and 11.7 shows the state responses for these designs for $\Delta k_1 = -0.25$ and $\Delta k_2 = 0.25$. Note that the robust nonlinear controller outperforms all other controllers in the sense of worst-case state

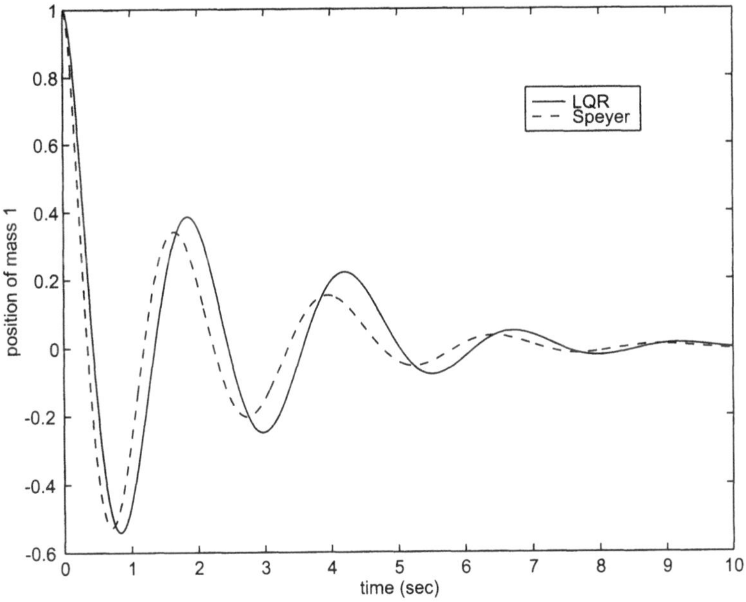

Figure 11.4 Comparison of LQR and Speyer controllers for the nominal system: Position of mass 1.

trajectory regulation. △

Example 11.8. Consider the pitch axis longitudinal dynamics model of the F-16 fighter aircraft system given in [394] for nominal flight conditions at 3000 ft and Mach number of 0.6. The nominal model is given by

$$\begin{bmatrix} \dot{x}_1(t) \\ \dot{x}_2(t) \\ \dot{x}_3(t) \end{bmatrix} = \begin{bmatrix} 0 & 1.00 & 0 \\ 0 & -0.87 & 43.22 \\ 0 & 0.99 & -1.34 \end{bmatrix} \begin{bmatrix} x_1(t) \\ x_2(t) \\ x_3(t) \end{bmatrix} + \begin{bmatrix} 0 & 0 \\ -17.25 & -1.58 \\ -0.17 & -0.25 \end{bmatrix} \begin{bmatrix} u_1(t) \\ u_2(t) \end{bmatrix}, \tag{11.199}$$

where x_1 is the pitch angle, x_2 is the pitch rate, x_3 is the angle of attack, u_1 is the elevator deflection, and u_2 is the flaperon deflection. Here, we consider uncertainty in the (2,2) and (3,3) entries of the dynamics matrix. Using the uncertainty structure given by (11.122), the actual dynamics are given by $A_{\text{actual}} = A + B_0 F C_0$, where

$$B_0 = \begin{bmatrix} 0 & 0 \\ 1 & 0 \\ 0 & 1 \end{bmatrix}, \quad F = \begin{bmatrix} f_1 & 0 \\ 0 & f_2 \end{bmatrix}, \quad C_0 = \begin{bmatrix} 0 & 1 & 0 \\ 0 & 0 & 1 \end{bmatrix}.$$

Using Proposition 11.18, with $R_1 = 3 \cdot I_3$, $R_2 = 0.001 \cdot I_2$, $\hat{R}_2 = 10 \cdot R_1$,

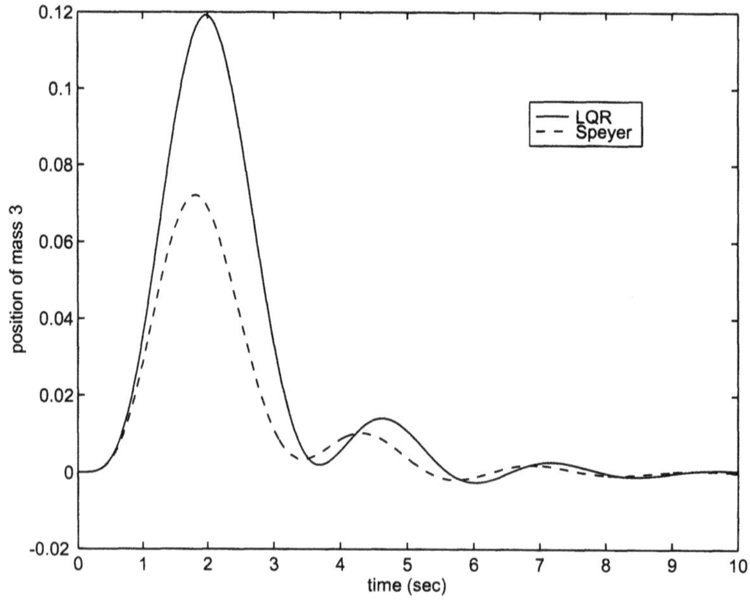

Figure 11.5 Comparison of LQR and Speyer controllers for the nominal system: Position of mass 3.

$|f_1| \leq 1$, $|f_2| \leq 5$, and $r = 2$, a robustified nonlinear controller was designed for the uncertain system. This controller is compared with the robustified linear controller given by (11.123), the Speyer [411] nonlinear controller, and the LQR controller. Figures 11.8 and 11.9 shows that for the case where $f_1 = -1$ and $f_2 = 5$ the LQR controller destabilizes the system while the nominal Speyer [411] controller maintains Lyapunov stability. This demonstrates the inherent robustness of the nominal (nonrobustified) nonlinear control in comparison to the nominal linear control. Furthermore, for the same uncertainty range Figures 11.10 and 11.11 shows the state response for the robustified nonlinear controller (Proposition 11.18) and the robustified linear controller obtained via (11.123). $\triangle$

11.9 Robust Nonlinear Controllers with Multilinear Performance Criteria

In this section, we specialize the results of Section 11.5 to linear uncertain systems controlled by nonlinear controllers that minimize a multilinear cost functional. Specifically, we assume $\mathcal{F}$ to be the set of uncertain linear systems given by (11.176). For the following result recall the definition of S and let $R_1 \in \mathbb{P}^n$, $R_2 \in \mathbb{P}^m$, and $\hat{R}_{2\nu} \in \mathcal{N}^{(2\nu,n)}$, $\nu = 2, \ldots, r$, be given where r is a given integer.

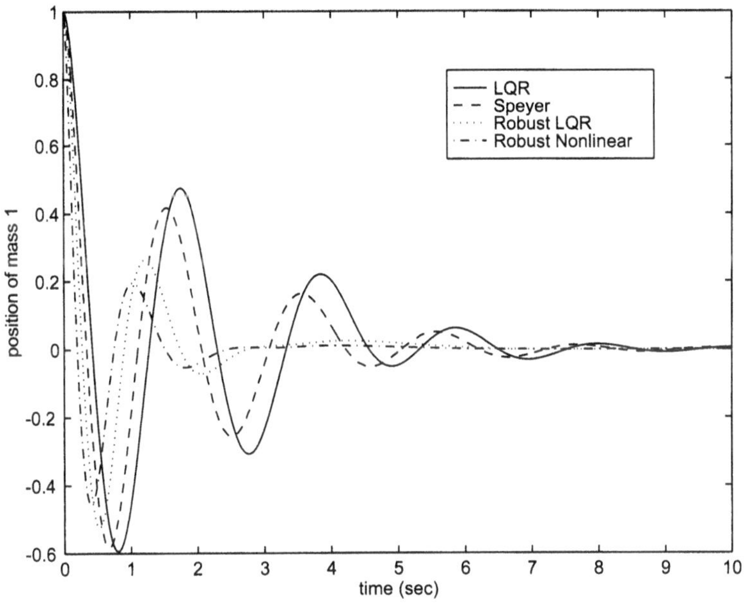

Figure 11.6 Comparison of LQR, Speyer, robust LQR, and robust nonlinear controllers for the uncertain system: Position of mass 1.

Theorem 11.6. Consider the linear uncertain controlled system

$$\dot{x}(t) = (A + \Delta A)x(t) + Bu, \qquad x(0) = x_0, \qquad t \geq 0, \tag{11.200}$$

where $u(\cdot) \in \mathcal{U}$ is admissible and $\Delta A \in \boldsymbol{\Delta}_A$. Assume there exist $\Omega : \mathbb{N}^n \to \mathbb{N}^n$, $P \in \mathbb{P}^n$, $\hat{\Omega}_\nu : \mathcal{N}^{(2\nu,n)} \to \mathcal{N}^{(2\nu,n)}$, and $\hat{P}_\nu \in \mathcal{N}^{(2\nu,n)}$, $\nu = 2, \ldots, r$, such that

$$\Delta A^{\mathrm{T}} P + P \Delta A \leq \Omega(P), \qquad \Delta A \in \boldsymbol{\Delta}_A, \tag{11.201}$$

$$\hat{\Omega}_\nu(\hat{P}_\nu) - \hat{P}_\nu(\overset{2\nu}{\oplus} \Delta A) \in \mathcal{N}^{(2\nu,n)}, \qquad \Delta A \in \boldsymbol{\Delta}_A, \quad \nu = 2, \ldots, r, \tag{11.202}$$

$$0 = A^{\mathrm{T}} P + PA + R_1 - PSP + \Omega(P), \tag{11.203}$$

and

$$0 = \hat{P}_\nu[\overset{2\nu}{\oplus} (A - SP)] + \hat{R}_{2\nu} + \hat{\Omega}_\nu(\hat{P}_\nu), \qquad \nu = 2, \ldots, r. \tag{11.204}$$

Then, with the feedback control $u = \phi(x) = -R_2^{-1} B^{\mathrm{T}}(Px + \frac{1}{2} g'^{\mathrm{T}}(x))$, where $g(x) \triangleq \sum_{\nu=2}^{r} \hat{P}_\nu x^{[2\nu]}$, the zero solution $x(t) \equiv 0$ of the uncertain system (11.200) is globally asymptotically stable for all $x_0 \in \mathbb{R}^n$ and $\Delta A \in \boldsymbol{\Delta}_A$, and the performance functional (11.127) satisfies

$$\sup_{\Delta A \in \boldsymbol{\Delta}_A} J_{\Delta A}(x_0, \phi(x(\cdot))) \leq \mathcal{J}(x_0, \phi(x(\cdot))) = x_0^{\mathrm{T}} P x_0 + \sum_{\nu=2}^{r} \hat{P}_\nu x_0^{[2\nu]}, \tag{11.205}$$

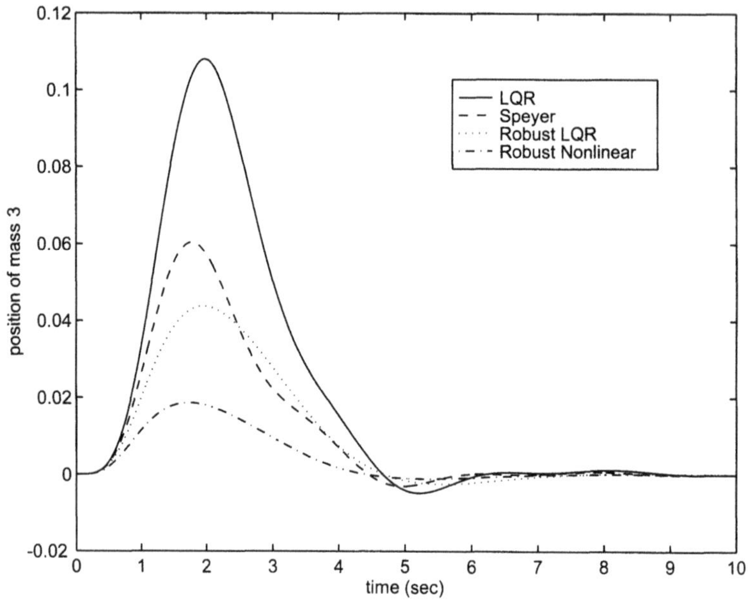

Figure 11.7 Comparison of LQR, Speyer, robust LQR, and robust nonlinear controllers for the uncertain system: Position of mass 3.

where

$$\mathcal{J}(x_0, u(\cdot)) \triangleq \int_0^\infty [L(x,u) + \Gamma(x,u)]\mathrm{d}t, \tag{11.206}$$

and where $u(\cdot)$ is admissible, and $x(t)$, $t \geq 0$, solves (11.200) with $\Delta A = 0$ and

$$\Gamma(x,u) = x^{\mathrm{T}}\Omega(P)x + \sum_{\nu=2}^{r} \hat{\Omega}_\nu(\hat{P}_\nu)x^{[2\nu]},$$

where $u(\cdot)$ is admissible and $\Delta A \in \mathbf{\Delta}_A$. In addition, the performance functional (11.127), with $R_2(x) = R_2$, $L_2(x) = 0$, and

$$L_1(x) = x^{\mathrm{T}}R_1x + \sum_{\nu=2}^{r} \hat{R}_{2\nu}x^{[2\nu]} + \frac{1}{4}g'(x)Sg'^{\mathrm{T}}(x)$$

is minimized in the sense that

$$\mathcal{J}(x_0, \phi(x(\cdot))) = \min_{u(\cdot)\in\mathcal{S}(x_0)} \mathcal{J}(x_0, u(\cdot)), \tag{11.207}$$

where $\mathcal{S}(x_0)$ is the set of regulation controllers for the nominal system and $x_0 \in \mathbb{R}^n$.

Proof. The result is a direct consequence of Corollary 11.4 with $f_0(x) = Ax$, $G_0(x) = B$, $\Delta f(x) = \Delta Ax$, $\Delta G(x) = 0$, $\mathbf{\Delta}_f = \mathbf{\Delta}_A$, and $V(x) = x^{\mathrm{T}}Px + \sum_{\nu=2}^{r}\hat{P}_\nu x^{[2\nu]}$. Specifically, conditions (11.128)–(11.162) are

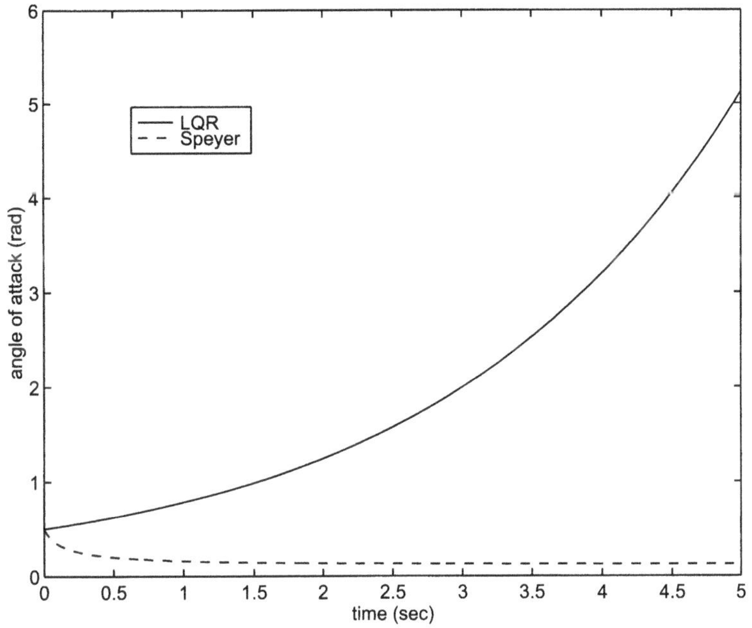

Figure 11.8 Comparison of LQR and Speyer controllers for the uncertain system: Angle of attack.

trivially satisfied. Now, it follows from (11.201) and (11.202) that

$$x^{\mathrm{T}}(\Delta A^{\mathrm{T}}P + P\Delta A - \Omega(P))x + \sum_{\nu=2}^{r}[\hat{P}_\nu(\overset{2\nu}{\oplus}\Delta A) - \hat{\Omega}_\nu(\hat{P}_\nu)]x^{[2\nu]} \le 0,\ x \in \mathbb{R}^n,$$

which implies (11.133) for all $\Delta A \in \mathbf{\Delta}_A$ so that all the conditions of Corollary 11.4 are satisfied. □

Note that since $g'(x)(A-SP)x = \sum_{\nu=2}^{r}\hat{P}_\nu[\overset{2\nu}{\oplus}(A-SP)]x^{[2\nu]}$ it follows that (11.204) can be equivalently written as

$$0 = g'(x)(A - SP)x + \sum_{\nu=2}^{r}[\hat{R}_{2\nu} + \hat{\Omega}_\nu(\hat{P}_\nu)]x^{[2\nu]},$$

for all $x \in \mathbb{R}^n$, and hence, it follows from Lemma 8.1 that there exists a unique $\hat{P}_\nu \in \mathcal{N}^{(2\nu,n)}$ such that (11.204) is satisfied. Theorem 11.6 generalizes the classical results of Bass and Webber [33] to robust nonlinear optimal control.

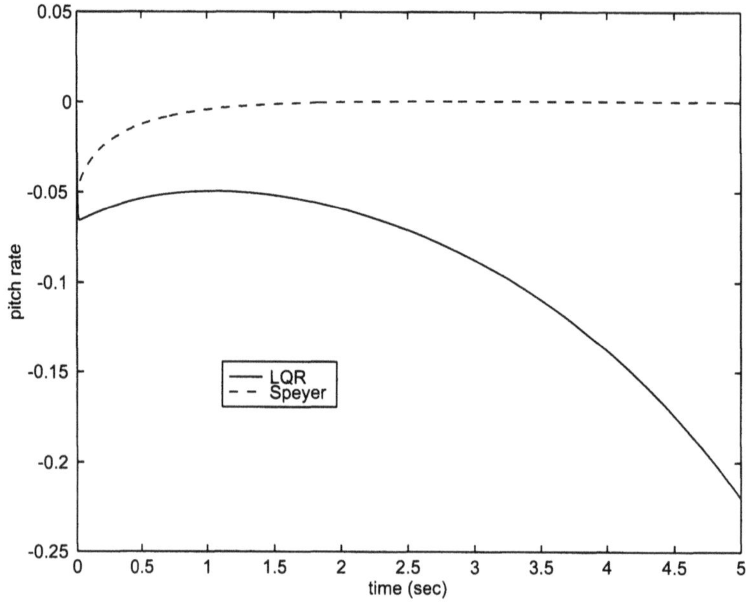

Figure 11.9 Comparison of LQR and Speyer controllers for the uncertain system: Pitch rate.

11.10 Problems

Problem 11.1. Consider the linear oscillator

$$\ddot{q}(t) + 2\zeta\omega_n\dot{q}(t) + \omega_n^2 q(t) = 0, \ q(0) = q_0, \ \dot{q}(0) = \dot{q}_0, \ t \geq 0, \tag{11.208}$$

where q, $\dot{q}$, $\ddot{q} \in \mathbb{R}$, and $\zeta \in [\zeta_{\rm l}, \zeta_{\rm u}]$ and ω_n denote the damping ratio and natural frequency, respectively, with $\zeta_{\rm l}$ and $\zeta_{\rm u}$ denoting the lower and upper bounds on ζ. Using $x(t) = [q(t), \ \dot{q}(t)]^{\rm T}$, represent (11.208) as (11.50) with $\mathbf{\Delta}_A$ given by (11.56).

Problem 11.2. Consider the linear matrix second-order uncertain dynamical system given by

$$M\ddot{q}(t) + (C_0 + \Delta C)\dot{q}(t) + (K_0 + \Delta K)q(t) = 0, \ q(0) = q_0, \ \dot{q}(0) = \dot{q}_0, \ t \geq 0, \tag{11.209}$$

where q, $\dot{q}$, $\ddot{q} \in \mathbb{R}^n$, M_0, C_0, and K_0 denote generalized inertia, damping, and stiffness matrices, respectively, and ΔC and ΔK denote damping and stiffness uncertainty such that $\sigma_{\max}(\Delta C) \leq \gamma_{\rm d}^{-1}$ and $\sigma_{\max}(\Delta K) \leq \gamma_{\rm s}^{-1}$. Using $x(t) = [q(t), \ \dot{q}(t)]^{\rm T}$, represent (11.209) as (11.50) with $\mathbf{\Delta}_A$ given by (11.27).

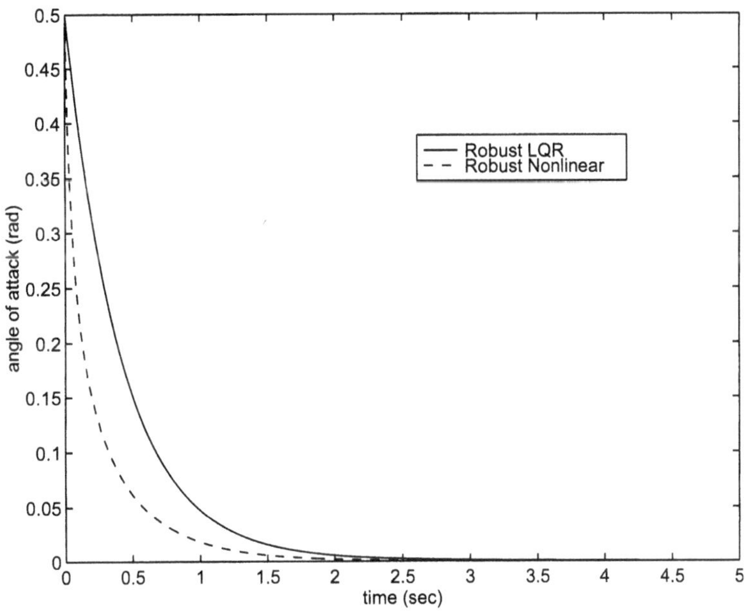

Figure 11.10 Comparison of robust LQR and robust nonlinear controllers for the uncertain system: Angle of attack.

Problem 11.3. Consider the uncertainty characterizations

$$\boldsymbol{\Delta}_1 \triangleq \{\Delta A \in \mathbb{R}^{n\times n} : \ \Delta A = \sum_{i=1}^{p} \delta_i A_i, \ |\delta_i| \le \gamma, \ i = 1, \ldots, p\}, \quad (11.210)$$

$$\boldsymbol{\Delta}_2 \triangleq \{\Delta A \in \mathbb{R}^{n\times n} : \ \Delta A = B_0 F C_0, \ \sigma_{\max}(F) \le \gamma\}, \quad (11.211)$$

where for $i = 1, \ldots, p$, $A_i \in \mathbb{R}^{n\times n}$, $B_0 \in \mathbb{R}^{n\times s}$, and $C_0 \in \mathbb{R}^{s\times n}$ are given, and $\gamma > 0$. Show that $\boldsymbol{\Delta}_1$ is equivalent to $\boldsymbol{\Delta}_2$ in the sense that an uncertainty set $\boldsymbol{\Delta}_1$ can always be written in the form of $\boldsymbol{\Delta}_2$ and vice versa.

Problem 11.4. Consider the linear uncertain system (11.50) where $\Delta A \in \boldsymbol{\Delta}_A$ and $\boldsymbol{\Delta}_A$ is given by

$$\boldsymbol{\Delta}_A \triangleq \{\Delta A \in \mathbb{R}^{n\times n} : \Delta A = \sum_{i=1}^{p} \delta_i A_i, \ |\delta_i| \le \gamma_i, \ i = 1, \ldots, p\}. \quad (11.212)$$

Show that the function

$$\Omega(P) = \sum_{i=1}^{p} \gamma_i |A_i^{\mathrm{T}} P + P A_i|, \quad (11.213)$$

where $|S|$ denotes $(S^2)^{1/2}$ for $S \in \mathbb{S}^n$ and $(\cdot)^{1/2}$ denotes the (unique) nonnegative-definite square root, satisfies (11.52) with $\boldsymbol{\Delta}_A$ given by (11.212).

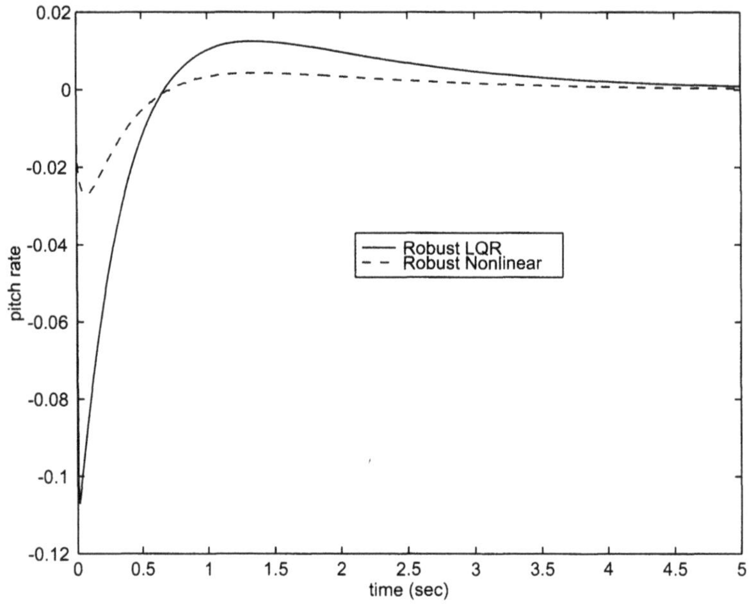

Figure 11.11 Comparison of robust LQR and robust nonlinear controllers for the uncertain system: Pitch rate.

Problem 11.5. Let $f : \mathbb{R} \to \mathbb{R}$ and define (with a minor abuse of notation) $f : \mathbb{S}^n \to \mathbb{S}^n$ by $f(S) \triangleq Uf(D)U^{\mathrm{T}}$, where $S = UDU^{\mathrm{T}}$, U is orthogonal, D is real diagonal, and $f(D)$ is the diagonal matrix obtained by applying f to each diagonal entry of D . Consider the linear uncertain system (11.50) where $\Delta A \in \boldsymbol{\Delta}_A$ and $\boldsymbol{\Delta}_A$ is given by (11.212). Let $f_i : \mathbb{R} \to \mathbb{R}$, $i = 1, \ldots, p$, be such that $f_i(x) \geq |x|$, $x \in \mathbb{R}$. Show that the function

$$\Omega(P) = \sum_{i=1}^{p} \gamma_i f_i(A_i^{\mathrm{T}} P + PA_i) \tag{11.214}$$

is an overbound for $\Omega(\cdot)$ given by (11.213) and, hence, satisfies (11.52) with $\boldsymbol{\Delta}_A$ given by (11.212). (**Hint:** Note if $f(x) = |x|$, then $f(S) = (S^2)^{1/2}$, where $(\cdot)^{1/2}$ denotes the (unique) nonnegative-definite square root.)

Problem 11.6. Consider the linear uncertain system (11.50) where $\Delta A \in \boldsymbol{\Delta}_A$ and $\boldsymbol{\Delta}_A$ is given by (11.212). Let $\beta_1, \ldots, \beta_p$ be arbitrary positive constants. Show that the function

$$\Omega(P) = \tfrac{1}{4} \sum_{i=1}^{p} \gamma_i \beta_i I_m + \sum_{i=1}^{p} (\tfrac{\gamma_i}{\beta_i})(A_i^{\mathrm{T}} P + PA_i)^2 \tag{11.215}$$

is an overbound for $\Omega(\cdot)$ given by (11.213) and, hence, satisfies (11.52) with $\boldsymbol{\Delta}_A$ given by (11.212).

Problem 11.7. Let $\mathbf{\Delta}_A$ given by (11.212) be defined by the positive constants $\gamma_1, \ldots, \gamma_p$ and let $\mathbf{\Delta}_A$ given by (11.56) be characterized by $\alpha_i = \left(\frac{\alpha\gamma_i}{\beta_i}\right)^{1/2}$, $i = 1, \ldots, p$, where $\alpha \triangleq \sum_{i=1}^{p} \gamma_i\beta_i$ and $\beta_1, \ldots, \beta_p$ are arbitrary positive constants. Show that the ellipse $\mathcal{E} \triangleq \{(\delta_1, \ldots, \delta_p) : \sum_{i=1}^{p} \frac{\delta_i^2}{\alpha_i^2} \le 1\}$ circumscribes the rectangle $\mathcal{R} \triangleq \{(\delta_1, \ldots, \delta_p) : |\delta_i| \le \gamma_i, \ i = 1, \ldots, p\}$, and hence, $\mathbf{\Delta}_A$ given by (11.56) contains $\mathbf{\Delta}_A$ given by (11.212).

Problem 11.8. Consider the linear uncertain system (11.50) where $\Delta A \in \mathbf{\Delta}_A$ and $\mathbf{\Delta}_A$ is given by (11.56). Let α be an arbitrary positive constant. Show that the function

$$\Omega(P) = \tfrac{\alpha}{4} I_n + \alpha^{-1} \sum_{i=1}^{p} \alpha_i^2 (A_i^{\mathrm{T}} P + P A_i)^2 \tag{11.216}$$

satisfies (11.52) with $\mathbf{\Delta}_A$ given by (11.56).

Problem 11.9. Consider the linear uncertain system (11.50) where $\Delta A \in \mathbf{\Delta}_A$ and $\mathbf{\Delta}_A$ is given by (11.56). Let α be an arbitrary positive constant. Show that for $P > 0$ the function

$$\Omega(P) = \tfrac{\alpha}{2} P + \tfrac{\alpha^{-1}}{2} \sum_{i=1}^{p} \alpha_i^2 [A_i^{2\mathrm{T}} P + A_i^{\mathrm{T}} P A_i + P A_i P^{-1} A_i^{\mathrm{T}} P + P A_i^2] \tag{11.217}$$

satisfies (11.52) with $\mathbf{\Delta}_A$ given by (11.56).

Problem 11.10. Consider the linear uncertain system (11.50) where $\Delta A \in \mathbf{\Delta}_A$ and $\mathbf{\Delta}_A$ is given by

$$\mathbf{\Delta}_A \triangleq \{\Delta A \in \mathbb{R}^{n\times n} : \ \Delta A = \sum_{i=1}^{p} \delta_i A_i, \ |\delta_i| \le \gamma^{-1}, \ i = 1, \ldots, p\}, \tag{11.218}$$

where $\gamma > 0$. For $i = 1, \ldots, p$, let $\alpha_i \in \mathbb{R}$, $\beta_i > 0$, $S_i \in \mathbb{R}^{n\times n}$ and define $Z_i \triangleq [(S_i + S_i^{\mathrm{T}})^2]^{1/2}$ and $\hat{I}_i \triangleq [S_i \ A_i^{\mathrm{T}}][S_i \ A_i^{\mathrm{T}}]^{\dagger}$. Show that the function

$$\Omega(P) = \sum_{i=1}^{p} [\gamma^{-2} (\alpha_i S_i + \beta_i A_i^{\mathrm{T}} P)^{\mathrm{T}} (\alpha_i S_i + \beta_i A_i^{\mathrm{T}} P) + \gamma^{-1} \beta_i^{-1} |\alpha_i| Z_i + \beta_i^2 \hat{I}_i \tag{11.219}$$

satisfies (11.52) with $\mathbf{\Delta}_A$ given by (11.218). (**Hint:** First show that $\hat{I}_i = \hat{I}_i^{\mathrm{T}} = \hat{I}_i^2$, $\hat{I}_i S_i = S_i$, and $A_i \hat{I}_i = A_i$.)

Problem 11.11. Consider the linear uncertain system (11.50) where $\Delta A \in \mathbf{\Delta}_A$ and $\mathbf{\Delta}_A$ is given by (11.56). Let α be an arbitrary positive constant and for each $P \in \mathbb{P}^n$, let $P_1 \in \mathbb{R}^{n\times m}$ and $P_2 \in \mathbb{R}^{m\times n}$ satisfy

$P = P_1P_2$. Show that the function

$$\Omega(P) \triangleq \alpha P_2^{\mathrm{T}}P_2 + \alpha^{-1}\sum_{i=1}^{p}\alpha_i^2 A_i^{\mathrm{T}}P_1P_1^{\mathrm{T}}A_i \tag{11.220}$$

satisfies (11.52) with $\mathbf{\Delta}_A$ given by (11.56). Using (11.220) show that the functions

$$\Omega(P) \triangleq \alpha I_n + \alpha^{-1}\sum_{i=1}^{p}\alpha_i^2 A_i^{\mathrm{T}}P^2A_i \tag{11.221}$$

and

$$\Omega(P) \triangleq \alpha P^2 + \alpha^{-1}\sum_{i=1}^{p}\alpha_i^2 A_i^{\mathrm{T}}A_i, \tag{11.222}$$

also satisfy (11.52) with $\mathbf{\Delta}_A$ given by (11.56).

Problem 11.12. Consider the linear uncertain system (11.50) where $\Delta A \in \mathbf{\Delta}_A$ and $\mathbf{\Delta}_A$ is given by (11.210). Let $\alpha > 0$, $N_i \in \mathbb{S}^n$, $i = 1,\ldots,p$, and define $\mathcal{N} \triangleq \{P \in \mathbb{P}^n : \ P - N_i \geq 0,\ i = 1,\ldots,p\}$. Show that the function

$$\Omega(P) = \sum_{i=1}^{p}[\alpha(P - N_i) + \tfrac{\gamma^2}{\alpha}A_i^{\mathrm{T}}(P - N_i)A_i + \gamma|A_i^{\mathrm{T}}N_i + N_iA_i|], \tag{11.223}$$

where $|S|$ denotes $(S^2)^{1/2}$, satisfies (11.52) with $\mathbf{\Delta}$ given by (11.210).

Problem 11.13. Consider the linear uncertain system (11.50) where $\Delta A \in \mathbf{\Delta}_A$ and $\mathbf{\Delta}_A$ is given by (11.210). Let $\alpha > 0$, $V_{1i}, V_{2i} \in \mathbb{R}^{n\times n}$, $N_i \in \mathbb{S}^n$, $i = 1,\ldots,p$, and define $\mathcal{N} \triangleq \{P \in \mathbb{P}^n : \ (P - N_i) > 0\}$, $i = 1,\ldots,p$. Show that the function

$$\begin{aligned}\Omega(P) \ = \ & \sum_{i=1}^{p}[\alpha(P - N_i) + \tfrac{\gamma}{2}|A_i^{\mathrm{T}}(V_{1i} + V_{2i}^{\mathrm{T}}) + (V_{2i} + V_{1i}^{\mathrm{T}})A_i|] \\ & + \tfrac{\gamma}{4\alpha}[A_i^{\mathrm{T}}(P - V_{1i}) + (P - V_{2i})A_i](P - N_i)^{-1} \\ & \cdot[A_i^{\mathrm{T}}(P - V_{1i}) + (P - V_{2i})A_i]^{\mathrm{T}},\end{aligned} \tag{11.224}$$

where $|S|$ denotes $(S^2)^{1/2}$, satisfies (11.52) with $\mathbf{\Delta}_A$ given by (11.210).

Problem 11.14. Consider the linear uncertain system (11.50) where $\Delta A \in \mathbf{\Delta}_A$ and $\mathbf{\Delta}_A$ is given by (11.59). Show that if there exists $P \in \mathbb{P}^n$ satisfying

$$0 = A^{\mathrm{T}}P + PA + \gamma^{-2}PB_0B_0^{\mathrm{T}}P + C_0^{\mathrm{T}}C_0 + R, \tag{11.225}$$

where $\gamma > 0$ and $R > 0$, then $A + \Delta A$ is asymptotically stable for all $\Delta A \in \mathbf{\Delta}_A$. Furthermore, show that there exists $P \in \mathbb{P}^n$ satisfying (11.225) if and only if $\|C_0(sI - A)^{-1}B_0\|_\infty < \gamma$.

Problem 11.15. Consider the linear uncertain controlled system (11.109) with $\Delta B = 0$, $\Delta A \in \mathbf{\Delta}_A$, where $\mathbf{\Delta}_A$ is given by (11.218), and performance functional (11.110). For $i = 1,\ldots,p$, let $\alpha_i \in \mathbb{R}$, $\beta_i > 0$, $S_i \in \mathbb{R}^{n\times n}$ and define $Z_i \triangleq [(S_i + S_i^{\mathrm{T}})^2]^{1/2}$ and $\hat{I}_i \triangleq [S_i\ A_i^{\mathrm{T}}][S_i\ A_i^{\mathrm{T}}]^{\dagger}$. Show that the zero solution $x(t) \equiv 0$ to (11.109) is globally asymptotically stable for all $\Delta A \subset \mathbf{\Delta}_A$ with the feedback control $\phi(x) = R_2^{-1}B^{\mathrm{T}}Px$, where $P > 0$ satisfies

$$0 = A_{\mathrm{s}\gamma}^{\mathrm{T}}P + PA_{\mathrm{s}\gamma} + R_1 + \sum_{i=1}^{p}[\gamma^{-2}(\alpha_i^2 S_i^{\mathrm{T}}S_i + \beta_i^2 PA_iA_i^{\mathrm{T}}P) + \gamma^{-1}\beta_i^{-1}|\alpha_i|Z_i$$
$$+\beta_i^2\hat{I}_i] - PBR_2^{-1}B^{\mathrm{T}}P, \tag{11.226}$$

and $A_{\mathrm{s}\gamma} \triangleq A + \gamma_i^2\sum_{i=1}^{p}\alpha_i\beta_iA_iS_i$.

Problem 11.16. Show that if $R \in \mathbb{P}^n$ and

$$\mathcal{A} \triangleq A_\alpha \oplus A_\alpha + \sum_{i=1}^{p}\frac{\alpha_i^2}{\alpha}A_i \otimes A_i$$

is Hurwitz, then there exists a unique $P \in \mathbb{R}^{n\times n}$ satisfying (11.64) and $P > 0$. Conversely, show that if for all $R \in \mathbb{P}^n$ there exists $P > 0$ satisfying (11.64), then $\mathcal{A}$, A_α, and A are Hurwitz. (**Hint:** Use the exponential product formula $e^{\mathcal{A}t} = \lim_{k\to\infty}\{\exp[\frac{1}{k}(A_\alpha \oplus A_\alpha)t]\exp[\frac{1}{k}\sum_{i=1}^{p}\frac{\alpha_i^2}{\alpha}(A_i \otimes A_i)]\}$.)

Problem 11.17. Let $\mathbf{\Delta}_A$ be given by (11.56) and let $\hat{\mathbf{\Delta}}_A \subseteq \mathbf{\Delta}_A$, where $\hat{\mathbf{\Delta}}_A$ is defined as in (11.56) with α_i replaced by $\hat{\alpha}_i \in [0,\alpha_i]$, $i = 1,\ldots,p$. Furthermore, let $R \in \mathbb{P}^n$, assume $\mathcal{A} \triangleq A_\alpha \oplus A_\alpha + \sum_{i=1}^{p}\frac{\alpha_i^2}{\alpha}A_i \otimes A_i$ is Hurwitz, and let $P \in \mathbb{P}^n$ satisfy (11.64). Show that there exists $\hat{P} \in \mathbb{P}^n$ satisfying

$$0 = A_\alpha^{\mathrm{T}}\hat{P} + \hat{P}A_\alpha + \alpha^{-1}\sum_{i=1}^{p}\hat{\alpha}_i^2A_i^{\mathrm{T}}\hat{P}A_i + R, \tag{11.227}$$

and $\hat{P} \leq P$.

Problem 11.18. Show that if

$$\left\|(A \oplus A)^{-1}\left(\alpha I_{n^2} + \alpha^{-1}\sum_{i=1}^{p}\alpha_i^2A_i \otimes A_i\right)\right\| < 1, \tag{11.228}$$

where $\|\cdot\|$ denotes an arbitrary submultiplicative norm on $\mathbb{R}^{n^2}$, then for all $R \in \mathbb{P}^n$ there exists $P \in \mathbb{P}^n$ satisfying (11.64).

Problem 11.19. Let $\kappa, \beta > 0$ satisfy $\|e^{At}\| \leq \kappa e^{-\beta t}$, $t \geq 0$, where A is Hurwitz and $\|\cdot\|$ denotes an arbitrary submultiplicative norm that is monotonic on $\mathbb{N}^n$. Show that if $R \in \mathbb{P}^n$ and $4\alpha\|B_0B_0^{\mathrm{T}}\|\ \|\alpha^{-1}C_0^{\mathrm{T}}NC_0 + R\| <$

ρ^2, where $\rho \triangleq 2\beta/\kappa$, then there exists $P \in \mathbb{P}^n$ satisfying (11.68).

Problem 11.20. Consider the linear uncertain system (11.50) where $\Delta A \in \boldsymbol{\Delta}_A$ and $\boldsymbol{\Delta}_A$ is given by

$$\boldsymbol{\Delta}_A \triangleq \{\Delta A \in \mathbb{R}^{n\times n} : \ \Delta A = B_0 F C_0, \ F \in \boldsymbol{\Delta}_{\rm bs}\}, \tag{11.229}$$

where $\boldsymbol{\Delta}_{\rm bs}$ denotes the set of block-diagonal matrices with possibly repeated blocks defined by

$$\begin{aligned}\boldsymbol{\Delta}_{\rm bs} \triangleq \{F \in \mathbb{R}^{s\times s} : \ & F = \text{block–diag}[I_{l_1} \otimes F_1, I_{l_2} \otimes F_2, \ldots, I_{l_p} \otimes F_p], \\ & F_i \in \mathbb{R}^{s_i \times s_i}, \ i = 1, \ldots, p\},\end{aligned} \tag{11.230}$$

and the dimension s_i of each block and the number of repetitions l_i of each block are given such that $\sum_{i=1}^p l_i s_i = s$. Furthermore, define the set of constant scaling matrices $\mathcal{D}$ by

$$\mathcal{D} \triangleq \{D \in \mathbb{R}^{s\times s} : \ D > 0, \ DF = FD, \ F \in \boldsymbol{\Delta}_{\rm bs}\}. \tag{11.231}$$

Show that if there exists $P \in \mathbb{P}^n$ satisfying

$$0 = A^{\rm T}P + PA + \gamma^{-2}PB_0D^{-2}B_0^{\rm T}P + C_0^{\rm T}D^2C_0 + R, \tag{11.232}$$

where $\gamma > 0$ and $R > 0$, then $A + \Delta A$ is asymptotically stable for all $\Delta A \in \boldsymbol{\Delta}_\gamma \triangleq \{F \in \boldsymbol{\Delta}_{\rm bs} : \ \sigma_{\max}(F) \leq \gamma^{-1}\}$. Furthermore, show that there exists $P \in \mathbb{P}^n$ satisfying (11.232) if and only if $|||DG(s)D^{-1}|||_\infty < \gamma$, where $G(s) = C_0(sI - A)^{-1}B_0$.

Problem 11.21. Consider the linear uncertain system (11.50) and (11.51) with $n = 2$,

$$A = \begin{bmatrix} -\nu & \omega \\ -\omega & -\nu \end{bmatrix}, \quad \nu > 0, \quad \omega \geq 0, \tag{11.233}$$

$V = \mathbb{E}[x_0 x_0^{\rm T}] = I_2$, $R = I_2$, and $\Delta A = \{\Delta A : \ \Delta A = \delta_1 A_1, \ \ |\delta_1| \leq \alpha_1\}$, where

$$A_1 = \begin{bmatrix} 0 & 1 \\ -1 & 0 \end{bmatrix}. \tag{11.234}$$

Show that (11.53) with $\Omega(P)$ given by (11.213) or (11.219) is nonconservative with respect to robust stability and performance. Alternatively, show that (11.213) and (11.217) give extremely conservative predictions, especially when ν is small.

Problem 11.22. Consider the linear dynamical system with state delay given by

$$\dot{x}(t) = Ax(t) + A_d x(t - \tau_d), \quad x(\theta) = \phi(\theta), \quad -\tau_{\rm d} \leq \theta \leq 0, \tag{11.235}$$

where $x(t) \in \mathbb{R}^n$, $A, A_d \in \mathbb{R}^{n\times n}$, and $\phi : [-\tau_d, 0] \to \mathbb{R}^n$ is a continuous vector-valued function specifying the initial state of the system. Show that the zero solution $x_t \equiv 0$ to (11.235) is globally asymptotically stable (in the sense of Problem 3.65) for all $\tau_d \geq 0$ if $\|DG(s)D^{-1}\|_\infty < 1$, where D is a positive-definite matrix and

$$G(s) \sim \left[\begin{array}{c|c} A & A_{\mathrm{d}} \\ \hline I & 0 \end{array}\right].$$

Furthermore, show that this problem can be represented as a feedback problem involving an uncertain diagonal operator $\Delta(s)$ satisfying $\|\Delta(s)\|_\infty \leq 1$.

Problem 11.23. Let $\mathcal{D} \subseteq \mathbb{R}^n$, $f : \mathcal{D} \to \mathbb{R}^n$ such that $f(0) = 0$, $G : \mathcal{D} \to \mathbb{R}^{n\times m}$, and $R_1 \in \mathbb{P}^n$. Consider the nonlinear uncertain system

$$\dot{x}(t) = f(x(t)) + G(x(t))\delta(x(t)) + G(x(t))u(t), \qquad x(0) = x_0, \qquad t \geq 0, \tag{11.236}$$

with performance functional

$$J_\delta(x_0, u(\cdot)) \triangleq \int_0^\infty x^{\mathrm{T}}(t)R_1x(t)\mathrm{d}t, \tag{11.237}$$

where $u(\cdot)$ is an admissible control and $\delta(\cdot) \in \boldsymbol{\Delta}_\delta \triangleq \{\delta : \mathcal{D} \to \mathbb{R}^m :\ \delta(0) = 0,\ \delta^{\mathrm{T}}(x)R_2\delta(x) \leq \delta_{\max}^2(x),\ x \in \mathcal{D}\}$, where $R_2 \in \mathbb{P}^m$ and $\delta_{\max} : \mathcal{D} \to \mathbb{R}$ such that $\delta(0) = 0$ are given. Assume that there exist a continuously differentiable function $V : \mathcal{D} \to \mathbb{R}$ and a control law $\phi(\cdot)$ such that (11.96)–(11.98) hold and

$$x^{\mathrm{T}}R_1x + V'(x)(f(x) + G(x)\phi(x)) + \delta_{\max}^2(x) + \phi^{\mathrm{T}}(x)R_2\phi(x) = 0,\ x \in \mathcal{D}. \tag{11.238}$$

Show that there exists a neighborhood of the origin $\mathcal{D}_0 \subseteq \mathcal{D}$ such that if $x_0 \in \mathcal{D}_0$, then the closed-loop system of (11.236) with the feedback control law $\phi(x) \triangleq -\frac{1}{2}R_2^{-1}G^{\mathrm{T}}(x)V'^{\mathrm{T}}(x)$ is locally asymptotically stable for all $\delta(\cdot) \in \boldsymbol{\Delta}_\delta$, and

$$\sup_{\delta(\cdot)\in\boldsymbol{\Delta}_\delta} J_\delta(x_0, \phi(x(\cdot))) \leq \mathcal{J}(x_0, \phi(x(\cdot))) = V(x_0), \tag{11.239}$$

where

$$\mathcal{J}(x_0, u(\cdot)) \triangleq \int_0^\infty [x^{\mathrm{T}}(t)R_1x(t) + u^{\mathrm{T}}(t)R_2u(t) + \delta_{\max}^2(x(t))]\mathrm{d}t, \tag{11.240}$$

and where $u(\cdot)$ is admissible and $x(t)$, $t \geq 0$ solves (11.236) with $\delta(x) \equiv 0$. Furthermore, if $x_0 \in \mathcal{D}_0$, show that

$$\mathcal{J}(x_0, \phi(x(\cdot))) = \min_{u(\cdot)\in\mathcal{S}(x_0)} \mathcal{J}(x_0, u(\cdot)), \tag{11.241}$$

where $\mathcal{S}(x_0)$ is the set of regulation controllers for the nominal system

$$\dot{x}(t) = f(x(t)) + G(x(t))u(t). \tag{11.242}$$

Problem 11.24. Let $\mathcal{D} \subseteq \mathbb{R}^n$, $f_0 : \mathcal{D} \to \mathbb{R}^n$ such that $f_0(0) = 0$, $G_\delta : \mathcal{D} \to \mathbb{R}^{n \times m_\delta}$, $G : \mathcal{D} \to \mathbb{R}^{n \times m}$, $R_1 \in \mathbb{P}^n$, and $R_2 \in \mathbb{P}^m$. Consider the nonlinear uncertain system

$$\dot{x}(t) = f_0(x(t)) + G_\delta(x(t))\delta(x(t)) + G(x(t))u(t), \quad x(0) = x_0, \quad t \geq 0, \tag{11.243}$$

with performance functional

$$J_\delta(x_0, u(\cdot)) \triangleq \int_0^\infty [x^{\mathrm{T}}(t)R_1 x(t) + u^{\mathrm{T}}(t)R_2 u(t)]\mathrm{d}t, \tag{11.244}$$

where $u(\cdot)$ is admissible control and $\delta(\cdot) \in \mathbf{\Delta}_\gamma \triangleq \{\delta : \mathcal{D} \to \mathbb{R}^{m_\delta} : \delta(0) = 0,\ \delta^{\mathrm{T}}(x)R_2\delta(x) \leq \delta_{\max}^2(x),\ \|R_2^{1/2}G^\dagger(x)G_\delta(x)\delta(x)\|_2^2 \leq \gamma_{\max}^2(x),\ x \in \mathcal{D}\}$, where $\delta_{\max}, \gamma_{\max} : \mathcal{D} \to \mathbb{R}$ such that $\delta_{\max}(0) = 0$, $\gamma_{\max}(0) = 0$, are given. Assume that there exist a continuously differentiable function $V : \mathcal{D} \to \mathbb{R}$ and control laws $\phi_1(\cdot)$, $\phi_2(\cdot)$ such that $\phi_1(0) = 0$, $\phi_2(0) = 0$, (11.96) and (11.97) hold, and

$$\begin{aligned} 0 = {} & x^{\mathrm{T}}R_1x + V'(x)[f_0(x) + G(x)\phi_1(x) + (I_n - G(x)G^\dagger(x))G_\delta(x)\phi_2(x)] \\ & +\delta_{\max}^2(x) + \gamma_{\max}^2(x) + \phi_1^{\mathrm{T}}R_2\phi_1(x) + \phi_2^{\mathrm{T}}\hat{R}_2\phi_2(x), \quad x \in \mathcal{D}, \end{aligned} \tag{11.245}$$

where $\hat{R}_2 = \rho R_2$, $\rho > 0$, $2\rho^2\|\phi_2(x)\|_2^2 \leq \lambda_{\min}^2(R_1)\|x\|_2^2$, $x \in \mathcal{D}$, and $\lambda_{\min}(R_1) < \rho$. Show that there exits a neighborhood of the origin $\mathcal{D}_0 \subseteq \mathcal{D}$ such that if $x_0 \in \mathcal{D}_0$, then the closed-loop system (11.243) with feedback control law $u = \phi_1(x) = -\frac{1}{2}R_2^{-1}G^{\mathrm{T}}(x)V'^{\mathrm{T}}(x)$ is locally asymptotically stable for all $\delta(\cdot) \in \mathbf{\Delta}_\delta$, and

$$\sup_{\delta(\cdot)\in\mathbf{\Delta}_\delta} J_\delta(x_0, \phi_1(x(\cdot)) \leq \mathcal{J}(x_0, \phi_1(\cdot), \phi_2(\cdot)) = V(x_0), \tag{11.246}$$

where

$$\begin{aligned} \mathcal{J}(x_0, u_1(\cdot), u_2(\cdot)) \triangleq {} & \int_0^\infty [x^{\mathrm{T}}(t)R_1x(t) + u_1^{\mathrm{T}}R_2u_1(t) + u_2^{\mathrm{T}}(t)\hat{R}_2u_2(t) \\ & +\rho^2\delta_{\max}^2(x(t)) + \gamma_{\max}(x(t))]\mathrm{d}t, \end{aligned} \tag{11.247}$$

and where $u_1(\cdot)$, $u_2(\cdot)$ are admissible and $x(t)$, $t \geq 0$, solves

$$\begin{aligned} \dot{x}(t) = {} & f_0(x(t)) + G(x(t))u_1(t) + [I_n - G(x(t))G^\dagger(x(t))]G_\delta(x(t))u_2(t), \\ & x(0) = x_0, \quad t \geq 0. \end{aligned} \tag{11.248}$$

(**Hint:** Decompose the uncertainty $G_\delta(x)\delta(x)$ into the sum of a matched component and an unmatched component by projecting $G_\delta(x)\delta(x)$ onto the range of $G(x)$, that is,

$$G_\delta(x)\delta(x) = G(x)G^\dagger(x)G_\delta(x)\delta(x) - (I_n - G(x)G^\dagger(x))G_\delta(x)\delta(x).) \tag{11.249}$$

Problem 11.25. Show that the robust stability and performance conditions given in Problem 11.24 still hold if $G(x(t))u(t)$ in (11.243) is replaced by $G(x(t))(u(t)+J_\delta(x(t))u(t))$, where $J_\delta : \mathcal{D} \to \mathbb{R}^{m\times m}$ is uncertain and satisfies $J_\delta(x) \geq 0$, $x \in \mathcal{D}$.

Problem 11.26. Consider the nonlinear uncertain dynamical system (11.125) and let $(\Delta f, \Delta G) \in \boldsymbol{\Delta}_f \times \boldsymbol{\Delta}_G$, where $\boldsymbol{\Delta}_f \times \boldsymbol{\Delta}_G$ is given by

$$\begin{aligned}\boldsymbol{\Delta}_f \times \boldsymbol{\Delta}_G \triangleq \{(\Delta f, \Delta G) : \Delta f(x) = G_\delta(x)\Delta(x)h_\delta(x),\\ \Delta G(x) = G_\delta(x)\Delta(x)J_\delta(x),\ x \in \mathbb{R}^n,\ \Delta(\cdot) \in \boldsymbol{\Delta}\},\end{aligned}$$

where $\boldsymbol{\Delta}$ satisfies

$$\boldsymbol{\Delta} = \{\Delta : \mathbb{R}^n \to \mathbb{R}^{m_\delta \times p_\delta} : \Delta^{\mathrm{T}}(x)\Delta(x) \leq M^{\mathrm{T}}(x)M(x),\ x \in \mathbb{R}^n\}, \quad (11.250)$$

and where $G_\delta : \mathbb{R}^n \to \mathbb{R}^{n\times m_\delta}$, $h_\delta : \mathbb{R}^n \to \mathbb{R}^{p_\delta}$, and $J_\delta : \mathbb{R}^n \to \mathbb{R}^{p_\delta \times m}$ are fixed functions denoting the structure of the uncertainty, $\Delta : \mathbb{R}^n \to \mathbb{R}^{m_\delta\times p_\delta}$ is an uncertain matrix function, and $M : \mathbb{R}^n \to \mathbb{R}^{m_\delta\times p_\delta}$ is a given matrix function. Show that condition (11.162) is satisfied with

$$\Gamma_{xx}(x) = \tfrac{1}{4}V'(x)G_\delta(x)g_\delta^{\mathrm{T}}(x)V'^{\mathrm{T}}(x) + h_\delta^{\mathrm{T}}(x)M^{\mathrm{T}}(x)M(x)h_\delta(x), \quad (11.251)$$

$$\Gamma_{xu}(x) = 2h_\delta^{\mathrm{T}}(x)M^{\mathrm{T}}(x)M(x)J_\delta(x), \quad (11.252)$$

$$\Gamma_{uu}(x) = J_\delta^{\mathrm{T}}(x)M^{\mathrm{T}}(x)M(x)J_\delta(x), \quad (11.253)$$

and the robust stabilizing control law (11.146) is given by

$$\phi(x) = -\tfrac{1}{2}R_{2a}^{-1}(x)[L_2(x) + V'(x)G_0(x) + 2h_\delta^{\mathrm{T}}(x)M^{\mathrm{T}}(x)M(x)J_\delta(x)]^{\mathrm{T}}, \quad (11.254)$$

where $R_{2a}(x) \triangleq R_2(x) + J_\delta^{\mathrm{T}}(x)M^{\mathrm{T}}(x)M(x)J_\delta(x)$.

Problem 11.27. Consider the nonlinear uncertain disturbed system

$$\dot{x}(t) = Ax(t) + \Delta f(x(t)) + Dw(t), \quad x(0) = x_0, \quad t \geq 0, \quad (11.255)$$

$$z(t) = Ex(t), \quad (11.256)$$

where $x \in \mathbb{R}^n$, $z \in \mathbb{R}^p$, $w(\cdot) \in \mathcal{L}_2$, $\Delta f(\cdot) \in \boldsymbol{\Delta}_f$, where

$$\boldsymbol{\Delta}_f \triangleq \{\Delta f : \mathbb{R}^n \to \mathbb{R}^n :\ \Delta f(x) = B_0\delta(C_0 x),\ x \in \mathbb{R}^n,\ \delta(\cdot) \in \boldsymbol{\Delta}\}, \quad (11.257)$$

and

$$\boldsymbol{\Delta} \triangleq \{\delta : \mathbb{R}^{l_0} \to \mathbb{R}^{m_0} :\ \delta(0) = 0,\ \|\delta(y_0)\|_2 \leq \gamma^{-1}\|y_0\|_2,\ y_0 = C_0 x \in \mathbb{R}^{l_0}\}, \quad (11.258)$$

where $B_0 \in \mathbb{R}^{n\times m_0}$ and $C_0 \in \mathbb{R}^{l_0\times n}$ are fixed matrices denoting the structure of the uncertainty, $\delta(\cdot)$ is an uncertain function, and $\gamma > 0$. Show that if there exist $n\times n$ matrices $P > 0$ and $R > 0$, and scalars $\gamma > 0$ and $\gamma_{\mathrm{d}} > 0$ such that

$$0 = A^{\mathrm{T}}P + PA + \gamma_{\mathrm{d}}^{-2}PDD^{\mathrm{T}}P + \gamma^{-2}PB_0B_0^{\mathrm{T}}P + C_0^{\mathrm{T}}C_0 + R, \quad (11.259)$$

then the zero solution $x(t) \equiv 0$ of the undisturbed ($w(t) \equiv 0$) system (11.255) is globally asymptotically stable for all $\Delta f(\cdot) \in \boldsymbol{\Delta}_f$. Furthermore, show that the solution $x(t)$, $t \geq 0$, of the disturbed system (11.255) satisfies the disturbance rejection constraint

$$\int_0^T z^{\mathrm{T}}(s)z(s)\mathrm{d}s < \gamma_{\mathrm{d}}^2 \int_0^T w^{\mathrm{T}}(s)w(s)\mathrm{d}s + x_0^{\mathrm{T}} P x_0, \quad w(\cdot) \in \mathcal{L}_2, \quad T \geq 0. \tag{11.260}$$

Problem 11.28. Consider the nonlinear uncertain cascade system

$$\begin{aligned} \dot{x}(t) &= f_0(x(t)) + \Delta f(x) + G_0(x(t))\hat{x}(t), \\ &\qquad x(0) = x_0, \ f_0 + \Delta f(\cdot) \in \mathcal{F}, \ t \geq 0, \qquad (11.261) \\ \dot{\hat{x}}(t) &= u(t), \qquad \hat{x}(0) = \hat{x}_0, \qquad (11.262) \end{aligned}$$

where $\mathcal{F}$ satisfies (11.23) and $\boldsymbol{\Delta}$ in (11.23) satisfies (11.24) with performance functional

$$J(x_0, \hat{x}_0, u(\cdot)) \triangleq \int_0^\infty L(x(t), \hat{x}(t), u(t))\mathrm{d}t, \tag{11.263}$$

where $u(\cdot)$ is admissible and $(x(t), \hat{x}(t))$, $t \geq 0$, solves (11.261) and (11.262) and where

$$L(x, \hat{x}, u) \triangleq L_1(x, \hat{x}) + L_2(x, \hat{x})u + u^{\mathrm{T}} R_2(x, \hat{x})u, \tag{11.264}$$

where $L_1 : \mathbb{R}^n \times \mathbb{R}^m \to \mathbb{R}$, $L_2 : \mathbb{R}^n \times \mathbb{R}^m \to \mathbb{R}^{1\times m}$, and $R_2 : \mathbb{R}^n \times \mathbb{R}^m \to \mathbb{P}^m$. Assume there exist continuously differentiable functions $\alpha : \mathbb{R}^n \to \mathbb{R}^m$ and $V_{\mathrm{sub}} : \mathbb{R}^n \to \mathbb{R}$ such that

$$\begin{aligned} \alpha(0) &= 0, && (11.265) \\ V_{\mathrm{sub}}(0) &= 0, && (11.266) \\ V_{\mathrm{sub}}(x) &> 0, \quad x \in \mathbb{R}^n, \quad x \neq 0, && (11.267) \end{aligned}$$

$$V'_{\mathrm{sub}}(x)[f_0(x) + G_0(x)\alpha(x)] + \tfrac{1}{4}V'_{\mathrm{sub}}(x)G_\delta(x)G_\delta^{\mathrm{T}}(x)V'^{\mathrm{T}}_{\mathrm{sub}}(x) + m^{\mathrm{T}}(h_\delta(x))m(h_\delta(x)) < 0, \quad x \in \mathbb{R}^n, \quad x \neq 0. \tag{11.268}$$

Furthermore, let $L_2 : \mathbb{R}^n \times \mathbb{R}^m \to \mathbb{R}^{1\times m}$ and $\hat{P} \in \mathbb{R}^{m\times m}$, $\hat{P} > 0$, be such that $L_2(0, 0) = 0$ and

$$\begin{aligned} &(\hat{x} - \alpha(x))^{\mathrm{T}}\hat{P}\Big[\hat{P}^{-1}G_0^{\mathrm{T}}(x)V'^{\mathrm{T}}_{\mathrm{sub}}(x) - 2\alpha'(x)(f_0(x) + G_0(x)\hat{x}) \\ &-R_2^{-1}(x, \hat{x})[2\hat{P}(\hat{x} - \alpha(x)) + L_2^{\mathrm{T}}(x, \hat{x})] - \alpha'(x)G_\delta(x)G_\delta^{\mathrm{T}}(x)V'^{\mathrm{T}}_{\mathrm{sub}}(x) \\ &+\alpha'(x)G_\delta(x)G_\delta^{\mathrm{T}}(x)\alpha'^{\mathrm{T}}(x)\hat{P}(\hat{x} - \alpha(x))\Big] < 0, \ (x, \hat{x}) \in \mathbb{R}^n \times \mathbb{R}^m, \ \hat{x} \neq \alpha(x). \end{aligned} \tag{11.269}$$

Show that the zero solution $(x(t), \hat{x}(t)) \equiv (0, 0)$ of the nonlinear uncertain cascade system (11.261) and (11.262) is globally asymptotically stable for all $\delta(\cdot) \in \mathbf{\Delta}$, where $\mathbf{\Delta}$ is given by (11.24), with feedback control law

$$\phi(x,\hat{x}) = -R_2^{-1}(x,\hat{x})[\tfrac{1}{2}L_2^{\mathrm{T}}(x,\hat{x}) + \hat{P}(\hat{x}-\alpha(x))]. \tag{11.270}$$

Furthermore, show that the performance functional (11.263) satisfies

$$\sup_{\delta(\cdot)\in\mathbf{\Delta}} J(x_0,\hat{x}_0,\phi(x,\hat{x})) \le \mathcal{J}(x_0,\hat{x}_0,\phi(x,\hat{x})) = V(x_0,\hat{x}_0),\ (x_0,\hat{x}_0) \in \mathbb{R}^n \times \mathbb{R}^m,$$

where

$$\begin{aligned}\mathcal{J}(x_0,\hat{x}_0,u(\cdot)) \triangleq \int_0^\infty &[L_1(x,\hat{x}) + L_2(x,\hat{x})u + u^{\mathrm{T}}R_2(x,\hat{x})u \\ &+m^{\mathrm{T}}(h_\delta(x))m(h_\delta(x)) \\ &+\tfrac{1}{4}[V'_{\mathrm{sub}}(x) - 2(\hat{x}-\alpha(x))^{\mathrm{T}}\hat{P}\alpha'(x)]G_\delta(x) \\ &\cdot G_\delta^{\mathrm{T}}(x)[V'_{\mathrm{sub}}(x) - 2(\hat{x}-\alpha(x))^{\mathrm{T}}\hat{P}\alpha'(x)]^{\mathrm{T}}]\mathrm{d}t,\end{aligned} \tag{11.271}$$

where $u(\cdot)$ is admissible, and $(x(t),\hat{x}(t))$, $t \ge 0$, solves (11.261) and (11.262) with $\delta(h_\delta(x)) \equiv 0$. In addition, show that the performance functional (11.271), with

$$\begin{aligned}L_1(x,\hat{x}) = {}& \phi^{\mathrm{T}}(x,\hat{x})R_2(x,\hat{x})\phi(x,\hat{x}) - [V'_{\mathrm{sub}}(x) - 2(\hat{x}-\alpha(x))\hat{P}\alpha'(x)] \\ &\cdot((f_0(x) + G_0(x)\hat{x})) - \tfrac{1}{4}[V'_{\mathrm{sub}}(x) - 2(\hat{x}-\alpha(x))^{\mathrm{T}}\hat{P}\alpha'(x)]G_\delta(x) \\ &\cdot G_\delta^{\mathrm{T}}(x)[V'_{\mathrm{sub}}(x) - 2(\hat{x}-\alpha(x))^{\mathrm{T}}\hat{P}\alpha'(x)]^{\mathrm{T}} - m^{\mathrm{T}}(h_\delta(x))m(h_\delta(x)),\end{aligned} \tag{11.272}$$

is minimized in the sense that

$$\mathcal{J}(x_0,\hat{x}_0,\phi(x(\cdot))) = \min_{u(\cdot)\in\mathcal{S}(x_0,\hat{x}_0)} \mathcal{J}(x_0,\hat{x}_0,u(\cdot)), \tag{11.273}$$

where $\mathcal{S}(x_0,\hat{x}_0)$ is the set of regulation controllers for the nonlinear system (11.261) and (11.262) with $\Delta f(x) \equiv 0$ by

$$\begin{aligned}\mathcal{S}(x_0,\hat{x}_0) \triangleq \{&u(\cdot) : u(\cdot) \text{ is admissible and } (x(\cdot),\hat{x}(\cdot)) \text{ given by (11.261) and} \\ &\text{(11.262) satisfies } (x(t),\hat{x}(t)) \to 0 \text{ as } t \to \infty \text{ with } \Delta f(x) \equiv 0\}.\end{aligned}$$

Finally, show that a particular choice of $L_2(x,\hat{x})$ satisfying (11.269) is given by

$$\begin{aligned}L_2(x,\hat{x}) = {}& \Big\{V'_{\mathrm{sub}}(x)G_0(x)\hat{P}^{-1} - 2[f_0(x) + G_0(x)\hat{x})]^{\mathrm{T}}\alpha'^{\mathrm{T}}(x) \\ &-V'_{\mathrm{sub}}(x)G_\delta(x)G_\delta^{\mathrm{T}}(x)\alpha'^{\mathrm{T}}(x) \\ &+(\hat{x}-\alpha(x))^{\mathrm{T}}\hat{P}\alpha'(x)G_\delta(x)G_\delta^{\mathrm{T}}(x)\alpha'^{\mathrm{T}}(x)\Big\}R_2(x,\hat{x}),\end{aligned} \tag{11.274}$$

yielding the feedback control law

$$\phi(x,\hat{x}) = -[R_2^{-1}(x,\hat{x}) + \tfrac{1}{2}\alpha'(x)G_\delta(x)g_\delta^{\mathrm{T}}(x)\alpha'^{\mathrm{T}}(x)]\hat{P}(\hat{x}-\alpha(x))$$

$$+\alpha'(x)[f_0(x)+G_0(x)\hat{x}]+\tfrac{1}{2}[\alpha'(x)G_\delta(x)g_\delta^{\mathrm{T}}(x)$$
$$-\hat{P}^{-1}G_0^{\mathrm{T}}(x)]V_{\mathrm{sub}}'^{\mathrm{T}}(x). \qquad (11.275)$$

Problem 11.29. Consider the nonlinear uncertain cascade system (11.261) and (11.262) where $\mathcal{F}$ satisfies (11.23) and $\boldsymbol{\Delta}$ in (11.23) satisfies (11.27), with performance functional (11.263). Assume there exist continuously differentiable functions $\alpha:\mathbb{R}^n\to\mathbb{R}^m$ and $V_{\mathrm{sub}}:\mathbb{R}^n\to\mathbb{R}$ such that (11.265)–(11.267) are satisfied and

$$V_{\mathrm{sub}}'(x)[f_0(x)+G_0(x)\alpha(x)]+V_{\mathrm{sub}}'(x)G_\delta(x)m_1(h_\delta(x))$$
$$+\tfrac{1}{4}[m(h_\delta(x))+G_\delta^{\mathrm{T}}(x)V_{\mathrm{sub}}'^{\mathrm{T}}(x)]^{\mathrm{T}}[m(h_\delta(x))+G_\delta^{\mathrm{T}}(x)V_{\mathrm{sub}}'^{\mathrm{T}}(x)]<0,\quad x\neq 0. \qquad (11.276)$$

Furthermore, let $L_2:\mathbb{R}^n\times\mathbb{R}^m\to\mathbb{R}^{1\times m}$ and $\hat{P}\in\mathbb{R}^{m\times m}$, $\hat{P}>0$, be such that $L_2(0,0)=0$ and

$$(\hat{x}-\alpha(x))^{\mathrm{T}}\hat{P}\Big[\hat{P}^{-1}G_0^{\mathrm{T}}(x)V_{\mathrm{sub}}'^{\mathrm{T}}(x)-2\alpha'(x)(f_0(x)+G_0(x)\hat{x})$$
$$-R_2^{-1}(x,\hat{x})[2\hat{P}(\hat{x}-\alpha(x))]+L_2^{\mathrm{T}}(x,\hat{x})$$
$$-\alpha'(x)G_\delta(x)[m_1(h_\delta(x))+m_2(h_\delta(x))+G_\delta^{\mathrm{T}}(x)V_{\mathrm{sub}}'^{\mathrm{T}}(x)]$$
$$+\alpha'(x)G_\delta(x)G_\delta^{\mathrm{T}}(x)\alpha'^{\mathrm{T}}(x)\hat{P}(\hat{x}-\alpha(x))\Big]<0,\quad (x,\hat{x})\in\mathbb{R}^n\times\mathbb{R}^m,\ \hat{x}\neq\alpha(x). \qquad (11.277)$$

Show that the zero solution $(x(t),\hat{x}(t))\equiv(0,0)$ of the nonlinear uncertain cascade system (11.261) and (11.262) is globally asymptotically stable for all $\delta(\cdot)\in\boldsymbol{\Delta}$, where $\boldsymbol{\Delta}$ is given by (11.27), with feedback control law

$$\phi(x,\hat{x})=-R_2^{-1}(x,\hat{x})[\tfrac{1}{2}L_2^{\mathrm{T}}(x,\hat{x})+\hat{P}(\hat{x}-\alpha(x))]. \qquad (11.278)$$

Furthermore, show that the performance functional (11.263) satisfies

$$\sup_{\delta(\cdot)\in\boldsymbol{\Delta}} J(x_0,\hat{x}_0,\phi(x,\hat{x}))\le\mathcal{J}(x_0,\hat{x}_0\phi(x,\hat{x}))$$
$$=V(x_0,\hat{x}_0),\quad (x_0,\hat{x}_0)\in\mathbb{R}^n\times\mathbb{R}^m, \qquad (11.279)$$

where

$$\mathcal{J}(x_0,\hat{x}_0,u(\cdot))\triangleq\int_0^\infty\Big[L_1(x,\hat{x})+L_2(x,\hat{x})u+u^{\mathrm{T}}R_2(x,\hat{x})u$$
$$+\Big\{[V_{\mathrm{sub}}'(x)-2(\hat{x}-\alpha(x))^{\mathrm{T}}\hat{P}\alpha'(x)]G_\delta(x)m_1(h_\delta(x))$$
$$+\tfrac{1}{4}[m(h_\delta(x))+G_\delta^{\mathrm{T}}(x)\{V_{\mathrm{sub}}'^{\mathrm{T}}(x)-2\alpha'^{\mathrm{T}}(x)P(\hat{x}-\alpha(x))\}]^{\mathrm{T}}$$
$$\cdot[m(h_\delta(x))+G_\delta^{\mathrm{T}}(x)\{V_{\mathrm{sub}}'^{\mathrm{T}}(x)-2\alpha'^{\mathrm{T}}(x)P(\hat{x}-\alpha(x))\}]\Big\}\Big]\mathrm{d}t, \qquad (11.280)$$

where $u(\cdot)$ is admissible, and $(x(t), \hat{x}(t))$, $t \geq 0$, solves (11.261) and (11.262) with $\delta(h_\delta(x)) \equiv 0$. In addition, show that the performance functional (11.280), with

$$\begin{aligned} L_1(x,\hat{x}) = {} & \phi^{\mathrm{T}}(x,\hat{x})R_2(x,\hat{x})\phi(x,\hat{x}) \\ & -[V'_{\mathrm{sub}}(x) - 2(\hat{x}-\alpha(x))\hat{P}\alpha'(x)](f_0(x)+G_0(x)\hat{x}) \\ & -\tfrac{1}{4}[m(h_\delta(x)) + G_\delta^{\mathrm{T}}(x)\{V'^{\mathrm{T}}_{\mathrm{sub}}(x) - 2\alpha'^{\mathrm{T}}(x)P(\hat{x}-\alpha(x))\}]^{\mathrm{T}} \\ & \cdot[m(h_\delta(x)) + G_\delta^{\mathrm{T}}(x)\{V'^{\mathrm{T}}_{\mathrm{sub}}(x) - 2\alpha'^{\mathrm{T}}(x)P(\hat{x}-\alpha(x))\}] \\ & -[V'_{\mathrm{sub}}(x) - 2(\hat{x}-\alpha(x))^{\mathrm{T}}\hat{P}\alpha'(x)]G_\delta(x)m_1(h_\delta(x)), \end{aligned} \tag{11.281}$$

is minimized in the sense that

$$\mathcal{J}(x_0,\hat{x}_0,\phi(x(\cdot))) = \min_{u(\cdot)\in\mathcal{S}(x_0,\hat{x}_0)} \mathcal{J}(x_0,\hat{x}_0,u(\cdot)). \tag{11.282}$$

Problem 11.30. Consider the nonlinear uncertain block cascade system

$$\dot{x}(t) = f_0(x(t)) + \Delta f(x(t)) + G(x(t))y(t), \quad x(0) = x_0, \quad t \geq 0, \tag{11.283}$$

$$\dot{\hat{x}}(t) = \hat{f}_0(\hat{x}(t)) + \Delta\hat{f}(\hat{x}(t)) + \hat{G}(\hat{x}(t))u(t), \quad \hat{x}(0) = \hat{x}_0, \tag{11.284}$$

$$y(t) = h(\hat{x}(t)), \tag{11.285}$$

where (11.128) has been augmented by a nonlinear input subsystem with $\Delta G(x) \equiv 0$, $\hat{x} \in \mathbb{R}^q$, $\hat{f} : \mathbb{R}^q \to \mathbb{R}^q$ satisfies $\hat{f}_0(0) = 0$, $h : \mathbb{R}^q \to \mathbb{R}^m$ satisfies $h(0) = 0$, $\hat{G} : \mathbb{R}^q \to \mathbb{R}^{q\times m}$, and $\hat{f}_0(\cdot)+\Delta\hat{f}(\cdot) \in \hat{\mathcal{F}}$, where $\hat{\mathcal{F}}$ is assumed to have the same form as $\mathcal{F}$ defined by (11.23). Assume that the input subsystem (11.284) and (11.285) is feedback strictly passive for all $\hat{f}_0(\cdot) + \Delta\hat{f}(\cdot) \in \hat{\mathcal{F}}$ such that there exist a positive-definite storage function $V_{\mathrm{s}} : \mathbb{R}^q \to \mathbb{R}$ and functions $\Gamma_{\mathrm{s}} : \mathbb{R}^q \to \mathbb{R}$ and $k : \mathbb{R}^q \to \mathbb{R}^m$ such that

$$0 \geq V'_{\mathrm{s}}(\hat{x})\Delta\hat{f}(\hat{x}) - \Gamma_{\mathrm{s}}(\hat{x}), \quad \hat{f}_0(\cdot) + \Delta\hat{f}(\cdot) \in \hat{\mathcal{F}}, \tag{11.286}$$

$$0 > V'_{\mathrm{s}}(\hat{x})\hat{f}_0(\hat{x}) + V'_{\mathrm{s}}(\hat{x})\hat{G}(\hat{x})k(\hat{x}) + \Gamma_{\mathrm{s}}(\hat{x}), \quad \hat{x} \in \mathbb{R}^n, \quad \hat{x} \neq 0, \tag{11.287}$$

$$0 = \tfrac{1}{2}\hat{G}^{\mathrm{T}}(\hat{x})V'^{\mathrm{T}}_{\mathrm{s}}(\hat{x}) - h(\hat{x}). \tag{11.288}$$

Also, assume that the subsystem (11.283) has a globally stable equilibrium at $x = 0$ with $y = 0$ for all $\Delta f(\cdot) \in \boldsymbol{\Delta}$ and Lyapunov function $V_{\mathrm{sub}}(x)$, $x \in \mathbb{R}^n$, so that

$$V'_{\mathrm{sub}}(x)\Delta f(x) \leq \Gamma_{\mathrm{sub}}(x), \qquad \Delta f(\cdot) \in \boldsymbol{\Delta}, \tag{11.289}$$

$$V'_{\mathrm{sub}}(x)f_0(x) + \Gamma_{\mathrm{sub}}(x) < 0, \qquad x \neq 0, \tag{11.290}$$

where $\Gamma_{\mathrm{sub}} : \mathbb{R}^n \to \mathbb{R}$ such that $\Gamma_{\mathrm{sub}}(0) = 0$. Furthermore, assume there exists a function $L_2 : \mathbb{R}^n \times \mathbb{R}^q \to \mathbb{R}^{1\times m}$ such that

$$L_2(0,0) = 0, \tag{11.291}$$

$$h^{\mathrm{T}}(\hat{x})\left\{G^{\mathrm{T}}(\hat{x})V'^{\mathrm{T}}_{\mathrm{sub}}(x) - \tfrac{1}{2}R_2^{-1}(x,\hat{x})L_2^{\mathrm{T}}(x,\hat{x}) - k(\hat{x})\right\} \leq 0. \tag{11.292}$$

Show that the zero solution $(x(t), \hat{x}(t)) \equiv (0,0)$ of the nonlinear uncertain block cascade system (11.283)–(11.285) is globally asymptotically stable for all $(f_0 + \Delta f, \hat{f}_0 + \Delta\hat{f}) \in \mathcal{F} \times \hat{\mathcal{F}}$ with the feedback control law

$$\phi(x, \hat{x}) = -\tfrac{1}{2}R_2^{-1}(x, \hat{x})\left[L_2^{\mathrm{T}}(x, \hat{x}) + h(\hat{x})\right]. \tag{11.293}$$

Furthermore, show that the performance functional (11.263) satisfies

$$J(x_0, \hat{x}_0, \phi(x(\cdot), \hat{x}(\cdot))) \le \mathcal{J}(x_0, \hat{x}_0, \phi(x(\cdot), \hat{x}(\cdot))) = V_{\mathrm{sub}}(x_0) + \hat{V}(\hat{x}_0), \tag{11.294}$$

where

$$\mathcal{J}(x_0, \hat{x}_0, u) \triangleq \int_0^\infty \left[L_1(x, \hat{x}) + L_2(x, \hat{x})u + u^{\mathrm{T}}R_2(x, \hat{x})u + \Gamma_{\mathrm{sub}}(x) + \Gamma_{\mathrm{s}}(\hat{x})\right]\mathrm{d}t, \tag{11.295}$$

where $u(\cdot)$ is admissible and $(x(t), \hat{x}(t))$, $t \ge 0$, solves (11.283) and (11.284) with $(\Delta f(x), \Delta\hat{f}(\hat{x}) \equiv (0,0)$. In addition, show that the performance functional (11.295) with

$$\begin{aligned} L_1(x, \hat{x}) &= \phi^{\mathrm{T}}(x, \hat{x})R_2(x, \hat{x})\phi(x, \hat{x}) - V_{\mathrm{sub}}'(x)[f_0(x) + G(x)h(\hat{x})] \\ &\quad - V_{\mathrm{s}}'(\hat{x})\hat{f}_0(\hat{x}) - \Gamma_{\mathrm{sub}}(x) - \Gamma_{\mathrm{s}}(\hat{x}), \end{aligned} \tag{11.296}$$

is minimized in the sense that

$$\mathcal{J}(x_0, \hat{x}_0, \phi(x(\cdot), \hat{x}(\cdot))) = \min_{u(\cdot) \in \mathcal{S}(x_0, \hat{x}_0)} \mathcal{J}(x_0, \hat{x}_0, u(\cdot)). \tag{11.297}$$

Problem 11.31. Consider the controlled nonlinear dynamical system

$$\dot{x}(t) = F(x(t), \sigma(u(t))), \qquad x(0) = x_0, \qquad t \ge 0, \tag{11.298}$$

where $u(\cdot) \in \mathcal{U}$ is an admissible input such that $u(t) \in U$, $t \ge 0$, where the control constraint set U is given such that $0 \in U$ and $\sigma(\cdot) \in \mathcal{M} \subset \{\sigma : U \to \mathbb{R}^m : \sigma(0) = 0\}$ denotes an input nonlinearity. Furthermore, consider the performance functional

$$J(x_0, u(\cdot)) \triangleq \int_0^\infty L(x(t), u(t))\mathrm{d}t, \tag{11.299}$$

where $L : \mathcal{D} \times U \to \mathbb{R}$. Assume there exist functions $V : \mathcal{D} \to \mathbb{R}$, $\Gamma : \mathcal{D} \times U \to \mathbb{R}$, and control law $\phi : \mathcal{D} \to U$, where $V(\cdot)$ is a continuously differentiable function, such that

$$V(0) = 0, \tag{11.300}$$

$$V(x) > 0, \quad x \in \mathcal{D}, \quad x \ne 0, \tag{11.301}$$

$$\phi(0) = 0, \tag{11.302}$$

$$V'(x)F(x, \sigma(\phi(x))) \le V'(x)F(x, \sigma_0(\phi(x))) + \Gamma(x, \phi(x)), \quad x \in \mathcal{D}, \quad \sigma(\cdot) \in \mathcal{M}, \tag{11.303}$$

$$V'(x)F(x, \sigma_0(\phi(x))) + \Gamma(x, \phi(x)) < 0, \quad x \in \mathcal{D}, \quad x \ne 0, \tag{11.304}$$

$$H(x,\phi(x))=0, \quad x\in\mathcal{D}, \tag{11.305}$$
$$H(x,u)\geq 0, \quad x\in\mathcal{D}, \quad u\in U, \tag{11.306}$$

where $\sigma_0(\cdot)\in\mathcal{M}$ is a given nominal input nonlinearity and

$$H(x,u)\triangleq L(x,u)+V'(x)F(x,\sigma_0(u))+\Gamma(x,u). \tag{11.307}$$

Show that, with the feedback control $u(\cdot)=\phi(x(\cdot))$, there exists a neighborhood $\mathcal{D}_0\subseteq\mathcal{D}$ of the origin such that if $x_0\in\mathcal{D}_0$, the zero solution $x(t)\equiv 0$ of the closed-loop system

$$\dot{x}(t)=F(x(t),\sigma(\phi(x(t)))), \qquad x(0)=x_0, \qquad t\geq 0, \tag{11.308}$$

is locally asymptotically stable for all $\sigma(\cdot)\in\mathcal{M}$. Furthermore, show that

$$\sup_{\sigma(\cdot)\in\mathcal{M}} J(x_0,\phi(x(\cdot)))\leq\mathcal{J}(x_0,\phi(x(\cdot)))=V(x_0), \tag{11.309}$$

where

$$\mathcal{J}(x_0,u(\cdot))\triangleq\int_0^\infty[L(x(t),u(t))+\Gamma(x(t),u(t))]\mathrm{d}t, \tag{11.310}$$

and where $u(\cdot)$ is admissible and $x(t)$, $t\geq 0$, solves (11.298) with $\sigma(u(t))=\sigma_0(u(t))$. In addition, if $x_0\in\mathcal{D}_0$ then show that the feedback control $u(\cdot)=\phi(x(\cdot))$ minimizes $\mathcal{J}(x_0,u(\cdot))$ in the sense that

$$\mathcal{J}(x_0,\phi(x(\cdot)))=\min_{u(\cdot)\in\mathcal{S}(x_0)}\mathcal{J}(x_0,u(\cdot)), \tag{11.311}$$

where

$$\begin{aligned}\mathcal{S}(x_0)\triangleq\{&u(\cdot):u(\cdot)\text{ is admissible and }x(\cdot)\text{ given by (11.298)}\\ &\text{satisfies }x(t)\to 0\text{ as }t\to\infty\text{ with }\sigma(\cdot)=\sigma_0(\cdot)\}.\end{aligned}$$

Finally, if $\mathcal{D}=\mathbb{R}^n$, $U=\mathbb{R}^m$, and

$$V(x)\to\infty\text{ as }\|x\|\to\infty,$$

show that the solution $x(t)=0$, $t\geq 0$, of the closed-loop system (11.308) is globally asymptotically stable.

Problem 11.32. Consider the controlled nonlinear dynamical system

$$\dot{x}(t)=f(x(t))+G(x(t))\sigma(u(t)), \quad x(0)=x_0, \qquad t\geq 0, \tag{11.312}$$

where $f:\mathbb{R}^n\to\mathbb{R}^n$ satisfies $f(0)=0$, $G:\mathbb{R}^n\to\mathbb{R}^{n\times m}$, $\mathcal{D}=\mathbb{R}^n$, and

$$\begin{aligned}\sigma(\cdot)\in\mathcal{M}\triangleq\{&\sigma:\mathbb{R}^m\to\mathbb{R}^m:\sigma(0)=0,\\ &[\sigma(u)-M_1u]^{\mathrm{T}}[\sigma(u)-M_2u]\leq 0, u\in\mathbb{R}^m\},\end{aligned} \tag{11.313}$$

where $M_1,M_2\in\mathbb{R}^{m\times m}$ are given diagonal matrices such that $M\triangleq M_2-M_1$ is positive definite, and $u(t)\in\mathbb{R}^m$ for all $t\geq 0$. Furthermore, consider the performance functional

$$J(x_0,u(\cdot)) = \int_0^\infty [L_1(x(t)) + L_2(x(t))u(t) + u^{\mathrm{T}}(t)R_2(x(t))u(t)]\mathrm{d}t, \quad (11.314)$$

where $L_1 : \mathbb{R}^n \to \mathbb{R}$, $L_2 : \mathbb{R}^n \to \mathbb{R}^{1\times m}$, and $R_2 : \mathbb{R}^n \to \mathbb{P}^m$. Assume there exist a continuously differentiable function $V : \mathbb{R}^n \to \mathbb{R}$ and a function $L_2 : \mathbb{R}^n \to \mathbb{R}^{1\times m}$ such that

$$V(0) = 0, \quad (11.315)$$

$$L_2(0) = 0, \quad (11.316)$$

$$V(x) > 0, \quad x \in \mathbb{R}^n, \quad x \neq 0, \quad (11.317)$$

$$\begin{aligned} &V'(x)[f(x) - \tfrac{1}{2}G(x)M_1R_{2a}^{-1}(x)V_a(x)] \\ &+\tfrac{1}{4}[-\tfrac{1}{2}MR_{2a}^{-1}(x)V_a(x) + G^{\mathrm{T}}(x)V'^{\mathrm{T}}(x)]^{\mathrm{T}} \\ &\cdot[-\tfrac{1}{2}MR_{2a}^{-1}(x)V_a(x) + G^{\mathrm{T}}(x)V'^{\mathrm{T}}(x)] < 0, \ x \in \mathbb{R}^n, \ x \neq 0, \end{aligned} \quad (11.318)$$

and

$$V(x) \to \infty \text{ as } \|x\| \to \infty, \quad (11.319)$$

where $R_{2a}(x) \triangleq R_2(x) + \frac{1}{4}M^{\mathrm{T}}M$ and $V_a(x) \triangleq [L_2(x) + \frac{1}{2}V'(x)G(x)(M_1 + M_2)]^{\mathrm{T}}$. Show that, with the feedback control law

$$\phi(x) = -\tfrac{1}{2}R_{2a}^{-1}(x)V_a(x), \quad (11.320)$$

the zero solution $x(t) \equiv 0$ of the nonlinear system (11.312) is globally asymptotically stable for all $\sigma(\cdot) \in \mathcal{M}$. Furthermore, show that the performance functional (11.314) satisfies

$$\sup_{\sigma(\cdot)\in\mathcal{M}} J(x_0,\phi(x(\cdot))) \leq \mathcal{J}(x_0,\phi(x(\cdot))) = V(x_0), \quad (11.321)$$

where

$$\begin{aligned}\mathcal{J}(x_0,u(\cdot)) \triangleq &\int_0^\infty [L_1(x) + L_2(x)u + u^{\mathrm{T}}R_2(x)u + \tfrac{1}{4}[Mu + G^{\mathrm{T}}(x)V'^{\mathrm{T}}(x)]^{\mathrm{T}} \\ &\cdot[Mu + G^{\mathrm{T}}(x)V'^{\mathrm{T}}(x)]\mathrm{d}t, \end{aligned} \quad (11.322)$$

where $u(\cdot)$ is admissible, and $x(t)$, $t \geq 0$, solves (11.312) with $\sigma(u) = M_1u$. In addition, show that the performance functional (11.322), with

$$L_1(x) = \phi^{\mathrm{T}}(x)R_{2a}(x)\phi(x) - V'(x)f(x) - \tfrac{1}{4}V'(x)G(x)G^{\mathrm{T}}(x)V'^{\mathrm{T}}(x), \quad (11.323)$$

is minimized in the sense that

$$\mathcal{J}(x_0,\phi(x(\cdot))) = \min_{u(\cdot)\in\mathcal{S}(x_0)} \mathcal{J}(x_0,u(\cdot)), \quad (11.324)$$

where

$$\mathcal{S}(x_0) \triangleq \{u(\cdot) : u(\cdot) \text{ is admissible and } x(\cdot) \text{ given by (11.312)}$$

satisfies $x(t) \to 0$ as $t \to \infty$ with $\sigma(u) = M_1 u\}$.

Problem 11.33. Consider the controlled dynamical system

$$\dot{x}(t) = Ax(t) + B\sigma(u(t)), \quad x(0) = x_0, \quad t \geq 0, \tag{11.325}$$

where $A \in \mathbb{R}^{n\times n}$, $B \in \mathbb{R}^{n\times m}$, and $\sigma(\cdot) \subset \mathcal{M}$, where $\mathcal{M}$ is given by (11.313), and $u(t) \in \mathbb{R}^m$ for all $t \geq 0$, with performance functional

$$J(x_0, u(\cdot)) = \int_0^\infty [x^{\mathrm{T}}(t)R_1 x(t) + 2x^{\mathrm{T}}(t)R_{12}u(t) + u^{\mathrm{T}}(t)R_2 u(t)]\mathrm{d}t, \tag{11.326}$$

where $R_1 \in \mathbb{R}^{n\times n}$, $R_{12} \in \mathbb{R}^{n\times m}$, and $R_2 \in \mathbb{R}^{m\times m}$ such that $R_1 > 0$, $R_2 > 0$, and $R_1 - R_{12}R_2^{-1}R_{12}^{\mathrm{T}} \geq 0$. Assume there exists an $n \times n$ positive-definite matrix P satisfying

$$0 = A^{\mathrm{T}}P + PA + R_1 + PBB^{\mathrm{T}}P - P_a^{\mathrm{T}}R_{2a}^{-1}P_a, \tag{11.327}$$

and let K be given by

$$K = -R_{2a}^{-1}P_a, \tag{11.328}$$

where $P_a \triangleq \frac{1}{2}(M_1+M_2)B^{\mathrm{T}}P+R_{12}^{\mathrm{T}}$ and $R_{2a} \triangleq R_2+\frac{1}{4}M^{\mathrm{T}}M$. Show that, with $u = Kx$, the zero solution $x(t) \equiv 0$ to (11.325) is globally asymptotically stable for all $\sigma(\cdot) \in \mathcal{M}$. If, in addition, $\sigma(\cdot) \in \mathcal{M}_{\mathrm{b}}$, where

$$\mathcal{M}_{\mathrm{b}} \triangleq \{\sigma : \mathbb{R}^m \to \mathbb{R}^m : M_{1_i}u_i^2 \leq \sigma_i(u)u_i \leq M_{2_i}u_i^2, \\ \underline{u_i} \leq u_i \leq \overline{u_i},\ i = 1, ..., m\}, \tag{11.329}$$

where $\underline{u_i} < 0$ and $\overline{u_i} > 0$, $i = 1, \ldots, m$, are given, then show that the zero solution $x(t) \equiv 0$ to (11.325) with $u = Kx$ is locally asymptotically stable, and

$$\mathcal{D}_{\mathrm{A}} \triangleq \{x \in \mathcal{B} :\ x^{\mathrm{T}}Px \leq V_\Gamma\}, \tag{11.330}$$

where $\mathcal{B} \triangleq \bigcap_{i=1}^m \mathcal{B}_i$, $\mathcal{B}_i \triangleq \{x \in \mathbb{R}^n : \underline{u_i} \leq \phi_i(x) \leq \overline{u_i}\}$ and

$$V_\Gamma = \min\left\{\min_{i=1,\ldots,m} \frac{\overline{u_i}^2}{\mathrm{row}_i(K)P^{-1}\mathrm{row}_i^{\mathrm{T}}(K)},\ \min_{i=1,\ldots,m} \frac{\underline{u_i}^2}{\mathrm{row}_i(K)P^{-1}\mathrm{row}_i^{\mathrm{T}}(K)}\right\}, \tag{11.331}$$

is a subset of the domain of attraction for (11.325). Furthermore, show that the performance functional (11.326) satisfies

$$\sup_{\sigma(\cdot)\in\mathcal{M}} J(x_0, \phi(x(\cdot))) \leq \mathcal{J}(x_0, \phi(x(\cdot))) = V(x_0), \tag{11.332}$$

where

$$\mathcal{J}(x_0, u(\cdot)) \triangleq \int_0^\infty [x^{\mathrm{T}}R_1 x + 2x^{\mathrm{T}}R_{12}u + u^{\mathrm{T}}R_2 u + (\tfrac{1}{2}Mu + B^{\mathrm{T}}Px)^{\mathrm{T}} \\ \cdot(\tfrac{1}{2}Mu + B^{\mathrm{T}}Px)]\mathrm{d}t, \tag{11.333}$$

where $u(\cdot)$ is admissible, and $x(t)$, $t \geq 0$, solves (11.325) with $\sigma(u) = M_1 u$. In addition, show that the performance functional (11.333) is minimized in the sense that

$$\mathcal{J}(x_0, \phi(x(\cdot))) = \min_{u(\cdot)\in\mathcal{S}(x_0)} \mathcal{J}(x_0, u(\cdot)), \tag{11.334}$$

where

$$\mathcal{S}(x_0) \triangleq \{u(\cdot) : u(\cdot) \text{ is admissible and } x(\cdot) \text{ given by (11.325)} \\ \text{satisfies } x(t) \to 0 \text{ as } t \to \infty \text{ with } \sigma(u) = M_1 u\}.$$

11.11 Notes and References

Quadratic Lyapunov functions form the basis for linear-quadratic control theory and have provided one of the principal tools for robust analysis and synthesis [25–27,29,41,59,83,99,115,131,194,195,200,233,237,243,266,267, 300,312,313,337,347,352,356,401,416,430,442,447,471,472,480]. Among the earliest Lyapunov function frameworks for linear robust control were those developed by Michael and Merriam [313], Chang and Peng [83], Horisberger and Belanger [200], Vinkler and Wood [447], and Leitmann [266, 267]. In contrast to the above cited literature, the work of Chang and Peng [83] additionally addressed bounds on worst-case quadratic performance within full-state feedback control design. Extensions of the guaranteed cost control approach of Chang and Peng to full- and reduced-order dynamic control were addressed by Bernstein [42], Bernstein and Haddad [48,51], and Haddad and Bernstein [146]. A systematic treatment of quadratic Lyapunov bounds is given by Bernstein and Haddad [50]. More recent results involving shifted quadratic guaranteed cost bounds for robust stability and performance are developed by Haddad, Chellaboina, and Bernstein [161], and Bernstein and Osburn [53].

Connections between absolute stability theory and robust control were first noted by Popov [363] where the notion of hyperstability was used to describe nonlinear robustness implicit in the inequalities of Lyapunov stability theory. The input-output functional analysis approach to absolute stability theory was identified by Zames [476] as a way of capturing system uncertainty. A more modern treatment of the connections of absolute stability and stability robustness was given by Safonov [377] using topological separation of graphs of feedback operators. More recently, explicit connections between absolute stability theory and robust stability and performance were given by Haddad and Bernstein [147, 148, 151] and Haddad, How, Hall, and Bernstein [172] using fixed and parameter-dependent Lyapunov functions.

Even though the theory of nonlinear robust control for nonlinear uncertain systems with parametric uncertainty remains relatively undeveloped in comparison to linear robust control, notable exceptions include the work of

Spong [412], Basar and Bernhard [31], van der Schaft [438,439,441], Freeman and Kokotović [128], Haddad, Chellaboina, and Fausz [164], and Haddad, Chellaboina, Fausz, and Leonessa [166]. The nonlinear-nonquadratic robust control framework presented in this chapter is adopted from Haddad, Chellaboina, and Fausz [164] and Haddad, Chellaboina, Fausz, and Leonessa [166].

Chapter Twelve

Structured Parametric Uncertainty and Parameter-Dependent Lyapunov Functions

12.1 Introduction

The analysis and synthesis of robust feedback controllers entails a fundamental distinction between parametric and nonparametric uncertainty. Parametric uncertainty refers to plant uncertainty that is modeled as constant real parameters, whereas nonparametric uncertainty refers to uncertain transfer function gains that may be modeled as complex frequency-dependent quantities or nonlinear dynamic operators. In the time domain, nonparametric uncertainty is manifested as uncertain real parameters that may be time varying.

The distinction between parametric and nonparametric uncertainty is critical to the achievable performance of feedback control systems. For example, in the problem of vibration suppression for flexible structures, if stiffness matrix uncertainty is modeled as nonparametric uncertainty, then perturbations to the damping matrix will inadvertently be allowed. Predictions of stability and performance for given feedback gains will consequently be extremely conservative, thus limiting achievable performance [52]. Alternatively, this problem can be viewed by considering the classical analysis of Hill's equation (e.g., the Mathieu equation) which shows that time-varying parameter variations can destabilize a system even when the parameter variations are confined to a region in which constant variations are *non*destabilizing. Consequently, a feedback controller designed for time-varying parameter variations will unnecessarily sacrifice performance when the uncertain real parameters are actually constant.

To further illuminate the above discussion consider the nonlinear uncertain system

$$\dot{x}(t) = f_0(x(t)) + \Delta f(x(t)), \qquad x(0) = x_0, \qquad t \geq 0, \tag{12.1}$$

where $x(t) \in \mathbb{R}^n$ is the system state vector, $f_0 : \mathbb{R}^n \to \mathbb{R}^n$ satisfies $f_0(0) = 0$, and $\Delta f(\cdot) \in \mathbf{\Delta} \subset \{\Delta f : \mathbb{R}^n \to \mathbb{R}^n : \Delta f(0) = 0\}$, where $\mathbf{\Delta}$ is a set of system perturbations. To determine whether the zero solution $x(t) \equiv 0$ to (12.1) remains stable, one can construct functions $V : \mathbb{R}^n \to \mathbb{R}$ and $\Gamma : \mathbb{R}^n \to \mathbb{R}$, where $V(\cdot)$ is continuously differentiable, such that $V(0) = 0$, $V(x) > 0$, $x \in \mathbb{R}^n$, $x \neq 0$, $V(x) \to \infty$ as $\|x\| \to \infty$,

$$V'(x)\Delta f(x) \le \Gamma(x), \quad x \in \mathbb{R}^n, \quad \Delta f(\cdot) \in \mathbf{\Delta}, \tag{12.2}$$

$$V'(x)f_0(x) + \Gamma(x) < 0, \quad x \in \mathbb{R}^n, \quad x \neq 0, \tag{12.3}$$

$$L(x) + V'(x)f_0(x) + \Gamma(x) = 0, \quad x \in \mathbb{R}^n, \tag{12.4}$$

where $L : \mathbb{R}^n \to \mathbb{R}$ and satisfies $L(x) > 0$, $x \in \mathbb{R}^n$, $x \neq 0$. Now, it follows from Corollary 11.1 that the zero solution $x(t) \equiv 0$ to (12.1) is globally asymptotically stable for all $\Delta f(\cdot) \in \mathbf{\Delta}$, and the performance functional

$$J_{\Delta f}(x_0) = \int_0^\infty L(x(t))\mathrm{d}t, \tag{12.5}$$

satisfies the bound $\sup_{\Delta f(\cdot) \in \mathbf{\Delta}} J_{\Delta f}(x_0) \le V(x_0)$, $x_0 \in \mathbb{R}^n$.

As shown in Section 11.3, although the Lyapunov-function-based framework discussed above applies to problems in which $f(x)$ is perturbed by an uncertain function $\Delta f(x)$, a reinterpretation of these bounds yields standard nonlinear system theoretic criteria. For example, the bounding function (11.26) forms the basis for nonlinear nonexpansivity theory while the bounding function (11.28) forms the basis for nonlinear passivity theory. Although not immediately evident, a defect of the above framework is the fact that stability is guaranteed even if Δf is an explicit function of t. This observation follows from the fact that the Lyapunov derivative $\dot{V}(x(t)) \triangleq V'(x(t))[f(x(t)) + \Delta f(t, x(t))]$ need only be negative for each fixed value of time t. Although this feature is desirable if Δf is time-varying, as discussed above, it leads to conservatism when Δf is actually time invariant. This defect, however, can be remedied as in the linear robust control literature [145, 147, 151] by utilizing an alternative approach based upon *parameter-dependent Lyapunov functions*. The idea behind parameter-dependent Lyapunov functions is to allow the Lyapunov function to be a function of the uncertainty Δf. In the usual case, $V(x)$ is a fixed function, whereas a parameter-dependent Lyapunov function represents a *family* of Lyapunov functions.

To demonstrate parameter-dependent Lyapunov functions for robust analysis of nonlinear systems in a manner consistent with the above discussion, consider the Lyapunov function

$$V(x) = V_{\mathrm{I}}(x) + V_{\Delta f}(x), \tag{12.6}$$

where $V : \mathbb{R}^n \to \mathbb{R}$ satisfies the above conditions with (12.2)–(12.4) replaced by

$$V_{\mathrm{I}}'(x)\Delta f(x) \le \Gamma(x) - V_{\Delta f}'(x)(f_0(x) + \Delta f(x)), \quad x \in \mathbb{R}^n, \quad \Delta f(\cdot) \in \boldsymbol{\Delta}, \tag{12.7}$$

$$V_{\mathrm{I}}'(x)f_0(x) + \Gamma(x) < 0, \quad x \in \mathbb{R}^n, \quad x \neq 0, \tag{12.8}$$

$$0 = L(x) + V_{\mathrm{I}}'(x)f_0(x) + \Gamma(x), \quad x \in \mathbb{R}^n, \tag{12.9}$$

respectively. In contrast to the framework developed in Chapter 11, the bounding function $\Gamma(\cdot)$ is *not* assumed to satisfy (12.2) and (12.3) but rather (12.7) and (12.8). Note that if $V_{\Delta f}(x)$ is identically zero, then (12.7) and (12.8) specialize to (12.2) and (12.3). The idea behind this framework is that only the fixed part of the Lyapunov function $V_{\mathrm{I}}(x)$ is a solution to the steady-state Hamilton-Jacobi-Bellman equation for the nominal system, while the overall Lyapunov function $V(x)$ is needed to establish robust stability.

For practical purposes the form of the parameter-dependent Lyapunov function $V(x)$ given by (12.6) is useful since the presence of Δf within (12.6) restricts the allowable time-varying uncertain parameters. That is, if $\Delta f(t, x(t))$ were permitted, then terms involving $\frac{\partial \Delta f(t,x)}{\partial t}$ would arise and potentially subvert the negative definiteness of $\dot{V}(x)$.

In this chapter, we extend the framework developed in Chapter 11 to address the problem of optimal nonlinear-nonquadratic robust feedback control via parameter-dependent Lyapunov functions. Specifically, we transform a robust nonlinear control problem into an optimal control problem. This is accomplished by properly modifying the cost functional to account for system uncertainty so that the solution of the modified optimal nonlinear control problem serves as the solution to the robust control problem. The present framework generalizes the linear guaranteed cost control approach via parameter-dependent Lyapunov functions for addressing robust stability and performance [145, 147, 151, 172] to nonlinear uncertain systems with nonlinear-nonquadratic performance functionals.

The main contribution of this chapter is a methodology for designing nonlinear controllers that provide both robust stability and robust performance over a prescribed range of nonlinear time-invariant, structured real parameter system uncertainty. The present framework is an extension of the parameter-dependent Lyapunov function approach developed in [145, 147, 151, 162] to nonlinear systems by utilizing a performance bound to provide robust performance in addition to robust stability. In particular, the performance bound can be evaluated in closed form as long as the nonlinear-nonquadratic cost functional considered is related in a specific way to the parameter-independent part of an underlying parameter-dependent Lyapunov function that is composed of a fixed (parameter-independent) and a variable (parameter-dependent) part that guarantees robust stability over a prescribed nonlinear time-invariant, real parameter uncertainty set. The fixed part of the Lyapunov function is shown to be the solution to the steady-state form of the Hamilton-Jacobi-Bellman equation for the nominal system and plays a key role in constructing the optimal nonlinear robust control law.

12.2 Linear Uncertain Systems and the Structured Singular Value

For completeness, in this section we redirect our attention to the problem of robust stability of *linear* systems with *structured* real and complex parameter uncertainty. A general framework for this problem is provided by mixed-μ theory [117] as a refinement of the complex structured singular value. The ability of the structured singular value to account for complex, real, and mixed uncertainty provides a powerful framework for robust stability and performance problems in both analysis and synthesis (see [110,117,344,381,475] and the references therein). Since exact computation of the structured singular value is, in general, an intractable problem, the development of practically implementable bounds remains a high priority in robust control research. Recent work in this area includes upper and lower bounds for mixed uncertainty [117, 150, 153, 160, 263, 475] as well as LMI-based computational techniques [66, 130].

An alternative approach to developing bounds for the structured singular value is to specialize absolute stability criteria for sector-bounded nonlinearities to the case of linear uncertainty. This approach, which has been explored by Chiang and Safonov [93], Haddad and Bernstein [145, 147, 148, 150, 151], How and Hall [206], and Haddad *et al.* [172], demonstrates the direct applicability of the classical theory of absolute stability to the modern structured singular value framework. In particular, the rich theory of multiplier-based absolute stability criteria due to Luré and Postnikov [5, 265, 331, 364], Popov [362], Yakubovich [469, 470], Zames and Falb [479], and numerous others can be seen to have a close, fundamental relationship with recently developed structured singular value bounds.

We start by stating and proving an absolute stability criterion for multivariable systems with generalized positive real frequency-dependent stability multipliers. This criterion involves a square nominal transfer function $G(s)$ in a negative feedback interconnection with a complex, square, uncertain matrix Δ as shown in Figure 12.1. Specifically, we consider the set of block-diagonal matrices with possibly repeated blocks defined by

$$\begin{aligned}\boldsymbol{\Delta}_{\mathrm{bs}} \triangleq \{\Delta \in \mathbb{C}^{m\times m} : \Delta = \text{block-diag}[I_{l_1} \otimes \Delta_1, I_{l_2} \otimes \Delta_2, \ldots, I_{l_{r+c}} \otimes \Delta_{r+c}],\\ \Delta_i \in \mathbb{R}^{m_i\times m_i}, i = 1, \ldots, r;\ \Delta_i \in \mathbb{C}^{m_i\times m_i},\ i = r+1, \ldots, r+c\},\end{aligned}$$

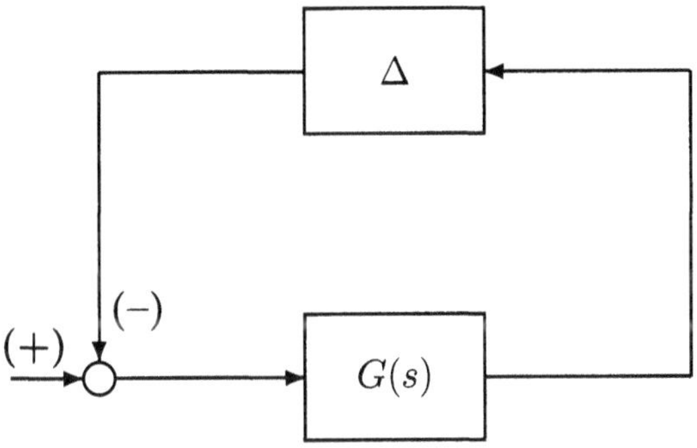

Figure 12.1 Interconnection of transfer function $G(s)$ with uncertain matrix Δ.

(12.10)

where the dimension m_i and the number of repetitions l_i of each block are given and $r+c \geq 1$. Furthermore, define the subset $\boldsymbol{\Delta} \subseteq \boldsymbol{\Delta}_{\mathrm{bs}}$ consisting of sector-bounded matrices

$$\boldsymbol{\Delta} \triangleq \{\Delta = \Delta^* \in \boldsymbol{\Delta}_{\mathrm{bs}} : \; M_1 \leq \Delta \leq M_2\}, \tag{12.11}$$

where $M_1, M_2 \in \boldsymbol{\Delta}_{\mathrm{bs}}$ are Hermitian matrices such that $M \triangleq M_2 - M_1$ is positive definite. Note that M_1 and M_2 are elements of $\boldsymbol{\Delta}$.

To prove the multivariable absolute stability criterion for sector-bounded uncertain matrices, define the sets $\mathcal{D}$ and $\mathcal{N}$ of Hermitian frequency-dependent scaling matrix functions by

$$\mathcal{D} \triangleq \{D : \jmath\mathbb{R} \to \mathbb{C}^{m\times m} : D(\jmath\omega) \geq 0, D(\jmath\omega)\Delta = \Delta D(\jmath\omega), \omega \in \mathbb{R}, \Delta \in \boldsymbol{\Delta}_{\mathrm{bs}}\}, \tag{12.12}$$

and

$$\mathcal{N} \triangleq \{N : \jmath\mathbb{R} \to \mathbb{C}^{m\times m} : N(\jmath\omega) = N^*(\jmath\omega), N(\jmath\omega)\Delta = \Delta N(\jmath\omega), \\ \omega \in \mathbb{R}, \Delta \in \boldsymbol{\Delta}_{\mathrm{bs}}\}. \tag{12.13}$$

Furthermore, define the set $\mathcal{Z}$ of complex multiplier matrix functions by

$$\mathcal{Z} \triangleq \{Z : \jmath\mathbb{R} \to \mathbb{C}^{m\times m} : Z(\jmath\omega) = D(\jmath\omega) - \jmath N(\jmath\omega), D(\cdot) \in \mathcal{D}, N(\cdot) \in \mathcal{N}\}. \tag{12.14}$$

Note that if $Z(\cdot) \in \mathcal{Z}$, $D(\cdot) \in \mathcal{D}$, and $N(\cdot) \in \mathcal{N}$, then $Z(\jmath\omega) = D(\jmath\omega) - \jmath N(\jmath\omega)$ if and only if $D(\jmath\omega) = \mathrm{He}\, Z(\jmath\omega)$ and $N(\jmath\omega) = \jmath\mathrm{Sh}\, Z(\jmath\omega)$. Hence, since $D(\jmath\omega) \geq 0$, $\omega \in \mathbb{R} \cup \infty$, $Z(\cdot) \in \mathcal{Z}$ consists of arbitrary *generalized* positive real functions [9]. For $\Delta \in \boldsymbol{\Delta}_{\mathrm{bs}}$, $\mathcal{D}$ and $\mathcal{N}$ are given by

$$\mathcal{D} = \{D : \jmath\mathbb{R} \to \mathbb{C}^{m\times m} : D = \text{block–diag}[D_1 \otimes I_{m_1}, D_2 \otimes I_{m_2}, \ldots,$$

$$D_{r+c} \otimes I_{m_{r+c}}],\ 0 \le D_i \in \mathbb{C}^{l_i \times l_i}, i = 1, \ldots, r+c\}, \qquad (12.15)$$

$$\mathcal{N} = \{N : \jmath\mathbb{R} \to \mathbb{C}^{m \times m} : N = \text{block-diag}[N_1 \otimes I_{m_1}, N_2 \otimes I_{m_2}, \ldots, N_{r+c} \otimes I_{m_{r+c}}], N_i = N_i^* \in \mathbb{C}^{l_i \times l_i}, N_i \otimes \Delta_i = N_i \otimes \Delta_i, i = 1, \ldots, r+c\}. \qquad (12.16)$$

Although the condition $D(\jmath\omega)\Delta = \Delta D(\jmath\omega)$ in $\mathcal{D}$ arises in complex and mixed-μ analysis [117], the condition $N(\jmath\omega)\Delta = \Delta N(\jmath\omega)$ in $\mathcal{N}$ has no counterpart in [117]. As shown in [153] this condition generalizes mixed-μ analysis to address nondiagonal real matrices which are not considered in standard mixed-μ theory. The condition $N(\jmath\omega)\Delta = \Delta N(\jmath\omega)$ is an extension of the condition used in [145] for Popov controller synthesis with constant real matrix uncertainty.

Next, we introduce the following key lemma.

Lemma 12.1. Let $Z(\cdot) \in \mathcal{Z}$, let $\omega \in \mathbb{R} \cup \infty$, and suppose $\det(I + G(\jmath\omega)M_1) \neq 0$. If

$$\text{He}\,[Z(\jmath\omega)(M^{-1} + (I + G(\jmath\omega)M_1)^{-1}G(\jmath\omega))] > 0, \qquad (12.17)$$

then $\det(I + G(\jmath\omega)\Delta) \neq 0$ for all $\Delta \in \mathbf{\Delta}$.

Proof. For notational convenience we write D for $D(\jmath\omega)$ and N for $N(\jmath\omega)$. Suppose that there exists $\Delta \in \mathbf{\Delta}$ such that $\det(I + G(\jmath\omega)\Delta) = 0$. Then there exists $x \in \mathbb{C}^m$, $x \neq 0$, such that $(I + \Delta G(\jmath\omega))x = 0$. Hence, $-x = \Delta G(\jmath\omega)x$ and $-x^* = x^*G^*(\jmath\omega)\Delta$.

Since $M_1 \le \Delta \le M_2$, it follows that

$$(\Delta - M_1)M^{-1}(\Delta - M_1) - (\Delta - M_1) \le 0,$$

or, equivalently,

$$\text{He}[\Delta M^{-1}\Delta - 2\Delta M^{-1}M_1 + M_1M^{-1}M_1 - \Delta + M_1] \le 0. \qquad (12.18)$$

Now, since $D(\cdot) \in \mathcal{D}$ and $M_1, M_2 \in \mathbf{\Delta}$ it follows that $DM_1 = M_1D$ and $DM^{-1} = M^{-1}D$. Next, forming D(12.18) yields

$$\text{He}[\Delta DM^{-1}\Delta - 2\Delta DM^{-1}M_1 + M_1DM^{-1}M_1 - \Delta D + DM_1] \le 0. \qquad (12.19)$$

Furthermore, forming $x^*G^*(\jmath\omega)(12.19)G(\jmath\omega)x$ yields

$$x^*\text{He}[DM^{-1} + 2DM^{-1}M_1G(\jmath\omega) + G^*(\jmath\omega)M_1DM^{-1}M_1G(\jmath\omega) + DG(\jmath\omega) + G^*(\jmath\omega)DM_1G(\jmath\omega)]x \le 0. \qquad (12.20)$$

Next, note that $\text{He}\,[Z(\jmath\omega)(M^{-1} + (I + G(\jmath\omega)M_1)^{-1}G(\jmath\omega))] > 0$ is equivalent to

$$\mathrm{He}[Z(\jmath\omega)(M^{-1}+G(\jmath\omega)(I+M_1G(\jmath\omega))^{-1})]>0.$$

Now, premultiplying and postmultiplying the above inequality by $I + G^*(\jmath\omega)M_1$ and $I + M_1G(\jmath\omega)$ yields

$$\mathrm{He}[(I+G^*(\jmath\omega)M_1)Z(\jmath\omega)(M^{-1}+M^{-1}M_1G(\jmath\omega)+G(\jmath\omega))]>0. \quad (12.21)$$

Since $N(\cdot)\in\mathcal{N}$ and $M_1, M_2\in\boldsymbol{\Delta}$ it follows that $NM_1 = M_1N$ and $NM_2 = M_2N$. Thus, $NM = MN$, and hence, $NM^{-1} = M^{-1}N$. Using these relations (12.21) simplifies to

$$\begin{aligned}\mathrm{He}[DM^{-1}&+2DM^{-1}M_1G(\jmath\omega)+G^*(\jmath\omega)M_1DM^{-1}M_1G(\jmath\omega)+DG(\jmath\omega)\\ &+G^*(\jmath\omega)DM_1G(\jmath\omega)]>\mathrm{He}[\jmath NG(\jmath\omega)]. \quad (12.22)\end{aligned}$$

Now, forming $x^*(12.22)x$ yields

$$\begin{aligned}x^*\mathrm{He}[DM^{-1}&+2DM^{-1}M_1G(\jmath\omega)+G^*(\jmath\omega)M_1DM^{-1}M_1G(\jmath\omega)\\ &+DG(\jmath\omega)+G^*(\jmath\omega)DM_1G(\jmath\omega)]x\\ &>\mathrm{He}[\jmath x^*NG(\jmath\omega)x]\\ &=-x^*G^*(\jmath\omega)\mathrm{He}[\jmath\Delta N]G(\jmath\omega)x.\end{aligned}$$

Since $N(\cdot)\in\mathcal{N}$, it follows that $\mathrm{He}[\jmath\Delta N]=\frac{\jmath}{2}(\Delta N - N\Delta)=0$, and hence,

$$\begin{aligned}x^*\mathrm{He}[DM^{-1}&+2DM^{-1}M_1G(\jmath\omega)+G^*(\jmath\omega)M_1DM^{-1}M_1G(\jmath\omega)\\ &+DG(\jmath\omega)+G^*(\jmath\omega)DM_1G(\jmath\omega)]x>0,\end{aligned}$$

which contradicts (12.20). Consequently, $\det(I+G(\jmath\omega)\Delta)\neq 0$ for all $\Delta\in\boldsymbol{\Delta}$. □

For the next result we assume that the feedback interconnection of $G(s)$ and Δ is well posed, that is, $\det[I+G(\infty)\Delta]\neq 0$ for all $\Delta\in\boldsymbol{\Delta}$.

Theorem 12.1. Suppose $G_{\mathrm{s}}(s)\triangleq(I+G(s)M_1)^{-1}G(s)$ is asymptotically stable. If there exists $Z(\cdot)\in\mathcal{Z}$ such that

$$\mathrm{He}\,[Z(s)(M^{-1}+G_{\mathrm{s}}(s))]>0, \quad (12.23)$$

for all $s=\jmath\omega$, $\omega\in\mathbb{R}\cup\infty$, then the negative feedback interconnection of $G(s)$ and Δ is asymptotically stable for all $\Delta\in\boldsymbol{\Delta}$.

Proof. Let $\Delta\in\boldsymbol{\Delta}$ and

$$G(s)\sim\left[\begin{array}{c|c}A & B\\ \hline C & D\end{array}\right]$$

be minimal so that the negative feedback interconnection of $G(s)$ and Δ is given by

$$(I+G(s)\Delta)^{-1}G(s) \sim \left[\begin{array}{c|c} A-B\Delta(I+D\Delta)^{-1}C & B-B\Delta(I+D\Delta)^{-1}D \\ \hline (I+D\Delta)^{-1}C & (I+D\Delta)^{-1}D \end{array}\right].$$

Suppose that, *ad absurdum*, $(I+G(s)\Delta)^{-1}G(s)$ is not asymptotically stable so that $A-B\Delta(I+D\Delta)^{-1}C$ is not Hurwitz. Since by assumption $G_{\mathrm{s}}(s)$ is asymptotically stable it follows that $A-BM_1(I+DM_1)^{-1}C$ is Hurwitz. Hence, there exists $\varepsilon \in (0,1)$ such that $A-B\Delta_\varepsilon(I+D\Delta_\varepsilon)^{-1}C$ has an eigenvalue $\jmath\hat{\omega}$ on the imaginary axis, where $\Delta_\varepsilon \triangleq \varepsilon\Delta+(1-\varepsilon)M_1$.

Next, note that

$$\begin{aligned}
&\det(I+G(\jmath\hat{\omega})\Delta_\varepsilon)\\
&\quad= \det[I+C(\jmath\hat{\omega}I-A)^{-1}B\Delta_\varepsilon+D\Delta_\varepsilon]\\
&\quad= \det(I+D\Delta_\varepsilon)^{-1}\det[I+(I+D\Delta_\varepsilon)^{-1}C(\jmath\hat{\omega}I-A)^{-1}B\Delta_\varepsilon]\\
&\quad= \det(I+D\Delta_\varepsilon)^{-1}\det(\jmath\hat{\omega}I-A)^{-1}\det[\jmath\hat{\omega}I-(A-B\Delta_\varepsilon(I+D\Delta_\varepsilon)^{-1}C]\\
&\quad= 0.
\end{aligned}$$

However, since $\Delta_\varepsilon \in \mathbf{\Delta}$ and $\det(I+G(\jmath\hat{\omega})M_1) \neq 0$, Lemma 12.1 with $\omega=\hat{\omega}$ implies that $\det(I+G(\jmath\hat{\omega})\Delta_\varepsilon) \neq 0$, which is a contradiction. □

Next, we specialize Theorem 12.1 to the case of norm-bounded uncertainty in order to draw connections with the structured singular value for real and complex block-structured uncertainty. Letting $M_1=-\gamma^{-1}I$ and $M_2=\gamma^{-1}I$, where $\gamma>0$, it follows that $M=2\gamma^{-1}I$ so that $M^{-1}=\frac{1}{2}\gamma I$. The set $\mathbf{\Delta}$ thus becomes

$$\mathbf{\Delta}_\gamma=\{\Delta\in\mathbf{\Delta}_{\mathrm{bs}}: -\gamma^{-1}I\le\Delta\le\gamma^{-1}I\}.$$

Now, $\Delta\in\mathbf{\Delta}_\gamma$ if and only if $\sigma_{\max}(\Delta)\le\gamma^{-1}$. Therefore, $\mathbf{\Delta}_\gamma$ is given by

$$\mathbf{\Delta}_\gamma=\{\Delta\in\mathbf{\Delta}_{\mathrm{bs}}: \sigma_{\max}(\Delta)\le\gamma^{-1}\}.$$

Alternatively, one can also consider the case where there may exist $\Delta\in\mathbf{\Delta}_\gamma$ such that $\Delta\neq\Delta^*$ and still arrive at $\Delta\in\mathbf{\Delta}_\gamma$ for the case where $M_1=-\gamma^{-1}I$ and $M_2=\gamma^{-1}I$. In this case, it can be shown that Lemma 12.1 also holds (see Problem 12.2). For the remainder of this section we consider uncertainties $\Delta\in\mathbf{\Delta}_\gamma$ such that $\Delta\neq\Delta^*$.

Now, we consider a special case of the sets $\mathbf{\Delta}_{\mathrm{bs}}$, $\mathcal{D}$, and $\mathcal{N}$ where $\Delta\in\mathbf{\Delta}_{\mathrm{bs}}$. In particular, let $\mathbf{\Delta}_{\mathrm{bs}}$ be the set of block-structured matrices with possibly repeated real scalar entries, complex scalar entries, and complex blocks given by

$$\begin{aligned}
\mathbf{\Delta}_{\mathrm{bs}}=\{\Delta\in\mathbb{C}^{m\times m}: \Delta=&\text{block-diag}[\delta_1^{\mathrm{r}}I_{l_1},\ldots,\delta_r^{\mathrm{r}}I_{l_r};\delta_{r+1}^{\mathrm{c}}I_{l_{r+1}},\ldots,\delta_{r+q}^{\mathrm{c}}I_{l_{r+q}};\\
&\Delta_{r+q+1}^{\mathrm{C}},\ldots,\Delta_{r+c}^{\mathrm{C}}],\delta_i^{\mathrm{r}}\in\mathbb{R}, i=1,\ldots,r;\delta_i^{\mathrm{c}}\in\mathbb{C}, i=r+1,\ldots,r+q;
\end{aligned}$$

$$\Delta_i^{\mathrm{C}} \in \mathbb{C}^{\hat{m}_i \times \hat{m}_i}, i = r+q+1, \ldots, r+c\}. \quad (12.24)$$

Then $\mathcal{D}$ and $\mathcal{N}$ are the sets of frequency-dependent positive-definite and Hermitian matrices, respectively, given by

$$\begin{aligned}\mathcal{D} &= \{D : \jmath\mathbb{R} \to \mathbb{C}^{m\times m} : D = \text{block-diag}[D_1, \ldots, D_{r+q}, d_{r+q+1}I_{\hat{m}_{r+q+1}}, \ldots, \\ &\quad d_{r+c}I_{\hat{m}_{r+c}}], 0 < D_i \in \mathbb{C}^{l_i \times l_i}, i = 1, \ldots, r+q; 0 < d_i \in \mathbb{R}, \\ &\quad i = r+q+1, \ldots, r+c\}, \quad (12.25)\\ \mathcal{N} &= \{N : \jmath\mathbb{R} \to \mathbb{C}^{m\times m} : N = \text{block-diag}[N_1, \ldots, N_r, 0_{r+1}, \ldots, 0_{r+c}], \\ &\qquad N_i = N_i^* \in \mathbb{C}^{l_i \times l_i}, i = 1, \ldots, r\}. \quad (12.26)\end{aligned}$$

Note that this special case is equivalent to the mixed-μ set considered in [117]. Furthermore, with $D(\cdot) \in \mathcal{D}$ and $N(\cdot) \in \mathcal{N}$ given by (12.25) and (12.26), respectively, the compatibility conditions required in $\mathcal{D}$ and $\mathcal{N}$ are automatically satisfied for $\boldsymbol{\Delta}_{\mathrm{bs}}$ given by (12.24).

Alternatively, let $\boldsymbol{\Delta}_{\mathrm{bs}}$ be given by (12.24) with the additional constraint that the complex blocks possess internal matrix structure. Then $\mathcal{D}$ and $\mathcal{N}$ are given by

$$\begin{aligned}\mathcal{D} &= \{D : \jmath\mathbb{R} \to \mathbb{C}^{m\times m} : D = \text{block-diag}[D_1, \ldots, D_{r+c}], 0 < D_i \in \mathbb{C}^{l_i \times l_i}, \\ &\quad i = 1 \ldots r+c; D_i\Delta_i^{\mathrm{C}} = \Delta_i^{\mathrm{C}} D_i, i = r+q+1, \ldots, r+c\}, \\ \mathcal{N} &= \{N : \jmath\mathbb{R} \to \mathbb{C}^{m\times m} : N = \text{block-diag}[N_1, \ldots, N_r, 0_{r+1}, \ldots, 0_{r+q}, \\ &\quad N_{r+q+1}, \ldots, N_{r+c}],\ N_i = N_i^* \in \mathbb{C}^{l_i \times l_i}, i = 1, \ldots, r+c; \\ &\quad N_i\Delta_i^{\mathrm{C}} = \Delta_i^{\mathrm{C}*} N_i, i = r+q+1, \ldots, r+c\}.\end{aligned}$$

For example, if $\Delta_i^{\mathrm{C}} = \Delta_i^{\mathrm{C}*} \in \boldsymbol{\Delta}_{\mathrm{bs}}$ then we can choose $D_i = d_i I, d_i \in \mathbb{R}$ and $N_i = n_i I, n_i \in \mathbb{R}, i = r+q+1, \ldots, r+c$.

Next, we present a key lemma which is important in connecting the absolute stability criterion given in Theorem 12.1 to the mixed-μ upper bounds.

Proposition 12.1. Let $Z(\cdot) \in \mathcal{Z}, D(\cdot) \in \mathcal{D}, N(\cdot) \in \mathcal{N}$, and $\omega \in \mathbb{R} \cup \infty$. Then the following statements are equivalent:

i) $\mathrm{He}\,[Z(\jmath\omega)(\frac{1}{2}\gamma I + G_{\mathrm{s}}(\jmath\omega))] > 0$.

ii) $G^*(\jmath\omega)DG(\jmath\omega) + \jmath\gamma(NG(\jmath\omega) - G^*(\jmath\omega)N) - \gamma^2 D < 0$.

Proof. Note that $\mathrm{He}\,[Z(\jmath\omega)(\frac{1}{2}\gamma I + G_{\mathrm{s}}(\jmath\omega))] > 0$ is equivalent to

$$\begin{aligned}&(D - \jmath N)(\tfrac{1}{2}\gamma I + G(\jmath\omega)(I - \gamma^{-1}G(\jmath\omega))^{-1}) \\ &\qquad + (\tfrac{1}{2}\gamma I + (I - \gamma^{-1}G^*(\jmath\omega))^{-1}G^*(\jmath\omega))(D + \jmath N) > 0.\end{aligned}$$

Now, premultiplying and postmultiplying the above inequality by $I - \gamma^{-1}G^*(\jmath\omega)$ and $I - \gamma^{-1}G(\jmath\omega)$, respectively, we obtain

$$\begin{aligned} &G^*(\jmath\omega)D(I - \gamma^{-1}G(\jmath\omega)) - \jmath(I - \gamma^{-1}G^*(\jmath\omega))NG(\jmath\omega) \\ &\quad +\jmath G^*(\jmath\omega)N(I - \gamma^{-1}G(\jmath\omega)) + \gamma(I - \gamma^{-1}G^*(\jmath\omega))D(I - \gamma^{-1}G(\jmath\omega)) \\ &\quad +(I - \gamma^{-1}G^*(\jmath\omega))DG(\jmath\omega) > 0, \end{aligned}$$

which, upon collecting terms, further simplifies to

$$\gamma D - \gamma^{-1}G^*(\jmath\omega)DG(\jmath\omega) - \jmath(NG(\jmath\omega) - G^*(\jmath\omega)N) > 0,$$

which is equivalent to (*ii*). □

We now obtain upper bounds for the structured singular value for real and complex multiple-block structured uncertainty. These bounds are based upon the absolute stability criterion of Theorem 12.1 for norm-bounded, block-structured uncertain matrices. The structured singular value [117] of a complex matrix $G(\jmath\omega)$ for mixed real and complex uncertainty is defined by $\mu(G(\jmath\omega)) \triangleq 0$ if $\det(I + G(\jmath\omega)\Delta) \neq 0$, $\Delta \in \boldsymbol{\Delta}_{\rm bs}$, and

$$\mu(G(\jmath\omega)) \triangleq \left(\min_{\Delta \in \boldsymbol{\Delta}_{\rm bs}} \{\sigma_{\max}(\Delta) : \det\left(I + G(\jmath\omega)\Delta\right) = 0\} \right)^{-1}, \tag{12.27}$$

otherwise. Hence, a necessary and sufficient condition for robust stability of the feedback interconnection of $G(s)$ and Δ is given by the following theorem. For the statement of this result we assume that the feedback interconnection of $G(s)$ and Δ is well posed for all $\Delta \in \boldsymbol{\Delta}_\gamma$.

Theorem 12.2. Let $\gamma > 0$ and suppose $G(s)$ is asymptotically stable. Then the negative feedback interconnection of $G(s)$ and Δ is asymptotically stable for all $\Delta \in \boldsymbol{\Delta}_\gamma$ if and only if

$$\mu(G(\jmath\omega)) < \gamma, \qquad \omega \in \mathbb{R} \cup \infty. \tag{12.28}$$

Proof. Let

$$G(s) \sim \left[\begin{array}{c|c} A & B \\ \hline C & D \end{array}\right],$$

where A is Hurwitz, and suppose the negative feedback interconnection of $G(s)$ and Δ given by

$$(I + G(s)\Delta)^{-1}G(s) \sim \left[\begin{array}{c|c} A - B\Delta(I + D\Delta)^{-1}C & B - B\Delta(I + D\Delta)^{-1}D \\ \hline (I + D\Delta)^{-1}C & (I + D\Delta)^{-1}D \end{array}\right]$$

is asymptotically stable for all $\Delta \in \boldsymbol{\Delta}_\gamma$. Next, note that, for all $\Delta \in \boldsymbol{\Delta}_\gamma$ and $\omega \in \mathbb{R} \cup \infty$

$$\begin{aligned} &\det[I + G(\jmath\omega)\Delta] \\ &\quad = \det[I + (C(\jmath\omega I - A)^{-1}B + D)\Delta] \end{aligned}$$

$$
\begin{aligned}
&= \det(I + D\Delta)\det[I + (\jmath\omega I - A)^{-1}B\Delta(I + D\Delta)^{-1}C] \\
&= \det(I + D\Delta)\det(\jmath\omega I - A)^{-1}\det[\jmath\omega I - (A - B\Delta(I + D\Delta)^{-1}C)] \\
&\neq 0.
\end{aligned}
$$

Hence, by definition $\mu(G(\jmath\omega)) < \gamma$ for all $\omega \in \mathbb{R} \cup \infty$.

Conversely, suppose $\mu(G(\jmath\omega)) < \gamma$ for all $\omega \in \mathbb{R} \cup \infty$ and assume that

$$
G(s) \sim \left[\begin{array}{c|c} A & B \\ \hline C & D \end{array}\right]
$$

is minimal. Then, by assumption, $\det(I + G(\infty)\Delta) = \det(I + D\Delta) \neq 0$ for all $\Delta \in \boldsymbol{\Delta}_\gamma$. Now, *ad absurdum*, suppose there exists $\Delta \in \boldsymbol{\Delta}_\gamma$ such that $(I + G(s)\Delta)^{-1}G(s)$ is not asymptotically stable, and hence, $A - B\Delta(I + D\Delta)^{-1}C$ is not Hurwitz. Since $G(s)$ is assumed to be asymptotically stable it follows that A is Hurwitz, and hence, there exists $\varepsilon \in (0, 1)$ such that $A - \varepsilon B\Delta(I + \varepsilon D\Delta)^{-1}C$ has an imaginary eigenvalue $\jmath\hat{\omega}$. Hence,

$$
\begin{aligned}
&\det[I + \varepsilon G(\jmath\hat{\omega})\Delta] \\
&\quad = \det(I + \varepsilon D\Delta)\det(\jmath\hat{\omega} I - A)^{-1}\det[\jmath\hat{\omega} I - (A - \varepsilon B\Delta(I + \varepsilon D\Delta)^{-1}C)] \\
&\quad = 0.
\end{aligned}
$$

However, since $\varepsilon\Delta \in \boldsymbol{\Delta}_\gamma$ it follows from the definition of $\mu(G(\jmath\omega))$ that $\det[I + \varepsilon G(\jmath\hat{\omega})\Delta] \neq 0$, which is a contradiction. □

Next, define $\overline{\mu}(G(\jmath\omega))$ by

$$
\overline{\mu}(G(\jmath\omega)) \triangleq \inf\{\gamma > 0 : \text{there exists } Z(\cdot) \in \mathcal{Z} \text{ such that } \operatorname{He}[Z(\jmath\omega)(\tfrac{1}{2}\gamma I + G_{\mathrm{s}}(\jmath\omega))] > 0\}, \quad (12.29)
$$

or, equivalently, using Proposition 12.1

$$
\overline{\mu}(G(\jmath\omega)) = \inf\{\gamma > 0 : \text{there exist } D(\cdot) \in \mathcal{D} \text{ and } N(\cdot) \in \mathcal{N} \text{ such that } G^*(\jmath\omega)DG(\jmath\omega) + \jmath\gamma(NG(\jmath\omega) - G^*(\jmath\omega)N) - \gamma^2 D < 0\}. \quad (12.30)
$$

To show that $\overline{\mu}(G(\jmath\omega))$ is an upper bound to $\mu(G(\jmath\omega))$, we require the following immediate result.

Lemma 12.2. Let $\omega \in \mathbb{R} \cup \infty$. If there exists $Z(\cdot) \in \mathcal{Z}$ such that

$$
\operatorname{He}[Z(\jmath\omega)(\tfrac{1}{2}\gamma I + G_{\mathrm{s}}(\jmath\omega))] > 0, \quad (12.31)
$$

then $\gamma \geq \overline{\mu}(G(\jmath\omega))$ and $\det(I + G(\jmath\omega)\Delta) \neq 0$ for all $\Delta \in \boldsymbol{\Delta}_\gamma$. Conversely, if $\gamma > \overline{\mu}(G(\jmath\omega))$ then there exists $Z(\cdot) \in \mathcal{Z}$ such that (12.31) holds and $\det(I + G(\jmath\omega)\Delta) \neq 0$ for all $\Delta \in \boldsymbol{\Delta}_\gamma$.

Proof. Suppose there exists $Z(\cdot) \in \mathcal{Z}$ such that (12.31) holds. Since

$\overline{\mu}(G(\jmath\omega))$ is the infimum over all γ such that there exists $Z(\cdot) \in \mathcal{Z}$ and (12.31) holds, it follows that $\gamma \geq \overline{\mu}(G(\jmath\omega))$. Conversely, suppose that $\gamma > \overline{\mu}(G(\jmath\omega))$. Then there exists $\hat{\gamma}$ satisfying $\overline{\mu}(G(\jmath\omega)) \leq \hat{\gamma} < \gamma$ and $Z(\cdot) \in \mathcal{Z}$ such that $\mathrm{He}\,[Z(\jmath\omega)(\frac{1}{2}\hat{\gamma}I + G_{\mathrm{s}}(\jmath\omega))] > 0$. Now, using the fact that $D(\jmath\omega) = \mathrm{He}\,Z(\jmath\omega) \geq 0$, it follows that

$$\begin{aligned}\mathrm{He}\,[Z(\jmath\omega)(\tfrac{1}{2}\gamma I + G_{\mathrm{s}}(\jmath\omega))] &= \tfrac{1}{2}(\gamma - \hat{\gamma})\mathrm{He}\,Z(\jmath\omega) + \mathrm{He}\,[Z(\jmath\omega)(\tfrac{1}{2}\hat{\gamma}I + G_{\mathrm{s}}(\jmath\omega))]\\ &> 0.\end{aligned}$$

Finally, applying Lemma 12.1 with $M^{-1} = \frac{1}{2}\gamma I$ and $\boldsymbol{\Delta} = \boldsymbol{\Delta}_\gamma$, it follows that $\det(I + G(\jmath\omega)\Delta) \neq 0$ for all $\Delta \in \boldsymbol{\Delta}_\gamma$. □

Theorem 12.3. Let $\omega \in \mathbb{R} \cup \infty$ and let $G(\jmath\omega)$ be a complex matrix. Then

$$\mu(G(\jmath\omega)) \leq \overline{\mu}(G(\jmath\omega)). \tag{12.32}$$

Proof. Suppose, *ad absurdum*, that $\overline{\mu}(G(\jmath\omega)) < \mu(G(\jmath\omega))$ and let $\gamma > 0$ satisfy $\overline{\mu}(G(\jmath\omega)) < \gamma \leq \mu(G(\jmath\omega))$. Then, from the definition of $\mu(G(\jmath\omega))$ it follows that $\min_{\Delta \in \boldsymbol{\Delta}_{\mathrm{bs}}}\{\sigma_{\max}(\Delta) : \det(I + G(\jmath\omega)\Delta) = 0\} \leq \gamma^{-1}$. It thus follows that there exists $\Delta \in \boldsymbol{\Delta}_\gamma$ such that $\det(I + G(\jmath\omega)\Delta) = 0$. However, using Lemma 12.2 we know if $\overline{\mu}(G(\jmath\omega)) < \gamma$, then $\det(I + G(\jmath\omega)\Delta) \neq 0$ for all $\Delta \in \boldsymbol{\Delta}_\gamma$, which is a contradiction. Hence, $\mu(G(\jmath\omega)) \leq \overline{\mu}(G(\jmath\omega))$. □

Next, in order to provide a systematic comparison of mixed-μ bounds for a fixed, internally block-structured uncertainty set $\boldsymbol{\Delta}_{\mathrm{bs}}$ define $\mu_i(G(\jmath\omega))$ by

$$\begin{aligned}\mu_i(G(\jmath\omega)) \triangleq \inf\{\gamma > 0 :\ &\text{there exist } D(\cdot) \in \mathcal{D}_i \text{ and } N(\cdot) \in \mathcal{N}_i \text{ such that}\\ &G^*(\jmath\omega)DG(\jmath\omega) + \jmath\gamma(NG(\jmath\omega) - G^*(\jmath\omega)N) - \gamma^2 D < 0\},\end{aligned} \tag{12.33}$$

where $\mathcal{D}_i$, $\mathcal{N}_i$ correspond to the pairs of frequency-dependent scaling matrix sets tailored to a fixed uncertainty structure $\boldsymbol{\Delta}_{\mathrm{bs}}$. The following result is immediate.

Proposition 12.2. Let $G(\jmath\omega)$ be a complex matrix and let $\mathcal{D}_i, \mathcal{N}_i$ and $\mathcal{D}_j, \mathcal{N}_j$ be frequency-dependent scaling matrix sets associated with a fixed uncertainty structure $\boldsymbol{\Delta}_{\mathrm{bs}}$. Suppose $\mathcal{D}_j \subseteq \mathcal{D}_i$ and $\mathcal{N}_j \subseteq \mathcal{N}_i$. Then

$$\mu_i(G(\jmath\omega)) \leq \mu_j(G(\jmath\omega)). \tag{12.34}$$

Proof. Suppose, *ad absurdum*, $\mu_j(G(\jmath\omega)) < \mu_i(G(\jmath\omega))$ and let some $\gamma > 0$ satisfy $\mu_j(G(\jmath\omega)) < \gamma \leq \mu_i(G(\jmath\omega))$. Then by definition of $\mu_j(G(\jmath\omega))$ there exist $D(\cdot) \in \mathcal{D}_j$ and $N(\cdot) \in \mathcal{N}_j$ such that $G^*(\jmath\omega)DG(\jmath\omega) + \jmath\gamma(NG(\jmath\omega) - G^*(\jmath\omega)N) - \gamma^2 D < 0$. Now, since $\mathcal{D}_j \subseteq \mathcal{D}_i$ and $\mathcal{N}_j \subseteq \mathcal{N}_i$, it follows that $D(\cdot) \in \mathcal{D}_i$ and $N(\cdot) \in \mathcal{N}_i$, and hence, $\mu_i(G(\jmath\omega) < \gamma$, which is a contradiction. Hence, $\mu_i(G(\jmath\omega)) \leq \mu_j(G(\jmath\omega))$. □

Finally, note that in the complex uncertainty case, that is, $r = 0$, we take $D > 0$ and $N = 0$ so that the complex-μ upper bound [110] given by

$$\overline{\mu}(G(\jmath\omega)) = \inf\{\gamma > 0 : \text{ there exists } D(\cdot) \in \mathcal{D} \text{ such that } G^*(\jmath\omega)DG(\jmath\omega) - \gamma^2 D < 0\},$$

or, equivalently,

$$\overline{\mu}(G(\jmath\omega)) = \inf_{D(\cdot)\in\mathcal{D}} \sigma_{\max}(D^{\frac{1}{2}}G(\jmath\omega)D^{-\frac{1}{2}}) \tag{12.35}$$

is recovered. In this case, the upper bound can be computed via a convex optimization problem [396].

12.3 Robust Stability Analysis of Nonlinear Uncertain Systems via Parameter-Dependent Lyapunov Functions

In this section, we present sufficient conditions for robust stability for a class of nonlinear uncertain systems. Specifically, we extend the framework of Chapter 11 in order to address stability of a class of nonlinear systems with *time-invariant structured* uncertainty. Here, we restrict our attention to time-invariant infinite-horizon systems. Once again, for the class of nonlinear uncertain systems considered we assume that the required properties for the existence and uniqueness of solutions are satisfied. For the following result, let $\mathcal{D} \subset \mathbb{R}^n$ be an open set, assume $0 \in \mathcal{D}$, let $L : \mathcal{D} \to \mathbb{R}$, and let $\mathcal{F} \subset \{f : \mathcal{D} \to \mathbb{R}^n : f(0) = 0\}$ denote the class of uncertain nonlinear systems with $f_0(\cdot) \in \mathcal{F}$ defining the nominal nonlinear system.

Theorem 12.4. Consider the nonlinear uncertain dynamical system

$$\dot{x}(t) = f(x(t)), \qquad x(0) = x_0, \qquad t \geq 0, \tag{12.36}$$

where $f(\cdot) \in \mathcal{F}$, with performance functional

$$J_f(x_0) \triangleq \int_0^\infty L(x(t))\mathrm{d}t. \tag{12.37}$$

Furthermore, assume there exist functions $\Gamma : \mathcal{D} \to \mathbb{R}$ and $V_{\mathrm{I}}, V_{\Delta f}, V : \mathcal{D} \to \mathbb{R}$, where $V_{\mathrm{I}}(\cdot)$ and $V_{\Delta f}(\cdot)$ are continuously differentiable functions such that $V_{\mathrm{I}}(x) + V_{\Delta f}(x) = V(x)$ for all $x \in \mathcal{D}$ and

$$V(0) = 0, \tag{12.38}$$

$$V(x) > 0, \quad x \in \mathcal{D}, \quad x \neq 0, \tag{12.39}$$

$$V'(x)f(x) \leq V_{\mathrm{I}}'(x)f_0(x) + \Gamma(x), \quad x \in \mathcal{D}, \quad f(\cdot) \in \mathcal{F}, \tag{12.40}$$

$$V_{\mathrm{I}}'(x)f_0(x) + \Gamma(x) < 0, \quad x \in \mathcal{D}, \quad x \neq 0, \tag{12.41}$$

$$L(x) + V_{\mathrm{I}}'(x)f_0(x) + \Gamma(x) = 0, \quad x \in \mathcal{D}, \tag{12.42}$$

where $f_0(\cdot) \in \mathcal{F}$ defines the nominal nonlinear system. Then the zero solution $x(t) \equiv 0$ to (12.36) is locally asymptotically stable for all $f(\cdot) \in \mathcal{F}$, and there exists a neighborhood of the origin $\mathcal{D}_0 \subseteq \mathcal{D}$ such that

$$J_f(x_0) \le \mathcal{J}(x_0) = V(x_0), \quad f(\cdot) \in \mathcal{F}, \quad x_0 \in \mathcal{D}_0, \tag{12.43}$$

where

$$\mathcal{J}(x_0) \triangleq \int_0^\infty [L(x(t)) + \Gamma(x(t)) - V'_{\Delta f}(x(t)) f_0(x(t))] \mathrm{d}t, \tag{12.44}$$

and where $x(t)$, $t \ge 0$, is the solution to (12.36) with $f(x(t)) = f_0(x(t))$. Finally, if $\mathcal{D} = \mathbb{R}^n$ and

$$V(x) \to \infty \text{ as } \|x\| \to \infty, \tag{12.45}$$

then the zero solution $x(t) \equiv 0$ to (12.36) is globally asymptotically stable for all $f(\cdot) \in \mathcal{F}$.

Proof. Let $f(\cdot) \in \mathcal{F}$ and $x(t)$, $t \ge 0$, satisfy (12.36). Then

$$\dot{V}(x(t)) \triangleq \frac{\mathrm{d}}{\mathrm{d}t} V(x(t)) = V'(x(t)) f(x(t)), \quad t \ge 0. \tag{12.46}$$

Hence, it follows from (12.40) and (12.41) that

$$\dot{V}(x(t)) < 0, \quad t \ge 0, \quad x(t) \ne 0. \tag{12.47}$$

Thus, from (12.38), (12.39), and (12.47) it follows that $V(\cdot)$ is a Lyapunov function for (12.36), which proves local asymptotic stability of the zero solution $x(t) \equiv 0$ to (12.36) for all $f(\cdot) \in \mathcal{F}$. Consequently, $x(t) \to 0$ as $t \to \infty$ for all initial conditions $x_0 \in \mathcal{D}_0$ for some neighborhood of the origin $\mathcal{D}_0 \subset \mathcal{D}$. Now, (12.46) implies that

$$0 = -\dot{V}(x(t)) + V'(x(t)) f(x(t)), \quad t \ge 0,$$

and hence, using (12.40) and (12.42),

$$\begin{aligned} L(x(t)) &= -\dot{V}(x(t)) + L(x(t)) + V'(x(t)) f(x(t)) \\ &\le -\dot{V}(x(t)) + L(x(t)) + V'_{\mathrm{I}}(x(t)) f_0(x(t)) + \Gamma(x(t)) \\ &= -\dot{V}(x(t)). \end{aligned}$$

Now, integrating over $[0, t)$ yields

$$\int_0^t L(x(s)) \mathrm{d}s \le -V(x(t)) + V(x_0).$$

Letting $t \to \infty$ and noting that $V(x(t)) \to 0$ for all $x_0 \in \mathcal{D}_0$ yields $J_f(x_0) \le V(x_0)$.

Next, let $x(t)$, $t \ge 0$, satisfy (12.36) with $f(x(t)) = f_0(x(t))$. Then it follows from (12.42) that

$$\begin{aligned} L(x(t)) + \Gamma(x(t)) - V'_{\Delta f}(x(t))f_0(x(t)) &= -\dot{V}(x(t)) + L(x(t)) \\ &\quad + V'_{\mathrm{I}}(x(t))f_0(x(t)) + \Gamma(x(t)) \\ &= -\dot{V}(x(t)). \end{aligned}$$

Integrating over $[0, t)$ yields

$$\int_0^t [L(x(s)) + \Gamma(x(s)) - V'_{\Delta f}(s)f_0(x(t))]\mathrm{d}s = -V(x(t)) + V(x_0).$$

Now, letting $t \to \infty$ yields $\mathcal{J}(x_0) = V(x_0)$. Finally, for $\mathcal{D} = \mathbb{R}^n$ global asymptotic stability of the solution $x(t) = 0$, $t \geq 0$, for all $f(\cdot) \in \mathcal{F}$ is a direct consequence of the radially unbounded condition (12.45) on $V(x)$. □

Theorem 12.4 is an extension of Theorem 8.1. Specifically, if $V_{\Delta f}(x) = 0$ and $\mathcal{F}$ consists of only the nominal nonlinear system $f_0(\cdot)$, then $\Gamma(x) = 0$ for all $x \in \mathcal{D}$ satisfies (12.40), and hence, $J_{f_0}(x_0) = J(x_0)$. In this case, Theorem 12.4 specializes to Theorem 8.1. Alternatively, setting $V_{\Delta f}(x) = 0$ and allowing $f(\cdot) \in \mathcal{F}$ we recover Theorem 11.1.

Next, we specialize Theorem 12.4 to nonlinear uncertain systems of the form

$$\dot{x}(t) = f_0(x(t)) + \Delta f(x(t)), \quad x(0) = x_0, \quad t \geq 0, \tag{12.48}$$

where $f_0 : \mathcal{D} \to \mathbb{R}^n$ satisfies $f_0(0) = 0$ and $f_0 + \Delta f \in \mathcal{F}$. Here, $\mathcal{F}$ is such that

$$\mathcal{F} \subset \{f_0 + \Delta f : \mathcal{D} \to \mathbb{R}^n : \ \Delta f \in \mathbf{\Delta}\}, \tag{12.49}$$

where $\mathbf{\Delta}$ is a given nonlinear uncertainty set of nonlinear perturbations Δf of the nominal system dynamics $f_0(\cdot) \in \mathcal{F}$. Since $\mathcal{F} \subset \{f : \mathcal{D} \to \mathbb{R}^n : f(0) = 0\}$ it follows that $\Delta f(0) = 0$ for all $\Delta f \in \mathbf{\Delta}$.

Corollary 12.1. Consider the dynamical system given by (12.48) where $\Delta f(\cdot) \in \mathbf{\Delta}$, with performance functional (12.37). Furthermore, assume there exist functions $\Gamma : \mathbb{R}^n \to \mathbb{R}$ and $V_{\mathrm{I}}, V_{\Delta f}, V : \mathbb{R}^n \to \mathbb{R}$, where $V_{\mathrm{I}}(\cdot)$ and $V_{\Delta f}(\cdot)$ are continuously differentiable functions such that $V_{\mathrm{I}}(x) + V_{\Delta f}(x) = V(x)$ for all $x \in \mathbb{R}^n$ and

$$V(0) = 0, \tag{12.50}$$

$$V(x) > 0, \quad x \in \mathbb{R}^n, \quad x \neq 0, \tag{12.51}$$

$$V'_{\mathrm{I}}(x)\Delta f(x) \leq \Gamma(x) - V'_{\Delta f}(x)(f_0(x) + \Delta f(x)), \quad x \in \mathbb{R}^n, \quad \Delta f(\cdot) \in \mathbf{\Delta}, \tag{12.52}$$

$$V'_{\mathrm{I}}(x)f_0(x) + \Gamma(x) < 0, \quad x \in \mathbb{R}^n, \quad x \neq 0, \tag{12.53}$$

$$L(x) + V'_{\mathrm{I}}(x)f_0(x) + \Gamma(x) = 0, \quad x \in \mathbb{R}^n, \tag{12.54}$$

and

$$V(x) \to \infty \text{ as } \|x\| \to \infty. \tag{12.55}$$

Then the zero solution $x(t) \equiv 0$ to (12.48) is globally asymptotically stable for all $\Delta f(\cdot) \in \mathbf{\Delta}$, and

$$J_{\Delta f}(x_0) \le \mathcal{J}(x_0) = V(x_0), \qquad \Delta f(\cdot) \in \mathbf{\Delta}, \tag{12.56}$$

where

$$\mathcal{J}(x_0) \triangleq \int_0^\infty [L(x(t)) + \Gamma(x(t)) - V'_{\Delta f}(x(t)) f_0(x(t))]\mathrm{d}t, \tag{12.57}$$

and where $x(t)$, $t \ge 0$, is the solution to (12.48) with $\Delta f(x) \equiv 0$.

Proof. The result is a direct consequence of Theorem 12.4 with $f(x) = f_0(x) + \Delta f(x)$. □

Having established the theoretical basis of our approach, we now assign explicit structure to the set $\mathbf{\Delta}$ and the bounding function $\Gamma(\cdot)$. Specifically, the uncertainty set $\mathbf{\Delta}$ is assumed to be of the form

$$\mathcal{F} = \{f_0(x) + \Delta f(x) : \Delta f(x) = G_\delta(x)\delta(h_\delta(x)),\ x \in \mathbb{R}^n,\ \delta(\cdot) \in \mathbf{\Delta}\}, \tag{12.58}$$

where $\mathbf{\Delta}$ satisfies

$$\mathbf{\Delta} = \{\delta(\cdot) \in \mathbf{\Delta}_{\mathrm{bs}} :\ [\delta(y) - m_1(y)]^{\mathrm{T}}[\delta(y) - m_2(y)] \le 0,\ y \in \mathbb{R}^{p_\delta}\}, \tag{12.59}$$

and where $m_1(\cdot), m_2(\cdot) \in \mathbf{\Delta}_{\mathrm{bs}}$ are such that $m_1^{\mathrm{T}}(y)m_2(y) \le 0$, $y \in \mathbb{R}^{p_\delta}$, $G_\delta : \mathbb{R}^n \to \mathbb{R}^{n \times m_\delta}$, $h_\delta : \mathbb{R}^{m_\delta} \to \mathbb{R}^{m_\delta}$ is continuously differentiable and satisfies $h_\delta(0) = 0$, and

$$\mathbf{\Delta}_{\mathrm{bs}} \subset \left\{\delta : \mathbb{R}^{m_\delta} \to \mathbb{R}^{m_\delta} :\ \delta(0) = 0,\ \delta(\cdot) \text{ is continuously differentiable, and } \frac{\mathrm{d}\delta(y)}{\mathrm{d}y} = \left[\frac{\mathrm{d}\delta(y)}{\mathrm{d}y}\right]^{\mathrm{T}}\right\}. \tag{12.60}$$

For the structure of $\mathbf{\Delta}$ as specified by (12.58), the bounding function $\Gamma(\cdot)$ satisfying (12.52) can now be given a concrete form. For the following result define the set of compatible scaling matrices $\mathcal{H}_{\mathrm{p}}$ and $\mathcal{N}_{\mathrm{s}}$ by

$$\mathcal{H}_{\mathrm{p}} \triangleq \{H \in \mathbb{P}^{m_\delta} :\ H\delta(y) = \delta(Hy),\ y \in \mathbb{R}^{m_\delta},\ \delta(\cdot) \in \mathbf{\Delta}_{\mathrm{bs}},\}, \tag{12.61}$$

$$\mathcal{N}_{\mathrm{s}} \triangleq \{N \in \mathbb{S}^{m_\delta} :\ N\delta(y) = \delta(Ny),\ N(\delta(y) - m_1(y)) \ge 0, y \in \mathbb{R}^{m_\delta},\ \delta(\cdot) \in \mathbf{\Delta}_{\mathrm{bs}},\}. \tag{12.62}$$

Furthermore, define $\tilde{h}(x) \triangleq H[m_1(h_\delta(x)) + m_2(h_\delta(x))] + Nh'_\delta(x)f_0(x)$ and $R_0(x) \triangleq 2H - Nh'_\delta(x)G_\delta(x) - G_\delta^{\mathrm{T}}(x)h'^{\mathrm{T}}_\delta(x)N$.

Proposition 12.3. Let $H \in \mathcal{H}_{\rm p}$ and $N \in \mathcal{N}_{\rm s}$ be such that $R_0(x) > 0$, $x \in \mathbb{R}^n$. Then the functions

$$\Gamma(x) = \tfrac{1}{2}[V_{\rm I}'(x)G_\delta(x) + \tilde{h}^{\rm T}(x)]R_0^{-1}(x)[V_{\rm I}'(x)G_\delta(x) + \tilde{h}^{\rm T}(x)]^{\rm T} - m_1^{\rm T}(h_\delta(x))Hm_2(h_\delta(x)), \quad (12.63)$$

$$V_{\Delta f}(x) = \int_0^{h_\delta(x)} \delta^{\rm T}(y)N{\rm d}y, \quad (12.64)$$

satisfy (12.52) with $\mathcal{F}$ given by (12.58) and $\boldsymbol{\Delta}$ given by (12.59).

Proof. First, note that if $H \in \mathcal{H}_{\rm p}$ then it follows that

$$[\delta(y) - m_1(y)]^{\rm T}H[\delta(y) - m_2(y)] \leq 0, \quad y \in \mathbb{R}^{m_\delta}, \quad \delta(\cdot) \in \boldsymbol{\Delta}_{\rm bs}.$$

Next, since $\delta^{\rm T}(y)N$ is a gradient vector of a real-value function the line integral $V_{\Delta f}(x) = \int_0^{h_\delta(x)} \delta^{\rm T}(y)N{\rm d}y$ is well defined, and hence, $V_{\Delta f}'(x) = \delta^{\rm T}(h_\delta(x))Nh_\delta'(x)$. Now,

$$\begin{aligned} 0 &\leq \tfrac{1}{2}[V_{\rm I}'(x)G_\delta(x) + \tilde{h}^{\rm T}(x) - R_0(x)\delta(h_\delta(x))]R_0^{-1}(x) \\ &\quad \cdot[V_{\rm I}'(x)G_\delta(x) + \tilde{h}^{\rm T}(x) - R_0(x)\delta(h_\delta(x))]^{\rm T} \\ &= \Gamma(x) + m_1^{\rm T}(h_\delta(x))Hm_2(h_\delta(x)) - V_{\rm I}'(x)G_\delta(x)\delta(h_\delta(x)) \\ &\quad -\tilde{h}^{\rm T}(x)\delta(h_\delta(x)) + \tfrac{1}{2}\delta^{\rm T}(h_\delta(x))R_0(x)\delta(h_\delta(x)) \\ &= \Gamma(x) - V_{\rm I}'(x)\Delta f(x) - \delta^{\rm T}(h_\delta(x))Nh_\delta'(x)[f_0(x) + G_\delta(x)\delta(h_\delta(x))] \\ &\quad +[\delta(h_\delta(x)) - m_1(h_\delta(x))]^{\rm T}H[\delta(h_\delta(x)) - m_2(h_\delta(x))] \\ &\leq \Gamma(x) - V_{\rm I}'(x)\Delta f(x) - V_{\Delta f}'(x)[f_0(x) + G_\delta(x)\delta(h_\delta(x))], \end{aligned}$$

which proves (12.52) with $\mathcal{F}$ given by (12.58) and $\boldsymbol{\Delta}$ given by (12.59). □

Next, consider the nonlinear uncertain dynamical system (12.48) and assume that the uncertainty set $\boldsymbol{\Delta}$ is given by (12.58) with $\boldsymbol{\Delta}$ given by

$$\boldsymbol{\Delta} = \{\delta(\cdot) \in \boldsymbol{\Delta}_{\rm bs} : \ \delta^{\rm T}(y)Q\delta(y) + 2\delta^{\rm T}(y)Sy + y^{\rm T}Ry \leq 0, \ y \in \mathbb{R}^{p_\delta}\}, \quad (12.65)$$

where $Q \in \mathbb{P}^{m_\delta}$, $-R \in \mathbb{N}^{p_\delta}$, and $S \in \mathbb{R}^{m_\delta \times p_\delta}$. For this uncertainty characterization, the bounding function $\Gamma(\cdot)$ satisfying (12.52) can be given a concrete form. For the following result define $\tilde{h}(x) \triangleq Nh_\delta'(x)f_0(x) - 2Sh_\delta(x)$ and $R_0 \triangleq 2Q - Nh_\delta'(x)G_\delta(x) - G_\delta^{\rm T}(x)h_\delta'^{\rm T}(x)N$.

Proposition 12.4. Let $N \in \mathcal{N}_{\rm s}$ be such that $R_0(x) > 0$, $x \in \mathbb{R}^n$. Then the functions

$$\Gamma(x) = \tfrac{1}{2}[V_{\rm I}'(x)G_\delta(x) + \tilde{h}^{\rm T}(x)]R_0^{-1}[V_{\rm I}'(x)G_\delta(x) + \tilde{h}^{\rm T}(x)]^{\rm T} - h_\delta^{\rm T}(x)Rh_\delta(x), \quad (12.66)$$

$$V_{\Delta f}(x) = \int_0^{h_\delta(x)} \delta^{\rm T}(y)N{\rm d}y, \quad (12.67)$$

satisfy (12.52) with $\mathcal{F}$ given by (12.58) and $\mathbf{\Delta}$ given by (12.65).

Proof. Since $\delta^{\mathrm{T}}(y)N$ is a gradient vector of a real-value function the line integral $V_{\Delta f}(x) = \int_0^{h_\delta(x)} \delta^{\mathrm{T}}(y)N\mathrm{d}y$ is well defined, and hence, $V'_{\Delta f}(x) = \delta^{\mathrm{T}}(h_\delta(x))Nh'_\delta(x)$. Now,

$$\begin{aligned}
0 &\le \tfrac{1}{2}[V_{\mathrm{I}}'(x)G_\delta(x) + \tilde{h}^{\mathrm{T}}(x) - R_0(x)\delta(h_\delta(x))]R_0^{-1}(x) \\
&\quad \cdot[V_{\mathrm{I}}'(x)G_\delta(x) + \tilde{h}^{\mathrm{T}}(x) - R_0(x)\delta(h_\delta(x))]^{\mathrm{T}} \\
&= \Gamma(x) + h_\delta^{\mathrm{T}}(x)Rh_\delta(x) - V_{\mathrm{I}}'(x)G_\delta(x)\delta(h_\delta(x)) \\
&\quad -\tilde{h}^{\mathrm{T}}(x)\delta(h_\delta(x)) + \tfrac{1}{2}\delta^{\mathrm{T}}(h_\delta(x))R_0(x)\delta(h_\delta(x)) \\
&= \Gamma(x) - V_{\mathrm{I}}'(x)\Delta f(x) - \delta^{\mathrm{T}}(h_\delta(x))Nh'_\delta(x)[f_0(x) + G_\delta(x)\delta(h_\delta(x))] \\
&\quad +\delta^{\mathrm{T}}(h_\delta(x))Q\delta(h_\delta(x)) + 2\delta^{\mathrm{T}}(h_\delta(x))Sh_\delta(x) + h_\delta^{\mathrm{T}}(x)Rh_\delta(x) \\
&\le \Gamma(x) - V_{\mathrm{I}}'(x)\Delta f(x) - V'_{\Delta f}(x)[f_0(x) + G_\delta(x)\delta(h_\delta(x))],
\end{aligned}$$

which proves (12.52) with $\mathcal{F}$ given by (12.58) and $\mathbf{\Delta}$ given by (12.65). □

We now combine the results of Corollary 12.1 and Propositions 12.3 and 12.4 to obtain conditions guaranteeing robust stability and performance for the nonlinear system (12.48). For the statement of the next result define $\tilde{h}(x) \triangleq H[m_1(h_\delta(x)) + m_2(h_\delta(x))] + Nh'_\delta(x)f_0(x)$ and $R_0(x) \triangleq 2H - Nh'_\delta(x)G_\delta(x) - G_\delta^{\mathrm{T}}(x)h'^{\mathrm{T}}_\delta(x)N$.

Proposition 12.5. Consider the nonlinear uncertain system (12.48). Let $L(x) > 0$, $x \in \mathbb{R}^n$, let $H \in \mathcal{H}_{\mathrm{p}}$, let $N \in \mathcal{N}_{\mathrm{s}}$, and let $R_0(x) > 0$, $x \in \mathbb{R}^n$. Assume that there exists a continuously differentiable function $V_{\mathrm{I}} : \mathbb{R}^n \to \mathbb{R}$ such that $V_{\mathrm{I}}(0) = 0$, $V_{\mathrm{I}}(x) > 0$, $x \in \mathbb{R}^n$, $x \neq 0$, $V_{\mathrm{I}}(x) \to \infty$ as $\|x\| \to \infty$, and

$$\begin{aligned}
0 &= V_{\mathrm{I}}'(x)f_0(x) + \tfrac{1}{2}[V_{\mathrm{I}}'(x)G_\delta(x) + \tilde{h}(x)]R_0^{-1}(x)[V_{\mathrm{I}}'(x)G_\delta(x) + \tilde{h}(x)]^{\mathrm{T}} \\
&\quad -m_1^{\mathrm{T}}(h_\delta(x))Hm_2(h_\delta(x)) + L(x), \quad x \in \mathbb{R}^n. \qquad (12.68)
\end{aligned}$$

Then the zero solution $x(t) \equiv 0$ to (12.48) is globally asymptotically stable for all $\Delta f(\cdot) \in \mathbf{\Delta}$, with $\mathbf{\Delta}$ given by (12.59), and

$$J_{\Delta f}(x_0) \le \mathcal{J}(x_0) = V(x_0), \qquad \Delta f(\cdot) \in \mathbf{\Delta}, \tag{12.69}$$

where $V(x) = V_{\mathrm{I}}(x) + V_{\Delta f}(x)$ with $V_{\Delta f}(x) = \int_0^{h_\delta(x)} \delta^{\mathrm{T}}(y)N\mathrm{d}y$, and

$$\mathcal{J}(x_0) \triangleq \int_0^\infty [L(x(t)) + \Gamma(x(t)) - V'_{\Delta f}(x(t))f_0(x(t))]\mathrm{d}t, \tag{12.70}$$

and where $\Gamma(x)$ is given by (12.63) and $x(t)$, $t \ge 0$, is the solution to (12.48) with $\Delta f(x) \equiv 0$.

For the following result define $\tilde{h}(x) \triangleq Nh'_\delta(x)f_0(x) - 2Sh_\delta(x)$ and $R_0 \triangleq 2Q - Nh'_\delta(x)G_\delta(x) - G_\delta^{\mathrm{T}}(x)h'^{\mathrm{T}}_\delta(x)N$.

Proposition 12.6. Consider the nonlinear uncertain system (12.48). Let $L(x) > 0$, $x \in \mathbb{R}^n$, let $N \in \mathcal{N}_{\rm s}$, and let $R_0(x) > 0$, $x \in \mathbb{R}^n$. Assume that there exists a continuously differentiable function $V_{\rm I} : \mathbb{R}^n \to \mathbb{R}$ such that $V_{\rm I}(0) = 0$, $V_{\rm I}(x) > 0$, $x \in \mathbb{R}^n$, $x \neq 0$, $V_{\rm I}(x) \to \infty$ as $\|x\| \to \infty$, and

$$\begin{aligned} 0 = {} & V_{\rm I}'(x) f_0(x) + \tfrac{1}{2}[V_{\rm I}'(x) G_\delta(x) + \tilde{h}(x)] R_0^{-1}(x) [V_{\rm I}'(x) G_\delta(x) + \tilde{h}(x)]^{\rm T} \\ & - h_\delta^{\rm T}(x) R h_\delta(x) + L(x), \quad x \in \mathbb{R}^n. \end{aligned} \tag{12.71}$$

Then the zero solution $x(t) \equiv 0$ to (12.48) is globally asymptotically stable for all $\Delta f(\cdot) \in \mathbf{\Delta}$, with $\mathbf{\Delta}$ given by (12.65), and

$$J_{\Delta f}(x_0) \le \mathcal{J}(x_0) = V(x_0), \qquad \Delta f(\cdot) \in \mathbf{\Delta}, \tag{12.72}$$

where $V(x) = V_{\rm I}(x) + V_{\Delta f}(x)$ with $V_{\Delta f}(x) = \int_0^{h_\delta(x)} \delta^{\rm T}(y) N {\rm d}y$, and

$$\mathcal{J}(x_0) \triangleq \int_0^\infty [L(x(t)) + \Gamma(x(t)) - V_{\Delta f}'(x(t)) f_0(x(t))] {\rm d}t, \tag{12.73}$$

and where $\Gamma(x)$ is given by (12.66) and $x(t)$, $t \ge 0$, is the solution to (12.48) with $\Delta f(x) \equiv 0$.

The following corollary specializes Theorem 12.4 to a class of linear uncertain systems which connects the framework of Theorem 12.4 to the parameter-dependent Lyapunov bounding framework of Haddad and Bernstein [145, 151]. Specifically, in this case we consider $\mathbf{\Delta}$ to be the set of uncertain linear systems given by

$$\mathcal{F} = \{(A + \Delta A)x : x \in \mathbb{R}^n,\ A \in \mathbb{R}^{n \times n},\ \Delta A \in \mathbf{\Delta}_A\},$$

where $\mathbf{\Delta}_A \subset \mathbb{R}^{n \times n}$ is a given bounded uncertainty set of uncertain perturbations ΔA of the nominal system matrix A such that $0 \in \mathbf{\Delta}_A$.

Corollary 12.2. Let $R \in \mathbb{P}^n$. Consider the linear uncertain dynamical system

$$\dot{x}(t) = (A + \Delta A)x(t), \qquad x(0) = x_0, \qquad t \ge 0, \tag{12.74}$$

with performance functional

$$J_{\Delta A}(x_0) \triangleq \int_0^\infty x^{\rm T}(t) R x(t) {\rm d}t, \tag{12.75}$$

where $\Delta A \in \mathbf{\Delta}_A$. Let $\Omega_0 : \mathcal{N} \subseteq \mathbb{S}^n \to \mathbb{S}^n$ and $P_0 : \mathbf{\Delta}_A \to \mathbb{S}^n$ be such that

$$\begin{aligned} \Delta A^{\rm T} P + P \Delta A \le {} & \Omega_0(P) - [(A + \Delta A)^{\rm T} P_0(\Delta A) + P_0(\Delta A)(A + \Delta A)], \\ & \Delta A \in \mathbf{\Delta}_A,\ P \in \mathcal{N}. \end{aligned} \tag{12.76}$$

Furthermore, suppose there exists $P \in \mathbb{P}^n$ satisfying

$$0 = A^{\rm T} P + PA + \Omega_0(P) + R \tag{12.77}$$

and such that $P+P_0(\Delta A) \in \mathbb{P}^n$, $\Delta A \in \boldsymbol{\Delta}_A$. Then the zero solution $x(t) \equiv 0$ to (12.74) is globally asymptotically stable for all $\Delta A \in \boldsymbol{\Delta}_A$, and

$$\sup_{\Delta A \in \boldsymbol{\Delta}_A} J_{\Delta A}(x_0) \leq \sup_{\Delta A \in \boldsymbol{\Delta}_A} \mathcal{J}(x_0) = x_0^{\mathrm{T}} P x_0 + \sup_{\Delta A \in \boldsymbol{\Delta}_A} x_0^{\mathrm{T}} P_0(\Delta A) x_0, \ x_0 \in \mathbb{R}^n, \tag{12.78}$$

where

$$\mathcal{J}(x_0) \triangleq \int_0^\infty x^{\mathrm{T}}(t)[R + \Omega_0(P) - A^{\mathrm{T}} P_0(\Delta A) - P_0(\Delta A) A] x(t) \mathrm{d}t, \tag{12.79}$$

and where $x(t)$, $t \geq 0$, solves (12.74) with $\Delta A = 0$. If, in addition, there exists $\overline{P}_0 \in \mathbb{S}^n$ such that $P_0(\Delta A) \leq \overline{P}_0$, $\Delta A \in \boldsymbol{\Delta}_A$, then

$$\sup_{\Delta A \in \boldsymbol{\Delta}_A} J_{\Delta A}(x_0) \leq x_0^{\mathrm{T}}(P + \overline{P}_0) x_0. \tag{12.80}$$

Proof. The result is a direct consequence of Theorem 12.4 with $f(x) = (A + \Delta A)x$, $f_0(x) = Ax$, $L(x) = x^{\mathrm{T}} R x$, $V_{\mathrm{I}}(x) = x^{\mathrm{T}} P x$, $V_{\Delta f}(x) = x^{\mathrm{T}} P_0(\Delta A) x$, $\Gamma(x) = x^{\mathrm{T}} \Omega_0(P) x$, and $\mathcal{D} = \mathbb{R}^n$. Specifically, conditions (12.38) and (12.39) are trivially satisfied. Now, $V'(x) f(x) = x^{\mathrm{T}}[(A + \Delta A)^{\mathrm{T}}(P + P_0(\Delta A)) + (P + P_0(\Delta A))(A + \Delta A)]x$, and hence, it follows from (12.76) that $V'(x) f(x) \leq V_{\mathrm{I}}'(x) f_0(x) + \Gamma(x) = x^{\mathrm{T}}(A^{\mathrm{T}} P + PA + \Omega_0(P))x$, for all $\Delta A \in \boldsymbol{\Delta}_A$. Furthermore, it follows from (12.77) that $L(x) + V_{\mathrm{I}}'(x) f_0(x) + \Gamma(x) = 0$, and hence, $V_{\mathrm{I}}'(x) f_0(x) + \Gamma(x) < 0$ for all $x \neq 0$, so that all the conditions of Theorem 12.4 are satisfied. Finally, since $V(x)$ is radially unbounded (12.74) is globally asymptotically stable for all $\Delta A \in \boldsymbol{\Delta}_A$. □

Corollary 12.2 is the deterministic version of Theorem 3.1 of [151] involving parameter-dependent Lyapunov functions for addressing robust stability and performance analysis of linear uncertain systems with constant real parameter uncertainty.

Next, we illustrate two bounding functions for two different uncertainty characterizations. Specifically, we assign explicit structure to the uncertainty set $\boldsymbol{\Delta}$ and the function $\Omega_0(\cdot)$ and $P_0(\cdot)$ satisfying (12.76). First, we assume that the uncertainty set $\boldsymbol{\Delta}_A$ is of the form

$$\boldsymbol{\Delta}_A = \{\Delta A : \ \Delta A = B_0 F C_0, \ F \in \boldsymbol{\Delta}\} \tag{12.81}$$

where $\boldsymbol{\Delta}$ satisfies

$$\boldsymbol{\Delta} \subset \boldsymbol{\Delta}_{\mathrm{bs}} = \{F \in \mathbb{S}^s : \ M_1 \leq F \leq M_2\}, \tag{12.82}$$

and where $B_0 \in \mathbb{R}^{n \times s}$, $C_0 \in \mathbb{R}^{s \times n}$, are fixed matrices denoting the structure of the uncertainty, $F \in \mathbb{S}^s$ is a symmetric uncertainty matrix, and $M_1, M_2 \in \boldsymbol{\Delta}_{\mathrm{bs}}$ are uncertainty bounds such that $M \triangleq M_2 - M_1 \in \mathbb{P}^s$. Note that $\boldsymbol{\Delta}$ may be equal to $\boldsymbol{\Delta}_{\mathrm{bs}}$, although, for generality, $\boldsymbol{\Delta}$ may be a specified proper subset of $\boldsymbol{\Delta}_{\rm bs}$. For example, $\boldsymbol{\Delta}_{\rm bs}$ may consist of block-structured matrices $F = \text{block-diag}[I_{\ell_1} \otimes F_1, \ldots, I_{\ell_r} \otimes F_r]$. Note that if $F = \text{block-diag}[F_1, \ldots, F_r]$, $M_1 = \text{block-diag}[M_{11}, \ldots, M_{1r}]$, and $M_2 = \text{block-diag}[M_{21}, \ldots, M_{2r}]$, then $M_{1i} \leq F_i \leq M_{2i}$, $i = 1, \ldots, r$. Furthermore, we assume that $0 \in \boldsymbol{\Delta}$ and $M_1, M_2 \in \boldsymbol{\Delta}$. Finally, we define the sets $\mathcal{H}_{\rm p}$, $\mathcal{N}_{\rm s}$, and $\mathcal{N}_{\rm nd}$ such that the product of every matrix in $\mathcal{H}_{\rm p}$ and $\mathcal{N}_{\rm s}$ and every matrix in $\boldsymbol{\Delta}_{\rm bs}$ is symmetric and the product of every matrix in $\mathcal{N}_{\rm nd}$ and $F - M_1$, where $F \in \boldsymbol{\Delta}_{\rm bs}$, is nonnegative definite by

$$\mathcal{H}_{\rm p} \triangleq \{H \in \mathbb{P}^s : FH = HF,\ F \in \boldsymbol{\Delta}_{\rm bs}\},$$

$$\mathcal{N}_{\rm s} \triangleq \{N \in \mathbb{S}^s : FN = NF,\ F \in \boldsymbol{\Delta}_{\rm bs}\},$$

and

$$\mathcal{N}_{\rm nd} \triangleq \{N \in \mathcal{N}_{\rm s} : (F - M_1)N \geq 0, F \in \boldsymbol{\Delta}_{\rm bs}\}.$$

The condition $FN = NF$, $F \in \boldsymbol{\Delta}_{\rm bs}$, is analogous to the commuting assumption between the D-scales and Δ uncertainty blocks in μ-analysis which accounts for structure in the uncertainty Δ. For the statement of the next result define $\tilde{C} \triangleq \frac{1}{2}[H(M_1 + M_2)C_0 + NC_0A]$ and $R_0 \triangleq H - \frac{1}{2}[NC_0B_0 - B_0^{\rm T}C_0^{\rm T}N]$.

Proposition 12.7. Let $M_1, M_2 \in \mathbb{S}^s$, $H \in \mathcal{H}_{\rm p}$, $N \in \mathcal{N}_{\rm s}$, and $R_0 > 0$. Then the functions

$$\Omega_0(P) = [\tilde{C} + B_0^{\rm T}P]^{\rm T}R_0^{-1}[\tilde{C} + B_0^{\rm T}P] - \tfrac{1}{2}C_0^{\rm T}(M_1HM_2 + M_2HM_1)C_0, \tag{12.83}$$

$$P_0(F) = C_0^{\rm T}FNC_0, \tag{12.84}$$

satisfy (12.76) with $\boldsymbol{\Delta}$ given by (12.82).

Proof. The proof is a direct consequence of Proposition 12.4. □

An alternative to Proposition 12.7 is given below. For the statement of the next result define $\tilde{C} \triangleq HC_0 + NC_0(A + B_0M_1C_0)$, $R_0 \triangleq (HM^{-1} - NC_0B_0) + (HM^{-1} - NC_0B_0)^{\rm T}$, and $F_{\rm s} \triangleq F - M_1$.

Proposition 12.8. Let $M_1, M_2 \in \mathbb{S}^s$, $H \in \mathcal{H}_{\rm p}$, $N \in \mathcal{N}_{\rm s}$, and $R_0 > 0$. Then the functions

$$\Omega_0(P) = [\tilde{C} + B_0^{\rm T}P]^{\rm T}R_0^{-1}[\tilde{C} + B_0^{\rm T}P] + PB_0M_1C_0 + C_0^{\rm T}M_1B_0^{\rm T}P, \tag{12.85}$$

$$P_0(F) = C_0^{\rm T}F_{\rm s}NC_0, \tag{12.86}$$

satisfy (12.76) with $\boldsymbol{\Delta}$ given by (12.82).

Proof. Since $R_0 > 0$ and by Problem 12.1 $F_{\rm s} - F_{\rm s}M^{-1}F_{\rm s} \geq 0$, it follows that

$$\begin{aligned}
0 &\le [\tilde{C} + B_0^{\mathrm{T}} P - R_0 F_{\mathrm{s}} C_0]^{\mathrm{T}} R_0^{-1} [\tilde{C} + B_0^{\mathrm{T}} P - R_0 F_{\mathrm{s}} C_0] \\
&\quad + 2 C_0^{\mathrm{T}} H [F_{\mathrm{s}} - F_{\mathrm{s}} M^{-1} F_{\mathrm{s}}] C_0 \\
&= \Omega_0(P) - P B_0 M_1 C_0 - C_0^{\mathrm{T}} M_1 B_0^{\mathrm{T}} P - [\tilde{C} + B_0^{\mathrm{T}} P]^{\mathrm{T}} F_{\mathrm{s}} C_0 \\
&\quad - C_0^{\mathrm{T}} F_{\mathrm{s}} [\tilde{C} + B_0^{\mathrm{T}} P] + C_0^{\mathrm{T}} F_{\mathrm{s}} R_0 F_{\mathrm{s}} C_0 \\
&\quad + 2 C_0^{\mathrm{T}} H [F_{\mathrm{s}} - F_{\mathrm{s}} M^{-1} F_{\mathrm{s}}] C_0 \\
&= \Omega_0(P) - A^{\mathrm{T}} C_0^{\mathrm{T}} N F_{\mathrm{s}} C_0 - C_0^{\mathrm{T}} M_1 B_0^{\mathrm{T}} C_0^{\mathrm{T}} N F_{\mathrm{s}} C_0 - P B_0 F C_0 \\
&\quad - C_0^{\mathrm{T}} F_{\mathrm{s}} N C_0 A - C_0^{\mathrm{T}} F_{\mathrm{s}} N C_0 B_0 M_1 C_0 - C_0^{\mathrm{T}} F B_0^{\mathrm{T}} P \\
&\quad - C_0^{\mathrm{T}} F_{\mathrm{s}} N C_0 B_0 F_{\mathrm{s}} C_0 - C_0^{\mathrm{T}} F_{\mathrm{s}} B_0^{\mathrm{T}} C_0^{\mathrm{T}} N F_{\mathrm{s}} C_0 \\
&= \Omega_0(P) - [(A + \Delta A)^{\mathrm{T}} P_0(F) + P_0(F)(A + \Delta A)] - [\Delta A^{\mathrm{T}} P + P \Delta A],
\end{aligned}$$

which proves (12.76) with $\boldsymbol{\Delta}$ given by (12.82). □

Next, assume the uncertainty set $\boldsymbol{\Delta}$ to be of the form

$$\boldsymbol{\Delta} \subset \boldsymbol{\Delta}_{\mathrm{bs}} = \{F \in \mathbb{R}^{m \times p} : \; F^{\mathrm{T}} Q F + F^{\mathrm{T}} S + S^{\mathrm{T}} F + R \le 0\}, \tag{12.87}$$

where $Q \in \mathbb{P}^m$, $S \in \mathbb{R}^{m \times p}$, and $-R \in \mathbb{N}^p$. For the statement of the next result define

$$\mathcal{N}_{\mathrm{s}} \triangleq \{N \in \mathbb{R}^{p \times m} : F^{\mathrm{T}} N = N^{\mathrm{T}} F \ge 0, \; F \in \boldsymbol{\Delta}_{\mathrm{bs}}\},$$

$\tilde{C} \triangleq N C_0 A - S C_0$, and $R_0 \triangleq Q - N C_0 B_0 - B_0^{\mathrm{T}} C_0^{\mathrm{T}} N^{\mathrm{T}}$.

Proposition 12.9. Let $N \in \mathcal{N}_{\mathrm{s}}$ and let $R_0 > 0$. Then the functions

$$\Omega_0(P) = [\tilde{C} + B_0^{\mathrm{T}} P]^{\mathrm{T}} R_0^{-1} [\tilde{C} + B_0^{\mathrm{T}} P] - C_0^{\mathrm{T}} R C_0, \tag{12.88}$$

$$P_0(F) = C_0^{\mathrm{T}} F^{\mathrm{T}} N C_0, \tag{12.89}$$

satisfy (12.76) with $\boldsymbol{\Delta}$ given by (12.89).

Proof. Since $R_0 > 0$ it follows that

$$\begin{aligned}
0 &\le [\tilde{C} + B_0^{\mathrm{T}} P - R_0 F C_0]^{\mathrm{T}} R_0^{-1} [\tilde{C} + B_0^{\mathrm{T}} P - R_0 F C_0] \\
&= \tilde{C}^{\mathrm{T}} R_0^{-1} \tilde{C} - \tilde{C}^{\mathrm{T}} F C_0 - C_0^{\mathrm{T}} F^{\mathrm{T}} \tilde{C} + C_0^{\mathrm{T}} F^{\mathrm{T}} R_0 F C_0 \\
&= \Omega_0(P) + C_0^{\mathrm{T}} R C_0 - A^{\mathrm{T}} C_0^{\mathrm{T}} N^{\mathrm{T}} F C_0 - C_0^{\mathrm{T}} S^{\mathrm{T}} F C_0 - P B_0 F C_0 \\
&\quad - C_0^{\mathrm{T}} F^{\mathrm{T}} N C_0 A - C_0^{\mathrm{T}} F^{\mathrm{T}} S C_0 - C_0^{\mathrm{T}} F^{\mathrm{T}} B_0^{\mathrm{T}} P + C_0^{\mathrm{T}} F^{\mathrm{T}} Q F C_0 \\
&\quad - C_0^{\mathrm{T}} F^{\mathrm{T}} N B_0 F C_0 - C_0^{\mathrm{T}} F^{\mathrm{T}} B_0^{\mathrm{T}} C_0^{\mathrm{T}} N^{\mathrm{T}} C_0 \\
&= \Omega(P) - [(A + \Delta A)^{\mathrm{T}} P_0(F) + P_0(F)(A + \Delta A)] - [\Delta A^{\mathrm{T}} P + P \Delta A] \\
&\quad + C_0^{\mathrm{T}} [F^{\mathrm{T}} Q F + F^{\mathrm{T}} S + S^{\mathrm{T}} F + R] C_0 \\
&= \Omega(P) - [(A + \Delta A)^{\mathrm{T}} P_0(F) + P_0(F)(A + \Delta A)] - [\Delta A^{\mathrm{T}} P + P \Delta A],
\end{aligned}$$

which proves (12.76) with $\boldsymbol{\Delta}$ given by (12.89). □

As in the nonlinear case, we now combine the results of Corollary 12.2 and Propositions 12.8 and 12.9 to obtain conditions guaranteeing robust stability and performance for the linear system (12.74). For the statement of the next result define $\tilde{C} \triangleq HC_0 + NC_0(A_0 + B_0M_1C_0)$, $R_0 \triangleq (HM^{-1} - NC_0B_0) + (HM^{-1} - NC_0B_0)^{\mathrm{T}}$, and $F_{\mathrm{s}} \triangleq F - M_1$.

Proposition 12.10. Consider the linear uncertain system (12.74). Let $R \in \mathbb{P}^n$, $M_1, M_2 \in \mathbb{S}^m$, $H \in \mathcal{H}_{\mathrm{p}}$, $N \in \mathcal{N}_{\mathrm{nd}}$, and $R_0 > 0$. Furthermore, suppose there exists a $P \in \mathbb{P}^n$ satisfying

$$0 = (A + B_0M_1C_0)^{\mathrm{T}}P + P(A + B_0M_1C_0) + [\tilde{C} + B_0^{\mathrm{T}}P]^{\mathrm{T}}R_0^{-1}[\tilde{C} + B_0^{\mathrm{T}}P] + R. \tag{12.90}$$

Then $A + \Delta A$ is asymptotically stable for all $\Delta A \in \mathbf{\Delta}_A$ with $\mathbf{\Delta}$ given by (12.82), and

$$\sup_{\Delta A \in \mathbf{\Delta}_A} J_{\Delta A}(x_0) \le x^{\mathrm{T}}Px + \sup_{\Delta A \in \mathbf{\Delta}} x_0^{\mathrm{T}}C_0^{\mathrm{T}}F_{\mathrm{s}}NC_0x_0 \le x_0^{\mathrm{T}}(P + \overline{P}_0)x_0, \quad x_0 \in \mathbb{R}^n, \tag{12.91}$$

where $\overline{P}_0 \in \mathbb{S}^n$ is such that $F_{\mathrm{s}}N \le \overline{P}_0$ for all $F_{\mathrm{s}} \in \mathbf{\Delta}$.

For the statement of the next result define $\tilde{C} \triangleq NC_0A - SC_0$ and $R_0 \triangleq Q - NC_0B_0 - B_0^{\mathrm{T}}C_0^{\mathrm{T}}N^{\mathrm{T}}$.

Proposition 12.11. Consider the linear uncertain system (12.74). Let $\hat{R} \in \mathbb{P}^n$, $N \in \mathcal{N}_{\mathrm{nd}}$, and $R_0 > 0$. Furthermore, suppose there exists a $P \in \mathbb{P}^n$ satisfying

$$0 = A^{\mathrm{T}}P + PA + [\tilde{C} + B_0^{\mathrm{T}}P]^{\mathrm{T}}R_0^{-1}[\tilde{C} + B_0^{\mathrm{T}}P] - C_0^{\mathrm{T}}RC_0 + \hat{R}. \tag{12.92}$$

Then $A + \Delta A$ is asymptotically stable for all $\Delta A \in \mathbf{\Delta}_A$ with $\mathbf{\Delta}$ given by (12.89), and

$$\sup_{\Delta A \in \mathbf{\Delta}_A} J_{\Delta A}(x_0) \le x^{\mathrm{T}}Px + \sup_{\Delta A \in \mathbf{\Delta}} x_0^{\mathrm{T}}C_0^{\mathrm{T}}F^{\mathrm{T}}NC_0x_0 \le x_0^{\mathrm{T}}(P + \overline{P}_0)x_0, \quad x_0 \in \mathbb{R}^n, \tag{12.93}$$

where $\overline{P}_0 \in \mathbb{S}^n$ is such that $F_{\mathrm{s}}N \le \overline{P}_0$ for all $F_{\mathrm{s}} \in \mathbf{\Delta}$.

Next, we use Theorem 12.2 and Corollary 12.2 to provide robust stability analysis tests for linear uncertain systems with structured uncertainty via an algebraic Riccati equation. Consider the linear uncertain dynamical system (12.74) with $\Delta A = B_0FC_0$, where $B_0 \in \mathbb{R}^{n\times p}$, $C_0 \in \mathbb{R}^{p\times n}$ denote the structure of the uncertainty, and $F \in \mathbf{\Delta}_{\mathrm{bs}} \subseteq \mathbb{R}^{p\times p}$ is the uncertain matrix. Here, $\mathbf{\Delta}_{\mathrm{bs}}$ is assumed to be the of the form

$$\begin{aligned}\mathbf{\Delta}_{\mathrm{bs}} = \{&F \in \mathbb{R}^{p\times p} : F = F^{\mathrm{T}},\ F = \text{block-diag}[\delta_1 I_{l_1}, \ldots, \delta_r I_{l_r}; \Delta_{r+1}, \ldots,\\ &\Delta_{r+q}], \delta_i \in \mathbb{R},\ i = 1, \ldots, r;\ \Delta_i \in \mathbb{R}^{p_i\times p_i},\ i = r+1, \ldots, r+q\},\end{aligned} \tag{12.94}$$

where the integers l_i, $i = 1, \ldots, r$, p_i, $i = r+1, \ldots, r+q$, are such that $\sum_{i=1}^{r} l_i + \sum_{i=r+1}^{r+q} p_i = p$. The following result provides a necessary and sufficient condition for robust stability of the feedback interconnection of $G(s) \triangleq C_0(sI - A)^{-1}B_0$ and F for all $F \in \mathbf{\Delta}_\gamma$, where

$$\mathbf{\Delta}_\gamma = \{F \in \mathbf{\Delta} : \ \sigma_{\max}(F) \le \gamma^{-1}\}, \tag{12.95}$$

and where $\gamma > 0$.

Theorem 12.5. Let $\gamma > 0$ and suppose A is Hurwitz. Then $A + B_0 F C_0$ is Hurwitz for all $F \in \mathbf{\Delta}_\gamma$ if and only if

$$\mu(G(\jmath\omega)) < \gamma, \qquad \omega \in \mathbb{R} \cup \infty. \tag{12.96}$$

Proof. The result is a direct consequence of Theorem 12.2 with $G(s) = C_0(sI - A)^{-1}B_0$ and $\Delta = F$. □

Next, we use Corollary 12.2 to provide an upper bound to the structured singular value via an algebraic Riccati equation. For the following result the set of compatible scaling matrices $\mathcal{H}_\mathrm{p}$ and $\mathcal{N}_\mathrm{s}$ are given by

$$\mathcal{H}_\mathrm{p} = \{H \in \mathbb{P}^m : \ HF = FH, \ F \in \mathbf{\Delta}_\gamma\}, \tag{12.97}$$

$$\mathcal{N}_\mathrm{s} = \{N \in \mathbb{S}^m : \ FN = NF, \ FN \ge \gamma^{-1}N, \ F \in \mathbf{\Delta}_\gamma\}, \tag{12.98}$$

and R_0 and $\tilde{C}$ are given by

$$R_0 = \gamma H - N C_0 B_0 - C_0^\mathrm{T} B_0^\mathrm{T} N, \qquad \tilde{C} = H C_0 + N C_0 (A - \gamma^{-1} B_0 C_0). \tag{12.99}$$

Theorem 12.6. Let $\gamma > 0$, $R \in \mathbb{P}^n$, $H \in \mathcal{H}_\mathrm{p}$, $N \in \mathcal{N}_\mathrm{s}$, and $R_0 > 0$. Consider the linear uncertain dynamical system (12.74) where $\Delta A = B_0 F C_0$, $F \in \mathbf{\Delta}_\gamma$, with performance functional (12.75). Furthermore, assume there exist $P \in \mathbb{P}^n$ such that

$$0 = (A - \gamma^{-1}B_0C_0)^\mathrm{T} P + P(A - \gamma^{-1}B_0C_0) + (\tilde{C} + B_0^\mathrm{T}P)^\mathrm{T} R_0^{-1} (\tilde{C} + B_0^\mathrm{T} P) + R. \tag{12.100}$$

Then the zero solution $x(t) \equiv 0$ to (12.74) is globally asymptotically stable for all $F \in \mathbf{\Delta}_\gamma$, and hence,

$$\mu(G(\jmath\omega)) < \gamma, \qquad \omega \in \mathbb{R} \cup \infty. \tag{12.101}$$

Furthermore,

$$\sup_{F \in \mathbf{\Delta}_\gamma} J_F(x_0) \le \sup_{F \in \mathbf{\Delta}_\gamma} \mathcal{J}(x_0) = x_0^\mathrm{T} P x_0 + \sup_{F \in \mathbf{\Delta}_\gamma} x_0^\mathrm{T} P_0(F) x_0, \qquad x_0 \in \mathbb{R}^n, \tag{12.102}$$

where

$$\mathcal{J}(x_0) \triangleq \int_0^\infty x^\mathrm{T}(t)(R + \Omega_0(P) - A^\mathrm{T} P_0(F) - P_0(F)A)x(t)\mathrm{d}t, \tag{12.103}$$

$P_0(F) \triangleq C_0^{\mathrm{T}}(F-\gamma^{-1}I)NC_0$, and $x(t)$, $t \geq 0$, solves (12.74) with $F=0$. If, in addition, there exists $\overline{P}_0 \in \mathbb{S}^n$ such that $P_0(F) \leq \overline{P}_0$, $F \in \mathbf{\Delta}_\gamma$, then

$$\sup_{F\in\mathbf{\Delta}_\gamma} J_F(x_0) \leq x_0^{\mathrm{T}}(P+\overline{P}_0)x_0. \tag{12.104}$$

Proof. The result is a direct consequence of Corollary 12.2 and Theorem 12.5. □

Setting $N=0$ and $H=I$ in (12.100) yields

$$0 = A^{\mathrm{T}}P + PA + \gamma^{-1}PB_0B_0^{\mathrm{T}}P + \gamma^{-1}C_0^{\mathrm{T}}C_0 + R, \tag{12.105}$$

or, equivalently,

$$0 = A^{\mathrm{T}}P + PA + \gamma^{-2}PB_0B_0^{\mathrm{T}}P + C_0^{\mathrm{T}}C_0 + R. \tag{12.106}$$

If only robust stability is of interest, then the weighting matrix R need not have physical significance. In this case, one can set $R=\varepsilon I$, where $\varepsilon>0$ is arbitrarily small. Hence, it follows from the results of Chapter 10 that if there exists a $P \in \mathbb{P}^n$ such that (12.106) is satisfied, then $|||G|||_\infty \leq \gamma$, where $G(s) = C_0(sI-A)^{-1}B_0$. Thus, it follows from Theorem 12.6 that $\mu(G(\jmath\omega)) \leq |||G|||_\infty$.

12.4 Robust Optimal Control for Nonlinear Systems via Parameter-Dependent Lyapunov Functions

In this section, we consider a control problem involving a notion of optimality with respect to an *auxiliary cost* which guarantees a bound on the worst-case value of a nonlinear-nonquadratic cost criterion over a prescribed uncertainty set. The optimal robust feedback controllers are derived as a direct consequence of Theorem 12.4. To address the robust optimal control problem let $\mathcal{D} \subset \mathbb{R}^n$ be an open set and let $U \subset \mathbb{R}^m$, where $0 \in \mathcal{D}$ and $0 \in U$. Furthermore, let $\mathcal{F} \subset \{F : \mathcal{D} \times U \to \mathbb{R}^n : F(0,0)=0\}$. Next, consider the controlled uncertain system

$$\dot{x}(t) = F(x(t),u(t)), \quad x(0)=x_0, \quad t \geq 0, \tag{12.107}$$

where $F(\cdot,\cdot) \in \mathcal{F}$ and the control $u(\cdot)$ is restricted to the class of admissible controls consisting of measurable functions $u(\cdot)$ such that $u(t) \in U$ for all $t \geq 0$, where the control constraint set U is given. We assume $0 \in U$. Given a control law $\phi(\cdot)$ and a feedback control $u(t) = \phi(x(t))$, the closed-loop system has the form

$$\dot{x}(t) = F(x(t),\phi(x(t))), \quad x(0)=x_0, \quad t \geq 0, \tag{12.108}$$

for all $F(\cdot,\cdot) \in \mathcal{F}$.

Next, we present an extension of Theorem 11.2 for characterizing robust feedback controllers that guarantee robust stability over a class of nonlinear uncertain systems and minimize an auxiliary performance functional. For the statement of this result let $L : \mathcal{D} \times U \to \mathbb{R}$ and define the set of regulation controllers for the nominal nonlinear system $F_0(\cdot,\cdot)$ by

$$\mathcal{S}(x_0) \triangleq \{u(\cdot) : u(\cdot) \text{ is admissible and } x(\cdot) \text{ given by (12.107)} \\ \text{satisfies } x(t) \to 0 \text{ as } t \to \infty \text{ with } F(\cdot,\cdot) = F_0(\cdot,\cdot)\}.$$

Theorem 12.7. Consider the nonlinear uncertain controlled system (12.107) with performance functional

$$J_F(x_0, u(\cdot)) \triangleq \int_0^\infty L(x(t), u(t))\mathrm{d}t, \tag{12.109}$$

where $F(\cdot,\cdot) \in \mathcal{F}$ and $u(\cdot)$ is an admissible control. Assume there exist functions $V_\mathrm{I}, V_{\Delta f}, V : \mathcal{D} \to \mathbb{R}$, $\Gamma : \mathcal{D} \times U \to \mathbb{R}$, and control law $\phi : \mathcal{D} \to U$, where $V_\mathrm{I}(\cdot)$ and $V_{\Delta f}(\cdot)$ are continuously differentiable functions such that $V_\mathrm{I}(x) + V_{\Delta f}(x) = V(x)$ for all $x \in \mathcal{D}$ and

$$V(0) = 0, \tag{12.110}$$
$$V(x) > 0, \quad x \in \mathcal{D}, \quad x \neq 0, \tag{12.111}$$
$$\phi(0) = 0, \tag{12.112}$$
$$V'(x)F(x,\phi(x)) \leq V_\mathrm{I}'(x)F_0(x,\phi(x)) + \Gamma(x,\phi(x)), \ x \in \mathcal{D}, \ F(\cdot,\cdot) \in \mathcal{F}, \tag{12.113}$$
$$V_\mathrm{I}'(x)F_0(x,\phi(x)) + \Gamma(x,\phi(x)) < 0, \quad x \in \mathcal{D}, \quad x \neq 0, \tag{12.114}$$
$$H(x,\phi(x)) = 0, \quad x \in \mathcal{D}, \tag{12.115}$$
$$H(x,u) \geq 0, \quad x \in \mathcal{D}, \quad u \in U, \tag{12.116}$$

where $F_0(\cdot,\cdot) \in \mathcal{F}$ defines the nominal system and

$$H(x,u) \triangleq L(x,u) + V_\mathrm{I}'(x)F_0(x,u) + \Gamma(x,u). \tag{12.117}$$

Then, with the feedback control $u(\cdot) = \phi(x(\cdot))$, there exists a neighborhood of the origin $\mathcal{D}_0 \subseteq \mathcal{D}$ such that if $x_0 \in \mathcal{D}_0$, the zero solution $x(t) \equiv 0$ of the closed-loop system (12.108) is locally asymptotically stable for all $F(\cdot,\cdot) \in \mathcal{F}$. Furthermore,

$$J_F(x_0, \phi(x(\cdot))) \leq \mathcal{J}(x_0, \phi(x(\cdot))) = V(x_0), \qquad F(\cdot,\cdot) \in \mathcal{F}, \tag{12.118}$$

where

$$\mathcal{J}(x_0, u(\cdot)) \triangleq \int_0^\infty [L(x(t),u(t)) + \Gamma(x(t),u(t)) - V_{\Delta f}'(x(t))F_0(x(t),u(t))]\mathrm{d}t, \tag{12.119}$$

and where $u(\cdot)$ is admissible and $x(t)$, $t \geq 0$, solves (12.107) with $F(x(t),u(t)) = F_0(x(t),u(t))$. In addition, if $x_0 \in \mathcal{D}_0$ then the feedback control $u(\cdot) = \phi(x(\cdot))$ minimizes $\mathcal{J}(x_0, u(\cdot))$ in the sense that

$$\mathcal{J}(x_0, \phi(x(\cdot))) = \min_{u(\cdot)\in\mathcal{S}(x_0)} \mathcal{J}(x_0, u(\cdot)). \tag{12.120}$$

Finally, if $\mathcal{D} = \mathbb{R}^n$, $U = \mathbb{R}^m$, and

$$V(x) \to \infty \text{ as } \|x\| \to \infty, \tag{12.121}$$

then the zero solution $x(t) \equiv 0$ of the closed-loop system (12.108) is globally asymptotically stable for all $F(\cdot) \in \mathcal{F}$.

Proof. Local and global asymptotic stability are a direct consequence of (12.110)–(12.114) and (12.121) by applying Theorem 12.4 to the closed-loop system (12.108). Furthermore, using (12.115), condition (12.118) is a restatement of (12.43) as applied to the closed-loop system. Next, let $u(\cdot) \in \mathcal{S}(x_0)$ and let $x(\cdot)$ be the solution of (12.107) with $F(\cdot,\cdot) = F_0(\cdot,\cdot)$. Then it follows that

$$0 = -\dot{V}(x(t)) + V'(x(t))F_0(x(t), u(t)).$$

Hence

$$\begin{aligned} L(x(t), u(t)) &+ \Gamma(x(t), u(t)) \\ &= -\dot{V}(x(t)) + L(x(t), u(t)) + V'(x(t))F_0(x(t), u(t)) + \Gamma(x(t), u(t)) \\ &= -\dot{V}(x(t)) + H(x(t), u(t)) + V'_{\Delta f}(x(t))F_0(x(t), u(t)). \end{aligned}$$

Now, using (12.117) and (12.119) and the fact that $u(\cdot) \in \mathcal{S}(x_0)$, it follows that

$$\begin{aligned} \mathcal{J}(x_0, u(\cdot)) &= \int_0^\infty [-\dot{V}(x(t)) + H(x(t), u(t))]\mathrm{d}t \\ &= -\lim_{t\to\infty} V(x(t)) + V(x_0) + \int_0^\infty H(x(t), u(t))\mathrm{d}t \\ &= V(x_0) + \int_0^\infty H(x(t), u(t))\mathrm{d}t \\ &\geq V(x_0) \\ &= \mathcal{J}(x_0, \phi(x(\cdot))), \end{aligned}$$

which yields (12.120). □

If $V_{\Delta f}(x) = 0$ and $\mathcal{F}$ consists of only the nominal nonlinear closed-loop system $F_0(\cdot,\cdot)$, then $\Gamma(x, u) = 0$ for all $x \in \mathcal{D}$ and $u \in U$ satisfies (12.113), and hence, $J_f(x_0, u(\cdot)) = J(x_0, u(\cdot))$. In this case, Theorem 12.7 specializes to Theorem 8.2. Alternatively, setting $V_{\Delta f}(x) = 0$ and allowing $F(\cdot,\cdot) \in \mathcal{F}$, Theorem 12.7 specializes to Theorem 11.2.

Next, we specialize Theorem 12.7 to linear uncertain systems and provide connections to the parameter-dependent Lyapunov bounding synthesis framework developed by Haddad and Bernstein [145, 151]. Specifically, in this case we consider $\mathcal{F}$ to be the set of uncertain linear systems given by

$$\mathcal{F} = \{(A + \Delta A)x + Bu : x \in \mathbb{R}^n, A \in \mathbb{R}^{n\times n}, B \in \mathbb{R}^{n\times m}, \Delta A \in \mathbf{\Delta}_A\},$$

where $\mathbf{\Delta}_A \subset \mathbb{R}^{n\times n}$ is a given bounded uncertainty set of the uncertain perturbation ΔA of the nominal system A such that $0 \in \mathbf{\Delta}_A$. For the following result let $R_1 \in \mathbb{P}^n$ and $R_2 \in \mathbb{P}^m$ be given.

Corollary 12.3. Consider the linear uncertain controlled dynamical system

$$\dot{x}(t) = (A + \Delta A)x(t) + Bu(t), \qquad x(0) = x_0, \qquad t \geq 0, \tag{12.122}$$

with performance functional

$$J_{\Delta A}(x_0, u(\cdot)) \triangleq \int_0^\infty [x^{\mathrm{T}}(t)R_1x(t) + u^{\mathrm{T}}(t)R_2u(t)]\mathrm{d}t, \tag{12.123}$$

where $u(\cdot)$ is admissible and $\Delta A \in \mathbf{\Delta}_A$. Furthermore, assume there exist $P \in \mathbb{P}^n$, $\Omega_{xx} : \mathbb{P}^n \to \mathbb{N}^n$, $\Omega_{xu} : \mathbb{P}^n \to \mathbb{R}^{n\times m}$, $\Omega_{uu} : \mathbb{P}^n \to \mathbb{N}^m$, and $P_0 : \mathbf{\Delta}_A \to \mathbb{S}^n$ such that

$$\begin{aligned}\Delta A^{\mathrm{T}}P + P\Delta A \leq\ & \Omega_{xx}(P) - [(A + \Delta A - BR_{2a}^{-1}P_a)^{\mathrm{T}}P_0(\Delta A)\\ & +P_0(\Delta A)(A + \Delta A - BR_{2a}^{-1}P_a)] - P_a^{\mathrm{T}}R_{2a}^{-1}\Omega_{xu}^{\mathrm{T}}(P)\\ & -\Omega_{xu}(P)R_{2a}^{-1}P_a + P_a^{\mathrm{T}}R_{2a}^{-1}\Omega_{uu}(P)R_{2a}^{-1}P_a, \end{aligned}\tag{12.124}$$

and

$$0 = A^{\mathrm{T}}P + PA + R_1 + \Omega_{xx}(P) - P_a^{\mathrm{T}}R_{2a}^{-1}P_a, \tag{12.125}$$

where $R_{2a} \triangleq R_2 + \Omega_{uu}(P)$ and $P_a \triangleq B^{\mathrm{T}}P + \Omega_{xu}^{\mathrm{T}}(P)$. Then the zero solution $x(t) \equiv 0$ to (12.122) is globally asymptotically stable for all $x_0 \in \mathbb{R}^n$ and $\Delta A \in \mathbf{\Delta}_A$, with the feedback control $u = \phi(x) \triangleq -R_{2a}^{-1}P_ax$, and

$$\begin{aligned}\sup_{\Delta A\in\mathbf{\Delta}_A} J_{\Delta A}(x_0, \phi(x(\cdot))) &\leq \sup_{\Delta A\in\mathbf{\Delta}_A} \mathcal{J}(x_0, \phi(x(\cdot)))\\ &= x_0^{\mathrm{T}}Px_0 + \sup_{\Delta A\in\mathbf{\Delta}_A} x_0^{\mathrm{T}}P_0(\Delta A)x_0,\end{aligned}\tag{12.126}$$

where

$$\begin{aligned}\mathcal{J}(x_0, u(\cdot)) \triangleq \int_0^\infty & [x^{\mathrm{T}}(t)(R_1 + \Omega_{xx}(P) - A^{\mathrm{T}}P_0(\Delta A) - P_0(\Delta A)A)x(t)\\ & +u^{\mathrm{T}}(t)R_2u(t)]\mathrm{d}t,\end{aligned}\tag{12.127}$$

and where $u(\cdot)$ is admissible and $x(t)$, $t \geq 0$, solves (12.122) with $\Delta A = 0$. Furthermore,

$$\mathcal{J}(x_0, \phi(x(\cdot)) = \min_{u(\cdot)\in\mathcal{S}(x_0)} \mathcal{J}(x_0, u(\cdot)), \tag{12.128}$$

where $\mathcal{S}(x_0)$ is the set of regulation controllers for the nominal system and $x_0 \in \mathbb{R}^n$. If, in addition, there exists $\overline{P}_0 \in \mathbb{S}^n$ such that $P_0(\Delta A) \leq \overline{P}_0$,

$\Delta A \in \boldsymbol{\Delta}_A$, then

$$\sup_{\Delta A \in \boldsymbol{\Delta}_A} J_{\Delta A}(x_0, \phi(x(\cdot))) \leq x_0^{\mathrm{T}}(P + \overline{P}_0)x_0. \tag{12.129}$$

Proof. The result is a direct consequence of Theorem 12.7 with $F(x,u) = (A+\Delta A)x+(B+\Delta B)u$, $F_0(x,u) = Ax+Bu$, $L(x,u) = x^{\mathrm{T}}R_1x+u^{\mathrm{T}}R_2u$, $V_{\mathrm{I}}(x) = x^{\mathrm{T}}Px$, $V_{\Delta f}(x) = x^{\mathrm{T}}P_0(\Delta A)x$, $\Gamma(x,u) = x^{\mathrm{T}}\Omega_{xx}(P)x + 2x^{\mathrm{T}}\Omega_{xu}(P)u + u^{\mathrm{T}}\Omega_{uu}(P)u$, $\mathcal{D} = \mathbb{R}^n$, and $U = \mathbb{R}^m$. Specifically, conditions (12.110) and (12.111) are trivially satisfied. Now, forming $x^{\mathrm{T}}(12.124)x$ it follows that, after some algebraic manipulation, $V'(x)F(x,\phi(x)) \leq V_{\mathrm{I}}'(x)F_0(x,\phi(x))+\Gamma(x,\phi(x))$, for all $\Delta A \in \boldsymbol{\Delta}_A$. Furthermore, it follows from (12.125) that $H(x,\phi(x)) = 0$, and hence, $V_{\mathrm{I}}'(x)F_0(x,\phi(x)) + \Gamma(x,\phi(x)) < 0$ for all $x \neq 0$. Thus, $H(x,u) = H(x,u) - H(x,\phi(x)) = [u - \phi(x)]^{\mathrm{T}}R_{2a}[u - \phi(x)] \geq 0$ so that all the conditions of Theorem 12.7 are satisfied. Finally, since $V(x)$ is radially unbounded (12.122), with $u(t) = \phi(x(t)) = -R_{2a}^{-1}P_ax(t)$, is globally asymptotically stable for all $\Delta A \in \boldsymbol{\Delta}_A$. □

The optimal feedback control law $\phi(x)$ in Corollary 12.3 is derived using the properties of $H(x,u)$ as defined in Theorem 12.7. Specifically, since $H(x,u) = x^{\mathrm{T}}(A^{\mathrm{T}}P + PA + R_1 + \Omega_{xx}(P))x + u^{\mathrm{T}}R_{2a}u + 2x^{\mathrm{T}}P_a^{\mathrm{T}}u$ it follows that $\partial^2 H/\partial u^2 = R_{2a} > 0$. Now, $\partial H/\partial u = 2R_{2a}u + 2P_ax = 0$ gives the unique global minimum of $H(x,u)$. Hence, since $\phi(x)$ minimizes $H(x,u)$ it follows that $\phi(x)$ satisfies $\partial H/\partial u = 0$ or, equivalently, $\phi(x) = -R_{2a}^{-1}P_ax$.

In order to make explicit connections with linear robust control, we now assign explicit structure to the set $\boldsymbol{\Delta}_A$ and the bounding functions $\Omega_{xx}(\cdot)$, $\Omega_{xu}(\cdot)$, and $\Omega_{uu}(\cdot)$. Specifically, the uncertainty set $\boldsymbol{\Delta}_A$ is assumed to be of the form

$$\boldsymbol{\Delta}_A \triangleq \{\Delta A : \Delta A = B_0FC_0,\ F \in \boldsymbol{\Delta}\}, \tag{12.130}$$

where $\boldsymbol{\Delta}$ satisfies

$$\boldsymbol{\Delta} \subseteq \boldsymbol{\Delta}_{\mathrm{bs}} = \{F \in \mathbb{R}^{s\times s} : M_1 \leq F \leq M_2\},$$

and where $B_0 \in \mathbb{R}^{n\times s}$, $C_0 \in \mathbb{R}^{s\times n}$, are fixed matrices denoting the structure of the uncertainty, $F \in \mathbb{S}^s$ is a symmetric uncertain matrix, and $M_1, M_2 \in \mathbb{S}^s$ are the uncertainty bounds such that $M \triangleq M_2 - M_1 \in \mathbb{P}^s$.

Next, let

$$\begin{aligned}
\Omega_{xx}(P) &= (\tilde{C} + B_0^{\mathrm{T}}P)^{\mathrm{T}}R_0^{-1}(\tilde{C} + B_0^{\mathrm{T}}P) + C_0^{\mathrm{T}}M_1B_0^{\mathrm{T}}P + PB_0M_1C_0,\\
\Omega_{xu}(P) &= (\tilde{C} + B_0^{\mathrm{T}}P)^{\mathrm{T}}R_0^{-1}NC_0B,\\
\Omega_{uu}(P) &= B^{\mathrm{T}}C_0^{\mathrm{T}}N^{\mathrm{T}}R_0^{-1}NC_0B,
\end{aligned}$$

where $\tilde{C} \triangleq HC_0 + NC_0(A + B_0M_1C_0)$ and $R_0 \triangleq (HM^{-1} - NC_0B_0) +$

$(HM^{-1} - NC_0B_0)^{\rm T}$. Furthermore, let $P_0(F) = C_0^{\rm T}F_{\rm s}NC_0$, where $F_{\rm s} \triangleq F - M_1$. Now, note that for all $H \in \mathcal{H}_{\rm p}$ and $N \in \mathcal{N}_{\rm s}$,

$$\begin{aligned}0 \leq{} & [(\tilde{C} + B_0^{\rm T}P - BR_{2a}^{-1}P_a) - R_0F_{\rm s}C_0]^{\rm T}R_0^{-1}[(\tilde{C} + B_0^{\rm T}P - BR_{2a}^{-1}P_a)\\ & -R_0F_{\rm s}C_0] + 2C_0^{\rm T}(F_{\rm s} - F_{\rm s}M^{-1}F_{\rm s})C_0,\end{aligned}$$

which further implies

$$\begin{aligned}0 \leq{} & \Omega_{xx}(P) - P_a^{\rm T}R_{2a}^{-1}\Omega_{xu}^{\rm T}(P) - \Omega_{xu}^{\rm T}(P)R_{2a}^{-1}P_a + P_a^{\rm T}R_{2a}^{-1}\Omega_{uu}(P)R_{2a}^{-1}P_a\\ & -A^{\rm T}C_0^{\rm T}N^{\rm T}F_{\rm s}C_0 - C_0^{\rm T}F_{\rm s}NC_0A - C_0^{\rm T}M_1B_0^{\rm T}C_0^{\rm T}N^{\rm T}F_{\rm s}C_0\\ & -C_0^{\rm T}F_{\rm s}NC_0B_0M_1C_0 + P_a^{\rm T}R_{2a}^{-1}B^{\rm T}C_0^{\rm T}N^{\rm T}F_{\rm s}C_0 + C_0^{\rm T}F_{\rm s}NC_0BR_{2a}^{-1}P_a\\ & -C_0^{\rm T}FB_0^{\rm T}P - PB_0FC_0 - C_0^{\rm T}F_{\rm s}NC_0B_0F_{\rm s}C_0 - C_0^{\rm T}F_{\rm s}B_0^{\rm T}C_0^{\rm T}N^{\rm T}F_{\rm s}C_0\\ ={} & \Omega_{xx}(P) - [(A + \Delta A - BR_{2a}^{-1}P_a)^{\rm T}P_0(F) + P_0(F)(A + \Delta A - BR_{2a}^{-1}P_a)]\\ & -P_a^{\rm T}R_{2a}^{-1}\Omega_{xu}^{\rm T}(P) - \Omega_{xu}(P)R_{2a}^{-1}P_a + P_a^{\rm T}R_{2a}^{-1}\Omega_{uu}(P)R_{2a}^{-1}P_a\\ & -(\Delta A^{\rm T}P + P\Delta A), \qquad (12.131)\end{aligned}$$

and hence, (12.124) holds. Furthermore, (12.125) specializes to

$$0 = A_P^{\rm T}P + PA_P + R_1 + \tilde{C}^{\rm T}R_0^{-1}\tilde{C} + PB_0R_0^{-1}B_0^{\rm T}P - P_a^{\rm T}R_{2a}^{-1}P_a, \quad (12.132)$$

where $A_P \triangleq A + B_0M_1C_0 + BR_0^{-1}\tilde{C}$ and P_a and R_{2a} specialize to $P_a = B^{\rm T}P + B^{\rm T}C_0^{\rm T}N^{\rm T}R_0^{-1}(\tilde{C} + B_0^{\rm T}P)$ and $R_{2a} = R_2 + B^{\rm T}C_0^{\rm T}N^{\rm T}R_0^{-1}NC_0B$, respectively. In this case, the optimal feedback law is given by $\phi(x) = -R_{2a}^{-1}P_ax$. This corresponds to the results obtained by Haddad and Bernstein [145, 151] for full-state feedback control.

12.5 Robust Control for Nonlinear Uncertain Affine Systems

In this section, we specialize Theorem 12.7 to affine (in the control) uncertain systems of the form

$$\dot{x}(t) = f_0(x(t)) + \Delta f(x(t)) + G_0(x(t))u(t), \quad x(0) = x_0, \qquad t \geq 0, \quad (12.133)$$

where $f_0 : \mathbb{R}^n \to \mathbb{R}^n$ satisfies $f_0(0) = 0$, $G_0 : \mathbb{R}^n \to \mathbb{R}^{n\times m}$, $\mathcal{D} = \mathbb{R}^n$, $U = \mathbb{R}^m$, and $\Delta f \in \boldsymbol{\Delta}$. Furthermore, we consider performance integrands $L(x, u)$ of the form

$$L(x, u) = L_1(x) + u^{\rm T}R_2(x)u, \quad (12.134)$$

where $L_1 : \mathbb{R}^n \to \mathbb{R}$ and $R_2 : \mathbb{R}^n \to \mathbb{P}^m$ so that (12.109) becomes

$$J(x_0, u(\cdot)) = \int_0^\infty [L_1(x(t)) + u^{\rm T}(t)R_2(x)u(t)]{\rm d}t. \quad (12.135)$$

Corollary 12.4. Consider the nonlinear uncertain controlled affine system (12.133) with performance functional (12.135). Assume there exist functions $V_{\rm I}, V_{\Delta f}, V : \mathbb{R}^n \to \mathbb{R}$, $\Gamma_{xx} : \mathbb{R}^n \to \mathbb{R}$, $\Gamma_{xu} : \mathbb{R}^n \to \mathbb{R}^{1\times m}$, and

$\Gamma_{uu} : \mathbb{R}^n \to \mathbb{N}^m$, where $V_{\mathrm{I}}(\cdot)$, $V_{\Delta f}(\cdot)$ are continuously differentiable such that $V_{\mathrm{I}}(x) + V_{\Delta f}(x) = V(x)$, $x \in \mathbb{R}^n$, and

$$V(0) = 0, \tag{12.136}$$

$$\Gamma_{xu}(0) = 0, \tag{12.137}$$

$$V(x) > 0, \quad x \in \mathbb{R}^n, \quad x \neq 0, \tag{12.138}$$

$$V'(x)\Delta f(x) - V'_{\Delta f}(x)[f_0(x) - \tfrac{1}{2}G(x)R_{2a}^{-1}(x)V_a(x)] \leq \Gamma_{xx}(x)$$
$$-\tfrac{1}{2}\Gamma_{xu}(x)R_{2a}^{-1}(x)V_a(x) + \tfrac{1}{4}V_a^{\mathrm{T}}(x)R_{2a}^{-1}(x)\Gamma_{uu}(x)R_{2a}^{-1}(x)V_a(x), \ \Delta f(\cdot) \in \boldsymbol{\Delta}, \tag{12.139}$$

$$V'_{\mathrm{I}}(x)[f_0(x) - \tfrac{1}{2}G(x)R_{2a}^{-1}(x)V_a(x)] + \Gamma_{xx}(x) - \tfrac{1}{2}\Gamma_{xu}(x)R_{2a}^{-1}(x)V_a(x)$$
$$+\tfrac{1}{4}V_a^{\mathrm{T}}(x)R_{2a}^{-1}(x)\Gamma_{uu}(x)R_{2a}^{-1}(x)V_a(x) < 0, \ x \in \mathbb{R}^n, \ x \neq 0, \tag{12.140}$$

and

$$V(x) \to \infty \text{ as } \|x\| \to \infty, \tag{12.141}$$

where $R_{2a}(x) \triangleq R_2(x) + \Gamma_{uu}(x)$ and $V_a(x) \triangleq [\Gamma_{xu}(x) + V'_{\mathrm{I}}(x)G(x)]^{\mathrm{T}}$. Then the zero solution $x(t) \equiv 0$ of the nonlinear uncertain system (12.133) is globally asymptotically stable for all $\Delta f(\cdot) \in \boldsymbol{\Delta}$ with the feedback control law

$$\phi(x) = -\tfrac{1}{2}R_{2a}^{-1}(x)V_a(x), \tag{12.142}$$

and the performance functional (12.135) satisfies

$$\sup_{\Delta f \in \boldsymbol{\Delta}} J(x_0, \phi(x(\cdot))) \leq \mathcal{J}(x_0, \phi(x(\cdot))) = V(x_0), \tag{12.143}$$

where

$$\mathcal{J}(x_0, u(\cdot)) \triangleq \int_0^\infty [L(x(t), u(t)) + \Gamma(x(t), u(t))$$
$$-V'_{\Delta f}(x(t))(f_0(x(t)) + G(x(t))u(t))]\mathrm{d}t, \tag{12.144}$$

and

$$\Gamma(x, u) = \Gamma_{xx}(x) + \Gamma_{xu}(x)u + u^{\mathrm{T}}\Gamma_{uu}(x)u, \tag{12.145}$$

where $u(\cdot)$ is admissible, and $x(t)$, $t \geq 0$, solves (12.133) with $\Delta f(x) = 0$. In addition, the performance functional (12.144), with

$$L_1(x) = \phi^{\mathrm{T}}(x)R_{2a}(x)\phi(x) - V'_{\mathrm{I}}(x)f_0(x) - \Gamma_{xx}(x), \tag{12.146}$$

is minimized in the sense that

$$\mathcal{J}(x_0, \phi(x(\cdot))) = \min_{u(\cdot) \in \mathcal{S}(x_0)} \mathcal{J}(x_0, u(\cdot)). \tag{12.147}$$

Proof. The result is a direct consequence of Theorem 12.7 with $\mathcal{D} = \mathbb{R}^n$, $U = \mathbb{R}^m$, $F_0(x, u) = f_0(x) + G_0(x)u$, $F(x, u) = f_0(x) + \Delta f(x) + G(x)u$, $L(x, u)$ given by (12.134), and $\Gamma(x, u)$ given by (12.145). Specifically, with

(12.133), (12.134), and (12.145), the Hamiltonian has the form

$$\begin{aligned} H(x,u) &= L_1(x) + u^{\mathrm{T}} R_2(x) u + V_{\mathrm{I}}'(x)(f_0(x) + G(x)u) \\ &\quad + \Gamma_{xx}(x) + \Gamma_{xu}(x) u + u^{\mathrm{T}} \Gamma_{uu}(x) u. \end{aligned}$$

Now, the proof follows as in the proof of Corollary 11.4. □

12.6 Robust Nonlinear Controllers with Polynomial Performance Criteria

In this section, we specialize the results of Section 12.4 to linear uncertain systems controlled by nonlinear controllers that minimize a polynomial cost functional. Specifically, assume $\mathcal{F}$ to be the set of uncertain systems given by

$$\mathcal{F} = \{(A+\Delta A)x + Bu : x \in \mathbb{R}^n, u \in \mathbb{R}^m, A \in \mathbb{R}^{n\times n}, B \in \mathbb{R}^{n\times m}, \Delta A \in \mathbf{\Delta}_A\},$$

where $\mathbf{\Delta}_A \subset \mathbb{R}^{n\times n}$ is a given bounded uncertainty set of the uncertain perturbation ΔA of the nominal system A such that $0 \in \mathbf{\Delta}_A$. For the following result let $R_1 \in \mathbb{P}^n$, $R_2 \in \mathbb{P}^m$, and $\hat{R}_k \in \mathbb{N}^n$, $k = 2, \ldots, r$, be given where r is a positive integer.

Theorem 12.8. Consider the linear uncertain controlled system

$$\dot{x}(t) = (A + \Delta A)x(t) + Bu(t), \qquad x(0) = x_0, \qquad t \geq 0, \tag{12.148}$$

where $u(\cdot)$ is admissible and $\Delta A \in \mathbf{\Delta}_A$. Assume there exist functions $\Omega : \mathbb{N}^n \to \mathbb{N}^n$, $\Omega_{xx} : \mathbb{P}^n \to \mathbb{N}^n$, $\Omega_{xu} : \mathbb{P}^n \to \mathbb{R}^{n\times m}$, $\Omega_{uu} : \mathbb{P}^n \to \mathbb{N}^m$, and $P_0 : \mathbf{\Delta}_A \to \mathbb{S}^n$, such that

$$\Delta A^{\mathrm{T}} P + P \Delta A \leq \Omega(P), \qquad \Delta A \in \mathbf{\Delta}_A, \quad P \in \mathbb{N}^n, \tag{12.149}$$

and assume there exist $P \in \mathbb{P}^n$ and $Y_k \in \mathbb{N}^n$, $k = 2, \ldots, r$, such that

$$\begin{aligned} \Delta A^{\mathrm{T}} P + P \Delta A &\leq \Omega_{xx}(P) - [(A + \Delta A - BR_{2a}^{-1} P_a)^{\mathrm{T}} P_0(\Delta A) \\ &\quad + P_0(\Delta A)(A + \Delta A - BR_{2a}^{-1} P_a)] - P_a^{\mathrm{T}} R_{2a}^{-1} \Omega_{xu}^{\mathrm{T}}(P) \\ &\quad - \Omega_{xu}(P) R_{2a}^{-1} P_a + P_a^{\mathrm{T}} R_{2a}^{-1} \Omega_{uu}(P) R_{2a}^{-1} P_a, \end{aligned} \tag{12.150}$$

$$0 = A^{\mathrm{T}} P + PA + R_1 + \Omega_{xx}(P) - P_a^{\mathrm{T}} R_{2a}^{-1} P_a, \tag{12.151}$$

and

$$0 = (A - BR_{2a}^{-1} P_a)^{\mathrm{T}} Y_k + Y_k (A - BR_{2a}^{-1} P_a) + \hat{R}_k + \Omega(Y_k), \quad k = 2, \ldots, r. \tag{12.152}$$

Then, with the feedback control law

$$u = \phi(x) \triangleq -R_{2a}^{-1} \left(P_a + \sum_{k=2}^{r} (x^{\mathrm{T}} Y_k x)^{k-1} B^{\mathrm{T}} Y_k \right) x,$$

the zero solution of the uncertain system (12.148) is globally asymptotically stable for all $x_0 \in \mathbb{R}^n$ and $\Delta A \in \mathbf{\Delta}_A$, and the performance functional (12.144) satisfies

$$\begin{aligned} J_{\Delta A}(x_0, \phi(x(\cdot))) &\leq \mathcal{J}(x_0, \phi(x(\cdot))) \\ &= x_0^{\mathrm{T}}(P + P_0(\Delta A))x_0 + \sum_{k=2}^{r} \tfrac{1}{k}(x_0^{\mathrm{T}} Y_k x_0)^k, \quad \Delta A \in \mathbf{\Delta}_A, \end{aligned} \tag{12.153}$$

where

$$\mathcal{J}(x_0, u(\cdot)) \triangleq \int_0^\infty [L(x,u) + \Gamma(x,u) - x^{\mathrm{T}}(t)(A^{\mathrm{T}} P_0(\Delta A) + P_0(\Delta A)A)x(t)]\mathrm{d}t, \tag{12.154}$$

and where $u(\cdot)$ is admissible, and $x(t)$, $t \geq 0$, solves (12.148) with $\Delta A = 0$ and

$$\begin{aligned} \Gamma(x,u) &= x^{\mathrm{T}} \left(\Omega_{xx}(P) + \sum_{k=2}^{r} (x^{\mathrm{T}} Y_k x)^{k-1} \Omega(Y_k) \right) x \\ &\quad + 2x^{\mathrm{T}} \Omega_{xu}(P) u + u^{\mathrm{T}} \Omega_{uu}(P) u, \end{aligned}$$

where $u(\cdot)$ is admissible and $\Delta A \in \mathbf{\Delta}_A$. In addition, the performance functional (12.144), with $R_2(x) = R_2$ and

$$\begin{aligned} L_1(x) &= x^{\mathrm{T}} \left(R_1 + \sum_{k=2}^{r} (x^{\mathrm{T}} Y_k x)^{k-1} \hat{R}_k + \left[\sum_{k=2}^{r} (x^{\mathrm{T}} Y_k x)^{k-1} B^{\mathrm{T}} Y_k \right]^{\mathrm{T}} R_{2a}^{-1} \right. \\ &\quad \left. \cdot \left[\sum_{k=2}^{r} (x^{\mathrm{T}} Y_k x)^{k-1} B^{\mathrm{T}} Y_k \right] \right) x, \end{aligned}$$

is minimized in the sense that

$$\mathcal{J}(x_0, \phi(x(\cdot))) = \min_{u(\cdot) \in \mathcal{S}(x_0)} \mathcal{J}(x_0, u(\cdot)), \tag{12.155}$$

where $\mathcal{S}(x_0)$ is the set of regulation controllers for the nominal system and $x_0 \in \mathbb{R}^n$.

Proof. The result is a direct consequence of Corollary 12.4 with $f_0(x) = Ax$, $\Delta f(x) = \Delta A x$, $G_0(x) = B$, $V_{\mathrm{I}}(x) = x^{\mathrm{T}} P x + \sum_{k=2}^{r} (\frac{1}{k})(x^{\mathrm{T}} Y_k x)^k$, and $V_{\Delta f}(x) = x^{\mathrm{T}} P_0(\Delta A) x$. □

The Lyapunov function that establishes robust stability in Theorem 12.8 is given by

$$V(x) = x^{\mathrm{T}} P x + \sum_{k=2}^{r} \tfrac{1}{k}(x^{\mathrm{T}} Y_k x)^k + x^{\mathrm{T}} P_0(\Delta A) x$$

and, hence, is explicitly dependent on the uncertain parameters ΔA. In the terminology of [145,151], this is a parameter-dependent Lyapunov function. As discussed in Section 12.1 and [145, 151] the ability of such Lyapunov functions to guarantee stability with respect to time-varying parameter variations is curtailed, thus reducing conservatism with respect to constant real parameter uncertainty.

Theorem 12.8 requires the solutions of $r-1$ modified Riccati equations in (12.152) to obtain the optimal robust controller. However, if $\hat{R}_k = \hat{R}_2$, $k = 3, \ldots, r$, then $Y_k = Y_2$, $k = 3, \ldots, r$, satisfies (12.152). In this case, we require the solution of one modified Riccati equation in (12.152). This special case is considered in Corollary 12.5 below.

As in Chapters 10 and 11 the performance functional can be written as

$$J_{\Delta A}(x_0, u(\cdot)) =$$
$$\int_0^\infty \left[x^{\mathrm{T}}(R_1 + \sum_{k=2}^{r} (x^{\mathrm{T}} Y_k x)^{k-1} \hat{R}_k) x + u^{\mathrm{T}} R_2 u + \phi_{\mathrm{NL}}^{\mathrm{T}}(x) R_{2a} \phi_{\mathrm{NL}}(x) \right] \mathrm{d}t,$$

where $\phi_{\mathrm{NL}}(x)$ is the nonlinear part of the optimal feedback control

$$\phi(x) = \phi_{\mathrm{L}}(x) + \phi_{\mathrm{NL}}(x),$$

where $\phi_{\mathrm{L}}(x) \triangleq -R_{2a}^{-1} P_a x$ and $\phi_{\mathrm{NL}} \triangleq -R_{2a}^{-1} B^{\mathrm{T}} \sum_{k=2}^{r} (x^{\mathrm{T}} Y_k x)^{k-1} Y_k x$.

Next, we consider the special case in which $r = 2$. In this case, note that if there exist $P \in \mathbb{P}^n$ and $Y_2 \in \mathbb{N}^n$ such that

$$0 = A^{\mathrm{T}} P + PA + R_1 + \Omega_{xx}(P) - P_a R_{2a}^{-1} P_a$$

and

$$0 = (A - BR_{2a}^{-1} P_a)^{\mathrm{T}} Y_2 + Y_2 (A - BR_{2a}^{-1} P_a) + \hat{R}_2 + \Omega(Y_2),$$

then (12.148), with the performance functional

$$\begin{aligned} J_{\Delta A}(x_0, u(\cdot)) = & \int_0^\infty [x^{\mathrm{T}} R_1 x + u^{\mathrm{T}} R_2 u + (x^{\mathrm{T}} Y_2 x)(x^{\mathrm{T}} \hat{R}_2 x) \\ & + (x^{\mathrm{T}} Y_2 x)^2 (x^{\mathrm{T}} Y_2 B R_{2a}^{-1} B^{\mathrm{T}} Y_2 x)] \mathrm{d}t, \end{aligned}$$

is globally asymptotically stable for all $x_0 \in \mathbb{R}^n$ with the feedback control law $u = \phi(x) = -R_{2a}^{-1}(P_a + (x^{\mathrm{T}} Y_2 x) B^{\mathrm{T}} Y_2) x$.

Finally, using the uncertainty characterization given by (12.130) we present a specialization of Theorem 12.8.

Corollary 12.5. Consider the uncertain controlled system

$$\dot{x}(t) = (A + \Delta A) x(t) + Bu(t), \qquad x(0) = x_0, \qquad t \geq 0, \tag{12.156}$$

where $u(\cdot)$ is admissible and $\Delta A \in \mathbf{\Delta}_A$, where $\mathbf{\Delta}_A$ is given by (12.130). Assume $H \in \mathcal{H}_{\rm p}$ and $N \in \mathcal{N}_{\rm nd}$ and suppose there exist $P \in \mathbb{P}^n$ and $Y_2 \in \mathbb{N}^n$ such that

$$0 = A_P^{\rm T}P + PA_P + R_1 + \tilde{C}^{\rm T}R_0^{-1}\tilde{C} + PB_0R_0^{-1}B_0^{\rm T}P - P_a^{\rm T}R_{2a}^{-1}P_a, \quad (12.157)$$

$$0 = (A_{Y_2} - BR_{2a}^{-1}P_u)^{\rm T}Y_2 + Y_2(A_{Y_2} - BR_{2a}^{-1}P_u) + \hat{R}_2 + \tfrac{1}{2}C_0^{\rm T}MC_0 + \tfrac{1}{2}Y_2B_0MB_0Y_2, \quad (12.158)$$

where $A_{Y_2} \triangleq A + \frac{1}{2}B_0(M_1 + M_2)C_0$. Then, with the feedback control law $u = \phi(x) = -R_2^{-1}B^{\rm T}(P + \sum_{k=2}^{r}(x^{\rm T}Y_2x)^{k-1}Y_2)x$, the zero solution $x(t) \equiv 0$ of the uncertain system (12.156) is globally asymptotically stable for all $x_0 \in \mathbb{R}^n$ and $\Delta A \in \mathbf{\Delta}_A$, and the performance functional (12.144) satisfies

$$J_{\Delta A}(x_0, \phi(x(\cdot))) \leq \mathcal{J}(x_0, \phi(x(\cdot))) = x_0^{\rm T}(P + P_0(\Delta A))x_0 + \sum_{k=2}^{r} \tfrac{1}{k}(x_0^{\rm T}Y_kx_0)^k, \qquad \Delta A \in \mathbf{\Delta}_A, \quad (12.159)$$

where

$$\mathcal{J}(x_0, u(\cdot)) \triangleq \int_0^\infty [L(x,u) + \Gamma(x,u) - x^{\rm T}(t)(A^{\rm T}P_0(\Delta A) + P_0(\Delta A)A)x(t)]{\rm d}t, \quad (12.160)$$

and where $u(\cdot)$ is admissible, and $x(t)$, $t \geq 0$, solves (12.156) with $\Delta A = 0$ and

$$\Gamma(x,u) = x^{\rm T}\Big[(\tilde{C} + B_0^{\rm T}P)^{\rm T}R_0^{-1}(\tilde{C} + B_0^{\rm T}P) + \tfrac{1}{2}\sum_{k=2}^{r}(x^{\rm T}Y_2x)^{k-1}(C_0 + B_0^{\rm T}Y_2)^{\rm T}M(C_0 + B_0^{\rm T}Y_2)\Big]x + 2x^{\rm T}B^{\rm T}C_0^{\rm T}N^{\rm T}R_0^{-1}(\tilde{C} + B_0^{\rm T}P)u + u^{\rm T}B_0^{\rm T}C_0^{\rm T}N^{\rm T}R_0^{-1}NC_0Bu.$$

In addition, the performance functional (12.144), with $R_2(x) = R_2$ and

$$L_1(x) = x^{\rm T}\left[R_1 + \sum_{k=2}^{r}(x^{\rm T}Y_2x)^{k-1}\hat{R}_2 + \left(\sum_{k=2}^{r}(x^{\rm T}Y_2x)^{k-1}\right)^2 Y_2BR_{2a}^{-1}B^{\rm T}Y_2\right]x,$$

is minimized in the sense that

$$\mathcal{J}(x_0, \phi(x(\cdot))) = \min_{u(\cdot)\in\mathcal{S}(x_0)} \mathcal{J}(x_0, u(\cdot)), \quad (12.161)$$

where $\mathcal{S}(x_0)$ is the set of regulation controllers for the nominal system and $x_0 \in \mathbb{R}^n$.

Proof. The result is a direct consequence of Theorem 12.8 with

$\Omega_{xx}(P) = (\tilde{C}+B_0^{\mathrm{T}}P)^{\mathrm{T}}R_0^{-1}(\tilde{C}+B_0^{\mathrm{T}}P)+C_0^{\mathrm{T}}M_1B_0^{\mathrm{T}}P+PB_0M_1C_0$, $\Omega_{xu}(P) = B^{\mathrm{T}}C_0^{\mathrm{T}}N^{\mathrm{T}}R_0^{-1}(\tilde{C}+B_0^{\mathrm{T}}P)$, $\Omega_{uu}(P) = B^{\mathrm{T}}C_0^{\mathrm{T}}N^{\mathrm{T}}R_0^{-1}NC_0B$, $\Omega(Y_2) = \frac{1}{2}(C_0 + B_0^{\mathrm{T}}Y_2)^{\mathrm{T}}M(C_0+B_0^{\mathrm{T}}Y_2)+C_0^{\mathrm{T}}M_1B_0^{\mathrm{T}}Y_2+Y_2B_0M_1C_0$, and $P_0(F) = C_0^{\mathrm{T}}F_{\mathrm{s}}NC_0$. In this case, as was shown in (12.131) and (12.150) holds. Furthermore, for all $\Delta A \in \boldsymbol{\Delta}_A$ it follows that

$$\begin{aligned}
0 &\le \tfrac{1}{2}[(C_0 + B_0^{\mathrm{T}}Y_2) - 2M^{-1}F_{\mathrm{s}}C_0]^{\mathrm{T}}M[(C_0 + B_0^{\mathrm{T}}Y_2) - 2M^{-1}F_{\mathrm{s}}C_0] \\
&\quad +2C_0^{\mathrm{T}}[F_{\mathrm{s}} - F_{\mathrm{s}}M^{-1}F_{\mathrm{s}}]C_0 \\
&= \Omega(Y_2) - C_0^{\mathrm{T}}FB_0^{\mathrm{T}}Y_2 - Y_2B_0FC_0 \\
&= \Omega(Y_2) - [\Delta A^{\mathrm{T}}Y_2 + Y_2\Delta A],
\end{aligned}$$

and hence, (12.149) holds. The result now follows as a direct consequence of Theorem 12.8. □

12.7 Robust Nonlinear Controllers with Multilinear Performance Criteria

In this section, we specialize the results of Section 12.5 to linear uncertain systems controlled by nonlinear controllers that minimize a multilinear cost functional. Specifically, we assume $\mathcal{F}$ to be the set of uncertain linear systems given by

$$\mathcal{F} = \{(A+\Delta A)x+Bu : x \in \mathbb{R}^n, u \in \mathbb{R}^m, A \in \mathbb{R}^{n\times n}, B \in \mathbb{R}^{n\times m}, \Delta A \in \boldsymbol{\Delta}_A\},$$

where $\boldsymbol{\Delta}_A \subset \mathbb{R}^{n\times n}$ is a given bounded uncertainty set of the uncertain perturbation ΔA of the nominal system A such that $0 \in \boldsymbol{\Delta}_A$. For the following let $R_1 \in \mathbb{P}^n$, $R_2 \in \mathbb{P}^m$, and $\hat{R}_{2\nu} \in \mathcal{N}^{(2\nu,n)}$, $\nu = 2,\ldots,r$, be given where r is a given integer.

Theorem 12.9. Consider the uncertain controlled system

$$\dot{x}(t) = (A+\Delta A)x(t) + Bu(t), \qquad x(0) = x_0, \qquad t \ge 0, \tag{12.162}$$

where $u(\cdot)$ is admissible and $\Delta A \in \boldsymbol{\Delta}_A$. Assume there exist $\Omega_{xx} : \mathbb{N}^n \to \mathbb{N}^n$, $\Omega_{xu} : \mathbb{N}^n \to \mathbb{R}^{n\times m}$, $\Omega_{uu} : \mathbb{N}^n \to \mathbb{N}^m$, $P_0 : \boldsymbol{\Delta}_A \to \mathbb{S}^n$, $P \in \mathbb{P}^n$, $\hat{\Omega}_\nu : \mathcal{N}^{(2\nu,n)} \to \mathcal{N}^{(2\nu,n)}$, and $\hat{P}_\nu \in \mathcal{N}^{(2\nu,n)}$, $\nu = 2,\ldots,r$, such that

$$\begin{aligned}
\Delta A^{\mathrm{T}}P + P\Delta A \le{}& \Omega_{xx}(P) - [(A+\Delta A - BR_{2a}^{-1}P_a)^{\mathrm{T}}P_0(\Delta A) + P_0(\Delta A) \\
&\cdot(A+\Delta A - BR_{2a}^{-1}P_a)] - P_a^{\mathrm{T}}R_{2a}^{-1}\Omega_{xu}^{\mathrm{T}}(P) \qquad (12.163)\\
&-\Omega_{xu}(P)R_{2a}^{-1}P_a + P_a^{\mathrm{T}}R_{2a}^{-1}\Omega_{uu}(P)R_{2a}^{-1}P_a, \quad \Delta A \in \boldsymbol{\Delta}_A,
\end{aligned}$$

$$\hat{\Omega}_\nu(\hat{P}_\nu) - \hat{P}_\nu(\overset{2\nu}{\oplus}\Delta A) \in \mathcal{N}^{(2\nu,n)}, \qquad \Delta A \in \boldsymbol{\Delta}_A, \quad \nu = 2,\ldots,r, \tag{12.164}$$

$$0 = A^{\mathrm{T}}P + PA + R_1 + \Omega_{xx}(P) - P_a^{\mathrm{T}}R_{2a}^{-1}P_a, \tag{12.165}$$

and

$$0 = \hat{P}_\nu[\overset{2\nu}{\oplus}(A - BR_{2a}^{-1}P_a)] + \hat{R}_{2\nu} + \hat{\Omega}_\nu(\hat{P}_\nu), \qquad \nu = 2,\ldots,r. \tag{12.166}$$

Then, with the feedback control law $u = \phi(x) \triangleq -R_{2a}^{-1}(P_a x + \frac{1}{2}B^{\mathrm{T}}g'^{\mathrm{T}}(x))$, where $g(x) = \sum_{\nu=2}^{r} \hat{P}_\nu x^{[2\nu]}$, the zero solution $x(t) \equiv 0$ of the uncertain system (12.162) is globally asymptotically stable for all $x_0 \in \mathbb{R}^n$ and $\Delta A \in \mathbf{\Delta}_A$, and the performance functional (12.144) satisfies

$$J_{\Delta A}(x_0, \phi(x(\cdot))) \leq \mathcal{J}(x_0, \phi(x(\cdot))) = x_0^{\mathrm{T}}(P + P_0(\Delta A)x_0 + \sum_{k=2}^{r} \hat{P}_\nu x_0^{[2\nu]},$$
$$\Delta A \in \mathbf{\Delta}_A, \qquad (12.167)$$

where

$$\mathcal{J}(x_0, u(\cdot)) \triangleq \int_0^\infty [L(x,u) + \Gamma(x,u) - x^{\mathrm{T}}(t)(A^{\mathrm{T}}P_0(\Delta A) + P_0(\Delta A)A)x(t)]\mathrm{d}t, \qquad (12.168)$$

and where $u(\cdot)$ is admissible, and $x(t)$, $t \geq 0$, solves (12.162) with $\Delta A = 0$ and

$$\Gamma(x,u) = x^{\mathrm{T}}\Omega_{xx}(P)x + 2x^{\mathrm{T}}\Omega_{xu}(P)u + u^{\mathrm{T}}\Omega_{uu}(P)u + \sum_{\nu=2}^{r} \hat{\Omega}_\nu(\hat{P}_\nu)x^{[2\nu]},$$

where $u(\cdot)$ is admissible, $\Delta A \in \mathbf{\Delta}_A$. In addition, the performance functional (12.144), with $R_2(x) = R_2$ and

$$L(x,u) = x^{\mathrm{T}}R_1 x + \sum_{\nu=2}^{r} \hat{R}_{2\nu}x^{[2\nu]} + \tfrac{1}{4}g'(x)BR_{2a}^{-1}B^{\mathrm{T}}g'^{\mathrm{T}}(x)$$

is minimized in the sense that

$$\mathcal{J}(x_0, \phi(x(\cdot)) = \min_{u(\cdot)\in\mathcal{S}(x_0)} \mathcal{J}(x_0, u(\cdot)), \qquad (12.169)$$

where $\mathcal{S}(x_0)$ is the set of regulation controllers for the nominal system and $x_0 \in \mathbb{R}^n$.

Proof. The result is a direct consequence of Corollary 12.4 with $f_0(x) = Ax$, $\Delta f(x) = \Delta Ax$, $G_0(x) = B$, $V_{\mathrm{I}}(x) = x^{\mathrm{T}}Px + \sum_{\nu=2}^{r} \hat{P}_\nu x^{[2\nu]}$, and $V_{\Delta f}(x) = x^{\mathrm{T}}P_0(\Delta A)x$. □

12.8 Problems

Problem 12.1. Consider the subset $\mathbf{\Delta} \subseteq \mathbf{\Delta}_{\mathrm{bs}}$ consisting of sector-bounded matrices

$$\mathbf{\Delta} \triangleq \{\Delta \in \mathbf{\Delta}_{\mathrm{bs}} : 2(\Delta - M_1)^*(M_2 - M_1)^{-1}(\Delta - M_1) \leq (\Delta - M_1) + (\Delta - M_1)^*\}, \qquad (12.170)$$

where $M_1, M_2 \in \mathbf{\Delta}_{\mathrm{bs}}$ are Hermitian matrices such that $M \triangleq M_2 - M_1$ is positive definite. Let $\Delta \in \mathbf{\Delta}_{\mathrm{bs}}$. Show that the following statements are equivalent:

i) $\Delta \in \boldsymbol{\Delta}$.

ii) $\begin{bmatrix} \mathrm{He}\,\Delta - M_1 & \Delta^* - M_1 \\ \Delta - M_1 & M \end{bmatrix} \geq 0.$

iii) $\begin{bmatrix} \mathrm{He}\,\Delta - M_1 & \Delta - M_1 \\ \Delta^* - M_1 & M \end{bmatrix} \geq 0.$

iv) $\begin{bmatrix} \mathrm{He}\,\Delta - M & \mathrm{Sh}\,\Delta \\ -\mathrm{Sh}\,\Delta & M_2 - \mathrm{He}\,\Delta \end{bmatrix} \geq 0.$

If, in addition, $\det(M_2 - \mathrm{He}\,\Delta) \neq 0$, show that

v) $M_1 - \mathrm{Sh}\,\Delta(M_2 - \mathrm{He}\,\Delta)^{-1}\mathrm{Sh}\,\Delta \leq \mathrm{He}\,\Delta \leq M_2$, is equivalent to *i)*–*iv)*.

Furthermore, show that if $\Delta \in \boldsymbol{\Delta}$, then the following statements hold:

vi) $M_1 \leq \mathrm{He}\,\Delta \leq M_2$.

vii) $\sigma_{\max}(\Delta) \leq \sigma_{\max}(M) + \sigma_{\max}(M_1)$.

Finally, show that $\Delta \in \boldsymbol{\Delta}$ if and only if $\Delta^* \in \boldsymbol{\Delta}$ and in the case $\Delta = \Delta^*$ for all $\Delta \in \boldsymbol{\Delta}$ then $\Delta \in \boldsymbol{\Delta}$ if and only if $M_1 \leq \Delta \leq M_2$.

Problem 12.2. Let $Z(\cdot) \in \mathcal{Z}$, let $\omega \in \mathbb{R} \cup \infty$, and suppose $\det(I + G(\jmath\omega)M_1) \neq 0$. Show that if (12.17) holds, then $\det(I + G(\jmath\omega)\Delta) \neq 0$ for all $\Delta \in \boldsymbol{\Delta}$, where $\boldsymbol{\Delta}$ is given by (12.170).

Problem 12.3. Consider $\boldsymbol{\Delta}_{\mathrm{bs}}$ given by (12.24) with $r = 0$. Let $G(\jmath\omega) \in \mathbb{C}^{m\times m}$ be such that $G(\jmath\omega)\Delta = \Delta G(\jmath\omega)$ for all $\Delta \in \boldsymbol{\Delta}_{\mathrm{bs}}$. Show that $\mu(G(\jmath\omega) = \rho(G(\jmath\omega))$, where $\rho(\cdot)$ denotes the spectral radius.

Problem 12.4. Let $G(\jmath\omega) \in \mathbb{C}^{m\times m}$. Show that

$$\rho(G(\jmath\omega)) = \inf_{D\in\mathcal{D}} \sigma_{\max}(DG(\jmath\omega)D^{-1}),$$

where $\mathcal{D} \triangleq \{D \in \mathbb{C}^{m\times m} : \det D \neq 0\}$. Using this result show that for $\boldsymbol{\Delta}_{\mathrm{bs}} = \{\Delta \in \mathbb{C}^{m\times m} : \Delta = \delta I_m,\ \delta \in \mathbb{C}\}$, $\overline{\mu}(G(\jmath\omega))$ given by (12.29) is nonconservative.

Problem 12.5. Let $G(\jmath\omega) \in \mathbb{C}^{m\times m}$ be such that $G(\jmath\omega) \geq 0$, $\omega \in \mathbb{R}$. Show that $\mu(G(\jmath\omega)) = \inf_{D\in\mathcal{D}} \sigma_{\max}(DG(\jmath\omega)D^{-1})$ for every $\mathcal{D}$.

Problem 12.6. Let $\omega \in \mathbb{R}$ and define

$$\rho_{\rm R}(G(\jmath\omega)) \triangleq \begin{cases} \max\{|\lambda| : \lambda \in \operatorname{spec}(G(\jmath\omega)) \cap \mathbb{R}\}, & \text{if } \operatorname{spec}(G(\jmath\omega)) \cap \mathbb{R} \neq \emptyset, \\ 0, & \text{otherwise.} \end{cases}$$

Show that $\rho_{\rm R}(G(\jmath\omega)) \le \mu(G(\jmath\omega))$. Furthermore, show that if $r = 0$ in $\mathbf{\Delta}_{\rm bs}$, then $\rho(G(\jmath\omega)) \le \mu(G(\jmath\omega))$.

Problem 12.7. Let $\Delta \in \mathbf{\Delta}_{\rm bs}$, where

$$\mathbf{\Delta}_{\rm bs} = \{\Delta \in \mathbb{R}^{m\times m} : \Delta = \operatorname{diag}[\delta_1, \delta_2, \ldots, \delta_m], \delta_i \in \mathbb{R}, i = 1, \ldots, m\}. \tag{12.171}$$

Define the frequency-dependent scaling matrix functions in $\mathcal{D}$ and $\mathcal{N}$ corresponding to small gain ($\mathcal{D}_{\rm sg}$, $\mathcal{N}_{\rm sg}$), Popov ($\mathcal{D}_{\rm P}$, $\mathcal{N}_{\rm P}$) [151, 172], monotonic ($\mathcal{D}_{\rm RC}$, $\mathcal{N}_{\rm RC}$) [172, 328, 331], odd-monotonic ($\mathcal{D}_{\rm RLC}$, $\mathcal{N}_{\rm RLC}$) [172, 328, 331], generalized odd-monotonic ($\mathcal{D}_{\rm GRLC}$, $\mathcal{N}_{\rm GRLC}$) [425, 428], and LC multipliers ($\mathcal{D}_{\rm LC}$, $\mathcal{N}_{\rm LC}$) [71], respectively, by

$$\begin{aligned}
\mathcal{D}_{\rm sg} &\triangleq \{D \in \mathbb{R}^{m\times m} : D = I\}, \\
\mathcal{N}_{\rm sg} &\triangleq \{N \in \mathbb{R}^{m\times m} : N = 0\}, \\
\mathcal{D}_{\rm P} &\triangleq \{D \in \mathbb{R}^{m\times m} : D = \operatorname{diag}(\alpha_{i0}), 0 < \alpha_{i0} \in \mathbb{R}, i = 1, \ldots, m\}, \\
\mathcal{N}_{\rm P} &\triangleq \{N : \mathbb{R} \to \mathbb{R}^{m\times m} : N(\omega) = \operatorname{diag}(-\omega\beta_{i0}), \beta_{i0} \in \mathbb{R}, i = 1, \ldots, m\}, \\
\mathcal{D}_{\rm RC} &\triangleq \{D : \mathbb{R} \to \mathbb{R}^{m\times m} : D(\omega) = D_{\rm P} + \operatorname{diag}\left(\textstyle\sum_{j=1}^{m_{i_1}} \alpha_{ij}(1 - \frac{\alpha_{ij}\eta_{ij}}{\beta_{ij}(\omega^2+\eta_{ij}^2)})\right), \\
&\qquad 0 \le \alpha_{ij}, \beta_{ij}, \eta_{ij} \in \mathbb{R}, \eta_{ij}\beta_{ij} - \alpha_{ij} \le 0, \\
&\qquad i = 1, \ldots, m;\ j = 1, \ldots, m_{i_1}, D_{\rm P} \in \mathcal{D}_{\rm P}\}, \\
\mathcal{N}_{\rm RC} &\triangleq \{N : \mathbb{R} \to \mathbb{R}^{m\times m} : N(\omega) = N_{\rm P}(\omega) + \operatorname{diag}\left(\textstyle\sum_{j=1}^{m_{i_1}} -\frac{\alpha_{ij}^2\omega}{\beta_{ij}(\omega^2+\eta_{ij}^2)}\right), \\
&\qquad 0 \le \alpha_{ij}, \beta_{ij}, \eta_{ij} \in \mathbb{R}, \eta_{ij}\beta_{ij} - \alpha_{ij} \le 0, \\
&\qquad i = 1, \ldots, m;\ j = 1, \ldots, m_{i_1}, N_{\rm P}(\cdot) \in \mathcal{N}_{\rm P}\}, \\
\mathcal{D}_{\rm RLC} &\triangleq \{D : \mathbb{R} \to \mathbb{R}^{m\times m} : D(\omega) = \operatorname{diag}\left(\textstyle\sum_{j=m_{i_1}+1}^{m_{i_2}} \alpha_{ij}(1 + \frac{\alpha_{ij}\eta_{ij}}{\beta_{ij}(\omega^2+\eta_{ij}^2)})\right) \\
&\qquad +D_{\rm RC}(\omega),\ 0 \le \alpha_{ij}, \beta_{ij}, \eta_{ij} \in \mathbb{R}, \\
&\qquad i = 1, \ldots, m;\ j = m_{i_1}+1, \ldots, m_{i_2}, D_{\rm RC}(\cdot) \in \mathcal{D}_{\rm RC}\}, \\
\mathcal{N}_{\rm RLC} &\triangleq \{N : \mathbb{R} \to \mathbb{R}^{m\times m} : N(\omega) = \operatorname{diag}\left(\textstyle\sum_{j=m_{i_1}+1}^{m_{i_2}} \frac{\alpha_{ij}^2\omega}{\beta_{ij}(\omega^2+\eta_{ij}^2)}\right), \\
&\qquad +N_{\rm RC}(\omega),\ 0 \le \alpha_{ij}, \beta_{ij}, \eta_{ij} \in \mathbb{R}, \\
&\qquad i = 1, \ldots, m;\ j = m_{i_1}+1, \ldots, m_{i_2}, N_{\rm RC}(\cdot) \in \mathcal{N}_{\rm RC}\}, \\
\mathcal{D}_{\rm GRLC} &\triangleq \{D : \mathbb{R} \to \mathbb{R}^{m\times m} : D(\omega) = D_{\rm RLC}(\omega)
\end{aligned}$$

$$
\begin{aligned}
&+\operatorname{diag}\left(\sum_{j=m_{i_2}+1}^{m_{i_3}} \alpha_{ij}\frac{\omega^4+\omega^2(a_{ij}\lambda_{ij}-b_{ij}-\eta_{ij})+b_{ij}\eta_{ij}}{(\eta_{ij}-\omega^2)^2+\lambda_{ij}^2\omega^2}\right),\\
&a_{ij}\in\mathbb{R}, 0\le \alpha_{ij},\beta_{ij},\eta_{ij}\in\mathbb{R}, a_{ij}\lambda_{ij}-b_{ij}-\eta_{ij}\ge 0,\\
&i=1,\ldots,m;\ j=m_{i_2}+1,\ldots,m_{i_3}, D_{\mathrm{RLC}}(\cdot)\in\mathcal{D}_{\mathrm{RLC}}\},\\
\mathcal{N}_{\mathrm{GRLC}} &\triangleq \{N:\mathbb{R}\to\mathbb{R}^{m\times m}: N(\omega)=N_{\mathrm{RLC}}(\omega)\\
&+\operatorname{diag}\left(\sum_{j=m_{i_2}+1}^{m_{i_3}} -\omega\alpha_{ij}\frac{\omega^2(a_{ij}-\lambda_{ij})+(a_{ij}\eta_{ij}-\lambda_{ij}b_{ij})}{(\eta_{ij}-\omega^2)^2+\lambda_{ij}^2\omega^2}\right),\\
&a_{ij}\in\mathbb{R}, 0\le \alpha_{ij},\beta_{ij},\eta_{ij}\in\mathbb{R}, a_{ij}\lambda_{ij}-b_{ij}-\eta_{ij}\ge 0,\\
&i=1,\ldots,m;\ j=m_{i_2}+1,\ldots,m_{i_3}, N_{\mathrm{RLC}}(\cdot)\in\mathcal{N}_{\mathrm{RLC}}\},\\
\mathcal{D}_{\mathrm{LC}} &\triangleq \{D\in\mathbb{R}^{m\times m}: D=0\},\\
\mathcal{N}_{\mathrm{LC}} &\triangleq \{N:\mathbb{R}\to\mathbb{R}^{m\times m}: N(\omega)=\operatorname{diag}(\mp\omega^{\pm 1}\tfrac{\Pi(\alpha_i^2-\omega^2)}{\Pi(\beta_i^2-\omega^2)}),\\
&\alpha_i,\beta_i\in\mathbb{R}, \beta_i\ne 0, i=1,\ldots,m\}.
\end{aligned}
$$

Show that for $G(\jmath\omega))\in\mathbb{C}^{m\times m}$

$$
\begin{aligned}
\mu(G(\jmath\omega)) &\le \overline{\mu}(G(\jmath\omega)) \le \mu_{\mathrm{GRLC}}(G(\jmath\omega)) \le \mu_{\mathrm{RLC}}(G(\jmath\omega))\\
&\le \mu_{\mathrm{RC}}(G(\jmath\omega)) \le \mu_{\mathrm{P}}(G(\jmath\omega)) \le \mu_{\mathrm{sg}}(G(\jmath\omega)), \qquad (12.172)
\end{aligned}
$$

where $\mu_i(G(\jmath\omega))$, for $i=$GRLC, RLC, P, and sg, corresponds to μ bounds predicated on fixed frequency-dependent scaling functions $\mathcal{D}_i$ and $\mathcal{N}_i$. What can you say about $\mu_{\mathrm{LC}}(G(\jmath\omega))$ with respect to the above ordering?

Problem 12.8. Let $\beta>0$, $G(\jmath\omega)\in\mathbb{C}^{m\times m}$, $D(\cdot)\in\mathcal{D}$, $N(\cdot)\in\mathcal{N}$, and define

$$
\begin{aligned}
\mathcal{X}(D,N,\beta) &\triangleq G^*(\jmath\omega)DG(\jmath\omega)+\jmath(NG(\jmath\omega)-G^*(\jmath\omega)N)-\beta D,\\
\Phi(D,N) &\triangleq \inf\{\eta>0:\mathcal{X}(D,N,\eta)<0\}.
\end{aligned}
$$

Show that there exist $D(\cdot)\in\mathcal{D}$ and $N(\cdot)\in\mathcal{N}$ such that $\Phi(D,N)<\eta$ if and only if $\mathcal{X}(D,N,\eta)<0$.

Problem 12.9. A functional f on a vector space $\mathcal{H}$ is *quasiconvex* if, for all $\alpha\in[0,1]$ and $H,\hat{H}\in\mathcal{H}$, $f(\alpha H+(1-\alpha)\hat{H})\le\max\{f(H),f(\hat{H})\}$. Show that $\Phi(D,N)$ defined by (12.173) is quasiconvex on $\mathcal{D}\times\mathcal{N}$.

Problem 12.10. Consider the linear uncertain system (12.74) where $\Delta A\in\boldsymbol{\Delta}_A$ and $\boldsymbol{\Delta}_A$ is given by

$$
\boldsymbol{\Delta}_A \triangleq \{\Delta A\in\mathbb{R}^{n\times n}:\Delta A=B_0FC_0,\ M_1\le F\le M_2\}. \qquad (12.173)
$$

Let $X\in\mathbb{R}^{p\times p}$ and $Y\in\mathbb{N}^n$ be such that

$$
B_0X^{\mathrm{T}}(F-M_1)C_0+C_0^{\mathrm{T}}(F-M_1)XB_0^{\mathrm{T}}\le Y, \qquad (12.174)
$$

and let $H \in \mathcal{H}_{\rm p}$ and $N \in \mathcal{N}_{\rm s}$ be such that

$$R_0 \triangleq [HM^{-1} - NC_0B_0] + [HM^{-1} - NC_0B_0]^{\rm T} > 0, \tag{12.175}$$

where $M \triangleq M_2 - M_1$. Show that the functions

$$\Omega_0(P) - [HC_0 + NC_0(A + B_0M_1C_0) + B_0^{\rm T}P - XB_0^{\rm T}]^{\rm T}R_0^{-1}[HC_0 + NC_0(A + B_0M_1C_0) + B_0^{\rm T}P - XB_0^{\rm T}] + PB_0M_1C_0 + C_0^{\rm T}M_1B_0^{\rm T}P + Y, \tag{12.176}$$

$$P_0(F) = C_0^{\rm T}(F - M_1)NC_0, \tag{12.177}$$

satisfy (12.76) with ΔA given by (12.173). Furthermore, show that

$$Y = B_0X^{\rm T}MXB_0^{\rm T} + C_0^{\rm T}MC_0, \tag{12.178}$$

satisfies (12.174) for all $X \in \mathbb{R}^{p\times p}$ and $F \in \mathbb{S}^p$ such that $M_1 \leq F \leq M_2$. Finally, for the special case of diagonal uncertainty F show that $Y = B_0X^{\rm T}XB_0^{\rm T} + C_0^{\rm T}M^2C_0$ also satisfies (12.174).

Problem 12.11. Consider the linear uncertain system (12.74) where $\Delta A \in \boldsymbol{\Delta}_A$ and $\boldsymbol{\Delta}_A$ is given by (12.173). Let $X \in \mathbb{R}^{p\times p}$, $Y \in \mathbb{N}^n$, and $H \in \mathcal{H}_{\rm p}$ be such that

$$B_0X^{\rm T}(F - M_1)HC_0 + C_0^{\rm T}H(F - M_1)XB_0^{\rm T} \leq Y, \tag{12.179}$$

and let $N \in \mathcal{N}_{\rm s}$ be such that (12.175) holds. Show that the functions

$$\begin{aligned}\Omega_0(P) &= [HC_0 + NC_0(A + B_0M_1C_0) + H^{-1}B_0^{\rm T}P - XB_0^{\rm T}]^{\rm T}R_0^{-1}[HC_0 \\ &\quad + NC_0(A + B_0M_1C_0) + H^{-1}B_0^{\rm T}P - XB_0^{\rm T}] \\ &\quad + PB_0M_1C_0 + C_0^{\rm T}M_1B_0^{\rm T}P + Y, &(12.180)\\ P_0(F) &= C_0^{\rm T}(F - M_1)NC_0, &(12.181)\end{aligned}$$

satisfy (12.76) with ΔA given by (12.173).

Problem 12.12. Consider the linear uncertain system (12.122) with $\Delta A \in \boldsymbol{\Delta}_A$, where $\boldsymbol{\Delta}_A$ is given by (12.173), and the performance functional (12.123). Let $H \in \mathcal{H}_{\rm p}$ and $N \in \mathcal{N}_{\rm nd}$ be such that (12.175) holds, and let $X \in \mathbb{R}^{p\times p}$ and $Y \in \mathbb{N}^n$ be such that (12.174) is satisfied. Show that the zero solution $x(t) \equiv 0$ to (12.122) is globally asymptotically stable for all $\Delta A \in \boldsymbol{\Delta}_A$ with the feedback control $\phi(x) = -R_{2a}^{-1}P_ax$, where

$$\begin{aligned}R_{2a} &\triangleq R_2 + B^{\rm T}C_0^{\rm T}NR_0^{-1}NC_0B, \\ P_a &\triangleq B^{\rm T}P + B^{\rm T}C_0^{\rm T}NR_0^{-1}[HC_0 + NC_0(A + B_0M_1C_0) + B_0^{\rm T}P - XB_0^{\rm T}],\end{aligned}$$

and $P > 0$ satisfies

$$\begin{aligned}0 = {}& A_P^{\rm T}P + PA_P + R_1 + Y + [HC_0 + NC_0(A + B_0M_1C_0) - XB_0^{\rm T}]^{\rm T}R_0^{-1} \\ & \cdot[HC_0 + NC_0(A + B_0M_1C_0) - XB_0^{\rm T}] + PB_0R_0^{-1}B_0^{\rm T}P - P_a^{\rm T}R_{2a}^{-1}P_a,\end{aligned}$$

where $A_P \triangleq A + B_0 M_1 C_0 + B_0 R_0^{-1}[HC_0 + NC_0(A + B_0 M_1 C_0) - XB_0^{\mathrm{T}}]$.

Problem 12.13. Consider the linear dynamical system with state delay given by

$$\dot{x}(t) = Ax(t) + A_{\mathrm{d}} x(t - \tau_{\mathrm{d}}), \quad x(\theta) = \phi(\theta), \quad -\tau_{\mathrm{d}} \le \theta \le 0, \tag{12.182}$$

where $x(t) \in \mathbb{R}^n$, $A \in \mathbb{R}^{n\times n}$, $A_{\mathrm{d}} \in \mathbb{R}^{n\times n}$, and $\phi : [-\tau_{\mathrm{d}}, 0] \to \mathbb{R}^n$ is a continuous vector-valued function specifying the initial state of the system. Show that the zero solution $x_t \equiv 0$ to (12.182) is globally asymptotically stable (in the sense of Problem 3.65) for all $\tau_{\mathrm{d}} \in [0, \overline{\tau}]$ if $|||DG(s)D^{-1}|||_\infty < 1$, where

$$G(s) \sim \left[\begin{array}{c|cc} A + QA_{\mathrm{d}} & \overline{\tau}Q & (I_n - Q)A_{\mathrm{d}} \\ \hline A_{\mathrm{d}}A & 0 & A_{\mathrm{d}}^2 \\ I_n & 0 & 0 \end{array}\right],$$

$D = \mathrm{diag}(D_1, D_2)$, $D_1 > 0$, $D_2 > 0$, and $Q \in \mathbb{R}^{n\times n}$. Show that the result also holds if $G(s)$ is replaced by

$$G(s) \sim \left[\begin{array}{c|cc} A + A_{\mathrm{d}} & \overline{\tau}A_{\mathrm{d}} & \overline{\tau}A_{\mathrm{d}} \\ \hline A & 0 & 0 \\ A_{\mathrm{d}} & 0 & 0 \end{array}\right].$$

Finally, show that in both cases the problem can be represented as a feedback problem involving an uncertain block-structured operator $\Delta(s) = \mathrm{diag}(\Delta_1(s), \Delta_2(s))$ satisfying $|||\Delta_i(s)|||_\infty \le 1$, $i = 1, 2$.

12.9 Notes and References

The efficacy of parameter-dependent Lyapunov functions for nonconservatively addressing real parameter uncertainty was first conjectured by Narendra [325] and later rigorously proven by Thathachar and Srinath [426]. Specifically, Thathachar and Srinath [426] constructed a parameter-dependent Lyapunov function for proving necessary and sufficient conditions for robust stability of a linear time-invariant system with a single constant uncertainty. The proof of this fundamental result stemmed from absolute stability theory [5, 64, 65, 67, 70, 71, 94, 326, 328–331, 361, 362, 364, 425, 427, 428, 458, 477, 479] and was based on the fact that if the uncertain system is robustly stable, then there always exists a plant-dependent inductor-capacitor multiplier such that the tandem connection of the nominal plant and multiplier is positive real. For further details on this fact see Narendra and Taylor [331] and How and Haddad [204]. Building on the wealth of knowledge of absolute stability theory, Haddad and Bernstein [147–149, 151, 152] unified and extended classical absolute stability theory for linear time-invariant dynamical systems with loop nonlinearities to address the modern-day robust analysis and synthesis problem via parameter-dependent Lyapunov functions. Potential advantages of parameter-dependent Lyapunov functions over "fixed" Lyapunov functions were also discussed by Barmish and DeMarco [27] and Leal and Gibson [261].

Connections between absolute stability theory and mixed-μ bounds for real parameter uncertainty were first reported by Haddad, How, Hall, and Bernstein [170–172] and How and Hall [206], with further extensions given by Safonov and Chiang [381] and Haddad, Bernstein, and Chellaboina [153]. A direct benefit of this unification resulted in new machinery for mixed-μ controller synthesis by providing an alternative to the standard multiplier-controller iteration and curve-fitting procedure [97, 203, 205, 207, 381].

Wong [465] was the first to consider the robustness problem with structured uncertainty, laying down the framework for multivariable stability and μ theory. Doyle [110] was the first to introduce the term structured singular value. Building on the work of Wong [465], Safonov and Athans [380] and Safonov [378] addressed stability margins of diagonally perturbed multivariable feedback systems and introduced the notion of excess stability margin; the reciprocal of the structured singular value. The mixed-μ bounds for *diagonal* real parameters are due to Fan, Tits, and Doyle [117] while the mixed-μ bounds for real and complex multiple-block uncertainty with internal matrix structure are due to Haddad, Bernstein, and Chellaboina [153]. For a complete historical perspective on the structured singular value the reader is referred to the editorial by Safonov and Fan [382].

Finally, the presentation of the structured singular value in this chapter is adopted from Haddad, Bernstein, and Chellaboina [153], while the presentation of the robust nonlinear-nonquadratic feedback control problem via parameter-dependent Lyapunov functions is adopted from Haddad and Chellaboina [156].

Chapter Thirteen

Stability and Dissipativity Theory for Discrete-Time Nonlinear Dynamical Systems

13.1 Introduction

In the first twelve chapters of this book we presented a thorough treatment of nonlinear analysis and control for continuous-time dynamical systems using a Lyapunov and dissipative dynamical systems approach. In the next two chapters we give a condensed review of the same theory for discrete-time systems. Since the theory for nonlinear discrete-time dynamical systems closely parallels the theory of nonlinear continuous-time dynamical systems, many of the results are similar. Hence, the contents in this and the next chapter are brief, except in those cases where the discrete-time results deviate markedly from their continuous-time counterparts. Even though many of the proofs are similar to the continuous-time proofs, for completeness of exposition we provide a self-contained treatment of the fundamental discrete-time results.

13.2 Discrete-Time Lyapunov Stability Theory

We begin by considering the general discrete-time nonlinear dynamical system

$$x(k+1) = f(x(k)), \qquad x(0) = x_0, \qquad k \in \overline{\mathbb{Z}}_+, \tag{13.1}$$

where $x(k) \in \mathcal{D} \subseteq \mathbb{R}^n$, $k \in \overline{\mathbb{Z}}_+$, is the system state vector, $\mathcal{D}$ is an open set, $0 \in \mathcal{D}$, $f : \mathcal{D} \to \mathcal{D}$, and $f(0) = 0$. We assume that $f(\cdot)$ is continuous on $\mathcal{D}$. Furthermore, we denote the solution to (13.1) with initial condition $x(0) = x_0$ by $s(\cdot, x_0)$, so that the map of the dynamical system given by $s : \overline{\mathbb{Z}}_+ \times \mathcal{D} \to \mathcal{D}$ is continuous on $\mathcal{D}$ and satisfies the consistency property $s(0, x_0) = x_0$ and the semigroup property $s(\kappa, s(k, x_0)) = s(k + \kappa, x_0)$, for all $x_0 \in \mathcal{D}$ and $k, \kappa \in \overline{\mathbb{Z}}_+$. For this and the next chapter we use the notation $s(k, x_0)$, $k \in \overline{\mathbb{Z}}_+$, and $x(k)$, $k \in \overline{\mathbb{Z}}_+$, interchangeably as the solution of the nonlinear discrete-time system (13.1) with initial condition $x(0) = x_0$.

Unlike continuous-time dynamical systems, establishing existence and uniqueness of solutions for discrete-time dynamical systems is straightforward. To see this, consider the discrete-time dynamical system (13.1) and let $\mathcal{I}_{x_0}^+ \subseteq \overline{\mathbb{Z}}_+$ be the maximal interval of existence for the solution $x(\cdot)$ of (13.1). Now, to construct a solution to (13.1) we can construct the *solution sequence* or *discrete trajectory* $x(k) = s(k, x_0)$ iteratively by setting $x(0) = x_0$ and using $f(\cdot)$ to define $x(k)$ recursively by $x(k+1) = f(x(k))$. Specifically,

$$\begin{aligned} s(0, x_0) &= x_0 \\ s(1, x_0) &= f(s(0, x_0)) = f(x_0) \\ s(2, x_0) &= f(s(1, x_0)) \\ &\vdots \\ s(k, x_0) &= f(s(k-1, x_0)). \end{aligned} \tag{13.2}$$

If $f(\cdot)$ is continuous, it follows that $f(s(k-1, \cdot)$ is also continuous since it is constructed as a composition of continuous functions. Hence, $s(k, \cdot)$ is continuous. If $f(\cdot)$ is such that $f : \mathbb{R}^n \to \mathbb{R}^n$, this iterative process can be continued indefinitely, and hence, a solution to (13.1) exists for all $k \geq 0$. Alternatively, if $f(\cdot)$ is such that $f : \mathcal{D} \to \mathbb{R}^n$, the solution may cease to exist at some point if $f(\cdot)$ maps $x(k)$ into some point $x(k+1)$ outside the domain of $f(\cdot)$. In this case, the solution sequence $x(k) = s(k, x_0)$ will be defined on the maximal interval of existence $x(k)$, $k \in \mathcal{I}_{x_0}^+ \subset \overline{\mathbb{Z}}_+$. Furthermore, note that the solution sequence $x(k)$, $k \in \mathcal{I}_{x_0}^+$, is uniquely defined for a given x_0 if $f(\cdot)$ is a continuous function. That is, any other solution sequence $y(k)$ starting from x_0 at $k = 0$ will take exactly the same values as $x(k)$ and can be continued to the same interval as $x(k)$. It is important to note that if $k \in \overline{\mathbb{Z}}_+$, uniqueness of solutions backward in time need not necessarily hold. This is due to the fact that $(k, x_0) = f^{-1}(s(k+1, x_0))$, $k \in \overline{\mathbb{Z}}_+$, and there is no guarantee that $f(\cdot)$ is invertible for all $k \in \overline{\mathbb{Z}}_+$. However, if $f : \mathcal{D} \to \mathcal{D}$ is a homeomorphism for all $k \in \overline{\mathbb{Z}}_+$, then the solution sequence is unique for all $k \in \mathbb{Z}$. Identical arguments can be used to establish existence and uniqueness of solutions for time-varying discrete-time systems. In light of the above discussion the following theorem is immediate.

Theorem 13.1. Consider the nonlinear dynamical system (13.1). Assume that $f : \mathcal{D} \to \mathcal{D}$ is continuous on $\mathcal{D}$. Then for every $x_0 \in \mathcal{D}$, there exists $\mathcal{I}_{x_0}^+ \subseteq \overline{\mathbb{Z}}_+$ such that (13.1) has a unique solution $x : \mathcal{I}_{x_0}^+ \to \mathbb{R}^n$. Moreover, for each $k \in \mathcal{I}_{x_0}^+$, the solution $s(k, \cdot)$ is continuous. If, in addition, $f(\cdot)$ is a homeomorphism of $\mathcal{D}$ onto $\mathbb{R}^n$, then the solution $x : \mathcal{I}_{x_0} \to \mathbb{R}^n$ is unique in all $\mathcal{I}_{x_0} \in \mathbb{Z}$ and $s(k, \cdot)$ is continuous for all $k \in \mathcal{I}_{x_0}$. Finally, if $\mathcal{D} = \mathbb{R}^n$, then $\mathcal{I}_{x_0} = \mathbb{Z}$.

An *equilibrium point* of (13.1) is a point $x \in \mathcal{D}$ satisfying $f(x) = x$ or, equivalently, $s(k, x) = x$ for all $k \in \overline{\mathbb{Z}}_+$. Unless otherwise stated, we assume $f(0) = 0$ for $\mathcal{I}_{x_0} = \overline{\mathbb{Z}}_+$. The following definition introduces several types of stability corresponding to the zero solution $x(k) \equiv 0$ of the discrete-time system (13.1) for $\mathcal{I}_{x_0} = \overline{\mathbb{Z}}_+$.

Definition 13.1. *i*) The zero solution $x(k) \equiv 0$ to (13.1) is *Lyapunov stable* if for all $\varepsilon > 0$ there exists $\delta = \delta(\varepsilon) > 0$ such that if $\|x(0)\| < \delta$, then $\|x(k)\| < \varepsilon,\ k \in \overline{\mathbb{Z}}_+$.

ii) The zero solution $x(k) \equiv 0$ to (13.1) is *asymptotically stable* if it is Lyapunov stable and there exists $\delta > 0$ such that if $\|x(0)\| < \delta$, then $\lim_{k\to\infty} x(k) = 0$.

iii) The zero solution $x(k) \equiv 0$ to (13.1) is *geometrically stable* if there exist positive constants $\alpha,\ \beta > 1$, and δ such that if $\|x(0)\| < \delta$, then $\|x(k)\| \le \alpha\|x(0)\|\beta^{-k}$, $k \in \overline{\mathbb{Z}}_+$.

iv) The zero solution $x(k) \equiv 0$ to (13.1) is *globally asymptotically stable* if it is Lyapunov stable and for all $x(0) \in \mathbb{R}^n$, $\lim_{k\to\infty} x(k) = 0$.

v) The zero solution $x(k) \equiv 0$ to (13.1) is *globally geometrically stable* if there exist positive constants α and $\beta > 1$ such that $\|x(k)\| \le \alpha\|x(0)\|\beta^{-k}$, $k \in \overline{\mathbb{Z}}_+$, for all $x(0) \in \mathbb{R}^n$.

vi) Finally, the zero solution $x(k) \equiv 0$ to (13.1) is *unstable* if it is not Lyapunov stable.

The following result, known as Lyapunov's direct method, gives sufficient conditions for Lyapunov and asymptotic stability of a discrete-time nonlinear dynamical system.

Theorem 13.2. Consider the discrete-time nonlinear dynamical system (13.1) and assume that there exists a continuous function $V : \mathcal{D} \to \mathbb{R}$ such that

$$V(0) = 0, \tag{13.3}$$

$$V(x) > 0, \quad x \in \mathcal{D}, \quad x \neq 0, \tag{13.4}$$

$$V(f(x)) - V(x) \le 0, \quad x \in \mathcal{D}. \tag{13.5}$$

Then the zero solution $x(k) \equiv 0$ to (13.1) is Lyapunov stable. If, in addition,

$$V(f(x)) - V(x) < 0, \quad x \in \mathcal{D}, \quad x \neq 0, \tag{13.6}$$

then the zero solution $x(k) \equiv 0$ to (13.1) is asymptotically stable. Alternatively, if there exist scalars $\alpha, \beta > 0$, $\rho > 1$, and $p \ge 1$, such that

$V : \mathcal{D} \to \mathbb{R}$ satisfies

$$\alpha\|x\|^p \le V(x) \le \beta\|x\|^p, \qquad x \in \mathcal{D}, \tag{13.7}$$

$$\rho V(f(x)) \le V(x), \qquad x \in \mathcal{D}, \tag{13.8}$$

then the zero solution $x(k) \equiv 0$ to (13.1) is geometrically stable. Finally, if $\mathcal{D} = \mathbb{R}^n$ and $V(\cdot)$ is such that

$$V(x) \to \infty \text{ as } \|x\| \to \infty \tag{13.9}$$

then (13.6) implies (respectively, (13.7) and (13.8) imply) that the zero solution $x(k) \equiv 0$ to (13.1) is globally asymptotically (respectively, globally geometrically) stable.

Proof. Let $\varepsilon > 0$ be such that $\mathcal{B}_\varepsilon(0) \subseteq \mathcal{D}$. Since $\overline{\mathcal{B}}_\varepsilon(0)$ is compact and $f(x)$, $x \in \mathcal{D}$, is continuous, it follows that

$$\eta \triangleq \max\left\{\varepsilon, \max_{x\in\overline{\mathcal{B}}_\varepsilon(0)} \|f(x)\|\right\} \tag{13.10}$$

exists. Next, let $\alpha \triangleq \min_{x\in\mathcal{D}:\ \varepsilon\le\|x\|\le\eta} V(x)$. Note $\alpha > 0$ since $0 \notin \partial\mathcal{B}_\varepsilon(0)$ and $V(x) > 0$, $x \in \mathcal{D}$, $x \neq 0$. Next, let $\beta \in (0, \alpha)$ and define $\mathcal{D}_\beta \triangleq \{x \in \mathcal{B}_\varepsilon(0) : V(x) \le \beta\}$. Now, for every $x \in \mathcal{D}_\beta$, it follows from (13.5) that $V(f(x)) \le V(x) \le \beta$, and hence, it follows from (13.10) that $\|f(x)\| \le \eta$, $x \in \mathcal{D}_\beta$. Next, suppose, *ad absurdum*, that there exists $x \in \mathcal{D}_\beta$ such that $\|f(x)\| \ge \varepsilon$. This implies $V(x) \ge \alpha$, which is a contradiction. Hence, for every $x \in \mathcal{D}_\beta$, it follows that $f(x) \in \mathcal{B}_\varepsilon(0) \subset \mathcal{D}_\beta$, which implies that $\mathcal{D}_\beta$ is a positively invariant set (see Definition 13.4) with respect to (13.1). Next, since $V(\cdot)$ is continuous and $V(0) = 0$, there exists $\delta = \delta(\varepsilon) \in (0, \varepsilon)$ such that $V(x) < \beta$, $x \in \mathcal{B}_\delta(0)$. Now, let $x(k)$, $k \in \overline{\mathbb{Z}}_+$, satisfy (13.1). Since $\mathcal{B}_\delta(0) \subset \mathcal{D}_\beta \subset \mathcal{B}_\varepsilon(0) \subseteq \mathcal{D}$ and $\mathcal{D}_\beta$ is positively invariant with respect to (13.1) it follows that for all $x(0) \in \mathcal{B}_\delta(0)$, $x(k) \in \mathcal{B}_\varepsilon(0)$, $k \in \overline{\mathbb{Z}}_+$. Hence, for all $\varepsilon > 0$ there exists $\delta = \delta(\varepsilon) > 0$ such that if $\|x(0)\| < \delta$, then $\|x(k)\| < \varepsilon$, $k \in \overline{\mathbb{Z}}_+$, which proves Lyapunov stability.

To prove asymptotic stability suppose that $V(f(x)) < V(x)$, $x \in \mathcal{D}$, $x \neq 0$, and $x(0) \in \mathcal{B}_\delta(0)$. Then it follows that $x(k) \in \mathcal{B}_\varepsilon(0)$, $k \in \overline{\mathbb{Z}}_+$. However, $V(x(k))$, $k \in \overline{\mathbb{Z}}_+$, is decreasing and bounded from below by zero. Now, suppose, *ad absurdum*, that $x(k)$, $k \in \overline{\mathbb{Z}}_+$, does not converge to zero. This implies that $V(x(k))$, $k \in \overline{\mathbb{Z}}_+$, is lower bounded by a positive number, that is, there exists $L > 0$ such that $V(x(k)) \ge L > 0$, $k \in \overline{\mathbb{Z}}_+$. Hence, by continuity of $V(x)$, $x \in \mathcal{D}$, there exists $\delta' > 0$ such that $V(x) < L$ for $x \in \mathcal{B}_{\delta'}(0)$, which further implies that $x(k) \notin \mathcal{B}_{\delta'}(0)$, $k \in \overline{\mathbb{Z}}_+$. Next, let $L_1 \triangleq \min_{\delta'\le\|x\|\le\varepsilon} [V(x) - V(f(x))]$. Now, (13.6) implies $V(x) - V(f(x)) \ge L_1$,

$\delta' \le \|x\| \le \varepsilon$, or, equivalently,

$$V(x(k)) - V(x(0)) = \sum_{i=0}^{k-1}[V(f(x(i))) - V(x(i))] \le -L_1 k,$$

and hence, for all $x(0) \in \mathcal{B}_\delta(0)$,

$$V(x(k)) \le V(x(0)) - L_1 k.$$

Letting $k > \frac{V(x(0))-L}{L_1}$, it follows that $V(x(k)) < L$, which is a contradiction. Hence, $x(k) \to 0$ as $k \to \infty$, establishing asymptotic stability.

Next, to show geometric stability note that (13.8) implies

$$V(x(k)) \le V(x(0))\rho^{-k}, \qquad k \in \overline{\mathbb{Z}}_+. \tag{13.11}$$

Now, since $V(x(0)) \le \beta\|x(0)\|^p$ and $\alpha\|x(k)\|^p \le V(x(k))$ it follows that

$$\alpha\|x(k)\|^p \le \beta\|x(0)\|^p\rho^{-k}, \qquad k \in \overline{\mathbb{Z}}_+, \tag{13.12}$$

which implies that

$$\|x(k)\| \le \left(\frac{\beta}{\alpha}\right)^{1/p}\|x(0)\|(\rho^{1/p})^{-k}, \tag{13.13}$$

establishing geometric stability.

Finally, to prove global asymptotic and geometric stability, let $x_0 \in \mathbb{R}^n$ and $\beta \triangleq V(x_0)$. Now, the radial unboundedness condition (13.9) implies that there exists $\varepsilon > 0$ such that $V(x) \ge \beta$ for $\|x\| \ge \varepsilon$, $x \in \mathbb{R}^n$. Hence, it follows from (13.6) that $V(x(k)) \le V(x(0)) = \beta$, $k \in \overline{\mathbb{Z}}_+$, which implies that $x(k) \in \mathcal{B}_\varepsilon(0)$, $k \in \overline{\mathbb{Z}}_+$. Now, the proof follows as in the proof of the local results. □

A continuous function $V(\cdot)$ satisfying (13.3) and (13.4) is called a *Lyapunov function candidate* for the discrete-time nonlinear dynamical system (13.1). If, additionally, $V(\cdot)$ satisfies (13.5), $V(\cdot)$ is called a *Lyapunov function* for the discrete-time nonlinear dynamical system (13.1). Next, we give a key definition involving the domain, or region, of attraction of the zero solution $x(k) \equiv 0$ of the discrete-time nonlinear dynamical system (13.1).

Definition 13.2. Suppose the zero solution $x(k) \equiv 0$ to (13.1) is asymptotically stable. Then the *domain of attraction* $\mathcal{D}_0$ of (13.1) is given by

$$\mathcal{D}_0 \triangleq \{x_0 \in \mathcal{D} : \text{ if } x(0) = x_0, \text{ then } \lim_{k\to\infty} x(k) = 0\}. \tag{13.14}$$

As in the continuous-time case, constructing the actual domain of attraction of a discrete-time nonlinear dynamical system is system trajectory dependent. To estimate a subset of the domain of attraction of the dynamical system (13.1) assume there exists a continuous function $V : \mathcal{D} \to \mathbb{R}$ such that (13.3), (13.4), and (13.6) are satisfied. Next, let $\mathcal{D}_\beta \triangleq \{x \in \mathcal{D} : V(x) \leq \beta\}$. Note that $\mathcal{D}_\beta \subseteq \mathcal{D}$ and every trajectory starting in $\mathcal{D}_\beta$ will move to an inner energy surface, and hence, cannot escape $\mathcal{D}_\beta$. Hence, $\mathcal{D}_\beta$ is an estimate of the domain of attraction of the nonlinear dynamical system (13.1). Now, to maximize this estimate of the domain of attraction we maximize β such that $\mathcal{D}_\beta \subseteq \mathcal{D}$. Hence, define $V_\Gamma \triangleq \max\{\beta > 0 : \mathcal{D}_\beta \subseteq \mathcal{D}\}$ so that

$$\mathcal{D}_{\mathrm{A}} \triangleq \{x \in \mathcal{D} : V(x) \leq V_\Gamma\}, \tag{13.15}$$

is a subset of the domain of attraction for (13.1) since $\Delta V(x) \triangleq V(f(x)) - V(x) < 0$ for all $x \in \mathcal{D}_{\mathrm{A}} \backslash \{0\} \subseteq \mathcal{D} \backslash \{0\}$.

13.3 Discrete-Time Invariant Set Stability Theorems

In this section, we use the discrete-time Barbashin-Krasovskii-LaSalle invariance principle to relax one of the conditions on the Lyapunov function $V(\cdot)$ in the theorems given in Section 13.2. In particular, the strict negative-definiteness condition on the Lyapunov difference can be relaxed while ensuring system asymptotic stability. To state the main results of this section several definitions and a key lemma analogous to the ones given in Section 2.12 are needed.

Definition 13.3. The trajectory $x(k)$, $k \in \overline{\mathbb{Z}}_+$, of (13.1) is *bounded* if there exists $\gamma > 0$ such that $\|x(k)\| < \gamma$, $k \in \overline{\mathbb{Z}}_+$.

Definition 13.4. A set $\mathcal{M} \subset \mathcal{D} \subseteq \mathbb{R}^n$ is a *positively invariant set* for the nonlinear dynamical system (13.1) if $s_k(\mathcal{M}) \subseteq \mathcal{M}$, for all $k \in \overline{\mathbb{Z}}_+$, where $s_k(\mathcal{M}) \triangleq \{s_k(x) : x \in \mathcal{M}\}$. A set $\mathcal{M} \subseteq \mathcal{D} \subseteq \mathbb{R}^n$ is an *invariant set* for the dynamical system (13.1) if $s_k(\mathcal{M}) = \mathcal{M}$ for all $k \in \overline{\mathbb{Z}}_+$.

Definition 13.5. A point $p \in \mathcal{D}$ is a *positive limit point* of the trajectory $x(k)$, $k \in \overline{\mathbb{Z}}_+$, of (13.1) if there exists a monotonic sequence $\{k_n\}_{n=0}^\infty$ of nonnegative numbers, with $k_n \to \infty$ as $n \to \infty$, such that $x(k_n) \to p$ as $n \to \infty$. The set of all positive limit points of $x(k)$, $k \in \overline{\mathbb{Z}}_+$, is the *positive limit set* $\omega(x_0)$ of $x(k)$, $k \in \overline{\mathbb{Z}}_+$.

Note that if $p \in \mathcal{D}$ is a positive limit point of the trajectory $x(\cdot)$, then for all $\varepsilon > 0$ and finite $K \in \mathbb{Z}_+$ there exists $k > K$ such that $\|x(k) - p\| < \varepsilon$. This follows from the fact that $\|x(k) - p\| < \varepsilon$ for all $\varepsilon > 0$ and some $k > K > 0$ is equivalent to the existence of a sequence of integers $\{k_n\}_{n=0}^\infty$, with $k_n \to \infty$ as $n \to \infty$, such that $x(k_n) \to p$ as $n \to \infty$.

Next, we state and prove a key lemma involving positive limit sets for discrete-time systems.

Lemma 13.1. Suppose the solution $x(k)$ to (13.1) corresponding to an initial condition $x(0) = x_0$ is bounded for all $k \in \overline{\mathbb{Z}}_+$. Then the positive limit set $\omega(x_0)$ of $x(k)$, $k \in \overline{\mathbb{Z}}_+$, is a nonempty, compact, invariant set. Furthermore, $x(k) \to \omega(x_0)$ as $t \to \infty$.

Proof. Let $x(k)$, $k \in \overline{\mathbb{Z}}_+$, denote the solution to (13.1) corresponding to the initial condition $x(0) = x_0$. Next, since $x(k)$ is bounded for all $k \in \overline{\mathbb{Z}}_+$, it follows from the Bolzano-Weierstrass theorem (Theorem 2.3) that every sequence in the positive orbit $\mathcal{O}_{x_0}^+ \triangleq \{s(k, x_0) : k \in \overline{\mathbb{Z}}_+\}$ has at least one accumulation point $p \in \mathcal{D}$ as $k \to \infty$, and hence, $\omega(x_0)$ is nonempty. Next, let $p \in \omega(x_0)$ so that there exists an increasing unbounded sequence $\{k_n\}_{n=0}^\infty$, with $k_0 = 0$, such that $\lim_{n\to\infty} x(k_n) = p$. Now, since $x(k_n)$ is uniformly bounded in n it follows that the limit point p is bounded, which implies that $\omega(x_0)$ is bounded. To show that $\omega(x_0)$ is closed let $\{p_i\}_{i=0}^\infty$ be a sequence contained in $\omega(x_0)$ such that $\lim_{i\to\infty} p_i = p$. Now, since $p_i \to p$ as $i \to \infty$ for every $\varepsilon > 0$, there exists i such that $\|p - p_i\| < \varepsilon/2$. Next, since $p_i \in \omega(x_0)$, there exists $k \geq K$, where $K \in \overline{\mathbb{Z}}_+$ is arbitrary and finite, such that $\|p_i - x(k)\| < \varepsilon/2$. Now, since $\|p - p_i\| < \varepsilon/2$ and $\|p_i - x(k)\| < \varepsilon/2$, $k \geq K$, it follows that $\|p - x(k)\| \leq \|p_i - x(k)\| + \|p - p_i\| < \varepsilon$. Thus, $p \in \omega(x_0)$. Hence, every accumulation point of $\omega(x_0)$ is an element of $\omega(x_0)$ so that $\omega(x_0)$ is closed. Thus, since $\omega(x_0)$ is closed and bounded it is compact.

To show positive invariance of $\omega(x_0)$ let $p \in \omega(x_0)$ so that there exists an increasing sequence $\{k_n\}_{n=0}^\infty$ such that $x(k_n) \to p$ as $n \to \infty$. Now, let $s(k_n, x_0)$ denote the solution $x(k_n)$ of (13.1) with initial condition $x(0) = x_0$ and note that since $f : \mathcal{D} \to \mathcal{D}$ in (13.1) is continuous, $x(k)$, $k \in \overline{\mathbb{Z}}_+$, is the unique solution to (13.1) so that $s(k+k_n, x_0) = s(k, s(k_n, x_0)) = s(k, x(k_n))$. Now, since $s(k, x_0)$, $k \in \overline{\mathbb{Z}}_+$, is continuous with respect to x_0, it follows that, for $k + k_n \geq 0$, $\lim_{n\to\infty} s(k + k_n, x_0) = \lim_{n\to\infty} s(k, x(k_n)) = s(k, p)$, and hence, $s(k, p) \in \omega(x_0)$. Hence, $s_k(\omega(x_0)) \subseteq \omega(x_0)$, $k \in \overline{\mathbb{Z}}_+$, establishing positive invariance of $\omega(x_0)$.

To show invariance of $\omega(x_0)$ let $y \in \omega(x_0)$ so that there exists an increasing unbounded sequence $\{k_n\}_{n=0}^\infty$ such that $s(k_n, x_0) \to y$ as $n \to \infty$. Next, let $k \in \overline{\mathbb{Z}}_+$ and note that there exists N such that $k_n > k$, $n \geq N$. Hence, it follows from the semigroup property that $s(k, s(k_n - k, x_0)) = s(k_n, x_0) \to y$ as $n \to \infty$. Now, it follows from the Bolzano-Lebesgue theorem (Theorem 2.4) that there exists a subsequence z_{n_i} of the sequence $z_n = s(k_n - k, x_0)$, $n = N, N+1, \ldots$, such that $z_{n_i} \to z \in \mathcal{D}$ and, by definition, $z \in \omega(x_0)$. Next, it follows from the continuous dependence property that $\lim_{i\to\infty} s(k, z_{n_i}) = s(k, \lim_{i\to\infty} z_{n_i})$, and hence, $y = s(t, z)$, which implies that $\omega(x_0) \subseteq s_k(\omega(x_0))$, $k \in \overline{\mathbb{Z}}_+$. Now, using positive invariance of $\omega(x_0)$ it follows that $s_k(\omega(x_0)) = \omega(x_0)$, $k \in \overline{\mathbb{Z}}_+$, establishing invariance of the positive limit set $\omega(x_0)$.

Finally, to show $x(k) \to \omega(x_0)$ as $k \to \infty$, suppose, *ad absurdum*, that $x(k) \not\to \omega(x_0)$ as $k \to \infty$. In this case, there exists a sequence $\{k_n\}_{n=0}^{\infty}$, with $k_n \to \infty$ as $n \to \infty$, such that

$$\inf_{p\in\omega(x_0)} \|x(k_n) - p\| > \varepsilon, \quad n \in \overline{\mathbb{Z}}_+. \tag{13.16}$$

However, since $x(k)$, $k \in \overline{\mathbb{Z}}_+$, is bounded, the bounded sequence $\{x(k_n)\}_{n=0}^{\infty}$ contains a convergent subsequence $\{x(k_n^*)\}_{n=0}^{\infty}$ such that $x(k_n^*) \to p^* \in \omega(x_0)$ as $n \to \infty$, which contradicts (13.16). Hence, $x(k) \to \omega(x_0)$ as $k \to \infty$. □

Next, we present a discrete-time version of the Barbashin-Krasovskii-LaSalle invariance principle.

Theorem 13.3. Consider the discrete-time nonlinear dynamical system (13.1), assume $\mathcal{D}_c$ is a compact invariant set with respect to (13.1), and assume that there exists a continuous function $V\colon \mathcal{D}_c \to \mathbb{R}$ such that $V(f(x)) - V(x) \le 0$, $x \in \mathcal{D}_c$. Let $\mathcal{R} \triangleq \{x \in \mathcal{D}_c : V(f(x)) = V(x)\}$ and let $\mathcal{M}$ denote the largest invariant set contained in $\mathcal{R}$. If $x(0) \in \mathcal{D}_c$, then $x(k) \to \mathcal{M}$ as $k \to \infty$.

Proof. Let $x(k)$, $k \in \overline{\mathbb{Z}}_+$, be a solution to (13.1) with $x(0) \in \mathcal{D}_c$. Since $V(f(x)) \le V(x)$, $x \in \mathcal{D}_c$, it follows that

$$V(x(k)) - V(x(\kappa)) = \sum_{i=\kappa}^{k-1} [V(f(x(i))) - V(x(i))] \le 0, \qquad k - 1 \ge \kappa,$$

and hence, $V(x(k)) \le V(x(\kappa))$, $k - 1 \ge \kappa$, which implies that $V(x(k))$ is a nonincreasing function of k. Next, since $V(x)$ is continuous on the compact set $\mathcal{D}_c$, there exists $L \ge 0$ such that $V(x) \ge L$, $x \in \mathcal{D}_c$. Hence, $\gamma_{x_0} \triangleq \lim_{k\to\infty} V(x(k))$ exists. Now, for all $p \in \omega(x_0)$ there exists an increasing unbounded sequence $\{k_n\}_{n=0}^{\infty}$, with $k_0 = 0$, such that $x(k_n) \to p$ as $n \to \infty$. Since $V(x)$, $x \in \mathcal{D}$, is continuous, $V(p) = V(\lim_{n\to\infty} x(k_n)) = \lim_{n\to\infty} V(x(k_n)) = \gamma_{x_0}$, and hence, $V(x) = \alpha$ on $\omega(x_0)$. Now, since $\mathcal{D}_c$ is compact and invariant it follows that $x(k)$, $k \in \overline{\mathbb{Z}}_+$, is bounded, and hence, it follows from Lemma 13.1 that $\omega(x_0)$ is a nonempty, compact invariant set. Hence, it follows that $V(f(x)) = V(x)$ on $\omega(x_0)$, and hence, $\omega(x_0) \subset \mathcal{M} \subset \mathcal{R} \subset \mathcal{D}_c$. Finally, since $x(k) \to \omega(x_0)$ as $k \to \infty$ it follows that $x(k) \to \mathcal{M}$ as $k \to \infty$. □

Now, using Theorem 13.3 we provide a generalization of Theorem 13.2

for local asymptotic stability of a nonlinear dynamical system.

Corollary 13.1. Consider the nonlinear dynamical system (13.1), assume $\mathcal{D}_c$ is a compact invariant set with respect to (13.1) such that $0 \in \overset{\circ}{\mathcal{D}}_c$, and assume that there exists a continuous function $V\colon \mathcal{D}_c \to \mathbb{R}$ such that $V(0) = 0$, $V(x) > 0$, $x \neq 0$, and $V(f(x)) - V(x) \leq 0$, $x \in \mathcal{D}_c$. Furthermore, assume that the set $\mathcal{R} \triangleq \{x \in \mathcal{D}_c\colon V(f(x)) = V(x)\}$ contains no invariant set other than the set $\{0\}$. Then the zero solution $x(k) \equiv 0$ to (13.1) is asymptotically stable and $\mathcal{D}_c$ is a subset of the domain of attraction of (13.1).

Proof. The result is a direct consequence of Theorem 13.3. □

In Theorem 13.3 and Corollary 13.1 we explicitly assumed that there exists a compact invariant set $\mathcal{D}_c \subset \mathcal{D}$ of (13.1). Next, we provide a result that does not require the existence of a compact invariant set $\mathcal{D}_c$.

Theorem 13.4. Consider the nonlinear dynamical system (13.1) and assume that there exists a continuous function $V\colon \mathbb{R}^n \to \mathbb{R}$ such that

$$V(0) = 0, \tag{13.17}$$
$$V(x) > 0, \quad x \in \mathbb{R}^n, \quad x \neq 0, \tag{13.18}$$
$$V(f(x)) - V(x) \leq 0, \quad x \in \mathbb{R}^n. \tag{13.19}$$

Let $\mathcal{R} \triangleq \{x \in \mathbb{R}^n\colon V(f(x)) = V(x)\}$ and let $\mathcal{M}$ be the largest invariant set contained in $\mathcal{R}$. Then all solutions $x(k)$, $k \in \overline{\mathbb{Z}}_+$, to (13.1) that are bounded approach $\mathcal{M}$ as $k \to \infty$.

Proof. Let $x \in \mathbb{R}^n$ be such that trajectory $s(k, x)$, $k \in \overline{\mathbb{Z}}_+$, of (13.1) is bounded. Now, with $\mathcal{D}_c = \overline{\mathcal{O}_x^+}$, it follows from Theorem 13.3 that $s(k, x) \to \mathcal{M}$ as $k \to \infty$. □

Next, we present the global invariant set theorem for guaranteeing global asymptotic stability of a discrete-time nonlinear dynamical system.

Theorem 13.5. Consider the nonlinear dynamical system (13.1) and assume that there exists a continuous function $V\colon \mathbb{R}^n \to \mathbb{R}$ such that

$$V(0) = 0 \tag{13.20}$$
$$V(x) > 0, \quad x \in \mathbb{R}^n, \quad x \neq 0, \tag{13.21}$$
$$V(f(x)) - V(x) \leq 0, \quad x \in \mathbb{R}^n, \tag{13.22}$$
$$V(x) \to \infty \text{ as } \|x\| \to \infty. \tag{13.23}$$

Furthermore, assume that the set $\mathcal{R} \triangleq \{x \in \mathbb{R}^n\colon V(f(x)) = V(x)\}$ contains no invariant set other than the set $\{0\}$. Then the zero solution $x(k) \equiv 0$ to

(13.1) is globally asymptotically stable.

Proof. Since (13.20)–(13.22) hold, it follows from Theorem 13.2 that the zero solution $x(k) \equiv 0$ to (13.1) is Lyapunov stable while the radial unboundedness condition (13.23) implies that all solutions to (13.1) are bounded. Now, Theorem 13.4 implies that $x(k) \to \mathcal{M}$ as $k \to \infty$. However, since $\mathcal{R}$ contains no invariant set other than the set $\{0\}$, the set $\mathcal{M}$ is $\{0\}$, and hence, global asymptotic stability is immediate. □

13.4 Converse Lyapunov Theorems for Discrete-Time Systems

In the previous sections the existence of a Lyapunov function is assumed while stability properties of a discrete-time nonlinear dynamical system are deduced. As in the continuous-time case, the existence of a lower semi-continuous and continuous time-invariant Lyapunov function for Lyapunov stable and asymptotically stable, respectively, discrete-time systems can be established. In order to state and prove the converse Lyapunov theorems for discrete-time systems the following key lemma is needed.

Lemma 13.2. Let $\sigma : \overline{\mathbb{Z}}_+ \to \overline{\mathbb{Z}}_+$ be a class $\mathcal{L}$ function. Then there exists a continuous class $\mathcal{K}$ function $\gamma : \overline{\mathbb{Z}}_+ \to \overline{\mathbb{Z}}_+$ such that $\sum_{k=0}^{\infty} \gamma[\sigma(k)] < \infty$.

Proof. The proof is similar to the proof of Lemma 3.1 and, hence, is omitted. □

Theorem 13.6. Assume that the zero solution $x(k) \equiv 0$ to (13.1) is Lyapunov (respectively, asymptotically) stable and $f : \mathcal{D} \to \mathcal{D}$ is continuous. Then there exist a set $\mathcal{D}_0 \subseteq \mathcal{D}$ with $0 \in \overset{\circ}{\mathcal{D}}_0$ and a lower semi-continuous (respectively, continuous) function $V : \mathcal{D}_0 \to \mathbb{R}^n$ such that $V(\cdot)$ is continuous at the origin, $V(0) = 0$, $V(x) > 0$, $x \in \mathcal{D}_0$, $x \neq 0$, and $V(f(x)) \leq V(x)$, $x \in \mathcal{D}_0$ (respectively, $V(f(x)) - V(x) < 0$, $x \in \mathcal{D}_0$, $x \neq 0$).

Proof. Let $\varepsilon > 0$. Since the zero solution $x(k) \equiv 0$ to (13.1) is Lyapunov stable it follows that there exists $\delta > 0$ such that if $x_0 \in \mathcal{B}_\delta(0)$, then $s(k, x_0) \in \mathcal{B}_\varepsilon(0)$, $k \in \overline{\mathbb{Z}}_+$. Now, let $\mathcal{D}_0 = \{y \in \mathcal{B}_\varepsilon(0) :$ there exist $k \in \overline{\mathbb{Z}}_+$ and $x_0 \in \mathcal{B}_\delta(0)$ such that $y = s(k, x_0)\}$, that is, $\mathcal{D}_0 = s_k(\mathcal{B}_\delta(0))$. Note that $\mathcal{D}_0 \subseteq \mathcal{B}_\varepsilon(0)$, $\mathcal{D}_0$ is positively invariant, and $\mathcal{B}_\delta(0) \subseteq \mathcal{D}_0$. Hence, $0 \in \overset{\circ}{\mathcal{D}}_0$. Next, define $V(x) \triangleq \sup_{k \in \overline{\mathbb{Z}}_+} \|s(k, x)\|$, $x \in \mathcal{D}_0$, and since $\mathcal{D}_0$ is positively invariant and bounded it follows that $V(\cdot)$ is well defined on $\mathcal{D}_0$. Now, $x = 0$ implies $s(k, x) \equiv 0$, and hence, $V(0) = 0$. Furthermore, $V(x) \geq \|s(0, x)\| = \|x\| > 0$, $x \in \mathcal{D}_0$, $x \neq 0$.

Next, since $f(\cdot)$ in (13.1) is continuous, it follows that for every $x \in \mathcal{D}_0$, $s(k,x)$, $k \in \overline{\mathbb{Z}}_+$, is the unique solution to (13.1) so that $s(k,x) = s(k-\kappa, s(\kappa,x))$, $0 \le \kappa \le k$, which implies that for every $k, \kappa \in \overline{\mathbb{Z}}_+$, such that $k \ge \kappa$,

$$\begin{aligned} V(s(\kappa,x)) &= \sup_{\theta\in\overline{\mathbb{Z}}_+} \|s(\theta, s(\kappa,x))\| \\ &= \sup_{\theta\in\overline{\mathbb{Z}}_+} \|s(\kappa+\theta, x)\| \\ &\ge \sup_{\theta\ge k-\kappa} \|s(\kappa+\theta, x)\| \\ &= \sup_{\theta\ge k-\kappa} \|s(\theta-(k-\kappa), s(k,x))\| \\ &= \sup_{\theta\in\overline{\mathbb{Z}}_+} \|s(\theta, s(k,x))\| \\ &= V(s(k,x)), \end{aligned} \tag{13.24}$$

which proves that $V(f(x)) - V(x) = V(s(1,x)) - V(s(0,x)) \le 0$.

Next, since the zero solution $x(k) \equiv 0$ to (13.1) is Lyapunov stable it follows that for every $\hat{\varepsilon} > 0$ there exists $\hat{\delta} > 0$ such that if $x_0 \in \mathcal{B}_{\hat{\delta}}(0)$, then $s(k,x_0) \in \mathcal{B}_{\hat{\varepsilon}/2}(0)$, $k \in \overline{\mathbb{Z}}_+$, which implies that $V(x_0) = \sup_{k\in\overline{\mathbb{Z}}_+} \|s(k,x_0)\| \le \hat{\varepsilon}/2$. Hence, for every $\hat{\varepsilon} > 0$ there exists $\hat{\delta} > 0$ such that if $x_0 \in \mathcal{B}_{\hat{\delta}}(0)$, then $V(x_0) < \hat{\varepsilon}$, establishing that $V(\cdot)$ is continuous at the origin. Finally, to show that $V(\cdot)$ is lower semicontinuous everywhere on $\mathcal{D}_0$, let $x \in \mathcal{D}_0$ and let $\hat{\varepsilon} > 0$, and since $V(x) = \sup_{k\in\overline{\mathbb{Z}}_+} \|s(k,x)\|$ there exists $K = K(x,\hat{\varepsilon}) > 0$ such that $V(x) - \|s(K,x)\| < \hat{\varepsilon}$. Now, consider a sequence $\{x_i\}_{i=1}^{\infty} \in \mathcal{D}_0$ such that $x_i \to x$ as $i \to \infty$. Next, since $f(\cdot)$ in (13.1) is continuous, it follows that for every $k \in \overline{\mathbb{Z}}_+$, $s(k,\cdot)$ is continuous. Hence, since $\|\cdot\| : \mathcal{D}_0 \to \mathbb{R}$ is continuous, $\|s(K,x)\| = \lim_{i\to\infty} \|s(K,x_i)\|$, which implies that

$$\begin{aligned} V(x) &< \|s(K,x)\| + \hat{\varepsilon} \\ &= \lim_{i\to\infty} \|s(K,x_i)\| + \hat{\varepsilon} \\ &\le \liminf_{i\to\infty} \sup_{k\in\overline{\mathbb{Z}}_+} \|s(k,x_i)\| + \hat{\varepsilon} \\ &= \liminf_{i\to\infty} V(x_i) + \hat{\varepsilon}, \end{aligned} \tag{13.25}$$

which, since $\hat{\varepsilon} > 0$ is arbitrary, further implies that $V(x) \le \liminf_{i\to\infty} V(x_i)$. Thus, since $\{x_i\}_{i=1}^{\infty}$ is arbitrary sequence converging to x, it follows that $V(\cdot)$ is lower semicontinuous on $\mathcal{D}_0$.

Next, assume that the zero solution $x(k) \equiv 0$ to (13.1) is asymptotically stable. It follows that (see Problem 3.76) there exist class $\mathcal{K}$ and $\mathcal{L}$ functions $\alpha(\cdot)$ and $\beta(\cdot)$, respectively, such that if $\|x_0\| < \delta$, for some $\delta > 0$, then $\|x(k)\| \le \alpha(\|x_0\|)\beta(k)$, $k \in \overline{\mathbb{Z}}_+$. Next, let $x \in \mathcal{B}_\delta(0)$ and let $s(k,x)$

denote the solution to (13.1) with the initial condition $x(0) = x$ so that for all $\|x\| < \delta$, $\|s(k,x)\| \le \alpha(\|x\|)\beta(k)$, $k \in \overline{\mathbb{Z}}_+$. Now, it follows from Lemma 13.2 that there exists a continuous class $\mathcal{K}$ function $\gamma : \overline{\mathbb{R}}_+ \to \overline{\mathbb{R}}_+$ such that $\sum_{k=0}^{\infty} \gamma(\alpha(\delta)\beta(k)) < \infty$. Since $\|s(k,x)\| \le \alpha(\|x\|)\beta(k) \le \alpha(\delta)\beta(k)$, $k \in \overline{\mathbb{Z}}_+$, it follows that the function

$$V(x) = \sum_{k=0}^{\infty} \gamma(\|s(k,x)\|),$$

is well defined on $\mathcal{D}$. Furthermore, note that $V(0) = 0$. Next, note that $V(x) = \sum_{k=0}^{\infty} \gamma(\ \|s(k,x)\|) > 0$, $x \in \mathcal{D}$, $x \neq 0$. Finally, note that since $f(\cdot)$ is continuous the trajectory $s(k,x)$, $k \in \overline{\mathbb{Z}}_+$, is unique, and hence,

$$\begin{aligned} V(s(k,x)) &= \sum_{\kappa=0}^{\infty} \gamma(\|s(\kappa, s(k,x))\|) \\ &= \sum_{\kappa=0}^{\infty} \gamma(\|s(\kappa + k, x)\|) \\ &= \sum_{\kappa=k}^{\infty} \gamma(\|s(\kappa, x)\|), \end{aligned}$$

which implies that $V(f(x)) - V(x) = V(s(1,x)) - V(s(0,x)) = -\gamma(\|x\|) < 0$, $x \in \mathcal{D}_0$, $x \neq 0$. The result is now immediate by noting that $V : \mathcal{D} \to \mathbb{R}$ defined above is continuous. □

The next result gives a converse Lyapunov theorem for geometric stability.

Theorem 13.7. Assume that the zero solution $x(k) \equiv 0$ to (13.1) is geometrically stable and $f : \mathcal{D} \to \mathcal{D}$ is continuous. Then, for every $p > 1$, there exist a set $\mathcal{D}_0 \subseteq \mathcal{D}$ with $0 \in \overset{\circ}{\mathcal{D}}_0$, a continuous function $V : \mathcal{D}_0 \to \mathbb{R}$, and scalars $\alpha, \beta > 0$, $\beta > \alpha$, and $\rho > 1$ such that

$$\alpha\|x\|^p \le V(x) \le \beta\|x\|^p, \quad x \in \mathcal{D}_0, \tag{13.26}$$

$$\rho V(f(x)) \le V(x), \quad x \in \mathcal{D}_0. \tag{13.27}$$

Proof. Since the zero solution $x(k) \equiv 0$ to (13.1) is, by assumption, geometrically stable it follows that there exist scalars $\alpha_1 \ge 1$, $\beta_1 > 1$, and $\delta > 0$ such that if $\|x_0\| < \delta$, then $\|x(k)\| \le \alpha_1 \|x_0\| \beta_1^{-k}$, $k \in \overline{\mathbb{Z}}_+$. Next, let $x \in \mathcal{B}_\delta(0)$ and let $s(k,x)$ denote the solution to (13.1) with the initial condition $x(0) = x$ so that for all $\|x\| < \delta$, $\|s(k,x)\| \le \alpha_1 \|x\| \beta_1^{-k}$, $k \in \overline{\mathbb{Z}}_+$. Now, using identical arguments as in Theorem 13.6 it follows that for $\alpha(\delta) = \alpha_1$, $\beta(k) = \beta_1^{-k}$, $\gamma(\sigma) = \sigma^p$,

$$\sum_{k=0}^{\infty} \alpha_1^p \beta_1^{-pk} < \infty. \tag{13.28}$$

Hence,

$$V(x) = \sum_{k=0}^{\infty} \|s(k, x)\|^p \tag{13.29}$$

is a continuous Lyapunov function candidate for (13.1).

To show that (13.26) holds note that

$$\begin{aligned} V(x) &= \sum_{k=0}^{\infty} \|s(k, x)\|^p \\ &\leq \sum_{k=0}^{\infty} \alpha_1^p \|x\|^p \beta_1^{-pk} \\ &= \frac{\alpha_1^p \beta_1^p}{\beta_1^p - 1} \|x\|^p, \quad x \in \mathcal{D}, \end{aligned} \tag{13.30}$$

which proves the upper bound in (13.26) with $\beta = \frac{\alpha_1^p \beta_1^p}{\beta_1^p - 1} > 1$. Next,

$$V(x) = \sum_{k=0}^{\infty} \|s(k, x)\|^p \geq \|s(0, x)\|^p = \|x\|^p, \quad x \in \mathcal{D}_0, \tag{13.31}$$

which proves the lower bound in (13.26) with $\alpha = 1$.

Finally, to show (13.27) note that with $\gamma(\sigma) = \sigma^p$ it follows from Theorem 13.6 that $V(f(x)) - V(x) = -\gamma(\|x\|) = -\|x\|^p$. Now, the result is immediate from (13.26) by noting that

$$V(f(x)) = V(x) - \|x\|^p \leq (1 - \tfrac{1}{\beta}) V(x), \quad x \in \mathcal{D}_0, \tag{13.32}$$

which proves (13.27) with $\rho = \frac{\beta}{\beta - 1}$. □

Next, we present a corollary to Theorem 13.7 that shows that p in Theorem 13.7 can be taken to be equal to 2 without loss of generality.

Corollary 13.2. Assume that the zero solution $x(k) \equiv 0$ to (13.1) is geometrically stable and $f : \mathcal{D} \to \mathcal{D}$ is continuous. Then there exist a set $\mathcal{D}_0 \subseteq \mathcal{D}$ with $0 \in \overset{\circ}{\mathcal{D}}_0$, a continuous function $V : \mathcal{D} \to \mathbb{R}$, and scalars α, $\beta > 1$, and $\rho > 1$, such that

$$\alpha \|x\|^2 \leq V(x) \leq \beta \|x\|^2, \quad x \in \mathcal{D}_0, \tag{13.33}$$

$$\rho V(f(x)) \leq V(x), \quad x \in \mathcal{D}_0. \tag{13.34}$$

Proof. The proof is a direct consequence of Theorem 13.7 with $p = 2$. □

Finally, we present a converse theorem for global geometric stability.

Theorem 13.8. Assume that the zero solution $x(k) \equiv 0$ to (13.1) is globally geometrically stable and $f : \mathbb{R}^n \to \mathbb{R}^n$ is continuous. Then there exist a continuous function $V : \mathbb{R}^n \to \mathbb{R}$ and scalars α, $\beta > 1$, and $\rho > 1$, such that

$$\alpha\|x\|^2 \leq V(x) \leq \beta\|x\|^2, \quad x \in \mathbb{R}^n, \tag{13.35}$$

$$\rho V(f(x)) \leq V(x), \quad x \in \mathbb{R}^n. \tag{13.36}$$

Proof. The proof is identical to the proof of Theorem 13.7 by replacing $\mathcal{D}_0$ with $\mathbb{R}^n$ and setting $p = 2$. □

13.5 Partial Stability of Discrete-Time Nonlinear Dynamical Systems

In this section, we present partial stability theorems for discrete-time nonlinear dynamical systems. Specifically, consider the discrete-time nonlinear autonomous dynamical system

$$x_1(k+1) = f_1(x_1(k), x_2(k)), \quad x_1(0) = x_{10}, \quad k \in \overline{\mathbb{Z}}_+, \tag{13.37}$$

$$x_2(k+1) = f_2(x_1(k), x_2(k)), \quad x_2(0) = x_{20}, \tag{13.38}$$

where $x_1 \in \mathcal{D} \subseteq \mathbb{R}^{n_1}$, $\mathcal{D}$ is an open set with $0 \in \mathcal{D}$, $x_2 \in \mathbb{R}^{n_2}$, $f_1 : \mathcal{D} \times \mathbb{R}^{n_2} \to \mathbb{R}^{n_1}$ is continuous and for every $x_2 \in \mathbb{R}^{n_2}$, $f_1(0, x_2) = 0$, and $f_2 : \mathcal{D} \times \mathbb{R}^{n_2} \to \mathbb{R}^{n_2}$ is continuous. Note that under the above assumptions the solution $(x_1(k), x_2(k))$ to (13.37) and (13.38) exists and is unique over $\overline{\mathbb{Z}}_+$. The following definition introduces eight types of partial stability, that is, stability with respect to x_1, for the nonlinear dynamical system (13.37) and (13.38).

Definition 13.6. *i*) The nonlinear dynamical system (13.37) and (13.38) is *Lyapunov stable with respect to* x_1 if, for every $\varepsilon > 0$ and $x_{20} \in \mathbb{R}^{n_2}$, there exists $\delta = \delta(\varepsilon, x_{20}) > 0$ such that $\|x_{10}\| < \delta$ implies that $\|x_1(k)\| < \varepsilon$ for all $k \in \overline{\mathbb{Z}}_+$.

ii) The nonlinear dynamical system (13.37) and (13.38) is *Lyapunov stable with respect to* x_1 *uniformly in* x_{20} if, for every $\varepsilon > 0$, there exists $\delta = \delta(\varepsilon) > 0$ such that $\|x_{10}\| < \delta$ implies that $\|x_1(k)\| < \varepsilon$ for all $k \in \overline{\mathbb{Z}}_+$ and for all $x_{20} \in \mathbb{R}^{n_2}$.

iii) The nonlinear dynamical system (13.37) and (13.38) is *asymptoti-*

cally stable with respect to x_1 if it is Lyapunov stable with respect to x_1 and, for every $x_{20} \in \mathbb{R}^{n_2}$, there exists $\delta = \delta(x_{20}) > 0$ such that $\|x_{10}\| < \delta$ implies that $\lim_{k\to\infty} x_1(k) = 0$.

iv) The nonlinear dynamical system (13.37) and (13.38) is *asymptotically stable with respect to* x_1 *uniformly in* x_{20} if it is Lyapunov stable with respect to x_1 uniformly in x_{20} and there exists $\delta > 0$ such that $\|x_{10}\| < \delta$ implies that $\lim_{k\to\infty} x_1(k) = 0$ uniformly in x_{10} and x_{20} for all $x_{20} \in \mathbb{R}^{n_2}$.

v) The nonlinear dynamical system (13.37) and (13.38) is *globally asymptotically stable with respect to* x_1 it is Lyapunov stable with respect to x_1 and $\lim_{k\to\infty} x_1(k) = 0$ for all $x_{10} \in \mathbb{R}^{n_1}$ and $x_{20} \in \mathbb{R}^{n_2}$.

vi) The nonlinear dynamical system (13.37) and (13.38) is *globally asymptotically stable with respect to* x_1 *uniformly in* x_{20} if it is Lyapunov stable with respect to x_1 uniformly in x_{20} and $\lim_{k\to\infty} x_1(k) = 0$ uniformly in x_{10} and x_{20} for all $x_{10} \in \mathbb{R}^{n_1}$ and $x_{20} \in \mathbb{R}^{n_2}$.

vii) The nonlinear dynamical system (13.37) and (13.38) is *geometrically stable with respect to* x_1 *uniformly in* x_{20} if there exist scalars $\alpha, \beta > 1$, and $\delta > 0$ such that $\|x_{10}\| < \delta$ implies that $\|x_1(k)\| \le \alpha\|x_{10}\|\beta^{-k}$, $k \in \overline{\mathbb{Z}}_+$, for all $x_{20} \in \mathbb{R}^{n_2}$.

viii) The nonlinear dynamical system (13.37) and (13.38) is *globally geometrically stable with respect to* x_1 *uniformly in* x_{20} if there exist scalars $\alpha, \beta > 1$ such that $\|x_1(k)\| \le \alpha\|x_{10}\|\beta^{-k}$, $k \in \overline{\mathbb{Z}}_+$, for all $x_{10} \in \mathbb{R}^{n_1}$ and $x_{20} \in \mathbb{R}^{n_2}$.

Next, we present sufficient conditions for partial stability of the nonlinear dynamical system (13.37) and (13.38). For the following result define $\Delta V(x_1, x_2) \triangleq V(f(x_1, x_2)) - V(x_1, x_2)$, where $f(x_1, x_2) \triangleq [f_1^{\mathrm{T}}(x_1, x_2)\ f_2^{\mathrm{T}}(x_1, x_2)]^{\mathrm{T}}$, for a given continuous function $V : \mathcal{D} \times \mathbb{R}^{n_2} \to \mathbb{R}$.

Theorem 13.9. Consider the nonlinear dynamical system (13.37) and (13.38). Then the following statements hold:

i) If there exist a continuous function $V : \mathcal{D} \times \mathbb{R}^{n_2} \to \mathbb{R}$ and a class $\mathcal{K}$ function $\alpha(\cdot)$ such that

$$V(0, x_2) = 0, \qquad x_2 \in \mathbb{R}^{n_2}, \tag{13.39}$$

$$\alpha(\|x_1\|) \le V(x_1, x_2), \qquad (x_1, x_2) \in \mathcal{D} \times \mathbb{R}^{n_2}, \tag{13.40}$$

$$V(f(x_1, x_2)) \le V(x_1, x_2), \qquad (x_1, x_2) \in \mathcal{D} \times \mathbb{R}^{n_2}, \tag{13.41}$$

then the nonlinear dynamical system given by (13.37) and (13.38) is Lyapunov stable with respect to x_1.

ii) If there exist a continuous function $V : \mathcal{D} \times \mathbb{R}^{n_2} \to \mathbb{R}$ and class $\mathcal{K}$ functions $\alpha(\cdot), \beta(\cdot)$ satisfying (13.40), (13.41), and

$$V(x_1, x_2) \leq \beta(\|x_1\|), \qquad (x_1, x_2) \in \mathcal{D} \times \mathbb{R}^{n_2}, \tag{13.42}$$

then the nonlinear dynamical system given by (13.37) and (13.38) is Lyapunov stable with respect to x_1 uniformly in x_{20}.

iii) If there exist a continuous function $V : \mathcal{D} \times \mathbb{R}^{n_2} \to \mathbb{R}$ and class $\mathcal{K}$ functions $\alpha(\cdot)$ and $\gamma(\cdot)$ satisfying (13.39), (13.40), and

$$\Delta V(x_1, x_2) \leq -\gamma(\|x_1\|), \qquad (x_1, x_2) \in \mathcal{D} \times \mathbb{R}^{n_2}, \tag{13.43}$$

then the nonlinear dynamical system given by (13.37) and (13.38) is asymptotically stable with respect to x_1.

iv) If there exist a continuous function $V : \mathcal{D} \times \mathbb{R}^{n_2} \to \mathbb{R}$ and class $\mathcal{K}$ functions $\alpha(\cdot), \beta(\cdot), \gamma(\cdot)$ satisfying (13.40), (13.42), and

$$\Delta V(x_1, x_2) \leq -\gamma(\|x_1\|), \qquad (x_1, x_2) \in \mathcal{D} \times \mathbb{R}^{n_2}, \tag{13.44}$$

then the nonlinear dynamical system given by (13.37) and (13.38) is asymptotically stable with respect to x_1 uniformly in x_{20}.

v) If $\mathcal{D} = \mathbb{R}^{n_1}$ and there exist a continuous function $V : \mathbb{R}^{n_1} \times \mathbb{R}^{n_2} \to \mathbb{R}$, a class $\mathcal{K}$ function $\gamma(\cdot)$, and a class $\mathcal{K}_\infty$ function $\alpha(\cdot)$ satisfying (13.40) and (13.43), then the nonlinear dynamical system given by (13.37) and (13.38) is globally asymptotically stable with respect to x_1.

vi) If $\mathcal{D} = \mathbb{R}^{n_1}$ and there exist a continuous function $V : \mathbb{R}^{n_1} \times \mathbb{R}^{n_2} \to \mathbb{R}$, a class $\mathcal{K}$ function $\gamma(\cdot)$, and class $\mathcal{K}_\infty$ functions $\alpha(\cdot), \beta(\cdot)$ satisfying (13.40), (13.42), and (13.44), then the nonlinear dynamical system given by (13.37) and (13.38) is globally asymptotically stable with respect to x_1 uniformly in x_{20}.

vii) If there exist a continuous function $V : \mathcal{D} \times \mathbb{R}^{n_2} \to \mathbb{R}$ and positive constants α, β, γ, p such that $\gamma < \beta$, $p \geq 1$, and

$$\alpha\|x_1\|^p \leq V(x_1, x_2) \leq \beta\|x_1\|^p, \qquad (x_1, x_2) \in \mathcal{D} \times \mathbb{R}^{n_2}, \tag{13.45}$$

$$\Delta V(x_1, x_2) \leq -\gamma\|x_1\|^p, \qquad (x_1, x_2) \in \mathcal{D} \times \mathbb{R}^{n_2}, \tag{13.46}$$

then the nonlinear dynamical system given by (13.37) and (13.38) is geometrically stable with respect to x_1 uniformly in x_{20}.

viii) If $\mathcal{D} = \mathbb{R}^{n_1}$ and there exist a continuous function $V : \mathbb{R}^{n_1} \times \mathbb{R}^{n_2} \to \mathbb{R}$ and positive constants α, β, γ, p such that $p \geq 1$ and (13.45) and (13.46) hold, then the nonlinear dynamical system given by (13.37) and (13.38) is globally geometrically stable with respect to x_1 uniformly in x_{20}.

Proof. *i*) Let $x_{20} \in \mathbb{R}^{n_2}$, let $\varepsilon > 0$ be such that $\mathcal{B}_\varepsilon(0) \triangleq \{x_1 \in \mathbb{R}^{n_1} : \|x_1\| < \varepsilon\} \subset \mathcal{D}$, define $\eta \triangleq \alpha(\varepsilon)$, and define $\mathcal{D}_\eta \triangleq \{x_1 \in \mathcal{B}_\varepsilon(0) : V(x_1, x_{20}) < \eta\}$. Since $V(\cdot,\cdot)$ is continuous and $V(0, x_{20}) = 0$, $x_{20} \in \mathbb{R}^{n_2}$, there exists $\delta = \delta(\varepsilon, x_{20}) > 0$ such that $V(x_1, x_{20}) < \eta$, $x_1 \in \mathcal{B}_\delta(0)$. Hence $\mathcal{B}_\delta(0) \subset \mathcal{D}_\eta$. Next, since $\Delta V(x_1, x_2) \leq 0$ it follows that $V(x_1(k), x_2(k))$ is a nonincreasing function of k, and hence, for every $x_{10} \in \mathcal{B}_\delta(0)$ it follows that

$$\alpha(\|x_1(k)\|) \leq V(x_1(k), x_2(k)) \leq V(x_{10}, x_{20}) < \eta = \alpha(\varepsilon).$$

Thus, for every $x_{10} \in \mathcal{B}_\delta(0)$, $x_1(k) \in \mathcal{B}_\varepsilon(0)$, $k \in \overline{\mathbb{Z}}_+$, establishing Lyapunov stability with respect to x_1.

ii) Let $\varepsilon > 0$ and let $\mathcal{B}_\varepsilon(0)$ and η be given as in the proof of *i*). Now, let $\delta = \delta(\varepsilon) > 0$ be such that $\beta(\delta) = \alpha(\varepsilon)$. Hence, it follows from (13.42) that for all $(x_{10}, x_{20}) \in \mathcal{B}_\delta(0) \times \mathbb{R}^{n_2}$,

$$\alpha(\|x_1(k)\|) \leq V(x_1(k), x_2(k)) \leq V(x_{10}, x_{20}) < \beta(\delta) = \alpha(\varepsilon),$$

and hence, $x_1(k) \in \mathcal{B}_\varepsilon(0)$, $k \in \overline{\mathbb{Z}}_+$.

iii) Lyapunov stability follows from *i*). To show asymptotic stability suppose, *ad absurdum*, that $\|x_1(k)\| \nrightarrow 0$ as $k \to \infty$ or, equivalently, $\limsup_{k\to\infty} \|x_1(k)\| > 0$. In addition, suppose $\liminf_{k\to\infty} \|x_1(k)\| > 0$, which implies that there exist constants $K > 0$ and $\theta > 0$ such that $\|x_1(k)\| \geq \theta$, $k \geq K$. Then it follows from (13.43) that $V(x_1(k), x_2(k)) \to -\infty$ as $k \to \infty$, which contradicts (13.40), and hence, $\liminf_{k\to\infty} \|x_1(k)\| = 0$. Now, since $\limsup_{k\to\infty} \|x_1(k)\| > 0$ and $\liminf_{k\to\infty} \|x_1(k)\| = 0$ it follows that there exist an increasing sequence $\{k_i\}_{i=1}^\infty$ and a constant $\theta > 0$ such that

$$\theta < \|x_1(k_i)\|, \quad i = 1, 2, \ldots. \tag{13.47}$$

Hence, $\sum_{i=1}^n \gamma(\|x_1(k_i)\|) > n\gamma(\theta)$. Now, using (13.43), it follows that

$$\begin{aligned} V(x_1(k_n), x_2(k_n)) &= V(x_{10}, x_{20}) + \sum_{k=1}^{k_n} \Delta V(x_1(k), x_2(k)) \\ &\leq V(x_{10}, x_{20}) + \sum_{i=1}^{n} \Delta V(x_1(k_i), x_2(k_i)) \\ &\leq V(x_{10}, x_{20}) - \sum_{i=1}^{n} \gamma(\|x_1(k_i)\|) \\ &\leq V(x_{10}, x_{20}) - n\gamma(\theta). \end{aligned} \tag{13.48}$$

Hence, for large enough n the right-hand side of (13.48) becomes negative, which contradicts (13.40). Hence, $x_1(k) \to 0$ as $k \to \infty$, proving asymptotic stability of (13.37) and (13.38) with respect to x_1.

iv) Lyapunov stability uniformly in x_{20} follows from *ii*). Next, let $\varepsilon > 0$ and $\delta = \delta(\varepsilon) > 0$ be such that for every $x_{10} \in \mathcal{B}_\delta(0)$, $x_1(k) \in \mathcal{B}_\varepsilon(0)$, $k \in \overline{\mathbb{Z}}_+$, (the existence of such a (δ, ε) pair follows from uniform Lyapunov stability) and assume that (13.44) holds. Since (13.44) implies (13.41) it follows that for every $x_{10} \in \mathcal{B}_\delta(0)$, $V(x_1(k), x_2(k))$ is a nonincreasing function of time and, since $V(\cdot,\cdot)$ is bounded from below, it follows from the monotone convergence theorem (Theorem 2.10) that there exists $L \geq 0$ such that $\lim_{k\to\infty} V(x_1(k), x_2(k)) = L$. Now, suppose for some $x_{10} \in \mathcal{B}_\delta(0)$, *ad absurdum*, that $L > 0$ so that $\mathcal{D}_L \triangleq \{x_1 \in \mathcal{B}_\varepsilon(0) : V(x_1, x_2) \leq L$ for all $x_2 \in \mathbb{R}^{n_2}\}$ is nonempty and $x_1(k) \notin \mathcal{D}_L$, $k \in \overline{\mathbb{Z}}_+$. Thus, as in the proof of *i*), there exists $\hat{\delta} > 0$ such that $\mathcal{B}_{\hat{\delta}}(0) \subset \mathcal{D}_L$. Hence, it follows from (13.44) that for the given $x_{10} \in \mathcal{B}_\delta(0)\backslash\mathcal{D}_L$ and $k \in \overline{\mathbb{Z}}_+$,

$$\begin{aligned} V(x_1(k), x_2(k)) &= V(x_{10}, x_{20}) + \sum_{i=0}^{k-1} \Delta V(x_1(i), x_2(i)) \\ &\leq V(x_{10}, x_{20}) - \sum_{i=0}^{k-1} \gamma(\|x_1(i)\|) \\ &\leq V(x_{10}, x_{20}) - \gamma(\hat{\delta})k. \end{aligned}$$

Letting $k > \frac{V(x_{10},x_{20})-L}{\gamma(\hat{\delta})}$, it follows that $V(x_1(k), x_2(k)) < L$, which is a contradiction. Hence, $L = 0$, and, since $x_0 \in \mathcal{B}_\delta(0)$ was chosen arbitrarily, it follows that $V(x_1(k), x_2(k)) \to 0$ as $k \to \infty$ for all $x_{10} \in \mathcal{B}_\delta(0)$. Now, since $V(x_1(k), x_2(k)) \geq \alpha(\|x_1(k)\|) \geq 0$ it follows that $\alpha(\|x_1(k)\|) \to 0$ or, equivalently, $x_1(k) \to 0$ $k \to \infty$, establishing asymptotic stability with respect to x_1 uniformly in x_{20}.

v) Let $\delta > 0$ be such that $\|x_{10}\| < \delta$. Since $\alpha(\cdot)$ is a class $\mathcal{K}_\infty$ function, it follows that there exists $\varepsilon > 0$ such that $V(x_{10}, x_{20}) < \alpha(\varepsilon)$. Now, (13.43) implies that $V(x_1(k), x_2(k))$ is a nonincreasing function of time, and hence, it follows that $\alpha(\|x_1(k)\|) \leq V(x_1(k), x_2(k)) \leq V(x_{10}, x_{20}) < \alpha(\varepsilon)$, $k \in \overline{\mathbb{Z}}_+$. Hence, $x_1(k) \in \mathcal{B}_\varepsilon(0)$, $k \in \overline{\mathbb{Z}}_+$. Now, the proof follows as in the proof of *iii*).

vi) Let $\delta > 0$ be such that $\|x_{10}\| < \delta$. Since $\alpha(\cdot)$ is a class $\mathcal{K}_\infty$ function it follows that there exists $\varepsilon > 0$ such that $\beta(\delta) \leq \alpha(\varepsilon)$. Now, (13.44) implies that $V(x_1(k), x_2(k))$ is a nonincreasing function of time, and hence, it follows from (13.42) that $\alpha(\|x_1(k)\|) \leq V(x_1(k), x_2(k)) \leq V(x_{10}, x_{20}) < \beta(\delta) < \alpha(\varepsilon)$, $k \in \overline{\mathbb{Z}}_+$. Hence, $x_1(k) \in \mathcal{B}_\varepsilon(0)$, $k \in \overline{\mathbb{Z}}_+$. Now, the proof follows as in the proof of *iv*).

vii) Let $\varepsilon > 0$ be given as in the proof of *i*) and let $\eta \triangleq \alpha\varepsilon$ and $\delta = \frac{\eta}{\beta}$. Now, (13.46) implies that $\Delta V(x_1, x_2) \leq 0$, and hence, it follows that $V(x_1(k), x_2(k))$ is a nonincreasing function of time. Hence, as in the proof of ii), it follows that for all $(x_{10}, x_{20}) \in \mathcal{B}_\delta(0) \times \mathbb{R}^{n_2}$, $x_1(k) \in \mathcal{B}_\varepsilon(0)$, $k \in \overline{\mathbb{Z}}_+$. Furthermore, it follows from (13.45) and (13.46) that for all $k \in \overline{\mathbb{Z}}_+$ and $(x_{10}, x_{20}) \in \mathcal{B}_\delta \times \mathbb{R}^{n_2}$,

$$\Delta V(x_1(k), x_2(k)) \leq -\gamma \|x_1(k)\|^p \leq -\frac{\gamma}{\beta} V(x_1(k), x_2(k)),$$

which implies that

$$V(x_1(k), x_2(k)) \leq V(x_{10}, x_{20})(1 - \tfrac{\gamma}{\beta})^k.$$

It now follows from (13.45) that

$$\begin{aligned} \alpha \|x_1(k)\|^p &\leq V(x_1(k), x_2(k)) \\ &\leq V(x_{10}, x_{20})(1 - \tfrac{\gamma}{\beta})^k \\ &\leq \beta \|x_{10}\|^p (1 - \tfrac{\gamma}{\beta})^k, \quad k \in \overline{\mathbb{Z}}_+, \end{aligned}$$

and hence,

$$\|x_1(k)\| \leq \left(\frac{\beta}{\alpha}\right)^{1/p} \|x_{10}\| (1 - \tfrac{\gamma}{\beta})^k, \quad k \in \overline{\mathbb{Z}}_+,$$

establishing geometric stability with respect to x_1 uniformly in x_{20}.

viii) The proof follows as in *vi*) and *vii*). □

By setting $n_1 = n$ and $n_2 = 0$, Theorem 13.9 specializes to the case of nonlinear autonomous systems of the form $x_1(k) = f_1(x_1(k))$. In this case, Lyapunov (respectively, asymptotic) stability with respect to x_1 and Lyapunov (respectively, asymptotic) stability with respect to x_1 uniformly in x_{20} are equivalent to the classical Lyapunov (respectively, asymptotic) stability of nonlinear autonomous systems presented in Section 13.2. In particular, note that it follows from Problems 3.75 and 3.76 that there exists a continuous function $V : \mathcal{D} \to \mathbb{R}$ such that (13.40), (13.42), and (13.44) hold if and only if $V(0) = 0$, $V(x_1) > 0$, $x_1 \neq 0$, and $V(f_1(x_1)) < V(x_1)$, $x_1 \neq 0$. In addition, if $\mathcal{D} = \mathbb{R}^{n_1}$ and there exist class $\mathcal{K}_\infty$ functions $\alpha(\cdot), \beta(\cdot)$ and a continuously differentiable function $V(\cdot)$ such that (13.40), (13.42), and (13.44) hold if and only if $V(0) = 0$, $V(x_1) > 0$, $x_1 \neq 0$, $V(f_1(x_1)) < V(x_1)$, $x_1 \neq 0$, and $V(x_1) \to \infty$ as $\|x_1\| \to \infty$. Hence, in this case, Theorem 13.9 collapses to the classical Lyapunov stability theorem for autonomous systems given in Section 13.2.

In the case of time-invariant systems the Barbashin-Krasovskii-LaSalle invariance theorem (Theorem 13.3) shows that bounded system trajectories of a nonlinear dynamical system approach the largest invariant set $\mathcal{M}$ characterized by the set of all points in a compact set $\mathcal{D}$ of the state space where the Lyapunov derivative identically vanishes. In the case of partially stable systems, however, it is not generally clear on how to define the set $\mathcal{M}$ since $\Delta V(x_1, x_2)$ is a function of both x_1 and x_2. However, if $\Delta V(x_1, x_2) \le -W(x_1) \le 0$, where $W : \mathcal{D} \to \mathbb{R}$ is continuous and nonnegative definite, then a set $\mathcal{R} \supset \mathcal{M}$ can be defined as the set of points where $W(x_1)$ identically vanishes, that is, $\mathcal{R} = \{x_1 \in \mathcal{D} : \ W(x_1) = 0\}$. In this case, as shown in the next theorem, the partial system trajectories $x_1(k)$ approach $\mathcal{R}$ as k tends to infinity. For this result, the following lemma is necessary.

Lemma 13.3. Let $\sigma : \overline{\mathbb{Z}}_+ \to \overline{\mathbb{R}}_+$ and suppose that $\sum_{k=0}^{\infty} \sigma(k)$ exists and is finite. Then, $\lim_{k\to\infty} \sigma(k) = 0$.

Proof. Suppose, *ad absurdum*, that $\limsup_{k\to\infty} \sigma(k) > 0$ and let $\alpha_1 \in \mathbb{R}$ be such that $0 < \alpha_1 < \limsup_{k\to\infty} \sigma(k)$. In this case, for every $K_1 > 0$, there exists $k_1 \ge K_1$ such that $\sigma(k_1) \ge \alpha_1$. Next, let $K_2 > k_1$ and using a similar argument as above it follows that there exists $k_2 \ge K_2$ such that $\sigma(k_2) \ge \alpha_1$. Hence, there exists an increasing unbounded sequence $\{k_i\}_{i=1}^{\infty}$ such that $\sigma(k_i) \ge \alpha_1$, and hence, $\sum_{i=1}^{\infty} \sigma(k_i) \ge \sum_{i=1}^{\infty} \alpha_1 = \infty$, which is a contradiction. □

Theorem 13.10. Consider the nonlinear dynamical system given by (13.37) and (13.38) and assume $\mathcal{D} \times \mathbb{R}^{n_2}$ is a positively invariant set with respect to (13.37) and (13.38). Furthermore, assume there exist functions $V : \mathcal{D} \times \mathbb{R}^{n_2} \to \mathbb{R}$, $W, W_1, W_2 : \mathcal{D} \to \mathbb{R}$ such that $V(\cdot,\cdot)$ is continuous, $W_1(\cdot)$ and $W_2(\cdot)$ are continuous and positive definite, $W(\cdot)$ is continuous and nonnegative definite, and, for all $(x_1, x_2) \in \mathcal{D} \times \mathbb{R}^{n_2}$,

$$W_1(x_1) \le V(x_1, x_2) \le W_2(x_1), \tag{13.49}$$

$$\Delta V(x_1, x_2) \le -W(x_1). \tag{13.50}$$

Then there exists $\mathcal{D}_0 \subseteq \mathcal{D}$ such that for all $(x_{10}, x_{20}) \in \mathcal{D}_0 \times \mathbb{R}^{n_2}$, $x_1(k) \to \mathcal{R} \triangleq \{x_1 \in \mathcal{D} : \ W(x_1) = 0\}$ as $k \to \infty$. If, in addition, $\mathcal{D} = \mathbb{R}^{n_1}$ and $W_1(\cdot)$ is radially unbounded, then for all $(x_{10}, x_{20}) \in \mathbb{R}^{n_1} \times \mathbb{R}^{n_2}$, $x_1(k) \to \mathcal{R} \triangleq \{x_1 \in \mathbb{R}^{n_1} : \ W(x_1) = 0\}$ as $k \to \infty$.

Proof. Assume (13.49) and (13.50) hold. Then it follows from Theorem 13.9 that the nonlinear dynamical system given by (13.37) and (13.38) is Lyapunov stable with respect to x_1 uniformly in x_{20}. Let $\varepsilon > 0$ be such that $\mathcal{B}_\varepsilon(0) \subset \mathcal{D}$ and let $\delta = \delta(\varepsilon) > 0$ be such that if $x_{10} \in \mathcal{B}_\delta(0)$, then $x_1(k) \in \mathcal{B}_\varepsilon(0)$, $k \in \overline{\mathbb{Z}}_+$. Now, since $V(x_1(k), x_2(k))$ is monotonically nonincreasing and bounded from below by zero, it follows from the monotone convergence theorem (Theorem 2.10) that $\lim_{k\to\infty} V(x_1(k), x_2(k))$ exists and is finite. Hence, since for every $k \in \overline{\mathbb{Z}}_+$,

$$\sum_{\kappa=0}^{k-1} W(x_1(\kappa)) \le -\sum_{\kappa=0}^{k-1} \Delta V(x_1(\kappa), x_2(\kappa)) = V(x_{10}, x_{20}) - V(x_1(k), x_2(k)),$$

it follows that $\sum_{\kappa=0}^{\infty} W(x_1(\kappa))$ exists and is finite. Hence, it follows from Lemma 13.3 that $W(x_1(k)) \to 0$ as $k \to \infty$. Finally, if, in addition, $\mathcal{D} = \mathbb{R}^{n_1}$ and $W_1(\cdot)$ is radially unbounded, then, as in the proof of *iv*) of Theorem 13.9, for every $x_{10} \in \mathbb{R}^{n_1}$ there exist $\varepsilon, \delta > 0$ such that $x_{10} \in \mathcal{B}_\delta(0)$ and $x_1(k) \in \mathcal{B}_\varepsilon(0)$, $k \in \overline{\mathbb{Z}}_+$. Now, the proof follows by repeating the above arguments. □

Theorem 13.10 shows that the partial system trajectories $x_1(k)$ approach $\mathcal{R}$ as k tends to infinity. However, since the positive limit set of the partial trajectory $x_1(k)$ is a subset of $\mathcal{R}$, Theorem 13.10 is a weaker result than the standard invariance principle, wherein one would conclude that the partial trajectory $x_1(k)$ approaches the *largest invariant set* $\mathcal{M}$ contained in $\mathcal{R}$. This is not true in general for partially stable systems since the positive limit set of a partial trajectory $x_1(k)$, $k \in \overline{\mathbb{Z}}_+$, is not an invariant set.

13.6 Stability Theory for Discrete-Time Nonlinear Time-Varying Systems

In this section, we use the results of Section 13.5 to extend Lyapunov's direct method to nonlinear time-varying systems thereby providing a unification between partial stability theory for autonomous systems and stability theory for time-varying systems. Specifically, we consider the nonlinear time-varying dynamical system

$$x(k+1) = f(k, x(k)), \quad x(k_0) = x_0, \quad k \ge k_0, \tag{13.51}$$

where $x(k) \in \mathcal{D}$, $k \ge k_0$, $\mathcal{D} \subseteq \mathbb{R}^n$ is an open set such that $0 \in \mathcal{D}$, $f : \{k_0, \ldots, k_1\} \times \mathcal{D} \to \mathbb{R}^n$ is such that $f(\cdot, \cdot)$ is continuous and, for every $k \in \{k_0, \ldots, k_1\}$, $f(k, 0) = 0$. Note that under the above assumptions the solution $x(k)$, $k \ge k_0$, to (13.51) exists and is unique over the interval $\{k_0, \ldots, k_1\}$. The following definition provides eight types of stability for the nonlinear time-varying dynamical system (13.51).

Definition 13.7. *i*) The nonlinear time-varying dynamical system (13.51) is *Lyapunov stable* if, for every $\varepsilon > 0$ and $k_0 \in \overline{\mathbb{Z}}_+$, there exists $\delta = \delta(\varepsilon, k_0) > 0$ such that $\|x_0\| < \delta$ implies that $\|x(k)\| < \varepsilon$ for all $k \ge k_0$.

ii) The nonlinear time-varying dynamical system (13.51) is *uniformly Lyapunov stable* if, for every $\varepsilon > 0$, there exists $\delta = \delta(\varepsilon) > 0$ such that $\|x_0\| < \delta$ implies that $\|x(k)\| < \varepsilon$ for all $k \ge k_0$ and for all $k_0 \in \overline{\mathbb{Z}}_+$.

iii) The nonlinear time-varying dynamical system (13.51) is *asymptotically stable* if it is Lyapunov stable and, for every $k_0 \in \overline{\mathbb{Z}}_+$, there exists $\delta = \delta(k_0) > 0$ such that $\|x_0\| < \delta$ implies that $\lim_{k\to\infty} x(k) = 0$.

iv) The nonlinear time-varying dynamical system (13.51) is *uniformly asymptotically stable* if it is uniformly Lyapunov stable and there exists $\delta > 0$ such that $\|x_0\| < \delta$ implies that $\lim_{k\to\infty} x(k) = 0$ uniformly in k_0 and x_0 for all $k_0 \in \overline{\mathbb{Z}}_+$.

v) The nonlinear time-varying dynamical system (13.51) is *globally asymptotically stable* if it is Lyapunov stable and $\lim_{k\to\infty} x(k) = 0$ for all $x_0 \in \mathbb{R}^n$ and $k_0 \in \overline{\mathbb{Z}}_+$.

vi) The nonlinear time-varying dynamical system (13.51) is *globally uniformly asymptotically stable* if it is uniformly Lyapunov stable and $\lim_{k\to\infty} x(k) = 0$ uniformly in k_0 and x_0 for all $x_0 \in \mathbb{R}^n$ and $k_0 \in \overline{\mathbb{Z}}_+$.

vii) The nonlinear time-varying dynamical system (13.51) is (*uniformly*) *geometrically stable* if there exist scalars $\alpha, \delta > 0$ and $\beta > 1$, such that $\|x_0\| < \delta$ implies that $\|x(k)\| \le \alpha\|x_0\|\beta^{-k}$, $k \ge k_0$ and $k_0 \in \overline{\mathbb{Z}}_+$.

viii) The nonlinear time-varying dynamical system (13.51) is *globally* (*uniformly*) *geometrically stable* if there exist scalars $\alpha > 0$ and $\beta > 1$, such that $\|x(k)\| \le \alpha\|x_0\|\beta^{-k}$, $k \ge k_0$, for all $x_0 \in \mathbb{R}^n$ and $k_0 \in \overline{\mathbb{Z}}_+$.

Next, using Theorem 13.9 we present sufficient conditions for stability of the nonlinear time-varying dynamical system (13.51). For the following result define

$$\Delta V(k,x) \triangleq V(k+1, f(k,x)) - V(k,x)$$

for a given continuous function $V : \overline{\mathbb{Z}}_+ \times \mathcal{D} \to \mathbb{R}$.

Theorem 13.11. Consider the nonlinear time-varying dynamical system (13.51). Then the following statements hold:

i) If there exist a continuous function $V : \overline{\mathbb{Z}}_+ \times \mathcal{D} \to \mathbb{R}$ and a class $\mathcal{K}$ function $\alpha(\cdot)$ such that

$$V(k,0) = 0, \qquad k \in \overline{\mathbb{Z}}_+, \tag{13.52}$$

$$\alpha(\|x\|) \le V(k,x), \qquad (k,x) \in \overline{\mathbb{Z}}_+ \times \mathcal{D}, \tag{13.53}$$

$$\Delta V(k,x) \le 0, \qquad (k,x) \in \overline{\mathbb{Z}}_+ \times \mathcal{D}, \tag{13.54}$$

then the nonlinear time-varying dynamical system given by (13.51) is Lyapunov stable.

ii) If there exist a continuous function $V : \overline{\mathbb{Z}}_+ \times \mathcal{D} \to \mathbb{R}$ and class $\mathcal{K}$ functions $\alpha(\cdot), \beta(\cdot)$ satisfying (13.53) and (13.54), and

$$V(k,x) \le \beta(\|x\|), \qquad (k,x) \in \overline{\mathbb{Z}}_+ \times \mathcal{D}, \tag{13.55}$$

then the nonlinear time-varying dynamical system given by (13.51) is uniformly Lyapunov stable.

iii) If there exist a continuous function $V : \overline{\mathbb{Z}}_+ \times \mathcal{D} \to \mathbb{R}$ and class $\mathcal{K}$ functions $\alpha(\cdot), \gamma(\cdot)$ satisfying (13.53) and

$$\Delta V(k,x) \le -\gamma(\|x\|), \qquad (k,x) \in \mathbb{Z}_+ \times \mathcal{D}, \tag{13.56}$$

then the nonlinear time-varying dynamical system given by (13.51) is asymptotically stable.

iv) If there exist a continuous function $V : \overline{\mathbb{Z}}_+ \times \mathcal{D} \to \mathbb{R}$ and class $\mathcal{K}$ functions $\alpha(\cdot), \beta(\cdot), \gamma(\cdot)$ satisfying (13.53), (13.55), and

$$\Delta V(k,x) \le -\gamma(\|x\|), \qquad (k,x) \in \overline{\mathbb{Z}}_+ \times \mathcal{D}, \tag{13.57}$$

then the nonlinear time-varying dynamical system given by (13.51) is uniformly asymptotically stable.

v) If $\mathcal{D} = \mathbb{R}^n$ and there exist a continuous function $V : \overline{\mathbb{Z}}_+ \times \mathbb{R}^n \to \mathbb{R}$, a class $\mathcal{K}$ function $\gamma(\cdot)$, class $\mathcal{K}_\infty$ functions $\alpha(\cdot)$ satisfying (13.53) and (13.56), then the nonlinear time-varying dynamical system given by (13.51) is globally asymptotically stable.

vi) If $\mathcal{D} = \mathbb{R}^n$ and there exist a continuous function $V : \overline{\mathbb{Z}}_+ \times \mathbb{R}^n \to \mathbb{R}$, a class $\mathcal{K}$ function $\gamma(\cdot)$, class $\mathcal{K}_\infty$ functions $\alpha(\cdot), \beta(\cdot)$ satisfying (13.53), (13.55), and (13.57), then the nonlinear time-varying dynamical system given by (13.51) is globally uniformly asymptotically stable.

vii) If there exist a continuous function $V : \overline{\mathbb{Z}}_+ \times \mathcal{D} \to \mathbb{R}$ and positive constants α, β, γ, p such that $p \ge 1$ and

$$\alpha\|x\|^p \le V(k,x) \le \beta\|x\|^p, \qquad (k,x) \in \overline{\mathbb{Z}}_+ \times \mathcal{D}, \tag{13.58}$$

$$\Delta V(k,x) \le -\gamma\|x\|^p, \qquad (k,x) \in \overline{\mathbb{Z}}_+ \times \mathcal{D}, \tag{13.59}$$

then the nonlinear time-varying dynamical system given by (13.51) is (uniformly) geometrically stable.

viii) If $\mathcal{D} = \mathbb{R}^n$ and there exist a continuous function $V : \overline{\mathbb{Z}}_+ \times \mathbb{R}^n \to \mathbb{R}$ and positive constants α, β, γ, p such that $p \ge 1$ and (13.58) and (13.59) hold, then the nonlinear time-varying dynamical system given by (13.51) is globally (uniformly) geometrically stable.

Proof. Let $n_1 = n$, $n_2 = 1$, $x_1(k-k_0) = x(k)$, $x_2(k-k_0) = k$, $f_1(x_1,x_2) = f(k,x)$, and $f_2(x_1,x_2) = x_2+1$. Now, note that with $\kappa = k-k_0$, the solution $x(k)$, $k \ge k_0$, to the nonlinear time-varying dynamical system (13.51) is equivalently characterized by the solution $x_1(\kappa)$, $\kappa \ge 0$, to the nonlinear autonomous dynamical system

$$x_1(\kappa+1) = f_1(x_1(\kappa), x_2(\kappa)), \quad x_1(0) = x_0, \quad \kappa \geq 0,$$
$$x_2(\kappa+1) = x_2(\kappa) + 1, \quad x_2(0) = k_0.$$

Now, the result is a direct consequence of Theorem 13.9. □

13.7 Lagrange Stability, Boundedness, and Ultimate Boundedness

In this section, we introduce the notions of *Lagrange stability, boundedness*, and *ultimate boundedness* and present Lyapunov-like theorems for boundedness and ultimate boundedness of nonlinear discrete-time dynamical systems.

Definition 13.8. *i*) The nonlinear dynamical system (13.37) and (13.38) is *Lagrange stable with respect to* x_1 if, for every $x_{10} \in \mathcal{D}$ and $x_{20} \in \mathbb{R}^{n_2}$, there exists $\varepsilon = \varepsilon(x_{10}, x_{20}) > 0$ such that $\|x_1(k)\| < \varepsilon$, $k \in \overline{\mathbb{Z}}_+$.

ii) The nonlinear dynamical system (13.37) and (13.38) is *bounded with respect to* x_1 *uniformly in* x_2 if there exists $\gamma > 0$ such that, for every $\delta \in (0, \gamma)$, there exists $\varepsilon = \varepsilon(\delta) > 0$ such that $\|x_{10}\| < \delta$ implies $\|x_1(k)\| < \varepsilon$, $k \in \overline{\mathbb{Z}}_+$. The nonlinear dynamical system (13.37) and (13.38) is *globally bounded with respect to* x_1 *uniformly in* x_2 if, for every $\delta \in (0, \infty)$, there exists $\varepsilon = \varepsilon(\delta) > 0$ such that $\|x_{10}\| < \delta$ implies $\|x_1(k)\| < \varepsilon$, $k \in \overline{\mathbb{Z}}_+$.

iii) The nonlinear dynamical system (13.37) and (13.38) is *ultimately bounded with respect to* x_1 *uniformly in* x_2 *with bound* ε if there exists $\gamma > 0$ such that, for every $\delta \in (0, \gamma)$, there exists $K = K(\delta, \varepsilon) > 0$ such that $\|x_{10}\| < \delta$ implies $\|x_1(k)\| < \varepsilon$, $k \geq K$. The nonlinear dynamical system (13.37) and (13.38) is *globally ultimately bounded with respect to* x_1 *uniformly in* x_2 *with bound* ε if, for every $\delta \in (0, \infty)$, there exists $K = K(\delta, \varepsilon) > 0$ such that $\|x_{10}\| < \delta$ implies $\|x_1(k)\| < \varepsilon$, $k \geq K$.

The following results present Lyapunov-like theorems for boundedness and ultimate boundedness. For these results define $\Delta V(x_1, x_2) \triangleq V(f(x_1, x_2)) - V(x_1, x_2)$, where $f(x_1, x_2) \triangleq [f_1^{\mathrm{T}}(x_1, x_2)\ f_2^{\mathrm{T}}(x_1, x_2)]^{\mathrm{T}}$ and $V : \mathcal{D} \times \mathbb{R}^{n_2} \to \mathbb{R}$ is a continuous function.

Theorem 13.12. Consider the nonlinear dynamical system (13.37) and (13.38). Assume there exist a continuous function $V : \mathcal{D} \times \mathbb{R}^{n_2} \to \mathbb{R}$ and class $\mathcal{K}$ functions $\alpha(\cdot)$, $\beta(\cdot)$ such that

$$\alpha(\|x_1\|) \leq V(x_1, x_2) \leq \beta(\|x_1\|), \quad x_1 \in \mathcal{D}, \quad x_2 \in \mathbb{R}^{n_2}, \tag{13.60}$$
$$\Delta V(x_1, x_2) \leq 0, \quad x_1 \in \mathcal{D}, \quad \|x_1\| \geq \mu, \quad x_2 \in \mathbb{R}^{n_2}, \tag{13.61}$$

where $\mu > 0$ is such that $\mathcal{B}_{\alpha^{-1}(\beta(\mu))}(0) \subset \mathcal{D}$. Furthermore, assume that $\sup_{(x_1,x_2)\in\overline{\mathcal{B}}_\mu(0)\times\mathbb{R}^{n_2}} V(f(x_1,x_2))$ exists. Then the nonlinear dynamical system (13.37) and (13.38) is bounded with respect to x_1 uniformly in x_2. Furthermore, for every $\delta \in (0,\gamma)$, $x_{10} \in \overline{\mathcal{B}}_\delta(0)$ implies that $\|x_1(k)\| \leq \varepsilon$, $k \in \overline{\mathbb{Z}}_+$, where

$$\varepsilon = \varepsilon(\delta) \triangleq \alpha^{-1}(\max\{\eta, \beta(\delta)\}), \tag{13.62}$$

$\eta \geq \max\{\beta(\mu), \sup_{(x_1,x_2)\in\overline{\mathcal{B}}_\mu(0)\times\mathbb{R}^{n_2}} V(f(x_1,x_2))\}$, and $\gamma \triangleq \sup\{r > 0 : \mathcal{B}_{\alpha^{-1}(\beta(r))}(0) \subset \mathcal{D}\}$. If, in addition, $\mathcal{D} = \mathbb{R}^{n_1}$ and $\alpha(\cdot)$ is a class $\mathcal{K}_\infty$ function, then the nonlinear dynamical system (13.37) and (13.38) is globally bounded with respect to x_1 uniformly in x_2 and for every $x_{10} \in \mathbb{R}^{n_1}$, $\|x_1(k)\| \leq \varepsilon$, $k \in \overline{\mathbb{Z}}_+$, where ε is given by (13.62) with $\delta = \|x_{10}\|$.

Proof. First, let $\delta \in (0,\mu]$ and assume $\|x_{10}\| \leq \delta$. If $\|x_1(k)\| \leq \mu$, $k \in \overline{\mathbb{Z}}_+$, then it follows from (13.60) that $\|x_1(k)\| \leq \mu \leq \alpha^{-1}(\beta(\mu)) \leq \alpha^{-1}(\eta)$, $k \in \overline{\mathbb{Z}}_+$. Alternatively, if there exists $K > 0$ such that $\|x_1(K)\| > \mu$, then, since $\|x_1(0)\| \leq \mu$, it follows that there exists $\kappa \leq K$ such that $\|x_1(\kappa-1)\| \leq \mu$ and $\|x_1(k)\| > \mu$, $k \in \{\kappa,\ldots,K\}$. Hence, it follows from (13.60) and (13.61) that

$$\begin{aligned}\alpha(\|x_1(K)\|) &\leq V(x_1(K), x_2(K))\\ &\leq V(x_1(\kappa), x_2(\kappa))\\ &= V(f(x_1(\kappa-1), x_2(\kappa-1)))\\ &\leq \eta,\end{aligned}$$

which implies that $\|x_1(K)\| \leq \alpha^{-1}(\eta)$. Next, let $\delta \in (\mu,\gamma)$ and assume $x_{10} \in \overline{\mathcal{B}}_\delta(0)$ and $\|x_{10}\| > \mu$. Now, for every $\hat{k} > 0$ such that $\|x_1(k)\| \geq \mu$, $k \in \{0,\ldots,\hat{k}\}$, it follows from (13.60) and (13.61) that

$$\alpha(\|x_1(k)\|) \leq V(x_1(k), x_2(k)) \leq V(x_{10}, x_{20}) \leq \beta(\delta),$$

which implies that $\|x_1(k)\| \leq \alpha^{-1}(\beta(\delta))$, $k \in \{0,\ldots,\hat{k}\}$. Next, if there exists $K > 0$ such that $\|x_1(K)\| \leq \mu$, then it follows as in the proof of the first case that $\|x_1(k)\| \leq \alpha^{-1}(\eta)$, $k \geq K$. Hence, if $x_{10} \in \mathcal{B}_\delta(0)$, then $\|x_1(k)\| \leq \alpha^{-1}(\max\{\eta,\beta(\delta)\})$, $k \in \overline{\mathbb{Z}}_+$. Finally, if $\mathcal{D} = \mathbb{R}^{n_1}$ and $\alpha(\cdot)$ is a class $\mathcal{K}_\infty$ function it follows that $\beta(\cdot)$ is a class $\mathcal{K}_\infty$ function, and hence, $\gamma = \infty$. Hence, the nonlinear dynamical system (13.37) and (13.38) is globally bounded with respect to x_1 uniformly in x_2. □

Theorem 13.13. Consider the nonlinear dynamical system (13.37) and (13.38). Assume there exist a continuous function $V : \mathcal{D} \times \mathbb{R}^{n_2} \to \mathbb{R}$ and class $\mathcal{K}$ functions $\alpha(\cdot)$, $\beta(\cdot)$ such that (13.60) holds. Furthermore, assume that there exists a continuous function $W : \mathcal{D} \to \mathbb{R}$ such that $W(x_1) > 0$, $\|x_1\| > \mu$, and

$$\Delta V(x_1,x_2) \leq -W(x_1), \quad x_1 \in \mathcal{D}, \quad \|x_1\| > \mu, \quad x_2 \in \mathbb{R}^{n_2}, \tag{13.63}$$

where $\mu > 0$ is such that $\mathcal{B}_{\alpha^{-1}(\beta(\mu))}(0) \subset \mathcal{D}$. Finally, assume that

$$\sup_{(x_1,x_2)\in\overline{\mathcal{B}}_\mu(0)\times\mathbb{R}^{n_2}} V(f(x_1,x_2))$$

exists. Then the nonlinear dynamical system (13.37) and (13.38) is ultimately bounded with respect to x_1 uniformly in x_2 with bound $\varepsilon \triangleq \alpha^{-1}(\eta)$, where $\eta > \max\{\beta(\mu), \sup_{(x_1,x_2)\in\overline{\mathcal{B}}_\mu(0)\times\mathbb{R}^{n_2}} V(f(x_1,x_2))\}$. Furthermore, $\limsup_{k\to\infty}\|x_1(k)\| \le \alpha^{-1}(\eta)$. If, in addition, $\mathcal{D} = \mathbb{R}^n$ and $\alpha(\cdot)$ is a class $\mathcal{K}_\infty$ function, then the nonlinear dynamical system (13.37) and (13.38) is globally ultimately bounded with respect to x_1 uniformly in x_2 with bound ε.

Proof. First, let $\delta \in (0,\mu]$ and assume $\|x_{10}\| \le \delta$. As in the proof of Theorem 13.12, it follows that $\|x_1(k)\| \le \alpha^{-1}(\eta) = \varepsilon$, $k \in \overline{\mathbb{Z}}_+$. Next, let $\delta \in (\mu,\gamma)$, where $\gamma \triangleq \sup\{r > 0 : \mathcal{B}_{\alpha^{-1}(\beta(r))}(0) \subset \mathcal{D}\}$ and assume $x_{10} \in \mathcal{B}_\delta(0)$ and $\|x_{10}\| > \mu$. In this case, it follows from Theorem 13.12 that $\|x_1(k)\| \le \alpha^{-1}(\max\{\eta, \beta(\delta)\})$, $k \in \overline{\mathbb{Z}}_+$. Suppose, *ad absurdum*, that $\|x_1(k)\| \ge \beta^{-1}(\eta)$, $k \in \overline{\mathbb{Z}}_+$, or, equivalently, $x_1(k) \in \mathcal{O} \triangleq \mathcal{B}_{\alpha^{-1}(\max\{\eta,\beta(\delta)\})}(0)\backslash\mathcal{B}_{\beta^{-1}(\eta)}(0)$, $k \in \overline{\mathbb{Z}}_+$. Since $\overline{\mathcal{O}}$ is compact and $W(\cdot)$ is continuous and $W(x_1) > 0$, $\|x_1\| \ge \beta^{-1}(\eta) > \mu$, it follows from Theorem 2.13 that $\theta \triangleq \min_{x_1\in\overline{\mathcal{O}}} W(x_1) > 0$ exists. Hence, it follows from (13.63) that

$$V(x_1(k), x_2(k)) \le V(x_{10}, x_{20}) - k\theta, \qquad k \in \overline{\mathbb{Z}}_+, \tag{13.64}$$

which implies that

$$\alpha(\|x_1(k)\|) \le \beta(\|x_{10}\|) - k\theta \le \beta(\delta) - k\theta, \qquad k \in \overline{\mathbb{Z}}_+. \tag{13.65}$$

Now, letting $k > \beta(\delta)/\theta$ it follows that $\alpha(\|x_1(k)\|) < 0$, which is a contradiction. Hence, there exists $K = K(\delta,\eta) > 0$ such that $\|x_1(K)\| < \beta^{-1}(\eta)$. Thus, it follows from Theorem 13.12 that $\|x_1(k)\| \le \alpha^{-1}(\eta)$, $k \ge K$, which proves that the nonlinear dynamical system (13.37) and (13.38) is ultimately bounded with respect to x_1 uniformly in x_2 with bound $\varepsilon = \alpha^{-1}(\eta)$. Furthermore, $\limsup_{k\to\infty}\|x_1(k)\| \le \alpha^{-1}(\eta)$. Finally, if $\mathcal{D} = \mathbb{R}^{n_1}$ and $\alpha(\cdot)$ is a class $\mathcal{K}_\infty$ function it follows that $\beta(\cdot)$ is a class $\mathcal{K}_\infty$ function, and hence, $\gamma = \infty$. Hence, the nonlinear dynamical system (13.37) and (13.38) is globally ultimately bounded with respect to x_1 uniformly in x_2 with bound ε. □

Corollary 13.3. Consider the nonlinear dynamical system (13.37) and (13.38). Assume there exist a continuous function $V : \mathcal{D}\times\mathbb{R}^{n_2} \to \mathbb{R}$ and class $\mathcal{K}$ functions $\alpha(\cdot)$, $\beta(\cdot)$ such that (13.60) holds. Furthermore, assume that there exists a class $\mathcal{K}$ function $\gamma : \mathcal{D} \to \mathbb{R}$ such that

$$\Delta V(x_1, x_2) \le -\gamma(\|x_1\|) + \gamma(\mu), \quad x_1 \in \mathcal{D}, \quad x_2 \in \mathbb{R}^{n_2}, \tag{13.66}$$

where $\mu > 0$ is such that $\mathcal{B}_{\alpha^{-1}(\beta(\mu))}(0) \subset \mathcal{D}$. Then the nonlinear dynamical system (13.37) and (13.38) is ultimately bounded with respect to x_1 uniformly in x_2 with bound $\varepsilon \triangleq \alpha^{-1}(\eta)$, where $\eta = \beta(\mu) + \gamma(\mu)$. Furthermore, $\limsup_{k\to\infty} \|x_1(k)\| \le \alpha^{-1}(\eta)$. If, in addition, $\mathcal{D} = \mathbb{R}^n$ and $\alpha(\cdot)$ is a class $\mathcal{K}_\infty$ function, then the nonlinear dynamical system (13.37) and (13.38) is globally ultimately bounded with respect to x_1 uniformly in x_2 with bound ε.

Proof. The result is a direct consequence of Theorem 13.13 with $W(x_1) = \gamma(\|x_1\|) - \gamma(\mu)$ and $\eta = \beta(\mu) + \gamma(\mu)$. Specifically, it follows from (13.66) that for every $(x_1, x_2) \in \mathcal{D} \times \mathbb{R}^{n_2}$, $\|x_1\| \le \mu$,

$$\begin{aligned} V(f(x_1,x_2)) &\le V(x_1,x_2) - \gamma(\|x_1\|) + \gamma(\mu) \\ &\le \beta(\|x_1\|) - \gamma(\|x_1\|) + \gamma(\mu) \\ &\le \beta(\mu) + \gamma(\mu). \end{aligned}$$

Hence, it follows that

$$\sup_{(x_1,x_2)\in\overline{\mathcal{B}}_\mu(0)\times\mathbb{R}^{n_2}} V(f(x_1,x_2)) \le \beta(\mu) + \gamma(\mu).$$

Finally, note that $\beta(\mu) \le \beta(\mu) + \gamma(\mu)$, and hence,

$$\eta \ge \max\left\{\beta(\mu), \sup_{(x_1,x_2)\in\overline{\mathcal{B}}_\mu(0)\times\mathbb{R}^{n_2}} V(f(x_1,x_2))\right\}$$

satisfying all the conditions of Theorem 13.13. □

Next, we specialize Theorems 13.12 and 13.13 to nonlinear time-varying dynamical systems. The following definition is needed for these results.

Definition 13.9. *i*) The nonlinear time-varying dynamical system (13.51) is *Lagrange stable* if, for every $x_0 \in \mathbb{R}^n$ and $k_0 \in \mathbb{Z}$, there exists $\varepsilon = \varepsilon(k_0, x_0) > 0$ such that $\|x(k)\| < \varepsilon$, $k \ge k_0$.

ii) The nonlinear time-varying dynamical system (13.51) is *uniformly bounded* if there exists $\gamma > 0$ such that, for every $\delta \in (0, \gamma)$, there exists $\varepsilon = \varepsilon(\delta) > 0$ such that $\|x_0\| < \delta$ implies $\|x(k)\| < \varepsilon$, $k \ge k_0$. The nonlinear time-varying dynamical system (13.51) is *globally uniformly bounded* if, for every $\delta \in (0, \infty)$, there exists $\varepsilon = \varepsilon(\delta) > 0$ such that $\|x_0\| < \delta$ implies $\|x(k)\| < \varepsilon$, $k \ge k_0$.

iii) The nonlinear time-varying dynamical system (13.51) is *uniformly ultimately bounded with bound* ε if there exists $\gamma > 0$ such that, for every $\delta \in (0, \gamma)$, there exists $K = K(\delta, \varepsilon) > 0$ such that $\|x_0\| < \delta$ implies $\|x(k)\| < \varepsilon$, $k \ge k_0 + K$. The nonlinear time-varying dynamical system (13.51) is *globally uniformly ultimately bounded with bound* ε if, for every $\delta \in (0, \infty)$, there exists $K = K(\delta, \varepsilon) > 0$ such that $\|x_0\| < \delta$ implies $\|x(k)\| < \varepsilon$, $k \geq k_0 + K$.

For the following result define

$$\Delta V(k,x) \triangleq V(k+1, f(k,x)) - V(k,x),$$

where $V : \mathbb{R} \times \mathcal{D} \to \mathbb{R}$ is a given continuous function.

Corollary 13.4. Consider the nonlinear time-varying dynamical system (13.51). Assume there exist a continuous function $V : \mathbb{R} \times \mathcal{D} \to \mathbb{R}$ and class $\mathcal{K}$ functions $\alpha(\cdot)$, $\beta(\cdot)$ such that

$$\alpha(\|x\|) \leq V(k,x) \leq \beta(\|x\|), \quad x \in \mathcal{D}, \quad k \in \mathbb{Z}, \tag{13.67}$$

$$\Delta V(k,x) \leq 0, \quad x \in \mathcal{D}, \quad \|x\| \geq \mu, \quad k \in \mathbb{Z}, \tag{13.68}$$

where $\mu > 0$ is such that $\mathcal{B}_{\alpha^{-1}(\beta(\mu))}(0) \subset \mathcal{D}$. Furthermore, assume that $\sup_{(k,x)\in\mathbb{Z}\times\overline{\mathcal{B}}_\mu(0)} V(k, f(k,x))$ exists. Then the nonlinear time-varying dynamical system (13.51) is uniformly bounded. If, in addition, $\mathcal{D} = \mathbb{R}^n$ and $\alpha(\cdot)$ is a class $\mathcal{K}_\infty$ function, then the nonlinear time-varying dynamical system (13.51) is globally uniformly bounded.

Proof. The result is a direct consequence of Theorem 13.12. Specifically, let $n_1 = n$, $n_2 = 1$, $x_1(k - k_0) = x(k)$, $x_2(k - k_0) = k$, $f_1(x_1, x_2) = f(k,x)$, and $f_2(x_1, x_2) = x_2 + 1$. Now, note that with $\kappa = k - k_0$, the solution $x(k)$, $k \geq k_0$, to the nonlinear time-varying dynamical system (13.51) is equivalently characterized by the solution $x_1(\kappa)$, $\kappa \geq 0$, to the nonlinear autonomous dynamical system

$$\begin{aligned} x_1(\kappa+1) &= f_1(x_1(\kappa), x_2(\kappa)), \quad x_1(0) = x_0, \quad \kappa \geq 0, \\ x_2(\kappa+1) &= x_2(\kappa) + 1, \quad x_2(0) = k_0. \end{aligned}$$

Now, the result is a direct consequence of Theorem 13.12. □

Corollary 13.5. Consider the nonlinear time-varying dynamical system (13.51). Assume there exist a continuous function $V : \mathbb{R} \times \mathcal{D} \to \mathbb{R}$ and class $\mathcal{K}$ functions $\alpha(\cdot)$, $\beta(\cdot)$ such that (13.67) holds. Furthermore, assume that there exists a continuous function $W : \mathcal{D} \to \mathbb{R}$ such that $W(x) > 0$, $\|x\| > \mu$, and

$$\Delta V(k,x) \leq -W(x), \quad x \in \mathcal{D}, \quad \|x\| > \mu, \quad k \in \mathbb{Z}, \tag{13.69}$$

where $\mu > 0$ is such that $\mathcal{B}_{\alpha^{-1}(\beta(\mu))}(0) \subset \mathcal{D}$. Finally, assume that $\sup_{(k,x)\in\mathbb{Z}\times\overline{\mathcal{B}}_\mu(0)} V(k, f(k,x))$ exists. Then the nonlinear time-varying dynamical system (13.51) is uniformly ultimately bounded with bound $\varepsilon \triangleq \alpha^{-1}(\eta)$ where $\eta \geq \max\{\beta(\mu), \sup_{(k,x)\in\mathbb{Z}\times\overline{\mathcal{B}}_{\mu(0)}} V(k, f(k,x))\}$. Furthermore, $\limsup_{k\to\infty} \|x(k)\| \leq \alpha^{-1}(\eta)$. If, in addition, $\mathcal{D} = \mathbb{R}^n$ and $\alpha(\cdot)$ is a class $\mathcal{K}_\infty$ function, then the nonlinear time-varying dynamical system (13.51) is globally uniformly ultimately bounded with bound ε.

Proof. The result is a direct consequence of Theorem 13.13 using similar arguments as in the proof of Corollary 13.4 and, hence, is omitted. □

Finally, we specialize Corollaries 13.4 and 13.5 to nonlinear time-invariant dynamical systems. For these results we need the following specialization of Definition 13.9.

Definition 13.10. *i*) The nonlinear dynamical system (13.1) is *Lagrange stable* if, for every $x_0 \in \mathbb{R}^n$, there exists $\varepsilon = \varepsilon(x_0) > 0$ such that $\|x(k)\| < \varepsilon$, $k \in \overline{\mathbb{Z}}_+$.

ii) The nonlinear dynamical system (13.1) is *bounded* if there exists $\gamma > 0$ such that, for every $\delta \in (0, \gamma)$, there exists $\varepsilon = \varepsilon(\delta) > 0$ such that $\|x_0\| < \delta$ implies $\|x(k)\| < \varepsilon$, $k \in \overline{\mathbb{Z}}_+$. The nonlinear dynamical system (13.1) is *globally bounded* if, for every $\delta \in (0, \infty)$, there exists $\varepsilon = \varepsilon(\delta) > 0$ such that $\|x_0\| < \delta$ implies $\|x(k)\| < \varepsilon$, $k \in \overline{\mathbb{Z}}_+$.

iii) The nonlinear dynamical system (13.1) is *ultimately bounded with bound* ε if there exists $\gamma > 0$ such that, for every $\delta \in (0, \gamma)$, there exists $K = K(\delta, \varepsilon) > 0$ such that $\|x_0\| < \delta$ implies $\|x(k)\| < \varepsilon$, $k \geq K$. The nonlinear dynamical system (13.1) is *globally ultimately bounded with bound* ε if, for every $\delta \in (0, \infty)$, there exists $K = K(\delta, \varepsilon) > 0$ such that $\|x_0\| < \delta$ implies $\|x(k)\| < \varepsilon$, $k \geq K$.

Corollary 13.6. Consider the nonlinear dynamical system (13.1). Assume there exist a continuous function $V : \mathcal{D} \to \mathbb{R}$ and class $\mathcal{K}$ functions $\alpha(\cdot)$ and $\beta(\cdot)$ such that

$$\alpha(\|x\|) \leq V(x) \leq \beta(\|x\|), \quad x \in \mathcal{D}, \tag{13.70}$$

$$V(f(x)) \leq V(x), \quad x \in \mathcal{D}, \quad \|x\| \geq \mu, \tag{13.71}$$

where $\mu > 0$ is such that $\mathcal{B}_{\alpha^{-1}(\eta)}(0) \subset \mathcal{D}$, with $\eta \geq \beta(\mu)$. Then the nonlinear dynamical system (13.1) is bounded. If, in addition, $\mathcal{D} = \mathbb{R}^n$ and $V(x) \to \infty$ as $\|x\| \to \infty$, then the nonlinear dynamical system (13.1) is globally bounded.

Proof. The result is a direct consequence of Corollary 13.4. □

Corollary 13.7. Consider the nonlinear dynamical system (13.1). Assume there exist a continuous function $V : \mathbb{R} \times \mathcal{D} \to \mathbb{R}$ and class $\mathcal{K}$ functions $\alpha(\cdot)$ and $\beta(\cdot)$ such that (13.70) holds and

$$V(f(x)) < V(x), \quad x \in \mathcal{D}, \quad \|x\| > \mu, \tag{13.72}$$

where $\mu > 0$ is such that $\mathcal{B}_{\alpha^{-1}(\eta)}(0) \subset \mathcal{D}$ with $\eta > \beta(\mu)$. Then the nonlinear dynamical system (13.1) is ultimately bounded with bound $\varepsilon \triangleq \alpha^{-1}(\eta)$. Furthermore, $\limsup_{k\to\infty} \|x(k)\| \le \alpha^{-1}(\eta)$. If, in addition, $\mathcal{D} = \mathbb{R}^n$ and $V(x) \to \infty$ as $\|x\| \to \infty$, then the nonlinear dynamical system (13.1) is globally ultimately bounded with bound ε.

Proof. The result is a direct consequence of Corollary 13.5. □

13.8 Stability Theory via Vector Lyapunov Functions

In this section, we consider the method of vector Lyapunov functions for stability analysis of discrete-time nonlinear dynamical systems. To develop the theory of vector Lyapunov functions, we first introduce some results on vector difference inequalities and the *vector comparison principle* for discrete-time systems. Specifically, consider the discrete-time nonlinear dynamical system given by

$$z(k+1) = w(z(k)), \quad z(k_0) = z_0, \quad k \in \mathcal{I}_{z_0}, \tag{13.73}$$

where $z(k) \in \mathcal{Q} \subseteq \mathbb{R}^q$, $k \in \mathcal{I}_{z_0}$, is the system state vector, $\mathcal{I}_{z_0} \subseteq \mathcal{T} \subseteq \mathbb{Z}$ is the maximal interval of existence of a solution $z(k)$ to (13.73), $\mathcal{Q}$ is an open set, $0 \in \mathcal{Q}$, and $w : \mathcal{Q} \to \mathbb{R}^q$ is a continuous function on $\mathcal{Q}$.

Theorem 13.14. Consider the discrete-time nonlinear dynamical system (13.73). Assume that the function $w : \mathcal{Q} \to \mathbb{R}^q$ is continuous and $w(\cdot)$ is of class $\mathcal{W}_\mathrm{d}$. If there exists a continuous vector function $V : \mathcal{I}_{z_0} \to \mathcal{Q}$ such that

$$V(k+1) - V(k) \le\le w(V(k)), \quad k \in \mathcal{I}_{z_0}, \tag{13.74}$$

then

$$V(k_0) \le\le z_0, \quad z_0 \in \mathcal{Q}, \tag{13.75}$$

implies

$$V(k) \le\le z(k), \quad k \in \mathcal{I}_{z_0}, \tag{13.76}$$

where $z(k)$, $k \in \mathcal{I}_{z_0}$, is the solution to (13.73).

Proof. It follows from (13.73), (13.74), (13.75), and the fact that $w(\cdot) \in \mathcal{W}_\mathrm{d}$ that

$$\begin{aligned} V(k_0+1) &= V(k_0) + V(k_0+1) - V(k_0) \\ &\le\le V(k_0) + w(V(k_0)) \\ &\le\le z_0 + w(z_0) \\ &= z(k_0+1). \end{aligned}$$

Hence, $V(k_0+1) \leq\leq z(k_0+1)$. Recursively repeating this procedure for all $k_i \in \mathcal{I}_{z_0}$ yields (13.76). □

Next, consider the discrete-time nonlinear dynamical system given by

$$x(k+1) = f(x(k)), \quad x(k_0) = x_0, \quad k \in \mathcal{I}_{x_0}, \tag{13.77}$$

where $x(k) \in \mathcal{D} \subseteq \mathbb{R}^n$, $k \in \mathcal{I}_{x_0}$, is the system state vector, $\mathcal{I}_{x_0} \subset \mathbb{Z}$ is the maximal interval of existence of a solution $x(k)$ to (13.77), $\mathcal{D}$ is an open set, $0 \in \mathcal{D}$, and $f(\cdot)$ is continuous on $\mathcal{D}$. The following result is a direct corollary of Theorem 13.14.

Corollary 13.8. Consider the discrete-time nonlinear dynamical system (13.77). Assume there exists a continuous vector function $V : \mathcal{D} \to \mathcal{Q} \subseteq \mathbb{R}^q$ such that

$$V(f(x)) - V(x) \leq\leq w(V(x)), \quad x \in \mathcal{D}, \tag{13.78}$$

where $w : \mathcal{Q} \to \mathbb{R}^q$ is a continuous function, $w(\cdot) \in \mathcal{W}_{\rm d}$, and the equation

$$z(k+1) = w(z(k)), \quad z(k_0) = z_0, \quad k \in \mathcal{I}_{z_0}, \tag{13.79}$$

has a unique solution $z(k)$, $k \in \mathcal{I}_{z_0}$. If $\{k_0, \ldots, k_0+\tau\} \subseteq \mathcal{I}_{x_0} \cap \mathcal{I}_{z_0}$, then

$$V(x_0) \leq\leq z_0, \quad z_0 \in \mathcal{Q}, \tag{13.80}$$

implies

$$V(x(k)) \leq\leq z(k), \quad k \in \{k_0, \ldots, k_0+\tau\}. \tag{13.81}$$

Proof. For every $x_0 \in \mathcal{D}$, the solution $x(k)$, $k \in \mathcal{I}_{x_0}$, to (13.77) is well defined. With $\eta(k) \triangleq V(x(k))$, $k \in \mathcal{I}_{x_0}$, it follows from (13.78) that

$$\eta(k+1) - \eta(k) \leq\leq w(\eta(k)), \quad k \in \mathcal{I}_{x_0}. \tag{13.82}$$

Moreover, if $\{k_0, \ldots, k_0+\tau\} \subseteq \mathcal{I}_{x_0} \cap \mathcal{I}_{z_0}$, then it follows from Theorem 13.14 that $V(x_0) = \eta(k_0) \leq\leq z_0$ implies

$$V(x(k)) = \eta(k) \leq\leq z(k), \quad k \in \{k_0, \ldots, k_0+\tau\}, \tag{13.83}$$

which establishes the result. □

Note that if the solutions to (13.77) and (13.79) are globally defined for all $x_0 \in \mathcal{D}$ and $z_0 \in \mathcal{Q}$, then the result of Corollary 13.8 holds for every $k \geq k_0$. For the remainder of this section we assume that the solutions to the systems (13.77) and (13.79) are defined for all $k \geq k_0$. Furthermore, consider the comparison discrete-time nonlinear dynamical system given by

$$z(k+1) = w(z(k)), \quad z(k_0) = z_0, \quad k \geq k_0, \tag{13.84}$$

and the nonlinear dynamical system

$$x(k+1) = f(x(k)), \quad x(k_0) = x_0, \tag{13.85}$$

where $z_0 \in \mathcal{Q} \subseteq \overline{\mathbb{R}}_+^q$, $0 \in \mathcal{Q}$, $x_0 \in \mathcal{D}$, $w : \mathcal{Q} \to \mathbb{R}^q$ is continuous, $w(\cdot) \in \mathcal{W}_\mathrm{d}$, $w(0) = 0$, $f : \mathcal{D} \to \mathbb{D}$ is continuous on $\mathcal{D}$, and $f(0) = 0$. Note that since $w(\cdot) \in \mathcal{W}_\mathrm{d}$ and $w(0) = 0$, then for every $x \in \mathcal{D}$ and $z \in \mathcal{Q} \cap \overline{\mathbb{R}}_+^q$ it follows that $w(z) \geq\geq 0$, which implies that for every $z_0 \in \mathcal{Q} \cap \overline{\mathbb{R}}_+^q$ the solution $z(k)$, $k \geq k_0$, remains in $\overline{\mathbb{R}}_+^q$ (see Problem 13.17).

Theorem 13.15. Consider the discrete-time nonlinear dynamical system (13.77). Assume that there exist a continuous function $V : \mathcal{D} \to \mathcal{Q} \cap \overline{\mathbb{R}}_+^q$ and a positive vector $p \in \mathbb{R}_+^q$ such that $V(0) = 0$, the scalar function $v : \mathcal{D} \to \overline{\mathbb{R}}_+$ defined by $v(x) \triangleq p^\mathrm{T} V(x)$, $x \in \mathcal{D}$, is such that $v(x) > 0$, $x \neq 0$, and

$$V(f(x)) - V(x) \leq\leq w(V(x)), \quad x \in \mathcal{D}, \tag{13.86}$$

where $w : \mathcal{Q} \to \mathbb{R}^q$ is continuous, $w(\cdot) \in \mathcal{W}_\mathrm{d}$, and $w(0) = 0$. Then the following statements hold:

i) If the zero solution $z(k) \equiv 0$ to (13.84) is Lyapunov stable, then the zero solution $x(k) \equiv 0$ to (13.77) is Lyapunov stable.

ii) If the zero solution $z(k) \equiv 0$ to (13.84) is asymptotically stable, then the zero solution $x(k) \equiv 0$ to (13.77) is asymptotically stable.

iii) If $\mathcal{D} = \mathbb{R}^n$, $\mathcal{Q} = \mathbb{R}^q$, $v : \mathbb{R}^n \to \overline{\mathbb{R}}_+$ is positive definite and radially unbounded, and the zero solution $z(k) \equiv 0$ to (13.84) is globally asymptotically stable, then the zero solution $x(k) \equiv 0$ to (13.77) is globally asymptotically stable.

iv) If there exist constants $\nu \geq 1$, $\alpha > 0$, and $\beta > 0$ such that $v : \mathcal{D} \to \overline{\mathbb{R}}_+$ satisfies

$$\alpha \|x\|^\nu \leq v(x) \leq \beta \|x\|^\nu, \quad x \in \mathcal{D}, \tag{13.87}$$

and the zero solution $z(k) \equiv 0$ to (13.84) is geometrically stable, then the zero solution $x(k) \equiv 0$ to (13.77) is geometrically stable.

v) If $\mathcal{D} = \mathbb{R}^n$, $\mathcal{Q} = \mathbb{R}^q$, there exist constants $\nu \geq 1$, $\alpha > 0$, and $\beta > 0$ such that $v : \mathcal{D} \to \overline{\mathbb{R}}_+$ satisfies (13.87), and the zero solution $z(k) \equiv 0$ to (13.84) is globally geometrically stable, then the zero solution $x(k) \equiv 0$ to (13.77) is globally geometrically stable.

Proof. Assume there exist a continuous function $V : \mathcal{D} \to \mathcal{Q} \cap \overline{\mathbb{R}}_+^q$ and a positive vector $p \in \mathbb{R}_+^q$ such that $v(x) = p^\mathrm{T} V(x)$, $x \in \mathcal{D}$, is positive definite, that is, $v(0) = 0$, $v(x) > 0$, $x \neq 0$. Note that $v(x) = p^{\mathrm{T}}V(x) \leq \max_{i=1,\ldots,q}\{p_i\}\mathbf{e}^{\mathrm{T}}V(x)$, $x \in \mathcal{D}$, and hence, the function $\mathbf{e}^{\mathrm{T}}V(x)$, $x \in \mathcal{D}$, is also positive definite. Thus, there exist $r > 0$ and class $\mathcal{K}$ functions $\alpha, \beta : [0, r] \to \overline{\mathbb{R}}_+$ such that $\mathcal{B}_r(0) \subset \mathcal{D}$ and

$$\alpha(\|x\|) \leq \mathbf{e}^{\mathrm{T}}V(x) \leq \beta(\|x\|), \quad x \in \mathcal{B}_r(0). \tag{13.88}$$

i) Let $\varepsilon > 0$ and choose $0 < \hat{\varepsilon} < \min\{\varepsilon, r\}$. It follows from Lyapunov stability of the nonlinear comparison system (13.84) that there exists $\mu = \mu(\hat{\varepsilon}) = \mu(\varepsilon) > 0$ such that if $\|z_0\|_1 < \mu$ and $z_0 \in \overline{\mathbb{R}}_+^q$, then $\|z(k)\|_1 < \alpha(\hat{\varepsilon})$ and $z(k) \in \overline{\mathbb{R}}_+^q$, $k \geq k_0$, for every $x_0 \in \mathcal{D}$. Now, choose $z_0 = V(x_0) \geq\geq 0$, $x_0 \in \mathcal{D}$. Since $V(x)$, $x \in \mathcal{D}$, is continuous, the function $\mathbf{e}^{\mathrm{T}}V(x)$, $x \in \mathcal{D}$, is also continuous. Hence, for $\mu = \mu(\hat{\varepsilon}) > 0$ there exists $\delta = \delta(\mu(\hat{\varepsilon})) = \delta(\varepsilon) > 0$ such that $\delta < \hat{\varepsilon}$ and if $\|x_0\| < \delta$, then $\mathbf{e}^{\mathrm{T}}V(x_0) = \mathbf{e}^{\mathrm{T}}z_0 < \mu$, which implies that $\mathbf{e}^{\mathrm{T}}z(k) = \|z(k)\|_1 < \alpha(\hat{\varepsilon})$, $k \geq k_0$. Now, with $z_0 = V(x_0) \geq\geq 0$, $x_0 \in \mathcal{D}$, and the assumption that $w(\cdot) \in \mathcal{W}_{\mathrm{d}}$, it follows from (13.86) and Corollary 13.8 that $0 \leq\leq V(x(k)) \leq\leq z(k)$ on any compact interval $\{k_0, \ldots, k_0 + \tau\}$, and hence, $\mathbf{e}^{\mathrm{T}}z(k) = \|z(k)\|_1$, $k \in \{k_0, \ldots, k_0 + \tau\}$. Let $\tau > k_0$ be such that $z(k) \in \mathcal{B}_r(0)$, $k \in \{k_0, \ldots, k_0 + \tau\}$, for all $x_0 \in \mathcal{B}_r(0)$. Thus, using (13.88), if $\|x_0\| < \delta$, then

$$\alpha(\|x(k)\|) \leq \mathbf{e}^{\mathrm{T}}V(x(k)) \leq \mathbf{e}^{\mathrm{T}}z(k) < \alpha(\hat{\varepsilon}), \quad k \geq k_0, \tag{13.89}$$

which implies $\|x(k)\| < \hat{\varepsilon} < \varepsilon$, $k \geq k_0$. Now, suppose, *ad absurdum*, that for some $x_0 \in \mathcal{B}_\delta(0)$ there exists $\hat{k} > k_0 + \tau$ such that $\|x(\hat{k})\| = \hat{\varepsilon}$. Then, for $z_0 = V(x_0)$ and the compact interval $\{k_0, \ldots, \hat{k}\}$ it follows from (13.86) and Corollary 13.8 that $V(x(\hat{k})) \leq\leq z(\hat{k})$, which implies that $\alpha(\hat{\varepsilon}) = \alpha(\|x(\hat{k})\|) \leq \mathbf{e}^{\mathrm{T}}V(x(\hat{k})) \leq \mathbf{e}^{\mathrm{T}}z(\hat{k}) < \alpha(\hat{\varepsilon})$. This is a contradiction, and hence, for a given $\varepsilon > 0$ there exists $\delta = \delta(\varepsilon) > 0$ such that for all $x_0 \in \mathcal{B}_\delta(0)$, $\|x(k)\| < \varepsilon$, $k \geq k_0$, which implies Lyapunov stability of the zero solution $x(k) \equiv 0$ to (13.77).

ii) It follows from *i)* and the asymptotic stability of the nonlinear comparison system (13.84) that the zero solution to (13.77) is Lyapunov stable and there exists $\mu > 0$ such that if $\|z_0\|_1 < \mu$ and $z_0 \in \overline{\mathbb{R}}_+^q$, then $\lim_{k\to\infty} z(k) = 0$ for every $x_0 \in \mathcal{D}$. As in *i)*, choose $z_0 = V(x_0) \geq\geq 0$, $x_0 \in \mathcal{D}$. It follows from Lyapunov stability of the zero solution to (13.77) and the continuity of $V : \mathcal{D} \to \mathcal{Q} \cap \overline{\mathbb{R}}_+^q$ that there exists $\delta = \delta(\mu) > 0$ such that if $\|x_0\| < \delta$, then $\|x(k)\| < r$, $k \geq k_0$, and $\mathbf{e}^{\mathrm{T}}V(x_0) = \mathbf{e}^{\mathrm{T}}z_0 = \|z_0\|_1 < \mu$. Thus, by asymptotic stability of (13.84), for every $\varepsilon > 0$ there exists $K = K(\varepsilon) > k_0$ such that $\mathbf{e}^{\mathrm{T}}z(k) = \|z(k)\|_1 < \alpha(\varepsilon)$, $k \geq K$. Thus, it follows from (13.86) and Corollary 13.8 that $V(x(k)) \leq\leq z(k)$, $k \geq K$, and hence, by (13.88),

$$\alpha(\|x(k)\|) \leq \mathbf{e}^{\mathrm{T}}V(x(k)) \leq \mathbf{e}^{\mathrm{T}}z(k) < \alpha(\varepsilon), \quad k \geq K. \tag{13.90}$$

Now, suppose, *ad absurdum*, that for some $x_0 \in \mathcal{B}_\delta(0)$, $\lim_{k\to\infty} x(k) \neq 0$, that is, there exists a sequence $\{k_n\}_{n=1}^{\infty}$, with $k_n \to \infty$ as $n \to \infty$, such that $\|x(k_n)\| \geq \hat{\varepsilon}$, $n \in \overline{\mathbb{Z}}_+$, for some $0 < \hat{\varepsilon} < r$. Choose $\varepsilon = \hat{\varepsilon}$ and the interval $\{k, \ldots, k+\tau\}$ such that at least one $k_n \in \{k, \ldots, k+\tau\}$. Then it follows from (13.89) that $\alpha(\varepsilon) \leq \alpha(\|x(k_n)\|) < \alpha(\varepsilon)$, which is a contradiction. Hence, there exists $\delta > 0$ such that for all $x_0 \in \mathcal{B}_\delta(0)$, $\lim_{k\to\infty} x(k) = 0$ which, along with Lyapunov stability, implies asymptotic stability of the zero solution $x(k) \equiv 0$ to (13.77).

iii) Suppose $\mathcal{D} = \mathbb{R}^n$, $\mathcal{Q} = \mathbb{R}^q$, $v : \mathbb{R}^n \to \overline{\mathbb{R}}_+$ is a positive definite, radially unbounded function, and the nonlinear comparison system (13.84) is globally asymptotically stable. In this case, for $V : \mathbb{R}^n \to \overline{\mathbb{R}}_+^q$ the inequality (13.88) holds for all $x \in \mathbb{R}^n$ where the functions $\alpha, \beta : \overline{\mathbb{R}}_+ \to \overline{\mathbb{R}}_+$ are of class $\mathcal{K}_\infty$. Furthermore, Lyapunov stability of the zero solution $x(k) \equiv 0$ to (13.77) follows from *i*). Next, for every $x_0 \in \mathbb{R}^n$ and $z_0 = V(x_0) \in \overline{\mathbb{R}}_+^q$ the identical arguments as in *ii*) can be used to show that $\lim_{k\to\infty} x(k) = 0$, which proves global asymptotic stability of the zero solution $x(k) \equiv 0$ to (13.77).

iv) Suppose (13.87) holds. Since $p \in \mathbb{R}_+^q$, then

$$\hat{\alpha}\|x\|^\nu \leq \mathbf{e}^{\mathrm{T}} V(x) \leq \hat{\beta}\|x\|^\nu, \quad x \in \mathcal{D}, \tag{13.91}$$

where $\hat{\alpha} \triangleq \alpha / \max_{i=1,\ldots,q}\{p_i\}$ and $\hat{\beta} \triangleq \beta / \min_{i=1,\ldots,q}\{p_i\}$. It follows from the geometric stability of the nonlinear comparison system (13.84) that there exist positive constants γ, μ, and $\eta > 1$ such that if $\|z_0\|_1 < \mu$ and $z_0 \in \overline{\mathbb{R}}_+^q$, then $z(k) \in \overline{\mathbb{R}}_+^q$, $k \geq k_0$, and

$$\|z(k)\|_1 \leq \gamma \|z_0\|_1 \eta^{-(k-k_0)}, \quad k \geq k_0, \tag{13.92}$$

for all $x_0 \in \mathcal{D}$. Choose $z_0 = V(x_0) \geq\geq 0$, $x_0 \in \mathcal{D}$. By continuity of $V : \mathcal{D} \to \mathcal{Q} \cap \overline{\mathbb{R}}_+^q$, there exists $\delta = \delta(\mu) > 0$ such that for all $x_0 \in \mathcal{B}_\delta(0)$, $\mathbf{e}^{\mathrm{T}} V(x_0) = \mathbf{e}^{\mathrm{T}} z_0 = \|z_0\|_1 < \mu$. Furthermore, it follows from (13.91) and (13.92), and Corollary 13.8 that for all $x_0 \in \mathcal{B}_\delta(0)$,

$$\hat{\alpha}\|x(k)\|^\nu \leq \mathbf{e}^{\mathrm{T}} V(x(k)) \leq \mathbf{e}^{\mathrm{T}} z(k) \leq \gamma \|z_0\|_1 \eta^{-(k-k_0)} \leq \gamma\hat{\beta}\|x_0\|^\nu \eta^{-(k-k_0)}, \quad k \geq k_0. \tag{13.93}$$

This in turn implies that for every $x_0 \in \mathcal{B}_\delta(0)$,

$$\|x(k)\| \leq \left(\frac{\gamma\hat{\beta}}{\hat{\alpha}}\right)^{\frac{1}{\nu}} \|x_0\| \eta^{-\frac{k-k_0}{\nu}}, \quad k \geq k_0, \tag{13.94}$$

which establishes geometric stability of the zero solution $x(k) \equiv 0$ to (13.77).

v) The proof is identical to the proof of *iv*). □

If $V : \mathcal{D} \to \mathcal{Q} \cap \overline{\mathbb{R}}_+^q$ satisfies the conditions of Theorem 13.15 we say that $V(x)$, $x \in \mathcal{D}$ is a *vector Lyapunov function.* Note that for stability analysis each component of a vector Lyapunov function need not be positive definite, nor does it need to have a negative definite time difference along the trajectories of (13.85). This provides more flexibility in searching for a vector Lyapunov function as compared to a scalar Lyapunov function for addressing the stability of discrete-time nonlinear dynamical systems.

Finally, we provide a time-varying extension of Theorem 13.15. In particular, we consider the discrete-time nonlinear time-varying dynamical system

$$x(k+1) = f(k, x(k)), \quad x(k_0) = x_0, \quad k \geq k_0, \tag{13.95}$$

where $x(k) \in \mathcal{D} \subseteq \mathbb{R}^n$, $0 \in \mathcal{D}$, $f : \{k_0, \ldots, k_1\} \times \mathcal{D} \to \mathbb{D}$ is such that $f(\cdot, \cdot)$ is continuous, and, for every $k \in \{k_0, \ldots, k_1\}$, $f(k, 0) = 0$.

Theorem 13.16. Consider the discrete-time nonlinear time-varying dynamical system (13.95). Assume that there exist a continuous vector function $V : \overline{\mathbb{Z}}_+ \times \mathcal{D} \to \mathcal{Q} \cap \overline{\mathbb{R}}_+^q$, a positive vector $p \in \mathbb{R}_+^q$, and class $\mathcal{K}$ functions $\alpha, \beta : [0, r] \to \overline{\mathbb{R}}_+$ such that $V(k, 0) = 0$, $k \in \overline{\mathbb{Z}}_+$, the scalar function $v : \overline{\mathbb{Z}}_+ \times \mathcal{D} \to \overline{\mathbb{R}}_+$ defined by $v(k, x) \triangleq p^{\mathrm{T}} V(k, x)$, $(k, x) \in \overline{\mathbb{Z}}_+ \times \mathcal{D}$, is such that

$$\alpha(\|x\|) \leq v(k, x) \leq \beta(\|x\|), \quad (k, x) \in \overline{\mathbb{Z}}_+ \times \mathcal{B}_r(0), \quad \mathcal{B}_r(0) \subseteq \mathcal{D}, \tag{13.96}$$

and

$$V(k+1, f(x)) - V(k, x) \leq\leq w(k, V(k, x)), \qquad x \in \mathcal{D}, \quad k \in \overline{\mathbb{Z}}_+, \tag{13.97}$$

where $w : \overline{\mathbb{Z}}_+ \times \mathcal{Q} \to \mathbb{R}^q$ is continuous, $w(k, \cdot) \in \mathcal{W}_{\mathrm{d}}$, and $w(k, 0) = 0$, $k \in \overline{\mathbb{Z}}_+$. Then the stability properties of the zero solution $z(k) \equiv 0$ to

$$z(k+1) = w(k, z(k)), \quad z(k_0) = z_0, \quad k \geq k_0, \tag{13.98}$$

where $z_0 \in \mathcal{Q} \cap \overline{\mathbb{R}}_+^q$, imply the corresponding stability properties of the zero solution $x(k) \equiv 0$ to (13.95). That is, if the zero solution $z(k) \equiv 0$ to (13.98) is uniformly Lyapunov (respectively, uniformly asymptotically) stable, then the zero solution $x(k) \equiv 0$ to (13.95) is uniformly Lyapunov (respectively, uniformly asymptotically) stable. If, in addition, $\mathcal{D} = \mathbb{R}^n$, $\mathcal{Q} = \mathbb{R}^q$, and $\alpha(\cdot), \beta(\cdot)$ are class $\mathcal{K}_\infty$ functions, then global uniform asymptotic stability of the zero solution $z(k) \equiv 0$ to (13.98) implies global uniform asymptotic stability of the zero solution $x(k) \equiv 0$ to (13.95). Moreover, if there exist constants $\nu \geq 1$, $\alpha > 0$, and $\beta > 0$ such that $v : \overline{\mathbb{Z}}_+ \times \mathcal{D} \to \overline{\mathbb{R}}_+$ satisfies

$$\alpha \|x\|^\nu \leq v(k, x) \leq \beta \|x\|^\nu, \quad (k, x) \in \overline{\mathbb{Z}}_+ \times \mathcal{D}, \tag{13.99}$$

then geometric stability of the zero solution $z(k) \equiv 0$ to (13.98) implies geometric stability of the zero solution $x(k) \equiv 0$ to (13.95). Finally, if $\mathcal{D} = \mathbb{R}^n$, $\mathcal{Q} = \mathbb{R}^q$, then there exist constants $\nu \geq 1$, $\alpha > 0$, and $\beta > 0$ such that $v : \overline{\mathbb{Z}}_+ \times \mathbb{R}^n \to \overline{\mathbb{R}}_+$ satisfies (13.99), then global geometric stability of the zero solution $z(k) \equiv 0$ to (13.98) implies global geometric stability of the zero solution $x(k) \equiv 0$ to (13.95).

Proof. The proof is similar to the proof of Theorem 13.15 and is left as an exercise for the reader. □

13.9 Dissipative and Geometrically Dissipative Discrete-Time Dynamical Systems

In this section, we consider the discrete-time nonlinear dynamical systems $\mathcal{G}$ of the form

$$x(k+1) = F(x(k), u(k)), \quad x(k_0) = x_0, \quad k \geq k_0, \tag{13.100}$$

$$y(k) = H(x(k), u(k)), \tag{13.101}$$

where $x \in \mathcal{D} \subseteq \mathbb{R}^n$, $\mathcal{D}$ is open with $0 \in \mathcal{D}$, $u(k) \in U \subseteq \mathbb{R}^m$, $y(k) \in Y \subseteq \mathbb{R}^l$, $F : \mathcal{D} \times U \to \mathbb{R}^n$, and $H : \mathcal{D} \times U \to Y$. We assume that $F(\cdot,\cdot)$ and $H(\cdot,\cdot)$ are continuous mappings and $F(\cdot,\cdot)$ has at least one equilibrium so that, without loss of generality, $F(0,0) = 0$ and $H(0,0) = 0$. Note that since all input-output pairs $u(\cdot) \in \mathcal{U}$, $y(\cdot) \in \mathcal{Y}$, of the discrete-time dynamical system $\mathcal{G}$ are defined on $\overline{\mathbb{Z}}_+$, the *supply rate* $r : U \times Y \to \mathbb{R}$ satisfying $r(0,0) = 0$ is locally summable for all input-output pairs satisfying (13.100) and (13.101), that is, $\sum_{k=k_1}^{k_2} |r(u(k), y(k))| < \infty$, $k_1, k_2 \in \overline{\mathbb{Z}}_+$.

Definition 13.11. A nonlinear discrete-time dynamical system $\mathcal{G}$ of the form (13.100) and (13.101) is *dissipative with respect to the supply rate* $r(u,y)$ if the *dissipation inequality*

$$0 \leq \sum_{i=k_0}^{k-1} r(u(i), y(i)) \tag{13.102}$$

is satisfied for all $k-1 \geq k_0$ and all $u(\cdot) \in \mathcal{U}$ with $x(k_0) = 0$ along the trajectories of $\mathcal{G}$. A discrete-time dynamical system $\mathcal{G}$ of the form (13.100) and (13.101) is *geometrically dissipative with respect to the supply rate* $r(u,y)$ if there exists a constant $\rho > 1$ such that the *geometric dissipation inequality*

$$0 \leq \sum_{i=k_0}^{k-1} \rho^{i+1} r(u(i), y(i)) \tag{13.103}$$

is satisfied for all $k-1 \geq k_0$ and all $u(\cdot) \in \mathcal{U}$ with $x(k_0) = 0$ along the trajectories of $\mathcal{G}$. A dynamical system $\mathcal{G}$ of the form (13.100) and (13.101) is *lossless with respect to the supply rate* $r(u,y)$ if $\mathcal{G}$ is dissipative with respect to the supply rate $r(u,y)$ and the dissipation inequality (13.102) is satisfied as an equality for all $k-1 \geq k_0$ and all $u(\cdot) \in \mathcal{U}$ with $x(k_0) = x(k) = 0$

along the trajectories of $\mathcal{G}$.

Next, define the *available storage* $V_{\rm a}(x_0)$ of the discrete-time nonlinear dynamical system $\mathcal{G}$ by

$$V_{\rm a}(x_0) \triangleq -\inf_{u(\cdot),\, K\geq 0} \sum_{k=0}^{K-1} r(u(k),y(k)) = \sup_{u(\cdot),\, K\geq 0} \left[-\sum_{k=0}^{K-1} r(u(k),y(k))\right], \quad (13.104)$$

where $x(k)$, $k \geq k_0$, is the solution to (13.100) with admissible input $u(\cdot) \in \mathcal{U}$. Note that $V_{\rm a}(x) \geq 0$ for all $x \in \mathcal{D}$ since $V_{\rm a}(x)$ is the supremum over a set of numbers containing the zero element ($K = 1$). Similarly, define the *available geometric storage* $V_{\rm a}(x_0)$ of the nonlinear dynamical system $\mathcal{G}$ by

$$V_{\rm a}(x_0) \triangleq -\inf_{u(\cdot),\, K\geq 0} \sum_{k=0}^{K-1} \rho^{k+1} r(u(k),y(k)), \quad (13.105)$$

where $x(k)$, $k \in \overline{\mathbb{Z}}_+$, is the solution to (13.100) with $x(0) = x_0$ and admissible input $u(\cdot) \in \mathcal{U}$. Note that if we define the available geometric storage as the time-varying function

$$\hat{V}_{\rm a}(x_0, k_0) = -\inf_{u(\cdot),\, K\geq k_0} \sum_{k=k_0}^{K-1} \rho^{k+1} r(u(k),y(k)), \quad (13.106)$$

where $x(k)$, $k \geq k_0$, is the solution to (13.100) with $x(k_0) = x_0$ and admissible input $u(\cdot) \in \mathcal{U}$, it follows that, since $\mathcal{G}$ is time invariant,

$$\hat{V}_{\rm a}(x_0, k_0) = -\rho^{k_0} \inf_{u(\cdot),\, K\geq 0} \sum_{k=0}^{K-1} \rho^{k+1} r(u(k),y(k)) = \rho^{k_0} V_{\rm a}(x_0). \quad (13.107)$$

Hence, an alternative expression for the available geometric storage function $V_{\rm a}(x_0)$ is given by

$$V_{\rm a}(x_0) = -\rho^{-k_0} \inf_{u(\cdot),\, K\geq k_0} \sum_{k=k_0}^{K-1} \rho^{k+1} r(u(k),y(k)). \quad (13.108)$$

Next, we show that the available storage is finite and zero at the origin if and only if $\mathcal{G}$ is dissipative. For this result we require the following three definitions.

Definition 13.12. A nonlinear dynamical system $\mathcal{G}$ is *completely reachable* if for all $x_0 \in \mathcal{D} \subseteq \mathbb{R}^n$ there exist a $k_{\rm i} < k_0$ and a square summable input $u(k)$ defined on $[k_{\rm i}, k_0]$ such that the state $x(k)$, $k \geq k_{\rm i}$, can be driven from $x(k_{\rm i}) = 0$ to $x(k_0) = x_0$. $\mathcal{G}$ is *completely null controllable* if for all $x_0 \in \mathcal{D} \subseteq \mathbb{R}^n$ there exist a finite time $k_{\rm f} > k_0$ and a square summable input $u(k)$ defined on $[k_0, k_{\rm f}]$ such that $x(k)$, $k \geq k_0$, can be driven from $x(k_0) = x_0$ to $x(k_{\rm f}) = 0$.

Definition 13.13. Consider the discrete-time nonlinear dynamical system $\mathcal{G}$ given by (13.100) and (13.101). A continuous nonnegative-definite function $V_{\rm s} : \mathcal{D} \to \mathbb{R}$ satisfying $V_{\rm s}(0) = 0$ and

$$V_{\rm s}(x(k)) \leq V_{\rm s}(x(k_0)) + \sum_{i=k_0}^{k-1} r(u(i), y(i)), \qquad k-1 \geq k_0, \tag{13.109}$$

for all $k_0, k \in \overline{\mathbb{Z}}_+$, where $x(k)$, $k \in \overline{\mathbb{Z}}_+$, is the solution of (13.100) with $u(\cdot) \in \mathcal{U}$, is called a *storage function* for $\mathcal{G}$.

Definition 13.14. Consider the discrete-time nonlinear dynamical system $\mathcal{G}$ given by (13.100) and (13.101). A continuous nonnegative-definite function $V_{\rm s} : \mathcal{D} \to \mathbb{R}$ satisfying $V_{\rm s}(0) = 0$ and

$$\rho^k V_{\rm s}(x(k)) \leq \rho^{k_0} V_{\rm s}(x(k_0)) + \sum_{i=k_0}^{k-1} \rho^{i+1} r(u(i), y(i)), \qquad k-1 \geq k_0, \tag{13.110}$$

for all $k_0, k \in \overline{\mathbb{Z}}_+$, where $x(k)$, $k \geq k_0$, is the solution of (13.100) with $u(\cdot) \in \mathcal{U}$, is called an *geometric storage function* for $\mathcal{G}$.

Theorem 13.17. Consider the discrete-time nonlinear dynamical system $\mathcal{G}$ given by (13.100) and (13.101), and assume that $\mathcal{G}$ is completely reachable. Then $\mathcal{G}$ is dissipative (respectively, geometrically dissipative) with respect to the supply rate $r(u, y)$ if and only if the available system storage $V_{\rm a}(x_0)$ given by (13.104) (respectively, the available geometric storage $V_{\rm a}(x)$ given by (13.105)) is finite for all $x_0 \in \mathcal{D}$ and $V_{\rm a}(0) = 0$. Moreover, if $V_{\rm a}(0) = 0$ and $V_{\rm a}(x_0)$ is finite for all $x_0 \in \mathcal{D}$, then $V_{\rm a}(x)$, $x \in \mathcal{D}$, is a storage function (respectively, geometric storage function) for $\mathcal{G}$. Finally, all storage functions (respectively, geometric storage functions) $V_{\rm s}(x)$, $x \in \mathcal{D}$, for $\mathcal{G}$ satisfy

$$0 \leq V_{\rm a}(x) \leq V_{\rm s}(x), \quad x \in \mathcal{D}. \tag{13.111}$$

Proof. Suppose $V_{\rm a}(x_0)$, $x_0 \in \mathcal{D}$, is finite. Now, it follows from (13.104) (with $K = 1$) that $V_{\rm a}(x_0) \geq 0$, $x_0 \in \mathcal{D}$. Next, let $x(k)$, $k \geq k_0$, satisfy (13.100) with admissible input $u(k)$, $k \in [k_0, K-1]$. Since $-V_{\rm a}(x_0)$, $x_0 \in \mathcal{D}$, is given by the infimum over all admissible inputs $u(\cdot) \in \mathcal{U}$ in (13.104), it follows that for all admissible inputs $u(\cdot) \in \mathcal{U}$ and $K-1 \geq k_0$,

$$\begin{aligned}
-V_{\mathrm{a}}(x(k_0)) &\leq \sum_{k=k_0}^{K-1} r(u(k), y(k)) \\
&= \sum_{k=k_0}^{k_{\mathrm{f}}-1} r(u(k), y(k)) + \sum_{k=k_{\mathrm{f}}}^{K-1} r(u(k), y(k)),
\end{aligned}$$

which implies

$$-V_{\mathrm{a}}(x(k_0)) - \sum_{k=k_0}^{k_{\mathrm{f}}-1} r(u(k), y(k)) \leq \sum_{k=k_{\mathrm{f}}}^{K-1} r(u(k), y(k)).$$

Hence,

$$\begin{aligned}
V_{\mathrm{a}}(x(k_0)) + \sum_{k=k_0}^{k_{\mathrm{f}}-1} r(u(k), y(k)) &\geq -\inf_{u(\cdot),\, K \geq k_{\mathrm{f}}} \sum_{k=k_{\mathrm{f}}}^{K-1} r(u(k), y(k)) \\
&= V_{\mathrm{a}}(x(k_{\mathrm{f}})) \\
&\geq 0, \qquad (13.112)
\end{aligned}$$

which implies that

$$\sum_{k=k_0}^{k_{\mathrm{f}}-1} r(u(k), y(k)) \geq -V_{\mathrm{a}}(x(k_0)). \qquad (13.113)$$

Hence, since by assumption $V_{\mathrm{a}}(0) = 0$, $\mathcal{G}$ is dissipative with respect to the supply rate $r(u, y)$. Furthermore, $V_{\mathrm{a}}(x)$, $x \in \mathcal{D}$, is a storage function for $\mathcal{G}$.

Conversely, suppose that $\mathcal{G}$ is dissipative with respect to the supply rate $r(u, y)$. Since $\mathcal{G}$ is completely reachable it follows that for every $x_0 \in \mathbb{R}^n$ such that $x(k_0) = x_0$, there exist $\hat{k} \leq k < k_0$ and an admissible input $u(\cdot) \in \mathcal{U}$ defined on $[\hat{k}, k_0]$ such that $x(\hat{k}) = 0$ and $x(k_0) = x_0$. Furthermore, since $\mathcal{G}$ is dissipative with respect to the supply rate $r(u, y)$ and $x(\hat{k}) = 0$ it follows that

$$\sum_{k=\hat{k}}^{K-1} r(u(k), y(k)) \geq 0, \qquad K - 1 \geq \hat{k}, \qquad (13.114)$$

or, equivalently,

$$\sum_{k=k_0}^{K-1} r(u(k), y(k)) \geq -\sum_{k=\hat{k}}^{k_0-1} r(u(k), y(k), \quad K - 1 \geq k_0, \qquad (13.115)$$

which implies that there exists a function $W : \mathbb{R}^n \to \mathbb{R}$ such that

$$\sum_{k=k_0}^{K-1} r(u(k), y(k)) \geq W(x_0) > -\infty, \qquad K-1 \geq k_0. \tag{13.116}$$

Now, it follows from (13.116) that for all $x \in \mathbb{R}^n$,

$$\begin{aligned} V_a(x) &= -\inf_{u(\cdot),\, K\geq 0} \sum_{k=0}^{K-1} r(u(k), y(k)) \\ &\leq -W(x), \end{aligned} \tag{13.117}$$

and hence, the available storage $V_a(x) < \infty$, $x \in \mathcal{D}$. Furthermore, with $x(k_0) = 0$ and for all admissible $u(k)$, $k \geq k_0$,

$$\sum_{k=k_0}^{K-1} r(u(k), y(k)) \geq 0, \qquad K-1 \geq k_0, \tag{13.118}$$

which implies that

$$\sup_{u(\cdot),\, K\geq k_0} \left[-\sum_{k=k_0}^{K-1} r(u(k), y(k)) \right] \leq 0, \tag{13.119}$$

or, equivalently, $V_a(x(k_0)) = V_a(0) \leq 0$. However, since $V_a(x) \geq 0$, $x \in \mathcal{D}$, it follows that $V_a(0) = 0$.

Moreover, if $V_a(x)$ is finite for all $x \in \mathcal{D}$, it follows from (13.112) that $V_a(x)$, $x \in \mathbb{R}^n$, is a storage function for $\mathcal{G}$. Next, if $V_s(x)$, $x \in \mathcal{D}$, is a storage function then it follows that, for all $K-1 \geq 0$ and $x_0 \in \mathcal{D}$,

$$\begin{aligned} V_s(x_0) &\geq V_s(x(K)) - \sum_{k=0}^{K-1} r(u(k), y(k)) \\ &\geq -\sum_{k=0}^{K-1} r(u(k), y(k)), \end{aligned}$$

which implies

$$\begin{aligned} V_s(x_0) &\geq -\inf_{u(\cdot),\, K\geq 0} \sum_{k=0}^{K-1} r(u(k), y(k)) \\ &= V_a(x_0). \end{aligned}$$

Finally, the proof for the geometrically dissipative case follows an identical construction and, hence, is omitted. □

For the remainder of this chapter, we assume that all storage functions of $\mathcal{G}$ are continuous on $\mathcal{D}$. Now, the following corollary is immediate from Theorem 13.17 and shows that a system $\mathcal{G}$ is dissipative (respectively, geometrically dissipative) with respect to the supply rate $r(\cdot,\cdot)$ if and only if there exists a continuous storage function $V_{\rm s}(\cdot)$ satisfying (13.109) (respectively, a continuous geometric storage function $V_{\rm s}(\cdot)$ satisfying (13.110)).

Corollary 13.9. Consider the discrete-time nonlinear dynamical system $\mathcal{G}$ given by (13.100) and (13.101) and assume that $\mathcal{G}$ is completely reachable. Then $\mathcal{G}$ is dissipative (respectively, geometrically dissipative) with respect to supply rate $r(\cdot,\cdot)$ if and only if there exists a continuous storage function (respectively, geometric storage function) $V_{\rm s}(x)$, $x \in \mathcal{D}$, satisfying (13.109) (respectively, (13.110)).

Proof. The result is immediate from Theorem 13.17 with $V_{\rm s}(x) = V_{\rm a}(x)$. □

The following theorem provides conditions for guaranteeing that all storage functions (respectively, geometric storage functions) of a given discrete-time dissipative (respectively, geometrically dissipative) nonlinear dynamical system are positive definite. For this result we require the following definition.

Definition 13.15. A dynamical system $\mathcal{G}$ is *zero state observable* if $u(k) \equiv 0$ and $y(k) \equiv 0$ implies $x(k) \equiv 0$.

Theorem 13.18. Consider the discrete-time nonlinear dynamical system $\mathcal{G}$ given by (13.100) and (13.101), and assume that $\mathcal{G}$ is completely reachable and zero-state observable. Furthermore, assume that $\mathcal{G}$ is dissipative (respectively, geometrically dissipative) with respect to supply rate $r(u,y)$ and there exists a function $\kappa : Y \to U$ such that $\kappa(0) = 0$ and $r(\kappa(y), y) < 0$, $y \neq 0$. Then all the storage functions (respectively, geometric storage functions) $V_{\rm s}(x)$, $x \in \mathcal{D}$, for $\mathcal{G}$ are positive definite, that is, $V_{\rm s}(0) = 0$ and $V_{\rm s}(x) > 0$, $x \in \mathcal{D}$, $x \neq 0$.

Proof. It follows from Theorem 13.17 that the available storage $V_{\rm a}(x)$, $x \in \mathcal{D}$, is a storage function for $\mathcal{G}$. Next, suppose there exists $x \in \mathcal{D}$ such that $V_{\rm a}(x) = 0$, which implies that $r(u(k), y(k)) = 0$, $k \in \overline{\mathbb{Z}}_+$, for all admissible inputs $u(\cdot) \in \mathcal{U}$. Since there exists a function $\kappa : Y \to U$ such that $r(\kappa(y), y) < 0$, $y \neq 0$, it follows that $y(k) = 0$, $k \in \overline{\mathbb{Z}}_+$. Now, since $\mathcal{G}$ is zero-state observable it follows that $x = 0$, and hence, $V_{\rm a}(x) = 0$ if and only if $x = 0$. The result now follows from (13.111). Finally, the proof for the geometrically dissipative case is identical and, hence, is omitted. □

An equivalent statement for dissipativity of $\mathcal{G}$ with respect to the supply rate $r(u, y)$ is

$$\Delta V_{\rm s}(x(k)) \le r(u(k), y(k)), \qquad k \in \overline{\mathbb{Z}}_+, \tag{13.120}$$

where $\Delta V_{\rm s}(x(k)) \triangleq V(x(k+1)) - V(x(k))$. Alternatively, an equivalent statement for geometric dissipativity of $\mathcal{G}$ with respect to the supply rate $r(u, y)$ is

$$\rho V_{\rm s}(x(k+1)) - V_{\rm s}(x(k)) \le \rho r(u(k), y(k)), \qquad k \in \overline{\mathbb{Z}}_+. \tag{13.121}$$

Furthermore, a system $\mathcal{G}$ with storage function $V_{\rm s}(\cdot)$ is *strictly dissipative* with respect to the supply rate $r(u, y)$ if and only if

$$V_{\rm s}(x(k)) < V_{\rm s}(x(k_0)) + \sum_{i=k_0}^{k-1} r(u(i), y(i)), \qquad k-1 \ge k_0. \tag{13.122}$$

Note that geometric dissipativity implies strict dissipativity, however, the converse is not necessarily true.

Next, we introduce the concept of a required supply of a discrete-time nonlinear dynamical system. Specifically, define the *required supply* $V_{\rm r}(x_0)$ of the discrete-time nonlinear dynamical system $\mathcal{G}$ by

$$V_{\rm r}(x_0) = \inf_{u(\cdot),\, K \ge 1} \sum_{k=-K}^{-1} r(u(k), y(k)), \tag{13.123}$$

where $x(k)$, $k \ge -K$, is the solution to (13.100) with $x(-K) = 0$ and $x(0) = x_0$. It follows from (13.123) that the required supply of a nonlinear dynamical system is the minimum amount of generalized energy that has to be delivered to the dynamical system in order to transfer it from an initial state $x(-K) = 0$ to a given state $x(0) = x_0$. Similarly, define the *required geometric supply* of the nonlinear dynamical system $\mathcal{G}$ by

$$V_{\rm r}(x_0) = \inf_{u(\cdot),\, K \ge 1} \sum_{k=-K}^{-1} \rho^{k+1} r(u(k), y(k)), \tag{13.124}$$

where $x(k)$, $k \ge -K$, is the solution to (13.100) with $x(-K) = 0$ and $x(0) = x_0$. Note that since, with $x(0) = 0$, the infimum in (13.123) is zero it follows that $V_{\rm r}(0) = 0$.

Next, using the notion of a required supply, we show that all storage functions are bounded from above by the required supply and bounded from below by the available storage, and hence, a dissipative discrete-time dynamical system can deliver to its surroundings only a fraction of its stored generalized energy and can store only a fraction of the generalized work done to it.

Theorem 13.19. Consider the discrete-time nonlinear dynamical system $\mathcal{G}$ given by (13.100) and (13.101), and assume that $\mathcal{G}$ is completely reachable. Then $\mathcal{G}$ is dissipative (respectively, geometrically dissipative) with respect to the supply rate $r(u,y)$ if and only if $0 \le V_{\rm r}(x) < \infty$, $x \in \mathcal{D}$. Moreover, if $V_{\rm r}(x)$ is finite and nonnegative for all $x \in \mathcal{D}$, then $V_{\rm r}(x)$, $x \in \mathcal{D}$, is a storage function (respectively, geometric storage function) for $\mathcal{G}$. Finally, all storage functions (respectively, geometric storage functions) $V_{\rm s}(x)$, $x \in \mathcal{D}$, for $\mathcal{G}$ satisfy

$$0 \le V_{\rm a}(x) \le V_{\rm s}(x) \le V_{\rm r}(x) < \infty, \qquad x \in \mathcal{D}. \tag{13.125}$$

Proof. Suppose $0 \le V_{\rm r}(x) < \infty$, $x \in \mathcal{D}$. Next, let $x(k)$, $k \in \overline{\mathbb{Z}}_+$, satisfy (13.100) and (13.101) with admissible inputs $u(\cdot) \in \mathcal{U}$ and $x(0) = x_0$. Since $V_{\rm r}(x)$, $x \in \mathcal{D}$, is given by the infimum over all admissible inputs $u(\cdot) \in \mathcal{U}$ and $K > 0$ in (13.123), it follows that for all admissible inputs $u(k)$ and $k \in [-K, 0]$,

$$V_{\rm r}(x_0) \le \sum_{i=-K}^{-1} r(u(i),y(i)) = \sum_{i=-K}^{k-1} r(u(i),y(i)) + \sum_{i=k}^{-1} r(u(i),y(i)),$$

and hence,

$$\begin{aligned} V_{\rm r}(x_0) &\le \inf_{u(\cdot),\,K\ge 1} \left[\sum_{i=-K}^{k-1} r(u(i),y(i))\right] + \sum_{i=k}^{-1} r(u(i),y(i)) \\ &= V_{\rm r}(x(k)) + \sum_{i=k}^{-1} r(u(i),y(i)), \end{aligned} \tag{13.126}$$

which shows that $V_{\rm r}(x)$, $x \in \mathcal{D}$, is a storage function for $\mathcal{G}$, and hence, $\mathcal{G}$ is dissipative.

Conversely, suppose $\mathcal{G}$ is dissipative with respect to the supply rate $r(u,y)$ and let $x_0 \in \mathcal{D}$. Since $\mathcal{G}$ is completely reachable it follows that there exist $K > 0$ and $u(k)$, $k \in [-K, 0]$, such that $x(-K) = 0$ and $x(0) = x_0$. Hence, since $\mathcal{G}$ is dissipative with respect to the supply rate $r(u,y)$ it follows that, for all $K \ge 1$,

$$0 \le \sum_{k=-K}^{-1} r(u(k),y(k)), \tag{13.127}$$

and hence,

$$0 \le \inf_{u(\cdot),\,K\ge 1} \left[\sum_{k=-K}^{-1} r(u(k),y(k))\right], \tag{13.128}$$

which implies that

$$0 \le V_{\rm r}(x_0) < \infty, \qquad x_0 \in \mathcal{D}. \tag{13.129}$$

Next, if $V_s(\cdot)$ is a storage function for $\mathcal{G}$, then it follows from Theorem 13.17 that

$$0 \le V_a(x) \le V_s(x), \qquad x \in \mathcal{D}. \tag{13.130}$$

Furthermore, for all $K \ge 1$ such that $x(-K) = 0$ it follows that

$$V_s(x_0) \le V_s(0) + \sum_{k=-K}^{-1} r(u(k), y(k)), \tag{13.131}$$

and hence,

$$V_s(x_0) \le \inf_{u(\cdot),\, K\ge 1} \left[\sum_{k=-K}^{-1} r(u(k), y(k)) \right] = V_r(x_0) < \infty, \tag{13.132}$$

which implies (13.125).

Finally, the proof for the geometrically dissipative case follows a similar construction and, hence, is omitted. □

In the light of Theorems 13.17 and 13.19 the following result on lossless discrete-time dynamical systems is immediate.

Theorem 13.20. Consider the discrete-time nonlinear dynamical system $\mathcal{G}$ given by (13.100) and (13.101) and assume $\mathcal{G}$ is completely reachable to and from the origin. Then $\mathcal{G}$ is lossless with respect to the supply rate $r(u, y)$ if and only if there exists a continuous storage function $V_s(x)$, $x \in \mathcal{D}$, satisfying (13.109) as an equality. Furthermore, if $\mathcal{G}$ is lossless with respect to the supply rate $r(u, y)$, then $V_a(x) = V_r(x)$, and hence, the storage function $V_s(x)$, $x \in \mathcal{D}$, is unique and is given by

$$V_s(x_0) = -\sum_{k=0}^{K_+-1} r(u(k), y(k)) = \sum_{k=-K_-}^{-1} r(u(k), y(k)), \tag{13.133}$$

where $x(k)$, $k \in \overline{\mathbb{Z}}_+$, is the solution to (13.100) with admissible $u(\cdot) \in \mathcal{U}$ and $x(0) = x_0$, $x_0 \in \mathcal{D}$, for every $K_-, K_+ > 0$ such that $x(-K_-) = 0$ and $x(K_+) = 0$.

Proof. Suppose $\mathcal{G}$ is lossless with respect to the supply rate $r(u, y)$. Since $\mathcal{G}$ is completely reachable to and from the origin it follows that, for every $x_0 \in \mathcal{D}$, there exist $K_-, K_+ > 0$ and $u(k) \in U$, $k \in [-K_-, K_+]$, such that $x(-K_-) = 0$, $x(K_+) = 0$, and $x(0) = x_0$. Now, it follows that

$$0 = \sum_{k=-K_-}^{K_+-1} r(u(k), y(k))$$

$$
\begin{aligned}
&= \sum_{k=-K_-}^{-1} r(u(k), y(k)) + \sum_{k=0}^{K_+-1} r(u(k), y(k)) \\
&\geq \inf_{u(\cdot),\, K\geq 1} \sum_{k=-K}^{-1} r(u(k), y(k)) + \inf_{u(\cdot),\, K>1} \sum_{k=0}^{K-1} r(u(k), y(k)) \\
&= V_{\mathrm{r}}(x_0) - V_{\mathrm{a}}(x_0), \qquad (13.134)
\end{aligned}
$$

which implies that $V_{\mathrm{r}}(x_0) \leq V_{\mathrm{a}}(x_0)$, $x_0 \in \mathcal{D}$. However, since by definition $\mathcal{G}$ is dissipative with respect to the supply rate $r(u, y)$ it follows from Theorem 13.19 that $V_{\mathrm{a}}(x_0) \leq V_{\mathrm{r}}(x_0)$, $x_0 \in \mathcal{D}$, and hence, every storage function $V_{\mathrm{s}}(x_0)$, $x_0 \in \mathcal{D}$, satisfies $V_{\mathrm{a}}(x_0) = V_{\mathrm{s}}(x_0) = V_{\mathrm{r}}(x_0)$. Furthermore, it follows that the inequality in (13.134) is indeed an equality, which implies (13.133).

Next, let $k_0, k, K-1 \geq 0$ be such that $k_0 < k < K$, $x(K) = 0$. Hence, it follows from (13.133) that

$$
\begin{aligned}
0 &= V_{\mathrm{s}}(x(k_0)) + \sum_{\hat{k}=k_0}^{K-1} r(u(\hat{k}), y(\hat{k})) \\
&= V_{\mathrm{s}}(x(k_0)) + \sum_{\hat{k}=k_0}^{k-1} r(u(\hat{k}), y(\hat{k})) + \sum_{\hat{k}=k}^{K-1} r(u(\hat{k}), y(\hat{k})) \\
&= V_{\mathrm{s}}(x(k_0)) + \sum_{\hat{k}=k_0}^{K-1} r(u(\hat{k}), y(\hat{k})) - V_{\mathrm{s}}(x(k)),
\end{aligned}
$$

which implies that (13.109) is satisfied as an equality.

Conversely, if there exists a storage function $V_{\mathrm{s}}(x)$, $x \in \mathcal{D}$, satisfying (13.109) as an equality, it follows from Corollary 13.9 that $\mathcal{G}$ is dissipative with respect to the supply rate $r(u, y)$. Furthermore, for every $u(k) \in U$, $k \in \overline{\mathbb{Z}}_+$, and $x(k_0) = x(k) = 0$, it follows from (13.109) (with an equality) that

$$
\sum_{\hat{k}=k_0}^{k-1} r(u(\hat{k}), y(\hat{k})) = 0,
$$

which implies that $\mathcal{G}$ is lossless with respect to the supply rate $r(u, y)$. □

13.10 Extended Kalman-Yakubovich-Popov Conditions for Discrete-Time Dynamical Systems

In this section, we show that dissipativity, geometric dissipativity, and losslessness of a discrete-time nonlinear affine dynamical system $\mathcal{G}$ of the form

$$x(k+1) = f(x(k)) + G(x(k))u(k), \quad x(k_0) = x_0, \quad k \geq k_0, \tag{13.135}$$
$$y(k) = h(x(k)) + J(x(k))u(k), \tag{13.136}$$

where $x \in \mathcal{D} \subseteq \mathbb{R}^n$, $\mathcal{D}$ is an open set with $0 \in \mathcal{D}$, $u(k) \in U \subseteq \mathbb{R}^m$, $y(k) \in Y \subseteq \mathbb{R}^l$, $f : \mathcal{D} \to \mathbb{R}^n$, $G : \mathcal{D} \to \mathbb{R}^{n\times m}$, $h : \mathcal{D} \to \mathbb{R}^l$, and $J : \mathcal{D} \to \mathbb{R}^{l\times m}$, can be characterized in terms of the system functions $f(\cdot)$, $G(\cdot)$, $h(\cdot)$, and $J(\cdot)$. Here, we assume that $f(\cdot)$, $G(\cdot)$, $h(\cdot)$, and $J(\cdot)$ are continuous mappings and $f(\cdot)$ has at least one equilibrium so that, without loss of generality, $f(0) = 0$ and $h(0) = 0$. For the following result we consider the special case of dissipative systems with quadratic supply rates. Specifically, set $\mathcal{D} = \mathbb{R}^n$, $U = \mathbb{R}^m$, $Y = \mathbb{R}^l$, let $Q \in \mathbb{S}^l$, $R \in \mathbb{S}^m$, and $S \in \mathbb{R}^{l\times m}$ be given, and assume $r(u, y) = y^{\mathrm{T}}Qy + 2y^{\mathrm{T}}Su + u^{\mathrm{T}}Ru$. Furthermore, we assume that there exists a function $\kappa : \mathbb{R}^l \to \mathbb{R}^m$ such that $\kappa(0) = 0$ and $r(\kappa(y), y) < 0$, $y \neq 0$.

Theorem 13.21. Let $Q \in \mathbb{S}^l$, $S \in \mathbb{R}^{l\times m}$, and $R \in \mathbb{S}^m$. If there exist functions $V_{\mathrm{s}} : \mathbb{R}^n \to \mathbb{R}$, $P_{1u} : \mathbb{R}^n \to \mathbb{R}^{1\times m}$, $P_{2u} : \mathbb{R}^n \to \mathbb{N}^m$, $\ell : \mathbb{R}^n \to \mathbb{R}^p$, and $\mathcal{W} : \mathbb{R}^n \to \mathbb{R}^{p\times m}$ such that $V_{\mathrm{s}}(\cdot)$ is continuous and positive definite, $V_{\mathrm{s}}(0) = 0$,

$$V_{\mathrm{s}}(f(x) + G(x)u) = V_{\mathrm{s}}(f(x)) + P_{1u}(x)u + u^{\mathrm{T}}P_{2u}(x)u, \quad x \in \mathbb{R}^n, \quad u \in \mathbb{R}^m, \tag{13.137}$$

and, for all $x \in \mathbb{R}^n$,

$$0 = V_{\mathrm{s}}(f(x)) - V_{\mathrm{s}}(x) - h^{\mathrm{T}}(x)Qh(x) + \ell^{\mathrm{T}}(x)\ell(x), \tag{13.138}$$
$$0 = \tfrac{1}{2}P_{1u}(x) - h^{\mathrm{T}}(x)(QJ(x) + S) + \ell^{\mathrm{T}}(x)\mathcal{W}(x), \tag{13.139}$$
$$0 = R + S^{\mathrm{T}}J(x) + J^{\mathrm{T}}(x)S + J^{\mathrm{T}}(x)QJ(x) - P_{2u}(x) - \mathcal{W}^{\mathrm{T}}(x)\mathcal{W}(x), \tag{13.140}$$

then $\mathcal{G}$ is dissipative with respect to the quadratic supply rate $r(u, y) = y^{\mathrm{T}}Qy + 2y^{\mathrm{T}}Su + u^{\mathrm{T}}Ru$. If, alternatively, there exist functions $V_{\mathrm{s}} : \mathbb{R}^n \to \mathbb{R}$, $P_{1u} : \mathbb{R}^n \to \mathbb{R}^{1\times m}$, and $P_{2u} : \mathbb{R}^n \to \mathbb{N}^m$ such that $V_{\mathrm{s}}(\cdot)$ is continuous and positive definite, and, for all $x \in \mathbb{R}^n$, (13.137) holds and

$$\mathcal{N}(x) \triangleq R + S^{\mathrm{T}}J(x) + J^{\mathrm{T}}(x)S + J^{\mathrm{T}}(x)QJ(x) - P_{2u}(x) > 0, \tag{13.141}$$
$$0 \geq V_{\mathrm{s}}(f(x)) - V_{\mathrm{s}}(x) - h^{\mathrm{T}}(x)Qh(x) + [\tfrac{1}{2}P_{1u}(x) - h^{\mathrm{T}}(x)(QJ(x) + S)] \cdot\mathcal{N}^{-1}(x)[\tfrac{1}{2}P_{1u}(x) - h^{\mathrm{T}}(x)(QJ(x) + S)]^{\mathrm{T}}, \tag{13.142}$$

then $\mathcal{G}$ is dissipative with respect to quadratic supply rate $r(u, y) = y^{\mathrm{T}}Qy + 2y^{\mathrm{T}}Su + u^{\mathrm{T}}Ru$.

Proof. Suppose that there exist functions $V_{\mathrm{s}} : \mathbb{R}^n \to \mathbb{R}$, $\ell : \mathbb{R}^n \to \mathbb{R}^p$, and $\mathcal{W} : \mathbb{R}^n \to \mathbb{R}^{p\times m}$ such that $V_{\mathrm{s}}(\cdot)$ is continuous and positive definite, and (13.137)–(13.140) are satisfied. Then, for every $u \in \mathbb{R}^m$ and $x \in \mathbb{R}^n$, it follows from (13.137)–(13.140) that

$$\begin{aligned} r(u,y) &= y^{\mathrm{T}}Qy + 2y^{\mathrm{T}}Su + u^{\mathrm{T}}Ru \\ &= h^{\mathrm{T}}(x)Qh(x) + 2h^{\mathrm{T}}(x)(S + QJ(x))u + u^{\mathrm{T}}(J^{\mathrm{T}}(x)QJ(x) \\ &\quad + S^{\mathrm{T}}J(x) + J^{\mathrm{T}}(x)S + R)u \\ &= V_{\mathrm{s}}(f(x)) - V_{\mathrm{s}}(x) + P_{1u}(x)u + \ell^{\mathrm{T}}(x)\ell(x) + 2\ell^{\mathrm{T}}(x)\mathcal{W}(x)u \\ &\quad + u^{\mathrm{T}}P_{2u}(x)u + u^{\mathrm{T}}\mathcal{W}^{\mathrm{T}}(x)\mathcal{W}(x)u \\ &= \Delta V_{\mathrm{s}}(x) + [\ell(x) + \mathcal{W}(x)u]^{\mathrm{T}}[\ell(x) + \mathcal{W}(x)u], \end{aligned}$$

and hence, for every admissible $u(\cdot) \in \mathcal{U}$, $k_1, k_2 \in \overline{\mathbb{Z}}_+$, $k_2 - 1 \geq k_1$,

$$\sum_{k=k_1}^{k_2-1} r(u(k), y(k)) \geq V_{\mathrm{s}}(x(k_2)) - V_{\mathrm{s}}(x(k_1)),$$

where $x(k)$, $k \in \overline{\mathbb{Z}}_+$, satisfies (13.135) and $\Delta V_{\mathrm{s}}(\cdot)$ denotes the total difference of the storage function along the trajectories $x(k)$, $k \in \overline{\mathbb{Z}}_+$ of (13.135). Now, the result is immediate from Corollary 13.9.

To show (13.142) note that (13.138)–(13.140) can be equivalently written as

$$\begin{bmatrix} \mathcal{A}(x) & \mathcal{B}(x) \\ \mathcal{B}^{\mathrm{T}}(x) & \mathcal{C}(x) \end{bmatrix} = -\begin{bmatrix} \ell^{\mathrm{T}}(x) \\ \mathcal{W}^{\mathrm{T}}(x) \end{bmatrix} \begin{bmatrix} \ell(x) & \mathcal{W}(x) \end{bmatrix} \leq 0, \quad x \in \mathbb{R}^n, \tag{13.143}$$

where $\mathcal{A}(x) \triangleq V_{\mathrm{s}}(f(x)) - V_{\mathrm{s}}(x) - h^{\mathrm{T}}(x)Qh(x)$, $\mathcal{B}(x) \triangleq \frac{1}{2}P_{1u}(x) - h^{\mathrm{T}}(x)(QJ(x) + S)$, and $\mathcal{C}(x) \triangleq -\mathcal{N}(x)$. Now, for all invertible $\mathcal{T} \in \mathbb{R}^{(m+1)\times(m+1)}$ (13.143) holds if and only if $\mathcal{T}^{\mathrm{T}}$(13.143)$\mathcal{T}$ holds. Hence, the equivalence of (13.138)–(13.140) to (13.142) in the case when (13.141) holds follows from the (1,1) block of $\mathcal{T}^{\mathrm{T}}$(13.143)$\mathcal{T}$, where

$$\mathcal{T} \triangleq \begin{bmatrix} 1 & 0 \\ -\mathcal{C}^{-1}(x)\mathcal{B}^{\mathrm{T}}(x) & I_n \end{bmatrix}.$$

This completes the proof. □

Next, we provide sufficient conditions for geometric dissipativity with respect to quadratic supply rates.

Theorem 13.22. Let $Q \in \mathbb{S}^l$, $S \in \mathbb{R}^{l\times m}$, and $R \in \mathbb{S}^m$. If there exist functions $V_{\mathrm{s}} : \mathbb{R}^n \to \mathbb{R}$, $P_{1u} : \mathbb{R}^n \to \mathbb{R}^{1\times m}$, $P_{2u} : \mathbb{R}^n \to \mathbb{N}^m$, $\ell : \mathbb{R}^n \to \mathbb{R}^p$, and $\mathcal{W} : \mathbb{R}^n \to \mathbb{R}^{p\times m}$, and a scalar $\rho > 1$, such that $V_{\mathrm{s}}(\cdot)$ is continuous and positive definite, $V_{\mathrm{s}}(0) = 0$, and, for all $x \in \mathbb{R}^n$, (13.137) holds and

$$0 = V_{\mathrm{s}}(f(x)) - \tfrac{1}{\rho}V_{\mathrm{s}}(x) - h^{\mathrm{T}}(x)Qh(x) + \ell^{\mathrm{T}}(x)\ell(x), \tag{13.144}$$

$$0 = \tfrac{1}{2}P_{1u}(x) - h^{\mathrm{T}}(x)(QJ(x) + S) + \ell^{\mathrm{T}}(x)\mathcal{W}(x), \tag{13.145}$$

$$0 = R + S^{\mathrm{T}}J(x) + J^{\mathrm{T}}(x)S + J^{\mathrm{T}}(x)QJ(x) - P_{2u}(x) - \mathcal{W}^{\mathrm{T}}(x)\mathcal{W}(x), \tag{13.146}$$

then $\mathcal{G}$ is geometrically dissipative with respect to the quadratic supply rate $r(u, y) = y^{\mathrm{T}}Qy + 2y^{\mathrm{T}}Su + u^{\mathrm{T}}Ru$. If, alternatively, there exist functions $V_{\mathrm{s}} : \mathbb{R}^n \to \mathbb{R}$, $P_{1u} : \mathbb{R}^n \to \mathbb{R}^{1\times m}$, and $P_{2u} : \mathbb{R}^n \to \mathbb{N}^m$, and a scalar $\rho > 1$, such that $V_{\mathrm{s}}(\cdot)$ is continuous and positive definite and, for all $x \in \mathbb{R}^n$, (13.137) holds and

$$0 < \mathcal{N}(x), \tag{13.147}$$
$$0 \geq V_{\mathrm{s}}(f(x)) - \tfrac{1}{\rho}V_{\mathrm{s}}(x) - h^{\mathrm{T}}(x)Qh(x) + [\tfrac{1}{2}P_{1u}(x) - h^{\mathrm{T}}(x)(QJ(x) + S)] \cdot\mathcal{N}(x)^{-1}[\tfrac{1}{2}P_{1u}(x) - h^{\mathrm{T}}(x)(QJ(x) + S)]^{\mathrm{T}}, \tag{13.148}$$

then $\mathcal{G}$ is geometrically dissipative with respect to the quadratic supply rate $r(u, y) = y^{\mathrm{T}}Qy + 2y^{\mathrm{T}}Su + u^{\mathrm{T}}Ru$.

Proof. The proof is analogous to the proof of Theorem 13.21. □

Next, we provide necessary and sufficient conditions for the case where $\mathcal{G}$ is lossless with respect to the quadratic supply rate $r(u, y)$.

Theorem 13.23. Assume $\mathcal{G}$ is zero-state observable and completely reachable. Let $Q \in \mathbb{S}^l$, $S \in \mathbb{R}^{l\times m}$, and $R \in \mathbb{S}^m$. Then $\mathcal{G}$ is lossless with respect to the quadratic supply rate $r(u, y) = y^{\mathrm{T}}Qy + 2y^{\mathrm{T}}Su + u^{\mathrm{T}}Ru$ if and only if there exist functions $V_{\mathrm{s}} : \mathbb{R}^n \to \mathbb{R}$, $P_{1u} : \mathbb{R}^n \to \mathbb{R}^{1\times m}$, $P_{2u} : \mathbb{R}^n \to \mathbb{N}^m$, $\ell : \mathbb{R}^n \to \mathbb{R}^p$, and $\mathcal{W} : \mathbb{R}^n \to \mathbb{R}^{p\times m}$ such that $V_{\mathrm{s}}(\cdot)$ is continuous and positive definite, $V_{\mathrm{s}}(0) = 0$, and, for all $x \in \mathbb{R}^n$, (13.137) holds and

$$0 = V_{\mathrm{s}}(f(x)) - V_{\mathrm{s}}(x) - h^{\mathrm{T}}(x)Qh(x), \tag{13.149}$$
$$0 = \tfrac{1}{2}P_{1u}(x) - h^{\mathrm{T}}(x)(QJ(x) + S), \tag{13.150}$$
$$0 = R + S^{\mathrm{T}}J(x) + J^{\mathrm{T}}(x)S + J^{\mathrm{T}}(x)QJ(x) - P_{2u}(x), \tag{13.151}$$

If, in addition, $V_{\mathrm{s}}(\cdot)$ is two-times continuously differentiable then

$$P_{1u}(x) = V_{\mathrm{s}}'(f(x))G(x), \tag{13.152}$$
$$P_{2u}(x) = \tfrac{1}{2}G^{\mathrm{T}}(x)V_{\mathrm{s}}''(f(x))G(x). \tag{13.153}$$

Proof. Sufficiency follows as in the proof of Theorem 13.21. To show necessity, suppose that $\mathcal{G}$ is lossless with respect to the quadratic supply rate $r(u, y)$. Then, it follows that there exists a continuous function $V_{\mathrm{s}} : \mathbb{R}^n \to \mathbb{R}^n$ such that

$$\begin{aligned} V_{\mathrm{s}}(f(x) + G(x)u) &= V_{\mathrm{s}}(x) + r(u, y), \\ &= V_{\mathrm{s}}(x) + y^{\mathrm{T}}Qy + 2y^{\mathrm{T}}Su + u^{\mathrm{T}}Ru \\ &= V_{\mathrm{s}}(x) + h^{\mathrm{T}}(x)Qh(x) + 2h^{\mathrm{T}}(x)(QJ(x) + S)u \end{aligned}$$

$$+u^{\mathrm{T}}(R+S^{\mathrm{T}}J(x)+J^{\mathrm{T}}(x)S+J^{\mathrm{T}}(x)QJ(x))u, \quad x\in\mathbb{R}^n,\ u\in\mathbb{R}^m. \quad (13.154)$$

Since the right-hand-side of (13.154) is quadratic in u it follows that $V_{\mathrm{s}}(f(x)+G(x)u)$ is quadratic in u, and hence, there exist $P_{1u}:\mathbb{R}^n\to\mathbb{R}^{1\times m}$ and $P_{2u}:\mathbb{R}^n\to\mathbb{N}^m$ such that

$$V_{\mathrm{s}}(f(x)+G(x)u)=V_{\mathrm{s}}(f(x))+P_{1u}(x)u+u^{\mathrm{T}}P_{2u}(x)u, \quad x\in\mathbb{R}^n, \quad u\in\mathbb{R}^m. \quad (13.155)$$

Now, using (13.155) and equating coefficients of equal powers in (13.154) yields (13.149)–(13.151). Finally, if $V_{\mathrm{s}}(\cdot)$ is two-times continuously differentiable, applying a Taylor series expansion on (13.155) about $u=0$ yields (13.152) and (13.153). □

Note that if $\mathcal{G}$ is dissipative with respect to the quadratic supply rate $r(u,y)=y^{\mathrm{T}}Qy+2y^{\mathrm{T}}Su+u^{\mathrm{T}}Ru$ with a continuous positive-definite radially unbounded storage function $V_{\mathrm{s}}(\cdot)$, and if $Q\le 0$ and $u(k)\equiv 0$, then it follows that

$$\Delta V_{\mathrm{s}}(x(k))\le y^{\mathrm{T}}(k)Qy(k)\le 0, \qquad k\in\overline{\mathbb{Z}}_+. \quad (13.156)$$

Hence, the zero solution $x(k)\equiv 0$ of the undisturbed ($u(k)\equiv 0$) nonlinear system (13.135) is Lyapunov stable. Alternatively, if $\mathcal{G}$ with a continuous positive-definite radially unbounded storage function, is geometrically dissipative with respect to the quadratic supply rate $r(u,y)=y^{\mathrm{T}}Qy+2y^{\mathrm{T}}Su+u^{\mathrm{T}}Ru$, and if $Q\le 0$ and $u(k)\equiv 0$, then it follows that

$$\begin{aligned}V_{\mathrm{s}}(f(x(k)))-V_{\mathrm{s}}(x(k)) &\le -\tfrac{\rho-1}{\rho}V_{\mathrm{s}}(x(k))+y^{\mathrm{T}}(k)Qy(k)\\ &\le -\tfrac{\rho-1}{\rho}V_{\mathrm{s}}(x(k)), \qquad k\in\overline{\mathbb{Z}}_+. \quad (13.157)\end{aligned}$$

Hence, the zero solution $x(k)\equiv 0$ of the undisturbed ($u(k)\equiv 0$) nonlinear system (13.135) is asymptotically stable. If, in addition, there exist scalars $\alpha,\beta>0$ and $p\ge 1$, such that

$$\alpha\|x\|^p\le V_{\mathrm{s}}(x)\le\beta\|x\|^p, \qquad x\in\mathbb{R}^n, \quad (13.158)$$

then the zero solution $x(k)\equiv 0$ of the undisturbed ($u(k)\equiv 0$) nonlinear dynamical system (13.135) is geometrically stable.

The following results present the discrete-time nonlinear versions of the Kalman-Yakubovich-Popov positive real lemma and the bounded real lemma.

Corollary 13.10. If there exist functions $V_{\mathrm{s}}:\mathbb{R}^n\to\mathbb{R}$, $P_{1u}:\mathbb{R}^n\to\mathbb{R}^{1\times m}$, $P_{2u}:\mathbb{R}^n\to\mathbb{N}^m$, $\ell:\mathbb{R}^n\to\mathbb{R}^p$, and $\mathcal{W}:\mathbb{R}^n\to\mathbb{R}^{p\times m}$ such that $V_{\mathrm{s}}(\cdot)$ is continuous and positive definite, $V_{\mathrm{s}}(0)=0$, and, for all $x\in\mathbb{R}^n$, (13.137) holds and

$$0=V_{\mathrm{s}}(f(x))-V_{\mathrm{s}}(x)+\ell^{\mathrm{T}}(x)\ell(x), \quad (13.159)$$

$$0 = \tfrac{1}{2}P_{1u}(x) - h^{\mathrm{T}}(x) + \ell^{\mathrm{T}}(x)\mathcal{W}(x), \tag{13.160}$$
$$0 = J(x) + J^{\mathrm{T}}(x) - P_{2u}(x) - \mathcal{W}^{\mathrm{T}}(x)\mathcal{W}(x), \tag{13.161}$$

then $\mathcal{G}$ is passive. If, alternatively, there exist functions $V_{\mathrm{s}} : \mathbb{R}^n \to \mathbb{R}$, $P_{1u} : \mathbb{R}^n \to \mathbb{R}^{1\times m}$, and $P_{2u} : \mathbb{R}^n \to \mathbb{N}^m$, such that $V_{\mathrm{s}}(\cdot)$ is continuous and positive definite and, for all $x \in \mathbb{R}^n$, (13.137) holds and

$$0 < J(x) + J^{\mathrm{T}}(x) - P_{2u}(x), \tag{13.162}$$
$$0 \geq V_{\mathrm{s}}(f(x)) - V_{\mathrm{s}}(x) + [\tfrac{1}{2}P_{1u}(x) - h^{\mathrm{T}}(x)] \cdot [J(x) + J^{\mathrm{T}}(x) - P_{2u}(x)]^{-1}[\tfrac{1}{2}P_{1u}(x) - h^{\mathrm{T}}(x)]^{\mathrm{T}}, \tag{13.163}$$

then $\mathcal{G}$ is passive.

Proof. The result is a direct consequence of Theorem 13.21 with $l = m$, $Q = 0$, $S = I_m$, and $R = 0$. □

Corollary 13.11. If there exist functions $V_{\mathrm{s}} : \mathbb{R}^n \to \mathbb{R}$, $P_{1u} : \mathbb{R}^n \to \mathbb{R}^{1\times m}$, $P_{2u} : \mathbb{R}^n \to \mathbb{N}^m$, $\ell : \mathbb{R}^n \to \mathbb{R}^p$, and $\mathcal{W} : \mathbb{R}^n \to \mathbb{R}^{p\times m}$ such that $V_{\mathrm{s}}(\cdot)$ is continuous and positive definite, $V_{\mathrm{s}}(0) = 0$, and, for all $x \in \mathbb{R}^n$, (13.137) holds and

$$0 = V_{\mathrm{s}}(f(x)) - V_{\mathrm{s}}(x) + h^{\mathrm{T}}(x)h(x) + \ell^{\mathrm{T}}(x)\ell(x), \tag{13.164}$$
$$0 = \tfrac{1}{2}P_{1u}(x) + h^{\mathrm{T}}(x)J(x) + \ell^{\mathrm{T}}(x)\mathcal{W}(x), \tag{13.165}$$
$$0 = \gamma^2 I_m - J^{\mathrm{T}}(x)J(x) - P_{2u}(x) - \mathcal{W}^{\mathrm{T}}(x)\mathcal{W}(x), \tag{13.166}$$

where $\gamma > 0$, then $\mathcal{G}$ is nonexpansive. If, alternatively, there exist functions $V_{\mathrm{s}} : \mathbb{R}^n \to \mathbb{R}$, $P_{1u} : \mathbb{R}^n \to \mathbb{R}^{1\times m}$, and $P_{2u} : \mathbb{R}^n \to \mathbb{N}^m$, such that $V_{\mathrm{s}}(\cdot)$ is continuous and positive definite, $V_{\mathrm{s}}(0) = 0$, and, for all $x \in \mathbb{R}^n$, (13.137) holds and

$$0 < \gamma^2 I_m - J^{\mathrm{T}}(x)J(x) - P_{2u}(x), \tag{13.167}$$
$$0 \geq V_{\mathrm{s}}(f(x)) - V_{\mathrm{s}}(x) + h^{\mathrm{T}}(x)h(x) + [\tfrac{1}{2}P_{1u}(x) + h^{\mathrm{T}}(x)J(x)] \cdot [\gamma^2 I_m - J^{\mathrm{T}}(x)J(x) - P_{2u}(x)]^{-1}[\tfrac{1}{2}P_{1u}(x) + h^{\mathrm{T}}(x)J(x)]^{\mathrm{T}}, \tag{13.168}$$

then $\mathcal{G}$ is nonexpansive.

Proof. The result is a direct consequence of Theorem 13.21 with $Q = -I_l$, $S = 0$, and $R = \gamma^2 I_m$. □

The following results present the discrete-time nonlinear versions of the Kalman-Yakubovich-Popov strict positive real lemma and strict bounded real lemma for geometrically passive and geometrically nonexpansive systems, respectively.

Corollary 13.12. If there exist functions $V_{\mathrm{s}} : \mathbb{R}^n \to \mathbb{R}$, $P_{1u} : \mathbb{R}^n \to \mathbb{R}^{1\times m}$, $P_{2u} : \mathbb{R}^n \to \mathbb{N}^m$, $\ell : \mathbb{R}^n \to \mathbb{R}^p$, and $\mathcal{W} : \mathbb{R}^n \to \mathbb{R}^{p\times m}$, and a scalar

$\rho > 1$, such that $V_s(\cdot)$ is continuous and positive definite, $V_s(0) = 0$, and, for all $x \in \mathbb{R}^n$, (13.137) holds and

$$0 = V_s(f(x)) - \tfrac{1}{\rho}V_s(x) + \ell^T(x)\ell(x), \tag{13.169}$$

$$0 = \tfrac{1}{2}P_{1u}(x) - h^T(x) + \ell^T(x)\mathcal{W}(x), \tag{13.170}$$

$$0 = J(x) + J^T(x) - P_{2u}(x) - \mathcal{W}^T(x)\mathcal{W}(x), \tag{13.171}$$

then $\mathcal{G}$ is geometrically passive. If, alternatively, there exist functions $V_s : \mathbb{R}^n \to \mathbb{R}$, $P_{1u} : \mathbb{R}^n \to \mathbb{R}^{1\times m}$, and $P_{2u} : \mathbb{R}^n \to \mathbb{N}^m$, and a scalar $\rho > 1$, such that $V_s(\cdot)$ is continuous and positive definite and, for all $x \in \mathbb{R}^n$, (13.137) holds and

$$0 < J(x) + J^T(x) - P_{2u}(x), \tag{13.172}$$

$$0 \geq V_s(f(x)) - \tfrac{1}{\rho}V_s(x) + [\tfrac{1}{2}P_{1u}(x) - h^T(x)] \cdot[J(x) + J^T(x) - P_{2u}(x)]^{-1}[\tfrac{1}{2}P_{1u}(x) - h^T(x)]^T, \tag{13.173}$$

then $\mathcal{G}$ is geometrically passive.

Proof. The result is a direct consequence of Theorem 13.22 with $l = m$, $Q = 0$, $S = I_m$, and $R = 0$. □

Corollary 13.13. If there exist functions $V_s : \mathbb{R}^n \to \mathbb{R}$, $P_{1u} : \mathbb{R}^n \to \mathbb{R}^{1\times m}$, $P_{2u} : \mathbb{R}^n \to \mathbb{N}^m$, $\ell : \mathbb{R}^n \to \mathbb{R}^p$, and $\mathcal{W} : \mathbb{R}^n \to \mathbb{R}^{p\times m}$, and a scalar $\rho > 1$, such that $V_s(\cdot)$ is continuous and positive definite, $V_s(0) = 0$, and, for all $x \in \mathbb{R}^n$, (13.137) holds and

$$0 = V_s(f(x)) - \tfrac{1}{\rho}V_s(x) + h^T(x)h(x) + \ell^T(x)\ell(x), \tag{13.174}$$

$$0 = \tfrac{1}{2}P_{1u}(x) + h^T(x)J(x) + \ell^T(x)\mathcal{W}(x), \tag{13.175}$$

$$0 = \gamma^2 I_m - J^T(x)J(x) - P_{2u}(x) - \mathcal{W}^T(x)\mathcal{W}(x), \tag{13.176}$$

where $\gamma > 0$, then $\mathcal{G}$ is geometrically nonexpansive. If, alternatively, there exist functions $V_s : \mathbb{R}^n \to \mathbb{R}$, $P_{1u} : \mathbb{R}^n \to \mathbb{R}^{1\times m}$, and $P_{2u} : \mathbb{R}^n \to \mathbb{N}^m$, and a scalar $\rho > 1$, such that $V_s(\cdot)$ is continuous and positive definite, $V_s(0) = 0$, and, for all $x \in \mathbb{R}^n$, (13.137) holds and

$$0 < \gamma^2 I_m - J^T(x)J(x) - P_{2u}(x), \tag{13.177}$$

$$0 \geq V_s(f(x)) - \tfrac{1}{\rho}V_s(x) + h^T(x)h(x) + [\tfrac{1}{2}P_{1u}(x) + h^T(x)J(x)] \cdot[\gamma^2 I_m - J^T(x)J(x) - P_{2u}(x)]^{-1}[\tfrac{1}{2}P_{1u}(x) + h^T(x)J(x)]^T, \tag{13.178}$$

then $\mathcal{G}$ is geometrically nonexpansive.

Proof. The result is a direct consequence of Theorem 13.22 with $Q = -I_l$, $S = 0$, and $R = \gamma^2 I_m$. □

Finally, we close this section by noting that Proposition 5.2, with appropriate modifications, is also valid for discrete-time systems.

13.11 Linearization of Dissipative Dynamical Systems

In this section, we present several key results on linearization of dissipative, geometrically dissipative, passive, geometrically passive, nonexpansive, and geometrically nonexpansive systems. For these results, we assume that there exists a function $\kappa : \mathbb{R}^l \to \mathbb{R}^m$ such that $\kappa(0) = 0$ and $r(\kappa(y), y) < 0$, $y \neq 0$, and the available storage function (respectively, geometrically storage function) $V_{\rm a}(x)$, $x \in \mathbb{R}^n$, is a three-times continuously differentiable function.

Theorem 13.24. Let $Q \in \mathbb{S}^l$, $S \in \mathbb{R}^{l\times m}$, $R \in \mathbb{S}^m$, and suppose $\mathcal{G}$ given by (13.135) and (13.136) is completely reachable and dissipative with respect to the quadratic supply rate $r(u,y) = y^{\rm T}Qy + 2y^{\rm T}Su + u^{\rm T}Ru$. Then, there exist matrices $P \in \mathbb{R}^{n\times n}$, $L \in \mathbb{R}^{p\times n}$, and $W \in \mathbb{R}^{p\times m}$, with P nonnegative definite, such that

$$P = A^{\rm T}PA - C^{\rm T}QC + L^{\rm T}L, \tag{13.179}$$
$$0 = A^{\rm T}PB - C^{\rm T}(QD + S) + L^{\rm T}W, \tag{13.180}$$
$$0 = R + S^{\rm T}D + D^{\rm T}S + D^{\rm T}QD - B^{\rm T}PB - W^{\rm T}W, \tag{13.181}$$

where

$$A = \left.\frac{\partial f}{\partial x}\right|_{x=0}, \quad B = G(0), \quad C = \left.\frac{\partial h}{\partial x}\right|_{x=0}, \quad D = J(0). \tag{13.182}$$

If, in addition, (A, C) is observable, then $P > 0$.

Proof. First note that since $\mathcal{G}$ is dissipative with respect to a quadratic supply rate there exists a continuous nonnegative function $V_{\rm s} : \mathbb{R}^n \to \mathbb{R}$ such that

$$V_{\rm s}(f(x) + G(x)u) - V_{\rm s}(x) \le r(u, y), \quad x \in \mathbb{R}^n, \quad u \in \mathbb{R}^m. \tag{13.183}$$

Next, it follows from (13.183) that there exists a three-times continuously differentiable function $d : \mathbb{R}^n \times \mathbb{R}^m \to \mathbb{R}$ such that $d(x,u) \ge 0$, $d(0,0) = 0$, and

$$0 = V_{\rm s}(f(x) + G(x)u) - V_{\rm s}(x) - r(u, h(x) + J(x)u) + d(x, u). \tag{13.184}$$

Now, expanding $V_{\rm s}(\cdot)$ and $d(\cdot,\cdot)$ via a Taylor series expansion about $x = 0$ and $u = 0$, and using the fact that $V_{\rm s}(\cdot)$ and $d(\cdot,\cdot)$ are nonnegative definite and $V_{\rm s}(0) = 0$, $d(0,0) = 0$, it follows that there exist matrices $P \in \mathbb{R}^{n\times n}$, $L \in \mathbb{R}^{p\times n}$, and $W \in \mathbb{R}^{p\times m}$, with P nonnegative definite, such that

$$V_{\rm s}(x) = x^{\rm T}Px + V_{\rm sr}(x), \tag{13.185}$$
$$d(x, u) = (Lx + Wu)^{\rm T}(Lx + Wu) + d_{\rm sr}(x, u), \tag{13.186}$$

where $V_{\rm sr} : \mathbb{R}^n \to \mathbb{R}$ and $d_{\rm sr} : \mathbb{R}^n \times \mathbb{R}^m \to \mathbb{R}$ contain higher-order terms of $V_{\rm s}(\cdot)$, $d(\cdot,\cdot)$, respectively.

Next, let $f(x) = Ax + f_{\rm r}(x)$ and $h(x) = Cx + h_{\rm r}(x)$, where $f_{\rm r}(x)$ and $h_{\rm r}(x)$ contain nonlinear terms of $f(x)$ and $h(x)$, respectively, and let $G(x) = B + G_{\rm r}(x)$ and $J(x) = D + J_{\rm r}(x)$, where $G_{\rm r}(x)$ and $J_{\rm r}(x)$ contain nonconstant terms of $G(x)$ and $J(x)$, respectively. Using (13.185) and (13.186), (13.184) can be written as

$$\begin{aligned} 0 = {} & (Ax+Bu)^{\rm T}P(Ax+Bu) - x^{\rm T}Px - (x^{\rm T}C^{\rm T}QCx + 2x^{\rm T}C^{\rm T}QDu \\ & + u^{\rm T}D^{\rm T}QDu + 2x^{\rm T}C^{\rm T}Su + 2u^{\rm T}D^{\rm T}Su + u^{\rm T}Ru) \\ & + (Lx+Wu)^{\rm T}(Lx+Wu) + \delta(x,u), \end{aligned} \tag{13.187}$$

where $\delta(x,u)$ is such that

$$\lim_{\|x\|^2+\|u\|^2\to 0} \frac{\delta(x,u)}{\|x\|^2+\|u\|^2} = 0.$$

Now, viewing (13.187) as the Taylor series expansion of (13.184) about $x = 0$ and $u = 0$ it follows that for all $x \in \mathbb{R}^n$ and $u \in \mathbb{R}^m$,

$$\begin{aligned} 0 = {} & x^{\rm T}(A^{\rm T}PA - P - C^{\rm T}QC + L^{\rm T}L)x \\ & + 2x^{\rm T}(A^{\rm T}PB - C^{\rm T}S - C^{\rm T}QD + L^{\rm T}W)u \\ & + u^{\rm T}(W^{\rm T}W - D^{\rm T}QD - D^{\rm T}S - S^{\rm T}D - R + B^{\rm T}PB)u. \end{aligned} \tag{13.188}$$

Now, equating coefficients of equal powers in (13.188) yields (13.179)–(13.181).

Finally, to show that $P > 0$ in the case where (A, C) is observable, note that it follows from Theorem 13.21 and (13.179)–(13.181) that the linearized system $\mathcal{G}$ with storage function $V_{\rm s}(x) = x^{\rm T}Px$ is dissipative with respect to the quadratic supply rate $r(u, y)$. Now, the positive definiteness of P follows from Theorem 13.18. □

The following corollaries are immediate from Theorem 13.24 and provide linearization results for passive and nonexpansive systems, respectively.

Corollary 13.14. Suppose the nonlinear dynamical system $\mathcal{G}$ given by (13.135) and (13.136) is completely reachable and passive. Then there exist matrices $P \in \mathbb{R}^{n\times n}$, $L \in \mathbb{R}^{p\times n}$, and $W \in \mathbb{R}^{p\times m}$, with P nonnegative definite, such that

$$P = A^{\rm T}PA + L^{\rm T}L, \tag{13.189}$$
$$0 = A^{\rm T}PB - C^{\rm T} + L^{\rm T}W, \tag{13.190}$$
$$0 = D + D^{\rm T} - B^{\rm T}PB - W^{\rm T}W, \tag{13.191}$$

where A, B, C, and D are given by (13.182). If, in addition, (A, C) is observable, then $P > 0$.

Corollary 13.15. Suppose the nonlinear dynamical system $\mathcal{G}$ given by (13.135) and (13.136) is completely reachable and nonexpansive. Then there exist matrices $P \in \mathbb{R}^{n\times n}$, $L \in \mathbb{R}^{p\times n}$, and $W \in \mathbb{R}^{p\times m}$, with P nonnegative definite, such that

$$P = A^{\mathrm{T}}PA + C^{\mathrm{T}}C + L^{\mathrm{T}}L, \tag{13.192}$$
$$0 = A^{\mathrm{T}}PB + C^{\mathrm{T}}D + L^{\mathrm{T}}W, \tag{13.193}$$
$$0 = \gamma^2 I_m - D^{\mathrm{T}}D - B^{\mathrm{T}}PB - W^{\mathrm{T}}W, \tag{13.194}$$

where A, B, C, and D are given by (13.182) and $\gamma > 0$. If, in addition, (A, C) is observable, then $P > 0$.

Next, we present a key linearization theorem for geometrically dissipative systems.

Theorem 13.25. Let $Q \in \mathbb{S}^l$, $S \in \mathbb{R}^{l\times m}$, $R \in \mathbb{S}^m$, and suppose $\mathcal{G}$ given by (13.135) and (13.136) is completely reachable and geometrically dissipative with respect to the quadratic supply rate $r(u, y) = y^{\mathrm{T}}Qy + 2y^{\mathrm{T}}Su + u^{\mathrm{T}}Ru$. Then, there exist matrices $P \in \mathbb{R}^{n\times n}$, $L \in \mathbb{R}^{p\times n}$, and $W \in \mathbb{R}^{p\times m}$, with P nonnegative definite, and a scalar $\rho > 1$, such that

$$\tfrac{1}{\rho}P = A^{\mathrm{T}}PA - C^{\mathrm{T}}QC + L^{\mathrm{T}}L, \tag{13.195}$$
$$0 = A^{\mathrm{T}}PB - C^{\mathrm{T}}(QD + S) + L^{\mathrm{T}}W, \tag{13.196}$$
$$0 = R + S^{\mathrm{T}}D + D^{\mathrm{T}}S + D^{\mathrm{T}}QD - B^{\mathrm{T}}PB - W^{\mathrm{T}}W, \tag{13.197}$$

where A, B, C, and D are given by (13.182). If, in addition, (A, C) is observable, then $P > 0$.

Proof. The proof is analogous to the proof of Theorem 13.24. □

Linearization results for geometrically passive systems and geometrically nonexpansive systems follow immediately from Theorem 13.25.

13.12 Positive Real and Bounded Real Discrete-Time Dynamical Systems

In this section, we specialize the results of Section 13.10 to the case of linear discrete-time systems and provide connections to the frequency domain versions of passivity, geometric passivity, nonexpansivity, and geometric nonexpansivity. Specifically, we consider linear systems

$$\mathcal{G} = G(z) \sim \left[\begin{array}{c|c} A & B \\ \hline C & D \end{array}\right]$$

with a state space representation

$$x(k+1) = Ax(k) + Bu(k), \qquad x(0) = 0, \qquad k \in \overline{\mathbb{Z}}_+, \quad (13.198)$$

$$y(k) = Cx(k) + Du(k), \quad (13.199)$$

where $x \in \mathbb{R}^n$, $u \in \mathbb{R}^m$, $y \in \mathbb{R}^l$, $A \in \mathbb{R}^{n\times n}$, $B \in \mathbb{R}^{n\times m}$, $C \in \mathbb{R}^{l\times n}$, and $D \in \mathbb{R}^{l\times m}$. To present the main results of this section we first give several key definitions.

Definition 13.16. A square transfer function $G(z)$ is *positive real* if *i*) all the entries of $G(z)$ are analytic in $|z| > 1$ and *ii*) He $G(z) \geq 0$, $|z| > 1$. A square transfer function $G(z)$ is *strictly positive real* if there exists $\rho > 1$ such that $G(z/\rho)$ is positive real.

Definition 13.17. A transfer function $G(z)$ is *bounded real* if *i*) all the entries of $G(z)$ are analytic in $|z| > 1$ and *ii*) $\gamma^2 I_m - G^*(z)G(z) \geq 0$, $|z| > 1$, where $\gamma > 0$. A transfer function $G(z)$ is *strictly bounded real* if there exists $\rho > 1$ such that $G(z/\rho)$ is bounded real.

As in the continuous case, note that *ii*) in Definition 13.17 implies that $G(z)$ is analytic in $|z| \geq 1$, and hence, a bounded real transfer function is asymptotically stable. To see this, note that $\gamma^2 I_m - G^*(z)G(z) \geq 0$, $|z| > 1$, implies that

$$[\gamma^2 I_m - G^*(z)G(z)]_{(i,i)} = \gamma^2 - \sum_{j=1}^{m} |G_{(j,i)}(z)|^2 \geq 0, \quad |z| > 1, \quad (13.200)$$

and hence, $|G_{(i,j)}(z)|$ is bounded by γ^2 at every point in $|z| > 1$. Hence, $G_{(i,j)}(z)$ cannot possess a pole in $|z| = 1$ since in this case, $|G_{(i,j)}(z)|$ would take on arbitrary large values in $|z| > 1$ in the vicinity of this pole. Hence, $G(e^{\jmath\theta}) = \lim_{\sigma\to 1, \sigma>1} G(\sigma e^{\jmath\theta})$ exists for all $\theta \in [0, 2\pi]$ and $\gamma^2 I_m - G^*(e^{\jmath\theta})G(e^{\jmath\theta}) \geq 0$, $\theta \in [0, 2\pi]$. Now, since $G^*(e^{\jmath\theta})G(e^{\jmath\theta}) \leq \gamma^2 I_m$, $\theta \in [0, 2\pi]$, is equivalent to $\sup_{\theta\in[0,2\pi]} \sigma_{\max}[G(e^{\jmath\theta})] \leq \gamma$, it follows that $G(z)$ is bounded real if and only if $G(z)$ is asymptotically stable and $|||G(z)|||_\infty \leq \gamma$. Similarly, it can be shown that strict bounded realness is equivalent to $G(z)$ asymptotically stable and $G^*(e^{\jmath\theta})G(e^{\jmath\theta}) < \gamma^2 I_m$, $\theta \in [0, 2\pi]$, or, equivalently, $|||G(z)|||_\infty < \gamma$.

The following theorem gives a frequency domain test for positive realness.

Theorem 13.26. Let $G(z)$ be a square, real rational transfer function. $G(z)$ is positive real if and only if the following conditions hold:

i) No entry of $G(z)$ has a pole in $|z| > 1$.

ii) He $G(e^{\jmath\theta}) \geq 0$ for all $\theta \in [0, 2\pi]$, with $e^{\jmath\theta}$ not a pole of any entry of $G(z)$.

iii) If $e^{\jmath\hat{\theta}}$ is a pole of any entry of $G(z)$ it is at most a simple pole, and the residue matrix $G_0 \triangleq \lim_{z\to e^{\jmath\hat{\theta}}}(z - e^{\jmath\hat{\theta}})G(z)$ is nonnegative definite.

Proof. The proof follows from the maximum modulus theorem of complex variable theory by forming a Nyquist-type closed contour Γ in $|z| > 1$ and analyzing the function $f(z) = x^* G(z^{-1})x$, $x \in \mathbb{C}^m$, on Γ. For details see [197]. □

Now, we present the key results of this section for characterizing positive realness, strict positive realness, bounded realness, and strict bounded realness of a linear discrete-time dynamical system in terms of the system matrices A, B, C, and D.

Theorem 13.27 (Positive Real Lemma). Consider the dynamical system

$$G(z) \overset{\min}{\sim} \left[\begin{array}{c|c} A & B \\ \hline C & D \end{array}\right]$$

with input $u(\cdot) \in \mathcal{U}$ and output $y(\cdot) \in \mathcal{Y}$. Then the following statements are equivalent:

i) $G(z)$ is positive real.

ii) $\sum_{k=0}^{K-1} u^{\mathrm{T}}(k)y(k) \geq 0, \quad K - 1 \geq 0.$

iii) There exist matrices $P \in \mathbb{R}^{n\times n}$, $L \in \mathbb{R}^{p\times n}$, and $W \in \mathbb{R}^{p\times m}$, with P positive definite, such that

$$P = A^{\mathrm{T}}PA + L^{\mathrm{T}}L, \tag{13.201}$$
$$0 = A^{\mathrm{T}}PB - C^{\mathrm{T}} + L^{\mathrm{T}}W, \tag{13.202}$$
$$0 = D + D^{\mathrm{T}} - B^{\mathrm{T}}PB - W^{\mathrm{T}}W. \tag{13.203}$$

If, alternatively, $D + D^{\mathrm{T}} - B^{\mathrm{T}}PB > 0$, then $G(z)$ is positive real if and only if there exists an $n \times n$ positive-definite matrix P such that

$$P \geq A^{\mathrm{T}}PA + (B^{\mathrm{T}}PA - C)^{\mathrm{T}}(D + D^{\mathrm{T}} - B^{\mathrm{T}}PB)^{-1}(B^{\mathrm{T}}PA - C). \tag{13.204}$$

Proof. First, we show that *i)* implies *ii)*. Suppose $G(z)$ is positive real. Then it follows from Parseval's theorem (see Problem 13.24) that, for all $K - 1 \geq 0$, and the truncated input function

$$u_K(k) = \begin{cases} u(k), & k = \{0, 1, \ldots, K-1\}, \\ 0, & \text{otherwise}, \end{cases} \tag{13.205}$$

$$\begin{aligned}\sum_{k=0}^{K-1} y^{\mathrm{T}}(k)u(k) &= \sum_{k=-\infty}^{\infty} y^{\mathrm{T}}(k)u_K(k)\\ &= \frac{1}{2\pi}\int_{-\pi}^{\pi} y^*(e^{\jmath\theta})u_K(e^{\jmath\theta})\mathrm{d}\theta\\ &= \frac{1}{4\pi}\int_{-\pi}^{\pi} u_K^*(e^{\jmath\theta})[G(e^{\jmath\theta})+G^*(e^{\jmath\theta})]u_K(e^{\jmath\theta})\mathrm{d}\theta\\ &\geq 0,\end{aligned}$$

which implies that $G(z)$ is passive.

Next, we show that *ii*) implies *iii*). If $G(z)$ is passive, then it follows from Corollary 13.14 with $f(x) = Ax$, $G(x) = B$, $h(x) = Cx$, $J(x) = D$, that there exist matrices $P \in \mathbb{R}^{n\times n}$, $L \in \mathbb{R}^{p\times n}$, and $W \in \mathbb{R}^{p\times m}$, with P positive definite, such that (13.201)–(13.203) are satisfied.

Next, to show that *iii*) implies *i*), note that if there exist matrices $P \in \mathbb{R}^{n\times n}$, $L \in \mathbb{R}^{p\times n}$, and $W \in \mathbb{R}^{p\times m}$, with P positive definite, such that (13.201)–(13.203) are satisfied, then, for all $|z| > 1$,

$$\begin{aligned}&G(z)+G^*(z)\\ &\quad= C(zI_n-A)^{-1}B+D+B^{\mathrm{T}}(z^*I_n-A)^{-\mathrm{T}}C^{\mathrm{T}}+D^{\mathrm{T}}\\ &\quad= W^{\mathrm{T}}W+B^{\mathrm{T}}PB+(B^{\mathrm{T}}PA+W^{\mathrm{T}}L)(zI_n-A)^{-1}B\\ &\qquad +B^{\mathrm{T}}(z^*I_n-A)^{-\mathrm{T}}(A^{\mathrm{T}}PB+L^{\mathrm{T}}W)\\ &\quad= W^{\mathrm{T}}W+W^{\mathrm{T}}L(zI_n-A)^{-1}B+B^{\mathrm{T}}(z^*I_n-A)^{-\mathrm{T}}L^{\mathrm{T}}W\\ &\qquad +B^{\mathrm{T}}(z^*I_n-A)^{-\mathrm{T}}[(z^*I_n-A)^{\mathrm{T}}P(zI_n-A)+(z^*I_n-A)^{\mathrm{T}}PA\\ &\qquad +A^{\mathrm{T}}P(zI_n-A)](zI_n-A)^{-1}B\\ &\quad= W^{\mathrm{T}}W+W^{\mathrm{T}}L(zI_n-A)^{-1}B+B^{\mathrm{T}}(z^*I_n-A)^{-\mathrm{T}}L^{\mathrm{T}}W\\ &\qquad +B^{\mathrm{T}}(z^*I_n-A)^{-\mathrm{T}}[L^{\mathrm{T}}L+(|z|-1)P](zI_n-A)^{-1}B\\ &\quad\geq [W+L(zI_n-A)^{-1}B]^*[W+L(zI_n-A)^{-1}B]\\ &\quad\geq 0.\end{aligned}$$

To show analyticity of the entries of $G(z)$ in $|z| > 1$ note that an entry of $G(z)$ will have a pole at $z = \lambda$ only if $\lambda \in \operatorname{spec}(A)$. Now, it follows from (13.201) and the fact that $P > 0$ that all eigenvalues of A have magnitude less than or equal to unity. Hence, $G(z)$ is analytic in $|z| > 1$, which implies that $G(z)$ is positive real. Finally, (13.204) follows from (13.163) with the linearization given above. □

Theorem 13.28 (Strict Positive Real Lemma). Consider the dynamical system

$$G(z) \overset{\min}{\sim} \left[\begin{array}{c|c} A & B \\ \hline C & D \end{array}\right]$$

with input $u(\cdot) \in \mathcal{U}$ and output $y(\cdot) \in \mathcal{Y}$. Then the following statements are equivalent:

i) $G(z)$ is strictly positive real.

ii) $\sum_{k=0}^{K-1} \rho^{k+1} u^{\mathrm{T}}(k) y(k) \geq 0, \qquad K-1 \geq 0.$

iii) There exist matrices $P \in \mathbb{R}^{n \times n}$, $L \in \mathbb{R}^{p \times n}$, and $W \in \mathbb{R}^{p \times m}$, with P positive definite, and a scalar $\rho > 1$ such that

$$\tfrac{1}{\rho}P = A^{\mathrm{T}}PA + L^{\mathrm{T}}L, \tag{13.206}$$

$$0 = A^{\mathrm{T}}PB - C^{\mathrm{T}} + L^{\mathrm{T}}W, \tag{13.207}$$

$$0 = D + D^{\mathrm{T}} - B^{\mathrm{T}}PB - W^{\mathrm{T}}W. \tag{13.208}$$

If, alternatively, $D + D^{\mathrm{T}} - B^{\mathrm{T}}PB > 0$, then $G(z)$ is strictly positive real if and only if there exist $n \times n$ positive-definite matrices P and R such that

$$\tfrac{1}{\rho}P = A^{\mathrm{T}}PA+(B^{\mathrm{T}}PA-C)^{\mathrm{T}}(D+D^{\mathrm{T}}-B^{\mathrm{T}}PB)^{-1}(B^{\mathrm{T}}PA-C)+R. \tag{13.209}$$

Proof. The equivalence of *i)* and *iii)* is a direct consequence of Theorem 13.27 by noting that $G(z)$ is strictly positive real if and only if there exists $\rho > 1$ such that

$$G(z/\sqrt{\rho}) \overset{\min}{\sim} \left[\begin{array}{c|c} \sqrt{\rho}A & \sqrt{\rho}B \\ \hline C & D \end{array}\right]$$

is positive real. The fact that *iii)* implies *ii)* follows from Corollary 13.12 with $f(x) = Ax$, $G(x) = B$, $h(x) = Cx$, $J(x) = D$, $V_{\mathrm{s}}(x) = x^{\mathrm{T}}Px$, $\ell(x) = Lx$, and $\mathcal{W}(x) = W$. To show that *ii)* implies *iii)*, note that if $G(z)$ is geometrically passive, then it follows from Theorem 13.25 with $f(x) = Ax$, $G(x) = B$, $h(x) = Cx$, $J(x) = D$, $Q = 0$, $S = I_m$, and $R = 0$, that there exist matrices $P \in \mathbb{R}^{n \times n}$, $L \in \mathbb{R}^{p \times n}$, and $W \in \mathbb{R}^{p \times m}$, with P positive definite, and a scalar $\rho > 1$ such that (13.206)–(13.208) are satisfied.

Finally, with the linearization given above, it follows from (13.173) that if $D + D^{\mathrm{T}} - B^{\mathrm{T}}PB > 0$, then $G(z)$ is strictly positive real if and only if there exist a scalar $\rho > 1$ and a positive-definite matrix $P \in \mathbb{R}^{n \times n}$ such that

$$\tfrac{1}{\rho}P \geq A^{\mathrm{T}}PA + (B^{\mathrm{T}}PA - C)^{\mathrm{T}}(D + D^{\mathrm{T}} - B^{\mathrm{T}}PB)^{-1}(B^{\mathrm{T}}PA - C). \tag{13.210}$$

Now, if there exist a scalar $\rho > 1$ and a positive-definite matrix $P \in \mathbb{R}^{n \times n}$ such that (13.210) is satisfied, then there exists an $n \times n$ positive-definite matrix R such that (13.209) is satisfied. Conversely, if $D + D^{\mathrm{T}} - B^{\mathrm{T}}PB > 0$ and there exists an $n \times n$ positive-definite matrix R such that (13.209) is satisfied, then, with $\rho = \sigma_{\min}(R)/\sigma_{\max}(P)$, (13.209) implies (13.210). Hence, if $D + D^{\mathrm{T}} - B^{\mathrm{T}}PB > 0$, then $G(z)$ is strictly positive real if and only if there exist $n \times n$ positive-definite matrices P and R such that (13.209) is satisfied. □

Next, we present analogous results for bounded real systems.

Theorem 13.29 (Bounded Real Lemma). Consider the dynamical system

$$G(z) \overset{\min}{\sim} \left[\begin{array}{c|c} A & B \\ \hline C & D \end{array}\right]$$

with input $u(\cdot) \in \mathcal{U}$ and output $y(\cdot) \in \mathcal{Y}$. Then the following statements are equivalent:

i) $G(z)$ is bounded real.

ii) $\sum_{k=0}^{K-1} y^{\mathrm{T}}(k)y(k) \leq \gamma^2 \sum_{k=0}^{K-1} u^{\mathrm{T}}(k)u(k), \quad K-1 \geq 0, \quad \gamma > 0.$

iii) There exist matrices $P \in \mathbb{R}^{n\times n}$, $L \in \mathbb{R}^{p\times n}$, and $W \in \mathbb{R}^{p\times m}$, with P positive definite, such that

$$P = A^{\mathrm{T}}PA + C^{\mathrm{T}}C + L^{\mathrm{T}}L, \tag{13.211}$$
$$0 = A^{\mathrm{T}}PB + C^{\mathrm{T}}D + L^{\mathrm{T}}W, \tag{13.212}$$
$$0 = \gamma^2 I_m - D^{\mathrm{T}}D - B^{\mathrm{T}}PB - W^{\mathrm{T}}W. \tag{13.213}$$

If, alternatively, $\gamma^2 I_m - D^{\mathrm{T}}D - B^{\mathrm{T}}PB > 0$, then $G(z)$ is bounded real if and only if there exists an $n \times n$ positive-definite matrix P such that

$$P \geq A^{\mathrm{T}}PA + C^{\mathrm{T}}C + (B^{\mathrm{T}}PA + D^{\mathrm{T}}C)^{\mathrm{T}}(\gamma^2 I_m - D^{\mathrm{T}}D - B^{\mathrm{T}}PB)^{-1} \cdot(B^{\mathrm{T}}PA + D^{\mathrm{T}}C). \tag{13.214}$$

Proof. First, we show that *i)* implies *ii)*. Suppose $G(z)$ is bounded real. Then it follows from Parseval's theorem (see Problem 13.24) that, for all $K-1 \geq 0$, and the truncated input function $u_K(\cdot)$ given by (13.205),

$$\begin{aligned}
\sum_{k=0}^{K-1} y^{\mathrm{T}}(k)y(k) &= \sum_{k=-\infty}^{\infty} y^{\mathrm{T}}(k)y(k) \\
&= \frac{1}{2\pi}\int_{-\pi}^{\pi} y^*(e^{\jmath\theta})y(e^{\jmath\theta})\mathrm{d}\theta \\
&= \frac{1}{2\pi}\int_{-\pi}^{\pi} u_K^*(e^{\jmath\theta})G^*(e^{\jmath\theta})G(e^{\jmath\theta})u_K(e^{\jmath\theta})\mathrm{d}\theta \\
&\leq \frac{\gamma^2}{2\pi}\int_{-\pi}^{\pi} u_K^*(e^{\jmath\theta})u_K(e^{\jmath\theta})\mathrm{d}\theta
\end{aligned}$$

$$= \gamma^2 \sum_{k=0}^{K-1} u^{\mathrm{T}}(k)u(k),$$

which implies that $G(z)$ is nonexpansive.

Next, we show that *ii*) implies *iii*). If $G(z)$ is nonexpansive, then it follows from Corollary 13.15 with $f(x) = Ax$, $G(x) = B$, $h(x) = Cx$, $J(x) = D$, that there exist matrices $P \in \mathbb{R}^{n\times n}$, $L \in \mathbb{R}^{p\times n}$, and $W \in \mathbb{R}^{p\times m}$, with P positive definite, such that (13.211)–(13.213) are satisfied.

Now, to show that *iii*) implies *i*), note that if there exist matrices $P \in \mathbb{R}^{n\times n}$, $L \in \mathbb{R}^{p\times n}$, and $W \in \mathbb{R}^{p\times m}$, with P positive definite, such that (13.211)–(13.213) are satisfied, then, for all $|z| \geq 1$,

$$\begin{aligned}
\gamma^2 I_m - G^*(z)G(z) \\
&= \gamma^2 I_m - [C(zI_n - A)^{-1}B + D]^*[C(zI_n - A)^{-1}B + D] \\
&= [\gamma^2 I_m - D^{\mathrm{T}}D] - [C(zI_n - A)^{-1}B]^*[C(zI_n - A)^{-1}B] \\
&\quad -[C(zI_n - A)^{-1}B]^*D - D^{\mathrm{T}}[C(zI_n - A)^{-1}B] \\
&= W^{\mathrm{T}}W + B^{\mathrm{T}}PB + [B^{\mathrm{T}}PA + W^{\mathrm{T}}L](zI_n - A)^{-1}B + B^{\mathrm{T}}(zI_n \\
&\quad -A)^{-*}[A^{\mathrm{T}}PB + L^{\mathrm{T}}W] - B^{\mathrm{T}}(zI_n - A)^{-*}C^{\mathrm{T}}C(zI_n - A)^{-1}B \\
&= W^{\mathrm{T}}W + B^{\mathrm{T}}PB + W^{\mathrm{T}}L(zI_n - A)^{-1}B + B^{\mathrm{T}}(zI_n - A)^{-*}L^{\mathrm{T}}W \\
&\quad -B^{\mathrm{T}}(zI_n - A)^{-*}C^{\mathrm{T}}C(zI_n - A)^{-1}B \\
&\quad +B^{\mathrm{T}}(zI_n - A)^{-*}[(zI_n - A)^*PA + A^{\mathrm{T}}P(zI_n - A)](zI_n - A)^{-1}B \\
&= W^{\mathrm{T}}W + W^{\mathrm{T}}L(zI_n - A)^{-1}B + B^{\mathrm{T}}(zI_n - A)^{-*}L^{\mathrm{T}}W \\
&\quad +B^{\mathrm{T}}(zI_n - A)^{-*}L^{\mathrm{T}}L(zI_n - A)^{-1}B \\
&\quad +(|z| - 1)B^{\mathrm{T}}(zI_n - A)^{-*}P(zI_n - A)^{-1}B \\
&\geq [W + L(zI_n - A)^{-1}B]^*[W + L(zI_n - A)^{-1}B] \\
&\geq 0.
\end{aligned}$$

To show analyticity of the entries of $G(z)$ in $|z| \geq 1$ note that it follows from (13.211) and the fact that $P > 0$ and (A, C) is observable that all the eigenvalues of A lie inside the unit disk. Hence, $G(z)$ is analytic in $|z| \geq 1$, which implies that $G(z)$ is bounded real. Finally, (13.214) follows from (13.168) with the linearization given above. □

Theorem 13.30 (Strict Bounded Real Lemma). Consider the dynamical system

$$G(z) \overset{\min}{\sim} \left[\begin{array}{c|c} A & B \\ \hline C & D \end{array}\right]$$

with input $u(\cdot) \in \mathcal{U}$ and output $y(\cdot) \in \mathcal{Y}$. Then the following statements are equivalent:

i) $G(z)$ is strictly bounded real.

ii) $\sum_{k=0}^{K-1} \rho^{k+1} y^{\mathrm{T}}(k) y(k) \leq \gamma^2 \sum_{k=0}^{K-1} \rho^{k+1} u^{\mathrm{T}}(k) u(k), \quad K-1 \geq 0, \quad \gamma > 0.$

iii) There exist matrices $P \in \mathbb{R}^{n\times n}$, $L \subset \mathbb{R}^{p\times n}$, and $W \in \mathbb{R}^{p\times m}$, with P positive definite, and a scalar $\rho > 1$ such that

$$\frac{1}{\rho}P = A^{\mathrm{T}}PA + C^{\mathrm{T}}C + L^{\mathrm{T}}L, \tag{13.215}$$

$$0 = A^{\mathrm{T}}PB + C^{\mathrm{T}}D + L^{\mathrm{T}}W, \tag{13.216}$$

$$0 = \gamma^2 I_m - D^{\mathrm{T}}D - B^{\mathrm{T}}PB - W^{\mathrm{T}}W. \tag{13.217}$$

If, alternatively, $\gamma^2 I_m - D^{\mathrm{T}}D - B^{\mathrm{T}}PB > 0$, then $G(z)$ is strictly positive real if and only if there exist $n \times n$ positive-definite matrices P and R such that

$$\frac{1}{\rho}P = A^{\mathrm{T}}PA + C^{\mathrm{T}}C + (B^{\mathrm{T}}PA + D^{\mathrm{T}}C)^{\mathrm{T}}(\gamma^2 I_m - D^{\mathrm{T}}D - B^{\mathrm{T}}PB)^{-1} \cdot (B^{\mathrm{T}}PA + D^{\mathrm{T}}C) + R. \tag{13.218}$$

Proof. The equivalence of *i)* and *iii)* is a direct consequence of Theorem 13.29 by noting that $G(z)$ is strictly bounded real if and only if there exists $\rho > 1$ such that

$$G(z/\sqrt{\rho}) \overset{\min}{\sim} \left[\begin{array}{c|c} \sqrt{\rho}A & \sqrt{\rho}B \\ \hline C & D \end{array}\right]$$

is bounded real. The fact that *iii)* implies *ii)* follows from Corollary 13.13 with $f(x) = Ax$, $G(x) = B$, $h(x) = Cx$, $J(x) = D$, $V_{\mathrm{s}}(x) = x^{\mathrm{T}}Px$, $\ell(x) = Lx$, and $\mathcal{W}(x) = W$. To show that *ii)* implies *iii)*, note that if $G(z)$ is geometrically nonexpansive, then it follows from Theorem 13.25 with $f(x) = Ax$, $G(x) = B$, $h(x) = Cx$, $J(x) = D$, $Q = -I_p$, $S = 0$, and $R = \gamma^2 I_m$, that there exist matrices $P \in \mathbb{R}^{n\times n}$, $L \in \mathbb{R}^{p\times n}$, and $W \in \mathbb{R}^{p\times m}$, with P positive definite, such that (13.215)–(13.217) are satisfied.

Finally, with the linearization given above, it follows from (13.178) that if $\gamma^2 I_m - D^{\mathrm{T}}D - B^{\mathrm{T}}PB > 0$, then $G(z)$ is strictly bounded real if and only if there exist a scalar $\varepsilon > 0$ and a positive-definite matrix $P \in \mathbb{R}^{n\times n}$ such that

$$P \geq \rho A^{\mathrm{T}}PA + (\rho B^{\mathrm{T}}PA - C)^{\mathrm{T}}(D + D^{\mathrm{T}} - \rho B^{\mathrm{T}}PB)^{-1}(\rho B^{\mathrm{T}}PA - C). \tag{13.219}$$

Now, if there exist a scalar $\rho > 1$ and a positive-definite matrix $P \in \mathbb{R}^{n\times n}$ such that (13.219) is satisfied, then there exists an $n \times n$ positive-definite matrix R such that (13.218) is satisfied. Conversely, if $\gamma^2 I_m - D^{\mathrm{T}}D - B^{\mathrm{T}}PB > 0$ and there exists an $n \times n$ positive-definite matrix R such that (13.218) is satisfied, then, with $\varepsilon = \sigma_{\min}(R)/\sigma_{\max}(P)$, (13.218) implies

(13.219). Hence, if $\gamma^2 I_m - D^{\mathrm{T}}D - B^{\mathrm{T}}PB > 0$, then $G(z)$ is strictly bounded real if and only if there exist $n \times n$ positive-definite matrices P and R such that (13.218) is satisfied. □

The following result characterizes necessary and sufficient conditions for a transfer function to be strictly positive real in terms of a frequency domain test.

Theorem 13.31. Let

$$G(z) \sim \left[\begin{array}{c|c} A & B \\ \hline C & D \end{array}\right]$$

be an $m \times m$ transfer function and suppose that $G(z)$ is not singular. Then, $G(z)$ is strictly positive real if and only if the following conditions hold:

i) No entry of $G(z)$ has a pole in $|z| \geq 1$.

ii) He $G(e^{\jmath\theta}) > 0$ for all $\theta \in [0, 2\pi]$.

Proof. Suppose *i)* and *ii)* hold. Let $\rho > 1$ and note that

$$\begin{aligned} G(\tfrac{z}{\rho}) &= C(\tfrac{z}{\rho}I - A)^{-1}B + D \\ &= C(zI - A)^{-1}(zI - A)(\tfrac{z}{\rho}I - A)^{-1}B + D \\ &= C(zI - A)^{-1}(\tfrac{z}{\rho}I - A + \tfrac{\rho-1}{\rho}zI))(\tfrac{z}{\rho}I - A)^{-1}B + D \\ &= G(z) + \tfrac{\rho-1}{\rho}G_\rho(z), \end{aligned} \tag{13.220}$$

where $G_\rho(z) \triangleq zC(zI - A)^{-1}(\tfrac{z}{\rho}I - A)^{-1}B$. Since A is Schur, $zI - A$ is nonsingular for all $z = e^{\jmath\theta}$ and there exists $\hat{\rho}$ such that $\tfrac{z}{\rho}I - A$ is nonsingular for all $\rho \in [1, \hat{\rho}]$ and for all $z = e^{\jmath\theta}$. Hence, $f(\theta, \rho) \triangleq \max_i |\lambda_i[\mathrm{He}\ G_\rho(e^{\jmath\theta})]|$ is finite for all $\rho \in [1, \hat{\rho}]$ and for all $\theta \in [0, 2\pi]$. Since $f(\theta, \cdot)$ is continuous it follows that there exists $k_1 > 0$ such that $f(\theta, \rho) < k_1$ for all $\rho \in [1, \hat{\rho}]$ and for all $\theta \in [0, 2\pi]$, so that $-k_1 I_m \leq \mathrm{He}\ G_\rho(z) \leq k_1 I_m$ for all $\rho \in [1, \hat{\rho}]$ and for all $z = e^{\jmath\theta}$. Now, since He $G(e^{\jmath\theta}) > 0$ for all $\theta \in [0, 2\pi]$, it follows that there exists $k_2 > 0$ such that He $G(\jmath\theta) \geq k_2 I_m > 0$, $\theta \in [0, 2\pi]$. Choosing $1 < \rho < \min\{\hat{\rho}, \frac{k_1}{k_1 - k_2}\}$, it follows that

$$\mathrm{He}\ G(\tfrac{z}{\rho}) = \mathrm{He}\ G(z) + \frac{\rho - 1}{\rho}\mathrm{He}\ G_\rho(z) \geq k_2 I_m - \frac{\rho - 1}{\rho}k_1 I_m > 0, \tag{13.221}$$

for all $z = e^{\jmath\theta}$. Hence, $G(\tfrac{z}{\rho})$ is positive real, and hence, by definition, $G(z)$ is strictly positive real.

Conversely, suppose $G(z)$ is strictly positive real and let $\rho > 1$ be such that $G(\tfrac{z}{\rho})$ is positive real. Note that $G(z)$ is asymptotically stable and positive real. Hence, He $G(e^{j\theta}) \geq 0$, $\theta \in [0, 2\pi]$. Now, let (A, B, C, D) be a minimal realization of $G(z)$. Using Theorem 13.28, it follows from (13.206)–(13.207) that

$$\begin{aligned}
G(z) + G^*(z) & \\
&= C(zI_n - A)^{-1}B + D + B^{\mathrm{T}}(z^*I_n - A)^{-\mathrm{T}}C^{\mathrm{T}} + D^{\mathrm{T}} \\
&= W^{\mathrm{T}}W + B^{\mathrm{T}}PB + (B^{\mathrm{T}}PA + W^{\mathrm{T}}L)(zI_n - A)^{-1}B \\
&\quad + B^{\mathrm{T}}(z^*I_n - A)^{-\mathrm{T}}(A^{\mathrm{T}}PB + L^{\mathrm{T}}W) \\
&= W^{\mathrm{T}}W + W^{\mathrm{T}}L(zI_n - A)^{-1}B + B^{\mathrm{T}}(z^*I_n - A)^{-\mathrm{T}}L^{\mathrm{T}}W \\
&\quad + B^{\mathrm{T}}(z^*I_n - A)^{-\mathrm{T}}[(z^*I_n - A)^{\mathrm{T}}P(zI_n - A) + (z^*I_n - A)^{\mathrm{T}}PA \\
&\quad + A^{\mathrm{T}}P(zI_n - A)](zI_n - A)^{-1}B \\
&= W^{\mathrm{T}}W + W^{\mathrm{T}}L(zI_n - A)^{-1}B + B^{\mathrm{T}}(z^*I_n - A)^{-\mathrm{T}}L^{\mathrm{T}}W \\
&\quad + B^{\mathrm{T}}(z^*I_n - A)^{-\mathrm{T}}[L^{\mathrm{T}}L + (|z| - \tfrac{1}{\rho})P](zI_n - A)^{-1}B. \qquad (13.222)
\end{aligned}$$

Now, suppose, *ad absurdum*, that He $G(e^{j\theta})$ is not positive definite for all $\theta \in [0, 2\pi]$. Then, for some $\theta = \hat{\theta}$ there exists $x \in \mathbb{C}^m$, $x \neq 0$, such that $x^*[\text{He } G(e^{j\theta})]x = 0$. In this case, it follows from (13.222) that $Bx = 0$ and $Wx = 0$. Hence, $x^*[\text{He } G(z)]x = 0$ for all $z \in \mathbb{C}$, and hence, $\det[\text{He } G(z)] \equiv 0$, which leads to a contradiction. Thus, He $G(e^{j\theta}) > 0$, $\theta \in [0, 2\pi]$, which proves the result. □

13.13 Feedback Interconnections of Dissipative Dynamical Systems

In this section, we consider feedback interconnections of dissipative dynamical systems. Specifically, using the notion of dissipative and geometrically dissipative dynamical systems, with appropriate storage functions and supply rates, we construct Lyapunov functions for interconnected dynamical systems by appropriately combining storage functions for each subsystem. We begin by considering the nonlinear discrete-time dynamical system $\mathcal{G}$ given by

$$x(k+1) = f(x(k)) + G(x(k))u(k), \quad x(0) = x_0, \quad k \in \overline{\mathbb{Z}}_+, \qquad (13.223)$$

$$y(k) = h(x(k)) + J(x(k))u(k), \qquad (13.224)$$

where $x \in \mathbb{R}^n$, $u \in \mathbb{R}^m$, $y \in \mathbb{R}^l$, $f : \mathbb{R}^n \to \mathbb{R}^n$ satisfies $f(0) = 0$, $G : \mathbb{R}^n \to \mathbb{R}^{n \times m}$, $h : \mathbb{R}^n \to \mathbb{R}^l$ satisfies $h(0) = 0$, and $J : \mathbb{R}^n \to \mathbb{R}^{l \times m}$, with the nonlinear feedback system $\mathcal{G}_c$ given by

$$x_c(k+1) = f_c(x_c(k)) + G_c(u_c(k), x_c(k))u_c(k), \quad x_c(0) = x_{c0}, \quad k \in \overline{\mathbb{Z}}_+, \qquad (13.225)$$

$$y_c(k) = h_c(u_c(k), x_c(k)) + J_c(u_c(k), x_c(k))u_c(k), \qquad (13.226)$$

where $x_c \in \mathbb{R}^{n_c}$, $u_c \in \mathbb{R}^l$, $y_c \in \mathbb{R}^m$, $f_c : \mathbb{R}^{n_c} \to \mathbb{R}^{n_c}$ satisfies $f_c(0) = 0$, $G_c : \mathbb{R}^l \times \mathbb{R}^{n_c} \to \mathbb{R}^{n_c \times l}$, $h_c : \mathbb{R}^l \times \mathbb{R}^{n_c} \to \mathbb{R}^m$ satisfies $h_c(0) = 0$, and $J_c : \mathbb{R}^l \times \mathbb{R}^{n_c} \to \mathbb{R}^{m \times l}$. Note that with the feedback interconnection given by Figure 13.1, $u_c = y$ and $y_c = -u$. The following results give sufficient conditions for Lyapunov, asymptotic, and geometric stability of the feedback interconnection given by Figure 13.1. For these results we assume that the negative feedback interconnection of $\mathcal{G}$ and $\mathcal{G}_c$ is well posed.

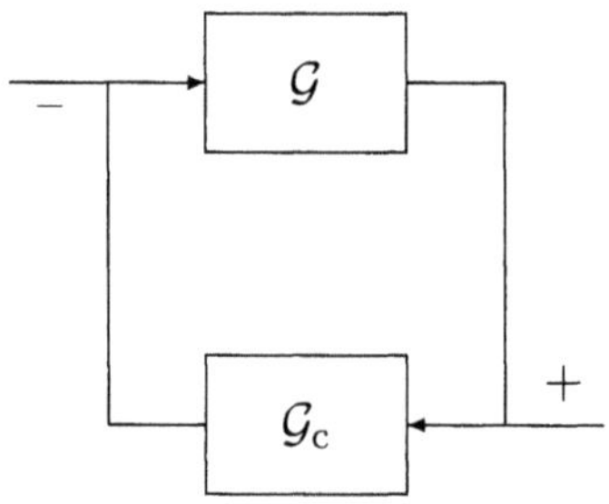

Figure 13.1 Feedback interconnection of $\mathcal{G}$ and $\mathcal{G}_c$.

Theorem 13.32. Consider the closed-loop system consisting of the nonlinear dynamical systems $\mathcal{G}$ given by (13.223) and (13.224) and $\mathcal{G}_c$ given by (13.225) and (13.226) with input-output pairs (u, y) and (u_c, y_c), respectively, and with $u_c = y$ and $y_c = -u$. Assume $\mathcal{G}$ and $\mathcal{G}_c$ are zero state observable and dissipative with respect to the supply rates $r(u, y)$ and $r_c(u_c, y_c)$ and with continuous positive definite radially unbounded storage functions $V_s(\cdot)$ and $V_{sc}(\cdot)$, respectively, such that $V_s(0) = 0$ and $V_{sc}(0) = 0$. Furthermore, assume there exists a scalar $\sigma > 0$ such that $r(u, y) + \sigma r_c(u_c, y_c) \leq 0$. Then the following statements hold:

i) The negative feedback interconnection of $\mathcal{G}$ and $\mathcal{G}_c$ is Lyapunov stable.

ii) If $\mathcal{G}_c$ is geometrically dissipative with respect to supply rate $r_c(u_c, y_c)$ and $\operatorname{rank}[G_c(u_c, 0)] = m$, $u_c \in \mathbb{R}^l$, then the negative feedback interconnection of $\mathcal{G}$ and $\mathcal{G}_c$ is globally asymptotically stable.

iii) If $\mathcal{G}$ and $\mathcal{G}_c$ are geometrically dissipative with respect to supply rates $r(u, y)$ and $r_c(u_c, y_c)$, respectively, $V_s(\cdot)$ and $V_{sc}(\cdot)$ are such that there exist constants $\alpha, \alpha_c, \beta, \beta_c > 0$ such that

$$\alpha \|x\|^2 \leq V_s(x) \leq \beta \|x\|^2, \qquad x \in \mathbb{R}^n, \tag{13.227}$$

$$\alpha_c \|x_c\|^2 \leq V_{sc}(x_c) \leq \beta_c \|x_c\|^2, \qquad x_c \in \mathbb{R}^{n_c}, \tag{13.228}$$

then the negative feedback interconnection of $\mathcal{G}$ and $\mathcal{G}_c$ is globally geometrically stable.

Proof. *i)* Consider the Lyapunov function candidate $V(x, x_c) = V_s(x) + \sigma V_{sc}(x_c)$. Now, the corresponding Lyapunov difference is given by

$$\Delta V(x, x_c) = \Delta V_s(x) + \sigma \Delta V_{sc}(x_c) \le r(u, y) + \sigma r_c(u_c, y_c) \le 0, \quad (x, x_c) \in \mathbb{R}^n \times \mathbb{R}^{n_c},$$

which implies that the negative feedback interconnection of $\mathcal{G}$ and $\mathcal{G}_c$ is Lyapunov stable.

ii) Next, consider the Lyapunov function candidate $V(x, x_c) = V_s(x) + \sigma V_{sc}(x_c)$. If $\mathcal{G}_c$ is geometrically dissipative it follows that for some scalar $\rho_c > 1$,

$$\begin{aligned} \Delta V(x, x_c) &= \Delta V_s(x) + \sigma \Delta V_{sc}(x_c) \\ &\le -\sigma \frac{\rho_c - 1}{\rho_c} V_{sc}(x_c) + r(u, y) + \sigma r_c(u_c, y_c) \\ &\le -\sigma \frac{\rho_c - 1}{\rho_c} V_{sc}(x_c), \qquad (x, x_c) \in \mathbb{R}^n \times \mathbb{R}^{n_c}. \end{aligned}$$

Next, let $\mathcal{R} \triangleq \{(x, x_c) \in \mathbb{R}^n \times \mathbb{R}^{n_c} : \Delta V(x, x_c) = 0\}$ and, since $V_{sc}(x_c)$ is positive definite, note that $\Delta V(x, x_c) = 0$ only if $x_c = 0$. Now, since $\operatorname{rank}[G_c(u_c, 0)] = m$, $u_c \in \mathbb{R}^l$, it follows that on every invariant set $\mathcal{M}$ contained in $\mathcal{R}$, $u_c(k) = y(k) \equiv 0$, and hence, $x(k+1) = f(x(k))$. Now, since $\mathcal{G}$ is zero-state observable it follows that $\mathcal{M} = \{(0, 0)\}$ is the largest invariant set contained in $\mathcal{R}$. Hence, it follows from Theorem 13.5 that $(x(k), x_c(k)) \to \mathcal{M} = \{0, 0\}$ as $k \to \infty$. Now, global asymptotic stability of the negative feedback interconnection of $\mathcal{G}$ and $\mathcal{G}_c$ follows from the fact that $V_s(\cdot)$ and $V_{sc}(\cdot)$ are, by assumption, radially unbounded.

iii) Finally, consider the Lyapunov function candidate $V(x, x_c) = V_s(x) + \sigma V_{sc}(x_c)$. Since $\mathcal{G}$ and $\mathcal{G}_c$ are geometrically dissipative it follows that

$$\begin{aligned} \Delta V(x, x_c) &= \Delta V_s(x) + \sigma \Delta V_{sc}(x_c) \\ &\le -\frac{\rho - 1}{\rho} V_s(x) - \sigma \frac{\rho_c - 1}{\rho_c} V_{sc}(x_c) + r(u, y) + \sigma r_c(u_c, y_c) \\ &\le -\min\left\{\frac{\rho - 1}{\rho}, \frac{\rho_c - 1}{\rho_c}\right\} V(x, x_c), \qquad (x, x_c) \in \mathbb{R}^n \times \mathbb{R}^{n_c}, \end{aligned}$$

which implies that the negative feedback interconnection of $\mathcal{G}$ and $\mathcal{G}_c$ is globally geometrically stable. □

The next result presents Lyapunov, asymptotic, and geometric stability of dissipative feedback systems with quadratic supply rates.

Theorem 13.33. Let $Q \in \mathbb{S}^l$, $S \in \mathbb{R}^{l \times m}$, $R \in \mathbb{S}^l$, $Q_c \in \mathbb{S}^m$, $S_c \in \mathbb{R}^{m \times l}$, and $R_{\mathrm{c}} \in \mathbb{S}^l$. Consider the closed-loop system consisting of the nonlinear dynamical systems $\mathcal{G}$ given by (13.223) and (13.224), and $\mathcal{G}_{\mathrm{c}}$ given by (13.225) and (13.226), and assume $\mathcal{G}$ and $\mathcal{G}_{\mathrm{c}}$ are zero-state observable. Furthermore, assume $\mathcal{G}$ is dissipative with respect to the quadratic supply rate $r(u,y) = y^{\mathrm{T}}Qy + 2y^{\mathrm{T}}Su + u^{\mathrm{T}}Ru$ and has a continuous radially unbounded storage function $V_{\mathrm{s}}(\cdot)$ and $\mathcal{G}_{\mathrm{c}}$ is dissipative with respect to the quadratic supply rate $r_{\mathrm{c}}(u_{\mathrm{c}}, y_{\mathrm{c}}) = y_{\mathrm{c}}^{\mathrm{T}}Q_{\mathrm{c}}y_{\mathrm{c}} + 2y_{\mathrm{c}}^{\mathrm{T}}S_{\mathrm{c}}u_{\mathrm{c}} + u_{\mathrm{c}}^{\mathrm{T}}R_{\mathrm{c}}u_{\mathrm{c}}$ and has a continuous radially unbounded storage function $V_{\mathrm{sc}}(\cdot)$. Finally, assume there exists $\sigma > 0$ such that

$$\hat{Q} \triangleq \begin{bmatrix} Q + \sigma R_{\mathrm{c}} & -S + \sigma S_{\mathrm{c}}^{\mathrm{T}} \\ -S^{\mathrm{T}} + \sigma S_{\mathrm{c}} & R + \sigma Q_{\mathrm{c}} \end{bmatrix} \leq 0. \tag{13.229}$$

Then the following statements hold:

i) The negative feedback interconnection of $\mathcal{G}$ and $\mathcal{G}_{\mathrm{c}}$ is Lyapunov stable.

ii) If $\mathcal{G}_{\mathrm{c}}$ is geometrically dissipative with respect to supply rate $r_{\mathrm{c}}(u_{\mathrm{c}}, y_{\mathrm{c}})$ and $\operatorname{rank}[G_{\mathrm{c}}(u_{\mathrm{c}}, 0)] = m$, $u_{\mathrm{c}} \in \mathbb{R}^l$, then the negative feedback interconnection of $\mathcal{G}$ and $\mathcal{G}_{\mathrm{c}}$ is globally asymptotically stable.

iii) If $\mathcal{G}$ and $\mathcal{G}_{\mathrm{c}}$ are geometrically dissipative with respect to supply rates $r(u,y)$ and $r_{\mathrm{c}}(u_{\mathrm{c}}, y_{\mathrm{c}})$ and there exist constants $\alpha, \beta, \alpha_{\mathrm{c}}, \beta_{\mathrm{c}} > 0$ such that (13.227) and (13.228) hold, then the negative feedback interconnection of $\mathcal{G}$ and $\mathcal{G}_{\mathrm{c}}$ is globally geometrically stable.

iv) If $\hat{Q} < 0$, then the negative feedback interconnection of $\mathcal{G}$ and $\mathcal{G}_{\mathrm{c}}$ is globally asymptotically stable.

Proof. Statements *i)*–*iii)* are a direct consequence of Theorem 13.32 by noting

$$r(u,y) + \sigma r_{\mathrm{c}}(u_{\mathrm{c}}, y_{\mathrm{c}}) = \begin{bmatrix} y \\ y_{\mathrm{c}} \end{bmatrix}^{\mathrm{T}} \hat{Q} \begin{bmatrix} y \\ y_{\mathrm{c}} \end{bmatrix},$$

and hence, $r(u,y) + \sigma r_{\mathrm{c}}(u_{\mathrm{c}}, y_{\mathrm{c}}) \leq 0$.

To show *iv)* consider the Lyapunov function candidate $V(x, x_{\mathrm{c}}) = V_{\mathrm{s}}(x) + \sigma V_{\mathrm{sc}}(x_{\mathrm{c}})$. Now, the corresponding Lyapunov difference is given by

$$\begin{aligned} \Delta V(x, x_{\mathrm{c}}) &= \Delta V_{\mathrm{s}}(x) + \sigma \Delta V_{\mathrm{sc}}(x_{\mathrm{c}}) \\ &\leq r(u,y) + \sigma r_{\mathrm{c}}(u_{\mathrm{c}}, y_{\mathrm{c}}) \\ &= y^{\mathrm{T}}Qy + 2y^{\mathrm{T}}Su + u^{\mathrm{T}}Ru + \sigma(y_{\mathrm{c}}^{\mathrm{T}}Q_{\mathrm{c}}y_{\mathrm{c}} + 2y_{\mathrm{c}}^{\mathrm{T}}S_{\mathrm{c}}u_{\mathrm{c}} + u_{\mathrm{c}}^{\mathrm{T}}R_{\mathrm{c}}u_{\mathrm{c}}) \\ &= \begin{bmatrix} y \\ y_{\mathrm{c}} \end{bmatrix}^{\mathrm{T}} \hat{Q} \begin{bmatrix} y \\ y_{\mathrm{c}} \end{bmatrix} \\ &\leq 0, \end{aligned}$$

which implies that the negative feedback interconnection of $\mathcal{G}$ and $\mathcal{G}_c$ is Lyapunov stable. Now, the proof of global asymptotic stability of the closed-loop system is identical to that of the continuous-time case given in Theorem 6.2 and, hence, is omitted. □

The following corollaries are a direct consequence of Theorem 13.33. For both results note that if a nonlinear dynamical system $\mathcal{G}$ is dissipative (respectively, geometrically dissipative) with respect to a supply rate $r(u, y) = u^{\mathrm{T}}y - \varepsilon u^{\mathrm{T}}u - \hat{\varepsilon}y^{\mathrm{T}}y$, where $\varepsilon, \hat{\varepsilon} \geq 0$, then with $\kappa(y) = ky$, where $k \in \mathbb{R}$ is such that $k(1 - \varepsilon k) < \hat{\varepsilon}$, $r(u, y) = [k(1 - \varepsilon k) - \hat{\varepsilon}]y^{\mathrm{T}}y < 0$, $y \neq 0$. Hence, if $\mathcal{G}$ is zero-state observable it follows from Theorem 13.21 that all storage functions (respectively, geometrically storage functions) of $\mathcal{G}$ are positive definite. For the next result, we assume that all storage functions of $\mathcal{G}$ and $\mathcal{G}_c$ are continuous.

Corollary 13.16. Consider the closed-loop system consisting of the nonlinear dynamical systems $\mathcal{G}$ given by (13.223) and (13.224) and $\mathcal{G}_c$ given by (13.225) and (13.226), and assume $\mathcal{G}$ and $\mathcal{G}_c$ are zero-state observable. Then the following statements hold:

i) If $\mathcal{G}$ is passive, $\mathcal{G}_c$ is geometrically passive, and $\operatorname{rank}[G_c(u_c, 0)] = m$, $u_c \in \mathbb{R}^l$, then the negative feedback interconnection of $\mathcal{G}$ and $\mathcal{G}_c$ is asymptotically stable.

ii) If $\mathcal{G}$ and $\mathcal{G}_c$ are geometrically passive with storage functions $V_s(\cdot)$ and $V_{sc}(\cdot)$, respectively, such that (13.227) and (13.228) hold, then the negative feedback interconnection of $\mathcal{G}$ and $\mathcal{G}_c$ is geometrically stable.

iii) If $\mathcal{G}$ is nonexpansive with gain $\gamma > 0$, $\mathcal{G}_c$ is geometrically nonexpansive with gain $\gamma_c > 0$, $\operatorname{rank}[G_c(u_c, 0)] = m$, and $\gamma\gamma_c \leq 1$, then the negative feedback interconnection of $\mathcal{G}$ and $\mathcal{G}_c$ is asymptotically stable.

iv) If $\mathcal{G}$ and $\mathcal{G}_c$ are geometrically nonexpansive with storage functions $V_s(\cdot)$ and $V_{sc}(\cdot)$, respectively, such that (13.227) and (13.228) hold, with gains $\gamma > 0$ and $\gamma_c > 0$, respectively, such that $\gamma\gamma_c \leq 1$, then the negative feedback interconnection of $\mathcal{G}$ and $\mathcal{G}_c$ is geometrically stable.

v) If $\mathcal{G}$ is passive and $\mathcal{G}_c$ is input-output strict passive then the negative feedback interconnection of $\mathcal{G}$ and $\mathcal{G}_c$ is asymptotically stable.

vi) If $\mathcal{G}$ and $\mathcal{G}_c$ are input strict passive then the negative feedback interconnection of $\mathcal{G}$ and $\mathcal{G}_c$ is asymptotically stable.

vii) If $\mathcal{G}$ and $\mathcal{G}_c$ are output strict passive then the negative feedback interconnection of $\mathcal{G}$ and $\mathcal{G}_c$ is asymptotically stable.

Proof. The proof is a direct consequence of Theorem 13.33. Specifically, *i*) and *ii*) follow from Theorem 13.33 with $Q = Q_c = 0$, $S = S_c = I_m$, and $R = R_c = 0$, while *iii*) and *iv*) follow from Theorem 13.33 with $Q = -I_l$, $S = 0$, $R = \gamma^2 I_m$, $Q_c = -I_{l_c}$, $S_c = 0$, and $R_c = \gamma_c^2 I_{m_c}$. Statement *v*) follows from Theorem 13.33 with $Q = 0$, $S = I_m$, $R = 0$, $Q_c = -\hat{\varepsilon} I_m$, $S_c = I_m$, and $R_c = -\varepsilon I_m$, where $\varepsilon, \hat{\varepsilon} > 0$. Statement *vi*) follows from Theorem 13.33 with $Q = 0$, $S = I_m$, $R = -\varepsilon I_m$, $Q_c = 0$, $S_c = \frac{1}{2} I_m$, and $R_c = -\hat{\varepsilon} I_m$, where $\varepsilon, \hat{\varepsilon} > 0$. Finally, *vii*) follows from Theorem 13.33 with $Q = -\varepsilon I_m$, $S = I_m$, $R = 0$, $Q_c = -\hat{\varepsilon} I_m$, $S_c = I_m$, and $R_c = 0$, where $\varepsilon, \hat{\varepsilon} > 0$. □

Corollary 13.17. Consider the closed-loop system consisting of the nonlinear dynamical systems $\mathcal{G}$ given by (13.223) and (13.224), and $\mathcal{G}_c$ given by (13.225) and (13.226). Let $a, b, a_c, b_c, \delta \in \mathbb{R}$ be such that $b > 0$, $0 < a+b$, $0 < 2\delta < b-a$, $a_c = a+\delta$, and $b_c = b+\delta$, let $M \in \mathbb{R}^{m\times m}$ be positive definite, and assume $\mathcal{G}$ and $\mathcal{G}_c$ are zero-state observable. If $\mathcal{G}$ is dissipative with respect to the supply rate $r(u,y) = u^{\mathrm{T}} M y + \frac{ab}{a+b} y^{\mathrm{T}} M y + \frac{1}{a+b} u^{\mathrm{T}} M u$ and has a continuous radially unbounded storage function and $\mathcal{G}_c$ is dissipative with respect to the supply rate $r_c(u_c, y_c) = u_c^{\mathrm{T}} M y_c - \frac{1}{a_c+b_c} y_c^{\mathrm{T}} M y_c - \frac{a_c b_c}{a_c+b_c} u_c^{\mathrm{T}} M u_c$ and has a continuous radially unbounded storage function, then the negative feedback interconnection of $\mathcal{G}$ and $\mathcal{G}_c$ is globally asymptotically stable.

Proof. The proof is identical to that of the continuous-time case given in Corollary 6.2 and, hence, is omitted. □

13.14 Stability Margins of Discrete Regulators

In this section, we develop sufficient conditions for gain, sector, and disk margin guarantees for discrete-time nonlinear systems controlled by nonlinear feedback regulators. To consider relative stability margins for nonlinear discrete-time regulators consider the nonlinear system $\mathcal{G}$ given by

$$x(k+1) = f(x(k)) + G(x(k))u(k), \quad x(0) = x_0, \quad k \in \overline{\mathbb{Z}}_+, \tag{13.230}$$
$$y(k) = -\phi(x(k)), \tag{13.231}$$

where $\phi : \mathbb{R}^n \to \mathbb{R}^m$ is such that $\mathcal{G}$ is asymptotically stable with $u = -y$. Furthermore, assume that the system $\mathcal{G}$ is zero-state observable. The following results provide algebraic sufficient conditions that guarantee disk margins for the nonlinear dynamical system $\mathcal{G}$ given by (13.230) and (13.231).

Theorem 13.34. Consider the nonlinear dynamical system $\mathcal{G}$ given by (13.230) and (13.231). Let $\alpha, \beta \in \mathbb{R}$ be such that $0 < \alpha \le 1 \le \beta < \infty$. Suppose there exists a positive-definite diagonal matrix $Z \in \mathbb{R}^{m\times m}$ and there exist functions $V_s : \mathbb{R}^n \to \mathbb{R}$, $P_{1u} : \mathbb{R}^n \to \mathbb{R}^{1\times m}$, $P_{2u} : \mathbb{R}^n \to \mathbb{N}^m$, $\ell : \mathbb{R}^n \to \mathbb{R}^p$, and $\mathcal{W} : \mathbb{R}^n \to \mathbb{R}^{p\times m}$ such that $V_s(\cdot)$ is continuous, $V_s(0) = 0$,

$V_s(x) > 0$, $x \in \mathbb{R}^n$, $x \neq 0$, and for all $x \in \mathbb{R}^n$, (13.137) holds and

$$0 = V_s(f(x)) - V_s(x) - \tfrac{\alpha\beta}{\alpha+\beta}\phi^T(x)Z\phi(x) + \ell^T(x)\ell(x), \quad (13.232)$$
$$0 = P_{1u}(x) + \phi^T(x)Z + 2\ell^T(x)\mathcal{W}(x), \quad (13.233)$$
$$0 = \tfrac{1}{\alpha+\beta}Z - P_{2u}(x) - \mathcal{W}^T(x)\mathcal{W}(x). \quad (13.234)$$

Then the nonlinear system $\mathcal{G}$ has a structured disk margin (α, β). Alternatively, if (13.232)–(13.234) are satisfied with $Z = I_m$, then the nonlinear system $\mathcal{G}$ has a disk margin (α, β).

Proof. The proof is identical to that of the continuous-time case given in Theorem 6.5 and, hence, is omitted. □

Corollary 13.18. Consider the nonlinear system $\mathcal{G}$ given by (13.230) and (13.231). Let $\alpha, \beta \in \mathbb{R}$ be such that $0 < \alpha \leq 1 \leq \beta < \infty$. Suppose there exist a positive-definite diagonal matrix $Z \in \mathbb{R}^{m\times m}$, functions $P_{1u} : \mathbb{R}^n \to \mathbb{R}^{1\times m}$, $P_{2u} : \mathbb{R}^n \to \mathbb{N}^m$, and a continuous function $V_s : \mathbb{R}^n \to \mathbb{R}$ such that $V_s(0) = 0$, $V_s(x) > 0$, $x \in \mathbb{R}^n$, $x \neq 0$, and for all $x \in \mathbb{R}^n$, (13.137) holds and

$$0 < \tfrac{1}{\alpha+\beta}Z - P_{2u}(x), \quad (13.235)$$
$$0 \geq V_s(f(x)) - V_s(x) - \tfrac{\alpha\beta}{\alpha+\beta}\phi^T(x)Z\phi(x)$$
$$+\tfrac{1}{4}(\phi^T(x)Z + P_{1u}(x))(\tfrac{1}{\alpha+\beta}Z - P_{2u}(x))^{-1}(\phi^T(x)Z + P_{1u}(x))^T. \quad (13.236)$$

Then the nonlinear system $\mathcal{G}$ has a structured disk margin (α, β). Alternatively, if (13.235) and (13.236) are satisfied with $Z = I_m$, then the nonlinear system $\mathcal{G}$ has a disk margin (α, β).

Proof. It follows from Theorem 13.21 that (13.232)–(13.234) are equivalent to (13.236). Now, the result is a direct consequence of Theorem 13.34. □

The following theorem gives the nonlinear version of the results of [262].

Theorem 13.35. Consider the nonlinear dynamical system $\mathcal{G}$ given by (13.230) and (13.231). Let $\alpha, \beta \in \mathbb{R}$ be such that $0 < \alpha \leq 1 \leq \beta < \infty$. Suppose there exist a positive-definite diagonal matrix $Z \in \mathbb{R}^{m\times m}$, a continuous function $V : \mathbb{R}^n \to \mathbb{R}$, $\hat{P}_{1u} : \mathbb{R}^n \to \mathbb{R}^{1\times m}$, $\hat{P}_{2u} : \mathbb{R}^n \to \mathbb{N}^m$, and a scalar $q > 0$, and for all $x \in \mathbb{R}^n$, (13.137) holds and

$$0 < qI - \hat{P}_{2u}(x), \quad (13.237)$$
$$0 \geq V(f(x)) - V(x) - \tfrac{\alpha\beta}{q(\alpha+\beta)^2}\hat{P}_{1u}(x)Z^{-1}\hat{P}_{1u}^T(x). \quad (13.238)$$

Then, with $\phi(x) = -\frac{1}{q(\alpha+\beta)}Z^{-1}G^{\mathrm{T}}(x)V'^{\mathrm{T}}(x)$, the nonlinear system $\mathcal{G}$ has a structured disk margin (α,β). Alternatively, if (13.238) is satisfied with $Z = I_m$, then the nonlinear system $\mathcal{G}$ has a disk margin (α,β).

Proof. The result is a direct consequence of Corollary 13.18 with $V_{\mathrm{s}}(x) = \frac{1}{q(\alpha+\beta)}V(x)$, $P_{1u}(x) = \frac{1}{q(\alpha+\beta)}\hat{P}_{1u}(x)$, and $P_{2u}(x) = \frac{1}{q(\alpha+\beta)}\hat{P}_{2u}(x)$. Specifically, since $\phi(x) = -Z^{-1}G^{\mathrm{T}}(x)\cdot\hat{P}_{1u}(x)$ it follows from (13.238) that for all $x \in \mathbb{R}^n$,

$$\begin{aligned}
0 &\geq V(f(x)) - V(x) - \tfrac{\alpha\beta}{q(\alpha+\beta)^2}P_{1u}(x)P_{1u}^{\mathrm{T}}(x)\\
&= q(\alpha+\beta)\left(V_{\mathrm{s}}(f(x)) - V_{\mathrm{s}}(x) - \tfrac{\alpha\beta}{\alpha+\beta}\phi^{\mathrm{T}}(x)\phi(x)\right)\\
&= q(\alpha+\beta)\Big(V_{\mathrm{s}}(f(x)) - V_{\mathrm{s}}(x) - \tfrac{\alpha\beta}{\alpha+\beta}\phi^{\mathrm{T}}(x)\phi(x)\\
&\quad +\tfrac{1}{4}(\phi(x)+\hat{P}_{1u}^{\mathrm{T}}(x))^{\mathrm{T}}(\tfrac{1}{\alpha+\beta}I - \hat{P}_{2u}(x))^{-1}(\phi(x)+\hat{P}_{1u}^{\mathrm{T}}(x))\Big),
\end{aligned}$$

which implies (13.236), so that all the conditions of Corollary 13.18 are satisfied. □

13.15 Control Lyapunov Functions for Discrete-Time Systems

In this section, we consider a feedback control problem and introduce the notion of control Lyapunov functions for discrete-time systems. Consider the controlled discrete-time nonlinear dynamical system given by

$$x(k+1) = F(x(k),u(k)), \quad x(t_0) = x_0, \quad k \in \overline{\mathbb{Z}}_+, \tag{13.239}$$

where $x(k) \in \mathcal{D} \subseteq \mathbb{R}^n$, $k \in \overline{\mathbb{Z}}_+$, is the state vector, $\mathcal{D}$ is an open set with $0 \in \mathcal{D}$, $u(k) \in U \subseteq \mathbb{R}^m$ is the control input, and $F : \mathcal{D}\times U \to \mathbb{R}^n$ satisfies $F(0,0) = 0$. We assume that the control input $u(\cdot) \in \mathcal{U}$ in (13.239) is restricted to the class of *admissible controls* consisting of measurable functions $u(\cdot) \in \mathcal{U}$ such that $u(k) \in U$ for all $k \in \overline{\mathbb{Z}}_+$, where the constraint set U is given with $0 \in U$. A mapping $\phi : \mathcal{D} \to U$ satisfying $\phi(0) = 0$ is called a *control law.* Furthermore, if $u(k) = \phi(x(k))$, where ϕ is a control law and $x(k)$, $k \in \overline{\mathbb{Z}}_+$, satisfies (13.239), then $u(\cdot) \in \mathcal{U}$ is called a *feedback control law.* With $u(k) = \phi(x(k))$ the closed-loop system is given by

$$x(k+1) = F(x(k),\phi(x(k))), \quad x(0) = x_0, \quad k \in \overline{\mathbb{Z}}_+. \tag{13.240}$$

The following two definitions are required for stating the results of this section.

Definition 13.18. Let $\phi : \mathcal{D} \to U$ be a mapping on $\mathcal{D}\backslash\{0\}$ with $\phi(0) = 0$. Then (13.239) is *feedback asymptotically stabilizable* if the zero solution $x(k) \equiv 0$ of the closed-loop system (13.240) is asymptotically stable.

Definition 13.19. Consider the controlled discrete-time nonlinear dynamical system given by (13.239). A continuous positive-definite function $V : \mathcal{D} \to \mathbb{R}$ satisfying

$$\inf_{u \in U} V(F(x,u)) < V(x), \quad x \in \mathcal{D}, \quad x \neq 0, \tag{13.241}$$

is called a *control Lyapunov function.*

Note that it follows from (13.241) that there exists a feedback control law $\phi : \mathcal{D} \to U$ such that $V(F(x,\phi(x))) - V(x) < 0$, $x \in \mathcal{D}$, $x \neq 0$, and hence, Theorem 13.2 implies that if there exists a control Lyapunov function for the nonlinear dynamical system (13.239), then there exists a feedback control law $\phi(x)$ such that the zero solution $x(k) \equiv 0$ of the closed-loop nonlinear dynamical system (13.239) is asymptotically stable. Conversely, if there exists a feedback control law $u = \phi(x)$ such that the zero solution $x(k) \equiv 0$ of the nonlinear dynamical system (13.239) is asymptotically stable, then it follows from Theorem 13.6 that there exists a continuous positive-definite function $V : \mathcal{D} \to \mathbb{R}$ such that $V(F(x,\phi(x))) - V(x) < 0$, $x \in \mathcal{D}$, $x \neq 0$, or, equivalently, there exists a control Lyapunov function for the nonlinear dynamical system (13.239). Hence, a given nonlinear dynamical system of the form (13.239) is feedback asymptotically stabilizable if and only if there exists a control Lyapunov function satisfying (13.241). Finally, in the case where $\mathcal{D} = \mathbb{R}^n$ and $U = \mathbb{R}^m$ the zero solution $x(k) \equiv 0$ to (13.239) is globally asymptotically stabilizable if and only if $V(x) \to \infty$ as $\|x\| \to \infty$.

Next, we consider the special case of nonlinear affine systems in the control and construct state feedback controllers that globally asymptotically stabilize the zero solution of the nonlinear dynamical system under the assumption that the system has a radially unbounded control Lyapunov function. Specifically, we consider nonlinear affine systems of the form

$$x(k+1) = f(x(k)) + G(x(k))u(k), \quad x(t_0) = x_0, \quad k \in \overline{\mathbb{Z}}_+, \tag{13.242}$$

where $f : \mathbb{R}^n \to \mathbb{R}^n$ is continuous and satisfies $f(0) = 0$, and $G : \mathbb{R}^n \to \mathbb{R}^{n \times m}$ is continuous. For the nonlinear system $\mathcal{G}$ we assume that there exists a two-times continuously differentiable control Lyapunov function $V(x)$ and there exist functions $P_{1u} : \mathbb{R}^n \to \mathbb{R}^{1 \times m}$ and $P_{2u} : \mathbb{R}^n \to \mathbb{N}^m$ such that $V(f(x) + G(x)u) = V(f(x)) + P_{1u}(x)u + u^{\mathrm{T}}P_{2u}(x)u$, $x \in \mathbb{R}^n$, $u \in \mathbb{R}^m$. Furthermore, we assume that

$$\left.\frac{\partial^2 V}{\partial x^2}\right|_{x=f(x)} > 0.$$

In this case, it can be shown using a Taylor series expansion about $x = f(x)$ that $P_{1u}(x) = V'(f(x))G(x)$, $P_{2u}(x) = G^{\mathrm{T}}(x) \left.\frac{\partial^2 V}{\partial x^2}\right|_{x=f(x)} G(x)$, and

$\operatorname{rank}[P_{2u}(x)\ P_{1u}^{\mathrm{T}}(x)] = \operatorname{rank} P_{2u}(x)$, $x \in \mathbb{R}^n$. Next, note that the function $\Gamma(x) \triangleq V(f(x)+G(x)u) - V(x) = V(f(x)) + P_{1u}(x)u + u^{\mathrm{T}}P_{2u}(x)u - V(x)$ is a convex function since $\frac{\partial^2\Gamma}{\partial u^2} \geq 0$, $u \in \mathbb{R}^m$, and hence, $\Gamma(x)$ has a global minimum. Now, setting $\frac{\partial\Gamma}{\partial u} = 0$ it follows that $u = -\frac{1}{2}P_{2u}^{\dagger}(x)P_{1u}^{\mathrm{T}}(x)$. Hence, $\inf_{u\in\mathbb{R}^m}\Gamma(x) = V(f(x)) - V(x) - \frac{1}{4}P_{1u}(x)P_{2u}^{\dagger}(x)P_{1u}^{\mathrm{T}}(x)$, which implies that $V(x)$ is a control Lyapunov function for $\mathcal{G}$ if and only if

$$V(f(x)) - V(x) - \tfrac{1}{4}P_{1u}(x)P_{2u}^{\dagger}(x)P_{1u}^{\mathrm{T}}(x) < 0, \quad x \in \mathbb{R}^n, \quad x \neq 0. \quad (13.243)$$

In this case, a stabilizing feedback control law is given by $\phi(x) = -\frac{1}{2}P_{2u}^{\dagger}(x)P_{1u}^{\mathrm{T}}(x)$.

13.16 Problems

Problem 13.1. Consider the dynamical system (13.1) with $f(x) = Ax$, where $A \in \mathbb{R}^{n\times n}$. Show that the following statements are equivalent:

i) The zero solution $x(k) \equiv 0$ to (13.1) is Lyapunov stable.

ii) For every initial condition $x_0 \in \mathbb{R}^n$, $x(k)$ is bounded for all $k \in \overline{\mathbb{Z}}_+$.

iii) If $\lambda \in \operatorname{spec}(A)$, then either $|\lambda| < 1$, or both $|\lambda| = 1$ and λ is semisimple.

iv) There exists $\alpha > 0$ such that $\|A^k\| < \alpha$, $k \in \overline{\mathbb{Z}}_+$.

Problem 13.2. Consider the dynamical system (13.1) with $f(x) = Ax$, where $A \in \mathbb{R}^{n\times n}$. Show that the following statements are equivalent:

i) The zero solution $x(k) \equiv 0$ to (13.1) is globally asymptotically stable.

ii) For every initial condition $x_0 \in \mathbb{R}^n$, $\lim_{k\to\infty} x(k) = 0$.

iii) If $\lambda \in \operatorname{spec}(A)$, then $|\lambda| < 1$.

iv) $\lim_{k\to\infty} A^k = 0$.

Problem 13.3. Show that the zero solution $x(k) \equiv 0$ to (13.1) with $f(x) = Ax$, where $A \in \mathbb{R}^{n\times n}$, is asymptotically stable if and only if it is geometrically stable.

Problem 13.4. Consider the dynamical system (13.1) with $f(x) = Ax$, where $A \in \mathbb{R}^{n\times n}$. Show that A is asymptotically stable if and only if there exists a positive-definite matrix $P \in \mathbb{R}^{n\times n}$ that satisfies the discrete-time Lyapunov equation

$$P = A^{\mathrm{T}}PA + R, \quad (13.244)$$

where $R \in \mathbb{R}^{n\times n}$ is a positive-definite matrix. Furthermore, show if A is asymptotically stable, then P is the unique solution to (13.244) and is given by

$$P = \sum_{k=0}^{\infty} A^{k\mathrm{T}} R A^k. \tag{13.245}$$

Problem 13.5. Consider the dynamical system (13.1) with $f(x) = Ax$, where $A \in \mathbb{R}^{n\times n}$ Let $R = C^{\mathrm{T}}C$, where $C \in \mathbb{R}^{l\times n}$, and assume (A, C) is observable. Show if there exists a positive-definite matrix $P \in \mathbb{R}^{n\times n}$ satisfying the discrete-time Lyapunov equation (13.244), then A is asymptotically stable.

Problem 13.6. Let $\mathcal{D}_0 \subset \mathcal{D}$ be a compact positively invariant set for the nonlinear dynamical system (13.1). Show that if there exists a continuous function $V : \mathcal{D} \to \mathbb{R}$ such that

$$V(x) = 0, \quad x \in \mathcal{D}_0, \tag{13.246}$$
$$V(x) > 0, \quad x \in \mathcal{D}, \quad x \notin \mathcal{D}_0, \tag{13.247}$$
$$V(f(x)) - V(x) \le 0, \quad x \in \mathcal{D}, \tag{13.248}$$

then $\mathcal{D}_0$ is Lyapunov stable (see Section 4.9). If, in addition,

$$V(f(x)) - V(x) < 0, \quad x \in \mathcal{D}, \quad x \notin \mathcal{D}_0, \tag{13.249}$$

show that $\mathcal{D}_0$ is asymptotically stable (see Section 4.9). Finally, if $\mathcal{D} = \mathbb{R}^n$ and $V(x) \to \infty$ as $\|x\| \to \infty$ show that $\mathcal{D}_0$ is globally asymptotically stable (see Section 4.9).

Problem 13.7. Consider the nonlinear dynamical system (13.1) and recall the definition of a semistable equilibrium point $x \in \mathcal{D}$ (see Problem 3.44). Suppose the orbit $\mathcal{O}_x$ of (13.1) is bounded for all $x \in \mathcal{D}$ and assume that there exists a continuous function $V : \mathcal{D} \to \mathbb{R}$ such that

$$V(f(x)) - V(x) \le 0, \quad x \in \mathcal{D}. \tag{13.250}$$

Furthermore, let $\mathcal{M}$ denote the largest invariant set contained in $\mathcal{R} \triangleq \{x \in \mathcal{D} : V(f(x)) - V(x) = 0\}$. Show that if $\mathcal{M} \subseteq \{x \in \mathcal{D} : f(x) = x$ and x is Lyapunov stable$\}$, then (13.1) is semistable.

Problem 13.8. Consider the dynamical system (13.1) with $f(x) = Ax$, where $A \in \mathbb{R}^{n\times n}$. Using the results of Problem 13.7 show that the following statements are equivalent:

i) The zero solution $x(k) \equiv 0$ to (13.1) is semistable.

ii) For every initial condition $x_0 \in \mathbb{R}^n$, $\lim_{k\to\infty} x(k)$ exists.

iii) If $\lambda \in \operatorname{spec}(A)$, then either $|\lambda| < 1$, or both $\lambda = 1$ and λ is semisimple.

iv) $\lim_{k\to\infty} A^k$ exists.

Problem 13.9. Consider the dynamical system (13.1) with $f(x) = Ax$, where $A \in \mathbb{R}^{n\times n}$. Show that if there exist $n \times n$ matrices $P \geq 0$ and $R \geq 0$ such that

$$P = A^{\mathrm{T}}PA + R, \tag{13.251}$$

$$\mathcal{N}\left(\begin{bmatrix} R \\ R(A - I_n) \\ \vdots \\ R(A - I_n)^{n-1} \end{bmatrix}\right) = \mathcal{N}(A - I_n), \tag{13.252}$$

then the zero solution $x(t) \equiv 0$ to (13.1) is semistable (see Problem 13.8). (**Hint:** First show that $\mathcal{N}(P) \subseteq \mathcal{N}(A - I_n) \subseteq \mathcal{N}(R)$ and $\mathcal{N}(A - I_n) \cap \mathcal{R}(A - I_n) = \{0\}$.)

Problem 13.10 (Lyapunov's Indirect Method). Let $x(k) \equiv 0$ be an equilibrium point for the nonlinear dynamical system

$$x(k+1) = f(x(k)), \quad x(0) = x_0, \quad k \in \overline{\mathbb{Z}}_+, \tag{13.253}$$

where $f : \mathcal{D} \to \mathcal{D}$ is continuously differentiable and $\mathcal{D}$ is an open set with $0 \in \mathcal{D}$. Furthermore, let

$$A = \left.\frac{\partial f}{\partial x}(x)\right|_{x=0}.$$

Show that:

i) If $|\lambda| < 1$, where $\lambda \in \operatorname{spec}(A)$, then the origin of the nonlinear dynamical system (13.253) is asymptotically stable.

ii) If there exists $\lambda \in \operatorname{spec}(A)$ such that $|\lambda| > 1$, where $\lambda \in \operatorname{spec}(A)$, then the origin of the nonlinear dynamical system (13.253) is unstable.

Problem 13.11. Let $A \in \mathbb{R}^{n\times n}$. A is *nonnegative* if $A_{(i,j)} \geq 0$, $i, j = 1, \ldots, n$. Consider the dynamical system (13.1) with $f(x) = Ax$, where $A \in \mathbb{R}^{n\times n}$. Show that $\overline{\mathbb{R}}_+^n$ is an invariant set with respect to (13.1) if and only if A is nonnegative.

Problem 13.12. Consider the nonlinear dynamical system

$$x_1(k+1) = \frac{\alpha x_2(k)}{1 + x_1^2(k)}, \quad x_1(0) = x_{10}, \quad k \in \overline{\mathbb{Z}}_+, \tag{13.254}$$

$$x_2(k+1) = \frac{\beta x_1(k)}{1 + x_2^2(k)}, \quad x_2(0) = x_{20}. \tag{13.255}$$

Analyze the stability of the zero solution $(x_1(k), x_2(k)) \equiv (0,0)$ to (13.254) and (13.255) using the Lyapunov function candidate $V(x_1, x_2) = x_1^2 + x_2^2$ for *i)* $\alpha^2 < 1$ and $\beta^2 < 1$, *ii)* $\alpha^2 \leq 1$, $\beta^2 \leq 1$, and $\alpha^2 + \beta^2 < 2$, *iii)* $\alpha^2 = \beta^2 = 1$, and *iv)* $\alpha^2 > 1$ and $\beta^2 > 1$.

Problem 13.13. The nonlinear dynamical system (13.1) is *nonnegative* if for every $x(0) \in \overline{\mathbb{R}}_+^n$, the solution $x(k)$, $k \in \overline{\mathbb{Z}}_+$, to (13.1) is nonnegative, that is, $x(k) \geq\geq 0$, $k \in \overline{\mathbb{Z}}_+$. The equilibrium solution $x(k) \equiv x_{\rm e}$ of the nonnegative dynamical system (13.1) is *Lyapunov stable* if for every $\varepsilon > 0$ there exists $\delta = \delta(\varepsilon) > 0$ such that if $x_0 \in \mathcal{B}_\delta(x_{\rm e}) \cap \overline{\mathbb{R}}_+^n$, then $x(k) \in \mathcal{B}_\varepsilon(x_{\rm e}) \cap \overline{\mathbb{R}}_+^n$, $k \in \overline{\mathbb{Z}}_+$. The equilibrium solution $x(k) \equiv x_{\rm e}$ of the nonnegative dynamical system (13.1) is *asymptotically stable* if it is Lyapunov stable and there exists $\delta > 0$ such that if $x_0 \in \mathcal{B}_\delta(x_{\rm e}) \cap \overline{\mathbb{R}}_+^n$, then $\lim_{k\to\infty} x(k) = x_{\rm e}$. Consider the dynamical system (13.1) with $f(x) = Ax$, where $A \in \mathbb{R}^{n\times n}$ is nonnegative (see Problem 13.11). Show that the following statements hold:

i) If there exist vectors $p, r \in \mathbb{R}^n$ such that $p >> 0$ and $r \geq\geq 0$ satisfy

$$p = A^{\rm T} p + r, \tag{13.256}$$

then A is Lyapunov stable.

ii) If there exist vectors $p, r \in \mathbb{R}^n$ such that $p \geq\geq 0$ and $r \geq\geq 0$ satisfy (13.256) and $(A, r^{\rm T})$ is observable, then $p >> 0$ and A is asymptotically stable.

Furthermore, show that the following statements are equivalent:

iii) A is asymptotically stable.

iv) There exist vectors $p, r \in \mathbb{R}^n$ such that $p >> 0$ and $r >> 0$ satisfy (13.256).

v) There exist vectors $p, r \in \mathbb{R}^n$ such that $p \geq\geq 0$ and $r >> 0$ satisfy (13.256).

vi) For every $r \in \mathbb{R}^n$ such that $r >> 0$, there exists $p \in \mathbb{R}^n$ such that $p >> 0$ satisfies (13.256).

(**Hint:** Use the Lyapunov function candidate $V(x) = p^{\rm T} x$ in your analysis and show that the Lyapunov stability theorems of Section 13.2 and the invariant set theorem of Section 13.3 can be used directly for nonnegative systems with the required sufficient conditions verified in $\overline{\mathbb{R}}_+^n$.)

Problem 13.14. Prove Lemma 13.2.

Problem 13.15. Prove Theorem 13.22.

Problem 13.16. Prove Theorem 13.25.

Problem 13.17. Let $f = [f_1, \ldots, f_n]^{\mathrm{T}} : \mathcal{D} \to \mathcal{D}$, where $\mathcal{D}$ is an open subset of $\mathbb{R}^n$ that contains $\overline{\mathbb{R}}_+^n$. Then f is *nonnegative* if $f_i(x) \geq 0$, for all $i = 1, \ldots, n$, and $x \in \overline{\mathbb{R}}_+^n$. Consider the nonlinear dynamical system (13.1). Show that the following statements hold:

i) Suppose $\overline{\mathbb{R}}_+^n \subset \mathcal{D}$. Then $\overline{\mathbb{R}}_+^n$ is an invariant set with respect to (13.1) if and only if $f : \mathcal{D} \to \mathcal{D}$ is nonnegative.

ii) Suppose $f(0) = 0$ and $f : \mathcal{D} \to \mathcal{D}$ is nonnegative and continuously differentiable on $\overline{\mathbb{R}}_+^n$. Then $A \triangleq \frac{\partial f}{\partial x}(x)\Big|_{x=0}$ is nonnegative (see Problem 13.11).

iii) If $f(x) = Ax$, where $A \in \mathbb{R}^{n \times n}$, then f is nonnegative if and only if A is nonnegative.

Problem 13.18. Consider a discrete epidemic model involving two distinct populations where infected members of one population can transmit a disease to a susceptible in the other population. Letting x_i denote the infected fraction of the ith population and $(1 - x_i)$ denote the fraction of susceptibles, the epidemic model can be characterized as

$$x_1(k+1) = a_1 x_2(k)[1 - x_1(k)] + (1 - b_1)x_1(k), \quad x_1(0) = x_{10}, \quad k \in \overline{\mathbb{Z}}_+, \tag{13.257}$$

$$x_2(k+1) = a_2 x_1(k)[1 - x_2(k)] + (1 - b_2)x_2(k), \quad x_2(0) = x_{20}, \tag{13.258}$$

where $a_i \in (0, 1)$ and $b_i \in (0, 1)$ for $i = 1, 2$. Characterize all the equilibria of (13.257) and (13.258). Using the Lyapunov function candidate $V(x) = p^{\mathrm{T}}x$, where $x = [x_1, \ x_2]^{\mathrm{T}}$ and $p \in \mathbb{R}_+^2$, analyze the stability of the equilibria of (13.257) and (13.258). Make sure to justify that $V(x) = p^{\mathrm{T}}x$ is a valid Lyapunov function. (**Hint:** See Problem 13.17.)

Problem 13.19. The nonlinear dynamical system $\mathcal{G}$ given by (13.135) and (13.136) is *nonnegative* if for every $x(0) \in \overline{\mathbb{R}}_+^n$ and $u(k) \geq\geq 0$, $k \in \overline{\mathbb{Z}}_+$, the solution $x(k)$, $k \in \overline{\mathbb{Z}}_+$, to (13.100) and the output $y(k)$, $k \in \overline{\mathbb{Z}}_+$, are nonnegative, that is, $x(k) \geq\geq 0$, $k \in \overline{\mathbb{Z}}_+$, and $y(k) \geq\geq 0$, $k \in \overline{\mathbb{Z}}_+$. Consider the nonlinear dynamical system $\mathcal{G}$ given by (13.135) and (13.136). Show that if $f : \mathcal{D} \to \mathbb{R}^n$ is nonnegative (see Problem 13.17), $G(x) \geq\geq 0$, $h(x) \geq\geq 0$, and $J(x) \geq\geq 0$, $x \in \overline{\mathbb{R}}_+^n$, then $\mathcal{G}$ is nonnegative.

Problem 13.20. Let $q \in \mathbb{R}^l$ and $r \in \mathbb{R}^m$. Consider the nonlinear nonnegative dynamical system $\mathcal{G}$ (see Problem 13.19) given by (13.135) and (13.136) where $f : \mathcal{D} \to \mathbb{R}^n$ is nonnegative (see Problem 13.17), $G(x) \geq\geq 0$, $h(x) \geq\geq 0$, and $J(x) \geq\geq 0$, $x \in \overline{\mathbb{R}}^n_+$. Show that if there exist functions $V_s : \overline{\mathbb{R}}^n_+ \to \overline{\mathbb{R}}_+$, $\ell : \overline{\mathbb{R}}^n_+ \to \overline{\mathbb{R}}_+$, $\mathcal{W} : \overline{\mathbb{R}}^n_+ \to \overline{\mathbb{R}}^m_+$, $P_{1u} : \mathbb{R}^n \to \mathbb{R}^{1\times m}$, and a scalar $\rho > 1$ (respectively, $\rho = 1$) such that $V_s(\cdot)$ is continuous and nonnegative definite, $V_s(0) = 0$,

$$V_s(f(x) + G(x)u) = V_s(f(x)) + P_{1u}(x)u, \quad x \in \overline{\mathbb{R}}^n_+, \quad u \in \overline{\mathbb{R}}^m_+, \tag{13.259}$$

and, for all $x \in \overline{\mathbb{R}}^n_+$,

$$0 = V_s(f(x)) - \frac{1}{\rho}V_s(x) - q^{\mathrm{T}}h(x) + \ell(x), \tag{13.260}$$

$$0 = P_{1u}(x) - q^{\mathrm{T}}J(x) - r^{\mathrm{T}} + \mathcal{W}^{\mathrm{T}}(x), \tag{13.261}$$

then $\mathcal{G}$ is geometrically dissipative (respectively, dissipative) with respect to the supply rate $s(u, y) = q^{\mathrm{T}}y + r^{\mathrm{T}}u$. (**Hint:** The definition of dissipativity and geometric dissipativity should be modified to reflect the fact that $x_0 \in \overline{\mathbb{R}}^n_+$, and $u(k)$, $k \in \overline{\mathbb{Z}}_+$, and $y(k)$, $k \in \overline{\mathbb{Z}}_+$, are nonnegative.)

Problem 13.21. Let $q \in \mathbb{R}^l$ and $r \in \mathbb{R}^m$. Consider the nonlinear nonnegative dynamical system $\mathcal{G}$ (see Problem 13.19) given by (13.135) and (13.136) where $f : \mathcal{D} \to \mathbb{R}^n$ is nonnegative (see Problem 13.17), $G(x) \geq\geq 0$, $h(x) \geq\geq 0$, and $J(x) \geq\geq 0$, $x \in \overline{\mathbb{R}}^n_+$. Show that $\mathcal{G}$ is lossless with respect to the supply rate $s(u, y) = q^{\mathrm{T}}y + r^{\mathrm{T}}u$, $u \in \overline{\mathbb{R}}^m_+$, if and only if there exist functions $V_s : \overline{\mathbb{R}}^n_+ \to \overline{\mathbb{R}}_+$ and $P_{1u} : \mathbb{R}^n \to \mathbb{R}^{1\times m}$ such that $V_s(\cdot)$ is continuous, $V_s(0) = 0$, and for all $x \in \overline{\mathbb{R}}^n_+$, (13.259) holds and

$$0 = V_s(f(x)) - V_s(x) - q^{\mathrm{T}}h(x), \tag{13.262}$$

$$0 = P_{1u}(x) - q^{\mathrm{T}}J(x) - r^{\mathrm{T}}. \tag{13.263}$$

If, in addition, $V_s(\cdot)$ is continuously differentiable show that

$$P_{1u}(x) = V_s'(f(x))G(x). \tag{13.264}$$

Problem 13.22. Let $q \in \mathbb{R}^l$ and $r \in \mathbb{R}^m$ and assume $\mathcal{G}$ given by (13.135) and (13.136) is such that $f : \mathcal{D} \to \mathbb{R}^n$ is nonnegative (see Problem 13.17), $G(x) \geq\geq 0$, $h(x) \geq\geq 0$, and $J(x) \geq\geq 0$, $x \in \overline{\mathbb{R}}^n_+$. Suppose $\mathcal{G}$ is geometrically dissipative (respectively, dissipative) with respect to the supply rate $s(u, y) = q^{\mathrm{T}}y + r^{\mathrm{T}}u$. Show that there exist $p \in \overline{\mathbb{R}}^n_+$, $l \in \overline{\mathbb{R}}^n_+$, and $w \in \overline{\mathbb{R}}^m_+$ and a scalar $\rho > 1$ (respectively, $\rho = 1$) such that

$$0 = A^{\mathrm{T}}p - \frac{1}{\rho}p - C^{\mathrm{T}}q + l, \tag{13.265}$$

$$0 = B^{\mathrm{T}}p - D^{\mathrm{T}}q - r + w, \tag{13.266}$$

where

$$A = \left.\frac{\partial f}{\partial x}\right|_{x=0}, \quad B = G(0), \quad C = \left.\frac{\partial h}{\partial x}\right|_{x=0}, \quad D = J(0). \tag{13.267}$$

If, in addition, (A, C) is observable, show that $p >> 0$.

Problem 13.23. Let $q \in \mathbb{R}^l$ and $r \in \mathbb{R}^m$. Consider the nonnegative dynamical system $\mathcal{G}$ given by (13.198) and (13.199) where $A \geq\geq 0$, $B \geq\geq 0$, $C \geq\geq 0$, and $D \geq\geq 0$. Show that $\mathcal{G}$ is geometrically dissipative (respectively, dissipative) with respect to the supply rate $s(u, y) = q^{\mathrm{T}}y + r^{\mathrm{T}}u$ if and only if there exist $p \in \overline{\mathbb{R}}_+^n$, $l \in \overline{\mathbb{R}}_+^n$, and $w \in \overline{\mathbb{R}}_+^m$, and a scalar $\rho > 1$ (respectively, $\rho = 1$) such that

$$0 = A^{\mathrm{T}}p - \frac{1}{\rho}p - C^{\mathrm{T}}q + l, \tag{13.268}$$

$$0 = B^{\mathrm{T}}p - D^{\mathrm{T}}q - r + w. \tag{13.269}$$

(**Hint:** Use Problems 13.21 and 13.22 to show the result.)

Problem 13.24 (Parseval's Theorem). Let $u : \overline{\mathbb{Z}}_+ \to \mathbb{R}^m$ and $y : \overline{\mathbb{Z}}_+ \to \mathbb{R}^l$ be in ℓ_p, $p \in [0, \infty)$, and let $u(z)$ and $y(z)$ denote their Z-transforms, respectively. Show that

$$\sum_{k=0}^{\infty} u^{\mathrm{T}}(k)y(k) = \frac{1}{2\pi}\int_{-\pi}^{\pi} u^*(e^{\jmath\theta})y(e^{\jmath\theta})\mathrm{d}\theta. \tag{13.270}$$

Problem 13.25 (Positivity Theorem). Consider the controllable and observable system

$$x(k+1) = Ax(k) + Bu(k), \quad x(0) = x_0, \quad k \in \overline{\mathbb{Z}}_+, \tag{13.271}$$

$$y(k) = Cx(k) + Du(k), \tag{13.272}$$

$$u(k) = -\sigma(y(k), k), \tag{13.273}$$

where $x \in \mathbb{R}^n$, $u, y \in \mathbb{R}^m$, and

$$\sigma(\cdot,\cdot) \in \Phi_{\mathrm{pr}} \triangleq \{\sigma : \mathbb{R}^m \times \overline{\mathbb{Z}}_+ \to \mathbb{R}^m : \sigma(0,\cdot) = 0, \\ \sigma^{\mathrm{T}}(y, k)y \geq 0, \ y \in \mathbb{R}^m, \ k \in \overline{\mathbb{Z}}_+\}.$$

Furthermore, suppose

$$G(z) \sim \left[\begin{array}{c|c} A & B \\ \hline C & D \end{array}\right]$$

is strictly positive real. Show that the negative feedback interconnection of (13.271)–(13.273) is globally uniformly asymptotically stable for all $\sigma(\cdot,\cdot) \in \Phi_{\mathrm{pr}}$.

Problem 13.26 (Small Gain Theorem). Consider the controllable and observable system

$$x(k+1) = Ax(k) + Bu(k), \quad x(0) = x_0, \quad k \in \overline{\mathbb{Z}}_+, \tag{13.274}$$
$$y(k) = Cx(k) + Du(k), \tag{13.275}$$
$$u(k) = \sigma(y(k), k), \tag{13.276}$$

where $x \in \mathbb{R}^n$, $u \in \mathbb{R}^m$, $y \in \mathbb{R}^l$, and

$$\sigma(\cdot,\cdot) \in \Phi_{\rm br} \triangleq \{\sigma : \mathbb{R}^l \times \overline{\mathbb{Z}}_+ \to \mathbb{R}^m : \ \sigma(0,\cdot) = 0, \ \|\sigma(y,k)\|_2 \le \gamma^{-1}\|y\|_2, \\ y \in \mathbb{R}^m, k \in \overline{\mathbb{Z}}_+\},$$

and where $\gamma > 0$. Furthermore, suppose

$$G(z) \sim \left[\begin{array}{c|c} A & B \\ \hline C & D \end{array}\right]$$

is strictly positive real. Show that the negative feedback interconnection of (13.274)–(13.276) is globally uniformly asymptotically stable for all $\sigma(\cdot,\cdot) \in \Phi_{\rm br}$.

Problem 13.27 (Circle Criterion). Consider the controllable and observable system

$$x(k+1) = Ax(k) + Bu(k), \quad x(0) = x_0, \quad k \in \overline{\mathbb{Z}}_+, \tag{13.277}$$
$$y(k) = Cx(k) + Du(k), \tag{13.278}$$
$$u(k) = -\sigma(y(k), k), \tag{13.279}$$

where $x \in \mathbb{R}^n$, $u \in \mathbb{R}^m$, $y \in \mathbb{R}^l$, and

$$\sigma(\cdot,\cdot) \in \Phi_{\rm c} \triangleq \{\sigma : \mathbb{R}^l \times \overline{\mathbb{Z}}_+ \to \mathbb{R}^m : \ \sigma(0,\cdot) = 0, \\ [\sigma(y,k) - M_1 y]^{\rm T}[\sigma(y,k) - M_2 y] \le 0, \ y \in \mathbb{R}^m, \ k \in \overline{\mathbb{Z}}_+\},$$

and where $M_1, M_2 \in \mathbb{R}^{m \times l}$. Furthermore, suppose $[I + M_2 G(z)][I + M_1 G(z)]^{-1}$ is strictly positive real, where

$$G(z) \sim \left[\begin{array}{c|c} A & B \\ \hline C & D \end{array}\right]$$

and $\det[I + M_1 G(x)] \ne 0$, $|z| \ge 1$. Show that the negative feedback interconnection of (13.277)–(13.279) is globally uniformly asymptotically stable for all $\sigma(\cdot,\cdot) \in \Phi_{\rm c}$.

Problem 13.28 (Szegö Criterion). Consider the controllable and observable system

$$x(k+1) = Ax(k) + Bu(k), \quad x(0) = x_0, \quad k \in \overline{\mathbb{Z}}_+, \tag{13.280}$$
$$y(k) = Cx(k), \tag{13.281}$$
$$u(k) = -\sigma(y(k)), \tag{13.282}$$

where $x \in \mathbb{R}^n$, $u, y \in \mathbb{R}^m$, and

$$\begin{aligned}\sigma(\cdot) \in \Phi_{\rm S} \triangleq \{\sigma : \mathbb{R}^m \to \mathbb{R}^m :\ & \sigma(0) = 0, \sigma^{\rm T}(y)[M^{-1}\sigma(y) - y] \le 0,\ y \in \mathbb{R}^m,\\ & \sigma(y) = [\sigma_1(y_1), \sigma_2(y_2), \ldots, \sigma_m(y_m)]^{\rm T},\ \text{and}\\ & 0 < \frac{\sigma_i(\nu) - \sigma_i(\hat{\nu})}{\nu - \hat{\nu}} < \mu_i,\ \nu, \hat{\nu} \in \mathbb{R},\ i = 1, \ldots, m\},\end{aligned}$$

and where $M \in \mathbb{R}^{m \times m}$ is a positive-definite matrix and $\nu \neq \hat{\nu}$. Furthermore, suppose there exists a positive-definite diagonal matrix N such that $M^{-1} + [I + (z-1)N]G(z) - \frac{1}{2}|z-1|^2 G^*(z)\mu N G(z)$ is strictly positive real, where

$$G(z) \sim \left[\begin{array}{c|c} A & B \\ \hline C & 0 \end{array}\right]$$

and $\mu = \mathrm{diag}[\mu_1, \mu_2, \ldots, \mu_m]$. Show that the negative feedback interconnection (13.280)–(13.282) is globally asymptotically stable for all $\sigma(\cdot) \in \Phi_{\rm S}$.

Problem 13.29. Show that the results of Problems 6.4–6.6 also hold in the case where $\mathcal{G}_1$ and $\mathcal{G}_2$ are discrete-time dynamical systems.

Problem 13.30. Consider the nonnegative dynamical system $\mathcal{G}$ (see Problem 13.19) given by (13.198) and (13.199), and assume that (A, C) is observable and $\mathcal{G}$ is geometrically dissipative with respect to the supply rate $s(u, y) = e^{\rm T}u - e^{\rm T}My$, where $M >> 0$. Show that the positive feedback interconnection of $\mathcal{G}$ and $\sigma(\cdot,\cdot)$ is globally asymptotically stable for all $\sigma(\cdot,\cdot) \in \Phi$, where

$$\begin{aligned}\Phi \triangleq \{\sigma : \overline{\mathbb{Z}}_+ \times \overline{\mathbb{R}}_+^l \to \overline{\mathbb{R}}_+^m :\ & \sigma(\cdot, 0) = 0,\ 0 \le\le \sigma(k, y) \le\le My,\\ & y \in \overline{\mathbb{R}}_+^l,\ k \in \overline{\mathbb{Z}}_+\},\end{aligned} \tag{13.283}$$

$M >> 0$, and $M \in \mathbb{R}^{m \times l}$. (**Hint:** Use Problem 13.23 to show that if $\mathcal{G}$ is geometrically dissipative with respect to the supply rate $s(u, y) = \mathbf{e}^{\rm T}u - \mathbf{e}^{\rm T}My$, then there exist $p \in \overline{\mathbb{R}}_+^n$, $l \in \overline{\mathbb{R}}_+^n$, and $w \in \overline{\mathbb{R}}_+^m$, and a scalar $\rho > 1$ such that

$$0 = A^{\rm T}p - \frac{1}{\rho}p + C^{\rm T}M^{\rm T}\mathbf{e} + l, \tag{13.284}$$

$$0 = B^{\rm T}p + D^{\rm T}M^{\rm T}\mathbf{e} - \mathbf{e} + w. \tag{13.285}$$

Now, use the Lyapunov function candidate $V(x) = p^{\rm T}x$.)

Problem 13.31. Let $q \in \mathbb{R}^l$, $r \in \mathbb{R}^m$, $q_{\rm c} \in \mathbb{R}^{l_{\rm c}}$, and $r_{\rm c} \in \mathbb{R}^{m_{\rm c}}$. Consider the nonlinear nonnegative dynamical systems $\mathcal{G}$ and $\mathcal{G}_{\rm c}$ (see Problem 13.19) given by (13.135) and (13.136), and (13.225) and (13.226), respectively, where $f : \mathcal{D} \to \mathbb{R}^n$ is nonnegative, $G(x) \ge\ge 0$, $h(x) \ge\ge 0$, $J(x) \ge\ge 0$, $x \in \overline{\mathbb{R}}_+^n$, $f_{\rm c} : \mathbb{R}^{n_{\rm c}} \to \mathbb{R}^{n_{\rm c}}$ is nonnegative, $G_{\rm c}(u_{\rm c}, x_{\rm c}) = G_{\rm c}(x_{\rm c}) \ge\ge 0$, $h_{\rm c}(u_{\rm c}, x_{\rm c}) = h_{\rm c}(x_{\rm c}) \ge\ge 0$, $x_{\rm c} \in \overline{\mathbb{R}}_+^{n_{\rm c}}$, and $J_{\rm c}(u_{\rm c}, x_{\rm c}) \equiv 0$. Assume that $\mathcal{G}$

is dissipative with respect to the linear supply rate $s(u,y) = q^{\mathrm{T}}y + r^{\mathrm{T}}u$ and with a continuous positive-definite storage function $V_{\mathrm{s}}(\cdot)$, and assume that $\mathcal{G}_{\mathrm{c}}$ is dissipative with respect to the linear supply rate $s_{\mathrm{c}}(u_{\mathrm{c}},y_{\mathrm{c}}) = q_{\mathrm{c}}^{\mathrm{T}}y_{\mathrm{c}} + r_{\mathrm{c}}^{\mathrm{T}}u_{\mathrm{c}}$ and with a continuous positive-definite storage function $V_{\mathrm{sc}}(\cdot)$. Show that the following statements hold:

i) If there exists a scalar $\sigma > 0$ such that $q + \sigma r_{\mathrm{c}} \leq\leq 0$ and $r + \sigma q_{\mathrm{c}} \leq\leq 0$, then the positive feedback interconnection of $\mathcal{G}$ and $\mathcal{G}_{\mathrm{c}}$ is Lyapunov stable.

ii) If $\mathcal{G}$ and $\mathcal{G}_{\mathrm{c}}$ are zero-state observable and there exists a scalar $\sigma > 0$ such that $q + \sigma r_{\mathrm{c}} << 0$ and $r + \sigma q_{\mathrm{c}} << 0$, then the positive feedback interconnection of $\mathcal{G}$ and $\mathcal{G}_{\mathrm{c}}$ is asymptotically stable.

iii) If $\mathcal{G}$ is zero-state observable, rank $G_{\mathrm{c}}(0) = m_{\mathrm{c}}$, $\mathcal{G}_{\mathrm{c}}$ is geometrically dissipative with respect to the supply rate $s_{\mathrm{c}}(u_{\mathrm{c}},y_{\mathrm{c}}) = q_{\mathrm{c}}^{\mathrm{T}}y_{\mathrm{c}} + r_{\mathrm{c}}^{\mathrm{T}}u_{\mathrm{c}}$, and there exists a scalar $\sigma > 0$ such that $q + \sigma r_{\mathrm{c}} \leq\leq 0$ and $r + \sigma q_{\mathrm{c}} \leq\leq 0$, then the positive feedback interconnection of $\mathcal{G}$ and $\mathcal{G}_{\mathrm{c}}$ is asymptotically stable.

iv) If $\mathcal{G}$ is geometrically dissipative with respect to the supply rate $s(u,y) = q^{\mathrm{T}}y + r^{\mathrm{T}}u$, $\mathcal{G}_{\mathrm{c}}$ is geometrically dissipative with respect to the supply rate $s_{\mathrm{c}}(u_{\mathrm{c}},y_{\mathrm{c}}) = q_{\mathrm{c}}^{\mathrm{T}}y_{\mathrm{c}} + r_{\mathrm{c}}^{\mathrm{T}}u_{\mathrm{c}}$, and there exists a scalar $\sigma > 0$ such that $q + \sigma r_{\mathrm{c}} \leq\leq 0$ and $r + \sigma q_{\mathrm{c}} \leq\leq 0$, then the positive feedback interconnection of $\mathcal{G}$ and $\mathcal{G}_{\mathrm{c}}$ is asymptotically stable.

(**Hint:** First show that the positive feedback interconnection of $\mathcal{G}$ and $\mathcal{G}_{\mathrm{c}}$ gives a nonnegative closed-loop system.)

13.17 Notes and References

In comparison to the stability theory of continuous-time dynamical systems, there is very little literature on the stability theory of discrete-time systems. This is due to the fact that discrete-time stability theory very closely parallels continuous-time stability theory and is often presented as a footnote to the continuous-time theory. Among the earliest papers on discrete-time stability theory is due to Li [276]. A self-contained summary of the application of Lyapunov stability theory to discrete-time systems is given by Hahn [177]. See also Kalman and Bertram [229]. The invariance principle and the invariant set theorems for discrete-time systems is due to LaSalle; see for example [259]. Nonlinear discrete-time extensions of passivity and losslessness are due to Byrnes and Lin [78] and Lin and Byrnes [281, 282]. Extended Kalman-Yakubovich-Popov equations for nonlinear discrete-time dissipative systems are given in Chellaboina and

Haddad [86]. The concepts of geometric dissipativity, geometric passivity, and geometric nonexpansivity are a discrete-time analog of exponential dissipativity introduced by Chellaboina and Haddad [88] for continuous-time systems. The discrete-time positive real and bounded real lemmas for single-input, single-output systems are due to Szegö and Kalman [422] and Szegö [421], respectively. Multivariable generalizations are given in Hitz and Anderson [197] and Vaidyanathan [437]. A textbook treatment of linear discrete-time passivity is given by Cains [80].

Gain, sector, and disk margins for nonlinear discrete-time systems are due to Chellaboina and Haddad [86] and can be viewed as a generalization of the work by Lee and Lee [262] on gain and phase margins for discrete-time linear systems. Finally, the concept of discrete-time control Lyapunov functions introduced in this chapter is a generalization of the single-input, discrete-time control Lyapunov function introduced by Amicucci, Monaco, and Normand-Cyrot [6] to multi-input systems.

Chapter Fourteen

Discrete-Time Optimal Nonlinear Feedback Control

14.1 Introduction

Since most physical processes evolve naturally in continuous time, it is not surprising that the bulk of nonlinear control theory has been developed for continuous-time systems. Nevertheless, it is the overwhelming trend to implement controllers digitally. Despite this fact, the development of nonlinear control theory for discrete-time systems has lagged its continuous-time counterpart. This is in part due to the fact that concepts such as zero dynamics, normal forms, and minimum phase are much more intricate for discrete-time systems. For example, in contrast to the continuous-time case, technicalities involving passivity analysis tools needed to prove global stability via smooth feedback controllers [282] as well as system relative degree requirements [78] are more involved in the discrete-time case.

In this chapter, we extend the framework developed in Chapters 8–11 to address the problem of optimal discrete-time nonlinear analysis and feedback control with nonlinear-nonquadratic performance criteria. Specifically, we consider discrete-time autonomous nonlinear regulation in feedback control problems on an infinite horizon involving nonlinear-nonquadratic performance functionals. As in the continuous-time case, the performance functional can be evaluated in closed form as long as the nonlinear-nonquadratic cost functional considered is related in a specific way to an underlying Lyapunov function that guarantees asymptotic stability of the nonlinear closed-loop system. This Lyapunov function is shown to be the solution of the discrete-time steady-state Bellman equation arising from the principle of optimality in dynamic programming and plays a key role in constructing the optimal nonlinear control law. The overall framework provides the foundation for extending discrete-time, linear-quadratic synthesis to nonlinear-nonquadratic problems.

14.2 Optimal Control and the Bellman Equation

In this section, we consider a discrete-time control problem involving a notion of optimality with respect to a nonlinear-nonquadratic cost functional. Specifically, we consider the following optimal control problem.

Optimal Control Problem. Consider the nonlinear controlled system given by

$$x(k+1) = F(x(k), u(k), k), \quad x(k_0) = x_0, \ x(k_f) = x_f, \ u(k) \in U, \ k \geq k_0, \tag{14.1}$$

where $x(k) \in \mathcal{D} \subseteq \mathbb{R}^n$, $k \in \overline{\mathbb{Z}}_+$, is the state vector, $\mathcal{D}$ is an open set with $0 \in \mathcal{D}$, $u(k) \in U \subseteq \mathbb{R}^m$, $k \in \overline{\mathbb{Z}}_+$, is the control input, $x(k_0) = x_0$ is given, $x(k_f) = x_f$ is fixed, and $F : \mathcal{D} \times U \times \mathbb{R} \to \mathbb{R}^n$ satisfies $F(0, 0, \cdot) = 0$. We assume that $u(\cdot)$ is restricted to the class of *admissible* controls $\mathcal{U}$ consisting of measurable functions $u(\cdot)$ such that $u(k) \in U$ for all $k \in \overline{\mathbb{Z}}_+$, where the constraint set U is given with $0 \in U$. Furthermore, we assume that $F(\cdot, \cdot, \cdot)$ is continuous. Then determine the control input $u(k) \in U$, $k \in [k_0, k_f]$, such that the cost functional

$$J(x_0, u(\cdot), k_0) = \sum_{k=k_0}^{k_f} L(x(k), u(k), k), \tag{14.2}$$

is minimized, where $L : \mathcal{D} \times U \times \mathbb{R} \to \mathbb{R}$ is given.

To solve the optimal control problem we present *Bellman's principle of optimality* which provides necessary and sufficient conditions, for a given control $u(k) \in U$, $k \in \overline{\mathbb{Z}}_+$, to minimize the cost functional (14.2).

Lemma 14.1. Let $u^*(\cdot) \in \mathcal{U}$ be an optimal control that generates the trajectory $x(k)$, $k \in [k_0, k_f]$, with $x(k_0) = x_0$. Then the trajectory $x(\cdot)$ from (k_0, x_0) to (k_f, x_f) is optimal if and only if for all $k_1, k_2 \in [k_0, k_f]$, the portion of the trajectory $x(\cdot)$ going from $(k_1, x(k_1))$ to $(k_2, x(k_2))$ optimizes the same cost functional over $[k_1, k_2]$, where $x(k_1) = x_1$ is a point on the optimal trajectory generated by $u^*(\cdot)$.

Proof. Let $u^*(\cdot) \in \mathcal{U}$ solve the optimal control problem and let $x(k)$, $k \in [k_0, k_f]$, be the solution to (14.1) generated by $u^*(\cdot)$. Next, suppose, *ad absurdum*, that there exist $k_1 \geq k_0$, $k_2 \leq k_f$, and $\hat{u}(k)$, $k \in [k_1, k_2]$, such that

$$\sum_{k=k_1}^{k_2} L(\hat{x}(k), \hat{u}(k), k) < \sum_{k=k_1}^{k_2} L(x(k), u^*(k), k),$$

where $\hat{x}(k)$ solves (14.1) for all $k \in [k_1, k_2]$ with $u(k) = \hat{u}(k)$, $\hat{x}(k_1) = x(k_1)$, and $\hat{x}(k_2) = x(k_2)$. Now, define

$$u_0(k) \triangleq \begin{cases} u^*(k), & k \in [k_0, k_1), \\ \hat{u}(k), & k \in [k_1, k_2], \\ u^*(k), & k \in (k_2, k_{\rm f}]. \end{cases}$$

Then,

$$\begin{aligned} & J(x_0, u_0(\cdot), k_0) \\ &= \sum_{k=k_0}^{k_{\rm f}} L(x(k), u_0(k), k) \\ &= \sum_{k=k_0}^{k_1} L(x(k), u^*(k), k) + \sum_{k=k_1}^{k_2} L(\hat{x}(k), \hat{u}(k), k) + \sum_{k=k_2}^{k_{\rm f}} L(x(k), u^*(k), k) \\ &< \sum_{k=k_0}^{k_1} L(x(k), u^*(k), k) + \sum_{k=k_1}^{k_2} L(x(k), u^*(k), k) + \sum_{k=k_2}^{k_{\rm f}} L(x(k), u^*(k), k) \\ &= J(x_0, u(\cdot), k_0), \end{aligned}$$

which is a contradiction.

Conversely, if $u^*(\cdot)$ minimizes $J(\cdot,\cdot,\cdot)$ over $[k_1, k_2]$ for all $k_1 \geq k_0$ and $k_2 \leq k_{\rm f}$, then it minimizes $J(\cdot,\cdot,\cdot)$ over $[k_0, k_{\rm f}]$. □

Lemma 14.1 states that $u^*(\cdot)$ solves the optimal control problem over the time interval $[k_0, k_{\rm f}]$ if and only if $u^*(\cdot)$ solves the optimal control problem over every subset of the time interval $[k_0, k_{\rm f}]$. Next, let $u^*(k)$, $k \in [k_0, k_{\rm f}]$, solve the optimal control problem and define the optimal cost $J^*(x_0, k_0) \triangleq J(x_0, u^*(\cdot), k_0)$. Furthermore, define, for $p : \mathbb{R}^n \times \overline{\mathbb{Z}}_+ \to [0, \infty)$, the Hamiltonian $H(x, u, p(x, k), k) \triangleq L(x, u, k) + p(F(x, u, k), k+1) - p(x, k)$. With these definitions we have the following result.

Theorem 14.1. Let $J^*(x, k)$ denote the minimal cost for the optimal control problem with $x_0 = x$ and $k_0 = k$. Then

$$0 = \min_{u(k) \in U} H(x(k), u(k), J^*(x(k), k), k). \tag{14.3}$$

Furthermore, if $u^*(\cdot)$ solves the optimal control problem, then

$$0 = H(x(k), u^*(k), J^*(x(k), k), k). \tag{14.4}$$

Proof. It follows from Lemma 14.1 that, for all $k_{\rm f} \geq k + 1$,

$$J^*(x(k), k) = \min_{u(\cdot) \in \mathcal{U}} \sum_{i=k}^{k_{\rm f}} L(x(i), u(i), i)$$

$$= \min_{u(\cdot)\in\mathcal{U}} \left\{ L(x(k),u(k),k) + \sum_{i=k+1}^{k_{\rm f}} L(x(i),u(i),i) \right\}$$
$$= \min_{u(k)\in U} \{L(x(k),u(k),k) + J^*(x(k+1),k+1)\},$$

or, equivalently,

$$0 = \min_{u(k)\in U} \{(J^*(x(k+1),k+1) - J^*(x(k),k)) + L(x(k),u(k),k)\},$$

which yields (14.3). Finally, (14.4) can be proved in a similar manner by replacing $u(\cdot)$ with $u^*(\cdot)$, where $u^*(\cdot)$ is the optimal control. □

Next, we provide the converse result to Theorem 14.1.

Theorem 14.2. Suppose there exist a continuous function $V : \mathcal{D}\times\mathbb{R} \to \mathbb{R}$ and an optimal control $u^*(\cdot)$ such that $V(x(k_{\rm f}+1),k_{\rm f}+1) = 0$,

$$0 = H(x,u^*(k),V(x(k),k),k), \quad x\in\mathcal{D}, \quad k\in\overline{\mathbb{Z}}_+, \tag{14.5}$$

and

$$H(x,u^*(k),V(x(k),k),k) \le H(x,u(k),V(x(k),k),k),$$
$$x\in\mathcal{D}, \quad u(k)\in U, \quad k\in\overline{\mathbb{Z}}_+. \tag{14.6}$$

Then $u^*(\cdot)$ solves the optimal control problem, that is,

$$J^*(x_0,k_0) = J(x_0,u^*(\cdot),k_0) \le J(x_0,u(\cdot),k_0), \qquad u(\cdot)\in\mathcal{U}, \tag{14.7}$$

and

$$J^*(x_0,k_0) = V(x_0,k_0). \tag{14.8}$$

Proof. Let $x(k)$, $k\in\overline{\mathbb{Z}}_+$, satisfy (14.1) and, for all $k\in[k_0,k_{\rm f}+1]$, define

$$\Delta V(x(k),k) \triangleq V(F(x(k),u(k)),k+1),k) - V(x(k),k). \tag{14.9}$$

Then, with $u(\cdot) = u^*(\cdot)$, it follows from (14.5) that

$$0 = \Delta V(x(k),k) + L(x(k),u^*(k),k).$$

Now, summing over $[k_0,k_{\rm f}]$ and noting that $V(x(k_{\rm f}+1),k_{\rm f}+1) = 0$ yields

$$V(x_0,k_0) = \sum_{k=k_0}^{k_{\rm f}} L(x(k),u^*(k),k) = J(x_0,u^*(\cdot),k_0) = J^*(x_0,k_0).$$

Next, for all $u(\cdot)\in\mathcal{U}$ it follows from (14.5) and (14.9) that

$$J(x_0,u(\cdot),k) = \sum_{k=k_0}^{k_{\rm f}} L(x(k),u(k),k)$$

$$
\begin{aligned}
&= \sum_{k=k_0}^{k_f} \{-\Delta V(x(k),k) + L(x(k),u(k),k) \\
&\quad + V(F(x(k),u(k),k),k) - V(x(k),k)\} \\
&\quad - \sum_{k=k_0}^{k_f} \{-\Delta V(x(k),k) + H(x(k),u(k),V(x(k),k),k)\} \\
&\geq \sum_{k=k_0}^{k_f} \{-\Delta V(x(k),k) + H(x(k),u^*(k),V(x(k),k),k)\} \\
&= \sum_{k=k_0}^{k_f} -\Delta V(x(k),k) \\
&= V(x_0,k_0) \\
&= J^*(x_0,k_0),
\end{aligned}
$$

which completes the proof. □

Note that (14.5) and (14.6) imply

$$0 = \min_{u(k)\in U} H(x(k),u(k),V(x(k),k),k), \tag{14.10}$$

which is known as the Bellman equation. It follows from Theorems 14.1 and 14.2 that the Bellman equation provides necessary and sufficient conditions for characterizing the optimal control for time-varying nonlinear dynamical systems over a finite time interval or the infinite horizon. In the infinite-horizon, time-invariant case, $V(\cdot)$ is independent of k so that the Bellman equation reduces to the time-invariant equation

$$0 = \min_{u\in U} H(x,u,V(x)), \qquad x \in \mathcal{D}. \tag{14.11}$$

14.3 Stability Analysis of Discrete-Time Nonlinear Systems

In this section, we present sufficient conditions for stability of nonlinear discrete-time systems. In particular, we consider the problem of evaluating a nonlinear-nonquadratic performance functional depending upon a nonlinear discrete-time difference equation. As in the continuous-time case, it is shown that the cost functional can be evaluated in closed form as long as the cost functional is related in a specific way to an underlying Lyapunov function that guarantees stability. Here, we restrict our attention to time-invariant infinite horizon systems. For the following result, let $\mathcal{D} \subset \mathbb{R}^n$ be an open set, assume $0 \in \mathcal{D}$, let $L : \mathcal{D} \to \mathbb{R}$, and let $f : \mathcal{D} \to \mathcal{D}$ be such that $f(0) = 0$.

Theorem 14.3. Consider the nonlinear discrete-time dynamical system

$$x(k+1) = f(x(k)), \qquad x(0) = x_0, \qquad k \in \overline{\mathbb{Z}}_+, \tag{14.12}$$

with nonlinear-nonquadratic performance functional

$$J(x_0) \triangleq \sum_{k=0}^{\infty} L(x(k)). \tag{14.13}$$

Furthermore, assume there exists a continuous function $V : \mathcal{D} \to \mathbb{R}$ such that

$$V(0) = 0, \tag{14.14}$$
$$V(x) > 0, \qquad x \in \mathcal{D}, \qquad x \neq 0, \tag{14.15}$$
$$V(f(x)) - V(x) < 0, \qquad x \in \mathcal{D}, \qquad x \neq 0, \tag{14.16}$$
$$L(x) + V(f(x)) - V(x) = 0, \qquad x \in \mathcal{D}. \tag{14.17}$$

Then then the zero solution $x(k) \equiv 0$ to (14.12) is a locally asymptotically stable and there exists a neighborhood of the origin $\mathcal{D}_0 \subseteq \mathcal{D}$ such that

$$J(x_0) = V(x_0), \quad x_0 \in \mathcal{D}_0. \tag{14.18}$$

Finally, if $\mathcal{D} = \mathbb{R}^n$ and

$$V(x) \to \infty \text{ as } \|x\| \to \infty, \tag{14.19}$$

then the zero solution $x(k) \equiv 0$ to (14.12) is globally asymptotically stable.

Proof. Let $x(k)$, $k \in \overline{\mathbb{Z}}_+$, satisfy (14.12). Then it follows from (14.16) that

$$\Delta V(x(k)) = V(f(x(k)) - V(x(k)) < 0, \quad k \in \overline{\mathbb{Z}}_+, \quad x(k) \neq 0. \tag{14.20}$$

Thus, from (14.14), (14.15), and (14.20) it follows that $V(\cdot)$ is a Lyapunov function for (14.12), which proves local asymptotic stability of the zero solution $x(k) \equiv 0$ to (14.12). Consequently, $x(k) \to 0$ as $k \to \infty$ for all initial conditions $x_0 \in \mathcal{D}_0$ for some neighborhood of the origin $\mathcal{D}_0 \subseteq \mathcal{D}$. Now, since

$$0 = -\Delta V(x(k)) + V(f(x(k))) - V(x(k)), \qquad k \in \overline{\mathbb{Z}}_+,$$

it follows from (14.17) that

$$L(x(k)) = -\Delta V(x(k)) + L(x(k)) + V(f(x(k))) - V(x(k)) = -\Delta V(x(k)).$$

Now, summing over $[0, N]$ yields

$$\sum_{k=0}^{N} L(x(k)) = -V(x(N)) + V(x_0).$$

Letting $N \to \infty$ and noting that $V(x(N)) \to 0$ for all $x_0 \in \mathcal{D}_0$ yields $J(x_0) = V(x_0)$. Finally, for $\mathcal{D} = \mathbb{R}^n$ global asymptotic stability is a direct consequence of the radially unbounded condition (14.19) on $V(x)$. □

Note that if (14.17) holds, then (14.16) is equivalent to $L(x) > 0$, $x \in \mathcal{D}$, $x \neq 0$. Theorem 14.3 is the discrete-time analog to Theorem 8.1. The key feature of Theorem 14.3 is that it provides sufficient conditions for stability of discrete-time nonlinear systems. Furthermore, the nonlinear-nonquadratic performance functional is given in terms of an underlying Lyapunov function that guarantees asymptotic stability.

The following corollary specializes Theorem 14.3 to discrete-time linear systems.

Corollary 14.1. Let $A \in \mathbb{R}^{n\times n}$ and $R \in \mathbb{P}^n$. Consider the linear system

$$x(k+1) = Ax(k), \qquad x(0) = x_0, \qquad k \in \overline{\mathbb{Z}}_+, \tag{14.21}$$

with performance functional

$$J(x_0) \triangleq \sum_{k=0}^{\infty} x^{\mathrm{T}}(k)Rx(k). \tag{14.22}$$

Furthermore, assume there exists $P \in \mathbb{P}^n$ such that

$$P = A^{\mathrm{T}}PA + R. \tag{14.23}$$

Then the zero solution $x(k) \equiv 0$ to (14.21) is globally asymptotically stable and

$$J(x_0) = x_0^{\mathrm{T}}Px_0. \tag{14.24}$$

Proof. The result is a direct consequence of Theorem 14.3 with $f(x) = Ax$, $L(x) = x^{\mathrm{T}}Rx$, $V(x) = x^{\mathrm{T}}Px$, and $\mathcal{D} = \mathbb{R}^n$. Specifically, conditions (14.14) and (14.15) are trivially satisfied. Now, $V(f(x)) - V(x) = x^{\mathrm{T}}(A^{\mathrm{T}}PA - P)x$, and hence, it follows from (14.23) that $L(x) + V(f(x)) - V(x) = 0$ so that all the conditions of Theorem 14.3 are satisfied. Finally, since $V(x)$ is radially unbounded, the zero solution $x(k) \equiv 0$ to (14.21) is globally asymptotically stable. □

It follows from Corollary 14.1 that Theorem 14.3 is an extension of the discrete-time $\mathcal{H}_2$ analysis framework to nonlinear systems. Recall that the $\mathcal{H}_2$ Hardy space consists of complex matrix-valued functions $G(z) \in \mathbb{C}^{l\times m}$ that are analytic outside the unit disk and satisfy

$$\sup_{\eta>0} \frac{1}{2\pi}\int_{-\pi}^{\pi} \|G(e^{\eta+\jmath\theta})\|_{\mathrm{F}}^2 \mathrm{d}\theta < \infty. \tag{14.25}$$

The norm of an $\mathcal{H}_2$ function $G(z)$ is defined by

$$|||G|||_2 \triangleq \left[\frac{1}{2\pi}\int_{-\pi}^{\pi} \|G(e^{\jmath\theta})\|_{\mathrm{F}}^2 \mathrm{d}\theta\right]^{1/2}. \tag{14.26}$$

Alternatively, using Parseval's theorem (see Problem 13.24) we can express the $\mathcal{H}_2$ norm as an ℓ_2 norm of the impulse response $H(k) = \frac{1}{2\pi}\int_{-\pi}^{\pi} G(e^{\jmath\theta})e^{\jmath\theta k}\mathrm{d}\theta$. In particular, the ℓ_2 norm of the matrix-valued impulse response function $H(k) \in \mathbb{R}^{l\times m}$, $k \in \overline{\mathbb{Z}}_+$, is defined by

$$|||H|||_2 \triangleq \left[\sum_{k=0}^{\infty} \|H(k)\|_{\mathrm{F}}^2\right]^{1/2}. \tag{14.27}$$

Now, letting $R = E^{\mathrm{T}}E$ and defining the free response $z(k) \triangleq Ex(k) = EA^k x_0$, $k \in \overline{\mathbb{Z}}_+$, it follows that the performance functional (14.22) can be written as

$$\begin{aligned} J(x_0) &= \sum_{k=0}^{\infty} z^{\mathrm{T}}(k)z(k) \\ &= \sum_{k=0}^{\infty} x_0^{\mathrm{T}} A^{k^{\mathrm{T}}} E^{\mathrm{T}} E A^k x_0 \\ &= x_0^{\mathrm{T}} P x_0 \\ &= \sum_{k=0}^{\infty} \|H(k)\|_{\mathrm{F}}^2 \\ &= \frac{1}{2\pi}\int_{-\pi}^{\pi} \|G(e^{\jmath\theta})\|_{\mathrm{F}}^2 \mathrm{d}\theta \\ &= |||G|||_2^2, \end{aligned} \tag{14.28}$$

where $H(k) = EA^k x_0$ and

$$G(z) \sim \left[\begin{array}{c|c} A & x_0 \\ \hline E & 0 \end{array}\right].$$

Alternatively, assuming $x_0 x_0^{\mathrm{T}}$ has an expected value V, that is, $\mathbb{E}[x_0 x_0^{\mathrm{T}}] = V$, where $\mathbb{E}$ denotes expectation, and letting $V = DD^{\mathrm{T}}$, it follows that the averaged performance functional is given by

$$\mathbb{E}[J(x_0)] = \mathbb{E}[x_0^{\mathrm{T}} P x_0] = \operatorname{tr} D^{\mathrm{T}} P D = |||G|||_2^2, \tag{14.29}$$

where P satisfies (14.23) and

$$G(z) \sim \left[\begin{array}{c|c} A & D \\ \hline E & 0 \end{array}\right].$$

Next, we specialize Theorem 14.3 to linear and nonlinear systems with multilinear cost functionals. The following lemma is needed. For this result define the *spectral radius* of $A \in \mathbb{R}^{n\times n}$ by

$$\rho(A) \triangleq \max\{|\lambda| : \lambda \in \operatorname{spec} A\}.$$

We say that the matrix $A \in \mathbb{R}^{n\times n}$ is *Schur* if and only if $\rho(A) < 1$.

Lemma 14.2. Let $A \in \mathbb{R}^{n\times n}$ be Schur and let $h : \mathbb{R}^n \to \mathbb{R}$ be a q-multilinear function. Then there exists a unique q-multilinear function $g : \mathbb{R}^n \to \mathbb{R}$ such that

$$0 = g(Ax) - g(x) + h(x), \qquad x \in \mathbb{R}^n. \tag{14.30}$$

Furthermore, if $h(x)$, $x \in \mathbb{R}^n$, is nonnegative (respectively, positive) definite, then $g(x)$, $x \in \mathbb{R}^n$, is nonnegative (respectively, positive) definite.

Proof. Let $h(x) = \Psi x^{[q]}$ and define $g(x) \triangleq \Gamma x^{[q]}$, where $\Gamma \triangleq -\Psi(A^{[q]} - I_n^{[q]})^{-1}$. Note that $A^{[q]} - I_n^{[q]}$ is invertible since A, and hence, $A^{[q]}$ is Schur. Now, note that for all $x \in \mathbb{R}^n$,

$$\begin{aligned} g(Ax) - g(x) &= \Gamma((Ax)^{[q]} - x^{[q]}) \\ &= \Gamma(A^{[k]}x^{[q]} - x^{[q]}) \\ &= \Gamma(A^{[q]} - I_n^{[q]})x^{[q]} \\ &= -\Psi x^{[q]} = -h(x). \end{aligned}$$

To prove uniqueness, suppose that $\hat{g}(x) = \hat{\Gamma}x^{[q]}$ also satisfies (14.30). Then it follows that

$$\Gamma(A^{[q]} - I_n^{[q]})x^{[q]} = \hat{\Gamma}(A^{[q]} - I_n^{[q]})x^{[q]}, \qquad x \in \mathbb{R}^n.$$

Since $A^{[q]}$ is Schur and $(A^{[q]})^j = (A^j)^{[q]}$, it follows that, for all $x \in \mathbb{R}^n$,

$$\begin{aligned} \Gamma x^{[q]} &= \Gamma(A^{[q]} - I_n^{[q]})(A^{[q]} - I_n^{[q]})^{-1}x^{[q]} \\ &= -\Gamma(A^{[q]} - I_n^{[q]})\sum_{j=1}^{\infty}(A^{[q]})^j x^{[q]} \\ &= -\Gamma(A^{[q]} - I_n^{[q]})\sum_{j=1}^{\infty}(A^j)^{[q]}x^{[q]} \\ &= -\sum_{j=1}^{\infty}\Gamma(A^{[q]} - I_n^{[q]})(A^j x)^{[q]} \\ &= -\sum_{j=1}^{\infty}\hat{\Gamma}(A^{[q]} - I_n^{[q]})(A^j x)^{[q]} \\ &= \hat{\Gamma}x^{[q]}, \end{aligned}$$

which shows that $g(x) = \hat{g}(x)$, $x \in \mathbb{R}^n$.

Finally, if $h(x)$, $x \in \mathbb{R}^n$, is nonnegative definite, then it follows that, for all $x \in \mathbb{R}^n$,

$$\begin{aligned} g(x) &= -\Psi(A^{[q]} - I_n^{[q]})^{-1} x^{[q]} \\ &= \Psi \sum_{j=1}^{\infty} (A^{[q]})^j x^{[q]} \\ &= \Psi \sum_{j=1}^{\infty} (A^j)^{[q]} x^{[q]} \\ &= \sum_{j=1}^{\infty} \Psi (A^j x)^{[q]} \\ &\geq 0. \end{aligned}$$

If, in addition, $h(x)$, $x \in \mathbb{R}^n$, is positive definite, then $g(x)$, $x \in \mathbb{R}^n$, is positive definite. □

Next, assume A is Schur, let P be given by (14.23), and consider the case in which $L(\cdot)$, $f(\cdot)$, and $V(\cdot)$ are given by

$$L(x) = x^{\mathrm{T}} R x + h(x), \tag{14.31}$$

$$f(x) = Ax + N(x), \tag{14.32}$$

$$V(x) = x^{\mathrm{T}} P x + g(x), \tag{14.33}$$

where $h : \mathcal{D} \to \mathbb{R}$ and $g : \mathcal{D} \to \mathbb{R}$ are nonquadratic and $N : \mathcal{D} \to \mathbb{R}^n$ is nonlinear. In this case, (14.17) holds if and only if

$$\begin{aligned} 0 = {} & x^{\mathrm{T}} R x + h(x) + x^{\mathrm{T}} A^{\mathrm{T}} P A x + N^{\mathrm{T}}(x) P N(x) + 2x^{\mathrm{T}} A^{\mathrm{T}} P N(x) - x^{\mathrm{T}} P x \\ & + g(Ax + N(x)) - g(x), \qquad x \in \mathcal{D}, \end{aligned} \tag{14.34}$$

or, equivalently,

$$\begin{aligned} 0 = {} & x^{\mathrm{T}}(A^{\mathrm{T}} P A - P + R)x + g(Ax + N(x)) - g(x) + h(x) + N^{\mathrm{T}}(x) P N(x) \\ & + 2x^{\mathrm{T}} A^{\mathrm{T}} P N(x), \qquad x \in \mathcal{D}. \end{aligned} \tag{14.35}$$

If A is Schur, then we can choose P to satisfy (14.23) as in the linear-quadratic case. Now, suppose $N(x) \equiv 0$ and let P satisfy (14.23). Then (14.35) specializes to

$$0 = g(Ax) - g(x) + h(x), \qquad x \in \mathcal{D}. \tag{14.36}$$

Next, given $h(\cdot)$, we determine the existence of a function $g(\cdot)$ satisfying (14.36). To this end, we focus our attention on multilinear functions for which (14.36) holds with $\mathcal{D} = \mathbb{R}^n$.

Consider the linear system (14.21) and let $h(x)$, $x \in \mathbb{R}^n$, be a positive-definite q-multilinear function, where q is necessarily even. Furthermore, let $g(x)$, $x \in \mathbb{R}^n$, be the positive-definite q-multilinear function given by Lemma 14.2. Then, since $g(Ax) - g(x) < 0$, $x \in \mathbb{R}^n$, $x \neq 0$, it follows that $g(x)$ is a

Lyapunov function for (14.21). Hence, Lemma 14.2 can be used to generate Lyapunov functions of specific structure.

Suppose now that $h(x)$ in (14.31) is of the more general form

$$h(x) = \sum_{\nu=2}^{r} h_{2\nu}(x), \tag{14.37}$$

where, for $\nu = 2, \ldots, r$, $h_{2\nu} : \mathbb{R}^n \to \mathbb{R}$ is a nonnegative-definite 2ν-multilinear function. Now, using Lemma 14.2, let $g_{2\nu} : \mathbb{R}^n \to \mathbb{R}$ be the nonnegative-definite 2ν-multilinear functions satisfying

$$0 = g_{2\nu}(Ax) - g_{2\nu}(x) + h_{2\nu}(x), \qquad x \in \mathbb{R}^n, \qquad \nu = 2, \ldots, r, \tag{14.38}$$

and define

$$g(x) \triangleq \sum_{\nu=2}^{r} g_{2\nu}(x). \tag{14.39}$$

Now, summing (14.38) over ν yields (14.30). Since (14.17) is satisfied with $L(x)$ and $V(x)$ given by (14.31) and (14.33), respectively, (14.18) implies that

$$J(x_0) = x_0^{\mathrm{T}} P x_0 + g(x_0). \tag{14.40}$$

As another illustration of condition (14.30), suppose that $V(x)$ is constrained to be of the form

$$V(x) = x^{\mathrm{T}} P x + (x^{\mathrm{T}} M x)^2, \tag{14.41}$$

where P satisfies (14.23) and M is an $n \times n$ symmetric matrix. In this case, $g(x) = (x^{\mathrm{T}} M x)^2$ is a nonnegative-definite 4-multilinear function. Then (14.30) yields

$$h(x) = -(x^{\mathrm{T}}(A^{\mathrm{T}} M A + M)x)(x^{\mathrm{T}}(A^{\mathrm{T}} M A - M)x). \tag{14.42}$$

If $\hat{R}$ is an $n \times n$ symmetric matrix and M is chosen to satisfy

$$M = A^{\mathrm{T}} M A + \hat{R}, \tag{14.43}$$

then (14.42) implies that $h(x)$ satisfying (14.30) is of the form

$$h(x) = (x^{\mathrm{T}}(A^{\mathrm{T}} M A + M)x)(x^{\mathrm{T}} \hat{R} x). \tag{14.44}$$

Note that if $\hat{R}$ is nonnegative definite, then M is also nonnegative definite, and hence, $h(x)$ is a nonnegative-definite 4-multilinear function. Thus, if $V(x)$ is of the form (14.41), and $L(x)$ is given by

$$L(x) = x^{\mathrm{T}} R x + (x^{\mathrm{T}}(A^{\mathrm{T}} M A + M)x)(x^{\mathrm{T}} \hat{R} x), \tag{14.45}$$

where M and $\hat{R}$ satisfy (14.43), then condition (14.30), and hence, (14.17) is satisfied. The following proposition generalizes the above results to general polynomial functionals.

Proposition 14.1. Let $A \in \mathbb{R}^{n\times n}$ be Schur, $R \in \mathbb{P}^n$, and $\hat{R}_q \in \mathbb{N}^n$, $q = 2,\ldots,r$. Consider the linear system (14.21) with performance functional

$$J(x_0) \triangleq \sum_{k=0}^{\infty} x^{\mathrm{T}}(k)Rx(k) + \sum_{q=2}^{r}\left[(x^{\mathrm{T}}(k)\hat{R}_q x(k))\sum_{j=1}^{q}(x^{\mathrm{T}}(k)M_q x(k))^{j-1}\right.$$
$$\left.\cdot(x^{\mathrm{T}}(k)A^{\mathrm{T}}M_q Ax(k))^{q-j}\right]. \tag{14.46}$$

Furthermore, assume that there exist $P \in \mathbb{P}^n$ and $M_q \in \mathbb{N}^n$, $q = 2,\ldots,r$, such that

$$P = A^{\mathrm{T}}PA + R, \tag{14.47}$$
$$M_q = A^{\mathrm{T}}M_q A + \hat{R}_q, \qquad q = 2,\ldots,r. \tag{14.48}$$

Then the zero solution $x(k) \equiv 0$ to (14.21) is globally asymptotically stable and

$$J(x_0) = x_0^{\mathrm{T}}Px_0 + \sum_{q=2}^{r}(x_0^{\mathrm{T}}M_q x_0)^q. \tag{14.49}$$

Proof. The result is direct consequence of Theorem 14.3 with $f(x) = Ax$, $L(x) = x^{\mathrm{T}}Rx + \sum_{q=2}^{r}\left[(x^{\mathrm{T}}\hat{R}_q x)\sum_{j=1}^{q}(x^{\mathrm{T}}M_q x)^{j-1}(x^{\mathrm{T}}A^{\mathrm{T}}M_q Ax)^{q-j}\right]$, $V(x) = x^{\mathrm{T}}Px + \sum_{q=2}^{r}(x^{\mathrm{T}}\ M_q x)^q$, and $\mathcal{D} = \mathbb{R}^n$. Specifically, conditions (14.14) and (14.15) are trivially satisfied. Now,

$$V(f(x)) - V(x) = x^{\mathrm{T}}(A^{\mathrm{T}}PA - P)x + \sum_{q=2}^{r}\left[x^{\mathrm{T}}(A^{\mathrm{T}}M_q A - M_q)x\right.$$
$$\left.\cdot\sum_{j=1}^{q}(x^{\mathrm{T}}M_q x)^{j-1}(x^{\mathrm{T}}A^{\mathrm{T}}M_q Ax)^{q-j}\right],$$

and hence, it follows from (14.47) and (14.48) that $L(x)+V(f(x))-V(x) = 0$ so that all the conditions of Theorem 14.3 are satisfied. Finally, since $V(x)$ is radially unbounded (14.21) is globally asymptotically stable. □

Proposition 14.1 requires the solutions of $r-1$ Lyapunov equations in (14.48) to obtain the nonquadratic cost. However, if $\hat{R}_q = \hat{R}_2$, $q = 3,\ldots,r$, then $M_q = M_2$, $q = 3,\ldots,r$, satisfies (14.48). In this case, we require the solution of one Lyapunov equation in (14.48).

14.4 Optimal Discrete-Time Nonlinear Control

In this section, we consider the discrete-time control problem involving a notion of optimality with respect to a nonlinear-nonquadratic cost criterion. The optimal feedback controllers are derived as a direct consequence of Theorem 14.3 and provide a discrete-time analog of the continuous-time Hamilton-Jacobi-Bellman conditions for time-invariant, infinite-horizon problems for addressing optimal controllers addressed in Section 8.3. To address the discrete-time optimal control problem let $\mathcal{D} \subset \mathbb{R}^n$ be an open set and let $U \subseteq \mathbb{R}^m$, where $0 \in \mathcal{D}$ and $0 \in U$, and let $F : \mathcal{D} \times U \to \mathbb{R}^n$ such that $F(0,0) = 0$. Next, consider the controlled system

$$x(k+1) = F(x(k), u(k)), \qquad x(0) = x_0, \qquad k \in \overline{\mathbb{Z}}_+, \tag{14.50}$$

where the control $u(\cdot)$ is restricted to the class of *admissible* controls consisting of measurable functions $u(\cdot)$ such that $u(k) \in U$ for all $k \in \overline{\mathbb{Z}}_+$ where the control constraint set U is given. Given a control law $\phi(\cdot)$ and a feedback control law $u(k) = \phi(x(k))$, the closed-loop system has the form

$$x(k+1) = F(x(k), \phi(x(k))), \qquad x(0) = x_0, \qquad k \in \overline{\mathbb{Z}}_+. \tag{14.51}$$

Next, we present a key theorem for characterizing discrete-time feedback controllers that guarantee stability for a nonlinear discrete-time system and minimize a nonlinear-nonquadratic performance functional. For the statement of this result let $L : \mathcal{D} \times U \to \mathbb{R}$ and define the set of regulation controllers for the nonlinear system $F(\cdot,\cdot)$ by

$$\mathcal{S}(x_0) \triangleq \{u(\cdot) : u(\cdot) \text{ is admissible and } x(\cdot) \text{ given by (14.50)} \\ \text{satisfies } x(k) \to 0 \text{ as } k \to \infty\}.$$

Theorem 14.4. Consider the nonlinear controlled system (14.50) with performance functional

$$J(x_0, u(\cdot)) \triangleq \sum_{k=0}^{\infty} L(x(k), u(k)), \tag{14.52}$$

where $u(\cdot)$ is an admissible control. Assume that there exist a continuous function $V : \mathcal{D} \to \mathbb{R}$ and control law $\phi : \mathcal{D} \to U$ such that

$$V(0) = 0, \tag{14.53}$$
$$V(x) > 0, \qquad x \in \mathcal{D}, \qquad x \neq 0, \tag{14.54}$$
$$\phi(0) = 0, \tag{14.55}$$
$$V(F(x,\phi(x))) - V(x) < 0, \qquad x \in \mathcal{D}, \qquad x \neq 0, \tag{14.56}$$
$$H(x,\phi(x)) = 0, \qquad x \in \mathcal{D}, \tag{14.57}$$
$$H(x,u) \geq 0, \qquad x \in \mathcal{D}, \qquad u \in U, \tag{14.58}$$

where

$$H(x,u) \triangleq L(x,u) + V(f(x,u)) - V(x). \tag{14.59}$$

Then, with the feedback control $u(\cdot) = \phi(x(\cdot))$, the zero solution $x(k) \equiv 0$ of the closed-loop system (14.51) is locally asymptotically stable and there exists a neighborhood of the origin $\mathcal{D}_0 \subseteq \mathcal{D}$ such that

$$J(x_0, \phi(x(\cdot))) = V(x_0), \quad x_0 \in \mathcal{D}_0. \tag{14.60}$$

In addition, if $x_0 \in \mathcal{D}_0$ then the feedback control $u(\cdot) = \phi(x(\cdot))$ minimizes $J(x_0, u(\cdot))$ in the sense that

$$J(x_0, \phi(x(\cdot))) = \min_{u(\cdot)\in\mathcal{S}(x_0)} J(x_0, u(\cdot)). \tag{14.61}$$

Finally, if $\mathcal{D} = \mathbb{R}^n$ and

$$V(x) \to \infty \text{ as } \|x\| \to \infty, \tag{14.62}$$

then the zero solution $x(k) \equiv$ of the closed-loop system (14.51) is globally asymptotically stable.

Proof. Local and global asymptotic stability is a direct consequence of (14.53)–(14.56) and (14.62) by applying Theorem 14.3 to the closed-loop system (14.51). Furthermore, using (14.57), condition (14.60) is a restatement of (14.18) as applied to the closed-loop system. Next, let $x_0 \in \mathcal{D}_0$, let $u(\cdot) \in \mathcal{S}(x_0)$, and let $x(\cdot)$ be the solution of (14.50). Then it follows that

$$0 = -\Delta V(x(k)) + V(F(x(k), u(k))) - V(x(k)).$$

Hence,

$$\begin{aligned} L(x(k), u(k)) &= -\Delta V(x(k)) + L(x(k), u(k)) + V(F(x(k), u(k))) - V(x(k)) \\ &= -\Delta V(x(k)) + H(x(k), u(k)). \end{aligned}$$

Now, using (14.59) and the fact that $u(\cdot) \in \mathcal{S}(x_0)$, it follows that

$$\begin{aligned} J(x_0, u(\cdot)) &= \sum_{k=0}^{\infty}[-\Delta V(x(k)) + H(x(k), u(k))] \\ &= -\lim_{k\to\infty} V(x(k)) + V(x_0) + \sum_{k=0}^{\infty} H(x(k), u(k)) \\ &= V(x_0) + \sum_{k=0}^{\infty} H(x(k), u(k)) \\ &\geq V(x_0) \\ &= J(x_0, \phi(x(\cdot)), \end{aligned}$$

which yields (14.61). □

Next, we specialize Theorem 14.4 to discrete-time linear systems and provide connections to the discrete-time, linear-quadratic-regulator problem.

For the following result let $A \in \mathbb{R}^{n\times n}$, $B \in \mathbb{R}^{n\times m}$, $R_1 \in \mathbb{P}^n$, and $R_2 \in \mathbb{P}^m$ be given.

Corollary 14.2. Consider the discrete-time linear controlled system

$$x(k+1) = Ax(k) + Bu(k), \qquad x(0) = x_0, \qquad k \in \overline{\mathbb{Z}}_+, \tag{14.63}$$

with performance functional

$$J(x_0, u(\cdot)) \triangleq \sum_{k=0}^{\infty} [x^{\mathrm{T}}(k)R_1x(k) + u^{\mathrm{T}}(k)R_2u(k)], \tag{14.64}$$

where $u(\cdot)$ is an admissible control. Furthermore, assume that there exists a positive-definite matrix $P \in \mathbb{R}^{n\times n}$ such that

$$P = A^{\mathrm{T}}PA + R_1 - A^{\mathrm{T}}PB(R_2 + B^{\mathrm{T}}PB)^{-1}B^{\mathrm{T}}PA. \tag{14.65}$$

Then, with the feedback control $u = \phi(x) \triangleq -(R_2 + B^{\mathrm{T}}PB)^{-1}B^{\mathrm{T}}PAx$, the zero solution $x(k) \equiv 0$ to (14.63) is globally asymptotically stable and

$$J(x_0, \phi(x(\cdot))) = x_0^{\mathrm{T}}Px_0. \tag{14.66}$$

Furthermore,

$$J(x_0, \phi(x(\cdot))) = \min_{u(\cdot)\in\mathcal{S}(x_0)} J(x_0, u(\cdot)), \tag{14.67}$$

where $\mathcal{S}(x_0)$ is the set of regulation controllers for (14.63) and $x_0 \in \mathbb{R}^n$.

Proof. The result is a direct consequence of Theorem 14.4 with $F(x,u) = Ax + Bu$, $L(x,u) = x^{\mathrm{T}}R_1x + u^{\mathrm{T}}R_2u$, $V(x) = x^{\mathrm{T}}Px$, $\mathcal{D} = \mathbb{R}^n$, and $U = \mathbb{R}^m$. Specifically, conditions (14.53) and (14.54) are trivially satisfied. Next, it follows from (14.65) that $H(x, \phi(x)) = 0$, and hence, $V(F(x,\phi(x))) - V(x) < 0$ for all $x \neq 0$. Thus, $H(x,u) = H(x,u) - H(x,\phi(x)) = [u - \phi(x)]^{\mathrm{T}}(R_2 + B^{\mathrm{T}}PB)[u - \phi(x)] \geq 0$ so that all the conditions of Theorem 14.4 are satisfied. Finally, since $V(x)$ is radially unbounded the solution $x(k) = 0$, $k \in \overline{\mathbb{Z}}_+$, of (14.63) with $u(k) = \phi(x(k)) = -(R_2 + B^{\mathrm{T}}PB)^{-1}B^{\mathrm{T}}PAx(k)$, is globally asymptotically stable. □

The optimal feedback control law $\phi(x)$ in Corollary 14.2 is derived using the properties of $H(x,u)$ as defined in Theorem 14.4. Specifically, since $H(x,u) = x^{\mathrm{T}}R_1x + u^{\mathrm{T}}R_2u + (Ax+Bu)^{\mathrm{T}}P(Ax+Bu) - x^{\mathrm{T}}Px$ it follows that $\partial^2H/\partial u^2 = R_2 + B^{\mathrm{T}}PB > 0$. Now, $\partial H/\partial u = 2(R_2 + B^{\mathrm{T}}PB)u + 2B^{\mathrm{T}}PAx = 0$ gives the unique global minimum of $H(x,u)$. Hence, since $\phi(x)$ minimizes $H(x,u)$ it follows that $\phi(x)$ satisfies $\partial H/\partial u = 0$ or, equivalently, $\phi(x) = -(R_2 + B^{\mathrm{T}}PB)^{-1}B^{\mathrm{T}}PAx$.

14.5 Inverse Optimal Control for Nonlinear Affine Systems

In this section, we specialize Theorem 14.4 to affine systems. As in the continuous-time case, in order to avoid the complexity in solving the Bellman equation, we consider an inverse optimal control problem. Consider the discrete-time nonlinear system given by

$$x(k+1) = f(x(k)) + G(x(k))u(k), \quad x(0) = x_0, \quad k \in \overline{\mathbb{Z}}_+, \tag{14.68}$$

where $f : \mathbb{R}^n \to \mathbb{R}^n$ satisfies $f(0) = 0$, $G : \mathbb{R}^n \to \mathbb{R}^{n\times m}$, $\mathcal{D} = \mathbb{R}^n$, and $U = \mathbb{R}^m$. Furthermore, we consider performance summands $L(x,u)$ of the form

$$L(x,u) = L_1(x) + L_2(x)u + u^{\mathrm{T}} R_2(x)u, \tag{14.69}$$

where $L_1 : \mathbb{R}^n \to \mathbb{R}, L_2 : \mathbb{R}^n \to \mathbb{R}^{1\times m}$, and $R_2 : \mathbb{R}^n \to \mathbb{P}^m$ so that (14.52) becomes

$$J(x_0, u(\cdot)) = \sum_{k=0}^{\infty} [L_1(x(k)) + L_2(x(k))u(k) + u^{\mathrm{T}}(k)R_2(x)u(k)]. \tag{14.70}$$

Theorem 14.5. Consider the discrete-time nonlinear controlled affine system (14.68) with performance functional (14.70). Assume that there exist functions $V : \mathbb{R}^n \to \mathbb{R}$, $L_2 : \mathbb{R}^n \to \mathbb{R}^{1\times m}$, $P_{12} : \mathbb{R}^n \to \mathbb{R}^{1\times m}$, and a nonnegative-definite function $P_2 : \mathbb{R}^n \to \mathbb{N}^m$ such that $V(\cdot)$ is continuous,

$$L_2(0) = 0, \tag{14.71}$$

$$P_{12}(0) = 0, \tag{14.72}$$

$$V(0) = 0, \tag{14.73}$$

$$V(x) > 0, \; x \in \mathbb{R}^n, \; x \neq 0, \tag{14.74}$$

$$V[f(x) - \tfrac{1}{2}G(x)(R_2(x) + P_2(x))^{-1}L_2^{\mathrm{T}}(x)] - V(x) < 0, \; x \in \mathbb{R}^n, \; x \neq 0, \tag{14.75}$$

$$V(f(x) + G(x)u) = V(f(x)) + P_{12}(x)u + u^{\mathrm{T}}P_2(x)u, \; x \in \mathbb{R}^n, \; u \in \mathbb{R}^m, \tag{14.76}$$

and

$$V(x) \to \infty \text{ as } \|x\| \to \infty. \tag{14.77}$$

Then the zero solution $x(k) \equiv 0$ of the closed-loop system

$$x(k+1) = f(x(k)) + G(x(k))\phi(x(k)), \quad x(0) = x_0, \quad k \in \overline{\mathbb{Z}}_+, \tag{14.78}$$

is globally asymptotically stable with the feedback control law

$$\phi(x) = -\tfrac{1}{2}(R_2(x) + P_2(x))^{-1}[L_2(x) + P_{12}(x)]^{\mathrm{T}}, \tag{14.79}$$

and the performance functional (14.70), with

$$L_1(x) = \phi^{\mathrm{T}}(x)(R_2(x) + P_2(x))\phi(x) - V(f(x)) + V(x), \tag{14.80}$$

is minimized in the sense that

$$J(x_0, \phi(x(\cdot))) = \min_{u(\cdot)\in\mathcal{S}(x_0)} J(x_0, u(\cdot)), \ x_0 \in \mathbb{R}^n. \tag{14.81}$$

Finally,

$$J(x_0, \phi(x(\cdot))) = V(x_0), \quad x_0 \in \mathbb{R}^n. \tag{14.82}$$

Proof. The result is a direct consequence of Theorem 14.4 with $F(x,u) = f(x)+G(x)u$, $L(x,u) = L_1(x)+L_2(x)u+u^{\mathrm{T}}R_2(x)u$, $\mathcal{D} = \mathbb{R}^n$, and $U = \mathbb{R}^m$. Specifically, with (14.68), (14.69), and (14.76) the Hamiltonian (14.59) has the form

$$H(x,u) = L_1(x)+V(f(x))+[L_2(x)+P_{12}(x)]u+u^{\mathrm{T}}(R_2(x)+P_2(x))u-V(x). \tag{14.83}$$

Now, the feedback control law (14.79) is obtained by setting $\frac{\partial H(x,u)}{\partial u} = 0$. With (14.79), it follows that (14.71)–(14.75) imply (14.53)–(14.56). Next, with $L_1(x)$ given by (14.80) and $\phi(x)$ given by (14.79) and (14.57) holds. Finally, since

$$H(x,u) = H(x,u) - H(x,\phi(x)) = (u-\phi(x))^{\mathrm{T}}(R_2(x)+P_2(x))(u-\phi(x)), \tag{14.84}$$

and $R_2(x) + P_2(x) > 0$, $x \in \mathbb{R}^n$, condition (14.58) holds. The result now follows as a direct consequence of Theorem 14.4. □

Note that (14.75) is equivalent to

$$\Delta V(x) \triangleq V(f(x)+G(x)\phi(x)) - V(x) < 0, \quad x \in \mathbb{R}^n, \quad x \neq 0, \tag{14.85}$$

with $\phi(x)$ given by (14.79). Furthermore, conditions (14.73), (14.74), and (14.85) ensure that $V(x)$ is a Lyapunov function for the closed-loop system (14.78). Furthermore, as in the continuous-time case, it is important to recognize that the function $L_2(x)$ which appears in the summand of the performance functional (14.69) is an arbitrary function of $x \in \mathbb{R}^n$ subject to conditions (14.71) and (14.75). Thus, $L_2(x)$ provides flexibility in choosing the control law.

With $L_1(x)$ given by (14.80) and $\phi(x)$ given by (14.79), $L(x,u)$ can be expressed as

$$\begin{aligned}
L(x,u)& \\
&= (u-\phi(x))^{\mathrm{T}}(R_2(x)+P_2(x))(u-\phi(x)) - V(f(x)+G(x)u)+V(x)\\
&= [u+\tfrac{1}{2}(R_2(x)+P_2(x))^{-1}L_2^{\mathrm{T}}(x)]^{\mathrm{T}}(R_2(x)+P_2(x))[u+\tfrac{1}{2}(R_2(x)\\
&\quad +P_2(x))^{-1}L_2^{\mathrm{T}}(x)] - [V(f(x)+G(x)\phi(x))-V(x)] + \phi^{\mathrm{T}}(x)P_2(x)\phi(x)\\
&\quad -\tfrac{1}{4}P_{12}(x)(R_2(x)+P_2(x))P_{12}^{\mathrm{T}}(x) - u^{\mathrm{T}}P_2(x)u.
\end{aligned} \tag{14.86}$$

Since $R_2(x)+P_2(x) \geq R_2(x) > 0$ for all $x \in \mathbb{R}^n$ the first and third terms of the right-hand side of (14.86) are nonnegative, while (14.85) implies that the second term is nonnegative. Thus, we have

$$L(x,u) \geq -\tfrac{1}{4}P_{12}(x)(R_2(x)+P_2(x))P_{12}^{\mathrm{T}}(x) - u^{\mathrm{T}}P_2(x)u, \tag{14.87}$$

which shows that $L(x,u)$ may be negative. As a result, there may exist a control input u for which the performance functional (14.70) is negative. However, if the control u is a regulation controller, that is, $u \in \mathcal{S}(x_0)$, then it follows from (14.81) and (14.82) that

$$J(x_0,u) \geq V(x_0) \geq 0, \quad x_0 \in \mathbb{R}^n, \quad u(\cdot) \in \mathcal{S}(x_0). \tag{14.88}$$

Furthermore, in this case substituting $u = \phi(x)$ into (14.86) yields

$$L(x,\phi(x)) = -[V(f(x)+G(x)\phi(x)) - V(x)], \tag{14.89}$$

which, by (14.85), is positive.

Next, we specialize Theorem 14.5 to the case of quadratic Lyapunov functions.

Corollary 14.3. Consider the discrete-time nonlinear controlled affine system (14.68) with performance functional (14.70). Assume that there exist a function $L_2 : \mathbb{R}^n \to \mathbb{R}^{1\times m}$ and a positive-definite matrix $P \in \mathbb{P}^m$ such that

$$L_2(0) = 0, \tag{14.90}$$

$$V[f(x) - \tfrac{1}{2}G(x)\hat{R}_2^{-1}(x)(L_2(x) + f^{\mathrm{T}}(x)PG(x))^{\mathrm{T}}] - V(x) < 0, \quad x \in \mathbb{R}^n, \ x \neq 0, \tag{14.91}$$

where

$$V(x) = x^{\mathrm{T}}Px \tag{14.92}$$

and $\hat{R}_2(x) \triangleq R_2(x) + G^{\mathrm{T}}(x)PG(x)$. Then the zero solution $x(k) \equiv 0$ of the closed-loop system (14.78) is globally asymptotically stable with the feedback control law

$$\phi(x) = -\tfrac{1}{2}\hat{R}_2^{-1}(x)\left[L_2(x) + 2f^{\mathrm{T}}(x)PG(x)\right]^{\mathrm{T}}, \tag{14.93}$$

and the performance functional (14.70), with

$$L_1(x) = \phi^{\mathrm{T}}(x)\hat{R}_2(x)\phi(x) - f^{\mathrm{T}}(x)Pf(x) + x^{\mathrm{T}}Px, \tag{14.94}$$

is minimized in the sense that

$$J(x_0,\phi(x(\cdot))) = \min_{u\in\mathcal{S}(x_0)} J(x_0,u(\cdot)), \quad x_0 \in \mathbb{R}^n. \tag{14.95}$$

Finally,

$$J(x_0,\phi(x(\cdot))) = V(x_0), \quad x_0 \in \mathbb{R}^n. \tag{14.96}$$

Proof. The result is a direct consequence of Theorem 14.5 with $V(x) = x^{\mathrm{T}}Px$, $P_{12}(x) = 2f^{\mathrm{T}}(x)PG(x)$, and $P_2(x) = G^{\mathrm{T}}(x)PG(x)$. Specifically, conditions (14.73), (14.74), (14.76), and (14.77) are trivially satisfied by (14.92). Next, conditions (14.90) and (14.91) imply (14.71) and (14.75). Now, (14.93) and (14.94) are immediate from Theorem 14.5. □

14.6 Gain, Sector, and Disk Margins of Discrete-Time Optimal Regulators

Gain and phase margins of continuous-time state feedback linear-quadratic optimal regulators were extensively discussed in Chapter 8. In particular, in terms of classical control relative stability notions, these controllers possess at least a $\pm 60^\circ$ phase margin, infinite gain margin, and 50 percent gain reduction for each control channel. Alternatively, in terms of absolute stability theory [10] these controllers guarantee sector margins in that the closed-loop system will remain asymptotically stable in the face of a memoryless static input nonlinearity contained in the conic sector $(\frac{1}{2}, \infty)$. In contrast, the stability margins of discrete-time linear-quadratic optimal regulators are not as well known and depend on the open-loop and closed-loop poles of the discrete-time dynamical system [397, 459].

Synthesis techniques for discrete-time linear state feedback control laws guaranteeing that the closed-loop system possesses prespecified sector, gain, and phase margins were developed in [262]. However, unlike continuous-time nonlinear-nonquadratic inverse optimal state feedback regulators possessing guaranteed sector and disk margins to component decoupled input nonlinearities in the conic sector $(\frac{1}{2}, \infty)$ [321] and dissipative dynamic input operators [395], sector and disk margin guarantees for discrete-time nonlinear-nonquadratic regulators have not been addressed in the literature.

In this section, we develop sufficient conditions for gain, sector, and disk margin guarantees for discrete-time nonlinear systems controlled by optimal and inverse optimal nonlinear regulators that minimize a nonlinear-nonquadratic performance criterion involving a nonlinear-nonquadratic function of the state and a quadratic function of the feedback control. In the case where we specialize our results to the linear-quadratic case, we recover the classical discrete-time linear-quadratic optimal regulator gain and phase margin guarantees obtained in [262, 459]. Specifically, we derive the relative stability margins for a discrete-time nonlinear optimal regulator that minimizes a nonlinear-nonquadratic performance criterion involving a nonlinear-nonquadratic function of the state and a quadratic function of the feedback control.

Consider the nonlinear system given by

$$x(k+1) = f(x(k)) + G(x(k))u(k), \quad x(0) = x_0, \quad k \in \overline{\mathbb{Z}}_+, \tag{14.97}$$
$$y(k) = -\phi(x(k)), \tag{14.98}$$

where $\phi : \mathbb{R}^n \to \mathbb{R}^m$, with a nonlinear-nonquadratic performance criterion

$$J(x_0, u(\cdot)) \triangleq \sum_{k=0}^{\infty} [L_1(x(k)) + u^{\mathrm{T}}(k) R_2(x(k)) u(k)], \tag{14.99}$$

where $L_1 : \mathbb{R}^n \to \mathbb{R}$ and $R_2 : \mathbb{R}^n \to \mathbb{R}^{m \times m}$ are given such that $L_1(x) \geq 0$ and $R_2(x) > 0$, $x \in \mathbb{R}^n$. In this case, the optimal nonlinear feedback controller $u = \phi(x)$ that minimizes the nonlinear-nonquadratic performance criterion (14.99) is given by the following result.

Theorem 14.6. Consider the nonlinear system (14.97) with performance functional (14.99). Assume there exist functions $V : \mathbb{R}^n \to \mathbb{R}$, $P_{1u} : \mathbb{R}^n \to \mathbb{R}^{1 \times m}$, and $P_{2u} : \mathbb{R}^n \to \mathbb{N}^m$, with $V(\cdot)$ continuous such that (14.73), (14.74), and (14.76) are satisfied and

$$V(f(x) - \tfrac{1}{2} G(x)(R_2(x) + P_{2u}(x))^{-1} P_{1u}^{\mathrm{T}}(x)) - V(x) < 0, \quad x \in \mathbb{R}^n, \quad x \neq 0, \tag{14.100}$$

$$0 = L_1(x) + V(f(x)) - V(x) - \tfrac{1}{4} P_{1u}(x)(R_2(x) + P_{2u}(x))^{-1} P_{1u}^{\mathrm{T}}(x), \quad x \in \mathbb{R}^n, \tag{14.101}$$

and

$$V(x) \to \infty \text{ as } \|x\| \to \infty. \tag{14.102}$$

Then the zero solution $x(k) \equiv 0$ of the closed-loop system

$$x(k+1) = f(x(k)) + G(x(k))\phi(x(k)), \quad x(0) = x_0, \quad k \in \overline{\mathbb{Z}}_+, \tag{14.103}$$

is globally asymptotically stable with the feedback control law

$$\phi(x) = -\tfrac{1}{2}(R_2(x) + P_{2u}(x))^{-1} P_{1u}^{\mathrm{T}}(x), \tag{14.104}$$

and the performance functional (14.99) is minimized in the sense that

$$J(x_0, \phi(x(\cdot))) = \min_{u(\cdot) \in \mathcal{S}(x_0)} J(x_0, u(\cdot)), \quad x_0 \in \mathbb{R}^n. \tag{14.105}$$

Finally,

$$J(x_0, \phi(x(\cdot))) = V(x_0), \quad x_0 \in \mathbb{R}^n. \tag{14.106}$$

Proof. The proof is identical to the proof of Theorem 14.5. □

The following key lemma is needed for developing the main result of this section.

Lemma 14.3. Consider the nonlinear system $\mathcal{G}$ given by (14.97) and (14.98), where $\phi(x)$ is a stabilizing feedback control law given by (14.104) and where $V(x)$, $P_{1u}(x)$, and $P_{2u}(x)$ satisfy (14.73), (14.74), (14.76), (14.100), and (14.101). Then for all $u(\cdot) \in \mathcal{U}$ and $k_1, k_2 \in \overline{\mathbb{Z}}_+$, $k_1 < k_2$, the solution $x(k)$, $k \in \overline{\mathbb{Z}}_+$, of the closed-loop system (14.103) satisfies

$$V(x(k_2)) \leq \sum_{k=k_1}^{k_2-1} \big[(u(k)+y(k))^{\mathrm{T}}(R_2(x(k))+P_{2u}(x(k)))(u(k)+y(k)) \\ -u^{\mathrm{T}}(k)R_2(x(k))u(k)\big] + V(x(k_1)). \tag{14.107}$$

Proof. Note that it follows from (14.101) that for all $u(\cdot) \in \mathcal{U}$ and $k \in \overline{\mathbb{Z}}_+$,

$$\begin{aligned} & u^{\mathrm{T}}(k)R_2(x(k))u(k) \\ &\quad \leq L_1(x(k)) + u^{\mathrm{T}}(k)R_2(x(k))u(k) \\ &\quad = -V(f(x(k)) + V(x(k)) + \tfrac{1}{4}P_{1u}(x)(R_2(x)+P_{2u}(x))^{-1}P_{1u}^{\mathrm{T}}(x) \\ &\qquad +u^{\mathrm{T}}(k)R_2(x(k))u(k) \\ &\quad = -V(f(x(k)) + G(x(k))u(k)) + V(x(k)) \\ &\qquad +\tfrac{1}{4}P_{1u}(x)(R_2(x)+P_{2u}(x))^{-1}P_{1u}^{\mathrm{T}}(x) + u^{\mathrm{T}}(k)R_2(x(k))u(k) \\ &\qquad +P_{1u}(x(k))u(k) + u^{\mathrm{T}}(k)P_{2u}(x(k))u(k) \\ &\quad = -V(f(x(k)) + G(x(k))u(k)) + V(x(k)) \\ &\qquad +(u(k)+y(k))^{\mathrm{T}}(R_2(x(k))+P_{2u}(x(k)))(u(k)+y(k)) \\ &\quad = -V(x(k+1)) + V(x(k)) + (u(k)+y(k))^{\mathrm{T}}(R_2(x(k)) \\ &\qquad +P_{2u}(x(k)))(u(k)+y(k)). \end{aligned}$$

Now, summing over $[k_1, k_2]$ yields (14.107). □

Note that with $R_2(x) \equiv I$ condition (14.107) is precisely the discrete-time counterpart of the return difference condition given by (8.120) for continuous-time systems. However, as shown in Theorem 8.5, in the continuous-time case a feedback law $\phi(x)$ satisfying the return difference condition is equivalent to the fact that a continuous-time nonlinear affine system with input u and output $y = -\phi(x)$ is dissipative with respect to the quadratic supply rate $[u+y]^{\mathrm{T}}[u+y] - u^{\mathrm{T}}u$. Hence, using the nonlinear Kalman-Yakubovich-Popov lemma one can show that a feedback control law $\phi(x)$ satisfies the return difference inequality if and only if $\phi(x)$ is optimal with respect to a performance criterion involving a nonnegative-definite weighting function on the state. Alternatively, in the discrete-time case (14.107) is not equivalent to the dissipativity property of (14.97) and (14.98) due to the presence of $P_{2u}(x)$. However, as will be shown below, (14.107) does imply that $\mathcal{G}$ is dissipative with respect to a quadratic supply rate.

Now, we present our main result, which provides disk margins for the nonlinear-nonquadratic optimal regulator given by Theorem 14.6. For the following result define

$$\overline{\gamma} \triangleq \sup_{x\in\mathbb{R}^n} \sigma_{\max}(R_2(x) + P_{2u}(x)), \qquad \underline{\gamma} \triangleq \inf_{x\in\mathbb{R}^n} \sigma_{\min}(R_2(x)). \tag{14.108}$$

Theorem 14.7. Consider the nonlinear system $\mathcal{G}$ given by (14.97) and (14.98) where $\phi(x)$ is a stabilizing feedback control law given by (14.104) and where $V(x)$, $P_{1u}(x)$, and $P_{2u}(x)$ satisfy (14.73), (14.74), (14.76), (14.100), and (14.101). Then the nonlinear system $\mathcal{G}$ has a disk margin $(\frac{1}{1+\theta}, \frac{1}{1-\theta})$, where $\theta \triangleq \sqrt{\underline{\gamma}/\overline{\gamma}}$.

Proof. Note that for all $u(\cdot) \in \mathcal{U}$ and $k_1, k_2 \in \overline{\mathbb{Z}}_+$, $k_1 < k_2$, it follows from Lemma 14.3 that the solution $x(k)$, $k \in \overline{\mathbb{Z}}_+$, of the closed-loop system (14.103) satisfies

$$\begin{aligned} V(x(k_2)) - V(x(k_1)) \leq \sum_{k=k_1}^{k_2-1} &\left[(u(k)+y(k))^{\mathrm{T}}(R_2(x(k))\right.\\ &\left.+P_{2u}(x(k)))(u(k)+y(k)) - u^{\mathrm{T}}(k)R_2(x(k))u(k)\right], \end{aligned}$$

which implies that

$$V(x(k_2)) - V(x(k_1)) \leq \sum_{k=k_1}^{k_2-1} [\overline{\gamma}(u(k)+y(k))^{\mathrm{T}}(u(k)+y(k)) - \underline{\gamma}u^{\mathrm{T}}(k)u(k)].$$

Hence, with the storage function $V_{\mathrm{s}}(x) = \frac{1}{2\overline{\gamma}}V(x)$, $\mathcal{G}$ is dissipative with respect to supply rate $r(u,y) = u^{\mathrm{T}}y + \frac{(1-\theta^2)}{2}u^{\mathrm{T}}u + \frac{1}{2}y^{\mathrm{T}}y$. Now, the result follows immediately from Corollary 13.17 and Definition 6.3 with $\alpha = \frac{1}{1+\theta}$ and $\beta = \frac{1}{1-\theta}$. □

14.7 Linear-Quadratic Optimal Regulators

In this section, we specialize Theorem 14.7 to the case of linear discrete-time systems. Specifically, consider the stabilizable linear system given by

$$x(k+1) = Ax(k) + Bu(k), \qquad x(0) = x_0, \qquad k \in \overline{\mathbb{Z}}_+, \tag{14.109}$$

$$y(k) = -Kx(k), \tag{14.110}$$

where $A \in \mathbb{R}^{n\times n}$, $B \in \mathbb{R}^{n\times m}$, and $K \in \mathbb{R}^{m\times n}$, and assume that (A, K) is detectable and the linear system (14.109) and (14.110) is asymptotically stable with the feedback $u = -y$. Furthermore, assume that K is an optimal regulator which minimizes the quadratic performance functional given by

$$J(x_0, u(\cdot)) = \sum_{k=0}^{\infty} [x^{\mathrm{T}}(k) R_1 x(k) + u^{\mathrm{T}}(k) R_2 u(k)], \tag{14.111}$$

where $R_1 \in \mathbb{R}^{n\times n}$ and $R_2 \in \mathbb{R}^{m\times m}$ are such that $R_1 \geq 0$, $R_2 > 0$, and (A, R_1) is observable. In this case, it follows from Theorem 14.6 with $f(x) = Ax$, $G(x) = B$, $L_1(x) = x^{\mathrm{T}} R_1 x$, $R_2(x) = R_2$, $\phi(x) = Kx$, and $V(x) = x^{\mathrm{T}} P x$ that the optimal control law K is given by $K = -(R_2 + B^{\mathrm{T}} P B)^{-1} B^{\mathrm{T}} P A$, where $P > 0$ is the solution to the discrete-time algebraic regulator Riccati equation given by

$$P = A^{\mathrm{T}} P A + R_1 - A^{\mathrm{T}} P B (R_2 + B^{\mathrm{T}} P B)^{-1} B^{\mathrm{T}} P A. \tag{14.112}$$

The following result provides guarantees of gain and sector margins for the linear system (14.109) and (14.110).

Corollary 14.4. Consider the linear system (14.109) and (14.110) with performance functional (14.111). Then, with $K = -(R_2 + B^{\mathrm{T}} P B)^{-1} B^{\mathrm{T}} P A$, where $P > 0$ solves (14.112), the linear system (14.109) and (14.110) has sector (and, hence, gain) margin $(\frac{1}{1+\theta}, \frac{1}{1-\theta})$, where

$$\theta = \left(\frac{\sigma_{\min}(R_2)}{\sigma_{\max}(R_2 + B^{\mathrm{T}} P B)} \right)^{1/2}. \tag{14.113}$$

Proof. The result is a direct consequence of Theorem 14.7 with $f(x) = Ax$, $G(x) = B$, $\phi(x) = Kx$, $V(x) = x^{\mathrm{T}} P x$, and $L_1(x) = x^{\mathrm{T}} R_1 x$. Specifically, (14.73), (14.74), and (14.76) are trivially satisfied and note that (14.112) is equivalent to (14.101). Finally, since (A, R_1) is observable (14.112) implies (14.100) so that all the conditions of Theorem 14.7 are satisfied. □

Note that for single-input systems

$$\begin{aligned}
\theta^2 &= \frac{R_2}{R_2 + B^{\mathrm{T}} P B} \\
&= 1 - (R_2 + B^{\mathrm{T}} P B)^{-1} B^{\mathrm{T}} P B \\
&= \det(I - B(R_2 + B^{\mathrm{T}} P B)^{-1} B^{\mathrm{T}} P) \\
&= \frac{\det(A - B(R_2 + B^{\mathrm{T}} P B)^{-1} B^{\mathrm{T}} P A)}{\det A} \\
&= \frac{\det(A + BK)}{\det A},
\end{aligned}$$

where $K = -(R_2 + B^{\mathrm{T}} P B)^{-1} B^{\mathrm{T}} P A$. In this case, the gain margins obtained in Corollary 14.4 are precisely the gain margins given in [459] for discrete-time linear-quadratic optimal regulators.

The following result specializes Lemma 13.35 to discrete-time linear systems and recovers Theorem 1 of [262].

Corollary 14.5. Consider the linear system (14.109) and (14.110) and let $\alpha, \beta \in \mathbb{R}$ be such that $0 \le \alpha < 1 < \beta < \infty$. Suppose there exist a positive-definite matrix $P \in \mathbb{R}^{n\times n}$ and a scalar $q > 0$ such that

$$0 < 2qI - B^{\mathrm{T}}PB, \tag{14.114}$$
$$0 \ge A^{\mathrm{T}}PA - P - \tfrac{2\alpha\beta}{q(\alpha+\beta)^2}A^{\mathrm{T}}PBB^{\mathrm{T}}PA. \tag{14.115}$$

Then, with $K = -\frac{1}{q(\alpha+\beta)}B^{\mathrm{T}}PA$, the linear system (14.109) and (14.110) has a disk margin (α, β).

Proof. The result is a direct consequence of Theorem 13.35 with $f(x) = Ax$, $G(x) = B$, $\phi(x) = Kx$, $V_{\mathrm{s}}(x) = x^{\mathrm{T}}Px$, $P_{1u}(x) = 2x^{\mathrm{T}}A^{\mathrm{T}}PB$, $P_{2u}(x) = B^{\mathrm{T}}PB$, and $Z = 2qI$. □

Note that since the controller in Corollary 14.5 guarantees a disk margin of (α, β) it also guarantees sector and gain margins of (α, β) and phase margin ϕ given by (see Problem 6.20)

$$\cos(\phi) = \frac{1+\alpha\beta}{\alpha+\beta}. \tag{14.116}$$

14.8 Stability Margins, Meaningful Inverse Optimality, and Control Lyapunov Functions

In this section, we give sufficient conditions that guarantee that a given nonlinear feedback controller of a specific form is inverse optimal and has certain disk, sector, and gain margins.

Theorem 14.8. Consider the nonlinear system $\mathcal{G}$ given by (14.97) and (14.98). Assume there exist functions $V : \mathbb{R}^n \to \mathbb{R}$, $P_{1u} : \mathbb{R}^n \to \mathbb{R}^{1\times m}$, $P_{2u} : \mathbb{R}^n \to \mathbb{N}^m$, and $\gamma : \mathbb{R}^n \to \mathbb{R}$ such that $V(\cdot)$ is continuous and satisfies (14.73), (14.74), (14.76), and for all $x \in \mathbb{R}^n$,

$$0 > V(f(x) - \tfrac{1}{2}\gamma^{-1}(x)G(x)P_{1u}^{\mathrm{T}}(x)) - V(x), \quad x \neq 0, \tag{14.117}$$
$$0 < \gamma(x)I - P_{2u}(x), \tag{14.118}$$
$$0 \ge V(f(x)) - V(x) - \tfrac{1}{4}\gamma^{-1}(x)P_{1u}(x)P_{1u}^{\mathrm{T}}(x), \tag{14.119}$$

and

$$V(x) \to \infty \text{ as } \|x\| \to \infty. \tag{14.120}$$

Then, with the feedback control law $\phi(x) = -\frac{1}{2}\gamma^{-1}(x)P_{1u}^{\mathrm{T}}(x)$, the nonlinear system $\mathcal{G}$ has a disk margin $(\frac{1}{1+\theta}, \frac{1}{1-\theta})$, where

$$\theta = \left(\frac{\inf_{x\in\mathbb{R}^n} \sigma_{\min}(\gamma(x)I - P_{2u}(x))}{\sup_{x\in\mathbb{R}^n} \gamma(x)} \right)^{1/2}. \tag{14.121}$$

Furthermore, with the feedback control law $\phi(x)$ the performance functional

$$\begin{aligned} J(x_0, u(\cdot)) = \sum_{k=0}^{\infty} [-V(f(x(k)) + G(x(k))u(k)) + V(x(k)) \\ +\gamma(x(k))(u(k) - \phi(x(k)))^{\mathrm{T}}(u(k) - \phi(x(k)))], \end{aligned} \tag{14.122}$$

is minimized in the sense that

$$J(x_0, \phi(x(\cdot))) = \min_{u(\cdot)\in\mathcal{S}(x_0)} J(x_0, u(\cdot)), \qquad x_0 \in \mathbb{R}^n. \tag{14.123}$$

Proof. The result is direct consequence of Theorem 14.6 and Theorem 14.7 with $R_2(x) = \gamma(x)I - P_{2u}(x)$ and $L_1(x) = -V(f(x)) + V(x) + \gamma(x)\phi^{\mathrm{T}}(x)\phi(x)$. □

Example 14.1. Consider the discrete-time nonlinear system given by

$$x_1(k+1) = \tfrac{1}{2}x_1(k), \quad x_1(0) = x_{10}, \quad k \in \overline{\mathbb{Z}}_+, \tag{14.124}$$

$$x_2(k+1) = \tfrac{1}{2}x_1^2(k) + u(k), \quad x_2(0) = x_{20}, \tag{14.125}$$

$$y(k) = 2(x_1^2(k) + u(k)). \tag{14.126}$$

With $x = [x_1,\ x_2]^{\mathrm{T}}$, $f(x) = [\frac{1}{2}x_1,\ \frac{1}{2}x_1^2]^{\mathrm{T}}$, $G(x) = [0,\ 1]^{\mathrm{T}}$, $h(x) = 2x_1^2$, $J(x) = 2$, $V_{\mathrm{s}}(x) = x_1^4 + 4x_2^2$, $\hat{P}_{1u}(x) = 4x_1^2$, $\hat{P}_{2u}(x) = 4$, $\ell(x) = [\frac{1}{4}x_1^2,\ 2x_2]^{\mathrm{T}}$, $\mathcal{W}(x) = 0$, $Q = R = 0$, and $S = I_2$, it follows from Theorem 13.21 that the discrete-time nonlinear system $\mathcal{G}$ is passive, that is, $\mathcal{G}$ is dissipative with respect to the supply rate $r(u, y) = 2yu$. Specifically, note that $V_{\mathrm{s}}(0) = 0$, $V_{\mathrm{s}}(x) > 0$, $x \neq 0$, and

$$\begin{aligned} V_{\mathrm{s}}(f(x) + G(x)u) &= \tfrac{17}{16}x_1^4 + 4x_1^2u + 4u^2 \\ &= V_{\mathrm{s}}(f(x)) + \hat{P}_{1u}(x)u + \hat{P}_{2u}(x)u^2, \end{aligned} \tag{14.127}$$

so that (13.137) is satisfied. Now, (13.138)–(13.140) can be easily verified to show the passivity of the discrete-time system $\mathcal{G}$.

Next, with $V(x) = x_1^4 + 4x_2^2$, $P_{1u}(x) = 4x_1^2$, $P_{2u}(x) = 4$, $L_1(x) = \frac{1}{8}x_1^4 + 4x_2^2$, and $R_2(x) = 60$, it follows from Theorem 14.7 that the feedback control law $\phi(x) = -\frac{1}{2}(R_2(x) + P_{2u}(x))^{-1}P_{1u}^{\mathrm{T}}(x) = -\frac{1}{32}x_1^2$ minimizes the performance functional (14.99) in the sense of (14.105). Furthermore, it follows that $\theta = \sqrt{15/16}$, and hence, the nonlinear system $\mathcal{G}$ has a disk margin $(0.5081, 31.4919)$. Finally, by choosing $\gamma(x)$ to be a constant such that $\gamma(x) > 4$, a whole class of inverse optimal controllers can be constructed using Theorem 14.8. In particular, setting $\gamma(x)$ to be very large we can obtain an inverse optimal controller that has a disk margin $(\frac{1}{1+\theta}, \frac{1}{1-\theta})$, where θ is arbitrarily close to 1 so as to maximize disk margin guarantees. △

Next, we give sufficient conditions that guarantee that a given nonlinear feedback controller of a specific form is inverse optimal and has *prespecified* stability margins.

Corollary 14.6. Consider the nonlinear system $\mathcal{G}$ given by (14.97) and (14.98). Let $\theta \in (0,1)$ and assume there exist functions $V : \mathbb{R}^n \to \mathbb{R}$, $P_{1u} : \mathbb{R}^n \to \mathbb{R}^{1\times m}$, $P_{2u} : \mathbb{R}^n \to \mathbb{N}^m$, and a scalar $q > 0$ such that $V(\cdot)$ is continuous and satisfies (14.73), (14.74), and (14.76), and for all $x \in \mathbb{R}^n$,

$$0 > V(f(x) - \tfrac{1-\theta^2}{2q} G(x) P_{1u}^{\mathrm{T}}(x)) - V(x), \quad x \neq 0, \tag{14.128}$$

$$0 < qI - P_{2u}(x), \tag{14.129}$$

$$0 \geq V(f(x)) - V(x) - \tfrac{1-\theta^2}{4q} P_{1u}(x) P_{1u}^{\mathrm{T}}(x), \tag{14.130}$$

and

$$V(x) \to \infty \text{ as } \|x\| \to \infty. \tag{14.131}$$

Then, with the feedback control law $\phi(x) = -\frac{1-\theta^2}{2q} P_{1u}^{\mathrm{T}}(x)$, the nonlinear system $\mathcal{G}$ has a disk margin $(\frac{1}{1+\theta}, \frac{1}{1-\theta})$. Furthermore, with the feedback control law $\phi(x)$ the performance functional

$$J(x_0, u(\cdot)) = \sum_{k=0}^{\infty} \Big[-V(f(x(k)) + G(x(k))u(k)) + V(x(k)) + \tfrac{q}{1-\theta^2}(u(k) - \phi(x(k)))^{\mathrm{T}}(u(k) - \phi(x(k)))\Big], \tag{14.132}$$

is minimized in the sense that

$$J(x_0, \phi(x(\cdot))) = \min_{u(\cdot)\in\mathcal{S}(x_0)} J(x_0, u(\cdot)), \qquad x_0 \in \mathbb{R}^n. \tag{14.133}$$

Proof. The result is a direct consequence of Theorem 14.8 with $\gamma(x) = \frac{q}{1-\theta^2} I$. Specifically, since $\theta \in (0,1)$, (14.129) implies (14.118) so that all the conditions of Theorem 14.8 are satisfied. □

Note that (14.129) implies $0 < \frac{q}{1-\theta^2} I - P_{2u}(x)$, $x \in \mathbb{R}^n$ and $\theta \in (0,1)$. Hence, if there exist functions $V : \mathbb{R}^n \to \mathbb{R}$, $P_{1u} : \mathbb{R}^n \to \mathbb{R}^{1\times m}$, $P_{2u} : \mathbb{R}^n \to \mathbb{N}^m$, and a scalar $q > 0$ such that (14.73), (14.74), (14.76) and (14.128)–(14.130) are satisfied, then it follows from Theorem 14.8 with $\gamma(x) = \frac{q}{1-\theta^2}$ that the nonlinear system $\mathcal{G}$ has a disk margin of $(\frac{1}{1+\hat{\theta}}, \frac{1}{1-\hat{\theta}})$, where

$$\hat{\theta} \triangleq \left(1 - \tfrac{1-\theta^2}{q} \sup_{x\in\mathbb{R}^n} \sigma_{\min}(P_{2u}(x))\right)^{1/2}.$$

In this case, since $0 < qI - P_{2u}(x)$, $x \in \mathbb{R}^n$, it follows that $\theta \le \hat{\theta}$ so that the disk margins provided by Theorem 14.8 are always greater than or equal to the disk margins provided by Corollary 14.6. However, in the latter case considerable numerical simplification in computing the disk margins is achieved in comparison to Theorem 14.8.

The following result specializes Corollary 14.6 to discrete-time linear systems and recovers Theorem 3 of [262].

Corollary 14.7. Consider the linear system (14.109) and (14.110), and let $\theta \in \mathbb{R}$ be such that $0 < \theta < 1$. Suppose there exist a positive-definite matrix $P \in \mathbb{R}^{n\times n}$, a nonnegative-definite matrix $R_1 \in \mathbb{R}^{n\times n}$, and a scalar $q > 0$ such that

$$0 < 2qI - B^{\mathrm{T}}PB, \tag{14.134}$$
$$0 = A^{\mathrm{T}}PA - P + R_1 - \tfrac{1-\theta^2}{2q}A^{\mathrm{T}}PBB^{\mathrm{T}}PA. \tag{14.135}$$

If (A, R_1) is observable then, with $K = -\frac{1-\theta^2}{2q}B^{\mathrm{T}}PA$, the linear system (14.109) and (14.110) has a disk margin $(\frac{1}{1+\theta}, \frac{1}{1-\theta})$. In addition, with the feedback control law $\phi(x) = Kx$, the performance functional

$$J(x_0, u(\cdot)) = \sum_{k=0}^{\infty}\left[x^{\mathrm{T}}(k)R_1x(k) + u^{\mathrm{T}}(k)(\tfrac{2q}{1-\theta^2}I - B^{\mathrm{T}}PB)u(k)\right], \tag{14.136}$$

is minimized in the sense that

$$J(x_0, \phi(x(\cdot))) = \min_{u(\cdot)\in\mathcal{S}(x_0)} J(x_0, u(\cdot)), \qquad x_0 \in \mathbb{R}^n. \tag{14.137}$$

Proof. The result is a direct consequence of Corollary 14.6 with $f(x) = Ax$, $G(x) = B$, $\phi(x) = Kx$, $V(x) = x^{\mathrm{T}}Px$, $P_{1u}(x) = 2x^{\mathrm{T}}A^{\mathrm{T}}PB$, $P_{2u}(x) = B^{\mathrm{T}}PB$, and q replaced by $2q$. □

Next, since $\phi(x)$ given by (13.243) has no a priori guarantees of stability margins we use Theorem 14.8 to obtain a feedback control law that has stability margins. Let $\gamma(x) = \sigma_{\max}(P_{2u}(x)) + \eta(x)$, where $\eta(x) > 0$, $x \in \mathbb{R}^n$, so that $\gamma(x)I > P_{2u}(x)$. In this case, it follows from Theorem 14.8 that if $\gamma(\cdot)$, $V(\cdot)$, and $P_{1u}(\cdot)$ satisfy (14.119) then the feedback control law $\phi(x) = -\frac{1}{2}\gamma^{-1}(x)P_{1u}^{\mathrm{T}}(x)$ is inverse optimal with respect to the performance functional (14.122) and the nonlinear system $\mathcal{G}$ possesses certain disk margins. Next, note that (14.119) is equivalent to

$$P_{1u}(x)P_{2u}^{\dagger}(x)P_{1u}^{\mathrm{T}}(x) - \gamma^{-1}(x)P_{1u}(x)P_{1u}^{\mathrm{T}}(x) \le \varepsilon(x), \qquad x \in \mathbb{R}^n, \tag{14.138}$$

where $\varepsilon(x) \triangleq 4(V(x) - V(f(x))) + P_{1u}(x)P_{2u}^{\dagger}(x)P_{1u}^{\mathrm{T}}(x) > 0$, $x \neq 0$, since $V(x)$ is a control Lyapunov function. Note that (14.138) can equivalently be written as

$$\eta(x)(P_{1u}(x)P_{2u}^{\dagger}(x)P_{1u}^{\mathrm{T}}(x)-\varepsilon(x)) \leq P_{1u}(x)P_{1u}^{\mathrm{T}}(x) \\ -\sigma_{\max}(P_{2u}(x))(P_{1u}(x)P_{2u}^{\dagger}(x)P_{1u}^{\mathrm{T}}(x)-\varepsilon(x)). \quad (14.139)$$

Now, if $P_{1u}(x)P_{2u}^{\dagger}(x)P_{1u}^{\mathrm{T}}(x)-\varepsilon(x)<0$, then $\eta(x)$ can be chosen to be an arbitrary positive-definite function. However, if $P_{1u}(x)P_{2u}^{\dagger}(x)P_{1u}^{\mathrm{T}}(x)-\varepsilon(x)>0$, then there may not exist a function $\eta(x)$ such that (14.139) is satisfied, which implies that the control laws obtained by using the above control Lyapunov function do not possess any disk margins. This is in contrast to the continuous-time case where it was shown in Section 8.8 that if there exists a control Lyapunov function, then there always exist feedback control laws that have guaranteed sector and gain margins of $(\frac{1}{2},\infty)$.

In order to demonstrate the above observations consider the linear discrete-time system given by

$$\begin{bmatrix} x_1(k+1) \\ x_2(k+1) \end{bmatrix} = \begin{bmatrix} 2 & 0 \\ 0 & 2 \end{bmatrix}\begin{bmatrix} x_1(k) \\ x_2(k) \end{bmatrix} + \begin{bmatrix} 1 & 0 \\ 0 & 0.1 \end{bmatrix}\begin{bmatrix} u_1(k) \\ u_2(k) \end{bmatrix}, \\ \begin{bmatrix} x_1(0) \\ x_2(0) \end{bmatrix} = \begin{bmatrix} x_{10} \\ x_{20} \end{bmatrix}, \quad (14.140)$$

with control Lyapunov function candidate $V(x)=x^{\mathrm{T}}x$, where $x \triangleq [x_1,x_2]^{\mathrm{T}}$. In this case, $f(x)=2x$, $G(x)=\mathrm{diag}[1,0.1]$, $u=[u_1,\ u_2]^{\mathrm{T}}$, $P_{1u}(x)=[4x_1,\ 0.4x_2]^{\mathrm{T}}$, and $P_{2u}(x)=\mathrm{diag}[1,\ 0.01]$. Hence,

$$V(f(x))-V(x)-\tfrac{1}{4}P_{1u}(x)P_{2u}^{\dagger}(x)P_{1u}^{\mathrm{T}}(x) = -x^{\mathrm{T}}x<0, \quad x\in\mathbb{R}^2, \quad x\neq 0,$$

which shows that $V(x)=x^{\mathrm{T}}x$ is a control Lyapunov function for (14.140). In this case, (14.139) becomes

$$\eta(x)[12x_1^2+12x_2^2] \leq 4x_1^2-11.84x_2^2. \quad (14.141)$$

Now, taking $x=[1,\ 1]^{\mathrm{T}}$, (14.141) is equivalent to

$$\eta(x)\leq -0.326<0, \quad (14.142)$$

which shows that it is not possible to construct $\eta(x)>0$ such that (14.139) is satisfied.

14.9 Nonlinear Discrete-Time Dynamical Systems with Bounded Disturbances

In this and the next two sections we extend the analysis results of Chapter 10 to discrete-time dynamical systems. Specifically, we present sufficient conditions for dissipativity for a class of nonlinear discrete-time systems with bounded energy and bounded amplitude disturbances. In addition, we consider the problem of evaluating a performance bound for a nonlinear-nonquadratic cost functional. The cost bound is evaluated in closed form by relating the cost functional to an underlying Lyapunov function that guarantees asymptotic stability of the nonlinear system. For the following result, let $\mathcal{D} \subset \mathbb{R}^n$ be an open set, assume $0 \in \mathcal{D}$, let $f : \mathcal{D} \to \mathcal{D}$ be such that $f(0) = 0$, $h : \mathcal{D} \to \mathbb{R}^p$ be such that $h(0) = 0$, $J_1 : \mathcal{D} \to \mathbb{R}^{n\times d}$, and $J_2 : \mathcal{D} \to \mathbb{R}^{p\times d}$. Finally, let $\mathcal{W} \subset \mathbb{R}^d$ and let $r : \mathbb{R}^p \times \mathbb{R}^d \to \mathbb{R}$ be a given function.

Lemma 14.4. Consider the nonlinear discrete-time system

$$x(k+1) = f(x(k)) + J_1(x(k))w(k), \quad x(0) = x_0, \quad k \in \overline{\mathbb{Z}}_+, \quad w(\cdot) \in \ell_2, \tag{14.143}$$

$$z(k) = h(x(k)) + J_2(x(k))w(k). \tag{14.144}$$

Furthermore, assume there exist functions $\Gamma : \mathcal{D} \to \mathbb{R}$, $P_{1w} : \mathcal{D} \to \mathbb{R}^{1\times d}$, $P_{2w} : \mathcal{D} \to \mathbb{N}^d$, and $V : \mathcal{D} \to \mathbb{R}$ such that $V(\cdot)$ is continuous and

$$P_{1w}(0) = 0, \tag{14.145}$$

$$V(0) = 0, \tag{14.146}$$

$$V(x) \geq 0, \quad x \in \mathcal{D}, \tag{14.147}$$

$$P_{1w}(x)w + w^{\mathrm{T}}P_{2w}(x)w \leq r(z,w) + \Gamma(x), \quad x \in \mathcal{D}, \quad w \in \mathcal{W}, \tag{14.148}$$

$$V(f(x) + J_1(x)w) = V(f(x)) + P_{1w}(x)w + w^{\mathrm{T}}P_{2w}(x)w, \quad x \in \mathcal{D}, \quad w \in \mathcal{W}, \tag{14.149}$$

$$V(f(x)) - V(x) + \Gamma(x) \leq 0, \quad x \in \mathcal{D}. \tag{14.150}$$

Then the solution $x(k)$, $k \in \overline{\mathbb{Z}}_+$, of (14.143) satisfies

$$V(x(k+1)) \leq \sum_{i=0}^{k} r(z(i), w(i)) + V(x_0), \qquad w(\cdot) \in \ell_2, \qquad k \in \overline{\mathbb{Z}}_+. \tag{14.151}$$

Proof. Let $x(k)$, $k \in \overline{\mathbb{Z}}_+$, satisfy (14.143) and let $w(\cdot) \in \ell_2$. Then it follows from (14.148), (14.149) and (14.150) that for all $k \in \overline{\mathbb{Z}}_+$,

$$\begin{aligned}
\Delta V(x(k)) &\triangleq V(x(k+1)) - V(x(k)) \\
&= V(f(x(k)) + J_1(x(k))w(k)) - V(x(k)) \\
&= V(f(x(k))) + P_{1w}(x(k))w(k) + w^{\mathrm{T}}(k)P_{2w}(x(k))w(k) - V(x(k)) \\
&= V(f(x(k))) - V(x(k)) + \Gamma(x(k)) + P_{1w}(x(k))w(k) \\
&\quad + w^{\mathrm{T}}(k)P_{2w}(x(k))w(k) - \Gamma(x(k)) \\
&\leq r(z(k), w(k)), \quad k \in \overline{\mathbb{Z}}_+
\end{aligned} \tag{14.152}$$

Now, summing over $[0,k]$ yields

$$V(x(k+1)) - V(x_0) \le \sum_{i=0}^{k} r(z(i),w(i)), \qquad k \in \overline{\mathbb{Z}}_+,$$

which proves the result. □

For the next result let $L : \mathcal{D} \to \mathbb{R}$ be given.

Theorem 14.9. Consider the nonlinear system given by (14.143) and (14.144) with performance functional

$$J(x_0) \triangleq \sum_{k=0}^{\infty} L(x(k)), \tag{14.153}$$

where $x(k)$, $k \in \overline{\mathbb{Z}}_+$, solves (14.143) with $w(k) \equiv 0$. Assume there exist functions $\Gamma : \mathcal{D} \to \mathbb{R}$, $P_{1w} : \mathcal{D} \to \mathbb{R}^{1\times d}$, $P_{2w} : \mathcal{D} \to \mathbb{N}^d$, and $V : \mathcal{D} \to \mathbb{R}$ such that $V(\cdot)$ is continuous and

$$P_{1w}(0) = 0, \tag{14.154}$$
$$V(0) = 0, \tag{14.155}$$
$$V(x) > 0, \quad x \in \mathcal{D}, \quad x \neq 0, \tag{14.156}$$
$$P_{1w}(x)w + w^{\mathrm{T}}P_{2w}(x)w \le r(z,w) + L(x) + \Gamma(x), \; x \in \mathcal{D}, \; w \in \mathcal{W}, \tag{14.157}$$
$$V(f(x)) - V(x) < 0, \quad x \in \mathcal{D}, \quad x \neq 0, \tag{14.158}$$
$$V(f(x) + J_1(x)w) = V(f(x)) + P_{1w}(x)w + w^{\mathrm{T}}P_{2w}(x)w, \quad x \in \mathcal{D}, \quad w \in \mathcal{W}, \tag{14.159}$$
$$L(x) + V(f(x)) - V(x) + \Gamma(x) = 0, \quad x \in \mathcal{D}. \tag{14.160}$$

Then the zero solution $x(k) \equiv 0$ of the undisturbed ($w(k) \equiv 0$) system (14.143) is locally asymptotically stable and there exists a neighborhood $\mathcal{D}_0 \subseteq \mathcal{D}$ of the origin such that if $\Gamma(x) \ge 0$, $x \in \mathcal{D}$, then

$$J(x_0) \le \mathcal{J}(x_0) = V(x_0), \quad x_0 \in \mathcal{D}_0, \tag{14.161}$$

where

$$\mathcal{J}(x_0) \triangleq \sum_{k=0}^{\infty} [L(x(k)) + \Gamma(x(k))] \tag{14.162}$$

and where $x(k)$, $k \in \overline{\mathbb{Z}}_+$, is a solution to (14.143) with $w(k) \equiv 0$. Furthermore, the solution $x(k)$, $k \in \overline{\mathbb{Z}}_+$, to (14.143) satisfies the dissipativity constraint

$$\sum_{i=0}^{k} r(z(i),w(i)) + V(x_0) \ge 0, \quad w(\cdot) \in \ell_2, \quad k \in \overline{\mathbb{Z}}_+. \tag{14.163}$$

Finally, if $\mathcal{D} = \mathbb{R}^n$, $w(k) \equiv 0$, and

$$V(x) \to \infty \text{ as } \|x\| \to \infty, \tag{14.164}$$

then the solution $x(k) = 0$, $k \in \overline{\mathbb{Z}}_+$, of (14.143) is globally asymptotically stable.

Proof. Let $x(k)$, $k \in \overline{\mathbb{Z}}_+$, satisfy (14.143). Then

$$\begin{aligned}\Delta V(x(k)) &\triangleq V(x(k+1)) - V(x(k)) \\ &= V(f(x(k))) - V(x(k)) + P_{1w}(x(k))w(k) \\ &\quad + w^{\mathrm{T}}(k)P_{2w}(x(k))w(k), \qquad k \in \overline{\mathbb{Z}}_+. \end{aligned} \tag{14.165}$$

Hence, with $w(k) \equiv 0$, it follows from (14.158) that

$$\Delta V(x(k)) < 0, \qquad k \in \overline{\mathbb{Z}}_+, \qquad x(k) \neq 0. \tag{14.166}$$

Thus, from (14.154), (14.156), and (14.166) it follows that $V(\cdot)$ is a Lyapunov function for (14.143), which proves local asymptotic stability of the solution $x(k) \equiv 0$ with $w(k) \equiv 0$. Consequently, $x(k) \to 0$ as $k \to \infty$ for all initial conditions $x_0 \in \mathcal{D}_0$ for some neighborhood $\mathcal{D}_0 \subseteq \mathcal{D}$ of the origin.

Next, if $\Gamma(x) \geq 0$, $x \in \mathcal{D}$, and $w(k) \equiv 0$, (14.160) implies

$$\begin{aligned} L(x(k)) &= -\Delta V(x(k)) + L(x(k)) + V(f(x(k))) - V(x(k)) \\ &\leq -\Delta V(x(k)) + L(x(k)) + V(f(x(k))) - V(x(k)) + \Gamma(x(k)) \\ &= -\Delta V(x(k)). \end{aligned}$$

Now, summing over $[0, k]$ yields

$$\sum_{i=0}^{k} L(x(i)) \leq -V(x(k+1)) + V(x_0).$$

Letting $k \to \infty$ and noting that $V(x(k)) \to 0$ as $k \to \infty$ for all $x_0 \in \mathcal{D}_0$ yields $J(x_0) \leq V(x_0)$.

Next, let $x(k)$, $k \in \overline{\mathbb{Z}}_+$, satisfy (14.143) with $w(k) \equiv 0$. Then it follows from (14.160) that

$$\begin{aligned} &L(x(k)) + \Gamma(x(k)) \\ &\quad = -\Delta V(x(k)) + L(x(k)) + V(f(x(k))) - V(x(k)) + \Gamma(x(k)) \\ &\quad = -\Delta V(x(k)). \end{aligned}$$

Summing over $[0, k]$ yields

$$\sum_{i=0}^{k} [L(x(i)) + \Gamma(x(i))] = -V(x(k+1)) + V(x_0).$$

Now, letting $k \to \infty$ yields $\mathcal{J}(x_0) = V(x_0)$. Finally, it follows that (14.154), (14.155), (14.157), (14.159), and (14.160) imply (14.145)–(14.150), and hence, with $\Gamma(x)$ replaced by $L(x)+\Gamma(x)$, Lemma 14.4 yields

$$V(x(k+1)) \leq \sum_{i=0}^{k} r(z(i), w(i)) + V(x_0), \qquad w(\cdot) \in \ell_2, \qquad k \in \overline{\mathbb{Z}}_+.$$

Now, (14.163) follows by noting that $V(x(k+1)) \geq 0$, $k \in \overline{\mathbb{Z}}_+$. Finally, for $\mathcal{D} = \mathbb{R}^n$ global asymptotic stability of the solution $x(k) = 0$, $k \in \overline{\mathbb{Z}}_+$, is a direct consequence of the radially unbounded condition (14.164) on $V(x)$, $x \in \mathbb{R}^n$. □

14.10 Specialization to Dissipative Systems with Quadratic Supply Rates

In this section, we consider the special case in which $r(z, w)$ is a quadratic functional. Specifically, let $h : \mathcal{D} \to \mathbb{R}^p$, $J_2 : \mathcal{D} \to \mathbb{R}^{p \times d}$, $\hat{Q} \in \mathbb{S}^p$, $\hat{S} \in \mathbb{R}^{p \times d}$, $\hat{R} \in \mathbb{S}^d$, and

$$r(z, w) = z^{\mathrm{T}} \hat{Q} z + 2 z^{\mathrm{T}} \hat{S} w + w^{\mathrm{T}} \hat{R} w, \tag{14.167}$$

such that

$$N(x) \triangleq J_2^{\mathrm{T}}(x) \hat{Q} J_2(x) + J_2^{\mathrm{T}}(x) \hat{S} + \hat{S}^{\mathrm{T}} J_2(x) + \hat{R} > P_{2w}(x), \qquad x \in \mathcal{D}.$$

Furthermore, let $L(x) \geq 0$, $x \in \mathcal{D}$. Then

$$\begin{aligned} \Gamma(x) = {} & \left[\tfrac{1}{2} P_{1w}^{\mathrm{T}}(x) - J_2^{\mathrm{T}}(x) \hat{Q} h(x) - \hat{S}^{\mathrm{T}} h(x)\right]^{\mathrm{T}} (N(x) - P_{2w}(x))^{-1} \\ & \cdot \left[\tfrac{1}{2} P_{1w}^{\mathrm{T}}(x) - J_2^{\mathrm{T}}(x) \hat{Q} h(x) - \hat{S}^{\mathrm{T}} h(x)\right] - h^{\mathrm{T}}(x) \hat{Q} h(x), \end{aligned}$$

satisfies (14.157) since in this case

$$\begin{aligned} & L(x) + \Gamma(x) + r(z, w) - P_{1w}(x) w - w^{\mathrm{T}} P_{2w}(x) w \\ & \quad = L(x) + [\tfrac{1}{2} P_{1w}^{\mathrm{T}}(x) - J_2^{\mathrm{T}}(x) \hat{Q}(x) h(x) - \hat{S}^{\mathrm{T}}(x) h(x) \\ & \qquad -(N(x) - P_{2w}(x)) w]^{\mathrm{T}} (N(x) - P_{2w}(x))^{-1} [\tfrac{1}{2} P_{1w}^{\mathrm{T}}(x) \\ & \qquad - J_2^{\mathrm{T}}(x) \hat{Q}(x) h(x) - \hat{S}^{\mathrm{T}}(x) h(x) - (N(x) - P_{2w}(x)) w] \\ & \quad \geq 0. \end{aligned} \tag{14.168}$$

Corollary 14.8. Let $L(x) \geq 0$, $x \in \mathcal{D}$, and consider the nonlinear system given by (14.143) and (14.144) with performance functional

$$J(x_0) \triangleq \sum_{k=0}^{\infty} L(x(k)), \tag{14.169}$$

where $x(k)$, $k \in \overline{\mathbb{Z}}_+$, solves (14.143) with $w(k) \equiv 0$. Assume there exist functions $P_{1w} : \mathcal{D} \to \mathbb{R}^{1 \times d}$, $P_{2w} : \mathcal{D} \to \mathbb{N}^d$, and $V : \mathcal{D} \to \mathbb{R}$ such that $V(\cdot)$ is continuous and

$$P_{1w}(0) = 0, \tag{14.170}$$
$$V(0) = 0, \tag{14.171}$$
$$V(x) > 0, \quad x \in \mathcal{D}, \quad x \neq 0, \tag{14.172}$$
$$V(f(x)) - V(x) < 0, \quad x \in \mathcal{D}, \quad x \neq 0, \tag{14.173}$$
$$V(f(x) + J_1(x)w) = V(f(x)) + P_{1w}(x)w + w^{\mathrm{T}}P_{2w}(x)w, \quad x \in \mathcal{D}, \quad w \in \mathcal{W}, \tag{14.174}$$
$$\gamma^2 I_d - J_2^{\mathrm{T}}(x)J_2(x) - P_{2w}(x) > 0, \quad x \in \mathcal{D}, \tag{14.175}$$
$$L(x) + V(f(x)) - V(x) + \Gamma(x) = 0, \quad x \in \mathcal{D}, \tag{14.176}$$

where $\gamma > 0$ and

$$\Gamma(x) = \left[\tfrac{1}{2}P_{1w}^{\mathrm{T}}(x) + J_2^{\mathrm{T}}(x)h(x)\right]^{\mathrm{T}}\left[\gamma^2 I_d - J_2^{\mathrm{T}}(x)J_2(x) - P_{2w}(x)\right]^{-1} \cdot \left[\tfrac{1}{2}P_{1w}^{\mathrm{T}}(x) + J_2^{\mathrm{T}}(x)h(x)\right] + h^{\mathrm{T}}(x)h(x). \tag{14.177}$$

Then the zero solution $x(k) \equiv 0$ to the undisturbed ($w(k) \equiv 0$) system (14.143) is locally asymptotically stable and there exists a neighborhood $\mathcal{D}_0 \subseteq \mathcal{D}$ of the origin such that

$$J(x_0) \leq \mathcal{J}(x_0) = V(x_0), \quad x_0 \in \mathcal{D}_0, \tag{14.178}$$

where

$$\mathcal{J}(x_0) \triangleq \sum_{k=0}^{\infty}[L(x(k)) + \Gamma(x(k))] \tag{14.179}$$

and where $x(k)$, $k \in \overline{\mathbb{Z}}_+$, solves (14.143) with $w(k) \equiv 0$. Furthermore, the solution $x(k)$, $k \in \overline{\mathbb{Z}}_+$, of (14.143) satisfies the nonexpansivity constraint

$$\sum_{i=0}^{k} z^{\mathrm{T}}(i)z(i) \leq \gamma^2 \sum_{i=0}^{k} w^{\mathrm{T}}(i)w(i) + V(x_0), \quad w(\cdot) \in \ell_2, \quad k \in \overline{\mathbb{Z}}_+. \tag{14.180}$$

Finally, if $\mathcal{D} = \mathbb{R}^n$, $w(k) \equiv 0$, and

$$V(x) \to \infty \text{ as } \|x\| \to \infty, \tag{14.181}$$

then the solution $x(k) = 0$, $k \in \overline{\mathbb{Z}}_+$, of (14.143) is globally asymptotically stable.

Proof. With $\hat{Q} = -I_p$, $\hat{S} = 0$, and $\hat{R} = \gamma^2 I_d$, it follows from (14.168) that $\Gamma(x)$ given by (14.177) satisfies (14.157). The result now follows as a direct consequence of Theorem 14.9. □

Note that if $L(x) = h^{\mathrm{T}}(x)h(x)$ in Corollary 14.8 then $\Gamma(x)$ can be chosen as

$$\Gamma(x) = \left[\tfrac{1}{2}P_{1w}^{\mathrm{T}}(x) + J_2^{\mathrm{T}}(x)h(x)\right]^{\mathrm{T}}\left[\gamma^2 I_d - J_2^{\mathrm{T}}(x)J_2(x) - P_{2w}(x)\right]^{-1}$$

$$\cdot \left[\tfrac{1}{2}P_{1w}^{\mathrm{T}}(x) + J_2^{\mathrm{T}}(x)h(x)\right].$$

The framework presented in Corollary 14.8 is an extension of the mixed-norm $\mathcal{H}_2/\mathcal{H}_\infty$ framework of Haddad *et al.* [154] to nonlinear systems. Specifically, letting $f(x) = Ax$, $J_1(x) = D$, $h(x) = Ex$, $J_2(x) = 0$, $L(x) = x^{\mathrm{T}}Rx$, and $V(x) = x^{\mathrm{T}}Px$, where $A \in \mathbb{R}^{n\times n}$, $D \in \mathbb{R}^{n\times d}$, $E \in \mathbb{R}^{p\times n}$, $R \triangleq E^{\mathrm{T}}E > 0$, and $P \in \mathbb{P}^n$ satisfies

$$P = A^{\mathrm{T}}PA + A^{\mathrm{T}}PD(\gamma^2 I_d - D^{\mathrm{T}}PD)^{-1}D^{\mathrm{T}}PA + R, \tag{14.182}$$

it follows from Corollary 14.8 with $L(x) = h^{\mathrm{T}}(x)h(x) = x^{\mathrm{T}}Rx$, $\Gamma(x) = x^{\mathrm{T}}A^{\mathrm{T}}PD(\gamma^2 I_d - D^{\mathrm{T}}PD)^{-1}D^{\mathrm{T}}PAx$, where $\gamma^2 I_d - D^{\mathrm{T}}PD > 0$, and $x_0 = 0$ that

$$\sum_{i=0}^{k} x^{\mathrm{T}}(i)Rx(i) \le \gamma^2 \sum_{i=0}^{k} w^{\mathrm{T}}(i)w(i), \qquad w(\cdot) \in \ell_2, \qquad k \in \overline{\mathbb{Z}}_+, \tag{14.183}$$

or, equivalently, the $\mathcal{H}_\infty$ norm of

$$G(z) \sim \left[\begin{array}{c|c} A & D \\ \hline E & 0 \end{array}\right]$$

satisfies

$$\|G\|_\infty \triangleq \sup_{\theta\in[0,2\pi]} \sigma_{\max}(G(e^{\jmath\theta})) \le \gamma, \tag{14.184}$$

where $\sigma_{\max}(\cdot)$ denotes the maximum singular value. Now, (14.178) implies

$$\begin{aligned}\sum_{k=0}^{\infty} x^{\mathrm{T}}(k)Rx(k) &\le \sum_{k=0}^{\infty} x^{\mathrm{T}}(k)[R + A^{\mathrm{T}}PD(\gamma^2 I_d - D^{\mathrm{T}}PD)^{-1}D^{\mathrm{T}}PA]x(k)\\ &= \sum_{k=0}^{\infty} x_0^{\mathrm{T}}(A^k)^{\mathrm{T}}[R + A^{\mathrm{T}}PD(\gamma^2 I_d - D^{\mathrm{T}}PD)^{-1}D^{\mathrm{T}}PA]A^k x_0,\end{aligned}$$

where $x(k)$, $k \in \overline{\mathbb{Z}}_+$, solves (14.143) with $w(k) \equiv 0$.

To eliminate the explicit dependence on the initial condition x_0 we assume $x_0x_0^{\mathrm{T}}$ has expected value V, that is, $\mathbb{E}[x_0x_0^{\mathrm{T}}] = V$, where $\mathbb{E}$ denotes expectation. Invoking this step leads to

$$\mathbb{E}\left[\sum_{k=0}^{\infty} x^{\mathrm{T}}(k)Rx(k)\right] = \mathbb{E}\left[\sum_{k=0}^{\infty} x_0^{\mathrm{T}}(A^k)^{\mathrm{T}}RA^k x_0\right] = \mathbb{E}[x_0^{\mathrm{T}}\hat{P}x_0] = \operatorname{tr}\,\hat{P}V,$$

where

$$\hat{P} = A^{\mathrm{T}}\hat{P}A + R$$

and

$$\mathbb{E}\left[\sum_{k=0}^{\infty} x^{\mathrm{T}}(k)(R + A^{\mathrm{T}}PD(\gamma^2 I_d - D^{\mathrm{T}}PD)^{-1}D^{\mathrm{T}}PA)x(k)\right]$$

$$= \mathbb{E}\left[\sum_{k=0}^{\infty} x_0^{\mathrm{T}}(A^k)^{\mathrm{T}}[R + A^{\mathrm{T}}PD(\gamma^2 I_d - D^{\mathrm{T}}PD)^{-1}D^{\mathrm{T}}PA]A^k x_0\right]$$
$$= \mathbb{E}[x_0^{\mathrm{T}}Px_0]$$
$$= \operatorname{tr} PV, \tag{14.185}$$

where P satisfies (14.182). Hence, $|||G|||_2^2 = \operatorname{tr} \hat{P}V \leq \operatorname{tr} PV$, which implies that $\mathcal{J}(x_0)$ given by (14.179) provides an upper bound to the $\mathcal{H}_2$ norm of $G(z)$.

Corollary 14.9. Let $L(x) \geq 0$, $x \in \mathcal{D}$, $p = d$, and consider the nonlinear system given by (14.143) and (14.144) with performance functional

$$J(x_0) \triangleq \sum_{k=0}^{\infty} L(x(k)), \tag{14.186}$$

where $x(k)$, $k \in \overline{\mathbb{Z}}_+$, solves (14.143) with $w(k) \equiv 0$. Assume there exist functions $P_{1w} : \mathcal{D} \to \mathbb{R}^{1\times d}$, $P_{2w} : \mathcal{D} \to \mathbb{N}^d$, and $V : \mathcal{D} \to \mathbb{R}$ such that $V(\cdot)$ is continuous and

$$P_{1w}(0) = 0, \tag{14.187}$$
$$V(0) = 0, \tag{14.188}$$
$$V(x) > 0, \quad x \in \mathcal{D}, \quad x \neq 0, \tag{14.189}$$
$$V(f(x)) - V(x) < 0, \quad x \in \mathcal{D}, \quad x \neq 0, \tag{14.190}$$
$$V(f(x) + J_1(x)w) = V(f(x)) + P_{1w}(x)w + w^{\mathrm{T}}P_{2w}(x)w, \quad x \in \mathcal{D}, \quad w \in \mathcal{W}, \tag{14.191}$$
$$J_2(x) + J_2^{\mathrm{T}}(x) - P_{2w}(x) > 0, \quad x \in \mathcal{D}, \tag{14.192}$$
$$L(x) + V(f(x)) - V(x) + \Gamma(x) = 0, \quad x \in \mathcal{D}, \tag{14.193}$$

where

$$\Gamma(x) = \left[\tfrac{1}{2}P_{1w}^{\mathrm{T}}(x) - h(x)\right]^{\mathrm{T}} \left[J_2(x) + J_2^{\mathrm{T}}(x) - P_{2w}(x)\right]^{-1} \left[\tfrac{1}{2}P_{1w}^{\mathrm{T}}(x) - h(x)\right]. \tag{14.194}$$

Then the zero solution $x(t) \equiv 0$ of the undisturbed ($w(k) \equiv 0$) system (14.143) is locally asymptotically stable and there exists a neighborhood $\mathcal{D}_0 \subseteq \mathcal{D}$ of the origin such that

$$J(x_0) \leq \mathcal{J}(x_0) = V(x_0), \quad x_0 \in \mathcal{D}_0, \tag{14.195}$$

where

$$\mathcal{J}(x_0) \triangleq \sum_{k=0}^{\infty}[L(x(k)) + \Gamma(x(k))] \tag{14.196}$$

and where $x(k)$, $k \in \overline{\mathbb{Z}}_+$, solves (14.143) with $w(k) \equiv 0$. Furthermore, the solution $x(k)$, $k \in \overline{\mathbb{Z}}_+$, of (14.143) satisfies the passivity constraint

$$2\sum_{i=0}^{k} z^{\mathrm{T}}(i)w(i) + V(x_0) \geq 0, \qquad w(\cdot) \in \ell_2, \qquad k \in \overline{\mathbb{Z}}_+. \tag{14.197}$$

Finally, if $\mathcal{D} = \mathbb{R}^n$, $w(k) \equiv 0$, and

$$V(x) \to \infty \text{ as } \|x\| \to \infty, \tag{14.198}$$

then the solution $x(k) = 0$, $k \in \overline{\mathbb{Z}}_+$, of (14.143) is globally asymptotically stable.

Proof. With $\hat{Q} = 0$, $\hat{S} = I$, and $\hat{R} = 0$, it follows from (14.168) that $\Gamma(x)$ given by (14.194) satisfies (14.157). The result now follows as a direct consequence of Theorem 14.9. □

The framework presented in Corollary 14.9 is an extension of the continuous-time $\mathcal{H}_2$/positivity framework of Haddad and Bernstein [146] to nonlinear discrete-time systems. Specifically, letting $f(x) = Ax$, $J_1(x) = D$, $h(x) = Ex$, $J_2(x) = E_\infty$, $L(x) = x^{\mathrm{T}}Rx$, $V(x) = x^{\mathrm{T}}Px$, and $\Gamma(x) = x^{\mathrm{T}}(D^{\mathrm{T}}PA - E)^{\mathrm{T}}(E_\infty + E_\infty^{\mathrm{T}} - D^{\mathrm{T}}PD)^{-1}(D^{\mathrm{T}}PA - E)x$, where $E_\infty + E_\infty^{\mathrm{T}} - D^{\mathrm{T}}PD > 0$ and where $A \in \mathbb{R}^{n\times n}$, $D \in \mathbb{R}^{n\times d}$, $E \in \mathbb{R}^{d\times n}$, $E_\infty \in \mathbb{S}^d$, $R \in \mathbb{P}^n$, and $P \in \mathbb{P}^n$ satisfies

$$P = A^{\mathrm{T}}PA + (D^{\mathrm{T}}PA - E)^{\mathrm{T}}(E_\infty + E_\infty^{\mathrm{T}} - D^{\mathrm{T}}PD)^{-1}(D^{\mathrm{T}}PA - E) + R, \tag{14.199}$$

it follows from Corollary 14.9, with $x_0 = 0$, that

$$\sum_{i=0}^{k} 2w^{\mathrm{T}}(i)z(i) \geq 0, \qquad w(\cdot) \in \ell_2, \qquad k \in \overline{\mathbb{Z}}_+, \tag{14.200}$$

or, equivalently,

$$G_\infty(z) + G_\infty^*(z) \geq 0, \qquad |z| > 1, \tag{14.201}$$

where

$$G_\infty(z) \sim \left[\begin{array}{c|c} A & D \\ \hline E & E_\infty \end{array}\right].$$

Now, using similar arguments as in the $\mathcal{H}_\infty$ case, (14.195) implies

$$\begin{aligned}\operatorname{tr} \hat{P}V &= \mathbb{E}\left[\sum_{k=0}^{\infty} x^{\mathrm{T}}(k)Rx(k)\right] \\ &\leq \mathbb{E}\Bigg[\sum_{k=0}^{\infty} x^{\mathrm{T}}(k)[R + (D^{\mathrm{T}}PA - E)^{\mathrm{T}}(E_\infty + E_\infty^{\mathrm{T}} - D^{\mathrm{T}}PD)^{-1} \\ &\qquad \cdot(D^{\mathrm{T}}PA - E)]x(k)\Bigg],\end{aligned}$$

or, equivalently, since

$$\begin{aligned}
&\mathbb{E}\left[\sum_{k=0}^{\infty} x^{\mathrm{T}}(k)[R+(D^{\mathrm{T}}PA-E)^{\mathrm{T}}(E_{\infty}+E_{\infty}^{\mathrm{T}}-D^{\mathrm{T}}PD)^{-1}\right.\\
&\qquad\left.\cdot(D^{\mathrm{T}}PA-E)]x(k)\right]\\
&=\mathbb{E}\left[\sum_{k=0}^{\infty} x_0^{\mathrm{T}}(A^k)^{\mathrm{T}}[R+(D^{\mathrm{T}}PA-E)^{\mathrm{T}}(E_{\infty}+E_{\infty}^{\mathrm{T}}-D^{\mathrm{T}}PD)^{-1}\right.\\
&\qquad\left.\cdot(D^{\mathrm{T}}PA-E)]A^k x_0\right]\\
&=\mathbb{E}[x_0^{\mathrm{T}}Px_0]\\
&=\operatorname{tr} PV,
\end{aligned}$$

where P satisfies (14.199). Hence, $\|G\|_2^2 = \operatorname{tr} \hat{P}V \leq \operatorname{tr} PV$, which implies that $\mathcal{J}(x_0)$ given by (14.196) provides an upper bound to the $\mathcal{H}_2$ norm of $G(z)$.

Next, define the subset of square-summable bounded disturbances

$$\mathcal{W}_{\beta} \triangleq \left\{ w(\cdot) \in \ell_2 : \sum_{i=0}^{\infty} w^{\mathrm{T}}(i)w(i) \leq \beta \right\}, \tag{14.202}$$

where $\beta > 0$. Furthermore, let $L : \mathcal{D} \to \mathbb{R}$ be given such that $L(x) \geq 0$, $x \in \mathcal{D}$.

Theorem 14.10. Let $\gamma > 0$, $L(x) \geq 0$, $x \in \mathcal{D}$, and consider the nonlinear system (14.143) with performance functional (14.153). Assume that there exist functions $P_{1w} : \mathcal{D} \to \mathbb{R}^{1\times d}$, $P_{2w} : \mathcal{D} \to \mathbb{N}^d$, and $V : \mathcal{D} \to \mathbb{R}$ such that $V(\cdot)$ is continuous and

$$P_{1w}(0) = 0, \tag{14.203}$$

$$V(0) = 0, \tag{14.204}$$

$$V(x) > 0, \qquad x \in \mathcal{D}, \qquad x \neq 0, \tag{14.205}$$

$$V(f(x)) - V(x) < 0, \qquad x \in \mathcal{D}, \qquad x \neq 0, \tag{14.206}$$

$$V(f(x) + J_1(x)w) = V(f(x)) + P_{1w}(x)w + w^{\mathrm{T}}P_{2w}(x)w, \; x \in \mathcal{D}, \; w \in \mathcal{W}, \tag{14.207}$$

$$\tfrac{\gamma}{\beta}I_d - P_{2w}(x) > 0, \qquad x \in \mathcal{D}, \tag{14.208}$$

$$L(x)+V(f(x))-V(x)+\tfrac{1}{4}P_{1w}(x)\left(\frac{\gamma}{\beta}I_d - P_{2w}(x)\right)^{-1}P_{1w}^{\mathrm{T}}(x) = 0, \qquad x \in \mathcal{D}. \tag{14.209}$$

Then the zero solution $x(k) \equiv 0$ of the undisturbed ($w(k) \equiv 0$) system (14.143) is locally asymptotically stable and there exists a neighborhood $\mathcal{D}_0 \subseteq \mathcal{D}$ of the origin such that

$$J(x_0) \le \mathcal{J}(x_0) = V(x_0), \quad x_0 \in \mathcal{D}_0, \tag{14.210}$$

where

$$\mathcal{J}(x_0) \triangleq \sum_{k=0}^{\infty}[L(x(k)) + \Gamma(x(k))], \tag{14.211}$$

$$\Gamma(x) = \tfrac{1}{4}P_{1w}(x)\left(\frac{\gamma}{\beta}I_d - P_{2w}(x)\right)^{-1}P_{1w}^{\mathrm{T}}(x), \tag{14.212}$$

and where $x(k)$, $k \in \overline{\mathbb{Z}}_+$, solves (14.143) with $w(k) \equiv 0$. Furthermore, if $x_0 = 0$ then the solution $x(k)$, $k \in \overline{\mathbb{Z}}_+$, of (14.143) satisfies

$$V(x(k)) \le \gamma, \qquad w(\cdot) \in \mathcal{W}_\beta, \qquad k \in \overline{\mathbb{Z}}_+. \tag{14.213}$$

Finally, if $\mathcal{D} = \mathbb{R}^n$, $w(k) \equiv 0$, and

$$V(x) \to \infty \text{ as } \|x\| \to \infty, \tag{14.214}$$

then the solution $x(k) = 0$, $k \in \overline{\mathbb{Z}}_+$, of (14.143) is globally asymptotically stable.

Proof. The proofs for local and global asymptotic stability and the performance bound (14.210) are identical to the proofs of local and global asymptotic stability given in Theorem 14.9 and the performance bound (14.161). Next, with $r(z,w) = \frac{\gamma}{\beta}w^{\mathrm{T}}w$ and $\Gamma(x)$ given by (14.212), it follows from Lemma 14.4 that

$$V(x(k)) \le \frac{\gamma}{\beta}\sum_{i=0}^{k-1} w^{\mathrm{T}}(i)w(i), \qquad k > 0, \qquad w(\cdot) \in \ell_2, \qquad k \in \overline{\mathbb{Z}}_+,$$

which yields (14.213). □

14.11 A Riccati Equation Characterization for Mixed $\mathcal{H}_2/\ell_1$ Performance

In this section, we consider the dynamical system (14.143) and (14.144) with $f(x) = Ax$, $J_1(x) = D$, $h(x) = Ex$, and $J_2(x) = E_\infty$, where $A \in \mathbb{R}^{n\times n}$, $D \in \mathbb{R}^{n\times d}$, $E \in \mathbb{R}^{p\times n}$, $E_\infty \in \mathbb{R}^{p\times d}$, and A is asymptotically stable so that

$$x(k+1) = Ax(k) + Dw(k), \; x(0) = 0, \; k \in \overline{\mathbb{Z}}_+, \; w(\cdot) \in \hat{\mathcal{W}}, \tag{14.215}$$

$$z(k) = Ex(k) + E_\infty w(k), \tag{14.216}$$

where $\hat{\mathcal{W}}$ consists of unit-peak input signals defined by

$$\hat{\mathcal{W}} \triangleq \left\{w(\cdot) : w^{\mathrm{T}}(k)w(k) \le 1, \; k \in \overline{\mathbb{Z}}_+\right\}. \tag{14.217}$$

The following result provides an upper bound to the ℓ_1 norm (ℓ_∞ equi-induced norm) of the convolution operator G of the linear time-invariant system (14.215) and (14.216) given by

$$|||G|||_1 \triangleq \sup_{w(\cdot)\in\hat{\mathcal{W}}} \left\{ \sup_{k\in\overline{\mathbb{Z}}_+} \|z(k)\| \right\},$$

where $\|\cdot\|$ denotes the Euclidean vector norm.

Theorem 14.11. Let $\alpha > 1$ and consider the linear system (14.215) and (14.216). Then

$$|||G|||_1 \le \sigma_{\max}^{\frac{1}{2}}(EP^{-1}E^{\mathrm{T}}) + \sigma_{\max}^{\frac{1}{2}}(E_\infty E_\infty^{\mathrm{T}}), \tag{14.218}$$

where $P > 0$ satisfies

$$(\alpha-1)I_d - \alpha D^{\mathrm{T}}PD > 0, \tag{14.219}$$

$$P \ge \alpha A^{\mathrm{T}}PA + \alpha^2 A^{\mathrm{T}}PD\left[(\alpha-1)I_d - \alpha D^{\mathrm{T}}PD\right]^{-1} D^{\mathrm{T}}PA. \tag{14.220}$$

Proof. Let $N > 0$, $N \in \overline{\mathbb{Z}}_+$, and consider the dilated linear system

$$\tilde{x}(k+1) = A_\alpha\tilde{x}(k) + D_\alpha v(k), \quad \tilde{x}(0) = 0, \quad k \in \overline{\mathbb{Z}}_+, \tag{14.221}$$

$$z(k) = \alpha^{-\frac{k-N}{2}}(E\tilde{x}(k) + E_\infty v(k)), \tag{14.222}$$

where $A_\alpha \triangleq \sqrt{\alpha}A$, $D_\alpha \triangleq \sqrt{\alpha}D$, $\tilde{x}(k) \triangleq \alpha^{\frac{k-N}{2}}x(k)$, and $v(k) \triangleq \alpha^{\frac{k-N}{2}}w(k)$. Note that (14.221) and (14.222) are equivalent to (14.215) and (14.216). Furthermore, note that if $w(\cdot) \in \hat{\mathcal{W}}$ then $v(\cdot) \in \mathcal{V}$, where $\mathcal{V} \triangleq \left\{v(\cdot) : \sum_{k=0}^{N-1} v^{\mathrm{T}}(k)v(k) \le \frac{1}{\alpha-1}\right\}$. Hence,

$$|||G|||_1 \le \sup_{N\ge 0}\sup_{v(\cdot)\in\mathcal{V}} \|z(N)\|.$$

Next, with $f(x) = A_\alpha\tilde{x}(k)$, $J_1(x) = D_\alpha$, $V(x) = \tilde{x}^{\mathrm{T}}P\tilde{x}$, $P_{1w}(x) = 2\tilde{x}^{\mathrm{T}}A_\alpha^{\mathrm{T}}PD_\alpha$, $P_{2w}(x) = D_\alpha^{\mathrm{T}}PD_\alpha$, $\mathcal{W} = \mathcal{V}$, $\beta = \frac{1}{\alpha-1}$, $\gamma = 1$, and $L(x) = \tilde{x}^{\mathrm{T}}R\tilde{x}$, where $R \in \mathbb{R}^{n\times n}$ is an arbitrary positive-definite matrix, it follows from Theorem 14.10 that if there exists $P > 0$ such that

$$P = \alpha A^{\mathrm{T}}PA + \alpha^2 A^{\mathrm{T}}PD\left[(\alpha-1)I_d - \alpha D^{\mathrm{T}}PD\right]^{-1} D^{\mathrm{T}}PA + R, \tag{14.223}$$

then

$$\tilde{x}^{\mathrm{T}}(N)P\tilde{x}(N) \le 1,$$

and hence, since $x(N) = \tilde{x}(N)$, for all $v(\cdot) \in \mathcal{V}$,

$$\|z(N)\| = \|Ex(N) + E_\infty w(N)\| \le \sigma_{\max}^{\frac{1}{2}}(EP^{-1}E^{\mathrm{T}}) + \sigma_{\max}^{\frac{1}{2}}(E_\infty E_\infty^{\mathrm{T}}).$$

The result is now immediate by noting that (14.223) is equivalent to (14.220).

□

Note that the Riccati inequality (14.220) is equivalent to the linear matrix inequality

$$\begin{bmatrix} P & 0 \\ 0 & (\alpha-1)I_d \end{bmatrix} \geq \alpha \begin{bmatrix} A^{\mathrm{T}} \\ D^{\mathrm{T}} \end{bmatrix} P \begin{bmatrix} A & D \end{bmatrix}, \tag{14.224}$$

or, equivalently,

$$\begin{bmatrix} P & 0 \\ 0 & (\alpha-1)I_d \end{bmatrix} \geq \alpha \begin{bmatrix} A & D \\ 0 & 0 \end{bmatrix}^{\mathrm{T}} \begin{bmatrix} P & 0 \\ 0 & (\alpha-1)I_d \end{bmatrix} \begin{bmatrix} A & D \\ 0 & 0 \end{bmatrix}. \tag{14.225}$$

Next, since $\alpha > 1$, using Schur complements, (14.225) is equivalent to

$$\begin{bmatrix} P^{-1} & 0 \\ 0 & \frac{1}{(\alpha-1)}I_d \end{bmatrix} \geq \alpha \begin{bmatrix} A & D \\ 0 & 0 \end{bmatrix} \begin{bmatrix} P^{-1} & 0 \\ 0 & \frac{1}{(\alpha-1)}I_d \end{bmatrix} \begin{bmatrix} A & D \\ 0 & 0 \end{bmatrix}^{\mathrm{T}}. \tag{14.226}$$

Now, letting $\mathcal{Q} = P^{-1}$ in (14.226) yields

$$\mathcal{Q} \geq \alpha A \mathcal{Q} A^{\mathrm{T}} + \frac{\alpha}{\alpha-1} D D^{\mathrm{T}}. \tag{14.227}$$

Hence, Theorem 14.11 yields

$$|\!|\!| G |\!|\!|_1 \leq \sigma_{\max}^{\frac{1}{2}}(E \mathcal{Q} E^{\mathrm{T}}) + \sigma_{\max}^{\frac{1}{2}}(E_\infty E_\infty^{\mathrm{T}}), \tag{14.228}$$

where $\mathcal{Q}$ satisfies (14.227). It is interesting to note that in the case where $E_\infty = 0$ the solution $\mathcal{Q}$ to (14.227) satisfies the bound

$$Q \leq \mathcal{Q}, \tag{14.229}$$

where Q satisfies

$$Q = AQA^{\mathrm{T}} + DD^{\mathrm{T}}, \tag{14.230}$$

and hence,

$$|\!|\!| G |\!|\!|_2^2 = \operatorname{tr}\, EQE^{\mathrm{T}} \leq \operatorname{tr}\, E\mathcal{Q}E^{\mathrm{T}}. \tag{14.231}$$

Thus, (14.227) can be used to provide a trade-off between $\mathcal{H}_2$ and mixed $\mathcal{H}_2/\ell_1$ performance. For further details see [157].

14.12 Robust Stability Analysis of Nonlinear Uncertain Discrete-Time Systems

In this section, we extend the analysis results of Chapter 11 to discrete-time uncertain dynamical systems. Specifically, we consider the problem of evaluating a performance bound for a nonlinear-nonquadratic cost functional depending upon a class of nonlinear uncertain systems. As in the continuous-time case, the cost bound can be evaluated in closed form as long as the cost functional is related in a specific way to an underlying Lyapunov function that guarantees robust stability over a prescribed uncertainty set. Hence, the overall framework provides for robust stability and performance where robust performance here refers to a guaranteed bound on the worst-case value of a nonlinear-nonquadratic cost criterion over a prescribed uncertainty set. For the following result, let $\mathcal{D} \in \mathbb{R}^n$ be an open set, assume $0 \in \mathcal{D}$, let $L : \mathcal{D} \to \mathbb{R}$, and let $\mathcal{F} \subset \{f : \mathcal{D} \to \mathcal{D} : f(0) = 0\}$ denote the class of uncertain nonlinear systems with $f_0(\cdot) \in \mathcal{F}$ defining the nominal nonlinear system.

Theorem 14.12. Consider the nonlinear uncertain system

$$x(k+1) = f(x(k)), \quad x(0) = x_0, \qquad k \in \overline{\mathbb{Z}}_+, \tag{14.232}$$

where $f(\cdot) \in \mathcal{F}$, with performance functional

$$J(x_0) \triangleq \sum_{k=0}^{\infty} L(x(k)). \tag{14.233}$$

Assume there exist functions $\Gamma : \mathcal{D} \to \mathbb{R}$ and $V : \mathcal{D} \to \mathbb{R}$ such that $V(\cdot)$ is continuous and

$$V(0) = 0, \tag{14.234}$$

$$V(x) > 0, \quad x \in \mathcal{D}, \quad x \neq 0, \tag{14.235}$$

$$V(f(x)) \leq V(f_0(x)) + \Gamma(x), \ x \in \mathcal{D}, \ f(\cdot) \in \mathcal{F}, \tag{14.236}$$

$$V(f_0(x)) - V(x) + \Gamma(x) < 0, \quad x \in \mathcal{D}, \quad x \neq 0, \tag{14.237}$$

$$L(x) + V(f_0(x)) - V(x) + \Gamma(x) = 0, \quad x \in \mathcal{D}, \tag{14.238}$$

where $f_0(\cdot) \in \mathcal{F}$ defines the nominal nonlinear system. Then the zero solution $x(k) \equiv 0$ of (14.232) is locally asymptotically stable for all $f(\cdot) \in \mathcal{F}$ and there exists a neighborhood $\mathcal{D}_0 \subseteq \mathcal{D}$ of the origin such that the performance functional (14.233) satisfies

$$\sup_{f(\cdot)\in\mathcal{F}} J(x_0) \leq \mathcal{J}(x_0) = V(x_0), \quad x_0 \in \mathcal{D}_0, \tag{14.239}$$

where

$$\mathcal{J}(x_0) \triangleq \sum_{k=0}^{\infty} [L(x(k)) + \Gamma(x(k))], \tag{14.240}$$

and where $x(k), k \in \overline{\mathbb{Z}}_+$, is the solution to (14.232) with $f(x(k)) = f_0(x(k))$. Finally, if $\mathcal{D} = \mathbb{R}^n$, and

$$V(x) \to \infty \text{ as } \|x\| \to \infty, \tag{14.241}$$

then the zero solution $x(k) = 0$, $k \in \overline{\mathbb{Z}}_+$, of (14.232) is globally asymptotically stable.

Proof. Let $f(\cdot) \in \mathcal{F}$ and $x(k)$, $k \in \overline{\mathbb{Z}}_+$, satisfy (14.232). Then

$$\Delta V(x(k)) \triangleq V(x(k+1)) - V(x(k)) = V(f(x(k))) - V(x(k)), \quad k \in \overline{\mathbb{Z}}_+. \tag{14.242}$$

Hence, from (14.236) and (14.237), it follows that

$$\Delta V(x(k)) < 0, \qquad k \in \overline{\mathbb{Z}}_+, \qquad x(k) \neq 0. \tag{14.243}$$

Thus, from (14.234) and (14.235), and (14.243) it follows that $V(\cdot)$ is a Lyapunov function for (14.244), which proves local asymptotic stability of the zero solution $x(k) \equiv 0$ for all $f(\cdot) \in \mathcal{F}$. Consequently, $x(k) \to 0$ as $k \to \infty$ for all initial conditions $x_0 \in \mathcal{D}_0$ for some neighborhood $\mathcal{D}_0 \subseteq \mathcal{D}$ of the origin.

Next, (14.242) implies that

$$0 = -\Delta V(x(k)) + V(f(x(k))) - V(x(k)), \quad k \in \overline{\mathbb{Z}}_+,$$

and hence, using (14.236) and (14.238),

$$\begin{aligned} L(x(k)) &= -\Delta V(x(k)) + L(x(k)) + V(f(x(k))) - V(x(k)) \\ &\leq -\Delta V(x(k)) + L(x(k)) + V(f_0(x(k))) - V(x(k)) + \Gamma(x(k)) \\ &= -\Delta V(x(k)). \end{aligned}$$

Now, summing over $[0, k]$ yields

$$\sum_{i=0}^{k} L(x(i)) \leq -V(x(k+1)) + V(x_0).$$

Letting $k \to \infty$ and noting that $V(x(k)) \to 0$ as $k \to \infty$ for all $x_0 \in \mathcal{D}_0$ and $f(\cdot) \in \mathcal{F}$ yields $J(x_0) \leq V(x_0)$.

Next, let $x(k)$, $k \in \overline{\mathbb{Z}}_+$, satisfy (14.232) with $f(x(k)) = f_0(x(k))$. Then, it follows from (14.238) that

$$\begin{aligned} &L(x(k)) + \Gamma(x(k)) \\ &\quad = -\Delta V(x(k)) + L(x(k)) + V(f_0(x(k))) - V(x(k)) + \Gamma(x(k)) \\ &\quad = -\Delta V(x(k)). \end{aligned}$$

Summing over $[0, k]$ yields

$$\sum_{i=0}^{k} [L(x(i)) + \Gamma(x(i))] = -V(x(k+1)) + V(x_0).$$

Now, letting $k \to \infty$ yields $\mathcal{J}(x_0) = V(x_0)$. Finally, for $\mathcal{D} = \mathbb{R}^n$ global asymptotic stability of the zero solution $x(k) = 0$, $k \in \overline{\mathbb{Z}}_+$, is a direct consequence of the radially unbounded condition (14.241) on $V(x)$, $x \in \mathbb{R}^n$.

□

Next, we specialize Theorem 14.12 to nonlinear uncertain systems of the form

$$x(k+1) = f_0(x(k)) + \Delta f(x(k)), \quad x(0) = x_0, \qquad k \in \overline{\mathbb{Z}}_+, \qquad (14.244)$$

where $f_0 : \mathcal{D} \to \mathbb{R}^n$ is such that $f_0(0) = 0$, and $f_0 + \Delta f \in \mathcal{F}$. Here, $\mathcal{F}$ is such that

$$\mathcal{F} \subset \{f_0 + \Delta f : \mathcal{D} \to \mathcal{D} : \Delta f \in \boldsymbol{\Delta}\},$$

where $\boldsymbol{\Delta}$ is a given nonlinear uncertainty set of nonlinear perturbations Δf of the nominal system dynamics $f_0(\cdot) \in \mathcal{F}$. Since $\mathcal{F} \subset \{f : \mathcal{D} \to \mathcal{D} : \ f(0) = 0\}$ it follows that $\Delta f(0) = 0$.

Corollary 14.10. Consider the nonlinear uncertain system given by (14.244) with performance functional (14.233). Assume there exist functions $\Gamma : \mathcal{D} \to \mathbb{R}$, $P_{1f} : \mathcal{D} \to \mathbb{R}^{1\times n}$, $P_{2f} : \mathcal{D} \to \mathbb{N}^n$, and $V : \mathcal{D} \to \mathbb{R}$ such that $V(\cdot)$ is continuous, (14.234), (14.235), (14.237), and (14.238) hold and

$$P_{1f}(0) = 0, \qquad (14.245)$$

$$\Delta f^{\mathrm{T}}(x)P_{1f}^{\mathrm{T}}(x) + P_{1f}(x)\Delta f(x) + \Delta f^{\mathrm{T}}(x)P_{2f}(x)\Delta f(x) \le \Gamma(x), \quad x \in \mathcal{D}, \quad \Delta f(\cdot) \in \boldsymbol{\Delta}, \qquad (14.246)$$

$$V(f_0(x) + \Delta f(x)) = V(f_0(x)) + \Delta f^{\mathrm{T}}(x)P_{1f}^{\mathrm{T}}(x) + P_{1f}(x)\Delta f(x) + \Delta f^{\mathrm{T}}(x)P_{2f}(x)\Delta f(x), \ x \in \mathcal{D}, \ \Delta f(\cdot) \in \boldsymbol{\Delta}. \qquad (14.247)$$

Then then the zero solution $x(k) \equiv 0$ of (14.244) is locally asymptotically stable for all $\Delta f(\cdot) \in \boldsymbol{\Delta}$ and there exists a neighborhood $\mathcal{D}_0 \subseteq \mathcal{D}$ of the origin such that the performance functional (14.233) satisfies

$$\sup_{\Delta f(\cdot)\in\boldsymbol{\Delta}} J(x_0) \le \mathcal{J}(x_0) = V(x_0), \quad x_0 \in \mathcal{D}_0, \qquad (14.248)$$

where

$$\mathcal{J}(x_0) \triangleq \sum_{k=0}^{\infty}[L(x(k)) + \Gamma(x(k))], \qquad (14.249)$$

where $x(k), k \in \overline{\mathbb{Z}}_+$, solves (14.244) with $\Delta f(k) \equiv 0$. Finally, if $\mathcal{D} = \mathbb{R}^n$, and $V(x), x \in \mathbb{R}^n$, satisfies (14.241), then the solution $x(k) = 0$, $k \in \overline{\mathbb{Z}}_+$, of (14.244) is globally asymptotically stable for all $\Delta f(\cdot) \in \boldsymbol{\Delta}$.

Proof. The result is a direct consequence of Theorem 14.12 with $f(x) = f_0(x) + \Delta f(x)$ and $V(f(x))$ given by (14.247). Specifically, in this case it follows from (14.246) and (14.247) that $V(f(x)) \le V(f_0(x)) + \Gamma(x)$ for all $x \in \mathbb{R}^n$ and $\Delta f(\cdot) \in \boldsymbol{\Delta}$. Hence, all conditions of Theorem 14.12 are satisfied. □

Having established the theoretical basis to our approach, we now give a concrete structure for the bounding function $\Gamma(x)$ for the structure $\mathcal{F}$ as specified by (11.23) with $\boldsymbol{\Delta}$ satisfying (11.24) and (11.27), respectively.

Proposition 14.2. Let $P_{1f} : \mathbb{R}^n \to \mathbb{R}^{1\times n}$ be such that $P_{1f}(0) = 0$ and let $P_{2f} : \mathbb{R}^n \to \mathbb{N}^n$ be such that

$$I_{m_\delta} - g_\delta^{\mathrm{T}}(x)P_{2f}(x)g_\delta(x) > 0, \quad x \in \mathbb{R}^n. \tag{14.250}$$

Then the function

$$\begin{aligned}\Gamma(x) &= P_{1f}(x)g_\delta(x)[I_{m_\delta} - g_\delta^{\mathrm{T}}(x)P_{2f}(x)g_\delta(x)]^{-1}g_\delta^{\mathrm{T}}(x)P_{1f}^{\mathrm{T}}(x)\\ &\quad +m^{\mathrm{T}}(h_\delta(x))m(h_\delta(x))\end{aligned} \tag{14.251}$$

satisfies (14.246) with $\mathcal{F}$ given by (11.23) and $\mathbf{\Delta}$ given by (11.24).

Proof. Note that for all $x \in \mathcal{D}$,

$$\begin{aligned}0 &\leq \{[P_{1f}(x)g_\delta(x) - \delta^{\mathrm{T}}(h_\delta(x))][I_{m_\delta} - g_\delta^{\mathrm{T}}(x)P_{2f}(x)g_\delta(x)]\}\\ &\quad \cdot[I_{m_\delta} - g_\delta^{\mathrm{T}}(x)P_{2f}(x)g_\delta(x)]^{-1}\{[I_{m_\delta} - g_\delta^{\mathrm{T}}(x)P_{2f}(x)g_\delta(x)]\\ &\quad \cdot[P_{1f}(x)g_\delta(x) - \delta^{\mathrm{T}}(h_\delta(x))]^{\mathrm{T}}\}\\ &= P_{1f}(x)(g_\delta(x)[I_{m_\delta} - g_\delta^{\mathrm{T}}(x)P_{2f}(x)g_\delta(x)]^{-1}g_\delta^{\mathrm{T}}(x)P_{1f}^{\mathrm{T}}(x)\\ &\quad -P_{1f}(x)g_\delta(x)\delta(h_\delta(x)) - \delta^{\mathrm{T}}(h_\delta(x))g_\delta^{\mathrm{T}}(x)P_{1f}^{\mathrm{T}}(x)\\ &\quad +\delta^{\mathrm{T}}(h_\delta(x))\delta(h_\delta(x)) - \delta^{\mathrm{T}}(h_\delta(x))g_\delta^{\mathrm{T}}(x)P_{2f}(x)g_\delta(x)\delta(h_\delta(x)\\ &\leq \Gamma(x) - [\Delta f^{\mathrm{T}}(x)P_{1f}^{\mathrm{T}}(x) + P_{1f}(x)\Delta f(x) + \Delta f^{\mathrm{T}}(x)P_{2f}(x)\Delta f(x)],\end{aligned}$$

which proves (14.246) with $\mathcal{F}$ given by (11.23) and $\mathbf{\Delta}$ given by (11.27). □

Proposition 14.3. Let $P_{1f} : \mathbb{R}^n \to \mathbb{R}^{1\times n}$ be such that $P_{1f}(0) = 0$ and let $P_{1f} : \mathbb{R}^n \to \mathbb{N}^n$ be such that

$$2I_{m_\delta} - g_\delta^{\mathrm{T}}(x)P_{2f}(x)g_\delta(x) > 0, \quad x \in \mathbb{R}^n. \tag{14.252}$$

Then the function

$$\begin{aligned}\Gamma(x) &= [P_{1f}(x)g_\delta(x) + m_1(h_\delta(x)) + m_2(h_\delta(x))][2I_{m_\delta} - g_\delta^{\mathrm{T}}(x)P_{2f}(x)g_\delta(x)]^{-1}\\ &\quad \cdot[P_{1f}(x)g_\delta(x) + m_1(h_\delta(x)) + m_2(h_\delta(x))]^{\mathrm{T}} + m_1^{\mathrm{T}}(h_\delta(x))m_2(h_\delta(x))\\ &\quad +m_2^{\mathrm{T}}(h_\delta(x))m_1(h_\delta(x))\end{aligned} \tag{14.253}$$

satisfies (14.246) with $\mathcal{F}$ given by (11.23) with $\mathbf{\Delta}$ given by (11.27).

Proof. The proof is identical to that of Proposition 14.2 and, hence, is left as an exercise for the reader. □

We now combine the results of Corollary 14.10 and Propositions 14.2 and 14.3 to obtain a series of sufficient conditions guaranteeing robust stability and performance for the nonlinear uncertain discrete-time system (14.244).

Proposition 14.4. Consider the nonlinear uncertain system (14.244).

Let $L(x) > 0$, $x \in \mathbb{R}^n$, and suppose there exist functions $P_{1f} : \mathbb{R}^n \to \mathbb{R}^{1\times n}$, $P_{2f} : \mathbb{R}^n \to \mathbb{N}^n$, and a continuous radially unbounded function $V : \mathbb{R}^n \to \mathbb{R}$ such that (14.234), (14.235), and (14.247) hold and

$$\begin{aligned} V(x) &= V(f_0(x)) + P_{1f}(x)g_\delta(x)[I_{m_\delta} - g_\delta^{\mathrm{T}}(x)P_{2f}(x)g_\delta(x)]^{-1}g_\delta^{\mathrm{T}}(x)P_{1f}^{\mathrm{T}}(x) \\ &\quad + m^{\mathrm{T}}(h_\delta(x))m(h_\delta(x)) + L(x), \quad x \in \mathbb{R}^n. \end{aligned} \tag{14.254}$$

Then the zero solution $x(t) \equiv 0$ to (14.244) is globally asymptotically stable for all $\Delta f(\cdot) \in \mathbf{\Delta}$ with $\mathbf{\Delta}$ given by (11.24), and the performance function (14.233) satisfies

$$\sup_{\Delta f(\cdot)\in\mathbf{\Delta}} J(x_0) \le \mathcal{J}(x_0) = V(x_0), \tag{14.255}$$

where

$$\begin{aligned} \mathcal{J}(x_0) &\triangleq \\ &\sum_{k=0}^{\infty}[L(x(k)) + P_{1f}(x(k))g_\delta(x(k))[I_{m_\delta} - g_\delta^{\mathrm{T}}(x(k))P_{2f}(x(k))g_\delta(x(k))]^{-1} \\ &\cdot g_\delta^{\mathrm{T}}(x(k))P_{1f}^{\mathrm{T}}(x(k)) + m^{\mathrm{T}}(h_\delta(x(k)))m(h_\delta(x(k))), \end{aligned} \tag{14.256}$$

where $x(k)$, $k \in \overline{\mathbb{Z}}_+$, is the solution to (14.244) with $\Delta f(x) \equiv 0$.

Proposition 14.5. Consider the nonlinear uncertain system (14.244). Let $L(x) > 0$, $x \in \mathbb{R}^n$, and suppose there exist functions $P_{1f} : \mathbb{R}^n \to \mathbb{R}^{1\times n}$, $P_{2f} : \mathbb{R}^n \to \mathbb{N}^n$, and a continuous radially unbounded function $V : \mathbb{R}^n \to \mathbb{R}$ such that (14.234), (14.235), and (14.247) hold and

$$\begin{aligned} V(x) &= V(f_0(x)) + [P_{1f}(x)g_\delta(x) + m_1(h_\delta(x)) + m_2(h_\delta(x))][2I_{m_\delta} \\ &\quad - g_\delta^{\mathrm{T}}(x)P_{2f}(x)g_\delta(x)]^{-1}[P_{1f}(x)g_\delta(x) + m_1(h_\delta(x)) + m_2(h_\delta(x))]^{\mathrm{T}} \\ &\quad + m_1^{\mathrm{T}}(h_\delta(x))m_2(h_\delta(x)) + m_2^{\mathrm{T}}(h_\delta(x))m_1(h_\delta(x)) + L(x), \quad x \in \mathbb{R}^n. \end{aligned} \tag{14.257}$$

Then the zero solution $x(t) \equiv 0$ to (14.244) is globally asymptotically stable for all $\Delta f(\cdot) \in \mathbf{\Delta}$ with $\mathbf{\Delta}$ given by (11.27), and the performance function (14.233) satisfies

$$\sup_{\Delta f(\cdot)\in\mathbf{\Delta}} J(x_0) \le \mathcal{J}(x_0) = V(x_0), \tag{14.258}$$

where

$$\begin{aligned} \mathcal{J}(x_0) &\triangleq \sum_{k=0}^{\infty}[L(x(k)) + [P_{1f}(x(k))g_\delta(x(k)) + m_1(h_\delta(x(k))) + m_2(h_\delta(x(k)))] \\ &\quad \cdot[2I_{m_\delta} - g_\delta^{\mathrm{T}}(x(k))P_{2f}(x(k))g_\delta(x(k))]^{-1}[P_{1f}(x(k))g_\delta(x(k)) \\ &\quad + m_1(h_\delta(x(k))) + m_2(h_\delta(x(k)))]^{\mathrm{T}} + m_1^{\mathrm{T}}(h_\delta(x(k)))m_2(h_\delta(x(k))) \\ &\quad + m_2^{\mathrm{T}}(h_\delta(x(k)))m_1(h_\delta(x(k))), \end{aligned} \tag{14.259}$$

where $x(k)$, $k \in \overline{\mathbb{Z}}_+$, is the solution to (14.244) with $\Delta f(x) \equiv 0$.

The following corollary specializes Theorem 14.12 to linear uncertain systems and connects the framework of Theorem 14.12 to the quadratic Lyapunov bounding framework of Haddad *et al.* [148]. Specifically, in this case we consider $\mathcal{F}$ to be the set of uncertain linear systems given by

$$\mathcal{F} = \{(A + \Delta A)x : x \in \mathbb{R}^n, A \in \mathbb{R}^{n\times n}, \Delta A \in \mathbf{\Delta}_A\},$$

where $\mathbf{\Delta}_A \subset \mathbb{R}^{n\times n}$ is a given bounded uncertainty set of uncertain perturbations ΔA of the nominal system matrix such that $0 \in \mathbf{\Delta}_A$. For the statement of this result let $R \in \mathbb{P}^n$.

Corollary 14.11. Consider the linear uncertain system

$$x(k+1) = (A + \Delta A)x(k), \qquad x(0) = x_0, \qquad k \in \overline{\mathbb{Z}}_+, \tag{14.260}$$

with performance functional

$$J_{\mathbf{\Delta}_A}(x_0) \triangleq \sum_{k=0}^{\infty} x^{\mathrm{T}}(k)Rx(k). \tag{14.261}$$

where $\Delta A \in \mathbf{\Delta}_A$. Let $\Omega : \mathcal{N} \subseteq \mathbb{S}^n \to \mathbb{N}^n$ be such that

$$A^{\mathrm{T}}P\Delta A + \Delta A^{\mathrm{T}}PA + \Delta A^{\mathrm{T}}P\Delta A \le \Omega(P), \quad \Delta A \in \mathbf{\Delta}_A, \quad P \in \mathcal{N}. \tag{14.262}$$

Furthermore, suppose there exists $P \in \mathbb{P}^n$ satisfying

$$P = A^{\mathrm{T}}PA + \Omega(P) + R. \tag{14.263}$$

Then the zero solution $x(k) \equiv 0$ of (14.260) is globally asymptotically stable for all $\Delta A \in \mathbf{\Delta}_A$, and

$$\sup_{\Delta A \in \mathbf{\Delta}_A} J_{\mathbf{\Delta}_A}(x_0) \le \mathcal{J}(x_0) = x_0^{\mathrm{T}}Px_0, \quad x_0 \in \mathbb{R}^n, \tag{14.264}$$

where

$$\mathcal{J}(x_0) = \sum_{k=0}^{\infty} x^{\mathrm{T}}(k)[R + \Omega(P)]x(k), \tag{14.265}$$

and where $x(k), k \in \overline{\mathbb{Z}}_+$, solves (14.260) with $\Delta A = 0$.

Proof. The result is a direct consequence of Theorem 14.12 with $f(x) = (A + \Delta A)x, f_0(x) = Ax, L(x) = x^{\mathrm{T}}Rx, V(x) = x^{\mathrm{T}}Px, \Gamma(x) = x^{\mathrm{T}}\Omega(P)x$, and $\mathcal{D} = \mathbb{R}^n$. Specifically, conditions (14.234) and (14.235) are trivially satisfied. Now, $V(f(x)) = x^{\mathrm{T}}A^{\mathrm{T}}PAx + x^{\mathrm{T}}(\Delta A^{\mathrm{T}}P + P\Delta A + \Delta A^{\mathrm{T}}P\Delta A)x$, and hence, it follows from (14.262) that $V(f(x)) \le V(f_0(x)) + \Gamma(x) = x^{\mathrm{T}}(A^{\mathrm{T}}PA + \Omega(P))x$, for all $x \in \mathbb{R}^n$ and $\Delta A \in \mathbf{\Delta}_A$. Furthermore, it follows from (14.263) that $L(x) + V(f_0(x)) - V(x) + \Gamma(x) = 0, x \in \mathbb{R}^n$, and hence, $V(f_0(x)) - V(x) + \Gamma(x) < 0$, for all $x \in \mathbb{R}^n, x \ne 0$, so that all conditions of Theorem 14.12 are satisfied. Finally, since $V(x)$ is radially unbounded the zero solution $x(k) \equiv 0$ of (14.260) is globally asymptotically stable for all $\Delta A \in \mathbf{\Delta}_A$. $\square$

Corollary 14.11 is the deterministic version of Theorem 4.1 of [173] involving quadratic Lyapunov bounds for addressing robust stability and performance analysis of linear uncertain discrete-time systems. To demonstrate the applicability of Corollary 14.11 consider the uncertain system (14.260) with $\mathbf{\Delta}_A$ given by (11.59). For this uncertainty characterization the bound $\Omega(\cdot)$ satisfying (14.262) can now be given a concrete form.

Proposition 14.6. Let $\mathcal{M} \subseteq \mathbb{S}^n$ denote the set of $P \in \mathbb{P}^n$ such that

$$\mathcal{M} \triangleq \{P \in \mathbb{N}^n : I_{m_\delta} - B_0^{\mathrm{T}} P B_0 > 0\}. \quad (14.266)$$

Then the function

$$\Omega(P) = A^{\mathrm{T}} P B_0 (I_{m_\delta} - B_0^{\mathrm{T}} P B_0)^{-1} B_0^{\mathrm{T}} P A + C_0^{\mathrm{T}} N C_0 \quad (14.267)$$

satisfies (14.262) with $\mathbf{\Delta}_A$ given by (11.59).

Proof. The proof is a direct consequence of Proposition 14.2 with $P_{1f}(x) = x^{\mathrm{T}} A^{\mathrm{T}} P$, $P_{2f}(x) = P$, and $m(h_\delta(x) = N^{1/2} C_0 x$. □

Finally, as in the nonlinear case, we now combine the results of Corollary 14.11 and Proposition 14.6 to obtain a sufficient condition for guaranteeing robust stability and performance for the uncertain discrete-time system (14.260).

Proposition 14.7. Consider the linear uncertain system (14.260). Let $R \in \mathbb{P}^m$, $N \in \mathbb{N}^n$, and suppose there exists $P \in \mathcal{M}$ satisfying

$$P = A^{\mathrm{T}} P A + A^{\mathrm{T}} P B_0 (I_{m_\delta} - B_0^{\mathrm{T}} P B_0)^{-1} B_0^{\mathrm{T}} P A + C_0^{\mathrm{T}} N C_0 + R. \quad (14.268)$$

Then $A + \Delta A$ is asymptotically stable for all $\Delta A \in \mathbf{\Delta}_A$ given by (11.59), and

$$\sup_{\Delta A \in \mathbf{\Delta}_A} J_{\Delta A}(x_0) \le x_0^{\mathrm{T}} P x_0, \quad x_0 \in \mathbb{R}^n. \quad (14.269)$$

14.13 Problems

Problem 14.1. Consider the system (14.1) and (14.2) with $F(x,u,k) = Ax + Bu$ and $L(x,u,k) = x^{\mathrm{T}} R_1 x + u^{\mathrm{T}} R_2 u$, where $A \in \mathbb{R}^{n\times n}$, $B \in \mathbb{R}^{n\times m}$, $R_1 \in \mathbb{R}^{n\times n}$, and $R_2 \in \mathbb{R}^{m\times m}$, such that $R_1 \ge 0$ and $R_2 > 0$. Show that the optimal control law characterized by the Bellman equation (14.10) is given by

$$u(k) = -[R_2 + B^{\mathrm{T}} P(k+1) B]^{-1} B^{\mathrm{T}} P(k+1) A x(k), \quad (14.270)$$

where $P(k_{\mathrm{f}} + 1) = 0$ and

$$P(k) = A^{\mathrm{T}} P(k+1) A + R_1 - A^{\mathrm{T}} P(k+1) B [R_2 + B^{\mathrm{T}} P(k+1) B]^{-1} B^{\mathrm{T}} P(k+1) A. \quad (14.271)$$

Problem 14.2. Consider the cascade system

$$x(k+1) = f(x(k)) + G(x(k))y(k), \quad x(0) = x_0, \quad k \in \overline{\mathbb{Z}}_+, \tag{14.272}$$
$$\hat{x}(k+1) = A\hat{x}(k) + Bu(k), \quad \hat{x}(0) = \hat{x}_0, \tag{14.273}$$
$$y(k) = C\hat{x}(k) + Du(k), \tag{14.274}$$

where $\hat{x} \in \mathbb{R}^q$, $u, y \in \mathbb{R}^m$, $A \in \mathbb{R}^{q\times q}$, $B \in \mathbb{R}^{q\times m}$, $C \in \mathbb{R}^{m\times q}$, and $D \in \mathbb{R}^{m\times m}$, with performance functional

$$J(x_0, \hat{x}_0, u) = \sum_{k=0}^{\infty} L(x(k), \hat{x}(k), u(k)), \tag{14.275}$$

where $(x(k), \hat{x}(k))$, $k \in \overline{\mathbb{Z}}_+$, solves (14.272) and (14.273), and $L(x, \hat{x}, u)$ is given by (14.276) and where

$$L(x, \hat{x}, u) \triangleq L_1(x, \hat{x}) + L_2(x, \hat{x})u + u^{\mathrm{T}} R_2(x) u. \tag{14.276}$$

Assume that (A, B, C, D) is strict feedback positive real, that is, there exist a scalar $\rho > 1$ and matrices $\hat{P} \in \mathbb{P}^q$, $L \in \mathbb{R}^{l\times q}$, $W \in \mathbb{R}^{l\times m}$, and $K \in \mathbb{R}^{m\times q}$ such that

$$\tfrac{1}{\rho}\hat{P} = (A + BK)^{\mathrm{T}}\hat{P}(A+BK) + L^{\mathrm{T}}L, \tag{14.277}$$
$$B^{\mathrm{T}}\hat{P}(A+BK) = C + DK - W^{\mathrm{T}}L, \tag{14.278}$$
$$D + D^{\mathrm{T}} = B^{\mathrm{T}}\hat{P}B + W^{\mathrm{T}}W, \tag{14.279}$$

and the nonlinear subsystem (14.272) has a globally stable equilibrium at $x(k) = 0$, $k \in \overline{\mathbb{Z}}_+$, and Lyapunov function $V_{\mathrm{sub}}(x)$ so that

$$V_{\mathrm{sub}}(f(x)) < V_{\mathrm{sub}}(x), \qquad x \in \mathbb{R}^n, \qquad x \neq 0. \tag{14.280}$$

Furthermore, assume there exist functions $L_2 : \mathbb{R}^n \times \mathbb{R}^q \to \mathbb{R}^{1\times m}$, $P_{12} : \mathbb{R}^n \to \mathbb{R}^{1\times m}$, and $P_2 : \mathbb{R}^n \to \mathbb{N}^m$ such that

$$L_2(0,0) = 0, \tag{14.281}$$
$$P_{12}(0) = 0, \tag{14.282}$$
$$V_{\mathrm{sub}}(f(x) + G(x)y) = V_{\mathrm{sub}}(f(x)) + P_{12}(x)y + y^{\mathrm{T}}P_2(x)y, \tag{14.283}$$
$$\begin{aligned} &y^{\mathrm{T}}\{P_{12}^{\mathrm{T}}(x) + P_2(x)C\hat{x} - 2K\hat{x} - (I_m + \tfrac{1}{2}P_2(x)D)R_2^{-1}(x) \\ &\cdot[2(B^{\mathrm{T}}\hat{P}A + D^{\mathrm{T}}P_2(x)C)\hat{x} + L_2^{\mathrm{T}}(x,\hat{x}) + D^{\mathrm{T}}P_{12}^{\mathrm{T}}(x)]\} \leq 0, \\ &\qquad (x, \hat{x}) \in \mathbb{R}^n \times \mathbb{R}^q, \end{aligned} \tag{14.284}$$

where $R_2(x) \triangleq R_2(x) + B^{\mathrm{T}}\hat{P}B + D^{\mathrm{T}}P_2(x)D$ and $\hat{P}$, K satisfy (14.277)–(14.279). Show that the zero solution $(x(k), \hat{x}(k)) \equiv (0,0)$ of the cascade system (14.272) and (14.273) is globally asymptotically stable with the feedback control law $u = \phi(x, \hat{x})$, where

$$\phi(x, \hat{x}) = -\tfrac{1}{2}R_2^{-1}(x)[2(B^{\mathrm{T}}\hat{P}A + D^{\mathrm{T}}P_2(x)C)\hat{x} + L_2^{\mathrm{T}}(x,\hat{x}) + D^{\mathrm{T}}P_{12}^{\mathrm{T}}(x)]. \tag{14.285}$$

Furthermore, show that for $(x_0, \hat{x}_0) \in \mathbb{R}^n \times \mathbb{R}^m$,

$$J(x_0, \hat{x}_0, \phi(x(\cdot), \hat{x}(\cdot))) = V(x_0, \hat{x}_0), \tag{14.286}$$

where

$$V(x, \hat{x}) = V_{\text{sub}}(x) + \hat{x}^{\mathrm{T}} \hat{P} \hat{x}, \tag{14.287}$$

and the performance functional (14.275), with

$$\begin{aligned} L_1(x, \hat{x}) &= \phi^{\mathrm{T}}(x, \hat{x}) R_2(x) \phi(x, \hat{x}) + V_{\text{sub}}(x) \\ &\quad - V_{\text{sub}}(f(x) + G(x)C\hat{x}) + \hat{x}^{\mathrm{T}}(\hat{P} - A^{\mathrm{T}} \hat{P} A)\hat{x}, \end{aligned} \tag{14.288}$$

is minimized in the sense that

$$J(x_0, \hat{x}_0, \phi(x(\cdot), \hat{x}(\cdot))) = \min_{u \in \mathcal{S}(x_0, \hat{x}_0)} J(x_0, \hat{x}_0, u(\cdot)), \tag{14.289}$$

where

$$\begin{aligned} \mathcal{S}(x_0, \hat{x}_0) \triangleq \{&u(\cdot) : u(\cdot) \in \mathcal{U} \text{ and } (x(\cdot), \hat{x}(\cdot)) \text{ given by} \\ &(14.272) \text{ and } (14.273) \text{ satisfies } (x(k), \hat{x}(k)) \to 0 \text{ as } k \to \infty\}. \end{aligned}$$

Problem 14.3. Consider the cascade system

$$x(k+1) = f(x(k)) + G(x(k))y(k), \quad x(0) = x_0, \quad k \in \overline{\mathbb{Z}}_+, \tag{14.290}$$

$$\hat{x}(k+1) = \hat{f}(\hat{x}(k)) + \hat{g}(\hat{x}(k))u(k), \quad \hat{x}(0) = \hat{x}_0, \tag{14.291}$$

$$y(k) = h(\hat{x}(k)) + J(\hat{x}(k))u(k), \tag{14.292}$$

where $\hat{x} \in \mathbb{R}^q$, $u, y \in \mathbb{R}^m$, $\hat{f} : \mathbb{R}^q \to \mathbb{R}^q$ satisfies $\hat{f}(0) = 0$, $\hat{g} : \mathbb{R}^q \to \mathbb{R}^{q \times m}$, $h : \mathbb{R}^q \to \mathbb{R}^m$ satisfies $h(0) = 0$, and $J : \mathbb{R}^q \to \mathbb{R}^{m \times m}$, with performance functional (14.275) where $(x(k), \hat{x}(k))$, $k \in \overline{\mathbb{Z}}_+$, solves (14.290) and (14.291), and $L(x, \hat{x}, u)$ is given by (14.276). Assume that the input subsystem (14.291) and (14.292) is feedback strictly passive and assume the subsystem (14.290) has a globally stable equilibrium at $x(k) = 0$, $k \in \overline{\mathbb{Z}}_+$, and Lyapunov function $V_{\text{sub}}(x)$ so that

$$V_{\text{sub}}(f(x)) < V_{\text{sub}}(x), \qquad x \in \mathbb{R}^n, \qquad x \neq 0. \tag{14.293}$$

Furthermore, assume there exist functions $L_2 : \mathbb{R}^n \times \mathbb{R}^q \to \mathbb{R}^{1 \times m}$, $P_{12} : \mathbb{R}^n \to \mathbb{R}^{1 \times m}$, and $P_2 : \mathbb{R}^n \to \mathbb{N}^m$ such that

$$L_2(0, 0) = 0, \tag{14.294}$$

$$P_{12}(0) = 0, \tag{14.295}$$

$$V_{\text{sub}}(f(x) + G(x)y) = V_{\text{sub}}(f(x)) + P_{12}(x)y + y^{\mathrm{T}} P_2(x) y, \tag{14.296}$$

$$\begin{aligned} y^{\mathrm{T}} \Big\{ &P_{12}^{\mathrm{T}}(x) + P_2(x)h(\hat{x}) - 2k(\hat{x}) - (I_m + \tfrac{1}{2}P_2(x)J(\hat{x}))R_{2a}^{-1}(x, \hat{x})[\hat{P}_{12}^{\mathrm{T}}(\hat{x}) \\ &+ 2J^{\mathrm{T}}(\hat{x})P_2(x)h(\hat{x}) + L_2^{\mathrm{T}}(x, \hat{x}) + J^{\mathrm{T}}(\hat{x})P_{12}^{\mathrm{T}}(x)] \Big\} \leq 0, \quad (x, \hat{x}) \in \mathbb{R}^n \times \mathbb{R}^q, \end{aligned} \tag{14.297}$$

where $R_{2a}(x,\hat{x}) \triangleq R_2(x,\hat{x}) + \hat{P}_2(\hat{x}) + J^{\mathrm{T}}(\hat{x})P_2(x)J(\hat{x})$ and $k(\hat{x})$ satisfies

$$0 > V_{\mathrm{s}}(\hat{f}(\hat{x}) + \hat{g}(\hat{x})k(\hat{x})) - V_{\mathrm{s}}(\hat{x}) + l^{\mathrm{T}}(\hat{x})l(\hat{x}), \quad \hat{x} \neq 0, \tag{14.298}$$

$$0 = \tfrac{1}{2}\hat{P}_{12}^{\mathrm{T}}(\hat{x}) + \mathcal{W}^{\mathrm{T}}(\hat{x})l(\hat{x}) - (h(\hat{x}) + J(\hat{x})k(\hat{x})) + \hat{P}_2(\hat{x})k(\hat{x}), \tag{14.299}$$

$$0 = \hat{P}_2(\hat{x}) + \mathcal{W}^{\mathrm{T}}(\hat{x})\mathcal{W}(\hat{x}) - (J(\hat{x}) + J^{\mathrm{T}}(\hat{x})). \tag{14.300}$$

Show that the zero solution $(x(k),\hat{x}(k)) \equiv (0,0)$ of the cascade system (14.290) and (14.291) is globally asymptotically stable with the feedback control law $u = \phi(x,\hat{x})$, where

$$\begin{aligned}\phi(x,\hat{x}) &= -\tfrac{1}{2}R_{2a}^{-1}(x,\hat{x})[\hat{P}_{12}^{\mathrm{T}}(\hat{x}) + 2J^{\mathrm{T}}(\hat{x})P_2(x)h(\hat{x}) + L_2^{\mathrm{T}}(x,\hat{x}) \\ &\quad + J^{\mathrm{T}}(\hat{x})P_{12}^{\mathrm{T}}(x)].\end{aligned} \tag{14.301}$$

Furthermore, show that for $(x_0,\hat{x}_0) \in \mathbb{R}^n \times \mathbb{R}^m$,

$$J(x_0,\hat{x}_0,\phi(x(\cdot),\hat{x}(\cdot))) = V(x_0,\hat{x}_0), \tag{14.302}$$

where

$$V(x,\hat{x}) = V_{\mathrm{sub}}(x) + V_{\mathrm{s}}(\hat{x}), \tag{14.303}$$

and the performance functional (14.275), with

$$\begin{aligned}L_1(x,\hat{x}) &= \phi^{\mathrm{T}}(x,\hat{x})R_{2a}(x,\hat{x})\phi(x,\hat{x}) + V_{\mathrm{sub}}(x) \\ &\quad - V_{\mathrm{sub}}(f(x) + G(x)h(\hat{x})) + V_{\mathrm{s}}(\hat{x}) - V_{\mathrm{s}}(\hat{f}(\hat{x})),\end{aligned} \tag{14.304}$$

is minimized in the sense that

$$J(x_0,\hat{x}_0,\phi(x(\cdot),\hat{x}(\cdot))) = \min_{u \in \mathcal{S}(x_0,\hat{x}_0)} J(x_0,\hat{x}_0,u(\cdot)). \tag{14.305}$$

Problem 14.4. Consider the discrete-time linear dynamical system

$$G(z) \sim \left[\begin{array}{c|c} A & D \\ \hline E & 0 \end{array}\right],$$

where $G \in \mathcal{RH}_\infty$. Define the *discrete entropy of G at infinity* by

$$I(G,\gamma) \triangleq \tfrac{-\gamma^2}{2\pi}\int_{-\pi}^{\pi} \ln\left|\det(I_m - \gamma^{-2}G^*(e^{\jmath\theta})G(e^{\jmath\theta}))\right| \mathrm{d}\theta, \tag{14.306}$$

and assume there exists a positive-definite matrix $P \in \mathbb{R}^{n \times n}$ satisfying

$$P = A^{\mathrm{T}}PA + A^{\mathrm{T}}PD(\gamma^2 I - D^{\mathrm{T}}PD)^{-1}D^{\mathrm{T}}PA + E^{\mathrm{T}}E, \tag{14.307}$$

where $\gamma > 0$. Show that the following statements hold:

i) The transfer function G satisfies $\|G\|_\infty \leq \gamma$.

ii) If $\|G\|_\infty < \gamma$, then $I(G,\gamma) \leq -\ln\det(\gamma^2 I - D^{\mathrm{T}}PD)$.

iii) The $\mathcal{H}_2$ norm of G satisfies $\|G\|_2 \leq I(G,\gamma)$.

iv) All real symmetric solutions to (14.307) are nonnegative definite.

v) There exists a (unique) minimal solution to (14.307) in the class of real symmetric solutions.

vi) P is the minimal solution to (14.307) if and only if $\rho(A + D(\gamma^2 I - D^{\mathrm{T}}PD)^{-1}D^{\mathrm{T}}PA) < 1$.

vii) $\|G\|_\infty < \gamma$ if and only if $A+D(\gamma^2 I - D^{\mathrm{T}}PD)^{-1}D^{\mathrm{T}}PA$ is Schur, where P is the minimal solution to (14.307).

viii) If P is the minimal solution to (14.307) and $\|G\|_\infty < \gamma$, then $I(G,\gamma) = -\ln\det(\gamma^2 I - D^{\mathrm{T}}PD)$.

Problem 14.5. Consider the controlled nonlinear dynamical system

$$x(k+1) = F(x(k),u(k)) + J_1(x(k))w(k), \quad x(0) = x_0, \quad w(\cdot) \in \ell_2, \quad k \in \overline{\mathbb{Z}}_+, \tag{14.308}$$

$$z(k) = h(x(k),u(k)) + J_2(x(k))w(k), \tag{14.309}$$

with performance functional

$$J(x_0,u(\cdot)) \triangleq \sum_{k=0}^{\infty} L(x(k),u(k)), \tag{14.310}$$

where $F : \mathcal{D} \times U \to \mathbb{R}^n$ is continuous and satisfies $F(0,0) = 0$, $J_1 : \mathcal{D} \to \mathbb{R}^{n\times d}$, $h : \mathcal{D} \times U \to \mathbb{R}^p$ satisfies $h(0,0) = 0$, $J_2 : \mathcal{D} \to \mathbb{R}^{p\times d}$, $L : \mathcal{D} \times U \to \mathbb{R}$, and $u(\cdot) \in \mathcal{U}$ is an admissible control. Assume there exist functions $V : \mathcal{D} \to \mathbb{R}$, $\Gamma : \mathcal{D} \times U \to \mathbb{R}$, $P_{1w} : \mathcal{D} \times U \to \mathbb{R}^{1\times d}$, $P_{2w} : \mathcal{D} \times U \to \mathbb{N}^d$, and a control law $\phi : \mathcal{D} \to U$ such that $V(\cdot)$ is continuous and

$$P_{1w}(0,0) = 0, \tag{14.311}$$

$$V(0) = 0, \tag{14.312}$$

$$V(x) > 0, \quad x \in \mathcal{D}, \quad x \neq 0, \tag{14.313}$$

$$\phi(0) = 0, \tag{14.314}$$

$$V(F(x,\phi(x))) - V(x) < 0, \quad x \in \mathcal{D}, \quad x \neq 0, \tag{14.315}$$

$$P_{1w}(x,\phi(x))w + w^{\mathrm{T}}P_{2w}(x,\phi(x))w \le r(z,w) + L(x,\phi(x)) + \Gamma(x,\phi(x)), \quad x \in \mathcal{D}, \quad w \in \mathcal{W}, \tag{14.316}$$

$$V(F(x,u) + J_1(x)w) = V(F(x,u)) + P_{1w}(x,u)w + w^{\mathrm{T}}P_{2w}(x,u)w, \quad x \in \mathcal{D}, \quad u \in U, \quad w \in \mathcal{W}, \tag{14.317}$$

$$H(x,\phi(x)) = 0, \quad x \in \mathcal{D}, \tag{14.318}$$

$$H(x,u) \ge 0, \quad x \in \mathcal{D}, \quad u \in U, \tag{14.319}$$

where

$$H(x,u) \triangleq V(F(x,u)) - V(x) + L(x,u) + \Gamma(x,u). \tag{14.320}$$

Show, with the feedback control $u(\cdot) = \phi(x(\cdot))$, that there exists a neighborhood $\mathcal{D}_0 \subseteq \mathcal{D}$ of the origin such that if $x_0 \in \mathcal{D}_0$ and $w(k) \equiv 0$, the solution zero $x(k) \equiv 0$ of the closed-loop system is locally asymptotically stable. If, in addition, $\Gamma(x, \phi(x)) \geq 0$, $x \in \mathcal{D}$, then show

$$J(x_0, \phi(x(\cdot))) \leq \mathcal{J}(x_0, \phi(x(\cdot))) = V(x_0), \tag{14.321}$$

where

$$\mathcal{J}(x_0, u(\cdot)) \triangleq \sum_{k=0}^{\infty} [L(x(k), u(k)) + \Gamma(x(k), u(k))] \tag{14.322}$$

and where $u(\cdot)$ is admissible and $x(k)$, $k \in \overline{\mathbb{Z}}_+$, solves (14.308) with $w(k) \equiv 0$. Furthermore, if $x_0 \in \mathcal{D}_0$ show that the feedback control $u(\cdot) = \phi(x(\cdot))$ minimizes $\mathcal{J}(x_0, u(\cdot))$ in the sense that

$$\mathcal{J}(x_0, \phi(x(\cdot))) = \min_{u(\cdot) \in \mathcal{S}(x_0)} \mathcal{J}(x_0, u(\cdot)), \tag{14.323}$$

where $\mathcal{S}(x_0)$ is the set of regulation controllers for the system (14.308) with $w(k) \equiv 0$ and $x_0 \in \mathcal{D}$. Finally, show that the solution $x(k)$, $k \in \overline{\mathbb{Z}}_+$, of (14.308) satisfies the dissipativity constraint

$$\sum_{i=0}^{k} r(z(i), w(i)) + V(x_0) \geq 0, \qquad w(\cdot) \in \ell_2, \qquad k \in \overline{\mathbb{Z}}_+. \tag{14.324}$$

Problem 14.6. Consider the linear discrete-time system

$$x(k+1) = Ax(k) + Bu(k) + Dw(k), \; x(0) = x_0, \; k \in \overline{\mathbb{Z}}_+, \; w(\cdot) \in \ell_2, \tag{14.325}$$

$$z(k) = E_1 x(k) + E_2 u(k), \tag{14.326}$$

with performance functional

$$J(x_0, u(\cdot)) = \sum_{k=0}^{\infty} [x^{\mathrm{T}}(k) R_1 x(k) + u^{\mathrm{T}}(k) R_2 u(k)], \tag{14.327}$$

where $A \in \mathbb{R}^{n \times n}$, $B \in \mathbb{R}^{n \times m}$, $D \in \mathbb{R}^{n \times d}$, $E_1 \in \mathbb{R}^{p \times n}$, $E_2 \in \mathbb{R}^{p \times m}$, $R_1 \triangleq E_1^{\mathrm{T}} E_1 > 0$, $R_2 \triangleq E_2^{\mathrm{T}} E_2 > 0$, and $u(\cdot)$ is admissible. Assume there exists $P \in \mathbb{P}^n$ such that

$$P = A^{\mathrm{T}} PA + R_1 + A^{\mathrm{T}} PD(\gamma^2 I_d - D^{\mathrm{T}} PD)^{-1} D^{\mathrm{T}} PA - P_a^{\mathrm{T}} R_{2a}^{-1} P_a, \tag{14.328}$$

where $\gamma > 0$ and

$$R_{2a} \triangleq R_2 + B^{\mathrm{T}} PB + B^{\mathrm{T}} PD(\gamma^2 I_d - D^{\mathrm{T}} PD)^{-1} D^{\mathrm{T}} PB,$$
$$P_a \triangleq B^{\mathrm{T}} PA + B^{\mathrm{T}} PD(\gamma^2 I_d - D^{\mathrm{T}} PD)^{-1} D^{\mathrm{T}} PA.$$

Show that, with the feedback control law $u = \phi(x) = -R_{2a}^{-1} P_a x$, the closed-loop undisturbed ($w(k) \equiv 0$) system (14.325) is globally asymptotically stable for all $x_0 \in \mathbb{R}^n$. Furthermore, show

$$J(x_0, \phi(x(\cdot))) \le \mathcal{J}(x_0, \phi(x(\cdot))) = x_0^{\mathrm{T}} P x_0, \tag{14.329}$$

where

$$\mathcal{J}(x_0, u(\cdot)) = \sum_{k=0}^{\infty} [(D^{\mathrm{T}}P(Ax(k) + Bu(k)))^{\mathrm{T}}(\gamma^2 I_d - D^{\mathrm{T}}PD)^{-1}(D^{\mathrm{T}}P(Ax(k) + Bu(k))) + x^{\mathrm{T}}(k)R_1x(k) + u^{\mathrm{T}}(k)R_2u(k)], \tag{14.330}$$

and where $u(\cdot)$ is admissible and $x(k)$, $k \in \overline{\mathbb{Z}}_+$, solves (14.325) with $w(k) \equiv 0$. In addition, show

$$\mathcal{J}(x_0, \phi(x(\cdot))) = \min_{u(\cdot) \in \mathcal{S}(x_0)} \mathcal{J}(x_0, u(\cdot)), \tag{14.331}$$

where $\mathcal{S}(x_0)$ is the set of regulation controllers for the system (14.325) with $w(k) \equiv 0$ and $x_0 \in \mathbb{R}^n$. Finally, if $x_0 = 0$ show that, with $u = \phi(x)$, the solution $x(k)$, $k \in \overline{\mathbb{Z}}_+$, of (14.325) satisfies the nonexpansivity constraint

$$\sum_{i=0}^{k} z^{\mathrm{T}}(i)z(i) \le \gamma^2 \sum_{i=0}^{k} w^{\mathrm{T}}(i)w(i), \qquad w(\cdot) \in \ell_2, \qquad k \in \overline{\mathbb{Z}}_+, \tag{14.332}$$

or, equivalently, $\|\tilde{G}(z)\|_\infty \le \gamma$, where $K \triangleq -R_{2a}^{-1}P_a$ and

$$\tilde{G}(z) \sim \left[\begin{array}{c|c} A + BK & D \\ \hline E_1 + E_2K & 0 \end{array}\right].$$

Problem 14.7. Let $p = d$ and consider the linear system (14.325) with performance variables

$$z(k) = E_1x(k) + E_2u(k) + E_\infty w(k), \tag{14.333}$$

and performance functional (14.327). Assume there exists $P \in \mathbb{P}^n$ such that

$$P = A^{\mathrm{T}}PA + R_1 + (D^{\mathrm{T}}PA - E_1)^{\mathrm{T}}R_0^{-1}(D^{\mathrm{T}}PA - E_1) - P_s^{\mathrm{T}}R_{2s}^{-1}P_s, \tag{14.334}$$

where

$$\begin{aligned} R_0 &\triangleq E_\infty + E_\infty^{\mathrm{T}} - D^{\mathrm{T}}PD, \\ R_{2s} &\triangleq B^{\mathrm{T}}PB + R_2 + (D^{\mathrm{T}}PB - E_2)^{\mathrm{T}}R_0^{-1}(D^{\mathrm{T}}PB - E_2), \\ P_s &\triangleq B^{\mathrm{T}}PA + (D^{\mathrm{T}}PB - E_2)^{\mathrm{T}}R_0^{-1}(D^{\mathrm{T}}PA - E_1). \end{aligned}$$

Show that, with the feedback control law $u = \phi(x) = -R_{2s}^{-1}P_sx$, the closed-loop undisturbed ($w(k) \equiv 0$) system (14.325) is globally asymptotically stable for all $x_0 \in \mathbb{R}^n$. Furthermore, show

$$J(x_0, \phi(x(\cdot))) \le \mathcal{J}(x_0, \phi(x(\cdot))) = x_0^{\mathrm{T}} P x_0, \tag{14.335}$$

where

$\mathcal{J}(x_0, u(\cdot))$

$$= \sum_{k=0}^{\infty} [x^{\mathrm{T}}(k)(R_1 + (D^{\mathrm{T}}PA - E_1)^{\mathrm{T}} R_0^{-1} (D^{\mathrm{T}}PA - E_1))x(k) + u^{\mathrm{T}}(k)$$
$$\cdot (R_{2\mathrm{s}} - B^{\mathrm{T}}PB)u(k) + 2x^{\mathrm{T}}(k)(D^{\mathrm{T}}PA - E_1)^{\mathrm{T}} R_0^{-1} (D^{\mathrm{T}}PB - E_2)u(k)], \tag{14.336}$$

and where $u(\cdot)$ is admissible and $x(k)$, $k \in \overline{\mathbb{Z}}_+$, solves (14.325) with $w(k) \equiv 0$. In addition, show

$$\mathcal{J}(x_0, \phi(x(\cdot))) = \min_{u(\cdot) \in \mathcal{S}(x_0)} \mathcal{J}(x_0, u(\cdot)), \tag{14.337}$$

where $\mathcal{S}(x_0)$ is the set of regulation controllers for the system (14.325) with $w(k) \equiv 0$ and $x_0 \in \mathbb{R}^n$. Finally, if $x_0 = 0$ show that, with $u = \phi(x)$, the solution $x(k)$, $k \in \overline{\mathbb{Z}}_+$, of (14.325) satisfies the passivity constraint

$$\sum_{i=0}^{k} 2z^{\mathrm{T}}(i)w(i) \geq 0, \qquad w(\cdot) \in \ell_2, \qquad k \in \overline{\mathbb{Z}}_+, \tag{14.338}$$

or, equivalently, $\tilde{G}_\infty(z) + \tilde{G}_\infty^*(z) \geq 0$, $|z| > 1$, where $K = -R_{2\mathrm{s}}^{-1} P_s$ and

$$\tilde{G}_\infty(z) \sim \left[\begin{array}{c|c} A + BK & D \\ \hline E_1 + E_2 K & E_\infty \end{array}\right].$$

Problem 14.8. Consider the controlled uncertain dynamical system

$$x(k+1) = F(x(k), u(k)), \quad x(0) = x_0, \quad k \in \overline{\mathbb{Z}}_+, \tag{14.339}$$

with performance functional

$$J(x_0, u(\cdot)) \triangleq \sum_{k=0}^{\infty} L(x(k), u(k)), \tag{14.340}$$

where $F(\cdot,\cdot) \in \mathcal{F} \subset \{F : \mathcal{D} \times U \to \mathbb{R}^n : F(0,0) = 0\}$, $\mathcal{F}$ denotes the class of uncertain closed-loop systems with $F_0(\cdot,\cdot) \in \mathcal{F}$ defining the nominal closed-loop system, $L : \mathcal{D} \times U \to \mathbb{R}$, and $u(\cdot)$ is an admissible control. Assume there exist functions $V : \mathcal{D} \to \mathbb{R}$, $\Gamma : \mathcal{D} \times U \to \mathbb{R}$, and a control law $\phi : \mathcal{D} \to U$ such that $V(\cdot)$ is continuous and

$$V(0) = 0, \tag{14.341}$$
$$V(x) > 0, \quad x \in \mathcal{D}, \quad x \neq 0, \tag{14.342}$$
$$\phi(0) = 0, \tag{14.343}$$
$$V(F(x, \phi(x))) \leq V(F_0(x, \phi(x))) + \Gamma(x, \phi(x)), \quad x \in \mathcal{D}, \quad F(\cdot,\cdot) \in \mathcal{F}, \tag{14.344}$$
$$V(F_0(x, \phi(x))) - V(x) + \Gamma(x, \phi(x)) < 0, \quad x \in \mathcal{D}, \quad x \neq 0, \tag{14.345}$$
$$H(x, \phi(x)) = 0, \quad x \in \mathcal{D}, \tag{14.346}$$
$$H(x, u) \geq 0, \quad x \in \mathcal{D}, \quad u \in U, \tag{14.347}$$

where

$$H(x,u) \triangleq L(x,u) + V(F_0(x,u)) - V(x) + \Gamma(x,u). \tag{14.348}$$

Show that, with the feedback control $u(\cdot) = \phi(x(\cdot))$, there exists a neighborhood $\mathcal{D}_0 \subseteq \mathcal{D}$ of the origin such that if $x_0 \in \mathcal{D}_0$, the zero solution $x(k) \equiv 0$ of the closed-loop system is locally asymptotically stable for all $F(\cdot,\cdot) \in \mathcal{F}$. Furthermore, show

$$\sup_{F(\cdot,\cdot)\in\mathcal{F}} J(x_0, \phi(x(\cdot))) \leq \mathcal{J}(x_0, \phi(x(\cdot))) = V(x_0), \tag{14.349}$$

where

$$\mathcal{J}(x_0, u(\cdot)) \triangleq \sum_{k=0}^{\infty}[L(x(k),u(k)) + \Gamma(x(k),u(k))], \tag{14.350}$$

and where $u(\cdot)$ is admissible and $x(k)$, $k \in \overline{\mathbb{Z}}_+$, solves (14.339) with $F(x(k),u(k)) = F_0(x(k),u(k))$. In addition, if $x_0 \in \mathcal{D}_0$ show that the feedback control $u(\cdot) = \phi(x(\cdot))$ minimizes $\mathcal{J}(x_0, u(\cdot))$ in the sense that

$$\mathcal{J}(x_0, \phi(x(\cdot))) = \min_{u(\cdot)\in\mathcal{S}(x_0)} \mathcal{J}(x_0, u(\cdot)), \tag{14.351}$$

where $\mathcal{S}(x_0)$ is the set of regulation controllers for (14.339) with $F(\cdot,\cdot) = F_0(\cdot,\cdot)$.

14.14 Notes and References

The principle of optimality is due to Bellman [36]. The Bellman equation was developed by Richard E. Bellman [36] in the late 1950s and is the functional equation of dynamic programming. For a modern textbook treatment, see Kirk [239] and Bryson and Ho [72].

The results in Sections 11.2–11.6 on stability analysis and optimal and inverse optimal control of nonlinear discrete-time systems are due to Haddad and Chellaboina [155] and Haddad, Chellaboina, Fausz, and Abdallah [165]. In particular, [155,165] give an overview of the discrete-time, nonlinear-nonquadratic control problem with polynomial and multilinear cost functionals. A brief treatment of nonlinear-nonquadratic discrete-time optimal control is also given by Jacobson [217]. For a treatment on stabilization of discrete-time nonlinear systems see Tsinias [432] and Byrnes, Lin, and Ghosh [79]. Gain, sector, and disk margins of nonlinear-nonquadratic optimal regulators are due to Chellaboina and Haddad [86]. The original work on gain and phase margins of linear-quadratic discrete-time optimal regulators is due to Willems and van de Voorde [459] and Shaked [397]. Later extensions were presented by Lee and Lee [262]. The discrete-time nonlinear disturbance rejection framework using dissipativity, passivity, and nonexpansivity concepts presented in Sections 11.10–11.12 is due to Haddad, Chellaboina, and Wu [168]. See also Lin and Byrnes [280]. The linear discrete-time mixed-norm $\mathcal{H}_2/\mathcal{H}_\infty$ problem was formulated by Haddad, Bernstein, and Mustafa [154]. A state space approach to linear discrete-time $\mathcal{H}_\infty$ can be found in Stoorvogel [418] and Iglesias and Glover [211].

Finally, the stability analysis of nonlinear uncertain discrete-time systems presented in Section 14.12 is a generalization of the linear discrete-time guaranteed cost control framework developed by Haddad, Huang, and Bernstein [173] and Haddad and Bernstein [148, 149].

Bibliography

[1] R. A. Adomaitis and E. H. Abed, "Bifurcation Analysis of Nonuniform Flow Patterns in Axial-Flow Gas Compressors," in *Proc. 1st World Congress of Nonlinear Analysts*, Tampa, FL, pp. 1597–1608, 1992.

[2] R. A. Adomaitis and E. H. Abed, "Local Nonlinear Control of Stall Inception in Axial Flow Compressors," in *29th Joint Propulsion Conference and Exhibit—Paper No. AIAA 93-2230*, Monterey, CA, 1993.

[3] D. Aeyels and J. Peutman, "A New Asymptotic Stability Criterion for Nonlinear Time-Variant Differential Equations," *IEEE Trans. Autom. Control*, vol. 43, pp. 968–971, 1998.

[4] R. P. Agarwal and V. Lakshmikantham, *Uniqueness and Nonuniqueness Criteria for Ordinary Differential Equations.* Singapore: World Scientific, 1993.

[5] M. A. Aizerman and F. R. Gantmacher, *Absolute Stability of Regulator Systems.* San Francisco, CA: Holden-Day, 1964.

[6] G. L. Amicucci, S. Monaco, and D. Normand-Cyrot, "Control Lyapunov Stabilization of Affine Discrete-Time Systems," in *Proc. IEEE Conf. Dec. Control*, San Diego, CA, pp. 923–924, 1997.

[7] B. D. O. Anderson, "Algebraic Description of Bounded Real Matrices," *Electron. Lett.*, vol. 2, pp. 464–465, 1966.

[8] B. D. O. Anderson, "A System Theory Criterion for Positive Real Matrices," *SIAM J. Control Optim.*, vol. 5, pp. 171–182, 1967.

[9] B. D. O. Anderson and J. B. Moore, "Algebraic Structure of Generalized Positive Real Matrices," *SIAM J. Control*, vol. 6, pp. 615–624, 1968.

[10] B. D. O. Anderson and J. B. Moore, *Optimal Control: Linear Quadratic Methods.* Englewood Cliffs, NJ: Prentice-Hall, 1991.

[11] B. D. O. Anderson and S. Vongpanitlerd, *Network Analysis and Synthesis: A Modern Systems Theory Approach.* Englewood Cliffs, NJ: Prentice-Hall, 1973.

[12] T. M. Apostol, *Mathematical Analysis.* Reading, MA: Addison-Wesley, 1957.

[13] Z. Artstein, "Stabilization with Relaxed Controls," *Nonlinear Anal. Theory, Methods Appl.*, vol. 7, pp. 1163–1173, 1983.

[14] S. J. Asseo, "Optimal Control of a Servo Derived from Nonquadratic Performance Criteria," *IEEE Trans. Autom. Control*, vol. 14, pp. 404–407, 1969.

[15] J.-P. Aubin, *Viability Theory.* Boston, MA: Birkhauser, 1991.

[16] E. Awad and F. E. C. Culick, "On the Existence and Stability of Limit Cycles for Longitudinal Acoustic Modes in a Combustion Chamber," *Combust. Sci. Technol.*, vol. 9, pp. 195–222, 1986.

[17] A. Bacciotti, *Local Stabilizability of Nonlinear Control Systems.* Series on Advances in Mathematics and Applied Sciences. Singapore: World Scientific, 1992.

[18] A. Back, J. Guckenheimer, and M. Myers, "A Dynamical Simulation Facility for Hybrid Systems," in *Hybrid Systems* (R. Grossman, A. Nerode, A. Ravn, and H. Rischel, eds.), pp. 255–267. New York: Springer, 1993.

[19] O. O. Badmus, S. Chowdhury, and C. N. Nett, "Nonlinear Control of Surge in Axial Compression Systems," *Automatica*, vol. 32, no. 1, pp. 59–70, 1996.

[20] D. D. Bainov and P. S. Simeonov, *Systems with Impulse Effect: Stability, Theory and Applications.* London: Ellis-Horwood, 1989.

[21] O. E. Balje, "A Contribution to the Problem of Designing Radial Turbomachines," *Trans. ASME*, vol. 74, pp. 451–472, 1952.

[22] J. A. Ball and J. W. Helton, "$\mathcal{H}_\infty$ Optimal Control for Nonlinear Plants: Connection with Differential Games," in *Proc. IEEE Conf. Dec. Control*, Tampa, FL, pp. 956–962, 1989.

[23] E. A. Barbashin and N. N. Krasovskii, "On the Stability of Motion in the Large," *Dokl. Akad. Nauk.*, vol. 86, pp. 453–456, 1952.

[24] B. R. Barmish, "Stabilization of Uncertain Systems via Linear Control," *IEEE Trans. Autom. Control*, vol. 28, pp. 848–850, 1983.

[25] B. R. Barmish, "Necessary and Sufficient Conditions for Quadratic Stabilizability of an Uncertain Linear System," *J. Optim. Theory Appl.*, vol. 46, pp. 399–408, 1985.

[26] B. R. Barmish, M. Corless, and G. Leitmann, "A New Class of Stabilizing Controllers for Uncertain Dynamical Systems," *SIAM J. Control Optim.*, vol. 21, pp. 246–255, 1983.

[27] B. R. Barmish and C. L. DeMarco, "A New Method for Improvement of Robustness Bounds for Linear State Equations," in *Proc. Conf. Infor. Sci. Syst.*, Princeton University, pp. 115–120, 1986.

[28] B. R. Barmish and A. R. Galimidi, "Robustness of Luenberger Observers: Linear Systems Stabilized via Non-linear Control," *Automatica*, vol. 22, pp. 413–423, 1986.

[29] B. R. Barmish, I. R. Petersen, and B. Feuer, "Linear Ultimate Boundedness Control of Uncertain Dynamic Systems," *Automatica*, vol. 19, pp. 523–532, 1983.

[30] R. G. Bartle, *The Elements of Real Analysis,* 2nd edition. New York: Wiley, 1976.

[31] T. Basar and P. Bernhard, *$\mathcal{H}_\infty$-Optimal Control and Related Minimax Design Problems.* Boston, MA: Birkhauser, 1995.

[32] T. Basar and G. J. Olsder, *Dynamic Noncooperative Game Theory.* New York: Academic Press, 1982.

[33] R. W. Bass and R. F. Webber, "Optimal Nonlinear Feedback Control Derived from Quartic and Higher-Order Performance Criteria," *IEEE Trans. Autom. Control*, vol. 11, pp. 448–454, 1966.

[34] W. T. Baumann and W. J. Rugh, "Feedback Control of Nonlinear Systems by Extended Linearization," *IEEE Trans. Autom. Control*, vol. 31, pp. 40–46, 1986.

[35] R. L. Behnken, R. D'Andrea, and R. M. Murray, "Control of Rotating Stall in a Low-Speed Axial Flow Compressor Using Pulsed Air," in *Proc. IEEE Conf. Dec. Control*, New Orleans, LA, pp. 3056–3061, 1995.

[36] R. Bellman, *Dynamic Programming.* Princeton, NJ: Princeton University Press, 1957.

[37] R. Bellman, *Introduction to Matrix Analysis.* New York: McGraw-Hill, 1960. Reprinted by SIAM, Philadelphia, 1995.

[38] R. Bellman, "Vector Lyapunov Functions," *SIAM J. Control*, vol. 1, pp. 32–34, 1962.

[39] I. Bendixson, "Sur les Courbes Définies par des Équations Différentielles," *Acta Math.*, vol. 24, pp. 1–88, 1901.

[40] A. Berman and R. J. Plemmons, *Nonnegative Matrices in the Mathematical Sciences.* New York: Academic Press, 1979.

[41] D. S. Bernstein, "Robust Static and Dynamic Output-Feedback Stabilization: Deterministic and Stochastic Perspectives," *IEEE Trans. Autom. Control*, vol. 32, pp. 1076–1084, 1987.

[42] D. S. Bernstein, "Robust Stability and Performance via Fixed-Order Dynamic Compensation," *SIAM J. Control Optim.*, vol. 27, pp. 389–406, 1989.

[43] D. S. Bernstein, "Nonquadratic Cost and Nonlinear Feedback Control," *Int. J. Robust Nonlinear Control*, vol. 3, pp. 211–229, 1993.

[44] D. S. Bernstein, "From Infancy to Potency: Lyapunov's Second Method and the Past, Present, and Future of Control Theory," plenary presentation at the *IEEE Conf. Dec. Control*, Orlando, FL, 2001.

[45] D. S. Bernstein, *Matrix Mathematics.* Princeton, NJ: Princeton University Press, 2005.

[46] D. S. Bernstein and S. P. Bhat, "Lyapunov Stability, Semistability, and Asymptotic Stability of Matrix Second-Order Systems," *ASME Trans. J. Vibr. Acoustics*, vol. 117, pp. 145–153, 1995.

[47] D. S. Bernstein and S. P. Bhat, "Nonnegativity, Reducibility, and Semistability of Mass Action Kinetics," in *Proc. IEEE Conf. Dec. Control*, Phoenix, AZ, pp. 2206–2211, 1999.

[48] D. S. Bernstein and W. M. Haddad, "The Optimal Projection Equations with Petersen-Hollot Bounds: Robust Stability and Performance via Fixed-Order Compensation for Systems with Structured Real-Valued Parameter Uncertainty," *IEEE Trans. Autom. Control*, vol. 33, pp. 578–582, 1988.

[49] D. S. Bernstein and W. M. Haddad, "LQG Control with $\mathcal{H}_\infty$ Performance Bound: A Riccati Equation Approach," *IEEE Trans. Autom. Control*, vol. 34, pp. 293–305, 1989.

[50] D. S. Bernstein and W. M. Haddad, "Robust Stability and Performance Analysis for State Space Systems via Quadratic

Lyapunov Bounds," *SIAM J. Matrix Anal. Appl.*, vol. 11, pp. 236–271, 1990.

[51] D. S. Bernstein and W. M. Haddad, "Robust Stability and Performance via Fixed-Order Dynamic Compensation with Guaranteed Cost Bounds," *Math. Control, Signals, Syst.*, vol. 3, pp. 136–163, 1990.

[52] D. S. Bernstein, W. M. Haddad, and D. C. Hyland, "Small Gain versus Positive Real Modelling of Real Parameter Uncertainty," *AIAA J. Guid. Control Dyn.*, vol. 15, pp. 538–540, 1992.

[53] D. S. Bernstein and S. L. Osburn, "Guaranteed Cost Inequalities for Robust Stability and Performance Analysis," *Int. J. Robust Nonlinear Control*, vol. 12, pp. 1275–1297, 2002.

[54] S. P. Bhat and D. S. Bernstein, "Lyapunov Analysis of Semistability," in *Proc. American Control Conf.*, San Diego, CA, pp. 1608–1612, 1999.

[55] S. P. Bhat and D. S. Bernstein, "Finite-time Stability of Continuous Autonomous Systems," *SIAM J. Control Optim.*, vol. 38, pp. 751–766, 2000.

[56] S. P. Bhat and D. S. Bernstein, "Arc-Length-Based Lyapunov Tests for Convergence and Stability in Systems Having a Continuum of Equilibria," in *Proc. American Control Conf.*, Denver, CO, pp. 2961–2966, 2003.

[57] S. P. Bhat and D. S. Bernstein, "Nontangency-Based Lyapunov Tests for Convergence and Stability in Systems Having a Continuum of Equilibra," *SIAM J. Control Optim.*, vol. 42, pp. 1745–1775, 2003.

[58] N. P. Bhatia and G. P. Szegö, *Stability Theory of Dynamical Systems*. Berlin: Springer-Verlag, 1970.

[59] R. M. Biernacki, H. Hwang, and S. P. Bhattacharya, "Robust Stability with Structured Real Parameter Perturbations," *IEEE Trans. Autom. Control*, vol. 32, pp. 495–506, 1987.

[60] G. D. Birkhoff, "Recent Advances in Dynamics," *Science*, vol. 51, pp. 51–55, 1920.

[61] G. D. Birkhoff, "Collected Mathematical Papers," *Am. Math. Soc.*, vol. 1,2,3, 1950.

[62] F. Blanchini, "Feedback Control for Linear Time-Invariant Systems with State and Control Bounds in the Presence of Disturbances," *IEEE Trans. Autom. Control*, vol. 35, pp. 1231–1234, 1990.

[63] A. M. Bloch and J. E. Marsden, "Stabilization of Rigid Body Dynamics by the Energy-Casimir Method," *Syst. Control Lett.*, vol. 14, pp. 341–346, 1990.

[64] J. J. Bongiorno, Jr., "An Extension of the Nyquist-Borkhausen Stability Criterion to Linear Lumped-Parameter Systems with Time-Varying Elements," *IEEE Trans. Autom. Control*, vol. 8, pp. 166–170, 1963.

[65] J. J. Bongiorno, Jr., "Real-Frequency Stability Criteria for Linear Time-Varying Systems," *Proc. IEEE*, vol. 52, pp. 832–841, 1964.

[66] S. P. Boyd, L. El Ghaoui, E. Feron, and V. Balakrishnan, *Linear Matrix Inequalities in System and Control Theory.* Philadelphia, PA: SIAM, 1994.

[67] R. W. Brockett, "The Status of Stability Theory for Deterministic Systems," *IEEE Trans. Autom. Control*, vol. 12, pp. 596–666, 1967.

[68] R. W. Brockett, "System Theory on Group Manifolds and Coset Spaces," *SIAM J. Control*, vol. 10, pp. 265–284, 1972.

[69] R. W. Brockett, "Feedback Invariants for Nonlinear Systems," *IFAC Congress*, vol. 7, pp. 1115–1120, 1978.

[70] R. W. Brockett and H. B. Lee, "Frequency-Domain Instability Criteria for Time-Varying and Nonlinear Systems," *Proc. IEEE*, vol. 55, pp. 604–619, 1967.

[71] R. W. Brockett and J. L. Willems, "Frequency Domain Stability Criteria, Parts I and II," *IEEE Trans. Autom. Control*, vol. 10, pp. 255–261, 407–413, 1965.

[72] A. E. Bryson and Y. C. Ho, *Applied Optimal Control: Optimization, Estimation and Control.* New York: Hemisphere, 1975.

[73] R. T. Bupp and D. S. Bernstein, "A Benchmark Problem for Nonlinear and Robust Control," *Int. J. Robust Nonlinear Control*, vol. 8, pp. 307–310, 1998.

[74] C. I. Byrnes and A. Isidori, "Local Stabilization of Minimum-Phase Nonlinear Systems," *Syst. Control Lett.*, vol. 11, pp. 9–17, 1988.

[75] C. I. Byrnes and A. Isidori, "New Results and Examples in Nonlinear Feedback Stabilization," *Syst. Control Lett.*, vol. 12, pp. 437–442, 1989.

[76] C. I. Byrnes and A. Isidori, "Asymptotic stabilization of minimum phase nonlinear systems," *IEEE Trans. Autom. Control*, vol. 36, pp. 1122–1137, 1991.

[77] C. I. Byrnes, A. Isidori, and J. C. Willems, "Passivity, Feedback Equivalence and the Global Stabilization of Minimum Phase Nonlinear Systems," *IEEE Trans. Autom. Control*, vol. 36, pp. 1228–1240, 1991.

[78] C. I. Byrnes and W. Lin, "Losslessness, Feedback Equivalence, and the Global Stabilization of Discrete-Time Nonlinear Systems," *IEEE Trans. Autom. Control*, vol. 39, pp. 83–98, 1994.

[79] C. I. Byrnes, W. Lin, and B. K. Ghosh, "Stabilization of Discrete-Time Nonlinear Systems by Smooth State Feedback," *Syst. Control Lett.*, vol. 21, pp. 255–263, 1993.

[80] P. E. Cains, *Linear Stochastic Systems*. New York: Wiley, 1989.

[81] S. L. Campbell and N. J. Rose, "Singular Perturbation of Autonomous Linear Systems," *SIAM J. Math. Anal.*, vol. 10, pp. 542–551, 1979.

[82] B. C. Chang and J. B. Pearson, "Optimal Disturbance Reduction in Linear Multivariable Systems," *IEEE Trans. Autom. Control*, vol. 29, pp. 880–997, 1984.

[83] S. S. L. Chang and T. K. C. Peng, "Adaptive Guaranteed Cost Control of Systems with Uncertain Parameters," *IEEE Trans. Autom. Control*, vol. 17, pp. 474–483, 1972.

[84] V. Chellaboina and W. M. Haddad, "Is the Frobenius Matrix Norm Induced?," *IEEE Trans. Autom. Control*, vol. 40, pp. 2137–2139, 1995.

[85] V. Chellaboina and W. M. Haddad, "Stability Margins of Nonlinear Optimal Regulators with Nonquadratic Performance Criteria Involving Cross-Weighting Terms," *Syst. Control Lett.*, vol. 39, pp. 71–78, 2000.

[86] V. Chellaboina and W. M. Haddad, "Stability Margins of Discrete-Time Nonlinear-Nonquadratic Optimal Regulators," *Int. J. Syst. Sci.*, vol. 33, pp. 577–584, 2002.

[87] V. Chellaboina and W. M. Haddad, "A Unification Between Partial Stability and Stability Theory for Time-Varying Systems," *IEEE Control Syst. Magn.*, vol. 22, pp. 66–75, 2002.

[88] V. Chellaboina and W. M. Haddad, "Exponentially Dissipative Systems: A Nonlinear Extension of Strict Positive Realness," *Math. Probl. Eng.*, vol. 1–2, pp. 24–45, 2003.

[89] V. Chellaboina, W. M. Haddad, D. S. Bernstein, and D. A. Wilson, "Induced Convolution Operator Norms for Linear Dynamical Systems," *Math. Control, Signals, Syst.*, vol. 13, pp. 216–239, 2000.

[90] V. Chellaboina, W. M. Haddad, and J. Oh, "Fixed-Order Dynamic Compensation for Linear Systems with Actuator Amplitude and Rate Saturation Constraints," *Int. J. Control*, vol. 73, pp. 1087–1103, 2000.

[91] V. Chellaboina, A. Leonessa, and W. M. Haddad, "Generalized Lyapunov and Invariant Set Theorems for Nonlinear Dynamical Systems," *Syst. Control Lett.*, vol. 38, pp. 289–295, 1999.

[92] N. G. Chetaev, *The Stability of Motion.* New York: Pergamon Press, 1961.

[93] R. Chiang and M. G. Safonov, "Real $K_{\rm m}$-Synthesis via Generalized Popov Multipliers," in *Proc. American Control Conf.*, Chicago, IL, pp. 2417–2418, 1992.

[94] Y. S. Cho and K. S. Narendra, "An Off-Axis Circle Criterion for the Stability of Feedback Systems with a Monotonic Nonlinearity," *IEEE Trans. Autom. Control*, vol. 13, pp. 413–416, 1968.

[95] D. Chung, T. Kang, and J.G. Lee, "Stability Robustness of LQ Optimal Regulators for the Performance Index with Cross-Product Terms," *IEEE Trans. Autom. Control*, vol. 39, pp. 1698–1702, 1994.

[96] E. A. Coddington and N. Levinson, *Theory of Ordinary Differential Equations.* New York: McGraw-Hill, 1955.

[97] E. G. Collins, Jr., W. M. Haddad, L. Watson, and D. Sadhukhan, "Probability-One Homotopy Algorithms for Robust Controller Synthesis with Fixed-Structure Multipliers," *Int. J. Robust Nonlinear Control*, vol. 7, pp. 165–185, 1997.

[98] W. A. Coppel, *Stability and Asymptotic Behavior of Differential Equations.* Boston, MA: D. C. Heath and Co., 1965.

[99] M. J. Corless and G. Leitmann, "Continuous State Feedback Guaranteeing Uniform Ultimate Boundedness for Uncertain Dynamic Systems," *IEEE Trans. Autom. Control*, vol. 26, pp. 1139–1144, 1981.

[100] J. B. Cruz, *Feedback Systems.* New York: McGraw-Hill, 1971.

[101] M. A. Dahleh and I. J. Diaz-Bobillo, *Control of Uncertain Systems: A Linear Programming Approach.* Englewood Cliffs, NJ: Prentice-Hall, 1995.

[102] E. J. Davison and E. M. Kurak, "A Computational Method for Determining Quadratic Lyapunov Functions for Nonlinear Systems," *Automatica*, vol. 7, pp. 627–636, 1971.

[103] B. de Jager, "Rotating Stall and Surge Control: A Survey," in *Proc. IEEE Conf. Dec. Control*, New Orleans, LA, pp. 1857–1862, 1995.

[104] C. A. Desoer and M. Vidyasagar, *Feedback Systems: Input-Output Properties.* New York: Academic Press, 1975.

[105] S. G. Dharmasanam, R. S. Erwin, D. S. Bernstein, and D. A. Wilson, "Synthesis of Optimal Generalized LQG and Hankel-Norm Controllers," in *Proc. American Control Conf.*, Albuquerque, NM, pp. 3078–3082, 1997.

[106] G. L. Dirichlet, "Note sur la Stabilité de l'Équilibre," *J. Math. Pures Appl.*, vol. 12, pp. 474–478, 1847.

[107] S. L. Dixon, *Fluid Mechanics, Thermodynamics of Turbomachinery.* Oxford: Pergamon Press, 1975.

[108] P. Dorato and R. F. Drenick, "Optimality, Insensitivity, and Game Theory," in *Sensitivity Methods in Control Theory* (L. Radanović, ed.), pp. 78–102. New York: Pergamon Press, 1966.

[109] J. C. Doyle, "Guaranteed Margins for LQG Regulators," *IEEE Trans. Autom. Control*, vol. 23, pp. 756–757, 1978.

[110] J. C. Doyle, "Analysis of Feedback Systems with Structured Uncertainties," *IEE Proc.*, vol. 129-D, pp. 242–250, 1982.

[111] J. C. Doyle, K. Glover, P. P. Khargonekar, and B. A. Francis, "State Space Solutions to the Standard $\mathcal{H}_2$ and $\mathcal{H}_\infty$ Control Problems," *IEEE Trans. Autom. Control*, vol. 34, pp. 831–846, 1989.

[112] J. Doyle, K. Zhou, K. Glover, and B. Bodenheimer, "Mixed $\mathcal{H}_2$ and $\mathcal{H}_\infty$ Performance Objectives II: Optimal Control," *IEEE Trans. Autom. Control*, vol. 39, pp. 1575–1587, 1994.

[113] J. L. Dussourd, G. W. Pfannebecker, and S. K. Singhania, "An Experimental Investigation of the Control of Surge in Radial Compressors Using Close Coupled Resistances," *J. Fluid Eng.*, vol. 99, pp. 64–76, 1977.

[114] C. H. Edwards, Jr., *Advanced Calculus of Several Variables.* New York: Academic Press, 1973.

[115] M. Eslami and D. L. Russell, "On Stability with Large Parameter Variations Stemming from the Direct Method of Lyapunov," *IEEE Trans. Autom. Control*, vol. 25, pp. 1231–1234, 1980.

[116] L. Euler, *Theoria Motuum Lunae.* Saint Petersburg, Russia: Acad. Imp. Sci. Petropolitanae, 1753.

[117] M. K. H. Fan, A. L. Tits, and J. C. Doyle, "Robustness in the Presence of Mixed Parametric Uncertainty and Unmodelled Dynamics," *IEEE Trans. Autom. Control*, vol. 36, pp. 25–38, 1991.

[118] A. F. Filippov, *Differential Equations with Discontinuous Right-Hand Sides,* Mathematics and Its Applications (Soviet Series). Dordrecht, The Netherlands: Kluwer Academic, 1988.

[119] D. A. Fink, N. A. Cumptsy, and E. M. Greitzer, "Surge Dynamics in a Free-Spool Centrifugal Compressor System," *J. Turbomach.*, vol. 114, pp. 321–332, 1992.

[120] W. Fleming, *Functions of Several Variables.* New York: Springer-Verlag, 1987.

[121] B. A. Francis, *A Course in $\mathcal{H}_\infty$ Control Theory.* New York: Springer-Verlag, 1987.

[122] B. A. Francis and J. C. Doyle, "Linear Control Theory with an $\mathcal{H}_\infty$ Optimality Criterion," *SIAM J. Control Optim.*, vol. 25, pp. 815–844, 1987.

[123] B. A. Francis, J. W. Helton, and G. Zames, "$\mathcal{H}_\infty$-Optimal Feedback Controllers for Linear Multivariable Systems," *IEEE Trans. Autom. Control*, vol. 29, pp. 888–900, 1984.

[124] P. M. Frank, *Introduction to System Sensitivity Theory.* New York: Academic Press, 1978.

[125] J. F. Frankena and R. Sivan, "A Nonlinear Optimal Control Law for Linear Systems," *Int. J. Control*, vol. 30, pp. 159–178, 1979.

[126] R. A. Freeman and P. V. Kokotović, "Optimal Nonlinear Controllers for Feedback Linearizable Systems," in *Proc. American Control Conf.*, Seattle, WA, pp. 2722–2726, 1995.

[127] R. A. Freeman and P. V. Kokotović, "Inverse Optimality in Robust Stabilization," *SIAM J. Control Optim.*, vol. 34, pp. 1365–1391, 1996.

[128] R.A. Freeman and P.V. Kokotović, *Robust Nonlinear Control Design: State-Space and Lyapunov Techniques.* Boston, MA: Birkhauser, 1996.

[129] R. A. Freeman and J. A. Primbs, "Control Lyapunov Functions: New Ideas from an Old Source," in *Proc. IEEE Conf. Dec. Control*, Kobe, Japan, pp. 3926–3931, 1996.

[130] P. Gahinet and A. Nemirovskii, *LMI Lab: A Package for Manipulation and Solving LMI's.* Paris: INRIA, 1993.

[131] A. R. Galimidi and B. R. Barmish, "The Constrained Lyapunov Problem and Its Application to Robust Output Feedback Stabilization," *IEEE Trans. Autom. Control*, vol. 31, pp. 410–419, 1986.

[132] F. Gantmacher, *Matrix Theory.* Providence, RI: Chelsea, 1959.

[133] W. L. Garrard and J. M. Jordan, "Design of Nonlinear Automatic Flight Control Systems," *Automatica*, vol. 13, pp. 497–505, 1977.

[134] J. P. Gauthier and G. Bornard, "Stabilisation des Systemes Nonlineaires," in *Outilset Methodes Mathematiques pour I'Autrmatique* (I. D. Landau, ed.), pp. 307–324. CNRS, 1981.

[135] S. T. Glad, "Robustness of Nonlinear State Feedback—A Survey," *Automatica*, vol. 23, pp. 425–435, 1987.

[136] K. Glover, "All Optimal Hankel-Norm Approximations of Linear Multivariable Systems and Their $\mathcal{L}_\infty$-Error Bounds," *Int. J. Control*, vol. 39, pp. 1115–1193, 1984.

[137] K. Glover and J. C. Doyle, "State-Space Formulae for all Stabilizing Controllers that Satisfy an $\mathcal{H}_\infty$ Bound and Relations to Risk Sensitivity," *Syst. Control Lett.*, vol. 11, pp. 167–172, 1988.

[138] J. T. Gravdahl and O. Egeland, "Speed and Surge Control for a Low Order Centrifugal Compressor Model," *Proc. IEEE Conf. Control Appl.*, pp. 344–349, 1997.

[139] L. M. Graves, *The Theory of Functions of Real Variables.* New York: McGraw-Hill, 1956.

[140] E. M. Greitzer, "Surge and Rotating Stall in Axial Flow Compressors: Part 1 and 2," *J. Eng. Power*, vol. 98, pp. 190–217, 1976.

[141] G. Gu, S. Banda, and A. Sparks, "An Overview of Rotating Stall and Surge Control for Axial Flow Compressors," *Proc. IEEE Conf. Dec. Control*, pp. 2786–2791, 1996.

[142] S. Gutman, "Uncertain Dynamical Systems—A Lyapunov Min-Max Approach," *IEEE Trans. Autom. Control*, vol. 24, pp. 437–443, 1979.

[143] D. L. Gysling and E. M. Greitzer, "Dynamic Control of Rotating Stall in Axial Flow Compressors using Aeromechanical Feedback," *Proc. ASME Int. Gas Turbines and Aeroengines Congress and Exposition*, 1994.

[144] W. M. Haddad and D. S. Bernstein, "Generalized Riccati Equations for Full- and Reduced-Order Mixed-Norm $\mathcal{H}_2/\mathcal{H}_\infty$ Standard Problem," *Syst. Control Lett.*, vol. 14, pp. 185–197, 1990.

[145] W. M. Haddad and D. S. Bernstein, "Parameter-Dependent Lyapunov Functions, Constant Real Parameter Uncertainty, and the Popov Criterion in Robust Analysis and Synthesis, Parts I and II," in *Proc. IEEE Conf. Dec. Control*, Brighton, U.K., pp. 2274–2279, 2632–2633, 1991.

[146] W. M. Haddad and D. S. Bernstein, "Robust Stabilization with Positive Real Uncertainty: Beyond the Small Gain Theorem," *Syst. Control Lett.*, vol. 17, pp. 191–208, 1991.

[147] W. M. Haddad and D. S. Bernstein, "Explicit Construction of Quadratic Lyapunov Functions for Small Gain, Positivity, Circle, and Popov Theorems and Their Application to Robust Stability. Part I: Continuous-Time Theory," *Int. J. Robust Nonlinear Control*, vol. 3, pp. 313–339, 1993.

[148] W. M. Haddad and D. S. Bernstein, "Explicit Construction of Quadratic Lyapunov Functions for Small Gain, Positivity, Circle, and Popov Theorems and Their Application to Robust Stability. Part II: Discrete-Time Theory," *Int. J. Robust Nonlinear Control*, vol. 4, pp. 249–265, 1994.

[149] W. M. Haddad and D. S. Bernstein, "Parameter-Dependent Lyapunov Functions and the Discrete-Time Popov Criterion for Robust Analysis," *Automatica*, vol. 30, pp. 1015–1021, 1994.

[150] W. M. Haddad and D. S. Bernstein, "The Octomorphic Criterion for Real Parameter Uncertainty: Real-μ Bounds Without Circles and D, N-Scales," *Syst. Control Lett.*, vol. 25, pp. 175–183, 1995.

[151] W. M. Haddad and D. S. Bernstein, "Parameter-Dependent Lyapunov Functions and the Popov Criterion in Robust Analysis and Synthesis," *IEEE Trans. Autom. Control*, vol. 40, pp. 536–543, 1995.

[152] W. M. Haddad and D. S. Bernstein, "The Multivariable Parabola Criterion for Robust Controller Synthesis," *J. Math. Syst. Est. Control*, vol. 6, pp. 79–103, 1996.

[153] W. M. Haddad, D. S. Bernstein, and V. Chellaboina, "Generalized Mixed-μ Bounds for Real and Complex Multiple-Block Uncertainty with Internal Matrix Structure," *Int. J. Control*, vol. 64, pp. 789–806, 1996.

[154] W. M. Haddad, D. S. Bernstein, and D. Mustafa, "Mixed-Norm $\mathcal{H}_2/\mathcal{H}_\infty$ Regulation and Estimation: The Discrete-Time Case," *Syst. Control Lett.*, vol. 16, pp. 235–247, 1991.

[155] W. M. Haddad and V. Chellaboina, "Discrete-Time Nonlinear Feedback Control with Nonquadratic Performance Criteria," *J. Franklin Inst.*, vol. 333B, pp. 849–860, 1996.

[156] W. M. Haddad and V. Chellaboina, "Robust Nonlinear-Nonquadratic Feedback Control via Parameter-Dependent Lyapunov Functions," *Int. J. Control*, vol. 66, pp. 843–861, 1997.

[157] W. M. Haddad and V. Chellaboina, "Mixed-Norm $\mathcal{H}_2/\mathcal{L}_1$ Controller Synthesis via Fixed-Order Dynamic Compensation: A Riccati Equation Approach," *Int. J. Control*, vol. 71, pp. 35–59, 1998.

[158] W. M. Haddad and V. Chellaboina, "Nonlinear Fixed-Order Dynamic Compensation for Passive Systems," *Int. J. Robust Nonlinear Control*, vol. 8, pp. 349–365, 1998.

[159] W. M. Haddad and V. Chellaboina, "Optimal Nonlinear-Nonquadratic Feedback Control with $\mathcal{L}_2$ and $\mathcal{L}_\infty$ Disturbances," *Nonlinear Anal. Theory, Methods Appl.*, vol. 34, pp. 229–255, 1998.

[160] W. M. Haddad, V. Chellaboina, and D. S. Bernstein, "Real-μ Bounds Based on Fixed Shapes in the Nyquist Plane: Parabolas, Hyperbolas, Cissoids, Nephroids, and Octomorphs," *Syst. Control Lett.*, vol. 27, pp. 55–66, 1996.

[161] W. M. Haddad, V. Chellaboina, and D. S. Bernstein, "An Implicit Small Gain Condition and an Upper Bound for the Real Structured Singular Value," *Syst. Control Lett.*, vol. 29, pp. 197–205, 1997.

[162] W. M. Haddad, V. Chellaboina, and J. L. Fausz, "Robust Nonlinear Feedback Control with Nonquadratic Performance Criteria," in *Proc. IEEE Conf. Dec. Control*, New Orleans, LA, pp. 1982–1987, 1995.

[163] W. M. Haddad, V. Chellaboina, and J. L. Fausz, "Optimal Nonlinear Disturbance Rejection Control for Nonlinear Cascade Systems," *Int. J. Control*, vol. 68, pp. 997–1018, 1997.

[164] W. M. Haddad, V. Chellaboina, and J. L. Fausz, "Robust Nonlinear Feedback Control for Uncertain Linear Systems with Nonquadratic Performance Criteria," *Syst. Control Lett.*, vol. 33, pp. 327–338, 1998.

[165] W. M. Haddad, V. Chellaboina, J. L. Fausz, and C. T. Abdallah, "Optimal Discrete-Time Control for Nonlinear Cascade Systems," *J. Franklin Inst.*, vol. 335B, pp. 827–839, 1998.

[166] W. M. Haddad, V. Chellaboina, J. L. Fausz, and A. Leonessa, "Optimal Nonlinear Robust Control for Nonlinear Uncertain Systems," *Int. J. Control*, vol. 73, pp. 329–342, 2000.

[167] W. M. Haddad, V. Chellaboina, and S. G. Nersesov, *Thermodynamics: A Dynamical Systems Approach.* Princeton, NJ: Princeton Univ. Press, 2005.

[168] W. M. Haddad, V. Chellaboina, and W. Wu, "Optimal Nonlinear-Nonquadratic Feedback Control for Nonlinear Discrete-Time Systems with ℓ_2 and ℓ_∞ Disturbances," *Nonlinear Anal. Theory, Methods Appl.*, vol. 41, pp. 287–312, 2000.

[169] W. M. Haddad, J. L. Fausz, V. Chellaboina, and C. T. Abdallah, "A Unification Between Nonlinear-Nonquadratic Optimal Control and Integrator Backstepping," *Int. J. Robust Nonlinear Control*, vol. 8, pp. 879–906, 1998.

[170] W. M. Haddad, J. P. How, S. R. Hall, and D. S. Bernstein, "Extensions of Mixed-μ Bounds to Monotonic and Odd Monotonic Nonlinearities Using Absolute Stability Theory: Part I," in *Proc. IEEE Conf. Dec. Control*, Tucson, AZ, pp. 2813–2819, 1992.

[171] W. M. Haddad, J. P. How, S. R. Hall, and D. S. Bernstein, "Extensions of Mixed-μ Bounds to Monotonic and Odd Monotonic Nonlinearities Using Absolute Stability Theory: Part II," in *Proc. IEEE Conf. Dec. Control*, Tucson, AZ, pp. 2820–2823, 1992.

[172] W. M. Haddad, J. P. How, S. R. Hall, and D. S. Bernstein, "Extensions of Mixed-μ Bounds to Monotonic and Odd Monotonic Nonlinearities Using Absolute Stability Theory," *Int. J. Control*, vol. 60, pp. 905–951, 1994.

[173] W. M. Haddad, H.-H. Huang, and D. S. Bernstein, "Robust Stability and Performance via Fixed-Order Dynamic Compensation: The Discrete-Time Case," *IEEE Trans. Autom. Control*, vol. 38, pp. 776–782, 1993.

[174] W. M. Haddad, V. Kapila, and V. S. Chellaboina, "Guaranteed Domains of Attraction for Multivariable Luré Systems via Open Lyapunov Surfaces," *Int. J. Robust Nonlinear Control*, vol. 7, pp. 935–949, 1997.

[175] W. M. Haddad, S. G. Nersesov, and V. Chellaboina, "A Lyapunov Function Proof of Poincaré's Theorem," *Int. J. Syst. Sci.*, vol. 35, pp. 287–292, 2004.

[176] T. Hagiwara and M. Araki, "Design of a Stable Feedback Controller Based on the Multirate Sampling of the Plant Output," *IEEE Trans. Autom. Control*, vol. 33, pp. 812–819, 1988.

[177] W. Hahn, "On the Application of the Method of Lyapunov to Difference Equations," *Math. Ann.*, vol. 136, pp. 403–441, 1958.

[178] W. Hahn, *Stability of Motion.* Berlin: Springer-Verlag, 1967.

[179] J. K. Hale, *Ordinary Differential Equations,* 2nd edition. New York: Wiley, 1980. Reprinted by Krieger, Malabar, 1991.

[180] J. K. Hale and H. Kocak, *Dynamics and Bifurcations.* New York: Springer-Verlag, 1991.

[181] J. K. Hale and S. M. Verduyn Lunel, *Introduction to Functional Differential Equations.* New York: Springer-Verlag, 1993.

[182] P. R. Halmos, *Finite Dimensional Vector Spaces.* Princeton, NJ: Princeton University Press, 1947.

[183] R. W. Hamilton, "On a General Method in Dynamics," *Philos. Trans. R. Soc., Part II*, pp. 247–308, 1834.

[184] K. E. Hansen, P. Jorgensen, and P. S. Larsen, "Experimental and Theoretical Study of Surge in a Small Centrifugal Compressor," *J. Fluid Eng.*, vol. 103, pp. 391–394, 1981.

[185] P. Hartman, *Ordinary Differential Equations.* New York: Wiley, 1964.

[186] C. A. Harvey and G. Stein, "Quadratic Weights for Asymptotic Regulator Properties," *IEEE Trans. Autom. Control*, vol. 23, pp. 378–387, 1978.

[187] J. M. Haynes, G. J. Hendricks, and A. H. Epstein, "Active Stabilization of Rotating Stall in a Three-Stage Axial Compressor," *J. Turbomach.*, vol. 116, pp. 226–239, 1994.

[188] D. J. Hill and P. J. Moylan, "The Stability of Nonlinear Dissipative Systems," *IEEE Trans. Autom. Control*, vol. 21, pp. 708–711, 1976.

[189] D. J. Hill and P. J. Moylan, "Stability Results for Nonlinear Feedback Systems," *Automatica*, vol. 13, pp. 377–382, 1977.

[190] D. J. Hill and P. J. Moylan, "Connections Between Finite-Gain and Asymptotic Stability," *IEEE Trans. Autom. Control*, vol. 25, pp. 931–936, 1980.

[191] D. J. Hill and P. J. Moylan, "Dissipative Dynamical Systems: Basic Input-Output and State Properties," *J. Franklin Inst.*, vol. 309, pp. 327–357, 1980.

[192] D. J. Hill and P. J. Moylan, "Dissipative Dynamical Systems: Basic Input-Output and State Properties," *J. Franklin Inst.*, vol. 309, pp. 327–357, 1980.

[193] P. G. Hill and C. R. Peterson, *Mechanics and Thermodynamics of Propulsion.* Reading, MA: Addison-Wesley, 1992.

[194] D. Hinrichsen and A. J. Pritchard, "Stability Radii of Linear Systems," *Syst. Control Lett.*, vol. 7, pp. 1–10, 1986.

[195] D. Hinrichsen and A. J. Pritchard, "Stability Radius for Structured Perturbations and the Algebraic Riccati Equation," *Syst. Control Lett.*, vol. 8, pp. 105–113, 1987.

[196] M. W. Hirsch and S. Smale, *Differential Equations, Dynamical Systems and Linear Algebra.* New York: Academic Press, 1974.

[197] L. Hitz and B. D. O. Anderson, "Discrete Positive-Real Functions and Their Application to System Stability," *Proc. IEE*, vol. 116, pp. 153–155, 1969.

[198] K. Hoffman, *Analysis in Euclidean Space.* Englewood Cliffs, NJ: Prentice-Hall, 1975.

[199] J. Hong, "Experimental Implementation of Fixed-Gain and Adaptive Control." Ph. D. thesis, The University of Michigan, 1998.

[200] H. P. Horisberger and P. R. Belanger, "Regulators for Linear Time-Invariant Plants with Uncertain Parameters," *IEEE Trans. Autom. Control*, vol. 21, pp. 705–708, 1976.

[201] R. A. Horn and R. C. Johnson, *Matrix Analysis.* Cambridge: Cambridge University Press, 1985.

[202] R. A. Horn and R. C. Johnson, *Topics in Matrix Analysis.* Cambridge: Cambridge University Press, 1991.

[203] J. P. How, E. G. Collins, Jr., and W. M. Haddad, "Optimal Popov Controller Analysis and Synthesis for Systems with Real Parameter Uncertainties," *IEEE Control Syst. Technol.*, vol. 4, pp. 200–207, 1996.

[204] J. P. How and W. M. Haddad, "Robust Stability and Performance Analysis Using LC Multipliers and the Nyquist Criterion," in *Proc. American Control Conf.*, Baltimore, MD, pp. 1418–1419, 1994.

[205] J. P. How, W. M. Haddad, and S. R. Hall, "Application of Popov Controller Synthesis to Benchmark Problems with Real Parameter Uncertainty," *AIAA J. Guid. Control Dyn.*, vol. 17, pp. 759–768, 1994.

[206] J. P. How and S. R. Hall, "Connections Between the Popov Stability Criterion and Bounds for Real Parametric Uncertainty," in *Proc. American Control Conf.*, San Francisco, CA, pp. 1084–1089, 1993.

[207] J. P. How, S. R. Hall, and W. M. Haddad, "Robust Controllers for the Middeck Active Control Experiment Using Popov Controller Synthesis," *IEEE Control Syst. Technol.*, vol. 2, pp. 73–87, 1994.

[208] Q. Hui, W. M. Haddad, and S. P. Bhat, "Finite-Time Semistability Theory with Applications to Consensus Protocols in Dynamical Networks," in *Proc. American Control Conf.*, New York, pp. 2411–2416, 2007.

[209] Q. Hui, W. M. Haddad, and S. P. Bhat, "Finite-Time Stability Theory with Applications to Consensus Protocols in Dynamical Networks," *IEEE Trans. Autom. Control*, submitted.

[210] L. R. Hunt, R. Su, and G. Meyer, "Design for Multi-Input Nonlinear Systems," *Differential Geometric Control Theory*, vol. 27, pp. 268–298, 1983.

[211] P. A. Iglesias and K. Glover, "State Space Approach to Discrete-Time $\mathcal{H}_\infty$ Control," *Int. J. Control*, vol. 54, pp. 1031–1073, 1991.

[212] A. Isidori, *Nonlinear Control Systems: An Introduction.* New York: Springer-Verlag, 1989.

[213] A. Isidori, "Dissipation Inequalities in Nonlinear $\mathcal{H}_\infty$ Control," in *Proc. IEEE Conf. Dec. Control*, Tucson, AZ, pp. 3265–3270, 1992.

[214] A. Isidori, "$\mathcal{H}_\infty$ Control via Measurement Feedback for Affine Nonlinear Systems," *Int. J. Robust Nonlinear Control*, vol. 4, pp. 552–574, 1994.

[215] A. Isidori and A. Astolfi, "Disturbance Attenuation and $\mathcal{H}_\infty$-Control via Measurement Feedback in Nonlinear Systems," *IEEE Trans. Autom. Control*, vol. 37, pp. 1283–1293, 1992.

[216] A. Isidori and A. Astolfi, "Nonlinear $\mathcal{H}_\infty$ Control via Measurement Feedback," *J. Math. Syst. Est. Control*, vol. 2, pp. 31–44, 1992.

[217] D. H. Jacobson, *Extensions of Linear-Quadratic Control Optimization and Matrix Theory*. New York: Academic Press, 1977.

[218] D. H. Jacobson, D. H. Martin, M. Pachter, and T. Geveci, *Extensions of Linear-Quadratic Control Theory*. Berlin: Springer-Verlag, 1980.

[219] J. A. Jacquez, *Compartmental Analysis in Biology and Medicine,* 2nd edition. Ann Arbor, MI: University of Michigan Press, 1985.

[220] J. A. Jacquez and C. P. Simon, "Qualitative Theory of Compartmental Systems," *SIAM Rev.*, vol. 35, pp. 43–79, 1993.

[221] B. N. Jain, "Guaranteed Error Estimation in Uncertain Systems," *IEEE Trans. Autom. Control*, vol. 20, pp. 230–232, 1975.

[222] S. Jain and F. Khorrami, "Application of a Decentralized Adaptive Output Feedback Based on Backstepping to Power Systems," in *Proc. IEEE Conf. Dec. Control*, New Orleans, LA, pp. 1585–1590, 1995.

[223] W. M. Jungovski, M. H. Weiss, and G. R. Price, "Pressure Oscillations Occuring in a Centrifugal Compressor System with and without Passive and Active Surge Control," *J. Turbomach.*, vol. 118, pp. 29–40, 1996.

[224] V. Jurdjevic and J. P. Quinn, "Controllability and Stability," *J. Differ. Equations*, vol. 28, pp. 381–389, 1978.

[225] W. Kahan, "Numerical Linear Algebra," *Can. Math. Bull.*, vol. 9, pp. 757–801, 1966.

[226] R. E. Kalman, "Lyapunov Functions for the Problem of Luré in Automatic Control," *Proc. Natl. Acad. Sci.*, vol. 49, pp. 201–205, 1963.

[227] R. E. Kalman, "When Is a Linear Control System Optimal?" *ASME J. Basic Eng.*, vol. 86 D, pp. 51–60, 1964.

[228] R. E. Kalman and J. E. Bertram, "Control System Analysis and Design via the Second Method of Lyapunov, Part I: Continuous-Time Systems," *ASME J. Basic Eng.*, vol. 82, pp. 371–393, 1960.

[229] R. E. Kalman and J. E. Bertram, "Control System Analysis and Design via the Second Method of Lyapunov, Part II: Discrete-Time Systems," *ASME J. Basic Eng.*, vol. 82, pp. 394–400, 1960.

[230] I. Kanellakopoulos, P. V. Kokotović, and A. S. Morse, "Systematic Design of Adaptive Controllers for Feedback Linearizable Systems," *IEEE Trans. Autom. Control*, vol. 36, pp. 1241–1253, 1991.

[231] I. Kanellakopoulos, P. V. Kokotović, and A. S. Morse, "A Toolkit for Nonlinear Feedback Design," *Syst. Control Lett.*, vol. 18, pp. 83–92, 1992.

[232] M. Kawski, "Stabilization of Nonlinear Systems in the Plane," *Syst. Control Lett.*, vol. 12, pp. 169–175, 1989.

[233] L. H. Keel, S. P. Bhattacharya, and J. W. Howze, "Robust Control with Structured Perturbations," *IEEE Trans. Autom. Control*, vol. 33, pp. 68–78, 1988.

[234] S. S. Keerthi and E. G. Gilbert, "Computation of Minimum-Time Feedback Control Laws for Discrete-Time Systems with State-Control Constraints," *IEEE Trans. Autom. Control*, vol. 32, pp. 432–435, 1987.

[235] H. K. Khalil, *Nonlinear Systems.* Upper Saddle River, NJ: Prentice-Hall, 1996.

[236] P. P. Khargonekar, I. R. Petersen, and M. A. Rotea, "$\mathcal{H}_\infty$-Optimal Control with State-Feedback," *IEEE Trans. Autom. Control*, vol. 33, pp. 786–788, 1988.

[237] P. P. Khargonekar, I. R. Petersen, and K. Zhou, "Robust Stabilization of Uncertain Linear Systems: Quadratic Stabilizability and $\mathcal{H}_\infty$ Control Theory," *IEEE Trans. Autom. Control*, vol. 35, pp. 356–361, 1990.

[238] P. P. Khargonekar and M. A. Rotea, "Mixed $\mathcal{H}_2/\mathcal{H}_\infty$ Control: A Convex Optimization Approach," *IEEE Trans. Autom. Control*, vol. 36, pp. 824–837, 1991.

[239] D. E. Kirk, *Optimal Control Theory: An Introduction.* Englewood Cliffs, NJ: Printice-Hall, 1970.

[240] D. E. Koditschek, "Adaptive Techniques for Mechanical Systems," in *Proc. 5th Yale Workshop on Adaptive Systems*, New Haven, CT, pp. 259–265, 1987.

[241] P. V. Kokotović and H. J. Sussmann, "A Positive Real Condition for Global Stabilization of Nonlinear Systems," *Syst. Control Lett.*, vol. 13, pp. 125–133, 1989.

[242] A. A. Kolesnikov, "Analytical Construction of Nonlinear Aggregated Regulators for a Given Set of Invariant Manifolds," *Electromechanika, Izv. VTU*, 1987. Part I, No. 3, pp. 100–109, Part II, No. 5, pp. 58–66.

[243] O. I. Kosmidou and P Bertrand, "Robust-Controller Design for Systems with Large Parameter Variations," *Int. J. Control*, vol. 45, pp. 1076–1084, 1987.

[244] R. L. Kosut, "Nonlinear Optimal Cue-Shaping Filters for Motion Base Simulators," *AIAA J. Guid. Control Dyn.*, vol. 2, pp. 486–490, 1979.

[245] N. N. Krasovskii, *Problems of the Theory of Stability of Motion.* Stanford, CA: Stanford University Press, 1959.

[246] M. Krstić, I. Kanellakopoulos, and P.V. Kokotović, "Adaptive Nonlinear Control without Overparameterization," *Syst. Control Lett.*, vol. 19, pp. 177–185, 1992.

[247] M. Krstić, I. Kanellakopoulos, and P.V. Kokotović, *Nonlinear and Adaptive Control Design.* New York: John Wiley and Sons, 1995.

[248] M. Krstić and P. V. Kokotović, "Lean Backstepping Design for Jet Engine Compressor Model," in *Proc. IEEE Conf. Control Appl.*, pp. 1047–1052, 1995.

[249] M. Krstić, J.M. Protz, J.D. Paduano, and P.V. Kokotović, "Backstepping Designs for Jet Engine Stall and Surge Control," in *Proc. IEEE Conf. Dec. Control*, New Orleans, LA, pp. 3049–3055, 1995.

[250] J. Kurzweil, "On the Inversion of Lyapunov's Second Theorem on Stability of Motion," *Tran. Am. Math. Soc. Transl.*, vol. 24, pp. 19–77, 1963, translated from *Czech. Math. J.*, vol. 81, pp. 217–259; 455–484, 1956.

[251] J. L. Lagrange, *Méchanique Analytique.* Paris: Mme Ve Courcier, 1787. See also Vols. 1 and 2, Gauthier-Villars, Paris, 1853.

[252] J. L. Lagrange, *Méchanique Analitique.* Paris: Desaint, 1788.

[253] V. Lakshmikantham and S. Leela, *Differential and Integral Inequalities: Theory and Applications.* New York: Academic Press, 1969.

[254] V. Lakshmikantham, S. Leela, and A. A. Martynyuk, *Stability Analysis of Nonlinear Systems.* New York: Marcel Dekker, 1969.

[255] V. Lakshmikantham, V. M. Matrosov, and S. Sivasundaram, *Vector Lyapunov Functions and Stability Analysis of Nonlinear Systems.* Dordrecht, The Netherlands: Kluwer Academic, 1991.

[256] P. Lancaster and M. Tismenetsky, *The Theory of Matrices.* New York: Academic Press, 1985.

[257] P. S. Laplace, *Oeuvres Complètes de Laplace.* Paris: Gauthier-Villars, 1895.

[258] J. P. LaSalle, "Some Extensions of Liapunov's Second Method," *IRE Trans. Circ. Theory*, vol. CT-7, pp. 520–527, 1960.

[259] J. P. LaSalle, *The Stability of Dynamical Systems.* Philadelphia, PA: SIAM, 1976.

[260] J. P. LaSalle and S. Lefschetz, *Stability by Lyapunov's Direct Method.* New York: Academic Press, 1961.

[261] M. A. Leal and J. S. Gibson, "A First-Order Lyapunov Robustness Method for Linear Systems with Uncertain Parameters," *IEEE Trans. Autom. Control*, vol. 35, pp. 1068–1070, 1990.

[262] T.-T. Lee and S.-H. Lee, "Gain and Phase Margins for Discrete-Time Systems," *Int. J. Control*, vol. 44, pp. 1415–1426, 1986.

[263] L. Lee and A. L. Tits, "Linear Fractional Transformations for the Approximation of Various Uncertainty Sets," in *Control Unc. Dyn. Sys* (S. Bhattacharyya and L. Keel, eds.), pp. 53–62. Boca Raton, FL: CRC Press, 1991.

[264] S. Lefschetz, *Differential Equations: Geometric Theory.* New York: Interscience, 1962.

[265] S. Lefschetz, *Stability of Nonlinear Control Systems.* New York: Academic Press, 1965.

[266] G. Leitmann, "Guaranteed Asymptotic Stability for a Class of Uncertain Linear Dynamical Systems," *J. Optim. Theory Appl.*, vol. 27, pp. 99–106, 1979.

[267] G. Leitmann, "Guaranteed Asymptotic Stability for Some Linear Systems with Bounded Uncertainties," *ASME J. Dyn. Syst. Meas. Control*, vol. 101, pp. 212–216, 1979.

[268] G. Leitmann, "On the Efficacy of Nonlinear Control in Uncertain Linear Systems," *ASME J. Dyn. Syst. Meas. Control*, vol. 102, pp. 95–102, 1981.

[269] A. Leonessa, V. Chellaboina, and W. M. Haddad, "Globally Stabilizing Controllers for Multi-Mode Axial Flow Compressors via Equilibria-Dependent Lyapunov Functions," in *Proc. American Control Conf.*, Albuquerque, NM, pp. 993–997, 1997.

[270] A. Leonessa, V. Chellaboina, and W. M. Haddad, "Robust Stabilization of Axial Flow Compressors with Uncertain Pressure-Flow Maps," in *Proc. IEEE Conf. Control Appl.*, Hartford, CT, pp. 671–676, 1997.

[271] A. Leonessa, W. M. Haddad, and V. Chellaboina, "Nonlinear System Stabilization via Equilibrium-Dependent Lyapunov Functions: Beyond Gain Scheduling Control," in *Proc. American Control Conf.*, Philadelphia, PA, pp. 1669–1673, 1998.

[272] A. Leonessa, W. M. Haddad, and V. Chellaboina, *Hierarchical Nonlinear Switching Control Design with Applications to Propulsion Systems*. New York: Springer-Verlag, 2000.

[273] A. Leonessa, W. M. Haddad, and V. Chellaboina, "Nonlinear System Stabilization via Hierarchical Switching Control," *IEEE Trans. Autom. Control*, vol. 46, pp. 17–28, 2001.

[274] W. S. Levine and M. Athans, "On the Determination of the Optimal Output Feedback Gains for Linear Multivariable Systems," *IEEE Trans. Autom. Control*, vol. 15, pp. 44–48, 1970.

[275] J. B. Lewis, "The Use of Nonlinear Feedback to Improve the Transient Response of a Servo-Mechanism," *Trans. AIEE, Part II. Appl. Ind.*, vol. 71, pp. 449–453, 1953.

[276] T. Li, "The Stability Problem in Difference Equations," *Acta Math.*, vol. 63, pp. 99–141, 1934.

[277] D. C. Liaw and E. H. Abed, "Stability Analysis and Control of Rotating Stall," *IFAC Nonlinear Control Syst.*, pp. 295–300, 1992.

[278] H. Li, A. Leonessa, and W. M. Haddad, "Globally Stabilizing Controllers for a Centrifugal Compressor Model with Spool Dynamics," in *Proc. American Control Conf.*, Philadelphia, PA, pp. 2160–2164, 1998.

[279] D. J. N. Limebeer, B. D. O. Anderson, and B. Hendel, "A Nash Game Approach to Mixed $\mathcal{H}_2/\mathcal{H}_\infty$ Control," *IEEE Trans. Autom. Control*, vol. 39, pp. 1564–1574, 1994.

[280] W. Lin and C. I. Byrnes, "Dissipativity, $\mathcal{L}_2$-Gain and $\mathcal{H}_\infty$ Control for Discrete-Time Nonlinear Systems," in *Proc. American Control Conf.*, Baltimore, MD, pp. 2257–2260, 1994.

[281] W. Lin and C.I. Byrnes, "KYP Lemma, State Feedback and Dynamic Output Feedback in Discrete-Time Bilinear Systems," *Syst. Control Lett.*, vol. 23, pp. 127–136, 1994.

[282] W. Lin and C. I. Byrnes, "Passivity and Absolute Stabilization of a Class of Discrete-Time Nonlinear Systems," *Automatica*, vol. 31, pp. 263–267, 1995.

[283] J. Liouville, "Formules Générales Relatives à la Question de la Stabilité de l'Équilibre d'Une Masse Liquide Homogène Douée d'un Mouvement de Rotation Autour d'un Axe," *J. Math. Pures Appl.*, vol. 20, pp. 164–184, 1855.

[284] X. Liu, "Stability Results for Impulsive Differential Systems with Applications to Population Growth Models," *Dyn. Stab. Syst.*, vol. 9, pp. 163–174, 1994.

[285] K. A. Loparo and G. L. Blankenship, "Estimating the Domain of Attraction of Nonlinear Feedback Systems," *IEEE Trans. Autom. Control*, vol. AC-23, no. 4, pp. 602–608, 1978.

[286] E. N. Lorentz, "Deterministic Non-Periodic Flow," *J. Atmos. Sci.*, vol. 20, pp. 130–141, 1963.

[287] R. Lozano, B. Brogliato, and I. D. Landau, "Passivity and Global Stabilization of Cascaded Systems," *IEEE Trans. Autom. Control*, vol. 37, pp. 1386–1388, 1992.

[288] D. Luenberger, *Optimization by Vector Space Methods.* New York: Wiley, 1969.

[289] K.-Y. Lum, D. S. Bernstein, and V. T. Coppola, "Global Stabilization of the Spinning Top with Mass Imbalance," *Dyn. Stab. Syst.*, vol. 10, pp. 339–365, 1995.

[290] A. I. Luré, "On the Problem of Stability of Control Systems," *Prikl. Mat. Mehk.*, vol. 15, 1951.

[291] A. I. Luré, *Some Nonlinear Problems in the Theory of Automatic Control.* Moscow: Gostekhizdat, 1951.

[292] A. I. Luré and V. N. Postnikov, "On the Theory of Stability of Control Systems," *Prikl. Mat. Mehk.*, vol. 8, 1944.

[293] A. M. Lyapunov, *The General Problem of the Stability of Motion.* Kharkov, Russia: Kharkov Mathematical Society, 1892.

[294] A. M. Lyapunov, " *"Problème Generale de la Stabilité du Mouvement," in* Annales de la Faculté Sciences de l'Université de Toulouse *(É. Davaux, ed.),*" vol. 9, pp. 203–474. 1907. Reprinted by Princeton University Press, Princeton, NJ, 1949.

[295] A. M. Lyapunov, *The General Problem of Stability of Motion* (A. T. Fuller, trans. and ed.). Washington, DC: Taylor and Francis, 1992.

[296] E. F. Mageirou and Y. C. Ho, "Decentralized Stabilization via Game Theoretic Methods," *Automatica*, vol. 13, pp. 888–896, 1977.

[297] I. G. Malkin, "On the Theory of Stability of Control of Systems," *Prikl. Mat. Mekh.*, vol. 15, 1951.

[298] I. G. Malkin, "On a Question of Reversibility of Liapunov's Theorem on Asymptotic Stability," *Prikl. Mat. Mekh.*, vol. 18, pp. 129–138, 1954. English translation found in *Nonlinear Systems: Stability Analysis*, pp. 161–170. (J. K. Agarwal and M. Vidyasagar, eds.), New York: Dowden, Hutchinson and Ross, 1977.

[299] J. Malmborg, B. Bernhardsson, and K. J. Astrom, "A Stabilizing Switching Scheme for Multi Controller Systems," in *Proc. 13th IFAC World Congress*, San Francisco, CA, pp. 229–234, 1996.

[300] J. M. Martin and B. A. Hewer, "Smallest Destabilizing Perturbations for Linear Systems," *Int. J. Control*, vol. 45, pp. 1495–1504, 1987.

[301] B. M. Maschke, R. Ortega, and A. J. van der Schaft, "Energy-Based Lyapunov Functions for Forced Hamiltonian Systems with Dissipation," in *Proc. IEEE Conf. Dec. Control*, Tampa, FL, pp. 3599–3604, 1998.

[302] B. M. Maschke, R. Ortega, A. J. van der Schaft, and G. Escobar, "An Energy-Based Derivation of Lyapunov Functions for Forced Systems with Application to Stabilizing Control," in *Proc. IFAC World Congress*, vol. E, pp. 409–414, 1999.

[303] B. M. Maschke and A. J. van der Schaft, "Port-Controlled Hamiltonian Systems: Modelling Origins and System-Theoretic Properties," in *Proc. 2nd IFAC NOLCOS*, Bordeaux, France, pp. 282–288, 1992.

[304] B. M. Maschke and A. J. van der Schaft, "System-Theoretic Properties of Port-Controlled Hamiltonian Systems," vol. II of *Systems and Networks: Mathematical Theory and Applications*, pp. 349–352. Berlin: Akademie-Verlag, 1994.

[305] B. M. Maschke, A. J. van der Schaft, and P. C. Breedveld, "An Intrinsic Hamiltonian Formulation of Network Dynamics: Non-standard Poisson Structures and Gyrators," *J. Franklin Inst.*, vol. 329, pp. 923–966, 1992.

[306] J. L. Massera, "On Liapunoff's Conditions of Stability," *Ann. Math.*, vol. 50, pp. 705–721, 1949.

[307] J. L. Massera, "Contributions to Stability Theory," *Ann. Math.*, vol. 64, pp. 182–206, 1956.

[308] V. M. Matrosov, "On the Theory of Stability of Motion," *Prikl. Mat. Mekh.*, vol. 26, 1962.

[309] J. D. Mattingly, *Elements of Gas Turbine Propulsion.* New York: McGraw-Hill, 1996.

[310] J. C. Maxwell, *On the Stability of the Motion of Saturn's Rings.* London: Macmillan, 1859.

[311] F. E. McCaughan, "Bifurcation Analysis of Axial Flow Compressor Stability," *SIAM J. Appl. Math.*, vol. 20, pp. 1232–1253, 1990.

[312] G. Menga and P. Dorato, "Observer-Feedback Design for Linear Systems with Large Parameter Uncertainties," in *Proc. IEEE Conf. Dec. Control*, Phoenix, AZ, pp. 872–878, 1974.

[313] G. S. Michael and C. W. Merriam, "Stability of Parametrically Disturbed Linear Optimal Control Systems," *J. Math. Anal. Appl.*, vol. 28, pp. 294–302, 1969.

[314] A. N. Michel, R. K. Miller, and B. H. Nam, "Stability Analysis of Interconnected Systems Using Computer Generated Lyapunov Functions," *Theor. Comput. Sci.*, vol. CAS-29, no. 7, pp. 431–440, 1982.

[315] R. K. Miller and A. N. Michel, *Ordinary Differential Equations.* New York: Academic Press, 1982.

[316] P. Molander and J. C. Willems, "Synthesis of State Feedback Control Laws with a Specified Gain and Phase Margin," *IEEE Trans. Autom. Control*, vol. 25, pp. 928–931, 1980.

[317] B. P. Molinari, "The Stable Regulator Problem and Its Inverse," *IEEE Trans. Autom. Control*, vol. 18, pp. 454–459, 1973.

[318] F. K. Moore, "A Theory of Rotating Stall of Multistage Axial Compressors: Part 1, 2, and 3," *J. Eng. Gas Turbines Power*, vol. 106, pp. 313–336, 1984.

[319] F. K. Moore and E. M. Greitzer, "A Theory of Post-Stall Transients in Axial Compression Systems: Part 1 and 2," *J. Eng. Gas Turbines Power*, vol. 108, pp. 68–76, 231–239, 1986.

[320] P. J. Moylan, "Implications of Passivity in a Class of Nonlinear Systems," *IEEE Trans. Autom. Control*, vol. 19, pp. 373–381, 1974.

[321] P. J. Moylan and B. D. O. Anderson, "Nonlinear Regulator Theory and an Inverse Optimal Control Problem," *IEEE Trans. Autom. Control*, vol. 18, pp. 460–465, 1973.

[322] P. J. Moylan and D. J. Hill, "Stability Criteria for Large-Scale Systems," *IEEE Trans. Autom. Control*, vol. 23, pp. 143–149, 1978.

[323] J. R. Munkres, *Topology: A First Course*. Englewood Cliffs, NJ: Prentice-Hall, 1975.

[324] K. Nagpal, J. Abedor, and K. Poolla, "An LMI Approach to Peak-to-Peak Gain Minimization: Filtering and Control," *Int. J. Robust Nonlinear Control*, vol. 6, pp. 899–927, 1996.

[325] K. S. Narendra, "A Conjecture on the Existence of a Quadratic Lyapunov Function for Linear Time-Invariant Feedback Systems," in *Proc. 4th Allerton Conf. on Circuit and System Theory*, University of Illinois, Urbana, 1966.

[326] K. S. Narendra and Y. S. Cho, "Stability of Feedback Systems Containing a Single Odd Monotonic Nonlinearity," *IEEE Trans. Autom. Control*, vol. 12, pp. 448–450, 1967.

[327] K. S. Narendra and R. M. Goldwyn, "A Geometrical Criterion for the Stability of Certain Nonlinear Nonautonomous Systems," *IEEE Trans. Circ. Theory*, vol. 11, pp. 406–407, 1964.

[328] K. S. Narendra and C. P. Neuman, "Stability of a Class of Differential Equations with a Single Monotone Nonlinearity," *SIAM J. Control*, vol. 4, pp. 295–308, 1966.

[329] K. S. Narendra and C. P. Neuman, "Stability of Continuous Time Dynamical Systems with m-Feedback Nonlinearities," *AIAA J.*, vol. 5, pp. 2021–2027, 1967.

[330] K. S. Narendra and J. H. Taylor, "Lyapunov Functions for Nonlinear Time-Varying Systems," *Info. Control*, vol. 12, pp. 378–393, 1968.

[331] K. S. Narendra and J. H. Taylor, *Frequency Domain Criteria for Absolute Stability*. New York: Academic Press, 1973.

[332] A. Naylor and G. R. Sell, *Linear Operator Theory in Engineering and Science*. Berlin: Springer-Verlag, 1986.

[333] V. V. Nemytskii and V. V. Stepanov, *Qualitative Theory of Differential Equations*. Princeton, NJ: Princeton University Press, 1960.

[334] S. G. Nersesov and W. M. Haddad, "On the Stability and Control of Nonlinear Dynamical Systems via Vector Lyapunov Functions," *IEEE Trans. Autom. Control*, vol. 51, pp. 203–215, 2006.

[335] I. Newton, *Philosophiae Naturalis Principia Mathematica*. London: Royal Society, 1687.

[336] H. Nijmeijer and A. J. van der Shaft, *Nonlinear Dynamical Control Systems,* 1st edition. Berlin: Springer-Verlag, 1990.

[337] E. Noldus, "Design of Robust State Feedback Laws," *Int. J. Control*, vol. 35, pp. 935–944, 1982.

[338] E. P. Odum, *Fundamentals of Ecology,* 3rd edition. Philadelphia, PA: Saunders, 1971.

[339] R. Ortega, "Passivity Properties for Stabilization of Cascaded Nonlinear Systems," *Automatica*, vol. 27, pp. 423–424, 1991.

[340] R. Ortega, A. Loria, R. Kelly, and L. Praly, "On Output Feedback Global stabilization of Euler-Lagrange systems," *Int. J. Control*, vol. 5, pp. 313–324, 1995.

[341] R. Ortega, A. Loria, P. J. Nicklasson, and H. Sira-Ramírez, *Passivity-Based Control of Euler-Lagrange Systems*. London: Springer-Verlag, 1998.

[342] R. Ortega and M. W. Spong, "Adaptive Motion Control of Rigid Robots: A Tutorial," *Automatica*, vol. 25, pp. 877–888, 1989.

[343] R. Ortega, A. van der Schaft, B. Maschke, and G. Escobar, "Interconnection and Damping Assignment Passivity-Based Control of Port-Controlled Hamiltonian Systems," *Automatica*, vol. 38, pp. 585–596, 2002.

[344] A. Packard and J. C. Doyle, "The Complex Structured Singular Value," *Automatica*, vol. 29, pp. 71–109, 1993.

[345] J. D. Paduano, L. Valavani, A. H. Epstein, E. M. Greitzer, and G. R. Guenette, "Modeling for Control of Rotating Stall," *Automatica*, vol. 30, no. 9, pp. 1357–1373, 1994.

[346] K. M. Passino, A. N. Michel, and P. J. Antsaklis, "Lyapunov Stability of a Class of Discrete Event Systems," *IEEE Trans. Autom. Control*, vol. 39, pp. 269–279, 1994.

[347] R. V. Patel and M. Toda, "Qualitative Measures of Robustness for Multivariable Systems," in *Proc. Joint Autom. Control Conf.*, San Francisco, CA, 1989.

[348] P. Peleties and R. DeCarlo, "Asymptotic Stability of m-Switched Systems using Lyapunov-Like Functions," in *Proc. American Control Conf.*, Boston, MA, pp. 1679–1684, 1991.

[349] L. Perko, *Differential Equations and Dynamical Systems.* New York: Springer-Verlag, 1996.

[350] K. P. Persidskii, "Uber die Stabilität Nach der Ersten Näherung," *Mat. Sbornik*, vol. 40, pp. 284–293, 1933.

[351] I. R. Petersen, "Nonlinear Versus Linear Control in the Direct Output Feedback Stabilization of Linear Systems," *IEEE Trans. Autom. Control*, vol. 30, pp. 799–802, 1985.

[352] I. R. Petersen, "A Riccati Equation Approach to the Design of Stabilizing Controllers and Observers for a Class of Uncertain Linear Systems," *IEEE Trans. Autom. Control*, vol. 30, pp. 399–408, 1985.

[353] I. R. Petersen, "Disturbance Attenuation and $\mathcal{H}_\infty$ Optimization: A Design Method Based on the Algebraic Riccati Equation," *IEEE Trans. Autom. Control*, vol. 32, pp. 427–429, 1987.

[354] I. R. Petersen, "A Stabilization Algorithm for a Class of Uncertain Linear Systems," *Syst. Control Lett.*, vol. 8, pp. 351–357, 1987.

[355] I. R. Petersen, "Complete Results for a Class of State Feedback Disturbance Attenuation Problems," *IEEE Trans. Autom. Control*, vol. 34, pp. 1196–1199, 1989.

[356] I. R. Petersen and C. V. Hollot, "A Riccati Equation Approach to the Stabilization of Uncertain Systems," *Automatica*, vol. 22, pp. 397–411, 1986.

[357] J. E. Pinsley, G. R. Guenette, and E. M. Greitzer, "Active Stabilization of Centrifugal Compressor Surge," *J. Turbomach.*, vol. 113, pp. 723–732, 1991.

[358] H. Poincaré, "Mémoire sur les Courbes Définies par une Equation Différentielle," *J. Mathématiques*, vol. 7, pp. 375–422, 1881. Oeuvre (1880–1890), Gauthier-Villar, Paris.

[359] H. Poincaré, "Sur les Equations de la Dynamique et le Probleme des Trois Corps," *Acta Math.*, vol. 13, pp. 1–270, 1890.

[360] H. Poincaré, "Sur les Proprietes des Fonctions Definies par les Equations aux Differences Partielles," in *Oeuvres*, vol. 1. Paris: Gauthier-Villars, 1929.

[361] V. M. Popov, "Absolute Stability of Nonlinear Systems of Automatic Control," *Automatika Telemekhanika*, vol. 22, pp. 961–978, 1961.

[362] V. M. Popov, "Absolute Stability of Nonlinear Systems of Automatic Control," *Autom. Remote Control*, vol. 22, pp. 857–875, 1962.

[363] V. M. Popov, "Hyperstability and Optimality of Automatic Systems with Several Control Functions," *Rev. Electrotech. Energie Acad. Rep. Pop. Rom.*, vol. 9, 1964.

[364] V. M. Popov, *Hyperstability of Control Systems.* New York: Springer, 1973.

[365] R. H. Rand, R. J. Kinsey, and D. L. Mingori, "Dynamics of Spinup Through Resonance," *Int. J. Non-Linear Mech.*, vol. 27, pp. 489–502, 1992.

[366] Z. V. Rekasius, "Suboptimal Design of Intentionally Nonlinear Controllers," *IEEE Trans. Autom. Control*, vol. 9, pp. 380–386, 1964.

[367] M. A. Rotea, "The Generalized $\mathcal{H}_2$ Control Problem," *Automatica*, vol. 29, pp. 373–385, 1993.

[368] N. Rouche, P. Habets, and M. Laloy, *Stability Theory by Liapunov's Direct Method.* New York: Springer, 1977.

[369] E. J. Routh, *A Treatise on the Stability of a Given State of Motion.* London: Macmillan, 1877.

[370] E. J. Routh, *The Advanced Part of a Treatise on the Dynamics of a System of Rigid Bodies.* New York: Dover, 1955.

[371] H. L. Royden, *Real Analysis.* New York: Macmillan, 1988.

[372] E. N. Rozenvasser, "On the Construction of a Lyapunov Function for a Class of Nonlinear Systems," *Tzevestia Akad. Nauk SSSR, OTN*, vol. 2, 1960.

[373] W. Rudin, *Principles of Mathematical Analysis,* 3rd edition. New York: McGraw-Hill, 1976.

[374] V. V. Rumyantsev, "On the Stability of Motion with Respect to Part of the variables," *Vestn. Mosk. Unive. Ser. Mat. Mekh. Fiz. Astron. Khim.*, vol. 4, pp. 9–16, 1957.

[375] E. P. Ryan, "Optimal Feedback Control of Saturating Systems," *Int. J. Control*, vol. 35, pp. 531–534, 1982.

[376] A. Saberi, P. V. Kokotović, and H. J. Sussmann, "Global Stabilization of Partially Linear Composite Systems," *SIAM J. Control Optim*, vol. 28, pp. 1491–1503, 1990.

[377] M. G. Safonov, *Stability and Robustness of Multivariable Feedback Systems.* Cambridge, MA: MIT Press, 1980.

[378] M. G. Safonov, "Stability Margins of Diagonally Perturbed Multivariable Feedback Systems," in *Proc. IEEE Conf. Dec. Control*, San Diego, CA, 1981.

[379] M. G. Safonov and M. Athans, "Gain and Phase Margin for Multiloop LQG Regulators," *IEEE Trans. Autom. Control*, vol. 22, pp. 173–179, 1977.

[380] M. G. Safonov and M. Athans, "A Multiloop Generalization of the Circle Criterion for Stability Margin Analysis," *IEEE Trans. Autom. Control*, vol. 26, pp. 415–422, 1981.

[381] M. G. Safonov and R. Y. Chiang, "Real/Complex K_{m}-Synthesis without Curve Fitting," *Control Dyn. Syst.*, vol. 56, Part 2, pp. 303–324, 1993.

[382] M. G. Safonov and K. H. M. Fan, "Editorial on the Multivariable Stability Margin," *Int. J. Robust Nonlinear Control*, vol. 7, pp. 97–103, 1997.

[383] M. G. Safonov, E. A. Jonckhere, M. Verma, and D. J. N. Limebeer, "Synthesis of Positive Real Multivariable Feedback Systems," *Int. J. Control*, vol. 45, pp. 817–842, 1987.

[384] S. V. Salehi and E. P. Ryan, "On Optimal Nonlinear Feedback Regulation of Linear Plants," *IEEE Trans. Autom. Control*, vol. 27, pp. 1260–1264, 1982.

[385] D. M. Salmon, "Minimax Controller Design," *IEEE Trans. Autom. Control*, vol. 13, pp. 369–376, 1968.

[386] C. Samson, "Robust Nonlinear Control of Robotic Manipulators," in *Proc. IEEE Conf. Dec. Control*, pp. 1211–1216, 1983.

[387] I. W. Sandberg, "A Frequency-Domain Condition for Stability of Feedback Systems Containing a Single Time-Varying Nonlinear Element," *Bell Syst. Tech. J.*, vol. 43, pp. 1601–1608, 1964.

[388] I. W. Sandberg, "On the $\mathcal{L}_2$-Boundedness of Solutions of Nonlinear Functional Equations," *Bell Syst. Tech. J.*, vol. 43, pp. 1581–1599, 1964.

[389] I. W. Sandberg, "On the Properties of Some Systems that Distort Signals," *Bell. Syst. Tech. J.*, vol. 4, pp. 91–112, 1964.

[390] I. W. Sandberg, "Some Results on the Theory of Physical Systems Governed by Nonlinear Functional Equations," *Bell Syst. Tech. J.*, vol. 44, pp. 871–898, 1964.

[391] J. Sandor and D. Williamson, "A Design of Nonlinear Regulators for Linear Plants," *IEEE Trans. Autom. Control*, vol. 22, pp. 47–50, 1977.

[392] S. S. Sastry and A. Isidori, "Adaptive Control of Linearizable Systems," *IEEE Trans. Autom. Control*, vol. 34, pp. 1123–1131, 1989.

[393] F. C. Schweppe, *Uncertain Dynamic Systems.* Englewood Cliffs, NJ: Prentice-Hall, 1973.

[394] W. E. Scmitendorf, "Stabilization via Dynamic Output Feedback: A Numerical Approach," *AIAA J. Guid. Control Dyn.*, vol. 14, pp. 1083–1086, 1991.

[395] R. Sepulchre, M. Jankovic, and P.V. Kokotović, *Constructive Nonlinear Control.* London: Springer, 1997.

[396] R. Sezginer and M. Overton, "The Largest Singular Value of $e^X A e^{-X}$ is Convex on Convex Sets of Commuting Matrices," *IEEE Trans. Autom. Control*, vol. 35, pp. 229–230, 1990.

[397] U. Shaked, "Guaranteed Stability Margins for the Discrete-Time Linear Quadratic Optimal Regulator," *IEEE Trans. Autom. Control*, vol. 31, pp. 162–165, 1986.

[398] J. S. Shamma and M. Athans, "Analysis of Gain Scheduled Control for Nonlinear Plants," *IEEE Trans. Autom. Control*, vol. 35, pp. 898–907, 1990.

[399] L. Shaw, "Nonlinear Control of Linear Multivariable Systems via State-Dependent Feedback Gains," *IEEE Trans. Autom. Control*, vol. 24, pp. 108–112, 1979.

[400] D. D. Siljak, *Large Scale Dynamic Systems: Stability and Structures.* New York: North-Holland, 1978.

[401] D. D. Siljak, "Parameter Space Methods for Robust Control Design: A Guided Tour," *IEEE Trans. Autom. Control*, vol. 34, pp. 674–688, 1989.

[402] D. D. Siljak, *Decentralized Control of Complex Systems.* Boston, MA: Academic Press, 1991.

[403] V. A. Sinitsyn, "On Stability of Solution in Inertial Navigation Problem," in *Certain Problems on Dynamics of Mechanical Systems*, Moscow, pp. 46–50, 1991.

[404] J. J. Slotine and S. S. Sastry, "Tracking Control of Nonlinear Systems Using Sliding Surfaces with Applications to Robot Manipulators," *Int. J. Control*, vol. 2, pp. 465–492, 1983.

[405] E. D. Sontag, "Smooth Stabilization Implies Coprime Factorization," *IEEE Trans. Autom. Control*, vol. 34, pp. 435–443, 1989.

[406] E. D. Sontag, "A Universal Construction of Artstein's Theorem on Nonlinear Stabilization," *Syst. Control Lett.*, vol. 13, pp. 117–123, 1989.

[407] E. D. Sontag, "On the Input-to-State Stability Property," *Eur. J. Control*, vol. 1, 1995.

[408] E. D. Sontag and H. J. Sussmann, "Further Comments on the Stabilizability of the Angular Velocity of a Rigid Body," *Syst. Control Lett.*, vol. 12, pp. 437–442, 1988.

[409] E. D. Sontag and Y. Wang, "On Characterizations of the Input-to-State Stability Property," *Syst. Control Lett.*, vol. 24, pp. 351–359, 1995.

[410] E. Soroka and U. Shaked, "On the Robustness of LQ Regulators," *IEEE Trans. Autom. Control*, vol. 29, pp. 664–665, 1984.

[411] J. L. Speyer, "A Nonlinear Control Law for a Stochastic Infinite Time Problem," *IEEE Trans. Autom. Control*, vol. 21, pp. 560–564, 1976.

[412] M. W. Spong, "Robust Stabilization for a Class of Nonlinear Systems," in *Theory and Application of Nonlinear Control Systems* (C. I. Byrnes and A. Lindquist, eds.). Amsterdam: North-Holland, 1986.

[413] M. W. Spong, J. S. Thorp, and J. M. Kleinwaks, "The Control of Robotic Manipulators with Bounded Control Part II: Robustness and Disturbance Rejection," in *Proc. IEEE Conf. Dec. Control*, Las Vegas, NW, pp. 1047–1052, 1984.

[414] M. W. Spong and M. Vidyasagar, "Robust Nonlinear Control of Robot Manipulators," in *Proc. IEEE Conf. Dec. Control*, pp. 1767–1772, 1985.

[415] M. W. Spong and M. Vidyasagar, *Robot Dynamics and Control.* New York: Wiley, 1989.

[416] H. L. Stalford and C. H. Chao, "A Necessary and Sufficient Condition in Lyapunov Robust Control," *J. Optim. Theory Appl.*, vol. 63, pp. 191–204, 1989.

[417] G. W. Stewart and J. Sun, *Matrix Perturbation Theory.* San Diego, CA: Academic Press, 1990.

[418] A. A. Stoorvogel, "The $\mathcal{H}_\infty$ Control Problem: A State Space Approach." Ph. D. thesis, Technical University Eidenhoven, The Netherlands, 1990.

[419] R. Su, "On the Linear Equivalents of Nonlinear Systems," *Syst. Control Lett.*, vol. 2, pp. 48–52, 1982.

[420] W. Sun, P. P. Khargonekar, and D. Shim, "Solution to the Positive Real Control Problem for Linear Time-Invariant Systems," *IEEE Trans. Autom. Control*, vol. 39, pp. 2034–2046, 1994.

[421] G. Szegö, "On the Absolute Stability of Sampled-Data Control Systems," *Proc. Natl. Acad. Sci.*, vol. 49, pp. 558–560, 1963.

[422] G. Szegö and R. E. Kalman, "Sur la Stabilite d'un System d'Équations aux Différences Finies," *C. R. Acad. Sci. Paris*, vol. 257, pp. 388–390, 1963.

[423] M. Sznaier, M. Holmes, and J. Bu, "Mixed $\mathcal{H}_2/\mathcal{L}_1$ Control with Low Order Controllers: A Linear Matrix Inequality Approach," in *Proc. IEEE Conf. Dec. Control*, New Orleans, LA, pp. 1352–1357, 1995.

[424] J. Terjeki and L. Hatvani, "On Asymptotic Halting in the Presence of Viscous Friction," *Prikl. Mat. Mekh.*, vol. 46, pp. 20–26, 1982.

[425] M. A. L. Thathachar and M. D. Srinath, "Some Aspects of the Luré Problem," *IEEE Trans. Autom. Control*, vol. 12, pp. 451–453, 1967.

[426] M. A. L. Thathachar and M. D. Srinath, "Stability of Linear Time-Invariant Systems," *IEEE Trans. Autom. Control*, vol. 12, pp. 335–336, 1967.

[427] M. A. L. Thathachar, M. D. Srinath, and G. Krishna, "Stability Analysis of Nonlinearity in a Sector," *IEEE Trans. Autom. Control*, vol. 11, pp. 311–312, 1966.

[428] M. A. L. Thathachar, M. D. Srinath, and H. K. Ramaprige, "On a Modified Luré Problem," *IEEE Trans. Autom. Control*, vol. 12, pp. 731–740, 1967.

[429] F. E. Thau, "Optimum Nonlinear Control of a Class of Randomly Excited Systems," *ASME J. Dyn. Syst. Meas. Control*, pp. 41–44, 1971.

[430] J. S. Thorp and B. R. Barmish, "On Guaranteed Stability of Uncertain Linear Systems via Linear Control," *J. Optim. Theory Appl.*, vol. 35, pp. 559–579, 1981.

[431] E. Torricelli, *Opera Geometrica.* Florence, Italy: Musse, 1644.

[432] J. Tsinias, "Stabilizability of Discrete-Time Nonlinear Systems," *IMA J. Math. Control Info.*, vol. 6, pp. 135–150, 1989.

[433] J. Tsinias, "Sufficient Lyapunov-like Conditions for Stabilization," *Math. Control, Signals, Syst.*, vol. 2, pp. 343–357, 1989.

[434] J. Tsinias, "Existence of Control Lyapunov Functions and Applications to State Feedback Stabilizability of Nonlinear Systems," *SIAM J. Control Optim.*, vol. 29, pp. 457–473, 1991.

[435] F. Tyan and D. S. Bernstein, "Shifted Quadratic Guaranteed Cost Bounds for Robust Controller Synthesis," in *Proc. 13th IFAC World Congress*, San Francisco, CA, pp. 5002–5007, 1996.

[436] F. Tyan, S. R. Hall, and D. S. Bernstein, "A Double-Commutator Guaranteed Cost Bound for Robust Stability and Performance," *Syst. Control Lett.*, vol. 25, pp. 125–129, 1995.

[437] P. P. Vaidyanathan, "The Discrete-Time Bounded-Real Lemma in Digital Filtering," *IEEE Trans. Circ. Syst.*, vol. 32, pp. 918–924, 1985.

[438] A. J. van der Schaft, "A State-Space Approach to Nonlinear $\mathcal{H}_\infty$ Control," *Syst. Control Lett.*, vol. 16, pp. 1–8, 1991.

[439] A. J. van der Schaft, "$\mathcal{L}_2$-Gain Analysis of Nonlinear Systems and Nonlinear $\mathcal{H}_\infty$ Control," *IEEE Trans. Autom. Control*, vol. 37, pp. 770–784, 1992.

[440] A. J. van der Schaft, "Nonlinear Control and Hamilton-Jacobi Inequalities," in *IFAC Nonlinear Control Systems Design*, Bordeaux, France, pp. 9–14, 1992.

[441] A. J. van der Schaft, *$\mathcal{L}_2$-Gain and Passivity Techniques in Nonlinear Control.* New York: Springer, 1996.

[442] C. van Loan, "How Near is a Stable Matrix to an Unstable Matrix?" *Contemp. Math.*, vol. 47, pp. 465–478, 1985.

[443] A. Vannelli and M. Vidyasagar, "Maximal Lyapunov Functions and Domains of Attraction for Autonomous Nonlinear Systems," *Automatica*, vol. 21, pp. 69–80, 1985.

[444] M. Vidyasagar, "Optimal Rejection of Persistent Bounded Disturbances," *IEEE Trans. Autom. Control*, vol. 31, pp. 527–535, 1986.

[445] M. Vidyasagar, *Nonlinear Systems Analysis.* Englewood Cliffs, NJ: Prentice-Hall, 1993.

[446] M. Vidyasagar and A. Vannelli, "New Relationships Between Input-Output and Lyapunov Stability," *IEEE Trans. Autom. Control*, vol. 27, pp. 481–483, 1982.

[447] A. Vinkler and L. J. Wood, "Multistep Guaranteed Cost Control of Linear Systems with Uncertain Parameters," *J. Guid. Control*, vol. 2, pp. 449–456, 1979.

[448] V. I. Vorotnikov, *Partial Stability and Control.* Boston, MA: Birkhäuser, 1998.

[449] C.J. Wan and D.S. Bernstein, "Nonlinear Feedback Control with Global Stabilization," *Dyn. Control*, vol. 5, pp. 321–346, 1995.

[450] C. J. Wan, V. T. Coppola, and D. S. Bernstein, "A Lyapunov Function for the Energy-Casimir Method," in *Proc. IEEE Conf. Dec. Control*, San Antonio, TX, pp. 3122–3123, 1993.

[451] H. O. Wang, R. A. Adomaitis, and E. H. Abed, "Nonlinear Analysis and Control of Rotating Stall in Axial Flow Compressors," in *Proc. American Control Conf.*, Baltimore, MD, pp. 2317–2321, 1994.

[452] J. Wang and W. J. Rugh, "Feedback Linearization Families for Nonlinear Systems," *IEEE Trans. Autom. Control*, vol. 32, pp. 935–941, 1987.

[453] N. Wiener, *Nonlinear Problems in Random Theory.* Cambridge, MA: MIT Press, 1958.

[454] S. Wiggins, *Introduction to Applied Nonlinear Dynamical Systems and Chaos.* New York: Springer-Verlag, 1990.

[455] J. C. Willems, "The Generation of Lyapunov Functions for Input-Output Systems," *SIAM J. Control Optim.*, vol. 9, pp. 105–134, 1971.

[456] J. C. Willems, "Dissipative Dynamical Systems Part I: General Theory," *Arch. Rat. Mech. Anal.*, vol. 45, pp. 321–351, 1972.

[457] J. C. Willems, "Dissipative Dynamical Systems Part II: Quadratic Supply Rates," *Arch. Rat. Mech. Anal.*, vol. 45, pp. 359–393, 1972.

[458] J. L. Willems, "The Circle Criterion and Quadratic Lyapunov Functions for Stability Analysis," *IEEE Trans. Autom. Control*, vol. 18, p. 184, 1973.

[459] J. L. Willems and H. van de Voorde, "The Return Difference for Discrete-Time Optimal Feedback Systems," *Automatica*, vol. 14, pp. 511–513, 1978.

[460] F. W. Wilson, "Smoothing Derivatives of Functions and Applications," *Trans. Am. Math. Soc.*, vol. 139, pp. 413–428, 1969.

[461] D. A. Wilson, "Convolution and Hankel Operator Norms of Linear Systems," *IEEE Trans. Autom. Control*, vol. 34, pp. 94–97, 1989.

[462] D. A. Wilson, "Induced Norms for Mixed $\mathcal{H}_2/\mathcal{H}_\infty$ and $\mathcal{H}_2/\mathcal{H}_\infty/\mathcal{L}_\infty$ Control," in *Proc. IEEE Conf. Dec. Control*, New Orleans, LA, pp. 3164–3168, 1995.

[463] D. A. Wilson, M. A. Nekou, and G. D. Halikias, "An LQR Weight Selection Approach to the Discrete Generalized $\mathcal{H}_2$ Control Problem," *Int. J. Control*, vol. 71, pp. 93–101, 1998.

[464] H. S. Witsenhausen, "Minimax Control of Uncertain Systems," Tech. Rep. ESL-R-269, MIT, Cambridge, MA, May 1966.

[465] P. K. Wong, "On the Interaction Structure of Linear Multi-Input Feedback Control Systems." Master's thesis, Electrical Engineering, MIT, 1975.

[466] P. K. Wong and M. Athans, "Closed-Loop Structural Stability for Linear-Quadratic Optimal Systems," *IEEE Trans. Autom. Control*, vol. 22, pp. 94–99, 1977.

[467] V. A. Yakubovich, "On Nonlinear Differential Equations for Control Systems with a Single Regulator," *Vestn. LGU*, vol. 2, 1960.

[468] V. A. Yakubovich, "Solution of Certain Matrix Inequalities Encountered in Nonlinear Control Theory," *Sov. Math. Dokl.*, vol. 5, pp. 652–656, 1964.

[469] V. A. Yakubovich, "Frequency Conditions for the Absolute Stability and Dissipativity of Control Systems with a Single Differentiable Nonlinearity," *Sov. Math.*, vol. 6, pp. 98–101, 1965.

[470] V. A. Yakubovich, "The Method of Matrix Inequalities in the Theory of the Stability of Nonlinear Systems II: Absolute Stability in a Class of Nonlinearities with a Condition on the Derivative," *Autom. Remote Control*, vol. 26, pp. 577–592, 1965.

[471] R. K. Yedavalli, "Improved Measures of Stability Robustness for Linear State Space Models," *IEEE Trans. Autom. Control*, vol. 30, pp. 577–579, 1985.

[472] R. K. Yedavalli, S. S. Banda, and D. B. Ridgley, "Time-Domain Stability Robustness Measures for Linear Regulators," *J. Guid. Control Dyn.*, vol. 8, pp. 520–524, 1985.

[473] H. Ye, A. N. Michel, and L. Hou, "Stability Theory for Hybrid Dynamical Systems," *IEEE Trans. Autom. Control*, vol. 43, pp. 461–474, 1998.

[474] T. Yoshizawa, *Stability Theory by Liapunov's Second Method.* Tokyo: The Mathematical Society of Japan, 1966.

[475] P. M. Young, M. P. Newlin, and J. C. Doyle, "μ Analysis with Real Parametric Uncertainty," in *Proc. IEEE Conf. Dec. Control*, Brighton, U.K., pp. 1251–1256, 1991.

[476] G. Zames, "On the Input-Output Stability of Time-Varying Nonlinear Feedback Systems, Part I: Conditions Derived Using Concepts of Loop Gain, Conicity, and Positivity," *IEEE Trans. Autom. Control*, vol. 11, pp. 228–238, 1966.

[477] G. Zames, "On the Input-Output Stability of Time-Varying Nonlinear Feedback Systems, Part II: Conditions Involving Circles in the

Frequency Plane and Sector Nonlinearities," *IEEE Trans. Autom. Control*, vol. 11, pp. 465–476, 1966.

[478] G. Zames, "Feedback and Optimal Sensitivity: Model Reference Transformations, Multiplicative Seminorms, and Approximate Inverses," *IEEE Trans. Autom. Control*, vol. 26, pp. 301–320, 1981.

[479] G. Zames and P. L. Falb, "Stability Conditions for Systems with Monotone and Slope-Restricted Nonlinearities," *SIAM J. Control Optim.*, vol. 4, pp. 89–108, 1968.

[480] K. Zhou and P. P. Khargonekar, "Stability Robustness Bounds for Linear State-Space Models with Structured Uncertainty," *IEEE Trans. Autom. Control*, vol. 32, pp. 621–623, 1987.

[481] V. I. Zubov, *Methods of A. M. Lyapunov and Their Application.* Groningen, The Netherlands: P. Noordhoff, 1964.

[482] V. I. Zubov, *The Dynamics of Controlled Systems.* Moscow: Vysshaya Shkola, 1982.

Index

T

U

GPSR Authorized Representative: Easy Access System Europe - Mustamäe tee 50, 10621 Tallinn, Estonia, gpsr.requests@easproject.com

www.ingramcontent.com/pod-product-compliance
Ingram Content Group UK Ltd.
Pitfield, Milton Keynes, MK11 3LW, UK
UKHW051843110826
473955UK00005B/66

* 9 7 8 0 6 9 1 1 3 3 2 9 4 *